Ch

To Our Son, David, We love you.
Mom & Dad

NEOTROPICAL ORNITHOLOGY

Cover drawing: Blue-crowned Motmot (*Momotus momota*). From a pen and ink drawing by Michel Kleinbaum.

ORNITHOLOGICAL MONOGRAPHS

This series, published by the American Ornithologists' Union, has been established for major papers too long for inclusion in the Union's journal, *The Auk*. Publication has been made possible through the generosity of the late Mrs. Carll Tucker and the Marcia Brady Tucker Foundation, Inc.

Correspondence concerning manuscripts for publication in the series should be addressed to the Editor, Dr. Mercedes S. Foster, USFWS, National Museum of Natural History, Washington, D.C. 20560.

Copies of *Ornithological Monographs* may be ordered from the Assistant to the Treasurer of the AOU, Frank R. Moore, Department of Biology, University of Southern Mississippi, Southern Station Box 5018, Hattiesburg, Mississippi 39406. (See list at the end of the volume).

Ornithological Monographs, No. 36, xii + 1041 pp.

Editor of AOU Monographs, Mercedes S. Foster

Editors, P. A. Buckley, U.S. National Park Service Cooperative Research Unit, Doolittle Hall, Rutgers University, New Brunswick, New Jersey 08903; Mercedes S. Foster, Museum Section, USFWS/NHB-378, National Museum of Natural History, Washington, D.C. 20560; Eugene S. Morton, Department of Zoological Research, National Zoological Park, Smithsonian Institution, Washington, D.C. 20008; Robert S. Ridgely, Division of Vertebrate Biology, Academy of Natural Sciences, Philadelphia, Pennsylvania 19103; Francine G. Buckley, Center for Coastal and Environmental Studies, Doolittle Hall, Rutgers University, New Brunswick, New Jersey 08903

Manuscript received 1 June 1983; accepted 2 January 1984; final revision submitted 16 July 1984
Issued July 30, 1985

Price $70.00

Library of Congress Catalogue Card Number 85-71524

Printed by Allen Press, Inc., Lawrence, Kansas 66044

It would be impossible to produce a volume of this size and complexity without the assistance of many individuals. In particular, the efforts made by those who reviewed the manuscripts considered for this publication must be recognized and their assistance most gratefully acknowledged. Their evaluations and constructive criticisms contributed significantly to the scientific and scholarly aspects of the volume. Their names are listed below (individuals who reviewed more than one manuscript are indicated with an asterisk): K. P. Able*, D. G. Ainley, D. W. Anderson, R. C. Banks*, L. F. Baptista, G. F. Barrowclough, B. M. Beehler, W. Belton, D. Boersma, J. Bond*, J. H. Brown, E. H. Burtt, Jr., G. S. Caldwell, S. L. Coats, J. L. Cracraft, A. Cruz, J. S. Denslow*, J. M. Diamond, R. W. Dickerman*, J. A. Endler, R. M. Erwin, D. S. Farner, A. Feduccia*, R. P. ffrench, J. W. Fitzpatrick*, R. M. Fraga, S. A. Gauthreaux, Jr.*, F. Gill, M. Gochfeld, J. Gradwohl, G. R. Graves*, R. S. Greenberg, J. H. Haffer, J. L. Hand, J. W. Hardy*, B. A. Harrington, H. A. Hespenheide, S. L. Hilty, L. C. Holcomb, T. R. Howell, P. S. Humphrey, D. J. T. Hussell, R. L. Hutto, D. H. Janzen, J. Jehl, N. K. Johnson, R. F. Johnston, P. M. Kahl, J. R. Karr, J. A. Keast, J. Kikkawa, W. B. King, J. C. Kricher, J. A. Kushlan, W. E. Lanyon*, C. F. Leck, S. Marchant, P. Mason, D. W. Mock*, B. L. Monroe, Jr., D. H. Morse, M. H. Moynihan, C. A. Munn, III*, B. G. Murray, Jr., J. P. Myers, M. Nores, R. J. O'Connor, R. D. Ohmart*, S. L. Olson, J. P. O'Neill, Y. Oniki, G. H. Orians, L. W. Oring, G. W. Page, T. A. Parker, III, K. C. Parkes*, R. F. Pasquier, J. Patton, R. B. Payne, D. L. Pearson*, O. P. Pearson, J. Picman, F. A. Pitelka, M. A. Plenge, W. Post, G. V. N. Powell*, D. M. Power, H. D. Pratt, R. J. Raikow*, J. V. Remsen, Jr.*, J. D. Rising, C. S. Robbins, T. Royama, S. M. Russell, D. Schneider, G. D. Schnell, T. S. Schulenberg, J. M. Sheppard, L. L. Short*, J. N. Smith, N. G. Smith*, R. G. Smith, W. J. Smith, D. W. Snow, D. W. Steadman, L. Stenzel, E. W. Stiles, G. F. Stiles, R. W. Storer, J. W. Terborgh, F. Vuilleumier, S. L. Warter, G. E. Watson, M. Weller, J. S. Weske, N. T. Wheelwright, J. A. Wiens, D. E. Willard, E. O. Willis*, L. L. Wolf*, D. S. Wood, A. Worthington, J. Wunderle, R. A. Zell, R. M. Zink*.

Ms. Susana Salas translated into Spanish all Abstracts except those by Cruz et al., Humphrey and Livezey, and Moermond and Denslow. Other valuable assistance was provided by C. J. Angle, R. B. Clapp, M. B. Hayes, M. B. Scott, and especially, R. C. Banks. In addition, I thank L. M. Wolfe, my secretary through much of the period during which the volume was produced and without whose help it would never have succeeded.

Finally, it is appropriate to thank the authors, some for adhering to deadlines, some for their patience with delays caused by reviewers and authors who did not, and all for their scientific contributions and their considerable efforts to make this volume a success.

MERCEDES S. FOSTER
Editor, Ornithological Monographs

FRONTISPIECE. Selected members of the *Thryothorus euophrys* complex (from top to bottom): *T. eisenmanni,* adult male and female; *T. euophrys schulenbergi,* adult; *T. e. atriceps,* adult; *T. e. longipes,* adult. From a mixed media painting by John P. O'Neill.

NEOTROPICAL ORNITHOLOGY

EDITORS:

P. A. BUCKLEY, MERCEDES S. FOSTER, EUGENE S. MORTON,
ROBERT S. RIDGELY, AND FRANCINE G. BUCKLEY

ORNITHOLOGICAL MONOGRAPHS NO. 36

PUBLISHED BY
THE AMERICAN ORNITHOLOGISTS' UNION
WASHINGTON, D.C.
1985

This volume is dedicated to the memory
of
Eugene Eisenmann
in recognition of his contributions
to neotropical ornithology

PREFACE

For historical reasons, most research on birds in the neotropics has been conducted by temperate zone residents of the Old or New World. Initially, their efforts were descriptive. Later, these individuals began to unravel the ecological and evolutionary workings of tropical environments, and, naturally, their interpretations of these processes were shaped by their temperate zone experiences. Especially in the last thirty years or so, the scientific community has become aware that tropical ecosystems are not only quantitatively, but also qualitatively different from those in temperate zones, and intense effort has been made to try to understand the limits of those differences.

At the forefront of tropical ecological research has been work with birds in Latin America, which possesses the richest avifauna in the world, and from which new species and even new genera are still being described. Many important ecological principles now widely accepted were first stated and explained using data from neotropical birds. Yet, despite a voluminous literature, few attempts have been made to collect into one volume papers that present a picture of those principles, that demonstrate how far along we are in understanding the ecological workings of neotropical avian communities, or that indicate the areas toward which ornithological research may profitably be directed in the future.

After Eugene Eisenmann died most unexpectedly in October 1981, several of us attending his memorial service pondered what would be a proper tribute in recognition of his infectious and generous support of students, young and old, of neotropical birds. It seemed that the most fitting memorial to him—in a lasting format he especially would have appreciated—would be the assembling of a series of papers on the general topic of neotropical ornithology, a task that was timely in its own right.

We invited authors, ranging from senior scientists long established in the field to graduate students, to prepare papers on a wide variety of topics. We sought a representative array from the various scientific specialities in which neotropical bird research is underway today, soliciting longer synthesizing review papers and shorter contributions reflecting recent research findings; where appropriate, we hoped for emphasis on ecological and evolutionary processes peculiar to, or more typical of, the New World tropics. As this was a memorial tribute to Eugene Eisenmann, we also hoped that the topics treated would reflect his own interests which, broad as they were, nevertheless centered on taxonomy and zoogeography; that objective, however, was secondary to the others.

Because this volume was not intended to be a textbook on neotropical ornithology, we did not edit to remove overlap or to achieve a common style of writing. This is a collection of papers on what has been and what is being done by researchers on birds in the neotropics. Wherever we have been able to induce authors to do so, we have sought comparisons between tropical and temperate zones (stressing the uniqueness of the former), and between paleotropical and neotropical birds and ecosystems.

From the beginning, we had as one of our goals the inclusion of a large number of contributions from Latin Americans, and we contacted many potential authors there during the volume's gestation. We are distressed—as we know Gene Eisenmann would have been—at how few Latin American authors are actually represented in the book. We likewise made great effort to obtain manuscripts on topics ultimately almost unrepresented (e.g., fossil birds, migration, breeding physiology, seabirds).

Thomas Howell sets the stage with a biographical sketch assessing Eugene Eisenmann's importance to Latin American ornithology, especially in encouraging students and Spanish-speaking ornithologists; Kenneth Parkes closes it with a "pre-publication review" that attempts to place the entire collection in perspective *inter se*; it would be useful to read this paper first. Various of our colleagues had planned to describe new taxa named in honor of Eugene Eisenmann, and they were persuaded to do so here. Several of the new taxa are depicted in the color plates we were fortunate to be able to secure for this work. John Fitzpatrick, Dana Gardner, Robert Mengel, John O'Neill, Roger Tory Peterson, and Arthur Singer donated their scientific skills and artistic talents to produce paintings for seven of the color plates that elucidate the papers they accompany. Morris Williams provided the photograph used for the eighth color plate.

Any work of this size and complexity is expensive to publish. Because we had as another of our operating maxima a desire to keep the cost of the book reasonable to ensure widespread dissemination among students and our English-reading Latin American colleagues, we embarked on a program of fund-raising. That you hold this volume in your hands is due to the generous financial support of the following institutions (in alphabetical order): American Museum of Natural History (Ornithology Department); American Ornithologists' Union (Ornithological Monographs' Fund); International Council for Bird Preservation (Pan American Section); Smithsonian Institution (Alexander Wetmore Fund); and Smithsonian Tropical Research Institute. Substantial individual support has come from two anonymous donors and Howard P. Brokaw, W. L. Brown, Betty Carnes, Crawford H. Greenwalt, A. J. Lauro, John S. McIlhenny, William H. Phelps, Jr., and Richard H. Pough.

At critical junctions, especially helpful assistance was provided by K. P. Able, S. Coats, T. R. Howell, and L. L. Short. Unstinting clerical support came from B. Dorn (USNPS, Boston), P. Eager and D. Rucci (Rutgers, New Brunswick), and L. M. Wolfe (USFWS, Washington, D.C.).

If this book acts as a stimulus for additional work on the birds of the neotropics, then we will be enormously satisfied and confident that our efforts of the past two and a half years have been well spent as have those of our authors, referees, artists, and patrons. We thank them all for their various contributions, and hope that they will agree that this volume on Neotropical Ornithology forms a fitting tribute to Gene Eisenmann and his love for the birds of the New World tropics.

May 1984

P. A. Buckley
Mercedes S. Foster
Eugene S. Morton
Robert S. Ridgely
Francine G. Buckley

TABLE OF CONTENTS

DEDICATION vi
PREFACE vii
TABLE OF CONTENTS ix
LIST OF COLOR ILLUSTRATIONS xi
EUGENE EISENMANN AND THE STUDY OF NEOTROPICAL BIRDS by
    Thomas R. Howell 1

NEW TAXA 7
    Parker, Theodore A., III, and John P. O'Neill. A New Species and a New
        Subspecies of *Thryothorus* Wren From Peru 9
    Delgado B., Francisco S. A New Subspecies of the Painted Parakeet (*Pyrrhura
        picta*) from Panama 17
    Stiles, F. Gary. Geographic Variation in the Fiery-throated Hummingbird, *Pan-
        terpe insignis* 23
    Storer, Robert W., and Thomas Getty. Geographic Variation in the Least Grebe
        (*Tachybaptus dominicus*) 31
    Weske, John S. A New Subspecies of Collared Inca Hummingbird (*Coeligena
        torquata*) from Peru 41

ZOOGEOGRAPHY AND DISTRIBUTION 47
    Cracraft, Joel. Historical Biogeography and Patterns of Differentiation within
        the South American Avifauna: Areas of Endemism 49
    Fjeldså, Jon. Origin, Evolution, and Status of the Avifauna of Andean Wet-
        lands 85
    Haffer, Jürgen. Avian Zoogeography of the Neotropical Lowlands 113
    Haffer, Jürgen, and John W. Fitzpatrick. Geographic Variation in Some Am-
        azonian Forest Birds 147
    Parker, Theodore A., III, Thomas S. Schulenberg, Gary R. Graves, and
        Michael J. Braun. The Avifauna of the Huancabamba Region, Northern
        Peru 169
    Robbins, Mark B., Theodore A. Parker III, and Susan E. Allen. The Avifauna
        of Cerro Pirre, Darién, Eastern Panama 198
    Sick, Helmut. Observations on the Andean-Patagonian Component of South-
        eastern Brazil's Avifauna 233
    Snow, David W. Affinities and Recent History of the Avifauna of Trinidad and
        Tobago 238
    Voous, K. H. Additions to the Avifauna of Aruba, Curaçao, and Bonaire, South
        Caribbean 247
    Vuilleumier, François. Forest Birds of Patagonia: Ecological Geography,
        Speciation, Endemism, and Faunal History 255
    Wiedenfeld, David A., Thomas S. Schulenberg, and Mark B. Robbins. Birds
        of a Tropical Deciduous Forest in Extreme Northwestern Peru 305

SYSTEMATICS 317
    Bock, Walter J. Is *Diglossa* (?Thraupinae) Monophyletic? 319
    Braun, Micahel J., and Theodore A. Parker III. Molecular, Morphological,
        and Behavioral Evidence Concerning the Taxonomic Relationships of
        "*Synallaxis*" *gularis* and Other Synallaxines 333
    Capparella, A. P., and Scott M. Lanyon. Biochemical and Morphometric Anal-
        yses of the Sympatric, Neotropical, Sibling Species, *Mionectes macconnelli*
        and *M. oleagineus* 347
    Dickerman, Robert W. Taxonomy of the Lesser Nighthawks (*Chordeiles acuti-
        pennis*) of North and Central America 356
    Lanyon, Wesley E. A Phylogeny of the Myiarchine Flycatchers 361
    Morony, John J., Jr. Systematic Relations of *Sericossypha albocristata* (Thrau-
        pinae) 383
    Schulenberg, Thomas S. An Intergeneric Hybrid Conebill (*Conirostrum* × *Oreo-
        manes*) from Peru 390

SIBLEY, CHARLES G., AND JON E. AHLQUIST. Phylogeny and Classification of New World Suboscine Passerine Birds (Passeriformes: Oligomyodi: Tyrannides) 396
TRAYLOR, MELVIN A., JR. Species Limits in the *Ochthoeca diadema* Species-group (Tyrannidae) 431

EVOLUTION 445
FITZPATRICK, JOHN W. Form, Foraging Behavior, and Adaptive Radiation in the Tyrannidae 447
GRANT, P. R. Climatic Fluctuations on the Galapagos Islands and their Influence on Darwin's Finches 471
JEHL, JOSEPH R., JR. Hybridization and Evolution of Oystercatchers on the Pacific Coast of Baja California 484
MURRAY, BERTRAM G., JR. Evolution of Clutch Size in Tropical Species of Birds 505
MYERS, J. P., J. L. MARON, AND MICHEL SALLABERRY. Going to Extremes: Why Do Sanderlings Migrate to the Neotropics? 520
ONIKI, YOSHIKA. Why Robin Eggs are Blue and Birds Build Nests: Statistical Tests for Amazonian Birds 536
SCHNEIDER, DAVID. Migratory Shorebirds: Resource Depletion in the Tropics? 546
SHORT, LESTER L. Neotropical-Afrotropical Barbet and Woodpecker Radiations: A Comparison 559
SKUTCH, ALEXANDER F. Clutch Size, Nesting Success, and Predation on Nests of Neotropical Birds, Reviewed 575
WUNDERLE, JOSEPH M., JR., AND KENNETH H. POLLOCK. The Bananaquit-Wasp Nesting Association and a Random Choice Model 595

COMMUNITY AND POPULATION ECOLOGY 605
CRUZ, ALEXANDER, TIMOTHY MANOLIS, AND JAMES W. WILEY. The Shiny Cowbird: A Brood Parasite Expanding its Range in the Caribbean Region 607
FAABORG, JOHN. Ecological Constraints on West Indian Bird Distributions 621
GOCHFELD, MICHAEL. Numerical Relationships Between Migrant and Resident Bird Species in Jamaican Woodlands 654
KUSHLAN, JAMES A., GONZALO MORALES, AND PAULA C. FROHRING. Foraging Niche Relations of Wading Birds in Tropical Wet Savannas 663
MUNN, CHARLES A. Permanent Canopy and Understory Flocks in Amazonia: Species Composition and Population Density 683
POWELL, GEORGE V. N. Sociobiology and Adaptive Significance of Interspecific Foraging Flocks in the Neotropics 713
REMSEN, J. V., JR. Community Organization and Ecology of Birds of High Elevation Humid Forest of the Bolivian Andes 733
STILES, F. GARY. Seasonal Patterns and Coevolution in the Hummingbird-Flower Community of a Costa Rican Subtropical Forest 757
WILLARD, DAVID E. Comparative Feeding Ecology of Twenty-two Tropical Piscivores 788
WRIGHT, S. JOSEPH, JOHN FAABORG, AND CLAUDIA J. CAMPBELL. Birds Form Tightly Structured Communities in the Pearl Archipelago, Panama 798

EVOLUTIONARY AND BEHAVIORAL ECOLOGY 815
FOSTER, MERCEDES S. Pre-nesting Cooperation in Birds: Another Form of Helping Behavior 817
FRAGA, ROSENDO M. Host-Parasite Interactions Between Chalk-browed Mockingbirds and Shiny Cowbirds 829
GREENBERG, RUSSELL, AND JUDY GRADWOHL. A Comparative Study of the Social Organization of Antwrens on Barro Colorado Island, Panama 845
HOUSTON, DAVID C. Evolutionary Ecology of Afrotropical and Neotropical Vultures in Forests 856
MOERMOND, TIMOTHY C., AND JULIE SLOAN DENSLOW. Neotropical Avian Frugivores: Patterns of Behavior, Morphology, and Nutrition, with Consequences for Fruit Selection 865
ROBINSON, SCOTT K. The Yellow-rumped Cacique and its Associated Nest Pirates 898

SHERRY, THOMAS W. Adaptation to a Novel Environment: Food, Foraging, and Morphology of the Cocos Island Flycatcher 908

THOMAS, BETSY TRENT. Coexistence and Behavior Differences Among the Three Western Hemisphere Storks 921

BREEDING BIOLOGY 933

ESCALANTE, RODOLFO. Taxonomy and Conservation of Austral-breeding Royal Terns 935

HUMPHREY, P. S., AND B. C. LIVEZEY. Nest, Eggs, and Downy Young of the White-headed Flightless Steamer-duck 945

MASON, PAUL. The Nesting Biology of Some Passerines of Buenos Aires, Argentina 954

CONSERVATION 975

COATS, SADIE, AND WILLIAM H. PHELPS, JR. The Venezuelan Red Siskin: Case History of an Endangered Species 977

FFRENCH, RICHARD. Changes in the Avifauna of Trinidad 986

GARRIDO, ORLANDO H. Cuban Endangered Birds 992

HILTY, STEVEN L. Distributional Changes in the Colombian Avifauna: A Preliminary Blue List 1000

RAPPOLE, JOHN H., AND EUGENE S. MORTON. Effects of Habitat Alteration on a Tropical Avian Forest Community 1013

OVERVIEW 1023

PARKES, KENNETH C. Neotropical Ornithology—An Overview 1025

INDEX 1038

LIST OF COLOR ILLUSTRATIONS

FRONTISPIECE: Selected members of the *Thryothorus euophrys* complex iv

PLATE I. Painted Parakeet (*Pyrrhura picta eisenmanni*) 16

II. Fiery-throated Hummingbird (*Panterpe insignis*) 22

III. Collared Inca Hummingbird (*Coeligena torquata eisenmanni*) 40

IV. Rufous Flycatcher (*Myiarchus semirufus*) and its nest and eggs 360

V. White-capped Tanager (*Sericossypha albocristata*) 382

VI. Selected taxa of the *Ochthoeca diadema* species-group 430

VII. Downy young of *Tachyeres* 944

EUGENE EISENMANN AND THE STUDY OF NEOTROPICAL BIRDS

THOMAS R. HOWELL

Department of Biology, University of California Los Angeles, Los Angeles, California 90024 USA

Eugene Eisenmann was born in Panama in 1906 and lived there for his first 10 years. His father was born in Philadelphia, his mother in Panama, and his family was prominent in Panamanian affairs; Gene always retained a deep interest and affection for his native country. He came to the United States for his early schooling, and after graduation from DeWitt Clinton High School in New York City, he entered Harvard College in 1923. He continued at Harvard Law School, and after receiving his degree, commenced legal practice with a prestigious New York firm with which he retained a long affiliation as Senior Partner. He was secretly proud of having argued successfully before the U.S. Supreme Court. In 1956 he retired from his law firm to assume a full-time career in ornithology as Research Associate at the American Museum of Natural History; this continued until his death on October 16, 1981.

Those are the minimal biographical facts about a remarkable person who has had a profound impact on ornithology in the Americas and influenced generations of investigators of the birds of that region. By his own account, Eugene Eisenmann's early and casual interest in birds did not take a more serious turn until 1919 at a summer camp in the Adirondacks, when he was only 13. Before a visit to Panama in 1923, he prepared his own field guide consisting of crayon drawings based on Ridgway's and Salvin and Godman's volumes and modelled after the pictures in Chester Reed's primitive booklets. His interest languished during his college and law school years, only to revive after he had settled into legal practice in the late 1930's. This revival was sparked by his participation in National Audubon Society field trips, acquaintances with John Bull, Allan Cruickshank, and Roger Tory Peterson, and especially membership in the Linnaean Society of New York, which held meetings at the American Museum. In the wartime 1940's, attendance at the Society's meetings was small but included ornithologists of great distinction, among them, members of the museum staff. In addition to those previously mentioned, the participants included Dean Amadon, James Chapin, Joseph J. Hickey, Robert Cushman Murphy, and Ernst Mayr, who was then editor of the Society's publications. Eisenmann received great intellectual stimulation from a talk by Mayr on bird speciation, based on Mayr's forthcoming book, *Systematics and the Origin of Species* (1942). He was also deeply impressed with Hickey's and Mayr's views on the importance of the contributions of amateurs to serious ornithological work, and was to become one of the most important of the serious "amateurs," i.e., those originally pursuing a professional career in a field other than the biological sciences.

Eisenmann continued to practice law with great success, but he devoted more and more time to ornithology and longed to do so on a full-time basis. He visited Panama almost annually, and in 1952 his first major publication, *Annotated List of Birds of Barro Colorado Island, Panama Canal Zone* was issued. The prosaic-sounding title perhaps does not suggest a landmark contribution, but the study proved exceptionally valuable.

Barro Colorado Island, formerly a large hill rising out of the lowlands, was formed by the gradual flooding of Gatun Lake from 1912 to 1914. The new island was designated a biological reserve in 1923. Its potential for ornithological study was soon recognized, but Eisenmann's 1952 paper was the first comprehensive account of the birds of the island. For 27 years his paper constituted the principal ornithological source for research by zoogeographers, population ecologists, environmental physiologists, and other ornithologists studying the avifauna of this unique biological preserve. Since 1979, a revised edition of the island's avifauna, co-authored with Edwin O. Willis, has become the standard reference. The two publications are of particular importance because they provide for the recognition and documentation of changes in species composition and individual abundance in a rich natural environment that became an island at a precisely known time, thus providing an excellent testing ground for theories of island biogeography and population dynamics. Eisenmann's summarizing accounts

1

have provided the basic data for studies of this kind. James R. Karr (1982a, b) reviewed much of this literature in his own recent contributions, which use use Willis and Eisenmann's 1979 check-list as a baseline for studies of population variability and extinction in the Barro Colorado Island avifauna.

Gene's interest in the systematics and distribution of neotropical birds was also linked with his concern for providing appropriate English names for them. Following World War II, as modern transportation opened Middle America to visitors, the need for a concise, accurate, comprehensive check-list became imperative. Furthermore, a number of awkward, artificial, and sometimes absurd English names in Ridgway's and Hellmayr's works needed to be revised or replaced. Eisenmann's *The Species of Middle American Birds* (1955) was a major achievement. It provided, for the first time, a brief but comprehensive summary of the taxonomy and distribution of Middle American birds that could otherwise be obtained only by the most laborious abstraction from multi-volume works of which the earliest parts were long outdated. Furthermore, Eisenmann's principles for the use of English names were exemplified fully for the first time. These principles are described in his introduction to that small volume and need not be repeated here. Despite some criticism, which Gene readily accepted when justified, his basic ideas have prevailed and most of the Central American bird names he proposed have become standard. He did essentially the same for all of South America when assisting Meyer de Schauensee with his books on the birds of that continent. English-language names for birds are still the subject of intense feeling and controversy, and space in this account is insufficient for a full discussion of the subject. Suffice it to say that Eisenmann's contributions were important and lasting, and set the standards by which others' efforts are still judged.

More important is the contribution of his 1955 publication to the awareness and understanding of the systematics and distribution of Middle American birds. It would be facile to say that this was really just a compilation of existing data that any knowledgeable person could have done. The point is that if there were indeed any such equally knowledgeable persons, none had done it. Often in the history of science, a deceptively simple summation of existing information not previously available in that form constitutes a catalyst and a stimulus for a series of major advances in the field. In my opinion, this scholarly work accomplished that goal.

Eisenmann was fluent in Spanish, and this fact and his Panamanian background brought him in close contact with students and established ornithologists of Latin America. To all of them he gave assistance and advice. Almost throughout that region, ornithological libraries are decidedly limited, and Gene made special efforts to bring his correspondents and visitors up to date on the literature in their research fields. Conversely, he frequently informed North American ornithologists of new data from Latin America that appeared in Spanish-language publications of limited circulation or which were simply reported to him personally. He gave unstinting editorial advice and helped to put countless manuscripts into idiomatic English. He was always especially pleased to hear from persons who were gathering new and interesting data on some taxon or in some Neotropical region that was previously little studied, or who were publishing their first works in ornithology.

Eisenmann's contributions to the ornithological literature of Latin America went far beyond his own publications. He was a principal adviser to Rodolphe Meyer de Schauensee while the latter was preparing his books on the birds of Colombia (1964) and on the species of birds of South America (1966). Other books to which he contributed valuable assistance include Meyer de Schauensee and Phelps' *A Guide to the Birds of Venezuela* (1978) and, especially, Ridgely's *A Guide to the Birds of Panama* (1976), where Eisenmann advised on English names and also on taxonomic matters and field identification. His generous assistance was warmly acknowledged in these and many other publications, and he not only provided information used in the texts of various books but also contributed to the illustrations. Because of his familiarity with a vast number of neotropical species in the field, he was a helpful and discerning critic to artists who sometimes had to paint birds that they had seen only as museum specimens and about which they were uncertain on details of carriage, proportions, and field appearance.

Although not academically trained in biological science, Eisenmann was thoroughly at home with systematics in the broadest sense. He was a knowledgeable proponent of the biological species concept and a moderate in questions of lumping and splitting. In technical matters of nomenclature and interpretation of the International Code, his legal knowledge was put to expert use, and he served on the prestigious International Commission on Zoological Nomenclature. He readily disclaimed detailed knowledge of disciplines such as molecular genetics, biochemical physiology, and theoretical ecology, but he kept abreast of the hypotheses relevant to ornithology that developed in these fields and evaluated them objectively from the perspective of his own knowledge and experience. Predictably, he relied mainly on traditional criteria of morphology, behavior, and presence or absence of interbreeding as indications of relationship, but he was always ready to consider other kinds of evidence. He was truly impatient only with proposals based on insufficient data or on field studies that he considered too hastily done or including dubious species identifications. To one researcher who opined that it did not matter if a small forest bird was identified to the species level as long as one knew it was different from others, Eisenmann replied, "But you don't know if it's really different unless you've identified it!" Yet, even those with whom he disagreed most vigorously admired the breadth of his knowledge and respected his views.

Over a long period of time, the Frank M. Chapman Fellows in the Ornithology Department of the American Museum used the office adjacent to his. They often sought advice and criticism from him, and he, in turn, sought to broaden his knowledge and circle of acquaintances through them. He particularly liked to chat and discuss matters with newly trained ornithologists, as if to update and expand his own conceptual framework. He was influential in the training of a long list of Chapman Fellows, including Charles Collins, Joel Cracraft, Julian Ford, William George, Richard Johnston, Robert Selander, Lester Short, Paul Slud, Gary Stiles, Hans Winkler, Edwin Willis, and Larry Wolf. They found him ever ready to listen, and ever a gentleman. Always he gave more of himself than he asked or expected of others.

His influence in neotropical ornithology is well known, but it would do his memory a disservice to think of him that restrictedly. Rarely put off by anyone, always a seeker of truth, he was repeatedly sought out at international meetings, whether of ornithologists or of conservationists or members of other disciplines. His broad knowledge of birds and his expertise in nomenclature, along with his personal charm, made him an attraction at many an ornithological gathering. At the 1969 Pan-African Ornithological Congress in Kruger Park, on his first and only trip to Africa, he became an instant celebrity as he sought out new birds with gusto, making comparisons with American counterparts, and entering into enthusiastic discussions with African colleagues.

Eugene Eisenmann's major ornithological papers were few, but his influence extended far beyond his publications. Everyone who has worked on the systematics, behavior, or distribution of New World birds benefitted from the sharing of his encyclopedic knowledge of these subjects and from the personally acquired information, the insights, the seminal ideas, and the always courteously phrased criticisms that he provided in correspondence. For those fortunate enough to have visited him in his office, the long discussions in which he enthusiastically participated were invaluable. The ornithological honors accorded him are many. Among the most significant were his offices in the Linnaean Society, his service on the Council of the American Ornithologists' Union, his editorship of The Auk in 1958–1959, Vice-Presidency of the A.O.U. in 1967–1969 (he declined to be put in nomination for the Presidency), and, especially, his chairmanship of the Committee on Classification and Nomenclature (better known as the A.O.U. Check-list Committee). As a member of that committee, I can testify to his leadership and particularly his ability to deal reasonably and fairly with the sometimes acrimonius differences of opinion that developed during our meetings. In these situations his legal training served him well, as did his warm and friendly personality and especially his wit and wisdom.

The numbers and kinds of papers in this volume speak more eloquently of his influence, his scholarship, and the personal esteem in which he was held than any introduction possibly could. I fear that we shall not see his like again, but we can rejoice in the privilege of having known him, and continue to profit from his lasting contributions to ornithology. Fittingly, his life and work have continued to contribute by providing the impetus for this landmark collection of papers on neotropical ornithology.

LITERATURE CITED

EISENMANN, E. 1952. Annotated list of birds of Barro Colorado Island, Panama Canal Zone. Smithson. Misc. Collect. 117:1–62.

EISENMANN, E. 1955. The species of Middle American birds. Trans. Linn. Soc. N.Y. 7:1–128.

KARR, J. R. 1982a. Avian extinctions on Barro Colorado Island, Panama: a reassessment. Am. Nat. 119:220–239.

KARR, J. R. 1982b. Population variability and extinction in the avifauna of a tropical land bridge island. Ecology 63:1975–1978.

MAYR, E. 1942. Systematics and the origin of species. Columbia University Press, New York.

MEYER DE SCHAUENSEE, R. 1964. The Birds of Colombia. Livingston Publ. Co., Narbeth, Pennsylvania.

MEYER DE SCHAUENSEE, R. 1966. The Species of Birds of South America and Their Distribution. Livingston Publ. Co., Narbeth, Pennsylvania.

MEYER DE SCHAUENSEE, R., AND W. H. PHELPS, JR. 1978. A Guide to the Birds of Venezuela. Princeton University Press, Princeton, New Jersey.

RIDGELY, R. S. 1976. A Guide to the Birds of Panama. Princeton University Press, Princeton, New Jersey.

WILLIS, E. O., AND E. EISENMANN. 1979. A revised list of the birds on Barro Colorado Island, Panama. Smithson. Contrib. Zool. 291:1–31.

NEW TAXA

PARKER, THEODORE A., III, AND JOHN P. O'NEILL. A New Species and a New Subspecies of *Thryothorus* Wren from Peru 9

DELGADO B., FRANCISCO S. A New Subspecies of the Painted Parakeet (*Pyrrhura picta*) from Panama 17

STILES, F. GARY. Geographic Variation in the Fiery-throated Hummingbird, *Panterpe insignis* 23

STORER, ROBERT W., AND THOMAS GETTY. Geographic Variation in the Least Grebe (*Tachybaptus dominicus*) 31

WESKE, JOHN S. A New Subspecies of Collared Inca Hummingbird (*Coeligena torquata*) from Peru 41

A NEW SPECIES AND A NEW SUBSPECIES OF *THRYOTHORUS* WREN FROM PERU

THEODORE A. PARKER, III AND JOHN P. O'NEILL

Museum of Zoology, Louisiana State University, Baton Rouge, Louisiana 70803-3216 USA

ABSTRACT. Field and museum work reveal that the *Thryothorus euophrys* complex includes two species, one of which is described here, and that there is an additional subspecies of *T. euophrys*, also described in this paper. Both new forms are from Peru.

RESUMEN. Trabajos llevados a cabo en el campo y en museos muestran que el complejo taxonómico de *Thryothorus euophrys* incluye dos especies, una de las cuales es descrita en el presente trabajo, y que existe una subespecie adicional de *T. euophrys*, la cual también se describe. Estas dos nuevas formas taxonómicas se encuentran en Perú.

Fieldwork in previously unexplored regions of the Peruvian Andes continues to reveal numerous undescribed species and subspecies of birds (O'Neill and Parker 1981; Schulenberg and Williams 1982; Graves et al. 1983). In this paper we report the discovery of a new species and a new subspecies of wren in the *Thryothorus euophrys* complex. As previously recognized, *T. euophrys* was represented by three subspecies, inhabitants of bamboo thickets in the high-elevation Andean cloud forests of southwestern Colombia, eastern Ecuador, and extreme northern Peru north of the arid Marañón River valley. The two new forms occur as disjunct, isolated populations in the high mountains to the south of this valley in the eastern cordilleras of Peru.

NEW SPECIES

In 1965, O'Neill and the late George H. Lowery, Jr., visited the Inca ruins of Macchu Picchu and repeatedly saw a rufous-backed, stripe-breasted *Thryothorus* wren that resembled *T. euophrys* of Colombia and Ecuador. Not until 1974 were specimens obtained. They were obviously strikingly different from *T. euophrys,* and clearly represented a new form. We have determined that this well-marked form from the Department of Cuzco should be named as a full species that we are pleased to call:

Thryothorus eisenmanni, new species
INCA WREN

Holotype.—Louisiana State University Museum of Zoology No. 78913, adult male from San Luís on Ollantaitambo-Quillabamba road, above Huyro, 13°06′S,72°25′W, elevation "9000 ft" [= 2744 m], Department of Cuzco, Peru, 4 August 1974; collected by John P. O'Neill, original number 4878.

Diagnosis.—A relatively small, strikingly marked, sexually dimorphic member of the *Thryothorus euophrys* complex most closely resembling *T. e. euophrys* of the western Andes of Colombia and Ecuador (frontispiece), but differing from that form by having the crown and nape black, rather than dull brown mottled and bordered with dusky black, by being somewhat brighter, more russet-brown above, by having ventral black markings in the form of broad streaks rather than barlike spots, by having the belly markings in the form of broad streaks rather than barlike spots, and by having the belly also streaked black and white rather than nearly concolor with the flanks and with only a few obsolete spots. *Thryothorus e. longipes* of the eastern Andes of Colombia and Ecuador is even less marked below and has an almost completely brown crown; *T. e. atriceps* of northern Peru and a new form of *T. euophrys* from south of the Marañón River in northern Peru, described later in this article, are both unmarked below.

Description of type.—Crown, nape, lores, moustachial streak, and upper half of ear coverts dull black; broad supercilium white; lower half of ear coverts white, finely streaked with black; mantle, scapulars, back, rump, and upper tail coverts bright russet-brown (nearest Hazel of

Ridgway 1912); tail of the same general color but slightly duller, and with obscure dusky barring crossing both webs of each rectrix; throat white; breast and all but lowest portion of belly white, boldly streaked with black, these streaks formed by feathers that are white with two large black spots on each web and a white tip; sides, flanks, and undertail coverts dull yellowish brown (nearest Tawny Olive of Ridgway 1912); tertials, secondaries, primaries, and greater primary covert blackish brown edged with russet-brown; greater, median, and lesser secondary coverts russet-brown; soft part colors (in life): maxilla, dark brown; mandible, silver-blue with blackish tip; feet and tarsi, grayish-horn; iris color not recorded; weight 27.0 g.

Measurements of type (mm). — Wing, unflattened 62.1; tail 56.8; tarsus 26.9; culmen from base 20.1.

Variation. — Birds of both sexes vary in three ways. The variation tends to be parallel, perhaps being age related. Males with the palest crowns (charcoal gray rather than almost pure, dull black, have the palest backs (more orange, less red), the most heavily barred tails, and have the whitest underparts (the black streaks on the breast are fairly wide, but those on the belly are thinner, much less obvious). The series contains only birds with fully pneumatized skulls or those lacking data on skull pneumatization, but none that are obviously immature. Only two females (LSUMZ 78916, 78919) have bellies that are obviously streaked; in "typical" females the bold black and white streaks are confined to the breast, with the belly a warm gray-brown spotted lightly with a mixture of dusky and dirty white. All females have charcoal gray crowns with nearly black foreheads and crown margins. Only one (LSUMZ 78916, skull 100% ossif.) has an obviously barred tail. Back color of females varies only slightly, being a bit more orange in some than others. All specimens with iris color data had "chestnut" or "reddish chestnut" eyes. The maxilla is described as "horn" or "black," and the mandible as "silver-blue," or "blue-gray," sometimes with a black tip. The feet and tarsi are described as "gray," "grayish horn," or "gray-black." For variation in weights and measurements see Table 1.

Range. — Known from 1830 to 3350 m elevation in subtropical and temperate forests of the valleys of the Río Urubamba (Machu Picchu), Río Santa María (San Luís, Bosque Aputinye above Huyro), and Río Mapitunari (Cordillera Vilcabamba), all in the Department of Cuzco; not found in similar habitats and elevations in the Departments of Ayacucho to the north or Puno to the south that have been extensively investigated.

Specimens examined. — Peru (all Depto. Cuzco). San Luís:3 ♂♂ (LSUMZ 78913–78915), 4 ♀♀ (LSUMZ 78619–78919); Bosque Aputinye, 2 ♂♂ (LSUMZ 78921; 78923), 2 ♀♀ (LSUMZ 78922; 78924); Hda. Huyro [= Bosque Aputinye], 1 ♂ (LSUMZ 78920); Machu Picchu, 1 ♂, 1 ♀ (LSUMZ 78912–78913); Cordillera Vilcabamba, 1 ♂ imm. (AMNH 820524).

Etymology. — It is a great pleasure to name this distinctive wren for the late Eugene Eisenmann in honor of his multitude of contributions to Neotropical ornithology. We deeply regret that he is not alive to receive this honor. It is appropriate that he did, however, see the bird at Machu Picchu on his trip to Peru with Robert Ridgely in 1969.

ECOLOGY AND BEHAVIOR

Along one well-studied elevational gradient in the Huayopata Valley (Bosque Aputinye, San Luís, Canchailloc), *Thryothorus eisenmanni* has been found from 1830 to 3350 m. To our knowledge, no other *Thryothorus* occurs as high as *eisenmanni* and *euophrys*, although the closely related *T. coraya* ranges up to 1850 m in the Carpish Mountains, Depto. Huánuco (pers. observ.), and to 2050 m at Santa Cruz, 9 km SE Oxapampa, Depto. Pasco (specimens, LSUMZ).

Forest within the elevational range of *T. eisenmanni* varies from tall (25–30 m canopy) at 1800–2500 m, to stunted (4–8 m canopy) at 3000–3500 m. Above 2000 m, arboreal epiphytic growth of mosses and bromeliads is particularly abundant. In all known localities *T. eisenmanni* is locally common in dense thickets of *Chusquea* bamboo on moderate to steep slopes in cloud forest. These wrens seem to avoid dense, well-shaded forest, and are usually noted in bamboo along road edges and natural landslides. *Chusquea* typically grows in monospecific stands that average 3 to 5 m tall at 2500 m; the wrens are most often noted from 0.5 to 1.5 m above ground in the nearly impenetrable lower branches of this vegetation. They move in pairs or groups of three to six individuals, and frequently probe clusters of stems at the internodes of bamboo stalks and also search dense foliage and curled, dead leaves trapped in the tangled crowns of thickets. The latter foraging behavior is typical of many *Thryothorus*

TABLE 1

SELECTED MEASUREMENTS OF MEMBERS OF THE *THRYOTHORUS EUOPHRYS* COMPLEX[1]

	Wing	Tail	Tarsus	Culmen base	Weight (g)
	Male				
T. euophrys euophrys	67.8	66.1	28.7	22.9	
	64.0–71.2	62.4–69.0	27.2–29.2	22.2–23.8	–
	(7)	(6)	(7)	(8)	
T. e. longipes	70.7	66.1	30.5	23.2	
	69.4–73.5	60.8–72.5	28.3–34.6	21.7–24.2	–
	(8)	(8)	(8)	(8)	
T. e. atriceps	68.2	68.2	29.2	22.1	27.5
	67.6–68.6	67.6–69.8	28.9–29.6	20.8–23.2	27.5–27.5
	(3)	(3)	(3)	(4)	(3)
T. e. schulenbergi	72.7	64.7	30.9	22.9	36.0
	72.6–72.8	63.3–66.0	30.6–31.2	22.5–23.2	34.0–38.0
	(2)	(2)	(2)	(2)	(2)
T. eisenmanni	63.5	59.2	26.1	20.9	24.6
	61.7–66.2	55.4–61.8	24.8–26.9	20.1–21.4	22.0–27.0
	(8)	(8)	(8)	(8)	(5)
	Female				
T. euophrys euophrys	66.5	63.8	27.8	22.3	
	64.1–73.4	58.4–68.6	26.8–28.6	21.1–22.5	–
	(14)	(12)	(13)	(14)	
T. e. longipes	68.3	62.7	28.7	22.3	
	63.2–71.3	61.2–64.2	27.9–29.2	21.4–23.5	–
	(4)	(4)	(4)	(3)	
T. e. atriceps	64.6	62.6	29.3	22.7	26.5
	63.6–65.6	61.5–63.6	28.6–29.9	22.5–22.8	26.0–27.0
	(2)	(2)	(2)	(2)	(2)
T. e. schulenbergi	67.2	62.7	28.8	23.4	31.5
	64.1–70.2	62.1–63.2	28.5–29.0	22.6–24.1	29.5–33.5
	(2)	(2)	(2)	(2)	(2)
T. eisenmanni	59.7	56.6	25.4	19.8	21.7
	57.8–62.4	53.6–59.8	22.9–27.9	19.0–20.7	19.0–23.0
	(6)	(6)	(6)	(6)	(4)

[1] Adults only; values given in descending order are means, ranges, and sample size (in parentheses).

species, especially *T. genibarbis* and *T. coraya,* which are closely related to the *T. euophrys* complex (Remsen and Parker 1984). Known prey items (from stomach contents and field observations) include small beetles (Coleoptera), caterpillars (Lepidoptera) and roaches (Blattoidea). As with other *Thryothorus* wrens (pers. observ.), *eisenmanni* is not a regular participant in mixed-species flocks.

This relatively large wren apparently has no close competitors, although the following green foliage and dead-leaf searchers occur in the same bamboo thickets: *Cranioleuca marcapatae, Schizoeaca helleri, Basileuterus luteoviridis, Hemispingus parodii, H. atropileus* and *Catamblyrhynchus diadema* (Parker and O'Neill 1980). The latter three, like *Thryothorus eisenmanni,* are almost entirely restricted to bamboo, as are many bird species in the Neotropics (Parker 1982). At the lower limit of the elevational range of *eisenmanni,* the smaller Gray-breasted Wood-Wren (*Henicorhina leucophrys*) is also common in forest undergrowth but shows no preference for bamboo.

During our June–September fieldwork in 1974 (Parker and O'Neill 1980), *Thryothorus eisenmanni* was not found in suitable bamboo habitat above 2800 m, but in October 1980, following large scale flowering and subsequent die-off of *Chusquea* that was curiously restricted to a 2450 to 2900 m elevational zone, the wren was conspicuous in higher elevation forest at 3350 m. The species apparently had abandoned the brown, nearly leafless thickets and had moved up (and perhaps down) into more verdant habitats. Such movements may explain why *T. eisenmanni* was not collected during the years of intensive biological study following the discovery of the Machu Picchu ruins (Chapman 1921). Bamboo also may not have been common there before the ruins were largely cleared of forest growth.

Vocalizations.—Pairs of *Thryothorus eisenmanni* perform well-synchronized, antiphonal duets that characterize nearly all *Thryothorus* (Farabaugh 1982). These duets consist of rapidly uttered series of whistles that rise and fall in a wide frequency range of 1 to 8 kH. One bird, probably the smaller female, gives a steady succession of whistles including an unusually high-pitched note that occurs at intervals of 1.5 sec, whereas the other sings a continuous series of closely spaced notes on a lower pitch. Song bouts typically last from 15 to 30 sec, with some continuing for up to 60 sec. Song is first heard shortly after dawn, and then sporadically through the day. One singing pair is almost instantly answered by several others on adjacent territories.

Songs of *T. eisenmanni* differ from those of *T. euophrys atriceps* and *T. e. longipes* in several ways. The highest notes of *eisenmanni* peak at ca. 8 kH, whereas the corresponding notes of *atriceps* peak at ca. 4 kH (Fig. 1). Typical *eisenmanni* song consists of 13 to 15 notes per 2.5 sec (including male and female contributions), whereas *atriceps* song consists of only 8 to 10 notes per 2.5 sec. There is more overlap between individual notes in *eisenmanni* song, and this undoubtedly contributes to the less synchronized sound of song in this species. In addition to being generally lower-pitched, songs of *T. e. atriceps* and *T. e. longipes* are slower and more rhythmic than those of *T. eisenmanni*. In terms of sound quality, songs of *T. euophrys* are more reminiscent of *T. coraya* and *T. genibarbis* than of *T. eisenmanni*.

Vocal differences between *T. eisenmanni* and other members of the *T. euophrys* complex may serve as reproductive isolating mechanisms. To test this hypothesis Parker and M. Braun carried out a series of playback experiments with *T. euophrys atriceps* in extreme northern Peru. On four mornings in July 1980 they tested the responses of one or two pairs of *atriceps* to tapes of *T. euophrys longipes* of eastern Ecuador and of *T. eisenmanni* of southern Peru. On the first and third mornings, 3 min of uninterrupted *T. e. longipes* song was played, followed by a pause of 3 min, and then 3 min of *eisenmanni* song. The order was reversed on the second and fourth mornings (not four consecutive days). Both members of *T. e. atriceps* pairs consistently sang within 10 to 30 sec of hearing *T. e. longipes* song, and only after 2.0 to 3.0 min of *eisenmanni* song, or not at all. Clearly, individuals of *atriceps* were able to differentiate between songs of the two forms, which supports our taxonomic decision to regard *eisenmanni* as a full species. Lanyon (1978) demonstrated that some closely related *Myiarchus* flycatchers respond weakly to playbacks of each others' songs. Parker was able to stimulate a low-level vocal response in *eisenmanni* by playing songs of the Gray-breasted Wood-Wren (*Henicorhina leucophrys*).

NEW SUBSPECIES

As mentioned above, Louisiana State University Museum of Zoology personnel have collected a series of specimens from two localities south of the arid Río Marañón valley, that closely resemble *T. euophrys atriceps,* but are easily separable by their overall dull coloration and large size. We are pleased to name them:

Thryothorus euophrys schulenbergi, new subspecies

Holotype.—Louisiana State University Museum of Zoology No. 88579, adult male from Cordillera Colán, SE La Peca, ca. 5°34'S, 78°19'W, elevation 2713 m, Department of Amazonas, Peru, 12 October 1978; collected by Thomas S. Schulenberg, original number 1196.

Diagnosis.—The largest and dullest member of the *Thryothorus euophrys* complex (Frontispiece); most similar to *T. e. atriceps,* from which it differs by its grayer crown and nape, pale gray rather than white superciliaries, grayish-white rather than pure white throat, much less russet dorsum and wings, and grayer, less brown posterior underparts.

Description of the type.—Crown, nape, lores, moustachial streak, and upper half of ear coverts dull grayish black; broad superciliaries and lower half of ear coverts dull medium gray; mantle, scapulars, back, rump, and upper tail coverts dull rusty olive-brown (nearest Cinnamon Brown of Ridgway 1912); tail of the same general color with the edges of the feathers slightly brighter than the centers and with no trace of barring; throat dull grayish white; breast and middle of upper belly dull brownish gray (nearest Drab Gray of Ridgway 1912); lower belly and flanks dull brown (nearest Saccardo's Umber of Ridgway 1912); tertials,

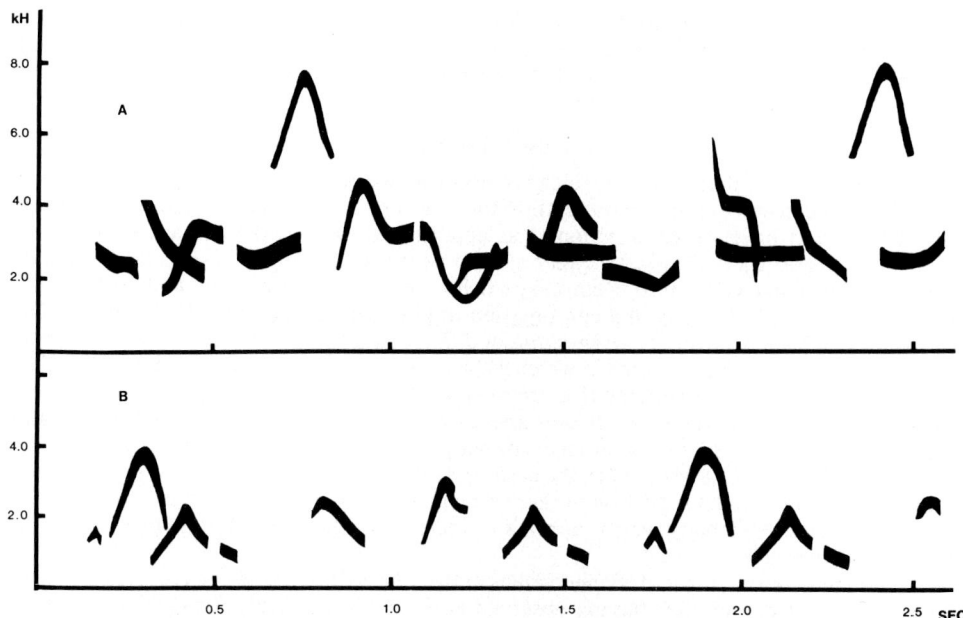

Fig. 1.   Sonograms of portions of songs of *Thryothorus eisenmanni* (A), and *T. euophrys atriceps* (B), to show differences in frequency and duration.

secondaries, primaries, and greater primary coverts blackish brown edged with rusty olive-brown; greater, median, and secondary coverts same rusty olive-brown but slightly brighter than mantle; soft part colors (in life); irides, brown; maxilla, black; mandible, blue-gray; feet and tarsi, pinkish brown; weight 38.0 g.

Measurements of type (mm). — Wing, unflattened 72.6; tail 63.3; tarsus 30.6; culmen from base 23.2.

Variation. — The four adults are all fairly uniform in color. One female has a clear gray crown with only the forehead and edges of the crown deep charcoal. The type (LSUMZ 88579) is somewhat browner on the lower belly than the others. The one immature bird (LSUMZ 104540) is described above. The iris color is described as "brown" or "reddish brown." The maxilla is described as "slate," "gray," or "black," the mandible is recorded as "blue-gray," or "silver-blue." The tarsi and feet are recorded as "pinkish brown," "horn," or "pale gray-white."

This very large wren is the only member of the *T. euophrys* complex to have upperparts that are not distinctly reddish or rusty in general coloration. Its plumage is characterized by having a dull or "dirty" overall tone. It lacks the clean white throat and clear gray upper breast of *T. e. atriceps* from which it is isolated by the arid Río Marañón valley, and is suffused with gray on the lower breast rather than with the clean buffy brown of *T. e. atriceps*. Females differ little from males except in averaging smaller, and one has the mid-crown and nape purer gray than the darker forehead and borders of males. One immature male (LSUMZ 104540) has the same overall pattern as an adult, but is even duller and has a dull grayish brown crown and nape, much browner anteriorly, a pale buff throat, and, in life, had a slate rather than black maxilla, a yellow rather than blue-gray mandible, and gray-pink rather than a pinkish brown or horn-colored tarsi and feet. It is also quite small (28.2 g; unflattened wing 63.1 mm).

Range. — Known only from the Cordillera Colán in Depto. Amazonas, and from Puerta del Monte in Depto. San Martín, Peru, from 2600 to 3250 m. At both localities it inhabits thickets of *Chusquea* bamboo.

Specimens examined. — Peru: Cordillera Colán, from 2623 to 2747 m SE of La Peca, Depto. Amazonas, 1 ♂ (LSUMZ 88579), 2 ♀♀ (LSUMZ 88577–88578); Puerta del Monte, ca. 30 km NE de Los Alisos, Depto. San Martín at 3250 m, 2 ♂♂ (LSUMZ 104540–104541).

Etymology.—It is our pleasure to name this well marked subspecies in honor of Thomas S. Schulenberg, whose efforts on the Cordillera Colán expedition, as well as in the overall program of research at LSU, have produced a tremendous amount of information on Peruvian birds.

DISCUSSION

There is little doubt that gene flow does not occur between *T. eisenmanni* and *T. euophrys schulenbergi,* the nearest *euophrys* population to the north. These taxa are separated by more than 650 km and have never been found at appropriate, well-worked localities along the Tayabamba-Ongón trail, Depto. La Libertad, the Carpish Mountains, Depto. Huánuco, or the Tambo-Apurímac valley trail, Depto. Ayacucho. *Thryothorus eisenmanni* differs from any member of the *euophrys* group in a combination of plumage, size, and vocal characters. The plumage similarities between *T. eisenmanni* and *T. e. euophrys* exemplify a fairly common pattern of variation in Andean birds in which geographically distant relatives are more similar-looking than those found in between (Remsen 1984). Both are heavily marked below, whereas geographically intermediate *T. e. atriceps* and *T. e. schulenbergi* are unmarked below, and the latter appears to be the largest in terms of wing length and body weight. Variation in the populations of *T. euophrys* is clinal in the eastern Andes (*longipes, atriceps,* and *schulenbergi*) with a north-south parallel reduction in the reddish coloration of the upperparts and spotting on the underparts. Also body weight appears to increase to the south, but a larger sample is needed to confirm this.

After having looked at most of the specimens of the *Thryothorus euophrys* complex in North American museums, and after having observed all forms in the field, except *T. e. euophrys* of the western Andes of Colombia and Ecuador, we conclude that the complex is best divided taxonomically into two species, one polytypic with four races, *T. euophrys euophrys, T. e. longipes, T. e. atriceps,* and *T. e. schulenbergi,* and monotypic *T. eisenmanni.* All members of the complex are inhabitants of high-elevation *Chusquea* bamboo thickets and do not occur syntopically with any other member of the genus.

OTHER SPECIMENS EXAMINED

T. e. euophrys.—*COLOMBIA:* Nariño; Mayasquer, 4 ♂♂ (ANSP 149994, 149996–149998), 3 ♀♀ (ANSP 149999–150001). *ECUADOR:* Liqui, 1 ♂ (ANSP 59702). Hda. Garzón, 2 ♀♀ (ANSP 59700–59701).

T. e. longipes.—*ECUADOR:* El Tablón, 1 ♂ (ANSP 164087); Sumaco arriba, 2 ♂♂ (ANSP 83610–83611); Planchas, 1 ♂ (LSUMZ 83496), 1 ♀ subad. (LSUMZ 83495).

T. e. atriceps.—*PERU:* Depto. Cajamarca; E. slope Cerro Chinguela, 3 ♂♂ (LSUMZ 97790–97791, 97793), 2 ♀♀ (LSUMZ 97789, 97792).

ACKNOWLEDGMENTS

A great number of people contributed to the success and completion of this project. We continue to be most grateful for the continued support and enthusiasm of J. S. McIlhenny, B. M. Odom, L. Schweppe, I. Schweppe, W. Carter, and others who have sponsored our field program. Manuel and Isabel Plenge, Arturo and Helen Koenig, and Gustavo del Solar Rojas have all continued to give us a tremendous amount of help. In the field we owe most of our efficiency to our able assistants, M. Sánchez S., K. Wehr, and R. Rivera A., who often work under the most trying of climatic circumstances. We are also most appreciative of the efforts of our field colleagues, especially M. Braun and T. S. Schulenberg, who have taken a special interest in this project both in the field and in the museum. We are grateful to M. S. Foster, J. W. Hardy, W. E. Lanyon, E. S. Morton, and J. V. Remsen, who reviewed the manuscript. We thank W. E. Lanyon, American Museum of Natural History (AMNH), and F. B. Gill, Academy of Natural Sciences of Philadelphia (ANSP), for the loan of specimens. Museum work included in this project was financed in part by grants from the Frank M. Chapman Memorial Fund to O'Neill in 1975 and Parker in 1979. We continue to appreciate the collaboration of the Dirección General Forestal y de Fauna, and of INFOR, both of the Ministerio de Agricultura in Lima, under whose permission and encouragement our work in Peru takes place.

LITERATURE CITED

CHAPMAN, F. M. 1921. The distribution of birdlife in the Urubamba valley of Peru. U.S. Natl. Mus. Bull. 117:1–138.

FARABAUGH, S. M. 1982. The ecological and social significance of duetting. Pp. 85–124, *In* D. E. Kroodsma and E. H. Miller (eds.), Acoustic Communication in Birds. Academic Press, New York.

GRAVES, G. R., J. P. O'NEILL, AND T. A. PARKER, III. 1983. *Grallaricula ochraceifrons,* a new species of antpitta from northern Peru. Wilson Bull. 95:1–6.

LANYON, W. E. 1978. Revision of the *Myiarchus* flycatchers of South America. Bull. Am. Mus. Nat. Hist. 161:429–627.

O'NEILL, J. P., AND T. A. PARKER, III. 1981. New subspecies of *Pipreola riefferii* and *Chlorospingus ophthalmicus* from Peru. Bull. Br. Ornithol. Club 101:294–299.

PARKER, T. A., III. 1982. Observations of some unusual rainforest and marsh birds in southeastern Peru. Wilson Bull. 94:477–493.

PARKER, T. A., III, AND J. P. O'NEILL. 1980. Notes on little known birds of the upper Urubamba valley, southern Peru. Auk 97:167–176.

REMSEN, J. V., JR. 1984. High incidence of "leapfrog" pattern of geographic variation in Andean birds: implications for the speciation process. Science 224:171–173.

REMSEN, J. V., JR., AND T. A. PARKER. 1984. Arboreal dead-leaf-searching birds of the neotropics. Condor 86:36–41.

RIDGWAY, R. 1912. Color Standards and Color Nomenclature. Published by the author, Washington, D.C.

SCHULENBERG, T. S., AND M. D. WILLIAMS. 1982. A new species of antpitta (*Grallaria*) from northern Peru. Wilson Bull. 94:105–113.

PLATE I.   Painted Parakeet (*Pyrrhura picta eisenmanni*). From a painting by Arthur Singer.

A NEW SUBSPECIES OF THE PAINTED PARAKEET (*PYRRHURA PICTA*) FROM PANAMA

FRANCISCO S. DELGADO B.

Central Regional Universitario de Veraguas, Santiago de Veraguas, Republica de Panamá

ABSTRACT. *Pyrrhura picta eisenmanni,* a new subspecies, is described from a series of 11 specimens taken in the southern corner of the Azuero Peninsula, Los Santos Province, Panama. It resembles most closely *P. picta subandina* of northern Colombia and represents an extension of this mainly South American species to these mountains of Panama.

RESUMEN. Una nueva subespecie, *Pyrrhura picta eisenmanni* se describe en base a 11 especímenes coleccionados en el extremo sur de la Península de Azuero, en la provincia de los Santos, Panamá. Ella se asemeja más a la subespecie *P. picta subandina* del norte de Colombia y el hallazgo de estos ejemplares representa una extensión en el rango de distribución de esta especie primeramente sudamericana hasta esas montañas panameñas.

In 1979 I collected a parakeet in the southwest portion of the Azuero Peninsula (7°18′N, 80°43′W), in an area whose avifauna is, perhaps, the least known in Panama. I tentatively identified the specimen as *Pyrrhura picta,* the Painted Parakeet, a South American species that had not been previously recorded from Panama. I sent that and additional specimens to Eugene Eisenmann who confirmed the identification and also agreed with me that they differed in plumage coloration from the South American forms.

Here I provide some behavioral and morphological data on this population, which I name:

Pyrrhura picta eisenmanni, new subspecies

Holotype.—American Museum of Natural History No. 824181, adult male from Los Piraguales, El Cortezo de Tonosí, Los Santos Province, Azuero Peninsula, Republic of Panama, elevation 1050 m, 26 February 1979; collected by Francisco Delgado, original number 4.

Diagnosis.—A typical *Pyrrhura* parakeet of the *picta-leucotis* complex (Plate I). Closest in appearance to *P. picta subandina* of the Río Sinú in northern Colombia, the geographically nearest population (Todd 1947), but differing most obviously in its mainly sooty (not blue) crown, and in having less red on the forecrown, lores, and orbital area. *Pyrrhura picta eisenmanni* is also generally duller than *P. picta subandina,* has less maroon on the sides of the throat, and has a notably larger culmen and longer tail. Differs from *P. picta caeruleiceps* of northern Colombia also in lacking the blue crown, and in having only a vestige of red on the shoulders. The recently described subspecies *pantchenkoi* from Sierra de Perijá in Venezuela and Colombia (Phelps 1977) resembles *caeruleiceps* but is darker on the breast and crown (more dusky, less blue), and resembles *eisenmanni* in the relatively large culmen.

Description of the holotype.—Upperparts generally dull green. Forehead, crown, and nape dull sooty, slightly suffused with indistinct blue on the forecrown, and with a narrow blue streak on the side of the forehead. Narrow red frontal band extending to the lores and ocular region, becoming duskier red on the cheeks. Ear-coverts buffy-white. Chin white, the feathers of the throat and chest dull blackish, broadly edged with white. Breast dark greenish-blue, the feathers also broadly edged white, and with narrow lateral yellow tipping, giving a scalloped appearance. Patch of the center of the belly dull red, remainder of the lower underparts pale green. Lower back to upper tail coverts dull red. Shoulder very narrowly fringed red. Tail above dull red, becoming green toward its base; below dull brownish-red.

Soft part colors (in life): iris pale ochre, tarsi grayish-black, bill dull black, bare orbital skin dull sooty.

Measurements of the holotype.—Wing (chord) 115.5 mm; tail 112 mm; culmen from cere 15.6 mm; exposed culmen 17.7 mm; tarsus 13.6 mm.

Variation.—Inspection of 11 specimens demonstrated that they are all much alike in color, pattern on the head, cheeks, breast, belly, and tail. All differ in the amount of red feathering on the shoulders and have some feathers with red color on only one side of the rachis. This

TABLE 1

MEASUREMENTS OF SELECTED RACES OF *PYRRHURA PICTA*[1]

Subspecies	Wing chord	Tail length	Exposed culmen length	Tarsus length
eisenmanni				
♂♂ (n = 7)	116.5	115.3	17.2	11.4
♀♀ (n = 4)	118.7	117.0	16.6	11.5
subandina				
♂♂ (n = 9)	115.1	104.1	14.9	13.3
♀♀ (n = 12)	116.3	103.2	15.0	12.8
caeruleiceps				
♂♂ (n = 3)	123.0	112.7	15.0	12.7
♀♀ (n = 3)	121.0	114.3	14.3	13.3

[1] Value given = means in mm. Measurements of South American races from Forshaw (1973).

variation in color pattern seems typical of South American races and is possibly age related. Topotypical female AMNH 824933 (original number FD 8) measures: wing 116 mm; tail 112 mm; culmen from cere 14.7 mm; exposed culmen 16.5 mm (bill tip broken, but attached); tarsus 12.5 mm.

Specimens examined.—The type and two topotypes of *Pyrrhura picta eisenmanni* were compared by Eugene Eisenmann and K. C. Parkes with the types of *Pyrrhura picta subandina* and *P. picta caeruleiceps*. Eisenmann also supplied me with descriptions of other races of *P. picta*. Nine additional specimens of the subspecies *eisenmanni* that I collected are deposited in the collections of the Central Regional University of Veraguas.

Etymology.—I am pleased to name this subspecies for the late Eugene Eisenmann, who was the first ornithologist to be born in the Republic of Panama, and who worked for almost 50 years on various aspects of the neotropical avifauna, especially that of the Panamanian region.

DISCUSSION

Taxonomy.—While it appears that *Pyrrhura picta eisenmanni* is most like *P. picta subandina* of northern Colombia and less like *P. picta caeruleiceps* and *P. picta pantchenkoi*, at least in coloration (Table 1), the possibility of a specific relationship between *P. picta* and the closely related *P. leucotis* in eastern South America remains. These two allopatric species are widespread in the tropical lowlands of South America. No intergradation between the forms has been shown. The Azuero birds are the species *picta*; nevertheless, they approach *leucotis* in their distinctly whitish ear-coverts and especially in the somewhat less pointed and more rounded scalloping of the breast feather edging (approaching a more or less scaled effect) as Eugene Eisenmann first pointed out (*in litt.*).

Although more information is needed, Eisemann felt (*in litt.*) that *P. picta* and *P. leucotis* were possibly conspecific as was recently suggested (Smith 1982).

Behavior.—These parakeets are gregarious and almost invariably are seen in groups of two or three, or occasionally, as many as 20. They usually remain high in the forest canopy. Their flight is swift and direct, and during flights, they utter a short *eek*. I noted other vocalizations, a single loud *peea* voiced by solitary birds as they attempted to relocate their flocks, and a harsh, guttural *kleek-kleek,* given when the birds were perched. Perched birds also may utter soft, barely perceptible calls, often while preening each other.

Breeding information.—The specimens obtained in January–February 1979 and in January 1980 appeared to be in breeding condition, and also show pre-breeding molt in the wings. Males had enlarged testes (5–7 mm). In January 1980, on Los Tres Cerros at the southern end of their range, a group of three birds entered two holes on the underside of a large branch of a tall leafless wild cashew (*Anacardium excelsum*); the birds peered out of each hole for a few moments and then flew into the forest. It seems likely that they were prospecting for a nest site. I never located an active nest, but countrymen of the region told me that the parakeets lay eggs in March, and that fledged young can be seen in late March and April (the beginning of the local rainy season).

Habitat.—*Pyrrhura picta eisenmanni* has been found exclusively in the humid forests of

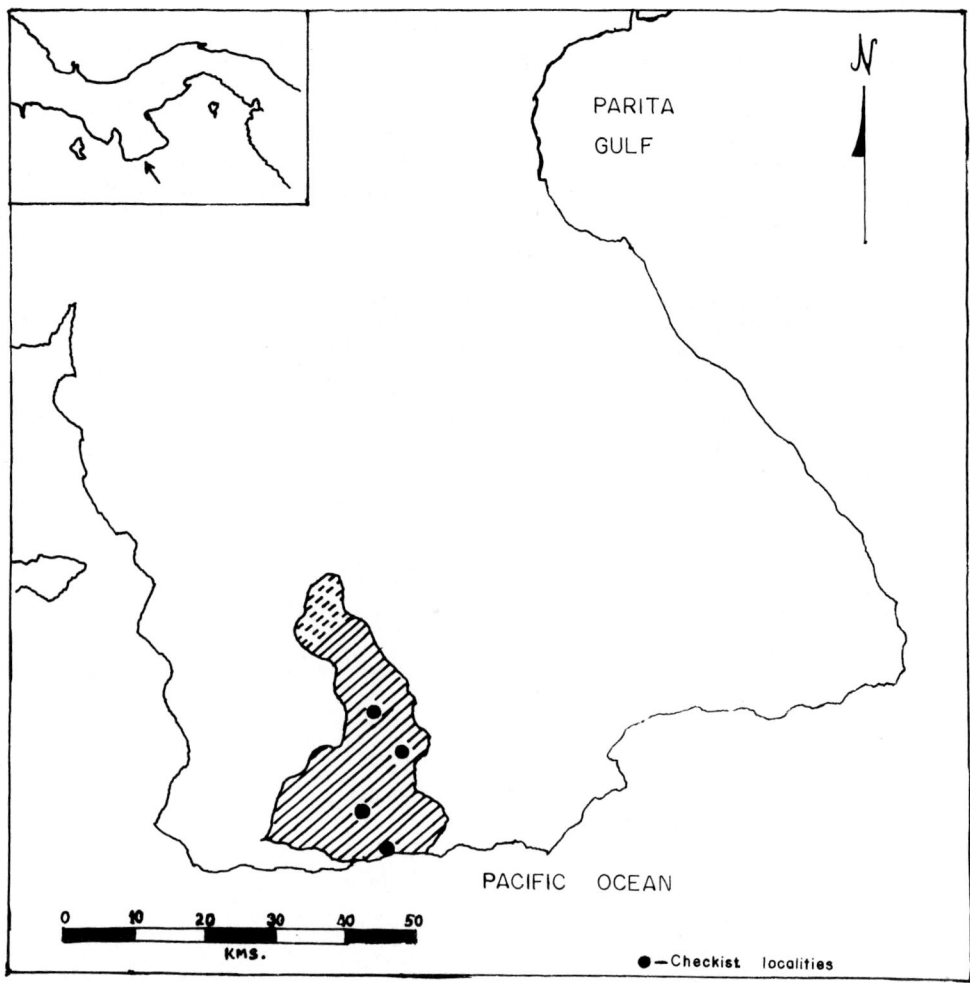

FIG. 1. Distribution of the Painted Parakeet (*Pyrrhura picta eisenmanni*) on the Azuero Peninsula (7°18′N, 80°43′W, Republic of Panama. Shaded region = known range of the species (dashed area = zone where good habitat remains). Dots indicate collecting localities.

the southwestern region of Azuero Peninsula, in Central Panama (Fig. 1). It regularly occurs at forest edge and, occasionally, in adjacent partly cleared areas. It may engage in limited elevational movements. Local residents at the lower end of its elevational range report that these birds are present only at certain seasons, and that they visit bean plantations. The overall range is probably increasingly restricted by deforestation.

Conservation. — The region where this parakeet is found still supports extensive forest, the last on the western Panamanian mainland south of the central mountains. The maximum elevation in the area is reached by Cerro Hoya (1660 m), also known as Los Tres Cerros. This area, while still little explored biologically, evidently supports a small endemic fauna of considerable interest (Myers 1969), of which this new parakeet is a good example.

In recent years the clearing of forest elsewhere on the Azuero Peninsula has proceeded rapidly, and many formerly forested areas have now been converted to pastures. Part of the range of *Pyrrhura picta eisenmanni*, especially along its northern and northeastern borders, has been recently cleared, and habitat destruction ultimately may threaten the existence of this form and many others. However, I am pleased to report that based on our field surveys, Panamanian authorities, by means of Law 74, passed on 2 October 1984, declared the Cerro Hoya region a national park.

ACKNOWLEDGMENTS

I am immensely grateful to the late Eugene Eisenmann who coordinated the comparison of my specimens with the holotypes and representative specimens of all the described subspecies of *Pyrrhura picta* in the United States. Thanks are also due to J. Forshaw, W. H. Phelps, Jr., K. C. Parkes, J. W. Aldrich, R. Laybourne, and R. S. Ridgely who examined specimens and compared them with the holotypes or topotypes. For assistance in putting the manuscript into acceptable scientific English, and for other aid in the description of this new form, I am particularly indebted to R. S. Ridgely, N. G. Smith, L. L. Short, and M. S. Foster.

LITERATURE CITED

FORSHAW, J. M. 1973. Parrots of the World. 1st ed. Doubleday, New York.

MYERS, C. W. 1969. The ecological geography of cloud forest in Panama. Am. Mus. Novit. No. 2396.

PHELPS, W. H., JR. 1977. Una nueva especie y dos nuevas subespecies de Aves (Psittacidae, Furnariidae) de la Sierra de Perija cerca de la divisoria Colombo-Venezolana. Bol. Soc. Venez. Cienc. Nat. 33(134):43–53.

SMITH, G. A. 1982. *Pyrrhura* conures. Parrot Soc. Mag. 16:365–372.

TODD, W. E. C. 1947. New South American parrots. Ann. Carnegie Mus. 30:331–338.

PLATE II.   Fiery-throated Hummingbird (*Panterpe insignis*). *Above: P. i. eisenmanni. Below: P. i. insignis.* From a watercolor painting by Dana Gardner.

GEOGRAPHIC VARIATION IN THE FIERY-THROATED HUMMINGBIRD, *PANTERPE INSIGNIS*

F. GARY STILES

Escuela de Biología, Universidad de Costa Rica, Ciudad Universitaria, Costa Rica, C.A.

ABSTRACT. Geographic variation in the measurements and coloration of *Panterpe insignis* is described, with particular reference to a newly discovered population on Volcán Miravalles in the Cordillera de Guanacaste of northwestern Costa Rica. This population, which extends the distribution of the species ca. 50 km to the northwest, has probably been isolated since the Pleistocene and is sufficiently differentiated to warrant recognition as *P. i. eisenmanni,* new subspecies. The ecology and annual cycle of the Volcán Miravalles population, and seasonal movements in other populations, are briefly described.

RESUMEN. Se describe la variación en las medidas y coloración de *Panterpe insignis* en relación con su distribución geográfica, refiriéndose especialmente a una población recientemente descubierta en el Volcán Miravalles de la Cordillera de Guanacaste en el noroeste costarricense. Esta población, cuyo descubrimiento extiende el rango de distribución de la especie aproximadamente 50 km hacia el noroeste, probablemente ha permanecido aislada desde el pleistoceno y se diferencia de manera suficiente como para garantizar el reconocimiento de *P. i. eisenmanni,* subesp. nova. Se describe brevemente la ecología y el ciclo anual de esta población del Volcán Miravalles, así como los movimientos estacionales de otras poblaciones.

The Fiery-throated Hummingbird, *Panterpe insignis,* is one of a number of bird species endemic to the high mountains of southern Central America (Slud 1964; Wolf 1976). The monotypic genus *Panterpe* is itself endemic, and although *P. insignis* shows a considerable resemblance in overall morphology and behavior to certain Andean genera such as *Metallura* and *Heliangelus* (R. L. Zusi, F. L. Ortiz-Crespo, pers. comm.; pers. observ.), its coloration is distinctive and its precise relationship obscure. Until recently the Fiery-throated Hummingbird was thought to be confined to the highest parts of the Cordillera Central of Costa Rica and the Cordillera de Talamanca of Costa Rica-Panamá (Slud 1964). However, it is now well known from the northern outlier of these ranges, the Cordillera de Tilarán, where it is rather sharply confined to elevations greater than 1550 m near the crest (Stiles and Hespenheide 1972; Law and Fogden 1981). The present report extends the range of *P. insignis* still farther to the northwest with the discovery of an isolated population on Volcán Miravalles in the Cordillera de Guanacaste.

The Cordillera de Guanacaste comprises four isolated volcanic massifs extending northwest in a line ca. 80 km long, from the Arenal Gap nearly to the Lago de Nicaragua (Fig. 1). Whereas the main mountain ranges of southern Central America are of Tertiary age, the Guanacaste volcanos date from the Quaternary (Dengo 1973); only Rincón de la Vieja is still active. Although many collections of birds and other organisms have been made on the lower Pacific slopes of these volcanos over the last century, their upper reaches have remained all but unknown biologically. Since 1976, I have made at least one ascent to the upper parts of each massif. Just how poorly known are the biotas of the tops of these volcanos is indicated by the fact that very limited collecting has already produced two new species of plants (Stiles 1980; Gómez and Gómez-Laurito 1982), one new butterfly (P. de Vries, pers. comm.), and numerous northward range extensions of highland birds (Stiles and Smith 1980). The range extension of *P. insignis* reported here is the first for which adequate specimen material indicates taxonomic differentiation in the species concerned.

Volcán Miravalles is the tallest massif in the Cordillera de Guanacaste (2020 m, Fig. 1), with the largest area of cloud forest about the peak [Rincón de la Vieja has an equivalent area in the highest life zone, Lower Montane Rain Forest (cf. Tosi 1969), but much of the vegetation has been severely modified by volcanic activity]. On the Caribbean (windward) slope of the massif, this cloud forest extends locally down to ca. 1400 m, as on Cerro Las Nubes, a tall hill (ca. 1850 m) just northeast of the main peak and connected to it by a high ridge. A

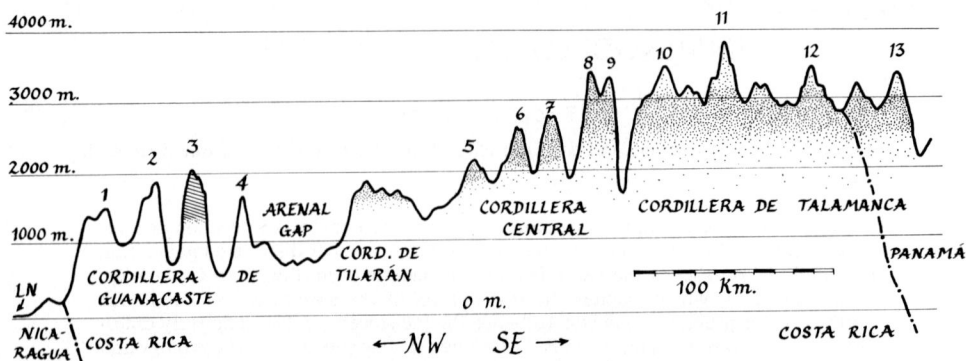

Fig. 1. Profile of the mountains of Costa Rica and extreme western Panamá, showing the known distribution of *Panterpe i. insignis* (stippled: heaviest stipple indicates elevation range of greatest abundance) and *P. i. eisenmanni* (cross-hatched). Abbreviations: LN = Lago de Nicaragua; 1 = Volcán Orosí; 2 = Volcán Rincón de la Vieja; 3 = Volcán Miravalles; 4 = Volcán Tenorio; 5 = Volcán Viejo; 6 = Volcán Poás; 7 = Volcán Barva; 8 = Volcán Irazú; 9 = Volcán Turrialba; 10 = Cerro de la Muerte; 11 = Cerro Chirripó; 12 = Cerro Kamuk; 13 = Volcán Barú (Volcán Chiriquí), Panamá.

specimen of the Fiery-throated Hummingbird collected on Cerro Las Nubes in April 1982 appeared to represent an undescribed race. Further collections in August 1983 confirmed the distinctness of the form, which I propose to name:

Panterpe insignis eisenmanni, new subspecies

Holotype. — No. 2741 of the Museo de Zoología, Universidad de Costa Rica, an adult male collected at an elevation of 1550 m on Cerro Las Nubes (ca. 8 km NW Bijagua, Provincia Alajuela) on 1 August 1983 by F. G. Stiles (original number F.G.S. 1963).

Diagnosis. — Distinguished from all other known populations of *P. insignis* by its much more extensively blue-violet (as opposed to green or blue-green) belly, breast, and upper tail-coverts, its much more extensive blackish "hood" extending over the upper back, and its much shorter bill (Plate II, Fig. 2, Tables 1, 2).

Description of holotype. — To facilitate comparisons, I have written this description in part with reference to the color system of Smithe (1974, 1981). I emphasize that it is often extremely difficult to obtain an unequivocal match between opaque color swatches and iridescent colors like those of *Panterpe,* but even in such cases an effort to standardize color nomenclature seems worthwhile.

Forehead and crown brilliant blue (varies with angle from near 168, Cobalt Blue, to 270, Ultramarine); sides of head and nape deep black; upper back black, faintly glossed green; mid-back bright green (near 62, Spectrum Green, in some lights approaching 163, Emerald Green) becoming more blue-green (near 65, Turquoise and Blue) on lower back, shading through blue (near 70, Smalt Blue) to violet (near 72, Spectrum Violet) on upper tail-coverts. Tail blue-black (near 173, Indigo Blue). Chin and throat brilliantly iridescent: medially copper-orange to copper-rose (near 106, Salmon, but brighter and pinker), becoming more golden-green laterally and posteriorly (near 158, Chartreuse), shading to brilliant green (62, Spectrum Green) on sides of neck. A large patch of violet (172, True Violet) covers most of lower breast, shading laterally through violet-blue to golden-green on sides; violet extends medially to upper belly. Medial belly feathers blue (near 69, Spectrum Blue) with green bases; lateral and posterior underparts dark blue-green (near 65, Turquoise Blue, but darker and duller). Shoulders, wing coverts bronzy-green (between 61, Apple Green, and 159, Lime Green); remiges blackish, glossed dull purplish (darker than 4, Deep Vinaceous). Iris dark brown, bill black with basal half of mandible pink (near 108 D, Rose-Pink, when fresh); feet blackish. Weight 6.3 grams, moderately fat, left testis 3.5 × 3.2 mm. Bill length 17.1 mm, wing chord 63.7 mm, tail length 40.5 mm.

Variation. — Four other males from the type locality are available for comparison; all agree closely with the type in coloration. In two the iridescence of the chin and throat is slightly more rosy in hue, and in one the longest upper tail-coverts are faintly glossed with a more

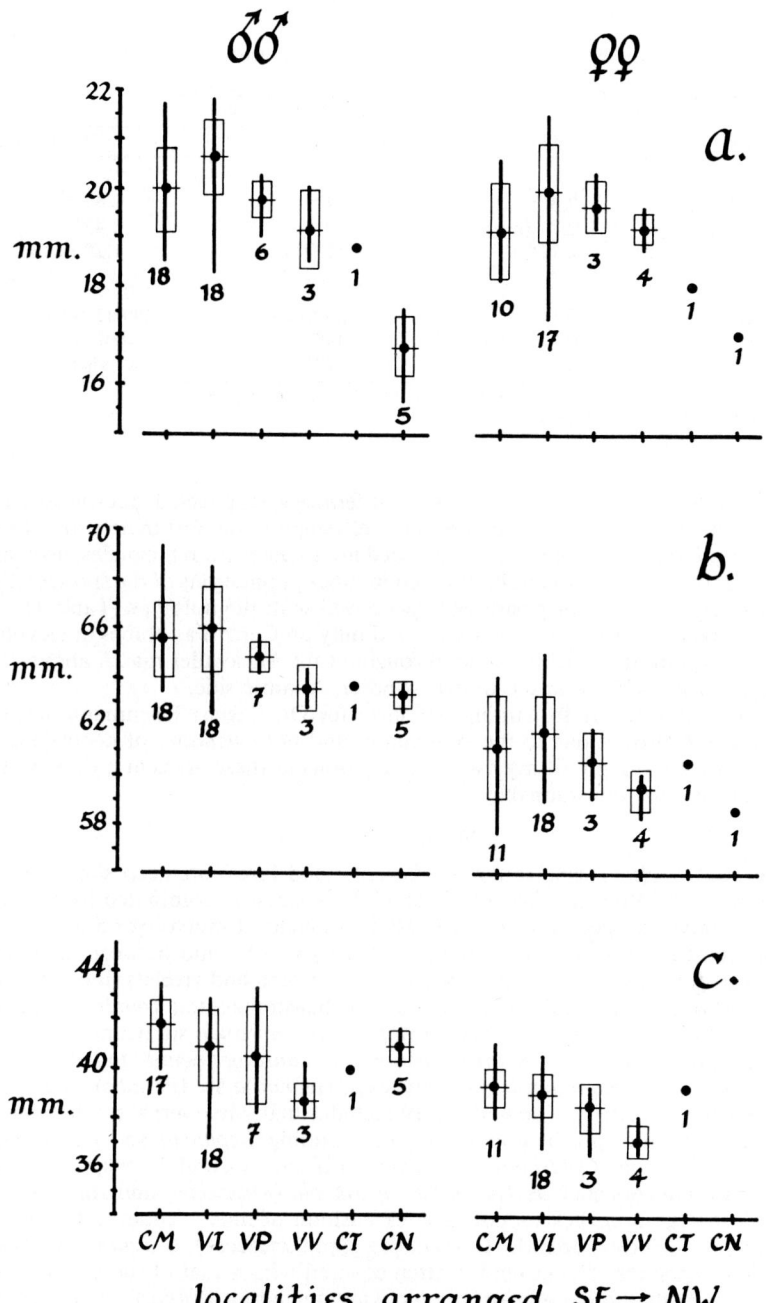

FIG. 2.   Measurements of *Panterpe insignis* from six localities; a. length of exposed culmen; b. wing chord; c. tail length. Box encloses 1 standard deviation on either side of the mean (= horizontal line); vertical line = range; number = sample size. Localities are: CM = Cerro de la Muerte; VI = Volcán Irazú; VP = Volcán Poás; VV = Volcán Viejo; CT = Cordillera de Tilarán; CN = Cerro Las Nubes.

TABLE 1

STATISTICAL COMPARISONS BETWEEN ADJACENT POPULATIONS OF *PANTERPE INSIGNIS*[1]

Sex and dimension	Populations compared[2]		
	CM vs. VI	VI vs. VP + VV + CT	CP + VV + CT vs. CN
Males			
Bill (exposed culmen)	2.36*	3.67**	8.47***
Wing chord	0.77 (n.s.)	2.92**	2.47*
Tail length	2.17*	1.55 (n.s.)	1.29 (n.s.)
Females			
Bill (exposed culmen)	2.16*	1.80 (n.s.)	(insufficient
Wing chord	0.81 (n.s.)	2.84*	data for
Tail length	2.20*	3.01**	comparisons)

[1] Numbers given = values of Student's t; n.s. = $P > .05$; * = $P < .05$; ** = $P < .01$; *** = $P < .001$.
[2] For abbreviations of localities, see legend to Figure 2.

greenish blue (near 65, Cyan) at the very tip. A single female was collected, but was so destroyed by shot that I was unable to prepare a specimen, although I was fortunately able to obtain bill and wing measurements. In plumage it appeared not to differ from the males; no consistent color difference between the sexes has been noted in other populations of the species (Ridgway 1911). In measurements, the other paratypes agree well with the holotype (Table 1).

Distribution.—To date this form has been found only on Cerro Las Nubes at elevations of 1400 to at least 1800 m; it probably occurs throughout the Miravalles massif above 1600 m, and regularly down to 1400 m on the wetter, windier, Atlantic side.

Etymology.—I take pleasure in naming this form for Dr. Eugene Eisenmann, not only in honor of his many contributions to the systematics and nomenclature of neotropical birds, but also for his encouragement of my own investigations in these areas at a time when such studies were increasingly unfashionable.

ECOLOGY

The habitat of *P. i. eisenmanni* is the windswept cloud forest on steep slopes and ridges near the summit of the Volcán Miravalles massif. This forest is dominated by *Clusia alata* (Guttiferae), the only tree species to regularly attain a height of much over 5 to 6 m. Above about 1600 m, the *Clusia* trees tend to be rather widely spaced, and between them a dense, even canopy 4 to 5 m high is formed by various tall shrubs and treelets, including *Blakea austin-smithii, Miconia* spp., and *Conostegia* spp. (Melastomaceae); *Dendropanax* sp. and *Oreopanax* spp. (Araliaceae); *Senecio megaphylla* and *Neomiranda* sp. (Compositae); *Weinmannia* spp. (Cunoniaceae); *Vaccinium poasanum, V. consanguineum,* and *Arctostaphylos* sp. (Ericaceae); *Drimys guatemalensis* (Winteraceae); *Geonoma* sp. (Palmae), and *Chusquea* sp. (Gramineae), among others. The understory includes many tree ferns; small palms (*Chamaedorea* spp., *Geonoma* sp.); bamboos; shrubs, including *Cephaelis* spp., *Palicourea* spp., *Psychotria* spp. (Rubiaceae); *Centropogon nubicola* (Lobeliaceae), and Piper spp. (Piperaceae); and large-leaved monocots such as *Heliconia vulcanicola* (Musaceae) and *Alpinia* sp. (Zingiberaceae). Most large branches support a thick cushion of moss, scattered through which are other epiphytes including orchids, *Cavendishia* spp. (Ericaceae), *Columnea* sp. (Gesneriaceae), but few bromeliads. The ground is often covered with a mat of roots and moss, and some plants grow both terrestrially or as epiphytes, including *Anthurium* spp. (Araceae), *Asplundia* sp. (Cyclanthaceae), and *Schefflera* sp. (Araliaceae). Hemiepiphytes, including *Clusia* spp. (Guttiferae), various Araliaceae, and *Hillia loranthoides* (Apocynaceae) were also numerous.

In April 1982, *P. insignis* activity was mostly concentrated about flowering *Clusia* trees; the flowers of *Columnea* sp. were also visited. The single male collected had small (1 mm) testes and was finishing molt. Suitable flowers for hummingbird visitation were more numerous in August 1983, especially those of *Cavendishia* spp. and *Vaccinium poasanum*. All males collected had enlarged testes (2.5–3.5 mm) and fairly worn plumage, and considerable deposits of fat in the furcular area and feather tracts. Some males appeared to be territorial at large

TABLE 2

VARIATION IN COLOR AMONG SEVERAL POPULATIONS OF *PANTERPE INSIGNIS*[1]

Plumage area	Population			
	Cerro de la Muerte-Volcán Irazú	Volcán Poás-Volcán Viejo	Cordillera de Tilarán (Monteverde)	Volcán Miravalles (Cerro Las Nubes)
Nape-hindneck	Black, faint purple-green sheen	Black, faint green sheen	Black, faint green sheen	Black, faint bluish purple sheen
Upper back	Bright green, at most faintly veiled blackish	Deep green sometimes tinged bluish, faintly veiled blackish	Deep green, faintly veiled blackish	Dusky-black, rather faintly glossed green
Lower back	Deep green, slight or no bluish tinge	Deep green, slight bluish tinge	Deep green, slight bluish tinge	Deep green, rather strongly tinged bluish
Rump, proximal tail-coverts	Deep green to blue-green	Deep blue-green	Deep blue-green	Blue to biolet-blue at most glossed turquoise
Longest tail-coverts	Turquoise with violet gloss to violet-blue with blue-green gloss	Deep violet-blue, tipped and more or less glossed blue-green	Violet-blue, tipped and glossed blue-green	Deep violet-blue with at most slight turquoise gloss at tip
Midbreast	Violet-blue spot covers up to ⅔ of midbreast (sometimes much less)	Violet-blue spot covers ½ to ¾ of midbreast	Violet-blue spot covers ½ to ¾ of midbreast	Violet-blue patch covers ⅔ or more of midbreast, extends medially into belly
Posterior underparts	Bright green golden highlights laterally, at most faint bluish tinge	Bright green, faint to moderate bluish tinge, golden highlights laterally	Bright green, faint to moderate bluish tinge, golden highlights laterally	Violet-blue medially to dark greenish-blue laterally
Lower tail-coverts	Green to dark blue-green	Dark blue-green	Dark blue-green	Violet-blue, edged blue-green

[1] Plumage areas showing little or no geographical color variation are not included.

clumps of flowers, but one male observed for ca. 30 min and finally collected seemed to allow a second bird into his territory. (This bird was also shot but fell into a deep ravine and could not be recovered.) If this second bird was a female, then the association resembled that seen in breeding *P. i. insignis* in the Cordillera de Talamanca (Wolf and Stiles 1970), and reinforces my impression from the gonad data, that the Cerro Las Nubes birds were breeding in August 1983.

I saw no other hummingbird species above ca. 1500 m on Cerro Las Nubes, but below this elevation, *P. i eisenmanni* was sympatric with the Green Hermit (*Phaethornis guy*) and the Purple-throated Mountain-gem (*Lampornis calolaema*) over both of which it appeared to be dominant at flowers. The only other nectarivore noted above 1550 m was the Slaty Flowerpiercer (*Diglossa plumbea*), which robbed flowers in *Panterpe* territories and elsewhere, as it does on the Cordillera de Talamanca (Wolf and Stiles 1970). Thus, allowing for the differences in habitat, many aspects of the ecology and behavior of *P. i. eisenmanni* appear similar to those of the relatively well-studied Cerro de la Muerte population of *P. i. insignis* (cf. Wolf et al. 1976).

GEOGRAPHIC VARIATION IN *PANTERPE INSIGNIS*

It now remains to integrate the features of *eisenmanni* into the context of variation in *Panterpe insignis* as a whole, insofar as possible with the material at hand. I do not have access to specimens from Volcán Barva or from any but the northernmost massif (Cerro de la Muerte) of the Cordillera de Talamanca. I analyzed geographical samples from the Cerro de la Muerte, Volcán Irazú (including Volcán Turrialba), Volcán Poás, Volcán Viejo, Cordillera de Tilarán (Monteverde), and Volcán Miravalles (Cerro Las Nubes). These mountains comprise an essentially linear array from southeast to northwest. I collected the last three samples and part of that from Cerro de la Muerte; other specimens included are from the collections of the Museo Nacional and Universidad de Costa Rica, except for four birds from Volcán Poás in the Louisiana State University Museum of Zoology, which were measured by J. P. O'Neill.

Measurements.—These data are summarized in Figure 2; results of statistical tests are given in Table 1. I compared adjacent populations by Student's *t* (save that I had to combine the small samples from Volcán Poás, Volcán Viejo, and Monteverde). I consider this procedure superior to a single nested ANOVA because it specifically takes into account the linear arrangement of the populations.

With respect to overall size (especially bill length), the largest birds occur on Volcán Irazú; those from Cerro de la Muerte are slightly but significantly shorter-billed (but longer-tailed). Birds from Volcán Poás average smaller in all dimensions, and those from Volcán Viejo are smaller still (although most differences are not significant). Measurements of the two birds from Monteverde are similar to those collected on Volcán Viejo, except for the shorter bill of the Monteverde female. However, this bird is a juvenile with corrugations in the maxillary ramphotheca, and its bill may not be full-grown. The bills of the Cerro Las Nubes birds are much shorter (especially notable in the good sample of males). The birds of this population differ only slightly in wing length from those to the south; the Monteverde and Cerro Las Nubes populations seem to reverse a trend towards shorter tails from Volcán Irazú north to Volcán Viejo. Statistically, each sample differs significantly in at least one dimension from adjacent populations. The notable differences between the Irazú and Poás-Viejo-Tilarán samples reflect both the necessity of combining the latter and, possibly, the lack of material from the geographically and elevationally intermediate Volcán Barva (cf. Fig. 1). Even so, the largest and most highly significant difference between the Cerro Las Nubes birds and those to the south is in male bill length (Table 1). Bill length, thus, provides the sole unequivocal mensural character for distinguishing *eisenmanni* (certainly in males; more data are required to confirm this in females) from *insignis*.

Coloration.—Perception of strongly iridescent colors like those of *Panterpe* is greatly affected by the viewing angle and feather angle (specifically, the degree to which the feathers are appressed to the skin), which varies with the manner of specimen preparation. The accumulation of dust in the plumage of old specimens often reduces considerably the brilliance of the iridescence, and this must also be taken into account when comparing skins of very different ages. Hence, I have relied whenever possible on specimens that I have collected myself in making these comparisons, as they are of uniform preparation and recently taken.

Birds from Volcán Irazú and Cerro de la Muerte are quite similar in coloration; any slight average color differences between these samples are far exceeded by individual variation. Progressing northwest from Volcán Irazú, several parts of the plumage tend to be increasingly bluish (or violet) and less greenish (Plate III, Table 2). As with bill length, this cline is decidedly "stepped" between Monteverde and Cerro Las Nubes. This is also the case with the degree of blackish "veiling" over the upper back. Although populations from the northern Cordillera Central and the Cordillera de Tilarán approach that of Volcán Miravalles in both color and measurements, I believe that all except the latter should be assigned to nominate *insignis* Cabanis and Heine.

DISTRIBUTION AND ALTITUDINAL MOVEMENTS

Over most of its range, the Fiery-throated Hummingbird's center of abundance is in the highland oak forests above 2000 to 2500 m elevation, roughly the upper Lower Montane and Montane belts of Holdridge (1967). Progressing northward from Volcán Irazú, each successive massif is lower (Fig. 1) and the Lower Montane belt extends to lower elevations, especially on the Caribbean (windward) slope (Tosi 1969). Thus, *P. insignis* breeds as low as 1800 m on Volcán Viejo, 1600 m on the Cordillera de Tilarán, and 1400 m on Volcán Miravalles. Particularly on the latter massif, the dominant trees of this life-zone are *Clusia* rather than oaks.

Below these elevations, *P. insignis* is an uncommon to sporadic visitor, mostly outside the breeding season, which lasts from July or August to about January in the Talamancas (Wolf et al. 1976) and probably on Cerro Las Nubes. I have twice seen the species around San José (ca. 1200 m) in March, and once in April at Tres Ríos (1300 m). The Museo Nacional has one specimen from Cartago (ca. 1330 m) taken in July, one from Escazú (ca. 1200? m) taken in April, and four May specimens from La Estrella (ca. 1600 m). One of the latter, plus one Escazú specimen in the Carnegie Museum (K. C. Parkes, pers. comm.) have notably short bills and may represent young birds. The lowest record I know for the species is a bird observed by A. F. Skutch at his farm in the El General valley (ca. 680 m) in May 1975.

Altitudinal wandering on this scale would certainly facilitate movements among all the mountain ranges inhabited by the nominate race, including the Cordillera de Tilarán. The Arenal Gap, with no mountains over 1000 m in a distance of nearly 30 km (Fig. 1) must represent a much more formidable barrier to dispersal and gene flow, which is evidently reflected in the differentiation of *P. i. eisenmanni*. How the Volcán Miravalles population originated is unknown. Birds may have dispersed across the Arenal Gap during a cooler Pleistocene glacial interlude, or the Miravalles population may represent a relict from a more continuous Pleistocene distribution. Given that the original differentiation of *P. insignis* from its (presumably) Andean ancestors most likely occurred in the Cordillera de Talamanca (Dengo 1973; Wolf 1976), I should emphasize that the two possibilities are by no means mutually exclusive.

SPECIMENS EXAMINED

Museo de Zoología, Universidad de Costa Rica: Cerro Las Nubes 5 ♂♂; Volcán Viejo 3 ♂♂, 4 ♀♀; Monteverde 1 ♂, 1 ♀; Cerro de la Muerte 10 ♂♂, 8 ♀♀; Volcán Irazú 1 ♂, 1 ♀. Museo Nacional de Costa Rica: Volcán Irazú 18 ♂♂, 17 ♀♀; 15 ? ; Escazú 1 ♂; La Estrella, 3 ♂♂, 1 ♀; Volcán Poás 4 ♂♂, 2 ♀♀; Cartago 1 ♂. Also measurements of 3 ♂♂, 1 ♀ from Volcán Poás, in Louisiana State Museum of Zoology, were included.

ACKNOWLEDGMENTS

I thank C. Gómez, I. Chacón, A. Solís, and R. Campos for field assistance, R. and C. Ramírez and G. Franco for hospitality during my visits to Cerro Las Nubes, and N. M. Rojas for typing the manuscript. J. A. Tosi granted permission to collect *Panterpe* at Monteverde. K. C. Parkes provided useful critical comments; J. P. O'Neill kindly provided measurements of four specimens.

LITERATURE CITED

DENGO, G. 1973. Estructura Geológica, Historia Tectónica, y Morfología de la América Central. Centro Regional Ayuda Técnica, México, D. F.

GÓMEZ, L. D., AND J. GÓMEZ-LAURITO. 1982. Plantae mesoamericanae novae V. Phytologia 51:474–478.

HOLDRIDGE, L. R. 1967. Life Zone Ecology. Tropical Science Center, San José, Costa Rica.

LAW, B. W., AND M. P. L. FOGDEN. 1981. The Birds of Monteverde. Pensión Quetzal, Monteverde, Costa Rica.

RIDGWAY, R. 1911. The birds of North and Middle America, Pt. 5. Bull. U.S. Natl. Mus. No. 50.

SLUD, P. 1964. The birds of Costa Rica: distribution and ecology. Bull. Am. Mus. Nat. Hist. 128:1–430.

SMITHE, F. B. 1974. The Naturalist's Color Guide and Supplement. American Museum Natural History, New York.

SMITHE, F. B. 1981. The Naturalist's Color Guide, Pt. III. American Museum Natural History, New York.

STILES, F. G. 1980. Further data on the genus *Heliconia* (Musaceae) in northern Costa Rica. Brenesia 18:147–154.

STILES, F. G., AND H. A. HESPENHEIDE. 1972. Observations on two rare Costa Rican finches. Condor 74:99–101.

STILES, F. G., AND S. M. SMITH. 1980. Notes on bird distribution in Costa Rica. Brenesia 17:137–156.

TOSI, J. A., JR. 1969. Mapa ecológico de Costa Rica. Tropical Science Center, San José, Costa Rica.

WOLF, L. L. 1976. Birds of the Cerro de la Muerte region, Costa Rica. Am. Mus. Novit. No. 2606.

WOLF, L. L., AND F. G. STILES. 1970. Evolution of pair cooperation in a tropical hummingbird. Evolution 24:759–773.

WOLF, L. L., F. G. STILES, AND F. R. HAINSWORTH. 1976. Ecological organization of a tropical, highland hummingbird community. J. Anim. Ecol. 32:349–379.

Note added in proof.—During recent trips to Cerro Las Nubes and Monteverde, I collected additional specimens of *Panterpe*. Data were obtained too late to be incorporated into this paper, but the small sample sizes presented in the latter induce me to present this information here (weight in grams, measurements in mm):

Locality	Date	Sex	Weight	Bill	Wing	Tail
Cerro Las Nubes	21 April 1984	♂	5.8	16.3	63.3	40.7
Cerro Las Nubes	21 April 1984	♂	6.1	16.8	63.3	39.3
Cerro Las Nubes	22 April 1984	♀	5.8	16.6	57.5	37.5
Monteverde	29 April 1984	♀	5.8	18.1	59.3	37.0

All measurements are comparable to those reported in the paper (cf. Fig. 2). The female from Cerro Las Nubes is the first of *eisenmanni* to be preserved; it differs from the males in being more greenish on the belly, approaching the condition in the Monteverde birds. Thus, the color characters of the new race may be most pronounced in males. All the Cerro Las Nubes birds were in nonbreeding condition and finishing molt, in accord with previous data for this population.

GEOGRAPHIC VARIATION IN THE LEAST GREBE (*TACHYBAPTUS DOMINICUS*)

ROBERT W. STORER[1] AND THOMAS GETTY[1,2]

[1]*Museum of Zoology and Division of Biological Sciences, The University of Michigan, Ann Arbor, Michigan 48109 USA;* [2]*present address: Zoology Department, University of Georgia, Athens, Georgia 30602 USA*

ABSTRACT. Analysis of geographic variation based on 686 specimens showed that the Least Grebes (*Tachybaptus dominicus*) of North and Central America, the Greater Antilles, and the Atlantic drainage of South America merit separation as the subspecies *brachypterus, dominicus,* and *brachyrhynchus,* respectively. Two smaller isolates, *T. d. bangsi* of Baja California, and a new subspecies from the Pacific slope of Ecuador are also recognized. Samples from Jamaica and Hispaniola, Cuba, and Puerto Rico differ significantly from each other, but the differences are insufficient to merit recognition as subspecies. Least Grebes from the Atlantic drainage of North America between latitudes 9° and 25°N averaged slightly larger than those from the Pacific drainage. No noticeable trends were noted within the range of the species in the Atlantic drainage of South America.

RESUMEN. Al analizar geográficamente las variaciones que presentan 686 especímenes de zambullidor chico (*Tachybaptus dominicus*) de América del Norte y Central, las Antillas Mayores y la cuenca atlántica de Sudamérica, se observa que merecen reconocerse subespecies de *brachypterus, dominicus* y *brachyrhynchus,* respectivamente. También deben reconocerse como subespecies, dos poblaciones mas pequeñas y aisladas: *T. d. bangsi,* de Baja California y una nueva subespecie en las vertientes del Pacífico de Ecuador. Los ejemplares de Jamaica, Española, Cuba y Puerto Rico, si bien se diferencían de manera significativa entre sí, no presentan una diferenciación suficiente como para que sean reconocidos como subespecies distintas. Los zambullidores chicos de la cuenca del Atlántico de América del Norte, entre las latitudes 9° y 25°N son estadísticamente más grandes que aquellos de la cuenca del Pacífico. No se ha notado que exista una tendencia en la variación dentro del intervalo de distribución para la especie en la cuenca del Atlántico sudamericano.

The Least Grebe (*Tachybaptus dominicus*) is one of several small, rather generalized grebes found throughout most of the world. Its range lies almost entirely between 30°N and 30°S latitude in the New World and includes the Bahamas and the Greater Antilles, as well as the mainland. In temperate South America, it is replaced by the White-tufted or Rolland's Grebe (*Rollandia rolland*) and in much of the Old World by the little grebes (*Tachybaptus ruficollis* and *T. novaehollandiae*), which range through the tropics and much of the temperate zone.

The species was originally described by Linnaeus (1766) from "Dominica" (= Santo Domingo, Dominican Republic, on the island of Hispaniola). In 1899, Chapman separated the South American populations as *Colymbus dominicus brachyrhynchus,* and the North American populations, as *C. d. brachypterus,* from the nominate form on the basis of the shorter bills or wings, respectively. In 1926, Wetmore applied the name *speciosus* to the South American populations of *dominicus,* but this name is a synonym of *Rollandia rolland chilensis* (Storer 1975). In 1937 van Rossem and Hachisuka described the birds from Baja California as *C. d. bangsi* on the basis of their coloration and bills, which are paler and shorter than those of the other North American populations. In the most recent revision of the species, Wetmore (1943), using 81 specimens, characterized three races of this grebe, *dominicus, brachypterus,* and *speciosus* (= *brachyrhynchus*) while withholding an opinion on *bangsi,* owing to a paucity of material. The recent availability of nearly 850 specimens of this bird, more than 10 times the number studied by Wetmore, has permitted us to make a far more detailed analysis of the geographic variation in this species than was possible earlier.

MATERIALS AND METHODS

Where possible, the following measurements in millimeters were taken on each specimen: wing length (arc), tarsus length, bill length (from the anterior edge of the nostril to the tip of the maxilla), and bill depth (at the posterior end of the nostril). Other data, including locality, sex, date of collection, weight, and miscellaneous information, such as the size of the gonads, stomach contents, and weight were also recorded. As is the case with other grebes, males are larger, especially in bill measurements, than females. Hence, the data for the sexes were analyzed separately. In a few instances, unsexed and obviously missexed birds were sexed by bill size, and special notations were made on the data sheets indicating those determinations. For the analysis, latitude, elevation, and drainage (Atlantic or Pacific) were obtained for each locality. All measurements were made by the senior author, using the same instruments.

Of the nearly 850 specimens examined, 686 adults were used for the analysis. These included material from the 33 collections listed in Appendix I. Birds were considered adult if no indication of the juvenal pattern remained on the head. Because Least Grebes are resident in most, if not all, areas where they occur regularly, the samples analyzed included specimens from all seasons. Specimens with noticeably worn wings were excluded from the analyses of wing length. The geographic distribution of the specimens used is listed in Appendix II.

Samples were first selected on the basis of the previously recognized subspecies, that is, Baja California, the rest of North and Middle America, the Bahamas and Greater Antilles, and South America. The two continental areas were then divided into their Atlantic and Pacific drainages because Least Grebes are usually found below 2000 m elevation and, therefore, rarely, if ever, cross highland barriers. The specimens from the Pacific drainage of South America were all from a small area in Ecuador and hence were considered a single sample. Variation in the other three continent-and-drainage groups was examined by first scatter-plotting each mensural character against latitude and then subdividing the groups into appropriate subsamples for further analysis. The West Indian population was divided by islands, except for the sample from the Bahamas, which was too small to be further subdivided into samples for use in a meaningful statistical analysis.

Geographic differences in the various measurements of size were tested for statistical significance with two-tailed Student's t-tests; the levels of significance for the pairwise tests are presented in Tables 1 and 2. The t-tests were corroborated with non-parametric Median Tests and Mann-Whitney U-tests.

Principal component analyses were used to summarize patterns of phenotypic variation among subsamples. Analyses were computed separately for males and females. The computations were based on sample means for the four measurements (wing length, tarsus length, bill length, and bill depth) for each of the 14 subsamples used. Principal components were extracted from the covariance matrix calculated from sample means of log-transformed measurements. The geographic areas included in each sample and the size of each sample are given in the legend for Figure 1. For both analyses, the first two principal components together summarize over 96 percent of the variance in the data sets. These vectors, displayed in Figure 1, provide a convenient visual summary of phenotypic similarity among the various geographic samples surveyed.

GEOGRAPHIC VARIATION

The sample from the West Indies was large enough (66 males, 40 females) to permit a preliminary analysis of inter-island variation. Of the Greater Antillean samples (Table 2), the Cuban one is the most distinct, differing significantly from each of the others (except Puerto Rico) in at least one character. In all measurements (except tarsus length and bill depth in the small sample of males from Puerto Rico) Cuban birds average larger than birds from the other West Indian samples. They are, thus, least like the mainland populations; this is contrary to expectation because Cuba is much nearer to the mainland (Yucatan) than is any one of the other Greater Antillean islands. Males from Jamaica and Hispaniola have significantly shorter wings and bills than those from Cuba. (The low levels of significance for these characters in females presumably result from the small sample sizes.) On the other hand, the birds from Jamaica and Hispaniola do not differ significantly from each other in any character. This is also unexpected because the water gap between them is approximately 180 km wide whereas each is closer to Cuba (ca. 135 and 90 km, respectively). Although the small size of the Puerto Rican sample does not permit a meaningful statistical comparison with samples from the other islands, bills of the Puerto Rican birds average shorter than those of Cuban birds, and the degree of sexual dimorphism in the Puerto Rican population may prove greater than those of other West Indian populations.

The Bahaman sample comes from five islands spanning a distance of 700 km (Table 3).

TABLE 1

MEASUREMENTS OF THE MAJOR SAMPLES OF LEAST GREBES, WITH MATRICES INDICATING THE DEGREE OF SIGNIFICANCE OF DIFFERENCES AMONG THEM[1]

	N	Range	Mean	s.d.	BC	NA	WI	SA(P)
Males: wing length								
BC	16	90.0–97.0	93.00	1.75				
NA	137	87.0–101.0	94.03	2.82	NS			
WI	66	93.0–104.0	99.14	2.35	**	**		
SA (P)	4	88.0–95.0	92.25	3.10	NS	NS	**	
SA (A)	69	93.0–107.0	98.97	3.17	**	**	NS	**
Males: tarsus length								
BC	16	29.7–32.8	31.41	0.92				
NA	146	28.7–35.2	32.11	1.21	*			
WI	66	30.1–35.0	32.91	1.29	**	**		
SA (P)	4	31.5–33.6	32.33	0.91	NS	NS	NS	
SA (A)	71	29.6–36.0	32.81	1.27	**	**	NS	NS
Males: bill length								
BC	16	12.1–14.4	13.32	0.60				
NA	142	11.6–16.9	14.32	1.02	**			
WI	66	13.5–18.6	15.84	1.16	**	**		
SA (P)	4	9.9–12.4	11.08	1.10	**	**	**	
SA (A)	69	11.4–15.4	13.28	0.92	NS	**	**	**
Males: bill depth								
BC	15	6.5–7.2	6.87	0.23				
NA	112	5.9–8.5	7.15	0.41	**			
WI	55	6.5–8.0	7.27	0.38	**	NS		
SA (P)	4	5.8–6.7	6.35	0.44	**	**	**	
SA (A)	55	6.2–8.2	7.09	0.41	*	NS	*	**
Females: wing length								
BC	16	87.0–93.0	89.81	1.80				
NA	116	85.0–98.0	91.60	2.75	*			
WI	40	91.0–102.0	95.80	2.59	**	**		
SA (P)	8	88.0–94.0	91.25	2.05	NS	NS	**	
SA (A)	72	90.0–104.0	96.03	3.28	**	**	NS	**
Females: tarsus length								
BC	15	28.4–31.5	30.48	0.80				
NA	121	27.6–35.1	30.99	1.07	NS			
WI	43	28.5–35.0	31.60	1.32	**	**		
SA (P)	8	29.4–33.5	31.30	1.36	NS	NS	NS	
SA (A)	72	28.2–34.5	31.64	1.35	**	**	NS	NS
Females: bill length								
BC	14	10.6–12.2	11.55	0.46				
NA	116	11.2–14.9	12.75	0.77	**			
WI	42	11.1–17.4	13.76	1.28	**	**		
SA (P)	8	9.8–10.8	10.38	0.33	**	**	**	
SA (A)	74	10.0–14.5	11.97	0.89	NS	**	**	**
Females: bill depth								
BC	12	5.9–7.1	6.33	0.29				
NA	106	5.7–7.5	6.53	0.36	NS			
WI	41	6.1–7.5	6.70	0.30	**	**		
SA (P)	8	6.0–6.7	6.33	0.21	NS	NS	**	
SA (A)	51	5.9–7.6	6.58	0.38	*	NS	NS	NS

[1] All measurements in mm. Abbreviations are BC = Baja California, NA = North (and Middle) America, WI = West Indies, SA(P) = Pacific drainage of South America, SA (A) = Atlantic drainage of South America. Significance levels: NS = not significant, * = $P \leq 0.05$, ** = $P \leq 0.01$; N = sample size; s.d. = standard deviation.

TABLE 2

MEASUREMENTS OF SAMPLES OF LEAST GREBES FROM THE BAHAMAS AND GREATER ANTILLES,
WITH MATRICES INDICATING THE DEGREE OF SIGNIFICANCE OF DIFFERENCES AMONG THEM[1]

	N	Range	Mean	s.d.	BA	CU	JA	HI
Males: wing length								
BA	7	97.0–101.0	99.29	1.50				
CU	24	97.0–103.0	100.5	1.69	NS			
JA	17	96.0–101.0	97.94	1.64	NS	**		
HI	11	93.0–101.0	97.09	2.55	NS	**	NS	
PR	6	96.0–104.0	100.33	3.01	NS	NS	*	*
Males: tarsus length								
BA	8	30.5–33.0	32.20	0.80				
CU	24	30.4–35.0	33.37	1.34	*			
JA	17	30.1–35.0	32.67	1.45	NS	NS		
HI	11	31.1–34.0	32.46	0.91	NS	*	NS	
PR	5	31.7–34.6	33.42	1.08	*	NS	NS	NS
Males: bill length								
BA	8	14.9–16.8	15.83	0.64				
CU	23	13.5–18.6	16.69	1.20	NS			
JA	17	13.8–16.9	15.34	0.83	NS	**		
HI	11	14.0–17.7	15.16	1.09	NS	**	NS	
PR	6	14.4–16.3	15.42	0.60	NS	*	NS	NS
Males: bill depth								
BA	4	6.6–7.4	7.03	0.35				
CU	21	6.5–8.0	7.24	0.41	NS			
JA	15	6.7–7.8	7.39	0.38	NS	NS		
HI	10	6.5–7.9	7.21	0.40	NS	NS	NS	
PR	4	7.2–7.7	7.43	0.21	NS	NS	NS	NS
Females: wing length								
BA	6	94.0–98.0	95.83	1.60				
CU	9	94.0–102.0	98.33	2.65	NS			
JA	8	93.0–98.0	95.13	2.10	NS	*		
HI	5	93.0–98.0	95.80	1.92	NS	NS	NS	
PR	8	92.0–99.0	94.63	2.83	NS	*	NS	NS
Females: tarsus length								
BA	6	29.5–33.5	31.05	1.67				
CU	11	30.4–35.0	32.52	1.23	NS			
JA	9	28.9–32.5	31.37	1.33	NS	NS		
HI	6	30.0–32.6	31.27	1.01	NS	NS	NS	
PR	7	28.5–32.5	31.36	1.34	NS	NS	NS	NS
Females: bill length								
BA	5	12.2–14.2	13.32	0.84				
CU	11	13.7–17.4	14.95	1.17	*			
JA	9	12.5–15.0	13.42	0.77	NS	**		
HI	6	13.1–14.6	13.73	0.61	NS	*	NS	
PR	7	12.9–16.0	13.89	1.05	NS	NS	NS	NS
Females: bill depth								
BA	5	6.4–7.2	6.76	0.38				
CU	11	6.6–7.5	6.89	0.29	NS			
JA	9	6.1–7.1	6.53	0.30	NS	NS		
HI	5	6.2–6.9	6.56	0.30	NS	NS	NS	
PR	7	6.5–7.3	6.71	0.27	NS	NS	NS	NS

[1] All measurements in mm. Abbreviations are BA = Bahamas, CU = Cuba, JA = Jamaica, HI = Hispaniola, PR = Puerto Rico. Significance levels: NS = not significant, * = $P \leq 0.05$, ** = $P \leq 0.01$; N = sample size; s.d. = standard deviation.

TABLE 3
Measurements of Bahaman Least Grebes[1]

Number	Island	Age	Wing	Tarsus	Bill
		Males			
AS 1298	Andros	ad.	101	32.7	16.3
AS 1297	Andros	ad.	101	32.2	15.7
FM 33207	Andros	im.	100	33.4	15.4
FM 33206	Eleuthera	ad.	—	33.0	15.4
MCZ 160921	Watling	ad.	99	32.6	16.4
NMNH 276333	Watling	ad.	100	30.5	15.3
NMNH 108003[2]	Watling	ad.	99	32.4	14.9
NMNH 108208	Rum Cay	ad.	97	32.6	15.8
AS 1380	Great Inagua	ad.	98	31.6	16.8
		Females			
AS 1320	Andros	ad.	97	32.3	13.9
NMNH 276334	Watling	ad.	94	29.5	13.6
CMP 30899	Watling	ad.	96	29.9	12.7
NMNH 108207	Rum Cay	ad.	96	29.5	12.2
AS 1379	Great Inagua	ad.	98	33.5	—
CMP 30849	Great Inagua	ad.	94	31.6	14.2

[1] All measurements in mm.
[2] Sexed by bill measurements.

The samples from the various islands are too small to permit statistical analysis, but comparisons of the measurements of individual birds with means and extremes of the Cuban and Hispaniolan samples provide some clues to the origin of the various Bahaman populations. Thus, the birds from Andros Island, which lies north of central Cuba are most like birds from Cuba. Those from Great Inagua Island, ca. 80 km north of Cuba and 65 km north of Hispaniola, appear somewhat more similar to birds from the latter island. The birds from San Salvador (Watling) Island, ca. 330 km from Cuba and 450 km from Hispaniola are more like the Cuban birds in wing length and like the Hispaniolan birds in other characters. It is, thus, likely that the Bahamas were colonized from both Cuba and Hispaniola, and that there was some subsequent interchange of birds within the Bahamas.

Wetmore (1943:231) studied eight Least Grebes from Cozumel Island off the coast of Yucatan and assigned them to the West Indian race on the basis of their "darker flanks and sides and . . . greater amount of fuscous on the breast." Paynter (1955:20–21) did not find his single specimen from Cozumel different from mainland birds and listed three possible explanations for the discrepancy between his and Wetmore's conclusions: that his Cozumel specimen was a migrant, that the mainland race is not a valid one, or that the Antillean form "once existed on Cozumal but has been replaced by immigrants from the mainland." Measurements made by R.W.S. (Table 4) of the sample Wetmore examined are much more similar to those of the North American sample than to those of the West Indian samples, particularly the Cuban one, which is geographically closest to Cozumel Island. In addition, we found considerable seasonal variation in the color of the underparts of Least Grebes, and with the exception of *bangsi*, few, if any constant color differences among the races. We believe, therefore, that colonization of Cozumel Island from the West Indies is unlikely.

Source of the West Indian populations.—It is unlikely that the Least Grebe originated in the West Indies and later spread to the mainland because its closest relatives, the little grebes, are widespread in the Old World. Hence, the ancestral stock presumably reached North America from the Old World and, thus, existed on the mainland of North America before the Old and New World populations diverged.

Assuming a mainland origin for the West Indian populations, two means of reaching the Antilles are possible. The vicariance model of Rosen (1976) could account for the pattern of variation shown by the Least Grebe, but if the separation of the Antilles from the mainland "probably occurred no later than the beginning of the Oligocene" (Baker and Genoways 1978: 71), one would expect far greater differences between the populations of the Antilles and those

TABLE 4

MEASUREMENTS OF LEAST GREBES FROM COZUMEL ISLAND[1]

	N	Range	Mean	s.d.
Males				
Wing length	3	92.0–95.0	93.67	1.53
Tarsus length	3	31.0–32.0	31.37	0.55
Bill length	3	12.8–15.3	13.87	1.29
Bill depth	2	6.8–7.7	7.25	0.64
Females				
Wing length	5	90.0–97.0	94.20	2.77
Tarsus length	5	30.0–32.5	30.98	1.08
Bill length	5	12.3–14.6	13.04	0.93
Bill depth	4	6.5–6.9	6.65	0.19

[1] All measurements in mm. N = sample size; s.d. = standard deviation.

of the mainland than now exist. Therefore, overwater dispersal seems more likely. One or more invasions could have come from Florida, the Yucatan Peninsula, Central America, or northern South America (either directly or by way of the Lesser Antilles). Although southern Florida is close to Cuba and the Bahamas, colonization from Florida is unlikely because the Least Grebe does not occur there now and, furthermore, is unknown from Pleistocene deposits in Florida in which the Pied-billed Grebe (*Podilymbus podiceps*) is common. Colonization via the Lesser Antilles is also unlikely because no relict populations exist on any of these islands east and south of the Virgin Islands.

The relative likelihood of colonization of the West Indies from North or South America was tested by comparing measurements of samples from these areas with those of the West Indian birds. The results were inconclusive. In wing length, the South American birds are between the Cuban and the Jamaican-Hispaniolan samples but nearer the latter; however, the South American birds are much shorter billed than all the West Indian samples although again nearest to the Jamaican-Hispaniolan sample, to which they are also geographically nearest. The North American sample, including Middle America, is also most like the Jamaican-Hispaniolan sample, although closest geographically to Cuba; however, the wings of the North American sample average considerably shorter than those of the West Indian birds.

If these comparisons of size reflect the source of the colonizing birds, which is by no means certain, Jamaica and/or Hispaniola were colonized by birds from North and/or South America that later spread to the other islands. In any event, the colonization(s) must have taken place early enough for differentiation among the West Indian populations to have occurred. Other possibilities, such as an early invasion followed by differentiation and further invasions cannot be ruled out.

Variation within North and Central America.—The measurements, especially of bill length, of the Baja California population (*bangsi*) are small (Table 1). Several other comparisons were made between the North and Central American samples. Scatter plots of the various measurements against latitude showed no evident trends in either the birds from the Atlantic or Pacific drainages (excluding Baja California and Sonora). Sonora birds are intermediate between those of Baja California and the rest of the Pacific drainage of Mexico. Birds from the Atlantic and Pacific drainages of Costa Rica and Panama did not differ significantly. A combined sample of these birds is slightly longer winged (males not significantly, females at the 0.05 level) than birds from southern Texas, thus approaching the birds from the Atlantic slope of South America in this character. In other measurements, however, birds from Texas and Costa Rica/Panama do not differ significantly. Birds from the Atlantic drainage of North and Central America between latitudes 9° and 26°N are significantly larger than those from the Pacific drainage of this area (significance levels, 0.01 for wing length and 0.05 for tarsus and bill length).

Variation within South America.—The distinctness of the Least Grebes from the Pacific slope of Ecuador is evident from Table 1. Scatter plots of the birds from the Atlantic slope indicated no latitudinal trends. For pairwise analyses, birds from this large area were divided into three subsamples: those from the equator north, those from the equator to 17°S, and

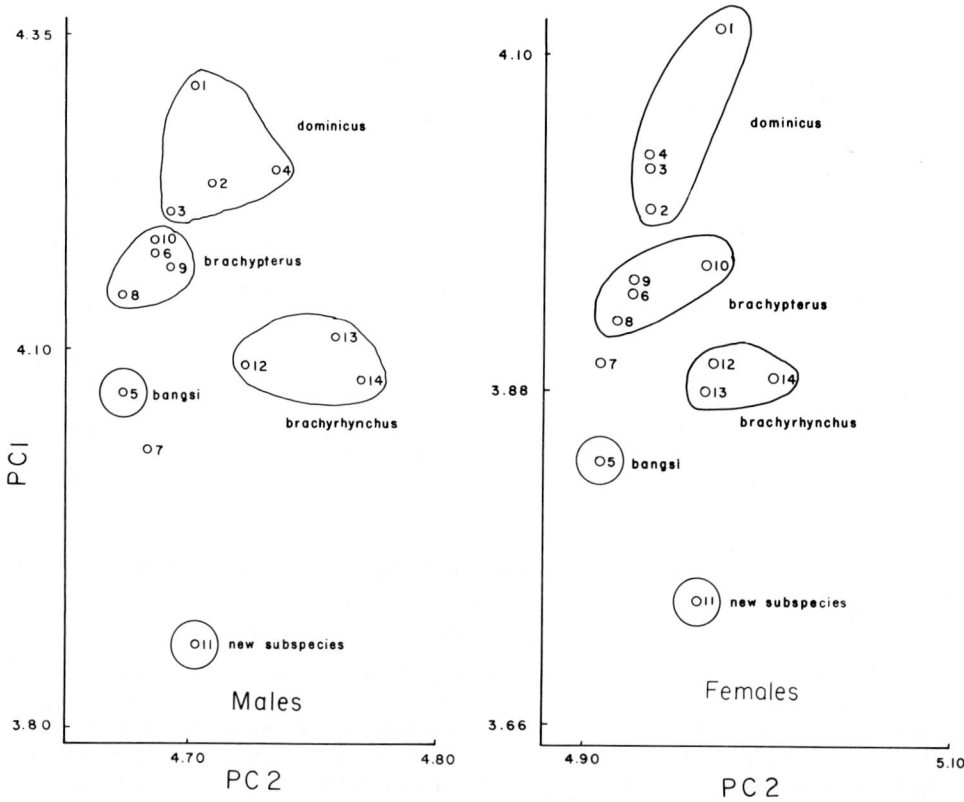

FIG. 1. Plot of first two principal components for 14 samples of Least Grebes grouped by recognized subspecies (Table 5). Samples: 1 = Cuba (24 males, 11 females); 2 = Jamaica (17 m, 9 f); 3 = Hispaniola (11 m, 6 f); 4 = Puerto Rico (6 m, 8 f); 5 = Baja California (16 m, 16 f); 6 = Southern Texas (54 m, 47 f); 7 = Sonora (4 m, 3 f); 8 = Pacific slope of Middle America between latitudes 7° and 26° (68 m, 57 f); 9 = Atlantic slope of Middle America between latitudes 7° and 26° (48 m, 44 f); 10 = Costa Rica and Panama (29 m, 24 f); 11 = Pacific slope of Ecuador (4 m, 8 f); 12 = Atlantic slope of South America north of the Equator (31 m, 33 f); 13 = Atlantic slope of South America from the Equator to 17°S (15 m, 15 f); 14 = Atlantic slope of South America south of latitude 17°S (27 m, 27 f).

those from 17°S to the southern limit of the range. The last subsample was selected because it corresponds closely to the area of sympatry with the White-tufted Grebe, which is similar in size to the Least Grebe and appears to replace it ecologically in most of temperate South America. No significant differences were found among these three subsamples, and hence there was no indication of possible character displacement in the area of sympatry.

Principal component analyses. — The results of these analyses for the two sexes are in general agreement (Table 5; Fig. 1). They also corroborate the pairwise analyses. The sample means cluster in groups corresponding to the recognized subspecies, with the exception of the samples from the Pacific slope of Ecuador, which are well separated from samples from the Atlantic drainage of South America with which they were formerly placed (Fig. 1). The principal component analyses also reflect the differences between the Cuban birds and the other Greater Antillean birds. The very small size of the Sonoran samples (4 males, 3 females) is probably the reason for the different relative positions of the samples for males and females of this population in Figure 1. Birds from this Mexican state are considered intermediate between *bangsi* and *brachypterus,* a determination corroborated by the sample of females.

TAXONOMIC CONCLUSIONS

Although statistically significant, the differences between the Cuban Least Grebes and those from the other Greater Antilles are not, in our opinion, sufficient to warrant subspecific

TABLE 5

RESULTS OF PRINCIPAL COMPONENT ANALYSTS FOR 14 POPULATION SAMPLES OF THE LEAST GREBE

Principal component	1	2	3	4
Males				
Percent of variance (cumulative)	90.78	96.75	99.10	100.0
Log wing length	0.190	0.854	0.213	0.435
Log tarsus length	0.072	0.237	0.577	−0.778
Log bill length	0.919	−0.318	0.191	0.130
Log bill depth	0.337	0.337	−0.765	−0.434
Females				
Percent of variance (cumulative)	88.45	96.23	99.29	100.0
Log wing length	0.177	0.711	0.468	0.494
Log tarsus length	0.090	0.448	0.199	−0.867
Log bill length	0.957	−0.277	0.085	−0.024
Log bill depth	0.212	0.465	−0.857	0.065

recognition. On the other hand, the birds from the Pacific slope of Ecuador are more distinct. These we propose to name:

Tachybaptus dominicus eisenmanni, new subspecies

Holotype.—University of Michigan Museum of Zoology No. 94684, adult female in winter (basic) plumage with a few black feathers growing in on the throat, from Santa Elena, Guayas, Ecuador, 23 December 1933; collected by Phillip Hershkovitz, original number A-4.

Diagnosis.—Smaller than *T. d. brachyrhynchus* in all measurements, especially wing length; shorter billed than *T. d. brachypterus* and *T. d. dominicus*; shorter billed and darker in color than *T. d. bangsi.*

Range.—Known only from the lowlands of western Ecuador.

Specimens examined.—(All localities listed in Paynter and Traylor 1977.) Ecuador: Manabí: Chone 5 (AMNH), Jama 1 (BM); Los Ríos: Río Palenque 1 (NMNH); Guayas: Santa Elena 1 (BM), 4 (UMMZ); El Oro: Santa Rosa, near Machala 1 (BM).

Etymology.—This form is named for Eugene Eisenmann in recognition of his studies of neotropical birds and his help and encouragement to others working in that area.

ACKNOWLEDGMENTS

This study could not have been completed without the friendly cooperation and assistance of the curators in charge of the many collections from which specimens were examined and the computing expertise of P. Myers and K. Creighton, who ran the principal component analyses. We are also grateful to R. C. Banks, T. R. Howell, N. K. Johnson, K. C. Parkes, B. L. Monroe, Jr., and L. L. Short, as well as the reviewers of the manuscript for their helpful comments. The two museums in Argentina were visited while the senior author was supported by a grant from the National Geographic Society.

LITERATURE CITED

BAKER, R. J., AND H. H. GENOWAYS. 1978. Zoogeography of Antillean bats. Zoogeography in the Caribbean. Spec. Publ. Acad. Nat. Sci. Phila. 13:53–97.

CHAPMAN, F. M. 1899. Description of two new subspecies of *Colymbus dominicus.* Bull. Am. Mus. Nat. Hist. 12:255–256.

LINNAEUS, C. 1766. Systema Naturae, ed. 12, Vol. 1.

PAYNTER, R. A., JR. 1955. The ornithogeography of the Yucatan Peninsula. Peabody Mus. Nat. Hist. Yale Univ. Bull. 9.

PAYNTER, R. A., JR., AND M. A. TRAYLOR, JR. 1977. Ornithological Gazetteer of Ecuador. Harvard College, Cambridge, Massachusetts.

ROSEN, D. E. 1976. A vicariance model of Caribbean biogeography. Syst. Zool. 24:431–464.

STORER, R. W. 1975. The status of the Least Grebe in Argentina. Bull. Br. Ornithol. Club 95:148–151.

VAN ROSSEM, A. J., AND THE MARQUESS HACHISUKA. 1937. A further report on birds from Sonora, Mexico, with descriptions of two new races. Trans. San Diego Soc. Nat. Hist. 8:321–336.

WETMORE, A. 1926. Observations on the birds of Argentina, Paraguay, Uruguay, and Chile. U.S. Natl. Mus. Bull. 133.

WETMORE, A. 1943. The birds of southern Veracruz, Mexico. Proc. U.S. Natl. Mus. 93:215–340.

APPENDIX I

COLLECTIONS FROM WHICH SPECIMENS WERE EXAMINED, WITH ACRONYMS FOR COLLECTIONS MENTIONED IN THE TEXT

Academy of Natural Sciences of Philadelphia; Albert Schwartz Collection, Miami Dade Junior College (AS); American Museum of Natural History, New York (AMNH); British Museum (Natural History), Tring (BM); Cambridge Museum, Cambridge University; Canadian National Museum, Ottawa; Carnegie Museum, Pittsburgh (CMP); Chicago Academy of Sciences; Cornell University, Ithaca, N.Y.; Delaware Museum of Natural History, Greenville; Field Museum of Natural History, Chicago (FM); George M. Sutton Collection, now at Delaware Museum of Natural History; James Ford Bell Museum, University of Minnesota; Los Angeles County Museum; Louisiana State University; Merseyside County Museum, Liverpool; Moore Laboratory of Zoology, Occidental College, Los Angeles; Museo Argentino de Ciencias Naturales "Bernardino Rivadavia," Buenos Aires; Museo de La Plata, La Plata, Argentina; Museo Zoologico de la Universidad Nacional de Asunción, San Lorenzo, Paraguay; Museum of Comparative Zoology, Harvard College (MCZ); Museum of Vertebrate Zoology, University of California, Berkeley; National Museum of Natural History, Washington (NMNH); Ohio State University, State Museum, Columbus; Peabody Museum, Yale University; Princeton Museum, Princeton University; Royal Ontario Museum, Toronto; Stanford University, Palo Alto, California; University of California at Los Angeles; University of Miami, Coral Gables, Florida; University of Michigan Museum of Zoology, Ann Arbor (UMMZ); Western Foundation of Vertebrate Zoology, Los Angeles; Winnipeg Museum of Man and Nature.

APPENDIX II

GEOGRAPHIC DISTRIBUTION OF SPECIMENS USED IN THE ANALYSIS

UNITED STATES: Texas 102. MEXICO: Baja California 32, Sonora 7, Nuevo Leon 5, Sinaloa 26, Zacatecas 1, San Luis Potosí 6, Nayarit 16, Jalisco 3, Colima 3, Michoacan 1, Veracruz 35, Guerrero 1, Oaxaca 28, Tabasco 1, Campeche 8, Yucatan 3, Quintana Roo 12, Chiapas 13. GUATEMALA 26. BELIZE 6. HONDURAS 7. EL SALVADOR 22. COSTA RICA 28. PANAMA 25. BAHAMAS 14. CUBA 35. JAMAICA 26. HISPANIOLA 17. PUERTO RICO 17. COLOMBIA: Cauca 3, Caquetá 3, Valle 3, Boyacá 1, Antioquia 2, Bolívar 1, Atlantico 3, Magdalena 6. VENEZUELA: Bolívar 4, Monagás 2, Sucre 1, Aragua 3, Carabobo 10, Mérida 2, Trujillo 1, Lara 2, Zulia 14. GUYANA 2. GUYANE 1. BRAZIL: Paraná 1, São Paulo 5, Rio de Janeiro 4, Minas Gerais 4, Goiás 4, Mato Grosso 12, Bahía 1, Marañhao 3, Pará 7. ECUADOR: Guayas 5, Ríos 1, Manabí 6. PERU: Cajamarca 6, Amazonas 2, Loreto 1. BOLIVIA: Tarija 1, Chuquisaca 1, Santa Cruz 2, Cochabamba 4. PARAGUAY 13. ARGENTINA: Salta 10, Formosa 3.

PLATE III.   A new subspecies of the Collared Inca (*Coeligena torquata eisenmanni*) from the Cordillera Vilcabamba, Peru. From a mixed media painting by John P. O'Neill.

A NEW SUBSPECIES OF COLLARED INCA HUMMINGBIRD (*COELIGENA TORQUATA*) FROM PERU

JOHN S. WESKE

Division of Birds, National Museum of Natural History, Smithsonian Institution, Washington, D.C. 20560 USA

ABSTRACT. *Coeligena torquata eisenmanni,* a new race of the Collared Inca hummingbird, is described from mid-elevation cloud forest in the northern Cordillera Vilcabamba, a front-range of the Andes in south-central Peru. Its plumage pattern and its geographic range provide a link between the northern, white-banded *torquata* subspecies group and the southern, rufous-banded *inca* group, confirming the conspecificity of the two groups. Lowlands of the Río Apurímac valley form a barrier between *C. t. eisenmanni* and the neighboring race to the west and northwest, *C. t. insectivora.*

RESUMEN. Se describe una nueva raza de picaflor inca de collar, *Coeligena torquata eisenmanni,* que ha sido encontrado en el bosque de neblinas de elevación mediana en el norte de la cordillera de Vilcabamba, la cual forma parte de la cordillera de los Andes en el centro-sur del Perú. Las características de su plumaje así como su rango de distribución la enlazan con las subespecies del grupo *torquata* del norte, de banda blanca, y con el grupo austral *inca,* de banda rojiza, confirmando de esta manera la conespecificidad ("conspecificity") de los dos grupos. Las tierras bajas del valle del río Apurímac forman la barrera entre *C. t. eisenmanni* y *C. t. insectivora,* la raza vecina hacia el oeste y noroeste.

During ornithological exploration of the northern Cordillera Vilcabamba in south-central Perú, my colleagues and I encountered a large, showy hummingbird representing *Coeligena torquata,* the Collared Inca, but conspicuously different from any previously known form of that species. I name it:

Coeligena torquata eisenmanni, new subspecies

Holotype.—American Museum of Natural History No. 820361, adult male from Cordillera Vilcabamba, 12°38'S, 73°36'W, elevation 2170 m, Provincia de La Convención, Departamento de Cuzco, Perú, 12 July 1967; collected by John S. Weske and John W. Terborgh; prepared by Weske, original number 1289.

Diagnosis.—A large, straight-billed hummingbird with a dark green body, black head, white pectoral band, and mostly white tail (Plate III). Differs from all other races of *C. torquata* in having shining coppery upper tail-coverts. Nearest to *C. t. omissa* of southeastern Peru and *C. t. inca* of northern Bolivia, but differs in both sexes by having a white (not orange-rufous) pectoral band, by the darker and clearer (less bronzy) shining green of back and belly, and by the bronzy (not greenish) hue of the median rectrices as seen from above; and in males by the presence of a golden green crown patch, by the near-absence of bronzy green on the centers of the black feathers of the throat, and by somewhat more extensive white and reduced dark tips in the tail; and in females by the paler, duller rufescence of the throat. Differs from *C. t. insectivora* of central Peru in both sexes by having the white pectoral band narrower and not extending forward onto the lower throat, and by bronzy (not dark green) median rectrices; and in males by the much larger, brilliant blue-green frontal plaque that is always present, by the shining dark green upper back (not velvety black glossed with green), by the reduced greenish tinge to the velvety black of the head, by the lack of a glittering green throat, and by more extensive white and much reduced dark terminal areas in the rectrices; and in females by the plain rufescent throat (not buffy white with green spots), the presence of a blue-green frontal patch, and more extensive white in the tail with the inner vanes of the outer two pairs of rectrices entirely white.

Measurements of holotype.—Chord of wing 80.6 mm, tail (from insertion of the two central rectrices to tip of the longest rectrix) 51.6 mm, exposed culmen 33.5 mm.

Range.—At elevations from about 2070 m to 2840 m on the slopes of the northern Cordillera

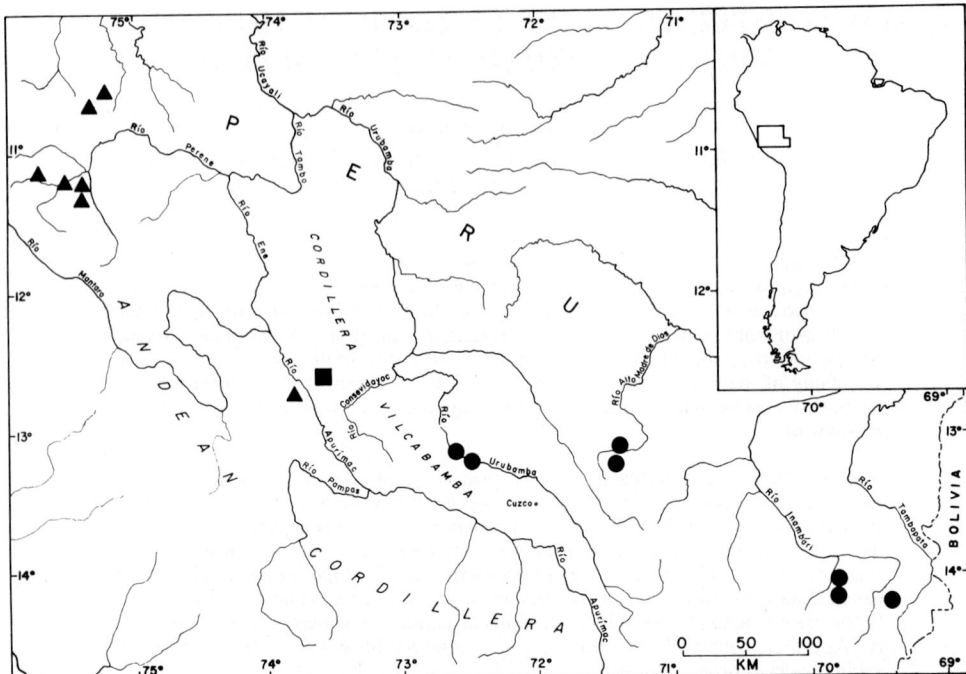

FIG. 1. Distribution of *Coeligena torquata insectivora* (triangles), *C. t. eisenmanni* (square), and *C. t. omissa* (circles). In some cases the symbol marks two or more nearby localities.

Vilcabamba in the Department of Cuzco, Peru. The type locality lies on the east side of the Río Apurímac valley in the watershed of its tributary the Río Mapitunari.

Etymology.—This race is named for the late Eugene Eisenmann in appreciation of his friendship, his hospitality, his helpfulness, and his intellect. In recent years, whenever I went to the American Museum to work on a manuscript, he always offered encouragement and was more than willing to discuss a problem or read and criticize the draft of a manuscript. Thus, as I prepare this paper at the same museum, I feel his loss with particular keenness.

REMARKS

Zimmer (1948) united the *Coeligena inca* populations of southeastern Peru and northern Bolivia, which had hitherto been considered a distinct species, with *C. torquata* of Venezuela,

TABLE 1

MEASUREMENTS OF THREE SUBSPECIES OF *COELIGENA TORQUATA*[1]

Race	Wing chord				Exposed culmen			
	N	Mean	s.d.	Range	N	Mean	s.d.	Range
Males								
C. t. insectivora	19	77.6	1.1	75.9–79.9	19	31.8	1.1	29.6–33.5
C. t. eisenmanni	5	79.7	1.5	78.0–81.4	7	32.7	0.6	31.9–33.5
C. t. omissa	12	76.4	1.5	74.3–79.2	8	32.2	1.2	30.2–33.3
Females								
C. t. insectivora	11	71.7	2.1	68.8–75.3	11	34.2	1.3	31.3–36.0
C. t. eisenmanni	4	72.4	1.1	71.1–73.5	5	33.3	1.4	31.7–35.3
C. t. omissa	9	70.7	0.9	69.0–72.1	8	33.6	1.4	31.4–35.2

[1] All measurements in millimeters.

TABLE 2

RESULTS OF STUDENT'S *t*-TEST COMPARING MEASUREMENTS OF *COELIGENA TORQUATA* POPULATIONS

Comparisons between races

Races compared	♂ Wing			♀ Wing			♂ Culmen			♀ Culmen		
	t	d.f.	*P*	*t*	d.f.	*P*	*t*	d.f.	*P*	*t*	d.f.	*P*
insectivora vs *eisenmanni*	3.46	22	<0.01	0.57	13	>0.5	1.93	24	>0.05	1.34	14	=0.2
eisenmanni vs *omissa*	3.99	15	<0.01	2.88	11	<0.02	0.98	13	>0.2	0.33	11	>0.5
insectivora vs *omissa*	2.55	29	<0.02	1.43	18	>0.1	0.83	25	>0.4	1.12	17	>0.2

Comparisons between sexes

Comparison	*C. t. insectivora*			*C. t. eisenmanni*			*C. t. omissa*		
	t	d.f.	*P*	*t*	d.f.	*P*	*t*	d.f.	*P*
♂ Wing vs ♀ Wing	10.38	28	<0.001	7.98	7	<0.001	10.05	19	<0.001
♂ Culmen vs ♀ Culmen	5.45	28	<0.001	1.09	10	>0.2	2.15	14	=0.05

Colombia, Ecuador, and northern and central Peru. This was a bold step, for white-banded birds of the *torquata* group and the rufous-banded *inca* differ as much from each other in appearance as do members of other hummingbird groups known to be specifically distinct. Collections at the time of Zimmer's study left a 370 km gap in Peru between the southeast-ernmost locality for a *torquata* representative (*C. t. insectivora*) in the vicinity of Maraynioc, Depto. Junín and the northwesternmost one for an *inca* representative (*C. t. omissa*) at Machu Picchu, Depto. Cuzco. The type locality of *C. t. eisenmanni* lies in this intervening area about 130 km west-northwest of Machu Picchu (Fig. 1). Moreover, recent field work by John Terborgh and myself and by a Louisiana State University field party has shown that *C. t. insectivora*'s range extends southeastward from Junín into the Department of Ayacucho. Individuals of this race occurred from 1900 m to 2600 m on the western slope of the valley of the Río Apurímac about 20 km in straight-line distance westward and directly opposite the known range of *C. t. eisenmanni*. The elevation at river level is 600 m, the valley apparently providing an effective barrier to contact between *insectivora* and *eisenmanni*. Nonetheless, *eisenmanni* shows morphologic features that link the *torquata* and *inca* groups, confirming Zimmer's decision to lump them as one species.

Most prominently, the pectoral band of *eisenmanni* is white as in *insectivora*, but narrower as in *omissa*. The adult male has the golden-green crown patch of *insectivora* (lacking in *omissa*) and the prominent blue-green frontal plaque of *omissa* (reduced or lacking in *insectivora*). The color of the back and belly is the darker, less bronzy green of *insectivora*, but the brightness of the upper back of the male recalls *omissa*. In lacking a glittering green gorget, *eisenmanni* differs from *insectivora* and resembles *omissa*. Female *eisenmanni* are nearer *omissa* in throat color and also show the sexual difference in tail pattern that occurs in *omissa* but not in *insectivora*. From both of its neighboring races *eisenmanni* stands apart by possessing shining coppery upper tail-coverts (dark green in *insectivora*, bronzy green in *omissa*), by having bronzy median rectrices (green in the others), and by having more white in the tail of the male than either of the others.

Measurements for the samples of specimens of the three races are given in Table 1, and the results of statistical comparisons using the Student's *t*-test are shown in Table 2. *Coeligena torquata eisenmanni* is significantly longer-winged than *omissa* in both sexes and is significantly longer-winged than *insectivora* in males but not in females. Culmen lengths do not differ significantly.

Within races, comparisons between sexes reveal that males are much longer-winged than females. On the other hand, females tend to have longer bills than males. This difference is significant for *insectivora* and *omissa*, but not for *eisenmanni*.

The northern Cordillera Vilcabamba is a sort of montane peninsula bordered by the Apurímac, Ene, Tambo, and Urubamba rivers. *Eisenmanni* can be expected to range throughout this area wherever elevations exceed ca. 2100 m. To the southeast, it is not clear whether *eisenmanni* and *omissa* intergrade rather narrowly or fail to come into contact. On maps, a tributary of the Río Urubamba, the Río Consevidayoc, separates most of the northern and southern portions of the Cordillera Vilcabamba, but an "isthmus" at moderate elevations seems to connect the two montane areas. The vegetational and geographic features of this saddle, and the status of *Coeligena torquata* in its vicinity, await investigation.

The type locality of *C. t. eisenmanni* is the same as that of the Cloud-forest Screech-Owl, *Otus marshalli*. The very moist, well-developed cloud forest habitat and some features of the local avifauna were discussed by Weske and Terborgh (1981). *Eisenmanni* was commonly mist-netted at elevations from 2070 m to 2260 m. It also occured less commonly in elfin forest along ridges up to 2840 m. Specimens were taken between 28 June and 11 August. None was in breeding condition. Three of four late-June birds were in final stages of wing and tail molt, with the outermost primaries and central rectrices sheathed basally and not quite full length.

SPECIMENS EXAMINED

All are from Peru; all are deposited in the AMNH unless noted otherwise:

Coeligena torquata eisenmanni.—The type series: 7 ♂♂, 5 ♀♀.

C. t. insectivora.—Depto. Pasco: Santa Cruz, 4 ♂♂ and 1 ♀ (LSUMZ); Cumbre de Ollón, 10 ♂♂ and 6 ♀♀ (LSUMZ). Depto. Junín: Huacapistana, 1 ♂ (ANSP); Culumachay, Maraynioc, 1 ♂; Chilpes, 1 ♂, 1 ♂?. Depto. Ayacucho: Yuraccyacu, 1 ♂ (LSUMZ), 1 ♀; Estera Rohuana, 1 ♂, 3 ♀♀.

C. t. omissa.—Depto. Cuzco: Santa Rita, Urubamba Canyon, 2 ♂♂; Urubamba Canyon, 1 ♂; Machu Picchu, 1 ♀; Torontoy, 1 ♂ (USNM); Huasampilla, 3 ♂♂ (incl. type), 1 ♀; Pillahuata, 1 ♀ (FMNH). Depto. Puno: Oconeque, 3 ♂♂ and 3 ♀♀ (ANSP), 1 ♀; Limbani, Carabaya, 2 ♂♂, 2 ♀♀.

ACKNOWLEDGMENTS

I thank my fellow expedition members J. W. Terborgh, W. C. Russell, J. S. Knox, and T. R. Dudley, who shared in the discovery of *C. t. eisenmanni.* J. P. O'Neill participated in field work on the Apurímac Valley's west slope and lent specimens from the Louisiana State University's Museum of Zoology (LSUMZ). I am grateful to him for this help and to F. B. Gill of the Academy of Natural Sciences of Philadelphia (ANSP) and J. W. Fitzpatrick of the Field Museum of Natural History (FMNH) for lending additional material. R. A. Paynter, Jr., of the Museum of Comparative Zoology and D. S. Peters of the Natur-Museum Senckenberg provided information on specimens in their care. J. R. Weske, my father, translated passages of 19th-century French-language literature that form part of the historical background for this study. J. W. Terborgh and reviewers R. C. Banks, P. A. Buckley, M. S. Foster, and J. W. Hardy critically read my manuscript and made helpful suggestions. W. E. Lanyon and L. L. Short, Jr., of the American Museum of Natural History (AMNH) have been generous in letting me use the collections and facilities in their care, and the Frank M. Chapman Memorial Fund provided partial support for my work in the Vilcabamba. Support from the National Geographic Society and the National Science Foundation (Grants GB-12378 and GB-20170) is likewise acknowledged.

LITERATURE CITED

WESKE, J. S., AND J. W. TERBORGH. 1981. *Otus marshalli,* a new species of screech-owl from Perú. Auk 98:1–7.
ZIMMER, J. T. 1948. Two new Peruvian hummingbirds of the genus *Coeligena.* Auk 65:410–416.

ZOOGEOGRAPHY AND DISTRIBUTION

CRACRAFT, JOEL. Historical Biogeography and Patterns of Differentiation within the South American Avifauna: Areas of Endemism 49

FJELDSÅ, JON. Origin, Evolution, and Status of the Avifauna of Andean Wetlands 85

HAFFER, JÜRGEN. Avian Zoogeography of the Neotropical Lowlands 113

HAFFER, JÜRGEN, AND JOHN W. FITZPATRICK. Geographic Variation in Some Amazonian Forest Birds 147

PARKER, THEODORE A., III, THOMAS S. SCHULENBERG, GARY R. GRAVES, AND MICHAEL J. BRAUN. The Avifauna of the Huancabamba Region, Northern Peru 169

ROBBINS, MARK B., THEODORE A. PARKER, III, AND SUSAN E. ALLEN. The Avifauna of Cerro Pirre, Darién, Eastern Panama 198

SICK, HELMUT. Observations on the Andean-Patagonian Component of Southeastern Brazil's Avifauna 233

SNOW, DAVID W. Affinities and Recent History of the Avifauna of Trinidad and Tobago 238

VOOUS, K. H. Additions to the Avifauna of Aruba, Curaçao, and Bonaire, South Caribbean 247

VUILLEUMIER, FRANÇOIS. Forest Birds of Patagonia: Ecological Geography, Speciation, Endemism, and Faunal History 255

WIEDENFELD, DAVID A., THOMAS S. SCHULENBERG, AND MARK B. ROBBINS. Birds of a Tropical Deciduous Forest in Extreme Northwestern Peru 305

HISTORICAL BIOGEOGRAPHY AND PATTERNS OF DIFFERENTIATION WITHIN THE SOUTH AMERICAN AVIFAUNA: AREAS OF ENDEMISM

JOEL CRACRAFT

Department of Anatomy, University of Illinois, Chicago, Illinois 60680 USA, and Division of Birds, Field Museum of Natural History, Chicago, Illinois 60605 USA

ABSTRACT. This study provides the background for examining the historical biogeography of the South American avifauna by postulating 33 areas of endemism and listing representative endemic taxa for each area. The results of this analysis suggest that most avian species are narrowly endemic rather than widely distributed, which implies that phylogenetic analysis of different avian lineages should provide important evidence about the histories of the areas of endemism. Generally speaking, previous discussions about South American biogeography have interpreted most areas of endemism as Pleistocene forest "refugia." An alternative possibility is proposed here, namely that many of these areas are considerably older than the Pleistocene and that hierarchical patterns of endemism were established prior to that epoch. This view implies that the physiographic evolution of South America was probably as crucial in forming patterns of avian differentiation as were habitat changes caused by cyclical Pleistocene climates.

RESUMEN. Este estudio brinda una base para examinar históricamente la biogeografía de la avifauna sudamericana, proponiendo 33 áreas de endemismo y nombrando taxa endémica representativa para cada área. Los resultados de este análisis sugieren que la mayoría de las especies de aves son endémicas en áreas muy pequeñas ("narrowly endemic") en lugar de ampliamente distribuídas, lo cual implica que los análisis filogenéticos de los distintos linajes de aves deben proveer evidencias de importancia para explicar la historia de las áreas de endemismo. Generalmente, en discusiones previas respecto de la biogeografía sudamericana, la mayoría de las áreas de endemismo se han interpretado como "refugios" de los bosques del pleistoceno. En el presente trabajo se propone una posibilidad alternativa; muchas de esa áreas son considerablemente más antiguas que el pleistoceno y los patrones jerárquicos de endemismo fueron establecidos previamente a esa época. Este punto de vista implica que la evolución fisiográfica de Sudamérica fue probablemente tan importante en la formación de modelos de diferenciación de avifauna como los cambios en los habitats causados por los ciclos climáticos del pleistoceno.

Two themes have dominated recent analyses of the historical biogeography of continental biotas. First is the idea of waves of intercontinental dispersal followed by "adaptive radiation" into the "empty niches" of the continent being invaded. That this view has been important, particularly when the time-scale being considered extends well-back into the Cenozoic, scarcely can be denied. One only has to read recent discussions about the history of continental biotas (including their avifaunas) to see how the dispersalist theme has shaped our thinking (*general papers:* Darlington 1957; Simpson 1965; Cracraft 1973; *North America:* Mayr 1946; *Central and South America:* Mayr 1964; *Australia:* Mayr 1944, 1972; Cracraft 1972, 1976; Keast 1972, 1981; Rich 1975).

A second dominant theme concerns the role played by "refuges" in the distribution and diversification of continental biotas (see summaries in Prance 1982a). In this case the time-scale is almost always restricted to the Pleistocene. Biotas are envisioned to have developed by repeated isolations of wet- and dry-forest habitats subjected to cyclical contraction and expansion during glacial and interglacial periods. Again, this theme has played a major role in our interpretations of avian continental biogeography, particularly in South America (Haffer 1967a, 1969, 1974, 1982; Vuilleumier 1969, 1970, 1980; Simpson and Haffer 1978; Vuilleumier and Simberloff 1980) and in Australia (Keast 1961, 1981).

In this paper I propose that the above themes can be considered components of a still larger problem within biogeography, the problem of areas of endemism and their historical interrelationships. This approach emphasizes that to reconstruct patterns of diversification, one must have knowledge of areas of endemism and their history (e.g., Platnick and Nelson 1978; Rosen 1978; Nelson and Platnick 1981).

The concept of "refuges" is closely related to, but is not identical with, the notion of areas of endemism. Hypotheses about refuges and their historical significance are only as good as our understanding of areas of endemism and their history, because the concept of refuges is a derivative of our knowledge about endemism. Refuges are regions of survival hypothesized to have existed following climatic reduction of a once more extensive habitat, and they are also considered to be areas of differentiation for some, but not necessarily all, of the taxa in them. Yet, scarcely any direct evidence exists, other than isoclines of species' distributions, to indicate where refuges might be located (Haffer 1974:143; 1980). Finally, refuges, as areas of differentiation, usually are postulated to be of Quaternary origin.

Areas of endemism, in contrast, are regions defined by distributional congruence of the constituent taxa. The simplest hypothesis is that areas of endemism represent common regions of biotic differentiation. Phylogenetic histories of taxa endemic in three or more areas provide evidence for their historical interrelationship, and it is these hypotheses regarding the history of areas that allow us to investigate patterns of differentiation in the component taxa (Rosen 1978; Cracraft 1982, 1983a, b). When undertaking such studies, no prior assumptions about the age of areas or their taxa are necessary or desirable.

Analysis of areas of endemism is important for three reasons. First, we are led to generate hypotheses about the history of geographic units and their biotas. Second, such analysis forms a basis for constructing hypotheses about the differentiation of component clades within those areas. And third, it provides a means of identifying those areas with unique biotas. Because of the rapid destruction of the continental biotas of the world, the analysis of areas of endemism can help determine priorities for their preservation (e.g., Lovejoy 1982; Myers 1982; Terborgh and Winter 1982).

Biogeographers certainly have not ignored the problem of defining areas of endemism within the South American avifauna. Müller (1972, 1973, 1976), for example, analyzed areas of endemism for the entire South American biota, and much of his classification was based on bird distributions. Haffer (1967a, b, 1970b, 1974, 1975, 1977, 1978), in particular, focused on the areas of endemism within the tropical forests of Amazonia and northwestern South America and is most responsible for the development of the refuge concept as it is applied to the South American biota. Other important ornithological works that have helped to define areas of endemism include Chapman (1917, 1926), Miller (1947, 1952), Phelps (1966), Mayr and Phelps (1967), and Short (1975). Despite this work, our understanding of the history of areas of endemism remains rudimentary. Not only do the areas themselves remain insufficiently well-known, but current methods of analysis have not been adequate to resolve historical relationships among the areas (see Discussion).

This paper has a limited purpose: to identify the major areas of endemism within the South American avifauna and to provide preliminary lists of the endemic taxa. This study represents one of the few attempts to define continental areas of endemism based on the avifauna (see Stegmann 1937; Müller 1973; Haffer 1978), and serves as a basis for subsequent studies whose focus will be the historical biogeography of the South American avifauna.

PRINCIPLES, DEFINITIONS, AND METHODS

Areas of endemism are defined by the congruence of species' ranges. Usually this congruence correlates with a physiographic or climatic barrier, yet congruence is rarely perfect owing to taxa being of different ages, ecologies, or dispersal abilities. Two or more taxa may have distributions that overlap relatively little and still be assigned to the same area of endemism: in such a case, the delimitation of areas depends on the patterns described by all the taxa in an area and by the pattern of relationships and distributions of close relatives.

The concepts of congruence and endemism are scale-dependent. If our sample region is the world, then we might speak of South America, the neotropics, or Laurasia (Holarctica) as areas of endemism. Taxa considered to be endemic in a region share some similarity in their distributions not shared by taxa inhabiting another area of endemism, and depending on the scale, we do not require, or necessarily expect, that these taxa will share perfect (or even near perfect) overlap in their ranges. One species may be North American, another European, yet both are endemic to the Holarctic. What this means, of course, is that two or more smaller areas of endemism may be nested within the larger one. Areas of endemism, therefore, appear to be organized hierarchically.

As the sample region becomes smaller, the areas of endemism within that region are defined by an increasingly higher degree of overlap (congruence) in species' distributions. Thus, the amount of congruence that is acceptable in defining endemism is a function of the scale employed.

The sample region of this paper is South America, hence the degree of overlap in species' distributions used to define areas of endemism must be judged *relative to South America as a whole*. As an example, Figure 1 depicts the ranges of ten of the taxa used to describe the South Amazonian Center of endemism (see methods and discussion below). At the small scale of the area itself, congruence might be judged imprecise, and each species is observed to differ in the details of its distribution. At a much larger scale, however, the degree of congruence increases. Thus, when all of South America is used as a basis of description, the ranges of these 10 taxa describe a well-marked area of endemism.

In recognizing areas of endemism in this paper, those proposed by Müller (1973) were first investigated. The validity of these areas was examined using distributional data for each species of South American bird compiled from the following primary sources: Peters (1931–1979), Meyer de Schauensee (1964, 1966, 1970), Phelps and Phelps (1950, 1958), Blake (1977), Meyer de Schauensee and Phelps (1978), and Parker et al. (1982). In addition, numerous monographs and papers were consulted (see papers cited in Introduction and in the following section), in particular Delacour and Amadon (1973), Short (1975), Vaurie (1980), and Snow (1982). The areas recognized by Müller, inasmuch as they were partially delimited by avian distributions, provided a convenient starting point for investigating avian endemism in more detail. This study confirmed that many of his areas are accurate descriptors of avian endemism, and his lists are expanded here. Other areas recognized in this paper are based on distributional congruence as revealed by the data collected from the sources cited above.

No map or written description can delineate a species' distribution accurately. The most accurate map would be that which shows reliable specimen and sight records, but an analysis of this type was not practical for this study and would require extensive resources, particularly of money and personnel, that are not available at this time. The distributions of most described taxa are still poorly known, especially in the Andes. Taxa were assigned to an area if their distributions were almost completely within that area. Occasionally, taxa were also assigned to an area even though recorded from outside that area: in these cases, outside distributions had to be restricted to adjacent areas and be interpretable as "recently" established or chance wanderings (e.g., ranges barely crossing a boundary, marginal distributions in adjacent biomes, or only scattered, isolated records). Thus, this study employs a more restricted definition of "endemic" than did Müller (1973) or Haffer (1978); I distinguish between taxa that are "characteristic" of an area and those that are endemic (restricted) to it.

Some of the areas postulated here undoubtedly will be divided into two or more distinct areas of endemism upon further study. This will be true particularly for the Andes where taxa are distributed in three dimensions and on a landscape that is complex topographically and ecologically. To give an example, the subtropical montane forest avifauna traditionally has been considered to be derived from the lowland tropical biota (Chapman 1917). Such a relationship may be corroborated eventually, but many taxa within the subtropical montane biota have sister-group relationships to forms within the montane forests that lie either to the north or south and not to the adjacent lowland tropical biotas (Cracraft, unpubl. observ.). This implies a distinct history for some of these montane and lowland taxa. Hence, the independent existence of an area of endemism that includes part or all of the montane biota will be determined once the phylogenetic patterns of the endemic species are established. The same argument extends to those species that inhabit "upper" subtropical and temperate montane forests and that, occasionally, also extend to the páramo-puna biome. Answers to the questions—Does the puna-páramo avifauna have a history distinct from the temperate avifauna? Does the temperate avifauna have a history distinct from that of the subtropical biome?—cannot be provided merely by mapping distributions and examining the ecology of the species; an area may be distinct historically and still contain an avifauna that is highly complex ecologically.

In order to avoid creating an area of endemism for each ecological "zone" within the Andes, I have, perhaps somewhat arbitrarily, included taxa that have "lower" subtropical montane

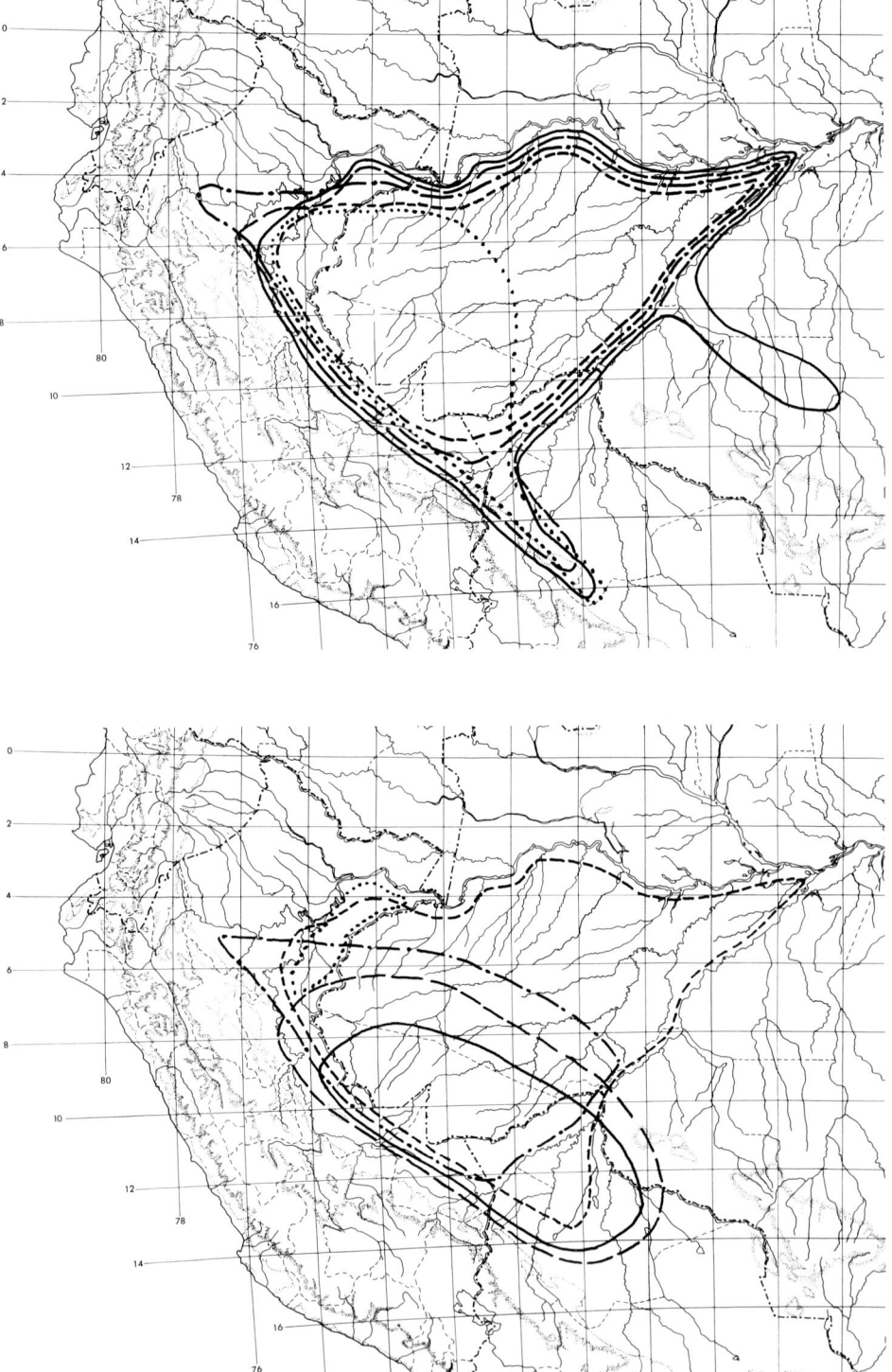

FIG. 1. Estimated distributions of 10 species of birds used to define the South Amazon (Inambari) Center of endemism (see Fig. 5, area 20). Note that not all species exhibit precise congruence even though they are all considered to be endemic in this area (see text). Data from Peters (1931–1979), Meyer de Schauensee (1966), and Haffer (1970b, 1974, 1978). Taxa include, top: *Pteroglossus beauharnaesii* (solid line), *P. mariae* (long dash), *Galbula cyanescens* (dot-dash), *Thamnomanes schistogynus* (dotted), and *Phaethornis philippi* (short dash); bottom: *Pyrrhura rupicola* (solid line), *Neopelma sulphureiventer* (long dash), *Muscisaxicola fluviatilis* (dot-dash), *Myrmoborus melanurus* (dotted), and *Psophia leucoptera* (Short dash).

distribution within some areas of endemism comprised predominately of lowland tropical species. For example, the Chocó Rainforest Center (area 1, below) is primarily lowland but also includes some more montane elements such as *Diglossa indigotica, Bangsia edwardsi,* and *B. aureocincta.* In doing this, I advance the hypothesis that these lowland tropical and "lower" subtropical species have shared a common history. Likewise, I have defined the montane centers to include taxa that may extend downward to "upper" subtropical or temperate forests. Whether, in fact, these taxa do share elements of history with puna-páramo taxa will have to await future systematic analyses.

Finally, I agree with Haffer (1978:42) that the use of taxa at the subspecific level often presents numerous difficulties in biogeographic analysis. Areas of endemism are defined by discrete (diagnosable), differentiated taxonomic units. Unfortunately, ornithologists have variously classified these as either species or subspecies, and all too frequently decisions about taxonomic ranking have been entirely subjective. To make matters worse, many subspecies are not discrete evolutionary taxa, but simply arbitrary descriptors of clinal variation. Such subspecies have little, if any, relevance for defining areas of endemism. In this study, I have provisionally accepted named taxa as given in the literature, and considering the scope of this paper, it has not proven possible to evaluate the status of each taxon as a differentiated taxonomic unit. Future analyses of individual areas or of specific taxonomic groups will have to verify the identity of these diagnosable taxa, because they constitute the basis for investigating taxonomic differentiation and the historical biogeography of areas (Cracraft 1983a).

Inclusion of subspecies creates other problems, particularly in introducing an imbalance in the composition of lists of endemics, as several of my colleagues have pointed out. Thus, areas of endemism in Venezuela and Colombia include a proportionately higher number of subspecies simply because recent checklists are available for these countries. Some colleagues have suggested that I eliminate subspecies altogether. Although this suggestion has merit, the potential for eliminating biogeographically important data strongly argues against it. If some subspecies on these lists represent ill-defined clinal variation, then no information is gained by including them, but neither is any gained by excluding them. If, however, some of these subspecies are diagnosable taxonomic units (which many are), then eliminating them from the lists does exclude relevant systematic data. I therefore have chosen to adopt a conservative approach and include subspecies even though this sometimes may give the appearance of imbalance. In adopting this policy, moreover, it should be clear that I have not included all the taxa that may be considered endemic to a given area. In this regard, species' lists for each of the recognized areas are reasonably complete, but the lists are less complete with respect to taxa of subspecific rank.

AREAS OF ENDEMISM

Chocó Rainforest Center

This area of endemism (Fig. 2: area 1) extends from the eastern Panamanian lowlands south along the west coast of Colombia to the east end of the Golfo de Guayaquil where the rainforest gives way to a region of more arid vegetation. The eastern border of the Chocó Center is determined by the upper subtropical-temperate biome of the Andes. The northern limits of the Chocó apparently are associated with the Pacific coast ranges (Serranía de Baudó, Serranía de los Saltos) that extend into Panamá, and with the broad delta of the Río Atrato (see Haffer 1967a, b, 1970a, 1975). Chapman (1917:106–117; 1926:54–62) discussed the Chocó (= his Colombian-Pacific) avifauna in detail.

Representative endemic taxa are:

Crypturellus kerriae
C. berlepschi
Leucopternis occidentalis
L. plumbea
Micraster plumbeus
Penelope ortoni
Odontophorus melanonotus
O. dialeucos
O. e. erythrops
Aramides wolfi
Leptotila pallida

Geotrygon goldmani
Columba goodsoni
C. subvinacea ruberrima
Ara ambigua guayaquilensis
Pionopsitta pulchra
Neomorphus radiolosus
Nyctibius aethereus chocoensis
Phaethornis yaruqui
Coeligena wilsoni
Eriocnemis mirabilis
Urochroa b. bougueri

Amazilia rosenbergi
Heliodoxa imperatrix
Goldmania violiceps
Urosticte b. benjamini
Aglaiocercus coelestis
Boissonneaua jardini
Heliangelus strophianus
Haplophaedia lugens
Philodice mitchellii
Acestrura berlepschi
Trogon massena australis
T. melanurus mesurus
Bucco noanamae
Capito squamatus
C. quinticolor
Semnornis ramphastinus
Andigena laminirostris
Pteroglossus erythropygius
Piculus leucolaemus litae
Xiphorhynchus erythropygius aequatorialis
Thripadectes ignobilis
Cranioleuca erythrops griseigularis
Sclerurus guatemalensis salvini
Margarornis stellatus
Formicarius nigricapillus destructus
Pittasoma rufopileatum
Gymnopithys leucaspis aequatorialis
Sipia rosenbergi
S. berlepschi
Anabacerthia variegaticeps temporalis
Xenornis setifrons
Myrmotherula brachyura ignota
Phaenostictus mcleannani pacificus
Myrmeciza exsul maculifer
Dysithamnus puncticeps flemmingi
D. o. occidentalis
Grallaria haplonota parambae
G. perspecillata periophthalmica
Lipaugus unirufus castaneotinctus
Carpodectes hopkei

Pipreola jucunda
Cephalopterus penduliger
Pachyramphus spodiurus
Machaeopterus deliciosus
Pipra mentalis minor
Rhynchocyclus brevirostris pacificus
Myiophobus phoenicomitra litae
Rhytipterna holoerythra rosenbergi
Mitrephanes phaeocercus berlepschi
Pseudotriccus pelzelni annectens
Cyanolyca pulchra
Campylorhynchus turdinus aenigmaticus
Thryothorus nigricapillus schottii
T. n. connectens
T. fasciatoventris albigularis
T. spadix
Cyphorhinus aradus phaeocephalus
C. a. chocoanus
Entomodestes coracinus
Psarocolius cassini
Basileuterus chrysogaster chlorophrys
Diglossa indigotica
Dacnis berlepschi
Tangara johannae
T. palmeri
T. rufigula
T. i. icterocephala
T. florida auriceps
T. l. lavinia
Erythrothlypis salmoni
Anisognathus notabilis
Bangsia aureocincta
B. edwardsi
B. rothschildi
Chlorothraupis stolzmanni
Chlorospingus semifuscus
Chlorochrysa phoenicotis
Chlorophonia flavirostris
Oreothraupis arremonops

Comments.—Some of the species endemic to this center range into the lowland forests of Panamá. The above list could have been expanded if those Chocó forms whose distributions include both Panamá and parts of Costa Rica were also included. Müller (1973) placed the Nechí Center as a subcenter of the Chocó, but I keep them separate here. When studied in detail, avian distributions may show distinct subdivisions within the Chocó, as, apparently, some plant groups do (Gentry 1982).

NECHÍ RAINFOREST CENTER

This area of endemism, lying north of the Andes, is closely associated with the Chocó Center (Fig. 2: area 2). Müller (1973:38) suggested that this center lies between the Río Sinú and Río Cauca, but very few avian taxa are so narrowly distributed. Instead, with regard to avian distributions, the Nechí area extends from the Serranía de Abibe and Río Sinú on the west to the lower Magdalena Valley and the eastern Andean cordillera. The northern limits of the center are ambiguous: many, perhaps most, of the endemics are restricted to the rainforest biome, but some taxa find their northern limits within the dry forest.

Representative endemic taxa are:

Ortalis garrula
Crax alberti

Leptotila c. cassinii
Phaeochroa cuvierii berlepschi

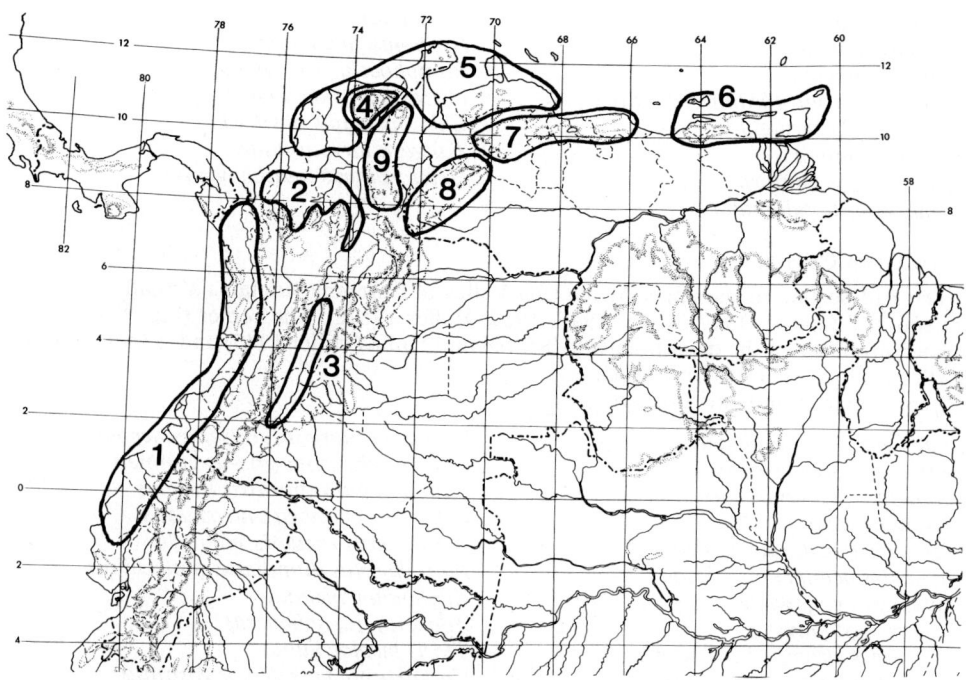

FIG. 2.   Postulated areas of endemism for birds in northern South America: 1, Chocó Rainforest Center; 2, Nechí Rainforest Center; 3, Magdalena Center; 4, Santa Marta Center; 5, Guajiran Center; 6, Parian Center; 7, Venezuelan Montane Center; 8, Meridan Montane Center; and 9, Perijan Montane Center.

Lepidopyga coeruleogularis
L. lilliae
Trogon melanurus macroura
Brachygalba salmoni
Malacoptila fulvogularis huilae
M. panamensis magdalenae
Capito hypoleucus
Synallaxis b. brachyura
Deconychura longicauda minor
Thamnophilus nigriceps magdalenae
T. n. nigriceps
Grallaria perspicillata pallidior
Myrmotherula fulviventris salmoni
Attila spadiceus caniceps

Laniocera rufescens griseigula
Aphanotriccus audax
Thryothorus f. fasciatoventris
T. rutilus interior
Campylorhynchus nuchalis pardus
C. zonatus brevirostris
Cyphorhinus aradus propinquus
Turdus grayi incomptus
Myadestes ralloides candelae
Gymnostinops quatimozinus
Habia fusicauda erythrolaema
H. gutturalis
Atlapetes atricapillus

Comments. — Some of the above species also extend into the rainforests of Panama. Because the Chocó and Nechí centers share many taxa, numerous workers have considered them to be closely related. Some of the taxa endemic in this combined area (some also range into Panamá include):

Geotrygon veraguensis
Damophila julie
Heliothryx b. barroti
Androdon aequatorialis
Trogon comptus
Notharchus pectoralis
Nystalus radiatus
Capito maculicoronatus
Pteroglossus sanguineus

Xiphorhynchus lachrymosus alarum
Xenerpestes minlosi
Phaenostictus mcleannani chocoanus
Myrmeciza exsul cassini
Dysithamnus puncticeps intensus
Gymnopithys leucaspis ruficeps
Hylophylax n. naevioides
Microrhopias quixensis consobrina
Myrmotherula f. fulviventris

Cotinga nattererii
Manacus vitellinus
Sapayoa aenigma
Campylorhynchus turdinus harterti
Polioptila schistaceigula
Hylophilus minor
Geothlypis s. semiflava
Dacnis venusta fuliginata

D. viguieri
Euphonia fulvicrissa
Ramphocelus icteronotus
Mitrospingus cassinii
Chlorothraupis olivacea
Tangara larvata fanny
Heterospingus xanthopygius

MAGDALENA CENTER

The arid portion of the upper Magdalena Valley of Colombia contains a small number of endemics (Chapman 1917; Miller 1947, 1952; Haffer 1967c). The center (Fig. 2: area 3) is located primarily in the departments of Tolima and Huila (Miller 1947) and is limited to the north by more mesic environments.

Representative endemic taxa are:

Ortalis guttata columbiana
Colinus cristatus leucotis
Columbina passerina parvula
Leptotila conoveri
Speotyto cunicularia tolimae
Chordeiles acutipennis crissalis
Lepidopyga g. goudoti
Nystalis r. radiatus
Melanerpes chrysauchen pulcher
Myrmeciza longipes boucardi
Chamaeza ruficauda turdina
Thamnophilus doliatus albicans
Manacus manacus flaveolus

Idioptilon margaritaceiventer septentrionalis
Campylorhynchus griseus zimmeri
Catharus aurantiirostris insignis
Polioptila plumbea anteocularis
Basileuterus rivularis motacilla
Ramphocelus dimidiatus molochinus
Euphonia concinna
Atlapetes fuscoolivaceus
Tiaris bicolor huilae
Coryphospingus pileatus rostratus
Arremon conirostris inexpectatus
Sporophila intermedia agustini

SANTA MARTA MONTANE CENTER

The Santa Marta Mountains of northeastern Colombia are isolated from neighboring mountain ranges by lowlands (Fig. 2: area 4; see Todd and Carriker 1922, for details about the avifauna).

Representative endemic taxa are:

Penelope argyrotis colombiana
Chamaepetes goudtii sanctaemarthae
Odontophorus a. atrifrons
Colinus cristatus littoralis
Geotrygon linearis infusca
Pionus sordidus saturatus
Pyrrhura viridicata
Caprimulgus parvulus heterurus
Campylopterus phainopeplus
Oxypogon guerinii cyanolaemus
Anthocephala f. floriceps
Coeligena phalerata
Ramphomicron dorsale
Trogon personatus sanctaemartae
Pharomachrus fuligidus festatus
Veniliornis fumigatus exsul
Piculus rubiginosus alleni
Lepidocolaptes affinis sanctaemartae
Asthenes wyatti sanctaemartae
Leptasthenura andicola extima
Xenops rutilans phelpsi
Sclerurus albigularis propinquus
Anabacerthia striaticollis anxia

Premnoplex brunnescens coloratus
Synallaxis fuscorufa
Cranioleuca hellmayri
Automolus rubiginosus rufipectus
Formicarius analis virescens
Grallaria bangsi
G. rufula spatiator
Scytalopus l. latebricola
S. femoralis sanctaemartae
Pipreola aureopectus decora
Zimmerius viridiflavus minimus
Mionectes olivaceus galbinus
Tyranniscus nigrocapillus flavimentum
Mecocerculus leucophrys montensis
Hemitriccus granadensis lehmanni
Pyrrhomyias cinnamomea assimilis
Ochthoeca diadema jesupi
O. rufipectoralis poliogastra
Attila spadiceus parvirostris
Myiotheretes pernix
Campylorhynchus zonatus curvirostris
Henicorhina leucophrys anachoreta
H. l. bangsi

Troglodytes solstitialis monticola
Turdus olivater sanctaemartae
T. fuscater cacozelus
Catharus fuscater sanctaemartae
C. aurantiirostris sierrae
Cinclus leucocephalus rivularis
Myioborus flavivertex
M. miniatus sanctaemartae
Basileuterus basilicus
B. culicivorus indignus
B. conspicillatus

Anisognathus melanogenys
Dubusia taeniata carrikeri
Thraupis cyanocephala margaritae
Chlorophonia cyanea psittacina
Tersina viridis grisescens
Catamenia oreophila
C. analis alpica
Atlapetes melanocephalus
A. torquatus basilicus
Carduelis spinescens capitanea

GUAJIRAN CENTER

This area of endemism includes the arid coastal avifauna of northern Colombia and Venezuela (Fig. 2: area 5). As used here, the center extends eastward from the region of Barranquilla and includes the lowlands around the Sierra Nevada de Santa Marta, the Península de Guajira, and northern Zulia, northern Lara, and Falcón, Venezuela (Hueck and Seibert 1981). Some of the taxa are distributed various distances eastward along the Venezuelan coast (see Haffer 1967c).

Representative endemic taxa are:

Columba corensis
Amazona b. barbadensis
Leucippus fallax
Picumnus c. cinnamomeus
P. c. perijanus
Melanerpes rubricapillus paraguanae
Hypnelus ruficollis decolor
Xiphorhynchus picus paraguanae
X. p. picirostris
Synallaxis albescens perpallida

Poecilurus candei venezuelensis
P. c. candei
Sakesphorus canadensis phainoleucus
Todirostrum viridanum
Inezia tenuirostris
Icterus icterus ridgwayi
Thraupis glaucocolpa
Cardinalis phoeniceus
Arremonops tocuyensis
Saltator orenocensis rufescens

PARIAN MONTANE CENTER

The Parian Montane Center of endemism (Fig. 2: area 6) includes the coastal mountain avifauna of northeastern Anzoátegui, northwestern Monagas, Sucre, and the Península de Paria. The western limits of this center extend to the lowlands of central Anzoátegui. The Península de Paria can also be considered to form a discrete subcenter.

Representative endemic taxa are:

Tinamus tao septentrionalis
Pionus sordidus antelius
Pyrrhura leucotis auricularis
Leptotila rufaxilla hellmayri
Columba subvinacea peninsularis
Chaetocercus j. jourdanii
Aglaiocercus kingi berlepschi
Thalurania furcata refulgens
Phaethornis g. guy
Campylopterus ensipennis
Hylonympha macrocerca
Aulacorhynchus sulcatus erythrognathus
Veniliornis k. kirkii
Piculus r. rubiginosus
Picumnus squamulatus obsoletus
Syndactyla guttulata pallida
Premnoplex brunnescens tatei
P. b. pariae
Synallaxis cinnamomea striatipectus
S. c. pariae

Grallaria haplonota pariae
Grallaricula nana cumanensis
G. n. pariae
Leptopogon superciliaris pariae
Zimmerius viridiflavus cumanensis
Phyllomyias virescens urichi
Mecocerculus leucophrys nigriceps
Myiophobus flavicans caripensis
Pyrrhomyias cinnamomea spadix
P. c. pariae
Pipreola formosa rubidior
P. f. pariae
Platycichla flavipes melanopleura
Catharus aurantiirostris birchalli
Turdus serranus cumanensis
Cyanocorax violaceus pallidus
Cyclarhis gujanensis flavipectus
Icterus nigrogularis trinitatis
Myioborus pariae
Basileuterus griseiceps

B. culicivorus olivascens
B. tristriatus pariae
Diglossa venezuelensis
Hemispingus frontalis iteratus
Tachyphonus luctuosus flaviventris

Habia rubica crissalis
Ramphocelus carbo capitalis
Thraupis cyanocephala subcinerea
T. c. buesingi
Chlorophonia cyanea miniscula

Comments.—Some of the above taxa range to one or more of the offshore islands north and east of the mainland center.

VENEZUELAN MONTANE CENTER

This area of endemism (Fig. 2: area 7) includes avifaunal elements found in the Venezuelan Coastal Forest and Venezuelan Montane Forest centers of Müller (1973:54–56). It is defined by the limits of the coastal mountains of northern Venezuela, extending from eastern Lara on the west through Miranda on the east.
Representative endemic taxa are:

Crypturellus obsoletus cerviniventris
Pyrrhura hoematotis
P. leucotis emma
Pionus sordidus
Aglaiocercus kingi margarethae
Metallura tyrianthina chloropogon
Ocreatus underwoodii polystictus
Coeligena coeligena zuloagae
C. c. coeligena
Heliodoxa l. leadbeateri
Adelomyia melanogenys aeneosticta
Amazilia tobaci monticola
Aulacorhynchus s. sulcatus
Thripadectes virgaticeps klagesi
Philydor rufus columbianus
Anabacerthia striaticollis venezuelana
Pseudocolaptes boissonneautii striaticeps
Synallaxis unirufa castanea
S. cinnamomea bolivari
Syndactyla g. guttulata
Premnoplex brunnescens rostratus
Lepidocolaptes affinis lafresnayi
Xiphorhynchus triangularis hylodromus
Chamaeza ruficauda chionogaster
Grallaria h. haplonota
G. ruficauda avilae

Grallaricula loricata
G. nana olivascens
Pipreola f. formosa
P. aureopectus festiva
Phylloscartes opthalmicus purus
P. venezuelanus
Pipromorpha oleaginea abdominalis
Acrochordopus burmeisteri viridiceps
Tyranniscus vilissimus petersi
Mecocerculus leucophrys palliditergum
Hemitriccus granadensis federalis
Ochthoeca diadema tovarensis
Henicorhina leucophrys venezuelensis
Psarcolius angustifrons oleagineus
Basileuterus tristriatus bessereri
Diglossa albilatera federalis
D. cyanea tovarensis
D. c. caerulescens
Conirostrum albifrons cyanopterum
Tangara rufigenis
Chlorospingus ophthalmicus falconensis
Hemispingus frontalis hanieli
Rhodinocichla r. rosea
Eucometis penicillata affinis
Thraupis cyanocephala olivicyanea
Chlorophonia cyanea frontalis

MERIDAN MONTANE CENTER

The Andes of western Venezuela constitute an area of endemism for species restricted to páramo and montane forest habitats (Fig. 2: area 8). The Táchira depression in the region of the Río Torbes and Río Quinimari (see Vuilleumier and Ewert 1978) defines the southern limits for most taxa. Some endemics extend slightly south to Cerro de Tamá, which is provisionally included in this center although its closest biogeographic affinities eventually may be shown to be with the eastern Colombian Andes. The northern limits of the Meridan Center appear to be defined by the lowlands of southern Lara, but this will have to be determined more accurately by mapping individual species' distributions.
Representative endemic taxa are:

Crypturellus obsoletus knoxi
Zenaida auriculata pentheria
Hapalopsitta amazonina theresae
Pyrrhura rhodocephala
Heliangelus spencei

Acestrura heliodor meridae
Oxypogon guerinii lindeni
Metallura tyrianthina oreopola
Ramphomicron microrhynchus andicolum
Coeligena torquata conradii

Lafresnaya lafresnayi greenwalti
Amazilia saucerrottei braccata
A. distans
Thalurania furcata rostrifera
Piculus rivolii meridae
Picumnus olivaceus tachirensis
Sittasomus griseicapillus tachirensis
Dendrocincla tyrannina hellmayri
Thripadectes flammulatus bricenoi
Asthenes wyatti mucuchiesi
Synallaxis gularis cinereiventris
S. unirufa meridana
Leptasthenura andicola certhia
Cinclodes fuscus heterurus
Schizoeaca coryi
Grallaria chthonia
G. g. griseonucha
G. g. tachirae
G. ruficapilla nigrolineata
Leptopogon rufipectus venezuelanus
Phyllomyias nigrocapillus aureus
Ochthoeca diadema meridana
O. cinnamomeiventris nigrita

O. fumicolor superciliosa
Myiotheretes fumigatus lugubris
Turdus serranus atrosericeus
Henicorhina leucophrys meridana
Cistothorus meridae
C. platensis tamae
Cyanocorax yncas andicolus
Cyanolyca viridicyana meridana
Basileuterus tristriatus meridanus
Myioborus albifrons
Diglossa carbonaria gloriosa
Conirostrum sitticolor intermedium
Chlorospingus ophthalmicus venezuelensis
Hemispingus goeringi
H. frontalis ignobilis
H. superciliaris chrysophrys
H. reyi
Rhodinocichla rosea beebei
Habia rubica coccinea
Atlapetes albofrenatus meridae
Pheucticus aureoventris meridensis
Sporophila obscura haplochroma

PERIJAN MONTANE CENTER

The Perijan Montane Center of endemism (Fig. 2: area 9) is recognized for the Serranía de Perijá which straddles the Venezuelan-Colombian border. It is isolated to the east, west, and north by lowlands. The southern limits are ambiguous and require more investigation. Many species extend their distributions into the Andes of Norte de Santander (Colombia) and Táchira (Venezuela). The valley of the Río de Oro in Norte de Santander may prove to be an important distributional barrier.

Representative endemic taxa are:

Crax pauxi gilliardi
Amazona autumnalis salvini
Pionus chalcopterus cyanescens
P. sordidus ponsi
Pyrrhura picta pantchenkoi
Metallura iracunda
Heliangelus amethysticollis violiceps
Coeligena bonapartei consita
C. coeligena zuliana
Picumnus olivaceus eisenmanni
Sittasomus griseicapillus perijanus
Campylorhamphus pusillus tachirensis
Synallaxis gularis brunneidorsalis
Asthenes wyatti perijanus
Schizoeaca fuliginosa perijana
Sclerurus albigularis kunanensis
Myrmeciza immaculata brunnea
Grallaria rufula saltuensis

G. ruficapilla perijana
Acrochordopus burmeisteri wetmorei
Myiopagis viridicata zuliae
Phylloscartes superciliaris griseocapillus
P. poecilotis pifanoi
Platyrinchus mystaceus perijanus
Myiobius villosus schaeferi
Ochthoeca diadema rubellula
O. rufipectoralis rubicundulus
Catharus aurantiirostris barbaritoi
Turdus fuscater clarus
Cinnycerthia unirufa chakei
Diglossa cyanea obscura
D. caerulescens ginesi
Chlorospingus ophthalmicus ponsi
Hemispingus frontalis flavidorsalis
Habia rubica perijana
Anisognathus lachrymosus pallididorsalis

PANTEPUI CENTER

Müller (1973:62–68) divided the Pantepui region (Mayr and Phelps 1967; Dickerman and Phelps 1982) into two areas of endemism, a Roraima fauna for the savannah forms of eastern Venezuela, and the Pantepui fauna for the montane regions of the entire Venezuelan highlands. The distribution patterns of birds suggest that a classification based on geographic areas rather than on ecological criteria may be best for deciphering the historical biogeography of the region (Fig. 3: area 10). Inasmuch as the relationships of the various endemics are uncertain,

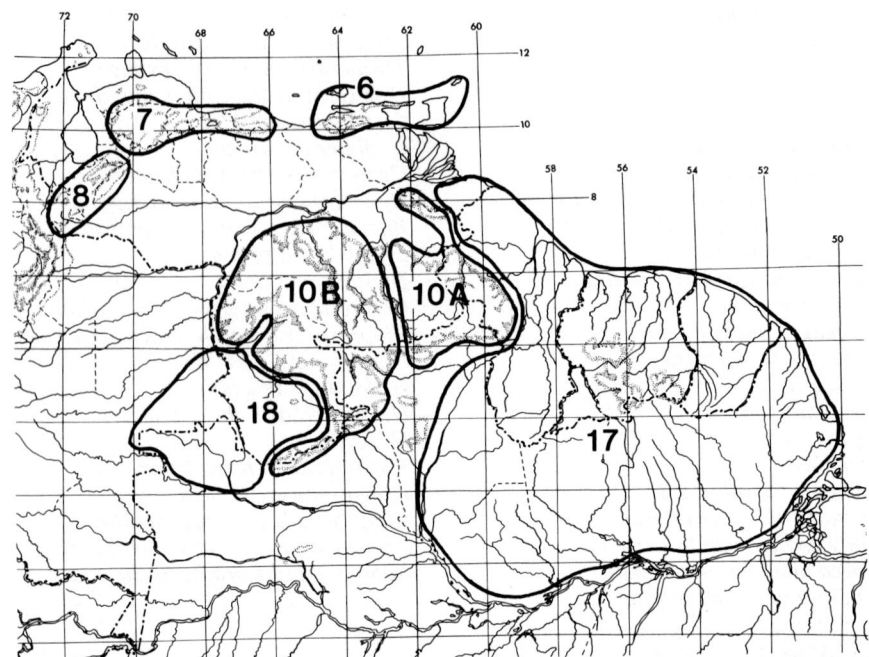

FIG. 3.   Postulated areas of endemism for the birds of northeastern South America: 6, Parian Montane Center; 7, Venezuelan Montane Center; 8, Meridan Montane Center; 10A, Gran Sabana Subcenter of Pantepui Center; 10B, Duida Subcenter of Pantepui Center; 17, Guyanan Center; and 18, Imeri Center.

exact subdivision of the Pantepui Center must remain tentative. Two moderately well-defined subcenters seem recognizable, although certainly each will be shown to be divisible further; the Gran Sabana Subcenter (Fig. 3: area 10A) includes the tepuis located east of the Río Caroní near the Venezuelan-Guyanan border, and the Duida Subcenter (Fig. 3: area 10B), those highland areas to the west of the Río Caroní. The historical unity of the Duida Subcenter is much less certain than that of the Gran Sabana.

Representative endemic taxa of the entire Pantepui region are:

Otus guatemalae roraimae
Caprimulgus whitelyi
Neomorphus rufipennis
Streptoprocne zonaris albicincta
Doryfera johannae guianensis
Polytmus milleri
Colibri coruscans germanus
Amazilia lactea zimmeri
A. versicolor hollandi
Heliodoxa xanthogonys
Trogon personatus roraimae
Veniliornis kirkii monticola
Synallaxis m. macconnelli
Chamaeza campanisona obscura
Grallaria guatimalensis roraimae
Phylloscartes nigrifrons
Contopus fumigatus duidae

Elaenia pallatangae olivina
Mecocerculus leucophrys roraimae
Tolmomyias suphurescens duidae
Platyrinchus mystaceus duidae
Knipolegus poecilurus salvini
Turdus ignobilis murinus
Microcerculus bambla caurensis
Hylophilus sclateri
Macroagelaius imthurni
Basileuterus bivittatus roraimae
Parula pitiayumi roraimae
Myioborus miniatus verticalis
Coereba flaveola roraimae
Chlorophonia cyanea roraimae
Piranga flava haemalea
Catamenia homochroa duncani

Representative endemic taxa of the Gran Sabana Subcenter are:

Crypturellus ptaritepui
Pyrrhura egregia

Nyctibius aethereus longicauda
Caprimulgus longirostris roraimae

Campylopterus hyperythrus
Phaethornis bourcieri whitelyi
Lophornis p. pavonina
L. p. punctigula
Amazilia viridigaster cupreicauda
Trogon personatus ptaritepui
Aulacorhynchus derbianus whitelianus
Piculus rubiginosus viridissimus
P. r. guianae
Lochmias nematura chimantae
L. n. castanonota
Cranioleuca d. demissa
Xiphocolaptes promeropirhynchus tenebrosus
Grallaricula nana kukenamensis
Percnostola leucostigma saturata
P. l. obscura
Chamaeza campanisona fulvescens
Margarornis a. adjusta
Automolus r. roraimae
Myrmotherula behni inornata
M. s. simplex
Herpsilochmus r. roraimae
Dysithamnus mentalis ptaritepui
Pachyramphus viridis griseigularis
Pipreola w. whitelyi
P. w. kathleenae
Lipaugus streptophorus
Pipra serena suavissima
Chloropipo u. uniformis
Acrochordopus burmeisteri bunites
Todirostrum russatum
Phylloscartes c. chapmani
Hemitriccus margaritaceiventer auyantepui

Elaenia d. dayi
E. d. auyantepui
Platyrinchus mystaceus imatacae
P. m. ptaritepui
Pipromorpha oleaginea dorsalis
Turdus olivater roraimae
Myadestes leucogenys gularis
Troglodytes r. rufulus
T. r. fulvigularis
Cistothorus platensis alticola
Microcerculus u. ustulatus
M. u. obscurus
Cyphorhinus aradus urbanoi
Thryothorus coraya obscurus
T. c. ridgwayi
Ramphocaenus melanurus albiventris
Hylophilus ochraceiceps luteifrons
Myioborus brunniceps castaneocapillus
Diglossa m. major
D. m. gilliardi
D. m. disjuncta
D. m. chimantae
Mitrospingus o. oleagineus
M. o. obscuripectus
Tangara g. guttata
T. cyanoptera whitelyi
Haplospiza rustica arcana
Atlapetes p. personatus
A. p. collaris
Zonotrichia capensis macconnelli
Spinus magellanicus longirostris
Spordiornis rusticus

Representative endemic taxa of the Duida Subcenter are:

Otus choliba duidae
Glaucidium brasilianum duidae
Phaethornis b. bourcieri
Lophornis pavonina duidae
Chlorostilbon mellisugus duidae
Colibri coruscans rostratus
Campylopterus d. duidae
C. d. guaiquinimae
Trogon personatus duidae
Aulacorhynchus derbianus duidae
Cranioleuca demissa cardonai
Synallaxis moesta yavii
Philydor hylobius
Xiphocolaptes promeropirhynchus neblinae
Glyphorhynchus spirurus coronobscurus
Margarornis adjusta mayri
M. a. duidae
M. a. obscurodorsalis
Automolus roraimae paraguensis
A. r. duidae
A. r. urutani
Taraba major duidae
Chamaeza campanisona yavii
C. c. huachamacarii

Myrmotherula behni yavi
M. b. camanii
M. simplex duidae
M. s. guaiquinimae
M. s. pacaraimae
Percnostola c. caurensis
P. c. australis
Herpsilochmus roraimae kathleenae
Dysithamnus mentalis spodionotus
Chloropipo uniformis duidae
Platyrinchus mystaceus ventralis
Knipolegus poecilurus paraguensis
Phylloscartes chapmani duidae
Hemitriccus margaritaceiventer duidae
H. m. breweri
Elaenia dayi tyleri
Mecocerculus leucophrys parui
Troglodytes rufulus yavii
T. r. duidae
T. r. wetmorei
Microcerculus ustulatus duidae
M. u. lunatipectus
Turdus olivater duidae
T. o. paraguensis

T. o. kemptoni
T. ignobilis arthuri
Ramphocaenus melanurus duidae
Myioborus brunniceps duidae
M. b. maguirei
M. cardonai
M. albifacies
Diglossa d. duidae
D. d. hitchcocki

Tangara guttata chrysophrys
Pipraeidea melanonota venezuelensis
Atlapetes personatus duidae
A. p. jugularis
A. p. parui
A. p. paraguensis
Emberizoides herbicola duidae
Zonotrichia capensis inaccessibilis

NORTH ANDEAN CENTER

Defining areas of endemism in the Andes is particulary difficult, in part because they span thousands of kilometers in distance and height. Not only are species' distributions often poorly known, but many species are confined to specific elevational-ecological zones whose geographic boundaries may differ latitudinally and longitudinally from those of zones located higher or lower. Furthermore, the complex topography and ecology of the Andes fragment species' ranges, making it unrealistic to expect that species will be distributed continuously. Eventually, areas of endemism within the Andes will have to be evaluated by plotting individual specimen localities. In this paper three major areas of endemism are identified: the North Andean Center, the Peruvian Andean Center (with three subcenters), and the Austral Andean Center. I fully expect that a detailed analysis of the Andean avifauna will reveal many additional subdivisions of these main areas.

The North Andean Center includes those species generally restricted to the upper subtropical, temperate, and páramo biomes of Colombia and Ecuador (Fig. 4: area 11). The northern limits of many species in this area stop short of the Cordillera de Mérida and the Serranía de Perijá. In the south, the distributions of many subtropical and temperate biome species terminate in northern Peru near the valley of the Río Marañón, which represents an important geographic boundary for avian distribution. Vuilleumier (1968, 1971, 1980) also has recognized the Northern Peruvian Low (Fig. 4), lying northwest of the Río Marañón, as an important barrier to birds of the páramo, puna, and montane forest.

Representative endemic taxa are:

Penelope barbata
P. cauca
Chamaepetes goudoti fagani
Phalcoboenus carunculatus
Odontophorus hyperythrus
O. strophium
Ognorhynchus icterotis
Bolborhynchus ferrugineifrons
Pyrrhura calliptera
Touit stictoptera
Haplopsittaca a. amazonina
H. fuertesi
H. pyrrhops
Pionus chalcopterus
Eriocnemis derbyi
E. nigrivestis
E. godini
E. mosquera
Heliangelus exortis
Coeligena lutetiae
C. prunellei
C. orina
C. helianthea
Metallura baroni
M. odomae
M. williami
Chalcostigma herrani

C. heteropogon
Picumnus granadensis
Cinclodes e. excelsior
Asthenes wyatti azuay
A. w. aequatorialis
Synallaxis subpudica
Schizoeaca griseomurina
Siptornis striaticollis
Grallaria milleri
G. gigantea
G. alleni
G. rufocinerea
G. nuchalis
G. q. quitensis
G. q. alticola
Grallaricula lineifrons
G. peruviana
Thamnophilus multistriatus
Pipreola lubomirskii
Lipaugus fuscocinereus
Pyroderus scutatus occidentalis
Uromyias agilis
Muscisaxicola maculirostris niceforoi
M. m. rufescens
Agriornis montana solitaria
Polystictus pectoralis bogotensis
Leptopogon rufipectus

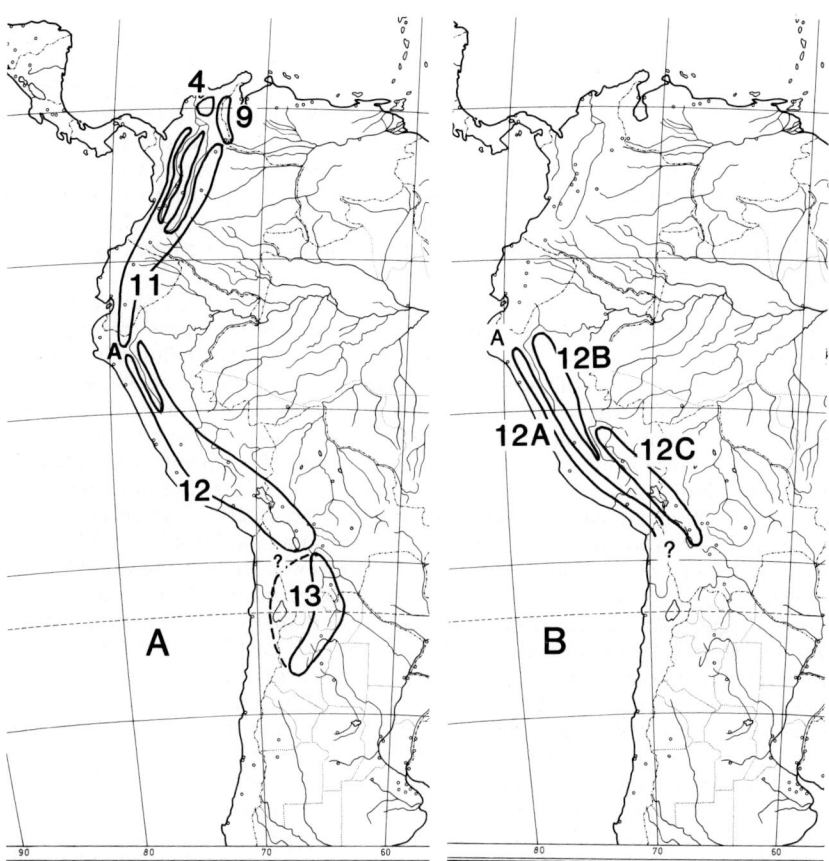

FIG. 4. Postulated areas of endemism for Andean birds: 4, Santa Marta Center; 9, Perijan Montane Center; 11, North Andean Center; 12, Peruvian Andean Center; 12A, West Peruvian Andean Subcenter; 12B, East Peruvian Andean Subcenter; 12C, South Peruvian Andean Subcenter; and 13, Austral Andean Center.

Myiophobus lintoni
Cyanolyca turcosa
Cistothorus apolinari
Thryothorus nicefori
Ramphocaenus melanurus badius
Cyclarhis nigrirostris
Hypopyrrhus pyrohypogaster
Macroagelaius subalaris
Basileuterus luteoviridis quindianus
B. l. richardsoni
Dacnis hartlaubi
Diglossa brunneiventris vuilleumieri
D. humeralis
D. aterrima
D. gloriosissima
Urothraupis stolzmanni

Iridosornis rufivertex
I. porphyrocephala
Chlorospingus canigularis conspicillatus
Chlorochrysa nitidissima
Buthraupis wetmorei
B. eximia
Bangsia melanochlamys
Habia cristata
Cnemoscopus r. rubrirostris
Hemispingus verticalis
Atlapetes pallidinucha
A. flaviceps
A. pallidiceps
A. s. schistaceus
A. l. leucopterus

Comments.—Some of the taxa in this center extend southward a short distance west of the Río Marañón. There, in the mountains of the departments of Cajamarca and Piura, broad river valleys (Quiroz, Chinchipe, Tabaconas, Huayllabamba, and Huancabamba rivers) depress the elevation below 2000 m, thus creating an effective barrier to the montane avifauna.

PERUVIAN ANDEAN CENTER

As defined here, this area of endemism includes the upper subtropical, temperate, and páramo-puna biomes extending from the valley of the Río Marañón in northern Peru south to central Bolivia (Fig. 4: area 12). Most of the avifauna resides on the eastern side of the Andes. The southern boundary of this center is uncertain, but many distributions seem to be limited by the dry valleys of the Grande, Chayanta, and Mizque rivers between Sucre and Cochabamba, Bolivia. In this area there is a marked north–south turnover in the avifauna, particularly in that restricted to the montane forests (see below; M. A. Traylor, pers. comm.). Hueck and Seibert (1981) indicated a substantial disruption of these forest biomes north and northwest of Sucre, Chuquisaca.

Although the Peruvian Andean Center probably constitutes a unified biogeographic entity relative to other centers of endemism within the Andes, it nevertheless can be subdivided into at least three subcenters. Two of these subcenters segregate the avifauna into eastern and western components. The first of these, termed here the West Peruvian Andean Subcenter (Fig. 4B: area 12A), lies to the south of the Northern Peruvian Low and to the west of the valley of the Río Marañón. Taxa found in this arid subtropical-temperate to puna biome (Parker et al. 1982) range south for variable distances along the Cordillera Occidental (some extend into Chile), and, consequently, the southern limits of this area are ill-defined. A second subcenter, the East Peruvian Andean Subcenter (Fig. 4B: area 12B), is delimited on the north and west by the Río Marañón, on the east by the tropical and subtropical lowlands, and on the south by the valley of the Río Apurímac. This subcenter defines the humid upper sub-tropical-temperate and puna biota within the Cordillera Central and Cordillera Oriental. Even though the Apurímac valley seems to constitute a barrier to many taxa, it appears less effective with increasing elevation. Distributional patterns within the East Peruvian Andean Subcenter are more complex than indicated here. The subcenter appears to be divisible into two or more, smaller areas of endemism (e.g., one between the Río Marañón and Río Huallaga, another between the latter and the Río Apurímac). The relationships of these areas to each other and to areas outside the East Peruvian Andean Subcenter are in need of much more study. Moreover, distributional patterns of species confined to the puna biome are often ambiguous with respect to their placement in the East Peruvian Andean or West Peruvian Andean subcenters. Finally, the third subcenter, the South Peruvian Andean Subcenter (Fig. 4B: area 12C), lies to the south of the Apurímac river valley and extends southward to Cochabamba, Bolivia (see above).

Representative endemic taxa, distributed in two or more subcenters, are:

Nothocercus nigrocapillus	*Grallaria andicola*
Bolborhynchus orbygnesius	*Pipreola intermedia*
Pionus tumultuosus	*P. frontalis*
Amazilia viridicauda	*Myiotheretes rufipennis*
Coeligena violifer	*M. fuscorufus*
Polyonymus caroli	*Myiophobus ochraceiventris*
Chalcostigma olivaceum	*Entomodestes leucotis*
Eubucco versicolor	*Psarcolius atrovirens*
Aulacorhynchus coeruleicinctis	*Conirostrum ferrugineiventre*
Geositta saxicolina	*Xenodacnis parina*
Cinclodes palliatus	*Iridosornis analis*
Leptasthenura striata	*Delothraupis castaneoventris*
L. yanacensis	*Hemispingus trifasciatus*
Asthenes humilis	*H. xanthophthalmus*
A. urubambensis	*Diuca speculifera*
Thripadectes scrutator	

Representative endemic taxa of the West Peruvian Andean Subcenter are:

Nothoprocta kalinowskii	*Chrysoptilus a. atricollis*
Otus robustus	*Geositta crassirostris*
Aglaeactis aliciae	*Upucerthia serrana*
Coeligena iris eva	*U. albigula*
Metallura phoebe	*Leptasthenura pileata*
Tephrolesbia griseiventris	*Synallaxis zimmeri*

Asthenes pudibunda
Automolus r. ruficollis
Zaratornis stresemanni
Ochthoeca piurae
O. pulchella jelskii
Anairetes reguloides
A. a. alpinus
Atlapetes rufigenis
A. seebohmi

A. nationi
Sicalis raimondii
Incaspiza pulchra
I. personata
I. ortizi
I. laeta
Poospiza alticola
P. rubecula

Representative endemic taxa of the East Peruvian Andean Subcenter are:

Nothoprocta taczanowskii
N. curvirostris peruviana
Podiceps taczanowskii
Hapalopsittaca melanotis peruviana
Oreotrochilus melanogaster
Aglaeactis castelnaudii
Metallura theresiae
M. eupogon
Oreonympha nobilis
Loddigesia mirabilis
Aulacorhynchus huallagae
Picumnus steindachneri
Cranioleuca albicapilla
Schizoeaca palprebalis
S. fuliginosa peruviana
Asthenes ottonis
A. virgata
A. wyatti graminicola
Thripophaga berlepschii

Grallaria przewalskii
G. capitalis
G. quitensis atuensis
Grallaricula flavirostris similis
Scytalopus macropus
Ampelion sclateri
Pipreola pulchra
Chloropipo unicolor
Ochthoeca pulchella similis
Anairetes agraphia squamigera
Leptopogon tacznowskii
Agriornis montana insolens
Basileuterus luteoviridis striaticeps
Tangara v. viridicollis
T. argyrofenges caeruleigularis
Iridosornis reinhardti
Thlypopsis pectoralis
Cnemoscopus rubrirostris chrysogaster

Representative endemic taxa of the South Peruvian Andean Subcenter are:

Odontophorus balliviani
Hapalopsittaca m. melanotis
Aglaeactis pamela
Metallura aeneocauda
Andigena cucullata
Leptasthenura xenothorax
Cinclodes excelsior aricomae
Cranioleuca marcapatae
C. albiceps
Schizoeaca helleri
S. harterti
Asthenes berlepschi
A. wyatti punensis
Thamnophilus aroyae
Terenura sharpei
Grallaria albigula
G. erythrotis
G. erythroleuca
G. andicola punensis

Grallaricula flavirostris boliviana
Conopophaga ardesiaca
Chirocylla uropygialis
Ochthoeca p. pulchella
Myiophobus inornatus
Pseudotriccus simplex
Hemitriccus rufigularis
H. spodiops
Anairetes a. agraphia
Mecocerculus hellmayri
Zimmerius bolivianus
Basileuterus luteoviridis euophrys
B. signatus
Diglossa c. carbonaria
D. mystacalis
Tangara a. argyrofenges
Iridosornis jelskii bolivianus
Creurgops dentata
Poospiza caesar

AUSTRAL ANDEAN CENTER

This area of endemism is defined for those species primarily restricted to the upper sub-tropical-temperate forest biome extending from central Chuquisaca and western Santa Cruz, Bolivia, south through the eastern Andes to Tucumán and southern Catamarca, Argentina (Fig. 4A: area 13). In addition, many high Andean forms have their southern limits in this region (or slightly farther south), which is apparently related to the end of the puna vegetation at about 26°S latitude (Hueck and Seibert 1981). The northern limits of this area of endemism, particularly in the western puna zone, are ambiguous, as many taxa are distributed north into the Andes of Peru (see below). Nevertheless, a distinct area of endemism for puna species does seem to begin in northwestern Bolivia, in the vicinity of Lake Titicaca, and to extend southward along the Andes. For convenience, both high montane forest species and those of the puna are included in this area of endemism; future systematic work will reveal whether this is valid biogeographically.

Representative endemic taxa are:

Penelope dabbenei	*Asthenes steinbachi*
P. obscura bridgesi	*Batara cinerea argentina*
Metriopelia aymara	*Scytalopus superciliaris*
M. morenoi	*Thamnophilus ruficapillus cochabambae*
Amazona tucumana	*T. r. subfasciatus*
Leptotila megalura	*Phylloscartes ventralis tucumanus*
Cypseloides major	*Todirostrum plumbeiceps viridiceps*
Eriocnemis glaucopoides	*T. p. obscurum*
Sappho sparganura	*Cinclus shultzi*
Microstilbon burmeisteri	*Mimus dorsalis*
Oreotrochilus leucopleurus	*Oreopsar bolivianus*
O. adela	*Basileuterus bivittatus argentinae*
Veniliornis frontalis	*Myioborus b. brunniceps*
Geositta rufipennis	*Atlapetes fulviceps*
G. isabellina	*A. citrinellus*
G. punensis	*Poospiza boliviana*
Synallaxis superciliosa	*P. hypochondria*
Upucerthia albigula	*P. erythrophrys*
U. ruficauda	*P. nigrorufa whitii*
U. andaecola	*Sicalis luteocephala*
U. harterti	*Phrygilus dorsalis*
Leptasthenura fuliginiceps	

Comments.—A substantial number of taxa are distributed over part or whole of both the Peruvian Andean Subcenter and the Austral Andean Center. Generally speaking, most of these taxa are restricted to the puna biome, but some occupy the forests of the temperate and upper subtropical biomes. Some of these endemics include:

Nothoprocta pentlandii	*M. juninensis*
N. ornata	*M. frontalis*
Tinamotis pentlandii	*M. albifrons*
Phalcoboenus megalopterus	*Anairetes flavirostris*
Aeronautes andecolus	*Petrochelidon andecola*
Picumnus dorbygnianus	*Turdus n. nigriceps*
Geositta tenuirostris	*Thlypopsis ruficeps*
Upucerthia validirostris	*Sicalis lutea*
Cinclodes atacamensis	*S. uropygialis*
Asthenes maculicauda	*S. olivascens*
Phacellodomus striaticeps	*Phrygilus atriceps*
Conopophaga ardesiaca	*Spinus crassirostris*
Kniplolegus signatus	*S. atratus*
Phyllomyias sclateri	*S. uropygialis*
Muscisaxicola rufivertex	

TUMBESAN CENTER

The Tumbesan Center, basically equivalent to the Ecuadorian Subcenter of Müller (1973: 100–101; see also Chapman 1926), lies directly to the south of the Chocó Center (Fig. 5: area 14). It includes an avifauna generally endemic to the narrow strip of dry forest that extends from slightly north of the Golfo de Guayaquil down the coast to Libertad, Peru (some species may occupy more humid habitats in this area). A small proportion of the taxa assigned to this center extend slightly north into the subtropical zone of the Chocó.

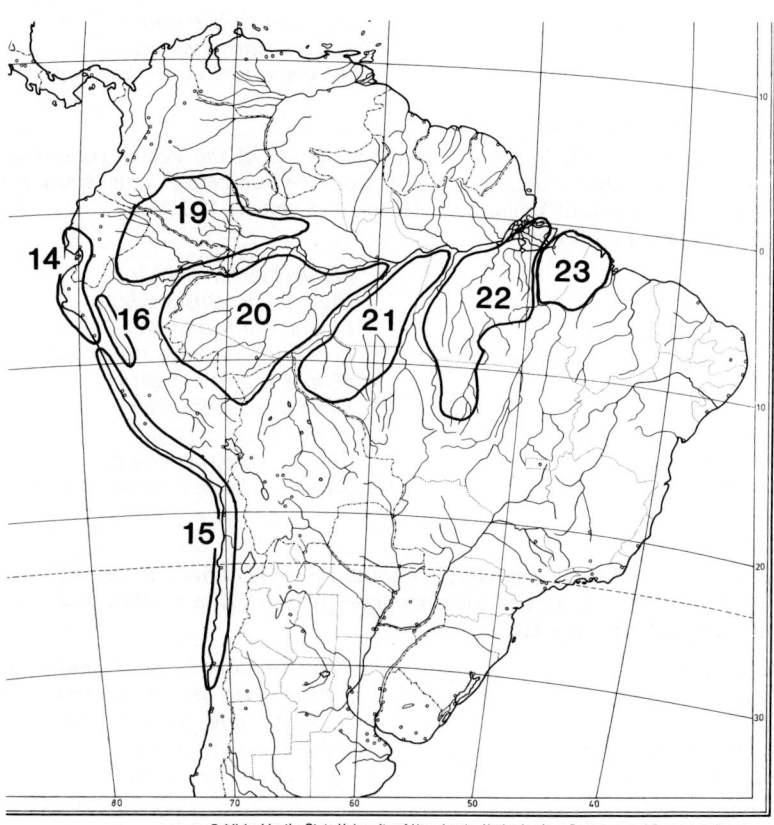

Published by the State University of Utrecht, the Netherlands   Department of Systematic Botany

FIG. 5. Postulated areas of endemism for the birds of central Amazonia and the western coast of South America:14, Tumbesan Center; 15, Peruvian Arid Coastal Center; 16, Marañón Center; 19, North Amazon (Napo) Center; 20, South Amazon (Inambari) Center; 21, Rondônia Center; 22, Pará Center; and 23, Belém (Maranhão) Center.

Representative endemic taxa are:

Crypturellus transfasciatus
Ortalis erythroptera
Penelope albipennis
Leptotila ochraceiventris
Columbina buckleyi
Aratinga erythrogenys
Forpus coelestis
Brotogeris pyrrhopterus
Leucippus baeri
Myrmia micrura
Phloeoceastes gayaquilensis
Picumnus sclateri
Synallaxis tithys
S. stictothorax
Automolus ruficollis
Hylocryptus erythrocephalus
Sakesphorus b. bernardi
S. b. piurae
S. b. cajamarcae
Myrmeciza griseiceps

Melanopareia elegans
Fluvicola nengeta atripennis
Muscigralla brevicauda
Tumbezia salvini
Empidonax griseipectus
Myiobius atricaudus portovelae
Myiodynastes bairdi
Myiopagis leucospodia
Petrochelidon fulva aequatorialis
Tachycineta albilinea stolzmanni
Cyanocorax mystacalis
Turdus reevei
T. nudigenis maculirostris
Thryothorus superciliaris
Mimus longicaudatus albogriseus
Icterus graceannae
Dives dives warszewiczi
Basileuterus fraseri
Saltator nigriceps
Piezorhina cinerea

Gnathospiza taczanowskii
Atlapetes albiceps
Arremon a. abeillei

Aimophila stolzmanni
Rhodospingus cruentus
Sporophila peruviana devronis

PERUVIAN ARID COASTAL CENTER

This center (Fig. 5: area 15) is defined by the endemics of the Pacific coastal desert and adjacent shoreline (arid tropical biome) extending from the Tumbesan Center south to northern Chile between Antofagasta and Valparaiso (Hueck and Seibert 1981).
Representative endemic taxa are:

Eulidia yarrellii
Rhodopis vesper
Leucippus taczanowskii
Thaumastura cora
Chrysoptilus a. atricollis
Geositta maritima
G. peruviana
Cinclodes taczanowskii
Asthenes cactorum

Myiarchus semirufus
Elaenia albiceps modesta
Phytotoma raimondii
Petrochelidon fulva rufocollaris
Mimus l. longicaudatus
Dives dives kalinowskii
Xenospingus concolor
Sporophila p. peruviana
Saltator albicollis immaculatus

MARAÑÓN CENTER

The Marañón Center (Fig. 5: area 16) includes those taxa confined to the arid valley of the upper Río Marañón (and the area of its immediate tributaries) in northcentral Peru.
Representative endemic taxa are:

Crypturellus tataupa inops
Columba oenops
Forpus xanthops
Aratinga wagleri minor
Chrysoptilus atricollis peruvianus
Phacellodomus dorsalis
Synallaxis stictothorax chinchipensis
Siptornopsis hypochondriacus
Melanopareia maranonica

Sakesphorus bernardi shumbae
Knipolegus aterrimus heterogyna
Mimus longicaudatus maranonicus
Turdus maranonicus
Thlypopsis inornata
Saltator albicollis peruvianus
Arremon abeillei nigriceps
Incaspiza watkinsi
Sporophila simplex

GUYANAN CENTER

This area of endemism primarily includes the lowland tropical rainforest of Guyana, Surinam, French Guiana, and Amapá, Brazil (Müller 1973; Fig. 3: area 17). Endemics are limited to the north by the delta of the Orinoco or by savannah vegetation (Hueck and Seibert 1981), but some species, nevertheless, are distributed north of the Orinoco, even as far as Trinidad. The center is bounded on the south by the Amazon River. To the north the highlands of western Guyana and southeastern Venezuela form the western edges of the Guyanan Center, but farther south species are variously distributed westward. For some, the campos region along the Brazil-Guyana-Surinam-French Guiana borders serves as a boundary; other species extend farther westward, some as far as the Río Negro and the southern border of the Venezuelan highlands.
Representative endemic taxa are:

Pionopsitta caica
Caprimulgus maculosus
Topaza p. pella
Lophornis ornata
Threnetes niger
T. loehkeni
Phaethornis malaris
Selenidera culik
Ramphastos v. vitellinus
Celeus u. undatus
Veniliornis sanguineus
Dendrexetastes r. rufigula

Myrmeciza f. ferruginea
Sakesphorus melanothorax
Terenura callinota guianensis
T. spodioptila elaopteryx
Gymnopithys r. rufigula
Percnostola r. rufifrons
Iodopleura fusca
Pachyramphus surinamus
Haematoderus militaris
Pipra s. serena
Contopus albogularis
Microcochlearius josephinae

Phylloscartes virescens
Euscarthmus rufomarginatus savannophilus
Polioptila g. guianensis
Euphonia finschi
E. cayennensis

Tangara v. velia
T. m. mexicana
Cyanicterus cyanicterus
Periporphyrus erythromelas

IMERI CENTER

This center of endemism is located in the vicinity of the Brazilian, Venezuelan, and Colombian frontiers (Fig. 3: area 18). It includes the lowlands of southern Amazonas, Venezuela, Brazil north of the upper Río Negro and Río Vaupés, and the eastern portions of Vaupés, Guainía, and possibly southeastern Vichada, Colombia. Its western border is ambiguous although it possibly lies near the boundary of the evergreen lowland forest of the North Amazon Center (see Hueck and Seibert 1981).

Representative endemic taxa are:

Crypturellus duidae
C. casiquiare
Nonnula amaurocephala
Selenidera nattereri
Picumnus pumilus
Hylexetastes stresemanni insignis
Myrmeciza pelzelni
M. disjuncta
Thripophaga cherriei

Myrmotherula ambigua
M. cherriei
Rhegmatorhina cristata
Gymnopithys leucaspis lateralis
Heterocercus flavivertex
Cyanocorax heilprini
Hylophilus b. brunneiceps
Dolospingus fringilloides

NORTH AMAZON (NAPO) CENTER

The North Amazon Center of endemism (Fig. 5: area 19) is defined by the Marañón and Amazon rivers on the south, the Andes on the west, and the limit of lowland rainforest to the north (near the Río Guaviare in central Colombia; Hueck and Seibert 1981). The eastern border of this center is more difficult to specify, but many taxa do not extend beyond the Río Negro.

Representative endemic taxa are:

Crax salvini
Pyrrhura albipectus
Neomorphus pucheranii
Campylopterus villaviscensio
Leucippus chlorocercus
Heliodoxa gularis
Taphrospilus h. hypostictus
Topaza pella pamprepta
P. pyra
Pharomachrus p. pavoninus
Nonnula ruficapilla rufipectus
Selenidera r. reinwardtii
Celeus s. spectabilis
Synallaxis moesta
Thamnophilus praecox
Myrmotherula s. sunensis
M. longicauda soderstromi
Herpsilochmus sticturus dugandi
Hypocnemis hypoxantha
Pithys castanea

Rhegmatorhina m. melanosticta
Phlegopsis barringeri
Gymnopithys leucaspis castanea
Grallaria dignissima
Hylopezus f. fulviventris
Pipra pipra discolor
P. coronata caquetae
Heterocercus aurantiivertex
Attila citriniventris
A. torridus
Contopus n. nigrescens
Mionectes oleagineus hauxwelli
Myiophobus cryptoxanthus
Todirostrum c. calopterum
Cyphorhinus aradus salvini
Microcerculus bambla albigularis
Hylophilus hypoxanthus fuscicapillus
Ocyalus latirostris
Cacicus sclateri

SOUTH AMAZON (INAMBARI) CENTER

The boundaries of the South Amazon Center are defined by the Rio Madeira to the east, by either the Río Madre de Dios or Río Beni (near the limit of lowland rainforest) to the south, by the Andes to the west, and by the Río Marañón (sometimes by the Río Ucayali) to the north (Figs. 1, 5: area 20).

Representative endemic taxa are:

Crypturellus bartletti
Psophia leucoptera
Pauxi unicornis
Ara couloni
Pyrrhura rupicola
Phaethornis stuarti
P. philippii
Amazilia lactea bartletti
Phlogophilus harterti
Hylocharis cyanus rostrata
Heliodoxa branickii
Taphrospilus hypostictus peruvianus
Brachygalba albogularis
Galbula cyanescens
Malacoptila semicincta
Nonnula sclateri
N. r. ruficapilla
Eubucco tucinkae
Pteroglossus mariae
P. beauharnaesii
Selenidera reinwardtii langsdorfii
Picumnus borbae juruanus
Celeus torquatus occidentalis
Hylexetastes stresemanni
Simoxenops ucayalae
S. striatus
Thamnomanes schistogynus
Myrmotherula iheringi heteroptera

Myrmoborus melanurus
Percnostola lophotes
Myrmeciza goeldii
Gymnopithys salvini
G. leucaspis peruana
Rhegmatorhina melanosticta purusiana
Formicarius rufifrons
Grallaria eludens
Conioptilon mcilhennyi
Pipra coronata exquisita
P. chloromeros
P. coeruleocapilla
Neopelma sulphureiventer
Pyroderus scutatus masoni
Muscisaxicola fluviatilis
Myiornis albiventris
Poecilotriccus tricolor (*albifacies*)
Todirostrum calopterum pulchellum
Mionectes oleagineus maynana
Myiobius barbatus amazonicus
Mionectes macconnelli peruanus
Lophotriccus eulophotes
Thryothorus griseus
Cyphorhinus aradus modulator
Hylophilus semicinereus juruanus
Cacicus koepckeae
Ramphocelus melanogaster
Tachyphonus rufiventer

Comments.—A number of the species listed above are narrowly distributed in the region of the upper Madeira, Purus, and Madre de Dios rivers. The relationship of this area of endemism to the remainder of the South Amazon Center is uncertain. Müller (1973:82–88) united the North and South Amazon centers into a single area of endemism. In fact, a sufficiently high number of taxa are restricted to this combined area to warrant listing them here. These endemic taxa (see also Haffer 1978:53–54) include:

Ortalis g. guttata
Nothocrax urumutum
Crax globulosa
Anurolimnas castaneiceps
Aramides calopterus
Micrastur buckleyi
Geotrygon saphirina
Odontophorus stellatus
Pulsatrix melanota
Aratinga weddellii
Pionopsitta barrabandi
Amazona f. festiva
Neomorphus geoffroyi aequatorialis
N. pucheranii
Chordeiles rupestris
Popelairia langsdorfi melanosternon
Heliodoxa schreibersii
Galbula tombacea
G. pastazae
G. leucogastra chalcothorax
Galbalcyrhynchus leucotis
Nystalus s. striolatus

Monasa flavirostris
Nonnula rubecula cineracea
N. brunnea
Eubucco richardsoni
Capito aurovirens
Pteroglossus inscriptus humboldti
Selenidera reinwardtii
Ramphastos vitellinus culminatus
Piculus l. leucolaemus
Picumnus castelnau
P. rufiventris
Dendrexetastes rufigula devillei
Furnarius minor
Synallaxis cherriei
S. albigularis
Sclerurus mexicanus peruvianus
Thripophaga fusciceps
Metopothrix aurantiacus
Philydor erythropterus
Automolus dorsalis
A. melanopezus
Frederickena unduligera

Myrmotherula obscura
M. erythrura
M. sunensis
M. assimilis
Terenura humeralis
Cercomacra serva
Percnostola schistacea
Myrmeciza melanoceps
M. fortis
M. hyperythra
Rhegmatorhina melanosticta
Phlegopsis erythroptera
Chamaeza nobilis
Neoctantes niger
Myrmochanes hemileucus
Conopophaga peruviana
Cotinga maynana
Porphyrolaema porphyrolaema
Pipreola chlorolepidota
Lipaugus subalaris

Phoenicircus nigricollis
Machaeropterus regulus striolatus
Chloropipo holochlora
Cnipodectes subbrunneus minor
Poecilotriccus capitale
Serpophaga h. hypoleuca
Tyranniscus cinereicapillus
Sirystes sibilator albocinereus
Clypicterus oseryi
Microbates cinereiventris peruvianus
Agelaius xanthophthalmus
Conirostrum margaritae
Ramphocelus nigrogularis
Euphonia laniirostris melanura
Tangara callophrys
T. schrankii
T. mexicana boliviana
T. x. xanthogastra
Calochaetes coccineus
Caryothraustes humeralis

RONDÔNIA CENTER

The Rondônia Center (Fig. 5: area 21) is one of the main areas of endemism within the southern Amazon basin (Haffer 1978:58–60). The Madeira and Beni rivers form the western boundary of this area and separate it from the South Amazon Center. The Rondônia Center is delimited to the north by the Rio Amazonas, to the east by the Rio Tapajós, although some forms extend eastward to the Rio Xingú, and by the limit of the tropical rainforest to the south.

Representative endemic taxa are:

Penelope pileata
Pipile pipile nattereri
Pyrrhura rhodogaster
Celeus torquatus angustus
Picumnus b. borbae
Dendrocolaptes hoffmannsi
Hylexetastes perrotii uniformis
Dendrexetastes rufigula moniliger
Myrmotherula i. iheringi
M. longipennis ochrogyna
M. leucophthalma phaeonota

Myrmeciza stictothorax
M. ferruginea elata
Rhegmatorhina berlepschi
R. hoffmannsi
Skutchia borbae
Tityra leucura
Pipra nattereri
Todirostrum senex
Odontorchilus cinereus
Hylophilus muscicapinus griseifrons

PARÁ CENTER

This southern Amazonian rainforest center extends from the Rio Amazonas on the north, and the Rio Tapajós on the west, east to the Rio Tocantins and south to the limits of the lowland rainforest (Müller 1973:75–79; Hueck and Seibert 1981; Fig. 5: area 22).

Representative endemic taxa are:

Ortalis motmot ruficeps
Pyrrhura perlata anerythra
Nonnula rubecula simplex
Pipra pipra separabilis
P. iris eucephala
P. vilasboasi
P. aureola aurantiicollis
Rhynchocyclus olivaceus sordidus
Dendrexetastes rufigula paraensis
Myrmotherula leucophthalma sordida

Rhegmatorhina gymnops
Hemitriccus m. minor
H. aenigma
H. striaticollis griseiceps
Serpophaga hypoleuca pallida
Myioborus barbatus insignis
Psarcolius bifasciatus
Hylophilus brunneiceps inornatus
Tangara velia signata
Sporophila leucoptera mexianae

BELÉM (MARANHÃO) CENTER

This is the easternmost of the southern Amazonian rainforest centers (Fig. 5: area 23). Its western border is the Rio Tocantins. Some forms extend eastward along the coast of Brazil and then south according to the distribution of the rainforest (Hueck and Seibert 1981).
Representative endemic taxa are:

Ortalis motmot superciliaris
Pyrrhura perlata lepida
P. p. coerulescens
Topaza pella microrhyncha

Brachygalba lugubris naumburgi
Nystalus striolatus torridus
Malacoptila striata minor
Pipra i. iris

SERRA DO MAR CENTER

The Serra do Mar region of southeastern Brazil is perhaps the most well-defined area of endemism for South American birds (Müller 1973:125–137). It includes a narrow strip of evergreen tropical rainforest along the coast and a somewhat broader zone of subtropical forest mixed with savannah vegetation just inland (Fig. 6: area 24). This area is located roughly from the Rio São Francisco in the north (Pernambuco) to the southern termination of the rainforest in Santa Catarina. Distributional limits inland are presumably determined by autecological factors relating to habitat and to physiological tolerance of more arid conditions. This center surely will be subdivided into smaller subcenters when bird distributions are examined in more detail (see Müller 1973:131–133; Kinzey 1982).
Representative endemic taxa are:

Leucopternis lacernulata
Penelope obscura bronzina
Crax blumenbachii
C. m. mitu
Ortalis guttata araucuan
Pyrrhura cruentata
P. l. leucotis
Brotogeris tirica
Touit melanota
T. surda
Amazona brasiliensis
A. dufresniana rhodocorythra
Neomorphus geoffroyi dulcis
N. g. maximiliani
Nyctibius l. leucopterus
Ramphodon dohrnii
Phaethornis squalidus
Hylocharis c. cyanus
Heliomaster squamosus
Popelairia l. langsdorfii
Amazilia l. lactea
Aphantochroa cirrhochloris
Malacoptila s. striata
Jacamaralcyon tridactyla
Baillonius bailloni
Piculus flavigula erythropis
Celeus torquatus tinnunculus
Veniliornis maculifrons
Oreophylax moreirae
Cranioleuca pallida
Cichlocolaptes leucophrys
Sclerurus caudacutus umbretta
S. mexicanus bahiae
Thripophaga macroura
Thamnomanes c. caesius
T. plumbeus

Myrmeciza loricata
M. ruficauda
Formicivora serrana
Cercomacra brasiliana
Dysithamnus stictothorax
D. xanthopterus
Conopophaga melanops
Drymophila squamata
D. genei
D. ochropyga
Myrmotherula axillaris luctuosus
M. erythronotos
M. unicolor
M. urosticta
Pyriglena atra
Rhopornis ardesiaca
Chamaeza r. ruficauda
Psilorhamphus guttatus
Merulaxis ater
M. stresemanni
Scytalopus indigoticus
S. speluncae
Laniisoma e. elegans
Tijuca atra
T. condita
Carpornis melanocephalus
Cotinga maculata
Xipholena atropurpurea
Iodopleura pipra
Calyptura cristata
Procnias nudicollis
Lipaugus lanioides
Pachyramphus m. marginatus
Piprites pileatus
Pipra pipra cephaleucos

Machaeropterus r. regulus
Ilicura militaris
Schiffornis t. turdinus
Neopelma aurifrons
Myiozetetes cayanensis erythropterus
Attila spadiceus uropygiatus
Rhynchocyclus o. olivaceus
Knipolegus nigerrimus
Polystictus superciliaris
Mionectes o. oleagineus
Myioborus atricaudus ridgwayi
M. barbatus mastacalis
Phylloscartes difficilis
P. paulistus
P. oustaleti
Hemitriccus obsoletus
H. nidipendulus
Platyrinchus leucoryphus
Ceratotriccus furcatus
Todirostrum poliocephalum
Rhytipterna s. simplex
Oreotriccus griseocapillus

Thryothorus l. longirostris
Neochelidon t. tibialis
Mimus gilvus antelius
Hylophilus t. thoracicus
Agelaius cyanopus atroolivaceus
Dacnis nigripes
Chlorophanes spiza axillaris
Thraupis ornata
Tangara peruviana
T. demaresti
T. velia cyanomelaena
T. mexicanus brasiliensis
T. fastuosa
Ramphocelus bresilius
Orchesticus abeillei
Hemithraupis flavicollis melanoxantha
H. ruficapilla
Orthogonys chloricterus
Sporophila ardesiaca
S. falcirostris
Oryzoborus m. maximiliani
Poospiza thoracica

Paraná Center

South of the Serra do Mar Center lies the somewhat elevated Paraná area of endemism (Fig. 6: area 25). Although this area is characterized by a forest of *Araucaria angustifolia* (Müller 1973:138–139; Hueck and Seibert 1981), not all avian endemics are restricted ecologically to that habitat. The northern boundary of this center lies at about the level of São Paulo, extends to the west nearly to the Rio Paraná, and has its southern limits at the Rio Jacuí (Müller 1973).

Representative endemic taxa are:

Ortalis guttata squamata
Penelope o. obscura
Amazona pretrei
Picumnus temminckii
P. nebulosus
Dryocopus galeatus

Clibanornis dendrocolaptoides
Leptasthenura setaria
L. striolata
Hemitriccus kaempferi
Cyanocorax caeruleus
Anthus nattereri

Comments.—Many taxa distributed in the tropical-subtropical forest of Serra do Mar also extend southward and eastward into the Paraná Center. Together these two centers comprise a well-defined area of endemism. Some of these endemics include:

Tinamus solitarius
Crypturellus noctivagus
Leucopternis polionota
Penelope superciliaris major
Pipile jacutinga
Odontophorus capueira
Claravis godefrida
Ara maracana
Pionopsitta pileata
Amazona vinacea
Triclaria malachitacea
Otus atricapillus
Pulsatrix koeniswaldiana
Strix hylophila
Aegolius harrisii iheringi
Macropsalis creagra
Ramphodon naevius

Melanotrochilus fuscus
Stephanoxis lalandi
Lophornis magnifica
Phaethornis eurynome
Leucochloris albicollis
Clytolaema rubricauda
Trogon surrucura
Notharchus macrorhynchos swainsoni
Selenidera m. maculirostris
Ramphastos discolorus
Piculus aurulentus
Automolus leucophthalmus
Anabazenops fuscus
Philydor r. rufus
P. atricapillus
Synallaxis cinerascens
Cranioleuca obsoleta

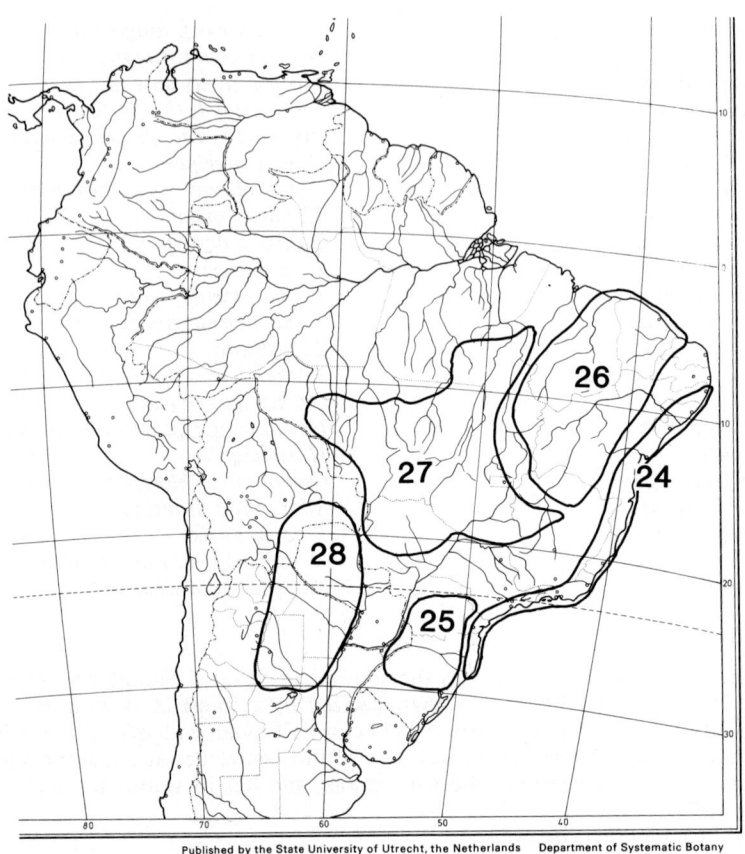

Published by the State University of Utrecht, the Netherlands   Department of Systematic Botany

FIG. 6.   Postulated areas of endemism for the birds of southeastern South America: 24, Serra do Mar Center; 25, Paraná Center; 26, Caatinga Center; 27, Campo Cerrado Center; and 28, Chaco Center.

Heliobletus contaminatus
Phacellodomus erythrophthalmus
Anabacerthia amaurotis
Sclerurus s. scansor
Hypoedaleus guttatus
Mackenziaena severa
M. leachii
Biatas nigropectus
Formicarius colma ruficeps
Batara c. cinerea
Pyriglena leucoptera
Myrmeciza squamosa
Drymophila malura
D. ferruginea
Dysithamnus m. mentalis
Herpsilochmus r. rufimarginatus
Terenura maculata
Grallaria varia imperator
Phibalura f. flavirostris
Pachyramphus c. castaneus
Pyroderus s. scutatus
Chiroxiphia caudata

Schiffornis virescens
Attila rufus
Knipolegus lophotes
Muscipipra vetula
Conopias t. trivirgata
Sirystes s. sibilator
Elaenia obscura sordida
E. mesoleuca
Ramphotrigon m. megacephala
Phylloscartes v. ventralis
Mionectes rufiventris
Hemitriccus orbitatus
H. diops
Todirostrum p. plumbeiceps
Myiornis a. auricularis
Pogonotriccus eximus
Leptotriccus sylviolus
Xanthomyias virescens
Turdus a. albicollis
T. nigriceps subalaris
Phaeothlypis r. rivularis
Thraupis cyanoptera

Tachyphonus coronatus
Pyrrhocoma ruficeps
Tangara c. cyanocephala
Euphonia pectoralis
Pitylus fuliginosus

Haplospiza unicolor
Saltator maxillosus
Sporophila frontalis
S. melanogaster

Caatinga Center

This area of endemism is located in the dry uplands of eastern Brazil in Ceará, Piauí, and western Paraíba and Pernambuco (Fig. 6: area 26). Its boundaries do not appear to be well-marked, but roughly, they exclude the more mesic vegetation zones to the west and northwest (more or less delimited by the Rio Paranaíba), the lowlands of coastal Brazil, and areas beyond the limits of the "Caatinga" vegetation zone (Hueck and Seibert 1981:36) near southern Bahia.
Representative endemic taxa are:

Penelope jacucaca
Anodorhynchus leari
Cyanopsitta spixii
Aratinga acuticaudata haemorrhous
A. cactorum
Pyrrhura leucotis griseipectus
Caprimulgus hirundinaceus
Phaethornis gounellei
Augastes lumachellus
Picumnus limae

P. pygmaeus
Xiphocolaptes flacirostris
Myrmorchilus s. strigilatus
Procnias a. averano
Todirostrum sylvia schulzi
Hemitriccus mirandae
Stigmatura budytoides gracilis
Thryothorus longirostris bahiae
Paroaria dominicana
Sporophila albogularis

Campo Cerrado Center

The campo cerrado savannah vegetation delineates an area of endemism in the uplands of south-central Brazil (Müller 1973:120–124). Its approximate boundaries include the Paraná and Paranaíba rivers to the south, the Chaco to the west and southwest, the tropical lowlands to the north and northwest, and the caatinga vegetation to the east (Hueck and Seibert 1981; Fig. 6: area 27).
Representative endemic taxa are:

Nothura minor
Taoniscus nanus
Penelope ochrogaster
Columbina cyanopis
Amazona xanthops
Chordeiles p. pusillus
Cypseloides senex
Phaethornis nattereri
Augastes scutatus
Heliactin cornuta
Geobates poecilopterus
Herpsilochmus longirostris
H. pectoralis
Formicivora iheringi
Cercomacra ferdinandi
Melanopareia torquata
Scytalopus novacapitalis

Antilophia galeata
Knipolegus aterrimus franciscanus
Phylloscartes roquettei
Polystictus superciliaris
Cyanocorax cristatellus
C. cyanopogon
Basileuterus leucophrys
Cypsnagra hirundinacea
Sericossypha loricata
Neothraupis fasciata
Conothraupis mesoleuca
Saltator atricollis
Paroaria baeri
Sporophila cinnamomea
Sicalis c. citrina
Poospiza cinerea
Embernagra longicauda

Chaco Center

The Chaco area of endemism consists of dry forest and plains in southwestern Brazil, southeastern Bolivia, western Paraguay, and north-central Argentina (Müller 1973; Hueck and Seibert 1981; see especially Short 1975). This center is bordered on the east by the Río Paraguay, on the north by more mesic vegetation of central Bolivia, the Andes on the west, and the more temperate dry forests of north-central Argentina (Fig. 6: area 28).

Representative endemic taxa are:

Nothura cinerascens (ranges slightly north)
N. boraguira
N. chacoensis
Eudromia f. formosa
E. f. mira
Ortalis canicollis
Chunga burmeisteri
Pyrrhura devillei
Strix rufipes chacoensis
Phaethornis subochraceus
Celeus lugubris
Colaptes melanochloros nigroviridis
Dryocopus schulzi
Campephilus leucopogon
Campylorhamphus trochilirostris lafresnayanus
C. t. hellmayri

Xiphocolaptes major (ranges slightly north)
Upucerthia certhiodes
Thripophaga baeri chacoensis
Pseudoseisura cristata unirufa
Thamnophilus caerulescens dinellii
T. c. paraguayensis
Myrmorchilus strigilatus suspicax
Xenopsaris a. albinucha
Xolmis i. irupero (ranges slightly beyond Chaco)
Thraupis sayaca obscura
Paroaria capitata
Sporophila nigrorufa
Lophospingus pusillus
Aimophila strigiceps
Embernagra platensis olivascens
Saltatricula multicolor

NOTHOFAGUS (CHILEAN ANDEAN) CENTER

This area of endemism describes the ranges of those taxa confined to the *Nothofagus* forests or adjacent, non-forested mountain and coastal areas of southern Chile and extreme western Argentina (Müller 1973:155–165; Vuilleumier 1985). The area extends from the region near Valdivia south to Tierra del Fuego (Fig. 7: area 29).

Representative endemic taxa are:

Nothoprocta perdicaria
Columba araucana
Enicognathus ferrugineus
E. leptorhynchus
Sephanoides sephanoides
Colaptes pitius
Picoides lignarius
Phloeoceastes magellanicus
Chilia melanura
Cinclodes oustaleti
Asthenes humicola

Pygarrhichas albogularis
Pteroptochos castaneus
P. megapodius
Scelorchilus albicollis
S. rubecula
Eugralla paradoxa
Agriornis livida
Pyrope pyrope
Phytotoma rara
Mimus thenca
Melanodera xanthogramma

PATAGONIAN CENTER

The Patagonian Center of endemism corresponds, at least in birds, to the extent of the patagonian steppe vegetation (Hueck and Seibert 1981). The center lies east of the Andes, has its northern limits near the Río Chubut, and extends southward through Tierra del Fuego (Fig. 7: area 30).

Representative endemic taxa are:

Pterocnemia p. pennata
Tinamotis ingoufi
Cyanoliseus patagonus
Geositta antarctica
Eremobius phoenicurus
Asthenes patagonica

Cinclodes antarcticus maculirostris
Neoxolmis rufiventris
Agriornis murina
A. microptera microptera
Sicalis lebruni
Melanodera melanodera princetoniana

Comments.—Some of the taxa in the *Nothofagus* and Patagonian centers also are disjunct in the Falkland Islands; in the above lists, this area is ignored for simplicity.

Many far southern taxa are found in open habitats and mountainous areas; thus, they are distributed in both the *Nothofagus* and Patagonian centers. Some are more widely distributed in one center than in the other. Some of these endemics include:

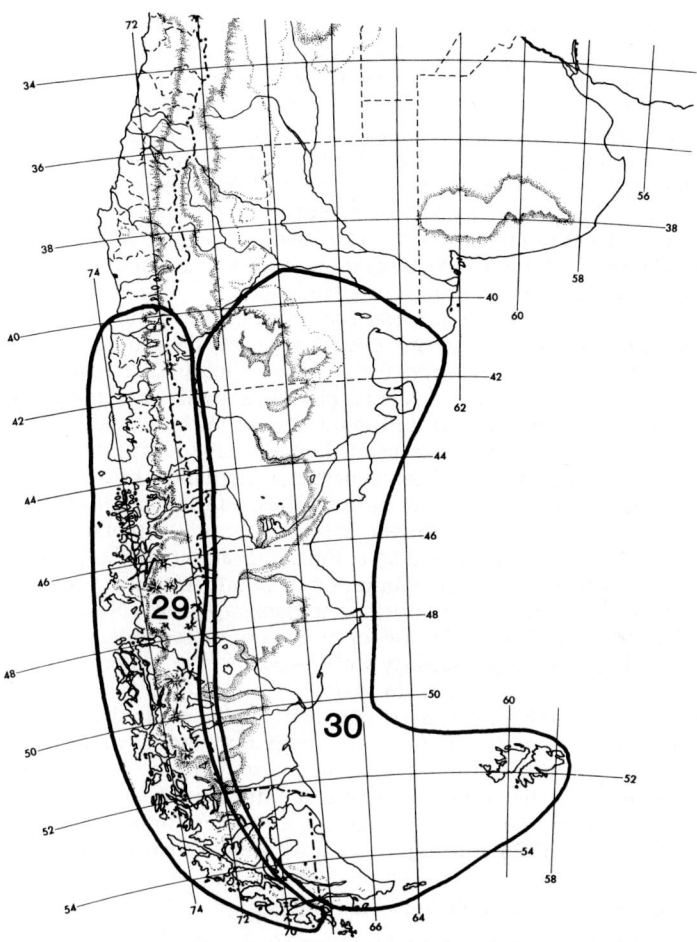

FIG. 7. Postulated areas of endemism for the birds of far southern South America: 29, *Nothofagus* (Chilean Andean) Center; and 30, Patagonian Center.

Buteo ventralis
Phalcoboenus albigularis
Falco kreyenborgi
Attagis malouinus
Glaucidium nanum
Cinclodes p. patagonicus
Sylviorthorhynchus desmursii
Aphrastura s. spinicauda

Pteroptochos tarnii
Scytalopus m. magellanicus
Agriornis livida
Ochthoeca parvirostris
Turdus falcklandii magellanicus
Curaeus curaeus
Phrygilus patagonicus
Spinus barbatus

DISCUSSION

SIGNIFICANCE OF AVIAN AREAS OF ENDEMISM

Because birds have powers of flight, it is often believed that they generally exhibit significant capabilities for long-distance dispersal. This being the case, birds are considered to be a particularly difficult group with which to decipher biogeographic pattern (e.g., Darlington 1957:236, 239, 242, 267). Data presented in this study suggest otherwise. The 33 areas of endemism recognized here (and actually many more will be added with further study) demonstrate that significant distributional pattern exists for birds. The areas also show that many taxa are very narrowly endemic. Just what proportion of the South American avifauna is narrowly endemic is difficult to assess. The exact percentage will remain in doubt because many currently recognized "species" that might be considered to be widespread are actually differentiated spatially, with well-marked "subspecies" themselves being narrowly endemic. A rough survey of the distributions of South American land bird species (using lists in Meyer de Schauensee 1966) suggests that a substantial proportion of the avifauna is narrowly endemic (considered here to be restricted to two, or rarely three, adjacent areas of endemism as recognized above). Moreover, that pattern would be significantly stronger if differentiated subspecies were the basis of tabulation. We can conclude that the narrow endemism of most birds provides the raw material for biogeographic studies and that birds are probably no less useful for this task than are other groups of organisms.

Present evidence suggests that avian areas of endemism exhibit considerable congruence with those determined for other organisms, even though very few groups have been investigated with the purpose of defining areas of endemism. Of the 31 South American centers of endemism identified by Müller (1973), only a few (e.g., his Barranquilla, Catatumbo, Caribbean, and Uruguayan centers) do not seem to be well-marked areas for birds. Other differences between his areas and those recognized here can be attributed to alternative interpretations of distribution patterns within the Andes and the Venezuelan highlands. Brown (1982a, b) described over 40 areas of endemism for neotropical butterflies, and taking into account that a single avian area may encompass two or more areas based on butterfly distributions, most of Brown's areas are congruent with those proposed here. Finally, Prance (1982b:150–156) postulated 26 centers of endemism for the lowland rainforests of South America. Compared to those for birds, most of these plant areas are more restricted geographically. Nevertheless, much agreement exists between his findings and those for birds as proposed in this paper and in the work of Haffer (1969, 1977, 1978), particularly with respect to the centers in the Amazon basin, the Chocó, and southeast Brazil (Serra do Mar).

One conclusion from these preliminary data, then, is that avian distribution patterns have significant generality: that is, distributions are highly nonrandom and congruent with those of other organisms. This does not imply that all species conform to these patterns, nor that all species assigned to a center will show identical patterns of distribution within the center. This generality does imply, however, that there is a problem to be investigated, namely, the origin and development of biotas as represented by these centers of endemism.

HYPOTHESES ABOUT THE ORIGIN OF AREAS OF ENDEMISM

Many systematists and biogeographers investigating distributional patterns within the South American biota, including those of the Amazon basin (Prance 1982a) and high Andes (Vuilleumier 1969; Haffer 1970c), have accepted the hypothesis that areas of endemism have arisen by vicariance of more widespread biotas. This hypothesis has been adopted whether the vicariance events are thought to have been paleogeographic changes, alterations in river systems, cyclical climatic events of the Pleistocene, or some combination thereof (Haffer 1982).

The evidence leading these workers to this conclusion is two-fold. First is the substantial congruence in the distributions of very different kinds of organisms to form well-defined areas. Although an area of endemism might be characterized as "lowland tropical," "humid montane," or perhaps "páramo" in a general ecological sense, component species of those areas often have distinctly different ecological-physiological requirements. Areas, one may therefore postulate, are primarily historical entities rather than simply manifestations of a single, coherent ecology. [This is sometimes a difficult point to convey and understand. Although avian endemics might be restricted to a particular type of forest habitat, the area of endemism for birds is not necessarily defined ecologically on the basis of that habitat. We still are compelled to ask why the plant species comprising that habitat are also endemic in this area.] Second, most areas are delimited by easily observable physiographic barriers and, thus, are geographic entities also. Summarizing, it is this combination of congruence in distribution correlated with geography that is so persuasive of vicariance.

Some biologists have rejected the above interpretation, suggesting instead that areas of endemism arise as a result of geographic gradients in "selection pressures" exhibiting a parallel influence on the taxa of a large area. According to this hypothesis, this process results in the borders of areas of endemism being defined by regions of steep change along a "selection gradient" (Endler 1977, 1982a, b; Benson 1982). Benson (1982:610) described the hypothesis in the following manner:

> *"An alternative model [to vicariance], based on population biology, involving a competition between alleles or species, accounts for the biogeographic patterns in terms of adaptation to environmental conditions that vary from one region to another (Endler 1977) . . . . The model produces essentially the same qualitative results as those attributed to refuges and documented in biogeographic studies . . . it is important to note that the parapatric differentiation model depends on the operation of opposing selection pressures between regions."*

Space does not permit a detailed examination of this hypothesis, but it does seem relevant to ask whether the "population biology model" (also termed the "parapatric model") is likely to be important in explaining large-scale patterns of endemism. The model can be criticized from three standpoints. The first concerns the pattern of endemism itself. An allopatric (vicariance) explanation would predict that areas of endemism should be delineated by major geographic or ecological *barriers*. In the Amazon basin, these barriers appear to be large river systems (e.g., the Amazon, Río Negro, and Napo rivers), whereas in the Andes extensive river valleys often are effective in defining areas (e.g., the Marañón and Apurímac river valleys in Peru). Many different kinds of organisms are stopped by these major barriers (see papers in Prance 1982a) but not by the many smaller river systems that exist within each area of endemism. The population biology model, on the other hand, would not predict boundaries of areas of endemism to be congruent with major geographic barriers unless the latter are correlated with the locations of well-defined ecological transitions. Theoretically, at least, these transitions are independent of the presence or size of geographic barriers. It is important to note that, under the population biology model, geographic barriers to gene flow are much less important relative to variance (between the areas) in the selection pressures said to cause differentiation.

The predictions of the population biology model do not appear to be fulfilled, at least as far as the lowland tropical and Andean biotas of South America are concerned. These areas of endemism are circumscribed by geographic or strong ecological barriers. The two models would be expected to yield the same boundaries for areas of endemism only when those regions are defined by major changes in ecology rather than by geographic barriers. Yet, even in this case, the models can be distinguished, and the reason for this constitutes the second criticism of the population biology model.

The population biology model is often said to produce the same biogeographic patterns as the vicariance model (Benson 1982; Endler 1982a, b; Turner 1982), yet on closer analysis that claim is seen to be false. The proponents of the model do not specify precisely which patterns would be predicted to be shared by the two models, but a pattern of distributional parapatry is usually implicit. And, as most systematists recognize, parapatric distributions may arise either directly by parapatric differentiation or by dispersal following allopatric differentiation. Used in this sense, however, parapatric distributions represent two-taxon statements and, therefore, are not particularly informative, either from the standpoint of systematics or biogeography. If, instead, we ask what *historical patterns* emerge with consideration of three or more taxa, each endemic to a different area, then the parapatric and allopatric models can be distinguished readily (Cracraft 1982). Parapatric differentiation cannot lead to congruent patterns of spatial differentiation from one clade to another (each with three or more taxa), whereas vicariance can. The extent to which endemic taxa participate in congruent historical patterns is a measure of the explanatory power of the vicariance model (Platnick and Nelson 1978; Rosen 1978; Nelson and Platnick 1981; Wiley 1981; Cracraft 1982, 1983a, b). Empirical data for South America are meagre at present, but patterns in frogs (Duellman 1982) and birds (Prum 1982; Cracraft unpubl. observ. on numerous clades) suggest that congruent patterns will be discovered once the relevant studies are undertaken (see, for example, Cracraft 1982, 1983a, for an evaluation of the parapatric versus allopatric origins of Australian areas of endemism).

A final criticism of the population biology model focuses on the likelihood that differentiation in numerous kinds of organisms can be correlated so precisely in terms of its spatial location. Even if one makes the assumption that a gradient in one or more environmental parameters will elicit a parallel gradient in selection pressure acting on each (or many) of the taxa distributed along the gradient, one must be able to relate the variance in selection (across the boundaries of two or more areas) directly to the phenotypic changes that take place in each taxon. Thus, a specific implication of the parapatric model is that differentiation between taxa in two adjacent areas is the result of selective responses to a common environmental gradient and that each "phenotypic response" can be related to directional natural selection associated with that gradient. As an example, the model would make the bold claim that changes in pattern and color of avian plumages, pattern and color of butterfly wings or beetle elytra, the shape of genitalia in worms, bugs, and spiders, or the shapes of leaves, are all attributable to the correlated responses of selection for the same (clinal) change in their environment. To my knowledge, a selective response such as this has not been demonstrated to define the boundaries of areas of endemism.

In summary, proponents of the parapatric differentiation model have failed to explain how that model can produce the major patterns of endemism we see within the South American biota. Most biologists agree that clinal environmental changes can result in clinal phenotypic responses in different kinds of organisms. This observation, however, does not lead to the conclusion that this phenomenon can produce a complex pattern of endemism. The allopatric (vicariance) model still remains the most parsimonious explanation for this pattern.

THE AGE OF AREAS AND THEIR BIOTAS

South American areas of endemism (often considered to be "refuges," see below) are considered by the majority of workers to be Pleistocene in age (e.g., Prance 1982a). Why this is the case is not entirely clear. The hypothesis seems to rest on the following evidence and suppositions: (1) the Pleistocene climatic fluctuations influenced the extent of lowland rainforest and dry forest vegetation in a cyclical manner, (2) the Andes experienced between 1000 and 3000 m rise in elevation during the Pleistocene, an event undoubtedly having a marked effect on the ecology and distribution patterns of the South American biota, (3) species distributions are nonrandomly clustered within areas, and (4) taxa whose differentiation is at the "subspecific" or "specific" levels are likely to be of recent origin (typically considered to be Pleistocene).

Taken together, these observations might lead one to infer that Pleistocene events probably have been important in shaping biogeographic and systematic patterns. They are insufficient, however, to demonstrate a Pleistocene origin for the areas of endemism. Any of the above observations or assumptions is equally compatible with a hypothesis that the areas originated prior to the Pleistocene. Indeed, one might easily argue that complex biogeographic patterns could not have been established so recently as the Pleistocene. This argument could even apply to those habitats said not to "exist" prior to the Pleistocene. The example most often cited, perhaps, is the high Andean puna-páramo biota (see, e.g., Simpson 1975:275, 291). Yet, much of the Andes may have been 2500 to 3500 m in elevation throughout much of the Cenozoic (Simpson 1975). Whereas this may not have been high enough to support "puna-páramo" conditions as we see them today, dry open habitats were surely available to support a "high" Andean biota, including perhaps the ancestral species of today's flora and fauna. Inasmuch as the Andes have not exhibited a history of stasis over the past 40 to 50 million years, there is no reason not to expect some of the presently observed biogeographic patterns to antedate the Pleistocene. This is equally true with respect to the lowland biotas. It is important to remember that climatic fluctuations were significant throughout most of the Cenozoic.

Much needs to be learned about the history of South American areas of endemism, and in the search for that history it would be prudent to keep an open mind and to reject a prioristic assumptions, especially when those assumptions are unnecessary. Thus, several groups of investigators of the tropical lowland vertebrate fauna (Heyer and Maxson 1982; Weitzman and Weitzman 1982) have been unable to interpret their systematic and biogeographic patterns within the framework of the Pleistocene, and yet they have not found it necessary to reject the notion of Pleistocene "refuges" altogether. If South American areas of endemism have complex hierarchical interrelationships similar to those postulated for the areas of Australia (Cracraft 1982), then certainly they have had a much longer history than currently accepted by many workers.

Undoubtedly the "age problem" of these areas can be traced to the assumption that species and subspecies of many organisms (particularly vertebrates) must be relatively young (see Mayr and Phelps 1967, as only one of many examples that could be cited). To adopt the view that the amount of differentiation is proportional to age requires acceptance of the hypothesis that the rate of change is equal for all characters and taxa. All biologists would no doubt see this as an unnecessary and unwarranted assumption. If so, then we must accept the possibility that some subspecies are older than some species, some species older than some genera, and so on. In fact, one must also entertain the possibility that disjunct, undifferentiated populations of some species may be as old as differentiated taxa of differing rank living in the same area.

THE INADEQUACY OF PHENETIC BIOGEOGRAPHY

Rejection of the equal-rate hypothesis also requires the rejection of "phenetic biogeography" (Nelson and Platnick 1978) in developing hypotheses about the history of areas of endemism. Phenetic biogeography is an expression of the assumption that areas having the most taxa in common are the most closely related to one another (Nelson and Platnick 1978). Although workers have not yet attempted to reconstruct the hierarchical relationships of South American areas of endemism, they have used the methods of phenetic biogeography to assess the interrelationships of avifaunas (Mayr and Phelps 1967; Müller 1973; Vuilleumier 1983).

With respect to the areas discussed in this paper, it is very easy to make phenetic comparisons and then to assume that they have some historical significance. Thus, the Chocó Center can be shown to share many taxa either with certain Central American centers, with the Nechí Center, or with the North and South Amazon centers, but the question remains, "To which one of these centers is the Chocó most closely related historically?" Similar questions present themselves when any of the South American areas are considered, but we cannot answer them using phenetic methods. This does not mean that comparisons of faunal similarity carry no information, only that they do not constitute a reliable method for revealing historical interrelationships among biotas (Nelson and Platnick 1978, 1981).

The interrelationships of the South American areas will be better understood once the phylogenetic relationships of endemic taxa are investigated (Nelson and Platnick 1978, 1981; Platnick and Nelson 1978; Rosen 1978; Cracraft 1982, 1983a, b). Preliminary studies suggest a pattern of congruence among the systematic and biogeographic relationships of some avian taxa (Prum 1982; Cracraft unpubl. observ.). Because many of these areas may be very old, and because it is likely that Pleistocene events had a role in shaping biotic distribution and endemism, deciphering the history of these areas will be difficult indeed. Nevertheless, given sufficient and reliable systematic data, patterns should emerge.

"REFUGES" OR AREAS OF ENDEMISM?

The "refuge concept" is based upon the assumption that during arid climatic periods, particularly during the Pleistocene (Haffer 1982), once broad tracts of rainforest were reduced to a small number of "refuges" in regions where mesic conditions still prevailed, and that these refuges then served as centers of taxonomic differentiation. The "reality" of these refuges has taken on paradigmatic proportions in the analysis of South American biogeography and taxonomic diversification (e.g., Prance 1982a). As noted earlier, however, there is still little direct evidence for the existence, extent, or location of specific refuges (Haffer 1980).

Even though indirect evidence strongly suggests that wet forests were restricted in their geographic extent during portions of the Pleistocene, one could still argue that the conceptual demands placed upon biogeographic analysis by the refuge concept may be counterproductive in the long run. As already mentioned, the use of the term "refuge" almost always implies a Pleistocene viewpoint, thereby de-emphasizing, if not eliminating, consideration of pre-Pleistocene biogeographic phenomena. Thus, many adherents of "refuge theory," while admitting pre-Pleistocene influences, nevertheless rarely, if ever, incorporate them into their analyses. Furthermore, the refuge theory places its emphasis almost entirely on habitat expansion and contraction, as mediated by cyclical climatic change (even in the Andes: Vuilleumier 1969; Haffer 1970c). Much less consideration has been given to the role of paleogeographic barriers in isolating biotas (particularly in discussions of the tropical lowland biota).

I suggest that the concept of areas of endemism is more general than the refuge concept. As they are currently understood and discussed, most, if not all, "refuges" are areas of endemism, but not all areas of endemism are "refuges." Areas of endemism are relatively discrete entities, at least to the extent that species' distributions are congruent with one another. "Refuges," however, are less discrete in that their status also depends upon an evaluation of where a particular habitat may have been located (e.g., Haffer 1982:10, calls "refuge" as "interpretive term"). Many aspects of South American biogeography—including postulated Pleistocene events—can be discussed within the framework of an "areas of endemism" approach, whereas the "refuge concept" is more restrictive in its explanatory power.

This discussion is not meant as a specific criticism of the biogeographic findings of the "refuge concept" school. On the contrary, those workers have contributed immeasurably to our understanding of South American biogeography. Instead, the above is an argument for seeing the hypothesis of Pleistocene refuges in the context of a larger problem, the origin and development of areas of endemism. It is also a plea for expanding our point of view and against interpreting South American biogeographic patterns from a narrow explanatory perspective.

ACKNOWLEDGMENTS

I am extremely grateful to J. Fitzpatrick, J. Haffer, T. A. Parker III, J. V. Remsen, R. S. Ridgely, T. S. Schulenberg, D. F. Stolz, M. A. Traylor, and F. Vuilleumier for their comments on the manuscript or providing me with unpublished distributional data. They improved the outcome immeasurably, though they cannot be held responsible for any remaining errors. This research was undertaken under grant DEB79-21492 from the National Science Foundation.

LITERATURE CITED

BENSON, W. W. 1982. Alternative models for infrageneric diversification in the humid tropics: tests with passion vine butterflies. Pp. 608–640, *In* G. T. Prance (ed.), Biological Diversification in the Tropics. Columbia University Press, New York.

BLAKE, E. R. 1977. Manual of Neotropical Birds, Vol. 1. University Chicago Press, Chicago, Illinois.

BROWN, K. S., JR. 1982a. Historical and ecological factors in the biogeography of aposematic neotropical butterflies. Am. Zool. 22:453–471.

BROWN, K. S., JR. 1982b. Paleoecology and regional patterns of evolution in neotropical forest butterflies. Pp. 255–308, *In* G. T. Prance (ed.), Biological Diversification in the Tropics. Columbia University Press, New York.

CHAPMAN, F. M. 1917. The distribution of bird-life in Colombia; a contribution to a biological survey of South America. Bull. Am. Mus. Nat. Hist. 36:1–729.

CHAPMAN, F. M. 1926. The distribution of bird-life in Ecuador; a contribution to a study of the origin of Andean bird-life. Bull. Am. Mus. Nat. Hist. 55:1–784.

CRACRAFT, J. 1972. Continental drift and Australian avian biogeography. Emu 72:171–174.

CRACRAFT, J. 1973. Continental drift, paleoclimatology, and the evolution and biogeography of birds. J. Zool. 169:455–545.

CRACRAFT, J. 1976. Avian evolution on southern continents: influences of palaeogeography and palaeoclimatology. Proc. 16th Int. Ornithol. Congr., pp. 40–52.

CRACRAFT, J. 1982. Geographic differentiation, cladistics, and vicariance biogeography: reconstructing the tempo and mode of evolution. Am. Zool. 22:411–424.

CRACRAFT, J. 1983a. Species concepts and speciation analysis. Current Ornithol. 1:159–187.

CRACRAFT, J. 1983b. Cladistic analysis and vicariance biogeography. Am. Sci. 71:273–281.

DARLINGTON, P. J., JR. 1957. Zoogeography: the Geographical Distribution of Animals. John Wiley, New York.

DELACOUR, J., AND D. AMADON. 1973. Curassows and Related Birds. American Museum Natural History, New York.

DICKERMAN, R. W., AND W. H. PHELPS, JR. 1982. An annotated list of the birds of Cerro Urutani on the border of Estado Bolivar, Venezuela, and Territorio Roraima, Brazil. Am. Mus. Novit. 2732: 1–20.

DUELLMAN, W. E. 1982. Quaternary climatic-ecological fluctuations in the lowland tropics: frogs and forests. Pp. 389–402, *In* G. T. Prance (ed.), Biological Diversification in the Tropics. Columbia University Press, New York.

ENDLER, J. A. 1977. Geographic Variation, Speciation, and Clines. Princeton University Press, Princeton, New Jersey.

ENDLER, J. A. 1982a. Problems in distinguishing historical from ecological factors in bioegography. Am. Zool. 22:441–452.

ENDLER, J. A. 1982b. Pleistocene forest refuges: fact or fancy? Pp. 641–657, *In* G. T. Prance (ed.), Biological Diversification in the Tropics. Columbia University Press, New York.

GENTRY, A. W. 1982. Phytogeographic patterns as evidence for a Chocó refuge. Pp. 112–136, *In* G. T. Prance (ed.), Biological Diversification in the Tropics. Columbia University Press, New York.

HAFFER, J. 1967a. Speciation in Colombian forest birds west of the Andes. Am. Mus. Novit. 2294: 1–57.

HAFFER, J. 1967b. Some allopatric species pairs of birds in northwestern Colombia. Auk 84:366–378.

HAFFER, J. 1967c. Zoogeographical notes on the "nonforest" lowland bird faunas of northwestern South America. El Hornero 10:315–333.

HAFFER, J. 1969. Speciation in Amazonian forest birds. Science 165:131–137.

HAFFER, J. 1970a. Geologic-climatic history and zoogeographic significance of the Uraba region in northwestern Colombia. Caldasia 10:603–636.

HAFFER, J. 1970b. Art-Entstehung bei einigen Waldvögeln Amazoniens. J. Ornithol. 111:285–331.
HAFFER, J. 1970c. Entstehung und Ausbreitung nord-Andiener Bergvögel. Zool. Jahrb. Abt. Syst. Oekol. Geogr. Tiere 97:301–337.
HAFFER, J. 1974. Avian speciation in tropical South America. Publ. Nuttall Ornithol. Club No. 14.
HAFFER, J. 1975. Avifauna of northwestern Colombia, South America. Bonn. zool. Monogr. No. 7.
HAFFER, J. 1977. Pleistocene speciation in Amazonian birds. Amazoniana 6:161–191.
HAFFER, J. 1978. Distribution of Amazonian forest birds. Bonn. zool. Beitr. 29:38–78.
HAFFER, J. 1980. Avian speciation patterns in upper Amazonia. Proc. 17th Int. Ornithol. Congr., pp. 1251–1255.
HAFFER, J. 1982. General aspects of the refuge theory. Pp. 6–24, In G. T. Prance (ed.), Biological Diversification in the Tropics. Columbia University Press, New York.
HEYER, W. R., AND L. R. MAXSON. 1982. Distributions, relationships, and zoogeography of lowland frogs. The Leptodactylus complex in South America, with special reference to Amazonia. Pp. 375–388, In G. T. Prance (ed.), Biological Diversification in the Tropics. Columbia University Press, New York.
HUECK, K., AND P. SEIBERT. 1981. Vegetationskarte von Sudamerika. Gustav Fischer Verlag, Stuttgart, Germany.
KEAST, A. 1961. Bird speciation on the Australian continent. Bull. Mus. Comp. Zool. 123:305–495.
KEAST, A. 1972. Faunal elements and evolutionary patterns: some comparisons between the continental avifaunas of Africa, South America, and Australia. Proc. 15th Int. Ornithol. Congr., pp. 594–622.
KEAST, A. 1981. The evolutionary biogeography of Australian birds. Pp. 1587–1635, In A. Keast (ed.), Ecological Biogeography of Australia. Dr. W. Junk, The Hague.
KINZEY, W. G. 1982. Distribution of primates and forest refuges. Pp. 455–482, In G. T. Prance (ed.), Biological Diversification in the Tropics. Columbia University Press, New York.
LOVEJOY, T. E. 1982. Designing refugia for tomorrow. Pp. 673–680, In G. T. Prance (ed.), Biological Diversification in the Tropics. Columbia University Press, New York.
MAYR, E. 1944. Timor and the colonization of Australia by birds. Emu 44:113–130.
MAYR, E. 1946. History of the North American bird fauna. Wilson Bull. 58:3–41.
MAYR, E. 1964. Inferences concerning the Tertiary American bird fauna. Proc. Natl. Acad. Sci. 51:280–288.
MAYR, E. 1972. Continental drift and the history of the Australian bird fauna. Emu 72:26–28.
MAYR, E., AND W. H. PHELPS, JR. 1967. The origin of the bird fauna of the south Venezuelan highlands. Bull. Am. Mus. Nat. Hist. 136:269–328.
MEYER DE SCHAUENSEE, R. 1964. The Birds of Colombia. Livingston Publ. Co., Narberth, Pennsylvania.
MEYER DE SCHAUENSEE, R. 1966. The Species of Birds of South America and Their Distribution. Livingston Publ. Co., Narberth, Pennsylvania.
MEYER DE SCHAUENSEE, R. 1970. A Guide to the Birds of South America. Livingston Publ. Co., Wynnewood, Pennsylvania.
MEYER DE SCHAUENSEE, R., AND W. H. PHELPS, JR. 1978. A Guide to the Birds of Venezuela. Princeton University Press, Princeton, New Jersey.
MILLER, A. H. 1947. The tropical avifauna of the upper Magdalena Valley, Colombia. Auk 64:351–381.
MILLER, A. H. 1952. Supplementary data on the tropical avifauna of the arid upper Magdalena Valley of Colombia. Auk 69:450–457.
MÜLLER, P. 1972. Centres of dispersal and evolution in the neotropical region. Stud. Neotrop. Fauna 7:173–185.
MÜLLER, P. 1973. The Dispersal Centres of Terrestrial Vertebrates in the Neotropical Realm. Dr. W. Junk, The Hague.
MÜLLER, P. 1976. Andean dispersal centers and their affinities. Biogeographica (Neotrop. Okosys.) 7:183–201.
MYERS, N. 1982. Forest refuges and conservation in Africa with some appraisal of survival prospects for tropical moist forests throughout the biome. Pp. 658–672, In G. T. Prance (ed.), Biological Diversification in the Tropics. Columbia University Press, New York.
NELSON, G., AND N. I. PLATNICK. 1978. The perils of plesiomorphy: widespread taxa, dispersal, and phenetic biogeography. Syst. Zool. 27:474–477.
NELSON, G., AND N. I. PLATNICK. 1981. Systematics and Biogeography: Cladistics and Vicariance. Columbia University Press, New York.
PARKER, T. A., III, S. A. PARKER, AND M. A. PLENGE. 1982. An Annotated Checklist of Peruvian Birds. Buteo Books, Vermilion, South Dakota.
PETERS, J. A. 1931–1979. Check-list of Birds of the World. Museum Comparative Zoology, Cambridge, Massachusetts.
PHELPS, W. H., AND W. H. PHELPS, JR. 1950. Lista de las aves de Venezuela con su distribucion. Parte 2. Passeriformes. Bol. Soc. Venez. Cienc. Nat. 12:1–427.
PHELPS, W. H., AND W. H. PHELPS, JR. 1958. Lista de las aves de Venezuela con su distribucion. Parte 1. No Passeriformes. Bol. Soc. Venez. Cienc. Nat. 19:1–317.
PHELPS, W. H., JR. 1966. Contribucion al analisis de los elementos que componen la avifauna subtropical de las cordilleras de la costa norte de Venezuela. Bol. Acad. Cienc. Fis., Mat. Nat. 26:14–43.

PLATNICK, N. I., AND G. NELSON. 1978. A method of analysis for historical biogeography. Syst. Zool. 17:1–16.

PRANCE, G. T. (ed.). 1982a. Biological Diversification in the Tropics. Columbia University Press, New York.

PRANCE, G. T. 1982b. Forest refuges: evidence from woody angiosperms. Pp. 137–158, In G. T. Prance (ed.), Biological Diversification in the Tropics. Columbia University Press, New York.

PRUM, R. O. 1982. Systematics and biogeography of the family Ramphastidae. Unpubl. Senior Honors Thesis, Harvard College, Cambridge, Massachusetts.

RICH, P. V. 1975. Antarctic dispersal routes, wandering continents and the origin of Australia's non-passerine avifauna. Mem. Nat. Mus. Victoria 36:63–125.

ROSEN, D. E. 1978. Vicariant patterns and historical explanation in biogeography. Syst. Zool. 27:159–188.

SHORT, L. L. 1975. A zoogeographic analysis of the South American Chaco avifauna. Bull. Am. Mus. Nat. Hist. 154:163–352.

SIMPSON, B. B. 1975. Pleistocene changes in the flora of the high tropical Andes. Paleobiology 1:273–294.

SIMPSON, B. B., AND J. HAFFER. 1978. Speciation patterns in the Amazonian forest biota. Annu. Rev. Ecol. Syst. 9:497–518.

SIMPSON, G. G. 1965. The Geography of Evolution. Chilton Books, Philadelphia, Pennsylvania.

SNOW, D. 1982. The Cotingas. Cornell University Press, Ithaca, New York.

STEGMANN, B. 1937. Grundzüge der ornithogeographischen Gliederung des palaarktischen Gebietes. Inst. Zool. Acad. Sci. U.S.S.R., New Ser. No. 19:1–157 (in Russian, extended German summary).

TERBORGH, J., AND B. WINTER. 1982. Evolutionary circumstances of species with small ranges. Pp. 587–600, In G. T. Prance (ed.), Biological Diversification in the Tropics. Columbia University Press, New York.

TODD, W. E. C., AND M. A. CARRIKER, JR., 1922. The birds of the Santa Marta region of Colombia. Ann. Carnegie Mus. 14:1–611.

TURNER, J. R. G. 1982. How do refuges produce diversity. Allopatry and parapatry, extinction and gene flow in mimetic butterflies. Pp. 309–335, In G. T. Prance (ed.), Biological Diversification in the Tropics. Columbia University Press, New York.

VAURIE, C. 1980. Taxonomy and geographical distribution of the Furnariidae (Aves, Passeriformes). Bull. Am. Mus. Nat. Hist. 166:1–357.

VUILLEUMIER, F. 1968. Population structure of the Asthenes flammulata superspecies (Aves: Furnariidae). Brevoria 297:1–21.

VUILLEUMIER, F. 1969. Pleistocene speciation in birds living in the high Andes. Nature 223:1179–1180.

VUILLEUMIER, F. 1970. Insular biogeography in continental regions. I. The northern Andes of South America. Am. Nat. 104:373–388.

VUILLEUMIER, F. 1971. Evolutionary relationships of some South American ground tyrants. Chapter 1. Generic relationships and speciation patterns in Ochthoeca, Myiotheretes, Xolmis, Neoxolmis, Agriornis, and Muscisaxicola. Bull. Mus. Comp. Zool. 141:181–232.

VUILLEUMIER, F. 1980. Speciation in birds of the high Andes. Proc. 17th Int. Ornithol. Congr. (Berlin), pp. 1256–1261.

VUILLEUMIER, F. 1983. The origin of high Andean birds. Nat. Hist. 90(7):50–57.

VUILLEUMIER, F. 1985. Forest birds of Patagonia: ecological geography, speciation, endemism, and faunal history. Pp. 255–304, In P. A. Buckley, M. S. Foster, E. S. Morton, R. S. Ridgely, and F. G. Buckley (eds.). Neotropical Ornithology. Ornithol. Monogr. No. 36.

VUILLEUMIER, F., AND D. EWERT. 1978. The distribution of birds in Venezuelan páramos. Bull. Am. Mus. Nat. Hist. 162:47–90.

VUILLEUMIER, F., AND D. SIMBERLOFF. 1980. Ecology versus history as determinants of patchy and insular distributions in high Andean birds. Evol. Biol. 12:235–379.

WEITZMAN, S. H., AND M. WEITZMAN. 1982. Biogeography and evolutionary diversification in neotropical freshwater fishes, with comments on the refuge theory. Pp. 403–422, In G. T. Prance (ed.), Biological Diversification in the Tropics. Columbia University Press, New York.

WILEY, E. O. 1981. Phylogenetics: the Theory and Practice of Phylogenetic Systematics. John Wiley, New York.

ORIGIN, EVOLUTION, AND STATUS OF THE AVIFAUNA OF ANDEAN WETLANDS

Jon Fjeldså

Zoological Museum, University of Copenhagen, Universitetsparken 15, DK-2100 Copenhagen Ø, Denmark

ABSTRACT. The distribution of Andean wetland habitats in the Pleistocene and present is reviewed and considered in relation to the known distribution of waterbirds. Analysis based on suture zones, disjunctions, core areas for endemic taxa, and phylogenetic relationships suggests evolutionary processes closely tied to events in the Pleistocene. The distributions of species that prefer barren habitats suggest a center of origin in the glacial Lake Michín on the Bolivian Altiplano. Other glacial refugia that gave rise to new taxa were Lake Atacama in Chile, Lake Junín in Peru, and the large lakes that previously were present in the Bogotá area, Colombia. Semi-open refugia along the Andean slopes had slight evolutionary significance. The inhabitants of Andean marsh habitats are generally poorly differentiated, morphologically, from lowland counterparts, and most probably they immigrated to the area in postglacial time. The colonization of the Andes was almost unidirectional, from the southern lowlands, which has resulted in a strong northward reduction of taxa adapted to barren habitats. The previous lakes of the Bogotá area received propagules from other directions, including from North America, but the many marsh birds in this area have not shown adaptive shifts in response to vacant niches in barren habitats.

RESUMEN. Se revee la distribución de los habitats pantanosos de los Andes tanto durante el pleistoceno como en la actualidad y se los considera en relación con la distribución conocidas de aves acuáticas. Análisis basados en zonas de sutura, disjunción, áreas centrales para grupos taxonómicos endémicos y relaciones filogenéticas, sugieren procesos de evolución cercanamente relacionados con sucesos del pleistoceno. La distribución de especies que prefieren habitats estériles hace suponer que el lago glacial Michín, en el altiplano boliviano, ha sido el centro de origen (de las mismas). Otros refugios glaciales, como el lago de Atacama en Chile, la laguna de Junín en Perú y los grandes lagos que se encontraban anteriormente en el área de Bogotá, Colombia, también brindaron origen a nuevos grupos taxonómicos. Refugios semidescubiertos en las laderas de los Andes tienen escaso significado a nivel de evolución. Los habitantes de los habitats pantanosos andinos generalmente presentan pocas diferencias morfológicas respecto a sus contrapartes en las tierras bajas, y probablemente han emigrado hacia el área luego de la época glacial. La colonización de los Andes fue casi unidireccional desde las tierras bajas del sur, lo cual ha determinado que hacia el norte se de una marcada disminución de taxa adaptada a los ambientes estériles. Los lagos que existían anteriormente en el área de Bogotá habían sido colonizados desde otras direcciones, incluyendo América del Norte, pero muchas de las aves acuáticas de esta area no muestran cambios adaptativos con respecto a los nichos vacantes en los habitats estériles.

Previous investigations of Andean waterbirds have been mainly qualitative, primarily delimiting general distributions. But, despite the tremendous recent interest in the wildlife of South America, the distribution of potential waterbird habitats is still known only in broad terms, and population data and detailed life history data are lacking for most species. Lack of sufficient data prevented for a long time the continuation of the analysis of speciation of Andean birds initiated by Chapman (1917, 1926). Recently, detailed patterns of speciation have been proposed for some groups of subtropical, temperate, and páramo landbirds (F. Vuilleumier 1968, 1969a, 1970a, 1971; Paynter 1972, 1978; Fitzpatrick 1973; Haffer 1974; Carpenter 1976; Graves 1982), and several scientists have examined in broad terms effects of Pleistocene climatic shifts on such patterns (Haffer 1967, 1969, 1974, 1979, 1981; Vuilleumier 1969b, 1970b, 1980, 1981; B. Vuilleumier 1971; Van der Hammen 1973; Simpson 1975; Vuilleumier and Simberloff 1980). However, waterbirds, which may be expected to show speciation patterns quite different from those of landbirds, have received hardly any attention.

In this paper I review the distribution patterns and systematic relations of waterbirds and against this background, attempt a general reconstruction of the patterns of immigration and evolution of the avifauna of the Andean wetlands. I shall also try to assess whether the fauna has reached equilibrium.

PREHISTORIC BACKGROUND

The following survey of the prehistoric scenery in which the Andean waterbirds evolved is based mainly on reviews by Auer (1960), Jeannel (1967), Putzer (1968), Van der Hammen (1973), Simpson (1975), and Mercer (1976).

GEOLOGICAL HISTORY OF POTENTIAL FOUNDING AREAS

By the start of the Tertiary, South America comprised three forested shields, all separated by shallow seas: Archi-Guyana, Archi-Brazil, and Archi-Plata, a forested prolongation of western Antarctica. The land that now forms the inter-Andean Altiplano had just been uplifted above sea level to the north of the latter shield.

Although sediments were being deposited between the shields, there were occasional marine ingressions into the Amazon area from the Pacific as well as the Atlantic. After the Ecuadorian and Peruvian Andes were connected in the mid-Tertiary, a definite eastward-directed drainage pattern was established from the sub-Andean basin, but, nevertheless, the area was periodically covered by large freshwater lakes or by Pleistocene marine waters (Fig. 1). These waters, presumed barriers to landbirds, may have facilitated the dispersal of waterbirds. We could, thus, expect a slow, phyletic evolution, resulting in a few widespread taxa. Speciation and formation of a more diverse fauna could, on the other hand, be expected in the savannas which covered vast areas in cold and dry periods but in warm periods became restricted to the llanos north and south of the rain forests.

The elevated shields were generally unsuited to waterbirds, but the plains formed by sedimentation in the La Plata basin certainly had numerous lakes and marshes and, periodically, marine wetlands. These plains changed as the Andes were elevated. Perhaps the greatest effect of these mountains was that they prevented humidity from the South Pacific Anticyclone from entering the Argentine plains. While wet rain forest existed in southern Chile, an elongate zone of rain shadow extended east of the Andes from Bolivia southward. The woodland savanna changed into cold shrub-steppes, which led to a complete isolation of the Chilean forest zone (Vuilleumier 1967).

During the Pleistocene glacial periods, the Argentine plains improved greatly as waterbird habitats, as meltwater led to formation of enormous wetlands all the way from the foot of the Andes to La Plata. The Salar de Pipanaco and Salinas Grandes in Santiago del Estero, and many other salinas in Mendoza are remains of these lakes. Meltwater lakes also appeared in the central valley of the Atacama desert.

DEVELOPMENT IN THE ANDES

During the Miocene, parts of the middle portion of the Andes locally reached 1500 to 2000 m, but still had a tropical climate and forested slopes. An enormous lake, Lake Ballivian, formed on the Altiplano as the Cordilleras were jammed against it. During series of fragmentations of the lake habitat, several groups of limnic organisms underwent "explosive speciation" (e.g., Willwock 1962), which resulted in a diverse food supply for birds in the vestige of Ballivian, known today as Lake Titicaca.

Areas with temperate climate formed on the mountain ridges from the end of the Pliocene, and this led to formation of a discontinuous dispersion corridor between the temperate zones of North America and southern South America. The uplift continued during the Pleistocene into recent times, as indicated by raised river terraces, tilted or faulted sediments, diluvial fan deposits and earthquakes. In the last half of the Pleistocene, the high parts of the Andes were subject to a series of glacial and interglacial periods, which caused an alternation between periods of dispersal and isolation (Vuilleumier and Simberloff 1980). Palynological data give a basis for judging the climatic changes in the northern part of the Andes, in Bolivia, and in the Fuegian zone, but for most of the Andes, the Pleistocene history is known only in broad terms.

The southern Andes, the Chilean fjordlands, and the Fuegian zone were completely glaciated, but ice-free areas with subantarctic moors and heaths existed on the outer Chilean coast as far south as the Guafo and Chiloé islands. The *Nothofagus* forest was concentrated between

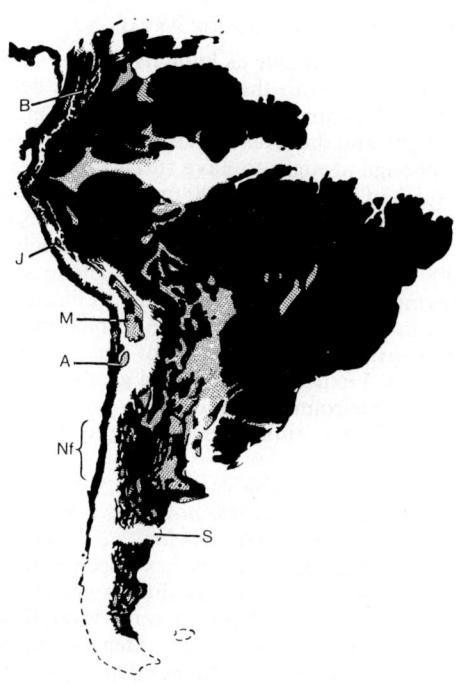

FIG. 1.   South America in the last glacial period. Land shown in black, large wetlands stippled, glaciers white. B = wetlands of the Bogotá and Ubaté Savannas; J = Lake Junín; M = Lake Michín; A = Lake Atacama; Nf = *Nothofagus* forest; S = Somuncura glacier.

30° and 36°S. The southern Patagonian steppes, with numerous wetlands fed by meltwater, were isolated from the pampas by the large Somuncura glacier which reached from the Andean ice to the Atlantic (Fig. 1).

The snowline rose northward, and large parts of the inter-Andean high plateau were ice-free, with large lakes isolated by glaciated cordilleras. Lake Michín on the Bolivian Altiplano connected the present lakes Titicaca and Poopó, and the salinas of Uyuni and Coipasá (Fig. 1; Lohman 1970). This area was isolated by the completely glaciated Western and Royal Cordilleras and to the north by ice-covered transverse ranges. A large freshwater lake in the present Salar de Atacama at 2400 m, 23°S, was only partly isolated to the west, and its fauna may, thus, have had some contact with that of the meltwater lakes in the Chilean central valley.

In western Peru, a high annual precipitation at high elevations led to development of continuous woods and many rather small wetlands. Glaciers were present above 3500 m, but some branches extended to much lower levels (Oppenheim 1945). The snouts of the glaciers dammed lakes in some valleys, e.g., in the Department of Cuzco, Peru. Two savanna plains at 2550 m near Cochabamba, Bolivia, suggest large Pleistocene wetlands. The inter-Andean plain at Junín, at 4080 m, 11°S, was ice-free at least in the last glacial period and comprised a well isolated area of level, waterlogged tundra, and a lake (Fjeldså 1981a).

The Andes of northern Peru, Ecuador, Colombia, and Venezuela had glaciers only locally, mainly above 3000 m. The climate 15,000–27,000 years before present (B.P.) was very dry and 6–7° colder than now, with alpine vegetation down to 2000 m (Van der Hammen 1974; Schubert 1974; Hastenrath 1981); thus, this vegetation was continuous except for small gaps at the Huancabamba deflection in northern Peru, Paso de Cruces in the eastern Andes of Colombia, and the Tachíra disjunction between Colombia and Venezuela. Large temperate wetlands may have existed in the inter-Andean basins in Ecuador, in the upper Cauca Valley, in the Central Andes, and on the present Bogotá and Ubaté savannas at 2600 m in Colombia. The latter savannas were formed by sedimentation into two large lakes, dammed by slightly tilted fault blocks.

ANDEAN GLACIAL LAKES AS POTENTIAL REFUGE AREAS

Whether the glacial lakes were suitable as habitats for birds can be judged by comparison with present Andean lakes. Even with the large oscillations between night-winter and day-summer conditions, modern highland lakes north of ca. 15°S are "tropical"; the annual variation in temperatures is slight, and the lakes are heated sufficiently by the intense solar radiation and conserve heat well enough at night to have surface temperatures constantly above 4°C. Usually, water temperatures of lakes at ca. 4000 m are constant between 9° and 12°C (Vellard 1952). In contrast to tropical lowland lakes, the diurnal variation in heat gain and heat loss prevents development of a stable stratification and promotes polymictic circulation; therefore, even small lakes at 5000 m never freeze. Large lakes create a local high pressure at night, which insures stable thermal conditions in the vicinity (Gilson and Holmes 1939; Monheim 1956). Studies by Lindroth (1965) suggest that the climate in closed glacial refuges was determined by the general Pleistocene deterioration of macroclimate and not by the proximity of the ice. For these reasons, I expect that the large glacial lakes in the puna zone were ice-free habitats for waterbirds year-round. Although they probably lacked tall marsh vegetation, I expect that the lakes had sufficient submergents rising to the surface for breeding grebes and coots.

In the most southern parts of the puna zone, some shallow lakes are known presently to freeze occasionally. Lakes with hot springs, however, remain open and act as refuges for waterbirds when night-time temperatures drop below −20°C (Johnson 1965). Because of strong Pleistocene volcanic activity, such springs may have been numerous in the glacial periods.

For the Bogotá and Ubaté wetlands, past conditions are well documented (Van Geel and Van der Hammen 1973). In cold periods with low water level, there was submerged vegetation of water milfoils (*Myriophyllum elatinoides*), and extensive marsh vegetation. From 3000 to 5000 years B.P., the areas were covered by rather open, typical *Potamogeton* lakes. This was followed by a short period in which *Typha* beds and floating *Hydrocotyle* vegetation spread, then a strong development of marsh vegetation, and disappearance of pondweeds.

PRESENT WATERBIRD HABITATS AND WATERBIRDS IN THE ANDES

TROPHIC CONDITIONS OF THE HABITATS

Aquatic plants, especially submergents, have less developed chemical and structural defenses against grazing than do terrestrial plants (reviewed in Hutchinson 1957). This apparently has made it easier for aquatic than for terrestrial birds to develop vegetarian habits. There are several reasons for the exceptional concentrations of bird life in some wetland habitats, but the most important appears to be that the birds can make direct use of primary production. The avifaunas of wetlands differ from those of most other habitats, above all by an exceptional density of large, herbivorous species (Reichholf 1975; Fjeldså 1980).

This trophic structure cannot, however, develop everywhere. In the tropics, particularly in the Amazon area, avifaunas are strongly influenced by complex fish faunas, whose species occupy many positions as consumers that otherwise might be available to birds (Reichholf 1975). Rain forest areas generally show small densities of waterbirds, and most species are piscivorous. In temperate zones, certain benthos-feeding and vegetarian fishes create turbid lakes with increased energy turnover but fewer resources available to birds (Andersson 1981). Diving ducks may directly avoid places with predation by fish (Eriksson 1983), and presence of certain fish types may well be expected to exclude, for instance, the occurrence of such social, plankton-feeding birds as flamingos. Probably as a result of such interference, the richness of waterbird species increases strongly from Amazon areas to areas with poor fish faunas, e.g., in the southern lowlands (Reichholf 1975), on the Falkland Islands (Weller 1972), on Tierra del Fuego (Weller 1975), and in much of the Andes (Fjeldså 1981c).

Another pertinent factor for waterbirds is related in the chemical dichotomy among freshwater plants (Lohammer 1938; Moyle 1945). Some are typical of soft water (e.g., moisture-influenced páramo habitats), others of hard water (limestone zones and arid regions). Plants specialized for high pH conditions (e.g., stonewarts, *Chara* spp., and narrow-leaved pondweeds, *Potamogeton, Zanichellia*, and *Ruppia* spp.) have mechanisms for using bicarbonate ions as a source of carbon and, as a result of consequent fast growth, constitute favorable food resources for vegetarian birds. Rather high pH conditions and dense submergents also permit the development of a rich invertebrate fauna. Finally, weeds reaching the water surface provide appropriate habitats for birds that have floating nests. Such habitats, sometimes found to

5000 m, may be particularly important in barren highland areas where nests placed on the shore are very vulnerable to predation. The stonewart vegetation in lakes in limestone zones is particularly valuable, because it often rises to the surface as large, floating carpets consolidated by precipitated marl.

DISTRIBUTION OF WETLAND HABITATS IN THE SOUTHERN ANDES

Waterbird habitats in lowlands surrounding the southern Andes have been described by Weller (1967, 1969, 1975), Reichholf (1975), Drouilly (1977), Aldridge (1981), and Bucher and Herrera (1981), but published information about mountain lakes is very fragmented.

The lowlands immediately to the east of the Southern Andes mostly have unsuitable habitats for waterbirds. The rich tule-marsh habitat of the pampas is separated from the Andes by wide zones of chaco and shrub-steppes, and the lakes of the Patagonian plains are deteriorating as bird-habitats owing to drought and intensive grazing by sheep. Tall waterside vegetation is almost entirely lacking, and the lakes become turbid and eventually silt in because of soil erosion. The numerous, shallow and unstable wind-influenced claypan lakes offer suitable habitats for few species, mainly shorebirds and plankton-feeders. Therefore, most waterbirds are restricted to large lakes near the foot of the Andes and small caldera lakes on the almost unexplored foothill mesetas. All wetlands on these windy plateaus lack emergent sedges. Many lakes are strongly turbid, but some produce much zooplankton, and there are numerous rather deep, stable, and clear lakes with dense submergents or dense carpets of water milfoils covering the surface. The meseta at 800 to 1200 m between Lakes Cardiel and Strobel, had more than 60,000 waterfowl in February 1984 (pers. observ.). Another important locality, Lake Blanca in western Chubut, has large breeding colonies of grebes, flamingos, and swans.

Owing to wind exposure, the Fuegian zone also has barren habitats. Precipitation, however, is more regular than in Patagonia, and the plains are densely dotted by lakes, weedy ponds, marshes, and wet meadows, with large numbers of geese, swans, and ducks. In the narrow Patagonian Cordillera, ice-capped ridges alternate with forested valleys with countless marshes, bogs and glacial lakes ranging from tarns and cirque lakes to deep fjord lakes, mainly below 800 m. This Valdivian rain forest zone has an oceanic, temperate, ever-humid climate. The lakes are mainly acidic and oligotrophic, with short bottom vegetation consisting primarily of quillworts (*Isoëtes*), but stable enough to have marginal reeds in sheltered parts. Some lakes, in areas with eutrophic, brown, forest soils, have dense pondweeds. The fish fauna is poor, but sufficient to support grebes and cormorants.

It is significant that these waterbird areas are close to mild marine wintering habitats, such as the tidal estuaries and salt-marshes in Ancud on the island of Chiloé and near Puerto Natales and Porvenir on Tierra del Fuego, and in the freshwater marshes along the Río Cruces in Valdivia.

WETLAND HABITATS IN THE PUNA ZONE

The central portion of the Andes is separated from adjacent lowlands by slopes with few waterbird habitats. Above 3500 m, this part of the Andes is characterized by rolling grassland, the puna. This habitat is unbroken today (unlike in glacial periods). There are no marked ecological barriers, although a certain longitudinal zonation exists, with desert-like conditions along the western edge grading into humid conditions in the east and north. The lake districts of the puna zone present some of the finest waterfowl spectacles of the continent. Although the grassland and semi-desert habitats are tediously uniform, the wetlands are manifold, including vast salt- and freshwater lakes, large marshes, wet meadows with meandering streams and temporarily inundated parts, highland bogs, and sterile lakes on very high levels.

The desert puna covers the slopes above the Atacama desert and the adjacent inter-Andean plain. Large, salt-encrusted depressions, remnants of Lake Michín, characterize the plains. The salt lakes are unstable. However, some very shallow, alkaline waters produce large quantities of big diatoms, protozoans and crustaceans, which are eaten by hundreds of thousands of flamingos and, for part of the year, Wilson's Phalaropes, *Phalaropus tricolor* (Hurlbert and Keith 1979; Hurlbert and Chang 1983; S. H. Hurlbert, *in litt.*). The majority of birds, however, requires the presence of fringing semi-bogland with freshwater springs supplied by water from the melting snow in the mountains. Because of the extreme climate, even well-watered places lack vegetation more than a few centimetres high. On the other hand, relatively deep and mildly salty lakes are filled with submergents, mostly *Ruppia* spp. Very high in the mountains, where the climate is extreme, glacial lakes with narrow weed-zones (Johnson 1965; McFarlane 1975) provide breeding habitat for the Giant Coot (*Fulica gigantea*).

Almost the full complement of Andean waterbirds, flamingos, geese, grebes, ducks, coots, and northern sandpipers, is found in complex wetlands such as Lake Pozuelos in Jujuy, Argentina (Correa 1973), the Parinacocha marshes, Lake Chugura, and the Surire salt lake in Arica, Chile, and Lake Lagunillas, southern Peru.

The wetter parts of the puna usually lack salt lakes, but most waters, except those very high up, are hard enough to have abundant weeds. Some areas with large amounts of accumulated glacial drift have numerous weedy kettles, which become breeding habitats for grebes and ducks in the rainy season. Characteristic of stable lakes in the humid puna are extensive tule and rush marshes. Tules cover a 12 km wide zone northwest of Lake Titicaca and a 2 to 5 km wide zone all the way around Lake Junín. In addition, fine tule-marsh habitats occur in a few places along the east side of the Bolivian Altiplano, in several lakes in Cuzco, and in Lake Parinacochas in Peru. Limnological conditions and vegetation are described by Gilson (1939) and Tutin (1940), whereas the birdlife is described by Niethammer (1953), Koepcke and Koepcke (1956), Dorst (1967), Dourojeanni et al. (1968), and Fjeldså (1981c, 1983a). These habitats are occupied by several marsh-birds unknown in the desert puna.

The original, almost continuous marsh vegetation, which was still widespread around Lake Titicaca in 1937 (Tutin 1940), had modest densities of birds. Since then, human overpopulation and intense harvesting of tules and weeds as food for man and cattle have created more open habitats. Particularly fine habitats for grebes have been formed where large tule marshes have been turned into mosaics of floating weeds with scattered tule beds, as, for example, in Lake Umayo near Puno, Peru. Altogether, human activity has altered the original zonation of the vegetation, but it has probably increased total species richness.

Possibly the very best breeding site for Andean waterbirds is Lake Junín, at 4080 m in central Peru. It is surrounded by 119 km² of strongly grazed meadows full of creeks and hummocky *Distichia* bogs, in many sections with the weedy zone teeming with waterfowl near the shore. Inside this zone lie 156 km² of marshes, both unbroken tule zones and areas of open water with soft marl bottoms and scattered patches of rushes, and, finally, 143 km² of open water with a dense growth of submerged stonewarts in shallow areas (Dourojeanni et al. 1968; Fjeldså 1983a). Dourojeanni et al. (1968) estimated a total of one million waterbirds, whereas recent estimates (Harris 1981; Fjeldså 1983a) of grebes, waterfowl, rails, and coots range to 100,000.

WETLAND HABITATS IN THE NORTHERN ANDES

In northern Peru, Ecuador, Colombia, and Venezuela, the Andes are covered with montane forest, except in the páramos above ca. 3200 m. The topography leads to a patchy and discontinuous distribution of open, alpine habitats, unlike the continuous habitats present during the cold parts of the Pleistocene (Vuilleumier and Simberloff 1980). The Ecuadorian páramos are very large and include some rather large lakes (Ortiz-Crespo 1983), but elsewhere, true lake districts exist only near "nevadas," e.g., in Sierra Nevada del Cocuy, northwest of Páramo de Santurban, and around Nevada del Ruiz.

This part of the Andes is generally under the influence of the intertropical low-pressure trough and is permanently humid. Many areas are strongly influenced by water-laden air rising upslope from the lowland savannas to the east, and are continuously wrapped in clouds and mists. Swampy terrain often has oligotrophic bogs with a thick layer of *Sphagnum* mosses, *Carex* sedges, dense fringing bamboo thickets (*Chusquea*) and moss-laden dwarf forest. Also, the lakes are oligotrophic, with clear, soft water. The vegetation, mainly sedges, submerged mosses, quillwarts, and *Ranunculus kunthianus,* is usually limited to estuarine shallows. Exceptions to this descriptions are, however, wetlands of rather dry intermontane areas, which resemble those of the humid puna.

Of particular importance are the wetlands at 2600 m in the Bogotá area and the 56 km² large Lake Tota, at 3020 m. Lake Tota is described by Borrero (1963) and Aguire and Rangel (1976), its ornithology by Adams and Espin (in press).

Most of the Bogotá and Ubaté plains were originally swamp interrupted by a multitude of lakes that probably provided excellent habitats for a wide variety of Andean marsh birds and wintering northern waterfowl. An important peripheral vegetation, the alder (*Alnus jorullensis*) swamp forest, has been removed. Another type of vegetation, still well represented in Lakes

Fuquene, Cucunuba, and Tota, is comprised of tall tules and cattails with an undergrowth of several herbs. Outside this zone there are usually a variety of floating plants, whereas in open water, generally, submergents are dense. The extensive marsh areas and weedy shallows around Lake Tota were turned into fields in the 1950's by lowering of the lake level, and the present weed zones consist almost solely of *Elodea,* which is becoming seriously polluted by filamentous algae (Adams and Espin, in press). Lakes Fuquene and Cucunuba have, since the 1940's, become very turbid because of soil erosion, and have lost their submerged vegetation; the rest of the wetlands has been cultivated or turned into pasture. Some small remaining marshes near Río Bogotá are now strongly polluted. The last fine bird lake on the savanna, Lake Herrera, was partly drained ca. 1973.

GENERAL AVIAN DISTRIBUTIONS

I have found it superfluous to review species distributions in detail, since such data already exist in numerous handbooks. However, recent information not incorporated in Hellmayr et al. (1918–1949), Meyer de Schauensee (1948–1952, 1966), Delacour (1954–1964), Olrog (1963), Johnson (1965), Humphrey et al. (1970), Kear and Duplaix-Hall (1975), Blake (1977), Ripley (1977), Meyer de Schauensee and Phelps (1978) and Johnsgard (1978, 1981) is given in Appendix I. A great deal of new data may be expected in the near future as a result of the survey of wetland birds and habitats recently launched by the International Waterfowl Research Bureau and the International Council for Bird Preservation. Summaries of selected distribution patterns are given below (Fig. 2).

The jagged, southern Andes have endemic, well-defined subspecies of Torrent Duck (*Merganetta armata*) and Cordilleran Snipe (*Gallinago stricklandi*). The foothill plateaus to the east side have another endemic, the Hooded Grebe (*Podiceps gallardoi;* Appendix I, Fig. 2A). Further, several taxa are endemic to the zone of Fuegian lowlands and adjacent mountains, a subspecies of Great Grebe (*Podiceps major;* Appendix I, Fig. 2K), Upland Goose (*Chloephaga picta;* two ecologically distinct subspecies or species *picta* and *dispar*), Flying Steamer-duck (*Tachyeres patachonicus;* Appendix I), Spectacled Duck (*Anas specularis*), Magellanic Oystercatcher (*Haematopus leucopodus;* Appendix I), Two-banded Plover (*Charadrius falklandicus;* Fig. 2F), Rufous-chested Dotterel (*Charadrius modestus*), and Magellanic Plover (*Pluvianellus socialis;* Appendix I). Taxa with a wide distribution in the southern lowlands and highlands are the nominate subspecies of the Silvery Grebe (*Podiceps occipitalis;* Appendix I, Fig. 2J), the nominate subspecies of the Black-faced Ibis (*Theristicus melanopis;* Fig. 2G), the Black-necked Swan (*Cygnus melanocoryphus*), the Red-gartered and White-winged Coots (*Fulica armillata, F. leucoptera*), and the nominate subspecies of the Rufous-backed Negrito (*Lessonia rufa*). Some species of the southern lowlands and cordilleras continue northward into the puna zone, where a few are differentiated into rather weak subspecies: White-tufted Grebe (*Rollandia rolland;* Appendix I, Fig. 2M), Chilean Flamingo (*Phoenicopterus chilensis*), Crested Duck (*Anas specularioides*), and Speckled Teal (*Anas flavirostris flavirostris* and *oxyptera;* Fig. 2N). Still others continue to the páramos: Yellow-billed Pintail (*Anas georgica*), Ruddy Duck (*Oxyura jamaicensis*), and Cinereous Harrier (*Circus cinereus*).

The puna zone has several endemic taxa of distinct appearance. Some of these are very local: Titicaca Flightless Grebe (*Rollandia microptera;* Titicaca area, Fig. 2A); Junín Flightless Grebe (*Podiceps taczanowskii;* Junín, Appendix I, Fig. 2A); subspecies *tuerosi* of the Black Rail (*Laterallus jamaicensis;* Junín, Appendix I); Horned Coot (*Fulica cornuta;* desert puna). Others are rather local: Andean and James' Flamingos (*Phoenicoparrus andinus, P. jamesi;* mainly desert puna), still others inhabit large parts of the zone: Puna Ibis (*Plegadis ridgwayi;* Fig. 2H), Puna Teal (*Anas puna;* Fig. 2I), Giant Coot, Puna Plover (*Charadrius alticola;* Appendix I, Fig. 2F), Andean Avocet (*Recurvirostra andina;* Fig. 2C), Puna Snipe (*Gallinago andina*), and subspecies *oreas* of the Rufous-backed Negrito. A few species extend their ranges to the high parts of the southern Andes: Andean Goose (*Chloephaga melanoptera*), Diademed Sandpiper-Plover (*Phegornis mitchellii*), Andean Gull (*Larus serranus*); others range to the páramos: subspecies *juninensis* of Silvery Grebe (Appendix I, Fig. 2J), subspecies *branickii* of Black-faced Ibis (Fig. 2G), Andean Coot (*Fulica ardesiaca;* Appendix I, Fig. 2D), and *leucogenys* complex of Torrent Duck. Widespread lowland species with moderately well-defined puna zone subspecies are: Cinnamon Teal (*Anas cyanoptera;* Appendix I, Fig. 2L), Plumbeous Rail (*Pardirallus sanguinolentus*), Common Moorhen (*Gallinula chloropus*), Wren-like Rush-bird (*Phleocryptes melanops*), Many-colored Rush-Tyrant (*Tachuris rubrigastra*),

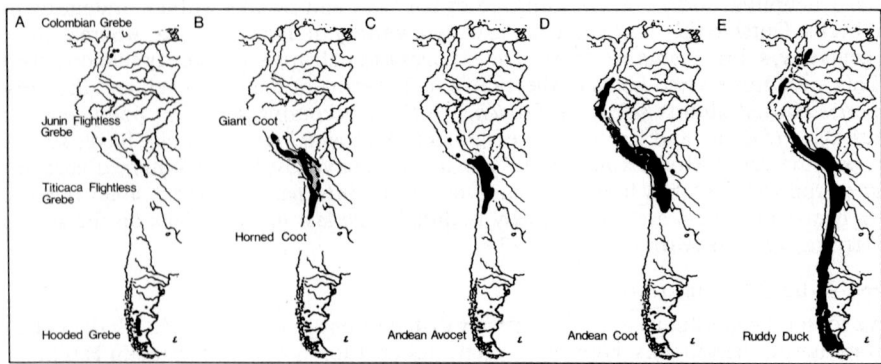

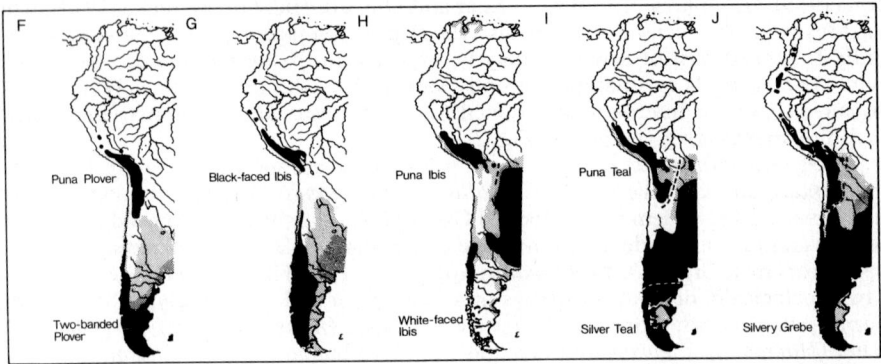

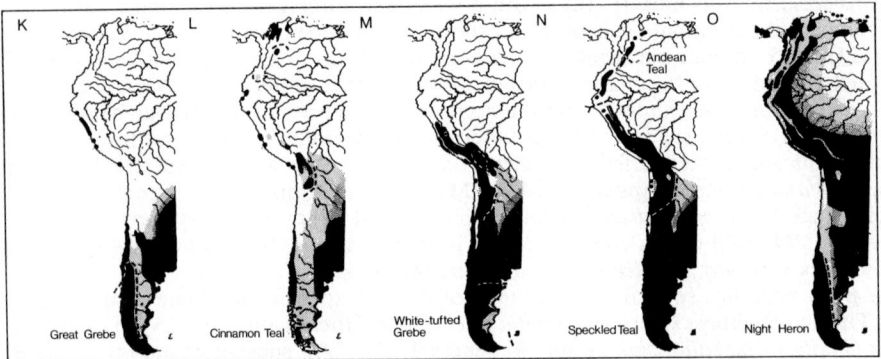

FIG. 2. Examples of distribution patterns. Black = areas of regular occurrence; stippled = areas of irregular or accidental occurrence; broken lines indicate sutures or narrow disjunctions between morphologically distinct populations. Top row = examples of Andean endemics. Middle row shows species and semispecies pairs with lowland and highland counterparts. Bottom row shows lowland species that ascend to the Andes at least locally.

Grass Wren (*Cistothorus platensis*), and Yellow-winged Blackbird (*Agelaius thilius*). No morphological differentiation has been noted, however, in the highland populations of Olivaceous Cormorant (*Phalacrocorax olivaceus*) or Black-crowned Night-heron (*Nycticorax nycticorax*; Fig. 2O).

Two páramo taxa occur also in boggy habitats in the middle portion of the Andes, the distinct subspecies *jamesoni* of the Cordilleran Snipe, and the Imperial Snipe (*Gallinago imperialis*). The páramos have a few endemic species or subspecies: Colombian Grebe (*Podiceps andinus*; Bogotá area, Appendix I, Fig. 2A), subspecies *colombiana* of Torrent Duck (Colombia, Venezuela), subspecies *niceforoi* of Yellow-billed Pintail (Bogotá area, Appendix

I), Andean Teal (*Anas flavirostris andinum* and *atlapetens*; widespread, Fig. 2N), Bogota Rail (*Rallus semiplumbeus*; Bogotá area, Appendix I), Noble Snipe (*Gallinago nobilis*; widespread), and Apolinar's Marsh Wren (*Cistothorus apolinari*; Bogotá area, Appendix I). Some species, mainly with lowland distribution, show well isolated subspecies in the Bogotá area: Least Bittern (*Ixobrychus exilis*; Appendix I), Cinnamon Teal (Appendix I, Fig. 2L), Ruddy Duck (Appendix I); Spot-flanked Gallinule (*Porphyriops melanops*), American Coot (*Fulica americana*; Appendix I), and Yellow-hooded Blackbird (*Agelaius icterocephalus*). The distribution and geographic variation of the Grass Wren in the northern Andes is not fully understood.

The peripheral occurrence of lowland birds in highland areas is important as our only clue to how Andean habitats may have been colonized in the past. Many southern lowland birds, which also occur in the southern Andes, have already been mentioned. In addition, several lowland species use habitats in the southern Andes as molting or staging areas, or they breed there occasionally: subspecies *obscurus* of Black-crowned Night-Heron (Fig. 2O), Coscoroba Swan (*Coscoroba coscoroba*), Chiloé Wigeon (*Anas sibilatrix*), Red Shoveler (*Anas platalea*; Appendix I), Southern Lapwing (*Vanellus chilensis*), Common Snipe (*Gallinago gallinago*), and Kelp Gull (*Larus dominicanus*), with 14 other species being rather casual. The marine Blue-eyed Cormorant (*Phalacrocorax atriceps*) breeds in two lakes in the Argentinian Andes.

Among widespread lowland species, Cattle Egret (*Bubulcus ibis*), Great Egret (*Egretta alba*), and Black-necked Stilt (*Himantopus mexicanus*) visit many Andean wetlands, sometimes in large numbers, and have recently nested in Lake Junín (Appendix I). Many lowland species reach the semiarid zone at 2550 m in Cochabamba, Bolivia, and Comb Duck (*Sarkidiornis melanotos*), White-cheeked Pintail (*Anas bahamensis*), Wattled Jacana (*Jacana jacana*), and Collared Plover (*Charadrius collaris*; breeding?) have been seen, some in considerable numbers, on some lakes in this region (Dott 1984; J. V. Remsen, *in litt.*; Appendix I). Altogether, 18 lowland species and some 30 Nearctic migrants are known to be regular or accidental visitors to temperate wetlands in the middle portion of the Andes.

As to the páramos, Virginia Rail (*Rallus limicola*) ascends to the temperate zone in Ecuador and southern Colombia; Pied-billed Grebe (*Podilymbus podiceps*) is common in the Eastern Andes of Colombia, and no less than 10 other lowland species nest(ed) or are suspected to nest more or less sporadically in the latter areas (Appendix I). Moreover, 20 lowland species and 25 Nearctic migrants are known to visit páramo wetlands.

ORIGIN AND EVOLUTION OF THE FAUNA

For Andean landbirds, speciation is favored wherever some kind of habitat discontinuity breaks up the range of a species (Vuilleumier 1980). Because of the instability of wetland habitats, and the great general mobility of waterbirds, these birds may be expected to exhibit phyletic evolution at a rate far slower than that of landbirds, unless isolation barriers of very large dimensions are involved. It is, therefore, not surprising that the waterbird fauna reviewed above and in Appendix I includes but few cases of complex geographic variation (Torrent Duck, Speckled and Andean Teal, Andean Coot). Notwithstanding these facts, and the general impossibility of testing putative speciation models, some useful clues for reconstructing the past exist.

SUTURE ZONES

Previously fragmented ranges of species may be revealed by the existence of interspecific suture zones, or zones within a species' range where the individual morphological variation is great, or where a stepped cline separates areas of uniform character expression (Fig. 3B). In order to judge to what extent sutures are determined by present ecological barriers, I also considered the positions of major intraspecific range disjunctions (Fig. 3A). Such major disjunctions coincide with the Andes of Colombia, Ecuador, and northern Peru, which form a main barrier at present for tropical waterbird populations, with the Andes of Argentina and Chile, which form a main barrier to waterbirds inhabiting the Argentine pampas and the lowlands of central Chile, and with the central Amazon area which is a main barrier to waterbirds inhabiting the savannas to the north and south (Fig. 3A). Other disjunctions are mainly associated with the Atacama Desert and the adjacent desert puna, and with discontinuities along the northern Andes.

The suture zones follow general patterns other than those associated with the intraspecific disjunctions, which suggests that most sutures reflect past rather than present-day barriers. The main suture zone follows the Central Cordillera in Bolivia or the edge of the high plateaus

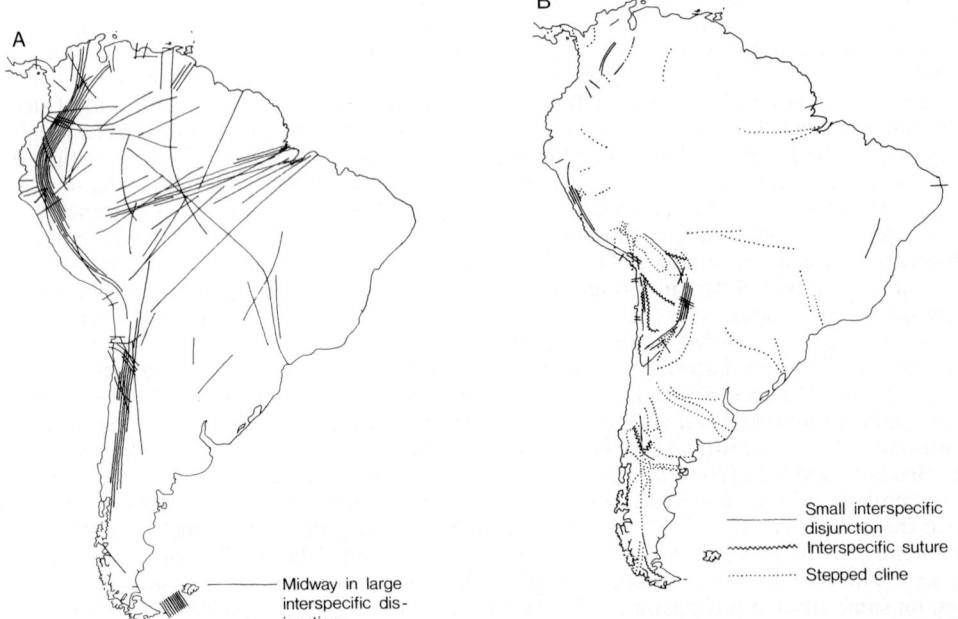

Fig. 3. The positions (A) of large intraspecific disjunctions and (B) of narrow interspecific disjunctions, suture zones, and stepped clines between adjacent populations of Neotropic waterbirds.

in Potosí, Jujuy, and Salta adjacent to the lowlands (Figs. 2I, L, M, N, and Fig. 3B). It must be remembered, however, that some narrow range disjunctions (e.g., between Puna and Silvery Teal, Fig. 2I, and between subspecies of Plumbeous Rail and Many-colored Rush-tyrant) possibly should be classified as major disjunctions instead. In general, the extent to which highland and lowland taxa are in direct contact is poorly known. The disjunction may be due to a paucity of wetlands on the Andean slopes, but might also reflect habitat differences between the forms, evolved during past isolation. Because these sutures follow a present-day ecotone, they yield suggestive rather than conclusive evidence about the past. However, the fact that some highland forms (e.g., Andean Coot and Speckled Teal; Figs. 2D, N) descend to the lowlands only to the west, from which their lowland counterparts are absent, suggests that the sutures in some cases reflect inability of the involved taxa to invade each other's range.

For a few taxa, such as Silvery Grebe, the suture may lie farther south (Catamarca, San Juan; Fig. 2J). The contact zones between Andean and White-winged Coots and between the subspecies (or species) of Rufous-backed Negrito may lie on the Chilean slope of the Altiplano.

Another suture zone, usually between subspecies, follows the eastern foothills of the Patagonian Andes, which are low and discontinuous and do not form a barrier (Fig. 3B; compare with Fig. 3A). For Great Grebe (Fig. 2K), the stepped cline follows the demarcation of areas with high summer precipitation so closely that it could reflect adaptations to present-day ecological conditions (Fjeldså, in press a). For other species, mainly lowland forms, the suture suggests, instead, the meeting of populations that expanded from the glacial forest refuge in central Chile (Fig. 1) and from the Argentine plains. White-tufted Grebe, Silver Teal, Southern Lapwing, and Grass Wren show stepped clines across Patagonia at ca. 42°S (Figs. 2, 3B). As this demarcation does not coincide directionally with present ecotones, it suggests the position of a past barrier, maybe the Somuncura glacier (Fig. 1).

No well marked suture zones occur within the puna. Giant and Horned Coots meet in only a few places along a line from northern Jujuy to Tarapacá (Fig. 2B). Slightly northwest of Lake Titicaca, Silvery Grebes change morphologically (Fig. 2J; Appendix I). Andean Coots show slight morphological variation south of this line but are polymorphic north of it (Fig. 2D; Appendix I). These changes and some zones of great individual variation in the Torrent

Duck in Peru do not coincide obviously with present ecotones and may represent sutures between populations once separated.

The Huancabamba deflection, northern Peru, separates four pairs of landbird allospecies (Schulenberg and Williams 1982), and also subspecies of Andean and Speckled Teal (Fig. 2N), Torrent Duck, White-capped Dipper, and perhaps Andean Coot (Fig. 2D). An apparent primary cline in Andean Coot in northern Ecuador and southern Colombia (Appendix I) suggests past character displacement as a result of selection against interbreeding at a past suture zone with American Coots (Fjeldså 1983b). The suture between two forms of the Andean Teal (Fig. 2N) lies slightly north of the Porculla Pass. Some gaps between subspecies on the Bogotá plateau and in the adjacent lowlands (Fig. 3B) coincide with present ecotones.

RANGE LIMITS AND LARGE DISTRIBUTION GAPS

Range limits that do not represent interspecific sutures and that do not coincide with obvious ecological ecotones may indicate that the populations in question have not had time to spread from a past refuge area to occupy their entire potential range. The most obviously unoccupied, apparently suitable, waterbird habitats in the Andes are in the north, where ibises, geese, many ducks, and gulls do not occur. Despite the presence of apparently excellent habitats, wigeons are absent from all the highlands, avocets from the southern lowlands, and dippers from all of the southern Andes.

Large gaps between the breeding ranges of the subspecies of Black-faced Ibis, and between Puna and White-faced Ibises (*Plegadis chihi*; Fig. 2H) do not appear to be due to discontinuities between structurally similar highland and lowland habitats. The absence of ibises from much of the southern puna zone may indicate that they differentiated when remotely separated, which in turn suggests that they were absent from the glacial Michín refuge. Alternatively, they may have disappeared from some areas due to the drought that sometimes ravages the wetland habitats of the Altiplano. Also the large gap between Puna and Two-banded Plovers (Fig. 2F) is noteworthy and suggests that representatives of this superspecies were absent from the large Pleistocene wetlands in Santiago del Estero, San Luis, and La Pampa. The absence of Cinnamon Teal from many apparently suitable habitats in the Andes between Cuzco and the Bogotá area (Fig. 2L) suggests colonization from the two ends of the Andes.

CORE AREAS

Following Haffer (1974), I mapped faunistic core areas by superimposing the range outline maps of relatively local taxa with a common main distribution. The area with the highest number of species in one distribution group is supposed to be a past refuge area or center of later dispersal. This conclusion requires that historical changes of the various habitats have consisted only of oscillations in their relative size, and not in their horizontal displacement. The method must, therefore, be used with care at high latitudes, where the vegetation zones have been subject to directional shifts. The method may also be misleading unless one considers whether the congruence of endemic distribution patterns simply reflects concentrations of particularly suitable habitats. In order to judge this possibility, I considered, in addition to maps of the individual local core areas, continent-wide maps of variation in relative distributions of endemic and widespread species.

A taxon was regarded as endemic if it inhabited only a part of one continuous biome. Taxa were analyzed according to membership in one of three crude guilds: (1) that associated with marsh and swamp habitats and not requiring large areas of open water (Fig. 4), (2) that associated with open water but requiring at least some marginal or patchy cover of reeds (Fig. 5), and (3) that associated with open wetland habitats and preferring rocky, sandy, or clay shores, or shore meadows with very short vegetation (Fig. 6).

Swamp and marsh birds have a complex zoogeography, with very large numbers of widespread as well as local species in the La Plata basin, Mato Grosso, and some tropical delta areas (Fig. 4). The distribution of endemic taxa in the lowlands of northern South America suggests a core area on the Colombian/Venezuelan llanos, another in the lowlands of northern Colombia, and yet another along the Colombian Pacific coast. The cores could be due partly to the distribution of marsh and swamp habitats and are not very apparent on the map of relative endemism. The recognition of core areas in this part of the continent is, nevertheless, fruitful, since it clearly suggests successful trans-Andean jumps in Colombia. Numerous instances of lowland species straggling high up in the Venezuelan and Colombian Andes are known, and this may have led to some permanent trans-Andean colonizations (Fig. 4). It is,

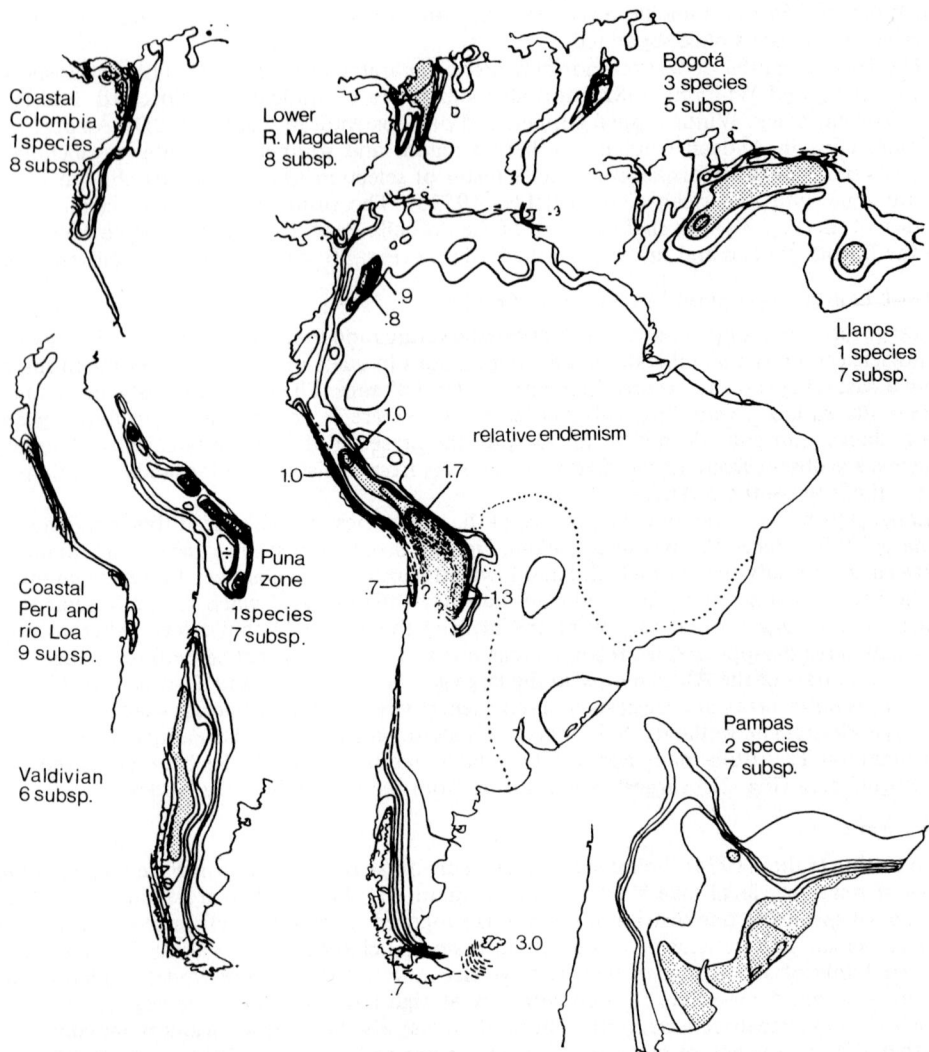

Fig. 4.   Core areas for endemic taxa (species, subspecies, or other morphologically distinct populations) associated with marsh and swamp habitats. Peripheral maps show numbers of taxa endemic to the same general area; the central map shows the continent-wide variation in number of endemic taxa relative to the number of widespread taxa.

therefore, not unexpected to find many swamp birds established in the temperate wetlands around Bogotá. However, the endemism (three species and five subspecies, or nearly as many as the number of non-endemic taxa) is remarkable and demonstrates an efficient isolation of these wetlands both in the past and at present.

A more southern core area with a high relative endemism at the subspecific level is found in littoral Peru. Very close to this lies an assemblage of three endemic subspecies in the isolated oasis of Río Loa in the Atacama desert. Some weakly defined core areas that somewhat resemble Haffer's (1974) Napo and Huallage refuges are indicated to the east of the Andes. Local subspecies also suggest past refuge areas for inhabitants of creeks and marshes in the foothills of Cochabamba, Bolivia, and Tucumán, Argentina. A large core area in the Pampa zone (with peak numbers of local taxa in the tule-marsh zone of Buenos Aires, in the marsh-lands of Paraguay, and near the coast of southeastern Brazil) can be explained, in part, by the present concentration of rich habitats. It is, therefore, not as conspicuous as might be expected

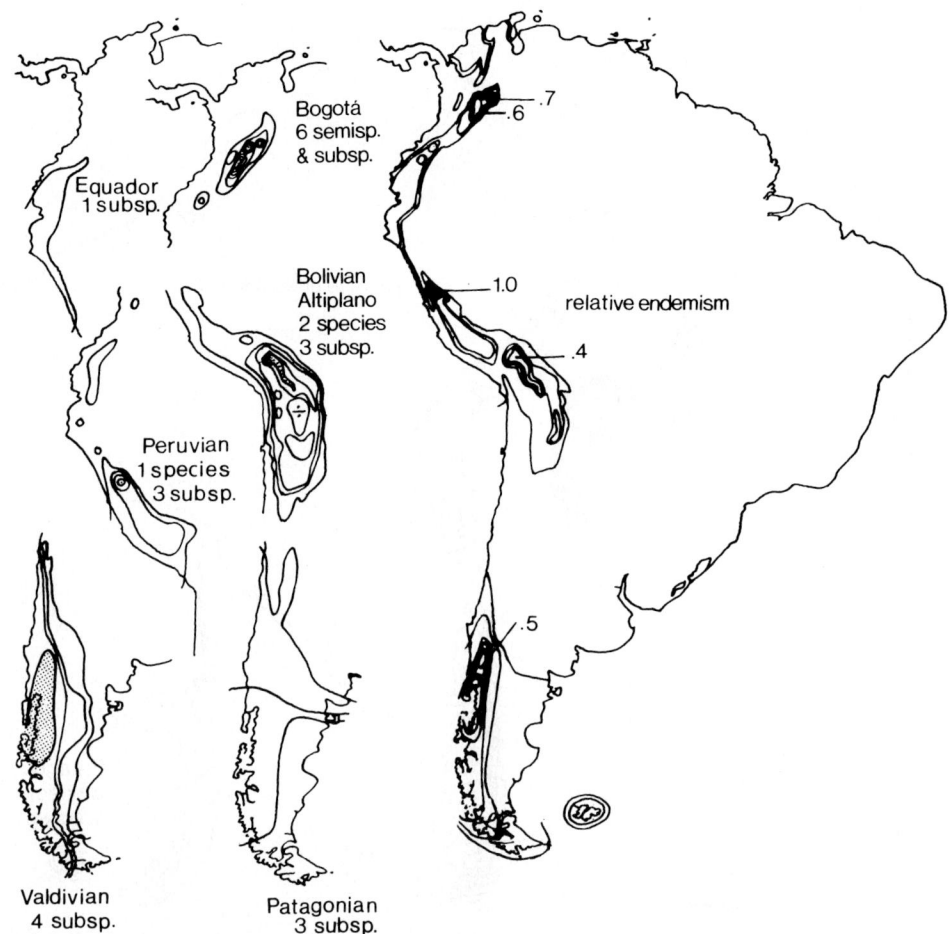

FIG. 5. Core areas for endemic taxa associated with open water, but normally requiring at least some marginal or patchy cover of reeds. Peripheral maps show numbers of taxa endemic to the same general area; the central map shows the continent-wide variation in number of endemic taxa relative to the number of widespread taxa.

on the central map of Figure 4. A last lowland core area lies in the Valdivian rain forest. Areas between the lowland core areas do not lack marsh and swamp birds, but have mainly widespread and/or opportunistic species.

A very high relative endemism is found in the puna zone, especially in the large marshes in the Junín and Titicaca areas. Except for Puna Ibis and Puna Snipe, this endemism exists, however, only at the subspecific level. Most Andean subspecies differ from lowland populations by being slightly larger, possibly a rapidly evolved adaptation to the climate and high elevations, and are maintained by lack of gene flow across the marked ecotones present along the Andean slopes.

The endemic taxa of waterbirds that prefer open lakes with some cover of tules or marginal rushes are restricted to the Andes, the lowlands of northern Colombia, and lowlands along the Pacific coast (Fig. 5). This does not correspond well with the distribution of widespread taxa, whose numbers peak on the Argentine plains. The highest percent of endemism is in Lake Tota, the Ubaté and Bogotá savannas, and Lake Junín, and a slightly lower proportion is found in Lake Titicaca and southward, east of the desert puna, and in the Valdivian forest zone.

Local endemism of waterbirds that prefer habitats without tall marginal vegetation occurs only along the Andes, with cores in southwestern Santa Cruz in Patagonia, and on the Bolivian

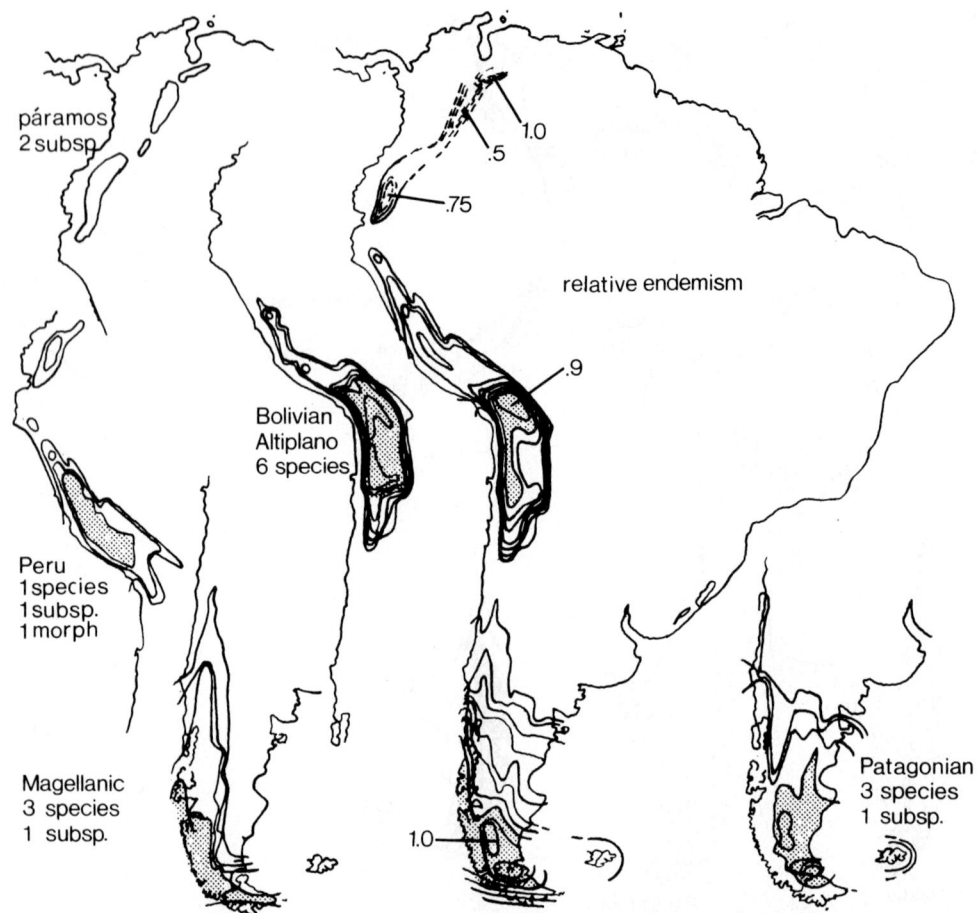

FIG. 6.   Core areas for endemic taxa associated with open wetlands that have barren shores. Peripheral maps show numbers of taxa endemic to the same general area; the central map shows the continent-wide variation in number of endemic taxa relative to the number of widespread taxa.

Altiplano (Fig. 6). This does not simply reflect a general suitability of these areas, as endemism in the Titicaca area and around the junction of the Argentine/Bolivian/Chilean territories near the Atacama salt lake is conspicuous both in absolute and relative terms (Fig. 6). As the core areas are not presently framed by marked ecotones in the north and south, I find it probable that they represent refuge areas, possibly Lake Michín and Atacama.

Also for the Patagonian core, marked framing ecotones are presently lacking to the north. The core area is partly inside previously glaciated areas, but a short postglacial displacement is likely to have occurred.

Areas in the páramos (Fig. 6) have an insufficient number of endemics to be recognized as core areas, perhaps because the total number of species requiring open habitat is very low.

RELATIONSHIPS BETWEEN ANDEAN AND LOWLAND FAUNAS

I connected the core area of each endemic Andean taxon to the faunal area(s) in which its presumed closest relative lives (Fig. 7). The strongest connection appears to exist between the puna (especially its southern part) and the biomes of the pampas and the southern part of the continent. Although a dispersal corridor appears to continue northward throughout the Andes, the connection with the southern lowlands becomes weaker as it passes into the páramos.

In contrast to this almost unambiguous pattern, the Bogotá area shows faunal connections in all directions, to the south (Andean Teal, Nicéforo's Pintail), to the Pampas (Spot-flanked

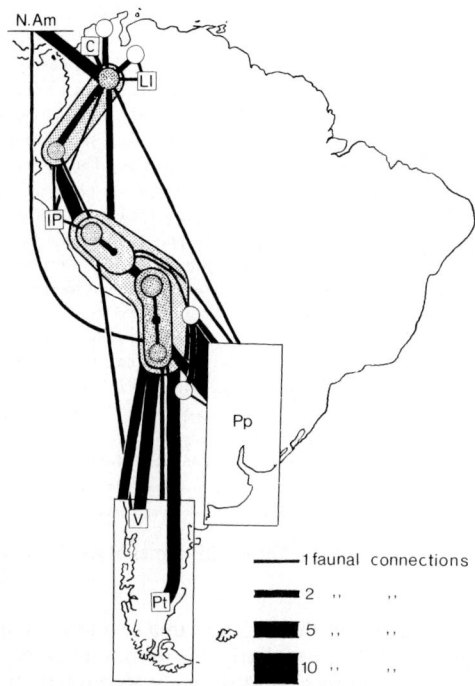

FIG. 7. Closest-relationship connections between core areas of endemic Andean taxa and the faunal areas of their nearest relatives. In the Andes, core areas are shown as circles or ellipses; in the lowlands as squares. N.Am = North America; C = northern Colombia; Ll = Llanos; IP = littoral Peru; Pp = Argentine pampas; Pt = Patagonia; V = Valdivian forest.

Gallinule), to Peru (Bogota Rail, assuming the presumed relationship with "*Rallus peruvianus*" is correct), to the Venezuelean highlands (Grass Wrens), to northern Colombia (Yellow-headed Blackbird), and to North America (Colombian Grebe, Least Bittern, American Coot). The Colombian Ruddy Duck may have had a hybrid origin, representing faunal connections both to the south and the north (see Appendix I).

Because the lowlands predate the Andes, it is probable that lowland species predate Andean ones, and that connections between core areas (Fig. 7), thus, reflect unidirectional dispersal from the south through the Andes. Massive entomological and botanical evidence suggests, however, that paleo-Antarctic taxa were "preserved" in the Andes (e.g., Brundin 1967; Jeannel 1967; Illies 1969), which means that dispersal also could have occurred from the Andes to the lowlands. To be certain, our speculations should be based on phylogenetic models. It is usually assumed that isolated species came from areas with many relatives; however, the relative distribution of taxa representing low and high branches on the phylogenetic tree should be considered as well (Cracraft 1983).

Phylogeny and zoogeography of grebes.—I developed an approximately rooted phylogenetic tree (Fig. 8) for all the grebes of the world, based on the classification of Storer (1979 and additional data given by Fjeldså 1981a, 1982a, 1984a, in press c).

The oldest known grebe fossil is from the Argentine Eocene, and southern South America has representatives of nearly all major clades on the tree, including species that show trans-Antarctic relationships. This suggests that the origin of the family was Antarctic or Archiplatean.

Two early lines (Pied-billed Grebe and Least Grebe, *Tachybaptus dominicus*) apparently spread to the subtropical parts of South America and reached North America but ascended to Andean wetlands only locally. The White-tufted Grebe, near the base of the phylogeny, remained mainly in the Archiplata area (Fig. 2M) and apparently displaced the ecologically similar Least Grebe from this zone (Fjeldså 1981a). Ethological data show that an ancestral form of the White-tufted Grebe theoretically could have been ancestral to most other grebes

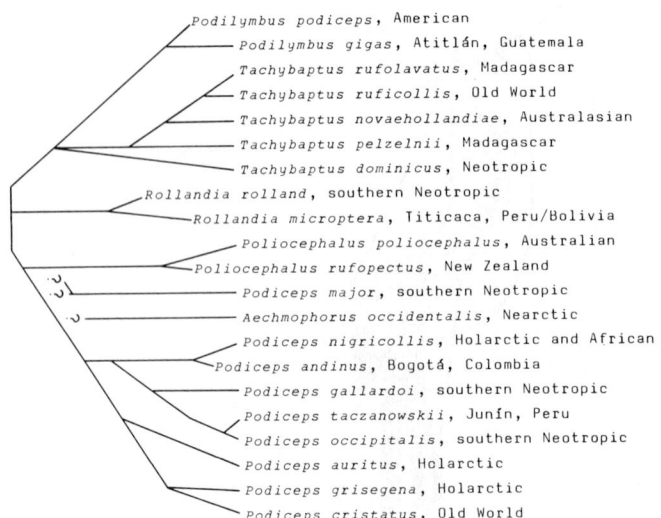

```
Podilymbus podiceps, American
    Podilymbus gigas, Atitlán, Guatemala
    Tachybaptus rufolavatus, Madagascar
    Tachybaptus ruficollis, Old World
    Tachybaptus novaehollandiae, Australasian
    Tachybaptus pelzelnii, Madagascar
    Tachybaptus dominicus, Neotropic
    Rollandia rolland, southern Neotropic
    Rollandia microptera, Titicaca, Peru/Bolivia
    Poliocephalus poliocephalus, Australian
    Poliocephalus rufopectus, New Zealand
    Podiceps major, southern Neotropic
    Aechmophorus occidentalis, Nearctic
    Podiceps nigricollis, Holarctic and African
    Podiceps andinus, Bogotá, Colombia
    Podiceps gallardoi, southern Neotropic
    Podiceps taczanowskii, Junín, Peru
    Podiceps occipitalis, southern Neotropic
    Podiceps auritus, Holarctic
    Podiceps grisegena, Holarctic
    Podiceps cristatus, Old World
```

FIG. 8.   A phylogenetic tree of the grebes of the world; names of species followed by species distributions.

(Fjeldså, in press c). The presence in the Titicaca area of a large congener with reduced wings, specialized feeding (Fjeldså 1981a), peculiarly adumbrated juvenal and natal plumages, and specialized displays (Fjeldså, in press c) is most likely accounted for by assuming that an early White-tufted Grebe type colonized the Titicaca area and became isolated there when the Cordilleras were ice-capped. After a second northward spread of White-tufted Grebe, sympatry with the derived Titicaca form was established, but with interference competition and some ecological displacement (Fjeldså 1981a, 1983f).

Great Grebe, another early line with trans-Antarctic relationships (Fjeldså 1984a) remained largely in the Archiplata area (Fig. 2K). It is not clear, however, whether the radiation among the more advanced, plumed *Podiceps* spp. started in southern South America or in the Northern Hemisphere. The second interpretation is not parsimonious since it requires one long distance dispersal to the north followed by one or two back to the south, and I, therefore, prefer the south to north explanation. Colombian Grebe, which inhabits the only Andean wetlands not colonized by the ecologically similar Silvery Grebe, looks primitive in relation to the Eared Grebe (*Podiceps nigricollis*; less melanistic and with less strongly elongated crown-feathers). It may, thus, represent a relatively primitive, relict population, but as reversal to a more primitive plumage type can be explained genetically, it is also possible that the population near Bogotá was established by southward displacement of Eared Grebes during a cold period. Hooded Grebe, which represents the next branch in the phylogeny, has a relict population within the Archiplata area (Fig. 2A), inside previously glaciated areas, but still close to a Pleistocene refuge (Fig. 1).

Silvery Grebe evidently shows its most primitive state in the Archiplata area (pigmented chin and throat, buff ear-plumes, conspicuous white area on wing), and changes stepwise as we pass northward in the Andes. This suggests that the colonization proceeded northward. Possibly, one population became isolated in the glacial Lake Junín during the last cold period. Under these conditions, with glaciers all around, it may have become flightless and behaved as a full species in relation to the renewed, postglacial influx of Silvery Grebes. This interpretation is strongly supported by anatomical evidence, which shows, for example, that the Junín Flightless Grebe has retained the precise jaw anatomy of the foliage-gleaning, insectivorous Silvery Grebe in spite of its present long bill and mainly piscivorous habits (Fjeldså 1981a). Lake Junín Silvery Grebes have a very small bill compared with allopatric founding populations. This anomaly appears to have become accentuated over the last century, and as small individual variations in bill dimensions have been shown to affect the interspecific dietary overlap, ecological character displacement appears to have been involved (Fjeldså 1981a, 1983f).

Phylogeny and zoogeography of coots. — The most primitive, gallinule-like coot, Red-fronted

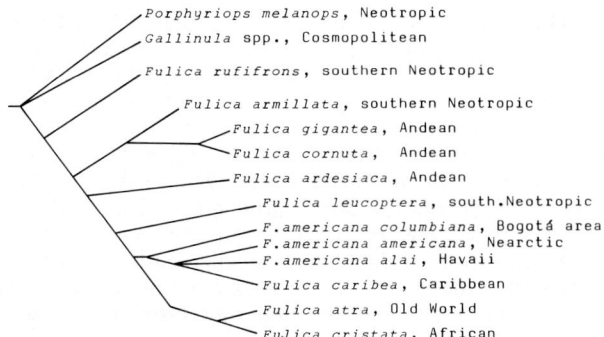

FIG. 9.   A phylogenetic tree for the coots of the world; names of species followed by species distributions.

Coot (Olson 1973), as well as other primitive species in a rooted phylogeny of the group (Fig. 9; based on Fjeldså 1982b, 1983b, d) occur in southern South America, which suggests that this is the center of origin of these birds. Presumably, an early stock related to the red-gartered species invaded the high Andes from southern lowlands, became isolated there, adapted strongly to high alpine conditions, and later differentiated into the present Giant and Horned Coots.

Above this level on the phylogeny, the pattern is more difficult to interpret, as the Andean Coot is primitive in some respects compared to the apparent allospecies that live to the south of it (White-winged Coot), north of it (Caribbean Coot, *Fulica caribea,* and American Coot), and in the Old World (Black and Crested Coots, *F. atra* and *cristata*). Possibly an early stock in this complex spread from the south to the puna, became isolated and specialized for highland conditions, and gave rise to the present Andean Coot which, after Pleistocene range fragmentations, acquired a complex geographical variation. Populations that remained in the south evolved the advanced character states of the White-winged Coot and of the populations that managed to spread to the Northern Hemisphere. Since coots mainly inhabit temperate biota and are absent from tropical rain forest, colonization along the Andes might seem most likely. Some alternatives exist, including long jumping or a trans-Equatorial spread along the Pacific coast, or across the savannas that covered the Amazon area in the cold periods. The result is, anyway, that highly evolved coots are absent from the Andean habitats presently occupied by Andean Coots, but exist south and north of it. The Colombian Coot approaches the Andean in size, which may be an independent adaptation to highland conditions. In other respects it resembles White-winged and Nearctic Coots (Fjeldså 1983d). This coot population may represent a relict from the period of northward dispersal, or it may have been established by North American birds during a southward range displacement in a cold period.

Other groups.—Opinions have been expressed on the systematic affinities of most Andean taxa, but the views have only rarely been expressed in a form that permits the construction of a directed phylogenetic tree. Numerous points regarding the phylogenetic history of waterfowl remain in doubt (Johnsgard 1978). The zoogeography of waterfowl may be particularly hard to reconstruct because of their very easy dispersal. It appears that most groups of neotropical waterfowl show African faunal connections. This is likely, as some extant species live on both sides of the Atlantic. The general distribution of stifftails, and Siegfried's (1976) suggestion that the Ruddy Duck, subspecies *ferruginea,* may be related to the Southern Hemisphere taxa *vittata, maccoa,* and *australis,* point to a southern origin, *viz.* that the Nearctic *jamaicensis* represents a derived evolutionary state. Among dabbling ducks, Silver and Puna Teal certainly have an African origin, with the first species most closely resembling the ancestral form. Red Shoveler and Chiloé Wigeon may have originated in the Northern Hemisphere, leaping the Equatorial region. The direction of dispersal of the rest of the group is difficult to determine.

LESSONS FROM THE RECORDS OF LOWLAND BIRDS IN THE HIGHLANDS

The Andes form a considerable barrier to the interchange of lowland faunas. However, judging from distribution patterns and from the records of strays in the highlands, some dispersal across the chain must occur. There are numerous indications that wetlands in the temperate parts of the northern Andes, particularly on the Bogotá Savanna, have been colonized from adjacent lowlands. It is noteworthy, however, that very widespread neotropical lowland species show a stronger tendency to congregate and to breed in highland habitats in Bolivia and Peru than in areas with milder climates in the northern Andes. Olivaceous Cormorant and Black-crowned Night-Heron (Fig. 2O) breed at high elevations in the puna zone, but are rare stragglers in the northern Andes. (The Pied-billed Grebe is, however, an example of the opposite [Appendix I].) Great Egret and Black-necked Stilt nest in Junín and may do so also in other puna zone wetlands, but are rare visitors in the northern Andes, although they are widespread in lowlands both in the south and the north. A noteworthy concentration of strays from the tropical chaco zone is found in the Cochabamba region. At the southern end of the continent, most lowland waterbirds occur also in the lakes of the adjacent mountain zones.

This difference between the north and the south of the continent is hard to explain. I see, for instance, no obvious reasons why Black-crowned Night-Herons can breed in southern but not in equatorial highlands. It may be decisive, however, that southern populations are partly migratory; because birds in these populations are adapted to rather unpredictable conditions, they may be more adaptable than tropical waterbirds. Irrespective of the reason, the described pattern suggests that also the past colonization of Andean wetlands was most likely to proceed from the south rather than from the north.

Andean birds stray mainly from the puna zone to the adjacent Pacific coast, which provides little support to ideas of colonization of southern lowlands from the southern Andes or from the puna zone.

ADAPTATIONS OF ANDEAN WATERBIRDS

Adaptive changes of Andean waterbirds to thin air, strong solar radiation, and extreme temperatures, will be mentioned only to the extent needed to judge how long various Andean populations have been isolated.

Endotherms generally evolve increased size in response to cold climate, both in the Arctic and in high mountains near the Equator. Thus, modest size increments in Andean birds are to be expected, and need not indicate long isolation. Large size increases, as shown by Titicaca Flightless Grebe, Andean and above all Giant and Horned Coots, are more significant. Also the reduced size, relative to low elevation counterparts, of James' Flamingo, Andean Lapwing, Puna Plover, and Puna Snipe are noteworthy. Both size changes suggest niche shifts relative to lowland forms.

Although the biological significance is still disputed, birds living in cold climates often develop a more or less completely white plumage. In the usually snow-free puna zone, only Titicaca Flightless Grebe, Andean Goose, and Puna Plover are markedly pale compared to lowland allies. A few species are slightly paler, whereas waterbirds from the north Andes show a slight tendency to be darker (compare Graves 1982). In addition, Titicaca Flightless Grebe, Andean Goose and Andean Avocet have acquired color patterns quite different from those of their nearest relatives. Silvery Grebe, three-toed flamingos, Puna Teal, Black Rail, the coots (soft parts), Andean Gull, and Rufous-backed Negrito have slightly different patterns. Also the snipes inhabiting páramo habitats are quite distinctly patterned.

One noteworthy adaptive change is the evolution of flightlessness. Although grebes adapted to semi-arid and unpredictable environments fly readily (Fjeldså 1984a), those of low latitude mountain lakes fly only exceptionally and have relatively short wings. Flightlessness appears to have evolved three times, and has involved a drastic anatomic change in Titicaca Flightless Grebe (Sanders 1967). The evolution of this state may have been related to high weight, resulting from large size and big fat reserves evolved as protection against cold (Niethammer 1953). In addition, because the environments were stable, it probably was not necessary to leave the large mountain lakes in winter; the potential fatal consequences of flying across surrounding barriers of Pleistocene ice may also have been important (Fjeldså 1981a).

Although juvenile Giant Coots certainly can fly and potentially can colonize new habitats, the adults appear to be so heavy as to be flightless (Fjeldså 1981c). This sedentary adulthood may involve certain advantages in the apparently rather stable environment. A pair, once established, runs little risk of losing its territory and, remaining for years in the same place, can build enormous, inaccessible, safe nesting isles in the open highland lakes. Whether this applies also to the Horned Coot is unknown. Puna species showing indications of niche-shifts in relation to lowland counterparts occupy exposed habitats without tall vegetation. Grebes will usually nest among tules, but are not strictly dependent on this tall vegetation cover. When tules are lacking, they nest open to view in areas with dense submergents that rise to the surface (Fjeldså 1981a), a vegetation found even in very barren habitats. No profound changes have been found in puna zone taxa strictly dependent on tall marsh vegetation, and this suggests that no such species have been isolated for a long time in the puna. It also makes it unlikely that the semi-open wetland refuges that existed on adjacent Andean slopes in the Pleistocene were isolated well enough to permit vicariance events. Only the northern part of the Andes has well differentiated taxa belonging to habitats with tall marsh and bog vegetation.

A General Hypothesis for the Origin and Evolution of the Fauna

Many kinds of evidence point to one quite general pattern of faunal immigration and speciation. The strong suture zones south of the Bolivian highlands, as well as patterns of relationships between inhabitants of different core areas, suggest that the nearest relatives of birds of Andean wetlands generally are found in the Argentine biota. Many extant species occur throughout the Argentina lowlands and into the southern foothills or into the puna zone, and lowland species with a continent-wide range usually invade highland habitats only in the puna zone, and appear only as accidentals in the páramo wetlands. Study of bird groups for which a phylogenetic model exists similarly suggests that the colonization went from southern lowlands into the puna, and not vice versa. Apparently, the adaptations of southern lowland birds to rather cold and unpredictable climate conditions and to open grassland environments functioned as preadaptations for life in the high Andes. Farther northward, dispersal may have occurred gradually and by jumping, as suggested by a present day northward reduction in the number of southern species. Also some large disjunctions between southern and North American taxa suggest long jumps or, alternatively, extinction in part of the range.

The general pattern of faunal connections is followed by several other animal groups. According to Mann (1968), it holds true for the puna fauna in general, and Simpson (1975) suggested its validity also for part of the páramo fauna. The result, however, does not agree with early ideas (Chapman 1917, 1926; Chardon 1938) that the biota in any part of the Andes, evolved as an adjacent lowland biota was gradually uplifted with the Andes. Among waterbirds, the influence of the tropical fauna is seen only in the unique fauna of the Bogotá area, but not throughout the northern Andes, as Mann (1968) suggested. The idea of faunal origin by gradual uplift of ancient lowland biota might hold true for such enigmatic birds as Torrent Duck and Diademed Sandpiper-Plover, but as most Andean waterbirds are only moderately differentiated from lowland taxa, most colonization probably resulted from rather recent, active dispersal from the lowlands.

Suture zones, core areas, and distribution gaps coincide only partly with present ecotones, but fit well with possible barriers and refuge areas of the glacial periods (Fig. 1). I therefore assume that these structures were the main obstacles to active dispersal and were widely responsible for the vicariance events in and near the Andes.

Possibly a variety of lowland birds colonized the Andes both before and between the glacial periods but failed to differentiate from the lowland stocks because of insufficient barriers. During cold periods, most of them had to retire to the wetlands which formed on the Argentine and Bolivian plains. Some, such as Yellow-billed Pintail, may have spread across the Amazonean savannas and into the continuous páramo habitats, whereas others were widespread in transient or semi-open refuges along Andean slopes. The ranges of some may have been fragmented by the glaciation of the Argentine/Chilean cordilleras, as suggested by differentiation of Buff-necked and Black-faced Ibises and subspecies of Black-crowned Night-Heron, Torrent Duck, Plumbeous Rail, Spot-flanked Gallinule, and Yellow-winged Blackbird, and for tundra species, by the separation of the two small species of sheldgeese (*Chloephaga poliocephala* and *rubidiceps*) and of two subspecies of the Upland Goose. Some differentiation on the subspecific level may have been due to the trans-Patagonian Somuncura glacier (as indicated by Fig. 3), but most Argentine waterbirds were possibly too mobile to have been affected by this low barrier.

The fact that well differentiated endemic species and semispecies of puna birds are found mainly in barren habitats suggests that only species occupying this type of habitat could survive within the glacial Lake Michín refuge. Already an early (the first?) glaciation may have led to the isolation of populations of *Rollandia,* a three-toed flamingo, a sheldgoose, a coot that evolved gigantic size, and an avocet. Unfortunately, DNA-DNA hybridizations with the nearest relatives, which could possibly date the isolations, have not yet been done (C. G. Sibley, *in litt.*). Silvery Grebe, Puna Teal, Andean Coot, Puna Plover, Puna Snipe, Andean Gull, and Rufous-backed Negrito also may have been differentiated from founding lowland stocks by isolation in Lake Michín.

Very few vicariance events have occurred within the puna zone. The ancestral three-toed flamingo and gigantic coot were split into two species-pairs after interglacial dispersal and subsequent range fragmentation. James' Flamingo and Giant Coot probably evolved in isolation in Lake Michín, and Andean Flamingo and Horned Coot in the Atacama Lake and maybe also in meltwater lakes in the Chilean Central Valley.

Silvery Grebes and Andean Coots possibly lived both south and north of the glaciated transverse ranges at 14° to 15°S. Silvery Grebes that became "locked" in the closed glacial refuge in Junín gave rise to the Junín Flightless Grebe (Fjeldså 1981a), and isolated white-fronted Andean Coots became adapted to very exposed habitats with a food supply mainly of stonewarts (Fjeldså 1982b).

Unlike the endemics of barren habitats, the big puna subspecies of some marsh birds may have evolved in semi-open marginal refuges, and some may not have started to differentiate before post-glacial time. The Junín subspecies of the White-tufted Grebe could possibly have evolved *in situ* (Fjeldså 1981a), but the Junín Black Rail (Appendix I) more probably immigrated from past wetlands on the Pacific slope, and may be limited to Lake Junín now in the absence of alternative habitats in the vicinity (Fjeldså 1983c).

The continuous, dry páramo vegetation of the glacial periods should have facilitated the northward dispersal of species already adapted to puna conditions. However, the northward diminution of species nevertheless suggests some short- and long-distance jumping, or it is a consequence of island biogeography principles operating after post-glacial range fragmentation. The fragmentation of páramo habitats in warm periods led to the separation of the Speckled and Andean Teal, of two subspecies of the latter, and of a pintail population in the Bogotá area. Rails and marsh wrens have also been isolated in the Bogotá area, maybe since the last interglacial period. The range of the northern Ruddy Duck shifted so far south in the glacial period that these birds hybridized with black-headed neotropical birds. Eared Grebes and Nearctic Coots may have established themselves near Bogotá during the cold period. A population of the Spot-flanked Gallinule may have become isolated here after extirpation from most lowland areas as a result of competition from the Common Moorhen. Other local subspecies in the Bogotá area may represent rather recent, possibly postglacial isolations.

IS THE ANDEAN WATERBIRD FAUNA COMPLETE?

METHOD OF ANALYSIS

A few, superficial descriptions of Andean waterbird communities exist (e.g., Dorst 1967), but no investigations of community structure are available. In order to group the species in guilds, we ideally need quantitative information on their distributions relative to habitat gradients and on their feeding habits. In the absence of such data, I studied the extent to which different species coexist in the same places, and compared this with the co-occurrence expected based on random distribution of all records of the involved taxa (Fjeldså 1981a, b, in press d). The matrices (Fig. 10) were arranged so that the strongest mutual associations of species were concentrated near the diagonals of the diagrams. This represents a maximal clustering of coexisting specis. The method of presentation has an advantage over traditional dendrograms based on euclidean distance-similarity coefficients, because, in addition to showing nearest neighbors, it gives a clear impression of the mutual association between any two taxa.

A representative of a community that shows strong coexistence with other members of its community but few associations with members of other communities I call an indicator species. Its presence in a locality is strongly indicative of presence of certain other species and probably also of some specific ambient conditions. These conditions may be revealed by comparing the habitat descriptions of all localities that have several indicator species, in order to see if they have common features that set them apart from places lacking these species.

Waterbird communities in the puna zone.—Birds recorded in 177 wetlands in the puna zone (Fig. 10A, adapted from Fjeldså 1981b) associate in one major and two minor communities. The small *Phleocryptes* community has herons, a grebe, two waterhens, and two (locally three) passerines. Areas inhabited by these species have extensive marsh vegetation with relatively small open spaces. The second small cluster, the *Podiceps* community, is characterized by

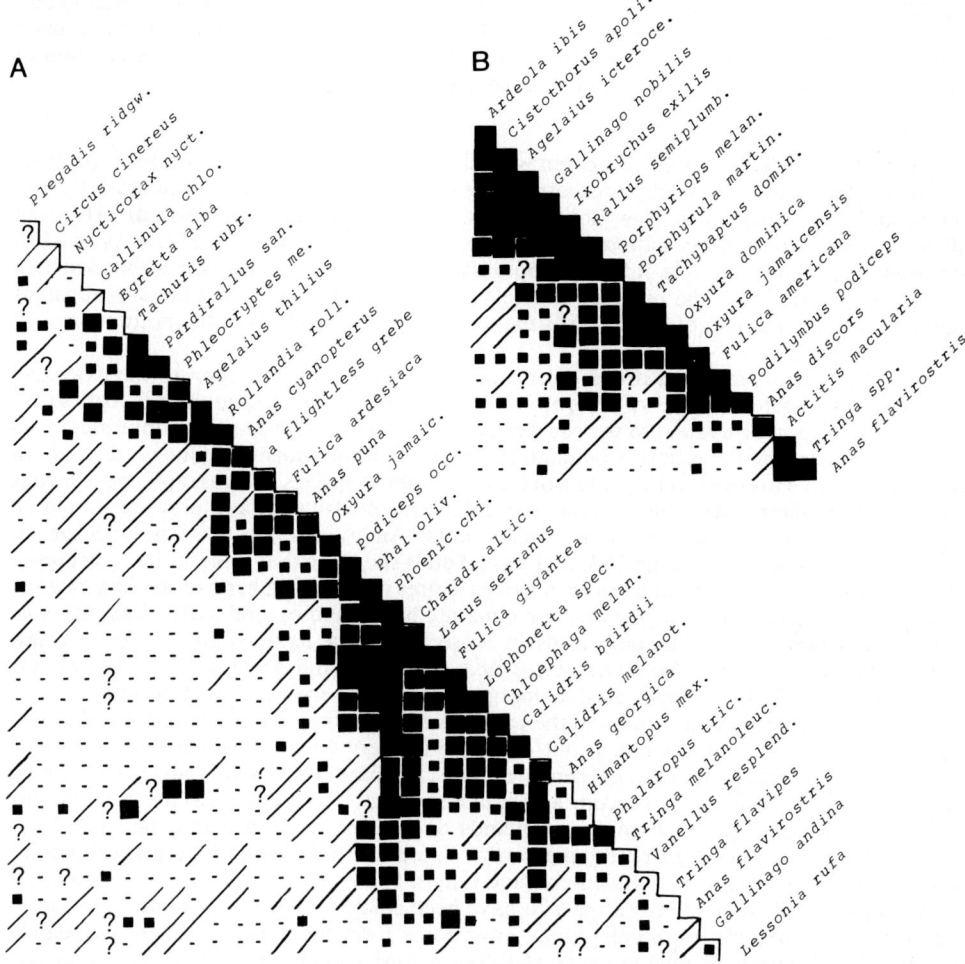

Fig. 10. Statistical coincidence between occurrences of waterbirds (omitting species recorded in fewer than 4% of the localities) in (A) 177 wetland sites in the puna zone, and (B) 77 wetlands in the Eastern Andes of Colombia. Large squares = species occurring in the same places more than 3 times as often as expected by chance distribution; medium squares = 2 to 3 times as often as expected; small squares = 1.5 to 2 times as often as expected; oblique lines = not significantly different from expected (χ^2 = <3.8, P > 0.05, 1 d.f.); − = coincidence rarer than expected.

three grebes, two dabbling ducks, one diving duck, and a coot, but these are often associated also with large numbers of ubiquitous species. Habitats occupied include extensive areas of open water with abundant submergents and a patchy cover of tules. The large *Phoenicopterus* community is characterized by a grebe, a cormorant, flamingos, a sheldgoose, two ducks, a large coot, a gull, and many shorebirds; in addition, most sites have many ubiquitous species. These indicator species occupy large, open areas of shallow, highly productive water surrounded by wide beaches and heavily grazed meadows. Poor, open habitats show impoverished communities with ubiquitous species, and do not have a specific guild.

  Waterbird communities in páramos.—I also considered associations among species in 77 localities in the Eastern Andes of Colombia (Fig. 10B). A compact *Agelaius* community includes two herons, a rail, a snipe, and two passerines. Their habitat is usually extensive marsh vegetation, often adjoining rushy pastures. A somewhat larger *Oxyura* community is comprised of two grebes (plus one now extinct), a dabbling duck (plus one extinct and one near extinction), two diving ducks, gallinules, and a coot. Actually, three subcommunities can be perceived, one associated with the border between areas with the *Agelaius* community and open water, another associated with water with dense floating leaves, and a third in open water with abundant weeds and a patchy or peripheral cover of tules. A small association of Speckled Teal and sandpipers is found on barren shores.

COMPARISON AND INTERPRETATION

Figures 10A and B have been compared with similar diagrams from 212 wetland localities visited in Magellanic Chile and southern Patagonia, 750 Scandinavian wetlands (Fjeldså 1981b) and 271 wetlands in New South Wales, Australia (Fjeldså, in press d). Although differences exist, community patterns like that of the puna zone appear to have evolved in all temperate grassland biomes. It further appears that the avifauna of southern South America has been transferred into the puna zone with very slight loss of species. The only apparent reduction is in the number of diving birds that pick tiny arthropods from the bottom or from the submerged foliage.

Much of the southern fauna drops out on the way north from the puna to the extensive wetlands in the Eastern Andes of Colombia. The *Phoenicopterus* community is lost, so we are left with a fauna equivalent to the southern *Phleocryptes* and *Podiceps* communities. This is partly explained by the lack of true flamingo habitats in the moisture-influenced northern part of the Andes. However, at least some of the open-habitat species, such as rapid, piscivorous divers of open lake habitat, ibises, geese, some dabbling ducks, gulls and plovers, should be able to find suitable conditions even in the Colombian Andes. Habitats for some such species occur in the open water along the 26 km of Lake Tota's shoreline that lacks marsh vegetation. There are numerous ephemeral lakes with barren shores in semi-arid parts of the Bogotá area, the large water-storage dams of the area have mainly barren shores, and barren habitats exist also in the páramo lakes. The only birds to be seen in these habitats are, however, a few northern sandpipers, Lesser Scaup (*Aythya affinis*; migrants), and Andean Teal. The habitats are only to a very limited degree used as marginal habitat by the many marsh birds.

A few southern species have reached the páramos of Ecuador and southernmost Colombia. Some of these prefer open habitats, but only two (Silvery Grebe, Andean Gull) are good indicator species (Fig. 10A). Most of them are almost as ubiquitous as the Andean Teal, which suggests that only species able to occupy a wide variety of habitats are successful in maintaining "insular populations" in small páramo wetlands or in using them as "stepping-stones" for reaching the Bogotá area. Apparently, the northern Andes with the Bogotá wetlands functions as a "biogeographic umbilicus" (Diamond and Gilpin 1983) with regard to southern species.

The Bogotá wetlands presently appear to have been colonized from the adjacent lowlands, although in cold periods, immigrants may have come from the Nearctic. These propagules all appear to have been adapted to aquatic habitats with rich vegetation cover, and the species remain in this kind of habitat. Presumably, time has not been sufficient for the evolution of adaptive shifts in response to the vacant niches that must exist in barren habitats. Apparently, then, a considerable relaxation period is needed in order to evolve such predictable patterns of maximal niche packing, as postulated by Cody (1974).

ACKNOWLEDGMENTS

Field studies and museum visits that served as the basis for this paper were funded by the Danish Natural Science Research Council (Grants 511-8136, 11-2250 and 11-4043), the Frank M. Chapman Memorial Fund, G. E. C. Gads Foundation, Knud Højgaard's Foundation, and Queen Margrethe and Prince Henrik's Foundation. For fine companionship and cooperation during various parts of the field work I wish to thank E. Bering, D. Boertmann, P. and S. Brehmer, M. Dunn, A. Johnson, N. Krabbe, G. L. Nuechterlein, O. H. Post, N. V. Urquiso Lopez, and J. E. Wasmuth. My gratitude is also due to all those persons mentioned in the text who provided information. B. Ree and U. Friis typed the final version of the manuscript.

LITERATURE CITED

ADAMS, J., AND P. J. ESPIN. In press. The waterfowl of Lake Tota, Colombia. Wildfowl.
AGUIRE, C. J., AND O. RANGEL CH. 1976. Contribucion al estudio ecologico y fitosociologico de las communidades acuaticas macroscopicas y continentales del Lago de Tota y alrededores. Unpubl. thesis. Universidad Nacional de Colombia, Bogotá.
ALDRIDGE, D. K. 1981. Aspectos ecologicos de patos silvestres en otoño para las provincias de Valdivia y Osorno, Chile. Unpubl. thesis. Lic. Med. Vet., Universidad Austral de Chile, Valdivia.

ANDERSSON, G. 1981. Fiskars inverkan på sjöfågel och fågelsjöar. Anser 20:21–34.

AUER, V. 1960. The quaternary history of Fuego-Patagonia. Proc. R. Soc. Lond., B (Biol. Sci.) 152: 507–516.

BLAKE, E. R. 1977. Manual of Neotropical Birds, Vol. I. Spheniscidae to Laridae. University Chicago Press, Chicago, Illinois.

BORRERO, J. I. 1963. El Lago de Tota. Rev. Fac. Nac. Agr. 23:1–15.

BRUNDIN, L. 1967. Transantarctic relationships and their significance, as evidenced by chironomid midges. Kungl. Sven. Vetenskapsakad. Handb. 4th ser. 11(1):1–472.

BUCHER, H., AND G. HERRERA. 1981. Communidades de aves acuáticas de la Laguna Mar Chiquita (Cordoba, Argentina). Ecosur 15:91–120.

CARPENTER, F. L. 1976. Ecology and evolution of an Andean hummingbird. Univ. Calif. Publ. Zool. 106:1–74.

CHAPMAN, F. M. 1917. The distribution of bird-life in Colombia: a contribution to a biological survey of South America. Bull. Am. Mus. Nat. Hist. 36:1–729.

CHAPMAN, F. M. 1926. The distribution of bird-life in Ecuador: a contribution to a study of the origin of Andean bird-life. Bull. Am. Mus. Nat. Hist. 55:1–784.

CHARDON, C. E. 1938. Apuntaciones sobre el origin de la vida en los Andes. Bol. Soc. Cienc. Nat. (Caracas, Venezuela) 5:1–47.

CODY, M. L. 1974. Competition and Structure of Bird Communities. Princeton University Press, Princeton, New Jersey.

CORREA, L. H. 1973. Laguna de Pozuelos (Jujuy), un importante refugio de aves. Inf. Interno Serv. Nac. Parq. Nac., Buenos Aires.

CRACRAFT, J. 1983. Cladistic analysis and vicariance biogeography. Am. Sci. 71:273–281.

DELACOUR, J. 1954–1964. The Waterfowl of the World. 4 vols., Country Life, London.

DIAMOND, J. M. AND M. E. GILPIN. 1983. Biogeographic umbilici and the origin of the Philippine avifuana. Oikos 41:307–321.

DORST, J. 1967. Considérations zoogéographiques et écologiques sur les oiseaux des Hautes Andes. Pp. 471–504, In C. D. Deboutteville and E. Rapoport (eds.), Biologie de l'Amérique Australe, Vol. III. Éditions Centre National Recherche Scientifique, Paris.

DOTT, H. E. M. 1984. Range extensions, one new record, and notes on winter breeding of birds in Bolivia. Bull. Br. Ornithol. Club 104:104–109.

DOUROJEANNI, M., R. HOFMAN, R. GARCIA, J. MALLEAUX, AND A. TOVAR. 1968. Observáciones preliminares para le manejo de las aves aquaticas del Lago Junín, Peru. Rev. For. Peru 2:3–52.

DROUILLY, P. 1977. Censo otoñal de patos (Anseriformes) realizado entre el río Limarí y el río Maule, provincia de Talca, durante marzo y abril de 1976. Medio Ambiente 2:102–106.

ERIKSSON, M. O. G. 1983. The rôle of fish in the selection of lakes by nonpiscivorous ducks: mallard, teal and goldeneye. Wildfowl 34:27–32.

FITZPATRICK, J. W. 1973. Speciation in the genus Ochthoeca (Aves: Tyrannidae). Breviora 402:1–13.

FJELDSÅ, J. 1980. Fuglebestanden in Regnemark Mose, Midtsjælland, med visse beregninger omkring fuglenes rolle i Mosens stofomsætning. Dan Ornithol. Foren. Tidsskr. 74:91–104.

FJELDSÅ, J. 1981a. Comparative ecology of Peruvian grebes. A study of the mechanisms of evolution of ecological isolation. Vidensk. Medd. Dan. Naturhist. Foren. 143:125–246.

FJELDSÅ, J. 1981b. Biological notes on the Giant Coot Fulica gigantea. Ibis 123:423–437.

FJELDSÅ, J. 1981c. A comparison of bird communities in temperate and subarctic wetlands in northern Europe and the Andes. Proc. 2nd Nordic Congr. Ornithol. 1979:101–108.

FJELDSÅ, J. 1981d. Podiceps taczanowskii (Aves, Podicipedidae), the endemic grebe of Lake Junín, Peru. A review. Steenstrupia 7:237–259.

FJELDSÅ, J. 1982a. Some behaviour patterns of four closely related grebes Podiceps nigricollis, P. gallardoi, P. occipitalis and P. taczanowskii, with reflections on phylogeny and adaptive aspects of the evolution of displays. Dan. Ornithol. Foren. Tidsskr. 76:37–68.

FJELDSÅ, J. 1982b. Biology and systematic relations of the Andean Coot "Fulica americana ardesiaca." Steenstrupia 8:1–21.

FJELDSÅ, J. 1983a. Vertebrates of the Junín area, central Peru. Steenstrupia 8:285–298.

FJELDSÅ, J. 1983b. Geographic variation in the Andean Coot Fulica ardesiaca. Bull. Br. Ornithol. Club 103:18–21.

FJELDSÅ, J. 1983c. A black rail from Junín, central Peru. Laterallus jamaicensis tuerosi ssp. nov. Steenstrupia 8:277–282.

FJELDSÅ, J. 1983d. Systematic and biological notes on the Colombian Coot Fulica americana columbiana (Aves, Rallidae). Steenstrupia 8:209–219.

FJELDSÅ, J. 1983f. Ecological character displacement and character release in grebes. Ibis 125:463–481.

FJELDSÅ, J. 1984a. Social behaviour and displays of the Hoary-headed Grebe Poliocephalus poliocephalus. Emu 83:129–140.

FJELDSÅ, J. 1984b. The Hooded Grebe—known since 1974 (in Danish). Nat. Verden 1984:131–142.

FJELDSÅ, J. In press a. Geographic variation in the Great Grebe Podiceps major (Aves: Podicipedidae). Steenstrupia.

FJELDSÅ, J. In press b. Color variation in the Ruddy Duck Oxyura jamaicensis andina. Wilson Bull.

FJELDSÅ, J. In press c. Displays of the two primitive grebes *Rollandia rolland* and *R. microptera,* and the origin of the complex courtship rituals of the *Podiceps* species (Aves: Podicipedidae). Steenstrupia.

FJELDSÅ, J. In press d. Classification of waterbird communities in south-eastern Australia. Emu.

GILL, F. B. 1964. The shield color and relationships of certain Andean coots. Condor 66:209–211.

GILSON, H. C. 1939. The Percy Sladen Trust Expedition to Lake Titicaca in 1937. I. Description of the expedition. Trans. Linn. Soc. Lond. 1939 7:1–20.

GILSON, H. C., AND P. F. HOLMES. 1939. The Percy Sladen Trust Expedition to Lake Titicaca in 1937. II. Meteorology. Trans. Linn. Soc. Lond. 1939 7:161–189.

GRAVES, G. R. 1982. Speciation in the Carbonated Flower-piercer (*Diglossa carbonaria*) complex of the Andes. Condor 84:1–14.

HAFFER, J. 1967. Speciation in Colombian forest birds west of the Andes. Am. Mus. Novit. No. 2294.

HAFFER, J. 1969. Speciation in Amazonian forest birds. Science 165:131–137.

HAFFER, J. 1974. Avian speciation in tropical South America. Publ. Nuttall Ornithol. Club No. 14.

HAFFER, J. 1979. Quaternary biogeography of tropical lowland South America. Pp. 107–140, *In* W. E. Duellman (ed.), The South American Herpetofauna. Its Origin, Evolution, and Dispersal. Univ. Kansas Mus. Nat. Hist. Monogr. 7.

HAFFER, J. 1981. Aspects of Neotropical bird speciation during the Cenozoic. Pp. 371–394, *In* G. Nelson and D. E. Rosen (eds.), Vicariance biogeography: a critique. Colombia University Press, New York.

HARRIS, M. P. 1981. The waterbirds of Lake Junín, central Peru. Wildfowl 32:137–145.

HASTENRATH, S. 1981. The Glaciation of the Ecuadorian Andes. Balkema, Rotterdam, Netherlands.

HELLMAYR, C., B. CONOVER, AND C. B. CORY. 1918–1949. Catalogue of Birds of the Americas, 11 vols. Field Mus. Nat. Hist., Chicago, Illinois.

HUMPHREY, P. S., D. BRIDGE, P. D. REYNOLDS, AND R. T. PETERSON. 1970. Birds of Isla Grande, Tierra del Fuego. Smithsonian Institution Press, Washington, D.C.

HUMPHREY, P. S., AND B. C. LIVEZEY. 1982. Molts and plumages of flying steamer-ducks (*Tachyeres patachonicus*). Occas. Pap. Mus. Pap. Mus. Nat. Hist., Univ. Kans. 103:1–30.

HURLBERT, S. H., AND J. O. KEITH. 1979. Distribution and spatial patterning of flamingos in the Andean Altiplano. Auk 96:328–342.

HURLBERT, S. H., AND C. C. Y. CHANG. 1983. Ornitholimnology: Effects of grazing by the Andean flamingo (*Phoenicoparrus andinus*). Proc. Natl. Acad. Sci. USA 80:4766–4769.

HUTCHINSON, G. E. 1957. A Treatise on Limnology, Vol. I. John Wiley and Sons, New York.

ILLIES, J. 1969. Biogeography and ecology of neotropical freshwater insects, especially those from running waters. Pp. 685–780, *In* J. Illies, H. Klinge, G. H. Schwabe, and H. Sioli (eds.), Biogeography and Ecology in South America. Dr. W. Junk, The Hague.

JEANNEL, R. 1967. Biogéographie de l'Amérique Australe. Pp. 401–460, *In* C. D. Deboutteville and E. Rapoport (eds.), Biologie de l'Amérique Australe, Vol. III. Éditions Centre National Recherche Scientifique, Paris.

JEHL, J. R., JR. 1975. *Pluvianellus socialis*: Biology, ecology, and relationships of an enigmatic Patagonian shorebird. Trans. San Diego Soc. Nat. Hist. 18:25–74.

JOHNSGARD, P. A. 1978. Ducks, Geese and Swans of the World. University Nebraska Press, Lincoln, Nebraska.

JOHNSGARD, P. A. 1981. The Plovers, Sandpipers, and Snipes of the World. University Nebraska Press, Lincoln, Nebraska.

JOHNSON, A. W. 1965. The Birds of Chile, Vol. I. Platt. Etabl. Gráficos, Buenos Aires.

KEAR, J., AND N. DUPLAIX-HALL (eds.). 1975. Flamingos. Poyser, Berkhamsted, England.

KOEPCKE, H. W., AND M. KOEPCKE. 1956. La cuenca del Lago Parinacochas, región ideal para parque nacional. Bol. Com. Nac. Protec. Nat. 15:1–10.

KING, W. B. (ed.). 1981. Aves. Red Data Book. International Union Conservation Nature, Morges, Switzerland.

LINDROTH, C. H. 1965. Skaftafell, Iceland—a living glacial refugium. Oikos (Suppl.):1–142.

LOHAMMER, G. 1938. Wasserchemie und höhere Vegetation schwedischer Seen. Symb. Bot. Uppsala 3: 1–252.

LOHMAN, H. H. 1970. Outline of tectonic history of Bolivian Andes. Bull. Am. Assoc. Pet. Geol. 54: 735–757.

MANN, G. 1968. Die Ökosysteme Südamerikas. Pp. 171–220, *In* E. J. Fittkau, J. Illies, H. Klinge, G. H. Schwabe, and H. Sioli (eds.), Biogeography and Ecology in South America. Dr. W. Junk, The Hague.

MCFARLANE, R. W. 1975. Notes on the Giant Coot (*Fulica gigantea*). Condor 77:324–327.

MERCER, J. H. 1976. Glacial history of southernmost South America. Quat. Res. 6:125–166.

MEYER DE SCHAUENSEE, R. 1948–1952. The birds of the Republic of Colombia. Their distribution and keys to their identification. Caldasia 5:251–1214.

MEYER DE SCHAUENSEE, R. 1966. The Species of Birds of South America and Their Distribution. Livingston, Narberth, Pennsylvania.

MEYER DE SCHAUENSEE, R., AND W. H. PHELPS, JR. 1978. A Guide to the Birds of Venezuela. Princeton University Press, Princeton, New Jersey.

MONHEIM, F. 1956. Beiträge zur Klimatologie und Hydrobiologie des Titicacabeckens. Selbstverlag Geogr. Inst. Univ. Heidelberg, pp. 1–152.

MOYLE, J. B. 1945. Some chemical factors influencing the distribution of aquatic plants in Minnesota. Am. Midl. Nat. 34:402–420.

NIETHAMMER, G. 1953. Zur Vogelwelt Boliviens. I. Bonn. Zool. Beitr. 4:195–303.

OLIVARES, A. 1969. Aves de Cundinamarca. Dirección Divulgación Cultural. Bogotá.

OLIVARES, A. 1973. Las Ciconiiformes Colombianos. Editorial Tercer Mundo, Bogotá.

OLROG, C. C. 1963. Lista y distribucion de las aves Argentinas. Univ. Noc. Tucuman, Instituto Miguel Lillo, Tucuman, Argentina.

OLSON, S. L. 1973. A classification of the Rallidae. Wilson Bull. 83:381–416.

OPPENHEIM, V. 1945. Las glaciaciones en el Peru. Bol. Soc. Geol. Perú 18:37–43.

ORTIZ CRESPO, F. I. 1983. Ecuadorean wetlands: past, present and future, with special mention of waterfowl. Proc. Int. Waterfowl Res. Bur. Symp. Edmonton 1982:127–132.

PAYNTER, T. A., JR. 1972. Biology and evolution of the *Atlapetes schistaceus* species-group (Aves: Emberizinae). Bull. Mus. Comp. Zool. 143:297–320.

PAYNTER, T. A., JR. 1978. Biology and evolution of the avian genus *Atlapetes* (Emberizinae). Bull. Mus. Comp. Zool. 148:323–369.

PUTZER, H. 1968. Überblick über die geologische Entwicklund Südamerikas. Pp. 1–24, *In* E. J. Fittkau, J. Illies, H. Klinge, G. H. Schwabe, and H. Sioli (eds.), Biogeography and Ecology in South America. Dr. W. Junk, The Hague.

REICHHOLF, J. 1975. Biogeographie und Ökologie der Wasservögel im subtropisch-tropischen Südamerika. Anz. Ornithol. Ges. Bayern 14:1–69.

REMSEN, J. V., AND R. S. RIDGELY. 1980. Additions to the avifauna of Bolivia. Condor 82:69–75.

RIPLEY, D. S. 1977. Rails of the World. Goldine, Boston, Massachusetts.

SANDERS, S. W. H. 1967. The osteology and myology of the pectoral appendage of grebes. Unpubl. Ph.D. dissert., University of Michigan, Ann Arbor.

SCHUBERT, C. 1974. Late Pleistocene Merida glaciation, Venezuelan Andes, Boreas 3:147–152.

SCHULENBERG, T. S., AND T. A. PARKER. 1981. Status and distribution of some northwest Peruvian birds. Condor 83:209–216.

SCHULENBERG, T. S., AND M. D. WILLIAMS. 1982. A new species of Antpitta (*Grallaria*) from northern Peru. Wilson Bull. 94:105–113.

SIEGFRIED, W. R. 1976. Social organization in Ruddy and Maccoa Ducks. Auk 93:560–570.

SIMPSON, B. 1975. Pleistocene changes in the flora of the high tropical Andes. Paleobiology 1:273–294.

STORER, R. W. 1976. The behavior and relationships of the Least Grebe. Trans. San Diego Soc. Nat. Hist. 18:113–126.

STORER, R. W. 1979. Order Podicipediformes. Pp. 140–155, *In* E. Mayr and G. W. Cottrel (eds.), Checklist of Birds of the World, 2nd ed., Vol. 1. Museum Comparative Zoology, Cambridge, Massachusetts.

STORER, R. W. 1982. The Hooded Grebe on Laguna de los Escarchados: ecology and behavior. Living Bird 19:51–67.

TUTIN, T. G. 1940. The Percy Sladen Trust Expedition to Lake Titicaca in 1937. X. The macrophytic vegetation of the lake. Trans. Linn. Soc. Lond. 1940:161–189.

VAN DER HAMMEN, T. 1973. The quaternary of Colombia: introduction to a research project and a series of publications. Palaeogeogr. Palaeoclimatol. Palaeoecol. 14:1–7.

VAN DER HAMMEN, T. 1974. The Pleistocene changes of vegetation and climate in tropical South America. J. Biogeogr. 1:3–26.

VAN GEEL, B., AND T. VAN DER HAMMEN. 1973. Upper Quarternary vegetational and climate sequences of the Fuquene Area (Eastern Cordillera, Colombia). Palaeogeogr. Palaeoclimatol. Palaeoecol. 14: 9–92.

VELLARD, J. 1952. Adaptation des Batraciens a la vie a grande hauteur dans les Andes. Bull. Soc. Zool. France 77:169–187.

VUILLEUMIER, B. S. 1971. Pleistocene changes in the fauna and flora of South America. Science 173: 771–780.

VUILLEUMIER, F. 1967. Phyletic evolution in modern birds of the Patagonian forest. Nature 215(5098): 247–248.

VUILLEUMIER, F. 1968. Population structure of the *Asthenes flammulata* superspecies (Aves: Furnariidae). Breviora 297:771–780.

VUILLEUMIER, F. 1969a. Pleistocene speciation in birds living in the high Andes. Nature 223:1179–1180.

VUILLEUMIER, F. 1969b. Systematics and evolution in *Diglossa* (Aves: Coerebidae). Am. Mus. Novit. No. 2381.

VUILLEUMIER, F. 1970a. Generic relations and speciation patterns in the caracaras (Aves: Falconidae). Breviora 355:1–29.

VUILLEUMIER, F. 1970b. Insular biogeography in continental regions. I. The Northern Andes of South America. Am. Nat. 104:373–388.

VUILLEUMIER, F. 1971. Generic relationships and speciation patterns in *Ochthoeca, Myiotheretes, Xolmis, Neoxolmis, Agriornis* and *Muscisaxicola.* Bull. Mus. Comp. Zool. 141:181–232.
VUILLEUMIER, F. 1980. Speciation in birds of the high Andes. Proc. XVII Congr. Int. Ornithol. Berlin, pp. 1256–1261.
VUILLEUMIER, F. 1981. The origin of the high Andean birds. Nat. Hist. 90:50–56.
VUILLEUMIER, F., AND D. N. EWERT. 1978. The distribution of birds in Venezuelan páramos. Bull. Am. Mus. Nat. Hist. 162:49–90.
VUILLEUMIER, F., AND D. SIMBERLOFF. 1980. Ecology vs. history as determinants of patchy and insular distributions in high Andean birds. Evol. Biol. 12:235–379.
WELLER, N. W. 1967. Notes on some marsh-birds of Cape San Antonio, Argentina. Ibis 109:391–411.
WELLER, N. W. 1969. Comments on waterfowl habitat and management. Wildfowl 20:126–130.
WELLER, N. W. 1972. Ecological studies of Falkland Islands' waterfowl. Wildfowl 23:25–44.
WELLER, N. W. 1975. Habitat selection by waterfowl of Argentine Isla Grande. Wilson Bull. 87:83–90.
WILLWOCK, W. 1962. Die Gattung *Orestias* (Pisces: Microcyprini) und die Frage der intralakustrinschen Speziation im Titicaca-Seen-gebiet. Verh. Deutscher Zool. Ges. Wien 54:610–624.

APPENDIX I

RECENT INFORMATION ON DISTRIBUTION, STATUS, AND GEOGRAPHICAL VARIATION OF BIRDS OF TEMPERATE, ANDEAN WETLANDS

White-tufted Grebe (*Rollandia rolland*). Fjeldså (1981a, 1983f) gives detailed accounts of the ecological requirements, and some new details of distribution and geographic variation. It is not known for certain whether the large, heavy-billed subspecies *morrisoni* is limited to Lake Junín, but this is probably the case. Bill-size is at a minimum in the Titicaca area. Birds breeding in Magellanic Chile and in Patagonia are short-billed, rather long-winged and long-legged, and have longer ornamental plumes than those living farther north. A stepped cline appears to exist at ca. 42°S, and another on the Andean slopes (Fig. 2M).

Pied-billed Grebe (*Podilymbus podiceps*). This lowland species occurs to 2550 m in Cochabamba, Bolivia (Remsen and Ridgely 1980). It may straggle to 4080 m in Peru (Fjeldså 1983a), and up to 3200 m in Venezuela (Vuilleumier and Ewert 1978). It is one of the commonest waterbirds breeding at 2500 to 3100 m in the Eastern Andes of Colombia (Adams and Espin, in press, pers. observ.).

Least Grebe (*Tachybaptus dominicus*). This tropical and subtropical species has been taken to 2500 m in the eastern parts of the Bolivian and Peruvian Andes (Storer 1976) and appears to breed in small numbers at least to 2700 m in the Eastern Andes of Colombia (pers. observ.). It is accidental in Venezuelan páramos (Vuilleumier and Ewert 1978).

Great Grebe (*Podiceps major*). Populations breeding in the Fuegian zone, in southern Chile to ca. 38°S, near the forest refuges in western Patagonia, and to ca. 800 m in the Patagonian Cordillera are recognizable as a separate subspecies (Fjeldså, in press d).

Colombian Grebe (*Podiceps andinus*). This close relative of the Eared Grebe (*P. nigricollis*) of other continents nested in rather open habitats on the Bogotá Savanna as late as the 1940's (Lakes Herrera, Parque la Florida, El Caro, Embalse de Muña), on the Ubaté Savanna at Lake Fuquene in the 1950's, possibly still later in Lake Cucunuba, and in Lake Tota, where it was common in the 1950's (Borrero 1963, *in litt.*; G. Cammacho, pers. comm.). It declined rapidly in Lake Tota because of environmental changes and hunting in the nesting colonies. Local people claim that the bird disappeared in the mid-1960's. R. S. Ridgely (*in litt.*) saw at least one bird in Feburary 1977, but none was found during my search for it in 1981 or ICBP's expedition in 1982 (Adams and Espin, in press).

Hooded Grebe (*Podiceps gallardoi*). Since its discovery in 1974 on Laguna de las Escarchados, Meseta de Viscachas, Santa Cruz, Argentina, this population has declined from 150 (Storer 1982) to 30 birds (1983–1984, pers. observ.). A population of 200 birds discovered in 1982 (A. Johnson, *in litt.*) north of Lake Viedma dropped to 45 in 1983–1984. A main breeding area, discovered in February 1984 (P. and S. Brehmer, N. Krabbe and I) at 750 to 1200 m on Meseta de Strobel between Lakes Cardiel and Strobel, probably has more than 1500 pairs. Because of extreme environmental conditions, and local absence and circumstantial overexploitation of the optimal food (*Lymnaea diaphana*; Gastropoda), the birds often fail to rear young. This may be a reason for frequent movements to other lakes (Fjeldså 1984b, unpubl. data). Potential wintering quarters occur within and around Meseta de Strobel.

Silvery Grebe (*Podiceps occipitalis*). Ethological differences and apparent lack of evidence of clinal transition at the suture between subspecies *occipitalis* and *juninensis* might suggest that two full species are involved (Fjeldså 1982a). Within the range of *juninensis*, the most southern populations have brass-colored auricular plumes that are replaced by drab-grey plumes northward of central Cuzco, Peru. Measurements peak around the Raya Range, drop in central Cuzco, and are at minimum in Lake Junín (Fjeldså 1981a). The species is very local and scarce, and probably declining, in the northern Andes (Ortiz Crespo 1983), although the known range has recently been extended to Lake Otun in Tolima, Colombia (G. Cammacho, pers. comm.).

Junín Flightless Grebe (*Podiceps taczanowskii*). The evolutionary history and ecological adaptations of this endemic species of Lake Junín have been outlined in some detail (Fjeldså 1981a, d). Because of sedimentation of iron oxides from mine-washing activities and fluctuating water-levels after a regulation of the lake, the population has declined to 300 birds (Fjeldså 1981d, Harris 1981).

Least Bittern (*Ixobrychus exilis*). Large subspecies *bogotensis* of the Bogotá and Ubaté savannas and Lake Tota, Eastern Andes of Colombia, appears now to be very rare on the savannas, but at least 25 pairs were present on Lake Tota in 1981–1982 (Olivares 1973; Adams and Espin, in press, pers. observ.).

Green-backed Heron (*Butorides striatus*). Known from many wetlands in the temperate parts of Peru, Ecuador, and Colombia. Breeding has never been documented, but is suspected in wetlands of the Eastern Andes of Colombia (Olivares 1973; Adams and Espin, in press, pers. observ.).

Great Egret (*Egretta alba*). Visits many puna zone wetlands regularly. Breeding attempts in recent years in Lake Junín, Peru have failed (F. T. Aldana, pers. comm.).

Cattle Egret (*Ardeola ibis*). Present on most large areas of rushy pastureland in the Andes. I have seen 1100 birds in a roost near Bogotá, and three nests with signs of completed breeding in Lake Junín, Peru (1983); I suspect breeding in some other highland areas.

Fulvous Whistling-Duck (*Dendrocygna bicolor*). This monotypic lowland species occasionally visits wetlands in the northern Andes, was rather common until the 1950's in Lake Fuquene near Bogotá, and previously occurred also in other places in this area (G. Cammacho, pers. comm.). It has nested once on Lake Junín (Fjeldså 1983a).

Flying Steamer-Duck (*Tachyeres patachonicus*). The plumage variation described by Humphrey and Livezey (1982) for coastal populations, does not apply well to Andean populations, at least not to those of the foothill mesetas at 50–51°S (pers. observ.). This suggests the existence of geographic variation in the species.

White-cheeked Pintail (*Anas bahamensis*). A tropical lowland species encountered accidentally at 4080 m in Junín, Peru (Fjeldså 1983a), and regularly at Lake Alalay, 2550 m, in Cochabamba, Bolivia (Dott 1984; J. V. Remsen, *in litt.*).

Yellow-billed Pintail (*Anas georgica*). The dark, short-tailed form *niceforoi* previously had a limited range in the Eastern Andes of Colombia, being found at Herrera, Fuquene until the 1940's, and at Tibabuyes and Lake Tota, where it was common, in the 1950's, but disappeared as the shallows outside the village Aquitania were drained and cultivated. It is probably extinct now (G. Cammacho, pers. comm.).

Cinnamon Teal (*Anas cyanoptera*). The previous picture of a highly disjunct distribution is less obvious now, as birds of yet unknown subspecies status have been found locally in littoral Peru (Schulenberg and Parker 1981), in Junín, and rather commonly in the Tungasuca area in Cuzco, Peru (pers. observ.). The subspecies *borreroi* once occurred from 2100 to 3600 m in the Eastern Andes and at Cocha in Nariño, Colombia, and in northern Ecuador (six unreported specimens from 1909 in the Zoological Museum of Copenhagen). It may now occur only in small numbers around Botogá (pers. observ.).

Red Shoveler (*Anas platalea*). This common inhabitant of southern lowland marshes breeds in small numbers and molts in tens of thousands at 700 to 1200 m on the Patagonian foothill plateaus (pers. observ.). There are two records of strays from the puna zone.

Southern Pochard (*Netta erythrophthalma*). A vanishing species throughout the neotropics (King, 1981). An important breeding area existed on the Bogotá Savanna (Herrera, Fontibon) and in Lake Fuquene until the submergents disappeared in the 1940's (G. Camacho pers. comm.). Known also from the intermontane Lake San Pablo, Ecuador, but likely to disappear from this country (Ortiz-Crespo 1983).

Rosybill Pochard (*Netta peposaca*). This pochard is known mainly as an inhabitant of duckweed-covered marshes on the Argentine plains and in central Chile. In Chile, it appears to increaese in abundance toward the south (Drouilly 1977), and it can now be seen in potential breeding habitats in Chilean Patagonia and as a visitor to the Andean foothill plateaus at 700 to 1100 m in Santa Cruz, Argentina (A. Johnson, pers. comm., pers. observ.).

Comb Duck (*Sarkidiornis melanotos*). A local forest duck, reaching the temperate zone in Colombia, Peru (accidentally), and the Cochabamba area, Bolivia (Dott 1984).

Ruddy Duck (*Oxyura jamaicensis*). The form *andinus* of the Central and Eastern Andes of Colombia probably numbers only a few hundred birds. It is usually described as intermediate between the Andean *ferruginea* and the northern *jamaicensis,* but actually most drakes are close to one or the other of these two subspecies. The situation suggests part sympatry and interbreeding, and may possibly be maintained by negative assortative mating (Fjeldså, in press b).

Masked Duck (*Oxyura dominica*). This inhabitant of placid, leaf-covered aquatic habitats in tropical and subtropical lowlands ascends to the temperate zone in the Eastern Andes of Colombia. It nested at Embalse de Muña near Bogotá at least in the 1940's (G. Cammacho, pers. comm.), and has been seen on several lakes from Bogotá to Lake Tota (pers. observ.).

Bogota Rail (*Rallus semiplumbeus*). Small numbers still persist in several marshes near Bogotá and Ubaté, and a few hundred live around Lake Tota. A nest record at 3300 m in Páramo de Chingaza may suggest a wider range (pers. observ.).

Black Rail (*Laterallus jamaicensis*). Well-marked subspecies *tuerosi* (Fjeldså 1983c) is probably endemic to Lake Junín, 4080 m, Peru.

Spot-flanked Gallinule (*Porphyriops melanops*). Poorly known subspecies *bogotensis* appears presently to be commonest waterbird within its small range at 2500 to 3020 m from the Bogotá and Ubaté Savannas to Lake Tota and Duitama, Colombia (pers. observ.).

Purple Gallinule (*Porphyrula martinica*). This lowland species is known as a visitor to several highland sites and, apparently, has a small, resident population on Lake Tota, 3020 m, Colombia (J. Adams and P. J. Espin, pers. comm.; pers. observ.).

American Coot (*Fulica americana*). Large race *columbiana* inhabits the Eastern Andes, Colombia, from 2000 to 3200 m, and has been taken accidentally as far away as Ibarra, northern Ecuador. The population is declining and may today amount to at most 2000 birds, with a concentration of 800 on Lake Tota (Fjeldså 1983c).

Andean Coot (*Fulica ardesiaca*). Since Gill (1964) reported interbreeding of the white-fronted *ardesiaca* and the red-fronted *F. americana peruviana,* they have been lumped under the name *F. americana.* Fjeldså (1982b) confirmed the existence of local interbreeding, although the mating was found to be negatively assortative, and ecological differences were found between the two morphs. Because displays and downy young of this polymorphic coot differ considerably from those of American Coot, it was recognized as a full species (Fjeldså 1982b, 1983b, c).

Red-fronted birds, showing minor individual variation, occur in the south of the range, north of Catamarca, Argentina, and Antofagasta, Chile (not from Tierra del Fuego, as proposed by Ripley 1977). Strong individual variation, with two main morphs, exists northward from the Mejia Lagoons and the Raya Range in southern Peru. North of this line, red-fronted birds occur alone near Tungasuca in Cuzco and in Huancavelica, Peru, and near Guayaquil, Ecuador, whereas white-fronted birds are most numerous at high elevations, 4500 to 4700 m in Junín, Peru, and occur alone in some barren habitats where they feed on stonewarts. The normally white undertail coverts, which in many coots function as a display signal, become suffused with black color in subspecies *atrura* (Fjeldså 1983b), which breeds along the coast of northern Peru and through Ecuador to Nariño, Colombia. The expression of the diagnostic character is strongest in the north, where past sympatry with the American Coot is suspected (Fjeldså 1983b).

Magellanic Oystercatcher (*Haematopus leucopodus*). Known mainly as an inhabitant of the Fuegian lowlands, but its range extends also to 900 m on the Andean foothills, at least to north of Lake Viedma (pers. observ.).

Black-necked Stilt (*Himantopus mexicanus*). Occurs in flocks in many puna zone wetlands, and small breeding colonies of the nominate subspecies occur around Lake Junín, Peru (F. Tueros Aldana, pers. comm.).

Collared Plover (*Charadrius collaris*). This tropical lowland species is quite common, and possibly breeds, at Lake Alalay at 2550 m in Cochabamba, Bolivia (Dott 1984; J. V. Remsen, *in litt.*).

Magellanic Plover (*Pluvianellus socialis*). In addition to the Fuegian lowland populations of ca. 1000 individuals (Jehl 1975; King 1981), several populations of breeding birds in the Andean foothills at 750 to 1200 m, between 48° and 51°S (e.g., more than 100 pairs on Meseta de Strobel; pers. observ.).

Apolinar's Marsh-Wren (*Cistothorus apolinari*). This inhabitant of tule-marsh vegetation between 2540 m and 4000 m in the Eastern Andes, Colombia, is disappearing. At present, it is positively known from five sites on the Bogotá and Ubaté Savannas, and has recently been netted also in Lake Tota (Adams and Espin, in press).

Paramo Wren (*Cistothorus meridae*). New localities for this Venezuelan endemic have been published by Vuilleumier (1970b) and Vuilleumier and Ewert (1978).

Yellow-hooded Blackbird (*Agelaius icterocephalus*). Large race *bogotensis* was previously abundant on the Bogotá and Ubaté Savannas, Colombia (Olivares 1969), but now numbers only a few hundred individuals (el Caro, Cucunuba, Fuquene, Herrera, Parque la Florida, San Ramon; pers. observ.).

AVIAN ZOOGEOGRAPHY OF THE NEOTROPICAL LOWLANDS

JÜRGEN HAFFER

Tommesweg 60, 4300 Essen-1, Federal Republic of Germany

ABSTRACT.   The neotropical avifauna, comprising approximately 3300 species, is the richest in the world. Most of the 1100 species of the suboscine Passeriformes (suborder Tyranni) are found in South America where they outnumber oscines (songbirds) two to one and locally three to one. The majority of suboscines is insectivorous, including many tyrant-flycatchers and furnariids adapted to open, and even terrestrial, environments. In South America the seed-eating niche is left to the oscines (finches, buntings, grosbeaks). From an historical perspective, the neotropical avifauna is comprised of (a) a Southern Hemisphere Element, (b) a Northern Element, (c) a South American Element, and (d) an Unanalyzed Element. The known fossil avifauna of the Cenozoic includes 60 families (9 extinct), 193 genera (51 extinct), and 274 species (94 extinct) of landbirds and freshwater birds. Most of these fossil South American birds are non-passeriform, only 10 families of Passeriformes being represented. Pliocene faunas are quite distinct from present avifaunas at both generic and specific levels. Faunal turnover rates in the Pleistocene are markedly increased as compared to the Late Tertiary and may reflect increased rates of allopatric speciation during the Pleistocene.

Numerous bird species are clustered in fairly restricted portions of the extensive neotropical lowlands, and characterize a series of areas of endemism (15 in the neotropical forests and seven in the South American nonforest regions) each with 10 to 50 species and many subspecies. Most contact zones of subspecies and closely related parapatric species cluster between the central portions of areas of endemism. Large Amazonian rivers delimit, along their wide lower portions, the ranges of many bird species. Ecological interactions of species, rather than the inability to cross the watercourses or to circumvent rivers in the headwater region, often appear to cause coincidence of species' range borders with rivers. Another conspicuous biogeographic phenomenon is the wide disjunction of numerous conspecific populations. Consistent and coterminous distribution patterns exist in diverse bird families that differ widely in their feeding preferences; examples include omnivorous toucans and trumpeters, insectivorous puffbirds, jacamars, woodpeckers, antbirds, flycatchers, and icterids, and predominantly frugivorous guans, parrots, cotingas, manakins, and tanagers.

An attempt is made to determine historical interrelationships of some areas of endemism on the basis of the phylogenetic patterns of selected endemic taxa and data on geological and climatic-vegetational changes (range fragmentations). Extensive vegetational fluctuations during the Pleistocene probably determined the present patterns of differentiation and distribution of most subspecies and species of neotropical birds. Analysis of the older patterns of speciation pertaining to the long time span of the Tertiary period will follow when more information becomes available on the geological and paleontological history of the neotropical avifauna.

RESUMEN.   La avifauna neotropical consta de aproximadamente 3.300 especies, siendo la más rica del mundo. La mayoría de las 1.100 especies de los Passeriformes suboscines (suborden Tyranni) se encuentran en Sudamérica, donde sobrepasan en dos a uno y localmente tres a uno a los oscines (aves cantoras). La mayoría de los suboscines son insectívoros, incluyendo muchos atrapa moscas y furnariidae adaptados a espacios abiertos e inclusive medios terrestres. En Sudamérica el nicho correspondiente a los semilleros ha sido ocupado por oscines (pinzones, trigueros y pepiteros). Si analizamos la avifauna neotropical a través de una perspectiva histórica, vemos que está compuesta por (a) un elemento del hemisferio sur; (b) un elemento del norte; (c) un elemento sudamericano y (d) un elemento aun no analizado. La avifauna fósil del cenozoico, actualmente conocida, incluye 60 familias (9 extinguidas), 193 géneros (51 extinguidos) y 274 especies (94 extinguidas) de aves acuáticas y terrestres. La mayoría de estas aves fósiles de Sudamérica son no-passeriformes, sólo 10 familias de Passeriformes están representadas. Las avifaunas del plioceno son muy diferentes a las actuales, a nivel de género y de especie. Los cambios acaecidos en la fauna durante el pleistoceno son notablemente mayores si los comparamos con el terciario tardío y podrían reflejar una mayor taza de especiación alopátrica en diferentes áreas durante el pleistoceno.

Numerosas especies de aves se agrupan en porciones relativamente reducidas de las extensas llanuras neotropicales, las cuales caracterizan una serie de áreas de endemismo (15 para los bosques neotropicales y siete para las regiones no boscosas de Sudamérica) cada una posee entre 10 y 50 especies y muchas subespecies. La mayoría de las zonas de contacto entre subespecies y también especies parapátricas cercanamente relacionadas se agrupan entre las partes centrales de las áreas de endemismo. Los grandes ríos amazónicos delimitan a lo largo de sus anchos cauces inferiores, los rangos de distribución de muchas especies de aves. Las interacciones ecológicas de las especies, más que la inhabilidad para cruzar el agua o salvar las nacientes de los ríos, a menudo parece determinar la coincidencia en los límites de los rangos de distribución de las especies, con los ríos. Otro fenómeno biogeográfico conspicuo es la amplia disyunción de numerosas poblaciones conespecíficas. Diversas familias de aves que difieren ampliamente en sus preferencias alimenticias, presentan modelos de distribución consistentes y limítrofes; por ejemplo omnívoros tales como tucanes y trompeteros; insectívoros como "puffbirds," "jacamars," carpinteros, "antbirds," atrapa moscas y de la familia Icteridae y de manera predominante frugívoros "guans," loros, cotingas, "manakins" y "tanagers."

Se intenta determinar de manera histórica la interrelación existente en algunas áreas de endemismo en base a modelos filogenéticos de determinados grupos taxonómicos endémicos e información en cambios geológicos y en la vegetación debidos a factores climáticos ("range fragmentation"). Las extensas fluctuaciones en la vegetación durante el pleistoceno, probablemente determinaron los modelos actuales de diferenciación y distribución para la mayoría de las especies y subespecies de aves neotropicales. Cuando se disponga de más información sobre geología y la historia paleontológica de la avifauna neotropical se llevarán a cabo análisis de modelos de especiación más antiguos, pertenecientes al largo período del Terciario.

The exceedingly varied avifauna of Middle and South America inhabits the most extensive and luxuriant rainforests in the world as well as widely different nonforest habitats of the lowlands and mountains. The distribution of neotropical birds is comparatively well known because naturalists have accumulated a large amount of data from many remote areas, but even today, some regions in Amazonia and along the slopes of the Andes have been sampled poorly or not at all. New species of birds are still being described from South America at a rate of several per year. During the period 1951 to 1983, 60 new species were discovered, 31 from Peru alone (including at least five species presently being described by specialists; see Haffer 1983). Only a few of these species are local representatives (allospecies) of widely distributed superspecies. Most of the recently discovered birds lack close relatives, and four of them, all from Perú, are isolated taxonomically to such a degree that they are placed in separate monotypic genera (*Xenoglaux, Conioptilon, Nephelornis, Wetmorethraupis;* "*Zaratornis*" is an *Ampelion*). *Nephelornis* is of uncertain familial affinity. All new species occupy remote and rather restricted areas and for this reason have remained undetected in the past. Additional localized species probably exist in South America and remain to be discovered.

For a number of neotropical birds, such as the Cayenne Nightjar (*Caprimulgus maculosus*), several terrestrial antpittas of montane forests in the Andes (*Grallaria chthonia, G. alleni, G. milleri*), and a number of Amazonian spinetails, antbirds, and flycatchers (*Thripophaga cherriei, Thamnophilus praecox, Myrmeciza disjuncta, M. stictothorax, Hemitriccus aenigma, Todirostrum senex*), no more is presently known than their names and the description of the type specimens collected many years ago. Further ornithological exploration may clarify details of the natural history and systematic relations of these rare birds as was the case with the Imperial Snipe (*Gallinago imperialis*), the White-winged Guan (*Penelope albipennis*), and the Indigo Macaw (*Anodorhynchus leari*), the Colombian Bronzed Cowbird (*Molothrus armenti*), and the Rufous-fronted Antthrush (*Formicarius rufifrons*) that had been known for decades from a few specimens until they were rediscovered recently (Terborgh and Weske 1972; Sick 1979; Sick and Teixeira 1980; Williams 1980; Dugand and Eisenmann 1983; Parker 1984).

Despite the gaps in our knowledge of the geographical distribution of neotropical birds, the basic distribution patterns of the avifauna can be analyzed. The excellent catalogues of Middle American birds by Eisenmann (1955) and of South American birds by Meyer de Schauensee (1966, 1970) summarize the available information, and several fieldguides and numerous regional lists provide detailed distributional data for individual countries.

At the level of descriptive zoogeography, several zoologists have attempted to subdivide the Neotropical Region into subregions and districts or provinces whose limits vary according to the taxonomic group studied (reviewed in Rapoport 1968; Udvardy 1975, in press; Vuilleumier 1975). The results, however, especially the detailed discussions of the delimitation of the various zoogeographical units recognized, seem to have little general biological relevance. In contrast, quantitative numerical techniques provide modern refinements to the study of faunal resemblance across a continent and may lead to a detailed hierarchical subdivision of faunal regions, subregions, and districts (e.g., Kikkawa and Pearse 1969). So far these methods have not been applied to the neotropical bird fauna.

Following the reasoning of Vuilleumier (1975) and earlier authors, the species and its distribution are the basic data of zoogeographical analysis which are preferably linked to ecological units such as vegetation formations. I summarize in this review several general zoogeographical aspects of the neotropical avifauna pertaining to an analysis of patterns and processes of biogeographical differentiation. I discuss the faunal composition and distribution of species whose ranges cluster in a number of restricted regions of Middle and South America, characterizing areas of endemism. Other biogeographical phenomena reviewed are the clustering of contact zones of subspecies and parapatric species between areas of endemism, and the occurrence of wide disjunctions between populations of certain bird species. The disjunct populations show various degrees of taxonomic differentiation. Historical interpretations of the zoogeographical phenomena are discussed following the presentation of various regional distribution patterns in the avifauna of forest and nonforest lowland regions.

METHODS

Illustrations of the distribution of individual species and superspecies are kept to a minimum in this review; numerous examples have been given by Blake (1977) for several families of the non-passeriforms, Vaurie (1968) and Delacour and Amadon (1973) for curassows, Forshaw (1973) for parrots, Chapman (1923) for motmots, Short (1982) for woodpeckers, Vaurie (1980) for spinetails and allies, Snow (1982) for cotingas, Lanyon (1978) and Fitzpatrick (1976, 1980) for several groups of flycatchers, Selander (1964) for wrens of the genus *Campylorhynchus*, Paynter (1972, 1978) for finches of the genus *Atlapetes*, Hardy (1969) for jays, Willis (1968, 1969) for selected antbirds, and Haffer (1967a, b, 1970, 1974, 1975, 1977, in press a) for several parrots, trogons, ground-cuckoos, jacamars, toucans, antbirds, cotingas, manakins, tanagers, and others. The early comprehensive publications by Chapman (1917, 1926) include illustrations and important discussions of the distribution of birds in the northwestern portion of the Neotropical Region.

A vegetation map (Fig. 1) illustrates the distribution of the main vegetation formations in the lowlands of South America and will serve as a basic frame of reference. In describing these formations, I use the informal terms *cis-Andean* and *trans-Andean* in the sense of "east of the Andes" and "west of the Andes," respectively. The cis-Andean lowlands include Amazonia, the llanos of Colombia and Venezuela, the llanos of Bolivia, the chaco and Patagonia. The trans-Andean lowlands include the Pacific lowlands of South America, the Caribbean lowlands of northern Colombia southward into the Magdalena Valley as well as all of the Middle American lowlands. It should be noted that Chapman (1917, 1926) sometimes referred to *cis-Andean* and *trans-Andean* in the opposite sense.

One of the main phytogeographical aspects of South America (Fig. 1) is the existence of four more or less isolated forest regions, (a) the trans-Andean rainforest region from western and northwestern Colombia into Middle America, (b) the Amazonian forests of central South America including the Guianan region north to the Orinoco delta, (c) the forests along the Atlantic coast of eastern South America and inland in southeastern Brazil, and (d) the widely isolated beech (*Nothofagus*) forests of southern Chile and Argentina. Extensive nonforest regions of South America are (a) the caatinga region of northeastern Brazil, (b) the campos cerrados of central Brazil, (c) the chaco region of western Paraguay, northern Argentina and adjacent parts of Bolivia and Rio Grande do Sul, Brazil, (d) the pampas of Uruguay and northeastern Argentina, and (e) the cold steppes of Patagonia. North of Amazonia, the llanos of eastern Colombia and Venezuela are extensive nonforest regions. The Pacific coast of western Peru, southwestern Ecuador, and northern Chile represent the most arid region of South America. The Neotropical Region ". . . is distinguished from all the other great Zoological divisions of the globe by the small proportion of its surface occupied by deserts, by the large proportion of its lowlands, and by the altogether unequalled extent and luxuriance of its tropical forests" (Wallace 1876, vol. 2, p. 3).

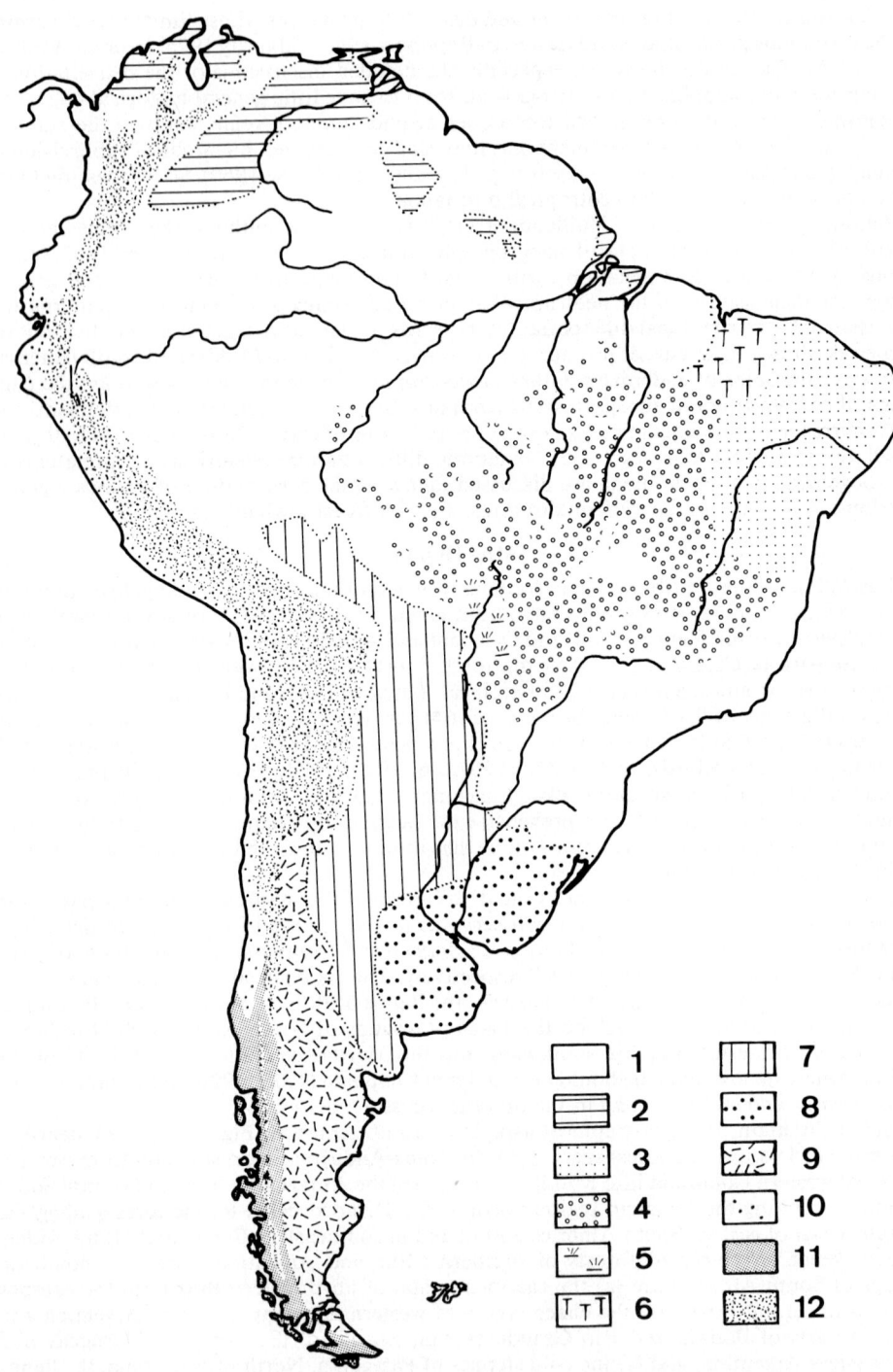

FIG. 1. Vegetation map of the South American lowlands (simplified from Hueck and Seibert 1972).
1 = evergreen and seasonal forests; 2 = nonforest vegetation of northern South America (grass plains of
Llanos and xerophytic vegetation near the Caribbean coast); 3 = caatinga; 4 = campos cerrados; 5 =
pantanal of Mato Grosso; 6 = palm forest (*Orbignya martiana*); 7 = chaco vegetation, savanna in eastern
Bolivia and transition to monte vegetation in Argentina; 8 = pampas; 9 = monte steppes and Patagonian
steppes; 10 = arid zone along the Pacific coast of Chile, Peru, and southwestern Ecuador; 11 = *Nothofagus*
forest; 12 = montane vegetation of the Andes.

ZOOGEOGRAPHIC ASPECTS OF THE
NEOTROPICAL AVIFAUNA

COMPOSITION

The Middle and South American avifauna is the richest in the world; approximately 3300 species of 95 bird families or more than one-third of the world's avifauna are native to or have been reported from these regions (Blake 1977). Fourteen families of nonpasseriforms (Rheidae 2 species; Tinamidae 46; Anhimidae 3; Opisthocomidae 1; Psophiidae 3; Eurypygidae 1; Cariamidae 2; Thinocoridae 4; Steatornithidae 1; Nyctibiidae 5; Momotidae 9; Galbulidae 17; Bucconidae 32; Ramphastidae 33) and eight families of Passeriformes [Dendrocolaptidae 52 species; Furnariidae 218; Formicariidae 238; Cotingidae 65; Pipridae 59; Rhinocryptidae 28; Oxyruncidae 1 (results of DNA hybridization studies indicate that this species is a cotingid; Sibley and Ahlquist 1984); Phytotomidae 3] are entirely or almost entirely endemic to Middle and South America. More widely distributed groups that occur in South and North America are the Trochilidae, Tyrannidae, Thraupinae, and Icterinae. The guans are a neotropical group today, but fossil members of this family have been found in North America. Most species of Troglodytidae, Mimidae, Vireonidae, and Parulinae occur in Middle America and southern North America. Among South American bird families that are represented in all tropical regions of the world are the Psittacidae, Trogonidae, Capitonidae, Columbidae, Phasianidae, Cuculidae, Corvidae, and Turdinae. Several other families represented in South America have worldwide ranges.

The ratio of Passeriformes to non-Passeriformes in the avifauna of tropical America is low (1.5:1) compared to that of the avifaunas of Temperate and Boreal North America (between 2.3:1 and 3:1), the difference being due to a relative increase of nonpasserine groups in tropical regions (Slud 1976). These additional groups are comprised of parrots, many hummingbirds, trogons, barbets, toucans, and, among Passeriformes, the manakins, many cotingas, and tanagers.

One of the most conspicuous features of the neotropical bird fauna is that it includes almost all of the 1100 species of suboscine Passeriformes (suborder Tyranni); only 50 species are known from the Old World. The number of suboscines increases southward in the Neotropical Region from 20 percent of the native passerines in Mexico to 60 percent in central South America. Native suboscine/oscine proportions in selected sample regions are 0.51—Guatemala, 0.73—Nicaragua, 0.96—Panama, 1.19—Colombia, 1.32—Venezuela, 1.36—Ecuador, 1.57—Peru, 1.34—Bolivia, 1.82—Brazil (Amazonian Brazil 1.98), 1.25—Paraguay, 1.35—Argentina, and 1.23—Chile (Slud 1976). The number of antbirds (Formicariidae), one of the characteristic suboscine families, increases gradually from nine species in Mexico, and 19 and 29 species in Honduras and Costa Rica, respectively, to 128 species in Colombia, and 144 in Ecuador-Perú, illustrating comparable trends in other neotropical families. In Amazonia, suboscine species outnumber oscine species by up to two to one, or locally, by even three to one.

The proportion of suboscines decreases in dry lowland regions (e.g., Caribbean Colombia, northeastern Venezuela, southwestern Ecuador, and coastal Perú) and also with increasing elevation in the mountains (Slud 1976).

Besides some frugivorous groups, especially in the Pipridae and Cotingidae, the majority of the suboscines is insectivorous, including many flycatchers and furnariids that have adapted to open, and even terrestrial, environments; no mainly granivorous groups have developed despite the wide extent of grassland in the neotropics since Early Tertiary time (Solbrig 1976; Webb 1978). As pointed out by Eisenmann (1961), the seed-eating niche is left to the oscines even in South America (189 species of finches, buntings, and grosbeaks). If the lack of granivorous suboscines is confirmed by future studies of fossil avifaunas, it may indicate a lesser adaptive potential of the suboscines as compared to the oscines which have radiated widely also as fruit-eaters in South America (180 species of Thraupinae). By contrast, the generally insectivorous oscine families are less diversified (only 33 warblers, 21 vireos, 57 blackbirds, 38 wrens, and 28 thrushes in South America; Traylor and Fitzpatrick 1982).

Water birds such as grebes, herons, ducks, geese, and others are often neglected in biogeographical analyses because of the wide distributions of their component species. Reichholf (1975) analyzed certain ecological aspects of their distribution in South America and compared geographical trends in the fish-eating herons and detritus-eating ducks. Whereas the number of species of herons (Ardeidae) increases toward the tropics, following the general pattern, the number of ducks (Anatidae) *decreases* from the high latitudes toward the tropics. Reichholf

(1975) speculated that ducks may be in competition for food with the species-rich tropical fish fauna (see Löffler 1977, for a critical discussion of this interpretation). The ratio of water bird species to land bird species decreases drastically with decreasing latitude, being lowest in the humid tropics (Slud 1976).

The Neotropical Region harbors numerous bird species with surprisingly localized distributions despite the fact that, in many cases, their habitat zones are fairly continuous and extensive. Future studies may show that in some, or even many, cases an ecological or physical barrier prevents further dispersal. A relative increase in number of localized species in Tropical versus Temperate Zone avifaunas is illustrated by a comparison of species numbers in small and large areas of the Temperate Zone with those in small and large areas of the tropics. Species numbers of extensive tropical areas are disproportionately large compared to those of small areas, indicating that relatively more geographically restricted species subdivide the large tropical areas among themselves (MacArthur 1969). The distribution patterns of numerous superspecies of neotropical birds whose component species (allospecies) replace one another geographically in uniform habitat zones with no or only limited hybridization along the zones of contact illustrate this. The geographical ranges of many superspecies resemble large scale mosaics composed of neatly interlocking patches formed by the ranges of their component species (see examples in Delacour and Amadon 1973; Haffer 1974, in press a; Snow 1982).

FAUNAL ELEMENTS AND FOSSILS

Early members of the endemic neotropical bird families have been assumed to have originated either within South America from immigrant ancestors or to have immigrated from North America or (via North America) from Eurasia (von Ihering 1927; Mayr 1946, 1964; Darlington 1957). Recent studies of the interrelationships of some of the neotropical families on a worldwide basis in connection with geological data on continental movements during the Late Cretaceous and Cenozoic indicate the need for reinterpretation of the early history of neotropical bird families. Cracraft (1973) hypothesized about the evolutionary history and dispersal of the ancestors of the South American bird families and assigned the latter to four categories (Table 1): (a) a *Southern Hemisphere Element,* including several families that almost certainly evolved and radiated within South America, but whose ancestors probably were more broadly distributed on southern lands; (b) a *Northern Element,* including several families that either arrived from North America or Eurasia or whose ancestors are of probable northern origin; (c) a *South American Element,* including 14 families that probably originated in South America; their interfamilial affinities, however, are not well enough known yet to permit assignment to one of the two previous categories; (d) an *Unanalyzed Element,* comprising several wide-ranging or cosmopolitan families. Future cladistic analyses of the genera within these families may reveal zoogeographic patterns that permit assignment of these families to one of the first two categroies.

The endemic families in South America apparently evolved there from ancestors that presumably had a wide distribution in western Gondwanaland, and that radiated after South America was isolated both from Africa (Middle Cretaceous) and from Antarctica (Early Tertiary). This is especially true for the species-rich suboscines, for several gruiform groups (cariamids, phororhacoids, probably aramids and eurypygids), and for several piciform groups, such as Galbulidae, Ramphastidae, and possibly Bucconidae (Cracraft 1973, 1976). The Tertiary history of neotropical birds probably is similar in principle to that of the mammals, which left a rich fossil record (Simpson 1969; Patterson and Pascual 1972). The record of South American fossil birds, on the other hand, is still too scanty to be useful in a reconstruction of the composition and distribution of Tertiary avifaunas (Vuilleumier, in press). Some extinct taxa indicate diversifications in the past that have left no trace in the Recent fauna, e.g., the large, carnivorous, flightless phororhacoids.

The fossil record of Cenozoic landbirds and freshwater birds includes 60 families (9 extinct), 193 genera (51 extinct), and 274 species (94 extinct). Most fossil South American birds are non-Passeriformes, only 10 families of passeriforms being represented. An important faunal turnover at the family level took place between the Oligocene and the Pliocene; a little less than half of the families present about 30 million years ago have disappeared since that time. Pliocene avifaunas are quite distinct from present faunas at both generic and specific levels (82% of the genera and 100% of the species extinct). Vuilleumier (in press) calculated markedly accelerated rates of faunal turnover, expressed as the average of origination and extinction of

TABLE 1
FAUNAL ELEMENTS OF THE SOUTH AMERICAN AVIFAUNA[1]

Southern Hemisphere Element[2]

Tinamidae (Pli-Monte), Rheidae (Mi-Santa), Spheniscidae, Columbidae (Late Plst), Psittacidae (Plst-Ense), Opisthocomidae (Mi-Fria), Cuculidae (Late Plst), Cracidae (Mi-Santa), Phasianidae (Late Plst), Dendrocolaptidae (Late Plst), Furnariidae (Plst-Ense), Formicariidae (Late Plst), Conopophagidae, Rhinocryptidae, Tyrannidae, Pipridae, Cotingidae, Phytotomidae.

Northern Element

Trochilidae (Late Plst), Momotidae (Late Plst), Alcedinidae, Cathartidae (Pli-Monte), Alaudidae, Motacillidae, Cinclidae, Troglodytidae, Mimidae (Late Plst), Turdinae, Sylviinae, Polioptilinae, Corvidae (Late Plst), Emberizidae (Plst-Ense), Fringillidae (Late Plst), Icteridae (Late Plst), Thraupinae (Late Plst), Vireonidae (Late Plst), Parulinae.

South American Element[3]

Anhimidae (Ho), Eurypygidae, Cariamidae (Pli-Monte), Phororhacoidea, 3 families (Ol-Desea to Pli-Monte), Psophiidae, Aramidae (Ol-Desea), Thinocoridae (Late Plst), Steatornithidae, Nyctibiidae (Late Plst), Galbulidae, Bucconidae (Late Plst), Ramphastidae (Late Plst), Telmabatidae, Presbyornithidae (Eo-Casa), Cladornithidae (Ol-Desea).

Unanalyzed Element[4]

Podicipedidae (Late Plst), Diomedeidae, Procellariidae, Oceanitidae, Pelecanoididae, Phaethontidae, Pelecanidae (Mi-Santa), Sulidae, Phalacrocoracidae (Late Plst), Anhingidae, Fregatidae, Ardeidae (Late Plst), Ciconiidae (Ol-Desea), Threskiornithidae (Mi-Santa), Phoenicopteridae (Ol-Desea), Anatidae (Ol-Desea), Accipitridae (Ol-Desea), Falconidae (Mi-Santa), Rallidae (Plst-Luja), Heliornithidae, Jacanidae (Late Plst), Haematopodidae, Charadriidae (Late Plst), Scolopacidae (Late Plst), Recurvirostridae (Late Plst), Burhinidae (Late Plst), Chionididae, Stercorariidae, Laridae (Plst-Luja), Rhynchopidae, Strigidae (Late Plst), Tytonidae, Caprimulgidae (Late Plst), Apodidae (Late Plst), Capitonidae, Trogonidae (Late Plst), Picidae (Late Plst), Hirundinidae (Late Plst).

---

[1] Elements according to classification of Cracraft (1973:519); fossil occurrence of landbirds and freshwater birds (in parentheses) after Vuilleumier (in press). Abbreviations: Ho = Holocene; Plst = Pleistocene; Pli = Pliocene; Mi = Miocene; Ol = Oligocene; Eo = Eocene; Cr = Cretaceous; Luja = Lujanian; Ense = Ensenadan; Monte = Montehermosan; Huay = Huayquerian; Fria = Friasian; Santa = Santacrucian; Desea = Deseadan; Casa = Casamayoran.

[2] Onychopterygidae (Eo-Casa) perhaps to be added to Southern Hemisphere Element.

[3] Baptornithidae (Cr) and Cunampaiidae (Ol-Desea) perhaps to be added to South American Element.

[4] Teratornithidae (Pli-Huay) perhaps to be added to Unanalyzed Element (but could also be added to South American Element).

---

genera per million years, in the Late Pleistocene (3.8–4.5) as compared to the Late Tertiary (2.0 in the Oligocene, 0.8 in the Miocene, 1.5–3.0 in the Pliocene). This he interpreted to reflect increased rates of allopatric speciation during the Late Pleistocene. Increased fragmentations of avian distributions as a result of Pleistocene climatic-vegetational fluctuations, as discussed below, may be the historical correlates of the patterns observed in the fossil avifauna. Older vicariance events during the Tertiary, both vegetational and paleogeographic changes, fragmented the ranges of the ancestors of those species that differentiated during the Pleistocene.

AVIAN DISTRIBUTION PATTERNS IN THE NEOTROPICAL LOWLANDS

AREAS OF ENDEMISM

Many bird species and well differentiated subspecies are clustered in fairly restricted regions of the extensive lowlands and characterize a number of *areas of endemism*. Other designations of such regions are *centers of endemism* or *distribution centers* (Haffer 1967a, b, 1969, 1974), *core areas* (Vanzolini and Williams 1970), *dispersal centers* (Müller 1973), and *centers of evolution* (Brown 1976, 1977a, b); the term *"area of endemism"* is probably the most bias-free descriptive term. Areas of endemism either coincide with ecological regions or, in the case of Amazonia and the Atlantic forest region of eastern Brazil, represent portions of ecological regions.

AREAS OF ENDEMISM IN FORESTED LOWLANDS

Trans-Andean Region (Figs. 2, 3).—Localized clusters of endemic species in the tropical lowlands west of the Andes are found (Haffer 1975) in Pacific Colombia (Chocó, 32 species),

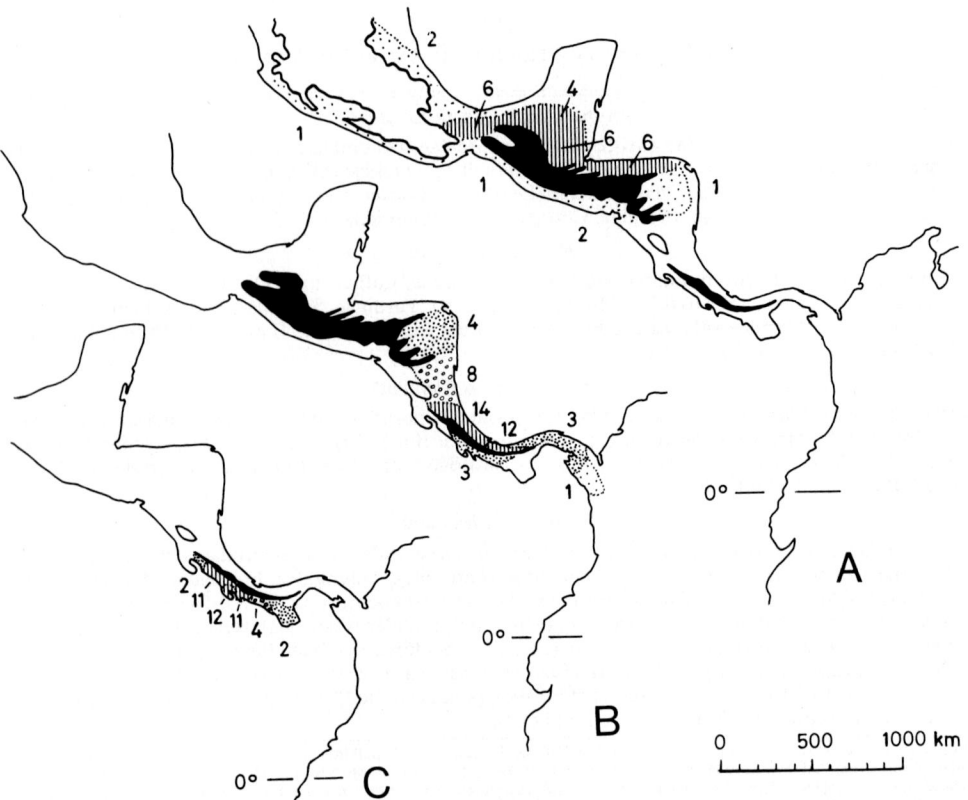

Fig. 2. Areas of endemism in the lowlands of Middle America. A: Caribbean northern Middle America; B: Caribbean southern Middle America; C: Pacific southern Middle America. Superimposed ranges of 7 endemic species in A, of 14 endemic species in B, and of 12 endemic species in C, as listed in Haffer (1975). Numbers and the respective shadings indicate totals of endemic species recorded for each of these groups. Mountains above 1000 m elevation are indicated in black.

in the Cauca-Magdalena (Nechí) region of northwestern Colombia (14 species), in Pacific southern Middle America (12 species), Caribbean southern Middle America (14 species), and in Caribbean northern Middle America (7 species). Several well differentiated subspecies characterize the Catatumbo forest in the southwestern portion of the Maracaibo basin. Most of the endemic species of the trans-Andean clusters are quite distinct, representing either independent species without close relatives or allospecies of more widespread superspecies. Rather few have ever been considered conspecific with other isolated trans-Andean populations or Amazonian species.

Other distribution patterns among endemic trans-Andean species include (a) Middle American (35 species) along the humid Caribbean lowlands from southeastern Mexico to Panamá (in some cases extending into humid Pacific Costa Rica or northern Colombia), (b) western Colombian and southern Middle American (54 species), including the ranges of several clusters of endemic birds, (c) trans-Andean (27 species) from Pacific Colombia north to Guatemala or southern Mexico. Seven species included in group (c) inhabit the Maracaibo basin of northwestern Venezuela, and one (*Penelope purpurascens*) is found east to the mouth of the Orinoco River. The lowland forests of northern and northwestern Venezuela zoogeographically represent a wide transition zone between the trans-Andean and cis-Andean avifaunas.

The groups of trans-Andean forest birds mentioned above total 195 endemic species. Between 60 (southeastern Mexico) and 112 (Pacific Colombia) of these species occur in any given local area representing 35 to 45 percent of the local avifaunas. The balance of more than half of the trans-Andean forest bird species (55–65% of local avifaunas) also inhabit the cis-Andean

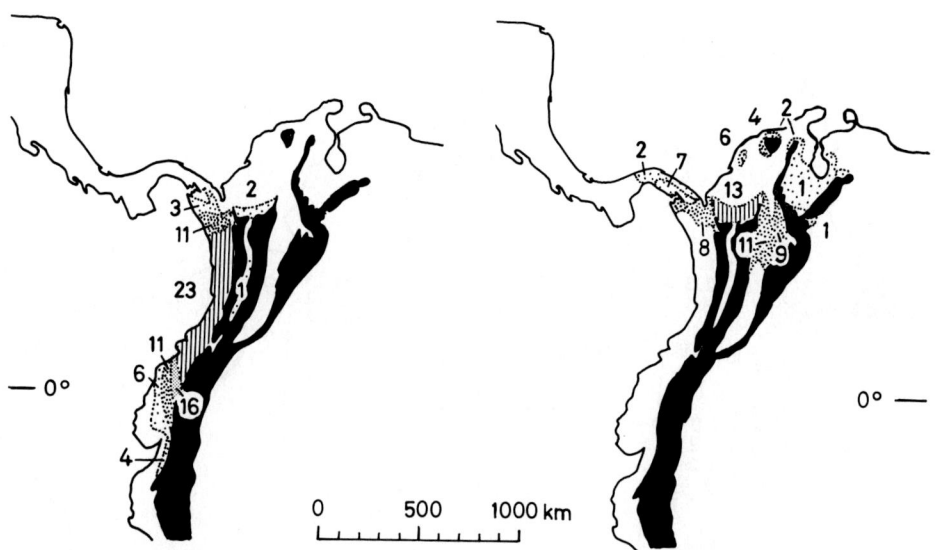

FIG. 3.   Trans-Andean areas of endemism in northwestern Colombia. Left: Pacific Colombia (Chocó); right: Cauca-Magdalena (Nechí). Superimposed ranges of 32 endemic species (left) and of 14 endemic species (right) as listed in Haffer (1975). Numbers and the respective shading indicate totals of endemic species recorded for each of these groups. Mountains above 1000 m elevation are indicated in black.

region where they are more or less widespread in Amazonia, the populations west of the Andes being either undifferentiated or, at most, subspecifically distinct. These numbers emphasize the close relationship between the forest bird faunas to the east and west of the Andes.

The most distinct, and perhaps oldest, elements of the trans-Andean forest avifauna are 30 species that belong to 20 or 21 endemic trans-Andean genera, 13 or 14 of which are monotypic. Examples of the latter are *Rhynchortyx, Androdon, Hylomanes, Clytoctantes, Xenornis, Gymnocichla, Phaenostictus,* and *Sapayoa* (Haffer 1975:40; *"Erythrothlypis"* and *"Zarhynchus"* now are included in *Chrysothlypis* and *Psarocolius,* respectively). By listing these names I emphasize the existence in the trans-Andean region of conspicuously differentiated species with no close relatives. (Whether or not all of them are correctly assigned to monotypic genera is a taxonomic problem and of little relevance in this zoogeographical context. The same applies to the monotypic genera in the avifauna of southern South America listed in Appendix I.) Chapman (1917), Slud (1960), and Howell (1969) discussed additional regional aspects of the trans-Andean avifauna.

Cis-Andean Region. —Conspicuous clusters of endemic birds characterize several areas of endemism in Amazonia (Fig. 4) and in the widely separated rainforests of southeastern Brazil (Fig. 5).

Within Amazonia, I described six areas of endemism, *Napo, Inambari, Imerí, Rondônia, Guiana,* and *Belém,* each characterized by 10 to 50 species (Haffer 1974, 1978). I mapped these areas of endemism by superimposing outlines of the ranges of the endemic species of each group and by contouring the number of sympatric species within each group. In this way areas of maximal overlap of breeding ranges of each particular species group are emphasized, and ill-defined range boundaries are deemphasized. The central portion of each of these group contour maps have been used to illustrate the location of the six Amazonian areas of endemism (Fig. 4). The number of regionally sympatric species in each group decreases in directions away from the central region resulting in steep to gentle gradients of species totals in several directions.

The six areas of endemism are located more or less peripherally in Amazonia (Fig. 4). These species clusters represent, together, a total of 150 avian species or 25 percent of the Amazon forest and forest-edge avifaunas. More widespread species inhabit increasingly extensive distributional areas comprising two or more of the above areas of endemism. Neighboring clusters

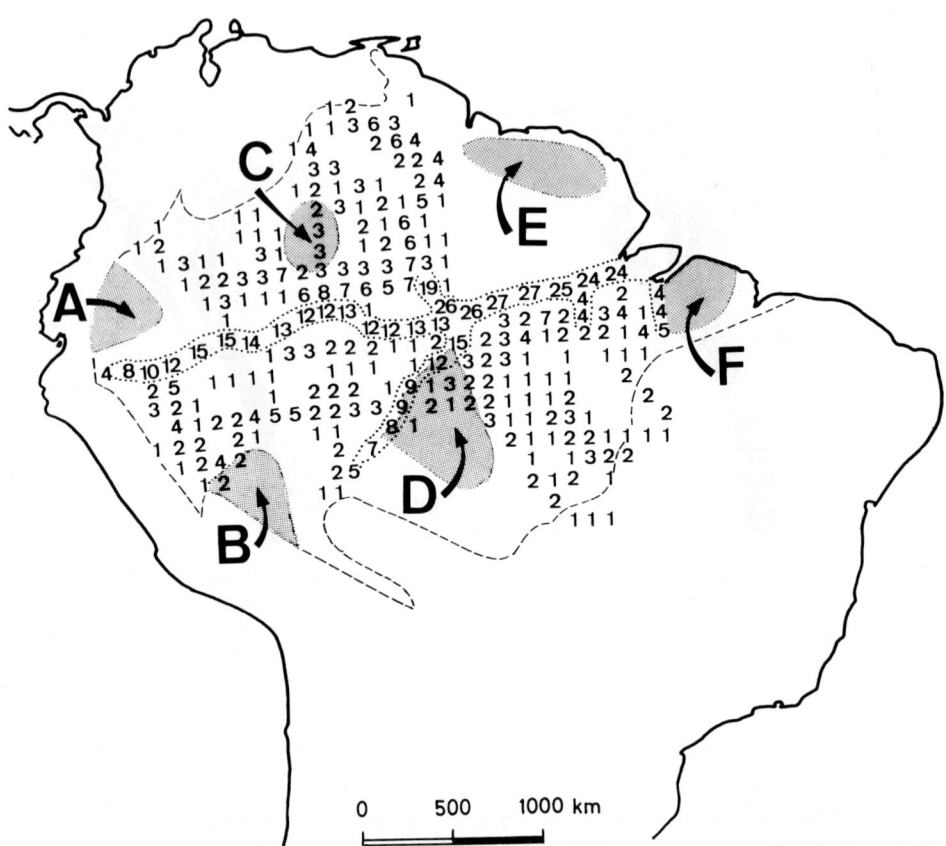

FIG. 4.   Distribution patterns of contact zones among Amazonian forest and forest-edge birds. Shaded regions indicate the central portions of six areas of endemism as mapped by Haffer (1978). A = Napo Area; B = Inambari Area; C = Imerí Area; D = Rondônia Area; E = Guiana Area; F = Belém Area. Amazonia indicated by dashed line. Figures indicate numbers of species and subspecies pairs in contact in 1° latitude-longitude blocks; a sample of 430 species is considered. Contact zones occur mostly between areas of endemism, and high numbers of contact zones occur along major rivers (e.g., areas within dotted lines).

of endemic birds overlap to some extent. It should be noted that the size and outline assigned to each area of endemism largely depends upon which contour line of the clusters is chosen to delimit the areas. The species of each area of endemism are listed in Haffer (1978).

  Napo Area (Fig. 4A): about 35 endemic species and well marked subspecies characterize this region in northwestern Amazonia centering around eastern Ecuador and the Río Napo. *Inambari Area* (Fig. 4B): the ranges of about 45 endemic species and subspecies characterize the forests of southwestern Amazonia centering in southeastern Peru in the region of the Río Inambari north to the hills in the headwater region of the Río Purús. Eight to 13 of these species reach the southern bank of the Rio Solimões without crossing this river. The Rio Madeira delimits the ranges of most species in the east. *Imerí Area* (Fig. 4C): the ranges of about 15 species cluster in the region between the upper Rio Negro and the upper Río Orinoco around the western end of the Sierra Imerí. *Rondônia Area* (Fig. 4D): southcentral Amazonia harbors a number of characteristic birds whose ranges center in the forests between the upper Rio Madeira and the Rio Tapajós. The lower Madeira and the Amazon rivers delimit the ranges of most of these species in the north. *Guiana Area* (Fig. 4E): the forest avifauna of northeastern Amazonia is quite distinctive due to a fairly large number of endemic species (about 50) that occur together in southern portions of the Guianas. Many of these species circumvent the Roraima massif in the north and/or reach the northern bank of the lower

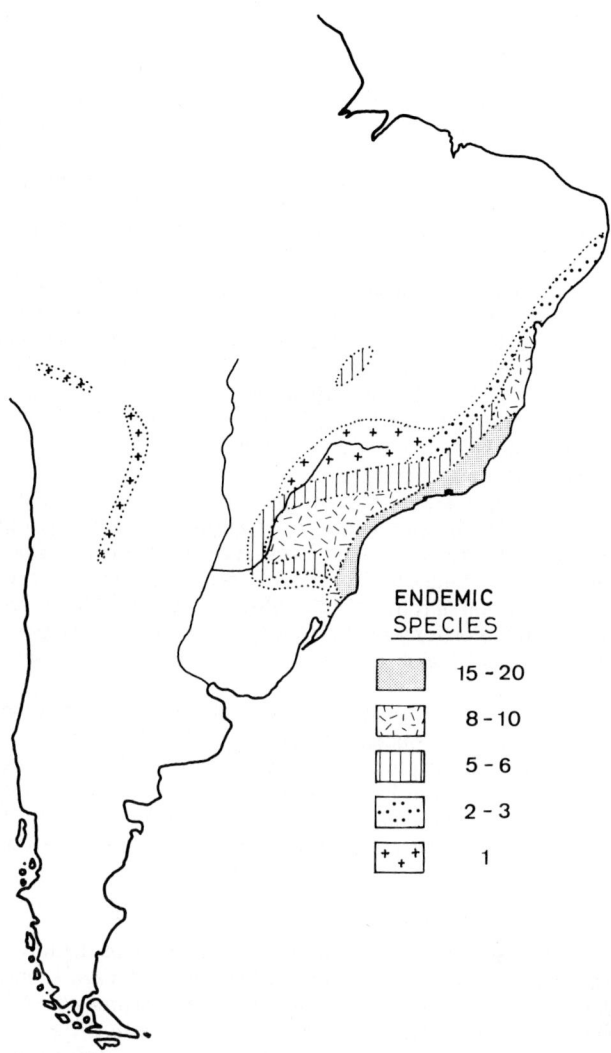

FIG. 5. Superimposed ranges of 21 bird species inhabiting the tropical forests of southeastern Brazil. Numbers and shading indicate totals of species in this group recorded. Names of the species are listed in Appendix I.

Amazon and the eastern bank of the lower Rio Negro. These rivers delimit the ranges of many Guianan species: only 14 of them have occupied small areas south of the lower Amazon or its mouth. *Belém Area* (Fig. 4F): a small number of species and subspecies are distinctive elements of the avifauna of the forests to the south of the mouth of the Amazon River. The ranges of some of them extend west to the Rio Tapajós or even to the lower Rio Madeira.

The majority of Amazonian forest and forest-edge birds occupy extensive portions of Amazonia including two or more of the above areas of endemism. Some of these species range across the entire, or almost the entire, Amazon region; others are restricted to western (upper) Amazonia, eastern (lower) Amazonia, to central portions of Amazonia, or to northern or southern Amazonia, respectively (Haffer 1978).

Numerous endemic bird species (about 140) inhabit the forests of *Southeastern Brazil* and make this region an important area of endemism (Fig. 5). Included as part of this area are the humid tropical forests extending north along the Atlantic coast from Rio Grande do Sul to Paraiba in the northeast and from the tropical to subtropical forests in the hilly region west of the coastal Serra do Mar into the Department of Misiones in Argentina and, beyond the Rio Paraná, into eastern Paraguay. About 30 of the endemic species are so distinct that they are placed in monotypic genera (Appendix I). Several workers (Müller 1973; Jackson 1978; Kinzey 1982) recognized three to four separate, smaller areas of endemism along coastal Brazil based on the distribution mainly of reptiles, birds, and monkeys. They have been named the *Pernambuco, Bahia, Rio Dôce,* and *Paulista Areas.*

Despite the distinct character of the southeast Brazilian forest avifauna, its rather close relationship with the Amazonian fauna is shown by the numerous species that are shared by both (30 non-passeriform species and 67 passeriform species; Müller 1973). With few exceptions, these species have widely disjunct and often subspecifically distinct populations in Amazonia and in the forests of southeastern Brazil. Large isolated forests in central Brazil (Mato Grosso de Goiás and forests in the headwater regions of several southern tributaries of the Amazon River; Hueck and Seibert 1972) with an Amazonian bird fauna suggest that the Amazon forest and the Atlantic forests were at one time more or less continuous. This is also indicated by even more isolated forests with an improvished Amazonian avifauna around several mountains in arid northeastern Brazil (Serrania Ibiapabá, Baturité, etc.) located between the Amazon forest of the Belém region and the Atlantic forests of Pernambuco (Pinto and Camargo 1961; see also Vanzolini 1981).

Extensive deserts and steppes separate the tropical to subtropical forests of central South America from the forests of Chile and southern Argentina which are dominated by several species of southern beech (*Nothofagus*). This forest region extends for 2000 km (20 degrees of latitude) along the southern Andes to Tierra del Fuego. It has a mediterranean subtropical character in the northern portion of central Chile, whereas the central forest region ("Valdivian rainforest") is more humid. Portions of the forests on the eastern slopes of the Andes in Patagonia and in the Magellan region are variously deciduous in contrast to the evergreen humid forest of the Pacific slopes.

This strongly isolated *Nothofagus* forest region is inhabited by a correspondingly distinctive avifauna characterized by a proportionately large number of endemic taxa (Vuilleumier 1967). Half of the 44 species of land birds (40 genera) are endemic to the *Nothofagus* forest region. Five to seven endemic species have been placed in monotypic genera, three of which are particularly distinct (*Sylviorthorhynchus, Aphrastura, Pygarrhichas*). Only four bird species seem to have speciated in the rather uniform *Nothofagus* region; most show little or no geographical variation.

AREAS OF ENDEMISM IN NONFORESTED LOWLANDS

The avifaunas of nonforest regions in South America differ considerably from those of the forests. Certain groups of spinetails and horneros, flycatchers, tanagers, and finches are particularly conspicuous among the birds of open habitats. Several areas of endemism are found in the nonforest regions of South America north of Amazonia in the grass plains (llanos) of eastern Colombia-Venezuela and along the Caribbean coast, as well as south of Amazonia from northeastern Brazil to the pampas and steppes of Argentina. The Pacific coasts of Peru and Chile are extensively arid. Moreover, many deep Andean valleys are predominantly dry and unforested, as the humidity is caught by the surrounding mountains. The cool winds that blow into these valleys from the highlands have an additional desiccating effect as they are warmed at the lower tropical elevations. Thorn scrub and cacti or xerophytic woods often are the only vegetation found in these areas.

Relatively few species characterize the *Colombian-Venezuelan Area* of endemism which is comprised of the dry Caribbean lowlands of Colombia and Venezuela and the Llanos plains to the south (Haffer 1967c). Endemic genera are lacking unless a monotypic genus *Hypnelus* is recognized for the puffbird, *Bucco ruficollis*. The avifauna is closely related to the Brazilian nonforest avifauna because both share a large number of species that are only subspecifically differentiated. Moreover, some of the endemic species are merely strongly differentiated representatives of Brazilian species with which they form superspecies (e.g., *Columba corensis/ C. picazuro, Aratinga pertinax/A. cactorum, Thraupis glaucocolpa/T. sayaca, Sporophila min-uta/S. hypoxantha*). Examples of species occurring as widely disjunct populations to the north and south of the Amazonian rainforest region include the Scaled Dove (*Scardafella squam-mata*), the Blue-crowned Parakeet (*Aratinga acuticaudata*), the flycatchers *Xenopsaris albin-ucha, Machetornis rixosus, Contopus cinereus, Hemitriccus margaritaceiventer, Elaenia cris-tata, Polystictus pectoralis,* the Flavescent Warbler (*Basileuterus flaveolus*), and the finches

FIG. 6.   Disjunct ranges of conspecific populations of South American nonforest birds (adapted from Short 1975). Range outlines of 1 = *Fluvicola nengeta*, 2 = *Coryphospingus pileatus*, 3 = *Sicalis flaveola*.

Sporophila plumbea, Sicalis flaveola, and *Coryphospingus pileatus* (Fig. 6). A fairly large number of the northern Colombian species also inhabits the unforested Cauca and upper Magdalena valleys between the Andean mountains (35 endemic forms are differentiated at the subspecific level), and 46 species recur on the dry Pacific slope of Middle America (Haffer 1967c).

Isolated savannas and islands of campina vegetation on white sandy soil exist in Amazonia, especially in the comparatively dry transverse zone across lower Amazonia extending from central Venezuela to northeastern Brazil. Birds inhabiting these patchy Amazonian nonforest

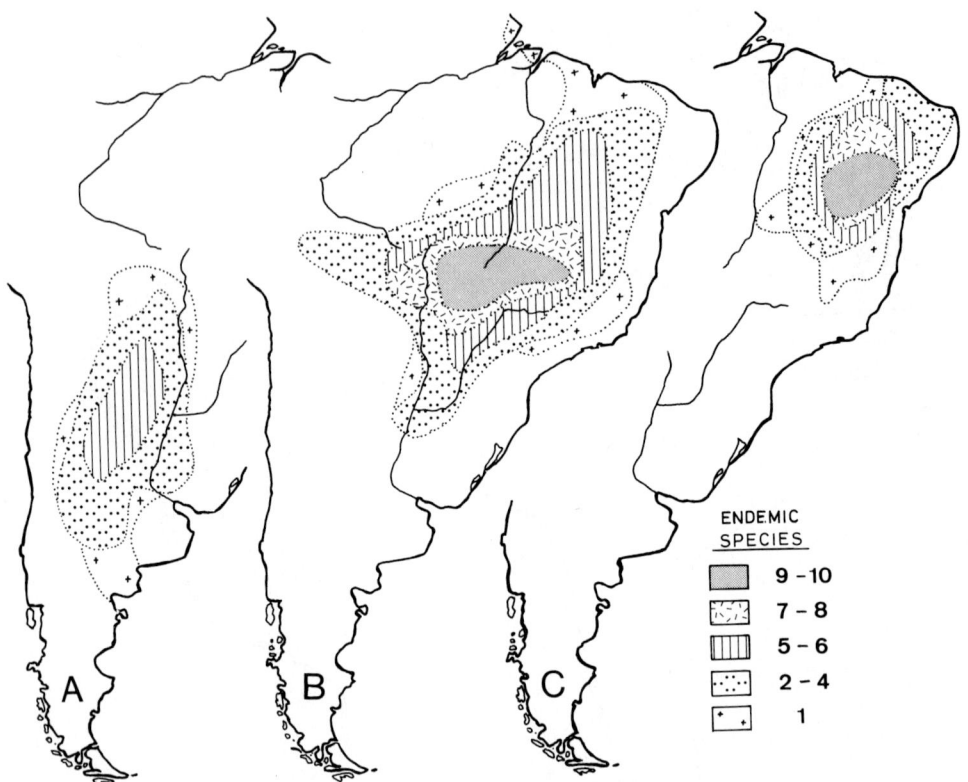

FIG. 7.   Superimposed ranges of bird species characteristic for A: the Chaco Area (7 species); B: the Campos Cerrados Area (10 species); C: the Caatinga Area (10 species). Species names are listed in Appendix I. Numbers and shading indicate totals of species for each group recorded.

habitats have their main distribution areas to the north and/or to the south of Amazonia. Some of the populations living on isolated Amazonian savannas are sufficiently well differentiated to be recognized as subspecies or even as species (see lists in Haffer 1974:111–112). Among the 111 species of the Amazonian campina avifauna, 21 have endemic forms in the upper Rio Negro region, the Moyobamba region, the Jamundá region, and in the Cachimbo region (Oren 1981).

Vast nonforest regions covered by distinct vegetation formations stretch diagonally across southcentral South America from northeastern and central Brazil to eastern Bolivia, Paraguay, Uruguay, and Argentina (Fig. 1). Widespread and ecologically tolerant birds found throughout this vast open region include the Rhea (*Rhea americana*), the Seriema (*Cariama cristata*), Guira Cuckoo (*Guira guira*), Suiriri flycatchers (*Suiriri suiriri, S. affinis*), Rufous and Ash-throated Casiornis (*Casiornis rufa, C. fusca*), Chopi Blackbird (*Gnorimopsar chopi*), and Saffron Finch (*Sicalis flaveola*). Several distinct avian assemblages inhabit the nonforest vegetation zones, caatinga, campos cerrados, chaco, pampas, and Patagonian steppes, extending from the northeast to the southwest, respectively.

In the *Caatinga Area,* the vegetation includes low xerophytic woods, deciduous thorny thickets, and cacti growing on poor soil. Semihumid, deciduous forests grow along the base and on the slopes of several isolated mountains in this region where the conditions of humidity are somewhat more favorable. A number of endemic bird species are restricted to this area (Appendix I). In constructing the contour map of this group of species (Fig. 7C), I superimposed their range outline maps.

The vegetation of the *Campos Cerrados Area* occupying the extensive central Brazilian tableland includes a ground cover of grasses and scattered low bushes and gnarled stunted trees. Pure grass plains (campos limpos) are present in certain parts of the cerrado region as

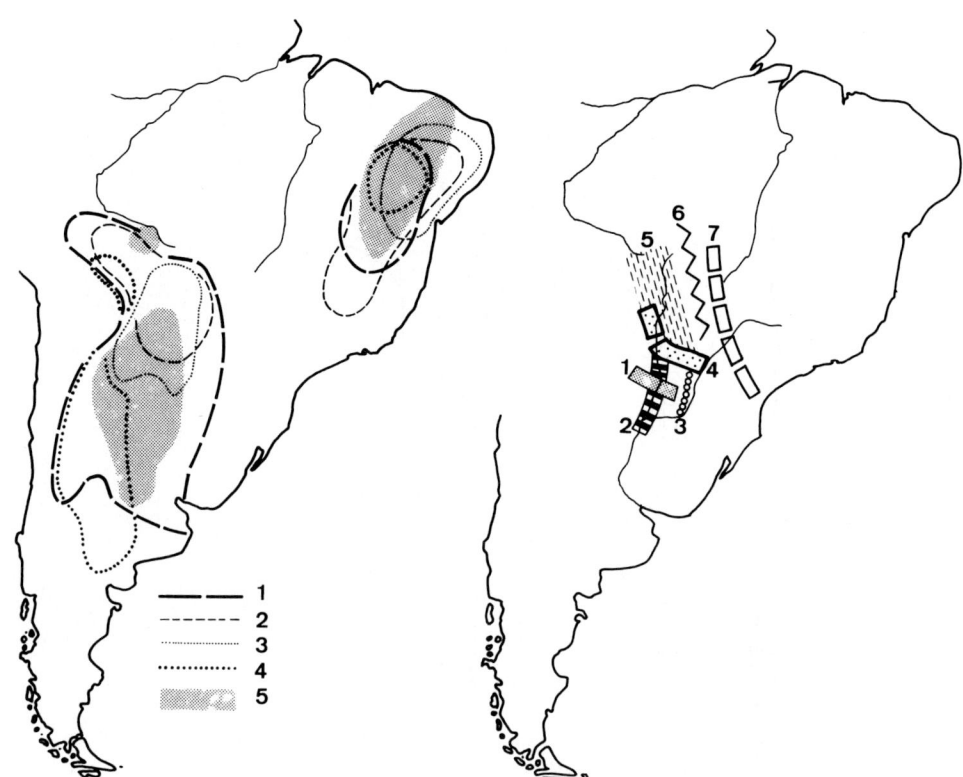

FIG. 8.   Distribution patterns of selected nonforest bird species in southcentral South America (adapted from Short 1975). Left: faunal relations of the Chaco and Caatinga areas are illustrated by species that have isolated populations in both areas (mostly subspecifically distinct) but are missing from the intervening campos cerrados of central Brazil. Superimposed ranges of 1 = *Aratinga acuticaudata*, 2 = *Pseudoseisura cristata*, 3 = *Myrmorchilus strigilatus*, 4 = *Stigmatura budytoides*, and 5 = *Xenopsaris albinucha*. Species 1 and 5 have isolated populations in northern South America. Right: clustered contact zones in southcentral Brazil between the following species and subspecies pairs: 1 = *Colaptes c. campestris* and *campestroides*; 2 = *Cyanocompsa cyanea argentina* and *sterea*; 3 = *Celeus lugubris* and *flavescens*; 4 = *Dendrocopus m. mixtus* and *cancellatus* groups, as well as *Suiriri suiriri* and *affinis*; 5 = *Phacellodomus rufifrons* ssp. and *P. r. sincipitalis* as well as *Sporophila collaris melanocephalus* and *S. c.* ssp.; 6 = *Nystalus m. maculatus* group and *striatipectus* group; 7 = *Sicalis flaveola pelzelni* and *brasiliensis*.

are forests that are quite extensive in certain areas (e.g., Mato Grosso de Goiás). Several characteristic birds of the campos cerrados (Appendix I) are taxonomically well isolated and are placed in monotypic genera (see also Sick 1965, 1966). I superimposed the ranges of the species in this group in constructing a contour map (Fig. 7B).

Dry woodlands, thorn forest, and shrub thickets characterize the *Chaco Area* of western Paraguay, northern Argentina, and adjacent parts of Bolivia and Rio Grande do Sul, Brazil. They grade southward into the xerophilous bushy steppes (monte) of central Argentina. The habitats of the Chaco Area also include grasslands and gallery forests along river courses. Physiognomically, the chaco vegetation resembles the caatinga vegetation of northeastern Brazil, and a number of bird species inhabit portions of both vegetation zones but are missing from the intervening campos cerrados (Fig. 8). These isolated populations show no or only slight racial differentiation, possibly indicating a rather recent connection of these vegetation formations in the geological past. A fairly large number of bird species is restricted in distribution to the Chaco Area (Appendix I). In constructing the contour map (Fig. 7A) I superimposed the ranges of seven characteristic species.

The extensive grass plains comprising the *Pampas Area* around the Río de la Plata are fairly humid near the coast, becoming drier inland where they grade into dry forests in the

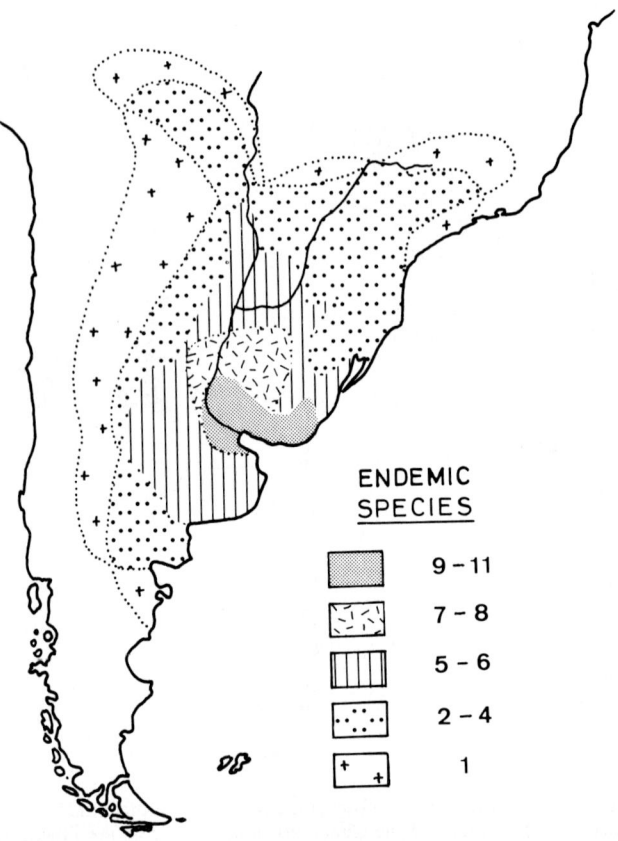

ENDEMIC
SPECIES

▨	9 – 11
▧	7 – 8
⦀	5 – 6
⬚	2 – 4
⊞	1

FIG. 9. Superimposed ranges of 11 bird species characteristic of the Pampas Area. Numbers and shading indicate totals of species in this group recorded; species names are listed in Appendix I.

north and into the Patagonian monte steppe in the west. Extensive grass plains that are periodically flooded occur north along the Paraná and Paraguay rivers into eastern Bolivia and the Gran Pantanal of southwestern Brazil (Mato Grosso). A number of birds characteristic of the Pampas Area, therefore, range far to the north of the La Plata River (Fig. 9).

The windswept, cold plains of the *Patagonian and Monte Steppe Area* of southern Argentina resemble semidesert areas on low mesa mountains; scattered trees grow locally along river beds. To the north, xerophilous, bushy steppes (monte) occur with scattered Algarrobo woods and small palm groves. Bird species characteristic of this area are listed in Appendix I.

Arid Pacific Coast Area includes the deserts of northern Chile and western Pacific Peru, which grade north into the semideserts, steppes, and dry deciduous woods of southwestern Ecuador. The avifauna of the latter region was analyzed in detail by Chapman (1926) who emphasized its small number of species (58), its high proportion of endemics (18 species, 7 genera), and its obvious relations to the Brazilian nonforest fauna. A few species, such as the White-winged Dove (*Zenaida asiatica*) of the United States and Middle America, and the Scrub Blackbird (*Dives warszewiczi*), which may be conspecific with the Melodious Blackbird (*Dives dives*) of the Pacific coast of Mexico to Nicaragua, indicate relations to the nonforest avifauna of northern South America, also. A number of widely separated dry pockets in deeply incised valleys in the eastern foothills of the Peruvian Andes form a discontinuous connection between the arid region of the Pacific coast, the dry upper Marañón Valley, and the Bolivian-Brazilian nonforest region. For example, the lower Río Urubamba Valley of eastern Peru, located halfway between the Río Marañón and the Bolivian savanna, is treeless and arid. Its distinctive avifauna ". . . has evidently been derived through western Brazil . . ." (Chapman 1921:28). Of a total of 66 species in the Urubamba Valley 38 are generally distributed in the tropics; 19 of the remaining 28 species are of Brazilian origin (Chapman 1921). The avifauna of the dry inter-Andean upper Marañón Valley in northern Peru resembles in many respects that of the arid Pacific coast from which it is separated by the rather low Porculla Pass, 2150 m (Dorst 1957). Notes on the birds of the dry inter-Andean valleys of Colombia have been given by Haffer (1967c).

CONTACT ZONES

Numerous well differentiated subspecies and closely related parapatric species of neotropical birds exclude each other along sharply defined contact zones in forest and nonforest regions (Delacour and Amadon 1973; Haffer 1974, in press b; Snow 1982). The contact zones often cluster in certain areas forming faunal "suture zones," as described for North America by Remington (1968). Most of the contact zones are located between the central portions of areas of endemism and along wide Amazonian rivers (Fig. 4). Mutual exclusion of areas of endemism and intervening zones of clustered contact zones is particularly conspicuous in northern Amazonia where, for example, no contact zones are known from the Napo Area or the extensive Guiana Area, each of which is inhabited by numerous endemic birds with restricted ranges. A few contact zones are located within portions of certain Amazonian areas of endemism (Imerí, Rondônia, Inambari areas) and the mutual exclusion, therefore, is locally less pronounced. Relatively low numbers of contact zones are also found in upper Amazonia south of the Rio Solimões and west of the Rio Madeira, as well as south of the lower Amazon River between the Rondônia and Belém areas (Fig. 4); however, no endemic species are restricted to these regions. In northwestern Colombia, a conspicuous cluster of contact zones is located between the Chocó Area, the Nechí Area, and the Caribbean southern Middle America Area (Haffer 1974, 1975).

Many unanswered questions remain with regard to contact zones between parapatric species in fairly uniform habitats. Still to be determined are (1) the mechanisms that assure reproductive isolation of competing species in contact, (2) whether behavioral responses or resource preemption by their respective partners prevent parapatric species from overlapping their ranges, (3) if the location of the contact zones is a result of historical or extant ecological factors, and (4) the reasons for parapatric species not penetrating each other's ranges. In instances of sympatry, the species might be expected to maintain interspecific territories or to occupy mutually exclusive patchy areas of varying size.

CONTACT ZONES IN FOREST BIRDS

In the trans-Andean lowlands, a number of contact zones are found in Honduras-Nicaragua and, especially, in Panama-northwestern Colombia (Haffer 1967a, b, 1974, 1975). Several conspecific or allospecific representatives of Middle and South America are separated by distributional gaps of varying width in central and eastern Panamá despite the continuity of the forests (e.g., *Cotinga, Carpodectes, Psarocolius*). In other similar cases a cis-Andean representative occupies all or part of the gap in Panama, establishing contact with the Middle American population of the trans-Andean representative as well as with the Pacific Colombian population. (Examples include the trans-Andean *Glaucis aenea*/cis-Andean *G. hirsuta*; *Rhynchocyclus brevirostris/R. olivaceus*; *Pipra mentalis/P. erythrocephala*; *Galbula r. melanogenia/G. r. ruficauda*). The Middle American and the Pacific Colombian populations of the first three are the same species, and in the last example, the same subspecies.

Many contact zones are clustered along the Amazon River and the wide lower portions of some of its tributaries (Rio Negro, Madeira, Tapajós, Tocantins). It has been known for more than one hundred years, from the reports of early naturalist explorers such as H. W. Bates (1862) and A. R. Wallace (1972), that large Amazonian rivers delimit, along their wide lower portions, the ranges of many bird species. Although these rivers probably represent absolute barriers to species unable to cross or to circumvent them, in other cases rivers serving as partial barriers to dispersal probably stabilized the equilibrium between competing species or between hybridizing subspecies. Somewhat different ecological conditions and/or competition with more distantly related species on an opposite river bank also may prevent the spread of some species despite the fact that a "bridge head" exists. These explanations suggest that species interactions rather than inability to cross or to circumvent a water course may cause the limit of a species range to coincide with a river in Amazonia. Ranges of the Razor-billed Curassow (*Mitu mitu*) and its allies (Fig. 10), and of the *Cercomacra tyrannina* superspecies (Fig. 11), for example (others are discussed in Haffer 1974, in press a), indicate that complex

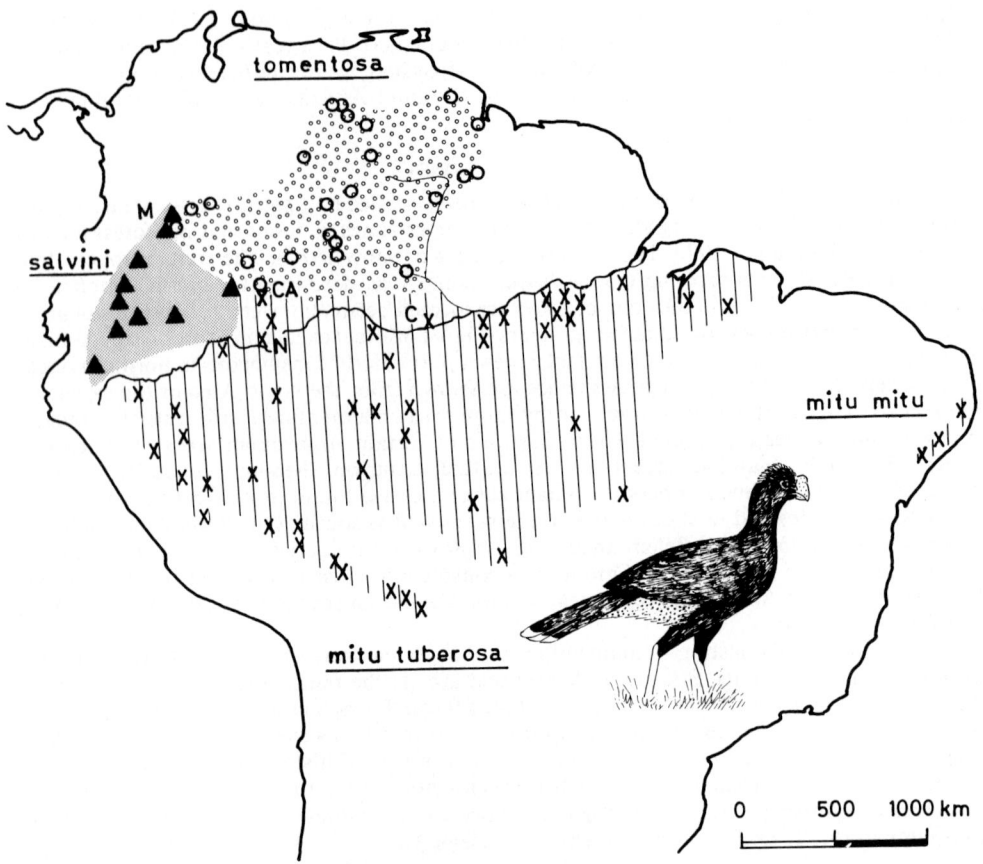

F<small>IG.</small> 10. Distribution of the Razor-billed Curassow and allies (*Mitu mitu* superspecies, Cracidae; adapted from Vaurie 1967). *M. mitu* (♂ illustrated): crosses and vertical hatching. *M. salvini*: closed triangles and shading. *M. tomentosa*: open circles and dots. Symbols denote locality records. C = Codajás, Brazil; N = Puerto Nariño, CA = Caquetá region, M = Macarena region, of Colombia (see text for discussion of these localities).

interspecific relations, rather than physiographic barriers alone, determine details of the species' distributions.

The allospecies of *Mitu* differ in the color of the undertail coverts (chestnut in *Mitu mitu* and *tomentosa*, white in *salvini*), of the tip of the tail (chestnut in *tomentosa*, white in *mitu* and *salvini*) as well as in the development of the crest and the shape of the bill (Delacour and Amadon 1973). The lower Amazon River does indeed delimit the range of the Razor-billed Curassow (*Mitu mitu*) in the north, but west of the Rio Negro, an extensive "bridge head" exists on the north bank of the Rio Solimões in central and upper Amazonia. Records of adult males from this region are available from Codajás, Brazil (C of Fig. 10, Gyldenstolpe 1951, but overlooked by subsequent authors), from Puerto Nariño, Colombia (N of Fig. 10; Nicéforo and Olivares 1965, considered "erroneous" by Vaurie 1967), and from several localities in the Caquetá region of southeastern Colombia (CA of Fig. 10, Scheuermann 1977). The location of the contact zone between *M. mitu* and *M. salvini* in the border region of southeastern Colombia and northeasternmost Peru is based on various kinds of indirect evidence such as information from local hunters (Scheuermann 1977). The three species of *Mitu* replace each other in featureless lowland forests traversed by many rivers of different size. Previous authors (Vuilleumier 1965, 1981; Vaurie 1967; Delacour and Amadon 1973) assumed that the Amazon River along its entire length prevented contact and range overlap of the southern and northern representatives. It remains to be seen whether in northern Perú *M. mitu* is confined to the

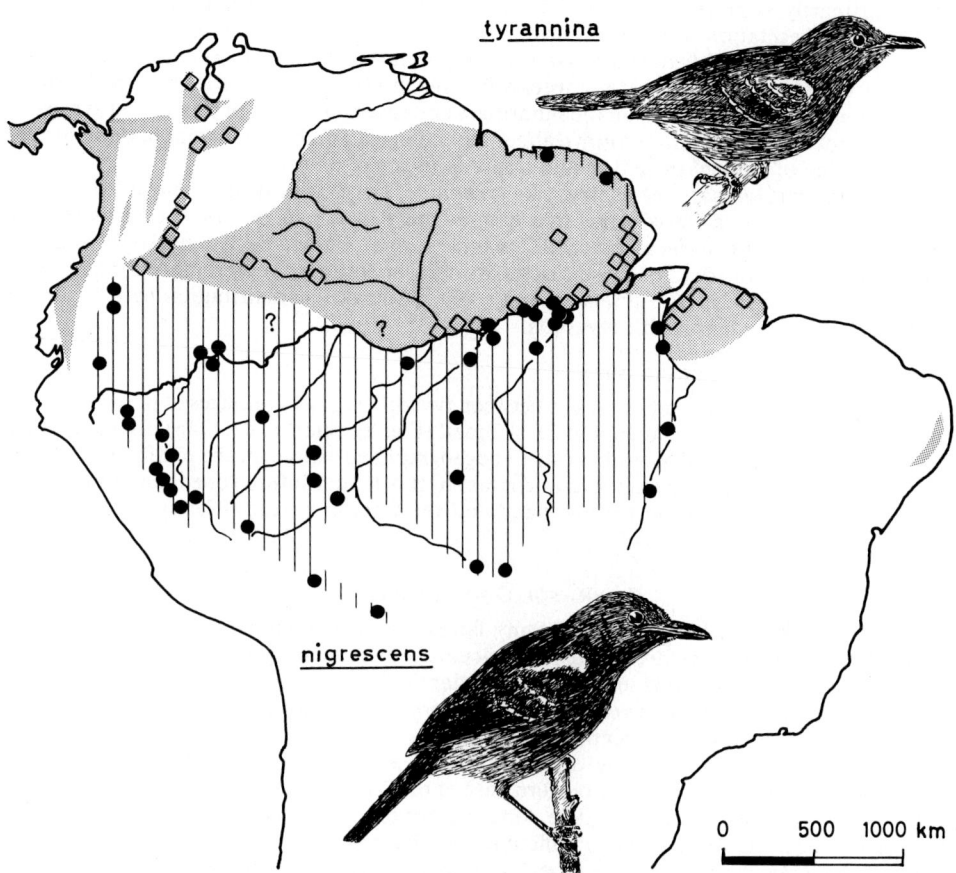

FIG. 11.   Distribution of the *Cercomacra tyrannina* superspecies (antbirds, Formicariidae). *Cercomacra tyrannina* (above): open squares and shading. Individual records for this widespread and common species are shown for selected regions only. *C. nigrescens* (below): solid circles and hatching. The birds illustrated are adult males. Symbols denote locality records.

region south of the Río Marañón. Neither *M. salvini* nor *M. mitu* is known from the lowlands just north of this river. Contact between *salvini* and *tomentosa* also takes place in the Macarena region of southeastern Colombia (M of Fig. 10; Haffer 1967b). Hybridization between these three curassow species is not known from any area of contact. The isolated population of *M. mitu mitu* in eastern Brazil is threatened by extinction due to deforestation (Sick 1980).

The ranges of the Dusky Antbird (*Cercomacra tyrannina*) and the Blackish Antbird (*Cercomacra nigrescens*) overlap narrowly along the lower Amazon River and in the coastal lowlands of the Guianas (Fig. 11). The males of *tyrannina* are slate-gray, whereas those of *nigrescens* are dark gray to blackish gray; males of both species have large concealed white interscapular patches, white bends of the wing and white-edged upper wing coverts. The females of both species are brown above (darker in *nigrescens*) and rufous below (more orange in *nigrescens* including the forehead). Both species inhabit borders of humid forest and second growth as well as thickets along river courses where they are encountered singly or in pairs. *Tyrannina* inhabits the entire trans-Andean forest region and northern Amazonia as well as small areas south of the mouth of the Amazon River and in northeastern Brazil (Pernambuco and Alagoas). Allospecies *nigrescens* is a bird of southern Amazonia but, in the west, also inhabits northeastern Peru and eastern Ecuador north of the Río Marañón from which *tyrannina* is missing. Both species are found along the northern Amazon's bank (Itacoatiara, Faro, Óbidos; Meyer de Schauensee 1966), but they may be ecologically separated. *Nigrescens* inhabits primarily swampy forests of the varzea, whereas *tyrannina* may prefer forest and second growth vegetation of the terra firme (Snethlage 1913:529); their ecological interrelationship requires detailed field study. *Nigrescens* also ranges along the Atlantic coastal plain northward where it inhabits swampy forests in French Guiana and Surinam. On the other hand, *tyrannina* occupies forests on the sandridges and savanna forests in this area of sympatry (Haverschmidt 1968). Nothing is known about the ecological interrelationships of these species along their zone of contact in central and upper Amazonia.

Other northern (Guiana) species that, like *tyrannina*, have crossed the Amazon River near its mouth and occupy a fairly large area east of the Rio Xingú or the Tocantins include *Ramphastos t. tucanus, Celeus undatus, Cotinga cotinga, Haematoderus militaris,* and *Euphonia cayennensis.* In some of these cases a western or southern representative exists (*R. t. cuvieri, Celeus grammicus, E. rufiventris*) and a lower Amazonian distribution pattern similar to that of the *Cercomacra*-antbirds results.

CONTACT ZONES IN NONFOREST BIRDS

A conspicuous cluster of contact zones between two different assemblages of nonforest birds occurs in eastern Paraguay and southcentral Brazil (Fig. 8). Short (1975:346) discussed most of these cases in some detail as well as other zoogeographical aspects of the nonforest avifauna of South America. The representatives in contact in this faunal suture zone inhabit the Chaco Area to the south and west on the one hand and the Campo Cerrado-Caatinga areas to the north and east on the other.

RANGE DISJUNCTIONS

Important zoogeographical features of any fauna are range disjunctions of its component species. Among neotropical forest birds, conspicuously disjunct ranges exist in those numerous species that inhabit both Amazonia and the Atlantic forests of southeastern Brazil but are absent from the extensive intervening campos region or here form additional disjunct populations in isolated forests of central and/or northeastern Brazil. Also, many Amazonian species recur on the western base of the Andes in Pacific Colombia with no connecting population in the forests along the northern base of the Andes. A large portion of these species range for varying distances into Middle America. Several of the isolated trans-Andean populations at or near the species level of differentiation have been variously treated as subspecies of the Amazonian species (in parentheses) or as a separate allospecies; examples include *Crypturellus (cinereus) berlepschi, Leucopternis (schistacea) plumbea, Heliothryx (aurita) barroti, Myrmornis (torquata) stictoptera,* and *Tangara (nigrocincta) larvata.*

Range disjunctions of nonforest species found north and south of Amazonia (Fig. 6) have been mentioned already. Most of these disjunct populations are differentiated at the level of subspecies, and some representatives are considered allospecies that together constitute a superspecies. Isolated populations of widely distributed cis-Andean species also occur in Pacific western Ecuador and northwestern Peru. One such species, the Masked Water-Tyrant (*Fluvicola nengeta*), has a widely disjunct population distributed from Maranhão to Bahia and Espíritu Santo in eastern Brazil (Fig. 6). The Yellow-faced Siskin (*Spinus yarellii*) is known from northeastern Brazil and from a single locality in northern Venezuela (where recorded only once).

Species with disjunct populations in South and Middle America include the Pearl Kite (*Gampsonyx swainsonii*), which inhabits the greater part of nonforested tropical South America and has an isolated population in Nicaragua, the Rusty-backed Spinetail (*Cranioleuca vulpina*) of Brazil and the Llanos region of Colombia and Venezuela, which has an isolated population on Coiba Island off the Pacific coast of Panama, and the Great-billed Seed-Finch (*Oryzoborus maximiliani*) of South America, which has isolated Caribbean populations from western Panamá to Nicaragua. The latter populations are considered by Stiles (1984) as an allospecies, *O. nuttingi.* Several localized isolates of northern Middle America (*Granatellus* spp., *Cyanocorax dickeyi, Campylorhynchus chiapensis*) are considered specifically distinct from their widely separated South American representatives (*G. pelzelni, Cyanocorax affinis, C. mystacalis,* and *Campylorhynchus griseus,* respectively).

HISTORICAL INTERPRETATION

In the above ornithogeographical analysis I emphasized regional differences in the composition of neotropical lowland avifaunas by delineating areas of endemism, contact zones, and range disjunctions. In an effort to trace the history and origin of these zoogeographical units, in this section I examine the interrelationships of selected species characterizing some areas of endemism and ascertain the degree of concordance between the resulting cladograms. In a second step, I consider geological data pertaining to the history of fragmentation of land areas and habitat zones in these regions and compare the results with those obtained from the study of the species interrelationships.

Concordant cladograms of various taxa imply common (allopatric) speciation events during faunal differentiation. In addition, congruence of area cladograms based on taxonomic data with area cladograms based on geological evidence indicates that the geological events that caused the fragmentation (vicariance) of land areas or habitat zones have caused the allopatric differentiation of the organisms considered (Platnick and Nelson 1978; Cracraft 1982, 1983a, b; Simberloff 1983). In other words, an attempt is made to determine historical interrelationships of the areas of endemism on the basis of the phylogenetic patterns of the respective endemic taxa. It should be noted, however, that common geological histories will not necessarily produce common phylogenetic patterns because barriers and corridors will affect different groups or taxa differentially (Endler 1982a). Cracraft (1982) suggested that lack of concordance among taxonomic cladograms of various species groups might indicate species differentiation through parapatric speciation. However, conflicting (non-concordant) cladograms among taxonomic groups may also result from allopatric differentiation if geological events have led to the more or less simultaneous fragmentation (on the geological time scale) of an area or a habitat zone into several isolated portions, such as ecological refugia inhabited by restricted animal populations.

The above method of zoogeographic analysis emphasizes aspects of endemism and vicariance. Its applicability to ornithological studies will depend upon the degree of confidence with which cladograms of species and genera can be constructed on the basis of morphology and plumage coloration, and upon the amount of detailed geological data available to resolve the historical sequence of fragmentation of geological or vegetational units. Aspects of dispersal are deemphasized in this analysis, although they also need to be considered in a more complete study. This is reflected in Mayr's (1965) statement that faunas are composites of several elements, each with a different history. A future refinement of methods may permit us to distinguish between the various vicariant and dispersal elements in the neotropical avifauna. In any case, a pluralistic approach to zoogeographic analysis is recommended in view of the diversity of phenomena causing the distribution patterns of any particular taxonomic group (Ball 1983; Mayr 1983).

PHYLOGENETIC PATTERNS OF SOME ENDEMIC TAXA

The group of araçari toucans consists of four cis-Andean species (*Pteroglossus aracari, P. castanotis, P. pluricinctus, P. beauharnaesius*) and two trans-Andean groups that hybridize and are conspecific (*P. torquatus,* incl. *sanguineus; P. t. frantzii* of southern Costa Rica and western Panamá is sometimes ranked as a separate species). Characters considered derived and used in the construction of the cladogram are (Fig. 12, squares and letters) the red belly band (x); color pattern of the bill (a; dark culmen stripe, light colored upper mandible), black throat and head (b), additional black breast band and belly band mixed red and black (c), breast band reduced to black chest spot (d), black longitudinal stripe along upper mandible (e), chestnut collar (f), body proportions (g; relatively short tail), curled crown feathers (h). In upper Amazonia *P. castanotis* is sympatric with *pluricinctus* and *beauharnaesius;* the latter two species exclude each other geographically north and south of the Río Marañón. The ancestor of the trans-Andean forms *sanguineus, torquatus,* and *frantzii* presumably separated from the upper Amazonian pre-*pluricinctus* relatively late during the differentiation of this group of toucans.

Toucans of the *Ramphastos vitellinus* species group include three cis-Andean and two trans-Andean species (Fig. 12). Characters considered derived and used in the construction of the cladogram of this group are (squares and letters, Fig. 12) the keeled bill (a), yellow or yellowish-green facial skin (b), yellow throat and upper breast (c), elaborate coloration of bill (d, e), lateral channelling of the upper mandible in adult birds (f), extensive black on bill (g), yellow culmen (h), basal blue or yellow on bill (i), and the broad and flattened culmen (j). Adults of *vitellinus* and *dicolorus* are channel-billed, as they possess a groove running lengthwise along both sides of the upper mandible below the culmen. The shape of this groove varies in the different forms from narrow and deep to shallow and broad. The bill of juveniles in these species is keeled as it is in adults of the keel-billed species *toco, brevis,* and *sulfuratus.* The

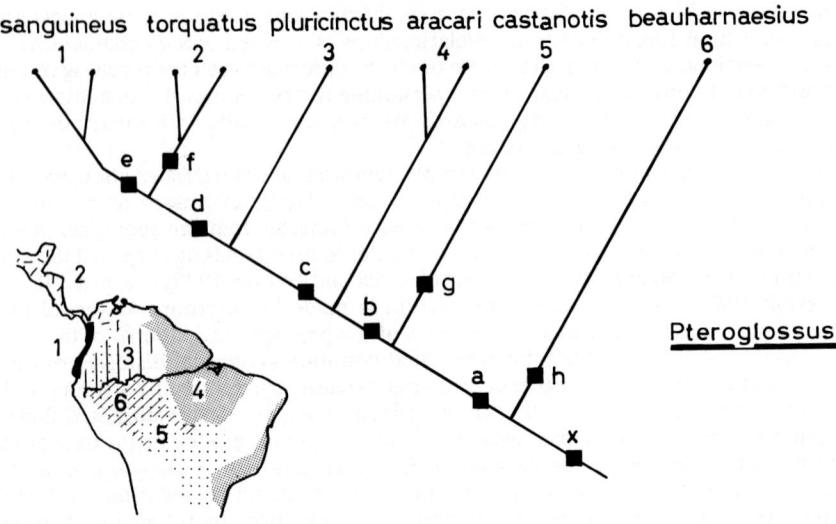

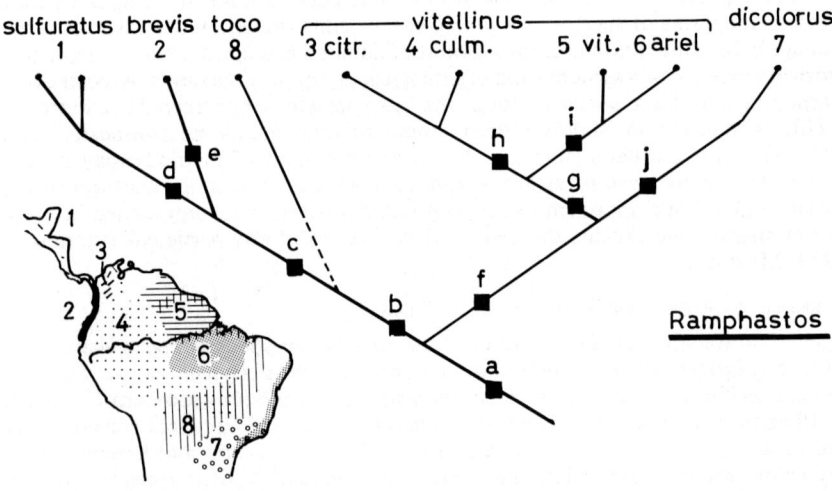

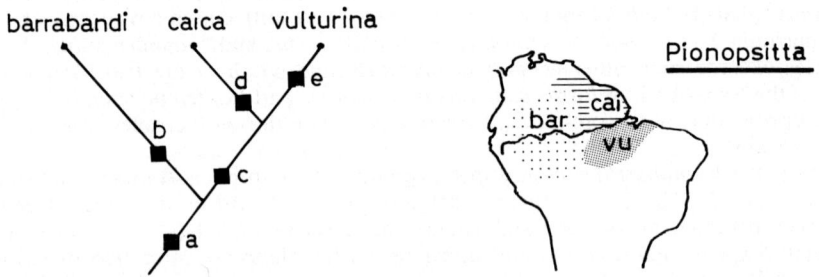

FIG. 12.   Distribution and postulated phylogenetic relationships of three species groups of neotropical forest birds. See text for derived characters (squares and letters) used in construction of cladograms. Above: toucans of the *Pteroglossus aracari* species group. Notice sympatry of *P. castanotis* with *pluricinctus* and *beauharnaesius* in upper Amazonia. Center: toucans of the *Ramphastos vitellimus* species group. Notice

channel-billed condition, therefore, is considered to be derived. The very similar croaking voice of the channel- and keel-billed toucans, as well as their parapatric distribution pattern in northwestern South America (where *vitellinus* and *sulfuratus* are in contact) further indicate their close relationship. Keel-billed *toco* inhabits open woodland and gallery forest in central South America, north to lower Amazonia; its relations with the other species of this group remain somewhat questionable. The ancestor of the trans-Andean species was apparently isolated from the cis-Andean group relatively early and prior to the differentiation of the channel-billed group in Amazonia. The Amazonian forms are ranked as subspecies because of close morphological resemblance and extensive hybridization along their zones of contact (Haffer 1974).

The sequence of differentiation in the *vitellinus* group contrasts markedly from that within the *Pteroglossus aracari* group (Fig. 12). The trans-Andean representatives of *Pteroglossus* apparently developed relatively late and after the differentiation of the Amazonian forms, whereas in the *Ramphastos vitellinus* group the trans-Andean forms seem to have originated relatively early and before the differentiation of the cis-Andean forms. East of the Andes, the separation of the ranges and the differentiation of forms to the north and south of the lower Amazon River appear to be relatively later events than the differentiation of western (upper) Amazonian and eastern (lower) Amazonian forms, as shown by both species groups.

The parrots of the *Pionopsitta caica* superspecies (Fig. 12) include two closely related lower Amazonian species (*caica* and *vulturina*) and one upper Amazonian species (*barrabandi*) illustrating a major difference between eastern and western Amazonian forms. Because of the naked head of adult birds, *P. vulturina* is frequently separated taxonomically as a monotypic genus "*Gypopsitta*" (Forshaw 1973) despite doubts about this taxonomic procedure raised by Hellmayr (1912), Griscom and Greenway (1941), and Haffer (1970). The separation of the western and eastern Amazonian forms is again in this group an earlier event than the differentiation of the species to the north and south of the lower Amazon River.

Many additional studies are required to verify the tentative statements regarding the species groups discussed above and possibly to establish concordance of the phylogenetic patterns among a larger number of species groups of neotropical birds and other sympatric organisms.

GEOLOGICAL HISTORY OF AREAS OF ENDEMISM

For about 100 million years South America, which had separated from Africa during the Early Cretaceous and from Antarctica during the Early Tertiary, was an island continent. During the Late Cretaceous and Tertiary, South America drifted westward (and slightly southward). Connection with North America was established across the Isthmus of Panama during the Late Pliocene only about 3 million years ago (Keigwin 1978, 1982; Marshall et al. 1982). Geological events in South America that led to the fragmentation of neotropical biotas and potentially to the differentiation of species on opposite sides of barriers include mountain building (orogenesis), differential erosion of uplifted areas, the formation of epicontinental seas, climatic-vegetational fluctuations, changes of world sea level, and the formation of river systems.

Mountain building. — The final uplift of the Andes led to the formation of an effective barrier to the interchange of trans- and cis-Andean lowland faunas. During the Quaternary, interchange was restricted to the north Colombian lowlands, to the area where the low Eastern Cordillera of Colombia joins the Central Cordillera at the head of the Río Magdalena Valley, and to the region of the low Porculla Pass (2150 m) in northern Peru (Chapman 1917; Haffer 1967a).

Differential erosion. — The differential destruction of uplifted areas by erosion continuing over millions of years can lead to the isolation of an upland fauna on mountain massifs that lag behind in the regional leveling process. This mode of isolation, possibly coupled with local uplift and subsidence, may have been relevant for the differentiation of certain birds inhabiting the upper levels of the tropical Andes and the high mesa mountains (tepuis) of southern Venezuela.

←

local sympatry of *R. vitellinus* and *dicolorus* in southeastern Brazil; *R. toco* inhabits savanna woodlands of south central South America and the lower Amazon valley. Below: parrots of the *Pionopsitta caica* superspecies.

Epicontinental seas.—Regional subsidence of land areas due to epeirogenic movements often led to the formation or change in extent of epicontinental seas and the concomitant separation of portions of a previously continuous biota. Such isolation of portions of the neotropical fauna occurred repeatedly during the Tertiary paleogeographic development of South America as mapped by Harrington (1962; see also Beurlen 1970; Bigarella 1973; Fittkau 1974). The extensive regions of the combined Guianan and Brazilian shields have been above sea level since Paleozoic times, and formed low-lying lands under the hot arid climate of the Mesozoic. Continued uplift of the shield areas occurred during the course of the Tertiary when the Amazon basin subsided and the southeast Brazilian mountains as well as the table mountains of southern Venezuela-Guyana (Pantepui) reached their present elevations of 1500–3000 m above sea level. Details about the paleogeographic history of central South America during the Tertiary, in particular the varying barrier effect of the Amazon valley for the northern and southern faunas are unknown because of the lack of marker horizons and index fossils in the strata of that period. Probably the sediments in the Amazon valley were deposited in large swamps and marshes and on vast flood plains that were connected to the west with the sub-Andean basin of the upper Amazon region and with the Atlantic Ocean to the east. Toward the end of the Tertiary (Pliocene), most of the Amazon valley was covered by the vast Belterra Lake whose clay deposits are presently found at 180 m above sea level in the lower Amazon valley.

Portions of the tropical Andes, as well as of the Venezuelan coastal mountains, were raised above sea level as narrow low-lying islands at the end of the Cretaceous and during the early Tertiary. Marine ingressions from the Pacific Ocean into the sub-Andean basin east of the Andes occurred until the middle Tertiary (Oligocene), after which time the rising Ecuadorian and Peruvian Andes were connected, thereby closing the last portal to the Pacific Ocean.

Climatic-vegetational fluctuations.—Studies of the Quaternary sediments and of their pollen content, as well as analyses of surface land forms in South America, have revealed dramatic and wide-ranging reversals in the climate at tropical latitudes; these changes probably caused vast shifts in vegetation types in the tropical lowlands during the last one to two million years. Forests expanded during humid periods and survived in variously extensive "refuge" areas during dry climatic periods (or at least one dry period); the opposite was true for savanna regions (reviewed by Livingstone and Van der Hammen 1978; Flenley 1979; Prance 1982; Haffer, in press b). In southern South America, the Chilean forests retreated some distance northward when the glaciers advanced and returned south during the interglacial periods (Mercer 1976; Simpson 1979). Geological, mineralogical, and micropaleontological studies of sea-bottom cores taken in the Caribbean Sea and off northeastern South America indicate climatic aridity during the glacial phases in adjacent onshore regions (Damuth and Fairbridge 1970; Bonatti and Gartner 1973; Parmenter and Folger 1974). Certain soil formations, dissected and gullied land surfaces, and crusts (stone lines) found in portions of Amazonia and in other humid lowland regions of tropical South America, indicate the previous existence of sparse vegetation in these areas pointing to at least one former period (and probably more) of drier climate than today (Garner 1974, 1975; Tricart 1974; Ab'Saber 1982; Klammer 1982). Palynological investigations of Quaternary sediment cores from several study sites in tropical South America permit an analysis of the changing vegetation at these localities through time (Wijmstra and Van der Hammen 1966; Wijmstra 1971; Eden 1974; Van der Hammen 1974, 1983). In contrast to the highlands of the northern Andes (for which a rich palynological documentation exists), pollen data from the neotropical lowlands, especially Amazonia, are still insufficient as far as the Pleistocene is concerned; most of the available lowland sections are Holocene (i.e., post-Pleistocene) in age. The vegetational fluctuations recognized for the neotropical lowlands on the basis of palynological data of the last 10,000 years, together with the results from a few Pleistocene pollen cores taken in southern Amazonia and in the coastal lowlands of the Guianas, as well as the results of geomorphological studies, support the notion of vast changes in the distribution of forest and nonforest vegetation during the Pleistocene (Garner 1974, 1975; Tricart 1974; Journaux 1975; Livingstone and Van der Hammen 1978; Brown and Ab'Saber 1979; Flenley 1979; Peterson et al. 1979; Ab'Saber 1982). Model studies of the July climate of 18,000 years B.P. also agree with this interpretation (Climap 1976; Gates 1976).

To be sure, the neotropical forest and nonforest vegetation is old, having originated during Cretaceous and Tertiary times. In the areas of their uninterrupted occurrence forests and savannas continuously offered ecologically fairly stable and rather equable conditions to the species of tropical animals. It is the overall geographical distribution of these vegetation types that has varied in response to the changing world climates of the Pleistocene and the Tertiary.

Changes of world sea level.—During Pleistocene glacial periods of lowered sea level, coastal lowlands were periodically enlarged, and islands on continental shelves became interconnected or connected with the mainland, e.g., some islands along the north coast of South America and along the coast of southeastern Brazil. Additional islands formed when portions of the present coastal lowlands were flooded during the height of the interglacial periods, when sea level was approximately 50 to 60 m higher than today (Flint 1971).

Formation of major river systems.—Some authors have assumed that the developing network of rivers in an extensive lowland region, such as Amazonia, toward the end of the Tertiary, fragmented a pre-existing continuous biota. The populations on opposite river banks presumably differentiated to the level of subspecies or even species. This theory has been invoked on various occasions to explain species differentiation in Amazonia, but the theory has never been formally proposed or quantitatively tested (reviewed by Haffer 1974).

Summarizing, the extensive South American shield areas, that is the major part of the continent east of the Andes south to Uruguay, formed stable land areas throughout most of the Mesozoic and Cenozoic, a period of more than 150 million years. Faunal differentiation in these regions probably took place in the absence of conspicuous tectonic-geological fragmentation (vicariance) events. An area cladogram for the South American lowlands based on geological data would indicate no more than a single split of the preexisting huge continent into a northern (Guiana Shield) and a southern portion (Brazilian Shield) during the Cretaceous. Continued differential movements during the Tertiary led to varying uplift of the mountains in southern Venezuela and southeastern Brazil to their present elevations; this uplift, coupled with differential erosion, caused increasing isolation of individual massifs. The history of the Andean region to the west is characterized by the appearance above sea level and continued uplift of extensive elongated island regions that became progressively interconnected as the mountain chains rose during the Tertiary. Paleogeographic details are insufficiently known to allow for construction of geological area cladograms for the Andean region.

Present knowledge of the Tertiary and, especially, the Quaternary history of South America indicates repeated complex formations and disappearances of ecological barriers over the entire continent through climatic-vegetational fluctuations. Available geoscientific data, however, are insufficient to allow for the mapping of changes in distributions of forest and nonforest vegetation during the various climatic periods and, in particular, for the tracing of the history and the location of areas of remnant forests and savannas that presumably served as refugia for animal populations. Therefore, no area cladograms for the Tertiary-Quaternary history of vegetational units in the neotropics can yet be constructed for a comparison with area cladograms derived from taxonomic studies of particular groups.

The areas of endemism in the neotropical lowlands, as mapped for birds (Haffer 1967a, 1969, 1974, 1978), reptiles (Vanzolini 1970; Vanzolini and Williams 1970), several groups of vertebrates (Müller 1973), butterflies (Brown et al. 1974; Descimon 1977; Brown 1982, in press a), and plants (Prance 1973), and summarized by Simpson and Haffer (1978), Prance (1982), and Whitmore (in press), may be linked historically to the formation of Quaternary ecological refugia and may indicate their general geographical location. Isolated populations of the refugia presumably differentiated as a result of selection and chance and developed, in part at least, into endemic species and subspecies. The early history of several or all areas of endemism probably will be traced back to climatic-vegetational and/or tectonic events of the preceding Tertiary period when those species that survived in the Pleistocene refugia more or less undifferentiated to the Recent, originated. In any case it should be remembered that the refugia are primarily geological, not biological, phenomena. The location and the extent of the refugia eventually must be traced by geoscientific data, especially from the fields of palynology and geomorphology, and not exclusively on the basis of distributional data of endemic plant and animal species. Nevertheless, there can be a valid interactive iterative process involving geological and biological data sets in tracing the history of the vegetation and faunas.

DISCUSSION

According to the traditional view held by many scientists during the last several decades, the adaptations and distribution patterns of tropical organisms are at least Tertiary in age and, in any case, much older than those of the North and South Temperate zone floras and faunas, which have been deeply influenced by the vicissitudes of the Pleistocene climatic and vegetational changes (Darlington 1957; Fittkau 1969:631; Schwabe 1969:126). Intensified fieldwork during the last twenty years, however, has proven the concept of environmental stability in the tropics during the Quaternary to be ill-founded and erroneous. Frequent barrier formation and concomitant range fragmentation probably led to extensive speciation and subspeciation in many taxa and to local extinction and adjustment of distribution patterns in most groups of neotropical organisms during the last two million years (Haffer 1967a, 1969; Vanzolini 1970; Dorst 1976). The point to be emphasized is that the differentiation of the neotropical fauna as we know it today probably was not complete at the end of the Tertiary but continued into the Quaternary when species and subspecies possibly originated in ecological refugia following the model of allopatric speciation (Mayr 1963). Of course, this does not mean in the least that all species of birds and other animals are Pleistocene in age or that speciation occurred in all cases in geographically isolated populations (although the latter assumption appears likely for birds). It will be the task of future ornithologists to estimate the number of bird species that originated during the early Quaternary, that survived more or less unchanged from the Tertiary, or that became extinct during these periods. Answers to these questions will contribute to a solution of the still unsolved problems of extinction and of the origin and age of tropical species diversity. In other words, attempts should now be made, through comparative cladistic studies of organisms and geological records, to place the biogeographic effects of the Quaternary events in proper perspective by also analyzing the older patterns of differentiation pertaining to the long time span of the Tertiary period (60 million years) when the ancestors of most modern avian groups and also a number of extant (especially non-passerine) species originated. The conspicuous disjunctions of the ranges of numerous lowland species discussed above may be explained in many cases on the basis of the Late Tertiary and Quaternary vegetational fluctuations that permitted range expansion of ancestral populations and, subsequently, caused range fragmentations through ecological shifts. For this reason, transportation by man (Haemig 1978, 1979) and tectonic-geological events (Croizat 1976) may be invoked to explain particular patterns, but are unlikely general causes to explain the disjunctions of so many conspecific and consuperspecific populations of neotropical birds and other animals.

The biogeographical phenomena that favor the above interpretation of a Tertiary-Quaternary origin of the patterns of differentiation and distribution include: (1) the clusters of endemic species and subspecies forming areas of endemism in superficially uniform habitats; (2) the numerous contact zones between subspecies and parapatric species that cluster between several areas of endemism, and, if interpreted as secondary contact zones, that can be taken to indicate the former existence of barrier zones that have since disappeared; (3) the existence of widely disjunct populations of representative subspecies and species with poor dispersal capabilities, and (4) the occurrence of terrestrial nonforest animals on isolated enclaves in Amazonia. These consistent and conterminous distribution patterns exist in diverse bird families that differ widely in their feeding preferences.

The avifaunas of the major habitat types in South America (forest; open woodland; prairies, steppes, and deserts) are conspicuously different at the specific and generic levels; their separation probably began correspondingly early during the Tertiary, continuing during later stages of faunal differentiation. Habitat shifts during relatively recent speciation occurred in several South American superspecies of birds that include allospecies confined to forest on the one hand and to nonforest habitats on the other (Vuilleumier 1981). The allospecies *Phrygilus patagonicus,* of the *Phrygilus gayi* superspecies (Fringillidae), has invaded the Chilean forest zone, as has *Phalcobaenus albogularis* of the *Phalcobaenus megalopterus* superspecies (Falconidae). All other members of the two superspecies (*Phrygilus atriceps* and *gayi*; *Phalcobaenus carunculatus* and *megalopterus,* respectively) inhabit various Andean steppe habitats. An ecological shift (forest to woodland or *vice versa*) also occurred during the differentiation of the *Thamnophilus caerulescens-amazonicus* superspecies (Formicariidae). The number of South American superspecies of passerines that have representatives in forest and nonforest zones probably is small, comprising less than 10 percent of the total; by comparison, this figure is nine percent in the Afrotropical avifauna (Hall 1972). Habitat specializations probably arise either during a phase of geographic separation, due to differential selection pressures, or, after secondary contact of former isolates is established, as a result of interspecific interactions between close relatives (Keast 1972, 1980). The historical mechanisms promoting the origin of such habitat shifts among closely related species during geographic separation remain obscure. The model of the "vanishing refuge" (Vanzolini 1981; Vanzolini and Williams 1981) implies that in areas affected by vegetational fluctuations, certain forest species confined to a refuge that eventually vanishes may become completely adapted to semiopen or even open nonforest habitat. The same might be assumed for nonforest species in the opposite sense.

The historical interpretation of avian differentiation in the neotropics reviewed above is based primarily on the occurrence of repeated range fragmentations of organisms (secondary disjunctions, vicariance events) followed in many cases by secondary contact and range overlap through biotic dispersal (diffusion) which were caused by geologic-tectonic events as well as climatic-vegetational fluctuations during the Cenozoic (Tertiary and Quaternary). Faunal differentiation due to active dispersal across barriers (founder effect; primary disjunctions) probably also occurred in several groups of birds and other organisms, but its relative importance remains unknown. This problem may be approached through comparative cladistic analyses of taxonomic groups with disjunct populations and of geographic-geological units (reviewed by Wiley 1981). An example of a neotropical avifauna that probably originated in part through active (jump) dispersal of some of its component species or their immediate ancestors is the montane avifauna of the strongly isolated table mountains of southern Venezuela (Mayr and Phelps 1967; Haffer 1981). Direct contact of this montane fauna with that of the Andes to the west or the Venezuelan mountains to the north during any time of the geological past is very unlikely. Those montane forms that cannot be derived from ancestors of the surrounding lowlands most probably have dispersed across the lowlands, in most cases from the Andes; low table mountains in southeastern Colombia probably provided a discontinuous "bridge" across the Amazonian lowlands.

Croizat (1958, 1976) claimed in his interpretations, first, that *all* extant patterns of faunal distribution and differentiation in South America originated through range fractionation (vicariance) and, second, that these vicariance events occurred exclusively through tectonic movements and erosional differentiation of the earth's relief during the Mesozoic and Tertiary. This latter suggestion is difficult to uphold in view of the relative tectonic stability of the extensive South American shield areas during the Tertiary and Quaternary. Furthermore, no geological data exist to support the idea of land areas off the northern, western, and eastern coasts of South America during the Tertiary; Croizat (1958, 1976) used such land areas to explain the previous dispersal and continuous range of ancestral avian populations, followed later by disjunctions through the presumed foundering of the postulated connecting land areas. Moreover, Croizat's views on evolution and an assumed old age of all extant bird species (about mid-Tertiary) are not in accordance with present consensus (Brodkorb 1971; Ballmann 1977; Vuilleumier, in press).

In view of the fact that fragmentation of species ranges through climatic-vegetational fluctuations also includes vicariance events, and that range expansion is assumed to have taken place as biotic dispersal (diffusion) through continuous habitat zones (not across barriers), the conceptual basis for the vigorous attacks by Croizat (1976) on the suggestions by Haffer and other authors regarding the differentiation of the neotropical fauna at low taxonomic levels during the Cenozoic remains unclear. As far as the Quaternary is concerned, the only reason seems to be Croizat's assumption of a Tertiary age of all species and even subspecies of birds. "Refuge biogeography is but a recent chapter of vicariance biogeography on a sub-continental scale, with more ecological emphasis, . . . and based on abundant and reliably recent geoscientific data-sets" (Brown, in press b).

Another interpretation of faunal differentiation in South America relates the observed biogeographical patterns of neotropical biotas to current ecological diversity. Thus, Endler (1977, 1982a, b) suggested that biotic differentiation across environmental gradients might lead to biogeographic patterns similar to those of coalesced refugia. In particular, he stated that partial barriers or ecological transition zones might produce "stepped clines" and "hybrid zones" which are indistinguishable from zones of secondary intergradation. This situation is said to lead frequently to parapatric speciation (which occurs whenever species evolve as contiguous populations in a continuous cline; Bush 1975). It remains to be confirmed that ecological gradients in the Neotropical Region are sufficiently pronounced to produce, through clinal variation, narrow zones of high variability ("hybrid zones"), not only in insects, but also in birds, without prior separation of the populations concerned. Furthermore, different taxonomic groups respond differently to environmental gradients and to the ecological diversity. In assessing the theoretical plausibility and available evidence for non-allopatric speciation, Futuyma and Mayer (1980:254, 269) concluded that "neither the theory nor the evidence are sufficient to justify a major departure from the traditional view: . . . speciation does indeed occur most often by the divergence of populations confined to small geographic areas."

In view of the implicit importance of ecological forces in an understanding of the past movements and present geographical distribution of bird species, I emphasize the urgent need for ecological surveys in the neotropics to test the validity of the assumed existence of several biotic interactions (Wiens 1983). Detailed studies should be carried out to determine the presence of limiting resources and the significance of competition and predation in local communities. Zones of purported secondary contact of parapatric species would be particularly suited for such studies and their results vital at this time. Many parapatric species exclude each other geographically in uniform habitat zones of South America, their abrupt replacement traceable over hundreds of kilometers. No field study has so far tested the hypothesis that ecological competition between these species when they are in contact prevents their coexistence and, thus, leads to geographical exclusion. An improved understanding of the ecological relations between species will contribute substantially to an evaluation of the various alternative theories of speciation in the neotropics and lead to a refined analysis of the biogeographical patterns observed in the lowlands of Middle and South America.

ACKNOWLEDGMENTS

The late Eugene Eisenmann gave me much encouraging advice in many friendly letters between 1958 and 1967 when I was stationed in Colombia. During my later repeated visits to the American Museum of Natural History, New York, he was always ready to discuss aspects of the distribution and speciation of neotropical birds with me and to review my manuscripts at various stages of completion. My debt of gratitude to him is very great indeed.

I thank the following persons who advised me of recent publications or furnished me with copies of them: M. LeCroy, B. B. Simpson, C. C. Olrog, M. D. F. Udvardy, P. E. Vanzolini, and F. Vuilleumier. I am also grateful to P. A. Buckley, M. S. Foster, J. P. O'Neill, and D. L. Pearson for critically reading the manuscript and for many helpful suggestions for its improvement.

LITERATURE CITED

Ab'Saber, A. N. 1982. The paleoclimate and paleoecology of Brazilian Amazonia. Pp. 41–59, In G. T. Prance (ed.), Biological Diversification in the Tropics. Columbia University Press, New York.

Ball, I. R. 1983. Planarians, plurality and biogeographical explanations. In R. W. Sims, J. H. Price, and P. E. S. Whalley (eds.), Evolution, Time and Space. The Emergence of the Biosphere. Syst. Assoc., Spec. Vol. 23:409–430.

Ballmann, P. 1977. Neuere Erkenntnisse über die zeitliche Entstehung der rezenten Vogelarten. Mitt. Bayer. Staatssamml. Palaeontol. Hist. Geol. 17:169–175.

Bates, H. W. 1862. The Naturalist on the River Amazons. J. Murray, London.

Beurlen, K. 1970. Geologie von Brasilien. Beitr. Regionalen Geol. Erde No. 9.

Bigarella, J. J. 1973. Geology of the Amazon and Parnaiba basins. Pp. 25–86, In A. E. M. Nairn and F. G. Stehli (eds.), The Ocean Basins and Margins, Vol. 1 (South Atlantic). Plenum Press, New York.

Blake, E. R. 1977. Manual of Neotropical Birds, Vol. 1. University Chicago Press, Chicago, Illinois.

Bonatti, E., and S. Gartner. 1973. Caribbean climate during Pleistocene ice ages. Nature 244:563–565.

Brodkorb, P. 1971. Origin and evolution of birds. Pp. 19–55, In D. S. Farner, J. R. King, and K. C. Parkes (eds.), Avian Biology, Vol. 1. Academic Press, New York.

Brown, K. S., Jr. 1976. Geographical patterns of evolution in neotropical Lepidoptera. Systematics and derivation of known and new Heliconiini (Nymphalidae: Nymphalinae). J. Entomol. Ser. B, 44:201–242.

Brown, K. S., Jr. 1977a. Geographical patterns of evolution in neotropical forest Lepidoptera (Nymphalidae: Ithomiinae-Heliconiini). Pp. 118–160, In H. Descimon (ed.), Biogéographie et evolution en Amérique tropicale. Publ. Lab. Zool. École Norm. Suppl. 9.

Brown, K. S., Jr. 1977b. Centros de evolução, refugios quaternarios, e conservação de patrimonios genéticos na região neotropical: padrões de diferenciação em Ithomiinae (Lepidoptera: Nymphalidae). Acta Amazônica 7:75–137.

Brown, K. S., Jr. 1982. Paleoecology and regional patterns of evolution in neotropical forest butterflies. Pp. 255–308, In G. T. Prance (ed.), Biological Diversification in the Tropics. Columbia University Press, New York.

BROWN, K. S., JR. In press a. Biogeography and evolution of neotropical butterflies. *In* T. C. Whitmore (ed.), Biogeography and Quaternary History in Tropical America. Clarendon Press, Oxford, England.

BROWN, K. S., JR. In press b. Conclusions, synthesis and alternative hypotheses. *In* T. C. Whitmore (ed.), Biogeography and Quaternary History in Tropical America. Clarendon Press, Oxford, England.

BROWN, K. S., JR., AND A. N. AB'SABER. 1979. Ice-age refuges and evolution in the Neotropics: correlation of paleoclimatological, geomorphological and pedological data with modern biological endemism. Univ. São Paulo, Inst. Geogr., Paleoclimas 5:1–30.

BROWN, K. S., JR., P. M. SHEPPARD, AND J. R. G. TURNER. 1974. Quaternary refugia in tropical America: evidence from race formation in *Heliconius* butterflies. Proc. R. Soc. Lond. B Biol. Sci. 187:369–378.

BUSH, G. L. 1975. Modes of animal speciation. Annu. Rev. Ecol. Syst. 6:339–364.

CHAPMAN, F. M. 1917. The distribution of bird-life in Colombia. Bull. Am. Mus. Nat. Hist. 36:1–729.

CHAPMAN, F. M. 1921. The distribution of bird life in the Urubamba Valley of Peru. Bull. U.S. Natl. Mus. 117:1–138.

CHAPMAN, F. M. 1923. The distribution of the motmots of the genus *Momotus*. Bull. Am. Mus. Nat. Hist. 48:27–59.

CHAPMAN, F. M. 1926. The distribution of bird-life in Ecuador, a contribution to the study of the origin of Andean bird life. Bull. Am. Mus. Nat. Hist. 55:1–784.

CLIMAP PROJECT MEMBERS. 1976. The surface of the ice-age Earth. Science 191:1131–1137.

CRACRAFT, J. 1973. Continental drift, paleoclimatology, and the evolution and biogeography of birds. J. Zool. (Lond.) 169:455–545.

CRACRAFT, J. 1976. Avian evolution on southern continents: influences of paleogeography and paleoclimatology. Pp. 40–52, *In* H. J. Frith and J. H. Calaby (eds.), Proc. 16th Int. Ornithol. Congr., Australian Academy Science, Canberra.

CRACRAFT, J. 1982. Geographic differentiation, cladistics, and vicariance biogeography: reconstructing the tempo and mode of evolution. Am. Zool. 22:411–424.

CRACRAFT, J. 1983a. Cladistic analysis and vicariance biogeography. Am. Sci. 71:273–281.

CRACRAFT, J. 1983b. Species concepts and speciation analysis. Current Ornithol. 1:159–187.

CROIZAT, L. 1958. Panbiogeography, Vol. 1 (The New World). Published by the author, Caracas.

CROIZAT, L. 1976. Biogeografia analítica y sintética ("Panbiogeografia") de las Américas. Bibl. Acad. Cien. Fís. Mat. Nat., Caracas 15–16:1–890.

DAMUTH, J. E., AND R. W. FAIRBRIDGE. 1970. Equatorial Atlantic deep-sea arkosic sands and ice-age aridity in tropical South America. Geol. Soc. Am. Bull. 81:189–206.

DARLINGTON, P. J. 1957. Zoogeography: the Geographical Distribution of Animals. Wiley, New York.

DELACOUR, J., AND D. AMADON. 1973. Curassows and Related Birds. American Museum Natural History, New York.

DESCIMON, H. (ed.). 1977. Biogéographie et evolution en Amérique tropicale. Publ. Lab. Zool. École Norm. Suppl. 9:1–344.

DORST, J. 1957. Contribution a l'étude écologique des oiseaux du haut Marañón (Pérou septentrional). Oiseau 27:235–269.

DORST, J. 1976. Historical factors influencing the richness and diversity of the South American avifauna. Pp. 17–35, *In* H. J. Frith and J. H. Calaby (eds.), Proc. 16th Int. Ornithol. Congr., Australian Academy Science, Canberra.

DUELLMAN, W. E. (ed.). 1979. The South American herpetofauna: its origin, evolution, and dispersal. Univ. Kansas Mus. Nat. Hist. Monogr. 7:1–485.

DUGAND, A., AND E. EISENMANN. 1983. Rediscovery of, and new data on, *Molothrus armenti* Cabanis. Auk 100:991–992.

EDEN, M. J. 1974. Palaeoclimatic influences and the development of savanna in southern Venezuela. J. Biogeogr. 1:95–109.

EISENMANN, E. 1955. The species of Middle American birds. Trans. Linn. Soc. N.Y. 7:1–128.

EISENMANN, E. 1961. Review of "The birds of Finca 'La Selva,' Costa Rica: A tropical wet forest locality" by P. Slud. Auk 78:283–284.

ENDLER, J. A. 1977. Geographic variation, speciation, and clines. Monogr. Popul. Biol. 10.

ENDLER, J. A. 1982a. Problems in distinguishing historical from ecological factors in biogeography. Am. Zool. 22:441–452.

ENDLER, J. A. 1982b. Pleistocene forest refuges: fact or fancy? Pp. 641–657, *In* G. T. Prance (ed.), Biological Diversification in the Tropics. Columbia University Press, New York.

FITTKAU, E. J. 1969. The fauna of South America. Pp. 624–658, *In* E. J. Fittkau, J. Illies, H. Klinge, G. H. Schwabe, and H. Sioli (eds.), Biogeography and Ecology in South America. Junk, The Hague.

FITTKAU, E. J. 1974. Zur ökologischen Gliederung Amazoniens, I. Die erdgeschichtliche Entwicklung Amazoniens. Amazoniana 5:77–134.

FITZPATRICK, J. W. 1976. Speciation and biogeography of the tyrannid genus *Todirostrum* and related genera (Aves). Bull. Mus. Comp. Zool. 147:435–463.

FITZPATRICK, J. W. 1980. Some aspects of speciation in South American flycatchers. Pp. 1273–1279, *In* R. Nöhring (ed.), Acta XVII Congr. Int. Ornithol., Deutsche Ornithologien-Gesellschaft, Berlin.

FLENLEY, R. J. 1979. The Equatorial Rain Forest. A Geological History. Butterworth, Boston, Massachusetts.

FLINT, R. F. 1971. Glacial and Quaternary Geology. J. Wiley, New York.

FORSHAW, J. M. 1973. Parrots of the World. Doubleday, New York.

FUTUYMA, D. J., AND G. C. MAYER. 1980. Non-allopatric speciation in animals. Syst. Zool. 29:254–271.

GARNER, J. F. 1974. The Origin of Landscapes. Oxford University Press, London.

GARNER, H. F. 1975. Rainforests, deserts and evolution. Anais Acad. Brasil. Cienc. 47(Supl.):127–133.

GATES, W. L. 1976. Modeling the ice-age climate. Science 191:1138–1144.

GRISCOM, L., AND J. C. GREENWAY, JR. 1941. Birds of lower Amazonia. Bull. Mus. Comp. Zool. 88:83–344.

GYLDENSTOLPE, N. 1951. The ornithology of the Rio Purús region in western Brazil. Ark. Zool. 2:1–320.

HAEMIG, P. D. 1978. Aztec emperor Auitzotl and the Great-tailed Grackle. Biotropica 10:11–17.

HAEMIG, P. D. 1979. Secret of the Painted Jay. Biotropica 11:81–87.

HAFFER, J. 1967a. Speciation in Colombian forest birds west of the Andes. Am. Mus. Novit. No. 2294.

HAFFER, J. 1967b. Some allopatric species pairs of birds in northwestern Colombia. Auk 84:343–365.

HAFFER, J. 1967c. Zoogeographical notes on the "nonforest" lowland bird faunas of northwestern South America. Hornero 10:315–333.

HAFFER, J. 1969. Speciation in Amazonian forest birds. Science 165:131–137.

HAFFER, J. 1970. Art-Entstehung bei einigen Waldvögeln Amazoniens. J. Ornithol. 111:285–331.

HAFFER, J. 1974. Avian speciation in tropical South America. With a systematic survey of the toucans (Ramphastidae) and jacamars (Galbulidae). Publ. Nuttall Ornithol. Club No. 14.

HAFFER, J. 1975. Avifauna of northwestern Colombia, South America. Bonn. Zool. Monogr. 7:1–182.

HAFFER, J. 1977. A systematic review of the neotropical ground-cuckoos (Aves, *Neomorphus*). Bonn. Zool. Beitr. 28:48–76.

HAFFER, J. 1978. Distribution of Amazon forest birds. Bonn. Zool. Beitr. 29:38–78.

HAFFER, J. 1981. Aspects of neotropical bird speciation during the Cenozoic. Pp. 371–394, *In* G. Nelson and D. E. Rosen (eds.), Vicariance Biogeography: A Critique. Columbia University Press, New York.

HAFFER, J. 1983. Ergebnisse moderner ornithologischer Forschung im tropischen Amerika. Spixiana, Suppl. 9:117–166.

HAFFER, J. In press a. Biogeography of neotropical birds. *In* T. C. Whitmore (ed.), Biogeography and Quaternary History in Tropical America. Clarendon Press, Oxford.

HAFFER, J. In press b. Quaternary history of tropical America. *In* T. C. Whitmore (ed.), Biogeography and Quaternary History in Tropical America. Clarendon Press, Oxford.

HALL, B. P. 1972. Causal ornithogeography of Africa. Pp. 585–593, *In* K. H. Voous (ed.), Proc. 15th Int. Ornithol. Congr. E. J. Brill, Leiden, Holland.

HARDY, J. W. 1969. A taxonomic revision of the New World jays. Condor 71:360–375.

HARRINGTON, H. J. 1962. Paleogeographic development of South America. Bull. Am. Assoc. Petr. Geol. 46:1773–1814.

HAVERSCHMIDT, F. 1968. Birds of Surinam. Oliver and Boyd, Edinburgh.

HELLMAYR, C. E. 1912. Zoologische Ergebnisse einer Reise in das Mündungsgebiet des Amazonas. II Vögel. Bayer. Akad. Wiss. Abh. Math.-Phys. Kl. 26:1–142.

HOWELL, T. R. 1969. Avian distribution in Central America. Auk 86:293–326.

HUECK, K., AND P. SEIBERT. 1972. Vegetationskarte von Südamerika (1 : 8 Mill.) mit Erläuterungen. Vegetationsmonographien der einzelnen Grossräume, Band IIa. Fischer, Stuttgart, Federal Republic of Germany.

IHERING, H. VON. 1927. The geographic origin of the birds of South America. Ibis, ser. 12:427–442.

JACKSON, J. F. 1978. Differentiation in the genera *Enyalius* and *Strobilurus* (Iguanidae): implications for Pleistocene climatic changes in eastern Brazil. Arq. Zool. (São Paulo) 30:1–79.

JOURNAUX, A. 1975. Recherches géomorphologiques en Amazonie brésilienne. Bull. Cent. Géomorph. Caen 20:1–67.

KEAST, A. 1972. Ecological opportunities and dominant families, as illustrated by the neotropical Tyrannidae (Aves). Evol. Biol. 5:229–277.

KEAST, A. 1980. The evolution of habitat specializations in space and time. Pp. 1025–1030, *In* R. Nöhring (ed.), Acta XVII Congr. Int. Ornithol., Deutsche Ornithologien-Gesellschaft, Berlin.

KEIGWIN, L. D., JR. 1978. Pliocene closing of the isthmus of Panama, based on biostratigraphic evidence from nearby Pacific Ocean and Caribbean Sea cores. Geology 6:630–634.

KEIGWIN, L. 1982. Isotopic paleoceanography of the Caribbean and east Pacific: role of Panama uplift in Late Neogene time. Science 217:350–353.

KIKKAWA, J., AND K. PEARSE. 1969. Geographical distribution of land birds in Australia—a numerical analysis. Aust. J. Zool. 17:821–840.

KINZEY, W. G. 1982. Distribution of primates and forest refuges. Pp. 455–482, *In* G. T. Prance (ed.), Biological Diversification in the Tropics. Columbia University Press, New York.

KLAMMER, G. 1982. Die Paläowüste des Pantanal von Mato Grosso und die pleistozäne Klimageschichte der brasilianischen Randtropen. Z. Geomorph. (N.F.) 26:393–416.

LANYON, W. E. 1978. Revision of the *Myiarchus* flycatchers of South America. Bull. Am. Mus. Nat. Hist. 161:427–628.

LIVINGSTONE, D. A., AND T. VAN DER HAMMEN. 1978. Paleogeography and paleoclimatology. Tropical forest ecosystems. A state-of-knowledge report. UNESCO/UNEP/FAO, Paris.

LÖFFLER, H. 1977. Beobachtungen zur Anatidenfauna der Bale-Berge (Äthiopien). Egretta 20:36–44.

MacARTHUR, R. H. 1969. Patterns of communities in the tropics. Biol. J. Linn. Soc. 1:19–30.

MARSHALL, L. G., S. D. WEBB, J. J. SEPKOSKI, JR., AND D. M. RAUP. 1982. Mammalian evolution and the great American interchange. Science 215:1351–1357.

MAYR, E. 1946. History of the North American bird fauna. Wilson Bull. 58:1–68.

MAYR, E. 1963. Animal Species and Evolution. Harvard University Press, Cambridge, Massachusetts.

MAYR, E. 1964. Inferences concerning the Tertiary American bird faunas. Proc. Nat. Acad. Sci. 51: 280–288.

MAYR, E. 1965. What is a fauna? Zool. Jahrb. Abt. Syst. Oekol. Geogr. Tiere 92:473–486.

MAYR, E. 1983. Introduction. Pp. 1–21, *In* A. H. Brush and G. A. Clark, Jr. (eds.), Perspectives in Ornithology. Cambridge University Press, Cambridge, England.

MAYR, E., AND W. H. PHELPS, JR. 1967. The origin of the bird fauna of the south Venezuelan highlands. Bull. Am. Mus. Nat. Hist. 136:269–328.

MERCER, J. H. 1976. Glacial history of southernmost South America. Quat. Res. (N.Y.) 6:125–166.

MEYER DE SCHAUENSEE, R. 1966. The Species of Birds of South America. Livingston Publ. Co., Narberth, Pennsylvania.

MEYER DE SCHAUENSEE, R. 1970. A Guide to the Birds of South America. Livingston Publ. Co., Narberth, Pennsylvania.

MÜLLER, P. 1973. Dispersal centers of terrestrial vertebrates in the Neotropical Realm. Biogeographica No. 2.

NICÉFORO, M., AND A. OLIVARES. 1965. Adiciones a la avifauna colombiana, II. Bol. Soc. Venez. Cienc. Nat. 26:36–58.

OREN, D. C. 1981. Zoogeographic analysis of the white sand campina avifauna of Amazonia. Unpubl. Ph.D. dissert., Harvard University, Cambridge, Massachusetts.

PARKER, T. A. 1984. Rediscovery of the Rufous-fronted Antthrush (*Formicarius rufifrons*) in southeastern Peru. Gerfaut 73:287–289 ("1983").

PARMENTER, C., AND D. W. FOLGER. 1974. Eolian biogenic detritus in deep sea sediments: a possible index of equatorial ice age aridity. Science 185:695–698.

PATTERSON, B., AND R. PASCUAL. 1972. The fossil mammal fauna of South America. Pp. 247–310, *In* A. Keast, F. C. Erk, and B. Glass (eds.), Evolution, Mammals and Southern Continents. State University New York Press, Albany, New York.

PAYNTER, R. A., JR. 1972. Biology and evolution of the *Atlapetes schistaceus* species-group (Aves: Emberizinae). Bull. Mus. Comp. Zool. 143:297–320.

PAYNTER, R. A., JR. 1978. Biology and evolution of the avian genus *Atlapetes* (Emberizinae). Bull. Mus. Comp. Zool. 148:323–369.

PETERSON, G. M., T. WEBB III, J. E. KUTZBACH, T. VAN DER HAMMEN, T. A. WIJMSTRA, AND F. A. STREET. 1979. The continental record of environmental conditions at 18,000 yr B.P.: an initial evaluation. Quat. Res. (N.Y.) 12:47–82.

PINTO, O. M., AND E. A. DE CAMARGO. 1961. Resultados Ornitologicos de quatro recentes expediçãos do Departamento de Zoologia ao nordeste do Brasil, com a descripção de seis novas subspecies. Arq. Zool. (São Paulo) 11:193–284.

PLATNICK, N. I., AND G. NELSON. 1978. A method of analysis for historical biogeography. Syst. Zool. 27:1–16.

PRANCE, G. T. 1973. Phytogeographic support for the theory of Pleistocene forest refuges in the Amazon basin, based on evidence from distribution patterns in Caryocaraceae, Chrysobalanaceae, Dichapetalaceae and Lecythidaceae. Acta Amazônica 3:5–28.

PRANCE, G. T. (ed.). 1982. Biological Diversification in the Tropics. Columbia University Press, New York.

RAPOPORT, E. H. 1968. Algunos problemas biogeográficos del Nuevo Mundo con especial referencia a la región neotropical. Pp. 53–110, *In* C. Delamare-Deboutteville and E. H. Rapoport (eds.), Biologie de l'Amérique Australe. Centre Nacional Recherche Scientifique (CNRS), Paris.

REICHHOLF, J. 1975. Biogeographie und Ökologie der Wasservögel im subtropisch-tropischen Südamerika. Anz. Ornithol. Ges. Bayern 14:1–69.

REMINGTON, C. L. 1968. Suture-zones of hybrid interaction between recently joined biotas. Evol. Biol. 2:321–428.

SCHEUERMANN, R. G. 1977. Hallazgos del Paujil *Crax mitu* (Aves: Cracidae) al norte del Rio Amazonas y notas sobre su distribución. Lozania 22:1–8.

SCHWABE, G. H. 1969. Towards an ecological characterisation of the South American continent. Pp. 113–136, *In* E. J. Fittkau, J. Illies, H. Klinge, G. H. Schwabe, and H. Sioli (eds.), Biogeography and Ecology in South America. Junk, The Hague.

SELANDER, R. K. 1964. Speciation in wrens of the genus *Campylorhynchus*. Univ. Calif. Publ. Zool. 74: 1–259.

SHORT, L. L. 1975. A zoogeographic analysis of the South American chaco avifauna. Bull. Am. Mus. Nat. Hist. 154:163–352.

SHORT, L. L. 1982. The woodpeckers of the world. Delaware Mus. Nat. Hist. Monogr. 4.

SIBLEY, C. G., AND J. AHLQUIST. 1984. The relationships of the Sharpbill (*Oxyruncus cristatus*). Condor 86:48–52.

SICK, H. 1965. A fauna do cerrado. Arq. Zool. (São Paulo) 12:71–93.

SICK, H. 1966. As aves do cerrado como fauna arborícola. Anais Acad. Bras. Cienc. 38:355–363.

SICK, H. 1979. Découverte de la patrie de l'Ara de Lear *Anodorhynchus leari*. Alauda 47:59–60.

SICK, H. 1980. Characteristics of the Razor-billed Curassow (*Mitu mitu mitu*). Condor 82:227–228.

SICK, H., AND D. M. TEIXEIRA. 1980. Discovery of the home of the Indigo Macaw in Brazil. Am. Birds 34:118–119, 212.

SIMBERLOFF, D. 1983. Biogeography: the unification and maturation of a science. Pp. 411–455, *In* A. H. Brush and G. A. Clark, Jr. (eds.), Perspectives in Ornithology. Cambridge University Press, Cambridge, England.

SIMPSON, B. B. 1979. Quaternary biogeography of the high montane regions of South America. Pp. 157–188, *In* W. E. Duellman (ed.), The South American Herpetofauna: Its Origin, Evolution, and Dispersal. Univ. Kansas Mus. Nat. Hist. Monogr. 7.

SIMPSON, B. B., AND J. HAFFER. 1978. Speciation patterns in the Amazonian forest biota. Annu. Rev. Ecol. Syst. 9:497–518.

SIMPSON, G. G. 1969. South American mammals. Pp. 879–909, *In* E. J. Fittkau, J. Illies, H. Klinge, G. H. Schwabe, and H. Sioli (eds.), Biogeography and Ecology in South America. Junk, The Hague.

SLUD, P. 1960. The birds of Finca 'La Selva,' Costa Rica: a tropical wet forest locality. Bull. Am. Mus. Nat. Hist. 121:49–148.

SLUD, P. 1976. Geographic and climatic relationships of avifaunas with special reference to comparative distribution in the neotropics. Smithson. Contr. Zool. No. 212.

SNETHLAGE, E. 1913. Über die Verbreitung der Vogelarten in Unteramazonien. J. Ornithol. 61:469–539.

SNOW, D. 1982. The Cotingas. British Museum (Natural History), Cornell University Press, Ithaca, New York.

SOLBRIG, O. T. 1976. The origin and floristic affinities of the South American temperate desert and semidesert regions. Pp. 7–49, *In* D. W. Goodall (ed.), Evolution of Desert Biota. University Texas Press, Austin, Texas.

STILES, F. G. 1984. The Nicaraguan Seed-Finch (*Oryzoborus nuttingi*) in Costa Rica. Condor 86:118–122.

TERBORGH, J., AND J. S. WESKE. 1972. Rediscovery of the Imperial Snipe in Peru. Auk 89:497–505.

TRAYLOR, M. A., JR., AND J. W. FITZPATRICK. 1982. A survey of the tyrant flycatchers. Living Bird 19: 7–45.

TRICART, J. 1974. Existence de périodes séches au Quaternaire en Amazonie et dans les régions voisines. Rev. Géomorph. Dynamique 4:145–158.

UDVARDY, M. D. F. 1975. A classification of the biogeographical provinces of the world. Int. Union Conserv. Nat. Resour., Occ. Pap. 18:1–48.

UDVARDY, M. D. F. In press. Biogeographical classification system for terrestrial environments. Proc. 3rd World National Parks Congress (1983).

VAN DER HAMMEN, T. 1974. The Pleistocene changes of vegetation and climate in tropical South America. J. Biogeogr. 1:3–26.

VAN DER HAMMEN, T. 1983. The paleoecology and paleogeography of savannas. *In* F. Bourlière (ed.), Tropical Savannas. Ecosystems of the World, 13. Elsevier, The Hague.

VANZOLINI, P. E. 1970. Zoologia sistematica, geografia e a origem das espécies. Inst. Georg. São Paulo, Ser. Téses Monogr. 3:1–56.

VANZOLINI, P. E. 1981. A quasi-historical approach to the natural history of the differentiation of reptiles in tropical geographic isolates. Pap. Avulsos Zool. (São Paulo) 34:189–204.

VANZOLINI, P. E., AND E. E. WILLIAMS. 1970. South American anoles: The geographic differentiation and evolution of the *Anolis chrysolepis* species group (Sauria, Iguanidae). Arq. Zool. (São Paulo) 19:1–298.

VANZOLINI, P. E., AND E. E. WILLIAMS. 1981. The vanishing refuge: A mechanism for ecogeographic speciation. Pap. Avulsos Zool. (São Paulo) 34:251–255.

VAURIE, C. 1967. Systematic notes on the bird family Cracidae. No. 10. The genera *Mitu* and *Pauxi* and the generic relationships of the Cracini. Am. Mus. Novit. No. 2307.

VAURIE, C. 1968. Taxonomy of the Cracidae (Aves). Bull. Am. Mus. Nat. Hist. 138:131–260.

VAURIE, C. 1980. Taxonomy and geographical distribution of the Furnariidae (Aves, Passeriformes). Bull. Am. Mus. Nat. Hist. 166:1–357.

VUILLEUMIER, F. 1965. Relationships and evolution within the Cracidae (Aves, Galliformes). Bull. Mus. Comp. Zool. 134:1–27.

VUILLEUMIER, F. 1967. Phyletic evolution in modern birds of the Patagonian forests. Nature 215:247–248.

VUILLEUMIER, F. 1975. Zoogeography. Pp. 421–496, In D. S. Farner, J. R. King, and K. C. Parkes (eds.), Avian Biology, Vol. 5. Academic Press, New York.

VUILLEUMIER, F. 1981. Ecological aspects of speciation in birds, with special reference to South American birds. Pp. 101–148, In O. Reig (ed.), Ecología y genética de la especiación animal. Universidad Simón Bolivar, Caracas.

VUILLEUMIER, F. In press. The development of avifaunas in South America: Estimates of faunal turnover from the fossil record. Proc. 18th Int. Ornithol. Congr.

WALLACE, A. R. 1876. The Geographical Distribution of Animals, 2 vols. Macmillan, London.

WALLACE, A. R. 1972. A Narrative of Travels on the Amazon and Rio Negro. Dover Reprint Co., New York (reprint of 1889 edition).

WEBB, S. D. 1978. A history of savanna vertebrates in the New World. Part II: South America and the great interchange. Annu. Rev. Ecol. Syst. 9:393–426.

WHITMORE, T. C. (ed.). In press. Biogeography and Quaternary History in Tropical America. Clarendon Press, Oxford, England.

WIENS, J. A. 1983. Avian community ecology: an iconoclastic view. Pp. 355–403, In A. H. Brush and G. A. Clark, Jr. (eds.), Perspectives in Ornithology. Cambridge University Press, Cambridge, England.

WIJMSTRA, T. A. 1971. The Palynology of the Guiana Coastal Basin. Akademisch Proefschrift, University Amsterdam, Holland.

WIJMSTRA, T. A., AND T. VAN DER HAMMEN. 1966. Palynological data on the history of tropical savannas in northern South America. Leidse Geol. Meded. 38:71–90.

WILEY, E. O. 1981. Phylogenetics. The Theory and Practice of Phylogenetic Systematics. J. Wiley and Sons, New York.

WILLIAMS, M. D. 1980. First description of the eggs of the White-winged Guan, *Penelope albipennis*, with notes on its nest. Auk 97:889–892.

WILLIS, E. O. 1968. Studies of the behavior of Lunulated and Salvin's Antbirds. Condor 70:128–148.

WILLIS, E. O. 1969. On the behavior of five species of *Rhegmatorhina*, ant-following antbirds of the Amazon basin. Wilson Bull. 81:363–395.

APPENDIX I

Selected Species Characteristic of Several Areas of Endemism in Southcentral and Southern South America[1]

Southeastern Brazil: *Orchesticus abeillei, Carpornis cucullatus + melanocephalus, Merulaxis ater + stresemanni, Cichlocolaptes leucophrys, Triclaria malachitacea, Macropsalis creagra, Ramphodon naevius, Ilicura militaris, Orthogonys chloricterus, Melanotrochilus fuscus, Pyrrhocoma ruficeps, Leucochloris albicollis, Leptotriccus sylviolus, Batara cinerea, Baillonius bailloni, Heliobletus contaminatus, Anabazenops fuscus, Hypoedalus guttatus,* and *Phibalura flavirostris.* The species of *Batara* and *Phibalura* form isolated populations along the base of the Bolivian and Argentinian Andes (Fig. 5).

Nothofagus forest of southern Chile and western Argentina: *Buteo ventralis, Pygarrhichas albogularis, Sephanoides sephaniodes, Sylviorthorhynchus desmursii, Aphrastura spinicauda, Enicognathus ferrugineus, E. leptorhynchus, Eugralla paradoxa, Coloramphus parvirostris, Pteroptochus castaneus, P. tarnii,* and *Scelorchilus rubecula.*

Caatinga Area (Fig. 7C): *Cyanopsitta spixi, Megaxenops parnaguae, Aratinga cactorum, Gyalophylax hellmayri, Sakesphorus cristatus, Cranioleuca semicinerea, Formicivora iheringi, Herpsilochmus pectoralis, Sericossypha loricata,* and *Paroaria dominicana.* Three of these endemic species have been placed in monotypic genera (*Cyanopsitta, Megaxenops, Gyalophylax*; the latter is very close to *Synallaxis*).

Campos Cerrados Area (Fig. 7B): *Uropelia campestris, Geobates poecilopterus, Antilophia galeata, Gubernetes yetapa, Alecturus tricolor, Culicivora caudacuta, Cypsnagra hirundinacea, Neothraupis fasciata, Porphyrospiza caerulescens,* and *Charitospiza eucosma.* These species are isolated taxonomically and are placed in monotypic genera.

Chaco Area (Fig. 7A): *Chunga burmeisteri, Nandayus nenday, Drymornis bridgesi, Coryphistera alaudina, Rhinocrypta lanceolata, Entotriccus striaticeps,* and *Saltatricula multicolor.* Except for *Nandayus,* which is close to *Aratinga,* these species are isolated taxonomically and are placed in monotypic genera. Additional endemic species of the Chaco Area include *Nothura chacoensis, Nothoprocta cinerascens, Eudromia elegans, Celeus lugubris, Dryocopus schulzi, Phloeoceastes leucopogon, Xiphocolaptes major, Upucerthia certhioides, Furnarius cristatus, Asthenes baeri, Phacellodomus sibilatrix, P. striaticollis, Pseudoseisura lophotes, Melanopareia maximiliani, Mimus triurus, Anthus chacoensis, Paroaria capitata, Lophospingus pusillus, Aimophila strigiceps, Poospiza ornata,* and *P. melanoleuca.*

Pampas Area (Fig. 9): *Anumbius anumbi, Spartanoica maluroides, Limnoctites rectirostris, L. curvirostris, Yetapa risoria, Amblyramphus holosericeus, Pezites defilippi, Xanthopsar flavus, Gubernatrix cristata, Donacospiza albifrons,* and *Embernagra platensis.* These forms are isolated taxonomically and, except for the species of *Pezites* and *Embernagra* are placed in monotypic genera.

Patagonia Area: *Pterocnemia pennata* (an isolated population inhabits the Andean puna plains), *Tinamotis ingoufi, Vultur gryphus* (north into the high Andes), *Geositta antarctica, Cinclodes fuscus* (north into the high Andes), *Eremobius phoenicurus, Cinclodes oustaleti, Asthenes pyrrholeuca, A. patagonica, Agriornis montana* (north into the high Andes), *Neoxolmis rufiventris, Xolmis coronata, X. murina, X. rubetra, Mimus patagonicus, Curaeus curaeus, Sicalis lebruni, Melanodera melanodera, M. xanthogramma,* and *Spinus barbatus.*

---

[1] Ranges of some of these species were used in the construction of contour maps (Figs. 5, 7, 9).

GEOGRAPHIC VARIATION IN SOME AMAZONIAN FOREST BIRDS

JÜRGEN HAFFER[1] AND JOHN W. FITZPATRICK[2]

[1]Tommesweg 60, 4300 Essen - 1, Federal Republic of Germany, and
[2]Field Museum of Natural History, Chicago, Illinois 60605 USA

ABSTRACT. Geographic variation in several common species of Amazon birds is studied through computer analyses and hand-contouring of mensural and color characters. Contour maps of individual characters illustrate regional gradients in character variation within Amazonia. These patterns permit meaningful application of the subspecies concept. Broad rivers are barriers to dispersal for some forest interior species, introducing steps or breaks in regional gradients of geographically variable characters. The extent of these breaks varies considerably between species, and they disappear entirely in the headwater regions of the Amazon tributaries. The strong-flying, canopy and forest-edge species show gentle clinal variation in Amazonia, with gradients not broken by major rivers. *Pitangus sulphuratus* shows symmetrical, parallel clines of increasing body size both north and south of the Amazon basin. In *Myrmoborus myotherinus*, character variation is pronounced, though quite different between males and females in central Amazonia. Broad regions of uniform character expression, connected by narrow transition zones, correspond to areas of species endemism, whose origins may lie in the historical climatic and vegetational fluctuations in the Amazon basin.

RESUMEN. Se estudió la variación geográfica de varias especies de aves amazónicas comunes por medio de análisis con computadora y perfiles hechos a mano de características de colores y medidas. Mapas de perfiles "contour" de variaciones para cada característica individual presentan gradientes regionales para las variaciones de carácteres ("character variation") en la región amazónica. Estos modelos permiten aplicaciones significativas del concepto de subespecie. Los ríos anchos representan barreras de dispersión para algunas especies que viven en pleno bosque, introduciendo pasos o brechas en los gradientes regionales de los carácteres de variación geográfica. La extensión de estas brechas varía considerablemente entre especies y desaparecen completamente en la zona de las nacientes de los ríos tributarios del Amazonas. Especies que tienen mucha resistencia volando, viven en el nivel tope del bosque y en los bordes de los bosques presentan una suave varición clinal en la región amazónica con gradientes que no son interrumpidos por grandes ríos. *Pitangus sulphuratus* presenta variaciones a lo largo de su rango de distribución ("clines"), aumentando las medidas de su cuerpo en forma simétrica y paralela tanto al norte como al sur de la cuenca amazónica. En *Myrmoborus myotherinus*, la variación de carácteres es pronunciada aunque levemente diferente entre machos y hembras en la región amazónica central. Amplias regiones con expresión de carácteres uniforme, conectadas por estrechas franjas de transición, corresponden a áreas de endemismo de especies cuyos orígenes pueden estar relacionado a fluctuaciones históricas de clima y vegetación en la cuenca amazónica.

Studies of geographic variation in neotropical birds traditionally have emphasized taxonomic aspects of population differences in measurements and plumage color, usually through the description of subspecies. This procedure does not facilitate the recognition and comparative analysis of regional patterns of variation. In contrast, quantitative comparisons of geographical variation using computers have increased considerably during recent years (Adams 1970; Selander 1971; Gould and Johnston 1972; Thorpe 1976; Pielou 1979; Johnson 1980). This technique is especially suited to populations distributed more or less continuously over large continental areas. With sufficient specimen material available over a large area, computers can analyze regional trends, statistically, and can generate isolines, contour maps, and trend-surface maps. In this way, regional patterns of character variation may be documented and analyzed quantitatively, without *a priori* reference to subspecies names.

Detailed descriptions of local populations and artificial delimitation of subspecies cannot depict accurately the complex patterns of geographic variation in many wide-ranging species.

In particular, the classic techniques do not effectively reflect cases of varying directions in different character gradients (clines), regionally varying amounts of character change, or different locations of steps in different clines.

Many previous authors have recommended the use of quantitative mapping techniques for the study of intraspecific variation in animals (Huxley 1942:226; Terentjew 1957; Mayr 1963; Beregovoy and Danilow 1967; Selander 1971). Only a few ornithologists, however, have published modern analyses, mostly on birds of the north temperate zone (Barth 1966, 1968; Johnston 1969; Barlow and Power 1970; James 1970; Johnston and Selander 1971; Mengel and Jackson 1977; Johnson 1980). For tropical birds, Crowe (1978) discussed variation in African guineafowl, Haffer (1974) analyzed hybridization in several South American toucans, and Handford (1983) depicted continental patterns of morphological variation in a South American sparrow (*Zonotrichia capensis*). Additional museum and field studies of intraspecific variation in birds are needed in order to document regional character changes along clines, across contact zones, and across various ecological gradients within the tropics. The results may contribute to a critical evaluation of the claim by some authors (e.g., Endler 1977) that, in contrast to the classical model of an allopatric origin of species (Mayr 1963), speciation frequently can occur parapatrically (species evolving as contiguous populations along a continuous cline).

In this paper we present introductory analyses of five species of widespread Amazonian birds and several closely related forms. Our primary goals are (1) to show how regional trends in variation can be depicted, both visually and quantitatively, within Amazonia as a whole, and (2) to point out examples of a few species for which traditional taxonomic and nomenclatural practices have obscured these trends. To add a dimension of ecological interest, we chose five species that roughly span the continuum between highly mobile, widely dispersing species (e.g., Blue-headed Parrot, *Pionus menstruus*), and sedentary, forest interior species that are incapable of much long distance dispersal, especially across large rivers (e.g., Black-faced Antbird, *Myrmoborus myotherinus*). As predicted from population genetics theory, the more sedentary forms show greater tendency toward local differentiation. Occasionally, this tendency produces some peculiar and confusing patterns.

Unfortunately, the quantity of specimen material available for this study still is scanty relative to the geographic area covered. Indeed, some remote parts of Amazonia have not yet been sampled at all. Despite these problems, general trends of geographic character variation are apparent and measurable in many species. Furthermore, despite a somewhat sparse data base, we introduce quantitative mapping techniques and quantitative analysis of regional patterns to the study of intraspecific variation in the neotropics even though the results are only preliminary.

METHODS

We assembled at the Field Museum of Natural History, Chicago, a total of nearly 1700 specimens of five bird species (*Aratinga leucophthalmus, Pionus menstruus, Myrmoborus myotherinus, Formicarius analis, Pitangus sulphuratus*) that are fairly common and widespread in Amazonia, and whose dispersal habits differ markedly. This material represents the major portion of the holdings of these birds in several large museums in the United States (see Acknowledgments). Using standard procedures, we measured the length of the wing (chord) and culmen (from the base and from the cere in parrots) in all five species, tail length in four species and tarsus length in three species. For a semiquantitative analysis of regionally varying plumage color in three species (*Aratinga leucophthalmus, Myrmoborus myotherinus, Formicarius analis*), we established for each portion of the plumage to be compared a scoring system on the basis of five to ten reference specimens, subdividing the total range of observed color variation into visually equivalent steps or stages (Appendix I). All specimens were scored for color by comparing their respective plumages to the reference series. Nomenclature and numbers of colors follow Smithe (1975). We did not apply quantitative methods of plumage color analysis (e.g., spectrophotometry) because of the varying ages and curatorial histories of the specimens used. We firmly believe that fresh and uniformly prepared specimens are required for accurate spectrophotometric work. We attempted to locate five to 10 specimens of each sex per sample locality for measurements and color comparison, but many localities are represented by fewer than five specimens. Males and females are treated separately in most analyses.

Automatic contouring of average values for population measurements and scored values for each locality were carried out for several characters using SURFACE II, a computer

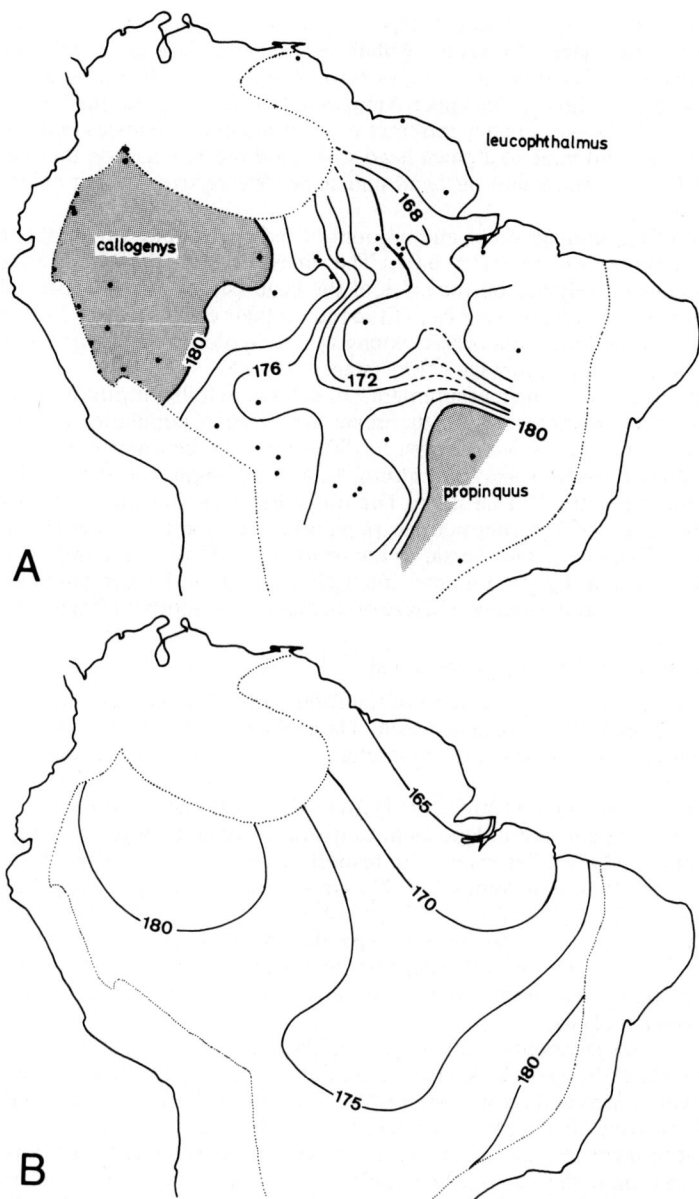

FIG. 3. Geographic variation in average wing length (mm) among male White-eyed Conures (*Aratinga leucophthalmus*). A = machine-contoured map; B = third order trend-surface map. Dots indicate sample locations. Dotted line indicates the approximate range of the species.

America from dusky "yellow green" to slightly bluish "parrot green." We distinguished five stages in this color series (Appendix I).

The most intensely colored populations ($\bar{X}$'s = 4.0–5.0) occur in upper Amazonia near the base of the Andes, in southeastern Colombia, eastern Ecuador, and eastern Peru. The populations with relatively small average measurements (wing size and culmen length) in the Guianas, near the mouth of the Amazon, and in central Brazil are paler green ($\bar{X}$'s = 1.0–2.8). A wide area in central Amazonia, eastern Peru, and eastern Bolivia is inhabited by populations with an intermediate green plumage color ($\bar{X}$'s = 3.0–4.0). Size (wing length) and color intensity are positively correlated ($\delta\delta$, $r = 0.5$, $P < 0.001$; $\circ\circ$, $r = 0.4$, $P < 0.01$), especially in westernmost and eastern Amazonia (inhabited by large/dark green and small/pale green populations, respectively). Birds from west-central Amazonia (Tefé, Purús, and lower Madeira Rivers) are as large as those from upper Amazonia but have an intermediate plumage color.

Red patches on head.—Character extremes in adult males and females range from an almost entirely green head and nape to a green head with solid red blotches on the cheeks and many scattered red feathers elsewhere on head and nape. We recognized ten color stages in this series (Appendix I).

Intensity of green plumage color and amount of red cheek color are weakly correlated ($\delta\delta$, $r = 0.34$, $P < 0.05$; $\circ\circ$, $r = 0.46$, $P < 0.01$). The cheeks of many dark green adults from upper Amazonia are extensively red on the sides of the head (stages 5–10). Most individuals with intermediate or pale green plumage have little red on their cheeks (stages 1–3), but exceptions are not rare in central Amazonia, where extensively red-colored cheeks are sometimes present in birds with intermediate plumage colors (stages 2.5–3.5).

Taxonomy.—Based on wing measurements of males, and following traditional arrangements (Gyldenstolpe 1951; Forshaw 1973), the upper Amazonian populations, consisting of large birds (*A. l. callogenys*, $\delta\delta$, $\bar{X}$ wing length > 180 mm), may be distinguished from the lower Amazonian populations (*A. l. leucophthalmus*, $\delta\delta$, $\bar{X}$ wing length < 170 mm), the latter ranging north to the mouth of the Río Orinoco. The transition zone in central Amazonia (widening southward) is inhabited by intermediate populations that may be designated *A. l. leucophthalmus* ⪎ *callogenys* or vice versa, as appropriate. Only the westernmost populations of the subspecies *callogenys* possess a more intensely colored, dark green plumage. Our data are insufficient to document the increase in size of this species in southern Brazil (*A. l. propinquus*).

BLUE-HEADED PARROT (*PIONUS MENSTRUUS*)

This species is a common inhabitant of the forested neotropical lowlands from Costa Rica to southeastern Brazil and is frequently seen in large flocks. Individuals and flocks range widely over the forest canopy, flying many kilometers and regularly crossing even the very largest rivers.

We studied 169 specimens (120 $\delta\delta$, 49 $\circ\circ$) from 50 Amazonian localities. No color variation is sufficiently evident and age-independent to warrant a detailed regional analysis of plumage pattern. We obtained a similar result with respect to geographic variation in size. Males from the foothills of the Peruvian Andes (n = 21) are significantly longer-winged than those from lower Amazon localities of Brazil (n = 24; $t = 4.6$, $P < 0.001$). This difference breaks down, however, when the areas compared are expanded. Wing length is rather variable in most places across Amazonia ($\bar{X}$'s = 180–190 mm), but is especially great, and possibly less variable, along the immediate base of the Andes in Peru. The number of female specimens was insufficient for formal analysis.

Taxonomy.—The geographically isolated populations of this parrot in the lowlands west of the Andes and in southeastern Brazil (which we did not study) have been designated as separate subspecies (*rubrigularis* and *reichenowi*, respectively) on the basis of certain details of plumage coloration. However, all populations inhabiting the huge area of Amazonia and Venezuela are combined to form the nominate race *P. m. menstruus*. This reflects a virtual absence of geographic variation in the cis-Andean populations, as confirmed by our own study. Evidently, the pronounced mobility of this bird leads to effective gene flow across large areas, resulting in uniform character expression in the populations across Amazonia.

BLACK-FACED ANTBIRD (*MYRMOBORUS MYOTHERINUS*)

This small species is a member of a group of four rather similarly colored Amazonian antbirds. This highly sedentary species inhabits the interior of mature rainforest, where pairs permanently occupy contiguous territories a few hundred meters across (Fitzpatrick, pers. observ.). In most areas of the species' range the male is medium gray above and light gray below, with a black throat and sides of the head. In central Amazonia males are dark gray below and blackish above. Females are generally cinnamon-buff below, this color becoming pale creamy white in a small area of central Amazonia. Throat color ranges from pure white to dark cinnamon-buff, bordered below by a necklace of black spots. The upperparts are olivaceous brown to umber brown.

The Black-faced Antbird is common in most of Amazonia, but it is inexplicably missing

TABLE 1
MEASUREMENTS AND POPULATIONAL MEANS OF FOUR SUBSPECIES OF *MYRMOBORUS MYOTHERINUS*

	elegans		*myotherinus*		*ardesiacus*		*ochrolaema*	
	♂ (48, 25)[1]	♀ (70, 19)	♂ (160, 30)	♀ (85, 29)	♂ (38, 8)	♀ (45, 6)	♂ (79, 17)	♀ (48, 13)
Wing chord[2]								
Mean	62.9	61.7	63.0	61.4	61.6	61.3	60.8	59.2
Range	59.0–67.0	58.0–66.0	58.5–69.0	56.5–65.0	58.0–64.0	56.0–63.5	56.0–66.5	55.0–62.5
s.d.	1.6	0.9	1.7	1.2	1.3	1.5	1.3	1.5
Tail length								
Mean	40.1	39.1	41.1	39.7	38.9	38.1	40.3	38.8
Range	36.5–44.5	36.4–43.9	37.3–45.9	36.0–45.8	34.2–43.9	33.5–40.9	36.8–44.0	36.5–43.5
s.d.	1.4	0.8	1.4	1.1	2.0	1.9	1.2	1.4
Culmen length[3]								
Mean	18.4	18.0	18.1	17.9	18.6	18.2	18.1	17.4
Range	17.0–19.8	16.6–19.6	16.9–20.0	16.1–19.4	17.1–19.3	16.8–19.0	15.9–19.4	15.9–19.1
s.d.	0.5	0.5	0.6	0.4	0.2	0.5	0.6	0.6
Tarsus length								
Mean	25.8	25.7	26.1	25.7	24.9	25.0	24.7	24.4
Range	24.4–27.0	24.2–27.4	24.3–27.8	24.1–27.2	23.1–26.1	23.5–26.1	23.2–26.4	22.5–25.6
s.d.	0.7	0.7	0.8	0.7	0.8	0.7	0.8	0.7

[1] Sample sizes in parentheses: (total specimens, total localities).
[2] Mean values represent averages of all locality means within each recognized subspecies; ranges show smallest and largest values among individual specimens; standard deviations (s.d.) show inter-locality variation in mean values.
[3] Culmens measured from base.

from the Guianan region. The lower Rio Negro and the Rio Amazon sharply delimit its range in the northeast.

We studied a total of 581 specimens of *M. myotherinus* (325 ♂♂, 256 ♀♀) from the entire Amazonian breeding range (106 localities). Overall variation in measurements is inconspicuous (Table 1), but plumage color varies appreciably, especially the color of the breast and belly of males and the color of the throat and breast of females. Geographical variation in this respect is considerably stronger in females than in males (an example of heterogynism, Hellmayr 1929). In our reference series for each plumage character, 10 roughly equal stages of color change were represented.

Color of breast and belly in males.—Broadly speaking, male breast and belly colors vary clinally from northern Peru eastward across southern Amazonia to the lowlands south of the mouth of the Amazon. However, the cline is interrupted abruptly in a localized area of central Amazonia (Fig. 4).

Breast and belly color averages between light and medium neutral gray ($\bar{X}$'s = 5.0–6.5) in populations from northwestern Amazonia eastward to southern Venezuela. In southwestern Amazonia, the gray breast and belly are slightly paler ($\bar{X}$ = 3.5–4.5), only occasionally deepening to stage 5 in some populations along the humid base of the Peruvian and Bolivian Andes. Very pale gray underparts ($\bar{X}$'s = 2.0–3.0) characterize the populations from southeastern Amazonia south to central Brazil. Individual variation in southeastern Amazonia is pronounced; both pale and dark gray birds (stages 1 through 5) occur in this region, where the coefficient of variation ($C.V.$) for this character ranges at different localities from 24 to 55. Individual variation within localities in upper Amazonia is less pronounced ($C.V.$ = 5–22).

Males from the areas between the lower Rio Negro and the Rio Solimões in central Amazonia, as well as between the mouth of the Rio Purús and that of the Rio Madeira are conspicuously different from those of all other populations, having extremely dark gray breasts and bellies. Transitions to the west and south apparently are abrupt, but no specimens from intermediate localities are available to estimate the width of these zones precisely. The width of the western transition zone on the south bank of the Rio Solimões also needs verification. The general location of this zone is indicated by a specimen from the west bank of the lower

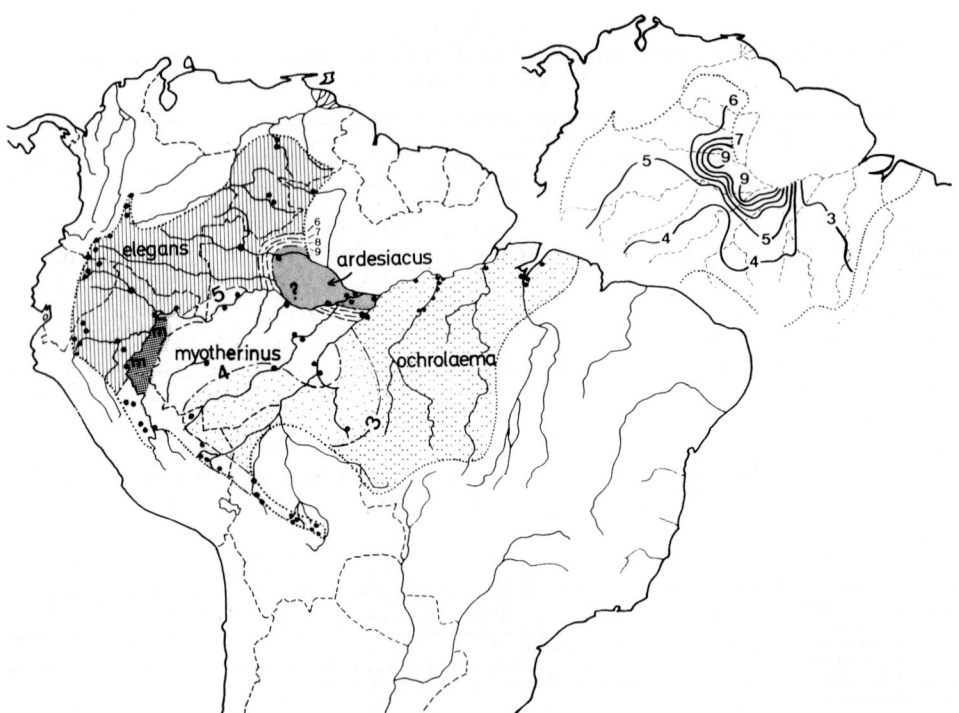

Fig. 4.   Geographic variation in breast and belly color among male Black-faced Antbirds (*Myrmoborus myotherinus*). Left: hand-contoured map; numbers represent isoclines for color-score values (see text for explanation of scores). Dots indicate sample locations, and dotted line illustrates the approximate range of the species; m = approximate range of *Myrmoborus melanurus*. Right: portion of a computer-generated contour map of the same data. Compare with Figure 5. Dashed lines show major rivers. Areas of different shading or stippling represent ranges of different subspecies or intermediate populations.

Rio Madeira (Santo Antonio de Guajará) which we assigned a 7.5 on our scale. By contrast, males from Borba on the east bank of this river average 2.3. Thus, the lower Rio Madeira separates two wide-ranging populations with conspicuously different plumage colors. This "river effect" disappears southward, where the influence of the darker western populations extends east and southeast across the Rio Madeira. No comparable effect is seen in the case of the other southern tributaries of the lower Amazon (Rios Tapajós, Xingú, Tocantins), whose eastern and western banks are inhabited by phenotypically similar or identical male populations.

   Breast color in females.—This color varies geographically (Fig. 5) from deep orange-buff or dark cinnamon-buff in birds from large portions of Amazonia to pale cinnamon-buff or creamy-white in some specimens from other regions. Ten color stages are recognized (Appendix I).

   Breast color averages deep cinnamon-buff to orange-buff ($\bar{X}$'s = 8.0–9.5) in northwestern Amazonia and in southeastern Amazonia ($\bar{X}$'s = 7.0–8.0, occasionally 9). In the intervening regions of southwestern Amazonia (E. Peru and W. Brazil), the breast is pale buff to pale cinnamon-buff ($\bar{X}$'s = 3.0–5.0), and creamy-white ($\bar{X}$'s = 1.0–2.0) in a relatively small region north of the Rio Solimões. This huge river has prevented the development of a transition zone between the pale population north of the river and the more intensely colored populations south of it. Specimens from São Paulo de Olivença on the south bank are similar (stage 4) to those from the Purús River. On the other hand, a fairly wide transition zone apparently occurs between the white-bellied population near the north bank of the Solimões River and the cinnamon-buff bellied population farther to the north. At three sample localities in this region, the breast color is intermediate ($\bar{X}$'s = 3.0–3.6) and highly variable (e.g., stages 1–5, average 3.4, in 22 specimens from Mancapurú on the north bank of the Rio Solimões, near the mouth

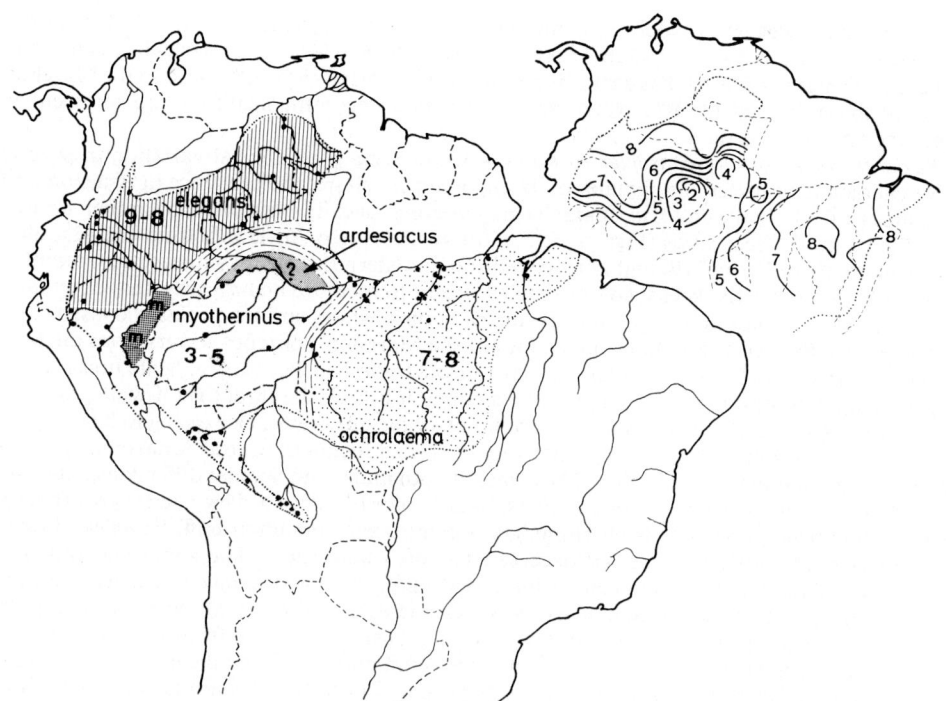

FIG. 5.   Geographic variation in breast color among female Black-faced Antbirds (*Myrmoborus myotherinus*). Left: hand-contoured map; numbers represent color-score values (see text for explanation of scores). Semicircular contour lines in north-central Amazonia refer to score values (from north to south of 8, 6, 4, 3, 2, and 1, respectively); contours in the narrow transition zone south of the Amazon represent score values 5 (west), 6, and 7 (east). Dots indicate sample locations, and dotted line illustrates the approximate range of the species; m = approximate range of *Myrmoborus melanurus*. Right: portion of a computer-generated contour map of the same data as at left. Dashed lines show major rivers.

of the Río Negro). South of the Amazon, the transition zone from the paler western to the more deeply colored eastern populations takes place between the Purús and Madeira Rivers. Based on the data available at present, no "river effect" is noticed in this region, which also applies to the wide lower portions of the southern tributaries of the Amazon River. Both banks of the Tapajós, Xingú, and Tocantins Rivers are inhabited by similarly colored females.

Throat color in females.—Throat color varies geographically from pure white to pale and dark cinnamon-buff, this latter color increasingly invading the white throat from the breast. Ten color stages are recognized (Appendix I).

The throat is pure or almost pure white ($\bar{X}$'s = 1.0–1.5) throughout northwestern Amazonia from the base of the Andes to southern Venezuela and to the mouth of the Río Negro. It is creamy off-white ($\bar{X}$ = 2.0) in southwestern Amazonia, and intermediate in color (pale to medium cinnamon-buff) in the Rio Purús-Madeira region of central Amazonia. East of the Madeira, the throat is uniformly dark connamon-buff ($\bar{X}$'s = 9.0–10.0) and only slightly paler than the breast and belly. A "river effect" in this character appears along one sharp boundary in central Amazonia, where the throat colors of the females from the east and west banks of the wide, lower Rio Madeira differ conspicuously (dark cinnamon to the east and pale creamy-white, variously invaded from the breast by cinnamon-buff, on the west bank). A similar situation exists in the females on opposite banks of the Rio Solimões just west of the mouth of the Río Negro ($\bar{X}$'s = 1.5 to the north, 3.3 and 3.8 at two sample locations to the south).

The necklace is best developed in birds from upper Amazonia near the base of the Andes, where it consists of 5 to 10 black spots in eastern Peru through Bolivia and no more than 4 to 6 spots in birds from eastern Ecuador to southeastern Colombia and southern Venezuela. The number of spots is reduced in central Amazonia (2 to 5 spots in the Purús region and north of the lower Solimões near the mouth of the Rio Negro). Between the mouth of the Rio

Purús and the lower Rio Madeira, the number of spots averages only 0.5 to 1. In birds from the extensive region of southeastern Amazonia east of the Rio Madeira, no necklace is present.

Female dorsum.—The back is dark olivaceous brown in eastern Ecuador east to the Marañón River at the mouth of the Río Napo, and near Pebas, becoming slightly more rufescent in other regions of Amazonia.

Regional correlation of character variation.—Comparing geographical variation in several color characters of males and females of *M. myotherinus* shows the existence in Amazonia of four areas of relatively uniform character expressions, usually connected by rather narrow transition zones. These areas (Figs. 4, 5) are, (a) northwestern Amazonia from the Marañón valley in northern Peru north and northeast through eastern Ecuador, southeastern Colombia and southern Venezuela. Breast color of males is light to medium neutral gray; that of females is deep cinnamon-buff to orange-buff, the throat is pure white, and the necklace consists of 4 to 8 spots. (b) Southwestern Amazonia south of the Marañón-Solimões Rivers from the base of the Peruvian and Bolivian Andes eastward through the Rio Purús area. Male breast color is pale to light neutral gray; that of females is cinnamon-buff; the throat is off-white, and the necklace consists of 5 to 10 spots in the populations near the Andes, decreasing to 3 to 5 spots in birds from the Purús region. (c) A relatively restricted region in central Amazonia between the Rio Negro and the Rio Solimões. The colors of both males and females differ conspicuously from those of other Amazonian populations; breast color of males is dark neutral gray (nearly black). The breast of females is off-white, grading into pale cinnamon-buff. Females' throats are almost pure white, and the necklace consists of 3 to 7 spots. The somewhat different geographical extent of this character region based upon male and female characters, respectively, is discussed below. (d) Southeastern Amazonia east of the Rio Madeira to the mouth of the Amazon and south into central Brazil. Male breast color ranges from very pale gray to grayish-white, breast and throat in females are deep cinnamon-buff, and the necklace is absent. Another fairly restricted "character region" may exist within *M. myotherinus* just east of the Rio Ucayali in northeastern Peru (Figs. 4, 5), although it is more likely that these populations represent a separate species (*M. melanurus;* see below).

An exact delimitation of the above character regions is not feasible, as the locations of the transition zones for various characters do not coincide precisely. Near the Andes, an artificial separation of the northwestern and southwestern regions may be drawn along the lower Río Ucayali (males, Fig. 4) or along the Río Marañón (females, Fig. 5). The small central Amazonian region could be delimited in the west using one of the isolines on the map (Fig. 5) illustrating female breast color. Curiously, in this region somewhat different patterns of variation in male and female plumage occur (Figs. 4, 5). Based on female breast color, this region extends considerably farther west than it does based on male breast color. Equally puzzling is the fact that among females, no influence of the northern plumage type is seen south of the Rio Solimões. In contrast, the extreme dark gray male plumage north of the Solimões does occur on the south bank of the Rio Solimões, near the mouth of the Purús and Madeira Rivers. We are at a loss to explain these anomalies. The lower Rio Madeira could serve to separate the character regions to the west and east (regions c and d) but, in the upper Madeira region, any separation between it and the Rio Roosevelt breaks down because the characters change fairly gradually. Genetic mixing presumably occurs more freely in this region than farther downstream.

Taxonomy.—We recommend designating the populations inhabiting the four character regions as subspecies, as follows: (a) *M. m. elegans* in NW Amazonia, (b) *M. m. myotherinus* in SW Amazonia, (c) *M. m. ardesiacus* in the restricted central Amazonian region, and (d) *M. m. ochrolaema* in SE Amazonia. No additional information is transmitted by maintaining the four additional races recognized by previous authors (Hellmayr 1910, 1929; Cory and Hellmayr 1924; Zimmer 1932). *Myrmoborus myotherinus "napensis"* of eastern and northern Peru was based on the somewhat darker upperparts of the females, which are only slightly more olivaceous brown and less rufescent above than the females from other portions of Amazonia; *M. m. "incanus"* described from Tocantins on the Rio Solimões refers to a population from the transition zone between *M. m. ardesiacus* and *M. m. myotherinus/ elegans,* composed of white-bellied females (as in *ardesiacus*) and light gray males (as in *M. m. myotherinus*). The females of *ardesiacus* at the type locality (Manacapurú, north bank of the Rio Solimões near the mouth of the Rio Negro) are not pure white below, indicating an influence of the female plumage color of northern and/or southern populations. The name

M. m. "proximus" refers to a population between the lower Purús and Madeira Rivers in which the male plumage agrees with that of *M. m. ardesiacus* to the north of the Rio Solimões, whereas the female plumage is intermediate between *M. m. myotherinus* and *ochrolaema*. *Myrmoborus myotherinus "sororius"* was described from the transition zone between *M. m. myotherinus* and *ochrolaema* in the upper Madeira region.

BLACK-TAILED ANTBIRD (*MYRMOBORUS MELANURUS*)

The rare, upper Amazonian *M. melanurus* is known from only eight specimens (of which we examined six) and represents a localized member of the *M. myotherinus* complex, as pointed out by Zimmer (1932). The black and dark gray male (wing coverts fringed white) and the white-bellied female of *melanurus* are suspiciously similar to the rather localized *M. myotherinus ardesiacus* (including *"incanus"*) farther downstream and north of the Rio Solimões.

Myrmoborus melanurus is known from a restricted area just east of the lower Río Ucayali in northeastern Peru (Figs. 4, 5). Collecting localities are Sarayacu on the lower Río Ucayali (AMNH, 2 ♂♂, 2 ♀♀); Cashiboya, ca. 90 km east of Sarayacu (British Mus. [Nat. Hist.], 1 ♂); Orosa, nearly opposite the mouth of the Río Napo (AMNH, 1 ♂); Yarinacocha on the middle Río Ucayali (FMNH, 1 ♂), and "lower" Ucayali (British Mus. [Nat. Hist.], 1 ♀). E. Bartlett collected the latter bird at an unspecified locality on the lower Ucayali River in 1865; he labeled it "upper Ucayali" probably because he felt after "a long and tedious journey up the Ucayali" from its mouth to Sarayacu that he was reaching the upper course of that river. This, however, was not the case (see his account and map in Sclater and Salvin 1873).

Head and throat of male *melanurus* are jet black; back, upper breast, and belly are dark slaty gray (between colors 82 and 83), and underwing coverts are white. A white interscapular patch is well developed in one adult male and lacking in two others. The female of *melanurus* resembles that of *ardesiacus/incanus* in the white lower surface and white fringes of the wing coverts but has brown sides of the head, a pale fulvous breast, and no "necklace."

Myrmoborus m. myotherinus so far has not been collected inside the range of *M. melanurus,* but both species have been taken along the Río Ucayali near Yarinacocha and Sarayacu as well as in the vicinity of Orosa on the right bank of the Río Marañón; two males of *M. m. myotherinus* in typical plumage were taken "40 miles east of Iquitos, Peru" (quoted from specimen labels, FMNH 281287, 281288), south of the Amazon River, and not far from Orosa where *M. melanurus* occurs. Whether these localities indicate areas of contact (where little or no hybridization takes place) between two parapatric species or whether *M. myotherinus* is sympatric with *M. melanurus* throughout the restricted range of the latter species remains unknown. Because of pronounced differences in plumage color and the fact that pure phenotypes of both forms have been collected in close geographic proximity, we consider *M. melanurus* as a separate species rather than as just another "character region" within *M. myotherinus,* although this possibility cannot be ruled out entirely.

ASH-BREASTED ANTBIRD (*MYRMOBORUS LUGUBRIS*)

Ecologically, this third species of *Myrmoborus* is a member of a small group of Amazonian birds that is restricted to riparian swamp forest and thickets on flood plains along the Amazon River and the lower portions of some of its tributaries.

We examined 48 specimens of *M. lugubris* from its entire range. The general coloration of this species is similar to that of its congeners *M. leucophrys* and *M. myotherinus.* Geographical variation of male plumage in *M. lugubris* is inconspicuous along its nearly linear range (Fig. 6) from the lower Río Napo and the Ucayali in northeastern Peru to the mouth of the Amazon. Compared to males from the lower Amazon (*M. l. lugubris*), those from the more humid upper Amazonian portion of the range (*M. l. berlepschi*) are somewhat darker gray. Female plumages differ more conspicuously than those of the males (another case of heterogynism, Hellmayr 1929). The front and sides of the head are light ferruginous in females from the lower Amazon (*M. l. lugubris*), whereas the sides of the head are black in the females of the western populations in upper Amazonia (*M. l. berlepschi*). Female plumages intermediate between those just mentioned occur in relatively restricted areas along the lower Rio Solimões/ Río Negro and along the lower Rio Madeira. These intermediate plumage types appear to result from secondary intergradation of the black-faced western (*berlepschi*) and the red-faced eastern (*lugubris*) populations. The intermediate female plumages have been described as two

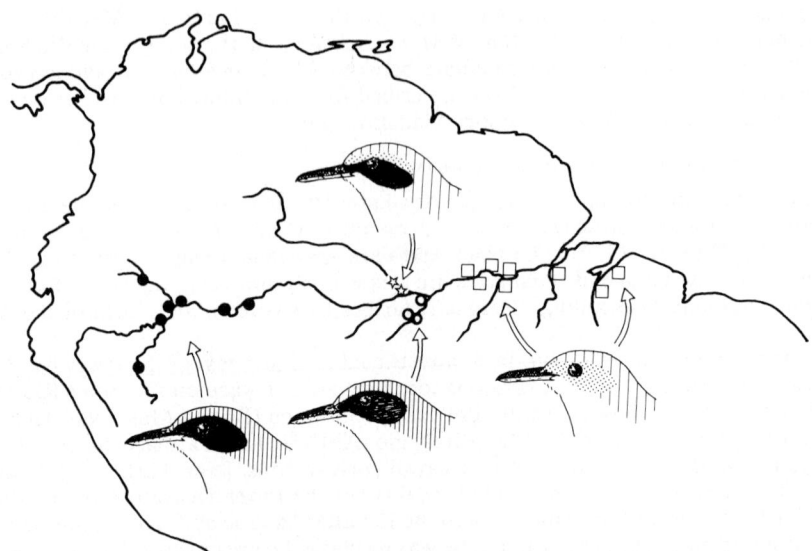

FIG. 6.   Range of the Ash-breasted Antbird (*Myrmoborus lugubris*). Solid circles = *M. l. berlepschi*; open circles = *M. l. femininus*; open stars = *M. l. stictopterus*; open rectangles = *M. l. lugubris*. Plumage colors of females as illustrated: solid = black; blank = white; broadly hatched = reddish cinnamon to rufescent sepia brown; narrowly hatched = russet brown; stippled = ferruginous.

separate subspecies, *M. l. "femininus"* from the lower Río Madeira and *M. l. "stictopterus"* from the north bank of the Rio Solimões (Manacapurú) and from the Rio Anavillhana, a left bank tributary of the lower Rio Negro.

Additional material from the middle course of the Amazon River is needed to verify the existence and regional extent of phenotypically stable intermediate plumage types, for which the application of separate subspecies names could be justified. In the more probable event that these intermediate populations are geographically restricted and phenotypically unstable (i.e., characterized by individually variable plumages), their taxonomic designations as *M. l. berlepschi* ≳ *lugubris* for *"femininus"* and *M. l. lugubris* ≳ *berlepschi* for *"stictopterus"* would be more meaningful biologically than designations as separate subspecies.

BLACK-FACED ANTTHRUSH (*FORMICARIUS ANALIS*)

This sedentary, terrestrial antbird inhabits the shaded, open floor of neotropical rainforests from Amazonia north into Middle America. Males and females have similar plumages, showing color variation in the breast, sides of neck, crissum, and dorsum as well as variously sized white loral spots. This widespread species appears to be absent from southern Venezuela and easternmost Colombia, for unknown reasons.

We compared 266 specimens of the Black-faced Antthrush from various parts of Amazonia (186 ♂♂, 80 ♀♀, from 89 localities). Males and females from humid upper Amazonia are darker and more richly colored and, locally, somewhat smaller than those from lower Amazonia and the Guianas. For example, 5 males from Limoncocha, eastern Ecuador have a wing length of 84.6 (X̄ in mm), tail 50.8, culmen 20.3, tarsus 29.3 compared to 13 males from Tamanoir, Cayenne with the following respective measurements: 91.7, 51.0, 22.7, 30.9. The more variable populations in central and southwestern Amazonia are intermediate between these extremes.

Color of breast and belly.—For regional analysis and mapping we recognize eight stages between the dark and light plumage color found in upper and lower Amazonia, respectively (Appendix I).

Contour and trend-surface maps of female breast color illustrate regional patterns of color variation (Figs. 7A, B). The darkest populations inhabit northwestern Amazonia north of the Marañón. Average values increase from 5.0 along this river to 6.0 to 7.0 near the base of the Ecuadorian and Colombian Andes, some individuals being as dark as blackish neutral gray (stage 8). Intermediate color stages around 4.0 occur in southwestern Amazonia, and in the

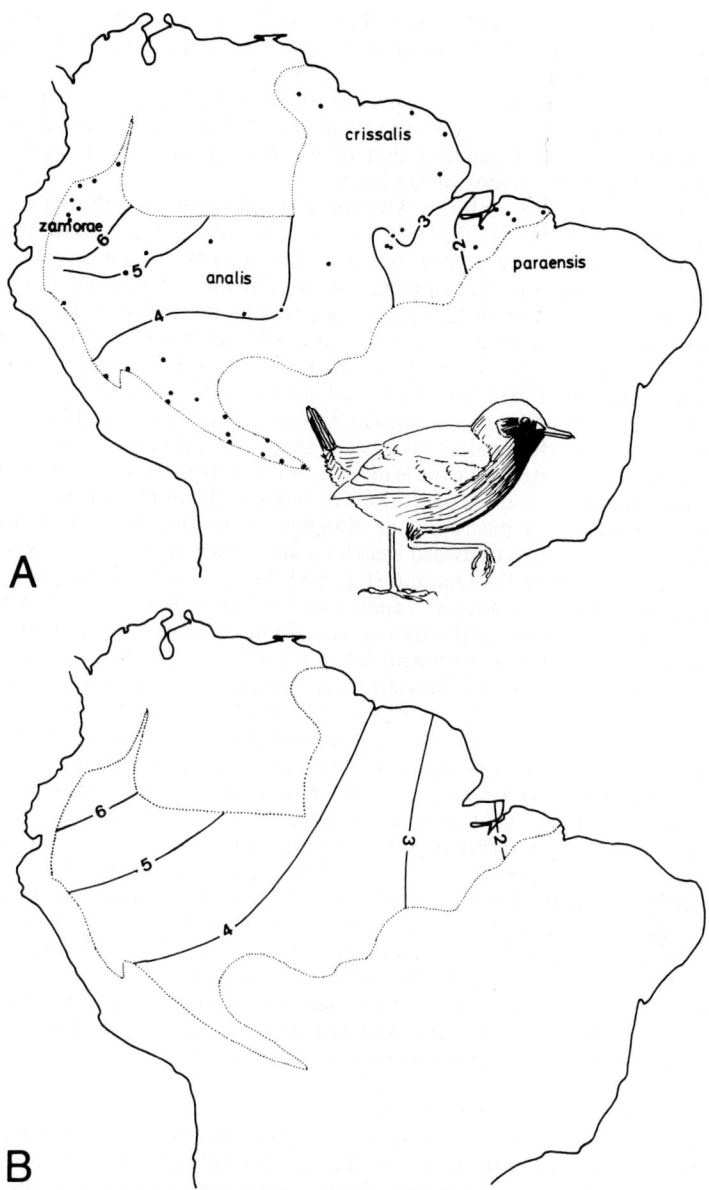

FIG. 7. Geographic variation in breast and belly color of female Black-faced Antthrush (*Formicarius analis*), Amazonian populations only. A = machine-contoured map; B = second order trend-surface map. Numbers represent color-score values (see text for explanation of scores). Dots indicate sample locations, and dotted line illustrates the approximate range of the species in Amazonia.

Guianas. The underparts are increasingly suffused with creamy white in southeastern Amazonia, the populations on both banks of the lower Rio Tapajós averaging stage 3, and this value decreasing to stage 2 south of the mouth of the Amazon. Some individuals from the Belém region have extensive creamy-white bellies (stage 1), the populations in this area being rather variable in this respect (*C.V.* = 15–20).

Color of crissum.—Eight color stages are recognized (Appendix I). In relatively pale specimens the crissum is slightly richer than cinnamon, and deepens in intermediately colored specimens through light and dark rufous to dark chestnut. Geographic variation of crissum color is correlated significantly with that of breast and belly color ($\delta\delta$, $r = 0.78$, $P < 0.001$; $\varphi\varphi$, $r = 0.87$, $P < 0.001$). Individual variation is rather pronounced, the $C.V.$ ranging between 12 and 44 in various populations.

Color of sides of neck.—Eight color stages are recognized (Appendix I). In birds from the Guianas the sides of the neck are grayish cinnamon. A light drab neck color decreases in intensity in stages 3 to 7 and matches that of the back (dark olive brown) in stage 8 in specimens collected near the Andes of Ecuador.

Geographic variation of neck color in Amazonia is correlated with that of breast color ($\delta\delta$, $r = 0.74$, $P < 0.001$; $\varphi\varphi$, $r = 0.77$, $P < 0.001$), i.e., pale colors predominate in the east. The lightest color stages (1–2), characterized by extensively grayish cinnamon on the sides of the neck, occur in populations from the Guianas, followed by the populations south of the mouth of the Amazon (stages 2–3). From this area, a gradual increase in population averages runs westward; Tapajós valley scores are 2.0 to 4.0, upper Madeira region, 4.0, Purús region, 5.0 to 5.7, base of the Andes in Bolivia and eastern Peru, 5.0 to 6.7, upper Solimões-Marañón region and eastern Ecuador, 6.0 to 8.0. Individual variation in these populations is pronounced ($C.V. = 12-16$) and is especially conspicuous in the lower Tapajós region ($C.V. = 30-50$) where different birds from the same localities represent stages 1.5 through 7.0. We could detect no "river effect" except at the mouth of the Amazon, which separates northern populations with grayish-cinnamon necks from those with drab or buffy necks to the south.

Loral spot.—The length of the round or elongate white loral spot increases from upper Amazonia ($\bar{X}$'s $= 1.5-2.5$ mm) eastward, reaching an average of 4.0 mm south of the mouth of the Amazon and 5.0 mm in the Guianas (Fig. 8A). The size of the loral spot is intermediate in central Amazonia (Purús, Madeira, Tapajós regions) as well as in southeastern Peru and eastern Bolivia. Variation in size of the spot in Amazonia is gradual. It is only weakly correlated with body size as measured by wing length ($\delta\delta$, $r = 0.40$, $P < 0.001$; $\varphi\varphi$, $r = 0.38$, $P < 0.05$).

Taxonomy.—The number and delimitation of Amazonian subspecies as currently recognized (Cory and Hellmayr 1924; Zimmer 1931; Novaes 1957), agree reasonably well with the results of our analysis of variation in selected plumage characters. The rather intensely colored populations of birds with small loral spots are found in upper Amazonia north of the Solimões-Marañón and in eastern Ecuador, and represent *F. a. zamorae*; the intermediate populations of eastern Peru, Bolivia, and central Amazonia represent the nominate form (*F. a. analis*), and the populations with conspicuously grayish cinnamon sides of the neck and a prominent loral spot have been designated *F. a. crissalis.* The populations south of the lower Amazon are substantially paler on the breast and belly than is any other population in Amazonia, and consequently are separated as *F. a. paraensis* (Novaes 1957). Note, however, that in neck color and size of the loral spot these populations approach *crissalis* of the Guianas. *Formicarius analis* "*olivaceus,*" described from the base of the Andes in northern Peru (Zimmer 1931:21), is characterized as ". . . having the upper parts darker and more olivaceous, less rufescent . . ." (than *zamorae*). On the basis of our analysis of regional trends in plumage color (Figs. 7, 8), we believe that recognition of this subspecies is unwarranted.

GREAT KISKADEE (*PITANGUS SULPHURATUS*)

This conspicuous, arboreal bird inhabits open vegetation formations, including gardens and parks of cities and villages from southern Texas, through Middle and South America, to Argentina. It occurs throughout Amazonia along lake margins and river courses. Its abundance may be increasing in this region as forests are extensively cleared along rivers and roads.

We measured 415 specimens (245 $\delta\delta$, 170 $\varphi\varphi$) of this species from Amazonia and the highlands of eastern Bolivia (wing, tail, culmen and tarsus; 122 localities). The contour and trend-surface maps of average wing length (Figs. 9A, B) in males illustrate zones of gradually increasing size in both northern and central South America, separated by an intervening zone of reduced measurements ($\bar{X}$'s $= 106-110$ mm) in the forested lowlands of Amazonia. The populations with smallest body size ($\bar{X}$ wing length ≥ 106 mm) inhabit upper Amazonia near the Andes. The conspicuous, steep gradient in the southwestern portion of the trend-surface map (Fig. 9B) represents a semi-isolated population with large measurements inhabiting the Andean valleys and highlands of eastern Bolivia; wing length in the impressively large birds of this population averages 128 mm in males and 124 mm in females. These values are about 12 percent higher than in the immediately adjacent populations in lowland Bolivia, and more than 20 percent higher than in west-central Amazonia. In addition, the Bolivian highland specimens all are conspicuously paler than the uniformly colored lowland populations through-

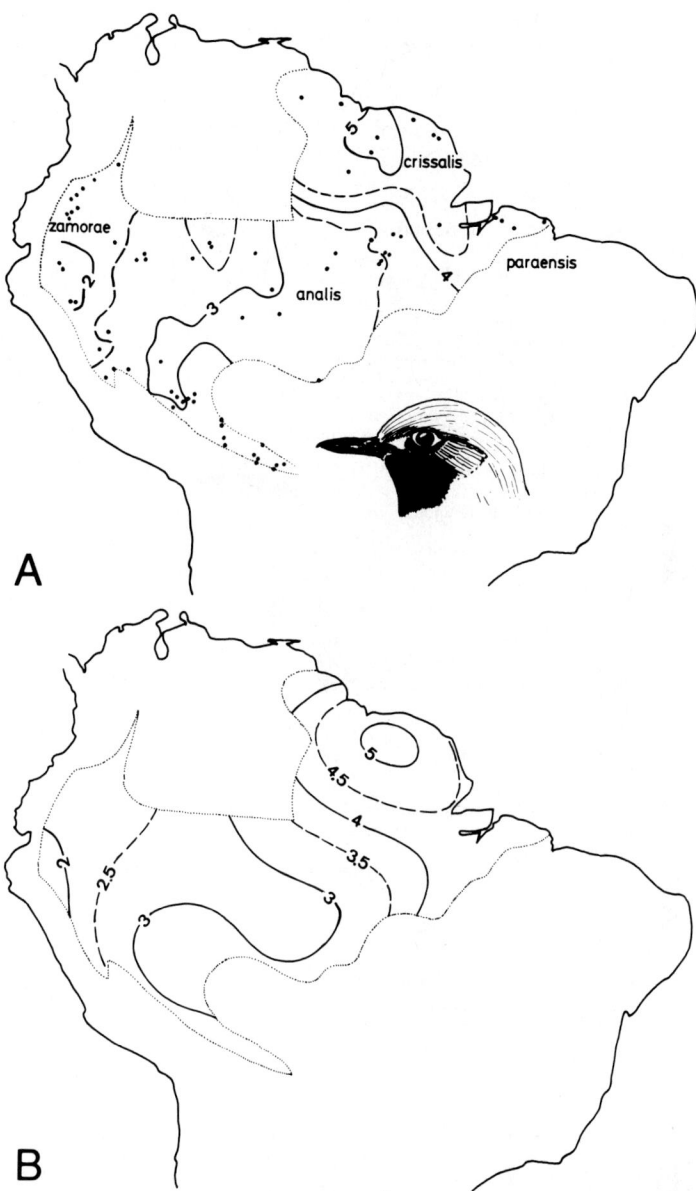

A

B

Fig. 8.   Geographic variation in the length (mm) of the white loral spot of male Black-faced Antthrush (*Formicarius analis*), Amazonian populations only. A = machine-contoured map; B = fourth order trend-surface map. Numbers represent average length of loral spot along isoclines. Dots indicate sample locations, and dotted line indicates the approximate range of the species in Amazonia.

out Amazonia. The contour map of average wing length in females depicts a similar pattern to that in males; average wing length is 102 mm or less in Amazonia, and increases to more than 106 mm both northward and southward.

Regional variations in tail, culmen, and tarsus lengths are closely correlated with wing length; the respective correlation coefficients, r, are 0.96, 0.50, and 0.93 in males, and 0.43, 0.86 and 0.74 in females ($P < 0.001$ for all values).

Except for the highland population, no clear subpopulations are evident from any of the

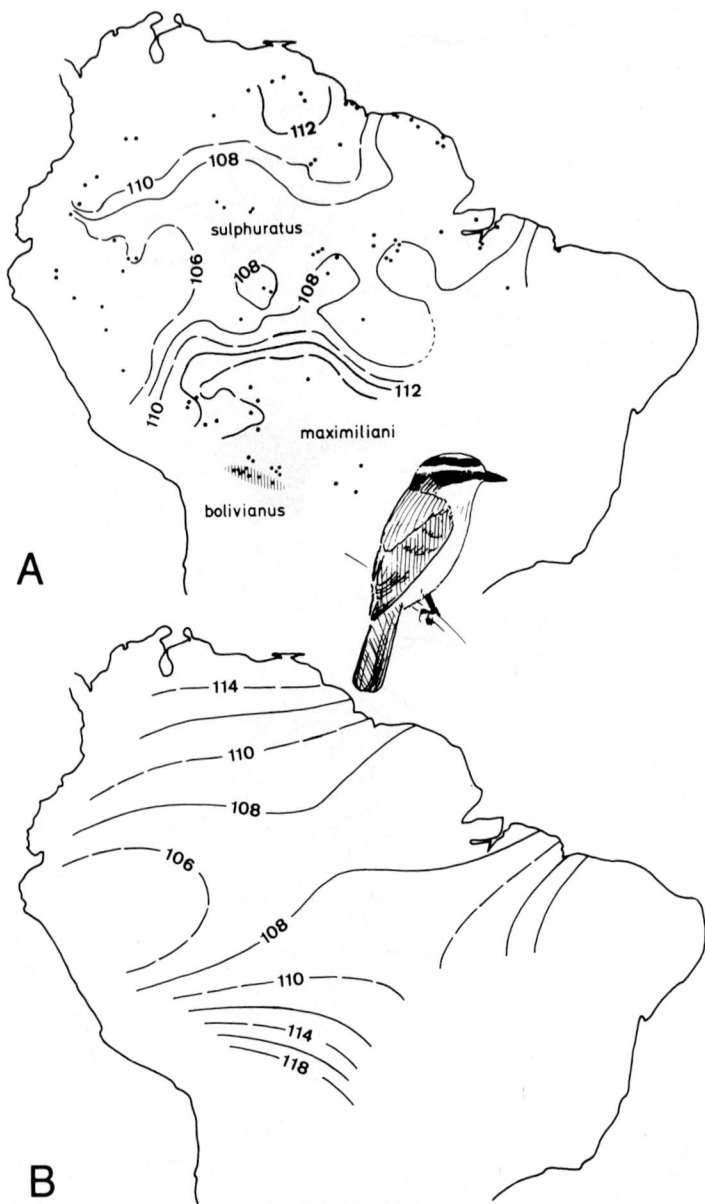

FIG. 9. Geographic variation of wing length (mm) in male Great Kiskadees (*Pitangus sulphuratus*), populations of central South America only. A = machine-contoured map; B = fourth order trend-surface map. Numbers indicate average wing lengths along isoclines. Dots indicate sample locations; vertically hatched area illustrates range of populations inhabiting eastern Bolivian highlands.

measurements we made. Furthermore, the gradual clines of increasing wing length northward and southward out of Amazonia are not correlated with any color changes. We find no evidence of abrupt character change across any of the major rivers.

   Taxonomy.—The currently recognized subspecies of the Great Kiskadee in tropical South America reasonably reflect the variation of this species in this region. The populations of Amazonia, including those showing slightly larger measurements in southeastern Colombia and the Guianas, are combined as the nominate form, *P. s. sulphuratus*. The larger birds of eastern Bolivia and southern Brazil, well south of Amazonia, have been designated *P. s.*

maximiliani, and the populations inhabiting the valleys and high plains of the eastern Bolivian Andes represent *P. s. bolivianus.*

DISCUSSION

CHARACTER VARIATION

Subspecies.—Contour maps of character variation in widespread bird species depict regionally varying gradients of character change. Areas of more or less uniform phenotypic character expression and intervening zones of sharp or gentle character change are readily identified. In our opinion this method of mapping geographic character variation is superior to wordy descriptions and rigid reference to subspecific names. A taxonomic, nomenclatural framework imposed upon continuous geographic variation is especially artificial where characters change gradually over extensive regions.

The subspecies concept clearly is most useful where applied to discrete, differentiated populations that are separated by distributional gaps ("evolutionary units" of Cracraft 1983). We advocate continued use of subspecies within *continuous* populations in only two situations: (1) at the extreme ends of steep clines, if and only if the two terminal populations show uniformity over a substantial portion of their ranges; and (2) where two or more wide-ranging populations show different, but in each case fairly uniform, character expression ("plateaus" on contour maps) connected by relatively narrow zones of character change. In this respect we concur with Johnson (1980). We stress that our goal should be to establish and study the underlying patterns of variation, not merely to differentiate subspecies (Storer 1982; Zusi 1982).

Among the species discussed in this paper, the geographic variation of *Aratinga leucophthalmus* illustrates gentle clines in both size (wing length) and coloration across Amazonia from west to east. The uniform populations near the Andes and in lower Amazonia may be recognized as subspecies, but their delimitation throughout central Amazonia is purely a matter of degree. The pattern of variation in the Black-faced Antbird (*Myrmoborus myotherinus*) is characterized by four wide-ranging Amazonian populations that exhibit rather uniform, correlated character expression in regions that are connected by relatively narrow transition zones. We apply subspecies names to these populations. Some differences in the extent and outline of certain character regions between males and females cause problems in the delimitation of certain central Amazonian subspecies of *M. myotherinus.* Nomenclaturally, we treat these cases by labelling the respective populations as "intermediate" and connecting the respective subspecies names with the symbol "≷." Mayr (1941, 1942: pp. 49, 83) discusses similar examples of independent geographic variation in males and females of birds from southeast Asia (*Microscelis leucocephalus*) and the southwest Pacific (*Petroica multicolor*). In those cases, isolated populations on mountains and islands, respectively, are involved. Zink (1983) reported different patterns of variation in males and females of the Fox Sparrow (*Passerella iliaca*). Continental examples of continuous distributions showing variation as complex as that in *Myrmoborus* probably are rare.

With climate.—Plumage color in several species may be generally correlated with climate, as indicated by geographic patterns of rainfall (Fig. 2). In *Aratinga leucophthalmus* and *Formicarius analis* the most richly colored populations occur near the Andes of Ecuador and Colombia. Similarly, in male *Myrmoborus myotherinus* the pale gray plumage of southeastern Amazonia becomes darker gray in western Amazonia. However, it is darkest (nearly black) in a fairly small region of central Amazonia, with less annual rainfall. In the females of this species, similarly intense coloration of the breast and belly occurs in northwestern and southeastern Amazonia, areas with widely different annual rainfall (although a rich, burnt orange-buff belly color is restricted to the very humid base of the Andes in southeastern Colombia and eastern Ecuador). Many additional patterns in other species need to be analyzed before a general explanation of these correlations and "anomalies" can be attempted. Historical factors, including climatic-vegetational changes in Amazonia during the geological past, also may have influenced or determined the origin of certain features mapped (see below).

Amazonian rivers.—Great variation exists in the degree to which Amazonian rivers represent effective barriers to gene flow in forest birds (Haffer 1974). In the cases illustrated here, no effect is seen among the strong flying and flocking parrots, *Pionus menstruus* and *Aratinga leucophthalmus* (Fig. 3). Body size of the latter species increases clinally in Amazonia from east to west without respect to river geography. Similarly, the remarkably symmetrical contour lines emanating northward and southward from Amazonia, reflecting size variation in the

Great Kiskadee (Fig. 9), are not influenced by the location of rivers. In all three of these cases, the species occupy open vegetation formations and the forest canopy. They are strong fliers, and their typical dispersal patterns undoubtedly include frequent crossing of even the largest rivers.

In contrast to the species just mentioned, the antbirds *Myrmoborus myotherinus* and *Formicarius analis* inhabit the interior of humid forests on or near the ground. They rarely venture into open habitats, their flight is weak, and their dispersal distances are known to be extremely short (Fitzpatrick and Terborgh, unpubl. data). Therefore, the gradual and gently clinal variation in various characters of the Black-faced Antthrush (*F. analis*) in Amazonia is rather surprising (Figs. 7, 8). The eastward increase in size of the loral spot in southern Amazonia, as well as the westward darkening of the plumage, seem not to be influenced by the location of rivers, except at the mouth of the Amazon.

On the other hand, the lower Amazon and the Río Negro delimit the entire northeasterly range of *Myrmoborus myotherinus*, clearly the most sedentary of the five species studied in detail. In central Amazonia, the Rio Solimões and Rio Madeira separate conspicuously different populations of this species on opposite river banks ("river effect"). Even in these cases, however, the barrier effect disappears in the headwater regions, where populations genetically separated by the rivers in their lower courses cross the river and intermingle in the forests on the respective opposite banks. No "river effect" is noted in this species in several other regions of Amazonia, where more or less identical populations occupy the areas on both sides of the rivers.

The lack of a barrier effect along most Amazonian rivers presumably reflects the fact that most rivers except the widest ones can change course quickly, especially in flood season, even locally shifting their main beds several kilometers overnight. In this way, even the most sedentary birds and other animals of the forest interior are passively transferred across the rivers and, thereby, continually reintroduce genes from one bank to the other.

Amazonian zoogeography.—The main "character regions" mapped for *Myrmoborus myotherinus* in northwestern, southwestern, and southeastern Amazonia center around areas inhabited by numerous endemic forest bird species, viz. Napo forest birds, Inambari forest birds, and Rondônia-Belém forest birds (Haffer 1978, 1985). As in many other Amazon forest species, the two upper Amazonian populations of *M. myotherinus* are closer to each other (white throat and necklace in females) than either is to the eastern form. Strongly differentiated populations restricted to the Amazonian forests between Rio Solimões and the lower Rio Negro, such as *Myrmoborus myotherinus ardesiacus*, are rare among Amazon birds. Some other examples include *Amazona autumnalis diadema, Pteroglossus flavirostris azara, Psophia crepitans ochroptera,* and *Nonnula amaurocephala.*

The origin of most of the geographical subspecies in *Myrmoborus myotherinus* may be related to that of the endemic Napo, Inambari, and Rondônia-Belém birds, respectively. It has been suggested (Haffer 1969, 1974) that many of these localized Amazonian species originated in comparatively restricted areas of forest when, during one or more glacial periods of the Pleistocene, the climate of South America was relatively dry and the Amazon rain forest was reduced to several forest "refugia." Animals and plants would have re-established contact during the intervening interglacial humid periods (as today), when the forests merged again. The same process presumably could give rise to geographic differentiation at the "subspecific" level, followed by a meeting of the various populations along zones of secondary contact. More detailed studies of other groups of birds and of other animals and plants are needed to ascertain whether a similar interpretation is likely for the localized forms between the Solimões and Negro Rivers.

Usefulness of Contour Maps

Contour lines drawn independently for different species and different characters can be used to test general ideas about geographic variation. For example, Wright's (1943) isolation-by-distance model for local differentiation predicts that appropriately scaled contour lines should show an average density (i.e., reflect a given degree of change with distance) approximately inversely proportional to dispersal distances, when compared between species. This prediction is loosely supported by our findings here, but a quantitative examination using more species would be desirable. Because trend-surface contour lines largely eliminate statistical noise in the raw data, they allow us to visualize the overall concordance or discordance between various characters, both within a species and between species. Measuring the angles between lines (90° angles represent perfect discordance) quantifies this feature. Such an exercise, performed among many species within Amazonia, for example, might point out centers of generally concordant variation separated by regions of discordance. We suspect that the western Amazonian region identified by Haffer (1974) as the Napo area of endemism would represent one such center, supported by the maps in this study. Such patterns would argue against random drift being responsible for major geographic patterns of variation, although they would not distinguish climatic and historic explanations.

Contour maps illustrating geographic variation of species inhabiting ornithologically incompletely known regions such as Amazonia may serve as "program maps" for future field work. Gaps of coverage as well as areas for profitable detailed sampling across zones of steep character gradients are easily identified. In the best example uncovered here, the location and width of the transition zones in males and females of the central and upper Amazonian forms of *Myrmoborus myotherinus,* and the contact zone with *M. melanurus,* require detailed study along the lower Río Ucayali, the Rio Yavarí, and both banks of the Rio Solimões.

We hope that quantitative studies of avian intraspecific variation eventually will lead to the publication of atlases illustrating contour maps of intraspecific variation of many or most species in continental regions. This would supplement the currently available species and subspecies checklists and greatly enhance their biological information content.

Summarizing the methodological aspects of this paper, we recommend that investigators studying widespread and continuously distributed species, (1) illustrate patterns of geographic variation of all sorts of characters or character combinations with the help of contour maps, (2) employ the computer as a tool for standardizing the mapping procedure, (3) compare patterns of geographic variation of various species with different ecological preferences and habits, and finally, (4) apply subspecies names only to geographic populations that show relatively uniform character-state "plateaus" as reflected on contour maps that quantify patterns and amounts of variation within continuous geographic distributions.

ACKNOWLEDGMENTS

We are grateful to the curators of the following museums for the loan of material used in this study: American Museum of Natural History (AMNH); Academy of Natural Sciences of Philadelphia (ANSP); Museum of Zoology, Louisiana State University (LSU); Carnegie Museum (CM); Museum of Comparative Zoology, Harvard University (MCZ); Natural History Museum of Los Angeles County (LACM); National Museum of Natural History (USNM); Museum of Zoology, University of Michigan (UMMZ); Alexander Koenig Museum, Bonn; Royal Natural History Museum, Stockholm. We thank A. Hofmann, who supervised the computer mapping of our data, F. Bairlein for statistical analysis of certain results, and E. Mayr for information on independent geographic variation in males and females of certain species of birds. We also thank M. S. Foster, D. L. Pearson, and one anonymous reviewer for their suggestions and critical reading of the manuscript.

The first author gratefully acknowledges travel grants received from the Deutsche Forschungsgemeinschaft (Bonn-Bad Godesberg) which enabled him to visit the Field Museum of Natural History and to present this paper at the 18th International Ornithological Congress (Moscow, August 1982).

LITERATURE CITED

ADAMS, R. P. 1970. Contour mapping and differential systematics of character variation. Syst. Zool. 19:385–390.

BARLOW, J. C., AND D. M. POWER. 1970. Analysis of character variation in Red-eyed and Philadelphia Vireos (Aves: Vireonidae) in Canada. Can. J. Zool. 48:673–694.

BARTH, E. K. 1966. Mantle colour as a taxonomic feature in *Larus argentatus* and *Larus fuscus.* Nytt Mag. Zool. 13:56–82.

BARTH, E. K. 1968. The circumpolar systematics of *Larus argentatus* and *Larus fuscus* with special reference to the Norwegian populations. Nytt Mag. Zool. 15 (supl. 1):1–50.

BEREGOVOY, V. E., AND N. N. DANILOV. 1967. Intraspecific variability in birds and phenography. *In* Intraspecific variability of land vertebrates and microevolution. Izd-vo Ural'skogo filial, Sverdlovsk.

CORY, C. B., AND C. E. HELLMAYR. 1924. Catalogue of birds of the Americas. Pt. 3. Field Mus. Nat. Hist., Zool. Ser. 13:1–369.

CRACRAFT, J. 1983. Species concepts and speciation analysis. Current Ornithol. 1:159–187.

CROWE, T. M. 1978. The evolution of guinea-fowl (Galliformes, Phasianidae, Numidinae). Taxonomy, phylogeny, speciation and biogeography. Ann. S. Af. Mus. 76:43–136.

DAVIS, J. C. 1973. Statistics and data analysis in geology. J. Wiley and Sons, New York.

ENDLER, J. A. 1977. Geographic variation, speciation, and clines. Monogr. Popul. Biol. No. 10.

FORSHAW, J. M. 1973. Parrots of the World. Lansdowne Press, Melbourne, Australia.

GOULD, S. J., AND R. F. JOHNSTON. 1972. Geographic variation. Annu. Rev. Ecol. Syst. 3:345–498.

GYLDENSTOLPE, N. 1951. The ornithology of the Rio Purus region in western Brazil. Arkiv Zool., Ser. 2, 2:1–320.

HAFFER, J. 1969. Speciation in Amazonian forest birds. Science 165:131–137.

HAFFER, J. 1974. Avian speciation in tropical South America. Publ. Nuttall Ornithol. Club No. 14.

HAFFER, J. 1978. Distribution of Amazon forest birds. Bonner Zool. Beitr. 29:38–78.

HAFFER, J. 1985. Avian zoogeography of the neotropical lowlands. Pp. 113–146, In P. A. Buckley, M. S. Foster, E. S. Morton, R. S. Ridgely, and F. G. Buckley (eds.), Neotropical Ornithology. Ornithol. Monogr. No. 36.

HANFORD, P. 1983. Continental patterns of morphological variation in a South American sparrow. Evolution 37:920–930.

HELLMAYR, C. E. 1910. Notes sur quelques oiseaux de l'Amérique tropicale. Rev. Fr. Ornith. 11:161–165.

HELLMAYR, C. E. 1929. On heterogynism in formicarian birds. J. Ornithol. 77 (Ergänz. II):41–70.

HUXLEY, J. 1942. Evolution. The Modern Synthesis. Allen and Unwin, London.

JAMES, F. 1970. Geographic size variation in birds and its relationship to climate. Ecology 51:365–390.

JOHNSON, N. K. 1980. Character variation and evolution of sibling species in the Empidonax difficilis-flavescens complex (Aves: Tyrannidae). Univ. Calif. Publ. Zool. 112:1–151.

JOHNSTON, R. F. 1969. Character variation and adaptation in European Sparrows. Syst. Zool. 18:201–231.

JOHNSTON, R. F., AND R. K. SELANDER. 1971. Evolution in the House Sparrow. II. Adaptive differentiation in North American populations. Evolution 25:1–28.

MAYR, E. 1941. Die geographische Variation der Färbungstypen von Microscelis leucocephalus. J. Ornithol. 89:377–392.

MAYR, E. 1942. Systematics and the Origin of Species. Columbia University Press, New York.

MAYR, E. 1963. Animal Species and Evolution. Harvard University Press, Cambridge, Massachusetts.

MEYER DE SCHAUENSEE, R. 1966. The Species of Birds of South America and Their Distribution. Livingston, Narberth, Pennsylvania.

MENGEL, R. M., AND J. A. JACKSON. 1977. Geographic variation of the Red-cockaded Woodpecker. Condor 79:349–355.

NOVAES, F. C. 1957. Notas de ornitologia Amazônica. 1. Generos Formicarius e Phlegopsis. Bol. Mus. Paraense E. Goeldi, Belem, Zool. No. 8.

PAES DE CAMARGO, A., A. R. REMO ALFONSI, H. S. PINTO, AND J. V. CHIARINI. 1977. Zoneamento da aptidao climática para culturas comerciais em areas de cerrado. In M. G. Ferri (ed.), IV Simposio Sôbre o Cerrado: Bases para Utilizaçao Agropecuaria. Itatiaia/EDUSP, Belo Horizonte, Brazil.

PIELOU, E. C. 1979. Biogeography. John Wiley and Sons, New York.

RATISBONA, L. R. 1976. The climate of Brazil. Pp. 219–269, In W. Schwerdtfeger (ed.), Climates of Central and South America. World Survey of Climatology 12. Elsevier, Amsterdam.

REINKE, R. 1962. Das Klima Amazoniens. Unpubl. Ph.D. dissert., University of Tübingen, Federal Republic of Germany.

SCLATER, P. L., AND O. SALVIN. 1873. On the birds of eastern Peru. Proc. Zool. Soc. Lond. 1873:252–311.

SELANDER, R. K. 1971. Systematics and speciation in birds. Pp. 57–147, In D. S. Farner, J. R. King, and K. C. Parkes (eds.), Avian Biology, Vol. 1. Academic Press, New York.

SMITHE, F. B. 1975. Naturalist's Color Guide. American Museum Natural History, New York.

STORER, R. W. 1982. Subspecies and the study of geographic variation. Auk 99:599–601.

TERENTJEW, P. W. 1957. Die Anwendbarkeit des Subspeziesbegriffes bei der Erforschung der innerartlichen Variabilität. Vestnik Leningr. Univ. 21:75–81 (transl. from Russian).

THORPE, R. S. 1976. Biometric analysis of geographic variation and racial affinities. Biol. Rev. 51:407–452.

WRIGHT, S. 1943. Isolation by distance. Genetics 28:114–138.

ZIMMER, J. T. 1931. Studies of Peruvian birds. I. New and other birds from Peru, Ecuador, and Brazil. Am. Mus. Novit. No. 500.

ZIMMER, J. T. 1932. Studies of Peruvian birds. VI. The formicarian genera Myrmoborus and Myrmeciza in Peru. Am. Mus. Novit. No. 545.

ZINK, R. M. 1983. Evolutionary and systematic significance of temporal variation in the Fox Sparrow. Syst. Zool. 32:223–238.

ZUSI, R. 1982. Infraspecific geographic variation and the subspecies concept. Auk 99:606–608.

APPENDIX I

Color Stages and Reference Specimens Used to Quantify Plumage Variation in Three Amazonian Bird Species[1]

Color	Specimen	Rank
Aratinga leucophthalmus: color of breast		
Yellow Green (slightly more dusky than color 58)	UMMZ 89127 (♂)	1
(Increasingly toward Parrot Green):	LSU 69355 (♂)	2
	USNM 121056 (♂)	3
	FMNH 251232 (♂)	4
Bluish Parrot Green (slightly more blue than color 60)	LSU 87253 (♂)	5
Aratinga leucophthalmus: red color on face		
All green or one small red feather	FMNH 257862 (♂)	1
2–5 red feathers on cheek	FMNH 262785 (♂)	2
4–10 red feathers on cheek	AMNH 474397 (♂)	3
(Increasingly large red patch on cheek, scattered red on nape, oculars, and breast):	FMNH 257894 (♂)	4
	AMNH 34561 (♂)	5
	AMNH 283368 (♂)	6
	AMNH 121079 (♂)	7
	LSU 51579 (♂)	8
	FMNH 286590 (♂)	9
Entire lower cheek solid red, abundant red on head and nape	CM 98776 (♂)	10
Myrmoborus myotherinus: breast and belly color of males		
Grayish white, or white suffused with gray	AMNH 429531	1
Slightly deeper grayish white	AMNH 137135	2
Pale Neutral Gray (color 86)	LSU 64203	3
Between Pale and Light Neutral Gray	AMNH 102163	4
Light Neutral Gray (color 85)	ANSP 120295	5
Between Light and Medium Neutral Gray	AMNH 255913	6
Medium Neutral Gray (color 84)	ANSP 165108	7
Between Medium and Dark Neutral Gray	none	8
Nearly Dark Neutral Gray	CM 98227	9
Dark Neutral Gray (color 83) to blackish	CM 98095	10
Myrmoborus myotherinus: breast and belly color of females		
Off-white	CM 98173	1
Pale buffy white	LSU 102160	2
Slightly paler than Buff (color 124)	LSU 64205	3
Pale cinnamon-buff	AMNH 137141	4
(Increasingly rich cinnamon-buff):	FMNH 180526	5
	CM 72441	6
(Increasingly dark cinnamon-buff):	AMNH 228597	7
	AMNH 286689	8
	AMNH 824579	9
Deep orange-buff	ANSP 165107	10
Myrmoborus myotherinus: throat color of females		
Pure white	FMNH 292868	1
Creamy white	AMNH 137144	2
(Creamy, invaded increasingly by cinnamon-buff from breast):	CM 99392	3
	CM 99355	4
	CM 86796	5
(Cinnamon-buff, increasingly dark):	none	6
	CM 72441	7
	AMNH 429526	8
	CM 77019	9
Dark cinnamon-buff (= stage 8 on breast-color scale)	AMNH 279577	10
Formicarius analis: breast and belly color		
Dark Neutral Gray (color 83) grading to creamy-white on belly	USNM 513335 (♂)	1
(Increasingly blackish gray, belly decreasingly suffused with creamy-white):	CM 69049 (♀)	2
	FMNH 264478 (♂)	3
	UMMZ 222934 (♂)	4

APPENDIX I
Continued

Color	Specimen	Rank
	CM 75364 (♂)	5
	AMNH 231922 (♂)	6
	FMNH 287156 (♀)	7
Blackish Neutral Gray (color 82), belly tinged olive	LSU 83336 (♂)	8
Formicarius analis: color of crissum		
Slightly richer than Cinnamon (color 123A)	AMNH 430878 (♀)	1
(Light to dark rufous):	CM 65513	2
	AMNH 277050	3
	CM 33806	4
	CM 70983	5
	LSU 92435	6
	AMNH 179365	7
Dark chestnut (near color 32)	CM 97891 (♂)	8
Formicarius analis: color of sides of neck		
Grayish cinnamon (between colors 139 and 132B)	CM 68433 (♀)	1
Dull grayish buff	LACM 31911 (♂)	2
(Light Drab (color 119C), decreasing in intensity):	CM 69881 (♂)	3
	AMNH 429541 (♂)	4
	AMNH 106826 (♂)	5
	AMNH 279588 (♂)	6
	AMNH 279594 (♂)	7
Dark olive brown, matching dorsum	AMNH 184401 (♂)	8

[1] Capitalized and numbered colors from Smithe (1975).

THE AVIFAUNA OF THE HUANCABAMBA REGION, NORTHERN PERU

Theodore A. Parker, III, Thomas S. Schulenberg,
Gary R. Graves, and Michael J. Braun

Museum of Zoology, Louisiana State University, Baton Rouge, Louisiana 70803-3216 USA

ABSTRACT. This paper reports on the distribution, behavior and ecology of 43 species from the Huancabamba region in the Andes of northern Peru. Included are 23 species new to Peru, and five recorded in that country for the second time. The rediscoveries of *Grallaricula peruviana, Myiophobus lintoni,* and *Incaspiza ortizi* are reported. The natural history of additional rare species, including *Gallinago imperialis, Nyctibius leucopterus, Hapaloptila castanea, Myornis senilis, Notiochelidon flavipes,* and *Buthraupis wetmorei,* is also discussed. Lists of species recorded along two elevational gradients (west slope of western cordillera and east slope of eastern cordillera) are presented, with an indication of elevational distribution, relative abundance, and social system. Twenty-six species are found to be at the southern limits of their distribution; most of these are replaced by closely related congeners south of the Río Marañón. The Río Marañón valley represents one of the most important barriers to montane forest birds along the eastern slope of the Andes. Its efficacy as a barrier is enhanced by the relative aridity of the western slope of the Andes, which increases as one moves south in Peru.

RESUMEN. En este trabajo se presenta información sobre la distribución, el comportamiento y la ecología de 43 especies de la región de Huancabamba en los Andes del norte peruano. Se incluyeron 23 especies nuevas para el Perú y cinco registradas por segunda vez en ese país. Se informa sobre el redescubrimiento de *Grallaricula peruviana, Myiophobus lintoni* y *Incaspiza ortizi.* También se discuten aspectos de la historia natural de especies raras, incluyendo *Gallinago imperialis, Nyctibius leucopterus, Hapaloptila castanea, Myornis senilis, Notiochelidon flavipes* y *Butharaupis wetmorei.* Se presentan listas de especies localizadas a lo largo de 2 gradientes elevacionales (la vertiente oeste de la cordillera occidental y la vertiente este de la corillera oriental) con indicaciones de distribución altitudinal, abundancia relativa y sistema social. Se encontró que 26 especies están en el límite sur de sus rangos de distribución, la mayoría de las cuales están reemplazadas al sur del Río Marañón por sus congéneres más cercanamente relacionadas. El valle del Río Marañón representa una de las barreras más importantes para aves de ceja de selva a lo largo de las vertientes orientales de los Andes. Su eficacia como barrera se ve realzada por la relativa aridez de las vertientes occidentales de los Andes, las cuales se tornan aún más áridas a medida que uno se dirige al sur del Perú.

Geographic barriers and faunal discontinuities have long been of central importance to the study of biogeography. Chapman (1926) was the first to recognize that the humid temperate avifauna of Ecuador and extreme northwest Peru was more closely related to that of Colombia and the northern Andes than to that of the central and southern Andes south of the Río Marañón in Peru and Bolivia. This north Andean component is generally thought to reach its southern limit north of the North Peruvian Low or Huancabamba Depression, which is characterized by relatively low mountains and the deep, arid, interandean basin of the Río Marañón (Fig. 1). The Marañón valley separates the eastern cordilleras of northwestern and central Peru and poses a major barrier to dispersal of birds restricted to humid montane forests. In this region humid forest extends southward along two roughly parallel spurs of the eastern cordillera of Ecuador, the Divisoria de Huancabamba and the Cordillera del Condor (east of the Río Chinchipe). South of the Divisoria de Huancabamba, montane forest is discontinuous and patchy and mountain passes are as low as 2000 m (Abra Porculla, 5°50'S, 79°31'W). East of the Cordillera del Condor, and south of the middle Río Marañón, humid montane forest is more or less continuous again along the eastern cordillera south to the Department of Santa Cruz, Bolivia.

Despite references in the literature attesting to the importance of the North Peruvian Low as a zoogeographic barrier (e.g., Vuilleumier 1969), the avifauna of this region remains poorly known. From 1974 to 1980, the Louisiana State University Museum of Zoology (LSUMZ) sent a series of expeditions to survey the avifaunas of the Divisoria de Huancabamba and adjacent ranges in the Departments of Piura and Cajamarca, Peru. The purpose of this paper is to document the southernmost occurrences of the north Andean forest avifauna, including 23 species new to Peru, and to report behavioral and ecological observations of poorly known species. Additionally, we include range extensions of four species north across the North Peruvian Low.

HISTORY OF EXPLORATION

G. K. Noble visited the vicinity of Huancabamba in August 1916, and also briefly visited more humid regions to the southeast of Tabaconas. He was primarily collecting reptiles and amphibians, but also made a small bird collection for the Museum of Comparative Zoology (Bangs and Noble 1918). H. Watkins, of the American Museum of Natural History, collected in the vicinity of Palambla, Huancabamba, and Chaupe (in humid forest east of Huancabamba at ca. 1850 m) in 1922–1923. M. A. Carriker, Jr. retraced the path of Watkins in 1933 while collecting for the Academy of Natural Sciences of Philadelphia. C. Kalinowski collected at high elevations southeast of Huancabamba for the Field Museum of Natural History in May 1954.

Fieldwork in the region by LSUMZ personnel began in 1974 when Parker and K. and R. Thomas worked above Canchaque (a few kilometers from Palambla) and at Cruz Blanca, on the upper west slope of the western cordillera, from 25 November to 10 December (Fig. 1). J. P. O'Neill and R. S. Kennedy returned to these localities for the period 19–25 August 1975. Parker, Braun, and L. Barkley collected at Cruz Blanca from 27 July–5 August 1980. The first LSUMZ party (Graves, Schulenberg, J. W. Eley) to collect on the humid, forested slopes northeast of Huancabamba visited Cerro Chinguela from 11–22 October 1977. Unfortunately, their collection was later lost in a robbery in Peru, but the initial results of this effort were so interesting that the LSUMZ sponsored two additional expeditions to Cerro Chinguela in 1978 and 1980. Graves, Parker, D. Hunter, J. W. Eley, and R. Semba collected in the paramo at Cerro Chinguela and in tall subtropical forest at Playón from 7 June–12 July 1978, and Parker, Braun, and L. Barkley worked again in the paramo and lower at Batán from 10 June–25 July 1980.

LSUMZ COLLECTING LOCALITIES

Cruz Blanca.—This locality (Fig. 1) lies at the crest of the western cordillera (3050 m) along the Canchaque-Huancabamba road, about 33 km southwest of the latter town, and only about 8 km northeast of Tambo, a well-known collecting locality (Chapman 1926). Our 1974, 1975, and 1980 camps were situated along the road at treeline on the west side of the ridge about 20 m below the crest. Between 2150 and 3050 m, the west slope was covered by a mixed evergreen forest (including *Clusia, Oreopanax, Podocarpus* and *Polylepis*). Unfortunately, human pressure on this habitat is increasing; the undergrowth is trampled and grazed by livestock in many places at all elevations, and forest is being cleared from above and below. Only scattered patches of forest exist below 2150 m. A mule trail descends about 1000 m from Cruz Blanca through mature forest, continuing downward through cleared lands to Canchaque. We worked the forest along this trail almost daily during our visits to the mountain. Treeline forest is dominated by *Polylepis,* with a canopy 4 to 6 m high. Taller trees cover the lower slopes and stream valleys. Arboreal bromeliads are conspicuous, but tree ferns are absent (or very scarce), and *Chusquea* bamboo is uncommon. In general, the area is much less humid than Cerro Chinguela, but foggy, cloudy weather occurred regularly on all three visits. Extensive grasslands nearly devoid of bushes cover the upper mountain slopes to the north and south of the road pass. Although these areas are intensively grazed by livestock, we still found a variety of puna/paramo bird species in them. Isolated *Polylepis* groves up to 2 ha in size survive, as do thickets of *Brachyotum, Gynoxys, Hesperomeles,* and *Lupinus.* The open habitats above Cruz Blanca are generally rockier and drier than those of Cerro Chinguela. The eastern slope of this cordillera is densely settled and under intensive cultivation. Only scattered hedgerows, orchards and small clumps of *Alnus* and *Polylepis* trees provide cover for woodland birds. The lower slopes of the west side of the Huancabamba valley support desert vegetation, including a variety of cacti, agaves, and terrestrial bromeliads. A cumulative list of the species recorded at Cruz Blanca with their elevational distribution, relative abundance, and social system, is presented in Appendix III.

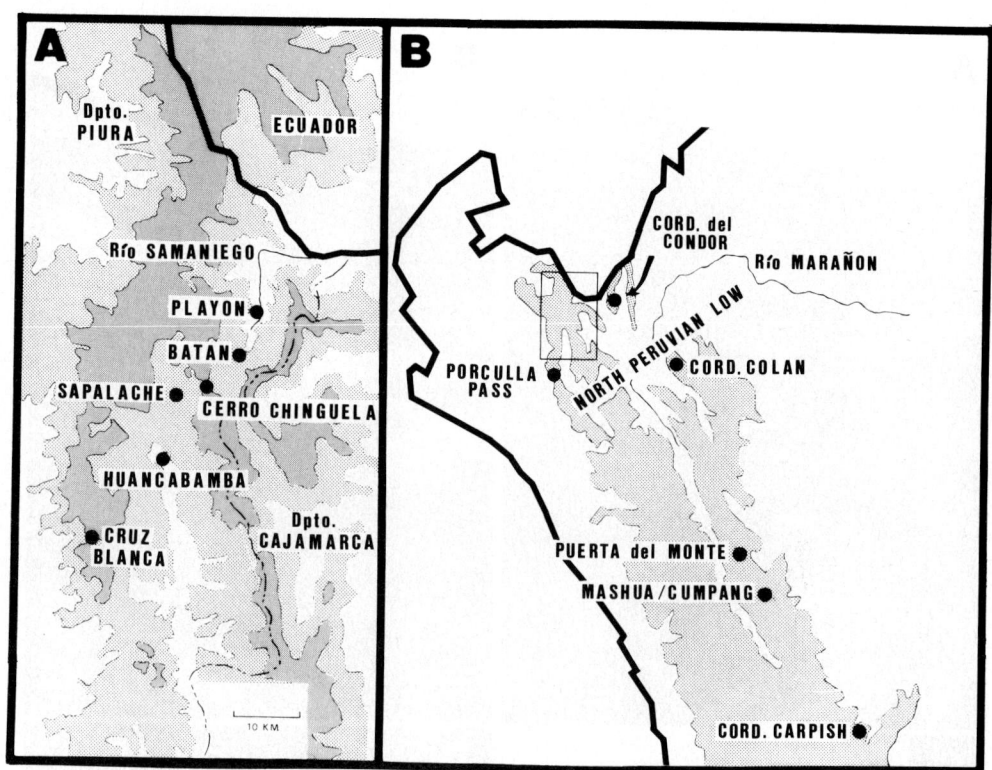

Fig. 1.   A. The Huancabamba region, shaded within the 2000 m (light stippling) and 3000 m (dark stippling) contours, showing main LSUMZ collecting localities. B. The Andes of northern Peru (shaded within the 2000 m contour), showing localities mentioned in the text. Rectangle in top center indicates area shown in detail in Map A.

Huancabamba Valley. — Huancabamba lies in a narrow rain-shadow valley at 1925 m (Fig. 1). Although this arid valley is densely settled and intensively cultivated, a few small remnants of desert scrub remain within walking distance of the city. From 11–13 June and on 27 July 1980 we visited an area of this habitat ca. 2 km NE Huancabamba on the road to Sapalache. The vegetation is composed of dense shrubbery and scattered *Acacia* trees and clumps of columnar cacti. The farmlands above this dry terrain eventually give way to humid forest at about 2150 m. A list of the species recorded during these visits is presented in Appendix II.

Cerro Chinguela. — This is a mountain north and east of Huancabamba and Sapalache (Fig. 1); it separates the relatively dry valley of the Río Huancabamba from the very humid, forested valley of the Río Samaniego (a tributary of the Río Canchis). The slopes are uniformly covered with cloudforest on the west side, from about 2150 m (the upper limit of agricultural and pasturelands) to ca. 2930 m (the beginning of paramo grassland). Undisturbed cloudforest on the east slope extends from treeline down at least to 1500 m (and lower away from the main trail). Our main camp was situated in the páramo at treeline just off the Sapalache-El Carmen mule trail that crosses the ridge at 3050 m. From there we worked the higher (to 3700 m) grassland and isolated forest groves to the north. We also penetrated the uppermost cloudforest (2400–3050 m) on both east and west slopes, especially along the main trail. This was typical high-elevation, montane forest; tree and shrub genera included *Clethra, Clusia, Gynoxys, Hesperomeles, Podocarpus, Shefflera,* and *Weinmannia.* Trees were thickly covered with bromeliads and mosses. *Chusquea* bamboo grew in luxuriant stands in and along the edges of forest at all elevations. In contrast to treeline woodland at Cruz Blanca, *Polylepis* was very scarce, if present at all. Forest abruptly gave way to grassland within several kilometers of the main trail (Fig. 2), but a shrubby corridor mixed with tall grasses characterized the paramo/forest ecotone in more remote areas away from the influence of man. From 1977 to 1980

Fig. 2.   A. East slope of Cerro Chinguela from 2100 to 3600 m; mule trail visible in center of photograph. B. Paramo grassland and woodland patches northeast of Cerro Chinguela camp, 3600 m. C. Epiphyte-laden forest along Río Samaniego near Machete, 2050 m. D. Typical forest with bamboo understory near Lúcuma, 2000 m.

only one man and a few dozen cows were noted on the mountaintop, but according to local people extensive areas of grassland are periodically burned to "improve the quality of the grass." This probably accounts for the limited shrubby growth away from the forest edge. High rainfall and frequent cloud cover may discourage human colonists. We experienced prolonged periods of foggy, rainy weather for nearly 75 percent of the time spent on Cerro Chinguela on three expeditions.

Río Samaniego Valley.—We worked from two additional camps on the east slope of Cerro Chinguela in the valley of the small Río Samaniego (Fig. 1). *Batán* is a small clearing and rest area for pack animals along a stream crossed by the main trail at 2250 m. From this site we worked primarily along the main trail, upslope to forest at treeline, and down to Machete (2050 m), the next stream crossing the trail to the east, and Lúcuma, a fairly level area of tall forest at 2000 m. At the latter locality some tall (to 30 m) *Podocarpus* were being cut by local people. Otherwise, except for a few clearings along the ridge separating Batán and Machete and around Lúcuma, the trail was bordered continuously by cloud forest from treeline on Cerro Chinguela down to Playón, a distance of about 10 km (Fig. 2).

Playón, at 1675 m, is the site of a few homesteads, small gardens, and a few hectares of pastureland on the north bank floodplain of the Río Samaniego; Playón was surrounded by cloudforest. We concentrated our efforts in forest on the slope just north of the clearings and along the trail back toward Lúcuma and Chinguela. This locality is in the Department of Piura (Stephens and Traylor 1983), not the Department of Cajamarca as previously reported (Eley et al. 1979; O'Neill and Schulenberg 1979; Parker and Parker 1980).

A list of the species recorded on the east slope of Cerro Chinguela, with their elevational distribution, relative abundance, and social system, is presented in Appendix I.

SPECIES ACCOUNTS

Phalcoboenus megalopterus. Mountain Caracara. From one to four of these striking birds were observed daily at Cruz Blanca (3050–3350 m) and on Cerro Chinguela (2900–3350 m). One was seen at 2290 m in the vicinity of Sapalache. Nearly all individuals observed were adults. Most were flying low over grassland and forest edge; occasionally one or two individuals were noted walking about in badly overgrazed or recently burned areas above treeline. Unfortunately these birds were wary, and we were unable to obtain specimens. Vuilleumier (1970) gave localities in the puna zone of central Cajamarca as the northernmost records for this species and mentioned the "low Andes of Northern Peru" as a probable barrier to its northward dispersal. Our records, however, document the occurrence of *Phalcoboenus megalopterus* north of the Porculla Pass in the western cordillera, and the Marañón Depression to the east.

Penelope barbata. Bearded Guan. Virtually nothing is known concerning the natural history of this species (Delacour and Amadon 1973). We found it on both slopes of Cerro Chinguela (2400–2900 m) and also on the west slope of the western cordillera below Cruz Blanca (1800–3000 m). In both areas the species was uncommon and probably decreasing due to hunting and habitat destruction. Displays observed were similar to those of other *Penelope*. In August 1980 at Cruz Blanca, individuals performed the "wing-drum" from 05:30 to 05:45, in the semi-darkness just before and at dawn. Preceding this display, which was repeated two to four times at intervals of 4–6 min, the birds uttered a peculiar whinnying call. Individuals or both members of pairs disturbed along trails gave two other types of vocalizations: a persistently-repeated, very loud yelping call, and a barely audible, ascending whistle. The latter call was also given by a female accompanied by two small young on the east slope of Cerro Chinguela on 29 July 1980. Most observations were of pairs or small groups of three or four individuals. The Chinguela records of *barbata* are the first for the eastern cordillera of Peru, where this species apparently replaces *Penelope montagnii,* which is not known on the east slope between southern Ecuador and the mountains east of Bagua, Depto. Amazonas, Peru.

Specimen data: LSUMZ 80343, 97567; 1 ♂, 1 ♀, no weights; iris medium brown; facial skin gray; bill dark horn; gular skin and legs carmen.

Gallinago andina. Puna Snipe. An individual collected by Parker on 3 August 1980 in a small bog surrounded by grassland at 3150 m above Cruz Blanca represents a northerly range extension of 150 km. The raspy flight call of this species is very much like that of *Gallinago gallinago.* The Puna Snipe was previously known north to central Cajamarca (Hellmayr and Conover 1948). This is the first record north of Porculla Pass.

Specimen data: LSUMZ 97569; ♀, 108 g; stomach: unidentified invertebrates.

Gallinago imperialis. Banded Snipe. Parker and Braun tape-recorded two or three different individuals of this large, very rare snipe just before dawn on several mornings in June and July 1980, high on the east slope of Cerro Chinguela. These birds were displaying over stunted forest between 2745 and 2960 m. Our observations agree closely with those of Terborgh and Weske (1972). On Cerro Chinguela *G. imperialis* was greatly outnumbered by the equally large *G. jamesoni*, a bird of páramo grassland and bogs above treeline. Both species could be heard simultaneously from the mule trail at 2900 m. On 17 July 1980 Parker flushed a large snipe, probably *Gallinago nobilis*, from the edge of a treeline bog on the eastern side of Cerro Chinguela at 3100 m. This is the first indication that *nobilis* may occur south of Ecuador. In addition to the above records of *imperialis*, LSUMZ personnel have the following recent records: several were heard almost daily by Parker, Schulenberg, and M. Robbins in October 1979 above Cumpang, on the Tayabamba-Ongón trail (ca. 2900 m), Depto. La Libertad; and a few were also heard regularly by Schulenberg and M. Williams in September 1978 near treeline (2930 m), on Cordillera Colán, NE Bagua, Depto. Amazonas. These records substantially augment the previously known distribution of *G. imperialis* (two "Bogota" trade skins, and one specimen from the Cordillera Vilcabamba, Depto. Cuzco, southern Peru [Terborgh and Weske 1972]). It is now apparent that *imperialis* is geographically widespread, but very local and difficult to collect (we did not acquire any specimens). With more thorough fieldwork, *Gallinago imperialis* will probably be found in other localities in the eastern Andes of Colombia, Ecuador, Peru, and perhaps even Bolivia. Tape-recordings made at Cerro Chinguela and Cumpang are deposited in the Library of Natural Sounds, Cornell University.

Hapalopsittaca amazonina. Rusty-faced Parrot. We have few records of this beautiful parrot from our several months of fieldwork on Cerro Chinguela. Schulenberg collected one of a pair on 16 October 1977 on the west slope at 2530 m; this specimen was later stolen. In June 1978 Parker twice saw pairs in mossy forest high on the west slope (2680–2870 m), and once on the east slope at 2960 m. The species is very rare in the region. This is the first report of this species for Peru. The nearest records are of the disjunct Ecuadorian population *H. a. pyrrhops*, known from only a few localities in the Provinces of Loja, Azuay, and Morona-Santiago (Chapman 1926; Ridgely 1980).

Specimen data: (T. S. Schulenberg 333, specimen now lost); 1 ♂, 97 g; iris chestnut-brown; bill whitish, cere gray; tarsi and feet gray.

Aegolius harrisii. Buff-fronted Owl. Two individuals of this rare owl were mist-netted on 5 and 6 December 1974 by K. and R. Thomas, 15 km by road east of Canchaque, Depto. Piura at 1740 m. Several others were heard there by Parker during the same period. The habitat in this locality, which is on the west slope of the western cordillera directly below Cruz Blanca, is semi-humid, rather open forest about 10 m tall. A third specimen was netted on 28 June 1978 by J. W. Eley, in very mossy, stunted forest near treeline (ca. 2960 m) on the east slope of Cerro Chinguela. The species is known from only a handful of curiously disjunct and widely separated localities in South America (Meyer de Schauensee 1966). Remsen and Traylor (1983) recently reported the first record of this species for Peru, two specimens (in the Field Museum of Natural History = FMNH) collected at Yurinaqui Alto, Depto. Junín in 1969. An additional unpublished record is of a male collected by J. W. Fitzpatrick (specimen in the Museum of Comparative Zoology = MCZ) on 16 July 1976 at 1950 m in the Cordillera del Condor, E San José de Lourdes, Depto. Cajamarca.

Specimen data: Cruz Blanca: LSUMZ 77995–96; Cerro Chinguela: 87296; Cordillera del Condor: MCZ 331073; 3 ♂♂, 113–117 g ($\bar{X}$ = 115 g); 1 ♀, 135 g; iris yellow; bill gray, maxilla tipped white; toes yellowish flesh; stomach: chitin fragments, mouse remains.

Nyctibius leucopterus. White-winged Potoo. On 30 June 1980 Braun collected a specimen of this little-known species at Lúcuma (2000 m). The bird was roosting atop a large, broken, dead branch protruding above the canopy of cloudforest about 25 m above ground. The specimen is of the form *maculosus*, known from fewer than 10 specimens from widely scattered localities in the Andes from Venezuela to northern Bolivia (records summarized in Schulenberg et al. 1984). This is the first record of *N. l. maculosus* for Peru. This form is almost certainly specifically distinct from nominate *leucopterus*, known from two specimens from northeastern Brazil (Schulenberg et al. 1984).

Specimen data: LSUMZ 97586; 1 ♀, 145 g; iris yellow; bill black; tarsi pink, feet grayish white; stomach: large Coleoptera.

Coeligena lutetiae. Buff-winged Starfrontlet. This large hummingbird was very common in forest on both slopes of Cerro Chinguela between 2600 and 3350 m. It was replaced on the lower west slope of Chinguela and at Cruz Blanca by the equally common *C. iris.* Three individuals (males?) of *lutetiae* were often seen pursuing one another high above the canopy; otherwise the species was noted singly at all heights in the forest. The flight call was a raspy *chhet,* which is rather unlike the sharp, clear *cheet* of closely related *C. violifer,* which inhabits similar high mountain forest on the east slope of the Andes from Depto. Amazonas, Peru south to Depto. Cochabamba, Bolivia (Meyer de Schauensee 1966). The two species, which are of similar size, color, and elevational distribution, should probably be regarded as members of a superspecies. This is the first record of *Coeligena lutetiae* for Peru. It was previously known from the Central Andes of Colombia, and both slopes of the Andes of Ecuador (Meyer de Schauensee 1966).

Specimen data: LSUMZ 87457–87464, 97605 (skins); 89779–97478 (skeletons); 89392–89398 (spirits); 5 ♂♂, 6.2–7.2 g ($\bar{X}$ = 6.8 g); 9 ♀♀, 6.3–7.7 g ($\bar{X}$ = 7.0 g); bill black; feet tan or brown; stomach: insects.

Eriocnemis vestitus. Glowing Puffleg. This was a common hummingbird of treeline forest edge and isolated patches of trees and shrubs in the paramo on Cerro Chinguela from 2900–3200 m; the species was also found in man-made clearings on the east slope of Cerro Chinguela from 2290–2590 m. It was frequently observed hanging on the flowers of *Brachyotum* shrubs in both of the above habitats. Interspecific disputes between *E. vestitus* and *Metallura odomae* were noted at treeline, and at lower elevations aggressive interactions were often observed between *E. vestitus* and *Heliangelus amethysticollis. Eriocnemis vestitus* was generally quiet and spent most of its non-foraging time perched within the foliage of trees and shrubs. This is the first report of this hummingbird for Peru; it was previously known from the Andes of Venezuela, Colombia, and Ecuador south to Prov. Loja (R. S. Ridgely, pers. comm.).

Specimen data: LSUMZ 87513–87522, 97614 (skins); 89819–89832, 97483–97486 (skeletons); 89412–89420 (spirits); 12 ♂♂, 4.1–5.0 g ($\bar{X}$ = 4.4 g); 10 ♀♀, 3.9–4.8 g ($\bar{X}$ = 4.4 g); bill black; feet black.

Metallura odomae. Neblina Metaltail. This species was recently described from seven specimens taken in the paramo on Cerro Chinguela (Graves 1980). At Cerro Chinguela in July 1980 we observed *Metallura odomae* in low densities in shrubby growth at forest edge and in patches of shrubs and short trees in the grassland. We recorded it daily in small numbers. Eight were seen during one walk of approximately four km in this habitat. Most individuals perched conspicuously atop shrubs and pursued intruding conspecifics and the occasional *Eriocnemis vestitus* that entered their territories. The only vocalization noted for *M. odomae* was a rather loud *seet-seet-seet-ti-tttt.* This is very like calls of the closely related *M. theresiae* of central Peru, and *M. aeneocauda* of southern Peru-northern Bolivia. *Metallura odomae* was observed flycatching, as well as feeding at the flowers of *Brachyotum* and a *Berberis*-like shrub.

Specimen data: LSUMZ 87545–87549, 97615–97617 (skins); 89835 (skeletons); 3 ♂♂, 5.0–5.4 g ($\bar{X}$ = 5.2 g); 4 ♀♀, 4.7–5.1 g ($\bar{X}$ = 4.8 g); bill black; feet black; stomach: insects.

Chalcostigma herrani. Rainbow-bearded Thornbill. On Cerro Chinguela this species was a scarce inhabitant of shrubby places in the paramo, especially on well-drained, rocky slopes with ferns (*Blechnum*), bromeliads (*Puya, Tillandsia*), and various shrubs (especially *Brachyotum*). This habitat appears to be maintained by fire and is very similar to that described for the species in Ecuador (Ridgely 1980). An occasional individual was also seen along the edges of bogs or forest patches in more humid parts of the paramo. The species hung on the flowers of *Brachyotum* and *Puya* while feeding. This is the first record of *Chalcostigma herrani* for Peru. Ridgely (1980) recently found this thornbill as far south as the Loja-Zamora road, Prov. Loja, Ecuador.

Specimen data: LSUMZ 87562–87569 (skins); 89427 (spirit); 4 ♂♂, 5.7–7.0 g ($\bar{X}$ = 6.2 g); 3 ♀♀, 5.1–6.3 g ($\bar{X}$ = 5.6 g); bill black; feet black.

Hapaloptila castanea. White-faced Nunbird. Only one record of this rare, peculiar bucconid was obtained. Until 1979, this species was known in Peru only from a specimen collected at Cumpang, Depto. La Libertad (Ménégaux 1910). In October–November 1979 an LSUMZ field party found the decidedly uncommon *Hapaloptila* at the latter locality, in cloudforest between 2590–2900 m. Then, on 3 July 1980, Braun collected one of a pair at 2150 m on the east slope of Cerro Chinguela. In addition to these Peruvian localities, *Hapaloptila castanea* is known from a few localities in the western cordilleras of Colombia and Ecuador (Meyer de Schauensee 1966).

At Cumpang, pairs of this nunbird frequented the canopy of forest on steep slopes. The birds perched quietly for long periods in the upper branches, occasionally sallying 3 to 9 m up into the air to catch large insects such as beetles and grasshoppers. The only vocalization noted for *H. castanea* was a mournful, downslurred *wuooooo*. In response to playbacks of this call, individuals approached to within 20 m and peered about with outstretched necks; occasionally they bobbed their heads and bowed forward in the direction of the played call. When particularly excited, they perched lengthwise on a branch in the manner of a nighthawk (*Chordeiles*). This position was held for only a few seconds at a time.

Specimen data: Machete: LSUMZ 97623; Cumpang: LSUMZ 92012–13, 92015 (skins), 92014 (skeleton and partial skin), 91493 (spirit); 3 ♂♂, 75–84 g ($\bar{X}$ = 80 g); 2 ♀♀, 81 g, 82 g; iris crimson; bill black; tarsi and feet gray; stomach: Coleoptera, Hymenoptera, Orthoptera, Lepidoptera (larvae).

Colaptes rupicola. Andean Flicker. This terrestrial woodpecker was found in small numbers at both Cruz Blanca and on Cerro Chinguela, in grassland and forest edge. Territorial birds sang from exposed perches on large rocks or in trees; otherwise the species spent most of its time on the ground. The song is a short, emphatic trill of 2.5 sec repeated for up to several minutes. Also given was a loud, rather high-pitched *kek*. This disturbance call was usually uttered several times in rapid succession as a bird flew up from the ground. Our records of *C. r. cinereicollis* are the northernmost for this common, widespread species. This information supplements that given by Short (1972).

Specimen data: LSUMZ 75175, 78190–91, 80414–15, 87652; 1 ♂, 190 g; 3 ♀♀, 175–190 g ($\bar{X}$ = 183 g); 1 sex ?, 161 g; iris golden yellow; bill slate; feet and tarsi yellowish flesh.

Campylorhamphus pusillus. Brown-billed Scythebill. One was netted at Playón (1675 m) on 3 July 1978. The LSUMZ also has specimens from the following Peruvian localities: Cordillera Colán (1675–2000 m), NE Bagua, Depto. Amazonas; 33 km NE Ingenio (2125 m), Depto. Amazonas; and 15 km NE Abra Patricia (1675 m), Depto. San Martín. In addition, J. W. Fitzpatrick collected a specimen (American Museum of Natural History) on 17 June 1975 in the Cordillera del Condor, E San José de Lourdes, Depto. Cajamarca. The species is an uncommon and difficult-to-observe inhabitant of cloudforest undergrowth and middlestory. These are the first published records for Peru.

Specimen data: Playón: LSUMZ 87717; Cordillera Colán: LSUMZ 87718–87723 (skins), 89433 (spirit); Ingenio: LSUMZ 81901; Abra Patricia: LSUMZ 81902; 8 ♂♂, 33–48 g ($\bar{X}$ = 39.9 g); 1 ♀, 37 g; iris brown; bill light brown, base usually darker; tarsi and feet olive.

Synallaxis gularis. White-browed Spinetail. This species was noted uncommonly in forest undergrowth on the upper east slope of Cerro Chinguela between 2625 and 2850 m. Solitary individuals and pairs were seen in dense bamboo thickets and other tangled vegetation from 0.5 to 3 m above ground. These birds crept along branches and probed clumps of mosses, bark, and the internodes of bamboo stalks. The infrequently heard song consists of an accelerating series of high-pitched notes (*cheet teet-teet-ti-titititi*) resembling, in pattern and quality, those of *Cranioleuca* spp. [but not including *Certhiaxis*, merged by Vaurie (1980) with *Cranioleuca*]. Vaurie (1980) listed but one record for Peru, a specimen taken at Maraynioc, Depto. Junín in 1891; this specimen, the type of *S. g. rufiventris*, is no longer extant. The LSUMZ has 15 specimens from Cerro Chinguela and 14 specimens from the following localities (all south of the Río Marañón): Cordillera Colán, Depto, Amazonas; Puerta del Monte, ca. 30 km NE Los Alisos, Depto. San Martín; Mashua, on the Tayabamba-Ongón trail, Depto. La Libertad; and the Cordillera Carpish, Depto. Huánuco. The habits and elevational distributions of *Synallaxis gularis* are similar at all five Peruvian localities. The series from south of the Río Marañón is easily distinguished from Cerro Chinguela specimens by the more brightly-colored, cinnamon underparts and more rufescent upperparts. In this respect, this series matches descriptions of the type of *S. g. rufiventris* (Cory and Hellmayr 1925:110), and we recommend that this well-marked form be recognized (contra Vaurie [1980]). See Braun and Parker (1985) for a discussion of the affinities of this species.

Specimen data: Cerro Chinguela: LSUMZ 87778–87788, 97649–97654 (skins); 89884–89886, 97493 (skeletons); 89434 (spirit); Cordillera Colán: LSUMZ 87788–87791 (skins); 89887 (skeleton); Puerta del Monte: LSUMZ 104472; Mashua: LSUMZ 92176–92178; Cordillera Carpish: LSUMZ 73994–73996, 79682–83, 80508 (skins); 75613, 81252 (skeletons); 79551 (spirit); 22 ♂♂, 10.5–14.5 g ($\bar{X}$ = 12.5 g); 7 ♀♀, 11–13 g ($\bar{X}$ = 12.2 g); iris brown; maxilla black; mandible flesh-pink; tarsi and feet olive or olive-brown; stomach: insects.

Schizoeaca griseomurina. Mouse-brown Thistletail. Although this species was previously thought to be endemic to southern Ecuador (Meyer de Schauensee 1966), we found it fairly common, although secretive, in forest edge shrubbery and bamboo of the paramo/forest ecotone on Cerro Chinguela, between 2960 and 3200 m. We normally encountered pairs in very thick growth within 1 m of the ground. These birds hopped deliberately and gleaned insects from small leaves and twigs. When disturbed, they flicked both wings simultaneously and raised and lowered their tails. They also uttered sharp *peent* or *feent* notes, which are quite like those of other members of the *S. fuliginosa* superspecies (*coryi, perijana, fuliginosa, griseomurina, palpebralis, vilcabambae, helleri,* and *harterti*; Vaurie 1980; Remsen 1981). The song was a rapidly intensifying trill of about 2 sec given at intervals of 6 sec.

Specimen data: LSUMZ 87758–87766; 3 ♂♂, 17.5–19 g ($\bar{X}$ = 18.5 g); 3 ♀♀, 15.5–17.5 g ($\bar{X}$ = 16.6 g); iris brown; maxilla dark gray; mandible paler, brown, gray, or flesh; tarsi and feet blue-gray.

Asthenes wyatti. Streak-backed Canastero. On 2 August 1980 Parker observed at least seven of these canasteros in grassland ca. 2 km N Cruz Blanca from 3100–3350 m. They were hopping on the ground amidst closely spaced clumps of grasses not far from scattered patches of bushes (especially *Gynoxys*). Occasionally, the birds flushed into the latter vegetation where they perched within the foliage and uttered disturbance calls (a dry *chik* or *chek*). Unfortunately, no specimens were collected. It is interesting to note the absence of *Asthenes flammulata* from seemingly suitable habitat at this locality. *Flammulata* is quite like *A. wyatti* behaviorally, although it appears to prefer more humid areas in Peru. *Asthenes wyatti* was not previously known in Peru north of central Cajamarca (Vuilleumier and Simberloff 1980), although it has been collected in southern Ecuador (Chapman 1926).

Asthenes flammulata. Many-striped Canastero. This species was fairly common in the paramo on Cerro Chinguela. Small numbers of solitary individuals were flushed daily from grassland with scattered shrubs, which were used for cover and song perches. The species was not recorded in open grassy areas devoid of shrubby growth, or in low, dense vegetation along forest edges where *Schizoeaca griseomurina* occurred. The song is an accelerating series of buzzy notes (*zhree-zree-ree-rrrr*) delivered from atop a bush or shrub. Disturbed individuals repeatedly utter a distinctive mewing call. Five specimens from the Huancabamba region resemble specimens of the nominate race but have more richly colored and well-defined gular patches [see Vuilleumier (1968) for a discussion of the geographic variation in the *Asthenes flammulata* superspecies]. *Asthenes flammulata* was previously unrecorded between Prov. El Oro, Ecuador (Chapman 1926) and central Depto. Cajamarca (Bond 1945) and southern Depto. Amazonas, Peru (LSUMZ 80530–80531, FMNH 65849, Museum of Vertebrate Zoology, University of California, Berkeley 156492), a gap of 300 to 400 km. Members of this superspecies may be more widespread than previously recognized, especially on the eastern slope of the eastern Andes in Peru. *A. flammulata* was common at Mashua, Depto. La Libertad and the Cordillera Carpish, Depto. Huánuco (specimens, LSUMZ), and *A. virgata* is now known from the Urubamba Valley, Depto. Cuzco (Parker and O'Neill 1980).

Specimen data: Cerro Chinguela: LSUMZ 87803–87805; FMNH 222316–222317; Mashua: LSUMZ 92202–92218; Cordillera Carpish: 74013–74016; 80529; 10 ♂♂, 21–27 g ($\bar{X}$ = 23 g); 9 ♀♀, 17–25.5 g ($\bar{X}$ = 22 g); iris brown; bill black, base of mandible blue-gray; tarsi olive; stomach: insect parts.

Automolus ruficollis. Rufous-necked Foliage-gleaner. A few individuals of this species were encountered in dense to moderately open forest understory (especially bamboo) and middlestory, on the west slope of the western cordillera below Cruz Blanca, between 1700–2900 m. These foliage-gleaners occurred singly. They probed arboreal bromeliads, ferns, and mosses on limbs as high as 10 m above ground, and hopped through dense tangles in bamboo, where their foraging movements were difficult to observe. Most sightings were of individuals 2 to 3 m above ground. Considerable trampling of the undergrowth by cattle and clearing of the bamboo by local people for food for pack animals pose a threat to this and other undergrowth inhabitants such as *Myrmeciza griseiceps* (primarily a bamboo bird). Unfortunately, these two species have restricted geographical and elevational ranges in densely settled regions of southwest Ecuador and northwest Peru.

Vaurie (1980) considered *Automolus ruficollis* to be a close relative of *A. ochrolaemus*. We see little morphological or behavioral similarity between *ruficollis* and any member of *Automolus.* The streaked underparts suggest a *Syndactyla.* The territorial song of *ruficollis* consists of an accelerating series (*chi chi-chi-chi-chchchchchch*) of notes; the call is an emphatic *chech*. Both vocalizations are strikingly similar in pattern and quality to homologous vocalizations of *Syndactyla rufosuperciliata,* another inhabitant of montane forest undergrowth (including bamboo thickets) in the Andes of Peru and Bolivia. *Automolus ruficollis* and *Syndactyla rufosuperciliata* share these types of vocalizations with at least one member of yet a third genus, *Simoxenops ucayalae* (Parker 1982). More thorough studies of every aspect of the morphology and natural history of these furnariids are needed.

Specimen data: LSUMZ 78408–10, 79686 (skins); 79799–79800 (skeletons); 5 ♀♀, 29–34 g ($\bar{X}$ = 31 g); iris brown; maxilla dark brown; mandible pale brown or pale gray; tarsi olive.

Thripadectes flammulatus. Flammulated Treehunter. Like many other forest-dwelling furnariids, this species was an inconspicuous inhabitant of dense undergrowth and trail edge thickets on the east slope of Cerro Chinguela, where four birds were mist-netted between 2135 and 2960 m. Our few sight records were of individuals in tangles of bamboo and other foliage within 2 m of the ground. The birds infrequently uttered emphatic *chek* notes. The territorial song is a loud, descending series of grating notes that accelerates toward the end. These two vocalizations are very similar to calls of *Thripadectes scrutator* of northern Peru to northern Bolivia (Remsen and Ridgely 1980). *T. scrutator* was previously known "only from about nine specimens taken in Junín, Ayacucho, and Cuzco in central Peru" (Vaurie 1980), but the LSUMZ has 14 specimens from the following additional Peruvian localities: Cordillera Colán (2440–2990 m), Depto. Amazonas; Puerta del Monte, ca. 30 km NE Los Alisos (3250 m), Depto. San Martín; Cumpang (2625–3000 m) on Tayabamba-Ongón trail, Depto. La Libertad; and several places in the Cordillera Carpish (2700–3050 m), Depto. Huánuco. *T. scrutator* forms a superspecies with *T. flammulatus* (see comments by Vuilleumier, in Vaurie 1980: 342). This is the first report of *T. flammulatus* for Peru. An additional unpublished record is of a bird collected by J. W. Fitzpatrick (specimen in the Museum of Comparative Zoology) on 28 July 1976 at 2450 m in the Cordillera del Condor, E San José de Lourdes, Depto. Cajamarca. The species was previously known from the Andes of Venezuela and Colombia south to Prov. Loja, southern Ecuador (Vaurie 1980).

Specimen data: Cerro Chinguela: LSUMZ 87896–98, 97674; Cordillera del Condor: MCZ 331072; 1 ♂, 53 g; 4 ♀♀, 50–62 g ($\bar{X}$ = 55.5 g); iris brown; bill black; tarsi and feet black or gray-brown; stomach: insects.

Grallaricula ferrugineipectus. Rusty-breasted Antpitta. A female (LSUMZ 78547) collected by Parker on the west slope of the western cordillera below Cruz Blanca at 3050 m and another female taken by M. Koepcke near Canchaque at 1750–1800 m represented the first records for the western cordillera of Peru (Schulenberg and Parker 1981). This species was previously known in Peru on the basis of one specimen, the type of *G. f. leymebambae* from Depto. Amazonas (Carriker 1933). At present, the LSUMZ has 45 additional specimens from the following Peruvian localities: Cordillera Colán, Depto. Amazonas (2000–2440 m); 33 km by road NE Ingenio (1830–2135 m), Depto. Amazonas; Puerta del Monte, ca. 30 km NE Los Alisos (3250 m), Depto. San Martín; Cumpang and Mashua (2450–3350 m), along Tayabamba-Ongón trail, Depto. La Libertad; and numerous localities in the Cordillera Carpish (2135–2800 m), Depto. Huánuco. The species was also listed for the Cordillera Vilcabamba, Depto. Cuzco (Weske 1972) and was recently discovered in Depto. La Paz, northern Bolivia (Schulenberg and Remsen 1982).

Specimen data: Cordillera Colán; LSUMZ 88074–88083 (skins); 89959–60 (skeletons); 89462 (spirit); Ingenio: 78548, 79687, 81995; Puerta del Monte:104493; Mashua and Cumpang: 92494–92501; Cordillera Carpish:74075–74090, 75244, 78550–79688, 80580–83 (skins); 74875–74878, 79811–79813, 81260 (skeletons); 75014, 81146–48 (spirit); 23 ♂♂, 15–21 g ($\bar{X}$ = 17.5 g); 20 ♀♀, 13–18 g ($\bar{X}$ = 16.8 g); iris brown; bill black, base of mandible white or pinkish white; tarsi and feet pink or pinkish gray; stomach: beetles, other insects.

Grallaricula nana. Slate-crowned Antpitta. This secretive antpitta was heard almost daily in small numbers on the east slope of Cerro Chinguela between 2400–2930 m. It was also recorded on the west slope at 2530 m in October 1977. The few individuals seen were hopping about in bamboo thickets and other vegetation within 1 m of the ground. These birds made short forward sallies to foliage and stems and occasionally attempted to hawk insects from the air within a few centimeters of a perch. They also regularly descended to the ground and foraged in the manner of a small thrush (*Catharus*), hopping several times and then stopping abruptly to capture small insects and arthropods on the forest floor. Like other *Grallaricula* this species frequently twitched both wings simultaneously. The song consisted of a soft, descending series of slightly buzzy whistles (*bzree-zree-zree-ree-ee-eeee*), which were delivered from perches about 0.5 m above ground. The song was given within the first hour of daylight on clear days and often until late morning on foggy days. An emphatic *chep* was often uttered in response to playbacks of songs. This is the first report of *Grallaricula nana* for Peru. Heretofore, the species was known from the Andes of Venezuela, Colombia, and Ecuador south to Prov. Napo (Chapman 1926). At Cruz Blanca and across the Marañón Valley on Cordillera Colán *G. nana* is replaced by *G. ferrugineipectus.*

Specimen data: LSUMZ 88084–88090, 97704 (skins); 86542 (partial skeleton); 89463 (spirit); 4 ♂♂, 18.5–20.5 g ($\bar{X}$ = 19.5 g); 4 ♀♀, 17.5–21.5 g ($\bar{X}$ = 19.8 g); iris brown; bill black, base of mandible pink or whitish; tarsi and feet gray.

Grallaricula peruviana. Peruvian Antpitta. Five individuals of this rare species were mistnetted in dense to moderately open undergrowth of forest at Playón, Lúcuma, and Machete (1680 to 2050 m). Despite intensive searches in the areas where individuals were netted, no free-flying birds were observed, and we consider the species to be genuinely uncommon. *Grallaricula peruviana* was known from only a single male specimen, the type, collected at Chaupe by H. Watkins on 3 March 1923 (Chapman 1923). The female, heretofore unknown, resembles the male except for the color of the crown. In one specimen the crown is nearly concolor with the back (olive-brown), but the forecrown is darker, almost black. In the other two specimens the crown is slightly rustier than the back, especially on the forecrown, but, nonetheless, is much closer to the color of the back than it is to the rust-orange color of the crown in the male. The only other sexually dimorphic *Grallaricula* is the recently described *G. ochraceifrons,* which is known from northern Peru south of the Río Marañón, and which may form a superspecies with *peruviana* (Graves et al. 1983).

Specimen data: LSUMZ 88091, 97706–97709; 2 ♂♂, 17 g, 17.5 g; 3 ♀♀, 19.5–21 g ($\bar{X}$ = 20 g); iris brown; maxilla black; mandible flesh, edged pale brown; tarsi and feet olive; stomach: beetles, other insects, fruit.

Grallaria quitensis. Tawny Antpitta. This species was known from Peru until now on the basis of two specimens collected by Watkins at El Tambo, Depto. Piura (Chapman 1926) and the type of *G. q. atuensis* collected at Atuén, Depto. Amazonas (Carriker 1933). Nominate *quitensis* was common at Cruz Blanca and on Cerro Chinguela. In the former locality it occurred at treeline and in patches of *Polylepis* and *Gynoxys* in the grassland from 3050–3200 m. On Cerro Chinguela *quitensis* was found in the edge of humid forest at treeline and also in isolated clumps of *Hesperomeles* shrubs bordering bogs in the paramo, from 2850 to 3300 m. Individuals of this antpitta regularly foraged up to 30 m from the cover of trees and shrubs in grassland, especially just after dawn or just before dusk, and throughout the day during foggy weather. Their loud, ringing song was a characteristic sound of the paramo; it consisted of three notes, the first higher and the last two on the same pitch (*uut—oo-oo*). The call note was a piercing *beeert.* Vocalizations were given from perches about 1 m above ground in dense cover at forest edge. In September–October 1979, *G. q. atuensis* was found to be common in similar treeline habitats at Mashua (between 3350–3500 m) on the Tayabamba-Ongón trail, Depto. La Libertad. The song of *atuensis* is like that of nominate *quitensis,* but the terminal syllable is distinctly slurred upward.

Specimen data: Cerro Chinguela and Cruz Blanca: LSUMZ 88070–71, 97683–97689; Puerta del Monte: 104492; Mashua: 92478–92494 (skins); 94019–20 (skeletons); 19 ♂♂, 57–83 g ($\bar{X}$ = 66 g); 6 ♀♀, 60–70 g ($\bar{X}$ = 64.7 g); iris brown; bill black, base of mandible of some specimens brown; tarsi and feet pale gray; stomach: caterpillars, beetles, dipteran larvae, other insects, leech, small bones (anuran?).

Grallaria hypoleuca. White-bellied Antpitta. A specimen collected by Graves at Playón (1680 m) on 7 July 1978 and eight unreported specimens [American Museum of Natural History (= AMNH) 181328–181335] taken nearby at Chaupe by Watkins between 31 January–26 March 1923 document the occurrence of this antpitta in Peru. This species was uncommon in tall, dense secondary forest along the edges of man-made clearings, trails, natural landslides, and in adjacent mature forest (especially bamboo) at Playón and Lúcuma (1675–2140 m). Throughout this elevational range, *G. hypoleuca* overlaps ecologically with the similarly-sized *G. ruficapilla.* The song of *hypoleuca* consists of three loud whistled notes (*too—tew-tew*); the last two are higher than the first and on the same pitch. Also uttered was a soft *whee-whee-whee-whee* series on one pitch; this is probably a disturbance call. *Grallaria h. castanea* was previously known from the Andes of Colombia and Ecuador south to Provs. Napo and Tungurahua (Chapman 1926).

Specimen data: AMNH 181328–181335; LSUMZ 88047; 1 ♀, 82 g with 8.3 g egg (fully-formed, shell light blue) in oviduct.

Grallaria nuchalis. Chestnut-naped Antpitta. This species was frequently heard in the high forest on the east slope of Cerro Chinguela between 2200 (rarely) and 2960 m. It was more or less restricted to the darkened recesses of bamboo thickets and adjacent forest undergrowth on steep slopes and in stream ravines. At least four territorial pairs of *G. nuchalis* were located along about 0.7 km of trail just below the paramo from 2750 to 2900 m, and many additional individuals were regularly heard singing farther away from the trail. The birds were extremely wary, and only with the aid of a tape-recorder were we able to obtain a series of specimens. The loud territorial song consisted of an accelerating, slightly ascending series of metallic notes that could be heard at a great distance (*tih—teh-teh-teh-teh-ti-ti-ti-tttttt*). This vocalization was given from dawn until mid-morning, or later during foggy weather. A response to song playback was a strident *chee-chee-chee-chee-chee*; this may have been uttered only by the female. This is the first report of *Grallaria nuchalis* for Peru. Our specimens are referable to the nominate race, previously known from Provs. Napo (Chapman 1926) and Zamora-Chinchipe (R. S. Ridgely, pers. comm.), Ecuador. Also observed in the immediate vicinity of *G. nuchalis* was the larger and rarer *G. squamigera,* and the smaller and commoner *G. rufula.* *Nuchalis* is replaced south of the Río Marañón by a recently described allospecies, *G. carrikeri* (Schulenberg and Williams 1982).

Specimen data: LSUMZ 88041, 97678–80; 2 ♂♂, 115 g, 122 g; 2 ♀♀, 110 g, 122 g; iris gray; bill black; tarsi and feet blue-gray; stomach: ants, beetles, other insect parts, millipede.

Chamaeza mollissima. Barred Antthrush. Parker saw a single individual of this species on 21 June 1980 on the ground under light undergrowth in tall cloudforest near a stream at Lúcuma (1830 m), and heard one or two others at Machete (ca. 2135 m) in early July 1980. The species has also been found in the Cordillera Vilcabamba, Depto. Cuzco (Weske 1972), and at Abra de Maruncunca and Valcón (2000–3000 m), Depto. Puno (specimens, LSUMZ). These Peruvian records fill a large gap in the distribution of this species, which was previously known only from the Andes of Colombia, Ecuador, and Bolivia (Meyer de Schauensee 1966). The distribution of *Chamaeza mollissima* along the eastern slopes of the Peruvian Andes is probably more continuous than the few records indicate.

Myornis senilis. Ash-colored Tapaculo. Almost nothing has been published on this monotypic genus. On Cerro Chinguela and at Cruz Blanca we found this secretive, solitary inhabitant of *Chusquea* bamboo thickets from 2685 to 3000 m. In contrast to similarly-plumaged, but more terrestrial members of *Scytalopus, Myornis senilis* was noted from 1 to 4 m above ground in thick foliage and tangles of bamboo stems. *Scytalopus latebricola* also occurs in *Chusquea* at these elevations on Cerro Chinguela (as does *S. unicolor* at Cruz Blanca), but it rarely ascends to more than 1 m above the ground. *Myornis senilis* was very difficult to observe, but its loud, distinctive vocalizations were heard daily at both of the above localities. The territorial song consists of a long series (95 sec) of about 15 emphatic *chuck* notes given at 10 sec intervals (decreasing to 5 sec toward the end), terminating in a rapid trill or pair of trills. The final few seconds of this vocalization may be phonetically transcribed as *chuk-chuk-chuchuk-chuchuk tititi-tititi-ttttit* etc. At times, only the well-spaced *chucks* were given. In response to playbacks of territorial song a persistent churring vocalization was uttered. Songs and calls were heard primarily only during the first hour of daylight (05:45–06:45). This is the first report of *Myornis senilis* in Peru. The LSUMZ has eight specimens from the following additional Peruvian localities: Cordillera Colán (2655 m), Depto. Amazonas; 33 km by road NE Ingenio (ca. 2300 m), Depto. Amazonas; Mashua and Cumpang (2745–3050 m), Depto. La Libertad. This tapaculo is rare in collections; it was previously known from only three localities in Colombia (Meyer de Schauensee 1966), and three in Ecuador (Chapman 1926).

Specimen data: Cerro Chinguela: LSUMZ 88105–07; Cordillera Colán: LSUMZ 88108; Ingenio: LSUMZ 82004 (skin), 84015 (skeleton); Mashua: LSUMZ 92510–11; 1 ♂, 23.5 g; 6 ♀♀, 18.1–24.5 g ($\bar{X}$ = 20.7 g); iris brown; maxilla dark gray; mandible gray; tarsi and feet gray-brown or olive-brown.

Scytalopus latebricola. Brown-rumped Tapaculo. This species was relatively common in the high-elevation forest on the east slope of Cerro Chinguela (2600–2900 m). Most individuals heard or seen were in bamboo, especially tangles of dead stalks and foliage within 1 m of the ground, but occasionally in branches and leaves as high as 2 m. *Scytalopus latebricola* shares the latter microhabitat with *Myornis senilis.* Three vocalizations were noted for *latebricola.* A territorial song given at dawn consisted of a low-pitched trill of about 10 sec. A shorter trill of about 2.5 sec was uttered throughout the day. An explosive *bzeek* was given primarily in response to playbacks of the previous two vocalizations. At higher elevations *S. latebricola* was replaced by *S. magellanicus* (primarily a treeline forest bird), and in lower, taller forest by *S. unicolor* and *S. femoralis*. *Scytalopus latebricola* was not previously known from Peru. It had been found as far south as Prov. Chimborazo, Ecuador (Peters 1951).

Specimen data: LSUMZ 97692–94; 2 ♂♂, 21 g, 24 g; 1 ♀, 23 g; iris brown, maxilla black; mandible brownish gray; tarsi brownish gray; stomach: beetles, other insects, plant material.

Acropternis orthonyx. Ocellated Tapaculo. One netted on 20 June 1978 in dense forest undergrowth near treeline (2960 m) on the upper east slope of Cerro Chinguela represents the first record of this beautiful species for Peru. In June–July 1980 it was noted irregularly in a narrow band of forest in the same general area. In October 1978 Schulenberg obtained three additional specimens on the upper west slope of Cordillera Colán, Depto. Amazonas. There the species was found in cloudforest undergrowth between 2290 and 2590 m.

In both localities *A. orthonyx* was recorded mainly by voice. The song, usually given within the first hour of daylight, consisted of a loud, mellow introductory whistle followed by several shorter whistles that were lower in volume and pitch; these were transcribed in field notes as *weéeoooooo-ooo-ooo-ooo* (Schulenberg) or *wheéeeeer-wheer-wheer* (Parker). Individuals walked on the forest floor in the manner of a rail and hopped deliberately through dense vegetation (especially dense bamboo) within 1 m of the ground. Like other forest-dwelling rhynocryptids, *Acropternis orthonyx* was very difficult to observe and has undoubtedly been overlooked in many places throughout its range. The species was previously known as far south as Provs. Pichincha (Chapman 1926) and Morona-Santiago (R. S. Ridgely, pers. comm.), Ecuador.

Specimen data: Cerro Chinguela: LSUMZ 88161, 97695; Cordillera Colán: LSUMZ 88162–64; 3 ♂♂, 90–100 g (X̄ = 95 g); 2 ♀♀, 81 g, 89 g; iris brown; maxilla gray; mandible dusky gray; tarsi gray-brown; stomach: beetles, spider, plant material.

Lipaugus fuscocinereus. Dusky Piha. We obtained only four records of this seemingly rare cotinga. One was collected by Parker on 16 June 1980 at Batán (2250 m) as it moved through forest canopy with a group of *Cacicus leucoramphus* (also scarce in the area). Single individuals were seen at the same locality on 14 July 1978 by D. Hunter and on 24 July 1980 by Parker and Braun, and another was observed below Batán at 2125 m by R. Rivera on 26 July 1980. These are the first Peruvian records of this cotinga, a species until now known only from the Andes of Colombia and Ecuador south to Prov. Zamora-Chinchipe (Snow 1979).

Specimen data: LSUMZ 97703; 1 ♀, 138 g; iris dark brown; bill black; tarsi dark gray; stomach: green fruit, 11 × 15 mm.

Myiophobus lintoni. Orange-banded Flycatcher. One or two small groups of these flycatchers were regularly observed in mixed-species flocks of other flycatchers, warblers, and tanagers in cloudforest canopy on the east slope of Cerro Chinguela between 2450–2750 m during June 1978 and June–July 1980. Probable sightings were also made on the west slope at 2530 m in October 1977. These birds perched upright and conspicuously atop leaves in the crowns of trees. Their foraging movements were mainly short, forward sally-gleans to upper leaf surfaces, and less frequently upward sallies to foliage from within the canopy. Individuals were very restless, seldom lingering in one tree for more than one minute, and calling almost constantly, most frequently uttering *chip* and *chep* notes. In its gregarious behavior and habit of perching atop canopy foliage *Myiophobus lintoni* is quite like *M. ochraceiventris*; the latter, not previously known north of Depto. Huánuco, Peru (Traylor 1979), was found in August–October 1978 on Cordillera Colán (2715–3050 m), Depto. Amazonas, and in September–October 1979 at Mashua and Cumpang (2800–3200 m), Depto. La Libertad, considerably reducing the gap between these two species. Morphological similarities also suggest that *lintoni* and *ochraceiventris* form a superspecies. Meyer de Schauensee (1951), however, who described *lintoni*, considered *M. pulcher* to be its closest relative. *Lintoni* was known previously from three specimens from two localities in the Andes of southern Ecuador (Meyer de Schauensee 1951). We obtained two adult males (LSUMZ 97736–97737) and one immature (88390) specimens. The adults agree with the type description, but the crown patch of both is Chrome Yellow of Ridgway (1912); the crown of the type is said to be Orange. Similar variation in crown color, apparently unrelated to age, sex, or locality, occurs in *M. ochraceiventris* (49 LSUMZ specimens). The immature lacks a crown patch, and its upperparts are browner than those of the adults, with little or no contrast between the back and sides of the crown. Also, the wing bars of the immature are a much deeper, rustier buff than are those of the adults.

Specimen data: LSUMZ 88390, 97736–37; 2 ♂♂, 9.5 g, 10 g; iris grayish yellow; maxilla blackish; mandible orange, tipped black; tarsi blackish; stomach contents: insects, small seeds.

Anairetes reguloides. Pied-crested Tit-Tyrant. One individual and a pair of this small flycatcher were observed by Parker on several days in *Polylepis* woodland and associated shrubbery at treeline (3050 m), 3 km SW of Cruz Blanca. These birds were territorial and very responsive to playbacks of their calls, including an explosive, rapid series (*wheek-tititittttttiti*) about 3 sec long, repeated at 40 to 60 sec intervals. A shorter *wheek-tic titititi* was also given. *Reguloides* made short, forward sallies to foliage and also perch-gleaned twigs and small leaves in the manner of other *Anairetes*. In the Cruz Blanca area *A. reguloides* was greatly outnumbered by *A. parulus*, which occurred in the same habitat. The above records of *reguloides* probably refer to the highland form *A. r. nigrocristatus*, previously known north to Chota, Depto. Cajamarca, about 150 km to the southeast (Traylor 1979). Primarily lowland *A. r. albiventris* inhabits riparian thickets in the foothills and along the coast of Peru north to Depto. Ancash, some 400 km south of Cruz Blanca.

Leptopogon rufipectus. Rufous-breasted Flycatcher. Pairs of this conspicuous flycatcher were frequently observed with mixed-species flocks from 1 to 20 m above ground in forest at Playón, Lúcuma, and Machete (1670–2140 m). In the undergrowth *L. rufipectus* regularly associated with *Basileuterus coronatus, B. tristriatus,* and *Hemispingus frontalis.* At mid-heights and in the lower canopy *rufipectus* followed larger flocks containing various *Tangara* spp. (*nigroviridis, parzudakii, vassorii,* and *xanthocephala*), *Iridosornis analis, Anisognathus flavinucha, Diglossa cyanea, D. caerulescens,* other small flycatchers, and furnariids. Like other members of *Leptopogon*, this species characteristically raised, opened, and flashed one wing at regular, short intervals; Ridgely and Gaulin (1980) noted similar behavior for *L. rufipectus* in Colombia. Foraging movements were primarily upward sallies to the undersides of leaves. The species frequently uttered an explosive *kweek*. This is the first report of *Leptopogon rufipectus* from Peru; the species was previously known only as far south as Prov. Napo and Tungurahua, Ecuador (Chapman 1926). *Leptopogon rufipectus* forms a superspecies with the very similar *L. taczanowskii*, found south of the Río Marañón valley from Depto. Amazonas to Depto. Cuzco, Peru (Meyer de Schauensee 1966).

Specimen data: LSUMZ 88507–88510, 97761–62 (skins); 97509 (skeleton); 97549 (spirit); 4 ♂♂, 11.5–18 g ($\bar{X}$ = 14.3 g); 2 ♀♀, 10 g, 12.8 g; iris brown; bill black or dark brown; tarsi and feet gray or bluish gray; stomach: beetles, other insects.

Notiochelidon flavipes. Pale-footed Swallow. Parker and O'Neill (1980) recently summarized the available information on the behavior and distribution of this cloudforest swallow. Since that publication, we have found the species in the following additional Peruvian localities, from north to south (specimens, LSUMZ): Cerro Chinguela, on east and west slopes (2440–2960 m; rarely as low as 1950 m), Depto. Piura; Cordillera Colán (2450–2930 m), Depto. Amazonas; Cumpang and Mashua (2450–3050 m), Depto. La Libertad; and Valcón, 5 km NNW Quiaca (3000 m), Depto. Puno. These records fill large distributional gaps in northern and extreme southern Peru. The species was also recently found in northern Bolivia (Parker et al. 1980). On Cerro Chinguela *N. flavipes* was seen daily in groups as large as 30 to 40 individuals. On a few occasions, mainly during inclement weather, they were observed as low as 2135 m over man-made clearings and forest edge in association with many *N. cyanoleuca* and (on 16 June 1980) one *Hirundo rustica*. Otherwise *N. flavipes* was restricted to the upper cloudforest. Braun and Parker saw several individuals entering and leaving a cavity in the underside of a huge moss clump on a large tree limb in dense forest at 2745 m. *Notiochelidon flavipes* may nest and roost in such sites. In August 1981, in the Cordillera Carpish, Depto. Huánuco, Parker watched two individuals of this swallow repeatedly land on a steep, moss-covered bank, gather slender strips of bamboo leaves in their bills, and fly off over the forest. The vocalizations and general behavior of *Notiochelidon flavipes* on Cerro Chinguela and at Cumpang were as described by Parker and O'Neill (1980).

Specimen data: Cerro Chinguela: LSUMZ 97770; Cordillera Colán: 88533, 88535–36 (skins); 88534 (skeleton and partial skin); Cumpang: 92930; Cordillera Carpish:79709; Valcón 98598–99; 5 ♂♂, 8.8–10.0 g ($\bar{X}$ = 9.3 g); 3 ♀♀, 9–10 g ($\bar{X}$ = 9.5 g); iris brown; bill black; tarsi and feet pale flesh.

Cinnycerthia unirufa. Rufous Wren. This was one of the common flocking birds of forest undergrowth and the shrubby forest/paramo ecotone on both slopes of Cerro Chinguela from 2440 to 3200 m. It almost always was encountered in groups of four to six individuals, which often formed the nucleus of mixed-species flocks that contained other inhabitants of dense undergrowth such as *Hemispingus atropileus, Catamblyrhynchus diadema, Atlapetes pallidinucha,* and *A. rufinucha.* These species were frequently observed in *Chusquea* bamboo thickets within forest. *Cinnycerthia unirufa* is a superb singer; duets performed by pairs within (family?) groups consisted of loud, melodious, canary-like whistles and trills. While one individual repeated a melodious *chu-woo* or *chi-wee* phrase, the other simultaneously rapidly trilled on one pitch. Song bouts were about 10 to 15 sec long. Singing birds perched upright and side by side with their bills almost touching. The call note, a raspy *chit,* was often uttered in unison by all members of a group. This is one of the few cloudforest species that respond vigorously to swishing or squeaking; whole parties approached closely and gave their loud, characteristic calls. On several occasions allopreening between two or three members of a group was observed. *Cinnycerthia unirufa* replaces its relative *C. peruana* at upper elevations on Cerro Chinguela. The latter, which is behaviorally and ecologically very similar to *unirufa,* was an uncommon species at Machete and Batán (2135–2350 m); it may narrowly overlap with *unirufa* at the upper limit of its elevational range. South of the Río Marañón *peruana* ranges as high as 3000 m, an apparent example of competitive release. This is the first report of *Cinnycerthia unirufa* for Peru. It was previously known as far south as Prov. Loja, Ecuador (Chapman 1926).

Specimen data: LSUMZ 88551–88561, 97772–97777 (skins); 90121–22, 90125, 97510 (skeletons); 89510–16 (spirits); 7 ♂♂, 23.5–28.5 g (X̄ = 26 g); 10 ♀♀, 21–29 g (X̄ = 24.1 g); iris gray-brown; bill black (dark brown with paler mandible in juveniles); tarsi and feet dark gray; stomach: beetles.

Thryothorus euophrys. Plain-tailed Wren. This species was previously known from Peru on the basis of a single specimen, the type of *T. e. atriceps* collected by Watkins at Chaupe ("6100 ft," Chapman 1924). We found these wrens to be uncommon at 2350 m, along the mule trail above Batán, on the east slope of Cerro Chinguela. Several pairs were located in dense *Chusquea* bamboo thickets along the edges of natural landslides within 75 m of the trail, and others were heard in bamboo along a knife-like ridge crest across a narrow valley to the north. The birds remained well-hidden within thickets from 0.5 to 1.5 m above ground. Their loud, chortling songs were heard daily just after dawn. The vocal behavior and taxonomy of the *T. euophrys* superspecies are discussed in Parker and O'Neill (1985).

Specimen data: LSUMZ 97789–97793; 3 ♂♂, all 27.5 g; 2 ♀♀, 26 g, 27 g; iris chestnut brown; maxilla blackish; mandible blue-gray; tarsi and feet gray or gray-brown; stomach: beetles, caterpillars, other insects.

Iridosornis rufivertex. Golden-crowned Tanager. Small numbers of this species (1–3 per flock) were observed daily in mixed-species flocks of tanagers and other birds in the upper cloudforest on both slopes of Cerro Chinguela, and in isolated groves of trees and shrubs in the paramo. The species was less commonly noted in taller forest as low as 2590 m. Treeline forest flocks usually contained the following species: *Margarornis squamiger, Pseudocolaptes boissonneauti, Mecocerculus leucophrys, Myioborus melanocephalus, Conirostrum sitticolor, Buthraupis eximia,* and *Hemispingus verticalis. Iridosornis rufivertex* tended to stay concealed within dense foliage, often of shrubby growth at the edge of wooded areas. The species was not very vocal during our extended visits. Individuals of *I. rufivertex* were observed taking various small fruits and also gleaning leaves. This is the first published report of this tanager for Peru. An additional specimen was collected by J. P. O'Neill on 19 July 1976 in the Cordillera del Condor, E San José de Lourdes, Depto. Cajamarca. The species was previously known from the Andes of Venezuela and Colombia, south to Prov. Loja, Ecuador (Chapman 1926).

Specimen data: Cerro Chinguela: LSUMZ 89014–89024, 97924–25 (skins); 90260 (skeleton); 89568 (spirit); Cordillera del Condor: LSUMZ 82270; 8 ♂♂, 19–25.5 g (X̄ = 21.9 g); 4 ♀♀, 21–27 g (X̄ = 23 g); iris brown or red-brown; maxilla black; mandible silver-blue, sometimes with black tip; tarsi black; stomach: seeds, insects.

Buthraupis eximia. Black-chested Mountain-Tanager. We observed this large tanager almost daily, usually at the upper limit of cloudforest on both slopes of Cerro Chinguela (2850–3300 m). *Buthraupis eximia* was noted in conspecific groups of three to five individuals, and in about one of four mixed-species flocks of other tanagers, wrens, flower-piercers, and brush-finches. The mountain-tanagers normally stayed within the dense foliage of trees and shrubs, searching foliage and epiphytic growth on branches. They were also noted in several varieties of fruiting trees, including melastomes and *Shefflera* spp. This species was observed on at least two occasions as low as 2745 m, where individuals were seen in canopy flocks that included the larger *B. montana. Buthraupis eximia* rarely vocalized, except for occasional *seep* notes; like its allospecies *B. aureodorsalis,* however, *eximia* is capable of producing a long, fairly elaborate song that continues for up to 30 sec. Parker transcribed the repeating element of one such song as follows: *tititi-turry-tititi-tee-ter-turry* etc. *Buthraupis eximia* was previously known to occur south to Prov. Pichincha, Ecuador (Meyer de Schauensee 1966).

Specimen data: LSUMZ 88964–88970, 97905–97912 (skins); 90230 (skeleton); 89562 (spirit); 10 ♂♂, 50–70 g (X̄ = 63 g); 5 ♀♀, 58–65 g (X̄ = 62 g); iris brown; bill black or dark brown; tarsi and feet black or dark brown; stomach: seeds, fruit.

Buthraupis wetmorei. Masked Mountain-Tanager. This rare, little-known species was observed only twice during more than two months of fieldwork on Cerro Chinguela. Two or three were seen by Parker on 13 July 1980, and one was collected from what was probably the same group the following day. In both instances *B. wetmorei* was with a varied assortment of other tanagers, including *Buthraupis eximia, Iridosornis rufivertex,* and *Hemispingus verticalis,* in the canopy of treeline forest by our campsite (2900 m). Thinking that *wetmorei* may be restricted to the uppermost patches of montane forest and, perhaps, to isolated tree groves in the upper paramo (3100–3400 m), we searched extensively for it there, but to no avail. *Buthraupis wetmorei* was known previously only from one locality in Colombia (Puracé, Depto. Cauca) and one in Ecuador (Mt. Sangay, Prov. Morona-Santiago [Meyer de Schauensee 1966]).

Specimen data: LSUMZ 97913; 1 ♀, 62 g; iris brown; maxilla blackish; mandible blue-gray tipped black; tarsi dark brown; stomach: seeds.

Hemispingus verticalis. Black-headed Hemispingus. This tanager was a common, conspicuous member of canopy mixed-species flocks in high-elevation cloudforest from 2680 to 3200 m (treeline) on both slopes of Cerro Chinguela, and in isolated tree groves in the paramo up to 3350 m. Three to five *H. verticalis* were usually seen per flock. One pair within each flock appeared to be territorial. These birds simultaneously sang a complex, unmusical jumble of notes lasting 5–15 sec, and repeated at short intervals, often for many minutes at a time. In terms of pattern, quality, and duration, these songs were quite like those of frequent flock associates *Basileuterus luteoviridis* (common below 2890 m), *Cnemoscopus rubrirostris, Hemispingus atropileus,* and *Catamblyrhynchus diadema.* Such flocks were very cohesive and seemed to be led by a pair of *Basileuterus luteoviridis.* At treeline *Hemispingus verticalis* may play a central role in flock leadership. This species behaved in the manner of *H. xanthophthalmus,* an allospecies found south of the Río Marañón (Parker and O'Neill 1980; Parker et al. 1980). They walked and hopped through the uppermost leaf clusters of shrubs and trees, often perching on top of leaves and gleaning their upper surfaces. They also fed on small fruits of several kinds, including melastomes. This is the first report of *verticalis* for Peru; it was previously known to occur south to Prov. Loja, southern Ecuador (Ridgely 1980).

Specimen data: LSUMZ 88928–30, 97894–95 (skins); 97520 (spirit); 4 ♂♂, 12–14 g (X̄ = 13.3 g); 2 ♀♀, 12.5 g, 13 g; iris yellow; bill black; tarsi and feet dark brown; stomach: beetles, caterpillars.

Atlapetes pallidinucha. Pale-naped Brush-Finch. This was one of the most conspicuous birds of the paramo/forest ecotone and upper cloudforest on both slopes of Cerro Chinguela. The species was noted in pairs or groups of three or four individuals, often in association with other undergrowth species, particularly *Cinnycerthia unirufa. Atlapetes pallidinucha* was most often seen within 1 m of the ground in dense shrubbery and low trees at the forest edge, but occurs also in the interior of temperate forest where it shows a preference for bamboo thickets. At the lower limit of its elevational range, *A. pallidinucha* narrowly overlaps with its similar congener *A. rufinucha* from 2800 to 2900 m, where both species were noted in the same mixed-species flocks. On Cerro Chinguela *rufinucha* inhabits both forest and edge situations below ca. 2900 m; at Cruz Blanca, in the absence of *pallidinucha, rufinucha* was common up to the uppermost shrubby vegetation (3250 m). *Atlapetes pallidinucha* gave at least three types of vocalizations on Cerro Chinguela: a musical, whistled *wheet-tew-tew-tew,* which is presumably a territorial or advertising song; a long, complex series of notes that drop in pitch and volume, that are given simultaneously by both members of a pair (the "chatter duet" of Paynter 1978), and that apparently serve to announce territorial boundaries; and an emphatic, often protracted *seee,* which is probably a contact call. *Atlapetes pallidinucha* was previously known from the Andes of Venezuela and Colombia, south to Prov. Azuay, Ecuador (Hellmayr 1938). This gregarious, high-elevation brush-finch apparently has no ecological equivalent in the paramo zone south of the Río Marañón.

Specimen data: LSUMZ 88764–73, 97827–29 (skins); 90178–79 (skeletons); 5 ♂♂, 31.0–40.0 g (X̄ = 37.1 g); 7 ♀♀, 31.0–38.0 g (X̄ = 34.4 g); iris chestnut brown; bill black; tarsi brown; stomach: seeds, other plant material, grit.

Incaspiza ortizi. Gray-winged Inca-Finch. On 13 June 1980 Parker, Braun, and J. W. Eley found a population of this rare finch in dense scrub 2 km by road NE Huancabamba at 2150

m. A small series was obtained, and several males were tape-recorded. Like other members of *Incaspiza,* this species was uncommon and difficult to approach. The birds foraged mainly on the ground under a dense cover of shrubs that averaged about 1.5 m high. Various cacti, including a large, columnar *Cereus*-like species and an *Opuntia* sp., and terrestrial bromeliads were common in this habitat. The singing males perched conspicuously atop bushes or short trees. Their song was comprised of three very high-pitched, insect-like notes. This vocalization was difficult to hear at a distance of more than 75 m. A female was seen carrying a caterpillar about 2 cm long, and we assume that young were nearby, since most of the bird species were breeding. *Incaspiza ortizi* was greatly outnumbered by other emberizids, including *Sporophila luctuosa, Phrygilus plebejus,* and *Zonotrichia capensis.* On 27 October 1980, conditions were much drier and most reproductive behavior had terminated. One male *ortizi* was in full song.

Incaspiza ortizi was known previously only from the type, a female collected on the Pacific slope of the western cordillera at La Esperanza, and an immature male from the Río Marañón drainage at Hacienda Limón, both in Depto. Cajamarca (Zimmer 1952). In early August 1976 the LSUMZ obtained 11 additional specimens at Hacienda Limón, including the first adult males. As in other species of *Incaspiza,* the sexes are similar in plumage; male *ortizi* differ from females only in having the gray of the breast and flanks purer, the center of the belly whiter, and the black of the face deeper and slightly more extensive. Both sexes differ from *I. pulchra* and *I. personata* by lacking all traces of reddish-brown on the wings and back. On the basis of vocal and morphological differences between *ortizi, pulchra,* and *personata,* and contact between some members of the group, we regard all three taxa as full species (Parker, unpubl. data).

Specimen data: Huancabamba: LSUMZ 97821–97826 (skins); Hacienda Limón: LSUMZ 81013, 81015, 81017–81024 (skins); 81353 (skeleton); 10 ♂♂, 29–38 g ($\bar{X}$ = 35 g); 7 ♀♀, 28–33 g ($\bar{X}$ = 31 g); iris chestnut; bill orange (dull yellow with horn-colored culmen in some ♀ specimens); tarsi and feet yellow or yellow-orange; stomach: plant and insect parts.

DISCUSSION

The Huancabamba region is one of the best studied areas in the Peruvian Andes (Chapman 1926; LSUMZ, 380+ worker-days). Zoogeographic patterns in this area are complex, however, and avifaunal analyses must be regarded as tentative pending better understanding of the forces governing bird distribution in the Andes.

Several preliminary conclusions can be reported. The addition of 23 north Andean species to the Peruvian avifauna suggests that there are few or no significant barriers to dispersal along the east slope from north-central Ecuador to northern Peru. It is probable that most montane forest species recorded south along the eastern Andean slope to Prov. Morona-Santiago, Ecuador (ca. 400 km NNE Huancabamba) will eventually be recorded in Peru. Similarly, montane forest species occurring north of the Río Marañón, but not yet recorded from Ecuador (e.g., *Gallinago imperialis, Metallura odomae, Heliangelus regalis, Campylorhamphus pusillus, Grallaricula peruviana, Henicorhina leucoptera*), probably range northward to Prov. Zamora-Chinchipe or Morona-Santiago.

Of the 250+ montane species recorded from the Huancabamba region, 26 apparently reach their southern limits at the North Peruvian Low. This area has two geographic components, the low Andean passes of the western cordillera, and the low arid valley of the Río Marañón. These two topographic features appear to differ in their effectiveness as barriers to bird distribution. As many as 18 pairs of allospecies are separated by the Río Marañón (Table 1). In this respect, the Río Marañón is by far the most important barrier to forest birds along the eastern slope of the Andes from Venezuela to Bolivia. As expected, high-elevation species are most affected. Only two allospecies pairs (*Grallaricula peruviana—G. ochraceifrons, Leptopogon rufipectus—L. taczanowskii*) listed in Table 1 regularly occur below 2000 m elevation. The arid floodplain of the Marañón Valley would seem to be an effective barrier to high-elevation species. Accordingly, the discovery of relatively undifferentiated populations of *Acropternis orthonyx* and *Myornis senilis* south of the Río Marañón is perplexing. Neither of these species is an *a priori* candidate for dispersal across an arid, low-elevation barrier. Either they are better dispersers than their ecologies suggest, or they have undergone little morphological divergence (cf. Table 1) since the initial vicariant event (disruption of forest by the eroding Marañón Valley). Other Huancabamba species with "equal or better" flight capabilities and no apparent taxonomic or ecological counterpart immediately east of the Río Marañón (Cordillera Colán) have not been recorded there (e.g., *Aegolius harrisii, Hapalopsittaca ama-*

TABLE 1
PROBABLE ALLOSPECIES PAIRS SEPARATED BY THE RÍO MARAÑÓN

Cerro Chinguela	Cordillera Colán
Nothocercus bonapartei[1]	Nothocercus nigrocapillus
Penelope barbata	Penelope montagnii
Coeligena lutetiae	Coeligena violifer
Metallura odomae	Metallura theresiae
Eubucco bourcierii[2]	Eubucco versicolor
Schizoeaca griseomurina	Schizoeaca fuliginosa
Thripadectes flammulatus	Thripadectes scrutator
Grallaria nuchalis	Grallaria carrikeri
Grallaricula nana	Grallaricula ferrugineipectus
Grallaricula peruviana	Grallaricula ochraceifrons
Ochthoeca diadema	Ochthoeca pulchella
Poecilotriccus ruficeps	Poecilotriccus sp. nov.
Myiophobus lintoni	Myiophobus ochraceiventris
Leptopogon rufipectus	Leptopogon taczanowskii
Diglossa lafresnayii	Diglossa mystacalis
Diglossa humeralis	Diglossa carbonaria
Iridosornis rufivertex	Iridosornis reinhardti
Hemispingus verticalis	Hemispingus xanthophthalmus

[1] Collected north of the North Peruvian Low at Chaupe (Meyer de Schauensee 1966).
[2] Collected north of the North Peruvian Low at San Ignacio (Meyer de Schauensee 1966).

zonina, Scytalopus latebricola, Lipaugus fuscocinereus, Cinnycerthia unirufa, Turdus fulviventris, Buthraupis wetmorei). Likewise, several species recorded on Cordillera Colán (e.g., Xenoglaux loweryi, Loddigesia mirabilis, Thripophaga berlepschi, Thamnophilus caerulescens, Tangara argyrofenges, Hemispingus rufosuperciliaris) have no counterparts west and north of the Río Marañón.

Some species or superspecies show puzzling range gaps in the Huancabamba region. For example, the Oreotrochilus estella superspecies has well-differentiated allospecies in the Andes of central Ecuador (O. chimborazo) and north-central Peru south of the North Peruvian Low. Apparently suitable habitat exists above 3600 m on the drier west slope of Cerro Chinguela just south of the Ecuadorian border. With additional fieldwork, we predict that a representative of Oreotrochilus will eventually be found there, but the taxon—chimborazo, estella or an undescribed allospecies—remains a question. On the other hand, some distributional gaps may be real. The gap in the distribution of the Anairetes agilis-agraphia superspecies in the Huancabamba region does not appear to be an artifact of incomplete sampling. Observers thoroughly familiar with the behavior of this species failed to find it in "suitable" habitat during three expeditions. Similar distributional gaps are known for other species in the Andes of Ecuador and Peru. Whether these represent true distributional patchiness or sampling artifacts can only be determined by additional fieldwork.

Our fieldwork near Huancabamba raises questions about the efficacy of the low passes of the western cordillera as a barrier to montane species, especially those inhabiting steppe-scrub. Four species previously unknown north of the North Peruvian Low (Phalcoboenus megalopterus, Gallinago andina, Colaptes rupicola, Anairetes reguloides) were found in the Huancabamba region. More significant are the several dozen species (e.g., Penelope barbata, Piculus rivolii, Ensifera ensifera, Hemispingus superciliaris) characteristic of montane forest that occur well south of the low passes of this area. Extensive pockets of such forest are found at least south to Chota (LSUMZ field party) and Taulis (Koepcke 1961), Depto. Cajamarca (130 km SE Porculla Pass). The distribution of the humid montane forest species appears to be limited more by increasing aridity southward along the western cordillera than by the low passes. Strictly speaking, there are no high-elevation allospecies pairs with species separated by low-elevation passes of the North Peruvian Low.

To date we have many more questions than answers about the mechanisms of geographic exclusion, the stability of "guild" structure across geographic barriers, and the speciation process in the Andes. We hope that continued fieldwork directed at the distribution and natural history of Andean birds will increase our understanding of these phenomena.

ACKNOWLEDGMENTS

Our field investigations in Peru were conducted with the support of the Dirección General Forestal y de Fauna of the Ministerio de Agricultura, Lima, Peru. We are most grateful to J. S. McIlhenny, H. I. and L. R. Schweppe, and B. M. Odom for their continued interest in and support of the LSUMZ fieldwork. We owe a special debt to A. and H. Koenig, M. and I. Plenge, and G. del Solar for their hospitality and logistical support in Peru. Our expeditions to the Huancabamba region would not have been possible without the aid of our field assistants, M. Sánchez S. and R. Rivera A., and our field companions, L. J. Barkley, J. W. Eley, D. H. Hunter, and R. D. Semba. P. A. Buckley, S. Coats, J. W. Fitzpatrick, M. S. Foster, J. V. Remsen, and R. S. Ridgely provided valuable comments on the manuscript.

LITERATURE CITED

BANGS, O. AND G. K. NOBLE. 1918. List of birds collected on the Harvard Peruvian Expedition of 1916. Auk 35:442–463.

BOND, J. 1945. Notes on Peruvian Furnariidae. Proc. Acad. Nat. Sci. Phila. 97:17–39.

BRAUN, M. J., AND T. A. PARKER, III. 1985. Molecular, morphological, and behavioral evidence concerning the taxonomic relationships of "*Synallaxis*" *gularis* and other synallaxines. Pp. 333–346. *In* P. A. Buckley, M. S. Foster, E. S. Morton, R. S. Ridgely, and F. G. Buckley (eds.), Neotropical Ornithology. Ornithol. Monogr. No. 36.

CARRIKER, M. A., JR. 1933. Descriptions of new birds from Peru, with notes on other little-known species. Proc. Acad. Nat. Sci. Phila. 85:1–38.

CHAPMAN, F. M. 1923. Descriptions of proposed new birds from Venezuela, Colombia, Ecuador, Peru, and Chile. Am. Mus. Novit. No. 96.

CHAPMAN, F. M. 1924. Descriptions of new birds from Ecuador, Colombia, Peru, and Bolivia. Am. Mus. Novit. No. 138.

CHAPMAN, F. M. 1926. The distribution of bird-life in Ecuador. Bull. Am. Mus. Nat. Hist. 55:1–784.

CORY, C. B., AND C. E. HELLMAYR. 1925. Catalogue of birds of the Americas. Part 4. Field Mus. Nat. Hist., Zool. Ser., 13(4):1–390.

DELACOUR, J., AND D. AMADON. 1973. Curassows and Related Birds. American Museum of Natural History, New York.

ELEY, J. W., G. R. GRAVES, T. A. PARKER, III, AND D. R. HUNTER. 1979. Notes on *Siptornis striaticollis* (Furnariidae) in Peru. Condor 81:319.

GRAVES, G. R. 1980. A new species of metaltail hummingbird from northern Peru. Wilson Bull. 92: 1–7.

GRAVES, G. R., J. P. O'NEILL, AND T. A. PARKER III. 1983. *Grallaricula ochraceifrons*, a new species of antpitta from northern Peru. Wilson Bull. 95:1–6.

HELLMAYR, C. E. 1938. Catalogue of birds of the Americas. Part 11. Field Mus. Nat. Hist., Zool. Ser., 13(11):1–662.

HELLMAYR, C. E., AND B. CONOVER. 1948. Catalogue of birds of the Americas. Part 1, No. 3. Field Mus. Nat. Hist., Zool. Ser., 13(1):1–383.

KOEPCKE, M. 1961. Birds of the western slope of the Andes of Peru. Am. Mus. Novit. No. 2028.

MÉNÉGAUX, A. 1910. Etude d'une collection d'oiseaux du Pérou. Bull. Mus. Hist. Nat. 16:359–367.

MEYER DE SCHAUENSEE, R. 1951. Notes on Ecuadorian birds. Notulae Naturae 234:1–11.

MEYER DE SCHAUENSEE, R. 1966. The Species of Birds of South America and Their Distribution. Livingston Publ. Co., Narberth, Pennsylvania.

O'NEILL, J. P., AND T. S. SCHULENBERG. 1979. Notes on the Masked Saltator, *Saltator cinctus*, in Peru. Auk 96:610–613.

PARKER, T. A., III. 1982. Observations of some unusual rainforest and marsh birds in southeastern Peru. Wilson Bull. 94:477–493.

PARKER, T. A., III, AND J. P. O'NEILL. 1980. Notes on little known birds of the upper Urubamba Valley, southern Peru. Auk 97:167–176.

PARKER, T. A., III, AND J. P. O'NEILL. 1985. A new species and a new subspecies of *Thryothorus* wren from Peru. Pp. 9–15, *In* P. A. Buckley, M. S. Foster, E. S. Morton, R. S. Ridgely, and F. G. Buckley (eds.), Neotropical Ornithology. Ornithol. Monogr. No. 36.

PARKER, T. A., III, AND S. A. PARKER. 1980. Rediscovery of *Xenerpestes singularis* (Furnariidae). Auk 97:203–205.

PARKER, T. A., III, J. V. REMSEN, JR., AND J. A. HEINDEL. 1980. Seven bird species new to Bolivia. Bull. Br. Ornithol. Club 100:160–162.

PAYNTER, R. A., JR. 1978. Biology and evolution of the avian genus *Atlapetes* (Emberizinae). Bull. Mus. Comp. Zool. 148:323–369.

PETERS, J. L. 1951. Check-list of Birds of the World, Vol. 7. Museum of Comparative Zoology, Cambridge, Massachusetts.

REMSEN, J. V., JR. 1981. A new subspecies of *Schizoeaca harterti* with notes on taxonomy and natural history of *Schizoeaca* (Aves: Furnariidae). Proc. Biol. Soc. Wash. 94:1068–1075.

REMSEN, J. V., JR., AND R. S. RIDGELY. 1980. Additions to the avifauna of Bolivia. Condor 82:69–75.

REMSEN, J. V., JR., AND M. A. TRAYLOR, JR. 1983. Additions to the avifauna of Bolivia, Part 2. Condor 85:95–98.

RIDGELY, R. S. 1980. Notes on some rare or previously unrecorded birds in Ecuador. Am. Birds 34: 242–248.

RIDGELY, R. S., AND S. J. C. GAULIN. 1980. The birds of Finca Merenberg, Huila Department, Colombia. Condor 82:379–391.

RIDGWAY, R. 1912. Color Standards and Color Nomenclature. Published by the author. Washington, D.C.

SCHULENBERG, T. S., S. E. ALLEN, D. F. STOTZ, AND D. A. WIEDENFELD. 1984. Distributional records from the Cordillera Yanachaga, central Peru. Gerfaut 74:57–70.

SCHULENBERG, T. S., AND T. A. PARKER, III. 1981. Status and distribution of some northwest Peruvian birds. Condor 83:209–216.

SCHULENBERG, T. S., AND J. V. REMSEN, JR. 1982. Eleven bird species new to Bolivia. Bull. Br. Ornithol. Club 102:52–57.

SCHULENBERG, T. S., AND M. D. WILLIAMS. 1982. A new species of antpitta (*Grallaria*) from northern Peru. Wilson Bull. 94:105–113.

SHORT, L. L. 1972. Systematics and behavior of South American flickers (Aves, *Colaptes*). Bull. Am. Mus. Nat. Hist. 149:1–110.

SNOW, D. L. 1979. Cotingidae. Pp. 281–308. *In* M. A. Traylor, Jr. (ed.), Check-list of Birds of the World, Vol. 8. Museum of Comparative Zoology, Cambridge, Massachusetts.

STEPHENS, L., AND M. A. TRAYLOR, JR. 1983. Ornithological Gazetteer of Peru. Museum of Comparative Zoology, Cambridge, Massachusetts.

TERBORGH, J. W., AND J. S. WESKE. 1972. Rediscovery of the Imperial Snipe in Peru. Auk 89:497–505.

TRAYLOR, M. A., JR. 1979. Tyrannidae. Pp. 1–245, *In* M. A. Traylor, Jr. (ed.), Check-list of Birds of the World, Vol. 8. Museum of Comparative Zoology, Cambridge, Massachusetts.

VAURIE, C. 1980. Taxonomy and geographical distribution of the Furnariidae (Aves, Passeriformes). Bull. Am. Mus. Nat. Hist. 166:1–357.

VUILLEUMIER, F. 1968. Population structure of the *Asthenes flammulata* superspecies (Aves: Furnariidae). Breviora 297:1–21.

VUILLEUMIER, F. 1969. Pleistocene speciation in birds living in the high Andes. Nature 223:1179–1180.

VUILLEUMIER, F. 1970. Generic relations and speciation patterns in the caracaras (Aves: Falconidae). Breviora 355:1–29.

VUILLEUMIER, F., AND D. SIMBERLOFF. 1980. Ecology versus history as determinants of patchy and insular distributions in high Andean birds. Evol. Biol. 12:235–379.

WESKE, J. S. 1972. The distribution of the avifauna in the Apurimac Valley of Peru with respect to environmental gradients, habitat, and related species. Unpubl. Ph.D. dissert., University of Oklahoma, Norman.

ZIMMER, J. T. 1952. A new finch from northern Perú. J. Washington Acad. Sci. 42:103–104.

APPENDIX I

BIRD SPECIES RECORDED ON THE EAST SLOPE OF CERRO CHINGUELA (1700–3350 M), WITH
ELEVATIONAL LIMITS, RELATIVE ABUNDANCE, AND SOCIAL SYSTEM[1]

	L[2] U[2]	RA[3]	SS[4]
Nothocercus julius	–(2900)	R	S
Crypturellus obsoletus†	–(1700)	R	S
Nothoprocta curvirostris	3050 – 3200	R	S
Tigrisoma fasciatum	– 2000	R	S
Merganetta armata†	1700 – 2500	U	S
Buteo polyosoma†	2150 – 2300	U	S
B. albigula†	1700 – 2450	R	S
B. magnirostris	– 2000	FC	S
Oroaetus isidori†	(1850)–	R	S
Phalcoboenus megalopterus†	2900 – 3350	FC	S, F
Falco femoralis†	2950 – 3050	R	S
F. sparverius†	– 3000	R	S
Penelope barbata	2400 – 2900	U	S
Chamaepetes goudotii	1700 – 2150	R	S
Gallinago jamesoni	2850 – 3200	FC	S
G. imperialis†	2750 – 2900	U	S
G. (nobilis)†	–(3100)	R	S
Columba fasciata	1750 – 2600	U	S, F
Geotrygon frenata†	–(2500)	R	S
Claravis mondetoura	–(2000)	R	S
Leptotila verreauxi†	– 1700	U	S
Bolborhynchus (lineola)†	2300 – 2450	U	F
*Hapalopsittaca amazonina†	2550 – 3000	R	S, F
Pionus sordidus†	– 1850	U	S, F
P. seniloides	2050 – 2300	FC	S, F
Amazona mercenaria†	2600 – 2750	R	F
Piaya cayana†	– 2300	U	S, MF
Otus ingens	–(1700)	R	S
O. sp. nov.	–(1700)	R	S
O. albogularis	2700 – 2800	U	S
Glaucidium jardinii	2550 – 2850	U	S
Ciccaba albitarsus	2250 – 2550	U	S
Aegolius harrisii	–(3000)	R	S
Nyctibius leucopterus	–(2000)	R	S
Lurocalis rufiventris†	1950 – 2600	U	S
Caprimulgus longirostris	2000 – 2950	U	S
Streptoprocne zonaris†	– 2900	U	F
Cypseloides rutilus	1950 – 2150	FC	F
Phaethornis syrmatophorus	– 1950	FC	S
Eutoxeres aquila	– 2000	U	S
Colibri thalassinus†	2000 – 2300	C	S
C. coruscans†	(1950)–	R	S
Adelomyia melanogenys	1700 – 2450	FC	S
Heliodoxa rubinoides	1700 – 1950	U	S
Lafresnaya lafresnayi	2300 – 2900	U	S
Pterophanes cyanopterus	2850 – 3200	U	S
Coeligena coeligena	– 1950	FC	S
C. torquata	1700 – 2300	FC	S
*C. lutetiae	2600 – 3350	C	S
C. iris†	– 2600 (W)	FC	S
Ensifera ensifera†	2600 – 2850	U	S
Boissonneaua matthewsii	1900 – 2300	U	S
Heliangelus amethysticollis	2050 – 2600	U	S
H. exortis	2400 – 2900	U	S
H. viola†	– 2800 (W)	U	S
*Eriocnemis vestitus	2300 – 3200	C	S
Ocreatus underwoodii	–(1700)	R	S
Metallura odomae	2850 – 3350	FC	S

APPENDIX I

CONTINUED

	L^2	U^2	RA^3	SS^4
M. tyrianthina	2400 –	3400	U	S
Chalcostigma ruficeps	2250 –	2300	U	S
*C. herrani	2750 –	3200	U	S
Aglaiocercus kingi		– 2600	U	S
Acestrura mulsant†	1900 –	2300	U	S
Pharomachrus auriceps†		– 2300	U	S
Trogon personatus		– 2350	U	S
Momotus momota		– 2700	U	S
Malacoptila fulvogularis		– 1700	R	S
Hapaloptila castanea		–(2150)	R	S
Aulacorhynchus prasinus		– 2000	FC	S, F
Andigena hypoglauca†	2300 –	2800	U	S
Colaptes rupicola	2900 –	3400	FC	S
Piculus rivolii†		– 2250	FC	S, MF
Veniliornis fumigatus		– 2050	U	MF
V. nigriceps	2600 –	2950	U	MF
Campephilus pollens†	2300 –		R	S
C. haematogaster		–(1700)	R	S
Dendrocincla tyrannina		– 2300	R	S, MF
Sittasomus griseicapillus†		– 1700	R	MF
Xiphocolaptes promeropirhynchus	2550 –	2850	R	S, MF
Xiphorhynchus triangularis		– 2000	FC	MF
Lepidocolapates affinis		– 2000	U	MF
Campylorhamphus pusillus		– 1700	R	MF
Cinclodes fuscus	2950 –	3200	U	S
Synallaxis azarae		– 2300	C	S
S. unirufa	2400 –	2800	U	S
S. gularis	2600 –	2850	U	S
Cranioleuca antisiensis†		– 2700 (W)	FC	S, MF
C. curtata		– 1700	FC	MF
*Schizoeaca griseomurina	2750 –	3200	FC	S
Asthenes flammulata	2800 –	3350	FC	S
*Siptornis striaticollis		– 1700	R	MF
Xenerpestes singularis†		– 1700	R	MF
Margarornis squamiger	1800 –	3250	C	MF
Premnornis guttuligera		– 2000	U	MF
Premnoplex brunnescens		– 2000	FC	S, MF
Pseudocolapates boissonneautii	2000 –	2900	FC	S, MF
Syndactyla rufosuperciliata	2250 –	2450	R	S
S. subalaris		– 1700	R	S
Anabacerthia striaticollis		– 1750	U	MF
*Thripadectes flammulatus	2150 –	2950	U	S, MF
T. holostictus†		–(1700)	R	S, MF
Xenops rutilans		– 1850	FC	MF
Sclerurus albigularis		–(1700)	R	S
Lochmias nematura		–(1700)	R	S
Thamnophilus (palliatus)†		–(1700)	U	S
T. unicolor		–(1700)	U	S, MF
Dysithamnus mentalis		–(1700)	U	MF
Pyriglena leuconota		–(1700)	U	S
Drymophila caudata†		–(1700)	FC	S
Terenura callinota		– 1950	FC	MF
Chamaeza mollisima†	1850 –	2150	U	S
Grallaria squamigera	2250 –	2800	U	S
G. ruficapilla	1700 –	2150	U	S
*G. nuchalis	2200 –	2950	FC	S
*G. hypoleuca	1700 –	2000	U	S
G. rufula	2600 –	2900	FC	S
G. quitensis	2850 –	3300	C	S

APPENDIX I
CONTINUED

	L[2]	U[2]	RA[3]	SS[4]
*Grallaricula nana	2400 – 2950		FC	S
*G. peruviana	1700 – 2050		R	S
Myornis senilis	2600 – 2850		FC	S
Scytalopus unicolor	(2000)–		U	S
S. femoralis	1700 – 1950		FC	S
*S. latebricola	2600 – 2900		FC	S
S. magellanicus	2600 – 3500		FC	S
Acropternis orthonyx	2600 – 2950		U	S
Ampelion rubrocristatus	2700 – 3300		U	S, F
Pipreola riefferii	1950 – 2350		U	S
P. arcuata	2250 – 2900		FC	S
*P. lubomirskii†	(1750)–		R	S
*Lipaugus fuscocinereus	2100 – 2250		R	S, MF
Rupicola peruviana	– 1700		R	S
Masius chrysopterus	– 1700		U	S
Phyllomyias plumbeiceps	– 1850		U	MF
P. nigrocapillus	2300 – 2950		FC	MF
P. cinereiceps	–(2050)		R	MF
Elaenia obscura†	– 2700		R	S, MF
E. pallatangae†	(2450)–		R	S, MF
Mecocerculus leucophrys	2800 – 3200		C	F, MF
M. poecilocercus	1700 – 2300		FC	MF
M. calopterus	1700 – 1950		R	MF
M. minor	1700 – 2000		FC	MF
M. stictopterus	2300 – 2450		R	MF
Serpophaga cinerea†	– 1950		R	S
Mionectes striaticollis	1700 – 2650		U	S, MF
*Leptopogon rufipectus	– 2250		FC	MF
Phylloscartes poecilotis	– 2000		U	MF
P. ophthalmicus	– 1800		FC	MF
Pseudotriccus ruficeps	2400 – 2900		U	S
P. pelzelni	– 2000		R	S
Lophotriccus pileatus	– 1900		U	S
*Poecilotriccus ruficeps	–(2450)		R	S
Hemitriccus granadensis	2400 – 2850		U	S
Todirostrum cinereum†	– 1700		U	S
Rhynchocyclus fulvipectus	–(1700)		R	S, MF
Platyrinchis mystaceus	– 1700		U	S
P. flavigularis	–(1700)		R	S
Myiophobus flavicans†	–(2000)		R	MF
*M. lintoni	2450 – 2750		FC	MF
Pyrrhomyias cinnamomea	– 2800		FC	S, MF
Mitrephanes olivaceus†	– 1700		U	S, MF
Contopus fumigatus†	– 2300		FC	S, MF
Sayornis nigricans†	– 2050		U	S
Ochthoeca cinnamomeiventris	1950 – 2450		U	S
O. diadema	2350 – 2800		FC	S
O. frontalis	2600 – 2900		FC	S
O. rufipectoralis	2250 – 2800		C	S, MF
O. fumicolor	2750 – 3350		C	S
Myiotheretes striaticollis†	1700 – 2700		R	S
M. erythropygius	2900 – 3500		R	S
M. fumigatus	2750 – 2950		U	S, MF
Muscisaxicola alpina†	2900 – 3350		R	S
Knipolegus poecilurus	2000 – 2150		FC	S
Colonia colonus†	– 1950		R	S
Hirundinea ferruginea†	– 2000		FC	S
Myiarchus tuberculifer†	– 2450		FC	S, MF
M. cephalotes	– 1700		U	S, MF

APPENDIX I
CONTINUED

	L[2]	U[2]	RA[3]	SS[4]
Tyrannus melancholicus†		– 1950	FC	S
Pachyramphus versicolor†	1700	– 2950	U	MF
P. albogriseus†		–(1700)	R	MF
Progne sp.		–(2250)	R	F
Notiochelidon murina	2750	– 3350	C	F
N. cyanoleuca		– 2600	FC	F
N. flavipes	1950	– 2950	C	F
Hirundo rustica†		– 2250	R	F
*Cyanolyca turcosa	2200	– 2800	FC	F, MF
Cyanocorax yncas		– 2000	FC	F, MF
Cinclus leucocephalus†		– 2300	U	S
Odontorchilus branickii†		– 1700	R	MF
*Cinnycerthia unirufa	2450	– 3200	C	F, MF
C. peruana	2000	– 2350	U	F, MF
Cistothorus platensis	2850	– 3700	FC	S
Thryothorus euophrys	2350	– 2450	U	S
T. maculipectus		–(1700)	R	S
Troglodytes aedon		– 2300	FC	S
T. solstitialis		– 2950	C	S, MF
Henicorhina leucophrys		– 2300	C	S
Myadestes ralloides	1700	– 2250	U	S
Catharus fuscater		– 2300	U	S
Platycichla leucops		–(2900)	R	S
Turdus fuscater	2400	– 3000	C	S, MF
T. serranus		– 2000	FC	S
T. nigriceps		–(1700)	R	S
*T. fulviventris	1700	– 2000	R	S
Anthus bogotensis	2900	– 3200	FC	S
Cyclarhis gujanensis		– 2000	U	S, MF
Vireo gilvus		– 2150	C	MF
Cacicus leucoramphus	2300	– 2600	R	F, MF
C. uropygialis		– 2000	U	F
C. holosericeus	2450	– 2900	FC	F, MF
Parula pitiayumi†		– 1850	U	MF
Dendroica fusca†		– ? (W)	FC	MF
Myioborus miniatus		– 2000	C	MF
M. melanocephalus	2000	– 3250	FC	MF
Basileuterus nigrocristatus	2000	– 3350	C	MF
B. luteoviridis	2600	– 3050	C	MF
B. tristriatus		– 1950	U	MF
B. coronatus		– 2600	FC	MF
Conirostrum sitticolor	2600	– 3050	FC	MF
C. albifrons		– 2250	FC	MF
Diglossa caerulescens		– 2450	FC	S, MF
*D. lafresnayii	2600	– 3350	C	S
D. humeralis	2850	– 2950	R	S
D. albilatera		– 2450	C	S, MF
D. cyanea	2000	– 2900	C	S, MF
Iridophanes pulcherrima†		– 2000	U	MF
Chlorophonia pyrrophrys†	1700	– 2300	R	S, MF
Euphonia xanthogaster†		– 1700	U	MF
Pipraeidea melanonota		– 1700	U	S, MF
Chlorochrysa calliparaea†		– 1700	U	MF
Tangara xanthocephala†		– 2300	C	MF
T. parzudakii		– 1950	C	MF
T. cyanotis†		– 1700	U	MF
T. labradorides		– 1700	U	MF
T. ruficervix		– 2000	U	MF
T. nigroviridis		– 2000	C	MF

APPENDIX I
CONTINUED

	L[2]	U[2]	RA[3]	SS[4]
T. vassorii†	2000 –		FC	MF
T. viridicollis	– 1700		U	MF
Iridosornis analis	– 2150		FC	MF
*I. rufivertex	2590 – 3250		FC	MF
Anisognathus igniventris	2750 – 3200		U	MF
A. lacrymosus	2600 – 3300		C	MF
A. flavinucha	– 2050		C	MF
Buthraupis montana	2300 – 2900		C	F, MF
*B. eximia	2850 – 3300		FC	F, MF
*B. wetmorei	(2900)–		R	MF
Dubusia taeniata	2600 – 2950		U	MF
Thraupis cyanocephala†	– 2600		C	MF
Piranga rubriceps	2000 – 2400		U	MF
P. leucoptera†	– 1700		U	MF
Creurgops verticalis	1950 – 2050		U	MF
Thlypopsis ornata†	– 2900		U	MF
Sericossypha albocristata	1700 – 2400		U	F
Chlorospingus ophthalmicus	– 2250		U	MF
C. parvirostris	1750 – 2250		C	F, MF
Cnemoscopus rubrirostris	2600 – 2900		FC	MF
Hemispingus atropileus	2450 – 2950		C	MF
H. superciliaris	– 2750 (W)		FC	MF
H. frontalis	– 2000		FC	MF
*H. verticalis	2350 – 3350		C	MF
Chlorornis riefferii†	1950 – 2600		FC	MF
Catamblyrhynchus diadema	2250 – 2950		FC	MF
Saltator cinctus	1700 – 2900		R	S, MF
Catamenia homochroa	2350 – 2900		R	S, F
Phrygilus unicolor	2900 – 3400		U	S
Haplospiza rustica	2600 – 2900		U	S
Atlapetes rufinucha	2000 – 2950		C	S, MF
*A. pallidinucha	2800 – 3350		C	MF
A. leucopterus	–(1700)		R	S
A. brunneinucha	– 2650		U	R
Lysurus castaniceps	– 1700		R	S
Zonotrichia capensis†	– 2950		C	S
Carduelis magellanica†	– 2400		U	S, F

[1] An asterisk denotes a species that reaches the southern limit of its distribution north of the North Peruvian Low. A dagger after the species name indicates a species not collected at the locality by the LSUMZ. Parentheses are placed around the specific epithet of a bird not positively identified to species.

[2] Lower (L) and Upper (U) elevational limits (in meters) are based on collected specimens and sight records. The lower limit for most lower montane species was not determinable due to the great amount of habitat disturbance below 1700 m and was left blank for all species for which we have records as low as that elevation. Elevations for species recorded only once are given in parentheses. A few species recorded only on the west slope of Cerro Chinguela, in remnant forest between Sapalache and the crest, are included in the list; these are indicated by (W) placed after their upper elevational limit.

[3] Relative abundance: C = seen or heard in moderate to large numbers daily (generally more than 10 individuals); FC = seen or heard in small numbers daily (fewer than 10 individuals); U = seen or heard on one out of three days (occurs in small numbers); R = seen or heard on one out of six days or less often (occurs in very small numbers).

[4] Social system: S = occurs singly, in pairs or small family groups; F = occurs in single-species flocks (usually more than 5 individuals); MF = occurs in mixed-species flocks (flocks of two or more species).

APPENDIX II
BIRD SPECIES RECORDED IN DESERT SCRUB NORTHEAST OF HUANCABAMBA WITH RELATIVE
ABUNDANCES

	RA[1]		RA
Buteo polyosoma	U	*Notiochelidon cyanoleuca*	C
Falco sparverius	FC	*Campylorhynchus fasciatus*	U
Zenaida auriculata	C	*Troglodytes aedon*	FC
Columbina cruziana	C	*Mimus longicaudatus*	U
Leptotila verreauxi	FC	*Turdus chiguanco*	U
Forpus coelestis	FC	*Polioptila plumbea*	FC
Coccyzus melacorhyphus	U	*Cyclarhis gujanensis*	FC
Crotophaga sulcirostris	FC	*Sturnella bellicosa*	FC
Glaucidium brasilianum	U	*Geothlypis aequinoctialis*	FC
Colibri coruscans	FC	*Conirostrum cinereum*	C
Leucippus taczanowskii	FC	*Thraupis bonariensis*	FC
Myrtis fanny	C	*Saltator albicollis*	U
Chrysoptilus atricollis	U	*Pheucticus chrysopeplus*	U
Furnarius leucopus	U	*Sporophila luctuosa*	C
Synallaxis azarae	FC	*S. nigricollis*	U
Cranioleuca antisiensis	FC	*Catamenia analis*	C
Pyrocephalus rubinus	FC	*Phrygilus plebejus*	C
Euscarthmus meloryphus	C	*Atlapetes torquatus*	U
Anairetes flavirostris	FC	*Incaspiza ortizi*	U
Elaenia sp.	R	*Zonotrichia capensis*	C
Camptostoma obsoletum	FC	*Carduelis magellanica*	U

[1] Relative abundance. Abbreviations as in footnote 3, Appendix I.

APPENDIX III

BIRD SPECIES RECORDED ON THE WEST SLOPE OF THE WESTERN CORDILLERA AT CRUZ BLANCA 1700–3350 M), WITH ELEVATIONAL LIMITS, RELATIVE ABUNDANCE, AND SOCIAL SYSTEM[1]

	L U	RA	SS
Nothoprocta curvirostris †	3100 – 3200	R	S
Vultur gryphus †	– 3200	R	S
Cathartes aura †	– 2150	U	S
Chondrohierax uncinatus †	– 1750	R	S
Accipiter striatus	– 2750	U	S
Geranoaetus melanoleucus †	– 3050	R	S
Buteo polyosoma †	– 3300	FC	S
B. albigula †	–(2450)	R	S
B. albonotatus †	–(1750)	R	S
B. magnirostris	– 2150	FC	S
B. leucorrhous †	–(1750)	R	S
Harpyhaliaetus solitarius †	–(1850)	R	S
Phalcoboenus megalopterus †	3000 – 3350	FC	S, F
Falco sparverius †	– 1750	U	S
Penelope barbata	1800 – 3000	U	S
Vanellus resplendens	3100 – 3200	U	S, F
Gallinago andina	–(3150)	R	S
G. jamesoni †	3050 – 3150	FC	S
Columba fasciata	2100 – 3000	U	S, F
Columbina cruziana †	– 1750	U	S, F
Leptotila verreauxi †	– 2450	U	S
Geotrygon (frenata) †	(2600)–	R	S
Aratinga wagleri	– 2450	U	S, F
Bolborhynchus sp.†	– 3050	R	F
Piaya cayana †	– 1850	U	S, MF
Crotophaga sulcirostris†	– 1850	U	F
Tyto alba †	– 3050	R	S
Otus albogularis	2300 – 3050	FC	S
Bubo virginianus	3000 – 3100	FC	S
Glaucidium jardinii	–(2600) (E)	R	S
Ciccaba albitarsus	1850 – 3000	U	S
Aegolius harrisii	1750 –	R	S
Caprimulgus longirostris †	2300 – 3050	U	
Nyctidromus albicollis †	– 2300	R	S
Streptoprocne zonaris †	– 3000	U	F
Chaetura brachyura †	– 1850	R	S, F
Aeronautes montivagus	– 2600	U	S, F
Colibri thalassinus	2100 – 2450	FC	S
C. coruscans	1750 – 2290	U	S
Adelomyia melanogenys	1750 – 2750	FC	S
Aglaeactis cupripennis	3000 – 3050	C	S
Lafresnaya lafresnayi	2450 – 3050	FC	S
Coeligena iris	1500 – 3050	C	S
Heliangelus viola	2150 – 3050	C	S
Lesbia victoriae †	3000 – 3200	U	S
L. nuna	2300 – 3050	FC	S
Metallura tyrianthina	2600 – 3200	C	S
Acestrura mulsant	– 1700	R	S
Pharomachrus auriceps	1650 – 2600	U	S
Trogon personatus	– 2900	U	S
Colaptes rupicola	3050 – 3300	FC	S
Piculus rivolii	2450 – 3050	FC	S, MF
Veniliornis fumigatus	1700 – 3050	FC	MF
Lepidocolaptes affinis	1750 –	U	MF
Cinclodes fuscus †	3050 – 3200	U	S
Synallaxis azarae	1700 – 3200	C	S
Cranioleuca antisiensis	1700 – 3150	C	MF
Asthenes wyatti †	3100 – 3350	FC	S

APPENDIX III
CONTINUED

	L U	RA	SS
Margarornis squamiger	2450 – 3050	FC	MF
Pseudocolaptes boissonneautii	2250 – 3050	FC	MF
Automolus ruficollis	1700 – 2900	U	S
Thamnophilus doliatus	– 1700	R	S
Myrmeciza griseiceps	1500 – 2900	R	S
Grallaria squamigera	2450 – 3050	R	S
G. ruficapilla	1700 – 2750	FC	S
G. rufula	2900 – 3050	FC	S
G. quitensis	3050 – 3200	C	S
Grallaricula ferrugineipectus	– 2450	R	S
Myornis senilis†	2900 – 3000	U	S
Scytalopus unicolor	1700 – 3200	C	S
Ampelion rubrocristatus	2450 – 3050	FC	S, F
Phyllomyias nigrocapillus†	1750 – 3050	U	MF
Zimmerius viridiflavus	1750 – 1850	FC	S, MF
Myiopagis subplacens	– 1750	R	
Phaeomyias murina†	– 1750	R	S
Elaenia albiceps	–(3000)	R	S
E. pallatangae	3000 – 3050	C	S
Mecocerculus leucophrys	2600 – 3200	C	F, MF
M. poecilocercus	1750 – 2100	FC	MF
M. stictopterus	2300 – 3050	FC	MF
Anairetes nigrocristatus†	– 2750	U	S
A. parulus	2750 – 3250	C	S, MF
Mionectes striaticollis	1700 – 2600	FC	S, MF
Contopus fumigatus	1700 – 2750	FC	S, MF
C. cinereus	– 2950	R	S
Ochthoeca pulchella	1750 – 2900	FC	S
O. fumicolor	3050 – 3250	C	S
O. rufipectoralis	2450 – 3050	FC	S
Myiotheretes striaticollis†	2950 – 3100	U	S
Agriornis montana†	2750 (E)– 3250	U	S
Muscisaxicola (alpina)†	3200 –	R	S
Myiarchus tuberculifer	1500 – 2300	FC	S, MF
Pachyramphus albogriseus†	– 3050	R	MF
Notiochelidon cyanoleuca†	– 2300	FC	F
N. murina	2600 – 3200	FC	F
Cyanolyca turcosa	1850 – 3100	FC	F, MF
Cistothorus platensis†	3050 – 3250	FC	S
Troglodytes aedon†	– 3050	FC	S
Catharus fuscater	– 1750	R	S
C. ustulatus	1500 – 2900	FC	S
Platycichla leucops†	–(1750)	R	S
Turdus (chiguanco)†	–(2500)	R	S
T. fuscater	1750 – 3250	C	S, MF
T. serranus	1750 –	R	S
T. nigriceps	– 1750	R	S
T. reevei	– 1850	U	S
Anthus bogotensis	3050 – 3300	FC	S
Cyclarhis gujanensis	– 2900	FC	S, MF
Vireo gilvus	1500 – 2150	FC	MF
Parula pitiayumi	– 2900	FC	MF
Dendroica fusca	1500 – 2900	FC	MF
Myioborus miniatus	– 2450	C	MF
M. melanocephalus	2300 – 3050	C	MF
Basileuterus nigrocristatus	1750 – 3050	C	MF
B. trifasciatus	1500 – 3050	C	MF
B. coronatus	1750 – 3050	FC	MF
B. fraseri	– 1850	R	S, MF

APPENDIX III
CONTINUED

	L	U	RA	SS
Conirostrum cinereum †	3000	3200	FC	S, MF
C. sitticolor	2600	3100	FC	MF
Diglossa humeralis	2750	3300	C	S
D. albilatera	1750	2900	FC	S, MF
D. cyanea †	2450	3050	FC	MF
D. baritula †		2450	R	S, MF
Euphonia musica †		2600	FC	S, MF
E. laniirostris		1850	R	S
Tangara vassorii †	2450	2900	FC	MF
T. viridicollis	1500	1850	FC	MF
Anisognathus igniventris	2900	3050 (E)	R	MF
A. lacrymosus	2400	3050	C	MF
Dubusia taeniata	2900		U	MF
Thraupis cyanocephala	1750	2900	FC	MF
Piranga flava	1750	3000	U	S, MF
Thlypopsis ornata	2200	3050	FC	MF
Hemispingus melanotis	1750	2900	FC	MF
H. superciliaris	2600	3050	FC	MF
Catamblyrhynchus diadema †	1750	2900	FC	MF
Saltator nigriceps †	1750	2600	R	S, MF
Pheucticus chrysogaster †	1500	1850	U	S, MF
Catamenia inornata	3000	3300	FC	S, F
C. homochroa		(3050)	R	S, F
Phrygilus alaudinus	2500	3000 (E)	FC	S
Haplospiza rustica	1850	2750	U	S
Atlapetes rufinucha	1850	3250	C	S, MF
A. leucopterus		1850	FC	S, F
A. torquatus	1750	2900	FC	S
Zonotrichia capensis		3200	C	S
Carduelis magellanica †		3050	U	S, F
C. psaltria		1750	R	S, F

[1] Symbols and abbreviations as in Appendix I, except that an (E) is placed after the upper elevational limit of species recorded only on the east slope.

THE AVIFAUNA OF CERRO PIRRE, DARIÉN, EASTERN PANAMA

Mark B. Robbins,[1,2] Theodore A. Parker, III,[1] and Susan E. Allen[1]

[1]Museum of Zoology, Louisiana State University, Baton Rouge, Louisiana 70803-3216 USA;
[2]present address: Academy of Natural Sciences, 19th and The Parkway,
Philadelphia, Pennsylvania 19103 USA

ABSTRACT. Recent Louisiana State University Museum of Zoology fieldwork on the east slope of Serranía a de Pirre at Cana is discussed. A list of bird species recorded by us along a 600–1550 m elevational gradient, with specimen data, and with an indication of relative abundance, is given. Information is presented on the status, behavior, and habitat preferences of 25 species of poorly known birds. Six are restricted to the highlands of Darién, Panama. We report the first specimen of Purple-throated Woodstar (*Calliphlox mitchellii*) for Central America with comments on the identification of female woodstars. We discuss the effects of long-termed human disturbance on a tropical forest fauna.

RESUMEN. La información obtenida recientemente sobre la vertiente oriental de la Serranía de Pirre en Cana, Darién por el Museo de Zoología de la Universidad del Estado de Luisiana es presentada en este trabajo. Se presenta una lista de especies de aves observadas a lo largo del gradiente de 600 a 1.550 m junto con información de cada ejemplar y una indicación sobre su abundancia relativa. También se presenta información sobre el estado biológico ("status"), comportamiento y preferencia de habitat para 25 especies de aves poco conocidas. Seis especies se encuentran restringidas a la zona montañosa del Darién, Panamá. Se presenta información sobre el primer espécimen del "Woodstar" de garganta purpura (*Calliphlox mitchellii*) para America Central con comentarios para la identificación de hembras de estas aves. Se discuten los efectos que han causado las perturbaciones producidas por grupos humanos durante largo tiempo, en fauna de un bosque tropical.

Serranía de Pirre on the Panamanian-Colombian border is of interest to zoogeographers because of its relatively high degree of faunal endemism. Analyses of the avifauna have offered cues about patterns of dispersal and speciation in this region (Haffer 1967). The high-elevation bird fauna is derived from the Andes (e.g., *Pharomachrus auriceps, Margarornis bellulus,* and *Scytalopus vicinior*) and Chiriquí Highlands to the west (e.g., *Otus clarkii, Basileuterus ignotus,* and *Tangara fucosa*). The high degree of endemism in montane birds of this mountain range led Haffer (1967) to suggest that a forest refugium existed on the slopes of Pirre during dry periods of the Pleistocene. This hypothesis is supported by herpetological discoveries of Myers (1966) and Wake et al. (1970). Haffer (1967) further demonstrated that the lowland avifauna of this region has closer affinities to northwestern and Amazonian South America than to Central America.

Pirre's endemism is a product of its geographic isolation from other montane areas. Nearly 50 km of lowland rainforest separate it from the nearest range of comparable elevation, Cerro Tacarcuna (which shares some unique taxa with Pirre), and 150 km of lowlands isolate Pirre from the Western Andes of northwestern Colombia (Fig. 1). The highest point in the Serranía de Pirre is Cerro Pirre at 1550 m. The range is approximately 50 km long and 25 km wide at the broadest point.

Although the composition of the avifauna of Pirre is relatively well known (Goldman 1920; Griscom 1929; Wetmore 1965, 1968, 1972), little has been published concerning the natural history of its endemics, nor of other birds of restricted distribution that occur in this area. In this paper we present specimen data and information on the status, behavior, voice, and habitat preferences of some of these poorly known species. We report the first record of *Calliphlox mitchellii* (Purple-throated Woodstar) in Central America. We also provide a list (Appendix I) of all bird species we encountered, along with some indication of their relative abundances along a 600 to 1550 m elevational gradient on the east slope of Cerro Pirre above Cana. Finally, we comment on the resiliency of this avifauna in the face of nearly 400 years of human disturbance.

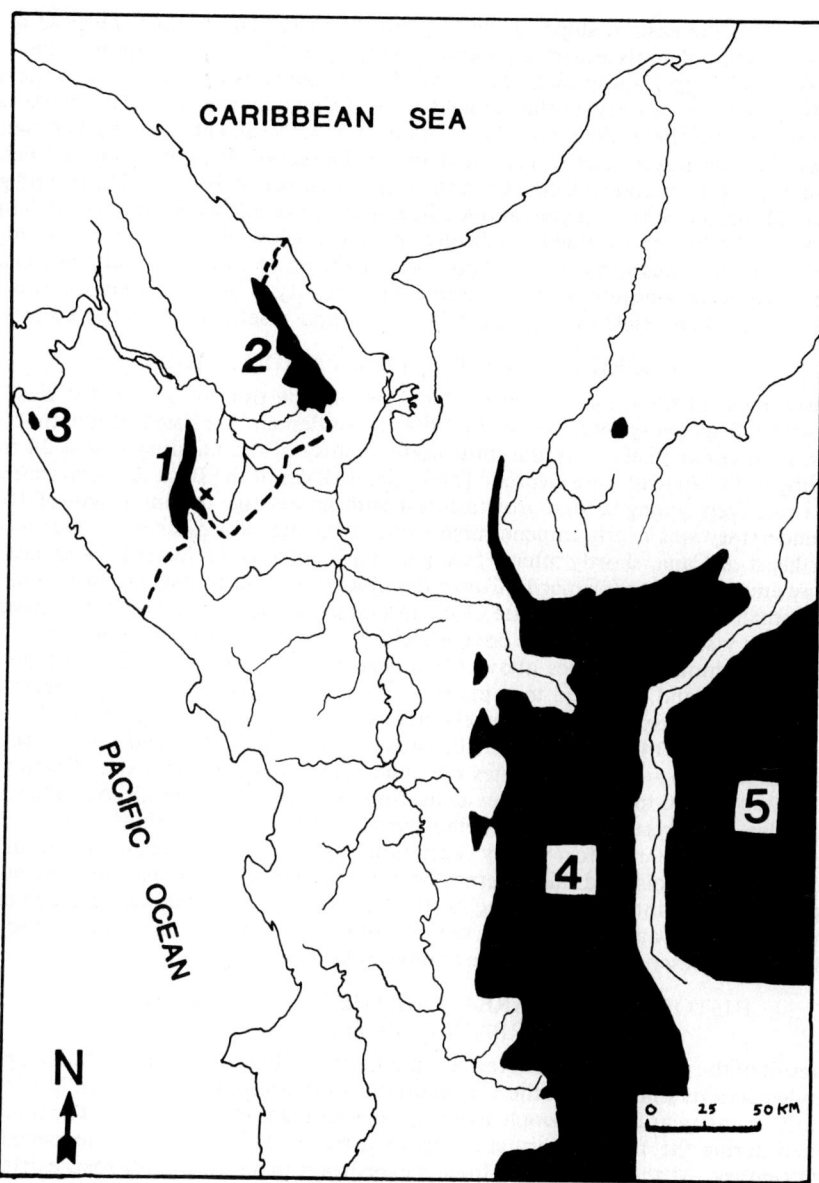

FIG. 1.   Map of eastern Panama and northwestern Colombia showing (1) Serranía de Pirre, (2) Serranía de Tacurcuna, (3) Cerro Sapo, (4) Western Andes, and (5) Central Andes. Areas in black indicate elevations above 1000 m. X denotes the approximate location of Cana. Modified from Haffer (1975).

The avifauna was first surveyed in 1912 (Goldman 1920). Nelson (1912) reported on Goldman's collection, describing a number of new genera, species, and subspecies. The lower slopes (ca. 820 m elev.) of this range were visited by Griscom in 1928, and he subsequently described 15 additional new subspecies and reported more than 30 species new to Panama (Griscom 1929). Wetmore's observations from a 1961 visit were included in his treatise on Panamanian birds (1965, 1968, 1972). Although additional fieldwork has resulted in the lumping of many of the taxa originally described, a relatively high degree of endemism is still evident (1 genus, 6 species, and numerous well-differentiated subspecies).

During 1982, two field parties from the Louisiana State University Museum of Zoology

(LSUMZ) visited the eastern slope of the Serranía de Pirre. The primary purpose of these expeditions was to obtain tissues of tyrannoid genera as part of an investigation of the familial boundaries of this superfamily by S. M. Lanyon. This paper is a by-product of the above investigation. Both parties flew to the old gold mine at Cana, located at 600 m on the eastern slope of Cerro Pirre. The initial group, M. J. Braun, J. P. O'Neill, and T. A. Parker, reopened an old trail that ascends a ridge to the backbone of the massif. Between 4 and 12 February they camped at 1480 m, and collected from the ridge top down to 1000 m. The second group, S. E. Allen, D. Braun, S. M. Lanyon, and M. B. Robbins, established a camp at 1000 m (Fig. 2) and between 18 July and 1 August, collected from just below this camp up to the ridge top at 1550 m. Fieldwork was also conducted around Cana from 15 to 17 July and 2 to 16 August; our efforts there were concentrated in the immediate vicinity of the gold mine settlement and along the first few kilometers of the trail to Boca de Cupe, a settlement on the Tuira River.

DESCRIPTION OF THE VEGETATION ZONES

Formerly, most of the Cana area was cleared, but at the time of our visit it was largely covered with tall second-growth forest. In 1982 the settlement consisted of five tin-covered buildings. During our visits, only the immediate vicinity of the buildings and a 20 m wide path leading to the airstrip were cleared. The initial 4–5 km of the Boca de Cupe trail passed through a relatively young second-growth forest with an average canopy height of 10 to 15 m. The understory was nearly impenetrable without the aid of a machete. About 5 km (by trail) northeast of Cana, shortly after crossing a major stream, the forest became less dense with many emergent trees (primarily *Cavanillesia*) as tall as 30 m. An extensive marsh fed by several small streams abuts the base of the mountain at the eastern end of the airstrip. At the time of our visits this marsh was covered with a dense mat of vegetation.

Forest on the mountain slopes above Cana was taller (averaging 15–20 m) than at the settlement clearing and appeared to support a much higher diversity of tree species. Palms were common and conspicuous at all levels, but especially along the ridge crests above 1450 m. From about 700 to 1400 m the understory was open. Between 1000 and 1200 m, the forest was more humid, and arboreal epiphytes were abundant. An extensive stand of bamboo was present at 1200 m, and tree ferns were evident from 1450 m to the summit. At 1480 m, along the main ridge, the forest was stunted ("elfin forest" in the accounts below), with an average canopy height of 8 to 10 m. Most woody vegetation was covered with mosses and bromeliads. The understory was moist and thick. Ferns and herbs comprised the most common elements. The ridge top was continuously shrouded in clouds. Measurable rainfall was recorded on three days during February and nearly every day at both camps in July and August. See Myers (1969) for further description of the forest above 1000 m.

HISTORY OF THE AREA AND THE EFFECTS OF HUMAN DISTURBANCE ON THE FAUNA

As a result of the gold mines, the forest around Cana has been under heavy human pressure for extended periods since at least the 16th century (summarized below from Goldman 1920). Apparently, as many as 20,000 people lived at Cana during the 17th century. The mines were abandoned during the 18th century and remained relatively inactive until reopened in the late 19th century. At the time of Goldman's expedition in 1912, a tram road existed from Boca de Cupe to within 10 km of Cana. These final 10 km were traversed by a mule-drawn car, indicating a major swath was cut through the forest. Photographs taken during the Goldman expedition (see Goldman 1920:14) clearly indicate that the area was extensively disturbed, although Goldman commented that heavily-forested mountain slopes were easily accessible. Since Goldman's visit, the mines have been periodically opened and closed, although there had not been any mining activity in the previous 2–3 years prior to our visit in 1982. At the time of our arrival only two families were living at Cana.

In view of the long disturbance, we anticipated that a number of the larger mammals (e.g., monkeys, tapir, jaguar) and birds (e.g., Harpy Eagle, cracids, and macaw) would be greatly reduced in numbers or extirpated. Surprisingly, not one of these fauna was absent.

One to three troops of Black Spider Monkeys (*Ateles paniscus*) were seen daily from our 1200 m camp up to 1500 m. Black Howler Monkeys (*Alouatta villosa*) were encountered daily from below Cana (Boca de Cupe trail) up to the elfin forest (1485 m). The White-faced Capuchin (*Cebus capucinus*) was observed less frequently, ranging from Cana up to 1000 m. Geoffrey's Marmoset (*Saguinus geoffroyi*) was seen regularly along the Boca de Cupe trail, and less

FIG. 2.   Eastern slope of Serranía de Pirre. View is to the north from our camp at 1000 m.

frequently on the lower slopes just above Cana. A Jaguar (*Felix onca*) was seen by one field assistant, and tapir (*Tapirus bairdi*) tracks were observed on the Boca de Cupe trail within 100 m of the Cana runway.

An adult Harpy Eagle (*Harpia harpyja*) was seen and heard calling on 21 July along a ridge opposite our 1000 m camp. Formerly, widespread throughout lowland rain forest from Mexico to northern Argentina, it has been extirpated from large areas of its range during the past few decades. *Harpia* occurs naturally in low densities. Thus, even limited human hunting pressure on this bird and its primary food source (monkeys), coupled with forest fragmentation, may dramatically affect its distribution. At least in some parts of its range, this eagle is shot for its massive talons and feathers.

Like *Harpia*, the Great Curassow (*Crax rubra*) is a good indicator of the intensity of human hunting pressure. Although none was observed along the Boca de Cupe trail and in the immediate area of Cana, where hunting pressure is more intense, one to two were observed infrequently above 1200 m. The Crested Guan (*Penelope purpurascens*) was common on the slopes above Cana. As many as a dozen were recorded in a day, with a few seen as high as 1550 m. We saw fewer along the Boca de Cupe trail.

Although the residents occasionally shoot the larger macaws, all four species were relatively common. Small numbers (2–10) of the Blue-and-yellow Macaw (*Ara ararauna*) were observed daily in forest bordering the large marsh. The Great Green Macaw (*A. ambigua*) was seen daily in small numbers (2–8) only in hill forest above Cana. The Red-and-green Macaw (*A. chloroptera*) was the commonest *Ara*. Groups up to 12 were seen from flat terrain below Cana to the ridge tops. Ten to twelve Chestnut-fronted Macaws (*A. severa*) were noted daily in forest bordering the marsh.

These findings are encouraging and point to the importance of corridors of forest as avenues for recolonization of formerly disturbed areas. It is likely that the extent of forest at Cana was greatly reduced when the mines were at their height of production during the 17th century. During this period, a "paved" road reputedly extended from the area of Real de Santa María, across the Setetule mountains to Cana. At this time montane and lowland forest faunas may have been bisected, with reservoirs of forest existing north and south of Cana. When Goldman visited about 300 years later, much of the area was again forested. This reforestation continues today.

SPECIES ACCOUNTS

Taxonomy follows the American Orntihologists' Union Check-list of North American Birds (1983) and Meyer de Schauensee (1970).

Gampsonyx swainsonii. Pearl Kite. A single bird was seen in the clearing at Cana on 16 July, and the same or another individual was seen again in the clearing on 3 August. The first record for Panama was obtained in the southwestern area of the Canal Zone on 12 June 1977 (Pujals et al. 1977), and there have been a number of subsequent reports from both slopes of central Panama, as well as one from Bocas del Toro coast (R. S. Ridgely, pers. comm.). Our sightings support Eisenmann's suggestion in the above paper that this species is invading Panama from Colombia and not from Nicaragua.

Geotrygon goldmani. Russet-crowned Quail-Dove. In February this species was heard and seen daily in small numbers from 1000 m up to the ridgetop elfin forest; in July it was seen on only two occasions in the same area. Solitary individuals were observed walking quietly and usually rapidly on the ground, and scratching in damp leaf litter. The birds were very vocal in February, when up to five were heard daily along ca. 1 km of trail. They sang from moss- and bromeliad-laden limbs and bamboo branches 3 to 6 m above the ground. Singing birds were extremely difficult to locate and were wary and hard to approach once seen. The song is a very low-pitched, downward-inflected *wooooo* characteristically repeated at 4–5 sec intervals for several minutes at a time (see Wetmore 1968:54). This vocalization may be mistaken for the similar but longer call of *Leptotila cassinii,* a fairly common terrestrial species on the lower slopes of Cerro Pirre around Cana.

Pionopsitta pyrilia. Saffron-headed Parrot. Small flocks of this beautiful parrot were encountered infrequently from Cana up to 1400 m. The largest group was a flock of 10 on 18 July at 1000 m. Birds were seen only in flight. *Pionopsitta haematotis* (Brown-hooded Parrot) was found to be syntopic throughout the same elevational range as *P. pyrilia,* but no information is available on how these two species coexist. The Saffron-headed Parrot was previously known in Panama from two specimens collected in the Río Tuira drainage not far from Cana (Wetmore 1968), but there are a number of other recent sight records from eastern Darién (R. S. Ridgely, pers. comm.). The species occurs only from extreme eastern Panama through northern Colombia to western Venezuela.

Neomorphus geoffroyi. Rufous-vented Ground-Cuckoo. Although this species has a broad distribution from Nicaragua to northern Bolivia and central Brazil, little is known of its natural history. Individuals were encountered from 550 m to 1000 m. In February M. Braun observed a single bird foraging on insects flushed by an army ant swarm at 1000 m. This bird frequently clacked its bill, and raised and lowered its crest as it foraged, making quick, brief dashes from logs into the ant swarm for insects. Similar behavior has been noted in Nicaragua (Howell 1957) and Panama (Ridgely 1976). On 10 August Robbins found a downy young perched about 2 m up on a horizontal limb in dense undergrowth along the trail to Boca de Cupe. Upon being disturbed, the bird erected its crest, leapt from the limb, and ran along the ground, clacking its bill. Imitation of the young bird's bill clacking induced an adult male to approach. This agrees with findings of Sick (in Roth 1981), who observed young accompanying an adult pair. The downy young proved to be a female, and the plumage closely fits that of a juvenal female described by Howell (in Haffer 1977).

Otus clarkii. Bare-shanked Screech-Owl. At least two pairs of this owl were noted in stunted ridge top forest near our camp at 1480 m in February. These birds were vocal early on clear moonlit evenings, especially from 19:00–21:00 hrs, and again just before dawn. The song of a male was much like that described by Wetmore (1968:157), a soft *coo coo-coo-coo*; the terminal series varied from 3 to 5 notes and tapered off in volume. Typically, songs were given about 10 seconds apart; occasionally another individual, presumably the female, sang simultaneously, adding a *cu* after each of the male's *coo*'s, producing what sounded like an *Otus trichopsis*-like duet; *coo coo-cu coo-cu coo-cu coo-cu.* One pair of *clarkii* closely approached our camp, where numerous large beetles and moths were attracted to kerosene lamps. *Otus clarkii* is fairly widespread, but apparently local.

Androdon aequatorialis. Tooth-billed Hummingbird. This strikingly-marked, but inconspicuous, hummingbird was encountered infrequently from Cana (600 m) to cloud forest at 1450 m. Most sightings were of individuals feeding at flowers from 3 to 7 m above ground in forest openings, but the species was heard also in the dark cloud forest interior and was seen there twice feeding at red flowers of a middlestory vine. Once an individual was flycatching about 3 m above ground in a shady forest clearing at 1300 m. The call note is a sharp, penetrating *cheet*, sometimes doubled *cheet-it.* A captive bird uttered a piercing, rapid series *chit-sit-sit-sit-sit.* Androdon was previously known in Panama from a few specimens from the Darién highlands (Wetmore 1968).

Goethalsia bella. Pirre Hummingbird. Although we recorded the species from secondary forest around Cana at 600 m to ridge top forest at 1450 m, it appeared to reach its greatest abundance from 1100 to 1275 m. At these latter elevations *Goethalsia* was observed daily in small numbers; individuals fed almost exclusively at the red and blue flowers of *Cephalus* shrubs in forest understory. In the upper forest above 1300 m these flowers were vigorously and successfully defended by individuals of *Haplophaedia aureliae,* which drove off intruding *Goethalsia.* The former species far outnumbered the latter and may restrict its upper distribution. Wetmore (1968:323) summarized information on *Goethalsia bella,* which had previously been known from fewer than 10 specimens; we obtained 11 birds.

Calliphlox mitchellii. Purple-throated Woodstar. A female was mist-netted on 11 August at the edge of secondary forest along the Boca de Cupe trail near Cana. The net was located under a flowering tree frequented by *Phaethornis guy, Colibri delphinae, Chalybura buffoni,* and *Heliodoxa jacula.* We never observed this species.

The only other record of a woodstar from Serranía de Pirre is of another female-plumaged bird taken at Cana, 13 April 1938 by Oliver Pearson, and identified by Wetmore as *Acestrura heliodor* (Wetmore 1968:373). Reexamination of the specimen (Academy of Natural Sciences, Philadelphia 131874) revealed that the bird is actually *C. mitchellii.* Wetmore had ascribed it to the nominate race of *heliodor,* an identification apparently based on a comparison with specimens at USNM. Wetmore was aware that the Cana specimen was atypical for nominate *heliodor* as he stated (1968:373), "It is whiter on the throat, foreneck, and upper breast than usual in this species . . . It is marked as male, which if accurate may indicate that it represents a local race, as the wing is longer than in immature males of typical *A. h. heliodor* from Colombia of similar stage of plumage."

Robbins examined the USNM series of *A. h. heliodor,* with which Wetmore compared the Cana bird, and found two females (USNM 392262, 392271) that appear to be incorrectly identified. Female nominate *heliodor* can be separated from other woodstars of Central America and northwestern South America by their distinctive rufous upper tail coverts and relatively short wings (Table 1). Neither of these birds has rufous upper tail coverts. Furthermore, they have relatively long wings (37.1 and 38.2 mm, respectively) and broader outer rectrices than

TABLE 1
WING LENGTHS OF FEMALE WOODSTARS (*TROCHILIDAE*)[1]

Species	Range	Mean (N)
Calliphlox bryantae[2]	41.3–43.7	42.5 (10)
Calliphlox mitchellii	36.0–38.3	36.8 (6)
Acestrura mulsant	39.0–41.4	40.3 (21)
Acestrura h. heliodor	31.8–34.5	33.0 (19)
Acestrura h. astreans	33.0–35.3	34.2 (10)
Chaetocercus jourdanii rosae	37.0–38.2	37.2 (7)

[1] Wing length measured = chord, in millimeters. Sample may include some missexed immature males. Males have shorter wings than females.
[2] Measurements taken from Wetmore (1968:369).

do female nominate *heliodor*. These two birds appear to be *Chaetocercus jourdanii rosae*. They were collected at the same locality (Buenos Aires, Depto. Santander, Colombia) as nominate *heliodor*. This misidentification is likely the explanation for the unusually large upper end of the range of wing length given for female nominate *heliodor* by Wetmore (1968: 372).

The significant distributional gap between the Cana birds and other woodstars, coupled with the extreme morphological similiarities shared by female woodstars, necessitated a critical comparison of the Cana birds with all other southern Central American and northwestern South American woodstars.

Our specimen and the Pearson bird can be separated from female *Chaetocercus jourdanii rosae* by dorsal color and bill length. Dorsally *C. mitchellii* is darker green with less iridescence than *jourdanii*. The bill (measured from the base) of *mitchellii* (17.9–19.0 mm; n = 6) is longer than that of *jourdanii* (16.5–17.1 mm; n = 7).

Females of the endemic Santa Marta race of *A. heliodor, astreans,* lack rufous upper tail coverts, but can be readily distinguished from all other female woodstars (including nominate *heliodor*) by their narrow, relatively pointed, outer rectrix. The outer rectrix in the others is broader and more rounded at the tip.

Female *mitchellii* can be separated from *Acestrura mulsant* and *Calliphlox bryantae* by wing length (Table 1).

Heretofore, *mitchellii* was unknown north of Cali, Depto. Valle, Colombia (Miller 1963). The Cana birds represent a range extension of ca. 500 km. The inclusion of *Acestrura heliodor* in the A.O.U. Check-list (1983) is based solely on Wetmore's identification of the Cana specimen. Deletion of *heliodor* from the check-list is now warranted.

It should be noted that differences in morphology between woodstar species (in both males and females) are relatively small. We question the allocation of the species in this group to separate genera (e.g., the five species discussed in this paper are assigned into three genera). Clearly, revision of this group at the generic level is needed.

Brachygalba salmoni. Dusky-backed Jacamar. Three or four were seen together in broken canopy of secondary forest along the Boca de Cupe trail and at the clearing edge in February, and single individuals were noted twice in similar habitat in August. All individuals made short sallies to air within several meters of their treetop perches. Their plaintive call, a clear whistled *sweet* or *feet* with an upward inflection, was uttered frequently. Two individuals were netted in young second growth at the Cana clearing edge on 12 August. This species was previously known in Panama from a handful of specimens from the Tuira River drainage, including Cana (Wetmore 1968).

Xenerpestes minlosi. Double-banded Graytail. Parker saw a single individual in a mixed-species flock at 1000 m on 5 February. One or two accompanied a large mixed-species flock in second growth forest along the Boca de Cupe trail at 600 m on 11–12 August. The species is very warbler-like in foraging habits, as noted by Griscom (1927). Birds typically maintain a nearly horizontal position reminiscent of *Mniotitla varia,* while foraging and resting. When foraging, birds are very active, working along the outer branches of trees, peering along and beneath leaves and twigs, and probing leaf clusters and dense blossoms located 5 to 15 m above the ground. The foraging habits of *X. minlosi* are very similar to those reported for *X. singularis* (Parker and Parker 1980).

Margarornis bellulus. Pearled Treerunner. This treerunner appears to be rare in cloud forest above 1400 m on Cerro Pirre. Two were seen in February, one in July. Single individuals were observed with three mixed-species flocks of insectivores including *Cranioleuca erythrops*, *Troglodytes ochraceus*, and *Myioborus miniatus*. The treerunners crept along branches and up vines, briefly pausing to probe epiphytic mosses and small bromeliads. A short trill was given by one individual (D. Wolf, pers. comm.)

Terenura callinota. Rufous-rumped Antwren. One to four were seen daily in tall forest near Cana and up to 1400 m. These birds often foraged among small leaves (<10 × 4 cm) and twigs in the outer extremities of trees. In tall forest along the Boca de Cupe trail individuals usually foraged in crowns of trees (>20 m above ground); nevertheless, on steep slopes above Cana, birds foraged as low as 6 m in trees at the edge of small clearings. Presumably, the phenology of the vegetation (as well as lighting and insect composition?) is similar between these two microhabitats. *Terenura* frequently accompanied mixed-species flocks; nonetheless, it often remained in trees foraging after a flock had moved on. In Colombia, this species associates with mixed-species flocks in middle and upper levels of forest; it "peers under branches or hangs upside-down" as it forages (Ridgely and Gaulin 1980). Remsen et al. (1982) reported similar behavior for *T. sharpei* in Bolivia. These are the first records for the Pirre range.

Pittasoma michleri. Black-crowned Antpitta. A few individuals of this large, terrestrial antpitta were found in disturbed forest around Cana. It preferred dense thickets around treefalls and forest edge. When disturbed, these antpittas uttered a startlingly loud, harsh, squirrel-like (*Sciurus* spp.) chatter *kwa-kwa-kwa-kwa-kwa-kwa* that slows towards the end. These call series were given at ca. 3-sec intervals for up to a minute at a time. *Pittasoma michleri* is apparently rare and local throughout its range from Costa Rica to northwestern Colombia.

Scytalopus vicinior. Narino Tapaculo. We encountered this secretive species only four times in February, but it was heard calling daily at 1200 m in July. Solitary individuals at 1150 to 1400 m were tape-recorded, and seen when they responded to playbacks. All remained in tangled undergrowth within a few centimeters of the ground. One hopped about on moss-covered branches of a fallen tree at the edge of a recently-made clearing. Another was noted on the steep slope of a densely forested ravine. The song is a rapidly uttered, deliberate series of notes *keh-keh-keh* etc. at least 30 sec long. In response to playbacks of this vocalization, and when alarmed by observers, *S. vicinior* also gave a loud, shorter series *thu-tu-tu-tu-tu-tutu* or a single, explosive *pzeert*. This species was previously known in Panama from two specimens taken at 1410 m and 1525 m respectively, on Cerro Pirre (Wetmore 1972).

Aphanotriccus audax. Black-billed Flycatcher. This flycatcher was found only in deeply shaded second-growth forest along the Boca de Cupe trail near Cana. Individuals perched quietly for relatively long periods from 3 to 4 m above ground. One bird uttered a three-note call, the final note of which had a distinctly buzzy quality. The species is found only from eastern Panama to northwestern Colombia.

Thryothorus spadix. Sooty-headed Wren. This relatively inconspicuous wren inhabits forest undergrowth from Cana up to 1200 m. It broadly overlaps *Henicorhina leucosticta*, elevationally and ecologically. *Thryothorus spadix* foraged in pairs from 1 to 10 m up in dense undergrowth, especially vine tangles. Several individuals probed dead, curled leaves trapped by vines; such dead-leaf-searching is known for several species of *Thryothorus* (Remsen and Parker 1984). The most frequent song was a rather high-pitched, sprightly *weechoaweeawee-chuwoo* given at about 12-second intervals. This wren is known in Panama only from Cerro Pirre.

Polioptila schistaceigula. Slate-throated Gnatcatcher. Parker saw one on 12 February with a large canopy mixed-species flock on the lower slope of Cerro Pirre just above Cana at 750 m. The bird was in the uppermost foliage of a 20 m tree. It gleaned small leaves and twice made short sallies. Other species in the flock included two *Hylophilus decurtatus*, two *Dacnis cayana*, three *Vermivora peregrina*, four *Dendroica castanea*, and two *Tangara icterocephala*. There are two previous records of this species in eastern Panama, but apparently none for Cerro Pirre (Ridgely 1976), although there is one recent sight record, 17 July 1975, at 600 m on Cerro Quía (R. S. Ridgely, pers. comm.). This species, as well as several others (e.g., *Vireolanius eximius*, *Heterospingus xanthopygius*, *Caryothraustes canadensis*), may be restricted to the lower slopes of the Pirre range, for individuals were encountered only from just above Cana (650 m) up to 1000 m.

Myadestes coloratus. Varied Solitaire. This species was one of the most frequently heard cloud forest birds from 1350 m to the ridge tops. At dawn, late in the afternoon, and during foggy periods throughout the day, the species sang from concealed perches in the middlestory and canopy. The song consists of short, flute-like phrases interspersed with less musical, often burry ones; phrases are given about 3–6 sec apart. In pattern and quality *M. coloratus* song closely resembles those of *M. ralloides* (Peruvian populations) and *M. melanops* (Parker, pers. obs.). The strong similarities in voice and plumage shared by these solitaires suggest that they should be treated as conspecific. *Myadestes coloratus* is restricted to the highlands of eastern Darién.

Vireolanius eximius. Yellow-browed Shrike-Vireo. In February single, singing individuals of this species were heard in three different canopy mixed-species flocks from 650 to 1000 m on Cerro Pirre. These birds moved sluggishly through foliage and were extremely difficult to observe. The song consisted of a three-note phrase, *peea-peea-peea* repeated at ca. 3-second intervals for up to several minutes. Thus, *eximius* is vocally similar to *V. pulchellus* (Parker, pers. observ.; and see Ridgely 1976). Cerro Pirre is one of only two known Panamanian localities for *eximius,* the other being nearby Cerro Quía, where R. S. Ridgely (pers. comm.) reports seeing two at 550 and 600 m on 17 July 1975.

Basileuterus ignotus. Pirre Warbler. This species was fairly common above 1400 m in elfin forest where 2 to 4 individuals were encountered at least three times per morning along ca. 1 km of trail. These birds were seen from 2 to 10 m above ground, usually above eye-level. Typically, they foraged apart from other species, but occasionally, a few associated with *Chlorospingus inornatus* and other species. No singing was noted during February or July; the distinctive call note, a penetrating *tseeut* or *tseeit,* was frequently heard. Slud (1964) noted a very restricted vocal repertoire in *ignotus'* sister species, *B. melanogenys,* in Costa Rica. Adults were seen feeding recently-fledged young in July. This species is restricted to the Pirre and Tacarcuna ranges and has been known previously from fewer than five specimens; we obtained eight.

Tangara palmeri. Gray-and-gold Tanager. One was seen on 10 February, and several others were heard in *Cecropia* trees at the edge of a small clearing in tall forest on Cerro Pirre at 1000 m in February. Chip notes were occasionally followed by a protracted, upslurred *seeep.* Despite intensive work in this area in July and August, no other records of this rare tanager were obtained. Heretofore, *Tangara palmeri* has been known from three localities in eastern Darién (Ridgely 1976); it is more common in western Colombia and northwestern Ecuador.

Tangara fucosa. Green-naped Tanager. Typically 3 to 4 were seen in about one-third of the mixed-species flocks encountered above 1400 m on Cerro Pirre. A few individuals were seen as low as 1000 m. These tanagers were regularly noted in fruiting *Miconia* at 1480 m; individuals in mixed-species flocks searched mosses on slender branches and small leaves. Two were observed feeding on berries of an epiphytic vine just below 1500 m in July. The call is a soft *tsip,* often given three or four times in rapid succession.

Chlorospingus inornatus. Pirre Bush-Tanager. This endemic was the most common and conspicuous elfin forest bird. It was found in small numbers in canopy of tall forest at elevations as low as 1200 m. During February, an average of 30 individuals was noted daily along about 1 km of trail; 3 to 6 individuals were seen in nearly every mixed-species flock encountered above 1400 m, and small conspecific groups were regularly observed in elfin forest. Frequent flock associates included *Basileuterus ignotus, Tangara fucosa,* and *Chrysothlypis chrysomelas.* The bush-tanagers were most often observed from canopy to mid-heights, where they searched epiphytic mosses on branches and gleaned leaves. In February, they also frequently visited fruiting *Miconia* at the 1480 m camp. Typical vocalizations of *C. inornatus* include a hoarse, rapidly uttered *chawee-chawee chawee-chawee* (pairs of notes often repeated four times), and a variety of sharp *chet* and *seet* notes. No *C. opthalmicus*-like territorial songs were heard (O'Neill and Parker 1981; Parker and Parker 1982). *Chlorospingus inornatus* is known only from Cerro Pirre, Cerro Sapo, and Cerro Nique (in adjacent Colombia).

Chrysothlypis chrysomelas. Black-and-yellow Tanager. This species was encountered daily, usually in monospecific flocks of 4 to 8 individuals, from 1150 m to the ridge top at 1575 m. They were very active and restless, seldom remaining in one tree for more than 2 minutes; they occasionally joined, but soon left, mixed-species flocks. Flock associates were primarily small insectivores, especially the parulines *Dendroica fusca, D. castanea* (in February), and *Parula pitiayumi,* and the tanagers *Tangara icterocephala, T. fucosa,* and *Euphonia xanthogaster.* The *Chrysothlypis* are insect foragers, gleaning leaves of all sizes, and often hover-gleaning surfaces of large leaves. They were also observed hanging and probing curled dead and dying leaves. One or more groups of these tanagers made several daily visits to clusters of fruiting *Miconia* at 1000 m (July) and 1480 m (February). These tanagers were quite vocal, uttering a near constant succession of *chips* and *seets,* especially as they moved from tree to tree.

Caryothraustes canadensis. Yellow-green Grosbeak. This gregarious, vocal species was common in forest canopy on the ridge above Cana in February. Two or three groups of 6 to 10 individuals were observed daily from 650 to 1000 m; they regularly joined canopy flocks of warblers, tanagers, foliage-gleaners, and woodcreepers, but also foraged away from mixed-species flocks. The grosbeaks moved quickly through upper, outer branches, searching bark and foliage, and were also noted in fruiting trees. Their loud, constantly uttered calls render the species almost impossible to overlook. Most characteristic of many vocalizations is an explosive buzzy *pzrrt* followed by a rapid succession of two to four clearer *tew* notes, *pzrrt tew-tew-tew;* all flock members seem to give this call, as well as a variety of shorter calls *pwee, pwit, see,* etc. On one occasion a melodious (advertising ?) song was heard: *churwee chip-chip-chip-chip.* The local race, *simulans,* is known only from slopes of Cerro Pirre, with an apparent hiatus in the species' range stretching to southeastern Colombia (Paynter 1970).

ACKNOWLEDGMENTS

The 1982 LSUMZ expeditions to the Serranía de Pirre could not have succeeded without the aid of a number of people. We are particularly grateful to our field companions, D. Braun, M. Braun, S. Lanyon, and J. O'Neill, who obtained much of the information presented in this paper. We wish to express our gratitude to Drs. E. Méndez, R. Watson, and P. Galindo of the Gorgas Memorial Laboratory, Panama City, for their assistance in making our field work in Panama possible. We are especially grateful to J. S. McIlhenny, B. M. Odom, H. Irving, and L. Schweppe for their continued interest in and support of our work in Latin America. This manuscript benefited from comments by S. L. Hilty, T. R. Howell, J. V. Remsen, Jr., R. S. Ridgely, and T. S. Schulenberg. R. S. Ridgely kindly provided information on a number of species. We thank G. E. Watson for access to and loan of specimens from the United States National Museum of Natural History (USNM) and G. S. Keith for loan of specimens from the American Museum of Natural History.

LITERATURE CITED

AMERICAN ORNITHOLOGISTS' UNION. 1983. Check-list of North American birds, 6th ed. American Ornithologists' Union, Washington, D.C.

GOLDMAN, E. E. 1920. Mammals of Panama. Smithson. Misc. Collect. 69:1–309.

GRISCOM, L. 1927. An ornithological reconnaissance in eastern Panama in 1927. Am. Mus. Novit. No. 282.

GRISCOM, L. 1929. A collection of birds from Cana, Darien. Bull. Mus. Comp. Zool. 69:148–190.

HAFFER, J. 1967. Speciation in Colombian forest birds west of the Andes. Am. Mus. Novit. No. 2294.

HAFFER, J. 1975. Avifauna of northwestern Colombia, South America. Bonn. Zool. Monogr. 7.

HAFFER, J. 1977. A systematic review of the neotropical ground-cuckoos (Aves: *Neomorphus*). Bonn. Zool. Beitr. 28:48–77.

HOWELL, T. R. 1957. Birds of a second-growth rain forest area of Nicaragua. Condor 59:73–111.

MEYER DE SCHAUENSEE, R. 1970. A Guide to the Birds of South America. Livingston Publ. Co., Wynnewood, Pennsylvania.

MILLER, A. H. 1963. Seasonal activity and ecology of the avifauna of an American equatorial cloud forest. Univ. Calif. Publ. Zool. 66:1–74.

MYERS, C. W. 1966. A new species of colubrid snake, genus *Coniophanes,* from Darien, Panama. Copeia 1966:665–668.

MYERS, C. W. 1969. The ecological geography of cloud forest in Panama. Am. Mus. Novit. No. 2396.

NELSON, E. W. 1912. Descriptions of new genera, species, and subspecies of birds from Panama, Colombia, and Ecuador. Smithson. Misc. Collect. 60:1–25.

O'NEILL, J.P., AND T. A. PARKER, III. 1981. New subspecies of *Pipreola riefferii* and *Chlorospingus ophthalmicus* from Peru. Bull. Br. Ornithol. Club. 101:294–299.

PARKER, T. A., III, AND S. A. PARKER. 1980. Rediscovery of *Xenerpestes singularis* (Furnariidae). Auk 97:203–205.

PARKER, T. A., III, AND S. A. PARKER. 1982. Behavioral and distributional notes on some unusual birds of a lower montane cloud forest in Peru. Bull. Br. Ornithol. Club 102:63–70.

PAYNTER, R. A., JR. 1970. Cardinalinae. Pp. 216–245, *In* R. A. Paynter (ed.), Check-list of Birds of the World. Vol. 13. Mus. Comp. Zool., Cambridge, Massachusetts.

PUJALS, J. J., J. W. WALL, AND D. W. WILCOVE. 1977. First record of the Pearl Kite in Panama. Am. Birds 31:1099–1100.
REMSEN, J. V., JR., T. A. PARKER III, AND R. S. RIDGELY. 1982. Natural history notes on some poorly known Bolivian birds. Gerfaut 72:77–87.
REMSEN, J. V., JR., AND T. A. PARKER III. 1984. Arboreal dead-leaf-searching birds of the neotropics. Condor 86:36–41.
RIDGELY, R. S. 1976. A Guide to the Birds of Panama. Princeton Univ. Press, Princeton, New Jersey.
RIDGELY, R. S., AND S. J. C. GAULIN. 1980. The birds of Finca Merenberg, Huila Department, Colombia. Condor 82:379–391.
ROTH, P. 1981. A nest of the Rufous-vented Ground-Cuckoo (*Neomorphus geoffroyi*). Condor 83:388.
SLUD, P. 1964. Birds of Costa Rica. Bull. Am. Mus. Nat. Hist. 128:1–430.
WAKE, D. B., A. H. BRAME, JR., AND C. W. MYERS. 1970. *Bolitoglossa taylori,* a new salamander from cloud forest of the Serranía de Pirre. Am. Mus. Novit. No. 2430.
WETMORE, A. 1965. The Birds of the Republic of Panama. Pt. 1. Smithson. Misc. Collect. 150(1).
WETMORE, A. 1968. The Birds of the Republic of Panama. Pt. 2. Smithson. Misc. Collect. 150(2).
WETMORE, A. 1972. The Birds of the Republic of Panama. Pt. 3. Smithson. Misc. Collect. 150(3).

APPENDIX I

RELATIVE ABUNDANCES AND SPECIMEN DATA FOR BIRD SPECIES RECORDED AT SERRANÍA DE PIRRE

	Abundance[1]			Weight[2]		Gonads[3]		Skull[4] pneuma.	Stomach[5] contents	Soft part colors
	CA	ES	RT	♂	♀	♂	♀			
Tinamidae										
Tinamus major	F	R								
Crypturellus soui	F									
C. (kerriae)	X	X								
Cathartidae										
Coragyps atratus	U									
Cathartes aura	F	U	U							
Sarcoramphus papa	U	R	R							
Accipitridae										
Elanoides forficatus	C	U	U							
Gampsonyx swainsonii	R									
Ictinia plumbea	C	U	U							
Accipiter superciliosus	R									
A. bicolor	R				430		C			facial skin: yellow-green; bill: black; cere: yellow-green; tarsi: yellow
Leucopterus princeps		F								
L. semiplumbea	U			250	325	B	C	f-1	5 (n = 1)	irides: yellow; bill: black; cere: yellow-orange; tarsi: orange
L. albicollis	R	U	U		650	B		f		irides: light yellow-green; cere: gray; tarsi: yellow
Buteogallus urubitinga	R	R			—		E	f	7	irides: dark; bill: black; cere: yellow; tarsi: yellow
Harpai harpyja		X								
Spizastur melanoleucus	R									
Spizaetus tyrannus	F	F	U							

APPENDIX I
CONTINUED

	Abundance[1]			Weight[2]		Gonads[3]		Skull[4] pneuma.	Stomach[5] contents	Soft part colors
	CA	ES	RT	♂	♀	♂	♀			
S. ornatus	R				950		D	f		irides: yellow bill: black cere: yellow tarsi: yellow
Falconidae										
Daptrius americanus	C									
Herpetotheres cachinnans	F									
Micrastur ruficollis	U	F								
M. mirandollei		R								
M. semitorquatus	U		R							
Falco rufigularis	U			140		B			1	bill: black facial skin, cere: yellow tarsi: orange
Cracidae										
Ortalis cinereiceps	F				490		E		3	irides: light brown bill: dark gray tarsi: gray
Crax rubra	U		R	2050		D	E-2	f-1	2 (n = 2) 4 (n = 1)	irides: red facial skin: blackish blue gular skin: orange-red bill: black tarsi: orange-red
Penelope purpurascens	U	F	U		2000–2150 (n = 2)					
Phasianidae										
Odontophorus erythrops	F				–		D	f		irides: brown facial skin: gray bill: black tarsi: gray
Rhynchortyx cinctus	U		R		54–150 (imm.)	A-1	D-1	b-1 f-1	1, 3 (n = 1)	irides: orange-brown bill: horn tarsi: blue-gray

APPENDIX I

CONTINUED

	Abundance[1]			Weight[2]		Gonads[3]		Skull[4] pneuma.	Stomach[5] contents	Soft part colors
	CA	ES	RT	♂	♀	♂	♀			
Rallidae										
Laterallus albigularis	C			43		B		b	1	irides: gray-brown bill: horn tarsi: greenish gray
Scolopacidae										
Actitis macularia	R									
Columbidae										
Columba cayennensis	U									
C. speciosa	U									
C. subvinacea		F								
C. nigrirostris		F								
Claravis pretiosa	X									
Leptotila verreauxi	U	F								
L. cassinii	C									
Geotrygon goldmani	U	F		—		D		f		irides: red-brown eyering, facial skin: red-violet bill: black tarsi: red-violet
Geotrygon montana	F		R		80		B	d	2	
Psittacidae										
Ara severa	C									
A. ambigua	U	F	U							
A. chloroptera	F	F	U							
A. ararauna	C									
Brotogeris jugularis	C			58	55	B	C	f-1	3 (n = 1)	irides: brown bill: creamy tan tarsi: creamy tan
Touit dilectissima	R	R	R							
Pionopsitta pyrilia	U	U								
P. haematotis	F	F	U		145–150 (n = 2)		C-2		2 (n = 1)	irides: green bill: yellow-horn tarsi: light gray

APPENDIX I
CONTINUED

	Abundance[1]			Weight[2]		Gonads[3]		Skull[4] pneuma.	Stomach[5] contents	Soft part colors
	CA	ES	RT	♂	♀	♂	♀			
Pionus menstruus	C	F								
Amazona autumnalis	R									
A. farinosa	C	F		620	644	B	E		2 (n = 1)	irides: red bill: max. horn; mand. light yellow tarsi: light yellow
Cuculidae										
Piaya cayana	F	F		110		B		f	1	irides: red bill: green-yellow tarsi: gray
P. minuta	U									
Neomorphus geoffroyi	R	R		350	400	C	C-1 D-1	f-3	1 (n = 2) 1,2,5 (n = 1)	irides: red (adults); brown (immature) bill: pale green (adults); black (immature) tarsi: gray
Crotophaga major	U									
C. ani	C									
Strigidae										
Otus guatemalae	R				125		A	f		irides: golden yellow bill: yellow tarsi: pale beige
O. clarkii	R	U		123		C		f		irides: golden yellow bill: pale green cere: orange tarsi: flesh-white
Pulsatrix perspicillata	U			750		B			1, 6	irides: yellow bill: greenish yellow
Glaucidium minutissimun	U				45		A			irides: yellow-orange bill: yellow tarsi: yellow
Ciccaba virgata	U	U								
Caprimulgidae										
Nyctidromus albicollis	C									

APPENDIX I
CONTINUED

	Abundance[1]			Weight[2]		Gonads[3]		Skull[4] pneuma.	Stomach[5] contents	Soft part colors
	CA	ES	RT	♂	♀	♂	♀			
Apodidae										
Streptoprocne zonaris	C	C	U							
Chaetura spinicauda	C	C	R							
Panyptila cayennensis	U			17.0	18.5			c-1	1	irides, bill, tarsi: dark gray
Trochilidae										
Glauis hirsuta	F			5.0–6.0 (n = 2)	5.5–7.0 (n = 3)	A-1 B-1	B-2 C-1	a-6	1 (n = 1)	irides: dark; bill: max. black; mand. yellow with black tip; tarsi: pink
Threnetes ruckeri	U									
Phaethornis guy	C	C	F	4.5–5.5 (n = 12)	4.0–5.5 (n = 11)	A-5 B-6	A-6 B-6	a-11 b-1	1 (n = 11)	irides: black; bill: max. black; mand. red; tarsi: pink
P. superciliosus	F	U		5.0	5.5–6.0 (n = 2)	B	A-1 B-1	1-2		irides: dark; bill: max. black; mand. yellow with black tip; tarsi: pinkish-gray
P. longuemareus	F	U		2.0 (n = 2)		B-2		a-1		irides: dark; bill: max. black; mand. yellow with black tip; tarsi: pink
Eutoxeres aquila	F	F	U	8.5–12.0 (n = 9)	9.0–9.5 (n = 5)	A-2 B-5 C-1	B-3 C-1	a-11 f-1		irides: brown; bill: max. black; mand. yellow; tarsi: pinkish gray
Androdon aequatorialis	R	U	U		5.5–6.0 (n = 2)		A-1 B-1 C-1	a-2	1 (n = 1)	irides: dark; bill: max. black; mand. black with pink base; tarsi: pink
Colibri delphinae	U			6.0 (n = 2)		B-2		a-2	1 (n = 1)	irides: brown; bill: gray; tarsi: gray
Klais guimeti	R									

APPENDIX I
CONTINUED

	Abundance[1]			Weight[2]		Gonads[3]		Skull[4] pneuma.	Stomach[5] contents	Soft part colors
	CA	ES	RT	♂	♀	♂	♀			
Lophornis delattrei	R									
Popelairia conversii	R	F	F							
Thalurania colombica				4.0–5.5 (n = 6)	3.5–4.0 (n = 2)	A-3 B-3	B-2	a-4 f-1		irides: dark bill: black tarsi: pinkish gray
Goethalsia bella	R	F	U	3.0–3.5 (n = 4)	3.0–4.0 (n = 4)	A-1 B-3	A-2 B-2 C-1	a-4 f-1		irides: black bill: max. black; mand. pink with black tip tarsi: pink
Amazilia amabilis	R			4.0		B	B	a		irides: brown bill: max. black; mand. coral tarsi: black
A. edward	U				4.5		B	a		irides: dark bill: max. black; mand. red with black tip tarsi: black
Chalybura buffoni	F			6.0–7.0 (n = 4)		B-4		a-3		irides: brown bill: black tarsi: pink
Helidoxa jacula	U	F	F	7.0–8.0 (n = 2)	6.0–8.0 (n = 9)	A-2	A-3 B-3 C-2	a-8 b-1		irides: dark bill: black tarsi: gray
Haplophaedia aureliae		U	C	5.0–6.0 (n = 5)	4.0–4.5 (n = 2)	A-2 B-4	B-1 C-1	a-5 f-2		irides: dark brown bill: black tarsi: pink
Heliothryx barroti		U	U	5.0 (n = 2)		A-1 B-1		a-1		irides, bill, tarsi: black
Heliomaster longirostris	R									
Calliphlox mitchellii	X				3.0		B	a		irides: brown bill & tarsi: black
Trogonidae										
Trogon viridis		F	U							

APPENDIX I
CONTINUED

	Abundance[1]			Weight[2]		Gonads[3]		Skull[4] pneuma.	Stomach[5] contents	Soft part colors
	CA	ES	RT	♂	♀	♂	♀			
T. collaris		F		57		B			2	irides: dark / eye-rings: red / bill: yellow / tarsi: yellow-green
T. rufus		U		50		B			2	irides: dark brown / eyerings: light blue / bill: yellow-green / tarsi: brown
T. melanurus		F								
T. massena		F	U		110		C	f	1, 2	irides: gray / bill: max. black with yellow base; mand. yellow / tarsi: yellow-gray
Pharomachrus auriceps		R	F							
Momotidae										
Hylomanes momotula		R	U	29.5–38 (n = 2)	27.5–34 (n = 6)	A-1 B-1	A-3 B-2	e-1 f-4	1 (n = 3)	irides: reddish brown / bill: black / tarsi: pale yellow
Electron platyrhynchum		U	R	53		A			1	irides: brown / bill: black / tarsi: black
Baryphthengus ruficapillus		U	R		155		C	f	1	irides: red / bill: black / tarsi: black
Alcedinidae										
Chlorceryle americana		U								
C. aenea		U								
Bucconidae										
Bucco radiatus		F	U		59		B		1	irides: pale yellow / bill: yellow-green with black commissure / tarsi: yellow-green

APPENDIX I
CONTINUED

	Abundance[1] CA ES RT	Weight[2] ♂	Weight[2] ♀	Gonads[3] ♂	Gonads[3] ♀	Skull[4] pneuma.	Stomach[5] contents	Soft part colors
B. macrorhynchos	R		100		C		1	irides: red bill, tarsi: black
B. tectus Malacoptila panamensis	U F R	38–46 (n = 4)	36–39 (n = 3)	B-4	B-2 C-1	f-4	1 (n = 5)	irides: red bill: max. black; mand. olive-yellow tarsi: olive-gray
Nonnula ruficapilla	U	15.0	14.5–17.5 (n = 6)	B	A-3 B-2 C-1	f-7	1 (n = 8)	irides: red-brown eye-rings: dusky red bill: gray tarsi: brownish gray
Monasa morphoeus	F		110–120 (n = 2)		C-2	f-2	1 (n = 2)	irides: brown bill: red tarsi: dark gray
Galbulidae Brachygalba salmoni	R	18.5	16.0	B	B	f-1	1 (n = 2)	irides: brown (♀♀); red (♂♂) bill: black tarsi: black
Jacamerops aurea	U U							
Capitonidae Capito maculicoronatus	U F	44–52 (n = 2)	47–48 (n = 2)	B-2	B-1 C-1	f-3	2 (n = 2) 3 (n = 1)	irides: dark red-brown bill: gray with dark tip tarsi: blue-gray
Eubucco bourcierii	U F F	34	31–35 (n = 2)	B	B-1 C-1	f-2	2 (n = 3)	irides: red bill: apple green tarsi: green
Ramphastidae Aulacorhynchus prasinus	F F	160–184 (n = 3)	155	B-2 C-1	D	f-4	2 (n = 3)	irides: dark bill: largely black and yellow-green tarsi: green
Pteroglossus torquatus	C							

APPENDIX I
Continued

	Abundance[1]			Weight[2]		Gonads[3]		Skull[4] pneuma.	Stomach[5] contents	Soft part colors
	CA	ES	RT	♂	♀	♂	♀			
Selenidera spectabilis	F			175–245 (n = 2)	200–240 (n = 3)	B-1 C-1	D-3	f-4	2 (n = 1)	irides: red orbital skin: aqua blue and yellow bill: chestnut with yellow culmen tarsi: blue
Ramphastos sulfuratus	C			540	275	C	A		2 (n = 2)	irides: olive-gold tarsi: blue-gray
R. swainsonii	C			625		D			2	irides: green bill: dark maroon with yellow culmen and tip tarsi: blue-gray
Picidae										
Picumnus olivaceus	U	U		10.5		A		f	1	irides: brown bill, tarsi: gray
Melanerpes pucherani	C	F		57	49	B	D			irides: brown bill: black tarsi: gray
Veniliornis kirkii	U	F								
Piculus leucolaemus	U	F		52	50	B	B	f-3	1 (n = 2)	irides: light gray (♀♀); dark (♂♂) bill: gray tarsi: blue-gray
Celeus loricatus	U			74		B		f		irides: red-brown bill: max. brown; mand. yellow tarsi: gray
Dyocopus lineatus	F	F								
Campephilus haematogaster	R	F		225–250 (n = 2)		B-1		f-1	1 (n = 2)	irides: brown bill, tarsi: black
C. melanoleucos	F				250–260 (n = 2)		C-2	f-1	1 (n = 1) 3 (n = 1)	bill: gray tarsi: brown
Furnariidae										
Synallaxis brachyura	C			17.0–21.0 (n = 5)	19.0–19.5 (n = 2)	A-1 B-2 C-2	A-1 C-1	a-2 c-1 f-2	1 (n = 2) 3 (n = 1)	irides: tan bill: gray (♀♀); black (♂♂) tarsi: olive-gray

APPENDIX I
CONTINUED

	Abundance[1]			Weight[2]		Gonads[3]		Skull[4] pneuma.	Stomach[5] contents	Soft part colors
	CA	ES	RT	♂	♀	♂	♀			
Cranioleuca erythrops	F			—	12.5–15.0 (n = 4)		B-4	b-1 c-1 f-1	1 (n = 4)	irides: red-brown bill: black with pink base of mand. tarsi: olive
Xenerpestes minlosi	R	R								
Premnoplex brunnescens		F				B				irides: dark bill, tarsi: black
Margarornis bellulus		R		18.5		A		e	1	irides: brown bill: max. brown; mand. flesh tarsi: gray
Syndactyla subalaris	F	F		27.5–34 (n = 2)	30–34 (n = 4)	A-1 D-1	B-2 C-2	f-5	1 (n = 5)	irides: brown bill: dark brown tarsi: olive-green
Philydor erythrocercus	F	U		25.0–27.5 (n = 2)		A-1		b-1		irides: brown bill: gray tarsi: gray-green
Automolus ochrolaemus		F			27.5–36 (n = 3)		B-1 C-2	f-3	1 (n = 1)	irides: dark bill: horn tarsi: greenish gray
A. rubiginosus		F		44	39	D-1	C-1	f-1	1 (n = 1)	irides: brown bill: max. black; mand. brown tarsi: brown
Xenops minutus	U	F		9.5–12.0 (n = 3)	9.0–11.5 (n = 3)	A-2 B-1	A-1 B-2	b-1 f-2	1 (n = 3)	irides: brown bill: light gray tarsi: gray
Sclerurus mexicanus		U		23.0	21.0–26.5 (n = 2)	C	C-2	f-1	1 (n = 1)	irides: dark brown bill: black tarsi: black
S. guatemalensis	R	U		31	36	B	C	f-1	1 (n = 2)	irides: black bill: black tarsi: gray
Dendrocolaptidae *Dendrocincla fuliginosa*	U	U		40 (n = 2)	34–38 (n = 4)	A-1 B-1	A-2 B-1 C-1	b-2 f-1	1 (n = 4)	irides: light brown bill: black tarsi: gray

APPENDIX I
CONTINUED

	Abundance[1]			Weight[2]		Gonads[3]		Skull[4] pneuma.	Stomach[5] contents	Soft part colors
	CA	ES	RT	♂	♀	♂	♀			
Sittasomus griseicapillus	F	F		13.0–15.5 (n = 3)		A-1 B-2		f-2	1 (n = 3)	irides: dark bill, tarsi: gray
Deconychura longicauda			U	21.0–24.0 (n = 3)	20	A-1 B-2	B-1	b-1 f-3	1 (n = 3)	irides: dark brown bill, tarsi: gray
Glyphorhynchus spirurus	F	F			12.0–14.5 (n = 6)		A-3 B-3	f-2	1 (n = 1)	irides: brown bill, tarsi: gray
Dendrocolaptes certhia			U	60–62 (n = 2)		B-1 C-1		f-2	1 (n = 1)	irides: brown bill: black
Xiphorhynchus guttatus	F	F		41–50 (n = 6)	41–48 (n = 2)	A-2 B-1 C-1 D-2	B-2	f-3	1 (n = 4)	tarsi: gray irides: brown bill: black tarsi: gray
X. lacrymosus	U	U		66		A		f	1	irides: brown bill: max. brown; mand. gray
X. erythropygius	F	U		40–54 (n = 3)	41–46 (n = 3)	B-2 C-1	B-2 C-1	f-5	1 (n = 3)	tarsi: gray irides: brown bill: max. black; mand. pinkish gray
Lepidocolaptes souleyetii	U				23		B	f	1	tarsi: gray irides, bill: brown
Campylorhamphus trocilirostris	U			36	29.5	B	C	f-2	1 (n = 1)	tarsi: gray irides: dark bill: red
C. pusillus		U	R	38		B				tarsi: olive-gray
Formicariidae										
Cymbilaimus lineatus	F			35–41 (n = 2)	—	B-2	C	e-1 f-1	1 (n = 1)	irides: red bill: black tarsi: blue-gray
Taraba major	C			66		B		f		irides: red bill: black tarsi: gray

APPENDIX I
CONTINUED

	Abundance[1]	Weight[2]		Gonads[3]		Skull[4] pneuma.	Stomach[3] contents	Soft part colors
	CA ES RT	♂	♀	♂	♀			
Thamnophilus punctatus	C	20.5	19.0–25.5 (n = 8)	B-2	A-1 B-5 C-1	b-2 f-2	1 (n = 3)	irides: brown / bill: max. black; mand. gray / tarsi: gray
Thamnistes anabatinus	F R	19.0–23.0 (n = 2)	19.0–20.0 (n = 2)	A-1 B-1	B-1 C-1	a-1 f-2	1 (n = 3)	irides: red / bill: max. slate; mand. bluish gray / tarsi: olive-gray
Dysithamnus mentalis	U F R	11.0–14.0 (n = 4)	11.0–15.5 (n = 6)	A-1 B-2	A-1 B-3 C-2	b-1 f-5	1 (n = 5)	irides: brown / bill, tarsi: gray
M. brachyura	F U							
M. surinamensis	U	8.5–9.0 (n = 2)	9.5	B-2	C-1	f-2	1 (n = 1)	irides: brown / bill: max. black; mand. pale gray / tarsi: bluish gray
M. fulviventris	C R	8.5–11.0 (n = 6)	8.5–9.5 (n = 4)	A-3 B-3	A-1 B-2	b-3 c-3 d-1 f-2	1 (n = 7)	irides: white / bill, tarsi: gray
M. schisticolor	C U	7.5–10.0 (n = 7)	7.5–9.0 (n = 7)	A-4 B-3	A-3 B-4	a-1 b-1 d-1 e-1 f-7	1 (n = 4)	irides: dark red-brown / bill: black / tarsi: gray
Herpsilochmus rufimarginatus	U U	11.5	10.0	B	A	a-1 f-1	1 (n = 2)	irides: brown / bill: max. black; mand. pale gray / tarsi: gray
Microrhopias quixensis	F F							
Terenura callinota	U F		7.0	B	B	e	1	irides: brown / bill: max. black; mand. pale gray / tarsi: gray
Cercomacra tyrannina	C	15.0–19.5 (n = 8)	13.0–18.0 (n = 5)	A-2 B-6	A-1 B-1 C-2 D-1	d-1 f-6	1 (n = 5)	irides: gray / bill: max. gray; mand. ivory / tarsi: gray

APPENDIX I
CONTINUED

	Abundance[1] CA ES RT	Weight[2] ♂	Weight[2] ♀	Gonads[3] ♂	Gonads[3] ♀	Skull[4] pneuma.	Stomach[5] contents	Soft part colors
C. nigricans	U		11.5		A	b	1	irides: gray bill: max. gray; mand. pale gray tarsi: bluish gray
Gymnocichla nudiceps	U	28.0–33 (n = 2)	27.0–33 (n = 3)	B-1 D-1	C-2 D-1	b-1 c-1 f-2	1 (n = 4)	irides: dark red (♂♂); brown (♀♀) crown and facial skin: blue bill: black
Myrmeciza exsul	C	23.5–26.0 (n = 4)	27.5–35 (n = 2)	B-4	C-1 D-1	f-3	1 (n = 2)	tarsi: gray irides: red orbital skin: blue bill: black
M. immaculata	R R	42		B		f		tarsi: gray irides: dark brown facial skin: pale blue bill, tarsi: black
Hylophylax naevoides	C	13.0–18.0 (n = 5)	14.0–16.5 (n = 4)	A-1 B-4	A-3 B-1 C-1	b-1 f-5	1 (n = 5)	irides, bill, tarsi: brown
Myrmornis torquata	F	41–48 (n = 4)	43–47 (n = 4)	B-4	C-3 D-1	f-4	1 (n = 1)	irides: brown facial skin: pale blue bill: black
Gymnopithys leucaspis	C C	22.0–31 (n = 9)	26.5–33 (n = 5)	B-1 C-7	B-2 C-2 D-1	f-5	1 (n = 5)	tarsi: gray irides: dark red-brown bill, tarsi: gray
Phaenostictus mcleannani	R F	44–56 (n = 4)	41	B-2 C-2	D	f-5	1 (n = 5)	irides: dark brown facial skin: blue bill: black
Formicarius analis	C F	42–53 (n = 2)	49–56 (n = 3)	C-1 D-1	B-2 D-1	f-2	1 (n = 4)	tarsi: pinkish gray irides: brown eyerings: pale blue bill: black tarsi: brown

APPENDIX I
CONTINUED

	Abundance[1] CA ES RT	Weight[2] ♂	Weight[2] ♀	Gonads[3] ♂	Gonads[3] ♀	Skull[4] pneuma.	Stomach[5] contents	Soft part colors
F. rufipectus	R F		75–82 (n = 2)		C-1 D-1	f-2	1 (n = 2)	irides: reddish brown; facial skin: grayish white; bill: black; tarsi: dark gray
Pittasoma michleri	U		—		C	f	1	irides: dark; bill: max. black; mand. black with white tip; tarsi: gray
Grallaria guatimalensis	R U		95		D		1	irides: brown; bill: black; tarsi: gray
Hylopezus perspicillata	R							
H. fulviventris	C	38–44 (n = 2)		B-1 C-1		f-1	1 (n = 1)	irides: brown; bill: max. black; mand. horn; tarsi: beige
Grallaricula flavirostris	U	14.0–17.0 (n = 3)		B-2		f-2	1 (n = 2)	irides: dark; bill: yellow with black culmen; tarsi: grayish green
Rhinocryptidae								
Scytalopus vicinior	U X							
Tyrannidae								
Zimmerius vilissimus	U	8.5		B		b	1,2	irides: pale; bill: black; tarsi: gray
Ornithion brunneicapillum	F							
Myiopagis caniceps	U							
M. viridicata	U							
Mionectes olivaceus	C C R	12.0–17.0 (n = 12)	12.0–17.0 (n = 11)	A-6 B-5	A-5 B-4 C-1	a-9 b-3 c-3	1 (n = 2) 2 (n = 8) 1,2 (n = 1)	irides: brown; bill: black; tarsi: pink
M. oleagineus	C	9.0–10.0 (n = 6)	8.0–11.0 (n = 12)	A-4 B-2	A-5 B-6 C-1	a-1 b-3 c-3 f-1	2 (n = 3) 1,2 (n = 1)	irides: brown; bill: max. black; mand. pink with black tip; tarsi: pinkish gray

APPENDIX I
CONTINUED

	Abundance[1] CA ES RT	Weight[2] ♂	♀	Gonads[3] ♂	♀	Skull[4] pneuma.	Stomach[5] contents	Soft part colors
Leptopogon superciliaris	U U	12.5–13.0 (n = 2)	12.5	B-1 C-1	B	a-1 b-1 f-1	1 (n = 2)	irides: dark brown bill: black tarsi: gray
Pseudotriccus pelzelni	R F	8.0	10.0	B	B	b-1 c-1		irides: reddish brown bill: black tarsi: olive
Myiornis atricapillus	F	6.0	5.5 (n = 2)	B-2	A-1 B-1	a-2 c-1	1 (n = 2)	irides: brown bill: black tarsi: pink
Lophotriccus pileatus	F	9.0	9.0	B	A	a-1 d-1	1 (n = 1)	irides: light gray bill: max. black; mand. flesh tarsi: gray
Oncostoma olivaceum	F	7.0–7.5 (n = 2)	7.0	B-2	B	c-1 d-1 f-1	1 (n = 2)	irides: cream bill: black tarsi: pale gray
Todirostrum cinereum	F	5.0–6.0 (n = 2)	6.0	A-1 B-1	B	a-1 b-1 e-1		irides: white bill: black tarsi: gray
T. nigriceps	U		6.5		A	f	1	irides: dark bill: max. black; mand. pink tarsi: gray
Cnipodectes subbrunneus	F	26.0–31 (n = 4)	20.0–23.0 (n = 2)	C-4	B-1 C-1	c-1 d-2 e-1 f-1	1 (n = 4)	irides: brown bill: max. black; mand. cream tarsi: horn
Rhynchocyclus brevirostris	F	25.0	20.0	A	B	a-1 b-1	1 (n = 2)	irides: brown bill: max. black; mand. pink tarsi: bluish gray
R. olivaceus	F	22.5	20.0–24.0 (n = 2)	A	B-2	b-1	1 (n = 3)	irides: brown bill: max. black; mand. yellow tarsi: bluish gray
Tolmomyias assimilis	U R	15.0		C		b	1	irides: brown bill: black tarsi: gray

APPENDIX I
CONTINUED

	Abundance[1] CA ES RT	Weight[2] ♂	Weight[2] ♀	Gonads[3] ♂	Gonads[3] ♀	Skull[4] pneuma.	Stomach[5] contents	Soft part colors
Platyrinchus mystaceus	U	10.0	9.0	B-1 C-1	—	f-1		irides: brown bill: max. black; mand. black with pink tip tarsi: pinkish gray
P. coronatus	F F	9.0–11.0 (n = 8)	8.0	A-4 B-3 C-2	A	a-2 c-3 d-1 e-3	1 (n = 5)	irides: brown bill: black tarsi: grayish brown
Onycorhynchus coronatus	U		16.5		A	e	1	irides: dark bill: max. black; mand. yellow with black tip tarsi: orange
Terenotriccus erythrurus	F U	7.0		A		f		irides: dark brown bill: max. black; mand. cream tarsi: light brown
Myiobius sulphureipygius	C	9.0–12.5 (n = 3)	9.5–11.0 (n = 2)	A-2 C-1	A-2	b-1 f-4	1 (n = 1)	irides: dark bill: max. slate; mand. slate with pink tip tarsi: pink
M. atricaudus	F U	8.0–11.0 (n = 3)	6.0–9.5 (n = 3)	A-2 C-1	A-2 B-1	c-1 d-2 f-1	1 (n = 2)	irides: dark bill: max. black; mand. pink with black tip tarsi: gray
Aphanotriccus audax	U	10.5–12.0 (n = 3)	11.0	A-2 B-1	C	c-1 f-2	1 (n = 3)	irides: brown bill: black tarsi: gray
Mitrephanes phaeocerus	F	8.0–9.5 (n = 2)		B-2		c-1 f-1	1 (n = 2)	irides: dark brown bill: max. black; mand. orange tarsi: black
Contopus borealis	U							
Empidonax spp.	F							
Colonia colonus	U U R							
Attila spadiceus	F U	38	35	C	A	c-1 f-1	1 (n = 1)	irides: reddish-brown bill: max. black; mand. pink with black tip tarsi: gray
Laniocera rufescens	R	50	56	—	C	c-1 f-1	1 (n = 1)	irides: brown bill: dark brown tarsi: gray

APPENDIX I
CONTINUED

	Abundance[1]			Weight[2]		Gonads[3]		Skull[4] pneuma.	Stomach[5] contents	Soft part colors
	CA	ES	RT	♂	♀	♂	♀			
Rhytipterna holerythra	F				39 (n = 2)		B-1 C-1	d-1	2 (n = 1)	irides: brown / bill: max. black; mand. light gray / tarsi: gray
Sirystes sibilator	U	F		36 (n = 2)		B-2		a-2	1 (n = 2)	irides: dark brown / bill, tarsi: black
Myiarchus tuberculifer	F	F	F							
M. crinitus	F	F								
Megarhynchus pitangua	U									
Myiozetetes cayanensis	F									
M. granadensis	F			25.5		A		b	1	irides: brown / bill, tarsi: black
Coryphotriccus albovittatus	U	F	U	24.0–26.0 (n = 2)	23.5–24.0 (n = 2)	B-2	B-1 C-1	c-1	1 (n = 1) 1,2 1 (n = 1)	irides: brown / bill, tarsi: black / irides: dark brown / bill: black / tarsi: dark gray
Myiodynastes chrysocephalus		U	F	43	39	B	B	f-1		
M. maculatus	U	U			43		B	f	1	irides: brown / bill: max. black; mand. pale brown with black tip / tarsi: gray
Legatus leucophaius	F									
Tyrannus melancholicus	R									
Pachyramphus cinnamomeus	F									
P. polychopterus	R									
P. homochrous	U			35		B			1	irides: brown / bill: max. black; mand. bluish gray / tarsi: gray
Tityra semifasciata	F	U								
Cotingidae										
Lipaugus unirufus	R	F		78		C			1	irides: brown / bill: gray / tarsi: black
Cotinga nattererii	U	R								
Carpodectes hopkei	R	R								
Querula purpurata	F			110	96	C-1 D-1	C	f-3	2 (n = 3)	

APPENDIX I
Continued

	Abundance[1] CA ES RT	Weight[2] ♂	Weight[2] ♀	Gonads[3] ♂	Gonads[3] ♀	Skull[4] pneuma.	Stomach[5] contents	Soft part colors
Pipridae								
Schiffornis turdinus	C C	33–40 (n = 7)	32–39 (n = 8)	A-3 B-4 D-1	A-1 B-6 C-1	c-2 d-1 e-1 f-7	1 (n = 1) 2 (n = 2) 4 (n = 1)	irides: brown bill, tarsi: gray
Sapayoa aenigma	R	23	21.0–21.5 (n = 2)	B	B-2	c-1 f-1	1 (n = 3)	irides: adult red; imm. tan bill: max. black; mand. gray tarsi: gray
Manacus vitellinus	C	15.5–19.0 (n = 11)	16.0–18.0 (n = 6)	A-4 B-6 C-1	A-2 B-1 C-2 D-1	b-3 c-2 f-7	2 (n = 6)	irides: brown bill: black tarsi: orange
Corapipo leucorrhoa	C R	9.0–10.0 (n = 35)	9.5–13.0 (n = 10)	A-13 B-14 C-2	A-2 B-1 C-5 D-1	a-7 b-9 c-6 f-1	2 (n = 6)	irides: dark brown bill, tarsi: gray
Pipra coronata	F	9.0	8.0–9.0 (n = 3)	B	A-1 B-1	c-2 f-2	2 (n = 1)	irides: dark bill, tarsi: black
P. erythrocephala	C C	10.0–13.5 (n = 13)	10.0–14.0 (n = 35)	A-3 B-5 C-5	A-15 B-12 C-6 D-1	a-3 b-8 c-6 d-2 e-1 f-8	2 (n = 16)	irides: brown (♀♀); white (♂♂) bill: brown (♀♀); yellow (♂♂) tarsi: pale brown
Oxyruncidae								
Oxyruncus cristatus	F	37–43 (n = 3)	41	B-2 C-1	C	b-1 f-1	2 (n = 2) 3 (n = 1)	irides: creamy yellow bill: max. dark gray; mand. pale gray tarsi: gray
Hirundinidae								
Stelgidopteryx ruficollis	F							
Corvidae								
Cyanocorax affinis	F	205	210	B	C	f-2	1,2 (n = 1) 3 (n = 1)	irides: yellow bill, tarsi: black

APPENDIX I

CONTINUED

	Abundance[1]			Weight[2]		Gonads[3]		Skull[4] pneuma.	Stomach[5] contents	Soft part colors
	CA	ES	RT	♂	♀	♂	♀			
Troglodytidae										
Donacobius atricapillus	U									
Campylorhynchus albobrunneus	F			27.5–39 (n = 3)	30	C-3	B	e-1 f-2	1 (n = 4)	irides: reddish brown bill: brownish gray tarsi: olive-gray
Thryothorus spadix		F								
T. fasciatoventris	F			25.0		B		f	1	irides: reddish brown bill, tarsi: dark gray
T. nigricapillus	C			20.5–24.5 (n = 5)	19.0–22.0 (n = 3)	B-4 C-1	A-2 C-1	c-1 d-1 f-6	1 (n = 5)	irides: brown bill: light gray tarsi: gray
Troglodytes aedon	F			8.5		A		a	1	irides: brown bill: dark brown tarsi: light brown
T. ochraceus	U	F			8.0		A		1	irides: reddish brown bill: max. slate; mand. horn tarsi: pale gray
Henicorhina leucosticta	C	C		14.0–15.5 (n = 6)	12.5–15.0 (n = 7)	A-2 B-4	A-3 B-3	a-1 f-8	1 (n = 5)	irides: brown bill: black tarsi: gray
H. leucophrys	U	C		17.0	15.0–17.0 (n = 2)	C	B-2	b-1		irides: light brown bill, tarsi: light gray
Cyphorhinus phaeocephalus	U	R								
Microcerculus marginatus	F	F		17.0–19.5 (n = 3)	15.5–18.5 (n = 3)	A-2 C-1	A-2 C-1	a-1 b-1 f-3	1 (n = 6)	irides: brown bill: black tarsi: gray
Muscicapidae										
Microbates cinereiventris	U	R		11.5	10.5	B	A	b-1 c-1	1 (n = 2)	irides: brown bill: max. black; mand. gray tarsi: gray
Polioptila schistaceigula	R									
Myadestes coloratus	C	C		26.0–34 (n = 5)	24.5–32 (n = 9)	A-3 B-2	A-3 B-1 C-4 D-1	a-3 b-2 f-7	2 (n = 5) 4 (n = 1)	irides: brown bill, tarsi: brown

APPENDIX I
CONTINUED

	Abundance[1] CA	Abundance[1] ES	Abundance[1] RT	Weight[2] ♂	Weight[2] ♀	Gonads[3] ♂	Gonads[3] ♀	Skull[4] pneuma.	Stomach[5] contents	Soft part colors
Catharus fuscater	F	F		28.5–33 (n = 4)	34	C-2 D-2	D	d-1 f-4		irides: white; bill, tarsi, eye-ring: orange
C. ustulatus	F									
Turdus obsoletus	F	C	R	45–58 (n = 4)	58–64 (n = 2)	A-1 B-2 C-2	A-3	a-2 b-1 f-1	2 (n = 3)	irides: brown; bill: black; tarsi: gray
T. assimilis	U			55		D		f		irides: light brown; eye-ring: yellow; bill: max. yellow; mand. black; tarsi: brown
Vireonidae										
Vireo olivaceus	F	F							1 (n = 1)	
Hylophilus ochraceiceps	U	U		11.0	12.5	B	A	f-1	2 (n = 1)	irides: white; bill: max. slate; mand. gray; tarsi: gray
H. decurtatus		U			—		A	a	1	irides: dark; bill: max. dark; mand. bluish gray; tarsi: bluish gray
Vireolanius eximius	F									
Emberizidae										
Vermivora chrysoptera	R	R								
V. peregrina	U	U								
Parula pitiayumi	C	F		7.5		B		f	1	irides: dark; bill: max. black; mand. yellow; tarsi: yellow
Dendroica virens	R									
D. fusca	F	F								
D. castanea	C	C								
Mniotilta varia	F	U								
Setophaga ruticilla	U	F								
Oporornis formosus	U									
O. philadelphia	U									
Wilsonia canadensis	F	U								
Myioborus miniatus				7.0		C		f		irides: brown; bill, tarsi: black

APPENDIX I
CONTINUED

	Abundance[1]			Weight[2]		Gonads[3]		Skull[4] pneuma.	Stomach[5] contents	Soft part colors
	CA	ES	RT	♂	♀	♂	♀			
Basileuterus ignotus			F	9.5–12.5 (n = 6)	10.5 (n = 2)	A-1 B-2 C-3	A-1 B-1	c-1 d-2 f-3	1 (n = 4)	irides: brown bill: max. brown; mand. horn tarsi: horn
Phaeothlypis fulvicauda	F		R	17.0	12.5–14.5 (n = 5)	B	B-5	a-3 f-3	1 (n = 6)	irides: brown bill: black tarsi: pinkish yellow
Coereba flaveola	F	U								
Conirostrum leucogenys	R				7.0		B	f	1	irides: brown bill: dark gray tarsi: gray
Tangara inornata	X	R								
T. palmeri		R								
T. florida	U	F		18.5	19.0–20.5 (n = 2)	C	B-1	f-2	2 (n = 1)	irides, bill: dark tarsi: bluish gray
T. icterocephala		F		24.5	20.0	C	C-1 C	f-2	3 (n = 1) 2 (n = 1)	irides: brown bill: black
T. guttata	F	F			20.0–21.0 (n = 2)		B-2	a-1	2 (n = 2)	tarsi: gray irides: dark brown bill: max. black; mand. bluish gray
T. gyrola	R	C			24.5–25.0 (n = 2)		C-2	f-1	2 (n = 2)	tarsi: gray irides: brown bill, tarsi: gray
T. larvata	F	U								
T. fucosa	U	F		21.5–23.0 (n = 3)	21.0–22.0 (n = 2)	B-1 C-2	B-1 C-1	f-5	2 (n = 1)	irides: dark brown bill: max. black; mand. silver with black tip
Dacnis venusta	U	U		16.0			B	f	2	tarsi: gray irides: gray irides: red bill: black tarsi: dark gray
D. cayana	F	U								
Chlorophanes spiza	F	U		21.5	19.0	B	B	a-1 f-1	3 (n = 1)	irides: brown (♀♀); red (♂♂) bill: black (♀♀); yellow with black culmen (♂♂) tarsi: gray-green (♀♀); yellow (♂♂)

APPENDIX I
CONTINUED

	Abundance[1] CA	ES	RT	Weight[2] ♂	♀	Gonads[3] ♂	♀	Skull[4] pneuma.	Stomach[5] contents	Soft part colors
Cyanerpes lucidis	F		U	12.0		A		f		irides: dark; bill: black; tarsi: yellow
Euphonia fulvicrissa	F			13.0		A		f		irides: brown; bill: max. black; mand. bluish gray; tarsi: gray
E. xanthogaster	F	F	U	12.0–14.5 (n = 3)	11.0–13.0 (n = 3)	A-1 B-1 C-1	A-2 B-1	a-2 b-1 f-2		irides: brown; bill: max. black; mand. silver with black tip; tarsi: gray
Thraupis episcopus	C				32–33 (n = 2)	B-1 C-1	B-1 C-1	f-2	2 (n = 1)	irides: dark; bill: max. black; mand. gray; tarsi: bluish gray
T. palmarum	C			37		A		f	2	bill, tarsi: dark gray
Chlorothraupis olivacea	U	U		37–43 (n = 3)		A B-1 C-1 D-1		f-1	1 (n = 1); 3 (n = 1)	irides: brown; bill: silver; tarsi: bluish gray
Heterospingus xanthopygius	R			—		C		f	4	bill: black; tarsi: gray
Tachyphonus luctuosus	C				15.0		C	f		irides: brown; bill: black; tarsi: gray
Habia spp.	X									
Piranga flava		F								
P. rubra		F								
Ramphocelus dimidiatus	F									
R. flammigerus	C			33–35 (n = 2)	31–33 (n = 2)	A-1 B-1	B-1 C-1	a-1 f-2		irides: brown; bill: bluish gray; tarsi: black
Mitrospingus cassinii	U	U			40–41 (n = 2)		B-2	f-1	2 (n = 2)	irides: gray; bill: max. black; mand. gray; tarsi: gray
Chlorospingus inornatus	U	C		30–36 (n = 6)	20.0–26.5 (n = 4)	A-2 B-3 C-1	A-1 B-1 C-2	b-1 f-9	1 (n = 1); 2 (n = 3); 1,2 (n = 3)	irides: creamy white; bill: black; tarsi: brownish gray

APPENDIX I
CONTINUED

	Abundance[1]			Weight[2]		Gonads[3]		Skull[4] pneuma.	Stomach[5] contents	Soft part colors
	CA	ES	RT	♂	♀	♂	♀			
Hemithraupis flavicollis	F									
Chrysothlypis chrysomelas		F	F	12.0–14.5 (n = 5)	11.5	A-3 B-2	A	a-1 f-2	2 (n = 5)	irides: brown bill: black tarsi: dark gray
Tersina viridis	C									
Saltator maximus	C	U		48		A		f		irides: brown bill: gray tarsi: brown
Pitylus grossus	F	U								
Caryothraustes canadensis	C			35	35	C	D		1,2 (n = 2)	irides: reddish brown bill; silver-blue with black tip tarsi: gray
Pheucticus ludovicianus	F	F								
Cyanocompsa cyanoides	F	F		26.5–32 (n = 5)	23.0–30 (n = 4)	B-1 C-4	A-1 B-1 C-1 D-1	a-1 d-1 f-7	1 (n = 2) 3 (n = 2) 4 (n = 1)	irides: reddish brown (♀♀); black (♂♂) bill, tarsi: black
Atlapetes brunneinucha	F	U		44	34–41 (n = 4)	B	B-1 C-2 D-1	f-5	1 (n = 1) 3 (n = 1)	irides: reddish brown bill: black tarsi: gray
A. atricapillus	U			47		C		f	1,3 (n = 1) 1	irides: brown bill: black tarsi: brown
Arremon aurantiirostris	C			23.0–32 (n = 9)	26.0–32 (n = 8)	A-3 B-2 C-3 D-1	A-2 B-4 C-2	a-4 b-2 e-1 f-6	1 (n = 2) 2 (n = 1) 3 (n = 1)	irides: dark bill: orange tarsi: light brown
Sporophila aurita	C			9.0–10.0 (n = 5)	9.0–11.5 (n = 9)	A-1 B-2 C-2	A-4 B-2 C-3	c-1 f-7	4 (n = 6) 3 (n = 7)	irides: brown (♀♀); black (♂♂) bill, tarsi: black
S. nigricollis	U									
S. minuta	U									
Oryzoborus funereus	C			11.0–12.0 (n = 2)		B-1 C-1				
Sturnella militaris	R							f		irides: dark brown bill, tarsi: black

APPENDIX I
CONTINUED

	Abundance[1]			Weight[2]		Gonads[3]		Skull[4] pneuma.	Stomach[5] contents	Soft part colors
	CA	ES	RT	♂	♀	♂	♀			
Icterus chrysater	F									
I. mesomelas	F			59		C				irides: brown bill: max. black; mand. gray tarsi: gray
Amblycercus holosericeus	C				65		D			irides, bill: yellow tarsi: gray
Cacicus cela	C									
Psarocolinus wagleri	C									

[1] Abundances are given for species present at the following localities: CA = Cana, vicinity of the gold mines at the base of Cerro Pirre, trail to Boca de Cupe, and lower east slope of Cerro Pirre; elevational range ca. 600–1000 m. ES = Eastern Slope, the east slope of the mountain along the trail from Cana at elevations of 1000–1400 m. RT = Ridge Top, 1400–1500 m. Abundances are: C = common, seen or heard daily in moderate to large numbers (≥10 individuals); F = fairly common, seen or heard daily in small numbers (<10 individuals); U = uncommon, seen or heard on one out of three days. Occurs in small numbers; R = rare, seen or heard on one of six days or less often, occurs in small numbers; X = present, not enough information to make a determination of abundance.

[2] Sample sizes for weight (g) ranges given in parentheses; absence of a number indicates a sample of one. Weights of specimens ≤30 g are rounded to the nearest 0.5 g. Weight data >30 g are rounded to the nearest gram.

[3] Letter designations code for the longest dimension (mm) of largest testis or of ovarian mass, as follows: A = ≤ 1; 1 < B ≤ 5; 5 < C ≤ 10; 10 < D ≤ 15; 15 < E ≤ 20. Numbers following the letters indicate sample sizes; absence of a number signifies a sample of one.

[4] Letter designations indicate the percent of pneumatization of the cranium, as follows: a = 0; 0 < b ≤ 25; 25 < c ≤ 50; 50 < d ≤ 75; 75 < e ≤ 100; f = 100. Numbers following the letters indicate sample sizes; absence of a number signifies sample of one.

[5] Numbers indicate diets as follows: 1 = insects; 2 = fruit; 3 = seeds; 4 = other plant matter; 5 = reptiles; 6 = mammals; 7 = birds. Sample sizes indicated in parentheses; absence of a number signifies sample of one.

OBSERVATIONS ON THE ANDEAN-PATAGONIAN COMPONENT OF SOUTHEASTERN BRAZIL'S AVIFAUNA

HELMUT SICK

Academia Brasileira de Ciencias, C. Postal 229, 20.000 Rio de Janeiro, Rio de Janeiro, Brazil

ABSTRACT. Several birds typical of southeastern Brazil's isolated and cold mountains have most probably been derived from Andean/Patagonian ancestors following Pleistocene glaciation. Typical species endemic to that area include Itatiaia Spinetail (*Schizoeaca moreirae*), Long-tailed Cinclodes (*Cinclodes pabsti*), and Mouse-colored Tapaculo (*Scytalopus speluncae*). Uniquely among Andean-Patagonian derivatives, Band-winged Nightjar (*Caprimulgus longirostris*) has also colonized low-lands, even reaching tropical, urban Rio de Janeiro. Secondary radiation has led to at least one species of Andean origin (Brasilia Tapaculo, *Scytalopus novacapitalis*) colonizing the central Brazil Plateau, apparently having followed the inland movement of coastal forests. Extension southward of Amazonian heat and northward of Patagonian aridity are advanced in explanation for the rupture of a formerly continuous band of vegetation linking the Andes and southeastern Brazil, with resulting isolation and speciation of landbirds.

RESUMEN. Varias aves típicas de las montañas aisladas y frías del sureste de Brasil han derivado muy probablemente de ancestros andino-patagónicos luego de las glaciaciones del pleistoceno. Especies endémicas típicas de esa área incluyen *Schizoeaca moreirae*, *Cinclodes pabsti*, y *Scytalopus speluncae*. La chotacabra serrana (*Caprimulgus longirostris*) es única entre los descendientes andinopatagónicos, ya que también ha colonizado las tierras bajas e incluso ha llegado a las áreas urbanas y tropicales en Río de Janeiro. La radiación secundaria ha hecho que al menos una especie de origen andino (*Scytalopus novacapitalis*) colonice la meseta central de Brasil, aparentemente habiendo seguido las extensiones tierra adentro del bosque atlántico. La extensión del calor amazónico hacia el sur y de la aridez patagónica hacia el norte brindan una hipótesis para explicar la ruptura de una banda de vegetación que en el pasado era continua y enlazaba los Andes con el sur de Brasil, dando como resultado la aislación y especiación de aves terrestres.

The Itatiaia Highlands of southeastern Brazil (Fig. 1) were recognized early in the 20th century as having a peculiar flora composed largely of endemics whose closest relatives were in the Bolivian Andes (Dusen 1903). Shortly thereafter, Miranda Ribeiro (1906) extended the analysis to the area's fauna, confirming its Andean affinities. He noted among other observations that the nearest relatives of certain catfishes (Siluroidei) and some birds were not in Brazil but in the Andes, a conclusion later reached for frogs and butterflies as well (Lutz 1951; Ebert 1960). In the present paper I offer observations on the taxonomic affinities of birds that occur generally in southeastern Brazil.

GEOLOGY AND CLIMATE

The Itatiaia Mountains appear to have been glaciated on several widely separated occasions (Ebert 1960), at least in the Wurm phase of the Pleistocene and possibly in pre-Devonian and Gondwanan times (Maack 1951), although not all experts concur that the area has ever been glaciated, and palynological evidence appears to be lacking (Oedman 1955; Beuerlen 1970). Nonetheless, most authors agree that U-shaped valleys 6–8 km long, typical glacial signs, are present, even though later water erosion has steepened their profiles to V-shapes. It is believed that vegetation covered the borders of the valleys, that the snow line was at ca. 1500 m, and that an ice-cap at ca. 2100–2600 m fed the glaciers (Ebert 1960).

Presumably, the same kinds of alternating warm-humid/cold-dry cycles that in the Andes facilitated the expansion and contraction of forests and other vegetation, with resulting speciation and range changes in birds (cf. Haffer 1967, 1981; Vuilleumier 1969a), also occurred in southeastern Brazil.

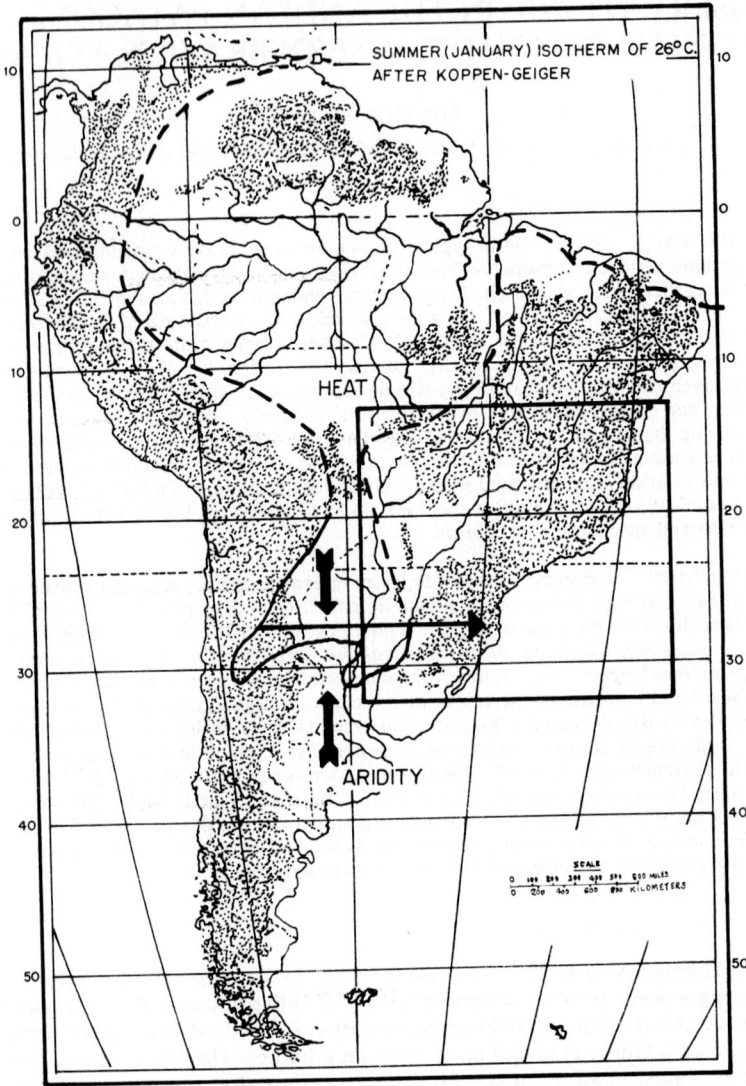

Fig. 1. Immigration of plants (horizontal arrow) from northern Argentina to southeast Brazil. Climate (heat, aridity; thick vertical arrows) and topography (stippled areas above 300 m) have been suggested as effecting disjunction of ranges between the Andes and southern Brazil. The square is the area shown in Figure 2. Adapted from Smith (1962).

REPRESENTATIVE TAXA

Schizoeaca moreirae. Itatiaia Spinetail. This very local furnariid most clearly demonstrates the faunal closeness of the Itatiaia Highlands and the Bolivian and Patagonian Andes. It is the most common species in the *Chusquea* bamboo above timberline (1800 m).

Hellmayr (1925) erected a new genus (*Oreophylax*) for this species, which is typically "orial" (=flora and fauna above timberline), in order to distinguish it from *Synallaxis,* but I believe that Vaurie (1971) was correct in allying it with the Andean thistletails (*Schizoeaca*). Its nest is a soft ellipsoid-shaped structure rather than an elongate one of rigid and spiny twigs (Sick 1970); this is typical of nests of *Schizoeaca* and in contrast to those of *Synallaxis.* Vuilleumier (1969b) described the nest of *Schizoeaca harterti* from Bolivia as very similar to that of *moreirae.* The vocalizations of the Itatiaia Spinetail are similar to those of tit-spinetails (*Leptasthenura*), also an Andean derivative, but not to those of *Synallaxis.*

Cinclodes pabsti. Long-tailed Cinclodes. This endemic, ground-dwelling furnariid is confined

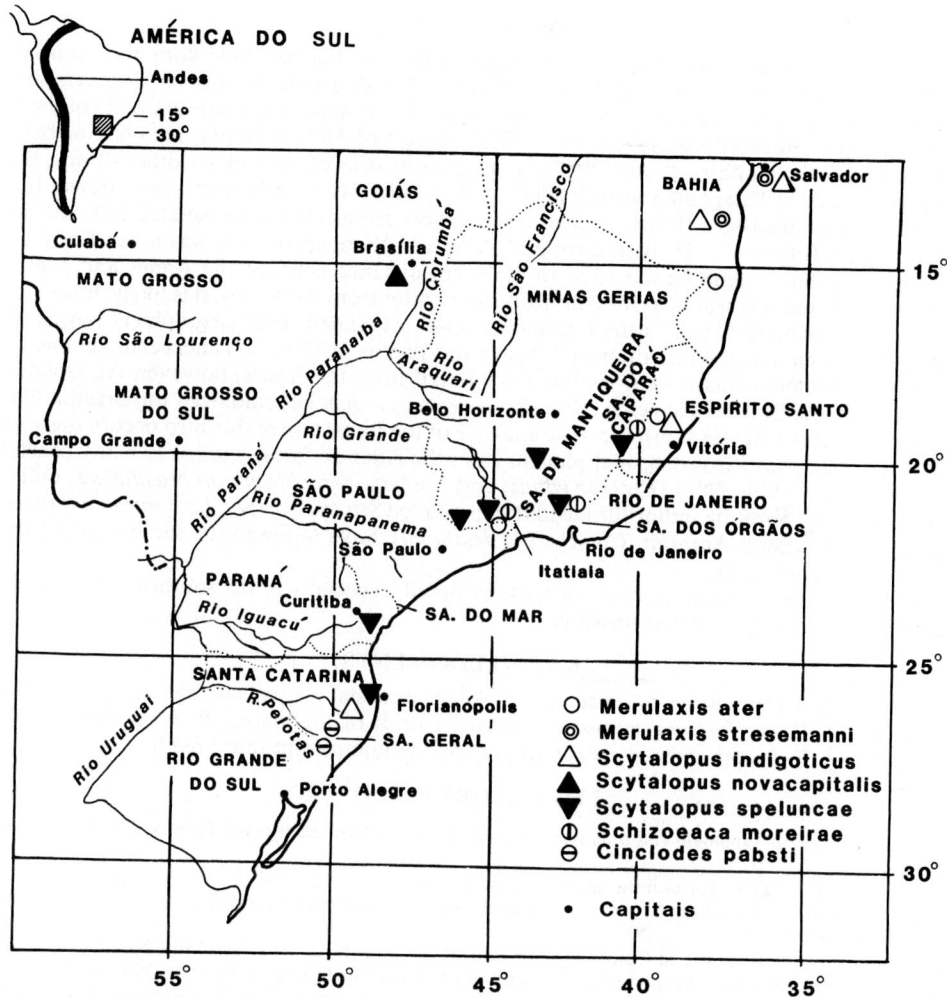

FIG. 2.   Distribution of some Andean tapaculos (Rhinocryptidae) and furnariids (Furnariidae) in Brazil showing colonization of southeast and central Brazil.

to the high plateau (e.g., 1700 m, Campo dos Padres) of northern Rio Grande do Sul and adjacent Santa Catarina, the coldest region of Brazil (Sick 1973). Only recently described (Sick 1969), it is allopatric with its nearest relative, Bar-winged Cinclodes (*C. fuscus*). This latter species is widespread throughout southern South America and in the Andes, although in winter it reaches lowland coastal areas in Rio Grande do Sul. *Cinclodes* is a well-defined genus, typical of the Andes and Patagonia, and occurs south to Tierra del Fuego.

Rhinocryptidae. Tapaculos. Tapaculos are widespread in the Andes and in Patagonia but have penetrated Brazil only in the southeastern mountains. *Scytalopus speluncae*, the Mouse-colored Tapaculo, is common on high, dense forest borders, even reaching peaks where, in *Chusquea*, it meets *Schizoeaca moreirae*. Most other tapaculos are scarce in southeastern Brazil.

Caprimulgus longirostris. Band-winged Nightjar. This typically Andean bird has invaded Brazil by way of the high, treeless parts of southeastern Brazil (Sick 1959). Unlike other montane derivatives, this species has continued its expansion in Brazil, moving into lower parts of Santa Catarina and Rio Grande do Sul. Most unexpectedly, and so far unique among Andean forms, Band-winged Nightjar has moved from the mountains to the tropical lowlands, and during the last 40 years has become a resident of the center city and suburbs of Rio de Janeiro (Sick 1963).

DISCUSSION

The most plausible explanation for the origin of the peculiar montane flora and fauna of southeastern Brazil is provided by Smith (1962), although a comprehensive biogeographic analysis of the area remains to be done. Smith argued that northern Argentina and southern Brazil were once linked floristically at ca. 27°S, the ranges of Andean plant genera having been constricted by the southward push of Amazonian heat (the 26° summer isotherm) and the northward push of Patagonian aridity (Fig. 1). This link has been broken many times. For example, the southeastern Brazil *Araucaria* forest now seems clearly an isolated relict of the Andean flora (Rambo 1951). Before rupture, however, there apparently was a band of continuous vegetation extending the 1500 to 2000 km distance between the Andes and southeastern Brazil, and this served as the colonization corridor from the Andes. It is likely, however, that Andean colonizers such as the Itatiaia Spinetail, expanded their geographical ranges by jumping between isolated mountains (cf. Mayr and Phelps 1967 for a Venezuelan analogue).

Not all movement was toward southeastern Brazil from the Andes, however. An isolated rhinocryptid, the Brasilia Tapaculo (*Scytalopus novacapitalis*), has colonized the Brasilia Plateau in Central Brazil (1000 m). Floristic analysis of the forest where this bird occurs suggests that it followed plants invading that plateau from the Atlantic forest; typical of many coastal plants there are edible palm (*Euterpe edulis*) and the large fern *Blechnum brasiliensis*. Other Brazilian coastal birds, including furnariids (Sharp-tailed Stream-creeper, *Lochmias nematura*) and tyrants (Southern Antpipit, *Corythopsis delalandi*), have followed the same route to the Brasilia Plateau (Fig. 2).

The unravelling of zoogeographic patterns in Brazil, especially in the southeast, has barely begun; the field is ripe for investigation.

ACKNOWLEDGMENTS

I thank C. T. Rizzini for information about Atlantic coastal plants that have invaded central Brazil, and J. Haffer for suggestions on an earlier draft of this paper. P. A. Buckley, L. L. Short, and M. S. Foster assisted greatly in putting the manuscript into final form.

LITERATURE CITED

BEUERLEN, K. 1970. Geologie von Brasilien. Beitrag zur regionalen Geologie der Erde, Vol. 9. Borntrager, Berlin.

DUSEN, D. K. H. 1903. Sur la flora de la Serra do Itatiaia. Arq. Mus. Nac. Rio de J. 13:1–119.

EBERT, H. 1960. Novas observacoes sobre a glaciacao pleistocenica na serra do Itatiaia. An. Acad. Bras. Cienc. 32:51–73.

HAFFER, J. 1967. Speciation in Colombian forest birds west of the Andes. Am. Mus. Novit. No. 2294.

HAFFER, J. 1981. Aspects of neotropical bird speciation during the Cenozoic. Pp. 371–394, *In* G. Nelson and D. E. Rosen (eds.), Vicariance Biogeography: A Critique. Columbia University Press, New York.

HELLMAYR, C. H. 1925. Catalogue of Birds of the Americas. Pt. IV. Field Mus. Nat. Hist. Publ. 234, Zool. Ser. Vol. 13.

LUTZ, B. 1951. Nota prévia sobre alguns anfibios anuros do Alto Itatiaia. Hospital: maio.

MAACK, R. 1951. Vestigios pre-devonianos de glacicao e a sequencia de camadas devonianos no Estado do Parana. Arq. Biol. Tecnol. (Curitiba) V–VI (16):197–230.

MAYR, E., AND W. H. PHELPS, JR. 1967. The origin of the bird fauna of South Venezuelan highlands. Bull. Am. Mus. Nat. Hist. 36:269–327.

OEDMAN, H. O. 1955. On the presumed glaciation in the Itatiaia mountain, Brazil. Engenh. Mineral Metal. Rio de J. 21:107–108.

RAMBO, B. 1951. O elemento andino no Pinhal Riograndesse. An. Bot. Herb. Barbosa Rodrigues 3: 7–39.

RIBEIRO, A. MIRANDA. 1906. Vertebrados do Itatiaia (Peixes, Serpentes, Saurios, Aves e Mamiferos). Arq. Mus. Nac. Rio de J. 13:165–190.

SICK, H. 1959. O redescobrimento no Brasil do Bacurau *Caprimulgus longirostris*. Bonaparte (Caprimulgidae, Aves). Bol. Mus. Nac., Rio de J. Zool. 204:1–15.

SICK, H. 1963. O bacurau *Caprimulgus longirostris* e outras aves noturnas do Estado da Guanabara. Vellozia I, 3:107–116.

SICK, H. 1969. Uber einige Topfervogel (Furnariidae) aus Rio Grande do Sul. Brasilien, mit Beschreibung eines neuen *Cinclodes*. Beitr. Neotrop. Fauna 6:63–79.

SICK, H. 1970. Der Strohschwanz, *Oreophylax moreirae*, andiner Furnariide Sudostbrasiliens. Bonn. Zool. Beitr. 21:251–268.

SICK, H. 1973. Nova Contribuicao ao conhecimento do *Cinclodes pabsti* Sick, 1969. Rev. Bras. Biol. 33:109–117.

SMITH, L. B. 1962. Origins of the flora of Southern Brazil. Contrib. U.S. Natl. Herb. 35:215–249.

VAURIE, C. 1971. Classification of the Ovenbirds (Furnariidae). Witherby, London.

VUILLEUMIER, F. 1969a. Pleistocene speciation in birds living in the high Andes. Nature 223:1179–1180.

VUILLEUMIER, F. 1969b. Field notes on some birds from the Bolivian Andes. Ibis 111:599–608.

AFFINITIES AND RECENT HISTORY OF THE AVIFAUNA OF TRINIDAD AND TOBAGO

DAVID W. SNOW

Subdepartment of Ornithology, British Museum (Natural History),
Tring, Hertfordshire HP23 6AP, United Kingdom

ABSTRACT.   Trinidad was last connected to the South American mainland about 11,000 years ago. Tobago has been separated for much longer (probably at least 2 million years), but during periods of lowered sea level, the water gap between Tobago and Trinidad may have been reduced to as little as 5 km. The avifauna of the two islands, excluding species that are widely distributed on the South American mainland, can be divided into four distributional categories, Andean, Northern Venezuelan, Caribbean, and Amazonian-Guianan. The Andean element is very closely related to the montane avifauna of the Cordillera de Caripe and Paria peninsula of northeastern Venezuela, more so than the latter is to the avifauna of the mountains farther west in Venezuela. The northern Venezuelan and Caribbean elements are small and undifferentiated, and their presence in the islands would be expected on the basis of their present distributions in adjacent areas. The Amazonian-Guianan element is more puzzling, as several of its members are geographically very isolated from the mainland populations; the degree of subspecific differentiation in this group is higher than that in the Andean element. The presence of the Amazonian-Guianan element in Trinidad and Tobago indicates a former extension of the Amazonian-Guianan avifauna up the east coast of Venezuela at a time when the sea level was low.

In spite of the fact that Tobago has been an island for about 200 times as long as Trinidad, its avifauna is only slightly more differentiated from that of the mainland. The most likely explanation of this apparent anomaly is that Tobago was so close to the mainland during periods of lowered sea level that it was effectively in biotic contact.

RESUMEN.   Trinidad ha estado en contacto por última vez con el continente sudamericano hace aproximadamente 11.000 años. Tobago ha estado separada por mucho más tiempo (probablemente dos millones de años por lo menos), pero durante períodos de descenso del nivel de los mares, el estrecho de agua entre Tobago y Trinidad puede haber sido reducido hasta 5 km de ancho. La avifauna de estas dos islas, excluyendo aquellas especies que se encuentran ampliamente distribuídas en el continente sudamericano, pueden ser divididas en 4 categorías de distribución: andinas, venezolanas del norte, caribeñas y amazónico-guyanas. La categoría andina está muy cercanamente relacionada con la avifauna de montañas de la cordillera de Caripe y la península de Paria en el noreste venezolano y está más relacionada que en el caso de la avifauna de la península con la avifauna de las montañas venezolanas, más al oeste. Los elementos del norte de Venezuela y del Caribe son pocos e indiferenciados y su presencia en las islas es lógica en base a su distribución actual en áreas adyacentes. El elemento amazónico-guyana es más enigmático, debido a que varios de sus miembros se encuentran muy aislados de las poblaciones continentales, el grado de diferenciación subespecifica en este grupo es mayor que en el elemento andino. La presencia del elemento amazónico-guyana en Trinidad y Tobago indica que en el pasado hubo una extensión de la fauna amazónico-guyana por la costa este de Venezuela durante el período en que el nivel del mar fue más bajo.

A pesar de que Tobago ha sido una isla por un período 200 veces mayor que Trinidad, la avifauna de la primera está sólo levemente más diferenciada de la del continente que la de la última. La explicación más apropiada para esta aparente anomalía sería que Tobago ha estado tan cercana al continente durante los períodos en que disminuyó el nivel del mar que existió contacto biológico ("biotic contact") entre ambos.

The composition of the avifauna of Trinidad and Tobago has not been recently analyzed. Chapman (1894) and Hellmayr (1906) discussed its affinities at a time when it was incompletely known and when the adjacent Venezuelan avifauna was very poorly known. Junge and Mees (1957) discussed them only briefly. It has long been recognized that, with a few exceptions, the birds of these two islands are of South American, not Antillean, origin and that Tobago, in accordance with its more isolated position, has a greater proportion of endemic forms. The degree of differentiation from mainland populations is, at most, slight in the great majority of species from both islands; no species is endemic. It is generally accepted that both islands have been connected with the mainland; both are on the continental shelf, Trinidad lying within the 100 m isobath and Tobago within the 200 m isobath. It also seems certain that Trinidad was separated comparatively recently, when the sea level rose at the end of the last glaciation—about 11,000 years ago (ffrench 1973). Tobago was probably separated much earlier—"early in the Pleistocene" (Lack 1976:161), i.e., about 2 million years ago, or "a few million years ago" (ffrench 1973). There seems to be no recent authoritative geological opinion, but it may be noted that a lowering of the sea level by about 100 m, as at the height of the last and other recent glaciations, would reduce the 25 km gap between the two islands to about 10 km, and perhaps to as little as 5 km. Greater precision is not possible given the present state of knowledge of the bathymetry of the area. Two deep channels separate the islands, so it is unlikely that they were connected in the Pleistocene (J. B. Saunders, pers. comm.).

Now that up-to-date compilations giving distributions of South American bird species (especially Meyer de Schauensee 1966) and recent handbooks on the birds of Trinidad and Tobago (ffrench 1973) and Venezuela (Meyer de Schauensee and Phelps 1978), as well as studies of the distribution and systematics of various bird groups (e.g., Delacour and Amadon 1973; Lanyon 1978), are available, it is possible to make a reasonably detailed analysis of the avifauna of Trinidad and Tobago. The island forms (species or subspecies) can be allocated to various distributional categories on the basis of the mainland distribution of the taxa to which they belong or, in the case of endemic Trinidad and Tobago subspecies, to which they are most closely related. This provides a reliable idea of the avifaunal groups to which the island birds belong. Analysis of the degree of differentiation of the island populations gives some basis for judging the length of time for which they have been separated from their mainland relatives. The combined results of these two approaches provide insight into the recent history of the avifauna of Trinidad and Tobago, and suggest hypotheses that may be tested as new data become available on past climatic changes and the alteration of sea levels and coast lines.

SPECIES INCLUDED IN THE ANALYSIS

For any zoogeographical analysis some species of birds are much less informative than others. In the case of continental islands close to the mainland, such as Trinidad and Tobago, the presence of wide-ranging, vagile species that might be expected to colonize the islands under present conditions tells one little or nothing about the past history of the avifauna. Many large nonforest birds, especially aquatic birds, are in this category, as are all migratory species. At the other extreme, birds that are strictly confined to forest, especially if they are poor fliers, are much more likely to give information about former connections to the mainland. Single cases may, of course, be misleading; thus, the largely terrestrial forest dove *Geotrygon montana* is a very successful colonist of islands (Bond 1979). If enough species are involved, however, the reliability of deductions made from them should be much increased. For these reasons I confine the analysis to landbirds that are not known to be migratory, giving most attention to forest species. In addition to waterbirds, I exclude non-forest diurnal raptors, as they tend to be very wide-ranging, and a few rare species, perhaps stragglers to Trinidad and Tobago and not known to breed.

The list of species for analysis numbers 204, of which 193 occur in Trinidad, 68 in Tobago, and 57 in both islands. This is 58 percent of the number of species (with the same exclusions of waterbirds, etc.) recorded from the adjacent mainland in the state of Sucre, an area of the same order of size as Trinidad and Tobago but with considerably higher mountains. I compared the Trinidad and Tobago avifauna with that of the immediately adjacent mainland state of Sucre, not Delta Amacuro which is about the same distance from Trinidad. Geographically Trinidad and Tobago are the final outliers of the Andean chain that runs through Sucre, and Sucre, like Trinidad, contains a wide variety of habitats. Delta Amacuro has no mountains except the Sierra de Imataca in the extreme south, which are part of the Guiana highlands; half of the state lies south of the Orinoco, a major faunal dividing line, and the other half consists largely of the mangrove and other swamps of the Orinoco delta.

TABLE 1

PERCENTAGES OF SPECIES IN DIFFERENT DISTRIBUTIONAL CATEGORIES[1,2]

	N	Andean	Amazonian-Guianan	Northern Venezuelan	Caribbean	Widespread
			Forest species			
Sucre	171	46	15	2	—	37
Trinidad	89	31	16	—	—	53
Tobago	32	38	9	3	3	47
			Non-forest species			
Sucre	152	6	8	12	1	73
Trinidad	104	3	12	2	3	81
Tobago	36	—	3	8	8	81

[1] All species breeding in Sucre, Trinidad, and Tobago, excluding water-birds, diurnal non-forest raptors, and migrants.
[2] Statistical note to Tables 1 and 2: The application of standard statistical tests to zoogeographical data is full of pitfalls (Mayr 1983), the main one of which is that the "units" involved (usually species) are almost never equivalent units subject to random sampling. Thus, the null hypothesis is extremely unlikely to be true, whether a formal test shows significance or not. If this objection is disregarded, the following differences in percentages in Tables 1 and 2 are significant, on the criterion that a difference in proportions must be more than three times its standard error. Table 1, forest species: the proportion of Andean species occurring in Trinidad and Tobago is less than the proportion occurring in Sucre, but not in Trinidad and Tobago; the proportion of widespread species occurring in Trinidad and Tobago is greater than the proportion occurring in Sucre, but not in Trinidad and Tobago. Table 2: the proportion of species characteristic of the three higher altitudinal levels in Trinidad and Tobago is lower than the proportion among species occurring in Sucre and not in Trinidad and Tobago.

The unit for analysis is the species, but in some cases the subspecific affinities of the island populations need to be taken into account. An example will illustrate the need (for full list, see Appendix I). The warbler *Basileuterus culicivorus* is widespread, ranging from Mexico through much of northern South America to southeastern Brazil. The Trinidad population (with the population of the adjacent part of Venezuela) belongs to the Amazonian-Guianan group of subspecies, not to the group that occurs farther west in the coastal mountains of Venezuela. The Trinidad population is therefore not classified as "widespread" (see next section) but is placed in the Amazonian-Guianan category.

I have been able to examine specimens of the great majority of the endemic Trinidad and Tobago subspecies, and have accepted the subspecific classification adopted by ffrench (1973) and Meyer de Schauensee and Phelps (1978), with two exceptions. I treat *Aburria (Pipile) pipile* as a well-marked subspecies (following Delacour and Amadon 1973; treated as an endemic species by ffrench) and have not recognized the slightly marked subspecies *Glaucis hirsuta insularum* (considered synonymous with *G. h. hirsuta* by Berlioz 1962, but recognized by ffrench).

DISTRIBUTIONAL CATEGORIES

Of the 204 species under consideration a large proportion (66%) occur widely in Venezuela, both north and south of the Orinoco, and farther to the west, south, and east. These species are designated as "widespread" in what follows. The remainder of the species fall into four additional distributional categories, Andean, Northern Venezuelan, Caribbean, and Amazonian-Guianan (Appendix I; Table 1). It is evident that Trinidad and Tobago have relatively (but not absolutely) more widespread species than does Sucre (Table 1). This is to be expected, as widespread species should be good colonizers and tolerant of a variety of ecological conditions, hence better able to reach and to survive on islands than less widespread species. Trinidad and Tobago have proportionately fewer Andean species than Sucre, as also would be expected given their lower elevation. Trinidad has a slightly greater and Tobago a smaller proportion of Amazonian-Guianan species than Sucre. The Caribbean element is most prominent in Tobago.

In general, the avifauna of Trinidad and Tobago is not very different from what would be expected if at the time of their separation from the mainland its avifauna had been just as it is now. This very broad analysis, however, fails to bring out a number of interesting points, as follows.

Andean element.—Included here are 32 species occurring in the northern serranias of Venezuela, many of them also farther west in the Andes, mostly in upper tropical and subtropical zones (e.g., *Sclerurus albigularis,* Fig.1). As mentioned, Trinidad and Tobago have propor-

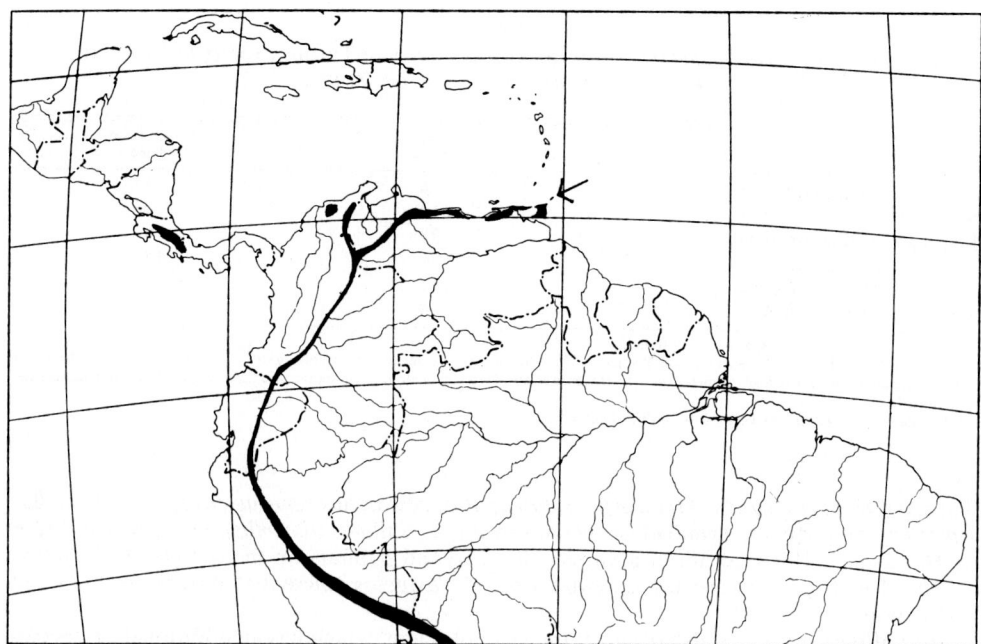

FIG. 1.   Distribution of the Gray-throated Leafscraper (*Sclerurus albigularis*), an Andean species occurring at all levels in Trinidad and Tobago but primarily at subtropical levels in the Andes. The Andean
range is shown as continuous, but records are patchy.

tionately fewer Andean species than has the adjacent state of Sucre. Those forms that occur
in Trinidad and Tobago are mainly from the lower montane zones in Venezuela (Table 2).
Thus, more than half of the forms occurring at lower elevations in Sucre (59%—bottom three
zones in Table 2) also occur in Trinidad and Tobago, but only 25 percent of the forms that
occur at higher elevations do so. Almost all of them, whether from the higher or the lower
zones in Venezuela, have to live at lower elevations in Trinidad and Tobago. Possibly climatic
and vegetational conditions typical of the upper tropical and subtropical zones in Venezuela
prevail at lower altitudes in the more equable, oceanic conditions of Trinidad and Tobago.
Even so, many of the higher-level montane birds of the Venezuelan Andes would surely find
conditions unsuitable in the lower mountains of Trinidad and Tobago. This alone seems
sufficient to account for the small size and composition of the montane element (Tables 1,
2).
    When the subspecific affinities of Andean species occurring in Trinidad and Tobago are
analyzed (Table 3), it is apparent that the number of subspecies endemic to Trinidad and/or
Tobago is considerably exceeded by the number of subspecies that are shared by Trinidad
and/or Tobago and the mountains of extreme northeastern Venezuela (the Caripe and Paria
cordilleras), different subspecies being found in the Venezuelan mountains farther west. This
is strikingly confirmed by the analysis made by Phelps (1966) of the subtropical element in
the avifauna of the northern cordilleras of Venezuela. Of the species occurring in the Cordillera
de Caripe, 85 percent are represented by a different subspecies in the Cordillera de Caracas
to the west, whereas of those occurring in the Cordillera de Caripe and Trinidad, only 45
percent are represented by different subspecies in the two areas. Affinities between the avifaunas
of the mountains of the Paria peninsula and Trinidad are even closer, only 30 percent of the
species in common being subspecifically distinct in the two areas. If a wider sample of species
is analyzed—the whole Andean element, not just the subtropical species included in Phelps'
analysis—the degree of differentiation is even less. Only 17 percent of the species occurring
in Trinidad and Tobago are represented by distinct subspecies in the Paria peninsula.
    Northern Venezuelan element.—This element includes six species occurring north of the
Orinoco, but not (or not primarily) in the mountains. Of these six species, four occur only on

TABLE 2

NUMBERS OF ANDEAN SPECIES OCCURRING IN SUCRE AND TRINIDAD AND TOBAGO THAT
OCCUPY VARIOUS ALTITUDINAL ZONES[1]

Zone[2]	Sucre	Trinidad and Tobago[3]
Subtropical-temperate	4	1
Subtropical	14	4
Upper tropical-subtropical	34	8
Tropical-subtropical	21	13
Tropical-lower subtropical	7	3
Tropical-upper tropical	4	3

[1] For statistical treatment, see Table 1: footnote 2.
[2] Zonation from Meyer de Schauensee and Phelps (1978). The category "tropical-lower subtropical" includes three species classed as "upper tropical-lower subtropical" or merely as "lower subtropical." Two very wide-ranging species (tropical or upper tropical to temperate) are omitted.
[3] All species in Trinidad and Tobago also occur in Sucre.

Tobago (*Ortalis ruficauda, Melanerpes rubricapillus, Myiarchus venezuelensis, Hylophilus flavipes*) and two only on Trinidad (*Euphonia trinitatis, Saltator albicollis*). With the exception of *H. flavipes*, which occurs in a variety of woodland habitats on the mainland, including forest, all are non-forest species. Likewise, only *H. flavipes* is differentiated from the mainland population.

Four of the six species also occur in the state of Sucre, which has an additional 18 species in this distributional category. They are mainly birds of strongly seasonal, in some cases arid, woodland and savanna. The poor representation of this group in Trinidad and Tobago (17% of the total, cf. 63% for the avifauna as a whole) may be due in part to the present paucity of suitable habitats in the islands, but it also suggests that even less suitable habitat was available at the time when Trinidad was isolated.

Caribbean element.—The five species in this small group have limited, mainly coastal distributions in Venezuela and adjacent parts of South America, with their centers of distribution in the West Indies. One species occurs on Trinidad only (*Quiscalus lugubris*), three on Tobago only (*Buteo platypterus, Progne dominicensis, Tiaris bicolor*), and one on both islands (*Tyrannus dominicensis*). Two of them (*Q. lugubris, T. bicolor*) also occur in Sucre, which has no other Caribbean element. This group thus represents the southern fringe of the Caribbean avifauna which, as might be expected, is more prominent in Tobago than in Trinidad. The Trinidad and Tobago populations are not subspecifically differentiated from the Antillean populations to the north.

Amazonian-Guianan element.—This element includes 27 species with Amazonian-Guianan distributions, some of which extend in a narrow belt up the east coast of Venezuela to, or a little beyond, the Orinoco delta region. As already mentioned (Table 1), Trinidad and Tobago combined have proportionately more Amazonian-Guianan species than has the adjacent state of Sucre. The actual number of species occurring in Trinidad and Tobago (forest = 15; non-forest = 13) is about three-quarters of the number occurring in Sucre (24 and 11, respectively). Of particular interest, as they occur in Trinidad and/or Tobago but not in Sucre or any other parts of Venezuela north of the Orinoco, are *Aburria pipile, Trogon violaceus, Momotus momota* (disregarding a doubtful Sucre record), *Myrmotherula axillaris, Manacus manacus,* and *Chiroxiphia pareola*. The degree of subspecific differentiation of the forest-living members of the Amazonian-Guianan element in Trinidad and Tobago is higher than that shown by the Andean element; thus, six of 15 (40%) are endemic subspecies, three of them well-marked. These three are members of the group of six species, listed above, that are geographically most isolated from conspecific populations south of the Orinoco.

The 13 non-forest species actually exceed the number recorded from Sucre, but all are recorded from adjacent parts of northeastern Venezuela north of the Orinoco (Monagas, Delta Amacuro). The lack of records from Sucre probably reflects the distribution of the kind of habitat in which they occur, especially mangroves and swampy savanna. In contrast to the forest species, none of the non-forest species in Trinidad and Tobago is subspecifically distinct from the nearest mainland population. In some cases the lack of differentiation may indicate

TABLE 3

SUBSPECIFIC AFFINITIES OF ANDEAN SPECIES OCCURRING ON TRINIDAD AND TOBAGO

Range of subspecies on mainland	Trinidad	Tobago
Endemic to Trinidad/Tobago	3	4
Caripe-Paria mountains	10	4
Venezuelan coastal cordilleras	5	1
Guiana highlands	1	—
Widespread	11	2

that the island and mainland populations are still in effective contact (for example, the macaws *Ara ararauna* and *A. manilata*).

SUBSPECIFIC DIFFERENTIATION OF TRINIDAD AND TOBAGO POPULATIONS

Of the 31 forest species that occur in Tobago, 42 percent are subspecifically distinct from mainland populations (cf. only 19% of non-forest species). Of the 89 forest species that occur in Trinidad, 18 percent are subspecifically distinct from mainland populations (cf. only 4% of non-forest species). Of the 25 forest species shared by Trinidad and Tobago, 28 percent are subspecifically distinct between the two islands (cf. only 11% of non-forest species). Thus, Trinidad populations have differentiated from Tobago populations more than they have from mainland populations, but not much more. Species that are undifferentiated both in Trinidad and in Tobago from mainland populations include a number of forest species that are probably very sedentary (e.g., *Veniliornis kirkii, Dendrocincla fuliginosa,* and *Platyrinchus mystaceus*).

Tobago, being a smaller island than Trinidad with less varied habitats, is presumably unable to support as large an avifauna, and selection should strongly favor adaptations to the simpler biotic conditions characteristic of small islands. Almost certainly such selection has operated. Of the 15 subspecies confined to Tobago, 12 are distinguished from the Trinidad subspecies or (in the case of three species that do not occur in Trinidad) from the mainland subspecies by their larger size. Of the 12 subspecies confined to Trinidad, only three are larger than the mainland subspecies, and one is smaller. Of the three subspecies shared by Trinidad and Tobago (all potentially good colonizers and hence likely to exchange genes between the two islands; *Amazona amazonica tobagensis, Chloroceryle americana croteta, Mimus gilvus tobagensis*), two are larger than the mainland subspecies. Evidently, selection has promoted increased size in Tobago populations, as has been found in some other isolated islands with small avifaunas. The adaptive basis of such changes in insular avifaunas is, however, far from being understood (Grant 1968).

DISCUSSION

The widespread element gives no clue to the past history of the Trinidad and Tobago avifauna; these species would be expected to occur on the islands, whatever the history of the avifauna of northern South America. The Caribbean and Northern Venezuelan elements also provide few clues; their presence in Trinidad and Tobago is expected, given their present distribution patterns. It is to the Andean and, especially, the Amazonian-Guianan elements that one must look for indications of the recent history and development of the avifauna of the two islands.

In attempting an historical interpretation it is tempting to assume that the degree of differentiation of an insular population of a species from its mainland relatives is related to the length of time for which it has been isolated and, hence, by implication, to the time of separation of the island from the mainland. There are, however, three possible complicating factors. (1) The rate of differentiation after isolation may be so variable that the degree of differentiation may give no clue to the length of time since separation. Recent research on avian phylogeny by DNA hybridization is providing evidence that rates of phenetic evolution are in fact extremely variable (Sibley and Ahlquist, in press). (2) Whatever the potential rate of differentiation in isolation may be, gene-flow from the mainland may reduce or suppress it. That this process has operated on the Trinidad and Tobago avifauna is strongly indicated by the relatively low level of subspecific differentiation in non-forest species, that is, those that are

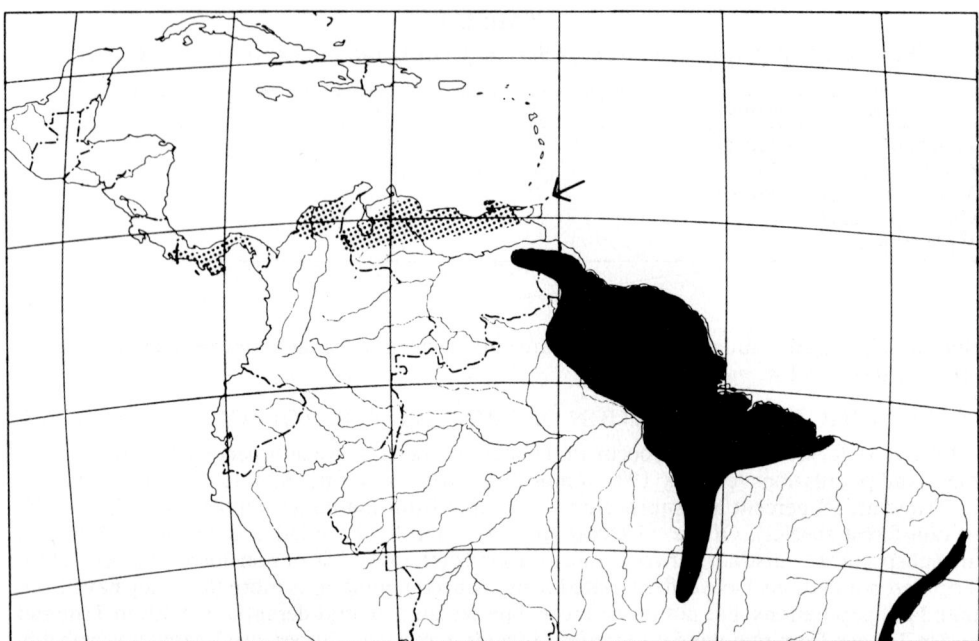

FIG. 2. Distribution of the Blue-backed Manakin (*Chiroxiphia pareola,* subspecies *pareola* [mainland, black area] and *atlantica* [Tobago, arrow]) and Lance-tailed Manakin (*C. lanceolata,* stippled area) in northern South America.

most tolerant of the open habitats that must be crossed in moving between island and mainland. (3) Finally, a very narrow water gap may allow more or less free interchange of birds to and from the adjacent mainland. This may be an especially important consideration in the case of Tobago. As has been noted, Tobago has probably been an island since the beginning of the Pleistocene or earlier, but during periods of lowered sea level it may have been separated from Trinidad (then part of the mainland) by channels no more than 5 km wide.

The preceding analysis shows that the Andean element of the Trinidad and Tobago avifauna has closer affinities with the montane avifauna of the Paria peninsula and Cordillera de Caripe of Venezuela than the latter has with the avifauna of the Venezuelan mountains farther west. This is in agreement with geological evidence (J. B. Saunders, per. comm.). The Caripe-Paria-Trinidad-Tobago mountains are geologically an island, and apparently have never been connected with the coastal mountains of Venezuela farther west, which are themselves continuous with the main chains of the Andes. Thus, the Trinidad and Tobago montane avifauna, like that of the adjacent mainland mountains, must have been derived by long-distance dispersal from the main Andean range. The process cannot be dated. It may be presumed, however, that at the height of the last glaciation, (ca. 13,000–21,000 years ago; Van der Hammen 1974), when the sea level and vegetation belts in northern South America were lower than now, the degree of physical isolation between the montane avifaunas must have been less than it is now, and interchange of montane birds between the Cordillera de Caripe and the mountains to the west may have been easier.

The presence of the Amazonian-Guianan element in Trinidad and Tobago is more puzzling. It is rather more differentiated than is the Andean element, but for the reasons given above, it may not be safe to conclude from this that it has been isolated longer from the mainland populations. The existence of this element, which includes several species most unlikely to cross extensive water gaps, suggests that during some past period when both Trinidad and Tobago were connected or (in the case of Tobago) very close to the mainland, forests containing an avifauna of the Amazonian-Guianan type extended up the coast of eastern Venezuela in the area of the present Orinoco delta. The available evidence indicates, however, that at times of lower sea level the climate of northern South America was cooler and drier than it is now

(Haffer 1974), so that such a forest extension seems unlikely. Clearly, further clarification must await a more detailed knowledge of the relative timing of sea-level and climatic changes. It is possible also that all the species of the Amazonian-Guianan element may not have been isolated for the same length of time.

Perhaps the most challenging member of the Amazonian-Guianan element is *Chiroxiphia pareola atlantica,* the only forest-living member of the group that is confined to Tobago. It is larger than the Guianan subspecies (*pareola*) but otherwise very similar, not only in plumage but also in vocalizations and display (Snow 1971). Thus, it is far more similar to the Guianan form than the latter is to the populations of western and central Amazonia (subspecies *regina, boliviana,* and *napensis*). The evolutionary history of these mainland forms has not been investigated, but they show about the same degree of differentiation as several of the super-species that Haffer (1974) has interpreted as products of isolation caused by the fragmentation of lowland forests during the last glaciation. It thus seems likely that the isolated population of *C. pareola* on Tobago dates from the last glaciation, when Tobago was separated from Trinidad by a very narrow water gap. It follows also that *C. pareola* must subsequently have become extinct on Trinidad and in adjacent parts of Venezuela. The most likely hypothesis suggests that this happened during a period of increased humidity, when the kind of seasonal forest that *Chiroxiphia* prefers (Snow 1977) was reduced in extent, and that when drier conditions supervened, the mainland adjacent to Trinidad was occupied by *C. lanceolata* spreading along the coast from the west (Fig. 2).

The evidence indicates that the present avifauna of Tobago, like that of Trinidad, is of rather recent origin. Not only is there no evidence, but also no likelihood, that any species now found on Tobago descended from a lineage that has been isolated for two million years or more. Although Tobago has been an island for this length of time, during glacial periods it was separated by such a narrow water gap that its avifauna was effectively in contact with the mainland.

ACKNOWLEDGMENTS

I am much indebted to J. B. Saunders for information on the geology of Trinidad and Tobago in advance of his forthcoming book on the subject, and to J. Haffer and T. Van der Hammen for help on certain points.

LITERATURE CITED

BERLIOZ, J. 1962. Notes critiques sur quelques espèces de Trochilidés. Oiseau Rev. Fr. Ornithol. 32: 135–144.
BOND, J. 1979. Derivations of Lesser Antillean birds. Proc. Acad. Nat. Sci. Phila. 131:89–103.
CHAPMAN, F. M. 1894. On the birds of the island of Trinidad. Bull. Am. Mus. Nat. Hist. 6:1–86.
DELACOUR, J., AND D. AMADON. 1973. Curassows and Related Birds. American Museum Natural History, New York.
FFRENCH, R. 1973. A Guide to the Birds of Trinidad and Tobago. Livingston Publ. Co., Wynnewood, Pennsylvania.
GRANT, P. R. 1968. Bill size, body size and the ecological adaptation of bird species to competitive situations on islands. Syst. Zool. 17:319–333.
HAFFER, J. 1974. Avian speciation in tropical South America. Publ. Nuttall Ornithol. Club No. 14.
HELLMAYR, C. E. 1906. On the birds of the island of Trinidad. Novit. Zool. 13:1–60.
JUNGE, G. C. A., AND G. F. MEES. 1957. The avifauna of Trinidad and Tobago. Zool. Verh. (Leiden) No. 37.
LACK, D. 1976. Island Biology. Blackwell, Oxford, England.
LANYON, W. E. 1978. Revision of the *Myiarchus* flycatchers of South America. Bull. Am. Mus. Nat. Hist. 161:427–628.
MAYR, E. 1983. Introduction. Pp. 1–21, *In* A. H. Brush, and G. A. Clark, Jr. (eds.), Perspectives in Ornithology. Cambridge University Press, Cambridge, England.
MEYER DE SCHAUENSEE, R. 1966. The Species of Birds of South America and Their Distribution. Livingston Publ. Co., Narberth, Pennsylvania.
MEYER DE SCHAUENSEE, R., AND W. H. PHELPS, JR. 1978. A Guide to the Birds of Venezuela. Princeton University Press, Princeton, New Jersey.
PHELPS, W. H., JR. 1966. Contribución al análisis de los elementos que componen la avifauna subtropical de las cordilleras de la costa norte de Venezuela. Bol. Acad. Cienc. Fís. Mat. Nat. (Caracas) 26: 14–34.
SIBLEY, C. G., AND J. E. AHLQUIST. In press. The phylogeny and classification of the passerine birds, based on comparisons of the genetic material, DNA. Proc. 18th Int. Ornithol. Congr.

Snow, D. W. 1971. Social organization of the Blue-backed Manakin. Wilson Bull. 83:35–38.
Snow, D. W. 1977. Duetting and other synchronised displays of the blue-backed manakins, *Chiroxiphia* spp. Pp. 239–251, *In* B. Stonehouse and C. Perrins (eds.), Evolutionary Ecology. Macmillan, London.
Van der Hammen, T. 1974. The Pleistocene changes of vegetation and climate in tropical South America. J. Biogeogr. 1:2–26.

APPENDIX I

SPECIES ON TRINIDAD AND TOBAGO INCLUDED IN THE ANDEAN AND AMAZONIAN-GUIANAN
DISTRIBUTIONAL CATEGORIES

ANDEAN

Forest species: *Columba fasciata, Geotrygon linearis, Touit batavica, Steatornis caripensis, Cypseloides rutilus, Phaethornis guy, Campylopterus ensippennis, Trogon collaris, Veniliornis kirkii, Dendrocincla fuliginosa, Sittasomus griseicapillus,* Synallaxis cinnamomea, Xenops rutilus, Sclerurus albigularis, Dysithamnus mentalis,* Grallaria guatimalensis, Procnias averano,* Leptopogon superciliaris, Mionectes olivaceus, Thryothorus rutilus, Catharus aurantiirostris, Platycichla flavipes, Turdus fumigatus,* Turdus albicollis,* Hylophilus aurantiifrons, Tangara chrysophrys, Tangara gyrola, Thraupis cyanocephala, Piranga flava, Habia rubica.*
Non-forest species: *Colibri delphinae, Tiaris fuliginosa, Spinus cucullatus.*

AMAZONIAN-GUIANAN

Forest species: *Aburria pipile, Leptotila rufaxilla, Chaetura spinicauda, Phaethornis longuemareus, Trogon violaceus,* Momotus momota, Ramphastos vitellinus, Celeus elegans, Myrmotherula axillaris, Manacus manacus, Chiroxiphia pareola, Myiarchus tuberculifer, Myiornis ecaudatus, Basileuterus culicivorus,* Euphonia violacea.*
Non-forest species: *Columbina minuta, Ara ararauna, Ara manilata, Anthracothorax viridigula, Lophornis ornata, Amazilia chionopectus, Sclateria naevia, Tyrannopsis sulphurea, Todirostrum maculatum, Icterus chrysocephalus, Tachyphonus luctuosus, Paroaria gularis, Sporophila americana.*

* Species allocated to distributional category on basis of subspecies occurring in Trinidad and/or Tobago.

ADDITIONS TO THE AVIFAUNA OF ARUBA, CURAÇAO, AND BONAIRE, SOUTH CARIBBEAN

K. H. Voous

Institute for Taxonomic Zoology, Zoological Museum, University of Amsterdam, The Netherlands

ABSTRACT. The islands of Aruba, Curaçao, and Bonaire, here called South Caribbean Islands, claim few indigenous land bird species (34) and even fewer waterbirds and seabirds (27). In contrast, the number of non-breeding visitors from North America (102 species) and other parts of the West Indies (1 species and 3 subspecies), regular visitors and stragglers from South America (2 and 24 species, respectively), and non-breeding seabirds (27 species), is remarkably large (159 species). The islands constitute real crossroads for birds moving across the Caribbean Sea in all directions. The birds originate from as far as Canada in the north and Argentina in the south. In the past fifty years, species have been added to the known avifauna whenever ornithologists have visited the islands. The present paper summarizes the occurrence of 77 species recorded in the islands for the first time since the last check-list (Voous 1965). These new data are discussed against faunistic and zoogeographical backgrounds. Patterns of migration and wandering are considered, as are the temporary appearance and nesting of fresh-water birds (freshwater bird nomadism), the occurrence of seabirds and introduced birds, and avian extinction and conservation.

RESUMEN. Las islas de Aruba, Curaçao y Bonaire, aquí llamadas Islas del sur del Caribe, poseen pocas especies de aves terrestres indígenas (34) y aün menos aves acuáticas y marinas (27). En contraste la cantidad de visitantes no anidadores desde América del Norte (102 especies) y otras partes de las Indias Occidentales (1 especie y 3 subspecies) y también visitantes regulares y rezagados de América del Sur (2 y 24 especies respectivamente) y aves marinas no anidadoras (27 especies) es notablemente grande (159 especies en total). Estas islas constituyen un verdadero cruce de rutas para aves que se mueven en todas direcciones a lo largo del mar Caribe. Las aves provienen desde tan lejos como Canadá al norte y Argentina al sur. Durante los últimos 50 años, cada vez que las islas fueron visitadas por ornitólogos, se aumentó el número de especies en la avifauna. En el presente trabajo se presenta un resumen sobre la presencia de 77 especies registradas en las islas por primera vez después la última lista de aves (Voous, 1965). Esta nueva información es discutida con respecto a los parámetros faunísticos y zoogeográficos. Se consideran los modelos migratorios y de dispersiones irregulares tales como la presencia temporaria y anidación de aves acuáticas ("fresh-water bird nomadism"), la aparición de aves marinas e introducidas y conservación y extinción de aves.

Apart from Margarita, Trinidad, and Tobago, the islands of Aruba, Curaçao, and Bonaire, here collectively called the South Caribbean Islands, are the largest situated off the north coast of South America. They are located about 12°N, 30 to 90 km from the mainland. Only Aruba forms part of the continental shelf; Curaçao and Bonaire are separated from Venezuela by sea depths reaching 1350 m and 1700 m, respectively. The islands are small (together ca. 925 km²), flat or slightly hilly, and covered with xerophytic scrub or woodland. Their leeward coasts and bays are fringed with mangroves (mainly *Rhizophora mangle*).

The last check-list of the birds of the islands of Aruba, Curaçao, and Bonaire, Netherlands Antilles, South Caribbean, listed 161 species (Voous 1965). Mainly through the enthusiastic cooperation of resident and visiting birders, the number of species recorded has now reached 238 (Voous 1983, and below). This represents an increase of 80 percent in the number of species known from the South Caribbean Islands since the first extensive survey was published about 30 years ago (131 species, Voous 1955, 1957) and of 47 percent in the last twenty years (161 species, Voous 1965). In view of the geographical position of the islands, further additions can be expected. In the present paper I summarize the faunistic novelties from these islands, many of which are of general interest (e.g., *Anser albifrons, Coccyzus lansbergi, Steatornis caripensis, Tachycineta leucopyga, Hirundo fulva, Oenanthe oenanthe*). All new data are classified and discussed against a general faunistic and zoogeographic background.

247

Some of the new data I collected myself during visits to the islands in 1977 and 1979, but most novelties are based either on specimens found incidentally and now preserved in the Zoological Museum of Amsterdam, or on color-slides made available by friends and correspondents. In a few cases I accepted reports of species new to the fauna on the basis of descriptions of observations. All sight records for which specimens are lacking have been carefully scrutinized, and quite a substantial number has been put aside to await future confirmation. In a number of difficult cases I invited and received generous help from the late Eugene Eisenmann, to whose memory this paper is gratefully dedicated.

LAND BIRDS

BREEDING

The number of species of land birds known to breed in the South Caribbean Islands is 34, of which 28 are known from Aruba, 30 from Curaçao, and 28 from Bonaire. No recent changes in the breeding status of species are known, apart from isolated cases of breeding by the Tropical Kingbird (*Tyrannus melancholicus chloronotus*) on Aruba (1971, 1972, 1973, 1974, 1978), Curaçao (1980), and Bonaire (1976, 1981), and possible breeding of Lesser Elaenia (*Elaenia chiriquensis*) on Bonaire (1974). The record of *E. chiriquensis* is based on the description by Brother Candidus of call notes different from those of the well-known Caribbean Elaenia (*E. martinica*) and on an unusually large nest in which the birds were incubating their clutch of two eggs. The nest was roughly built of fine twigs and branches, and was lined with fine plant stems. It strongly resembled that of a Tropical Mockingbird (*Mimus gilvus*) and, hence, differed markedly from the minute, neat cup of thin plant fibers and hair of *E. martinica* often found by Brother Candidus in the same locality. Specimens of *Elaenia chiriquensis albivertex* have been collected previously on Curaçao (27 October 1951) and on Bonaire (6 November 1951). In addition, at least four sight-records from Bonaire and one from Curaçao are tentatively referred to this species. Some of these may have been migrants from temperate South America.

NON-PASSERINE MIGRANTS FROM NORTH AMERICA

Species of non-passerine migrants from North America are few, numbering no more than 10. Among these, Yellow-billed Cuckoos (*Coccyzus americanus*) can be abundant from mid-October to early November when their tameness and exhausted condition make them conspicuous. Species not previously recorded are *Caprimulgus carolinensis* (Aruba [A], specimen, 11 December 1979; Curaçao [C], 16 and 21 October 1966; Bonaire [B], ca. 20 November 1981) and *Chaetura pelagica* (A, 24–25 October 1978; B, 28 October 1979).

PASSERINE MIGRANTS FROM NORTH AMERICA

The number of North American songbird species recorded passing through or wintering in the islands is increasing steadily and presently stands at 53, with 32 known from Aruba, 37 from Curaçao, and 38 from Bonaire. Included are the families Tyrannidae (3 spp.), Hirundinidae (6 spp.), Turdidae (5 spp.), Mimidae (1 sp.), Bombycillidae (1 sp.), Sturnidae (1 sp.), Vireonidae (2 spp.), Parulidae (25 spp.), Thraupidae (2 spp.), Emberizidae (5 spp.), and Icteridae (2 spp.).

Species not previously reported are *Contopus virens* (Aruba, Bonaire), *Tyrannus tyrannus* (Aruba, Bonaire), *Stelgidopteryx serripennis* (formerly *ruficollis,* Bonaire), *Hirundo fulva* (Curaçao, specimen), *Bombycilla cedrorum* (Aruba, specimen), *Sturnus vulgaris* (Aruba, Bonaire), *Vermivora chrysoptera* (19 October 1983, color-slide, Bonaire), *Dendroica pensylvanica* (Aruba, Bonaire), *D. caerulescens* (Aruba; Bonaire, specimen), *D. discolor* (Aruba, Curaçao), *D. cerulea* (Bonaire, specimen), *Seiurus motacilla* (Aruba, Curaçao; Bonaire, specimen), *Oporornis formosus* (Aruba, Bonaire), *Geothlypis trichas* (Aruba), *Wilsonia citrina* (Aruba, Bonaire), and *Pheucticus melanocephalus* (Curaçao).

The number of North American warbler species recorded on the South Caribbean Islands is astonishingly large. Before 1950 only two species (*Setophaga ruticilla, Seiurus noveboracensis*) were known. It was only after specimens had been collected and modern field-guides had come into wide use that it became apparent how important the islands are as a first landfall after hazardous trans-Caribbean flights. When in some years the passage of warblers is especially strong in October and early November, Blackpoll Warblers (*Dendroica striata*) dominate and can be seen almost everywhere, hopping even in rocky court yards and foraging in gutters along the streets. Those that have been found dead from exhaustion are literally "skin and bone;" those collected for study skins weighed less than 9 grams. The total list of parulid species recorded, apart from the resident Yellow Warbler (*Dendroica petechia rufopileata*), is given below. However, northern parulid warblers are not at all numerous each year and few stay throughout the winter. Only *Setophaga ruticilla, Seiurus aurocapillus,* and *S. noveboracensis* are regular winter residents.

North American Warblers recorded from Aruba (A), Curaçao (C), and Bonaire (B) (** = including winter period, 15 December–15 February; * = study skin available) are *Vermivora chrysoptera* (B), *V. peregrina* (C*, B), *Parula americana* ** (C*, B), *Dendroica pensylvanica* (A, B), *D. tigrina* ** (A, C, B), *D. caerulescens* (A, B*), *D. coronata* ** (C*, B), *D. virens* (A, B), *D. fusca* (A, C*, B*), *D. discolor* (A, C), *D. palmarum* (A, C*), *D. castanea* (C, B*), *D. striata* (A*, C*, B*), *D. cerulea* (B*), *Mniotilta varia* ** (A, C*, B), *Setophaga ruticilla* ** (A*, C*, B*), *Protonotaria citrea* ** (A*, C*, B), *Seiurus aurocapillus* ** (A*, C*, B*), *S. noveboracensis* ** (A, C*, B*), *S. motacilla* ** (A, C, B*), *Oporornis formosus* (A, B), *O. agilis* (A*, C*, B*), *O. philadelphia* (C), *Geothlypis trichas* ** (A), *Wilsonia citrina* ** (A, B).

Some North American migrants reach the southern limit of their winter ranges in the South Caribbean Islands. Whether their appearance here is exceptional or whether they have passed so far unnoticed is unknown. Species included are *Oenanthe oenanthe leucorrhoa* (C, 4 November 1962; B, 18 December 1975), *Hylocichla mustelina* (C*, 30 October 1951), *Toxostoma r. rufum* (C*, 2 October 1957), *Sturnus v. vulgaris* (A, 18 November 1977–2 January 1978; B, 10 November 1980), *Dendroica tigrina* (A, C, B, at least 5 records), *D. c. caerulescens* (A, B*, at least 4 records), *D. p. palmarum* (A, C*, winter 1956–1957 only), *Geothlypis trichas* (C, 26 January 1979), and *Zonotrichia albicollis* (A*, January 1964, likely by "assisted passage" on board a ship).

The occurrence of other species is less unexpected, although these are thought to migrate mostly through Central America rather than across the Caribbean, and some rarely migrate this far south. The species include *Tyrannus tyrannus, Hirundo pyrrhonota**, *H. fulva pallida* (C*, 6 October 1952), *Bombycilla cedrorum* (A*, 22 February 1979), *Catharus m. minimus**, *Vireo flavifrons* (C*, 21 March 1957), *V. o. olivaceus**, *Vermivora chrysoptera* (B, 19 October 1982), *V. peregrina**, *Dendroica pensylvanica, D. virens, D. fusca**, *Oporornis philadelphia, Wilsonia citrina, Piranga olivacea**, and *Pheucticus melanocephalus* (C*, 17 December 1978).

MIGRANTS FROM OTHER CARIBBEAN ISLANDS

Migrants from other Caribbean islands and the warmer parts of Florida have been dealt with elsewhere (Voous 1982). They include *Coccyzus minor maynardi* (A, 21 April 1930; C*, 19 January 1952), *Chordeiles (minor) gundlachii* (C*, 19 April and 17 September 1955), *Tyrannus d. dominicensis* (e.g., A, 16–23 September 1979; B*, 24 October 1979), *Progne (dominicensis) cryptoleuca* (C*, 8 September and 6 October 1955, 30 September 1956), *P. d. dominicensis* (C*, 26 May and 6 October 1955), *Vireo altiloquus barbatulus* (B*, 5 November 1951), *V. a. altiloquus* (A*, mid-October 1978), and *V. a. barbadensis* (B, 13 October 1979).

MIGRANTS FROM SOUTH AMERICA

Migrant land birds from South America recorded on the islands are considered to have originated from the temperate or cold southern part of the continent. Apart from *Muscivora tyrannus* (*Tyrannus savana* of American Ornithologists' Union 1983), they generally remain unnoticed, and it is hard to judge whether the occurrence of the few species reported is regular or rather exceptional, with individuals "overshooting" their normal wintering grounds. Included are *Elaenia chiriquensis albivertex* (C*, B*; see, however, "Breeding Birds"), *Elaenia parvirostris* (A*, 6 May 1908), *Tyrannus m. melancholicus* (B, 22 May 1979), *Muscivora tyrannus* (A, C*, B, May–early October), and *Tachycineta leucopyga* (C, 15 May 1977).

STRAGGLERS FROM SOUTH AMERICA

The number of species usually considered resident in northern South America and recorded probably as stragglers in the South Caribbean Islands is increasing steadily and at present stands at 20. All except one (*Chloroceryle amazona*) were reported by Voous (1982), who stressed the theoretical importance of the study of continental birds straggling to islands. Species included are *Elanus caeruleus* (sive *leucurus*; A), *Milvago chimachima* (C), *Columbina (Columbigallina) talpacoti* (B), *Coccyzus m. minor* (A*), *Coccyzus lansbergi* (B*), *Steatornis caripensis* (A*), *Chordeiles acutipennis* (C, B), *Glaucis hirsuta* (C*), *Florisuga mellivora* (A*, C), *Chloroceryle amazona* (A), *Pyrocephalus rubinus* (A), *Muscivora tyrannus monachus* (C*),

Tachycineta albiventer (C), *Cyanerpes cyaneus eximius* (B*), *Volatinia jacarina splendens* (C, B*), *Agelaius i. icterocephalus* (C, B), *Sturnella magna* (probably subsp. *paralios*; B), *Quiscalus lugubris* (probably subsp. *lugubris*; A,B).

A specimen of Guira Cuckoo (*Guira guira*) from Curaçao (12 June 1954) is most exceptional, as this species does not occur in adjacent parts of South America and is generally considered a strict resident. It is in an immaculately fresh plumage.

MARSH BIRDS AND SHOREBIRDS

BREEDING

In spite of the presence and abundance of suitable habitats in the form of quiet inland bays, mangroves, lagoons, mud flats, and muddy shores, the number (13) of marsh and shorebird species breeding in the islands is small, as, except for the Greater Flamingo (*Phoenicopterus r. ruber*), is the number of individuals. For the most part, man probably is responsible for the paucity of these birds, but irregular rainfall and consequent reduction of available food is another factor that may limit breeding efforts of these birds. This group includes the Greater Flamingo, which nests on Bonaire, and six species of heron, including Cattle Egret (*Bubulcus ibis*; nesting on Curaçao since 1967, on Aruba since 1980), Reddish Egret (*Egretta rufescens*) on Bonaire (Spaans 1974), and probably also Great Egret (*Casmerodius albus*) on Bonaire (1980). It also includes six species of shorebirds, among which Wilson's Plover (*Charadrius wilsonia cinnamominus*) and Snowy Plover (*Charadrius alexandrinus nivosus*) are the most regular, and the Killdeer (*Charadrius vociferus*; nesting on Aruba 1979–1982), the most unexpected. In 1968 the American Oystercatcher (*Haematopus p. palliatus*) was found breeding for the first time on the South Caribbean Islands (Bonaire; also in 1971, 1979).

Payne (1974) claimed that the resident Green Herons are mainly of the striated type and, therefore, should be referred to *Butorides s. striatus,* but all 15 adult specimens collected on the islands that I examined are of the *virescens* type (rich reddish brown neck, small size). They are listed in the Aruba–Curaçao–Bonaire Check-list (Voous 1965) as *B. virescens curacensis.* Not fully adult specimens with characters somewhat intermediate between *virescens* and *striatus* were collected on Aruba (22 June 1930) and Bonaire (17 October 1979), but only four of 29 birds that I scrutinized carefully in the field on Bonaire in October 1979 showed signs of intermediate plumage. So far, I have observed only one bird in striated plumage indistinguishable from adult South American *B. s. striatus* on the islands (Bonaire, 3 October 1979). Whether this was a member of the breeding population or a straggler from the nearby continental coast is unknown, but I list it below as a straggler.

MIGRANTS FROM NORTH AMERICA

Members of this large group (47 species) are prominent on the shores of bays, in lagoons, in saltpans, on mud flats, and in mangroves. They include a cormorant, herons (7 species, in addition to *Butorides striatus virescens*), geese and ducks (8 species), rails (2 species), and shorebirds (29 species). Not all of these are of certain North American origin, for some of the herons, at least, may have come from nearby continental South America (e.g., *Nycticorax nycticorax, Egretta caerulea, Casmerodius albus*). Species (and one subspecies) not previously recorded are *Phalacrocorax auritus* (B), *Ardea herodias wardi* (B), *Anser albifrons* (A, one adult, early June 1980), *Anas platyrhynchos* (B, 19 September–October 1983, one female among White-cheeked Pintail and Blue-winged Teal; C. van der Linden, pers. comm.), *A. acuta* (B), *A. clypeata* (A, B), *Aythya collaris* (B), *Charadrius melodus* (B), *Recurvirostra americana* (B), *Calidris bairdii* (A, C, B), *Tryngites subruficollis* (B), *Limnodromus scolopaceus* (A, C, B), *Phalaropus tricolor* (B), and *P. lobatus* (B).

STRAGGLERS FROM SOUTH AMERICA

This category includes marsh, fresh-water, and shorebirds of presumed continental South American origin. Those previously unreported are *Botaurus pinnatus* (A, 18 January 1972), *Cochlearius c. cochlearius* (B, 7 October 1972), *Butorides s. striatus* (B, 3 October 1979), *Mycteria americana* (A, 16 February 1977–August 1982), *Aramus g. guarauna* (A, February 1975), *Jacana jacana intermedia* (B*, 1971), *Belonopterus* (*Vanellus*) *chilensis* (A, 6 June 1979).

FRESH-WATER BIRDS

IRREGULARITY OF BREEDING

Recently I dealt in some detail (Voous 1982) with the phenomenon of fresh-water bird nomadism in the South Caribbean Islands. Aruba, Curaçao, and Bonaire have no permanent rivers and the few fresh water seeps are small. In the usually arid hills and valleys covered with cactus and thorn scrub, the presence of water birds, especially nesting ones, seems unlikely. Still, at irregular intervals rainfall is sufficient to fill reservoirs and valleys; then, grebes, rails, coots and ducks appear as if by magic and hurriedly start breeding, continuing to do so until the water is gone. After some months, the water subsides by evaporation and seepage, and the area regains its arid appearance with soil cracked under the burning sun. Then, the birds disappear as suddenly as they arrived, leaving behind numbers of partly incubated eggs and starving and starved young. They do not reappear until the next rainy period, for which they sometimes must wait five years or more. The species involved (with years of breeding) are *Tachybaptus dominicus brachyrhynchus* (tropical South American race) 1967, 1971, 1972 (C), 1981 (B); *Podilymbus podiceps antarcticus* (South American race) 1953 (B), 1954, 1955, 1956, 1971, 1972 (C), suspected nesting 1979 (A); *Porphyrula martinica* suspected nesting 1956 (C); *Gallinula chloropus cerceris* (Caribbean race) 1956, 1971 (C); *Fulica americana caribaea* (Caribbean race) 1956, 1971 (C), 1976, 1979 (A), 1981 (B). *Tachybaptus dominicus* had not been recorded nesting in the South Caribbean Islands before.

I consider the coot occurring on these islands to be a representative of *Fulica americana caribaea,* although most authors and the American Ornithologists' Union (1983) treat it as a full species. It has an expanded white frontal shield with corrugated surface and usually lacks the basal knob characteristic of North American *americana.* In a film made by W. Bokma of the breeding birds of Malpais, Curaçao, in 1971, one coot at least showed a thick, reddish brown knob at the frontal base of the shield. In another film made by E. van Campen in the same locality a coot with a prominent knob relieves its mate, which has a *caribaea* type of bill, at a nest with eggs. Coots with bill knobs were also seen by P. A. Buckley and F. G. Buckley among "Caribbean" Coots on Aruba and Bonaire in January 1971 (in litt.). Whether these coots were wintering North American birds or not is unknown. There is, however, sufficient reason to agree with those authors who consider Caribbean coots to be a polymorphic race of the widespread North American Coot with knobless birds predominating (Gill 1964; see also Bond 1976:13).

Interesting information on birds occupying the temporarily flooded valley of Malpais, Curaçao, between 17 January to 13 December 1971 was provided by W. Bokma and E. van Campen (pers. comm.). This valley, which is between low hills, occupies a relatively small depression in which water is held by a dam. Heavy rains fell in the area in mid-December 1970; more rain fell on several occasions in the first half of 1971. The water level in front of the dam ultimately reached a height of about 3 m. Water did not rise higher than 50 cm behind the dam before gradually seeping into the mud flats of the coastal lagoon. After June 1971, the water level dropped noticeably, and the first mud flats had dried by the end of July. Observations, unfortunately, were not continued after 13 December 1971. By March 1972 the whole valley had practically dried up. A summary of the numbers of breeding birds and the young reared follows. Total numbers include young birds.

Tachybaptus dominicus. Least Grebe. Recorded on each of 26 visits from 9 February 1971 onward. The first nest with eggs was found 30 March. Average numbers varied between 3 in February to April, 23 in May to September, and 26 in October to December, with a maximum of 46 on 19 September. Nests were built in dense stands of *Echinodorus cordifolius* (Alismataceae).

Podilymbus podiceps. Pied-billed Grebe. Recorded on each of 25 visits from 11 February 1971 onward. The first young attended by parents were observed 10 April. Average numbers varied between 2 in February to April, 7 in May to September, and 7 in October to December, with a maximum of 11 on 18 June. Nests were hidden in suddenly developed ephemeral clumps of burhead (*Echinodorus cordifolius*) and consisted of heaps of stonewort (*Chara*).

Gallinula chloropus. Common Gallinule. Recorded on each of 21 visits from 13 March 1971 onward. Nests were found from April onward. Average numbers varied from 2 in February to April, and 8 in May to September, to 7 in October to December, with a maximum of 11 on 19 September.

Fulica americana. American Coot. Recorded on 21 of 26 visits between 9 February and 13 December 1971. Nests were found from April onward, and, throughout the season, numbered 25, at least, producing ca. 120 young, many of which survived. Average numbers varied from 3 in February to April, and 36 in May to September, to 54 in October to December, with a maximum of 85 on 16 October. Nests were built among the branches of submerged trees and shrubs near water level, but they became more and more conspicuous as the water fell.

SPECIES APPEARING IRREGULARLY

Malpais Valley, Curaçao.—Fresh-water bird nomadism described above did not always lead to irregularly occurring local nesting. Some species appeared to take advantage only of an unexpected possibility to feed on a temporarily rich food supply between breeding periods elsewhere. This category included various species of herons, ibises, and ducks, among which *Sarkidiornis melanotos* (30 March 1971) had not been recorded previously from the islands. *Plegadis falcinellus* was present in the Malpais valley from 16 October to 19 November 1971 and was known previously only from one specimen taken on Aruba in early 1965.

Sewage ponds at Bubali, Aruba.—A large variety of fresh-water birds has turned up at the sewage ponds operating since 1973 behind the luxury beach hotels at Bubali. These ponds, stocked with cichlids and other fast-growing fishes, surrounded by a border of marsh vegetation, and temporarily overgrown with water-hyacinths (*Eichornia crassipes*), form an exotic habitat on the otherwise arid island. Several South American species have been recorded in the South Caribbean Islands, only here. These include *Eudocimus albus* (apart from one record from Curaçao around 1925), *Plegadis chihi* (June 1978, color-slide), *Mycteria americana,* and *Belonopterus chilensis* (6 June 1979, color-slide).

The Wood Stork (*Mycteria americana*) of which only one individual was seen, was first recorded on 16 February 1977 when still in juvenile plumage. It was observed and photographed on numerous occasions; it was still present in October 1984 as an adult. Although never found outside the area of the sewage ponds, it has been observed frequently soaring high in the air and circling over Aruba as if to leave the island, but it has always returned. Probably the sea with its lack of constant thermal air currents prevents it from returning to its native colony on the South American mainland even after so many years of solitude. Apparently, the sea passage of 30 km to the Paraguaná Peninsula is too wide to allow large soaring birds an easy crossing. This also applies to two Black Vultures (*Coragyps atratus*) taken from Colombia to Aruba in the early 1970's. They easily soar over the mangrove coast where they are fed by local fishermen, but they have never left the island.

BREEDING SEASONS IN VENEZUELA

Several species of large marsh and fresh-water birds, which nest in large colonies in Venezuela during the rainy period (August–November) when the llanos are flooded, leave in the subsequent dry season. Some of these birds are now observed in the Aruban sewage ponds at Bubali, particularly from February to June, in numbers previously unknown. Included are Olivaceous Cormorant (*Phalacrocorax olivaceus*), Great Egret (*Casmerodius albus,* hundreds), Snowy Egret (*Egretta thula,* numerous), Black-crowned Night Heron (*Nycticorax nycticorax,* most probably now nesting), White Ibis (*Eudocimus albus*), Scarlet Ibis (*E. ruber*), and Glossy Ibis (*Plegadis falcinellus*).

Since February 1977, White and Scarlet Ibises in juvenile and adult plumages have been observed, usually together, in the Bubali ponds on many occasions in numbers not exceeding a total of six. White Ibises were the more frequent of the two and probably originated from breeding sites in westernmost Venezuela. Usually one fiery-red Scarlet Ibis was present; from February 1977 to February 1978 it was often accompanied by a bird in a remarkably striated, white, pink, and red plumage. The latter may have been an aberrant immature, but probably was a hybrid between the White and the Scarlet species.

Black Skimmers (*Rynchops nigra cinerascens*) previously unrecorded in the islands, now also appear regularly on Aruba (since 1977; also observed on Bonaire, April 1979). They are present during the period of flooding (July–November) of the sand banks and sand beaches bordering the lowland rivers of the Orinoco and other northern South American river systems, which provide their exclusive breeding habitat.

Two other fresh-water species, either forced to leave their continental breeding grounds because of seasonal droughts, or induced by copious rainfall to disperse to new nesting areas, have been recorded for the first time in the islands. *Dendrocygna autumnalis* has been noted on Aruba (September 1977, July 1981) and Curaçao (January 1981), and *Oxyura dominica*, on Bonaire (February 1981).

SEABIRDS

BREEDING

As with marsh birds and shorebirds, the number of seabirds breeding on the South Caribbean Islands is remarkably small (9 species) considering the presence of what seem to be potentially favorable habitats all along coasts and in bays. On all islands, man always has been and still is a major destructive force. Breeding colonies of terns (7 species) are robbed of their eggs regularly, although this is illegal. In view of the extent of human depredation the young of nesting boobies and frigatebirds would have little chance of survival. In this respect it is remarkable that since 1966 some 20 pairs of Brown Pelicans (*Pelecanus occidentalis*) have managed to breed annually in the fly- and mosquito-infested strip of rhizophore mangrove on Aruba's reef wall. Audubon's Shearwater (*Puffinus lherminieri lherminieri*, not *loyemilleri*, see Voous 1983:32) may occasionally nest on Bonaire, but proof is lacking. Nocturnal courtship songs have been heard, however. No other recent additions of breeding seabirds are known.

NON-BREEDING, PELAGIC, AND STRAGGLERS

As a result of the ornithological contributions of R. van Halewijn (1970–1971) and D. M. C. Poppe (1972) to the internationally sponsored Cooperative Investigations of the Caribbean and Adjacent Regions Project (CICAR), the number of non-breeding seabird species recorded in the South Caribbean has increased to 27. Those observed on the coast and in territorial waters (declared since 1978) include the following species not recorded by Voous (1965): *Pterodroma hasitata, Bulweria bulwerii, Puffinus gravis, Oceanites oceanicus, Oceanodroma leucorhoa, Phaethon lepturus, Sula dactylatra, Stercorarius pomarinus, Catharacta (Stercorarius) skua* (or rather *maccormicki*), *Larus pipixcan, L. ridibundus, L. delawarensis, L. argentatus, L. fuscus, L. marinus, Sterna caspia*. Some of these may occur regularly in the Caribbean Basin and offshore waters. Others, such as the gulls, seem to concentrate on the relatively shallow leeward coast of Aruba (Voous 1977). Increase in numbers and frequency of occurrence of "northern gulls" has also been observed on the east coast of North America, the Florida and Gulf coasts, the northern Greater and Lesser Antilles, even Panama, Trinidad, French Guiana, and Brazil. The European Black-headed (*Larus ridibundus*) and Lesser Black-backed (*L. fuscus*) Gulls most probably followed a North American east coast route to the islands rather than crossing the mid-Atlantic.

INTRODUCED BIRDS

Fortunately, few species have been introduced in the islands. Of these, the House Sparrow (*Passer d. domesticus*) is now established on Curaçao (since 1953) and Aruba (first recorded 1978). The Saffron Finch (*Sicalis flaveola*), now found in luxuriant gardens of suburban Willemstad, Curaçao, was probably introduced from Venezuela (escaped pets, early or mid-1970's). The same may apply to the Carib Grackle (*Quiscalus lugubris*), now known from the gardens of Oranjestad, Aruba (first recorded 1982), but a record from Bonaire (1980) seems to refer to a genuine wild straggler. The Green-rumped Parrotlet (*Forpus passerinus viridissimus*), allegedly introduced from Colombia around 1940, apparently has not survived, but records of this species from the previous century (e.g., 1868) probably referred to stragglers from the continent. In contrast, the Old World Rose-winged Parakeet (*Psittacula krameri*) has established itself on Curaçao (e.g., Plantation Raphael), whereas other, unidentified species of *Psittacula* are sometimes seen as well. Other parrots, including the Red-lored Amazon (*Amazona autumnalis salvini*), have been recorded from time to time on Curaçao flying around, visiting fruit gardens and nesting in a wild state (e.g., 31 May 1977, nest with 2 eggs in hollow tree at border of mangroves, Groot Sint Joris).

Other noteworthy introductions include Rufous-collared Sparrow (*Zonotrichia capensis insularis*, mid-1950's) and Troupial (*Icterus icterus ridgwayi*, 1973, 1975) from Curaçao to Bonaire. Only the Troupial seems to have survived, but because of similarly aggressive and almost identical feeding habits it may ultimately clash with the indigenous Pearly-eyed Thrasher (*Margarops fuscatus*).

Recently (1982, 1983) Village Weavers (*Ploceus cucullatus*), African cage-birds that have escaped and become established in many parts of the world including the West Indies (Hispaniola, Bond 1971:225), have been found nesting in the gardens of one of the residential suburbs of Willemstad, Curaçao (Brother Yvo Nijsten, *in litt.*).

EXTINCTION AND PRESERVATION

No species has become extinct on the South Caribbean Islands. Nevertheless, the Yellow-shouldered Parrot (*Amazona b. barbadensis*) disappeared from Aruba around 1947. Another subspecies, *A. b. rothschildi,* survives on Bonaire. The species may have occurred on Curaçao too, at least until the second half of the 18th century.

A few species still occurring widely on the other islands, have not been found on Aruba, the most arid of the islands, for about half a century. These include Scaly-naped Pigeon (*Columba squamosa*) and Caribbean Elaenia (*Elaenia martinica*); they may have been extirpated by now. Other species on Aruba, such as the Scrub and Brown-crested Flycatchers (*Sublegatus modestus, Myiarchus tyrannulus*) are extremely localized.

Enforcement of nature conservation and bird preservation measures is greatly needed and essential for the survival of the Yellow-shouldered Parrot on Bonaire and for the continued nesting of terns and other seabirds. The protection of Greater Flamingoes on Bonaire is exemplary, but Crested Caracaras (*Polyborus plancus cheriway*) and White-tailed Hawks (*Buteo albicaudatus colonus*) are heavily persecuted and their numbers are diminishing rapidly. The same applies to the Barn Owl (*Tyto alba bargei*) of which a marginal population of probably less than 20 pairs survives only on Curaçao. The population (10–20 pairs) of the Burrowing Owl (*Athene cunicularia arubensis*) is likewise extremely vulnerable.

ACKNOWLEDGMENTS

I take great pleasure in acknowledging help from J. Bond, J. R. Jehl, Jr. (on dowitchers), and A. R. Phillips (mainly on swallows). A. R. Phillips and E. Eisenmann identified the *Progne* specimens. Friends and correspondents who provided additional assistance are too numerous to mention by name, but I thank all of them most warmly. M. S. Foster helped to put the manuscript in final form.

LITERATURE CITED

American Ornithologists' Union. 1983. Check-list of North American Birds. 6th ed. American Ornithologists' Union, Washington, D.C.

Bond, J. 1971. Birds of the West Indies. 2nd ed. Collins, London.

Bond, J. 1976. Twentieth supplement to the check-list of birds of the West Indies (1956). Academy Natural Sciences, Philadelphia, Pennsylvania.

Gill, F. B. 1964. The shield color and relationships of certain Andean Coots. Condor 66:209–211.

Payne, R. B. 1974. Species limits and variations of the new world Green Herons *Butorides virescens* and Striated Herons *B. striatus*. Bull. Br. Ornithol. Club 94:81–88.

Spaans, A. L. 1974. Some bird records from Bonaire. Ardea 62:236–238.

Voous, K. H. 1955. De vogels van de Nederlandse Antillen. Natuurwet. Werkgroep Ned. Antillen, Curaçao.

Voous, K. H. 1957. The birds of Aruba, Curaçao, and Bonaire. Stud. Fauna Curaçao Other Caribb. Isl. 7:1–260.

Voous, K. H. 1965. Check-list of the birds of Aruba, Curaçao, and Bonaire. Ardea 53:205–234.

Voous, K. H. 1977. Northern gulls in Aruba, Netherlands Antilles. Ardea 65:80–82.

Voous, K. H. 1982. Straggling to islands—South American birds in the islands of Aruba, Curaçao, and Bonaire, South Caribbean. J. Yamashina Inst. Ornithol. 14:171–178.

Voous, K. H. 1983. Birds of the Netherlands Antilles. Walburg Pers, Zutphen, The Netherlands.

FOREST BIRDS OF PATAGONIA: ECOLOGICAL GEOGRAPHY, SPECIATION, ENDEMISM, AND FAUNAL HISTORY

FRANÇOIS VUILLEUMIER

Department of Ornithology, The American Museum of Natural History, Central Park West at 79th Street,
New York, New York 10024 USA

ABSTRACT. The beech (*Nothofagus*) forests of Patagonia (southern South America) are isolated from other South American forests by 1100 (Andean forests of northwestern Argentina) to 1400 (lowland forests of southern Brazil) km of nonforest vegetation (including matorral, steppe, monte, espinal, pampa grassland, chaco woodland). This isolation makes the *Nothofagus* forest biome an island-like biota. Its avifauna includes 46 species of 40 genera, in 21 families of breeding landbirds. The patterns of ecological, elevational, and latitudinal distribution of these forest-inhabiting species of birds suggest that they have a wide habitat-niche, a pattern characteristic of truly insular taxa. Isolate formation has taken place in populations of a few species living on outlying islands but has led to insular speciation in three cases only, in the genera *Aphrastura, Anairetes,* and *Sephanoides.* The majority of forest species show no geographical variation on the continent. Only two cases of actual or potential continental speciation were detected, in the genera *Pteroptochos* and *Scytalopus.* Both are centered on the valley of the Rio Bío-Bío (central Chile), which may act as a barrier to gene flow. Several forest-dwelling allospecies are members of superspecies including one or more other allospecies that occupy forest or nonforest vegetation elsewhere in South America or in North America. Endemism is high in the Patagonian forest avifauna; about 41 percent of the species and allospecies, and 10 percent of the genera, are restricted in their distribution to the *Nothofagus* forest region. Levels of endemism were analyzed taking into account three different classifications of these birds, in order to assess the effects of changes in taxonomic practice on our evaluation of endemism. Levels of endemism in Patagonian forest birds (inhabiting a habitat island) are intermediate between the lower levels found in Tasmanian forest birds (a continental island) and the high levels found in New Zealand forest birds (an oceanic island group). Endemism in Patagonian forest birds is concentrated among the species that belong to families now distributed in South America, or to families that have been attributed to old autochthonous faunal elements (especially the Furnariidae and the Rhinocryptidae). The implications of these findings are discussed in terms of the Cenozoic history of the Patagonian forest region. The cycles of glaciation and deglaciation of the Pleistocene have played an important role in the development of the present avifauna. Niche shifts have been a biologically significant element in the evolution of several forest birds whose ancestors probably came from open vegetation farther north in the Andes or east in the lowlands of southern South America. The present fauna is a mixture of old relicts and of several strata of more or less recent immigrants.

RESUMEN. Los bosques de hayas (*Nothofagus*) de Patagonia (la porción más austral de Sudamérica) están aislados de otros bosques sudamericanos por 1.100 km (bosques andinos del noroeste argentino) hasta 1.400 km (bosques de galerías del sur de Brazil) por vegetación no boscosa (incluyendo matorral, estepa, monte, espinal, pampas y monte chaqueño). Esta aislación convierte a los bosques de *Nothofagus* en una biota de tipo insular. Su avifauna incluye 46 especies en 40 géneros, en 21 familias de aves terrestres anidadoras. Los modelos de distribución ecológica, altitudinal y latitudinal de las especies habitantes de estos bosques sugieren que poseen un amplio nicho-habitat, situación característica de taxa verdaderamente insulares. Se ha producido diferenciación ("isolate formation") en poblaciones de pocas especies que viven en islas periféricas, pero se ha producido especiación insular en sólo tres casos, en los géneros *Aphrastura, Anairetes* y *Sephanoides.* La mayoría de las especies de bosque no presentan diferenciación geográfica en el continente. Sólo dos casos de especiación actual o potencial han sido detectados en situaciones continentales: en los géneros *Pteroptochos* y *Scytalopus.* Ambos estan centrados en el valle del río Bío-Bío (región central de Chile), que puede actuar como barrera para el flujo genético. Varias alloespecies de habitats boscosos son miembros de superespecies las cuales incluyen una o más alloespecies que ocupan vegetación boscosa y no boscosa en otros sitios en

América del Sur y del Norte. El nivel de endemismo en Patagonia es alto, alrededor del 41 por ciento de las especies y alloespecies y el 10 por ciento de los géneros tienen su distribución restringida a los bosques de *Nothofagus* de la región. Los niveles de endemismo se analizaron tomando en consideración 3 diferentes clasificaciones para estas aves, de manera de preever los efectos de cambios en la práctica taxonómica en nuestro evaluación de endemismo. Los niveles de endemismo en las aves de los bosques patagónicos (habitando un habitat insular) son intermedios entre los niveles bajos que se encuentran en las aves de bosques de Tasmania (una isla continental) y los niveles altos de las aves de los bosques de Nueva Zelandia (un grupo de islas oceánicas). El endemismo en las especies de aves del bosque patagónico se concentra alrededor de especies pertenecientes a familias actualmente distribuidas en Sudamérica o familias que han estado atribuídas a elementos faunísticos autóctonos muy antiguos ("old autochthonous") (especialmente los Furnariidae y los Rhinocryptidae). Las implicaciones producto de estos hallazgos son discutidas en relación con la historia del cenozoico de los bosques de la región patagónica. Los ciclos de avance y retroceso de los glaciares durante el Pleistoceno han jugado un papel importante en el desarrollo de la avifauna actual. El cambio de nichos ha sido un elemento biológico significativo en la evolución de varias aves de bosques cuyos antecesores provienen probablemente de la vegetación de campo abierto más al norte en los Andes o en las llanuras orientales del sur de Sudamérica. La fauna actual es una mezcla de los vestigios antiguos y de inmigrantes mas o menos recientes de varios estratos.

Patagonia is defined as the southern part of temperate South America, approximately south of a line running southeast across the continent from the mouth of the Rio Maule in Chile to the mouth of the Rio Colorado in Argentina (see Aubert de la Rüe 1959). The two main vegetation formations of Patagonia are forests dominated by southern beeches (*Nothofagus*, Fagaceae) and semi-desertic steppes. The mountainous western side of the continent, a thin ribbon more than 2000 km long but only 100 to 200 km wide, is covered with forests, while steppes stretch across the much wider upland and lowland areas east of the Andes to the shores of the Atlantic Ocean. A very narrow ecotone separates these formations along the eastern foothills of the Andes. The floras of forest and steppe differ sharply from one another in taxonomic composition and physiognomy. Their climates are also dramatically different.

The *Nothofagus* forests of Patagonia are approximately 1100 km from the nearest montane forests in northwestern Argentina and 1400 km from the nearest lowland forests in southern Brazil, northeastern Argentina, and Paraguay (Fig. 1). The intermediate areas have a dry to arid climate (Knoch 1930; Sorge 1930) and are covered with nonforest vegetation, including pampa grassland, semi-desert scrub, monte, matorral, espinal, and savanna (Cabrera 1953; Schmithüsen 1956; Hueck and Seibert 1972; Solbrig 1976). The presence of an island of forest vegetation at the southern tip of South America immediately suggests a series of questions: Does the composition of the avifauna of Patagonian forests reflect the isolation of their environment? Is this fauna related to the faunas of distant forests elsewhere or to steppe faunas nearby? To what extent is this fauna tied ecologically to the forest environment? What are the ecological and evolutionary consequences of the ecological isolation? What is the history of this fauna? In an effort to answer these questions, I describe in this paper the composition of the Patagonian forest avifauna, discuss the ecological preferences of these birds, analyze their speciation, and examine their levels of endemism and their distribution patterns.

Similar questions have attracted the attention of biologists ever since Darwin visited Patagonia. One fact, especially, has whetted the curiosity of biogeographers. Forests dominated by *Nothofagus* are found in other parts of the southern hemisphere besides South America: Tasmania, Australia, New Zealand, New Caledonia, and New Guinea. Biotic relationships among the plants (Skottsberg 1910, 1916, 1960; Couper 1960; Godley 1960; Cranwell 1963, 1964; Schmithüsen 1964; Cerceau-Larrival 1968; Raven and Axelrod 1972, 1974, 1975) and animals (Simpson 1940, 1966; Darlington 1957, 1960, 1965; Brundin 1965, 1966; Rapoport 1968, 1971; Reig 1968; Hoffstetter 1970; Keast 1972) of these disjunct areas have been discussed in many papers.

Whereas several arthropod taxa found in Patagonian forests show clearcut affinities with those of New Zealand, Tasmania or Australia (Kuschel 1960; Brundin 1965, 1966; Darlington 1965), this is not the case for the birds. Fifteen species of the passerine suborder Tyranni (5 species of Furnariidae, 5 Rhinocryptidae, 4 Tyrannidae, and 1 Phytotomidae) comprise 33 percent of the 46 land bird species. Mayr (1964) classified the Furnariidae and Rhinocryptidae

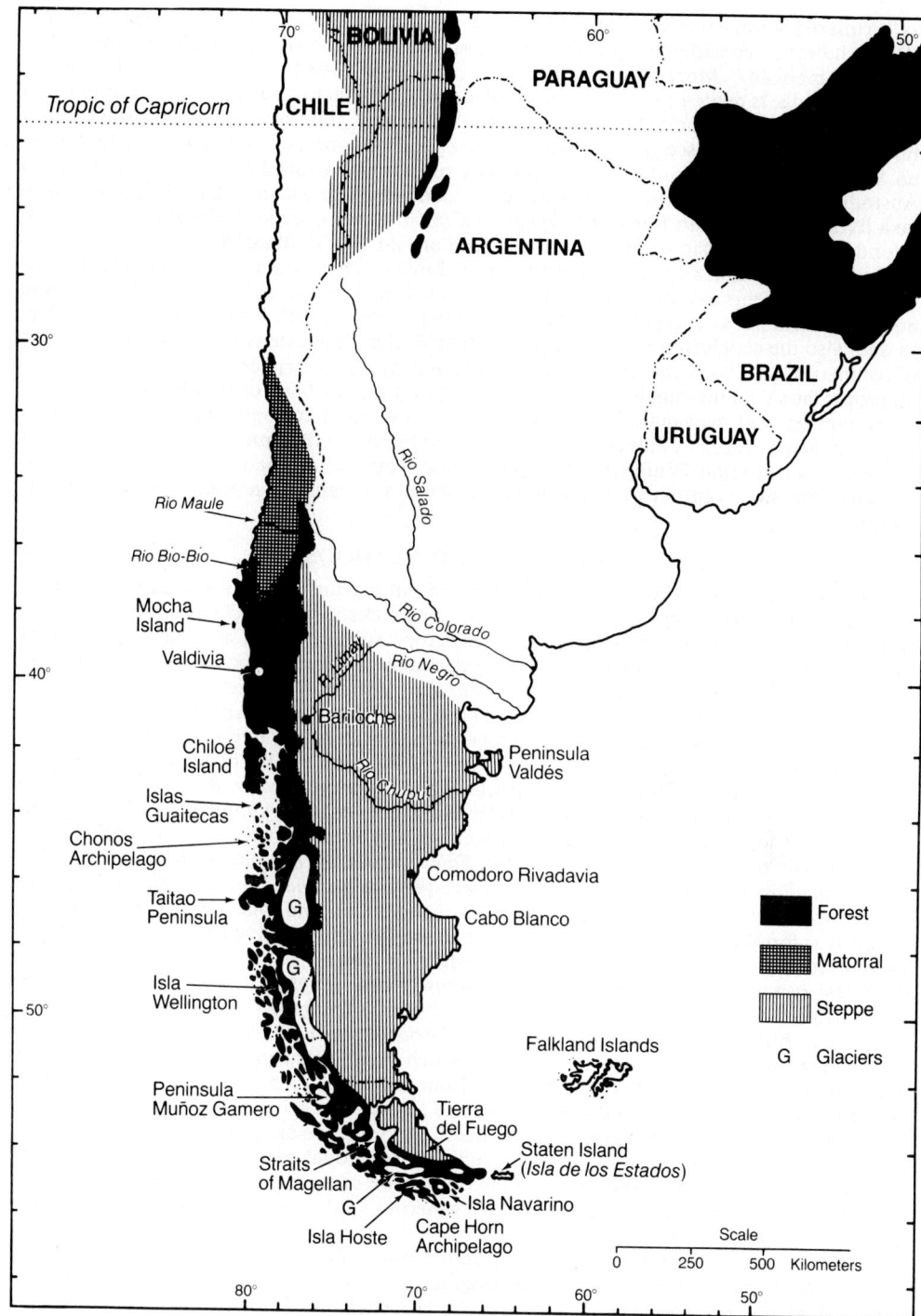

FIG. 1. Schematic map showing location of Patagonian forests in southwestern South America, of Andean montane forests in northwestern Argentina, and of lowland forests in southern Brazil. Puna steppes of Bolivia, Argentina, and Chile are disjunct from Patagonian steppes of Argentina. Other nonforest vegetation types (only matorral identified on map) occur between the three forested areas.

as "Primarily South American," and the Tyrannidae as "Expanding South American." Cracraft (1973), however, considered the suborder Tyranni to be "Southern Hemisphere" rather than "South American." More recently, Feduccia and Olson (1982) suggested that the family Rhinocryptidae is related to the Menurae of Australia. Whether these taxa are primarily South American in origin or have broader southern hemisphere affinities are clearly problems for further research. The discovery of a late Eocene or early Oligocene marsupial (Polydolopidae) on Seymour Island, Antarctic Peninsula, does suggest the possibility of exchange between Australia and South America via Antarctica (Woodburne and Zinsmeister 1983). Other bird taxa living in Patagonian forests (*Accipiter, Columba, Strix, Colaptes, Picoides, Troglodytes,* Icteridae, and Emberizidae) have Pan-American or Old World affinities.

The relative complexity of the affinities of the birds of Patagonian forests suggests that the development of this avifauna was more complicated than just an influx of immigrants from northerly latitudes, as suggested by Darlington (1957, 1960, 1965). This view needs revision, as does also the conclusion of Hershkovitz (1969:58) that "the extensive beech-conifer forest of southern Chile, bordered by sea on one side and by scrub steppes on the other side, has no proper fauna" of mammals (see data in Baker 1967). On the basis of data from amphibians, birds, reptiles, and mammals, Müller (1973) considered the Patagonian forest region to be a dispersal center. Birds (Vuilleumier 1967a), amphibians (Vuilleumier 1968; Formas 1979), and perhaps mammals (Vuilleumier, unpubl. data) seem to have had rather similar patterns of biogeographic evolution in Patagonian forests, but apparently reptiles did not (Formas 1979).

MATERIALS AND METHODS

Only the landbirds are considered in this paper. All marine birds, as well as the following freshwater birds occurring in or around Patagonian forests, were excluded: grebes (Podicipedidae), herons (Ardeidae), cormorants (Phalacrocoracidae), swans and ducks (Anatidae), rails and coots (Rallidae), gulls (Laridae), oystercatchers (Haematopodidae), snipes (Scolopacidae), and plovers (Charadriidae, except *Vanellus* [*chilensis*] *chilensis*).

There are 89 genera and 138 species of land birds in Patagonia (Appendix I), as I define this region. Of these, 40 genera and 46 species are considered to live in Patagonian forests (marked F and B in Appendix I; Appendix II). The decision to classify a given species as a forest bird was based on personal field experience and literature reports. Besides the 46 species accepted as forest birds, 11 others occur locally along (or even in) *Nothofagus* forests, especially clearings, but were excluded from the list because they breed in open country either in the Andes above timberline or east of the Andes: *Falco sparverius, Zenaida auriculata, Tyto alba, Upucerthia dumetaria, Cinclodes oustaleti, Leptasthenura aegithaloides, Cistothorus platensis, Mimus thenca, Mimus patagonicus, Molothrus bonariensis,* and *Diuca diuca.* In contrast, I included as forest birds several species that are found in steppes and open woodlands but that also commonly occur and breed along forest edges, in forest openings, or in parkland within the forest region. These are *Theristicus caudatus, Vultur gryphus, Coragyps atratus, Buteo* [*albicaudatus*] *polyosoma, Polyborus chimango, Vanellus* [*chilensis*] *chilensis, Cinclodes* [*fuscus*] *fuscus, Notiochelidon cyanoleuca,* and *Zonotrichia capensis.* Finally, I included *Ceryle* [*alcyon*] *torquata* and *Cinclodes patagonicus,* which inhabit stream and lake shores, but are widespread within the forest region. My definition of a forest bird is, thus, broader than that of Moreau (1966:81–82), who did not include as forest birds species that occur in clearings. Some ornithologists may disagree with my decision to include as forest birds several species that occur essentially in open habitats. Excluding these would reduce the number of species considered, but would probably increase the proportion of ecologically restricted or endemic taxa. If I have erred, I have done so on the "generous" side and, thus, have not biased the list "in favor" of forest taxa.

In Appendix II I have summarized the information obtained from my examination of museum skins and from the literature, on geographic variation, species limits, and taxonomic relationships for each species of forest bird; the nomenclature followed in this paper is also indicated. I used the superspecies concept and its formalization by means of brackets (Amadon 1966). I adopted Mayr's (1959) concept of population structure of species in the study of speciation.

Information on geographical and ecological distribution comes from my personal field work in and around *Nothofagus* forests of Argentina and Chile between February and April 1965 (Vuilleumier 1967b, 1972, and unpubl. data), supplemented by the literature (Crawshay 1907;

Pässler 1922; Peters 1923b; Wetmore 1926a, 1926b; Reynolds 1932, 1934, 1935; Radboone 1935; Morrison 1940; Trimble 1943; Behn 1947; Olrog 1948, 1950; Ripley 1950; Krieg 1951; Goodall et al. 1951, 1957, 1964; Philippi et al. 1954; Cawkell and Hamilton 1961; Horváth and Topál 1963; Bernath 1965; Johnson 1965, 1967, 1972; Johansen 1966; Short 1969a, 1969b, 1970a, 1970b, 1971, 1972, 1982; Humphrey et al. 1970; Keith 1970; Texera 1972; Forshaw and Cooper 1973; Araya et al. 1974; Schlatter 1976; Parmelee and MacDonald 1975; Jehl and Rumboll 1976; Venegas 1976, 1981, 1982a, 1982b; Sielfeld 1977; Contreras et al. 1980; Yáñez et al. 1982).

For the measurement of faunal resemblance I used Simpson's (1960:301) index no. 2:

$$\frac{\text{number of taxa common to both faunas}}{\text{total taxa in smaller fauna}} \times 100$$

since in most cases one fauna is richer than the other.

NOTHOFAGUS FORESTS OF PATAGONIA

Although these forests grow entirely at temperate latitudes, they are tall and luxuriant, and resemble tropical or subtropical rainforests in their evergreenness, structure (trees, epiphytes), and physiognomy. Reynolds (1935:66) described the vegetation found near the southernmost localities where forest grows, in the Cape Horn area, at ca. 56°S: "In sheltered places, such as the proximity of a lake at the northern side of Freycinet, precipitous ravines are submerged by evergreen forest of tropical aspect. So thick is the canopy that a chaos of decay exists in the semi-darkness prevailing beneath, and a small maidenhair fern combines with a deep layer of spongy moss to cover everything."

The resemblance to tropical rainforests does not extend to the taxonomic composition of the flora, however. As a result, some botanists have called the southern Patagonian forests "subantarctic" (Young 1972). Patagonian forests are floristically related to forests in New Zealand or Tasmania (Godley 1960), but some floral elements occur also in other South American forests (southern Brazil, Smith 1962; Andes, Hueck 1966).

Nothofagus forests stretch along the Andes from about 35°S south to Tierra del Fuego (Isla Grande) and the islands and islets of the Cape Horn Archipelago at about 56°S (Fig. 1). A few forest fragments are found north of 35°S; the northernmost, the Fray Jorge and Talinay woodlands at about 30°S, contain genera of plants with southern affinities (Muñoz and Pisano 1947; Schmithüsen 1956) but lack *Nothofagus*. Between 31° and 35°S, the patches of forest and woodland are still composed chiefly of trees belonging to genera other than *Nothofagus*. Only from about 36°S southward is *Nothofagus* the common or dominant taxon.

The northernmost species appears to be *Nothofagus obliqua*, found at middle elevations along the Chilean Andes in the Provinces of Santiago and Aconcagua. It associates with *N. procera* farther south, where both deciduous species comprise lowland forests which occur mostly south to the provinces of Valdivia, Osorno, and Llanquihue (Chile) along the western slope of the Andes, and in Neuquén Province (Argentina) along the eastern slope. Moving up into the Andes, these two species are progressively replaced altitudinally by *N. antarctica* and *N. pumilio* (both deciduous), which become more and more shrubby toward timberline. Timberline is found as high as 1800 to 2000 m in the northern *Nothofagus* forests.

In the Valdivian region, between 39° to 40°S and 44° to 45°S, forests are generally luxuriant and resemble tropical rainforests. On the whole, the forests are tallest and most humid in Chilean territory because of the influence of the moisture-carrying westerly winds. The forests decrease in density and floral diversity from west to east along a gradient of decreasing rainfall that corresponds to the lessened maritime influence of winds from the Pacific coast across the Andes and down to the eastern Andean foothills. The forests also decrease in diversity along a north to south gradient of increasing yearly temperature fluctuations.

Valdivian forests are dominated by *Nothofagus dombeyi,* an evergreen species found with the conifers *Saxegothea conspicua* (Podocarpaceae), *Fitzroya cupressoides* (Cupressaceae), *Pilgerodendron uviferum* (Cupressaceae), and *Podocarpus nubigena* (Podocarpaceae). Another conifer, *Araucaria araucana* (Araucariaceae) grows between 37° and 40°S on Andean slopes from about 1000 to 1600 meters. Together with *N. dombeyi,* it forms a montane forest belt. *Nothofagus pumilio* and *N. antarctica* are found up to 1800 m or even higher. A schematic representation of an elevational vegetation sequence on the west slope of the Andes at 41°S can be found in Heusser (1981: fig. 2).

South of 41°S the altitude of timberline decreases (Godley 1960). Between 47° to 48°S and

TABLE 1
Faunal Resemblance Among Avifaunas[1]

	Nothofagus forest	Patagonian nonforest	Andean forest	Puna	Californian coastal forest
Nothofagus forest	—	53	30	45	30
Patagonian nonforest	43	—	11	81	16
Andean forest	15	4	—	12	18
Puna	33	50	5	—	14
Californian coastal forest	11	9	7	7	—

[1] Values given are index 2 of Simpson (1960:301), in percent. Values to right of diagonal = generic level resemblance. Values to left = specific and superspecific level resemblance.

51° to 52°S two extensive ice caps (Fig. 1) reduce the area available for forests, which are only found in a thin band along the Andes. Furthermore, the southwestern fringe of the area, including many of the outlying islands, is covered in part with an open vegetation called Magellanic moorland (Godley 1960; Moore 1979). Forests occur again in large and more continuous tracts south of about 52°S.

The southernmost forests, often called Magellanic or Subantarctic (Young 1972), do not have the species diversity and height of Valdivian forests, and consist mostly of deciduous *Nothofagus pumilio* and *N. antarctica,* and evergreen *N. betuloides.* The boundary between the rich evergreen Valdivian rainforest and the poorer, largely deciduous Subantarctic forest lies between 46° and 48°S. However, the southernmost forests do not differ much in aspect from the Valdivian rainforests (Darwin 1906:199–200; Reynolds 1935).

In southern South America the Andes do not form an uninterrupted high mountain chain as they do farther north. Consequently, the forests do not occur in two separate tracts along the eastern and western slopes, respectively. Rather, the southern Andes can be visualized as a long series of mountain peaks and hills. Treeless alpine vegetation occurs on their tops whereas extensive forests clothe their slopes with little interruption from east to west across numerous relatively low passes.

More information on the vegetation of Patagonian forests can be found in Dusén (1903–1906), Reiche (1907), Skottsberg (1910, 1916), Hauman-Merck (1913), Ljungner (1939), Kalela (1941), Cabrera (1953), Schmithüsen (1956, 1960), Auer (1958), Thomasson (1959), Godley (1960), Dimitri (1962), Ward (1965), Hueck (1966), Eskuche (1968), Pisano (1974, 1977, 1981), Fernández (1976), Veblen et al. (1977, 1980, 1981), Tomaselli (1981), and Veblen (1982). Ecological descriptions were published by Kuschel (1960), di Castri (1968), and Formas (1979).

AFFINITIES OF THE PATAGONIAN FOREST AVIFAUNA

Faunal resemblance.—The Patagonian forest avifauna is closest geographically to the fauna living in the ecologically dissimilar Patagonian steppes. It is, thus, of interest to investigate faunal resemblances between these two avifaunas. Because the *Nothofagus* forests are distributed along the Andes, it is reasonable to inquire whether their avifauna resembles the fauna of other Andean forests, the montane forests much farther north in northwestern Argentina (Fig. 1). Similarly, because the Patagonian steppe has many taxa in common with the high Andean puna farther north, one wonders about resemblances between Patagonian forest and puna faunas. Finally, in view of the well-known resemblances between Pacific North American and Pacific South American biota (Constance 1963) one wonders whether the avifaunas of Patagonian forests and equivalent forests in Pacific North America resemble each other (for mammals, see Baker 1967).

Faunal resemblances (Table 1) were determined on the basis of species lists in Appendices I, III, IV, and V. Both specific (and allospecific) and generic levels of faunal resemblances were considered.

At the generic level, the *Nothofagus* forest avifauna is about equally similar to the neighboring (Fig. 1) Patagonian nonforest fauna (53%) and to the very distant puna fauna (45%). This is not unexpected in view of the high generic level resemblance (81%) between Patagonian nonforest and puna avifaunas. The *Nothofagus* forest fauna shares 30 percent of its genera with the fauna of Andean forests (distant) as it does also with the fauna of Californian forests

TABLE 2
NUMBERS OF GENERA AND SPECIES IN VARIOUS AVIFAUNAS

Avifauna	No. of		Species : genus ratio
	Genera	Species	
Patagonian forest	40	46	1.15
Patagonian nonforest	70	110	1.57
Andean forest	65	79	1.22
Puna	52	98	1.88
California forest	49	57	1.16
		Mean =	1.40
		Mean without puna =	1.28

(very distant). This is not surprising in view of the fact that Andean and Californian forests share 18 percent of their genera.

At the specific level the indices of similarity are, of course, lower, but the same patterns obtain, at least in part. The resemblance between Patagonian nonforest and puna avifaunas (50%) is reflected in the 43 percent resemblance between Patagonian forest and Patagonian nonforest avifaunas, as well as the 33 percent resemblance between Patagonian forest and puna avifaunas. The Patagonian forest fauna is much less similar to either the Andean forest fauna (15%) or the Californian forest fauna (11%).

Conclusions.—The Patagonian forest avifauna is mixed, sharing elements with Patagonian nonforest and puna avifaunas, and with Andean and North American forest avifaunas. Interestingly, the Patagonian forest avifauna is more closely related to nonforest than to other forest avifaunas.

Among the five faunas discussed, the Patagonian forest avifauna is the poorest in numbers of genera and species (Table 2). Its generic and specific diversities are most comparable to those of the Californian forest fauna. The Patagonian nonforest fauna is the richest, but also occupies the largest area, about three times that of Patagonian forests. The difference in faunal diversity is correlated with area (110 vs. 46 species, or a 2.4:1 ratio; if the 20 species common to Patagonian nonforest and forest faunas are excluded, the ratio is 90 vs. 26 species, or about 3.5:1). The discrepancy between the Andean forest fauna and the Patagonian forest fauna (79 vs. 46 species, or a 1.7:1 ratio) is also of interest because the area of Andean forests is less than half that of Patagonian forests (Fig. 1). Part of the richness in taxa of Argentine Andean forests can be attributed to the presence of a large reservoir of taxa in the rich tropical montane forests of the Peruvian and Bolivian Andes. Finally, the puna avifauna of the southern altiplano, occupying an area about half that of Patagonian nonforest, is only slightly poorer in number of species than the Patagonian steppe avifauna (98 vs. 110) but has proportionately more genera. The 1.88 species:genus ratio in the puna is the highest of all five faunas. This strongly suggests that speciation within the puna has played a role in species enrichment (Vuilleumier 1969, 1980; Vuilleumier and Simberloff 1980), whereas that phenomenon has been either negligible or of lesser importance in the other faunas. In conclusion it appears that both Patagonian faunas are depauperate. They are located at the temperate southern tip of South America and are rather distant from the rich potential sources of taxa at more tropical latitudes. Species numbers show clearly that Patagonian forests constitute a habitat island.

ECOLOGICAL GEOGRAPHY OF THE PATAGONIAN FOREST AVIFAUNA

Habitat preferences.—The 46 species exhibit varying degrees of ecological dependence on forest vegetation. Some species are rather strictly limited to dense forest and do not venture into more open types of vegetation. Other species, however, occur in forest as well as in other vegetation types, such as matorral (scrublands). Furthermore, some species appear to be restricted to forest during the breeding season, migrating north out of the Patagonian forest region, and into more open vegetation, during the southern hemisphere winter (Appendices I, II).

Most species (about 75%) occur in several different kinds of forest during the breeding season, and as many as 30 (65%) even enter the shrublands around the periphery of the forest region. Fewer than half also inhabit the steppe or semi-desert scrub (19 or 41%). The greatest

TABLE 3

ECOLOGICAL DISTRIBUTION OF LAND BIRDS OF PATAGONIAN FORESTS

Species or allospecies	Rain forest	Mesophytic forest	Montane forest	Parkland	Openings within forest	Forest/ steppe ecotone	Shrublands (matorral)	Steppe or semidesert scrub	Alpine scrub	Streams and lake shores	Moorland
Theristicus caudatus	—	—	—	—	X	X	X	X	—	X	X
Cathartes aura	?	X	X	X	X	X	X	X	—	—	—
Coragyps atratus	—	X	—	X	X	X	X	X	—	—	—
Vultur gryphus	—	X	X	X	X	X	X	X	X	—	—
Accipiter [bicolor] bicolor	X	X	X	X	X	X	—	—	—	—	—
Buteo [albicaudatus] polyosoma	—	X	X	X	X	X	X	X	X	—	—
B. [jamaicensis] ventralis	?	X	—	X	X	X	—	—	—	—	—
Geranoaetus melanoleucus	X	X	X	X	X	X	X	X	X	—	—
Polyborus chimango	—	—	—	X	X	X	X	X	—	X	X
P. [megalopterus] albogularis	?	X	X	X	X	X	X	?	?	—	—
P. plancus	—	—	—	X	X	X	X	X	—	—	—
Vanellus [chilensis] chilensis	—	—	—	X	X	—	—	X	—	X	X
Columba [fasciata] araucana	X	X	—	X	X	?	—	—	—	—	—
Enicognathus ferrugineus	X	X	X	X	X	X	X	—	—	—	—
E. leptorhynchus	—	X	—	X	X	X	—	—	—	—	—
Bubo [bubo] virginianus	?	X	X	X	X	X	X	—	—	—	—
Glaucidium [brasilianum] nanum	?	X	X	X	X	X	X	—	—	—	—
Strix rufipes	X	X	—	X	X	X	—	—	—	—	—
Sephanoides sephaniodes	X	X	X	X	X	X	X	—	—	—	—
Ceryle [alcyon] torquata	—	—	—	—	X	—	—	—	—	X	—
Picoides [mixtus] lignarius	X	X	X	X	X	X	X	—	—	—	—
Colaptes pitius	X	X	X	X	X	X	X	—	—	—	—
Campephilus magellanicus	X	X	X	X	—	—	—	—	—	X	—
Cinclodes patagonicus	—	—	—	—	X	X	—	—	—	X	—
C. [fuscus] fuscus	—	—	—	—	X	X	X	X	X	X	X
Sylviorthorhynchus desmursii	?	X	—	X	X	—	X	—	—	—	—
Aphrastura [spinicauda] spinicauda	X	X	X	X	X	X	X	—	—	—	—
Pygarrhichas albogularis	X	X	X	X	—	—	—	—	—	—	—
Pteroptochos [tarnii] tarnii	X	X	X	X	X	—	—	—	—	—	—
P. [tarnii] castaneus	X	X	X	X	—	—	—	—	—	—	—
Scelorchilus [rubecula] rubecula	X	X	X	X	—	—	—	—	—	—	—
Eugralla paradoxa	X	X	—	X	—	—	—	—	—	—	—
Scytalopus magellanicus	X	X	X	X	X	X	X	—	—	—	—
Elaenia albiceps	X	X	X	X	X	X	X	—	—	—	—
Anairetes parulus	—	—	—	X	X	X	X	X	—	—	—
Ochthoeca parvirostris	X	X	X	X	—	—	—	—	—	—	—
Xolmis pyrope	X	X	X	X	X	X	X	—	—	—	—
Phytotoma rara	—	—	—	X	X	X	X	X	—	—	—
Notiochelidon cyanoleuca	—	—	—	X	X	X	X	X	X	X	X
Tachycineta [leucorrhoa] leucopyga	X	X	X	X	X	X	—	—	—	X	—

TABLE 3
CONTINUED

Species or allospecies	Rain forest	Meso-phytic forest	Mon-tane forest	Park-land	Open-ings within forest	Forest/steppe ecotone	Shrub-lands (matorral)	Steppe or semi-desert scrub	Alpine scrub	Streams and lake shores	Moor-land
Troglodytes aedon	X	X	X	X	X	X	X	X	—	—	—
Turdus falcklandii	X	X	X	X	X	X	X	?	—	—	—
Curaeus curaeus	X	X	—	X	X	X	X	?	—	—	—
Phrygilus [gayi] pata-gonicus	X	X	X	X	X	X	?	—	—	—	—
Zonotrichia capensis	—	—	—	X	X	X	X	X	?	—	—
Carduelis [barbata] barbata	X	X	X	X	X	X	X	—	—	—	—
Totals (N = 46 species)	25–31	35	27	42	39	35–36	29–30	16–19	5–7	8	5

diversity is found in parklands and openings within the forest (39–42 species, or 85–91%; Table 3). In pure forest, the greatest diversity seems to be found in mesophytic forests (35 species or 76%, vs. 31 or 67% in rainforest, and 27 or 59% in montane forest). These results confirm what I described previously (Vuilleumier 1972), that dense rainforest has a somewhat less diverse avifauna than mesophytic forest (see Pearson and Pearson 1982, for a comparable analysis in mammals). As a whole, Patagonian forest birds appear to show little specialization to different forest types; they are found in most kinds of forest.

Elevational distribution. — All 46 species in the northern part of the Patagonian forests occur up to about 1000 meters; nearly 80 percent are found up to 1500, but only about 40 percent up to 2000 meters (Table 4). Most species (60%) thus occupy a broad elevational range, and only a few (40%) appear to fail to reach subalpine forests. The data, however, are still limited. Further field work may well show that more species occur in the 1500 to 2000 meter range. Patagonian forest species are clearly tolerant of a wide range of conditions along elevational gradients.

Latitudinal distribution. — The center of diversity of the 46 species during the breeding season (44–45 species, or 96–98%) occurs between latitudes 35°S and 44°S, an area that corresponds in part to the Valdivian rainforest (Table 5). Farther north, the number of species decreases rather abruptly but is still high (80% to latitude 33°S). Farther south, species numbers decrease progressively, south of 44°S: 87 percent at 44–47°S, 85 percent at 47–50°S, 78 percent at 50–53°S, and 74 percent at 53–56°S. It is interesting that the substantial reduction in forest area due to ice-caps (Fig. 1) and moorlands between about 47°S and 52°S seems to result in only a small decrease in bird species diversity. Furthermore, the poorer Magellanic or Subantarctic forests of higher latitudes still harbor 74 percent or more of the total species diversity. Birds appear to be relatively evenly distributed along the north–south eco-climatic gradient.

Conclusions. — Patagonian forest birds are rather widely distributed over forest types, as well as elevationally and latitudinally (Tables 3, 4, 5). This sort of broad niche-distribution is the one usually found in island birds (Blondel 1979). Cody (1970) noted that Chilean birds as a whole, not just Patagonian forest birds, have the following insular characteristics, (1) high within-habitat diversity (i.e., "individuals are more equitably distributed among species"), (2) extremely low species turnover between habitats in the same latitudinal zone, and (3) limited occupation of areas by bird species, which are replaced by others only with major shifts in vegetation or latitude. Kikkawa (1974) compared feeding and foraging niches of birds in *Nothofagus* forests of Patagonia, Tasmania, New Zealand, Australia, and New Guinea. He described some similarities, as well as differences, in the patterns of resource exploitation, but did not discuss the problem of niche width associated with the various degrees of insularity shown in these faunas. The problem of niche width in birds of Patagonian forests (a habitat island) is worth further investigation.

SPECIATION PHENOMENA IN PATAGONIAN FOREST BIRDS

The species: genus ratio is low (1.15) in Patagonian forest birds (Table 2). Of 40 genera, only 5 (*Buteo, Polyborus, Enicognathus, Cinclodes,* and *Pteroptochos*) or 12.5 percent have more than one species in the *Nothofagus* forest region. Speciation, therefore, does not seem

TABLE 4

ELEVATIONAL DISTRIBUTIONS OF BIRDS IN NORTHERN PATAGONIAN FORESTS BETWEEN 37°S AND 43°S[1]

Species or allospecies	Elevation (meters)			
	0–500	500–1000	1000–1500	1500–2000
Theristicus caudatus	X	X	X	—
Cathartes aura	X	X	—	—
Coragyps atratus	X	X	X	—
Vultur gryphus	—	X	X	X
Accipiter [bicolor] bicolor	X	X	—	—
Buteo [albicaudatus] polyosoma	X	X	X	X
B. [jamaicensis] ventralis	X	X	—	—
Geranoaetus melanoleucus	X	X	X	X
Polyborus chimango	X	X	X	—
P. [megalopterus] albogularis	X	X	X	—
P. plancus	X	X	X	—
Vanellus [chilensis] chilensis	X	X	—	—
Columba [fasciata] araucana	X	X	X	X
Enicognathus ferrugineus	X	X	—	—
E. leptorhynchus	X	X	X	X
Bubo [bubo] virginianus	X	X	X	—
Glaucidium [brasilianum] nanum	X	X	—	—
Strix rufipes	X	X	X	X
Sephanoides sephaniodes	X	X	—	—
Ceryle [alcyon] torquata	X	X	X	—
Picoides [mixtus] lignarius	X	X	X	—
Colaptes pitius	X	X	X	X
Campephilus magellanicus	X	X	X	—
Cinclodes patagonicus	X	X	X	X
C. [fuscus] fuscus	X	X	—	—
Sylviorthorhynchus desmursii	X	X	X	X
Aphrastura [spinicauda] spinicauda	X	X	X	X
Pygarrhichas albogularis	X	X	X	—
Pteroptochos [tarnii] tarnii	X	X	X	—
P. [tarnii] castaneus	X	X	X	—
Scelorchilus [rubecula] rubecula	X	X	—	—
Eugralla paradoxa	X	X	X	X
Scytalopus magellanicus	X	X	X	X
Elaenia albiceps	X	X	—	—
Anairetes parulus	X	X	X	—
Ochthoeca parvirostris	X	X	X	—
Xolmis pyrope	X	X	X	—
Phytotoma rara	X	X	X	X
Notiochelidon cyanoleuca	X	X	X	X
Tachycineta [leucorrhoa] leucopyga	X	X	X	X
Troglodytes aedon	X	X	X	X
Turdus falcklandii	X	X	X	—
Curaeus curaeus	X	X	X	X
Phrygilus [gayi] patagonicus	X	X	X	—
Zonotrichia capensis	X	X	X	X
Carduelis [barbata] barbata	X	X	X	—
Totals (N = 46 species)	45	46	36	18

[1] Breeding range only.

to be—or to have been—very active in Patagonian forests. It is necessary, nevertheless, to analyze speciation phenomena in detail to determine the possible reasons for the lack of active species formation. The analysis will be done in terms of isolate formation, superspecies, and old speciation patterns (Table 6).

POPULATION STRUCTURE AND ISOLATE FORMATION

As Mayr (1959) pointed out, population structure is of great interest from the speciation viewpoint, when the species is broken down into morphologically (and genetically) differen-

TABLE 5

LATITUDINAL DISTRIBUTIONS OF THE BREEDING RANGES OF PATAGONIAN FOREST BIRDS

Species or allospecies	Degrees S							
	33–35	35–38	38–41	41–44	44–47	47–50	50–53	53–56
Theristicus caudatus	x	x	x	x	x	x	x	x
Cathartes aura	x	x	x	x	x	x	x	x
Coragyps atratus	x	x	x	x	—	—	—	—
Vultur gryphus	x	x	x	x	x	x	x	x
Accipiter [bicolor] bicolor	x	x	x	x	x	x	x	x
Buteo [albicaudatus] polyosoma	x	x	x	x	x	x	x	x
B. [jamaicensis] ventralis	—	x	x	x	x	x	x	x
Geranoaetus melanoleucus	x	x	x	x	x	x	x	x
Polyborus chimango	x	x	x	x	x	x	x	x
P. [megalopterus] albogularis	—	x	x	x	x	x	x	x
P. plancus	x	x	x	x	x	x	x	x
Vanellus [chilensis] chilensis	x	x	x	x	x	x	x	x
Columba [fasciata] araucana	—	x	x	x	x	—	—	—
Enicognathus ferrugineus	—	x	x	x	x	x	x	x
E. leptorhynchus	x	x	x	x	—	—	—	—
Bubo [bubo] virginianus	x	x	x	x	x	x	x	x
Glaucidium [brasilianum] nanum	x	x	x	x	x	x	x	x
Strix rufipes	x	x	x	x	x	x	x	x
Sephanoides sephaniodes	x	x	x	x	x	x	x	x
Ceryle [alcyon] torquata	x	x	x	x	x	x	x	x
Picoides [mixtus] lignarius	x	x	x	x	x	x	x	—
Colaptes pitius	x	x	x	x	x	x	x	—
Campephilus magellanicus	—	x	x	x	x	x	x	x
Cinclodes patagonicus	x	x	x	x	x	x	x	x
C. [fuscus] fuscus	x	x	x	x	x	x	x	x
Sylviorthorhynchus desmursii	x	x	x	x	x	x	—	—
Aphrastura [spinicauda] spinicauda	x	x	x	x	x	x	x	x
Pygarrhichas albogularis	x	x	x	x	x	x	x	x
Pteroptochos [tarnii] tarnii	x	x	—	—	—	—	—	—
P. [tarnii] castaneus	—	—	x	x	x	x	—	—
Scelorchilus [rubecula] rubecula	x	x	x	x	x	x	—	—
Eugralla paradoxa	—	x	x	x	—	—	—	—
Scytalopus magellanicus	x	x	x	x	x	x	x	x
Elaenia albiceps	x	x	x	x	x	x	x	x
Anairetes parulus	x	x	x	x	x	x	—	—
Ochthoeca parvirostris	—	—	x	x	x	x	x	x
Xolmis pyrope	x	x	x	x	x	x	x	x
Phytotoma rara	x	x	x	x	x	x	—	—
Notiochelidon cyanoleuca	x	x	x	x	x	x	x	x
Tachycineta [leucorrhoa] leucopyga	x	x	x	x	x	x	x	x
Troglodytes aedon	x	x	x	x	x	x	x	x
Turdus falcklandii	x	x	x	x	x	x	x	x
Curaeus curaeus	x	x	x	x	x	x	x	x
Phrygilus [gayi] patagonicus	—	x	x	x	x	x	x	x
Zonotrichia capensis	x	x	x	x	x	x	x	x
Carduelis [barbata] barbata	x	x	x	x	x	x	x	x
Totals (N = 46 species)	37	44	45	45	40	39	36	34

tiated and geographically (and ecologically) isolated populations. Each isolate constitutes a potential or incipient species. I analyze this phenomenon separately for the continental and for the insular populations of forest species (Table 6).

Mainland populations.—Thirty-six of the 46 forest species (78%) show no geographical variation in the continental part of their ranges (however, some of these species also range beyond the forest belt and vary geographically outside of it). Of the 10 species (22%) that show geographical variation, nine (20%) vary gradually or perhaps clinally. Trends of geographical variation include a north to south increase in size (as measured by wing-length or bill-length) and a parallel north to south increase in the intensity or "saturation" of pigmen-

TABLE 6
Population Structure and Speciation Phenomena of Land Birds of Patagonian Forests[1]

Species or allospecies	Within mainland populations		Differentiation of island from mainland populations		Belongs to a superspecies
	Slight to moderate	Marked	Slight to moderate	Marked	
Theristicus caudatus	—	—	—	—	—
Cathartes aura	(falklandica, jota)	—	—	—	—
Coragyps atratus	—	—	—	—	—
Vultur gryphus	—	—	—	—	—
Accipiter [bicolor] bicolor	—	—	—	—	x
Buteo [albicaudatus] polyosoma	—	—	—	x	x
B.[jamaicensis] ventralis	—	—	—	—	x
Geranoaetus melanoleucus	—	—	—	—	—
Polyborus chimango	(temucoensis, chimango)	—	x	—	(ex-superspecies with P. chimachima)
P. [megalopterus] albogularis	—	—	—	—	x
P. plancus	—	—	—	—	—
Vanellus [chilensis] chilensis	—	—	—	—	x
Columba [fasciata] araucana	—	—	—	—	x
Enicognathus ferrugineus	(ferrugineus, minor)	—	—	—	—
E. leptorhynchus	—	—	—	—	—
Bubo [bubo] virginianus	—	—	—	—	x
Glaucidium [brasilianum] nanum	(nanum, vafrum)	—	—	—	x
Strix rufipes	—	—	x	—	—
Sephanoides sephaniodes	—	—	—	—	—
Ceryle [alcyon] torquata	—	—	—	—	x
Picoides [mixtus] lignarius	—	—	—	—	x
Colaptes pitius	(cachinnans, pitius)	—	x	—	(ex-superspecies with C. rupicola, C. campestris)
Campephilus magellanicus	—	—	—	—	(ex-superspecies with C. leucopogon)
Cinclodes patagonicus	(chilensis, patagonicus)	—	—	—	—
C. [fuscus] fuscus	—	—	—	—	x
Sylviorthorhynchus desmursii	—	—	—	—	—
Aphrastura [spinicauda] spinicauda	—	—	x	x	x
Pygarrhichas albogularis	—	—	—	—	—
Pteroptochos [tarnii] tarnii	—	—	—	—	(ex-superspecies with P. megapodius)
P. [tarnii] castaneus	—	—	—	—	
Scelorchilus [rubecula] rubecula	—	—	x	—	x
Eugralla paradoxa	—	—	—	—	—
Scytalopus magellanicus	—	(magellanicus, fuscus)	—	—	—
Elaenia albiceps	—	—	—	—	(ex-superspecies with E. parvirostris)
Anairetes parulus	(lippus, patagonicus)	—	—	—	(x?)
Ochthoeca parvirostris	—	—	—	—	—

TABLE 6

Continued

Species or allospecies	Within mainland populations		Differentiation of island from mainland populations		Belongs to a superspecies
	Slight to moderate	Marked	Slight to moderate	Marked	
Xolmis pyrope	—	—	x	—	—
Phytotoma rara	—	—	—	—	(ex-superspecies with *P. rutila, P. raimondii*)
Notiochelidon cyanoleuca	—	—	—	—	—
Tachycineta [leucorrhoa] leucopyga	—	—	—	—	x
Troglodytes aedon	—	—	—	x	—
Turdus falcklandii	(*magellanicus, pembertoni*)	—	x	—	—
Curaeus curaeus	—	—	x	—	—
Phrygilus [gayi] patagonicus	—	—	—	—	x
Zonotrichia capensis	(*australis, chilensis*)	—	—	—	—
Carduelis [barbata] barbata	—	—	—	—	x
Totals (N = 46 species)	9	1	8	3	18
(%)	20%	2%	17%	7%	39%

¹ For detailed treatment see Appendix II.

tation (or increase in barrings giving a darker tone to the general coloration). In all of the nine latter species geographical variation has been recognized by taxonomists who have described subspecies (Table 6). That very few species vary geographically was confirmed by Texera (1972), who stated that only three of a total of 122 land bird species in Magallanes Province (Chile) had subspecies; unfortunately, he did not list them.

In only one forest species, *Scytalopus magellanicus,* are the ranges of the continental populations disjunct. This species has two well-marked subspecies, one (*magellanicus*) south, and the other (*fuscus*) north of the Rio Bío-Bío. It is at present unclear whether the Bío-Bío river valley does indeed separate these forms or whether they actually overlap. Differences in morphology are said to parallel differences in vocalizations (Goodall et al. 1957; Johnson 1967). This species merits detailed field investigation.

Insular populations.—Forty of the 46 species have populations living on one or more of the islands, such as the distant Juán Fernández and Falkland Islands, and the closer Mocha, Chiloé, Guaitecas, Navarino, Hoste, and Cape Horn Islands, lying off southern South America (Fig. 1). Eleven of the 40 species (27.5%) have morphologically differentiated insular populations (Table 6). Eight species have slightly to moderately differentiated isolates (subspecies have been described for seven of these species), and three species have markedly differentiated subspecies (Table 7). In three additional species the insular isolate has clearly reached species or allospecies status (Table 7). Two of the three well-marked insular subspecies are from distant islands (Juán Fernández, Falklands), and all three cases of completed speciation involve the distant Juán Fernández archipelago.

One of the three species-level insular isolates, *Aphrastura [spinicauda] masafuerae,* is clearly an allospecies of the mainland taxon. A second, *Anairetes fernandezianus,* is related to a mainland taxon, but it is not clear to which one of two possible species (*A. reguloides* or *A. parulus*). The third (*Sephanoides fernandensis*) is sympatric with its mainland relative on one island, thus showing an instance of double invasion.

In conclusion, the probability of isolate formation seems to be extremely low in continental populations (1 of 46 species or 2%) but markedly higher in island populations (12 of 40 species, or 30%, including all instances in Table 7). If all isolates, rather than species, are counted then the probability is about 15 in 40 species or 37.5%. Bird species living in Patagonian forests can form morphologically differentiated isolates, as is shown by the insular populations of some of them. The fact that virtually no isolates occur within the continental forest belt

TABLE 7
INSULAR ISOLATES AMONG LANDBIRDS OF PATAGONIAN FORESTS

Species or allospecies on mainland	Insular subspecies or allospecies	Islands
A. Slightly to moderately differentiated isolates (subspecies)		
Polyborus chimango	*fuegiensis*	Isla Grande (Tierra del Fuego)
Strix rufipes	*sanborni*	Chiloé
Colaptes pitius	none named	Chiloé and Guaitecas
Aphrastura [spinicauda] spinicauda	*bullocki*	Mocha
Scelorchilus [rubecula] rubecula	*mochae*	Mocha
Xolmis pyrope	*fortis*	Chiloé
Turdus falcklandii	*falcklandii*	Falklands
Curaeus curaeus	*reynoldsi*	Tierra del Fuego, Cape Horn Archipelago, Navarino, Hoste
	recurvirostris	Isla Riesco
B. Markedly differentiated isolates (subspecies)		
Buteo [albicaudatus] polyosoma	*exsul*	Juán Fernández
Aphrastura [spinicauda] spinicauda	*fulva*	Chiloé
Troglodytes aedon	*cobbi*	Falklands
C. Insular isolates with species status		
Aphrastura [spinicauda] spinicauda	[*s.*] *masafuerae*	Másafuera (Juán Fernández)
Anairetes [?parulus] [?reguloides]	*fernandezianus*	Másafuera (Juán Fernández)
Sephanoides sephaniodes	*fernandensis*	Juán Fernández (sympatric on Másafuera with *S. sephaniodes*: double invasion)

suggests either that the continental area is so uniform geographically that a considerable amount of gene flow occurs within the species living in it, or that selection pressures from this uniform environment produce much phenetic uniformity, or both. Of course, former isolates may have gone extinct. Another possible reason for the virtual absence of isolate formation within the continental Patagonian forest region is that the present area of forest resulted from a late Pleistocene event (see Discussion and Conclusions).

SUPERSPECIES PHENOMENA

Superspecies with all allospecies within the forest region.—Eighteen of the 46 (39%) Patagonian forest species are members (allospecies) of superspecies, but only one of these superspecies occurs strictly within the forest region (Table 6). Its allospecies are the tapaculos, *Pteroptochos* [*tarnii*] *tarnii* and *P.* [*tarnii*] *castaneus*, which, in the absence of detailed field work, seem to be isolated by the Rio Bío-Bío (Behn 1944).

Superspecies with one allospecies in Patagonian forest and one or more allospecies outside the forest.—Fifteen of 40 forest genera (37.5%) and 16 of 46 forest species (34.8%) exhibit this pattern (Table 8). In all of these cases it is reasonably certain that the allospecies involved have reached or are close to reaching species status (Appendix II).

Distributionally, the most interesting superspecies are those nine (20%) in which the allospecies found in *Nothofagus* forests is restricted (or virtually restricted) to it. Ecologically, these superspecies can be classified into two categories. In the first, one allospecies is found in *Nothofagus* forests, while the other or others occur in nonforest vegetation elsewhere. Superspecies included are *Polyborus [megalopterus]*, *Picoides [mixtus]*, *Scelorchilus [rubecula]*, *Tachycineta [leucorrhoa]*, and *Phrygilus [gayi]*. *Polyborus*, *Scelorchilus*, and *Phrygilus* are essentially Andean superspecies, whereas the other two (*Picoides*, *Tachycineta*) are southern South American.

In the second category all allospecies are found in forests or woodlands; included are *Buteo [jamaicensis]*, *Columba [fasciata]*, *Aphrastura [spinicauda]*, and *Carduelis [barbata]*. *Columba* and *Carduelis* are essentially Andean, whereas the other two are southern South American (*Aphrastura*), or cosmopolitan (*Buteo*).

Ex-superspecies.—In the "ex-superspecies" category (Table 6) I include pairs or triplets of

TABLE 8

SUPERSPECIES WITH REPRESENTATIVES AMONG LAND BIRDS OF PATAGONIAN FORESTS[1]

Genus	Allospecies in Patagonian forests	Allospecies outside Patagonian forests (habitat and range)
Accipiter	*bicolor*	*cooperi* (woodlands, North America), *gundlachi* (woodlands, Cuba)
Buteo	*polyosoma*	*poecilochrous* (puna, high Andes), *albicaudatus* (open country, N and NE South America), *galapagoensis* (open country, Galapagos)
Buteo	*ventralis**	*buteo* (woodlands, Palearctic), *oreophilus* (woodlands, Africa), *jamaicensis* (woodlands, North America)
Polyborus	*albogularis**	*carunculatus* (páramo, high Andes), *megalopterus* (puna, high Andes)
Vanellus	*chilensis*	*resplendens* (puna, high Andes)
Columba	*araucana**	*fasciata* (woodlands, western North America; forests, Andes), *caribaea* (forests, West Indies)
Bubo	*virginianus*	*bubo* (various habitats, Palearctic), *africanus* (various habitats, Africa)
Glaucidium	*nanum*	*jardinii* (forests, high Andes), *brasilianum* (woodlands, forests, Central and South America)
Ceryle	*torquata*	*alcyon* (North America), *lugubris* (Asia), *maxima* (Africa)
Picoides	*lignarius**	*mixtus* (woodlands, southcentral South America)
Cinclodes	*fuscus*	*pabsti* (open areas, southern Brazil), ?*comechingonus* (open areas, Córdoba, Argentina)
Aphrastura	*spinicauda**	*masafuerae* (forests, Juán Fernández Islands)
Scelorchilus	*rubecula**	*albicollis* (matorral, central Chile)
Tachycineta	*leucopyga**	*leucorrhoa* (open habitats, southern South America)
Phrygilus	*patagonicus**	*gayi* (puna, Andes and steppes, Patagonia), *atriceps* (puna, high Andes), *chloronotus* (puna, high Andes)
Carduelis	*barbata**	*magellanica* (various habitats, incl. woodlands, Andes, and southern South America)

[1] * indicates superspecies with *Nothofagus* allospecies restricted to this region.

obviously related species that are either largely sympatric or else allopatric but too divergent morphologically to be grouped as a superspecies (e.g., *Polyborus chimango* and *P. chimachima*, see Vuilleumier 1970; Olson 1976). All exhibit southern South American and Andean patterns of distribution.

Conclusions.—Only one superspecies has been detected wholly within the Patagonian forest region (genus *Pteroptochos*); there is a similar pattern of variation, but at the subspecies level, in the genus *Scytalopus*. In both cases the apparent barrier is the Rio Bío-Bío. All other cases of superspecies or ex-superspecies involve one Patagonian forest allospecies or species and one or more non-Patagonian allospecies or species. There appear to be different ecological and distributional patterns of allospecific variation, including forest-forest and forest-nonforest, and Andean and southern South American. The barriers (if any) that separate the various allospecies at present are far from clear. Each one of these cases deserves careful study, especially those involving forest-nonforest pairs (or triplets) of allospecies, because this ecological pattern strongly implies an important niche shift during speciation (Vuilleumier, unpubl. data).

LONG-COMPLETED SPECIATION

Only one genus of 40 (2.5%) shows a case of long-completed speciation. In the parrot genus *Enicognathus* the two species (*ferrugineus* and *leptorhynchus*) are widely sympatric. The extent of morphological similarity between these two species (the only ones in the genus, Peters and Blake 1948) makes it likely that they are derived from a common ancestor. The extent of sympatry, however, does not allow one to infer speciational history. It is, therefore, not possible to tell whether the splitting of the original *Enicognathus* stock took place within the forest, with subsequent development of sympatry there, or whether the two species achieved sympatry as a result of a double invasion of the *Nothofagus* region from some other region.

TABLE 9

LEVELS OF ENDEMISM IN PATAGONIAN FOREST BIRDS ACCORDING TO DIFFERENT TAXONOMIC
TREATMENTS

	Treatment[1]		
Level of endemism	Hellmayr	Philippi	This paper
A. Subspecific and specific levels			
No. of non-endemic species without endemic subspecies	18 (40%)	19 (41%)	22 (48%)
No. of non-endemic species with endemic subspecies	5 (11%)	6 (13%)	5 (11%)
No. of endemic species or allospecies	22 (49%)	21 (46%)	19 (41%)
Total species	45 (100%)	46 (100%)	46 (100%)
B. Generic level			
No. of endemic genera	8 (19%)	7 (17%)	4 (10%)
Total genera	43	42	40

[1] Data for Hellmayr in Appendix VI. Data for Philippi in Appendix VII. Data for treatment in this paper in Appendix VIII.

ENDEMISM IN PATAGONIAN FOREST BIRDS

I define an endemic taxon here as a taxon (of any rank) living exclusively in the Patagonian forest region of southern South America. The 46 species of forest birds can be assigned to one of the following categories:

Non-endemic species without endemic subspecies.—These species live in *Nothofagus* forests as well as elsewhere. Their populations inside Patagonian forests are not differentiated at the subspecies level from subspecies living elsewhere. An example is *Bubo virginianus magellanicus,* a species widespread in North and South America and a subspecies widespread in the Andes and southern South America (Traylor 1958).

Non-endemic species with endemic subspecies.—These species live inside and outside Patagonian forests, but the subspecies inhabiting the forest region is restricted to that region. An example is *Accipiter bicolor chilensis.* The species is widespread in South America, but the subspecies is restricted to Patagonian forests (Stresemann and Amadon 1979).

Endemic species.—These are species that are restricted to Patagonian forests. An example is *Campephilus magellanicus* (Short 1970b).

Endemic genus.—Genera whose species are found only in the Patagonian forests are considered to be endemic, for example *Pygarrhichas* (Vaurie 1980).

In order to determine the influence of different taxonomic points of view on the perceived levels of endemism, I calculated levels of endemism (Table 9) based on the taxonomic treatments of Hellmayr (1932; Appendix VI), Philippi (1964; Appendix VII), and the one adopted in this paper (Appendices II, VIII).

As our knowledge of systematic affinities between species, geographical variation within species, and intergeneric relationships has progressed from Hellmayr's days to the present, our concepts of species and genus have been modified. As a result, we have: (1) an increased number of non-endemic species without endemic subspecies, (2) a concomitant decreased number of endemic species, and (3) a decreased number of endemic genera. Important as these changes may be in single instances, they do not appreciably modify the structure of endemism in the Patagonian forest fauna (Table 9). The important point is, thus, that the Patagonian forest avifauna has 40–48 percent non-endemic species, 11–13 percent endemic subspecies, and 41–49 percent endemic species. Several of the endemic species belong to endemic genera (10–19%). Different taxonomic interpretations of the taxa in the fauna do not influence markedly the results of an analysis of endemicity.

In order to place these levels of endemism in a broader perspective, I analyzed endemism in forest birds of Patagonia, Tasmania, and New Zealand (Table 10; Appendices VIII, IX, X).

Tasmania is a continental or landbridge island that has only been separated from Australia for about 12,000 years (Ridpath and Moreau 1966; Thomas 1974). Endemism should be relatively low, either because isolation is recent and has allowed little time for allopatric speciation, or because isolation is incomplete and allows gene flow to prevent or retard

TABLE 10
Comparison of Endemism in Forest Birds of Tasmania, Patagonia, and New Zealand[1]

Level of endemism	Tasmania	Patagonia	New Zealand
A. *Subspecific and specific level*			
No. of non-endemic species without endemic subspecies	36 (55%)	22 (48%)	3 (9.5%)
No. of non-endemic species with endemic subspecies	16 (25%)	5 (11%)	2 (6%)
No. of endemic species or allospecies	13 (20%)	19 (41%)	27 (84.5%)
Total species	65 (100%)	46 (100%)	32 (100%)
B. *Generic level*			
No. of endemic genera	2 (4%)	4 (10%)	16 (67%)
Total genera	47	40	24

[1] Tasmania = a continental or landbridge island; Patagonia = an ecological island; New Zealand = an oceanic island.

speciation; both factors could have played roles in the differentiation of the Tasmanian avifauna (see also Thomas 1974:350–356). The two oceanic islands forming New Zealand have been isolated from other land masses for about 50 million years (Fleming 1979). Endemism in them should, therefore, be very high. The Patagonian forest region is isolated from other forests in South America at present by more than 1000 km of nonforest vegetation. It may have been an ecological island for about 10 million years (Vuilleumier 1967a). It may, thus, be supposed to be intermediate in its degree of isolation between continental Tasmania and oceanic New Zealand. Levels of endemism in Patagonian forests may be expected to be intermediate, and this is what the data show (Table 10). Tasmania (least isolated) has the most species; New Zealand (most isolated) has the fewest species. Tasmanian forests have far more (80%) non-endemic species (with or without endemic subspecies) than either Patagonian (59%) or New Zealand forests (15.5%). But New Zealand forests have about twice (84.5%) as many endemic species as Patagonian forests (41%), and about four times as many endemic species as Tasmanian forests (20%). Finally, generic endemism is very high in New Zealand (67%) as compared to Patagonia (10%) or Tasmania (4%).

GEOGRAPHICAL DISTRIBUTION OF PATAGONIAN FOREST BIRDS

To gain insight into the geographical origins of the taxa living in Patagonian forests today one can analyze the present distribution of species and genera (Tables 11, 12). The majority of species and allospecies (19, 41.3%) are endemics; smaller numbers of species have southern South American (8, 17.4%), Andean (8, 17.4%), widespread South American (6, 13%), or Pan-American (5, 10.9%) distributions. (No species or allospecies has a Cosmopolitan distribution.) This suggests that the origins of the fauna are, potentially, quite diverse, although the great majority of species and allospecies (41, 89.1%) are South American.

An analysis of genera shows that half (20) belong in the pooled category Pan-American plus Cosmopolitan (Table 12). Other, smaller numbers of genera are distributed in various ways in South America (50% of the genera are South American by pooling all these categories).

In summary, the great majority of Patagonian forest species are South American (89%), the rest are Pan-American. Half the genera are South American, 20 percent are Pan-American, and a rather high percentage (30%) are Cosmopolitan. Only 15 percent of both genera and species are Andean. This mixture of geographical distribution patterns could lead one to conclude that the *Nothofagus* avifauna is composed of autochthonous elements (old or young) and of others that are immigrants (again, more or less recent).

CORRELATION BETWEEN ENDEMISM AND GEOGRAPHICAL DISTRIBUTION

Von Ihering (1927), Mayr (1946, 1964), and Cracraft (1973) speculated on the geographical origins of South American birds. Von Ihering (1927:433) called Archinotis a "large continent, which has for the most part disappeared," and supposed that "in the Cretaceous and in the beginning of the Tertiary it connected Patagonia with Australia and New Zealand." He assumed that "unusual families of birds in South America which cannot be ascribed to later immigration, must have come from Archinotis." Among these birds, he included the tracheo-

TABLE 11

GEOGRAPHICAL DISTRIBUTION OF SPECIES AND ALLOSPECIES OF LAND BIRDS OF PATAGONIAN FORESTS

Species or allospecies	Endemic	Southern South American	Andean	Widespread South American	Pan-American	Cosmo-politan
Theristicus caudatus	—	—	—	X	—	—
Cathartes aura	—	—	—	—	X	—
Coragyps atratus	—	—	—	—	X	—
Vultur gryphus	—	—	X	—	—	—
Accipiter [bicolor] bicolor	(endemic subspecies *chilensis*)	—	—	X	—	—
Buteo [albicaudatus] polyosoma	—	—	X	—	—	—
B. [jamaicensis] ventralis	X	—	—	—	—	—
Geranoaetus melanoleucus	—	—	X	—	—	—
Polyborus chimango	(endemic subspecies *temucoensis* and *fuegiensis*)	X	—	—	—	—
P. [megalopterus] albogularis	X	—	—	—	—	—
P. plancus	—	—	—	—	X	—
Vanellus [chilensis] chilensis	—	—	—	X	—	—
Columba [fasciata] araucana	X	—	—	—	—	—
Enicognathus ferrugineus	X	—	—	—	—	—
E. leptorhynchus	X	—	—	—	—	—
Bubo [bubo] virginianus	—	—	—	—	X	—
Glaucidium [brasilianum] nanum	(endemic subspecies *nanum* and *vafrum*)	—	X	—	—	—
Strix rufipes	(endemic subspecies *rufipes* and *sanborni*)	X	—	—	—	—
Sephanoides sephaniodes	—	X	—	—	—	—
Ceryle [alcyon] torquata	(endemic subspecies *stellata*)	—	—	X	—	—
Picoides [mixtus] lignarius	—	X	—	—	—	—
Colaptes pitius	—	X	—	—	—	—
Campephilus magellanicus	X	—	—	—	—	—
Cinclodes patagonicus	—	—	X	—	—	—
C. [fuscus] fuscus	—	—	X	—	—	—
Sylviorthorhynchus desmursii	X	—	—	—	—	—
Aphrastura [spinicauda] spinicauda	X	—	—	—	—	—
Pygarrhichas albogularis	X	—	—	—	—	—
Pteroptochos [tarnii] tarnii	X	—	—	—	—	—
P. [tarnii] castaneus	X	—	—	—	—	—
Scelorchilus [rubecula] rubecula	X	—	—	—	—	—
Eugralla paradoxa	X	—	—	—	—	—
Scytalopus magellanicus	—	—	X	—	—	—
Elaenia albiceps	—	X	—	—	—	—
Anairetes parulus	—	—	X	—	—	—
Ochthoeca parvirostris	X	—	—	—	—	—
Xolmis pyrope	X	—	—	—	—	—
Phytotoma rara	—	X	—	—	—	—
Notiochelidon cyanoleuca	—	—	—	X	—	—
Tachycineta [leucorrhoa] leucopyga	X	—	—	—	—	—
Troglodytes aedon	—	—	—	—	X	—
Turdus falcklandii	—	X	—	—	—	—
Curaeus curaeus	X	—	—	—	—	—
Phrygilus [gayi] patagonicus	X	—	—	—	—	—
Zonotrichia capensis	—	—	—	X	—	—
Carduelis [barbata] barbata	X	—	—	—	—	—
Totals (N = 46 species)	19	8	8	6	5	0
(%)	(41.3)	(17.4)	(17.4)	(13.0)	(10.9)	(0)

TABLE 12
GEOGRAPHICAL DISTRIBUTION OF GENERA OF LAND BIRDS OF PATAGONIAN FORESTS

Genus	Endemic	Southern South American	Andean	Widespread South American	Pan-American	Cosmo-politan
Theristicus	—	—	—	x	—	—
Cathartes	—	—	—	—	x	—
Coragyps	—	—	—	—	x	—
Vultur	—	—	—	—	x	—
Accipiter	—	—	—	—	—	x
Buteo	—	—	—	—	—	x
Geranoaetus	—	—	x	—	—	—
Polyborus	—	—	—	—	x	—
Vanellus	—	—	—	—	—	x
Columba	—	—	—	—	—	x
Enicognathus	x	—	—	—	—	—
Bubo	—	—	—	—	—	x
Glaucidium	—	—	—	—	—	x
Strix	—	—	—	—	—	x
Sephanoides	—	x	—	—	—	—
Ceryle	—	—	—	—	—	x
Picoides	—	—	—	—	—	x
Colaptes	—	—	—	—	x	—
Campephilus	—	—	—	—	x	—
Cinclodes	—	—	x	—	—	—
Sylviorthorhynchus	x	—	—	—	—	—
Aphrastura	—	x	—	—	—	—
Pygarrhichas	x	—	—	—	—	—
Pteroptichos	—	x	—	—	—	—
Scelorchilus	—	x	—	—	—	—
Eugralla	x	—	—	—	—	—
Scytalopus	—	—	x	—	—	—
Elaenia	—	—	—	x	—	—
Anairetes	—	—	x	—	—	—
Ochthoeca	—	—	x	—	—	—
Xolmis	—	—	—	x	—	—
Phytotoma	—	—	—	x	—	—
Notiochelidon	—	—	—	x	—	—
Tachycineta	—	—	—	—	x	—
Troglodytes	—	—	—	—	—	x
Turdus	—	—	—	—	—	x
Curaeus	—	x	—	—	—	—
Phrygilus	—	—	x	—	—	—
Zonotrichia	—	—	—	—	x	—
Carduelis	—	—	—	—	—	x
Totals (N = 40 genera)	4	5	6	5	8	12
(%)	(10)	(12.5)	(15)	(12.5)	(20)	(30)

phone Passeres, which "survive only in South America" but "must, however, have had a wider distribution in earlier times which could have only extended across Archinotis" (p. 439).

Mayr (1946) classified the faunas of the Americas in terms of faunal elements (1946: fig. 2), thus following Dunn's (1922, 1931) analysis of reptiles. Mayr's (1946) South American Element included three families of tracheophones now living in Patagonian forests, the Furnariidae, Rhinocryptidae, and Phytotomidae. In Mayr's (1946:25) opinion, "there can be no doubt about their South American origin."

Using somewhat different criteria in his 1964 paper Mayr classified the American faunas into faunal elements once again. He wrote (1964:283) that "The South American origin of the Suboscines can hardly be questioned." He included the Furnariidae and the Rhinocryptidae among the Primary South American families.

Cracraft (1973) re-evaluated these faunal elements on the basis of advances in our knowledge of plate tectonics and avian systematics, and provided a modified classification. He placed the Furnariidae, Rhinocryptidae, Phytotomidae, together with other families, such as other members of the Tyranni, in the Southern Hemisphere Element, which "includes those families with probable Southern affinities and which lack close relationships to taxa in the north" (p. 519). Cracraft (1973:519) stated that the Southern Hemisphere Element "contains birds that may have evolved in South America proper, but [whose] ancestors are considered to have had a broader distribution on southern lands." Cracraft (1973) thus reverted to the much earlier thinking of von Ihering (1927).

If the South American Element of Mayr (1946, 1964) or the Southern Hemisphere Element of Cracraft (1973) indeed represents an old autochthonous group of taxa, then their representatives in the now ecologically isolated forests of Patagonia, which are as much southern hemisphere as Neotropical in their floristic affinities, should be the ones with the greatest amount of differentiation as judged by their specific- or generic-level endemism.

I categorized the species of Patagonian forest birds as non-endemic, or endemic (Table 13). I also considered endemic genera. I then analyzed endemism in terms of the present geographical distribution of families (Table 13A), Mayr's (1946) faunal elements (Table 13B), and Cracraft's (1973) faunal elements (Table 13C). Because of the small number of families involved, I pooled Old World and North American elements (as Mayr did in his 1964 paper), and pooled Unanalyzed and Pantropical (= Cosmopolitan Tropical) because most Unanalyzed taxa are also Cosmopolitan or virtually so.

The majority of birds that now have a predominantly South American distribution are endemic species (Table 13A; 7 of 11 species, or 64%, are endemic; 3 of 4 genera, or 75%, are endemic). The majority of birds that are members of Mayr's South American Element are endemic (Table 13B; 10 of 17 species, or 59%; 3 of 4 genera, or 75%). Finally, the majority of birds in Cracraft's Southern Hemisphere Element are endemic (Table 13C; 12 of 18 species, or 67%; 4 genera, or 100%). Note that no Patagonian forest species belongs to his South American Element.

The majority of non-endemic species belong to Cosmopolitan families (Table 13A; 17 of 27, or 63%). Many non-endemic species belong to Old World and North American families (Table 13B; 11 of 27, or 41%), or to Northern Element families (30%). Nevertheless 9 of 19 endemic species or allospecies (47%, Table 13A) belong to Cosmopolitan families, 4 of them (21%) are members of Mayr's Old World and North American Element (Table 13B), and 3 of them (16%) are members of Cracraft's Northern Element (Table 13C).

In conclusion, endemism is clearly concentrated in families that are considered to be old and autochthonous (whether ultimately of South American or Southern Hemisphere origin). Of the 19 endemic species and allospecies, seven are shared among the three categories of South American distribution (Table 13A), South American Element (Table 13B), and Southern Hemisphere Element (Table 13C): *Sylviorthorhynchus desmursii, Aphrastura* [*spinicauda*] *spinicauda, Pygarrhichas albogularis, Pteroptochos* [*tarnii*] *castaneus, P.* [*t.*] *tarnii, Scelorchilus* [*rubecula*] *rubecula,* and *Eugralla paradoxa.* The first three are Furnariidae of uncertain affinities (Vaurie 1980); the others are Rhinocryptidae (claimed by Feduccia and Olson 1982, to be related to the Menurae of Australia). Three of the five endemic genera, *Sylviorthorhynchus, Pygarrhichas* and *Eugralla,* are similarly shared. The first two of these genera are probably the most differentiated taxa of the Patagonian forest fauna today.

DISCUSSION AND CONCLUSIONS

CHARACTERISTICS OF PATAGONIAN FOREST BIRDS

The 46 species of land birds breeding in Patagonian forests show four main characteristics.

(1) They resemble more, as a faunal assemblage, the avifaunas of the Patagonian steppes and of the high Andean puna than the avifauna of montane Andean forests, but they also show some resemblance to the avifauna of coastal forests in western North America. The *Nothofagus* avifauna can, thus, be termed Andean and southern South American with North American affinities.

(2) Most species living in *Nothofagus* forests range widely across habitat types and have a broad elevational and latitudinal range. The species may, thus, have wide habitat-niches, a phenomenon characteristic of insular birds.

TABLE 13
ENDEMISM AND GEOGRAPHICAL DISTRIBUTION OF LAND BIRDS OF PATAGONIAN FORESTS

	Numbers of			
	Non-endemic species	Endemic species and allospecies	Total species	Endemic genera
A. *Present geographical distribution of families*[1]				
Cosmopolitan	17	9	26	1
Pan-American	6	3	9	0
Tropical American and temperate South American (essentially South American)	4	7	11	3
Totals	27	19	46	4
B. *Mayr's (1946) faunal elements*[2]				
Old World and North American	11	4	15	0
South American	7	10	17	3
Unanalyzed and Pantropical	9	5	14	1
Totals	27	19	46	4
C. *Cracraft's (1973) faunal elements*[3]				
Northern Element	8	3	11	0
South American Element	—	—	—	—
Southern Hemisphere Element	6	12	18	4
Unanalyzed Element	13	4	17	0
Totals	27	19	46	4

[1] Cosmopolitan families: Threskiornithidae, Accipitridae, Falconidae, Charadriidae, Strigidae, Columbidae, Psittacidae, Alcedinidae, Picidae, Hirundinidae, Troglodytidae, Turdidae, Emberizidae, Carduelidae. Pan-American families: Cathartidae, Trochilidae, Tyrannidae, Icteridae. South American families: Furnariidae, Rhinocryptidae, Phytotomidae.

[2] Old World and North American Element: Cathartidae, Columbidae, Strigidae, Alcedinidae, Hirundinidae, Troglodytidae, Turdidae, Emberizidae, Carduelidae. South American Element: Trochilidae, Furnariidae, Rhinocryptidae, Tyrannidae, Phytotomidae, Icteridae. Unanalyzed Element and Pantropical: Threskiornithidae, Accipitridae, Falconidae, Charadriidae, Psittacidae, Picidae.

[3] Northern Element: Cathartidae, Trochilidae, Alcedinidae, Troglodytidae, Turdidae, Emberizidae, Carduelidae, Icteridae. Southern Hemisphere Element: Columbidae, Psittacidae, Furnariidae, Rhinocryptidae, Tyrannidae, Phytotomidae. Unanalyzed Element: Threskiornithidae, Accipitridae, Falconidae, Charadriidae, Strigidae, Picidae, Hirundinidae.

(3) Active or recent speciation within the *Nothofagus* forest region is limited to one instance of isolation within a species (*Scytalopus magellanicus*) and one case of isolation within a superspecies (*Pteroptochos [tarnii]*), both occurring across the Rio Bío-Bío, an area also noted as a barrier for plants (Simpson 1973, 1979). A pair of now largely sympatric parrots (*Enicognathus ferrugineus* and *E. leptorhynchus*) probably represents a case of older speciation within the forest region. Thus, intra-forest speciation is relatively unimportant, although the potential for differentiation in geographical isolation does exist in the fauna, as shown by several instances of insular isolates (on Mocha, Chiloé, the Guaitecas, the Juán Fernández, and other islands). In contrast, speciation involving one allospecies in the *Nothofagus* forest area and one or more allospecies in other areas in South or in North America (and even in the Old World) is frequent and involves 16 of 46 species (35%). In nine instances (29%) one allospecies is endemic to the Patagonian forest region, and one or more other allospecies occur elsewhere (southern South America, Andes, or North America). Speciation appears to involve niche shifts, as exemplified in the genus *Phrygilus*. All species in this genus occur in nonforest habitat, except *P. patagonicus,* an allospecies of the *gayi* superspecies living in *Nothofagus* forests. The three other members of the *gayi* superspecies live in Patagonian steppes or in high Andean scrub. "Invasion" of the forest area from steppes has thus been followed by species-level differentiation.

(4) Levels of endemism of Patagonian forest birds are intermediate between those of Tasmanian and New Zealand forest birds, thus conforming to the insular nature of their environment. Several taxa of endemics are without obvious taxonomic affinities with other genera of South American birds. They include *Sylviorthorhynchus* and *Pygarrhichas* (Furnariidae). Little doubt exists that these taxa are among the "primitive" Furnarioidea that belong to old groups, either Primarily South American (Mayr 1964), or Southern Hemisphere (Cracraft 1973) in origin.

RECONSTRUCTING THE HISTORY OF PATAGONIAN FOREST BIRDS

A possible model (scenario) for the evolution of the Patagonian forest fauna, taking into account the characteristics of Patagonian forest birds described above, might involve four stages.

(1) In the early to middle Cenozoic existed an autochthonous fauna of which only a few relicts survive today. These are *Enicognathus* (Psittacidae), *Sylviorthorhynchus, Aphrastura, Pygarrhichas* (Furnariidae), *Pteroptochos, Scelorchilus,* and *Eugralla* (Rhinocryptidae).

(2) During the middle to late Cenozoic, stocks present in the Andes or southern South America (or widespread in South America) invaded Patagonian forests where they left representatives, which are now relatively differentiated taxonomically from other members of their genera. Examples of these old South American immigrants are *Colaptes pitius, Campephilus magellanicus,* and *Ochthoeca parvirostris.* The mechanism of invasion and differentiation is unknown.

(3) During the Pleistocene, other South American or North American stocks penetrated the *Nothofagus* forests and left isolates that are now allospecific members of superspecies. These younger immigrants include *Buteo [jamaicensis] ventralis, Polyborus [megalopterus] albogularis, Columba [fasciata] araucana, Picoides [mixtus] lignarius, Tachycineta [leucorrhoa] leucopyga,* and *Phrygilus [gayi] patagonicus.*

(4) The process of invasion is still active and involves most of the remaining species (newcomers). An example is *Accipiter [bicolor],* with the subspecies *chilensis* in Patagonian forests.

For the four faunal components postulated above, the lack of intraforest speciation (except in *Enicognathus*) is explained if the very large area covered with forest today represents a southward expansion in the late Pleistocene of forest that was restricted to a small area north of the huge ice caps during the latest glacial maximum. If the old endemics are also the birds confined to (or commoner in) dense rainforest, whereas the younger non-endemics are commoner in other, more marginal or peripheral kinds of forest, a continental analog to Wilson's (1961) taxon cycle in insular faunas might have taken place in *Nothofagus* forest birds. Preliminary results (Vuilleumier 1972) suggested that this kind of forest occupancy might be correct. Hence, "older elements of the forest fauna . . . might be distributed throughout forest types, but in higher densities in rain forest, so that later colonists . . . were prevented from attaining higher densities there through competition" (Vuilleumier 1972:270). Interestingly, a parallel can be made with the avifauna of New Zealand forests studied by Kikkawa (1966) and Williams (1981). Cody (1970:459), however, suggested that both the earlier residents and later arrivals "occupy many habitat types." The solution of this problem will necessitate further field work. Pearson and Pearson (1982) studied the mammal fauna of Patagonian forests and reached conclusions somewhat similar to mine (Vuilleumier 1972).

POSSIBLE HISTORY OF PATAGONIAN FORESTS AND ITS RELEVANCE TO THE HISTORY OF THE FOREST AVIFAUNA

Previous studies.—A few years ago, I (Vuilleumier 1967a) hypothesized that the low speciation and high endemism in forest-inhabiting landbirds could be explained if their evolution had been phyletic during a rather long period of the Cenozoic. Simpson (1973) analyzed the distribution and differentiation patterns of Compositae of the genus *Perezia* and found that some taxa in this genus have evolutionary features compatible with the hypothesis of a phyletic mode.

Humphrey and Péfaur (1979) reconstructed the species diversity of birds on several outlying islands on the basis of predictions from equilibrium biogeography. They assumed that some islands (Isla Grande of Tierra del Fuego, Los Estados, Hoste, and Wollaston in the Cape Horn Archipelago) had been fully glaciated during the last major advance of the Patagonian ice cap (Llanquihue Glaciation, about 13,000 years BP) and that their avian populations had been eliminated. As a contrast, they studied the bird diversity of several unglaciated or partially glaciated islands (Mocha, Chiloé, Guaitecas). They concluded that species numbers were: (1) above equilibrium for two partially glaciated land-bridge islands (Chiloé, Guaitecas), (2) at equilibrium for an unglaciated island (Mocha), and (3) at equilibrium for the four glaciated islands. Humphrey and Péfaur (1979) included in their species lists all birds except the seabirds (Procellariiformes and Sphenisciformes). Thus, their conclusions may not apply entirely to the forest avifauna as defined in this paper.

In the remainder of this section I review the information available on past climates and environments in Patagonia and relate it to the possible history of the forest avifauna.

Cenozoic history of Patagonia.—To my knowledge no student with a thorough grasp of stratigraphy, fossil floras, and fossil faunas has attempted to write a synthetic review of the rich literature on fossil vertebrates, fossil plants, geology, and glaciology of Patagonia. In the meantime I present below a summary based upon the reviews of geology by Groeber (1936) and Harrington (1962), and of vertebrate fossil communities by Pascual and Odreman Rivas (1971), together with evidence from fossil plants (Berry 1937a, b; Gerth 1941; Couper 1960; Cranwell 1963, 1964; Cerceau-Larrival 1968; Menéndez 1971; Moore 1978).

At times during the early Cenozoic (Eocene), what is now covered with forest was under seawater. Marine transgressions occurred again in the Oligocene and Miocene. Between periods of transgression and volcanic activity, continental sedimentation occurred. The continental areas of Patagonia were warm and moist in the early Tertiary. Since the Andes were still low, humid air from the Pacific probably influenced the climate much farther east than it does today. Phases of orogeny took place during the Tertiary, but it is only from the Pliocene onward that the Andean uplift was well-marked, leading eventually to the present Patagonian geography and, in part, vegetation and climate.

The Patagonian climate during much of the period from the Eocene to the Pliocene must have been more benign and more moist than it is today. Vegetation included forests and woodlands, as shown by plant remains that belong either to the same taxa that occur in Patagonia today or their precursors, as well as a variety of more or less open types of landscape, including savannas and grassy steppes. The mammalian fossil assemblages appear to have included preponderantly open-country forms at times, and at other times, a mixture including forest or woodland forms. Other assemblages suggest the presence of marshes and ponds under warm temperate or even subtropical conditions.

Thus, Patagonia has had forest and nonforest vegetation formations. These vegetation types must have varied in location and area at various times, forming either a mosaic or zones, the extents and the patterns of which depended in part on the relation between sea (transgressions) and land (volcanic activity) in the first half of the Tertiary. No doubt exists that at times the forest extended considerably farther east than it does today. The primeval forest avifauna of the first half of the Tertiary, whatever its taxonomic composition may have been, lived partially where it is found today and partially where steppes now occur. The restriction of forests to western (Andean) Patagonia is a phenomenon associated with the desertification of eastern Patagonia and is largely correlated with the late Tertiary uplift phases of the Andes. This orogeny created a sharp moisture gradient along the eastern Andean slopes, starting some time in the Pliocene. One can, therefore, postulate that the remaining taxa of the old forest avifauna are relict in their geographical distribution, just as their habitat is.

In the late Cenozoic the climatic deterioration led to intense episodes of glaciation. Glaciation began about 3.5 million years ago in Patagonia and continued in numerous cycles of glacial advance and retreat until about 11,000 years BP, although some minor advances took place more recently still, for example 4600 to 4200 years BP and even 2700 to 2000 years BP. The onset of aridity in eastern Patagonia thus coincided not only with the uplift of the Andes but also with severe deterioration of the climate in the late Pliocene. These climatic changes began to influence the mammalian faunas as early as the Friasian (Miocene). Some of the warm elements of the mammalian fauna disappeared from Patagonia between the Friasian and the Chasicoan (Pliocene), although others persisted longer. If the primeval forest avifauna was tropical or subtropical, such a climatic deterioration was probably felt at the same time as it was for mammals elsewhere in Patagonia. One may suppose that several forest taxa of birds either became extinct or moved out of the present range of the forest. Others still might have adapted to the changing conditions and survived.

The Pleistocene of Patagonia has been extensively studied geologically, glaciologically, climatologically, and palynologically (Caldenius 1932; Auer 1946, 1958, 1960, 1970; Brüggen 1948, 1950; Flint 1959; Heusser 1960, 1961, 1966, 1974, 1981, 1982; Polanski 1965; Mercer 1972, 1976; Mercer et al. 1975; Heusser and Flint 1977; Paskoff 1977; Heusser and Streeter 1980; Heusser et al. 1981; Markgraf 1980, 1983; Porter 1981; Markgraf and Bradbury 1982). The effects of Pleistocene events have been studied from a biological point of view by Báez and Scillato Yané (1979), Formas (1979), B. S. Vuilleumier (1971) and Simpson (1979). The review is based on this literature.

Characteristically, during a glacial advance, the ice cap occupied the western coast of Pat-

FIG. 2.   Schematic map showing location of late Wisconsin (17,000 to 21,000 BP) ice margins in southern South America, simplified and slightly modified from Hollin and Schilling (1981). Estimated area of late Wisconsin Patagonian ice-cap about 478,000 km^2; today about 19,500 km^2.

agonia and the adjacent Andean backbone, and extended some distance eastward into the steppe area. Depending on the severity of the episode, some or all of the outlying islands were covered with ice (e.g., Fig. 2; Auer 1960; B. S. Vuilleumier 1971; Hollin and Schilling 1981). Isostatic lowering of sea level during a glacial period extended the continental margin considerably eastward (see reconstruction in B. S. Vuilleumier 1971). The Falkland Islands had cirque glaciers (Clapperton and Sugden 1976) and, insofar as present evidence shows, no trees during the late Pleistocene (Barrow 1978). Thus, although largely free of ice and occupying an area slightly larger than that occupied today, they probably could not have served as a refugium for forest birds.

In a reconstruction of the avifaunal history at the level of resolution attempted here, whether some of the outlying islands were glaciated or not during some glacial advances is not very important. It is likely that all southernmost islands were covered with ice during at least one episode of glacial expansion. Chiloé, somewhat farther north, apparently remained partially ice free. Markgraf (pers. comm.) suggested that forest elements survived locally in Tierra del Fuego, even though they were perhaps very restricted in their area. If forest refugia existed on one or more of the islands, they were apparently not large enough and/or of sufficient duration to have produced species level isolates among the birds. Within the continental area, however, the valley of the Bío-Bío River, which now apparently serves to isolate subspecies of *Scytalopus magellanicus,* and allospecies of the *Pteroptochos [tarnii]* superspecies, probably acted as a barrier during a glacial episode (Simpson 1973, 1979). The barrier is now near the northern limit of the *Nothofagus* forest region (Fig. 1). This location suggests that its main effects were felt in the past, when the forest belt was displaced northward.

One of the most important aspects of the Pleistocene history of forested Patagonia is the repetition of phases of glaciation and deglaciation, because they resulted in several episodes of retreat northward and of later re-advance southward of the forest vegetation along the Andean axis. Simpson (1973, 1979; see also B. S. Vuilleumier 1971) published a diagram showing the latitudinal contraction of vegetation northward during a glacial advance, as compared to the present (interglacial) distribution. During glacial maxima, the forest vegetation probably retreated northward to the area corresponding roughly to the mediterranean zone of Chile, today covered with matorral (Fig. 1), and was restricted to the western slopes of the Andes because mountain glaciers covered the crests (Fig. 2). Nevertheless, as pointed out by Markgraf (pers. comm.), some forests elements persisted in the far south during glacial maxima, so that not all plant movements involved constriction during glaciation, and recolonization after deglaciation. The latest southward re-advance of forest took place shortly after the last main advance of the ice cap about 13,000 years BP. Forest probably occupied its present area about 11,000 years BP (but see Heusser 1983), but shrank again somewhat during re-advances of glaciers on two or more occasions since that time. The repetition of cycles of forest retreat and re-expansion closely following the repetition of glaciation and deglaciation would not have been conducive to allopatric speciation because the forest area probably was not fragmented into blocks that could hold isolates or incipient species. Also, if small forest refugia existed (Markgraf, pers. comm.), they were probably too small either to support viable populations of some species, or they did not last long enough for the isolate within to become differentiated. Thus, the Pleistocene did not lead to repeated episodes of speciation in forest birds in Patagonia, as it did in puna birds of the high Andes (Vuilleumier 1969; B. S. Vuilleumier 1971; Vuilleumier and Simberloff 1980).

Old stocks of forest birds that did not become extinct as a consequence of the climatic deterioration in the Pliocene-Pleistocene may well have become extinct during the episodes of forest shrinkage of the Pleistocene, thus leading to a considerable impoverishment of the fauna. The surviving stocks did not multiply by allopatric speciation (possible exception *Enicognathus*). Faunal enrichment did occur, however, as is clearly evidenced in patterns of speciation involving forest and nonforest allospecies today (e.g., *Phrygilus [gayi]* superspecies). Steppe-inhabiting stocks were evidently able to adapt to the forest, most likely during periods of forest reduction corresponding to glacial advances. A stock that was in some way preadapted to the forest/steppe ecotone and that found itself "caught" along the edge of the forest as the latter shrank either adapted to forest conditions in the refuge area or became extinct. Vanzolini and Williams (1981) published a model (the vanishing refuge) to help account for speciation of forest forms that became adapted to open vegetation formations during dry-wet cycles in Amazonia. The reverse situation probably occurred in Patagonia: adaptation of steppe forms to forest and their subsequent speciation. The role of ecotones is very important in this scenario of speciation involving such marked niche shifts (Vuilleumier, unpubl. data).

ACKNOWLEDGMENTS

I thank E. Mayr, the late E. Eisenmann, the late B. Patterson, E. E. Williams, K. F. Koopman, O. A. Reig, and B. B. Simpson for help during this work. Very helpful comments were made on the manuscript by A. Keast, O. Pearson, M. Foster, V. Markgraf, and an anonymous referee. I am grateful to R. A. Paynter, Jr. (Museum of Comparative Zoology, Harvard), V. Aellen (Museum d'Histoire Naturelle, Geneva), the late R. A. Philippi B. (Museo Nacional de Historia Natural, Santiago), and the late F. Behn (Concepción) for permitting me to work in the collections in their care. For hospitality and assistance during field work, I thank the late F. Behn and M. A. Ricardi. Field work in Patagonia was made possible by grants from the National Science Foundation to the Department of Biology, Harvard University (Evolutionary Biology Fund), from the Frank M. Chapman Memorial Fund of the American Museum of Natural History, and from the Society of Sigma Xi. No synthetic paper on Patagonian birds could be attempted without the basic work of J. L. Peters, A. Wetmore, C. E. Hellmayr, C. C. Olrog, J. D. Goodall, A. W. Johnson, F. Behn, and R. A. Philippi B. All students of southern South American birds owe them much. V. Morales drew the illustrations. I am grateful to J. Drobnick and M. Ardagna for typing the manuscript.

LITERATURE CITED

AMADON, D. 1964. Taxonomic notes on birds of prey. Am. Mus. Novit. No. 2166.

AMADON, D. 1966. The superspecies concept. Syst. Zool. 15:245–249.

AMADON, D. 1982. A revision of the sub-buteonine hawks (Accipitridae, Aves). Am. Mus. Novit. No. 2741.

AMADON, D., AND L. L. SHORT. 1976. Treatment of subspecies approaching species status. Syst. Zool. 25:161–167.

AMERICAN ORNITHOLOGISTS' UNION. 1983. Check-list of North American birds, 6th ed.

ARAYA M., B., G. MILLIE H., AND O. MAGUERE B. 1974. Aves del Parque Nacional "Vicente Pérez Rosales." An. Mus. Hist. Nat. Valparaiso 7:311–316.

AUBERT DE LA RÜE, E. 1959. Quelques observations sur la biogéographie de la Patagonie chilienne et de la Terre de Feu. C. R. Somm. Séances Soc. Biogéogr. Paris, Nos. 314–316:61–65.

AUER, V. 1946. The Pleistocene and postglacial period in Fuego-Patagonia. Acta Soc. Sci. Fenn. Ser. B 2:1–20.

AUER, V. 1958. The Pleistocene of Fuego-Patagonia. Part II. The history of the flora and vegetation. Ann. Acad. Sci. Fenn. Ser. A III 50:1–239.

AUER, V. 1960. The Quaternary history of Fuego-Patagonia. Proc. R. Soc. Lond. B Biol. Sci. 152:507–516.

AUER, V. 1970. The Pleistocene of Fuego-Patagonia. Part V. Quaternary problems of southern South America. Ann. Acad. Sci. Fenn. Ser. A III 100:1–95.

BÁEZ, A. M., AND G. J. SCILLATO YANÉ. 1979. Late Cenozoic environmental changes in temperate Argentina. Pp. 141–156, In W. E. Duellman (ed.), The South American Herpetofauna: Its Origin, Evolution, and Dispersal. Mus. Nat. Hist., Univ. Kansas, Monogr. No. 7.

BAKER, R. M. 1967. Distribution of Recent mammals along the Pacific coastal lowlands of the Western Hemisphere. Syst. Zool. 16:28–37.

BARROW, C. J. 1978. Postglacial pollen diagrams from South Georgia (sub-Antarctic) and West Falkland Island (South Atlantic). J. Biogeogr. 5:251–274.

BEHN, K. F. 1944. Contribución al estudio del Pteroptochos castaneus Philippi et Landbeck. Hornero 8:464–470.

BEHN, K. F. 1947. Contribución al estudio del Buteo ventralis Gould. Bol. Soc. Biol. Concepción 22:3–5.

BEHN, K. F. 1957. Columba araucana in Chile durch die Newcastle Krankheit dezimiert. J. Ornithol. 98:124.

BERLEPSCH, H. VON. 1907. Studien über Tyranniden. Ornis 14:463–493.

BERLIOZ, J. 1955. Notes critiques sur les pics du genre Campephilus. Oiseau Rev. Fr. Ornithol. 25:27–39.

BERNATH, E. L. 1965. Observations in southern Chile in the southern hemisphere autumn. Auk 82:95–101.

BERRY, E. W. 1937a. A Paleocene flora from Patagonia. Johns Hopkins Univ. Stud. Geol. No. 12:33–50.

BERRY, E. W. 1937b. Eocene plants from Rio Turbio in the Territory of Santa Cruz, Patagonia. Johns Hopkins Univ. Stud. Geol. No. 12:91–97.

BLAKE, E. R. 1968. Family Icteridae. Pp. 138–202, In R. A. Paynter, Jr. (ed.), Check-list of Birds of the World, Vol. 14. Museum Comparative Zoology, Cambridge, Massachusetts.

BLAKE, E. R. 1977. Manual of Neotropical Birds. Vol. 1. Spheniscidae (Penguins) to Laridae (Gulls and Allies). University Chicago Press, Chicago, Illinois.

BLONDEL, J. 1979. Biogéographie et Écologie. Masson, Paris.

BOCK, W. J. 1958. A generic review of the plovers (Charadriinae, Aves). Bull. Mus. Comp. Zool. 118: 27–97.

BOND, J., AND R. MEYER DE SCHAUENSEE. 1943. The birds of Bolivia Part II. Proc. Acad. Nat. Sci. Phila. 95:167–221.

BRODKORB, P. 1934. Geographical variation in *Belonopterus chilensis* (Molina). Occas. Pap. Mus. Zool. Univ. Mich. 293:1–13.

BROWN, L., AND D. AMADON. 1968. Eagles, Hawks and Falcons of the World. Vols. I, II. McGraw-Hill, New York.

BRÜGGEN, J. 1948. La expansión del bosque en el sur de Chile en la época post-glacial. Rev. Univ. (Univ. Católica de Chile, Santiago) 33:105–114.

BRÜGGEN, J. 1950. Fundamentos de la geología de Chile. Instituto Geográfico Militar, Santiago.

BRUNDIN, L. 1965. On the real nature of transantarctic relationships. Evolution 19:496–505.

BRUNDIN, L. 1966. Transantarctic relationships and their significance, as evidenced by chironomid midges; with a monograph of the subfamilies Podonominae, Aphroteniinae and the austral Heptagyiae. Kungl. Svensk. Vetenskapsadad Handl. 4:1–472.

CABRERA, A. L. 1953. Esquema fitogeográfico de la República Argentina. Rev. Mus. La Plata, Bot. 8: 87–168.

CALDENIUS, C. C. 1932. Las glaciaciones cuaternarias en la Patagonia y Tierra del Fuego. Geogr. Ann. 14:1–164.

CAWKELL, E. M., AND J. E. HAMILTON. 1961. The birds of the Falkland Islands. Ibis 103a:1–27.

CERCEAU-LARRIVAL, T. 1968. Contribution palynologique et biogéographique à l'étude biologique de l'Amérique australe. Pp. 111–197, In C. Delamare-Deboutteville, and E. Rapoport (eds.), Biologie de l'Amérique Australe, Vol. 4. Centre National Recherche Scientifique, Paris.

CHAPMAN, F. M. 1919. Descriptions of proposed new birds from Peru, Bolivia, Argentina, and Chile. Bull. Am. Mus. Nat. Hist. 41:323–333.

CHAPMAN, F. M. 1922a. The distribution of the swallows of the genus *Pygochelidon*. Am. Mus. Novit. No. 30.

CHAPMAN, F. M. 1922b. Descriptions of apparently new birds from Colombia, Ecuador, and Argentina. Am. Mus. Novit. No. 31.

CHAPMAN, F. M. 1929. Descriptions of new birds from Mt. Duida, Venezuela. Am. Mus. Novit. No. 380.

CHAPMAN, F. M. 1934. Descriptions of new birds from Mocha Island, Chile, and the Falkland Islands, with comments on their bird life and that of the Juan Fernandez Islands, and Chiloe Island, Chile. Am. Mus. Novit. No. 762.

CHAPMAN, F. M. 1940. The post-glacial history of *Zonotrichia capensis*. Bull. Am. Mus. Nat. Hist. 77: 381–438.

CHAPMAN, F. M., AND L. GRISCOM. 1924. The House Wrens of the genus *Troglodytes*. Bull. Am. Mus. Nat. Hist. 50:279–304.

CHECKLIST COMMITTEE ORNITHOLOGICAL SOCIETY OF NEW ZEALAND. Checklist of New Zealand Birds. 1953. A. H. and A. W. Reed for the Ornithological Society New Zealand, Wellington.

CLAPPERTON, C. M., AND D. E. SUGDEN. 1976. The maximum extent of glaciers in part of West Falkland. J. Glaciol. 17:73–77.

CODY, M. L. 1970. Chilean bird distribution. Ecology 51:455–464.

CONSTANCE, L. 1963. Amphitropical relationships in the herbaceous flora of Pacific coast of North and South America: a symposium. Q. Rev. Biol. 38:109–116.

CONTRERAS, J. R. 1980. Avifauna Mendocina. III. El carpintero chico Picoides mixtus en Mendoza: subespecies presente y comentarios biogeográficos (Picidae). Hist. Nat., Mendoza 1:85–90.

CONTRERAS, J. R., V. G. ROIG, AND A. G. GIAI. 1980. La avifauna de la cuenca del Río Manso Superior y la orilla sur del Lago Mascardi, Parque Nacional Nahuel Huapi, Provincia de Río Negro. Hist. Nat., Mendoza 1:41–48.

COUPER, R. A. 1960. Southern hemisphere Mesozoic and Tertiary Podocarpaceae and Fagaceae and their palaeogeographic significance. Proc. R. Soc. Lond., B Biol. Sci. 152:491–500.

CRACRAFT, J. 1973. Continental drift, paleoclimatology, and the evolution and biogeography of birds. J. Zool. (Lond.) 169:455–545.

CRANWELL, L. M. 1963. *Nothofagus*: living and fossil. Pp. 387–400, In L. J. Gressitt (ed.), Pacific Basin Biogeography: a Symposium. University Hawaii Press, Honolulu, Hawaii.

CRANWELL, L. M. 1964. Antarctica: cradle or grave for its *Nothofagus*? Pp. 87–93, In L. M. Cranwell (ed.), Ancient Pacific Floras. University Hawaii Press, Honolulu, Hawaii.

CRAWSHAY, R. 1907. The Birds of Tierra del Fuego. Bernard Quaritch, London.

CUBILLOS, A., R. SCHLATTER, AND V. CUBILLOS. 1979. Diftero-viruela aviar en Torcaza (*Columba araucana*, Lesson) del Sur de Chile. Zentralbl. Veterinärmed. B 26:430–432.

DARLINGTON, P. J., JR. 1957. Zoogeography: The Geographical Distribution of Animals. John Wiley and Sons, New York.

DARLINGTON, P. J., JR. 1960. The zoogeography of the southern cold temperate zone. Proc. R. Soc. Lond. B Biol. Sci. 152:659–668.

DARLINGTON, P. J., JR. 1965. Biogeography of the Southern End of the World. Harvard University Press, Cambridge, Massachusetts.

DARWIN, C. 1906. The Voyage of the Beagle. E. P. Dutton, New York.

DI CASTRI, F. 1968. Esquisse écologique du Chili. Pp. 7–52, In C. Delamare-Deboutteville, and E. Rapoport (eds.), Biologie de l'Amérique australe, Vol. 4. Centre National Recherche Scientifique, Paris.

DIMITRI, M. J. 1962. La flora andino-patagónica. An. Parques Nac. 9:1–115.

DUNN, E. R. 1922. A suggestion to zoogeographers. Science 56:336–338.

DUNN, E. R. 1931. The herpetological fauna of the Americas. Copeia 1931:106–119.

DUSÉN, P. 1903–1906. The vegetation of western Patagonia. Rep. Princeton Univ. Exped. to Patagonia 1896–99 8:1–33.

ESKUCHE, V. 1968. Fisionomía y sociología de los bosques de Nothofagus dombeyi en la región de Nahuel Huapi. Vegetatio 19:264–285.

FEDUCCIA, A., AND S. L. OLSON. 1982. Morphological similarities between the Menurae and the Rhinocryptidae, relict passerine birds of the Southern Hemisphere. Smithson. Contr. Zool. No. 366: 1–22.

FERNÁNDEZ, J. 1976. El límite boreal de los bosques andino-patagónicos. Bol. Soc. Arg. Bot. 12:307–314.

FLEMING, C. A. 1979. The Geological History of New Zealand and Its Life. Auckland University Press, Auckland.

FLINT, R. F. 1959. La glaciación pleistocena y las gravas tehuelches. Holmbergia 6:87–92.

FORMAS, J. R. 1979. La herpetofauna de los bosques temperados de Sudamérica. Pp. 341–369, In W. E. Duellman (ed.), The South American Herpetofauna: Its Origin, Evolution, and Dispersal. Mus. Nat. Hist., Univ. Kansas, Monogr. No. 7.

FORSHAW, J. M., AND W. T. COOPER. 1973. Parrots of the World. Landsdowne Press, Melbourne, Australia.

FRY, C. M. 1980. The evolutionary biology of Kingfishers (Alcedinidae). Living Bird 18:113–160.

GERTH, H. 1941. Die Tertiärfloren des südlichen Südamerika und die angebliche Verlagerung des Südpols während dieser Periode. Geol. Rundschau 32:321–336.

GODLEY, E. J. 1960. The botany of southern Chile in relation to New Zealand and the Subantarctic. Proc. R. Soc. Lond. B Biol. Sci. 152:457–475.

GOODALL, J. D., A. W. JOHNSON, AND R. A. PHILIPPI B. 1951. Las Aves de Chile, Vol. II. Platt Establecimientos Gráficos, S.A., Buenos Aires.

GOODALL, J. D., A. W. JOHNSON, AND R. A. PHILIPPI B. 1957. Las Aves de Chile, Vol. I. 2nd ed. Platt Establecimientos Gráficos, S.A., Buenos Aires.

GOODALL, J. D., A. W. JOHNSON, AND R. A. PHILIPPI B. 1964. Secundo suplemento de las Aves de Chile. Platt Establecimientos Gráficos, S.A., Buenos Aires.

GOODWIN, D. 1959. Taxonomy of the genus Columba. Bull. Br. Mus. (Nat. Hist.), Zool. 6:1–23.

GOODWIN, D. 1970. Pigeons and Doves of the World, 2nd ed. Trustees British Museum (Nat. Hist.), London.

GRIGERA, D. E. 1976. Ecología alimentaria de cuatro especies de Fringillidae frecuentes en la zona del Nahuel Huapi. Physis C 35(91):279–292.

GROEBER, P. 1936. Oscilaciones de clima en la Argentina desde el Plioceno. Holmbergia 1:71–84.

HANDFORD, P. 1983. Continental patterns of morphological variation in a South American sparrow. Evolution 37:920–930.

HAUMAN-MERCK, L. 1913. La forêt valdivienne et ses limites. Rec. Inst. Bot. Leo Errera 9:346–408.

HARRINGTON, H. J. 1962. Paleogeographic development of South America. Bull. Am. Assoc. Petrol. Geol. 46:1773–1814.

HELLMAYR, C. E. 1927. Catalogue of birds of the Americas. Part V. Field Mus. Nat. Hist., Zool. Ser. 13:1–517.

HELLMAYR, C. E. 1932. The birds of Chile. Field Mus. Nat. Hist., Zool. Ser. 19:1–472.

HELLMAYR, C. E., AND B. CONOVER. 1949. Catalogue of birds of the Americas. Part I, No. 4. Field Mus. Nat. Hist., Publ. 634, Zool. Ser. 13:1–358.

HERSHKOVITZ, P. 1969. The recent mammals of the Neotropical Region: a zoogeographic and ecological review. Q. Rev. Biol. 44:1–70.

HEUSSER, C. J. 1960. Late-Pleistocene environments of the laguna de San Rafael area, Chile. Geogr. Rev. 50:555–577.

HEUSSER, C. J. 1961. Some comparisons between climatic changes in northwestern North America and Patagonia. Ann. N.Y. Acad. Sci. 95:642–657.

HEUSSER, C. J. 1966. Late-Pleistocene pollen diagrams from the province of Llanquihue, southern Chile. Proc. Am. Philos. Soc. 110:269–305.

HEUSSER, C. J. 1974. Vegetation and climate of the southern Chilean Lake District during and since the last interglaciation. Quat. Res. (N.Y.) 4:290–315.

HEUSSER, C. J. 1981. Palynology of the last interglacial-glacial cycle in midlatitudes of southern Chile. Quat. Res. (N.Y.) 16:293–321.

HEUSSER, C. J. 1982. Palynology of cushion bogs of the Cordillera Pelada, Province of Valdivia, Chile. Quat. Res. (N.Y.) 17:71–92.

HEUSSER, C. J. 1983. Quaternary pollen record from Laguna Tagua Tagua, Chile. Science 219:1429–1432.

HEUSSER, C. J., AND R. F. FLINT. 1977. Quaternary glaciations and environments of northern Isla Chiloe, Chile. Geology 5:305–308.

HEUSSER, C. J., AND S. S. STREETER. 1980. A temperature and precipitation record of the past 16,000 years in southern Chile. Science 210:1345–1347.

HEUSSER, C. J., S. S. STREETER, AND M. STRIVER. 1981. Temperature and precipitation record in southern Chile extended to ~43,000 yr ago. Nature 294:65–67.

HOFFSTETTER, R. 1970. Le peuplement mammalien de l'Amérique du Sud: rôle des continents austraux comme centres d'origine, de diversification et de dispersion pour certains groupes mammaliens. An. Acad. Brasil. Cienc. 43:125–144.

HOLLIN, J. T., AND D. M. SCHILLING. 1981. Late Wisconsin-Weichselian mountain glaciers and small ice caps. Pp. 179–206, In G. M. Denton and T. J. Hughes (eds.), The Last Great Ice Sheets. John Wiley and Sons, New York.

HORVÁTH, L., AND G. TOPÁL. 1963. The zoological results of G. Topal's collectings in South Argentina. 9. Aves. An. Hist.-Nat. Mus. Natl. Hung., Zool. 55:531–542.

HOWELL, T. R., R. A. PAYNTER, JR., AND A. L. RAND. 1968. Subfamily Carduelinae. Pp. 207–306, In R. A. Paynter, Jr. (ed.), Check-list of Birds of the World, Vol. 14. Museum Comparative Zoology, Cambridge, Massachusetts.

HUECK, K. 1966. Die Wälder Südamerikas: Ökologie, Zusammensetzung und wirtschaftliche Bedeutung. G. Fischer, Stuttgart, West Germany.

HUECK, K., AND P. SEIBERT. 1972. Vegetationskarte von Südamerika. G. Fischer, Stuttgart, West Germany.

HUMPHREY, P. S., D. BRIDGE, P. W. REYNOLDS, AND R. T. PETERSON. 1970. Birds of Isla Grande (Tierra del Fuego). Prelim. Smithson. Manual, University Kansas Museum Natural History, Lawrence, Kansas.

HUMPHREY, P. S., AND J. E. PÉFAUR. 1979. Glaciation and species richness of birds on austral South American islands. Occ. Pap. Mus. Nat. Hist., Univ. Kansas, No. 80:1–9.

IHERING, H. VON. 1927. The geographic origin of the birds of South America. Ibis, July 1927:427–442.

JEHL, J. R., JR., AND M. A. E. RUMBOLL. 1976. Notes on the avifauna of Isla Grande and Patagonia, Argentina. Trans. San Diego Soc. Nat. Hist. 18:145–154.

JOHANSEN, H. 1966. Die Vögel Feuerlands (Tierra del Fuego). Vidensk. Medd. Dan. Naturhist. Foren. 129:215–260.

JOHNSON, A. W. 1965. The Birds of Chile and Adjacent Regions of Argentina, Bolivia and Peru, Vol. I. Platt Establecimientos Gráficos, S. A., Buenos Aires.

JOHNSON, A. W. 1967. The Birds of Chile and Adjacent Regions of Argentina, Bolivia and Peru, Vol. II. Platt Establecimientos Gráficos, S. A., Buenos Aires.

JOHNSON, A. W. 1972. Supplement to the Birds of Chile and Adjacent Regions of Argentina, Bolivia and Peru. Platt Establecimientos Gráficos, S. A., Buenos Aires.

JOHNSON, A. W., AND F. BEHN. 1957. Milvago chimango fuegiensis subsp. nov. Pp. 353–354, In J. D. Goodall, A. W. Johnson, and R. A. Philippi (eds.), Las Aves de Chile: Su Conocimiento y sus Costumbres, Suplemento. Platt Establecimientos Gráficos, S. A., Buenos Aires.

JOHNSTON, R. F. 1962. The taxonomy of pigeons. Condor 64:69–74.

KALELA, E. K. 1941. Ueber die Holzarten und die durch die klimatische Verhältnisse verursachten Holzartenwechsel in den Wäldern Ostpatagoniens. Ann. Acad. Sci. Fenn. Ser. A IV:1–151.

KEAST, J. A. 1972. Continental drift and evolution of the biota on southern continents. Pp. 23–87, In A. Keast, F. C. Erk, and B. Glass (eds.), Evolution, Mammals, and Southern Continents. State University New York Press, Albany, New York.

KEITH, A. R. 1970. Bird observations from Tierra del Fuego. Condor 72:361–363.

KIKKAWA, J. 1966. Population distribution of land birds in temperate rainforest of sourthern New Zealand. Trans. R. Soc. N. Z. 7:215–277.

KIKKAWA, J. 1974. Niches of birds in Nothofagus forests. Abstracts of Papers, 16th Int. Ornithol. Congr., Canberra, pp. 90–91.

KNOCH, K. 1930. Klimakunde von Südamerika. Handbuch der Klimatologie, II, Teil G. Gebrüder Borntraeger, Berlin.

KRIEG, H. 1951. Als Zoologe in Steppen und Wäldern Patagoniens. Bayerischer Landwirtschaftsverlag, Munich, West Germany.

KUSCHEL, G. 1960. Terrestrial zoology in southern Chile. Proc. R. Soc. London B Biol. Sci. 152:540–550.

LJUNGNER, E. 1939. A forest section through the Andes of northern Patagonia. Sven. Bot. Tidskr. 33:321–337.

MANN, G. 1960. Regiones biogeográficas de Chile. Invest. Zool. Chil. 6:15–49.

MARKGRAF, V. 1980. New data on the late and postglacial vegetational history of "La Mision," Tierra del Fuego, Argentina. IV Int. Palynol. Conf., Lucknow (1976–1977) 3:68–74.

MARKGRAF, V. 1983. Late and postglacial vegetational and paleoclimatic changes during the last 15,000 years in subantarctic and arid environments in Argentina (South America). Palynology 7:43–70.

MARKGRAF, V., AND J. P. BRADBURY. 1982. Climatic history of South America. Striae 16:40–45.

MARKHAM, B. J. 1971. Descripción de una nueva subespecie de "tordo" Curaeus curaeus recurvirostris, subsp. nov. An. Inst. Patagonia Punta Arenas, 2:158–159.

MAYR, E. 1946. History of the North American bird fauna. Wilson Bull. 58:1–41.

MAYR, E. 1959. Trends in avian systematics. Ibis 101:293–302.

MAYR, E. 1964. Inferences concerning the Tertiary American bird faunas. Proc. Natl. Acad. Sci. 51: 280–288.

MAYR, E., AND L. L. SHORT. 1970. Species taxa of North American birds. Cambridge, Mass., Publ. Nuttall Ornithol. Club No. 9.

MENÉNDEZ, C. A. 1971. Floras terciarias de la Argentina. Ameghiniana 8:357–371.

MERCER, J. H. 1972. Chilean glacial chronology 20,000 to 11,000 Carbon-14 years ago: some global comparisons. Science 176:1118–1120.

MERCER, J. H. 1976. Glacial history of southernmost South America. Quat. Res. (N.Y.) 6:125–166.

MERCER, J. H., R. J. FLECK, E. A. MANKINEN, AND W. SANDER. 1975. Southern Patagonia: glacial events between 4 m.y. and 1 m.y. ago. Pp. 223–230, In R. P. Suggate and M. M. Cresswell (eds.), Quaternary Studies. Royal Society New Zealand, Wellington.

MEYER DE SCHAUENSEE, R. 1966. The Species of Birds of South America and Their Distribution. Livingston Publ. Co., Narberth, Pennsylvania.

MILLER, A. 1951. An analysis of the distribution of the birds of California. Univ. Calif. Publ. Zool. 50: 531–644.

MILLER, W. DE W. 1912. A revision of the classification of the kingfishers. Bull. Am. Mus. Nat. Hist. 31:239–311.

MILLER, W. DE W. 1920. The genera of ceryline kingfishers. Auk 27:422–429.

MOORE, D. M. 1978. Post-glacial vegetation in the South Patagonian territory of the giant ground sloth, Mylodon. Bot. J. Linn. Soc. 77:177–202.

MOORE, D. M. 1979. Southern oceanic wet-heathlands (including Magellanic moorland). Pp. 489–497, In R. L. Specht (ed.), Heathlands and Related Shrublands of the World, A. Descriptive Studies. Elsevier Publ. Co., Amsterdam.

MOREAU, R. E. 1966. The Bird Faunas of Africa and Its Islands. Academic Press, New York.

MORRISON, A. 1940. Brief notes on the birds of south Chile. Ibis, April 1940:248–256.

MÜLLER, P. 1973. The Dispersal Centres of Terrestrial Vertebrates in the Neotropical Realm. Junk, The Hague.

MUÑOZ, C., AND E. PISANO. 1947. Estudio de la vegetación y flora de los Parques Nacionales de Fray Jorge y Talinay. Agric. Téc. (Santiago) 7:71–190.

NOTTEBOHM, F. 1975. Continental patterns of song variability in Zonotrichia capensis: some possible ecological correlates. Am. Nat. 109:605–624.

OLROG, C. C. 1948. Observaciones sobre la avifauna de Tierra del Fuego y Chile. Acta Zool. Lilloana 5:437–531.

OLROG, C. C. 1949. La avifauna del Aconquija. Acta Zool. Lilloana 7:139–159.

OLROG, C. C. 1950. Notas sobre mamíferos y aves del Archipiélago de Cabo de Hornos. Acta Zool. Lilloana 9:505–532.

OLROG, C. C. 1962. Notas ornitológicas sobre la colección del Instituto Miguel Lillo (Tucumán). VI. Acta Zool. Lilloana 28:111–120.

OLROG, C. C. 1963. Lista y distribución de las aves Argentinas. Opera Lilloana 9:1–377.

OLROG, C. C. 1979. Nueva lista de la avifauna Argentina. Opera Lilloana 27:1–324.

OLSON, S. L. 1976. A new species of Milvago from Hispaniola, with notes on other fossil caracaras from the West Indies (Aves: Falconidae). Proc. Biol. Soc. Wash. 88:355–366.

PARMELEE, D. F., AND S. D. MACDONALD. 1975. Recent observations on the birds of Isla Contramaestre and Isla Magdalena, Straits of Magellan. Condor 77:218–220.

PASCUAL, R., AND O. E. ODREMAN RIVAS. 1971. Evolución de las comunidades de los vertebrados del Terciario argentino. Los aspectos paleozoogeográficos y paleoclimáticos relacionados. Ameghiniana 7:372–412.

PASKOFF, R. P. 1977. Quaternary of Chile: the state of research. Quat. Res. (N.Y.) 8:2–31.

PÄSSLER, R. 1922. In der Umgebung Coronel's (Chile) beobachtete Vögel. J. Ornithol. 70:430–482.

PAYNTER, R. A., JR. 1957. Taxonomic notes on the New World forms of Troglodytes. Breviora No. 71.

PAYNTER, R. A. 1960. Troglodytidae. Pp. 379–440, In E. Mayr and J. C. Greenway, Jr. (eds.), Check-list of Birds of the World, Vol. 9. Museum Comparative Zoology, Cambridge, Massachusetts.

PEARSON, O. P., AND A. K. PEARSON. 1982. Ecology and biogeography of the southern rainforests of Argentina. Spec. Publ. Pymatuning Lab. Ecol. No. 6:129–142.

PEÑA, L. E. 1964. Notas ornitológicas. Un nuevo género de ave para Chile y nuevos records de distri-bución geográfica para Aysén y Magallanes. Rev. Chil. Hist. Nat. 55:115–121.

PETERS, J. L. 1923a. A new babbler from Argentina. Proc. New England Zool. Club 8:45–46.
PETERS, J. L. 1923b. Notes on some summer birds of northern Patagonia. Bull. Mus. Comp. Zool. 65: 277–337.
PETERS, J. L. 1948. Check-list of Birds of the World, Vol. 6. Harvard University Press, Cambridge, Massachusetts.
PETERS, J. L., AND E. R. BLAKE. 1948. *Microsittace* not different generically from *Enicognathus.* Auk 65:288–289.
PHILIPPI B., R. A. 1964. Catálogo de las aves chilenas con su distribución geográfica. Inv. Zool. Chil. 11:1–179.
PHILIPPI B., R. A., A. W. JOHNSON, J. D. GOODALL, AND F. BEHN. 1954. Notas sobre aves de Magallanes y Tierra del Fuego. Bol. Mus. Nac. Hist. Nat. Santiago 26(3):1–55.
PISANO V., E. 1974. Estudio ecológico de la región continental sur del área Andino-patagónica. II. Contribución a la fitogeografía de la zona del Parque Nacional "Torres del Paine." An. Inst. Patagonia Punta Arenas 5:59–104.
PISANO V., E. 1977. Fitogeografía de Fuego-Patagonia chilena. I. Comunidades vegetales entre las latitudes 52° y 56° S. An. Inst. Patagonia Punta Arenas 8:121–250.
PISANO V., E. 1981. Bosquejo fitogeográfico de Fuego-Patagonia. An. Inst. Patagonia Punta Arenas 12: 159–171.
POLANSKI, J. 1965. The maximum glaciation in the Argentine cordillera. Geol. Soc. Am. Spec. Pap. 84: 453–472.
PORTER, S. C. 1981. Pleistocene glaciation in the southern lake district of Chile. Quat. Res. (N.Y.) 16: 263–292.
RADBOONE, S. 1935. Notas sobre algunas aves del Lago San Martín (Santa Cruz). Hornero 6:99–100.
RAPOPORT, E. H. 1968. Algunos problemas biogeográficos del nuevo mundo con especial referencia a la región neotropical. Pp. 54–110, *In* C. Delamare-Deboutteville and E. H. Rapoport (eds.), Biologie de l'Amérique Australe, Vol. 4. Centre National Recherche Scientifique, Paris.
RAPOPORT, E. H. 1971. The geographical distribution of Neotropical and antarctic collembola. Pac. Insects Monogr. 25:99–118.
RAVEN, P. H., AND D. AXELROD. 1972. Plate tectonics and Australasian paleobiogeography. Science 176:1379–1386.
RAVEN, P. H., AND D. AXELROD. 1974. Angiosperm biogeography and past continental movements. Ann. Mo. Bot. Gard. 61:539–673.
RAVEN, P. H., AND D. AXELROD. 1975. History of the flora and fauna of Latin America. Am. Sci. 63: 420–429.
REICHE, R. 1907. Grundzüge der Pflanzenverbreitung in Chile. Die Vegetation der Erde VIII. W. Engelmann, Leipzig, East Germany.
REIG, O. 1968. Peuplement en vertébrés tétrapodes de l'Amérique du Sud. Pp. 215–260, *In* C. Delamare-Deboutteville and E. M. Rapoport (eds.), Biologie de l'Amérique Australe, Vol. 4. Centre National Recherche Scientifique, Paris.
REYNOLDS, P. W. 1932. Notes on the birds of Snipe, and the Woodcock Islands, in the Beagle Channel. Ibis, Ser. 13, 2:34–39.
REYNOLDS, P. W. 1934. Apuntes sobre aves de Tierra del Fuego. Hornero 5:339–353.
REYNOLDS, P. W. 1935. Notes on the birds of Cape Horn. Ibis, Ser. 3, 5:65–101.
RIDPATH, M. G., AND R. E. MOREAU. 1966. The birds of Tasmania: ecology and evolution. Ibis 108: 348–393.
RIPLEY, S. D. 1950. A small collection of birds from Argentine Tierra del Fuego. Postilla No. 3.
RIPLEY, S. D. 1964. Turdinae. Pp. 13–227, *In* E. Mayr and R. A. Paynter, Jr. (eds.), Check-list of Birds of the World, Vol. 10. Museum Comparative Zoology, Cambridge, Massachusetts.
SCHLATTER, R. P. 1976. Aves observadas en un sector del Lago Rinihue, Provincia de Valdivia, con alcances sobre su ecología. Bol. Soc. Biol. Concepción 50:133–143.
SCHLATTER, R. P., G. REINHARDT, AND L. BURCHARD. 1978. Estudio del Jote (*Coragyps atratus foetens* Lichtenstein) en Valdivia: etología carronera y rol en diseminación de agentes patogenos. Arch. Med. Vet. 10(2):111–127.
SCHLATTER, R. P., J. L. YÁÑEZ, AND F. M. JAKSIĆ. 1980. Food-niche relationships between Chilean Eagles and Red-backed Buzzards in central Chile. Auk 97:897–898.
SCHMITHÜSEN, J. 1956. Die räumliche Ordnung der chilenischen Vegetation. Pp. 1–86, *In* Forschungen in Chile, Bonn. Geogr. Abh. No. 17.
SCHMITHÜSEN, J. 1960. Die Nadelhölzer in den Waldgesellschaften der südlichen Anden. Vegetatio 9:313–327.
SCHMITHÜSEN, J. 1964. Problems of vegetation history in Chile and New Zealand. Proc. 20th Int. Geogr. Congr., pp. 189–206.
SCLATER, P. L. 1890. Catalogue of the Birds in the British Museum. Vol. XV. Taylor and Francis, London.
SCLATER, W. L. 1939. A note on some American orioles of the family Icteridae. Ibis 1939:140–145.
SHORT, L. L. 1969a. Observations on three sympatric species of Tapaculos (Rhinocryptidae) in Argentina. Ibis 111:239–240.

SHORT, L. L. 1969b. Observations of the nuthatch-like White-throated Treerunner (*Pygarrhichas albogularis*) in Argentina. Condor 71:438–439.

SHORT, L. L. 1970a. Notes on the habits of some Argentine and Peruvian woodpeckers (Aves, Picidae). Am. Mus. Novit. No. 2413.

SHORT, L. L. 1970b. The habits and relationships of the Magellanic Woodpecker. Wilson Bull. 82:115–129.

SHORT, L. L. 1971. Systematics and behavior of some North American woodpeckers, genus *Picoides* (Aves). Bull. Am. Mus. Nat. Hist. 145:1–118.

SHORT, L. L. 1972. Systematics and behavior of South American flickers (Aves, *Colaptes*). Bull. Am. Mus. Nat. Hist. 149:1–110.

SHORT, L. L. 1975. A zoogeographic analysis of the South American chaco avifauna. Bull. Am. Mus. Nat. Hist. 154:163–352.

SHORT, L. L. 1980. Speciation in South American woodpeckers. Actis XVII Congr. Int. Ornithol., Berlin 1978. Vol II:1268–1272.

SHORT, L. L. 1982. Woodpeckers of the World. Delaware Mus. Nat. Hist. Monogr. Ser. No. 4.

SICK, H. 1960. Zur Systematik und Biologie der Bürzelstelzer (Rhinocryptidae), speziell Brasiliens. J. Ornithol. 101:141–174.

SIELFELD K., W. H. 1977. Reconocimiento macrofaunístico terrestre en el area de Seno Ponsonby (Isla Hoste). An. Inst. Patagonia Punta Arenas 8:275–296.

SIMPSON, B. B. 1973. Contrasting modes of evolution in two groups of Perezia (Mutisiae; Compositae) of southern South America. Taxon 22:525–536.

SIMPSON, B. B. 1979. Quaternary biogeography of the high montane regions of South America. Pp. 157–188, *In* W. E. Duellman (ed.), The South American Herpetofauna: Its Origin, Evolution, and Dispersal. Mus. Nat. Hist., University Kansas, Monogr. No. 7.

SIMPSON, G. G. 1940. Antarctica as a faunal migration route. Proc. 6th Pac. Sci. Congr. 2:755–768.

SIMPSON, G. G. 1960. Notes on the measurement of faunal resemblance. Am. J. Sci. 258-A:300–311.

SIMPSON, G. G. 1966. Mammalian evolution on the southern continents. Neues Jahrb. Geol. Palaeont. Abh. 125:1–18.

SKOTTSBERG, C. 1910. Botanische Ergebnisse der schwedischen Expedition nach Patagonien und dem Feuerlande 1907–1909. I. Uebersicht uber die wichstigsten Pflanzenformationen Südamerikas S. von 41°S, ihre geographische Verbreitung und Beziehungen zum Klima. Kungl. Sven. Vetenskapsakad. Handl. 46:1–28.

SKOTTSBERG, C. 1916. Die Vegetationsverhältnisse längs der Cordillera de los Andes S. von 41°S. Kungl. Sven. Vetenskapsakad. Handl. 56:1–366.

SKOTTSBERG, C. 1960. Remarks on the plant geography of the southern cold temperate zone. Proc. R. Soc. Lond. B Biol. Sci. 152:447–457.

SMITH, L. B. 1962. Origins of the flora of southern Brazil. Contrib. U.S. Natl. Herb. 35:215–249.

SMITH, W. J. 1971. Behavior of *Muscisaxicola* and related genera. Bull. Mus. Comp. Zool. 141:233–268.

SOLBRIG, O. T. 1976. The origin and floristic affinities of the South American temperate desert and semidesert regions. Pp. 7–14, *In* D. W. Goodall (ed.), Evolution of Desert Biota. University Texas Press, Austin, Texas.

SORGE, E. 1930. Die Trockengrenze Südamerikas. Zeitschr. Ges. Erdk. Berlin 1930:277–287.

STEINBACHER, J. 1979. Family Threskiornithidae. Pp. 253–268, *In* E. Mayr and G. W. Cottrell (eds.), Check-list of Birds of the World, Vol. 1, 2nd ed. Museum Comparative Zoology, Cambridge, Massachusetts.

STRESEMANN, E., AND D. AMADON. 1979. Order Falconiformes. Pp. 271–425, *In* E. Mayr and G. W. Cottrell (eds.), Check-list of Birds of the World, Vol. 1, 2nd ed. Museum Comparative Zoology, Cambridge, Massachusetts.

TEXERA, W. A. 1972. Distribución y diversidad de mamíferos y aves en la Provincia de Magallanes. 1 Analisis preliminar de la diversidad ecológica y variación taxonómica. An. Inst. Patagonia Punta Arenas 3:171–200.

THOMAS, D. G. 1974. Some problems associated with the avifauna. Pp. 339–365, *In* W. D. Williams (ed.), Biogeography and Ecology in Tasmania. Junk, The Hague.

THOMASSON, K. 1959. Nahuel Huapi: plankton of some lakes in an Argentine National Park, with notes on terrestrial vegetation. Acta Phytogeogr. Suec. 42:1–83.

TODD, W. E. C. 1926. A study of the Neotropical finches of the genus *Spinus*. Ann. Carnegie Mus. 17:11–82.

TOMASELLI, R. 1981. The longitudinal zoning of vegetation in the southern sector of the Andes. Studi Trentini Sci. Nat. Ser. B Biol. 58:471–484.

TRAYLOR, M. A., JR. 1958. Variation in South American Great Horned Owls. Auk 75:143–149.

TRAYLOR, M. A., JR. 1977. A classification of the tyrant flycatchers (Tyrannidae). Bull. Mus. Comp. Zool. 148:129–184.

TRAYLOR, M. A., JR. 1979. Family Tyrannidae. Pp. 3–245, *In* M. A. Traylor, Jr. (ed.), Check-list of Birds of the World, Vol. 8. Museum Comparative Zoology, Cambridge, Massachusetts.

TRAYLOR, M. A., JR. 1982. Notes on tyrant flycatchers (Aves: Tyrannidae). Fieldiana, Zool., New Ser. No. 13:1–22.

TRAYLOR, M. A., JR., AND J. W. FITZPATRICK. 1982. A survey of the tyrant flycatchers. Living Bird 19: 7–50.

TRIMBLE, R. 1943. Birds collected during two cruises of the "Vagabondia" to the west coast of South America. Ann. Carnegie Mus. 29:409–441.

VANZOLINI, P. E., AND E. E. WILLIAMS. 1981. The vanishing refuge: a mechanism for ecogeographic speciation. Pap. Avulsos Zool. (São Paulo) 34:251–255.

VAURIE, C. 1962. A systematic study of the red-backed hawks of South America. Condor 64:277–290.

VAURIE, C. 1980. Taxonomy and geographical distribution of the Furnariidae (Aves, Passeriformes). Bull. Am. Mus. Nat. Hist. 166:1–357.

VEBLEN, T. T. 1982. Regeneration patterns in *Araucaria araucana* forests in Chile. J. Biogeogr. 9:11–28.

VEBLEN, T. T., D. M. ASHTON, F. M. SCHLEGEL, AND A. T. VEBLEN. 1977. Plant succession in a timberline depressed by vulcanism in south-central Chile. J. Biogeogr. 4:275–294.

VEBLEN, T. T., F. M. SCHLEGEL, AND B. ESCOBAR R. 1980. Structure and dynamics of old-growth *Nothofagus* forests in the Valdivian Andes, Chile. J. Ecol. 68:1–31.

VEBLEN, T. T., Z. DONOSO, F. M. SCHLEGEL, AND B. ESCOBAR R. 1981. Forest dynamics in south-central Chile. J. Biogeogr. 8:211–247.

VENEGAS C., C. 1976. Observaciones ornitológicas en la tundra magallanica. 1. Recuento descriptivo del área y de las observaciones aviales entre los paralelos 51°31'S y 52°09'S. An. Inst. Patagonia Punta Arenas 7:171–184.

VENEGAS C., C. 1981. Aves de las Islas Wollaston y Bayly, Archiepiélago del Cabo de Hornos. An. Inst. Patagonia Punta Arenas 12:213–219.

VENEGAS C., C. 1982a. Nuevos registros ornitológicos en Magallanes. An. Inst. Patagonia Punta Arenas 13:183–187.

VENEGAS C., C. 1982b. Suplemento a la guía de campo para las aves de Magallanes. An. Inst. Patagonia Punta Arenas 13:189–206.

VOOUS, K. H. 1947. On the history of the distribution of the genus *Dendrocopos*. Limosa 20:1–142.

VUILLEUMIER, B. S. 1971. Pleistocene changes in the fauna and flora of South America. Science 173: 771–780.

VUILLEUMIER, F. 1967a. Phyletic evolution in modern birds of the Patagonian forests. Nature 215:247–248.

VUILLEUMIER, F. 1967b. Mixed species flocks in Patagonian forests, with remarks on interspecies flock formation. Condor 69:400–404.

VUILLEUMIER, F. 1968. Origin of frogs of Patagonian forests. Nature 219:87–89.

VUILLEUMIER, F. 1969. Pleistocene speciation in birds living in the high Andes. Nature 223:1179–1180.

VUILLEUMIER, F. 1970. Generic relations and speciation patterns in the caracaras (Aves: Falconidae). Breviora No. 355.

VUILLEUMIER, F. 1971. Generic relationships and speciation patterns in *Ochthoeca, Myiotheretes, Xolmis, Neoxolmis, Agriornis*, and *Muscisaxicola*. Bull. Mus. Comp. Zool. 141:181–232.

VUILLEUMIER, F. 1972. Bird species diversity in Patagonia (temperate South America). Am. Nat. 106: 266–271.

VUILLEUMIER, F. 1980. Speciation in birds of the high Andes. Acta XVII Congr. Int. Ornithol., Berlin 1978, Vol. II:1296–1301.

VUILLEUMIER, F., AND D. SIMBERLOFF. 1980. Ecology versus history as determinants of patchy and insular distributions in high Andean birds. Evol. Biol. 12:235–379.

WARD, R. T. 1965. Beech (*Nothofagus*) forests in the Andes of southwestern Argentina. Am. Midl. Nat. 74:50–56.

WATTEL, J. 1973. Geographical differentiation in the genus *Accipiter*. Publ. Nuttall Ornithol. Club No. 13.

WETMORE, A. 1922. New forms of neotropical birds. J. Wash. Acad. Sci. 12:323–328.

WETMORE, A. 1923. New subspecies of birds from Patagonia. Univ. Calif. Publ. Zool. 21:333–337.

WETMORE, A. 1926a. Observations on the birds of Argentina, Paraguay, Uruguay, and Chile. U.S. Natl. Mus. Bull. 133:1–448.

WETMORE, A. 1926b. Report on a collection of birds made by J. R. Pemberton in Patagonia. Univ. Calif. Publ. Zool. 24:395–474.

WETMORE, A. 1962. Systematic notes concerned with the avifauna of Panama. Smithson. Misc. Collect. 145(1):1–14.

WETMORE, A. 1964. A revision of the American vultures of the genus *Cathartes*. Smithson. Misc. Collect. 146(6):1–18.

WETMORE, A., AND J. L. PETERS. 1922. A new genus and four new subspecies of American birds. Proc. Biol. Soc. Wash. 35:41–46.

WHEELER, J. 1938. A new wood owl from Chile. Field Mus. Nat. Hist., Zool. Ser. 20(37):479–482.

WILLIAMS, G. R. 1981. Aspects of avian island biogeography in New Zealand. J. Biogeogr. 8:439–456.

WILSON, E. O. 1961. The nature of the taxon cycle in the Melanesian ant fauna. Am. Nat. 95:169–193.

WOODBURNE, M. O., AND W. J. ZINSMEISTER. 1983. Fossil land mammal from Antarctica. Science 218: 284–286.

YÁÑEZ, J. L., H. NÚÑEZ, AND F. M. JAKSIĆ. 1982. Food habits and weight of Chimango Caracaras in central Chile. Auk 99:170–171.

YOUNG, S. B. 1972. Subantarctic rain forest of Magellanic Chile: distribution, composition, and age and growth rate studies of common forest trees. Antarct. Res. Ser. 20:307–322.

ZIMMER, J. T. 1939. Studies of Peruvian birds. No. 32. The genus *Scytalopus*. Am. Mus. Novit. No. 1044.

ZIMMER, J. T. 1940. Studies of Peruvian birds. No. 35. Notes on the genera *Phylloscartes, Euscarthmus, Pseudocolopteryx, Tachuris, Spizitornis, Yanacea, Uromyias, Stigmatura, Serpophaga*, and *Mecocerculus*. Am. Mus. Novit. No. 1095.

ZIMMER, J. T. 1941. Studies of Peruvian birds. No. 36. The genera *Elaenia* and *Myiopagis*. Am. Mus. Novit. No. 1108.

APPENDIX I
LAND BIRDS OF PATAGONIA[1, 2]

Rheidae. *Pterocnemia pennata* (N).
Tinamidae. *Nothura darwinii* (N); *N. maculosa* (N); *N. [perdicaria] perdicaria* (N); *Eudromia elegans* (N); *Tinamotis [pentlandii] ingoufi* (N).
Threskiornithidae. *Theristicus caudatus* (B; M).
Cathartidae. *Cathartes aura* (B; P); *Coragyps atratus* (B; R); *Vultur gryphus* (B; R).
Accipitridae. *Circus [cyaneus] cinereus* (N); *Elanus [caeruleus] leucurus* (N); *Accipiter [bicolor] bicolor* (F; M); *Parabuteo unicinctus* (N); *Buteo [?brachyurus] albigula* (A); *B. [albicaudatus] polyosoma* (B; P?); *B. [jamaicensis] ventralis* (F; R); *Geranoaetus melanoleucus* (B; R).
Falconidae. *Polyborus chimango* (B; M); *P. [megalopterus] megalopterus* (A); *P. [megalopterus] albogularis* (B; R); *P. australis* (N); *P. plancus* (B; R); *Falco peregrinus* (N).
Charadriidae. *Vanellus [chilensis] chilensis* (B; P); *Eudromias ruficollis* (N); *Charadrius modestus* (N).
Thinocoridae. *Attagis gayi* (A); *A. malouinus* (A); *Thinocorus orbignyianus* (N); *T. rumicivorus* (N).
Columbidae. *Columba [fasciata] araucana* (F; R); *C. maculosa* (N); *Zenaida [macroura] auriculata* (N); *Metriopelia melanoptera* (N); *Columbina picui* (N).
Psittacidae. *Cyanoliseus patagonus* (N); *Enicognathus ferrugineus* (F; R?); *E. leptorhynchus* (F; R?).
Tytonidae. *Tyto alba* (N).
Strigidae. *Bubo [bubo] virginianus* (B; R); *Glaucidium [brasilianum] nanum* (F; P); *Athene cunicularia* (N); *Strix rufipes* (F; R); *Asio flammeus* (N).
Caprimulgidae. *Caprimulgus longirostris* (N).
Trochilidae. *Sephanoides sephaniodes* (F; P?); *Oreotrochilus [estella] leucopleurus* (A).
Alcedinidae. *Ceryle [alcyon] torquata* (B).
Picidae. *Picoides [mixtus] mixtus* (N); *P. [mixtus] lignarius* (F; P); *Colaptes pitius* (F; R); *Campephilus magellanicus* (F; R).
Furnariidae. *Geositta rufipennis* (A); *G. [cunicularia] cunicularia* (N); *G. [?cunicularia] antarctica* (N); *Upucerthia dumetaria* (N); *U. ruficauda* (A); *Eremobius phoenicurus* (N); *Cinclodes antarcticus* (N); *C. patagonicus* (B; P); *C. oustaleti* (A); *C. [fuscus] fuscus* (B; M); *C. [nigrofumosus] nigrofumosus* (N); *Sylviorthorhynchus desmursii* (F; R); *Aphrastura spinicauda* (F; P?); *Phleocryptes melanops* (N); *Leptasthenura aegithaloides* (N); *L. platensis* (N); *Asthenes pyrrholeuca* (N); *A. patagonica* (N); *A. humicola* (N); *A. modesta* (A); *A. [anthoides] anthoides* (N); *A. hudsoni* (N); *Anumbius annumbi* (N); *Pseudoseisura gutturalis* (N); *Pygarrhichas albogularis* (F; R).
Rhinocryptidae. *Pteroptochos [tarnii] tarnii* (F; R); *P. [tarnii] castaneus* (F; R); *Scelorchilus [rubecula] rubecula* (F; R); *Rhinocrypta lanceolata* (N); *Teledromas fuscus* (N); *Scytalopus magellanicus* (F; R); *Eugralla paradoxa* (F; R).
Tyrannidae. *Elaenia albiceps* (F; M); *Anairetes parulus* (B; P); *Tachuris rubrigastra* (N); *Pseudocolopteryx flaviventris* (N); *Ochthoeca parvirostris* (F; M); *Xolmis pyrope* (B; M); *Neoxolmis rubetra* (N); *N. rufiventris* (N); *Agriornis livida* (N); *A. microptera* (N); *A. montana* (A); *A. murina* (N); *Muscisaxicola [albilora] albilora* (A); *M. [albifrons] flavinucha* (A); *M. capistrata* (N); *M. macloviana* (N); *M. maculirostris* (N); *Lessonia [rufa] rufa* (N); *Knipolegus aterrimus* (N); *K. hudsoni* (N); *Hymenops perspicillata* (N).
Phytotomidae. *Phytotoma rara* (B; P).
Hirundinidae. *Progne [subis] modesta* (N); *Notiochelidon cyanoleuca* (B; M); *Tachycineta [leucorrhoa] leucopyga* (F; M).
Troglodytidae. *Cistothorus [platensis] platensis* (N); *Troglodytes aedon* (B; M).
Mimidae. *Mimus [saturninus] patagonicus* (N); *M. [thenca] thenca* (N); *Mimus [triurus] triurus* (N).
Turdidae. *Turdus falcklandii* (F; R?).
Motacillidae. *Anthus correndera* (N); *A. hellmayri* (N).
Icteridae. *Molothrus bonariensis* (N); *Agelaius thilius* (N); *Curaeus curaeus* (F; R); *Sturnella loyca* (N).
Emberizidae. *Sicalis auriventris* (A); *S. lebruni* (N); *S. luteola* (N); *Diuca [diuca] diuca* (N); *Phrygilus [gayi] patagonicus* (F; M); *P. [gayi] gayi* (N); *P. fruticeti* (N); *P. unicolor* (A); *P. carbonarius* (N); *Melanodera melanodera* (N); *M. xanthogramma* (A); *Zonotrichia capensis* (B; M).
Carduelidae. *Carduelis uropygialis* (A); *C. [barbata] barbata* (F; P).

[1] Compiled from various sources, especially Philippi (1964), Olrog (1963, 1979), and Meyer de Schauensee (1966). Species (e.g., *Spiziapteryx circumcinctus, Xolmis coronata, X. irupero, Pitangus sulphuratus*) that occur only at the northern margin of Patagonia, and whose distribution is basically extra-Patagonian, are not included.
[2] Habitat occupancy: F = species occurring in forest only; N = nonforest only; A = Andean nonforest vegetation; B = both forest and nonforest. Migratory status of forest birds (F and B): M = migrant (species moves north of forest range during austral winter); P = partial migrant (some, usually southern, populations of species move north during austral winter; some populations also have local or elevational movements during non-breeding season); R = year-round resident.

APPENDIX II

Landbirds of Patagonian Forests: Taxonomy and Distribution

Theristicus caudatus. Widespread South American (Blake 1977). Marked geographic variation: high Andean populations (subspecies *branickii*), and southern South American ones (subspecies *melanopis*) morphologically differentiated from lowland ones, and at times treated as a species (*T. melanopis*), distinct from *T. caudatus* (tropical lowlands). Steinbacher (1979) included both species in a superspecies. Lowland forms (*caudatus* and *hyperorius*), and highland (*branickii*) and Patagonian (*melanopis*) taxa here considered to be members of a single species. Populations in *Nothofagus* region do not vary geographically; all are *T. c. melanopis*. Populations of *T. c. melanopis* on small islands off southern Chile not differentiated morphologically from mainland populations. *T. caudatus* found in openings and clearings within *Nothofagus* belt. Migrant.

Cathartes aura. In *Nothofagus* region, continental populations of this Pan-American species (Blake 1977) show geographic variation, but insular populations do not. Wetmore (1964:10) recognized subspecies *jota* (Andes: southern Colombia to southern Chile) and *falklandica* [Falkland Islands, and "north along the western coast of South America to Chile (Isla Mocha, Penco), Peru (Talara, Islas de Chincha), and Ecuador (Isla Jambelí, Isla La Plata)"]. Stresemann and Amadon (1979) adopted same treatment. According to Wetmore (1964:11) populations under name *falklandica* "differ definitely in smaller size" from *jota*. He added: "It may be supposed that [*C. aura falklandica*] is adapted to maritime conditions influenced by colder oceanic waters." Migrant in southernmost part of its range (Humphrey et al. 1970).

Coragyps atratus. Pan-American species (Blake 1977) inhabits only northern section of *Nothofagus* region. Apparently, no insular populations. Wetmore (1962) distinguished under name *Coragyps atratus foetens* all Andean populations from Ecuador to southern Chile, and lowland populations from Paraguay to southern Argentina. Stresemann and Amadon (1979) did not recognize subspecies, but Blake (1977) admitted *foetens*. Schlatter et al. (1978) discussed pathogenic role of *Coragyps atratus* in forest near Valdivia. Appears resident.

Vultur gryphus. Chiefly Andean in northern part of its range, but in Patagonia occurs also in lowlands, including *Nothofagus* region (Blake 1977). No geographic variation (Stresemann and Amadon 1979); populations on islands off southern Chile undifferentiated morphologically. Appears resident.

Accipiter [bicolor] bicolor. Pan-American *Accipiter bicolor* (Wattel 1973; Blake 1977) varies geographically outside *Nothofagus* region. Birds living in *Nothofagus* region morphologically distinct (taxon *chilensis*). Hellmayr (1932:279) treated *chilensis* as a species, but remarked, "is probably subspecifically related to *A. bicolor*." Amadon (1964) kept *chilensis* as a subspecies of *A. bicolor*. Brown and Amadon (1968:533) treated *chilensis* as a subspecies of *bicolor* but wrote "Some of the forms [of *A. bicolor*] . . . have been regarded as separate species and *A. b. chilensis* may well be [one]." Wattel (1973:121) considered *chilensis* a subspecies of *A. bicolor*. He thought that "the resemblance (of *A. b. guttifer*) to *A. b. chilensis* is purely coincidental rather than an indication of recent contact between *A. b. guttifer* and *A. b. pileatus*." Wattel (1973) considered *bicolor, cooperi* and *gundlachii* closely related to each other, but did not include them in a superspecies. Stresemann and Amadon (1979) treated *chilensis* as a "megasubspecies" (*sensu* Amadon and Short 1976) of *A. bicolor*; they included *bicolor, cooperi* and *gundlachi* in a superspecies (see also Mayr and Short 1970). I follow their treatment here. Apparently, no geographic variation within *Nothofagus* region (taxon *chilensis*); no differentiation among insular populations. *A. bicolor chilensis* migrates to northwestern Argentina. Olrog (1949) mentioned a sight record of a bird he thought was *chilensis* in small *Alnus* forest near Tafí del Valle, Tucumán. Olrog believed *A. b. chilensis* might follow southern migrants, for example *Zonotrichia capensis australis*.

Buteo [albicaudatus] polyosoma. *B. polyosoma* (Andean; Vaurie 1962; Blake 1977), *B. poecilochrous* and perhaps also *B. galapagoensis* and *B. albicaudatus* included in superspecies by Stresemann and Amadon (1979), treatment followed here. Probably most closely related to *Buteo poecilochrous*, with which partly sympatric (Vaurie 1962). Coastal Ecuadorian and Peruvian birds have relatively short wings, Andean ones have relatively long wings (Vaurie 1962). Apparently no geographic variation within *Nothofagus* region. Vaurie (1962:286) considered "the population . . . from the Juan Fernandez Islands . . . [to be] very distinct" (*B. polyosoma exsul*). Taxon *exsul* is considered a megasubspecies by Stresemann and Amadon (1979). Food of *B. [albicaudatus] polyosoma* in central Chile north of forest region in Schlatter et al. (1980). Species appears resident, but partial migration possible (see Humphrey et al. 1970).

Buteo [jamaicensis] ventralis. Hellmayr and Conover (1949) and Amadon (1964) considered *B. ventralis* quite similar to North American *B. jamaicensis*. Stresemann and Amadon (1979:371) placed *ventralis* near *jamaicensis*, and remarked that "*B. buteo, oreophilus*, and *brachypterus* form a superspecies, to which, perhaps, *jamaicensis* and *ventralis* should be added." Mayr and Short (1970:38) think that to consider *buteo* and *ventralis* conspecific with *jamaicensis* "may prove to be correct." See also Brown and Amadon (1968). In this paper I consider *ventralis* and *jamaicensis* as allospecies. Behn (1947) described several phases of *B. ventralis*. Species seems restricted to *Nothofagus* region (Blake 1977), but Olrog (1949) identified (with some doubt) one specimen from Tucumán, NW Argentina, as *B. ventralis*. Bernath (1965) saw one near Rio Chico, where he reported forest birds (e.g., *Aphrastura spinicauda, Pygarrhichas albogularis*), but according to Humphrey et al. (1970), it is found in open country in Tierra del Fuego. No geographic variation. Appears resident.

APPENDIX II

CONTINUED

Geranoaetus melanoleucus. Andean (Blake 1977), shows only minor geographic variation. I follow Amadon (1964, 1982) in assignment of species to monotypic genus *Geranoaetus.* Patagonian populations belong to *G. m. australis* (Stresemann and Amadon 1979). *Nothofagus* region populations do not appear to vary geographically. Food of *G. melanoleucus* in central Chile north of forest region described by Schlatter et al. (1980). Species appears resident.

Polyborus chimango. Southern South American (Blake 1977), usually placed in genus *Milvago* (Stresemann and Amadon 1979), but here placed in *Polyborus* (Vuilleumier 1970; Short 1975; Olrog 1979). Brown and Amadon (1968) and Short (1975) placed *P. chimango* and *P. chimachima* in a superspecies, but Vuilleumier (1970) suggested such an allocation was incorrect because of large overlap between them (see Olson 1976). Geographic variation apparently correlated with ecological and climatic factors. Birds from dry or open habitats north and east of *Nothofagus* region generally paler than birds near or in forest (edges, clearings) in wetter localities. Intergradation occurs between two kinds of populations. Paler birds recognized as subspecies *chimango*; darker ones as *temucoensis.* Birds from Isla Grande (Tierra del Fuego) weakly differentiated (somewhat larger size) from mainland ones (subspecies *fuegiensis,* Johnson and Behn 1957:353). Stresemann and Amadon (1979) recognized only two subspecies, the form *temucoensis* being the Patagonian form (*fuegiensis* was not mentioned). See Olrog (1962) for geographic variation and migration. Yáñez et al. (1982) published on the food of *P. chimango* in central Chile north of forest region. Southern populations (*fuegiensis* and *temucoensis*) migratory.

Polyborus [megalopterus] albogularis. Generic allocation, see F. Vuilleumier (1971), Short (1975). Populations in *Nothofagus* region (*albogularis*) assigned by Hellmayr and Conover (1949) to a subspecies of *megalopterus,* and by Amadon (1964; Brown and Amadon 1968) to a distinct species. Stresemann and Amadon (1979) treated *albogularis* as a megasubspecies. Maps of *megalopterus* group in Blake (1977) and Vuilleumier (1970). Painting of *albogularis* by Louis Agassiz Fuertes in Wetmore (1926b). I (Vuilleumier 1970) suggested that *albogularis* is a subspecies of *megalopterus,* but would now prefer to treat it as an allospecies. Northernmost part of range of *albogularis* and southernmost of *megalopterus,* where two taxa may come in contact, is poorly known ornithologically (Johnson 1972:93). No known geographic variation within *albogularis.* In the south of its range, at least, it is a forest bird (Olrog 1948, 1950), but in the north (Rio Negro), it occurs also above timberline (pers. observ.) and in open vegetation east of the Andes (Peters 1923b). Bernath (1965:98) saw it "in mountain beechwood at the Payne Cordillera, 120 km. north of Puerto Natales, February, 1945." Appears resident.

Polyborus plancus. Pan-American species (Blake 1977). In Patagonia in steppes and open woodlands. In southern Patagonia occurs also in open forest and parkland. No geographical variation in forest region (including Tierra del Fuego and other islands), inhabited by subspecies *plancus.* Appears resident.

Vanellus [chilensis] chilensis. Lowland *chilensis* and Andean *resplendens* form a superspecies (Short 1975; Blake 1977). Bock (1958:64) thought *chilensis* and *resplendens* to be closely related to *V. vanellus,* and to constitute an "invasion of South America separate from [*Vanellus*] *cayanus,*" the third species of the genus in South America. Brodkorb's (1934) subspecies *fetensis* for Patagonian birds retained by Blake (1977). No geographic variation in *Nothofagus* region. Insular birds not morphologically differentiated. Common in openings and clearings within forest. Southern populations migratory (Humphrey et al. 1970).

Columba [fasciata] araucana. C. [*fasciata*] *araucana* in a superspecies with *fasciata* (western North and South America), and *caribaea* (Jamaica) (Goodwin 1959, 1970; Johnston 1962). Restricted to *Nothofagus* region (map in Goodwin 1970). Quite rare some time ago probably as a result of decimation by Newcastle virus in 1953–1955 (Behn 1957; Goodall et al. 1957:385; Goodall et al. 1964:513–514). Has slowly recovered since 1960, at least in Chile. In Argentina (Rio Negro), Contreras et al. (1980) stated species was extremely abundant until mid 1950's, declined until about 1957–58, and was not seen after 1960. Cubillos et al. (1979) reported avian pox in this species on the basis of a specimen captured in 1977 in Osorno and suggested that the earlier diagnosis of Newcastle virus infection was perhaps mistaken, or else both diseases have been operating. They thought that the episodic mortalities registered in this species recently could be due to avian pox. *C. fasciata* has no geographic variation. Appears resident.

Enicognathus ferrugineus. The two species of parakeets in (and restricted to) *Nothofagus* region are congeneric (Peters and Blake 1948) and broadly sympatric in south-central Chile. *Enicognathus ferrugineus* varies geographically: populations on island of Tierra del Fuego and on adjacent mainland are somewhat larger (subspecies *ferruginea*) than those on mainland farther north (subspecies *minor,* Chapman 1919). Populations on Guaitecas and Chiloé not differentiated from mainland ones. Map in Forshaw and Cooper (1973). Local movements possible (Forshaw and Cooper 1973).

Enicognathus leptorhynchus. Differs from *E. ferrugineus* in color, larger body size, and in elongated and decurved upper mandible (Peters and Blake 1948). More restricted range than *E. ferrugineus* (Forshaw and Cooper 1973). No geographic variation. Long billed *E. leptorhynchus* eats the cones of *Araucaria araucana* (Goodall et al. 1957:411; Forshaw and Cooper 1973), whereas small billed *E. ferrugineus* may not be able to open them. *Enicognathus leptorhynchus* apparently affected by Newcastle disease in the early 1950's (references under *Columba [fasciata] araucana*). Appears resident, but elevational movements possible (Forshaw and Cooper 1973).

APPENDIX II

CONTINUED

Bubo [bubo] virginianus. Pan-American. Mayr and Short (1970:51) stated: "We consider *bubo, virginianus* and *africanus* closely related but not conspecific," and, "indeed it is not entirely clear whether they comprise a superspecies or a species group" (see also Short 1975). Considered here to belong in *bubo* superspecies. Populations of *virginianus* in *Nothofagus* region assigned to subspecies *magellanicus,* which ranges north in Andes and east in open Patagonian vegetation (Traylor 1958). Birds on islands off southern Chile not differentiated morphologically. Appears resident.

Glaucidium [brasilianum] nanum. Taxon *nanum,* considered a species (e.g., Hellmayr 1932:266), or a subspecies of Pan-American *G. brasilianum* (e.g., Olrog 1963:172) is disjunct, with populations in *Nothofagus* region and in northwestern Argentina. Wetmore (1926a:200–201), Hellmayr (1932:268), and Mayr and Short (1970:51) thought *brasilianum* and *nanum* to be closely related. Short (1975) treated *jardinii, brasilianum,* and *nanum* as allospecies of a superspecies (map p. 234), a procedure adopted here, even though *nanum* (subspecies *tucumanum*) and *brasilianum* overlap in Argentina. Geographic variation in *nanum* led Wetmore (1922) to describe subspecies *vafrum,* but Hellmayr (1932:267) thought that "hardly more than 40 percent [of *vafrum* specimens] are distinguishable." See Olrog (1948) on geographical variation. Migratory (at least in Argentina: Olrog 1963, 1979). American Museum of Natural History Specimen 170333, 21 June 1920, from Moquegua, southern Peru, identified as *nanum vafrum* by Chapman (1922b:5; 1929) may be a migrant.

Strix rufipes. Southern South American. Short (1975) discussed relationships of *S. rufipes rufipes (Nothofagus* region) and *S. r. chacoensis* (chaco and monte vegetation). Mayr and Short (1970:52) wrote that "South American *S. rufipes* and *S. hylophila* are nearer [*varia*] than are Old World species." They included *S. varia* and *S. occidentalis* in the *S.* [*varia*] superspecies. Short (1975:235) stated that *S. rufipes* was "related to allopatric *hylophila,* but not so closely as to form a superspecies." Mainland and Tierra del Fuego birds referred to subspecies *rufipes.* Chiloé Island birds, described by Wheeler (1938) as subspecies *sanborni,* somewhat darker and smaller than mainland birds. Appears resident.

Sephanoides sephaniodes. Genus has one species (*S. fernandensis*) in Juán Fernández Islands, and another (*S. sephaniodes*) in *Nothofagus* region as well as Juán Fernández Islands, apparently a case of speciation by double invasion. *S. sephaniodes* shows no geographic variation, either on mainland, or on islands of Tierra del Fuego, Chiloé, Mocha, and even Más a Tierra of the Juán Fernández, thus suggesting recent colonization of that distant island. Species apparently not migratory although has local movements (Johnson 1967).

Ceryle [alcyon] torquata. Central and South American. *C. [alcyon] torquata* occurs along lakes and rivers in *Nothofagus* region. *C. [alcyon] torquata* "appears to be a close relative of *C. alcyon*" according to Mayr and Short (1970:54; see also Miller 1912, 1920). More recently, Fry (1980) suggested that *alcyon, torquata* (both New World), *lugubris,* and *maxima* (Old World) were all allospecies of a single superspecies. Short (1975:242) considered that *torquata* and *alcyon* form a superspecies, and that both "have their closest relatives in the tropical Old World (*C. maxima, C. lugubris*), and they most likely reached the New World from Africa." Populations in *Nothofagus* region not geographically variable (all belong to subspecies *stellata*). Populations on islands not morphologically differentiated. Migratory or at least partial migrant.

Picoides [mixtus] lignarius. Two species of South American *Picoides* form a superspecies (Short 1975). *Picoides lignarius* occurs in Cochabamba and Santa Cruz, Bolivia (Bond and Meyer de Schauensee 1943: 220), and in *Nothofagus* region, whereas *P. mixtus* occurs from south-central Brazil to southern Bolivia, Paraguay, northern and central Argentina, and Uruguay (Short 1970a, 1975, 1982). The two species very similar morphologically. Data insufficient to reach definite conclusions about partial overlap (Short 1982). Sympatry, or at least contact, between *P. mixtus* and *P. lignarius* appears possible (1) in southern Bolivia, and (2) in northern Patagonia, but two species seem to have somewhat different ecological requirements (Voous 1947; Short 1975). I agree with Voous (1947:98) that the separation of an ancestral stock into two is "recent" (i.e., Pleistocene) but doubt that his explanation of the event is correct. He postulated that "the Pleistocene transgression of a part of South America [caused] an oceanic penetration into the region of the Rio de la Plata, which again separated the Brazilian and the Patagonian parts of the continent." He wrote further: "In the Patagonian island *lignarius* developed, whereas *mixtus* was isolated either in the S. Brazilian island or in the temporarily connecting archipelago between the two parts of the continent." Although fluctuations in sea level occurred in southern South America during the Pleistocene, it seems doubtful that rises caused transgressions of such a magnitude to separate whole islands in the sense visualized by Voous. It seems more plausible that a rise in sea level in the Rio de la Plata area would have merely extended the estuary upstream.

Hellmayr (1932:252) mentioned that specimens of *P. lignarius* "from Valdivia and south are more heavily streaked underneath, and the Guaitecas bird, besides having a stronger bill, certainly is more coarsely marked than any other example seen by me." Geographical variation in *P. lignarius* should be studied with large series (see also Wetmore 1926b). Short (1971:108–109) published a phylogeny of New World species of *Picoides.* "The South American form reached the false beech (*Nothofagus*) forests of southern South America, where an isolate evolved into the still closely similar *P. lignarius*—the two

APPENDIX II
CONTINUED

supposed species *lignarius* and *mixtus* do not meet." Short (1980:1269) postulated a barrier "between the Andean slopes of Bolivia-Argentina and the chaco" to help account for speciation in *Picoides* [*mixtus*]. Contreras (1980) also discussed speciation in *Picoides* [*mixtus*] superspecies. Migratory in Argentina (Meyer de Schauensee 1966; Short, 1982) but apparently resident in Chile (Johnson 1967).

Colaptes pitius. Belongs in group of flickers including Andean *C. rupicola* and lowland *C. campestris* (Short 1972). Morphologically, *pitius* resembles more closely *rupicola* than *campestris*. Wetmore and Peters (1922) described subspecies *cachinnans* from northern Patagonia along Argentine slope of the Andes (shorter billed, and more heavy barrings of underparts; see also Goodall et al. 1964:454). Hellmayr (1932:249) thought *cachinnans* may not be valid; Short (1982) did not recognize it. According to Hellmayr (1932), birds from Chiloé and Guaitecas Islands are slightly differentiated.

Short (1970a, 1972, 1982) published detailed information on *Colaptes pitius*. He stated (1972:105): "In South America, speciation gave rise to the ancestor of *campestris* and of *pitius-rupicola*." And further (1972:105–106): "The ancestor of *pitius* and *rupicola* may have invaded the northern Andes along the forest fringes of the uplands, after which the northern and southern population became disjunct." In 1982 Short wrote (p. 383) that *C. pitius* was "Related somewhat to *C. rupicola,* but these differ too much to be considered allospecies of a superspecies, although their ranges might be considered complementary." From a zoogeographical point of view they might be called an "ex-superspecies." A painting by Louis Agassiz Fuertes is reproduced in Wetmore (1926b). Species apparently resident.

Campephilus magellanicus. Occurs only in *Nothofagus* region, where it shows no geographic variation. Formerly placed in monotypic genus *Ipocrantor* (Hellmayr 1932), but Peters (1948) put it in *Campephilus*. Berlioz (1955) included *Phloeoceastes* in *Campephilus* (see also Short 1970b). In some characters *magellanicus* appears closer to some tropical species than to *principalis* and *imperialis* of Central America, Florida, and West Indies (Short 1970b, 1982). *Campephilus magellanicus* may be related to primarily tropical ivory-billed woodpecker stock. This stock could have split when there was either a direct connection between beech forests and other forests farther north in South America, or when these forest tracts were closer than they are now. According to Short (1982:448): "A common ancestor of *Campephilus leucopogon* (which approaches *magellanicus* most closely in its range—the chaco and scrub of western Argentina) and of the *C. melanoleucos* group most probably gave rise to *magellanicus* (Short 1970[b]), which is monotypic." If the dry belt across southern South America (which isolates forests today) formed sometime in the early Pliocene or earlier, then invasion of beech forests by proto-*magellanicus* ancestor took place earlier. Short (1980:1270) suggested that a putative former barrier isolated the ancestors of *C. leucopogon* and *C. magellanicus* "in the eastern fringes of the chaco or slightly to the east, perhaps through inundation of the Parana-Paraguay river region, isolating . . . the ancestor of *Campephilus leucopogon-C. magellanicus* to the west and their corresponding relatives . . . the ancestor of *C.* [*melanoleucos*] and *C. robustus,* in the east." Apparently resident.

Cinclodes patagonicus. Southern South American, found along streams and lake shores within forest (Vaurie 1980). Birds of Tierra del Fuego somewhat larger and with more white along shaft of breast and abdomen feathers than birds farther north. Geographic variation appears clinal. Hellmayr (1932:177) stated that "Birds from Valdivia, Chiloe, and the Guaitecas Islands . . . are variously intermediate [assigned to subspecies *chilensis*], and certain specimens hardly differ in coloration from typical *patagonicus*" [subspecies in Straits of Magellan and Tierra del Fuego]. Vaurie (1980:39) wrote: "birds from the northern part of its range tend to be slightly less streaked on the breast." Southernmost populations leave breeding areas in austral winter (Humphrey et al. 1970); Chilean populations, however, appear resident (Johnson 1967).

Cinclodes [*fuscus*] *fuscus.* Andean and southern South American superspecies (Vaurie 1980) found along water or marshy areas in *Nothofagus* region. No geographic variation in continental forest region (subspecies *fuscus*). Island populations (e.g., Tierra del Fuego) not differentiated. Migratory in southern South America (Vaurie 1980).

Sylviorthorhynchus desmursii. Monotypic genus without close relatives among living Furnariidae, found exclusively in *Nothofagus* region (map in Vaurie 1980). Long ago Sclater (1890:31) noticed similarity between *Sylviorthorhynchus* and Australian malurine genus *Stipiturus*. Vaurie (1980) discussed resemblances further and concluded that they were due to convergence. Immature *Sylviorthorhynchus* similar to adults in general coloration, but lacks rufous on forehead, and shows obsolescent bars on underparts. Similar barrings on underparts found in several species of genus *Asthenes*. Nest of *Sylviorthorhynchus* (Goodall et al. 1957:264) rather similar to nests of genera *Synallaxis* and *Asthenes*. Evidence available indicates *Sylviorthorhynchus* is a Furnariidae. Additional kinds of evidence, especially biochemical, needed to shed more light on relationships. Not on Tierra del Fuego, but occurs on Guaitecas and Chiloé islands. No geographic variation either in mainland or in insular populations. Resident.

Aphrastura [*spinicauda*] *spinicauda.* *Aphrastura* perhaps closely related to monotypic *Phleocryptes melanops,* a bird living in reed-beds. *Aphrastura* and *Phleocryptes* show resemblances in plumage pattern (but parts of pattern found in other furnariid genera), but differ conspicuously in vocalizations, ecological preferences, and nest structure. Vaurie (1980) suggested that *Aphrastura* is related to *Leptasthenura,* and

Phleocryptes to *Synallaxis*. Occurs from Talinay and Fray Jorge woodlands south to Tierra del Fuego and Cape Horn Archipelago (Olrog 1950; Goodall et al. 1957). Continental and Tierra del Fuego populations of *A. spinicauda* (Vaurie 1980) show no geographic variation. Birds from Mocha Island slightly larger and more buff or ochraceous on underparts than mainland birds; described by Chapman (1934) as subspecies *bullocki* on the basis of two specimens. Population of Chiloé Island (*fulva*) differentiated from mainland birds in being more saturated in color (Hellmayr 1932:193). Vaurie (1980:59) wrote that "It is possible that birds from the islands farther south (Guaitecas and Chonos Archipelago) are also similar to those of Chiloé, as one specimen taken on Melchor Island in the Chonos has been identified as *fulva* by Trimble (1943)." Jehl (in Johnson 1972:100–101) wrote: "It appears that the dark-bellied spinetails (rayaditos) pertaining to the *A. s. fulva* or Chiloe form, may be irregularly distributed on nearshore islands over more than 400 miles of the Chilean coast. This distributional pattern suggests that the island populations were derived from separate invasions from the mainland and that the dark ventral coloration represents a parallel response to a common environmental condition, such as increased humidity. It would not be surprising to find dark-bellied populations on many unstudied islands. Exceptions, such as the white-bellied Guaitecas island birds, may represent recent invasions from the mainland." Population on Más Afuera of Juán Fernández Islands well-marked, larger in wing, tail, culmen, and tarsus lengths, and with less distinct patterning and more uniform dark brownish-gray coloration (*masafuerae*). *Aphrastura masafuerae* seen in 1928, but not in 1955 (Johnson 1967), and may be extinct. Más Afuera birds traditionally considered as a distinct species (e.g., Vaurie 1980). I include both forms in a superspecies (footnote no. 43 in Vaurie 1980). Lack of geographic variation on mainland, contrasted with slight to marked variation in insular populations, is notable. *Aphrastura spinicauda* social during non-breeding season (Vuilleumier 1967b). Flocking behavior may be partially responsible for morphological uniformity found on continent, by either favoring greater panmixia than in non-flocking species, or because advantages accruing to birds because of flocking selectively favor morphological unity. The above explanation may be pertinent to existing insular differentiation. If many island isolates are the result of invasion by separate founder populations, then selective factors due to flocking may reinforce the effects of the founder principle. A careful study of continental versus insular populations of *A. spinicauda* with large series might be worthwhile. *Aphrastura spinicauda* is convergent on *Parus* and *Certhia* (Johansen 1966; Vuilleumier 1967b; Humphrey et al. 1970). Resident, but may be partial migrant in southernmost part of its range (Humphrey et al. 1970).

Pygarrhichas albogularis. Well-marked monotypic genus living exclusively in *Nothofagus* region (Vaurie 1980). *Pygarrhichas* resembles several other genera in subfamily Phylidorinae in structural characters (bill shape), and in habit of digging its nest-hole in soft wood (Goodall et al. 1957). Nevertheless, *Pygarrhichas* has no close relatives among living Furnariidae. Convergence on *Sitta* mentioned by Johnson (1967), Short (1969b), Vuilleumier (1967b), Johansen (1966). Notes on behavior and habitat in Short (1969b) and Vuilleumier (1967b, 1972). Geographic variation apparently nil in continental, as well as insular (Tierra del Fuego), populations. Resident.

Pteroptochos [*tarnii*] *castaneus* and *P.* [*tarnii*] *tarnii*. Genus *Pteroptochos* contains one species (*megapodius*) in steppes, monte, and matorral in central Chile, and two others (*castaneus* and *tarnii*) replacing each other geographically in *Nothofagus* region. Hellmayr (1932) treated the last two as species. According to Behn (1944), *castaneus* and *tarnii* are similar in immature plumage, ecological requirements, and vocalizations, and can only be distinguished by color of the underparts in adult specimens: *tarnii* has gray chin and breast, while *castaneus* has chestnut chin and breast. Bío-Bío River separates the two forms. In Concepción area, where river is over 1 km in width, *tarnii* occurs only south, and *castaneus* only north. In upper Bío-Bío valley situation not so clear because Bío-Bío does not seem broad enough to act as a barrier in the same way it does near Concepción. Furthermore, there seems to be a gap in collecting records (see Behn's 1944 map). It is therefore not possible to say whether *tarnii* and *castaneus* hybridize, occur side by side without interbreeding, or overlap slightly. Apparently only a single immature specimen collected at Los Angeles, in the critical zone (see Behn's map); its immaturity prevents clear-cut identification. Behn (1944:465) thought *castaneus* and *tarnii* were subspecies of the same species, and added: "nada de raro sería que en los alrededores de Los Angeles, donde las barreras naturales disminuyen en importancia, exista intergradación." I consider these taxa as allospecies of a superspecies (see also Sick 1960:171). There might be slight geographical variation in *tarnii* (see Wetmore 1926a). Short (1969a) gave information on vocalization, behavior, and habitat of *Pteroptochos* [*tarnii*] *tarnii* that complements that in Johnson (1967). Both allospecies in *tarnii* superspecies resident.

Scelorchilus [*rubecula*] *rubecula*. Genus *Scelorchilus* has two allopatric, morphologically well-marked, and ecologically differentiated species. *Scelorchilus albicollis* occupies open shrub, matorral, and steppe, whereas *S. rubecula* lives in *Nothofagus* region. I consider *S. rubecula* and *S. albicollis* members of a superspecies. *Scelorchilus rubecula* does not occur on Tierra del Fuego. Mainland populations exhibit no apparent geographic variation although Peters (1923a) described subspecies *hylonympha* for birds living on Argentine side of forests in Rio Negro. Wetmore (1926b:445) stated that *hylonympha* and *nemorivaga* (which he described himself) were "based on postmortem changes in coloration and so must stand as

APPENDIX II
CONTINUED

synonyms of *rubecula*." He discussed faint variations in mainland birds from Argentina and Chile and "faintly darker" birds from Chiloé Island than from central Chile. Birds from Mocha Island, a little larger in wing- and tail-length than neighboring continental birds, were described as subspecies *mochae* by Chapman (1934). For notes on habitat and behavior of *S.* [*rubecula*] *rubecula* see Short (1969a), complementing data in Johnson (1965). Resident.

Eugralla paradoxa. Monotypic genus occurring only in *Nothofagus* region. Monotypic genera *Eugralla* and *Myornis* (*senilis*) present similarities with, and differences from, polytypic genus *Scytalopus*. Similarities are ecological, all three genera inhabiting dense, wet (normally montane) forests or thickets, and morphological, the adults being entirely gray (some species of *Scytalopus* and *Myornis*), or largely gray with brownish to rufescent lower belly (other species of *Scytalopus* and *Eugralla*). Differences among three genera include: immature plumage of *Myornis* appears to lack barrings present in *Scytalopus* and *Eugralla*; feet of adult *Eugralla* pale yellowish, whereas they are dark in *Myornis* and in most, if not all *Scytalopus*; finally, casque most developed in *Eugralla*, less so in *Myornis*, and least in *Scytalopus* (however, some species of *Scytalopus* have a culmen ridge only a trifle less prominent than that found in *Myornis senilis*. This difference may be size-related, since *Eugralla* is the largest, *Myornis* intermediate, and *Scytalopus* the smallest. Generic limits between *Eugralla*, *Myornis*, *Scytalopus* may be hard to draw. No geographic variation on mainland or islands of Chiloé and Mocha. *Eugralla paradoxa* is resident.

Scytalopus magellanicus. Tapaculos of genus *Scytalopus* offer difficult taxonomic problems because of morphological uniformity throughout genus. Birds in *Nothofagus* region belong to *S. magellanicus*, a species which, as understood by Zimmer (1939), occurs along Andes from Venezuela to Tierra del Fuego. Birds from Tierra del Fuego not differentiated from neighboring mainland ones. From the Straits of Magellan north to about region of Concepción, no apparent geographic variation exists although there is individual variation, especially presence of whitish on forehead (Hellmayr 1932:223). From about the Concepción area north to Atacama, birds are, as a whole, different morphologically from the ones occurring south of Concepción. Northern birds (taxon *fuscus*) are larger (tail length, bill length, leg size) and darker, less gray, than southern birds (taxon *magellanicus*). Exact ranges of the two forms insufficiently known, especially where they meet, near Rio Bío-Bío (Johnson 1967). Goodall et al. (1957:410) indicated that the two forms have distinct songs, and that their ranges may overlap slightly. Southern birds (*magellanicus*) apparently forest birds, whereas northern birds (*fuscus*) occur in more open situations. A specimen of *fuscus* I collected on 25 March 1965 at 2400 m in the Valle del Yeso, Cordillera de Santiago, was one of several birds in a treeless area, at the foot of a rocky cliff where cover was provided by large boulders. Much farther south, near the Laguna de la Laja (BíoBío/Ñuble boundary) at 1400 m, I heard *Scytalopus* in treeless areas with huge lava blocks that offered cover. Either *magellanicus*, or *fuscus*, or both, occur in this region, but, unfortunately, I was not able to collect specimens. In Cape Horn Archipelago, Olrog (1950) found *S. magellanicus* in forest and open habitats (high grassland; shrubby vegetation of *Berberis* and *Ribes*). For field notes on *Scytalopus magellanicus* see Short (1969a), complementing the information in Johnson (1965). Resident.

Elaenia albiceps. Widespread South American. Traylor (1979) recognized six subspecies, but apparently no geographic variation in *Nothofagus* region. Populations of Tierra del Fuego not differentiated from mainland ones. Birds of *Nothofagus* region all belong to subspecies *chilensis* (Hellmayr 1927, 1932). *Elaenia a. chilensis* ranges from Bolivia through Tierra del Fuego. (For an earlier treatment of *E. albiceps* see Zimmer 1941). Short (1975) stated that *E. albiceps* and *E. parvirostris* form a superspecies. Traylor (1982) discussed these two species in detail and published a partial map of their distribution. The situation described by Traylor is not simple. "In Argentina there is positive evidence that *E. parvirostris* and *E. albiceps chilensis* behave as distinct species" (Traylor 1982:13). But in Bolivia "at intermediate altitudes . . . populations are found that are intergrades between *E. parvirostris* and *E. albiceps*, suggesting that the two species hybridize in this area" (Traylor 1982:15). Traylor (1982) concluded that "They seem to be good sympatric species that, for some reason, hybridize in a limited part of their range." As in *Polyborus chimango* and *P. chimachima*, the broad overlap in ranges precludes one from concluding that *E. albiceps* and *E. parvirostris* are allospecies of a superspecies. *Elaenia albiceps* is migratory.

Anairetes parulus. Andean. Populations in *Nothofagus* belt, including those on islands off southern Chile, are not, or at best very slightly geographically variable (but subspecific names exist: *curatus* Wetmore and Peters 1922; *lippus* Wetmore 1923). Zimmer (1940) kept *lippus* but not *curatus*. Traylor (1979) recognized subspecies *patagonicus* from southern Mendoza to northern Santa Cruz in Argentine Patagonia and *parulus* in Chile and southwestern Argentina (area of *Nothofagus* forests). Birds from Más a Tierra in Juán Fernández group (*A. fernandezianus*) differ from mainland *A. parulus*, usually considered a species (e.g., by Chapman 1934; Traylor 1979; for remarks on the differences, see Hellmayr 1932:141). Differences between *fernandezianus* and any population of *parulus* are as large or larger than differences between *A. parulus* and another Andean species *A. reguloides*, and definitely larger than those between *A. parulus* and *A. flavirostris*, which are sympatric. Taxon *fernandezianus* has apparently reached species status. *Anairetes fernandezianus* may not be derived from *parulus*-like stock. Chapman (1934:6) noted, "in its heavily streaked throat and breast it [*fernandezianus*] more nearly resembles [*A.*] *reguloides albiventris*

APPENDIX II
CONTINUED

Chapman of the central Peruvian coast." Traylor and Fitzpatrick (1982:25) stated that *A. fernandezianus* has not been found on Más a Tíerra since 1917 "and may now be the only modern flycatcher to have become extinct." In *Nothofagus* region it is an edge bird, rather than a true forest bird. *Anairetes parulus* appears to be at least a partial (or local) migrant (see Johnson 1967; Humphrey et al. 1970).

Ochthoeca parvirostris. Found only in *Nothofagus* region. Populations of Tierra del Fuego not differentiated from mainland ones; continental populations do not exhibit geographic variation. In some checklists (Hellmayr 1927:400; Meyer de Schauensee 1966:379) this species was placed in the genus *Colorhamphus* immediately after *Mecocerculus*. As early as 1907, however, von Berlepsch included *parvirostris* in *Ochthoeca*. Traylor (1979) placed *O. parvirostris* between *O. oenanthoides* and *O. leucophrys*. *Parvirostris* forages in typical *Ochthoeca* fashion and has quick wing flicks (pers. observ.), also typical of the genus (Smith 1971). Its calls, a long, high-pitched whistle *piiii* . . . or *pee* . . . , emitted singly, are very reminiscent of a similar call emitted by *Ochthoeca cinnamomeiventris*. Migratory.

Xolmis pyrope. Placed by Meyer de Schauensee (1966:335) in monotypic genus *Pyrope,* but by F. Vuilleumier (1971) and Traylor (1977, 1979) in *Xolmis.* Birds from Tierra del Fuego and mainland do not show geographic variation. Birds from Chiloé Island larger than neighboring mainland birds recognized as subspecies *fortis.* Traylor (1979) recognized *fortis* (but not *ignea,* previously described by Wetmore 1923). Migratory.

Phytotoma rara. Phytotomidae have only three, rather closely related, allopatric species. *Phytotoma rara* occurs in matorral of central Chile and around *Nothofagus* belt. It clearly lives in wetter habitats than either *P. rutila* or *P. raimondii.* *Phytotoma rara* does not show any geographic variation. Short (1975) included *P. raimondii* and *P. rutila,* but not *P. rara,* in the same superspecies. Apparently resident in northern part of its range but migratory in south (Johnson 1967).

Notiochelidon cyanoleuca. Widespread South American; populations breeding in *Nothofagus* region belong to subspecies *patagonica* (Chapman 1922a:12). No geographic variation in *Nothofagus* region. Migratory.

Tachycineta [leucorrhoa] leucopyga. Forms a superspecies with allopatric *leucorrhoa* (Short 1975). *Tachycineta [leucorrhoa] leucopyga* occurs in *Nothofagus* region, whereas *T. [leucorrhoa] leucorrhoa* is found in more open country from Bolivia and southeastern Brazil to central Argentina. No geographic variation. Migratory.

Troglodytes aedon. Pan-american (Paynter 1957). Birds inhabiting *Nothofagus* region belong to subspecies *chilensis* (Paynter 1960). For an earlier treatment see Chapman and Griscom (1924:304). Tierra del Fuego populations identical with mainland ones, but Falkland Islands birds were considered sufficiently differentiated by Chapman (1934:8) to be called "a specifically distinct representative." The Falkland birds (*cobbi*) are more robust than mainland birds (longer and thicker bill, thicker legs, and heavier feet; but wing- and tail-lengths almost identical), and duller-colored than mainland birds (well-marked insular subspecies of *T. aedon,* Paynter 1960). Migratory.

Turdus falcklandii. Occurs in *Nothofagus* region, and on Juán Fernández and Falkland Islands (Ripley 1964). Continental populations show no or very slight geographic variation. Wetmore (1923) described subspecies *pembertoni* from northern part of forested Patagonia. Otherwise, all Patagonian mainland birds belong to subspecies *magellanicus* (Ripley 1964). Birds on islands of Tierra del Fuego, Chiloé, the Guaitecas, and distant Juán Fernández not differentiated from mainland birds. Birds from Mocha island described by Chapman (1934) as subspecies *mochae* because they have "a consistently larger bill" (but he examined only three specimens); *mochae* not retained by Ripley (1964). Birds from Falklands have slightly larger bills and are slightly more ochraceous on underparts and lighter on crown than mainland birds (subspecies *falcklandii*). Apparently resident.

Curaeus curaeus. Blackbird genus found in *Nothofagus* region and Brazil (Blake 1968). Affinities of *Curaeus* not clear. *Curaeus* may be more closely related to *Agelaius* than to other blackbird genera. According to Wetmore (1926a:378), *C. curaeus* has "more or less agelaiine songs," and resembles *Pseudoleistes* in general behavior. Blake (1968) placed *Curaeus* between *Hypopyrrhus* and *Gnorimopsar,* not far from *Sturnella* and *Quiscalus,* but rather removed from *Agelaius.* No geographic variation within continental populations. Birds of Tierra del Fuego and other islands (Cape Horn Archipelago, Navarino, Hoste) slightly larger (body size, bill length) than mainland ones, have been recognized taxonomically (subspecies *reynoldsi*). Subspecies *recurvirostris* from Riesco Island (Markham 1971) intermediate in size between *curaeus* and *reynoldsi,* but distinct in bill shape. Blake (1968) stated that according to Sclater (1939:143), birds "from Gray Harbour and Tom Bay, both on the long channel between Wellington Island and the mainland in southernmost Chile," show some evidence of intergradation with *reynoldsi.* Apparently resident.

Phrygilus [gayi] patagonicus. Only *Phrygilus* in *Nothofagus* region. *P. [gayi] patagonicus* allopatric with its nearest neighbor *P. [gayi] gayi,* but contact (and possibly slight sympatry) between both forms may take place in Tierra del Fuego (Isla Grande) and on continent in Magallanes, Santa Cruz, and Chubut. *Phrygilus [gayi] patagonicus* forms a superspecies with *gayi* (southern South America), and high Andean *atriceps* and *chloronotus* (Vuilleumier, unpubl. data). No geographic variation in mainland or insular

APPENDIX II

CONTINUED

populations. According to Grigera (1976), eats vegetal matter during austral winter, and insects as well as vegetal matter in austral summer. Migratory.

Zonotrichia capensis. Widespread Andean and South American; variable geographically (Chapman 1940; Handford 1983). According to Nottebohm (1975) birds in forested areas sing at higher frequencies than those in arid scrub. Populations from Patagonia have been assigned to several subspecies; those from *Nothofagus* region are *chilensis* (essentially along Chilean side of the Andes) and *australis* (along Argentine side). Migratory.

Carduelis [barbata] barbata. *Carduelis [barbata] barbata,* found exclusively in *Nothofagus* region, is closer to *C. magellanica* than to any other form of the genus, and I include both in a superspecies (see also Short 1975). However, interspecific relationships within South American *Carduelis* are poorly understood, so I am not yet prepared to accept Short's (1975) suggestion to include *notata* and *siemiradzkii* in this superspecies. Howell et al. (1968) placed *C. barbata* in the following sequence: *C. atriceps, C. spinescens, C. yarrellii, C. cucullata, C. crassirostris, C. magellanica, C. dominicensis, C. siemiradzkii, C. olivacea, C. notata, C. xanthogastra, C. atrata, C. barbata, C. tristis, C. psaltria, C. lawrencei.* If this sequence is an indication of relationship, then Howell et al. (1968) view *barbata* as close to *atrata* and *tristis.* Shows no geographic variation on the mainland or islands off southern Chile and southernmost South America (Todd 1926). Grigera (1976) gave data on food preferences. *Carduelis [barbata] barbata* is largely resident but has local movements.

APPENDIX III

LAND BIRDS OF MONTANE (ANDEAN) FORESTS OF ARGENTINA[1]

Tinamidae. *Crypturellus tataupa.*
Cathartidae. *Cathartes aura;* ?*Sarcoramphus papa.*
Accipitridae. *Accipiter [bicolor] bicolor; Buteo leucorhous; Oroaetus isidori.*
Falconidae. *Micrastur ruficollis.*
Cracidae. *Penelope obscura; P. dabbenei; P. montagnii.*
Columbidae. *Columba [fasciata] fasciata; Leptotila megalura; Geotrygon frenata.*
Psittacidae. *Aratinga mitrata; Pyrrhura molinae; Pionus maximiliani; Amazona tucumana.*
Cuculidae. *Piaya cayana.*
Strigidae. *Bubo [bubo] virginianus; Pulsatrix perspicillata; Glaucidium [brasilianum] brasilianum; Glaucidium [brasilianum] nanum.*
Apodidae. *Streptoprocne zonaris; Cypseloides rothschildi.*
Trochilidae. *Colibri thalassinus; Adelomyia melanogenys; Eriocnemis glaucopoides; Microstilbon burmeisteri.*
Trogonidae. *Trogon curucui.*
Picidae. *Picumnus cirratus; Piculus rubiginosus; Veniliornis fumigatus; V. frontalis; Campephilus leucopogon.*
Furnariidae. *Synallaxis superciliosa; Philydor rufosuperciliatus; Xenops rutilans; Lochmias nematura.*
Formicariidae. *Thamnophilus connectens; T. ruficapillus; Grallaria albigula.*
Rhinocryptidae. *Scytalopus superciliaris.*
Cotingidae. *Platypsaris rufus.*
Tyrannidae. *Phyllomyias sclateri; Elaenia spectabilis; E. albiceps; E. strepera; E. obscura; Mecocerculus leucophrys; M. hellmayri; Phylloscartes ventralis; Pyrrhomyias cinnamomea; Contopus cinereus; C. [fumigatus] fumigatus; Knipolegus cabanisi; Myiarchus tuberculifer.*
Corvidae. *Cyanocorax chrysops.*
Hirundinidae. *Notiochelidon cyanoleuca.*
Cinclidae. *Cinclus schultzi.*
Troglodytidae. *Troglodytes solstitialis.*
Turdidae. *Catharus dryas; Turdus chiguanco; T. serranus; T. nigriceps.*
Parulidae. *Myioborus brunniceps; Basileuterus signatus.*
Coerebidae. *Diglossa baritula.*
Thraupidae. *Pipraeidea melanonota; Hemithraupis guira; Thlypopsis ruficeps; Chlorospingus ophthalmicus.*
Catamblyrhynchidae. *Catamblyrhynchus diadema.*
Emberizidae. *Saltator rufiventris; Pheucticus aureoventris; Atlapetes fulviceps; A. citrinellus; A. torquatus; Arremon flavirostris; Poospiza erythrophrys.*

[1] Compiled chiefly from Olrog (1949, 1963, 1979) and Vuilleumier (unpubl. data).

APPENDIX IV

LAND BIRDS IN THE PUNA OF ARGENTINA, CHILE, AND BOLIVIA (SOUTHERN ALTIPLANO)[1]

Rheidae. *Pterocnemia pennata.*
Tinamidae. *Nothura ornata; Tinamotis [pentlandii] pentlandii.*
Threskiornithidae. *Theristicus caudatus.*
Cathartidae. *Cathartes aura: Vultur gryphus.*
Accipitridae. *Circus [cyaneus] cinereus; Buteo [albicaudatus] polyosoma; B. poecilochrous; Geranoaetus melanoleucus.*
Falconidae. *Polyborus [megalopterus] megalopterus; Falco femoralis; F. [tinnunculus] sparverius.*
Charadriidae. *Vanellus [chilensis] resplendens; Eudromias ruficollis.*
Thinocoridae. *Attagis gayi; Thinocorus orbigynianus; T. rumicivorus.*
Columbidae. *Metriopelia [ceciliae] ceciliae; M. [ceciliae] morenoi; M. aymara; M. melanoptera.*
Psittacidae. *Bolborhynchus aurifrons.*
Strigidae. *Bubo [bubo] virginianus; Athene cunicularia; Asio flammeus.*
Caprimulgidae. *Caprimulgus longirostris.*
Trochilidae. *Oreotrochilus [estella] estella; O. [estella] leucopleurus; Patagona gigas.*
Picidae. *Colaptes rupicola.*
Furnariidae. *Geositta isabellina; G. rufipennis; G. punensis; G. [cunicularia] cunicularia; G. tenuirostris; Upucerthia dumetaria; U. [validirostris] albigula; U. [validirostris] validirostris; U. andaecola; U. ruficauda; Cinclodes [fuscus] fuscus; C. atacamensis; Phleocryptes melanops; Leptasthenura andicola; L. striata; L. aegithaloides; Asthenes dorbignyi; A. modesta; A. [anthoides] wyatti; A. humilis; A. [flammulata] maculicauda; Phacellodonus striaticeps.*
Tyrannidae. *Anairetes parulus; A. flavirostris; Tachuris rubrigastra; Ochthoeca oenanthoides; O. leucophrys; Myiotheretes rufipennis; Agriornis microptera; A. montana; A. albicauda; Muscisaxicola rufivertex; M. [albilora] juninensis; M. [albifrons] albifrons; M. [alpina] cinerea; M. maculirostris; Lessonia rufa.*
Hirundinidae. *Notiochelidon murina; N. cyanoleuca; Petrochelidon andecola.*
Troglodytidae. *Troglodytes aedon.*
Mimidae. *Mimus [triurus] dorsalis.*
Turdidae. *Turdus chiguanco.*
Motacillidae. *Anthus furcatus; A. hellmayri; A. correndera; A. bogotensis.*
Icteridae. *Agelaius thilius.*
Emberizidae. *Catamenia inornata; Sicalis lutea; S. uropygialis; S. olivascens; Diuca [diuca] speculifera; Idiopsar brachyurus; Phrygilus [gayi] atriceps; P. [gayi] gayi; P. fruticeti; P. unicolor; P. [erythronotus] dorsalis; P. [erythronotus] erythronotus; P. plebejus; P. alaudinus; Zonotrichia capensis.*
Carduelidae. *Carduelis crassirostris; C. atrata; C. uropygialis; C. [barbata] magellanica.*

[1] Compiled from various sources, especially Meyer de Schauensee (1966), Vuilleumier and Simberloff (1980), Philippi (1964), Olrog (1963, 1979), and Vuilleumier (unpubl. data).

APPENDIX V

LAND BIRDS IN COASTAL FORESTS OF CALIFORNIA[1]

Cathartidae. *Cathartes aura.*
Accipitridae. *Accipiter [bicolor] cooperii; A. [nisus] striatus.*
Falconidae. *Falco [tinnunculus] sparverius.*
Tetraonidae. *Dendragapus obscurus* (S).
Phasianidae. *Callipepla picta.*
Columbidae. *Columba [fasciata] fasciata.*
Strigidae. *Otus asio* (S); *Bubo [bubo] virginianus; Glaucidium [gnoma] gnoma* (E); *Strix [varia] occidentalis* (E).
Apodidae. *Chaetura [pelagica] vauxi* (E).
Trochilidae. *Selasphorus [rufus] sasin.*
Picidae. *Picoides pubescens; P. villosus* (S); *Melanerpes formicivorus; Sphyrapicus [varius] varius; Colaptes auratus* (S); *Dryocopus pileatus.*
Tyrannidae. *Empidonax difficilis* (E); *Contopus borealis; C. [virens] sordidulus.*
Hirundinidae. *Tachycineta thalassina; T. bicolor; Progne [subis] subis* (E).
Corvidae. *Perisoreus [infaustus] canadensis* (E); *Cyanocitta stelleri* (S); *Corvus [brachyrhynchos] brachyrhynchos.*
Paridae. *Parus rufescens* (E).
Sittidae. *Sitta pusilla* (S).
Certhiidae. *Certhia familiaris* (E).
Troglodytidae. *Troglodytes aedon; T. troglodytes* (E).
Muscicapidae:
    Sylviinae. *Regulus [regulus] satrapa* (E).
    Turdinae. *Turdus migratorius; Zoothera naevia* (E); *Catharus guttatus* (S); *C. ustulatus; Sialia mexicana.*
Bombycillidae. *Bombycilla cedrorum* (E).
Vireonidae. *Vireo huttoni.*
Emberizidae:
    Parulinae. *Vermivora celata; Dendroica coronata; D. nigrescens; D. [virens] occidentalis; Oporornis philadelphia; Wilsonia pusilla.*
    Thraupinae. *Piranga [olivacea] ludoviciana.*
    Cardinalinae. *Pheucticus [ludovicianus] melanocephalus.*
    Icterinae. *Euphagus cyanocephalus.*
    Emberizinae. *Pipilo erythrophthalmus; Spizella passerina; Junco [hyemalis] hyemalis.*
Fringillidae:
    Carduelinae. *Carpodacus [erythrynus] purpureus* (E); *Loxia curvirostra* (S); *Carduelis [spinus] pinus; C. psaltria; Coccothraustes vespertinus.*

---

[1] Compiled from Miller (1951); sequence of families and subfamilies after American Ornithologists' Union (1983) check-list; nomenclature of genera, species, and allospecies after Mayr and Short (1970). Abbreviations: E = "exclusive and principal adherent" species (Miller 1951); S = species "with other principal affinities which as races are endemic to the coastal forest" (Miller 1951).

APPENDIX VI

HELLMAYR'S TREATMENT OF PATAGONIAN FOREST BIRDS[1]

Species	Endemism			
	None	Subspecies	Species	Genus
Theristicus caudatus	x	—	—	—
Cathartes aura	x	—	—	—
Coragyps atratus	x	—	—	—
Vultur gryphus	x	—	—	—
Accipiter chilensis	—	—	x	—
Buteo polyosoma	x	—	—	—
Geranoaetus melanoleucus	—	x	—	—
Milvago chimango	—	x	—	—
Phalcoboenus albogularis	—	—	x	—
Polyborus plancus	x	—	—	—
Belonopterus cayennensis	x	—	—	—
Columba araucana	—	—	x	—
Microsittace ferruginea	—	—	x	x
Enicognathus leptorhynchus	—	—	x	x
Bubo virginianus	x	—	—	—
Glaucidium nanum	—	—	x	—
Strix rufipes	—	x	—	—
Sephanoides sephaniodes	x	—	—	—
Megaceryle torquata	—	x	—	—
Dyctiopicus lignarius	—	—	x	—
Colaptes pitius	x	—	—	—
Ipocrantor magellanicus	—	—	x	x
Cinclodes patagonicus	x	—	—	—
C. fuscus	x	—	—	—
Sylviorthorhynchus desmursii	—	—	x	x
Aphrastura spinicauda	—	—	x	—
Pygarrhicus albogularis	—	—	x	x
Pteroptochos tarnii	—	—	x	—
P. castaneus	—	—	x	—
Scelorchilus rubecula	—	—	x	—
Eugralla paradoxa	—	—	x	x
Scytalopus magellanicus	—	—	x	—
Elaenia albiceps	—	x	—	—
Spizitornis parulus	x	—	—	—
Colorhamphus parvirostris	—	—	x	x
Xolmis pyrope	—	—	x	—
Phytotoma rara	x	—	—	—
Pygochelidon patagonica	x	—	—	—
Iridoprocne leucopyga	—	—	x	—
Troglodytes musculus	x	—	—	—
Turdus falcklandii	x	—	—	—
Notiopsar curaeus	—	—	x	x
Phrygilus patagonicus	—	—	x	—
Zonotrichia capensis	x	—	—	—
Spinus barbatus	—	—	x	—
Totals (N = 45 species)	18	5	22	8

[1] Nomenclature as in Hellmayr (1932), but sequence as in Appendix VIII.

APPENDIX VII

PHILIPPI'S TREATMENT OF PATAGONIAN FOREST BIRDS[1]

Species	Endemism			
	None	Subspecies	Species	Genus
Theristicus caudatus	x	—	—	—
Cathartes aura	x	—	—	—
Coragyps atratus	x	—	—	—
Vultur gryphus	x	—	—	—
Accipiter chilensis	—	—	x	—
Buteo polyosoma	x	—	—	—
B. ventralis	—	—	x	—
B. fuscescens	x	—	—	—
Milvago chimango	—	x	—	—
Phalcoboenus albogularis	—	—	x	—
Caracara plancus	x	—	—	—
Belonopterus chilensis	x	—	—	—
Columba araucana	—	—	x	—
Microsittace ferruginea	—	—	x	x
Enicognathus leptorhynchus	—	—	x	x
Bubo virginianus	x	—	—	—
Glaucidium nanum	—	—	x	—
Strix rufipes	—	x	—	—
Sephanoides sephaniodes	x	—	—	—
Ceryle torquata	—	x	—	—
Dendrocopos lignarius	x	—	—	—
Colaptes pitius	—	x	—	—
Campephilus magellanicus	—	—	x	—
Cinclodes patagonicus	x	—	—	—
C. fuscus	x	—	—	—
Sylviorthorhynchus desmursii	—	—	x	x
Aphrastura spinicauda	—	—	x	—
Pygarrhichas albogularis	—	—	x	x
Pteroptochos tarnii	—	—	x	—
P. castaneus	—	—	x	—
Scelorchilus rubecula	—	—	x	—
Eugralla paradoxa	—	—	x	x
Scytalopus magellanicus	—	x	—	—
Elaenia albiceps	x	—	—	—
Spizitornis parulus	—	x	—	—
Colorhamphus parvirostris	—	—	x	x
Xolmis pyrope	—	—	x	—
Phytotoma rara	x	—	—	—
Notiochelidon cyanoleuca	x	—	—	—
Tachycineta leucopyga	—	—	x	—
Troglodytes aedon	x	—	—	—
Turdus falcklandii	x	—	—	—
Notiopsar curaeus	—	—	x	x
Phrygilus patagonicus	—	—	x	—
Zonotrichia capensis	x	—	—	—
Spinus barbatus	—	—	x	—
Totals (N = 46 species)	19	6	21	7

[1] Nomenclature as in Philippi (1964), but sequence as in Appendix VIII.

APPENDIX VIII
Treatment of Patagonian Forest Birds in This Paper[1]

Species or allospecies	Endemism			
	None	Subspecies	Species or allospecies	Genus
Theristicus caudatus	x	—	—	—
Cathartes aura	x	—	—	—
Coragyps atratus	x	—	—	—
Vultur gryphus	x	—	—	—
Accipiter [bicolor] bicolor	—	(chilensis)	—	—
Buteo [albicaudatus] polyosoma	x	—	—	—
B. [jamaicensis] ventralis	—	—	x	—
Geranoaetus melanoleucus	x	—	—	—
Polyborus chimango	—	(temucoensis and fuegiensis)	—	—
P. [megalopterus] albogularis	—	—	x	—
P. plancus	x	—	—	—
Vanellus [chilensis] chilensis	x	—	—	—
Columba [fasciata] araucana	—	—	x	—
Enicognathus ferrugineus	—	—	x	x
E. leptorhynchus	—	—	x	—
Bubo [bubo] virginianus	x	—	—	—
Glaucidium [brasilianum] nanum	—	(nanum and vafrum)	—	—
Strix rufipes	—	(rufipes and sanborni)	—	—
Sephanoides sephaniodes	x	—	—	—
Ceryle [alcyon] torquata	—	(stellata)	—	—
Picoides [mixtus] lignarius	x	—	—	—
Colaptes pitius	x	—	—	—
Campephilus magellanicus	—	—	x	—
Cinclodes patagonicus	x	—	—	—
C. [fuscus] fuscus	x	—	—	—
Sylviorthorhynchus desmursii	—	—	x	x
Aphrastura [spinicauda] spinicauda	—	—	x	—
Pygarrhichas albogularis	—	—	x	x
Pteroptochos [tarnii] tarnii	—	—	x	—
P. [tarnii] castaneus	—	—	x	—
Scelorchilus [rubecula] rubecula	—	—	x	—
Eugralla paradoxa	—	—	x	x
Scytalopus magellanicus	—	—	—	—
Elaenia albiceps	x	—	—	—
Anairetes parulus	x	—	—	—
Ochthoeca parvirostris	—	—	x	—
Xolmis pyrope	—	—	x	—
Phytotoma rara	x	—	—	—
Notiochelidon cyanoleuca	x	—	—	—
Tachycineta [leucorrhoa] leucopyga	—	—	x	—
Troglodytes aedon	x	—	—	—
Turdus falcklandii	x	—	—	—
Curaeus curaeus	—	—	x	—
Phrygilus [gayi] patagonicus	—	—	x	—
Zonotrichia capensis	x	—	—	—
Carduelis [barbata] barbata	—	—	x	—
Totals (N = 46 species)	22	5	19	4

[1] For details see Appendix II.

APPENDIX IX
ENDEMISM IN LAND BIRDS OF TASMANIAN FOREST[1]

Species	Endemism			
	None	Subspecies	Species	Genus
Accipiter cirrocephalus	x	—	—	—
A. fasciatus	x	—	—	—
A. novaehollandiae	x	—	—	—
Aquila audax	—	x	—	—
Aegotheles cristata	x	—	—	—
Phaps chalcoptera	x	—	—	—
P. elegans	x	—	—	—
Cacomantis pyrrhophanus	x	—	—	—
Chalcites basalis	x	—	—	—
C. plagosus	x	—	—	—
Cuculus pallidus	x	—	—	—
Falco berigora	x	—	—	—
F. peregrinus	x	—	—	—
Synoicus australis	—	x	—	—
Podargus strigoides	—	x	—	—
Calyptorhynchus funereus	x	—	—	—
Glossopsitta concinna	x	—	—	—
Kakatoe galerita	x	—	—	—
Lathamus discolor	—	—	x	x
Platycercus caledonicus	—	—	x	—
P. eximius	—	x	—	—
Ninox novaeseelandiae	—	x	—	—
Turnix varia	x	—	—	—
Tyto alba	x	—	—	—
T. novaehollandiae	—	x	—	—
Artamus cyanopterus	x	—	—	—
Coracina novaehollandiae	—	x	—	—
Corvus coronoides	x	—	—	—
Cracticus torquatus	—	x	—	—
Strepera arguta	—	—	x	—
S. fuliginosa	—	—	x	—
Pardalotus punctatus	x	—	—	—
P. quadragintus	—	—	x	—
P. striatus	x	—	—	—
Zonaeginthus bellus	x	—	—	—
Hirundo neoxena	x	—	—	—
Hylochelidon nigricans	x	—	—	—
Acanthorhynchus tenuirostris	—	x	—	—
Anthochera chrysoptera	x	—	—	—
A. paradoxa	—	—	x	—
Meliornis novaehollandiae	x	—	—	—
Meliphaga flavicollis	—	—	x	—
Melithreptus affinis	—	—	x	—
M. validirostris	—	—	x	—
Myzantha melanocephala	—	x	—	—
Phylidonyris pyrrhoptera	x	—	—	—
Collurincincla harmonica	—	x	—	—
Myiagra cyanoleuca	x	—	—	—
Pachycephala olivacea	x	—	—	—
P. pectoralis	—	x	—	—
Petroica multicolor	x	—	—	—
P. phoenicea	x	—	—	—
P. rodinogaster	x	—	—	—
P. vittata	—	—	x	—
Rhipidura fuliginosa	—	x	—	—
Acanthiza ewingii	—	—	x	—
A. chrysorrhoa	x	—	—	—
A. pusilla	x	—	—	—

APPENDIX IX
CONTINUED

Species	Endemism			
	None	Subspecies	Species	Genus
Acanthornis magnus	—	—	x	x
Calamanthus fuliginosus	x	—	—	—
Malurus cyaneus	—	x	—	—
Sericornis humilis	—	—	x	—
Cinclosoma punctatum	—	x	—	—
Zoothera dauma	x	—	—	—
Zosterops lateralis	—	x	—	—
Totals (N = 65 species)	36	16	13	2

[1] Data from Ridpath and Moreau (1966).

APPENDIX X
ENDEMISM IN LAND BIRDS OF NEW ZEALAND FORESTS[1]

Species	Endemism			
	None	Subspecies	Species	Genus
Apteryx australis	—	—	x	—
A. oweni	—	—	x	—
A. haasti	—	—	x	x
Hemiphaga novaeseelandiae	—	—	x	x
Strigops habroptilus	—	—	x	x
Nestor meridionalis	—	—	x	x
Cyanorhamphus novaezelandiae	—	—	x	—
C. malherbi	—	—	x	—
Chalcites lucidus	x	—	—	—
Eudynamis taitensis	x	—	—	—
Ninox novaeseelandiae	—	x	—	—
Acanthisitta chloris	—	—	x	x
Xenicus longipes	—	—	x	—
X. gilviventris	—	—	x	x
X. lyalli	—	—	x	—
Rhipidura fuliginosa	—	x	—	—
Petroica macrocephala	—	—	x	—
P. australis	—	—	x	—
P. traversi	—	—	x	—
Bowdleria punctata	—	—	x	x
Finschia novaeseelandiae	—	—	x	x
Mohoua ochrocephala	—	—	x	x
Gerygone igata	—	—	x	—
G. albofrontata	—	—	x	—
Notiomystis cincta	—	—	x	x
Anthornis melanura	—	—	x	x
Prosthemadera novaeseelandiae	—	—	x	x
Zosterops lateralis	x	—	—	—
Philesturnus carunculatus	—	—	x	x
Heteralocha acutirostris	—	—	x	x
Callaeas cinerea	—	—	x	x
Turnagra capensis	—	—	x	x
Totals (N = 32 species)	3	2	27	16

[1] Data from Checklist of New Zealand Birds (Checklist Committee Ornithological Society of New Zealand 1953).

BIRDS OF A TROPICAL DECIDUOUS FOREST IN EXTREME NORTHWESTERN PERU

DAVID A. WIEDENFELD,[1,2] THOMAS S. SCHULENBERG,[1] AND MARK B. ROBBINS[1,3]

[1]*Museum of Zoology, Louisiana State University, Baton Rouge, Louisiana 70803 USA;*
[2]*present address: Department of Biology, Florida State University, Tallahassee, Florida 32306 USA;*
[3]*present address: Academy of Natural Science, 19th and The Parkway,*
Philadelphia, Pennsylvania 19103 USA

ABSTRACT. From 14 June to 5 July 1979 we studied the avifauna at two localities in Depto. Tumbes, Peru, near the Ecuadorean border. We obtained specimens of six species and 15 subspecies new for Peru, and collected specimens and natural history information on such poorly known taxa as *Ortalis erythroptera, Synallaxis tithys, Automolus erythrocephalus,* and *Thamnophilus zarumae.*

RESUMEN. Entre el 14 de junio y el 5 de julio de 1979 estudiamos la avifauna de dos localidades cercanas a la frontera con Ecuador en el Departamento de Tumbes, Perú. Obtuvimos especímenes de seis especies y 15 subspecies nuevas para el Perú y coleccionamos especímenes e información sobre la historia natural de taxa poco conocida como: *Ortalis erythroptera, Synallaxis tithys, Automolus erythrocephalus* y *Thamnophilus zarumae.*

In South America west of the Andes, the ranges of many humid tropical forest bird species extend from Colombia south to western Ecuador. The climate of this narrow, coastal corridor becomes much drier in southwestern Ecuador, and many species found in the more humid forests to the north are absent. The dry area extends from northwestern Peru in the Deptos. of Piura and Tumbes into west-central Ecuador, with a narrow arm extending into coastal northwestern Ecuador, and is bounded on the east by the Andean Cordillera. Noting the high endemism of birds, as well as other organisms, in the region, Chapman (1926) designated them as the "Equatorial Arid Fauna."

This region has been studied very little by ornithologists in recent years. Jelski and Stolzmann visited Tumbes in 1876, going on to collect at Lechugal, and in nearby Ecuador at "Palmal" (= Palmales; Taczanowski 1877). Collecting in the region was not extensive, however, until just after the turn of the century when several expeditions from the American Museum of Natural History worked in the area of Ecuador just north and east of what is now the Depto. Tumbes, Peru. Collectors from the American Museum spent about 200 field-days in the area between 1913 and 1922 (Chapman 1926). M. A. Carriker, Jr., later collected at La Laja, Peru, now near the Ecuadorean border (Bond 1945).

From 14 June to 5 July 1979 a field party (the authors, J. W. Eley, and Manuel Sánchez S.) from the Louisiana State University Museum of Zoology (LSUMZ) studied the birds of northeastern Depto. Tumbes near the Ecuadorean border. The aim of our collecting was to clarify the ranges of the forms that reach their southern limit in this region, which is believed to be at or near the southern terminus of Chapman's "Colombian-Pacific Fauna," to gather information on the poorly known species of the area, and to provide the Peruvian government with an initial faunal inventory of the Bosque Nacional de Tumbes. Our camp was ca. 2 km east of the Peruvian military post at El Caucho, on Quebrada Faical, about 24 km southeast of Pampa de Hospital (3°49'S, 80°17'W; Fig. 1), at about 400 m elevation. During this period, we also visited Campo Verde (3°51'S, 80°11'W), another military outpost ca. 11 km east-southeast of El Caucho and at 750 m elevation (30 June; 1, 3, and 4 July). At these localities we collected six bird species not previously recorded in Peru, and recorded 15 subspecies new for the country.

Because of the extensive earlier collections in nearby Ecuador, most species that we recorded new for Peru had previously been taken within 100 km of our El Caucho camp. Thus, most records are not major range extensions, and in most cases extend the range westward, not southward.

The area around El Caucho was a dry, deciduous forest. We were there during the early part of the dry season, and most trees were losing or had recently lost their leaves. Quebrada

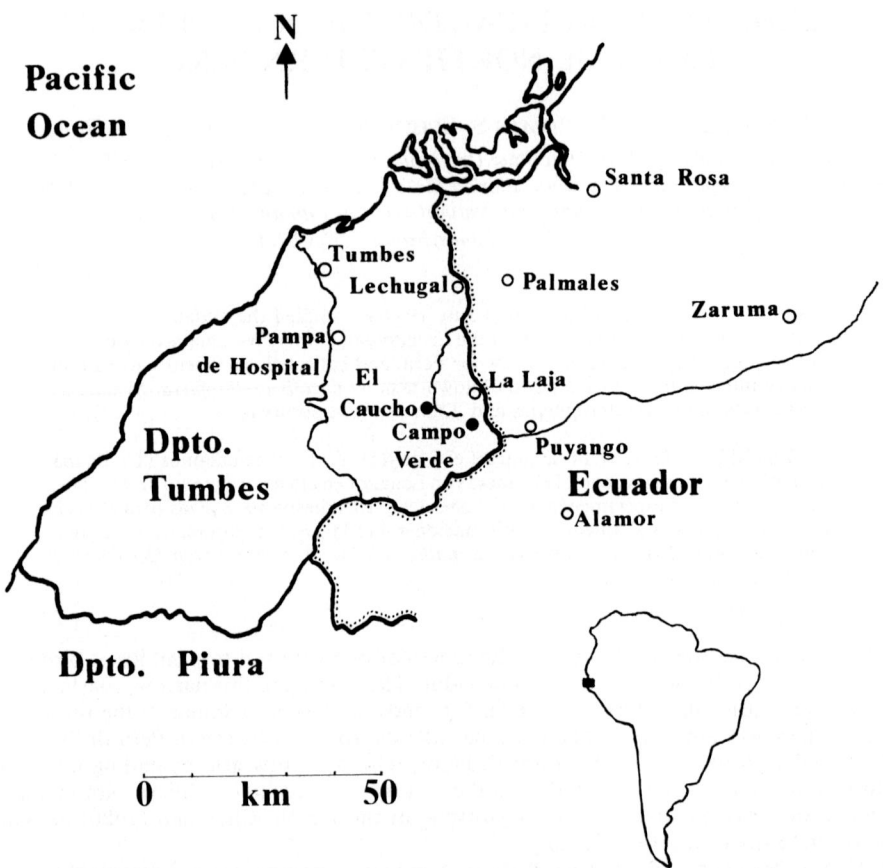

FIG. 1. Map of the Department of Tumbes, Peru, and adjacent Ecuador, showing our collecting localities (closed circles) and those of other ornithologists (open circles; see text).

Faical probably flows as a small stream during the wet season (January to May), but by mid-June the stream was reduced to a series of isolated pools.

The topography of the El Caucho area was hilly, with narrow canyons and steep slopes rising to 600 m elevation. The forest was composed primarily of *Bombax* spp., often reaching 20 m in height, but acacias were also found in the more open, level areas. Epiphytes, especially *Tillandsia* spp., were common. The forest understory was relatively open, but in the narrower canyons vine tangles were common. Few plants were in flower, and herbaceous plants were mainly brown and dry (Fig. 2).

The topography of the Campo Verde area was similar to that of El Caucho. The area was considerably more humid, however, and its forest much less deciduous. At the time of our visits most trees still had leaves. *Tillandsia* spp. were more common, and the forest understory was much more dense and tangled.

Even with the short amount of time we spent at Campo Verde, the locality proved to have a very interesting avifauna. Two of the six species new for the country were found only at Campo Verde, and 10 of the 15 subspecies new for the country were found only there.

At neither locality was breeding activity very evident. We heard little singing by most species, and in virtually all specimens the gonads were in regressed condition. In this region breeding probably occurs in the wet season (Marchant 1958).

Because of the seasonality of rainfall, the composition of the avifauna in this area probably changes during the year. As a result, and because bird lists from similar localities in Ecuador are incomplete and mist nets were not used there, it is premature to make comparisons between this and other studies.

The following are species accounts of the species and subspecies new for Peru, along with

Fig. 2. View of the forest near El Caucho. Many of the trees have lost some of their leaves. Elevation at the road is approximately 400 m.

notes on behavior and vocalizations for several species about which little is known. All species we recorded at El Caucho and Campo Verde, and body weights when available, are listed in Appendix I.

SPECIES ACCOUNTS

Crypturellus transfasciatus. Pale-browed Tinamou. This species is known from Prov. Manabí, Ecuador (Chapman 1926) south to Depto. Lambayeque, Peru (Schulenberg and Parker 1981). We collected nine specimens (LSUMZ 91592–91596, skins; 91463, alcoholic; 93847–93849, skeletons), all at El Caucho. Generally, iris colors were brown or olive-brown; the maxilla gray, dark gray, or black; mandible pinkish, white, off-white, or pale yellow; and the tarsi orange or pale orange. One specimen's crop was filled with seeds, and other specimens' stomachs contained vegetable material and seeds, or seeds and grit. One to 10 of these tinamous were encountered daily, both at El Caucho and Campo Verde. We never heard tinamous calling, but we often located the birds by the rustling sound they made while foraging in newly fallen leaves.

Elanoides forficatus. Swallow-tailed Kite. This distinctive species was seen at both El Caucho and Campo Verde, although we collected no specimens. This kite was not previously known from western Peru, although it is widely distributed in South America and was known from western Ecuador. On 4 July we located a nest with a single young about 15 m above ground in the top of a bombacaceous tree at Campo Verde. Four adults were observed in the area of the nest. The young bird was nearly full-grown and could be seen testing its wings.

Buteo albonotatus. Zone-tailed Hawk. The range of the species in South America seems to be poorly known. The only records from west of the Andes are one from Depto. Lima, Peru (Hellmayr and Conover 1949) and several sight records from Provs. Manabí and Guayas, Ecuador (Ridgely 1980). This hawk was seen several times at both El Caucho and Campo Verde. Also, J. P. O'Neill, T. A. Parker, III, and Schulenberg have seen Zone-tailed Hawks on several occasions in Depto. Lambayeque, Peru. Thus, this species should probably be considered a rare resident of western Ecuador and northwestern Peru south at least to Depto. Lambayeque.

Leucopternis occidentalis. Gray-backed Hawk. This rare species was not previously known from Peru, although it has been collected near the Tumbes border at Río Pullango and Alamor, Ecuador (Chapman 1926). Although we were unable to obtain specimens, we believe we saw this distinctive species at both El Caucho and Campo Verde, and three others in the hilly forest along the road from Pampa de Hospital to El Caucho on 14 June. On 25 June Robbins located one giving a single-noted scream as it perched near a troop of Howler Monkeys (probably *Allouatta palliata*). We almost always saw this species perched quietly in the middle level of open trees, as is typical for other members of the genus.

Spizaetus tyrannus. Black Hawk-Eagle. A single adult was soaring very high and calling over Campo Verde on 4 July. Blake (1977) reported this species west of the Andes only in Colombia (Baudó Mountains). Ridgely (1980), however, saw one at Tandayapa, Prov. Pichincha, Ecuador, in February 1980.

Ortalis erythroptera. Rufous-headed Chachalaca. We saw this poorly known species on several occasions at both localities. Nevertheless, the chachalacas were uncommon. They were seen usually in small flocks of two to six birds, on or within 2 to 3 m of the ground; on one occasion a flock of six was seen moving through the upper level of the forest. We never heard the chachalacas calling. Two birds were collected at El Caucho on 28 June. The male (LSUMZ 91613) had testes 10 × 6 mm; brown irides; blue-gray bill; gray tarsi; gular skin pinkish orange; orbital skin gray; skull ossified; tracheal loop 100 mm from top of keel to end of loop; and it had leaves in its stomach. The female (LSUMZ 91612) had an ovary 17 × 9 mm; soft parts same as male except orbital skin dark brown; skull ossified; no tracheal loop; crop contained small green leaves.

Chaetura cinereiventris. Gray-rumped Swift. This swift was fairly common at El Caucho, where it formed mixed flocks of up to 50 birds with *Chaetura brachyura*. We collected three individuals (LSUMZ 91711–91713), the first of this species from Peru west of the Andes. All were fat with small gonads and had insects in their stomachs.

Phaethornis superciliosus baroni. Long-tailed Hermit. Our three specimens (LSUMZ 91717–91719, all females) and sight records of this species from El Caucho and Campo Verde constitute the first records of the subspecies *baroni* from Peru. One bird had a black bill, with the basal two-thirds of the mandible dull red, and pink tarsi. No lekking activity was noted during our visit.

Damophila julie feliciana. Violet-bellied Hummingbird. The specimens taken at our two localities are the first records of the species for Peru. However, Chapman (1926) reported specimens from Cebollal, Las Piñas, and La Puente, Ecuador, all within about 30 km of El Caucho. We obtained eight specimens (LSUMZ 91733–91737, skins; 93878–79, skeletons; 91477, alcoholic). One typical male had a black maxilla, pinkish mandible with a black tip, and black tarsi. These hummingbirds usually fed 2 to 4 m above ground at the forest edge, but they were seen occasionally at canopy flowers.

Chalybura buffonii intermedia. White-vented Plumeleteer. Our five specimens (LSUMZ 91747–91750, skins; 93884, skeleton) from Campo Verde are the first from Peru. Again,

Chapman (1926) reported specimens from nearby at Río Pullango (= Puyango), Ecuador, only about 15 km from Campo Verde. Soft part colors of a typical specimen (LSUMZ 91748) were maxilla black, mandible flesh-colored tipped black, and tarsi flesh-colored, although a female specimen had the base of the mandible grayish and feet pinkish-brown. The flesh-colored base of the mandible agrees with Chapman's (1926) description of the subspecies. This hummingbird was seen feeding only in the forest undergrowth.

Trogon violaceus concinnus. Violaceous Trogon. Taczanowski (1877) reported a specimen of *Trogon "caligatus"* taken at Lechugal, Depto. Tumbes, Peru, and Zimmer (1948) concluded that the specimen is probably *T. violaceus concinnus.* This record, however, seems to have been overlooked by Peters (1945) and Meyer de Schauensee (1966), because the latter authors do not mention the species occurring in northwestern Peru. Zimmer (1948) also reported records of the subspecies in Ecuador at Portovelo (near Zaruma) and northward to Esmeraldas. We obtained a female and a male (LSUMZ 91983 and 91984, respectively) of this subspecies at El Caucho. The male had a gray bill and yellow-orange orbital ring. The female, with a 75 percent ossified skull, had a bluish gray bill with a dark gray culmen. Its stomach contained katydids and seeds. Both birds had brown irides and gray tarsi.

Veniliornis kirkii cecilii. Red-rumped Woodpecker. The single specimen obtained at Campo Verde represents the first record of the species for Peru. Schulenberg observed another individual in a mixed-species flock at El Caucho. Chapman (1926) reported a specimen from Pullango (= Puyango), only about 15 km away in Ecuador. Our specimen, a female (LSUMZ 92074), had dark gray maxilla, blue-gray mandible, olive-gray tarsi, and brown irides. Its stomach contained a large grub.

Dendrocincla fuliginosa ridgwayi. Plain-brown Woodcreeper. A specimen obtained at Campo Verde is the first for this subspecies in Peru. The specimen, a male (LSUMZ 92093), had black bill and tarsi, light gray irides, and skull completely ossified. Its stomach contained insect parts.

Synallaxis brachyura chapmani. Slaty Spinetail. We collected three females and one unsexed bird (LSUMZ 92154–92157) of this species at Campo Verde; these are the first records of the species from Peru. One female with full data (LSUMZ 92155) had reddish-brown irides, black maxilla, blue-gray mandible with black tip, and blue-gray tarsi. All of our specimens had the skull less than 25 percent ossified.

Hellmayr (1925) found little basis for separating *S. b. chapmani* (= *S. b. griseonuchus*) from the Central American *S. b. nigrofumosa.* He did acknowledge, however, that birds from southwestern Ecuador are "lighter," and proposed that if *S. b. chapmani* were retained, its range be restricted to southwestern Ecuador.

Our Tumbes specimens agree with Chapman's (1923) description of *S. b. chapmani,* and are distinct from specimens taken at Santo Domingo de los Colorados, Prov. Pichincha, Ecuador, at La Guayacana, Nariño, Colombia, and *S. b. nigrofumosa* specimens from Costa Rica (all specimens in LSUMZ). Our specimens differ from these in being paler and grayer on the forecrown and nape, paler olive-brown on the back; having a paler gray lower throat and breast; and a very pale gray, almost white, center of the abdomen. The specimens from Prov. Pichincha, Ecuador, and Nariño, Colombia, resemble the Costa Rican *S. b. nigrofumosa.*

We favor retention of the subspecies *S. b. chapmani,* with its range confined to southwestern Ecuador and extreme northwestern Peru, following the recommendation of Hellmayr (1925). We consider northwestern Ecuadorean and western Colombian birds to be *S. b. nigrofumosa.*

Synallaxis tithys. Blackish-headed Spinetail. This poorly known species was common at both El Caucho and Campo Verde. We prepared 15 skins (LSUMZ 92159–92173), three skeletons (LSUMZ 93968–93970), and two alcoholics (LSUMZ 91503–91504). A male specimen with a 60 percent ossified skull (LSUMZ 92167) had reddish-brown irides, black maxilla, blue-gray mandible with dusky tip, and blue-gray tarsi. The iris colors of the two individuals with unossified skulls were described as dark gray and light gray-brown. For several individuals with skulls more than 25 percent ossified, the iris was described as dark red, reddish, or reddish-brown. The stomach contents of the specimens were generally insect material. One stomach contained the head of a grasshopper (Orthoptera) 7 mm in diameter; another contained the head of a wasp (Hymenoptera) 3 mm in diameter; a third contained a "whole black beetle 8 × 3 mm" and other beetle elytra.

As is typical of most spinetails, the *Synallaxis tithys* were usually seen in thicket areas, and almost always within 2 or 3 m of the ground. In contrast to most *Synallaxis,* however, they spent a large portion of their time on the ground, searching the leaf litter. The birds were usually in pairs, although three birds were seen together on several occasions, and rarely a lone bird was seen. They were encountered occasionally with mixed-species flocks. The song of these spinetails consists of four notes and is similar to the songs of several other *Synallaxis* (Braun and Parker 1985).

Automolus e. erythrocephalus. Henna-hooded Foliage-Gleaner. We encountered this species at both El Caucho and Campo Verde, and our records are the first for this subspecies in Peru. We obtained 12 specimens (LSUMZ 92294–92303, skins; 93985, skeleton; and 91507, alcoholic). Our descriptions of iris colors ranged from gray-brown to brown and gray, but no individuals had "yellow" irides as reported for one specimen from Ecuador by Paynter (1972). We described the maxilla as gray or blackish-gray, and the mandible was generally paler, being "whitish, shading dusky distally" or horn. The tarsi were olive-gray or gray. Stomachs of eight specimens contained insect parts, and one also contained terrestrial isopods. Several of the foliage-gleaners we collected were infested with up to several hundred minute ticks. Some specimens of other ground-foraging birds, such as *Crypturellus transfasciatus* and *Synallaxis tithys,* were similarly infested.

The skull of one specimen was noted as only 30 percent ossified. The plumage of this bird differs in some respects from that of the adult. Feathers on the forehead and crown are edged with black; the throat feathers have a fine black fringe that becomes more prominent on the lower throat and breast, giving a markedly scalloped appearance to this area; similar, but less prominent, markings extend onto the belly. Chapman (1919) noted similar markings on an immature female. Two other specimens, both with the skulls noted as fully ossified, have similar, but less conspicuous markings, and probably also represent immature birds.

Foraging foliage-gleaners were invariably seen on or near the ground, often tossing dead leaves aside and probing into leaf litter, low vine tangles, and leaf clusters. They were often conspicuous because of the loud rustling noise they made when foraging in this manner. The birds hop, rather than walk, on the ground. Once an individual clung to and hitched up a tree trunk in dendrocolaptid fashion. When alarmed while feeding on the ground, foliage-gleaners assumed an upright posture with the head erect, and flicked the wings, occasionally making a low *chuck chuck chuck.*

Thamnophilus z. zarumae. Chapman's Antshrike. Our single specimen from Campo Verde proved to be the first record of this subspecies for Peru. The adult male (skull ossified; LSUMZ 92327) had yellow-white irides, black maxilla, blue-gray mandible, and blue-gray tarsi. Chapman (1926) reported the species from several localities nearby in Ecuador, including Milagros (listed as Milagros, Peru, in Chapman 1926 and Zimmer 1933).

Pyriglena leuconota pacifica. White-backed Fire-Eye. We collected a single female at Campo Verde. The specimen agrees with Zimmer's (1931) description of this form, extending the subspecies' range into northwestern Peru. The specimen (LSUMZ 92408) had an unossified skull, red irides, black maxilla, blue-gray mandible, and gray tarsi.

Myrmeciza griseiceps. Gray-headed Antbird. We obtained two specimens of this species (LSUMZ 94013, skeleton; 91510, alcoholic) at Campo Verde. The skeleton, a male by plumage, had brown irides, black bill, and blue-gray tarsi.

A specimen from 2.8 km by road SW Porculla Pass, Depto. Piura, Peru (LSUMZ 84912), differs from all other specimens in the LSUMZ collection in lacking all traces of the white interscapular patch. Otherwise, the plumage of the specimen matches fairly well the immature male described by Bond (1950), although the Porculla Pass bird was a female with a fully ossified skull.

Schiffornis turdinus rosenbergi. Thrush-like Manakin. Our specimen from Campo Verde and sightings at both camps are the first records of the subspecies for Peru. The female specimen (LSUMZ 92645) had an incompletely ossified skull. Its bill was black and tarsi gray.

Manacus manacus maximus. White-bearded Manakin. We collected a female of this species at Campo Verde, the first record of the subspecies from Peru. The specimen, an adult (skull ossified; LSUMZ 92637), had brown irides, black bill, and orange-yellow tarsi. Its stomach contained seeds from purple fruit.

Lophotriccus pileatus squamaecristata. Scale-crested Pygmy-Tyrant. Our five specimens (LSUMZ 92797–92801), all from El Caucho, are the first for this subspecies from Peru. One male (LSUMZ 92800; skull unossified) had white irides, black bill, and gray tarsi. The iris color of the specimens varied from yellow (specimen with skull ossified), to light tan (specimen with skull 50% ossified), to light gray-brown.

Mionectes oleagineus pacificus. Ochre-bellied Flycatcher. This species was seen at Campo Verde, and three specimens (LSUMZ 92919–92921) were collected at El Caucho. These are the first records for the subspecies in Peru. A female with typical soft part colors (LSUMZ

92920; skull 30% ossified) had brown irides, black maxilla, pink mandible with black tip, and gray tarsi.

Turdus nudigenis maculirostris. Bare-eyed Thrush. Although Ripley (1964) listed this species from northwestern Peru, the only previous record was from Milagros, a locality now in Ecuador. The species was fairly common at both Campo Verde and El Caucho, and we obtained four specimens (LSUMZ 93089–93091, skins; 91544, alcoholic). They had orange-brown irides, orange-yellow eye-ring, dull olive-yellow bill with olive base, and pale blue-gray tarsi.

Ramphocaenus melanurus rufiventris. Long-billed Gnatwren. Paynter (1964) and Meyer de Schauensee (1966) mentioned this species occurring west of the Andes only in Colombia, although Chapman (1926) recorded several individuals from Esmeraldas to Alamor in western Ecuador. The subspecies *rufiventris* had not been previously recorded from Peru, but we obtained two specimens at Campo Verde (LSUMZ 93092–93093). The specimens had gray-brown irides, gray maxillae, white or pale pinkish-gray mandibles, and blue-gray tarsi. Gnatwrens were usually observed foraging in vine tangles within a few meters of the ground.

Hylophilus minor. Lesser Greenlet. Two specimens (LSUMZ 93813–93814) obtained at El Caucho are the first of this species in Peru. Irides were brown, maxillae, dark gray, mandibles, blue-gray, and tarsi, blue-gray. Greenlets were typically observed gleaning insects in middle-story mixed-species flocks. Common flock associates were *Sittasomus griseicapillus, Dysithamnus mentalis, Tolmomyias sulphurescens, Parula pitiayumia,* and *Basileuterus fraseri.*

Euphonia xanthogaster quitensis. Orange-bellied Euphonia. Our two specimens (LSUMZ 93430–93431) of this subspecies are the first for Peru. The male, with ossified skull, had brown irides, black maxilla, blue-gray mandible with black tip, and dark gray tarsi. These birds were only found at Campo Verde.

Tangara gyrola nupera. Bay-headed Tanager. We collected a single specimen (LSUMZ 93456) of this species at Campo Verde for the first record of this subspecies for Peru. The male bird had its skull 95 percent ossified. Its bill and tarsi were dark brown.

Chlorospingus canigularis paulus. Ash-throated Bush-Tanager. Zimmer (1947) predicted that this subspecies would be found in Peru. Our three specimens (LSUMZ 93234–93236) from Campo Verde are the first records for the country. A male with ossified skull had dark gray maxilla, light gray mandible tipped dark, and gray tarsi.

Sporophila nigricollis vivida. Yellow-bellied Seedeater. At Campo Verde we collected a male (LSUMZ 93129) of this subspecies, a new record for Peru. The bird had a fully ossified skull, and its bill was pale blue and its tarsi black. South of Porculla Pass, Depto. Piura, *Sporophila nigricollis* is represented on the western slope of the Andes by *S. n. inconspicua* (Koepcke 1961), which is primarily a bird of the eastern slope of the Andes.

Arremon aurantiirostris santarosae. Orange-billed Sparrow. A single specimen (LSUMZ 93145) collected at Campo Verde adds this subspecies to the forms known from Peru. Its irides were brown, bill orange, and tarsi flesh-colored. When compared with *A. a. occidentalis,* the Campo Verde bird has a slight greenish tinge to the gray vertical stripe, and its underparts are mottled with gray, giving it a "dirty" appearance.

Carduelis sp. Small groups of siskins, including adult males, were occasionally noted at El Caucho. Two immature specimens (a female, LSUMZ 93838, and an unsexed bird, LSUMZ 93839) and an adult male (LSUMZ 91586, alcoholic) were obtained. The specimens remain unidentified, but their small size (wing chord 55.5 mm, 56.2 mm and 58.0 mm, respectively) suggests that they may be the very poorly-known *C. siemiradzkii* of western Ecuador. Stolzmann reported *C. siemiradzkii* from Tumbes, on the basis of sight observations (Taczanowski 1886), but this taxon is so similar to *C. magellanicus,* even in the hand, that this early record may be considered doubtful.

ACKNOWLEDGMENTS

We gratefully acknowledge the continued support of our Peruvian fieldwork by J. S. McIlhenny, B. M. Odom, L. Schweppe, H. I. Schweppe, and E. W. Mudge. We wish to thank E. Cardich B., S. Moller-H., A. Brack E., and R. Bustamante M. of the Dirección General Forestal y de Fauna, Ministerio de Agricultura, Lima, Peru, for their continued assistance. We also appreciate the kind loan of facilities at Olmos, Peru, by G. del Solar. The manuscript benefitted from the criticisms of J. V. Remsen, Jr.

LITERATURE CITED

BLAKE, E. R. 1977. Manual of Neotropical Birds, Vol. 1. University Chicago Press, Chicago, Illinois.

BOND, J. 1945. Notes on Peruvian Furnariidae. Proc. Acad. Nat. Sci. Phila. 97:17–39.

BOND, J. 1950. Notes on Peruvian Formicariidae. Proc. Acad. Nat. Sci. Phila. 102:1–26.

BRAUN, M. J., AND T. A. PARKER, III. 1985. Molecular, morphological, and behavioral evidence concerning the taxonomic relationships of "*Synallaxis*" *gularis* and other synallaxines. Pp. 333–346, *In* P. A. Buckley, M. S. Foster, E. S. Morton, R. S. Ridgely, and F. G. Buckley (eds.), Neotropical Ornithology. Ornithol. Monogr. No. 36.

CHAPMAN, F. M. 1919. Descriptions of proposed new birds from Peru, Bolivia, Brazil, and Colombia. Proc. Biol. Soc. Wash. 32:253–268.

CHAPMAN, F. M. 1923. Descriptions of proposed new Formicariidae and Dendrocolaptidae. Am. Mus. Novit. No. 86.

CHAPMAN, F. M. 1926. The distribution of bird-life in Ecuador. Bull. Am. Mus. Nat. Hist. 60:1–784.

HELLMAYR, C. E. 1925. Catalogue of birds of the Americas. Part IV. Field Mus. Nat. Hist. Publ., Zool. Ser., 13(4):1–390.

HELLMAYR, C. E., AND B. CONOVER. 1949. Catalogue of birds of the Americas. Part I, no. 4. Field Mus. Nat. Hist. Publ., Zool. Ser., 13(1):1–358.

KOEPCKE, M. 1961. Birds of the western slope of the Andes. Am. Mus. Novit. No. 2028.

MARCHANT, S. 1958. The birds of the Santa Elena Peninsula, S.W. Ecuador. Ibis 100:349–387.

MEYER DE SCHAUENSEE, R. M. 1966. The Species of Birds of South America and Their Distribution. Academy Natural Sciences Philadelphia, Livingston Publ. Co., Narberth, Pennsylvania.

PAYNTER, R. A., JR. 1964. Polioptilinae. Pp. 443–455, *In* E. Mayr and R. A. Paynter, Jr. (eds.), Check-list of Birds of the World, Vol. 10. Museum Comparative Zoology, Cambridge, Massachusetts.

PAYNTER, R. A., JR. 1972. Notes on the furnariid *Automolus* (*Hylocryptus*) *erythrocephalus*. Bull. Br. Ornithol. Club 92:154–155.

PETERS, J. L. 1945. Check-list of Birds of the World, Vol. 5. Harvard University Press, Cambridge, Massachusetts.

RIDGELY, R. S. 1980. Notes on some rare or previously unrecorded birds in Ecuador. Am. Birds 34: 242–248.

RIPLEY, S. D. 1964. Turdinae. Pp. 13–227, *In* E. Mayr and R. A. Paynter, Jr. (eds.), Check-list of Birds of the World, Vol. 10. Museum Comparative Zoology, Cambridge, Massachusetts.

SCHULENBERG, T. S., AND T. A. PARKER, III. 1981. Status and distribution of some northwest Peruvian birds. Condor 83:209–216.

TACZANOWSKI, L. 1877. Liste des oiseaux recueillis en 1876 au nord de Pérou occidental par MM. Jelski et Stolzmann. Proc. Zool. Soc. Lond. 1877:319–333.

TACZANOWSKI, L. 1886. Ornithologie du Pérou. Vol. 3. Oberthur, Paris.

ZIMMER, J. T. 1931. Studies of Peruvian birds. II. Peruvian forms of the genera *Microbates*, *Ramphocaenus*, *Sclateria*, *Pyriglena*, *Pithys*, *Drymophila*, and *Liosceles*. Am. Mus. Novit. No. 509.

ZIMMER, J. T. 1933. Studies of Peruvian birds. IX. The formicarian genus *Thamnophilus*. Part I. Am. Mus. Novit. No. 646.

ZIMMER, J. T. 1947. Studies of Peruvian birds. No. 52. The genera *Sericossypha*, *Chlorospingus*, *Cnemoscopus*, *Hemispingus*, *Conothraupis*, *Chlorornis*, *Lamprospiza*, *Cissopis*, and *Schistochlamys*. Am. Mus. Novit. No. 1367.

ZIMMER, J. T. 1948. Studies of Peruvian birds. No. 53. The family Trogonidae. Am. Mus. Novit. No. 1380.

APPENDIX I

SPECIES RECORDED AT EL CAUCHO AND CAMPO AND CAND VERDE, DEPTO. TUMBES, PERU, 14 JUNE TO 5 JULY 1979[1]

	El Caucho[2]	Campo Verde	Mean body weight (g) ♂♂ (N)	♀♀ (N)
Crypturellus transfasciatus	FC	x	273 (3)	293 (3)
*Ardea cocoi	R			
*Mycteria americana	R			
*Sarcoramphus papa	FC	x		
*Coragyps atratus	C	x		
*Cathartes aura	C	x		
*Elanoides forficatus	R	x		
*Accipter bicolor	R			
*A. striatus	R			
*Buteo albonotatus	R	x		
*B. brachyura		x		
*Parabuteo unicinctus	R			
*Leucopternis occidentalis	R	x		
Buteogallus urubitinga	R			
*Spizaetus tyrannus		x		
Geranospiza coerulescens	R			
*Herpetotheres cachinnans	R	x		
*Polyborus plancus	R			
Ortalis erythroptera	R	x	645 (1)	620 (1)
*Columba plumbea		x		
*Zenaida auriculata	R			
*Columbina buckleyi	FC			
C. cruziana	C	x		
Claravis pretiosa	R	x	58.0 (1)	
*Leptotila verrauxii	FC	x		
*Aratinga erythrogenys	U	x		
Forpus coelestis	U		27.9 (2)	24.5 (2)
Brotogeris pyrrhopterus	FC		68.0 (1)	60.0 (2)
Pionus chalcopterus	R	x	210 (1)	
Piaya cayana	U	x	107.0 (1)	
*Crotophaga sulcirostris	R	x		
Otus roboratus	U		74.0 (1)	
*Pulsatrix perspicillata	U			
Glaucidium brasilianum	U	x		
Nyctidromus albicollis	U		44.0 (1)	
*Streptoprocne zonaris	R			
Chaetura brachyura	FC	x		
C. cinereiventris	FC		18.0 (1)	17.5 (2)
*Panyptila cayennensis	R			
Phaethornis superciliosus	FC	x	5.6 (1)	5.5 (3)
P. griseogularis	U			2.5 (2)
Damophila julie	R	x	3.4 (4)	3.0 (3)
Amazilia amazilia	C	x	4.4 (3)	
Chalybura buffonii		x	6.0 (1)	5.5 (3)
Heliomaster longirostris	U		6.8 (1)	
*Trogon melanurus	FC			
T. violaceus	R		50.0 (1)	49.0 (1)
Chloroceryle americana	U			28.0 (1)
Momotus momota	FC		96.0 (1)	96.0 (2)
Picumnus sclateri	FC	x	9.8 (2)	12.4 (1)
Piculus rubiginosus	U		48.0 (1)	
Dryocopus lineatus	R	x		
Veniliornis callonotus	FC	x	25.0 (2)	
V. kirkii	R	x		30.0 (1)
*Phloeoceastes gayaquilensis	R			
Dendrocincla fuliginosa		x	38.0 (1)	
Sittasomus griseicapillus	FC	x	11.3 (3)	10.4 (5)

APPENDIX I

CONTINUED

	El Caucho[2]	Campo Verde	♂♂ (N)	♀♀ (N)
			Mean body weight (g)	
Xiphocolaptes promeropirhynchus	R			110.0 (1)
Lepidocolaptes souleyettii	C	x	25.8 (5)	25.6 (5)
Campylorhamphus trochilirostris	U	x	46.0 (1)	40.0 (2)
Furnarius leucopus	FC	x	45.5 (2)	
Synallaxis brachyura		x		16.8 (3)
S. tithys	FC	x	15.9 (10)	15.6 (5)
S. azarae		x	12.0 (1)	
Automolus ruficollis	R	x		35.0 (2)
A. erythrocephalus	U	x	44.8 (5)	48.2 (5)
Xenops rutilans	U	x	10.5 (1)	10.2 (4)
*Taraba major		x		
Sakesphorus bernardi	FC	x	32.0 (1)	28.9 (1)
Thamnophilus zarumae		x	22.0 (1)	
Dysithamnus mentalis	FC	x	10.5 (6)	11.0 (2)
Pyriglena leuconota		x		26.0 (1)
Myrmeciza griseiceps		x		
Grallaria ruficapilla	FC	x	64.2 (5)	63.0 (3)
Melanopareia elegans	R			16.0 (1)
Pachyramphus spodiurus	R	x	18.1 (1)	18.3 (3)
P. albogriseus	R	x		
Manacus manacus		x		19.0 (1)
Schiffornis turdinus	R	x		20.0 (1)
Pyrocephalus rubinus	FC		10.5 (1)	
Megarhynchus pitangua	R	x		
Myiodynastes bairdii	FC			45.0 (1)
*M. maculatus	R			
Myiarchus phaeocephalus	U	x	27.2 (1)	
M. tuberculifer		x	21.0 (1)	
Contopus cinereus	FC	x		9.3 (3)
C. fumigatus	U	x		19.5 (1)
Myiobius atricaudus	U	x		
Myiophobus fasciatus	U	x	10.0 (1)	9.5 (1)
Onychorhynchus coronatus	R			14.2 (1)
Tolmomyias sulphurescens	FC	x	12.0 (1)	12.6 (3)
Todirostrum cinereum	FC	x	5.9 (3)	
Lophotriccus pileatus	U	x	6.7 (3)	
Euscarthmus melorhyphus	FC	x	5.3 (1)	
Mecocerculus calopterus		x		7.0 (3)
Myiopagis subplacens	FC		14.9 (1)	16.9 (4)
Camptostoma obsoletum	FC	x		
Tyranniscus viridiflavus		x	10.0 (1)	
Mionectes oleagineus	R	x	11.8 (1)	10.3 (2)
*Progne chalybea	U			
Campylorhynchus fasciatus	C	x		24.9 (1)
Thryothorus maculipectus	FC	x	12.7 (3)	10.0 (1)
Troglodytes aedon	FC	x		
*Henicorhina leucophrys		x		
*Mimus longicaudatus	R			
Catharus dryas	R	x	29.0 (1)	27.5 (1)
Turdus reevei	FC	x	61.0 (1)	
T. nudigenis	U	x	65.0 (2)	72.0 (2)
Ramphocaenus melanurus		x		9.0 (1)
Polioptila plumbea	FC	x		
Cyclarhis gujanensis	FC	x	26.0 (1)	26.0 (2)
Vireo olivaceus	R	x		13.0 (1)
Hylophilus minor	R	x	8.0 (2)	
*Cacicus cela	R	x		
C. holosericeus	FC	x	54.0 (1)	48.0 (2)

APPENDIX I

CONTINUED

	El Caucho[2]	Campo Verde	♂♂ (N)	♀♀ (N)
			Mean body weight (g)	
*Icterus graceannae	U			
I. mesomelas	FC	x	33.4 (5)	
Parula pitiayumi	FC	x	6.6 (1)	5.7 (2)
Geothlypis aequinoctialis	U	x	8.2 (1)	10.0 (1)
*Myioborus miniatus		x		
*Basileuterus trifasciatus		x		
B. fraseri	FC	x	12.0 (12)	11.0 (6)
Coereba flaveola	U	x		7.6 (1)
Euphonia xanthogaster		x	17.0 (1)	14.0 (1)
E. laniirostris	FC	x	15.1 (2)	13.0 (2)
Tangara gyrola		x	25.0 (1)	
Thraupis episcopus	FC	x	28.0 (1)	27.0 (1)
Piranga flava	U	x	29.0 (2)	
Chlorospingus canigularis		x	15.8 (2)	14.5 (1)
*Saltator maximus		x		
S. albicollis	FC	x	43.5 (2)	
Pheucticus chrysopeplus	U	x	54.0 (1)	
*Cyanocompsa cyanoides		x		
Sporophila americana	C	x		10.0 (3)
S. nigricollis		x	9.0 (1)	
S. obscura	R	x		
S. telasco	R	x	8.4 (1)	
Phrygilus plebejus	U		13.6 (1)	
Rhodospingus cruentus	U	x	10.5 (1)	11.0 (1)
Atlapetes leucopterus	C	x	18.6 (3)	19.8 (4)
*A. albiceps	R			
A. torquatus	U	x	36.0 (2)	
Arremon aurantiirostris		x		
A. abeillei	C	x	26.0 (7)	25.8 (3)
Carduelis sp.	R	x	10.2 (1)	7.9 (1)

[1] The sequence of species follows Meyer de Schauensee (1966). Species for which we have only sight records are indicated by asterisks.
[2] Relative abundance is indicated for each species at El Caucho; we did not attempt an abundance estimate at Campo Verde. C = common, seen daily in moderate to large numbers (≥10/day); FC = fairly common, seen daily in small numbers (<10/day); U = uncommon, seen every few days; R = rare, only a few records during our stay.

SYSTEMATICS

Bock, Walter J. Is *Diglossa* (?Thraupinae) Monophyletic? ⸺ 319

Braun, Michael J., and Theodore A. Parker, III. Molecular, Morphological, and Behavioral Evidence Concerning the Taxonomic Relationships of "*Synallaxis*" *gularis* and Other Synallaxines ⸺ 333

Capparella, A. P., and Scott M. Lanyon. Biochemical and Morphometric Analyses of the Sympatric, Neotropical, Sibling Species, *Mionectes macconnelli* and *M. oleagineus* ⸺ 347

Dickerman, Robert W. Taxonomy of the Lesser Nighthawks (*Chordeiles acutipennis*) of North and Central America ⸺ 356

Lanyon, Wesley E. A Phylogeny of the Myiarchine Flycatchers ⸺ 361

Morony, John J., Jr. Systematic Relations of *Sericossypha albocristata* (Thraupinae) 383

Schulenberg, Thomas S. An Intergeneric Hybrid Conebill (*Conirostrum* × *Oreomanes*) from Peru ⸺ 390

Sibley, Charles G., and Jon E. Ahlquist. Phylogeny and Classification of New World Suboscine Passerine Birds (Passeriformes: Oligomyodi: Tyrannides) ⸺ 396

Traylor, Melvin A., Jr. Species Limits in the *Ochthoeca diadema* Species-group (Tyrannidae) ⸺ 431

IS *DIGLOSSA* (?THRAUPINAE) MONOPHYLETIC?.

WALTER J. BOCK

Department of Biological Sciences, Columbia University, New York, New York 10027 USA, and
Department of Ornithology, American Museum of Natural History, New York, New York 10024 USA

ABSTRACT. The currently recognized genus of Neotropical flowerpiercers, *Diglossa* (?Thraupinae), is polyphyletic as shown primarily by the structure of the corneous tongue. The species *caerulescens, cyanea* and *glauca* constitute the genus *Diglossopis* Sclater 1856, whereas the remaining species form the restricted genus *Diglossa* Wagler 1832. These genera are not closely related to one another within the New World nine-primaried oscines and have acquired their frilled tongue and other nectar feeding adaptations independently. Similarities in the flower-piercing nectarivorous habits and the bill structures of the two genera are examples of convergent evolution.

RESUMEN. El género actualmente reconocido para mieleros, *Diglossa* (?Thraupinae), tiene su origen en varios linajes ("polyphyletic") tal como lo muestra principalmente la estructura de la lengua cornea. Las especies *caerulescens, cyanea* y *glauca* constituyen el género *Diglossopis* Sclater 1856, mientras que las restantes especies, el género restringido *Diglossa* Wagler 1832. Estos géneros no están cercanamente relacionados entre sí con los oscines de nueve primarias del Nuevo Mundo y han adquirido su lengua escarolada y otras adaptaciones para su alimentación de néctar, de manera independiente. Las similaridades en los hábitos nectarívoros de picar flores y en la estructura del pico de los dos géneros son ejemplos de evolución convergente.

The genus *Diglossa* as recognized today (Vuilleumier 1969; Storer 1970) is a group of between 10 and 17 species, depending on the treatment of allopatric taxa, of New World nine-primaried oscines found in mountain forests and brushlands from southern Mexico to northern Argentina. They are nectar feeders with the specialized habit of holding on to the side of a flower with their hooked upper jaw and cutting through the corolla wall with their mandibular tip, hence the English name "flowerpiercers" for the group. *Diglossa* was included in the Coerebidae until that family had been shown to be unnatural (Beecher 1951) and broken up. Currently the flowerpiercers are included in the subfamily Thraupinae of the Emberizidae as delimited in Peters's "Check-list" (Paynter and Storer 1970), but their allocation to that group is regarded as tentative by many systematists because of uncertainties about the classification of the entire complex of New World nine-primaried oscines. During the course of a study of the taxonomic affinities of *Acanthidops bairdii* (? Emberizinae; Bock and Johnson, unpubl. data) which is believed by some ornithologists to be related to *Diglossa*, evidence accumulated which suggested that *Diglossa* is not a homogeneous, tight-knit genus as believed by ornithologists. Hence it was necessary to review the whole genus *Diglossa* and to inquire into the question of whether or not it is monophyletic (in the broad sense used in evolutionary classification, see Bock 1973, 1977), which is the goal of the current paper.

For the purposes of this analysis, I will accept the species classification proposed by Vuilleumier (1969), which represents the most recent detailed review of the genus. As he pointed out, considerable question exists on the recognition of many allopatric taxa as subspecies or as allospecies because *Diglossa* is an actively speciating group. Recognition of these geographic forms as species or as subspecies will not affect the conclusions of this study.

Vuilleumier reviewed the taxonomic history of this genus which had been divided into a number of genera with most allopatric taxa recognized as species during the last century. However, 100 years ago, Cassin (1864) and Sclater (1875) recognized only two genera, *Diglossa* and *Diglossopis* (for *caerulescens* only); they considered these genera to be closely related. Subsequently, Berlepsch (1884) stated that he knew no justification for maintaining *Diglossopis* as a separate genus. This conclusion was not accepted by all workers until Hellmayr (1935) recognized only *Diglossa*. Even workers separating these birds into two genera did not question the close relationship between them and usually pointed out that several species of *Diglossa* (e.g., *glauca* and *cyanea*) are intermediate between *Diglossopis caerulescens* and the typical species of *Diglossa*. The characteristic upturned hooked bill of the flowerpiercers was considered to be the unique hallmark of this genus. Basically the close relationships of these birds, i.e., the monophyly of all flowerpiercers, has never been doubted by avian taxonomists ever since the description of the first species and of the genus by Wagler in 1832.

TAXONOMIC HYPOTHESES

During the course of a comparative study of *Diglossa*, in connection with the analysis of *Acanthidops*, several features of cranial anatomy were discovered which suggested that the genus *Diglossa* is not only comprised of two distinct groups, but that these two groups may not be closely related to one another within the New World nine-primaried oscines. Based on these features, several taxonomic and phylogenetic hypotheses about groups (see Bock 1977, 1981) may be offered at the outset. These are:

(1) That the genus *Diglossa*, as currently recognized, is comprised of two groups that can be considered as distinct genera,
    (a) *Diglossa* Wagler 1832. Type: *Diglossa baritula* Wagler 1832. Includes: *D. indigotica, major, gloriosissima, lafresnayii, mystacalis, humeralis, carbonaria, duidae, albilatera, venezuelensis, baritula, plumbea,* and *sittoides.*
    (b) *Diglossopis* Sclater 1856. Type: *Diglossopis caerulescens* Sclater 1856. Includes: *D. caerulescens, cyanea,* and *glauca.*
(2) That the genera *Diglossa* and *Diglossopis* are not closely related to one another within the New World nine-primaried oscines, e.g., within the Thraupinae.
(3) That the genera *Diglossa* and *Diglossopis* became nectarivorous and evolved their specialized nectar-feeding adaptations independently.

These group hypotheses will be tested against several hypotheses about the taxonomic properties of features of the feeding apparatus. The major character hypotheses are:

(1) That the dissimilar frilled, nectarivorous, corneous tongues of *Diglossa* and of *Diglossopis* are not homologous as a tongue adapted for nectar feeding.
(2) That the corneous tongues present in *Diglossa* and in *Diglossopis* represent two different adaptive solutions (i.e., different paradaptations) to selection forces associated with nectar feeding.
(3) That the similar horny bills of *Diglossa* and of *Diglossopis* are not homologous as a bill specialized for the flower-piercing method of nectarivory, but rather that they are convergent as a result of selection forces associated with this feeding method.
(4) That the dissimilar features of the bony palate (e.g., the transpalatine process, the interpalatine process and the maxillopalatine) and of the mandible of *Diglossa* and of *Diglossopis* are not horizontally homologous (i.e., both are not traceable to the same feature in the common ancestor of the two groups).

These and other hypotheses about taxonomic properties of features will be tested against empirical evidence provided by a comparative analysis of morphological features only; these will be structures of the horny bill (rhamphotheca), of the skull, and of the corneous tongue. Sufficient comparative information about the behavior and ecology, including feeding methods and food, of enough species of flowerpiercers is not available to permit use of these attributes as taxonomic data on which to test the above group hypotheses. I was able to examine study skins of all species of flowerpiercers, and spirit and skeletal specimens of all species groups and of all biogeographic species recognized by Vuilleumier (1969) except *Diglossa indigotica* for which anatomical specimens were not available.

MATERIALS AND METHODS

A large series of study skins (200+) of all species of *Diglossa* (in the former sense) as recognized by Vuilleumier (1969) were examined in the American Museum of Natural History. Greatest attention was given to the external morphology of the bill (rhamphotheca) which was examined with the assistance of a dissecting stereomicroscope. Skeletons and spirit specimens were examined in the collections of the AMNH and the Museum of Zoology, Louisiana State University (MZ-LSU). These included members of all species groups and of all biogeographic species recognized by Vuilleumier (1969) except *Diglossa indigotica,* for which anatomical specimens are not available. Study of anatomical specimens concentrated on the structure of the skull and of the corneous tongue. Dissections were made with the help of a Wild M 5 stereomicroscope. Illustrations were drawn with a drawing tube (camera lucida) attached to the microscope (the corneous tongue) or traced from projections of 35 mm trans-parencies (the skulls).

The following anatomical specimens were examined:

(a) Spirit specimens: *Diglossopis caerulescens,* AMNH 4627; *D. cyanea,* AMNH 7936, 7937, 1755, 1756; *D. glauca,* MZ-LSU 107707; *Diglossa major,* AMNH 1753; *D. lafesnayii,* AMNH 1754, 1755, 7935; *D. carbonaria,* AMNH 4756, 4757, 4628, 3995, 3996, 3549, 3550, 3551, 7932, 7933; *D. albilatera,* MZ-LSU 83899; AMHH 122857 (skin with dried tongue attached); *D. baritula,* AMNH 1754; *D. duidae* (AMNH uncataloged; S. Coats 631, 632).

(b) Skeletons: *Diglossopis caerulescens,* AMNH 9501, MZ-LSU 84133, 84134, 90303, 75643; *D. cyanea* AMNH 6003, MZ-LSU 86666, 86716, 90304, 90305, 90306, 90307, 90308, 90309, 81330, 81332, 74927, 74928, 74929, 74931, 74932, 65050, 94180, 101629, 101630, *D. glauca,* 90302, 86438, 99533, 107524, 70271, 70272, 79988, 79890, 79891; *Diglossa major,* AMNH 9502, 9504; *D. lafresnayii,* AMNH 6952, 9505, MZ-LSU 81338, 90278, 90280, 90283, 90284, 90287, 90288, 90298, 104439; *D. carbonaria,* AMNH 7220, 7221, 8254, MZ-LSU 74938, 81333, 84128, 94165, 94173, 94175, 94176, 94177, 94178, 94179; *D. albilatera,* MZ-LSU 79892, 84129, 84130, 84131, 90301, 90481, 90482, 97530; *D. baritula,* AMNH 7219, 12702, 12707, 12708, 12695, 12696, MZ-LSU 51414, 48533, 48619, 48620, 48621, 48974, 62798.

John Morony kindly examined skeletons of *Diglossa* and of *Diglossopis* in the Museum of Zoology, Louisiana State University and the spirit specimen of *Diglossa albilatera.*

MORPHOLOGICAL FEATURES

Interspecific variation in the morphological features described below falls into two distinct categories. In each case, I first describe the condition seen in the "typical" species of flow-erpiercers (= *Diglossa*), followed by that seen in the "anomalous" species (= *Diglossopis*). No significance should be given to the adjectives typical and anomalous, which are used only as a mnemonic aid; the two groups could be labelled A and B.

BILL STRUCTURE

The rhamphotheca of the maxilla (upper jaw) of the "typical" species sweeps slightly upwards from its base and ends in a distinct strong hook (see Vuilleumier 1969:fig. 1 for illustrations of the bill of several species of *Diglossa* and of *Diglossopis*). Immediately behind the hook is a notch in the maxillary tomium which is followed by 2 to 4 serrations. Sometimes slight ridges and grooves begin at these serrations and extend backward along the lateral surface of the maxillary rhamphotheca. These serrations are weak (in *albilatera* and *major*) or almost absent (in *indigotica*) in some species. The rhamphotheca of the mandible (lower jaw) is upturned with both the mandibular tomium and the mandibular gonys sweeping upward from the base of the lower jaw; this structure of the mandible is most unusual for passerine birds. The anterior tip of the mandible is flat and slightly rounded (viewed dorsally) to form a distinct cutting edge. In the anterior third of the mandible, the edges of the paired tomia curl inward so that the anterior part of the lower jaw forms a partial tube; this tube is opened dorsally at the tip of the mandible. The completeness of this tube varies greatly in the typical species of flowerpiercers, but it is difficult to determine how much of the inward curling of the mandibular tomia is an artifact of drying. A narrow medial ridge is present on the dorsal surface of the mandibular tip inside the mouth. This ridge becomes lower and finally disappears as it extends backward along the midline.

The corneous tongue fits inside the mandibular tube and projects out of the anterior opening as it is protracted forward. However, the tongue is not confined to the mandibular tube, but may be lifted dorsal to it. The medial ridge may serve as a guide for the paired tubes of the tongue and may separate them slightly as the tongue moves forward out of the mouth.

This bill structure is present in all typical species of *Diglossa,* with the greatest variations being in the serrations along the maxillary tomium and in the completeness of the mandibular tube.

The rhamphotheca of the upper jaw of the three "anomalous" species (*caerulescens, cyanea* and *glauca*) extends straight from its base. It is hooked at its tip (only very slightly in *caeru-lescens*) with a tomial notch but no serrations along the maxillary tomium. The rhamphotheca of the mandible (lower jaw) is not upturned, but has a straight tomium; the gonys curves upward as is typical in many passerine birds. The anterior tip of the mandible is flat and slightly rounded (in dorsal view) to form a cutting edge as in the typical flowerpiercers. The edges of the paired tomia curl inwards in the anterior third of the mandible to form a partial tube; the tube is opened dorsally at the tip of the mandible. This mandibular tube appears to be almost complete in the anomalous flowerpiercers, but again it is difficult to say how much of the inward curling of the tomia is artificial. A narrow medial ridge is present on the dorsal surface of the mandibular rhamphotheca inside the mouth, starting at its tip. This ridge becomes lower and finally disappears as it extends posteriorly along the midline.

As in the typical species of flowerpiercers, the corneous tongue fits inside, but is not confined to, the mandibular tube and projects out of the anterior opening as it is protracted forward.

This bill structure is present in the three anomalous species. It appears very similar to that of the typical species of *Diglossa*, especially when the mandible is fitted into the maxilla in a well prepared skin. However, the shape of the straight mandible in the anomalous species is distinctly different from the upswept mandible of the typical flowerpiercers; the latter is most unusual and may be unique for passerine birds. The straight upper jaw of the anomalous flower-piercers also differs from the upturned upper jaw of the typical ones, but this difference is not as striking as that seen in the mandible.

The most interesting structure of the bill is the partial tube formed in the anterior third of the mandible by the inward curling of the paired mandibular tomia. This incomplete tube can constrain movement of the tongue so that it projects out of the mouth just above the flattened mandibular tip. As the tongue moves forward out of the mouth, it enters the slit made in the corolla by the tip of the lower jaw (see Skutch 1954:422–424; Moynihan 1963: 328–329; Vuilleumier 1969:5, for details of feeding). Thus, the hook and serrations of the upper jaw, the flat cutting edge of the mandibular tip, and the incomplete mandibular tube formed by the inwardly curling tomia are all parts of a specialized adaptive complex associated with the unusual method of nectar feeding of the flowerpiercers.

The structure of the horny covering of the bill, except for the shape of the mandibular tomium, of *Diglossa* and *Diglossopis* is so similar that it would be most reasonable to conclude that the rhamphotheca of these two genera is homologous as a bill specialized for the flower-piercing method of nectarivary. Indeed, avian taxonomists have accepted this conclusion for more than 100 years because bill structure has been the major characteristic used for placing all flowerpiercers in a single genus. Nevertheless for reasons presented below (see Discussion), the similar morphology of the horny bill of the two groups of flowerpiercers is, with high probability, convergent and therefore not homologous as a specialized flower-piercing bill.

SKULL

The cranial osteology could be compared in most species of flowerpiercers as skulls were available for all species groups and for most biogeographic species recognized by Vuilleumier; of the latter, only *Diglossa indigotica* and *D. duidae* were not available for study. A detailed comparison of all features of cranial osteology will not be attempted because in the absence of a broad comparison I was not certain which of the observed differences represent individual variation or specific traits and which may be indicative of relationships within the New World nine-primaried oscines. Attention will focus on features of the bony palate.

The skulls of typical flowerpiercers (Fig. 1) are characterized by the lack of an elongated, narrow transpalatine process at the posterolateral corner of the palatine. Instead the prepalatine bars usually continue in a smooth curve into the mediopalatine processes. Sometimes a short, broad, flat nubbin of bone is found at the posterolateral corner of the palatine at the site where the transpalatine process would be found. A strong tendon arises from the posterolateral corner of the palatine; this tendon may be partly ossified in some specimens but would be lost in prepared skulls. This tendon, like the transpalatine process, is associated with the origin of the M. pterygoideus lateralis (Bock, unpubl. data). Specimens of the species of *Diglossa* that lack a transpalatine process appear to be full adults from the condition of skull ossification and information available on the labels. Knowledge of the age of these specimens is important because the transpalatine process develops ontogenetically by ossification of the tendon of origin of the M. pterygoideus lateralis which subsequently fuses to the posterolateral corner of the palatine (Jollie 1957:421–2, 1958:28, see fig. 2, his posterolateral process). Complete or almost complete lack of the transpalatine process appears to be a normal condition in the adult of the typical species of flowerpiercers and not a consequence of possessing only young specimens in which the transpalatine process was still unossified and unfused to the palatine.

The interpalatine process (choanal process of Jollie) is very short, scarcely projecting anteriorly from the thin palatine shelf.

Most striking are the maxillopalatines (palatine processes of the maxillae), which end in large, swollen scroll-like terminal plates that fill most of the space between the two prepalatine bars. Shortness of the interpalatine processes is correlated with the large size of these terminal plates. The pedicel of the maxillopalatine arises from the maxilla well posterior to the junction of the prepalatine bar with the premaxilla.

The mandible of the typical flowerpiercers is characterized by an abrupt ventral step of the gonys at the mandibular symphysis (Fig. 2A) and a slightly concave tomium that corresponds to the upward curvature of the tomium of the mandibular rhamphotheca. The internal process of the mandible (Fig. 2B) extends almost vertically from its base, has a uniformly narrow width, and possesses a large pneumatic foramen at its base (the size and exact position of this fossa was difficult to show in the illustration because of the foreshortened perspective of the process). The anterior edge of the internal process starts far posterior along the medial articular facet resulting in a distinct corner (the intermandibular flange). At its dorsal tip, the internal process appears to curve forward slightly. A deep hollow separates the internal process and the retroarticular process. The medial articular facet is a distinct groove with a steep ventral slope from anterior to posterior, i.e., it drops sharply into the hollow between the internal process and the retroarticular process. The lateral articular surface begins on the dorsal surface of the retroarticular process and continues forward in a smooth crescent to end against the lateral edge of the medial facet.

The skulls of the anomalous species of flowerpiercers (Fig. 3) are characterized by the presence of a straight, thin transpalatine process projecting posteriorly at a slight angle to the palatine. The length of this process varies somewhat, which may reflect differential degrees of ossification (due partly to age differences?) or damage during preparation.

The interpalatine processes are longer than those in the typical species, and generally longer than those illustrated (Fig. 3). The longer length appears to be correlated with the smaller size of the maxillopalatines.

The maxillopalatines are much smaller in the anomalous species with small, flatter and less swollen terminal plates that do not extend as far posteriorly. These plates do not occupy the whole of the space between the prepalatine bars, nor do they cover the vomer as completely as in the typical species. The pedicel of the maxillopalatine arises from the maxilla at the common junction with the prepalatine process.

The differences in the structure of the bony palate, including the maxillopalatines in the two groups of flower piercers is striking; however, they should be interpreted with some care. Most important may be the large difference seen in the maxillopalatines which would be significant among any groups of passerine birds. Although more obvious, the difference in the transpalatine processes may be less important because this process is simply the ossified tendon of the M. pterygoideus lateralis fused to the body of the palatine. Hence, lack of ossification, or the failure of the ossified tendon to fuse with the palatine would produce the observed difference in the transpalatine processes of the two groups. Presence or absence of the transpalatine process almost certainly affects the functional and adaptive properties of the M. pterygoideus lateralis. But its presence or absence is a minor difference in terms of ontogenetical development; all that is needed for disappearance of the transpalatine process is suppression of ossification of the tendon.

The mandible of the anomalous species has a smooth gonys and a straight or slightly convex tomium (Fig. 4A). The internal process of the mandible extends dorsally at an angle of about 45° from its base, has a broad base and narrows gradually toward its distal tip, and possesses a smaller pneumatic foramen somewhat above its base (Fig. 4B). The anterior edge of the internal process is almost continuous with the anterior edge of the medial articular facet. The distal tip of the internal process does not deviate from the longitudinal orientation of the process and points backwards. As in the typical species, a deep hollow separates the internal process and the retroarticular process. The medial articular facet is a distinct groove that slopes sharply into the hollow between the internal process and the retroarticular process, but this groove is less sharply delimited than that in the typical flowerpiercers. (A corresponding difference may exist in the medial condyle of the quadrate in the two groups of flowerpiercers, but this difference is difficult to describe; the medial condyle in the typical species appears to be more elongated along its antero-posterior axis.) The lateral edge of the medial articular facet appears as a distinct lip which results largely from the separation of this facet from the lateral articular facet. This lateral articular surface starts on the dorsal surface of the retroar-

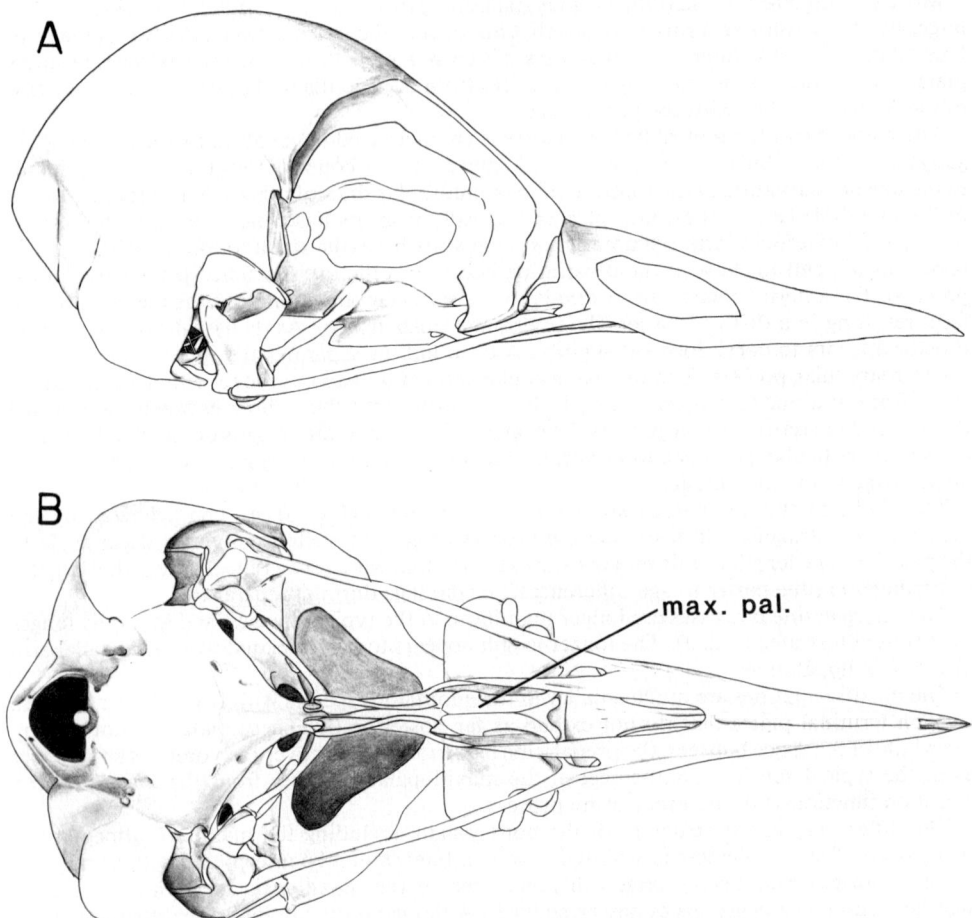

Fig. 1. Skull of *Diglossa lafresnayii* (AMNH 6972). (A) Lateral view. (B) Ventral view. Note the large maxillo-palatine (max. pal.) and the absence of the transpalatine process.

ticular process and continues straight forward. A distinct groove separates the anterior end of the lateral articular facet from the posterior end of the dorsal edge of the mandibular ramus and from the lateral border of the medial articular facet.

Because of the striking dissimilarities in the morphology of the transpalatine process, and of the maxillopalatine of *Diglossa* and of *Diglossopis,* I would conclude that the conditions of these features in the two genera are not horizontally homologous. This conclusion, however, cannot be supported strongly. It is possible to derive the palatal structure seen in *Diglossa* from that present in *Diglossopis,* or vice versa, but again neither conclusion can be supported strongly. The differences in the shape of the mandibular ramus reflect the differences in the shape of the mandibular rhamphotheca of the two groups. The differences in the structure of the mandibular articulation and shape of the internal process may not be homologous and may represent different paradaptations associated with the movement of the mandible during flower piercing, but these last conclusions cannot be supported strongly.

CORNEOUS TONGUE

The only description of the corneous tongue known to me is the brief one of Vuilleumier (1969:4–5) for an unspecified species of *Diglossa,* but clearly for one of the typical species. Because flowerpiercers are nectar feeders (all species?) tongue morphology is of particular interest to the question of affinity of these birds to one another.

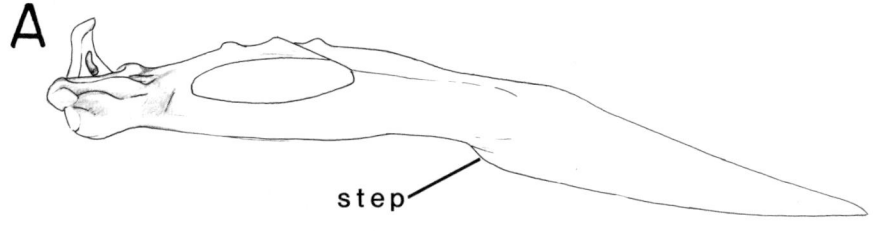

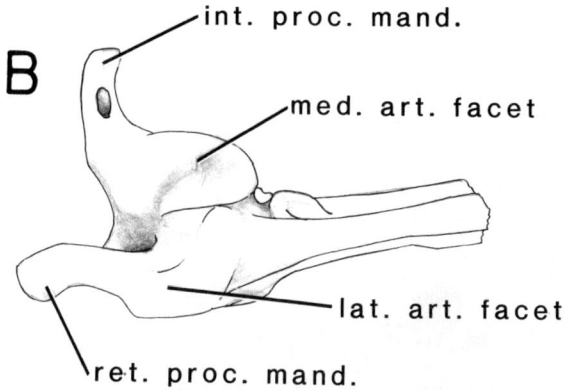

FIG. 2.   Mandible of *Diglossa lafresnayii* (AMNH 6972). (A) Lateral view. (B) Dorsal view (enlarged). Note the step in the mandibular gonys. Abbreviations are: int. proc. mand. = internal process of the mandible; ret. proc. mand. = retromandibular process; med. art. facet = medial articular facet; lat. art. facet = lateral articular facet.

The corneous tongue was examined in all species-groups and in all biogeographic species recognized by Vuilleumier with the exception of *D. indigotica*.

The tongue of the typical species of *Diglossa* (Fig. 5) is characterized by a broad flattish base that narrows abruptly about one-third of the way to the tip. The narrow anterior portion is split by a deep cleft which is about half the length of the corneous tongue (12 mm in a tongue 24 mm long) into two narrow halves, each only 0.5 mm wide in a tongue that is 3 mm wide at its base. Each half consists of a semitube that has a narrow medial opening. Each semitube opens ventrally at its posterior base and is continuous with the deep, ventrally facing groove present in the broad posterior half of the tongue. The ventral groove becomes increasingly shallower posteriorly and finally disappears just before the tongue base is reached. The walls of the two semitubes are composed of paper-thin sheaths. The tubes terminate anteriorly in a short (3 mm) frill. This consists of a series of 12 to 15 laciniae (fringes) which are formed by slits of the thin tubular sheath starting close to its ventromedial edge and curving dorsolaterally and then anteriorly. All of the fringes end in an approximately straight line across the anterior tip of the tube. Because the laciniae are so thin and colorless, they can be difficult to see under intense illumination with a dissecting microscope. The terminal fringes seen in these species of *Diglossa* are most similar to the terminal fringes of the tongue of most nectariniid species. The corneous tongue of the typical flowerpiercers is unusual for passerine nectarivorous birds and is most similar to that present in the Nectariniidae (Moller 1930; Bock, unpubl. data). The ventral groove in the base of the tongue with the paired semitubes opening into it is, to my knowledge, unique for nectar-feeding passerine birds and probably for all birds.

This tongue structure is present with no significant variation in all typical species of *Diglossa* examined.

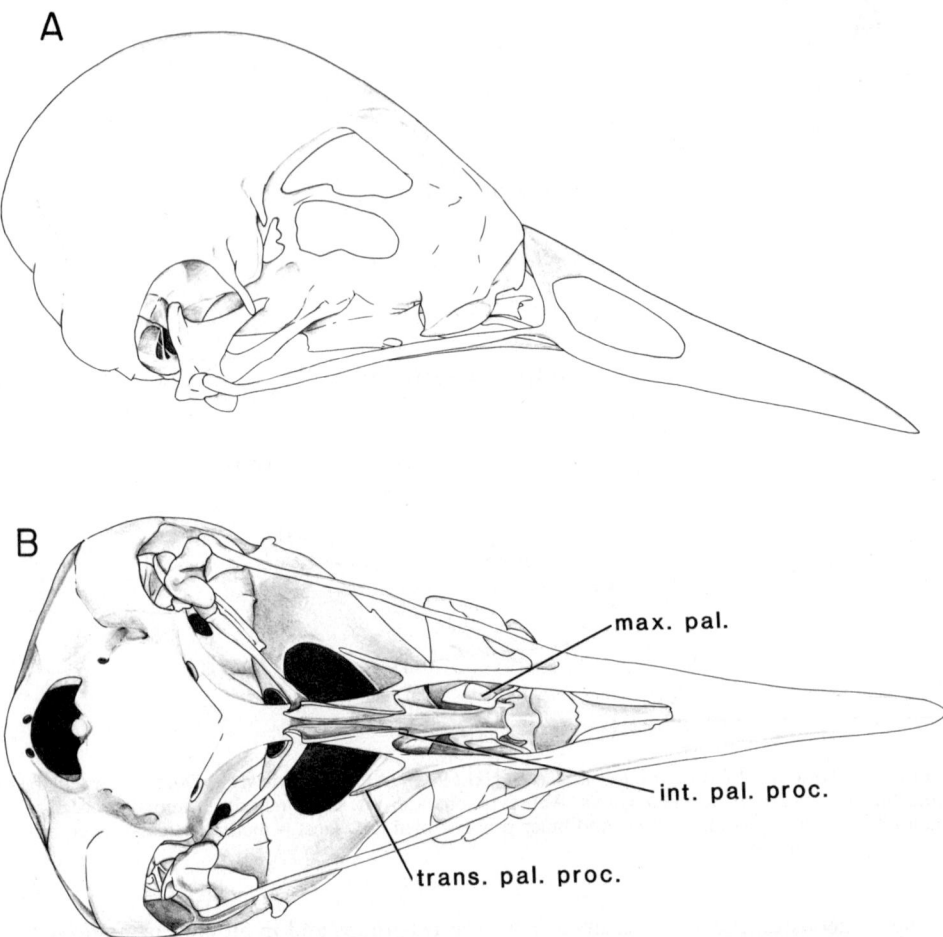

Fig. 3.   Skull of *Diglossopis cyanea* (AMNH 6003). (A) Lateral view. (B) Ventral view. Note the elongated, spine-like transpalatine process (trans. pal. proc.), the large interpalatine process (int. pal. proc.) and the small maxillo-palatine (max. pal.).

The tongue of the anomalous species of flowerpiercers (Fig. 6) has an equally broad base, but is deeper than that in the typical species. The tongue is deep throughout and narrows gradually toward the anterior tip from about the midpoint of the tongue. The tongue is solid throughout with a flat dorsal surface and rounded lateral and ventral surfaces. A very shallow ventral groove is present in the anterior half of the tongue. This groove disappears well before it reaches the tongue base. The tongue is split into two halves by a shallow cleft (4–5 mm long in a tongue 19–20 mm long). The dorsolateral edges of the tongue bear a series of heavy, black laciniae. These start at the midpoint (approximately) of the tongue and continue forward. They point anteriorly and medially. The posterior laciniae are quite short, but these fringes become longer towards the anterior tip of the tongue. A second series is present at the anterior tip of each tongue half. These project forward and curl slightly upward along the lateral and medial edges of each cleft. The individual laciniae are rather thick and coarse, especially along the dorsolateral edges of the tongue. Those at the anterior tip are longer, thinner, and finer in structure. Because of the dark color of the laciniae, the tongue appears pale colored with a heavy dark fringe.

This tongue structure is present without significant variation in the three anomalous species. It is a frilled tongue similar to that seen in a number of passerine species in diverse families and is most similar to that of *Acanthidops* (Bock and Johnson, unpubl. data). If anything, the

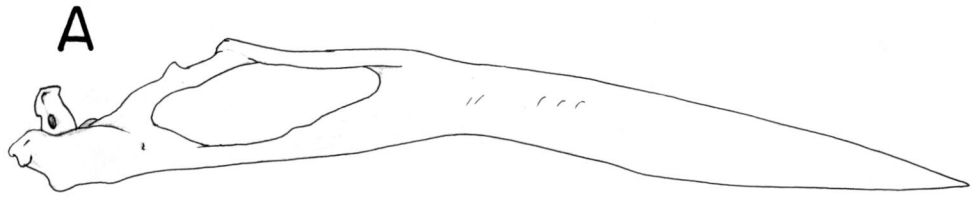

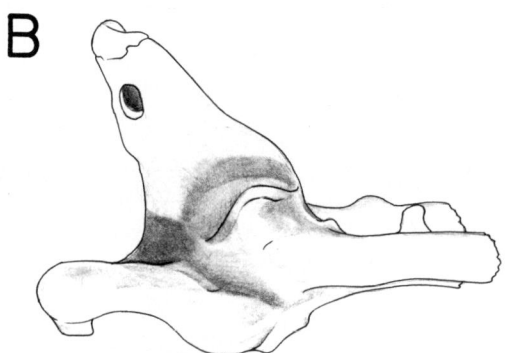

FIG. 4. Mandible of *Diglossopis cyanea* (AMNH 6003). (A) Lateral view. (B) Dorsal view (enlarged).

frilled tongue of *Acanthidops* appears to be more specialized for nectar feeding than those of the three anomalous species of flowerpiercers.

Without going into details of a proper analysis, I would conclude that each tongue type of the two groups of flowerpiercers is adapted for nectarivory. Yet the morphologies of these two tongue types are sharply distinct from one another in numerous details. Therefore I conclude that the frilled tubular tongue of the typical flowerpiercers is not homologous to the frilled tongue of the anomalous species, as a split frilled tongue. These two tongue types represent different paradaptations—different adaptive answers—to the selection forces associated with nectar feeding. Regardless of which of these two tongue types is assumed to be the ancestral one, it is not possible to derive either tongue type from the other if the birds were adapted to feeding on nectar throughout the whole sequence of evolutionary changes. The morphologies of the tongue fringes and of the cross-sections of the tongue—solid versus a pair of tubes—are so different that evolutionary change from one tongue type to the other would not be possible without passing through a intermediate nonfrilled stage. Such a nonfrilled tongue would be less adapted for feeding on nectar than either of the frilled tongues seen in the two groups of flowerpiercers. Because all species of flowerpiercers feed on nectar (no claim is made or needed that nectar is the only food of these birds—all nectar feeding birds take other food, usually insects and/or fruit), no basis exists to postulate a non-nectarivorous stage in a presumed evolution of one tongue type from the other. Both frilled, nectar adapted tongue types found in the flowerpiercers can be derived from a nonfrilled ancestral tongue, that is, a generalized passerine tongue probably subdivided into two short terminal halves by a short medial slit and with few or no terminal fringes. The two flowerpiercer tongue types are a classical case of multiple pathways of adaptive evolution which frequently provide very valuable taxonomic evidence (Bock 1967; 1973).

DISCUSSION

Having presented the results of a comparative survey of the several features of the feeding apparatus and having reached conclusions on the hypotheses about taxonomic properties of these features, I will now consider the several group hypotheses presented at the onset of this paper.

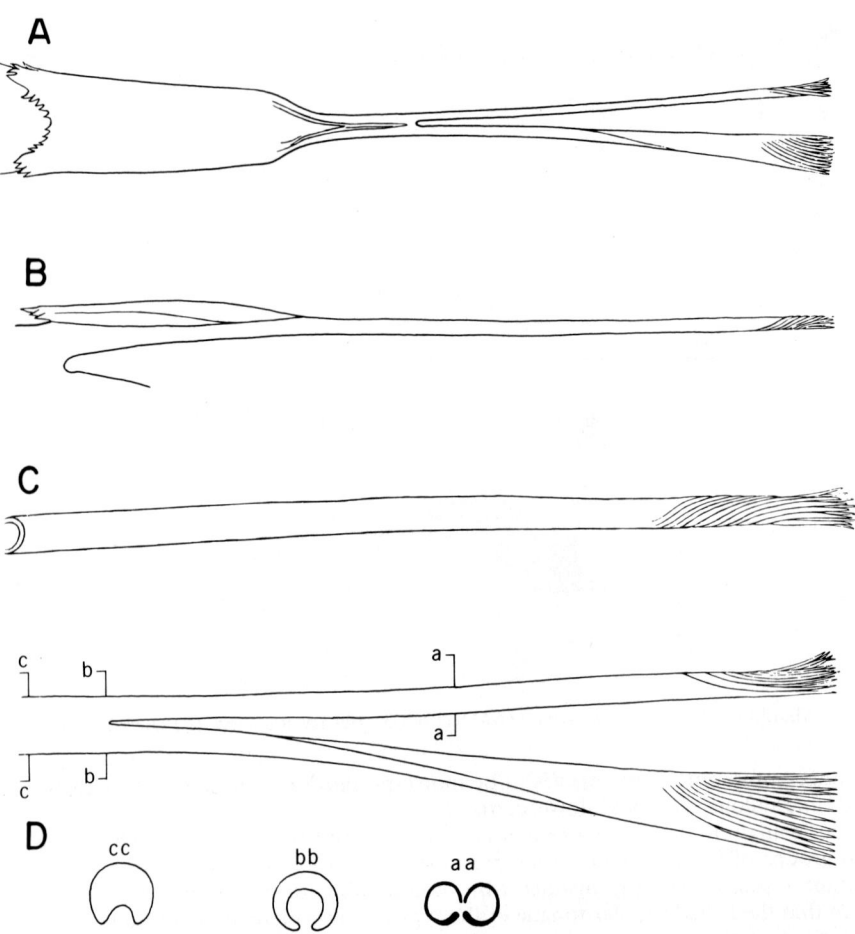

Fig. 5. Corneous tongue of *Diglossa major*. (A) Dorsal view with the anterior part of the right tube unrolled. (B) Lateral view. (C) Enlarged lateral view. (D) Enlarged dorsal view with the anterior part of the right tube unrolled and with three inserts showing cross-sections of the tongue. Section a-a is at the midpoint of the tubes showing the cross-section of both tubes in their normal position. Section b-b is at the anterior end of the undivided segment of the tongue. Section c-c is at the posterior end of the narrowed portion of the tongue.

The features described above divide the flowerpiercers into two distinct series of species with no intermediate forms and with neither group of species tending toward the other in any feature. These groups are most sharply separated by the morphology of the corneous tongue, followed by that of the bony palate, the maxillo-palatines and the mandible, and lastly by the structure of the rhamphotheca. I conclude that the conditions of these features in the two groups are nonhomologous, although I cannot provide a strong argument for the nonhomology of the cranial features. Moreover, the two corneous tongue types represent two different paradaptations for nectarivory. Thus, I accept the hypothesis that the genus *Diglossa* as currently recognized (Vuilleumier 1969; Storer 1970) should be divided into two genera, *Diglossa* and *Diglossopis,* each of which includes the species listed above (group hypothesis 1).

Anatomical specimens of *Diglossa indigotica* were not available, but this species can be assigned to the restricted genus *Diglossa,* with little question, on the basis of its bill structure, especially that of the mandibular rhamphotheca. The shape of the mandible in the two groups of flowerpiercers is clearly distinct from each other and is correlated with the cranial and lingual morphologies of those species for which anatomical material is available. Therefore

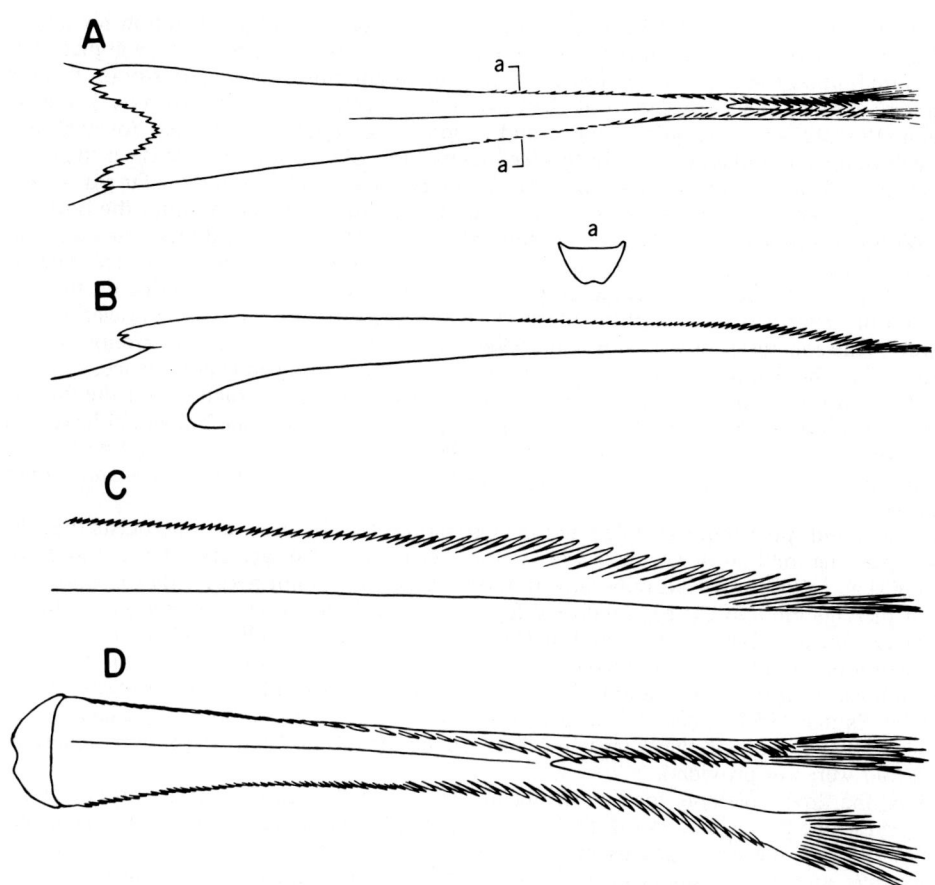

FIG. 6.   Corneous tongue of *Diglossopis cyanea*. (A) Dorsal view with insert showing the cross-section at the midpoint of the tongue. (B) Lateral view. (C) Enlarged lateral view. (D) Enlarged dorsal view.

until anatomical specimens of this species become available, it is most reasonable to retain *indigotica* in the typical flowerpiercers (*Diglossa*).

Second, I accept the hypothesis that the genera *Diglossa* and *Diglossopis* are not closely related to one another within the New World nine-primaried oscines. If they prove to be members of the same family group taxon, e.g., the Thraupinae, which is possible, then they are most probably not closely related within that taxon. This conclusion can be accepted with reasonable certainty without undertaking a broad comparative survey of possible relatives of these two genera because of the structure of their corneous tongue. As argued above, these two tongue types are best interpreted as adaptations for nectar feeding. Further, they are, with high probability, two different multiple adaptive solutions to the same general selection force arising from the demands of feeding on nectar. Structures that represent different adaptive answers to the same selection force can serve as strong phylogenetic tests especially of group hypotheses that two taxa are not closely related (Bock 1967, 1969, 1973). Groups of organisms possessing different adaptive answers (i.e., paradaptations) to the same selection force are with strong probability not monophyletic. The different paradaptations suggest the existence of different complexes of genetically based variation that have interacted with the same selection force (e.g., for nectar feeding) to produce the different adaptive answers (e.g., the two types of frilled tongues found in the flowerpiercers).

A counter argument could be presented that *Diglossa* and *Diglossopis* are actually closely related and had descended from a common ancestor which was characterized by a hooked upper jaw, cutting edge at the mandibular tip and partial mandibular tube. The morphology of the corneous tongue would have diverged as a result of further specialization for nectar feeding at a subsequent stage in the evolution of the group. It can be reasonably argued that the rhamphothecal features of the flowerpiercers are adaptations for nectarivory on flowers with long corollas. The description of *Diglossa baritula* feeding on such blossoms, given by Skutch (1954:422–424), is quite detailed and supports the conclusion that the horny sheath of the flowerpiercer bill is adapted for this feeding method. Skutch showed that the bird grasps the slender corolla cross-wise in its bill. The maxillary hook is held around the far side of the corolla to prevent its slipping; bruises are found on this side of the flower after the bird has fed. With the corolla held in place, a longitudinal (= vertical) slit is cut into its near wall with the mandibular tip; the tongue is protruded through this slit into the flower to obtain nectar. These slits were found on flowers after *Diglossa baritula* were observed feeding on them. I have no objections to this adaptive interpretation of the bill structure of the flowerpiercer. However, the conclusion that *Diglossa* and *Diglossopis* are closely related (e.g., monophyletic) requires that the features of the bill in the two genera are homologous and, hence, synapomorphous, not just apomorphous, as a complex of rhamphothecal structures for the flower-piercing method of nectar feeding. Thus, this complex of features of the bill would have had to evolve *once* and *early* in the history of nectar feeding in the flowerpiercers, and before any modifications in the tongue for nectarivory. This argument suffers from two flaws and cannot be accepted.

The first and most important flaw is the tacit assumption by most ornithologists that the flower-piercing method of nectar feeding is limited to the flowerpiercers (*Diglossa* and *Diglossopis*) and that it evolved only once in these birds. This is not necessarily the case. The flower-piercing method has been reported for *Diglossa lafresnayii, D. carbonaria* (Moynihan 1963), *D. baritula* (Skutch 1954), and in *Diglossopis cyanea* (Moynihan 1963). Moreover, it has been reported in *Coereba flaveola* (? Parulinae; Skutch 1954), in *Conirostrum cinereum* (? Parulinae; Moynihan 1963) and in *Phaethornis lonquemareus* and *Hylocharis leucotis* (Trochilidae; Skutch 1954). These other species were observed to peck a hole in the base of the corolla or to pierce it with the sharp tip of their bill, but details of the methods used in piercing the flower were not provided.

Thus, the flower-piercing method of nectar feeding on flowers with deep corollas is not restricted to the flowerpiercers (in the broadest sense), but is found in two additional genera of New World nine-primaried oscines, which may not be closely related to *Diglossa* and *Diglossopis,* and in two genera of hummingbirds. Lack of detailed observations on feeding methods used by other nectarivorous birds precludes knowing how many other groups may use this method. If the flower-piercing method of nectarivory evolved independently in several groups of birds, it is reasonable to conclude that it could have evolved independently in *Diglossa* and *Diglossopis.* If this is the case, the several rhamphothecal features as the hooked upper jaw, sharp tip of the mandible, and semitubular shape of the mandible evolved independently in these two genera as specializations for the flower-piercing method of nectarivory. Moreover, this complex of rhamphothecal features may be the only possible adaptive specializations for flower-piercing other than the bird holding the flower steady with its feet. No strong basis exists to support the conclusion that the similarities in bill structure of *Diglossa* and *Diglossopis* are homologous in view of the taxonomically widespread occurrence of the flower-piercing method of nectarivory. It is quite possible that these similarities are convergences (nonhomologous apomorphies) resulting from the action of the same set of selective demands associated with this method of nectar feeding. The differences in the morphology of the mandible in the two genera supports the argument of independent evolutionary origins and convergence of their similar flower-piercing bill.

The second flaw in the argument that *Diglossa* and *Diglossopis* are closely related is associated with the time of appearance of their diverse nectarivorous adaptations. This argument assumes that the adaptations of the rhamphotheca for the flower-piercing method of nectar feeding evolved fully in the common ancestor of *Diglossa* and *Diglossopis,* but that the corneous tongue was still unchanged from the generalized passerine condition—that is, the tongue possessed no modifications for nectarivory. This assumption is necessary because of the previous conclusion that the tongues in the two genera of flowerpiercers are different multiple adaptive answers for nectar feeding. Even accepting the basic ideas of mosaic evolution, this pattern of evolutionary change is highly unlikely because the bill and corneous tongue comprise a single functional complex for nectar feeding in *Diglossa* and *Diglossopis.* It is not reasonable to assume that the bill became well specialized for this feeding method before any evolutionary modification occurred in tongue structure. Consequently it is reasonable to assume that the corneous tongue of these genera was at least partly specialized for nectarivory at the time that rhamphothecal specializations started to evolve. Therefore, the specialized adaptations in the bill of the flowerpiercers appeared independently in the two groups and converged.

Lastly, I accept the hypothesis that *Diglossa* and *Diglossopis* acquired all of their feeding adaptations, both in the structure of the rhamphotheca and of the corneous tongue, independently. This hypothesis is closely related to the last and is supported by the argument presented above that the corneous tongues represent multiple pathways of adaptation and that the similarities in the bill are most likely convergences.

It should be stressed that all previous workers have concluded that the Neotropical flowerpiercers constitute a monophyletic group even when they were arranged in two (or more) genera, e.g., as advocated by Cassin (1864) and by Sclater (1875) in the last century. Moreover, it must be stressed that the limits of the genera *Diglossa* and *Diglossopis* recognized in this paper are different from the limits of these genera recognized by early workers, prior to Hellmayr (1935). No one had suggested previously that the species *cyanea* and *glauca* should be placed in *Diglossopis* with *caerulescens* rather than in the genus *Diglossa*. Inclusion of *indigotica* in the presumed primitive *caerulesceus* species group by Vuilleumier (1969:13–14) was based mainly on the blue plumage, black facial mask, "non-scaly"forehead, and poorly marked sexual dimorphism. The plumage similarity between *indigotica* and the species of *Diglossopis* may be the result of convergence stemming from social mimicry in feeding flocks (see Moynihan 1968). The proposed arrangement of the flowerpiercers into *Diglossa* and the unrelated *Diglossopis* should have no important effect on analyses of species relationships or patterns of speciation in the flowerpiercers (Vuilleumier 1969).

The conclusion that the "diglossids" are polyphyletic provides a taxonomic lesson that must be taken seriously. The flowerpiercers are a well known group of Neotropical birds that have been revised by a number of taxonomists over the past century without any suspicion that they might be polyphyletic. All earlier studies were based on comparisons of external features. Only a fortuitous examination of some aspects of the internal morphology in several species suggested that these birds might not be a tightly knit group as all workers had assumed. However success of the analysis depended absolutely on functional-adaptive considerations of the key taxonomic features (e.g., the rhamphotheca and the corneous tongue). Without a reasonable functional-adative analysis, as I have advocated in a number of earlier papers (Bock 1967, 1969, 1977, 1981), it would not have been possible to demonstrate that the corneous tongues in the two genera are distinct paradaptations and that most features of the rhamphotheca are most likely convergent adaptations to a similar method of nectar feeding that appeared independently in a number of nectarivorous birds. Quite likely a nonfunctional-adaptive taxonomic analysis, be it a conventional, phenetic, or cladistic one, would have led to the conclusion that the flowerpiercers are a monophyletic genus with two distinct subgroups. Avian taxonomists cannot assume that genera of birds, even apparently homogeneous ones, are really uniform and, therefore, that detailed morphological or other investigations of single species is sufficient. In the study that necessitated the present analysis, the question of whether *Acanthidops bairdii* is closely related to *Diglossa* (in the former sense) would have been answered "definitely no" or "probably yes" depending on the species of flowerpiercer chosen for comparison.

ACKNOWLEDGMENTS

I am most grateful to J. V. Remsen, Jr. and J. Morony who examined specimens of *Diglossa* in the Museum of Zoology, LSU and loaned me critical spirit specimens. F. Vuilleumier, J. Morony, and D. Homberger read the manuscript and provided many useful suggestions for which I am most thankful. I wish especially to thank F. Vuilleumier with whom I had many valuable discussions on the biology, systematics, and evolutionary history of the flowerpiercers. I wish to thank S. Coats for permitting me to examine specimens of *Diglossa duidae* which she collected after this study was completed.

LITERATURE CITED

BEECHER, W. J. 1951. Convergence in the Coerebidae. Wilson Bull. 63:274–287.
BERLEPSCH, H. v. 1884. Untersuchungen über die Vögel der Umgegend von Bucaramanga in Neu-Granada. J. Ornithol. 32:273–320.

BOCK, W. J. 1967. The use of adaptive characters in avian classification. Prov. XIV Int. Ornith. Cong., pp. 61–74.

BOCK, W. J. 1969. Comparative morphology in systematics. Pp. 411–448, *In* Systematic Biology. National Academy Sciences, Washington, D.C..

BOCK, W. J. 1973. Philosophical foundations of classical evolutionary classification. Syst. Zool. 22:375–392.

BOCK, W. J. 1977. Foundation and methods of evolutionary classification. Pp. 851–895, *In* M. K. Hecht, P. C. Goody, and B. M. Hecht (eds.), Major Patterns in Vertebrate Evolution. Plenum Press, New York.

BOCK, W. J. 1981. Functional-adaptive analysis in evolutionary classification. Am. Zool. 21:5–20.

CASSIN, J. 1864. Notes of an examination of the birds of the subfamily Coerebinae. Proc. Acad. Nat. Sci. Phila. 16:265–275.

HELLMAYR, C. E. 1935. Catalogue of birds of the Americas. Pt. VIII. Field Mus. Nat. Hist. Publ., Zool. Ser. 13:1–542.

JOLLIE, M. T. 1957. The head skeleton of the chicken and remarks on the anatomy of this region in other birds. J. Morphol. 100:389–436.

JOLLIE, M. T. 1958. Comments on the phylogeny and skull of the Passeriformes. Auk 75:26–35.

MOLLER, W. 1930. Ueber die Schnabel und Zungenmechankik blutenbesuchender Vögel. I. Biol. Gen. 6:651–726.

MOYNIHAN, M. 1963. Inter-specific relations between some Andean birds. Ibis 105:327–339.

MOYNIHAN, M. 1968. Social mimicry: character convergence versus character displacement. Evolution 22:315–331.

SCLATER, P. L. 1875. Synopsis of the species of the subfamily Diglossinae. Ibis, Ser. 3, 5:204–221.

SKUTCH, A. F. 1954. Life histories of Central American birds, families Fringillidae, Thraupidae, Icteridae, Parulidae and Coerebidae. Pac. Coast Avif. 31.

STORER, R. W. 1970. Subfamily Thraupinae, Tanagers. Pp. 246–408, *In* R. A. Paynter, Jr. and R. W. Storer (eds.), Check-list of Birds of the World, Vol. XIII. Museum Comparative Zoology, Cambridge, Massachusetts.

VUILLEUMIER, F. 1969. Systematics and evolution in *Diglossa* (Aves, Coerebidae). Am. Mus. Novit. No. 2381.

MOLECULAR, MORPHOLOGICAL, AND BEHAVIORAL EVIDENCE CONCERNING THE TAXONOMIC RELATIONSHIPS OF *"SYNALLAXIS" GULARIS* AND OTHER SYNALLAXINES

MICHAEL J. BRAUN[1] AND THEODORE A. PARKER, III[2]

[1]*Department of Biochemistry, Lousiana State University Medical Center, 1901 Perdido St., New Orleans, Louisiana 70112 USA, and*
[2]*Museum of Zoology, Louisiana State University, Baton Rouge, Louisiana 70893 USA*

ABSTRACT. Based on similarities in vocalizations, behavior, and morphology, we hypothesized that the White-browed Spinetail (*"Synallaxis" gularis*) is more closely related to species in the genus *Cranioleuca* than to other species of *Synallaxis*. We tested this hypothesis by using starch gel electrophoresis and specific staining techniques to compare the protein products from thirty genetic loci of *gularis* to those of four other species of *Synallaxis* and five species of *Cranioleuca*. Two species of *Schizoeaca* were included in the study as an outgroup. Vocalizations of selected taxa were compared by spectrographic analysis. Both the molecular and vocal evidence support our hypothesis. The mean unbiased genetic distance (D of Nei 1978) from *gularis* to *Cranioleuca* species was 0.29, whereas the mean distance to *Synallaxis* species was 0.52. The supposed outgroup, *Schizoeaca*, fell so much closer to *Cranioleuca* than to *Synallaxis* (mean D of 0.31 vs 0.47) that no cladistic analysis was attempted. We recommend resurrection of the genus *Hellmayrea* for *gularis*, and suggest possible relationships with several other synallaxine genera. The protein data also indicate that *Schizoeaca helleri* and *S. harterti*, recently lumped with *S. fuliginosa*, are quite well differentiated at the molecular level (D = 0.13) and probably should be retained as full species.

RESUMEN. Basandonos en similaridades de vocalizaciones, comportamiento y morfología, proponemos como hipótesis que el trepador de frente blanca (*"Synallaxis" gularis*) está más cercanamente relacionado a especies del género *Cranioleuca* que a otras especies del género *Synallaxis*. Para comprobar esta hipótesis utilizamos electroforesis por medio de gelatina de fécula y técnicas de tinte específicas para comprobar los productos proteínicos de 30 "loci" genéticos de *gularis* con aquellos de otras cuatro especies de *Synallaxis* y cinco especies de *Cranioleuca*. Se incluyeron dos especies de *Schizoeaca* como un grupo raíz comparativo "outgroup." Se compararon, por medio de análisis espectrográfico vocalizaciones de taxas específicas. Los resultados de ambos estudios, molecular y vocal, respaldan nuestra hipótesis. La distancia genética imparcial promedio ("mean unbiased genetic distance") (D en Nei 1978) desde *gularis* hasta especies *Cranioleuca* fue 0.29, mientras que la distancia promedio hasta especies de *Synallaxis* fue 0.52. El supuesto "outgroup," *Schizoeaca*, estuvo mucho más cercano a *Cranioleuca* que a *Synallaxis* (promedio D de 0.31 versus 0.47) por lo que no se intentó realizar ningún análisis cladístico. Recomendamos usar el género *Hellmayrea* para *gularis* y sugerimos relaciones posibles con varios otros géneros synallaxines. Los datos de proteínas también indican que *Schizoeaca helleri* y *S. harterti*, recientemente englobados con *S. fuliginosa*, están suficientemente diferenciados a nivel molecular (D = 0.13) y probablemente deben ser mantenidos como especies independientes.

The family Furnariidae represents a major adaptive radiation of suboscine passerine birds in South America. Its more than 200 species occupy every available habitat on the continent, from rocky seacoasts and barren deserts to freshwater swamps and marshes, from high Andean grasslands to Amazonian rainforests. The family is usually divided into three subfamilies, the largest of which, Synallaxinae, is comprised of more than 100 species, including the 12 in the present study. Very little information is available on many furnariid species, and the evolutionary relationships within the family are just beginning to be investigated (Vaurie 1971, 1980; Feduccia 1973). Because questions about the validity of some of Vaurie's generic grouping (Fitzpatrick 1982) have been raised, we use the generic names of Meyer de Schauensee (1966) which are probably most familiar to Neotropical ornithologists.

The White-browed Spinetail ("*Synallaxis*" *gularis* Lafresnaye 1843) is a poorly known furnariid species that inhabits the undergrowth of high elevation Andean cloud forest from Venezuela south to central Peru. Sztolcman (1926) created the monotypic genus *Hellmayrea* for *gularis* on the basis of bill and tail morphology (see Comparative Morphology below). Peters (1951) followed this treatment, but Meyer de Schauensee (1966) merged *Hellmayrea* with *Synallaxis*. Parker et al. (1985) summarized information on the behavior and distribution of this species. On the basis of vocalizations and behavior, we suspected that *gularis* is actually more closely related to species of *Cranioleuca*, another large genus of spinetails, than to species of *Synallaxis*. In this paper we report on our initial efforts to test that hypothesis, using the technique of starch gel electrophoresis to compare proteins of these birds. Although electrophoretic studies of birds that examine many specifically identified protein loci have been relatively few, the broad applicability of the technique is confirmed by the hundreds of studies of other organisms in which it has been employed (Nevo 1978).

MATERIALS AND METHODS

MORPHOLOGICAL ANALYSIS

Statements on comparative morphology are based on examination of the synallaxine material in the Louisiana State University Museum of Zoology (LSUMZ). This collection contains examples of all genera mentioned herein, and 78 of the 97 species in these genera.

ELECTROPHORESIS

Specimens included in the electrophoretic analysis (Table 2) were collected by LSUMZ personnel between June 1980 and August 1982 at various localities in Peru, Bolivia, and Panama (Appendix I). Voucher specimens are deposited at LSUMZ. Tissue samples were taken within four hours of collection. Samples were wrapped in aluminum foil and frozen in liquid nitrogen for one to twelve weeks before transport to the laboratory, where they were stored at $-60°C$ until used.

Samples of breast muscle were homogenized with an equal volume of a 2 mM $MgCl_2$, 0.2 mM dithiothreitol, 0.25 M sucrose solution and clarified by centrifugation. Vertical electrophoresis was carried out on 10% starch gels at 6 V/cm for 16–21 hrs. Buffer systems used were: Tris-citrate pH 7.5 (electrode baths—0.1 M tris (hydroxymethyl) aminomethane (Tris), 0.036 M citric acid; gel—4 fold dilution of the electrode stock); Phosphate-citrate pH 6.0 (electrode baths—0.1 M citric acid, 0.032 M Na_2HPO_4; gel—5 fold dilution of the electrode stock); Tris-EDTA-Borate pH 8.1 (electrode baths—0.175 M boric acid, 0.05 M NaOH, pH 8.6; gel—0.033 M Tris, 0.17 mM ethylenediaminetetraacetic acid (EDTA), adjusted to pH 8.1 with a saturated solution of boric acid. The following 30 protein loci were scored in the indicated buffer system: *Tris-citrate pH 7.5*—isocitrate dehydrogenase (ICD-1,-2), glutamate dehydrogenase (GDH), glycerol-3-phosphate dehydrogenase (GPD), adenylate kinase (AK), creatine kinase (CK), valyl-leucine peptidase (VL), leucylglycyl-glycine peptidase (LGG-1,-2), phenylalanyl-proline peptidase (PP-1,-2), umbelliferyl acetate esterase (EST-3,-4), phosphoglucomutase (PGM); *Phosphate-citrate pH 6.0*—malate dehydrogenase (MDH-1,-2), lactate dehydrogenase (heart type—LDH-1; muscle type—LDH-2), acid phosphatase (ACP), glutamate-pyruvate transaminase (GPT), glutamate-oxaloacetate transaminase (GOT-1,-2), phosphoglucoisomerase (PGI), myoglobin (Mb), hemoglobin (Hb); *Tris-EDTA-Borate pH 8.1*—superoxide dismutase (SOD-1,-2), adenosine deaminase (ADA), mannose-6-phosphate isomerase (MPI), 6-phosphogluconate dehydrogenase (6-PGD). When more than one locus with a particular enzymatic activity was present, loci were numbered beginning with the most anodally migrating one. Similarly, multiple alleles at a locus were designated alphabetically, beginning with the most anodal one. Staining procedures used to identify specific enzymes were similar to those given by Harris and Hopkinson (1976). GDH was localized using the method of Brewer (1970). Globins were stained by the benzidine test for their peroxidase activity.

We now give additional information on certain loci to facilitate identification of homologous proteins in other studies. *Esterases*—Four banding regions of umbelliferyl acetate esterase activity are commonly produced by bird muscle extracts. In this study, only the third and fourth most anodal regions could be interpreted as discrete loci. On the basis of heterozygote banding pattern, EST-3 was dimeric and EST-4 was monomeric. *GPT*—Bird muscle extracts usually produce two regions of GPT activity: a weak, anodal one and a strong, less anodal one, which is the one reported herein. In these spinetails and many other Neotropical suboscine

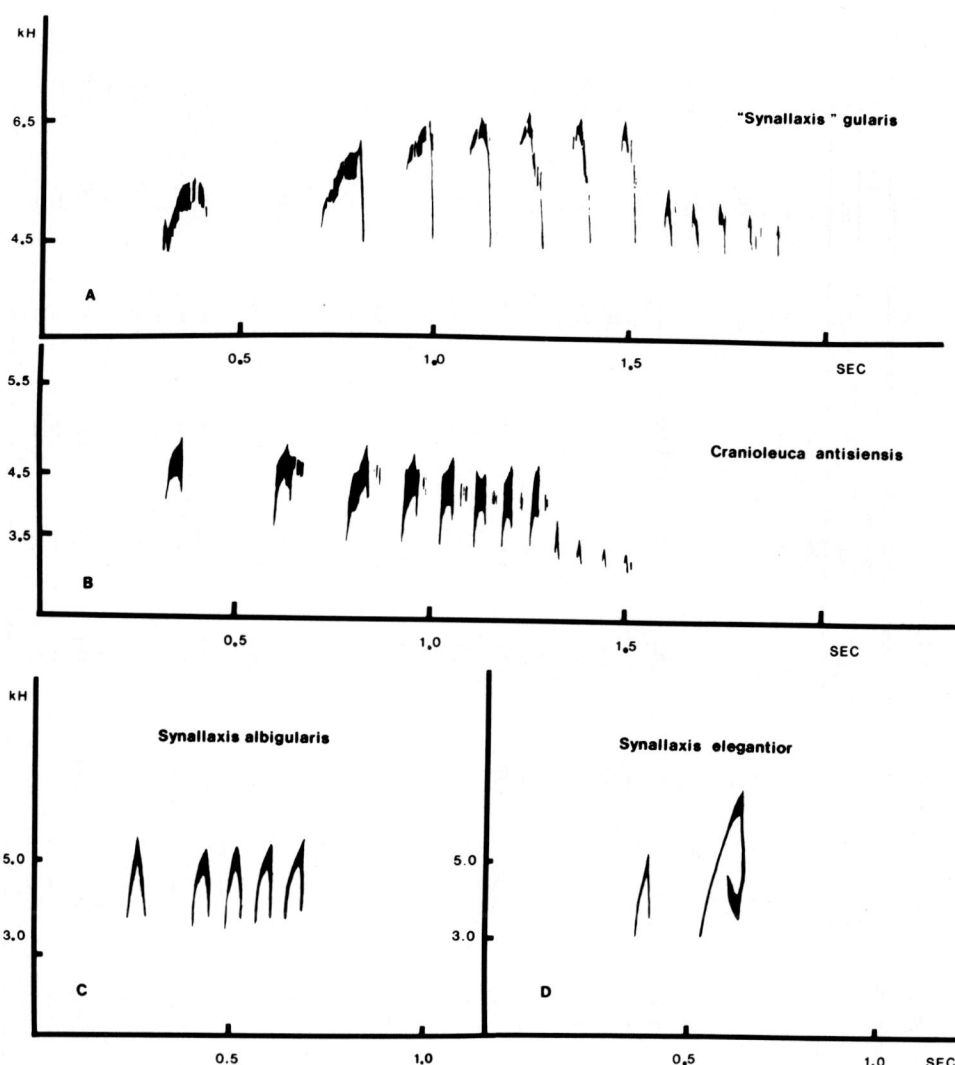

Fig. 1.   Sonograms of songs of four furnariid species.

species, every individual examined shows at least three bands in this region (Braun, pers. observ.). This pattern suggests a gene duplication has occurred, but, since conclusive evidence is lacking, the region was scored as a single locus. *Globins*—Of the four regions of peroxidase activity observed in spinetail muscle extracts, we tentatively identified the second and fourth most cathodal regions as Hb and Mb respectively, based on tissue distribution studies in other birds. Hb was scored as a single locus since it was impossible to determine which of the component polypeptides was responsible for a particular electrophoretic variant without further biochemical analysis.

ANALYSIS OF ELECTROPHORETIC DATA

Electrophoretic data were entered as single individual genotype scores into the BIOSYS-1 computer program of Swofford and Selander (1981) which computed allelic frequencies, heterozygosity estimates, and genetic distance measures. The program was also used to construct a phenogram by the unweighted pair group method (UPGMA; Sneath and Sokal 1970) and phylogenetic trees by the distance Wagner procedure (Farris 1972). Rogers' D (1972), as modified by Wright (1978), was used as the distance metric in both procedures. The Wagner

TABLE 1
ALLELIC FREQUENCIES AT VARIABLE LOCI

Locus	Allele	Synallaxis				"Synallaxis"			Cranioleuca			Schizoeaca	
		aza[1]	uni	sti	rut	gularis	ery	gut	ant	alb	vul	hel	har
GPT[2]	a	—	—	—	—	—	—	1.00	1.00	1.00	1.00	—	1.00
	b	1.00	1.00	1.00	—	—	—	—	—	—	—	—	—
	c	—	—	—	1.00	—	—	—	—	—	—	1.00	—
	d	—	—	—	—	1.00	—	—	—	—	—	—	—
	e	—	—	—	—	—	1.00	—	—	—	—	—	—
LDH-1	a	—	—	—	—	—	—	—	—	—	—	1.00	—
	b	1.00	1.00	1.00	1.00	1.00	1.00	1.00	1.00	1.00	1.00	—	1.00
GOT-2	a	1.00	1.00	1.00	1.00	1.00	1.00	1.00	1.00	1.00	1.00	1.00	1.00
	b	—	—	—	—	—	—	—	—	—	—	—	—
CK	a	1.00	1.00	1.00	—	—	—	—	—	—	—	—	1.00
	b	—	—	—	1.00	1.00	1.00	1.00	1.00	1.00	1.00	1.00	—
SOD-1	a	1.00	1.00	1.00	1.00	1.00	1.00	1.00	1.00	1.00	1.00	1.00	1.00
	b	—	—	—	—	—	—	—	—	—	—	—	—
EST-3	a	0.08	—	—	—	—	0.83	—	—	1.00	—	—	0.75
	b	0.92	1.00	1.00	1.00	0.75	0.17	0.13	0.50	—	1.00	1.00	0.25
	c	—	—	—	—	0.25	—	0.87	0.50	—	—	—	—
EST-4	a	—	—	—	0.10	—	—	0.13	0.25	1.00	1.00	1.00	1.00
	b	1.00	1.00	1.00	0.90	—	1.00	0.87	0.75	—	—	—	—
ADA	a	1.00	—	—	—	—	—	—	—	—	—	—	—
	b	—	0.17	—	—	—	—	—	—	—	—	—	—
	c	—	—	—	1.00	0.25	0.67	0.38	0.25	1.00	—	—	1.00
	d	—	0.83	1.00	—	0.75	0.33	0.62	0.75	—	1.00	1.00	—
	e	—	—	—	—	—	—	—	—	—	—	—	—
ICD-1	a	—	—	—	—	—	0.17	1.00	0.25	1.00	—	—	0.50
	b	1.00	1.00	—	—	—	—	—	—	—	—	—	—
	c	—	—	—	—	—	—	—	—	—	—	—	—
	d	—	—	1.00	—	—	0.83	—	0.75	—	—	—	0.50
	e	—	—	—	1.00	—	—	—	—	—	—	—	—
ICD-2	a	1.00	—	—	—	—	—	—	1.00	—	—	—	—
	b	—	—	—	1.00	0.25	1.00	1.00	—	1.00	1.00	1.00	1.00
	c	—	—	1.00	—	0.75	—	—	—	—	—	—	—
	d	—	1.00	—	—	—	—	—	—	—	—	—	—

TABLE 1
Continued

Locus	Allele	Synallaxis				"Synallaxis"			Cranioleuca			Schizoeaca	
		aza[1]	uni	sti	rut	gularis	ery	gut	ant	alb	vul	hel	har
PGI	a	0.92	1.00	—	1.00	1.00	1.00	1.00	1.00	1.00	1.00	1.00	1.00
	b	—	—	1.00	—	—	—	—	—	—	—	—	—
	c	0.08	—	—	—	—	—	—	—	—	—	—	—
PGM	a	1.00	—	—	0.90	0.25	—	—	1.00	1.00	1.00	1.00	1.00
	b	—	1.00	1.00	0.10	0.75	1.00	1.00	—	—	—	—	—
VL	a	—	—	—	—	—	—	0.13	—	—	—	—	—
	b	0.83	1.00	1.00	—	0.75	1.00	0.87	1.00	1.00	1.00	1.00	1.00
	c	0.08	—	—	1.00	0.25	—	—	—	—	—	—	—
	d	0.08	—	—	—	—	—	—	—	—	—	—	—
LGG-1	a	1.00	1.00	1.00	1.00	1.00	1.00	1.00	1.00	1.00	1.00	0.50	1.00
	b	—	—	—	—	—	—	—	—	—	—	0.50	—
LGG-2	a	1.00	1.00	1.00	—	0.25	—	—	—	—	—	—	—
	b	—	—	—	0.90	—	—	—	—	—	—	—	—
	c	—	—	—	—	0.75	—	—	—	—	—	—	—
	d	—	—	—	0.10	—	—	—	—	—	—	—	—
	e	—	—	—	—	—	0.83	1.00	1.00	1.00	1.00	1.00	0.50
	f	—	—	—	—	—	0.17	—	—	—	—	—	0.50
PP-1	a	—	—	—	—	0.25	—	—	—	—	—	—	—
	b	—	—	—	0.10	—	—	—	0.25	—	—	—	—
	c	1.00	1.00	1.00	0.80	0.75	1.00	1.00	0.75	1.00	1.00	1.00	1.00
	d	—	—	—	0.10	—	—	—	—	—	—	—	—
PP-2	a	—	—	—	—	—	—	—	—	—	—	—	—
	b	0.42	1.00	1.00	—	1.00	0.33	0.50	0.50	—	1.00	1.00	0.25
	c	—	—	—	—	—	0.67	—	—	—	—	—	0.75
	d	0.58	—	—	1.00	—	—	0.25	0.50	1.00	—	—	—
	e	—	—	—	—	—	—	0.25	—	—	—	—	—
Mb	a	1.00	1.00	1.00	1.00	1.00	1.00	1.00	1.00	1.00	1.00	1.00	1.00
	b	—	—	—	—	—	—	—	—	—	—	—	—
Hb	a	—	—	—	—	1.00	1.00	1.00	1.00	1.00	1.00	1.00	1.00
	b	1.00	1.00	1.00	1.00	—	—	—	—	—	—	—	—

TABLE 1
CONTINUED

Locus[2]	Allele	Synallaxis				"Synallaxis"		Cranioleuca				Schizoeaca	
		aza	uni	sti	rut	gularis	ery	gut	ant	alb	vul	hel	har
6-PGD	a	0.17	—	—	—	1.00	1.00	0.75	—	1.00	1.00	1.00	1.00
	b	0.83	—	0.50	0.60	—	—	0.25	1.00	—	—	—	—
	c	—	—	0.50	—	—	—	—	—	—	—	—	—
	d	—	1.00	—	0.40	—	—	—	—	—	—	—	—
GPD	a	—	—	—	0.20	—	—	—	—	—	—	—	—
	b	1.00	—	1.00	0.80	—	1.00	1.00	1.00	1.00	1.00	1.00	1.00
	c	—	1.00	—	—	—	—	—	—	—	—	—	—
	d	—	—	—	—	1.00	—	—	—	—	—	—	—
MPI	a	—	—	—	—	0.25	—	—	—	—	—	—	—
	b	—	—	—	0.40	0.75	—	—	—	—	—	—	—
	c	0.83	1.00	1.00	0.60	—	—	—	—	—	—	1.00	1.00
	d	—	—	—	—	—	1.00	1.00	1.00	1.00	1.00	—	—
	e	0.17	—	—	—	—	—	—	—	—	—	—	—

[1] Species abbreviations are: Synallaxis azarae (aza), S. unirufa (uni), S. stictothorax (sti), S. rutilans (rut), Cranioleuca erythrops (ery), C. gutturata (gut), C. antisiensis (ant), C. albiceps (alb), C. vulpina (vul), Schizoeaca helleri (hel), S. harterti (har). All species were monomorphic at ACP, MDH-1, MDH-2, LDH-2, GOT-1, AK, SOD-2, GDH. All specimens were scored at every locus.

[2] Locus abbreviations may be found under Materials and Methods.

trees were constructed using the multiple addition criterion algorithm of Swofford (1981). Five partial networks were saved for each successive step in the distance Wagner procedure. The percent standard deviation (% s.d.; Fitch and Margoliash 1967) was used in selecting partial networks, and the trees were rooted at the midpoint of the greatest patristic distance.

ANALYSIS OF VOCALIZATIONS

Field recordings of spinetail vocalizations were made on a Nagra IV-B tape recorder at 7.5 inches per second using an Elektra Voice shotgun microphone. These recordings were played back on the same machine into a Kay Sonograph (Model 6061A) set to analyze signals with the broad-band, 300 Hz filter and linear display. Copies of the original recordings are on file at the Library of Natural Sounds of the Cornell Laboratory of Ornithology.

RESULTS AND DISCUSSION

COMPARATIVE MORPHOLOGY

Vaurie (1980) published the most recent revision of the Furnariidae. He recognized four species groups within his genus *Synallaxis*. Three of these groups (29 species) form a homogeneous assemblage with respect to overall body morphology, having relatively short, rounded wings, long tails, black throat patches, and contrasting rufous wings or wing coverts. The fourth group, consisting of seven species including *gularis*, "cannot be characterized satisfactorily as it is more or less heterogeneous, although about half its species are probably related" (Vaurie 1980:94).

Of Vaurie's Group Four species, only *erythrothorax* and *cinnamomea* appear to be typical *Synallaxis* species in morphology and vocalizations (Vaurie and Schwartz 1972). The three *Poecilurus* species, first merged with *Synallaxis* by Vaurie, are clearly close to each other, but their relationship to *Synallaxis* remains uncertain. Except for its small size, *Synallaxis stictothorax* is strikingly similar in color and pattern to *Siptornopsis hypochondriacus*, a fact first pointed out by Hellmayr (1925:132). Both species have blunt-tipped rectrices (Peters 1951), both inhabit desert scrub in zoogeographically complementary areas in Peru and Ecuador (*stictothorax* along the coast and in the valleys of Cajamarca, Peru; *hypochondriacus* in the upper Marañon valley), and both build huge, roofed stick nests like those of *Phacellodomus* (Parker, pers. observ.).

Finally, we are left with *gularis*. Vaurie (1980:121) stated that *gularis* "is the most aberrant species in Group Four and in the entire genus because it lacks a rufous area on the upper surface of the wing and its black gular patch has virtually vanished." Additionally, *gularis* differs from all other *Synallaxis* in having well defined white superciliary lines and a white throat devoid of dark coloration, save the inconspicuous black tips found on some feathers of the lower throat. This pattern is shown by several *Cranioleuca* species, especially *antisiensis* and *pyrrhopia*, and to some extent by several *Leptasthenura* species, especially *xenothorax* and *striata*. *Gularis* also shares its uniform buffy orange body color with some *Leptasthenura*, especially *yanacensis*, and with *Silviorthorhynchus*, whereas nearly all *Cranioleuca* and *Synallaxis* have chestnut-brown wings that contrast boldly with the upper and underparts.

Gularis is also aberrant among *Synallaxis* spinetails in having a short tail for its size. Vaurie (1980:132) considered the ratio of tail length to wing length an important defining character of his genus *Certhiaxis*, which included *Cranioleuca*. He gave the mean and range of this ratio for his genera *Certhiaxis* (mean = 1.00; range = 0.88–1.08) and *Synallaxis* (mean = 1.27; range = 0.98–1.78). Apparently Vaurie did not include *gularis* in this analysis, because from his measurements (1980:92) it has a tail/wing ratio of 0.93 and, thus, would appear allied to *Cranioleuca* on the basis of this character.

Finally, *gularis* has a slender, rather straight bill, which separates it from all *Cranioleuca* and *Synallaxis*. Sztolcman (1926) used this character in recognizing *Hellmayrea*. Again, *gularis* resembles *Silviorthorhynchus* and some members of *Leptasthenura* in bill shape. Although none of the above morphological features is decisive taxonomically, in concert they indicate that *gularis* is not well accommodated in the genus *Synallaxis*.

VOCALIZATIONS

Vaurie and Schwartz (1972), in comparing the vocalizations of *Synallaxis unirufa* and *S. castanea*, provided acoustical information for 11 additional *Synallaxis* species. These authors stated (Vaurie and Schwartz 1972:6) that the primary song of "*Synallaxis* is, generally speaking, a constant repetition of a stereotyped phrase that consists of two or three figures in the

TABLE 2

Measures of Intraspecific Protein Variability

Species	N	% of loci polymorphic[1]	Mean heterozygosity ± s.e.	
			Direct count	Hardy-Weinberg expected[2]
Synallaxis				
azarae	6	20	0.05 ± 0.02	0.06 ± 0.02
unirufa	3	3	0.01 ± 0.01	0.01 ± 0.01
stictothorax	1	3	0.03	0.03
rutilans	5	23	0.08 ± 0.04	0.08 ± 0.03
"*Synallaxis*"				
gularis	2	27	0.13 ± 0.04	0.13 ± 0.04
Cranioleuca				
erythrops	3	17	0.08 ± 0.04	0.07 ± 0.03
gutturata	4	20	0.10 ± 0.04	0.08 ± 0.03
antisiensis	2	20	0.13 ± 0.05	0.11 ± 0.04
albiceps	1	0	0.00	0.00
vulpina	1	0	0.00	0.00
Schizoeaca				
helleri	1	3	0.03	0.03
harterti	2	13	0.07 ± 0.04	0.08 ± 0.04
Mean	2.6	13	0.06 ± 0.05	0.06 ± 0.04

[1] A locus was considered polymorphic if more than one allele was recorded at that locus in the species in question.
[2] Unbiased estimate of Nei (1978).

majority of species studied." Our own field observations and recordings when combined with a survey of the *Synallaxis* song material in the Cornell Library of Natural Sounds revealed that the song type described by Vaurie and Schwartz characterizes at least 14 members of the genus (e.g., *S. elegantior,* Fig. 1D). Three additional species produce a four to six note phrase (e.g., *S. albigularis,* Fig. 1C).

The song of *gularis* differs from all *Synallaxis* songs known to us in being structurally more complex. A typical territorial song phrase of *gularis* is comprised of an introductory note followed by a pause of 0.3 seconds, a series of six higher pitched notes, and a terminal component of five lower pitched notes (Fig. 1A). In pattern, this song closely resembles the song of *Cranioleuca antisiensis* (Fig. 1B), which also consists of 12 notes, including a terminal component which is similar to that of *gularis,* but which is unknown in any other *Synallaxis.* Of the 11 *Cranioleuca* species whose vocalizations are known, all give similarly complex territorial songs (Parker, pers. observ.). Thus, the song of *gularis* is more similar to those of species of *Cranioleuca* than to those of species of *Synallaxis,* both genera being homogeneous with respect to song.

INTRASPECIFIC PROTEIN VARIABILITY

Of the 30 loci examined, eight were monomorphic in all species studied (Table 1). Estimates of percentage of polymorphic loci and average heterozygosity (H) must be considered very imprecise due to the small sample sizes (Table 2). However, the values presented do indicate that the protein variability of these spinetails is not grossly different from that reported for other passerine bird species (Barrowclough and Corbin 1978; Yang and Patton 1981; Zink 1982) or vertebrates in general (Nevo 1978). That values of H are not inordinately large is important to help establish the validity of the genetic distance estimates and dendrograms presented below (Nei 1978).

INTERSPECIFIC PROTEIN DIFFERENTIATION

The average genetic distance separating *Cranioleuca* species (Tables 3, 4) is about the same as previously reported for many congeneric passerine species (Smith and Zimmerman 1976; Yang and Patton 1981; Zink 1982). The average genetic distance separating *Synallaxis* species (Tables 3, 4) is somewhat greater, indicating that more protein differentiation has occurred in this genus. However, these distance values still fall within the range reported for some avian genera (Avise et al. 1980a, b).

TABLE 3

Matrix of Nei's[1] and Rogers'[2] Genetic Distance Coefficients

	1	2	3	4	5	6	7	8	9	10	11	12
1. *Synallaxis azarae*	—	0.19	0.20	0.12	0.51	0.42	0.38	0.37	0.55	0.41	0.43	0.37
2. *S. unirufa*	0.41	—	0.16	0.20	0.48	0.44	0.46	0.42	0.56	0.40	0.52	0.49
3. *S. stictothorax*	0.43	0.39	—	0.28	0.57	0.45	0.47	0.42	0.61	0.44	0.51	0.48
4. *S. rutilans*	0.34	0.42	0.49	—	0.50	0.49	0.49	0.44	0.59	0.48	0.52	0.47
5. *"Synallaxis" gularis*	0.62	0.61	0.65	0.61	—	0.32	0.28	0.33	0.25	0.27	0.41	0.39
6. *Cranioleuca erythrops*	0.57	0.59	0.60	0.61	0.52	—	0.09	0.05	0.14	0.05	0.28	0.24
7. *C. gutturata*	0.55	0.60	0.61	0.60	0.49	0.30	—	0.07	0.08	0.08	0.36	0.28
8. *C. antisiensis*	0.54	0.58	0.59	0.58	0.53	0.25	0.29	—	0.10	0.05	0.33	0.27
9. *C. albiceps*	0.64	0.66	0.68	0.66	0.47	0.36	0.29	0.33	—	0.11	0.41	0.38
10. *C. vulpina*	0.58	0.57	0.60	0.61	0.49	0.23	0.29	0.24	0.32	—	0.27	0.25
11. *Schizoeaca helleri*	0.59	0.64	0.63	0.63	0.58	0.50	0.55	0.53	0.59	0.49	—	0.13
12. *S. harterti*	0.55	0.62	0.62	0.60	0.56	0.46	0.49	0.49	0.56	0.47	0.37	—

[1] Nei (1978); values above diagonal.
[2] Rogers (1972) as modified by Wright (1978); values below diagonal.

TABLE 4

MEAN GENETIC DISTANCE BETWEEN VARIOUS SYNALLAXINE TAXA

Taxa compared	Number of comparisons	Nei's distance[1] ± s.d.	Rogers' distance[2] ± s.d.
Within genera			
Synallaxis-Synallaxis	6	0.19 ± 0.05	0.41 ± 0.05
Cranioleuca-Cranioleuca	10	0.08 ± 0.03	0.29 ± 0.04
Schizoeaca-Schizoeaca	1	0.13	0.36
Between genera			
Synallaxis-Cranioleuca	20	0.47 ± 0.07	0.60 ± 0.03
Synallaxis-Schizoeaca	8	0.47 ± 0.05	0.61 ± 0.03
Cranioleuca-Schizoeaca	10	0.31 ± 0.06	0.51 ± 0.04
"*S.*" *gularis*—Other genera			
"*S.*" *gularis-Synallaxis*	4	0.52 ± 0.04	0.62 ± 0.02
"*S.*" *gularis-Cranioleuca*	5	0.29 ± 0.03	0.50 ± 0.02
"*S.*" *gularis-Schizoeaca*	2	0.40 ± 0.02	0.57 ± 0.01

[1] Nei (1978).
[2] Rogers (1972) as modified by Wright (1978).

For *gularis,* the average genetic distance to the *Cranioleuca* species studied was considerably less than that to the *Synallaxis* species studied (Table 4). Direct examination of the raw allelic frequency data (Table 1) shows that this difference is substantive. "*Synallaxis*" *gularis* shares alleles with *Cranioleuca* at six loci that contribute to the differentiation between *Cranioleuca* and *Synallaxis* (SOD-1, Hb, ICD-1, ICD-2, 6-PGD, CK) whereas it shares alleles with *Synallaxis* at only one such locus (ADA). Moreover, ADA is a rather variable protein in these and other birds (Braun, pers. observ.), which suggests it is prone to back mutation and, therefore, less reliable as a taxonomic character than more conservative proteins such as SOD-1 and Hb. Thus, at the protein loci studied, *gularis* is more similar to the *Cranioleuca* than to the *Synallaxis* species examined.

We initially included *Schizoeaca* in the analysis as an outgroup to try to establish primitive character states for the loci that differentiate *Cranioleuca* and *Synallaxis.* However, *Schizoeaca* groups strongly with *Cranioleuca* at most of the critical loci (SOD-1, Hb, GPD, 6-PGD, ICD-1, ICD-2; Table 1). This resemblance is reflected in the average genetic distances, which are less to *Cranioleuca* species than to *Synallaxis* species (Tables 3, 4). Because we do not know of any characters that clearly define *Synallaxis* and *Cranioleuca* as a clade with respect to *Schizoeaca* (see cladogram in Fitzpatrick 1982), we now consider its status as an outgroup questionable and do not attempt to establish primitive character states from the protein evidence currently available.

Although both "*S.*" *gularis* and *Schizoeaca* share more protein alleles with *Cranioleuca* than *Synallaxis,* they are, nevertheless, easily distinguished from the *Cranioleuca* species studied. This is due to the presence of autapomorphic alleles at several loci in each taxon (GPT, LGG-2, GPD and Mb for *gularis;* GPT, LGG-2, and LDH-1 for *Schizoeaca*). The distinctiveness of *gularis* and *Schizoeaca* from *Cranioleuca* and from each other is readily apparent from the genetic distances (Tables 3, 4).

Rogers' distances (Table 3) were used to construct the branching diagrams in Figure 2. On the whole, the topologies of the dendrograms are very similar. The *Synallaxis* species (excluding *gularis*) form one major grouping, whereas *Cranioleuca, Schizoeaca,* and *gularis* form another. The only difference in the branching sequence comes at the split between the latter three groups. In the UPGMA phenogram (Fig. 2A), *gularis* clusters with *Cranioleuca,* whereas *Schizoeaca* is somewhat more divergent. In the Wagner tree (Fig. 2B), *gularis* splits off before *Cranioleuca* and *Schizoeaca* diverge. After branch length optimization (Swofford 1981) of the Wagner tree, however, *Cranioleuca* and *Schizoeaca* are joined by a negative branch length (tree not shown). Clearly, ambiguity exists in this region, and the trichotomy must be considered unresolved by the electrophoretic data.

A word of caution is in order here. Small sample sizes surely affect the accuracy of our genetic distance estimates. However, when average heterozygosities are low and the number of loci examined is large, the error inherent in estimating genetic distance from a small number

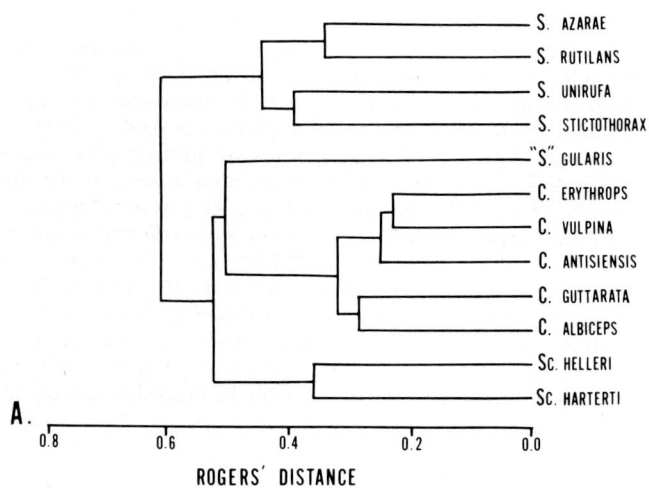

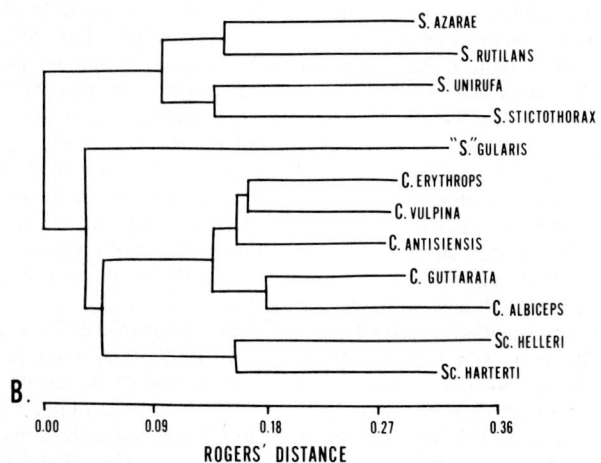

FIG. 2.   Branching diagrams derived from modified Rogers' distances (Table 3). A. Phenogram based on UPGMA clustering (% s.d. = 6.0; cophenetic correlation = 0.97). B. Distance Wagner tree rooted at midpoint of greatest patristic distance (% s.d. = 6.36; cophenetic correlation = 0.98). S = *Synallaxis*, C = *Cranioleuca*, and Sc. = *Schizoeaca*.

of individuals decreases as genetic distance increases (Nei 1978). Likewise, when a dendrogram is constructed from genetic distance estimates, the reliability of the branching sequence depends on the differences in genetic distance between different groups of species. When these differences are large, a single individual may be sufficient for obtaining a reliable branching sequence (Gorman and Renzi 1979). In the present case, average heterozygosities are moderate, the number of loci is reasonably large, and genetic distances are sizeable. Also, differences in genetic distance between different hierarchical levels are reasonably large. Thus, all conditions are good for obtaining the correct branching sequence of the dendrograms from the number of individuals studied.

TAXONOMIC AND BIOGEOGRAPHIC INTERPRETATION

SPECIES LIMITS

The electrophoretic data presented here have some bearing on an important species level problem. *Schizoeaca helleri* and *S. harterti* are part of a highly polytypic series of allopatric populations that had always been regarded as separate species until lumped into *S. fuliginosa* by Vaurie (1971). Remsen (1981) argued convincingly for the retention of full species status for all forms. The taxonomy of these birds is of interest because they show a pattern of distribution, common among Andean bird genera, in which several well differentiated allopatric forms are separated by major geographic features of the Andes. The various isolates of many genera are strikingly different, yet their taxonomic status has been open to question. Molecular studies offer the opportunity for objective comparison of the degree of genetic differentiation of these populations to that of populations known to represent separate species.

The electrophoretic data suggest that divergence between *helleri* and *harterti* has occurred at three loci (GPT, GOT-2, PP-1; Table 1). This leads to genetic distance estimates (Tables 3, 4) as large as those separating any of the *Cranioleuca* species in this study, and much larger than those separating many sympatrically breeding bird species (for example, *Catharus* thrushes or *Dendroica* warblers; Barrowclough and Corbin 1978; Avise et al. 1980a, c). If these genetic distance estimates prove accurate, it will indicate that extensive molecular differentiation exists between these forms, and it will probably be desirable to treat them as separate species.

GENERIC LIMITS

All available evidence indicates that the *Synallaxis* (excluding *gularis*) and *Cranioleuca* species considered here form natural groups. "*Synallaxis*" *gularis* appears more closely related to *Cranioleuca* than to *Synallaxis* on the basis of protein and song similarities. Whether or not *gularis* is included in *Cranioleuca* depends in part on the treatment of *Schizoeaca*, which shows a level of molecular similarity to *Cranioleuca* close to that of *gularis*. Several taxonomic treatments are tenable in light of the current study. First, *gularis* could be placed in *Cranioleuca*, and *Schizoeaca* and *Synallaxis* retained as separate genera. Second, *Schizoeaca* and *gularis* could be merged with *Cranioleuca*, and *Synallaxis* retained as a separate genus. Third, the monotypic genus *Hellmayrea* could be resurrected for *gularis*, retaining *Synallaxis*, *Cranioleuca*, and *Schizoeaca* as separate genera. Which treatment is adopted is, at present, largely a matter of personal preference. Until electrophoretic or other evidence becomes available that will allow an objective decision on generic limits in this group, we favor the final option. Separating *gularis* from other spinetails in the genus *Hellmayrea* requires no judgment about the *Cranioleuca-Schizoeaca-gularis* trichotomy and reflects the relatively large protein divergence of *gularis* and *Schizoeaca* from *Cranioleuca*.

The previously unsuspected similarities between *Cranioleuca* and *Schizoeaca* lead one to consider the evolutionary relationships of these and other synallaxine genera. For example, the relationships of *Schizoeaca* to *Asthenes* and *Leptasthenura* should be investigated. These genera are similar morphologically and behaviorally and occur mainly in the puna zone of the high Andes. Chapman (1926) suggested that such highland genera originated in the open country of southern South America. *Cranioleuca* species, on the other hand, occur mainly in humid montane forests of northern South America, from which one might surmise that they originated in adjacent Amazonia. The apparent relationship between *Cranioleuca* and *Schizoeaca*, however, makes these traditional zoogeographical interpretations questionable. Moreover, it is at odds with the phylogeny envisioned by Vaurie (1980, as interpreted by Fitzpatrick 1982). Final resolution of the relationships of these genera will be of considerable interest in the study of South American zoogeography.

ACKNOWLEDGMENTS

We dedicate this paper to the memory of Eugene Eisenmann, whose friendly smile and kind words were a great encouragement to us as neophyte ornithologists. For taking the time from their own research to collect tissue specimens we thank J. V. Remsen, Jr., J. P. O'Neill, T. S. Schulenberg, M. B. Robbins, S. M. Lanyon, A. P. Capparella, S. A. Stotz, D. Wiedenfield, G. H. Rosenberg, and T. J. Davis. The help of Drs. E. Mendez and P. Galindo in Panama and G. Bejarano B. in Bolivia made our fieldwork incalculably easier. Scientific permits were granted by the Direccion General Forestal y de la Fauna of Peru and by DICYT and the Academia Nacional de Ciencias of Bolivia. E. Tobey graciously made her expertise and the sound analysis equipment of the Kresge Speech and Hearing Laboratory available to us. Critical readings by H. C. Dessauer, J. V. Remsen, Jr., T. S. Schulenberg, R. M. Zink, M. S. Foster, and one anonymous reviewer strengthened the final manuscript. We especially wish to acknowledge the generous support of B. Odom, J. S. McIlhenny, H. I. Schweppe, and L. Schweppe, which made all our fieldwork possible.

LITERATURE CITED

AVISE, J. C., J. C. PATTON, AND C. F. AQUADRO. 1980a. Evolutionary genetics of birds. Comparative molecular evolution in New World warblers and rodents. J. Hered. 71:303–310.

AVISE, J. C., J. C. PATTON, AND C. F. AQUADRO. 1980b. Evolutionary genetics of birds. II. Conservative protein evolution in North American sparrows and relatives. Syst. Zool. 29:323–334.

AVISE, J. C., J. C. PATTON, AND C. F. AQUADRO. 1980c. Evolutionary genetics of birds. I. Relationships among North American thrushes and allies. Auk 97:135–147.

BARROWCLOUGH, G. F., AND K. W. CORBIN. 1978. Genetic variation and differentiation in the Parulidae. Auk 95:691–702.

BREWER, G. J. 1970. An Introduction to Isozyme Techniques. Academic Press, New York.

CHAPMAN, F. M. 1926. The distribution of birdlife in Ecuador. A contribution to a study of the origin of Andean birdlife. Bull. Am. Mus. Nat. Hist. No. 55.

FARRIS, J. S. 1972. Estimating phylogenetic trees from distance matrices. Am. Nat. 106:645–668.

FEDUCCIA, A. 1973. Evolutionary trends in the neotropical ovenbirds and woodhewers. Ornithol. Monogr. No. 13.

FITCH, W. M., AND E. MARGOLIASH. 1967. Construction of phylogenetic trees. Science 155:279–284.

FITZPATRICK, J. W. 1982. Review of "Taxonomy and geographical distribution of the Furnariidae" by C. Vaurie. Auk 99:810–813.

GORMAN, G. C., AND J. R. RENZI. 1979. Genetic distance and heterozygosity estimates in electrophoretic studies: effects of sample size. Copeia 1979:242–249.

HARRIS, H., AND D. A. HOPKINSON. 1976. Handbook of Enzyme Electrophoresis in Human Genetics. North Holland/Elsevier, Amsterdam.

HELLMAYR, C. E. 1925. Catalogue of the birds of the Americas, Pt. 4. Field Mus. Nat. Hist. Publ. Zool. Ser., Vol. 13.

MEYER DE SCHAUENSEE, R. 1966. The Species of Birds of South America. Livingston Publ. Co., Narberth, Pennsylvania.

NEI, M. 1978. Estimation of average heterozygosity and genetic distance from a small number of individuals. Genetics 89:583–590.

NEVO, E. 1978. Genetic variation in natural populations: patterns and theory. Theor. Pop. Biol. 13: 121–177.

PARKER, T. A., T. S. SCHULENBERG, G. GRAVES, AND M. J. BRAUN. 1985. The avifauna of the Huancabamba region, northern Peru. Pp. 169–197, In P. A. Buckley, M. S. Foster, E. S. Morton, R. S. Ridgely, and F. G. Buckley (eds.), Neotropical Ornithology. Ornithol. Monogr. No. 36.

PETERS, J. L. 1951. Check-list of the Birds of the World, Vol. 7. Museum Comparative Zoology, Cambridge, Massachusetts.

REMSEN, J. V. JR. A new subspecies of *Schizoeaca harterti* with notes on taxonomy and natural history of *Schizoeaca* (Aves: Furnariidae). Proc. Biol. Soc. Wash. 94:1068–1075.

ROGERS, J. S. 1972. Measures of genetic similarity and genetic distance. Stud. Genet. 7213:145–153.

SMITH, J. K., AND E. G. ZIMMERMAN. 1976. Biochemical genetics and evolution of North American blackbirds, family Icteridae. Comp. Biochem. Physiol. B. Comp. Biochem. 53:319–324.

SNEATH, P. H. A., AND R. R. SOKAL. 1973. Numerical Taxonomy. W. H. Freeman, San Francisco, California.

SWOFFORD, D. L. 1981. On the utility of the distance Wagner procedure. Pp. 25–43, In V. A. Funk and D. R. Brooks (eds.), Advances in Cladistics: Proceedings of the First Meeting of the Willi Hennig Society. New York Botanical Garden, New York.

SWOFFORD, D. L., AND R. B. SELANDER. 1981. BIOSYS-1: a FORTRAN program for the comprehensive analysis of electrophoretic data in population genetics and systematics. J. Hered. 72:281–283.

SZTOLCMAN, J. 1926. Etude des collections ornithologiques de Parana. Ann. Zool. Mus. Polonici Hist. Nat. 5:107–196.

VAURIE, C. 1971. Classification of the Ovenbirds (Furnariidae). Witherby, London.

VAURIE, C. 1980. Taxonomy and geographical distribution of the Furnariidae (Aves, Passeriformes). Bull. Am. Mus. Nat. Hist. 166:1–357.

VAURIE, C., AND P. SCHWARTZ. 1972. Morphology and vocalizations of *Synallaxis unirufa* and *Synallaxis castanea* (Furnariidae, Aves) with comments on other *Synallaxis*. Am. Mus. Novit. No. 2483.

WRIGHT, S. 1978. Evolution and the Genetics of Populations, Vol. 4. Variability Within and Among Natural Populations. University Chicago Press, Chicago, Illinois.

YANG, S. Y., AND J. L. PATTON. 1981. Genic variability and differentiation in the Galapagos finches. Auk 98:230–242.

ZINK, R. 1982. Patterns of genic and morphologic variation among sparrows in the genera *Zonotrichia, Melospiza, Junco,* and *Passerella.* Auk 99:632–649.

APPENDIX I
GENERAL COLLECTING LOCALITIES FOR THE TISSUE SPECIMENS ANALYZED[1]

Synallaxis azarae—PERU: Depto. Cajamarca, 97640; Depto. Pasco, 105891; Depto. Puno, 98243, 98247. BOLIVIA: Depto. La Paz, 101957, 101312.

S. unirufa—PERU: Depto. Cajamarca, 97641, 97643; Depto. Pasco, 105899. *S. stictothorax*—PERU: Depto. Lambayeque, 97644.

S. rutilans—PERU: Depto. Loreto, 109750, 109751, 111253, 111254. BOLIVIA: Depto. La Paz, 101969.

"Synallaxis" gularis—PERU: Depto. Piura-Cajamarca, 97649, 97650.

Cranioleuca erythrops—PANAMA: Prov. Darien, 104679, 104680, 108288.

C. guttarata—PERU: Depto. Pasco, 105905. BOLIVIA: Depto. La Paz, 101318, 101984, 101986.

C. antisiensis—PERU: Depto. Piura, 97655, 97656.

C. albiceps—BOLIVIA: Depto. La Paz, 101973.

C. vulpina—PERU: Depto. Loreto, 111250.

Schizoeaca helleri—PERU: Depto. Puno, 98233.

S. harterti—BOLIVIA: Depto. La Paz, 101950, 101952.

[1] All numbers are LSUMZ catalog numbers. Detailed locality data are available from LSUMZ.

BIOCHEMICAL AND MORPHOMETRIC ANALYSES OF THE SYMPATRIC, NEOTROPICAL, SIBLING SPECIES, *MIONECTES MACCONNELLI* AND *M. OLEAGINEUS*

A. P. Capparella and Scott M. Lanyon

Museum of Zoology and Department of Zoology and Physiology, Louisiana State University, Baton Rouge, Louisiana 70803 USA

ABSTRACT. Protein electrophoresis and discriminant function analysis (DFA) were used to analyze differences between members of a sibling species pair, *Mionectes oleagineus* and *M. macconnelli* (Tyrannidae). Three populations were examined: *oleagineus* from Bolivia, where it is sympatric with *macconnelli*, *macconnelli* from Bolivia, where it is sympatric with *oleagineus*, and *oleagineus* from northeastern Peru, where *macconnelli* does not occur. In spite of the extreme similarity in external morphology of the two *Mionectes*, genetic distance measures indicate that these species have diverged as much as other avian congeners. Percentage of polymorphic loci and average heterozygosity values are typical for birds, and no fixed differences were detected. DFA of morphological characters showed that the two species could be correctly separated 91.2% of the time. Contrary to predictions of competition theory, sympatric *oleagineus* and *macconnelli* are morphologically more similar to each other than either is to allopatric *oleagineus*.

RESUMEN. Se usaron electroforesis de proteínas y análisis de función discriminatoria (DFA) para analizar las diferencias entre miembros de un par de especies gemelas, *Mionectes oleagineus* y *M. macconnelli* (Tyrannidae). Tres poblaciones fueron examinadas: *oleagineus* de Bolivia, donde esta especie se encuentra junto con ("simpatrica") *macconnelli*; *macconnelli* de Bolivia donde se encuentra junto con *oleagineus*; y *oleagineus* del noreste de Perú donde no se encuentra *macconnelli*. A pesar de las similaridades morfológicas externas de ambas especies de *Mionectes*, las medidas de distancias genéticas indican que estas especies han divergido tanto entre si como con respecto a otras especies de aves congenéricas. Porcentajes de "loci" polimórficos y valores promedios de heterocigoticidad, son los típicos para aves, y no se detectaron diferencias fijas. DFA de carácteres morfológicos muestran que las dos especies pueden ser separadas correctamente el 91,2% de las veces. Contrariamente con las predicciones en la teoría de competición, las especies "simpatricas" *oleaginus* y *macconnelli* morfologicamente son más similares entre si que cualquiera de ellas lo es con la alopátrica *oleagineus*.

The degree of concordance in evolutionary change of different suites of characters is an area of increasing interest to evolutionary biologists. Recent investigations of the variation in evolutionary rates have concentrated on comparisons between genetic and morphological characters among taxa that are well-differentiated morphologically (e.g., Zink 1982). The degree of concordance between these two suites of characters in morphologically very similar species, "sibling species" *sensu* Mayr (1963), has not been investigated.

One hypothesis proposed to explain the morphological similarity between sibling species pairs is that they are younger and, therefore, of more recent origin than morphologically dissimilar sets of sister species. This assumes that the rate of morphological change over time is roughly equivalent within the taxonomic unit encompassing the compared sets of species, e.g., a family of birds. With a constant rate of change, the degree of morphological differentiation among sister species within a taxon should be an index of their recency of origin and, consequently, sibling species would be of more recent origin. On the other hand, if the rate of morphological change varies over the course of evolution, then morphological similarity may tell us relatively little concerning the age of the taxa concerned.

Analysis of biochemical characters provides a means of determining the constancy of morphological change. For example, electrophoretic analysis of protein variation enables us to quantify the degree of genetic differentiation between species. In theory, genetic distance provides an approximate measure of the average number of electrophoretically detectable codon substitutions per locus between two species since their separation (Nei 1972). If this rate is roughly constant through evolutionary time (the molecular clock hypothesis; Wilson et al. 1977), then a comparison of genetic distances will enable us to test the prediction that sibling species are the result of more recent speciation events than are non-sibling species (but see Avise and Aquadro 1982 for a critical discussion of the molecular clock hypothesis).

To investigate this prediction we studied the McConnell's Flycatcher, *Mionectes macconnelli*, and two forms of the Ochre-bellied Flycatcher, *Mionectes oleagineus* [these taxa were removed from the genus *Pipromorpha* by Traylor (1979)]. These two widely-distributed neotropical tyrannids occur sympatrically from the Guianas, through Amazonian Brazil, and into southeastern Peru and northern Bolivia (Meyer de Schauensee 1966). *Macconnelli* was first described as a subspecies of *oleagineus* (Chubb 1919), but Todd (1921) recognized that two distinct species are actually involved, differing in the presence of buffy edges to the inner secondaries and wing coverts in *oleagineus*.

Although extremely similar morphologically, the two *Mionectes* have different vocalizations and behavioral repertoires (Willis et al. 1978). The evidence is less clear regarding habitat segregation. Near Manaus, Brazil, *macconnelli* was found primarily in mature forest whereas *oleagineus* occurred principally in second growth and forest edge (Willis et al. 1978). Near Belem, Brazil, *macconnelli* was restricted to seasonally flooded forest (*varzea*) while *oleagineus* ranged from high ground (*terra firme*) forest to *varzea* (Lovejoy 1974). On Trinidad, where *macconnelli* does not occur, *oleagineus* was found in both secondary and primary forest (Snow and Snow 1979). At our study site in Bolivia, both species were netted primarily in hilly upland forest, with smaller numbers of each also occurring in adjacent river-edge second growth. All individuals were in non-breeding condition (48 *macconnelli*, 39 *oleagineus*); thus, we do not know if habitat separation occurs during the breeding season. For this reason, in the remainder of this paper we refer to the two Bolivian populations as sympatric to reflect our uncertainty concerning their syntopic status.

We studied the degrees of allozymic and morphological similarity between samples of *oleagineus* and *macconnelli* taken from their area of sympatry at our study site in Bolivia and between these samples and those of *oleagineus* taken in northeastern Peru where *macconnelli* does not occur. We used electrophoretic and discriminant function analyses to determine the concordance between the two character sets, compared these results to those from studies of other congeneric species, and evaluated the recent origin hypothesis of sibling species.

STUDY SITES

We compared specimens from three populations, (1) *Mionectes macconnelli amazonus* from site I in Bolivia; (2) *Mionectes oleagineus chloronota* from site I in Bolivia; and (3) *Mionectes oleagineus hauxwelli* from two pooled localities (sites II and III) in Peru. No form of *Mionectes macconnelli* was found at sites II and III. The location and habitat at each site are:

SITE I: Bolivia, Depto. La Paz, Río Beni, ca. 20 km by river N Puerto Linares, 600 m; 15°20'S, 67°40'W; slightly-disturbed hilly tropical forest and adjacent river edge second-growth.

SITE II: Peru, Depto. Loreto, Río Napo, 157 km by river NNE Iquitos, 110 m; 3°40'S, 71°50'W; disturbed lowland high ground (*terra firme*) tropical forest, approximately one kilometer inland and north of the Río Napo and its seasonally flooded (*varzea*) forest.

SITE III: Peru, Depto. Loreto, Río Napo, 80 km N Iquitos, 120 m; 3°0'S, 73°20'W; slightly-disturbed lowland *terra firme* tropical forest, approximately one kilometer inland and south of the Río Napo and its *varzea*, with a finger of *varzea* extending near the site along a stream.

METHODS

BIOCHEMICAL ANALYSIS

Birds were collected primarily with mist nets. Tissue samples, taken 0–4 hours after death, were stored in liquid nitrogen before transport to the laboratory, where they were stored at −60°C until used. Samples of breast muscle, heart, liver, and kidney were homogenized with an equal volume of a 2 mM $MgCl_2$, 0.2 mM dithiothreitol or 0.2 mM mercaptoethanol, 0.25 M sucrose solution and clarified by centrifugation. Vertical electrophoresis was carried out on 10% starch gels at 6 V/cm for 14–21 hours, depending on the mobility of the protein to be assayed. Buffer systems used were: tris-citrate (pH 7.5), phosphate-citrate (pH 6.0), tris-EDTA-borate (pH 8.1), and tris-maleate (pH 7.5).

TABLE 1

ALLELIC FREQUENCIES AT VARIABLE LOCI[1,2]

Locus	Allele	Mionectes macconnelli (Bolivia)	Mionectes oleagineus (Bolivia)	Mionectes oleagineus (Peru)
adenosine deaminase	a	—	0.10	—
	b	1.00	0.90	1.00
umbelliferyl acetate esterase 2	a	—	0.10	—
	b	1.00	0.90	1.00
glycerol-3-phosphate dehydrogenase	a	0.17	0.83	0.25
	b	0.83	0.17	0.75
hexokinase 2[3]	a	1.00	0.88	0.75
	b	—	0.13	0.25
leucyl-alanine peptidase 1	a	1.00	0.60	0.64
	b	—	0.40	0.36
leucyl-alanine peptidase 2	a	0.33	1.00	1.00
	b	0.67	—	—
phenylalanyl-proline peptidase 1	a	—	1.00	0.93
	b	1.00	—	0.07
phenylalanyl-proline peptidase 2	a	0.43	1.00	1.00
	b	0.57	—	—
leucyl-glycl-glycine peptidase 2	a	—	0.25	—
	b	—	—	0.10
	c	—	0.13	0.40
	d	1.00	0.63	0.20
	e	—	—	0.30

[1] Both species were fixed for the same allele at the following 23 loci: acid phosphatase; adenylate kinase, creatine kinase 1 and 2, ethanol dehydrogenase, globins, glutamate dehydrogenase, glutamate-oxaloacetate transaminase (= aspartate aminotransferase) 1 and 2, glutamate-pyruvate transaminase (= alanine aminotransferase), glyceraldehyde-phosphate dehydrogenase, lactate dehydrogenase, leucyl-glycl-glycine peptidase 1, malate dehydrogenase 1 and 2, mannose-6-phosphate isomerase, octanol dehydrogenase, phosphoglucomutase, 6-phosphogluconate deaminase, phosphoglucose isomerase, superoxide dismutase 1 and 2, and umbelliferyl acetate esterase 1.

[2] Not all individuals were scorable at every locus.

[3] Hexokinase 1 and 3 were not scorable.

When two or more loci with a particular enzymatic activity were scored, they were numbered in sequence from the most anodally migrating to the least. Similarly, multiple alleles at a locus were designated alphabetically, beginning with the most anodal one. Staining procedures used to identify specific enzymes were similar to those given by Harris and Hopkinson (1976). Glutamate dehydrogenase was localized using the method of Brewer (1970). Globins were stained by the benzidine test for their peroxidase activity: 25 mg benzidine dihydrochloride, 0.3 ml 30% hydrogen peroxide, and 0.2 ml glacial acetic acid dissolved in 40 ml water.

Four banding regions of umbelliferyl acetate esterase (EST) have been found in birds (Braun and Parker 1985), but only three were detected in this study, and EST-3 was not scorable. On the basis of heterozygote banding pattern, EST-2 was dimeric. The three anodal bands of glutamate-pyruvate transaminase found in many neotropical suboscine species (Braun and Parker 1985) were seen in this study, but lack of polymorphism does not shed light on whether or not this is the result of a gene duplication event; therefore, the region was scored as a single locus. Creatine kinase (CK) showed three anodal bands in heart and kidney, with the center band equidistant between the fast and slow bands. Although only two loci in birds have been confirmed, one predominately expressed in brain and heart tissue and the other in skeletal muscle tissue (Eppenberger et al. 1967), other investigators have noted three bands in heart tissue (Avise et al. 1980; Braun, pers. comm.). Whether this is caused by a third locus, a heterodimer, or other factor remains to be determined. In this paper, only the fast and slow bands were scored (CK-1 and CK-2), and the center band was considered to be a heterodimer.

Electrophoretic data were entered as individual genotypic scores into the BIOSYS-1 computer program of Swofford and Selander (1981), which computes allelic frequencies, heterozygosity estimates, and genetic distance measures.

MORPHOMETRIC ANALYSIS

Specimens were collected by Louisiana State University Museum of Zoology personnel in June–July 1981 in Bolivia and May–August 1982 in Peru; reference study skins are deposited

TABLE 2

MEASURES OF PROTEIN VARIABILITY FOR THREE *MIONECTES* POPULATIONS

Species	Sample size	Mean sample size per locus ± s.e.	Percentage of loci polymorphic[1]	Mean heterozygosity ± s.e.	
				Direct count	Hardy-Weinberg expected[2]
M. macconnelli (Bolivia)	7	6.6 ± 0.2	9.4	0.04 ± 0.03	0.04 ± 0.03
M. oleagineus (Bolivia)	5	4.6 ± 0.2	18.8	0.07 ± 0.03	0.07 ± 0.03
M. oleagineus (Peru)	7	4.3 ± 0.2	15.6	0.07 ± 0.03	0.08 ± 0.04

[1] A locus is considered polymorphic if the frequency of the most common allele does not exceed 0.95.
[2] Unbiased estimate of Nei (1978).

in the Museum. Culmen length (from anterior edge of nares), culmen depth (at anterior edge of nares), culmen width (at anterior edge of nares), tarsus length, wing length (chord), and tail length of study skins from the three study sites, including those specimens used in the biochemical analysis, were measured to the nearest 0.1 mm with dial calipers. In addition, sex and body weight were recorded from the specimen labels. To investigate statistical relationships independent of body size, all variables were divided by the cubed root of weight. Data were analyzed with the discriminant function analysis program of SPSS (Nie et al. 1975), using direct entry of variables rather than the stepwise method.

RESULTS AND DISCUSSION

BIOCHEMISTRY

Allelic frequencies at variable loci are presented in Table 1. No fixed allelic differences were found between the two species in the protein loci sampled, although the phenylalanyl-proline peptidase 1 locus almost shows fixation for alternate alleles. This absence of fixed differences is not unusual in congeneric species of birds. For example, Avise et al. (1980) found no fixed differences between four species of *Catharus* thrushes. However, other congeneric species do exhibit fixed differences (e.g., species of *Vireo*; Avise et al. 1982).

The percentage of loci that are polymorphic and the mean heterozygosity (H̄) for each population are presented in Table 2. The concordance between observed and expected H̄ values implies that the three populations are in Hardy-Weinberg equilibrium and, therefore, are panmictic. Although small sample size precludes statistical testing, the large number of loci sampled (32), and the empirical analysis by Gorman and Renzi (1979) regarding the negligible effects of small sample size on genetic distance and heterozygosity estimates give us confidence in our results. The H̄ values fall within the range (0.007–0.147) reported for other birds (Barrowclough 1980), and they are close to the means for birds, H̄ = 0.049 and H̄ = 0.053, calculated by Nevo (1978) and Barrowclough (1983), respectively.

The genetic distance between *macconnelli* and the sympatric Bolivian population of *oleagineus* is 0.08 (Nei's unbiased distance; Nei 1978) and 0.28 (Rogers' distance as modified by Wright 1978). This is virtually identical to that between *macconnelli* and the Peruvian population of *oleagineus* which is not sympatric with *macconnelli*, those values being 0.07 (Nei's) and 0.27 (Rogers'). The genetic distance between the two *oleagineus* populations is 0.01 (Nei's) and 0.14 (Rogers'). The genetic distances, as measured by two different methods, are 50 percent greater between *macconnelli* and either population of *oleagineus* than between the two *oleagineus* populations.

Comparison of these genetic distance measures to other studies is difficult because of differences in the numbers and types of loci chosen. However, Barrowclough (1980) has examined degree of genetic differentiation as a function of taxonomic rank using values culled from eight avian studies. He found a mean genetic distance (Nei's unbiased distance) of 0.0440 ± 0.0221 (s.d.) between congeneric species and 0.0048 ± 0.0049 (s.d.) between subspecies, about an order of magnitude difference in the means. Equivalent values for congeneric species have been found in some subsequent studies (all values average Nei's unbiased distance, D̄), e.g., among three species of *Melospiza* (D̄ = 0.059 ± 0.026 [s.d.]; Zink 1982), among five species of *Cranioleuca* (D̄ = 0.08 ± 0.03 [s.d.]; Braun and Parker 1985), but not all, e.g., among five species of *Zonotrichia* (D̄ = 0.118 ± 0.073 [s.d.]; Zink 1982). The mean values (D̄) of 0.075 ± 0.01 (s.d.) between the two *Mionectes* species and 0.01 between the two *oleagineus* populations (considered by taxonomists to be different subspecies) fall well within the wide ranges ascribed

TABLE 3

RESULTS OF CLASSIFICATION OF BOLIVIAN *MIONECTES OLEAGINEUS* AND *M. macconnelli* USING DISCRIMINANT FUNCTIONS PROVIDED BY DFA

		M. oleagineus ♂	*M. oleagineus* ♀	*M. macconnelli* ♂	*M. macconnelli* ♀
M. oleagineus ♂		83.3%	5.6%	0.0%	11.1%
N	18	(15)	(1)	(0)	(2)
M. oleagineus ♀		27.3%	72.7%	0.0%	0.0%
N	11	(3)	(8)	(0)	(0)
M. macconnelli ♂		4.3%	0.0%	87.0%	8.7%
N	23	(1)	(0)	(20)	(2)
M. macconnelli ♀		18.8%	0.0%	6.3%	75.0%
N	16	(3)	(0)	(1)	(12)

by Barrowclough (1980) to species (0.0078–0.1267) and subspecies (−0.0014–0.02), respectively. Therefore, the two *Mionectes* are as well-differentiated as other congeneric species and, if the molecular clock hypothesis is valid, have been separated for a length of time roughly equivalent to other bird species.

MORPHOMETRICS

Discriminant function analysis (DFA) of the sympatric populations of *oleagineus* and *macconnelli* identified three functions that result in 80.9 percent correct classification of individuals to sex and species, and 91.2 percent correct classification to species (Table 3). These values are similar to the values of 86.8 percent and 90.4 percent, respectively, found by Rising and Schueler (1980) in their study of a pair of sibling species, the Eastern and Western Wood Pewees (*Contopus virens* and *C. sordidulus*). Similarly, in another study (Rohwer 1972) of cryptic species using DFA, Eastern and Western Meadowlarks (*Sturnella magna* and *S. neglecta*) were correctly discriminated to species 100 percent of the time.

We believe that the ability of DFA to separate these various cryptic species is so highly influenced by, or dependent upon, the nature of the variables used that comparisons between studies based upon different characters are very difficult. Nevertheless, two of the studies of sibling species noted above resulted in greater than 0 percent misclassification; this would be a surprising result if found in a non-sibling species pair. Furthermore, the estimates of misclassification in these two studies are conservative; the estimates were obtained from the same individuals that were used to generate the functions initially (Hand 1981).

Few sibling species pairs, including those mentioned above, exhibit the degree of sympatry demonstrated by *oleagineus* and *macconnelli*. Consequently, it is important to ask how these extremely similar species coexist when competition theory (Grant 1975) predicts a considerable potential for interspecific competition. In light of this evolutionary consideration, the variables used in the DFA were analyzed using two different morphological distance measures for possible expression of morphological character displacement in the Bolivian population of *oleagineus* that occurs sympatrically with *macconnelli*. This assumes that differences in morphology between taxa reflect ecological differences, an assumption supported by numerous studies (see Hespenheide 1973, and references therein), although Wiens and Rotenberry (1980) have demonstrated that this correlation between ecology and morphology can be "tenuous."

TABLE 4

MORPHOMETRIC DISTANCE MATRIX FOR THREE *MIONECTES* POPULATIONS[1]

	M. macconnelli (Bolivia)	*M. oleagineus* (Bolivia)	*M. oleagineus* (Peru)
M. macconnelli (Bolivia)	—	2.66	3.83
M. oleagineus (Bolivia)	2.16	—	2.83
M. oleagineus (Peru)	1.64	1.34	—

[1] Manhattan's distance below diagonal, proportional distance above diagonal.

TABLE 5

UNSTANDARDIZED MORPHOLOGICAL MEASUREMENTS OF *MIONECTES OLEAGINEUS* AND *M. MACCONNELLI*

Variable[3]	I *M. oleagineus* (Peru; N = 47–51)[1]			II *M. oleagineus* (Bolivia; N = 31–32)			III *M. macconnelli* (Bolivia; N = 39–40)			Comparison of group means[2]		
	$\bar{X}$	s.d.	Range	$\bar{X}$	s.d.	Range	$\bar{X}$	s.d.	Range	I vs II	I vs III	II vs III
Culmen depth	3.1	.15	2.8–3.4	3.1	.15	2.9–3.4	3.2	.12	2.9–3.5	NS	**	**
Culmen length	7.5	.29	6.9–8.1	7.7	.27	7.2–8.3	7.5	.33	6.7–8.2	**	NS	**
Culmen width	3.6	.24	3.2–4.1	3.7	.20	3.4–4.1	3.8	.18	3.4–4.2	*	***	*
Wing length	58.6	2.59	54.4–64.1	62.3	2.33	56.9–67.0	63.8	3.02	58.8–69.9	***	***	*
Tail length	44.2	2.59	40.8–50.4	48.1	2.05	44.1–52.1	47.7	2.62	43.0–51.3	***	***	NS
Tarsus length	13.6	.39	12.9–14.6	14.5	.37	13.9–15.2	15.7	.52	14.7–16.5	***	***	***
Body weight	9.6	.67	7.5–10.8	12.0	1.15	10.0–14.8	13.5	1.32	10.0–16.0	***	***	***

[1] Sample sizes differ for the different variables because of damaged or molting specimens.
[2] Paired Student's *t*-test comparison of means for the three populations. NS = not significant; * = $P < .05$; ** = $P < .01$; *** = $P < .001$.
[3] Measurements in mm, body weights in grams.

TABLE 6
SIZE-STANDARDIZED MORPHOLOGICAL MEASUREMENTS OF *MIONECTES OLEAGINEUS* AND *M. MACCONNELLI*

Variable[3]	I M. oleagineus (Peru; N = 47–51)[1]			II M. oleagineus (Bolivia; N = 31–32)			III M. macconnelli (Bolivia; N = 39–40)			Comparison of group means[2]		
	X̄	s.d.	Range	X̄	s.d.	Range	X̄	s.d.	Range	I vs II	I vs III	II vs III
Culmen depth/$^3\sqrt{wt}$	1.44	.08	1.32–1.65	1.38	.08	1.22–1.49	1.36	.05	1.26–1.45	**	***	NS
Culmen length/$^3\sqrt{wt}$	3.51	.15	3.22–3.85	3.35	.16	2.99–3.79	3.19	.17	2.85–3.55	***	***	***
Culmen width/$^3\sqrt{wt}$	1.69	.11	1.43–1.89	1.64	.10	1.43–1.86	1.61	.08	1.45–1.82	*	***	NS
Wing length/$^3\sqrt{wt}$	27.64	1.28	25.21–30.59	27.29	1.01	25.77–29.34	26.9	1.1	25.07–31.19	*	***	*
Tail length/$^3\sqrt{wt}$	20.9	1.25	18.94–24.05	21.03	.92	19.35–23.25	19.62	2.96	18.42–23.58	NS	**	*
Tarsus length/$^3\sqrt{wt}$	6.34	.54	5.89–6.80	6.34	.19	5.98–6.68	6.61	.25	6.12–7.12	NS	**	***

[1] Sample sizes differ for the different variables because of damaged or molting specimens.
[2] Paired Student's t-test comparison of means for the three populations. NS = not significant; * = $P < .05$; ** = $P < .01$; *** = $P < .001$.
[3] Measurements in mm, divided by the cube root of body weight in grams.

If character displacement has occurred, the Bolivian population of *oleagineus* will be less like *macconnelli* than is Peruvian *oleagineus*.

The predicted relationship is demonstrated by Manhattan's distance, but the reverse is indicated by the proportional distance measure (Table 4). This finding does little to answer the question that was posed concerning character displacement, although it does raise additional questions about the merit of different morphological distance measures.

The data were further analyzed univariately using a Student's *t*-test to examine the differences between means (Table 5). The predicted relationship was not found. For all variables, with the exception of culmen depth, the mean values of *macconnelli* and Bolivian *oleagineus* are closer than are the mean values of *macconnelli* and Peruvian *oleagineus*. The two *oleagineus* populations do not differ statistically in culmen depth.

Rather than indicating active "character convergence" (Cody 1973) in sympatry, these data may illustrate that the geographic relationship is not the only factor influencing the evolution of morphological characters between congeners (Wiens and Rotenberry 1980). The marked difference in latitude of the two *oleagineus* populations supports this. Grant (1975) demonstrated that size differences between different populations, at first interpreted as character displacement, need not be a result of competitive pressures but rather a result of geographic variation.

To investigate the possibility that the observed morphological similarity of *macconnelli* and Bolivian *oleagineus* is a result of other evolutionary factors causing an overall increase in size, we standardized each variable for body size by dividing it by the cubed root of body weight (Table 6). As with the unstandardized data, no character displacement was observed. The two populations of *oleagineus* did not differ in relative tail and tarsus lengths, but did differ in the remaining four variables. In terms of these latter four variables, Bolivian *oleagineus* is different from the Peruvian population and is more like *macconnelli* with which it is sympatric.

Interpretation of the ecological importance of the morphological differences observed between the two *oleagineus* populations is extremely difficult for two reasons: (1) the increased size of Bolivian *oleagineus* could reflect a shift to larger prey size or adjustment to colder temperatures at higher latitudes; and (2) the differences in relative bill shape and wing length could reflect subtle differences in foraging behavior. The elucidation of how such similar species coexist must await further field work. At present, we conclude that our data do not demonstrate character displacement.

CONCLUSIONS

Comparison of the genetic distance between *Mionectes oleagineus* and *M. macconnelli* with those between other congeneric birds indicates that this sibling species pair is equally well-differentiated genetically. The implication under the molecular clock hypothesis is that *oleagineus* and *macconnelli* are not of especially recent origin relative to other avian congeners and, therefore, that they show very conservative morphological evolution. Morphologically they can be separated only by the presence or absence of buffy edges on the inner secondaries and wing coverts or by multivariate analysis of several morphological characters (and even then, 100% discrimination is not achieved). We know of no other extensively sympatric species that are so similar morphologically as are these two species of flycatcher. This lack of concordance between the levels of genetic and morphological differentiation provides additional evidence for the variation in evolutionary rates in different suites of characters.

ACKNOWLEDGMENTS

We thank the following individuals who collected specimens used in this study: S. W. Cardiff, T. J. Davis, J. V. Remsen, G. H. Rosenberg, M. Sánchez S., T. S. Schulenberg, and D. Wiedenfeld. Permits were granted by the Dirección General Forestal y de Fauna of the Ministerio de Agricultura, Lima, Peru and by DICYT in La Paz, Bolivia (through Gastón Bejarano). The assistance of the Academia Nacional de Ciencias of Bolivia is appreciated. For laboratory facilities, advice, and encouragement we thank H. C. Dessauer and M. J. Braun. For assistance with the BIOSYS-1 program we thank R. Lawson. For computer funds we thank the Department of Zoology and Physiology, Louisiana State University. Critical reading by J. V. Remsen, R. M. Zink, and members of the LSU Museum of Zoology Ornithology Discussion Group was greatly appreciated. We also wish to acknowledge the generous support of A. W. Geier, J. S. McIlhenny, B. M. Odom, H. I. Schweppe, and L. R. Schweppe which made fieldwork possible.

LITERATURE CITED

AVISE, J. C., AND C. F. AQUADRO. 1982. A comparative summary of genetic distances in the vertebrates: patterns and correlations. Evol. Biol. 15:151–185.

AVISE, J. C., C. F. AQUADRO, AND J. C. PATTON. 1982. Evolutionary genetics of birds. V. Genetic distances within Mimidae (mimic thrushes) and Vireonidae (vireos). Biochem. Genet. 20:95–104.

AVISE, J. C., J. C. PATTON, AND C. F. AQUADRO. 1980. Evolutionary genetics of birds. I. Relationships among North American thrushes and allies. Auk 97:135–147.

BARROWCLOUGH, G. F. 1980. Genetic and phenotypic differentiation in a wood warbler (Genus Dendroica) hybrid zone. Auk 97:655–668.

BARROWCLOUGH, G. F. 1983. Biochemical studies of microevolutionary processes. Pp. 223–261, In A. H. Brush and G. A. Clark, Jr. (eds.), Perspectives in Ornithology. Cambridge University Press, New York.

BRAUN, M. J., AND T. A. PARKER, III. 1985. Molecular, morphological, and behavioral evidence concerning the taxonomic relationships of "Synallaxis" gularis and other synallaxines. Pp. 333–346, In P. A. Buckley, M. S. Foster, E. S. Morton, R. S. Ridgely, and F. G. Buckley (eds.), Neotropical Ornithology. Ornithol. Monogr. No. 36.

BREWER, G. J. 1970. An Introduction to Isozyme Techniques. Academic Press, New York.

CHUBB, C. 1919. New forms of South-American birds. Ann. Mag. Nat. Hist. (Ser. 9) No. 4:301–304.

CODY, M. L. 1973. Character convergence. Annu. Rev. Ecol. Syst. 4:187–211.

EPPENBERGER, H. M., D. M. DAWSON, AND N. O. KAPLAN. 1967. The comparative enzymology of creatine kinases. I. Isolation and characterization from chicken and rabbit tissues. J. Biol. Chem. 242:204–209.

GORMAN, G. C., AND J. RENZI, JR. 1979. Genetic distance and heterozygosity estimates in electrophoretic studies: effects of sample size. Copeia 1979:242–249.

GRANT, P. R. 1975. The classical case of character displacement. Evol. Biol. 8:237–337.

HAND, D. I. 1981. Discrimination and Classification. Wiley and Sons, London.

HARRIS, H., AND D. A. HOPKINSON. 1976. Handbook of Enzyme Electrophoresis in Human Genetics. North Holland/Elsevier, Amsterdam, Netherlands.

HESPENHEIDE, H. A. 1973. Ecological inferences from morphological data. Annu. Rev. Ecol. Syst. 4:213–229.

LOVEJOY, T. E. 1974. Bird diversity and abundance in Amazon forest communities. Living Bird 13:127–191.

MAYR, E. 1963. Animal Species and Evolution. Belknap Press of Harvard University Press, Cambridge, Massachusetts.

MEYER DE SCHAUENSEE, R. 1966. The Species of Birds of South America. Livingston Publ. Co., Wynnewood, Pennsylvania.

NEI, M. 1972. Genetic distance between populations. Am. Natur. 106:283–292.

NEI, M. 1978. Estimation of average heterozygosity and genetic distance from a small number of individuals. Genetics 89:583–590.

NEVO, E. 1978. Genetic variation in natural populations: pattern and theory. Theor. Pop. Biol. 13:121–177.

NIE, N. H., C. H. HULL, J. G. JENKINS, K. STEINBRENNER, AND D. H. BENT. 1975. Statistical Package for the Social Sciences (SPSS). McGraw-Hill, New York.

RISING, J. D., AND F. W. SCHUELER. 1980. Identification and status of wood pewees (Contopus) from the Great Plains: what are sibling species? Condor 82:301–308.

ROHWER, S. A. 1972. A multivariate assessment of interbreeding between the meadowlarks, Sturnella. Syst. Zool. 21:313–338.

SNOW, B. K., AND D. W. SNOW. 1979. The Ochre-bellied Flycatcher and the evolution of lek behavior. Condor 81:286–292.

SWOFFORD, D. L., AND R. B. SELANDER. 1981. BIOSYS-1: a FORTRAN program for the comprehensive analysis of electrophoretic data in population genetics and systematics. J. Hered. 72:281–283.

TODD, W. E. C. 1921. Studies in the Tyrannidae. I. A revision of the genus Pipromorpha. Proc. Biol. Soc. Wash. 34:173–192.

TRAYLOR, M. A., JR. 1979. Tyrannidae. Pp. 1–245, In M. A. Traylor, Jr. (ed.), Check-list of Birds of the World, Vol. VIII. Museum Comparative Zoology, Cambridge, Massachusetts.

WIENS, J. A., AND J. T. ROTENBERRY. 1980. Patterns of morphology and ecology in grassland and shrubsteppe bird populations. Ecol. Monogr. 50:287–308.

WILLIS, E. O., D. WECHSLER, AND Y. ONIKI. 1978. On behavior and nesting of McConnell's Flycatcher (Pipromorpha macconnelli): does female rejection lead to male promiscuity? Auk 95:1–8.

WILSON, A. C., S. S. CARLSON, AND T. J. WHITE. 1977. Biochemical evolution. Annu. Rev. Biochem. 46:573–639.

WRIGHT, S. 1978. Evolution and the Genetics of Populations, Vol. 4. Variability, Within and Among Natural Populations. University Chicago Press, Chicago, Illinois.

ZINK, R. M. 1982. Patterns of genic and morphologic variation among sparrows in the genera Zonotrichia, Melospiza, Junco, and Passerella. Auk 99:632–649.

TAXONOMY OF THE LESSER NIGHTHAWKS (*CHORDEILES ACUTIPENNIS*) OF NORTH AND CENTRAL AMERICA

ROBERT W. DICKERMAN

Department of Microbiology, Cornell University Medical College, New York, New York 10021 USA

ABSTRACT. Three subspecies of the Lesser Nighthawk (*Chordeiles acutipennis*) are recognized in southwestern United States and in Middle America; *texensis* (with *inferior* as a synonym), Texas to California, south to the northern arid regions of Mexico, a large pale form with a pale and finely vermiculated juvenal plumage; *micromeris*, northern arid zone of Yucatan Peninsula, a small pale form with gray juvenal plumage with coarser dark markings; an *littoralis*, central Mexico south to Costa Rica (separating the ranges of *texensis* and *micromeris*), a medium-sized richly colored form with juveniles a darker gray with a generally brownish cast.

RESUMEN. En el sudoeste de los Estados Unidos y en América Central se reconocen tres subspecies del halcón nocturno (*Chordeiles acutipennis*): *texensis* (con *inferior* como sinónimo), desde Texas hasta California, y hacia el sur hasta las regiones áridas del norte de México, es grande y pálido con un plumaje juvenil pálido y suavemente vermiculado; *micromeris*, de la zona árida del norte de la penínensula de Yucatán, es pálido y pequeño y los juveniles presentan un plumaje grisaseo con marcas gruesas oscuras y *littoralis* que se encuentra desde el centro de México hacia el sur hasta Costa Rica (separando las áreas de distribución de *texensis* y *micromeris*), siendo de tamaño mediano con vistosa coloración, los juveniles son gris oscuro, generalmente, con tonos amarronados.

In attempting to identify specimens of the Lesser Nighthawk (*Chordeiles acutipennis*) collected in Guatemala, I found it necessary to reevaluate the validity of the several names available for the North and Middle American populations. Oberholser (1914) revised the species, and later Eisenmann (1962) provided a lucid analysis of the geographic variation and of the nomenclatural opinions that have been expressed regarding the Middle American populations. In brief, four names have been proposed. Two refer to northern, pale subspecies, i.e., *texensis* (larger) Texas to southern California, and *inferior* (described as averaging smaller) of Baja California. *Micromeris* was proposed for a small, pale subspecies of the northern, arid region of the Yucatan Peninsula; and *littoralis* was proposed for a small, more darkly-colored subspecies in the Pacific lowlands of Chiapas. The Baja California, Yucatan and Chiapas subspecies have been variously combined with each other, but without sufficient consideration of mensural and color characters, and populations as far northwest as Sonora and as far south as Panama have been considered to be *micromeris* (van Rossem 1942; Wetmore 1968).

METHODS

I obtained copies of original measurement sheets compiled by Drs. Oberholser and Wetmore from the files of the U.S. Fish and Wildlife Service and the U.S. National Museum, respectively. My measurements of a sample of the same specimens were very close to theirs, and, thus, I was able to combine their data with mine. I measured additional specimens from Baja California, Arizona, and Texas, and all available nesting season (May–July when measurable) specimens from the Yucatan Peninsula. Amadeo Rea kindly measured specimens of females from Baja California and the southern tier of counties of California in the San Diego Natural History Museum. Measurements taken were standard wing chord and tail length, rounded to the nearest millimeter. It is amazing, and discouraging, to find that the majority of specimens from Yucatan was collected by G. F. Gaumer between 1884 and 1910, and that exceedingly little recent material is available. Because the number of specimens measured from the northern portion of the range (subspecies *texensis*) was large, totals are cited by state only. Nesting season (May–July) specimens from Mexico and Central America are cited by locality; juveniles cited earlier (Dickerman 1981) are not included here.

TABLE 1

MEASUREMENTS OF WING CHORD AND TAIL FOR POPULATIONS OF *CHORDEILES ACUTIPENNIS*[1]

	Males		Females	
	Wing chord	Tail	Wing chord	Tail
A. c. texensis	173–191 (183.1) n = 34, s.d. 5.3	104–118 (110.8) n = 33, s.d. 4.0	168–180 (174.9) n = 28, s.d. 3.9	92–112 (105.1) n = 30, s.d. 4.2
A. c. "inferior"	165–185 (177.0) n = 38, s.d. 3.8	94–115 (102.2) n = 38, s.d. 4.6	162–177 (169.5) n = 24, s.d. 4.2	93–105 (99.9) n = 29, s.d. 3.7
A. c. micromeris	159–175 (167.0) n = 16, s.d. 3.8	91–103 (98.2) n = 16, s.d. 3.7	151–171 (162.5) n = 18, s.d. 4.8	90–103 (94.9) n = 18, s.d. 4.2
A. c. littoralis				
Brodkorb (1942)	166–180 (173.9) n = 7	91–101 (95.4) n = 7		
Other collections	166–188 (175.4) n = 13, s.d. 7.2	93–105 (101.2) n = 10, s.d. 4.0	158–180 (167.6) n = 15, s.d. 6.7	91–106 (99.4) n = 10, s.d. 5.6
Guatemala (AMNH)	174, 176, 179 n = 3	100, 102, 103 n = 3	164–178 (169.7) n = 6	95–104 (100.5) n = 6

[1] Values given are ranges followed by means in parentheses, sample size (n), and standard deviation (s.d.).

TAXONOMIC CONCLUSIONS

On the basis of measurements and color patterns of adults and of plumage of juveniles (Dickerman 1981), I recognize three subspecies:

Chordeiles acutipennis texensis
Chordeiles texensis Lawrence, 1856, Ann. Lyc. Nat. Hist. New York, 6:167.
Chordeiles acutipennis inferior Oberholser, 1914, Bull. U.S. Natl. Mus. 86:109.

Diagnosis.—Largest of the subspecies and pale; juvenal plumage palest, more cinnamon on tips of dorsal feathers, and with fine vermiculations.

Range.—Baja California and Southern California east to Texas and south through central Mexico.

Discussion.—*Chordeiles texensis* Lawrence was described from two specimens without data, presented to Lawrence by Captain J. P. McCown (Greenway 1978). Oberholser (1914:104) restricted the type locality to "Ringgold Barracks, near Rio Grande City," Texas. Van Rossem (1947), however, demonstrated that specimens of the Bridled Titmouse (*Parus wollweberi*) supposedly from that locality were probably collected farther south in Mexico. Such may be the case with one of the cotypes of *texensis*. A female (AMNH 43851) shows little wear and is large (wing chord 176 mm, tail 105 mm); the other (AMNH 43852), a male by plumage, is somewhat more worn and is smaller (wing chord 168 mm, tail 97 mm). Both are pale and match comparably-plumaged specimens of *texensis*. The wing measurement of the male falls outside the range of 33 male *texensis* (*sensu stricto*) and is considerably below the mean wing length of "*inferior*"; it is near the mean for the Yucatan population *micromeris* (Table 1). Thus, I designate the female cotype (AMNH 43851) as the lectotype of *C. texensis* Lawrence. Although no evidence exists that the female was actually taken in Texas, to avoid further uncertainty I follow Oberholser and recognize "Ringgold Barracks, near Rio Grande City," Texas, as the type locality of *C. texensis* Lawrence.

Oberholser described *inferior* as similar to *texensis,* but smaller. The means of wing and tail measurements of specimens from southern Baja California are significantly smaller than those of specimens of *texensis*. The means of the wing chord in males were 177 mm vs 183 mm (Student's $t = 5.656$, d.f. = 70, $P < .01$); in females 170 mm vs 176 mm ($t = 4.804$, d.f. = 50, $P < .01$). Differences between means of tail measurements were slightly greater ($t = 8.342$, d.f. = 69, $P < .01$ for males, $t = 5.040$, d.f. = 50, $P < .01$ for females). However, measurement overlap is extensive (Table 1). The species range is continuous, but specimens from northern Baja California, where birds intermediate in size should occur, are limited. Three males from Baja California collected 19 April (Carrizo Valley) and 23 April (Santa Cruz) have wings measuring 185, 187, and 190 mm and probably represent migrants from farther north.

Juveniles and definitively plumaged specimens from Texas to Baja California show no color or pattern differences. Van Rossem (1942) found specimens from Baja California and northwestern Mexico similar in color and pattern but used the name *micromeris* for them! I recommend that *C. a. inferior* be considered a junior synonym of *C. a. texensis*.

Areas of intergradation between *texensis* and *littoralis* are discussed under the latter subspecies.

Specimens examined.—149. UNITED STATES: *Texas* 23 (including lectotype); *Arizona* 20; *California* 22. MEXICO: *Baja California* 68, *Sonora* 3, *Chihuahua* 4, *Nuevo Leon* 2, *Tamaulipas* 2, *Durango* 5.

Chordeiles acutipennis micromeris Oberholser, 1914, Bull. U.S. Natl. Mus., 86:100.

Diagnosis.—Similar in color to *texensis* but smaller.

Range.—Arid northern portion of the Yucatan Peninsula.

Discussion.—In describing *micromeris,* Oberholser compared it to the nominate subspecies from northern South America rather than to the more similar northern forms, in keeping with his use of a south to north taxonomic arrangement. The single specimen of *micromeris* in juvenal plumage differs from series of juvenile *texensis* in having coarser dorsal markings (Dickerman 1981). The geographic range of *micromeris* is separated from the range of *texensis* by that of the dark subspecies *littoralis* of the humid tropical lowlands of southern Mexico. Specimens from northern Belize, an area in which intergrades between *micromeris* and *littoralis* would be expected, were not examined.

Specimens examined.—34. MEXICO: *Yucatan* Merida 1, Progreso 4, Ixmal 4, Santa Clara 6, Chichen Itzá 1, Dzidzantun 4, Temax 3, Xbac 4, Río Lagartos 6; *Quintana Roo* Chetumal 1.

Chordeiles acutipennis littoralis Brodkorb, 1940, Auk 57:543.

Diagnosis.—Similar in size to *micromeris* but darker and more richly colored, the spotting being more ochraceous, less buffy than in the paler populations; darker and smaller than *texensis.*

Range.—South-central Mexico (Colima, Irapuato, Puebla, Oaxaca, and Chiapas) south to Costa Rica and possibly adjacent Panama.

Discussion.—The juvenal plumage of *littoralis* is darker gray dorsally than in juvenal *texensis* and *micromeris* with a generally browner cast (Dickerman 1981). Faded and worn nesting adults are difficult to identify. A nesting season male from "15 miles" east of Oaxaca City is large (wing 188 mm, tail 111 mm) and is inseparable from *texensis*; however, two less worn females are smaller and darker and are typical *littoralis*. Six adults from Sinaloa are large (wings of five males 181–185 mm, one female 175 mm) but are darker than four nesting specimens from Chihuahua. Thus, *texensis* and *littoralis* intergrade, in size at least, across a broad zone of central Mexico, probably related to the intermediate area of rainfall between the arid north and humid Middle America.

Specimens examined.—44. MEXICO: *Colima* 3 mi. E. Cuyutlan 1; *Guanajuato* Irapuato 1; *Morelos* Tutepec 1, Acatlipa 1; *Guerrero* Ajuchitlan 3, Zirandaro 2; *Puebla* Izucar de Matamoros 1; *Oaxaca* Chivela 4, 10–15 mi. W. Cd. Oaxaca 3; *Chiapas* Arriaga (type locality) 7, 40 mi. W. Arriaga 1. BELIZE: All Pines 1. GUATEMALA: *San Marcos* Ocos 1; *Santa Rosa* La Avellana 5, Mouth Río Esclavos 1; *Zacapa* Santa Lucia 2. HONDURAS: *Olancho* San Esteban 1; *Yoro* Coyoles 2. NICARAGUA: *unspecified* 1. COSTA RICA: *Guanacaste* El Pelon 3; *Punta Arenas* near Rio Palo Seco 2.

ACKNOWLEDGMENTS

I wish to express my appreciation to the curators at the following museums for loan of specimens used in this study: Academy of Natural Sciences, Philadelphia; American Museum of Natural History (AMNH); Carnegie Museum of Natural History; Field Museum of Natural History; Delaware Museum of Natural History; Louisiana State University Museum of Zoology; Moore Laboratory of Zoology, Occidental College; Museum of Comparative Zoology, Harvard University; Peabody Museum of Natural History, Yale University; San Diego Museum of Natural History; U.S. National Museum of Natural History; University of Kansas Museum of Natural History; University of Michigan Museum of Zoology; and the Western Foundation of Vertebrate Zoology.

The specimens from Guatemala were collected during the course of studies on the ecology of arboviruses under permits from the Ministerio de Agricultura de Guatemala. That research was supported in part by U.S. Public Health Service Research Grant AI-06248. Sylvia Hope kindly calculated the Student's *t*-test statistics.

LITERATURE CITED

DICKERMAN, R. W. 1981. Geographic variation in the juvenal plumage of the Lesser Nighthawk (*Chordeiles acutipennis*). Auk 98:619–621.

EISENMANN, E. 1962. Notes on nighthawks of the genus *Chordeiles* in southern Middle America, with a description of a new race of *Chordeiles minor* breeding in Panama. Am. Mus. Novit. No. 2094.

GREENWAY, J. C., JR. 1978. Type specimens of birds in the American Museum of Natural History, Part 2. Bull. Am. Mus. Nat. Hist. 161:1–305.

OBERHOLSER, H. C. 1914. A monograph of the genus *Chordeiles* Swainson, type of a new family of goatsuckers. Bull. U. S. Natl. Mus. 86:1–123.

VAN ROSSEM, A. J. 1942. The Lower California nighthawk not a recognizable race. Condor 44:73–74.

VAN ROSSEM, A. J. 1947. Two races of the Bridled Titmouse. Fieldiana (Zool.) 31:87–92.

WETMORE, A. 1968. The birds of the Republic of Panama. Pt. 2. Smithson. Misc. Collect. 150(2).

PLATE IV.   Nests of the Rufous Flycatcher (*Myiarchus semirufus*). These nests illustrate one of the shared derived characters that define the myiarchine flycatchers: obligatory nesting in tree-cavities. (A) These eggs, photographed 12 May 1981, are less heavily blotched and lined than most eggs of *Myiarchus*. Note the presence of a soft nest lining of fur and feathers. (B) An adult carrying food to its four nestlings pauses at the cavity entrance, 24 April 1981. Photographs taken by M. D. Williams at Las Pampas, Lambayeque, Peru.

A PHYLOGENY OF THE MYIARCHINE FLYCATCHERS

WESLEY E. LANYON

American Museum of Natural History, New York, New York 10024 USA

ABSTRACT. The genus *Myiarchus* is diagnosed on the basis of shared derived character states of two morphological complexes, the nasal capsule of the skull and the supporting elements of the syrinx, as well as by a shared derived pattern of nesting behavior. Equivocal species having "aberrant" external morphology (*validus, magnirostris, semirufus, antillarum* and *sagrae*) are reexamined with respect to this diagnosis, and their assignment to *Myiarchus* justified.

Laniocera, Lipaugus, and *Nesotriccus* are examined and found to be unrelated to *Myiarchus* and inappropriately assigned to the subfamily Tyranninae, whereas *Sirystes, Casiornis, Rhytipterna,* and *Deltarhynchus* share with *Myiarchus* a derived character state of the nasal capsule, nest in tree-cavities, so far as is known, and are properly considered near relatives. An unsuspected near relative of *Myiarchus, Ramphotrigon,* formerly allied with the "flatbills," is found to share the same derived state of the nasal capsule and is now known to nest in tree-cavities. These latter six genera constitute the myiarchine assemblage within the subfamily Tyranninae.

Two branches of myiarchine flycatchers are defined by differences in syringeal morphology and possibly by differences in the materials used to line the nest-cavities. *Deltarhynchus* and *Ramphotrigon* are each other's closest relative, share a unique syringeal morphology, and nest in tree-cavities not lined with fur and feathers. A second branch, consisting of *Rhytipterna, Sirystes, Casiornis,* and *Myiarchus,* share an equally unique syringeal morphology and, so far as is known, nest in tree-cavities lined with fur and feathers.

Attila does not share the diagnostic characters of the myiarchine flycatchers but may be the sister group of that assemblage.

RESUMEN. El género *Myiarchus* se diagnostica sobre la base de patrones de caracteres derivados compartidos ("shared derived character states") de dos complejos morfológicos, la cápsula nasal del cráneo y los elementos de soporte de la siringe, así como también por un patrón derivado compartido de comportamiento de anidación ("shared derived pattern of nesting behavior"). Se reexaminan especies ambiguas que poseen morfología externa "aberrante" (*validus, magnirostris, semirufus, antillarum* y *sagrae*) con respecto a este diagnóstico y se justifica la inclusión de estas especies en el género *Myiarchus.*

Se examinaron *Laniocera, Lipaugus* y *Nesotriccus* y se encontró que no están relacionados con *Myiarchus* así como que han sido asignados a la subfamilia Tyranninae indebidamente, mientras que *Sirystes, Casiornis, Rhytipterna* y *Deltarhynchus* comparten con *Myiarchus* un patrón de caracteres derivados ("derived character state") de las cápsulas nasales, por lo que se sabe, anidan en cavidades en los árboles y están considerados apropiadamente como parientes cercanos. *Ramphotrigon,* un insospechado pariente cercano de *Myiarchus,* anteriormente asociado con los picos chatos (*Tolmomyias* spp. y *Rhynchocyclus* spp.), se encontró que comparte el mismo patrón derivado de cápsula nasal y ahora se sabe que anida en huecos en los árboles. Estos últimos seis géneros constituyen el ensamble "myiarchine" dentro de la subfamilia Tyranninae.

Dos ramas de atrapa moscas del grupo myiarchine son definidas por las diferencias en la morfología de la siringe y posiblemente por las diferencias respecto a los materiales utilizados para forrar la cavidad donde anidan. *Deltarhynchus* y *Ramphotrigon* son los parientes más cercanos para ambos, comparten una morfología de siringe única y anidan en cavidades de árboles que no están forradas con piel o plumas. Una segunda rama compuesta por *Rhytipterna, Sirystes, Casiornis* y *Myiarchus,* comparten así mismo una morfología de siringe única, y por lo que se sabe, anidan en cavidades de árboles forradas con piel y plumas.

Attila no comparte los caracteres diagnósticos de los atrapa moscas del grupo myiarchine, pero podría ser un grupo hermano ("sister group") de ese ensamble.

Upon completion of my revision (1978) of South American taxa of the genus *Myiarchus*, I was reasonably satisfied with my recognition of 22 species in this most successful and widespread of all flycatcher (Tyrannidae) genera. I had based the generic assignment of these species solely upon considerations of external morphology and foraging and nesting behavior. In some instances, however, external morphology was equivocal on the matter of generic affinity with the core of archetypical *Myiarchus,* and it was possible that patterns of foraging and nesting behavior might transcend the conventional limits of the genus. Moreover, the diagnostic vocal characters that had facilitated interpretation of external morphology and species' limits were of questionable value at the generic level. Therefore, it seemed prudent to examine internal morphological characters to strengthen my diagnosis of the genus, to reevaluate the equivocal cases among those species I had assigned to *Myiarchus,* and to redefine the limits of the genus where warranted. A comparative study of internal morphology should also provide the means for evaluating the affinities of putative near relatives as well as for identifying taxa heretofore unsuspected of being related to *Myiarchus.* Ultimately, such an approach should permit me to hypothesize about phylogenetic relationships among *Myiarchus,* its near relatives, and other tyrant flycatchers.

METHODOLOGY

I based my conclusions upon a concordance of shared derived character states of two internal morphological complexes as well as the sharing of derived patterns of nesting behavior. This approach already has proven effective in diagnosing several genera of tyrant flycatchers (Lanyon 1982, 1984; Lanyon and Fitzpatrick 1983) and holds promise for the eventual determination of a meaningful phylogeny of the family Tyrannidae (Traylor 1977; Traylor and Fitzpatrick 1982).

The two morphological complexes used in this study are the nasal capsule of the skull and the supporting elements of the syrinx, as demonstrated in landmark studies by Warter (1965) and Ames (1971), respectively. I have been influenced greatly by both of these works, and I follow the terminology of these authors. References to Warter and Ames are exclusively to these studies and, since numerous, are made hereafter without the usual literature citations. Fortunately, I have had access to a greater number and array of anatomical specimens than was available to either of them. In addition, all syringes that I examined were double-stained with alcian blue for cartilage and alizarin red for ossified bone (after Dingerkus and Uhler 1977) to facilitate the study of the supporting elements and to differentiate between bone and cartilage. The syringes studied by Ames were not stained. Specifically, I examined (1) the degree to which the nasal septum is ossified and is buttressed with internal and/or transverse supporting elements, and (2) the number, shape, and position of the bony and cartilaginous supporting elements in the syrinx, especially the internal cartilages.

As Warter reported, the nasal capsule in most birds is essentially unossified; little remains for study in the cleaned skull maintained in museum collections. The widespread occurrence of this unossified state suggests the primitive condition; the ontogenetic transformation from membrane, through cartilage, to bone could also be interpreted as support for the hypothesis that the primitive state is the unossified one. Fortunately for students of the Tyrannidae, tyrant flycatchers exhibit a great array of character states of the nasal capsule. These range from the presumed primitive condition in "flatbills" and "tody-tyrants," in which the nasal septum is unossified or represented only by a heavily buttressed dorsal remnant, to virtually complete septa replete with internal supporting structures and/or transverse trabecular plates (Lanyon, unpubl. data). Within this spectrum are shared derived character states that offer great promise for indicating generic and suprageneric relationships.

Tyrant flycatchers vary greatly with respect to the morphology of the syrinx (Ames). Whether it will ever be possible to recognize morphoclines or transformation series in the general configuration of the syrinx, the degree of fusion of the A-elements, and the shape and position of the internal cartilages and the A- and B-elements remains to be demonstrated. At present, derived character states of the syrinx are best identified by their uniqueness at the generic and suprageneric level, and evidence for such evolutionary novelties within the spectrum of tyrannid syringes is ample.

I examined the skulls of 88 of the 90 genera of tyrant flycatchers (*sensu* Traylor 1979), and the syringes of 90, and have data on nesting behavior of 81. In addition to the anatomical collections at the American Museum of Natural History (AMNH), I used material on loan from the Carnegie Museum of Natural History (CM), the Field Museum of Natural History (FMNH), the Museu Parense Emilio Goeldi in Belem, Brazil (MPEG), the Museum of Zoology, Louisiana State University (LSU), the Museum of Zoology, University of Michigan (UMMZ), the National Museum of Natural History (USNM), the Peabody Museum of Natural History, Yale University (PMNH), and the Western Foundation of Vertebrate Zoology (WFVZ). Specimens cited here are identified to collection by the abbreviations given above.

DIAGNOSIS OF THE GENUS *MYIARCHUS*

NESTING BEHAVIOR

Nesting behavior has remained "one of the more conservative features of flycatcher biology" during the "explosive ecological and morphological radiation" of this family (Traylor and Fitzpatrick 1982) and traditionally has been given considerable weight in tyrannid classifications. I now have data on the nesting habits of 20 of the 22 species of *Myiarchus* (*venezuelensis* and *yucatanensis* lacking). All are obligatory tree-cavity nesters, and the cavities are typically deep enough to hide the incubating or brooding female from without (Plate IV). The nests have a soft lining of fur and feathers, in addition to vegetable fibers, and usually, though not always, include fragments of shed reptilian skin, paper or plastic.

The eggs have been described for roughly half the species of *Myiarchus* and are rather uniformly creamy buff or pinkish-white, and heavily blotched and lined (scratchy appearance) with brown, purple, or lavender over the entire surface (Plate IV).

NASAL CAPSULE

I have examined the skulls of 21 of the 22 species of *Myiarchus* (*nugator* lacking). The nasal capsule is virtually fully ossified, including the alinasal walls and turbinals (Fig. 1). The nasal septum is well developed but lacks a transverse trabecular plate. A conspicuous internal supporting rod runs diagonally through the *Myiarchus* septum, from the roof of the capsule to the ventral edge. When viewed from below, this rod appears as a swelling in the otherwise comparatively thin, knife-like ventral edge of the septum.

SYRINX

I have examined 36 syringes taken from 18 of the 22 species of *Myiarchus* (*apicalis, sagrae, validus,* and *yucatanensis* lacking). Diagnostic features of the *Myiarchus* syrinx, as discussed below, are illustrated in Figure 2. As is characteristic of all genera in the subfamily Tyranninae (*sensu* Traylor 1977), the A-elements are not fused into a drum. The tracheo-bronchial junction has no bony pessulus, but rather consists of multiple bands of cartilage, the most prominent of which are located laterad to the junction of the two bronchi and are the cartilaginous segments of the complete A2 and/or A3 supporting rings. Only the A2s are involved in most instances, but sometimes the prominent lateral bands involve the A2 on one side and the A3 in the other bronchus, or only the A3s. The remainder of the junction of the two bronchi consists of smaller, less prominent, medially located cartilaginous bands, more variable in number and attachment (generally to the A3 or A4 rings dorsally, but sometimes showing no obvious attachment dorsally or ventrally). Two pairs of internal cartilages lie within the internal tympaniform membranes, (1) a large dorsal pair, each member of which is L-shaped and connected anteriorly by membrane or cartilage to the medio-ventral segment of the prominent cartilaginous bands that complete the A2 or A3 supporting rings, and (2) a much smaller ventral pair of variable size and shape, that has a membranous connection to the ventral ends of the large L-shaped cartilages and another connection to the ventral ends of the B2 or B3 bronchial elements. The B1 and B2 supporting elements of the bronchi have a consistent and characteristic configuration (Fig. 2: syrinx 1); B1 is narrow throughout except for a very broad dorsal end, whereas B2 is narrow throughout except for a broad triangular ventral end that is barely attached to the ventral end of the B1.

REEVALUATION OF THE EQUIVOCAL SPECIES OF *MYIARCHUS*

I retained (Lanyon 1978) three species in *Myiarchus,* "*Muscifur*" *semirufus,* "*Eribates*" *magnirostris,* and "*Hylonax*" *validus,* that have been assigned to monotypic genera at one time or another, on the basis of aberrant external morphology. Now that my diagnosis of the genus has been strengthened with a consideration of internal morphology, it is appropriate to reexamine these equivocal species to determine if any character contradictions exist.

Bangs and Penard (1921) assigned *semirufus* to a new genus, *Muscifur,* because of the unique rufous coloration of the plumage. Subsequent authors retained this coastal Peruvian endemic in *Myiarchus* (Todd 1922; Hellmayr 1927; Zimmer 1938), and my field observations on behavior support this view (Lanyon 1975). Since my report on the first known nest of this

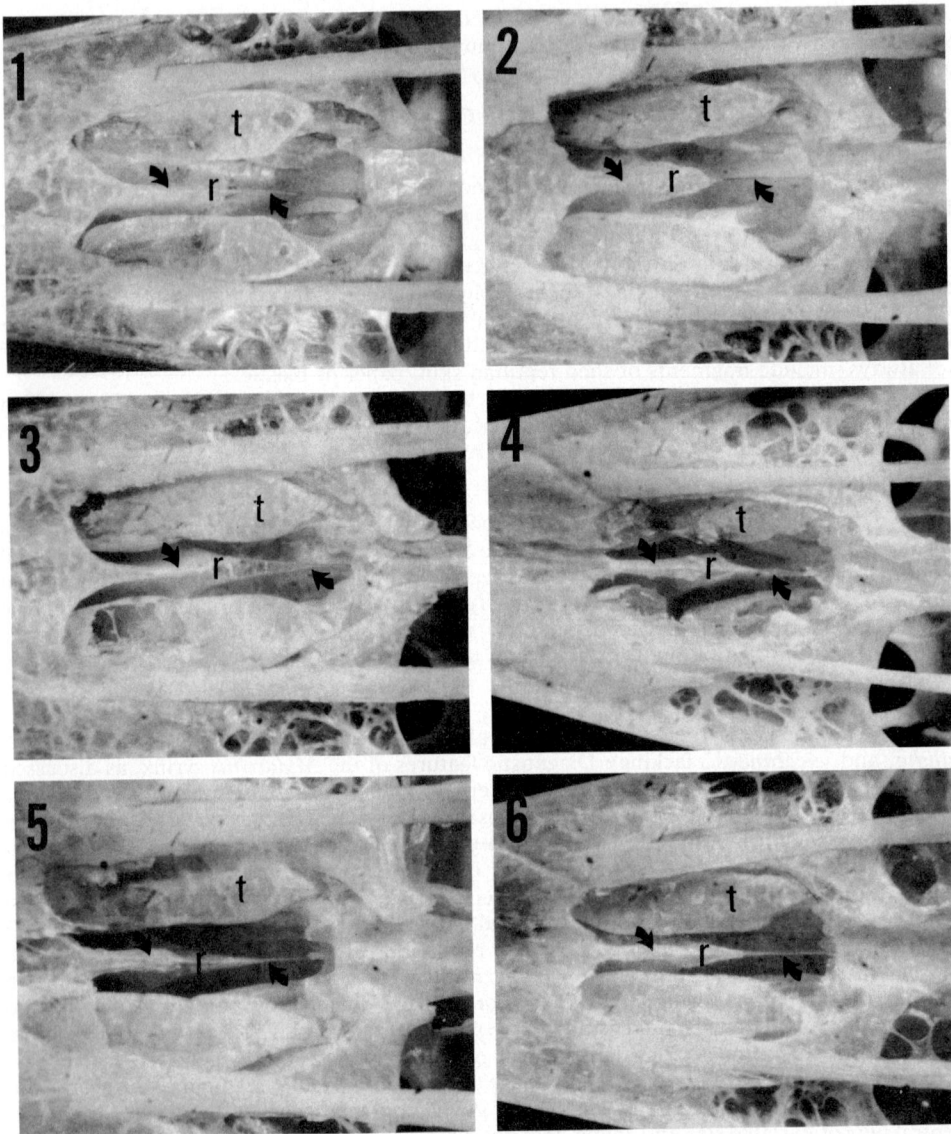

FIG. 1. Photographs, taken through a dissecting microscope (magnification = 7×), of the ventral aspect of the nasal region of various species of *Myiarchus* (anterior end of skull to left). Extremes in prominence of the internal supporting rod, r, within the genus are illustrated by the capsules of (1) *Myiarchus cephalotes,* AMNH 9248, and (2) *M. apicalis,* AMNH 9242. Other species exhibiting this character state are: (3) *M. semirufus,* AMNH 9062, (4) *M. magnirostris,* USNM 321065, (5) *M. validus,* USNM 502770, and (6) *M. antillarum,* USNM 554895. Arrows indicate ventral edge of nasal septum, which lacks a trabecular plate; t = alinasal turbinal, left side.

species (Lanyon 1975), T. A. Parker and M. D. Williams have located a number of additional nests, all in tree-cavities and lined with fur and feathers in typical myiarchine fashion (Plate IV). Recent examination of both the nasal capsule (Fig. 1: skull 3) and the syrinx (Fig. 2: syrinx 5) supports the view that *semirufus* belongs in *Myiarchus.*

The Galapagos endemic, *magnirostris,* was placed in its own genus, *Eribates,* by Ridgway (1893) in recognition of its longer tarsi and its bill shape. Later workers agreed that it should be retained in *Myiarchus* (Swarth 1931; Lanyon 1978; J. T. Zimmer, unpubl. ms.). I can now

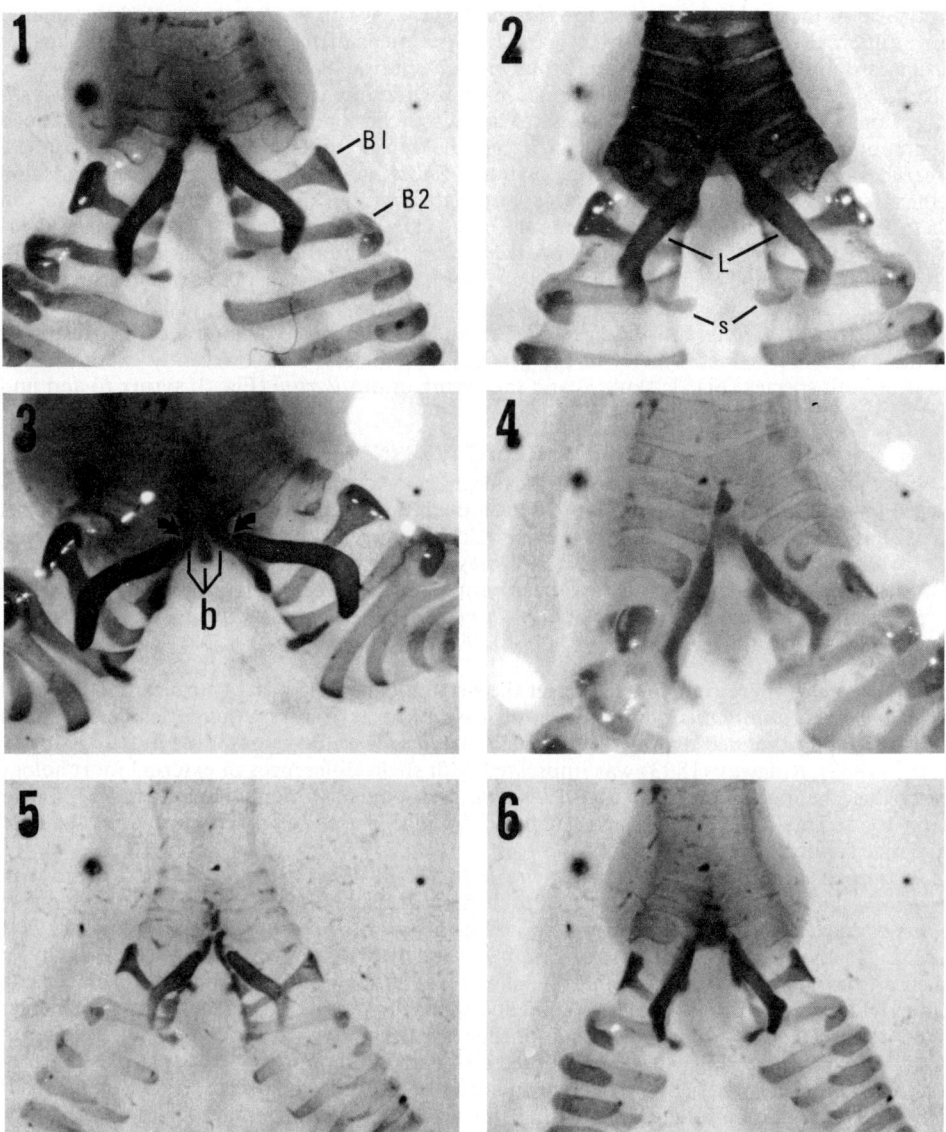

Fig. 2. Photographs, taken through a dissecting microscope (magnification = 10×), of the dorsal aspect of the syringes of various species of *Myiarchus*: (1) *Myiarchus crinitus*, PMNH 10225, (2) *M. venezuelensis*, AMNH 8316, (3) *M. tyrannulus*, PMNH 9028, (4) *M. magnirostris*, AMNH 6678, (5) *M. semirufus*, AMNH 4503, (6) *M. antillarum*, CM 827. The B1 and B2 = supporting elements, labeled in specimen 1. Large L-shaped dorsal cartilages, L, and small ventral cartilages, s, labeled only in specimen 2. Cartilaginous bands forming the tracheo-bronchial junction are darkly stained and labeled b, in specimen 3; arrows indicate the attachments of the large L-shaped cartilages to the medio-ventral segments of the lateral pair of these bands.

add internal morphology to the arguments for retention in *Myiarchus*. The nasal capsule is typically myiarchine, as described above (Fig. 1: skull 4). The two available syringes (AMNH 6678, FMNH 106203) differ in detail from those of *Myiarchus*; the two lateral cartilaginous bands contributing to the tracheo-bronchial junction are not nearly as prominent and are not attached dorsally to the A-elements. This means that the A1 through A3 supporting elements are incomplete around each bronchus. In all other respects the syringes of these specimens are typical of those in my sample of *Myiarchus* (Fig. 2: syrinx 4). Ames reported the smaller pair of internal cartilages to be absent from his specimens of *magnirostris,* but these smaller cartilages stained well and are conspicuous in both specimens that I examined.

Ridgway (1905) transferred *validus,* a Jamaican endemic, to a monotypic genus *Hylonax* in the family Cotingidae, largely because of tarsal scutellation. Hellmayr (1927:187) retained the monotypic genus but was unwilling to accept the transfer to the cotingids "until anatomical researches have decided its systematic position." Subsequent workers, aware of the variation in the tarsal envelope within the tyrannids, restored *validus* to *Myiarchus* (Bond 1956; Lanyon 1967; Traylor 1977; J. T. Zimmer, unpubl. ms.). The only new data to supplement those given in my revision of the West Indian *Myiarchus* (1967) are that the nasal capsule of *validus* is typically myiarchine (Fig. 1: skull 5). I have not had an opportunity to examine the syrinx.

Two other West Indian species, *antillarum* of Puerto Rico and the Virgin Islands and *sagrae* of Cuba, Grand Cayman and the Bahamas, might be categorized as equivocal because they are the only white-bellied species in the genus. Nevertheless, I have recently examined the skulls of both species (Fig. 1: skull 6) and the syrinx of *antillarum* (Fig. 2: syrinx 6) and find the characters typically myiarchine.

PUTATIVE NEAR RELATIVES OF *MYIARCHUS*

With the genus rigorously diagnosed, I can now consider the affinities of those flycatchers alleged to be the nearest relatives of it to determine if they exhibit any of the derived character states that define *Myiarchus,* and if they should be merged with that genus. If they do not exhibit these evolutionary novelties, it still may be possible to hypothesize about their real affinities. The tyrant genera that have been proposed as the near relatives of *Myiarchus* are *Deltarhynchus, Nesotriccus, Sirystes, Attila, Laniocera, Lipaugus, Casiornis,* and *Rhytipterna.*

DELTARHYNCHUS

I reported elsewhere (Lanyon 1982) on the nesting behavior, nasal capsule, and syrinx of *Deltarhynchus flammulatus,* which is endemic to southwestern and southern Mexico. Although it was originally assigned by Lawrence (1875) to *Myiarchus* and maintained in that genus by Sclater (1888), Ridgway (1893) was impressed with slight differences in external morphology and created the monotypic genus *Deltarhynchus.* Subsequent workers (Hellmayr 1927; Eisenmann 1955; Traylor 1977) followed Ridgway, but placed *Deltarhynchus* close to *Myiarchus,* a practice supported by my studies.

I examined three skulls of *Deltarhynchus* (AMNH 11390, 11391, 12078). All share with *Myiarchus* the derived state of the nasal capsule described above (Fig. 3: skull 5). Consideration of this character complex alone would argue for merger of *Deltarhynchus* with *Myiarchus.*

Although the single known nest of *Deltarhynchus* was located in a natural tree-cavity, it differed from *Myiarchus* nests in that the cavity was shallow and not lined with soft fur and feathers (Lanyon 1982: figs. 4, 5). A second and more convincing argument against the merger of these genera comes from comparison of their syringes. My two syringes of *Deltarhynchus* (AMNH 7761, 8128; Fig. 4: syrinx 5) have B1 and B2 supporting elements like those of *Myiarchus,* but are significantly different in a number of other respects. The A3 elements of both specimens are complete and ossified throughout, and the remainder of the tracheo-bronchial junction, though largely cartilaginous, is at least partially ossified. One specimen has only a single pair of internal cartilages. The other specimen has a suggestion of a second smaller pair of cartilages located in the same position as the smaller cartilages of *Myiarchus,* but not nearly as prominent. It is in the configuration and position of the large dorsal cartilages, present in both specimens, that one sees the greatest departure from the syrinx of *Myiarchus.* These cartilages are straight, not L-shaped, and are expanded distally. Moreover, they give the appearance of being twisted so that the proximal two-thirds of each is flattened at right angles to the internal tympaniform membrane, whereas the distal third is flattened within the plane of that membrane. The attachment of these cartilages is farther dorsad than in *Myiarchus,* for they are connected to the dorsal ends of the incomplete A2 elements and to the dorso-medial segments of the complete A3 elements.

These contradictions in the nesting biology and syringeal morphology of the two genera suggest that *Deltarhynchus* should be kept generically distinct, while cranial morphology and a proclivity toward nesting in tree-cavities argue for a close relationship between them.

NESOTRICCUS

Nesotriccus ridgwayi, a Cocos Island endemic, historically has been considered a near relative of *Myiarchus* (Ridgway 1907; Hellmayr 1927; Eisenmann 1955). The anatomical studies of

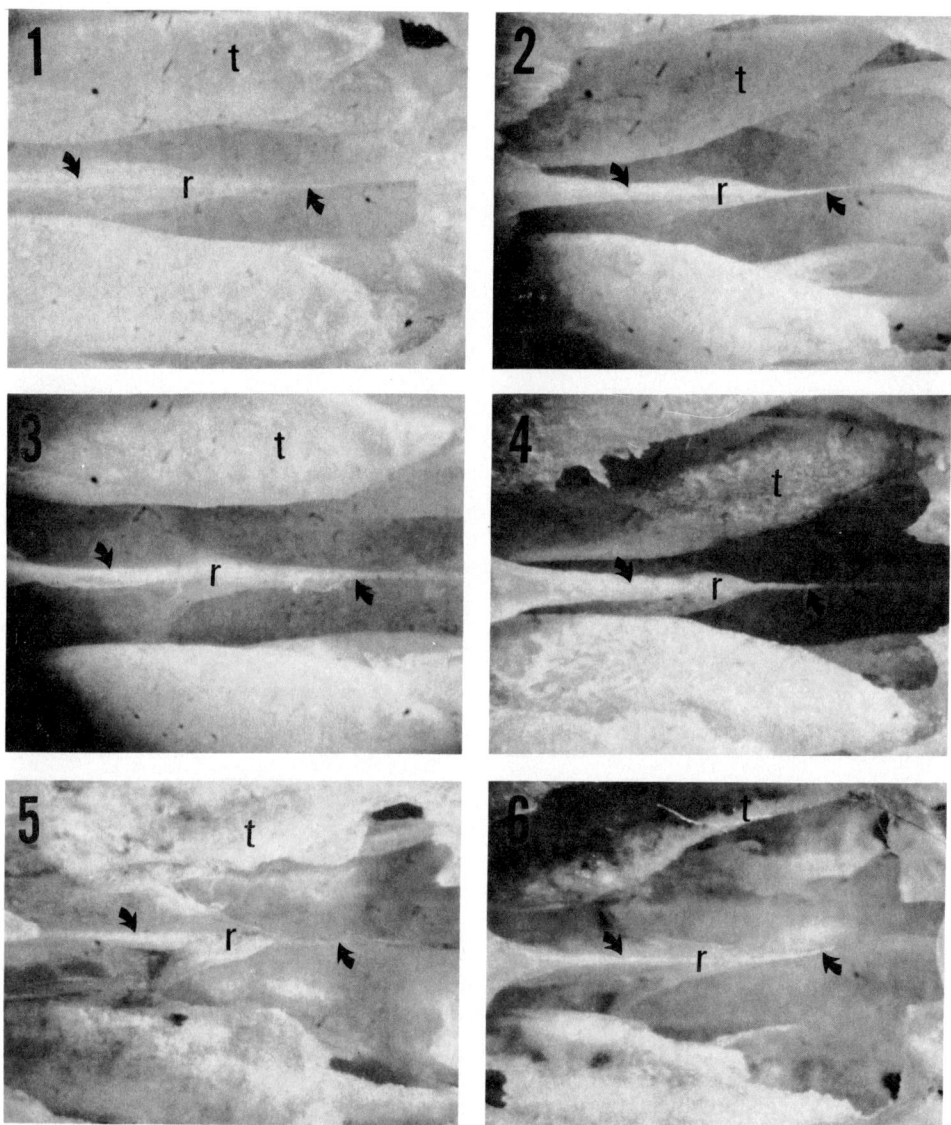

FIG. 3. Photographs, taken through a dissecting microscope (magnification = 13×), of the ventral aspect of the nasal region of the six genera of myiarchine flycatchers (anterior end of skull to left): (1) *Myiarchus tuberculifer*, AMNH 7175, (2) *Casiornis rufa*, AMNH 6697, (3) *Sirystes sibilator*, USNM 347723, (4) *Rhytipterna immunda*, AMNH 11617, (5) *Deltarhynchus flammulatus*, AMNH 11391, (6) *Ramphotrigon fuscicauda*, LSU 101510. Symbols as in Figure 1.

Warter and Ames raised serious questions about this alleged affinity, but these authors were equivocal as to how *Nesotriccus* should be classified. Traylor (1977) removed *Nesotriccus* from its traditional position near *Myiarchus* and placed it in another subfamily, the Fluvicolinae, on the basis of plumage characteristics.

I have argued in detail elsewhere (Lanyon 1984) that no morphological evidence supports the contention of a close relationship between *Nesotriccus* and *Myiarchus*. That the nests of *Nesotriccus* are cup-shaped and located near the distal tips of vegetation (T. S. Sherry, pers. comm.) is further confirmation. *Nesotriccus* shares its nasal capsule and syringeal character states, (Fig. 5: skull 1, Fig. 6: syrinx 2), which are very different from those of *Myiarchus*, with only two other monotypic genera, *Phaeomyias* and *Capsiempis*. Presumably, all three

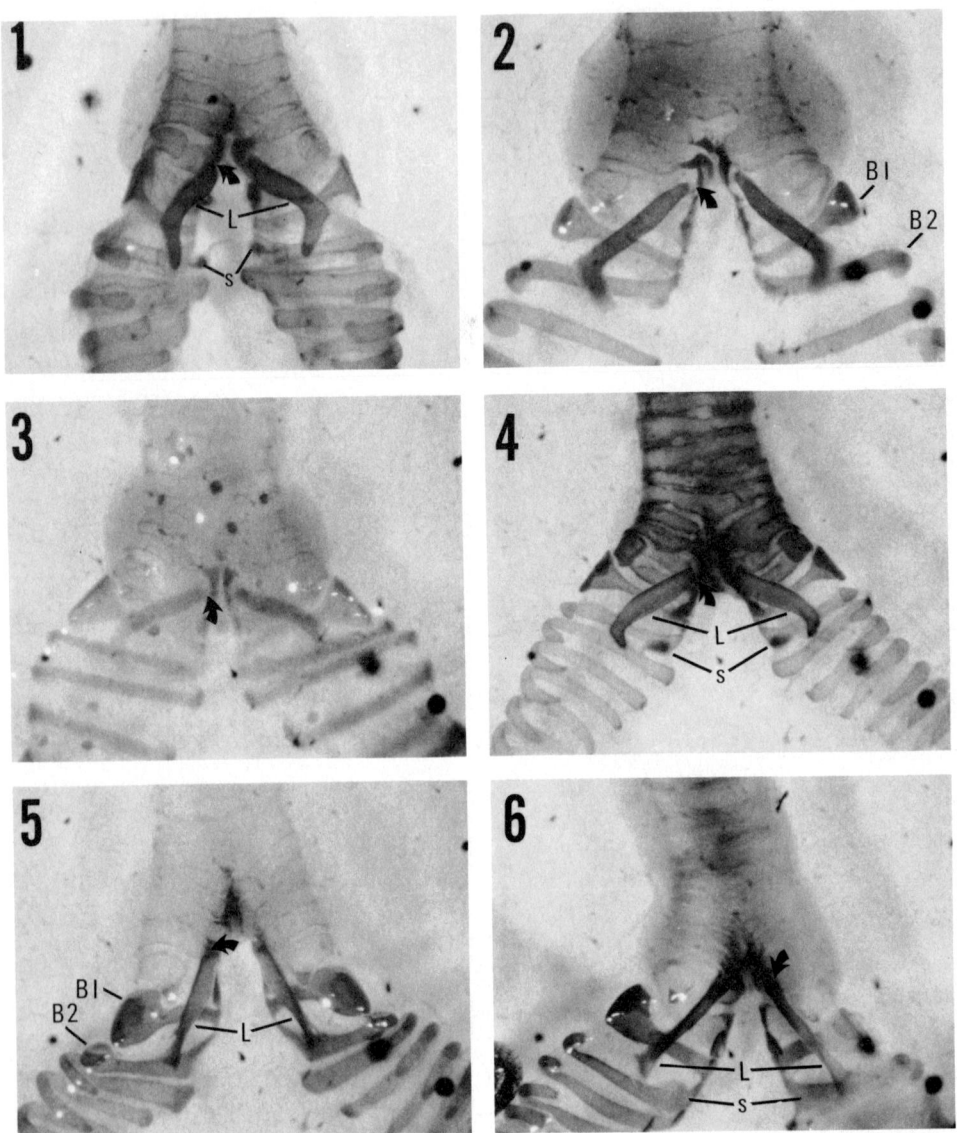

Fig. 4. Photographs, taken through a dissecting microscope (magnification = 10×), of the dorsal aspect of the syringes of the six genera of myiarchine flycatchers: (1) *Myiarchus nuttingi*, PMNH 9032, (2) *Sirystes sibilator*, PMNH 2791, (3) *Casiornis rufa*, USNM 227189, (4) *Rhytipterna immunda*, AMNH 8109, (5) *Deltarhynchus flammulatus*, AMNH 7761, (6) *Ramphotrigon fuscicauda*, LSU 102608. Arrows indicate attachments of large dorsal cartilages to the dorsal ends of the A2 supporting elements (in *Deltarhynchus*, 5, and *Ramphotrigon*, 6) or to the ventral portion of the tracheo-bronchial junction (in the other four genera). Other symbols as in figure 2; large dorsal cartilages, L, are straight and expanded distally in *Deltarhynchus*, 5, and *Ramphotrigon*, 6, and L-shaped in the other four genera; small cartilages lacking in *Deltarhynchus*.

of these genera belong somewhere within the subfamily Elaeniinae as constituted by Traylor (1977), but their nearest relatives remain obscure.

SIRYSTES

Although Ridgway (1907) placed *Sirystes* in the Cotingidae, on the basis of the tarsal envelope, this wide-ranging (Panama to northern Argentina), monotypic genus has generally

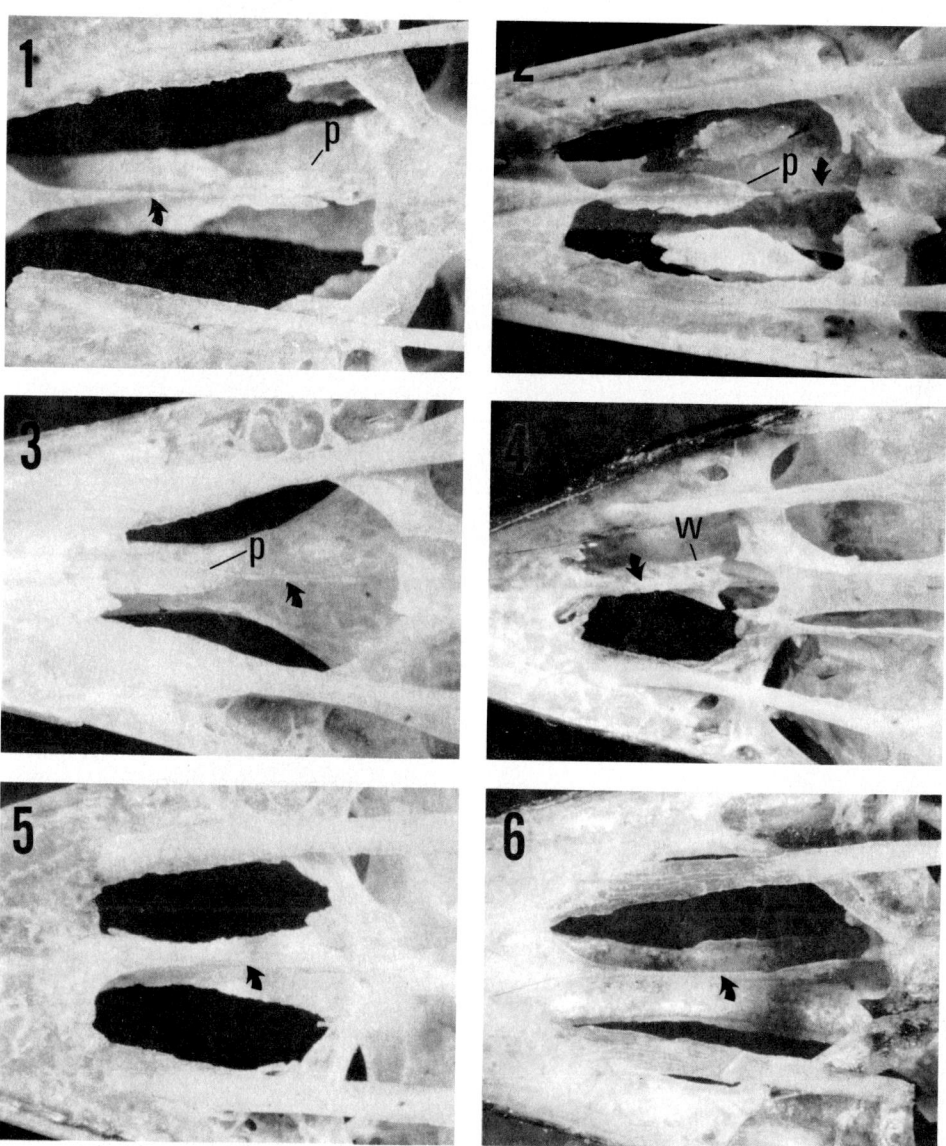

Fɪɢ. 5. Photographs, taken through a dissecting microscope (magnification = 7×), of the ventral aspect of the nasal region of various non-myiarchine flycatchers (anterior end of skull to left): (1) *Nesotriccus ridgwayi,* AMNH 9210, (2) *Machetornis rixosus,* USNM 499048, (3) *Laniocera hypopyrra,* AMNH 11448, (4) *Legatus leucophaius,* AMNH 11599, (5) *Empidonomus aurantioatrocristatus,* AMNH 6933, (6) *Attila cinnamomeus,* AMNH 11615. Arrows indicate ventral edge of nasal septum; p = trabecular plate; w = unique "wings" projecting from internal supporting rod in *Legatus.*

been considered a tyrant flycatcher. Hellmayr (1927) included it with *Tyrannus* and its allies. It was not until Warter's study of cranial anatomy that there was an indication of a closer relationship of *Sirystes* to *Myiarchus,* a suggestion later followed by Wetmore (1972) and Traylor (1977). The behavior and morphology of *Sirystes* also indicate unequivocally that it is a close relative of *Myiarchus* (Lanyon and Fitzpatrick 1983). The only remaining question is whether *Sirystes* should be merged with *Myiarchus.*

The single known nest of *Sirystes* was located in a natural tree-cavity, too high to permit documentation of the nest lining (Lanyon and Fitzpatrick 1983: fig. 1). The nasal capsules of

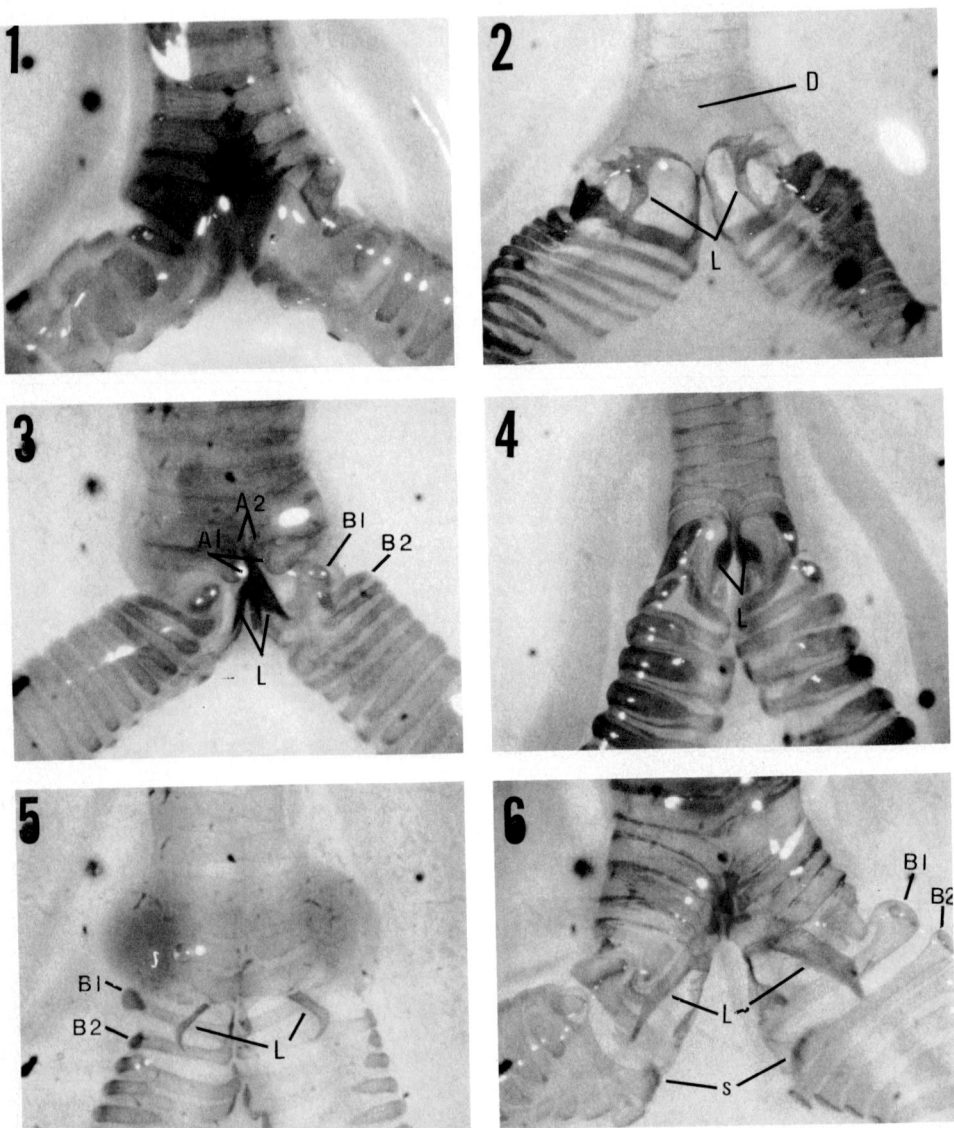

FIG. 6. Photographs, taken through a dissecting microscope (magnification = 10×, except 7× for *Lipaugus*), of the dorsal aspect of the syringes of various non-myiarchine flycatchers and of *Lipaugus*: (1) *Lipaugus subalaris,* FMNH 290407, (2) *Nesotriccus ridgwayi,* AMNH 8067, (3) *Laniocera rufescens,* LSU 108461, (4) *Machetornis rixosus,* PMNH 2705, (5) *Tyrannus melancholicus,* AMNH 6828, (6) *Attila spadiceus,* AMNH 8124. Note that the internal cartilages (absent from *Lipaugus,* 1) are shaped differently from those of the myiarchine flycatchers; fusion of the A elements into a drum, D, is characteristic of *Nesotriccus,* 2, and most of the subfamily Elaeniinae; the dorsal ends of the A1 and A2 elements of each bronchus are much closer to each other in *Laniocera,* 3, than in the syringes of any of the myiarchines; other symbols as in figure 2.

both *Sirystes* skulls examined (USNM 346048, 347723) fall within the range of variation of the derived state found in *Myiarchus*; they would not be separable from a series of skulls of *Myiarchus* species of comparable size (Fig. 3: skull 3). The supporting elements of the four syringes examined (AMNH 6803; PMNH 2790, 2791; LSU 108463) exhibit the derived state found in *Myiarchus,* with the exception that the cartilaginous bands completing the A2 elements are shorter, narrower, and less prominent, and the large internal L-shaped cartilages attach to the ventral ends of the A2s rather than to the medio-ventral segments of the cartilaginous bands completing the A2 rings (Fig. 4: syrinx 2).

ATTILA

Attila traditionally has been classified with the Cotingidae, on the basis of tarsal scutellation, but not without suspicion that it may more properly belong in the Tyrannidae (Sclater 1888; Ridgway 1907; Hellmayr 1927; Zimmer 1936). The cranial studies of Warter persuaded Meyer de Schauensee (1970) to transfer this successful and widespread genus (seven species; western Mexico to northeastern Argentina) to the Tyrannidae. Warter recommended such a transfer, but also believed the cranial differences merited separate subfamily status. Ames supported the transfer, but went further, reporting the syrinx of *Attila* to be like that of *Myiarchus*. Ames' findings persuaded Snow (1973) and Traylor (1977) to argue for recognition of *Attila* as a near relative of *Myiarchus*.

The species of *Attila* are known to nest in tree cavities but are more likely to nest in more exposed niches or crevices. They certainly could not be characterized as being obligatory tree-cavity nesters, fur and feathers are not used in the nest lining, and the eggs are like those of *Tyrannus* and its allies rather than those of *Myiarchus*.

I have examined the skulls of *Attila spadiceus* (AMNH 11438, 11447, 11593, 11608), *A. cinnamomeus* (AMNH 11615, 11616), and *A. bolivianus* (USNM 346062), the same species that were available to Warter. The nasal capsules of all three are like those of *Tyrannus* and its allies (Fig. 5: skull 6); the nasal septum is well ossified, but lacks a trabecular plate and a conspicuous internal supporting rod, whereas the remainder of the capsule is poorly ossified, with little or no remnant of the alinasal turbinals or walls. No characteristics argue against *Attila* being a member of the subfamily Tyranninae (*sensu* Traylor 1977). *Attila* differs from *Myiarchus* in the notable lack of the conspicous internal supporting rod in the nasal septum and in the poorly ossified condition of the capsule as a whole.

The syringes of *Attila spadiceus* (AMNH 8124, 8125, 8243; LSU SML 130), *A. cinnamomeus* (AMNH 8126, 8127), and *A. bolivianus* (FMNH 290391) differ from those of *Myiarchus* species. The two large internal cartilages (Fig. 6: syrinx 6) are basically straight and broad, and are shaped either like a long triangle (single specimen of *bolivianus*) or a trapezium (all other specimens). They are attached to the dorso-medial portions of the medial segments of the complete A2 supporting elements. The degree of calcification of the tracheo-bronchial junction varies considerably in my sample; the junction may be nearly fully ossified (including a well developed pessulus) or nearly completely cartilaginous. Most of my specimens have a smaller ventral pair of internal cartilages located as in *Myiarchus* but less conspicuous. The configuration of the B1 and B2 supporting elements is like that of *Myiarchus,* however.

In summary I consider *Attila* to be closer to *Myiarchus* than did Warter, certainly within the same tyrant subfamily, but somewhat farther removed from *Myiarchus* than did Ames.

LANIOCERA

The two species of *Laniocera* are poorly known in life, and I have found no reference to nesting behavior. The genus (ranging from southern Mexico to eastern Brazil) has been variously assigned to the Cotingidae (Sclater 1888; Hellmayr 1927), the Pipridae (Ridgway 1907), and the Tyrannidae (Meyer de Schauensee 1970; Wetmore 1972; Traylor 1977).

Warter did not have access to a skull of *Laniocera*; if he had, his findings almost surely would have staved off the subsequent move to relate this genus to *Myiarchus*. I have examined the skulls of *L. hypopyrra* (AMNH 11448) and *L. rufescens* (UMMZ 218410) and find both very different from *Myiarchus* (Fig. 5: skull 3). In these skulls the nasal septum is poorly ossified, there is no conspicuous internal supporting rod, and a well developed, narrow trabecular plate is located basally along the anterior half of the septum. A diagnostic feature of all tyrannine flycatchers is the absence of a trabecular plate on the nasal septum.

Ames reported that the syrinx of *Laniocera rufescens* is sufficiently like that of the other "cotingids" having syringes like *Myiarchus* (*Attila, Casiornis,* and *Rhytipterna*) to warrant including *Laniocera* in his *Myiarchus* group. It was this that persuaded Meyer de Schauensee (1970), Wetmore (1972), Snow (1973), and Traylor (1977) to transfer *Laniocera* more formally from the Cotingidae to the Tyrannidae, near *Myiarchus*.

I borrowed the specimen of *rufescens* available to Ames (PMNH 986), and in addition examined two other *rufescens* (LSU 108460, 108461) and three specimens of *L. hypopyrra* (AMNH 8077; LSU 102562, 102564). All of these specimens are very different from *Myiarchus* and from all genera in the tyrant subfamily Tyranninae (Fig. 6: syrinx 3). The dorsal ends of the A1 and A2 supporting elements in each bronchus are close to one another medially, so that the tracheo-bronchial junction is not as easily studied as it is in all tyrannines. When the bronchi are spread apart for study, the tracheo-bronchial junction can be seen to consist of a single calcified band or pessulus. There is a large pair of short, triangular internal cartilages; the anterior edge of each is attached to the dorsal half of the pessulus. Because the tracheo-bronchial junction is not as broad (fewer different supporting structures) as in *Attila* and *Myiarchus*, the large internal cartilages of *Laniocera* nearly meet medially. A pair of smaller internal cartilages of variable shape is located ventrally in the internal tympaniform membranes, in roughly the same position as in *Myiarchus*. The B1 and B2 supporting elements are not as in *Myiarchus* and *Attila*, but have a configuration as illustrated in syrinx 3, Figure 6.

I have complete confidence in concluding from these descriptions of the nasal capsule and syrinx that *Laniocera* is not a near relative of *Myiarchus* and that it does not belong in the subfamily Tyranninae. Its position within the Tyrannidae must await a better understanding of the phylogeny of the family as a whole.

LIPAUGUS

Not until Ames suggested that *Lipaugus* might belong to his *Myiarchus* group was there any serious question that this genus (seven species; Mexico to southern Brazil) belongs in the Cotingidae. Ames had difficulty interpreting his single damaged syrinx of *L. unirufus*, however, and was less certain about recommending its transfer to the Tyrannidae than the transfer of the other *Myiarchus*-like "contingids." Earlier Warter had found the skull of *Lipaugus* to be cotingid rather than tyrannid-like. Wetmore (1972) alone followed Ames' tentative suggestion, whereas Meyer de Schauensee (1970), Snow (1973), and Traylor (1977) retained *Lipaugus* in the Cotingidae.

I examined skulls of *Lipaugus vociferans* (AMNH 5074, 11450) and *L. unirufus* (UMMZ 153347), and I agree with Warter that they are not myiarchine. I have also concluded that syringes from *L. vociferans* (AMNH 2385, 7701; LSU 102270, 102273), *L. subalaris* (FMNH 290407), and *L. unirufus* (AMNH 6655) are not myiarchine (Fig. 6: syrinx 1). In referring to the nest of *Lipaugus*, Snow (1973:8–9) cited references to "a minute stick platform such as other medium-large cotingas (*Xipholena, Procnias*) build, but no tyrannids so far as known."

Lipaugus is not a myiarchine flycatcher, but I am unable to suggest even a tentative position for it.

CASIORNIS

The two allopatric and possibly conspecific forms of *Casiornis* (D. W. Snow, *in* Traylor 1979:191) are little known birds of the campos and caatinga of central and eastern South America. Traditionally the genus has been placed in the Cotingidae close to *Attila, Laniocera,* and *Rhytipterna*. Ames included *Casiornis* among his group of "cotingids" with a *Myiarchus*-like syrinx, whose transfer he recommended to the Tyrannidae. Ames' recommendation was followed by Meyer de Schauensee (1970), Snow (1973), and Traylor (1977).

Helmut Sick (pers. comm.) found *Casiornis rufa* nesting in a "hollow in a tree," about 1.5 m above ground, but was unable to examine the contents of the cavity. The only other reference (called to my attention by Sick) to nesting behavior in *Casiornis* is the report (Eisentraut 1935) of *rufa* using a knothole, 1.2 m above ground, but the eggs and nest are not described.

Warter did not examine a skull of *Casiornis*. I examined one skull each of *rufa* (AMNH 6697) and of *fusca* (AMNH 12665) and find that both share the derived state of the nasal capsule characteristic of *Myiarchus* (Fig. 3: skull 2). It would be impossible to separate these two skulls from a series of skulls of *Myiarchus* species of comparable size.

I borrowed the syrinx of *Casiornis rufa* (USNM 227189) that was the basis for Ames' recommendation and, in addition, examined two syringes from *C. fusca* (AMNH 12665; MPEG 34416). I concur with Ames that the syrinx of *Casiornis* is remarkably similar to that of *Myiarchus* (Fig. 4: syrinx 3). The syrinx of *rufa* would not be separable from a series of syringes of *Myiarchus* species of comparable size. In both of the *fusca* syringes, the medial sections of the complete A2 (one specimen) or A3 (the other specimen) elements, to which the large internal L-shaped cartilages attach, are completely calcified, rather than being cartilaginous as in the single specimen of *rufa* and in *Myiarchus*. Otherwise, they too would be inseparable from my sample of *Myiarchus* syringes.

Undoubtedly, *Casiornis* is a very near relative of *Myiarchus*. The only remaining question concerns the possible merger of the two genera.

RHYTIPTERNA

Another genus for which there is little information from life is *Rhytipterna*, which is distributed throughout the tropical zones of Central and South America. One of the three species, *R. immunda*, can be mistaken easily for the species of *Myiarchus* with which it is sympatric (Lanyon and Fry 1973). Historically the position of *Rhytipterna* within the Cotingidae, close to *Attila*, *Lipaugus*, and *Casiornis*, was accepted. Then Warter reported (p. 137) that the skulls of *Rhytipterna holerythra* and *R. simplex* are "virtually indistinguishable from those of *Sirystes*, *Myiarchus*, and *Eribates*" and recommended that the genus "be transferred to the Tyrannidae and placed near *Myiarchus*, which it most resembles." Ames reached a similar conclusion from examination of syringes of *R. holerythra*, and the recommendation of transfer was followed by Meyer de Schauensee (1970), Wetmore (1972), Snow (1973), and Traylor (1977).

I have examined the skulls of all three species of *Rhytipterna* (*holerythra*, UMMZ 153350; *simplex*, AMNH 10485, 11439, 11597; and *immunda*, AMNH 11452, 11617, 12080, 12081) and concur with Warter's conclusion. As with *Sirystes* and *Casiornis*, it would be impossible to separate any skulls of *Rhytipterna* that I examined from a series of skulls of *Myiarchus* species of comparable size (Fig. 3: skull 4).

My sample of 15 *Rhytipterna* syringes included all three species, *holerythra* (AMNH 6720, 8227, 8260; PMNH 1074, 1092; LSU 108458), *simplex* (AMNH 8106, 8107, 8129), and *immunda* (AMNH 8105, 8108, 8109, 8110, 8130; MPEG 34422). I agree with Ames that these syringes are remarkably similar to those of *Myiarchus*; some would not be separable from a series of syringes of *Myiarchus* species of comparable size (Fig. 4; syrinx 4). As noted in *Casiornis*, the species of *Rhytipterna* vary with respect to the medial sections of the complete A2 or A3 elements, to which the large internal L-shaped cartilages attach; in the three specimens of *simplex* these structures are completely cartilaginous (as in *Casiornis rufa* and my entire sample of *Myiarchus*), whereas in the six specimens of *immunda*, they are fully or mostly calcified (as in *Casiornis fusca*). In four specimens of *holerythra* they were completely cartilaginous and in two other specimens they are partly calcified. In all other respects the syringes of *Rhytipterna* are not separable from my series taken from *Casiornis* and *Myiarchus*.

UNSUSPECTED NEAR RELATIVES OF *MYIARCHUS*

In the course of my examination of tyrant flycatcher skulls and syringes, I looked for evidence of the independent development in unrelated lineages of the same derived character states that I have come to associate with *Myiarchus* and its near relatives. I wished to consider whether these character states were truly unique or appeared in other tyrant taxa as a result of convergence or parallelism.

My first indication that the derived character state that I described for the nasal capsule of *Myiarchus* and near relatives might not be unique to those genera came with a routine examination of the skull of *Ramphotrigon fuscicauda* (FMNH 290293). I was surprised to find that its nasal capsule had all the myiarchine characteristics. In the literature it had been linked only with the "flatbills" (*Platyrinchus*, *Rhynchocyclus*, *Tolmomyias*, and *Cnipodectes*) in a different subfamily (Hellmayr 1927; Traylor 1979). Shortly thereafter I received information from T. A. Parker (pers. comm.) about nests of all three species of *Ramphotrigon*. The nests were located in natural tree-cavities, and he described the eggs of one species (*fuscicauda*) as being like those of *Myiarchus*. None of the nests had a soft lining of fur and feathers.

This information from Parker rekindled my interest in *Ramphotrigon*, and I borrowed other skulls of *fuscicauda* (LSU 101509, 101510, 101511) as well as skulls of *ruficauda* (LSU 62765, 86571), and *megacephala* (LSU 101512). All three species have a myiarchine nasal capsule (Fig. 3: skull 6). That *megacephala* has a nasal capsule like those of the other two species is of special significance because Zimmer (1939) had transferred *megacephala* from *Tolmomyias* on the basis of rather equivocal considerations of external morphology. However, the nasal capsule of *Tolmomyias* (*flaviventris*, AMNH 9138; *sulphurescens*, AMNH 10307), as in the other genera of "flatbills," is of the presumed primitive type (i.e., unossified, with only a dorsal remnant of the nasal septum), and quite unlike that of *Ramphotrigon*. Warter did not examine *Ramphotrigon*.

Next I examined the syringes of *Ramphotrigon fuscicauda* (LSU 102603, 102604, 102608),

R. megacephala (LSU 102610; FMNH 291696) and *R. ruficauda* (LSU TJD 948). In these specimens the B1 and B2 elements are like those of *Myiarchus*; the large dorsal cartilages are straight, not L-shaped as in *Myiarchus,* and expanded distally (Fig. 4: syrinx 6). The proximal half of each of these dorsal cartilages is flattened at right angles to the internal tympaniform membrane, whereas the distal half is flattened within the plane of that membrane, lending the cartilage a somewhat twisted aspect. The dorsal cartilages are attached to the dorsal ends of the incomplete A2 elements and to the dorso-medial segments of the A3 elements, which are complete and ossified throughout. The remainder of the supporting elements of the tracheo-bronchial junction are largely ossified as well. A second pair of internal cartilages is located ventrally in the internal tympaniform membrane between the distal ends of the large cartilages and the ventral ends of the B2 and B3 elements, as in *Myiarchus.* Only the degree of ossification of the elements of the tracheo-bronchial junction varies, ranging from no to partial ossification. The only other species of flycatcher with a syrinx like this is *Deltarhynchus flammulatus.* The similarity in the syringes of *Ramphotrigon* and *Deltarhynchus* is remarkable (Fig. 4: syringes 5, 6), the only apparent difference being the absence or inconspicuousness of the smaller pair of ventral cartilages in *Deltarhynchus.*

Ames did not examine a syrinx of *Deltarhynchus,* but he commented on the striking differences in the syringes of *Tolmomyias sulphurescens* and *T. megacephalus,* offering further support for Zimmer's (1939) transfer of *megacephalus* out of *Tolmomyias,* at least. I examined the syringes of a number of the so-called "flatbills" to which *Ramphotrigon* has formally been allied (*Platyrinchus cancrominus, P. mystaceus, P. coronatus, P. platyrhynchos, Rhynchocyclus brevirostris, R. olivaceus, Tolmomyias sulphurescens, T. assimilis, T. poliocephalus, T. flaviventris,* and *Cnipodectes subbrunneus*) and found none with a syrinx remotely like that of *Ramphotrigon.*

I am completely confident that *Ramphotrigon* belongs with *Myiarchus* and its near relatives, on the basis of the shared derived character state of the nasal septum (unique to this assemblage) and the derived behavioral pattern of nesting in tree-cavities.

No other cases of heretofore unsuspected relatives of *Myiarchus* were uncovered in my general examination of tyrant flycatcher skulls and syringes.

PHYLOGENETIC RELATIONSHIPS OF *MYIARCHUS* AND NEAR RELATIVES

In constructing a working hypothesis of the relationships of *Myiarchus* and its near relatives, I omitted consideration of *Laniocera, Lipaugus,* and *Nesotriccus.* For reasons expressed above, we can conclude that none of these genera is a close relative of *Myiarchus.* The reader is encouraged to refer to Table 1 and to the phylogenetic diagram in Fig. 7 while the hypothesis is developed below.

Five genera of tyrant flycatchers, *Sirystes, Casiornis, Rhytipterna, Ramphotrigon,* and *Deltarhynchus,* possess the derived character state of the nasal septum that I identified as part of the generic diagnosis for *Myiarchus* (character state A in Table 1, Fig. 7; Figs. 1, 3). Although I have yet to examine the skulls of two of the 90 tyrant flycatcher genera and seven of the 42 cotingid and piprid genera, it seems highly unlikely that any of these unexamined genera will be found to possess this character state. Individual specimens in other flycatcher genera, notably in some whose species have a poorly ossified septum, exhibit a suggestion of an internal supporting rod, indicating a proclivity toward the development of such a structure. In a few genera with a fully ossified septum bearing a conspicuous transverse trabecular plate, the internal supporting rod may be nearly as prominent as in *Myiarchus* and its near relatives. What distinguishes the latter group from all other flycatchers is the presence of the conspicuous internal supporting rod *in the absence* of any kind of trabecular plate, and the well ossified septum in a virtually fully ossified nasal capsule, including the alinasal turbinals and walls.

The uniqueness of character state A is my principal argument for considering these six genera (Fig. 7: bold face) a monophyletic group. They can be regarded collectively as the myiarchines, because *Myiarchus* is the oldest name and is the largest genus.

My secondary reason for believing the myiarchines to be a natural assemblage is that they are obligatory tree-cavity nesters (character state B), as far as is presently known. This argument is weakened by the lack of information on the nesting behavior of *Rhytipterna* and by the small sample sizes for all of the remaining genera except *Myiarchus.* I believe that ultimately all of the myiarchines will be found to nest in tree-cavities.

What is the evidence for regarding nesting in tree-cavities as a derived behavioral pattern within the Tyrannidae? Other than among the myiarchine flycatchers, nesting in tree-cavities

TABLE 1
Distribution of Character States

	Symbol used in Fig. 7	Tyranninae											Other subfamilies
		Myiarchines						Other tyrannines					
		Myiarchus	*Casiornis*	*Sirystes*	*Rhytipterna*	*Ramphotrigon*	*Deltarhynchus*	*Attila*	*Legatus*	*Myiodynastes*	*Conopias*	Other genera	
Skull													
Nasal septum													
Poorly ossified													Flatbills, tody-tyrants and allies
Moderately ossified, no trabecular plate													Tityrinae
Fully ossified, no trabecular plate, no conspicuous internal supporting rod	L	X	X	X	X	X	X	X	X	X	X	X	
with supporting rod	A	X	X	X	X	X	X						
Fully ossified, with trabecular plate													Most genera, including *Laniocera, Nesotriccus* and *Machetornis*
Medial frontal ridge	M							X	X	X	X		
Syrinx													
Large internal cartilages													
L-shaped, attached to ventral trach.-bron. jct.	D	X	X	X	X								
Straight, expanded distally, attached to dorsal trach.-bron. jct.	E					X	X						
B1, B2 elements as in *Myiarchus*	N	X	X	X	X	X	X	X	X				Several genera
Nesting behavior													
Obligatory tree-cavity nests	B	X	X	X	?[1]	X	X						*Tityra, Colonia*
Niche, crevice or tree-cavity								X		X	X		*Machetornis, Xolmis*
Fur and feathers in lining of tree-cavity nest	F?[1]	X	?	?	?								
Eggs like those of *Myiarchus*	C?	X	?	?	?	X	X			X			*Machetornis*

[1] Question marks indicate uncertainties due to inadequate or contradictory data.

has been reported for only seven genera of flycatchers. Three of these (*Attila, Myiodynastes,* and *Conopias*) currently are classified in the subfamily Tyranninae to which the myiarchines belong (Traylor 1977), but they can not be described as obligatory tree-cavity nesters. The single species of *Conopias* known to nest in tree-cavities, *parva,* may also use abandoned cacique (Icteridae) nests (Traylor and Fitzpatrick 1982). Nesting behavior is known for only one of the remaining species of *Conopias*; *inornata* builds a cup-shaped nest, saddled on a horizontal fork of tree branches (Thomas 1979). The species of *Myiodynastes* may nest in shallow holes in trees but are also known to locate their nests in niches or crevices.

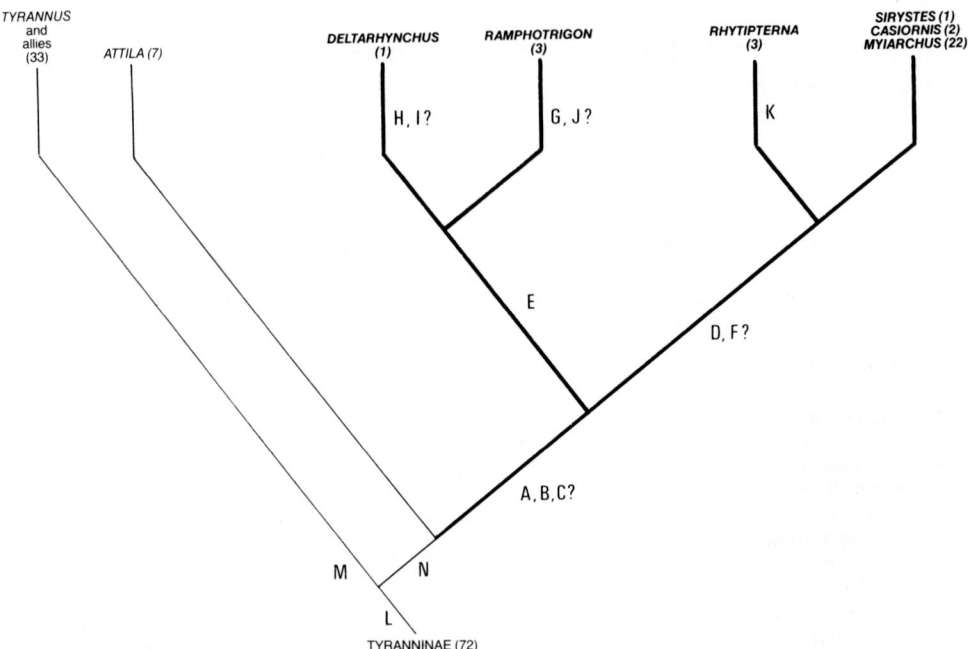

Fig. 7. Phylogenetic relationships of the genera of myiarchine flycatchers (bold face, thick lines) and their positions within the subfamily Tyranninae. Capital letters identify diagnostic character states described in the text and listed in Table 1. Numbers of species indicated in parentheses. Question marks indicate uncertainties due to inadequate or contradictory data.

The remaining four of these genera (*Machetornis, Xolmis, Colonia,* and *Tityra*) are assigned to other subfamilies within the Tyrannidae; clearly their hole-nesting behavior can be attributed to convergent or parallel evolution. Only *Colonia* and *Tityra* are obligatory tree-cavity nesters. Both *Machetornis* and *Xolmis* may use tree-cavities, but are more likely to use the enclosed mud or stick nests built by ovenbirds (Furnariidae) or to place their nests in crevices.

It is possible that egg coloration and pattern (character state C?) may be useful in diagnosing the myiarchines, but data are insufficient and contradictory. Outside of the genus *Myiarchus*, we currently have information on only one set of *Deltarhynchus* eggs and one set of eggs of *Ramphotrigon fuscicauda*; both of these egg sets were like those of *Myiarchus*. Of the seven genera of non-myiarchine hole-nesting tyrants identified above, only *Myiodynastes* and *Machetornis* have eggs colored and patterned as in *Myiarchus*. It could be argued for *Myiodynastes,* a member of the same subfamily as the myiarchines, that the tendency toward nesting in tree-cavities and the similarity in the eggs can be attributed to an older relationship between that genus and the myiarchines, prior to the development of the derived state of the nasal septum of the latter. The eggs of *Machetornis* are remarkably similar to those of *Myiarchus* (Naumburg 1930:264; AMNH egg set 13896), and Fitzpatrick (1978) argued that this monotypic South American genus should be placed in the subfamily Tyranninae (near the kingbirds), on the basis of behavior and external morphology. Nevertheless, the nasal septum of *Machetornis* has a well developed trabecular plate (Fig. 5: skull 2), and the internal cartilages of the syrinx are very different from those of any member of the subfamily Tyranninae (Fig. 6: syrinx 4). I would argue that convergent or parallel evolution best explains this character contradiction.

The morphology of the syrinx of tyrannids is less conservative than the morphology of the nasal capsule and shows more variation among closely related genera than does the latter. This is fortunate, because within the myiarchine assemblage, as defined by the nasal capsule, two unique suites of syringeal characters suggest a fundamental dichotomy. The genera *Myiarchus, Casiornis, Sirystes,* and *Rhytipterna* constitute one branch and are characterized by having large dorsal internal cartilages that are L-shaped and connected anteriorly to the medioventral segment of the tracheo-bronchial junction (character state D; Fig. 4: top and middle rows). The other branch of myiarchines, consisting of *Deltarhynchus* and *Ramphotrigon*, has large internal cartilages that are straight, expanded distally, twisted, and connected to the dorso-medial segment of the tracheo-bronchial junction and to the dorsal ends of the incomplete A2 elements (character state E; Fig. 4: bottom row). The contention that *Deltarhynchus* and *Ramphotrigon* are each other's closest relative is also supported by external morphology. Their bills are very similar, which is not surprising considering the etymology of the two generic names, and the flammulations on the breast of *Deltarhynchus* are even more pronounced in two of the three species of *Ramphotrigon*.

The nature of the materials used to line the nest-cavity may provide a second basis for this dichotomy within the myiarchines and lend credence to the division defined by syringeal morphology. None of the four cavity-nests reported for *Deltarhynchus* and *Ramphotrigon* had a soft lining of fur and feathers, in marked contrast to the nests of *Myiarchus* (character state F?). No data on nest lining of *Rhytipterna*, *Sirystes*, or *Casiornis* are, as yet, available.

The limits of some genera within the myiarchine assemblage need further clarification. Differences in the prominence of the second pair of internal cartilages in the syrinx are sufficient grounds for maintaining *Deltarhynchus* and *Ramphotrigon* as separate genera; these small cartilages in *Ramphotrigon* are as conspicuous as they are in *Myiarchus* (character state G), whereas in *Deltarhynchus* they are barely discernible or absent (character state H). Further field work may reveal consistent differences in the material used by these two genera to line their nest-cavities, but present data are insufficient and contradictory. My single nest of *Deltarhynchus* consisted entirely of vegetable fibers and small fragments of dried leaves and shredded bark (character state I?), whereas Parker reported (pers. comm.) that *Ramphotrigon* nests "consisted entirely of shiny black and blonde mammal hairs," in one instance, and of vegetable fibers and twigs in another (character state J?).

Within the other branch of the assemblage, an argument could be made for keeping *Rhytipterna* generically separate from *Myiarchus* on the basis of a unique tarsal scutellation; the recurved margins of the plantar scutes give the upper portion of the posterior edge of the tarsus a serrated appearance (character state K). Merger of the single species of *Sirystes* and the two species of *Casiornis* with *Myiarchus*, acceptable on the basis of shared derived syringeal morphology, must await determination of the material with which these three species line their nest-cavities.

The sister group of the myiarchine assemblage may be *Attila*, although this suggestion must remain tentative until we have more conclusive evidence of the monophyly of the subfamily Tyranninae (*sensu* Traylor 1977). *Attila* does not share with the myiarchines the derived state of the nasal septum (compare Fig. 5: skull 6 with the skulls in Fig. 3) or the obligatory tree-cavity nesting behavior of that group, its skull does appear to be tyrannine. Warter (p. 141) described it as "essentially tyrannine" but preferred to maintain it in its own subfamily because of the less completely ossified interorbital septum and the incomplete (less ossified) nasal septum. In my sample of *Attila* skulls I saw no evidence that the nasal septum is less ossified than in the skulls of *Tyrannus* and its allies. I have not discussed the interorbital septum as a useful skull character for it is much more variable than Warter's study indicated. Warter characterized the "tyranno-myiarchine" skull as having the Type 1 interorbital septum, whereas in my survey I found the Type 2 of Warter (less ossified than Type 1) occasionally in *Pitangus*, *Legatus*, *Myiarchus*, and other tyrannine genera. All of the *Deltarhynchus* and *Ramphotrigon* skulls that I examined had interorbital septa that would be classified as Type 2 (Warter lacked these genera). The *Attila* skulls in my sample have Type 2 interorbital septa, but in no specimen is the septum less ossified than in any of the *Deltarhynchus* or *Ramphotrigon* skulls. I conclude from this that the state of the interorbital septum is too variable to be useful in defining the limits of the "tyranno-myiarchine" skull in any precise manner, and that no reason exists to exclude *Attila* from the subfamily Tyranninae.

Monophyly of the tyrannines appears to rest on a derived state of the nasal septum in which the septum is fully ossified and lacks a transverse trabecular plate (character state L; Figs. 1, 3, 5: skulls 4, 5, 6). The two major lineages within the subfamily lead to the kingbirds on one hand and to the myiarchines on the other. All of the genera in the kingbird lineage share a derived cranial character: the presence of a conspicuous medial ridge in the frontal region of the skull (Lanyon, unpubl. data; character state M; Fig. 8: skulls 1, 2). *Attila*, the myiarchines, and all other flycatchers lack a medial frontal ridge (Fig. 8: skulls 3–6). Providing the subfamily is monophyletic, the absence of this ridge in *Attila* argues against its alliance with *Tyrannus* and allies and for its placement at the base of the myiarchine lineage.

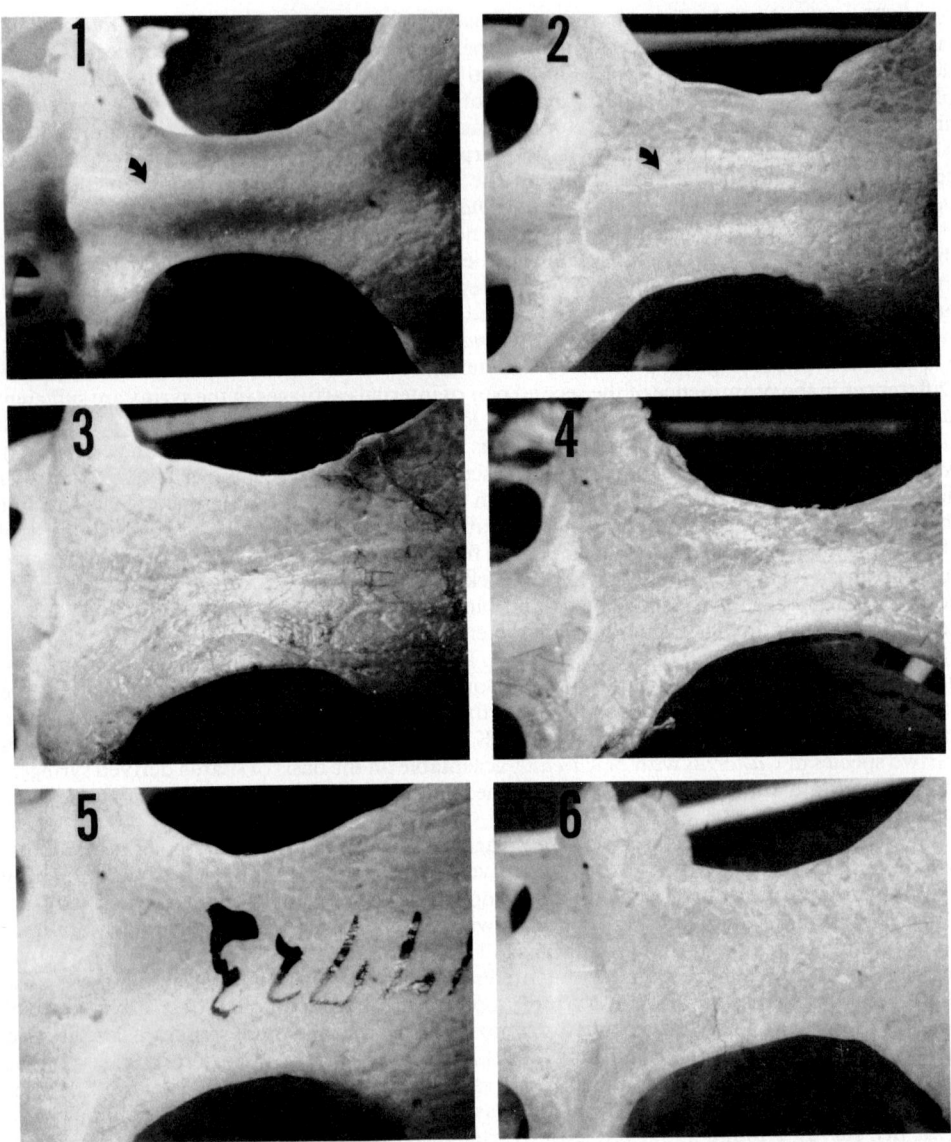

Fig. 8. Photographs taken through a dissecting microscope (magnification = 6×), of the frontal region of the skulls of various tyrannine flycatchers (anterior end of skull to left). The kingbirds and their allies have a conspicuous medial ridge, indicated by arrows: (1) *Myiozetetes similis*, AMNH 7173, (2) *Myiodynastes luteiventris*, AMNH 11388. *Attila* and the myiarchine flycatchers lack this ridge: (3) *Attila spadiceus*, AMNH 11447, (4) *Deltarhynchus flammulatus*, AMNH 11391, (5) *Sirystes sibilator*, USNM 347723, (6) *Myiarchus tuberculifer*, AMNH 7175.

That the affinities of *Attila* are with the myiarchines rather than with the kingbirds is suggested also by the configuration of the B1 and B2 supporting elements of the syrinx. In *Attila* these elements are shaped as they are in the myiarchines (character state N; Figs. 2, 4, 6: syrinx 6), not as they are in the kingbirds and their relatives (Fig. 6: syrinx 5). Caution must be exercised in the use of this syringeal character, however, for B1 and B2 elements with similar configurations have evolved independently within the various tyrannid subfamilies. Its value would appear to be limited to a consideration of relationships within lineages established by unique character states. The only contradiction of this character among the tyrannines is that *Legatus,* an ally of the kingbirds on the basis of the medial frontal ridge, has B1 and B2 elements like those of *Attila* and the myiarchines. *Legatus* is of special interest for other reasons, including the unique modification of the tyrannine nasal septum, in which "wings" project laterally from the septum (Fig. 5: skull 4), and will be discussed in more detail elsewhere (Lanyon, unpubl. data).

The data base used here to develop a phylogeny of the myiarchine flycatchers is limited, in that only two morphological complexes and one behavioral pattern have been examined in detail. The anatomical data are adequate and convincing, in my opinion, but still lacking or scarce are observations of nesting behavior and eggs for most of the genera. It remains to be seen whether my hypotheses of relationships within this assemblage, and between it and other tyrannines, will be supported by other data.

ACKNOWLEDGMENTS

I am grateful to the following individuals for arranging loan of specimens under their care: J. W. Fitzpatrick, L. F. Kiff, F. C. Novaes, S. L. Olson, K. C. Parkes, J. V. Remsen, C. G. Sibley, and R. W. Storer. Special thanks go to members of the field parties from the Museum of Zoology, Louisiana State University, for valuable specimens obtained in Bolivia, Peru, and Panama, to P. L. Ames for the loan of critical specimens, and to B. T. Thomas for obtaining needed specimens.

T. A. Parker, H. Sick, J. W. Fitzpatrick, and M. D. Williams shared with me their knowledge of these flycatchers, and Williams kindly permitted reproduction of one of his outstanding photographs. My wife Vicky has assisted me in countless ways, including with much of the field work in Middle and South America. I have benefited from the suggestions of G. F. Barrowclough, M. K. Hecht, S. M. Lanyon, M. C. McKitrick, K. C. Parkes, and S. L. Warter who critically read earlier drafts of the manuscript. The photographs of skulls and syringes were made with equipment in the laboratory of C. W. Myers.

I regret that Eugene Eisenmann did not live to share the genuine satisfaction that came with constructing a hypothesis of relationships among these neotropical flycatchers. As a colleague "just down the hall," Gene was a continual source of information, inspiration, and constructive criticism during most of this research; I missed him dearly during the last years of the study.

LITERATURE CITED

AMES, P. L. 1971. The morphology of the syrinx in passerine birds. Peabody Mus. Nat. Hist. Yale Univ. Bull. 37:1–194.

BANGS, O., and T. E. PENARD. 1921. Notes on some American birds, chiefly neotropical. Bull. Mus. Comp. Zool. 64:365–397.

BOND, J. 1956. Check-list of Birds of the West Indies. 4th ed. Academy Natural Sciences, Philadelphia, Pennsylvania.

DINGERKUS, G., AND L. D. UHLER. 1977. Enzyme clearing of alcian blue stained whole small vertebrates for demonstration of cartilage. Stain Technol. 52:229–232.

EISENMANN, E. 1955. The species of Middle American birds. Trans. Linn. Soc. N.Y. 7:1–128.

EISENTRAUT, M. 1935. Biologische Studien im bolivianischen Chaco. VI. Beitrag zur Biologie der Vogelfauna. Mitt. Zool. Mus. Berl. 20:367–443.

FITZPATRICK, J. W. 1978. Foraging behavior and adaptive radiation in the avian family Tyrannidae. Unpubl. Ph.D. dissert. Princeton University, Princeton, New Jersey.

HELLMAYR, C. E. 1927. Catalogue of birds of the Americas and the adjacent islands, part 5. Field Mus. Nat. Hist. Publ. Zool. Ser., Vol. 13.

LANYON, W. E. 1967. Revision and probable evolution of the *Myiarchus* flycatchers of the West Indies. Bull. Am. Mus. Nat. Hist. 136:329–370.

LANYON, W. E. 1975. Behavior and generic status of the Rufous Flycatcher of Peru. Wilson Bull. 87: 441–455.

LANYON, W. E. 1978. Revision of the *Myiarchus* flycatchers of South America. Bull. Am. Mus. Nat. Hist. 161:427–628.

LANYON, W. E. 1982. Behavior, morphology, and systematics of the Flammulated Flycatcher of Mexico. Auk 99:414–423.

LANYON, W. E. 1984. The systematic position of the Cocos Island Flycatcher. Condor 86:42–47.

LANYON, W. E., AND J. W. FITZPATRICK. 1983. Behavior, morphology, and systematics of *Sirystes sibilator* (Tyrannidae). Auk 100:98–104.

LANYON, W. E., AND C. H. FRY. 1973. Range and affinity of the Pale-bellied Mourner (*Rhytipterna immunda*). Auk 90:672–674.

LAWRENCE, G. N. 1875. Descriptions of two new species of birds of the families Tanagridae and Tyrannidae. Ann. Lyc. Nat. Hist. N.Y. 11 July 1874, pp. 70–72.

MEYER DE SCHAUENSEE, R. 1970. A Guide to the Birds of South America. Livingston Publ. Co., Wynnewood, Pennsylvania.

NAUMBURG, E. M. B. 1930. The birds of Matto Grosso, Brazil. Bull. Am. Mus. Nat. Hist. 60:1–432.

RIDGWAY, R. 1893. Remarks on the avian genus *Myiarchus*, with special reference to *M. yucatanensis* Lawrence. Proc. U.S. Natl. Mus. 16:605–608.

RIDGWAY, R. 1905. Descriptions of some new genera of Tyrannidae, Pipridae, and Cotingidae. Proc. Biol. Soc. Wash. 18:207–210.

RIDGWAY, R. 1907. The birds of North and Middle America, pt. IV. Bull. U.S. Natl. Mus. 50.

SCLATER, P. L. 1888. Catalogue of birds in the British Museum, vol. 14. British Mus. (Nat. Hist.), London.

SNOW, D. W. 1973. The classification of the Cotingidae (Aves). Breviora No. 409.

SWARTH, H. S. 1931. The avifauna of the Galapagos Islands. Occas. Pap. Calif. Acad. Sci. 18.

THOMAS, B. T. 1979. Behavior and breeding of the White-bearded Flycatcher (*Conopias inornata*). Auk 96:767–775.

TODD, W. E. C. 1922. Studies in the Tyrannidae. III. The South American forms of *Myiarchus*. Proc. Biol. Soc. Wash. 35:181–218.

TRAYLOR, M. A. JR. 1977. A classification of the tyrant flycatchers (Tyrannidac). Bull. Mus. Comp. Zool. 148:129–184.

TRAYLOR, M. A. JR. 1979. Tyrannidae. Pp. 1–228, *In* M. A. Traylor, Jr. (ed.), Peters' Check-list of Birds of the World, Vol. 8. Museum Comparative Zoology, Cambridge, Massachusetts.

TRAYLOR, M. A. JR., AND J. W. FITZPATRICK. 1982. A survey of the tyrant flycatchers. Living Bird 19: 7–50.

WARTER, S. L. 1965. The cranial osteology of the New World Tyrannoidea and its taxonomic implications. Ph.D. dissertation, Louisiana State University. Univ. Microfilms, Ann Arbor, Michigan, order no. 66-761.

WETMORE, A. 1972. The birds of the Republic of Panama, Pt. 3. Smithson. Misc. Collect. 150(3).

ZIMMER, J. T. 1936. Studies of Peruvian birds. No. 23. Notes on *Doliornis, Pipreola, Attila, Laniocera, Rhytipterna*, and *Lipaugus*. Am. Mus. Novit. No. 893.

ZIMMER, J. T. 1938. Studies of Peruvian birds. No. 29. The genera *Myiarchus, Mitrephanes*, and *Cnemotriccus*. Am. Mus. Novit. No. 994.

ZIMMER, J. T. 1939. Studies of Peruvian birds. No. 33. The genera *Tolmomyias* and *Rhynchocyclus* with further notes on *Ramphotrigon*. Am. Mus. Novit. No. 1045.

ZIMMER, J. T. Unpubl. ms. Notes on the Tyrannidae. Files Dept. Ornithology, Am. Mus. Nat. Hist., New York.

PLATE V.   White-capped Tanager (*Sericossypha albocristata*). From a painting by Roger Tory Peterson.

SYSTEMATIC RELATIONS OF
SERICOSSYPHA ALBOCRISTATA
(THRAUPINAE)

JOHN J. MORONY, JR.

Museum of Zoology, Louisiana State University, Baton Rouge, Louisiana 70893-3216 USA;
present address: Wildlife and Fisheries Commission, 400 Royal Street,
New Orleans, Louisiana 70130 USA

ABSTRACT. *Sericossypha albocristata* has consistently been placed in the Thraupinae, a taxon of New World nine-primaried oscines, in all ornithological works concerned with South America. Doubt regarding this arrangement has persisted, however, with opinion generally favoring placement in the Cotingidae, a suboscine taxa. I analyzed six skeletal characters, five cranial and one postcranial, of *S. albocristata* in a survey of the Cotingidae and Thraupinae. In all instances *S. albocristata* possessed structural characters similar to the Thraupinae but not the Cotingidae, and no characters shared with cotingas were not likewise shared with tanagers. The nine-primaried condition of *S. albocristata* is an additional character shared with all tanagers and differing from the contingas, which possess a functional tenth primary. While this analysis establishes the relationship of *S. albocristata* to the New World nine-primaried oscines, it does not provide a basis for determining its relationship to the tanagers. A new bony structure, the Processus manubrii, of the sternum that was not found in any other species surveyed is described for *S. albocristata*.

RESUMEN. En todos los trabajos ornitológicos concernientes a Sudamérica se ha considerado repitadamente a *Sericossypha albocristata* miembro de la subfamilia Thraupinae, un taxón del Nuevo Mundo de los oscines de nueve primarias. Sin embargo han persistido dudas respecto a este arreglo, con opiniones generalmente favorables a su colocación taxonómica con Cotingidae, una taxa suboscine. He analizado seis caracteres óseos, cinco craniales y uno postcranial, en *S. albocristata* en un reconocimiento de los Cotingidae y Thraupinae. En todas las instancias *S. albocristata* poseeyó carácteres estructurales similares a los de Thraupinae, pero no a los de Cotingidae, y no hubo carácteres compartidos con las cotingas que no fueran también compartidos con las tanagaras. La condición de nueve primarias que se encuentra en *S. albocristata* es un carácter adicional compartido con todas las tangaras y diferenciante de las cotingas las cuales que poseen una decima primaria funcional. Si bién este análisis establece la relación de *S. albocristata,* con los oscines de nueve primarias del Nuevo Mundo, no provee una base para su relación con las tanagaras. Una nueva estructura ósea del esternón, el Processus Manubrii, que no se encuentra en ninguna otra especie inspeccionada, es descrito para *S. albocristata*.

The White-capped Tanager, *Sericossypha albocristata,* first described in 1884 by Lafresnaye, has been consistently placed among the thraupines in all major ornithological works on South America including that of Sclater (1886). Doubt regarding this arrangement has persisted, however, with opinion generally favoring placement of *S. albocristata* in the Cotingidae. The nine-primaried condition of *S. albocristata* is the principal reason given for retaining this species in the Thraupinae; all suboscines, including the Cotingidae, have the outermost or tenth primary feather functional.

Storer (1970) listed *S. albocristata* as a monotypic species and genus adjacent to *Compsothraupis loricata.* Some authorities, for example Meyer de Schauensee (1966), regarded these two species as congeneric despite marked differences in plumage pattern and color.

Zimmer (1947:1) contended that *S. albocristata* ". . . is in need of internal examination which may confirm its place in the Thraupidae or show that it has other affinities elsewhere." Zimmer mentioned some similarities to the Icteridae, but he also mentioned that general appearance is cotinga-like. Meyer de Schauensee (1966:492) also contended that *Sericossypha* is in need of investigation being "Most unlike a tanager in appearances; a position among cotingas, which they rather resemble in pattern, has been suggested . . . ." Storer (1970:253) stated that "Allocation [of *Sericossypha*] to the tanagers requires confirmation." Raikow (1978: 30) in his study of the appendicular myology and relationships of the New World nine-primaried oscines commented that *S. albocristata* was the most aberrant of the tanagers he dissected. He concluded that the ". . . pattern of peculiar features does not resemble any other form studied, and so its relationship remains obscure."

What is most striking about the appearance of the plumage of *S. albocristata* is the color pattern rather than the color *per se* (Plate V). The brilliant white cap, found in no tanager or cotinga, contrasts markedly with the dark maroon throat and breast and the otherwise black plumage. The wings and tail are glossy contrasting with the duller black of the contour feathers. Aside from the white cap, this plumage color and pattern is quite similar to that of *Ramphocelus carbo*. This maroon color of *S. albocristata* is found among tanagers only in *R. carbo* and is not characteristic of any suboscine.

Examination of a skeleton of *S. albocristata* has provided much additional information regarding its proper taxonomic placement. Currently, one skeletal and two spirit specimens are available, all deposited in the Museum of Zoology, Louisiana State University. The specimen data for the skeleton records a male with slightly enlarged testes. The skull appears completely ossified. This species has a restricted distribution in South America and has proven difficult to collect because of its elusive behavior and the difficult nature of the terrain in which it occurs.

METHODS AND MATERIALS

Skeletons of 43 of the 57 genera of Thraupinae listed by Storer (1970) were surveyed: *Schistochlamys* (*melanopis*), *Conothraupis* (*speculigera*), *Cissopis* (*leveriana*), *Chlorornis* (*riefferii*), *Sericossypha* (*albocristata*), *Chlorospingus* (*ophthalmicus, inornatus, zeledoni, pileatus, parvirostris, flavigularis*), *Cnemoscopus* (*rubrirostris*), *Hemispingus* (*atropileus, calophrys, superciliaris, frontalis, melanotis, verticalis, xanthophthalmus, trifasciatus*), *Thlypopsis* (*ornata, pectoralis, sordida, ruficeps*), *Hemithraupis* (*flavicollis*), *Chrysothlypis* (*chrysomelas*), *Nemosia* (*pileata*), *Mitrospingus* (*cassinii*), *Chlorothraupis* (*carmioli, olivacea, stolzmanni*), *Eucometis* (*penicillata*), *Lanio* (*fulvus, versicolor, aurantius*), *Creurgops* (*verticalis*), *Tachyphonus* (*cristatus, rufiventer, surinamus, luctuosus, delatrii, coronatus, rufus, phoenicius*), *Trichothraupis* (*melanops*), *Habia* (*rubica, fuscicauda, atrimaxillaris*), *Piranga* (*bidentata, flava, rubra, olivacea, ludoviciana, leucoptera*), *Calochaetes* (*coccineus*), *Ramphocelus* (*sanguinolentus, nigrogularis, dimidiatus, melanogaster, carbo, bresilius, passerinii, flammigerus*), *Spindalis* (*zena*), *Thraupis* (*episcopus, sayaca, palmarum, cyanocephala, bonariensis*), *Buthraupis* (*montana, eximia*), *Wetmorethraupis* (*sterrhopteron*), *Anisognathus* (*lacrymosus, igniventris, flavinuchus*), *Stephanophorus* (*diadematus*), *Iridosornis* (*analis, jelskii, rufivertex*), *Dubusia* (*taeniata*), *Delothraupis* (*castaneoventris*), *Pipraeidea* (*melanonota*), *Euphonia* (*affinis, luteicapilla, chlorotica, violacea, laniirostris, hirundinacea, chrysopasta, mesochrysa, minuta, anneae, xanthogaster, rufiventris*), *Chlorophonia* (*cyanea, occipitalis*), *Chlorochrysa* (*calliparaea*), *Tangara* (*mexicana, chilensis, seledon, schrankii, florida, arthus, icterocephala, xanthocephala, chrysotis, parzudakii, xanthogastra, punctata, guttata, gyrola, cayana, ruficervix, cyanotis, cyanicollis, larvata, nigrocincta, dowii, nigroviridis, vassorii, viridicollis, argyrofenges, pulcherrima, velia, callophrys*), *Dacnis* (*lineata, venusta, cayana*), *Chlorophanes* (*spiza*), *Cyanerpes* (*nitidus, caeruleus, cyaneus*), *Xenodacnis* (*parina*), *Oreomanes* (*fraseri*), *Diglossa* (*plumbea, baritula, lafresnayii, carbonaria, albilatera, glauca, caerulescens, cyanea*), *Tersina* (*viridis*).

Skeletons of 17 of the 27 genera of Cotingidae listed by Snow (1979) were surveyed: *Phoenicircus* (*nigricollis*), *Ampelion* (*rubrocristata, rufaxilla, sclateri, stresemanni*), *Pipreola* (*riefferii, intermedia, arcuata, aureopectus, frontalis, chlorolepidota*), *Ampelioides* (*tschudii*), *Iodopleura* (*isabellae*), *Lipaugus* (*subalaris, cryptolophus, vociferans, unirufus*), *Pachyramphus* (*viridis, versicolor, spodiurus, castaneus, homochrous, cinnamomeus, polychopterus, marginatus, albogriseus, minor*), *Tityra* (*cayana, semifasciata, inquisitor*), *Porphyrolaema* (*porphyrolaema*), *Cotinga* (*cayana*), *Conioptilon* (*mcilhennyi*), *Gymnoderus* (*foetidus*), *Querula* (*purpurata*), *Pyroderus* (*scutatus*), *Cephalopterus* (*ornatus*), *Rupicola* (*peruviana*).

No skeletal material of *Compsothraupis loricata*, the presumed nearest relative of *S. albocristata*, exists in collections.

Illustrations were drawn by the author using a Wild dissecting scope and camera lucida.

SKELETAL ANATOMY

The term "thraupine-like," where employed herein, is not intended to imply that such characters are restricted to the Thraupinae. For the most part they are shared by other members of the New World nine-primaried oscines, especially the Cardinalinae, Emberizinae, and Parulinae of the Emberizidae (*sensu* Storer 1970).

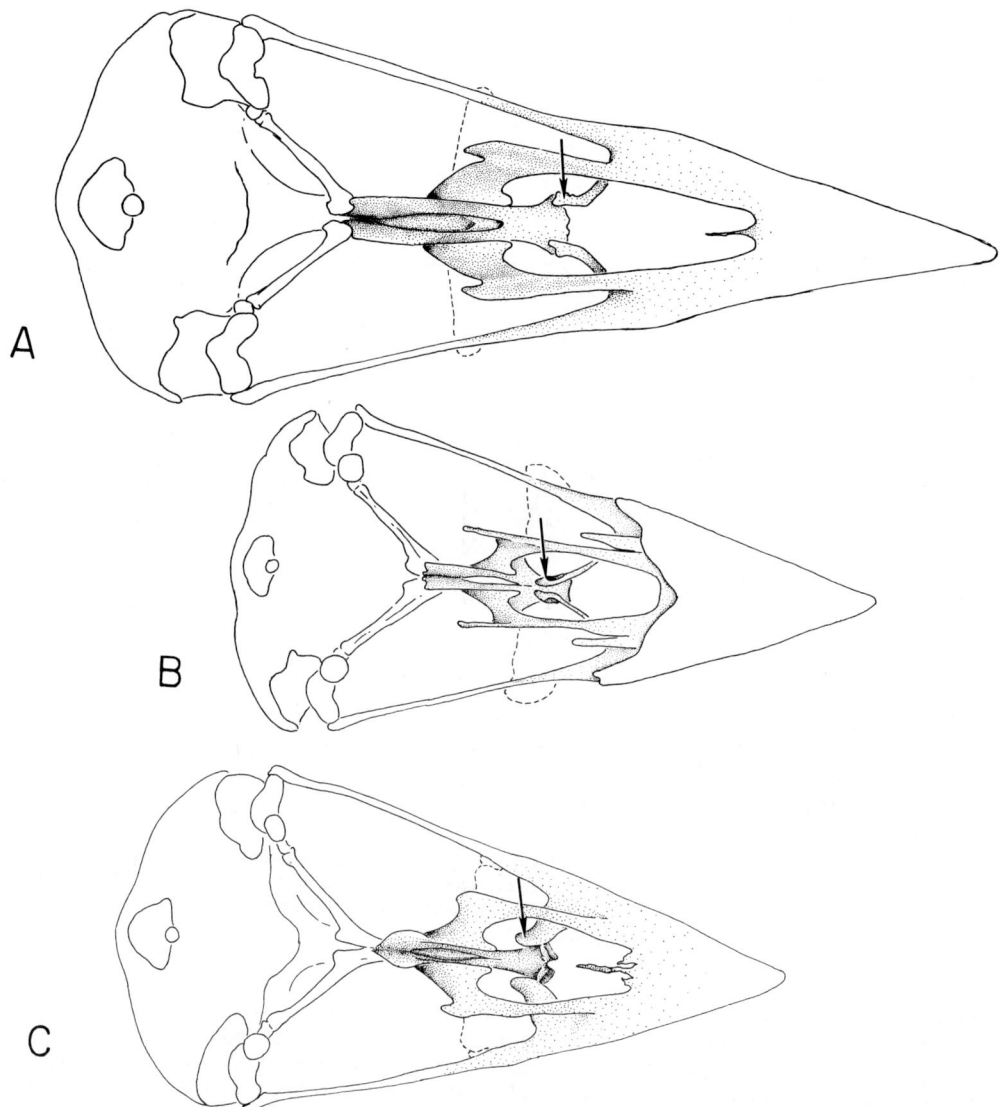

FIG. 1.   Ventral aspect of the nasal chamber illustrating variations in the form of the maxillo-palatines (indicated by arrows). Dotted line indicates the ventral margin of the ectethmoid plate, which constitutes the posterior wall of the nasal chamber. A. Semi-atrophied condition of the palatine process of the maxillo-palatines as found in *S. albocristata*. B. Typical inflated palatine process of the maxillo-palatines as exhibited by *Piranga rubra* (Thraupinae). C. Maxillo-palatines of the cotinga *Ampelion stresemanni*, illustrating the plate-like condition of the maxillo palatines and the absence of a palatine-process. Note the articular surfaces at the anterior margins of the vomer (between the maxillo-palatines), site of articulation with an ossified nasal concha.

The skeletal characters surveyed in this study include the basihyale of the hyoid apparatus, lacrimal process, maxillo-palatines, suprameatic and subtemporal fossae, foramen venae occipitale externum (foramen of a vein), and the pneumatic fossa of the humerus. A unique skeletal structure of the sternum is also described. The palatine process of the premaxilla was not included in this study because of its limited potential as an indicator of systematic relationships (Bock 1960).

Basihyale.—In a survey of the oscine hyoid apparatus, George (1962) distinguished two

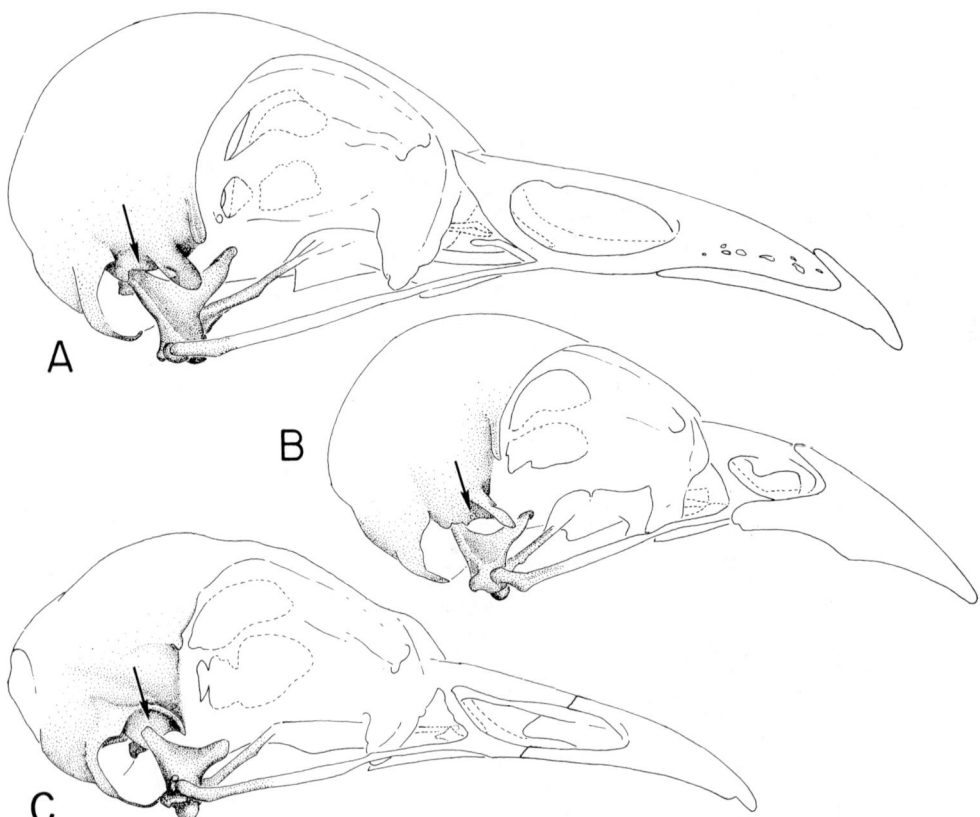

FIG. 2. A and B. Subtemporal fossa constituting the origin of the M. adductor mandibulae externus caudalis in *S. albocristata* and *Piranga rubra*, respectively. Suprameatic fossa too reduced in size to be discernible. C. Suprameatic fossa in a cotinga, *Ampelion sclateri*, in which there is no muscle attachment of any kind. Subtemporal fossa essentially lacking.

basic conditions of the basihyale, cylindrical and laterally compressed. He regarded these variations of the basihyale as reflections of certain functional and morphological attributes of the M. hypoglossus obliquus. The cylindrical condition predominates in the ten-primaried oscines with the exception of the Certhiidae. The laterally compressed condition characterizes the Certhiidae and the New World nine-primaried oscines with the exception of the Vireonidae, *Tersina* (generally regarded as related to the Thraupinae), and *Peucedramus* (Parulinae), all of which exhibit the cylindrical condition. I surveyed the condition of the basihyale in the Cotingidae and found that it is highly variable. In general, the basihyale in the Cotingidae is more truncated than in the oscines, whether laterally compressed or cylindrical. It also tends to be flattened dorsally and convex ventrally, sometimes with a lateral flange, as in *Lipaugus subalaris*. The basihyale in *Pipreola arcuata* is somewhat laterally compressed, resembling the condition in the oscines. This resemblance is rather superficial, however, because in overall appearance, the basihyale of *P. arcuata* is still more similar to that of other cotingas than to those of any nine-primaried oscines.

The basihyale in *S. albocristata* exhibits the typical laterally compressed condition characterizing the New World nine-primaried oscines.

Lacrimal bone.—A well developed lacrimal bone is always present in the Cotingidae. It generally is lacking from the Thraupinae, but is present in genera such as *Schistochlamys, Cissopis, Cnemoscopus, Buthraupis, Anisognathus, Stephanophorus, Iridosornis, Dubusia,* and *Delothraupis.* It is especially well developed in *Chlorornis riefferii* and occurs in *Hemispingus superciliaris* but not elsewhere in that genus (*H. goeringi* and *H. reyi* were not examined).

There is no indication of a lacrimal process in *S. albocristata*. Although the taxonomic

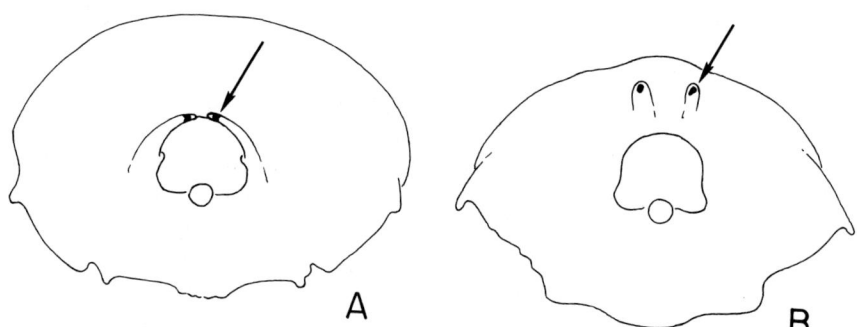

Fig. 3.   A. Typical arrangement of the Foramen venae occipitale externum in the New World nine-primaried oscines as exhibited by *S. albocristata*. B. Typical arrangement of the foramina in Cotingidae as exhibited by *Pipreola arcuata*.

significance of the lacrimal process in passerines has not been assessed, its absence from *S. albocristata* and its universal presence in the Cotingidae is none the less of interest.

Maxillo-palatines. — The maxillo-palatines of the Cotingidae may be readily distinguished from those of the Thraupinae (Fig. 1). In the Cotingidae, the maxillo-palatines are more or less flattened, plate-like structures that in most instances are attached to the anterior end of the vomer. The maxillo-palatines have no terminal processes and may be "pneumatic" or hollow, as in *Conioptilon*. The Thraupinae, as well as other members of the nine-primaried assemblage and most ten-primaried oscines, exhibit distinctively bulbous, pneumatic-like processes at the ends of the maxillo-palatines that are termed the palatine processes of the maxillo-palatines. These processes are readily observed in the posterior portion of the nasal chamber, and in the Thraupinae they do not articulate directly with the vomer. In some instances, ossification of the palatine processes may be incomplete, and all that may remain in a prepared skull are flattened ossified remnants. The palatine processes in this instance may be said to be atrophied. Nevertheless, those flattened, atrophied or semi-atrophied processes retain the general configuration characteristic of the maxillo-palatines of the Thraupinae and unlike those of the Cotingidae.

The maxillo-palatines of *S. albocristata* are semi-atrophied and quite indistinguishable from those found in some other species of Thraupinae.

Suprameatic and subtemporal fossae. — In the New World nine-primaried oscines, ventral to the zygomatic process there is a relatively large subtemporal fossa that serves as the site of the origin of the M. adductor mandibulae externus caudalis (Fig. 2A, B). This muscle is relatively large in the nine-primaried oscines compared to its condition in the Cotingidae and other suboscines. In the Cotingidae this muscle originates from the ventral edge of the zygomatic process so that the subtemporal fossa is essentially lacking. What is clearly discernible, however, is a large suprameatic fossa that remains free of any muscle attachment (Fig. 2C). In the illustrations of *S. albocristata* and *Piranga* in Figure 2, the suprameatic fossae are too small to be seen clearly. The fossa is easily revealed on a spirit specimen, however, by reflecting back the skin in the temporal area.

Although the subtemporal fossa in *S. albocristata* is relatively smaller than in many tanagers, it is nevertheless entirely occupied by the origin of the M. adductor mandibulae externus caudalis. The subtemporal fossa of *S. albocristata* may thus be said to exhibit the typical thraupinae condition, which is notably different from the condition in the suboscines.

Foramen venae occipitale externum. — These paired foramina (terminology follows Baumel et al. 1979) are associated with a pair of veins and are located dorsally on either side of the Foramen magnum. Although their placement is consistent within the Cotingidae and Thraupinae, it differs significantly between these two taxa (Fig. 3). The position of the Foramen venae occipitale externum in *S. albocristata* is typical of the nine-primaried oscines.

Pneumatic fossa of the humerus. — With respect to the pneumatic fossa of the humerus (Bock 1962), *S. albocristata* exhibits the typical two-fossae condition shared by all tanagers and all but a few members of the nine-primaried oscines assemblage. All cotingas, as well as all suboscines surveyed, have a single fossa.

Sternum. — A unique skeletal structure was found in *S. albocristata* in the form of an unpaired

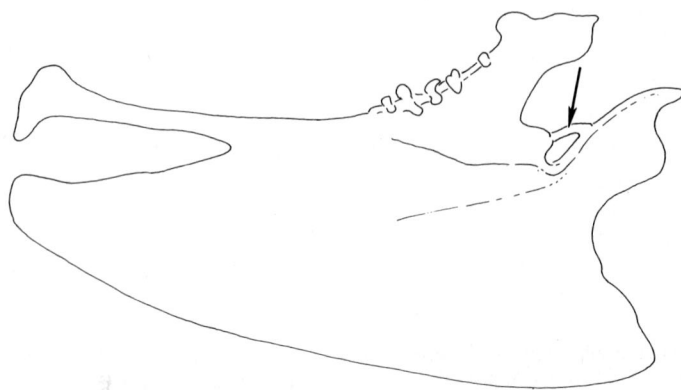

Fig. 4.   Lateral view of the sternum of *S. albocristata* illustrating the Processus manubrii.

ossified bridge-like structure of the sternum extending from the ventral manubrial spine to the dorsal spine (Fig. 4). Using the terminology of Baumel et al. (1979) this structure extends from the Spina interna to the Spina externa of the Rostrum sterni. I have named this structure the Processus manubrii or the manubrial process.

As only one skeleton of *S. albocristata* is available for study, the possibility that the manubrial process is simply an aberration of the specimen cannot be excluded. However, the structure appears to be substantial. Additional study of the structure does not appear to justify the extensive damage to the soft tissue that would result from dissection of the sternum of a spirit specimen. Rather, such a dissection should be combined with a dissection of the syrinx because both structures lie relatively close to one another.

An extensive survey of sterna of skeletal specimens of the Icteridae, Cardinalinae, Emberizidae, Ploceidae, and Fringillidae was conducted without encountering any trace of a manubrial process. For a list of species surveyed see Wood et al. 1982 for skeletal specimens of the appropriate families in the skeletal collection of the Museum of Zoology, Louisiana State University, Baton Rouge, Louisiana.

CONCLUSION

The present survey of anatomical characters supports the hypothesis that *S. albocristata* is a member of the New World nine-primaried oscines and not a cotinga. *S. albocristata* exhibits several structural characters shared with the nine-primaried oscine assemblage, but not the cotingas, but none shared with the latter and not likewise shared with the former. That *S. albocristata* exhibits a typical nine-primaried condition in contrast to the ten-primaried condition of cotingas, further supports the above conclusion.

The placement of *S. albocristata* in the Thraupinae, rather than some other taxa of the New World nine-primaried oscines, has been based on the bill shape and configuration which are decidedly thraupine-like. The plumage pattern and color are also not at variance with the range of variation of such in the tanagers.

Nothing can be said of *Compsothraupis loricata*, the supposed nearest relative of *S. albocristata*, based on this study. Anatomical material of the former species is needed for a fuller assessment of its taxonomic status which must remain in doubt.

ACKNOWLEDGMENTS

I take special pleasure in publishing this paper in a volume dedicated to the memory of the late Eugene Eisenmann. I well recall walking into Gene's office with a study skin of *S. albocristata* and his jovial, but unsuccessful, efforts to allay my concern that this supposed tanager was very probably a cotinga and had simply lost its outermost primary. What he knew almost instinctively about neotropical birds requires detailed analysis by others.

I am grateful to D. G. Homberger and J. V. Remsen for critical review of an early draft of the manuscript and W. J. Bock for numerous helpful suggestions and comments regarding the study. I thank A. J. Fogg for typing the manuscript.

LITERATURE CITED

BAUMEL, J. J., A. S. KING, A. M. LUCAS, J. E. BREAZILLE, AND H. E. EVANS (eds.). 1979. Nomina Anatomica Avium. Academic Press, London.

BOCK, W. J. 1960. The palatine process of the premaxilla in the Passeres. Bull. Mus. Comp. Zool. 122: 361–488.

BOCK, W. J. 1962. The pneumatic fossa of the humerus in the Passeres. Auk 79:425–443.

GEORGE, W. G. 1962. The classification of the Olive Warbler, *Peucedramus taeniatus*. Am. Mus. Novit. No. 2103.

MEYER DE SCHAUENSEE, R. 1966. The Species of Birds of South America and Their Distributions. Livingston Publishing Company, Narberth, Pennsylvania.

RAIKOW, R. J. 1978. Appendicular myology and relationship of New World nine-primaried oscines (Aves: Passeriformes). Bull. Carnegie Mus. Nat. Hist. 7:1–43.

SCLATER, P. L. 1886. Catalogue of Birds in the British Museum. Vol. 11. British Museum (Natural History), London.

SNOW, D. W. 1979. Cotingidae. Pp. 281–308, *In* M. A. Traylor, Jr. (ed.), Check-list of Birds of the World, Vol. 8. Museum Comparative Zoology, Cambridge, Massachusetts.

STORER, R. W. 1970. Thraupinae. Pp. 246–408, *In* R. A. Paynter, Jr. (ed.). Check-list of Birds of the World. Vol. 13. Museum Comparative Zoology, Cambridge, Massachusetts.

WOOD, S., R. L. ZUSI, AND M. A. JENKINSON. 1982. World inventory of avian skeletal specimens. American Ornithologists' Union and Oklahoma Biological Survey, Norman, Oklahoma.

ZIMMER, J. T. 1947. Studies of Peruvian Birds No. 52. Am. Mus. Novit. No. 1367.

AN INTERGENERIC HYBRID CONEBILL (*CONIROSTRUM* × *OREOMANES*) FROM PERU

THOMAS S. SCHULENBERG

Museum of Zoology, Louisiana State University, Baton Rouge, Louisiana 70803-3216 USA

ABSTRACT. A conebill collected in October 1980 in Dpto. Puno, Peru is considered a hybrid between *Conirostrum ferrugineiventre* and *Oreomanes fraseri*. Both genera are former members of the "Coerebidae," a family now not usually recognized, as it is thought to be polyphyletic. *Conirostrum* and *Oreomanes* are now usually regarded as belonging to different phyletic lineages. Hybridization between these genera, coupled with their basic similarities in plumage coloration and pattern and in general morphology, suggests that *Conirostrum* and *Oreomanes* are more closely related than was previously recognized, but leaves unanswered the question of their relationships to other nine-primaried oscines.

RESUMEN. En octubre de 1980 se coleccionó un pico de cono en el Departamento de Puno, Perú, el cual es considerado un híbrido entre *Conirostrum ferrugineiventre* y *Oreomanes fraseri*. Ambos géneros fueron miembros del "Coerebidae", una familia que en la actualidad no es usualmente reconocida, pues se la considera compuesta por diversos linajes ("polyphyletic"). *Conirostrum* y *Oreomanes* actualmente son considerados como pertenecientes a diferentes linajes de filos. La hibridación entre esos dos géneros, unida a sus similaridades básicas en coloración y modelo de plumajes y morfología en general, sugieren que *Conirostrum* y *Oreomanes* están más estrechamente relacionados que lo que se reconocía anteriormente, pero queda sin respuesta la cuestión de sus relaciones con los otros oscines de nueve primarias.

Intergeneric hybridization of birds in natural populations is rare, especially between species that are not highly sexually dimorphic. Intergeneric hybrids have nonetheless exerted a disproportionate influence on avian systematics. Study of such hybrids has often led to reassessment of the taxonomic affinities of the parental taxa and related groups (e.g., Banks and Johnson 1961; Parkes 1961, 1978; Short and Phillips 1966). Herein I report a new intergeneric hybrid that is of exceptional interest because the parental taxa are currently assigned by many systematists to different subfamilies.

The bird in question is a conebill collected in October 1980 during a Louisiana State University Museum of Zoology (LSUMZ) expedition to the high cloud forests of extreme southeastern Peru. The site is known locally as Valcón, and lies ca. 15 km southeast of Sandia, Departamento de Puno, at 3000 m elevation. On 25 October, near the upper extent of elfin forest at this locality, at 3325 m, Laurence C. Binford saw an unfamiliar bird which, with its chestnut underparts and superciliary, and pointed bill, superficially resembled the Rufous-browed Conebill *Conirostrum rufum*. The following day, 26 October, I located the bird in the same area. The bird was a member of a small mixed-species flock that also contained two or three each of *Margarornis squamiger* (Pearled Treerunner), *Mecocerculus leucophrys* (White-throated Tyrannulet), and *Thlypopsis ruficeps* (Rust-and-yellow Tanager), and a single *Conirostrum ferrugineiventre* (White-browed Conebill). I collected both conebills from the flock; both had incompletely pneumatized skulls, were undergoing heavy body molt, and had caterpillars and other insects in their stomachs. The *C. ferrugineiventre* (LSUMZ 98859) was a female, without enlarged follicles. The odd conebill (LSUMZ 98860) was a male (testes 3 × 2 mm). This specimen was larger and longer-billed than any *Conirostrum* species; in color it suggested *Oreomanes fraseri* (Giant Conebill), but it was smaller than *fraseri* and lacked that species' white cheek patch.

Study of this bird at the LSUMZ led me to conclude that it was a hybrid between *C. ferrugineiventre* and *Oreomanes fraseri* (Fig. 1). On close inspection both the plumage pattern and colors of the parental species prove to be similar to each other. In general, both color and pattern of various parts of the hybrid resemble one or the other of the parental species, although color of the crown and superciliary is intermediate.

FIG. 1. Dorsal (left), lateral (center) and ventral (right) aspects of specimens of conebills. A. *Oreomanes fraseri*. B. Hybrid. C. *Conirostrum ferrugineiventre*.

MORPHOLOGY

COLOR AND PATTERN

Oreomanes fraseri displays slight geographic variation (Vuilleumier 1984). The following descriptions of *O. fraseri* are based on specimens from Peru and may not apply in some details to other populations. *Conirostrum ferrugineiventre* does not vary geographically.

Crown.—C. ferrugineiventre: dark slate gray to black, darker anteriorly. Feathers of mid- and hind-crown of some individuals edged blue-gray. *O. fraseri*: forecrown white or silvery-gray, but black bases showing through, giving grizzled appearance. Posteriorly, black bases are extensive and feather tips light blue-gray, so that the mid-crown appears black-spotted. Black disappears from feathers of the hind-crown, which appears uniformly blue-gray (concolor with the back). *Hybrid*: similar to *C. ferrugineiventre*, but feathers of forecrown lightly fringed with blue-gray.

Face pattern.—C. ferrugineiventre: a conspicuous pale superciliary from the base of the maxilla to the nape, usually grayish-white but occasionally washed with light ochraceous-buff. Lores and area immediately behind eye dark slate gray to black. Auriculars gray with very narrow white shaft streaks. Malar area mottled gray and white or buff, sometimes mixed with black, forming a moustachial streak on the side of the throat. *O. fraseri*: a chestnut superciliary from the base of the maxilla to behind and surrounding the eye, sometimes bordered above with narrow white line. Lores gray to black. A white patch on the side of the face, extending from the base of the mandible to below the auriculars, and including the lower auriculars. Upper auriculars usually gray to blue-gray, but relative extent of the white variable, and on some birds almost the entire auriculars are white. *Hybrid*: superciliary predominately white before the eye, with small amounts of chestnut on the ventral border. Above and behind the eye, the superciliary is entirely chestnut, and extends to the nape. Lores and post-ocular spot black. Auriculars gray with white shaft streaks, as in *C. ferrugineiventre*. Malar area mottled white and chestnut, some feathers with black tips.

Back, rump, and upper tail coverts.—C. ferrugineiventre: entire upperparts blue-gray, sometimes slightly bluer on the rump and upper tail coverts. Some individual variation: the bluest individuals are males, although some are as gray as females. *O. fraseri*: similar to *ferrugineiventre*, but typically somewhat grayer, and never as blue as some male *ferrugineiventre*. *Hybrid*: closest to *ferrugineiventre*, being slightly bluer than most *fraseri*.

*Tail.—*The tail pattern is the same in *C. ferrugineiventre*, *O. fraseri*, and the hybrid. The tail is dark gray ventrally; dorsally the feathers are black with blue-gray edges to the outer webs, these edges more extensive basally.

*Wing.—*No significant differences between *C. ferrugineiventre*, *O. fraseri*, and the hybrid. Tertials and remiges dark gray to black. Almost the entire outer web of the tertials is blue-gray, with the amount of blue-gray edging decreasing out the wing to the primaries, the outer webs of which are only narrowly gray-edged. Outer webs of upperwing coverts bluish-gray.

*Underparts.—*Some individual variation in color intensity in both species. *C. ferrugineiven-*

TABLE 1

MEASUREMENTS AND WEIGHTS OF MALE *CONIROSTRUM FERRUGINEIVENTRE*, *OREOMANES FRASERI*, AND HYBRID

Measurement (mm)	*Conirostrum ferrugineiventre*				Hybrid	*Oreomanes fraseri*			
	N	Range	Mean	s.d.		N	Range	Mean	s.d.
Wing (chord)	10	65.3–71.5	69.1	2.0	77.5	10	87.9–93.7	90.1	2.0
Tail	10	50.5–55.4	53.0	1.7	55.8	10	61.3–65.4	63.4	1.3
Bill from anterior edge of nostril	10	6.8–7.8	7.3	0.4	10.6	10	12.7–15.0	14.0	0.7
Tarsus	10	17.6–19.2	18.6	0.5	20.7	10	21.3–22.8	21.9	0.5
Chord of hind claw	10	5.5–6.4	6.0	0.3	7.0	10	8.4–10.0	9.1	0.5
Middle toe and claw	10	14.5–16.5	15.7	0.7	18.4	10	20.1–22.4	21.4	0.7
Weight (g)	3	11.0–12.0	11.5		–	10	22.0–27.0	24.2	1.7

tre: entire underparts ochraceous-buff, paler (almost white) on the upper throat and chin. Tibio-tarsal feathers white. *O. fraseri:* entire underparts chestnut, matching the color of the superciliary. Tibio-tarsal feathers white. *Hybrid:* as in *fraseri*.

Soft part colors.—*C. ferrugineiventre:* iris brown; bill dark slate to black; tarsi dark brown, black. *O. fraseri:* iris brown; bill silvery-gray, dark gray, or black, mandible sometimes described as paler than maxilla; tarsi gray-brown, dark gray, black. *Hybrid:* iris not recorded; bill black; tarsi brown.

MEASUREMENTS

Oreomanes fraseri is larger than *Conirostrum ferrugineiventre*, and weighs twice as much (Table 1). The two taxa are of slightly different proportions, as *O. fraseri* has a relatively shorter tail and tarsus. The hybrid is intermediate in the two mensural characters, wing and bill length, that most clearly separate the parental taxa, although the wing length of the hybrid is a little closer to that of *C. ferrugineiventre* than to that of *O. fraseri*. Tail length of the hybrid, although greater than the mean of *C. ferrugineiventre*, is barely outside the range of even this small sample of that species. The tarsal lengths of *C. ferrugineiventre* and *O. fraseri* show surprisingly little difference. The tarsal length of the hybrid is, nonetheless, intermediate, despite the small gap separating the parental taxa. The hybrid is also intermediate in two measurements of the foot, chord of hind claw and length of middle toe plus claw.

DISTRIBUTION

Despite broad overlap in the geographic ranges of *Conirostrum ferrugineiventre* and *O. fraseri*, these two conebills probably come into contact only infrequently. *Conirostrum ferrugineiventre* is a bird of humid treeline forest and forest edge on the east slope of the Andes from Dptos. San Martín (sight records, S. Allen and D. Stotz), La Libertad (sight records, T. A. Parker, III and T. S. Schulenberg) and Huánuco (Hellmayr 1935; specimens LSUMZ), Peru, south to Cochabamba, Bolivia (Meyer de Schauensee 1966). *Oreomanes fraseri* has a more extensive range, from southern Colombia south to western Bolivia (Meyer de Schauensee 1966). Within its range *O. fraseri* is more patchily distributed than *C. ferrugineiventre*, as it is confined to *Polylepis* woodlands which are apparently restricted to areas where specialized ecological requirements are met (Simpson 1979). *Oreomanes fraseri* is typically found at somewhat higher elevations than *C. ferrugineiventre* (which reflects the distribution of *Polylepis* forests). Collecting localities in Peru range from 3000 to 3600 m for *C. ferrugineiventre*, and from 3500 to 4550 m for *O. fraseri*. At present the two species are known to be syntopic at only a single locality. In 1974, LSUMZ personnel collected *C. ferrugineiventre* at Canchaillo, 3257 m, Dpto. Cuzco, Peru, and collected *O. fraseri* in *Polylepis* forest directly above Canchaillo at 3950 m. *Conirostrum ferrugineiventre* was seen on one occasion by T. A. Parker, III in these *Polylepis* (Parker and O'Neill 1980). *Oreomanes* has been collected at several other localities in Dptos. Cuzco and Puno (Vuilleumier 1984), and there is no reason to think that the complex topography of the high Andes of southern Peru and Bolivia does not bring *O. fraseri* and *C. ferrugineiventre* into contact in other areas as well.

FEEDING BEHAVIOR

Despite the specialized tongues of some coerebids, probably only a few (e.g., *Coereba*, some *Diglossa*) rely on flowers to such an extent that they may legitimately be considered to be specialized nectarivores. Both *Conirostrum* and *Oreomanes* are primarily insectivorous. Despite the general similarity in their diets, they differ significantly in their foraging techniques.

Conirostrum ferrugineiventre gleans insects from the surfaces of leaves and twigs (Parker and O'Neill 1980; pers. observ.) and, consequently, moves through the foliage of bushes and trees. The diet and foraging behavior is similar to that of at least six other species of *Conirostrum*, including *C. leucogenys* (Robbins et al. 1985), *C. bicolor* (ffrench 1976), *C. cinereum* (Johnson 1967; Koepcke 1970; McFarlane and Loo 1974; pers. observ.), *C. tamarugensis* (McFarlane and Loo 1974; Tallman et al. 1978), *C. sitticolor* (Remsen 1985; pers. observ.), and *C. albifrons* (pers. observ.). *Oreomanes fraseri* searches for insects found under the scaly bark of *Polylepis* trees and bushes (George 1964; Johnson 1967; Koepcke 1970; Parker and O'Neill 1980; pers. observ.), and hitches along tree limbs while foraging. Observations were brief, but the actions of the hybrid conebill foraging in the foliage of *Gynoxys* and other shrubs were similar to those of the accompanying *C. ferrugineiventre*.

DISCUSSION

Both *Conirostrum* and *Oreomanes* have traditionally been regarded as members of the Coerebidae (Sclater 1886). Ridgway (1902:376–377) expressed doubts that this family was monophyletic. He also believed, on the basis of "external characters," that *Conirostrum* belonged to the Parulidae and was "inclined to believe" that *Oreomanes* was also a parulid, related to *Conirostrum*. Hellmayr (1935) later returned both genera to the Coerebidae without further comment.

Beecher (1951) argued strongly that the Coerebidae were polyphyletic. On the basis of variations in jaw musculature and the structure of the palate, he placed *Coereba* and *Conirostrum* in the Parulidae, and the remaining coerebid genera in the Thraupinae. Beecher's extensive dissections of jaw musculature eventually led to a completely original phylogeny of the oscines (Beecher 1953). The phylogenetic interpretations Beecher drew from his dissections were later criticized (Mayr 1955; Bock 1960) and all but ignored, with the important exception of his division of the Coerebidae, which was quickly and widely adopted (Van Tyne and Berger 1959; Lowery and Monroe 1968; Storer 1970; but cf. Meyer de Schauensee 1966).

Skutch (1962) and Morony (1968) argued that *Coereba* was not a parulid, but the broader issue of the affinities of the entire coerebid assemblage has received surprisingly little attention. Raikow (1978) proposed a cladogram of the nine-primaried oscines based in part on dissections of the appendicular myology, suggesting corroboration of Beecher's division of the Coerebidae. All of the coerebid genera dissected by Raikow, however, including both "paruline" and "thraupine" genera, have similar appendicular myologies and, therefore, cannot be used to divide this assemblage. Raikow (1978:35) noted that alternative phylogenies are possible and that additional evidence is needed to resolve the affinities of the Coerebidae.

Thus, in the more than 30 years since Beecher published his paper on the relationship of the Coerebidae, we have scarcely progressed in our understanding of the affinities of these genera. In this context, one should exercise caution in deriving any taxonomic implications from the hybridization of *Conirostrum* and *Oreomanes*. It is worth noting, however, that Beecher (1951) did not have anatomical specimens of two genera of coerebid, *Oreomanes* and *Xenodacnis*. Beecher "presumed" that both of these genera belong to his "thraupine" group on the basis of plumage color, although he did not specify the color similarities that he saw between *Oreomanes* and the tanagers. Actually, the plumage color of *Oreomanes* more closely resembles that of *Conirostrum*, especially *C. ferrugineiventre* (see above); furthermore, as Ridgway (1902) noted, the external morphology of these two genera is similar. External characters (color, pattern, and shape) and the intergeneric hybrid reported here suggest that *Oreomanes* is most closely related to *Conirostrum*, but leave open the question of the affinities of these genera to other nine-primaried oscines. The status of *Oreomanes* as a monotypic genus also remains undetermined. *Oreomanes* may be a sister group of all *Conirostrum*, or it may be more closely related to some Andean species of *Conirostrum* than to other species and, hence, phylogenetically merely a specialized *Conirostrum*.

No intra-generic *Conirostrum* hybrids are known. Although the opportunity for interspecific contact is reduced by the elevational segregation of *Conirostrum* species, as many as two or three may be found in the same feeding flocks in the Andes of Peru and Bolivia (Weske 1972; pers. observ.). Thus, that the first known *Conirostrum* hybrid involves a species of another genus is consistent with the observation that inter-generic hybrids in several avian families may be more common than intra-generic hybrids (Banks and Johnson 1961; Parkes 1961, 1978), presumably because "selection pressures favoring the evolution of reproductive isolating mechanisms would be strongest among closely related (= congeneric) sympatric species" (Parkes 1978:688).

ACKNOWLEDGMENTS

L. C. Binford, L. Campos L., M. Sánchez S., and A. Urbay T. participated in the 1980 LSUMZ expedition to Puno. The manuscript was criticized by S. Allen, R. C. Banks, K. C. Parkes, J. V. Remsen, Jr., L. L. Short, and F. Vuileumier. The Dirección General Forestal y de Fauna continues its support of LSUMZ field studies. I am most grateful to J. S. McIlhenny, H. I. and L. R. Schweppe, and B. M. Odom for their continued interest in and support of the field work of the LSUMZ.

LITERATURE CITED

BANKS, R. C., AND N. K. JOHNSON. 1961. A review of North American hybrid hummingbirds. Condor 63:3–28.
BEECHER, W. J. 1951. Convergence in the Coerebidae. Wilson Bull. 63:274–287.
BEECHER, W. J. 1953. A phylogeny of the Oscines. Auk 70:270–333.
BOCK, W. J. 1960. The palatine process of the premaxilla in the Passeres. Bull. Mus. Comp. Zool. 122: 361–488.
FFRENCH, R. 1976. A Guide to the Birds of Trinidad and Tobago. Harrowood, Valley Forge, Pennsylvania.
GEORGE, W. G. 1964. Rarely seen songbirds of Peru's high Andes. Nat. Hist. 38:26–29.
HELLMAYR, C. E. 1935. Catalogue of birds of the Americas. Field Mus. Nat. Hist. Publ. Zool. Ser. 13, Pt. 8.
JOHNSON, A. W. 1967. The Birds of Chile and Adjacent Regions of Argentina, Bolivia, and Peru. Vol. 2. Platt Establ. Gráf. S.A., Buenos Aires.
KOEPCKE, M. 1970. The Birds of the Department of Lima, Peru. Livingston Publ. Co., Wynnewood, Pennsylvania.
LOWERY, G. H., JR., AND B. L. MONROE, JR. 1968. Parulidae. Pp. 3–93, *In* Paynter, R. A., Jr. (ed.). Check-list of Birds of the World. Vol. 14. Museum Comparative Zoology, Cambridge, Massachusetts.
MAYR, E. 1955. Comments on some recent studies of song bird phylogeny. Wilson Bull. 67:33–44.
MCFARLANE, R. W., AND E. LOO P. 1974. Food habits of some birds in Tarapacá. Idesia 3:163–166.
MEYER DE SCHAUENSEE, R. 1966. The Species of Birds of South America and Their Distribution. Livingston Publ. Co., Narberth, Pennsylvania.
MORONY, J. J., JR. 1968. A survey of certain skull structures of the family Parulidae (Class Aves). Unpubl. Master's Thesis, Louisiana State University, Baton Rouge.
PARKER, T. A., III, AND J. P. O'NEILL. 1980. Notes on little known birds of the upper Urubamba Valley, southern Peru. Auk 97:167–176.
PARKES, K. C. 1961. Intergeneric hybrids in the family Pipridae. Condor 63:345–350.
PARKES, K. C. 1978. Still another parulid intergeneric hybrid (*Mniotilta* × *Dendroica*) and its taxonomic and evolutionary implications. Auk 95:682–690.
RAIKOW, R. J. 1978. Appendicular myology and relationships of the New World nine-primaried oscines (Aves: Passeriformes). Bull. Carnegie Mus. Nat. Hist. 7:1–43.
REMSEN, J. V., JR. 1985. Community organization and ecology of birds of high elevation humid forest of the Bolivian Andes. Pp. 733–756, *In* P. A. Buckley, M. S. Foster, E. S. Morton, R. S. Ridgely, and F. G. Buckley (eds.), Neotropical Ornithology. Ornithol. Monogr. No. 36.
RIDGWAY, R. 1902. The birds of North and Middle America. Part 2. U.S. Natl. Mus. Bull. 50, Part 2.
ROBBINS, M. B., T. A. PARKER, III, AND S. E. ALLEN. 1985. The Avifauna of Cerro Pirre, Darién, eastern Panama. Pp. 198–232, *In* P. A. Buckley, M. S. Foster, E. S. Morton, R. S. Ridgely, and F. G. Buckley (eds.), Neotropical Ornithology. Ornithol. Monogr. No. 36.
SCLATER, P. L. 1886. Catalogue of the birds in the British Museum. Volume 11. Br. Mus. (Nat. Hist.), London.
SHORT, L. L., AND A. R. PHILLIPS. 1966. More hybrid hummingbirds from the United States. Auk 83: 253–265.
SIMPSON, B. B. 1979. A revision of the genus *Polylepis* (Rosaceae: Sanguisorbeae). Smithson. Contrib. Bot. 43.
SKUTCH, A. F. 1962. Life histories of honeycreepers. Condor 64:92–116.

STORER, R. W. 1970. Thraupinae. Pp. 246–408, *In* R. A. Paynter, Jr. (ed.). Check-list of Birds of the World. Vol. 13. Museum Comparative Zoology, Cambridge, Massachusetts.

TALLMAN, D. A., T. A. PARKER, III, G. D. LESTER, AND R. A. HUGHES. 1978. Notes on 2 species of birds previously unreported from Peru. Wilson Bull. 90:444–446.

VAN TYNE, J., AND A. J. BERGER. 1959. Fundamentals of Ornithology. John Wiley and Sons, New York.

VUILLEUMIER, F. 1984. Patchy distribution and systematics of *Oreomanes fraseri* (Aves, ?Coerebidae) of Andean *Polylepis* woodlands. Am. Mus. Novit. No. 2777.

WESKE, J. S. 1972. The distribution of the avifauna in the Apurimac Valley of Peru with respect to environmental gradients, habitat, and related species. Unpubl. Ph.D. dissert. University of Oklahoma, Norman.

PHYLOGENY AND CLASSIFICATION OF NEW WORLD SUBOSCINE PASSERINE BIRDS (PASSERIFORMES: OLIGOMYODI: TYRANNIDES)

CHARLES G. SIBLEY AND JON E. AHLQUIST

Peabody Museum of Natural History and Department of Biology, Yale University, New Haven, Connecticut 06511 USA

ABSTRACT. During the past century the classifications of the suboscine passerine birds have included the following groups, as in Wetmore (1960): Superfamily Furnarioidea—Dendrocolaptidae, Furnariidae, Formicariidae, Conopophagidae, Rhinocryptidae; and Superfamily Tyrannoidea—Cotingidae, Pipridae, Tyrannidae, Oxyruncidae, Phytotomidae, Pittidae, Acanthisittidae, Philepittidae.

DNA-DNA hybridization data and congruent morphological characters show that a revision of this arrangement is indicated, as follows: (1) Woodcreepers (Dendrocolaptinae) and ovenbirds (Furnariinae) are subfamilies of the Furnariidae. (2) The Formicariidae of Wetmore (1960) is a diphyletic assemblage divisible into ground antbirds (Formicariidae), which are closely related to the gnateaters (Conopophagidae) and tapaculos (Rhinocryptidae), and typical antbirds (Thamnophilidae), which diverged from the lineage leading to the furnariids, formicariids, conopophagids, and rhinocryptids before these groups diverged from one another. (3) The Tyrannoidea includes only the New World taxa of Wetmore (1960). Pittidae is the sister group of the Eurylaimidae, and the New Zealand wrens (Acanthisittidae) are descendants of the oldest known branch in the suboscine phylogeny (Sibley et al. 1982). Affinities of the Philepittidae are uncertain, but the family is presumed to belong to the Old World suboscine infraorder Eurylaimides. (4) The Tyrannidae of Wetmore (1960) is divided into the Mionectidae, which consists of several genera previously thought to be tyrannids, and the Tyrannidae, which includes the Tyranninae, Tityrinae, Cotinginae, and Piprinae. (5) The Tyranninae includes most of the genera of tyrant flycatchers. (6) The Tityrinae (becards, tityras, *Schiffornis*) is the sister group of the Tyranninae. (7) The Cotinginae includes the cotingas, plantcutters (*Phytotoma*), and the Sharpbill (*Oxyruncus*) (Sibley et al. 1984). (8) The Piprinae includes the manakins, except *Schiffornis* which, tentatively, is the descendant of an early branch from the tityrine lineage.

RESUMEN. Durante el siglo pasado la clasificación de aves passeriformes suboscines ha incluído, tal como aparece en Wetmore (1960), los siguientes grupos: Superfamilia Furnarioidea—Dendrocolaptidae, Furnariidae, Formicariidae, Conopophagidae, Rhinocryptidae; y Superfamilia Tyrannoidea—Cotingidae, Pipridae, Tyrannidae, Oxyruncidae, Phytotomidae, Pittidae, Acanthisittidae, Philepittidae.

Datos de hibridación de ADN-ADN y caracteres morfológicos congruentes muestran que sería necesaria una revisión de este arreglo de la siguiente manera: (1) Los trepadores (Dendrocolaptinae) y horneros (Furnariinae) son subfamilias de los Furnariidae. (2) El Formicariidae de Wetmore (1960) es una unión de dos filos ("diphyletic") divisible entre los hormigueros terrestres (Formicariidae), (que están cercanamente relacionados con las perlitas (Conopophagidae) y los tapaculos (Rhinocryptidae) y típicos hormigueros (Thamnophilidae) que se apartaron del linaje que formó a los furnariidos, formicariidos, conopophagidos y rhinocryptidos, antes de que esos grupos se separasen entre si. (3) El Tyrannoidea incluye únicamente la taxa de Wetmore (1960) para el Nuevo Mundo. Pittidae es el grupo hermano ("sister group") de los Eurylaimidae, y los salta paredes (Acanthisittidae) de Nueva Zelanda son descendientes de la rama más antigua conocida en la filogenia de los suboscines (Sibley et al. 1982). Son inciertas las afinidades de los Philepittidae, pero se presume que la familia pertenece al infraorden suboscine Eurylaimides del Viejo Mundo. (4) El Tyrannidae de Wetmore (1960) se divide en Mionectidae, formado por varios géneros previamente considerados como tyrannidos y la Tyrannidae, que incluye Tyranninae, Tityrinae, Cotinginae y Piprinae. (5) El Tyranninae incluye a la mayoría de los géneros de atrapa moscas tiranos. (6) El Tityrinae (cabezones, rechinadores, *Schiffornis*) es el grupo hermano del Tyranninae. (7) El Cotinginae incluye las cotingas, los cortaplantas (*Phytotoma*), y los picoafilados (*Oxyruncus*) (Sibley et al. 1984). (8) El Piprinae incluye las matracas, excepto *Schiffornis,* el cual es considerado tentativamente descendiente de una rama más antigua del linaje de los tityrines.

The history of the classification of the suboscine passerines has been reviewed by Sibley (1970:23–31) and Ames (1971:127–129, 153–164). The classifications of Mayr and Amadon (1951) and Wetmore (1930, 1934, 1940, 1951, 1960) recognize four suborders in the Passeriformes and differ only in minor ways at the lower taxonomic levels. Wetmore's (1960) classification of the suboscines follows:

Order Passeriformes
    Suborder Eurylaimi
        Family Eurylaimidae, Broadbills
    Suborder Tyranni
      Superfamily Furnarioidea
        Family Dendrocolaptidae, Woodhewers
            Furnariidae, Ovenbirds
            Formicariidae, Antbirds
            Conopophagidae, Antpipits, Gnateaters
            Rhinocryptidae, Tapaculos
      Superfamily Tyrannoidea
        Family Cotingidae, Cotingas
            Pipridae, Manakins
            Tyrannidae, Tyrant Flycatchers
            Oxyruncidae, Sharpbills
            Phytotomidae, Plantcutters
            Pittidae, Pittas
            Acanthisittidae, New Zealand Wrens
            Philepittidae, Asities, False Sunbirds
    Suborder Menurae, Lyrebirds and Scrub-birds
    Suborder Passeres, Songbirds (Oscines)

"The work of Hans Gadow" (1893) was "taken as the starting point" by Wetmore (1930) for the first of the five editions of his classification. The arrangement proposed by Wetmore (1930) remained unchanged through the last edition of 1960. Thus, Wetmore's classification, although widely accepted, reflects the opinions developed by Gadow a century ago.

Various modifications of Wetmore's classification have been proposed, some of which are noted below. However, a complete revision of the classification of the suboscine passerines has not been presented and the proposals for partial changes have not achieved a consensus.

To reconstruct the phylogeny of the suboscines we have used the technique of DNA-DNA hybridization. This method measures genealogical distances by determining the degrees of base pair complementarity between the homologous DNA sequences of different taxa. The present study is based on the data from 967 DNA-DNA hybrids (Appendices I–XVI). These data are congruent with the morphological evidence of relationships described by Heimerdinger and Ames (1967), Ames et al. (1968), and Ames (1971). The DNA-DNA measurements provide the branching pattern of the suboscine lineages and the approximate datings of divergence nodes (Fig. 1). We also present a new classification based on the phylogeny revealed by the DNA comparisons.

Ames (1971:153) recognized the Furnarii as a fifth suborder including members of the Furnarioidea of Mayr and Amadon (1951). Ames (1971:155) also suggested that the Tyranni "should probably be restricted to the five New World families of Wetmore's (1960) superfamily Tyrannoidea: Cotingidae, Phytotomidae, Pipridae, Tyrannidae, and Oxyruncidae." Ames viewed the positions of the Pittidae and Acanthisittidae as uncertain and as showing "no clear relationship to any New World tyrannoid group." Sibley (1970:37) came to the same conclusion. Sibley et al. (1982) have shown that the Acanthisittidae are the descendants of the oldest known branch in the suboscine phylogeny (Fig. 1).

Olson (1971) recognized only three suborders of Passerines: Tyranni, Menurae, and Oscines. He placed the Eurylaimidae in the Tyranni because he believed them to be closely related to the Cotingidae. Snow (1973) suggested that the morphological similarities between cotingas and broadbills are due to convergence, not close relationship. Our DNA comparisons support Snow's suggestion.

Olson (1971) also noted that *Menura* has a quadrato-jugal articulation like that of the oscines and unlike that of the suboscines. Although he retained the Menurae as a suborder, he believed that the lyrebirds "will ultimately be found to be closer to the Oscines than to the Tyranni" (Olson 1971:514).

Sibley (1974) and Feduccia (1975) both concluded that *Menura* is oscinine, but Feduccia and Olson (1982:16), from comparisons of skeletal elements, decided that the "Menurae are much more similar to the Rhinocryptidae than to any other passerine group." However, (p. 17) they considered it "premature . . . to propose a phylogeny or any suggestions for classification, other than disallowing any association between the Menurae and the bowerbird assemblage."

As we have shown (Sibley and Ahlquist 1985, in press), the Menuridae (lyrebirds, scrubbirds) and the Ptilonorhynchidae (bowerbirds) are sister groups in the oscine suborder Corvi, superfamily Menuroidea. The divergence between the lyrebird and bowerbird lineages occurred ca. 40–45 MYA, but the suboscine-oscine dichotomy occurred ca. 85–90 MYA. Thus, the specialized characters of the lyrebirds (e.g., the syrinx) are derived, not primitive, and the similarities between the lyrebirds and the rhinocryptids are due to convergence, not recent common ancestry.

METHODS

Our procedures are based primarily on those of Britten and Kohne (1968), Kohne (1970), Kohne and Britten (1971) and Britten et al. (1974). Brief descriptions (Sibley and Ahlquist 1982a–f) and more complete versions (Sibley and Ahlquist 1981, 1983) of our methods are published elsewhere.

The thermal dissociation curve of a DNA-DNA hybrid is a plot of the different rates of nucleotide substitution that have occurred in different sequences (Grula et al. 1982). Those hybrid duplexes composed of homologous sequences that have diverged most rapidly, and hence are most different, will contain the largest number of mismatched base pairs and will, therefore, melt at the lower temperatures. The DNA hybrids that do not melt until exposed to the highest temperatures are composed of sequences that have evolved slowly and contain few, if any, mismatched bases.

To obtain a measure of the genealogical distance between taxa we calculate the $T_{50}H$ statistic which is the temperature in degrees Celsius at which 50 percent of the DNA duplexes in a given hybrid have melted. The difference, in degrees Celsius, between the $T_{50}H$ of a homoduplex hybrid (= Taxon A × Taxon A) and the $T_{50}H$ of a heteroduplex hybrid (= Taxon A × Taxon B) is the delta $T_{50}H$ between the two taxa (A and B) forming the heteroduplex hybrid. Thus, the delta $T_{50}H$ measures the average *amount* of DNA sequence divergence which is a product of the average *rate* of divergence. The delta $T_{50}H$ measures the average (or median) difference between the two genomes composing a DNA-DNA hybrid. In the calculation of $T_{50}H$ it is assumed that all of the single-copy sequences in the genomes of the two species being compared have homologs in the other species, that all single-copy sequences potentially can hybridize with their homologs, and that all degrees of divergence can be detected. However, the percentage of hybridization declines as the amount of divergence increases and the thermal stability curve is progressively truncated by the effects of the experimental conditions. For DNA-DNA hybrids between more diverged taxa whose melting curves are entirely above the 50 percent level, it is necessary to extrapolate the most linear portion of the curve to its intercept with the 50 percent level. This is done by fitting a cumulative distribution function to the data to find the intercept. Thus the $T_{50}H$ values incorporate the percentage of hybridization and the measure of thermal stability, in a single number. Kohne (1970:352) and Hall et al. (1980) discussed this subject. Kohne (1970) used "$T_{50}R$" for the same calculation as $T_{50}H$, and Hall et al. (1980) designated it as the "median sequence divergence." As Hall et al. (1980:101) noted "the median divergence takes into account the extent of reaction [= % hybridization] and therefore is a more accurate estimate of total divergence than the T_m difference for cases in which the interspecies reassociation is less than intraspecies reassociation." The T_m is the temperature in degrees Celsius at the median of a melting curve without taking into account the percentage of hybridization.

To convert the delta $T_{50}H$ values into a phylogeny requires a procedure to obtain a hierarchical clustering of taxa. We use the "average linkage" unweighted pair group procedure which begins by clustering the closest pair or pairs of taxa. The next step links the taxa that have the smallest average distance to any existing cluster. This procedure continues until all taxa are linked. The use of the average linkage procedure to reconstruct phylogenies from DNA-DNA hybridization data is justified because the *same average rate* of base substitution has been shown to occur in all avian lineages, thus all branch lengths are equal (Sibley and Ahlquist 1983, and see below).

The experimental error in our data has been measured and a single delta $T_{50}H$ value should be assumed to have a possible error of ± 1.0 (Sibley and Ahlquist 1983:270–275). We have found it possible to compensate for all sources of experimental error by using five or more species and/or replicates for each pairwise comparison between monophyletic clusters so that an average delta $T_{50}H$, its standard error (s.e.), and standard deviation (s.d.) can be calculated for each branch node in a phylogeny. Because of the incremental summation of values in the average linkage procedure, the older nodes in the phylogenies are often the averages of 10, 20, or more delta $T_{50}H$ measurements. In Figure 1 there are nodes which are the averages of 32, 36, 47, 48, 92, 107, and 116 delta values.

THE UNIFORM AVERAGE RATE OF DNA EVOLUTION

The time dimension of the DNA hybridization data derives from the observed evidence that the same *average* rate of sequence evolution occurs in all lineages of birds. We have some evidence that the same average rate may also occur in mammals (Sibley and Ahlquist 1984). This should not be surprising because the Uniform Average Rate (UAR) is nothing more (or less) than the inevitable statistical result of averaging over billions of nucleotides and millions of years. Different sequences evolve at many different rates but the *average* rate is the same in all lineages because the genome is so large compared with the range in the rates of different sequences. Sibley and Ahlquist (1983:270–275) provide additional details about the UAR.

Because the genomes of all birds evolve at the same *average* rate, the delta $T_{50}H$ values are measures of *relative* time. They may, therefore, be used to reconstruct the branching pattern of a phylogeny. To convert the delta values into *absolute* time it is necessary to calibrate them against an external dating source, viz., fossils or geological events that have caused phyletic dichotomies. Using this procedure, we have obtained an estimate of the calibration such that delta $T_{50}H$ 1.0 = ca. 4.5 million years (MY) (Sibley and Ahlquist 1981, 1983, 1984, in press b; Sibley et al. 1982; Diamond 1983). This constant of proportionality is tentative and subject to correction, but we will use it to provide approximate dates for divergence nodes. See below under "The History of the New World Suboscine Passerines" for additional comments.

PHYLOGENY AND CLASSIFICATION

The phylogeny of the suboscines (Fig. 1) was developed by average linkage from a 16×16 matrix based on the data in Tables 1–16. Some data from previous studies (Sibley et al. 1982, 1984; Sibley and Ahlquist, in press) were also used, especially for the older divergence nodes.

At first glance the series of delta $T_{50}H$ values in the tables appear to have no breaks in the increasing magnitude of the numbers, and the values for different groups ("Group Index") overlap or interdigitate. The experimental error causes the overlap in delta values between groups that branched close together in time, and interdigitation results when two different, but closely related, groups branched via the same node relative to the lineage of the tracer species. The delta $T_{50}H$ values within a table are between the tracer species and each of the "driver" species, but *not* between the driver species. Also, because of the UAR of DNA evolution, all taxa on opposite sides of the same node are equidistant from one another and are, therefore, members of sister groups.

The derivation of a classification from a phylogeny may take various forms, depending upon the opinions of the classifier. We believe that a classification should reflect the branching pattern of the phylogeny, that taxonomic rank should be determined by the age of origin of taxa, and that sister groups should be of coordinate rank. We also believe that a classification derived from DNA hybridization data will express objective measurements of genealogical divergence, although it may not reflect the subjectively evaluated degrees of morphological specialization ("grades") as judged by the human eye. We thus follow Hennig (1966) concerning the relationship between phylogeny and classification.

A classification of the suboscines was derived from the phylogeny (Fig. 1) using the calculated times of origin (i.e., divergence nodes) to assign categorical ranks. Because the DNA-based phylogenies contain so many dichotomies, we have found it impractical to assign different names to every pair of sister groups. Instead, as a compromise, categorical levels were determined by dividing the delta $T_{50}H$ scale into segments and assigning the same category to all lineages branching within a given segment. To increase the congruence between the phylogeny and the classification we used Nelson's (1973) principles of "subordination and sequencing."

Thus, the groups in the classification are subordinated according to their times of origin, and the groups assigned to the same categorical level are listed in the sequence that reflects the age of origin of their lineage. This makes it possible to reconstitute the branching pattern of the phylogeny from the classification, but to avoid the proliferation of names that would result from a strict application of the Hennigian rule that every branch requires new categorical names for the resultant sister groups. For example, we assign the Infraorder category to three groups: Acanthisittides, Eurylaimides, and Tyrannides, although the phylogeny (Fig. 1) shows that the first is the sister group of the other two. Similarly, we have three Parvorders with the same pattern. This is not ideal, but we view it as a reasonable solution to the problem. The classification of the suboscines is relatively simple, but if we applied the strict sister group rule to the entire Order Passeriformes (Sibley and Ahlquist 1985, in press), the number of categories would be excessive by any standard.

The assignment of categories in our classification is based on the following average divergence node values in delta $T_{50}H$ units: Delta 0–4 = genera; 4–7 = tribes; 7–9 = subfamilies; 9–11 = families; 11–13 = superfamilies; 13–15 = parvorders; 15–17 = infraorders; 17–20 = suborders. We found it necessary to treat these boundaries as guidelines, not as hard barriers. The resulting classification reflects the phylogeny, but we recognize that there are discrepancies. This is a stage in the development of a classification of birds from genealogical distance data, not necessarily the final product. It is also among the first attempts to apply the principle of categorical equivalence (Sibley and Ahlquist 1982d).

In the following classification the bracketed taxa are sister groups.

Classification of the Suboscine Passeriformes

Order Passeriformes
  Suborder Oligomyodi
    Infraorder Acanthisittides
        Family Acanthisittidae, New Zealand Wrens
    ┌Infraorder Eurylaimides
    │   Superfamily Pittoidea
    │   Family Pittidae, Pittas
    │   Family Eurylaimidae, Broadbills
    │   Family *inc. sedis* Philepittidae, Asities
    └Infraorder Tyrannides
      Parvorder Tyranni
        Superfamily Tyrannoidea
        Family Tyrannidae
          ┌Subfamily Tyranninae, Tyrant Flycatchers
          └Subfamily Tityrinae
            Tribe Schiffornithini, *Schiffornis*
            Tribe Tityrini, Tityras, Becards
          ┌Subfamily Cotinginae, Cotingas, Sharpbills, Plantcutters
          └Subfamily Piprinae, Manakins
        Family Mionectidae, Mionectid Flycatchers
    ┌Parvorder Furnarii
    │ Superfamily Furnarioidea
    │ Family Furnariidae
    │   Subfamily Furnariinae, Ovenbirds
    │   Subfamily Dendrocolaptinae, Woodcreepers
    │ Superfamily Formicarioidea
    │   Family Formicariidae, Ground Antbirds
    │   ┌Family Rhinocryptidae, Tapaculos
    │   └Family Conopophagidae, Gnateaters
    └Parvorder Thamnophili
        Family Thamnophilidae, Typical Antbirds
  Suborder Passeres, Oscines

RESULTS AND DISCUSSION

The DNA-DNA hybridization comparisons show that the New World suboscine groups, which we place in the Infraorder Tyrannides, are more closely related to one another than

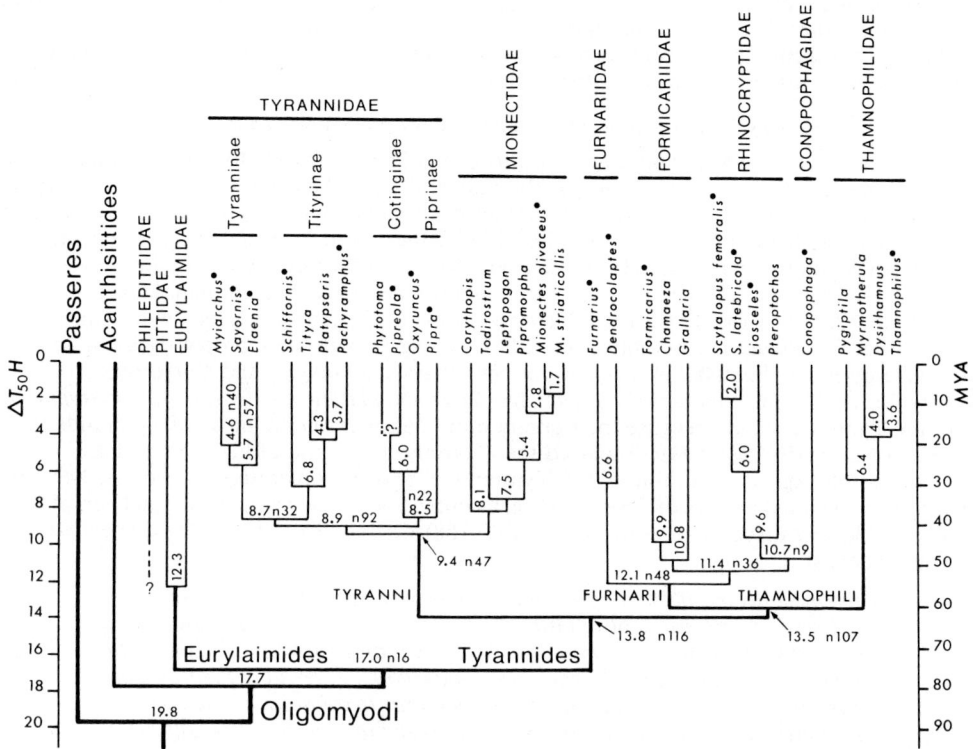

Fig. 1. Phylogeny of the suboscine passerine birds. Delta $T_{50}H$ values for divergence nodes were computed from the data in Tables 1–16. Black dots, e.g., *Myiarchus* ●, identify the radioiodine-labeled taxa. MYA = millions of years ago.

any one of them is to any Old World group. We have diagrammed the phylogeny of the living suboscine groups (Fig. 1), except for the Philepittidae, of which we lack DNA.

PARVORDER TYRANNI, SUPERFAMILY TYRANNOIDEA

We agree with Ames (1971:155) that this group should be restricted to the New World members of Wetmore's (1960) Tyrannoidea. The Old World Pittidae, Philepittidae, and Acanthisittidae are not closely related to the New World suboscines.

The syringeal characters of the tyrannoids are exceptionally variable, and Ames (1971:152) concluded that "the Tyrannoidea cannot be defined on the basis of the syrinx. Certain syringeal characters . . . appear to be useful at about the family and subfamily levels in the New World Tyrannoidea."

All tyrannoids have a two-notched sternum with little variation (Heimerdinger and Ames 1967:20).

Sclater (1888) and Ridgway (1907) defined some of the tyrannoid groups on the basis of foot structure and tarsal scutellation which Snow (1975:21) summarized as follows:

"Pipridae   Tarsus exaspidean; second phalanx of outer toe wholly united with that of middle toe.

Cotingidae   Tarsal scutellation of various types, but never exaspidean; second phalanx of outer and middle toes not united.

Tyrannidae   Tarsus exaspidean; neither outer nor inner toe united to middle toe by more than the basal part of the first phalanx."

Snow then noted that "the difficulty has been that a too rigid application of these criteria leads to the probable misplacement of a few genera; but no better criteria have been suggested, and recent anatomical studies have shown great variability within all three families, so that it is unlikely that clear-cut morphological criteria will be found."

Snow's conclusion apparently means that we cannot trust these characters to define the monophyletic clusters within the Tyrannoidea.

FAMILY TYRANNIDAE, SUBFAMILIES TYRANNINAE AND TITYRINAE

The tyrant flycatchers (Tyranninae) range from Alaska and Canada to Central America, the Caribbean islands, all of South America, the Falklands, and the Galapagos islands. Including the taxa that may belong to the Mionectidae (see below) Traylor (1977) recognized 374 species of tyrant flycatchers in 88 genera, and he divided them into three subfamilies: Elaeniinae, Fluvicolinae, and Tyranninae. He transferred *Attila, Pseudattila, Casiornis, Laniocera,* and *Rhytipterna* from the Cotingidae to the Tyrannidae on the basis of recommendations by Ames (1971) and Snow (1973). Traylor left *Lipaugus* in the Cotingidae but agreed with Ames et al. (1968) that *Corythopis* is a tyrannid. Our DNA data agree with these proposals for those taxa we have been able to study, namely, *Attila, Rhytipterna, Lipaugus,* and *Corythopis.*

Traylor (1977:136) "tentatively allied" the Tityrinae (*Tityra, Pachyramphus, Platypsaris*) "to the Tyrannidae only because their crania more nearly resemble those of Tyrannids than those of Cotingids." Snow (1979) placed the Tityrinae at the end of his Tyrannidae.

The morphological definition of the Tyrannidae, and the separation of tyrants, becards, cotingas, and manakins, are difficult because, as noted above, the characters usually employed are variable and discordant. Ames (1971:157–158) described the tyrant flycatchers (= our Tyranninae) as:

> "*the most diverse of suboscine families. Although the 'typical' tyrannid could be described as a small olive bird with a flat, slightly hooked bill and strong rictal bristles, there are more atypical forms than typical ones. Usually the tarsus is exaspidean and the toes lack syndactyly. . . . In all of their external characters the Tyrannidae are so variable that taxonomic boundaries and relationships within the family and with other families are often difficult to determine.*
>
> *With few exceptions the tyrannid syrinx is characterized by the presence of an intrinsic muscle, M. obliquus ventralis, and internal cartilages. Rarely, a second intrinsic muscle, M. obliquus lateralis, is present. Outside the traditional limits of the Tyrannidae (as revised by Hellmayr 1927) M. obliquus ventralis and internal cartilages occur in the New World Tyranni, separately or together, only in Oxyruncus, Corythopis, the manakin Ilicura and a few members of the Cotingidae.*"

Ames (1971:158–163) distinguished seven groups "each with a high degree of syringeal homogenity and with certain features not found elsewhere in the family." An eighth group, with six subgroups, contains genera of uncertain relationships, including *Mionectes, Pipromorpha,* and *Corythopis.* The becards (*Pachyramphus, Platypsaris*) were noted by Ames (1971: 163) as having "several tyrannid features of the syrinx that are not found in the more typical members of the Cotingidae." Ames thought it likely that the becards are "tyrannids" rather than "cotingids."

Warter (1965) found that the skulls of the becards are like those of tyrannine flycatchers and proposed that the Tityrinae be made a subfamily of the Tyrannidae, a conclusion that was accepted by Traylor (1977) and adopted by Snow (1979).

The DNA hybridization data support the morphological evidence of a fairly close relationship between the becards (Tityrinae) and the tyrants (Tyranninae) and the phylogram (Fig. 1) shows that they are sister groups with a divergence node at delta $T_{50}H$ 8.7. This indicates a divergence time ca. 35–40 MYA, in the late Eocene.

The genus *Schiffornis,* usually placed with the manakins, clusters with the Tityrinae on the basis of DNA comparisons. Its lineage branched from the becard clade at delta $T_{50}H$ 6.8, indicating a divergence in the Oligocene, ca. 30 MYA. This is a tentative allocation because we had only a small amount of *Schiffornis* DNA, and the delta values to some of the older divergence nodes are not concordant with the data from other taxa. These discrepancies can be seen in the tables containing DNA hybrids involving *Schiffornis.* However, note that the delta $T_{50}H$ values between *Schiffornis* and the becards in Appendices IV and V demonstrate the relationship between them.

FAMILY TYRANNIDAE, SUBFAMILY COTINGINAE

Snow (1973) recognized 27 genera and 79 species in his Cotingidae, but in 1979 he transferred the becards to the Tyrannidae, as the subfamily Tityrinae. Later Snow (1982) recognized 25 genera and 65 species of cotingas. Sibley et al. (1984) added *Oxyruncus* and the three species of *Phytotoma* to the Cotinginae.

The cotingas range from Mexico through Central and South America to northeastern Argentina and Paraguay. Cotingas vary in size from that of a kinglet (*Regulus*) to that of a crow (*Corvus*); they are exceptionally variable in color and external appearance with such extreme types as the umbrella birds (*Cephalopterus*), bellbirds (*Procnias*), and cocks-of-the-rock (*Rupicola*). The tiny Kinglet Calyptura (*Calyptura cristata*) has the size and coloration of a small manakin, but "its allocation to the Cotingidae is based on its tarsal scutellation and foot structure which are typical for cotingas: its tarsus is pycnaspidean . . . and its toes are free . . . . Its allocation to the cotingas has always been accepted, but its remoteness from the rest of the family in nearly all external characters supports the view that the cotingas may be an association of diverse phylogenetic lines that arose from the primitive tyranniform stock before the main lines leading to the flycatchers and manakins were recognizable" (Snow 1982:39). This discussion, and the uncertainties expressed in it, reflect the difficulties involved in determining genealogical relationships from morphological characters. The book by Snow (1982) provides a review of the cotingas, their characters, taxonomic problems, life histories, and distribution.

The DNA-DNA data (Appendices VI, VII; Fig. 1) confirm the inclusion in the Cotinginae of such morphologically diverse genera as *Pipreola*, *Lipaugus*, *Rupicola*, *Cephalopterus*, *Procnias*, and *Ampelion*. The Sharpbill (*Oxyruncus cristatus*) has usually been placed in the Tyrannidae or in a monotypic family. Sibley et al. (1984) compared the radioiodine-labeled single-copy DNA of *Oxyruncus* with the DNAs of other tyrannoids and found that it is clearly a cotinga, not a tyrant flycatcher. We include it in the subfamily Cotinginae.

Since 1888, when Sclater defined the tyrannoid groups, the cotingas have been viewed as having various types of tarsal scutellation, but never exaspidean, as in *Oxyruncus*. That the Sharpbill is a cotinga is supported by several characters in addition to the DNA hybridization evidence. *Oxyruncus* thus provides another example of the lack of congruence between tarsal scutellation and phylogenetic relationships in the suboscines.

Stiles and Whitney (1983) presented their observations on the behavior of the Sharpbill without coming to any conclusion concerning its relationships.

The plantcutters (*Phytotoma*) usually have been placed in a monotypic family, although clearly related to the cotingas. Küchler (1936) studied the pterylosis, external morphology, skeleton, syrinx, and alimentary tract of *P. rara* and concluded that *Phytotoma* is a cotinga. It has a pycnaspidean tarsal envelope, as do many cotingas, and the main artery of the leg is the sciatic, as in *Rupicola*. Ames (1971:156) noted the "evident similarity of the syrinx of *Phytotoma* to that of some cotingas (particularly *Heliochera*). . . ." (= *Ampelion*).

Although we do not yet have DNA hybridization data for the plantcutters, it seems clear that they are members of the cotingine radiation. Morphologically *Phytotoma* is even more like typical cotingas than is *Oxyruncus*, which is clearly a cotinga. We therefore include *Phytotoma* in the Cotinginae.

FAMILY TYRANNIDAE, SUBFAMILY PIPRINAE

Snow (1975, 1979) recognized 17 genera and 51 species of manakins. They occur from southeastern Mexico to southeastern Brazil, northeastern Argentina, southeastern Paraguay, Bolivia, and Peru. Most manakins are small, short-winged, short-tailed birds with small bills which are broad at the base and have a terminal hook on the maxilla. Most are sexually dimorphic, and the males of some species display in "courts." Only the females incubate.

Ames (1971:157) found considerable variation in the syringes of seven genera and noted that *Piprites* has a syrinx similar to that of the small tyrannids *Myiobius* and *Terenotriccus*. Other manakins show little similarity to either tyrannids or cotingas. Ames found that *Schiffornis*, the large, so-called "Thrush-like Manakin," has a syrinx "quite like that of the cotinga *Lipaugus* . . . *Schiffornis* is unique among manakins . . . in possessing internal cartilages, but the manakins are so heterogeneous in syringeal structure that I have come to expect almost anything."

W. E. Lanyon (pers. comm.) has found that, in addition to *Schiffornis*, the genera *Sapayoa*, *Piprites*, *Tyranneutes*, and *Neopelma* have internal cartilages. The other, "typical" manakins, lack these cartilages. All tyrannines have internal cartilages, and their presence in *Schiffornis*, which we find is most closely related to the becards, suggest that these other genera, which also have been considered to be manakins, may be either tyrannines or tityrines. Lanyon has examined 18 of the 25 genera of cotingas recognized by Snow (1982) and has found no internal cartilages in their syringes. Ames (1971:37) reported "a pair of very small . . . internal cartilages" in *Iodopleura*, a genus considered by Snow (1982) to be a cotinga. However, Ames (1971:156) noted that "The cotinga genus *Iodopleura* would appear from syringeal structure to be allied to the Tyrannidae. . . ."

Snow (1975:21) noted that *Manacus, Pipra, Chiroxiphia*, and a few smaller genera "are a well-defined group" but that *Schiffornis, Sapayoa*, and *Piprites* are "especially problematical" and (p. 24) their "allocation to the Pipridae [is] uncertain . . . all show some affinity to the cotingas or tyrant-flycatchers and are unspecialized in the direction of the typical manakins." Thus, in addition to *Schiffornis*, there are other "manakins" that may belong in the Tityrinae or the Tyranninae.

Tables 6 and 7 contain 24 delta $T_{50}H$ values between manakins and cotingas which average 8.5, s.e. = 0.04, s.d. = 0.2. We therefore place the two groups as sister subfamilies in the Tyrannidae.

FAMILY MIONECTIDAE

The DNA comparisons have revealed some new facets of the phylogeny of the tyrannoids and have confirmed certain proposals made by others. The most unexpected discovery was the evidence (Appendix VIII, Fig. 1) that a lineage of "flycatchers" branched from the rest of the tyrannoids before the radiation of the typical tyrants, becards, cotingas, and manakins began. This group, which we have designated as the family Mionectidae, includes, at least, *Mionectes* (including *Pipromorpha*), *Leptopogon, Pseudotriccus, Corythopis, Hemitriccus* (including *Idioptilon*), and *Todirostrum*. These genera are among those placed in the Elaeniinae by Traylor (1977:177). Traylor included *Pipromorpha* in *Mionectes*, but we separate them in Figure 1 to demonstrate their close, undoubtedly congeneric, relationship. The DNA evidence (Appendices I–III) indicates that the other genera in Traylor's Elaeniinae for which we have DNA data are indeed closer to *Elaenia* than to either *Sayornis* or *Myiarchus*, which represent Traylor's Fluvicolinae and Tyranninae, respectively. These include *Camptostoma, Phaeomyias, Suiriri, Tyrannulus, Myiopagis, Mecocerculus, Anairetes*, and *Euscarthmus*. The DNA data are equivocal about several other genera in Traylor's Elaeniinae, viz., *Lophotriccus, Atalotriccus, Cnipodectes, Rhynchocyclus*, and *Platyrinchus*. However, the DNA data agree completely with the genera Traylor assigned to his Fluvicolinae and Tyranninae, except for some for which our data do not indicate a clear choice. Thus, it seems likely that the Mionectidae are so convergently similar to the true elaeniines that they are difficult to separate by the characters used by Traylor. However, Ames (1971:162) noted that *Mionectes* and *Pipromorpha* have a "peculiar syringeal structure," and Traylor (1977:151–159) discussed some of the genera that we assign to the Mionectidae and noted several characters that seem to set them apart from the typical elaeniine tyrants. For example, *Leptopogon* and *Mionectes* build nests "unlike the nests of any Elaeniine flycatcher" and (Monroe 1968:277) they share the unusual behavioral trait of "single-wing flicking." As Traylor noted, these characters cannot be fully evaluated because they are unknown for many genera. The same problem afflicts the DNA evidence. We lack material of many genera, and it would be necessary to "label" the DNAs of several more to clarify the boundaries of the Mionectidae. The divergence of the Mionectidae and the Tyrannidae at delta $T_{50}H$ 9.4 is based on 47 DNA hybrids and we predict that the genera listed above as mionectids will be found to share additional morphological characters. To some degree they do share the "Type I" nasal septum defined by Warter (1965; see Traylor 1977:182), and most of them belong to the group that Warter (1965) called the "primitive tyrannids." W. E. Lanyon (pers. comm.) has told us that according to his examination of a large number of genera, the "primitive nasal capsule" is present in *Corythopis, Myiornis, Tachuris, Tolmomyias*, some species of *Myiopagis, Todirostrum, Platyrinchus, Onychorhynchus, Oncostoma, Rhynchocyclus, Hemitriccus, Lophotriccus, Poecilotriccus, Cnipodectes*, and, possibly *Leptopogon*, but not *Mionectes*!

We conclude that the Mionectidae is a valid taxon but that the determination of its boundaries and defining morphological characters will require additional study.

Parvorders Furnarii and Thamnophili

The DNA comparisions have revealed that the "Furnarioidea" of Wetmore (1960) is more complex than previously thought (see Fig. 1 and our classification). The "typical" antbirds (= our Thamnophilidae) are the descendants of one branch of a dichotomy at delta $T_{50}H$ 13.5, which indicates that this divergence occurred in the Paleocene, ca. 60 MYA. The sister branch produced the living groups we place in the Furnarioidea and Formicarioidea. The latter includes the "ground antbirds" (Formicariidae), the gnateaters (Conopophagidae), and the tapaculos (Rhinocryptidae). As we will show, the DNA hybridization data for these groups are congruent with the variations in the sternal notches described by Heimerdinger and Ames (1967), the characters of *Conopophaga* and *Corythopis* (Ames et al. 1968), and the syringeal characters studied by Ames (1971).

Parvorder Furnarii, Superfamily Furnarioidea

Ames (1971:153) found that the members of his suborder Furnarii (= our parvorder Furnarii) "share an elaborate syringeal form not found elsewhere. No structures even remotely resembling the Membrana trachealis and Processus vocalis are found in other passerines."

Family Furnariidae, Subfamily Furnariinae

Vaurie (1980) recognized 34 genera and ca. 214 species of ovenbirds. They range from central Mexico south over all of Central and South America to Tierra del Fuego and the Falkland Islands.

The ovenbirds may be morphologically and ecologically the most diverse avian subfamily yet recognized. They occur in all habitats from sea level to the puna zone in the Andes. Their adaptive radiation has produced morphotypes that converge on larks, wheatears, nuthatches, thrushes, thrashers, wrens, and dippers. Most of the species are small to medium in size among passerines, and they tend to be dull colored, often in shades of brown.

The syringes of the ovenbirds and woodcreepers (Furnarioidea) have two pairs of intrinsic muscles and, thus, differ from the members of the Formicarioidea (ground antbirds, gnateaters, tapaculos), which have either one pair, or no intrinsic muscles.

The syringes of the ovenbirds (except *Geositta*) differ from those of the woodcreepers in their lack of "horns" on the Processi vocales (Ames 1971:154).

Garrod (1877) described the nares in the ovenbirds as "schizorhinal," a reference to the shape of the long nasal opening that extends posteriad from the naso-frontal hinge. Feduccia (1973) found considerable variation in the bony nares in ovenbirds and termed them "pseudo-schizorhinal" in most species because they are not as deeply slit as the "true" schizorhinal condition in the Charadriiformes. Feduccia noted that the nares of some philydorine ovenbirds tend toward holorhiny, as in the woodcreepers.

Family Furnariidae, Subfamily Dendrocolaptinae

The 13 genera and ca. 47 species of woodcreepers (or woodhewers) range from northern Mexico to northern Argentina, Bolivia, and Peru. They tend to be larger than the ovenbirds, from 20 to 40 cm long, and are mostly olive or rufous brown, often streaked or barred below. The shafts of the rectrices are usually stiffened, and most species forage in a manner convergently similar to the Northern creepers (*Certhia*) or woodpeckers (Marchant 1964).

The woodcreepers also differ from the ovenbirds by having "horns" on the Processi vocales of the syrinx (Ames 1971:154), and holorhinal nares. However, *Geositta*, an otherwise typical ovenbird, has large "horns" on the Processi vocales.

Feduccia (1973) compared several anatomical characters and the electrophoretic patterns of the hemoglobins of ovenbirds and woodcreepers. He concluded that they are "very closely related groups" and that "the group of ovenbirds most closely related to the woodhewers is the subfamily Philydorinae. . . ." Feduccia found four genera that he considered to be "intermediates" between ovenbirds and woodcreepers.

The DNA hybridization data (Appendices IX, X) agree with the morphological evidence of the close relationship between ovenbirds and woodcreepers. The two groups diverged at delta $T_{50}H$ 6.6, which equates with ca. 30 MYA, in the Oligocene. They could be considered to be either tribes or subfamilies in the family Furnariidae.

PARVORDER FURNARII, SUPERFAMILY FORMICARIOIDEA

The three families we recognize in the Formicarioidea share two derived morphological structures, namely, a distinctive syrinx and a four-notched sternum. These synapomorphies are congruent with the DNA hybridization data. As noted above, the traditional antbird family Formicariidae of Wetmore (1960) and others, was a diphyletic composite of the "ground antbirds" and the "typical antbirds."

FAMILY FORMICARIIDAE, GROUND ANTBIRDS

The number and patterns of the notches and fenestra in the posterior border of the sternum in the suboscines were studied by Heimerdinger and Ames (1967). Two principal types, two-notched and four-notched, were found, with considerable variation in each. Most "antbirds" have a two-notched sternum, but the "ground-antbirds," including *Pittasoma, Grallaricula,* and *Grallaria,* have a four-notched sternum, or a tendency toward that condition. Other than these only the sternum of *Conopophaga* tends to be four-notched, and those of the rhinocryptids are consistently so. All other suboscines have two-notched sterna. [Feduccia and Olson (1982:19–20) commented on these characters.] Ames (1971:154) discovered that the "antbirds may be divided into two groups on the basis of syringeal morphology. The majority . . . "typical antbirds," are distinguished by having one pair of intrinsic syringeal muscles, a very small Processus, and M. sternotrachealis bifurcate near its insertion. Examples . . . are *Taraba, Dysithamnus, Thamnophilus,* and *Myrmotherula.* The second group, the "ground antbirds," is characterized by the absence of intrinsic syringeal muscles, a large Processus, and a simple M. sternotrachealis. To this group belong *Grallaria, Chamaeza, Formicarius,* and *Conopophaga.* Long-legged terrestrial birds, they appear to be intermediate between the Formicariidae [= our Thamnophilidae] and the Rhinocryptidae. Such intermediacy is suggested by the presence of a four-notch metasternum, classically a rhinocryptid character, in some species of *Grallaria* and in *Pittasoma* (Heimerdinger and Ames 1967)."

The morphological characters and DNA-DNA measurements are congruent in several respects (Appendices XI–XVI, Fig. 1). The delta $T_{50}H$ distances from *Formicarius* to *Chamaeza* (9.7, 10.1) and *Grallaria* (10.7, 10.9) could support the recognition of a subfamily or even another family. We prefer to leave the classification as presented above until additional genera can be "labeled" to delineate the boundaries of clusters within the formicarioid families. As we presently understand the Formicariidae it includes, at least, *Formicarius, Chamaeza, Grallaria, Grallaricula,* and *Pittasoma.* Feduccia and Olson (1982:19–20) came to virtually the same conclusion, based on morphological characters. They noted variations in some of the skeletal characters. It is possible that *Myrmornis, Myrmothera, Hylopezus,* and *Thamnocharis* are also members of the "ground antbird" group, which would then contain ca. 50 species. The other ca. 44 genera of antbirds are apparently members of the Thamnophilidae.

FAMILY CONOPOPHAGIDAE

The nine species of gnateaters in the genus *Conopophaga* have often been placed, with *Corythopis,* in the Conopophagidae. The gnateaters are small, long-legged, short-tailed birds with short, rounded wings. Most of the species have a post-ocular stripe of elongate, usually white, feathers. They are insectivorous ground-dwellers in the tropical and temperate forests of South America from Colombia, Surinam, and Brazil south to northeastern Argentina, southern Paraguay, Bolivia, and Peru.

Ames et al. (1968) reviewed the taxonomic history of the group and, primarily on the basis of sternal and syringeal characters, recommended that *Conopophaga* be transferred to the Formicariidae and placed near *Grallaria.* The genus *Corythopis,* which had long been associated with *Conopophaga,* was moved to the "Tyrannidae."

The DNA data (Appendix XII, Fig. 1) indicate that the rhinocryptid-conopophagid clade branched from the formicariid lineage at delta $T_{50}H$ 11.4 (ca. 50 MYA), and the conopophagids branched from the rhinocryptids at delta 10.7 (ca. 48 MYA). We recognize the Conopophagidae and Rhinocryptidae as families since they diverged between delta $T_{50}H$ 9 and 11. However, we have not placed *Grallaria* in a family apart from the Formicariidae, although its lineage diverged from that of *Formicarius* at delta $T_{50}H$ 10.8. This is an inconsistency in the application of our criteria for defining categories, but until the DNAs of additional formicarioid genera can be compared, we think it best to refrain from further splitting of the Formicariidae. The DNA data testify to the long history of these taxa and explain the distribution of their distinctive morphological characters.

FAMILY RHINOCRYPTIDAE

The 26 to 28 species of tapaculos are placed in 11 or 12 genera (Peters 1951; Meyer de Schauensee 1966). Most of the rhinocryptids are ground-dwellers, and they occur in a wide range of habitats from Costa Rica to Patagonia, and from sea level to ca. 4000 m in the Andes. They vary in size from that of a small wren to that of a large thrush and tend to run, rather than to fly (Sick 1964).

The opercula that cover the nostrils are the basis for the name of the type genus, and all rhinocryptids have a four-notched sternum (Heimerdinger and Ames 1967).

"In their syrinx the rhinocryptids possess a simple M. sternotrachealis and a dorsally originating intrinsic muscle. One genus, *Telodromas,* lacks the intrinsic muscle, its syrinx being much like those of *Formicarius* and *Grallaria.* The tapaculos are outwardly similar to the short-tailed, long-legged, ground-living antbirds which they resemble in syringeal structure . . . ." (Ames 1971:154).

Ames et al. (1968:24) noted that the pterylosis of the ventral tract in the rhinocryptids is "unique within passerines; the one exception to this is *Melanopareia maranonicus* whose pterylosis is very similar to several genera of Formicariidae . . . ."

The DNA hybridization data (Appendices 13–15, Fig. 1) show that the tapaculos are formicarioids and the sister group of the Conopophagidae. The excellent congruence between the morphological characters and the DNA comparisons is apparent.

PARVORDER THAMNOPHILI, FAMILY THAMNOPHILIDAE

The ca. 44 genera and 175 species of "typical antbirds" occur from southern Mexico to northern Argentina, Bolivia, and Peru. Some species associate with army ants, but most do not.

As noted above, the thamnophilids have a two-notched sternum and a syrinx with one pair of intrinsic muscles, a small Processus, and a bifurcate M. sternotrachealis.

The Furnarii and the Thamnophili are separated by a large phylogenetic gap (Appendices XI–XVI, Fig. 1). The node at delta $T_{50}H$ 13.5 is the average for 107 DNA-DNA hybrids and indicates that the dichotomy occurred in the Paleocene, ca. 60 MYA.

Heimerdinger and Ames (1967) and Ames (1971) discovered the morphological distinctions between the ground antbirds and the typical antbirds, and correctly associated the former with *Conopophaga* and the rhinocryptids. The DNA data support their conclusions and, because of the uniform average rate of DNA evolution, provide the time dimension that makes it possible to reconstruct the phylogeny. The congruence between the morphological and molecular evidence argues persuasively for the correctness of the phylogeny presented herein.

HISTORY OF THE NEW WORLD SUBOSCINE PASSERINES

The phylogeny of the suboscines (Fig. 1) provides the basis for a partial reconstruction of their historical biogeography, and for the calibration of the delta $T_{50}H$ measurements in absolute time.

The divergence between the Eurylaimides and the Tyrannides (delta $T_{50}H$ 17.0) presumably began when South America became isolated from the Old World portions of Gondwanaland as Africa and South America drifted apart during the Cretaceous. The last emergent land was probably across the South Atlantic between the Walvis Ridge and the Rio Grande Rise (Linden 1980; Reyment 1980). The submergence of this connection occurred ca. 80 MYA, although intermittent islands may have persisted into the early Tertiary (Bonatti and Chermak 1981).

Until ca. 80 MYA there was a fairly direct pathway between Africa and Australia via Madagascar, India, and such continental fragments as Kerguelen Island and the Seychelles, all of which were then far south of their present positions. Australia was 3000 km south of Asia and was connected to South America via Antarctica, which had a temperate climate (Hamilton 1979; Webb and Tracey 1981).

Thus, both of the two possible connections between Africa and South America were probably severed ca. 80 MYA. It is, therefore, reasonable to assume that the divergence between the Eurylaimides and the Tyrannides occurred ca. 80 MYA. This yields a calibration constant of delta $T_{50}H$ 1.0 = 4.7 MY (80/17.0).

In our study of the ratite birds (Sibley and Ahlquist 1981:322) we assumed that the common ancestor of the Ostrich (*Struthio camelus*) and the Common Rhea (*Rhea americana*) was separated into African and South American populations ca. 80 MYA by the opening of the

Atlantic, as described above. The delta $T_{50}H$ for the Ostrich-Rhea divergence is 17.4 (see revised calculation in Diamond 1983), thus the calibration constant is 4.6 (80/17.4).

The New Zealand wrens (Acanthisittides in Fig. 1) diverged from the other suboscines at delta $T_{50}H$ 17.7, presumably when the Tasman Sea opened between Australia and New Zealand, ca. 80 MYA, thus, the calibration constant is 4.5 (80/17.7) (Sibley et al. 1982).

In a study of the hominoid primates the divergence of the Orangutan (*Pongo pygmaeus*) lineage was dated from fossils at 16 or 17 MYA, and the delta $T_{50}H$ for the branch node was measured as 3.7. Thus, the constant is 4.3 or 4.6 (Sibley and Ahlquist 1984:13). The 17 MYA divergence was based on a recently discovered fossil from Kenya (Alan Walker, pers. comm.), and is probably closer to the true divergence time than the 16 MYA fossils from Pakistan described by Pilbeam (1983).

Thus, we have five calibration constants (4.3, 4.5, 4.6, 4.6, 4.7), with an average of 4.5, which we now use to calculate divergence times. When the three avian constants near 80 MYA, and the two hominoid constants at 16–17 MYA, are plotted on the regression between the delta $T_{50}H$ scale and absolute time, they fall on a straight line that passes through zero. As a result,

> *"We may be close to the correct calibration, but not as close as the available techniques permit. Furthermore, there are uncertainties in all of the fossil and geological datings and additional calibration points should be obtained before concluding that the dating problem is solved.... Comparisions between vicariant taxon pairs that began their divergences when the southern continents drifted apart will ... contribute to the solution of the calibration problem."*
>
> C. G. Sibley and J. E. Ahlquist 1984 (p. 13).

The triple branching at delta $T_{50}H$ 13.5–13.8 that produced the lineages leading to the Tyranni, Furnarii, and Thamnophili (Fig. 1), occurred ca. 60–62 MYA, in the Paleocene.

South America was an island continent that moved slowly northward during the Cenozoic, resulting in the emergence of the Panamanian land bridge, ca. 3 MYA (Marshall et al. 1982). The long isolation of South America provided the opportunities for adaptive radiations that produced the present diversity of the suboscines and other groups, including the oscine tanagers (Thraupini) and troupials (Icterini) (Sibley and Ahlquist, in press). The members of many South American groups that now occur north of the Isthmus of Panama presumably have expanded their ranges since the Pliocene.

The living "old endemic" passerines of Australia and New Guinea are ten-primaried oscines that began their radiation from a single ancestral taxon ca. 55–60 MYA, thus they were not derived from South American suboscines or New World nine-primaried oscines (Sibley and Ahlquist 1985). However, the Acanthisittides of New Zealand may be the only survivors of an earlier radiation that became extinct in Australia between 80 and 60 MYA. This is a reasonable hypothesis because the Acanthisittides branched from the other suboscines ca. 79–80 MYA (delta 17.7 × 4.5), thus 3–4 MY before the Eurylaimides and Tyrannides diverged from one another 76–77 MYA (17.0 × 4.5). The origin of the Australo-Papuan old endemics must have been from an African or Asian ten-primaried ancestor.

In summary, the passeriformes originated from an as-yet-unidentified non-passerine ancestor more than 90 MYA, and split into the two lineages that evolved into the suboscines (Oligomyodi) and oscines (Passeres) ca. 90 MYA (19.8 × 4.5). The next known divergence produced the Acanthisittides, of which only the New Zealand wrens (Acanthisittidae) survive, but this group may have been present in Australia, and perhaps Antarctica, before ca. 60 MYA.

The Eurylaimides and Tyrannides diverged ca. 76–77 MYA, when the Atlantic became impassable, and the adaptive radiation of the New World suboscines occurred in South America during the late Mesozoic and Cenozoic.

ACKNOWLEDGMENTS

We thank L. Wallace, R. Gatter, E. Nowicki, M. Mattie, and F. C. Sibley for laboratory assistance. For other help we are indebted to M. Avedillo, W. Belton, A. H. Bledsoe, J. I. Borrero, J. duPont, S. Furniss, A. Garza de Leon, P. Gertler, G. Graves, S. M. Lanyon, W. E. Lanyon, T. O. Lemke, C. O'Neill, J. O'Neill, T. Parker, H. D. Pratt, R. J. Raikow, J. V. Remsen, T. Schulenberg, R. Semba, F. C. Sibley, T. F. Smith, J. Spendelow, D. Tallman, N. and E. Wheelwright, A. Wilson, and D. Wysham.

Yale University and the National Science Foundation (DEB-79-26746) supported the laboratory work.

LITERATURE CITED

AMES, P. L. 1971. The morphology of the syrinx in passerine birds. Bull. Peabody Mus. Nat. Hist. 37: 1–194.

AMES, P. L., M. A. HEIMERDINGER, AND S. L. WARTER. 1968. The anatomy and systematic position of the antpipits *Conopophaga* and *Corythopis*. Postilla 114:1–32.

BONATTI, E., AND A. CHERMAK. 1981. Formerly emerging crustal blocks in the equatorial Atlantic. Tectonophysics 72:165–180.

BRITTEN, R. J., AND D. E. KOHNE. 1968. Repeated sequences in DNA. Science 161:529–540.

BRITTEN, R. J., D. E. GRAHAM, AND B. R. NEUFELD. 1974. Analysis of repeating DNA sequences by reassociation. Methods Enzymology 29:363–418.

DIAMOND, J. M. 1983. Taxonomy by nucleotides. Nature 305:17–18.

FEDUCCIA, A. 1973. Evolutionary trends in the Neotropical ovenbirds and woodhewers. Ornithol. Monogr. No. 13.

FEDUCCIA, A. 1975. Morphology of the bony stapes in the Menuridae and Acanthisittidae: evidence for oscine affinities. Wilson Bull. 87:418–420.

FEDUCCIA, A., AND S. L. OLSON. 1982. Morphological similarities between the Menurae and the Rhinocryptidae, relict passerine birds of the southern hemisphere. Smithson. Contrib. Zool. No. 366.

GADOW, H. 1893. Vögel, II. Systematischer Theil. *In* H. G. Bronn, Klassen und Ordnungen des Thierreichs, Vol. 6(4), 303 pp. C. F. Winter, Leipzig.

GARROD, A. H. 1877. Notes on the anatomy of passerine birds. Part 2. Proc. Zool. Soc. Lond. 1877: 447–452.

GRULA, J. W., T. J. HALL, J. A. HUNT, T. D. GIUGNI, G. J. GRAHAM, E. H. DAVIDSON, AND R. J. BRITTEN. 1982. Sea urchin DNA sequence variation and reduced interspecies differences of the less variable sequences. Evolution 36:665–676.

HALL, T. J., J. W. GRULA, E. H. DAVIDSON, AND R. J. BRITTEN. 1980. Evolution of sea urchin non-repetitive DNA. J. Mol. Evol. 16:95–110.

HAMILTON, W. 1979. Tectonics of the Indonesian region. U.S. Geol. Survey Prof. Paper 1078:1–345.

HEIMERDINGER, M. A., AND P. L. AMES. 1967. Variation in the sternal notches of suboscine passerine birds. Postilla 105:1–44.

HELLMAYR, C. E. 1927. Catalogue of birds of the Americas. Pt. 5. Field Mus. Nat. Hist. Publ. Zool. Ser. No. 242, 13:1–517.

HENNIG, W. 1966. Phylogenetic Systematics. University Illinois Press, Urbana, Illinois.

KOHNE, D. E. 1970. Evolution of higher-organism DNA. Q. Rev. Biophysics 33:327–375.

KOHNE, D. E., AND R. J. BRITTEN. 1971. Hydroxyapatite techniques for nucleic acid reassociation. Procedures Nucleic Acid Res. 2:500–512.

KÜCHLER, W. 1936. Anatomische Untersuchungen an *Phytotoma rara* Mol. J. Ornithol. 84:352–362.

LINDEN, W. J. M. VAN DER 1980. Walvis Ridge, a piece of Africa? Geology 8:417–421.

MARCHANT, S. 1964. Ovenbird. Pp. 571–574, *In* A. L. Thomson (ed)., A New Dictionary of Birds. Thos. Nelson, London.

MARSHALL, L. G., S. D. WEBB, J. J. SEPKOSKI, JR., AND D. M. RAUP. 1982. Mammalian evolution and the great American interchange. Science 215:1315–1357.

MAYR, E., AND D. AMADON. 1951. A classification of Recent birds. Am. Mus. Novit. No. 1496.

MEYER DE SCHAUENSEE, R. 1966. The Species of Birds of South America. Adademy Natural Sciences Philadelphia, Livingston Publ. Co., Narberth, Pennsylvania.

MONROE, B. L., JR. 1968. A distributional survey of the birds of Honduras. Ornithol. Monogr. No. 7.

NELSON, G. J. 1973. Classification as an expression of phylogenetic relationships. Syst. Zool. 22:344–359.

OLSON, S. L. 1971. Taxonomic comments on the Eurylaimidae. Ibis 113:507–516.

PETERS, J. L. 1951. Check-list of Birds of the World. Vol. 7. Museum Comparative Zoology, Cambridge, Massachusetts.

PILBEAM, D. 1983. Hominoid evolution and hominid origins. Pp. 43–61, *In* C. Chagas (ed.), Recent advances in the evolution of the primates. Pontif. Acad. Scient. Scripta Varia 50.

REYMENT, R. A. 1980. Mid-Cretaceous events. Nature and Resources 16:28–34.

RIDGWAY, R. 1907. The birds of North and Middle America. Pt. 4. U.S. Natl. Mus. Bull. 50:1–973.

SCLATER, P. L. 1888. Catalogue of the Birds in the British Museum. Vol. 14. Trustees, British Museum, London.

SIBLEY, C. G. 1970. A comparative study of the egg-white proteins of passerine birds. Yale Univ. Bull. Peabody Mus. Nat. Hist. 32:1–131.

SIBLEY, C. G. 1974. The relationships of the lyrebirds. Emu 74:65–79.

SIBLEY, C. G., AND J. E. AHLQUIST. 1981. The phylogeny and relationships of the ratite birds as indicated by DNA-DNA hybridization. Pp. 303–335, *In* G. G. E. Scudder and J. L. Reveal (eds.), Evolution Today. Proc. 2nd Intl. Congr. Syst. Evol. Biol., Hunt Inst. Bot. Doc., Carnegie-Mellon University, Pittsburgh, Pennsylvania.

SIBLEY, C. G., AND J. E. AHLQUIST. 1982a. The relationships of the Hawaiian honeycreepers (Drepaninini) as indicated by DNA-DNA hybridization. Auk 99:130–140.

SIBLEY, C. G., AND J. E. AHLQUIST. 1982b. The relationships of the Wrentit (*Chamaea fasciata*) as indicated by DNA-DNA hybridization. Condor 84:40–44.

SIBLEY, C. G., AND J. E. AHLQUIST. 1982c. The relationships of the vireos (Vireoninae) as indicated by DNA-DNA hybridization. Wilson Bull. 94:114–128.

SIBLEY, C. G., AND J. E. AHLQUIST. 1982d. The relationships of the Yellow-breasted Chat (*Icteria virens*), and the alleged "slow-down" in the rate of macromolecular evolution in birds. Postilla 187:1–19.

SIBLEY, C. G., AND J. E. AHLQUIST. 1982e. The relationships of the Australo-Papuan scrub-robins *Drymodes* as indicated by DNA-DNA hybridization. Emu 82:101–105.

SIBLEY, C. G., AND J. E. AHLQUIST. 1982f. The relationships of the swallows. (Hirundinidae). J. Yamashina Inst. Ornithol. 14:122–130.

SIBLEY, C. G., AND J. E. AHLQUIST. 1983. The phylogeny and classification of birds, based on the data of DNA-DNA hybridization. Current Ornithol. 1:245–292.

SIBLEY, C. G., AND J. E. AHLQUIST. 1984. The phylogeny of the hominoid primates, as indicated by DNA-DNA hybridization. J. Mol. Evol. 1984:2–15.

SIBLEY, C. G., AND J. E. AHLQUIST. 1985. The phylogeny and classification of the Australo-Papuan passerine birds. Emu 85:1–14.

SIBLEY, C. G., AND J. E. AHLQUIST. In press. The phylogeny and classification of the passerine birds, based on comparisons of the genetic material, DNA. Proc. 18th Int. Ornithol. Congr. V. D. Ilyichev (ed.). Nauka Publ., Moscow.

SIBLEY, C. G., G. R. WILLIAMS, AND J. E. AHLQUIST. 1982. The relationships of the New Zealand wrens (Acanthisittidae) as indicated by DNA-DNA hybridization. Notornis 29:113–130.

SIBLEY, C. G., S. M. LANYON, AND J. E. AHLQUIST. 1984. The relationships of the Sharpbill (*Oxyruncus cristatus*). Condor 86:48–52.

SICK, H. 1964. Tapaculo. Pp. 805–806, *In* A. L. Thomson (ed.), A New Dictionary of Birds. Thos. Nelson, London.

SNOW, D. W. 1973. The classification of the Cotingidae (Aves). Breviora 409:1–27.

SNOW, D. W. 1975. The classification of the manakins. Bull. Br. Ornithol. Club 95:20–27.

SNOW, D. W. 1979. Family Cotingidae. Pp. 281–308, *In* M. A. Traylor, Jr. (ed.), Check-list of Birds of the World. Vol. 8. Museum Comparative Zoology, Cambridge, Massachusetts.

SNOW, D. W. 1982. The Cotingas. Cornell University Press, Ithaca, New York.

STILES, F. G., AND B. WHITNEY. 1983. Notes on the behavior of the Costa Rican Sharpbill (*Oxyruncus cristatus frater*). Auk 100:117–125.

TRAYLOR, M. A. 1977. A classification of the tyrant flycatchers (Tyrannidae). Bull. Mus. Comp. Zool. 148:129–184.

VAURIE, C. 1980. Taxonomy and geographical distribution of the Furnariidae (Aves, Passeriformes). Bull. Am. Mus. Nat. Hist. 166:1–357.

WARTER, S. L. 1965. The cranial osteology of the New World Tyrannoidea and its taxonomic implications. Unpubl. Ph.D. dissert., Louisiana State University, Baton Rouge.

WEBB, L. J., AND J. G. TRACEY. 1981. Australian rainforests: patterns and change. Pp. 605–694, *In* A. Keast (ed.), Ecological Biogeography of Australia. W. Junk, The Hague.

WETMORE, A. 1930. A systematic classification for the birds of the world. Proc. U.S. Natl. Mus. 76(24): 1–8.

WETMORE, A. 1934. A systematic classification for the birds of the world, revised and amended. Smithson. Misc. Collect. 89(13):1–11.

WETMORE, A. 1940. A systematic classification of the birds of the world. Smithson. Misc. Collect. 99(7): 1–11.

WETMORE, A. 1951. A revised classification for the birds of the world. Smithson. Misc. Collect. 117(4): 1–22.

WETMORE, A. 1960. A classification for the birds of the world. Smithson. Misc. Collect. 139(11):1–37.

APPENDIX I

DNA-DNA Hybridization Values Between the Brown-crested Flycatcher (*Myiarchus Tyrannulus*) and Other Passerine Birds

Common name	Scientific name		Delta T₅₀H	Group index[1]
Brown-crested Flycatcher	*Myiarchus tyrannulus*		0.0	T
Ash-throated Flycatcher	*Myiarchus cinerascens*		0.2	T
Pale-edged Flycatcher	*Myiarchus cephalotes*		0.8	T
Short-crested Flycatcher	*Myiarchus ferox*	(3)[2]	0.9	T
Dusky-capped Flycatcher	*Myiarchus tuberculifer*		0.9	T
Apical Flycatcher	*Myiarchus apicalis*		1.4	T
Boat-billed Flycatcher	*Megarhynchus pitangua*		2.5	T
Rufous Mourner	*Rhytipterna holerythra*	(2)	2.5	T
Social Flycatcher	*Myiozetetes similis*	(2)	2.9	T
Bright-rumped Attila	*Attila spadiceus*	(2)	3.1	T
Cinnamon Attila	*Attila cinnamomeus*		3.2	T
Streaked Flycatcher	*Myiodynastes maculatus*		3.4	T
Great Kiskadee	*Pitangus sulphuratus*	(3)	3.5	T
Eastern Wood-Pewee	*Contopus virens*		3.9	T
Eastern Kingbird	*Tyrannus tyrannus*		3.9	T
Tropical Kingbird	*Tyrannus melancholicus*		4.1	T
Black-billed Shrike-Tyrant	*Agriornis montana*		4.2	T
White-headed Marsh-Tyrant	*Arundinicola leucocephala*		4.2	T
Pied Water-Tyrant	*Fluvicola pica*		4.2	T
Cattle Flycatcher	*Machetornis rixosus*		4.2	T
Fork-tailed Flycatcher	*Tyrannus savana*		4.2	T
Slaty-backed Chat-Tyrant	*Ochthoeca cinnamomeiventris*		4.2	T
Brown-backed Chat-Tyrant	*Ochthoeca fumicolor*		4.2	T
Fuscous Flycatcher	*Cnemotriccus fuscatus*		4.2	T
Acadian Flycatcher	*Empidonax virescens*		4.3	T
Rufous-tailed Tyrant	*Knipolegus poecilurus*		4.3	T
Puna Ground-Tyrant	*Muscisaxicola juninensis*		4.3	T
Flavescent Flycatcher	*Myiophobus flavicans*		4.3	T
Crowned Chat-Tyrant	*Ochthoeca frontalis*		4.3	T
White-tailed Shrike-Tyrant	*Agriornis albicauda*		4.4	T
Streamer-tailed Tyrant	*Gubernetes yetapa*		4.4	T
White-winged Black-Tyrant	*Knipolegus aterrimus*		4.4	T
Eastern Phoebe	*Sayornis phoebe*	(3)	4.5	T
Willow Flycatcher	*Empidonax traillii*		4.6	T
Olive-sided Flycatcher	*Contopus borealis*		4.6	T
Vermilion Flycatcher	*Pyrocephalus rubinus*		4.7	T
Least Flycatcher	*Empidonax minimus*		5.1	T
Lesser Elaenia	*Elaenia chiriquensis*		5.1	T
Tawny-crowned Pigmy-Tyrant	*Euscarthmus meloryphus*		5.1	T
Scrub Flycatcher	*Sublegatus arenarum*		5.2	T
White-crested Elaenia	*Elaenia albiceps*	(2)	5.3	T
Yellow-bellied Elaenia	*Elaenia flavogaster*		5.4	T
Mountain Elaenia	*Elaenia frantzii*	(2)	5.4	T
White-throated Tyrannulet	*Mecocerculus leucophrys*		5.5	T
Cinnamon Flycatcher	*Pyrrhomyias cinnamomea*		5.5	T
Yellow Tyrannulet	*Capsiempis flaveola*		5.8	T
Mouse-colored Tyrannulet	*Phaeomyias murina*		5.8	T
Greenish Elaenia	*Myiopagis viridicata*		6.2	T
White-crested Spadebill	*Platyrinchus platyrhynchos*		6.4	T
Yellow-crowned Tyrannulet	*Tyrannulus elatus*		6.5	T
Brownish Flycatcher	*Cnipodectes subbrunneus*		7.5	?
Scaly-crested Pygmy-Tyrant	*Lophotriccus pileatus*		7.5	?
Barred Becard	*Pachyramphus versicolor*		7.6	Ti
Golden-headed Manakin	*Pipra erythrocephala*	(2)	7.6	P
Band-tailed Manakin	*Pipra fasciicauda*		7.7	P
Rufous-headed Pygmy-Tyrant	*Pseudotriccus ruficeps*		7.7	?
Pale-eyed Pygmy-Tyrant	*Atalotriccus pilaris*		7.8	?
Swallow-tailed Manakin	*Chiroxiphia caudata*		7.8	P
Blue-backed Manakin	*Chiroxiphia pareola*		7.8	P

APPENDIX I

CONTINUED

Common name	Scientific name		Delta T$_{50}$H	Group index[1]
White-winged Becard	*Pachyramphus polychopterus*		7.8	Ti
Olivaceous Piha	*Lipaugus cryptolophus*	(2)	7.8	C
Blue-crowned Manakin	*Pipra coronata*		7.9	P
Pink-throated Becard	*Platypsaris minor*		7.9	Ti
Black-tailed Tityra	*Tityra cayana*		7.9	Ti
Andean Cock-of-the-rock	*Rupicola peruviana*		7.9	C
Ruddy-tailed Flycatcher	*Terenotriccus erythrurus*		8.0	?
Masked Tityra	*Tityra semifasciata*		8.0	Ti
Plum-throated Cotinga	*Cotinga maynana*		8.0	C
Black-crowned Tityra	*Tityra inquisitor*		8.1	Ti
Red-capped Manakin	*Pipra mentalis*		8.1	P
Jet Manakin	*Chloropipo unicolor*		8.1	P
Yellow-crowned Manakin	*Heterocercus flavivertex*		8.1	P
Striped Manakin	*Machaeropterus regulus*		8.2	P
Golden-breasted Fruiteater	*Pipreola aureopectus*		8.2	P
Royal Flycatcher	*Onychorhynchus coronatus*		8.2	?
Fulvous-breasted Flatbill	*Rhynchocyclus fulvipectus*		8.3	?
Green Manakin	*Chloropipo holochroa*		8.3	P
Sulphur-rumped Flycatcher	*Myiobius barbatus*		8.4	?
Wire-tailed Manakin	*Pipra filicauda*		8.5	P
Tawny-breasted Flycatcher	*Myiobius villosus*		8.5	?
Amazonian Umbrellabird	*Cephalopterus ornatus*		8.6	C
Golden-winged Manakin	*Masius chrysopterus*		8.7	P
Barred Fruiteater	*Pipreola arcuata*	(4)	8.7	C
Thrushlike Manakin	*Schiffornis turdinus*		8.7	Ti
Green-and-black Fruiteater	*Pipreola riefferii*		8.8	C
Red-crested Cotinga	*Ampelion rubrocristatus*		8.8	C
White-collared Manakin	*Manacus candei*		8.8	P
Bare-necked Fruitcrow	*Gymnoderus foetidus*		8.9	C
Stripe-necked Tody-Tyrant	*Idioptilon striaticolle*		8.9	M
Common Tody-Flycatcher	*Todirostrum cinereum*	(2)	9.1	M
Streak-necked Flycatcher	*Mionectes striaticollis*		9.1	M
Bare-throated Bellbird	*Procnias nudicollis*		9.2	C
Purple-throated Fruitcrow	*Querula purpurata*		9.2	C
Sepia-capped Flycatcher	*Leptopogon amaurocephalus*	(2)	9.3	M
Olive-striped Flycatcher	*Mionectes olivaceus*		9.4	M
Slaty-capped Flycatcher	*Leptopogon superciliaris*		9.4	M
Ringed Antpipit	*Corythopis torquata*		9.5	M
Ochre-bellied Flycatcher	*Mionectes oleagineus*		9.9	M
Brown-rumped Tapaculo	*Scytalopus latebricola*		13.0	R
White-shouldered Antbird	*Mymeciza melanoceps*		13.0	Th
Rufous-vented Tapaculo	*Scytalopus femoralis*		13.1	R
Black-billed Treehunter	*Thripadectes melanorhynchus*		13.2	F
Warbling Antbird	*Hypocnemis cantator*		13.2	Th
Rufous Gnateater	*Conopophaga lineata*	(2)	13.4	Co
Spotted Barbtail	*Premnoplex brunnescens*		13.6	F
White-chinned Thistletail	*Schizoeaca fuliginosa*		13.6	F
Plain Xenops	*Xenops minutus*		13.6	F
White-flanked Antwren	*Myrmotherula axillaris*		13.8	Th
Spotted-crowned Woodcreeper	*Lepidocolaptes affinis*		14.0	F
Spot-winged Antbird	*Percnostola leucostigma*		14.0	Th
Plain-brown Woodcreeper	*Dendrocincla fuliginosa*		14.1	F
Chestnut-crowned Gnateater	*Conopophaga castaneiceps*		14.2	Co
Spot-winged Antshrike	*Pygiptila stellaris*		14.2	Th
Rufous Hornero	*Furnarius rufus*		14.3	F
Curve-billed Scythebill	*Campylorhamphus procurvoides*		14.4	F
Tawny Antpitta	*Grallaria quitensis*		14.5	Fo
Black-capped Antshrike	*Thamnophilus schistaceus*		14.5	Th
Plain Antvireo	*Dysithamnus mentalis*		14.8	Th
Azara's Spinetail	*Synallaxis azarae*		14.9	F

APPENDIX I
CONTINUED

Common name	Scientific name	Delta $T_{50}H$	Group index[1]
Green Broadbill	*Calyptomena viridis*	16.3	E
New Zealand Rifleman	*Acanthisitta chloris*	17.5	A
Clay-colored Robin	*Turdus grayi*	19.3	O
Little Raven	*Corvus mellori*	19.4	O

[1] T = Tyranninae, P = Piprinae, Ti = Tityrinae, C = Cotinginae, M = Mionectidae, R = Rhinocryptidae, Th = Thamnophilidae, Co = Conopophagidae, F = Furnariidae, Fo = Formicariidae, E = Eurylaimidae, A = Acanthisittidae, O = Oscine, ? = uncertain affinities.
[2] Numbers in parentheses indicate that more than one DNA hybrid was averaged.

APPENDIX II
DNA-DNA HYBRIDIZATION VALUES BETWEEN THE EASTERN PHOEBE
(*SAYORNIS PHOEBE*) AND OTHER PASSERINE BIRDS

Common name	Scientific name		Delta $T_{50}H$	Group index[1]
Eastern Phoebe	*Sayornis phoebe*		0.0	T
Olive-sided Flycatcher	*Contopus borealis*		2.1	T
Eastern Wood-Pewee	*Contopus virens*		2.3	T
Least Flycatcher	*Empidonax minimus*		2.3	T
Acadian Flycatcher	*Empidonax virescens*		2.4	T
Slaty-backed Chat-Tyrant	*Ochthoeca cinnamomeiventris*		3.4	T
Crowned Chat-Tyrant	*Ochthoeca frontalis*		3.5	T
White-tailed Shrike-Tyrant	*Agriornis albicauda*		3.5	T
Puna Ground Tyrant	*Muscisaxicola juninensis*		3.6	T
Brown-backed Chat-Tyrant	*Ochthoeca fumicolor*		3.8	T
Smoky Bush-Tyrant	*Myiotheretes fumigatus*		3.8	T
Black-billed Shrike-Tyrant	*Agriornis montana*		3.9	T
Streamer-tailed Tyrant	*Gubernetes yetapa*		4.0	T
White-winged Black-Tyrant	*Knipolegus aterrimus*		4.1	T
Cattle Flycatcher	*Machetornis rixosus*		4.2	T
Rufous-tailed Tyrant	*Knipolegus poecilurus*		4.4	T
Short-tailed Field-Tyrant	*Muscigralla brevicauda*		4.4	T
Vermilion Flycatcher	*Pyrocephalus rubinus*		4.4	T
Scrub-Flycatcher	*Sublegatus arenarum*		4.4	T
White-headed Marsh-Tyrant	*Arundinicola leucocephala*		4.5	T
Pied Water-Tyrant	*Fluvicola pica*		4.5	T
Dusky-capped Flycatcher	*Myiarchus tuberculifer*		4.5	T
Brown-crested Flycatcher	*Myiarchus tyrannulus*	(2)[2]	4.6	T
Boat-billed Flycatcher	*Megarhynchus pitangua*		4.7	T
Rufous Mourner	*Rhytipterna holerythra*		4.7	T
Eastern Kingbird	*Tyrannus tyrannus*		4.7	T
Bright-rumped Attila	*Attila spadiceus*	(4)	4.8	T
Cinnamon Attila	*Attila cinnamomeus*		4.8	T
Grayish Mourner	*Rhytipterna simplex*	(4)	4.9	T
Streaked Flycatcher	*Myiodynastes maculatus*		4.9	T
Mouse-colored Tyrannulet	*Phaeomyias murina*		5.0	T
Yellow-crowned Tyrannulet	*Tyrannulus elatus*		5.0	T
Tawny-crowned Pygmy-Tyrant	*Euscarthmus meloryphus*		5.2	T
Mountain Elaenia	*Elaenia frantzii*		5.3	T
Scissor-tailed Flycatcher	*Tyrannus forficatus*		5.4	T
Greenish Elaenia	*Myiopagis viridicata*		5.4	T
Social Flycatcher	*Myiozetetes similis*		5.4	T
Great Kiskadee	*Pitangus sulphuratus*		5.4	T
White-throated Tyrannulet	*Mecocerculus leucophrys*		5.5	T

APPENDIX II
CONTINUED

Common name	Scientific name		Delta T$_{50}$H	Group index[1]
Lesser Elaenia	*Elaenia chiriquensis*		5.6	T
Tufted Tit-Tyrant	*Anairetes parulus*		5.6	T
Sulphury Flycatcher	*Tyrannopsis sulphurea*		5.7	T
White-crested Elaenia	*Elaenia albiceps*		5.7	T
Fuscous Flycatcher	*Cnemotriccus fuscatus*		6.0	T
Suiriri Flycatcher	*Suiriri suiriri*		6.2	T
Cinnamon Flycatcher	*Pyrrhomyias cinnamomea*		6.7	T
White-throated Spadebill	*Platyrinchus mystaceus*		7.9	?
Blue-backed Manakin	*Chiroxiphia pareola*		8.3	P
Golden-headed Manakin	*Pipra erythrocephala*	(4)	8.4	P
Andean Cock-of-the-rock	*Rupicola peruviana*		8.4	C
Amazonian Umbrellabird	*Cephalopterus ornatus*		8.5	C
Sepia-capped Flycatcher	*Leptopogon amaurocephalus*		8.5	M
Masked Tityra	*Tityra semifasciata*		8.5	Ti
Black-tailed Tityra	*Tityra cayana*	(4)	8.6	Ti
Green Manakin	*Chloropipo holochroa*		8.6	P
Barred Fruiteater	*Pipreola arcuata*	(3)	8.7	C
Scale-crested Pygmy-Tyrant	*Lophotriccus pileatus*		8.8	?
Stripe-necked Tody-Tyrant	*Idioptilon striaticolle*		8.8	M
Olivaceus Piha	*Lipaugus cryptolophus*	(4)	8.9	C
Common Tody-Flycatcher	*Todirostrum cinereum*		9.1	M
White-winged Becard	*Pachyramphus polychopterus*		9.2	C
Olive-striped Flycatcher	*Mionectes olivaceus*		9.2	M
Ochre-bellied Flycatcher	*Mionectes oleagineus*		9.2	M
Sulphur-rumped Flycatcher	*Myiobius barbatus*		9.4	?
Barred Becard	*Pachyramphus versicolor*		9.4	Ti
Royal Flycatcher	*Onychorhynchus coronatus*		9.4	?
Fulvous-breasted Flatbill	*Rhynchocyclus fulvipectus*		9.5	?
Rufous-headed Pygmy-Tyrant	*Pseudotriccus ruficeps*		9.7	M
Scale-crested Pygmy-Tyrant	*Lophotriccus pileatus*		9.9	?
Thrushlike Manakin	*Schiffornis turdinus*		10.6	Ti

[1] T = Tyranninae, P = Piprinae, C = Cotinginae, Ti = Tityrinae, M = Mionectidae, ? = uncertain affinities.
[2] Numbers in parentheses indicate that more than one DNA hybrid was averaged.

APPENDIX III
DNA-DNA HYBRIDIZATION VALUES BETWEEN THE MOUNTAIN ELAENIA (*ELAENIA FRANTZII*)
AND OTHER PASSERINE BIRDS

Common name	Scientific name	Delta T$_{50}$H	Group index[1]
Mountain Elaenia	*Elaenia frantzii*	0.0	T
White-crested Elaenia	*Elaenia albiceps*	2.2	T
Lesser Elaenia	*Elaenia chiriquensis*	2.3	T
Yellow-bellied Elaenia	*Elaenia flavogaster*	3.7	T
White-throated Tyrannulet	*Mecocerculus leucophrys*	3.9	T
Mouse-colored Tyrannulet	*Phaeomyias murina*	4.0	T
Yellow Tyrannulet	*Capsiempis flaveola*	4.2	T
Southern Beardless Flycatcher	*Camptostoma obsoletum*	4.5	T
Tufted Tit-Tyrant	*Anairetes parulus*	4.8	T
Greenish Elaenia	*Myiopagis viridicata*	5.0	T
Fuscous Flycatcher	*Cnemotriccus fuscatus*	5.1	T

APPENDIX III
CONTINUED

Common name	Scientific name		Delta $T_{50}H$	Group index[1]
Yellow-crowned Tyrannulet	*Tyrannulus elatus*		5.1	T
White-headed Marsh-Tyrant	*Arundinicola leucocephala*		5.3	T
White-winged Black-Tyrant	*Knipolegus aterrimus*		5.4	T
Smoky Bush-Tyrant	*Myiotheretes fumigatus*		5.4	T
Tawny-crowned Pygmy-Tyrant	*Euscarthmus meloryphus*		5.5	T
Eastern Wood-Pewee	*Contopus virens*		5.5	T
Rufous-tailed Tyrant	*Knipolegus poecilurus*		5.5	T
Streamer-tailed Tyrant	*Gubernetes yetapa*		5.6	T
Acadian Flycatcher	*Empidonax virescens*		5.6	T
Suiriri Flycatcher	*Suiriri suiriri*		5.6	T
Crowned Chat-Tyrant	*Ochthoeca frontalis*		5.7	T
Brown-backed Chat-Tyrant	*Ochthoeca fumicolor*		5.8	T
Bright-rumped Attila	*Attila spadiceus*		5.8	T
Short-tailed Field-Tyrant	*Muscigralla brevicauda*		5.8	T
Eastern Kingbird	*Tyrannus tyrannus*		5.8	T
Eastern Phoebe	*Sayornis phoebe*	(3)[2]	5.9	T
Puna Ground-Tyrant	*Muscisaxicola juninensis*		5.9	T
Olive-sided Flycatcher	*Contopus borealis*		5.9	T
Great Kiskadee	*Pitangus sulphuratus*		5.9	T
Cinnamon Attila	*Attila cinnamomeus*		6.0	T
Scissor-tailed Flycatcher	*Tyrannus forficatus*		6.0	T
Streaked Flycatcher	*Myiodynastes maculatus*		6.0	T
Grayish Mourner	*Rhytipterna simplex*		6.0	T
Cattle Flycatcher	*Machetornis rixosus*		6.1	T
Boat-billed Flycatcher	*Megarhynchus pitangua*		6.1	T
Ornate Flycatcher	*Myiotriccus ornatus*		6.1	T
Black-billed Shrike-Tyrant	*Agriornis montana*		6.3	T
Pied Water-Tyrant	*Fluvicola pica*		6.3	T
Social Flycatcher	*Myiozetetes similis*		6.3	T
Brown-crested Flycatcher	*Myiarchus tyrannulus*	(2)	6.5	T
Short-crested Flycatcher	*Myiarchus ferox*		6.5	T
Scale-crested Pygmy-Tyrant	*Lophotriccus pileatus*		7.1	M?
Green Manakin	*Chloropipo holochroa*		8.4	P
White-throated Spadebill	*Platyrinchus mystaceus*		8.5	M?
Barred Becard	*Pachyramphus versicolor*		8.5	Ti
Stripe-necked Tody-Tyrant	*Idioptilon striaticolle*		8.6	M
Black-tailed Tityra	*Tityra cayana*		8.7	Ti
Golden-headed Manakin	*Pipra erythrocephala*		8.8	P
Olivaceous Piha	*Lipaugus cryptolophus*		8.9	C
Masked Tityra	*Tityra semifasciata*		8.9	Ti
Barred Fruiteater	*Pipreola arcuata*		8.9	C
White-winged Becard	*Pachyramphus polychopterus*		9.0	Ti
Common Tody-Flycatcher	*Todirostrum cinereum*		9.0	M
Ruddy-tailed Flycatcher	*Terenotriccus erythrurus*		9.1	M?
Thrushlike Manakin	*Schiffornis turdinus*		9.1	Ti
Sepia-capped Flycatcher	*Leptopogon amaurocephalus*		9.1	M
Ochre-bellied Flycatcher	*Mionectes oleagineus*		9.2	M
Andean Cock-of-the-rock	*Rupicola peruviana*		9.2	C
Blue-backed Manakin	*Chiroxiphia pareola*		9.3	P
Royal Flycatcher	*Onychorhynchus coronatus*		9.5	M?
Slaty-capped Flycatcher	*Leptopogon superciliaris*		9.8	M
Olive-striped Flycatcher	*Mionectes olivaceus*	(3)	10.0	M
Fulvous-breasted Flatbill	*Rhynchocyclus fulvipectus*		10.0	M?
Streak-necked Flycatcher	*Mionectes striaticollis*		10.3	M

[1] T = Tyranninae, M = Mionectidae, M? = probably Mionectidae, P = Piprinae, Ti = Tityrinae, C = Cotinginae.
[2] Numbers in parentheses indicate that more than one DNA hybrid was averaged.

APPENDIX IV

DNA-DNA Hybridization Values Between the Thrushlike Manakin (*Schiffornis turdinus*) and Other Passerine Birds

Common name	Scientific name		Delta $T_{50}H$	Group index[1]
Thrushlike Manakin	*Schiffornis turdinus*		0.0	Ti
Black-tailed Tityra	*Tityra cayana*		6.5	Ti
Barred Becard	*Pachyramphus versicolor*		6.6	Ti
White-winged Becard	*Pachyramphus polychopterus*	(2)[2]	6.7	Ti
Pink-throated Becard	*Platypsaris minor*		6.8	Ti
Masked Tityra	*Tityra semifasciata*		6.9	Ti
Flavescent Flycatcher	*Myiophobus flavicans*		7.8	?
Lance-tailed Manakin	*Chiroxiphia lanceolata*		8.2	P
Blue-backed Manakin	*Chiroxiphia pareola*		8.3	P
Jet Manakin	*Chloropipo unicolor*		8.4	P
Green Manakin	*Chloropipo holochroa*		8.5	P
Golden-headed Manakin	*Pipra erythrocephala*		8.5	P
White-collared Manakin	*Manacus candei*	(2)	8.6	P
White-bearded Manakin	*Manacus manacus*		8.6	P
Golden-winged Manakin	*Masius chrysopterus*		8.6	P
Band-tailed Manakin	*Pipra fasciicauda*	(2)	9.0	P
Yellow-crowned Manakin	*Heterocercus flavivertex*		9.1	P
Amazonian Umbrellabird	*Cephalopterus ornatus*		9.1	C
Sulphur-rumped Flycatcher	*Myiobius sulphureipygius*		9.2	T
Green-and-black Fruiteater	*Pipreola riefferii*		9.2	C
Blue-crowned Manakin	*Pipra coronata*		9.3	P
Red-capped Manakin	*Pipra mentalis*		9.3	P
Greenish Elaenia	*Myiopagis viridicata*		9.3	T
Bare-necked Fruitcrow	*Gymnoderus foetidus*		9.4	C
Bare-throated Bellbird	*Procnias nudicollis*		9.4	C
Striped Manakin	*Machaeropterus regulus*		9.5	P
Golden-breasted Fruiteater	*Pipreola aureopectus*	(2)	9.6	C
Crowned Chat-Tyrant	*Ochthoeca frontalis*		9.6	T
Ringed Antpipit	*Corythopis torquata*		9.7	M
Rufous Mourner	*Rhytipterna holerythra*		9.7	T
Andean Cock-of-the-rock	*Rupicola peruviana*		9.9	C
Slaty-capped Flycatcher	*Leptopogon superciliaris*		9.9	M
White-winged Black-Tyrant	*Knipolegus aterrimus*		10.3	T
Short-crested Flycatcher	*Myiarchus ferox*		10.9	T
Bright-rumped Attila	*Attila spadiceus*		11.1	T
Ochre-bellied Flycatcher	*Mionectes oleagineus*		11.2	M
Olive-striped Flycatcher	*Mionectes olivaceus*		11.5	M
Olivaceous Woodcreeper	*Sittasomus griseicapillus*		12.5	F
Chestnut-crowned Gnateater	*Conopophaga castaneiceps*		12.9	Co
White-chinned Thistletail	*Schizoeaca fuliginosa*		13.4	F
Plumbeous Antbird	*Myrmeciza hyperythra*		13.5	Th
Rufous-vented Tapaculo	*Scytalopus femoralis*		14.3	R

[1] Ti = Tityrinae, P = Piprinae, C = Cotinginae, T = Tyranninae, M = Mionectidae, F = Furnariidae, Co = Conopophagidae, Th = Thamnophilidae, R = Rhinocryptidae, ? = uncertain affinities.
[2] Numbers in parentheses indicate that more than one DNA hybrid was averaged.

APPENDIX V

DNA-DNA Hybridization Values Between the White-winged Becard (*Pachyramphus polychopterus*) and Other Passerine Birds

Common name	Scientific name	Delta T$_{50}$H	Group index[1]
White-winged Becard	*Pachyramphus polychopterus*	0.0	Ti
Barred Becard	*Pachyramphus versicolor*	3.7	Ti
Pink-throated Becard	*Platypsaris minor*	3.7	Ti
Masked Tityra	*Tityra semifasciata*	4.2	Ti
Black-tailed Tityra	*Tityra cayana*	4.3	Ti
Thrushlike Manakin	*Schiffornis turdinus*	7.7	Ti
White-throated Tyrannulet	*Mecocerculus leucophrys*	8.0	T
Amazonian Umbrellabird	*Cephalopterus ornatus*	8.2	C
Green Manakin	*Chloropipo holochroa*	8.2	P
Sharpbill	*Oxyruncus cristatus*	8.3	C
Golden-headed Manakin	*Pipra erythrocephala*	8.3	P
Red-crested Cotinga	*Ampelion rubrocristatus*	8.4	C
Yellow-crowned Manakin	*Heterocercus flavivertex*	8.4	P
White-collared Manakin	*Manacus candei*	8.5	P
Purple-throated Fruitcrow	*Querula purpurata*	8.5	C
Andean Cock-of-the-rock	*Rupicola peruviana*	8.5	C
Blue-backed Manakin	*Chiroxiphia pareola*	8.7	P
Eastern Wood-Pewee	*Contopus virens*	8.7	T
Mountain Elaenia	*Elaenia frantzii*	8.7	T
Bare-necked Fruitcrow	*Gymnoderus foetidus*	8.7	C
Bare-throated Bellbird	*Procnias nudicollis*	8.7	C
Grayish Mourner	*Rhytipterna simplex*	8.7	T
Bright-rumped Attila	*Attila spadiceus*	8.8	T
Golden-breasted Fruiteater	*Pipreola aureopectus*	8.8	C
Cinnamon Attila	*Attila cinnamomeus*	8.9	T
Olivaceous Piha	*Lipaugus cryptolophus*	9.0	C
Puna Ground-Tyrant	*Muscisaxicola juninensis*	9.0	T
Short-crested Flycatcher	*Myiarchus ferox*	9.0	T
Brown-backed Chat-Tyrant	*Ochthoeca fumicolor*	9.0	T
Green-and-black Fruiteater	*Pipreola riefferii* (2)[2]	9.1	C
Rufous Mourner	*Rhytipterna holerythra*	9.1	T
Striped Manakin	*Machaeropterus regulus*	9.1	P
Slaty-capped Flycatcher	*Leptopogon superciliaris*	9.1	M
Eastern Kingbird	*Tyrannus tyrannus*	9.2	T
Sepia-capped Flycatcher	*Leptopogon amaurocephalus*	9.3	M
Eastern Phoebe	*Sayornis phoebe* (2)	9.5	T
Mouse-colored Tyrannulet	*Phaeomyias murina*	9.5	T
Greenish Elaenia	*Myiopagis viridicata*	9.6	T
Common Tody-Flycatcher	*Todirostrum cinereum*	9.7	M
Olive-striped Flycatcher	*Mionectes olivaceus* (2)	9.8	M
Streak-necked Flycatcher	*Mionectes striaticollis*	10.1	M
Ochre-bellied Flycatcher	*Mionectes oleagineus*	10.3	M
Rufous-breasted Antthrush	*Formicarius rufipectus*	11.8	Fo
White-chinned Thistletail	*Schizoeca fuliginosa*	13.7	F
Barred Antshrike	*Thamnophilus doliatus*	14.9	Th

[1] Ti = Tityrinae, T = Tyranninae, C = Cotinginae, P = Piprinae, M = Mionectidae, Fo = Formicariidae, F = Furnariidae, Th = Thamnophilidae.

[2] Numbers in parentheses indicate that more than one DNA hybrid was averaged.

DNA-DNA HYBRIDIZATION VALUES BETWEEN THE BARRED FRUITEATER (*PIPREOLA ARCUATA*) AND OTHER PASSERINE BIRDS

Common name	Scientific name	Delta $T_{50}H$	Group index[1]
Barred Fruiteater	*Pipreola arcuata*	0.0	C
Green-and-black Fruiteater	*Pipreola riefferii*	1.5	C
Golden-breasted Fruiteater	*Pipreola aureopectus*	2.5	C
Olivaceous Piha	*Lipaugus cryptolophus*	6.6	C
Plum-throated Cotinga	*Cotinga maynana* (3)[2]	7.1	C
Purple-throated Fruitcrow	*Querula purpurata*	7.2	C
Andean Cock-of-the-rock	*Rupicola peruviana*	7.3	C
Amazonian Umbrellabird	*Cephalopterus ornatus*	7.4	C
Bare-throated Bellbird	*Procnias nudicollis*	7.7	C
Bare-necked Fruitcrow	*Gymnoderus foetidus*	7.7	C
Red-crested Cotinga	*Ampelion rubrocristatus*	8.0	C
Yellow-crowned Manakin	*Heterocercus flavivertex*	8.4	P
Golden-winged Manakin	*Masius chrysopterus*	8.5	P
White-collared Manakin	*Manacus candei*	8.5	P
Wire-tailed Manakin	*Pipra filicauda*	8.5	P
Red-capped Manakin	*Pipra mentalis*	8.5	P
Golden-headed Manakin	*Pipra erythrocephala* (4)	8.6	P
Swallow-tailed Manakin	*Chiroxiphia caudata*	8.8	P
Bright-rumped Attila	*Attila spadiceus*	8.9	T
Scissor-tailed Flycatcher	*Tyrannus forficatus*	8.9	T
Brown-crested Flycatcher	*Myiarchus tyrannulus*	8.9	T
Green Manakin	*Chloropipo holochlora*	9.0	P
Cattle Flycatcher	*Machetornis rixosus*	9.0	T
White-crested Spadebill	*Platyrinchus platyrhynchos*	9.0	T
Rufous Mourner	*Rhytipterna holerythra*	9.2	T
Willow Flycatcher	*Empidonax traillii*	9.3	T
Brown-backed Chat-Tyrant	*Ochthoeca fumicolor*	9.3	T
Rufous-tailed Tyrant	*Knipolegus poecilurus*	9.4	T
Rusty-margined Flycatcher	*Myiozetetes cayanensis*	9.4	T
Great Kiskadee	*Pitangus sulphuratus* (3)	9.4	T
Short-crested Flycatcher	*Myiarchus ferox*	9.5	T
Vermilion Flycatcher	*Pyrocephalus rubinus*	9.5	T
Short-tailed Field-Tyrant	*Muscigralla brevicauda*	9.7	T
White-crested Elaenia	*Elaenia albiceps*	9.7	T
Eastern Phoebe	*Sayornis phoebe* (2)	9.8	T
Lesser Elaenia	*Elaenia chiriquensis*	9.8	T
Yellow-bellied Elaenia	*Elaenia flavogaster*	9.8	T
Black-masked Tityra	*Tityra cayana*	9.8	Ti
Barred Becard	*Pachyramphus versicolor*	9.9	Ti
White-winged Becard	*Pachyramphus polychopterus*	10.0	Ti
White-throated Tyrannulet	*Mecocerculus leucophrys*	10.0	T
Tufted Tit-Tyrant	*Anairetes parulus*	10.1	T
Thrushlike Manakin	*Schiffornis turdinus*	10.2	Ti
Southern Beardless Flycatcher	*Camptostoma obsoletum*	10.2	T
Common Tody-Flycatcher	*Todirostrum cinereum*	10.2	M
Sepia-capped Flycatcher	*Leptopogon amaurocephalus*	10.3	M
Ringed Antpipit	*Corythopis torquata*	10.4	M
Pied Water-Tyrant	*Fluvicola pica*	10.5	T
Streak-necked Flycatcher	*Mionectes striaticollis*	10.6	M
Ochre-bellied Flycatcher	*Mionectes oleagineus*	10.6	M
Rufous-vented Tapaculo	*Scytalopus femoralis*	13.2	R
Chestnut-crowned Gnateater	*Conopophaga castaneiceps*	13.4	Co
Rufous Hornero	*Furnarius rufus*	14.2	F
Azara's Spinetail	*Synallaxis azarae* (2)	14.3	F
Barred Woodcreeper	*Dendrocolaptes certhia*	14.4	F
Rufous-capped Antthrush	*Formicarius colma*	14.5	Fo
Green Broadbill	*Calyptomena viridis*	17.6	E
Noisy Pitta	*Pitta versicolor*	17.9	Pt
American Robin	*Turdus migratorius*	20.2	O

[1] C = Cotinginae, P = Piprinae, M = Mionectidae, Ti = Tityrinae, T = Tyranninae, R = Rhinocryptidae, Co = Conopophagidae, F = Furnariidae, Fo = Formicariidae, E = Eurylaimidae, P = Pittidae, O = Oscine, ? = uncertain affinities.

[2] Numbers in parentheses indicate that more than one DNA hybrid was averaged.

APPENDIX VII

DNA-DNA Hybridization Values Between the Golden-headed Manakin (*Pipra erythrocephala*) and Other Passerine Birds

Common name	Scientific name		Delta T$_{50}$H	Group index[1]
Golden-headed Manakin	*Pipra erythrocephala*		0.0	P
Yellow-crowned Manakin	*Heterocercus flavivertex*		2.1	P
Red-capped Manakin	*Pipra mentalis*		2.2	P
Swallow-tailed Manakin	*Chiroxiphia caudata*		2.3	P
White-bearded Manakin	*Manacus manacus*		2.4	P
Golden-collared Manakin	*Manacus vitellinus*		2.4	P
Blue-crowned Manakin	*Pipra coronata*		2.5	P
Blue-backed Manakin	*Chiroxiphia pareola*		2.5	P
Green Manakin	*Chloropipo holochroa*		2.5	P
Striped Manakin	*Machaeropterus regulus*		2.5	P
White-collared Manakin	*Manacus candei*		2.6	P
Wire-tailed Manakin	*Pipra filicauda*		2.7	P
Jet Manakin	*Chloropipo unicolor*		2.8	P
Golden-winged Manakin	*Masius chrysopterus*		3.2	P
Plum-throated Cotinga	*Cotinga maynana*		8.2	C
Olivaceous Piha	*Lipaugus cryptolophus*		8.2	C
Andean Cock-of-the-rock	*Rupicola peruviana*		8.2	C
Red-crested Cotinga	*Ampelion rubrocristatus*		8.3	C
Amazonian Umbrellabird	*Cephalopterus ornatus*		8.3	C
Bare-necked Fruitcrow	*Gymnoderus foetidus*		8.3	C
Barred Fruiteater	*Pipreola arcuata*	(3)[2]	8.4	C
Pink-throated Becard	*Platypsaris minor*		8.4	Ti
Masked Tityra	*Tityra semifasciata*		8.4	Ti
Golden-breasted Fruiteater	*Pipreola aureopectus*		8.5	C
Green-and-black Fruiteater	*Pipreola riefferii*		8.6	C
Bare-throated Bellbird	*Procnias nudicollis*		8.6	C
Black-tailed Tityra	*Tityra cayana*		8.7	Ti
Purple-throated Fruitcrow	*Querula purpurata*		8.7	C
White-throated Tyrannulet	*Mecocerculus leucophrys*		8.7	T
Flavescent Flycatcher	*Myiophobus flavicans*		8.7	T
Rufous Mourner	*Rhytipterna holerythra*		8.8	T
Grayish Mourner	*Rhytipterna simplex*		8.8	T
Bright-rumped Attila	*Attila spadiceus*		8.8	T
White-winged Becard	*Pachyramphus polychopterus*		8.8	Ti
Cinnamon Attila	*Attila cinnamomeus*		9.0	T
Lesser Elaenia	*Elaenia chiriquensis*		9.0	T
Thrushlike Manakin	*Schiffornis turdinus*	(2)	9.0	Ti
Mountain Elaenia	*Elaenia frantzii*		9.1	T
Social Flycatcher	*Myiozetetes similis*		9.2	T
Tawny-crowned Pigmy-Tyrant	*Euscarthmus meloryphus*		9.2	T
Crowned Chat-Tyrant	*Ochthoeca frontalis*		9.3	T
Barred Becard	*Pachyramphus versicolor*		9.3	Ti
Eastern Kingbird	*Tyrannus tyrannus*		9.3	T
Black-billed Shrike-Tyrant	*Agriornis montana*		9.4	T
Southern Beardless Flycatcher	*Camptostoma obsoletum*		9.4	T
Acadian Flycatcher	*Empidonax virescens*		9.4	T
Brown-crested Flycatcher	*Myiarchus tyrannulus*	(3)	9.4	T
Royal Flycatcher	*Onychorhynchus coronatus*		9.5	M?
Sulphur-rumped Flycatcher	*Myiobius barbatus*	(2)	9.7	T
Suiriri Flycatcher	*Suiriri suiriri*		9.8	T
Slaty-capped Flycatcher	*Leptopogon superciliaris*		9.8	M
Sepia-capped Flycatcher	*Leptopogon amaurocephalus*		9.9	M
Fulvous-breasted Flatbill	*Rhynchocyclus fulvipectus*		10.1	M?
Ringed Antpipit	*Corythopis torquata*		10.2	M
Olive-striped Flycatcher	*Mionectes olivaceus*		10.2	M
Ochre-bellied Flycatcher	*Mionectes oleagineus*		10.3	M
Rufous Hornero	*Furnarius rufus*		14.2	F
Rufous-vented Tapaculo	*Scytalopus femoralis*		14.3	R
Barred Woodcreeper	*Dendrocolaptes certhia*		14.4	F
Rufous Gnateater	*Conopophaga lineata*		14.6	Co
Black-capped Antshrike	*Thamnophilus schistaceus*		14.9	Th

[1] P = Piprinae, C = Cotinginae, Ti = Tityrinae, T = Tyranninae, M = Mionectidae, F = Furnariidae, R = Rhinocryptidae, Co = Conopophagidae, Th = Thamnophilidae, ? = uncertain affinities.
[2] Numbers in parentheses indicate that more than one DNA hybrid was averaged.

DNA-DNA Hybridization Values Between the Olive-striped Flycatcher (*Mionectes olivaceus*) and Other Passerine Birds

Common name	Scientific name		Delta T₅₀H	Group index[1]
Olive-striped Flycatcher	*Mionectes olivaceus*		0.0	M
Streak-necked Flycatcher	*Mionectes striaticollis*		1.7	M
Ochre-bellied Flycatcher	*Mionectes oleagineus*		2.8	M
Slaty-capped Flycatcher	*Leptopogon superciliaris*	(3)[2]	5.3	M
Sepia-capped Flycatcher	*Leptopogon amaurocephalus*		5.6	M
Rufous-headed Pygmy-Tyrant	*Pseudotriccus ruficeps*		6.7	M
Stripe-necked Tody-Tyrant	*Idioptilon striaticolle*		7.4	M
Common Tody-Flycatcher	*Todirostrum cinereum*		7.5	M
Ringed Antpipit	*Corythopis torquata*	(2)	8.1	M
Scale-crested Pygmy-Tyrant	*Lophotriccus pileatus*		8.6	M?
White-throated Tyrannulet	*Mecocerculus leucophrys*		8.7	T
Fulvous-breasted Flatbill	*Rhynchocyclus fulvipectus*		8.8	T
Tufted Tit-Tyrant	*Anairetes parulus*		8.8	T
Rufous Mourner	*Rhytipterna holerythra*		8.8	T
White-winged Black-Tyrant	*Knipolegus aterrimus*		8.9	T
Greenish Elaenia	*Myiopagis viridicata*		9.0	T
Least Flycatcher	*Empidonax minimus*		9.1	T
Mouse-colored Tyrannulet	*Phaeomyias murina*		9.1	T
Social Flycatcher	*Myiozetetes similis*		9.2	T
Andean Cock-of-the-rock	*Rupicola peruviana*		9.2	C
Golden-headed Manakin	*Pipra erythrocephala*	(2)	9.2	P
Black-tailed Tityra	*Tityra cayana*		9.3	Ti
Southern Beardless Flycatcher	*Camptostoma obsoletum*		9.4	T
Crowned Chat-Tyrant	*Ochthoeca frontalis*		9.4	T
Green Manakin	*Chloropipo holochroa*		9.5	P
Mountain Elaenia	*Elaenia frantzii*	(2)	9.5	T
Scissor-tailed Flycatcher	*Tyrannus forficatus*		9.5	T
White-throated Spadebill	*Platyrinchus mystaceus*		9.5	T
White-tailed Shrike-Tyrant	*Agriornis albicauda*		9.6	T
Bright-rumped Attila	*Attila spadiceus*		9.6	T
Swallow-tailed Manakin	*Chiroxiphia caudata*		9.6	P
Eastern Phoebe	*Sayornis phoebe*	(2)	9.7	T
Dusky-capped Flycatcher	*Myiarchus tuberculifer*	(2)	9.7	T
Short-crested Flycatcher	*Myiarchus ferox*		9.8	T
Ornate Flycatcher	*Myiotriccus ornatus*		9.9	T
White-winged Becard	*Pachyramphus polychopterus*		9.9	Ti
Green-and-black Fruiteater	*Pipreola riefferii*	(2)	10.0	C
Yellow-crowned Tyrannulet	*Tyrannulus elatus*		10.1	T
Flavescent Flycatcher	*Myiophobus flavicans*		10.2	T
Fuscous Flycatcher	*Cnemotriccus fuscatus*		10.3	T
Royal Flycatcher	*Onychorhynchus coronatus*		10.3	T
Ruddy-tailed Flycatcher	*Terenotriccus erythrurus*		10.6	T
Thrushlike Manakin	*Schiffornis turdinus*		10.7	Ti
Moustached Turca	*Pteroptochos megapodius*		12.3	R
Olivaceous Woodcreeper	*Sittasomus griseicapillus*		12.5	D
Rufous-breasted Antthrush	*Formicarius rufipectus*		12.5	Fo
Short-tailed Antthrush	*Chamaeza campanisona*		13.4	Fo
Rufous Hornero	*Furnarius rufus*		13.9	F
Azara's Spinetail	*Synallaxis azarae*		14.1	F
Spotted Barbtail	*Premnoplex brunnescens*		14.5	F
Black-capped Antshrike	*Thamnophilus schistaceus*		14.6	Th
Black-breasted Antwren	*Formicivora grisea*		14.7	Th
Curve-billed Scythebill	*Campylorhamphus procurvoides*		14.8	D
Chestnut-crowned Antpitta	*Grallaria ruficapilla*		14.8	Fo
Rufous Gnateater	*Conopophaga lineata*		14.9	Co
White-flanked Antwren	*Myrmotherula axillaris*		14.9	Th
Garnet Pitta	*Pitta granatina*		16.8	Pt
Black-and-red Broadbill	*Cymbirhynchus macrorhynchos*		17.1	E
Green Broadbill	*Calyptomena viridis*		17.3	E
Hooded Pitta	*Pitta sordida*		18.3	Pt
Long-billed Thrasher	*Toxostoma longirostre*		19.2	O
Superb Lyrebird	*Menura novaehollandiae*		19.5	O

[1] M = Mionectidae, T = Tyranninae, Ti = Tityrinae, C = Cotinginae, P = Piprinae, R = Rhinocryptidae, F = Furnariinae, D = Dendrocolaptinae, Fo = Formicariidae, Co = Conopophagidae, Th = Thamnophilidae, Pt = Pittidae, E = Eurylaimidae, O = Oscine.

[2] Numbers in parentheses indicate that more than one DNA hybrid was averaged.

APPENDIX IX

DNA-DNA Hybridization Values Between the Rufous Hornero (*Furnarius rufus*) and Other Passerine Birds

Common name	Scientific name		Delta $T_{50}H$	Group index[1]
Rufous Hornero	*Furnarius rufus*		0.0	F
Pale-legged Hornero	*Furnarius leucopus*		2.9	F
Stout-billed Cinclodes	*Cinclodes excelsior*		3.4	F
Spotted Barbtail	*Premnoplex brunnescens*		4.3	F
Streaked Tuftedcheek	*Pseudocolaptes boissonneautii*	(2)[2]	4.4	F
Cinnamon-rumped Foliage-gleaner	*Philydor pyrrhodes*		4.7	F
White-chinned Thistletail	*Schizoeca fuliginosa*		4.7	F
Olive-backed Foliage-gleaner	*Automolus infuscatus*		4.9	F
Chestnut-crowned Foliage-gleaner	*Automolus rufipileatus*		4.9	F
Black-billed Treehunter	*Thripadectes melanorhynchus*		5.0	F
Azara's Spinetail	*Synallaxis azarae*	(2)	5.0	F
Dark-breasted Spinetail	*Synallaxis albigularis*		5.0	F
Rusty-backed Spinetail	*Cranioleuca vulpina*		5.3	F
Striped Leaf-gleaner	*Hyloctistes subulatus*		5.7	F
Plain Xenops	*Xenops minutus*		5.8	F
Black-tailed Leafscraper	*Sclerurus caudacutus*		6.0	F
Olive-backed Woodcreeper	*Xiphorhynchus triangularis*		6.3	D
Spotted Woodcreeper	*Xiphorhynchus erythropygius*		6.4	D
Plain-brown Woodcreeper	*Dendrocincla fuliginosa*		6.5	D
Spot-crowned Woodcreeper	*Lepidocolaptes affinis*		6.5	D
Giant Woodcreeper	*Xiphocolaptes promeropirhynchus*		6.6	D
Curve-billed Scythebill	*Campylorhamphus procurvoides*		6.6	D
Straight-billed Woodcreeper	*Xiphorhynchus picus*		6.6	D
White-chinned Woodcreeper	*Dendrocincla merula*		6.7	D
Ocellated Woodcreeper	*Xiphorhynchus ocellatus*		6.8	D
Olivaceous Woodcreeper	*Sittasomus griseicapillus*		7.0	D
Barred Woodcreeper	*Dendrocolaptes certhia*	(2)	7.1	D
Rufous-vented Tapaculo	*Scytalopus femoralis*		10.5	R
Striated Antthrush	*Chamaeza nobilis*		12.3	Fo
Rufous Gnateater	*Conopophaga lineata*		12.4	Co
White-flanked Antwren	*Myrmotherula axillaris*		12.5	Th
Spot-backed Antbird	*Hylophylax naevia*		13.1	Th
Black-capped Antshrike	*Thamnophilus schistaceus*		13.3	Th
Black-spotted Bare-eye	*Phlegopsis nigromaculata*		13.6	Th
Golden-headed Manakin	*Pipra erythrocephala*		13.6	P
Red-crested Cotinga	*Ampelion rubrocristatus*		14.0	C
Black-breasted Antwren	*Formicivora grisea*		14.0	Th
Jet Manakin	*Chloropipo unicolor*		14.0	P
Brown-crested Flycatcher	*Myiarchus tyrannulus*		14.1	T
Spot-winged Antbird	*Percnostoloa leucostigma*		14.1	Th
Barred Fruiteater	*Pipreola arcuata*		14.3	C
Eastern Kingbird	*Tyrannus tyrannus*		14.4	T
Lesser Elaenia	*Elaenia chiriquensis*		14.7	T

[1] F = Furnariinae, D = Dendrocolaptinae, R = Rhinocryptidae, Co = Conopophagidae, Th = Thamnophilidae, P = Piprinae, C = Cotinginae, T = Tyranninae, Fo = Formicariidae.
[2] Numbers in parentheses indicate that more than one DNA hybrid was averaged.

APPENDIX X

DNA-DNA HYBRIDIZATION VALUES BETWEEN THE BARRED WOODCREEPER (*DENDROCOLAPTES CERTHIA*) AND OTHER PASSERINE BIRDS

Common name	Scientific name	Delta $T_{50}H$	Group index[1]
Barred Woodcreeper	*Dendrocolaptes certhia*	0.0	D
Spot-crowned Woodcreeper	*Lepidocolaptes affinis*	2.9	D
Giant Woodcreeper	*Xiphocolaptes promeropirhynchus*	3.0	D
Ocellated Woodcreeper	*Xiphorhynchus ocellatus*	3.1	D
Olive-backed Woodcreeper	*Xiphorhynchus triangularis*	3.2	D
Spotted Woodcreeper	*Xiphorhynchus erythropygius*	3.3	D
Straight-billed Woodcreeper	*Xiphorhynchus picus*	3.4	D
Curved-billed Scythebill	*Campylorhamphus procurvoides*	3.6	D
White-chinned Woodcreeper	*Dendrocincla merula*	4.1	D
Plain-brown Woodcreeper	*Dendrocincla fuliginosa*	4.2	D
Olivaceous Woodcreeper	*Sittasomus griseicapillus*	4.2	D
Wedge-billed Woodcreeper	*Glyphorhynchus spirurus* (2)[2]	4.6	D
Spotted Barbtail	*Premnoplex brunnescens*	5.9	F
Streaked Tuftedcheek	*Pseudocolaptes boissonneautii*	5.9	F
Plain Xenops	*Xenops minutus*	6.0	F
Dark-breasted Spinetail	*Synallaxis albigularis*	6.1	F
Azara's Spinetail	*Synallaxis azarae* (2)	6.3	F
Black-billed Treehunter	*Thripadectes melanorhynchus*	6.4	F
Olive-backed Foliage-gleaner	*Automolus infuscatus*	6.5	F
Cinnamon-rumped Foliage-gleaner	*Philydor pyrrhodes*	6.5	F
Chestnut-crowned Foliage-gleaner	*Automolus rufipileatus*	6.6	F
White-chinned Thistletail	*Schizoeaca fuliginosa*	6.6	F
Rusty-backed Spinetail	*Cranioleuca vulpina*	6.7	F
Stout-billed Cinclodes	*Cinclodes excelsior*	6.9	F
Pale-legged Hornero	*Furnarius leucopus*	6.9	F
Rufous Hornero	*Furnarius rufus* (2)	7.0	F
Black-tailed Leafscraper	*Sclerurus caudacutus*	7.3	F
Striped Leaf-gleaner	*Hyloctistes subulatus*	7.4	F
Rufous-vented Tapaculo	*Scytalopus femoralis*	9.4	R
Striated Antthrush	*Chamaeza nobilis*	10.1	Fo
Rusty-belted Tapaculo	*Liosceles thoracicus*	11.2	R
Black-spotted Bare-eye	*Phlegopsis nigromaculata*	12.6	Th
Lesser Elaenia	*Elaenia chiriquensis*	13.2	T
Jet Manakin	*Chloropipo unicolor*	13.3	P
Spot-backed Antbird	*Hylophylax naevia*	13.4	Th
Red-crested Cotinga	*Ampelion rubrocristatus*	13.5	C
Barred Fruiteater	*Pipreola arcuata*	13.5	C
Eastern Kingbird	*Tyrannus tyrannus*	13.5	T
Black-capped Antshrike	*Thamnophilus schistaceus*	13.6	Th
White-flanked Antwren	*Myrmotherula axillaris*	13.8	Th
Golden-headed Manakin	*Pipra erythrocephala*	13.9	P
Brown-crested Flycatcher	*Myiarchus tyrannulus*	14.0	T
Ringed Antpipit	*Corythopis torquata*	14.3	M

[1] D = Dendrocolaptinae, F = Furnariinae, R = Rhinocryptidae, Fo = Formicariidae, Th = Thamnophilidae, P = Piprinae, C = Cotinginae, T = Tyranninae, M = Mionectidae.

[2] Numbers in parentheses indicate that more than one DNA hybrid was averaged.

APPENDIX XI

DNA-DNA Hybridization Values Between the Rufous-breasted Antthrush (*Formicarius rufipectus*) and Other Passerine Birds

Common name	Scientific name		Delta $T_{50}H$	Group index[1]
Rufous-breasted Antthrush	*Formicarius rufipectus*		0.0	Fo
Black-headed Antthrush	*Formicarius nigricapillus*	(2)[2]	2.6	Fo
Black-faced Antthrush	*Formicarius analis*	(2)	3.1	Fo
Rufous-capped Antthrush	*Formicarius colma*	(2)	3.6	Fo
Short-tailed Antthrush	*Chamaeza campanisona*		9.7	Fo
Striated Antthrush	*Chamaeza nobilis*		10.1	Fo
Chestnut-crowned Antpitta	*Grallaria ruficapilla*		10.7	Fo
Tawny Antpitta	*Grallaria quitensis*		10.9	Fo
Rufous-vented Tapaculo	*Scytalopus femoralis*	(3)	11.2	R
Brown-rumped Tapaculo	*Scytalopus latebricola*		11.2	R
Chestnut-crowned Gnateater	*Conopophaga castaneiceps*	(3)	11.3	R
Rusty-belted Tapaculo	*Liosceles thoracicus*		11.3	R
Slate-crowned Antpitta	*Grallaricula nana*		11.5	Fo
Rufous Gnateater	*Conopophaga lineata*		11.8	R
Moustached Turca	*Pteroptochos megapodius*	(2)	11.9	R
Thrush-like Antpitta	*Myrmothera campanisona*		12.0	Fo?
Rufous Hornero	*Furnarius rufus*		12.1	F
Azara's Spinetail	*Synallaxis azarae*		12.1	F
Curve-billed Scythebill	*Campylorhamphus procurvoides*		12.4	F
Olivaceous Woodcreeper	*Sittasomus griseicapillus*		12.4	F
White-chinned Thistletail	*Schizoeaca fuliginosa*		12.6	F
Spotted Barbtail	*Premnoplex brunnescens*		12.7	F
Barred Woodcreeper	*Dendrocolaptes certhia*		12.8	F
Stout-billed Cinclodes	*Cinclodes excelsior*		12.9	F
Streaked Tuftedcheek	*Pseudocolaptes boissonneautii*		12.9	F
Plain Xenops	*Xenops minutus*		12.9	F
Yellow-bellied Elaenia	*Elaenia flavogaster*		13.3	T
Golden-headed Manakin	*Pipra eyrthrocephala*		13.3	P
Andean Cock-of-the-rock	*Rupicola peruviana*		13.4	C
Dusky-capped Flycatcher	*Myiarchus tuberculifer*		13.5	T
White-cheeked Antbird	*Gymnopithys leucaspis*		13.5	Th
Jet Antbird	*Cercomacra nigricans*		13.5	Th
Plain Antvireo	*Dysithamnus mentalis*		13.5	Th
Spot-winged Antbird	*Percnostola leucostigma*		13.5	Th
Warbling Antbird	*Hypocnemis cantator*		13.6	Th
Black Bushbird	*Neoctantes niger*		13.6	Th
Spot-winged Antshrike	*Pygiptila stellaris*		13.6	Th
Scaly-backed Antbird	*Hylophylax poecilonota*		13.6	Th
Eastern Phoebe	*Sayornis phoebe*		13.7	T
Black-spotted Bare-eye	*Phlegopsis nigromaculata*		13.7	Th
Dusky-throated Antshrike	*Thamnomanes ardesiacus*		13.7	Th
Jet Manakin	*Chloropipo unicolor*		13.8	P
White-flanked Antwren	*Myrmotherula axillaris*		13.8	Th
Reddish-winged Bare-eye	*Phlegopsis erythroptera*		13.8	Th
Silvered Antbird	*Sclateria naevia*		13.9	Th
White-backed Fire-eye	*Pyriglena leuconota*		13.9	Th
Green-and-black Fruiteater	*Pipreola riefferii*		13.9	C
Spot-backed Antbird	*Hylophylax naevia*		14.0	Th
Ocellated Antbird	*Phaenostictus mcleannani*		14.0	Th
White-shouldered Antbird	*Myrmeciza melanoceps*		14.1	Th
Barred Becard	*Pachyramphus versicolor*		14.1	Th
Black-capped Antshrike	*Thamnophilus schistaceus*	(2)	14.1	Th
Barred Antshrike	*Thamnophilus doliatus*		14.2	Th
Blackish-gray Antshrike	*Thamnophilus nigrocinereus*		14.2	Th
Black-faced Antbird	*Myrmoborus myotherinus*		14.3	Th
Black-breasted Antwren	*Formicivora grisea*		14.3	Th
Undulated Antshrike	*Frederickena unduligera*		14.3	Th
Olive-striped Flycatcher	*Mionectes olivaceus*		14.4	M
Sooty Antbird	*Myrmeciza fortis*		14.4	Th
Plumbeous Antbird	*Myrmeciza hyperythra*		14.5	Th
Collared Antshrike	*Sakesphorus bernardi*		14.8	Th

[1] Fo = Formicariidae, R = Rhinocryptidae, F = Furnariidae, T = Tyranninae, P = Piprinae, C = Cotinginae, Th = Thamnophilidae.
[2] Numbers in parentheses indicate that more than one DNA hybrid was averaged.

423

APPENDIX XII

DNA-DNA HYBRIDIZATION VALUES BETWEEN THE CHESTNUT-CROWNED GNATEATER
(*CONOPOPHAGA CASTANEICEPS*) AND OTHER PASSERINE BIRDS

Common name	Scientific name	Delta $T_{50}H$	Group index[1]
Chestnut-crowned Gnateater	*Conopophaga castaneiceps*	0.0	Co
Rufous Gnateater	*Conopophaga lineata*	1.9	Co
Rufous-vented Tapaculo	*Scytalopus femoralis*	10.2	R
Rufous-capped Antthrush	*Formicarius colma*	10.8	Fo
Tawny Antpitta	*Grallaria quitensis*	11.1	Fo
Plain Xenops	*Xenops minutus*	11.7	F
Spotted Barbtail	*Premnoplex brunnescens*	11.8	F
Azara's Spinetail	*Synallaxis azarae*	12.0	F
Curve-billed Scythebill	*Campylorhamphus procurvoides*	12.1	D
White-chinned Woodcreeper	*Dendrocincla merula*	12.1	D
Cinnamon-rumped Foliage-gleaner	*Philydor pyrrhodes*	12.2	F
Barred Woodcreeper	*Dendrocolaptes certhia*	12.4	D
White-shouldered Antbird	*Myrmeciza melanoceps*	12.8	Th
Plain Antvireo	*Dysithamnus mentalis*	12.9	Th
Rufous Mourner	*Rhytipterna holerythra*	13.0	T
Spot-winged Antbird	*Percnostola leucostigma*	13.2	Th
Common Tody-Flycatcher	*Todirostrum cinereum*	13.3	M
Sepia-capped Flycatcher	*Leptopogon amaurocephalus*	13.3	M
Ringed Antpipit	*Corythopis torquata*	13.4	M
Black-breasted Antwren	*Formicivora grisea*	13.4	Th
Black-capped Antshrike	*Thamnophilus schistaceus*	13.4	Th
Golden-headed Manakin	*Pipra erythrocephala*	13.6	P
Plum-throated Cotinga	*Cotinga maynana*	13.7	C
Rufous-tailed Tyrant	*Knipolegus poecilurus*	13.7	T

[1] Co = Conopophagidae, R = Rhinocryptidae, Fo = Formicariidae, F = Furnariinae, D = Dendrocolaptinae, Th = Thamnophilidae, M = Mionectidae, P = Piprinae, Co = Cotinginae, T = Tyranninae.

APPENDIX XIII

DNA-DNA HYBRIDIZATION VALUES BETWEEN THE BROWN-RUMPED TAPACULO (SCYTALOPUS LATEBRICOLA) AND OTHER PASSERINE BIRDS

Common name	Scientific name		Delta $T_{50}H$	Group index[1]
Brown-rumped Tapaculo	Scytalopus latebricola		0.0	R
Rufous-vented Tapaculo	Scytalopus femoralis		1.8	R
Rusty-belted Tapaculo	Liosceles thoracicus	(2)[2]	5.8	R
Moustached Turca	Pteroptochos megapodius		10.0	R
Rufous Gnateater	Conopophaga lineata		10.6	Co
Black-headed Antthrush	Formicarius nigricapillus		10.7	Fo
Rufous-breasted Antthrush	Formicarius rufipectus	(2)	10.8	Fo
Chestnut-crowned Gnateater	Conopophaga castaneiceps		10.9	Co
Chestnut-crowned Antpitta	Grallaria ruficapilla		11.2	Fo
Striated Antthrush	Chamaeza nobilis		11.6	Fo
Black-faced Antthrush	Formicarius analis		11.7	Fo
Tawny Antpitta	Grallaria quitensis		11.7	Fo
Slaty-crowned Antpitta	Grallaricula nana		11.7	Fo
Rufous-capped Antthrush	Formicarius colma		11.8	Fo
Short-tailed Antthrush	Chamaeza campanisona		11.8	Fo
Olivaceous Woodcreeper	Sittasomus griseicapillus		11.8	F
Plain Xenops	Xenops minutus		11.8	F
Stout-billed Cinclodes	Cinclodes excelsior		11.9	F
White-chinned Woodcreeper	Dendrocincla merula		11.9	F
Spotted Barbtail	Premnoplex brunnescens		12.2	F
Dark-breasted Spinetail	Synallaxis albigularis		12.2	F
Straight-billed Woodcreeper	Xiphorhynchus picus		12.3	F
Rufous Hornero	Furnarius rufus		12.4	F
Thrush-like Antpitta	Myrmothera campanisona		12.4	?
Jet Antbird	Cercomacra nigricans		12.9	Th
Plain Antvireo	Dysithamnus mentalis		12.9	Th
Spot-winged Antshrike	Pygiptila stellaris		13.0	Th
Spot-winged Antbird	Percnostola leucostigma		13.1	Th
Warbling Antbird	Hypocnemis cantator		13.2	Th
White-cheeked Antbird	Gymnopithys leucaspis		13.2	Th
White-flanked Antwren	Myrmotherula axillaris		13.2	Th
Eastern Phoebe	Sayornis phoebe		13.2	T
White-shouldered Antbird	Myrmeciza melanoceps		13.4	Th
Undulated Antshrike	Frederickena unduligera		13.5	Th
Black-spotted Bare-eye	Phlegopsis nigromaculata		13.5	Th
Collared Antshrike	Sakesphorus bernardi		13.5	Th
Spot-backed Antbird	Hylophylax naevia		13.6	Th
Barred Antshrike	Thamnophilus doliatus	(2)	13.7	Th
White-backed Fire-eye	Pyriglena leuconota		13.8	Th
Golden-headed Manakin	Pipra erythrocephala		13.9	P
Green-and-black Fruiteater	Pipreola riefferii		14.2	C
Ocellated Antbird	Phaenostictus mcleannani		14.4	Th
White-winged Becard	Pachyramphus polychopterus		14.4	Ti
Olive-striped Flycatcher	Mionectes olivaceus		14.5	M
Black-breasted Antwren	Formicivora grisea		14.7	Th

[1] R = Rhinocryptidae, Co = Conopophagidae, Fo = Formicariidae, F = Furnariidae, Th = Thamnophilidae, T = Tyranninae, P = Piprinae, C = Cotinginae, Ti = Tityrinae, M = Mionectidae.
[2] Numbers in parentheses indicate that more than one DNA hybrid was averaged.

APPENDIX XIV

DNA-DNA HYBRIDIZATION VALUES BETWEEN THE RUFOUS-VENTED TAPACULO
(*SCYTALOPUS FEMORALIS*) AND OTHER PASSERINE BIRDS

Common name	Scientific name		Delta $T_{50}H$	Group index[1]
Rufous-vented Tapaculo	*Scytalopus femoralis*		0.0	R
Brown-rumped Tapaculo	*Scytalopus latebricola*	(2)[2]	2.0	R
Rusty-belted Tapaculo	*Liosceles thoracicus*		5.8	R
Chestnut-crowned Gnateater	*Conopophaga castaneiceps*	(3)	10.6	Co
Rufous Gnateater	*Conopophaga lineata*		10.8	Co
Rufous-capped Antthrush	*Formicarius colma*	(2)	11.5	Fo
Rufous Hornero	*Furnarius rufus*		12.0	F
Spotted Barbtail	*Premnoplex brunnescens*		12.1	F
Azara's Spinetail	*Synallaxis azarae*		12.1	F
Tawny Antpitta	*Grallaria quitensis*		12.2	Fo
White-chinned Thistletail	*Schizoeaca fuliginosa*		12.2	F
Curve-billed Scythebill	*Campylorhamphus procurvoides*		12.3	D
Plain-brown Woodcreeper	*Dendrocincla fuliginosa*	(2)	12.3	D
Plain Xenops	*Xenops minutus*		12.3	F
Spotted-crowned Woodcreeper	*Lepidocolaptes affinis*		12.5	D
Green Manakin	*Chloropipo holochroa*		12.6	P
Ringed Antpipit	*Corythopis torquata*		12.7	M
Black-billed Treehunter	*Thripadectes melanorhynchus*		12.7	F
Spot-winged Antshrike	*Pygiptila stellaris*		12.8	Th
Warbling Antbird	*Hypocnemis cantator*		12.9	Th
White-flanked Antwren	*Myrmotherula axillaris*		13.1	Th
White-collared Manakin	*Manacus candei*		13.1	P
White-shouldered Antbird	*Myrmeciza melanoceps*		13.2	Th
Golden-headed Manakin	*Pipra erythrocephala*		13.3	P
Barred Fruiteater	*Pipreola arcuata*		13.4	C
Eastern Phoebe	*Sayornis phoebe*		13.4	T
Lesser Elaenia	*Elaenia chiriquensis*		13.4	T
Black-breasted Antwren	*Formicivora grisea*		13.4	Th
Spot-winged Antbird	*Percnostola leucostigma*		13.4	Th
Black-capped Antshrike	*Thamnophilus schistaceus*	(2)	13.6	Th
Amazonian Umbrellabird	*Cephalopterus ornatus*		13.7	C
Plain Antvireo	*Dysithamnus mentalis*		13.9	Th
White-throated Tyrannulet	*Mecocerculus leucophrys*		14.1	T
Barred Becard	*Pachyramphus versicolor*		14.3	Ti
Brown-crested Flycatcher	*Myiarchus tyrannulus*		14.5	T
Blue-headed Pitta	*Pitta baudii*		16.1	Pt
Banded Pitta	*Pitta guajana*		16.2	Pt
Green Broadbill	*Calyptomena viridis*		16.7	E
Black-and-red Broadbill	*Cymbirhynchus macrorhynchos*		16.9	E
New Zealand Rifleman	*Acanthisitta chloris*		18.2	A
Little Raven	*Corvus mellori*		19.8	O
American Robin	*Turdus migratorius*		20.0	O

[1] R = Rhinocryptidae, Co = Conopophagidae, Fo = Formicariidae, F = Furnariinae, D = Dendrocolaptinae, P = Piprinae, C = Cotinginae, M = Mionectidae, Th = Thamnophilidae, T = Tyranninae, Ti = Tityrinae, Pt = Pittidae, E = Eurylaimidae, A = Acanthisittidae, O = Oscine.

[2] Numbers in parentheses indicate that more than one DNA hybrid was averaged.

APPENDIX XV

DNA-DNA Hybridization Values Between the Rusty-belted Tapaculo (*Liosceles thoracicus*) and Other Passerine Birds

Common name	Scientific name		Delta $T_{50}H$	Group index[1]
Rusty-belted Tapaculo	*Liosceles thoracicus*		0.0	R
Brown-rumped Tapaculo	*Scytalopus latebricola*		6.1	R
Rufous-vented Tapaculo	*Scytalopus femoralis*		6.3	R
Moustached Turca	*Pteroptochos megapodius*		9.2	R
Chestnut-crowned Gnateater	*Conopophaga castaneiceps*		11.2	Co
Rufous Gnateater	*Conopophaga lineata*		11.2	Co
Striated Antthrush	*Chamaeza nobilis*		11.2	Fo
Black-headed Antthrush	*Formicarius nigricapillus*		11.4	Fo
Rufous-breasted Antthrush	*Formicarius rufipectus*	(2)[2]	11.4	Fo
Rufous-capped Antthrush	*Formicarius colma*		11.5	Fo
Short-tailed Antthrush	*Chamaeza campanisona*		11.5	Fo
Chestnut-crowned Antpitta	*Grallaria ruficapilla*		11.5	Fo
Black-faced Antthrush	*Formicarius analis*		11.7	Fo
Tawny Antpitta	*Grallaria quitensis*		11.7	Fo
Slate-crowned Antpitta	*Grallaricula nana*		12.0	Fo
Stout-billed Cinclodes	*Cinclodes excelsior*		12.0	F
Plain Xenops	*Xenops minutus*		12.0	F
Dark-breasted Spinetail	*Synallaxis albigularis*		12.1	F
Rufous Hornero	*Furnarius rufus*		12.2	F
White-chinned Woodcreeper	*Dendrocincla merula*		12.2	D
Spotted Barbtail	*Premnoplex brunnescens*		12.3	F
Olivaceous Woodcreeper	*Sittasomus griseicapillus*		12.3	D
Straight-billed Woodcreeper	*Xiphorhynchus picus*		12.4	D
Thrush-like Antpitta	*Myrmothera campanisona*		13.1	?
Spot-backed Antbird	*Hylophylax naevia*		13.1	Th
Warbling Antbird	*Hypocnemis cantator*		13.1	Th
Jet Antbird	*Cercomacra nigricans*		13.2	Th
White-cheeked Antbird	*Gymnopithys leucaspis*		13.2	Th
Plain Antvireo	*Dysithamnus mentalis*		13.4	Th
White-shouldered Antbird	*Myrmeciza melanoceps*		13.4	Th
Undulated Antshrike	*Frederickena unduligera*		13.5	Th
Spot-winged Antbird	*Percnostola leucostigma*		13.5	Th
Spot-winged Antshrike	*Pygiptila stellaris*		13.6	Th
Collared Antshrike	*Sakesphorus bernardi*		13.7	Th
Golden-headed Manakin	*Pipra erythrocephala*		13.8	P
Black-spotted Bare-eye	*Phlegopsis nigromaculata*		13.9	Th
Barred Antshrike	*Thamnophilus doliatus*	(2)	13.9	Th
White-backed Fire-eye	*Pyriglena leuconota*		14.2	Th
Black-breasted Antwren	*Formicivora grisea*		14.3	Th
Olive-striped Flycatcher	*Mionectes olivaceus*		14.5	M
White-winged Becard	*Pachyramphus polychopterus*		14.6	Ti
Ocellated Antbird	*Phaenostictus mcleannani*		14.7	Th
Green-and-black Fruiteater	*Pipreola riefferii*		14.7	C
Eastern Phoebe	*Sayornis phoebe*		14.7	T

[1] R = Rhinocryptidae, Co = Conopophagidae, Fo = Formicariidae, F = Furnariinae, D = Dendrocolaptinae, Th = Thamnophilidae, M = Mionectidae, Ti = Tityrinae, C = Cotinginae, T = Tyranninae.

[2] Numbers in parentheses indicate that more than one DNA hybrid was averaged.

APPENDIX XVI

DNA-DNA Hybridization Values Between the Black-capped Antshrike
(*Thamnophilus schistaceus*) and Other Passerine Birds

Common name	Scientific name		Delta T$_{50}$H	Group index[1]
Black-capped Antshrike	*Thamnophilus schistaceus*		0.0	Th
Blackish-gray Antshrike	*Thamnophilus nigrocinereus*		2.1	Th
Barred Antshrike	*Thamnophilus doliatus*	(2)[2]	3.0	Th
Plain Antvireo	*Dysithamnus mentalis*		3.6	Th
White-flanked Antwren	*Myrmotherula axillaris*		4.0	Th
Plain-throated Antwren	*Myrmotherula hauxwelli*		4.0	Th
Black-breasted Antwren	*Formicivora grisea*		4.1	Th
Black-spotted Bare-eye	*Phlegopsis nigromaculata*		4.5	Th
Dusky-throated Antshrike	*Thamnomanes ardesiacus*		4.6	Th
Spot-backed Antbird	*Hylophylax naevia*		4.8	Th
White-cheeked Antbird	*Gymnopithys leucaspis*		4.9	Th
Black-faced Antbird	*Myrmoborus myotherinus*		5.0	Th
Sooty Antbird	*Myrmeciza fortis*		5.2	Th
Spot-winged Antbird	*Percnostola leucostigma*		5.8	Th
Warbling Antbird	*Hypocnemis cantator*		6.0	Th
Spot-winged Antshrike	*Pygiptila stellaris*		6.4	Th
Striated Antthrush	*Chamaeza nobilis*		12.2	Fo
Curve-billed Scythebill	*Campylorhamphus procurvoides*		12.5	D
Chestnut-crowned Gnateater	*Conopophaga castaneiceps*		12.5	Co
Azara's Spinetail	*Synallaxis azarae*	(2)	12.6	F
Rufous Gnateater	*Conopophaga lineata*		12.6	Co
Rufous-vented Tapaculo	*Scytalopus femoralis*	(2)	12.6	R
Rusty-belted Tapaculo	*Liosceles thoracicus*		12.8	R
White-collared Manakin	*Manacus candei*		12.8	P
White-chinned Thistletail	*Schizoeaca fuliginosa*		12.8	F
Black-billed Treehunter	*Thripadectes melanorhynchus*		12.8	F
Golden-headed Manakin	*Pipra erythrocephala*		12.9	P
Green Manakin	*Chloropipo holochroa*		12.9	P
Rufous Hornero	*Furnarius rufus*		12.9	F
Plain-brown Woodcreeper	*Dendrocincla fuliginosa*		12.9	D
Brown-rumped Tapaculo	*Scytalopus latebricola*		12.9	R
Tawny Antpitta	*Grallaria quitensis*	(2)	12.9	Fo
Barred Fruiteater	*Pipreola arcuata*		13.0	C
White-throated Tyrannulet	*Mecocerculus leucophrys*		13.0	T
Brown-crested Flycatcher	*Myiarchus tyrannulus*		13.1	T
Streaked Tuftedcheek	*Pseudocolaptes boissonneautii*		13.1	F
Cinnamon-rumped Foliage-gleaner	*Philydor pyrrhodes*		13.3	F
Plain Xenops	*Xenops minutus*		13.3	F
Spot-crowned Woodcreeper	*Lepidocolaptes affinis*		13.3	D
Eastern Phoebe	*Sayornis phoebe*		13.3	T
Ringed Antpipit	*Corythopis torquata*		13.4	M
Black-faced Antthrush	*Formicarius analis*		13.4	Fo
Rufous-capped Antthrush	*Formicarius colma*		13.4	Fo
Spotted Barbtail	*Premnoplex brunnescens*		13.4	F
Barred Becard	*Pachyramphus versicolor*		13.4	Ti
Amazonian Umbrellabird	*Cephalopterus ornatus*		13.5	C
Lesser Elaenia	*Elaenia chiriquensis*		13.5	T
Noisy Pitta	*Pitta versicolor*		16.1	Pt
Blue-headed Pitta	*Pitta baudii*		16.5	Pt
Banded Pitta	*Pitta guajana*		16.6	Pt
Green Broadbill	*Calyptomena viridis*		17.0	E
Black-and-red Broadbill	*Cymbirhynchus macrorhynchos*		17.8	E
Little Raven	*Corvus mellori*		19.8	O
American Robin	*Turdus migratorius*		20.2	O

[1] Th = Thamnophilidae, Fo = Formicariidae, D = Dendrocolaptinae, F = Furnariinae, Co = Conopophagidae, R = Rhinocryptidae, P = Piprinae, C = Cotinginae, T = Tyranninae, M = Mionectidae, Ti = Tityrinae, Pt = Pittidae, E = Eurylaimidae, O = Oscine.
[2] Numbers in parentheses indicate that more than one DNA hybrid was averaged.

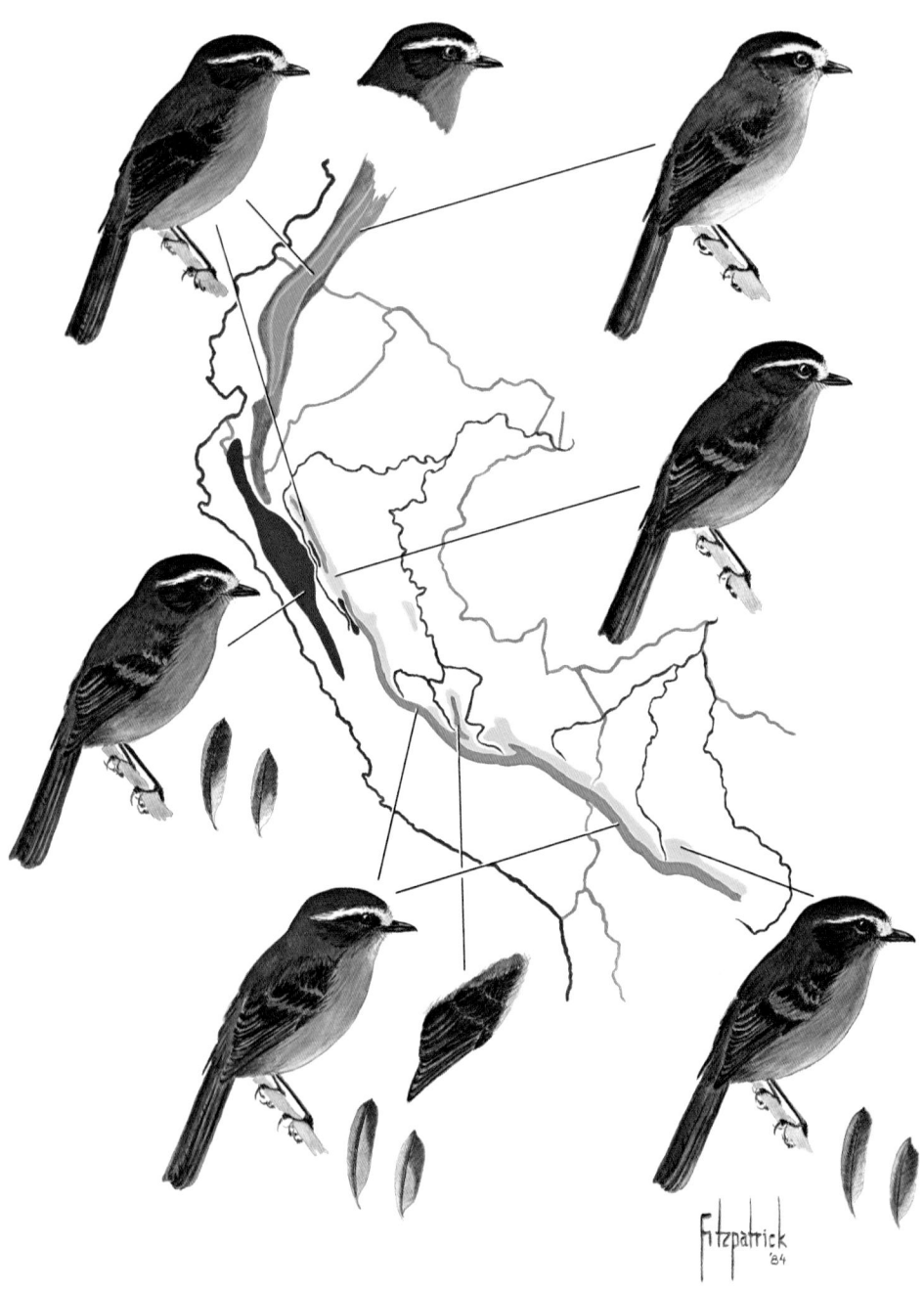

PLATE VI.   Taxa of the *Ochthoeca diadema* species-group encountered from southern Colombia to central Bolivia. The taxa of the *frontalis* superspecies are in the left hand column, from top to bottom, head of *O. f. albidiadema* (from Bogota, not on map), *O. f. frontalis, O. jelskii,* and *O. f. spondionota.* The inset wing belongs to the population of western Cuzco and adjoining Ayacucho. The birds in the right hand column are in the *diadema* superspecies, from top to bottom, *O. diadema gratiosa, O. pulchella similis,* and *O. p. pulchella.* The general aspect of the birds is shown, as well as the pattern of the tertials (secondaries 7–9), which is diagnostic for *spodionota.* The map shows the range of each taxon. From a painting by John W. Fitzpatrick.

SPECIES LIMITS IN THE *OCHTHOECA DIADEMA* SPECIES-GROUP (TYRANNIDAE)

MELVIN A. TRAYLOR, JR.

Field Museum of Natural History, Chicago, Illinois 60605-2496 USA

ABSTRACT. Recent classifications of the *Ochthoeca diadema* species-group recognize three species, *diadema, frontalis,* and *pulchella,* with the taxon *jelskii* considered a subspecies of *pulchella.* The present study shows that the species-group is composed of two superspecies, *diadema,* which includes *pulchella,* and *frontalis,* which includes *jelskii. Frontalis* is sympatric with *diadema* or *pulchella* throughout its range from central Colombia to central Bolivia. *Jelskii* is parapatric with the three other species.

RESUMEN. Recientes clasificaciones del grupo-especie ("species-group") *Ochthoeca diadema* reconocen tres especies, *diadema, frontalis* y *pulchella,* con el taxón *jelskii* considerado como subespecie de *pulchella.* El presente estudio muestra que el grupo-especie está compuesto por 2 superespecies, *diadema* que incluye a *pulchella* y *frontalis* que incluye a *jelskii. Frontalis* se encuentra junto con ("sympatric") *diadema* o *pulchella* en todo su rango de distribución desde el centro de Colombia hasta el centro de Bolivia. *Jelskii* se encuentra en distintas áreas ("parapatric") que las otras tres especies.

Within the Fluvicoline flycatchers of South America, the *Ochthoeca diadema* species-group, as recognized by Smith and Vuilleumier (1971) and Fitzpatrick (1973), is composed of three species: *diadema, frontalis,* and *pulchella.* The taxa of this group are confined to Andean subtropical and humid temperate forests from Venezuela to Bolivia. Within the group, *diadema* is a distinctive species usually treated as separate from the apparently more closely related *frontalis* and *pulchella.* The latter two are close phenotypically and were considered conspecific by Hellmayr (1927). The taxon *jelskii* of northwestern Peru has been assigned as a subspecies to both *frontalis* and *pulchella* at various times, even by the same author (Zimmer 1937, and unpubl. ms.), and its true position has not yet been satisfactorily determined. In the present study I show that *pulchella* is more closely related to *diadema* than to *frontalis.* Two superspecies comprise the species group: *diadema* and *pulchella,* and *frontalis* and *jelskii,* respectively. Unfortunately, this is not the classification of *Peters' Check-list* (Traylor 1979), but recently collected specimens have caused me to revise my former conclusions.

The classification derived here is presented below. Taken together with Plate VI and the distribution maps (Figs. 1, 2), it will permit the reader to recognize the diagnostic characters of the various taxa, and to follow the ensuing discussion more easily. Only those taxa occurring in the area of sympatry from southern Colombia to Bolivia are shown. There are five races of *diadema* in northeastern Colombia and northern Venezuela, and one race of *frontalis* in the Bogotá area that are not relevant to the present discussion.

diadema species-group
  frontalis superspecies
    Ochthoeca frontalis frontalis (syn. *orientalis*)
                  spodionota (syn. *boliviana*)
    Ochthoeca jelskii
  diadema superspecies
    Ochthoeca diadema gratiosa
                  cajamarcae (not on Plate VI)
    Ochthoeca pulchella similis
            pulchella

The only character that distinguishes the two superspecies is the color of the superciliary, white in *frontalis* and yellow in *diadema.* Within the *frontalis* superspecies, the subspecies *frontalis* is characterized by absence of wing bars and by dull colored tertials. Most populations of *spodionota* have wing bars, although they are apparently secondarily lost in those populations from the Vilcabamba range of western Cuzco and from adjoining Ayacucho. However, *spodionota* has a brighter, more sharply defined pattern on the tertials. *Jelskii* differs from *spodionota* by having duller colored tertials, and from *frontalis* by the presence of wing bars. It has a more rusty dorsum than either of the others.

Within the *diadema* superspecies, *diadema* is characterized by its yellow underparts. *Pulchella* has gray underparts, but the northern race, *similis,* has a distinctive olive wash that separates it from nominate *pulchella* which is gray below and grayish brown above. All representatives of the *diadema* superspecies discussed here have distinct wing bars.

DISTRIBUTION AND SPECIES RELATIONSHIPS

With the exception of *diadema,* the taxa described above are all similar in appearance, and Hellmayr (1927) placed them in one species. Later, Carriker (1933) placed them in two, and subsequently (1935) in three species, whereas Zimmer (1937) arranged them in two species but with different components from Carriker's. Zimmer (1937) used the color of the superciliary stripe as his species character, white in *frontalis,* in which he included *jelskii* and *spodionota,* and yellow in *pulchella* with which he included *similis.* However, in 1955, in an unpublished manuscript of the Tyrannidae for the *Check-list of Birds of the World,* Zimmer placed *jelskii* in *pulchella.* This was the classification used by Meyer de Schauensee (1966) and Traylor (1979).

Zimmer (1937) did not discuss *diadema* as a close relative of *frontalis* and *pulchella,* although he did note (p. 21) that the olivaceous wash and yellow superciliaries of *O. p. similis* suggested a possible relationship with *diadema.* The first paper to include *diadema* with *frontalis* and *pulchella* in a species-group was that of Smith and Vuilleumier (1971). Within the group they treated *diadema* as a sister species of the *frontalis* superspecies which included *frontalis* and *pulchella.* Fitzpatrick (1973) reviewed the genus and recognized the *diadema* species-group, but treated the three species *diadema, frontalis,* and *pulchella* as equal hierarchically.

Earlier authors were handicapped by a lack of material of these birds, which are quiet and unobtrusive and quite difficult to shoot. The use of mist nets, however, and recent intensive collecting in Peru and Bolivia, primarily by parties from Louisiana State University Museum of Zoology (LSU), have increased the number of specimens available and broadened our understanding of the abundance and distribution of these species. Zimmer, as late as the early 1950's, had only 17 specimens from Peru, whereas I have been able to examine 87 from Peru and 264 altogether. The Peruvian and Bolivian localities on the maps (Figs. 1, 2) all refer to specimens I personally examined (see Appendix I). Colombian and Ecuadorian localities are from both specimens and published reports. The ranges of all taxa appear to be essentially continuous, given the presence of suitable habitat.

In almost every locality, except within the range of *jelskii* in northwestern Peru, two taxa occur sympatrically (Fig. 1). In southern Colombia and northern Ecuador *gratiosa* and *frontalis* occur together; this sympatric distribution extends north through the Eastern Andes to Bogotá, where the northernmost race of *frontalis, albidiadema,* occurs. To the south *gratiosa* and *cajamarcae* are found in southern Ecuador and in northern Peru in Piura and Cajamarca, north and west of the Marañón, where they have been taken near *jelskii,* although not at the same localities. South and east of the Marañón, *frontalis* occurs along the eastern Andes of Amazonas and La Libertad, where it is regularly sympatric with *similis,* the northern race of *pulchella* (Fig. 2). South of La Libertad *frontalis* is replaced by *spodionota,* which is also sympatric with *similis* in Huánuco and Junín and then regularly with *pulchella* from western Cuzco south to central Bolivia (Fig. 1).

The taxon *jelskii* is unusual in not having been taken at the same locality with any other taxon in the group. This in part accounts for its assignment to both the species *frontalis* and *pulchella* at different times. As shown (Fig. 2), it is essentially parapatric with all three of the other species. In eastern Piura it is found within 40 km of *gratiosa* and in central Cajamarca within about 30 km of *cajamarcae.* In the former case *jelskii* is on the Pacific drainage and *gratiosa* on the Marañón drainage, but in the latter case, both are on the Marañón drainage, although on different river systems. Although *jelskii* is primarily a bird of the Pacific slope of the Andes in the northern part of its range, it occurs in the left bank Marañón drainage in central Cajamarca, as noted above, and has crossed to the right bank drainage at Pataz, La Libertad. Farther south, at Quebrada La Caldera about 10 km west of Mashua, it is found at tree line on the Marañón slope of the eastern Andes. It has also been taken at Quebrada Huamash, Huánuco, on the right bank of the Marañón, but at Quebrada Shugush, ca. 30 km

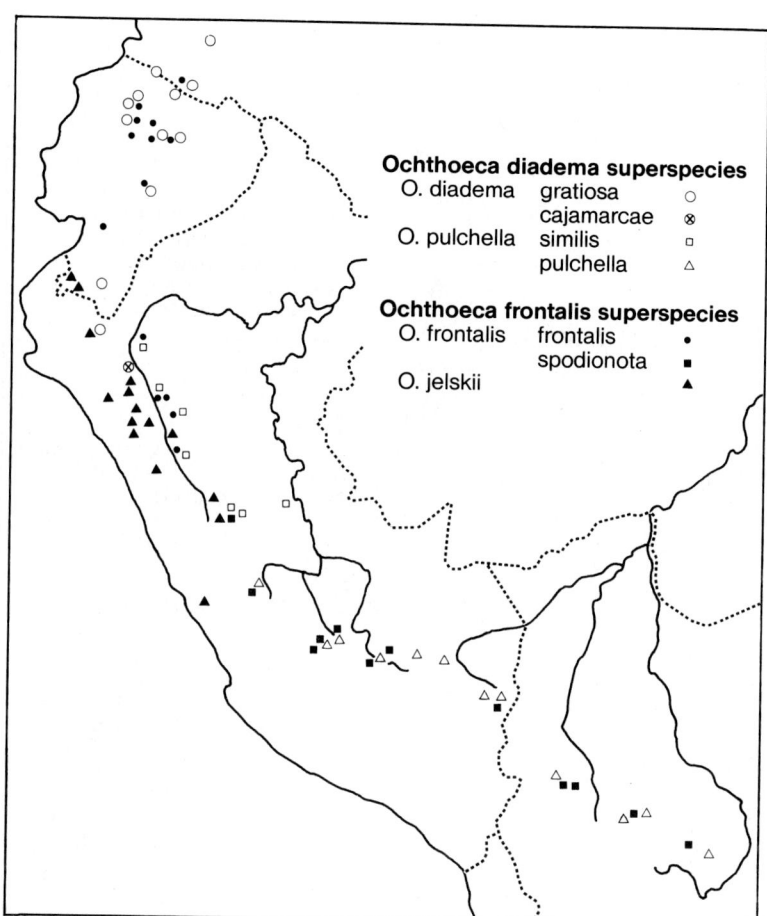

Ochthoeca diadema superspecies
O. diadema   gratiosa   ○
              cajamarcae   ⊗
O. pulchella similis   □
              pulchella   △

Ochthoeca frontalis superspecies
O. frontalis frontalis   •
              spodionota   ■
O. jelskii   ▲

FIG. 1. Distribution of the four species of the *Ochthoeca diadema* species-group, from southern Colombia to central Bolivia. *Frontalis* is found north to Bogotá and Antioquia, Colombia, and *diadema* is found north to Santa Marta, Colombia, and to the mountains of northern Venezuela.

west of the city of Huánuco, it has crossed the crest of the Andes to the Huallaga drainage. The population of *jelskii* at Quebrada La Caldera is only about 10 km from both *frontalis* and *similis* at Mashua on the eastern slope of the Andes, but that there is direct contact is doubtful because the intervening terrain is grassland, unsuitable for a forest-haunting bird. The Quebrada Shugush population, however, is only about 45 km from *frontalis* and *similis* in the Cordillera Carpish, and since they are on the same drainage, they probably meet somewhere between. In this area *jelskii* must certainly be considered parapatric with *frontalis* and *similis*.

It is evident that *diadema, similis,* and *pulchella* replace each other geographically from north to south, and in their association with *frontalis. Diadema* and *pulchella* are each others' closest relatives and form a superspecies within the species-group. This close relationship is not evident at first glance, particularly when specimens are viewed in a museum tray where the yellowish underparts of *diadema* are in sharp contrast to the gray underparts of *pulchella.* However, the northern *O. p. similis* is distinguished from *O. p. pulchella* by having an olive wash on the body plumage, and when *diadema* and *similis* are viewed from above, they are hard to separate. Both are slightly reddish brown on the back and darker on the crown, all with an olive wash, and they have rufous wing-bars and edgings on the secondaries and yellow superciliaries. The differences between *similis* and *pulchella* are much less, and there is no reason not to consider them races of a single species. At present *diadema* and *pulchella* are not in contact, the former being found west of the Marañón and the latter to the east of it.

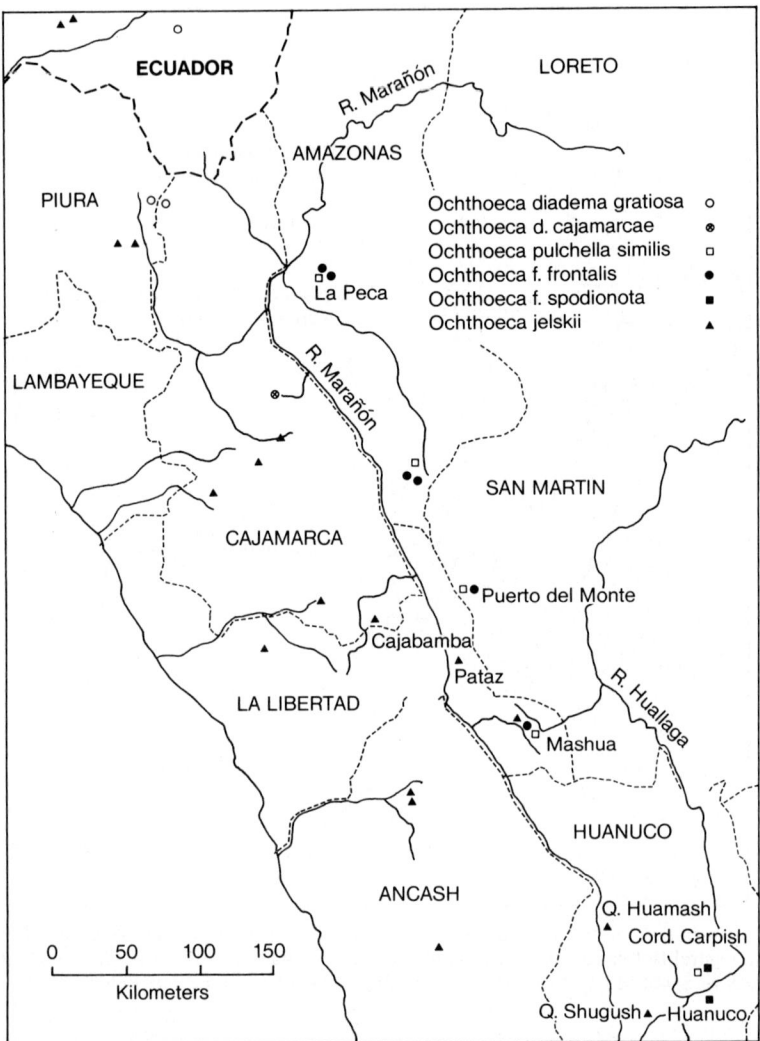

Fig. 2.   Map of northwestern Peru showing the distribution of *Ochthoeca jelskii* in relation to those of *diadema, frontalis,* and *pulchella.* Although *jelskii* has not been taken at the same localities as any of the others, it is parapatric with all three.

In the same way that *diadema* taxa replace each other geographically from north to south, so do *frontalis* and *spodionota.* Although they differ markedly in the presence (in *spodionota*) or absence (in nominate *frontalis*) of wing-bars and in the patterning of the tertials, they are clearly each others' closest relatives. At present there is a gap of about 150 km between the southernmost *frontalis* at Mashua, La Libertad, and the northernmost *spodionota* in the Cordillera Carpish, Huánuco, and we cannot be sure how they would behave if they met. However, until this gap is closed, they are best treated as a single species.

As noted above, *jelskii* is essentially parapatric with all three of the other species and cannot be made conspecific with any one of them. There is really nothing in its present range that suggests a closer relationship to either *frontalis* or *pulchella,* the two species it most nearly resembles. In southern Huánuco, the one area in which contact might be expected, it is equidistant from both. It could be a western representative of either one. It does share a white superciliary with *frontalis,* however, and this is the one character that consistently separates *frontalis* from *pulchella,* which has a yellow superciliary. Therefore, *jelskii* should be considered a separate species, probably in a superspecies with *frontalis.*

TABLE 1
COLLECTING ELEVATIONS OF *PULCHELLA* AND *FRONTALIS* IN AREAS OF SYMPATRY

Locality[1]	pulchella[2]	frontalis[2]
Peru, Amazonas, Cordillera Colán northeast of La Peca (LSU)	2440–2660 (7)	2880–3295 (10)
Peru, La Libertad, Mashua and 7 km east (LSU)	3150–3350 (3) 2600–2625 (10)	3350 (8)
Peru, Huánuco, Cordillera Carpish (LSU)	2350–2820 (17)	2530–3355 (7)
Peru, Cuzco, Cordillera Vilcabamba (Weske in litt.)	2070–2660 (50)	2820–3525 (47)
Peru, southern Cuzco/Puno (various)	2000–2900 (5)	3000–3660 (11)
Bolivia, La Paz (various)	2074–3050 (7)	3050–3300 (68)
Bolivia, Cochabamba (various)	1100–1595 (10)	2200–3350 (31)
Bolivia, Santa Cruz (various)	1680 (1)	2500 (1)

[1] The first four localities are all single sites from which good series of specimens are available. The last four cover broader geographic areas, as indicated. The source of the data is given in parentheses.
[2] Numbers in parentheses = number of specimens.

ELEVATIONAL RELATIONSHIPS

The relative elevational distributions of representatives of the two superspecies, wherever they occur together, also support the close relationship of *pulchella* to *diadema* rather than *frontalis*. Although the two superspecies are sympatric geographically, they are for the most part segregated elevationally throughout their ranges. This is not evident from published accounts in which both species are recorded from the same locality with only a single elevation listed, but it is apparent when good series with carefully labeled specimens are available or when experienced observers have studied both species in the field. In all cases the *frontalis* representative occurs at higher elevations than the *diadema* representative, although wandering individuals of one may occur within the range of the other.

Published records for Colombia, suggest overlap in elevational range, with *diadema* subspecies occurring from 1225 m to 3150 m, and *frontalis* subspecies from 2817 m to 3875 m. However, S. L. Hilty, who has extensive field experience in Colombia, writes (in litt.), "I just don't think these two taxa would ever occur together, even in the rare instances where their ranges may interdigitate. Based on my observations in the field, *O. frontalis* is most numerous from 3000 m to tree-line. It favors dense, short temperate forest . . . *O. diadema* is chiefly a subtropical species, most numerous about 1800–2500 m." Presumably, the same situation holds in northern Ecuador, where the same subspecies, *O. d. gratiosa* and *O. f. frontalis,* are found. Unfortunately, no recent collections are available from there. R. S. Ridgely (in litt.), however, says, "It does hold in southern Ecuador (Sangay area, east slope) where in October 1976 I found *frontalis* numerous from about 3400 m up to treeline (3700 m where we were), with no *diadema* there; and down below *diadema* was common in the 2800–2900 m range where we worked, with no *frontalis* there—again we were netting extensively."

Fine series of recently collected material show that the same elevational segregation found between *diadema* and *frontalis* in Colombia and Ecuador occurs between *pulchella* and *frontalis* in Peru and Bolivia (Table 1). The figures for Mashua, La Libertad, and the Cordillera Carpish, Huánuco, show that the elevational ranges of *pulchella* and *frontalis* occasionally overlap, but at Cordillera Colán, Amazonas, and the Cordillera Vilcabamba, Cuzco, elevational segregation is complete. The latter figures are particularly impressive, because they are from a series of net lines on a transect of the cordillera, that were operated over a period of four years. Nevertheless, the gap between *pulchella* at 2660 m and *frontalis* at 2820 m was actually a gap in the net line, so it is possible that their ranges meet in between.

The figures for more general areas in southern Peru and Bolivia show that elevational segregation extends to the southern limit of the species' ranges. The only apparent area of overlap is in La Paz at 3050 m, but the series from that elevation included only one *pulchella* versus 39 *frontalis*; the former was probably a wanderer from 2575 m where only *pulchella* was taken.

Although it is clear from elevational evidence that *diadema* and *pulchella* are subtropical representative species, segregated partially or wholly from the temperate forest *frontalis,* the status of the allopatric *jelskii* is not so clear. In the north of its range, in southwestern Ecuador and Piura, Peru, its elevational range is from 1220 to 2867 m, similar to those of the subtropical *diadema* and *pulchella*. From Cajamarca south, however, it ranges from 2640 to 3350 m, more like the temperate *frontalis*. This is particularly evident in the areas where *jelskii* most nearly approaches *frontalis* and *pulchella*. The elevation of Quebrada La Caldera was 3350 m, similar to that of Mashua where *frontalis* predominated; Quebrada Shugush, at 3100 m, is within the range of *frontalis* but not *pulchella* on the Cordillera Carpish. Although not conclusive, the elevational evidence supports the inclusion of *jelskii* in the *frontalis* superspecies.

MENSURAL DATA

Measurements tell us very little about relationships within the species-group. Variation in either absolute size or in proportions is slight (Figs. 3, 4). Nevertheless, whenever sympatric populations of the two superspecies are compared, the *frontalis* representative consistently is larger than the *diadema* representative. The actual differences are small, only occasionally being statistically significant at the .05 level, but they are significant in always being in the same direction. If these differences were intraspecific rather than interspecific, one would probably ascribe them to increase in size with elevation (Traylor 1950). There is some geographic variation, but it appears more erratic than clinal.

Because *jelskii* is not sympatric with any of the others, it cannot be directly compared to them, but in absolute size it fits in the middle of the species-group. *Jelskii* has one difference in proportions, however, that sets it apart. Its tail length is 88 percent of wing length, whereas no population of the other species has a tail length greater than 85 percent of wing length. While this difference is statistically significant at almost any level, and reinforces our conclusion that *jelskii* is a distinct species, it tells nothing about its relationship to *frontalis* or *pulchella*.

PLUMAGE CONVERGENCE AND SPECIATION PATTERNS

One of the most confusing aspects of the relationship between the two superspecies is the degree of resemblance between sympatric populations. This resemblance varies clinally from almost complete in Bolivia to slight in Colombia (Plate VI). In Cochabamba and Santa Cruz, Bolivia, nominate *pulchella* differs from *O. f. spodionota* only in having a yellow instead of white superciliary and rather duller patterned tertials. The two species have often been confused in the past. Proceeding north into Peru, *pulchella* develops a slight olive wash, which becomes accentuated in Junín and Huánuco. Farther north, in La Libertad, *spodionota* is replaced by nominate *frontalis,* which differs even more strikingly from *similis* in its lack of wing bars; these two are sympatric through northern Peru east of the Marañón. In Ecuador and northern Peru west of the Marañón *similis* is replaced by *O. diadema gratiosa,* which with its green and yellow underparts is completely unlike the gray and brown *frontalis.*

The precursor of the *diadema* species-group must have looked much like the present day *spodionota* or *jelskii*. The congeners most nearly related to this group are *O. cinnamomeiventris* and *rufipectoralis,* both small-sized species characteristic of the humid temperate zone. *Cinnamomeiventris* is dark slate above, with white forehead and superciliaries and completely unpatterned wings; the underparts are blackish with varying amounts of deep chestnut. Nominate *rufipectoralis* of Bolivia is dull brown above with white forehead and superciliaries and unpatterned wings; the underparts are rufous on throat and breast and whitish on the abdomen. However, *rufopectus,* the race of *rufipectoralis* from northern Peru to Colombia, has a single chestnut wing bar and pale patches on the tertials like nominate *frontalis.* The yellow forehead and two wing-bars are derived characters in the *diadema* species-group compared to their nearest relatives. Within the group, the yellow superciliaries of the *diadema* superspecies and

→

FIG. 3. Measurements of the taxa of the *Ochthoeca diadema* species-group discussed here. Horizontal lines (solid for the *frontalis* superspecies, dashed for the *diadema* superspecies) represent the observed range for each character; the vertical line is the mean. Measurements are grouped by geographical area where the pairs are sympatric; *jelskii* stands alone since it is not sympatric with any of the others. Numbers in parentheses = numbers of specimens. Col./N. Ec. = Central and Western Andes of Colombia, and Ecuador south through Pichincha; N. Peru = Amazonas, San Martín, and La Libertad; C. Peru = Huánuco and Junín; Vilcabamba = Cordillera Vilcabamba of western Cuzco, and adjoining Ayacucho; Cuzco/Puno = the rest of Cuzco and Puno; La Paz and Coch./S. C. = Bolivian departments of La Paz, Cochabamba, and Santa Cruz, respectively.

Length in millimeters

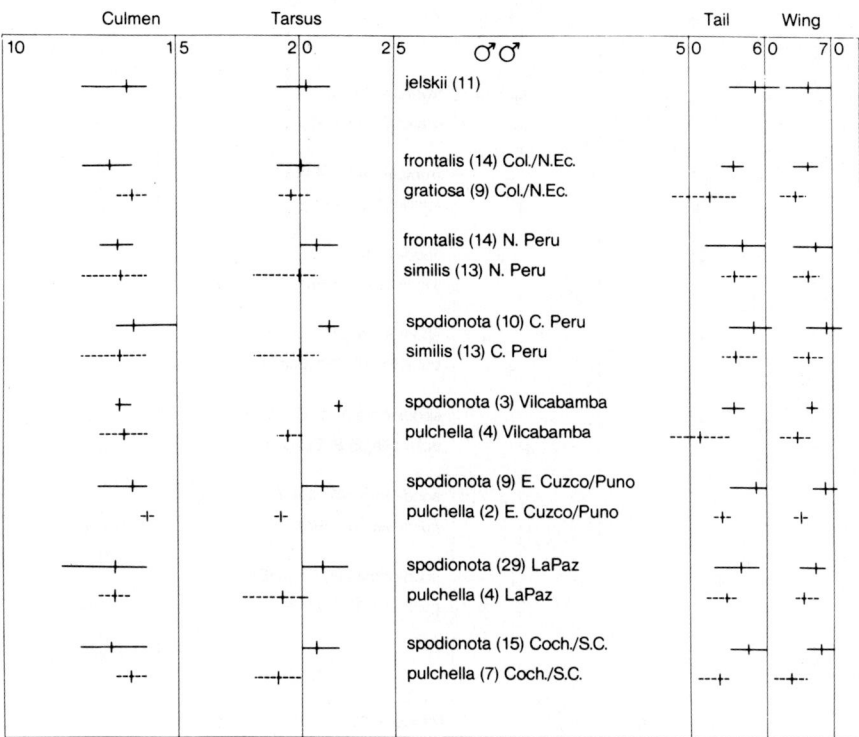

Length in millimeters

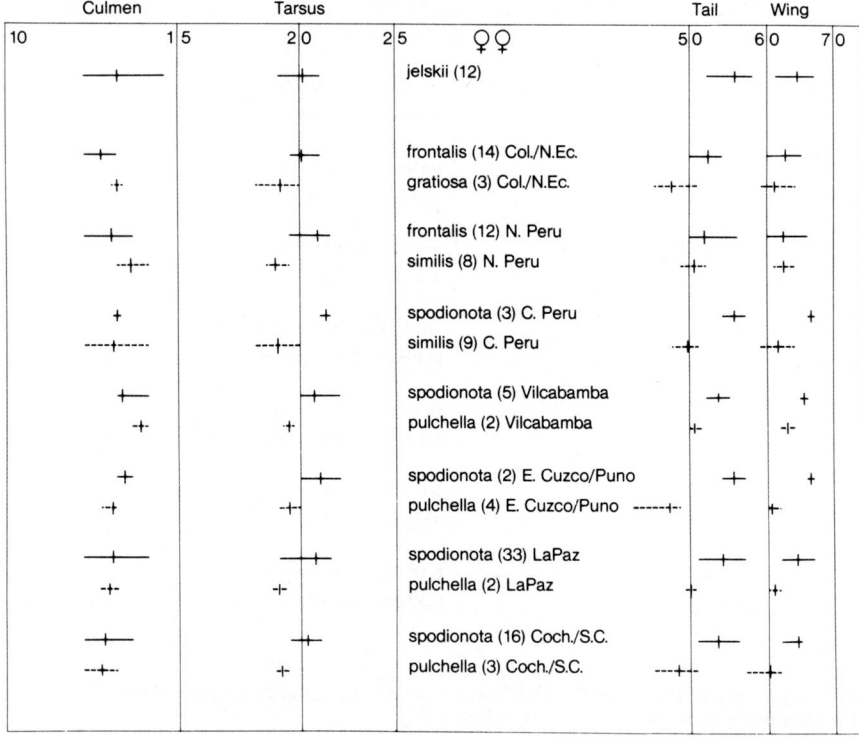

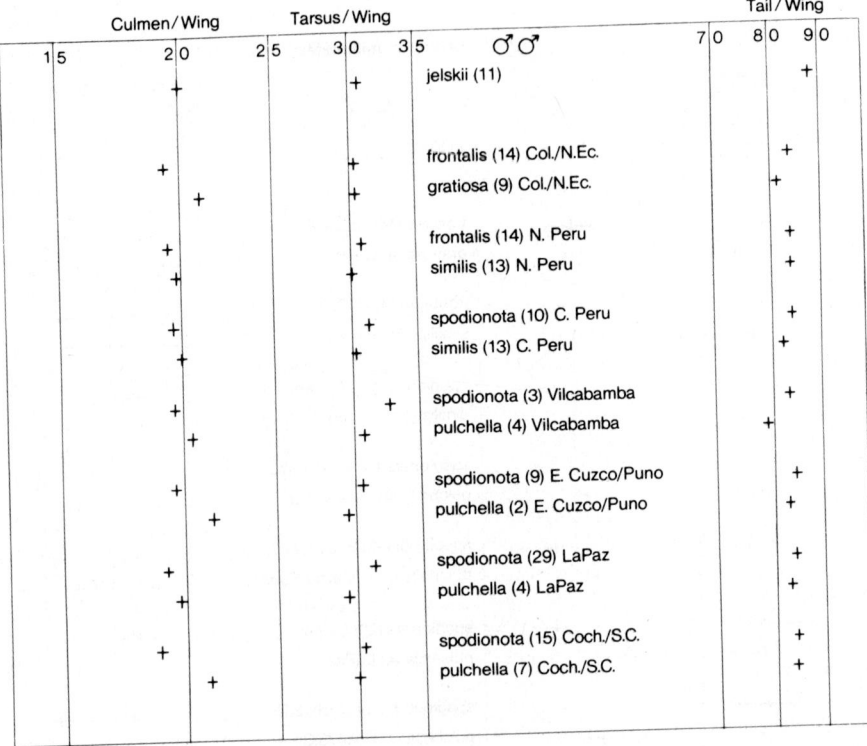

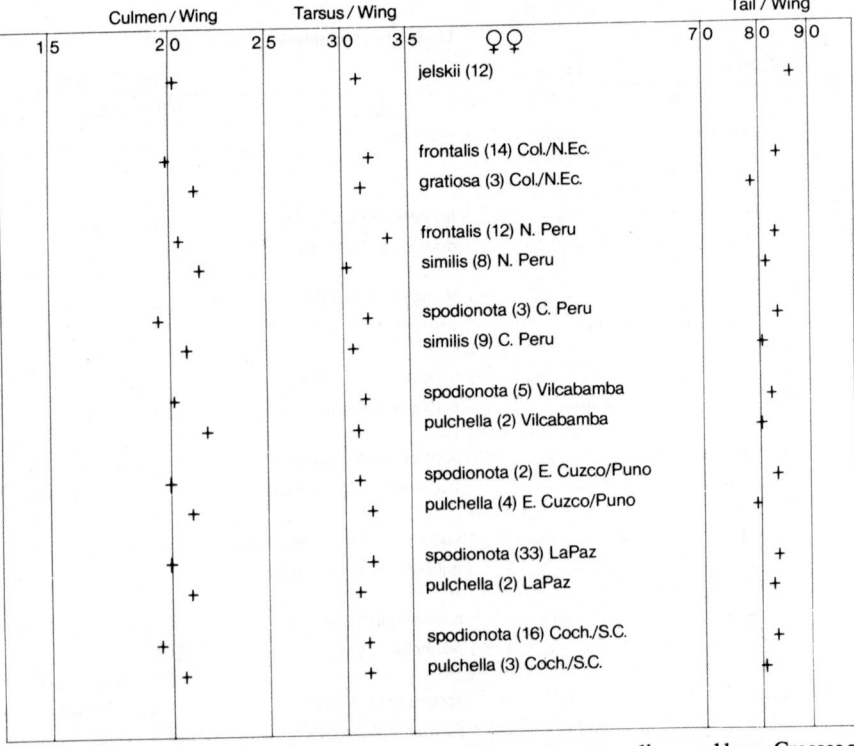

FIG. 4. Proportions of the taxa of the *Ochthoeca diadema* species-group discussed here. Crosses average values for each population. Ratios are grouped as in Figure 3.

the yellow underparts of *diadema* are certainly derived since they are found nowhere else in the genus. The presence or absence of wing-bars apparently has no phylogenetic significance since in *O. rufipectorialis* and *leucophrys*, as well as *frontalis* and *diadema*, these markings are found in some races but not others. Since they occur in some races of all four species of the species-group, however, they were probably found in the ancestral form.

Fitzpatrick (1973) discussed speciation in *Ochthoeca*, in relation to the changing climatic regimes in the Andes, cool and humid during glacial periods, and warm and dry during interglacials. As Haffer (1974) pointed out, during the glacial periods, the temperature drop was greater at higher elevations than in the tropics, so that the various montane vegetation zones were both lowered in elevation and compressed. They also became more continuous, particularly the upper zones, such as the humid temperate. It was during these cool periods that humid temperate species were able to disperse along the Andes. During the warm, dry interglacials, the montane zones were pushed to higher elevations on mountain peaks where they became fragmented by the drier river valleys. During these periods, speciation occurred in many isolated populations. In Fitzpatrick's reconstruction of speciation in *Ochthoeca*, the precursor of the *diadema* species-group had a continuous range along the Andes from western Venezuela to central Peru during the second, Mindel, glaciation. This seems a reasonable assumption considering the probable subsequent speciation. In the following warm, dry interglacial, the range would have become fragmented, with at least the dry Marañón valley forming a major barrier to dispersal. During this period of isolation, proto-*diadema* evolved its green coloration and yellow superciliaries in the population to the north of the Marañón, while proto-*frontalis*, to the south of the Marañón, remained, phenotypically at least, much as it had been. At the same time, proto-*diadema* became better adapted to subtropical rather than temperate forest, possibly because of the lesser extent of temperate forest in Colombia compared to Peru.

During the third glacial period, the Riss glaciation, the montane forest zones again became more or less continuous, and both species were able to disperse along the axis of the Andes; their adaptation to different elevational zones permitted almost complete geographical sympatry. Proto-*frontalis* extended north to central Colombia, and proto-*diadema* penetrated south into Peru, and, in conjunction with proto-*frontalis*, into Bolivia. The only area south of central Colombia in which the two were not sympatric was northwestern Peru west of the Marañón where only proto-*frontalis* occurred.

In the warm dry period following the Riss glaciation, the ranges of the two species would again have been fragmented. In the case of proto-*frontalis*, four populations were apparently isolated, one east of the Magdalena River in the eastern Andes of Colombia, one in southern Colombia and northern Ecuador, a third in northwestern Peru west of the Marañón, and the last in Peru east of the Marañón and south to Bolivia. During the period of isolation, the first population evolved a white instead of yellow forehead, characteristic of the race *O. f. albididiadema* (Plate VI). The population of southern Colombia lost its wing-bars, becoming the plain-winged race *frontalis*, while the population of Peru and Bolivia developed the bright patterning of the tertials characteristic of *O. f. spodionota*. The population of northwestern Peru did not apparently undergo much phenotypic change, but must have evolved behavioral characters that prevented it from interbreeding with *frontalis* when they subsequently met; this is the species *jelskii*.

During the last, Würm, glaciation, the montane forests again became more continuous, and plain-winged nominate *frontalis* evidently extended its range across the Marañón into northern Peru. The abrupt change from *frontalis* to *spodionota* in central Peru, with no intermediates known, suggests that any meeting between them would be secondary. Also, the two populations of *frontalis* in Colombia and northern Ecuador and in northern Peru, respectively, are identical except for a small size difference. During this same glacial time, *jelskii* extended its range across the Marañón and even into the Huallaga drainage where it is parapatric with *spodionota*.

Proto-*diadema* also underwent speciation during the warm, dry postglacial period. One or more populations were isolated east of the Magdalena in Colombia and Venezuela, another was isolated in southern Colombia and northern Ecuador, and the last was isolated in Peru south and east of the Marañón and in Bolivia. Those east of the Magdalena lost their wing-bars, and either evolved into or were precursors of the present series of subspecies in northeastern Colombia and Venezuela. The populations of southern Colombia apparently underwent no phenotypic change and are the present *O. d. gratiosa*. The populations of Peru and Bolivia, however, proto-*pulchella*, seem to have undergone a regressive evolution, losing the yellowish coloration of *diadema*, except for the superciliary, and converging on *spodionota* phenotypically. The loss of yellow was complete in Bolivian nominate *pulchella*, while the northern *similis* retains an olive wash. Why selection should favor convergence in Boliva and Peru, but not in Ecuador or Colombia is not clear. The degree of elevational segregation between the *frontalis* and *diadema* representatives seems almost complete in all areas, and no reason for convergence to be more adaptive in Bolivia than in Ecuador is now evident.

During the last glaciation, *O. d. gratiosa* extended its range south to Cajamarca in north-western Peru, but not across the Marañón or into the range of *jelskii*. *O. pulchella* extended its range north to the Marañón where it is sympatric with nominate *frontalis* instead of *O. f. spodionota*.

SUBSPECIES

In the description of the *diadema* species-group (p. 431), I discuss several subspecies; some of these forms are quite constant, and others show much clinal variation. The following discussion summarizes subspeciation within each species.

Diadema.—The seven races of *diadema*, as listed in the *Check-list of Birds of the World* (Traylor 1979), are based on varying intensities and distributions of general color and the presence or absence of wing-bars, neither of which forms any particular geographic pattern. Only three are sympatric with *frontalis*: *diadema* in the eastern Andes of Colombia, *gratiosa* from central Colombia to northern Cajamarca, Peru, and *cajamarcae* in central Cajamarca. At the time of the preparation of the manuscript for the *Check-list*, the only known specimens of *O. diadema* from Peru were the type series of the race *cajamarcae*, one male and three females from Chira. The population nearest to Chira was in northern Ecuador. Since then, I have examined a male from southern Ecuador in the Museum of Comparative Zoology and five specimens from northeast of Sapalache, on or near the Piura/Cajamarca boundary, north-western Peru. These recent specimens are all typical *gratiosa* of Colombia and northern Ecuador, which is characterized by a well marked, dark olive-green breast band. This would make the validity of *cajamarcae* suspect, but I have examined the three original females of *cajamarcae*, and they are, as stated in the original description, virtually without a breast-band. Pending the collection of additional topotypical material, *cajamarcae* must be recognized, but the range of *gratiosa* is extended to northern Cajamarca.

Pulchella.—There are two well-marked races of *pulchella*: *similis*, found from central Amazonas to Junín, and *pulchella*, occurring from western Cuzco and adjoining Ayacucho to western Santa Cruz, Bolivia. The former is a uniform race, characterized by a distinct olive wash above and below, and by a more reddish, less grayish brown underlying color on the back. The nominate race shows some clinal variation in the direction of *similis*, from central Bolivia to central Peru. Birds from Cochabamba are grayish-brown on the back, but as one progresses northward through southern Peru, the back becomes a warmer brown with a slight olive wash. It is not until Junín, however, that a strong olive wash appears on the underparts, and I place the *pulchella/similis* boundary somewhere between western Cuzco and central Junín.

Frontalis.—The two northern races of *frontalis* stand apart from the southernmost, *spodionota*, because of their unbarred wings and dull patterned tertials. Among them, *albidiadema* is distinguished from all other taxa in the species-group by having a white rather than yellow forehead; it is found in the Bogotá region. Nominate *frontalis* is found from the Central and Western Andes of Colombia to La Libertad, northern Peru, east of the Marañón. The race *orientalis*, from northeastern Ecuador to northern Peru was accepted in *Peters' Check-list* (Traylor 1979) because of a dearth of fresh material, but with good series from northern Peru and southern Colombia before me now, I fail to find any characters that distinguish this form. The remaining race, *spodionota*, is highly variable, and it was separated into two subspecies, *spodionota* and *boliviana*, in the check-list. Birds from the northernmost populations of *spodionota*, in Huánuco, have a broad chestnut wing-bar on the greater coverts, and a variable, narrow, darker one on the median coverts; occasionally the second bar is lacking. A single topotype of *spodionota* from Junín has only a single bar, but it is well marked. The most distinctive population is that from the Cordillera Vilcabamba of western Cuzco, and from adjoining Ayacucho. A few specimens have a vestige of the bar on the greater coverts and none on the lesser coverts, but others have none at all. Farther east in Cuzco the situation is like that in Junín, with only a single bar, but in Puno and even more in La Paz, Bolivia, the second bar is usually found and may be as bright as the first. Even in La Paz, near the type locality of *boliviana,* however, the second bar may be lacking. Farther east, in Cochabamba, the second bar is more regular and, often, as bright as the first, though narrower. In the *Check-list of Birds of the World* (Traylor 1979), I restricted the name *spodionota* to the populations of Junín and western Cuzco/Ayacucho, with *boliviana* having a divided range in Huánuco and from eastern Cuzco to Bolivia. The variation takes the form of a double cline, however, with many specimens from Huánuco and from eastern Cuzco to La Paz having only a single wing-bar like topotypical *spodionota*; thus, any line drawn between *boliviana* and *spodionota* seems artificial. The population from western Cuzco/Ayacucho with virtually no wing bars could be described as a distinct subspecies, but this would entail a divided range for *spodionota*. I prefer to recognize only a single, variable subspecies.

ACKNOWLEDGMENTS

Studies of this type would be impossible without access to all relevant material. For their kindness in lending me their specimens or allowing me to visit their collections, I thank J. Bond and F. B. Gill (Academy of Natural Sciences of Philadelphia), W. E. Lanyon and L. L. Short (American Museum of Natural History), K. C. Parkes (Carnegie Museum of Natural History), the late G. H. Lowery, J. P. O'Neill, and J. V. Remsen, Jr. (Museum of Zoology, Louisiana State University, Baton Rouge), R. A. Paynter, Jr. (Museum of Comparative Zoology, Harvard University), R. L. Zusi (National Museum of Natural History), and J. S. Weske (personal collection now in the American Museum of Natural History). K. L. Pruitt and F. Vuilleumier listed for me all Colombian and Ecuadorian specimens in the National Museum of Natural History and American Museum of Natural History, respectively. I also thank those who have shared their field experience of these birds with me: S. L. Hilty, J. V. Remsen, Jr., T. S. Schulenberg, D. F. Stotz, and J. S. Weske. Finally, I am particularly grateful to my colleague at Field Museum, J. W. Fitzpatrick, who carefully followed the evolution of my ideas and tried to clarify my thinking. All errors, however, are my own.

LITERATURE CITED

CARRIKER, M. A., JR. 1933. Descriptions of new birds from Peru, with notes on other little known species. Proc. Acad. Nat. Sci. Phila. 85:1–38.
CARRIKER, M. A., JR. 1935. Descriptions of new birds from Bolivia, with notes on other little known species. Proc. Acad. Nat. Sci. Phila. 87:313–341.
FITZPATRICK, J. W. 1973. Speciation in the genus *Ochthoeca* (Aves: Tyrannidae). Breviora, No. 402.
HAFFER, J. 1974. Avian speciation in tropical South America. Publ. Nuttall Ornithol. Club No. 14.
HELLMAYR, C. E. 1927. Catalogue of birds of the Americas. Field Mus. Nat. Hist., Zool. Ser. 13, Pt. 5.
MEYER DE SCHAUENSEE, R. 1966. The Species of Birds of South America with Their Distribution. Academy Natural Sciences Philadelphia, Livingston Publ. Co., Narbeth, Pennsylvania.
SMITH, W. J., AND F. VUILLEUMIER. 1971. Evolutionary relationships of some South American ground-tyrants. Bull. Mus. Comp. Zool. 141:179–268.
TRAYLOR, M. A., JR. 1950. Altitudinal variation in Bolivian birds. Condor 52:123–126.
TRAYLOR, M. A., JR. 1979. Tyrannidae. Pp. 3–245, *In* M. A. Traylor, Jr. (ed.), Check-list of Birds of the World. Vol. 8. Museum Comparative Zoology, Cambridge, Massachusetts.
ZIMMER, J. T. 1937. Studies of Peruvian birds, no. xxvi. Am. Mus. Novit. No. 930.

APPENDIX I
Specimens Examined[1]

O. d. gratiosa—COLOMBIA: Cauca: Cerro Munchique (5, FMNH); Nariño: Llorente (8, FMNH). ECUADOR: Cajanuma Divide (1,MCZ). PERU: Piura: NE of Sapalache (5, LSU).

O. d. cajamarcae—PERU: Cajamarca: Chira (3, ANS).

O. p. similis—PERU: Amazonas: Cordillera Colán (8, LSU); Leimebamba (4, ANS); La Libertad: Mashua (3, LSU); Cumpang (10, LSU); San Martín: Puerto del Monte (1, LSU); Huánuco; Cordillera Carpish (8, LSU); Huaylaspampa (3, LSU); above Acomayo (6, LSU; 2, FMNH); Cerros del Sira (4, AMNH); Junín: Rumicruz (1, AMNH).

O. p. pulchella—PERU: Ayacucho: Yuraccyacu (1, LSU; 1, AMNH); Cuzco: Cordillera Vilcabamba (4, AMNH); Pillahuate (1, FMNH); Puno: below Limbani (1, AMNH); Oconeque (3, ANS). BOLIVIA: La Paz: Sandillani (1, ANS); Chuspipata (1, LSU); Sacramento Alto (5,LSU); Cochabamba: El Limbo (1, LSU); Alto Palmar (1, LSU); El Palmar (2, FMNH); San Cristóbal (1, ANS); Incachaca (1, ANS; 3, CM); Santa Cruz: Samaipata (1, ANS).

O. f. albidiadema—COLOMBIA: Cundinamarca: Tocaimito (1, AMNH).

O. f. frontalis—COLOMBIA: Riseralda: Santa Isabel (3, AMNH); Caldas: La Leonera (2, ANS); Cauca: Purace (2, FMNH); Nariño: Puerres (1, ANS). ECUADOR: Canzocoto (1, AMNH); Intag (2, AMNH); Yanacocha (1, AMNH); Milligalli (2, AMNH); Mindo (1, AMNH); Oyacachi (4, AMNH; 1, ANS); Sumaco Arriba (5, AMNH); below Sumaco (2, ANS); Papallacta (1, AMNH). PERU: Amazonas: Llui (2, ANS); Atuen (2, ANS); N.E. of La Peca (11, LSU); La Libertad: Mashua (8, LSU); San Martín: Puerto del Monte (3, LSU).

O. f. spodionota—PERU: Huánuco: Huaylaspampa (4, LSU); above Acomayo (3, LSU); Bosque Unchog (4, LSU); Bosque Potrero (1, LSU); Junín: Rumicruz (1, AMNH); Ayacucho: Puncu (1, LSU); Tambo (1, AMNH); Cuzco: Cordillera Vilcabamba (6, AMNH); Cedrobamba (1, NMNH); N.E. of Abra Malaga (4, LSU); Puno: Valcón (6, LSU). BOLIVIA: La Paz: Hichuloma (2, ANS); Cotopata (22, LSU); Chuspipata (44, LSU); Cochabamba: Aduana (1, LSU; 1, FMNH); Chapare, Km. 104 (24, LSU); Incachaca (2, ANS; 2, CM); Santa Cruz: Comarapa (1, FMNH).

O. jelskii—ECUADOR: Guachanoma (1, AMNH); Celica (2, AMNH); San Bartolo (1, AMNH). PERU: Piura: Palambla (1, AMNH; 2 ANS); El Tambo (1, AMNH); Cajamarca: Sunchubamba (1, FMNH); Chugur (3, AMNH); Taulis (4, AMNH); Cajabamba (1, ANS); La Libertad: Pataz (1, ANS); Quebrada La Caldera (1, LSU); Huánuco: Quebrada Huamash (1, LSU); Quebrada Shugush (1, LSU); Lima: Zarate (2, LSU; 1, AMNH).

[1] Abbreviations of museums housing specimens examined are: AMNH = American Museum of Natural History; ANS = Academy of Natural Sciences, Philadelphia; CM = Carnegie Museum of Natural History; FMNH = Field Museum of Natural History; LSU = Louisiana State University Museum of Zoology; MCZ = Museum of Comparative Zoology, Harvard University; NMNH = U.S. National Museum of Natural History.

EVOLUTION

FITZPATRICK, JOHN W. Form, Foraging Behavior, and Adaptive Radiation in the
Tyrannidae 447

GRANT, P. R. Climatic Fluctuations on the Galapagos Islands and Their Influence on
Darwin's Finches 471

JEHL, JOSEPH R., JR. Hybridization and Evolution of Oystercatchers on the Pacific Coast
of Baja California 484

MURRAY, BERTRAM G., JR. Evolution of Clutch Size in Tropical Species of Birds 505

MYERS, J. P., J. L. MARON, AND MICHEL SALLABERRY. Going to Extremes: Why Do
Sanderlings Migrate to the Neotropics? 520

ONIKI, YOSHIKA. Why Robin Eggs are Blue and Birds Build Nests: Statistical Tests for
Amazonian Birds 536

SCHNEIDER, DAVID. Migratory Shorebirds: Resource Depletion in the Tropics? 546

SHORT, LESTER L. Neotropical-Afrotropical Barbet and Woodpecker Radiations:
A Comparison 559

SKUTCH, ALEXANDER F. Clutch Size, Nesting Success, and Predation on Nests of Neo-
tropical Birds, Reviewed 575

WUNDERLE, JOSEPH M., JR., AND KENNETH H. POLLOCK. The Bananaquit-Wasp Nesting
Association and a Random Choice Model 595

EVOLUTION

FORM, FORAGING BEHAVIOR, AND ADAPTIVE RADIATION IN THE TYRANNIDAE

JOHN W. FITZPATRICK

Division of Birds, Field Museum of Natural History, Chicago, Illinois 60605 USA

ABSTRACT. Relationships between foraging behavior and external measurements of the bill, wings, tarsus, and tail are analyzed for 94 representative species of tyrant flycatchers (one quarter of the family). Results are presented in three parts: (1) Allometric relationships are summarized for external measurements, followed by general correlations between measurements (corrected for body size) and foraging mode. Upward-strikers have wide bills, short wings, long tarsi, and reduced tails; perch-gleaners and ground-foragers both have narrow bills, and with the latter also showing substantially longer legs; aerial-hawkers have triangular bills, extremely long wings (as do long-distance migrants), and short legs. Generalists, using many prey capture techniques, have intermediate structural features. (2) Using three sample analyses I show that relative sizes of certain characters vary continuously among species along a quantitative scale of foraging behavior. The morphology-behavior functions suggest adaptive interpretations for much of the observed structural variation. Evolutionary change in structure appears to accompany fine-scale changes along behavioral spectra. Extreme structural peculiarity characterizes the most behaviorally specialized forms, and even could limit evolutionary flexibility at the end-points. These relationships support the notion that evolution proceeds most easily from generalized form and behavior, through intermediate conditions, to the most extreme behavioral and morphological specializations. (3) I speculate on the directions along which increasingly specialized descendants evolved from generalized ancestors in the Tyrannidae, suggesting that these pathways are still visible in present-day forms. The great diversity within the Tyrannidae results, in part, from the unusually competitor-free environment within which the nonoscine lineages may have evolved. This environment allowed various flycatcher species to stop along certain routes toward specialization, remaining or even proliferating at stable intermediate points while certain others continued to specialize further. In this way, today's family Tyrannidae may illustrate pathways of morphological and behavioral specialization long since obliterated in most other modern bird families.

RESUMEN. Se analiza la relación entre el comportamiento de alimentación y medidas externas del pico, alas, tarso y cola para 94 especies representativas de atrapa moscas tirano (un cuarto de la familia). Los resultados se presentan en tres partes: (1) Se resumen las relaciones alométricas para medidas externas, seguido por una correlación entre medidas (corregidas para el tamaño del cuerpo) y el modelo empleado para alimentarse. Las especies que picotean hacia arriba ("upward-strikers") tienen picos anchos, alas cortas, tarsos largos y cola reducida; aquellas aves que picotean posadas en ramas ("perch-gleaners") así como las que se alimentan en el suelo ("ground-foragers") tienen picos angostos y estas últimas aves muestran patas singularmente largas; las que cazan en el aire ("aerial-hawkers") tienen picos triangulares, alas extremadamente largas (tal como las que migran grandes distancias) y patas cortas. Aquellas aves que usan técnicas diferentes para la captura de sus presas ("generalists") presentan componentes intermedios. (2) Utilizo análisis de tres ejemplos para mostrar que los tamaños relativos de ciertos caracteres varían continuamente entre especies a lo largo de una escala cuantitativa de comportamiento de forraje. Las funciones de comportamiento-morfología sugieren interpretaciones adaptativas para la mayoría de las variaciones estructurales observadas. Los cambios evolucionarios en la estructura, parecerían acompañar cambios muy pequeños en los espectros de comportamiento. La presencia de peculiaridades estructurales externa caracteriza a las formas con comportamientos más especializados e inclusive podrían limitar la flexibilidad evolucionaria en los puntos de los espectros. Estas relaciones soportan la idea de que evolución avanza más fácilmente desde formas y comportamientos generalizados a través de las condiciones intermedias hacia las especializaciones más extremas de comportamiento y morfología. (3) Especulo sobre las direcciones a lo largo de las cuales evolucionaron los descendientes mayormente especializados de ancestros generalizados en los Tyrannidae, sugiriendo que esas direcciones aún son visibles en formas de la actualidad. La gran diversidad en Tyrannidae resulta en parte por la inusual falta de competidores en el medio en el cual es posible que hayan evolucionado los linajes de no-oscines. Este medio permitió que varias especies de atrapa moscas se detuvieran a lo largo de determinadas rutas de especialización permaneciendo o inclusive proliferando en puntos intermedios estables, mientras que otras continuaron especializandose aún más. De esta manera, la familia actual de Tyrannidae puede ilustrar rutas de especialización de comportamiento y morfología que hace mucho han sido obliteradas en otras familias de aves actuales.

Morphological radiation and gradually increasing ecological diversity within certain successful lineages are principal hallmarks of evolution. We would learn a great deal about the process of adaptation if only we could study through time the pathways by which such radiations occur. Such a feat is impossible, of course, because proliferation of many descendant species from a common ancestor occurs too slowly to observe directly. However, an alternative exists; available for study are numerous examples of diverse phylogenetic lineages that we can study in "cross-section" in present-day time. Many lineages currently are at revealing points along their individual histories of proliferation or decay. Comparative analysis of structure and function within such groups gives us a picture, albeit an indirect one, that allows us to infer some of the pathways and processes of adaptive radiation. In this context, I began studying the relationships between body form and foraging habits in the enormously diverse family Tyrannidae, the tyrant flycatchers.

The 375 tyrant flycatcher species are largely neotropical, and they span a range of ecological roles and morphological forms as great as that of any bird family. They form the subject of several preliminary studies on radiation, both at the family-wide level (Keast 1972) and within several important lineages within the family (e.g., Johnson 1963, 1980; Smith 1966; Lanyon 1967, 1978; Smith and Vuilleumier 1971). An overview of the tyrannid radiation is provided in Traylor and Fitzpatrick (1982).

In prior papers I described and classified the patterns of foraging behavior found in the family (Fitzpatrick 1980), and analyzed some quantitative aspects of flycatcher foraging strategies (Fitzpatrick 1981a). In the present paper I present a preliminary analysis of the external shapes of flycatchers as they relate to foraging. Patterns that are suggested by this study indicate that much of the variation in tyrannid body form is closely related to variation in foraging techniques. Despite the caution advised by Gould and Lewontin (e.g., 1979) in such matters, I interpret close correlations between features of form and behavior within a variable group of close relatives as evidence for *adaptive* radiation within the group.

Demonstration that form and function vary together among related species remains a crucial first step in any careful analysis of adaptation. In this regard, I emphasize that the present study is only a beginning. I focus on the broad patterns of functional correspondence between form and behavior within the Tyrannidae. These overall relationships can be illustrated rather clearly with visually comprehensible diagrams and bivariate statistics. However, such methods only crudely assess the degree to which morphological features vary together with one another and with behavior to form true adaptive character-suites. More sophisticated statistical treatments of this question, using a multi-variate approach and an expanded set of morphologic and behavioral data, currently are in progress.

METHODS

This paper combines field measurements of flycatcher foraging behavior with physical measurements of museum specimens representing the same species. Field sites and procedures are described in detail elsewhere (Fitzpatrick 1980, 1981a) and will be summarized only briefly here. To date, I have observed more than 200 flycatcher species in the wild, principally in Peru, Venezuela, and Brazil. I have quantitative foraging data for about 120 of these species. Data recorded for a foraging tyrannid include overall habitat and micro-habitat descriptions and the following variables for each perch used during a continuous foraging bout, usually lasting as long as the bird can be kept in sight: perch height, distance to nearest leaf above the bird, search time at perch (minus any handling time for captured prey), distance moved to next perch (after a sally or a give-up), sally angle, sally distance, and sally type. These data are organized into a "foraging mode profile" for each species (Fitzpatrick 1980). This permits each species to be classified on a gross level as a specialist or a generalist. Specialists use predominantly one prey-capture technique; generalists use several or many techniques with

TABLE 1
FORAGING MODES AND THEIR DISTRIBUTION AMONG GENERA IN THREE SUBFAMILIES OF TYRANT FLYCATCHERS[1]

	Elaeniinae (37)	Fluvicolinae (34)	Tyranninae (17)
Foliage-gleaning			
Generalized modes:			
Outward hover-gleaners	1		7
Frugivorous aerial-hawkers	*2		2
Frugivorous upward hover-gleaners	7		
Specialized modes:			
Perch-gleaners	13		
Upward-strikers	15	2	1
Aerial-foraging			
Generalized mode:			
Enclosed-perch-hawkers		7	*
Specialized mode:			
Aerial-hawkers		8	3
Ground-foraging			
Generalized mode:			
Near-ground generalists		5	1
Specialized modes:			
Perch-to-ground specialists		4	
Terrestrial specialists		5	1
Unknown	1	3	2

[1] Subfamilies defined in Traylor (1977) and Traylor and Fitzpatrick (1982). Numbers in parentheses show total number of currently recognized genera within each subfamily.
[2] Asterisk (*) shows foraging modes of secondary importance to one or more genera within subfamily. From Fitzpatrick (1980).

equal frequency. The actual frequencies of use of perching positions, sally types, etc. provide data for numerous comparisons along continuously varying behavioral scales.

Linear measurements used in this paper (mm) are averages from two typical, but otherwise randomly chosen males for each species, collected from the geographic regions where I studied them. I measured 94 species (one-quarter of the family), including members of 72 of the 87 genera recognized by Traylor (1977, 1979) and representing essentially the entire range of body form variation within the family. I measured wing chord, length of innermost and outermost rectrices, tarsus length, length of hallux and claw, culmen length from base and from anterior edge of nostril opening, bill width at base and half-way to the tip, and bill depth at the nostril.

Body weights were obtained primarily from birds mist-netted in Peru and Venezuela. For species I never caught, I used weights listed in Weske (1972), ffrench (1976), or recorded on specimen labels at the American Museum of Natural History, Field Museum of Natural History, and Louisiana State University Museum of Zoology.

Aspect ratios were calculated from wing tracings made from live or recently collected specimens in the field, with the anterior edge of the wing held as nearly straight (extended) as possible. To correct for variability, I made two separate tracings for each individual, and traced as many individuals as possible (N = 1 to 19). Wing areas were measured from these tracings with a compensating polar planimeter; aspect ratios were calculated as the square of the total length of the extended wing divided by wing area (i.e., wing length/mean aerodynamic width). These measures are proportional, but not identical, to the aspect ratios used in technical aerodynamic equations (e.g., Greenewalt 1975), because I measured only one wing instead of total wing area including the body.

To compare the sizes and shapes of external body parts across numerous species I relied principally upon bivariate statistics and bivariate graphs. Of special interest are the three body parts most directly associated with food-gathering: bill, wing, and tarsus. In all figures and

TABLE 2

SIZES AND ALLOMETRIC RELATIONSHIPS IN THE TYRANNIDAE[1]

	Wing	Tail	Tarsus	Bill Length	Bill Width	Bill Half-width	Body weight
Median	69.5	57.0	18.0	14.7	8.5	4.7	13.3
Minimum	30.5	14.0	11.7	8.9	4.0	2.5	4.7
Maximum	138.0	280.0	37.3	31.6	16.6	11.2	72.0

Allometric regressions:

Untransformed linear measurements vs log_{10} body weight

Slope	77.23	86.27	7.96	13.38	7.48	3.58	—
Intercept	−17.33	−34.58	9.51	0.46	−0.06	0.67	—
r^2	0.86	0.38	0.27	0.67	0.70	0.39	—

Untransformed linear measurements vs $\sqrt[3]{body\ weight}$

Slope	37.89	42.77	3.95	6.77	3.68	1.77	—
Intercept	−22.35	−41.32	8.87	−0.94	−0.57	0.41	—
r^2	0.85	0.38	0.28	0.70	0.70	0.39	—

Log_{10} measurements vs log_{10} body weight

Slope	0.45	0.52	0.17	0.33	0.37	0.32	—
Intercept	1.31	1.16	1.07	0.81	0.49	0.30	—
r^2	0.87	0.58*	0.28	0.67	0.69	0.43	—

[1] Based on measurements of 94 representative species. Measurements in millimeters, body weights in grams. Asterisk (*) marks the only case in which one regression shows a substantially better fit than the others in its column, thereby favoring one of the transformations over the other two.

graphs, each plotted point represents a species; its position on the graph reflects that species' morphological or behavioral position relative to the other species measured. Species are symbolized on some graphs according to their foraging modes, using the behavioral categories described in detail in Fitzpatrick (1980, 1981a). Those foraging modes and their frequencies within the three tyrannid subfamilies of Traylor (1977)—the Elaeniinae, Fluvicolinae, and Tyranninae—are listed in Table 1.

To test for correspondence between foraging mode and morphology I used two approaches. In the first ("Character Variation") I searched bivariate graphs for non-random clustering among species that share a foraging mode. The null hypothesis in these examples is that species with similar foraging characteristics plot randomly with respect to one another inside the two-dimensional space mapped by the family as a whole. Put conversely, I searched for statistical correlations between size-correct mensural traits and certain discrete categories of foraging behavior. In the second approach ("Morphology as a Function of Behavior") I plotted selected morphological measurements directly against continuously varying behavioral axes. Positions of species along these axes were determined from frequency distributions of their respective use of various prey capture maneuvers. The resulting correlations are examined largely by eye in this paper.

In one case (see Fig. 8), values for relative tarsus length represent residuals, based upon a tarsus-length versus body-weight regression calculated earlier (Fig. 5). The values were derived by subtracting 9.40 mm from the actual length and dividing by log body weight. The Y-intercept of the Figure 5 regression is 9.40. Thus, a tarsus of exactly average length for its body weight produces a value of 8.05 by this formula, which is the slope of the regression in Figure 5.

CHARACTER ANALYSES AND INTERPRETATIONS

ALLOMETRIC RELATIONSHIPS

In any analysis of body form variation, the effects of size upon shape must be considered. The sizes of individual characters tend to be strongly correlated with overall body mass, but the correlations are neither perfect nor similar to one another across characters. To make meaningful functional analyses of shape, we must factor out the overriding effects of size from our measurements. (Functional analyses of size per se also can be informative, but shall not be attempted here.)

The ranges of sizes and shapes contained within the Tyrannidae are summarized in Table 2. The results of a series of bivariate regression analyses indicate how each of five external characters varies against body weight across the family. Not surprisingly, all regressions show highly significant correlations between mensural characters and body weight (lowest correlation coefficient, r, is .52, d.f. = 92, $P < 0.01$ throughout), although the amount of variation "explained" by body weight alone (r^2) varies considerably among characters. Three types of allometric analyses are shown for the same set of measurements, using different transformations. Untransformed linear measurements were plotted against $\log_{10}$ body weight and against cube root of body weight, and $\log_{10}$ transformed measurements were plotted against $\log_{10}$ body weight. Not shown are regression results using untransformed body weight data, as these yielded substantially poorer fits. In general, the linear regression model fits all three data transformations equally well. This can be seen by comparing the three r^2 values for the three separate regressions within each column.

If the values of r^2 in Table 2 were 1.0, or close to it, then this paper would end here. Differences in shape would strictly reflect differences in body size. The remainder of this report analyzes the various components of variation in shape not explained by size alone. To do this, size is factored out of the mensural data, usually by dividing them by $\log_{10}$ body weight.

CHARACTER VARIATION

Bill shape.—Perhaps the most ecologically revealing morphological feature of a bird is its bill, the tool with which potential food is handled and ingested (e.g., Schoener 1965; Pulliam 1975). Variations, specializations and convergences in bill structure are widely known to reflect gross ecological roles (Storer 1971), and even small behavioral differences related to foraging tactics (e.g., Engels 1940) or food choice (reviewed by Hespenheide 1973; Abbott et al. 1975; Karr and James 1975).

Among flycatchers, most of which are predominantly or entirely insectivorous, bill structure varies only in subtleties of length and width measurements, rather than in over-all form (Traylor and Fitzpatrick 1982). Whereas Hespenheide (1971) showed prey sizes to be correlated at least grossly with bill size among tyrannids (in his species, larger birds have larger bills, but prey sizes are more closely correlated with body sizes), he also stressed the effects that subtle differences in foraging style can have on prey choice and community assembly among grossly similar species (Hespenheide 1975). It is, therefore, of ecological as well as evolutionary interest to determine how foraging style differences *themselves* relate to bill size and shape.

I plotted relative bill widths and lengths of 94 representative species of tyrant flycatchers in Figure 1. Body size effects were removed from this analysis by dividing both bill width and bill length by $\log_{10}$ body weight.

Seven foraging mode distinctions are made among the 94 species I considered. Bill shapes of the species using these modes fall within distinct subsets of the total morphological space defined by the family as a whole (Fig. 1). Bills of upward-strikers are relatively long and wide, whereas those of perch-gleaners are nearly as long, but relatively narrow. The highly frugivorous tyrannids (including species in all three subfamilies) have short, stubby bills relative to body size. Ground-foragers and perch-to-ground specialists show almost complete overlap with perch-gleaners in possessing relatively long, narrow bills, slightly wider at the base. Aerial-hawkers show wide variation in bill length, but uniformly intermediate bill widths, reflecting bill shapes that are all minor variants upon a broad isosceles triangle, providing a relatively wide gape.

Some simple functional interpretations of the above general pattern are available. Upward-strikers forage with highly stereotyped, explosive sallies upward toward leaf undersurfaces, clearly relying on surprising unwary insects. Their spatulate bills would seem to provide the necessary room for error during split-second contact with the prey substrate, during which the insect is literally scooped from the leaf. In contrast, both directional and temporal precision are smaller problems for a perch-gleaner, because no sally is made at all. In these species, the thin, pointed bill permits accurate picking and probing as the bird searches vegetation within its reach, in much the same way that tweezers are used by humans.

As a sidelight to the above, well-developed rictal bristles characterize the upward-strikers, and are conspicuously absent from tyrannid perch-gleaners. The role of these bristles in prey capture remains unclear (Stettenheim 1974); they may serve in part as protective devices about the eyes and face of a sallying flycatcher (Conover and Miller 1980), as well as expanding the effective area swept by an open bill in motion. The observation that bristles are best

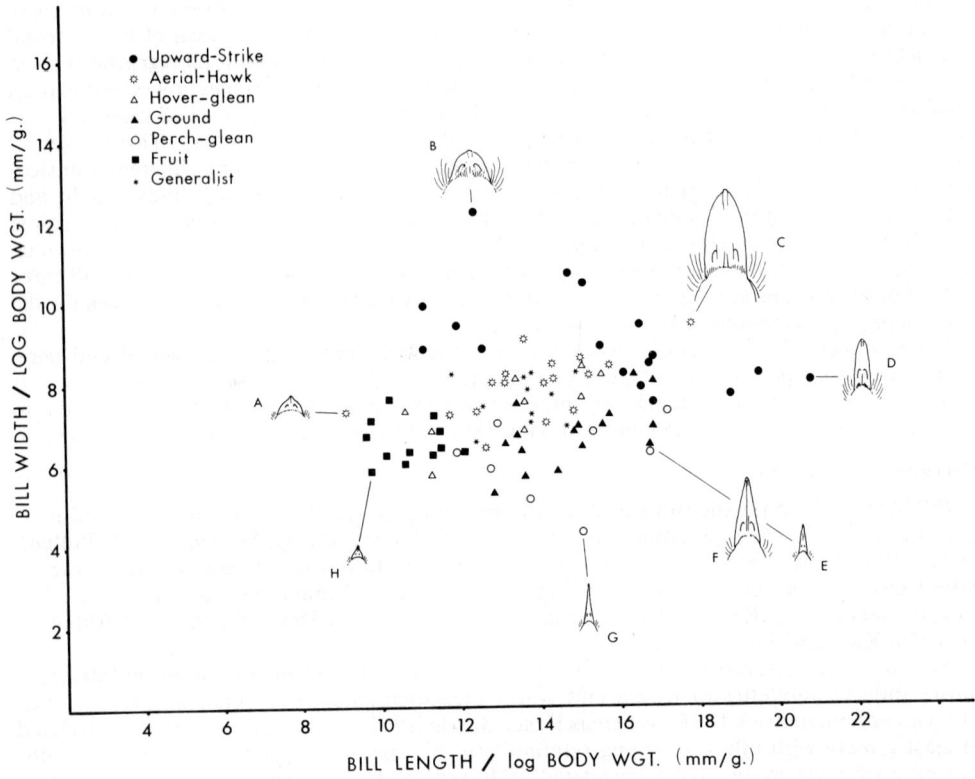

FIG. 1. Relative bill width at base plotted against relative bill length (from base) for 94 species of tyrant flycatchers, each assigned to one of seven foraging mode categories (upper left). Extreme bill shapes, drawn to scale: A = *Colonia colonus*, B = *Platyrinchus platyrhynchos*, C = *Megarynchus pitangua*, D = *Todirostrum cinereum*, E = *Anairetes paulus*, F = *Agriornis montana* (closed triangle), G = *Tachuris rubrigastra*, and H = *Tyrannulus elatus*.

developed among upward-strikers, and least among perch-gleaners, is consistent with both functions.

The bill-shape dichotomy just discussed suggests why most perch-gleaners, when they do sally after a prey item, usually do so with an upward-hover-glean (Fitzpatrick 1980). With a narrow, pointed bill, individuals presumably must become virtually stationary to use the bill tip accurately. During a sally, this can be accomplished only by hovering. Furthermore, because hovering presumably gives some kinds of insects a chance to escape, this restriction may have implications regarding the range of prey items usable by perch-gleaners versus upward-strikers. This point was stressed by Sherry (1982), who documented just such differences between the diets of these two ecological groups.

Method of prey capture used by ground-foraging flycatchers usually is similar to that of perch-gleaners. The simple picking motion is slower and more easily directed than in the sally-gleaners. The fact that many perch-to-ground salliers are large, and feed on relatively large prey items, probably explains the relatively wide gapes among bills of ground-foragers, whose bills otherwise closely match those of perch-gleaners (Fig. 2).

Aerial-hawkers in a number of avian groups show relatively wide gapes, presumably affording a means of sweeping a wider area with the open bill (Storer 1971). Possibly, this constraint leaves only bill *length* to vary functionally with prey size (Fig. 2). This interpretation predicts that bill length is a better indicator of prey size among aerial-hawkers than among other tyrannid foraging modes less constrained to a wide gape.

MacArthur (1971) pointed out that one "fundamental law of functional morphology"—that jacks-of-all-trades are masters of none—still remains to be "clearly enunciated." In this regard

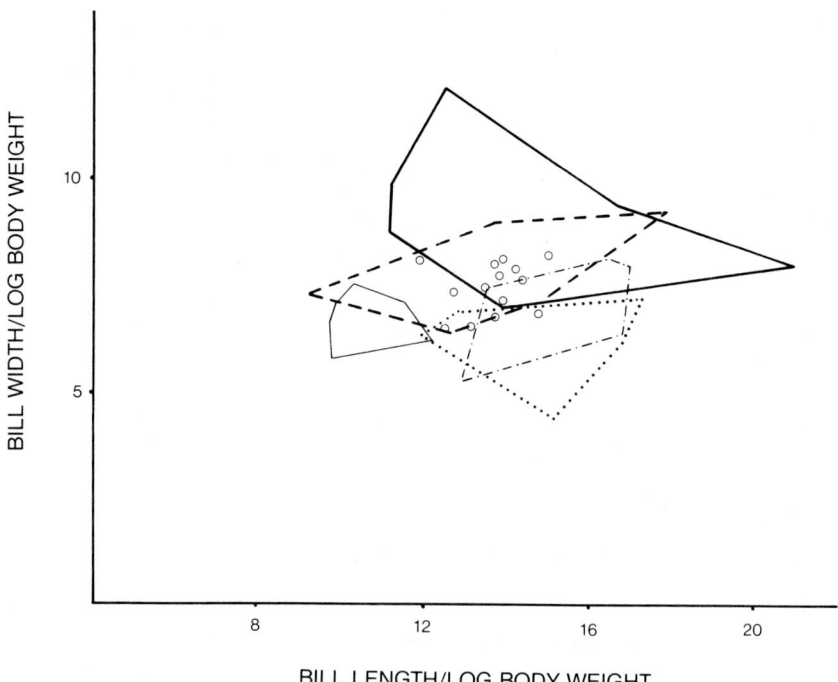

FIG. 2. Relative bill shapes of tyrant flycatchers, based on the data plotted in Figure 1. Lines enclose all values within the following foraging mode categories: upward-strike (thick solid line), aerial-hawk (thick dashed line), ground-related foraging (dot-and-dashed line), perch-glean (dotted line), frugivorous generalists (thin solid line). Foraging mode generalists are plotted as open circles.

it seems worth emphasizing that the tyrannid behavioral generalists (those species that use several sally techniques) do, in fact, fall in the very middle of their family's bill-shape spectrum (Fig. 2). Generalists appear to have settled on a "morphologically average" bill shape, in which the differing structural requirements of distinct behavioral traits combine to produce a master-of-none-of-them. This morphological intermediacy among behavioral generalists reappears throughout the analyses discussed in this paper.

   Wing shape.—Severe aerodynamic constraints have resulted in remarkable uniformity in wing form within any taxonomic group of birds. Indeed, regressions of wing length, area, and aspect ratio against body weight show that a single power function defines wing structures within the entire class Aves, spanning four orders of magnitude in body weight (Greenewalt 1962, 1975). Because of aerodynamic constraints, variation in wing shape within the Tyrannidae is comparatively slight. This is immediately apparent from the fact that body size alone accounts for much more of the variation in the tyrannid wing length than it does for any other character I measured (Table 2). Nevertheless, the family still shows more variation in wing design than does any other passerine family (Hartman 1961). This variation principally relates to migratory habits and reliance upon aerial foraging techniques.

   The Tyrannidae, like other passerines, show minor variations on the elliptical wing, a design that provides maximum lift and maneuverability at low air speed, and allows for rapid acceleration during a sally (Savile 1957). However, maximum power during open-air, high velocity flight is provided by longer, more pointed wings. One ecological subset of flycatchers, aerial-hawking species (N = 19) and species of other foraging modes known to migrate great distances each year (N = 11; Kipp 1958), approaches this latter design in having relatively long wings compared to the rest of the family (Fig. 3). Analyzed separately, the hawkers and the migrants show virtually identical regressions; pooled, their regression line shows the same slope but is significantly above that of the remainder of species (Fig. 3; F = 7.08; $P < 0.01$).

   The regression for upward-strikers (Fig. 3, open triangles) shows a trend toward relatively

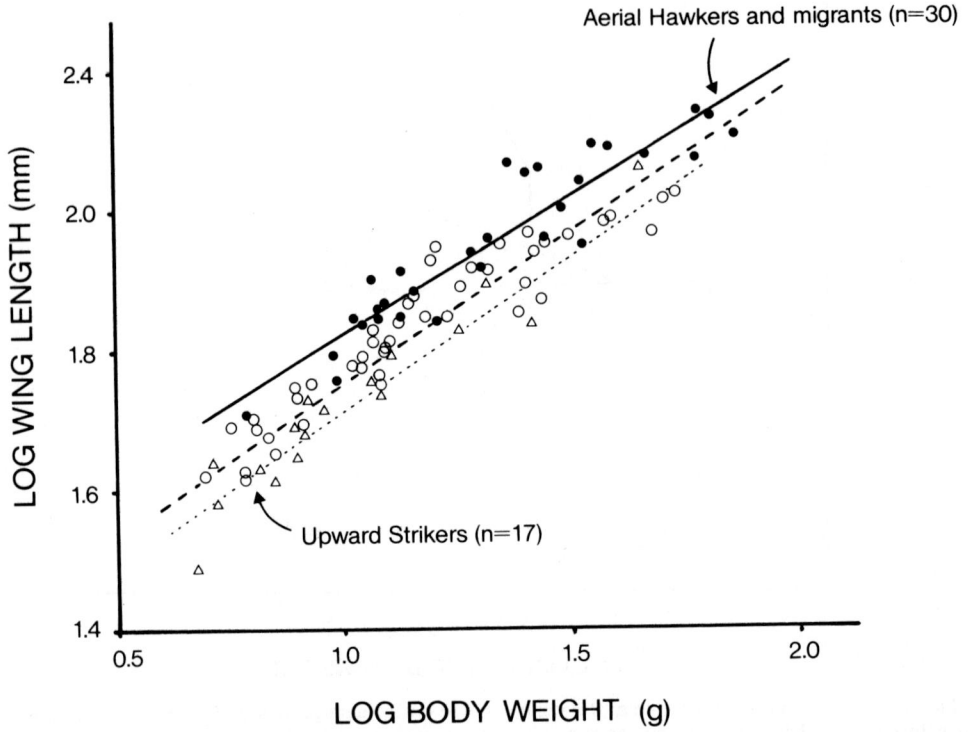

FIG. 3.   Log-log plot of wing length versus body weight in 94 species of tyrant flycatchers, categorized as aerial-hawkers or long distance migrants (closed circles), upward-strikers (open triangles), or others (open circles). Regression lines are shown for all three categories.

short wings compared to other tyrannids. Although not statistically significant, this trend is consistent with Savile's (1957) model for species that require rapid take-offs in dense foliage.

Wing designs differ more in aspect ratio than in simple wing length (Fig. 4). Even among the limited set of species for which I obtained both wing tracings and foraging data, some clear patterns emerge. The four aerial-hawking specialists have substantially more elongate wings than the other species (Fig. 4). Species that capture aerial prey between 20 and 80 percent of the time I consider to be facultative aerial-hawkers (Fig. 4, open stars). Just as generalists exhibit intermediate bill shapes (preceding section), the aspect ratios of facultative hawkers fall below those of aerial-hawking specialists but, on average, above those of species that rarely hawk.

The four aerial-hawking specialists plotted in Fig. 4 belong to two genera, *Contopus* (left-hand pair) and *Tyrannus* (right-hand pair). These pairs illustrate the cumulative effects of foraging strategy and migratory habits; each pair contains a resident species (*C. cinereus* and *T. melancholicus*) and a long-distance migrant (*C. virens* and *T. tyrannus*). In both genera the migrant, as predicted, has a substantially greater aspect ratio, despite being nearly identical in body weight.

Functional interpretations of these results are straightforward and parallel the discussion by Savile (1957). Longer wings generate power more efficiently during sustained, high speed flight, while shorter, rounded wings provide better lift at take-off and low velocities, and also permit greater mobility through dense vegetation. As expected for these mechanical reasons, aerial-hawkers and migrants show the relatively longest wings for their body size, while foliage-gleaning specialists show the shortest.

Tarsus length.—In contrast to wing form, which for aerodynamic reasons is restricted to minor variation in the family, tyrannid tarsi span a wide morphological range. Tarsus length and body weight are only weakly correlated across the family as a whole (Fig. 5, Table 2). A major portion of the variation is related to the use of the ground as a foraging substrate, a

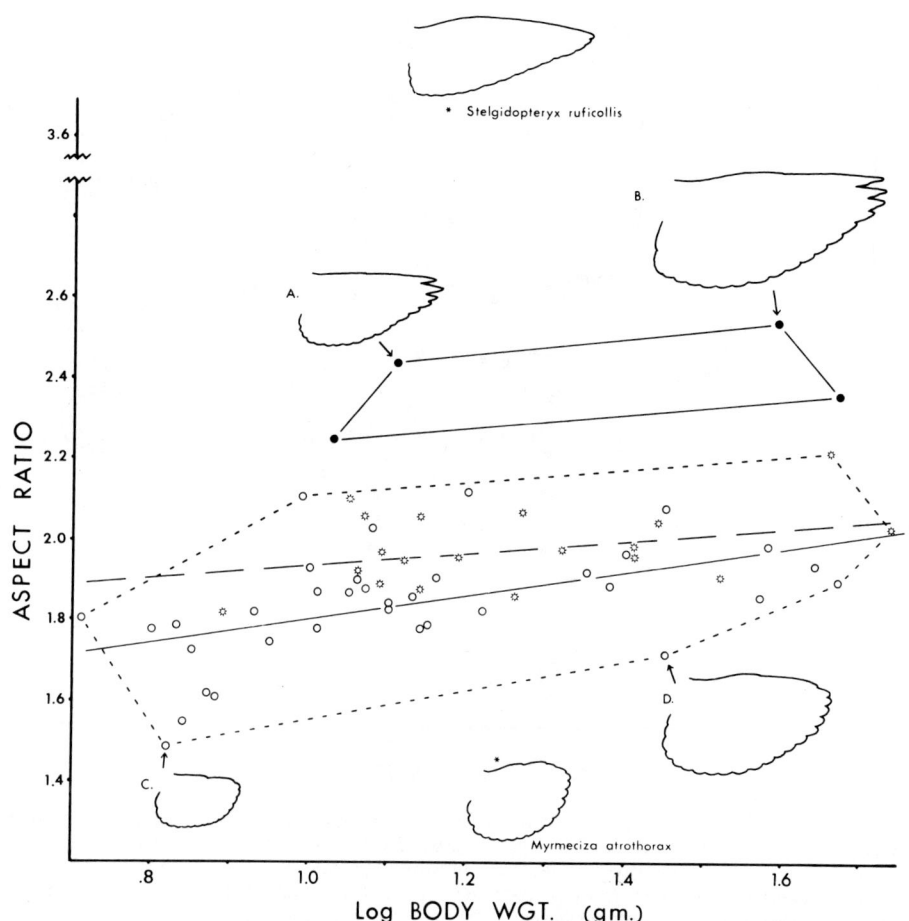

FIG. 4.   Aspect ratio plotted against log body weight for some representative species of tyrant flycatchers, categorized as aerial-hawkers (closed circles, solid lines), facultative aerial-hawkers (20–80% use of aerial-hawk; open stars), and others (open circles), the latter two groups enclosed by broken lines. Regression lines are shown for facultative hawkers (dashed line) and all non-hawkers (solid line). Wing shapes illustrated: an aerial-foraging swallow (*Stelgidopteryx ruficollis*); a terrestrial formicariid (*Myrmeciza atrothorax*); tyrannids: A = *Contopus virens,* B = *Tyrannus tyrannus,* both aerial-hawkers; C = *Todirostrum cinereum,* an upward-striker; D = *Myiarchus ferox,* an outward hover-gleaning generalist.

habit that is related to the evolution of long legs and long hind claws in a number of avian orders (e.g., Engels 1940; Berger 1952; Storer 1971). Tyrannid species that habitually forage upon, or sally to, the ground have significantly longer legs than the remaining species (Fig. 6). More important, the tarsus-length versus body-weight regression shows a much steeper slope among this group. Not only do ground-foraging species exhibit longer (and thicker) tarsi than do equivalently-sized species that perch on twigs in the typical passerine style (Schaffer 1903), but the legs of the heavier-bodied species proportionately are even longer than those of the lighter-bodied ground-users.

In relative tarsus length, the "perch-and-wait" aerial-hawkers represent the opposite of ground-foragers (Figs. 5, 6). Their legs are shorter, and their allometric increase is substantially shallower than among species that habitually use the ground (note slopes of two solid lines in Fig. 6).

Once again, species that combine the two behavioral extremes are morphologically inter-mediate. Five species (closed triangles, Fig. 6) are aerial-hawking specialists, but either they do so largely from the ground or rock surfaces over streams (*Serpophaga cinerea, Sayornis*

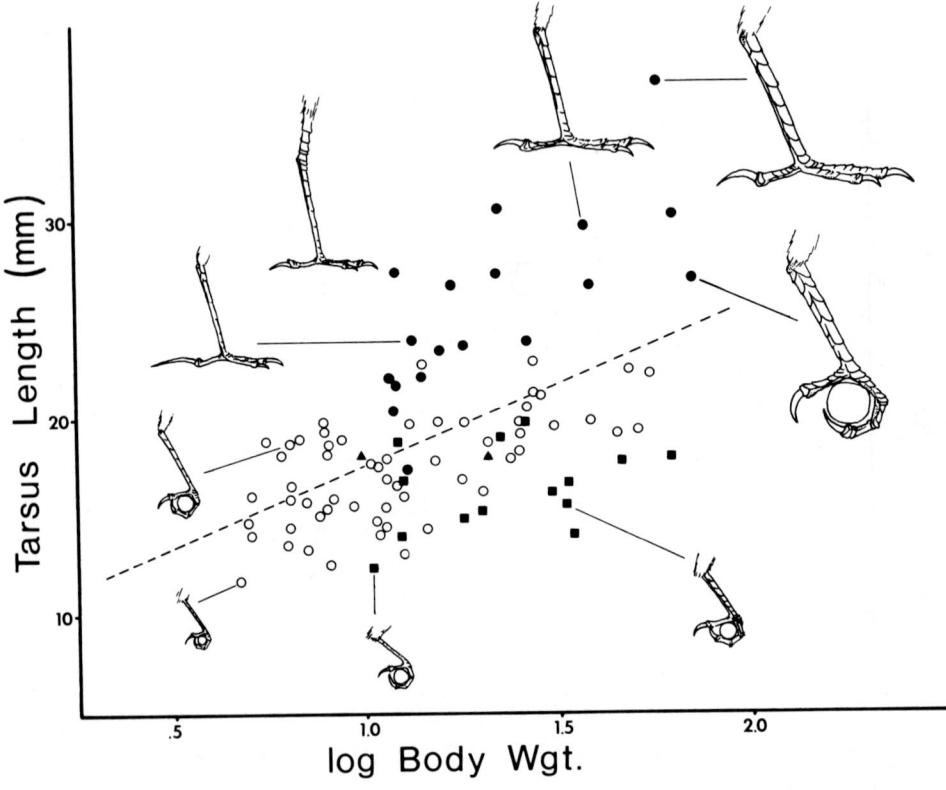

Fig. 5. Absolute tarsus length plotted against log body weight for 94 species of tyrant flycatchers. Regression line = dashed line. Ecological categories: ground-related foragers = closed circles; aerial-hawkers = closed squares; all others = open circles. Tarsi illustrated to scale, clockwise from lower left: *Myiornis ecuadatus, Todirostrum cinereum, Lessonia rufa, Muscigralla brevicauda, Machetornis rixosus, Agriornis montana, Gubernetes yetapa, Contopus borealis, Pyrrhomyias cinnamomea.*

nigricans), or they sally to the ground regularly (*Pyrocephalus rubinus, Knipolegus lophotes, Gubernetes yetapa*). The regression for tarsus-length versus body-weight among these five species has the same slope as that of ground-users, but its values fall between the lines for the two groups of behavioral specialists (Fig. 6).

In the case of intermediate behavior just described, it seems that the opposing physical demands for long legs, associated with ground-use, and for short ones, associated with aerial-hawking habits, result in tarsus lengths that approach the average for the respective body weights (Fig. 6). (In a subsequent section, I explore the extent to which this kind of morphological fine tuning exists.) What are these opposing physical demands? The functional morphology of long versus short legs among birds has received remarkably little attention for being such a clear dichotomy (but see Osterhaus 1962; Orians and Horn 1969). At present I can only speculate on the reasons for the differences. Among the ground-users, long legs presumably provide greater running speed, a higher viewing position, stronger physical support because of increased cross-sectional area, greater area for muscle attachment, and greater flexing capability during landing than would short legs. Among aerial-hawkers, short legs insure a low center of gravity upon the perch, favoring balance and stability during long, nearly motionless search periods. In addition, a shorter tarsus requires a less acute flexing angle at the heel joint to accomplish a given amount of passive flexing of the foot while perching (see Schaffer 1903). This provides a firmer hold for the perching, "sit-and-wait" aerial-hawkers than a long-legged bird could attain sitting in the same position. Both considerations would also hold for many other, ecologically similar bird groups that have unusually short tarsi (e.g., swifts, swallows, hummingbirds, kingfishers, jacamars).

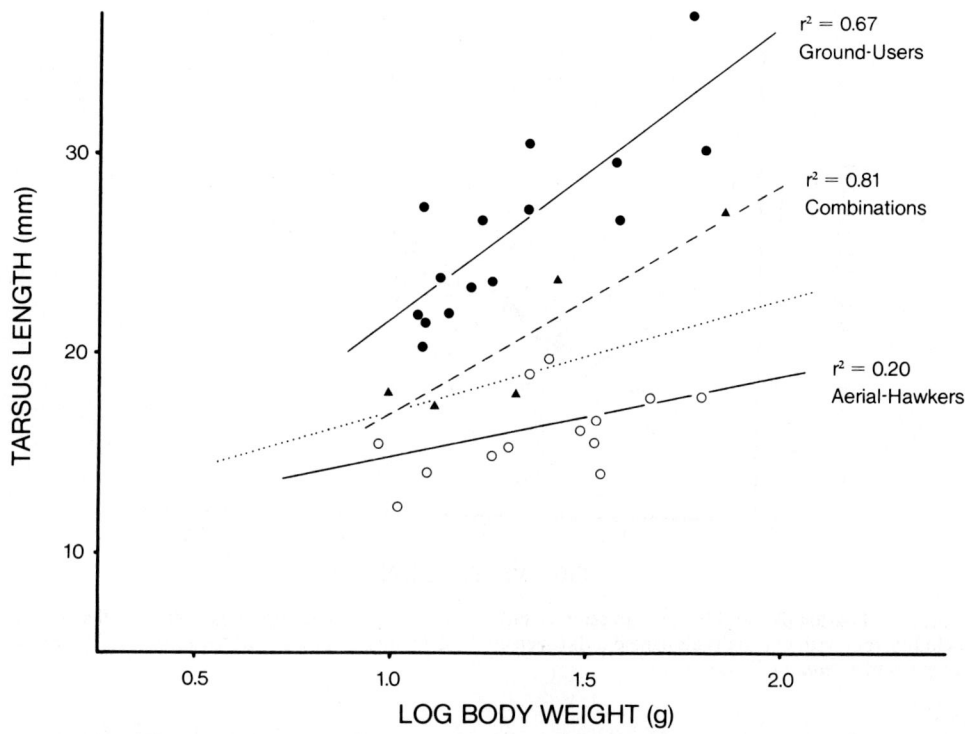

FIG. 6.   Tarsus length plotted against log body weight for species in three ecological categories, showing regression lines for each group and for the family as a whole (dotted line). Ground-users = closed circles, solid line; aerial-hawkers = open circles, solid line, "combination" species, or hawkers that use the ground extensively = closed triangles, dashed line.

Upward-strikers also possess relatively long legs, especially in the smallest size classes. Because these species habitually bob, lean, and peer while searching leaf surfaces, I surmise that their long legs may afford them increased viewing area in dense vegetation, where undersurfaces often are obstructed (diagrammed in Fitzpatrick 1978). Sherry (1982) suggested in addition that the need for explosively rapid take-off during sallies selects for longer legs that provide maximum spring. Both interpretations are supported by the proportionately large muscle mass associated with the hind limbs of many upward-strikers (pers. observ.).

Two of the more bizarre adaptations for ground foraging in the family are the extended hind claw of *Lessonia rufa* and the partially scuted tibiotarsus of *Muscigralla brevicauda* (Fig. 5). Both seem related to peculiar ecological features. *Lessonia* forages with fast ground-chases and fluttering aerial pursuits on moist, boggy alpine meadows, floating mats of vegetation along Andean lake margins, and on south temperate wet grasslands. Elongated hallux claws have arisen among many ground-dwelling bird groups (e.g., longspurs, longclaws, larks, pipits). They are best developed among species that use the hind claw for support while walking on soft, wet ground or floating vegetation, as in the jacanas (Storer 1971). *Lessonia* appears to provide a tyrannid example of this adaptation. *Muscigralla,* probably the most exclusively terrestrial, cursorial tyrant flycatcher, walks and runs on the dry earth of the arid Peruvian coast. Its legs are proportionately the longest in the family, comparable to those of flycatcher species two to three times its weight (Fig. 5). These oversized legs presumably provide added height and increased running speed, both associated with the species' habit of standing atop small mounds of earth and scanning, before darting after prey along the ground. The tremendously elongated tibiotarsus of this species has been accompanied by development of scutellation above the heel joint—a unique feature in the family.

Tail form.—Unlike variation in bill, wing, and tarsus features, many of the peculiar tail modifications within the Tyrannidae have social functions, placing them outside the scope of

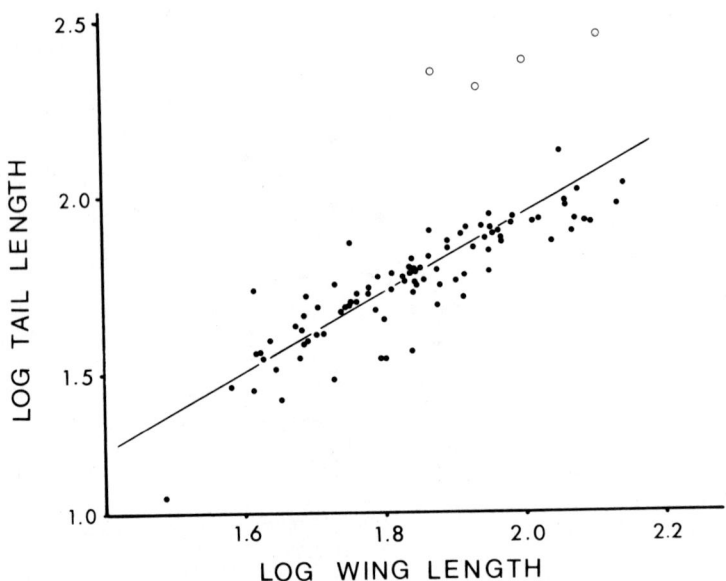

FIG. 7. Log-log plot and linear regression of tail length versus wing length in 94 species of flycatchers, including four with unusually elongated tails (open circles): *Tyrannus savana, Alectrurus risora, Gubernetes yetapa,* and *Colonia colonus.*

this paper. The typical tyrannid tail is of medium length with a small central notch. Tail length and wing length are strongly correlated (Fig. 7), and the relationship is even stronger if a few of the species with elaborate, greatly elongated tails are removed from the analysis (four open circles in Fig. 7; r^2 changes from .58 to .66). The closeness of this relationship supports the general assertion (e.g., Hartman 1961) that in most flycatchers the tail serves predominantly aerodynamic functions as in other typical passerines.

The most notable elaboration in tail form is the independent development of elongated outer tail feathers among several groups of species. This clearly represents an adaptation for maneuverability during aerial prey capture, where laterally extended outer rectrices provide a rudder against which the body can be sharply turned. Thus, aerial-hawkers show swallow-like gradations from deeply forked tails (especially in genera *Contopus, Pyrrhomyias, Hirundinea, Tyrannus*) to greatly exaggerated outer rectrices (*Muscipipra, Gubernetes, Tyrannus forficatus* and *savana*). In at least one genus (*Alectrurus*), containing marsh-dwelling, apparently polygynous species (Fitzpatrick, unpubl. data), this adaptation appears to have developed into a secondary sexual characteristic.

A few genera show reduced tail lengths. Most notable are *Myiornis* and *Muscigralla*. The latter species, discussed above, is so cursorial that it apparently has lost the need for the extra glide area provided by a tail of normal length. The genus *Myiornis*, which contains the smallest passerine bird (*M. ecaudatus,* Short-tailed Pygmy Tyrant, 4.5 g), uses an explosive sally for its upward-strike foraging technique, and a peculiar, hovering flight while moving amidst dense vine tangles high in the forest. Short, rounded wings provide all the lift in this form of flight (Greenewalt 1975). *Myiornis* performs no complicated aerial acrobatics, and, therefore, has little use for a typical, elongated passerine tail. Indeed, it may have lost the tail as a means of reducing drag (Sherry 1982); in these respects the genus seems to have converged upon the form and flight behavior of medium-sized beetles.

Graduated tails, in which the innermost rectrices are longest, occur in three unrelated genera: *Todirostrum (cinereum* species group only, Fitzpatrick 1976), *Fluvicola (nengeta* only), and *Stigmatura.* In all three cases the graduated rectrices are broadly tipped with white. The tail of *Todirostrum* is frequently cocked over the back during a peculiar, stereotyped, intraspecific display. *Stigmatura budytoides* apparently jerks and cocks the tail during normal foraging and in display (Wetmore 1926; Smith 1971). The combined features of conspicuous white rectrices and tail-cocking habits (both also present in *Inezia subflava,* in which the tail shows rudi-

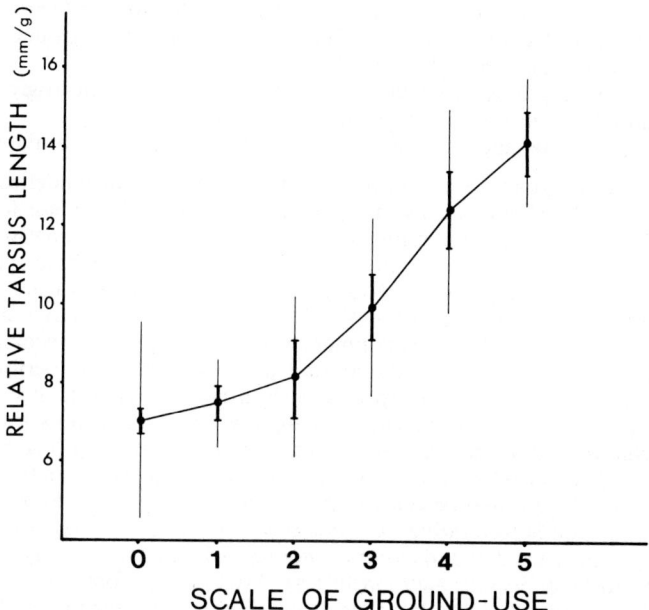

Fig. 8.   Relative tarsus lengths of flycatchers using the ground to different degrees: 0 = no ground-use; 5 = exclusively terrestrial species; categories defined in text. Mean relative tarsus length (dots) plus and minus one standard error (thick bars) and one standard deviation (thin lines) are shown for species in each ground-use category. See text for method of calculating relative tarsus length.

mentary graduation) strongly suggest that graduated rectrices may have secondary social functions in the Tyrannidae.

MORPHOLOGY AS A FUNCTION OF BEHAVIOR

In the preceding section I identified groups of species representing various ecological categories and showed, in broad terms, some morphological correlates of each behavioral specialization. "Upward-strikers" have wide bills, "aerial-hawkers" have long wings, "ground-foragers" have long legs, and so on. A much more revealing question now may be asked: To what extent do external morphological features vary *continuously* between species that show a gradient in behavioral specialization? The existence of a smooth continuum, in which morphology and behavior track one another, would imply at least two points relevant to adaptive radiation. (1) Directly functional, adaptive interpretations of the morphological variation are supported if the designs of various species are shown to be fine-tuned to behavior, especially if convergence among unrelated groups is implied; (2) morphological and behavioral change can occur relatively easily, at an evolutionary pace, if a small average change in a certain behavior pattern requires only a correspondingly small change in morphology to facilitate it.

Some evidence for such a morphological continuity principle has been mentioned already, in that behaviorally intermediate species also tend to be morphologically intermediate. I examine the question more directly here, using three sample analyses in which morphological measures are plotted directly against behavioral indices.

Tarsus length versus ground-use.—Figure 5 shows that tarsus length is proportionately longest among ground-foraging flycatchers. However, even within the ground-foragers, significant variation exists. Furthermore, other species not classified as "ground-foragers" do occasionally sally to the ground. To test more critically the relationship between tarsus length and ground-use I plotted relative lengths of flycatcher tarsi against degree of ground-use (Fig. 8), separating six categories of behavior as follows:

0 = strictly arboreal, essentially never perches on or sallies to the ground (n = 64 species);
1 = rarely sallies to ground, no searching from ground (n = 6);

2 = 10 to 30 percent of sallies to ground, no searching from ground (n = 4);
3 = 30 to 60 percent of sallies to ground, frequently searches from ground, but uses other foraging techniques extensively (n = 7);
4 = ground-foraging specialist, 60 to 90 percent of sallies either perch-to-ground or initiated from ground, but still uses other techniques (n = 7);
5 = exclusively terrestrial and cursorial, all perches and sallies occur at the ground (n = 4).

Average tarsus length does vary continuously with degree of dependence upon ground-related foraging (Fig. 8). This confirms the intuitive impression that tarsus length is an evolutionarily flexible feature among tyrannids, and it even suggests a scenario through which members of a lineage may become ground-foraging specialists. The difference in tarsus length, between species in behavioral category 0 (no ground use) and those in 1 (rare, facultative use) is slight, and not significant (Student's $t = .42$; $P > 0.05$). This implies that facultative use of the ground requires no special tarsal development. Given the appropriate ecological conditions or setting, then, occasional sallies to the ground physically could be performed and even adopted as a foraging pattern, by any species. Such a behavioral shift, if it persists, would cause selection to favor those individuals with slightly longer tarsi, thus leading to gradual morphological change in those populations (cf. categories 0, 1, and 2 in Fig. 8). This, in turn, would facilitate additional reliance upon the ground-related foraging tactics. Shifts in both behavior and morphology would be gradual at this "generalized" end of the spectrum, where use of the technique still is facultative. At some point on the way up the scale of ground-use (around category 2) physical demands for morphological specialization may become stronger, causing selection to favor increasingly longer legs. True ground-foraging specialists seem to require proportionately much longer legs than do perching species (see Fig. 6). Thus, I speculate that a species that becomes a ground-foraging specialist (categories 3 and 4 in Fig. 8) may be under intense pressure to develop extremely long legs. At this point, the slope of the curve steepens, and its overall asymmetry becomes important. Possibly, complete specialization (categories 4 and 5) could lead to the evolution of such long legs that a species becomes less capable of departing from its specialized foraging mode. Its body form, now well adapted for one mode, is less adequate for other styles of foraging, rendering the specialist at a competitive disadvantage in any situation that favors more generalized behavior. This suggests that, in evolutionary terms, it may be easier for generalized species to begin the ascent up a curve toward specialization than for a specialist to descend back down the curve after nearing the top (Fig. 8.).

It is important to note that the above scenario depends upon the accuracy of the ground-use scale. The steepening curve in Figure 8 could simply result from lumping different degrees of specialization into fewer categories at the upper end of the curve. Whether the non-linearity of the curve is an artifact of the scale remains to be investigated with more detailed, quantitative study of the species involved. In any case, a morphological continuum clearly seems to hold across a variety of generalized to specialized ground-foragers.

Bill width versus upward-striking.—We have already seen that among foliage-gleaning tyrannids, upward-strikers have proportionately wider bills than do perch-gleaners and hover-gleaners (Fig. 1). To test how the tendency toward spatulate-shaped bills varies along a continuum of gleaning techniques, I plotted relative half-widths (bill width half-way from base to tip, divided by the logarithm of body weight) against a behavioral index obtained for each species by subtracting its percentages of perch-gleans and upward hover-gleans from its percentage of strikes (Fig. 9). Thus, the index for a complete perch-gleaning or upward hover-gleaning specialist is -1.0, that for an upward-strike specialist is $+1.0$, and the index for a 50:50 generalist is 0. The "strike index" used here makes the simplifying assumption that the opposing effects of perch- or hover-gleaning versus striking combine additively. (Among other possible complications, for example, weighting each technique equally could squeeze fine ecological subdivisions closer together at the ends than at the middle of the continuum.) At least at the gross level, however, the change in morphology along the behavioral continuum is gradual and monotonic (Fig. 9). Specialized perch-gleaners and hover-gleaners have narrow bills, upward-strikers have the widest bills, and the generalists (few, in this case) are intermediate.

Once again, at a certain level of specialization the curve appears to steepen. Upward strike specialists ($\geq 80\%$ use) have proportionately much wider bills than birds at any other position along the spectrum.

Wing/tarsus ratio versus aerial-hawking.—In this example two morphological features are

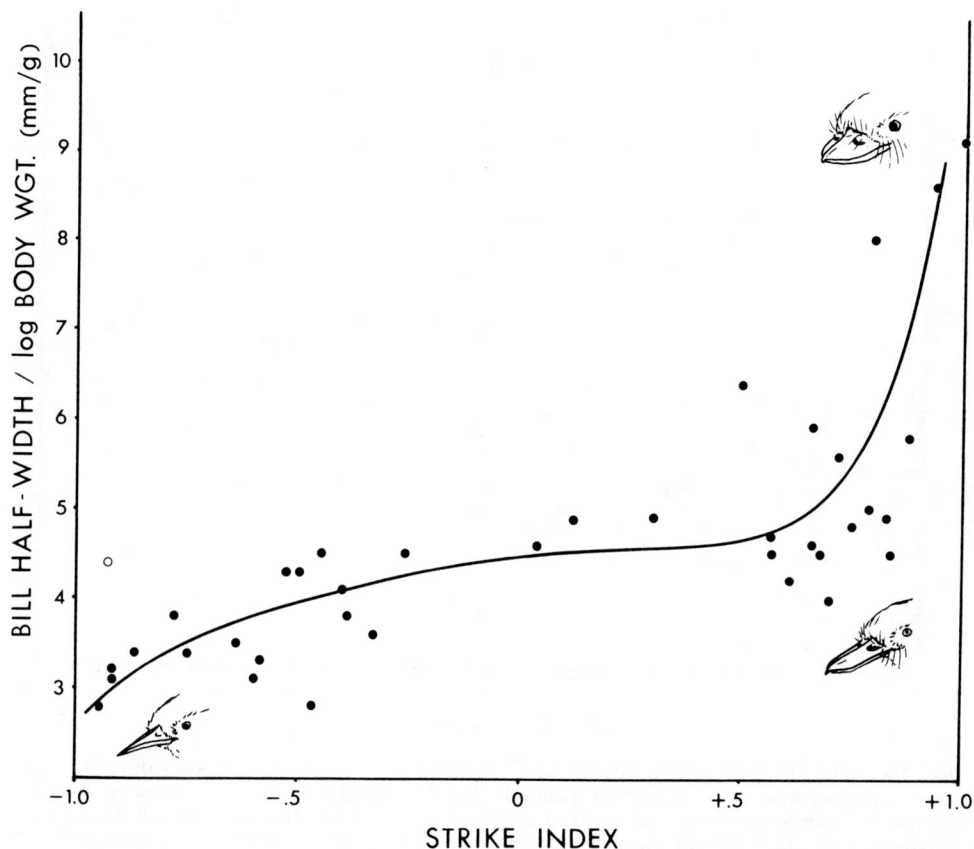

FIG. 9. Width of bill half-way from base to tip, divided by log body weight, plotted against the strike index (i.e., the proportion of use of upward-strikes, minus the summed proportions for perch-gleans and hover-gleans) for 38 foliage-gleaning flycatchers. A strike index of 0 indicates equal use of these techniques; +1.0 and −1.0 indicate pure specialists. Bill shape of one perch-gleaner (*Suiriri suriri*, open circle) shows extreme individual variation, for unknown reasons. The value shown is at the upper extreme for the species. The line is fitted by eye. Composite sketches illustrate bill shapes of three specialized genera: *Tachuris* (left-most), *Platyrinchus* (uppermost), and *Todirostrum* (lower right).

considered together in relation to a single, continuous behavioral variable. I have already shown (Figs. 4, 5) that aerial-hawking flycatchers possess relatively long wings and short tarsi. Here I combine these features and examine whether the ratio between wing and tarsus lengths varies continuously with degree of dependence upon aerial-hawking (Fig. 10).

As in the previous two examples, where only single characters were examined, a continuous relationship is apparent, in this case between a pair of seemingly independent characters and the use of a specialized behavioral trait. Again the data support the interpretation that long wings and short legs are adaptively favored features for aerial-hawkers. The continuum suggests that even when two characters are involved, flycatchers can evolve toward either behavioral or morphological extreme relatively easily.

Non-linearities in this relationship (Fig. 10) may occur at either end of the spectrum. Certain species (mostly upward-strikers) have proportionately shorter wings and longer tarsi than the rest, leaving them well below the central axis of the curve. For such species (black area in lower left of inset, Fig. 10), the changes in body form that would appear to adapt them to increased aerial-hawking are relatively greater than those that would be required of species closer to the central axis. This suggests that such non-users (e.g., upward-strikers) are less likely than other species to make successful ecological shifts of this sort. Extremely long wings and short tarsi of certain aerial-hawking specialists place them well *above* the central axis (Fig.

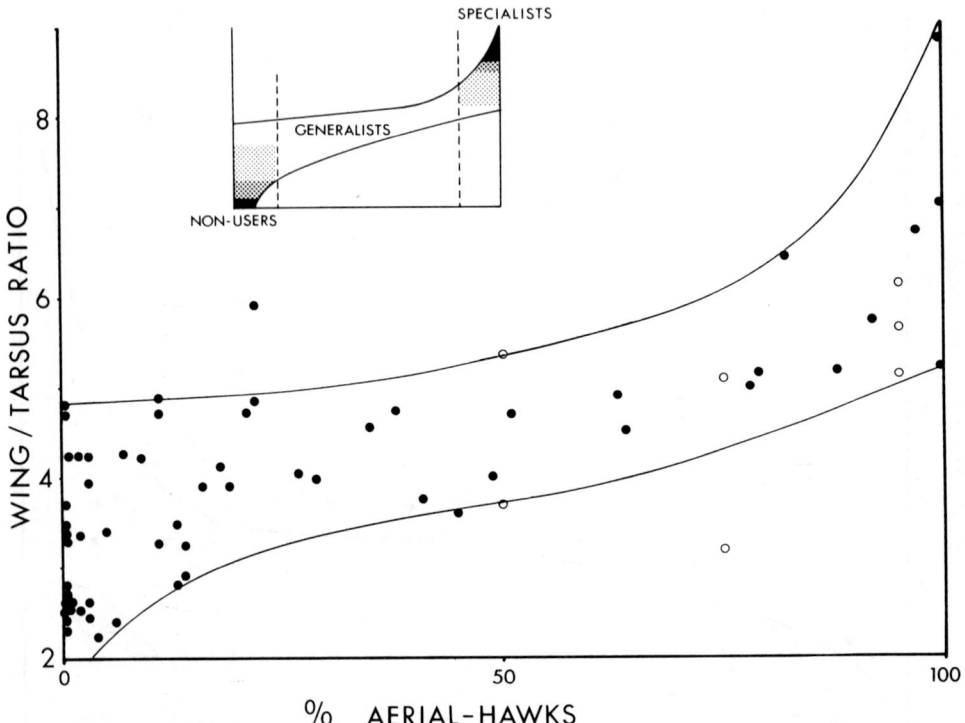

FIG. 10.   Wing chord divided by tarsus length, plotted against the percent frequency of aerial-hawks in the foraging repertoire of 67 species of tyrannids. Aerial-hawking percentages are based on field data with sample sizes of at least 30 sallies (closed circles), or less (open circles). Inset schematically summarizes the pattern, emphasizing the extreme differences between the wing-tarsus ratios of certain species (shaded areas) at either end of the spectrum compared to the ratios of generalists.

10). These species also may be behaviorally more restricted than the others. The body form of *Hirundinea ferruginea*, for example (extreme upper point in Fig. 10) is so specialized for aerial-hawking that it appears unsuitable for most other kinds of capture techniques. I doubt that *Hirundinea* could shift easily to any other foraging mode without passing through a phase of much decreased efficiency, in which its behavior would become generalized while its body remained specialized.

Most species fall along the central axis of the relationship (Fig. 10). These appear morphologically more free to shift one way or the other behaviorally, without drastic change in body form. This relationship suggests that facultative aerial-hawking is a foraging technique still available to most tyrannid groups.

PATTERNS OF ADAPTIVE RADIATION

In this section I indulge in speculation about the evolution of modern flycatchers, by combining my own findings with the independently derived phylogenetic inferences made by Traylor (1977, 1979). Traylor recognized three subfamilies, which constitute the major lineages of the family. Within them are several clearly defined assemblages of related genera (summarized in Traylor and Fitzpatrick 1982). I emphasize a crucial methodological point: Traylor arrived at his systematic picture of the Tyrannidae principally by integrating the anatomical studies of Warter (1965) and Ames (1971), published data on nest forms, and his own data on plumage patterns. External structural peculiarities played only a minor role, and behavior (except nest-contruction) was not involved at all. Thus, I was able to assess Traylor's proposed superstructure for the family independently, from an ecological and behavioral view. The extent to which Traylor's assemblages fit a logical, parsimonius pattern in terms of behavioral relationships is indeed striking.

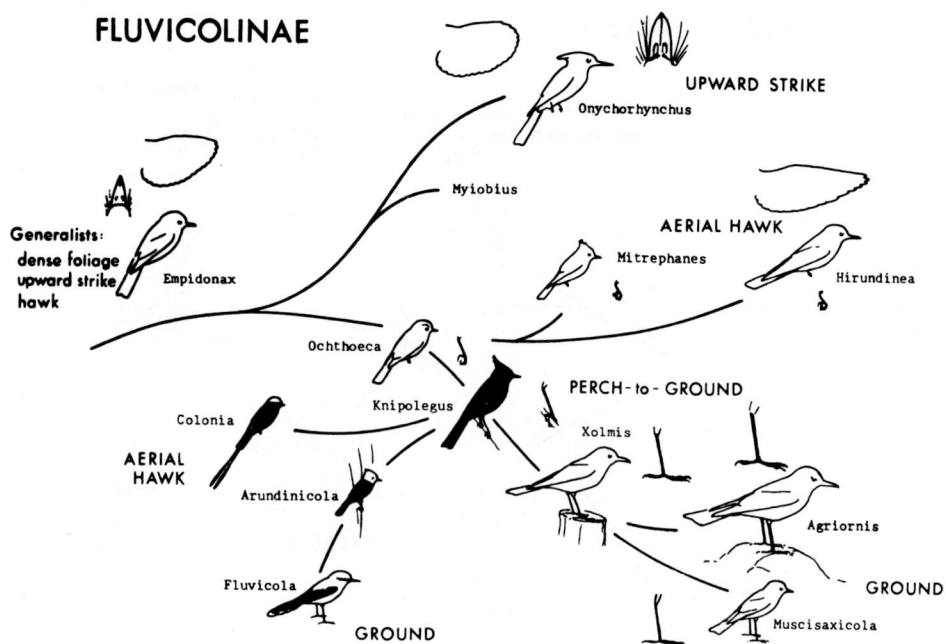

FLUVICOLINAE

UPWARD STRIKE
Onychorhynchus

Myiobius

AERIAL HAWK

Generalists:
dense foliage
upward strike
hawk

Empidonax

Mitrephanes

Hirundinea

Ochthoeca

PERCH-to-GROUND

Knipolegus

Colonia

Xolmis

AERIAL
HAWK

Arundinicola

Agriornis

Fluvicola

GROUND

GROUND

Muscisaxicola

FIG. 11. The major pathways of specialization in the subfamily Fluvicolinae, following the phylogenetic groupings of Traylor (1977). Important modern-day genera that reflect generalized (lower case lettering), intermediate, and specialized (capital lettering) foraging habits are shown along hypothesized evolutionary lines of increasing specialization. Selected features of external morphology illustrate the structural trends.

Earlier (Fitzpatrick 1980) I showed that within each of the three tyrannid subfamilies occur from one to three generalized foraging modes as well as a corresponding set of one or two highly specialized modes seemingly derived from the generalized ones. In the present paper I have emphasized some conspicuous relationships between those foraging modes and certain external features of flycatchers. With these, and Traylor's (1977) lineages, as background, I speculate on a few of the pathways followed during the evolutionary radiation of the tyrant flycatchers. These speculations are based on the following observations: (1) each major lineage contains its own unique set of generalists and specialists, spanning only a certain subset of the behavioral and morphological gradations described in the preceding sections; (2) by reason of parsimony, the morphology-behavior relationships considered (Figs. 8–10) represent some of the specific pathways of evolution between generalists and specialists; (3) many of these pathways we can still see preserved among closely related species, as if certain groups (e.g., ancestors of modern genera) became successful at intermediate positions while other forms continued toward more specialized behavior and form; (4) I interpret the non-linear relationships between morphology and foraging behavior (Figs. 8–10) to suggest that evolutionary trends in general proceed more readily up a scale toward specialization, than in both directions equally. Once again, I stress that this interpretation depends upon the assumptions that the behavioral and morphological factors are scaled in biologically meaningful ways, and that decreased efficiency at a novel foraging technique hampers specialists more than generalists. This admittedly controversial point is not critical to the present discussion, however, and will be developed more fully in a subsequent paper. The important evolutionary trends within the three flycatcher subfamilies are illustrated in Figures 11, 12, and 13.

GROUND-RELATED FORAGING

Independent adaptations to terrestrial foraging have occurred in four lineages, representing at least two subfamilies and possibly all three. Smith and Vuilleumier (1971) discussed relationships and speciation patterns in the clearest case, that of the fluvicoline ground-tyrants (Fig. 11). The key transition (Fig. 11, lower right) lies in facultative perch-to-ground sallying by generalists prone to aerial-hawking in open country and forest edge (e.g., the present-day

ELAENIINAE

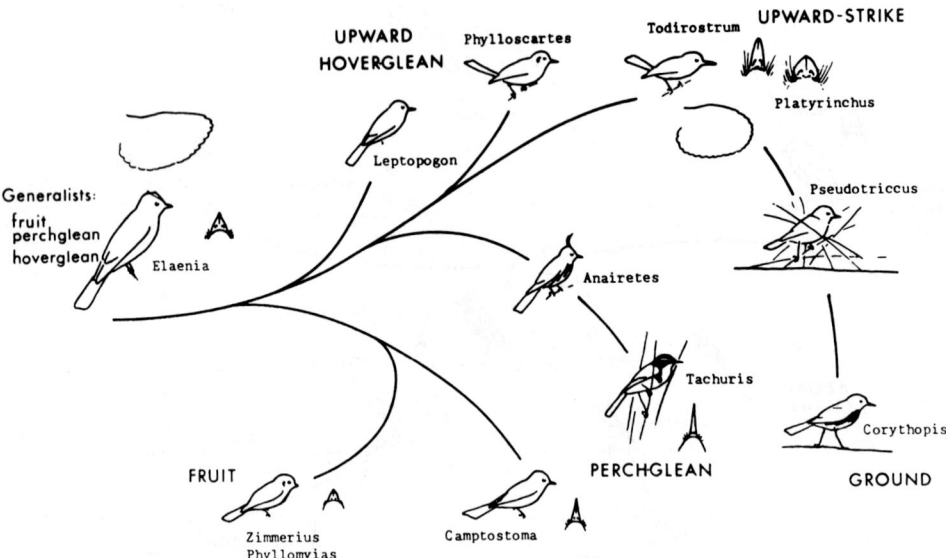

FIG. 12. The major pathways of specialization in the subfamily Elaeniinae, following the phylogenetic groupings of Traylor (1977). Schematic format as in Figure 11.

genera *Pyrocephalus, Sayornis, Ochthoeca*). Facultative perch-to-ground sallying gave rise to specialists (e.g., *Neoxolmis, Muscisaxicola, Muscigralla*). The pattern, both behavioral and morphological, seems precisely illustrated by Figure 8. Many of the intermediate positions shown in Figure 8 are represented by modern forms in this assemblage of closely related species.

A second line to terrestrial habits is illustrated by the black, sexually dimorphic fluvicolines *Knipolegus, Hymenops, Alectrurus, Arundinicola,* and *Fluvicola.* Most members of this group share a propensity for marsh and lake edge vegetations.

Machetornis contains one highly terrestrial species of uncertain affinities (Traylor 1977). It frequently performs perch-to-ground sallies (often from the backs of livestock), but more often walks in grassy pastures picking up ground prey as do meadowlarks (icterid genus *Sturnella*). The species shows striking similarities in plumage, voice, display, and attentuated primaries, to the kingbirds (*Tyrannus*) which are almost exclusively aerial-hawkers. I suggest that *Machetornis* may be a ground-adapted kingbird, having proceeded to its peculiar behavioral and morphological state along exactly the same route as did the fluvicolines (see Lanyon 1984 for an alternate view).

The Elaeniinae is a subfamily comprised of foliage-gleaners and generalist frugivores, with specialists in perch-gleaning and upward-striking (Fitzpatrick 1980). Here, too, a single genus of uncertain affinities has become terrestiral (*Corythopis*; Fig. 12). In this case, however, the route presumably was not through perch-to-ground intermediates. *Corythopis* is an upward-striker, typically sallying from the ground while walking deliberately along the forest floor. Many upward-strikers forage in extremely dense foliage near the ground, and their legs are unusually long for birds their size. Thus, they are pre-adapted for shifting entirely to the ground, where they can retain the specialized foraging technique amidst ground story vegetation of the forest floor.

FOLIAGE-GLEANERS

The most stereotyped foraging specialization among foliage-gleaners is upward-striking, a mode that carries with it numerous elaborations in body form: broad bill, short and rounded wings, long legs, and well developed rictal bristles. Upward-strike specialists are most prevalent in the Elaeniinae (Table 2, Fig. 12; see also Fitzpatrick 1980), with several genera in the

TYRANNINAE

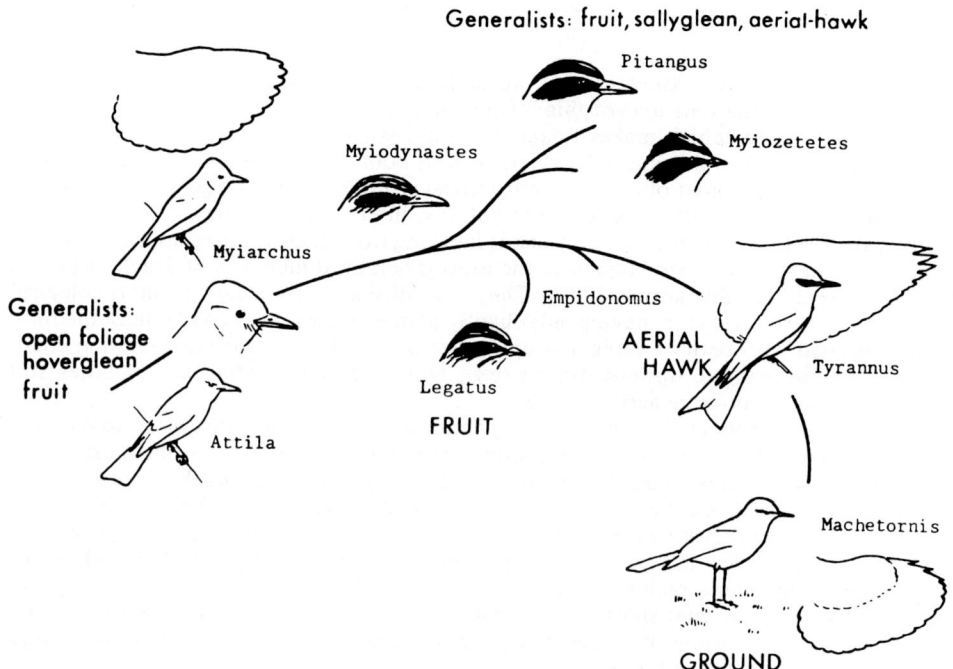

Generalists: fruit, sallyglean, aerial-hawk

Pitangus

Myiodynastes

Myiozetetes

Myiarchus

Generalists:
open foliage
hoverglean
fruit

Empidonomus

AERIAL
HAWK

Tyrannus

Legatus

FRUIT

Attila

Machetornis

GROUND

FIG. 13.   The major pathways of specialization in the subfamily Tyranninae, following the phylogenetic groupings of Traylor (1977). Schematic format as in Figure 11.

Fluvicolinae and one (*Myiodynastes*) in the Tyranninae. In the transition from generalized hover-gleaning intermediates to upward-strikers a wide bill and short wings become favored as upward hover-gleaners increasingly rely on explosive sallies toward leaf undersides (Fig. 9). Sherry (1982) suggested that marked specialization in diet accompanies such a morphological shift, because the shift permits increased ability to capture fast-moving prey through surprise attack. A clear, present-day intermediate genus is *Phylloscartes* (the bristle-tyrants), an assemblage that displays considerable variation in foraging styles. Most are narrow-billed upward hover-gleaners, but a few are wider billed and use the upward-strike as often as the hover-glean.

Adaptive radiation within the huge assemblage of elaeniine upward-strikers has been remarkably conservative. This is perhaps the most stereotyped major group within the family, both in foraging behavior and in morphology. Coexistence of numerous species is accompanied by unusually fine microhabitat segregation (e.g., Slud 1964; Terborgh and Weske 1969) and unusually sharp geographic replacement between closely related species (Fitzpatrick 1976, 1981b). These observations parallel the notion that the upward-strikers are so specialized, with a unique suite of morphological and behavioral peculiarities, that they have been largely unsuccessful at evolving out of their specialty. *Corythopis,* however, may represent a new macro-evolutionary step in this group, as proposed earlier (Fig. 12).

Besides the distinctive upward-strikers, the Elaeniinae include a homogeneous group of genera that combine the related techniques of perch-gleaning and upward hover-gleaning with varying amounts of frugivory. Small bills, associated with this combination of foraging styles, may restrict their use of aerial-hawking, a technique used regularly only by the relatively wider-billed genera *Sublegatus, Suriri,* and *Elaenia.* Many elaeniine genera specialize in perch-gleaning. The only external modification associated with this speciality is the long, narrow bill (Fig. 1). In *Pseudocolopteryx* and *Tachuris* this feature is extreme and is associated with wren-like gleaning habits in tall reeds and marsh grass. Otherwise, perch-gleaners have rather generalized morphology, hence they may not be as evolutionarily constrained as are other kinds of specialists.

AERIAL-HAWKING

With only one exception known to me, all obligate aerial-hawkers belong to one of two large, independent radiations toward this "typical flycatcher" foraging mode. The exception is *Serpophaga cinerea,* which makes frequent use of aerial-hawking (see Smith 1971 for a discussion of behavior in this and related species). In both the fluvicoline and the tyrannine radiations, potential avenues of change toward aerial specialization from generalized foliage-gleaning are represented fairly closely by present-day, intermediate species.

Certain species that frequently employ aerial-hawking from enclosed perches (*Myiophobus, Empidonax, Cnemotriccus,* etc.) represent the most generalized members of Traylor's (1977) Fluvicolinae (Fig. 11; Fitzpatrick 1980). They also display early stages in morphological adaptations for aerial-hawking, having only slightly pointed wings. They have bills of medium width, associated with both hawking and upward-striking. This generalized condition could give rise to round-winged, upward-strikers (e.g., *Onychorhynchus, Myiobius;* affinities still uncertain), as well as obligate aerial-hawkers.

One group of fluvicolines (*Sayornis, Pyrocephalus, Knipolegus,* etc.), discussed above, represents a transition toward open country, ground-related foraging. *Colonia* is the most stereotyped aerial-hawker of this group, with long wings and tiny tarsi, reflecting this specialization.

When enclosed-perch-hawkers forage along open edges of vegetation, they shift to almost exclusive use of aerial-hawking (pers. observ.). Closely related to the near-ground-generalists just mentioned are a few genera (e.g., *Mitrephanes, Pyrrhomyias*) that are habitual aerial-hawkers, restricting their foraging to openings and forest edge microhabitats. Their wings are relatively long and their tarsi short, in accordance with this shift. The most specialized end of this lineage consists primarily of the large genus *Contopus,* with its greatly elongated wings (Fig. 4) and notably short and weak tarsi.

I suggest that the monotypic genus *Hirundinea,* the most strikingly specialized aerial-hawker in the Tyrannidae, is a member of this group. The proportions of this problematic species are indeed extreme. It has the greatest wing-to-tarsus ratio in the entire family (Fig. 10). This should not be surprising, however, because relatively long wings combined with short tarsi characterize virtually all aerial-hawking specialists (Fig. 10). In behavior, habitat, overall plumage pattern, and even in technical aspects of voice (W. J. Smith, pers. comm.), *Hirundinea* is reminiscent of an enlarged *Pyrrhomyias,* and its bill shows a similar broad base and pinched tip. In my view, the evidence indicates that *Hirundinea* represents the morphological culmination of this fluvicoline lineage of aerial-hawkers (Fig. 11).

Two groups of genera comprise Traylor's Tyranninae (Fig. 13). One is a homogeneous assemblage of hole-nesting foliage-gleaners (*Myiarchus, Sirystes, Attila, Casiornis,* etc.). Their generalized foraging repertoires are characterized by a predominance of outward hover-gleaning, often for very large prey, along with some frugivory and occasional aerial-hawking. The tendency for these species to perch in the open and survey the upper surfaces of vegetation, their slightly oversized bills, and the relatively long wings required for hover-gleaning, all provide behavioral and body form pre-adaptations for more habitual aerial-hawking.

The genus *Myiodynastes* (also a hole-nester, along with *Conopias*) seemingly links the former group of generalists with the remaining tyrannine genera (Fig. 13; Lanyon 1984). These remaining genera show shortened tarsi, medium to long wings, restriction to open habitats, and an array of foraging habits unparalleled in any other flycatcher group. Many of these generalists use aerial-hawking extensively, and the genus *Tyrannus* has specialized in this mode, marking the endpoint of an adaptive lineage comparable to that of the aerial-hawking fluvicolines. Species of *Tyrannus* have extremely long wings, very short tarsi, and show a gradation in tail forms from deeply notched to long and forked.

One morphological transition in the Tyranninae that is not well illustrated by present-day species is that between *Empidonomus* and *Tyrannus* (Fig. 13). The former genus indeed relies heavily upon aerial-hawking technique, but does not bear a close physical resemblance to *Tyrannus,* even though a "species" originally described in the genus *Laphyctes* appears to be a hybrid between *Tyrannus melancholicus* and *Empidonomus varius* (Meise 1949).

GENERAL DISCUSSION

It comes as no surprise to evolutionary biologists that the foraging behavior and the external morphology of animals are closely related. Storer (1971) presented an overview of the major evolutionary trends along these lines among birds. With few exceptions, however, examinations of radiation in major avian lineages have been largely descriptive. The anatomic variation usually is described rather thoroughly (e.g., Engels 1940; Lack 1947; Amadon 1950; Beecher 1951; Bowman 1961; Osterhaus 1962; Feduccia 1973), but the behavioral end of the story usually is left unquantified, thereby providing only half the picture. A notable exception is the landmark study of geospizine finches currently underway on the Galapagos Islands (e.g., Abbott et al. 1975, 1977; Boag and Grant 1981). In those investigations the ecological and behavioral variables are being measured and compared quantitatively with the anatomical ones. Only with this procedure can we expect to reveal more than the most superficial patterns by which evolutionary divergence takes place.

I have shown here that a general pattern of morphological adaptation is apparent within a large continental radiation of birds. In the Tyrannidae, body form varies closely with foraging mode, and the two are related through various, *continuous* functions that change gradually across most of their ranges of values. This conforms to the idea that form and function are plastic in the evolutionary sense. The two can be modified together in response to ecological pressures or opportunities.

At certain endpoints of these functions—representing nearly complete (and measurable) specialization in a single behavioral mode—drastic modifications in body form seem to accompany relatively small increments of behavioral change. In this way, physical attributes presumably allow for more effective or efficient performance of the behavioral speciality than a generalized body form would allow. Such modifications, however, may arise at the expense of evolutionary flexibility, principally because behavioral shifts by a specialist toward a more generalized foraging mode probably entail a greater reduction in efficiency than do ones coming from the opposite direction.

The best tyrannid example of possible loss in flexibility may be the upward-strike foraging mode, characterizing the broad-billed elaeniine assemblage. This mode carries with it numerous external specializations (extremely wide bill, short and rounded wing, long and slender tarsus, reduced tail). Presumably, this suite of characters makes possible the odd behavior of "surprise attack" foraging (Sherry 1982). Speciation, however, seems to have occurred largely *within* this mode once it was achieved (discussed above). Ground-foraging may be almost as limiting, for the fluvicoline ground-tyrants show almost the same degree of habitat and geographic subdivision as the elaeniine upward-strikers (Smith and Vuilleumier 1971; Fitzpatrick 1981b). Aerial-hawking also carries with it extreme morphological specialization (long wing, short tarsus, wide and triangular bill), but it can develop toward open country ground-foraging (e.g., the more primitive ground tyrants, and possibly *Machetornis*), especially among species in which the tarsus is sufficiently long to facilitate facultative use of the ground (see Fig. 8).

Directions of behavioral change are not random. In its broadest sense this observation is trivial: a woodpecker cannot begin foraging like a heron without a long, perhaps nearly prohibitive period of evolution through intermediate stages. Even within the narrow range of behavior characterizing insectivorous passerines, however, directions of change follow discrete pathways. Perch-gleaners, for example, can evolve into upward-strikers by passing through a transition stage in which upward hover-gleaning is used (Fig. 9). Sallying foliage-gleaners can become aerial-hawkers through the intermediate stage of facultative enclosed-perch hawking, a stage that favors morphological changes presumably rendering perch-gleaning and hover-gleaning more difficult. True specialists in upward-striking, ground-foraging, and even aerial-hawking may have more limited potential for changing back toward generalization, compared to the potential that generalists have for becoming specialized.

These directions of radiation, and presumably the extent to which the pathways become expressed within a given taxonomic group, depend largely upon three variables: (1) opportunities for repeated speciation and secondary contact, giving rise to ecological interactions that cause related species to diverge in behavior and morphology; (2) the degree to which competition, from outside the group and within it, restricts or directs this divergence; (3) the extent to which morphology itself constrains behavioral divergence.

In the case of the Tyrannidae, climatic and orogenic events during the Pliocene and Pleistocene of South America appear to have provided massive opportunity for repeated fragmentation, differentiation, and dispersal of species (Vanzolini and Williams 1970; Haffer 1974). A paucity of passerine groups on that long-isolated continent permitted the few non-oscine groups to evolve in an environment relatively free of competitors (Keast 1972). As lineages produced species that evolved toward increasing specialization, it appears today as if certain forms "settled" at intermediate stages—perhaps thereby forcing other forms to specialize further. In turn, of course, the specialists may have competitively prohibited other species or lineages from converging upon their specialities. These radiating pathways thereby might be remarkably well preserved by modern-day forms which persisted at various stages along the continua. In this view, groups like the flycatchers present us with an unusually complete picture of the pathways themselves.

Precisely how close together along pathways of adaptive radiation can related species persist? The interplay between this question and that of competition forms an important, unsolved ecological and biogeographic problem. As a group radiates, continued competition from less specialized forms (which also can proliferate through time) may reinforce any canalizing effects of morphological specialization. The persistence of intermediate species, more effective over a wider range of behavior, may itself prohibit the specialists from giving rise to species that return toward more generalized behavior and body form. The relative strengths of the morphological versus these historical and ecological constraints remain a matter of conjecture in need of further study.

I must emphasize in conclusion that the scenario just painted—the notion that the Tyrannidae contains "preserved phylogenetic lineages"—is a hypothetical one, perhaps unduly speculative. Ecological and morphological intermediacy, even among close relatives, does not necessarily imply genetic and phylogenetic intermediacy. Necessary tests for the evolutionary models and pathways discussed in the preceding section require sound phylogenetic reconstructions both within and among the various monophyletic lineages of flycatchers. Such reconstructions using various, independent characters are being developed at present. Comparisons between detailed phylogenetic reconstructions and the relationships between form and foraging behavior discussed in this paper will constitute the next major step in evaluating the adaptive radiation of flycatchers.

ACKNOWLEDGMENTS

Much of this research was carried out during my graduate years with the Population Biology group at Princeton University. I owe my greatest debts to J. W. Terborgh and H. S. Horn for their guidance and criticisms during that stage of my flycatcher studies. I thank the Ministerio de Agricultura of Peru, and its Dirección General Forestal y de Fauna, for continued permission to work in their country. For logistical help in their respective countries, I thank P. Vanzolini, the late P. Schwartz, and M. Plenge. Assistance in the field has been provided by numerous colleagues, and I especially thank R. Kiltie, D. K. Moskovits, and D. E. Willard in this regard. I am grateful to curators at the following museums for permitting my use of specimens under their care: American Museum of Natural History, Louisiana State University Museum of Natural Science, Museum of Comparative Zoology, and Museo de Zoología du Universidade de São Paulo. Field work for this project was supported by funds from the National Science Foundation, Chapman Memorial Fund, Princeton University, and the Conover Fund of the Field Museum of Natural History. I am indebted to these institutions for their help. Finally, I thank M. S. Foster and two anonymous reviewers, whose criticisms vastly improved this manuscript.

LITERATURE CITED

ABBOTT, I., L. K. ABBOTT, AND P. R. GRANT. 1975. Seed selection and handling ability of four species of Darwin's finches. Condor 77:332–335.

ABBOTT, I., L. K. ABBOTT, AND P. R. GRANT. 1977. Comparative ecology of Galapagos Finches (*Geospiza* Gould); evaluation of the importance of floristic diversity and intraspecific competition. Ecol. Monogr. 47:151–184.

AMADON, D. 1950. The Hawaiian honeycreepers (Aves, Drepaniidae). Bull. Am. Mus. Nat. Hist. 95: 151–262.

AMES, P. L. 1971. The morphology of the syrinx in passerine birds. Bull. Peabody Mus. Nat. Hist. No. 37.

BEECHER, W. J. 1951. Adaptations for food-getting in the American blackbirds. Auk 68:411–440.

BERGER, A. J. 1952. The comparative functional morphology of the pelvic appendage in three genera of Cuculidae. Am. Midl. Nat. 47:513–605.

BOAG, P. T., AND P. R. GRANT. 1981. Intense natural selection in a population of Darwin's Finches (Geospizinae) in the Galapagos. Science 214:82–85.

BOWMAN, R. I. 1961. Morphological differentiation and adaptation in the Galapagos Finches. Univ. Calif. Publ. Zool. 58:1–326.

CONOVER, M. R., AND D. E. MILLER. 1980. Rictal bristle function in Willow Flycatcher. Condor 82: 469–471.

ENGELS, W. L. 1940. Structural adaptations in thrashers (Mimidae: genus *Toxostoma*) with comments on interspecific relationships. Univ. Calif. Publ. Zool. 42:341–400.

FEDUCCIA, A. 1973. Evolutionary trends in the Neotropical ovenbirds and woodhewers. Ornithol. Monogr. No. 13.

FITZPATRICK, J. W. 1976. Systematics and biogeography of the Tyrannid genus *Todirostrum* and related genera (Aves). Bull. Mus. Comp. Zool. 147:436–563.

FITZPATRICK, J. W. 1978. Foraging behavior and adaptive radiation in the avian family Tyrannidae. Unpubl. Ph.D. dissert., Princeton University, Princeton, New Jersey.

FITZPATRICK, J. W. 1980. Foraging behavior of Neotropical tyrant flycatchers. Condor 82:43–57.

FITZPATRICK, J. W. 1981a. Search strategies of tyrant flycatchers. Anim. Behav. 29:810–821.

FITZPATRICK, J. W. 1981b. Some aspects of speciation in South American flycatchers. Proc. XVI Int. Ornithol. Congr., pp. 1273–1279.

FFRENCH, R. 1976. A Guide to the Birds of Trinidad and Tobago. Asa Wright Nature Center, Publ. no. 1. Harrowood, Valley Forge, Pennsylvania.

GOULD, S. J., AND R. C. LEWONTIN. 1979. The spandrels of San Marco and the Panglossian paradigm: a critique of the adaptationist programme. Proc. R. Soc. Lond. B Biol. Sci. 205:581–598.

GREENEWALT, C. H. 1962. Dimensional relationships for flying animals. Smithson. Misc. Collect. no. 144.

GREENEWALT, C. H. 1975. The flight of birds. Trans. Am. Phil. Soc. 65:1–67.

HAFFER, J. 1974. Avian speciation in tropical South America. Publ. Nuttall Ornithol. Club No. 14.

HARTMAN, F. A. 1961. Locomotor mechanisms of birds. Smithson. Misc. Collect. 143:1–91.

HESPENHEIDE, H. A. 1971. Food preference and the extent of overlap in some insectivorous birds, with special reference to the Tyrannidae. Ibis 113:59–72.

HESPENHEIDE, H. A. 1973. Ecological inferences from morphological data. Annu. Rev. Ecol. Syst. 4: 213–229.

HESPENHEIDE, H. A. 1975. Prey characteristics and predator niche width. Pp. 158–180, *In* M. L. Cody and J. M. Diamond (eds.), Ecology and Evolution of Communities. Belknap, Cambridge, Massachusetts.

JOHNSON, N. K. 1963. Biosystematics of sibling species of flycatchers in the *Empidonax hammondii-oberholseri-wrightii* complex. Univ. Calif. Publ. Zool. 66:79–238.

JOHNSON, N. K. 1980. Character variation and evolution of sibling species in the *Empidonax difficilis-flavescens* complex (Aves: Tyrannidae). Univ. Calif. Publ. Zool. 112:1–151.

KARR, J. R., AND F. C. JAMES. 1975. Eco-morphological configurations and convergent evolution in species and communities. Pp. 258–291, *In* M. L. Cody and J. M. Diamond (eds.), Ecology and Evolution of Communities. Belknap, Cambridge, Massachusetts.

KEAST, A. 1972. Ecological opportunities and dominant families, as illustrated by the Neotropical Tyrannidae (Aves). Evol. Biol. 5:229–277.

KIPP, F. A. 1958. Zur Geschichte des Vogelzuges auf der Grundlage der Flügelanpassungen. Vogelwarte 19:233–242.

LACK, D. 1947. Darwin's Finches. Cambridge University Press, London.

LANYON, W. E. 1967. Revision and probable evolution of the *Myiarchus* flycatchers of the West Indies. Bull. Am. Mus. Nat. Hist. 136:329–370.

LANYON, W. E. 1978. Revision of the *Myiarchus* flycatchers of South America. Bull. Am. Mus. Nat. Hist. 161:427–628.

LANYON, W. E. 1984. A phylogeny of the kingbirds and their allies. Am. Mus. Novit. No. 2797.

MACARTHUR, R. H. 1971. Patterns of terrestrial bird communities. Pp. 189–221, *In* D. S. Farner and J. R. King (eds.), Avian Biology, Vol. 1. Academic Press, New York.

MEISE, W. 1949. Über einen Gattungsbastard und eine Zwillingsart der Tyrannen nebst Bemerkungen über Zuweite und Flügelform. Pp. 61–83, *In* E. Mayr and E. Schütz (eds.), Ornithologie als biologische Wissenschaft. Carl Winter Universitätsverlag, Heidelberg, Federal Republic of Germany.

ORIANS, G. H., AND H. S. HORN. 1969. Overlap in foods of four species of blackbirds in the potholes of central Washington. Ecology 50:930–938.

OSTERHAUS, M. B. 1962. Adaptive modifications in the leg structure of some North American warblers. Am. Midl. Nat. 68:474–486.

PULLIAM, H. R. 1975. Coexistence of sparrows: a test of community theory. Science 189:474–476.

SAVILE, D. B. O. 1957. Adaptive evolution in the avian wing. Evolution 11:212–224.

SCHAFFER, J. 1903. Über die Sperrvorrichtung an den Zehen der Vögel. Z. Wiss. Zool. 73:377–428.

SCHOENER, T. W. 1965. The evolution of bill size differences among sympatric congeneric species of birds. Evolution 19:189–213.

SHERRY, T. W. 1982. Ecological and evolutionary inferences from morphology, foraging behavior, and diet of sympatric insectivorous neotropical flycatchers (Tyrannidae). Unpubl. Ph.D. dissert., University of California, Los Angeles.

SLUD, P. 1964. The birds of Costa Rica: distribution and ecology. Bull. Am. Mus. Nat. Hist. 128:1–430.

SMITH, W. J. 1966. Communications and relationships in the genus *Tyrannus*. Publ. Nuttall Ornithol. Club No. 6.

SMITH, W. J. 1971. Behavioral characteristics of Serpophaginine tyrannids. Condor 73:259–286.

SMITH, W. J., AND F. VUILLEUMIER. 1971. Evolutionary relationships of some South American ground tyrants. Bull. Mus. Comp. Zool. 141:181–232.

STETTENHEIM, P. 1974. The bristles of birds. Living Bird 12:201–234.

STORER, R. 1971. Adaptive radiation of birds. Pp. 149–188, *In* D. S. Farner and J. R. King (eds.), Avian Biology, Vol. 1. Academic Press, New York.

TERBORGH, J., AND J. S. WESKE. 1969. Colonization of secondary habitats by Peruvian birds. Ecology 50:765–782.

TRAYLOR, M. A., JR. 1977. A classification of the tyrant flycatchers (Tyrannidae). Bull. Mus. Comp. Zool. 148:129–184.

TRAYLOR, M. A., JR. (ed.). 1979. Peters' Check-list of Birds of the World, Vol. 8. Museum of Comparative Zoology, Cambridge, Massachusetts.

TRAYLOR, M. A., JR., AND J. W. FITZPATRICK. 1982. A survey of the tyrant flycatchers. Living Bird 19:7–50.

VANZOLINI, P. E., AND E. E. WILLIAMS. 1970. South American anoles: the geographic differentiation and evolution of the *Anolis chrysolepis* species group (Sauria, Iguanidae). Arq. Zool. (São Paulo) 19:1–298.

WARTER, S. L. 1965. The cranial osteology of the New World Tyrannoidea and its taxonomic implications. Unpubl. Ph.D. dissert., Louisiana State University, Baton Rouge.

WESKE, J. S. 1972. The distribution of the avifauna in the Apurimac Valley of Peru with respect to environmental gradients, habitat, and related species. Unpubl. Ph.D. dissert., University of Oklahoma, Norman.

WETMORE, A. 1926. Observations on the birds of Argentina, Paraguay, Uruguay, and Chile. Bull. U.S. Natl. Mus. 122:1–448.

CLIMATIC FLUCTUATIONS ON THE GALÁPAGOS ISLANDS AND THEIR INFLUENCE ON DARWIN'S FINCHES

P. R. GRANT

Division of Biological Sciences, University of Michigan, Ann Arbor, Michigan 48109 USA

ABSTRACT. The purpose of this article is to explore the connection between past and present environments and evolutionary forces operating on Darwin's finches. The finches diversified from a single ancestral stock on the modern Galápagos Islands probably between 0.5 and 1.5 million years before present. During this time, they were probably exposed to fluctuating climates, and some species are likely to have gone extinct. Climatic conditions have varied from present conditions to much drier ones in the last 50,000 years, during which time all modern species of finches were present. Currently, some finch populations are subjected to intense directional selection on morphological variation in times of drought when food supply is low; possibly selection operates also under extreme wet conditions, but in an opposite direction. Unusually dry conditions and unusually wet conditions each occur once every decade, on average. Given a potential life span of 15 to 20 years, an individual finch may experience at least one drought and one wet year in its life-time. The main implication of these results and suggestions is that natural selection, both in the past and currently, is responsible for at least some and perhaps most of the differences between species. Modern analyses of selection can be extrapolated to yield estimates of the forces and conditions necessary to transform one species into another in the past, which is the key evolutionary process in the adaptive radiation.

RESUMEN. La finalidad de este artículo es explorar la conección entre el ambiente actual y pasado y las fuerzas evolucionarias operante en los pinzones de Darwin. Los pinzones se han diversificado a partir de un único grupo ancestral en las modernas islas Galápagos probablemente entre los últimos 0,5 y 1,5 millones de años. Es muy probable que durante ese tiempo los pinzones se vieran expuestos a climas fluctuantes y quizás algunas especies se extinguieran. Las condiciones climáticas han variado desde las condiciones actuales a otras mucho más secas en los últimos 50.000 años, tiempo en el cual todas las especies modernas de pinzones, ya existían. Actualmente, algunas poblaciones de pinzones están sujetas a una intensa selección direccional en variaciones morfológicas en tiempos de sequía cuando hay poca disponibilidad de comida; posiblemente la selección también se produce en condiciones de extrema humedad, pero en dirección contraria. Como promedio, una vez en cada década se dan condiciones de inusual sequía y así como también de humedad extraordinaria. Si consideramos que los pinzones tienen un lapso potencial de vida de entre 15 y 20 años, un individuo puede experimentar al menos una vez en su vida un año de inusual sequía y otro de inusual humedad. La principal implicación de estos resultados y sugerencias es que la selección natural, tanto en el pasado como actualmente, es responsable de al menos algunas y quizás la mayoría, de las diferencias entre especies. Se pueden extrapolar modernos análisis de selección para producir estimaciones de las fuerzas y condiciones necesarias en el pasado, para transformar una especie en otra, lo cual es la llave de los procesos evolucionarios en radiación adaptativa.

Darwin's Finches are a textbook example of the adaptive differentiation of species from a single ancestral stock because, being closely related and living in an isolated archipelago, they are unusually suitable for studies of phylogenetic reconstruction and interpretation. Such studies include assessing phylogenetic relationships by considering morphological and molecular features, deducing past environmental conditions under which the organisms of interest lived and evolved, and inferring past evolutionary processes from ecological studies of extant forms.

Lack's (1947) tentative phylogeny for Darwin's finches has been confirmed in broad outline by the electrophoretic analysis of protein polymorphisms (Yang and Patton 1981), although the phylogenetic and geographical orgins of the finches still remain uncertain (see Bowman 1961; Steadman 1982a). In this article I pursue the other two lines of enquiry mentioned above, by exploring the connection between past and present environments and evolutionary forces operating on the finches. First, I consider the age of the islands, how long the finches have been present, and the changes in climate they must have experienced. Then I discuss modern evidence for the influence of climatic fluctuations on finch populations and assess the frequency with which these fluctuations may have strong effects upon the finches. Finally, I point out ways in which links between studies of historical and contemporary evolutionary processes need to be strengthened in order to deepen our understanding of the adaptive radiation.

HISTORY OF THE ISLANDS AND FINCHES

ORIGINS AND AGES OF THE ISLANDS

The first task is to define the period of interest, i.e., how long have the islands existed, and for how long have finches been present. On biological grounds the islands were considered previously to be geologically very old and, at one time, connected to the mainland (Agassiz 1892; Baur 1897; Beebe 1924; Gulick 1932; Swarth 1934). The connections were thought to be the Carnegie and Cocos ridges, now submarine, that extend toward the Galápagos from South and Central America, respectively. In fact, the islands are relatively young and have never been connected to the mainland.

The islands are the product of outpourings of mantle material from the Galápagos hotspot (Holden and Dietz 1972). According to the hotspot theory of Wilson (1963a, b) and Morgan (1971), the hotspot is fixed, but the rigid Cocos and Nazca plates have moved over the hotspot in a general eastward direction toward the continent of South America. Changes in relative plate positions and motions have occurred (Hey et al. 1977), but by about 5 million years before present (m.y.B.P.) the hotspot was, for the first time, entirely under the Nazca plate (Hey 1977). The islands lie entirely on the Nazca plate (Hey 1977). Therefore, 5 m.y.B.P. seems a reasonable upper limit to the age of the islands. This is confirmed by potassium-argon dating which has shown, in conjunction with geomorphological data (Hall 1983), that the oldest rocks on the islands were formed no more than 3 to 5 m.y.B.P. (Cox and Dalrymple 1966; Bailey 1976; Cox 1983). All the current major islands had been formed by about 0.5 – 1 m.y.B.P. (Bailey 1976; Cox 1983).

The Cocos and Carnegie ridges are probably hotspot traces (Hey 1977). In the last 20 to 25 m.y.B.P. they have moved toward the continent on the Cocos and Nazca plates, respectively. Even with maximal lowering of sea level in recent times, they would not have constituted a land bridge (Bowman 1961), although if some land had been above sea level as islands (Holden and Dietz 1972), it could have provided stepping stones (MacArthur and Wilson 1967) for the colonization of the Galápagos by the finches. The finches could have undergone adaptive differentiations on these stepping stones. Rosen (1976) followed Holden and Dietz (1972) in suggesting that the modern islands were colonized by organisms crossing short water gaps separating the South American mainland from ancestral islands (since disappeared), and separating ancestral islands from each other and from modern ones. However, no guyots (submerged islands) are known from either the Carnegie or Cocos ridges (Holden and Dietz 1972), and the youth of the islands and their inhabitants (see below) makes it unlikely that numerous and well spaced guyots did exist but have since been subducted at the continental margin. Thus, the ancestral Darwin's Finches probably colonized the islands by extended overwater flight, as Lack (1940, 1945, 1947) argued, and subsequent differentiation took place on islands close to their current geographical positions approximately 1000 km west of Ecuador on the equator.

OCCUPATION OF THE ISLANDS BY THE FINCHES

Holden and Dietz (1972) speculated that biological evolution in general on the Galápagos proceeded slowly over a long time ("millions of years") on a succession of islands as they were sequentially formed. Hey (1977) offered the alternative suggestion of relatively rapid and recent evolution.

A direct answer using datable fossils cannot be given because Darwin's Finch remains of known age are less than 2500 years old (Steadman 1981, 1982b). An indirect answer can be given by estimating the times of divergence between taxa through the application of Nei's (1975) method of dating to protein polymorphisms. The method of estimation is based on several uncertain assumptions, including approximate regularity in the stochastic process of amino acid substitution in protein molecules (e.g., Thorpe 1982), and, therefore, it is not very precise. Using this method, Yang and Patton (1981) estimated that the earliest divergence in Darwin's Finches (the splitting of *Certhidea olivacea* from the ancestral stock) took place about 570,000 y.B.P. If the differentiation of the group started fairly soon after the arrival of the ancestral group (Lack 1947), Darwin's Finches colonized the Galápagos no more than about 600,000 y.B.P. Nei's (1975) method, however, may underestimate divergence times (Yang and Patton 1981; Thorpe 1982). For example, hybridization, which is known to occur (Boag 1981; Grant and Price 1981), tends to result in decreased estimates of divergence times. A more conservative estimate of the date at which the initial divergence took place is ca. 1.5 m.y.B.P. (Yang and Patton 1981). Other endemic groups such as the reptiles (Marlow and Patton 1981) have differentiated more recently.

These biological data support Hey's suggestion of relatively rapid evolution. Darwin's Finch fossil data (Steadman 1981, 1982b) point to the same conclusion. I conclude that the Finches have been present for a relatively small portion of the archipelago's history, from 0.5 to 1.5 million years.

CLIMATIC FLUCTUATIONS IN HISTORICAL TIMES

The most recent divergence of Darwin's Finch taxa, that of *Cactospiza* and *Platyspiza*, took place about 62,500 y.B.P. (Yang and Patton 1981). Unfortunately, no climatic data for the Galápagos before that time exist. The available data characterize the climate after all modern species had been formed.

For the last 50,000 years the Galápagos climate has not remained stable. This is known from an analysis of particles and plant products in cores from the El Junco lake on the summit of San Cristóbal (Colinvaux 1972). The oldest sediment is dated at more than 48,000 y.B.P. Changes in composition throughout the core reflect changes in climatic conditions at the time of deposition.

The most important discovery from the cores, apart from the climatic fluctuations themselves, is that the Galápagos experienced much drier climates than now, and for a long time, but have not had wetter climates than now. Before 34,000 y.b.p., the climate was approximately the same as now, as far as can be determined from the scarce data. The period 10,000–34,000 y.B.P. was dry with little precipitation or evaporation. Since this corresponds with a period of advance and retreat of glaciers in the north, it was also a period of fluctuating sea levels. The greatest extent of arid zone vegetation would have occurred at this time, during the glacials (Quinn 1971), partly because of the dry conditions, and partly because at times of low sea level, more land would have been exposed. The lowering of sea level, however, did not dramatically affect the size and isolation of the major islands (Simpson 1974; Connor and Simberloff 1978). Its principal effect was to connect the small islands and islets to nearby large islands.

In the last 10,000 years there have been two periods of climate similar to our own, from 8000 to 6200 y.B.P. and from 3000 to the present; the remaining two periods were drier, and possibly hotter, but not as dry as in the period before 10,000 y.B.P. These are only coarse categories. It is evident from the core profiles of *Azolla microphylla* massulae, or spore masses (Colinvaux 1972: fig. 3), that the lake level and, hence, presumably climate, fluctuated within each of the relatively dry periods in the last 10,000 years.

IMPLICATION OF RECENT CLIMATIC FLUCTUATIONS

During the driest period in the last 50,000 years, the area of mesic habitat on some high elevation islands, such as San Cristóbal, would have been reduced or absent. A consistent observation is the low proportion of endemic plant species in moist high elevation habitats now. This has been attributed by Johnson and Raven (1973) to the relative youth of the current habitat as a result of climatic fluctuations. It therefore seems highly probable that some finch species adapted to mesic habitat went extinct in dry periods. For example, early branches of the finch radiation that evolved under conditions similar to present ones (> 34,000 y.B.P.) probably went extinct in the succeeding dry period (34,000–10,000 y.B.P.).

More specifically, tree finch species, which generally occur now more commonly in mesic habitats at high elevation than in arid habitat at low elevation, and presumably always did, would have been more adversely affected than ground finch species in dry periods, and some may have gone extinct. There is no evidence for the origin of new species in the last 50,000 years, so a net reduction in the number of Darwin Finch species may have occurred during this period. The present Darwin's Finch fauna is basically an arid zone one. All species occur in arid zone habitat at low elevations, even though some, especially the tree finches, reach their maximum density in mesic habitat at moderately high elevations (Lack 1947; Schluter 1982). Possibly before 50,000 y.B.P. an opposite trend occurred during a wetter period than now, with perhaps the extinction of ground finch species at a time when arid zone habitat was much less extensive than now.

These suggested compositional changes in Galápagos finch faunas have a counterpart in the finch faunas of the West Indies where there is fossil evidence for climate-associated extinctions (Olson and McKitrick 1981; Pregill and Olson 1981). Compositional changes on the Galápagos were possibly contemporaneous with changes in avifaunas on the South American continent which experienced similar cycles of climatic and vegetational change (Simpson and Haffer 1978).

FINCH ECOLOGY

RAINFALL AND FINCH DEMOGRAPHY

The cores from El Junco provide a better record of precipitation than of thermal conditions. This is fortunate and useful because modern ecological studies on the islands suggest that finch demography is more affected by fluctuations in precipitation than by fluctuations in temperature, at least at low elevations. Hence a link can be established between modern and historical environments.

Variations in primary production are related to variations in precipitation. Secondary and tertiary production, including the breeding of finches, is intimately related to primary production. Thus, rainfall governs the breeding of finches (Lack 1950). In the strongly seasonal environment of the Galápagos, a season of heavy rainfall extends usually from January to April or May, occasionally from December to June (Grant and Boag 1980). *Geospiza* finches usually do not breed until the first rainfall of about 20 mm or more (Boag 1981; Grant and Grant 1983; Boag and Grant 1984). The occurrence and frequency of rainfall is highly unpredictable in this season (Grant and Boag 1980). The breeding of finches varies similarly (Grant and Grant 1983; Boag and Grant 1984; Millington and Grant 1984).

In the dry and cooler season from June to December precipitation is more frequent, particularly at high elevations, but it is usually not very heavy, and birds typically do not breed. This absence of breeding may reflect a dependence of their food supply (e.g., caterpillars) on temperature.

Superimposed on the seasonal cycle is a large degree of annual variation in precipitation. Correspondingly, finch populations fluctuate in size. In wet years breeding is extensive and successful. Individual pairs may produce five broods or more (Grant and Grant 1980a). Survival of both young and adults through the ensuing dry season can be as much as 85 percent or higher (Grant et al. 1975). In dry years breeding may not occur at all and mortality of adults and subadults combined may be as high as 85 percent (Boag and Grant 1981, 1984).

ECOLOGICAL AND EVOLUTIONARY CONSEQUENCES OF A FLUCTUATING CLIMATE

The dependence of finches on rainfall has both ecological and evolutionary implications for the understanding of the adaptive radiation of the finches. In the drought year of 1977 the numbers of females on Isla Daphne Major were reduced to 35 *Geospiza fortis* and 25 *G. scandens,* and by early 1979 the numbers of breeding adult females were further reduced to 28 and 21 respectively (Price et al. 1984). The two populations increased in size as a result of successful breeding in 1978, but successive drought years (two? three?) could have led to the extinction of one or both. Thus, some populations, especially those on small islands, are precarious.

A second, and related, effect is shown at the 'community' level. In the example above, *G. fortis* suffered 85 percent mortality whereas *G. scandens* suffered 60 percent mortality (Grant and Grant 1980b; Boag 1981; Boag and Grant 1984). The proportional representation of *G. scandens* increased from 25 percent to 50 percent. Therefore, climatic extremes can have selective effects at the community level.

These two ecological consequences suggest that dry periods in the climatic history of the Galápagos had unequal effects upon the distribution and abundance of different finch species. This applies to species within the genus *Geospiza,* as well as to the sub-family as a whole as pointed out in the previous discussion of tree finches.

The micro-evolutionary consequence of the drought on Isla Daphne was a selective shift in the phenotypic characters of both of the species. *G. fortis* experienced strong directional selection on body size and bill size (Boag and Grant 1981), and *G. scandens* experienced stabilizing selection (Boag 1981; Grant and Price 1981). These results have three important implications.

First, finch populations, particularly on small islands, should not be viewed as morphologically static entities. Rather, their morphological attributes are subject to change under contemporary selective pressures in relatively short time periods. Second, random drift may occur in these populations since their size can be drastically reduced. The importance of drift is difficult to assess, however, because gene flow between populations and natural selection, both of which do occur (Boag 1981; Grant and Price 1981), will tend to counteract it; of these two forces, selection has been clearly the more powerful. Third, the strong response of *G. fortis* to selection argues against the importance of the founder effect or genetic drift as an explanation for the small average body size of this population (see also Yang and Patton 1981).

In summary, the general implication is that natural selection, both historically and currently, is responsible for at least some, and perhaps most, of the morphological properties by which the species differ. This is a powerful justification, in addition to the usual arguments, for referring to the radiation of the finches as adaptive. Natural selection operated on the Daphne population of *G. fortis* strongly during a drought. It is likely that selection would also operate in unusually wet years but in a different manner (Grant et al. 1976; Boag and Grant 1981). For example, because large birds were favored in the drought, it is possible that small birds are favored in especially wet years, through either enhanced survival or enhanced reproduction (e.g., see Price and Grant 1984). To provide a better perspective of the frequency of strong selection events it is important, therefore, to document the frequency of especially dry and wet years.

CLIMATIC FLUCTUATIONS

The Frequency of Extreme Climatic Conditions

How often do Darwin's Finches experience droughts and usually wet conditions, and undergo microevolutionary shifts in response to selection? Our study on Isla Daphne Major cannot answer this question because it has continued for only nine years. We know, however, that selection occurred strongly in a year (1977) in which only 24 mm of rain fell, and less strongly in years of more rainfall and greater food supply (Price et al. 1984). The populations were not studied intensively in the last year of heavy rainfall (1975).

We can use annual rainfall records from other islands (Table 1) to provide broader perspective and an indirect answer to the question. Unfortunately, we do not know the relationship between rainfall and demography of any finch species on these islands, nor do we know the relationship between rainfall and food supply. Relationships may be different from those documented on I. Daphne Major. Furthermore, finches can escape the effects of a drought to some extent by seasonally migrating to more mesic environments at higher elevations on large islands such as I. San Cristóbal and I. Santa Cruz (Lack 1947; Downhower 1976). These higher elevations receive some rain when low elevations receive none. For example, *Geospiza fuliginosa* breed at moderately high elevations on the south side of Santa Cruz only in dry years (A. Kastdalen, pers. comm.); in other years their breeding distribution is restricted to lower elevations. Finally, as stated earlier, different species are likely to be affected to different extents by droughts, especially among the diverse assemblage of ground-finch and tree-finch species on these two islands. Nevertheless, rainfall records can provide a guide to the frequency of extreme conditions.

It is difficult to classify conditions as extreme without being arbitrary. I consider unusually wet conditions to be those associated with El Niño events (e.g., see Quinn and Burt 1970), that is unusually warm surface water along the coasts of Peru and Ecuador which usually extends westward to the Galápagos (Halpern et al. 1983). I consider unusually dry conditions to prevail when less than 30 mm of rain falls in the wet season.

I. San Cristóbal.—On this island, drought conditions comparable to the most extreme on I. Daphne did not occur in any of the 24 years for which data are available (Table 1). In two years, however, only slightly more than 30 mm of rain fell. The driest year was 1950; 33 mm fell in the wet season. By Daphne standards this was insufficient for successful breeding, and directional selection may have occurred moderately strongly. The next driest year was 1970; 39 mm fell in the wet season. By Daphne standards this would have generated a largely unsuccessful breeding response, and again directional selection may have occurred. Thus, in only two of 24 years was the precipitation sufficiently low that moderately strong selection

TABLE 1

TOTAL PRECIPITATION IN THE WET SEASON, JANUARY THROUGH MAY, AT COASTAL
RECORDING STATIONS ON THE GALÁPAGOS[1]

Year	San Cristóbal	Santa Cruz	Isabela	Baltra	Daphne	Genovesa
1950	33					
1951	369					
1952	88					
1953	1380					
1954	145					
1955	305					
1956	647					
1957	826					
1958	261					
1959	453					
1960	56					
1961	285					
1962	150					
1963	69					
1964	130			44		
1965	455	489	477	134		
1966	108	112	190	106		
1967	318	195	111	0		
1968	>236[2]	68	29	23		
1969	737	389	262	165		
1970	39	27	63	0		
1971		198	148	113		
1972		440	552	≥234[3]		
1973		423	517	174		
1974		113	147	29		
1975		864	594	>0[4]		
1976		410	104	≥35[5]	135	
1977	242	191	29		23	
1978	>346[6]	321	195		139	>153[7]
1979		182	161		69	
1980	153	232			53	164
1981		302			79	117
1982		62			51	69

[1] Values in mm. Sources of data: Grant and Boag (1980), Boag (1981) and Grant and Grant (1983). Monthly precipitation in the dry season, June to December, was generally low and a small fraction of the annual precipitation, except as follows: (1) San Cristóbal, 17 mm in June 1951, all other Junes received <10 mm; 75 mm in July 1951, all other Julys received <20 mm; 75 mm in December 1957, all other Decembers received <20 mm; (2) Santa Cruz, 124 mm in June 1972, all other Junes received <40 mm; 75 mm in December 1972 and 65 mm in December 1978, all other Decembers received <40 mm; (3) Isabela, 50 mm in June 1972, all other Junes received <40 mm; 106 mm in December 1972, all other Decembers received <20 mm.

[2] Data missing for April and May.
[3] Data missing for January.
[4] Data missing for January and March.
[5] Data missing for January, March–May.
[6] Data missing for May.
[7] Data missing for 1 to 15 of January.

may have occurred. This conclusion is subject to two caveats. On this high island the effects of dry conditions at low elevations may have been avoided by migration of some birds to higher elevations, and the effects may not have been experienced in some lowland parts of the island, if the spatial distribution of rainfall was patchy (cf. Grant and Boag 1980).

In the same period there were five years of heavy rainfall. Three were associated with El Niño events; in just the wet season, 1380 mm fell in 1953, 826 mm fell in 1957, and 485 mm in 1965. 1956 and 1969 were also wet years; 647 mm and 737 mm, respectively, fell in the wet season. If strong selection occurs in very wet years, the record for this island suggests that its frequency is two to three times greater than selection under drought conditions.

I. Santa Cruz.—In the 18 year period from 1965 to 1982 drought conditions occurred once (1970). The drought year on I. Daphne, 1977, was not a drought year on this island.

Considerable rain fell in the four El Niño years of 1965, 1972, 1973, and 1975. The heaviest fell in 1975, a year in which oceanographic conditions justified the appellation of only a weak El Niño event (Wyrtki et al. 1976).

I. Isabela.—Especially dry conditions prevailed in 1977, the year of the drought on I. Daphne, but they also occurred in 1968, i.e., twice in 16 years (Table 1). Especially wet conditions occurred in the same years as on I. Santa Cruz.

I. Baltra.—In a 13 year period no rain fell in two, possibly three years, and less than 30 mm fell in two additional years. This frequency of roughly one drought in three years is the highest recorded for any Galápagos locality; although it may be overestimated because rainfall records were not kept as carefully as at other Galápagos stations. It is also noteworthy that dry conditions prevailed in two successive years, twice. There are, in addition to the records in Table 1, Alpert's (1963) records of 39 mm rain in the wet season of 1943, 112 mm in 1944, and 107 mm in 1945. I. Baltra lies in the rain shadow of I. Santa Cruz, and sufficiently close so that in a dry year some finches may cross the water gap, flying to the highlands of Santa Cruz, and later returning. I. Baltra is also near Daphne.

In summary, the frequency of especially dry conditions in the usually wet season varies roughly from one in three years (Baltra) to one in twelve years (San Cristóbal) or more (Santa Cruz). Not surprisingly, droughts occur more frequently on low islands (Daphne, Baltra) than at low elevations on high islands (San Cristóbal, Santa Cruz, Isabela). Droughts are rarer than especially wet conditions, although this conclusion depends on an arbitrary classification. In the 33 year period from 1950 to 1982, especially wet conditions associated with El Niño events have occurred widely in the archipelago six times: 1953, 1957, 1965, 1972, 1973, and 1975. Their frequency is, thus, roughly one in five years.

EXTENDING THE RECORDS

To obtain better estimates of the frequency of unusual climatic conditions, I sought indirect, but longer, records. I first tried to use tree ring data from *Bursera graveolens* and *B. malacophylla,* in the hope that the widths of the rings would be strongly correlated with the amount of annual rainfall. There were no demonstrable correlations (Grant 1981b).

As an alternative, I used rainfall data for the years 1925 to 1980, from the coastal locality of Ancón on the Santa Elena peninsula of the Ecuadorian mainland (Fig. 1). Data were obtained from the Instituto Nacional de Meteorología e Hidrología, Quito, and from the Anglo-Ecuadorian Oil Co. An association between rainfall on the mainland and rainfall on the islands is to be expected for two reasons. First, amounts of precipitation are positively associated with sea surface temperatures at Galápagos (Houvenaghel 1974) and mainland localities (Murphy 1926; Sheppard 1933; Bjerknes 1961; Quinn and Burt 1970). Second, sea surface temperatures at Galápagos and mainland localities are positively associated in the early months of the year (Houvenaghel 1974).

In the period 1950 to 1970 together with 1977 and 1980, when records were kept at both Ancón and I. San Cristóbal, amounts of rainfall varied in parallel ($r = 0.698$, N = 25, $P < 0.005$). Significant positive correlations also exist between rainfall at Ancón and the other islands over shorter periods. There are some differences, however. For example, since 1950 most rain (>300 mm/y) fell at Ancón in the El Niño years of 1953, 1957, and 1972, which were years of extensive rainfall on the Galápagos, but Ancón received relatively little rain in two other years, 1965 and 1973, which were particularly wet on the Galápagos. The frequency of dry years (<30 mm/y) at Ancón (1 in 6) is comparable to the frequency of dry years on I. Isabela and I. Daphne (one in eight).

Given the correspondence between mainland and island rainfall patterns since 1950, we can use the full 56 year record of rainfall from Ancón as an approximate indication of long-term fluctuations in rainfall on the Galápagos Islands. Unusually heavy rain (>300 mm) fell in 14 years: 1925, 1926, 1929, 1932, 1933, 1939, 1941, 1943, 1949, 1953, 1957, 1972, 1975, and 1976. Thus, the observed frequency of very wet years is one in four. The frequency of dry years (<30 mm/y) is lower at one in eight. By my classification of wet and dry years, which is based partly on observed responses of finches to rainfall, this analysis does not confirm the supposition of Svenson (1946) that wet and dry periods even out (see also Quinn et al. 1969), although the classification is to some extent arbitrary.

Observations made on Galápagos in this early period, 1925 to 1950, are consistent with the Ancón pattern. The unusually wet conditions of 1925 were witnessed and recorded by Beebe (1926) on a visit to the islands. I. Santa Cruz was settled in 1926, and Svenson (1946) reported statements of settlers that 1929 was wet but 1930 was very dry; at Ancón the rainfall

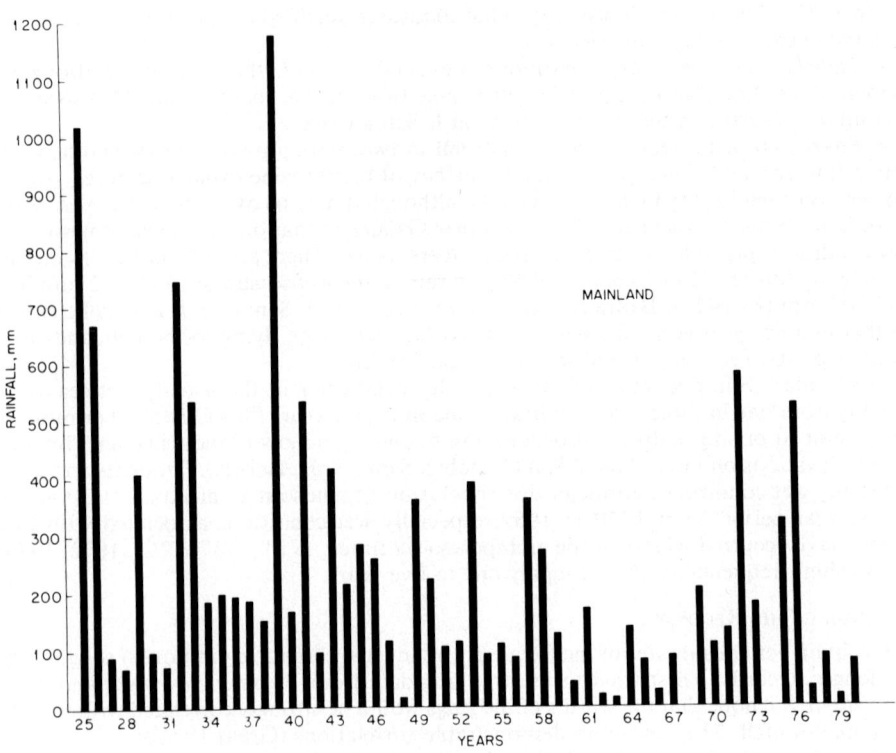

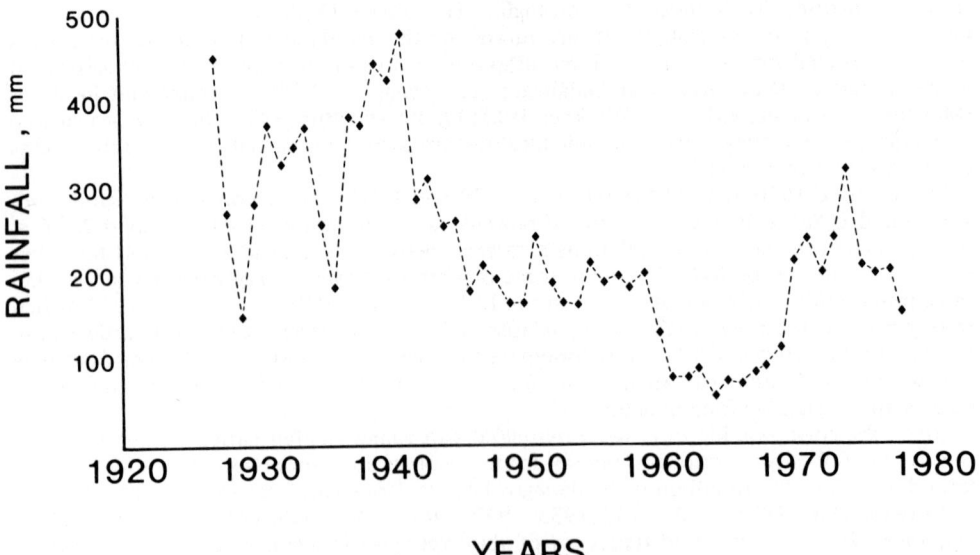

FIG. 1.   Annual precipitation at Ancón, Santa Elena peninsula, Ecuador. The lower figure gives five-year moving averages; the smoothing of annual variation makes it clear that recent rainfall, especially in the 1960's, has been relatively low.

totals were 408 mm and 104 mm for these two years, respectively. Kastdalen (1982) observed that the wettest year on the island after his arrival in 1935 was 1939, an El Niño year and also the year of Lack's (1945) visit. At Ancón, 1939 was the wettest year on record. Kastdalen noted also that 1953 was very wet, and for this observation there is independent evidence from San Cristóbal (Table 1). He remembered, but did not record (pers. comm.), that dry conditions prevailed in two successive years in the late 1940's; possibly these years were 1947 and 1948, in which 120 mm and 15 mm of rain, respectively, were recorded at Ancón.

Ideally, these estimates of frequency of extreme conditions would be based on rainfall records over a longer period so as to minimize the influence of short-term fluctuations. For example there is reason to believe that conditions are generally drier now than 40 years ago, both on the mainland (Fig. 1) and on Galápagos. At Ancón there were nine wet years (>300 mm) in the first 25 years, and only three in the second 25 years. There were no droughts in the first 20 years and only one in the first 25 years, but there were five in the next 25 years. Since 1975 there has been one more drought year, and rainfall has not exceeded 100 mm in any year except 1976.

Rainfall at Ancón may have diminished as a result of the cutting of woody vegetation, especially from the early 1940's on (Svenson 1946), but the fact that all of the early settlers on I. Santa Cruz believe that conditions were wetter in the 1930's and 1940's than now (A. Kastdalen, pers. comm.) argues for a general change in climate, because the same type of forest removal did not occur on that island. Some natural changes in vegetation on the uninhabited and undisturbed island of Daphne can also be interpreted as a response to drier conditions (Grant 1984).

A final point worth emphasizing is the extreme fluctuation shown by the Ancón rainfall. The coefficient of variation for annual rainfall, 104.16, is unusually large (Grant and Boag 1980).

Further Extension of the Records

The frequency of especially dry years prior to 1925 is unknown. Unusually heavy rains associated with El Niño phenomena have been recorded, however. The El Niño phenomenon is the occurrence of unusually warm water of low salinity that typically first appears off the coast of Peru in the December to April period (Quinn and Burt 1970; Philander 1983). It usually lasts from two to four months but may recur after an interval of several months (Miller and Laurs 1975). No universal agreement exists on what deviations from normal conditions warrant the El Niño classification (Bjerknes 1961, 1966; Quinn and Burt 1970; Miller and Laurs 1975; Wyrtki 1975; Wyrtki et al. 1976; but see footnote 1 in Halpern et al. 1983). The uncertainty arises first because the origins and characteristics of the El Niño phenomena vary among years (Bjerknes 1961, 1966; Philander 1983), and second because locally on the coast of South America the amount of associated rainfall varies from none (Idyll 1973) to an enormous amount (Murphy 1926; see also Ancón rain in Fig. 1). As a result, the effects of El Niño vary. They were particularly severe along coastal Peru and Ecuador in 1891 and 1925 (Murphy 1926); their effects were also present, but of unknown severity, on the Galápagos in those years (Agassiz 1892; Beebe 1926; Murphy 1926).

Considering all the data, I have arrived at a somewhat subjective classification of El Niño years (see also Hutchinson 1950; Quinn and Burt 1970, Miller and Laurs 1975). In the 192 year period since 1791 there have been 27 such years. This gives a frequency of about 14 per century, or one every seven years. These estimates based on sketchy long term data are similar to those based on recorded rainfall for a shorter period at Ancón and much shorter periods on the Galápagos. I conclude that the frequency of occurrence of wet years recorded on the Galápagos in the last 30 years is typical of the last two centuries.*

Summary of modern rainfall and its effect on finches. — Overlapping but non-congruent sets of rainfall data have been analyzed to assess the frequency of extreme conditions experienced by the finches. Modern studies show that finches may be subjected to strong stabilizing or directional selection in drought years (Boag and Grant 1981; Grant and Price 1981). In the

---

* Data concerning the recent 1982–1983 El Niño event, which has been described as the strongest oceanographic warming trend off the coast of South America in this century (Cane 1983; Kerr 1983), appeared too late to be included in this paper. The causes of the phenomenon are better understood now (Cane 1983; Philander 1983; Rasmusson and Wallace 1983; Smith 1983; Weare 1983). Biological effects are well documented in the marine environment (Barber and Chavez 1983), but less well documented in terrestrial environments including Galápagos.

last 30 years low islands and their finch populations have experienced droughts as frequently as one in three years (Baltra), or one in seven years (Daphne). Droughts occur less frequently on higher islands. They have occurred more frequently since 1950 than in the preceding 25 years.

The frequency of occurrence of very wet years is known better. Several estimates, including an indirect one from mainland data over a roughly 200 year period, agree on a figure of about one in seven years. Wet years occurred more frequently in the period 1925 to 1950 than in the following 25 years. The possibility of directional selection occuring in very wet years has never been investigated, and although it remains a speculative idea it seems plausible.

The life-span of finches in nature is not known. In captivity finches may live for 10 to 20 years (Bowman 1983). Banding studies on Daphne have shown that annual adult mortality rates may be less than 10 percent and that finches can live 10 years or more (Grant, unpubl. data). *Geospiza fortis* and *Geospiza scandens* individuals banded in 1973, when at least one year old, were still alive in 1983. These data show that those individual finches that survive their first year may experience at least one drought and one set of unusually wet conditions in their life-times. However, they may experience droughts more frequently and wet conditions less frequently now than their relatives did in the second quarter of this century.

CONCLUSIONS

Modern climate oscillates about a long term average between wet and dry conditions, and populations respond to extremes with microevolutionary shifts. Over millenia the long term climatic average has changed. It is reasonable to infer that this produced selective effects upon finch populations. However, the effect of climatic fluctuations on speciation events is unknown. We can speculate that such fluctuations probably caused changes in vegetation. These resulted in cycles of extinction, colonization, and adaptation of finches, with speciation the outcome in come cases.

A deeper understanding of the adaptive radiation of Darwin's Finches as an historical process requires an integration of several types of investigation. In the future, analysis of fossil material has the potential of illuminating evolutionary changes that have occurred in the last few hundred or even thousands of years (Steadman 1981, 1982b, pers. comm.). Paleobotanical studies could help in the interpretations of these changes by defining past environments more precisely, not only in terms of climatic conditions but in terms of the composition of the food plant species of the finches. For example, it is important to know the period during which modern island floras were established, and the interaction between finch diversification and the dynamics of colonization of the islands by the plant species from the mainland. Modern analyses of selection operating on the finches, combined with quantitative genetics theory (e.g., Lande and Arnold 1983) should enable us to estimate the conditions and forces necessary to transform one species into another (Price et al. 1984). Several recent ecological studies have thrown light on the question of why different islands have different communities of finches (Abbott et al. 1977; Smith et al. 1978; Grant 1981a; Grant and Schluter 1984). These studies have concentrated on the six species in the ground finch genus *Geospiza* and their food supply, and they should now be extended to the tree finches. Finally, the further application of biochemical techniques to Darwin's Finch material holds the promise of refining our knowledge of their relationships, both among their members and between them and continential relatives, as well as providing an improved chronology of their diversification.

ACKNOWLEDGMENTS

I thank S. Marchant, N. Lindner, A. Litchfield, and R. J. Norris for help with the Ancón rainfall records taken by members of the Anglo-Ecuadorian Oil Co., and the office of Instituto Nacional de Meteorología e Hidrología in Quito for supplying me with the most recent records. Rainfall records from the Galápagos were kindly supplied by T. de Vries, H. Hoeck, F. Köster, and C. MacFarland of the Charles Darwin Research Station. P. T. Boag gave much help in a variety of ways, and criticized the manuscript as did M. Foster, B. R. Grant, T. D. Price, D. Schluter, J. N. M. Smith, and a reviewer.

LITERATURE CITED

ABBOTT, I., L. K. ABBOTT, AND P. R. GRANT. 1977. Comparative ecology of Galápagos ground finches (*Geospiza* Gould): evaluation of the importance of florisitic diversity and interspecific competition. Ecol. Monogr. 47:151–184.

AGASSIZ, A. 1892. General sketch of the expedition of the *Albatross* from February to May, 1891. Bull. Mus. Comp. Zool. 23:1–89.

ALPERT, L. 1963. The climate of the Galápagos Islands. Occ. Pap. Calif. Acad. Sci. 44:21–44.

BAILEY, K. 1976. Potassium-argon ages from the Galápagos Islands. Science 192:465–467.

BARBER, R. T., AND F. P. CHAVEZ. 1983. Biological consequences of El Niño. Science 222:1203–1210.

BAUR, G. 1897. On the origin of the Galápagos Islands. Am. Nat. 31:661–680, 894–896.

BEEBE, W. 1924. Galápagos: World's End. Putnam's, New York.

BEEBE, W. 1926. The Arcturus oceanographic expedition. Zoologica 8:1–45.

BJERKNES, J. 1961. El Niño study based on analysis of ocean surface temperatures, 1935–1957. Bull. Inter-Am. Trop. Tuna Comm. 5:217–303.

BJERKNES, J. 1966. Survey of El Niño 1957–58 in its relation to tropical Pacific meteorology. Bull. Inter-Am. Trop. Tuna Comm. 12:25–86.

BOAG, P. T. 1981. Morphological variation in the Darwin's finches (Geospizinae) of Daphne Major Island, Galapagos. Unpubl. Ph.D. dissert., McGill University, Montreal.

BOAG, P. T., AND P. R. GRANT. 1981. Intense natural selection in a population of Darwin's Finches (Geospizinae) in the Galápagos. Science 214:82–85.

BOAG, P. T., AND P. R. GRANT. 1984. The Darwin's Finches (*Geospiza*) on Isla Daphne Major, Galápagos: breeding and feeding ecology in a climatically variable environment. Ecol. Monogr. 54:463–489.

BOWMAN, R. I. 1961. Morphological differentiation and adaptation in the Galápagos finches. Univ. Calif. Publ. Zool. 58:1–302.

BOWMAN, R. I. 1983. The evolution of song in Darwin's finches. Pp. 237–537, *In* R. I. Bowman, M. Berson, and A. Leviton (eds.), Patterns of Evolution in Galápagos Organisms. Spec. Publ., A.A.A.S., Pacific Division, San Francisco, California.

CANE, M. A. 1983. Oceanographic events during El Niño. Science 222:1189–1195.

COLINVAUX, P. A. 1972. Climate and the Galápagos Islands. Nature 240:17–20.

CONNOR, E. F., AND D. S. SIMBERLOFF. 1978. Species number and compositional similarity of the Galápagos flora and avifauna. Ecol. Monogr. 48:219–248.

COX, A. 1983. Ages of the Galápagos Islands. Pp. 11–23, *In* R. I. Bowman, M. Berson, and A. Leviton (eds.), Patterns of Evolution in Galápagos Organisms. Spec. Publ., A.A.A.S., Pacific Division, San Francisco, California.

COX, A., AND G. B. DALRYMPLE. 1966. Paleomagnetism and potassium-argon ages of some volcanic rocks from the Galápagos Islands. Nature 209:776–777.

DOWNHOWER, J. F. 1976. Darwin's Finches and the evolution of sexual dimorphism in body size. Nature 263:558–563.

GRANT, B. R., AND P. R. GRANT. 1983. Fission and fusion in a population of Darwin's Finches: an example of the value of studying individuals in ecology. Oikos 41:530–547.

GRANT, P. R. 1981a. Speciation and the adaptive radiation of Darwin's Finches. Am. Sci. 69:653–663.

GRANT, P. R. 1981b. Population fluctuations, tree rings and climate. Noticias de Galápagos. No. 33: 12–16.

GRANT, P. R. 1984. The endemic landbirds. Pp. 175–189, *In* R. L. Perry (ed.), The Galápagos. Key Environment Series, Vol. 1. Pergamon, Oxford, England.

GRANT, P. R., AND P. T. BOAG. 1980. Rainfall on the Galápagos and the demography of Darwin's Finches. Auk 97:227–244.

GRANT, P. R., AND B. R. GRANT. 1980a. The breeding and feeding characteristics of Darwin's Finches on Isla Genovesa, Galápagos. Ecol. Monogr. 50:381–410.

GRANT, P. R., AND B. R. GRANT. 1980b. Annual variation in finch numbers, foraging and food supply on Isla Daphne Major, Galápagos. Oecologia 46:55–62.

GRANT, P. R., B. R. GRANT, J. N. M. SMITH, I. J. ABBOTT, AND L. K. ABBOTT. 1976. Darwin's Finches: Population variation and natural selection. Proc. Nat. Acad. Sci. USA 73:257–261.

GRANT, P. R., AND T. D. PRICE. 1981. Population variation in continuously varying traits as an ecological genetics problem. Am. Zool. 21:795–811.

GRANT, P. R., AND D. SCHLUTER. 1984. Interspecific competition inferred from patterns of guild structure. Pp. 201–233, *In* D. R. Strong, D. Simberloff, L. G. Abele, and A. B. Thistle (eds.), Ecological Communities: Conceptual Issues and the Evidence. Princeton University Press, Princeton, New Jersey.

GRANT, P. R., J. N. M. SMITH, B. R. GRANT, I. J. ABBOTT, AND L. K. ABBOTT. 1975. Finch numbers, owl predation and plant dispersal on Isla Daphne Major, Galápagos. Oecologia 19:239–257.

GULICK, A. 1932. Biological peculiarities of oceanic islands. Q. Rev. Biol. 7:405–427.

HALL, M. L. 1983. Origin of Española Island and the age of terrestrial life on the Galápagos Islands. Science 221:545–547.

HALPERN, D., S. P. HAYES, A. LEETMAA, D. V. HANSEN, AND S. G. H. PHILANDER. 1983. Oceanographic observations of the 1982 warming of the tropical eastern Pacific. Science 221:1173–1175.

HEY, R. 1977. Tectonic evolution of the Cocos-Nazca spreading center. Bull. Geol. Soc. Am. 88:1404–1420.

HEY, R., G. L. JOHNSON, AND A. LOWRIE. 1977. Recent plate motions in the Galápagos area. Bull. Geol. Soc. Am. 88:1385–1403.

HOLDEN, J. C., AND R. S. DIETZ. 1972. Galápagos gore, NazCoPac triple junction and Carnegie/Cocos ridges. Nature 235:266–269.

HOUVENAGHEL, G. T. 1974. Equatorial undercurrent and climate in the Galápagos Islands. Nature 250: 565–566.

HUTCHINSON, G. E. 1950. Survey of contemporary knowledge of biogeochemistry. 3. The biogeochemistry of vertebrate excretion. Bull. Am. Mus. Nat. Hist. 96:1–554.

IDYLL, C. P. 1973. The Anchovy crises. Sci. Am. 228(6):22–29.

JOHNSON, M. P., AND P. H. RAVEN. 1973. Species number and endemism: the Galápagos archipelago revisited. Science 179:893–895.

KASTDALEN, A. 1982. Changes in the biology of Santa Cruz Island between 1935 and 1965. Noticias de Galápagos No. 35:7–12.

KERR, R. A. 1983. Fading El Niño broadening Scientists' view. Science 221:940–941.

LACK, D. 1940. Evolution of the Galápagos finches. Nature 146:324–327.

LACK, D. 1945. The Galápagos finches (Geospizinae): a study in variation. Occ. Pap. Calif. Acad. Sci. 21:1–159.

LACK, D. 1947. Darwin's Finches. Cambridge University Press, Cambridge, England.

LACK, D. 1950. Breeding seasons in the Galápagos. Ibis 92:268–278.

LANDE, R., AND S. J. ARNOLD. 1983. The measurement of selection on correlated characters. Evolution 37:1210–1226.

MACARTHUR, R. H., AND E. O. WILSON. 1967. The Theory of Island Biogeography. Princeton University Press, Princeton, New Jersey.

MARLOW, R. W., AND J. L. PATTON. 1981. Biochemical relationships of the Galápagos Giant tortoises. J. Zool. Lond. 195:413–422.

MILLER, F. R., AND R. M. LAURS. 1975. The El Niño of 1972–1973 in the eastern Tropical Pacific Ocean. Bull. Inter-Am. Trop. Tuna Comm. 16:403–448.

MILLINGTON, S. J., AND P. R. GRANT. 1984. The breeding ecology of the Cactus Finch *Geospiza scandens* on Isla Daphne Major, Galápagos. Ardea 72:177–188.

MORGAN, W. J. 1971. Convection plumes in the lower mantle. Nature 230:42–43.

MURPHY, R. C. 1926. Oceanic and climatic phenomena along the west coast of South America during 1925. Geogr. Rev. 16:26–54.

NEI, M. 1975. Molecular Population Genetics and Evolution. North-Holland, Amsterdam, The Netherlands.

OLSON, S. L., AND M. C. MCKITRICK. 1981. A new genus and species of emberizine finch from Pleistocene cave deposits in Puerto Rico (Aves: Passeriformes). J. Vert. Paleontol. 1:279–283.

PHILANDER, S. G. H. 1983. El Niño Southern Oscillation phenomena. Nature 302:295–301.

PREGILL, G. K., AND S. L. OLSON. 1981. Zoogeography of West Indian vertebrates in relation to Pleistocene climatic cycles. Ann. Rev. Ecol. Syst. 12:75–98.

PRICE, T. D., P. R. GRANT, AND P. T. BOAG. 1984. Genetic changes in the morphological differentiation of Darwin's Ground Finches. Pp. 49–66, *In* K. Wöhrmann and V. Löschcke (eds.), Population Biology and Evolution. Springer, Berlin.

PRICE, T. D., AND P. R. GRANT. 1984. Life history traits and natural selection for small body size in a population of Darwin's Finches. Evolution 37:in press.

QUINN, W. H. 1971. Late quaternary meteorological and oceanographic developments in the equatorial Pacific. Nature 229:330–331.

QUINN, W. H., AND W. V. BURT. 1970. Prediction of abnormally heavy precipitation over the equatorial Pacific dry zone. J. Appl. Meteor. 9:20–28.

QUINN, W. H., W. V. BURT, AND W. M. PAWLEY. 1969. A study of several approaches to computing surface insolation over tropical oceans. J. Appl. Meteor. 8:205–212.

RASMUSSON, E. M., AND J. M. WALLACE. 1983. Meteorological aspects of the El Niño/Southern Oscillation. Science 222:1195–1202.

ROSEN, D. E. 1976. A vicariance model of Caribbean biogeography. Syst. Zool. 24:431–464.

SCHLUTER, D. 1982. Distributions of Galápagos ground finches along an altitudinal gradient: the importance of food supply. Ecology 63:1106–1120.

SHEPPARD, G. 1933. The rainy season of 1932 in southwestern Ecuador. Geogr. Rev. 23:210–216.

SIMPSON, B. B. 1974. Glacial migrations of plants: island biogeographical evidence. Science 185:698–700.

SIMPSON, B. B., AND J. HAFFER. 1978. Speciation patterns in the Amazonian forest biota. Annu. Rev. Ecol. Syst. 9:497–518.

SMITH, J. N. M., P. R. GRANT, B. R. GRANT, I. J. ABBOTT, AND L. K. ABBOTT. 1978. Seasonal variation in feeding habits of Darwin's Ground Finches. Ecology 59:1137–1150.

SMITH, R. L. 1983. Peru coastal currents during El Niño: 1976 and 1982. Science 221:1397–1399.

STEADMAN, D. W. 1981. Vertebrate fossils in lava tubes in the Galápagos Islands. Pp. 549–550, *In* B. F. Beck (ed.), Proc. 8th Int. Congr. Speleol.

STEADMAN, D. W. 1982a. The origin of Darwin's finches (Fringillidae, Passeriformes). Trans. S. Diego Soc. Nat. Hist. 19:279–296.

STEADMAN, D. W. 1982b. Fossil birds, reptiles, and mammals from Isla Floreana, Galápagos Archipelago. Unpubl. Ph.D. dissert., Univ. of Arizona, Tuscon.

SVENSON, H. K. 1946. Vegetation of the coast of Ecuador and Peru and its relation to the Galápagos archipelago. Am. J. Bot. 33:394–426.

SWARTH, H. S. 1934. The bird fauna of the Galápagos Islands in relation to species formation. Biol. Rev. 9:213–234.

THORPE, J. P. 1982. The molecular clock hypothesis: biochemical evolution, genetic differentiation and systematics. Annu. Rev. Ecol. Syst. 13:139–168.

WEARE, B. C. 1983. The possible link between net surface heating and El Niño. Science 221:947–949.

WILSON, J. T. 1963a. A possible origin of the Hawaiian Islands. Canad. J. Physics 41:863–870.

WILSON, J. T. 1963b. Continental drift. Sci. Am. 208:86–100.

WYRTKI, K. 1975. El Niño—the dynamic response of the equatorial Pacific Ocean to atmospheric forcing. J. Phys. Oceanog. 5:572–584.

WYRTKI, K., E. STROUP, W. PATZERT, R. WILLIAMS, AND W. QUINN. 1976. Predicting and observing El Niño. Science 191:343–346.

YANG, S. Y., AND J. L. PATTON. 1981. Genic variability and differentiation in Galápagos finches. Auk 98:230–242.

HYBRIDIZATION AND EVOLUTION OF OYSTERCATCHERS ON THE PACIFIC COAST OF BAJA CALIFORNIA

JOSEPH R. JEHL, JR.

Hubbs-Sea World Research Institute, 1700 South Shores Road, San Diego, California 92109 USA

ABSTRACT. The ranges of the Black Oystercatcher (*Haematopus bachmani*) and the American Oystercatcher (*Haematopus palliatus frazari*) overlap for approximately 480 km along the Pacific coast of Baja California. Phenotypes are variable, especially in coloration, and hybridization has long been known. Yet, at most localities parental morphs predominate, hybrids are uncommon, and assortative mating occurs when possible.

The historical record is sufficiently detailed to allow demonstration of changes in the hybrid zone. At the turn of the century—or whenever they were first studied—most populations were composed largely of parental morphs. After intensive collecting, resulting in the virtual extermination of some populations, recolonization occurred from both north and south. The reconstituted populations differed from the original ones, and by the 1920s and 1930s the frequency of hybridization was high. Recently, however, many populations have returned to their original composition.

The prevalence of parental forms, a demonstration of assortative mating, and the resumption of stability in the zone of hybridization after a period of dynamic change, all show that there is selection against hybridization in this zone of secondary contact and that the two forms are specifically distinct. The nature of the selective forces remains to be determined.

RESUMEN. Los rangos de distribución del ostrero negro (*Haematopus bachmani*) y el ostrero americano (*Haematopus palliatus frazari*) se superponen por aproximadamente 480 km a lo largo de la costa del Pacífico de Baja California. Los fenotipos son variables, especialmente en coloración y desde hace mucho se conocen casos de hibridación. Todavía, en la mayoría de las localidades las formas parentales ("parental morphs") son las predominantes, los híbridos son poco comunes y los apareamientos se dan entre iguales cuando es posible.

Los registros históricos son suficientemente detallados como para permitir demostrar cambios en la zona de híbridos. A fines del siglo pasado o cuando se llevaron a cabo los primeros estudios, la mayoría de las poblaciones estaban compuestas comúnmente por "parental morphs." Después de la intensa colección de especímenes, que produjo la exterminación "virtual" de algunas poblaciones, ocurrió una recolonización tanto desde el norte como el sur. Las poblaciones reconstituídas fueron diferentes de aquellas originales y en las décadas de 1920 y 1930 la frecuencia de híbridos fue alta. Sin embargo recientemente muchas poblaciones han regresado a su composición original.

La prevalecencia de "parental forms," una demostración de apareamiento selectivo ("assortative mating") y una reasunción de la estabilidad en zonas de hibridación luego de un período de cambios dinámicos—todo indica que existe una selección en contra de los híbridos en estas zonas de contacto secundario y que las dos formas de ostreros son distintas a nivel de especie. Las características de las fuerzas de selección esperan ser determinadas.

"Oystercatchers are dimorphic through much of their world range, being either pied or wholly black. The melanistic forms have been particularly troublesome to taxonomists . . . especially since they do not often occur sympatrically with pied forms" (Baker 1973:330). One area of sympatry is Baja California.

For more than 70 years ornithologists have known that the American Black Oystercatcher (*Haematopus bachmani*) and the American Oystercatcher (*H. palliatus*) interbreed where their ranges overlap, mainly on small islands on the Pacific side of the Baja California peninsula (Fig. 1). Willett (1913) was the first to recognize that hybridization was occurring, pointing out that few birds on Cedros Island were typical of either form. At Scammon's Lagoon, Bancroft (1927) discovered another variable population, which included pied birds, black birds, and one intermediate, and he reported (p. 51) that C. C. Lamb had found a larger percentage of birds with "mixed underparts" at Natividad Island. Bancroft also collected an

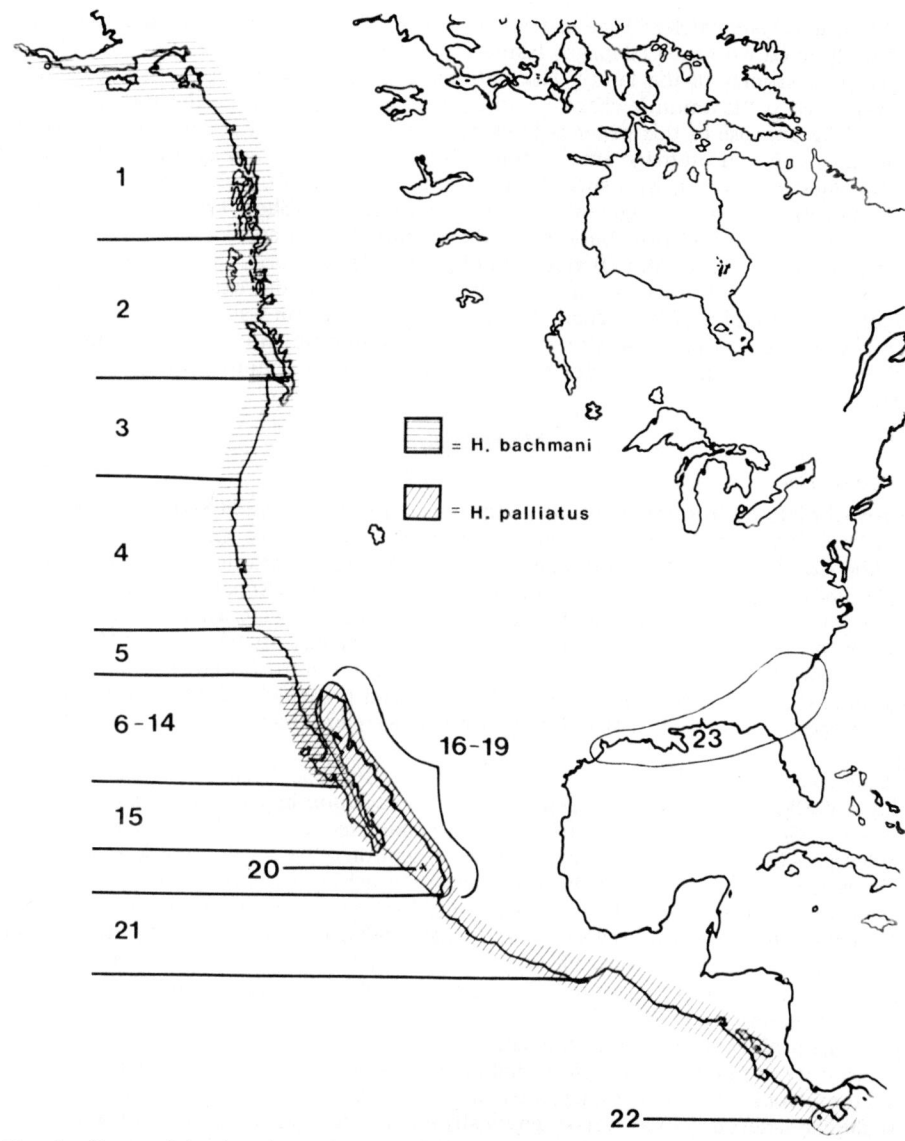

FIG. 1. Range of the American Black (*H. bachmani*) and American (*H. palliatus*) oystercatchers on the west coast of North and Middle America, and the zone of overlap in Baja California. Numbers 1–23 refer to populations analyzed in this paper (Appendix II). Localities in Baja California are shown in Figure 4.

intermediate at San Gerónimo Island. Kenyon (1949) provided further information for some localities in the zone of overlap.

Defining the species limits of North American birds was a task that dominated New World ornithology through the first half of the 20th century. It is surprising, therefore, that this situation remained unstudied, because the nature and extent of the hybridization bears directly on the taxonomic status of the two forms, a point on which there has been some disagreement. The American Ornithologists' Union Check-list Committee (American Ornithologists' Union 1983) and Murphy (1925) considered *bachmani* and *palliatus* as distinct species, whereas most other authors (e.g., Stresemann 1927; Peters 1934; Mayr and Short 1970; Heppleston 1973) considered them as a single species, *H. palliatus,* or lumped both with the Old World *H. ostralegus.* The population of *H. palliatus* in Mexico and in the zone of overlap (*H. p.*

frazari, "Frazar's Oystercatcher") has also been considered specifically distinct (Brewster 1888) or has been lumped with the Galápagos Island form, *H. galapagensis* (e.g., Gifford 1913).

The limits of species in the family Haematopodidae have always been disputed [Baker (1975:338) called it "taxonomic chaos"] and debates are not unique to the species in Baja California. The purpose of this paper is to clarify species limits by documenting the nature of variation in size and plumage among populations of oystercatchers on the west coast of North and Middle America, with special reference to the zone of overlap. Information necessary to render a taxonomic judgment in areas of hybridization (Short 1969) includes knowledge of the occurrence and distribution of hybrids, distribution and habits of the parental forms and of the hybrids, relative frequencies of parental and hybrid phenotypes in the area of hybridization, type of crossing that is occurring, occurrence and extent of introgression, and population dynamics of the forms involved. This information was not available to early taxonomists, who based their ideas on remarks in the literature and on casual examination of specimens. The data are now sufficiently complete to permit a more thorough assessment of the situation.

METHODS

MUSEUM STUDIES

I borrowed material from many of the major museums in the United States (see Acknowledgments) and examined 561 oystercatcher specimens collected along the west coast of North and Middle America, from Alaska to Panamá, as well as a small sample of *palliatus* from the eastern United States, far from any potential area of hybridization (Appendix I). Only "adults" (i.e., birds showing no traces of immature plumage or bill color) were used because juveniles are smaller and vary slightly in coloration. Standard measurements taken were: (1) exposed culmen to nearest 0.1 mm, (2) flattened wing to nearest 1 mm, (3) greatest depth of bill (approximately, depth at gonys) to nearest 0.1 mm, and (4) length of tarsus to nearest 0.1 mm. Worn specimens and those with bill deformities were excluded from size (but not color) analysis.

To study color variation, I established a Character Index (e.g., Rising 1970; Hybrid Index of Sibley and Short 1964) and classified the range of variation in 10 characters into five (0–4) or seven (0–6) character states (Table 1). These characters account for virtually all of the color differences between the two forms, with the exception of back color (black in *bachmani,* brown in *palliatus*), which I found to be too subject to bleaching and foxing to make evaluation reliable. In establishing the Character Index, I used reference specimens obtained far from the area of overlap (*bachmani,* Alaska; *palliatus,* eastern North America) to obviate problems in interpretation resulting from possible introgression. Thus, if there were no variability within each form, a "pure" *bachmani* would receive a score of "0," and a "pure" *palliatus* from eastern North America a score of "42."

Some of the problems associated with using a Character Index have been discussed (Rising 1970). A tacit assumption, for example, is that F_1 hybrids will have scores intermediate between those of the parental forms. Although this may often be the case, it is not necessarily so, because the characters used may not be genetically independent and because dominance effects or modifying genes may mask underlying genetic differences. Furthermore, there is no certainty that differences in ranking either within or between character states represent equal intervals, so that a score of "2" might represent an F_1 in one character and an F_2 in another. Nevertheless, Character Indices have proven useful in analyzing hybrid zones in many avian taxa and are used here.

In the oystercatcher literature black-and-white phenotypes are referred to as "pied," and melanistic phenotypes as "black"; intermediates may also be recognized. These terms are used for convenience and without implication regarding taxonomic status.

The genetics of oystercatchers are little known. Baker (1973:330) argued that differences between the pied and black forms of the Variable Oystercatcher (*H. unicolor*) of New Zealand could be explained by the action of a single major gene "whose dominance is modified only in the heterozygous condition," and that the pied form is dominant to the melanistic. He confined his analysis, however, to gross phenotypes ("pied," "intermediate," "black") and did not consider the expression of differences in single characters.

In analyzing size variation, I treated the sexes independently, as sexual dimorphism is considerable, females averaging larger than males especially in bill length; however, bills of

TABLE 1
Synopsis of Color Characters in Oystercatchers

Character	Character state and score
Upper tail coverts	0. Black, as in *H. bachmani* 1. Black, a few white mottlings 2. Nearly equally black and white 3. White, a few black mottlings 4. White, as in *H. palliatus*
Tail	0. Black, as in *H. bachmani* 1. Mainly black, trace of white at base of vanes 2. Basal ¼ of rectrices white 3. Basal ⅓ of rectrices white 4. Basal half of rectrices white, as in *H. palliatus*
Chest	0. Black, with black chest band extending smoothly onto mid-belly, as in *H. bachmani* 1. Black chest band extending onto upper ⅓ of belly 2. Black chest band extending onto upper ¼ of belly 3. Black chest band bordered by ragged edge on upper breast 4. Black chest band sharply delimited from white of upper chest, as in *H. palliatus*
Belly	0. Blackish, as in *H. bachmani* 1. Blackish, with traces of white on a few feathers 2. Blackish, white area around crissum 3. ¾ black, ¼ white 4. Nearly equally black and white 5. ¾ white, ¼ black 6. Entirely white, as in *H. palliatus*
Under tail coverts	0. Entirely black, as in *H. bachmani* 1. Mainly black with slight white mottling 2. Nearly equally black and white 3. Mainly white 4. Entirely white, as in *H. palliatus*
Thighs	0. Entirely black, as in *H. bachmani* 1. Black with grayish underdown, not noticeable externally 2. Puffs of grayish down noticeable 3. Mainly white 4. Entirely white, as in *H. palliatus*
Greater secondary coverts (width of white edging in folded wing)	0. Lacking, as in *H. bachmani* 1. <2 mm 2. 2–5 mm 3. 6–15 mm 4. >15 mm
Extent of white wing stripe	0. Lacking, as in *H. bachmani* 1. White markings confined to inner ½ of secondaries 2. White markings extend to outer secondaries, but not onto primaries 3. White present on some or all of inner 5 primaries 4. White present on at least one of primaries 6–10
Underwing coverts	0. Entirely black, as in *H. bachmani* 1. Mainly black, some white mottling 2. Nearly equally black and white 3. Mainly white 4. White, as in *H. palliatus*
Axillars	0. Black, as in *H. bachmani* 1. Mainly black, some white mottling 2. Nearly equally black and white 3. Mainly white 4. White, as in *H. palliatus*

males are much stouter than those of females (Appendix II). I found no sexual differences in color pattern, and the data for the two sexes are combined.

FIELD WORK

Since 1967 I have made more than a dozen trips to islands on the Pacific coast of Baja California and have visited all but San Roque at least once, the majority on several occasions. Most field work was accomplished in 1969–1972. At each island I attempted to determine (1) population size, by walking the entire periphery of the island, if possible, (2) character scores of as many individuals as possible, and (3) composition of mated pairs. Many birds were photographed. A few specimens were collected (deposited in the collections of The San Diego Natural History Museum), and those were mainly selected to include presumed hybrids or mated pairs.

Color differences between the pied and black forms are sufficiently distinct to allow for the determination of character scores with high precision (± 2 units) in the field. This means that variable populations can be studied and monitored without collecting large numbers of specimens. This is important because oystercatcher populations tend to be small, and collecting can have important effects on the subsequent composition of the population, as shown below.

RESULTS

ALLOPATRIC POPULATIONS

To determine whether variation in the zone of overlap represents geographic variation in the parental species forms or is attributable to hybridization, it is first necessary to examine morphological variation in areas of allopatry, where the potential for hybridization and introgression is low.

American Black Oystercatcher.—The American Black Oystercatcher is a resident of rocky coasts from the Aleutian Islands and southern Alaska to central Baja California. There is, apparently, only one record from the Gulf of California (Russell and Lamm 1978), where it does not breed. Typically it occurs on islands and isolated headlands and exists in small, disjunct populations comprised of only a few pairs (Table 2). In the area of allopatry (populations 1–5), which extends from the Aleutians to southern California, the American Black Oystercatcher shows no obvious geographic variation in size; mean values for all characters are similar, and variability does not change as the zone of overlap is neared (Appendix II). Therefore, mensural data for these populations were pooled (Fig. 2). Even though the northernmost populations reside in very cold areas, morphological variation predicted by well-known "biogeographic rules" is not evident. Coloration does vary geographically (Fig. 3). Birds from northern areas tend to be entirely black, occasionally with a few white feathers or feather edgings on the venter. More southerly populations show a variable but increasing amount of white in many areas, and a reduction in the extent to which the black of the chest extends onto the abdomen. Birds from the southern part of the range may have a brownish cast to the abdomen, which results in an area of contrast at the chest-belly border.

American Oystercatcher.—The American Oystercatcher is widespread on both coasts of North and South America. In México, as elsewhere in the species' range, *H. p. frazari* is typically a bird of sandy beaches and coastal mudflats, although it does occur along with *bachmani* on rocky islands on the west coast of Baja California. This race is common in the Gulf of California, locally common on the mainland coasts of Sonora and Sinaloa, and spottily distributed south to the Isthmus of Tehuantepec. Farther south it is very rare (Griscom 1933; Slud 1964), and when the species reappears in Costa Rica and Panamá it is as the nominate race (*palliatus*) of the Atlantic coasts (Murphy 1925; see also below). *Frazari* also occurs on the Tres Marías, but not on the Revillagigedos (Brattstrom and Howell 1956; *contra* Friedmann et al. 1950).

Frazari is larger than *bachmani* and, as befits a bird of its ecological preferences, has a longer and slimmer bill. Size does not vary geographically between the Gulf of California and southern México (populations 16–21, Appendix II), and the morphological data have been pooled (Fig. 2). (Data from Panamá are few and probably include missexed specimens.)

"Frazar's Oystercatcher" (Brewster 1888) was described from specimens collected in the Gulf of California. Birds there differ from *H. p. palliatus* in being slightly browner dorsally and having the white areas invaded to a variable degree by dark smudging, especially on the rump, under tail coverts, wing lining, and margin of the chest band; in addition, the amount

TABLE 2

Size Estimates of Oystercatcher Populations from Alaska to Panamá

Population	Size	Basis for estimate[1]
1. Alaska	Unknown, minimum several 1000 birds	Sowls et al. 1978; Jehl, est.
2. British Columbia	Unknown, perhaps 1000 pairs	Hatler et al. 1978; Jehl, est.
3. Washington	350–400 birds	Census, 1982 (S. L. Speich, pers. comm.)
Oregon	350 birds	Census, 1981 (R. L. Pitman, pers. comm.)
4. Northern California	650 birds	Census, 1979–80 (Sowls et al. 1980)
5. Southern California	350 birds	Census, 1979–80 (Sowls et al. 1980)
6. Northern Baja California		
Los Coronados Islands	5–8 pairs	This paper
Todos Santos Islands	Late 19th century common, 5–10 pairs; 1978–81, no more than 4 birds	Kaeding 1897 W. Everett, pers. comm.
7. San Martín Island	1896, 20 birds?; 1946–present, no sightings	This paper
8. San Gerónimo Island	1905–present, ca. 15–20 birds	This paper
9. San Benito Islands	1969–74, 14–20 pairs	This paper
10. Cedros Island	ca. 20 pairs	This paper
11. Scammon's Lagoon	1926, 200–300 birds; 1970, ca. 100 birds	This paper
12. Natividad Island	Late 19th century, 10–15 pairs; 1969, 5–10 pairs?	This paper
13. Turtle Bay	Several pairs	This paper
14. San Roque/Asunción islands	Original population ca. 30–40 birds, now much smaller	This paper
15. SW Baja California	100 pairs?	Jehl, est.
16–18. Gulf of California	1000–1500 pairs	D. W. Anderson, pers. comm.; Jehl, est.
19. Sonora/Sinaloa	Several hundred pairs?	Jehl, est.
20. Tres Marías Islands	Breeds on all four islands but evidently very uncommon; 10–20 pairs?	Stager 1957; Grant 1964
21. Southern México (Nayarit to Chiapas)	Rare and local. Under 100 pairs?	Jehl, est.
Guatemala/Costa Rica	Essentially absent	Griscom 1933; Slud 1964
22. Panamá	Rare and very local on Pacific coast	Ridgely 1976

[1] For many areas population sizes are known with good accuracy, either from historical records (Baja California) or recent censuses (Washington, Oregon, California). For other areas data are scanty, and I have estimated numbers from the regional literature, information from colleagues, and my own assessment of habitat availability.

of white in the primaries is reduced (*cf.* Murphy 1925). Although Gulf populations are fairly uniform, coloration varies geographically. Birds from southern México approach nominate *H. p. palliatus* in showing increasing amounts of white in several areas (e.g., upper tail coverts, under tail coverts, greater secondary coverts, primaries) and a more sharply-defined chest band; birds on the outer coast of Baja California are much more variable and average darker.

In summary, *bachmani* and *frazari* show no appreciable geographic variation in size within their allopatric ranges. Toward the zone of overlap both exhibit gradual and clinal changes in size and coloration; character scores begin to converge and variability increases. Within the zone of overlap populations seem intermediate and highly variable (Fig. 3). This suggests that the changes have resulted from hybridization and introgression (but see Clarke 1966). Both events have certainly occurred. However, in most of the zone of overlap (e.g., San Benito Islands) the populations are bimodal in coloration and consist largely of "pure" phenotypes (see below). The apparent intermediate values and high variability, then, are largely a mathematical consequence of treating the local populations as a unit. Character scores also increase

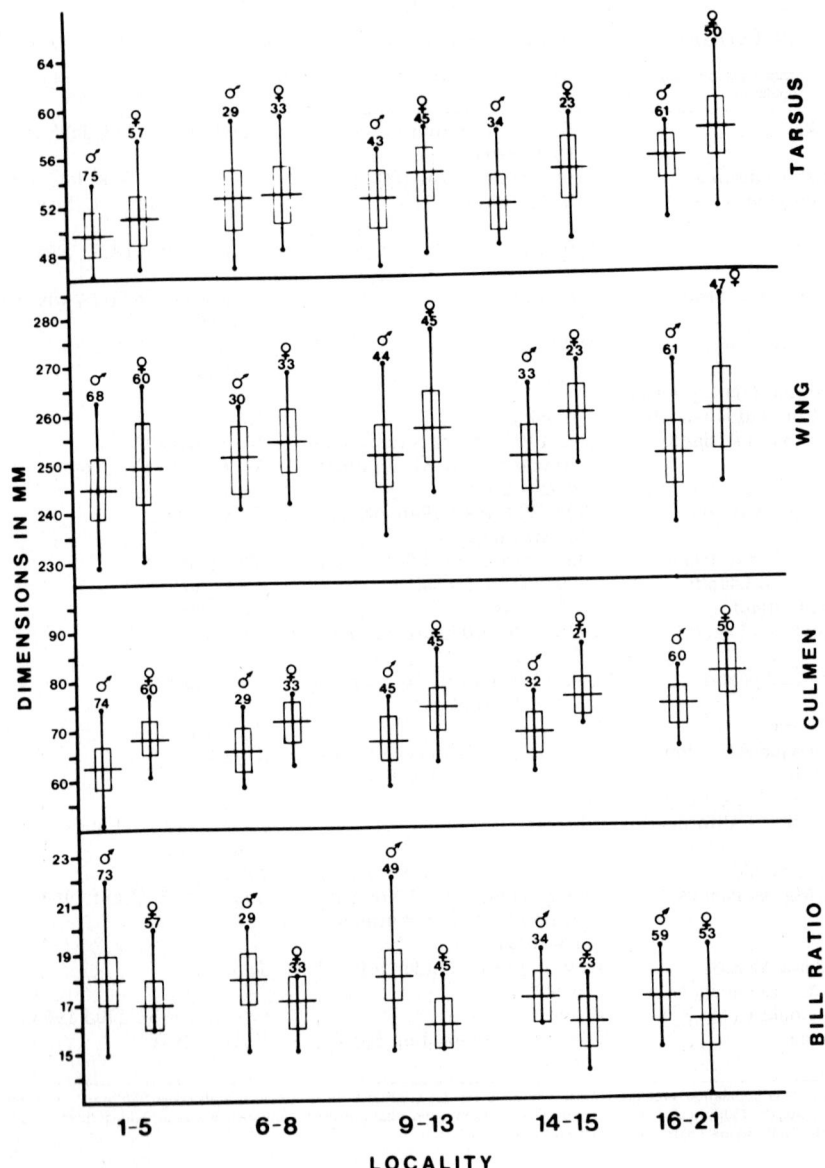

Fig. 2. Measurements of oystercatchers. The horizontal line represents the mean, the vertical line the range, and the rectangle one standard deviation on each side of the mean. Numbers above ranges are sample sizes. Localities are shown in Figures 1 and 4.

in southern México (and Costa Rica, Griscom 1933), probably as a result of introgression of *frazari* with *palliatus*.

In order to evaluate variation in the zone of overlap it is necessary to judge the degree to which geographic variation in allopatric populations may be the result of hybridization. I think that populations of *bachmani* from Oregon northward (localities 1–3) are sufficiently isolated to consider their character scores (0–9, mean 1.8, s.d. 1.9) as representative of normal variation. Similarly, the populations of *frazari* in the northern Gulf of California and on the coasts of Sonora and Sinaloa (localities 18–19) show little variation and are remote from potential areas of hybridization (both on the west coast of Baja California and in southern

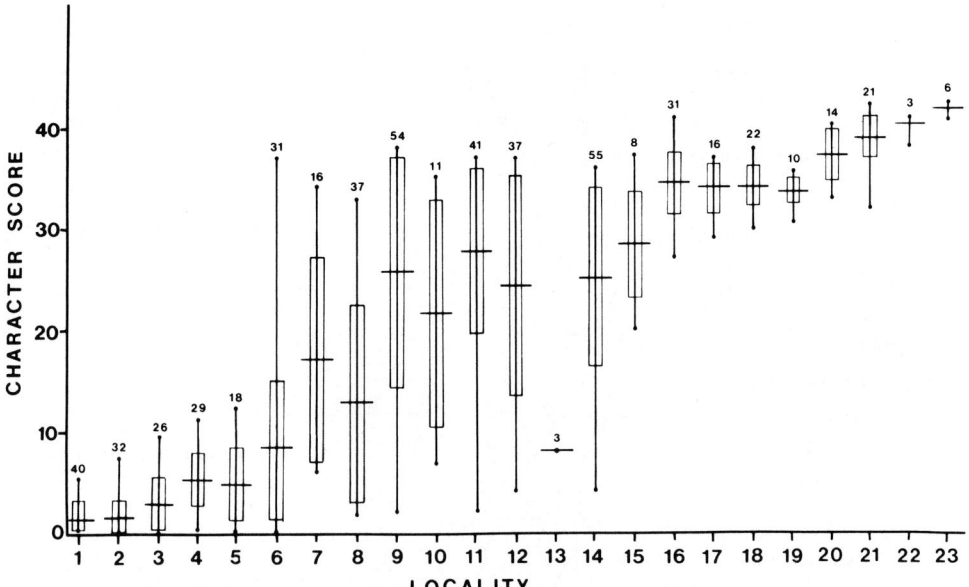

FIG. 3.   Character scores of color variation in oystercatcher populations. The horizontal line represents the mean, the vertical line the range, and the rectangle one standard deviation on each side of the mean. Numbers above the range are sample sizes. Localities are shown in Figures 1 and 4.

México). Therefore, I consider character scores of 30–38 (mean 33.6, s.d. 1.9) as typical of *frazari*. For the purposes of this paper, any individual with a score of 10 to 29 is considered to be a hybrid.

POPULATIONS IN THE ZONE OF OVERLAP

Midway along the outer coast of Baja California there is a transition between the Warm-temperate North Pacific and the Tropical Eastern Pacific biogeographic provinces (Vermeij 1978). In general, cold water faunas extend south to Punta Eugenia. However, "alternating areas of warm and cold water cause a discontinuous distribution of warm and cold water faunas, rather than a gradual transition" (Durham and Allison 1960:65). Thus, the boundary between these provinces is arbitrary (see also Hubbs 1960), but is commonly placed in the vicinity of Magdalena Bay. Also, midway along the peninsula the predominantly rocky shores and islands of western North America give way to the light colored sandy beaches of southern Baja California and the Gulf of California. It is in this region that the ranges of the two oystercatchers overlap.

The area in which both forms are consistently found together is broadly that extending from Los Coronados to San Roque and Asunción islands (Fig. 4), a distance of 480 km. Here the birds are mostly restricted to a few islands, although in the southern two-thirds of the region small numbers occur on isolated headlands in sparsely populated regions, and substantial numbers nest on islands in the larger lagoons. South of San Roque Island there are no offshore islands until the Gulf of California is reached, and oystercatcher populations here are probably small and local.

On many islands (Todos Santos, San Martín, San Gerónimo, Asunción, San Roque, Natividad) oystercatcher populations have declined in recent years as a result of human disturbance. Lamb (1927:67) recognized this problem on Natividad, but noted that the fishing camp there "is only temporary and before long the island will be uninhabited except for an occasional fisherman." He could not have anticipated that within several decades each of the islands would house a permanent fishing camp and that some would be visited by hundreds of tourists each year.

In general the west coast Baja California populations exhibit a gradual shift in size (Fig. 2), but a rapid shift in coloration (Fig. 3) from the *bachmani* to the *frazari* phenotype. Coloration

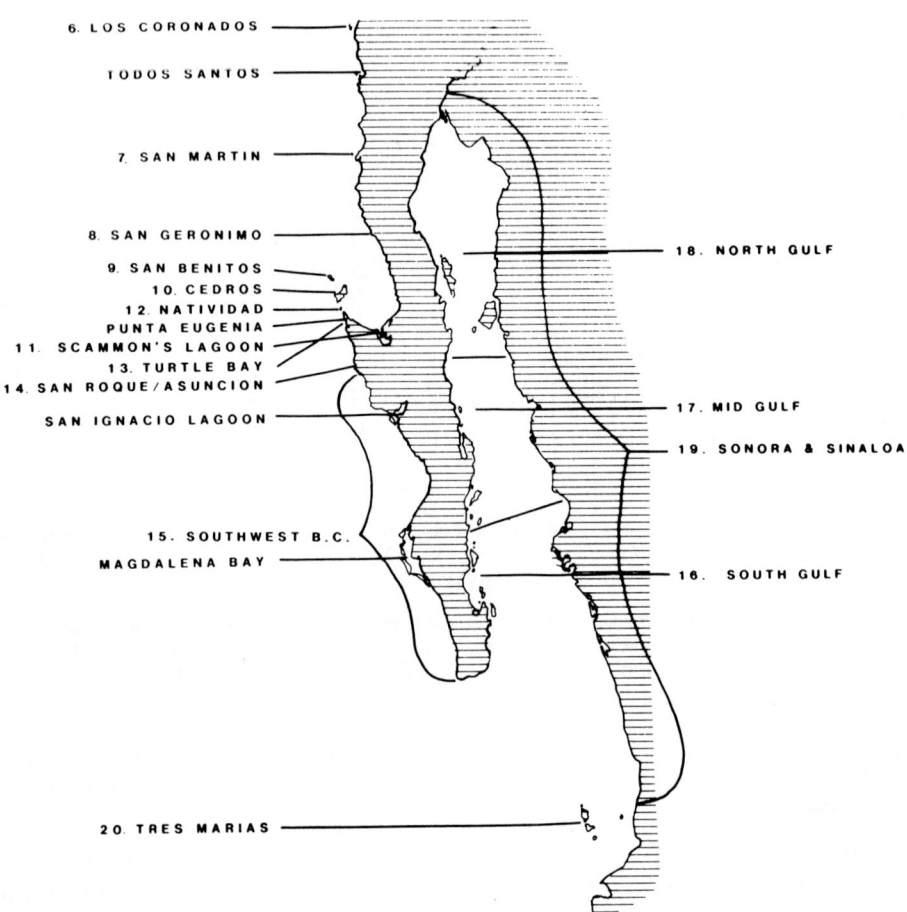

FIG. 4.   Baja California. Numbers refer to localities of populations sampled (Appendix II).

in each of the local populations (6–15), with the exception of Turtle Bay (13) is highly variable (Fig. 3), and it would be easy to conclude that random mating is the rule. However, analysis of a good historical record from many localities leads to a somewhat different interpretation.

Los Coronados (Area 6).—Historically, this population consisted of black birds. Occasionally pied birds have appeared, but they evidently have not nested. In 1967–1974 the population consisted of up to 19 individuals including five to eight pairs. I suspect that population size has remained stable in this century, because the number of nesting sites is limited. The birds are rather uniform in appearance, being dark-bodied with a grayish cast to the underwing and with scattered white feathers on the lower belly and thighs; a few show an indistinct wing stripe. In 1968, I estimated character scores at 5 to 12 (one "14" in 1974). All of the several *frazari* I have seen there had scores of more than 30. Phenotypes in this population have shown no change in nearly a century (Fig. 5).

Todos Santos Islands (Area 6).—Late in the 19th century both forms were present, for Kaeding (1897:109) reported collecting "a basket full of Heerman's Gulls, Black Turnstones, Black and Frazar's Oyster-catchers." Howell (1912) also found both common in 1910 and noted that *bachmani* outnumbered *frazari* by two to one. In 1912, *frazari* was not seen (Willett 1913), and in 1923 Van Denburgh (1924) noted only a "few pairs" of *bachmani* on South Island. In 1969 I estimated the population at four pairs, all *bachmani,* of which three were on the South Island. On four trips between 1978 and 1981, W. T. Everett (pers. comm.) never saw more than four individuals on South Island, including a single *frazari* in 1980.

San Martín Island (Area 7).—Oystercatchers were common in 1896 (13 collected in one

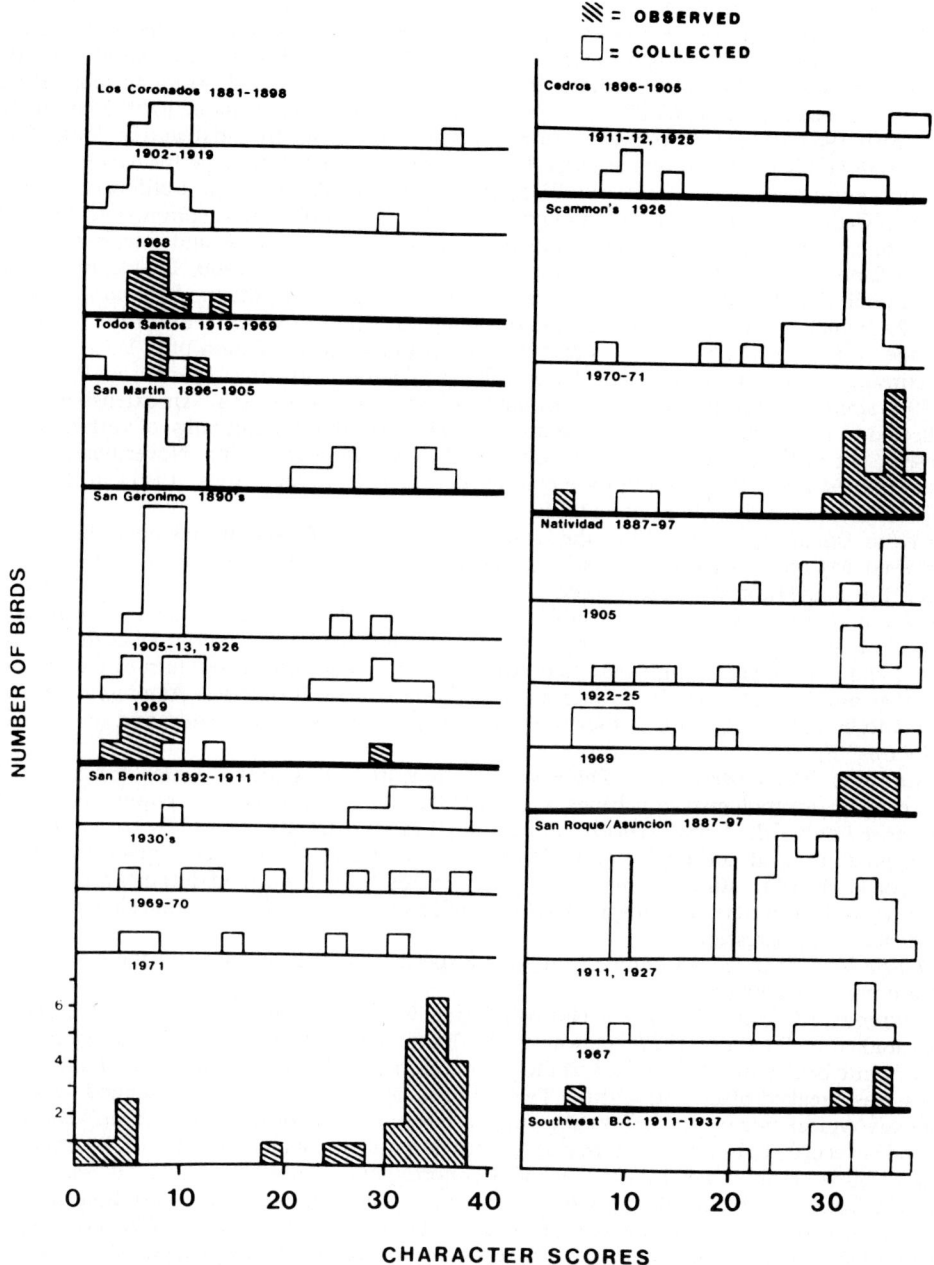

FIG. 5.   Temporal changes in character scores of oystercatchers in the zone of overlap on the west coast of Baja California.

day), and some were present until at least 1913 (Kaeding 1897; Willett 1913; Wright 1913). By 1946 they were gone (Kenyon 1949), and none has been seen since, despite dozens of visits by ornithologists. Intensive collecting probably caused their extirpation. Yet, despite the establishment of a permanent fishing camp, it is puzzling that the birds have not recolonized, because young birds wander to new locations, and because prey populations seem adequate to support at least one pair (pers. observ.). Early in the century *bachmani* outnumbered *frazari* by three to one, but hybrids were present and perhaps slightly commoner than *frazari*.

San Gerónimo Island (Area 8).—San Gerónimo is now the northernmost island on which pied birds are resident. In 1905 Gifford (1913) reported the population as 12 blacks and eight pied, but at least a few hybrids were present (Fig. 5). Willett (1913) found *bachmani* common in 1912 and collected an intermediate. In 1946, Kenyon (1949:193–194) found four blacks and two "intermediate or pied" birds, one of the latter showing signs of hybridization. In December 1967 12 birds were present (nine black, one pied, two hybrids), and in April 1969 there were 16 birds, all but two apparently being within the normal range for *bachmani*.

Punta Falsa area.—This location (28°55'N) is on the mainland of Baja California, slightly north of Cedros Island. Along 16 km of coast in 1963, G. S. Suffel (pers. comm.) encountered 14 oystercatchers, which he designated as *bachmani*-8, intermediate-5, and *frazari*-1.

San Benito Islands (Area 9).—Both forms were present (Anthony 1900; Thayer and Bangs 1907; Gifford 1913) at the turn of the century. Gifford gave the *frazari*:*bachmani* ratio as 10–12:1, which is slightly greater than the specimen data indicate (Fig. 5). From 1967 to 1974 I made 10 trips to the San Benitos and in 1971 estimated a population of at least 14 pairs (maximum 20) which included 21 *frazari*, five *bachmani*, and three hybrids; in 1969 and 1970 a similar composition was determined in less thorough surveys. An extensive series collected in 1930 and five birds collected 1969–1970 show that the full range of variation was present, but the latter series, at least, was biased by selective collecting. Nevertheless, even when the 1969 and 1970 samples are combined with 1971 observations, it is clear that this population consists largely of parental forms.

Cedros Island (Area 10).—This, the largest of the west coast islands, has never been fully surveyed. My impression is that oystercatchers are rarer here than on nearby islands and that the entire population does not exceed 20 pairs. Thayer and Bangs (1907) collected several birds and commented on their variability, and Willett (1913:22) noted that few birds were typical of either *bachmani* or *frazari*, most of them showing hybridization. Data from specimens and Gifford (1913) hint that pied birds were more common at the turn of the century but that numbers of black birds increased in the next several decades. At present, *frazari* seems to be in the majority; of four birds seen in 1971, three were *frazari*, and one was an "obvious" F_1.

Natividad Island (Area 12).—There are few data from this island, which has not been surveyed by ornithologists in decades. I estimate that the late 19th century population, which comprised only *frazari* and hybrids, numbered 10 to 15 pairs. By the 1920's *bachmani* may have predominated, although Lamb (1927) wrote that both forms were equally common. Kenyon (1949:194) estimated four to six pairs, all birds "showing a good deal of white." When I surveyed the southern half of the island in 1969, I saw only six *frazari*, none showing much evidence of introgression.

Turtle Bay (Area 13).—I know of only three specimens from this area on the outer coast; all are typical *bachmani*.

Scammon's Lagoon (Area 11).—This population, which nests on small islands in the lagoon, was studied by Bancroft (1927:51), who found 200 to 300 birds in 1926. Of these, 90 percent had "white bellies and the rest had all their underparts black, with the exception of one whose belly was streaked black and white." Two decades later Kenyon (1949) estimated 49 pairs, "mostly typical *frazari*, the others showing variable amounts of dark coloration." On Shell Island he recorded 21 pairs of *frazari* and two mixed pairs, which included the only "completely black" birds he observed. In 1957, R. M. Gilmore (pers. comm.) estimated 50 *frazari* and four *bachmani* there but, like previous observers, did not study plumage variation. In 1970 on Shell Island I found only one pair of *frazari* and one pair of *bachmani* (collected) in which one bird showed minor evidence of introgression (score 10). On Rocky Island, where Kenyon reported four pairs of *frazari* in 1946, I also found four pairs of oystercatchers in 1970 and 1971; in 1970 all but two and 1971 all but one of the mated birds were typical *frazari* (see below). Most of the birds collected in 1926 and both of those taken in 1946 show evidence of hybridization, whereas my sightings and photographs of 1970–1971 show that hybrids occurred in low frequency.

San Roque and Asunción islands (Area 14).—These small islands, only a few kilometers offshore, once supported large populations; in 1897, 11 birds were taken in one day on Asunción and 20 on San Roque. The original populations were highly variable and contained a high percentage of hybrids, but pied birds predominated. Three decades later Huey (1927a) found only several pairs of *frazari* on San Roque. In December 1967 I photographed one *bachmani* and three *frazari* on Asunción, but in June 1974 not a single oystercatcher was

FIG. 6. A mixed pair of oystercatchers (*H. p. frazari,* left; *H. bachmani,* right) at Scammon's Lagoon, 1971.

present. Apparently pied birds have always predominated here but the incidence of hybridization was far higher than elsewhere, probably because of the juxtaposition of rocky islands and the sandy beaches of the mainland.

Southwestern Baja California (Area 15).—There are no offshore islands south of Asunción Island, and the entire population from there to the Gulf of California is scattered along sandy beaches of the mainland and in coastal lagoons. Numbers are probably not large, but there are no reliable estimates. Pond Lagoon has not been studied in many years; Huey (1927b) observed "half a dozen" *frazari* there. Oystercatchers are very common in San Ignacio Lagoon, upwards of 100 *frazari* occurring in a single flock in winter (W. T. Everett, pers. comm.). Character scores in southwestern Baja California range from 20 to 36, most birds showing evidence of hybridization.

PAIR COMPOSITION

The composition of pairs in the zone of overlap and the extent to which back-crossing is occurring bear directly on the question of species limits. Such information is not easy to obtain, most obviously because oystercatchers engage in group displays and, therefore, birds seen together may not be paired. Other complications exist as well.

First, the number of nesting sites on some islands is limited, oystercatchers are long-lived, and pairs remain intact for many seasons, perhaps for life (Harris 1967). Thus, it is probably difficult for young birds to gain a territory and to enter the breeding population near their natal site.

Second, because the reproductive life of an oystercatcher is very long, it follows that annual production of breeders or survivorship of nonbreeders is low. And, because pair bonds are persistent, generation times are long, the opportunities for backcrossing and introgression are reduced, and selection to establish or reinforce potential isolating mechanisms acts slowly.

Finally, because populations are small and stable, an unmated bird may have to accept (not select) any bird of the opposite sex that becomes available if it is to enter the breeding population. Thus, the existence of mixed pairs is not necessarily evidence for random mating

TABLE 3
Composition of Mated Pairs of Oystercatchers on the West Coast of Baja California[1]

Location	Date	Pair composition	Reference
Los Coronados	1969–74	*bachmani* × *bachmani*—5–8 pairs, *bachmani* × *frazari*—1 pair.	This study
San Gerónimo Island	1913	The single female "*frazari*" present (score 28) was mated to a Black.	Willett 1913
	1946	6 birds present. "Two of the dark birds were mated, while each of the other two pairs included a dark and a pied bird."	Kenyon 1949:194
	1969	Population composition: *bachmani*—15, hybrid—1. Two pairs determined: *bachmani* × *bachmani* (2 × 7), and *bachmani* male × female hybrid (7 × 29).	Jehl, pers. observ.
Punta Falsa	1963	*bachmani* × *bachmani*—3 pairs, *bachmani* × *frazari*—1 pair, hybrid × hybrid—2 pairs.	S. Suffel, pers. comm.
San Benito Islands	1930	Male *bachmani* × female hybrid (4 × 23).	Museum specimens
	1969	Population composition: *frazari*—17, *bachmani*—3, hybrid—1 (incomplete census). Pairs determined: *frazari* × *frazari*—3, *bachmani* × *bachmani*—1?, *bachmani* male × hybrid female (5 × 25).	This study
	1970	Population composition: *frazari*—20, *bachmani*—1, hybrid—2 (incomplete census). Pairs determined: *frazari* × *frazari*—9; apparently unmated: *frazari*—2, *bachmani*—1, hybrid—2.	This study
	1971	Population composition: *frazari*—21, *bachmani*—5, hybrid—3 (full census). Pairs determined: *frazari* × *frazari*—8, *frazari* × hybrid—1, *bachmani* × hybrid—1, *frazari* × *bachmani*—4. Unmated: hybrid—1.	This study
Scammon's Lagoon	1926	"There was only one case I observed of a black bird paired with another black; all other blacks had white-bellied mates."	Bancroft 1927:51
	1946	Population estimated at 47 pairs: 23 pairs on Shell Island included *frazari* × *frazari*—21, *frazari* × *bachmani*—2.	Kenyon 1949
	1970	Shell Island. Pairs determined: *frazari* × *frazari*—1, *bachmani* × *bachmani*—1. Rocky Island: *frazari* × *frazari*—2, *frazari* × hybrid (32 × 29), hybrid male × *frazari* female (21 × 37).	This study
	1971	Rocky Island. Pairs determined: *frazari* × *frazari*—3, *frazari* × *bachmani*—1.	This study
Cedros Island	1913	Mixed pair collected, male "almost typical *frazari*," female "almost typical *bachmani*."	Willett 1913:22
Natividad Island	1969	Six birds seen, including 2 *frazari* × *frazari* pairs.	Jehl, pers. observ.

[1] Interpretations of phenotypes are those of the original author, except when character scores are given, the latter having been determined from analysis of museum specimens.

496

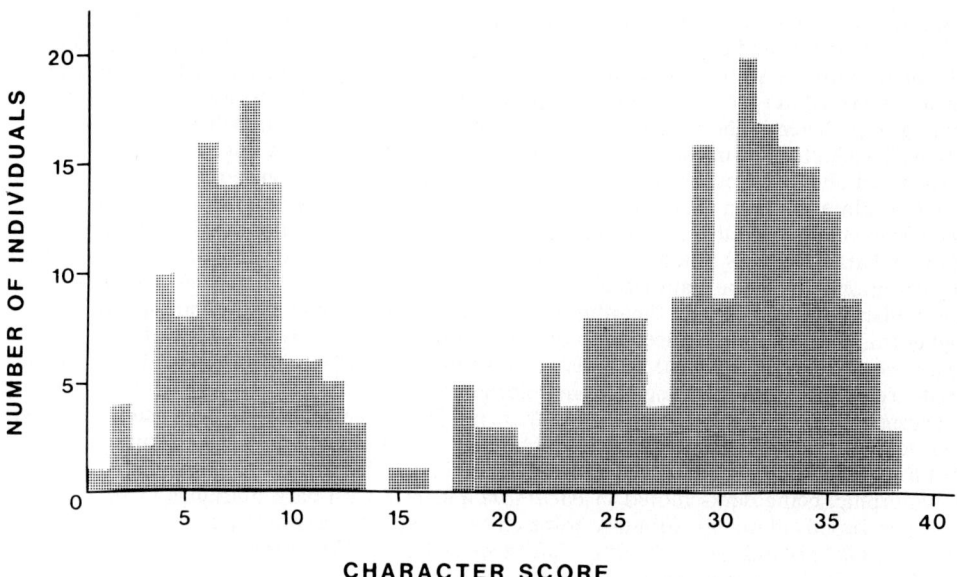

FIG. 7.   Character scores of oystercatchers from the west coast of Baja California (Localities 6–15), based on museum specimens and observational data. Data as in Figure 5.

or, more precisely, evidence against a preference for assortative mating. Observations from Los Coronados and Scammon's Lagoon illustrate some of these points.

Los Coronados.—This population is comprised of five to eight pairs of *bachmani,* and seems limited by the availability of nest sites. In early 1968 a single *frazari* appeared and remained there unmated, at least until March 1970, when I saw a pair of *bachmani* driving it away from North Island. In November 1970 it began to be seen regularly with a *bachmani* on Middle Island, and by April 1971 they seemed to be on territory. There was no evidence of nesting. The "pair" was seen at Middle Island into November 1971, when an apparent pair of *frazari* also appeared for a brief period. Evidently the mixed pair did not persist because in April 1972 the *frazari* was alone, and a pair of *bachmani* nested in its former territory. In May–June 1973 a single *frazari* (identity unknown) was again observed being driven off North Island. Two *frazari* (a pair?) appeared on Middle Island in March 1974 but could not be found two months later. In summary, *frazari* wander to Los Coronados regularly and may reside there for long periods apparently without nesting.

Scammon's Lagoon.—Kenyon (1949) found four pairs of *frazari* on Rocky Island in 1946. In 1970 it held two pairs of *frazari,* two *frazari* × hybrid pairs (scores 37 × 21, 32 × 29), two unmated *frazari,* and one unmated *bachmani.* The latter showed no sign of introgression and was continually chased by pied birds; one of the hybrids (score 21) was collected. In 1971 four pairs were still present on the periphery of the island. Territorial boundaries were as in 1970, but the territory from which the hybrid had been removed was now occupied by a *frazari* × *bachmani* pair (Fig. 6), the latter almost certainly being the unmated bird of 1970.

The data on pair composition in the literature, as in my observations, are biased in favor of mixed pairs (Table 3). In the absence of complementary information confirming the composition of pairs, and without knowledge of the age, sex, breeding status, and phenotypic frequency in the several island populations, it is not valid to compare pairing frequency expected from random mating with that actually observed. (Apparently-unmated birds, for example, may be pre-breeders rather than adults unable to obtain mates; or, a bird of breeding age may be unable to secure a territory; or, no potential mates may be available). Nevertheless, despite selective reporting, like pairs outnumber mixed pairs.

The compositions of six mated pairs for which sexes are known are *bachmani* male × hybrid female—four, *bachmani* male × *frazari* female—one, *frazari* male × *bachmani* female—one.

Indirect evidence on mate preference is provided by character scores in the zone of overlap.

If matings occurred at random, the result would be a hybrid swarm, and the distribution of character scores would approximate a normal distribution. The opposite is the case (Fig. 7). The distribution is strongly bimodal, clustering near "pure" scores for each of the parental forms; intermediate scores are poorly represented. I interpret this to mean that assortative mating is prevalent. The alternative interpretation—that immigration from parental populations is so high as to mask evidence of massive hybridization—is unsupported.

As noted above, juveniles were excluded from this analysis because they differ slightly in size and coloration from the adults; in addition, the number of specimens was relatively small. Data from young birds are critical to a full understanding of interactions in the zone of overlap. If their character scores differed from those of the adults, it would indicate that mixed pairs are commoner than the data on adults indicate, and, therefore, that pre-mating isolating mechanisms are weak. If, on the other hand, hybrid young are at a disadvantage (because of higher loss to predators, reduced viability, low fertility, etc.) a low frequency of intermediate scores would not be relevant to inferences about mate selection, but would indicate the existence of strong post-mating isolating mechanisms.

I suspect that the latter is more likely. There is a general correlation between substrate color and adult phenotypes, although it is obscured on such islands as San Gerónimo and San Benitos, where sandy beaches are intermingled with dark lava rocks. Oystercatcher chicks are polymorphic, being dark-colored in *bachmani* and pale with dark markings in *frazari* (Jehl 1968). In broods of mixed parentage some chicks will not be cryptically colored (see Bancroft 1927:52, Fig. 27) and the probability that those will be quickly discovered by Western Gulls (*Larus occidentalis*) seems high.

DISCUSSION

HISTORICAL FACTORS

In analyzing variation in the zone of overlap it is essential to acknowledge that oystercatchers were once collected so intensively that some populations were annihilated, or nearly so. Sampling could hardly have been very biased. Consequently, museum collections probably represent the prevailing conditions accurately.

Early in the century, or whenever they were first sampled, most local populations showed little variation (Fig. 5). At Los Coronados and Todos Santos, populations were comprised entirely of "pure" *bachmani* (except for wandering *frazari*). At San Martín, the variation was bimodal, *bachmani* predominating; parental types (both forms) comprised 56 percent of the population. *Bachmani* also predominated on San Gerónimo in the 1890's, with 80 percent of the birds being "pure," as well as on the rocky mainland coast south to Punta Falsa and Turtle Bay. On the other hand, at the San Benitos Islands and Cedros Island, and from there south, pied forms were in the majority. The percentage of parental types at those localities decreased southward, from 77 percent at the San Benitos, 67 percent at Cedros, and 60 percent at Scammon's Lagoon, to 57 percent at Natividad. Only at San Roque/Asunción (38%) and in southwestern Baja California (37.5%) were hybrids in the majority.

Apparently these conditions were greatly disrupted by collecting. Local populations were depleted, opportunities for colonization arose, and the islands were recolonized from both north and south by "founder" populations that differed from those that preceded them. This reshuffling changed the composition of the hybrid zone and evidently resulted in increased hybridization, as shown in the samples taken in the 1920's and 1930's. I know of no ecological changes at that time that could have selected for so dramatic a shift in phenotypes.

Today, at every locality for which adequate samples exist, the percentage of parental forms is greater than in the 1920–1930's, and at Los Coronados, San Gerónimo, San Benitos, Scammon's Lagoon, San Roque/Asunción, and probably Natividad, they comprise more than 50 percent of the population, which means that hybrids have been selected against. Of greater interest is the observation that, at each locality except San Roque/Asunción, character scores in the current population approximate those of the "original" population.

ZONES OF SECONDARY CONTACT

As a point of departure for analyzing variation in the zone of overlap, it was necessary to study the degree of phenotypic variation in *frazari*. I considered populations in the northern Gulf of California and the adjacent mainland of México as "typical" because variation was low and because those areas are distant from possible areas of hybridization. If one considers the population between southwestern Baja California and southern México as a whole, however, there is much variation; mean character scores vary from 28.3 to 38.6 (over-all range 20–42), suggestive of hybridization and introgression with *bachmani* to the north and *palliatus* to the south. The color variation is expressed in the degree to which dark coloration invades the white areas and, most conspicuously, in the occurrence of a ragged border to the chest band. Murphy (1925:11) argued that "the mottling of the breast, which is so strongly typical of [*frazari*], appears to be carried by a genetic factor deeply rooted in the species as a whole. . . ," and he noted that *frazari*-like phenotypes occur in other parts of the world as well. Indeed, some examples of *H. unicolor* (New Zealand), *H. p. durnfordi* (Argentina, see Jehl et al. 1973: 62, Fig. 3) and *H. p. galapagensis* (Galápagos Is.) may be distinguishable from *frazari* only on the basis of locality.

What does not seem to have been recognized, however, is that all of those forms are highly variable and that the ranges of all but *galapagensis* are near or between the ranges of a typical pied and a black form. Thus, I suggest that *frazari, durnfordi,* and *unicolor* are not classical subspecies but originated as hybrid swarms in zones of secondary contact. If so, Murphy's "deeply rooted genetic factor" is nothing more profound than hybrid phenotypes. Ostensibly *galapagensis* would seem to be an exception, because it is the only oystercatcher on the Galápagos and does not occur in a contact zone. However, it is so similar to *frazari* (they were once lumped) that we must entertain the possibility that they shared a common ancestor and that the Galápagos residents were derived from a hybrid population from Middle America.

CONCLUSION

It appears that the distribution of oystercatcher phenotypes on the Pacific coast of Baja California existed in a stable condition at the turn of the century and that in most localities the pied and black forms acted as separate species. At San Roque/Asunción and on the mainland coast south to Cabo San Lucas, however, the small populations may have been interbreeding freely. The situation was disrupted by intensive collecting, but local populations have since become reestablished at or near their original composition. From this I conclude that selection promoting the presence or absence of a particular phenotype at a specific location is strong. The resumption of the original condition, and the demonstration of a high frequency of assortative mating in current (and, by inference, in original) populations, as well as of a low frequency of hybrids in both original and current populations show that *bachmani* and *palliatus* are independent evolutionary units and must be considered as specifically distinct.

ACKNOWLEDGMENTS

Specimens used in this research were made available through: Academy of Natural Sciences, Philadelphia; American Museum of Natural History; Australian Museum, Sydney; California Academy of Sciences; Carnegie Museum of Natural History; Dominion Museum, Wellington, New Zealand; Field Museum of Natural History; Los Angeles County Museum of Natural History; Museum of Vertebrate Zoology, University of California, Berkeley; Oregon State University; San Diego Natural History Museum; United States National Museum of Natural History; University of California, Los Angeles; Western Foundation of Vertebrate Zoology. I am indebted to the curators of these collections for their generosity.

Many people over the years have contributed to this research, by assisting in the field work, providing transportation to the islands, or granting permission to carry out the studies. To all of these, too numerous to name, my thanks.

R. L. Pitman, D. W. Anderson, W. T. Everett, S. M. Speich, R. M. Gilmore, T. R. Wahl, G. S. Suffel, and G. McCaskie kindly shared with me their knowledge of local oystercatcher populations and provided unpublished information. Special thanks are due S. I. Bond, A. Bowles, A. Jackson and D. Kent for help in compiling and analyzing the data. D. G. Ainley, P. A. Buckley, E. S. Morton, and J. Rising provided helpful comments on the manuscript.

My debt to Eugene Eisenmann, who did so much to stimulate my early interests in evolution and systematics, is beyond expression.

LITERATURE CITED

AMERICAN ORNITHOLOGISTS' UNION. 1983. Check-list of North American Birds. 6th ed. American Ornithologists' Union, Washington, D.C.

ANTHONY, A. W. 1900. A night on land. Condor 2:28–29.

BAKER, A. J. 1973. Genetics of plumage variability in the Variable Oystercatcher (*Haematopus unicolor*). Notornis 20:330–345.

BAKER, A. J. 1975. Morphological variation, hybridization and systematics of New Zealand oyster-catchers (Charadriiformes: Haematopodidae). J. Zool. (Lond) 175:357–390.

BANCROFT, C. 1927. Breeding birds of Scammon's Lagoon, Lower California. Condor 29:29–57.

BRATTSTROM, B. H., AND T. R. HOWELL. 1956. The birds of the Revilla Gigedo Islands, Mexico. Condor 58:107–120.

BREWSTER, W. 1888. Descriptions of supposed new birds from Lower California, Sonora, and Chihuahua, Mexico and the Bahamas. Auk 5:82–95.

CLARKE, B. 1966. The evolution of morph-rates clines. Am. Nat. 100:389–402.

DURHAM, J. W., AND E. C. ALLISON. 1960. The geologic history of Baja California and its marine faunas. Syst. Zool. 9:47–91.

FRIEDMANN, H., L. GRISCOM, AND R. T. MOORE. 1950. Distributional check-list of the birds of Mexico. Part I. Pac. Coast Avif. No. 29.

GIFFORD, E. W. 1913. Expedition of the California Academy of Sciences to the Galapagos Islands, 1905–1906. VIII. The birds of the Galapagos Islands, with observations on the birds of Cocos and Clipperton Islands. Proc. Calif. Acad. Sci., ser. 2, 2:1–132.

GRANT, P. R. 1964. Birds of the Tres Marietas Islands, Nayarit, Mexico. Auk 81:514–519.

GRISCOM, L. 1933. Notes on the Havemeyer collection of Central American Birds. Auk 50:297–308.

HARRIS, M. P. 1967. The biology of oystercatchers Haematopus ostralegus on Skokholm Island, S. Wales. Ibis 109:180–193.

HATLER, D. F., R. W. CAMPBELL, AND A. DORST. 1978. Birds of the Pacific Rim National Park. Occas. Pap. B. C. Prov. Mus. No. 20.

HEPPLESTON, P. B. 1973. The distribution and taxonomy of oystercatchers. Notornis 20:102–112.

HOWELL, A. B. 1912. Notes from Todos Santos Islands. Condor 14:187–191.

HUBBS, C. L. 1960. The marine vertebrates of the outer coast. Syst. Zool 9:134–147.

HUEY, L. M. 1927a. Northernmost breeding station of the Heermann Gull on the Pacific Ocean, and other notes from San Roque Island, Lower California. Condor 29:205–206.

HUEY, L. M. 1927b. The bird life of San Ignacio and Pond Lagoons on the western coast of Lower California. Condor 29:239–243.

JEHL, J. R., JR. 1968. Relationships in the Charadrii (shorebirds): A taxonomic study based on color patterns of the downy young. San Diego Soc. Nat. Hist. Mem. 3.

JEHL, J. R., JR., M. A. E. RUMBOLL, AND J. P. WINTER. 1973. Winter bird populations of Golfo San Jose, Argentina. Bull. Br. Ornithol. Club. 93:56–63.

KAEDING, H. B. 1897. In Mexican waters. Success of the Anthony-Kaeding-McGregor expedition. Nidologist, 4 May, p. 109.

KENYON, K. W. 1949. Observations on behavior and populations of oystercatchers in Lower California. Condor 51:193–199.

LAMB, C. A. 1927. The birds of Natividad Island, Lower California. Condor 29:67–70.

MAYR, E., AND L. L. SHORT. 1970. Species taxa of North American Birds. Publ. Nuttall Ornith. Club No. 9.

MURPHY, R. C. 1925. Notes on certain species and races of oyster-catchers. Am. Mus. Novit. No. 194.

PETERS, J. L. 1934. Check-list of Birds of the World, Vol. 2. Harvard University Press, Cambridge, Massachusetts.

RIDGELY, R. S. 1976. A Guide to the Birds of Panama. Princeton University Press, Princeton, New Jersey.

RISING, J. D. 1970. Morphological variation and evolution in some North American orioles. Syst. Zool. 19:315–351.

RUSSELL, S. L., AND D. W. LAMM. 1978. Notes on the distribution of birds in Sonora, Mexico. Wilson Bull. 90:123–131.

SHORT, L. L. 1969. Taxonomic aspects of avian hybridization. Auk 86:84–105.

SIBLEY, C. G., AND L. L. SHORT, JR. 1964. Hybridization in the orioles of the Great Plains. Condor 66:130–150.

SLUD, P. 1964. The birds of Costa Rica. Bull. Am. Mus. Nat. Hist. Vol. 128.

SOWLS, A. L., S. A. HATCH, AND C. J. LENSINK. 1978. Catalog of Alaskan Seabird Colonies. U.S. Dept. Interior, Fish and Wildlife Service, Biological Services Program. FWS/OBS-78/78.

SOWLS, A. L., A. R. DEGANGE, J. W. NELSON, AND G. S. LESTER. 1980. Catalog of California Seabird Colonies. U.S. Dept. of Interior, Fish and Wildlife Service, Biological Services Program. FWS/OBS-37/80.

STAGER, K. E. 1957. The Avifauna of the Tres Marias Islands, Mexico. Auk 74:413–432.

STRESEMANN, E. 1927. Die schwarzen Austernfischer (Haematopus). Ornithol. Monatsber. 35:71–77.

THAYER, J. E., AND O. BANGS. 1907. Birds collected by W. W. Brown, Jr., on Cerros, San Benito and Natividad Islands in the spring of 1906, with notes on the biota of the islands. Condor 9:77–81.

VAN DENBURGH, J. 1924. The birds of the Todos Santos Islands. Condor 26:67–71.

VERMEIJ, G. J. 1978. Biogeography and Adaptation. Harvard University Press, Cambridge, Massachusetts.

WILLETT, G. 1913. Bird notes from the coast of northern Lower California. Condor 15:19–24.

WRIGHT, H. W. 1913. The birds of San Martin Island. Condor 15:207–210.

APPENDIX I

SPECIMENS EXAMINED

ALASKA: Prince William Sound—6, Aleutian Is.—7, Alaskan Peninsula—4, SE Alaska—18, Bering Sea—2, Wolf Rocks—2, unknown—1.

BRITISH COLUMBIA: Vancouver Is. area—20, Queen Charlotte Is.—6, Fort Simpson—2, unknown—5.

WASHINGTON: Jefferson Co.—5, Clallum Co.—4, Granville Is.—1, San Juan—1, unknown—4.

OREGON: Tillamook Co.—7, Curry Co.—5.

NORTHERN CALIFORNIA: Del Norte Co.—2, Humboldt Co.—1, Marin Co.—6, Monterey Co.—19, San Luis Obispo Co.—2.

SOUTHERN CALIFORNIA: Anacapa Is.—6, Santa Cruz Is.—3, San Miguel Is.—8, Santa Barbara Is.—1, San Diego—2, unknown—1.

NORTHERN BAJA CALIFORNIA (West Coast): Los Coronados—19, Todos Santos Is.—2, San Martín Is.—17, San Gerónimo Is.—31, Cedros Is.—11, San Benito Is.—26, Natividad Is.—31, San Roque Is.—38, Asunción Is.—7, Scammon's Lagoon—25, Turtle Bay—3, "Hole in the Wall"—1.

SOUTHERN BAJA CALIFORNIA (West Coast): Magdalena Bay-San Ignacio Lagoon—8.

GULF OF CALIFORNIA (South): San Jose Is.—14, San Francisco Is.—3, Espiritu Santo Is.—6, La Paz—3, Cerralvo Is.—4, San Juan—1.

GULF OF CALIFORNIA (Middle): Concepción Bay—3, Coronados Is.—8, Carmen Is.—4, Ildefonso Is.—1.

GULF OF CALIFORNIA (North): San Esteban Is.—3, Raza Is.—4, San Lorenzo Is.—1, Angel de la Guarda—4, Mejia Is.—1, Pond Is.—2, Encantada Is.—6, Montague Is.—1, Tres Marías Is., Isabella Is.—14.

MÉXICO (Northern Mainland): Sonora—8, Sinaloa—2.

MÉXICO (Southern Mainland): Colima—1, Guerrero—8, Oaxaca—3, Chiapas—2, Cayacall (Guerrero?)—3.

PANAMÁ: 3.

EAST AND GULF COASTS: South Carolina—4, Texas—1, Louisiana—1.

APPENDIX II

MEASUREMENTS AND CHARACTER SCORES OF WEST COAST OYSTERCATCHERS

Locality	Sex	Bill length		Wing		Tarsus		Bill ratio[1]		Character score[2]	
		N	Mean ± s.d. (range)	N	Mean ± s.d. (range)	N	Mean ± s.d. (range)	N	Mean ± s.d. (range)	N	Mean ± s.d. (range)
1. Alaska	F	14	72.8 ± 4.8 (65.2–82.5)	14	246.4 ± 8.0 (231–258)	14	51.6 ± 1.9 (48.5–56.0)	13	0.17 ± 0.01 (0.16–0.19)	40	1.5 ± 1.4 (0–5)
	M	24	68.3 ± 4.6 (56.4–78.0)	24	246.1 ± 5.2 (235–256)	25	51.5 ± 2.0 (48.2–55.0)	23	0.18 ± 0.02 (0.15–0.22)		
2. British Columbia	F	8	75.9 ± 2.5 (71.4–79.0)	9	253.2 ± 6.3 (244–265)	9	53.2 ± 3.1 (48.5–59.2)	9	0.17 ± 0.01 (0.16–0.18)	32	1.7 ± 1.6 (0–7)
	M	18	67.4 ± 3.7 (61.3–75.8)	18	242.6 ± 7.1 (228–257)	18	51.4 ± 1.8 (49.1–55.0)	18	0.18 ± 0.01 (0.16–0.20)		
3. Washington/Oregon	F	12	75.4 ± 2.8 (71.0–81.2)	12	248.2 ± 7.8 (230–260)	12	53.2 ± 1.5 (51.2–55.5)	12	0.17 ± 0.01 (0.16–0.18)	26	2.5 ± 2.7 (0–9)
	M	14	70.6 ± 3.5 (66.0–79.0)	11	244.8 ± 6.8 (234–256)	14	53.0 ± 1.5 (50.9–55.9)	14	0.18 ± 0.01 (0.16–0.19)		
4. Northern California	F	18	73.1 ± 1.5 (70.7–76.0)	18	248.1 ± 6.3 (237–260)	17	54.0 ± 1.4 (52.0–56.7)	16	0.17 ± 0.01 (0.16–0.18)	29	4.9 ± 2.7 (0–11)
	M	11	67.4 ± 3.6 (61.0–73.5)	9	246.1 ± 4.3 (241–265)	11	52.3 ± 1.5 (50.6–55.5)	11	0.18 ± 0.01 (0.17–0.20)		
5. Southern California	F	8	71.6 ± 3.0 (67.5–75.4)	7	249.0 ± 5.1 (242–258)	8	53.3 ± 1.2 (50.9–55.2)	8	0.17 ± 0.01 (0.16–0.18)	18	4.3 ± 3.6 (0–12)
	M	7	67.0 ± 4.9 (62.5–76.5)	6	246.3 ± 7.3 (238–256)	7	51.8 ± 1.7 (50.6–55.0)	7	0.18 ± 0.01 (0.16–0.20)		
6. Northern Baja (Los Coronados, Todos Santos Islands)	F	11	76.0 ± 3.9 (68.2–80.7)	10	254.5 ± 4.7 (246–263)	11	55.0 ± 2.4 (51.5–58.6)	11	0.17 ± 0.01 (0.16–0.18)	31	8.2 ± 7.2 (0–37)
	M	8	69.9 ± 6.3 (64.0–79.4)	8	250.6 ± 4.7 (243–259)	7	55.2 ± 3.5 (49.0–59.6)	8	0.18 ± 0.01 (0.16–0.20)		
7. San Martín Island	F	7	77.0 ± 2.7 (71.8–80.0)	8	252.3 ± 5.2 (244–261)	7	54.7 ± 1.6 (52.9–57.0)	7	0.17 ± 0.01 (0.16–0.17)	16	16.8 ± 10.2 (6–34)
	M	8	73.9 ± 3.3 (67.0–78.2)	8	254.9 ± 4.7 (248–262)	8	55.9 ± 2.6 (52.0–60.8)	8	0.17 ± 0.01 (0.15–0.18)		
8. San Gerónimo Island	F	15	77.2 ± 3.3 (72.0–82.1)	15	255.3 ± 7.0 (242–268)	15	54.5 ± 3.0 (50.2–61.0)	15	0.17 ± 0.01 (0.15–0.18)	37	12.9 ± 9.5 (2–33)
	M	13	70.1 ± 3.0 (64.0–76.1)	14	250.1 ± 7.1 (241–262)	14	53.3 ± 2.3 (49.8–56.9)	13	0.18 ± 0.01 (0.16–0.20)		

APPENDIX II
Continued

Locality	Sex	Bill length		Wing		Tarsus		Bill ratio[1]		Character score[2]	
		N	Mean ± s.d. (range)	N	Mean ± s.d. (range)	N	Mean ± s.d. (range)	N	Mean ± s.d. (range)	N	Mean ± s.d. (range)
9. San Benito Islands	F	13	78.4 ± 5.0 (70.2–88.2)	13	258.6 ± 7.2 (247–269)	13	56.5 ± 2.3 (53.1–59.6)	13	0.17 ± 0.01 (0.15–0.18)	54	25.9 ± 1.7 (2–38)
	M	9	70.8 ± 2.0 (68.2–74.5)	8	252.4 ± 5.4 (244–260)	8	55.6 ± 2.1 (51.8–54.7)	7	0.17 ± 0.01 (0.16–0.19)		
10. Cedros Island	F	7	77.8 ± 5.1 (68.5–86.0)	7	255.0 ± 6.1 (247–261)	7	56.2 ± 2.6 (50.8–58.9)	6	0.17 ± 0.01 (0.16–0.18)	11	21.6 ± 11.1 (7–35)
	M	4	71.2 (63.7–76.5)	3	242.7 (235–249)	4	54.9 (52.0–57.0)	4	0.19 (0.17–0.20)		
11. Scammon's Lagoon	F	11	79.8 ± 3.3 (75.5–86.0)	11	255.0 ± 8.7 (245–276)	11	55.8 ± 2.6 (49.7–60.0)	11	0.16 ± 0.01 (0.15–0.17)	41	27.8 ± 8.2 (2–37)
	M	14	73.5 ± 4.5 (63.5–81.0)	14	254.4 ± 6.6 (244–264)	14	54.3 ± 2.1 (50.9–58.1)	14	0.17 ± 0.01 (0.15–0.20)		
12. Natividad Island	F	12	80.7 ± 6.0 (72.5–90.1)	12	258.7 ± 7.3 (244–269)	12	56.1 ± 1.9 (52.0–59.0)	10	0.16 ± 0.01 (0.15–0.18)	37	24.1 ± 11.3 (4–37)
	M	17	72.6 ± 4.5 (64.5–80.5)	18	249.8 ± 6.3 (240–270)	18	53.4 ± 2.5 (48.5–57.5)	17	0.18 ± 0.01 (0.16–0.20)		
13. Turtle Bay	F	2	73.8 (72.3–75)	2	260.0 (250–270)	2	55.8 (55.1–56.5)	2	0.16 (0.16–0.17)	3	8.0
	M	1	67.0	1	252	1	53.1	1	0.19		
14. San Roque/Asunción Islands	F	19	80.6 ± 2.3 (74.5–84.0)	21	259.3 ± 5.3 (250–270)	21	56.2 ± 2.6 (51.0–61.0)	19	0.16 ± 0.01 (0.15–0.18)	55	24.8 ± 8.7 (4–36)
	M	26	72.7 ± 3.9 (65.7–81.5)	27	251.2 ± 6.0 (240–265)	28	53.5 ± 2.0 (50.5–59.4)	26	0.17 ± 0.01 (0.16–0.19)		
15. SW Baja California	F	2	86.2 (80.7–91.7)	2	257.0 (249–265)	2	58.1 (55.0–61.2)	2	0.15 (0.14–0.16)	8	28.3 ± 4.9 (20–36)
	M	6	78.3 ± 3.3 (74.8–82.7)	6	253.7 ± 6.2 (248–265)	6	55.8 ± 1.6 (53.5–57.8)	5	0.17 ± 0.01 (0.16–0.18)		
16. Southern Gulf of California	F	12	83.8 ± 5.8 (69.5–91.5)	12	261.8 ± 9.2 (248–275)	12	58.8 ± 2.6 (53.4–61.6)	12	0.16 ± 0.01 (0.15–0.19)	31	34.1 ± 3.0 (27–41)
	M	18	79.8 ± 4.0 (72.0–86.8)	19	251.7 ± 5.5 (239–260)	19	57.2 ± 2.1 (52.5–60.2)	18	0.17 ± 0.01 (0.15–0.18)		

APPENDIX II
Continued

Locality	Sex	Bill length		Wing		Tarsus		Bill ratio[1]		Character score[2]	
		N	Mean ± s.d. (range)	N	Mean ± s.d. (range)	N	Mean ± s.d. (range)	N	Mean ± s.d. (range)	N	Mean ± s.d. (range)
17. Middle Gulf	F	9	86.6 ± 2.4 (83.6–89.6)	9	260.1 ± 7.2 (250–270)	9	59.1 ± 1.6 (57.0–62.5)	9	0.15 ± 0.01 (0.13–0.16)	16	33.6 ± 2.3 (29–37)
	M	7	77.4 ± 3.0 (74.2–82.4)	7	250.1 ± 4.4 (245–258)	7	57.5 ± 1.6 (54.5–59.2)	7	0.17 ± 0.01 (0.16–0.18)		
18. Northern Gulf	F	8	84.9 ± 5.1 (78.0–95.2)	8	263.2 ± 9.0 (253–276)	8	58.4 ± 1.9 (56.5–62.0)	7	0.16 ± 0.01 (0.15–0.17)	22	33.6 ± 2.1 (30–38)
	M	14	79.1 ± 2.7 (75.4–83.0)	14	255.6 ± 7.1 (242–270)	14	57.7 ± 1.4 (55.4–60.5)	13	0.17 ± 0.01 (0.16–0.19)		
19. Sonora/Sinoloa	F	4	87.1 (82.0–89.2)	4	264.3 (253–283)	4	59.0 (55.8–62.5)	4	0.15 (0.15–0.16)	10	33.5 ± 1.7 (31–35)
	M	6	82.4 ± 3.8 (77.5–86.2)	6	250.2 ± 8.7 (237–262)	6	58.8 ± 0.7 (58.1–60.0)	6	0.16 ± 0.01 (0.16–0.17)		
20. Tres Marias Islands	F	4	84.6 (79.9–89.1)	4	262.0 (259–267)	4	61.6 (59.4–62.5)	4	0.17 (0.16–0.18)	14	37.3 ± 2.5 (33–40)
	M	9	75.6 ± 4.3 (70.2–82.5)	9	249.8 ± 5.5 (242–257)	9	54.5 ± 2.1 (59.3–60.5)	9	0.18 ± 0.01 (0.16–0.18)		
21. Southern México	F	13	86.6 ± 3.8 (83.0–95.0)	10	256.0 ± 7.3 (245–268)	13	61.0 ± 2.2 (57.5–66.3)	13	0.15 ± 0.01 (0.13–0.16)	21	38.6 ± 2.2 (34–42)
	M	6	72.3 ± 2.4 (75.0–82.0)	6	247 ± 5.0 (240–253)	6	57.2 ± 2.1 (55.0–59.6)	6	0.17 ± 0.01 (0.16–0.18)		
22. Panamá	F	1	75.1	1	250.0	1	58.5	1	0.18	3	40.0 ± 1.7 (38–41)
	M	2	76.4 (75.5–77.2)	2	254.5 (246–263)	2	59.2 (59.0–59.5)	1	0.18		
23. East and Gulf coasts	F	4	89.6 (84.0–95.2)	4	258.8 (253–264)	4	60.9 (60.0–61.8)	4	0.16 (0.15–0.17)	6	41.7 ± 0.5 (41–42)
	M	2	83.9 (82.4–85.5)	2	246.5 (246–247)	2	59.0 (58.2–59.9)	2	0.16 (0.16–0.17)		

[1] Sexes combined.
[2] Depth/bill length.

EVOLUTION OF CLUTCH SIZE IN TROPICAL SPECIES OF BIRDS

BERTRAM G. MURRAY, JR.

Department of Biological Sciences, Rutgers University, New Brunswick, New Jersey 08903 USA

ABSTRACT. The small size of clutches of tropical species of birds when compared with clutch sizes of related species at higher latitudes has been known for a long time. Several contending theories and their variations have been proposed. The most prominent suggest that clutch size is determined by the amount of energy available to the parents for reproduction. Others propose that reproduction is adjusted to the mortality rate, small clutches occurring in populations whose members are favored with long life expectancies. The theory supported in this paper was first proposed by R. E. Moreau. It states that individuals may produce more offspring in a season or in a lifetime by reducing the number of chicks in a brood relative to the maximum number of young that could be reared under prevailing conditions. The parents of larger broods have shorter life expectancies and, thus, produce fewer broods and, in the long run, fewer young.

The development of these and several other minor hypotheses is reviewed, and the theories are evaluated with respect to the data and arguments offered in their support. My version of Moreau's theory is most fully discussed and is applied to single-brooded species with clutch sizes that vary geographically (e.g., phasianids), single-brooded species without geographic variation (e.g., procellariiforms), multibrooded species with geographic variation (e.g., many passerine species), and multibrooded species without geographic variation (e.g., doves and hummingbirds). The patterns of variation in these groups are consistent with Moreau's theory, although available information is scanty, especially with regard to the life expectancies of birds and to the number of broods produced in a breeding season.

The theory seems to offer explanations for several lingering problems: (a) Ricklefs' observation that clutch size variation is surprisingly similar among species having widely different foraging behaviors and occupying habitats varying as much as 10-fold in productivity, (b) the similarity in clutch sizes in species with uniparental care and related species with biparental care, (c) the greater clutch sizes at higher latitudes of populations of nocturnal species, and (d) the evolution of populations whose clutch sizes are contrary to the general trend.

The theoretical argument suggests that patterns of clutch size variation cannot be understood without considering differences in survivorship schedules, ages of first and last reproduction, and the number of broods reared by parents during the breeding season. The need for research on these life history parameters is urged.

RESUMEN. Por largo tiempo se ha sabido que las nidadas ("clutch") de las especies de aves tropicales son más pequeña que las nidadas de las especies relacionadas de latitudes más elevadas. Se han propuesto varias teorías opuestas y sus variaciones. La teoría más destacada sugiere que el tamaño de la nidida está determinado por la cantidad de energía de la cual podrán disponer los padres para reprodicir. Otros proponen que la reproducción se ajusta a la tasa de mortalidad, por lo que se dan "clutches" pequeños en poblaciones cuyos individuos tienen una expectación de vida larga. La teoría sostenida en este trabajo fue primero propuesta por R. E. Moreau. Según esta teoría los individuos podrán producir más descendientes en una temporada de reproducción o durante su vida, si reducen el número de polluelos en la nidada en relación con el número máximo de polluelos que podrán ser criados bajo las condiciones reinantes. Los padres de nidadas más grandes tendrán expectativas de vida más corta y por ende producirán menos nidadas y a largo plazo menos polluelos.

Se revee el desarrollo de estas y otras varias hipótesis menores y se evaluan las teorías en relación a los datos y argumentos ofrecidos para su demostración. Mi versión de la teoría de Moreau es discutida más detalladamente y es aplicada a una especie que produce una sola nidada estacional y en la que el tomaño del "clutch" varía de acuerdo con su ubicación geográfica (por ej. phasianids), una especie que produce una sola nidada por temporada y no presenta variación geográficas (por ej. procellariiforms), una especie que produce mutliples nidadas por temporada con variaciones geográficas (por ej. muchas especies de passeriformes) y una especie con nidadas múltiples durante la estación sin variación geográfica (por ej. palomas y colibríes). Los patrones de variación de estos grupos son consistentes con la teoría de Moreau, si bién la información disponible es escasa, especialmente en lo que se refiere a espectativas de duración de vida y el número de nidadas producidas durante una temporada reproductiva.

La teoría parece ofrecer explacaciones para varios problemas que existen desde hace bastante tiempo: (a) la observación de Ricklefs en la que la variación en el tamaño de los "clutches" es, sorprendentemente, constante entre especies que tienen patrones de forraje ampliamente diferentes y ocupan habitats que varían hasta 10 veces en productividad; (b) la similaridad en el tamaño de los "clutches" en aquellas especies en que sólo uno de los padres cría los polluelos y especies relacionadas en las que ambos padres se encargan de la cría de los polluelos; (c) el mayor tamaño en los "clutches" en especies de hábitos nocturnos en poblaciones de latitudes más elevadas y (d) evolución de poblaciones en las que los tamaños de los "clutches" son contrarios a las normas generales.

Los argumentos teóricos sugieren que los modelos de variación en el tamaño de los "clutches" no pueden ser entendidos sin considerar las diferencias en niveles de sobrevivencia, edad en la primera y última reproducción y el número de nidadas criadas por los padres durante la temporada reproductiva. Se recomienda que se investiguen esos parámetros en la historia de vida.

Tropical birds are noted for having smaller clutch sizes than birds of higher latitudes (Lack 1947, 1948, 1954, 1968; Cody 1966, 1971; Klomp 1970; von Haartman 1971). This is especially marked in multibrooded passerines, which often have clutches of two or three eggs in the tropics, four or five eggs at middle latitudes, and five to seven eggs in the Arctic (Ricklefs 1969a). For example, clutch sizes of populations in South Africa average about half an egg larger than those of populations of the same species in equatorial Africa, and British populations of closely related species have even larger clutches (Moreau 1944). In a comparison of species within subfamilies and families of passerines, clutch sizes in tropical Africa were about half those in middle Europe (Lack 1968). A similar difference has been found in several New World families (Cody 1966, 1971). Some notable exceptions include all the procellariiforms, which have one-egg clutches, and many hummingbirds, nightjars, and pigeons, which have two-egg clutches throughout their extensive ranges.

The search for an explanation for the latitudinal gradient in clutch sizes and especially for the small clutches in the tropics has produced a variety of hypotheses. These are reviewed in this paper.

THEORIES ON THE EVOLUTION OF CLUTCH SIZE

EARLY HYPOTHESES

The small clutch sizes of tropical species of birds have been known since the early 1800s (Prince Wied 1830 and Schomburgk 1848, cited in Wagner 1957). Explanations came early in this century. Hesse (1923, cited in Klomp 1970) and Hesse et al. (1937) suggested that latitudinal variation in clutch size was probably a consequence of differences in day length during the breeding season, the longer days at higher latitudes allowing parents to find more food and produce more eggs and rear more young than birds in the tropics. This idea was later developed in detail by Lack (1947, 1948).

Other hypotheses proposed that reproduction is in some way adjusted to mortality rates. Large clutch sizes evolve in populations at higher latitudes in response to high mortality, owing to severe climatic conditions (Stresemann 1927–1934, cited in Moreau 1944; Rensch 1938). Moreau (1944) proposed instead that parents with small broods may be favored over parents with larger broods because their reduced effort allows them to rear more young during a season and during a longer lifetime. This is essentially the view I supported earlier (Murray 1979) and develop in further detail with respect to tropical species in this paper.

Thus, the major contending interpretations of variations in clutch size were proposed prior to the surge of research following the publication of Lack's (1947, 1948) papers.

LACK'S HYPOTHESIS

After thoroughly reviewing clutch size variation in birds, Lack (1947, 1948) further developed the theory of Hesse. Lack's work has influenced further theoretical developments and

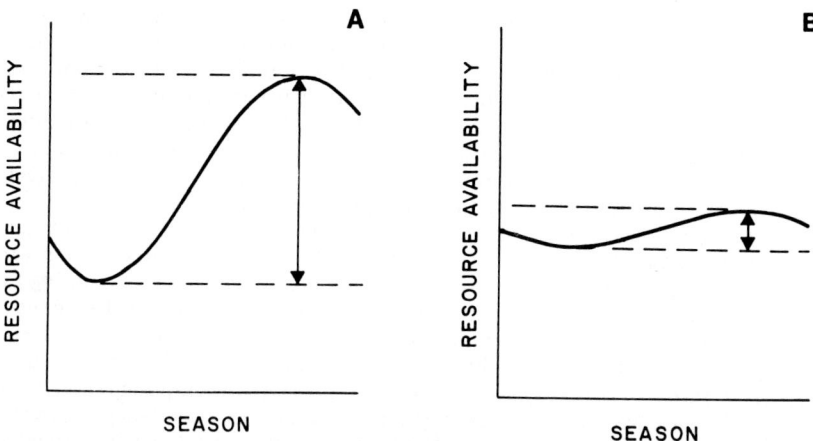

FIG. 1. Effect of seasonal variation in food supply on reproduction. A. Temperate (high latitude) regions. B. Tropical (low latitude) regions. The solid curve shows the seasonal variation in resource availability. The lower dashed line represents the resource level limiting the population's size during the non-breeding season. The upper dashed line represents the resource level during the breeding season. The difference between the lower and upper dashed lines, indicated by the arrow, represents the amount of resources available for reproduction. (Redrawn from Ricklefs 1980). In tropical regions variation in food supply is smaller than in temperate regions. Thus, there are fewer resources available for reproduction and smaller clutches in the tropics.

research until the present time. His hypothesis appears in many forms, but perhaps most representative is the following: "The number of eggs in the clutch has been evolved to correspond with that from which, on the average, the most young are raised. In nidicolous species, the limit is set by the amount of food which the parents can bring for their young, and in nidifugous species by the average amount of food available for the laying female, modified by the size of the egg" (Lack 1968:5). With respect to the latitudinal gradient in clutch size, Lack argued that birds at higher latitudes have more hours per day during the breeding season to gather food for their young than have birds in the tropics. Thus, birds at higher latitudes can feed larger broods, and, therefore, they have been able to evolve larger clutches.

Ashmole (1961), in a frequently cited (e.g., Lack and Moreau 1965; Lack 1968; Ricklefs 1980) but unpublished discussion (not seen by me) of Lack's hypothesis, observed that day length at mid-latitudes was 1.5 times the tropical day length, but the clutches in many groups of birds at mid-latitudes were about twice those of tropical species. He reasoned that some factor other than day length must be involved and suggested that temperate and tropical populations were limited by food in different ways. In temperate regions, the seasonal variation in food supply is great, and the lows in late winter result in high mortality. The population at the beginning of the breeding season would then be small relative to the available food supply, and the birds could rear large broods (Fig. 1A). In tropical regions, where seasonal variation in food supply is presumably slight, seasonal variation in mortality would be small, and population size would always be near that allowed by the available food supply during the breeding season (Fig. 1B). Thus, tropical species would be limited to rearing small broods and, therefore, to having small clutches. Skutch (1949, 1967, 1976) proposed a similar hypothesis for explaining the large clutch sizes at higher latitudes, but he did not accept it as an explanation for small clutch sizes of tropical species (see below).

Lack and Moreau (1965) attempted to test Ashmole's hypothesis by comparing the clutch sizes of equatorial African species inhabiting forest and savanna habitats. The species of the more seasonal savanna habitats had a clutch size about half an egg larger than the clutches of forest species, lending support to Ashmole's hypothesis.

Ricklefs (1980) examined Ashmole's hypothesis in a different way. Because measures of food availability are not only scarce but difficult to obtain, he compared clutch size to seasonal variations in the actual evapotranspiration (AE), which he assumed was proportional to primary and to overall productivity. In a comparison of data from thirteen localities from the tropics to the Arctic, Ricklefs found that clutch size varied directly with the ratio of summer AE to winter AE, inversely with winter AE, and not at all with summer AE. These results are consistent with expectations from Ashmole's hypothesis. Furthermore, in comparing data from five localities from Costa Rica to Alaska, Ricklefs found that clutch size varied directly with the ratio of summer AE to population density, inversely with population density, and not at all with breeding season AE. As a result of his analysis, Ricklefs (1980: 47) concluded that "variation in the seasonality of resources experienced by a population is the single most important cause of geographical patterns in clutch size."

Owen (1977) was concerned with the increasing clutch sizes of owls (von Haartman 1971) and litter sizes of nocturnal mammals (Lord 1960) with increasing latitude. This variation seems contrary to Lack's hypothesis because during the breeding season the hunting time for nocturnal animals decreases with increasing latitude. Presumably, nocturnal animals should be rearing smaller broods at higher latitudes than at lower latitudes, but they do not. Owen proposed, as an amendment to Lack's hypothesis, that differences between the tropics and higher latitudes in resource diversity play an important role in determining food availability. He argued that because species diversity is greater and species abundance is lower in the tropics than at higher latitudes, both diurnal and nocturnal tropical species should have a more difficult time in developing a search image for appropriate food items than temperate species. As a result, tropical species should have a more difficult time in obtaining food for their young, and, thus, they should have a smaller clutch size. According to this hypothesis, then, owls at higher latitudes could obtain more food during their short nights than tropical owls could during their longer nights because the reduced diversity and greater abundance of their prey allowed them to find, capture and deliver more prey to their young.

Regardless of the differences in emphasis, these authors hypothesize that the small clutches of tropical species result from limits set by the parents' abilities to obtain food for their young.

SKUTCH'S HYPOTHESIS

Skutch (1949) immediately challenged Lack regarding the factors leading to small clutches in tropical species, although he did accept Lack's hypothesis as an adequate explanation for the larger clutch sizes of temperate species (see also Skutch 1967, 1976). Skutch contended that parent birds in the tropics were not working as hard as they could to rear their broods. He provided anecdotal evidence of parents substantially increasing their rate of delivering food after a prolonged absence from the nest or when artificially provided with an enlarged brood, a point subsequently supported by Wagner (1957) and Morel (1967). Most compelling was his comparison of the clutch sizes and parental care of several species of tropical flycatchers in which the clutch sizes of species with uniparental care and species with biparental care did not differ. A comparison of temperate European species yielded similar results (von Haartman 1955).

Seemingly, if Lack's hypothesis were correct, species with biparental care should be able to rear larger broods than those with uniparental care. Although Skutch (1949, 1967, 1976) repeatedly pointed this out, I know of only three attempts to account for this discrepancy. First, Lack (1949) responded (a) that food of species in which only the female feeds the young may be easier to find, (b) that the male may be mated to another female, and (c) that the brood size may not be limited by food in those instances. In other words, anything is possible, making the testing of theory a genuine challenge. I consider Lack's alternative explanations an unsatisfactory solution to the problem.

Second, Royama (1966, 1969), working with Great Tits (*Parus major*), suggested that because of increasing heat loss per nestling with decreasing clutch size, a female would not gain energetically even if she reduced her clutch by half. Thus, one should not expect lone females to have smaller clutches than females being assisted by their mates. But, Royama has the argument backwards. The question is not why the clutches of uniparental species are not smaller but why the clutches of biparental species are not larger. If a lone female can rear a brood of two or three, why cannot a pair rear a brood of four or six?

Third, in a rather speculative discussion considering the adaptation of predators and prey to each other, Ricklefs (1970) proposed that if one considered the family as the exploiting unit, then uniparental families provide less intensive predation pressure on the prey population than biparental families, thus sufficiently relaxing prey adaptation to the predator and making it possible for a single parent to supply a brood equal in size to those of biparental species.

Among the many assumptions inherent in this discussion is the requirement that "all similar species employ the same unbalanced division of labor" (Ricklefs 1970:600), but, as pointed out by Skutch (1949, 1967, 1976) and von Haartman (1955), similar species do employ different patterns of division of labor between the sexes and yet have similar clutch sizes.

Having rejected Lack's hypothesis, Skutch (1949) proposed that the small clutch size of tropical species was an adaptation that reduced predation on the nestlings. With one or two young in the nest, parents could visit the nest less frequently than parents of larger broods, presumably lowering the likelihood that the parents' activities would draw attention of predators to the nest's location. Unfortunately, predators of young in nests probably do not find nests by observing the parents (Lack 1966), and in the few studies so far undertaken eggs were taken by predators about as frequently as nestlings (Lack 1966; Ricklefs 1969b).

Skutch's hypothesis, however, challenged Lack at a more fundamental level. Lack proposed not only that the clutch size represented the number of eggs from which the greatest number of young could be reared but that the excess production was removed by subsequent density-dependent mortality (Lack 1954). Skutch (1949, 1967, 1976), as well as others (Stresemann 1927–1934; Rensch 1938; Wynne-Edwards 1962), took the alternative view that clutch size was in one way or another affected by mortality rates. Although Skutch (1949) considered that this relationship was achieved by natural selection, the hypotheses generally referred to as "adjusted reproduction" seem more consistent with the theory of group selection (Wynne-Edwards 1962). Indeed, Skutch (1967) later explicitly expressed his theory of adjusted reproduction in terms of group selection, which for many evolutionary biologists is an inadequate theory of evolution (e.g., Lack 1966; Wiens 1966; Williams 1966a). Judging from his later writing, I think Skutch (1976) now agrees. The theories of adjusted reproduction, however, seem to be associated with the theory of group selection in the minds of evolutionists, and, thus, they have not been given the attention they deserve.

CODY'S HYPOTHESIS

Although Lack (1954) certainly recognized the influence of "modifying factors" such as predation rates, growth rates of nestlings, egg size, and differences between the sexes in parental care on the evolution of clutch size, he never attempted to integrate these factors into a general model of clutch size. The first to do so was Cody (1966), who suggested that individual reproductive success might be increased if birds reduced their clutch sizes and used the energy gained to avoid predators and to increase their competitive abilities. It was worth formalizing the notion that individuals of different populations might distribute their energy differently toward survival and reproduction depending upon local environmental situations and that this distribution could affect clutch sizes, but Cody predicted from his theory that clutch sizes should be smaller in the more climatically stable habitats, such as occur in the tropics, at lower elevations, and on oceanic islands. This prediction seems contrary to fact because it does not explain, for example, why the clutch sizes of birds of the climatically variable Central American highlands are no larger than those of species from the more stable lowlands (Skutch 1976). It should be noted that Cody's hypothesis is only a variation of Lack's, for it implicitly suggests that clutch size is limited by the amount of energy the parents can afford to devote to reproduction.

MISCELLANEOUS HYPOTHESES

B. K. Snow (1970), Lill (1974), and D. W. Snow (1978) suggested that clutch size in some tropical species could be limited by selection favoring small, inconspicuous nests. This hypothesis cannot serve as a general explanation for small clutch size in the tropics because other species with small clutches nest on the ground or on substantial platforms. Furthermore, and a more serious criticism of this idea, selection cannot favor nests that cannot accommodate a clutch or brood of at least replacement size. If it did, selection for small size of nests would result in extinction. No doubt, selection favors small, inconspicuous nests in many cases, but the minimum size is determined by the clutch or brood size, not the other way around.

Other hypotheses include the notions (a) that the small clutch size of nidicolous, tropical species is determined by the physical condition of females at time of laying (Wagner 1957), (b) that small clutch sizes allow overlap of breeding activities with molt, resulting in an extended breeding season (Foster 1974), and (c) that small clutches in the tropics result from the difficulty prey species have in adapting to the diversity of their predators (in this case,

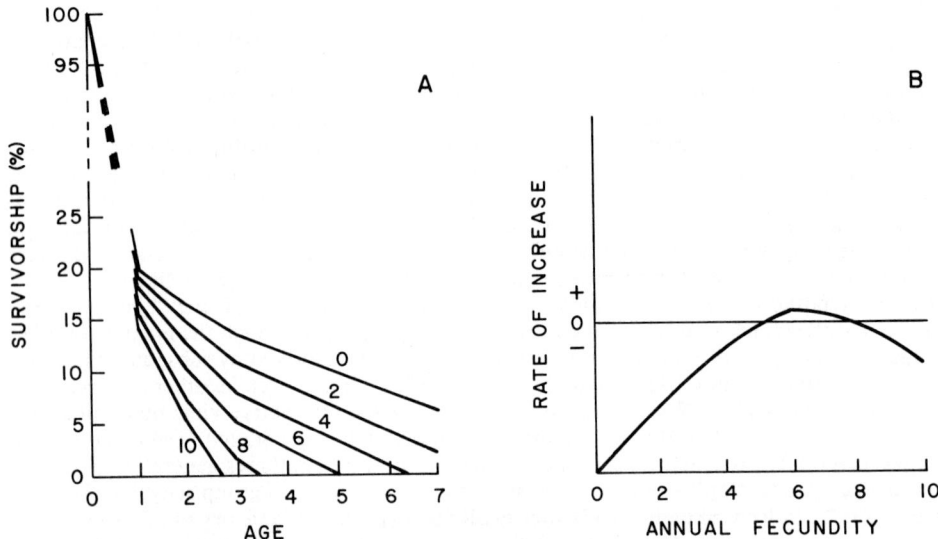

Fig. 2. A graphical model of the evolution of annual fecundity. A. Partial survivorship curves of birds with different gene-determined differences in annual fecundity: 0, 2, 4, 6, 8, 10 (odd number of eggs are not included in order to reduce clutter). Curves for annual fecundity of 0 and 2 (and 1) actually extend beyond age 7. These curves show the costs in survivorship of differences in annual fecundity. B. Annual rate of increase (ρ) of birds with different annual fecundities. The shape of the curve may be somewhat different but always increases to a maximum at the smallest annual fecundity that results in replacement. See text for fuller explanation.

birds) and, thus, prey suffer a high mortality and low population size, causing the predators to have smaller clutch sizes (Ricklefs 1970). None of these ideas has been applied to clutch size variation in any systematic way.

MURRAY'S THEORY

In a general model of the evolution of clutch size (Murray 1979), I proposed that females maximized their lifetime reproduction by minimizing reproductive effort during each breeding cycle. The least effort is achieved by producing clutches that will just assure replacement. In this way, females increase their probability of producing additional clutches, either in the same season or in later breeding seasons. Moreau (1944) first proposed this hypothesis, but he was ignored by later investigators. My development of this theory was put in hypothetico-deductive form (the assumptions are presented somewhat differently here).

ASSUMPTIONS

Assumption 1.—In a genetically diverse population, natural selection increases the frequencies of the traits of those individuals whose probabilities of survival and reproduction result in the greatest rate of increase, ρ, under prevailing conditions, as calculated from the equation,

$$1 = \Sigma \lambda_x \mu_x e^{-\rho x},$$

where λ_x is the probability that individuals with a particular trait have of surviving from birth to age class x, μ_x is the average number of offspring born to or sired by these individuals when members of age class x, e is the base of the natural logarithms, and x is the age class (Murray and Gårding 1984).

This assumption allows us to calculate the rate of change of alternative traits (gene-determined differences in phenotype) within a population. Thus, if the individuals with two-egg clutches and others with three-egg clutches have different probabilities of survival, we can calculate which phenotype has the greater rate of increase. Individuals possessing the phenotype with the greatest ρ are favored by natural selection.

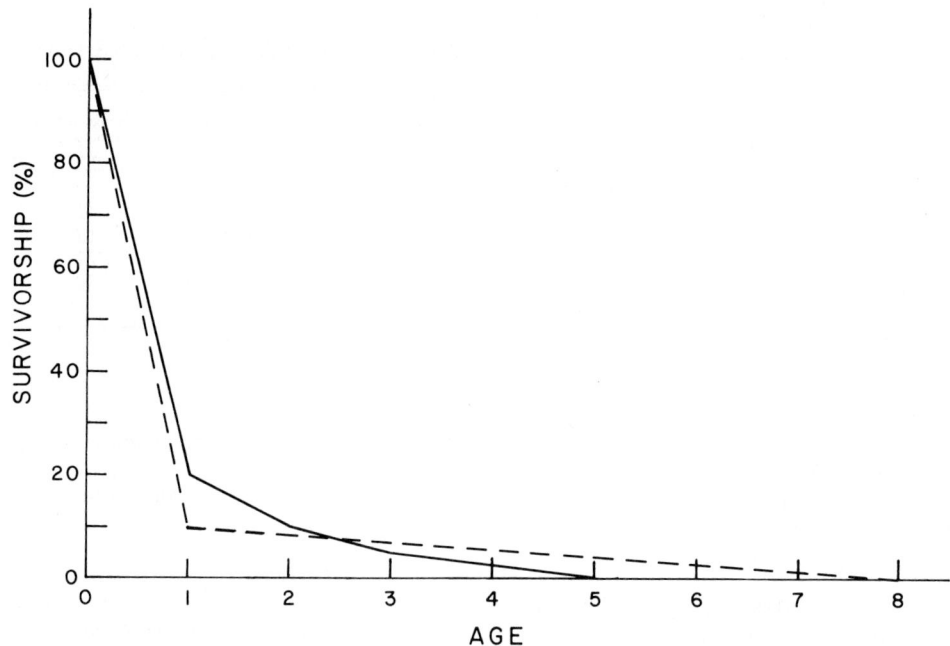

FIG. 3. Survivorship of two hypothetical populations. The population representing high latitude species (solid line) has higher nesting success and poorer adult survivorship than the population representing low latitude species (dashed line). Nevertheless, if the age of first breeding is 1 in both populations, the average annual fecundity is the same (cf. Tables 1, 2). If both populations have the same number of broods, they should have the same clutch size; if they differ in the number of broods they rear, then the population with the greater number of broods should have the smaller average clutch size.

Assumption 2.—As clutch size increases, an individual's probability of surviving to breed again decreases, or the probability of its offspring surviving to adulthood decreases, or both.

This assumption simply states that there is a cost to reproduction that is best measured in terms of survivorship. A consequence of this assumption is a series of survivorship curves associated with differences in annual fecundity (Fig. 2A).

Assumption 3.—A female's annual fecundity represents the fewest eggs required to replace herself because additional eggs are not worth their cost in reduced survivorship (Fig. 2B). Her annual fecundity is a function of her survivorship, ages of first and last reproduction, and the sex ratio.

When the primary sex ratio is 1:1, a female's annual replacement fecundity (μ_r), as given by Murray (1979), is

$$\mu_r = 2/\sum_{\alpha}^{\omega} \lambda_x,$$

where α is the age of first reproduction, and ω is the age of last reproduction.

This assumption implies a trade-off between survivorship and reproduction (Lack 1954; Cody 1966; Williams 1966a, 1966b; Charnov and Krebs 1974; Ricklefs 1977, 1983; Pianka 1978; Haukioja and Hakala 1979). Presumably, parents may be more successful if they expend some energy in defending their young and themselves from predators, contending with competitors, maintaining their own body condition, and so on, instead of producing additional eggs, and the young from smaller broods may have a better chance of surviving, because they obtain a greater share of parental care.

Assumption 4.—A female divides her annual fecundity among two or more clutches if she is able to rear two or more broods per season.

TABLE 1

LIFE HISTORY PARAMETERS OF A HYPOTHETICAL POPULATION OF A HIGH LATITUDE SPECIES

Age	Age class (x)	Probability of surviving to age class x (λ_x)	Age class at first reproduction (α)	Average annual replacement fecundity (μ_r)	Average annual fecundity (μ_a)
0	1	1.000	1	1.455	2
1	2	0.200	2	5.333	6
2	3	0.100	3	11.429	12
3	4	0.050	4	26.667	27
4	5	0.025	5	80.000	80
5	6	0.000	—	—	—

Assumption 5.—The annual variations in mortality and fecundity tend toward average values over longer periods of time.

From these assumptions we may make some deductions with regard to the evolution of clutch size in the tropics (others can be found in Murray 1979).

PREDICTIONS

Prediction 1.—Assuming that a favored trait eventually characterizes all but mutant individuals in a population or that alternative traits reach a state of balanced polymorphism, we should expect an inverse relationship between survivorship and fecundity. Species or populations with long life expectancies should have small annual fecundities, and those with short life expectancies should have large annual fecundities.

In evaluating the effects of life expectancy, we must consider survival at all ages. For example, it is generally believed that nesting success is lower in the tropics than in the temperate zone (Marchant 1960; Willis 1961, 1974; Snow and Snow 1964; Skutch 1966; Ricklefs 1969b; Fogden 1972), although Oniki (1979) has challenged this notion, at least with respect to birds in the Brazilian equatorial rain forest. It is also generally believed that adult survivorship is greater in the tropics (Willis 1961; Snow 1962; Cody 1971; von Haartman 1971; Fogden 1972; Ricklefs 1973a; Snow and Lill 1974). Thus, the survivorship schedules of tropical and temperate species are probably often different. A difference in survivorship schedules, however, need not be associated with a difference in annual fecundity (Fig. 3; Tables 1, 2).

Prediction 2.—Species or populations with similar annual fecundities should have similar average clutch sizes, unless they have breeding seasons of different lengths. Those with long breeding seasons relative to the time it takes to rear the young to independence should divide their annual fecundity into two or more clutches. Thus, because, in general, breeding seasons are longer in the tropics, we should expect that birds at low latitudes should have more clutches of smaller size than birds of the same or similar species at high latitudes.

Interpreting the significance of clutch size does not involve a simple, straightforward comparison of clutch size with some environmental factor only, but requires consideration of survivorship schedules, ages of first and last reproduction, and the number of broods that can be reared during the breeding season. Variation in the relationship between latitude and clutch size probably results from local conditions that affect survivorship, ages at first and last reproduction, and the number of broods that can be reared (see Murray 1979 for a discussion of local variations in the clutch size of the Great Tit).

CLUTCH SIZE IN TROPICAL BIRDS

Single-brooded species.—Some single-brooded species show geographical variation in clutch size (e.g., phasianids, rallids, larids), and others do not (e.g., procellariiforms). These groups must be considered separately.

The implication of the hypothesis is that $\sum_{\alpha}^{\omega} \lambda_x$ is greater for tropical single-brooded species with small clutches than for those temperate single-brooded species with large clutches. This seems to be consistent with evidence that adult mortality in birds in general is lower in tropical species (Willis 1961; Snow 1962; Fogden 1972; Snow and Lill 1974), although data on this

TABLE 2

LIFE HISTORY PARAMETERS OF A HYPOTHETICAL POPULATION OF A LOW LATITUDE SPECIES

Age	Age class (x)	Probability of surviving to age class x (λ_x)	Age class at first reproduction (α)	Average annual replacement fecundity (μ_r)	Average annual fecundity (μ_a)
0	1	1.000	1	1.444	2
1	2	0.100	2	5.195	6
2	3	0.085	3	7.018	8
3	4	0.070	4	10.000	10
4	5	0.055	5	15.385	16
5	6	0.040	6	26.667	27
6	7	0.025	7	57.143	58
7	8	0.010	8	200.000	200
8	9	0.000	—	—	—

point are woefully few; as we have seen (Fig. 3; Tables 1, 2), however, high mortality of eggs and chicks can counter the benefits of the longer life expectancy of survivors. From demographic considerations, then, the smaller clutch size in the tropics among single-brooded species means that the greater adult survivorship more than offsets their high nesting mortality, giving them a greater $\sum_\alpha^\omega \lambda_x$ than that for temperate species.

The pre-eminent examples of single-brooded species without geographic variation are members of the Procellariiformes. All species are characterized by 1-egg clutches, low mortality rates, and late ages of first breeding. The small clutch size has been attributed to the birds' difficulty in finding sufficient food for rearing two chicks or for forming a second egg (Ashmole 1963; Lack 1966; Perrins et al. 1973), which is a reiteration of Lack's hypothesis. A consideration, however, of the survivorship schedule of the Laysan Albatross (*Diomedea immutabilis*; Fisher 1975) indicates that a female could replace herself even if she produced a single egg per year starting at age 15 (Murray, unpubl. data). She would have to lay a clutch of two in every year of reproductive life if she began breeding between ages 16 and 21. In fact, female Laysan Albatrosses begin breeding on average at age 8.9 (Fisher 1975; Van Ryzin and Fisher 1976) and produce less than one egg per year. I interpreted this to mean that female Laysan Albatrosses could afford to wait for a monogamous mating relationship. I think the small clutch size is also a consequence of the albatross's long life expectancy. I think this is the case for other procellariiforms as well. Differences in survivorship between small (e.g., storm-petrels) and large (e.g., albatrosses) species are adjusted by varying their age of first breeding rather than by varying their clutch size.

Multibrooded species with geographic variation.—If the breeding season is long with respect to the time required for rearing young, then birds should divide their annual fecundity into two or more clutches, but if the breeding season is short, then parents will be forced to rear all their offspring for the year in a single brood. Because the length of the breeding season roughly varies with latitude, we should expect to find a relationship between latitude (more correctly, the length of the breeding season), clutch size, and the number of broods in multibrooded species with similar survivorship schedules.

Although clutch sizes are well known, few data are available on the number of clutches produced or broods reared in a season (Cody 1971; von Haartman 1971; Ricklefs 1973a). From the length of each nesting cycle and the length of the breeding seasons, Ricklefs (1973a) calculated that tropical birds might attempt as many as six or more clutches a season, whereas temperate zone species might try only three or four. Excluding replacement clutches, however, individuals probably attempt fewer than are theoretically possible (Skutch 1950, 1967).

Data suitable for testing this interpretation are almost nonexistent. Although the largest number of broods reared successfully is reported from the tropics (Ricklefs 1973a; Skutch 1976), the relationship between clutch size, number of broods, and length of the breeding season is not at all clear. Indeed, both Rensch (1938) and Moreau (1944) contended that the small clutch sizes in the tropics were not compensated for by more broods. The chief problem is a lack of data on the number of broods usually reared by females in a population, mainly because of the difficulty in determining this. We may, however, consider a few cases.

Skutch (1953, 1976) found that the Southern House-Wren (*Troglodytes musculus*) successfully reared as many as four broods at 900 m in Costa Rica during the breeding season from January to September, but only one at 2600 m in Guatemala during a shorter breeding season from April to May. Because the clutch size is usually four in both populations, the implication is that mortality must be lower in the highlands of Guatemala than at lower elevations in Costa Rica. This may be the case, at least for birds in general. For 28 highland species in Guatemala, 37 of 67 (55%) nests produced one or more fledglings, whereas nesting success in the lowlands of Costa Rica was rarely as high as 40 percent in three of four seasons (overall success, 85 of 208 nests of 37 species), presumably because predatory snakes were more common in the lowlands than in the highlands (Skutch 1966). However, 17 of 26 (65%) Southern House-Wren nests successfully produced at least one fledgling in Costa Rica (Skutch [1953], 33 of 45 [73%] in Skutch [1966]). Unfortunately, data on nest success for this species in the highlands of Guatemala are not available.

The closely related Northern House-Wren (*Troglodytes aedon*) usually rears only two broods from clutches averaging six eggs (range, 4–8) in Illinois (Kendeigh et al. 1956). In this population 5351 young fledged from 6773 eggs (79%) (Kendeigh 1942), although other studies with much smaller samples showed a range from 48.3 to 83.7 percent (Nice 1957). In the Southern House-Wren 51 of 90 eggs (57%) produced fledged young (Skutch 1953). This is in the expected direction. Again, unfortunately, no data seem available for differences in adult mortality in these species. Greater adult mortality in the Northern House-Wren is expected, and this relationship is consistent with the available data on mortality rates of temperate and tropical species.

In another comparison, Skutch (1976) pointed out that tropical finches are not known to start more than three clutches of two or three eggs, whereas Song Sparrows (*Melospiza melodia*) in Ohio (Nice 1937) and White-crowned Sparrows (*Zonotrichia leucophrys*) in California (Blanchard 1941) have been known to lay four clutches of three to five eggs. The relationship between clutch size, number of broods, and length of breeding season seems inconsistent with the proposed theory. It is important, however, to distinguish between the maximum number and average number of broods reared by members of a population and between the number of broods that can be reared and the number of clutches that can be produced. Females that have sufficient time to rear only two broods could produce as many or more replacement clutches as birds that have time to rear three or four broods. According to my hypothesis, given a particular annual fecundity, it is the potential number of broods that can be reared rather than the number of clutches that can be produced that determines the clutch size. (This relationship between the number of broods that can be reared and the number of clutches laid is more complex than indicated here, but that is a subject to be addressed at a later time.) In order to compare tropical and temperate populations, then, we need to know how many broods are, in fact, reared in both areas, or, more meaningfully, how frequently new broods are started after successful breeding. In fact, only a small proportion of both Song Sparrows and White-crowned Sparrows rears more than two broods.

Perhaps, the best data relating the length of the breeding season, number of broods, and clutch size are those on three species of thrushes, studied by Snow and Snow (1963) in Trinidad. The Cocoa Thrush (*Turdus fumigatus*) and White-necked Thrush (*T. albicollis*) are forest species with lengthy breeding seasons extending from October or November until the following August. The Bare-eyed Thrush (*T. nudigenis*), a bird of semi-open country, has a short breeding season of five months beginning in April. Both forest species produce three or four broods, whereas the Bare-eyed Thrush rarely produces more than two. The average clutch size of the White-necked Thrush was 2.08 and that of the Bare-eyed Thrush was 2.96, consistent with the expectations suggested by the theory. The average clutch size of the Cocoa Thrush, however, was 2.80, only slightly less than that of the Bare-eyed Thrush. Differences in nesting success do not account for the relatively large clutch of the Cocoa Thrush. It could be the result of low adult survivorship, but adequate data on survivorship of these species are not available.

Clearly, many data still need to be collected on the number of broods that can be reared by birds in different ecological conditions.

Multibrooded species without geographic variation.—Pigeons are multibrooded species whose clutches of one or two eggs do not vary with latitude. Lack (1947) contended that pigeons were exceptions to his hypothesis because parents feed their young with "milk" formed in their crops rather than food collected from their environment. He assumed that the volume of crop milk produced by parents was independent of day length and, therefore, latitude. Two unstated assumptions seem to be (a) that pigeons cannot feed more young during the longer days at higher latitudes or that tropical pigeons can feed their young at night as well as during the day and (b) that the amount of crop milk produced is independent of the amount of food the parents can collect for themselves during the day. Neither of these assumptions seems likely to me. Furthermore, crop milk is augmented by other foods after the first few days of nestling life (Murton 1964). That pigeons feed their young with crop milk seems an inadequate explanation for the lack of geographic variation in clutch size. Indeed, Lack does not seem to have repeated this idea in his later publications.

Pigeons, however, have lengthy breeding seasons, even in temperate regions. In Iowa, the average pair of Mourning Doves (*Zenaidura macroura*) made six nesting attempts and reared three broods between March and October (McClure 1942). Some Mourning Doves rear as many as five or six broods a year (cited in Nice 1957). In England, the average pair of Wood Pigeons (*Columba palumbus*) seems to rear three to four broods between May and November (Murton and Isaacson 1964). Also, in England the Rock Dove (*Columba livia*) lays eggs in every month of the year (Lees 1946; Murton and Clarke 1968). Furthermore, Wood Pigeons, Stock Doves (*Columba oenas*), and Rock Doves may be incubating eggs while still tending young from their previous brood (Murton and Isaacson 1964; Murton and Clarke 1968).

The lengthy breeding season, short nestling period, and almost (at least, sometimes) nonexistent interval between successive nests may well allow pigeons in the temperate zone to rear as many or more broods during the breeding season as many tropical species do. I find it hard to believe that the food given to young pigeons is so uniformly distributed throughout the year that parents could never rear more than two young or produce more than two eggs at *any* time during the year. I suggest that the lack of geographic variation in the clutch size of pigeons is a result of long breeding seasons in the temperate zone as well as in the tropics.

Hummingbirds have clutches of two eggs throughout the range of this large family from Alaska to Tierra del Fuego, and although hummingbirds may breed throughout the year in tropical lowlands, they certainly do not either at high latitudes or in tropical highlands (Skutch 1976). Unfortunately, little is known about the number of broods that can be reared by hummingbirds, probably because hummingbirds are generally small and difficult to mark individually and because females are not restricted to building their nests within the territories of their polygynous mates. As in pigeons, however, consecutive nesting attempts may overlap. Female Black-chinned Hummingbirds (*Archilochus alexandri*) frequently build one nest while attending another (Cogswell 1949). This has also been observed in the Ruby-throated Hummingbird (*Archilochus colubris*; Nickell 1948). Skutch (1976) has not observed any tropical hummingbird tending two nests simultaneously, but he did see a female White-eared Hummingbird (*Hylocharis leucotis*) feeding a well-grown juvenile between periods of sitting on eggs in a second nest.

If tropical hummingbirds produce more clutches than temperate hummingbirds, then the implication is that tropical hummingbirds suffer greater mortality from predators, for example, than some species of temperate hummingbirds suffer on their lengthy migrations. I think the demography of hummingbirds, when it becomes known, will prove to be most interesting.

DISCUSSION

A diversity of hypotheses has been proposed to account for the evolution of small clutch sizes in tropical birds. Most theories on the evolution of clutch size, however, can be grouped into two fundamentally different views. One group was first developed in detail by Lack (1947, 1948, 1954, 1968) and his followers. It is the notion that birds are reproducing up to the maximum allowed by the energy that parents can devote to laying eggs and rearing young. Excess reproduction is eliminated by subsequent density-dependent mortality.

A second group comprises the ideas often referred to as "adjusted reproduction," according to which the clutch size is adjusted to the mortality rate, species with long life expectancies having small clutches, and species with short life expectancies having large clutches. There is, however, little agreement on how this adjustment evolves (Stresemann 1927–1934; Rensch 1938; Moreau 1944; Skutch 1949, 1967, 1976; Wynne-Edwards 1962; Murray 1979).

The two groups of theories have quite different evolutionary consequences. The first implies that life expectancy is a consequence of density-dependent mortality, birds with large clutches suffering high mortality and having short life, whereas birds with small clutches have low mortality and long life. The second suggests explicitly that the clutch size is affected by life expectancy, such that an increase in life expectancy resulting from, for example, reduced predation should result in the evolution of a smaller clutch size, whereas a decrease in life expectancy should result in the evolution of a larger clutch size. In the first, then, life expectancy adjusts to changes in clutch size, and in the second, clutch size adjusts to changes in life expectancy.

Alternative hypotheses may be evaluated by how well each resolves problems raised by variations in the phenomenon they are intended to explain. Any hypothesis claiming to be a general explanation of clutch size variations must explicitly confront several issues.

(1) As pointed out by Ricklefs (1970), what is most striking about clutch size variation is its similarity among species having widely different foraging behaviors and diets and occupying habitats that differ as much as 10-fold in primary productivity. It seems incredible, at least to me, that food availability could be distributed in such a way that diverse species in a variety of habitats would be limited to similar clutch sizes. Should we not expect greater variation in clutch size among similar species in different habitats or diverse species in the same habitat than we find, if clutch size is determined by the amount of food that parents can bring to their young?

Instead, one can hypothesize that, because crude longevity is related to body size (Lindstedt and Calder 1976), species of similar size might be expected to have clutches of similar size, regardless of their differences in foraging ecology. This is generally so, considering variability in clutch size caused by differences in ages of first reproduction and the number of broods that can be reared during the breeding season. We should expect variations in the clutch sizes of birds of similar size to be smaller than those expected in birds exhibiting great diversity in foraging habits and habitats.

(2) There is the problem of the similar clutch size in birds with uniparental care and others with biparental care (Skutch 1949, 1967, 1976; von Haartman 1955). If clutch size were limited by the amount of food that can be delivered to the young by parents, then we should expect that species with biparental care should have larger clutches. The three subsidiary hypotheses attempting to account for the lack of differences seem weak (reviewed above).

Alternatively, if uniparental and biparental species of similar size have similar survivorship schedules, ages of first breeding, and number of broods per season, then clutches of similar size are expected. Why in some species males assist their mates and in other species they do not is probably a consequence of their probability of obtaining second mates (Murray 1984).

(3) The increasing litter or clutch sizes of nocturnal animals with increasing latitude (Lord 1960; von Haartman 1971) are inconsistent with the notions that clutch size is limited by the amount of food parents can obtain for their young as determined by day length (reviewed above). But, if the environments of diurnal and nocturnal animals vary in the same way with latitude, such that both groups have lower nesting success, greater adult survivorship, and longer breeding seasons in the tropics, then we should expect by the theory of adjusted reproduction that latitudinal variation in clutch size would be similar in both diurnal and nocturnal animals.

(4) Finally, there is the nature of the evidence used to support various ideas about the evolution of clutch size. The data are usually average differences in clutch size among species, subfamilies, or familes living in different habitats or in different geographical regions. In these comparisons, however, some species show no difference or even a reversal of the average trend. Lack's hypothesis and others, such as Cody's, provide no mechanism to account for exceptions to the general trends, but the clutch size of each population or species must be accounted for because evolution occurs at the population level rather than at the species, subfamily, or family level. The theory of adjusted reproduction does provide a mechanism within itself that allows for variations to general trends. According to this theory, clutch size variation is a function of differences in survivorship, ages of first breeding, and the length of the breeding season relative to the time required to rear a brood to independence, and these may differ between populations, not only with latitude and habitat but within habitats as well.

THE ROLE OF FOOD

In contrast to Lack's hypothesis and others, which tie clutch size to the amount of energy that parents have to devote to reproduction, the theory of adjusted reproduction implies that food supply plays no direct role in determining the *average* clutch size. This does not mean that food supply does not affect the actual clutch size of birds. For example, in species in which individuals compete for food, the food supply will affect survivorship, which in turn will affect clutch size, as outlined in the theory of adjusted reproduction. Note, however, that as food supply decreases, competition for food increases, life expectancies shorten, and the average clutch size must increase, just the reverse of what would be expected from Lack's hypothesis. I suspect that birds compete for food only infrequently (Murray 1979) and that, therefore, food supply does not act as an ultimate factor in the evolution of average clutch size.

Annual and seasonal variations in food supply do play a more direct role in determining variations in clutch size. As a proximate factor, when food is abundant, clutch size may be *greater than average,* and when sparse, clutch size can be *smaller than average,* but the food supply does not determine the *average* (Murray 1979).

LIFE IN THE TROPICS

The small clutch sizes, the long intervals between consecutive nesting attempts (Ricklefs 1973a), long incubation and nestling times (Ricklefs 1969a), and slow growth rates of nestlings (Ricklefs 1973b) of tropical zone species, compared with those of temperate zones, indicate to me that tropical birds have a more relaxed breeding schedule because of their longer life expectancies and longer breeding seasons. If the annual fecundity, which will result in replacement, is six eggs, then birds in the tropics with long breeding seasons can divide these six eggs among as many clutches as they have time to rear. After all, it can hardly be adaptive to put all one's eggs into one nest, especially in a region where nest predation is high. Thus, birds in the tropics can afford to work at a more leisurely pace in rearing their young than birds at higher latitudes where parents often must produce all their young in a single brood.

The fact that "twinning" experiments in many species (reviewed in Murray 1979) indicate that parents often do not rear any more young in experimental broods than found in their normal-sized broods does not mean that parents are rearing as many young as allowed by conditions because, as Cody (1971) and Hussell (1972) pointed out, foraging rates, abilities to find food items, and other behaviors associated with rearing young need not evolve to have a greater efficiency than necessary to provide for their normal brood.

The alternative notions, which are explicit or implicit in many discussions about the evolution of clutch size, that tropical birds are rearing as many young as circumstances allow and that the small clutch sizes and slow growth rates are consequences of limited food supplies, have additional implications. For example, one must count it only as convenient that species with small clutches in the tropics have long breeding seasons so they may produce enough eggs to keep from going extinct. Also, we must assume that longer life is an adaptation to limited food supplies available for rearing young.

All in all, I think the evidence and logic favor the notion that a species' rate of reproduction, as reflected by its clutch size, is adjusted by natural selection to the species' mortality rates, ages of first and last breeding, and length of the breeding season.

ACKNOWLEDGMENTS

I am grateful to the editors of this volume for inviting me to contribute a paper in memory of Eugene Eisenmann. He was always supportive, even if not always convinced, and I am going to miss him. I am also grateful to D. J. T. Hussell, J. R. Jehl, Jr., C. Leck, and W. M. Shields for commenting on the manuscript.

LITERATURE CITED

ASHMOLE, N. P. 1961. The biology of certain terns. Unpubl. Ph.D. dissert., Oxford University, Oxford, England. [Not seen; cited in Lack 1968 and others.]
ASHMOLE, N. P. 1963. The regulation of numbers of tropical oceanic birds. Ibis 103b:458–473.
BLANCHARD, B. D. 1941. The White-crowned Sparrows (*Zonotrichia leucophrys*) of the Pacific seaboard: environment and annual cycle. Univ. Calif. Publ. Zool. 46:1–178.
CHARNOV, E. L., AND J. R. KREBS. 1974. On clutch-size and fitness. Ibis 116:217–219.
CODY, M. L. 1966. A general theory of clutch size. Evolution 20:174–184.
CODY, M. L. 1971. Ecological aspects of reproduction. Pp. 461–512, *In* D. S. Farner and J. R. King (eds.), Avian Biology, Vol. 1. Academic Press, New York.
COGSWELL, H. L. 1949. Alternate care of two nests in the Black-chinned Hummingbird. Condor 51: 176–178.

FISHER, H. I. 1975. The relationship between deferred breeding and mortality in the Laysan Albatross. Auk 92:433–441.

FOGDEN, M. P. L. 1972. The seasonality and population dynamics of equatorial forest birds in Sarawak. Ibis 114:307–343.

FOSTER, M. S. 1974. A model to explain molt-breeding overlap and clutch size in some tropical birds. Evolution 28:182–190.

HAARTMAN, L. VON. 1955. Clutch size in polygamous species. Acta XI Int. Ornithol. Congr., pp. 450–453.

HAARTMAN, L. VON. 1971. Population dynamics. Pp. 391–459, In D. S. Farner and J. R. King (eds.), Avian Biology, Vol. 1. Academic Press, New York.

HAUKIOJA, E., AND T. HAKALA. 1979. On the relationship between avian clutch size and life span. Ornis Fenn. 56:45–55.

HESSE, R. 1923. Die Bedeutung der Tagesdauer für die Vögel. Sitzungsber. Nathist. Preuss. Rheinl., Bonn, pp. 13–17. [Not seen; cited in Klomp 1970.]

HESSE, R., W. C. ALLEE, AND K. P. SCHMIDT. 1937. Ecological Animal Geography. Wiley, New York.

HUSSELL, D. J. T. 1972. Factors affecting clutch size in arctic passerines. Ecol. Monogr. 42:317–364.

KENDEIGH, S. C. 1942. Analysis of losses in the nesting of birds. J. Wildl. Manage. 6:19–26.

KENDEIGH, S. C., T. C. KRAMER, AND F. HAMMERSTROM. 1956. Variations in egg characteristics of the House Wren. Auk 73:42–65.

KLOMP, H. 1970. A determination of clutch-size in birds. A review. Ardea 58:1–124.

LACK, D. 1947. The significance of clutch-size. Parts I and II. Ibis 89:302–352.

LACK, D. 1948. The significance of clutch-size. Part III. Ibis 90:25–45.

LACK, D. 1949. Comments on Mr Skutch's paper on clutch-size. Ibis 91:455–458.

LACK, D. 1954. The Natural Regulation of Animal Numbers. Oxford University Press, London.

LACK, D. 1966. Population Studies of Birds. Clarendon Press, Oxford, England.

LACK, D. 1968. Ecological Adaptations for Breeding in Birds. Methuen, London.

LACK, D., AND R. E. MOREAU. 1965. Clutch-size in tropical passerine birds of forest and savanna. Oiseau Rev. Fr. Ornithol. 35 (special number):76–89.

LEES, J. 1946. All the year breeding of the Rock-Dove. Br. Birds 39:136–141.

LILL, A. 1974. The evolution of clutch size and male "chauvinism" in the White-bearded Manakin. Living Bird 13:211–231.

LINDSTEDT, S. L., AND W. A. CALDER. 1976. Body size and longevity in birds. Condor 78:91–94.

LORD, R. D., JR. 1960. Litter size and latitude in North American mammals. Am. Midl. Nat. 64:488–499.

MARCHANT, S. 1960. The breeding of some S. W. Ecuadorian birds. Ibis 102:349–382, 584–599.

McCLURE, H. E. 1942. Mourning Dove production in southwestern Iowa. Auk 59:64–75.

MOREAU, R. E. 1944. Clutch-size: a comparative study, with special reference to African birds. Ibis 86:286–347.

MOREL, M.-Y. 1967. Les oiseaux tropicaux elevent-ils autant de jeunes qu'ils peuvent en nourrir? Le cas de Lagonosticta senegala. Terre Vie 1:77–82.

MURRAY, B. G., JR. 1979. Population Dynamics: Alternative Models. Academic Press, New York.

MURRAY, B. G., JR. 1984. A demographic theory on the evolution of mating systems as exemplified by birds. Evol. Biol. 18:71–140.

MURRAY, B. G., JR., AND L. GÅRDING. 1984. On the meaning of parameter x of Lotka's discrete equations. Oikos 42:323–326.

MURTON, R. K. 1964. Pigeon milk. Pp. 472–473, In A. Landsborough Thompson (ed.), A New Dictionary of Birds. McGraw-Hill, New York.

MURTON, R. K., AND S. P. CLARKE. 1968. Breeding biology of Rock Doves. Br. Birds 61:429–448.

MURTON, R. K., AND A. J. ISAACSON. 1964. Productivity and egg predation in the Woodpigeon. Ardea 52:30–47.

NICE, M. M. 1937. Studies in the life history of the Song Sparrow. 1. A population study of the Song Sparrow. Trans. Linn. Soc. N.Y. 4:1–247.

NICE, M. M. 1957. Nesting success in altricial birds. Auk 74:305–321.

NICKELL, W. P. 1948. Alternate care of two nests by a Ruby-throated Hummingbird. Wilson Bull. 60:242–243.

ONIKI, Y. 1979. Is nesting success of birds low in the tropics? Biotropica 11:60–69.

OWEN, D. F. 1977. Latitudinal gradients in clutch size: an extension of David Lack's theory. Pp. 171–179, In B. Stonehouse and C. Perrins (eds.), Evolutionary Ecology. University Park Press, Baltimore, Maryland.

PERRINS, C. M., M. P. HARRIS, AND C. K. BRITTON. 1973. Survival of Manx Shearwaters Puffinus puffinus. Ibis 115:535–548.

PIANKA, E. R. 1978. Evolutionary Ecology. 2nd ed. Harper and Row, New York.

RENSCH, B. 1938. Einwirkung des Klimas bei der Ausprägung von Vogelrassen, mit besonderer Berüksichtigung der Flügelform und der Eizahl. Proc. VIII Int. Ornithol. Congr., pp. 285–311.

RICKLEFS, R. E. 1969a. The nesting cycle of songbirds in tropical and temperate regions. Living Bird 8:165–175.

RICKLEFS, R. E. 1969b. An analysis of nesting mortality in birds. Smithson. Contrib. Zool. 9:1–48.

RICKLEFS, R. E. 1970. Clutch size in birds: outcome of opposing predator and prey adaptations. Science 168:599–600.

RICKLEFS, R. E. 1973a. Fecundity, mortality, and avian demography. Pp. 366–435, In D. S. Farner (ed.), Breeding Biology of Birds. National Academy Sciences, Washington, D.C.

RICKLEFS, R. E. 1973b. Patterns of growth in birds. II. Growth rate and mode of development. Ibis 115:177–201.

RICKLEFS, R. E. 1977. On the evolution of reproductive strategies in birds: reproductive effort. Am. Nat. 111:453–478.

RICKLEFS, R. E. 1980. Geographical variation in clutch size among passerine birds: Ashmole's hypothesis. Auk 97:38–49.

RICKLEFS, R. E. 1983. Comparative avian demography. Current Ornithol. 1:1–32.

ROYAMA, T. 1966. Factors governing feeding rate, food requirement and brood size of nestling Great Tits Parus major. Ibis 108:313–347.

ROYAMA, T. 1969. A model for the global variation of clutch size in birds. Oikos 20:562–567.

SCHOMBURGK, M. R. 1848. Reisen in British-Guiana in den Jahren 1840–1844. J. J. Weber, Leipzig, German Democratic Republic. [Not seen; cited in Wagner 1957.]

SKUTCH, A. F. 1949. Do tropical birds rear as many birds as they can nourish? Ibis 91:430–455.

SKUTCH, A. F. 1950. The nesting seasons of Central American birds in relation to climate and food supply. Ibis 92:185–222.

SKUTCH, A. F. 1953. Life history of the Southern House Wren. Condor 55:121–149.

SKUTCH, A. F. 1966. A breeding bird census and nesting success in Central America. Ibis 108:1–16.

SKUTCH, A. F. 1967. Adaptive limitation of the reproductive rate of birds. Ibis 109:579–599.

SKUTCH, A. F. 1976. Parent Birds and Their Young. University Texas Press, Austin, Texas.

SNOW, B. K. 1970. A field study of the Bearded Bellbird in Trinidad. Ibis 112:299–329.

SNOW, D. W. 1962. A field study of the Black and White Manakin, Manacus manacus, in Trinidad. Zoologica 47:65–104.

SNOW, D. W. 1978. The nest as a factor determining clutch-size in tropical birds. J. Ornithol. 119:227–230.

SNOW D. W., AND A. LILL. 1974. Longevity records for some neotropical land birds. Condor 76:262–267.

SNOW, D. W., AND B. K. SNOW. 1963. Breeding and the annual cycle in three Trinidad thrushes. Wilson Bull. 75:27–41.

SNOW, D. W., AND B. K. SNOW. 1964. Breeding seasons and annual cycles of Trinidad land birds. Zoologica 49:1–39.

STRESEMANN, E. 1927–1934. Aves. In W. Kückenthal and T. Krumbach (eds.), Handbuch der Zoologie. Vol. 7, Pt. 2. De Gruyter, Berlin. [Not seen; cited in Moreau 1944.]

VAN RYZIN, M. T., AND H. I. FISHER. 1976. The age of Laysan Albatrosses, Diomedea immutabilis, at first breeding. Condor 78:1–9.

WAGNER, H. O. 1957. Variation in clutch size at different latitudes. Auk 74:243–250.

WIED, MAXIMILIAN PRINZ ZU. 1830. Beiträge zur Naturgeschichte von Brasilien. Landes-Industrie-Comptoirs, Weimar, German Democratic Republic. [Not seen; cited in Wagner 1957.]

WIENS, J. A. 1966. On group selection and Wynne-Edwards' hypothesis. Am. Sci. 54:273–287.

WILLIAMS, G. C. 1966a. Adaptation and Natural Selection. Princeton University Press, Princeton, New Jersey.

WILLIAMS, G. C. 1966b. Natural selection, the costs of reproduction, and a refinement of Lack's principle. Am. Nat. 100:687–690.

WILLIS, E. O. 1961. A study of nesting ant-tanagers in British Honduras. Condor 63:479–503.

WILLIS, E. O. 1974. Populations and local extinctions of birds on Barro Colorado Island, Panama. Ecol. Monogr. 44:153–169.

WYNNE-EDWARDS, V. C. 1962. Animal Dispersion in Relation to Social Behaviour. Hafner, New York.

GOING TO EXTREMES: WHY DO SANDERLINGS MIGRATE TO THE NEOTROPICS?

J. P. Myers[1,2,3], J. L. Maron[3,4], and Michel Sallaberry[1,2,5]

[1]Academy of Natural Sciences, 19th and the Parkway, Philadelphia, Pennsylvania 19103 USA;
[2]Department of Biology, University of Pennsylvania, Philadelphia, Pennsylvania 19104 USA;
[3]Bodega Marine Laboratory, P.O. Box 247, Bodega Bay, California 94923 USA;
[4]Biology Department, University of North Dakota, Grand Forks, North Dakota 58202 USA; and
[5]Museo Nacional de Historia Natural, Casilla 787, Santiago, Chile

ABSTRACT. Sanderlings (*Calidris alba*) breed within a small latitudinal range in the arctic while spreading in winter virtually throughout temperate and tropical marine beaches of the world. This paper examines spatial variation in Sanderlings nonbreeding density across the New World, documents annual cycle differences between populations wintering in California and those wintering in Peru and Chile, and then explores demographic and ecological factors underlying Sanderlings migration to different wintering grounds.

Densities during the nonbreeding season are higher on the Pacific coast than on the Atlantic at all censused latitudes in the New World, and reach a peak in southwestern Peru and northwestern Chile adjacent to the Humboldt Current. Populations wintering in California spend a larger fraction of the year on the wintering site than do those wintering in Peru and Chile. Adults replace primaries during prebasic molt in both regions, as do first-winter Sanderlings in Peru and Chile. First-winter birds in California do not molt primaries.

Comparisons of weight and time-activity budgets near the northern and southern ends of the winter distributions along the Pacific coast of the western hemisphere indicate that resource conditions are more favorable for Sanderlings in the south.

RESUMEN. Los playeros (*Calidris alba*) se reproducen en un rango de distribución latitudinal muy circunscripto en el ártico, mientras que en invierno se dispersan hacia las playas templadas y tropicales de los mares del mundo. Este trabajo examina la variación espacial en la densidad de playeros no anidadores en el Nuevo Mundo, presenta información sobre diferencias en los ciclos anuales entre las poblaciones que invernan en California y aquellas que invernan en Perú y Chile y así explorar factores demográficos y ecológicos fundamentando las migraciones de playeros a diferentes áreas de invernación.

Se observa que para todas las latides censadas en el Nuevo Mundo las densidades durante la estación no reproductiva son mayores en la costa del Pacífico que en la del Atlántico, teniendo su pico máximo en el suroeste de Perú y el noroeste de Chile, adyacentes a la corriente de Humbolt. Las poblaciones que invernan en California pasan una mayor parte del año en el sitio de invernada que las poblaciones que invernan en Perú y Chile. Los adultos reemplazan las primarias durante la muda prebásica en ambas regiones tal como lo hacen los playeros jóvenes ("first-winter") que pasan su primera invernada en Perú y Chile. Los playeros que pasan la primera invernada en California no mudan las primarias.

Comparaciones del peso y patrones de actividad cerca de los límites norte y sur de las áreas de invernada a lo largo de la costa del Pacífico del hemisferio occidental indican que las condiciones de los recursos en el sur son mas favorables para los playeros.

Two salient features distinguish the distribution of Sanderlings (*Calidris alba*, Scolopacidae) in the New World. First, Sanderlings breed only in the high arctic. Second, during the nonbreeding season Sanderlings occur on temperate and tropical sandy beaches throughout the western hemisphere. Their broad winter range spans more than 100° latitude and is among the widest of all migrant birds' nonbreeding ranges.

In this paper we consider two questions posed by these distributions. How does Sanderling density vary across their nonbreeding distribution, and why do some Sanderlings migrate much farther than others?

We begin by describing spatial and temporal features of Sanderling distribution in the west, including original data from the United States, Mexico, Ecuador, Peru, and Chile. We then develop a demographic and ecological framework for dissecting hypotheses concerning the evolution of long-distance migration. Finally, we present data on ecological benefits gained by long-distance migrant Sanderlings in Chile compared to California.

This essay is neither a traditional review nor a research paper. Instead, it lies between the two, synthesizing current knowledge about the migration and annual cycle of one species that is now the subject of intense investigation. We emphasize that these studies are as yet incomplete, and offer this paper here to provoke parallel work with other species and to stimulate alternative approaches to the study of the evolution of migration.

The approach we take is reductionistic. Moreover, it concentrates on one species in order to avoid two pitfalls we believe have retarded progress in studies on the ecology and evolution of migration. The first is a confused intertwining of demographical and ecological hypotheses. The second is the unrealistic supposition that the behavior of all migratory species represents a cohesive evolutionary response to the same ecological factors. These pitfalls make it unlikely that a holistic approach to an entire migration system will yield anything but murky waters and untestable—even if appealing—assertions. We need to begin by understanding the migration of one species.

PART I. THE GEOGRAPHY AND TIMING OF SANDERLING MIGRATION

METHODS

CENSUSES

Sanderlings restrict their activities almost exclusively to sandy beaches and sandflats. The latter are covered by incoming tides. It is thus possible to determine local density in a simple fashion by making counts during a period of the tidal cycle when all members of the local population are forced by high water onto the beach (Connors et al. 1981).

Most censuses summarized in this paper were conducted on foot along a predetermined length of beach by tallying Sanderlings encountered. Birds flying past the observer from behind (i.e., from a section of the beach already censused) were subtracted from the total. Birds flying in the opposite direction were added. Censuses conducted by car were made by driving along the beach with two or more observers, one or more counting while one drove.

Censuses at Bodega Bay, California, were made over the years 1976 to 1982, with one to four censuses made each month in a given year. We distinguished adult from first-winter birds during counts in early fall using plumage characters (Prater et al. 1977). After October, these groups cannot be distinguished consistently in the field. Other sites along the United States Pacific Coast were censused by a network of 31 collaborating volunteers during mid-January 1982. Sites on the Pacific coast in South America were censused by the authors and assistants in February–March 1982 and December 1982. C. T. Schick, T. Johnson, and M. Kunde censused beaches in Baja California in December 1983. Myers censused beaches near Veracruz, Mexico in December 1983. Census data for January–February 1982 in North Carolina were obtained by J. R. Walters (1984). B. R. Chapman (1984) censused Texas beaches during mid-winter of 1980 and 1981. R. I. G. Morrison (1984) provided data from aerial censuses along the South American Atlantic and Caribbean coasts during January–February 1982.

Lengths of beach censuses were, North America: Washington, 96 km; Oregon, 71 km; Northern California, 103 km; Southern California, 26 km; North Carolina, 9 km; Texas, 190 km; Baja California, 543 km; Veracruz, Mexico, 35 km; and Neotropical America: Ecuador, 8 km; Peru, 55.4 km; Northern Chile, 31 km; North Central Chile, 11.4 km; Central Chile, 16.9 km; Southern Chile, 9 km; Tierra del Fuego, 15 km; Peninsula Valdez, Argentina, 19 km; Bahia Blanca, Argentina, 83 km; Porto Alegre, Brazil, 127 km; San Luis, Brazil, 34 km; mouth of Amazon, Brazil, 55 km; Surinam, 27 km; Venezuela, 280 km.

MOLT

Primary molt scores were obtained for all birds captured for banding. Each primary on the right wing was examined and assigned a score based on the proportion of growth of new primaries: 0 if old or missing, 0.1 to 0.9 for one-tenth to nine/tenths or more grown, 10 if fully grown. A feather was considered old if it was one of the primaries on the wing during the previous southward migration. A molt score for the bird varying between 0 and 10 was then calculated by summing the scores for each primary. The goal of this simplified molt

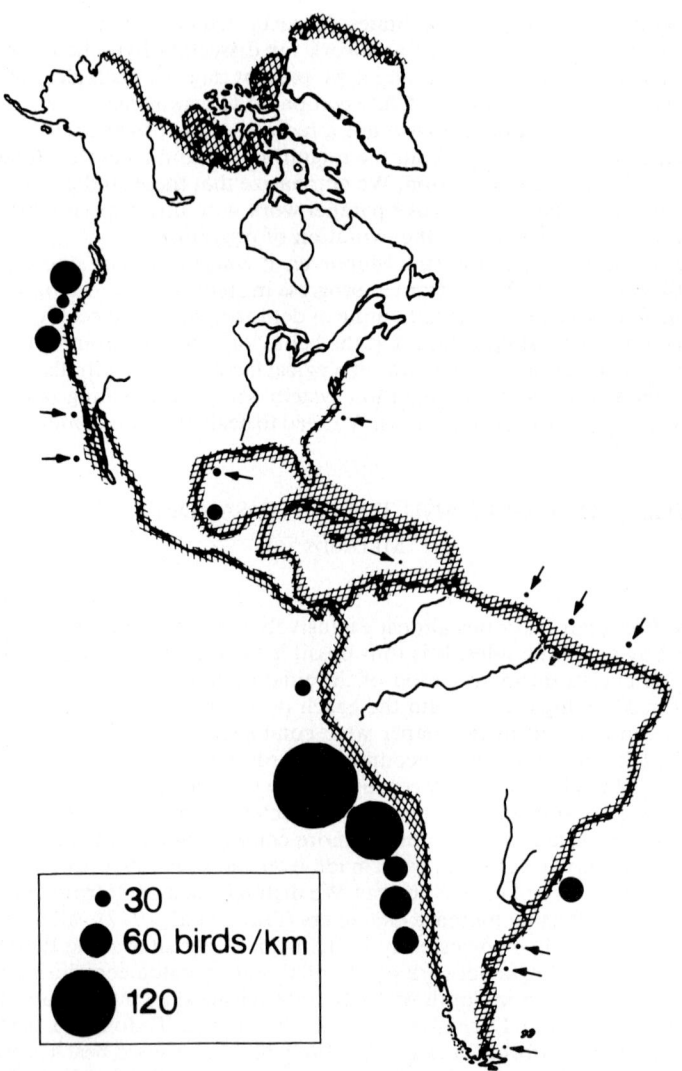

Fig. 1.   Sanderlings distribution in the Western Hemisphere. Hatching in Arctic: breeding. Hatching along temperate and tropical coastlines: nonbreeding. Nonbreeding density (birds/km) on censused outer coast sandy beaches proportional to the diameter of circle adjacent to the coastal census site. Arrows point to census results that might otherwise be missed because of low density. Coastal sectors without adjacent circles were not censused.

score method was merely to document when primary molt occurred in different age classes at different locations. Sites from Peru and Chile were pooled for comparison with California.

RESULTS

GEOGRAPHY

Sanderlings breed throughout the Holarctic Region chiefly north of 73°N (Bent 1927; Glutz von Blotzheim et al. 1975; Cramp 1983). Within the western hemisphere (Fig. 1) their breeding distribution is limited mostly to the Canadian arctic islands and Greenland, although they do breed at low densities in Alaska (Godfrey 1966; Pitelka 1974). During the nonbreeding period, they are found along most temperate and tropical marine beaches of the world, from 50°N on the Pacific Coast and 35°N on the Atlantic south to 50°S. This species' winter distribution thus spans some 80° to 100° latitude, in contrast to its narrow latitudinal breeding range.

How breeding populations array themselves on the wintering ground is virtually unknown, save migration by Greenland-breeding Sanderlings to the Old World (Branson 1979). Whether significant numbers from Siberian breeding grounds reach the western hemisphere in winter is uncertain. Individuals from some Siberian breeding populations migrate to Europe (Glutz von Blotzheim et al. 1975; Branson 1979). Preliminary data (Myers et al. 1984) on the movement of marked individuals indicate that populations wintering along the Pacific Coast of South America move north through the United States chiefly via two routes, along the Pacific coasts of California, Oregon, and Washington, or along the gulf coast of Texas. A few travel via the Atlantic coast.

Midwinter population densities within the Western Hemisphere are consistently higher along the Pacific than along the Atlantic coast at all latitudes, and higher in the southern hemisphere than in the Northern (Fig. 1). The largest known wintering populations are found in coastal Peru and northern Chile, adjacent to the Humboldt Current. Unfortunately, data are lacking for most of Central America.

Throughout the nonbreeding range of Sanderlings, local populations contain adults and juveniles, males and females. Myers (1981a) reported that sex and age classes do not segregate latitudinally, save for a slight but significant tendency for first-winter males to winter farther south than first-winter females. This separation is small compared to the overall latitudinal spread of the species during the nonbreeding season.

Important spring staging areas have been identified along both coasts in North America. On the Pacific, two sites are known. The more southern of these includes beaches within the Oregon Dunes National Recreation area and then beaches from northern Oregon to Gray's Harbor in southern Washington, especially around the mouth of the Columbia River (Myers et al. 1984). At least 30,000 Sanderlings passed through these areas in spring 1983. The second Pacific staging site lies to the north in the Copper River Delta of Alaska, where single flocks of 10,000+ Sanderlings have been observed (Isleib 1979). On the Atlantic coast, Sanderlings spread widely along the outer coast beaches from North Carolina to New Jersey, with a major concentration in Delaware Bay. At least 50,000 Sanderlings stage here in spring migration (Dunne et al. 1982).

No fall staging areas of comparable magnitude have been reported.

SEASONALITY

Numbers.—Adult Sanderlings begin returning to northern wintering grounds in central coastal California by mid-July (Fig. 2). Among the first adults to return are individuals that will remain throughout the nonbreeding season. For example, on 17 July 1982, of 17 Sanderlings present at Bodega Bay, four had been color-banded at Bodega Bay during previous years. These banded individuals then remained throughout the 1982–1983 winter in the area (although see Myers 1984 on regional vagility of the Bodega Bay population). The ~1:4 ratio of banded to unbanded birds is typical of the ratio prevailing at Bodega during the winter, suggesting that all 17 could have been local winter residents.

The major influx of adults begins in early August, and by October most have returned. Increases in adult population size at Bodega Bay during fall and early winter reflect two processes, revealed by studies of individually color-marked birds (Myers 1980a, 1984). First, newly returning adults continue to appear through December, although at much lower rates than in September and October. In many instances, these new returnees actually had been sighted at nearby estuaries in central California for a month or more, but simply had not reappeared at Bodega Bay. Second, in late summer and early fall, returned adults wander broadly along the central California coastline, staying for a few days or weeks at Bodega Bay, moving away for a period, and then returning (Myers 1984). Through September and October, the proportion of time spent away decreases, with the result that a greater fraction of the Bodega Bay adult population is present at a given time during late fall and winter.

The first juveniles begin trickling in by late August (Fig. 2). Here, as with adults, among the first birds are individuals that will remain until spring departure (Myers, unpubl. data). Numbers build rapidly in September. Beyond mid-October it is impossible to differentiate adults and juveniles during a census.

The combined adult/first-winter population builds through November, declines slowly through the winter, and then drops rapidly in spring. During some years in March, large portions of the Bodega Bay population move to an outer beach approximately 20 km south, out of the census area but still in central coastal California. The early spring decline (Fig. 2)

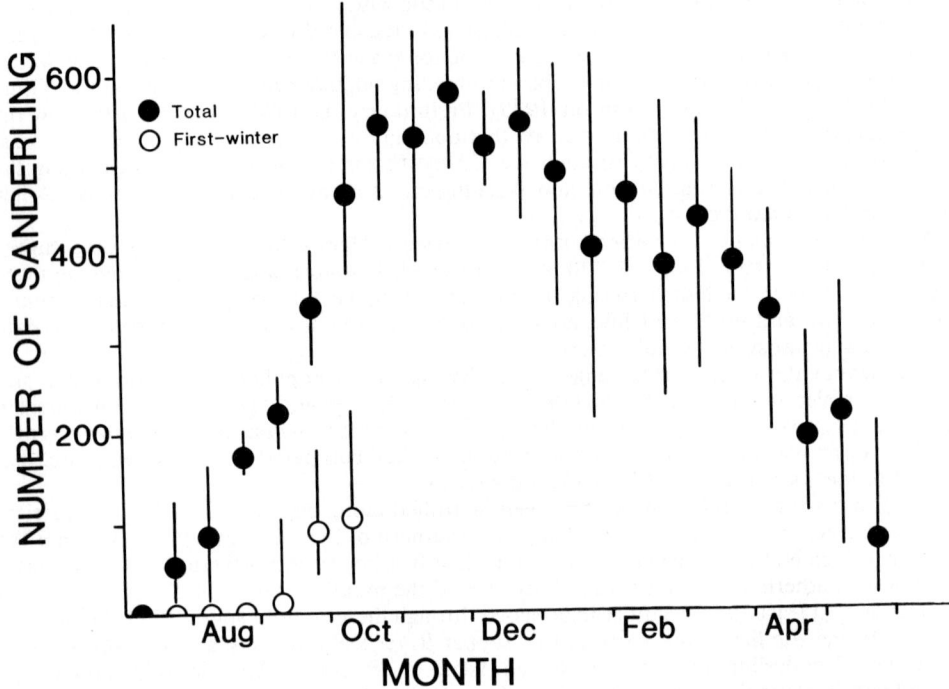

FIG. 2. Mean number and range of Sanderlings at Bodega Bay, California during the years 1976 to 1983. Solid circles: total population. Open circles: juveniles. Adults and juveniles cannot be distinguished reliably in the field after mid-October.

therefore, does not reflect migration. Movement north begins in April or early May, and virtually all individuals are gone by late May. Detailed records of departure times of individually color-banded birds at Bodega Bay reveal no striking age difference in departure schedule (Myers, unpubl. data).

The period of winter residency for many individuals at Bodega Bay thus lasts nine months and for some reaches ten.

Far less information is available on seasonality in southern populations, but patterns differ in at least two respects. First, arrival on the wintering grounds is later. Second, a large number of birds remain on South American wintering grounds during the breeding season, June through mid-July. These points are amplified below.

Adults begin arriving by early September whereas juveniles first appear in early October. Preliminary observations during autumn 1983 of birds color-marked in Peru and Chile during the nonbreeding season of 1982–1983 indicate that birds wintering in South America may spend up to a month or more in stopovers along the United States east coast. Individuals banded in Peru and Chile were seen as late as early September at sites between Florida and Prince Edward Island, Newfoundland. Moreover, birds banded the previous winter in northern Chile have been detected at stopovers in Peru on their way south the following fall as late as October (Myers, unpubl. data).

Local populations in Peru remain high through the end of March. Censuses near Lima begun in April 1983 indicate that numbers fall steadily through April, and that most individuals have left by the end of the month (G. Castro and E. Ortiz, pers. comm.).

Many Sanderlings in the southern hemisphere oversummer, failing to migrate north for the arctic breeding season. Such behavior is common in many shorebird species, particularly Southern Hemisphere migrants (Eisenmann 1951; Johnson 1979; Myers 1981b). R. A. Hughes (pers. comm.) reported flocks of Sanderlings at Mollendo, Peru during each northern summer, and also that several thousand Sanderlings oversummered at Tacna on the southern Peruvian coast in June 1982. William Belton (pers. comm.) reported similar observations for south-

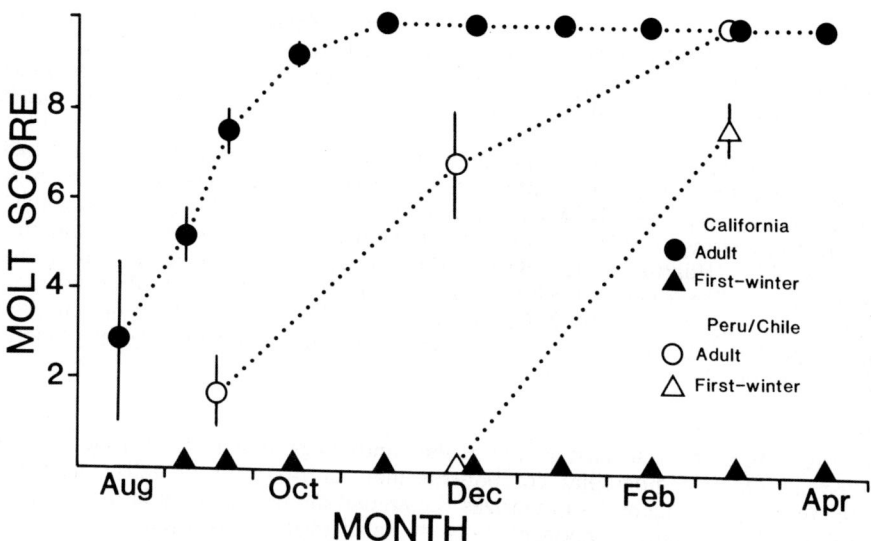

Fig. 3. Primary molt of Sanderlings in California and northwest Chile/southwest Peru.

eastern Brazil. Judging from work with other long-distance shorebird migrants (Elliot et al. 1976; Johnson and Johnson 1983), many Sanderlings are not likely to return north in their first summer.

Molt.—At Bodega Bay, post-nuptial body molt begins immediately after a bird reaches the wintering site (Myers, unpubl. data). By early September it is completed in all adults. Primary molt starts shortly after a bird returns to Bodega Bay (Fig. 3). By late September virtually all adults have begun primary molt, and many have finished. A few individuals drag the process through to December, but most complete primary molt by mid-October. Juveniles at Bodega Bay commence their first-winter molt shortly after arrival, changing most body feathers by early December. Juveniles do not molt their remiges during their first winter (Fig. 3) and retain some wing and upper tail coverts until the following summer or fall.

Along the coast of southern Peru and northern Chile, adult primary molt takes place in late fall and early winter, with a few completing it as late as mid-March (Fig. 3). By mid-March most have begun molting contour feathers to alternate plumage. Most juveniles are in body and primary molt in March, when they still retain a mixture of old and new wing and tail coverts. Some replace all their primaries while a low proportion (<1%) replace only the outer three. The body molt of almost all first-winter birds is to basic plumage.

PART II. WHY DO SOME SANDERLINGS MIGRATE FARTHER THAN OTHERS?

The combination of a narrow breeding latitudinal range and a broad winter range means that some Sanderlings migrate much farther than others. Birds wintering in central Chile may travel 7500 km beyond potential wintering sites in the northern hemisphere. The round-trip costs of that extra distance at 65 km/hr involve 230 hrs of flight and an estimated 1242 kilocalories (after McNeil and Cadieaux 1972), or almost 6000 10-mm long sandcrabs (*Emerita analoga*, Hippidae), each worth about 300 calories (Myers and Smith, unpubl. data), assuming 70 percent assimilation efficiency. These calories are roughly equivalent to a month's existence costs for a 50 gm Sanderling in mid-winter in California (calculated after Kendeigh et al. 1977). This is surely a simplistic assessment, as it assumes similar metabolism for migrants and winter residents and does not take into account non-caloric costs. Moreover, it implies that the calories are of consequence to the migrating individual, or in other words, that energy is one of the factors limiting migration. As a starting point, nonetheless, it conveys the magnitude of additional costs imposed by migration to southern wintering areas.

Why do so many Sanderlings bother to go to the Neotropics? In essence, studies of the evolution of migration focus on two broadly intertwined questions, (1) why migrate and, (2) if you migrate, where should you go? Clearly, the range of possibilities for (2) will influence (1), as formalized by Lack (1954) and later by Cohen (1967), Southwood (1977), Baker (1978), Dingle (1980), and others. Unfortunately, the field languishes far behind studies on migratory orientation, remaining poor in data and rife with imprecise theory (Dingle 1980; Gauthreaux 1982).

Modern theoretical approaches to the study of evolution of avian migration have had two components. The first is the demographic consequence of a given migratory option: mortality costs of increasing migration distance and productivity or survivorship gains that result from remaining in or moving to different wintering or breeding areas (Lack 1954, 1968; Salomonsen 1955; Cohen 1967; Haartman 1968; Baker 1978; Fretwell 1980; Greenberg 1980; Ketterson and Nolan 1982). The second is the ecological basis for the demographic effect, for example, resource availability (Blondel 1969; Gauthreaux 1978; Fretwell 1980), predation (Fretwell 1980), competition (Salomonsen 1955; Cox 1968) or environmental predictability (Alerstrom and Enckell 1979).

DEMOGRAPHIC MODELS

For birds with a life cycle such as that of the Sanderlings, migratory behavior potentially can affect a suite of demographic components that contribute to an individual's lifetime reproductive success. Figure 4 summarizes the annual demographic cycle of a Sanderling, focusing explicitly on those components liable to be affected by migration:

(1) survivorship from the time of arrival on the wintering ground to the time of departure, hereafter referred to as within-winter survivorship, S_w, at a particular wintering site for birds of a given age and sex;
(2) age, sex and distance-specific survivorship during northward migration, S_n;
(3) the age, sex and distance-specific probability of obtaining a breeding opportunity, B;
(4) the age, sex and location-specific effect that winter resources may have on breeding condition, E;
(5) age, sex and location-specific seasonal productivity, in terms of the number of progeny surviving to southward migration, P;
(6) age, sex and location-specific survivorship on the breeding ground for a given level of reproductive effort, S_b;
(7) age, sex and location-specific survivorship when oversummering south of the breeding ground, S_o;
(8) age, sex and distance-specific survivorship during southward migration, S_s.

Beginning when individuals reach the wintering ground, within-winter survivorship, S_w, sets the probability that an individual will survive through the winter. Come spring, surviving individuals may either migrate north, thus accepting the risks of migration plus the potential benefits of breeding, or they may remain south of the breeding ground. In such a decision, whether made in ecological or evolutionary time, the likelihood of successful reproduction (given S_n, E, B, S_b, P, and S_s specific to their age, sex, winter location and migration distance) must be weighed against the S_o and the probabilities for future reproductive opportunities (i.e., survival through another winter and migration plus successful reproduction).

S_n, B, and E together determine the probability and quality of reproductive effort. The latter two require additional comment. B refers to the effect that distance between wintering and breeding grounds may have on productivity by affecting time of arrival and thereby obtaining essential breeding resources such as a territory or a mate (Myers 1981a). The magnitude of B is unknown and its potential importance debated (Ketterson and Nolan 1983). E refers to the enhancement of breeding effort engendered by energy or nutritional reserves stored when on the wintering ground and mobilized during the breeding season. This phenomenon is well known in geese (Drent and Daan 1980) and may contribute to the differential attractiveness of different wintering areas. For Sanderlings in particular it seems unlikely to be important, as migration consumes more calories than can be stored on the wintering ground (see above). Thus, an individual could not carry stored reserves northward. Nutritional factors other than calories may be important, even though for Sanderlings it seems implausible because of the broad similarity of their diets—cirolanid isopods, amphipods, *Emerita*—throughout their winter distribution.

Demographic models of avian migration have considered two types of tradeoffs (Fig. 4), migration survivorship ($S_n \times S_s$) played against winter survivorship (S_w) and against produc-

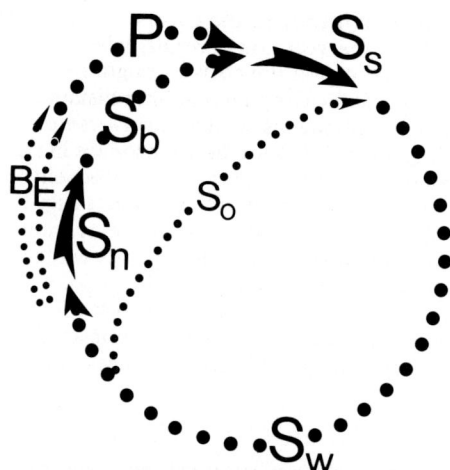

Fig. 4. The annual cycle of Sanderling demography. See text p. 526 for explanation of symbols.

tivity (P) (Cohen 1967; Haartmann 1968; Lack 1968; Fretwell 1980; Greenberg 1980; Ketterson and Nolan 1982). One assumption underlies virtually all interpretations and models, that S_w increases monotonically with distance, or increases to some peak determined by the geographic distribution of optimum breeding or wintering grounds. Most interpretations also assume that S_n and S_s decrease with migration distance. P may increase or decrease with distance, depending upon the model and the pattern being explained; viewing the breeding grounds as the starting point, P may decrease with distance if, by wintering farther from the breeding area, an individual lessens its chance of obtaining a breeding site (Haartman 1968) or has a shorter breeding season (Greenberg 1980, see below). Viewing the wintering grounds as the starting point, then P may increase with distance migrated if, by migrating farther, individuals obtain more resources essential for breeding or reduce nest predation (Cox 1968; Fretwell 1980). Alternatively, P may decrease at greater distances from wintering sites if individuals lose winter site dominance (Fretwell 1980). Depending upon the taxon involved, either starting point may be plausible. For Sanderlings, viewing the arctic breeding grounds as the starting point is appropriate, as all 24 members of the calidridine sandpipers breed in the arctic (Myers 1981b), and the vast majority of scolopacids nest in the northern hemisphere (Myers 1980b).

Despite the fundamental importance to theory of these tradeoffs, few studies address them empirically. The best data are those of Ketterson and Nolan (1982), indicating (1) that S_w of southern wintering populations of Dark-eyed Juncos (*Junco hyemalis*) exceeds S_w of northern wintering populations and (2) that annual survivorship ($S_w \times S_n \times S_b \times S_s$) of southern wintering populations equals that of northern populations. This finding implies higher mortality for the southern birds during another period of the year. Ketterson and Nolan propose that the southern birds have higher mortality during migration, in other words, that S_n or S_s or both decrease with migration distance. Ketterson and Nolan's work is the first empirical migration study making a concerted empirical effort to partition survivorship into its annual components (Fig. 4). Their data, however, are limited and are poor at distinguishing dispersal from mortality. Moreover, the results imply a perfect tradeoff between survivorship during migration ($S_n \times S_s$) and winter survivorship (S_w), which seems implausible.

Greenberg (1980) showed that the average North American migrant passerine has higher adult survivorship and lower productivity than does the average resident. This conflicts with O'Connor's (1981) demographic analysis of British migrants and residents. O'Connor found no difference in either survivorship or productivity, yet argued that migrants are r-selected while residents are k-selected. Both studies are marred by statistical problems. Greenberg's sample is geographically heterogeneous, while O'Connor's sample is taxonomically so, and neither O'Connor nor Greenberg controls for a positive relationship between body size and survivorship (e.g., Boyd 1962). Greenberg's results would probably be strengthened by this and O'Connor's weakened, because in both samples the residents are larger on average than the migrants. Finally, and most critically in the current context, neither study can partition survivorship among periods of the year. As a result, whether migration entails a tradeoff between within-winter survivorship and the costs of migration remains unresolved.

Greenberg argued through a graphical model that long-distance migrants have an enhanced S_w not simply because their instantaneous winter survivorship rate is higher. He proposed they have longer nonbreeding seasons because they reproduce in regions with shorter breeding seasons, and that as a result, they have more time to benefit cumulatively from a high instantaneous survivorship rate. This retrospectively predicts leap-frog migration patterns (e.g., Swarth 1920; Myers 1981b). Data from Sanderlings, however, contradict this aspect of Greenberg's model; as developed in the first section, individuals wintering in Peru spend less time on the wintering ground than those wintering in California.

Baker (1978) provides an ambitious logical framework for considering the evolution of migration. At an empirical level, however, his largely untestable general framework reduces to questions about the ultimate demographic consequences of juveniles settling in particular areas.

ECOLOGICAL FACTORS

Given that some demographic differences must exist among populations with different migratory tendencies, the question becomes what ecological factors underlie the demography. This question has been approached from a number of directions.

MacArthur (1959) and others (e.g., Willson 1976; Herrera 1978; Alerstrom and Ecknell 1979) have examined the geographic distribution of migrants in relation to ecological factors. These reviews address the first question above, i.e., why migrate, and the results clearly implicate seasonal patterns in the abundance and predictability of resources. Tundra insectivorous birds, such as Sanderlings, may leave the arctic during winter because of the disappearance of their food supply. But the question is more complex in regions with mixtures of migrant and resident species (e.g., MacArthur 1959), or for species with some resident and some migratory populations (Swarth 1920; Haartman 1968), or populations with both resident and migrant fractions (Nice 1937; Haartman 1968). Suddenly the self-evident conclusions disappear and the generalizations such as Lack's (1968), that "migration occurs in those species which survive in greater numbers if they leave, than if they remain in, their breeding grounds for the nonbreeding season," beg empirical documentation.

More focused ecological papers typically have argued for the primacy of one or a few factors. Their conclusions have rested more upon logic, untested assumptions and theory than upon data. The majority ultimately derive from seasonal patterns of resource availability and several interpretations seem plausible. Resource-based interpretations include hypotheses on competition (Cox 1968), behavioral dominance (Haartman 1968; Fretwell 1969, 1980; Gauthreaux 1978, 1982), and resource phenology (Klopfer et al. 1974). One less frequently mentioned alternative is predation, with geographic gradients in nest predation risk favoring migration from the optimum (determined by resources) wintering grounds (Fretwell 1980).

As attractive as these ideas are, the work to date fails to discriminate among competing ecological hypotheses that would provide a connection to demographic models. It assumes a common cause for migration across many taxonomic groups whose geography, ecology, and history differ radically. We are left unable to specify the immediate ecological costs or benefits of reaching alternative breeding or wintering grounds, because little has been done to test the empirical importance of ecological variables underlying choices between alternative wintering sites.

ECOLOGICAL DIFFERENCES AMONG SANDERLING
WINTERING SITES

For Sanderlings along the Pacific coast we can list four hypotheses that predict differences among sites in S_w: (1) milder physical conditions result in reduced physiological stress for birds in South America; (2) predation rates are lower in South America; (3) resource availability is higher in South America and therefore mortality due to starvation is lower; and (4) resources are more stable in South America resulting in lower starvation rates.

The first, while relevant to Sanderlings' global distribution in winter, probably is not important to the immediate question: why do so many Sanderlings migrate to the Neotropics? Many geographic locations north of the Neotropics have environments within Sanderlings' physiological limits; 70 percent of the adult Sanderlings present at Bodega Bay, California during the fall remain there during winter and return again the next fall (Myers 1980a). Thus, the birds wintering in Chile almost certainly could survive the environmental conditions of a California winter.

The other hypotheses cannot be rejected given current data. Raptor predation on shorebirds is heavy in central coastal California (Page and Whitacre 1975) and predation risk may vary geographically because of the distribution of Sanderling predators. In fact, one of the primary Sanderling predators of the northern hemisphere, the Merlin (*Falco columbarius*), rarely migrates south beyond Panama (R. S. Ridgely, pers. comm.), and there is no clear ecological equivalent beyond Panama. On the other hand, another major predator of Sanderlings in the northern winter, the Peregrine Falcon (*Falco peregrinus*), migrates throughout South America in winter and is conspicuously common for this species along the Pacific shore adjacent to the Humboldt Current, where it regularly hunts Sanderlings and other northern migrants (Myers et al. 1984). Distribution patterns alone will not resolve the issue. Rather, we need data on mortality due to predation and the risk run by individuals in different wintering areas.

The remaining hypotheses focus on whether resource conditions are more favorable in the south, and if so, why? The third refers to a difference in resource availability. This might come about because of higher resource abundance in the south, because of latitudinal variation in other factors mediating resource availability independent of abundance (see Myers et al. 1980), because of lower competitor density and, thus, even with equivalent resource abundance, a higher per capita resource level, or some combination of them all. Competitors could be hetero- or conspecific.

The fourth hypothesis proposes that resource levels are less variable in Chile and Peru than those in California. The benefit for traveling to the Neotropics thus results from the reliability of food resources, even if average food availability is equal to or lower than that farther north (i.e., hypotheses 3 and 4 are distinct). Such differences in resource reliability are plausible, as winter storms beset California beaches frequently while coastal sites in southwest Peru and northwest Chile have not experienced a single storm during the last 30 years (R. A. Hughes, pers. comm.). Storms, especially with strong winds, disrupt shorebird feeding (Davidson 1981a; Dugan et al. 1981). The waves that accompany them also radically alter beaches and, thus, Sanderling prey distributions (Myers et al. 1981). On the other hand, every six to seven years an El Niño Current wreaks profound changes in the Peruvian and northern Chilean coastal ecosystem. In early 1983 coastal water temperatures throughout Peru exceeded normal by 5° to 7°C, an increase accompanied by massive die-offs in different invertebrate species, most notably the clam *Mesodesma donacium,* a dominant member of the sandy beach invertebrate fauna (R. A. Hughes, pers. comm.). The magnitude of effect on sandy beach crustaceans such as *Emerita* is unknown.

Clearly, none of the four hypotheses has been tested adequately for Sanderlings, nor for any other migrant. We present here preliminary data that indicate that food availability is higher in coastal Chile and Peru than in California, even though potential competitors on sandy beaches are more abundant.

METHODS

Allocation of time to foraging.—In northern Chile (December 1982) and in central coastal California (January 1983), we measured the proportion of birds foraging versus roosting during censuses taken throughout the day at two-hour intervals. In both areas, we covered all occupied Sanderling habitat during these censuses. In Chile (near Hornitos, 22°S) one observer sampled three separate beaches totaling 12 km while riding a 3-wheel all-terrain vehicle. In California (Bodega Bay, 35° N) four observers sampled two beaches (7 km), a slough (~1 ha) and a sandflat (~200 ha) on foot and by spotting scope from a car. Censuses were run throughout the day on two different days at each site: one day with high tide in the morning, the other with high tide in afternoon.

Comparison of weights.—Sanderlings were captured (Myers and Sallaberry 1984) from night roosts (Chile and California) and from daytime foraging areas (Chile), weighed to the nearest 0.5 g with Pesola scales, and measured for tarsus, flattened wing chord, and bill length (proximal end of nares to tip). Weights were corrected for weight loss between time of capture and weighing using data from Schick (1983) and unpublished data from Bodega Bay. The California sample included all Sanderlings caught for banding at Bodega Bay in mid-winter from 1976 to 1983. Chilean birds were caught during December 1982.

Captured birds were grouped on the basis of capture site: natural feeding sites in California,

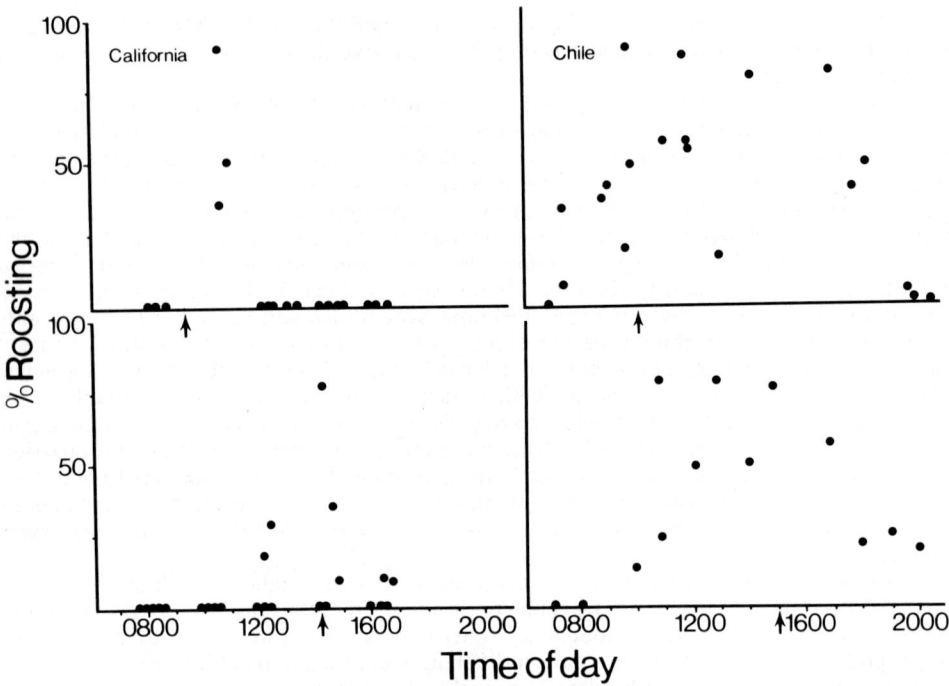

FIG. 5. Diurnal roosting pattern of Sanderlings in two different wintering populations. Graphs show the percentage of birds that were roosting during each census. Time of high tide indicated by arrow. Left: Bodega Bay, California; upper, 12 January 1983; lower, 4 January 1983. Right: Hornitos, Chile; upper, 16–17 December 1982; lower 22 December 1982.

natural feeding sites in Chile, and artificial feeding sites in Chile. The latter birds were all feeding on fish meal spilled in the yard of a fish processing plant near Mejillones (22°S), approximately 25 km south of the natural feeding site (Hornitos). They had access to food *ad libitum,* and alternated between feeding on piles of spilled fish meal in the yard and bathing and roosting on the shore nearby.

To compare the weights of birds in these groups we performed an analysis of covariance to control for the relationship between size and weight and possible body size differences between California and Chile. Tarsus length was the covariate. Differences between weights were then tested using a Student-Newman-Keuls test. All statistical tests were run at The University of Pennsylvania Computing Center using the statistical package SAS.

RESULTS

ALLOCATION OF TIME TO FORAGING

Figure 5 shows the proportion of birds roosting at different portions of the tidal cycle and at different times of day in California (January 1983) and in northern Chile (December 1982). The Chilean population sampled was feeding on natural prey, not at the fish meal factory. The Sanderlings in California roosted only around high tide, independent of the time of day. The Chilean birds, by contrast, began roosting within an hour after sun-up, and at any given time throughout the day a large fraction of the local population was roosting. This fraction decreased appreciably only just before dusk. They did not adjust to the changing tide schedule.

No quantitative samples were obtained at either location on night-time foraging. However, throughout these periods we attempted to capture birds at night roosts. During December 1982 in Chile, we searched over 35 km of beach during the night with a 300,000 candlepower spotlight. Less than one percent of the daytime population was observed on the beach at night, and most of the birds were roosting. Fieldwork in December 1983 revealed that they roost several hundred meters inland. At Bodega Bay, we regularly patrolled the beaches at night with a 300,000 candlepower spotlight. Here, most individuals within the local population

TABLE 1
WEIGHTS OF SANDERLINGS FEEDING UNDER DIFFERENT CONDITIONS[1]

	Mean	s.e.	N
	Adult		
California	48.7	0.4	118
Chile—natural	51.5	1.2	7
Chile—*ad libitum*	55.6	1.1	39
	First-winter		
California	48.5	0.5	91
Chile—natural	54.1	0.9	19
Chile—*ad libitum*	57.3	0.4	262

[1] Weights in grams; s.e. = standard error.

roost during mid- and high tide levels at night, often all in one single flock. During low tides at night, however, the roost is much more difficult to locate, and a portion of the population can be found feeding on exposed sandflats.

Taken together, the data for daylight foraging and the observations during night hours clearly show that Chilean birds spend far less time feeding than do individuals in California.

COMPARISON OF WEIGHTS

We compared weights of naturally-feeding Sanderlings in California and Chile and Chilean birds feeding at an artificial, superabundant food resource (Table 1). For first-winter birds, after removing tarsus as a covariate, the overall F for location was 86.1 ($P < 0.0001$). The Student-Newman-Keuls test revealed that California first-winter birds were lighter than both sets of Chilean first-winter birds. Chilean birds feeding naturally, in turn, were lighter than those feeding *ad libitum*. For adults, again after removing tarsus as a covariate, the overall F for location was 33.7 ($P < 0.0001$). Among the adult birds, those in California did not differ from naturally-feeding birds in Chile but both of these were lighter than the birds feeding on fish meal. Unfortunately, the sample of naturally-feeding adults in Chile is too small to accept this conclusion without reservation. The trends in weights for adults follow those for juveniles.

DISCUSSION

We interpret these results to indicate that resource conditions during the mid-boreal winter are more favorable in Chile than in California for both adults and juveniles. Even though weights of the naturally-feeding adults in the two areas did not differ statistically, the California birds required more foraging time to achieve those weights. Furthermore, the *ad libitum* data imply that natural food is not superabundant in Chile. Given free food, both adults and juveniles weigh more.

A suite of untested assumptions underlies the above interpretations, the most important of which is that relative body weight indicates resource conditions. This can be true, but it is not a simple issue (e.g., Blem 1981). The weights of the *ad libitum* birds are consistent with this assumption, in that they are greater than those of birds without superabundant food; experiments with caged shorebirds yielded similar results (Goss-Custard et al. 1981; Myers and Williams, unpubl. data). Weights of European waders in the wild decrease during periods of food stress (Davidson 1981a, 1982; Dugan et al. 1981). Alternatively, it has been argued that shorebirds are more likely to store fat when resource conditions are unpredictable and the likelihood of periods of starvation is high (Evans and Smith 1975; Davidson 1981b; but see above re resource predictability).

If the bird weights do reflect different resource conditions, then these differences could result from geographic differences in resource availability, in competition, or both. These two factors can be distinguished, although they are not mutually exclusive. The question is whether either of them, in itself, is both necessary and sufficient to explain the observed patterns. Differences in resource availability do not necessarily invoke a notion of Sanderling density. To date we lack direct comparative measurements of prey availability in the two locations, and they will be difficult to obtain because of problems inherent in measuring food availability from the birds' point of view (Myers et al. 1980).

TABLE 2

LINEAR DENSITIES OF BIRDS TAKING INTERTIDAL INVERTEBRATES ON SANDY BEACHES DURING
MID-BOREAL WINTER

Species	California[1]	SW Peru[2]	NW Chile[3]	C Chile[4]
Haematopus palliatus	0[5]	0.4	5	3.5
H. ater	0	0.5	0.2	0
Pluvialis squatarola	2.9	7.5	0.1	0
Charadrius alexandrinus	3.9	0.6	0.5	1.6
Numenius phaeopus	0.2	2.7	0.5	5.6
Limosa fedoa	3.2	0	0	0
Arenaria interpres	0.1	6.7	0	0
A. melanocephala	1.3	0	0	0
Catoptrophorus semipalmatus	4.5	0	0.1	0
Calidris alba	51	159	109	53
Larus modestus	0	220	110	29
Total	67	397	225	93

[1] Average density in 12 censuses of 1.5 km of sandy beach at Bodega Bay; sampled Decembers, 1978–1980.
[2] Average density on 8 samples of sandy beach totaling 36 censused km between Lima and Mollendo; March 1982.
[3] Average density on 6 samples of sandy beach near Hornitos totaling 31 censused km; December 1982.
[4] Average density on 2 samples of sandy beach near Rio Huasco totaling 11.4 censused km; December 1982.
[5] Birds/km.

The competition factor does not require any underlying geographic variation in the baseline abundance of food independent of bird density. Interspecifically there are no clear patterns in this case. The number of species of sympatrically wintering Nearctic shorebirds declines sharply south of the equator in the western hemisphere (Pitelka 1979), but within sandy beach habitats if there is any change, it is the opposite of that predicted (Table 2). Of those we considered (Table 2), several are known to share prey species with Sanderlings, particularly the Willet (*Catoptrophorus semipalmatus*), the Whimbrel (*Numenius phaeopus*), and the Grey Gull (*Larus modestus*) Reeder 1951; Koepke and Koepke 1952; Connors et al. 1981; Myers, unpubl. data. These three species and Sanderlings prey heavily upon the sandcrab *Emerita analoga* when feeding on open coast sandy beaches. The northern hemisphere has no form comparable to the Grey Gull, an abundant species along the Humboldt Current coast and a sandcrab specialist.

These comparisons tell us little, however, about competition per se. Not only are the data sparse, but more importantly, we lack baseline information on the availability of food and on the effect of bird density on foraging success at locally prevailing food abundances. In the northern hemisphere data now point toward strong competitive interactions: shorebirds deplete their prey through the winter (Evans et al. 1979; Goss-Custard 1980; Quammen 1980; Myers et al., unpubl. data). Duffy et al. (1981) have argued that shorebirds in Peru do not compete for food, but their results are empirically weak (Myers and McCaffery 1984). Puttick (1980) suggests that Curlew Sandpipers (*C. ferruginea*) in South Africa have superabundant resources. If these suggestions can be confirmed and generalized, then a good case may be built for reduced competition being a benefit for long-distance migrants. Densities of sympatric wintering shorebirds (Table 2) indicate, however, that a reduction in competition is not due simply to patterns in bird distribution; some geographic difference in resource abundance must be involved. In other words, reduction in competition may be a necessary factor, but it is not a sufficient one.

CONCLUSIONS

Sanderling nonbreeding distribution in the New World is not the homogeneous pattern one might expect from range maps or from casual excursions along a few sandy beaches. Sanderlings occur in most suitable habitat within their extraordinarily broad range, but in only a few regions do they become abundant. These areas of concentration lie along the Pacific Coast, especially in southwest Peru and northern Chile. Elsewhere, their apparent commonness may be more a result of the ease with which they can be detected in their habitat, than of numerical abundance.

We cannot yet resolve the demographic and ecological bases underlying this distribution, nor those of the long-distance migration it requires. Resource conditions for Sanderlings appear to be more favorable on Chile's northern coast than in California. The data remain sketchy, however, and demand amplification with detailed ecological and demographic comparisons at these sites, near the northern and southern ends of Sanderlings' winter distribution.

ACKNOWLEDGMENTS

F. A. Pitelka inspired this work, and his influence runs (wild) throughout. Grants to him from the National Science Foundation funded much of the basic work at Bodega Bay. South American field work was supported by The World Wildlife Fund-US and the International Council for Bird Preservation-Panamerican Section. We thank R. Lee and P.-J. Subaru for the use of a field vehicle. Volunteers in the Sanderlings Project, led by coordinators C. Hohenberger and T. Schick, obtained much of the census data along the United States Pacific Coast. A. Amos, R. I. G. Morrison and J. R. Walters provided unpublished censuses. The Bodega Marine Laboratory, particularly P. Siri, greatly facilitated our field efforts. L. Myers contributed editorial skills and tolerance, and F. Gill gave at least the former. M. S. Foster clarified our writing style. R. Ricklefs gave his and his computer's time generously. In the field we have been helped by B. McCaffery, M. Cikutovits, P. G. Connors, R. Culver, K. Havelock, E. Hernandez, C. Maizels, E. Ortiz, V. Pulido, J. Walters and S. Williams.

LITERATURE CITED

ALERSTROM, T., AND P. H. ENCKELL. 1979. Unpredictable habitats and the evolution of bird migration. Oikos 33:228–232.
BAKER, R. R. 1978. The Evolutionary Ecology of Animal Migration. Holmes and Meier, New York.
BENT, A. C. 1927. Life histories of North American shorebirds. Pt. I. U.S. Nat. Mus. Bull. 142.
BLEM, C. R. 1981. Geographic variation in mid-winter body composition of starlings. Condor 83:370–376.
BLONDEL, J. 1969. Les migrations transcontinentales d'oiseaux vues sous l'angle ecologique. Bull. Soc. Zool. Fr. 94:577–598.
BOYD, H. 1962. Mortality and fertility of European charadrii. Ibis 104:368–387.
BRANSON, M. J. B. A. (ED.). 1979. Wash Wader Ringing Group. Report 1977–1978. Cambridge, England.
CHAPMAN, B. R. 1984. Seasonal abundance and habitat use patterns of coastal bird populations on Padre and Mustang Island barrier beaches. U.S. Fish Wildl. Serv. Div. Biol. Serv., Washington, D.C. FWS/BS-84/02.
CRAMP, S. (SR. ED.). 1983. Handbook of the birds of Europe the Middle East and North Africa. Vol. III. Waders to Gulls. Oxford University Press, Oxford, England.
COHEN, D. 1967. Optimization of seasonal migratory behavior. Am. Nat. 101:5–18.
CONNORS, P. G., J. P. MYERS, C. S. W. CONNORS, AND F. A. PITELKA. 1981. Interhabitat movements by Sanderlings in relation to foraging profitability and the tidal cycle. Auk 98:49–64.
COX, G. W. 1968. The role of competition in the evolution of migration. Evolution 22:180–192.
DAVIDSON, N. C. 1981a. Survival of shorebirds (Charadrii) during severe weather: the role of nutritional reserves. Marine Science 15:231–240.
DAVIDSON, N. C. 1981b. Seasonal changes in nutritional condition of shorebirds (Charadrii) during the non-breeding season. Unpubl. Ph.D. dissert., University of Durham, Durham, England.
DAVIDSON, N. C. 1982. Changes in the body-condition of redshanks during mild winters: an inability to regulate reserves? Ringing Migr. 4:51–62.
DINGLE, H. 1980. Ecology and evolution of migration. Pp. 1–101, In S. A. Gauthreaux (ed.), Animal Migration, Orientation, and Navigation. Academic Press, New York.
DRENT, R. H., AND S. DAAN. 1980. The prudent parent: energetic adjustments in avian breeding. Ardea 68:225–252.
DUFFY, D. C., N. ATKINS, AND D. C. SCHNEIDER. 1981. Do shorebirds compete on their wintering grounds? Auk 98:215–229.
DUGAN, P. J., P. R. EVANS, L. R. GOODYEAR, AND N. C. DAVIDSON. 1981. Winter fat reserves in shorebirds: disturbance of regulated levels by severe weather conditions. Ibis 123:359–363.
DUNNE, P., D. SIBLEY, C. SUTTON, AND W. WANDER. 1982. Aerial surveys in Delaware Bay: confirming an enormous spring staging area for shorebirds. Wader Study Group Bull. 35:32–33.
EISENMANN, E. 1951. Northern birds summering in Panama. Wilson Bull. 63:181–185.
ELLIOT, C. C. H., M. WALTNER, L. G. UNDERHILL, J. S. PRINGLE, AND W. J. A. DICK. 1976. The migration system of the Curlew Sandpiper Calidris ferruginea in Africa. Ostrich 47:191–213.
EVANS, P. R., D. M. HERDSON, P. J. KNIGHTS, AND M. W. PIENKOWSKI. 1979. Short term effects of reclamation of part of Seal Sands, Teesmouth, on wintering waders and Shelduck. I. Shorebird

diets, invertebrate densities and the impact of predation on the invertebrates. Oecologia 41:183–206.

EVANS, P. R., AND P. C. SMITH. 1975. Studies of shorebirds at Lindisfarne, Northumberland, 2. Fat and pectoral muscle as indicators of body condition in the Bar-tailed Godwit. Wildfowl 26:64–76.

FRETWELL, S. D. 1969. Dominance behavior and winter habitat distribution in junco (*Junco hyemalis*). Bird-Banding 39:293–306.

FRETWELL, S. D. 1980. Evolution of migration in relation to factors regulating bird numbers. Pp. 493–504, *In* A. Keast and E. S. Morton (eds.), Migrant Birds in the Neotropics. Smithsonian Institution Press, Washington, D.C.

GAUTHREAUX, S. A. 1978. The ecological significance of behavioral dominance. Pp. 17–54, *In* P. P. G. Bateson and P. H. Klopfer (eds.), Perspectives in Ethology. Plenum, New York.

GAUTHREAUX, S. A. 1982. The ecology and evolution of avian migration systems. Pp. 93–168, *In* D. S. Farner, J. R. King, and K. C. Parkes (eds.), Avian Biology, Vol. 6. Academic Press, New York.

GLUTZ VON BLOTZHEIM, U. N., K. M. BAUER, AND E. BEZZEL. 1975. Handbuch der Vogel Mitteleuropas 6.

GODFREY, W. E. 1966. The birds of Canada. Natl. Mus. Canada Bull. No. 203, Biol. Ser. No. 73.

GOSS-CUSTARD, J. D. 1980. Competition for food and interference among waders. Ardea 68:31–52.

GOSS-CUSTARD, J. D., R. E. JONES, AND L. HARRISON. 1981. Weights of Knot in captivity. Wader Study Group Bull. 32:34–35.

GREENBERG, R. S. 1980. Demographic aspects of long-distance migration. Pp. 493–504, *In* A. Keast and E. S. Morton (eds.), Migrant Birds in the Neotropics. Smithsonian Institution Press, Washington, D.C.

HAARTMANN, L. VON. 1968. The evolution of resident versus migratory habit in birds. Some considerations. Ornis Fenn. 45:1–7.

HERRERA, C. M. 1978. On the breeding distribution pattern of European migrant birds: MacArthur's theme reexamined. Auk 95:496–509.

ISLEIB, P. 1979. Migratory shorebird populations on the Copper River Delta and Eastern Prince William Sound, Alaska. Stud. Avian Biol. 2:125–130.

JOHNSON, O. W. 1979. Biology of shorebirds summering on Enewetak Atoll. Stud. Avian Biol. 2:193–205.

JOHNSON, O. W., AND P. W. JOHNSON. 1983. Plumage-molt-age relationships in "over-summering" and migratory Lesser Golden-Plovers. Condor 85:406–419.

KENDEIGH, S. C., V. R. DOL'NIK, AND V. M. GAVRILOV. 1977. Avian energetics. Pp. 129–204, *In* J. Pinowski and S. C. Kendeigh (eds.), Inter. Biol. Prog. 12: Granivorous birds in ecosystems.

KETTERSON, E. D., AND V. NOLAN, JR. 1982. The role of migration and winter mortality in the life history of a temperate-zone migrant, the dark-eyed junco, as determined from demographic analyses of winter populations. Auk 99:243–259.

KETTERSON, E. D., AND V. NOLAN, JR. 1983. A review of hypotheses to account for the evolution of differential bird migration, with particular reference to the dark-eyed junco (*Junco h. hyemalis*). Current Ornithol. 1:357–402.

KLOPFER, P. H., D. I. RUBENSTEIN, R. S. RIDGELY, AND R. J. BARNETT. 1974. Migration and species diversity in the tropics. Proc. Natl. Acad. Sci. USA 71:339–340.

KOEPKE, H. W., AND M. KOEPKE. 1952. Sobre el proceso de transformación de la matéria organica en las playas arenosas marinas del Peru. Publ. Mus. Hist. Nat. Javier Prado, Peru. Ser. A. Zool. 8:1–25.

LACK, D. 1954. The Natural Regulation of Animal Numbers. Oxford University Press, London.

LACK, D. 1968. Bird migration and natural selection. Oikos 19:1–9.

MACARTHUR, R. H. 1959. On the breeding distribution patterns of North American migrant birds. Auk 76:318–325.

MCNEIL, R., AND F. CADIEAUX. 1972. Fat content and flight range capabilities of some adult spring and fall migrant North American shorebirds in relation to migration routes on the Atlantic Coast. Nat. Can. 99:589–606.

MORRISON, R. I. G. 1984. Migration patterns of shorebirds in the New World. In press, *In* J. Burger and O. Bolla (eds.), The behavior of Marine Organisms, Vol. 6.

MYERS, J. P. 1980a. Sanderlings *Calidris alba* at Bodega Bay: facts, inferences and shameless speculations. Wader Study Group Bull. 30:26–32.

MYERS, J. P. 1980b. The Pampas shorebird community: interactions between breeding and nonbreeding members. Pp. 37–49, *In* A. Keast and E. S. Morton (eds.), Migrant Birds in the Neotropics. Smithsonian Institution Press, Washington, D.C.

MYERS, J. P. 1981a. A test of three hypotheses for latitudinal segregation of the sexes in wintering birds. Can. J. Zool. 59:1527–1534.

MYERS, J. P. 1981b. Cross-seasonal interactions in the evolution of sandpiper social organization. Behav. Ecol. Sociobiol. 8:195–202.

MYERS, J. P. 1984. Spacing behavior of nonbreeding shorebirds. Behav. Marine Organ. 6:273–323.

MYERS, J. P., G. CASTRO, B. HARRINGTON, M. HOWE, J. MARON, E. ORTIZ, M. SALLABERRY, C. T. SCHICK, AND E. TABILO. 1984. The Pan American shorebird program: a progress report. Wader Study Group Bull. 42:26–31.

MYERS, J. P., P. G. CONNORS AND F. A. PITELKA. 1981. Optimal territory size and the Sanderlings: compromises in a variable environment. Pp. 135–158, In A. C. Kamil and T. D. Sargent (eds.), Mechanisms of Foraging Behavior. Garland Press, New York.

MYERS, J. P., AND B. J. McCAFFERY. 1984. Paracas revisited: do shorebirds compete on their wintering ground? Auk 101:197–199.

MYERS, J. P., AND M. SALLABERRY. 1984. Como capturar y anillar Calidris alba. Volante Migratorio 2:30–37.

MYERS, J. P., C. T. SCHICK, AND C. J. HOHENBERGER. 1984. Notes on the 1983 distribution of Sanderlings along the United States' Pacific Coast. Wader Study Group Bull. 40:22–26.

MYERS, J. P., S. L. WILLIAMS, AND F. A. PITELKA. 1980. An experimental analysis of prey availability for Sanderlings Calidris alba Pallas feeding on sandy beach crustaceans. Can. J. Zool. 58:1564–1574.

NICE, M. M. 1937. Studies in the life history of the Song Sparrow. I. Trans. Linn. Soc. 4:1–247.

O'CONNOR, R. J. 1981. Comparison between migrant and non-migrant birds in Britain. Pp. 167–196, In D. J. Aidley (ed.), Animal Migration. Soc. Exper. Biol., Sem. Ser. 13.

PAGE, G., AND D. F. WHITACRE. 1975. Raptor predation on wintering shorebirds. Condor 77:73–83.

PITELKA, F. A. 1974. An avifaunal review for the Barrow region and north slope of arctic Alaska. Arct. Alp. Res. 6:161–184.

PITELKA, F. A. 1979. Introduction: the Pacific Coast shorebird scene. Stud. Avian Biol. 2:1–11.

PRATER, A. J., J. H. MARCHANT, AND J. VUORINEN. 1977. Guide to the identification and ageing of holarctic waders. British Trust for Ornithology Guide 17.

PUTTICK, G. M. 1980. Energy budgets of Curlew Sandpipers at Langebaan Lagoon, South Africa. Estuarine Coast. Mar. Sci. 11:207–215.

QUAMMEN, M. L. 1980. The impact of predation by shorebirds, benthic feeding fish and a crab on shallow living invertebrates in intertidal mudflats of two southern California lagoons. Unpubl. Ph.D. dissert., University of California, Irvine.

REEDER, W. G. 1951. Stomach analysis of a group of shorebirds. Condor 53:43–45.

SALOMONSEN, F. 1955. The evolutionary significance of bird migration. Danske Viden. Sels. Biolog. Medd. 22:1–62.

SCHICK, C. T. 1983. Weight loss in Sanderlings Calidris alba after capture. Wader Study Group Bull. 38:33–34.

SOUTHWOOD, T. R. E. 1977. Habitat, the templet for ecological strategies? J. Anim. Ecol. 46:337–365.

SWARTH, H. S. 1920. Revision of the avian genus Paserella with special reference to the distribution and migration of the races in California. Univ. Calif. Publ. Zool. 21:75–224.

WILLSON, M. F. 1976. The breeding distribution of North American migrant birds: a critique of MacArthur (1959). Wilson Bull. 88:582–587.

WALTERS, J. R. 1984. Sanderlings in North Carolina. Wader Study Group Bull. 40:27–29.

WHY ROBIN EGGS ARE BLUE AND BIRDS BUILD NESTS: STATISTICAL TESTS FOR AMAZONIAN BIRDS

YOSHIKA ONIKI

Departamento de Zoologia, Universidade Estadual Paulista, 13.500 Rio Claro, São Paulo, Brasil

ABSTRACT. Data on egg color, nest type, and nest location for bird species nesting at the Amazonian localities of Belem and Manaus, Brazil were examined for possible adaptations against predation and other negative environmental factors. White egg color and spotting was not related to sunlight, except in the negative sense that concealed eggs in any habitat lacked pigments. White eggs, often with spots, were also associated with thin nests in which eggs were visible from below; such eggs may resemble transparent holes in the canopy, in the fashion of transparent wings of certain understory butterflies. Blue eggs normally were found in thick, dark cup nests in dappled sunlight, apparently imitating spots of light on dark foliage in the manner of black-and-blue butterflies of similar zones. Buff (pink) color was most common among eggs placed against dull substrates, and when combined with spotting, resembles the "tiger" pattern known for certain butterflies of dull substrates.

Oven-shaped nests were always thick, but evidently not for protection against weather; pouch nests, open above and to the weather, were also thick. More likely, such large nests had to be thick to discourage predator entry or detection of contents. Pouch nests were not necessary on thin supports (although a deep nest may secondarily keep eggs and young from rolling out); cup nests also occurred in such locations. Pale nests were more common on pale supports and in open areas, dark nests on dark supports and in dark locations, but hypotheses of heat gain or loss in relation to nest color were not supported. Small nests tended to be concealed from above, but none were oven-shaped; protection was against predators, not against sun or rain. Cavity nests were rare within 1 m of the ground, another fact explicable only if nests provide protection against predators rather than against sun or rain. A number of environmental factors, such as human interference (especially with ground or cavity nests) or low vegetation (grass, palms, second growth) limited nest sites. Limitation of nest sites may be as important as limitations of foraging niches in areas with limited foliage height diversity.

No postulated anti-weather function for nests was demonstrated; my data supported only anti-predator functions as reasons why birds build nests or have pigmented (rather than white) eggs.

RESUMEN. Se examinó información respecto al color de los huevos, tipo de nido y ubicación de los mismos, para especies de aves que anidan en Belén y Manaos, en la región amazónica de Brasil, para identificar posibles adaptaciones en contra de la depredacion y otros factores ambientales negativos. Ni los huevos de color blanco y ni los moteados, estuvieron relacionados con luz solar, excepto en el sentido negativo que los ocultaba en cualquier ambiente carente de pigmentación. Los huevos blancos, generalmente moteados, también estuvieron asociados a nidos muy delgados en los que los huevos podían ser vistos desde abajo, esos huevos pueden asemejarse a agujeros transparentes en una bóveda o techo, algo similar a las alas transparentes de ciertas mariposas del nivel bajo del bosque. Los huevos azules se encontraron normalmente en nidos gruesos oscuros con forma de copa salpicados de luz solar aparentemente imitando manchas de luz en follaje oscuro de la misma manera que las mariposas azul y negro de áreas similares. El color ante (rosado) fue más común entre los huevos puestos contra substratos poco coloridos y cuando estuvieron combinados con manchas se asemejaron al modelo "tigre," conocido en ciertas mariposas de substratos apagados.

Los nidos en forma de horno, siempre fueron gruesos, pero evidentemente no para protección climática, nidos en forma de bolsa, abiertos en su extremo superior y a los efectos climáticos, también fueron gruesos. Es más probable que esos nidos grandes sean gruesos para desalentar el ingreso de los depredadores o que el contenido sea detectado. Los nidos en forma de bolsa no estuvieron necesariamente fijados a elementos delgados (aunque un nido profundo puede secundariamente prevenir que los huevos y los polluelos se caígan del nido); también se encontraron en esa posicion nidos en forma de copa. Nidos palidos fueron más comunes en soportes pálidos y en áreas abiertas, nidos oscuros en soportes oscuros y áreas oscuras, pero estas hipótesis de ganancia o pérdida de temperatura en su relación con el color del nido no fueron soportadas. Los nidos pequeños suelen estar ocultos por arriba pero ninguno tuvo forma de horno, la protección fue contra los depredadores, pero no contra el sol o la lluvia. Fue raro encontrar nidos en huecos en el primer metro desde el nivel del suelo, otro factor sólo explicable si los nidos proveen protección contra depredadores, no contra el sol o la lluvia. Una serie de factores ambientales, tales como interferencia humana (especialmente con nidos en el suelo o en cavidades) o vegetación baja (pastos, palmeras, vegetación secundaria) limita los sitios para anidar. El disponer de limitados sitios de anidación puede ser tan importante como la limitación de nichos de forraje enáreas con limitada diversidad de follaje de altura.

No se demostró ningún postulado anti-clima en la función de los nidos; mis datos sólo indican que las aves construyen los nidos o ponen huevos pigmentados (en lugar de blancos) como funcion anti-depredadora.

Eggs and nests of birds have fascinated humans for centuries (Skutch 1976). Until recently nest or egg collecting was a popular hobby of Americans and Europeans, but they paid more attention to aberrant clutches or nests than to the obvious questions of why robin eggs are blue or why birds use nests. It was generally recognized that some egg colors provided camouflage, and that enclosed nests protect clutches from weather or predators. Lack (1958), however, is almost the only person to have examined carefully the relationships between egg color, nest type, and environment. Working only with Turdidae, he suggested a few potential functions for different egg colors and nests of various types.

Success rates for nests I studied at Belem and Manaus, in equatorial, Amazonian Brazil, were higher than expected (Oniki 1979a), probably because of protection from predators provided by egg color and certain nest types (Oniki 1979b). Here I examine statistically these and other Amazonian data on egg color and nest type, testing various hypotheses that suggest that these factors may help prevent predation, protect against excessive sunlight (Montevecchi 1976), or reduce breakage in dim light (Holyoak 1969).

STUDY AREAS AND METHODS

From June 1966 to May 1968, and from March 1972 to May 1973, I studied nests in and near the Agronomic Institute or the Museu Goeldi at Belem, Para, Brazil. From July 1973 to August 1974, I studied nests at Reserva Ducke and at a nearby weekend cottage, Sitio I-iruçanga (entrances a few meters apart at 2°55′S, 59°59′W), near kilometer 26 of the highway from Manaus to Itacoatiara, Amazonas, Brazil.

Habitats at Belem are described by Pires et al. (1953), Novaes (1970), Crump (1971), and Oniki (1972), and at Reserva Ducke and Sitio I-iruçanga by Takeuchi (1961), Willis (1977), and Oniki (1979a). Briefly, Belem study areas ranged from upland forest or second growth to nearby tidally flooded várzea forests, plantations, and houses; Manaus habitats ranged from upland forests to forest plantations, cleared areas, and houses.

For this study I gathered information on egg colors, nests, and locations of nests. In ground color, eggs were white, blue, or buff (pink); they were spotted or plain. Nests were cups, pouches, or enclosed "ovens," or were placed in cavities. Some nests were thick, others thin; they might be dark, or pale; they could be small (diameter ≤7 cm) or large (>7 cm). The nest interior was either lighted or dark. The eggs were visible, or not, from below or above. Nests were placed in dense foliage or not, in well or poorly lighted spots, in isolated bushes or not, in green, brown (e.g., dead leaves, ground), or dark (thick foliage) sites, on pale or dark supports, on slender (diameter 0.5 cm), or thick twigs, and on or under leaves, or not. Nests were built on the ground, in herbs or epiphytes, in dead trees, in live palms, or in dicotyledonous bushes or trees. Nests were near wasps, humans, protected by spiny trunks, or unprotected. Nest heights above ground were noted in meters. Nest habitats were either closed (forest, closed second growth) or semiopen (bushy pastures, yards). The nest habitats were isolated by water or not, and nest habitats were isolated on 1, 2, 3, 4 or no sides.

These data, along with those reported by Pinto (1953) for nests from Belem, were used to test several hypotheses, using Chi-square contingency tables. Original data and species for my nests are given in Oniki (1979b).

TABLE 1

TABLE 1

χ^2 CONTINGENCY-TABLE TESTS OF HYPOTHESES CONCERNING WHITE EGG COLOR IN AMAZONIAN BIRDS

| Test | Row A/B | Column C/D | Number of species | | | | χ^{2a} |
			AC	AD	BC	BD	
1	Habitat closed/open	Egg white/not	69[b]	20	35[b]	31	10.34**
2	Eggs exposed/concealed	Egg white/not	37	37	60[b]	13	16.88**
3	Habitat closed/open	Exposed egg spotted/not	40	20[b]	33	7[b]	1.66 n.s.
4	Habitat closed/open	Large egg spotted/not	40	9[b]	33	3[b]	1.75 n.s.
5	Nest interior dark/lighted	Egg white/not	54	13	45[b]	33	8.82**
6	Small or concealed egg/not	Egg white/not	60	14	13[b]	21	19.61**
7	White egg concealed/not	White egg spotted/not	10	65	21	2[b]	49.24**
8	Egg visible below/not	Exposed egg white/not	13	4	19	24	5.02*

[a] ** = highly significant, $P < 0.01$; * = significant, $P < 0.05$; l.s. = less significant, $P < 0.1$; n.s. = not significant, $P > 0.1$.

[b] The white eggs of *Ortalis superciliosus*, with an open nest in open or closed habitats, become dirty rapidly and, therefore, are doubtfully included in this category.

RESULTS

EGG COLOR

White egg color (Table 1: tests 1–8).—Montevecchi (1976) suggested that white eggs reflect light, preventing over-heating in nests subject to direct sun. Both he and Lack (1958) found, however, that fewer species in open areas than in forests lay white eggs, contrary to the hypothesis. At the Amazonian sites, fewer species in open habitats lay white eggs than species in closed habitats, also contrary to the hypothesis (test 1). More directly, I compared concealed eggs (including tiny ones) with large, exposed eggs and found that the latter are less often white (test 2).

One could suggest that eggs that are not white occur in open areas so as to gain more energy from sunlight (black or heavily spotted eggs would be best for this purpose, but such eggs are not common). If this were true, then for open nests (in which eggs can be heated by the sun), spotted eggs should be more frequent in open habitats. Nevertheless, differences in spotting of eggs between open and closed habitats are not significant (test 3). This was somewhat unexpected, as Lack (1958) reported eggs of open-habitat turdids to be more heavily spotted than eggs of forest turdids, apparently for camouflage reasons. I reasoned that white hummingbird eggs, from tiny nests, might be too small for predators to detect, and that removal of these eggs from the sample might show that spotted, large eggs were indeed more common in open habitats. This was not the case (test 4), however, and it is evident that spotted eggs occur even in forests in my samples. I found no evidence, therefore, to support the hypothesis that spotted eggs are favored in open areas to increase heat storage or for any other reason.

Holyoak (1969) painted eggs of cavity-nesting Jackdaws (*Corvus monedula*) and found that parents broke darkened eggs more often than normal white eggs. As a result, he predicted that more white eggs would be found in nests with dark interiors than in those with light interiors. My data support his hypothesis (test 5). An alternate hypothesis, that birds whose eggs cannot be seen by predators no longer waste energy or materials on egg coloring, predicts this same result. Enclosed nests with non-white eggs in the Amazonian samples are often oven-shaped with short entrance tunnels, supporting either hypothesis; dove eggs (constantly covered) and hummingbird eggs (tiny, hence hard to detect) are often white, facts not explained by Holyoak's hypothesis. Exposed large eggs are less frequently white than are small or concealed eggs (test 6), with a larger χ^2 than in the test for lighted versus dark nests. The difference in χ^2 indicates that lighted versus dark interiors (the Holyoak hypothesis) explains the data less well than does the alternate hypothesis.

The 13 species with white eggs in cup nests that I considered in test 6 have spotted rather than plain white eggs, except for *Ortalis superciliosus* whose white eggs soon become dirty. Pinto (1953) listed 10 other species with white eggs in exposed nest cups, and all eggs but those of *Piaya cayana* (which nests in dense foliage or vines) are spotted. Unconcealed white eggs are indeed spotted more often than are concealed ones (test 7). In nests where eggs are concealed from view, most species with spotted eggs belong to families in which exposed nests and spotted eggs are the rule (Formicariidae, Emberizidae).

TABLE 2

χ^2 Contingency-Table Tests of Hypotheses Concerning Blue Egg Color in Amazonian Birds

			Number of species				
Test	Row A/B	Column C/D	AC	AD	BC	BD	χ^{2a}
1	Habitat closed/open	Exposed egg blue/not	3	75	13	52	9.22**
2	Cup thick/thin	Exposed egg blue/not	10	15	2	21	6.26*
3	Leafy green site/not	Exposed egg blue/not	14	22	2	23	7.40*
4	Isolated bush/not	Exposed egg blue/not	3	50	8	21	7.71*
5	Egg visible below/not	Exposed egg white/blue	13	0	19	13	6.64*
6	Nest dark/pale	Exposed egg blue/not	4	0	1	12	12.25**

[a] As in footnote a, Table 1.

If a nest is thin and the eggs are visible from below, the eggs tend to be white (often with spots) rather than blue or buff (test 8). This is further investigated below; I would have expected eggs in such thin nests to be blue, so as to mimic the blue sky or green canopy.

Blue egg color (Table 2: tests 1–6). — Preliminary inspection of my data and those of Pinto (1953) indicated that blue eggs are often characteristic of species that build dark, thick cup nests in isolated bushes in open areas. I tested for the significance of these patterns of occurrence.

Blue eggs are much more common among species that nest outside the forest shade (test 1). The only species nesting in the forest shade that have blue eggs are *Tinamus guttatus*, which covers its eggs with leaf litter on departure, and *Tachyphonus surinamus*, a tanager related to canopy and open-area birds and commonest in well-lighted, semi-open, sand-ridge woodlands. *Cochlearius cochlearius* of the forest edge has pale blue eggs and is doubtfully counted as a bird of the forest shade.

Secondly, blue eggs occur significantly more often in species with thick cup nests than in those with thin ones (test 2). Small nests and those that conceal eggs were excluded from this comparison, as one cannot expect hidden eggs to be blue (most are white). Blue eggs are, therefore, more visible from above than from below. Blue egg-color and thick nests are common in thrushes and robins, the former trait leading to the common expression "robin's-egg blue." *Turdus fumigatus*, however, a bird of shaded várzea forests at Belem, often has spotted buff eggs, whereas the forest species *Turdus albicollis* has spotted white eggs. Blue eggs occur in thick cup nests of other groups occupying open areas in Amazonia, notably anis (*Crotophaga* spp., Cuculidae) and tanagers (Thraupinae).

I rejected the null hypothesis that blue eggs occur in dull or dark sites as often as in green and leafy sites, for blue eggs are significantly associated with green leaves (test 3). Concealed or small eggs were excluded from the test, for the reason given above. Blue eggs are found significantly more often in isolated, sunny bushes than in other sites (test 4).

Eggs that can be seen from below, through a thin nest, tend to be white rather than blue or buff (Table 1). In addition, a related test (test 5) showed that blue eggs were not visible from below, and hence that blue eggs cannot be mimicking green leaves or the blue sky from below. Blue eggs are usually in dark nests (test 6) even though dark nests are uncommon in well-lighted semiopen areas where blue eggs normally occur (see below).

Buff (pink) egg color (Table 3: tests 1–3). — Exposed buff (pink) eggs were often next to dead leaves on the ground, in pouch and oven nests with short entrances, or on trunks or limbs. In other words, buff eggs were close to dull-colored, often non-living material (test 1). Lack (1958) and other authors have often suggested that buff egg colors provide camouflage, and my data seem to confirm this.

The same test was performed omitting 12 species that have oven nests with short entrances (*Laterallus viridis*, *L. exilis*, *Myrmotherula gutturalis*, *Pyriglena leuconota*, *Pachyramphus rufus*, *Platypsaris minor*, *Myiozetetes cayanensis*, *Pitangus sulphuratus*, *Coereba flaveola*, *Arremon taciturnus*, and *Myospiza aurifrons*); this increased the χ^2 value slightly (test 2). Seven of these 12 species have white eggs rather than buff eggs, even though the eggs are slightly exposed to light and to predators. This type of nest evidently selects for both buff and white eggs. Many of these species have eggs of an intermediate color, either spotted white or pale

TABLE 3

χ^2 Contingency-Table Tests of Hypotheses Concerning Buff Egg Color in
Amazonian Birds

| Test | Row A/B | Column C/D | Number of species | | | | χ^{2a} |
			AC	AD	BC	BD	
1	Background brown/not	Exposed egg buff/not	25	11	4	36	28.56**
2	Background brown/not	Very exposed egg buff/not	20	4	4	36	34.42**
3	Egg visible below/not	Exposed egg white/buff	13	4[b]	19	11	0.83 n.s.

[a] As in footnote a, Table 1.
[b] There is doubt about the color of these eggs, as they are almost white (see text).

buff. Only *Tolmomyias poliocephalus* in the Amazonian samples has buff eggs that are placed within an oven nest in which eggs cannot be seen from the entrance; several species of Tyrannidae (*Rhynchocyclus olivaceus, Todirostrum maculatum, Mionectes macconnelli*) that build oven nests with long entrances have white eggs. Tyrannidae often have spotted pink eggs in cup nests set on broad and dull-colored limbs; species with such eggs in oven nests may be derived from relatives with open nests (Willis 1962).

Buff eggs are sometimes visible from below, and visibility does not differ significantly from that of white eggs when compared with them alone (test 3). The sample size is small, however, and the designation of color of eggs of the four species with thin nests and "buff" eggs is uncertain. These four species (*Eucometis penicillata, Chiroxiphia pareola, Pipra erythrocephala, P. pipra*) have very pale pink, almost white, eggs. Moreover, these nests are not especially thin, and it is not certain that the eggs could normally be seen by a ground predator below the nest. Most buff eggs cannot be seen against green or sky backgrounds regardless of direction of view.

NEST TYPES AND LOCATIONS

Birds expend energy in selecting nest sites and in building nests, so it is likely that they gain from their efforts by avoiding predation or excessive insolation, or from other factors. Koepcke (1972) and Brosset (1974) examined nests and nest sites of tropical birds as adaptations to reduce mortality. Here I perform similar analyses for nests at Belem and Manaus, using χ^2 contingency-table tests.

Protection against weather, predators, or wasps (Table 4: tests 1–10).—Oven-shaped nests were never thin, even though many cup nests were (test 1). Perhaps oven nests have to be thick to be structurally sound, but one still has to consider why birds with such nests do not use energetically and materially less expensive cup nests. Two likely reasons are the protection against weather (including preservation of convective heat inside the nest) or predators provided by the thick material of the nest. In this regard, it is useful to note that a pouch nest is essentially an oven nest turned on its side. Pouch nests were thick, much like oven nests, and hence, did not represent savings in materials (test 2). The only thin pouch was that of *Icterus chrysocephalus*, sewn and completely concealed under a large folded leaf of the buriti palm (*Mauritia flexuosa*). Oven-shaped, but thick, pouches open to sun and rain above (and to heat loss) suggest that protection from weather is not essential. Radiant and conductive heat may be preserved by thick pouch and oven nests, however. I have no data relevant to this possibility, except that oven and pouch nests are singularly rare in temperate and arctic zones where temperatures are low. This is the reverse of what one would expect on a thermodynamic basis.

Pouch nests are often located at the tips of flexible limbs or palm leaves; hence, pouches may keep eggs or young from falling out of the nest as well as minimizing the probability of discovery or attack by predators (because of both their thick walls and their facilitation of positioning at the tips of slender limbs, where large predators cannot go). Pouch nests are found significantly more often on thin supports than on thick ones, when compared to cup nests (test 3). If falling eggs or young were the primary reason for pouch nests, however, there should be no cup nests on slender supports, which is not the case (test 3, Category "BC"). Moreover, pouch nests sometimes fall from their supports (nests of *Cacicus cela* at Belem, pers. observ.) and thus negate somewhat the support value of a deep nest.

TABLE 4

χ^2 CONTINGENCY-TABLE TESTS OF HYPOTHESES CONCERNING NEST STRUCTURE IN
AMAZONIAN BIRDS

			Number of species				
Test	Row A/B	Column C/D	AC	AD	BC	BD	χ^{2a}
1	Nest thin/thick	Large cup/oven	35	0	26	26	25.13**
2	Nest thin/thick	Large cup/pouch	25	1	26	7	5.81*
3	Nest pouch/cup	Thin support/not	7	1	37	36	4.07*
4	Nest cup/oven	Thin support/not	37	36	11	14	0.31 n.s.
5	Nest pouch/oven	Thin support/not	7	1	11	14	4.50*
6	Nest cup/pouch	Habitat open/closed	43	51	6	3	1.41 n.s.
7	Nest cup/oven	Habitat open/closed	43	51	16	15	0.34 n.s.
8	Nest thin/thick	Nest small/large	0	36	15	57	8.71**
9	Forest cup thin/thick	Nest small/large	0	14	12	7	13.94**
10	Nest thin/thick	Near wasps/not	0	37	7	69	3.65 l.s.

[a] As in footnote a, Table 1.

Cup and oven nests are found on slender or thick supports about equally (test 4), and pouch nests are thus on slender supports significantly more often than are oven nests (test 5). Young or eggs should not fall from broad supports, so the effort expended in building a large nest (pouch or oven) is not needed for protection from falling. These tests add to the previous tests in indicating that weather (including wind) is not important in determining whether birds build a small cup or large oven or pouch; birds build thin cups exposed to wind, rain, and sun.

If oven or pouch nests only serve to protect against wind, rain, and sun, they should be rare inside the forest shade. If oven and pouch nests serve to conserve heat, they should be common in areas of cool forest shade. The data reject the hypotheses that oven and pouch nests are significantly less or more common in forests than in open areas, despite a slight prevalence of pouch nests in open areas (tests 6, 7).

Tiny nests, mainly of hummingbirds, are thick (tests 8, 9) even when thick pouch and oven nests or nests in open areas are eliminated from tallies. This correlation could be expected if (1) small, inconspicuous nests need not be thin in order to avoid detection by predators, (2) a thick nest slows heat loss (serious in small birds because of high surface area relative to volume), or (3) small, thin nests might fall apart as young grow. If heat loss were the problem, small nests in the forest should be thick more often than are small nests in open areas; but this is not true. Either of the other hypotheses may be true, but I have so far been unable to devise a test that would discriminate between the possibilities.

Nests near wasps and bees were always thick, as reported also by Skutch (1976), although this relationship (test 10) is significant only at the level of $P < 0.1$, presumably because of low sample size.

Nest coloration (Table 5: tests 1–10). — Nests often resemble leafy or other backgrounds, and some nests are unusually dark or pale. The null hypothesis that dark and pale nests are equally likely to occur in open or closed habitats is rejected (test 1). Pale nests are more often in well-lighted open habitats, and dark nests in dark forests, as would be expected for concealing coloration. Another comparison (test 2) shows that nests on pale leaves, grass, or limbs tend to be pale rather than dark, contrasted to a prevalence of dark nests on other and darker types of supports. When one omits nests on grass and limbs and considers only nests on leaves contrasted with nests on dark supports (test 3), differences are also significant. Since many pale nests on leaves are hummingbird nests in dark forests, the possible alternative hypothesis, that pale material is not available in forests, is not supported. Nests on grass and pale limbs are also pale significantly more often than they are dark, even when one omits nests on leaves (test 4). Pale nests are sometimes placed on pale trunks and leaves of palms, but some birds place dark nests in dark palm crowns (test 5, significance only at the level $P < 0.1$). This comparison was repeated (test 6) omitting pale or dark nests on leaves and trunks or in other species of trees since such nests tend to be pale (tests 3, 4). Pale nests were significantly more common in palms ($P < 0.05$).

Oven-shaped nests tend to be pale more often than large cup and pouch nests (test 7). This

TABLE 5
χ^2 CONTINGENCY-TABLE TESTS OF HYPOTHESES CONCERNING NEST COLORATION IN AMAZONIAN BIRDS

Test	Row A/B	Column C/D	AC	AD	BC	BD	χ^{2a}
1	Nest dark/pale	Habitat open/closed	5	18	16	11	7.30*
2	Nest dark/pale	Support pale/dark	5	18	19	6	14.11**
3	Nest dark/pale	Support leaf/dark	1	18	11	6	14.08**
4	Nest dark/pale	Pale support nonleaf/dark	4	18	8	6	5.72*
5	Nest dark/pale	Support palm/not	5	20	12	17	2.90 l.s.
6	Nest dark/pale	Support palm/other	5	18	12	9	5.85*
7	Nest pale/not	Large oven/cup, pouch	8	2	16	25	5.44*
8	Nest pale, dark/medium	Ground/not	4	22	1	37	3.61 l.s.
9	Large pale, dark/medium	<0.5 m/not	10	15	4	22	3.79 l.s.
10	Nest pale/not	Nest thin, tiny/large	15	2	27	25	7.23*

Number of species (spanning AC, AD, BC, BD columns)

[a] As in footnote a, Table 1.

was expected since oven-shaped nests often sit on pale limbs or trunks, while pouch nests tend to be pendent in dark parts of a tree, and cup nests are often dark with blue eggs. Reflection of excess sunlight by pale oven nests is an alternate explanation for this observation testable by seeing if pouch-shaped nests are also pale. Since pouch nests are rarely pale, the hypothesis is not supported. However, pouch nests would not keep sunlight off eggs in the middle of the day.

I also examined whether nests are protectively pale (in open zones) or dark (in closed zones) more often on the ground than above the ground (tests 8, 9). The hypothesis here is that ground nests are more accessible to predators (i.e., terrestrial as well as arboreal), and, thereby, selection favors concealing coloration. The results are significant in the expected direction, but only at the level of $P < 0.1$.

Finally, thin or tiny nests are often pale (test 10), supporting the original hypothesis that small nests can easily be placed on pale leaves and limbs and, thus, should be pale. Since small or thin nests are pale, and none is oven-shaped, the hypothesis (above) that pale nests serve to reflect heat becomes even less likely.

Small nests (Table 6: tests 1–6).—Small nests, although pale, are often located under leaves (test 1). Either a leaf reflects excess heat from sunlight, or it hides small nests from predators (large nests could rarely be hidden in this fashion). If small nests are placed under leaves as a means of gaining protection from rain or sun, one would expect oven-shaped small nests to be a suitable alternative, but no oven-shaped small nests were found (test 2). Such nests may attract more predator attention than do small cup nests. They might, however, simply be difficult to keep small.

Small nests are often located in palms (test 3), where they are hidden under large leaves (test 4) or associated with spiny trunks (test 5). No small nests were found on the ground, but, in fact, few species placed nests on the ground regardless of size (not significant, test 5). It also may be that small birds would not be able to defend ground nests against predators. Inundation could be a great problem for tiny nests on the ground, as it could for ground nests deep in burrows or for oven-shaped nests with entrances at ground level; no ground nest had raised sides to protect it from inundation.

Cavity nests (Table 7: tests 1–4).—Loss rates of eggs and young in nests in cavities (including woodpecker holes) were lower for those in cup nests (Oniki 1979a). Cavity nests were seldom on or near the ground (test 1). This is inconsistent with a hypothesis that tree cavities are used only because they provide shade or protection from rain, for then tree cavity nests would occur frequently below 1 m. Results are consistent with a hypothesis that ground predators, especially snakes, would raid tree cavities below 1 m, or attack birds digging cavities. Ground burrows are used for nesting, but because of horizontal entrances are not easily seen. Since adults digging burrows are subject to predation, protection of adults seems less likely to cause avoidance of low tree cavity nests than does protection of nest contents. Cavity nests do not occur in herbaceous species near the ground (test 2), even where masses of herbs provide room for such nests. Cavity nests do occur in bromeliads and similar epiphyte masses well above the ground.

TABLE 6

χ^2 CONTINGENCY-TABLE TESTS OF HYPOTHESES CONCERNING SMALL NESTS OF AMAZONIAN BIRDS

Test	Row A/B	Column C/D	Number of species				χ^{2a}
			AC	AD	BC	BD	
1	Nest small/large	Under leaf, etc./not	7	7	22	76	4.94*
2	Nest small/large	Oven/not	0	29	26	144	5.11*
3	Nest small/large	Palm/not	7	9	24	104	5.44*
4	On leaf/not	Palm/not	22	2	9	111	83.23**
5	Spiny trunk/not	Palm/not	21	8	14	107	40.07**
6	Nest small/large	Ground/not	0	17	16	155	1.96 n.s.

[a] As in footnote a, Table 1.

Cavity nests were seldom in dense or dark places (tests 3, 4). Possibly, the Amazonian cavity-nesting birds avoided sites where adults or young might be caught by predators. Difficulty in seeing inside the nest seems an unlikely reason especially for dense sites; there, foliage is generally present in response to light. Avoidance of cavities in open (sunlit) sites and in dark (shady) sites could be avoidance of both dry and moist microclimates. My data do not distinguish between these alternatives.

Human and other limitations of nest sites (Table 8: tests 1–10).—Cavity nests (including burrows) rarely occur near houses, if one omits nests inside houses (test 1). Nests in herbaceous vegetation and on the ground also are infrequent near houses (test 2). In both cases, humans and their commensals (dogs, rats) probably interfere with nesting.

Herbaceous vegetation, in addition to restricting nest height and being unsuitable near humans, is disadvantageous in that it never provides dark sites for nests. Herbs and epiphytes require light to survive and therefore more data would likely produce a significant negative relationship between dark areas and nests in herbaceous vegetation (test 3).

Cavity nests did not occur in live palms (test 4), although some were found in rotting palm stubs. Perhaps live palm trunks are difficult to excavate. Also, nests in palms were seldom more than 10 m up, since palms often are short in stature (test 5).

In the forest interior, certain types of nest location were unavailable or unused. Dense sites were rarely used in forests (test 6), and sites isolated on two to four sides were relatively seldom used (test 7). However, sites of both types are known to be used in the canopy, which was poorly sampled.

Nests on leaves, nests with pendent leaves, and nests isolated on islands or in water were rarely above 5 m up (tests 8–10). Wind probably would cause problems above 5 m in these cases, although predators might gain access to nests well above the water in tall vegetation. If so, these were the only significant effects of weather on this set of nests, and it is notable that these are not causes of nest-building but things that prevent nest-building in certain places.

DISCUSSION

WHY DO BIRDS BUILD NESTS?

No weather-related or thermal function postulated for the nests or eggs was supported by the evidence from the Amazonian samples. Many birds nest in the hot sun, without overhead

TABLE 7

χ^2 CONTINGENCY-TABLE TESTS OF HYPOTHESES CONCERNING CAVITY NESTS OF AMAZONIAN BIRDS

Test	Row A/B	Column C/D	Number of species				χ^{2a}
			AC	AD	BC	BD	
1	Tree cavity nest/not	<1 m/not	3	33	45	90	8.79**
2	Cavity nest/not	Ground herb/not	0	44	13	123	4.58*
3	Cavity nest/not	Dense place/not	2	31	35	62	10.92**
4	Cavity nest/not	Dark place/not	0	33	15	81	5.75*

[a] As in footnote a, Table 1.

TABLE 8

χ^2 CONTINGENCY-TABLE TESTS OF HYPOTHESES CONCERNING NEST-SITE AVAILABILITY

			Number of species				
Test	Row A/B	Column C/D	AC	AD	BC	BD	χ^{2a}
1	Cavity nest/not	Near houses/not	1	29	20	93	4.24*
2	Ground, herb/not	Near houses/not	0	24	26	136	4.56*
3	Dark/not	Herb, epiphyte/not	0	16	18	109	2.57 n.s.
4	Cavity nest/not	Live palm/not	0	31	32	84	10.82**
5	Palm/other tree	<10 m/not	31	4	54	31	7.48*
6	Site dense/not	Forest/edge, open	2	16	43	35	11.25**
7	Isolated 1 side/2–4	Habitat closed/open	35	19	9	57	33.50**
8	Nest on leaf/not	<5 m/not	23	3	74	54	8.65**
9	Nest pendent/not	<5 m/not	8	0	66	38	4.39*
10	Water around/not	<5 m/not	8	0	72	57	5.99*

ᵃ As in footnote a, Table 1.

cover. Unfavorable climatic conditions may have limited nest placement, explaining the absence of nests atop leaves or of pendent nests from presumably windy zones more than 5 m above ground, but it seems doubtful that birds of Belem and Manaus had to build nests to avoid rain, sun, or wind, or to gain some advantage from insolation. Statistical tests of factors related to nest predation were often significant, in contrast to tests related to weather. Obviously, some birds have secondarily come to use nest materials to heat eggs (megapodes), while others arrange nests to retain heat (eiders) or to avoid extreme heat (Ricklefs and Hainsworth 1969) or flooding. The basic reason that birds do not just incubate their eggs on the ground, however, is probably predation pressure, which forces them to build at the tips of limbs, inside tree cavities, and the like.

WHY ARE ROBIN EGGS BLUE?

The main function of egg color in the birds I studied seems to be protection against predation. Eggs not exposed to predators are white, so females do not expend energy or material on coloring. It is sometimes suggested that egg colors are excretory waste products, but it is difficult to believe that related birds with eggs of different colors are really different in excretory physiology, or that unrelated birds with similar egg colors are similar in physiology. There was no evidence in my data for either avoidance or retention of heat by egg color. Holyoak's (1969) suggestion that birds would break nonwhite eggs deep in cavities seems questionable. His jackdaws were probably not expecting dark eggs and may even have interpreted dark eggs as spoiled or already broken. Birds with blue eggs in cavities (*Sialia sialis* in North America, *Phoeniculus purpureus* in Africa) are not reported to break their eggs frequently.

Color may prevent breakage or burial of eggs of tinamous, however. The bright blue or glossy brown of tinamou eggs probably allows these birds to cover their eggs with leaf litter, then find every egg on return to the nest. The bright blue eggs of *Tinamus guttatus* were in deep forest shade, unlike all other blue eggs in the study, and, perhaps, have some special function.

All other blue eggs in this study were in thick, dark nests in isolated bushes in sunny areas. Since these eggs cannot be seen against the sky or canopy, they may be imitating spots of light on green leaves against a dark background (Oniki 1979b). Where no sunflecked green leaves are near the nest, blue eggs do not occur. Papageorgis (1975) suggested that a blue-and-black pattern in butterflies provides protective coloration in contrastingly-lighted, semi-open zones, such as the subcanopy and semiopen areas, and I interpret that pattern also as imitation of spots of light on dark green leaves. I would expect birds nesting in the subcanopy of tropical forests to have blue eggs, although I was unable to check this in my study. Robin's-egg blue, therefore, is normally a protective coloration in situations of contrasting light on green foliage.

Spotting and buff coloration tend to be associated with two other types of lighting. Buff eggs tend to occur against leaf litter and other dull but well-lighted substrates, especially near the ground. They correspond to the "tiger" pattern of butterfly color (Papageorgis 1975) in similar situations. Spotted white eggs tend to occur in thin nests in sparse foliage, especially where the eggs can be seen both from below and above. I suspect that spotted white eggs correspond to the "transparent" butterfly pattern (Papageorgis 1975), also of the forest interior or uniformly poorly lighted sites (such as grass tangles), and that in both cases the pale but spotted color tends to vanish against a speckled background.

ARE NEST SITES AND AVIAN SPECIES DIVERSITY LIMITED BY VEGETATION?

It is clear from several analyses of my data and from general considerations that diversity of bird egg-colors and nests is limited in habitats outside the forest. Humans limit the distribution of cavity and ground nests and of those in herbaceous vegetation, at least in areas of intensive use, while such low habitats as fields and palm groves limit nest heights and distribution of pouch nests. Not only are birds limited in species diversity by lack of differing foraging niches outside forest (MacArthur and MacArthur 1961), but also they are limited by lack of nesting niches and by consequent susceptibility to nest predation. Foliage height diversity was thought to restrict avian species diversity mainly by restricting bird foraging niches (MacArthur and MacArthur 1961), but the present study indicates that low foliage height diversity can directly reduce the number of nest niches and, hence, affect species diversity.

ACKNOWLEDGMENTS

Field work was conducted under grant GB-32921 from the National Science Foundation to R. H. MacArthur and E. O. Willis, and grant TC 6998-71 from the Conselho Nacional de Pesquisas of Brazil to Y. Oniki. I am grateful to the directors of the Agronomic Institute and Museu Goeldi at Belem and Instituto Nacional de Pesquisas de Amazonia at Manaus and to their field workers, who helped me to find nests. D. Wechsler helped with studies at Manaus. M. Corofino, owner of Sitio I-iruçanga, kindly permitted the study of nests on his property. I appreciate comments by M. S. Foster, E. H. Burtt, Jr., and an anonymous referee. E. O. Willis helped in the field and with the manuscript.

LITERATURE CITED

BROSSET, A. 1974. La nidification des oiseaux en forêt Gabonaise: architecture, situation des nids, et predation. Terre Vie 28:579–610.

CRUMP, M. L. 1971. Quantitative analysis of the ecological distribution of a tropical herpetofauna. Occas. Pap. Mus. Nat. Hist. Univ. Kans. 3:1–62.

HOLYOAK, D. 1969. The function of the pale egg colour of the Jackdaw. Bull. Br. Ornithol. Club 89: 159.

KOEPCKE, M. 1972. Über die Resistenzformen der Vogel-nester in einem begrenzten Gebiet des troposchen Regenwaldes in Peru. J. Ornithol. 113:138–160.

LACK, D. 1958. The significance of the colour of turdine eggs. Ibis 100:145–166.

MACARTHUR, R. H., AND J. MACARTHUR. 1961. On bird species diversity. Ecology 42:594–598.

MONTEVECCHI, W. A. 1976. Field experiments on the adaptive significance of avian eggshell pigmentation. Behaviour 58:26–39.

NOVAES, F. C. 1970. Distribuição e abundância das aves em um trecho da mata do Baixo Rio Guamá (Estado do Pará). Bol. Mus. Para. Emilio Goeldi. Nova Sér. Zool. 71:1–54.

ONIKI, Y. 1972. Studies of the guild of ant-following birds at Belém, Brazil. Acta Amazonica 2:59–79.

ONIKI, Y. 1979a. Is nesting success of birds low in the tropics? Biotropica 11:60–69.

ONIKI, Y. 1979b. Nest-egg combinations: possible antipredatory adaptations in Amazonian birds. Rev. Bras. Biol. 39:747–767.

PAPAGEORGIS, C. 1975. Mimicry in neotropical butterflies. Am. Sci. 63:522–532.

PINTO, O. 1953. Sobre a coleção Carlos Estevão de peles, ninhos e ovos das aves de Belém (Pará). Pap. Avulsos Zool. (São Paulo) 11:113–224.

PIRES, J. M., T. DOBZHANSKY, AND G. A. BLACK. 1953. An estimate of the number of species of trees in an Amazonian forest community. Bot. Gaz. 114:467–477.

RICKLEFS, R. E., AND F. R. HAINSWORTH. 1969. Temperature regulation in nestling Cactus Wrens: the nest environment. Condor 71:32–37.

SKUTCH, A. F. 1976. Parent Birds and Their Young. University Texas Press, Austin, Texas.

TAKEUCHI, M. 1961. The structure of the Amazonian vegetation. II, tropical rain forest. J. Fac. Sci. Univ. Tokyo, Sect. III Bot. 8:1–26.

WILLIS, E. O. 1962. Another nest of Pitangus lictor. Auk 79:111.

WILLIS, E. O. 1977. Lista preliminar das aves da parte noroeste e áreas vizinhas da Reserva Ducke, Amazonas, Brasil. Rev. Bras. Biol. 37:585–601.

MIGRATORY SHOREBIRDS: RESOURCE DEPLETION IN THE TROPICS?

DAVID SCHNEIDER

Smithsonian Tropical Research Institute, Balboa, Panama;
present address: Newfoundland Institute of Cold Ocean Science, Memorial University,
St. John's, Newfoundland A1B 3X7, Canada

ABSTRACT. Frequent reports of prey depletion by shorebird populations at temperate latitudes suggest that competition for food may be an active agent of natural selection on timing of migration. Yet rates of change in prey populations in coastal habitats used by migratory shorebirds in the tropics have not been considered. Therefore, I measured invertebrate densities in four intertidal feeding areas on the Pacific coast of Panama. Marked sites in each area were sampled in mid-January, late February or early March, and late April of 1978. Sites in one area, La Boca, were sampled again in mid-January and late February, 1979. Prey sizes were determined from fecal castings collected in the four areas and from a published analysis of the stomach contents of plovers collected in one of the study areas. During the 7-week period prior to the departure of wintering birds in April, the density of all invertebrate macrofauna decreased by 12%. The density of prey taxa increased by 10% during this same period. Prey patchiness, within study areas, decreased significantly during this 7-week period. The standing stock of polychaetes, within the size range taken by shorebirds, increased from 14.1 cm³/m² in early March to 26.8 cm³/m² in late April. The increase in standing stock was accompanied by a decrease in the density of large individuals and a decrease in the average length of worms. Changes in the density and patchiness of prey taxa were not consistent from year to year. The observed lack of consistency points toward disruptive rather than synchronizing selection on timing of migration in the neotropics.

RESUMEN. Noticias frecuentes sobre el agotamiento de presas producidos por poblaciones de aves costeras en latitudes templadas, sugieren que la competencia por comida puede ser un agente activo de selección natural del tiempo de migración. Sin embargo hasta ahora no se han considerado los cambios occurridos en las poblaciones de presas de los habitats costeros utilizados por aves migratorias costeras en lois trópicos. Por ello medí la densidad de invertebrados en cuatro áreas litorales de alimentación en la costa del Pacífico en Panamà. Se tomaron muestras en cada sitio marcado a mediados de enero, fines de febrero o principios de marzo y fines de abril de 1978. En el área La Boca se tomaron muestras de los sitios marcados a mediados de enero y fines de febrero de 1979. El tamaño de las presas fue determinado por el contenido de las feces juntado en las cuatro áreas y por datos publicados sobre análisis de contenidos estomacales de chorlitos, coleccionados en una de las áreas de estudio. Durante el período de siete semanas previo a la partida en abril de las aves invernantes, la densidad de toda la macrofauna de invertebrados disminuyó un 12%. La densidad de la taxa de las presas incremento un 10% durante ese mismo período. Concentraciones dispersas de presas ("prey patchiness") en el área de estudio, disminuyeron de manera significativa durante este período de siete semanas. La cantidad de lombrices, dentro del rango de tamaños comidos por los chorlitos, se incremento de 14,1 cm³/m² a comienzos de marzo hasta 26,8 cm³/m² a fines de abril. Este incremento en la cantidad de poliquetos (lombrices) fue acompañado por una disminución en la densidad de individuos grandes y una disminución en el tamaño promedio de las lombrices. Los cambios en la densidad y distribución de concetración ("patchiness") de la taxa presa no fueron consistentes de un año a otro. La ausencia de consistencia observada indica una selección disruptiva en lugar de sincronizante en el tiempo de migración en los trópicos del Nuevo Mundo.

Substantial numbers of North American sandpipers (Scolopacidae) and plovers (Charadriidae) spend the greater part of their lives in coastal habitats where they feed on a variety of intertidal invertebrates. Seasonal depletion of intertidal prey populations is a potentially important agent of synchronizing selection on timing of migration; the available evidence shows that prey depletion occurs in temperate and subtropical foraging areas (O'Connor and

Brown 1977; Evans et al. 1979; Quammen 1980; Schneider and Harrington 1981). Little is known about depletion rates of prey populations in the neotropics, where substantial numbers of shorebirds spend the non-breeding season (Pitelka 1979). I investigated depletion rates of prey populations in four intertidal areas used by foraging shorebirds in the Bay of Panama (8°56′N, 79°32′W). Change in the density, patchiness, and standing stock of prey taxa were measured over a three-month period prior to the departure of wintering shorebirds from the area (Schneider and Mallory 1982).

The Bay of Panama was chosen for the investigation because of extensive intertidal flats along its northern boundary. Tides in the Bay of Panama expose more than 1 km of intertidal flat along 150 km of coastline east of Panama City. Extensive intertidal flats also occur in the northwest corner of the Bay, near Aguadulce. Study areas were chosen because of their proximity to laboratory facilities at the Naos Marine Station operated by the Smithsonian Tropical Research Institute. Dexter (1972) reported a diverse community of intertidal invertebrates in the Naos area. Strauch and Abele (1979) showed that plovers (*Charadrius* spp.) wintering in the Naos area fed on a variety of intertidal invertebrates within definite size ranges.

LOCATION AND METHODS

Bird counts and invertebrate samples were taken in four areas near the south entrance of the Panama Canal (Fig. 1). All four areas were exposed twice a day by a semi-diurnal tide with a range of 6 m. The four areas differed in wave exposure and sediment type. The beach on Culebra Island (Fig. 1) was exposed to waves from the south; it consisted of well-sorted sand mixed with shell fragments. Invertebrate zonation down this beach was described by Dexter (1972). The beach at Amador (Fig. 1) was less exposed to wave action. The beach consisted of a flat of silty sand near mean low water, a field of cobbles and mud at mid-tide level, and a berm of coarse sand running up to extreme high water. Prey sizes of *Charadrius* plovers from this beach were described by Strauch and Abele (1979). The sheltered flat at Punta Mala (Fig. 1) was bounded on the north by Panama City and to the south by Fort Amador. The flats ranged from silty sand to fine mud with high water content. Area exposed at low tide was approximately 80 ha. The sheltered flats at La Boca (Fig. 1) consisted of fine sand with silt, grading to mud in meandering channels. Area exposed at low tide was approximately 90 ha.

Shorebird counts were made in all four areas from mid-March until late May, 1978. Counts at Punta Mala were made approximately 2 hr after low tide, as the flats were covered by the rising tide, and birds were concentrated toward the west end of the cove (Fig. 1). Counts at La Boca were made approximately 2.5 hr after low tide, as the rising tide concentrated birds toward the western margin of the flat. Counts at Amador were made approximately 3 hr after low tide; counts at Culebra Beach were made a few minutes later.

Shorebird counts were made during brief visits to other coastal locations, including Farfan (Fig. 1), Venado Beach (south of Farfan), Puerto Caimito (30 km W Panama City), Punta Chame (70 km SW Panama City), Aguadulce (150 km SW Panama City), and Juan Diaz (15 km E Panama City).

Wooden stakes were used to mark invertebrate sampling sites in each of the four study areas in mid-January 1978. Three sites were established on a transect up the beach at Culebra. The lowest site was exposed for ca. 2 hr per tidal cycle, the highest site was exposed for ca. 6 hr per tidal cycle. Two more transects of three sites each were established on the beach at Fort Amador. The two lowest sites were exposed for less than 2 hr at low tide, the two mid-beach sites were exposed for 3 to 4 hr per tidal cycle, and the two highest sites were exposed 6 to 8 hr per tidal cycle.

At Punta Mala, five sites were established at 50 m intervals along a transect extending from the southwest side of the cove (exposed 7–8 hr per tidal cycle) northeastward toward the middle of the cove (exposed 5–6 hr per tidal cycle). A sixth site was established near the south side of the cove; exposure was 5 to 6 hr per tidal cycle. At La Boca, five sites were set up at 50 m intervals along a transect that extended from the foot of a mangrove tree (exposed 8 hr per tidal cycle) eastward across a flat of silty sand, to a site exposed about 5 hr per tidal cycle. A sixth site (exposed about 8 hr per tidal cycle) was set on a muddy substrate at the base of a second mangrove tree.

Replicate cores were taken at each site by pushing a 10 cm diameter tube into the substrate to a depth of 10 cm. The tube was tipped to free the core, which was washed on a 1 mm mesh sieve. Organisms retained on the sieve were removed to the laboratory, where they were

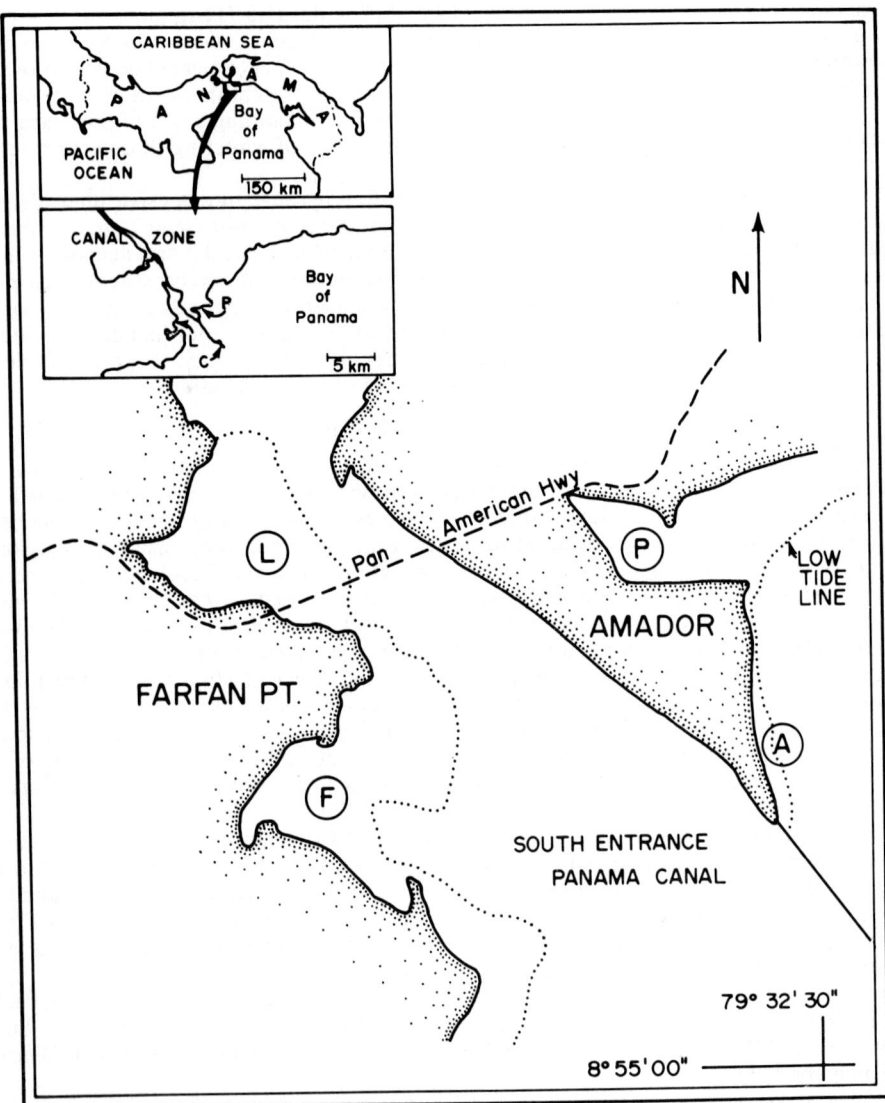

FIG. 1.   Study areas in Bay of Panama. A = Amador. C = Culebra Beach. F = Farfan. L = La Boca. P = Punta Mala.

sorted on trays. Published keys were used to identify organisms to the lowest taxonomic level. Not all keys were to the species level, so a catalogue was created, with descriptions of features normally used in differentiating genera and species within a family. The greatest dimension of each organism was recorded. Worm length was measured to the nearest centimeter. Worm diameter was also recorded for comparison with the work of Strauch and Abele (1979). Worm volume was the product of length and average cross-sectional area within each length class.

Two cores were collected at each of the 21 sites in mid-January; four cores were collected at each site in late February or early March; eight cores were collected at each site in late April. This sampling regime, based on increasing replication, was chosen because it is more efficient than equal replication for detecting decreases in density. A small number of cores at the outset also reduces the chance that sampling itself will alter invertebrate densities. Cores were taken at the six La Boca sites in mid-January and late February, 1979. All cores at a site were taken within a 2 m diameter area.

An attempt was made in January to exclude birds from a part of each sampling site. Four

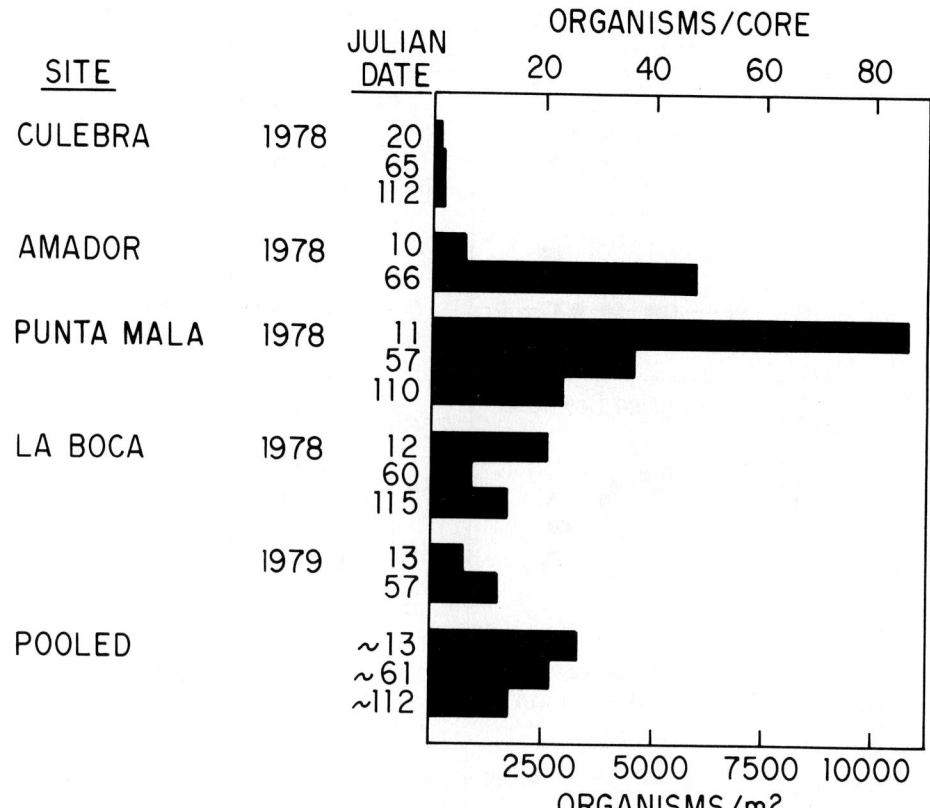

Fig. 2. Change in average density of all invertebrates collected in core samples. Pooled values are for all samples, regardless of year or location. For Julian Date, 1 = 1 January, 32 = 1 February, etc.

stakes were set in a 1 m square at each site. Plastic rope was strung around the tops of the stakes. Shorebirds walked through the roped areas, so wire canopies were placed over the stakes at two sites in early March. Canopies were checked for the presence of birds on subsequent visits to the areas. The canopies were open to lateral entrance by fish, crabs, and other high tide predators. The possibility that lateral movement of invertebrates might obscure cage effects (Goss-Custard et al. 1977) was checked by comparing rates of change in mobile and sedentary groups.

Prey size was determined by matching whole organisms with parts found in fecal castings collected in the four study areas. Regurgitated pellets were not found and were not used in the analysis. Fecal castings from Whimbrel (*Numenius phaeopus*) were easily distinguished from other castings, so the prey of Whimbrel were analyzed separately, Prey sizes of small plovers (*Charadrius wilsonia* and *Charadrius semipalmatus*) were obtained from Strauch and Abele (1979).

Invertebrate parts used in the analysis of prey size included crustacean legs and chelae, polychaete jaws and setae, and clam hinges. Nemerteans and phoronids, which lack hard parts, were commonly encountered in core samples. Nemerteans in excess of 2 cm were added to the list of prey taxa, based on the size of nemerteans found in the stomachs of Black-bellied Plover (*Pluvialis squatarola*) collected at Plymouth, Massachusetts (Schneider and Harrington 1981). Phoronids did not exceed 2 cm in length, and have not been reported as shorebird prey, and so were not added to the list of prey taxa. Based on the analysis of prey size, invertebrates were assigned to one or more of the following categories: animals taken by Whimbrel, animals taken by small plovers, animals taken by any shorebird, and animals not taken by any bird. The last two categories were mutually exclusive; the third category included the first two categories.

The invertebrate data did not meet the assumptions for analysis of variance, so G-statistics

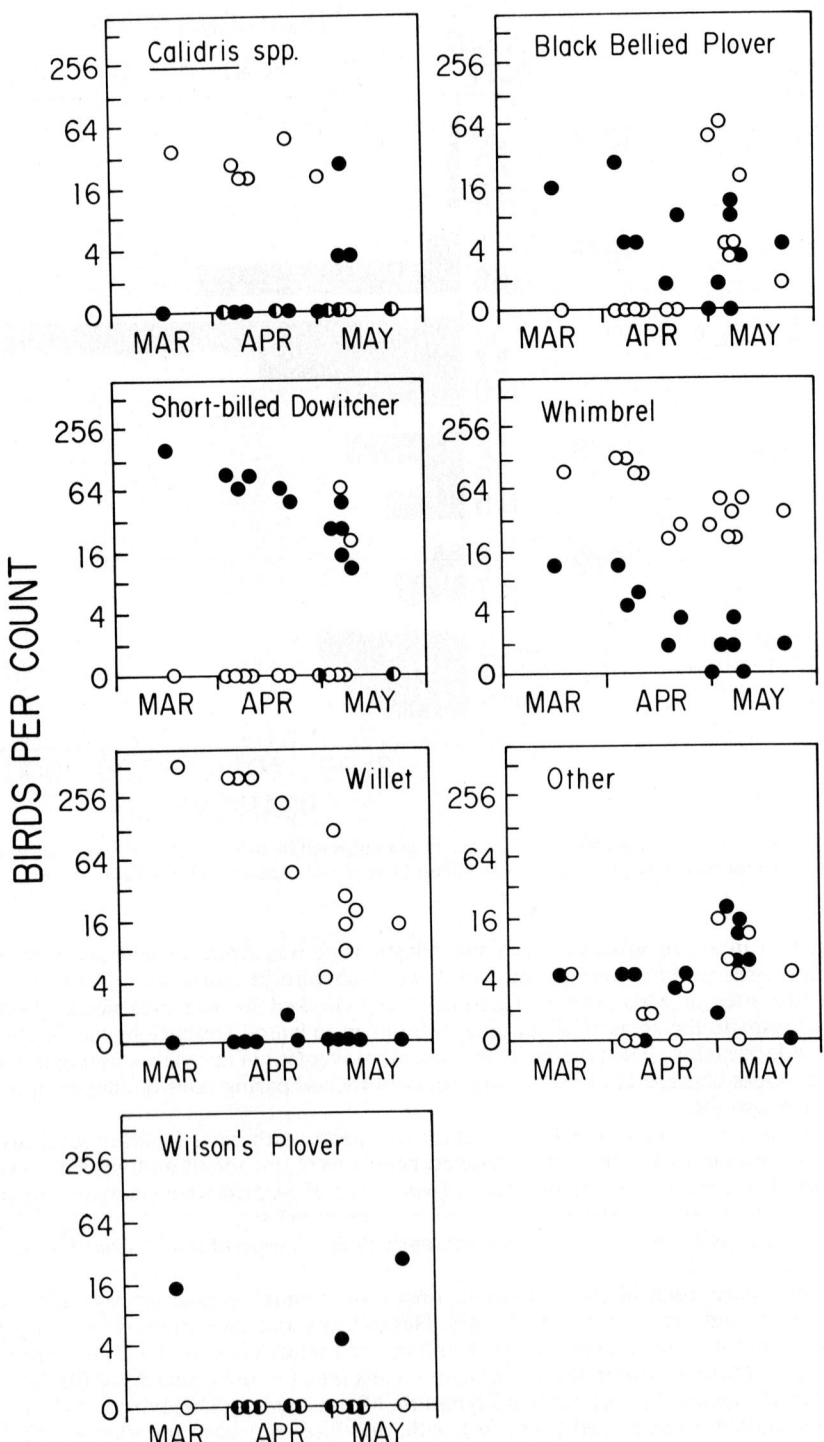

FIG. 3. Abundance of migratory shorebirds at La Boca (solid circles) and at Punta Mala (hollow circles). Half solid circle indicates the same count at both locations.

TABLE 1

SIZE OF INVERTEBRATES FOUND IN FECAL CASTINGS AND IN STOMACH CONTENTS OF
MIGRATORY SHOREBIRDS IN THE BAY OF PANAMA[1]

	Fecal castings		Plovers' stomach contents[2]
	Size[3]	Location	Size[3]
Whimbrel			
Crabs	5–15	A, C, L, P	
All charadriids and scolopacids			
Polychaetes (Annelida)			
Nereids	>20	A, L, P	>10 (0.5–2.5 wide)
Glycerids			>20 (1–2 wide)
Sabellids	>20	L	
Pelecypoda (Mollusca)			
Tellinids	>3	L	
Veneriids	5–10	A, P	
Gastropoda (Mollusca)			1–4
Crustacea			
Peracarids	3–10	L	2–6
Crabs	5–15	A, C, L, P	1–11
Other decapods	3–30	A, C, L	1–6
Brachiopoda			5–7

[1] Collections at Amador Beach (A), Culebra Beach (C), La Boca (L), and Punta Mala (P).
[2] Data from Strauch and Abele (1979).
[3] Maximum dimensions, in mm.

(Sokal and Rohlf 1969) were used to test whether changes in density were significant. Under the null hypothesis, no change in density, the number of organisms collected in an area was expected to be proportional to the number of cores collected. For example 12 cores were collected at Punta Mala in mid-January, and another 24 cores were collected at the same sites in early March. One would expect twice as many organisms in the early March collection, if there were no change in density. G-statistics were computed according to the following formula,

$$G = 2(n \ln n - n \ln \hat{n} + m \ln m - m \ln \hat{m}),$$

where $\hat{n} = p(n + m)$, $\hat{m} = (1 - p)(n + m)$, n = number of organisms collected on a given date, p = proportion of cores collected on that date, m = number of organisms collected on a subsequent date, and $1 - p$ = proportion of cores collected on the subsequent date.

The number of organisms collected in the canopied areas was expected to be proportional to the number of cores collected, and was also expected to be independent of where cores were collected (inside versus outside the canopy) if there were no significant canopy effect. The formula for a two-way contingency test (two dates vs. two treatments) was used to calculate the G-statistic. Exact probabilities were computed from the area under a Chi-square distribution with a single degree of freedom.

Patchiness within each area was measured as the variance in average density among sites on each date. Average density was the number of organisms collected at a site, divided by the number of cores from the site. An F-ratio was used to test whether variability within an area on one date differed from variability in that area on the previous date. This test does not involve a partitioning of variance (ANOVA) and its attendant assumptions.

RESULTS

SPATIAL VARIATION IN ABUNDANCES OF INVERTEBRATES AND SHOREBIRDS

The four study areas differed in density and relative abundance of both invertebrate and shorebird faunas. The lowest average invertebrate density was found at Culebra Beach (Fig. 2). A total of 16 types of invertebrate was collected, and of these, nine were polychaetes. Numerically dominant taxa included mole crabs (*Emerita* sp.), isopods (*Cirolana* sp.), and a

TABLE 2

DENSITY OF INVERTEBRATES WITHIN SIZE RANGES TAKEN BY SHOREBIRDS

	January			February–March			April	
	Organisms per core[1]	No. of cores	P^2	Organisms per core	No. of cores	P^2	Organisms per core	No. of cores
Taxa taken by any shorebird[3]								
1978								
Culebra	0.0	6	.004	0.83	12	.02	0.25	24
Amador	2.6	10	.94	2.65	20			
Punta Mala	10.5	12	.14	8.75	20	.89	8.98	43
La Boca	5.8	12	<.001	2.3	24	.06	3.1	48
1979								
La Boca	2.5	12	.03	3.9	24			
Pooled (1978)	6.5	30	<.001	4.3	56	.74	4.7	115
Taxa taken by plovers (*Charadrius* spp.)[4]								
1978								
Amador	1.1	10	.14	1.8	20			
La Boca	5.7	12	<.001	1.5	24	.09	2.0	48
1979								
La Boca	1.7	12	.16	2.4	24			

[1] All cores = 10 cm in diameter, 10 cm deep.
[2] Based on *G*-statistics (see text).
[3] Based on size ranges in Table 1.
[4] Based on sizes from Strauch and Abele (1979).

glycerid polychaete. The beach was visited by an occasional Whimbrel. *Emerita* fragments were found in Whimbrel fecal castings from this beach.

Invertebrate density was higher at Amador Beach than at Culebra Beach (Fig. 2). A total of 64 types of invertebrate were collected, and of these, 38 were polychaetes. The invertebrate faunas at the two low beach sites were dominated by tubiculous worms—onuphid polychaetes, sabellid polychaetes, and phoronids. The fauna at the mid-beach sites was dominated by crevice seeking forms—crabs and errant polychaetes. The high beach sites were inhabited by a small number of worms, mostly phoronids and polychaetes. Shorebird counts were highly variable at Amador, which was frequented by people. Maximum counts of foraging birds were 127 *Calidris* sandpipers (mostly *C. minutilla*), 56 Wilson's Plover (*Charadrius wilsonia*), 34 Willet (*Catoptrophorus semipalmatus*), 26 Ruddy Turnstone (*Arenaria interpres*) and six Whimbrel. Flocks of several hundred small *Charadrius* plovers were observed roosting on the beach after the flats were covered by the tide. Foraging Whimbrel were evenly distributed along the beach, whereas other species foraged in flocks over the middle and lower reaches of the beach.

Whimbrel castings from Amador contained brachyuran crab remains, hippid (mole) crab remains, and veneriid clam fragments. Willet castings contained brachyuran crab remains, clam remains, polychaete setae, and snail fragments. Identifiable groups included veneriids, sabellids, and a naticid. Castings from *Calidris* sandpipers contained crustacean, polychaete, and insect remains. Turnstone castings were not found, but birds were observed eating crabs at this beach.

Invertebrate density at Punta Mala was higher than at the other three areas (Fig. 2). A total of 70 types of invertebrate was collected at Punta Mala and of these, 36 were polychaetes. Nereids, spionids, and cirratulids were the numerically dominant polychaete families. Clams (*Felaniella, Corbula, Mactra, Tellina, Cyclinella, Tagelus*) were abundant at wetter sites, as were crustaceans (callianassids, xanthid crabs, and ocypodid crabs). Willet, Whimbrel, and Western Sandpipers (*Calidris mauri*) were the most abundant wintering birds (Fig. 3). Black-bellied Plover and Short-billed Dowitchers (*Limnodromus griseus*) were briefly abundant during spring passage in early May (Fig. 3). Willet castings from Punta Mala contained crab, clam, polychaete, and snail remains. Identifiable groups included ocypodids and xanthids (crabs), veneriids (clams), nereids and sabellids (polychaetes). Crab remains were found in smaller fecal castings, which probably came from small sandpipers and plovers.

TABLE 3
Patchiness of Invertebrates within Size Ranges Taken by Shorebirds

	January			February–March			April	
	Variance among sites	No. of sites	P^1	Variance among sites	No. of sites	P^1	Variance among sites	No. of sites
Taxa taken by any shorebird[2]								
1978								
Culebra	0.0	3		0.75	3	.33	0.88	3
Amador	9.35	5	.43	6.08	5			
Punta Mala	137.4	6	.35	192.8	6	.01	24.8	6
La Boca	178.9	6	<.001	0.97	6	.004	11.5	6
1979								
La Boca	4.0	6	.054	16.6	6			
Pooled (1978)	126.5	15	.22	77.7	15	.002	14.7	15
Taxa taken by plovers (*Charadrius* spp.)[3]								
1978								
Amador	3.11	5	.35	1.63	5			
La Boca	234.2	6	<.001	1.24	6	.10	3.78	6
1979								
La Boca	3.88	6	.10	11.98	6			

[1] Based on *F*-statistics (see text).
[2] Based on size ranges in Table 1.
[3] Based on sizes from Strauch and Abele (1979).

Invertebrate density at La Boca was lower than at Punta Mala (Fig. 2). A total of 112 types of organism was collected, and of these, 68 were polychaetes. Nereids were the most abundant organism, but did not dominate the samples as they did at Punta Mala. Clams, sipunculids, nemerteans, amphipods, brachyuran crabs, and a brachiopod (*Glottidea* sp.) were also abundant. Short-billed Dowitchers, Black-bellied Plovers, and Wilson's Plovers were the most abundant foragers in the winter (Fig. 3). Other species foraging in this area were Whimbrel, Willet, Spotted Sandpipers (*Actitis macularia*), Semipalmated Plovers, and Greater Yellowlegs (*Tringa melanoleuca*). *Calidris* sandpipers were briefly abundant during spring passage (Fig. 3). Foraging Whimbrel at La Boca defended territories (Mallory 1981). Fecal castings from Whimbrel contained crab remains and an occasional tellinid fragment. Dowitchers foraged in moving flocks. Fecal castings of Dowitchers contained setae from nereids and sabellids, and an occasional crab or mollusc fragment. Crab remains were found in the fecal castings of Wilson's and Black-bellied Plovers.

Temporal Variation in Abundances of Invertebrates

Density of macrofaunal invertebrates, averaged over all four study areas, decreased 21 percent between mid-January and early March, then decreased by 33 percent between early March and late April (Fig. 2). This reduction in density was due primarily to the reduction in density at Punta Mala, the area with the highest density of invertebrates in January. Densities at La Boca and Amador increased after early March (Fig. 2). The direction of change in density was not consistent from year to year at the La Boca sites. During 1978 densities decreased, but during the same period in 1979 densities increased (Fig. 2).

The size range of taxa preyed on by shorebirds is listed in Table 1. During the 7-week period prior to late April the density of prey taxa within these size ranges did not decrease (Table 2). Significant decrease in the density of taxa taken by shorebirds occurred before early March, rather than afterward (Table 2). Significant decrease in the density of taxa taken by small plovers occurred at La Boca before early March, but did not occur after early March (Table 2). Prey taxa taken by Whimbrel increased from 0.13 organisms/core in January to 0.33 organisms/core in early March, with no change afterward (cf. Mallory 1981).

During the seven-week period from early March to late April, patchiness of prey taxa decreased at Punta Mala and at La Boca (Table 3). Patchiness in these two areas increased before early March. Change in patchiness was not consistent from year to year. The direction

TABLE 4

LENGTH AND VOLUME OF WORM TAXA IN SIZE RANGES TAKEN BY SHOREBIRDS

	January	February–March	April
Identified in fecal castings			
All areas 1978			
Number of worms	97	185	451
Density of worms (no./m^2)	412	421	499
Average length (cm)	1.7	2.3	2.0
Variance in length (cm^2)	.99	2.17	2.11
Volume (cm^3/m^2)	4.8	14.1	26.8
Identified in plover stomachs[1]			
La Boca 1978			
Number of worms	5	21	66
Density of worms (no./m^2)	53	111	178
Average length (cm)	2.2	2.1	1.6
Variance in length (cm^2)	1.2	.95	2.4
Volume (cm^3/m^2)	.35	1.2	4.6
La Boca 1979			
Number of worms	13	41	
Density of worms (no./m^2)	138	218	
Average length (cm)	1.9	1.9	
Variance in length (cm^2)	.91	1.2	
Volume (cm^3/m^2)	.70	2.3	

[1] Size ranges taken from Strauch and Abele (1979).

of change was negative during the early part of 1978, and positive during the same period in 1979 (Table 3). Change in patchiness of prey taken by small plovers was not consistent from year to year. Patchiness decreased during early 1978 and increased during the same period in 1979 (Table 3).

Worm volume, an index of standing stock, increased in prey taxa from 4.8 cm^3/m^2 in mid-January to 26.8 cm^3/m^2 in late April (Table 4). The increase in volume prior to early March was due to an increase in the density of large worms, especially worms longer than 6 cm (Fig. 4). The increase in worm volume after early March was due to an increase in small (less than 6 cm) worms. The density of large worms decreased after early March (Fig. 4).

Worm volume, in those taxa taken by small plovers (Table 1), increased from 0.35 cm^3/m^2 in mid-January to 4.6 cm^3/m^2 in late April (Table 4). This change was due to an increase in density rather than an increase in average length (Table 4). Average length decreased due to reduced density of large worms and an increased density of small worms. The direction of change in worm volume was consistent from year to year (Table 4).

The canopy at La Boca reduced usage by shorebirds. Dowitchers ceased walking through the roped area at La Boca after the wire canopy was set over the ropes. No birds were seen walking under the wire canopy at La Boca, or under the canopy at Punta Mala. Not all shorebirds were excluded by the canopies. Footprints smaller than those of Dowitchers were observed on a few visits to the La Boca site. The footprints may have been those of Spotted Sandpipers, which were observed in the vicinity of the canopied area at La Boca.

Canopies had no significant effect on the rate of change in the density of all invertebrates (Table 5). The observed direction of change in the density of prey taxa increased beneath the canopy at La Boca, but decreased beneath the canopy at Punta Mala (Table 5). The contribution of lateral movement of invertebrates to this result was tested by comparing a sedentary prey group (clams), with mobile prey taxa (crustaceans and errant polychaetes). Treatment effects were greater for clams than for more mobile groups (Table 5). However, the number of clams collected was too small to permit computation of a G-statistic.

DISCUSSION

The density of prey taxa, averaged over all four study areas, did not decrease during the 7-week period prior to the departure of wintering birds, in mid to late April. Patchiness of

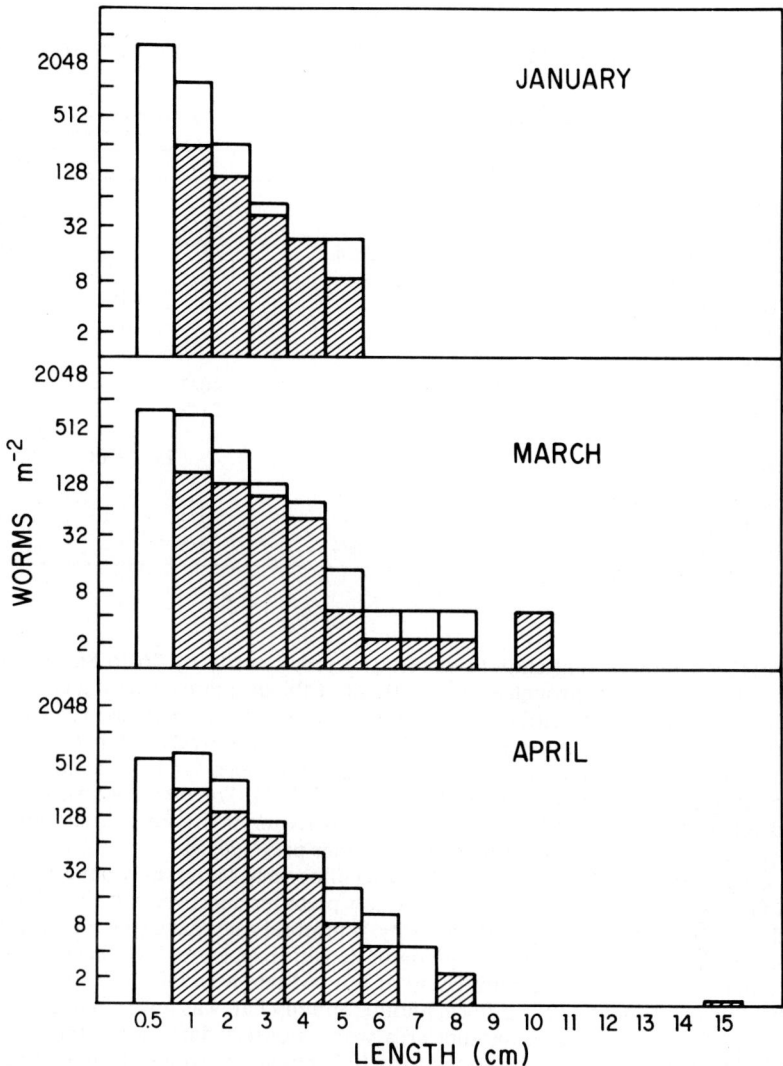

Fig. 4. Distribution of worm lengths, by collection date. Height of bar is frequency of all worms. Shaded portion is fraction identified as prey from fecal castings of shorebirds.

prey taxa decreased significantly during this same period. At least two mechanisms could have caused the reduction in patchiness without reducing average density. One is lateral movement of organisms, including diffusion of organisms away from high density sites. A second mechanism is differential removal from high density sites, balanced by recruitment and growth of larval invertebrates at low density sites. The second mechanism seems more likely in a community composed of infaunal clams and polychaetes, with generally low mobility.

A change in patchiness without any accompanying change in average density is consistent with the result of Duffy et al. (1981), who reported spatially heterogeneous rates of change in prey abundance over a one-month period in Peru (13°S latitude). The average rate of change in Panama, as in Peru (Duffy et al. 1981), was below those reported for temperate and subtropical foraging areas. In England (54°N) Evans et al. (1979) reported a 90 percent decrease in the biomass of benthic prey during an 8-month period. In northern Ireland (55°N) O'Connor and Brown (1977) reported a 71 percent decrease in the density of preferred prey during a 3-month period. At Plymouth, Massachusetts (42°N) Schneider and Harrington (1981) reported depletion rates ranging up to 90 percent over a 6-week period. In California (32°N)

TABLE 5

CHANGE IN NUMBER OF INVERTEBRATES BENEATH TWO CANOPIES, COMPARED TO CHANGES IN
NUMBERS OUTSIDE CANOPIES

	La Boca		Punta Mala	
	In	Out	In	Out
All macrofauna				
Initial (number/core)	27/2	31/2	134/2	248/2
Final (number/core)	78/4	78/4	29/3	75/4
				(=56.25/3)
Probability (one-tailed)[1]	.33		.43	
Combined tests		$P = .14$		
Identified in fecal castings				
Initial (number/core)	2/2	5/2	40/2	43/2
Final (number/core)	19/4	21/4	21/3	41/4
				(=30.75/3)
Probability (one-tailed)[1]	.17		In < Out	
Clam taxa in fecal castings				
Initial (number/core)	1/2	3/2	2/2	1/2
Final (number/core)	3/4	2/4	3/3	1/4

[1] Alternative hypothesis H_A: in (final/initial) > out (final/initial).

Quammen (1980) showed that shorebirds reduced prey density by 26 to 80 percent during the winter season (October through April), with all of the decrease occurring before January. For a similar latitude in the southern hemisphere Puttick (1980) reported removal of 34 to 50 percent of prey stocks during one year, with a maximum removal of 10 percent per month. Grant (1981) reported no decrease in the density of one prey species (*Acanthohaustorius millsi*) during a 57-day period in the fall, in South Carolina (32°N). The prey species, an amphipod, accounted for no more than 10 percent of the food requirements of one shorebird species during this period, and Grant did not report on other prey species.

The lack of reduction in prey density in Panama after early March may have been the result of several factors, including avian territoriality, low predator densities, low prey densities, and high production rates. Defense of territories by Whimbrel may have prevented depletion of its food resources, primarily crabs. Low predator densities appear to have been a contributing factor in other species. The maximum count at Punta Mala was 550 birds in an 80 ha area, or roughly 7 birds/ha. Density at La Boca, another muddy flat, was lower than at Punta Mala. Maximum density at Aguadulce, another area with extensive flats, was 8 foraging birds/ha along a 1 km transect perpendicular to the shoreline. Higher densities of foraging birds could not be found during visits to mudflats at Puerto Caimito, Farfan, Venado, and Juan Diaz. These densities are low relative to those reported at higher latitudes in the new world (Quammen 1980; Schneider and Harrington 1981). Low densities were not due to a lack of birds. Over 50,000 shorebirds were counted at Juan Diaz on 13 March 1979 as the birds flew westward toward Panama City on a rising tide. Foraging density and residence time of these birds could not be determined from the single visit.

Human disturbance at Culebra and Amador was frequent and may have contributed to the low densities of birds at these beaches (Strauch and Abele 1979). At a relatively isolated beach, Punta Chame, the density of foraging birds in a 20 m wide zone along 1 km of beach was 50 birds/ha, a density comparable to that reported at a migratory stopover with high rates of prey depletion (Schneider and Harrington 1981). Human disturbance did not appear to be a factor contributing to low shorebird densities on extensive mudflats such as Punta Mala and La Boca.

Low densities of foraging shorebirds, especially on mudflats, cannot be the sole reason for low rate of prey depletion, for prey densities were also low relative to higher latitudes. The highest prey densities occurred at Punta Mala, but this area was grossly polluted, and densities of all invertebrates in this area were higher than in the other three areas. Invertebrate densities at La Boca appeared to be typical of other mudflats along the coast, based on spot sampling at Aguadulce, Puerto Caimito, Venado, and Farfan. Density of prey at La Boca was 742 organisms/m² in January 1978. At higher latitudes, Evans et al. (1979) reported prey densities of 3400 organisms/m²; Schneider and Harrington (1981) reported densities of individual prey species ranging from 2 to 1540 organisms/m².

High rates of productivity may also contribute to low rates of prey depletion. Benthic productivity appears to have been high during the study, with heavy cropping by high tide predators, although this was not tested directly. Standing stock of worms increased between January and early March. The increase was due to growth and recruitment of juvenile organisms, as indicated by increases in the density of smaller worms. Worm production after early March evidently exceeded consumption by predators, since volume increased while density of large worms decreased. Lee (1978) reported high rates of recruitment and predation of shallow subtidal benthic populations near Amador during the dry season (January through April), a period of seasonal upwelling in the Bay of Panama (Schott 1931).

Recruitment and growth of intertidal populations during the dry season would seem to favor a seasonal influx of shorebirds into the Bay of Panama. However, the direction of change in the density of prey taxa was not consistent from year to year at La Boca. If shorebird prey populations at low latitudes are less seasonally predictable than at high latitudes, then "hard" selection (Wallace 1968) may maintain variability in migration schedules within shorebird populations wintering at low latitudes. Annual consistency in the prey supplies of shorebirds needs to be investigated more thoroughly, as do the effects of change in prey supply on the energy balance and survivorship of migratory shorebirds.

ACKNOWLEDGMENTS

I thank J. Lubchenco for canopy material, and E. P. Mallory for setting up the study sites, for making the January collection, and for permission to use data on bird numbers. I thank J. D. Goss-Custard, E. P. Mallory, J. P. Myers, P. Petraitis, and J. Strauch for comments on earlier drafts. Special thanks go to N. Smith for discussion and encouragement. The work was supported by the Smithsonian Tropical Research Institute with funds provided by the John and Edward Nobel Foundations.

LITERATURE CITED

DEXTER, D. M. 1972. Comparison of the community structures of a Pacific and an Atlantic Panamanian sandy beach. Bull. Mar. Sci. 22:449–462.

DUFFY, D. C., N. ATKINS, AND D. SCHNEIDER. 1981. Do shorebirds compete in the tropics? Auk 98: 215–229.

EVANS, P. R., D. M. HERDSON, P. J. KNIGHT, AND M. W. PIENKOWSKI. 1979. Short-term effects of reclamation of part of seal sands, Teesmouth on wintering waders and shelduck. I. Shorebird diets, invertebrate densities, and the impact of predation on the invertebrates. Oecologia (Berl.) 41:183–206.

GOSS-CUSTARD, J. D., R. A. JENYON, R. E. JONES, P. E. NEWBERRY, AND R. B. WILLIAMS. 1977. The ecology of The Wash. II. Seasonal variation in the feeding conditions of wading birds. J. Appl. Ecol. 14:701–719.

GRANT, J. 1981. A bioenergetic model of shorebird predation on infaunal amphipods. Oikos 37:53–62.

LEE, H. 1978. Predation and opportunism in tropical soft-bottom communities. Unpubl. Ph.D. dissert., University of North Carolina, Chapel Hill.

MALLORY, E. P. 1981. Ecological, behavioral and morphological adaptations of a shorebird (the Whimbrel, *Numenius phaeopus hudsonicus*) to its different migratory environments. Unpubl. Ph.D. dissert., Dartmouth College, Dartmouth, New Hampshire.

O'CONNOR, R. J., AND R. A. BROWN. 1977. Prey depletion and foraging strategy in the Oystercatcher *Haematopus ostralegus*. Oecologia. (Berl.) 27:75–92.

PITELKA, F. A. 1979. Introduction: The Pacific coast shorebird scene. Stud. Avian Biol. 2:1–11.

PUTTICK, G. M. 1980. Energy budgets of curlew sandpipers at Langebaan Lagoon, South Africa. Estuarine Coastal Mar. Sci. 11:207–215.

QUAMMEN, M. L. 1980. The impact of predation by shorebirds, benthic feeding fish, and a crab on the shallow living invertebrates in intertidal mudflats of two southern California lagoons. Unpubl. Ph.D. dissert., University of California, Irvine.

SCHNEIDER, D., AND B. A. HARRINGTON. 1981. Timing of shorebird migration in relation to prey depletion. Auk 98:801–811.

SCHNEIDER, D., AND E. P. MALLORY. 1982. Spring migration of shorebirds in Panama. Condor 84:344–345.

SCHOTT, G. 1931. Der Peru-strom und seine nordlichen Nachbargebiete in normaler und anormaler Ausbildung. Ann. Hydrogr. Marit. Meteorol. 59:161–169, 200–213, 240–252.

SOKAL, R. R., AND F. J. ROHLF. 1969. Biometry. W. H. Freeman, San Francisco, California.

STRAUCH, J. G., AND L. G. ABELE. 1979. Feeding ecology of three species of plovers wintering on the Bay of Panama, South America. Stud. Avian Biol. 2:217–230.

WALLACE, B. 1968. Topics in Population Genetics. Norton, New York.

NEOTROPICAL-AFROTROPICAL BARBET AND WOODPECKER RADIATIONS: A COMPARISON

LESTER L. SHORT

Ornithology Department, American Museum of Natural History, New York, New York 10024 USA

ABSTRACT. Afrotropical Piciformes include the barbets (Capitonidae), honey-guides (Indicatoridae), and woodpeckers (Picidae), and neotropical Piciformes include barbets, toucans (Ramphastidae, here regarded as an offshoot of, and closely related to, barbets), and woodpeckers. The relationship of the other extant supposed pici-forms, i.e., jacamars (Galbulidae) and puffbirds (Bucconidae), to the families just mentioned is controversial, and they are not considered here.

Many complex factors affect adaptive radiation of bird families and orders. Access to continental areas from without, opportunities presented for speciation, relatively small sample sizes, and uniqueness of associated biotic and climatic events on different continents are some of these. The paucity of fossils also makes comparisons difficult.

The taxa discussed are of uncertain origin. The pantropical Capitonidae are diverse and numerous in the Afrotropics (seven genera, 42 species), and much less so in the neotropics (three genera, seven species), unless toucans are included (six genera, 33 species), whereupon the diversity and numbers become more comparable. Honey-guides are primarily Afrotropical (four genera, 15 species) with penetration of southern Asia by one genus (two species). Other than the one genus of honeyguide (*Indicator*) reaching Asia, all other genera of neotropical and Afrotropical barbets, toucans, and honeyguides are endemic to those regions. Afrotropical picid genera number only six, and of the 26 species, 23 represent the three endemic genera, two of which (*Campethera*, 10 species, *Dendropicos*, 12 species) show the only radiations of picids in that region. In contrast the neotropics have 95 species of picids in 11 genera, four of which are endemic.

Climatic and other historical factors affecting distribution of forests, but more particularly of woodlands, have played a major role in the radiation of Afrotropical barbets, honeyguides, and the two genera of woodpeckers. Honeyguides themselves may have influenced the radiation of the other two groups, upon which they are nest parasites. Surprisingly few piciforms have entered the Afrotropics from without (only the picid genera *Sasia* from Asia and *Picoides* from Eurasia), or left the region (only the two honeyguides, perhaps one *Jynx,* and probably the ancestor of *Picoides*). In the neotropics opportunities for speciation have been greater because of alternating wet-dry periods in Amazonia and other areas, and because of the Andean uplift and its associated biotic effects. There has also been some immigration of picids into the neotropics from the north. *Picoides* is definitely an Old World derivative that entered North, and then South America. Middle America and the West Indies have been important secondary centers of radiation in the neotropics, with emigration of species of *Melanerpes, Colaptes,* and *Campephilus* to the nearctic, of *Dryocopus* to the nearc-tic, Eurasia, and Southern Asia, and, remarkably, of *Picumnus* and *Celeus* to (or conceivably but not likely from) southern Asia. Radiation in the genera of these families is discussed.

Understanding the patterns of radiation poses great problems because of a lack of data on factors (insect faunas, predators, non-piciform and even non-avian compet-itors, and floral differences) that could have affected them. Nest parasitism by hon-eyguides in the Afrotropics, and the usurpation of nest cavities in both the Afrotropics and neotropics by diverse other birds, mammals, and other animals are apt to have had a major influence on woodpecker radiation, as is the competition of bark foragers (this may be more severe in the Afrotropics, where even some barbets are likely competitors). The factors having the most important influence on the nature of and differences in radiation patterns discerned are probably the historical ones mentioned above.

RESUMEN. Los Piciformes de Africa tropical incluyen chaboclos (Capitonidae), mieleros ("honeyguides") (Indicatoridae), y carpinteros (Picidae); los Piciformes neo-tropicales incluyen chaboclos, tucanes (Ramphastidae, aquí considerados como de-rivados de y cercanamente relacionado a, chaboclos) y carpinteros. Es controversial la relación de las familias anteriormente mencionadas con las otras existentes, supuestamente Piciformes, por ejemplo jacares (Galbulidae) y chacurues (Bucconidae), por lo que en este trabajo no se los considera.

Muchos factores complejos afectan la radiacción adaptativa de las familias y órdenes de aves. Algunos de ellos son: acceso a áreas continentales desde afuera, oportunidad presentada para especiación, cantidad de ejemplares relativamente pequeña y singularidad de eventos bióticos y climáticos asociados en diferentes continentes. También se hacen difíciles las comparaciones debido a la escasez de fósiles.

Las taxas discutidas son de origen incierto. Los capitonidae pantropicales son diversos y numerosos en Africa tropical (siete géneros, 42 especies) y muchos menos en el neotrópico (tres géneros y siete especies) a no ser que se incluyen los tucanes (seis géneros y 33 especies), de manera que las diversidad y los números se hacen más comparables. "Honeyguides" son principalmente afrotropicales, (cuatro géneros y 15 especies) con penetración al sur de Asia por un género (dos especies). Con excepción del género de "honeyguide" (*Indicator*) que llega a Asia, todos los otros géneros de chaboclos, tucanes y "honeyguides" son endémicos de esas regiones. Los picidos de Africa tropical comprenden solamente seis géneros y de las 26 especies, 23 representan los tres géneros endémicos, dos de los cuales (*Campethera*, 10 especies, *Dendropicos*, 12 especies) muestran la única radiación de picidos en esa región. En contraste en la región neotropical hay 95 especies de picidos en 11 géneros, cuatro de los cuales son endémicos.

Factores climáticos e históricos que afectaron la distribución de los bosques, pero más particularmente de las zonas arbustivas; han jugado un rol importante en la radiación de chaboclos, "honeyguides" y dos géneros de carpinteros del Africa tropical. Los propios "honeyguides" pueden haber influído en la radiación de los otros 2 grupos, en cuyos nidos desovan como aves parásitas. Sorpresivamente, pocos piciformes han entrado a los tropicos africanos desde afuera (sólo el género *Sasia* de Asia y el *Picoides* de Eurasia) o partido de la región (sólo los dos "honeyguides," quizás un *Jynx* y probablemente el antecesor de *Picodies*). En los neotropicos las oportunidades de especiación han sido mayores, debido a los períodos de sequía y de humedad de la Amazonia y otras áreas y la elevación de los Andes y sus efectos bióticos asociados. También ha habido cierta inmigración de picidos desde el norte hacia los neotrópicos. *Picodies* es definitivamente una forma del Viejo Mundo que primero ingreso a América del Norte y luego a Sudamérica. América Central y las Antillas han sido secundariamente importantes como centros de radiación en la región neotropical, con inmigración de especies de *Melanerpes, Colaptes* y *Campephilus* hacia el neoártico y de *Dryocopus* hacia el neoártico, Eurasia y Sudasia y de manera remarcable *Picumnus* y *Celeus* hacia (o concebible pero no probable desde) Sudasia. Se discute la radiación en los géneros de estas familias.

Al tratar de entender los patrones de radiación se presentan bastantes problemas debido a la falta de datos sobre factores que pueden haber afectado dicha radiación (fauna de insectos, depredadores no-picidos e inclusive competidores que no son aves). La reproducción parásita de los "honeyguides" en el Africa tropical y la usurpación de huecos de anidación de diversas aves del Africa tropical y los neotrópicos por aves, mamíferos y otros animales pueden haber influído en la radiación de carpinteros, tanto como la competencia de los comedores en cortezas (esto puede ser más serio en Africa tropical donde algunos de los chaboclos son posibles competidores). Es posible que los factores históricos mencionados anteriormente hayan tenido la mayor influencia en la naturaleza de, y diferencia en, los patrones de radiación discernidos.

The purpose of this review is to compare the radiations of pantropical piciform birds in the neotropics with those in the Afrotropics. No controversey exists over the relationships among the piciform families Capitonidae, Ramphastidae, Indicatoridae and Picidae, whatever their connections with the questionably piciform Bucconidae and Galbulidae (Sibley and Ahlquist 1972; Short 1974b; Simpson and Cracraft 1981; Swierczewski and Raikow 1981). The family status of the toucans (Ramphastidae) appears to be questionable, and they can be regarded as specialized derivatives of neotropical barbets (Capitonidae), perhaps more closely related to them than the latter are to either Asian or Afrotropical barbets (P. J. Burton, *in litt.*; the Toucan Barbet *Semnornis ramphastinus* morphologically is close to being intermediate between neotropical barbets and toucans). Phylogenetically close interrelationships among the barbets, toucans, honeyguides (Indicatoridae) and woodpeckers (Picidae) are evident in (1) their visual and vocal displays (e.g., see Short and Horne 1979), (2) the fact that Afrotropical barbets and woodpeckers are the major hosts of most honeyguides, (3) the strong year-round behavioral interactions among barbets, woodpeckers, and toucans, all of which roost, as well as nest, in holes, and even of honeyguides with barbets and woodpeckers (Short and Horne 1983), and (4) their morphological resemblances and shared derived characters (Simpson and Cracraft 1981; Swierczewski and Raikow 1981). Based upon unique, derived character states and habits, the honeyguides are specialized derivatives of the capitonid-picid line, uncertainly closer to either, and the Picidae probably evolved from ancestral barbets. These families are distributed as follows: Capitonidae, pantropical; Ramphastidae, neotropical; Indicatoridae, paleotropical; and Picidae, virtually cosmopolitan. Various publications provide additional background information on honeyguides (Friedmann 1955, 1970), toucans (Haffer 1974), barbets (Ripley 1945; Goodwin 1964), Afrotropical woodpeckers (Short 1971a, 1980a; Short and Tarboton 1978), neotropical woodpeckers (Short 1975, 1980c), and woodpeckers generally (Short 1982). In particular, the ranges of various species discussed below, and only sketchily summarized herein, are described in detail or mapped in these publications.

This review of adaptive radiation in these birds is based upon two decades of research largely concentrated on tropical piciform birds. What makes the comparisons so appealing is that these taxa are closely interrelated, yet show diverse foraging specializations. One can compare their phylogenetic diversity, adaptive radiation, and speciation both within and between groups, given that all are related, and relatively specialized for aboreal existence centered about their use of holes in trees. The occurrence of two or three families in the Afrotropics and the neotropics, of course, permits and invites such comparisons.

There are many problems inherent in comparing faunas of different continents, given the varied continental histories and floral and faunal associations. One can reduce such problems by considering only limited groups, such as related taxa on the different continents. Even within an accepted monophyletic group, such as that of the Piciformes discussed here, it is clear that radiation reflects historical events generally unique to a region. Also, these events, with their related topographic and biotic effects, differentially affect dispersal, isolation, and speciation, even among closely related subgroups of piciform birds. Clear examples of these differing effects are that (1) the neotropical picids include genera represented on all the continents except Australasia and Antarctica, (2) some Afrotropical genera of Picidae (and Indicatoridae) are represented in tropical Asia, and in one case (*Picoides*), in Europe and the Americas, (3) Afrotropical and neotropical genera of Capitonidae are entirely endemic, and (4) the Ramphastidae are endemic to the neotropics. The radiation of a group may be assumed, in the absence or uncertainty of fossil data, to have been intrinsic to a region and to have been affected by intrinsic events, if all the taxa of that group are endemic there. Should any of the taxa, say genera, be represented elsewhere, however, then there is the chance that external radiation may have played a role in the continent being considered. Even when taxa are endemic to a continent, external events, particularly radiation of related (or in some cases, even unrelated) groups outside that continent, may influence the course of radiation within the continent in question. For example, events outside the continent may preclude or, on the other hand, enhance the opportunity for invasion of that continent by potential competitors, or nest parasites of the endemic taxa.

A general problem in avian zoogeographic comparisons is the relatively small sample sizes with which ornithologists must work. Many invertebrate groups have thousands of species even within one widely distributed family, whereas worldwide we have but 9200 or so avian species in about 150 families. The relatively small bird families and genera with their limited numbers of species are apt to show effects of unique events involving one or two species, or one genus. Within the piciform groups treated here are 34 genera, representing four ecologically distinct arrays of species (the families). Nevertheless, all share structural features and habits that are apt to be influenced similarly by the environment, and, because of their similarities, birds of the different piciform families may impinge upon each other ecologically.

The piciform groups under consideration are of uncertain origin in time and location, with few fossils definitely ascribed to them (Brodkorb 1971). Generally, the pantropical capitonids are most numerous in the Afrotropics, with 42 species of seven endemic African genera (Table 1); they are less numerous in the neotropics (13 species of three endemic genera) and southern Asia (30 species of three endemic genera, including *Megalaima,* with 28 species the largest capitonid genus). The endemic neotropical toucans have 33 species in seven genera. There are 15 Afrotropical honeyguides of four genera, three of which are Afrotropical endemics. The Picidae on the various continents have been enumerated elsewhere (Short 1982); essentially, the woodpecker faunas in tropical Asia and the neotropics are rich, and in the Afrotropics, depauperate, even more so than in the nearctic.

TABLE 1
Genera of Neotropical and Afrotropical Piciformes[1]

Neotropics	Afrotropics
Capitonidae—13 + 0	Capitonidae—42 + 0
(*Capito*)—7	(*Gymnobucco*)—4
(*Eubucco*)—4	(*Stactolaema*)—4
(*Semnornis*)—2	(*Buccanodon*)—1
	(*Pogoniulus*)—10
(Ramphastidae)—33 + 0	(*Tricholaema*)—6
(*Aulacorhynchus*)—6	(*Lybius*)—12
(*Pteroglossus*)—9	(*Trachyphonus*)—5
(*Selenidera*)—6	
(*Baillonius*)—1	Indicatoridae—15 + 2
(*Andigena*)—4	(*Prodotiscus*)—3
(*Ramphastos*)—7	(*Melignomon*)—2
	Indicator—9 + 2
Picidae—95 + 39	(*Melichneutes*)—1
Picumnus—22 + 1	
(*Nesoctites*)—1	Picidae—26 + 35
Melanerpes—18 + 3	*Jynx*—1 + 1
(*Xiphidiopicus*)—1	*Sasia*—1 + 2
Picoides—3 + 30	(*Campethera*)—10
(*Veniliornis*)—12	(*Geocolaptes*)—1
(*Piculus*)—7	(*Dendropicos*)—12
Colaptes—8	*Picoides*—1 + 32
Celeus—10 + 1	
Dryocopus—3 + 3	
Campephilus—10 + 1	

[1] Parentheses indicate endemic taxa; numbers refer to species within the given region with the plus followed by the number of extralimital congeneric species.

In my discussion of speciation herein, I take into account the refugium hypotheses of Haffer (1969, 1982), and also the objections to it of such workers as Livingstone (1982) and Endler (1982). Whether or not refugia need be completely isolated and, indeed, whether or not allopatric speciation requires complete geographic isolation have been questioned (Endler 1982), but the point is one of *effective* isolation, complete or not.

In characterizing piciform species below I use "tiny" for birds weighing less than 20 g, "small" for those 20 to 50 g, "medium" for 50 to 100 g, "large" for 100 to 200 g, and "very large" (only certain woodpeckers and many toucans attain such size) for more than 200 g.

CAPITONIDAE

The barbets are usually colorful, heavy-bodied birds with a broad, deep, pointed bill. Knowledge of the biology of neotropical capitonids is poor compared with that of Afrotropical (and Asian) species. Neotropical barbets are forest-dwelling small- to medium-sized birds. Eleven species comprise the closely related genera *Eubucco* and *Capito,* and two species, the genus *Semnornis.* In contrast there are seven Afrotropical genera of tiny (*Pogoniulus,* the tinkerbirds) to medium-sized (*Lybius* in part) barbets, most species of which occupy open woodland, water courses in dry country, and scrublands. The Afrotropical barbets include at ecological extremes the little tinkerbirds that feed on mistletoe berries, other fruits, and some insects, and the omnivorous ground barbets (*Trachyphonus*), neither of which have neotropical (or Asian) counterparts among barbets. All neotropical barbets inhabit forests and forest-edges. None ranges into temperate areas or even to the temperate border of the subtropics. This is in strong contrast to Afrotropical barbets that range from deserts to wet montane forests, and into temperate southern Africa.

The presumably more ancient radiation reflected by the modern Afrotropical genera is difficult to analyze. All the modern genera have at least one forest-inhabiting species, and two (*Gymnobucco, Buccanodon*) are confined to forest. I consider *Stactolaema* to have been derived from a *Gymnobucco*-like ancestor, as probably was *Pogoniulus. Buccanodon* and *Tricholaema* appear to connect *Pogoniulus* with *Lybius* and *Trachyphonus.* These relationships and such a phylogeny suggest a forest origin of these barbets. Earlier radiation involving forests largely has been masked by presumably more recent speciation and radiation, for example, of *Lybius, Tricholaema, Pogoniulus,* and *Trachyphonus,* in woodlands and bushlands outside of the forest (which may have diminished greatly in the Pleistocene, Livingstone 1982). All of the neotropical genera appear to have originated in forest, and, based on the distribution of their ancestors, so too do the toucans.

Certainly, all of the recent neotropical barbet radiation has involved speciation in relation to geographic isolation in forest areas, very likely reflecting a series of Amazonian forest refugia (Haffer 1969, 1974, 1982). In contrast, forest refugia seem to have been of little importance in the recent Afrotropical radiation of barbets, except perhaps in *Gymnobucco* (four forest species) and *Pogoniulus* (about four forest species). Most Afrotropical patterns of speciation seem to reflect dispersal through, and isolation in, woodland.

Afrotropical, as well as neotropical barbets are largely frugivorous. However, the tinkerbirds and some of the larger barbets are partly insectivorous, most barbets feed at least some invertebrates to their young, and birds of the genus *Trachyphonus* are omnivorous. Dependence upon a supply of fruits throughout the year probably is a major factor limiting the distribution of the family. The range of each Afrotropical barbet (also woodpeckers, honeyguides) is mapped in Snow (1978).

As a group, the four species of Afrotropical *Gymnobucco* occupy all major forest areas; two (*calvus, peli*) basically are western in distribution, corresponding to Grubb's (1982) Western and West Central refugia, and the other two (*sladeni, bonapartei*) are more eastern (in Grubb's East Central refugium). Speciation and radiation in *Gymnobucco* seem attributable to isolation of forest blocks, differentiation, and dispersal between these blocks in successive bursts. As many as three (*calvus, peli, bonapartei*) are now sympatric in Cameroun. Closely related *Stactolaema* is a forest-fringe and woodland array of four eastern to southern African species that essentially replace one another, although *olivacea* and *leucotis* show considerable range overlap. The very closely related western *anchietae* and eastern *whytii* barely meet in central Zambia. Species of both these genera are highly social, frugivorous and insectivorous, dull in color, and of similar size and proportions. *Buccanodon duchaillui* is a little known, mainly frugivorous barbet of the forest canopy. *Tricholaema* generally occupies eastern and southern African woodland and bushland, with three closely related allospecies of the *diademata* group (superspecies) plus *lachrymosa* of more moist woodlands, and the scrubland *melanocephala.* The one forest species, *hirsuta,* is probably fully allopatric with congeners. Species of this genus in East Africa may be narrowly sympatric but tend to be ecologically separated. All have a "tooth" or hook on the sides of the bill tomia, and feed upon a mixture of fruit and insects.

Lybius is a large genus comprised of several subgroups. The closely knit *L. bidentatus* group consists of three large-sized species with one in forest-edge (*bidentatus*) and two, western (*dubius*) and eastern (*rolleti*), in wooded parts of the Sahel. Peculiar and little known *undatus,* possibly connecting *Lybius* to *Tricholaema,* is endemic in Ethiopia. A closely knit group of social, duetting species that are mainly allopatric largely replace one another in woodlands fringing the forest block from West Africa (*vieillotii*) to eastern and southern Africa (*rubrifacies, guifsobalito, torquatus*). The large-billed *leucocephalus* has distinctive races fringing the core area of the forest from nothern Cameroun east to East Africa and southward and westward disjunctly to Angola, with a likely relative, *chaplini,* isolated south of it in Zambia. Finally there are the allopatric, social *melanopterus* of eastern Africa, and *minor* of the forest fringe to the south, from Angola and Zambia eastward and northward to Burundi. All of these are tooth-billed and fruit-dependent, although taking insects as well, especially when feeding young.

The ground barbets of the genus *Trachyphonus* number five species. One inhabits the forest understory (*purpuratus*) from West Africa to western Kenya, with a distinctive West African isolate (*goffinii*), and four occupy open woodland, bushland, and xeric streamside vegetation from West Africa well north of the forests, around them to the east, thence south to southern and southwestern Africa. The southern *vaillantii* is variously intermediate between forest-dwelling *purpuratus* and the other three species, which are the true ground barbets. Like *purpuratus, vaillantii* nests in trees. Long-billed *margaritatus* of the Sahel barely meets its East African allospecies *erythrocephalus* in parts of Ethiopia and Somalia. Smaller *darnaudii* is partly sympatric with *erythrocephalus,* occurring from Ethiopia to Tanzania, with four distinct races of rather limited range. The last three species nest and roost in holes in the ground, in termite mounds, or in earthen banks.

The large genus *Pogoniulus* occupies diverse habitats. Only two species, *pusillus* and *chrysoconus*, which form a superspecies widespread in western, north-central, eastern and southern Africa, can be considered birds of dry woodland or brushland. Forested central Africa may have up to four or possibly five sympatric species (*bilineatus, subsulphureus, atroflavus, scolopaceus*, and perhaps *coryphaeus* in part). These are distinct, lacking superspecies relationships, except for upland *coryphaeus* that forms a superspecies with highland East African *leucomystax* and coastal (to foothills) East African *simplex*. Widespread *bilineatus* essentially is a forest-edge species with hybridizing West-Central African (*leucolaima*) and eastern and southern African (*bilineatus*) subspecies groups (Prigogine 1980). The food of tinkerbirds includes fruits, especially mistletoe berries, and insects. In eastern and southern Africa the species tend to replace one another ecologically, although *bilineatus* is broadly sympatric with both *leucomystax* and *simplex*. The forest species are little known.

The 13 much less known neotropical barbets comprise only three genera, two of which (*Capito, Eubucco*) are poorly defined, the other being the distinctive genus *Semnornis* of forest slopes (Table 1). *Semnornis* contains two allospecies of the highlands of Central America (*frantzii*) and of western Colombia-Ecuador (*ramphastinus*). Most neotropical barbets are sexually dichromatic, in contrast to Afrotropical barbets, of which all but a few (e.g., *Trachyphonus erythrocephalus* and *T. margaritatus, Tricholaema hirsuta*, and *T. lachrymosa*, the last only in eye color) are sexually monochromatic. *Eubucco* and *Capito* inhabit forests in the Amazon and northwestern South America, but sympatry within each genus is very limited. Of the four *Eubucco* species, *richardsoni* occurs in western Amazonia. The eastern Andean front of Peru to Bolivia is occupied by *versicolor*. A relict species with a tiny distribution, *tucinckae*, inhabits central-eastern Peru. The fourth species, *bourcierii*, occurs disjunctly in the Andean slopes of western Ecuador to Colombia, eastern Ecuador to northeastern Peru, central Colombia, and northwestern Colombia to the Costa Rican highlands. Only the last species, *Capito maculicoronatus*, and *Semnornis frantzii*, of the 13 neotropical barbets, reach Central America, and none extends north of Costa Rica. The systematics and biology of these barbets is little known. However, the outlying *bourcierii* and *versicolor* have color patterns, including their sexual dichromatism, that suggest they are the most derived. Their distributions readily fit into several schemes drawn from Haffer's (1969, 1982; Simpson and Haffer 1978) refugium concept.

Among the seven species of *Capito* are *niger*, which is widely distributed in the Guianas, northwestern, west-central and southwestern Amazonia, allopatric *dayi*, of the major Amazonian tributary region south of the Amazon, and west-Amazonian *aurovirens*. There are also four allospecies of a single superspecies with adjacent ranges, *squamatus* of northwestern Ecuador to southwestern Colombia, *quinticolor* of southwestern Colombia, *maculicoronatus* of northwestern Colombia to eastern Panama, and *hypoleucus* of interior Colombia in the middle Magdalena and Cauca drainage. Terborgh and Winter (1982) ascribed *squamatus, quinticolor*, and *maculicoronatus* to an origin associated with the Choco Refuge, and *hypoleucus* to the Nechi Refuge. *Capito dayi* of the Amazonian group seems nearest the *maculicoronatus* complex.

Among congeners, only *C. niger* and *aurovirens* overlap broadly. Sympatry appears to involve no more than three species of neotropical barbets at one time, for example, *C. niger, C. aurovirens*, and *E. richardsoni* in west-central Amazonia, and *E. bourcierii* and *S. ramphastinus* plus *C. squamatus, C. quinticolor*, or *C. maculicoronatus* in northwestern Colombia. We have yet to learn how congeneric *aurovirens* and *niger* coexist, and ecological comparisons of *Eubucco* and *Capito* are lacking. In Kenya J. F. M. Horne and I have found as many as seven capitonid species together (indeed in the same tree), and it seems likely that as many could be sympatric in Gabon and Cameroun.

The largely frugivorous neotropical barbets are concentrated in northern South America. Their distributions fit well with the various forest refuges that have been proposed (Haffer 1982). It is unlikely that these barbets would traverse woodland or open grassland, being true forest species. In contrast, Afrotropical forest barbets seem better able to exist in fragmented habitats far from main tracts of forest, as in western Kenya (streams east of Kisumu); *Gymnobucco bonapartei* and *Trachyphonus purpuratus* occur there with such woodland species as *Pogoniulus chrysoconus, Tricholaema lachrymosa* and *Lybius guifsobalito* (pers. observ.). The origins of the Afrotropical forest barbets seem likely to relate to refuges, particularly in the cases of species of *Gymnobucco*, and probably some of the species of *Pogoniulus*. *Pogoniulus coryphaeus, leucomystax*, and *simplex* could have originated by simple fragmentation of their once common range into several refugia (for forest contacts and refugia in Africa see Diamond and Hamilton 1980; Grubb 1982). However, the main factors facilitating radiation in Afrotropical barbets doubtless have been the shifting of woodland zones coupled with adaptation of some forest barbets to existence in open country. The predominanatly woodland and scrubland genus *Tricholaema* and the extreme open-country *Trachyphonus* each have a single forest species, and so have either secondarily invaded forest or, more likely (Short and Horne, unpubl. data), were derived from forest-dwelling ancestors. At any rate only in Africa do we have a balanced barbet fauna, with some forest species and many others in woodland-scrubland. The origin of many woodland barbets in Africa probably lies in fragmentation of their ranges in the past by either forest intrusions (as to the East African coast), or xeric gaps (Grubb 1982). Such speciation has many parallels in other woodland birds (e.g., Picidae), but historical factors have not yet been studied to the extent that they have for forest avifaunas.

Without morphological and other studies it is difficult to establish definite relationships between Afrotropical and neotropical barbets. At least superficially the neotropical barbets show some resemblances to Afrotropical barbets such as *Tricholaema hirsuta*, which is sexually dichromatic, as are most neotropical barbets. The Afrotropical genera *Gymnobucco*, *Stactolaema*, and *Tricholaema* show similarities to both neotropical (*Tricholaema* with *Eubucco*, *Stactolaema* possibly with *Semnornis*) and southern Asian (*Gymnobucco* with *Colorhamphus*; possibly *Stactolaema*, through *olivacea*, with *Megalaima*) barbets. Afrotropical radiations of *Lybius*, *Pogoniulus*, and *Trachyphonus* have no equivalent in the neotropics (unless toucans are viewed as the *Lybius* counterpart) or in Asia.

RAMPHASTIDAE

Toucans effectively are large, specialized, toothed-billed barbets. Asian barbets lack teeth, and so do neotropical barbets other than species of *Semnornis*. The Afrotropical *Tricholaema* and *Lybius* have a toothed bill, and, in fact, the large species of the *bidentatus* group of *Lybius* have a complex pattern of "teeth" and grooves in the bill (these species also resemble toucans in some displays and vocalizations, Short and Horne, unpubl. data). Afrotropical *Gymnobucco* and, to a lesser extent *Stactolaema*, have a keel on the culmen, recalling those on toucans' bills. Since barbets and toucans feed on fruit, they are potential competitors; toucans also compete with larger barbets for tree cavities (Skutch 1983). Perhaps more than some woodpeckers (Short 1982) and barbets of the genus *Trachyphonus* (Short and Horne in press), toucans eat some animal food, particularly seizing nestling birds (Short 1974b; Skutch 1983).

Haffer (1974) treated speciation in the toucans in great detail, and I provide only a brief summary here. Toucans are much larger than barbets and can fly longer distances to forage, even visiting areas, such as cultivated areas, in which they cannot breed.

Haffer grouped the 33 toucans into 14 species and superspecies, among six genera. *Selenidera* effectively is monotypic (i.e., consists of a single zoogeographic species, see Mayr and Short 1970), and *Baillonius* has but one species. Species of the genera *Aulacorhynchus* and *Andigena* are highland toucans, occurring on the slopes of the Andes. *Aulacorhynchus prasinus* is widespread, and the genus otherwise consists of two superspecies that segregate altitudinally, *sulcatus* (allospecies *sulcatus*, *derbianus*) at lower elevations and *haematopygus* (*haematopygus*, *huallagae*, *coeruleicinctus*) in more upland forest. *Andigena* is composed of the allospecies *hypoglauca*, *laminirostris*, and *cucullata* of the superspecies *hypoglauca*, and of *A. nigrirostris* (see Haffer 1974 for maps and details). Both genera probably arose *in situ* from separate ancestors that happened to exist in the area of geologically recent Andean uplift. Thus, the chance event of the uplift and presence of toucans at the site led in these cases to successful adaptation to forests on mountain slopes. Glacioclimatic effects led to gaps in ranges, isolation, and speciation, plus further refinement of adaptations to sub-Andean conditions.

Lowland *Pteroglossus* includes the species *beauharnaesii* and three superspecies, *viridis* (allospecies *viridis*, *inscriptus*), *bitorquatus* (*bitorquatus*, *flavirostris*), and *aracari* (*aracari*, *castanotis*, *pluricinctus* and *torquatus*). These show an Amazonian-northern South American distribution generally, but some extend far into Central America and to southern Brazil and Argentina (for unknown reasons toucans extend farther from the tropical core of their distribution than do neotropical barbets). *Selenidera* contains one widespread superspecies, *maculirostris*, with the allospecies *maculirostris*, *gouldii*, *reinwardtii*, *nattereri*, *culik*, and *spectabilis*. Their distributions reflect well the refuge scheme advanced by Haffer (1974). *Baillonius bailloni* is widespread in southeastern South America, and its structure and coloration show the likely relationship of this genus to *Selenidera*.

The genus *Ramphastos* contains the largest species of toucans, the distinctive *toco*, the superspecies *dicolorus* (allospecies *dicolorus, vitellinus, brevis, sulfuratus*) and the superspecies *tucanus* (*tucanus, ambiguus*). Sympatry is limited to species of the different superspecies, and *toco*. Two of the species, *sulfuratus* and *ambiguus*, extend to northern Middle America (Honduras, Mexico). We lack studies showing food differences among sympatric genera of toucans. The fact that sympatry occurs between species having bills of different shape and size, often representing different genera, suggests that differences in diet exist. The evolution of toucans in the direction of greater size with a toothed bill makes them unique in the Piciformes. The failure of Afrotropical and Asian barbets to evolve along similar lines may be due to the toucan-like hornbills (Bucerotidae) of those areas closing the opportunity to barbets.

INDICATORIDAE

The honeyguides are so little known systematically, and so specialized in their wax-eating, honey-guiding and nest-parasitic habits as to defy meaningful comparison with the other groups treated here. However, they have to be considered briefly because of their direct effects on Afrotropical barbets and woodpeckers. Of the 17 species in four genera (Table 1), only two species of the genus *Indicator* occur outside Africa (in Asia). The Afrotropical endemic genus *Prodotiscus*, alone among the piciforms, is not associated with cavity-nesting in any way. The majority of honeyguides are nest parasites. Most are parasites of barbets and woodpeckers, but other hole-nesters (bee-eaters, kingfishers) and a number of non-hole nesting species, are also parasitized. Widespread *Indicator minor* attends paired barbets, provoking intense aggressive interactions even outside the breeding season (Short and Horne 1979, 1983). Because barbets and woodpeckers use their holes all year for roosting, honeyguides constantly interrupt their hosts' activities. These points and the fact that parasitism inevitably causes death of the host's young likely have made honeyguides important in relation to the evolution and radiation of Afrotropical barbets and woodpeckers. Honeyguides that parasitize barbets and woodpeckers include species of the large genus *Indicator*, at least one species of which occurs within the range of every Afrotropical barbet and woodpecker. Indicatorids are lacking and have no counterparts in the neotropics.

PICIDAE

Patterns of speciation of Afrotropical (Short 1971a, 1980a) and South American (Short 1975, 1980c) woodpeckers have been treated in detail. All species discussed here are listed with their classification in Short (1982:49–53). The species of Afrotropical Picidae are few (Table 1), and three of the six genera (containing 23 of the total of 26 species) are endemic. Further, size and structural diversity are not great, although a tiny piculet (*Sasia africana*) and one terrestrial species (*Geocolaptes olivaceus*) are included. In contrast neotropical picid species are numerous (95) and vary greatly in size (tiny *Picumnus* spp. to very large *Campephilus imperialis*). They also are diverse in habits. Departures from typical picid insectivorous and arboreal habits are omnivory (*Melanerpes* spp.), specialization for ant-foraging (e.g., *Celeus* spp.), excavation and scaling of tree bark by very sturdy-billed species (e.g., *Campephilus* spp.), and full terrestriality (*Colaptes rupicola, C. campestris*).

Many studies and hypotheses have dealt with the history of the forest biotas and forest refuges in the Americas (e.g., Haffer 1974; Simpson and Haffer 1978) and in Africa (e.g., Hall and Moreau 1970; Snow 1978; Diamond and Hamilton 1980; Grubb 1982; Livingstone 1982), and have compared Afrotropical and neotropical forest faunas (e.g., Meggers et al. 1973; Haffer 1974). Most Afrotropical picids are not forest birds, but inhabit woodlands and wooded grasslands, intertropical comparisons of which are few (see Fry 1980; Short 1980a, 1980b).

Because woodpeckers are specialists in their foraging, nesting, and roosting, in that they excavate trees (Short 1982), competition between them is apt to be sufficiently severe to affect their distributions (Short 1971b, 1973). The diversity of woodpeckers in tropical Asia and in the neotropics has permitted sympatric existence of 12 to 13 species in certain forest habitats (Short 1978). Most temperate regions show sympatry of four or five, or up to seven picid species. Within the Afrotropics picid sympatry regularly involves three to four species, with possibly a maximum of six at the junction of different habitats (e.g., highland-lowland forests as at Mount Cameroun or in gallery forests of Uganda, where woodland and forest species meet, see maps in Short and Tarboton 1978).

The Afrotropical woodpecker radiation was but little influenced by geologically recent events in woodpecker evolution outside the Afrotropics, for only three Afrotropical species have congeneric relatives outside of Africa. The one Afrotropical wryneck (*Jynx ruficollis*) has an Eurasian relative, the lone piculet (*Sasia africana*) is related to forms in southern Asia, and the single pied woodpecker (*Picoides obsoletus*) is the northern Afrotropical representative of a nearly cosmopolitan genus (Winkler and Short 1978). The remaining 23 Afrotropical picids represent three endemic genera with no close affinities (except *Picoides* derived from *Dendropicos*) to European or Asian picids (Short 1971a, 1982). The larger picid fauna of the neotropics includes four endemic genera, five genera with strong nearctic connections, and two essentially neotropical genera, *Picumnus* and *Celeus,* each having one species in tropical Asia (Short 1980a, 1982). It is possible that some neotropical picids resulted from radiations of *Melanerpes, Colaptes,* and perhaps *Dryocopus* in Middle America. These three genera may even have originated in Middle America. Endemic neotropical *Nesoctites* exists only in Middle America (Hispaniola). *Picoides,* with an endemic neotropical superspecies (*mixtus-lignarius*), seems to have entered the neotropics only very recently from the nearctic (and thence from Eurasia, Short 1971b, 1980a, 1982). *Picoides* is the only genus shared with the Afrotropics.

The nearctic infusion of picids into the neotropics, the presence of a somewhat distinct center of radiation in Middle America, the evolution and radiation of larger picids of the genera *Colaptes, Celeus, Dryocopus,* and *Campephilus,* and the extensive radiation of *Picumnus* in large part are responsible for the differences in numbers of Afrotropical and neotropical picids. Why the less isolated Afrotropics have had so little invasion of picids from without, and why there have been no similar radiations of large and very small Afrotropical picids remain unexplained.

The Picumninae are well distributed in the neotropical region, especially species of *Picumnus.* Monotypic *Nesoctites* is an unusual piculet, if really allied to the other members of the subfamily (Short 1974a). The Afrotropics have but one piculet, *Sasia africana,* with relatives only in tropical Asia. Since there are but three genera of Picumninae, one in Afro-Asia, one in the neotropics, and one in both southern Asia and the neotropics, it is difficult to establish a possible center of origin. Remarkable is the considerable radiation of *Picumnus* in the neotropics, whereas Africa has but a single, lowland forest inhabiting piculet. The ancestor of *Sasia africana* probably invaded Africa from Asia, either recently (hence no time for radiation), or at a more remote time (with failure to radiate).

The neotropical species of *Picumnus* are little known and include several of very restricted distribution. The number of species and the taxonomy of *Picumnus* are in need of investigation (Short 1982). All species inhabit forest or dense woodland, and none extends out of the subtropics. Only one species (*olivaceus*) occurs in Middle America, the genus having a markedly Amazonian center of distribution. Those species outside, or on the fringes of Amazonia and the Orinoco are *limae, fulvescens,* and *pygmaeus,* centered in northeastern Brazil, *nebulosus,* in southeastern Brazil, arid zone *sclateri* of western Ecuador and northwestern Peru, *granadensis* of interior Colombian valleys, and *cinnamomeus* and *squamulatus* of northern Colombia and Venezuela. The remaining 14 species occur in the Amazonian-Orinocan drainage, and some of them undoubtedly are related directly to the extralimital species just mentioned. Patterns of speciation have not yet been determined, and species complexes remain to be defined (Short 1982 attempted this with the *cirratus* complex). The important points are that the radiation of *Picumnus* is considerable, and it is centered in Amazonia. The distribution and radiation of piculets suggest that this subfamily has undergone replacement by more recent, small picids, especially species of *Picoides.* However, there are small woodpeckers in the Afrotropics, which has but one piculet, and in the neotropics, which has many piculets.

Two lines of woodpeckers represent restricted radiations in the Afrotropics. Together the three endemic genera of these two lines include 23 of the 26 picids, and 23 of the 24 picine species in the region. These genera, *Campethera, Geocolaptes* and *Dendropicos,* are closely interrelated, their only other relatives being the New World Colaptini, and *Picoides,* which is a derivative of *Dendropicos* (Short 1982). Monotypic *Geocolaptes* is isolated in South Africa, and evolved from an ancestral species of *Campethera* (Short 1971a, 1980a).

Species of *Campethera* are relatively unspecialized for woodpecking, are mainly ant-foraging, and have radiated largely in the woodland-grassland belt surrounding the forests. Five species form two forest-fringing superspecies, *nubica,* with allospecies *punctuligera, nubica,* and *bennettii,* and *abingoni,* with allospecies *abingoni* and *notata.* The woodland and forest groups are connected by the superspecies *maculosa,* with the western allospecies *maculosa* and eastern allospecies *cailliautii* that contains subpopulations in forest and in woodland (Short 1980a). Truly forest-dwelling are the very distinct *tullbergi, nivosa,* and *caroli* (Short

1982). Of these *tullbergi* is interesting in having disjunct montane megasubspecies (Amadon and Short 1976; Short 1982).

Dendropicos contains 12 species of which five comprise a tight-knit *fuscescens* group (*gabonensis, fuscescens, abyssinicus, poecilolaemus, elachus*), three form a group of especially "woodpecking" species (*namaquus, pyrrhogaster, xantholophus*), two form a widespread forest-woodland superspecies (*goertae, griseocephalus*), and two are rather isolated with no very close relatives (*stierlingi, elliotii*). The *fuscescens* group occupies all types of habitat that contain trees, from open grassland and desert washes, to lowland wet forest. Within the *fuscescens* group there seem to have been repeated speciation events involving forest-woodland disjunctions, with resulting sympatry of some of the species in forests; the sympatry is equivalent to that seen elsewhere, in such genera as *Picoides,* but not frequent in the Afrotropics. Connecting the *fuscescens* group and the *namaquus* group, and limited to brachystegia woodland of southeastern Africa is *stierlingi* (Short and Horne 1981). The *namaquus* group contains *namaquus* of woodlands with large trees in eastern and southern Africa. Within the Afrotropical forest belt *pyrrhogaster* of western Africa and *xantholophus* of central Africa form a superspecies; they appear to have evolved in separate forest blocks. Woodpeckers of the *namaquus* group drum loudly (unlike most Afrotropical picids) in signal communication, and they excavate and tap loudly while foraging in trees. The superspecies *goertae* includes woodland-inhabiting, northern and eastern African *goertae,* and southern forest-dwelling *griseocephalus.* The final species, *elliotii,* occupies forest and has highland and lowland forest megasubspecies, the only likely example of lowland-highland speciation of piciform taxa in Africa.

All the picid radiation in the Afrotropics is encompassed within *Campethera* and *Dendropicos,* including *Geocolaptes* as a derivative of *Campethera.* Neither genus shows marked or enhanced divergence compared with picid genera elsewhere, including the neotropics. Neotropical *Piculus, Veniliornis, Colaptes,* and *Melanerpes,* that are approximately as speciose as *Campethera* and *Dendropicos,* show about the same intrageneric range of morphological divergence. It is unclear why there are no large Afrotropical picids and why piculets have not radiated there (unless *Sasia* is a very recent entrant into Africa). Terrestrial woodpeckers are few in the Afrotropics probably because African grasslands have been less extensive than in the neotropics and have been associated closely with woodland throughout most of their history. It is a moot point whether wrynecks were once more numerous. Extant species may be relicts from a past radiation in Africa or Eurasia. *Picoides obsoletus* probably is not an ancient Afrotropical picid and if, as is likely, *Picoides* is derived from *Dendropicos,* then failure of radiation of *Picoides* in the presence of speciose *Dendropicos* is understandable (Short 1982). Both Eurasian *Picus* and the "red-bellied" array of advanced Eurasian *Picoides* (Short 1971b, 1982; Winkler and Short 1978) are represented in North Africa but have failed to penetrate the Afrotropics, or for that matter, the Americas. Hence radiation within the Afrotropics is restricted to its three endemic genera.

The neotropical picids include the endemic and essentially endemic (i.e., those barely extending into the nearctic) genera *Veniliornis, Piculus, Colaptes,* and *Campephilus,* and the piculet genus *Nesoctites* (Table 1). The first three of these comprise the tribe Colaptini, and parallel Afrotropical *Campethera, Dendropicos,* and *Geocolaptes. Piculus* contains seven species of relatively non-specialized tapping-foraging and ant-foraging picids, somewhat comparable to *Campethera.* The 12 species of *Veniliornis* are more woodpecking specialists in habits, and *Colaptes* is a terrestrial derivative of *Piculus* containing eight species and reaching temperate regions of North and South America.

Piculus has a distribution generally centering about the Amazon basin, although one subgroup fringes the Andes (*rubiginosus, rivolii*) and has an isolate far north in northwestern Mexico (*auricularis,* forming a superspecies with *rubiginosus*). *Leucolaemus* shows incipient speciation (megasubspecies) between Middle and South America. Widespread *chrysochloros* has an allospecies (*aurulentus*) in southeastern South America. Picid competitors for food, nesting sites and roosting sites are of the genera *Celeus, Melanerpes,* and *Colaptes* (Short 1972b, 1982). Species of *Piculus* are larger than those of *Campethera,* despite a lack of unspecialized small picid competitors.

Veniliornis is more Amazonian-centered than is *Piculus,* and but one species extends far into Middle America (*fumigatus*). Most species are similar in size, averaging smaller than *Piculus* and showing less morphological differentiation than do species of their Afrotropical counterpart *Dendropicos,* probably because of strong competition for nesting and roosting sites, and in foraging, from numerous species of six genera. An Andean slope element is distinguishable in *fumigatus, dignis,* and *nigriceps.* There is a west coastal species (*callonotus*), a southeastern South American forest species (*spilogaster*), and an isolated Guianan species (*sanguineus*). The "core" of the genus is composed of the superspecies *passerinus* with allospecies *passerinus* from Venezuela to northeastern Argentina, and southern *frontalis,* and *affinis,* including *cassini, kirkii, affinis,* and *maculifrons* in and bordering the Amazonian-Orinocan region. Isolation of populations in Amazonian refugia and within areas along the Andean slopes can account for virtually all speciation in this genus.

Colaptes may have originated in South America, Middle America, or even North America. An endemic Cuban species (*fernandinae*) has its closest relatives in South America. *Colaptes auratus* occurs in the nearctic, and south to Cuba, Grand Cayman, and highland Middle America. Both of these species forage on the ground and nest in trees. The forest flickers connect *Piculus* with *Colaptes.* Amazonia to Panama is the range of *punctigula.* Forests, woodlands, and wooded grasslands of Brazil to Patagonia are occupied by *melanochloros,* which forms a superspecies with *punctigula,* and itself includes a more arboreal megasubspecies, *melanochloros,* and more terrestrial megasubspecies, *melanolaimus* (Short 1972b). Arid woodlands of the Peruvian Andes and the west coast of South America form the range of *atricollis.* A terrestrial South American group occupies grasslands as far north as Suriname (*campestris*), the high Andes (*rupicola,* with northern and southern megasubspecies), and the Fuegian forest fringes (somewhat arboreal *pitius*). Radiation of these semi-terrestrial and terrestrial species probably reflects the more extensive pure grasslands of the neotropics compared with those of the Afrotropics. Also there likely were greater possibilities for speciation associated with forest-grassland advances and retractions in lowlands (Short 1971c), and with the rise and isolation of parts of the Andean chain (Haffer 1982). Both among and within species, birds of the more tropical, forest-dwelling taxa are smaller, more nearly the size of Afrotropical *Geocolaptes,* while more temperate, upland taxa tend to be larger and less arboreal.

The 11 species of the large-sized, highly specialized genus *Campephilus* include a temperate North American species *principalis* (reaching Cuba) that forms a superspecies with the giant *imperialis* of highland Mexico (both near extinction or possibly extinct), and a temperate Fuegian forest species, *magellanicus.* These three species of temperate areas are the largest of the genus, *imperialis* being the world's largest picid. The northern superspecies and *magellanicus* appear to have evolved independently from neotropical ancestors (Short 1970). Other species are widely distributed, with only two showing an Amazonian (*rubricollis*) or expanded Amazonian (*melanoleucos*) distribution. The latter species is part of the superspecies *guatemalensis,* also including *guatemalensis* of Middle America, and an isolate west of the Andes, *gayaquilensis. Pollens* and *haematogaster* occur in forests along the slopes of the northern Andes, *robustus* is confined to southeastern Brazil and adjacent areas, and *leucopogon* occurs from eastern Bolivia and western Brazil to northern Argentina. Only in Amazonia are two species (*rubricollis, melanoleucos*) found regularly and extensively in sympatry.

Monotypic Cuban *Xiphidiopicus,* related to *Melanerpes,* was derived from continental North or Middle America, or from elsewhere in the West Indies (Short 1982).

Dryocopus, likely allied closely to both *Campephilus* and *Celeus,* has one species (*D. galeatus*) almost morphologically intermediate between *Celeus* and *Dryocopus* (Short 1970, 1976). *Dryocopus galeatus* occurs in southeastern South America. The superspecies *pileatus* occupies most of the Americas (*schulzi* of the South American chaco, *lineatus* from tropical Mexico to Brazil and Argentina, and nearctic *pileatus*). The other two species, closely related to the *pileatus* superspecies, are Eurasian *martius* and eastern and southern Asia *javensis.* The northern and Old World *pileatus, martius,* and *javensis* appear to be derivatives of the neotropics that have become specialized in the absence of *Campephilus* or any other very large, woodpecking picids (Short 1982). The only sympatry in the genus occurs between weak billed *galeatus* and more specialized *lineatus* in southeastern South America, and *martius* and *javensis* in Korea (where, perhaps as a result of character displacement, they show morphological and ecological divergence, Short 1982).

The distribution of *Celeus,* like *Picumnus,* centers about Amazonia, with two species extending well into Middle America, but one species also occurs in southern Asia (*brachyurus*). Neotropical ancestors of *Celeus* gave rise to *Dryocopus* (thence to *Campephilus*) and probably to *Picus* (*P. miniaceus* clearly shows features of *Celeus*) and, thus, its Asian relatives (Short 1982). Amazonian species include small Guianan *undatus* and Amazonian *grammicus* (which form a superspecies), medium, unspecialized *flavus,* large *torquatus,* the most woodpecking-specialized member of the genus, and *elegans.* The rare, specialized *spectabilis* occurs along the eastern slope of the Andes, with an isolate in eastern Brazil. *Loricatus* is an essentially Middle American (to northwestern Ecuador) generalized species, which with the *undatus* and *elegans* groups, most nearly resembles Asian *brachyurus*. The moderately large, but relatively unspecialized (for woodpecking) superspecies *elegans* consists of Amazonain *elegans, lugubris* of central South America, *flavescens* in eastern and southeastern South America, and Middle American *castaneus* (Short 1972a). Most species of *Celeus* appear to be ant-foragers, and some nest in carton-like arboreal ant nests, as does *brachyurus* in Asia. There is remarkable uniformity of coloration but great morphological variation in bill structure among species of *Celeus. Flavus* and *undatus*, and *torquatus* represent, respectively, "weak-billed" and "strong-billed" extremes. Behaviorally, and morphologically *Celeus* resembles *Colaptes* (Short 1982).

Melanerpine woodpeckers are morphologically and behaviorally distinctive, with three genera, monotypic *Xiphidiopicus*, the strictly nearctic *Sphyrapicus* (four species), and *Melanerpes* (Short 1982). The last is widespread in the Americas, having highly derived species in North America (e.g., *lewis, erythocephalus*), the West Indies (e.g., *striatus, portoricensis*), Middle America, (e.g., *chrysogenys*), and South America (e.g., *cactorum, candidus*). *Melanerpes* has two subgroups. The first subgroup of species, well isolated from one another, is widespread, generalized, and omnivorous. This, the *Melanerpes* subgroup, reflects an ancient pattern of distribution with especially distinct species in North America (*lewis*), the West Indies (*herminieri*), and South America (*candidus*). The second subgroup is of Middle American origin, and is actively undergoing radiation (Short 1982). The latter "*Centurus*" subgroup of somewhat more specialized (woodpecking) species, has relict, presumably older, species in the West Indies (*striatus, radiolatus*), South America (*cactorum*, possibly), and Middle America (*chrysogenys, hypopolius*). More advanced members of the *Centurus* subgroup include *rubricapillus*, which has reached northernmost South America, and the strictly Middle-North American superspecies *carolinus* (nearctic *carolinus*, West Indian *superciliaris*, Middle to North American *uropygialis*, Middle American *hoffmannii*, and Middle to North American *aurifrons*).

The four strictly South American species of *Melanerpes* are honey-eating *candidus*, which occurs from the Amazon to Argentina, *cactorum* of dry forest and woodland from Peru to northern Argentina, *cruentatus*, widespread in northern South America, and *flavifrons* of eastern Brazil to Paraguay. Of these *cruentatus* and *flavifrons* form a superspecies with *chrysauchen* and *pucherani*, Middle American species of the *Melanerpes* subgroup that reach northwestern South America. Otherwise only *rubricapillus*, basically Middle American, and nut-storing *formicivorus* (*Melanerpes* subgroup) reach northern South America. Eight species thus occur in South America, four of which essentially are Middle American. The latter four and the endemic South American species occur in northern South America, and around the fringes of Amazonia. The nine mainland Middle American species include two distinctive endemics (*chrysogenys, hypopolius*), three that reach North America (*uropygialis, aurifrons, formicivorus*), and four that reach South America. The two endemics, three of the species reaching South America, and two reaching North America can be considered Middle American in origin (Short 1982). *Melanerpes formicivorus*, reaching both South America and widespread in North America, probably also originated in Middle America. All five West Indian species are endemic; *superciliaris* probably arrived recently from North or Middle America, whereas the others (*portoricensis, herminieri, radiolatus, striatus*) lack close relatives and probably are older. Sympatry is common between species of the *Melanerpes* and *Centurus* subgroups, but uncommon within each subgroup.

The distributions of *Sphyrapicus, Xiphidiopicus,* and *Melanerpes* suggest that this tribe, the Melanerpini, likely arose in Middle America or even the West Indies. Failure of the melanerpines to reach the Old World suggests a neotropical rather than nearctic origin.

The final genus, *Picoides*, is cosmopolitan, and derived from Afrotropical *Dendropicos* (Short 1971a, 1971b, 1982). An extensive radiation centering on Eurasia (where it may have started) has reached the Afrotropics (*obsoletus*), the nearctic (where there was a secondary radiation), and, ultimately, the neotropics. The New World invasion preceded the major Eurasian radiation of "red-belled" species now widespread within the palearctic and southern Asia (Short 1982). No American species inhabits tropical forests. Entry of *Picoides* into the neotropics has been along the Middle American temperate-montane "backbone" (e.g., *villosus*), and in both pinelands and xeric scrub (*scalaris*). Speciation in South America gave rise to the campo-chaco *mixtus* and its allospecies *lignarius*, which shows a curious disjunct distribution in the arid upland valleys of Bolivia and in the Fuegian forests (Short 1980a, 1982). No northward connections persist, and there is a gap between Middle American *scalaris* and Bolivian *lignarius*. These specialized species of *Picoides* either have reached South America too recently to radiate, have an intolerance for the tropics, or are unable to compete successfully with neotropical *Veniliornis*. Thus, they occupy only the southern fringes of the New World tropics and the south temperate zone.

Within the neotropics picid genera are broadly sympatric, with intrageneric sympatry usually involving distinct groups within a genus. Three congeneric sympatric species seem to be the usual maximum, except at the adjoining borders of ranges of essentially allopatric species or where diverse habitats come together. This is true of southern Asia (*Picus*, Short 1978), the Afrotropics (*Dendropicos, Campethera*, pers. observ.), Europe (*Picoides*, pers. observ.), the nearctic (*Picoides*, Short 1971b), and the neotropics (*Picumnus, Celeus*, perhaps *Veniliornis*, Short 1982).

It was shown that sympatric congeners in southern Asia tend to differ in size, foraging habits, or both (Short 1973, 1978). As a corollary, woodpeckers of the same size tend to be of different genera. Specialized woodpecking species tend to differ markedly in size if sympatric (Short 1978). This holds in Africa among species of *Dendropicos* (e.g., *namaquus, elliotii, gabonensis* in Cameroun, the specialized woodpeckers all being congeneric in most of the picid-depauperate Afrotropics). It also is true of the neotropics, where specialized *Campephilus*, somewhat specialized *Dryocopus*, the specialized species of *Celeus*, specialized *Veniliornis* and *Picoides*, and somewhat specialized *Picumnus* form a gapped size gradient. Species of all the neotropical genera just mentioned, except *Picoides*, are sympatric in many parts of South America. In ecologically less diverse areas, such as the Fuegian forests of temperate South America size gradation is particularly steep among sympatric picids, and includes both specialized and generalized woodpeckers (large specialized *Campehilus magellanicus*, medium-sized, generalized *Colaptes pitius* and small, specialized *Picoides lignarius*, Short 1982). The same phenomenon holds in such areas as peninsular Baja California, with *Colaptes auratus, Melanerpes uropygialis*, and *Picoides scalaris* grading from medium-large to small (Short 1982).

CONCLUSIONS

The three families of piciform birds considered here (Ramphastidae are considered confamilial with Capitonidae) show differential radiation in the Afrotropics and the neotropics. The Afrotropics are marked by two major, one moderate, and three minor barbet radiations, major (*Indicator*) and minor (*Prodotiscus*) radiations of honeyguides, and two major radiations within one tribe (Campetherini) of woodpeckers. The Afrotropical barbet genera are all endemic, and highly diverse. The neotropics are marked by one moderate and one minor radiations of barbets, one major and three moderate radiations of toucans, and six major and one minor radiations of woodpeckers. By treating toucans as barbets the number of Afrotropical and neotropical barbets becomes comparable (42 and 46, respectively), as do the numbers of genera (seven and nine), and the total numbers of radiations (six each). Of course, toucans have such habits and are of such a large size as to exceed the extreme of the Afrotropical barbet spectrum of variation tending toward toucans (*Lybius*). The Afrotropical barbets of medium and small size, however, are much more diverse than are their neotropical relatives. It has been noted that neotropical woodpeckers far exceed the spectrum of diversity of Afrotropical woodpeckers.

Reasons for the differential radiations are not easy to understand and are undoubtedly multiple. Historical factors and dispersal of the piciforms has been mentioned above. Factors involving differences between the Afrotropics and neotropics, such as the nature of the vegetation, the arboreal predators, the particular insects available as food, and fruit sources have not been fully addressed. Hole-nesting and hole-roosting have many advantages but also disadvantages (Short 1979). Freshly excavated cavities are sanitary but gradually develop a microbiota that may make secondary or long-term use less advantegous. Certainly nest-hole competitors vary greatly between the Afrotropics and the neotropics. Some Afrotropical falcons and owls, hornbills, parrots, wood-hoopoes, hoopoes, rollers, trogons, and starlings are hole-nesting, and often aggressive birds of generally medium to large size. These, with a large number of mammals (e.g., monkeys, squirrels) often take over holes excavated by piciforms. Among these competitors for holes, only the mammals compete all year. The piciform honeyguides parasitize many barbets and woodpeckers. One effect of parasitism and piracy of nests is to delay breeding of the victims, perhaps causing them to select poorer, secondary nest sites, and to nest at less advantageous times of the year (these probably account for some of the unusually long breeding seasons reported for many tropical picids).

Within the neotropics no nest parasites favor piciforms, but an array of secondary hole-nesters includes some falcons, owls, parrots, trogons, woodcreepers, ovenbirds, cotingas and tyrant-flycatchers, and many of these compete during their breeding seasons for piciform-excavated cavities. The toucans do not excavate, but use old picid holes that they enlarge, or other cavities, and to some degree they do compete with barbets and woodpeckers for cavities (Skutch 1983).

Competition is apt to be most severe between related taxa that share some morphological and behavioral features. This competition is for food and for nesting and roosting sites. Unlike most hole-nesters, barbets and woodpeckers require, use, and defend holes for roosting all year; hence, they are apt to compete more severely than other, breeding season hole-seekers. J. F. M. Horne and I have many times observed barbets, such as *Gymnobucco bonapartei* and *Tricholaema lachrymosa*, drive Cardinal Woodpeckers (*Dendropicos fuscescens*) from trees in which the latter sought only to feed. Toucans may or may not compete for fruit with barbets, but Skutch (1983) cited cases of both attempted predation of barbets by toucans, and nest-hole piracy by the latter. Within families there is intense competition both for holes, and in foraging. The White-headed Barbet (*Lybius leucocephalus*) attacks tinkerbirds in feeding trees, and *Gymnobucco bonapartei* chases *Tricholaema hirsuta* from where it feeds (pers. observ.). Among woodpeckers, dense populations of *Picoides nuttallii* prevent pairs of *Picoides villosus* from establishing a territory (Short 1971b). *Picoides macei* actively restricts smaller *P. canicapillus* to feeding in the outer edges of tall trees, away from trunks preferred by the former (Short 1973). *Melanerpes erythrocephalus* attacks *M. carolinus* in more open woods, restricting its feeding (L. Saul, and Short, pers. observ., Florida, Georgia). *Dryocopus javensis* enlarges holes made by smaller *Picus puniceus*, rendering them unusable (Short 1973), and *M. erythrocephalus* does the same at some nesting holes of *Picoides borealis* (fide J. Jackson). The different feeding sites and foraging modes of congeneric sympatric species within *Picus* and *Meiglyptes* in Asia imply strong competition (Short 1973, 1978), as do the precise separation of sizes and occurrence only of species representing different genera among wood-pecking specialists that are sympatric (Short 1978).

Non-piciform foraging competitors also differ in the two regions. It is often assumed that woodpeckers, by excavating into bark, are such specialists as to avoid competition generally, but in fact many woodpeckers glean or otherwise forage on, rather than in the bark, so other bark foragers are apt to offer some competition. Certainly barbets have numerous competitors, both when foraging for insects and for fruit, especially at the end of the fruiting periods of various trees (pers. observ.). Afrotropical frugivorous barbets (and barbets generally) are very aggressive toward other frugivores, even of larger size. There are apt to be intercontinental differences in fruits and insects, and in the array of fruit- or insect-foraging animals. Toucans are primarily frugivorous, but it is unclear whether their origin and radiation reflect something unique about the fruits available in the neotropics or whether they have an effective equivalent, the hornbills, in the Afrotropics, that preclude evolution of toucan-like Afrotropical barbets. Honeyguides have a unique foraging mode and must have little avian competition.

Even comparing Afrotropical and neotropical radiations of bark-foraging woodpeckers in terms of possible restriction due to other bark foragers is difficult because of many opportunist bark foragers, and lack of knowledge of the relative availability of insects in the bark of trees in the two regions. There are many bark-foraging birds in both areas. Afrotropical examples include especially some barbets (e.g., *Pogoniulus*, *Tricholaema*), wood-hoopoes, a creeper (*Salpornis*), tits (Paridae), some bulbuls, some babblers, some warblers (Sylviinae), and a group of weavers (Ploceidae). Neotropical bark foragers include the woodcreepers, some oven-birds, some antbirds, gnatcatchers (Sylviinae), vireos, some wood-warblers and a scattering of other birds. Since none of these actually excavates into the bark, their effects would be maximal on the generalized, less woodpecking picids. Many generalized woodpeckers (e.g., Afrotropical *Campethera*; neotropical *Colaptes*, many *Piculus* and *Celeus*, some *Melanerpes*) are ant-foragers, eating insects avoided by many potential competitors. The neotropics have some specialized picids (*Dryocopus*, some *Celeus*) that excavate deep into trees for ants. *Melanerpes* includes species that forage for fruits and nuts, even storing the latter in special ways (e.g., *M. formicivorus*).

Even such specialists in feeding as barbets and woodpeckers readily become opportunistic when suitable food is superabundant. J. F. M. Horne and I have observed many barbets and woodpeckers flycatching ineptly, but with some success, for winged termites and ants, even with flycatchers (Muscicapidae or Tyrannidae) about them. These piciforms even prolong their feeding until well after normal roosting time to avail themselves of this food source (pers. observ.).

One could argue that events favoring speciation have been responsible for the greater number of piciforms in the neotropics. However, whereas woodpeckers have not radiated greatly in Africa, barbets have, and, whereas woodpeckers show great radiation in the neotropics, barbets do not. Undoubtedly for certain groups (e.g., *Picumnus* and toucans in the neotropics; *Pogoniulus* and *Lybius* in the Afrotropics) there have been opportunities for speciation not available to other piciforms because of special feeding techniques or unique features. Constraints such as parasitism by honeyguides may be lessened or removed from species of *Pogoniulus* due to the small size of the latter.

We are left with the facts of the radiation as documented, and little in the way of explanations for them. I suspect that historical events (especially those allowing entry to or exit from the neotropics and Afrotropics, and those facilitating speciation), and the impingement on one another of the piciform groups are the major factors responsible for the initiation and extent of the radiations that have occurred. But, until basic data on the nature and availability of food sources, and on the identification and behaviors of competitors and predators are gathered and analyzed, we will grope for a full understanding of the observed patterns of radiation.

ACKNOWLEDGMENTS

Many persons and agencies have supported my research upon which this review largely is based; these are listed in Short (1982:xv–xvi). Here I especially wish to thank the authorities of the American Museum of Natural History where my analyses have been conducted, the National Science Foundation for support of some of my neotropical field studies, and the Leonard C. Sanford Fund and Gerald and May Ellen Ritter-Eugene Eisenmann Fund of the American Museum of Natural History that provided funding for African field work. Jennifer F. M. Horne has assisted me greatly with the African research, as well as imparting an understanding of the eastern Afrotropics. I thank M. S. Foster, and several referees for comments benefitting the manuscript. I recollect that my close friend and valued colleague, the late Eugene Eisenmann, shared with me my first field experiences in the Afrotropics in 1969.

LITERATURE CITED

AMADON, D., AND L. L. SHORT. 1976. Treatment of subspecies approaching species status. Syst. Zool. 25:161–167.

BRODKORB, P. 1971. Catalogue of fossil birds: Part 4 (Columbiformes through Piciformes). Bull. Fl. State Mus. 15:163–266.

DIAMOND, A. W., AND A. C. HAMILTON. 1980. The distribution of forest passerine birds and Quaternary climatic change in tropical Africa. J. Zool. (Lond.) 191:379–402.

ENDLER, J. A. 1982. Pleistocene forest refuges: fact or fancy? Pp. 641–657, *In* G. T. Prance (ed.), Biological Diversification in the Tropics. Columbia University Press, New York.

FRIEDMANN, H. 1955. The honey-guides. U.S. Natl. Mus. Bull. no. 208.

FRIEDMANN, H. 1970. Further information on the breeding biology of the honeyguides. Los Ang. Cty. Mus. Contrib. Sci. no. 205.

FRY, C. H. 1980. An analysis of the avifauna of African northern tropical woodlands. Proc. 4th Pan-Afr. Ornithol. Congr., pp. 77–87.

GOODWIN, D. 1964. Some aspects of taxonomy and relationships of barbets (Capitonidae). Ibis 106: 198–220.

GRUBB, P. 1982. Refuges and dispersal in the speciation of African forest animals. Pp. 537–553, *In* G. T. Prance (ed.), Biological Diversification in the Tropics. Columbia University Press, New York.

HAFFER, J. 1969. Speciation in Amazonian forest birds. Science 165:131–137.

HAFFER, J. 1974. Avian speciation in tropical South America. Publ. Nuttall Ornithol. Club no. 14.

HAFFER, J. 1982. General aspects of the refuge theory. Pp. 6–24, *In* G. T. Prance (ed.), Biological Diversification in the Tropics. Columbia University Press, New York.

HALL, B. P., AND R. E. MOREAU. 1970. Atlas of Speciation in African Passerine Birds. British Museum (Natural History), London.

LIVINGSTONE, D. A. 1982. Quaternary geography of Africa and the refuge theory. Pp. 523–536, *In* G. T. Prance (ed.), Biological Diversification in the Tropics. Columbia University Press, New York.

MAYR, E., AND L. L. SHORT. 1970. Species taxa of North American birds. Publ. Nuttall Ornithol. Club No. 9.

MEGGERS, B. J., E. S. AYENSU, AND W. D. DUCKWORTH (eds.). 1973. Tropical Forest Ecosystems in Africa and South America: A Comparative Review. Smithsonian Institution Press, Washington, D.C.

PRIGOGINE, A. 1980. Hybridization entre les Barbions *Pogoniulus bilineatus* et *Pogoniulus leucolaima* au Rwanda et au Burundi. Gerfaut 70:73–91.

RIPLEY, S. D. 1945. The barbets. Auk 62:542–563.

SHORT, L. L. 1970. The habits and relationships of the Magellanic Woodpecker. Wilson Bull. 82:115–129.

SHORT, L. L. 1971a. The affinity of African with neotropical woodpeckers. Ostrich Suppl. 8:35–40.

SHORT, L. L. 1971b. Systematics and behavior of some North American woodpeckers, genus *Picoides* (Aves). Bull. Am. Mus. Nat. Hist. 145:1–118.

SHORT, L. L. 1971c. The evolution of terrestrial woodpeckers. Am. Mus. Novit. No. 2467.

SHORT, L. L. 1972a. Relationships among the four species of the superspecies *Celeus elegans* (Aves, Picidae). Am. Mus. Novit. No. 2487.

SHORT, L. L. 1972b. Systematics and behavior of South American flickers (Aves, *Colaptes*). Bull. Am. Mus. Nat. Hist. 149:1–109.

SHORT, L. L. 1973. Habits of some Asian woodpeckers (Aves, Picidae). Bull. Am. Mus. Nat. Hist. 152:253–364.

SHORT, L. L. 1974a. Habits of three endemic West Indian woodpeckers (Aves, Picidae). Am. Mus. Novit. no. 2549.

SHORT, L. L. 1974b. Piciformes. Encyclopaedia Britannica, 15th ed., 14:447–452.

SHORT, L. L. 1975. A zoogeographic analysis of the South American chaco avifauna. Bull. Am. Mus. Nat. Hist. 154:163–352.

SHORT, L. L. 1976. The contribution of external morphology to avian classification. Proc. 16th Int. Ornithol. Congr., pp. 185–195.

SHORT, L. L. 1978. Sympatry in woodpeckers of lowland Malayan forest. Biotropica 10:122–133.

SHORT, L. L. 1979. Burdens of the picid hole-nesting habit. Wilson Bull. 91:16–28.

SHORT, L. L. 1980a. Speciation in African woodpeckers. Proc. 4th Pan-Afr. Ornithol. Congr., pp. 1–8.

SHORT, L. L. 1980b. Chaco woodland birds of South America—some African comparisons. Proc. 4th Pan-Afr. Ornithol. Congr., pp. 147–158.

SHORT, L. L. 1980c. Speciation in South American woodpeckers. Acta XVII Congr. Int. Ornithol., pp. 1268–1272.

SHORT, L. L. 1982. Woodpeckers of the world. Del. Mus. Nat. Hist. Monogr. 4.

SHORT, L. L., AND J. F. M. HORNE. 1979. Vocal displays and some interactions of Kenyan honeyguides (Indicatoridae) with barbets (Capitonidae). Am. Mus. Novit. No. 2684.

SHORT, L. L., AND J. F. M. HORNE. 1981. Vocal and other behaviour of Stierling's Woodpecker. Scopus 5:5–13.

SHORT, L. L., AND J. F. M. HORNE. 1983. The relationships of male Lesser Honeyguides *Indicator minor* with duetting barbet pairs. Bull. Br. Ornithol. Club 103:25–32.

SHORT, L. L., AND J. F. M. HORNE. In press. Aspects of duetting in some ground barbets. Proc. 5th Pan-Afr. Ornithol. Congr.

SHORT, L. L., AND W. TARBOTON. 1978. Picidae. Pp. 359–393, *In* D. W. Snow (ed.), An Atlas of Speciation in African Non-passerine Birds. British Museum (Natural History), London.

SIBLEY, C. G., AND J. E. AHLQUIST. 1972. A comparative study of the egg white proteins of non-passerine birds. Peabody Mus. Nat. Hist. Yale Univ. Bull. No. 39.

SIMPSON, B. B., AND J. HAFFER. 1978. Speciation patterns in the Amazonian forest biota. Annu. Rev. Ecol. Syst. 9:497–518.

SIMPSON, S. F., AND J. CRACRAFT. 1981. The phylogenetic relationships of the Piciformes (Class Aves). Auk 98:481–494.

SKUTCH, A. F. 1983. Birds of Tropical America. University Texas Press, Austin, Texas.

SNOW, D. W. (ed.). 1978. An Atlas of Speciation in African Non-passerine Birds. British Museum (Natural History), London.

SWIERCZEWSKI, E. W., AND R. J. RAIKOW. 1981. Hind limb morphology, phylogeny, and classification of the Piciformes. Auk 98:466–480.

TERBORGH, J., AND B. WINTER. 1982. Evolutionary circumstances of species with small ranges. Pp. 587–600, *In* G. T. Prance (ed.), Biological Diversification in the Tropics. Columbia University Press, New York.

WINKLER, H., AND L. L. SHORT. 1978. A comparative analysis of acoustical signals in pied woodpeckers (Aves, *Picoides*). Bull. Am. Mus. Nat. Hist. 160:1–110.

CLUTCH SIZE, NESTING SUCCESS, AND PREDATION ON NESTS OF NEOTROPICAL BIRDS, REVIEWED

ALEXANDER F. SKUTCH

San Isidro de El General, Costa Rica

ABSTRACT. A survey of the clutch size of 217 species of passerines of the humid neotropics shows that two is the prevailing number of eggs, sets of one and three are less frequent, and larger sets are rare. Contrary to what we should expect from the theory of maximum reproduction—that birds rear as many young as they can adequately nourish—unaided females commonly have broods as large as those attended by both parents, sometimes with helpers. Failure to find consistent correlation between clutch size and number of nest attendants, diet, habitat, or type of nest (other than the well-known tendency of hole-nesters to rear larger broods) leads us to seek some factor, or factors, that profoundly influence the reproduction of most birds of the humid neotropics. Not to be neglected is the high percentage of nest failures, greater in forest than in neighboring clearings and plantations, and greater at low than at high elevations. Available evidence leaves the effect of human visits on nest losses uncertain; hatching failure due to infertility, faulty incubation, or other intrinsic factors appears to be no greater in the tropics than at higher latitudes; predation is certainly responsible for most losses. The major factor responsible for the small clutches of tropical birds of many kinds appears to be, as Cody and Ricklefs have argued, the less strongly contrasting seasons of the humid tropics—a measure of which is the annual march of evapotranspiration—as compared with northern lands. The restrained reproductive effort of tropical birds is adjusted to their low annual mortality in a climate that does not force birds to confront a season of scarcity and stress unless they undertake hazardous migrations. Moreover, the high incidence of predation on nests makes it advantageous to limit the energy expended on a brood, so that, if this fails, strength remains for repeated trials. Also, the smaller the brood, the fewer the feeding visits that may reveal the nest's location to predators. Because ornithology was born in the north temperate zone where broods tend to be large, we ask why the broods of tropical birds are so small. If more ornithologists had grown up in the tropics, we would be asking why birds at high latitudes lay so many eggs—a question easier to answer.

RESUMEN. Un estudio del tamaño de las nidadas de 217 especies de passeri-formes de las regiones húmedas neotropicales, muestra que el número prevaleciente de huevos en una nidada es dos; siendo menos frecuentes nidadas de uno o tres huevos y son raras las nidadas más grandes. Contrariamente a lo que deberíamos esperar, si consideramos la teoría de máxima reproducción—que las aves crían tantos polluelos como les es posible alimentar—las hembras que no tienen ayuda cuídan nidadas tan grandes como aquellas nidadas que son atendidas por ambos padres, que algunas veces tienen ayudantes. El no encontrar una correlación consistente para la relación entre el tamaño de la nidada y el número de encargados del nido, dieta, habitat o tipo de nido (otra que la tendencia conocida para los anidadores en huecos que crían grandes nidadas), nos hace considerar ciertos factores que influyen profundamente la reproducción de la mayoría de las aves de las regiones húmedas de los neotrópicos. Algo que no debe ser descuidado es el alto procentaje de fracasos de anidación, los cuales son mayores en el bosque que en las zonas abiertas o plantaciones cercanas y mayor a baja que a altas elevaciones. No está claro que efecto tienen las visitas humanas en las pérdidas de nidos; fracasos de eclosión debido a infertilidad, incubación defectuosa, u otros factores intrínsecos que parecen no ser más importantes en los trópicos que en otras latitudes más elevadas; la depredación es por cierto la mayor responsable en la mayoría de las pérdidas. El factor mayormente responsable por el tamaño pequeño de las nidadas de las aves neotropicales de cualquier tipo parece ser, tal como lo discutiesen Cody y Ricklefs, el menor contraste entre las estaciones en los trópicos húmedos—lo cual puede ser medido por la marcha anual de evapo-transpiración—si se compara con tierras septentrionales. El esfuerzo reproductivo moderado de las aves tropicales, se ajusta a la baja mortalidad anual en un clima que no fuerza a las aves a enfrentar una estación de escasez y "stress," a no ser que participen en migraciones riesgosas. Mas aún la gran incidencia de depredación en los nidos hace ventajoso limitar el gasto de energía en una nidada, de manera que si falla, aún quedará con fuerzas suficientes para intentarlo nuevamente. Así mismo, cuanto más pequeña sea la nidada, será menor cantidad de visitas para alimentación que podrá revelar la posición del nido a los depredadores. Debido a que los estudios ornithológicos nacieron en las zonas templadas del norte, donde las nidadas tienden a ser grandes, nos preguntamos, porque las nidadas de aves tropicales son pequeñas. Si más ornitólogos hubiesen crecido en los trópicos, nos estaríamos preguntando porque las aves de latitudes más elevadas ponen tantos huevos—una pregunta más fácil de responder.

We know that the stream of soldiers, priests, colonists, and administrators that Spain and Portugal sent to their recently acquired possessions in tropical America included men interested in nature, such as Gonzalo Fernández de Oviedo y Valdés (1478–1557), author of the celebrated *Historia General y Natural de las Indias*. It would be surprising if none of these people noticed that the birds' nests around them held fewer eggs than those in their homeland in the Iberian Peninsula. It appears, however, that German explorers in South America, such as Prince Maximilian Wied (1825–1833) and M. R. Schomburgk (1847–1848), first brought this striking difference in the number of eggs laid in the tropics and the north temperate zone to the attention of the scientific world. Hesse (1922) was apparently the first to offer an explanation of the difference, so unexpected when one reflects upon the abundance of fruits and insects that the fecund tropics offer to parent birds feeding young. He suggested that tropical birds rear smaller families than do birds at higher latitudes because the shorter days in the breeding season give them less time to hunt food for their young.

The differences in the clutch sizes of birds in the tropics and at higher latitudes are evident in nearly all families, passerines and nonpasserines, ground-nesters and tree-nesters, nocturnal and diurnal birds. Notable exceptions are pigeons and hummingbirds, whose maximum clutch size of two eggs appears to be determined by factors independent of latitude, such as the size of their nests or their method of feeding their young. In this paper, however, I shall give attention chiefly to passerines, because the data for them are so much more abundant than those for nonpasserines, and the conclusions drawn from their study are equally applicable to nonpasserine land birds.

CLUTCH SIZE

I considered clutch size, type of nest, number of nest attendants, and principal foods of 217 species of neotropical passerines (Table 1). The data are chiefly from the northern tropics in Suriname (Haverschmidt 1968), Trinidad (ffrench 1973), and from Central America, mostly Costa Rica (my own published and unpublished records). They cover a range of latitude of about 12°, from 4° to 16°N. To confine this discussion to the humid tropics, I have omitted Marchant's (1960) abundant records from very arid southwestern Ecuador, where clutch sizes tend to be larger and nesting success substantially higher than in tropical regions with more abundant rainfall—the contrasts between the rainy tropics and so dry a region parallel those between the wetter tropics and the temperate zones. I have likewise excluded records from farther north, in Belize and Mexico, because at this distance from the equator the latitude effect, already evident at 16°N, becomes stronger. I have included southern Guatemala because from this country I have most records of the clutch sizes of birds that nest above 2500 m. Although clutch size increases with latitude, it appears not to do so with elevation, up, at least, to the limits of heavy forests at 3000 or 3500 m.

I would have liked to include mean clutch size as well as the extremes, but my sources did not always give the information needed for this. Moreover, clutch size appears to be smaller in Suriname than in Trinidad and Costa Rica, about 5° farther north; this may be attributed to the latitude effect. I cannot explain why some species occasionally lay more eggs in Trinidad than in southern Costa Rica, as these two regions are at about the same latitude and have rather similar climates. Although an island, Trinidad is only a slightly detached fragment of the South American continent.

Despite great differences among neotropical birds in habitat, diet, type of nest, and taxonomic position, two is by far their most frequent brood size (Table 1). A number of species usually or occasionally lay three eggs, but larger sets are decidedly rare. Only swallows, jays, certain wrens, and euphonias frequently lay larger sets. The House-Wren (scientific names given in Table 1 or 3), like the swallows, nests in crannies, holes, and burrows. It breeds from sea level to high in the mountains; for years it has been invading new areas as woodland, which it avoids, is replaced by farms and buildings where it flourishes. The Band-backed Wren is a bird of great ecological tolerance, thriving from lowlands up to 3000 m, in clearings as well as wet forests. The tiny euphonias, which build nests with side entrances, consistently lay more eggs than larger tanagers with open nests. The entry (Table 1) of five eggs for the Yellow-throated Euphonia is based on only two records, one from Guatemala and one from Costa Rica.

Aside from the well-known tendency of hole-nesting birds to lay larger sets of eggs than open-nesters do, clutch size does not appear to be correlated with any factor (Table 1). Insectivorous flycatchers lay sets of about the same size as those of the mainly frugivorous tanagers. Large thrushes are no more prolific than small wood warblers and vireos. Small flycatchers with pendent nests lay the same number of eggs as those that build open nests.

Most surprising of all, and most difficult to reconcile with the widely held view that birds everywhere rear as many young as they can adequately nourish (Lack 1947, 1948) is the absence of correlation between clutch size and number of nest attendants. Unaided, female manakins, like female hummingbirds, attend broods as large as those of the great majority of their monogamous neighbors, whose young are fed by two parents. The failure of clutch size to be correlated with number of attendants is most impressive when we compare birds of similar habitats and diets in the same family, as in the woodcreepers, cotingas, and flycatchers. Few nests of the secretive woodcreepers have been found, and fewer have been carefully studied, but we know that in two species of *Dendrocincla* and one of *Xiphorhynchus* the female attends the nest alone (Willis 1972a; Skutch 1981). In two species of *Lepidocolaptes* and in *Glyphorhynchus spirurus* both sexes incubate and feed the young (Skutch 1969, 1981), yet all these woodcreepers usually or always lay two eggs.

Much remains to be learned about the nesting of cotingas. Females of the larger species, including the Rufous Piha, Bearded Bellbird, and Calfbird, rear a single nestling without a male's help. At the only nest of the Purple-throated Fruitcrow that has been well studied, both parents and their helpers likewise attended a brood of one (Snow 1982). Although Snow removed the tityras and becards from the cotinga family, I include them here because I believe that if separated from the Cotingidae, they should be given familial status rather than included in the Tyrannidae.

The Tyrannidae provide the most abundant material for comparing the reproductive effort of solitary females and mated pairs. If we confine our attention to the smaller flycatchers of wholly tropical genera (e.g., *Terenotriccus*, *Zimmerius*, Table 1), we notice that all, except *Mionectes*, regularly or usually lay two eggs, whether their nest be a compact open cup or a less accessible pensile structure, whether they live in forest or clearings, whether the female is solitary or helped by a mate. It is particularly instructive to compare *Mionectes* with *Zimmerius*. Both build mossy nests with a side entrance, that of the Ochre-bellied Flycatcher (*M. oleaginus*) attached to a dangling vine or aerial root, the smaller structure of the Paltry Tyrannulet often tucked into a tuft of mosses or liverworts growing beneath a slender branch. Both are largely frugivorous; *Zimmerius* prefers the berries of mistletoes (Loranthaceae), while *Mionectes* seeks a greater variety of berries and arillate seeds. Both parents regularly attend the two nestlings of *Zimmerius*, but the solitary female *Mionectes* lays three eggs more often than two and, apparently very rarely, a larger number.

When we contrast the uniformly small clutch size of these birds of the more humid regions of tropical America with the great diversity of their habitats, mating habits, diet, and nidification, we are disinclined to seek particular reasons, such as whether one species is limited by its ability to nourish its young, another by the smallness of its nest, and so forth, to explain why they do not lay more eggs. Instead, we must look for some general, widely effective feature of the American tropics that has profoundly influenced the reproductive efforts of all, or most, of its so diverse feathered inhabitants. If the number of a bird's progeny depended solely upon how many eggs it lays, we might ask at once what this feature might be. However, since the bird's success in reproduction depends so greatly upon the hazards to which nests and young are exposed, it appears best to defer the investigation of this question until we have examined the magnitude of these hazards.

NESTING SUCCESS

To obtain a fair measure of the nesting success of any species accurate to within 5 or 10 percent requires a far larger number of records than we are likely to collect, especially amid the heavy vegetation of the humid tropics. Success varies from place to place, and from month

TABLE 1

Clutch Size, Nest Type, Number of Attendants, and Food of Passerine Birds in the Humid Neotropics[1]

Species	Nest type	Clutch size[2]	No. attendants[3]	Diet[4]
Furnariidae				
Pale-breasted Spinetail *Synallaxis albescens*	Bulky, enclosed	2–3	2	I
Slaty Spinetail, *S. brachyura*	Bulky, enclosed	2	2	I
Rufous-breasted Spinetail *S. erythrothorax*	Bulky, enclosed	3, 4	2	I
Buff-throated Foliage-gleaner, *Automolus ochrolaemus*	Burrow	2–3	2	I, L
Xenops, *Xenops* 2 spp.	Tree hole	2	2	I
Scaly-throated Leaftosser, *Sclerurus guatemalensis*	Burrow	2	2	I
Rufous-fronted Thornbird, *Phacellodomus rufifrons*	Bulky, pendent	3	2	I
Dendrocolaptidae				
Woodcreepers, *Dendrocincla* 2 spp.	Tree hole	2	1	I
Wedge-billed Woodcreeper, *Glyphorhynchus spirurus*	Tree hole	2 (3)	2	I
Buff-throated Woodcreeper, *Xiphorhynchus guttatus*	Tree hole	2	1	I
Woodcreepers, *Lepidocolaptes* 2 spp.	Tree hole	2	2	I
Formicariidae				
Antbirds, 12 genera, 17 spp.	Open cup or pouch	2 [1–3]	2	I
Antbirds, 2 genera, 2 spp.	Hollow stub	2	2	I
Tyrannidae				
Paltry Tyrannulet, *Zimmerius vilissimus*	Globe, side entrance	2	2	F, I
Southern Beardless Tyrannulet, *Camptostoma obsoletum*	Globe, side entrance	2	2	I, F
Mouse-colored Tyrannulet, *Phaeomyias murina*	Cup	2	2	F, I
Elaenias, *Myiopagis* 2 spp.	Cup	2	2	I, F
Elaenias, *Elaenia* 3 spp.	Compact cup	2 [1–3]	2	I, F
Torrent Flycatcher, *Serpophaga cinerea*	Mossy cup	2	2	I
Ochre-bellied Flycatcher, *Mionectes oleaginea*	Pendent, side entrance	3–2 [4, 5]	1	F, I
Yellow Flycatcher, *Capsiempis flaveola*	Cup	2	2	I
Bentbills, *Oncostoma* 2 spp.	Pendent, side entrance	1–2	1	I
Tody-Flycatchers, *Todirostrum* 3 spp.	Pendent, side entrance	2–3	2	I
Eye-ringed Flatbill, *Rhynchocyclus brevirostris*	Retort-shaped	2 (1)	1	I
Flycatchers, *Tolmomyias* 3 spp.	Pendent retort	2–3	2	I, F
Golden-crowned Spadebill, *Platyrinchus coronatus*	Cup	2	2	I
Royal Flycatchers, *Onychorhynchus* 2 spp.	Pendent, side entrance	2 (1)	1	I
Ruddy-tailed Flycatcher, *Terenotriccus erythrurus*	Pendent, side entrance	2	1	I
Flycatchers, *Myiobius* 2 spp.	Pendent, side entrance	2	1	I
Bran-colored Flycatcher, *Myiophobus fasciatus*	Deep cup	2	2	I (F)
Tropical Pewee, *Contopus cinereus*	Cup	2–3	2	I
Flycatchers, *Empidonax* 2 spp.	Cup	2–3	2	I
Black Phoebe, *Sayornis nigricans*	Mud cup	2–3	2	I
Pied Water-Tyrant, *Fluvicola pica*	Oval, side entrance	2–3	2	I
White-headed Marsh-Tyrant, *F. leucocephala*	Oval, side entrance	2–3 (4)	2	I
Bright-rumped Attila, *Attila spadiceus*	Niche in tree or bank	3–4 (2)	2	I, L, F
Flycatchers, *Myiarchus* 3 spp.	Cavity in tree	2–4	2	I, F
Great Kiskadee, *Pitangus sulphuratus*	Bulky, side entrance	3, 2, 4	2	O
Boat-billed Flycatcher, *Megarynchus pitangua*	Broad cup	2–3	2	I, F
Flycatchers, *Myiozetetes* 3 spp.	Bulky, side entrance	3, 2–4	2	I, F
White-bearded Flycatcher, *Conopias inornata*	Compact Cup	2	2-H	I, F

TABLE 1
CONTINUED

Species	Nest type	Clutch size[2]	No. atten- dants[3]	Diet[4]
Streaked Flycatcher, *Myiodynastes maculatus*	Cavity in tree	2–3	2	I, F
Piratic Flycatcher, *Legatus leucophaius*	Various, stolen	2, 3 (4)	2	I, F
Tropical Kingbird, *Tyrannus melancholicus*	Shallow cup	2, 3 (4)	2	I, F
Cotingidae				
Green-and-black Fruiteater, *Pipreola riefferii*	Mossy cup	2	2	F
Rufous Piha, *Lipaugus unirufus*	Tiny platform	1	1	F, I
Blue Cotingas, *Cotinga* 2 spp.	Shallow cup	2	1	F
White-winged Becard, *Pachyramphus polychopterus*	Bulky, side entrance	3, 2–4	2	I, F
Tityras, *Tityra* 2 spp.	Tree hole	2–3	2	I, F
Purple-throated Fruitcrow, *Querula purpurata*	Loose cup	1	2-H	F, I
Calfbird, *Perissocephalus tricolor*	Slight cup	1	1	F, I
Bearded Bellbird, *Procnias averano*	Slight cup	1	1	F
Cocks-of-the-rock, *Rupicola* 2 sp.	Mud cup on cliff	2	1	F
Pipridae				
Manakins, 4 genera, 9 spp.	Slight cup	2 (1)	1	F, I
Hirundinidae				
Gray-breasted Martin, *Progne chalybea*	Cranny	2–4 (5)	2	I
Blue-and-white Swallow, *Pygochelidon cyanoleuca*	Cranny or burrow	2–4	2	I
Southern Rough-winged Swallow, *Stelgidopteryx ruficollis*	Burrow	4–5 (3–6)	2	I
Corvidae				
Green Jay, *Cyanocorax yncas*	Bulky cup	4	2-H	I, F
Brown Jay, *C. morio*	Bulky cup	2–3	2-H	I, F
Bushy-crested Jay, *C. melanocyaneus*	Bulky cup	3–4	2-H	I, F
Troglodytidae				
Black-capped Donacobius, *Donacobius atricapillus*	Bulky cup	2–3	2	I
Band-backed Wren, *Campylorhynchus zonatus*	Bulky, side entrance	5, 3	2-H	I
Riverside Wren, *Thryothorus semibadius*	Globe, side entrance	2	2	I
Rufous-breasted Wren, *T. rutilus*	Globe, side entrance	3 (2–4)	2	I
Plain Wren, *T. modestus*	Globe, side entrance	2 (3)	2	I
House-Wren, *Troglodytes aedon*	Cranny	4, 3 (5, 6)	2	I
Rufous-browed Wren, *T. rufociliatus*	Cranny	3	2	I
Wood-wrens, *Henicorhina* 2 spp.	Globe, side entrance	2	2	I
Song Wren, *Cyphorhinus phaeocephalus*	Elbow-shaped	2	2	I
Musicapidae				
Long-billed Gnatwren, *Ramphocaenus melanurus*	Deep cup	2	2	I
Tropical Gnatcatcher, *Polioptila plumbea*	Small cup	3–2	2	I
Black-faced Solitaire, *Myadestes melanops*	Cup in niche	2–3	2	F, I
Nightingale-Thrushes, *Catharus* 3 spp.	Mossy cup	2 [3]	2	F, I
Yellow-legged Thrush, *Platycichla flavipes*	Shallow cup	2	2	F, I
White-necked Thrush, *Turdus albicollis*	Bulky cup	2 [3, 4]	2	I, F
Thrushes (Robins), *Turdus* 6 spp.	Bulky cup	3, 2 (4)	2	I, F
Mimidae				
Tropical Mockingbird, *Mimus gilvus*	Loose cup	3, 2–4	2	I, F
Ptilogonatidae				
Long-tailed Silky-flycatcher, *Ptilogonys caudatus*	Bulky cup	2	2	F, I

TABLE 1

CONTINUED

Species	Nest type	Clutch size[2]	No. atten- dants[3]	Diet[4]
Vireonidae				
Vireos, *Vireo* 2 spp.	Cup	3, 2	2	I, F
Greenlets, *Hylophilus* 5 spp.	Cup	2–3	2	I (F)
Rufous-browed Peppershrike, *Cyclarhis gujanensis*	Cup	2–3	2	I
Emberizidae				
Tropical Parula, *Parula pitiayumi*	Side entrance	2–3	2	I, F
Crescent-chested Warbler, *P. superciliosa*	Bulky cup	2–3	2	I
Flame-throated Warbler, *P. gutturalis*	Bulky cup	2	2	I
Masked Yellowthroat, *Geothlypis aequinoctialis*	Deep cup	2	2	I
Pink-headed Warbler, *Ergaticus versicolor*	Roofed, on bank	2–4	2	I
Redstarts, *Myioborus* 2 spp.	Side entrance	3, 2	2	I
Golden-crowned Warbler, *Basileuterus culicivorus*	Side entrance	2–4	2	I
Rufous-capped Warbler, *B. rufifrons*	Side entrance	2–3	2	I
Buff-rumped Warbler, *Phaeothlypis fulvicauda*	Roofed, on bank	2	2	I
Bananaquit, *Coereba flaveola*	Globe, side entrance	2, 3	2	N, I
Turquoise Tanager, *Tangara mexicana*	Cup	3	2-H	F, I
Small tanagers, *Tangara* 7 spp.	Cup	2	2	F, I
Green Honeycreeper, *Chlorophanes spiza*	Shallow cup	2	2	F, N, I
Red-legged Honeycreeper, *Cyanerpes cyaneus*	Shallow cup	2	2	F, N, I
Blue-crowned Chlorophonia, *Chlorophonia occipitalis*	Side entrance	3	2	F
Yellow-crowned Euphonia, *Euphonia luteicapilla*	Roofed, in cranny	3, 2–4	2	F
Violaceous Euphonia, *E. violacea*	Roofed, in cranny	3–4	2	F, I
Yellow-throated Euphonia, *E. hirundinacea*	Roofed, in cranny	5	2	I, F
Spot-crowned Euphonia, *E. imitans*	Roofed, in cranny	3, 2	2	F, I
White-vented Euphonia, *Euphonia minuta*	Roofed, in cranny	3	2	I, F
Blue-gray Tanager, *Thraupis episcopus*	Cup	2 [1–3]	2	F, I
Palm Tanager, *T. palmarum*	Cup in cranny	2 (3)	2	F, I
Gray-headed Tanager, *Eucometis penicillata*	Thin cup	2 (3)	2	I, F
White-lined Tanager, *Tachyphonus rufus*	Cup	2–3 (1)	2	F, I
Red-crowned Ant-Tanager, *Habia rubica*	Cup	2 (1–3)	2	F, I
Larger tanagers, *Ramphocelus* 3 spp.	Cup	2 [1–3]	2	F, I
Common Bush-Tanager, *Chlorospingus opthalmicus*	Bulky cup	2	2	F, I
Swallow-Tanager, *Tersina viridis*	Hole or burrow	3, 2	2	F, I
Saltators, *Saltator* 3 spp.	Bulky cup	2 (3)	2	F, I
Blue-black Grosbeak, *Cyanocompsa cyanoides*	Slight cup	2	2	S, F
Yellow-thighed Finch, *Pselliophorus tibialis*	Cup	2 (1)	2	I, F
Yellow-throated Brush-Finch, *Atlapetes gutturalis*	Cup	2–3	2	I, F
Chestnut-capped Brush-Finch, *A. brunneinucha*	Bulky cup	2 (1)	2	I
Striped-headed Brush-Finch, *A. torquatus*	Bulky cup	2	2	I, F
Orange-billed Sparrow, *Arremon aurantiirostris*	Roofed, on ground	2	2	I
Black-striped Sparrow, *Arremonops conirostris*	Bulky, side entrance	2 [3]	2	I, F, S
Blue-black Grassquit, *Volatinia jacarina*	Slight cup	2–3	2	S
Seedeaters, *Sporophila* 9 spp.	Slight cup	2 (3)	2	S, F
Thick-billed Seed-Finches, *Oryzoborus* 3 spp.	Cup	2–3	2	S, I
Yellow-faced Grassquit, *Tiaris olivacea*	Side entrance	2, 3 [4]	2	S, F, I

TABLE 1
CONTINUED

Species	Nest type	Clutch size[2]	No. attendants[3]	Diet[4]
Black-faced Grassquit, *T. bicolor*	Side entrance	2–3	2	S, F, I
Sooty Grassquit, *T. fuliginosa*	Side entrance	2–3 (4)	2	S, F
Flowerpiercers, *Diglossa* 3 spp.	Cup	2	2	N, I
Saffron Finch, *Sicalis flaveola*	Cup in cranny	2–4	2	S
Grassland Sparrow, *Ammodramus humeralis*	Cup, on ground	2	2	S
Rufous-collared Sparrow, *Zonotrichia capensis*	Cup	2, 3	2	S, I
Melodious Blackbird, *Dives dives*	Bulky cup	3	2	I, N
Carib Grackle, *Quiscalus lugubris*	Deep cup	2–4	1	I, F
Great-tailed Grackle, *Q. mexicanus*	Bulky cup	3, 2 [4]	1	O
Yellow Oriole, *Icterus nigrogularis*	Long pouch	3 (2–4)	2	I, F
Yellow-rumped Cacique, *Cacicus cela*	Long pouch	2 (3)	1	F, I
Red-rumped Cacique, *C. haemorrhous*	Long pouch	2	2	F, I
Crested Oropendola, *Psarocolius decumanus*	Long pouch	1–2	1	F, N, I
Montezuma Oropendola, *P. montezuma*	Long pouch	2	1	F, N, I
Fringillidae				
Lesser Goldfinch, *Carduelis psaltria*	Cup	3, 4, 2	2	S, I

[1] Sources: data for the cotingas are from Snow (1982); White-bearded Flycatcher from Thomas (1979); Green Jay from Alvarez (1975); Bushy-crested Jay from Hardy (1976). I have omitted records of Lawton and Guindon (1981) for the Brown Jay in Costa Rica because it appears that more than one female laid in a nest; included are mine from Guatemala. For all other species I have, where available, combined data from Haverschmidt (1968), ffrench (1973), and Skutch (1954, 1960, 1967a, 1969, 1972, 1981, unpubl. data).
[2] The most frequent clutch size is given first, separated by a comma from the next most frequent. When one size is not clearly more frequent than another, a dash separates the numbers. Infrequent clutches in parentheses; very rare, in brackets.
[3] H = helpers.
[4] F = fruits; I = insects; and other invertebrates; L = lizards; N = nectar; O = omnivorous; S = seeds.

to month and year to year in the same small area, according to the weather, the presence or absence of certain predators, or the magnitude of human disturbance (Table 2). Thus, in Summit Gardens on the Isthmus of Panama, Morton (1971) found that 15 of 21 nests (71%) of the Garden Thrush, or Clay-colored Robin, occupied before the rains returned were successful, but after the rainy season began and predators increased, only three of 17 nests (18%) yielded fledglings.

We can never be sure that we have found the nests most likely to escape disaster. We do not know whether those so well hidden in dense thickets scarcely penetrable by man that they remain unfound are the most or the least successful; their immunity from certain kinds of predators may be counterbalanced by their vulnerability to the snakes that abound in such

TABLE 2
NESTING SUCCESS IN OPEN HABITATS IN THE VALLEY OF EL GENERAL, COSTA RICA[1]

Year[2]	No. of species	No. of nests	Successful nests	
			No.	%
1942–1943	20	45	17	38
1943–1944	20	76	30	39
1944–1945	21	55	21	38
1960–1961	16	32	17	53
1981–1982	15	29	9	31
Total	42	237[3]	94	39.7

[1] Study area: 1.5 hectares of shady garden and pasture.
[2] Each year began 1 September and continued to the following 31 August.
[3] Includes 15 nests in holes and crannies.

TABLE 3
NESTING SUCCESS OF BIRDS OF THE HUMID TROPICS[1]

Species	Locality	Nest type	No. nests	Suc- cessful no.	Nests %
Ruddy Ground-Dove, *Columbina talpacoti*	Costa Rica	Saucer	22	5	22.7
Blue Ground-Dove, *Claravis pretiosa*	Costa Rica	Flimsy saucer	15	5	33.3
White-tipped Dove, *Leptotila verreauxi*	Costa Rica	Saucer	18	5	27.8
Ruddy Quail-Dove, *Geotrygon montana*	Costa Rica	Platform	17	5	29.4
Oilbird, *Steatornis carpipensis*	Trinidad	Ledge in cave	68	31	45.6
Bronzy Hermit, *Glaucis aenea*	Costa Rica	Beneath leaf	13	4	30.8
Rufous-breasted Hermit, *G. hirsuta*	Trinidad	Beneath leaf	185	32	17.3
Band-tailed Barbthroat, *Threnetes ruckeri*	Costa Rica	Beneath leaf	20	8	40.0
Little Hermit, *Phaethornis longuemareus*	Costa Rica	Beneath leaf	18	6	33.3
Scaly-breasted Hummingbird, *Phaeochroa cuvierii*	Costa Rica	Compact cup	20	10	50.0
Rufous-tailed Hummingbird, *Amazilia tzacatl*	Costa Rica	Compact cup	19	6	31.6
Rufous-tailed Hummingbird	Panama		13	3	23.1
Golden-naped Woodpecker, *Melanerpes chrysauchen*	Costa Rica	Tree hole	31	17	54.8
Spotted Antbird, *Hylophylax naevioides*	Panama	Bulky cup	84	16	19.0
Bicolored Antbird, *Gymnopithys leucaspis*	Panama	Low hollow stump	77	9	11.7
Paltry Tyrannulet, *Zimmerius vilissimus*	Costa Rica	Mossy globe	31	11	36.0
Yellow-bellied Elaenia, *Elaenia flavogaster*	Costa Rica	Compact cup	36	15	41.5
Lesser Elaenia, *E. chiriquensis*	Costa Rica	Compact cup	39	11	28.2
Ochre-bellied Flycatcher, *Mionectes oleagineus*	Trinidad	Pensile, side entrance	33	4	12.1
Ochre-bellied Flycatcher	Costa Rica		11	4	36.3
Yellow-olive Flycatcher, *Tolmomyias sulphurescens*	Costa Rica	Pensile retort	18	8	44.4
Sulphur-rumped Flycatcher, *Myiobius sulphureipygius*	Costa Rica	Pensile, roofed	10	5	50.0
Boat-billed Flycatcher, *Megarynchus pitangua*	Costa Rica	Bulky cup	19	8	42.1
Gray-capped Flycatcher, *Myiozetetes granadensis*	Costa Rica	Side entrance	40	18	45.0
Tropical Kingbird, *Tyrannus melancholicus*	Costa Rica	Cup	18	5	27.8
White-bearded Manakin, *Manacus manacus*	Trinidad	Tiny cup	227	44	19.4
Orange-collared Manakin, *M. aurantiacus*	Costa Rica	Tiny cup	78	20	25.5
House-Wren, *Troglodytes aedon*	Costa Rica	Cup in cranny	45	33	73.3
Orange-billed Nightingale-Thrush, *Catharus aurantiirostris*	Costa Rica	Mossy cup	27	17	63.0
Cocoa Thrush, *Turdus fumigatus*	Trinidad	Bulky cup	57	19	33.3
Clay-colored Robin, *Turdus grayi*	Costa Rica	Bulky cup	61	22	36.1
Clay-colored Robin	Panama		56	18	32.1
Bare-eyed Thrush, *T. nudigenis*	Trinidad	Bulky cup	21	7	33.3
White-necked Thrush, *T. albicollis*	Trinidad	Bulky cup	35	7	20.0
Long-tailed Silky-flycatcher, *Ptilogonys caudatus*	Costa Rica	Bulky cup	15	4	26.7
Buff-rumped Warbler, *Phaeothlypis fulvicauda*	Costa Rica	Roofed, on bank	29	9	31.0
Silver-throated Tanager, *Tangara icterocephala*	Costa Rica	Compact cup	35	19	54.3
Golden-masked Tanager, *T. larvata*	Costa Rica	Compact cup	35	10	28.6
Blue-gray Tanager, *Thraupis episcopus*	Costa Rica	Compact cup	47	22	46.8
Gray-headed Tanager, *Eucometis penicillata*	Costa Rica	Thin cup	24	9	37.5
Ant-Tanagers, *Habia* 2 spp.	Belize	Thin cup	53	8	15.1
Scarlet-rumped Tanager, *Ramphocelus passerinii*	Costa Rica	Cup	163	63	38.6

TABLE 3
CONTINUED

Species	Locality	Nest type	No. nests	Successful no.	Nests %
Buff-throated Saltator, *Saltator maximus*	Costa Rica	Bulky cup	47	15	31.9
Black-striped Sparrow, *Arremonops conirostris*	Costa Rica	Bulky, roofed	56	22	39.3
Variable Seedeater, *Sporophila aurita*	Costa Rica	Slight cup	44	24	54.5
Yellow-faced Grassquit, *Tiaris olivacea*	Costa Rica	Side entrance	82	29	35.4

[1] Data from Trinidad are from Snow (1962a, b) and Snow and Snow (1963, 1973, 1979); for the Spotted and Bicolored antbirds, from Willis (1967, 1972b); for the ant-tanagers, from Willis (1961); for the Clay-colored Robin in Panama, from Morton (1971). The remaining data are the author's.

thickets. Moreover, we are uncertain how our visits of inspection, no matter how circumspectly made, affect the outcome of nests.

EFFECT OF HUMAN VISITS

This question of the effect of man's visits on the outcome of nesting attempts has received a good deal of attention of late. Lenington (1979), writing of the "uncertainty principle" in field biology, showed that when the same colony of Red-winged Blackbirds (*Agelaius phoeniceus*) or Common Grackles (*Quiscalus quiscula*) was studied in successive years, the proportion of successful nests was likely to decrease, sometimes greatly, from year to year. She attributed this to the probability that predatory mammals and snakes learn to profit by scent trails or other signs left by visiting humans. Bart (1977), in a statistical analysis of a large number of records in the North American Nest Record Card Program, found that nests were much more likely to be pillaged during the first day after a human visit than on subsequent days, which implied that predators promptly followed fresh signs left by a visitor. However, nests that Gottfried and Thompson (1978) visited daily did not suffer significantly higher predation than those that remained unvisited for six days. Willis (1973), who could tell by the behavior of Bicolored Antbirds at swarms of army ants in a Panamanian forest whether they had eggs or nestlings, concluded that nests which he inspected were no less successful than those that he never saw. Thus, we are left with conflicting evidence about the effect of our visits upon birds' nests.

Another question that we must answer is whether, in order to gauge the success of nests, it is better to inspect them daily or as infrequently as is compatible with the acquisition of the information that we seek. Frequent or daily visits to nests, as recommended by Mayfield (1961, 1975) and now widely practiced, yield vast amounts of data and reveal the mean daily rate of nest losses, but do they provide the information that we most need to assess recruitment in avian populations? Parent birds whose young survive only until nearly ready to fly are no better off than those that lose a set of newly laid eggs. Indeed, they are worse off, for they have fruitlessly spent more energy and have lost more time that they might profitably have given to a replacement nest. In view of the uncertainty surrounding the effect of our visits to nests, it might be better to inspect them no more frequently than is necessary to learn when eggs are laid, when they hatch, and whether the nestlings survive until they are ready to depart.

STAGE AT WHICH NESTS ARE FOUND

A final question is whether we should base our estimate of success on all occupied nests that we find, at whatever stage, or only upon those discovered before or during the period of laying. Because some eggs are lost soon after they are laid, and nests are emptied at all stages, those that we find on any day after the first egg is laid are already a selected set that has survived perils to which neighboring nests may have succumbed. Accordingly, if we include all nests that we find, at whatever stage, our estimate of success is likely to be too high.

In an earlier paper (Skutch 1966) I compared the success of nests found before the last egg was laid with that of all nests whose outcome I knew. In some species the sample including all nests at whatever stage they were found was more successful, whereas in other species the smaller sample of nests found early did better—an unexpected conclusion. These contrary tendencies nearly balanced each other when I averaged the success of the two categories for

TABLE 4

Variation of Nesting Success of Altricial Birds with Region, Habitat, and Type of Nest[1]

Region	Habitats	Nest type[2]	No. species	Nests		Eggs	
				No.	% success[3]	No. laid	% success[4]
1. Wet tropics, Costa Rica	All	Open	49	885[5]	34.8		
2. Wet tropics, Costa Rica	All	Open	42	483[6]	33.3	978	29.5
3. Dry tropics, SW Ecuador	All	Open	14	1538	50.1	3618	45.3
4. N. Temperate	All	Open	24[7]	7788	49.3		
5. N. Temperate	All	Open	29[7]			21,951	45.9
6. Wet tropics, Costa Rica	All	Hole	16	145[5]	60.6		
7. N. Temperate	All	Hole	13			94,400	66.0
8. Wet tropics, Costa Rica	Rainforest	Open	28	129[5]	24.0		
9. Wet tropics, Costa Rica	Nearby garden and pasture	Open	38	222[5]	37.4		

[1] Sources: data for North Temperate Zones, lines 4, 5, 7, from Nice (1957); for southwestern Ecuador, line 3, from Marchant (1960); for Costa Rica, lines 1, 2, 6, 8, and 9, recalculated from Skutch (1966).
[2] Open = all nests not in holes, including cups, roofed, and pensile structures amid vegetation. Hole = nests in cavities in trees and termitaries, bird houses, burrows, etc.
[3] A nest is considered successful if at least one nestling survived to the usual time for fledging.
[4] Eggs that produced young who survived to the time of fledging.
[5] All nests of known outcome, at whatever stage they were found.
[6] Nests found before last egg was laid.
[7] Number of studies, not species.

all species (see Table 4, lines 1 and 2). Accordingly, in compiling Table 3, I included all nests for which I knew the outcome, at whatever stage they were found. This course permitted comparison of nesting success in the humid tropics with that of other regions for which much of the available information appears to have been based upon the total number of nests of known outcome. Similarly, Ricklefs (1969a), unable to learn how most of his data were collected, treated his large compilation "as if nests were found before the initiation of laying." Except as indicated in the footnotes of Table 3, all the values for nesting success are from my own records. Nearly all were gathered in the Valley of El General, on the Pacific slope of southern Costa Rica between 600 and 900 m elevation.

The success of nests in the wet tropics is substantially lower than that found by Marchant (1960) in southwestern Ecuador (Table 4, line 3). Nesting success in this region of light and undependable rains compares favorably with that of open nests in the north temperate zone. The data in lines 4, 5, and 7, from a compilation by Nice (1957), differ only slightly from those obtained by Lack (1954). Using a somewhat different set of studies, Lack calculated that on average about 45 percent of the eggs of altricial birds in open nests yield flying young, whereas about 67 percent of those laid in holes and nest boxes do so. Likewise, Ricklefs (1969a) found the nesting success of 12 temperate zone passerines with open nests to be 55.4 percent; the egg success of 13 species was 46.6 percent. In the humid neotropics as in the north temperate zone, nests in enclosed spaces are much more successful than those built in the open, and the advantage of occupying a cavity of some sort appears to be greater in the tropics than in the north.

In the north temperate zone as well as in the humid tropics, nests fare better in plantations, gardens, and other types of open country, than in woodland. In the forests of Trinidad, 20 percent of 35 nests of the White-necked Thrush and 23 percent of 22 nests of the Cocoa Thrush yielded fledglings, but on plantations 40 percent of 35 nests of the latter, and 33 percent of 21 nests of the Bare-eyed Thrush were successful. Thus, the Cocoa Thrush did nearly twice as well in plantations as in forest. It did very much better than Eurasian Blackbirds (*Turdus merula*) in English woodland, and nearly as well as Blackbirds and American Robins (*T.*

TABLE 5

VARIATION OF NESTING SUCCESS WITH ELEVATION IN SIX CENTRAL AMERICAN LOCALITIES

Locality	Elev. (m)	Period	No. nests found[1]	No. species	Nests of known outcome	Successful nests No.	Successful nests %
Barro Colorado Island, Panama	25–150	Feb.–June, 1935	83	38	62	13	21
Motagua Valley, Gautemala	60–240	Feb.–June, 1932	96	41	68	29	43
El General, Costa Rica	610–700	Jan.–June, 1939	136	61	85	28	33
Montaña Azul, Costa Rica	1525–1830	July, 1937– Aug., 1938	123	47	80	42	53
Los Cartagos, Costa Rica	1980–2285	Feb.–July, 1963	81	27	41	18	44
Sierra de Tecpán, Guatemala	2440–3050	Jan.–Dec., 1933	82	28	67	37	55

[1] Omits nests of colonial nesting icterids, found only on Barro Colorado, in the Motagua Valley, and at Montaña Azul.

migratorius) in northern gardens and parkland (Snow and Snow 1963: table 8). The Snows suggested that it may be necessary to reexamine the generalization that nesting success in tropical forests is exceptionally low. Low it certainly is, but this may be true of forests at all latitudes. The real contrast may be, not between tropical and temperate regions, but between wild woodland, where predators abound, and man-made habitats, where predation is much reduced. Seeking greater safety, forest birds frequently emerge to build their nests in neighboring gardens, plantations, or pastures, even when they fly back to the woods to find food for their young. Open country birds, however, hardly ever enter the forest, even a little way, to nest, although, like Clay-colored Robins, Scarlet-rumped Tanagers, and others, they may venture well into the forest in search of food.

In the Oriental forests of Sarawak, parent birds are in no better plight, and possibly worse, than in tropical America. Of 167 nests, excluding those in holes, as many as 144, or 86 percent, suffered predation (Fogden 1972). In contrast to other studies, Oniki (1979) found that forest nests in equatorial localities at Belém and Manaus on the Amazon, were as successful as, or more successful than, nests in open areas. She believed that the low success reported for localities such as Trinidad, Barro Colorado Island, the Valley of El General, and Sarawak was due to the fact that they are islands, or contained isolated tracts of forest amid farmlands. In such situations large predators, which may reduce the numbers of small, nest-robbing animals, tend to be rare or absent. This idea merits further investigation. In our 50 hectares or so of old forest, from which great raptors, large felines, and other predators not likely to be interested in small birds' nests long ago vanished, nest-robbing squirrels, and possibly small nocturnal marsupials that destroy nests, appear to have become more abundant. In Iowa, many fewer nests were raided by fairly large mammals such as racoons (*Procyon lotor*) and opossums (*Didelphis virginianus*) than by snakes, small mammals, and birds (Best and Stauffer 1980).

I also compared the success of nests at different elevations in Central America (Table 5). Of the six localities considered, Barro Colorado Island can be compared with Montaña Azul; in both places, many of the nests were in narrow clearings amid extensive forest. The Sierra de Tecpán most resembled El General in 1939; both had tracts of forest amid pastures and cultivation. In the Motagua Valley, I studied birds at "Alsacia" plantation, where pastures and great banana groves were distant from the high forest. This can be compared with Los Cartagos, Costa Rica, where pastures covered broad ridges between deep forested ravines on the massif of Volcán Barva (formerly Barba). An important factor in the increase of nesting success with elevation is the relative paucity of snakes in cool highlands. I saw few in the localities above 1500 m, but they abound on Barro Colorado Island and in the Valley of El General.

CAUSES OF NEST FAILURE

Despite great variation in the magnitude of nest failure between species and in the same species in different years and places, available evidence indicates that, on average, birds of the humid neotropics lose about two-thirds or more of their open nests, whereas in the north temperate zone loss of only half is more usual. In both regions, the causes of loss are the same, but of different weight. Everywhere, some nests fail because they were badly situated and fall. Those attached to growing monocotyledonous leaves, such as palm fronds or cattail leaves, may be so strongly tilted by the different rates of elongation of the supporting leaves that eggs or nestlings spill out [e.g., Red-winged Blackbirds in cattails, (*Typha*), and Great-tailed Grackles in Coconut Palms (*Cocos nucifera*)]. Falling branches and trees destroy some nests, especially of birds nesting in holes in decaying trunks or limbs. A few are pulled apart by birds seeking materials for their own nests. Birds, such as the Yellow-olive Flycatcher, whose pensile nests are made largely of fine, tough strands excellent for nest lining and hang conspicuously from the tips of branches, often suffer such depredations. Nests are more likely to be pulled apart while they are being built than after they are finished and contain eggs or nestlings. Such pilfering seems more often to delay laying than to destroy broods.

HATCHING FAILURE

Some eggs do not hatch because they are infertile, inadequately incubated, contain a genetically defective embryo, or contain a chick that fails to break out of the shell, sometimes because its head is in the narrow end of the egg. When the number of eggs in a clutch is almost invariable, the presence in a nest of this number of nestlings, plus the absence of an unhatched egg, is twofold evidence that every egg has produced a nestling. With the exception of woodpeckers, birds tend to leave unhatched eggs in the nest beside or beneath their growing young. Doubt arises when less than the full complement of young lie in a nest with no egg. Because a hatchable egg or a nestling could have been lost from such a nest, I have excluded these nests from the analysis.

I obtained data on the number of eggs that failed to hatch from intrinsic causes (infertility or death of embryos) in 30 species of birds (Table 6). Of 1779 eggs incubated for the full term, only 107, or 6 percent, failed to hatch. This is substantially less than the 8.1 percent of 3226 eggs of six species in temperate North America that remained unhatched at the end of the incubation period (Ricklefs 1969a). In my sample of 868 nests, only 11 whole clutches that were incubated for the full period, or usually longer, failed to hatch. Three were in nests of the Ochre-bellied Flycatcher in Trinidad, which also had the highest proportion of hatching failures (Table 6); none of 23 eggs of this species in Central America was infertile or otherwise defective. I surmise that so many eggs of the Band-tailed Barbthroat remained unhatched because they were infertile; this hummingbird was rare in the vicinity, and no known courtship assembly of males was near. Probably Clay-colored Robins failed to hatch two clutches that were incubated beyond the full term because they often flew from their nests when people passed.

Aside from the Band-tailed Barbthroat and the Ochre-bellied Flycatcher in Trinidad, the tropical birds with the highest proportion of unhatched eggs may be compared with related species in the north (data from Ricklefs 1969a). Of 175 eggs of the American Robin 10.3 percent were defective. The 10.5 percent failure of the Lesser Elaenia was hardly different from the 10.9 percent failure of 129 eggs of the Willow Flycatcher (*Empidonax traillii*), a bird of about the same size. The Yellow-faced Grassquit's hatching failure of 10.6 percent was smaller than that of the Yellow Warbler (*Dendroica petechia*) at 11.2 percent of 135 eggs. These data do not corroborate the conclusion of Koenig (1982) that hatchability increases slightly over 1 percent for every 10° increase in latitude.

OTHER MINOR CAUSES OF FAILURE

Except at high elevations in the tropics, where during prolonged cold rains whole broods may die of starvation or exposure, I have found few dead nestlings in nests. These probably died because a parent was killed or was frightened from her nest on a rainy night. Except for woodpeckers, which lay as many as four eggs, tropical birds seldom appear to have difficulty nourishing all members of their small broods. A more frequent cause of loss appears to be the falling of eggs or tiny nestlings that probably were brushed from a shallow nest by an alarmed parent leaving hastily.

At El General, where nearly all the data were recorded (Table 6), I have never found a parasitic egg or nestling. The Striped Cuckoo (*Tapera naevia*), which intrudes only into closed nests such as those of spinetails (*Synallaxis*) and wrens, is a relatively recent immigrant. The very few Bronzed Cowbirds (*Molothrus aeneus*) that I have seen were transients. In a different

TABLE 6
INFERTILE EGGS OR DEAD EMBRYOS[1]

Species	No. eggs[2]	Eggs failing to hatch	
		No.	%
White-tipped Dove	20	0	0
Ruddy Ground-Dove	32	0	0
Ruddy Quail-Dove	28	3	10.7
Band-tailed Barbthroat	26	5	19.2
Scaly-breasted Hummingbird	28	2	7.1
Rufous-tailed Hummingbird	42	2	4.8
Rufous-tailed Jacamar	30	0	0
Orange-collared Manakin	86	6	7.0
Blue-crowned Manakin	20	0	0
Gray-capped Flycatcher	75	7	9.3
Yellow-bellied Elaenia	44	3	6.8
Lesser Elaenia	38	4	10.5
Paltry Tyrannulet	37	1	2.7
Ochre-bellied Flycatcher	36	10	27.8
Ochre-bellied Flycatcher	23	0	0
House-Wren	87	7	8.0
Orange-billed Nightingale-Thrush	66	2	3.0
Clay-colored Robin	115	14	12.2
Bananaquit	58	2	3.4
Buff-rumped Warbler	44	2	4.5
Silver-throated Tanager	43	1	2.3
Golden-masked Tanger	35	3	8.6
Blue-gray Tanager	62	0	0
Gray-headed Tanager	45	2	4.4
Scarlet-rumped Tanager	240	5	2.1
Buff-throated Saltator	69	3	4.3
Blue-black Grosbeak	26	2	7.7
Orange-billed Sparrow	26	0	0
Black-striped Sparrow	85	4	4.7
Variable Seedeater	109	6	5.5
Yellow-face Grassquit	104	11	10.6
Total	1779	107	6.0

[1] All entries from the Valley of El General in Costa Rica except the first entry for the Ochre-bellied Flycatcher, which is for Trinidad (data from Snow and Snow 1979).
[2] Includes only eggs incubated for the full term.

category is the Piratic Flycatcher, which throws eggs and more rarely nestlings from the roofed nests of Gray-capped Flycatchers, the pensile retorts of Yellow-olive Flycatchers, and a variety of other covered structures. It incubates its eggs and rears its young in the stolen receptacle.

PREDATION

By far the greatest cause of nesting failures in the humid tropics is predation. I surmise that snakes destroy more nests in tropical America than all other predators together; certainly they are the creatures that I have most often seen plundering nests. Their long, slender, sinuous form enables them to creep along slender branches, pass easily through dense grass or thickets, climb branchless trunks, and intrude into narrow crannies. No type of nest, underground, on the surface, or high in trees, appears to be wholly immune from them. The snake that I have most frequently caught pillaging nests is *Spilotes pullatus*, which may reach more than 2 m in length. Like a number of other snakes, it attacks nests both day and night. It sometimes hides in woven pouch nests of oropendolas or caciques (Icterinae) by day, to sally under cover of darkness to plunder others in the same treetop colony (Skutch 1954). Most nest-robbing snakes are nonvenomous; an exception is the small *Bothrops schlegelii*, which I have twice found swallowing eggs. Unlike mammalian predators, which often leave fragments of shell in a nest, snakes swallow eggs whole. If one is in doubt whether nestlings almost ready to leave have departed spontaneously or been carried off by a predator, he can often tell by examining the nest's lining. If it is pulled up, the young have probably been torn away while clinging to it; if it lies flat, the nestlings have probably emerged alive.

As in the north, squirrels destroy many nests in tropical woodlands, parks, and gardens. The poor success of nests around our garden in 1981 and 1982 was caused by depredations of squirrels as well as snakes. I have seen the Tayra (*Eira barbara*) attack a nest of the Laughing Falcon (*Herpetotheres cachinnans*); a dreaded enemy of the farmer's chickens, this big black weasel is probably chiefly interested in larger birds. The omnivorous Coati (*Nasua nasua*) certainly eats any eggs or young found in trees or on the ground, but I have never witnessed its depredations. I once saw White-faced Monkeys (*Cebus capucinus*) take the eggs of a heron (*Butorides* sp.) on the wooded shore of a lake. The opossums, marmosas, and other marsupials large and small, operate chiefly by night, when they undoubtedly rob many birds' nests. Mice often make themselves snug in a small bird's nest, especially an enclosed structure such as Yellow-faced Grassquits build, by shredding the lining, leaving the egg intact beneath themselves. I strongly suspect that bats, some of which are known to eat birds, took nestlings from nests of Bronzy Hermits and Band-tailed Barbthroats fastened beneath banana leaves too slippery for snakes to climb and hardly accessible to any predator incapable of flight.

American Swallow-tailed Kites (*Elanoides forficatus*), although largely insectivorous in this region, occasionally attack nests at almost any height that they can reach while hovering; sometimes they carry the nest high in the air before removing the nestlings and dropping it. Double-toothed Kites (*Harpagus bidentatus*) and Black Hawk-Eagles (*Spizaetus tyrannus*) sometimes pillage nests, and once I saw a White Hawk (*Leucopternis albicollis*) capture a fledgling Collared Araçari (*Pteroglossus torquatus*) newly emerged from a high hole. Toucans (*Ramphastos, Pteroglossus, Aulacorhynchus*) vary their mainly frugivorous diets with eggs and nestling birds. Jays are still absent from El General, but in other parts of Costa Rica Brown Jays have become a serious menace to nesting birds.

Among insects, fire ants (*Solenopsis*) are, in my experience, the worst enemies of nesting birds. I have known them to attack nests in burrows underground, in shrubbery, and high in dead trunks. Although birds pluck ants and other insects from their nests, they cannot stem a massive invasion of these stinging ants. Army ants (*Eciton, Labidus*) appear rarely to harm eggs or nestlings. I have on several occasions watched their swarms flow close by a terrestrial or arboreal nest where a parent was incubating or brooding, or where nestlings rested alone, even sending scouts to the nest without injuring parents or young.

The foregoing appear to be the chief predators on birds' nests in the regions where I have worked in tropical America. What can the birds do to diminish this menace? The very diversity of their nests may help them to survive. It may be difficult for a predator to become familiar with all their forms and sites, thinly scattered as most of them are in tropical forests, or to develop a specific search image that would aid in their discovery. Some birds, especially certain cotingas and manakins, have reduced the size of their nests to the very minimum capable of holding two eggs, or only one. When a Rufous Piha sits on her arboreal nest with one big egg, the fantastically slight mat is hidden beneath her.

With such a great variety of nests, some of which are cunningly concealed, it would seem that at least diurnal predators must depend in large measure upon the movements of the birds themselves to reveal their position—as I do. If this be true, it should be advantageous to parent birds to approach and leave their nests as infrequently as possible. In fact, incubating birds of tropical forests take long sessions on their eggs, and their young are fed at rather long intervals, often with items about as big as the nestlings can swallow. Other birds feed their young by regurgitation, a method that reduces the frequency of meals. Unless well concealed, one may watch until his patience is exhausted without seeing some of the birds of lowland tropical forests come near their nests. Yet while they approach with the utmost caution, some of them, like Blue-black Grosbeaks and ant-tanagers, reveal their nervous tension by calling loudly, and they call or sing while sitting—all of which suggests that the predators they chiefly fear are insensitive to sound, as snakes seem to be. In contrast to the situation in lowland forests, I have found birds of remote highland forests surprisingly unconcerned about my presence. In this respect I have noticed much variation among species and individuals in the same woodland.

DIRECT DEFENSE OF NESTS

I have seen little effective defense of nests by parent birds. A parent Laughing Falcon (*Herpetotheres cachinnans*) timidly threatened but did not attack a Tayra that killed her downy nestling in a cavity 30 m up in a great tree (Skutch 1971). By spirited threats, Clay-colored Robins make maurading squirrels flee. Toucans intent upon plundering nests hold enraged parents at bay by menacing them with huge, vividly colored beaks. Although I have watched Riverside Wrens in Costa Rica and Rufous-fronted Thornbirds in Venezuela repeatedly peck snakes well over a meter long, I have never known a bird to save its nest from a snake. A Scarlet-rumped Tanager that tried to do so was caught and hung by one wing, until I rescued her, from the jaws of a long, green snake that seemed too slender to swallow her. A white-tipped Dove tried to drive a snake from her nest by striking it with a wing, but soon thereafter she lost her nestlings, probably taken by the snake. In Michigan, however, a pair of Veeries (*Catharus fuscescens*) kept a 75-cm Garter Snake (*Thamnophis sirtalis*) away from their nest by repeatedly striking it with wings and feet. Their success in saving the brood may owe much to the fact that the nearly fledged Brown-headed Cowbird (*Molothrus ater*), whom the snake seized by the neck and afterward by a wing, clung so tightly to the nest that the predator could not detach it (Pettingill 1976). One wonders how often such clinging, widespread among nestlings, saves their lives.

My impression, from what I have seen and read, is that small northern birds tend to be bolder in the defense of their nests than tropical birds, who in general seem milder-tempered. The reason for this appears to be that, being mostly permanently resident, they can settle conflicts over mates and territories in a more leisurely and pacific manner than do migratory or wandering birds with less time to arrange these matters.

Many tropical birds have given distraction displays to me, and once I watched a Black-striped Sparrow lure a small ground snake. However, I have rarely seen parents save eggs or young by this ruse. A pair of Common Pauraques (*Nyctidromus albicollis*) cooperated to lure an opossum (*Didelphis marsupialis*) from their eggs, which soon afterward vanished (Skutch 1972). In a seaside park, a Black-striped Sparrow very effectively lured a dog from a recently emerged fledgling.

REGULATION OF THE RATE OF REPRODUCTION OF BIRDS IN THE HUMID TROPICS

The foregoing survey leaves us with the picture of an extremely varied avifauna that rears small broods, usually of two nestlings, without relation to habitat, diet, type of nest, or number of nest attendants, and which, moreover, must contend with very high predation on eggs and young. The chief exceptions are some of the hole-nesters, who, for reasons given elsewhere (Skutch 1976), tend to lay larger clutches and suffer fewer losses. The difficulty of correlating clutch size with habitats or habits prompts us to seek some factor that affects all of these so diverse birds of the humid tropics. However they may differ, all live in much the same climate, which is likely to be the factor that we seek.

EFFECTS OF STABLE ENVIRONMENTS

Cody (1966) adduced evidence that all stable environments, including not only the tropics but also extratropical islands and coasts with maritime climates, favor reduced clutch size, whereas continental climates with great seasonal extremes promote larger clutch size. The reason for this difference appears to be that the more uniform climates support more stable populations of birds that remain more constantly near the limit of the capacity of their habitats to nourish them, and, accordingly, not only have less need for annual recruitment but in the breeding season have a smaller surplus of food for a new generation. Applying the principle of the allocation of time and energy, Cody postulated that in the temperate zones most energy would be used to increase the reproductive rate, whereas in the tropics, and in the more stable climates in general, more would be applied to self-maintenance, including avoidance of predators and competition for resources.

Ashmole (1963) suggested that clutch size should vary in direct proportion to the degree of seasonal fluctuation of the resources available to a population. Populations are regulated during the season when food is least abundant; clutch size is determined by the resources during the breeding season relative to population density. Hence, the greater the seasonal variation in resources, the larger the average clutch size. Ricklefs (1980) tested this hypothesis indirectly by relating clutch size to seasonal variation in actual evapotranspiration, which is proportional to primary production of organic matter by plants. He concluded that clutch size is directly related to the ratio between summer and winter evapotranspiration. It is inversely related to winter evapotranspiration but independent of its magnitude in summer. This is consistent with the hypothesis that the number of breeding individuals is regulated in winter, or, in general, during the leanest months, and that population size exerts a density-dependent effect on the resources available to each individual for reproduction in the more favorable season. Geographic trends in clutch size are caused primarily by factors that limit populations during the nonreproductive period rather than abundance of resources during the breeding season.

The analyses of both Cody and Ricklefs relate the small clutch size of tropical birds to the comparatively slight seasonal fluctuation in abundance of food in the humid tropics and the relatively low mortality of birds in the leanest months. Long ago (Skutch 1940), distressed by the early loss of so many of the nests that I found by much searching through tropical woodlands and wished to study, I was consoled by the thought that birds who reproduce with such great difficulty must live a long while. This deduction from observed nest losses has since been confirmed by studies of banded birds. In Trinidad, the annual survival of adult male White-bearded Manakins was 89 percent, and that of females was also very high (Snow 1962a). Many of the birds banded by the Snows in Trinidad between 1957 and 1961 were recaptured by Lill a decade later (Snow and Lill 1974). A number of these small birds were more than 10 years old. For 182 individuals of 15 species that were recorded again, Snow and Lill (1974) calculated an annual survival rate of 82 percent. This corresponds closely with the minimum survival rate of 86 percent of a population of forest birds in Sarawak (Fogden 1972). These values are very much higher than the 40 to 60 percent survival usual among small passerines in the north temperate zone (Lack 1954). Very exceptional among tropical birds was the population of Long-tailed Hermits (*Phaethornis superciliosus*) in the wet forests of northeastern Costa Rica, for which Stiles (1980) estimated an annual mortality of nearly 50 percent.

FOOD AVAILABILITY AND CLUTCH SIZE

Although we have growing evidence that the small clutch size of tropical birds is related to their longevity, it is not obvious that this is the causal factor directly responsible for their restrained reproduction. We must still seek the links in the causal chain. If we follow Lack's (1947, 1948) view that birds everywhere rear broods as large as they can adequately nourish, we must conclude that avian populations in the tropics are continuously so near the carrying capacity of their habitats that parents cannot find food for more than two or three (rarely more) nestlings in their relatively short working days.

The poor correlation between clutch size and number of attendants (Table 1) prompts me to ask why, if so many unaided females can rear two or even three nestlings, so few pairs rear more than this. They should be able to collect twice as much food. Furthermore, the correlation between brood size and length of working day, which varies with latitude, is poor. We cannot proceed so simply as to say that if two parents with a working day of 18 hours can supply six nestlings with food, two parents with a 12-hour day should be able to feed four; this loses sight of the fact that the adults also must eat. We probably shall not be far wrong if we assume that active parents need about as much nourishment as their inactive nestlings when the latter are growing most rapidly. Then, with a working day of 17 hours, as in the high forties or fifties of latitude near the summer solstice, two parents with a brood of six, not unusual at these latitudes, have 34 bird-hours for supplying eight individuals, or 4.25 hours for each. With the same allowance, a pair of birds at 10° latitude should be able to feed four nestlings in an active day that in the main breeding season hardly exceeds 12.75 hours ($6 \times 4.25 = 2 \times 12.75$). Since they so rarely do, we must conclude either that food is less abundant or perhaps harder to find, or that the parents do not work as hard as they might. The heavy but usually brief showers early in the rainy season when nesting is at its peak may interrupt foraging and feeding of young by parent birds, but it is not certain that they lose enough time to substantially reduce the number of nestlings they can nourish (Foster 1974a).

Because seasonal contrasts in the humid tropics are so much less striking than those in temperate regions, northerners tend to underestimate the magnitude of the changes that do occur. Caused by the alternation of wet and dry seasons rather than by great fluctuations of temperature, they are by no means negligible. Evapotranspiration may reveal the rate of primary production throughout the year, but it fails to indicate whether this is applied chiefly to vegetative growth, to flowering that provides nourishment to many nectar-drinkers, or to the production of fruits for frugivores. In the Valley of El General at 9° north, as in Trinidad at about the same latitude, the abundance of fruits available to birds increases in March or April, as the dry season yields to the rainy season, and reaches a peak in May or June (Skutch 1950, 1980; Snow and Snow 1964). At the same time, insects become obviously more abundant. Moreover, the vernal exodus of many wintering migrants (especially from Central America) has in one respect the same effect on the total avian population as high winter mortality would have: it increases the food available to permanent residents without, however, increasing their need of recruitment. All these factors cause a profusion of resources in the main breeding season, which begins at this time, comparable to that in spring and early summer beyond the tropics. Just at the time when nesting reaches its peak, and one would expect the birds to come more frequently to our feeder to supply themselves and their young, attendance declines sharply because the birds find so much wild fruit. If they reared as many young as this profusion of resources could nourish, many more would die during the leaner months toward the end of the long rainy season and the dry season in the first quarter of the year, when our feeder attracts more birds.

I have repeatedly been impressed by the fact that parents who feed their nestlings at long intervals can greatly accelerate their rate when necessary. Since I have given examples of this elsewhere (Skutch 1949), here I shall relate only that which I most recently witnessed. I passed a morning watching a nest of Riverside Wrens. At 05:24, when I first heard the male sing, the female immediately flew from the nest where she had brooded two nestlings becoming feathered. Even in the dim light of dawn, the parents foraged so effectively that in only four minutes both arrived with insects for their young. Together they fed 17 times in the first hour of the morning, 11 times in the second, but only three to six times in each of the following four hours. Evidently they were not exerting themselves strenuously.

When I increased a brood of Scarlet-rumped Tanagers from two to three, the two parents augmented their rate of feeding by 41 percent. At this nest, changes in the number of similarly aged nestlings from one to three caused changes in the feeding rate of only 7.1 to 7.5 meals per nestling per hour (Skutch 1981: table 6). By combining broods of House Finches (*Carpodacus mexicanus*) and Yellow-eyed Juncos (*Junco phaeonotus*) in the Mexican highlands, Wagner (1957) demonstrated that the parents could nourish twice their usual number of nestlings.

ADAPTIVE LIMITATION OF FECUNDITY OR MAXIMUM REPRODUCTION?

The established facts that we must consider to form a comprehensive theory of clutch size are that in climates without great seasonal contrasts birds that survive the first vulnerable months of their lives have a relatively long life expectancy and lay few eggs. Their reproductive effort is somehow related to their low mortality. How this is effected is far from clear and, doubtless, much more subtle than current theories recognize. To postulate for a population that remains near the limit of the carrying capacity of the habitat in the lean season that paucity of food available for nestlings and fledglings in the season of greatest abundance restricts their number is too simple a solution, contradicted by strong evidence.

Among the factors to be considered in constructing a theory of the clutch size of tropical birds are: (1) the high incidence of predation on their nests and the advantage of spending only a moderate amount of energy on one nesting, so that, if this fails, strength remains for repeated trials; (2) the probability that the risk of attracting certain kinds of predators to nests is reduced by small broods that require fewer visits with food; and (3) the long-term advantage of keeping a population in balance with its resources, as Wynne-Edwards (1962) maintained, although perhaps reproductive effort is restrained largely by means less direct than he suggested.

Wherever risks are great, investments, whether of money or time and energy, should be small and diversified. Birds that must try again and again to rear a brood do well to lay small clutches, thereby saving their resources to renest repeatedly, in different situations, until they succeed or the breeding season terminates. Four or five nesting attempts, possibly all unsuccessful, are not rare among tropical birds; Spotted Antbirds in Panama may renest up to 10 times during each rainy season from April to November (Willis 1972b). By overlapping nesting and molting, some tropical birds can prolong their breeding season (Foster 1974b).

As Snow (1978) pointed out, the small nests of hummingbirds, manakins, some cotingas, and others could not hold more than the two (less often, one) eggs or nestlings that these birds attend. But when we recall that many tropical birds with quite bulky nests lay no more than two eggs, we must conclude that some birds build tiny nests because they lay few eggs, not that they lay few eggs because they make nests that will hold no more.

In certain families of tropical brids, evolutionary developments have locked the clutch size at levels lower than it might otherwise attain (Skutch 1967b). This is notably true of manakins. Because the females were capable of rearing, unaided, the usual brood of two, the males, emancipated from domestic chores, could develop conspicuous plumage and elaborate court-ship displays. By adding a third nestling to broods, Lill (1974) demonstrated the female White-bearded Manakin's inability to rear so many young, who in any case were overcrowded in the diminutive nests. Evidently, in these small manakins, a mutation for greater fertility would be promptly eliminated by natural selection; to prevail, it would have to be supported by a highly improbable reversal of the manakin's social system that would put the males back to work, as well as by the construction of more capacious nests. Similar developments have occurred among hummingbirds, certain cotingas, and some flycatchers (*Myiobius, Mionectes, Lophotriccus, Terenotriccus*). For reasons unexplained, the emancipated males of these fly-catchers continue to resemble the females, and their courtship, where known, remains ex-tremely simple.

In other families, notably among the ovenbirds (Furnariidae), pairs remain intact and lavish upon the construction and maintenance of elaborate and relatively huge nests time and energy that might be applied to rearing larger families. Whether these well-enclosed but conspicuous nests enjoy greater freedom from predation is difficult to learn, for one can hardly see what they contain without altering them and thereby probably increasing their vulnerability; but I know that many of them are pillaged. Here, mutations for greater fertility would have to complete with the deeply rooted tendency to devote to structures energy that might be used for rearing progeny. In these and similar cases, we can understand why the mutations for greater fertility that probably arise occasionally in most species do not take firm hold, but it is not so evident why many tropical species do not raise their clutch size to the limit of the parents' ability adequately to nourish their young. Clutch sizes below the maximum are difficult to account for by evolutionary principles that often appeal by their simplicity rather than by their profundity (Kipp 1948).

An outstanding problem in tropical ornithology is why, in the face of extremely high predation on nests, the incubation periods of many species remain so long. Ricklefs' (1969b) suggestion that retardation of embryonic development, followed by slower nestling growth, would in certain cases enable the parents to rear one more young, will appeal to those who accept the theory of maximum reproduction without reservation. Predation pressure appears to be the factor that has made the incubation period of the big ostrich egg, lying exposed on the ground in predator-infested Africa, no longer than that of the much smaller egg of a petrel, in a safe burrow on an island. Nevertheless, the tiny eggs of manakins (*Manacus, Pipra*) need 18 or 19 days to hatch, which is about 50 percent longer than the incubation periods of certain tanagers and thrushes in the same region. The manakins' nestling periods of 13 or 14 days are likewise long for such small birds. The incubation periods of antbirds (Formicariidae), 14 to 18 days for those with open nests, are also long for birds of their size, but their nestling periods are notably short. These and other long incubation periods of tropical birds may be related to generally slower reproductive schedules. On average, tropical birds take longer to build their nests, delay longer to lay in them, have longer incubation and nestling periods, and wait longer to lay again after rearing a brood or losing a nest, than do northern birds (Ricklefs 1969b). And, as a rule, they continue much longer to care for their fledged young (Fogden 1972; Skutch 1976), as though the juvenile that has survived so many perils were an investment too precious to neglect. An advantage of a brood of only two is that each parent can give his whole attention to a single young during the very vulnerable interval between its departure from the nest and attainment of independence.

Because ornithology was born, and until quite recently nearly all ornithologists grew up, in the North Temperate Zone where clutches tend to be large, it is natural to ask why tropical birds lay so few eggs. If the tropics had given birth to ornithology and most ornithologists, we would probably be asking why northern birds lay so many eggs. To answer this question would be much easier than to answer that which has occupied us in this article. Whether they remain to endure winter's cold and dearth or undertake long and perilous migrations to escape it, northern birds have survival rates substantially lower than those of continuously resident tropical birds of similar size. At intervals, an exceptionally severe winter or a disaster on migration drastically reduces the population of a species. This is followed by a period of free expansion, as in a pioneer community, when the individuals that produce the greatest number of offspring contribute most to the rapid recovery of the population. The occasional repetitions of such episodes of decimation and recovery should, unless strong contrary pressures are operative, raise the breeding rate to the maximum (Skutch 1949). Birds of the humid tropics rarely, if ever, suffer such widespread disasters but maintain a more constant population density, from year to year, by adjusting their reproductive rates to their needs of recruitment. This is the great difference between tropical and extratropical birds, responsible not only for differences in the number of eggs that they lay but likewise for other aspects of their life histories.

ACKNOWLEDGMENTS

I am grateful to M. S. Foster, B. G. Murray, Jr., and R. J. O'connor for helpful comments on this article.

LITERATURE CITED

ALVAREZ, H. 1975. The social system of the Green Jay in Colombia. Living Bird 14:5–55.

ASHMOLE, N. P. 1963. The regulation of numbers of tropical oceanic birds. Ibis 103b:458–473.

BART, J. 1977. Impact of human visitations on avian nesting success. Living Bird 16:187–192.

BEST, L. B., AND D. F. STAUFFER. 1980. Factors affecting nesting success in riparian bird communities. Condor 82:149–158.

CODY, M. L. 1966. A general theory of clutch size. Evolution 20:174–184.

FFRENCH, R. 1973. A Guide to the Birds of Trinidad and Tobago. Livingston Publ. Co., Wynnewood, Pennsylvania.

FOGDEN, M. P. L. 1972. The seasonality and population dynamics of equatorial forest birds in Sarawak. Ibis 114:307–343.

FOSTER, M. S. 1974a. Rain, feeding behavior, and clutch size in tropical birds. Auk 91:722–726.

FOSTER, M. S. 1974b. A model to explain molt-breeding overlap and clutch size in some tropical birds. Evolution 28:182–190.

GOTTFRIED, B. M., AND C. F. THOMPSON. 1978. Experimental analysis of nest predation in an old field habitat. Auk 95:304–312.

HARDY, J. W. 1976. Comparative breeding behavior and ecology of the Bushy-crested and Nelson San Blas jays. Wilson Bull. 88:96–120.

HAVERSCHMIDT, F. 1968. Birds of Surinam. Oliver and Boyd, Edinburgh.

HESSE, R. 1922. Die Bedeutung der Tagesdauer für die Vögel. Sitzungsber. Nathist. Preuss. Rheinl. Bonn, 1922–1923, pp. A 13–17.

KIPP, F. A. 1948. Ueber die Eierzahl der Vögel. Biol. Zentralbl. 67:250–267.

KOENIG, W. D. 1982. Ecological and social factors affecting hatchability of eggs. Auk 99:526–536.

LACK, D. 1947. The significance of clutch-size. Ibis 89:302–352.

LACK, D. 1948. The significance of clutch-size. Ibis 90:25–45.

LACK, D. 1954. The Natural Regulation of Animal Numbers. Clarendon Press, Oxford, England.

LAWTON, M. F., AND C. F. GUINDON. 1981. Flock composition, breeding success, and learning in the Brown Jay. Condor 83:27–33.

LENINGTON, S. 1979. Predators and blackbirds: the "uncertainty principle" in field biology. Auk 96:190–192.

LILL, A. 1974. The evolution of clutch-size and male "chauvinism" in the White-bearded Manakin. Living Bird 13:211–231.

MARCHANT, S. 1960. The breeding of some S. W. Ecuadorian birds. Ibis 102:349–382, 584–599.

MAYFIELD, H. F. 1961. Nesting success calculated from exposure. Wilson Bull. 73:255–261.

MAYFIELD, H. F. 1975. Suggestions for calculating nest success. Wilson Bull. 87:456–466.

MORTON, E. S. 1971. Nest predation affecting the breeding season of the Clay-colored Robin, a tropical song bird. Science 171:920–921.

NICE, M. M. 1957. Nesting success in altricial birds. Auk 74:305–321.

ONIKI, Y. 1979. Is nesting success of birds low in the tropics? Biotropica 11:60–69.

OVIEDO Y VALDÉS, G. F. DE. 1851–1855. Historia General y Natural de las Indias, Islas y Tierra Firme del Mar Océano. 4 vols. Impr. Real Academia Historia, Madrid.

PETTINGILL, O. S., JR. 1976. Observed acts of predation on birds in northern lower Michigan. Living Bird 15:33–41.

RICKLEFS, R. E. 1969a. An analysis of nesting mortality in birds. Smithson. Contrib. Zool. 9:1–48.

RICKLEFS, R. E. 1969b. The nesting cycle of songbirds in tropical and temperate regions. Living Bird 8:165–175.

RICKLEFS, R. E. 1980. Geographical variation in clutch size among passerine birds: Ashmole's hypothesis. Auk 97:38–49.

SCHOMBURGK, M. R. 1847–1848. Reisen in British-Guiana in den Jahren 1840–1844. J. J. Weber, Leipzig, Germany.

SKUTCH, A. F. 1940. Some aspects of Central American bird life. Sci. Monthly 51:409–418, 500–511.

SKUTCH, A. F. 1949. Do tropical birds rear as many young as they can nourish? Ibis 91:430–455.

SKUTCH, A. F. 1950. The nesting seasons of Central American birds in relation to climate and food supply. Ibis 92:185–222.

SKUTCH, A. F. 1954. Life histories of Central American birds. Pac. Coast Avif. No. 31.

SKUTCH, A. F. 1960. Life histories of Central American birds II. Pac. Coast Avif. No. 34.

SKUTCH, A. F. 1966. A breeding bird census and nesting success in Central America. Ibis 108:1–16.

SKUTCH, A. F. 1967a. Life histories of Central American highland birds. Publ. Nuttall Ornithol. Club No. 7.

SKUTCH, A. F. 1967b. Adaptive limitation of the reproductive rate of birds. Ibis 109:579–599.

SKUTCH, A. F. 1969. Life histories of Central American birds III. Pac. Coast Avif. No. 35.

SKUTCH, A. F. 1971. A Naturalist in Costa Rica. University Florida Press, Gainesville, Florida.

SKUTCH, A. F. 1972. Studies of tropical American birds. Publ. Nuttall Ornithol. Club No. 10.

SKUTCH, A. F. 1976. Parent Birds and Their Young. University Texas Press, Austin, Texas.

SKUTCH, A. F. 1980. A Naturalist on a Tropical Farm. University California Press, Berkeley, California.

SKUTCH, A. F. 1981. New studies of tropical American birds. Publ. Nuttall Ornithol. Club No. 19.

SNOW, D. W. 1962a. A field study of the Black and White Manakin, *Manacus manacus*, in Trinidad. Zoologica 47:65–104.

SNOW, D. W. 1962b. The natural history of the Oilbird, *Steatornis caripensis*, in Trinidad, W. I. Part 2. Population, breeding ecology and food. Zoologica 47:199–221.

SNOW, D. W. 1978. The nest as a factor determining clutch-size in tropical birds. J. Ornithol. 119:227–230.

SNOW, D. 1982. The Cotingas. British Museum (Natural History), London.

SNOW, D. W., AND A. LILL. 1974. Longevity records for some neotropical land birds. Condor 76:262–267.

SNOW, D. W., AND B. K. SNOW. 1963. Breeding and the annual cycle in three Trinidad thrushes. Wilson Bulll 75:27–41.

SNOW, D. W., AND B. K. SNOW. 1964. Breeding seasons and annual cycles of Trinidad land-birds. Zoologica 49:1–39.

SNOW, D. W., AND B. K. SNOW. 1973. The breeding of the Hairy Hermit *Glaucis hirsuta* in Trinidad. Ardea 61:106–122.

SNOW, B. K., AND D. W. SNOW. 1979. The Ochre-bellied Flycatcher and the evolution of lek behavior. Condor 81:286–292.

STILES, F. G. 1980. The annual cycle of a tropical wet forest hummingbird community. Ibis 122:322–343.

THOMAS, B. T. 1979. Behavior and breeding of the White-bearded Flycatcher (*Conopias inornata*). Auk 96:767–775.

WAGNER, H. O. 1957. Variation in clutch size at different latitudes. Auk 75:243–250.

WIED, MAXIMILIAN PRINZ ZU. 1825–1833. Beiträge zur Naturgeschichte Brasilien. Landes-Industrie-Comptoirs, Weimar.

WILLIS, E. 1961. A study of nesting ant-tanagers in British Honduras. Condor 63:479–503.

WILLIS, E. O. 1967. The behavior of Bicolored Antbirds. Univ. Calif. Publ. Zool. 79:1–127.

WILLIS, E. O. 1972a. The behavior of Plain-brown Woodcreepers, *Dendrocincla fuliginosa*. Wilson Bull. 84:377–420.

WILLIS, E. O. 1972b. The behavior of Spotted Antbirds. Ornithol. Monogr. No. 10.

WILLIS, E. O. 1973. Survival rates for visited and unvisited nests of Bicolored Antbirds. Auk 90:263–267.

WYNNE-EDWARDS, V. C. 1962. Animal Dispersion in Relation to Social Behaviour. Oliver and Boyd, Edinburgh.

THE BANANAQUIT-WASP NESTING ASSOCIATION AND A RANDOM CHOICE MODEL

JOSEPH M. WUNDERLE, JR.[1] AND KENNETH H. POLLOCK[2]

[1]Departamento de Biología, Colegio Universitario, Universidad de Puerto Rico,
Cayey, Puerto Rico 00633 USA, and
[2]Department of Statistics, North Carolina State University, Raleigh, North Carolina 27650 USA

ABSTRACT. Bananaquits nesting in association (≤ 1 m) with wasp (*Polybia*) nests had significantly lower nest predation rates than neighbors with nests not associated with wasp nests. Males mated to females nesting near wasp nests were more likely to retain their mates through the breeding season than males mated to females having nests unassociated with wasp nests. A survey of Bananaquit and *Polybia* nests indicated that trees containing nests of both were more frequent than expected by chance. Within the same tree, the distance between Bananaquit nests and the nearest wasp nest averaged 82.3 cm $\pm$ 10.3 (s.e.). Females nesting close to wasp nests one year did not usually nest with wasps in the following year. Also, removal of Bananaquit nests near wasp nests did not cause the females to re-nest with wasps.

A probability model is presented as a null hypothesis to explain how spatial association between bird and wasp nests may arise. It is shown that a bird's random choice of a nest site in conjunction with increased nest success for nests close to wasp nests can (during several breeding attempts) produce a bird-wasp nesting association.

RESUMEN. Los reinitas que anidaron en asociación (=1 m) con nidos de avispas (*Polybia*) tenían una tasa de depredación significativamente más baja que sus vecinos anidando sin asociación con nidos de avispas. Los machos que se aparearon con hembras que anidaban cerca de nidos de avispas fueron más propensos a retener sus parejas a lo largo de la estación reproductiva, que los machos que se aparearon con hembras que anidaron sin asociarse con nidos de avispas. Un estudio de nidos de colibrí bananero y *Polybia* indicó que fue más frecuente encontrar árboles que contuviesen nidos de ambos que lo que se podía esperar por chance. En un mismo árbol, la distancia promdeio entre nidos de colibríes y el nido de avispa más cercano, fue de 82,3 cm$\pm$10,3 cm (DS). Las hembras que anidaron un año cerca de un nido de avispa, no lo hicieron al año siguiente. Así mismo, el remover un nido de colibrí cercano a un nido de avispa no hizo que la hembra volviese a anidar cerca del nido de avispas.

Se presenta un modelo de probabilidades como hipótesis nula, que explica como puede originarse la asociación espacial entre aves y avispas. Se muestra que la elección al azar de un sitio de anidación en conjunto con el aumento de éxito de anidación para nidos cercanos a nidos de avispas puede (durante varios intentos reproductivos) producir una asocciación de anidación entre aves y avispas.

Tropical regions are often noted for their many intricate biological interactions such as mutualism, commensalism, parasitism, and predation. The complexity and diversity of such tropical symbiotic relationships is partly a reflection of the high species richness of the tropics. One of these tropical interactions is the relationship between colonial nesting icterids and associated hymenopterans, several avian brood parasites, and an array of ectoparasites as studied in detail by Smith (1968, 1980). In his studies, Smith found that those colonial icterid nests associated with colonial hymenopterans are protected from ectoparasitic flies and vertebrate predators. This bird-hymenopteran nesting association appears to be an example of commensalism or possibly mutualism, for the nesting birds obtain a benefit through reduced nest predation, and some nesting birds may drive off predators that feed on bee or wasp larvae.

Historically, interest in bird-insect nesting associations has been considerable, as is evidenced by the early literature which contains numerous anecdotal observations of such associations. Reviews of this literature (e.g., Myers 1929, 1935; Moreau 1936; Maclaren 1950; Chisholm 1952; Hindwood 1955) indicate that a variety of bird species will nest in proximity to the nests of ants, wasps, or bees. Avian species in families as diverse as Cuculidae, Furnariidae, Tyrannidae, Corvidae, Muscicapidae, Nectariniidae, Dicaeidae, Emberizidae, Fringillidae, Estrildidae, and Ploceidae have been recorded nesting in the immediate neighborhood of hymenopteran nests. Most of these examples of bird-hymenopteran nesting associations are from tropical regions.

It is generally accepted (e.g., Moreau 1936) that nesting birds seek the association of hymenopteran colonies because the stinging or biting behavior of the ants, bees, or wasps deters nest predators. Except for Smith (1968) and Janzen (1969), however, no one has compared the fledging success of bird nests associated with hymenopteran nests to that of birds not associated with insect colonies. In addition, few of the early workers provide the data necessary to determine statistically whether a nesting association actually exists between the birds and colonial hymenopterans. Maclaren (1950) provided the only statistical demonstration of a bird-ant nesting association.

Although nesting by birds in the proximity of colonial hymenopterans, is well documented, many questions remain. What predators of avian eggs and nestlings are inhibited by stinging or biting insects? Does nesting in association with hymenopteran nests actually increase fledging success? How does this nesting association evolve? What is the role of stochastic events in the evolution of this relationship? How do factors such as the nest predation rate, female abandonment of nest site, and length of the breeding season affect the evolution of this association? In this paper we examine these questions with regard to the nesting association between Bananaquits (*Coereba flaveola*) and wasps (*Polybia occidentalis*). We then provide a model to explain the evolution of such nesting associations.

The Bananaquit is an abundant species widespread throughout the Caribbean, Central, and South America. On many of the Caribbean islands it is the commonest species in almost all terrestrial habitats from sea level to the tops of the highest mountains (MacArthur and Wilson 1967). Both males and females build their own woven, globular, domed nests as individual sleeping dormitories. Frequently, the male and female build a separate breeding nest which is indistinguishable from the dormitories. However, sometimes the dormitories are used for breeding. These nests are placed in almost every conceivable location with heights above the ground ranging from less than one to 30 meters (Skutch 1954; Biaggi 1955; Gross 1958). Bananaquits have been found nesting near the nests of *Polistes canadensis* and other wasp species in Trinidad, Jamaica, Haiti, and Venezuela (Myers 1935). Because not all Bananaquits nest near wasp nests, it is possible to compare the success of individuals nesting in association with those that nest away from wasp nests.

METHODS

This study was conducted on the island of Grenada at the southern end of the Lesser Antilles as part of a larger study of Bananaquit breeding biology (Wunderle 1980, 1981, 1982, 1984). Data were collected over five breeding seasons as follows: March–August 1975, March–September 1976, March–September 1977, March–October 1978, and May–August 1981. The study was done in southwestern Grenada, from Point Salines on the south to Grand Anse on the north, and to Mt. Hartman estate on the east, covering an area of approximately 900 hectares. Although the area has been extensively disturbed for agricultural purposes, remnants of a dry deciduous scrub woodland (described in Beard 1949) can be found.

We compare the fledging success of newly initiated Bananaquit nests associated with wasp nests with nearest-neighbor nests initiated during the same week that were not associated with wasp nests. Both members of a pair of Bananaquits were mist-netted and color-banded in 1976 to 1978. Pairing of data (associated versus non-associated nests) was necessary to control for changes in nest predation over time during the breeding season. Because nesting periods are synchronized with periods of rainfall, it was possible to pair nests to the exact week of initiation.

To determine if a nesting association occurred between Bananaquits and wasps we asked whether Bananaquit and wasp nests were found together within the same tree more frequently than predicted by chance. To answer this question Wunderle censused all the trees >2.0 m tall in three plots in different habitats. All trees with wasp nests only, wasp nests and Bananaquit nests, Bananaquit nests only, and trees with neither wasp nor Bananaquit nests were recorded. It was not possible to discriminate between Bananaquit roosting and breeding nests, so all bird nests were pooled. The first habitat surveyed (10 June and 9 November 1981) consisted of a mixed assemblage of *Acacia, Coccoloba, Cocos nucifera,* and *Hippomane macinella* located directly behind the beach at Morne Rouge Bay. This plot was ca. 0.25 ha with trees ranging from 5 to 10 m tall. The second location, surveyed on 21 June and 9 November 1981, consisted of ornamental plantings and open lawns of homes between Pinquin and Madame Jardin. This 1.1 ha site contained a wide diversity of trees and shrubs (4–17 m tall) such as *Tamarindus indica, Gliricidia sepium, Nerium oleander, Bursera simaruba,* and *Citharexylum berlandieri.* The third sample site (ca. 0.87 ha), at True Blue, was located in an open *Acacia* savanna with trees 4 to 8 m in height. It was sampled on 24 June and 1 November 1981.

The distance between Bananaquit nests and the nearest wasp nest in trees containing both nests was measured. In addition, Wunderle estimated the height of the nest tree and measured the maximum and minimum crown diameters using a tape placed on the ground under the tree. The length and maximum width of the wasp nests associated with Bananaquit nests were also measured.

To determine if females nesting in association (≤ 1 m) with a wasp nest would subsequently re-build in association the following year, Wunderle noted the location of nests of color-banded females in successive years. In addition, he removed the active nests of eight color-banded females and then located the replacement nest to see if females originally nesting near (≤ 1 m) a wasp nest would re-nest in association with a wasp nest in the same breeding season. Only females in areas with numerous *Polybia* nests were used, so that wasp nests were, presumably, available. Nests were collected only during the nestling stage.

As part of another study (Wunderle 1980), 507 Bananaquits were color-banded in southern Grenada. Wunderle mapped the locations of all territories and nests in three different study sites and noted if a female's first or replacement breeding nests (i.e., following nest predation) were associated with wasp nests.

RESULTS

When undisturbed, *Polybia occidentalis* is a docile species that will permit human approach to within several centimeters of the nest. However, during daylight hours a sudden jarring of a branch supporting a *Polybia* nest causes the wasps to swarm in the air around their nest for a variable distance (1–2 m) that may depend upon the size of the colony and the stage of the nest (N. G. Smith, pers. comm.). On dark nights the wasps swarmed over the surface of the nest and onto the supporting branch while producing an audible buzzing sound, but they rarely flew. Disturbance of the wasp nests on bright, moon-lit nights did cause some wasps to swarm around their nest. They readily stung humans and presumably would sting other animals that shook the nest or its supporting branch. The wasp sting and its swarming behavior may deter the major predators of Bananaquit eggs and nestlings such as snakes (*Boa enhydris*), grackles (*Quiscalus lugubris*), and the introduced rat (*Rattus rattus*).

Only 13.0 percent of the 23 Bananaquit nests within 1 m and on the same branch as a wasp nest were destroyed by predators compared with 43.5 percent of 23 nearby nests initiated within the same week, but not associated with a wasp nest (*G*-Test, Sokal and Rohlf 1981: 696; $G = 3.976, P < 0.05$). The three wasp-nest associated Bananaquit nests whose contents had been removed, showed no outward sign of predation as is characteristic of predation by snakes (Wunderle 1982). Of the eight females that nested 1 to 2 m from a *Polybia* nest, only three (37.5%) successfully fledged young whereas half of the control nests (N=8) not associated with wasp nests fledged young ($G = 0.002, P > 0.5$). Thus, the degree of protection provided by a wasp nest decreases with distance so that beyond approximately 1 m, nest predation rates are equivalent to those of nests not associated with wasp nests.

Wunderle (1984) demonstrated that females will frequently abandon both their nest sites and mates following nest predation. Therefore, females nesting in areas or sites with low nest predation rates should remain with their original mates for a longer period. Ninety percent of all males (N = 10) with mates nesting within 1 m of a *Polybia* nest had the same mate at the end of the breeding season, while only 36.4 percent of 11 males with mates having breeding nests not associated with wasp nests had the same mate at the season's end ($G = 4.57, P < 0.05$). Thus, nesting in association with a wasp nest reduces nest predation losses, and, as a result, reduces the probability of female abandonment.

The possible advantage to Bananaquits nesting near *Polybia* nests suggests that Bananaquit nests may frequently be associated with wasp nests. Trees (>2.0 m) containing both Bananaquit nests (either roosting or breeding) and *Polybia* nests were more frequently encountered than expected by chance within the Pinquin study site in June ($\chi^2 = 18.108, P < 0.005,$ d.f. = 1, Table 1). At the two other sites the number of trees with both Bananaquit and wasp nests was small (N = 2, N = 3), so significant Chi-square values (Morne Rouge $\chi^2 = 5.735, P < 0.025$; True Blue $\chi^2 = 4.641, P < 0.05$) may be inaccurate (see Zar 1974:50, for a discussion

TABLE 1

DISTRIBUTION OF BANANAQUIT AND *POLYBIA* WASP NESTS IN ORNAMENTAL TREES AND
SHRUBS AT PINQUIN, GRENADA

	Trees without Bananaquit nests	Trees with Bananaquit nests
Trees without *Polybia* nests	294* (289.36)†	12 (16.64)
Trees with 1 or more *Polybia* nests	19 (23.64)	6 (1.36)

* Observed
† Expected $\chi^2 = 18.108, P < 0.005$

of the problems using Chi-square with small sample sizes). The distance between the Bananaquit nest and the nearest *Polybia* nest within the same tree averaged 82.3 cm ± 10.3 (s.e.)
(Fig. 1). This distance averaged 15.6% ± 1.6 (s.e.) of the maximum horizontal distance through
the tree crown (Fig. 2).

The above evidence suggests that a nesting association exists between Bananaquits and
wasps, but does not show which animal initiates the association. Over the five years of field
work, Wunderle observed seven Bananaquit females as they initiated nest building (of breeding
nests only) near *Polybia* nests, but never found wasps building near an existing bird nest
(although he did find 19 wasp nests as they were being established in 1981). The average
length (16.7 cm ± 1.3 s.e.) and average width (9.8 cm ± 0.5 s.e.) of 34 *Polybia* nests associated
with Bananaquit nests did not differ from the average length (17.9 cm ± 1.1 s.e.) and width
(10.7 cm ± 0.6 s.e.) of 54 randomly chosen *Polybia* nests not associated with bird nests (length,
t-Test, *t* = 0.434, *P* > 0.5; width *t*-Test, *t* = 1.043, *P* > 0.3).

To determine if individual *Polybia* nests tended to survive (e.g., remain active) longer than
individual Bananaquit nests during the summer months, I recorded all active Bananaquit
nests associated (≤1.0 m) with active *Polybia* nests in May and June and returned in late
September (1981) to determine which nest of the pair was still active. Of the 29 pairs of wasp
and bird nests reexamined in September, three (10.3%) had both nests active, 17 (58.6%) had
an active wasp nest and a missing bird nest, four (13.8%) had an abandoned or missing wasp
nest with an active bird nest (breeding or roosting), and in five (17.2%) both nests had been
destroyed (cause unknown) or abandoned. *Polybia* nests remained active for a longer time
(*G* = 10.29, *P* < 0.005) than did the associated Bananaquit nest. This suggests that wasp nests
remain active long enough to provide predator deterence during the span of Bananaquit nesting.

No association was found between the placement of the Bananaquit nest relative to the
wasp nest and to the tree trunk. Eighteen Bananaquit nests were farther from the main trunk
than the wasp nest, whereas 16 bird nests were closer to the trunk in 1981. In addition, no
trees in June and July (1981) in the three sites surveyed had both wasp and bird nests together
in November (nonbreeding period), despite the presence of numerous roosting nests. These
findings suggest that no association exists between Bananaquit nest placement relative to wasp
nests and the tree trunk during the breeding season, and in the nonbreeding season, placements
of Bananaquit and wasp nests are not associated.

Female Bananaquits that nested in association with wasps in one breeding season did not
usually nest in association with them in the following year. Of 15 color-banded females nesting
in association with wasps, only one (6.6%) nested near (1.2 m) a wasp nest in the following
year. Nest removal experiments were run to determine if some females would always nest
near wasp nests during the same breeding season. All females used for this experiment had
several (range 9–25) *Polybia* nests available within their territories. After nest removal, one
of the females abandoned her territory, while the remaining seven females re-nested on their
territories which contained at least nine *Polybia* nests. At the time of nest removal, the eight
breeding nests were located an average of 0.65 m ± 0.09 (s.e.) from the nearest wasp nest;
replacement nests averaged 6.47 m ± 1.27 (s.e.) from the nearest wasp nest. This difference
was significant (Mann-Whitney *U*-test, Sokal and Rohlf 1981:432; *U* = 49, *P* < 0.001). The
inverse relationship was observed in four color-banded females that built their original breeding nests in trees without wasp nests (mean distance to nearest wasp nest = 8.43 m ± 2.02

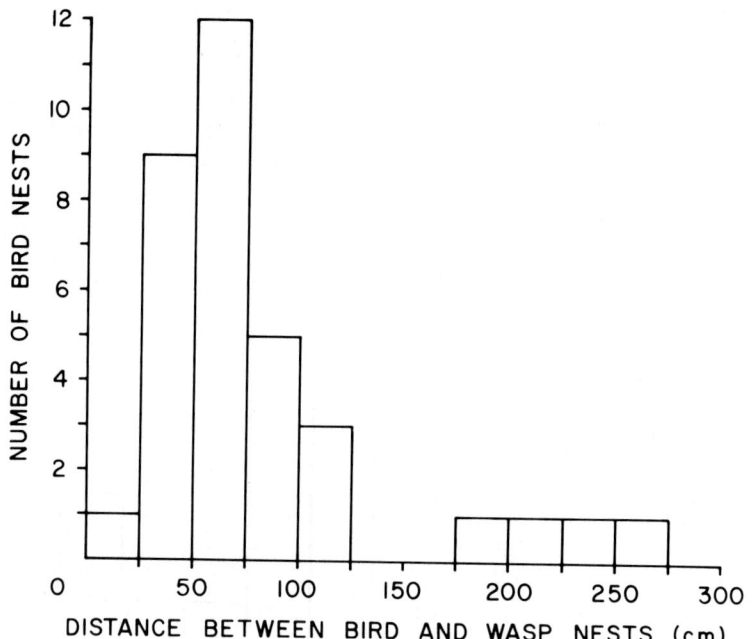

Fɪɢ. 1. The distance between Bananaquit and *Polybia* wasp nests in the same tree. (N = 34).

(s.e.), but after their nests were destroyed, re-nested an average of 0.69 m ± 0.07 (s.e.) from a wasp nest.

DISCUSSION

Bananaquit nests within 1 m of a *Polybia* nest had a lower predation rate than nearby nests which were not associated with wasp nests. However, those Bananaquit nests that were more distantly associated with wasp nests (<1 m but >2 m) had predation rates equivalent to nearby nests not associated with wasp nests. These findings are consistent with the hypothesis that stinging insects may deter potential nest predators and, thus, that birds nesting in association with them benefit by having lower nest predation rates. On Grenada, grackles and rats may be reluctant to disturb wasp nests. However, some large birds with heavy feather coverings may not be deterred by *Polybia* stings as illustrated by the observations of Gray-headed Kite (*Leptodon cayanensis*) predation upon a wasp brood in Costa Rica (Windsor 1976). It is also possible that some snakes may be able to thwart the *Polybia* defenses of associated bird nests by feeding at night and with slow movements that are unlikely to disturb the wasp colony. In this study, three Bananaquit broods in nests less than 1 m from a wasp nest were destroyed by snakes (possibly at night).

The Bananaquit-*Polybia* nesting association is probably a case of commensalism rather than mutualism. The *Polybia* wasps do not appear to benefit from the neighboring Bananaquits. No evidence suggests that Bananaquits deter the predators of wasp larvae, as has been suggested for tyrannids nesting near wasp nests (Smith 1980), nor is it likely that the bird nest is the more conspicuous component of this association, thus reducing the risk of accidental damage to the fragile wasp nest as discussed by Myers (1935). In this case, the bird nest may advertise the presence of a wasp nest and, thus, provide a warning to foraging animals.

The *Polybia* wasps showed only a mild response (e.g., swarmed over the surface of their nest, few flew in the air) to female Bananaquits as they began nest building near the colony. With time the wasps habituated to the activity of the Bananaquits. A similar situation occurs in ants that habituate to the activities of birds nesting in ant acacias (Janzen 1969). Whereas the light, regular movements of the nesting Bananaquits were ignored by wasps, a sudden jarring of the wasp nest or its associated branch produced an immediate swarming response

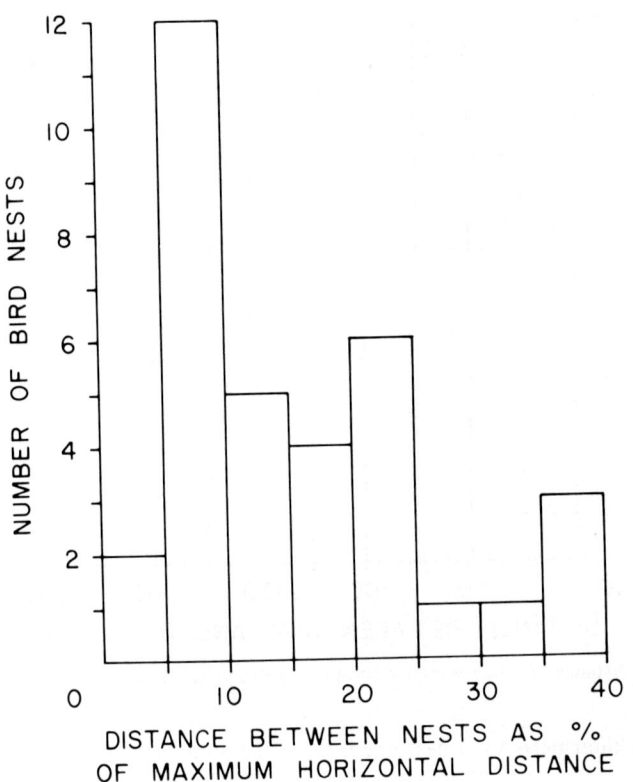

Fig. 2. The distance between Bananaquit and wasp nests as a percentage of the maximum horizontal distance through the tree crown. (N = 34).

by the wasps. Presumably, the domed Bananaquit nest protects the nestlings from the swarming, stinging wasps. Domed or tightly woven pendant nests that protect nestlings from stinging or biting insects may be a pre-adaptation necessary for the evolution of this type of association; such nests are characteristic of birds associated with aggressive insects (Myers 1929, 1935; Moreau 1936; Maclaren 1950; N. G. Smith, pers. comm.).

Male Bananaquits mated to females nesting in proximity to *Polybia* nests within their territories are more likely to retain their original mates through the breeding season than those males with nests not associated with wasp nests. As discussed elsewhere (Wunderle 1984), female disappearance following nest predation is a result of abandonment of nest site and mate and *not* a result of mortality due to the nest predator. An abandoning female will leave the territory of her original mate and re-nest with a nearby male at a new location. The original male may be without a mate for several weeks or even for the remainder of the breeding season. This could result in the loss of valuable breeding time for both sexes. In addition, a delay in breeding could also result in reduced fledging success because nest predation rates increase as the breeding season proceeds (Wunderle 1982, 1984). Thus, Bananaquits nesting near a *Polybia* nest will have a higher fledging success per clutch and possibly more successive clutches per breeding season with the same mate.

Female Bananaquits appear to choose the actual nest site (Wunderle, unpubl. observ.) and, hence, they, rather than males, are responsible for the selection of a site near a *Polybia* nest. On Grenada, females do not always nest in association with *Polybia* nests even when several "appear" to be available within the territory. The results of the nest removal experiments confirm this, or the nest removal (i.e., an act of predation) may have served as a negative conditioning agent that encouraged females to nest away from wasp nests. That females did not always nest in association with wasps in successive seasons (even though *Polybia* appeared to be available) suggests that other factors may be important in nest site selection.

It is unlikely that the Bananaquit-*Polybia* nesting association occurs because both species are attracted to a particular species of tree for their nests. In southern Grenada, Bananaquits were found nesting in 34 species of trees and shrubs whereas *Polybia* nests were found in 22 species. Bananaquits may have greater flexibility in nest site selection than wasps. No *Polybia* nests were found in plant species not used by Bananaquits. *Polybia* wasps require a horizontal, or nearly horizontal, branch from which to suspend their nest. Bananaquits will use a variety of structures for nesting as long as there is a fork to which they can weave their nest (Skutch 1954; Biaggi 1955; Gross 1958). Neither species will nest on the vertical trunk of a tree. However, both species can probably find numerous acceptable sites for nest attachment often within close proximity to each other.

Female Bananaquits always abandon their nests and often their territories following nest predation and re-nest elsewhere. They will make as many as six re-nesting attempts in a breeding season. If they are successful at one nest location, they remain there and produce up to four successful clutches unless the nest is destroyed (Wunderle 1980). This pattern may increase the likelihood of a female finding a safe nesting site, especially in close to a wasp nest. Once associated with the wasp nest, the female may remain and produce several successful clutches until her nest finally deteriorates. The observed statistical association between Bananaquit nests and wasp nests may represent the nest distribution resulting from the female abandonment and relocation behavior. Such females will eventually locate a predation "free" site where they may remain for several successive clutches. If most of these safe sites are near a wasp nest, we may expect to find more females nesting in association with wasps than predicted by chance. The association would not be dependent upon the female cuing on a wasp nest as part of her nest site selection behavior. Thus, as nest predation rates increase, the proportion of females nesting near wasp nests may increase.

This simple nest predation response of female site abandonment and relocation until a safe site is found could explain the evolution of some other bird-hymenopteran nesting associations. Many biologists have previously assumed that the existence of bird-hymenopteran nest spatial associations indicates that site selection by the birds is nonrandom. For example, Myers (1935) suggested that nestlings reared close to a hymenopteran nest may later prefer such an association for their own nest sites. We offer an alternative explanation as a null hypothesis, a random choice model to explain the evolution of similar nesting associations.

A Random Choice Model

We assume that a bird has a large number of potential nest sites near to and away from wasp nests so that we can ignore the problems of sampling without replacement. We further denote the proportion of potential safe sites that are "close to" a wasp nest by a parameter, P. Assuming that any bird makes a random choice of nest site from those available, the probability that the bird chooses a site close to a wasp nest is also given by P.

We also assume that there is a fixed "generation" time during which either a bird succeeds in rearing a brood or fails. It the bird is successful, it attempts to rear another brood in the same nest. If the bird fails, it moves to a new site and builds another nest for the next generation. We assume that there is a probability (ϕ_w) for success if the nest is close to a wasp nest and another lower probability (ϕ) for success if the nest is not close to a wasp nest.

We now derive the probability of a bird having a nest site close to a wasp nest after i generations in terms of the three parameters of the model (p, ϕ_w, ϕ). Using conditional probability arguments, we can easily show that the probability of a nest site being chosen close to a wasp nest at time t_1 is given by

$$P_1 = P[\phi_w + (1 - \phi_w)P] + (1 - P)[(1 - \phi)P]$$

As each generation flows to the next in the same manner, we have for the next generation

$$P_2 = P_1[\phi_w + (1 - \phi_w)P] + (1 - P_1)[(1 - \phi)P].$$

Using inductive reasoning, we have established the recurrence relation

$$P_i = P_{i-1}[\phi_w + (1 - \phi_w)P] + (1 - P_{i-1})](1 - \phi)P] \qquad (1)$$

Computations based on this equation (1) for a range of model parameter values are shown in Table 2. We consider wasp nests to be sparse (P = 0.2), abundant (P = 0.5) or super abundant (P = 0.8). We allow the degree of protection of wasp nests to range from small ($\phi_w = 0.8$, $\phi = 0.7$) to very large ($\phi_w = 0.8$, $\phi = 0$).

TABLE 2

PROBABILITY OF A BIRD NESTING NEAR A WASP NEST ON SUCCESSIVE TRIALS

Trial	Model parameters			Probabilities in successive generations						
	P	ϕ_w	ϕ	P_1	P_2	P_3	P_4	P_5	P_6	P_7
1	0.2	0.8	0.7	0.216	0.228	0.238	0.246	0.252	0.256	0.260
2	0.2	0.8	0.0	0.328	0.410	0.462	0.496	0.517	0.531	0.540
3	0.5	0.8	0.7	0.525	0.544	0.558	0.568	0.576	0.582	0.587
4	0.5	0.8	0.0	0.700	0.780	0.812	0.824	0.830	0.832	0.833
5	0.8	0.8	0.7	0.816	0.828	0.836	0.842	0.846	0.849	0.851
6	0.8	0.8	0.0	0.928	0.948	0.952	0.952	0.952	0.952	0.952

If the wasp nests are abundant and offer substantial protection, then it is clear that a bird-hymenopteran nest spatial association can develop very quickly. The rapid evolution of this nesting association would be encouraged by factors such as learning or nonrandom choice, a shortage of nest sites, restriction of wasp and bird nests to very similar nesting sites, and the possibility that unsuccessful birds have a shorter generation time than successful individuals.

In our model we have deliberately ignored the role of learning or nonrandom choice in an effort to illustrate the effects of random choice. It is apparent that random choice of nest sites alone can have some effect upon the evolution of this nesting association and may be important in the initial stages of its development. With the addition of learning (e.g., a female associates a wasp nest with previous nesting success and hence seeks wasp nests for future nest sites), this nesting association will develop at a faster rate. A final stage might be the genetic incorporation of a "mental image" of a wasp nest which serves as a cue for nest site selection. Thus, random choice could start and initially maintain the nesting association until the more powerful mechanism of behavioral choice (e.g., cuing on wasp nests for nest sites) finally controls nest site selection. As a result, we might find differences between bird species or even populations in the mechanism (i.e., random choice or learning) maintaining the nesting association, depending upon ecological circumstances.

As an island study, our results and suggested mechanism may not be applicable to the evolution of bird-wasp nesting associations in some mainland areas. For example, high population densities of both nesting birds and wasps, reduced diversity of predators (on both bird and wasp nests), and disturbed vegetation are common on some islands, but rare in mainland rainforests. However, these characteristics are frequently found in xeric areas or savannas of continental regions for which a large number of bird-insect nesting associations are also documented (e.g., Myers 1929, 1935; Maclaren 1950; Chisholm 1952). For example, wasp nesting densities on Grenada appear to be similar to densities in Guanacaste, Costa Rica (Wunderle, pers. observ.). In addition, some mainland bird species nesting with insects have very high densities similar to those of island Bananaquits (Maclaren 1950; Contino 1968). While Grenada does have a low diversity of nest predators (four predators of Bananaquit nests, Wunderle 1980), nest predation rates are as high as mainland rates (Wunderle 1982). Hence, our random choice model may be useful for understanding the evolution of bird-wasp nesting associations in xeric regions of the tropical mainland, but of limited utility in rainforest regions where predator diversity is higher, wasp and bird densities lower, and habitat disturbance less frequent.

Several factors may explain the prevalence of bird-hymenopteran nesting associations in the tropics relative to temperate regions. Obvious factors are the higher nest predation rates on tropical birds (Ricklefs 1969) and, possibly the greater number of colonial hymenopterans available to provide protection. In addition, nests that protect nestlings from insect attacks are an essential pre-adaptation for this association, and such nests may be more characteristic of tropical birds. Less obvious factors are the longer life spans of tropical birds (e.g., Snow and Lill 1974; Fry 1977) and a longer season available for re-nesting. As shown in our model, an increase in the number of nesting generations increases the probability that an individual will nest in association with a hymenopteran colony. A tropical species (e.g., Bananaquits) may build more replacement nests (because of higher nest predation rates) in a life-time than similar temperate zone species. Eventually some of these re-nesting attempts may bring the tropical species within close proximity of a protective hymenopteran nest.

ACKNOWLEDGMENTS

This paper benefited from the suggestions of R. Breitwisch, M. Foster, M. Melampy, N. Smith, and R. Waide. The senior author is grateful to J. Andersen, P. Arnold, D. and H. McNeill, and the Wren family for their hospitality on Grenada. The wasps were kindly identified by T. Nuhn. Dr. J. G. Vandenbergh kindly provided facilities of the Department of Zoology, North Carolina State University for the senior author. This work was funded in part by grants from the Dayton Natural History Fund and the Wilkie Fund of the James Ford Bell Museum of Natural History, University of Minnesota, the Frank M. Chapman Memorial Fund of the American Museum of Natural History, the Josselyn Van Tyne Fund of the American Ornithologists Union, and the National Geographic Society.

LITERATURE CITED

BEARD, J. S. 1949. The Natural Vegetation of the Windward and Leeward Islands. Clarendon Press, Oxford, England.

BIAGGI, V. 1955. The Puerto Rican honeycreeper (Reinita) *Coereba flaveola* (Bryant) Agric. Exp. Stn. Spec. Publ. Univ. Puerto Rico, 61 pp.

CHISHOLM, A. H. 1952. Bird-insect nesting associations in Australia. Ibis 94:395–405.

CONTINO, F. 1968. Observations on the nesting of *Sporophila obscura* in association with wasps. Auk 85:137–138.

FRY, C. H. 1977. The evolutionary significance of cooperative breeding in birds. Pp. 127–136. *In* B. Stonehouse and C. M. Perrins (eds.), Evolutionary Ecology. Macmillan, London.

GROSS, A. O. 1958. Life history of the Bananaquit of Tobago Island. Wilson Bull. 70:257–279.

HINDWOOD, K. A. 1955. Bird/wasp nesting associations. Emu 55:263–274.

JANZEN, D. H. 1969. Birds and the ant × acacia interaction in Central America, with notes on birds and other myrmecophytes. Condor 71:240–256.

MACARTHUR, R., AND E. O. WILSON. 1967. Island Biogeography. Princeton University Press, Princeton, New Jersey.

MACLAREN, P. I. R. 1950. Bird-ant nesting associations. Ibis 92:564–566.

MOREAU, R. E. 1936. Bird-insect nesting associations. Ibis 78:460–471.

MYERS, J. G. 1929. The nesting-together of birds, wasps, and ants. Proc. Entomol. Soc. Lond. 4:80–88.

MYERS, J. G. 1935. Nesting associations of birds with social insects. Trans. R. Entomol. Soc. Lond. 83:11–23.

RICKLEFS, R. E. 1969. An analysis of nesting mortality in birds. Smithson. Contrib. Zool. 9:1–48.

SKUTCH, A. F. 1954. Life histories of Central American birds. Pac. Coast Avif. No. 31.

SMITH, N. G. 1968. The advantage of being parasitized. Nature 219:690–694.

SMITH, N. G. 1980. Some evolutionary, ecological and behavioral correlates of communal nesting by birds with wasps or bees. Proc. 17th Int. Ornithol. Congr., pp. 1199–1205.

SNOW, D. W., AND A. LILL. 1974. Longevity records for some neotropical birds. Condor 76:262–267.

SOKAL, R. R., AND F. J. ROHLF. 1981. Biometry. W. H. Freeman, San Francisco, California.

WINDSOR, D. M. 1976. Birds as predators on the brood of *Polybia* wasps (Hymenoptera:Vespidae: Polistinae) in a Costa Rican deciduous forest. Biotropica 8:111–116.

WUNDERLE, J. M. 1980. The breeding ecology of the Bananaquit (*Coereba flaveola*) on Grenada. Unpubl. Ph.D. dissert., University of Minnesota, Minneapolis.

WUNDERLE, J. M. 1981. An analysis of a morph ratio cline in the Bananaquit (*Coereba flaveola*) on Grenada, W.I. Evolution 35:333–344.

WUNDERLE, J. M. 1982. The timing of the breeding season in the Bananaquit (*Coereba flaveola*) on the island of Grenada, W.I. Biotropica 14:124–131.

WUNDERLE, J. M. 1984. Mate switching and a seasonal increase in polygyny in the Bananaquit. Behaviour 88:123–144.

ZAR, J. H. 1974. Biostatistical Analysis. Prentice-Hall, Englewood Cliffs, New Jersey.

COMMUNITY AND POPULATION ECOLOGY

CRUZ, ALEXANDER, TIMOTHY MANOLIS, AND JAMES W. WILEY. The Shiny Cowbird: A Brood Parasite Expanding its Range in the Caribbean Region 607

FAABORG, JOHN. Ecological Constraints on West Indian Bird Distributions 621

GOCHFELD, MICHAEL. Numerical Relationships Between Migrant and Resident Bird Species in Jamaican Woodlands 654

KUSHLAN, JAMES A., GONZALO MORALES, AND PAULA C. FROHRING. Foraging Niche Relations of Wading Birds in Tropical Wet Savannas 663

MUNN, CHARLES A. Permanent Canopy and Understory Flocks in Amazonia: Species Composition and Population Density 683

POWELL, GEORGE V. N. Sociobiology and Adaptive Significance of Interspecific Foraging Flocks in the Neotropics 713

REMSEN, J. V., JR. Community Organization and Ecology of Birds of High Elevation Humid Forest of the Bolivian Andes 733

STILES, F. GARY. Seasonal Patterns and Coevolution in the Hummingbird-Flower Community of a Costa Rican Subtropical Forest 757

WILLARD, DAVID E. Comparative Feeding Ecology of Twenty-two Tropical Piscivores 788

WRIGHT, S. JOSEPH, JOHN FAABORG, AND CLAUDIA J. CAMPBELL. Birds Form Tightly Structured Communities in the Pearl Archipelago, Panama 798

THE SHINY COWBIRD: A BROOD PARASITE EXPANDING ITS RANGE IN THE CARIBBEAN REGION

*Alexander Cruz,[1] Timothy Manolis,[1] and James W. Wiley[2]

[1]Department of Environmental, Population and Organismic Biology, B-334, University of Colorado, Boulder, Colorado 80309 USA, and
[2]Puerto Rico Field Station, United States Fish and Wildlife Service, Box 21, Palmer, Puerto Rico 00721 USA

* Authors listed alphabetically.

ABSTRACT. The Shiny Cowbird (*Molothrus bonariensis*), an avian brood parasite, was originally confined to South America, Trinidad, and Tobago. During the last 86 years it has rapidly spread into the West Indies. The relationship between the Shiny Cowbird and the Yellow-hooded Blackbird (*Agelaius icterocephalus*) in Trinidad, an area of long parasite-host interaction, is compared with the relationship between the cowbird and the Yellow-shouldered Blackbird (*A. xanthomus*) in Puerto Rico, an area of recent cowbird-host interaction. The level of parasitism suffered by the Yellow-shouldered Blackbird (94.2% of nests) is significantly greater than that sustained by the Yellow-hooded Blackbird (44.8% of nests). This difference is related to dissimilarities between the two blackbird species in nest placement and form of nest defense.

In mangrove study areas in Puerto Rico 61 percent (11 of 18) of the resident passerine species were parasitized by the Shiny Cowbird. Cowbird parasitism reduced host productivity, depressing their nest success rate below that of non-parasitized nests. Parasitized pairs had fewer eggs and fledged fewer of their own chicks than did nonparasitized pairs. The high rates of parasitism observed for some species in Puerto Rico probably reflect the cowbird's recent arrival and exploitation of an island avifauna presumably lacking effective anti-parasite defenses. Some nesting species, however, have effective anti-parasite strategies, including alien egg rejection and nest guarding. The species of the recently exposed Puerto Rican avifauna are classified as acceptors and rejectors of alien eggs, behavioral categories also found in populations long associated with brood parasites. The rate of parasitism was negatively correlated with degree of a species' aggressive response toward the Shiny Cowbird, with regularly parasitized species showing less aggression toward cowbirds than non-parasitized species. Behavior effective in avoiding parasitism is similar to that used by certain birds in evading nest predation, and we suggest that anti-predator behavior is a preadaptation for countering cowbird parasitism.

RESUMEN. El Tordo Lustroso (*Molothrus bonariensis*), una ave parasitíca, se encontraba originalmente confinada a Sur America, Trinidad, y Tobago. Durante los últimos 86 años, esta ave se ha expandido rapidamente en la region Caribeña. Una comparación importante es entre el Turpial de Agua (*Agelaius icterocephalus*) en Trinidad, una región donde el Tordo Lustroso es endémico, y la Mariquita de Puerto Rico (*A. xanthomus*), una región donde el Tordo Lustroso ha reciente llegado. El nivel del parasitismo (94.2%) sostenido por La Mariquita es significativamente más grande que el sostenido por el Turpial de Agua (44.8%). Esta diferencia está relacionado con diferentes sitios de anidación e el grado de defensa de los nidos.

En areas de estudio de Puerto Rico, 61 por ciento (11 de 18) de las especies paserinas residente estaban siendo parasitados. El parasitismo del tordo efecta al hospedero por reducir la productivida del hospedero, disminuyendo el éxito del periodo de anidación en comparación con los nidos no parasitados. Parejas parasitadas producen menos huevos y pajaros volantones que parejas no-parasitados. Los altos indices de parasitismo observado para algunos especies en Puerto Rico están probablemente relacionados a la reciente llegado del tordo y a la explotación por parte de este de la fauna isleña que presumablemente carece de defensas antiparasiticas efecivas, incluyendo el rechazo de huevos extraños y vigilancia del nido. Encontramos que, al igual que en poblaciones de aves que tienen largos periodos coevolutivos con este tipo de parasitísmo de crianza, las recientemente expuesta poblaciones de aves anidando en Puerto Rico fueron dividas en categorias discretas de aceptor y no-aceptor de huevos extraños. El indice de parasitismo resulto estar negativamente correlacionado con la respuesta agresiva de los especies hacia el tordo, siendo las especies parasitados las que demuestran menos agresión hacia el tordo que las especies no parasitados. El comportamiento efectivo para evitar el parasitismo es similar al el usado por ciertos aves para evadir los depredadores de nidos, y nosotros sugerimos que el comportamiento de un antidepredador es preadaptivo contra el parasitismo del Tordos lustroso.

Recent changes in the range of the Shiny Cowbird (*Molothrus bonariensis*), a brood parasite, have brought it into contact with avian populations that have no previous exposure to brood parasitism (Fig. 1). Originally confined to South America, Trinidad, and Tobago, the northern race (*M. b. minimus*), aided by the destruction of forests and perhaps by man-made introductions, has spread through the West Indies during the last 86 years (Bond 1966, 1971, 1973; Ricklefs and Cox 1972; Post and Wiley 1977a). In 1899 it was recorded on Carriacou, Grenadines (Bond 1956). The first Shiny Cowbird was collected in Hispaniola in October 1972 (Bond 1973), and it is now common in many lowland areas on the island. O. H. Garrido (pers. comm.) found it at Cardenas, west Cuba in 1982.

Several workers have described the biology of the Shiny Cowbird in South America (Hudson 1920; Friedmann 1929; Sick 1958; Hoy and Ottow 1964; Selander 1964; Friedmann et al. 1977). Sick and Ottow (1958), King (1973), and Fraga (1978) made quantitative studies of the Shiny Cowbird's relationship with one of its main South American hosts, the Rufous-collared Sparrow (*Zonotrichia capensis*), in Brazil and Argentina, and Gochfeld (1979a, b) studied the interactions among Shiny Cowbirds and two meadowlark species [Long-tailed (*Sturnella loyca*) and Lesser Red-Breasted (*S. defillippi*)] in Argentina. Wiley and Wiley (1980) studied the effects of Shiny Cowbird parasitism on the Yellow-hooded Blackbird (*Agelaius icterocephalus*) in Trinidad, Surinam, and Venezuela. These studies and others of the Brown-headed Cowbird (*Molothrus ater*) of North America (Friedmann 1929, 1963; Bent 1958; Young 1963; Mayfield 1965, 1977a, b; Rothstein 1975a; Payne 1976, 1977; Friedmann et al. 1977; Elliot 1978; Linz and Bolin 1982) showed that cowbirds lower host reproductive output by egg stealing, egg-puncturing, and reducing egg hatchability and fledgling success. Ample evidence shows that cowbird parasitism reduces host success, although in most reported situations, parasites and hosts have coevolved; i.e., brood parasitism has constituted a selective pressure favoring the evolution of antiparasite defenses by the host. This is not the case in the West Indies where there is no evidence of cowbirds in fossil or subfossil deposits (Olson 1978; P. Brodkorb, pers. comm.). Consequently, for many species in this region little time has been available for the evolution of mechanisms to cope with cowbird parasitism.

Both the frequency of nest parasitism and the relative reduction in host nesting success due to parasitism should be inversely related to the length of time the Shiny Cowbird has been in contact with host populations. Host and parasite populations in regions where the cowbird is indigenous should exhibit stable reproductive relationships, whereas in areas recently invaded by the cowbird, non-equilibrium conditions should occur. One can predict that the Shiny Cowbird will have a more detrimental effect on West Indian species than on Trinidad and mainland species. On Puerto Rico, the Shiny Cowbird has been implicated as the main causative agent for the decline of the Yellow-shouldered Blackbird (*Agelaius xanthomus*; Post and Wiley 1976, 1977a, b; Post 1981). It has also been implicated as the cause of extinction or decline of several island bird populations, for example, endemic races of the House Wren (*Troglodytes aedon*) on Grenada (Bond 1971), the Yellow Warbler (*Dendroica petechia*) on Barbados (Bond 1966), and the endemic Martinique Oriole (*Icterus bonana*; M. Bon Saint Come, pers. comm.). However, with the exception of the Yellow-shouldered Blackbird, field studies on these species have not been undertaken. The habitat change that has permitted cowbird range expansion could also have caused the decline of some of these species independently of cowbird parasitism. Cowbirds prefer disturbed habitats, usually in association with agriculture and livestock.

Our studies of the Shiny Cowbird in the Caribbean region began in 1973. The purpose of this work was to document the spread of the Shiny Cowbird in the region, to determine its impact on the native avifauna, and to examine adaptations present in the avifauna against brood parasitism. Here we report on the cowbird-host relationships in an area of long parasite-host interactions (Trinidad) and in areas where the cowbird has only recently arrived (Puerto Rico and Hispaniola). We present data on levels of parasitism and on the effect of parasitism

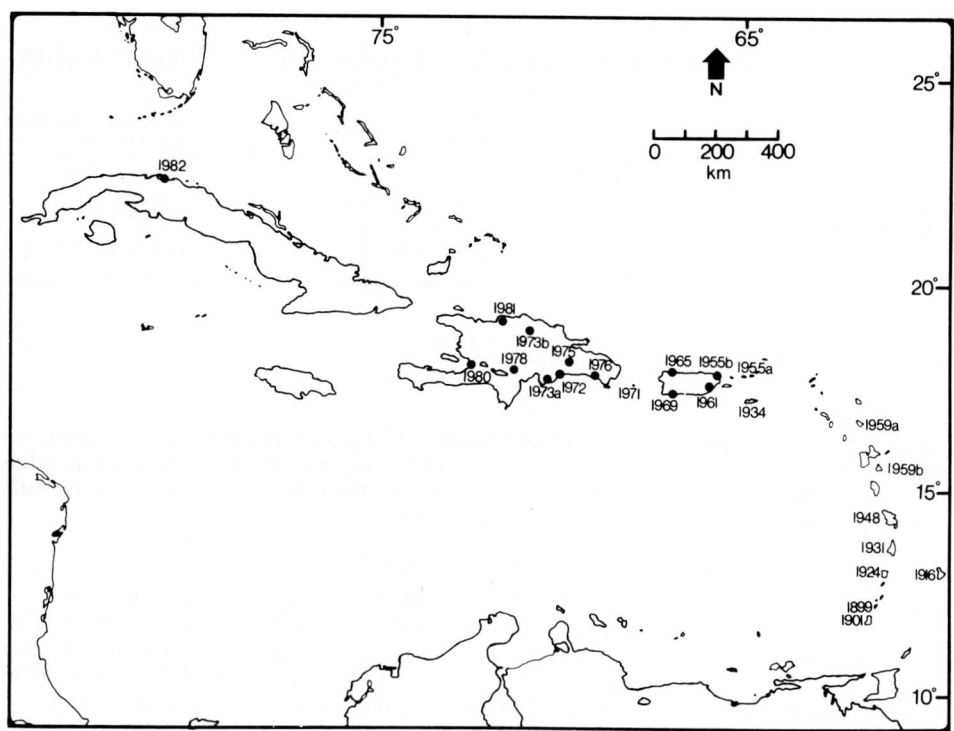

Fig. 1.   Range expansion of *Molothrus bonariensis* in the West Indies (dates represent first report of appearance of these localities). 1899: Carriacou, Grenadines (Bond 1956)—male and female collected. 1901: Grenada (Bond 1956). 1916: Barbados (Bond 1956). 1924: St. Vincent (Bond 1956). 1931: St. Lucia (Danforth 1935). 1934: St. Croix (Bond 1956). 1948: Martinique (Pinchon 1963). 1955: St. John (Robertson 1962). 1955: Cabezas de San Juan, Puerto Rico (Grayce 1957). 1959: Antiqua (Pinchon 1963). 1959: Marie-Galante (Pinchon 1963). 1961: Yabucoa, Puerto Rico (Biaggi 1963). 1965: Guajataca Cliffs, Puerto Rico (Buckley and Buckley 1970). 1969: Guánica, Puerto Rico (Kepler and Kepler 1970). 1971: Mona Island (Bond 1973). 1972: Hato Nuevo, 5 km SW Santo Domingo, Dominican Republic (Bond 1973). 1973: Najayo Abajo, 22 km SW Santo Domingo (A. Dod, pers. comm.). 1973: Santiago, Dominican Republic (A. Dod, pers. comm.). 1975: Finca Estrella, 10 km N Santo Domingo (pers. observ.) 1976: La Romana, Dominican Republic (pers. observ.) 1978: 10 km SE Neiba, Dominican Republic (Arendt and Vargas 1984). 1980: Port-au-Prince, Haiti (C. Mitchell, pers. comm.). 1981: Monte Criste, Dominican Republic (Arendt and Vargas 1984). 1982: Cardenas, Cuba (O. H. Garrido, pers. comm.).

on productivity of hosts in bird communities in Puerto Rico. We compare aspects of behavioral ecologies of the Yellow-shouldered Blackbird in Puerto Rico and the Yellow-hooded Blackbird in Trinidad, that may reduce risk of parasitism. We also report on experimental studies of host aggression toward cowbirds and cowbird egg rejection.

Robertson and Norman (1977) and others hypothesized that aggressiveness functions as a defense that reduces parasitism, and, hence, that aggressive individuals are favored by selection in proportion to the incidence of parasitism. This aggression-defense prediction assumes that other factors do not greatly affect the aggressiveness of hosts tested. A corollary of the aggression defense prediction is that host aggressiveness will increase as long as these selection pressures operate in a host. Range expansion of the Shiny Cowbird has resulted in a mosaic of populations that have been parasitized for varying periods of time, thus providing an excellent opportunity to test this prediction, i.e., that hosts in areas where the cowbird has been resident longer (e.g., Trinidad) will show stronger aggressive responses to cowbirds close to their nests than will hosts in areas that the cowbird has recently colonized (e.g., Puerto Rico).

Rothstein (1975a, b) made detailed studies of egg rejection behavior of a large number of North American species and found that species were geographically uniform in either having

TABLE 1
REPRODUCTIVE SUCCESS OF PARASITIZED AND UNPARASITIZED YELLOW-HOODED BLACKBIRD PAIRS, TRINIDAD, 1979–1981

Reproductive component[1]	Parasitized nests	Unparasitized nests
Nest success[2]	0.17 (7/41)	0.11 (7/63)
Hatching success[3]	0.32 (14/44)	0.29 (18/63)
Nests with fledgling/nests with hatchling[4]	0.58 (7/12)	0.39 (7/18)
Fledgling/egg	0.08 (8/95)	0.06 (8/127)

[1] Differences non-significant ($P > 0.5$) for nest success (d.f. = 1, $G = 0.33$), hatching success (d.f. = 1, $G = 0.02$), nests with fledgling/nests with hatching (d.f. = 1, $G = 0.04$), and fledgling/egg (d.f. = 1, $G = 0.11$).
[2] Proportion of nests fledgling at least one blackbird.
[3] Proportion of nests hatching at least one young blackbird and/or cowbird.
[4] Proportion of nests in which eggs hatched, that fledged at least one blackbird.

or lacking the defense. He predicted a rapid fixation of rejection behavior, once it occurs in a population, because of a strong positive selection pressure. He also noted that rejection has no adaptive value until a species is parasitized. As Puerto Rican birds have only been recently exposed to cowbird parasitism, we expected egg rejection to be rare.

STUDY AREAS AND METHODS

Study areas.—We did field work in Puerto Rico between 1973 and 1982 in two mangrove habitats, Roosevelt Roads Naval Station in eastern Puerto Rico, and Boquerón Commonwealth Forest in southwestern Puerto Rico. Investigations in Trinidad were carried out from 1979 through 1981, and in the Dominican Republic, between 1974 and 1978, and in 1981 and 1982. Detailed information on the study areas is provided by Post and Wiley (1977b), Post (1981), Manolis (1982), and Wiley (1982).

Methods.—We gathered information on the breeding biology of the cowbird and its potential and actual host species, the frequency of use of host species, and the effects of brood parasitism on host and parasite breeding success. Nests were located by regular searches through the study areas. Each nest was marked with a coded tag (inconspicuously placed) and plotted on field maps. At each visit to the study area (usually every other day), we inspected nests to determine number of host and, if present, parasite eggs and chicks. Each egg was inconspicuously marked with ink. Causes of nest failures were determined whenever possible. We defined a nest as active when the resident laid at least one egg or, if the nest was parasitized and held no host eggs, when the host incubated the cowbird egg(s). Egg success was defined as the proportion of eggs hatched/eggs laid, and nest success as proportion of nests fledging at least one young.

Species that reject cowbird eggs are generally not detectable from observations in the field. We followed Rothstein's (1971, 1975a, b) technique of artificial parasitism, using both real and artificial cowbird eggs. Real eggs were obtained from nests in other parts of the study areas where experiments were not being performed. Artificial cowbird eggs were constructed of wood and coated with paint to simulate the cowbird egg pattern. Size and weight of the artificial eggs closely approximated real eggs. Nests were artificially parasitized within two hours after dawn. The Shiny Cowbird lays its eggs in the morning until about noon (Hoy and Ottow 1964; pers. observ.). After nests were parasitized they were monitored to determine host response. Eggs were considered rejected if they were ejected or damaged, but we also considered desertion and nest build-overs as rejections.

We determined residents' aggressive responses to territory intruders by watching nesting birds from blinds and towers. Most observations were made during the pre-incubation and early incubation stages. We recorded the species of the intruder and the resident's reaction to the alien when the two were separated by eight predetermined distances (0.0–0.4, 0.5–0.9, 1.0–1.9, 2.0–2.9, 3.0–4.9, 5.0–9.9, 10.0–19.9, ≥20 m). The overt responses of the residents were scored on data sheets according to the following predetermined categories arranged in order of increasing aggressiveness: 0 = no detectable reaction by resident at alien's presence; 1 = resident orients toward intruder, gives low-intensity calls; 2 = resident leaves nest, moves in direction of intruder, gives alarm calls; 3 = resident flies at intruder, supplants, and chases it from territory; 4 = resident strikes intruder, grapples, plucks intruder's feathers. For further

TABLE 2

CLUTCH SIZES OF YELLOW-HOODED BLACKBIRDS IN PARASITIZED AND UNPARASITIZED NESTS,
TRINIDAD, 1979–1981

Nest status	Blackbird clutch size			
	1	2	3	Mean ± s.d.[1]
Unparasitized (N = 23)	1	6	16	2.65 ± 0.57
Parasitized (N = 20)	1	7	12	2.55 ± 0.60
All nests (N = 43)	2	13	28	2.60 ± 0.58

[1] Differences in clutch sizes between unparasitized and parasitized nests not significant (G-test, $P > 0.05$).

information on methods see Post and Wiley (1977b), Post (1981), Manolis (1982), and Wiley (1982).

RESULTS AND DISCUSSION

BLACKBIRD REPRODUCTIVE SUCCESS

All 44 Yellow-shouldered Blackbird nests examined in 1982 were parasitized; from 1975 to 1981, 152 of 164 (93%) nests examined in various parts of Puerto Rico were parasitized. In contrast, only 82 of 183 (44.8%) nests of the Yellow-hooded Blackbird examined in Trinidad were parasitized. The level of parasitism on Yellow-shouldered Blackbirds is significantly greater than that incurred by the Yellow-hooded Blackbird ($P < 0.001$, d.f. = 1, $G = 75.94$).

The negative effects of parasitism on host reproductive success are greater for Yellow-shouldered Blackbirds than for Yellow-hooded Blackbirds. In southwestern Puerto Rico, nest and egg successes were 40 percent and 30 percent, respectively, for parasitized nests, and 63 percent and 36 percent, respectively, for unparasitized nests (Post and Wiley 1977b). In 1982 nest and egg successes were 29 percent and 12 percent in southwestern Puerto Rico. The number of Yellow-shouldered Blackbird fledglings per nest was 0.34, significantly lower ($P < 0.001$, d.f. = 1, $G = 11.2$) than the 1975 figure (0.77) for the same region (Post and Wiley 1977b). This low fledgling success is mainly due to Shiny Cowbird parasitism. Host egg puncturing and breakage by Shiny Cowbirds are the main factors in the decline in number of blackbird eggs hatched (0.37/nest in 1975; 0.25/nest in 1982).

The reproductive successes of parasitized and unparasitized Yellow-hooded Blackbirds did not differ significantly ($P > 0.5$, G Test of Independence, Table 1). Blackbird clutch sizes in parasitized and unparasitized nests were not significantly different ($P > 0.05$, d.f. = 1, $G = 0.34$, Table 2). Parasitized Yellow-hooded Blackbird nests in Trinidad contained an average of 1.49 cowbird eggs (N = 61 nests) and suspected cases of egg-puncturing and egg removal were few (Manolis 1982). That cowbirds rarely damage eggs in parasitized nests is supported by data on hatching success. Parasitized and unparasitized nests for which complete life histories through hatching were available did not differ in hatching success ($P > 0.05$, d.f. = 1, $G = 0.94$); 82.9 percent of 40 blackbird eggs in 16 parasitized nests hatched compared to 71.4 percent of 42 eggs in 16 unparasitized nests. Although blackbirds did not experience decreased hatching or nest success as a result of cowbird parasitism, unparasitized nests exhibited a slightly greater tendency than parasitized nests to fledge two or more blackbirds, although the difference is not statistically significant ($P > 0.05$, d.f. = 1, $G = 0.43$, Table 3).

NEST DEFENSE BY BLACKBIRDS

Yellow-hooded Blackbirds appear to suffer less from cowbird parasitism than Yellow-shouldered Blackbirds because of nest defense and different nest spacing and placement. Male Yellow-hooded Blackbirds were rarely absent from their territories and vigorously chased cowbirds from their nest areas. Male Yellow-hooded Blackbirds were present during 45 of 47 (96%) intrusions by cowbirds into blackbird territories that we observed, and chased the cowbirds on 36 (80%) of these occasions. In contrast, at one Yellow-shouldered Blackbird nest the male blackbird was present during only 46 of 85 (54%) cowbird intrusions and challenged cowbirds on only 13 (28%) occasions. During the pre-egg-laying and egg-laying periods, Yellow-shouldered Blackbird nests (N = 7) were left unguarded for up to 86 percent of the total observation time (12 hrs; Post 1981).

Male Yellow-hooded Blackbirds are particularly responsive to the presence of female cow-

TABLE 3

FLEDGLING PRODUCTION BY SUCCESSFUL YELLOW-HOODED BLACKBIRD PAIRS AT PARASITIZED
AND UNPARASITIZED NESTS, TRINIDAD, 1979–1981[1]

	Fledglings/nest			
	1	2	3	Mean ± s.d.
Parasitized nests (N = 11):				
Blackbirds only	8	2		1.2 ± 0.42
Cowbirds only	8			1.0 ± 0
Blackbirds and cowbirds	4	5	2	1.8 ± 0.75
Unparasitized nests (N = 13):				
Blackbirds only	9	3	1	1.4 ± 0.65
All nests (N = 24)				
Blackbirds and cowbirds	13	8	3	1.6 ± 0.72

[1] Differences in numbers of blackbirds fledged from parasitized and unparasitized nests not statistically significant ($P > 0.05$, G-test).

birds near nests. Eighteen male blackbirds on territories were significantly more aggressive and responded more quickly (Wilcoxon's signed rank test, $P < 0.05$, $t = 4.66$, N = 18) to models of female cowbirds than to control models (Bananaquit [*Coereba flaveola*] placed near nests). Yellow-hooded Blackbirds apparently recognized female cowbirds as a greater threat than male cowbirds; blackbirds chased female cowbirds from their territories on 32 of 39 (82%) occasions, and male cowbirds on 4 of 8 (50%) occasions ($P < 0.05$, d.f. = 1, $G = 5.8$). Yellow-shouldered Blackbirds were less aggressive toward cowbirds than toward other species close to or moderately far from (0–9.9 m) nests (significantly different overall, $P < 0.02$, $\chi^2 = 17.88$, Friedman two-way Anova; Table 4). The Shiny Cowbird apparently is not recognized by the Yellow-shouldered Blackbird as a potential threat, and, therefore, the blackbird does not attack it with the same intensity that it directs toward other species that it may recognize as competitors or nest predators. The data presented support the prediction that hosts (Yellow-hooded Blackbird) in areas where the cowbird is native will show stronger aggressive responses toward cowbirds close to their nests than will hosts (Yellow-shouldered Blackbird) in areas to which the cowbird has recently spread.

Placement of blackbird nests.—Both Yellow-hooded and Yellow-shouldered Blackbirds often breed in colonies, and colonial breeding can effectively deter cowbird parasitism (Wiley and Wiley 1980). Differences between the two species in mating systems and nesting habitats, however, influence the susceptibility of their nests to parasitism. Yellow-hooded Blackbirds are primarily polygynous (Wiley and Wiley 1980), whereas Yellow-shouldered Blackbirds are monogamous (Post 1981). Yellow-shouldered Blackbirds form pair bonds about two months before egg deposition. Birds pair at or near the nest site, and before mating, males follow and guard females only while they are on the nesting grounds. After laying the first egg, the male defends the nest site and the female. At this time, however, the male also guards the female when she leaves the nesting grounds, and during the two- to three-day egg deposition period, the nest is often unguarded. Yellow-shoulder pairs usually aggregate in loose colonies, and each pair defends a radius of 3 to 4 m about its nest.

Wiley and Wiley (1980) suspected that colonial nesting in the Yellow-hooded Blackbird helped deter parasitism, but lacked the data to test this. At one study site in 1980 [see Manolis (1982) for maps of nests at the sites], nests in isolated territories more than 40 m from the center of a nearby colony were more frequently parasitized ($P < 0.05$, d.f. = 1, $G = 4.7$) than were nests in three, discrete clumps of tall, dense sedge (*Cyperus articulatus*) less than 20 m from the colony's center. Nests in peripheral territories, between 20 m and 40 m from the colony's center, were parasitized at an intermediate level (Table 5). These data suggest that territory clumping reduces cowbird parasitism, most importantly by providing colonial males with a protective "shield" of surrounding territorial males, making it more difficult for female cowbirds to enter host nesting areas. The first or only nests in Yellow-hooded territories were more frequently parasitized than were subsequent nests ($P < 0.05$, d.f. = 1, $G = 5.14$). As territories, particularly those in colonies, filled with nests, cowbirds appeared to have an increasingly difficult time monitoring active nests and synchronizing egg-laying with host

TABLE 4

AGGRESSIVE RESPONSES OF YELLOW-SHOULDERED BLACKBIRDS TOWARD INTRUDERS NEAR THEIR NESTS[1]

Intruder	Distance of intruder from resident's nest (m)							
	0.0–0.4	0.5–0.9	1.0–1.9	2.0–2.9	3.0–4.9	5.0–9.9	10.0–19.9	>20
All species	2.9	2.9	2.5	2.3	1.9	0.9	0.7	0.1
Non-cowbird	3.3	3.2	2.7	2.7	2.2	1.0	0.8	0.1
Cowbird	2.5	2.1	1.0	1.4	1.3	1.0	0.1	0.0

[1] Roosevelt Roads Naval Station, Puerto Rico, 1979–1981. Values given are means of Aggressive Indices (see Methods).

clutches. This temporal decline in parasitism probably did not result from any significant changes in the number of laying cowbirds as much as it may have reflected the seasonal accumulation of old, unused, or unsuccessful nests, which made it increasingly difficult for female cowbirds to find and keep track of new nests.

Yellow-shouldered Blackbirds mainly nest arboreally, frequently in mid-seral stages of mangroves, whereas Yellow-hooded Blackbirds usually nest in low, dense stands of emergent marsh plants. In the former habitat, cowbirds can monitor host nesting activities while perched quietly and inconspicuously in the trees. Cowbirds can also use tree foliage as a screen to conceal their approach to host nests. Conversely, in wide expanses of dense marsh vegetation, cowbirds can monitor host nests only by slowly flying over, or dropping directly to, the nests. These latter approaches render the cowbird vulnerable to attack by male blackbirds; thus, Yellow-hooded Blackbird nests at the center of large marsh colonies, far from trees and surrounded by other territorial male blackbirds, are particularly difficult for cowbirds to parasitize.

EFFECTS OF PARASITISM ON PUERTO RICAN AVIAN COMMUNITIES

In the Puerto Rican mangrove study areas, 61 percent (11/18) of the resident passerine species were parasitized (Table 6) with cowbird eggs or nestlings. Some species were parasitized only occasionally (≤20% of nests examined), and others were regularly parasitized (52–100% of nests). The proportion of passerines parasitized in Puerto Rico is much greater than in areas where the cowbird is endemic. Only 15 percent (3/20) of the breeding passerines on our Trinidad Study sites were parasitized. In Argentina, Gochfeld (1979b) found 21.4 percent (6/28) and Fraga (1985) 40 percent (14/35) of the passerines in their respective study areas parasitized. In the llanos of Venezuela, 22.9 percent (8/35) of the resident passerines were parasitized (Cruz and Andrews, unpubl. data).

Host nest success (proportion of nests hatching at least one host young) was generally greater at non-parasitized nests than at parasitized nests (Table 7); nest success at non-parasitized nests (9 species) averaged 41 percent above that at parasitized nests. Examined collectively, nest success of parasitized pairs was less than that of non-parasitized pairs for 11 species in both study areas ($P < 0.001$, $Max D = 0.155$, Kolmogorov-Smirnov test, N = 674). The single parasitized nests of the Red-legged Thrush (*Turdus plumbeus*), Bronze Mannikin (*Lonchura cucullata*), and Gray Kingbird (*Tyrannus dominicensis*) all failed, whereas the only parasitized

TABLE 5

INCIDENCE OF PARASITISM, NEST SUCCESS, AND FLEDGLING SUCCESS OF YELLOW-HOODED BLACKBIRD NESTS IN COLONIAL AND ISOLATED TERRITORIES[1]

Nest dispersion	Parasitized nests	Successful nests[2]	Fledging success[3]	
			Blackbirds	Cowbirds
Isolated territories	76 (16/21)	16.7 (3/18)	0.67 (2/3)	1.0 (3/3)
Colonial territories (peripheral)	50 (13/26)	13.7 (3/22)	1.0 (3/3)	—
Colonial territories (central)	24 (5/21)	10.0 (2/20)	2.5 (5/2)	0.5 (1/2)

[1] Caroni Swamp, Trinidad, 1980. See Manolis (1982) for detailed maps of nest dispersion. Values given = percents (sample/total).
[2] Nests fledging at least one blackbird or cowbird.
[3] Number of fledglings/successful nest.

TABLE 6

PREVALENCE OF BROOD PARASITISM IN TWO MANGROVE STUDY SITES[1]

Host species[2]	No. nests examined	No. nests parasitized	Percent parasitized
Puerto Rican Flycatcher	25	13	52.0
Gray Kingbird	92	1	1.1
Red-legged Thrush	27	1	3.7
Caribbean Martin	5	0	0
Northern Mockingbird	52	1	2.0
Pearly-eyed Thrasher	63	0	0
Black-whiskered Vireo	15	13	86.7
Yellow Warbler	144	92	63.9
Bananaquit	106	0	0
Stripe-headed Tanager	19	0	0
Black-faced Grassquit	30	0	0
Puerto Rican Bullfinch	0	0	0
Greater Antillean Grackle	232	23	9.9
Black-cowled Oriole	14	14	100.0
Yellow-shouldered Blackbird	208	186	94.2
Troupial	5	5	100.0
Bronze Mannikin	10	2	20.0
Nutmeg Mannikin	62	0	0

[1] Study sites: Roosevelt Roads Naval Station, Eastern Puerto Rico, and Boquerón Forest, Southwestern Puerto Rico, 1975–1982.
[2] Scientific names of species not mentioned in the text are, Caribbean Martin (*Progne dominicensis*), Pearly-eyed Thrasher (*Margarops fuscatus*), Stripe-headed Tanager (*Spindalis zena*), Black-faced Grassquit (*Tiaris bicolor*), Puerto Rican Bullfinch (*Loxigilla portoricensis*), Nutmeg Mannikin (*Lonchura punctulata*).

Northern Mockingbird (*Mimus polyglottos*) nest parasitized fledged young (although the cowbird chick did not fledge).

Host clutches in parasitized nests averaged 12 percent smaller than those in non-parasitized nests (Table 8). The number of cowbird eggs per nest was greater than the number of host eggs for some species (Table 9). The mean number of cowbird eggs exceeded the mean number of host eggs at Yellow-shouldered Blackbird (21% more cowbird than host eggs) and Black-cowled Oriole (*Icterus dominicensis*; 89%) nests. In contrast, host eggs outnumbered cowbird eggs at Greater Antillean Grackle (*Quiscalus niger*), Troupial, (*Icterus icterus*), and Yellow Warbler nests. For most parasitized species there was a negative relationship between the number of cowbird eggs deposited in nests and the number of host eggs laid. Regularly-

TABLE 7

NEST SUCCESS AND DEGREE OF SHINY COWBIRD PARASITISM AT NESTS OF ELEVEN HOST SPECIES[1]

Reproductive component	Host species[2]										
	YW	YsBb	GAG	Tr	Mb	GKb	PRF	BwV	BcO	BM	RIT
Total no. active nests	107	164	218	5	44	76	13	11	12	6	25
No. successful nests	44	63	138	4	25	56	9	7	10	4	12
(%)	(41)	(38)	(63)	(80)	(57)	(74)	(69)	(64)	(83)	(67)	(48)
No. nests parasitized	81	152	23	5	1	1	11	9	12	1	1
(%)	(76)	(93)	(11)	(100)	(2)	(1)	(85)	(82)	(100)	(17)	(4)
Mean nest success[3]	0.4	0.4	0.6	0.8	0.6	0.7	0.7	0.6	0.8	0.7	0.5
Mean nest success of parasitized nests	0.4	0.4	0.5	0.8	1.0	0.0	0.6	0.6	0.8	0.0	0.0
Mean nest success of nests not parasitized	0.3	0.2	0.7		0.6	0.8	1.0	1.0		0.8	0.5

[1] Roosevelt Roads Naval Station and Boquerón Forest, Puerto Rico, 1975–1981.
[2] Common names: YW = Yellow Warbler, YsBb = Yellow-shouldered Blackbird, GAG = Greater Antillean Grackle, Tr = Troupial, Mb = Northern Mockingbird, GKb = Gray Kingbird, PRF = Puerto Rican Flycatcher, BwV = Black-whiskered Vireo, BcO = Black-cowled Oriole, BM = Bronze Mannikin, RIT = Red-legged Thrush.
[3] Proportion of nests fledging at least one host young.

TABLE 8

CLUTCH SIZES OF SHINY COWBIRD HOST SPECIES AT PARASITIZED AND NON-PARASITIZED NESTS[1]

| Host species | Parasitized[2] | | Non-parasitized[2] | | |
	N	X̄ ± s.d.	N	X̄ ± s.d.	Significance[3]
Puerto Rican Flycatcher	11	2.9 ± 0.70	2	4.5 ± 0.71	$P < 0.0025$
Red-legged Thrush	1	3.0	24	3.0 ± 0.51	n.s.
Northern Mockingbird	1	3.0	12	2.9 ± 0.29	n.s.
Black-whiskered Vireo	9	2.2 ± 0.82	2	3.0 ± 0.0	$P < 0.001$
Yellow Warbler	54	2.1 ± 0.83	15	2.6 ± 0.49	$P < 0.0005$
Greater Antillean Grackle	23	3.1 ± 0.56	195	3.2 ± 0.82	n.s.
Black-cowled Oriole	14	2.1 ± 1.07	0		
Yellow-shouldered Blackbird	93	2.6 ± 0.95	10	2.7 ± 0.35	n.s.
Troupial	5	2.4 ± 0.98	0		
Bronze Mannikin	1	5.0	5	5.4 ± 0.55	n.s.

[1] Roosevelt Roads Naval Station and Boquerón Forest, Puerto Rico, 1975–1981.
[2] N = number of clutches; X̄ = mean clutch size.
[3] Mann-Whitney two-sample test; 1-tailed.

parasitized species (Puerto Rican Flycatcher [*Myiarchus antillarum*], Black-whiskered Vireo [*Vireo altiloquus*], Yellow-shouldered Blackbird, and Yellow Warbler) experienced an average reduction in clutch size at parasitized nests that ranged from 13 percent (at nests where only 1 cowbird egg was added) to 74 percent at nests that had the maximum number of cowbird eggs (8 at 3 *Agelaius* nests). On the average, parasitized pairs fledged 67 percent fewer host chicks than non-parasitized pairs (0.5 host chicks/nest vs. 1.5 chicks/nest at non-parasitized nests for 5 species; Table 10).

ANTI-PARASITE BEHAVIOR OF NESTING BIRDS

Although naive because of the absence of prior contract with brood parasites, some bird species in Puerto Rico have preadaptations that are effective in countering cowbird parasitism. These include (1) nest concealment, (2) egg rejection, and (3) nest attendance and aggressiveness.

Concealment.—Concealing nests from predators is a preadaptation that reduces incidence of brood parasitism. Wiley (1982) analyzed the nest site characteristics of Yellow Warblers and Yellow-shouldered Blackbirds in Puerto Rican mangrove habitats. Habitat features that better concealed nests from predators improved nest success. Some warblers apparently were able to avoid parasitism at sites where certain vegetation components obscured their nests (e.g., nests placed low in small trees within dense stands). Yellow-shouldered Blackbirds nesting in dense shrubs were parasitized to a lesser degree than those nesting in more open sites. Because non-parasitized pairs typically fledged more host chicks than parasitized pairs, selection of sites that reduce the chance of brood parasitism should be strongly favored by selection, particularly in populations where a high proportion of the individuals are parasitized.

TABLE 9

NUMBERS OF HOST AND SHINY COWBIRD EGGS IN PARASITIZED NESTS[1]

| Host species | Mean no. eggs/parasitized nest | | | |
	N[2]	Host	Cowbird	Significance[3]
Puerto Rican Flycatcher	11	2.91	3.18	$P > 0.05$
Black-whiskered Vireo	9	2.22	2.56	$P > 0.05$
Yellow Warbler	54	2.06	1.89	$P = 0.05$
Greater Antillean Grackle	23	3.09	1.13	$P = 0.025$
Black-cowled Oriole	14	2.07	3.92	$P < 0.0025$
Yellow-shouldered Blackbird	93	2.62	3.08	$P = 0.0025$
Troupial	5	2.40	1.60	$P > 0.05$

[1] Roosevelt Roads Naval Station and Boquerón Forest, Puerto Rico, 1975–1981.
[2] N = number of nests.
[3] Mann-Whitney two-sample test; 1-tailed.

TABLE 10

HOST CHICKS FLEDGED FROM PARASITIZED AND NON-PARASITIZED NESTS[1]

	Mean number of host chicks fledged per nest				
	Parasitized		Non-parasitized		
Species	N[2]	Mean ± s.d.	N[2]	Mean ± s.d.	Significance[3]
Puerto Rican Flycatcher	11	0.64 ± 1.29	2	3.00 ± 0.0	P < 0.05
Black-whiskered Vireo	9	0.33 ± 0.05	2	2.00 ± 0.0	P = 0.025
Yellow Warbler	81	0.14 ± 0.47	26	0.58 ± 0.95	P < 0.05
Greater Antillean Grackle	22	1.36 ± 1.29	195	1.48 ± 1.29	n.s.
Black-cowled Oriole	12	0.33 ± 0.49	0		
Yellow-shouldered Blackbird	152	0.25 ± 0.65	12	0.33 ± 0.78	n.s.
Troupial	5	0.80 ± 0.84	0		

[1] Roosevelt Roads Naval Station and Boquerón Forest, Puerto Rico, 1975–1981.
[2] N = number of nests.
[3] Mann-Whitney two-sample test; 1-tailed. H_0: number of host chicks fledged from parasitized nests = number fledged from non-parasitized nests.

Egg rejection.—Rothstein (1975a) suggested that nest desertion and egg burial may not be anti-parasite adaptions, but by-products of standard avian behavior patterns. However, he considered egg rejection to be interpreted most reasonably as an evolved anti-parasite defense. Presumably, in species that have not been exposed to brood parasitism, birds have not been subject to selection to enhance discrimination between their own and parasite eggs (Hamilton and Orians 1965). Therefore, ejection behavior would not be expected in a population recently exposed to parasitism (e.g., on Puerto Rico). Contrary to this prediction, we found that several species did exhibit rejection responses (Table 11). Regularly parasitized species were characteristically acceptors, but those species for which a low proportion of nests were parasitized were rejectors.

Judging from locations of recorded sightings of cowbirds in Puerto Rico, population trends of one of its major hosts (Yellow-shouldered Blackbird), and cowbird invasion patterns on other islands, Post and Wiley (1975b) estimated that the Shiny Cowbird arrived in Puerto Rico in the 1940's and 1950's. Rothstein (1975b) calculated that populations parasitized by Brown-headed Cowbirds would require from 20 to 100 years to go from 80 percent acceptance to 80 percent rejection of cowbird eggs. Presumably, there must also be a period when the rejection rate is zero before the trait makes its initial appearance in the parasitized populations.

Even though populations of nesting birds in Puerto Rico may meet the minimum number of years of exposure (30+) to cowbirds required to achieve the high rejection rates we observed, other circumstances need to be considered. Each of the several species that do exhibit rejection behavior are presently experiencing low, or no parasitism. This may be related to the species' rejection behavior, but a number of other characteristics of these birds reduce the chances of their nests being parasitized. For example, some species have diets different from that of the cowbird, making them poor choices of hosts. Some are extremely aggressive at their nests or exhibit very high attendance rates making it less likely that cowbirds can penetrate their defenses to lay eggs in the nests. Therefore, selection on the host as a result of parasite-mediated losses may not be very great. Evolution of a rejection trait within an irregularly-parasitized population may take considerably longer than in a host population experiencing extreme selective pressures (Wiley 1982).

It is possible that a low level of intraspecific nest parasitism (egg dumping) has always existed and that selection may have favored those inidividuals that reacted to a new or different egg added to their clutches (J. M. Wunderle, pers. comm.).

Nest defense—Nest attentiveness and aggressiveness toward predators are preadaptations for avoidance of parasitism. It is adaptive to chase certain egg or chick predators from one's nest and a non-discriminatory, aggressive individual may defend its territory against all intruders, including cowbirds. Such indiscriminant chasing by nesting birds may reduce cowbird visits to nests and, thereby reduce parasitism of those nests (Wiley 1982). In Puerto Rico we found that most aggressive species or individuals had the lowest rates of parasitism. Even among species that were regularly or occasionally parasitized, there was a moderate negative association of degree of aggression and the rate of parasitism. Whereas some non-parasitized

TABLE 11

RESPONSES OF SPECIES OF NESTING BIRDS TO EXPERIMENTAL PARASITISM OF NESTS[1]

Species	No. parasitized nests	Nests with rejection[2]		Natural parasitism, % of nests
		No.	%	
Puerto Rican Flycatcher	10	0	0	52.0
Gray Kingbird	21	18	85.7	1.1
Red-legged Thrush	3	3	100.0	3.7
Northern Mockingbird	9	7	77.7	2.0
Pearly-eyed Thrasher	21	17	81.0	0
Black-whiskered Vireo	3	1	33.3	86.7
Yellow Warbler	20	0	0	63.9
Bananaquit	14	9[2]	64.3	0
Stripe-headed Tanager	2	2	100.0	0
Black-faced Grassquit	10	0	0	0
Puerto Rican Bullfinch	1	0	0	0
Greater Antillean Grackle	36	32	88.8	9.9
Black-cowled Oriole	7	0	0	100.0
Yellow-shouldered Blackbird	11	1	9.1	94.1
Nutmeg Mannikin	15	0	0	0

[1] Roosevelt Roads Naval Station and Boquerón Forest, Puerto Rico, 1975–1982.
[2] Rejection includes not only ejected or damaged eggs, but also nest desertions and build-overs.

species did discriminate between cowbirds and other species and were more hostile toward the parasite, some parasitized species showed less aggression toward cowbirds than toward other intruders.

THE SHINY COWBIRD AS A GENERALIZED PARASITE

Although it has been suggested (reviewed in Payne 1977) that the cowbird's generalist strategy of using many host populations to maintain parasite loads more easily, some cowbird populations have significantly reduced the reproductive success of certain host species [e.g., Brown-headed Cowbird on Kirtland's Warbler (*Dendroica kirtlandii*; Mayfield 1972; Shake and Mattsson 1975; Kelly and DeCapita 1982); Shiny Cowbird on Rufous-collared Sparrow in Argentina (King 1973; Fraga 1978)]. The Shiny Cowbird and Brown-headed Cowbird are non-specialized parasites (Mayfield 1972; Friedmann et al. 1977; pers. observ.) although within different parts of their ranges, they may heavily parasitize a few species, while only occasionally parasitizing others (Marchant 1960; Friedmann et al. 1977; Post and Wiley 1977b; Fraga 1978; Gochfeld 1979b; Mason 1980; Manolis 1982; Wiley 1982). Even if a primary host species becomes rare in a community, however, and cowbirds switch to other hosts, the remaining individuals of the primary host are still parasitized whenever they are encountered.

While it is tempting to think of generalist parasites as primitive, the use of a wide variety of hosts can be adaptive, allowing opportunistic exploitation of many species encountered (Gochfeld 1979b). This is demonstrated by the rapid range expansion and increase in abundance of both Brown-headed and Shiny Cowbirds in the past century (Friedmann 1963; Mayfield 1965; Post and Wiley 1977a). About 130 species have been successfully parasitized by the Brown-headed Cowbird and about 180 species by the Shiny Cowbird (Friedmann et al. 1977; Manolis 1982; Wiley 1982). As the Shiny Cowbird expands into new regions with new avifaunas, new hosts are continually added.

COWBIRD PARASITISM ON HISPANIOLA

The first Shiny Cowbird was collected on Hispaniola in October 1972 (Bond 1973). A. Dod (pers. comm.) found this species as far west as Santiago in north-central Dominican Republic. From 1976 to 1978 Arendt and Vargas (1984) observed cowbirds in 13 widely-scattered geographical areas in eastern and central Dominican Republic, including Saona Island (off southeastern Hispaniola). By the 1980's the Shiny Cowbird had become established in most lowland sites throughout the island. In the Dominican Republic, cowbirds are common in lowland disturbed habitats, usually in association with livestock and agriculture (rice and maize).

Our studies in the Dominican Republic began in 1974 just as the Shiny Cowbird arrived

TABLE 12

Shiny Cowbird Parasitism at the Nests of Five Host Species in the Dominican Republic[1]

	1974–1977			1982		
		Parasitized			Parasitized	
Host species	Total nests[2]	No.	%	Total nests[2]	No.	%
Palmchat	243 (24)	13	5.3	62 (6)	16	25.8
Black-whiskered Vireo	14	2	14.3	9	6	66.7
Yellow Warbler	19	2	10.5	12	10	83.3
Black-cowled Oriole	24	7	29.2	6	6	100
Village Weaver	936 (78)	12	1.3	134 (11)	21	15.7

[1] All data collected in areas of the Dominican Republic where the Shiny Cowbird is known to occur.
[2] Number of colonies sampled given in parentheses.

there. We have concentrated our investigation of cowbird hosts on several native species and one introduced bird, the Village Weaver (*Ploceus cucullatus*; Table 12). Since we began our studies, the mean incidence of parasitism for five host species has increased by 46.14 percent, ranging from 14.4 percent for the Village Weaver to 72.8 percent for the Yellow Warbler. In addition to finding eggs and chicks in these hosts' nests, we have seen Black-cowled Orioles, Black-whiskered Vireos, Palmchats (*Dulus dominicus*), and Village Weavers feeding fledgling cowbirds. On separate occasions juvenile cowbirds have been observed foraging among flocks of Village Weavers. In Puerto Rico, the Black-cowled Oriole, Yellow Warbler, and Black-whiskered Vireo are heavily parasitized by the Shiny Cowbird. The Village Weaver, introduced from Africa, and the Palmchat, an island endemic, are not present on Puerto Rico.

As our work progresses in the Dominican Republic, we will be able to learn more about the interactions between the cowbird and local host species. We anticipate the Tawny-shouldered Blackbird (*Agelaius humeralis*) of northwestern Haiti and Cuba, and the Rufous-collared Sparrow found in the montane regions of Hispaniola, will be parasitized when the cowbird encounters them. In Puerto Rico, Trinidad, and on the South American mainland, *Agelaius* blackbirds are a common host species. In South America the Rufous-collared Sparrow overlaps extensively with the Shiny Cowbird, and throughout their common range, it is an important host species (Friedmann 1963; King 1973; Fraga 1978).

ACKNOWLEDGMENTS

This work was supported by NSF Grant PRM-8112194 to the University of Colorado, A. Cruz, principal investigator; by the U. S. Fish and Wildlife Service Office of Endangered Species and Patuxent Widlife Research Center; the New York Zoological Society; University of Colorado Grant-in-Aid; the Institute of Tropical Forestry; Caribbean Islands National Wildlife Refuge; the Frank M. Chapman Fund of the American Museum of Natural History; the Alexander Bache Fund of the National Academy of Sciences, and by the Graduate Schools of the University of Colorado and the University of Miami.

For their assistance in the field and laboratory, we thank W. Arendt, A. Arendt, C. Belitsky, J. Cardona, J. Colón, S. Corbett, C. Delannoy, J. DiTomaso, S. Furniss, E. Litovich, A. Nethery, E. Santana, D. Smith, J. Taapken, R. Johnson, T. Nakamura, J. Wunderle, W. Post, E. C. Phoebus, A. Valido, J. Blankenship, B. Wiley, B. Ramdial, S. Ramdeen, G. Ramdeen, T. Keeler-Wolf, G. Keeler-Wolf, R. ffrench, and B. Mohan.

We received additional logistical support from the U.S Navy and from Sean Furniss, Caribbean Islands National Widlife Refuge Manager. We thank Mr. Furniss for providing assistance in the form of YACC and YCC summer employees and U.S. Fish and Wildlife Service personnel. We also thank the Department of Natural Resources, Commonwealth of Puerto Rico, for allowing us to undertake investigations in the Commonwealth Boquerón Forest. In Trinidad, we received assistance from the University of West Indies, the Forestry Division of the Trinidad and Tobago Ministry of Agriculture, the Simla Research Station, and the Caribbean Epidemiology Center.

LITERATURE CITED

Arendt, W., and T. Vargas. 1984. Range expansion by the Shiny Cowbird in the Dominican Republic. J. Field Ornithol. 55:104–107.

BENT, A. C. 1958. Life histories of North American blackbirds, orioles, tanagers, and allies. U.S. Natl. Mus. Bull. No. 211.

BIAGGI, V., JR. 1963. Record of the White Pelican and additional information on the Glossy Cowbird from Puerto Rico. Auk 80:198.

BOND, J. 1956. Check-list of birds of the West Indies. 4th ed. Academy Natural Sciences, Philadelphia, Pennsylvania.

BOND, J. 1966. Eleventh supplement to the Check-list of birds of the West Indies (1956). Academy Natural Sciences, Philadelphia, Pennsylvania.

BOND, J. 1971. Sixteenth supplement to the Check-list of birds of the West Indies (1956). Academy Natural Sciences, Philadelphia, Pennsylvania.

BOND, J. 1973. Eighteenth supplement to the Check-list of birds of the West Indies (1956). Academy Natural Sciences, Philadelphia, Pennsylvania.

BUCKLEY, P. A., AND F. G. BUCKLEY. 1970. Notes on the distribution of some Puerto Rican birds and the courtship behavior of White-tailed Tropicbirds. Condor 72:83–486.

DANFORTH, S. T. 1935. The birds of Saint Lucia. Monogr. Univ. Puerto Rico, Phys. Biol. Sci., Ser. B. No. 3.

ELLIOTT, P. F. 1978. Cowbird parasitism in the Kansas tallgrass prairie. Auk 95:161–167.

FRAGA, R. M. 1978. The Rufous-collared Sparrow as host of the Shiny Cowbird. Wilson Bull. 90:271–284.

FRAGA, R. M. 1985. Host-parasite interactions between Chalk-browed Mockingbirds and Shiny Cowbirds. Pp. 829–844, In P. A. Buckley, M. S. Foster, E. S. Morton, R. S. Ridgely, and F. G. Buckley (eds.), Neotropical Ornithology. Ornithol. Monogr. No. 36.

FRIEDMANN, H. 1929. The cowbirds. A study in the biology of social parasitism. C. C. Thomas, Springfield, Illinois.

FRIEDMANN, H. 1963. Host relations of the parasitic cowbirds. U.S. Natl. Mus. Bull. No. 233.

FRIEDMANN, H., L. F. KIFF, AND S. I. ROTHSTEIN. 1977. A further contribution to the knowledge of host relations of the parasitic cowbirds. Smithson. Contrib. Zool. No. 235.

GOCHFELD, M. 1979a. Begging by nestling Shiny Cowbirds—adaptive or maladaptive? Living Bird 17:40–50.

GOCHFELD, M. 1979b. Brood parasite and host coevolution: Interaction between Shiny Cowbirds and two species of meadowlark. Am. Nat. 113:855–870.

GRAYCE, R. L. 1957. Range extensions in Puerto Rico. Auk 74:106.

HAMILTON, W. J., III, AND G. H. ORIANS. 1965. Evolution of brood parasitism in altricial birds. Condor 67:361–382.

HOY, G., AND J. OTTOW. 1964. Biological and oological studies of the molothrine cowbirds (Icteridae) of Argentina. Auk 82:186–203.

HUDSON, W. H. 1920. Birds of La Plata. E. P. Dutton, New York.

KELLY, S. T., AND M. E. DECAPITA. 1982. Cowbird control and its effect on Kirtland's Warbler reproductive success. Wilson Bull. 94:363–365.

KEPLER, C. B., AND A. K. KEPLER. 1970. Preliminary comparison of bird species diversity and density in Luquillo and Guanica Forests, Pp. E183–186, In H. Odum (ed.), A Tropical Rain Forest. U.S. Atomic Energy Comm., Oak Ridge, Tennessee.

KING, J. R. 1973. Reproductive relationships of the Rufous-collared Sparrow and the Shiny Cowbird. Auk 90:19–34.

LINZ, G. M., AND S. B. BOLIN. 1982. Incidence of Brown-headed Cowbird parasitism on Red-winged Blackbirds. Wilson Bull. 94:93–95.

MANOLIS, T. D. 1982. Host relationships and reproductive strategies of the Shiny Cowbird in Trinidad and Tobago. Unpubl. Ph.D. dissert., University of Colorado, Boulder.

MARCHANT, S. 1960. The breeding of some southwestern Ecuadorian birds. Part 2. Ibis 102:584–599.

MASON, P. 1980. Ecological and evolutionary aspects of host selection in cowbirds. Unpubl. Ph.D. dissert., University of Texas, Austin.

MAYFIELD, H. 1965. The Brown-headed Cowbird with old and new hosts. Living Bird 4:13–28.

MAYFIELD, H. 1972. Third decennial census of Kirtland's Warbler. Auk 89:263–268.

MAYFIELD, H. 1977a. Brown-headed Cowbirds: agent of extermination? Am. Birds 31:107–113.

MAYFIELD, H. 1977b. Brood parasitism—reducing interactions between Kirtland's Warblers and Brown-headed Cowbirds. Pp. 89–91, In S. A. Temple (ed.), Endangered Birds. University of Wisconsin Press, Madison, Wisconsin.

OLSON, S. L. 1978. A paleontological perspective of West Indian birds and mammals. Pp. 97–117, In F. B. Gill (ed.), Zoogeography in the Caribbean. Acad. Nat. Sci. Phila., Spec. Publ. 12.

PAYNE, R. B. 1976. The clutch size and numbers of eggs of Brown-headed Cowbirds: effects of latitude and breeding season. Condor 78:337–342.

PAYNE, R. B. 1977. The ecology of brood parasitism in birds. Annu. Rev. Ecol. Syst. 8:1–28.

PINCHON, R. 1963. Faune des Antilles Francaises. Les Oiseaux. Fort-de-France, Martinique.

POST, W. 1981. Biology of the Yellow-shouldered Blackbird Agelaius xanthomus on a tropical island. Bull. Fla. State Mus., Biol. Sci. 26:125–202.

POST, W., AND J. W. WILEY. 1976. The Yellow-shouldered Blackbird—present and future. Am. Birds. 30:12–20.

POST, W., AND J. W. WILEY. 1977a. The Shiny Cowbird in the West Indies. Condor 79:119–121.

POST, W., AND J. W. WILEY. 1977b. Reproductive interactions of the Shiny Cowbird and the Yellow-shouldered Blackbird. Condor 79:176–184.

RICKLEFS, R. E., AND G. W. COX. 1972. Taxon cycles in the West Indian avifauna. Am. Nat. 106:195–219.

ROBERTSON, W. B., JR. 1962. Observations on the birds of St. John, Virgin Islands. Auk 79:44–76.

ROBERTSON, R. J., AND R. F. NORMAN. 1977. The function and evolution of aggressive behavior towards the Brown-headed Cowbird (*Molothrus ater*). Can. J. Zool. 35:508–515.

ROTHSTEIN, S. I. 1971. Observations and experiments in the analysis of interactions between brood parasites and their hosts. Am. Nat. 105:71–74.

ROTHSTEIN, S. I. 1975a. An experimental and teleonomic investigation of avian brood parasitism. Condor 77:250–271.

ROTHSTEIN, S. I. 1975b. Evolutionary rates and host defenses against avian brood parasitism. Am. Nat. 190:161–176.

SELANDER, R. K. 1964. Behavior of captive South American cowbirds. Auk 81:394–402.

SHAKE, W. F., AND J. P. MATTSSON. 1975. Three years of cowbird control: an effort to save the Kirtland's Warbler. Jack-Pine Warbler 53:48–53.

SICK, H. 1958. Notas biologicas sobre o Guaderio, "*Molothrus bonariensis*" (Gmelin) (Icteridae, Aves). Rev. Bras. Biol. 18:417–431.

SICK, H., AND J. OTTOW. 1958. Vom brasilianischen kuhvogel, *Molothrus bonariensis,* und seinen Wirten, besonders dem Amerfinken, *Zonotrichia capensis.* Bonn Zool. Beitr. 1/9:40–62.

WILEY, J. W. 1982. Ecology of avian brood parasitism at an early interfacing of host and parasite populations. Unpubl. Ph.D. dissert., University of Miami, Coral Gables, Florida.

WILEY, R. H., AND M. S. WILEY. 1980. Spacing and timing in the nesting ecology of a tropical blackbird; comparison of populations in different environments. Ecol. Monogr. 50:153–178.

YOUNG, H. F. 1963. Breeding success of the cowbird. Wilson Bull. 75:115–121.

ECOLOGICAL CONSTRAINTS ON WEST INDIAN BIRD DISTRIBUTIONS

JOHN FAABORG

Division of Biological Sciences, University of Missouri–Columbia, Columbia, Missouri 65211 USA

ABSTRACT. In an attempt to understand ecological constraints on bird species distributions, the only major neotropical island system, the West Indies were studied. Discussed among previously-described patterns of structure from a 12 island study are: incidence regressions for guild membership on islands, using four simple food/foraging guilds (frugivores, gleaning insectivores, flycatching insectivores, and nectarivores); saturation curves for total species and guild membership within habitats on large islands; and size differences between coexisting members of insectivorous and nectarivorous guilds in all island habitats and between coexisting frugivores on small and medium islands. The existence of size shifts and other responses have shown these to be real patterns and not the result of either redundancy in species lists or chance colonizations.

Here this approach is expanded where possible to include 26 West Indian islands (excluding the Bahamas and extralimital islands) using published species distributions. The correlation between guild membership and total number of land-bird species on an island holds for frugivores and insectivores throughout the West Indies. Patterns of guild membership saturation and size differences between coexisting guild members (except large island frugivores) seem to apply to all islands but Jamaica and Grand Cayman; on these islands some habitats appear to be supersaturated (by West Indian standards) and guild members of unusually similar size coexist. The unusual patterns of structure on Jamaica may reflect resource differences from other islands, its proximity to source areas, or both. The variations found on Cuba and Grand Cayman may be the results of climatic constraints on the resources of those islands, which are regularly exposed to temperate weather patterns.

The distributional patterns of West Indian bird species are next discussed in the context of these apparent ecological constraints. The existence of ecological counterparts on different islands, habitat or size shifts due to varying competition, and other such patterns are noted. Situations where the ecological rules seem to be strained by present bird distributions are also indicated. In many cases the interactions between the ecological variation of a species and its taxonomic distinctions are readily apparent; further analyses at the subspecies level would yield further examples of this interaction.

In conclusion, West Indian patterns of bird community structure are compared with those on eight neotropical land-bridge islands. The regressions for guild membership in relation to total land bird species in the West Indies very closely describe the same relationships for insectivorous and frugivorous guilds on land-bridge islands. Only for nectarivores does the West Indian pattern differ greatly from the land-bridge island pattern. Habitat saturation patterns differ, and size separations within guilds rarely occur on the land-bridge islands, probably because of species-area relationships very different from those of the West Indies. Nevertheless, the convergence of at least some of the structural patterns on all neotropical islands suggests that someday the constraints of community assembly for communities as large as the 200 species of Trinidad may be understood. With this knowledge, it should be easier to understand the ecological constraints at work in the diverse tropical mainland avifaunas.

RESUMEN. En un intento por entender los constreñimientos ecológicos en la distribución de aves se estudió el único gran sistema insular neotropical—las Antillas. A lo largo de estudios realizados previamente fueron discutidos patrones de estructuras de 12 islas, tales como: las correlaciones de incidencia de miembros de grupos ecológicos ("incidence regressions for guild membership") en islas usando cuatro grupos simples de comida/forrajeo (frugívoros, espigadores ("gleaning") de insectos, atrapa-moscas, y nectarívoros); curvas de saturación para el total de especies y socios de grupos ecológicos ("guild membership") en habitats de islas grandes; y diferencias de tamaño entre miembros coexistentes de grupos ecológicos de insectívoros y nectarívoros en todos los habitats de islas y entre frugívoros coexistentes en islas pequeñas y medianas. La existencia de cambios de medidas u otras respuestas han mostrado que estos son verdaderos patrones y no el resultado de ya sea redundancia en la lista de especies o colonización al azar.

Este intento es extendido, cuando es posible, hasta incluír 26 islas antillanas (excluyendo Bahamas e islas extralimítrofes) usando distribuciones de especies publicadas. La correlación entre los socios de grupos ecológicos y el número total de especies de aves terrestres en una isla se mantiene para frugívoros e insectívoros, a lo largo de todas las Antillas. Parecen aplicarse patrones de saturación de socios de grupos ecológicos y diferencias en tamaños entre miembros coexistentes del mismo grupo (excepto para grandes fugívoros insulares) para todas las islas con excepción de Jamaica y Gran Cayman; en esas islas algunos habitats parecen estar supersaturados (para lo normal en las Antillas) y coexisten miembros de un grupo ecológico de medidas inusualmente similares. Los inusuales patrones de estructura en Jamaica pueden reflejar diferencias de recursos con respecto a otras islas, su proximidad a áreas de origen, o ambos. Las variaciones encontradas en Cuba y Gran Cayman, pueden ser los resultados de constreñimientos climáticos en los recursos de esas islas, que están expuestos regularmente a patrones de clima templado.

A continuación se discuten los patrones de distribución de las especies de aves de las Indias Occidentales en el contexto de esos constreñimientos ecológicos. Se describe la existencia de contrapartes ecológicos en diferentes islas, habitat o cambio de tamaño debido a competencia variable y otros patrones similares. También se indican las situaciones donde las reglas ecológicas parecen haber sido presionadas por las distribuciones actuales de aves. En muchos casos las interacciones entre variacion ecológica de una especie y sus distribuciones taxonómicas son fácilmente aparentes; más análisis al nivel de subespecie producirá ejemplos de esta interacción.

En conclusión se comparan los patrones de estructura de la comunidad de aves de las Indias Occidentales con aquellos en ocho islas neotropicales unidas al continente ("land bridge islands"). Las regresiones para socios de grupos ecológicos en relación al total de especies de aves terrestres de las Indias Occidentales describe muy cercanamente las mismas relaciones para socios insectívoros y frugívoros en islas unidas al continente. El patrón de las Indias Occidentales sólo difiere notablemente del de las islas unidas al continente en el caso de los nectarivoros. Los patrones de saturación de habitats difieren y las separaciones entre socios se puede presentar raramente en las islas unidas al continente, probablemente porque la relación especie-área es muy diferente de aquellas en las Indias Occidentales. Sim embargo la convergencia de al menos algunos de los patrones estructurales de todas las islas neotropicales sugiere que algún día podrán entenderse los constreñimientos del ensamble de la comunidad para comunidades tan grandes como la de las 200 especies de Trinidad. Con este conocimiento deberá ser fácil más entender los constreñimientos ecológicos que trabajan en las diversas avifaunas tropicales continentales.

Evolutionary ecologists often attempt to determine how and why various distributions of species occur. Quite clearly, this cannot be done until information is complete on the taxonomic relationships and distributional patterns of species in the region of concern. Thanks to people like Eugene Eisenmann, we have a good knowledge of the taxonomy and distribution of many neotropical birds, but, with nearly 3000 species, the step to understanding the ecological and evolutionary factors determining neotropical avian diversity is a large one.

In an attempt to answer questions about ecological controls while avoiding this sometimes overwhelming diversity, I have examined patterns of distribution of birds on the major neotropical island system, the West Indies. As with most islands, a generally clear relationship exists between the number of species each of these islands possesses and its area. My goal has been to understand the ecological constraints that, acting over time, have selected certain repeatable sets of birds on these islands. To date, I have focused on patterns in the ecological types of birds that may be found together. While these islands are perhaps exceptional in their simplicity, previous success in finding patterns of structure justifies such an approach. The extent to which West Indian patterns can be applied to land-bridge islands or mainland avifaunas will tell us about the existence of similar ecological controls in more diverse faunas.

This approach would not be possible without a basic knowledge of the taxonomy and distribution of West Indian birds, a knowledge provided by the life-long studies of James Bond. In this paper, I hope to explain the zoogeographic distributions he has presented by showing how each species' ecological traits fit a set of ecological rules structuring each community. The distributions and, to some extent, the taxonomic classifications should then make sense in ecological terms. I finish by comparing these West Indian patterns to bird distributions on neotropical land-bridge islands.

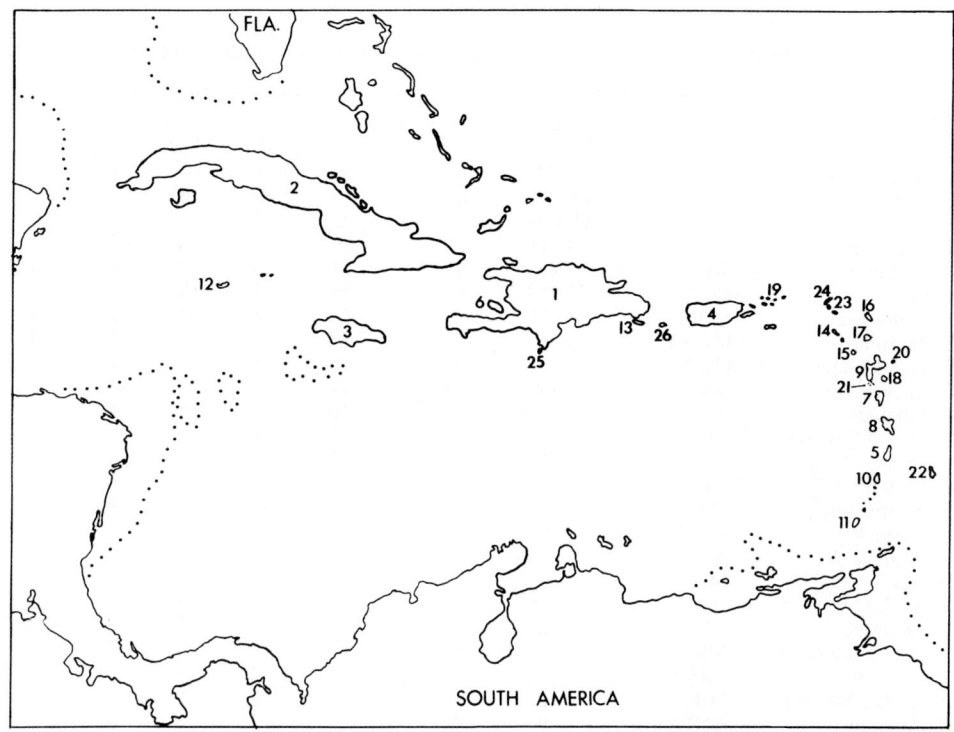

Fig. 1. Map of the West Indies. Islands included in this study are identified by numbers from Table 1. Dotted lines show approximate land contours in locations where Pleistocene ocean levels reduced island-mainland distances.

Even with the relative simplicity of the West Indian avifauna, some constraints must be put on the species to be considered. First, I limit my examination to the Greater and Lesser Antilles and exclude the Bahamas and various extralimital islands considered part of the West Indies by Bond (1971). I deal solely with native land bird species that breed on the islands, but exclude raptors. This approach is similar to the pigeons through passerines list used by Lack (1976) and in my previous studies. I will not discuss species that are only winter residents, although these are an important element of the West Indian avifauna (see Faaborg and Terborgh 1980; Terborgh and Faaborg 1980a). I count most recent extinctions, but exclude introduced species (most of which live in disturbed habitats). While our knowledge of fossil West Indian birds has increased dramatically in recent years (Olson 1978), these are beyond the scope of this paper. I follow the 6th edition of the American Ornithologists' Union's *Check-list of North American Birds* (1983), although conflicts between it and the taxonomy of Bond (1956, 1971) are sometimes noted.

THE ANTILLEAN ENVIRONMENT

Analyses of island biogeography have tended to focus on three criteria of the islands involved, area, location (particularly the degree of isolation from other island and/or mainland sources of colonists), and the degree of habitat complexity of each island. Hamilton et al. (1964) suggested that 93 percent of the variation in the numbers of West Indian bird species per island was explained by island area, but since area is also often correlated with topographic variation and, thus, habitat variation, the effect of area and habitat are difficult to separate. Within the Antilles, the role of isolation on total species per island is generally small, although sometimes important.

The Antilles consist of two arcs of islands of variable size (Fig. 1, Table 1), at different distances from mainland source areas, and composed of varying mosaics of habitat types. The generally small Lesser Antilles run north and south with the southernmost island, Grenada,

TABLE 1

CHARACTERISTICS OF ANTILLEAN ISLANDS, INCLUDING SUMMARIES OF GUILD COMPOSITION

				Guild				
Island	Abbreviation	Area (km²)	Total species	F	FI	GI	N	M
1. Hispaniola		76,000	69	30	7	14	4	14
2. Cuba		115,000	65	23	6	18	3	15
3. Jamaica		11,400	63	26	10	13	5	9
4. Puerto Rico		8900	51	20	6	12	6	7
5. St. Lucia	StL	600	41	21	4	9	4	3
6. Gonave	GON	800	40	18	4	10	3	5
7. Dominica	DOM	790	39	19	4	7	5	4
8. Martinique	MART	1000	37	17	4	7	5	4
9. Guadeloupe	GUA	1510	35	15	4	7	4	5
10. St. Vincent	StV	350	35	17	4	7	4	3
11. Grenada		310	34	16	6	6	4	2
12. Grand Cayman	GC	185	27	11	4	8	1	3
13. Saona	SAO	110	24	9	2	7	3	3
14. St. Kitts	StK	176	23	10	3	4	4	2
15. Montserrat	MTS	98	23	11	2	5	4	1
16. Barbuda	BRBU	160	24	10	3	4	4	3
17. Antigua	ANT	280	20	10	2	3	4	1
18. Marie Galante	MG	155	20	10	2	4	3	1
19. St. Johns	StJ	50	18	8	2	4	3	1
20. Desirade	DES	21	17	8	2	3	3	1
21. Isles des Saintes	IdS	12	16	6	2	4	3	1
22. Barbados	BRBA	430	16	7	2	3	3	1
23. St. Bartholomew	StB	21	16	8	2	3	3	0
24. Anguilla	ANG	91	15	7	2	3	3	0
25. Beata		47	15	6	1	4	2	2
26. Mona		52	11	8	1	2	0	0

¹ Guild composition from Appendix I. Values given = numbers of species in each guild. F = frugivore, FI = flycatching insectivore, GI = gleaning insectivore, N = nectarivore, and M = miscellaneous species.

about 150 km from the South American mainland. These islands vary from low, coral and limestone islands to those with volcanic peaks nearing 1500 m. The greater Antilles consist of four large and several small islands running generally east and west. The large islands are all quite complex topographically, with mountains exceeding 3000 m on Hispaniola and 2000 m on Cuba and Jamaica. Jamaica is presently separated from Central America by about 600 km, whereas Cuba is about 200 km from mainland Florida and the Yucatan Peninsula.

Two major vegetation types cover virtually all of the small Antilles (including all the Lesser Antilles) and most of the larger islands. Seasonally dry forest (also called sclerophyll scrub, tropical deciduous forest, etc.) covers the lowlands of most islands and may cover all of low, flat islands. It is particularly common in rain shadow zones where it may extend to moderate elevations. This forest type seems highly consistent in composition throughout the Antilles and, in fact, throughout the Caribbean basin. Beard (1948) pointed out that 40 percent of the native tree species in this vegetation are of wide tropical distribution, another 28 percent range at least from Puerto Rico through the Lesser Antilles, and few endemics exist.

The second dominant forest type is rainforest, which is sometimes split into lowland and montane rainforest, but which I consider a single unit because few Antillean bird species are specialized to only one of these forest subtypes. Rainforest occurs in the lowland areas on the windward sides of most islands and extends, in shorter form, to the summits of most mountains, excluding the high elevations of Cuba and Hispaniola. The Lesser Antillean rainforest contains about 40 percent West Indian endemic species (Beard 1948) and only about 20 percent species with widespread distributions in the tropics. Yet, these endemics are widespread within the Antilles such that island-to-island variation in this forest type is small.

These two forest types dominate the Lesser Antilles and small islands of the Greater Antilles. Only on some of the large Greater Antilles are there tracts of mangrove, savannah, pine forest, or other vegetation large enough to be ecologically meaningful to the bird species under consideration. While these two forest types cover roughly equal areas today (barring human interference), this was not always the case. Pregill and Olson (1981) showed how Pleistocene climatic cycles changed island characteristics and the distributions of these habitats. Lower water levels exposed shallow banks, reducing the water gaps between some islands and connecting others. The arid conditions of this period probably pushed dry forests to higher elevations, increasing the amount of dry forest and adding large expanses of savannah on larger islands. Wet forests decreased in size and became more isolated from one another. Thus, in recent history these islands were larger and drier, a factor critical for understanding some of the bird patterns seen today.

Pleistocene climatic shifts also significantly affected the water gaps between West Indian islands and mainland source areas and, thus, today's patterns of distribution (Fig. 1). The gap between Grenada and South America was reduced to about 40 km. The Cuba-Florida gap was not greatly changed, but the Bahamas, greatly enlarged at this time, could have served as stepping stones between Florida and Hispaniola or Cuba. Low water levels extended Honduras to the east and exposed several islands that served as potential stepping stones to Jamaica.

One can consider these islands as a string of mosaics of different habitat types that birds could use for survival. While each habitat type seems highly similar in vegetative structure and composition from island to island, we know little about island variation in the actual resources available to birds. Seeds available in dry scrub on Mona Island Puerto Rico show a similar size range and mean (Faaborg and Kennedy, unpubl.), but no published measurements of inter-island variation in resources exist.

METHODS

My goal is to show ecological patterns of community structure that may explain the patterns of distribution of individual species for the Antilles. My actual field work to date has been confined to a subset of the Antilles from Hispaniola eastward and Dominica northward. I chose these islands because they are removed from mainland sources of avian colonists. If there are consistent ecological controls on species distributions, they should show best on these islands; the existence of fairly recent colonists whose fates are unresolved might obscure patterns on islands closer to source faunas. This subset was also chosen because of its range of island sizes and species densities, from Hispaniola (77,600 km^2; 69 species) to Mona (65 km^2; 11 species). Within this arc of islands, I conducted extensive studies in wet or dry forests at 16 locations on 12 islands. All other habitat types were at least surveyed on these islands, as well as on five others. While I later consider patterns that seem to hold for all West Indian islands, I am most certain about my findings on these 12 that I shall refer to as the "study islands." The descriptions of other scientists of islands I have not studied (e.g., Lack is [1976] of Jamaica), although helpful, are not often directly adaptable to my study approach.

My basic tool is a list of the species breeding in the various major habitat types of each island. Initially, my own species lists were made for each study site where I mist-netted for at least three days and made four to six days of associated observations. Censuses and netting were done within the most homogeneous tracts of native forest accessible and away from human disturbance when possible. Species that occur only in disturbed areas, second growth, or around homes are recorded as such and not as part of native forest communities if they are not found in the forest itself. It should be noted that many dry forest species also occur in second growth, even in rainforest. This observation was often critical for understanding ecological patterns. Because these are islands occupied by relatively few species, but at high densities, these censuses appear to be complete; nevertheless, the available literature was searched and any missed species added to the list. Each bird netted was weighed with a Pesola scale to the nearest 0.1 g, and its wing chord was measured to the nearest 0.5 mm. Weights for species not netted are from the literature or are estimates made from similar birds of similar wing length. Lists and observations for the other West Indian islands are from the literature.

To search meaningfully for patterns within the West Indian avifauna, I assigned the species to ecological sets of guilds. These are defined as groups of species with the same diets and/or foraging techniques that have the potential to affect each other's distribution. Most West Indian forest species readily fall into one of four guilds: frugivores (including fruit and seed eaters), gleaning insectivores, flycatching insectivores, and nectarivores. Guild designation was made on the basis of published food habits (e.g., Wetmore 1916) and observed foods and foraging behavior. Species that potentially belong to more than one guild were assigned to the guild they were likely to use during severe conditions such as drought. For example, I

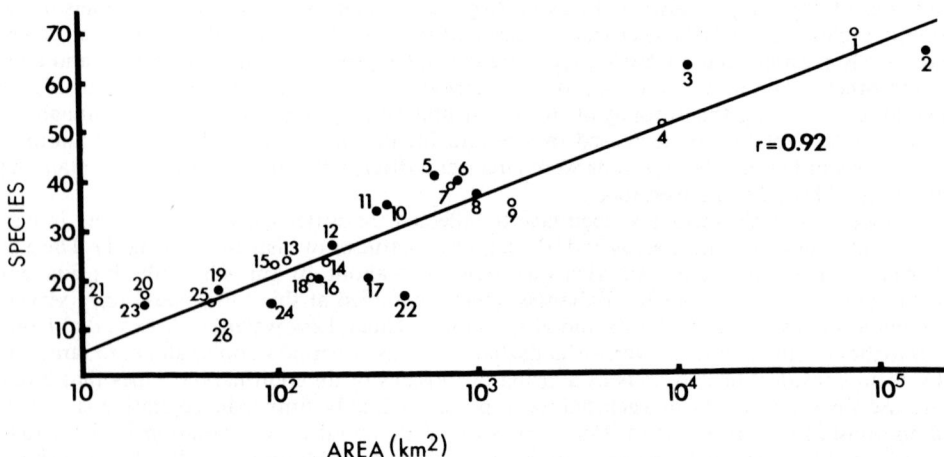

Fig. 2. Relationship of the number of land bird species to island area for the West Indian islands in this study. "Study islands" are shown by open circles, islands I have not visited, by solid circles. Numbers identifying the islands come from Table 1.

considered *Elaenia martinica* a flycatcher, because I saw it hawking insects during severe drought conditions in the summer of 1973, even though it regularly eats fruits when they are available. Diamond (1973) also found it primarily an insectivore on St. Lucia, but Johnston (1975) and Crowell (1968) considered it a frugivore. The forest bird species that did not fit into one of these four guilds included swifts (6 species) swallows (4 species), goatsuckers (7 species), woodpeckers (10 species), crows (4 species), and a single species of kingfisher. As mentioned earlier, I do not include owls (7 species) or hawks (10 species).

Basically, then, I am concerned with patterns in the membership of guilds in habitats on islands of different size. A clear species-area relationship exists for the West Indian Islands (Fig. 2); a similar relationship also exists between the number of species in a guild and island area. In both cases most of the remaining scatter is the result of isolation. For example, small islands close to larger islands (such as the Isles des Saintes near Guadeloupe) hold more total species than area alone would predict, while those more isolated than most (i.e., Barbados) hold fewer than area alone would predict. Working from the premise that these islands have an equilibrium species number that is a function of area and isolation and that this equilibrium reflects the survival of an ecologically coadapted set of species over time, I also have examined the number of species in a guild on an island as a function of total number of land bird species on that island. When a pattern appears, it says simply that as species number changes, guild membership changes in some orderly fashion. This reduces the need to deal with isolation apart from the ways it causes lower or higher species numbers than area alone suggests or apart from its affects on compositional patterns. The following figures relating species numbers by guild to total species numbers would show nearly identical patterns if guild species numbers were plotted against island areas.

PATTERNS ON NORTHEAST ANTILLEAN STUDY ISLANDS

When the species-area relationships for the West Indies are considered (Fig. 2), Mona Island, Desirade, the Isles des Saintes, and Dominica are the study islands most removed from the regression. Mona has few species than area would predict, perhaps because its low relief makes it difficult for colonists to find, perhaps because it is an arid limestone plateau with little soil, or perhaps because I do not include two species recorded once in the distant past but not seen since. The Saintes and Desirade have more species than area alone would predict because they are very close to the species-rich land of Guadeloupe. No obvious reason exists for the fact that Dominica has slightly more species than area suggests it should.

When plotting total species per guild against area, similar species-area curves appear, but tighter relationships are shown when the number of species in a guild is plotted against total

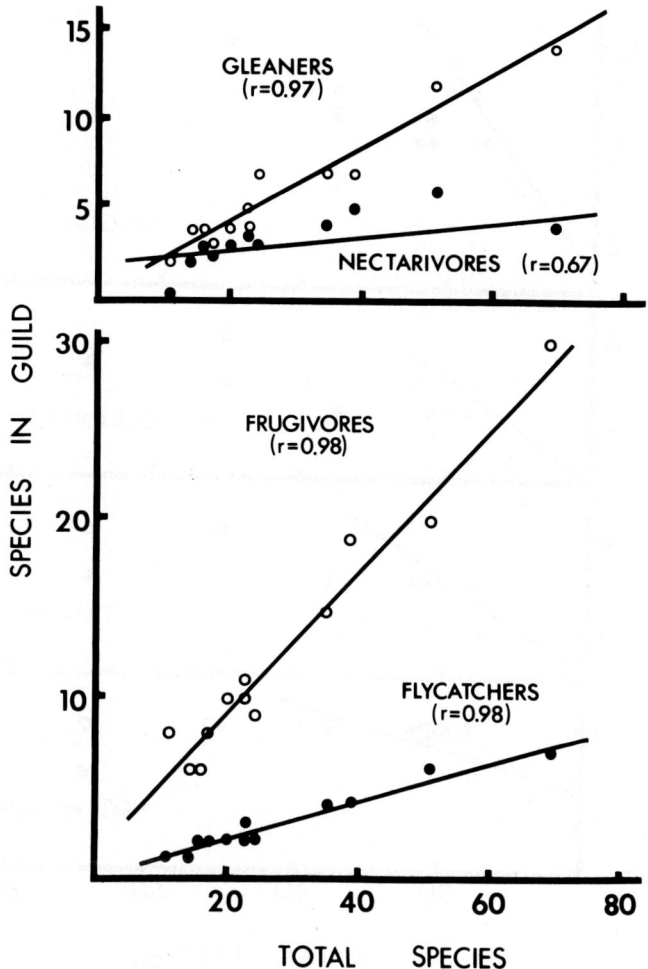

Fig. 3. The relationship of number of species within a guild to total number of species on an island for the four guilds on the 12 study islands.

species on an island (Fig. 3). This shows quite clearly that as species are added to an island, they are added in an orderly fashion with regard to trophic structure.

These patterns by themselves might suggest that large islands have 30 syntopic frugivores, 12 gleaners, etc. Certainly this many guild members coexist in mainland avifaunas. What we have discovered, however, is that larger islands with diverse species pools do not have complex sets of coexisting species but instead show simply structured communities whose compositions vary most between habitats (Terborgh and Faaborg 1980b; Faaborg 1982). We first saw that the number of species found in a habitat (alpha diversity) was constant on islands above about 30 species. Saturation levels were about 30 species for dry forest and 20 species for wet forest. (This characteristic runs counter to arguments correlating foliage diversity and bird species diversity and may have an historic explanation.) Not surprisingly, when one divides these species totals by guild, consistent patterns of composition appear (Fig. 4). A saturated community on these study islands contains 7 to 9 species of frugivores, 4 to 5 gleaning insectivores, 2 to 3 flycatchers, and 2 to 4 nectarivores, depending on the habitat involved. The difference between the total number of members of all guilds found in a particular habitat and the total number of species on the island reflects differences in the compositions of the communities of different habitats (beta diversity). Large islands support larger total guilds than do small

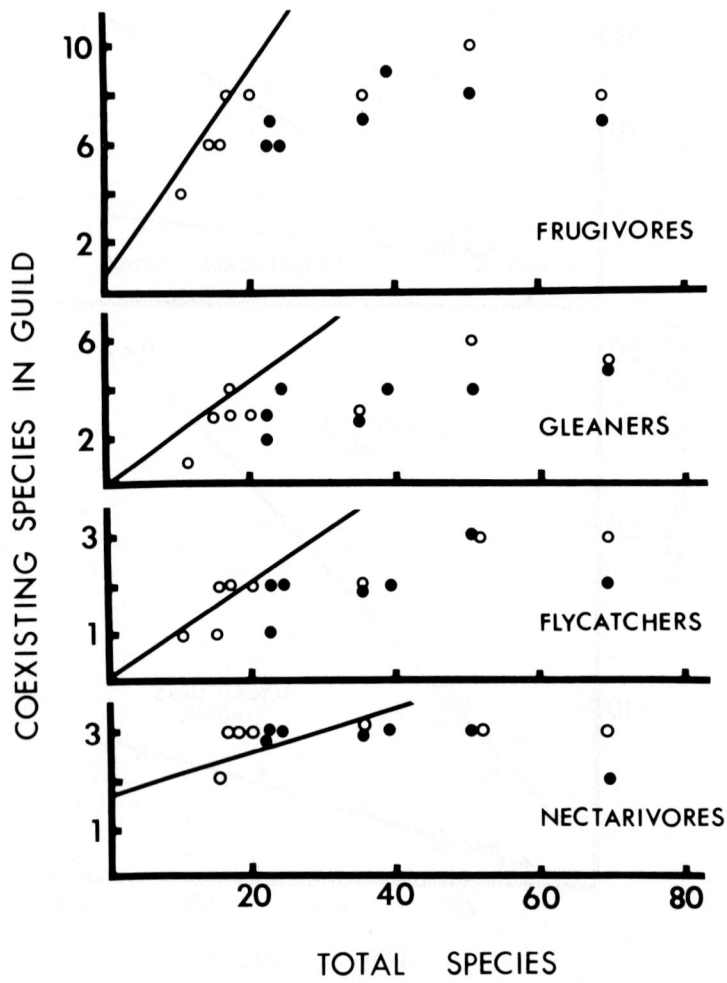

FIG. 4. The relationship of number of species coexisting within a guild to total number of species on an island for dry forest sites (open circles) and wet forest sites (solid circles) on the study islands. Straight lines = regressions from Figure 3 to emphasize the occurrence of habitat saturation on these islands.

islands both because species turnover between habitats is greater and because more habitat types are available.

Given this consistency in the number of guild members per habitats and per island, I looked for patterns of composition within guilds on the study islands. In nearly all instances coexisting guild members differ by weight (Figs. 5, 7–10). Coexisting gleaners, flycatchers, and nectarivores virtually always show these patterns; coexisting frugivores showed it on islands the size of Guadeloupe or smaller. On the largest islands, coexisting frugivores were more similar in size, although the most similar-sized frugivores were generally from different families. When size differences appeared to be the dominant factor separating coexisting guild members, the ratio of the weight of a species to the weight of its smaller guild associate was usually about 2.0 for birds less than 50 g [the "Hutchinsonian ratio" (Hutchinson 1959)] and about 1.3 for birds larger than 50 g [the ratio important in the guild of fruit pigeons studied by Diamond (1975)]. On smaller islands, the differences in weight between coexisting species of the same guild generally are greater, and smaller-sized guild-members are often absent. Only the largest member of a guild may occur on the smallest islands.

A few exceptions to this size separation pattern exist on the study islands. In a few cases

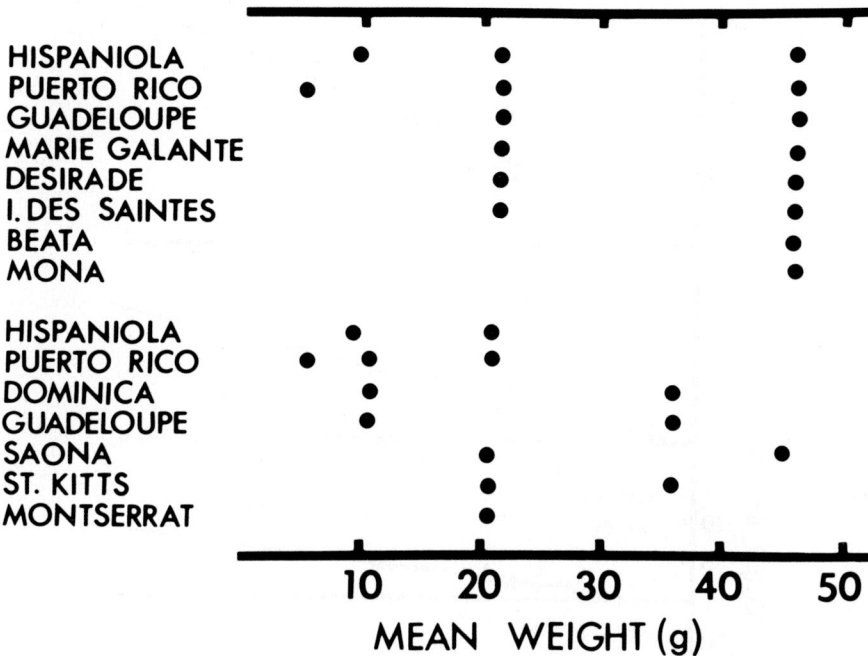

FIG. 5. Mean weights of species of flycatching insectivores coexisting in dry forest (upper) and wet forest (lower) study sites on West Indian islands.

(usually among large guild members), coexisting guild members are of similar size, but they show distinct within-habitat foraging differences. One species is usually terrestrial whereas the other is highly arboreal (see examples below). The Greater Antilles have two congeneric cuckoos of similar size, but one is only a summer resident. How these species coexist is unknown. That these are, in fact, repeatable size rules and not the result of the occurrence of the same species sets from place to place is supported by species that shift in size or habitat use on islands with differing sets of competitors. I have documented a number of these (Faaborg 1982; Case et al. 1983), including shifts from 32 to 52 g in *Loxigilla*, 37 to 52 g in *Saltator*, and 37 to 27 g in *Columbina*. Several cases of size shifts in *Myiarchus* species were negated by recent taxonomic revisions that split the species, generally in accord with size differences. More examples of these shifts occur in the material below.

The bird communities on the study islands appear to have been assembled according to a set of rules. Not surprisingly, the species on small islands, with small communities, show higher densities than do those in larger communities (Terborgh and Faaborg 1973; Faaborg 1980), further suggesting a response to changing competitive pressures. All of the evidence available underscores the importance of interspecific interactions in determining the types of species that can live in these communities. I am aware of arguments suggesting that island patterns may be the result of stochastic processes (e.g., Simberloff 1980), but our own tests (Case et al. 1983) have shown that chance cannot have led to these structural patterns. Chance does have a role is species distributions, but only affects which species plays a particular ecological role in a particular habitat on a West Indian island.

PATTERNS OF STRUCTURE FOR ALL THE ANTILLES

I next considered whether the patterns of structure found on 12 islands may extend to virtually all West Indian islands. In examining this larger set of islands, it should first be noted that the largest island I considered (Cuba) has, in fact, the second largest species total. This is only two species greater than Jamaica, an island roughly one-tenth its size (Table 1). If one added hawks and owls to this study, Cuba would take its "appropriate" place, for it supports 10 hawks and six owls, whereas Hispaniola supports five raptors and Jamaica only three. Explanations for these differences between islands are not readily apparent, although some

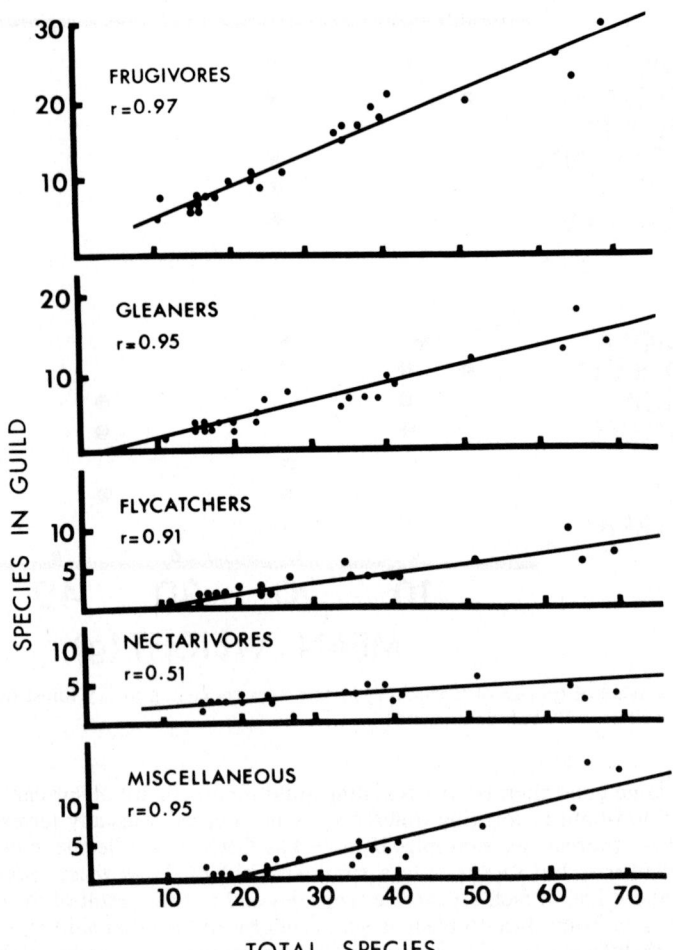

Fig. 6. Numbers of species assigned to the major guilds in relation to total land bird species on the 26 islands examined. Based on data in Appendix I, summarized in Table 1.

will be discussed later. Other variations in the species-area curve can be associated with isolation; isolated islands (Anguilla, Barbados) may have relatively few species, while close islands, such as Gonave, are sometimes species rich.

Despite this species-area variation, the relationship between species in a guild and total species on an island is highly significant for frugivores and both insectivore guilds in the 26 island sample (Fig. 6). While miscellaneous species occur on these islands in orderly fashion relative to total species, there is virtually no relationship between nectarivore diversity and total island species.

Further analysis is difficult because community-wide censuses on the islands I have not studied are few and often difficult to interpret within my approach. In the insectivorous guilds, it appears that species saturation within habitats and size separations among syntopic species occur in all Lesser Antillean habitats and on Cuba, but the Jamaican situation is decidedly more complex. Lack (1976) weighed and measured birds and summarized the elevational range of each species on Jamaica. While he found more species in dry forests than wet, he suggested that little habitat segregation occurred between congeners and showed that like-sized gleaners and flycatchers coexisted. Censuses by Ricklefs (1970) also show the coexistence of insectivores more similar in size than those in my study islands, and Cruz (1980) recorded three vireos and a warbler in one study location. Grand Cayman also has a surprisingly diverse and similar-sized set of insectivores.

Size differences between coexisting nectarivores seem to occur on all islands, but once again the differences between species are less on Jamaica than elsewhere. This guild is unusually small on both Cuba and Grand Cayman.

The frugivore guild is the most complex guild in the West Indies and lacks simple rules for guild structure in Greater Antillean habitats. Coexisting frugivores in the southern Lesser Antilles differ in size, but wet forests on Dominica, St. Lucia, and Martinique support an unusual complex of mimids and thrushes of relatively similar sizes. The most similar of these tend to differ by foraging location. Dry forests on these islands have the typical Lesser Antillean pattern of structure shown above. Cuba is somewhat low in frugivores relative to the other Greater Antilles. While it seems to have a full complement of pigeons and parrots, it has only one tanager and is the only large island without a mistletoe specialist (*Euphonia*).

It appears that in all West Indian bird communities the total number of species that can occur in a guild, given a certain total community size, is controlled. It also appears that on all islands other than Jamaica the number of guild members that can coexist in a habitat is limited. Within these limits on species number, coexisting members of the insectivore and nectarivore guilds, respectively, tend to be of different sizes on all islands. Sizes of frugivores coexisting on small islands differ, but on larger islands other factors (bill shape, subtle dietary preferences) become important, and similar-sized guild members will occur together.

One wonders why Jamaica appears to be so different and, in particular, why Cuba and Jamaica differ in so many ways. In part, differences may be related to proximity to sources of colonization. As noted earlier, during the Pleistocene, Jamaica was separated from tropical Central America by only a few narrow water gaps. The distance between Cuba and the extreme end of the Yucatan Peninsula was little affected. Equally important at this time may have been the effect of Pleistocene climatic cycles on both island and source area habitats. Much of Cuba consists of lowlands that were covered by grasslands or savannahs during dry periods. These and the potential source area of Yucatan were undoubtedly more exposed to temperate weather patterns than today. In contrast, Jamaica is a uniform block of fairly high elevation that probably remained forested throughout these dry cycles and that had more tropical ocean to buffer it from cold weather. This and other factors may mean that Jamaica is simply more tropical in nature than Cuba, and my earlier assumptions about constant resource levels on these islands may not apply. The occurrence of resource constraints on Cuba is indicated by the fact it has only two hummingbird species, one honeycreeper of restricted distribution, one other tanager, no large *Hyetornis* cuckoos, no mistletoe specialist, no bananaquit, and no partially frugivorous flycatcher of the genus *Elaenia*. It does have the only species of wren and gnatcatcher found in the Greater Antilles, five woodpeckers (Hispaniola has two), several blackbirds, and 16 raptors. Some of these factors may also apply to Grand Cayman, which supports several unusual combinations of species. The relative diversity of Jamaica is indicated by the structural patterns examined earlier and by the fact that it is the only West Indian island to support two species of large, non-hummingbird nectarivores and a member of the tropical subfamily of becards. If one supports Bond's (1978) arguments that the West Indian avifauna is of tropical North American derivation, it appears that Jamaica may have served as the main filter for nearly all colonization into the Antilles.

CONFORMING TO THE RULES: DISTRIBUTIONAL ECOLOGY OF WEST INDIAN BIRDS

We have seen that the number of species in the bird communities of most habitats in the West Indies and the acceptable sizes of coexisting guild members follow regular patterns. There are two basic mechanisms by which these patterns may have been achieved. They may have been produced by the selection of colonist species of appropriate size and habitat preference from a pool of colonist species of varying sizes. Alternatively, the patterns may have been produced by the evolutionary transformation of colonists that attained the proper sizes for coexistence after arrival on the island and interaction with the species present. Both patterns seem to have occurred, although species that show size shifts are fewer than are species that show little size variation but have multi-island distributions.

Attempts to explain the ecological controls of species distributions for whole avifaunas on large archipelagoes are few. Diamond (1975) did this for birds on Southwest Pacific islands both by looking at distributional patterns of select guilds (for which he designed "assembly rules" very similar to those shown above for West Indian guilds) and by looking at patterns of incidence of single species. His "incidence functions" plot the occurrence of a species relative to the total number of species on an island. Some species (termed "supertramps") were found on small, often isolated islands with low species numbers but were absent from larger, species-dense islands. These were considered to be generalized species that were good dispersers but poor competitors. At the other end of the scale were species that occurred only in species-rich bird communities (termed "high S" species). These were presumably poor colonizers or too specialized for small islands but excellent competitors able to survive in complex avifaunas.

The West Indian avifauna has a few supertramps and some endemics confined to only large islands. Among the former, *Margarops fuscatus* is distinctive. It is a dominant member of small northern Lesser Antillean communities and those of the Virgin Islands, Mona, Beata, and the southern Bahamas. It is less common and more irregularly distributed on mid-sized islands like Dominica and Puerto Rico and absent from Hispaniola, Jamaica, and Cuba despite a range that nearly surrounds Hispaniola. *Elaenia martinica* and to a lesser extent *Coereba flaveola* also may fit the supertramp category. While the larger islands have many endemics, these are apparently not diverse enough to competitively exclude anything but the few supertramps. In most cases in the West Indies, small island species are also found on large islands in at least some habitats. This fact and the regular occurrence of size-shifts within species, which are apparently rare in the Southwest Pacific avifauna, suggest that the West Indian and Southwest Pacific island systems differ considerably in dynamics of community organization. Part of this may reflect the relative simplicity of the West Indian avifauna, which has a maximum of 70 species per island, while the Solomons and Bismarcks of Diamond's work hold as many as 120 and 140 species, respectively. It also appears that colonization pressures in the Southwest Pacific are greater, such that the appropriate species from an available pool of sizes will colonize an island frequently enough to prevent the undoubtedly slow process of size alteration.

Keeping in mind both the constraints on the number and size of coexisting guild members in West Indian habitats and the mechanisms for achieving this structure, let us survey the distribution of West Indian bird species. Once again, this shall be done by guild with a focus on the major habitat types.

NECTARIVORES

This guild is the simplest in most habitats, consisting of the bananaquit (*Coereba flaveola*) and two hummingbirds (one big, one small) in nearly all cases (Fig. 7). It is so simple, in fact, that at least two authors have seen the regularity of structure of the hummingbird component of this guild (Lack 1973; Brown et al. 1978). The Lesser Antillean nectarivore community is highly consistent, composed of *Coereba* and the small hummingbird *Orthorhynchus cristatus* in nearly all habitats plus the large hummers *Eulampis jugularis* in wet forest and *E. holosericeus* in dry. Low dry islands generally lack *jugularis*, although Marie Galante is an exception, having *jugularis* and not *holosericeus*. *Eulampis jugularis* is only a vagrant on Grenada where is apparently is replaced by the South American *Glaucis hirsuta*. *Cyanophaia bicolor*, at 4.5 g almost twice as large as *Orthorhynchus*, also occurs in mid-elevation wet forest on Dominica and Martinique. These species coexist with *Eulampis jugularis* in some forests, but only *Eulampis* and *Cyanophaia* frequent high-elevation forests. The Lesser Antillean dry forest hummingbirds are also found on the Virgin Islands, despite the recent connection of these islands with Puerto Rico. Both species have also become established in parts of northeastern Puerto Rico, although their ranges do not seem to be expanding.

The Greater Antillean nectarivore community follows similar structural rules but is noteworthy for its small diversity on large islands. Ignoring the two Lesser Antillean invaders, Puerto Rico has a small widespread hummingbird (*Chlorostilbon maugaeus*), *Coereba*, and two large *Anthracothorax* hummers, *A. dominicus* in the lowlands and *A. viridis* in wet forests. *Anthracothorax dominicus* is also widespread on Hispaniola and its satellites, where it usually exists with the tiny *Mellisuga minima* and *Coereba*. In high elevation, temperate-climate habitats, the lowland hummers are replaced by *Chlorostilbon swainsonii*, and *Coereba* is rare or absent. Jamaica is unusual in occasionally having three coexisting hummers, the tiny *M. minima*, medium *Trochilus polytmus*, and large *Anthracothorax mango*. Lack (1976) suggested that the two larger hummingbirds are most abundant in habitats where they live alone, with *A. mango* more common in the lowlands and *Trochilus* in the highlands. Jamaica is also unusual in having, in addition to *Coereba*, the Orangequit (*Euneoris campestris*), a bird of questionable ancestry that feeds on nectar. One here could conceivably get up to five nectar-

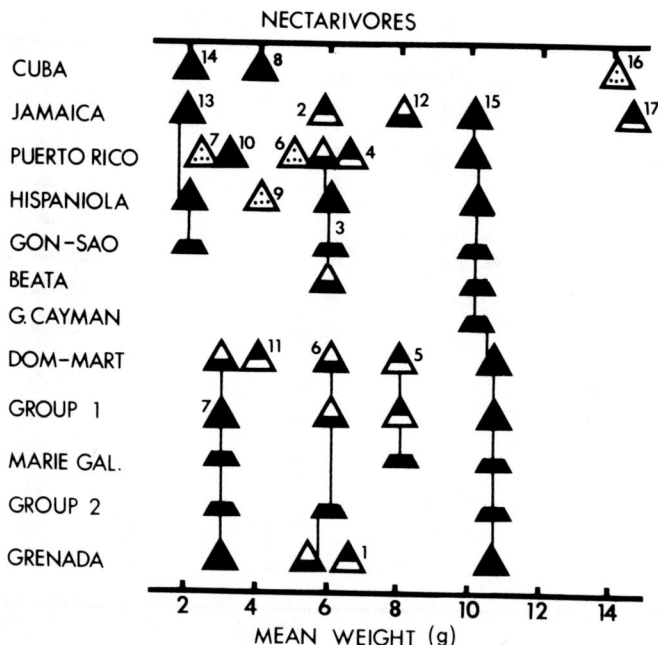

FIG. 7. Weights and general habitat distributions of nectarivores considered in this study. Location of a symbol shows the approximate mean weight of a species (numbers identify species according to the list in Appendix I) on an island or island group (abbreviations from Table 1). Group 1 includes Antigua, St. Kitts, Montserrat, Guadeloupe, St. Lucia, and St. Vincent; Group 2 includes St. Johns, Anguilla, St. Bartholomew, Barbuda, Barbados, Isles des Saintes, and Desirade. Triangles indicate high islands with both wet and dry forests, trapezoids, low islands with only one forest type (usually dry). Areas shaded black show a species' range in dry (lower) or wet (upper) forests. Stippled areas signify occupation of habitats other than wet or dry forest. Symbols of species with multi-island ranges are connected.

ivores in one site, weighing 2, 5, 8, 10, and 16 g. In sharp contrast, Cuba has but two hummers, the tiny *Mellisuga helenae* (reported to be the smallest bird in the world) and the mid-sized *Chlorostilbon ricordii*, and no *Coereba*. The honeycreeper *Cyanerepes cyaneus* exists in a few Cuban locales, but most habitats seem to lack a nectarivore larger than 4 g. Small Greater Antillean islands may lack the smallest hummingbird (Beata), both hummingbirds (Grand Cayman), or have no nectarivores at all (Mona).

The absence of *Coereba* from Cuba was discussed by MacArthur and Wilson (1967). Given its widespread distribution on cays surrounding Cuba and the restricted distribution of *Cyanerpes*, it is hard to believe that the latter competitively excludes *Coereba*. *Coereba* is generally less abundant in more diverse communities and also weighs less, decreasing from nearly 12 g on some Lesser Antillean islands to 9 g or less on some Greater Antillean islands. Yet, Cuba is no more diverse in any habitat than Puerto Rico or Jamaica. If some form of diffuse competition is excluding *Coereba*, perhaps it is with the winter resident species. These are reported to be very abundant on Cuba, and some (such as *Dendroica tigrina*) will feed on nectar. Cuba may also be abnormally low in flowers due to climatic factors, such that a big hummingbird and bananaquit are not able to coexist there. Further work is needed to solve this biogeographic puzzle.

FLYCATCHING INSECTIVORES

This guild is composed of members of the Tyrannidae and Todidae. All are distinguished by their ability to catch insects on the wing or while hovering, but many will occasionally eat fruit, and elaenias (*Elaenia* and *Myiopagis*) are difficult to assign to a guild because of this.

The richest Lesser Antillean flycatcher guild (found on Dominica, Guadeloupe, Martinique, and St. Lucia, Fig. 8) includes *Tyrannus dominicensis* (45 g) and *Elaenia martinica* (22 g) generally in dry forests and *Myiarchus oberi* (37 g) and *Contopus latirostris* (11 g) in wet

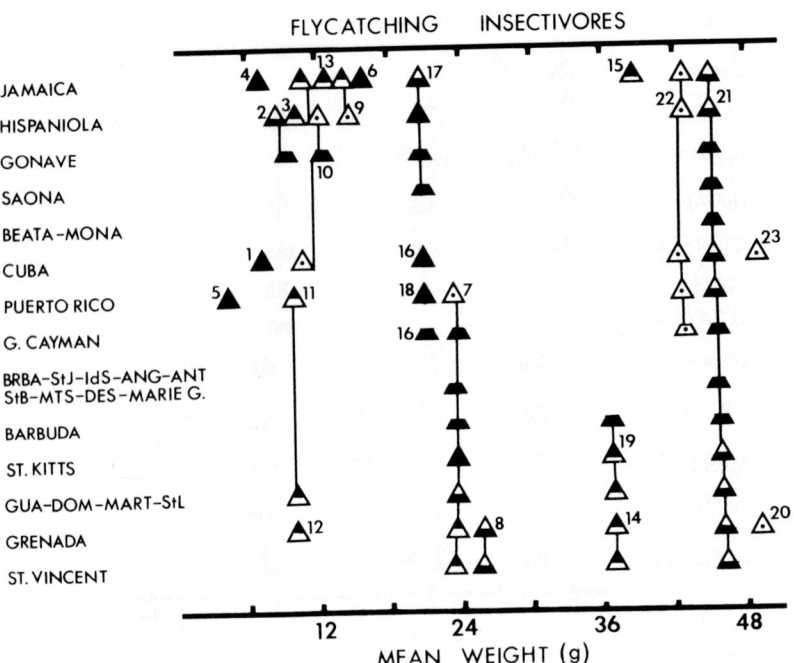

FIG. 8. Weights and general habitat distributions of all flycatching insectivores considered in this study, expect *Pachyramphus niger* from Jamaica. Location of a symbol shows the approximate mean weight of a species (numbers identify species according to the list in Appendix I) on an island or island group (abbreviations from Table 1). Triangles indicate high elevation islands with both wet and dry forests, trapezoids, low islands with only one forest type (usually dry). Areas shaded black show a species' range in dry (lower) or wet (upper) forests. Symbols with dots signify occupation of habitats other than wet or dry forest. Symbols of species with multi-island ranges are connected.

forests. On Dominica and, perhaps, other of the southern Lesser Antilles, greater habitat overlap may occur between the species in these pairs. Small, dry islands often have just *Tyrannus* and *Elaenia,* although these two and *Myiarchus* occur on Barbuda and St. Kitts. On the latter, *Myiarchus* and *Tyrannus* are similar in size (38 g vs 45 g) and apparently do not coexist; on Barbuda they do coexist but use different microhabitats and differ more in size (26 g vs 45 g). On islands with wet forest but only two or three flycatcher species, *Elaenia* invades the wet forest, but often at low densities (Terborgh et al. 1978). In the southern Lesser Antilles, the situation is somewhat more complex, due in part to recent invaders from South America. On Grenada, *Tyrannus melancholicus* occurs in disturbed habitats, the rare (if not extinct) *Empidonax euleri* may fill the *Contopus* niche in wet forests, *Myiarchus nugator* replaces *oberi,* and *Elaenia flavogaster* has apparently pushed *E. martinica* out of some habitats (Crowell 1968). *Tyrannus dominicensis, Myiarchus nugator,* and the two *Elaenia* also occur on St. Vincent.

Orderly assemblages of flycatchers occur on Greater Antillean islands, with size differences between coexisting species pronounced on all islands but Jamaica and Grand Cayman. *Tyrannus dominicensis* is found in all dry habitats and often along forest edge on all islands, and it is the only flycatcher remaining on small islands such as *Mona.* In certain sub-habitats it is replaced by *T. caudifasciatus* and in Cuban pines by the bigger-billed *T. cubensis.* Only one *Tyrannus* species is found in each habitat. In most habitats on Puerto Rico, Cuba, and Hispaniola, species of *Myiarchus* weighing 20 to 22 g coexist with a flycatcher weighing approximately 10 g and, rarely, with one weighing 5 g. On Puerto Rico this sequence may include *Todus mexicanus* (5.5 g), *Contopus latirostris* (11 g), and *M. antillarum,* although only *Todus* and *Myiarchus* occur in dry forest. *Todus multicolor* (6–7 g) and *C. caribaeus* (11 g) may coexist with *M. sagrae* on Cuba. Hispaniola supports similar-sized *C. caribaeus, T.*

subulatus, and *T. angustirostris,* but these seem to separate by habitat, with *Contopus* at high elevations, *subulatus* in the lowlands, and *angustirostris* in between. These coexist with *M. stolidus. Elaenia fallax* also occurs on Hispaniola, in temperate environments in the highlands. Puerto Rico is inhabited by isolated populations of *E. martinica*; its recent attempt at invading the Guanica Forest appears to have failed (Faaborg and W. J. Arendt, unpubl. data). This species also occurs on Grand Cayman, as befits a "supertramp" species.

This consistency and simplicity of structure is broken on Jamaica which, in addition to *Contopus* and two *Tyrannus,* has three *Myiarchus, Elaenia fallax,* and the elaenia-like *Myiopagis cotta.* The *Myiarchus* are of different sizes and show some habitat separation, and the elaenias show some elevational separation (Lack 1976), but the Jamaican situation is obviously more complex than elsewhere in the Antilles. Grand Cayman also is unusual in that *Elaenia martinica* and *Myiarchus sagrae* apparently coexist despite being the same size. Johnston (1975) suggested this was possible because *Elaenia* eats more fruit, but such an explanation apparently does not work where *Myiarchus* and *Elaenia* co-occur on other islands.

GLEANING INSECTIVORES

Virtually all West Indian forest habitats contain three classes of gleaning insectivores, a large cuckoo (usually *Coccyzus minor,* 65–70 g), a large vireo (usually *Vireo altiloquus,* 20 g), and an 11 g warbler or vireo. Most of the complexity occurs in the between-habitat replacements in this smallest category. Larger islands may have an oriole (35 g) and one or two larger cuckoos whose diets include many lizards. Several islands have a smaller (6–7 g) gleaner, as well, while the whole guild is more complex in Jamaica than elsewhere (Fig. 9).

In the standard Lesser Antillean situation, *C. minor* and *V. altiloquus* occur in all forested habitats. In the Virgin Islands and northern Lesser Antilles, the migratory *C. americanus* coexists with or replaces *C. minor.* This *Vireo-Coccyzus* pair often coexists with *Dendroica petechia,* a widespread Lesser Antillean species that is confined to mangroves on the Greater Antilles. In the wet forests (although occasionally occurring in drier forest) a variety of endemic forms fill the small insectivore niche. On Guadeloupe and Dominica, the 11 g *Dendroica plumbea* is abundant, while *Leucopeza semperi* occurs in scattered parts of St. Lucia and *Catharopeza bishopi* on St. Vincent. *Dendroica adelaidae* also occurs in St. Lucia forest, but it is not known if this species actually coexists with the larger *Leucopeza. Dendroica adelaidae* is the smallest of the warblers (6–7 g), which may allow its coexistence with 11 g or larger warblers on these islands. On the southern Lesser Antilles these forest warblers may coexist with an 11 g wren (*Troglodytes aedon*), which generally feeds lower than the warblers and which is confined to the lowlands on St. Lucia (Diamond 1973). Nowhere else do like-sized guild members coexist. On three islands, these warbler-vireo-cuckoo sets coexist with a 35 g oriole (*Icterus oberi* on Montserrat, *I. laudabilis* on St. Lucia; *I. bonana* on Martinique). Edge and disturbed habitats on these islands support the large *Crotophaga ani* and *Quiscalus lugubris.*

The Greater Antillean situation is more complex due primarily to habitat-replacement patterns, although occasionally a more complex sequence of species occurs. For example, dry forest in Puerto Rico holds the 7 g *Dendroica adelaidae,* 11 g *Vireo latimeri,* 20 g *Vireo altiloquus,* and 35 g *Icterus dominicensis.* In high wet forest *V. latimeri* is replaced by the recently discovered *D. angelae* (Kepler and Parks 1972), whereas *adelaidae* is absent. *Dendroica petechia* is restricted to mangroves, as it is throughout the Greater Antilles. On Hispaniola, simple sets generally exist. *Vireo nanus* fills the 11 g slot in some lowland habitats, while the 13 g *Microligea palustris* occurs in others. In mountain pines we find *Dendroica pinus* in the tree tops and *Xenoligea montana* (13 g) in the underbrush. Similar patterns seem to occur on Cuba, with *Vireo gundlachii* in scrubby broadleafed forest, *Dendroica pityophila* in pines, the allopatric *Teretistris fernandinae* and *T. fornsi* in the forest and sometimes scrub, a wren *Ferminia cerverai* in the Zapata Swamp, and a tiny gnatcatcher (*Polioptila lembeyei,* 5 g) in scrub. As usual, the most complex situation with small gleaning insectivores occurs on Jamaica, where three vireo species may coexist with a warbler (*Vireo modestus,* 10 g, *V. altiloquus* 19 g, and *V. osburni,* 20 g, with *Dendroica pharetra,* 11 g; Cruz 1980). While these species differ in size and show different overall habitat preferences, the situation is much more complex than elsewhere in the Antilles. *Vireo altiloquus* is not found on Mona, where no small gleaners exist, or on Grand Cayman, where it is replaced by *V. magister.* Grand Cayman also has *V. crassirostris, Dendroica petechia,* and *D. vitellina,* which gives it an unusual set of

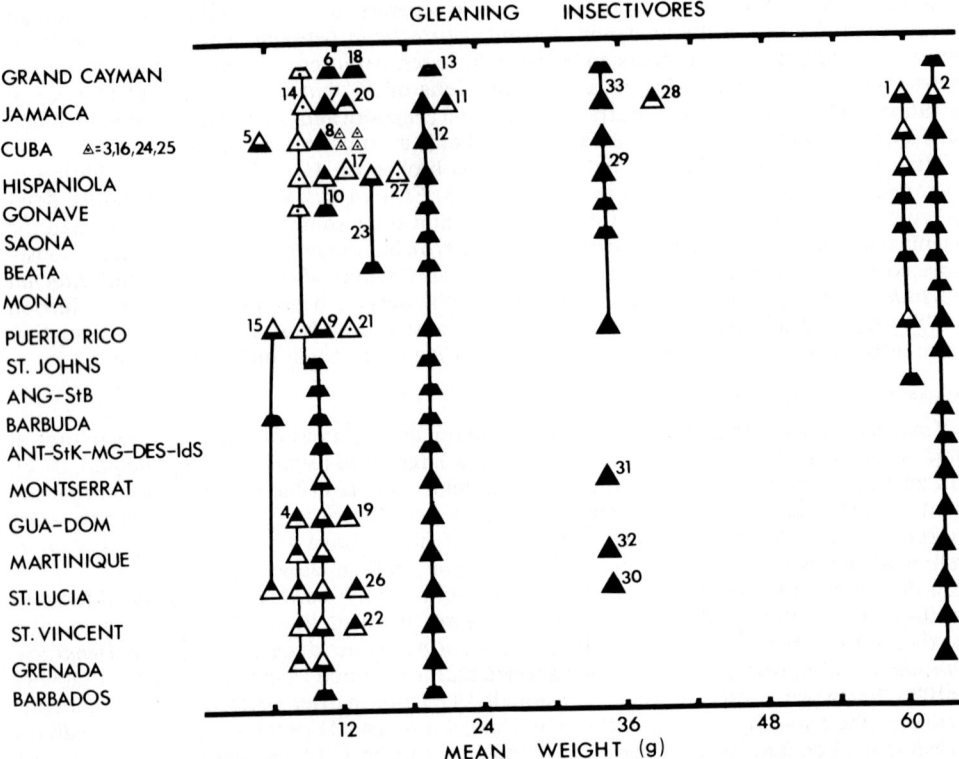

FIG. 9.   Weights and general habitat distributions of forest-dwelling gleaning insectivores considered in this study. All large cuckoos (*Saurothera, Hyetornis*) and insectivores restricted to grasslands, savannahs, swamps, or disturbed habitat (*Crotophaga, Quiscalus, Dives, Agelaius, Sturnella*) are excluded. Location of a symbol shows the approximate mean weight of a species (numbers identify species according to the list in Appendix I) on an island or island group (abbreviations from Table 1). Triangles indicate high elevation islands with both wet and dry forests, trapezoids, low islands with only one forest type (usually dry). Areas shaded black show a species' range in dry (lower) or wet (upper) forest. Symbols with dots signify occupation of forests other than wet and dry. Symbols of species with multi-island ranges are connected.

insectivorous birds for a small, dry island (Johnston et al. 1971). While like-sized members of this guild tend to occupy different habitats on this island (Johnston 1975), the situation requires further study.

At the larger end of the gleaning insectivore scale, the Greater Antilles support three general size classes. The most common cuckoo is the 65–70 g *Coccyzus minor,* which is found on nearly all islands, although it is apparently rare on Cuba. On the larger islands, it may coexist in the breeding season with the like-sized *C. americanus.* Although only some *minor* are migratory, all *americanus* are, and the ecological interactions between these species need further examination. On the four main islands, plus Gonave and Saona, the *Coccyzus* group coexists with the even larger *Hyetornis.* While *Saurothera* is often abundant, *Hyetornis* is thinly distributed on Jamaica (Lack 1976) and rare and local on Hispaniola. Few published weights exist for these species, especially *Hyetornis,* but it appears *Saurothera* weighs about 100 g and *Hyetornis* about 150 g.

On all the Greater Antilles, the gap between small gleaners and cuckoos is filled by a 35 g oriole (*Icterus dominicensis* on Hispaniola, Cuba, and Puerto Rico, *I. leucopteryx* on Jamaica and Grand Cayman). Jamaica is unusual in supporting a forest blackbird (*Nesopsar nigerrimus*) that may compete with the oriole and the woodpecker (Cruz 1978). In open habitats and mangroves, the larger Greater Antilles support a grackle (*Quiscalus niger*). Hispaniola has scattered populations of a blackbird (*Agelaius humeralis*), while the Puerto Rican form (*A.*

xanthomus) is more widespread in open habitats. Cuba supports three open country blackbirds (*Dives atroviolacea, Agelaius humeralis,* and *A. phoeniceus*) plus the meadowlark (*Sturnella magna*). Disturbed habitats on all these islands also support the large ani (*Crotophaga ani*).

The relative simplicity of the insectivore communities on these islands may be due in part to the fact that the Greater Antilles support great densities of small winter resident warblers (Terborgh and Faaborg 1980a). The effect of these non-breeders on resident insectivore densities and diversities may be profound.

FRUGIVORES

This most complex of West Indian guilds can best be examined by dividing it into passerine and nonpasserine components in addition to the regional division. The nonpasserine group includes pigeons and parrots, the largest of the West Indian frugivores, while the passerine component varies from 9 g grassquits through 100 g thrashers. Only *Columbina passerina* of the nonpasserine group falls within the passerine size range.

The parrots of the West Indies seem to be an old group with an almost relictual distribution in the Lesser Antilles (Olson 1978). Thus, St. Vincent and St. Lucia each support single species (*Amazona guildingii* and *A. versicolor,* respectively), Dominica has two (*A. imperialis* and *A. arausiaca*), but none of the other Lesser Antilles has any. Puerto Rico (*A. vittata*), Hispaniola and Gonave (*A. ventralis*), and Cuba and Grand Cayman (*A. leucocephala*) all support one parrot species, but Jamaica has two (*A. collaria* and *A. agilis*). Several islands have smaller parakeets (e.g., Cuba, *Aratinga euops*; Jamaica, *A. nana*, and Hispaniola, *A. chloroptera*). This latter species was extirpated from Mona Island around 1900 and may have occurred on Puerto Rico. Where they occur, *Amazona* is often widespread, whereas *Aratinga* may be confined to highland areas. Cuba once had a macaw (*Ara tricolor*), giving it psittacids of three sizes. Where two *Amazona* coexist, they tend to separate elevationally, but much overlap may occur.

The pigeons present a confusing situation on the Greater Antilles, but distinct patterns on the Lesser Antilles. Lesser Antillean dry forests support a large aboreal pigeon (usually *Columba leucocephala*, 250 g) and three sizes of ground dwellers, a small ground dove (*Columbina passerina*, 37 g), medium pigeon (usually *Zenaida aurita*, 150 g), and a large quail dove (*Geotrygon mystacea*, 230 g). In wet forests, *Columba leucocephala* is usually replaced by *C. squamosa, G. mystacea* is usually replaced by *G. montana*, and neither *Columbina* or *Zenaida* occur. Where *G. montana* is not found, *G. mystacea* invades wet forest. This simple Lesser Antillean picture is confused by two species in the south, *Zenaida auriculata* (100 g), which is apparently a fairly recent invader on St. Lucia, St. Vincent, Barbados, and Grenada, and *Leptotila rufaxilla*, a South American species found in dry forest on Grenada. *Geotrygon mystacea* is absent from those forests with *Leptotila* and two *Zenaida*.

The elements of this structural pattern tend to extent to the Greater Antillean pigeons, but much overlap exists, and one can find confusing sets of possible coexisting species. Generally, only one *Columba* breeds in a given habitat, with *squamosa* common in wet forest and *leucocephala* in dry. How *inornata* and *caribaea* fit in is unknown. *Zenaida aurita* is the usual dry forest and edge pigeon, but *Z. macroura* and *Z. asiatica* have invaded some habitats, and in certain locations all three can be found. This includes Mona Island in the breeding season, which helps to account for its unusually high number of frugivores. Jamaica and Grand Cayman also contain *Leptotila jamaicensis,* which further confuses the ecological picture with regard to mid-sized pigeons. The dry forests of all islands contain the small *Columbina*, whose density varies greatly depending on the number of other frugivorous species (Terbourgh and Faaborg 1973).

The ground-dwelling quail doves of the Greater Antilles also tend to separate by habitat, but occasionally overlap. Thus, *Geotrygon chrysia* is usually found in dry forests and *G. montana* in wet forest on islands where both occur (Puerto Rico, Hispaniola, Cuba). Both of these are smaller than their Lesser Antillean counterparts, and they may coexist with larger forms on Cuba and Hispaniola, including the 210 g *G. caniceps* on both islands, and the even larger (weight unknown) *Starnoenas cyanocephala* on Cuba. Jamaica has only two quail doves, and there *montana* is generally found at lower elevations than the large *G. versicolor.*

The last of the nonpasserine frugivores are from the Trogonidae. Hispaniola and Cuba each support a species (*Priotelus roseigaster* and *P. temnurus,* respectively). These are primarily highland forms but will wander to the lowlands.

The passerine component of West Indian frugivores is characterized by size differences between most coexisting guild members on smaller islands, although several pairs of like-sized coexisting species occur and appear to differ in foraging location. The Greater Antillean frugivore guild has many similar-sized forms (Fig. 10). The largest passerine frugivores are from the Mimidae and Muscicapidae (Turdinae) and are a dominant part of the Lesser Antillean avifauna, where up to eight species may occur on an island. Four of these are thrasher species that occur only rarely outside the Lesser Antilles. The dominant form is *Margarops fuscatus*, a 100 g thrasher found in most habitats from St. Lucia through Puerto Rico, Mona, Beata and the southern Bahamas (but not Hispaniola). Through most of its range, it coexists with the 75 g *Margarops fuscus*, and on high islands with wet forests these coexist with the wren-like trembler (*Cinclocerthia ruficauda*, 50 g). On Martinique and St. Lucia, *Cinclocerthia* may have been replaced in dry forests by the thrasher *Ramphocinclus brachyurus*, although this species has a rather patchy distribution today. These four, plus the mockingbird (*Mimus gilvus*) in edge habitats, give these islands an unusual diversity of mimids. Perhaps due to this diversity, *Cinclocerthia* and *M. fuscatus* on St. Lucia are unusually large (75 and 120 g, respectively). This thrasher group declines in density and diversity to the south and east, such that the southern Lesser Antilles are inhabited by only one of the forest thrashers (*M. fuscus*; Lack and Lack 1973; Lack et al. 1973) and the biggest of the Greater Antilles support only a mockingbird. This is usually *Mimus polyglottos* from North America, but Jamaica also has an isolated population of the Bahaman *M. gundlachii*, a large mockingbird at 75 g. The Turdinae show two areas of strength in the Antilles, the southern Lesser Antilles and the large Greater Antilles. In the central Lesser Antilles, the small solitaire, *Myadestes genibarbis*, (27 g) occupies wet forest. On St. Lucia and Dominica, the 100 g thrush *Cichlherminia lherminieri* also occurs in wet forests, leading to a coexisting thrush-thrasher group weighing 27, 50, 75, 100, and 100 g. The two 100 g species tend to forage differently, with the thrush primarily terrestrial and the thrasher arboreal. Dry forest on Dominica also is occupied by the 75 g *Turdus plumbeus*, which probably differs from the coexisting thrasher *Margarops fuscus* by foraging on the ground. *Cichlherminia* also occurs in wet forest on Montserrat and Guadeloupe, where it is the only thrush. Two South American robins (*Turdus fumigatus*, 73 g, and *T. nudigenis*, 54 g) have moved into the southern Lesser Antilles, which may explain the absence of some of the West Indian thrashers there. Both robins exist on Grenada and St. Vincent, with *fumigatus* in wet forest and *nudigenis* in dry. In the last 30 years *T. nudigenis* has also invaded edge habitats on Martinique and St. Lucia.

Thrushes are absent from the Northern Lesser Antilles and Virgin Islands. The most widespread of the Greater Antillean thrushes is *Turdus plumbeus* (75 g) of Puerto Rico, Saona, Gonave, and most habitats on Cuba and Hispaniola. In the higher mountains of southwestern Hispaniola it is replaced by *Turdus swalesi*. *Turdus ravidus* of Grand Cayman is now extinct. Jamaica has two totally different thrushes, *Turdus jamaicensis* (66 g) and *T. aurantius* (81 g), which coexist at higher elevations. Highland thrushes often live with a solitaire, *Myadestes elisabeth* on Cuba and *M. genibarbis* on Jamaica and Hispaniola.

The other passerine frugivores are members of the Emberizidae and Fringillidae and range from 9 to 52 g. Thus, they overlap in size only with *Mimus* and *Myadestes* from the thrush-thrasher group and *Columbina* of the pigeons. The Lesser Antillean component of species is very simple in structure on all but the southernmost islands. As noted earlier, size separations are common. The only tanager present on most islands is the mistletoe specialist, *Euphonia musica* (16 g), although Grenada and St. Vincent have the small *Tangara cucullata* (20 g). All Lesser Antillean islands have dry forests with *Tiaris bicolor* (9 g) and *Loxigilla noctis* (18 g) coexisting with *Columbina* (37 g). Dry forests on Guadeloupe, Dominica, Martinique, and St. Lucia also contain a large (52 g) race of the South American *Saltator albicollis*; on St. Kitts the 52 g dry-forest frugivore used to be a race of *Loxigilla portoricensis* from Puerto Rico, now extinct. Of these small frugivores, only *Loxigilla noctis* is common in wet forest. St. Lucia has an endemic finch, *Melanospiza richardsonii*, which is apparently fairly widespread, but uncommon. At 20 g it is larger than *Loxigilla*, which is smaller on St. Lucia than elsewhere. These species also differ in foraging height (Diamond 1973). Edge and field habitats on Grenada support *Sporophila nigricollis* and *Volatina jacarina*, both apparently recent colonists from South America.

The greater Antillean group of finches and tanagers does not align itself as clearly. All of the large islands have two small (9–10 g) grassquits. On Puerto Rico, Gonave, Hispaniola, and Jamaica, these are *Tiaris bicolor* and *T. olivacea*. Generally, *bicolor* lives in brushier vegetation and dry forest, whereas *olivacea* lives in open country and grasslands. On mainland Cuba *T. bicolor* is replaced by *T. canora*, although *bicolor* occurs on some of the northern

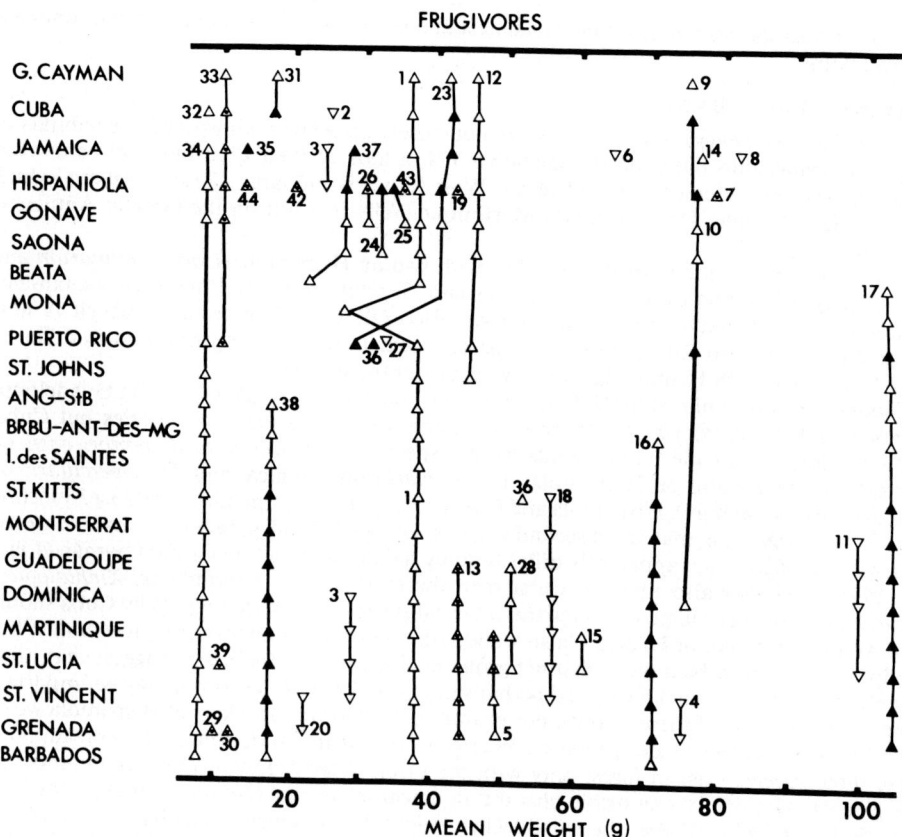

FIG. 10. Weights and general habitat distributions for small forest-dwelling frugivores considered in this study. All Psittacidae, all Columbidae, except *Columbina passerina*, and mistletoe (*Euphonia*) and grassland (*Ammodramus*) specialists are excluded. Location of a symbol shows the approximate mean weight of a species (numbers identify species according to the list in Appendix I) on an island or island group (abbreviations from Table 1). Species confined to lowland dry forest are shown by open triangles, those confined to rainforest, by inverted triangles, and those using both forest types, by triangles shaded black. Symbols with dots denote species occurring in other forest types. Symbols of birds with multi-island ranges are connected.

cays. Only *olivacea* is found on Grand Cayman. All major islands also have at least one larger finch. This category includes the 17 g *Melopyrrha nigra* on Cuba and Grand Cayman, the 22 g *Loxigilla violacea* on Beata, the 28 g *L. violacea* on Jamaica, Hispaniola, and Saona, and the 32 g *Loxigila portoricensis* on Puerto Rico. Jamaica has a third finch in some forests, the 12 g *Loxipasser anoxanthus*, which is more of a forest interior species than either *Tiaris* (Lack 1976). The only other West Indian finches are found in grasslands (*Ammodramus savannarum* on Jamaica, Hispaniola, and Puerto Rico), remnant swamps or deserts (*Torreornis inexpecta* on Cuba), or high mountain pine forests (*Carduelis dominicensis*, *Loxia leucoptera*, and *Zonotrichia capensis* on Hispaniola).

The tanager group of the Greater Antilles is small except on Hispaniola. The major islands except Cuba have a mistletoe specialist (*Euphonia musica* on Puerto Rico and Hispaniola, *E. jamaica* on Jamaica). They all also have the tanager *Spindalis zena*, although this bird does not occur in all habitats and varies in weight from 28 g on Puerto Rico to 43 g on Jamaica. Wet forest in Puerto Rico has a second tanager (*Nesospingus speculiferus*) at 35 g, whereas Hispaniola has three endemics. Two of these (*Phaenicophilus palmarum* and *P. poliocephalus*) are superspecies that divide the island geographically, while the third (*Calyptophilus frugivorus*) is rather patchily distributed and poorly know ecologically. Hispaniola is also home to the unusual frugivore, the palmchat (*Dulus dominicus*) of the family Dulidae.

This widespread species builds a communal nest in open forests; it shows a particular fondness for royal palm.

OTHER WEST INDIAN BIRDS

As noted earlier, several species either are not closely associated with particular habitats or do not fit cleanly into the guild designations. Yet, a look at their distributional patterns is revealing. The only kingfisher found in the West Indies is the large *Ceryle torquata*, which occurs on Guadeloupe, Dominica, and Martinique. Why it is not on the Greater Antilles is unknown.

Four crows occur on the Greater Antilles, with two on Hispaniola (*Corvus palmarum* and *C. leucognaphalus*), two on Cuba (*C. nasicus* and *C. palmarum*), and one each on Jamaica (*C. jamaicensis*), Gonave, Saona, and Puerto Rico (all *C. leucognaphalus,* which is now extirpated from Puerto Rico). When two species coexist on an island, they tend to differ by habitat, although both Hispaniolan species occur in pine forests.

Woodpeckers are confined to the Greater Antilles except for a single species on Guadeloupe (*Melanerpes herminieri*, 100 g). There is no island with more than two species but Cuba, which has five. All of the main islands have a species of *Melanerpes* (*M. portoricensis* on Puerto Rico; *M. striatus* on Hispaniola; *M. radiolatus* on Jamaica; and *M. superciliaris* on Cuba and Grand Cayman). Hispaniola and Gonave have the only "piculet," the 33 g *Nesoctites micromegas*. Grand Cayman has a second woodpecker, a small (88 g) race of *Colaptes auratus*. *Colaptes* and *Melanerpes* apparently differ both by habitat and diet on Grand Cayman (Johnston 1975). *Colaptes* also lives on Cuba with the endemics *C. fernandinae, Xiphidiopicus percussus,* and the very large *Campephilus principalis* (perhaps extirpated). Why Cuba should be so woodpecker rich, or Hispaniola so woodpecker poor, is an unanswered question. While woodpeckers seem to be a fairly distinct group ecologically, shifts in size suggest that other bark gleaners (orioles, blackbirds, and perhaps the trembler, *Cinclocerthia*) may be important competitors on some of these islands. For example, the two woodpeckers of Hispaniola weigh 33 and 65 g, while the single species on Puerto Rico is at the midpoint of this spread, 50 g. These three species exist in forest only with the oriole *Icterus dominicensis* as a competing bark gleaner. Jamaica has an oriole plus the Jamaican blackbird (*Nesopsar*) that gleans bark regularly (Cruz 1978). There, the woodpecker weighs 101 g. Guadeloupe has no oriole, but the trembler, although a thrasher, does much bark gleaning (Zusi 1969). The Guadeloupe woodpecker also weighs 100 g.

The swifts and swallows are wide-ranging aerial insectivores that are difficult to associate with distinct vegetation types. A large martin (*Progne*) is found on the large Greater Antilles and all but the northernmost Lesser Antilles. This is *P. cryptoleuca* on Cuba and *P. dominicensis* elsewhere. Hispaniola and Jamaica support the swallow *Tachycineta euchrysea* at high elevations, while Puerto Rico, Gonave, Hispaniola, Jamaica, and Cuba support colonies of *Hirundo fulva* in caves and cliffs. The Greater Antillean swifts come in three sizes, the small *Tachornis phoenicobia* (11 g), *Cypseloides niger* (33 g), and the large *Streptoprocne zonaris* (weight unknown). All three live on Cuba, Hispaniola, and Jamaica, with the larger two species common in the mountains and the small species in the lowlands. *Tachornis* is also found on some of the Hispaniolan satellite islands. Puerto Rico has only *Cypseloides,* which is also found on St. Kitts, Guadeloupe, Dominica, Martinique, and St. Lucia. On all of these islands but St. Kitts *Cypseloides* coexists with a smaller swift, *Chaetura martinica*. St. Vincent has *C. brachyura* and *C. cinereiventris,* like-sized swifts that tend to differ in habitat. Only *cinereiventris* occurs on Grenada.

The last of the miscellaneous groups includes nocturnal insectivores other than owls. In the Lesser Antilles these consist of the endemic *Caprimulgus otiosus* on St. Lucia and, on Martinique, *C. cayennensis* a form that is widespread on the South American mainland. Virtually all Greater Antillean islands support the nighthawk *Chordeiles gundlachii* during the summer. Whip-poor-will forms occur on Puerto Rico (*Caprimulgus noctitherus*), Hispaniola, and Cuba (both with *C. cubanensis*). A tiny paraque (*Siphonorhis brewsteri*) lives on Hispaniola and Gonave and has been extirpated from Jamaica. The large potoo (*Nyctibius griseus*) is found on Jamaica, Hispaniola, and Gonave. Where multiple species coexist, they differ either in size or foraging zone. The most common situation has *Chordeiles* feeding high in the air and *Caprimulgus* near the ground. Many of these species are local in distribution and poorly understood ecologically.

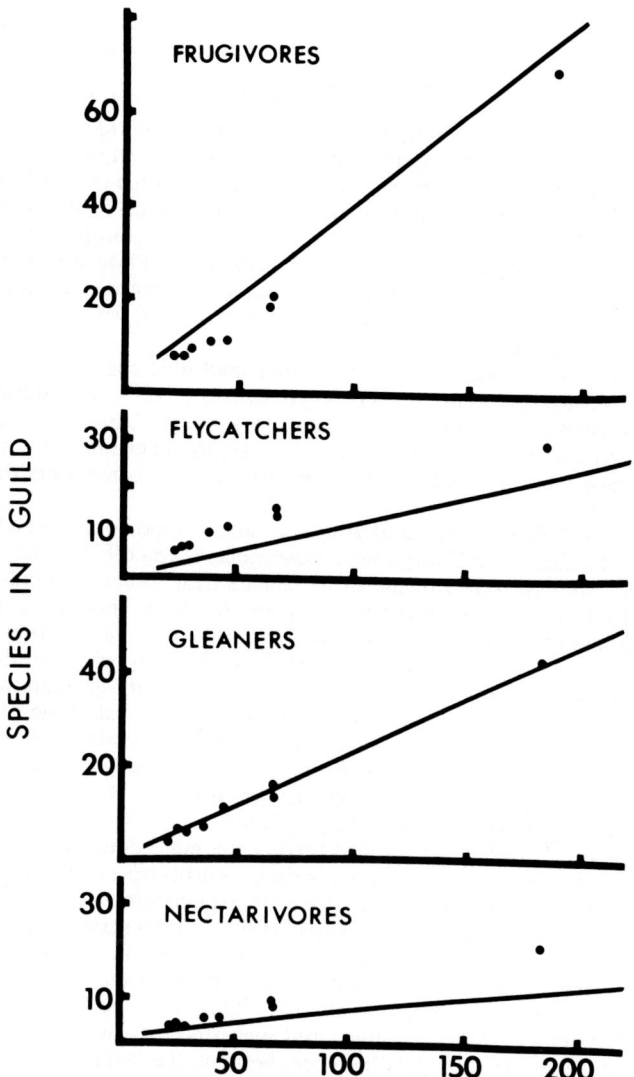

FIG. 11.   The relationship between the number of species in a guild and the total number of island bird species for eight neotropical land-bridge islands (points) compared to the relationship for West Indian islands (regression lines; from Fig. 6). Data for the land-bridge islands from ffrench (1973), Wetmore (1957), and Wright et al. (1985).

In the above material, I have combined the taxonomic and distributional work of Bond (1971) and others with the apparent ecological constraints that influence community formation. Where the ecological "rules" seem to be broken, one should examine such factors as resource base, island isolation, etc. The examination of replacements in ecological positions on these islands gives us a view of the role of chance colonization in determining distributions, the possible sequence of colonization on an island, and some measure of the dynamism of these islands. In many cases, extreme ecological shifts are closely tied to taxonomic disputes, such as is found in the *Myiarchus* flycatchers. Closer examination of the possible effects of ecological constraints on subspecies' distributions may be of interest, for in many cases these differences seem to correlate with size or other shifts associated with differing competitive environments. For example, the seven races of *Spindalis zena* may reflect its existence in different communities. The varying ecological pressures in these communities may have led to its range of sizes (28 to 43 g). Further inquiry at this level would possibly identify more clearly the interactions between ecological pressures and taxonomic distinctions.

THE GENERALITY OF WEST INDIAN STRUCTURAL PATTERNS

Given that I have shown a fairly simple set of patterns of structure of West Indian bird communities of varying size, one wonders to what extent any or all of these patterns apply to larger bird communities. The consistency of some of the patterns over a large variance in community size leads me to believe in the existence of some ecological controls on the assembly of communities. A major problem is that in large communities, birds may cope with the constraints present by means of many alternative responses. While minimum acceptable differences between competing species may always be found, one cannot hope for clear patterns of interactions for whole guilds as can be seen in the West Indies, yet the consistency of the West Indian patterns is encouraging.

On the basis of guild composition relative to total land-bird species, one can use West Indian regressions to make predictions about guild size in larger communities. One could simply extend the regression for guild membership to any mainland community size. The problem is that mainland areas and species lists are usually determined by political rather than ecological or geographic boundaries. Who knows the area constraints that affect neotropical mainland birds?

To circumvent this problem, I looked at guild size and composition on eight neotropical land-bridge islands (Trinidad and Tobago near Venezuela, Coiba off Panama, and San Jose, Rey, Pacheca, Canas, and Bayoneta of the Pearl Islands near Panama). Guild designations were done largely by taxonomic distinctions, following the West Indian groupings but recognizing that this is less meaningful with mainland species. What can be seen in most cases (Fig. 11) is a strong convergence in pattern between the guild composition of the West Indies and these land-bridge islands. This is particularly true for flycatchers, gleaners, and frugivores. If one did a detailed categorization of the land-bridge species, several of the Tyrannidae that are primarily fruit-eaters (such as *Mionectes oleaginea*) could be transferred from the flycatchers to the frugivores. This would tighten the fit of the land-bridge island points to the West Indian, oceanic-island regressions. Only for nectarivores does the West Indian regression do a poor job of predicting guild size on land-bridge islands.

The strength with which Indian regressions predict the guild composition of neotropical land-bridge islands certainly suggests some overriding controls on avian community organization. This does not mean that these islands share identical assembly rules. The areas of the islands involved vary greatly between the two island groups. Coiba supports a bird community nearly the size of the large Greater Antilles, yet it is smaller than Gonave. Bird species communities of the Pearl Islands, many of which consist of 20 or more bird species, exist on islands of a few square kilometers (Wright et al. 1985). Only with the Pearl Island flycatchers is any form of size separation within a whole guild suggested (MacArthur et al. 1972).

Further analysis of the similarities and differences between the avifaunas of these two island groups is underway. The fact that similar-sized bird communities, some formed by colonization over water, others through the extinction of a larger fauna, show such similarity in the sizes of their various guilds is significant. It reinforces the idea that there are ecological controls on bird community structure, and it suggests how these controls may vary with diet or foraging method. Increased knowledge of the ecology of birds on all neotropical islands should allow us eventually to understand the factors that determine which bird species are present on Trinidad. With a better appreciation of the ecological factors at work on this 200 species island, it will be but a small step to a fuller understanding of the constraints on mainland bird communities and how these affect species distributions.

ACKNOWLEDGMENTS

Anyone who has worked in the West Indies owes a debt to James Bond, and for this paper in particular that debt is substantial. He and J. Diamond provided many constructive comments on this manuscript, although any remaining deficiencies are my own. M. S. Foster provided exceptional editorial assistance. Some of the work reported here was done with J. W. Terborgh, who with H. Horn provided early inspiration and guidance. Many people have helped in the field; special thanks go to S. J. Wright, D. J. Howell, D. Willard, W. J. Arendt, and J. W. Faaborg. Financial support has been provided by NSF (GB-38325), the Frank M.

Chapman Fund of the American Museum of Natural History, and the Research Council of the Graduate School, University of Missouri–Columbia.

LITERATURE CITED

AMERICAN ORNITHOLOGISTS' UNION. 1983. Check-list of North American Birds. 6th ed. American Ornithologists' Union, Washington, D.C.

BEARD, J. S. 1948. The natural vegetation of the Windward and Leeward Islands. Oxford For. Mem. No. 21.

BOND, J. 1956. Check-list of the Birds of the West Indies. Academy Natural Sciences, Philadelphia, Pennsylvania.

BOND, J. 1971. Birds of the West Indies. Houghton Mifflin, Boston, Massachusetts.

BOND, J. 1978. Derivations and continental affinities of Antillean birds. Spec. Publ. Acad. Nat. Sci. Phila. 13:119–128.

BROWN, J. H., W. A. CALDER III, AND A. KODRIC-BROWN. 1978. Correlates and consequences of body size in nectar-feeding birds. Am. Zool. 18:687–700.

CASE, T., J. FAABORG, AND R. SIDELL. 1983. The role of body size in the assembly of West Indian bird communities. Evolution 37:1062–1074.

CROWELL, K. L. 1968. Competition between two West Indian flycatchers *Elaenia*. Auk 85:265–286.

CRUZ, A. 1978. Adaptive evolution in the Jamaican Blackbird *Nesopsar nigerrimus*. Ornis Scand. 9:130–137.

CRUZ, A. 1980. Feeding ecology of the Black-Whiskered Vireo and associated gleaning birds in Jamaica. Wilson Bull. 92:40–52.

DIAMOND, A. W. 1973. Habitats and feeding stations of St. Lucia forest birds. Ibis 115:313–329.

DIAMOND, J. M. 1975. Assembly of species communities. Pp. 342–444, *In* M. L. Cody and J. M. Diamond (eds.), Ecology and Evolution of Communities. Kelknap, Cambridge, Massachusetts.

FAABORG, J. 1980. Further observations on ecological release in Mona Island birds. Auk 97:624–627.

FAABORG, J. 1982. Trophic and size structure of West Indian bird communities. Proc. Natl. Acad. Sci. USA 79:1563–1567.

FAABORG, J. R., AND J. W. TERBORGH. 1980. Patterns of migration in the West Indies. Pp. 157–163, *In* A. Keast and E. S. Morton (eds.), Migrant Birds in the Neotropics. Smithsonian Institution Press, Washington, D.C.

FFRENCH, R. 1973. A Guide to the Birds of Trinidad and Tobago. Livingston, Wynnewood, Pennsylvania.

HAMILTON, T. H., R. H. BARTH, AND I. RUBINOFF. 1964. The environmental control of insular variation in bird species abundance. Proc. Natl. Acad. Sci. USA 52:132–140.

HUTCHINSON, G. E. 1959. Homage to Santa Rosalia, or why are there so many kinds of animals? Am. Nat. 93:145–159.

JOHNSTON, D. W. 1975. Ecological analysis of the Cayman Island avifauna. Bull. Fl. State Mus., Biol. Sci. 19:235–300.

JOHNSTON, D. W., C. H. BLAKE, AND D. W. BUDEN. 1971. Avifauna of the Cayman Islands. Q. J. Fl. Acad. Sci. 34:141–156.

KEPLER, C. B., AND K. C. PARKES. 1972. A new species of warbler (Parulidae) from Puerto Rico. Auk 89:1–18.

LACK, D. 1973. The number of species of hummingbirds in the West Indies. Evolution 27:326–337.

LACK, D. 1976. Island Biology. University of California Press, Berkeley, California.

LACK, D., AND A. LACK. 1973. Birds on Grenada. Ibis 115:53–59.

LACK, D., E. LACK, P. LACK, AND A. LACK. 1973. Birds on St. Vincent. Ibis 115:46–52.

MACARTHUR, R. H., AND E. O. WILSON. 1967. The Theory of Island Biogeography. Princeton University Press, Princeton, New Jersey.

MACARTHUR, R. H., J. M. DIAMOND, AND J. R. KARR. 1972. Density compensation in island faunas. Ecology 53:330–342.

OLSON, S. L. 1978. A paleontological perspective of West Indian birds and mammals. Spec. Publ. Acad. Nat. Sci. Phila. 13:99–117.

PREGILL, G. K., AND S. L. OLSON. 1981. Zoogeography of West Indian vertebrates in relation to Pleistocene climatic cycles. Annu. Rev. Ecol. Syst. 12:75–98.

RICKLEFS, R. E. 1970. Stage of taxon cycle and distribution of birds on Jamaica, Greater Antilles. Evolution 24:475–477.

SIMBERLOFF, D. 1980. Dynamic equilibrium island biogeography: the second stage. Acta XVII Congr. Int. Ornithol. 2:1289–1295.

TERBORGH, J., AND J. FAABORG. 1973. Turnover and ecological release in the avifauna of Mona Island, Puerto Rico. Auk 90:759–779.

TERBORGH, J., J. FAABORG, AND H. J. BROCKMANN. 1978. Island colonization by Lesser Antillean birds. Auk 95:59–72.

TERBORGH, J., AND J. FAABORG. 1980a. Factors affecting the distribution and abundance of North

American migrants in the eastern Caribbean region. Pp. 145–155, *In* A. Keast and E. S. Morton (eds.), Migrant Birds in the Neotropics. Smithsonian Institution Press, Washington, D.C.

TERBORGH, J. AND J. FAABORG. 1980b. Saturation of bird communities in the West Indies. Am. Nat. 116:178–195.

WETMORE, A. 1916. Birds of Puerto Rico. U.S. Dep. Agric. Bull. 326:1–140.

WETMORE, A. 1957. The birds of Isla Coiba, Panama. Smithson. Misc. Collect. 134(9).

WRIGHT, S. J., J. FAABORG, AND C. J. CAMPBELL. 1985. Birds form tightly structured communities in the Pearl Island Archipelago, Panama. Pp. 798–812, *In* P. A. Buckley, M. S. Foster, E. S. Morton, R. S. Ridgely, and F. G. Buckley (eds.), Neotropical Ornithology. Ornithol. Monogr. No. 36.

ZUSI, R. L. 1969. Ecology and adaptations of the Trembler on the Island of Dominica. Living Bird 8:137–164.

APPENDIX I
ISLAND DISTRIBUTIONS OF SPECIES CONSIDERED IN THIS STUDY[1]

Species	X̄ weight (g)	Cuba	Grand Cayman	Jamaica	Hispaniola	Gonave	Beata	Saona	Mona	Puerto Rico	St. Johns	Anguilla	St. Barts	Barbuda	Antigua	St. Kitts	Montserrat	Guadeloupe	Desirade	Marie Galante	Isles Des Saintes	Dominica	Martinique	St. Lucia	St. Vincent	Barbados	Grenada
Frugivores																											
Columbidae																											
Columba squamosa	250	R	D	R	R	R	D	D	D	R	D	D	D	D	D	R	R	R	D	D	D	R	R	R	R	D	R
C. leucocephala	250	D	D	D	D	D	D	D	D	D	D	D	D	D	D	D	R	D	D	D	D	D	D	D		D	
C. inornata	250	M		M	M	M			M	M																	
C. caribaea	250			R																							
Zenaida asiatica	130	M	M	M	M	M	M	M	M	M	D	D	D	D	D	D	D	D	D	D	D	D	D	D	D	D	D
Z. aurita	150	D	D	D	D	D	D	D	D	D	D	D	D	D	D	D	D	D	D	D	D	D	D	D	D	D	D
Z. auriculata	110																										
Z. macroura	150	M	D	M	M	M	M	M	M	M	D	D	D	D	D	D	D	D	D	D	D	D	D	D	D	D	D
1. Columbina passerina	27–37	D	D	D	D	D	D	D	D	D	D	D	D	D	D	D	D	D	D	D	D	D	D	D	D	D	D
Leptotila rufaxilla	?																										
L. jamaicensis	162		D	W																							
Geotrygon chrysia	145	D			D	D				D																	
G. mystacea	230									R	D	D	D	D	D	W	W	D	D	D	D	D	D	D			
G. caniceps	210	M			M																						
G. montana	140–180	R		W	R					R	D	D	D	D	D	D	W	D	D	D	D	R	R	R	R		R
G. versicolor	225			R																							
Starnoenas cyanocephala	?	M																									
Psittacidae																											
Aratinga chloroptera	?			M			E																				
A. euops	?	M	R																								
A. nana	?			M																							
Ara tricolor	?	E																										
Amazona leucocephala	?	W	D																								
A. collaria	?			R																							
A. ventralis	?				W	D																					

APPENDIX I
CONTINUED[1]

Species	X̄ weight (g)	Cuba	Grand Cayman	Jamaica	Hispaniola	Gonave	Beata	Saona	Mona	Puerto Rico	St. Johns	Anguilla	St. Barts	Barbuda	Antigua	St. Kitts	Montserrat	Guadeloupe	Desirade	Marie Galante	Isles Des Saintes	Dominica	Martinique	St. Lucia	St. Vincent	Barbados	Grenada
A. vittata	?									R																	
A. agilis	?			R																							
A. arausiaca	?																					R					
A. versicolor	?																							R			
A. guildingii	?																								R		
A. imperialis	?																					R					
Trogonidae																											
Priotelus temnurus	?	M																									
P. roseigaster	?				M																						
Muscicapidae (Turdinae)																											
2. Myadestes elisabeth	27	M																									
3. M. genibarbis	27			R	M																	R	R	R	R		R
4. Turdus fumigatus	73																						M	M	D		D
5. T. nudigenis	54																										
6. T. jamaicensis	66			R																							
7. T. swalesi	75				M																						
8. T. aurantius	81			R																							
9. T. ravidus	?		E																								
10. T. plumbeus	75	W			W	W				W												R					
11. Cichlherminia lherminieri	100																R	R				D		R			
Mimidae																											
12. Mimus polyglottos	45	D	D		D	D	D	D		D	D																
13. M. gilvus	45										D							D									D
14. M. gundlachii	75			D																							
15. Ramphocinclus brachyurus	50?																						D	D			
16. Margarops fuscus	75													D	D	W	W	W	D	D		W	W	W	W	D	D
17. M. fuscatus	100						D			W	W			D	D	W	W	W	D	D		W	W	W	E		R

APPENDIX I
Continued[1]

	X weight (g)	Cuba	Grand Cayman	Jamaica	Hispaniola	Gonave	Beata	Saona	Mona	Puerto Rico	St. Johns	Anguilla	St. Barts	Barbuda	Antigua	St. Kitts	Montserrat	Guadeloupe	Desirade	Marie Galante	Isles Des Saintes	Dominica	Martinique	St. Lucia	St. Vincent	Barbados	Grenada
18. *Cinclocerthia ruficauda*	50															R	R	R				R	R	R	R		R
Dulidae																											
19. *Dulus dominicus*	42				M	M		M																			
Emberizidae (Thraupinae)																											
20. *Tangara cucullata*	20																								R		W
21. *Euphonia jamaica*	16			W																							
22. *E. musica*	16	W	D	R	W	W				W							W	W		D	D	W	W	W	W		W
23. *Spindalis zena*	28–43	W	D	R	W	D				W																	
24. *Phaenicophilus palmarum*	31				W	D		W																			
25. *P. poliocephalus*	31				W	D																					
26. *Calyptophilus frugivorus*	30				M	D																					
27. *Nesospingus speculiferus*	35									R																	
Emberizidae (Cardinalinae)																											
28. *Saltator albicollis*	52																	D				D	D	D			
Emberizidae (Emberizinae)																											
29. *Volatinia jacarina*	11	W																								M	M
30. *Sporophila nigricollis*	11	D																								M	M
31. *Melopyrrha nigra*	17?	M	D																								
32. *Tiaris canora*	10	M	D																								
33. *T. olivacea*	9	M	D	M	M	M				M																	
34. *T. bicolor*	10	M	D	D	M	D				D	D	D	D	D	D	D	D	D	D	D	D	D	D	D	D	D	D
35. *Loxipasser anoxanthus*	12			W																							
36. *Loxigilla portoricensis*	32									W																	
37. *L. violacea*	22–29		W		W	D	D	W		W						E											
38. *L. noctis*	18											W	W	W	W	W	W	W	W	W	W	W	W	W	W	W	W
39. *Melanospiza richardsonii*	11																							W			
40. *Torreornis inexpecta*	?	M																									

APPENDIX I
CONTINUED[1]

	X̄ weight (g)	Cuba	Grand Cayman	Jamaica	Hispaniola	Gonave	Beata	Saona	Mona	Puerto Rico	St. Johns	Anguilla	St. Barts	Barbuda	Antigua	St. Kitts	Montserrat	Guadeloupe	Desirade	Marie Galante	Isles Des Saintes	Dominica	Martinique	St. Lucia	St. Vincent	Barbados	Grenada
41. *Ammodramus savannarum*	17				M					M																	
42. *Zonotrichia capensis*	21				M																						
Fringillidae																											
43. *Loxia leucoptera*	28				M																						
44. *Carduelis dominicensis*	9				M																						
FLYCATCHING INSECTIVORES																											
Todidae																											
1. *Todus multicolor*	?	W																									
2. *T. subulatus*	9				D	D																					
3. *T. angustirostris*	9				R	D																					
4. *T. todus*	7			W																							
5. *T. mexicanus*	6									W																	
Tyrannidae																											
6. *Myiopagis cotta*	13			W																							
7. *Elaenia martinica*	22									M	D	D	D	D	D	D	W	D	D	D	D	D	D	D	M	D	D
8. *E. flavogaster*	26	M		M	M	D																			M		M
9. *E. fallax*	14			R	M	D																					
10. *Contopus caribaeus*	10	M								R																	
11. *C. latirostris*	11									R								R				R	R	R			
12. *Empidonax euleri*	11			R																							
13. *Myiarchus barbirostris*	12			R																							
14. *M. nugator*	37																								R		R
15. *M. validus*	39			R																							
16. *M. sagrae*	21	W	W																								
17. *M. stolidus*	20		D	D		D		W																			
18. *M. antillarum*	20									W																	

APPENDIX I
CONTINUED[1]

	Grenada	Barbados	St. Vincent	St. Lucia	Martinique	Dominica	Isles Des Saintes	Marie Galante	Desirade	Guadeloupe	Montserrat	St. Kitts	Antigua	Barbuda	St. Barts	Anguilla	St. Johns	Puerto Rico	Mona	Saona	Beata	Gonave	Hispaniola	Jamaica	Grand Cayman	Cuba	X̄ weight (g)
19. *M. oberi*														D													37
20. *Tyrannus melancholicus*	M			R	R	R				R		R															45
21. *T. dominicensis*	D	D	D	D	D	D	D	D	D	D	D	D	D	D	D	D	D	D	D	D	D	D	D	D	D	M	45
22. *T. caudifasciatus*																		M					M	M	M	M	39–45
23. *T. cubensis*																											?
Pachyramphus niger																								R			?
GLEANING INSECTIVORES																											
Cuculidae																											
1. *Coccyzus americanus*		D	D	D			D	D	D						D	D	D	W	D	D	D	D	D		D	D	65
2. *C. minor*	W		W	W	W	W	D	D	D	W	W	W						W	D	W			W	W	D	W	65
Saurothera merlini																									D	W	?
S. vieilloti																		W									80
S. longirostris																				W		W	W				110
S. vetula																								W			?
Hyetornis pluvialis																								R			163
H. rufigularis																						D	W				?
Crotophaga ani	M						M	M		M							M	M		M	M	M	M	M	M	M	110
Troglodytidae																											
3. *Ferminia cerverai*																										M	?
4. *Troglodytes aedon*	R			R	R	R				R																	11
Muscicapidae (Sylviinae)																											
5. *Polioptila lembeyei*																										D	6
Vireonidae																											
6. *Vireo crassirostris*																									D		?
7. *V. modestus*																									W		10
8. *V. gundlachii*																										W	?

APPENDIX I
Continued[1]

Species	X̄ weight (g)	Cuba	Grand Cayman	Jamaica	Hispaniola	Gonave	Beata	Saona	Mona	Puerto Rico	St. Johns	Anguilla	St. Barts	Barbuda	Antigua	St. Kitts	Montserrat	Guadeloupe	Desirade	Marie Galante	Isles Des Saintes	Dominica	Martinique	St. Lucia	St. Vincent	Barbados	Grenada
9. *V. latimeri*	11									W																	
10. *V. nanus*	11				R	D																					
11. *V. osburni*	20			R																							
12. *V. altiloquus*	20	W	D	W	W	W		W		W	W	D	D	D	D	W	W	W	D	D	D	W	W	W	W	W	W
13. *V. magister*	?		D																								
Emberizidae (Parulinae)																											
14. *Dendroica petechia*	11	M	D	M	M	M		M		M	M	D	D	D	D	D	D	D	D	D	D	D	D	D	D	D	
15. *D. adelaidae*	7									D	D			D													
16. *D. pityophila*	?	M																									
17. *D. pinus*	11	M			M																						
18. *D. vitellina*	11		D																								
19. *D. plumbea*	11																	R				R					
20. *D. pharetra*	11			R																							
21. *D. angelae*	11									M																	
22. *Catharopeza bishopi*	?																								R		
23. *Microligea palustris*	13				W		D																				
24. *Teretistris fernandinae*	?	M																									
25. *T. fornsi*	?	M																									
26. *Leucopeza semperi*	?																							R			
27. *Xenoligea montana*	?				M																						
Emberizidae (Icterinae)																											
Agelaius phoeniceus	?	M																									
A. humeralis	?	M			M																						
A. xanthomus	?								M	M																	
28. *Nesopsar nigerrimus*	39			R																							
Sturnella magna	?	M	M																								
Dives atroviolacea	?	M																									
Quiscalus niger	?	M	M					M		M																	

APPENDIX I
CONTINUED[1]

Species	X̄ weight (g)	Cuba	Grand Cayman	Jamaica	Hispaniola	Gonave	Beata	Saona	Mona	Puerto Rico	St. Johns	Anguilla	St. Barts	Barbuda	Antigua	St. Kitts	Montserrat	Guadeloupe	Desirade	Marie Galante	Isles Des Saintes	Dominica	Martinique	St. Lucia	St. Vincent	Barbados	Grenada
Q. lugubris	?																M	M				M	M	M	M	M	
29. Icterus dominicensis	35	W			W	W		W		W																	
30. I. laudabilis	35																							R			
31. I. oberi	35																R										
32. I. bonana	35																						R				
33. I. leucopteryx	35		D	W																							
NECTARIVORES																											
Trochilidae																											
1. Glaucis hirsuta	7																										R
2. Anthracothorax mango	8			D																							
3. A. dominicus	6				W	D	D	W		D																	
4. A. viridis	6									R																	
5. Eulampis jugularis	9															D	D	D				D	D	D	D		
6. E. holosericeus	6									M	D	D	D	D	D	D	D	R	D	D	D	D	R	R	R	D	D
7. Orthorhynchus cristatus	3									M	D	D	D	D	D	W	W	W	D	D	D	D	D	D	W	D	D
8. Chlorostilbon ricordii	4	W	D																								
9. C. swainsonii	4				M																						
10. C. maugaeus	3									W																	
11. Cyanophaia bicolor	4																					R	R				
12. Trochilus polytmus	6			W																							
13. Mellisuga minima	2			W	W	W		W																			
14. M. helenae	2	W																									
Emberizidae (Coerebinae)																											
15. Coereba flaveola	8–12	W	D	W	W	D	D	W		W	W	D	D	D	D	D	W	W	D	D	W	W	W	W	D	W	W
Emberizidae (Thraupinae)																											
16. Cyanerpes cyaneus	14	M																									
Emberizidae (Emberizinae)																											
17. Euneornis campestris	16			R																							

APPENDIX I
CONTINUED[1]

MISCELLANEOUS SPECIES

Species	X̄ weight (g)	Cuba	Grand Cayman	Jamaica	Hispaniola	Gonave	Beata	Saona	Mona	Puerto Rico	St. Johns	Anguilla	St. Barts	Barbuda	Antigua	St. Kitts	Montserrat	Guadeloupe	Desirade	Marie Galante	Isles Des Saintes	Dominica	Martinique	St. Lucia	St. Vincent	Barbados	Grenada
Caprimulgidae																											
Chordeiles gundlachii	?	W	D	D	W	D	D	W		W																	
Siphonorhis brewsteri	?			E	M	M																					
Caprimulgus otiosus	?																							M			
C. cubanensis	?	M			M																						
C. noctitherus	?									D																	
C. cayennensis	?																						?				
Nyctibiidae																											
Nyctibius griseus	?			M	M	M																					
Apodidae																											
Cypseloides niger	31	M		M	M					M					M			M				M	M	M	M		
Streptoprocne zonaris	?	M		M	M																				M		M
Chaetura brachyura	?																										
C. cinereiventris	?																										
C. martinica	?																	M				M	M	M			
Tachornis phoenicobia	11	M		M	M			M		M																	
Alcedinidae																											
Ceryle torquata	?																	M				M	M				
Picidae																											
Nesoctites micromegas	33				W																						
Melanerpes herminieri	100																	W									
M. portoricensis	50									W																	
M. striatus	65				W																						
M. radiolatus	101	W																									
M. superciliaris	?	W	W																								

APPENDIX I
CONTINUED[1]

	X̄ weight (g)	Cuba	Grand Cayman	Jamaica	Hispaniola	Gonave	Beata	Saona	Mona	Puerto Rico	St. Johns	Anguilla	St. Barts	Barbuda	Antigua	St. Kitts	Montserrat	Guadeloupe	Desirade	Marie Galante	Isles Des Saintes	Dominica	Martinique	St. Lucia	St. Vincent	Barbados	Grenada
Xiphidiopicus percussus	?	W																									
Colaptes auratus	88–125	W	D																								
C. fernandinae	?	M																									
Campephilus principalis	?	M																									
Hirundinidae																											
Progne cryptoleuca	?	M																									
P. dominicensis	?			M	M					M	M						M	M	M	M	M	M	M	M	M	M	M
Tachycineta euchrysea	?			M	M																						
Hirundo fulva	?	M		M	M	M				M																	
Corvidae																											
Corvus palmarum	?	W			W																						
C. nasicus	?	W																									
C. leucognaphalus	?				W	W		W		E																	
C. jamaicensis	?			R																							

[1] Nomenclature follows the American Ornithologists' Union Check-list (1983). Habitat preferences are generalized, with D = dry forest, R = rainforest, W = widespread, E = extirpated, and M = miscellaneous habitats, including open forest, disturbed sites, and specialized habitats such as pine or mangrove. Weights are generalized values for several islands, or estimates when specific values are unknown.

NUMERICAL RELATIONSHIPS BETWEEN MIGRANT AND RESIDENT BIRD SPECIES IN JAMAICAN WOODLANDS

MICHAEL GOCHFELD

Department of Environmental and Community Medicine, U.M.D.N.J.—Rutgers Medical School, Piscataway, New Jersey 08854 USA

ABSTRACT. Using censuses from three forest types on Jamaica, West Indies, I compared the proportions of migrant wood warblers (Parulinae) and resident nine-primaried oscines, to determine whether statistical evidence for an impact of the migrants on the resident species exists. These two assemblages contributed equally to the total numbers of birds in the main community sampled (Southern Riverine Forest). Migrants accounted for 20 to 55 percent of individuals in Southern Riverine forest, 12 to 19 percent in Montane Humid Forest, and 30 to 41 percent in Mangrove woodland. Numbers of migrants and residents were not correlated across 12 riverine forest censuses, indicating neither repulsion nor positive covariation. Results are compared with data reported by Lack (1976). In summer when migrant warblers are absent, resident populations are augmented by arrival of three abundant breeding migrants. The case of the Yellow Warbler, *Dendroica petechia*, cited by Lack as evidence for competitive displacement by migrants, is reexamined and an alternative explanation presented. Compared with resident species, the assemblage of migrant warblers showed comparable species richness and variability across censuses, but showed greater species diversity, despite greater morphologic uniformity. The results of this study and Lack's obtained six years apart show good concordance.

RESUMEN. He comparado la proporción de gorjeadores migratorios (Parulinae) y oscines residentes de nueve primarias, utilizando censos de tres tipos de bosques en la isla de Jamaica en las Antillas, para determinar si existe una evidencia estadística sobre el impacto que las aves migratorias causan a las residentes. Estos dos grupos contribuyen de igual manera al número total de aves en la principal comunidad estudiada (bosque riverino del sur). La cantidad de aves migratorias fue de 20 a 55 por ciento de los individuos en el bosque riverino del sur, 12 a 19 por ciento en bosque montañoso húmedo y 30 a 40 por ciento en bosques de manglares.

Los números de aves migratorias y residentes no tuvieron correlación a lo largo de 12 bosques riverinos censados, indicando que no existe ni repulsión, ni covariación positiva. Los resultados son comparados con información que presentara Lack (1976). En el verano cuando los gorjeadores migratorios están ausentes, las poblaciones residentes aumentan con la llegada de tres especies migratorias anidadoras abundantes. Se reexamina y presenta una explicación alternativa respecto al caso del gorjeador amarillo, *Dendroica petechia*, citado por Lack como evidencia de desplazamiento competitivo por una especie migratoria. Comparados con especies residentes, el grupo de gorjeadores migratorios presentaba una riqueza y variedad de especies comparable a lo largo de los censos, pero se observa una mayor diversidad de especies, a pesar de la mayor uniformidad morfológica. Los resultados de este trabajo y los de Lack, hace 6 años, muestran una buena concordancia.

Ethnocentrism of ecologists who reside in the north temperate zone and migrate southward to study avian ecology, may have distorted our understanding of the role of wintering migrants in tropical bird communities. Emphasis on the "adverse impact" of migrants as competitors (Miller 1963), or on the relegation of migrants to peripheral habitats (see Chipley 1976), ignores the alternative that migrants are integral components of tropical ecosystems, and emigrate to the temperate zone only for breeding purposes (see papers in Keast and Morton 1980). Historically, the wood warblers (Parulinae) have played an important role in communities of the circum-Caribbean region, for the group probably evolved there, even though most speciation occurred in temperate and boreal North America (Mengel 1964).

In this paper I examine bird communities, in three types of Jamaican forest. I compare migrant wood warblers and resident species in terms of species richness and diversity, and compare my data with those obtained by Lack (1976). My study was prompted by the recognition that migrant birds often reach great abundance in some tropical habitats, and, thus, may have great impact on populations of resident species.

METHODS

I studied birds on Jamaica from 7 November 1964 to 28 January 1965, censusing birds in three types of forests (see Asprey and Robbins 1953). The main study area comprised 4 ha of mesic woodland on the south side of the Mona Reservoir (elevation 250 m) in a northeastern suburb of Kingston. This study site, adjacent to the University of the West Indies campus, has been discussed by Smith (1971), Lack (1976), and Diamond et al. (1977), and is termed Southern Riverine Forest (Lack 1976). The mangrove site was around the perimeter of Great Salt Pond, southwest of the Hellshire Hills. The highland forest site was at Hardwar Gap (elevation 1200 m).

Eugene Eisenmann (pers. comm.) was a stern critic of many studies of species diversity and populations done in the Neotropics, arguing that inexperienced field observers inevitably misidentified some species, overlooked many others that are inconspicuous, and failed to take into account seasonal changes in vocalizations and other habits, (see also Karr 1976). Because many of these problems could be avoided in Jamaica, it offered a favorable locale in which to examine a tropical bird community. For one thing, the avifauna of the Greater Antilles, which is believed (Bond 1978) to have originated mainly from tropical North American elements, is impoverished compared to mainland Neotropical faunas. Moreover, excellent field guides exist, so that with experience one can identify confidently almost every bird seen and most heard. This is in marked contrast to Central or South America where non-identification or mis-identification may significantly bias results. Moreover, migrants from North America are abundant on Jamaica (Lack 1976).

The main census techniques employed for tropical bird studies have been mist-net sampling and variations of the strip census techniques (Emlen 1971). Waide et al. (1980) found that these methods can produce somewhat different estimates, although for migrant wood warblers in Puerto Rico, Post (1978) found no significant difference. I used a strip transect method comparable to that employed by the Lacks (Lack and Lack 1972; Lack 1976). At Mona I established parallel transects 50 m apart and walked slowly along the paths counting all birds seen or heard within 25 m of my path. In the mangroves and montane forest I walked slowly along pre-existing paths for periods of 2 to 3 hours beginning at sunrise. All censuses were conducted in early morning and were completed by 3 hours after sunrise. I made 12 censuses (23 hrs) at Mona, eight censuses (18 hrs) in the mangroves, and eight censuses (21 hrs) at Hardwar Gap. The sites differed in configuration, and I walked single long transects in mangroves and mountains, but parallel transects at Mona, hence one must be careful in comparing the results across my sites.

For each census I estimated the number of individuals encountered of each species, and calculated the species diversity index using both the Shannon-Weiner equation ($H' = -\sum p_i \log_e p_i$) and an alternative measure ($1/\sum p_i^2$), where p_i is the proportion of all individuals encountered belonging to the ith species. Species richness is simply the total number of species encountered. Data were compared for all resident species of the nine-primaried oscine assemblage (Emberizidae: tanagers, orioles, wood warblers, honey creepers, and finches), and for the migrant wood warblers. The two assemblages are only partly comparable in that the nine-primaried oscines show greater morphological diversity. However, the bulk of those species resident in Jamaican forests are actually quite comparable to the parulinae, including in addition to resident wood warblers, two small to medium tanagers, and two honey creepers. The main differences lie in the inclusion of a large oriole and two small grassquits (*Tiaris*).

For three habitat types I was able to compare my data with that obtained by Lack (1976). I then compared Lack's (1976) data for summer and winter censuses across many habitats.

RESULTS

SOUTHERN RIVERINE FOREST

On 12 complete censuses (23 hrs) in the Mona Woods I observed 52 species, of which 55 to 87 percent ($\bar{X} \pm$ s.d. = 76.4 ± 10.9%) were passerines. Table 1 shows the relative abundance and diversity of all species, of migrant wood warblers (Parulinae), and of Resident Nine-primaried Oscines (RNPO), and gives the total daily counts for all species on the 12 censuses. Both assemblages accounted for about one-third of the 52 species found. Migrant warblers comprised 11 to 46 percent of the daily counts (30.2 ± 10.9%). Lack's Lowland Riverine

TABLE 1

RELATIVE ABUNDANCE AND DIVERSITY OF MIGRANT WARBLER AND RESIDENT NINE-PRIMARIED OSCINE ASSEMBLAGES IN MONA WOODS, JAMAICA[a]

Census	Total daily census for all species					Migrant Warblers					Resident nine-primaried Oscines				
	No. sp.	No. indiv.[b]	H'	$1/\Sigma p^2$	Eq.	No. sp.	No. (%) indiv.	H'	$1/\Sigma p^2$	Eq.	No. sp.	No. (%) indiv.	H'	$1/\Sigma p^2$	Eq.
1.	28	88	3.08	18.00	.65	9	42 (48)	1.96	6.25	.69	8	20 (23)	1.99	6.90	.86
2.	30	135	2.77	9.28	.31	10	34 (25)	2.13	7.60	.76	10	63 (47)	1.55	2.92	.29
3.	38	254	3.03	13.89	.37	10	28 (11)	2.21	8.52	.85	12	83 (33)	1.85	4.63	.39
4.	28	132	2.73	7.63	.26	9	25 (19)	2.10	7.71	.83	7	32 (24)	1.71	4.57	.65
5.	28	64	2.99	16.92	.71	7	13 (20)	1.88	6.26	.89	6	20 (31)	1.72	5.26	.88
6.	24	68	3.06	18.49	.71	11	25 (37)	2.21	7.91	.72	8	24 (35)	1.95	6.40	.80
7.	26	113	3.34	24.70	.75	10	35 (31)	2.18	8.22	.82	10	40 (35)	2.14	7.34	.73
8.	33	65	3.17	20.81	.74	9	30 (46)	2.09	7.50	.83	8	15 (23)	1.96	6.43	.80
9.	28	158	3.13	16.15	.47	10	42 (27)	2.04	6.73	.67	9	39 (25)	1.97	6.11	.68
10.	34	112	3.07	18.39	.68	8	38 (34)	1.83	5.16	.64	8	39 (35)	1.99	6.82	.85
11.	27	96	3.13	18.51	.62	10	35 (36)	2.15	7.80	.78	10	37 (38)	1.99	5.68	.57
12.	30	100	2.92	13.93	.52	9	29 (29)	1.99	6.52	.72	9	50 (50)	1.81	4.92	.55
Mean	29.4	115	3.03	16.93	.56	9.3	31 (30)	2.06	7.18	.78	8.7	39 (33)	1.89	5.70	.67
s.d.	3.89	52.7	.17	4.71	.18	1.07	8.2 (10.9)	.13	.99	.08	1.60	19.6 (8.9)	.16	1.36	.19
CV[d]	13.2	45.6	5.5	28.7	30.9	11.5	26 (36)	6.1	13.8	10.3	18.4	51 (27)	8.5	22.1	28.4

[a] Habitat type is Southern Riverine Forests; census period = November 1964 to January 1965.
[b] Number of individuals observed.
[c] H' = −Σ p $\log_e$ p, the Shannon-Wiener Measure of Diversity; $1/\Sigma p^2$ is the Simpson Index of Diversity; Eq = equitability = H'/H_{max} where H_{max} is maximum value of H' if all species are equally represented in population.
[d] CV = Coefficient of Variation ([SD/mean] × 100) for the 12 censuses.

TABLE 2
LANDBIRD CENSUSES IN THREE JAMAICAN FOREST HABITATS SHOWING CONTRIBUTION OF
MIGRANT PARULINAE[1]

	Southern Riverine Forest		Northern Riverine Forest		Mangrove Woodland	
	Present study	Lack (1976)	Present study	Lack (1976)	Present study	Lack (1976)
Species found	52	54	35	45	41	28
Mean no. individuals found						
All landbirds	520.2	nc	217.5	186	408.0	273
Migrant Parulids	159.3	193	27.3	35	124.5	113
Percent Parulids	30.2	25.8	12.6	18.8	30.5	41.4
Percent of species in common[2]	81%		69%		63%	
Rank correlation[3]	+0.64**		+0.69**		+0.72**	

[1] Compares data from present study with Lack (1976); nc = not counted.
[2] Percent of species on combined list of those recorded by Lack (1976) and in the present study.
[3] Kendall tau correlation between numbers of individuals for species recorded by both Lack (1976) and present study.
** Correlation significant at 0.01 level.

Forests data include an unspecified proportion of observations from the Ferry River so that our results are not strictly comparable (Table 2). I found three species not listed by Lack (1976), two flycatchers (Tyrannidae), and the Golden-winged Warbler (*Vermivora chrysoptera*), then the first record for Jamaica (Gochfeld 1974). Lack recorded five species I did not see, but some of these may have been at Ferry River. For the 46 species encountered by both of us, the numbers of individuals recorded per 10 hours observation across all species were significantly positively correlated ($r = 0.64$, $P < 0.01$) indicating an overall similarity in the way we sampled the faunas.

In the Mona Woods the migrant warblers had a slightly greater species richness than the RNPO assemblage ($\bar{X} = 9.3$ vs 8.7, NS), and slightly lower variability (coefficient of variation across censuses = 11.5 vs 18.4; Table 1). Species diversity of the migrant assemblage was significantly greater (Mann Whitney U = 24, $P < 0.02$) than that of the RNPO assemblage. However, the latter assemblage is morphologically and taxonomically more diverse including finches and as well as wood warblers, bananaquits, and small tanagers. I found no correlation ($r = 0.03$) between the numbers of individuals in the two assemblages over the 12 censuses.

MANGROVES

I made eight censuses (18 hrs) in mangroves over a three-month period, while Lack (1976) published results of 14.5 hours over a six-month period (Table 2). Of the 41 species I observed, Lack listed only 24 for mangroves, while he recorded four I did not find. These differences reflect, in part, the different configurations of the mangrove woodlands, for this forest forms a relatively narrow band around much of Jamaica, and at Great Salt Pond, birds from the adjacent arid woodlands frequently appeared in the mangroves. Of the 24 species common to both lists, the numbers of birds showed a significant positive correlation ($r = +0.69$, $P < 0.01$). In my sample 30.5 percent of all birds were migrant parulids.

MONTANE FOREST

On eight censuses (18 hrs) over a three-month period, I recorded 35 species of which 12 were North American migrants. Of the 33 species common to Lack's (1976) and my lists, the numbers of individuals recorded per 10 hours observation were strongly correlated ($r = +0.72$, $P < 0.01$). Lack worked the montane forest for 50 hours over a six-month period. I saw only three species not on Lack's list, including the Lincoln Sparrow (*Melospiza lincolnii*; Gochfeld 1969). Lack, however, found 13 species I did not see, almost all encountered at a rate of less than one individual per 10 hours. Of the 48 species on our combined list, 24 probably breed near Hardwar Gap; most others are winter visitors. One of the species, *Elaenia fallax,* is a migrant that winters in unknown quarters and returns to Jamaica to breed.

Among the migrant Parulinae, the Yellow-rumped (Myrtle; *Dendroica coronata*), Yellow-throated (*D. dominica*), and Black-throated Blue (*D. caerulescens*) Warblers were seen on most censuses the former two foraging in taller pine trees. These were among the parulids that Terborgh and Faaborg (1980) found predominant in highland pine forests.

DISCUSSION

Contrary to the traditional view of migrants simply as seasonal visitors to tropical communities, Emlen (1980), Rappole and Warner (1980), and Schwartz (1980), among others, argued that the winter communities of the tropics are well integrated, including both resident and migrant species, and that, in general, when the migrants depart in the spring, the residents do not expand their niches. Migrants are often territorial (Rappole and Warner 1980; Schwartz 1980). Chipley (1976) presented evidence that once migrants are present in the tropics, migrant-migrant, resident-resident, and migrant-resident associations in flocks do not differ from random.

The posthumous appearance of Lack's book (1976) provides an important base of comparison for my own data obtained six years prior to Lack's visit. Although I have grave concerns about generalizing distributional trends and relative abundances, much less competition and community evolution, from a single season of field work (see Karr 1976), it is reassuring to note the remarkable concordance between the results of the two studies. This stands in marked contrast, for example, to a comparison made by Vuilleumier (1978) for a wet tropical forest area in Ecuador, where there was only about 37 percent overlap in the species recorded by two observers.

PROPORTION OF MIGRANTS

In my study the proportion of migrants in Southern Riverine Forest averaged 38 percent. This compares with contributions for migrants (all species) of 30 percent in Jamaica (Lack and Lack 1972), 8 to 45 percent in Colombia (Chipley 1976; Hilty 1980), 14 percent in Panama (Willis 1980), 27 percent in Yucatan (Waide et al. 1980), 46 percent in Bahamas, and 69 percent in northern Florida (Emlen 1980). A strong gradient is apparent as one moves east and south from Jamaica through the Lesser Antilles (Terborgh and Faaborg 1980), with the proportion of migrants decreasing to only 1 percent.

HABITAT USE BY MIGRANTS

Many authors have described the occupation by migrants of marginal, dry, open, or disturbed habitat (e.g., Willis 1966, 1980; Chipley 1976; Fitzpatrick 1980). Migrants are considered characteristic of both wetter habitats (Waide et al. 1980) and dryer ones (Hespenheide 1980). They seem to dwell in the understory of Yucatan forests (Waide et al. 1980), but shun the undergrowth on the Bahamas (Emlen 1980). Theoretically, they may occupy seasonal or unpredictable habitat or those newly available that have not yet been colonized by resident species (Waide 1980). On the Bahamas migrants occupy open country and canopy, rather than forest understory (Emlen 1980). On Jamaica I found migrant species in all habitats surveyed. They actually comprised a lower proportion of the montane community, although other studies, for example, those of Hilty (1980, for Colombia) and Terborgh and Faaborg (1980, eastern Caribbean) report a greater proportion of migrant species as elevation increased.

MIGRANT PARULINAE ON JAMAICA

Lack's (1976) studies emphasized the role of competition in shaping bird communities. Spatial shifts (Lack and Lack 1972), temporal shifts in breeding (Miller 1963; Emlen 1980), and niche shifts (Chipley 1976) have been attributed to increased competitive pressure from wintering species. Karr (1976a) pointed out that in both the Old and New Worlds, migrants tend to exploit superabundant or temporally patchy food resources on their tropical wintering grounds. Post (1978) arrived at the same conclusion from his study of parulids on Puerto Rico. Although I made no attempt to infer feeding "niches" by examining foraging behavior, I estimated the impact of the wintering migrants by a numerical comparison of abundance, and by comparison of my data with Lack's (Lack and Lack 1972; Lack 1976).

Although some species of warblers show marked fluctuations in abundance from year to year, others may show directional trends. Comparisons of studies made at a substantial interval may clarify such trends, but two studies cannot distinguish fluctuations from trends.

Two species, the Arrow-headed (*Dendroica phaetra*) and Yellow (*D. petechia*) Warblers, breed on Jamaica. The latter is abundant in mangroves where I recorded 41.8 and Lack recorded 62 individuals per 10 hours of observation. Lack and Lack (1972) stated that in summer the Yellow Warbler expands its range from the mangroves, where it is seemingly confined in winter, into lowland forests in which it breeds. They implied that it is excluded from the forests by the large populations of wintering warblers. This hypothesis certainly merits further study on Jamaica, but some additional evidence can be adduced to clarify the situation. Even on the Lesser Antilles, where migrant parulids are much less numerous, the resident race of Yellow Warbler mainly occupies the mangroves (Bond 1971). It is hardly likely that these southern races are excluded from forests by the occasional migrant parulid; it seems likely that the apparent confinement of Yellow Warblers to Jamaican mangroves reflects the habitat preference of the species and its main ecological requirements, rather than competitive exclusion. Yellow Warblers are a mangrove species not only in the Antilles, but in Central America as well (E. Eisenmann, pers. comm.).

On Jamaica migrant warblers form a smaller proportion of the montane community than of the lowland forest community. Thus, although parulids comprised 33 percent of the species seen near Hardwar Gap, they accounted for only 13 percent of the individuals compared with about 38 percent of the individuals in the Mona Woods. In addition to the migrant warblers recorded in this study as regular or vagrant migrants or winter residents, Lack (1976) mentioned 12 rare transient species. One of these, the Black-poll (*Dendroica straita*) is clearly a migrant wintering mainly in South America (Bond 1971; ffrench 1973). The other species winter mainly in Central America and are probably best treated as vagrants (Gochfeld 1979).

Several authors have emphasized the possible role of migrants in interfering with residents. Thus, Miller (1963) and Emlen (1980) indicated that residents delay breeding until migrants leave. Waide et al. (1980), however, found that in Yucatan residents began breeding while migrants were still present, although young were being fed after the bulk of the migrants had gone. In the Mona Woods there was no correlation across censuses between the numbers of individuals in the wood warbler and the resident nine-primaried oscine assemblages. Thus, there was no evidence of short-term repulsion, nor were wood warblers increasing and decreasing as the RNPO increased and decreased. Such short-term adjustments, however, might not be expected to play an important role in alleviating competition.

To examine longer-term adjustments I compared (Table 3) summer and winter populations on Jamaica from data in appendices I and II of Lack's (1976) book. If the Jamaican avifauna is stressed by the presence of North American migrants, this should be reflected in much higher winter than summer populations. For both the Southern Riverine and the Montane Wooded Cultivation (partly cleared highland forest, see Lack 1976) habitats, differences between summer and winter populations are approximately equal to the influx of the wood warblers. Thus, the resident populations did not decrease when the migrants were present, nor was there an increase after their departure. In other habitats, however, the combined resident and migrant populations in winter were actually lower than the total summer population (residents plus summer migrants; Wilcoxon Matched-Pair Test, $P < 0.05$, one-tailed). Unfortunately, this difference is due partly to the fact that the winter data include only birds seen, whereas the summer data include birds seen or heard, and to the fact that birds may be more conspicuous in one season than in another. However, the data provide no basis for inferring a "release" for the residents when the migrants depart for the north.

Breeding of residents might be curtailed, however, if the migrant populations offered severe competition. Most resident species begin nesting in the spring, as the migrants are leaving, and rear their young in the absence of the migrant warblers. If one examines the contribution of the three breeding migrants, Gray Kingbird (*Tyrannus dominicensis*), Greater Antillean Elaenia (*Elaenia fallax*), and Black-whiskered Vireo (*Vireo altiloquus*), however, it can be seen (Table 3) that the populations of these birds, particularly the abundant vireo, augment the resident population to a level far above that experienced in winter. Yet this great increase in total population occurs at the time when the resident species are feeding young. Unfortunately, Lack's (1976) data do not allow a species by species examination. The departure of these three species in winter could be related to the arrival of the northern migrants, although the potential competition between these two groups is unknown. Because the two groups are not present at the same time on Jamaica, it is difficult to compare their requirements directly. Emlen (1980) showed that in both Florida and the Bahamas the number of birds present increased more than 200 percent, after the wintering migrants had arrived.

If the two groups of birds competed intensely for scarce resources, one would predict a negative correlation in numbers. If both groups responded similarly to favorable resources, a positive correlation might result. R. L. Hutto (pers. comm.) pointed out that the outcome

TABLE 3

Comparisons of Summer and Winter Bird Counts for Eleven Jamaican Habitats[1]

	Man-groves	Strand wood-land	Sea level	Arid lime	Arid ruinate	Bo-tanic garden	Rich second-ary	South-ern riv-erine	North-ern riv-erine	Hard-ward gap	Mon-tane culti-vation
				Winter							
Migrants	14	6	15	5	1	10	10	21	4	3	5
Residents	16	41	43	25	41	79	37	59	28	16	47
Total	30	47	58	30	42	89	47	78	32	19	52
				Summer[2]							
Migrants	6	0	30[v]	16[v]	10	28[vt]	19[v]	19[v]	15[v]	9	10
Residents	19	73	68	36	73	81	39	39	17	24[m]	32
Total	25	73	98	52	83	109	58	58	32	32	42
Seasonal change[3]	−5	+26	+40	+22	+41	+20	+11	−20	0	+13	−10

[1] All data are in birds seen per hour; see Lack (1976: Appendices I and II) for descriptions of habitat.
[2] m = *Myadestes genibarbis* is abundant and vocal during summer, hence more likely to be seen in summer than in winter. t = *Tyrannus dominicensis*: an abundant species migrating away from Jamaica after the breeding season; comprises >10 birds/hour. v = *Vireo altiloquus*: an abundant species migrating away from Jamaica after breeding season; comprises >10 birds/hour.
[3] Summer total − winter total.

depends on the scale for which the correlation is done. Terborgh and Faaborg (1980) found a nonsignificant negative correlation ($r = -.18$) for eight eastern Caribbean habitats. Hutto (1980) reported a stronger ($r = -.43$) negative correlation between habitats. Waide (1980) and Willis (1980) found a positive correlation between numbers of migrants and residents at various sites. Stiles (1980) found a positive correlation along an elevational gradient, whereas Emlen's (1980) data show a positive association over 10 guilds and 13 locations (my calculations). In the present study I found no significant correlation in numbers of individuals or of species between the migrant parulidae and the RNPO across 12 censuses at a single site, suggesting that neither repulsion nor attraction occurred.

INTERYEAR VARIATION

Two striking examples of interyear variation in numbers involve the Black-throated Green (*Dendroica virens*) and Yellow-rumped Warblers. In 1964–1965 all but one of the Black-throated Green Warblers I saw were feeding in pines on the edge of the montane forests. Lack (1976), however, found the species fairly common in lowland forests, but I could hardly have overlooked such a conspicuous and familiar species in my lowland forest tracts.

On a visit to Jamaica in April, 1963 I found Yellow-rumped Warbler among the most numerous of parulids in lowland forest and mangroves, while in 1964–1965 it was conspicuously absent, with only a few individuals being noted on the edge of montane forest. A partial explanation may derive from the fact that Yellow-rumped Warblers may not reach the Antilles until mid-winter (Terborgh and Faaborg 1980), and may not have distributed themselves over the lowland habitats when I departed in February 1965. Alternatively, in some years they may not reach the Antilles at all. I conclude that the data on these two species represent real differences in abundance or intra-island distribution in different years. By contrast Waide et al. (1980) found low interyear variability.

CONCLUSION

One can hardly expect to reconstruct historical evolutionary pressures directly from current events. Although one may find little current evidence for competition or exclusion in a particular study, this may reflect the outcome of previous competitive interactions. Moreover, there may be relatively rare seasons when because of population fluctuations, migrational influxes, or environmental conditions, competition may be temporarily intense.

Annual censuses providing documentation of species abundances over many years would be particularly valuable. Thus, Lack's (1976) data play an important role in documenting population size, providing a basis for future comparisons and provoking thoughtful examination of bird distribution, behavior, and population trends. Lack's emphasis on the role of competition and exclusion in shaping Antillean avifaunas remains open to question, but the basis for more detailed, refined, and long-term studies has been set.

Although Miller (1963) suggested a strong inhibitory impact of migrants on the breeding of residents, it is likely that residents' breeding cycles respond to other factors (e.g., local food cycles) rather than to migration schedules, whereas the migrants' annual cycles reflect factors operating remotely (e.g., food availability on their breeding grounds). Assuming these factors to be completely independent, chance alone could produce circumstances in which resident breeding followed migrant departure. Various authors (e.g., Skutch 1950; Willis 1966; Howell 1971) concluded that such an inhibitory impact is not prominent, and various studies showing no negative correlation in numbers of migrants and residents across space or time suggest that such inhibition does not often operate.

ACKNOWLEDGMENTS

R. Smith introduced me to the Mona Woods and its birds. E. Kidd and C. B. Lewis were extremely helpful in directing me to other forest locations suitable for study. Eugene Eisenmann and E. O. Willis provided valuable stimulation for studying neotropical birds, and useful discussion in studying tropical avian ecology. R. L. Hutto provided useful comments on the manuscript.

LITERATURE CITED

ASPREY, G. F., AND R. G. ROBBINS. 1953. The vegetation of Jamaica. Ecol. Monogr. 23:359–412.

BOND, J. 1971. Birds of the West Indies. Houghton Mifflin, Boston, Massachusetts.

BOND, J. 1978. Derivations and continental affinities of Antillean birds. Spec. Publ. Acad. Nat. Sci. Phila. 13:119–128.

CHIPLEY, R. 1976. The impact of winter migrant wood warblers on insectivorous passerines in a subtropical Colombian oak woodland. Living Bird 15:119–141.

DIAMOND, A. W., P. LACK, AND R. W. SMITH. 1977. Weights and fat condition of some migrant warblers in Jamaica. Wilson Bull. 89:456–465.

EMLEN, J. T. 1971. Population densities of birds derived from transect counts. Auk 99:323–342.

EMLEN, J. T. 1980. Interactions of migrant and resident land birds in Florida and Bahama pinelands. Pp. 133–144, In A. Keast and E. S. Morton (eds.), Migrant Birds in the Neotropics. Smithsonian Institution Press, Washington, D.C.

FFRENCH, R. 1973. A Guide to the Birds of Trinidad and Tobago. Livingston Publ. Co., Wynnewood, Pennsylvania.

FITZPATRICK, J. W. 1980. Wintering of North American Tyrant Flycatchers in the Neotropics. Pp. 67–78, In A. Keast and E. S. Morton (eds.), Migrant Birds in the Neotropics. Smithsonian Institution Press, Washington, D.C.

GOCHFELD, M. 1969. Status of Lincoln's Sparrow in Jamaica, West Indies. Wilson Bull. 81:219–220.

GOCHFELD, M. 1974. Status of the genus Vermivora (Aves, Parulidae) in the Greater Antilles with new records from Jamaica and Puerto Rico. Caribb. J. Sci. 14:177–181.

GOCHFELD, M. 1979. Wintering ranges of migrant warblers of eastern North America. Am. Birds 33: 742–745.

HESPENHEIDE, H. A. 1980. Bird community structure in two Panama forests: residents, migrants, and seasonality during the non-breeding season. Pp. 227–237, In A. Keast and E. S. Morton (eds.), Migrant Birds in the Neotropics. Smithsonian Institution Press, Washington, D.C.

HILTY, S. L. 1980. Relative abundance of north temperate zone breeding migrants in western Colombia and their impact at fruiting trees. Pp. 265–271, In A. Keast and E. S. Morton (eds.), Migrant Birds in the Neotropics. Smithsonian Institution Press, Washington, D. C.

HOWELL, T. R. 1971. An ecological study of the birds of the lowland pine savanna and adjacent rain forest in northeastern Nicaragua. Living Bird 10:185–242.

HUTTO, R. L. 1980. Winter habitat distribution of migratory land birds in western Mexico, with special reference to small foliage-gleaning insectivores. Pp. 181–203, In A. Keast and E. S. Morton (eds.), Migrant Birds in the Neotropics. Smithsonian Institution Press, Washington D.C.

KARR, J. R. 1976. Within and between avian habitat diversity in African and Neotropical lowland habitats. Ecol. Monogr. 46:457–481.

KEAST, A., AND E. S. MORTON. (eds.). 1980. Migrant Birds in the Neotropics. Smithsonian Institution Press, Washington, D.C.

LACK, D. 1976. Island Biology Illustrated by the Land Birds of Jamaica. University California Press, Berkeley, California.

LACK, D., AND P. LACK. 1972. Wintering warblers in Jamaica. Living Bird. 11:129–153.

MENGEL, R. M. 1964. The probable history of species formation in some northern wood warblers (Parulidae). Living Bird 3:9–44.

MILLER, A. H. 1963. Seasonal activity and ecology of the avifauna of an American equatorial cloud forest. Univ. Calif. Publ. Zool. 66:1–78.

POST, W. 1978. Social and foraging behavior of warblers wintering in Puerto Rican coastal scrub. Wilson Bull. 90:197–214.

RAPPOLE, J. H., AND D. W. WARNER. 1980. Ecological aspects of migrant bird behavior in Veracruz, Mexico. Pp. 353–393, In A. Keast and E. S. Morton (eds.), Migrant Birds in the Neotropics. Smithsonian Institution Press, Washington, D.C.

SCHWARTZ, P. 1980. Some considerations on migratory birds. Pp. 31–34, In A. Keast and E. S. Morton (eds.), Migrant Birds in the Neotropics. Smithsonian Institution Press, Washington, D.C.

SKUTCH, A. 1950. The nesting season of Central American birds in relation to climate and food supply. Ibis 92:185–222.

SMITH, R. W. 1971. Some results of the banding of North American wood warblers (Parulidae) in the Mona Woods. Gosse Bird Club Broadsheet 17:15–17.

STILES, F. G. 1980. Evolutionary implications of habitat relations between permanent and winter resident landbirds in Costa Rica. Pp. 421–435, In A. Keast and E. S. Morton (eds.), Migrant Birds in the Neotropics. Smithsonian Institution Press, Washington, D.C.

TERBORGH, J. W., AND J. R. FAABORG. 1980. Factors affecting the distribution and abundance of North American migrants in the Eastern Caribbean region. Pp. 145–155, In A. Keast and E. S. Morton (eds.), Migrant Birds in the Neotropics. Smithsonian Institution Press, Washington, D.C.

VUILLEUMIER, F. 1978. Remarques sur l'echantillonnage d'une riche avifaune de l'oeust de l'Ecuador. Oiseau Rev. Fr. Ornithol. 48:27–36.

WAIDE, R. B. 1980. Resource use between migrant and resident birds: the use of irregular resources. Pp. 337–352, In A. Keast and E. S. Morton (eds.), Migrant Birds in the Neotropics. Smithsonian Institution Press, Washington, D.C.

WAIDE, R. B., J. T. EMLEN, AND E. J. TRAMER. 1980. Distribution of migrant birds in the Yucatan peninsula: a survey. Pp. 165–180, In A. Keast and E. S. Morton (eds.), Migrant Birds in the Neotropics. Smithsonian Institution Press, Washington, D.C.

WILLIS, E. O. 1966. The role of migrant birds at swarms of army ants. Living Bird 5:187–231.

WILLIS, E. O. 1980. Ecological roles of migratory and resident birds on Barro Colorado Island, Panama. Pp. 205–225, In A. Keast and E. S. Morton (eds.), Migrant Birds in the Neotropics. Smithsonian Institution Press, Washington, D.C.

FORAGING NICHE RELATIONS OF WADING BIRDS IN TROPICAL WET SAVANNAS

James A. Kushlan,[1] Gonzalo Morales,[2] and Paula C. Frohring[1]

[1]Department of Biology, East Texas State University, Commerce, Texas 75428 USA,
and [2]Instituto de Zoología Tropical, Universidad Central de Venezuela,
Caracas, Venezuela

ABSTRACT. The foraging niche relations of ciconiiform wading birds in hyper-seasonal wet savannas are influenced by seasonal fluctuations in rainfall, water depth, and food abundance and availability. The ciconiiform wading bird community of the Venezuelan Llanos includes 22 species, that of the Florida Everglades includes 15. Both faunas contain more species of herons than ibises, spoonbills, or storks. The Llanos supports terrestrial ibises, very large storks, and species that live in mated pairs year-round, types of wading birds not found in the Everglades. The wading bird community of the Everglades is numerically dominated by a single species, whereas abundances are more equitably distributed among species in the Llanos. Seasonal cycles of prey availability, bird migration, and fluctuating water depth influence packing of species in the wading bird community of both savannas. The Llanos includes relatively more terrestrial habitat than does the Everglades and supports a greater variety and amount of potential prey. Thus the Llanos provides more diverse foraging opportunities. Wading bird species found in both areas feed on analogous types of prey, although the diets of some species in the Llanos include types of prey with no analogue in the Everglades, the fish fauna of which is derived from temperate North America.

RESUMEN. Las relaciones de los nichos de forraje de la comunidad de aves zancudas ciconiformes en savanas excesivamente húmedas están influenciendas por las fluctuaciones estacionales de lluvias, profundidad del agua y abundancia y disponibilidad de comida. La comunidad de aves zancudas ciconiformes de los llanos venezolanos incluye 22 especies, mientras que la de los Everglades en Florida, E.E.U.U. incluye 15. Ambas avifaunas contienen mayor cantidad de especies de garzas que de ibis, picos de espátula o cuchara y cigueñas. En los llanos se encuentran ibis terrestres, cigueñas muy grandes y especies que viven en pareja duranta todo el año, tipos de aves zancudas que no se encuentran en los Everglades. La comunidad de aves zancudas de los Everglades está dominada en el aspecto numérico por una sola especie, mientras que la abundancia de especies de los llanos está distribuída de manera más equitativa. Ciclos estacionales de disponibilidad de presas, migración de aves y fluctuación de la profundidad de las aguas son quienes influyen en el compactamiento de las especies de la comunidad de aves zancudas de ambas savanas. Los llanos incluyen relativamente mayor cantidad de habitats terrestres que los Everglades y poseen una mayor variedad y cantidad de posibles presas. Así los llanos preveen diversidad de oportunidades de forrajeo. Las aves zancudas encontradas en ambas áreas se alimentan en tipos de presas análogas, aunque las dietas de algunas especies de los llanos incluyen presas que no tienen análogos en los Everglades, la ictiofauna que se deriva del clima templado de América del Norte.

Wet savannas are among the most extensive of tropical ecosystems (Beard 1953; Cole 1960; Blydenstein 1967; Walter 1969; Sarmiento and Monasterio 1975). Although tropical and subtropical in character, these systems are far from climatically stable in that rainfall, generally over 1000 mm per year is highly seasonal, concentrated in a five to eight month period. In the western hemisphere, such wet savannas occur from southern North America, through para-equatorial regions, to southern South America. Their faunal components are determined by both ecological constraints and biogeographical considerations. Ecologically, the dramatic climatic fluctuations could be expected to impose fundamental constraints on the invasion and persistence of animal populations. These savannas are hyperseasonal marshes, in the terminology of Sarmiento and Monasterio (1975), and are characterized by the presence of plant and animal populations that are adapted to the dual stresses of flood and drought. Biogeographically, the fauna is constrained by the availability of suitably-adapted species in

Fig. 1.   Habitats of the Everglades (left) and Llanos (right). From top to bottom: Everglades wet prairie—Llanos estero; Everglades tree islands—Llanos wooded banco, with cattle; Everglades pond in the dry season—Llanos caño in the dry season. Llanos photographs courtesy of Dr. Mauricio Ramia.

that these savannas are relatively young in their current incarnation. In South America, glacial refugia undoubtedly provided sources for aquatic species having a long history of adaptation to tropical marshes, whereas in North America nonvolant species are derived from a temperate fauna.

Two of the largest Neotropical wet savannas are the South American Llanos, situated in the Orinoco River Basin of Venezuela (Ramia 1967), and the North American Everglades, situated on the southern tip of the Florida peninsula. The ecological and biogeographic similarities and differences of these two marshes provide a natural experiment in the assembly of animal communities. In such comparisons species may be found that exhibit ecological equivalency. A similar comparison of entire communities may reveal both equivalencies and divergences that result from environmental or biogeographic constraints peculiar to each system, and so can further our understanding of system effects and community responses.

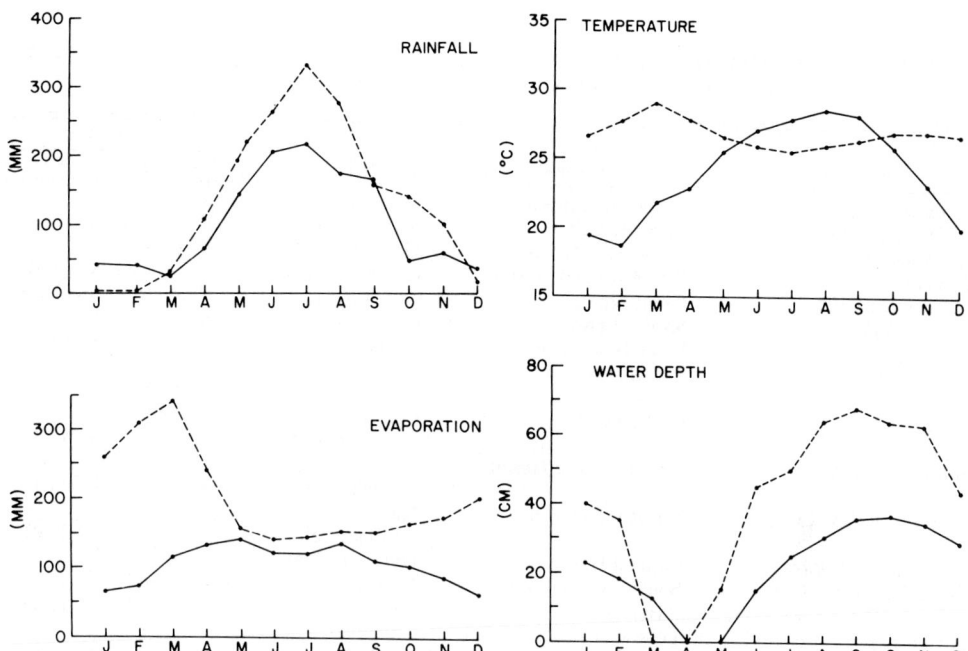

FIG. 2.   Seasonal variation of environmental conditions in the Everglades (solid line) and Llanos (dashed line). Data plotted are monthly averages for 1972–1980, except for water depth in the Everglades, which includes data from 1962–1982.

Ciconiiform wading birds are useful for such a comparison. They are large and abundant species that are especially suited to make use of the seasonal variability in prey populations that occurs in wet savannas (Kushlan 1978a, 1979a, b, 1981a; Pinowski et al. 1980) and may be influential in the flow of energy and materials within the system (Kushlan 1976a; Morales et al. 1981). Wading birds are a community of morpho-ecologically similar species, some of which occur in both North and South America. In this paper we compare the wading bird communities of the Florida Everglades and Venezuelan Llanos with respect to species composition, abundance structure, and foraging niche relations, and we consider particularly how differences and similarities are the result of biotic, environmental, and biogeographic constraints. Our goal is to understand the patterns of community composition and to suggest explanations for the patterns observed.

STUDY AREAS

The Llanos is in South America in the states of Apure, Barinas, Guarico, and Bolivar, Venezuela, at 6–9°N latitude. The Everglades is in North America in the state of Florida, United States, 25–27°N latitude. We examined the wading bird communities of study areas as representative of the two ecosystems. The Everglades study area was the fresh-water marshes west of Miami, Florida, 25°N, 81°W. The Llanos study area was in the low Llanos near Mantical, Apure, 7°N, 69°W (Ramia 1974). The seasonal hydrological pattern was wetter than the more extensive high Llanos and therefore more similar to that of the Everglades.

The landscapes (Fig. 1) of both savannas are mosaics of marsh and woody vegetation (Loveless 1959; Tamayo 1961; Ramia 1967; Troth 1979). The highest ground, occupied by trees, palms, and shrubs, is not normally flooded. In the Llanos, such high ground, called bancos, is variable in size and shape ranging from small islands and river banks to extensive, continuous sandy ridges (matas). High ground in the Everglades is restricted to small tree islands of fewer than 2 ha that occupy discontinuous rock outcrops. Slightly lower elevations are marshes (bajíos) dominated by sawgrass (*Mariscus*) in the Everglades and by *Panicum* and *Leersia* in the Llanos. Even lower elevations support wet prairies (esteros) dominated by aquatic sedges and grasses, *Eleocharis, Panicum,* and *Rhynchospora* in the Everglades and

TABLE 1

Ciconiiform Wading Bird Faunas of the Florida Everglades and Venezuelan Llanos and the Relative Abundance of Various Species

Scientific name	English name	Spanish name	Occurrence (rel. abund.)[2] Everglades	Occurrence (rel. abund.)[2] Llanos
Botaurus lentiginosus	American Bittern	Mirasol	X[1]	
Ixobrychus exilis	Least Bittern	Garza Enana	X	
Ardea herodias	Great Blue Heron	Garzón Cenizo	X (8.6)	
A. cocoi	White-necked Heron	Garza Morena		X (4.3)
Casmerodius albus	Great Egret	Garza Real	X (5.8)	X (17.2)
Egretta thula	Snowy Egret	Garza Chusmita	X	X (2.7)
E. caerulea	Little Blue Heron	Garcita Azul	X (2.1)	X (3.1)
E. tricolor	Tricolor Heron	Garza Pechiblanca	X (1.2)	
Bubulcus ibis	Cattle Egret	Garza Garrapatera	X	X (32.3)
Butorides striatus	Green-backed Heron	Chicuaco Cuello Gris	X	X (3.1)
Syrigma sibilatrix	Whistling Heron	Garza Silbadora		X (0.4)
Pilherodius pileatus	Capped Heron	Garciolo Real	X	
Nycticorax nycticorax	Black-crowned Night-Heron	Guaco	X	X (3.1)
N. violacea	Yellow-crowned Night-Heron	Chicuaco Enmascarado	X	X
Tigrisoma lineatum	Rufescent Tiger-Heron	Pájaro Vaco		X (2.1)
Theristicus caudatus	Buff-necked Ibis	Tautaco		X
Cercibis oxycerca	Sharp-tailed Ibis	Tarotaro		X
Mesembrenibis cayennensis	Green Ibis	Corocora Negra		X (12.6)
Phimosus infuscatus	Bare-faced Ibis	Tara		X (2.8)
Eudocimus albus	White Ibis	Corocora Blanca	X (78.3)	X (0.2)
E. ruber	Scarlet Ibis	Corocora Roja		X (10.3)
Pelgadis falcinellus	Glossy Ibis	Corocora Castaña	X	X (0.6)
Ajaia ajaja	Roseate Spoonbill	Garza Paleta	X[1]	X (0.2)
Jabiru mycteria	Jabiru	Garzon Soldado		X (1.5)
Mycteria americana	Wood-Stork	Gabán	X (3.7)	X (2.2)
Euxenura maguari	Maguari Stork	Gabán Peonío		X (1.4)

[1] Present but not nesting.
[2] Relative abundance is the percent of all birds present. Species for which no relative abundance is given were present but in relatively small numbers and were not adequately censused.

Hymenachne, Eleocharis, and *Luziola* in the Llanos. Floating and submerged species occur in deeper areas. In the Everglades deep pockets contain ponds, maintained by alligators. In the Llanos small seasonal streams, called caños, provide deep water habitats. In both areas, canals and borrow pits (préstamos) have been excavated for drainage, road fill, or levees. In some situations these hold water after natural habitats have dried.

Soils in the Everglades are peat and marl overlying highly permeable limestone. When the limestone is saturated, the ground water level rises above the surface, flooding the marshes. In the Llanos, the soil is a heavy clay that impedes vertical drainage, leading to surface flooding from rainfall and from overflow of permanent and seasonal rivers.

Rainfall in both areas exceeds 1200 mm per year and is heavily seasonal, peaking in July (Fig. 2). Both areas are, therefore, covered in the boreal summer and fall by moderately deep water, which recedes in the winter to a low point in spring. The low Llanos differs from the Everglades by being wetter in the wet season. Higher wet season water levels in the low Llanos result from a 25 percent greater annual rainfall, 1540 mm per year compared to 1225 mm per year in the Everglades. In the dry season the Llanos experiences less rainfall and higher rates of evaporation owing to higher temperatures.

TABLE 2
FAUNAL SIMILARITY OF THE WADING BIRD COMMUNITIES OF THE EVERGLADES AND VENEZUELAN LLANOS

	Bitterns	Herons	Tiger-Herons	Ibises	Spoonbills	Storks
Nesting in Everglades	1	9	0	2	0	1
Present but not nesting in Everglades	1	0	0	0	1	0
Total in Everglades	2	9	0	2	1	1
Present and nesting in Llanos	0	10	1	7	1	3
Common faunal elements	0	7	0	2	1	1
Faunal similarity[1]	0	0.58	0	0.29	1.00	0.33

[1] Jaccard similarity coefficient.

METHODS

Faunal lists were constructed by considering only those species that occur in the freshwater marshes of each geographic area. Species abundances were determined using monthly censusing along transects conducted by car and by airplane. Census results are expressed as the percent of the annual total number of each species occurring in each month. Observations of birds at feeding sites were made using cars as hides. Qualitative descriptions of foraging behavior, using the terminology of Kushlan (1978a), foraging sociality, interactions, and habitat use as presented in this paper are based on quantitative 1 or 5 min observations of randomly selected individuals. Food habits were quantified by collecting specimens and examining regurgitates of adults and nestlings, which had the same diets (Kushlan 1979a; Morales et al. 1981). Fish standing crop was sampled using enclosure traps, net traps at culverts, and poisoning (Kushlan 1981b; Ramos et al. 1981). Size of birds was indexed by one measurement of tarsal length: the distance from the point of the joint between the tibia and metatarsus to the point of the joint at the base of the middle toe in front.

Faunal similarity was calculated using the Jaccard similarity coefficient, $S_j = a/(a + b + c)$, where a = the number of common faunal elements, b = those only in Everglades, c = those only in the Llanos. Diversity was calculated using the Simpson index, $D = 1/\Sigma\ p_i^2$, where p_i is the proportion of individuals in the ith species. Climatological and hydrological data were obtained at permanent stations within each study area.

RESULTS

SPECIES RICHNESS

In the Everglades, the wading bird community consists of 15 species (Table 1). The Reddish Egret (*Egretta rufescens*) is not included in the faunal list because it is rarely found in the inland freshwater marshes, occurring more commonly in the south Florida marine habitats. Although also primarily a marine species in south Florida, the Roseate Spoonbill is included because it occurs regularly in the Everglades marshes in small numbers. The American Bittern is included because it winters there, although it does not nest in the Everglades.

Twenty-two species of wading birds are found in the Llanos (Table 1), and all of these species nest there although not necessarily in the study area. The Pinnated Bittern (*Botaurus pinnatus*) occurs in the high Llanos (M. Gochfeld, pers. comm.) but is not considered in this paper because it is not recorded from the study area.

Wading birds comprise a larger proportion of the total bird fauna in the Llanos (12%) than in the Everglades (4%) (Kushlan and Morales, unpubl. data). The species richness of wading birds in the Llanos exceeds that of the Everglades by seven species, and its nesting fauna is greater by nine species. Of six higher taxa (families and subfamilies) occurring in the two areas (Table 2), herons dominate both faunas, comprising 60 percent of the species in the Everglades and 45 percent of those in the Llanos. The seven extra species of ibises and storks in the Llanos account for much of the difference in species richness between the wading bird communities of the two areas.

Despite these differences, over 40 percent of the total number of wading bird species found in the Everglades and Llanos are common to both areas (Jaccard's coefficient = 0.42). The

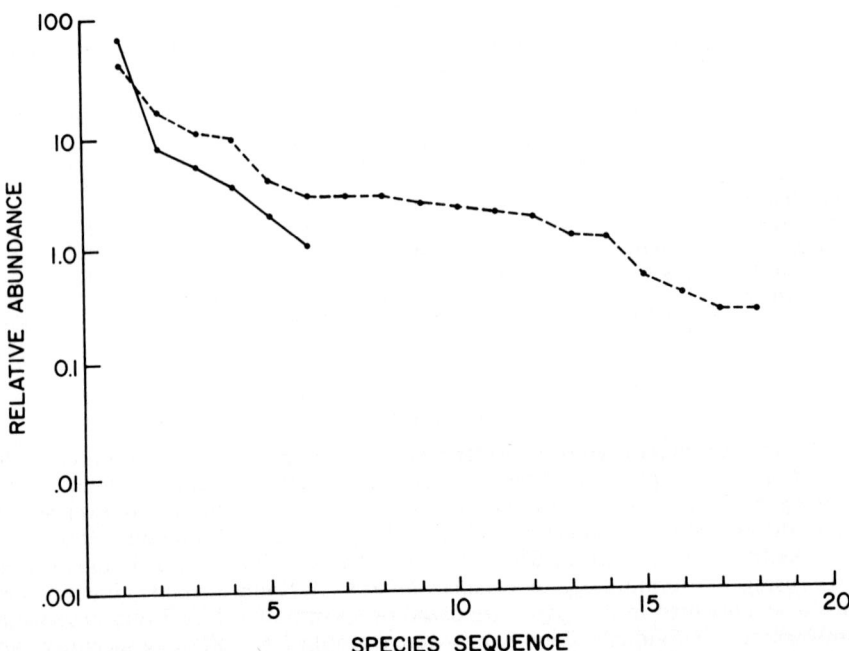

Fig. 3.   Relative abundances of wading bird species in the Everglades (solid line) and Llanos (dotted line). Relative abundance is the proportion of the combined wading birds population represented by each species, plotted in sequence of species from the most to the least abundant. The steeper the curve, the greater the dominance of one or a few species.

ecological commonality of the two areas is understated by recognized taxonomy (Hancock and Kushlan 1984). The Great Blue Heron and the White-necked (Cocoi) Heron are closely-related species that are geographic replacements, and probably ecological equivalents. Similarly, the White and Scarlet Ibises, treated as separate species in our analyses, are closely related and are thought to be conspecific (B. Busto and C. Ramo, pers. comm.).

COMMUNITY STRUCTURE

The Llanos and the Everglades differ in the relative abundances of wading bird species comprising the respective communities (Table 1). In the Llanos, the herons are the most abundant wading birds, while in the Everglades a single species, the White Ibis, constitutes more than 75 percent of the total wading bird fauna, a situation that apparently has always been the case (Kushlan and Frohring, unpubl. data). Cattle Egrets, which comprise half the heron fauna in the Llanos, are scarce in the Everglades marsh although abundant in terrestrial habitats of southern Florida (Kushlan and White 1977).

The distribution of relative abundances among species implies differences in community structure that may be fundamental (Fig. 3). The low species diversity of the Everglades fauna (Simpson's Index = 1.60) reflects the strong dominance of the single species, whereas the higher diversity of the Llanos fauna (Simpson's Index = 5.98) reflects a more even representation of species. The distribution of relative abundances in the Everglades, the relatively rapid fall of the curve in Figure 3, is similar to that expected of a non-equilibrial community in contrast to the more equitable situation suggested by the relatively slowly falling curve (Fig. 3) for the Llanos fauna. (See May [1976] for a discussion of such curves.) Thus the Llanos wading bird community may be closer to an equilibrium condition, with an equitable distribution of abundances spread out over a greater number of species.

SEASONALITY OF PREY AND FORAGING HABITAT

In the wet season, surface water in both the Llanos and the Everglades stands over all of the savanna except the tree islands (bancos). Wet season (June–December) water depths in

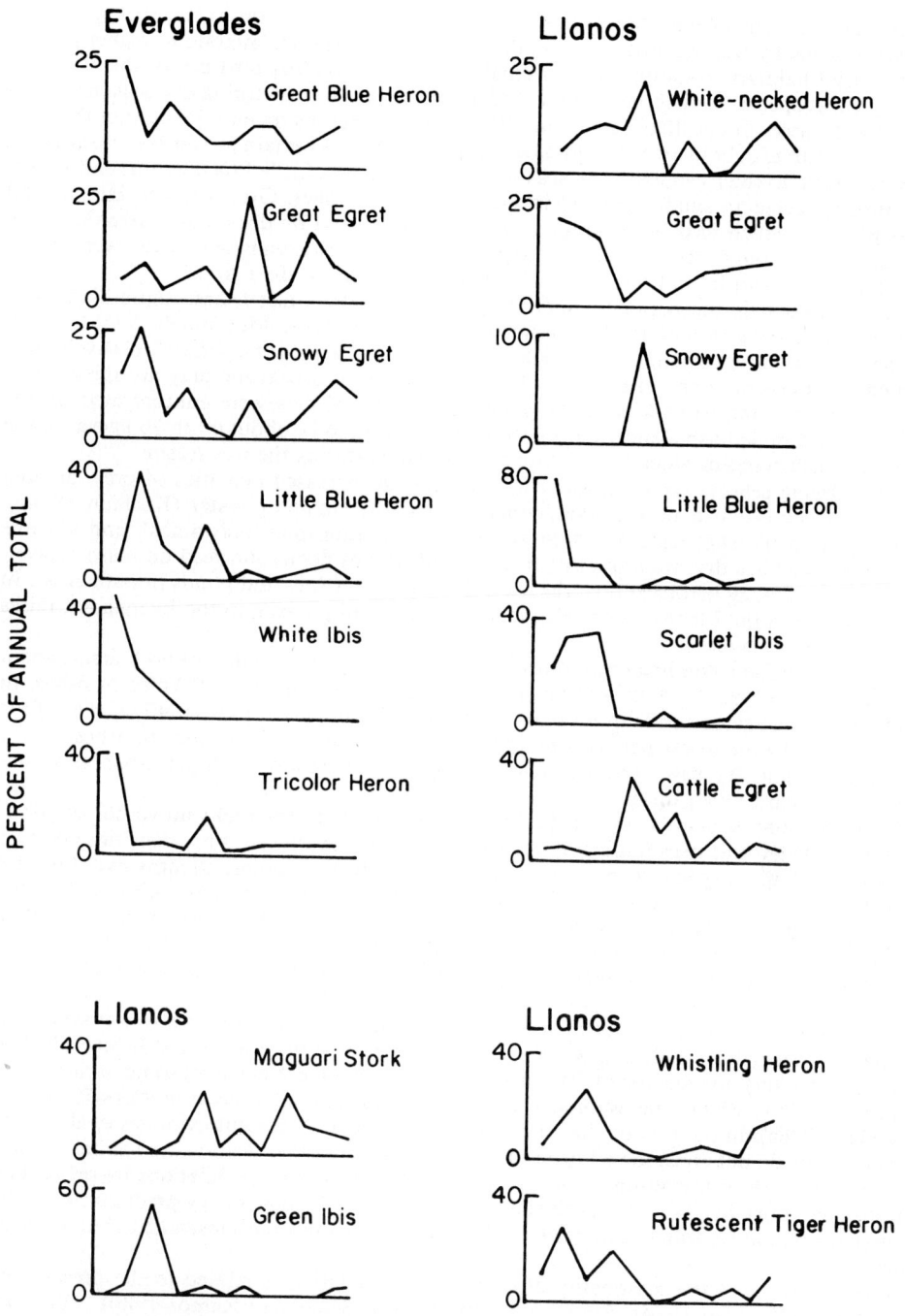

FIG. 4. Seasonal variation in abundance of wading bird species in the Everglades and Llanos. Abundance is expressed as the monthly percentage of the yearly total count for each species (based on monthly censuses), beginning in January and ending in December.

the marshes (38 cm in the Llanos and 24 cm in the Everglades, on average) (Fig. 2), can restrict foraging by wading birds. During this season many species migrate to higher ground or to coastal habitats, resulting in a seasonal variation in wading bird use of the savannas (Fig. 4). Most species tend to be more abundant in the late winter–spring dry season than in the fall wet season. Great Blue Herons and White-necked Herons remain throughout the year, as do Great Egrets in both areas. In the Llanos, Maguari Storks remain in numbers throughout the wet season. In the Everglades wet season, Tricolor Herons, Little Blue Herons, and American Bitterns occur in small numbers in the shallower marshes. Green-backed Herons and Least Bitterns, which feed by perching and are less restricted by deep water, remain in the Everglades year-round. Ibises typically migrate out of both savannas during the wet season, except for the territorial Buff-necked and Sharp-tailed Ibises in the Llanos.

The wet season is the period of reproductive activity for most fishes, amphibians, and aquatic invertebrates in both the Llanos and the Everglades (Lowe-McConnell 1975; Kushlan and Kushlan 1980; Kushlan, unpubl. data). Thus, the wet season is a time of relatively high abundance of potential prey. Ecological density, however, is low as the prey are spread over large expanses of marsh making them less available than if they were concentrated. In the Llanos, we estimated fish standing stock in the wet season to be about 60 to 70 kg/ha. In the Everglades fish standing stock is less, 50 kg/ha of marsh during the wet season.

In the drying season, receding water levels result in increased densities of prey, as prey become concentrated in progressively smaller areas of remaining water (Kushlan 1976a), including deep marshes (esteros), natural ponds, excavation pits (préstamos), and streams (caños). As marshes dry, wading birds form mixed-species flocks and feed on these concentrated prey. Flocking herons and storks use still-flooded marshes and ponds in both areas. In the study area in the Llanos, nearly-dry marshes are used by ibises; in the Everglades, ibises feed in wetter marshes.

The drying season standing crop of fishes is similarly high in marshes of both areas, about 90 kg/ha. In the Llanos, shallow excavation pits can contain up to 1300 kg/ha of fishes. In south Florida, alligator ponds in most years may contain about 500 kg/ha, 40 percent of the Llanos value. In the driest period wading birds seldom use dry Everglades marshes. In the Llanos, damp and dry areas continue to be used by Cattle Egrets, Whistling Herons, and Buff-necked and Sharp-tailed Ibises.

When the rainy season begins in the Everglades, remnant fish stocks move out of ponds via surface water that rises uniformly throughout the marsh system. Prey densities decrease rapidly, and birds disperse to more elevated habitats. In the Llanos, streams overflow and flood the savanna, carrying fishes from their dry-season refugia. Wading birds follow the flooding, initially using newly flooded marshes. These birds include, especially, Whistling Herons, Little Blue Herons, Maguari Storks, and ibises. As the savannas flood further, the wading bird community continues its adjustment toward the high water conditions noted previously.

The seasonal interrelationships of wading birds to their prey are complex in wet savannas. Deeper and permanent bodies of water serve as dry-season refugia for fishes. In sites shallow enough for wading, the feeding of wading birds can regulate the survival of some of these prey populations. In south Florida, wading birds can reduce prey populations as much as 76 percent (Kushlan 1976a); in the Llanos the reduction is more than 50 percent (Morales et al. 1981). Because birds do not consume all fishes present, if ponds or other deep-water habitats do not dry completely, bird predation prevents local extirpation of prey populations by preventing catastrophic mortality of the fish (Kushlan 1974b). In the Llanos, heavy predation by birds on juvenile predatory fishes (*Serrasalmus, Hoplias*) may reduce the subsequent abundance of adults.

Another aspect of the seasonality of wading bird predation is related to the presence of crocodilians (Kushlan 1974a; Staton and Dixon 1977; Seijas and Ramos 1980). Alligators (*Alligator mississippiensis*) in the Everglades and Spectacled Caiman (*Caiman crocodilus*) in the Llanos, also feed on aquatic prey and so may affect fish diversity. In the Everglades, alligators dig and maintain the ponds in which wading birds feed and in which the fish populations persist through dry seasons.

FORAGING NICHES

The foraging niches of wading birds may be interpreted to include axes related to food, habitat, and behavior. In our analysis of the wading bird communities of seasonal savannas, it is useful to consider macrohabitat, feeding strategy, size of bird, microhabitat, including

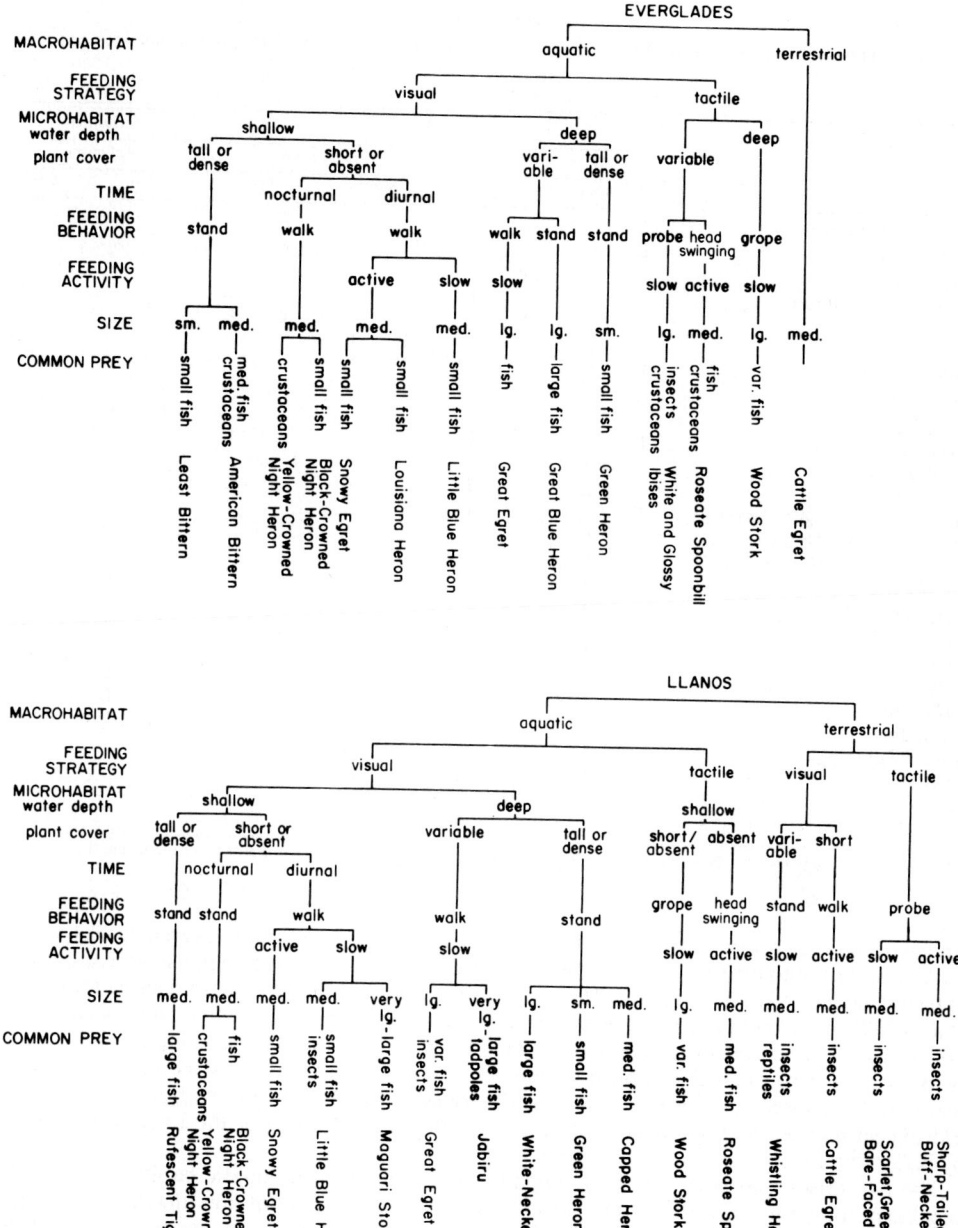

FIG. 5. Difference in niche characteristics within wading bird communities in the Llanos and Everglades.

water depth and plant cover, time of foraging, feeding behavior and activity, and prey consumed. We have qualitatively evaluated the differences among species with respect to these aspects of foraging niche in both the Everglades and Llanos (Fig. 5).

Macrohabitat.—Wading birds may forage in aquatic or terrestrial habitats, but most are primarily aquatic (Table 3). Terrestrial foraging could be an effective strategy in wet savannas only if topographically high areas or the length of the dry season were sufficiently extensive.

TABLE 3

CHARACTERISTICS OF FEEDING SITES USED BY WADING BIRDS IN THE EVERGLADES AND THE LLANOS[1]

	Aquatic				Terrestrial	
	Water depth		Plant cover		Plant cover	
	Shallow	Deep	Short	Tall	Short	Tall
American Bittern	E			E		
Least Bittern	E			E		
Great Blue Heron		E	E	e		
White-necked Heron		L	L	L		
Great Egret	E	E L	E L	e L		
Snowy Egret	E L		E L			
Little Blue Heron	E L		E L			
Tricolor Heron	E		E			
Cattle Egret					E	L
Green-backed Heron	E L	E L		E L	L	L
Whistling Heron	l			l		
Capped Heron	L	L	L	L		
Black-crowned Night Heron	E L		E L			
Yellow-crowned Night Heron	L			L		
Rufuscent Tiger-Heron	l					L
Buff-necked Ibis	l			L		L
Sharp-tailed Ibis	l			L		L
Green Ibis	L			L		L
Bare-faced Ibis	L			L		L
White Ibis	E L		E L		e	L
Scarlet Ibis	L			L		L
Glossy Ibis	E L		E L		e	L
Roseate Spoonbill	L			L		
Jabiru		L		L		
Wood Stork	e L	E L	E L	L		
Maguari Stork	L					

[1] Capital letters indicate primary feeding sites, small letters indicate secondary feeding sites in Everglades (E, e) and Llanos (L, l).

The two savannas appear to differ in this regard in that dry ground in the Everglades is limited both seasonally and areally. The wading bird communities reflect the differences in availability of terrestrial habitat. The Whistling Heron, the typical heron of neotropical savannas (Kushlan et al. 1982), feeds on ground that is dry or very shallowly covered by water. This species does not occur in the Everglades. The Cattle Egret, a dry-ground forager occurring in both areas, is the typical heron of paleotropical savannas, where it originated. It is the most abundant heron in the Llanos, but it rarely feeds in the Everglades marsh. That it is common in southern Florida on higher ground suggests that it is excluded from feeding in the Everglades marsh by water depths.

The two ibises that occur in the Everglades feed primarily in aquatic habitats. Although capable of foraging terrestrially, they do so only when aquatic habitats are unavailable. In the Llanos the same species, although often foraging in wet marshes, seem also to be terrestrial, especially in the drying season, choosing the drier sites of those available. The Buff-necked, Bare-faced, and Sharp-tailed Ibises characteristically feed on dry or damp ground, the White, Scarlet, and Glossy Ibises somewhat less so. Thus, in the Llanos, although ibises forage in both aquatic and dry habitats, they can especially use the latter in the drying season. It appears, then, that the permanent and facultative terrestrial component of the fauna is expanded in the Llanos; the addition of terrestrial foraging opportunities affects both heron and ibis species found there.

Feeding strategy.—Wading birds can be partitioned dichotomously into those locating prey primarily visually and those locating prey by tactile mechanisms (Kushlan 1978a). Herons and bitterns are primarily visual hunters; ibises and spoonbills are primarily tactile. Heavy cover and seasonally high densities of food enhance the effectiveness of tactile foraging in wet savannas. Four tactile foragers (two ibises, Roseate Spoonbill, and Wood Stork) occur in the

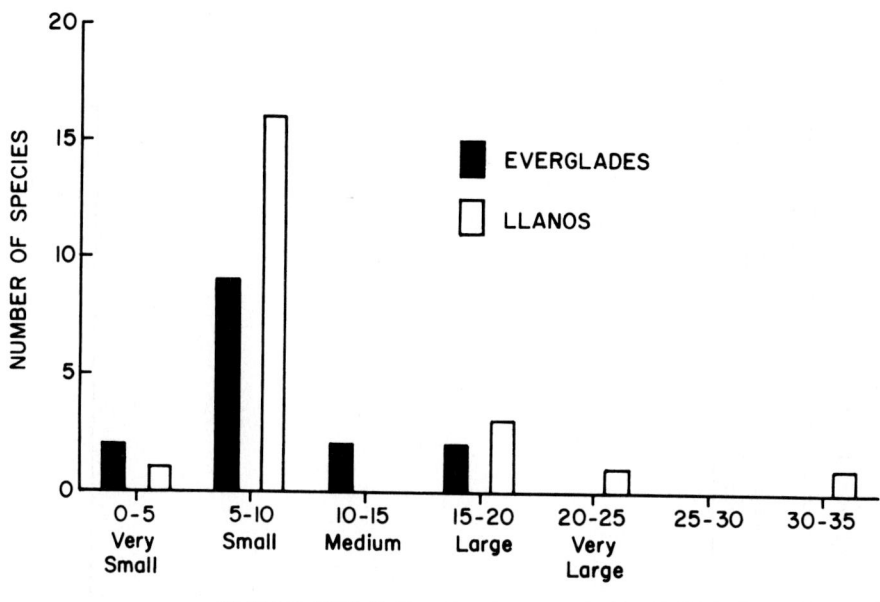

FIG. 6. Distribution by size of wading bird species in the Everglades and Llanos. Leg length is used as an index of size.

Everglades, but the Llanos supports five additional tactile-foraging species of ibises. The addition of these tactile-foraging ibises accounts for five of the seven species added to the Llanos fauna over that of the Everglades.

Size of the bird.—Body size is an important aspect of wading bird niche breadth, in that it is correlated with feeding strategy (Kushlan 1978a). Leg length appears to be the most crucial morphological feature in determining foraging sites available to the various species (Kushlan 1976a; Custer and Osborn 1977). Geographically and seasonally variable water depths in wet savannas provide conditions suitable for use by birds of different leg lengths. Based on this characteristic, we divided the savanna wading bird community into five classes of very small, small, medium, large, and very large species (Fig. 6). The species present in the Llanos, but not in the Everglades, are mostly small (the ibises) but also include the very large Jabiru and Maguari Storks, the latter two species having no equivalent in the Everglades.

Some size differences occur in similar species found in the two areas. Great Egrets in the Everglades are smaller than those in the Llanos (14.1 vs 16.1 cm tarsus length), and Great Blue Herons in the Everglades are larger than White-necked Herons in the Llanos (18.9 vs 17.9 cm). As a result, the Great Blue Heron and Great Egret are more dissimilar in size in the Everglades than are the White-necked Heron and Great Egret in the Llanos. The Great Egret in the Everglades is, in fact, situated along the leg-length gradient midway between the Great Blue Heron and the three small day herons (Little Blue Heron 9.6 cm, Snowy Egret 9.6 cm, Louisiana Heron 9.9 cm). This comparison suggests that size compensation has occurred in the Everglades. That the Great Blue Heron in the Everglades has not become even larger, and that the Wood Stork, the other large species, has not diverged at all from its size in the Llanos (19.8 cm in both areas) suggest the existence of minimal selective pressure to fill a niche space occupied by very large birds in the Everglades wading bird community.

Microhabitat.—Niche distinctions can be drawn with respect to characteristics of feeding sites, particularly water depth and plant cover (Table 3). Primary habitats are those used by a species most of the time, and those in which a species is expected to be found. All primarily aquatic species use shallow water. Very small species (Least Bittern, Green-backed Heron) and some medium-sized species are able to stand on vegetation and so can use deep water as well, as can larger species. The very large species that occur in the Llanos (Fig. 6) are capable of using the greater water depths found there (Fig. 2). In both areas, the cryptic bittern-like

TABLE 4

Foraging Behaviors Used by Wading Birds in the Everglades (E) and Llanos (L)[1]

	Number of behaviors recorded		Standing		Walking slowly		Walking quickly		Probing		Head swinging		Foot stirring		Gleaning		Groping	
	E	L	E	L	E	L	E	L	E	L	E	L	E	L	E	L	E	L
American Bittern	4	—	X		(X)													
Least Bittern	4	—	X		(X)													
Great Blue Heron	7	—	X		X				(X)									
White-necked Heron	—	2		X		X									(X)	(X)		
Great Egret	10	7	X	X	(X)		X	X		(X)								
Snowy Egret	20	7	X		X	X	X						X	X		X		
Little Blue Heron	13	6	(X)		X	X	(X)			(X)								
Tricolor Heron	12	—	X		(X)		X								X	X		
Cattle Egret	12	7	(X)		(X)		X		(X)				(X)					
Green-backed Heron	9	2	X	X	(X)			(X)										
Whistling Heron	—	6		X		X												
Capped Heron	—	2		X		(X)												
Black-crowned Night-Heron	4	1	X	(X)														
Yellow-crowned Night-Heron	5	1	X	X														
Rufescent Tiger-Heron	—	3		X		(X)												
Buff-necked Ibis	—	2								X								
Sharp-tailed Ibis	—	2								X								
Green Ibis	—	3								X								
Bare-faced Ibis	—	3								X								
White Ibis	5	3	(X)		(X)				X	X	(X)							
Scarlet Ibis	—	3								X								
Glossy Ibis	5	2	(X)		(X)				X	X	(X)							
Roseate Spoonbill	7	3	(X)					(X)			X	X						
Jabiru	—	2		(X)		X							X				X	
Wood Stork	7	6	(X)	(X)	X	X							X	X			X	X
Maguari Stork	—	2		(X)		X												

[1] X indicates the behavior used predominantly by a species in an area; (X) indicates a behavior used less frequently.

TABLE 5

FORAGING SOCIALITY OF WADING BIRDS IN EVERGLADES AND LLANOS[1]

	Flocking		Paired		Single	
	E	L	E	L	E	L
American Bittern					X	
Least Bittern					X	
Great Blue Heron					X	
White-necked Heron						X
Great Egret	X	X				
Snowy Egret	X	X				
Little Blue Heron	X				X	X
Tricolor Heron					X	
Cattle Egret	X	X				
Green-backed Heron					X	X
Whistling Heron				X		
Capped Heron				X		
Black-crowned Night Heron					X	X
Yellow-crowned Night Heron					X	X
Rufescent Tiger-Heron						X
Buff-necked Ibis				X		
Sharp-tailed Ibis				X		
Green Ibis				X		
Bare-faced Ibis		X				
White Ibis	X	X				
Scarlet Ibis		X				
Glossy Ibis	X	X				
Roseate Spoonbill	X	X				
Jabiru		X				
Wood Stork	X	X				
Maguari Stork						X

[1] Species categorized by their predominant behavior.

birds, i.e., American and Least Bitterns, Rufescent Tiger-Heron, and Green-backed Heron, prefer dense plant cover. Other species tend to feed in the open marsh and pools typical of savannas.

Time of foraging.—Most wading birds are diurnal, and some can feed nocturnally as well. The latter include the Great Blue Heron and Wood Stork. Some species, however, typically forage at night. In both areas, these include the two night-herons. This activity pattern distinguishes the night-herons from similarly sized herons that forage during the day.

Feeding behavior.—The behavioral aspects of the foraging niche of wading birds in savannas are complex because each species can be flexible in its behavior and activity level in response to habitat and prey availability. One aspect of a bird's response is its behavioral repertoire and the relative frequency of use of each behavior. We recorded the behaviors we have observed in each area and determined which were used most frequently (Table 4).

It appears that wading birds have a larger repertoire of foraging behaviors in the Everglades than in the Llanos. On average, herons had repertoires of 8.0 behaviors per species in the Everglades and 4.1 behaviors per species in the Llanos. For ibises the comparison is 5.0 behaviors per species in the Everglades versus 2.6 in the Llanos. For storks it is 7.0 versus 3.3. Such differences between the two areas also hold when individual species that occur in both areas are compared. The known behaviors of various species in the Everglades and Llanos respectively are: Great Egrets, 10 and 7; Snowy Egrets, 20 and 7; Little Blue Herons, 13 and 6; Cattle Egrets, 13 and 7; Roseate Spoonbill, 7 and 3. Such differences in known repertoires are not a function of observation effort, which was about the same in both areas, but rather seem to reflect differences in the diversity of behaviors needed to forage in each area.

The predominant behaviors used by species that occur in both areas were similar (Table 4), and most of these were the relatively inexpensive techniques that involve standing or walking slowly. It appears that birds in the Everglades have additional, rarer and usually more

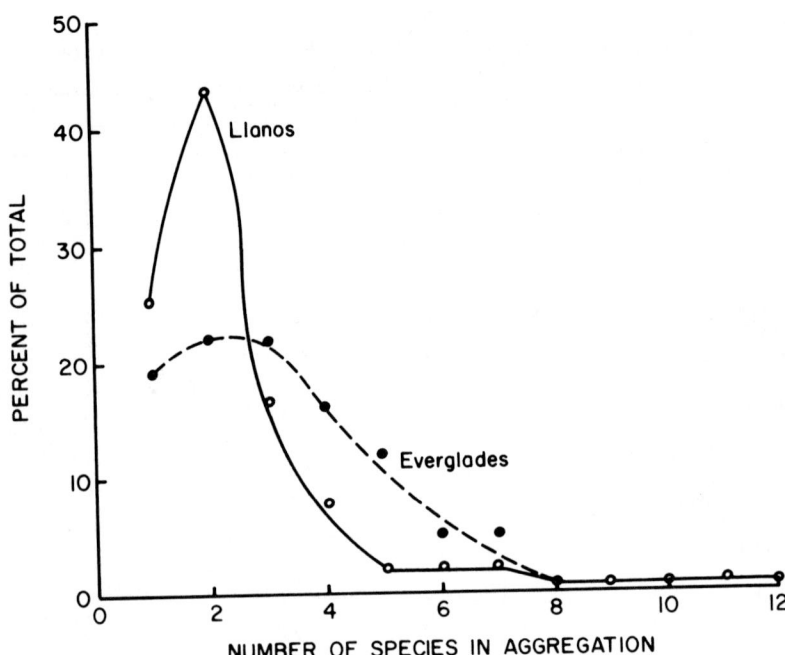

FIG. 7.   The percentage of all aggregations censused in the Florida Everglades (solid circles) and Venezuelan Llanos (hollow circles), respectively, that contained specific numbers of species.

energy-demanding behaviors. The higher abundance of prey in the Llanos may make a prey item less difficult to catch and, therefore, specialized behavior unnecessary.

Wading birds typically forage in tropical savannas and similar areas by aggregating at locations of high food availability (Kushlan 1976a; Caldwell 1981). A previous comparison of aggregation structure in a temperate marsh and in the Everglades demonstrated that the maximum and mean number of species in aggregations was higher in the Everglades with its greater overall wading bird species richness (Kushlan 1978a). This led to the hypothesis that aggregation diversity would continue to increase with increasing species richness. The results of the current study do not support this expectation, entirely; mean species richness of aggregations in the Llanos, 2.7 species per aggregation, did not surpass that of aggregations in the Everglades, 3.1 species per aggregation (Fig. 7). This result was due to the addition, in the Llanos, of species that occur primarily in mated pairs (Table 5), tallied as single-species aggregations. Such species, which appear to hold large, generally exclusive feeding territories and remain in pairs year round, include two herons and two ibises and represent a strategy that does not occur in the Everglades.

Maximum species richness, however, did increase in Llanos aggregations (Fig. 7), and one may hypothesize that interspecific interactions would be more common and complicated in such a species-rich community, thereby in part determining feeding behavior. For one such behavior, prey robbing (Kushlan 1978b; Caldwell 1980), we were able to obtain comparable information in both areas. However, contrary to expectations, prey robbing interactions appeared to be more complex in the Everglades than in the Llanos (Fig. 8). The species involved do not appear to be a factor contributing to the differences observed between areas, in that all the species shown to interact in the Everglades are also present in Llanos (if Great Blue Heron = Cocoi Heron). Rather, it seems likely that differences in prey robbing may be related to food availability. We suggest that the higher prey availability in the Llanos makes such interactions unnecessary. Thus, direct competition among species in the Llanos may be mitigated for much of the year by higher prey levels there.

Prey.—The prey consumed by a wading bird is a crucial component of its foraging niche. If one considers higher taxonomic categories of prey, some remarkable similarities in prey

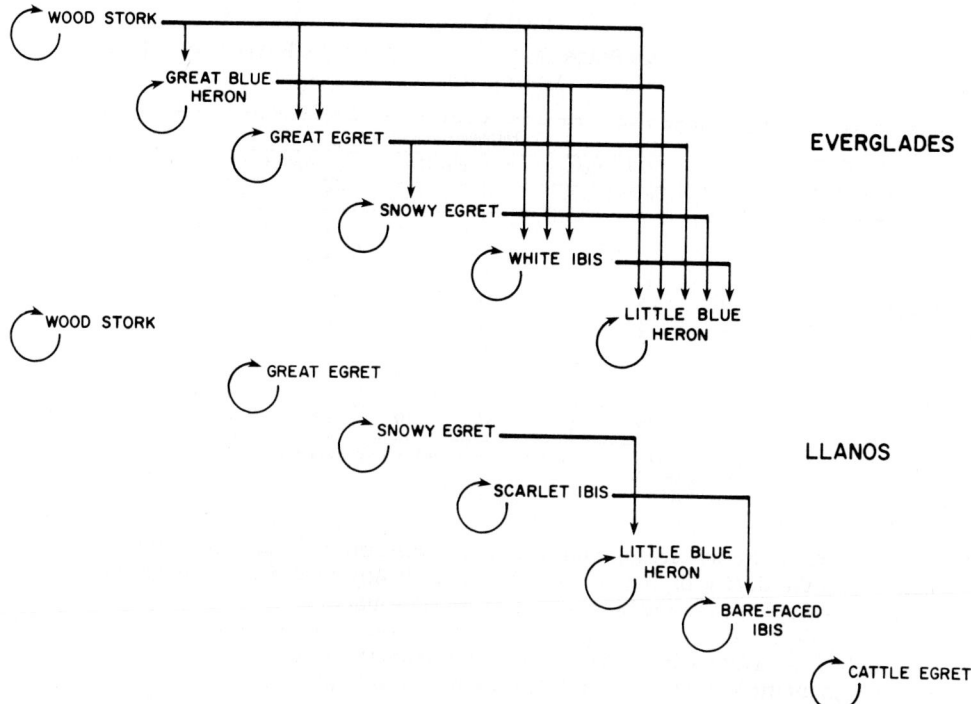

FIG. 8. Prey robbing interactions in the Everglades and Llanos. Arrows show direction of prey robbing from robber to victim.

consumption exist for wading birds that occur in both areas (Table 6). Wood Storks in both areas consume fish almost exclusively. Fish predominate in the diets of Great Egrets, Snowy Egrets, and Little Blue Herons in both areas. White Ibis exhibited the greatest difference in prey selection taking insects in the Llanos and crustaceans and fish in the Everglades. This difference seems to support the interpretation that ibises forage more terrestrially in the Llanos and more aquatically in the Everglades (Table 3).

In that the primary prey of most herons and storks is fishes, it is useful to examine the fish faunas of the two savannas. The Everglades fish fauna of 80 species is less than 30 percent as rich as the Llanos fauna of 286 species, and the species compositions of the two faunas are completely different (Mago-Leccia 1967, 1970; Loftus and Kushlan in press), meaning that the species of fish consumed by wading birds in the two areas by necessity differed taxonomically (Table 7).

Nonetheless the diets of wading birds in the two areas often contain ecologically-similar prey. For example, South American cichlids and North American centrarchids (sunfishes) are to a considerable extent ecological equivalents and are consumed by Great Egrets and Wood Storks where they occur. Similarly, cyprinodontid killifishes and poeciliid livebearers in the Everglades are similar to tetras (*Hyphessobrycon*; characins) and lebiasinids in the Llanos. These fishes are eaten by Great Egrets and Snowy Egrets in their respective locations. Wood Storks consume catfishes in both areas, callichtyid armored catfish (*Hoplosternum littorale*) in the Llanos and ictalurid bullheads in the Everglades.

Some dietary components in the Llanos have no analogue in the Everglades. For example, Great Egrets consume eel-like gymnotid electric fish (*Gymnotus carapo*) in the Llanos, and Snowy Egrets consume deep-bodied (*Markiana nigripinnis*) and compressed-bodied (*Ctenobrycon spirulus*) characins. There are no similar species in the Everglades.

Part of the dietary differences may owe to differing adaptations of the fish species. In the Llanos Snowy Egrets consume curimatas, a group similar to *Notemigonus*, which occurs in south Florida. However, *Notemigonus* is not eaten in the Everglades, because it occurs in

TABLE 6

DIETS OF FIVE SPECIES OF WADING BIRDS OCCURRING IN BOTH THE EVERGLADES (E) AND THE LLANOS (L)

	Percentage of total diet[1]									
	Great Egret		Snowy Egret		Little Blue Heron		White Ibis		Wood Stork	
	E	L	E	L	E	L	E	L	E	L
Worms							T			
Millipedes							T			
Spiders			T		T					
Crustaceans	T	14	15	5	21		52		T	T
Insects	T	8	4	6	4	21	14	100		
Snails	T				T		7			
Amphibians	3	3	T		11		2			
Reptiles	T	2								
Fishes	94	72	79	89	64	79	19		99	100

[1] Percentage of prey items consumed, except for White Ibis, which is expressed as the percentage biomass consumed because of the high degree of masceration. T = less than 2 percent of diet.

deep open water in the wet season and, being very susceptible to deoxygenation, is among the first fish to die during drying periods (Kushlan 1974b). Such factors make these fishes relatively unavailable to wading birds in the Everglades marsh.

The way in which fishes accommodate to the dry season may differ significantly between the two study areas. Llanos fishes generally possess adaptations for survival in deoxygenated waters; such adaptations are rare in Everglades fishes. As a result, dry season conditions affect large fishes in the Everglades more severely than in the Llanos (Kushlan 1976b). This difference influences the standing stocks and sizes of fishes available to birds in the two areas.

Other than for the Wood Stork, large fishes play an unimportant role in the diets of Everglades wading birds. Even the Wood Stork, capable of eating fish in excess of 22 cm long, consumed prey that averaged only 2.6 cm long (Ogden et al. 1976) in the Everglades. In contrast, large fishes in the Llanos are eaten by several birds including White-necked Herons, Great Egrets, and Jabiru, Maguari, and Wood Storks. These birds consume electric fishes, river eels (*Synbranchus marmoratus*), armored catfish, trahiras, caribe (*Serrasalmus*) (a characin), and large cichlids. The only equivalently-sized species in the Everglades marsh, bass (*Micropturus salmoides*), bowfin (*Amia calva*), and gar (*Lepisosteus platyrhincus*), are rare and not often taken by wading birds. Thus, the abundant and diverse large-item component of the fish fauna of the Llanos is not available to wading birds in the Everglades.

DISCUSSION

The Everglades and Llanos share a major portion of their wading bird fauna. However, as might be expected of the more tropically-situated ecosystem, the Llanos supports more species. Although the species compositions of the wading bird communities of both areas are dominated by herons, much of the increase in the Llanos is due to the terrestrial ibises and two very large storks, in addition to the Whistling Heron, a characteristic neotropical savanna species, and the Capped Heron, an Amazonian species.

Differences in the wading bird faunas in these two wetland ecosystems may in part be attributable to historical and biogeographic factors. An impoverishment of the vertebrate fauna in general exists in both the Everglades and the Llanos, and in the Everglades this in part reflects its location at the end of a temperate continental cline of decreasing species richness (Simpson 1964; Robertson and Kushlan 1974; Tramer 1974). Both savannas are also relatively young, perhaps less than 10,000 years old. South American savannas, however, were probably readily re-invaded by tropical species that had persisted in glacial refugia, and as a result the species there have had a longer time to adapt to conditions in fluctuating marshes. This would apply to birds directly as well as to their prey species. In contrast, the Everglades is isolated from South American refugia by a barrier of unsuitable marine habitat, inhibiting the invasion of many species characteristic of neotropical inland marshes, especially those unable to use intervening island and marine habitats of the West Indies. Thus the wading bird community of the Everglades is only a selection of that of South America, and the prey

TABLE 7
Comparison of the Fishes Eaten by Five Wading Birds in the Everglades (E) and Llanos (L)

	Great Egret		Snowy Egret		Little Blue Heron		White Ibis		Wood Stork	
	E	L	E	L	E	L	E	L	E	L
Gar	T[1]								T	
Freshwater eels	T									
Characins		19		21						
Electric-fishes		28								
Suckers	T		T							
Bullheads	T								2	
Killifishes	26		28		12		T		75	
Livebearers	47		50		49		2		13	
Sunfishes	13		T		2				9	
Cichlids	T	19	T	2						83
Sleepers					T					
Lebiasinids		6		32						
Armored catfish										17
Curimatas				34						
Trahiras						79				
Other fishes	8		T		T		16			

[1] T indicates less than 2 percent of diet.

fauna is entirely derived from temperate forms, poorly adapted to fluctuating tropical wetlands. Differing periods of evolutionary time and access, therefore, may be responsible for differing species richness in these systems.

Ecological factors must also have played an important role in structuring the wading bird communities in the Llanos and Everglades, which differ not only in numbers of species, but in their relative abundances as well. The dominance of one species, the White Ibis, in the Everglades and the greater equitability of species in the Llanos may reflect differing responses to resource availability. Higher equitability may suggest that a larger proportion of the wading bird community has accommodated to conditions in the Llanos as a result of niche compression or more likely because of more abundant and diverse resources.

In both the Everglades and the Llanos, differences among species can be characterized along several axes of their foraging niches, including foraging behavior, prey selectivity, and habitat use. In the Everglades a greater range of foraging behaviors are used and the larger herons are more widely separated along a morphological niche dimension. It is possible that direct competition is of greater consequence in structuring this community, an idea supported by the greater complexity of robbing interactions and the higher average number of species that feed together in aggregations. We hypothesize that a lower prey abundance and diversity in the Everglades may decrease the long-term effectiveness of foraging there as contrasted with the Llanos.

Prey selectivity by wading birds results primarily from the size and behavior of their prey. Thus the diets of wading birds in both savannas show a surprising degree of ecological similarity despite taxonomic differences in the prey taken. Although characteristics of the fish faunas converge in the two systems, a greater species richness, including types of fishes not found in the Everglades, occurs in the Llanos, as in many tropical freshwater systems (Lowe-McConnell 1975). This may provide more opportunities for predators and to some extent account for the greater richness of wading bird species in the Llanos. The presence of very large storks, for example, may in part be due to the opportunity to take advantage of the large-sized prey available there.

Another factor that may affect species composition is habitat diversity, which would increase the types of feeding areas available while decreasing the need for niche overlap. The savanna specialists, such as the Buff-necked Ibis, Sharp-tailed Ibis, and Whistling Herons, occur primarily in very shallow to dry habitats near high ground, which is more extensive and continuous in the Llanos than in the Everglades. These species are permanent residents, maintaining pair-based social and territorial systems not found in the Everglades. Such shallow and dry habitats provide foraging sites for small-sized tactile foraging ibises in the Llanos. These birds often migrate out of the area during the wet season when such habitats are no longer available.

An exception to the pattern of taking advantage of increased foraging opportunities in the Llanos can be seen in the ibises. These species overlap along niche dimensions and in the drying season feed together using similar behaviors and feeding sites. Although additional study of these species is needed (S. Ramos and B. Busto, pers. comm.), we hypothesize that during the dry season prey overlap should be high. However, because prey availability will also be high during these periods, competition may be minor, if it occurs at all. It is possible that differing patterns of habitat use in the wet season may separate the ibis species in the Llanos.

The ecology of hyperseasonal savannas is dominated by the fluctuation in water depths, which results in the seasonal alternation of floods and droughts. Greater fluctuations in water depth occur in the Llanos than in the Everglades, a result of higher rainfall, temperatures, and evapotranspiration, with the more tropical system therefore being the more variable. With a more diverse fish community in the Llanos, prey availability is higher there during the dry season. Because of their adaptations to stressful conditions, more fish species are available to birds in the Llanos later in the dry season, and more species and individuals survive to repopulate the marshes during the following wet season. In general, the maintenance of prey species diversity in tropical savannas depends on the complex interaction of biotic factors and fluctuating water levels. In both the Llanos and the Everglades, fish species richness may be affected by wading birds and crocodilians. During the dry seasons of both areas, deep-water areas provide refugia for fish populations while predation pressure is concomitantly increased.

Wading bird niche relations must therefore be elaborated within a context of seasonal variability of prey and habitat. Many characteristics of wading birds adapt them to such seasonal water depth cycles. Their relatively long legs permit them to use shallowly-flooded marshes, and some species are also able to forage terrestrially. Another important adaptation is seasonal migrations which take birds out of parts of the savannas during the wet season when water is too deep for wading. In the drying season, shorter movements permit foraging at sites where water depths are temporally suitable and prey are locally abundant. These ephemeral but superabundant food supplies, in addition to terrestrial habitat availability, support wading bird populations in the dry season.

Amid the apparently stressful cycles of prey availability, wading bird communities reflect specific accommodations of the species to the savanna environment. That the Llanos and the Everglades differ in the structure of their wading bird communities may be another instance of the general pattern of higher diversity and productivity seen in many tropical ecosystems as contrasted with their more temperate counterparts. In both systems, and probably in tropical savannas in general, it seems clear that cyclical and random fluctuations of environmental factors, together with biogeographic constraints, play primary roles in structuring constituent animal communities.

ACKNOWLEDGMENTS

We thank our many associates, especially C. Casler, K. Dobrowolski, F. Gómez, N. J. León, M. Madriz, C. F. Naranjo, O. T. Owre, J. Pacheco, and J. Pinowski. L. Bulla, W. Loftus, A. Machado, R. Martinez, R. Martinez-E., G. Pereira, D. Rada, and C. R. Robins provided information on prey species. Members of the Grupo de Ecología of the Instituto de Zoología Tropical were always encouraging and supportive, and we especially thank J. Pacheco for making our collaborations possible. E. Szeplaki made critical translations. M. Ramia supplied the Llanos photographs. Financial support was provided by the Maytag Chair of Ornithology at the University of Miami, Consejo de Desarrollo Científico y Humanístico of Universidad Central de Venezuela (Projects of C. 05.02/78 and C. 05.4/78), and CONICIT (Projecto Pima 7). We thank P. A. Buckley, G. S. Caldwell, M. S. Foster, M. Gochfeld, T. Jacobsen, C. S. Luthin, and D. Mock for comments and suggestions on the paper.

LITERATURE CITED

BEARD, J. S. 1953. The savanna vegetation of northern tropical America. Ecol. Monogr. 23:149–215.
BLYDENSTEIN, J. 1967. Tropical savanna vegetation of the Llanos of Columbia. Ecology 48:2–15.

CALDWELL, G. S. 1980. Underlying benefits of foraging aggression in egrets. Ecology 61:996–997.

CALDWELL, G. S. 1981. Attraction to tropical mixed-species heron flocks: proximate mechanisms and consequences. Behav. Ecol. Sociobiol. 8:99–103.

COLE, M. M. 1960. Cerrado, caatinga and pantanal: the distribution and origin of the savanna vegetation of Brazil. Georgr. J. 126:168–179.

CUSTER, T. W., AND R. G. OSBORN. 1977. Feeding-site description of three heron species near Beaufort, North Carolina. Pp. 355–360, In A. Sprunt, J. C. Ogden, and S. Winckler (eds.), Wading Birds. Natl. Audubon Soc. Res. Rep. No. 7.

HANCOCK, J., AND J. KUSHLAN. 1984. The Herons Handbook. Harper and Row, New York.

KUSHLAN, J. A. 1974a. Observations on the role of the American alligator (Alligator mississippiensis) in the southern Florida wetlands. Copeia 1974:993–996.

KUSHLAN, J. A. 1974b. Effect of a natural fish kill on the water quality, plankton, and fish population of a pond in the Big Cypress Swamp, Florida. Trans. Am. Fish. Soc. 2:235–243.

KUSHLAN, J. A. 1976a. Wading bird predation in a seasonally-fluctuating pond. Auk 93:464–476.

KUSHLAN, J. A. 1976b. Environmental stability and fish community diversity. Ecology 57:821–825.

KUSHLAN, J. A. 1978a. Feeding ecology of wading birds. Pp. 249–296, In A. Sprunt, J. C. Ogden, and S. Winckler (eds.), Wading Birds. Natl. Audubon Soc. Res. Rep. No. 7.

KUSHLAN, J. A. 1978b. Nonrigorous foraging by robbing egrets. Ecology 59:649–653.

KUSHLAN, J. A. 1979a. Foraging ecology and prey selection in the White Ibis. Condor 81:376–389.

KUSHLAN, J. A. 1979b. Prey choice by tactile-foraging wading birds. Proc. Col. Waterbird Grp. 3:133–142.

KUSHLAN, J. A. 1981a. Resource use strategies in wading birds. Wilson Bull. 93:145–163.

KUSHLAN, J. A. 1981b. Sampling characteristics of enclosure fish traps. Trans. Am. Fish. Soc. 110:557–562.

KUSHLAN, J. A., AND M. S. KUSHLAN. 1980. Population fluctuation of the prawn, Palaemonetes paludosus, in the Everglades. Am. Midl. Nat. 103:401–403.

KUSHLAN, J. A., AND D. A. WHITE. 1977. Nesting wading bird population in southern Florida. Fla. Sci. 40:65–72.

KUSHLAN, J. A., J. A. HANCOCK, J. PINOWSKI, AND B. PINOWSKA. 1982. Behavior of Whistling and Capped Herons in the seasonal savannas of Venezuela and Argentina. Condor 84:255–260.

LOVELESS, C. M. 1959. A study of the vegetation of the Florida Everglades. Ecology 40:1–9.

LOFTUS, W. L., AND J. A. KUSHLAN. In press. The fresh-water fishes of southern Florida. Bull. Fla. State Mus. Biol. Sci.

LOWE-MCCONNELL, R. 1975. Fish Communities in Tropical Fresh-waters. Longman, New York.

LOWE-MCCONNELL, R. H. 1969. Speciation in tropical freshwater fishes. Biol. J. Linn. Soc. 1:51–75.

MAGO-LECCIA, F. 1967. Notas preliminares sobre los peces de los Llanos de Venezuela. Bol. Soc. Venez. Cienc. Nat. 27:237–263.

MAGO-LECCIA, F. 1970. Estudio preliminares sobre la ecologia de los peces de los Llanos de Venezuela. Acta Biol. Venez. 7:71–102.

MAY, R. M. 1976. Patterns in multi-species communities. Pp. 142–162, In R. M. May (ed.), Theoretical Ecology, Principles and Applications. W. B. Saunders, Philadelphia, Pennsylvania.

MORALES, G., J. PINOWSKI, J. PACHECO, M. MADRIZ, AND F. GOMEZ. 1981. Densidades poblacionales, flujo de energía y hábitos alimentarios de las aves ictiófagas de los módulos de Apure, Venezuela. Acta Biol. Venez. 11:1–45.

OGDEN, J. C., J. A. KUSHLAN, AND J. T. TILMANT. 1976. Prey selectivity by the Wood Stork. Condor 78:324–330.

PIANKA, E. R. 1974. Evolutionary Ecology. Harper and Row, New York.

PINOWSKI, J., L. G. MORALES, J. PACHECO, K. A. DOBROWOLSKI, AND B. PINOWSKA. 1980. Estimation of the food consumption of fish-eating birds in the seasonally-flooded savannas (Llanos) of Alto Apure, Venezuela. Bull. Acad. Pol. Sci. Ser. Sci. Biol. 28:163–170.

RAMIA, M. 1967. Tipos de sabanas en los llanos de Venezuela. Bol. Soc. Venez. Cienc. Nat. 27:264–288.

RAMIA, M. 1974. Estudo ecologico del Módulo Experimental de Montecal (Alto Apure). Bol. Soc. Venez. Cien. Nat. 31:117–142.

RAMOS, S., S. DANIELEWSI, AND G. COLOMINE. 1981. Contribución a la ecología de los vertebrados acuáticos en esteros y bajíos de sabanas moduladas. Bol. Soc. Venez. Cienc. Nat. 35:79–103.

ROBERTSON, W. B., JR., AND J. A. KUSHLAN. 1974. The southern Florida avifauna. Miami Geol. Soc. Mem. 2:414–452.

SARMIENTO, G., AND M. MONASTERIO. 1975. A critical consideration of the environmental conditions associated with the occurrence of savanna ecosystems in tropical America. Pp. 223–250, In F. Golley and E. Medina (eds.), Tropical Ecological Systems. Springer-Verlag, New York.

SEIJAS, A. E., AND S. RAMOS. 1980. Características de la dieta de la baba (Caiman crocodilus) durante la estación seca en las sabanas moduladas del Estado Apure, Venezuela. Acta Biol. Venez. 10:373–389.

SIMPSON, G. G. 1964. Species density of North America recent mammals. Syst. Zool. 13:57–73.

STATON, M. A., AND J. R. DIXON. 1977. The herpetofauna of the central Llanos of Venezuela; noteworthy records, a tenative checklist and ecological notes. J. Herpetol. 11:17–24.

TAMAYO, F. 1961. Los Llanos de Venezuela. Instituto Pedagógico, Dirección de Cultura, Caracas.

TRAMER, E. J. 1974. Latitudinal gradients in avian diversity. Condor 76:123–130.

TROTH, R. G. 1979. Vegetational types on a ranch in the central Llanos of Venezuela. Pp. 17–30, In J. F. Eisenberg (ed.), Vertebrate Ecology in the Northern Neotropics. Smithsonian Institution Press, Washington, D.C.

WALTER, H. 1969. El problema de la sabana. Bol. Soc. Venez. Cienc. Nat. 28:123–144.

PERMANENT CANOPY AND UNDERSTORY FLOCKS IN AMAZONIA: SPECIES COMPOSITION AND POPULATION DENSITY

Charles A. Munn

*Department of Biology, Princeton University, Princeton, New Jersey 08544 USA;
present address: Wildlife Conservation International, New York
Zoological Society, Bronx, New York 10460 USA*

ABSTRACT. Two kinds of permanent, multi-species flocks inhabit the lowland forest of Amazonian Peru, canopy and understory flocks. Each flock type has a core composed of 5 to 10 different species of largely insectivorous birds, and each species is represented by a single bird, a mated pair, or a family group. Besides these core species, the flocks are joined by individuals of an additional 80 species with varying degrees of regularity. Since 1976, color-marking and long-term following of hundreds of individuals of flock species has demonstrated that the individuals of the core species generally spend their entire lives in one flock. Species that have higher population densities than the core species enter and leave flocks as flocks pass through their territories, whereas species with lower population densities than the core species alternate between several flocks on a daily or hourly basis. Based on 1500 h and 250 h of following individually-recognizable understory flocks and canopy flocks, respectively, I present territory maps for the 25 understory and seven canopy flocks occurring on a 1.8 km² study plot. I also present population density estimates for 78 flock species and detail their use of the 32 different study flocks. The population density estimates for some understory flock species are several times lower than Karr's (1971) estimates for the same species in Panama. Density compensation is hypothesized to explain this difference.

RESUMEN. Dos tipos de bandadas permanentes compuestas por varias especies conjuntas habitan las zonas bajas del bosque amazónico peruano, las bandadas que viven en las copas de los árboles ("canopy") y las que están en la vegetación inferior ("understory"). Cada tipo de bandada tiene un componente básico ("core") que varía entre 5 o 10 especies diferentes, mayormente aves insectívoras y cada especies está representada por: un único individuo, una pareja o un grupo familiar. Ademas del "core" se unen a la bandada, con varios grado de regularidad, individuos de 80 especies adicionales. Desde 1976, por medio de marcación con colores y seguimientos por largos períodos de cientos de individuos de especies participantes en bandadas, se ha demostrado que los individuos de las especies del "core" pasan generalmente todas sus vidas en una sola bandada. Especies que tienen una mayor densidad de población que las especies del "core" entran y salen de las bandadas a medida que estas pasan por sus territorios, mientras que aquellas especies cuya densidad es menor que la de las especies del "core" de la bandada alternan entre diversas bandadas en base a un patrón horario o diario. Presento mapas de los territorios para 25 bandadas de vegetación inferior y 7 bandadas de las copas de los árboles que se presentaron en el área de estudio de 1,8 km², basandome en 1500 h y 250 h de seguimientos de bandadas de vegetación inferior y de las copas de los árboles respectivamente, que eran reconocibles individualmente. También presento estimaciones de la densidad de población de 78 especies de bandadas y detallo su presencia en las 32 diferentes bandadas estudiadas. Las estimaciones de densidad de población para algunas especies de bandadas de la vegetación inferior son varias veces menores que las estimaciones que brindase Karr (1971) para las mismas especies en Panamá. Para explicar estas diferencias se sugieren compensaciones de densidad.

At least since Bates (1863), biologists traveling in the humid tropics have marveled at the large mixed-species flocks of insectivorous birds typical of lowland and mid-elevation forests. Bates himself reported (1863:334–335),

> "... *one may pass several days without seeing many birds; but now and then the surrounding bushes and trees appear suddenly to swarm with them. There are scores, probably hundreds of birds, all moving about with the*

greatest activity . . . . The bustling crowd loses no time, and although moving in concert, each bird is occupied, on its own account, in searching bark or leaf or twig . . . . In a few minutes the host is gone, and the forest path remains deserted and silent as before. I became, in course of time, so accustomed to this habit of birds in the woods near Ega, that I could generally find the flock of associated marauders whenever I wanted it. There appeared to be only one of these flocks in each small district; and, as it traversed chiefly a limited tract of woods and second growth, I used to try different paths until I came up with it."

In temperate latitudes, mixed flocks of parids and warblers form in the nonbreeding season (Morse 1970), whereas the mixed flocks of tropical forests are in many cases year-round phenomena (Moynihan 1962; Fogden 1972; Croxall 1976; Partridge and Ashcroft 1976; Jones 1978; Powell 1979; Morton 1980; Wiley 1980). Munn and Terborgh (1979) and Gradwohl and Greenberg (1980) color-marked and studied understory mixed-species flocks in southeastern Peru and Panama, respectively; in each study, the understory flocks were composed of individuals, mated pairs, or small family groups of each of the species, with the same individuals consorting together daily in a particular flock for their entire lives. Also, the individual birds in both the simple flocks in Panama and the more complex flocks in Peru jointly held and defended territories against neighboring understory flocks. In both locations, individual birds defended their flock positions by displaying toward and chasing away intruders of the same sex and species.

Munn and Terborgh (1979) briefly discussed multi-species canopy flocks at their study site, but since they were unable to mark canopy flock species, they only speculated about the possible structure of these groups. McClure's (1967) knowledge of the multi-species canopy flocks of southeast Asia was equally fragmentary.

I show here for the first time that, like the understory flocks, each multi-species canopy flock at my study site in Amazonian Peru is composed of a permanent core of mated pairs or families of birds, each representing a different species. Birds of many other species join these flocks as well, but, unlike the core species, do not spend the majority of their time in just one flock.

I began my study of the canopy flocks as a test of whether, like the understory flocks, they were constructed of a permanent core of mated pairs, each of a different species. These canopy flocks are dominated by tanagers, flycatchers, and vireos, and, thus, share few species with the understory flocks, which are dominated by antbirds, ovenbirds, and woodcreepers. It seemed unlikely, therefore, that similar flock structures would result from taxonomic similarity, but rather, that they would reflect similar selective forces operating on both understory and canopy birds. Furthermore, because canopy flocks occur in the same forest, sometimes even in the same vegetation layer, as the understory flocks and often travel together with understory flocks for several hours at a time, these two types of flocks should be experiencing somewhat similar foraging environments and predation pressures.

My observations of marked birds in canopy flocks showed that these flocks are similar in structure to understory flocks. Herein, I discuss these similarities, present maps of the territories of 25 understory and seven canopy flocks, and give the species composition of each of these flocks. Finally, based on these species compositions, I calculate or estimate the population sizes of 78 of the 101 different species that join these flocks.

STUDY AREA AND METHODS

The study site is situated at 71°19′W, 11°51′S (elevation ca. 400 m) in undisturbed rainforest within 1 km of the Manu River at Cocha Cashu Biological Station in Manu National Park. The Park includes portions of the departments of Cuzco and Madre de Dios in southeastern Peru.

The study area (Fig. 1) is flat except for gentle depressions and rises left from ancient changes in the course of the river. In the rainy season, flooding occurs in low areas near the river and in shallow depressions in the forest. The study area is primary tropical forest. Occasional very tall trees ("emergents") rise to 60 m, while the more-or-less continuous canopy of tall trees is at 30 to 45 m. Below the canopy are several tree layers (Richards 1952; Terborgh 1983). I could walk easily through the open understory except where trees had fallen.

I studied flock birds for a total of 29 months from 1976 to 1982 (August–November 1976, July–August 1978, August–December 1979, July–November 1980, June–January 1981, and

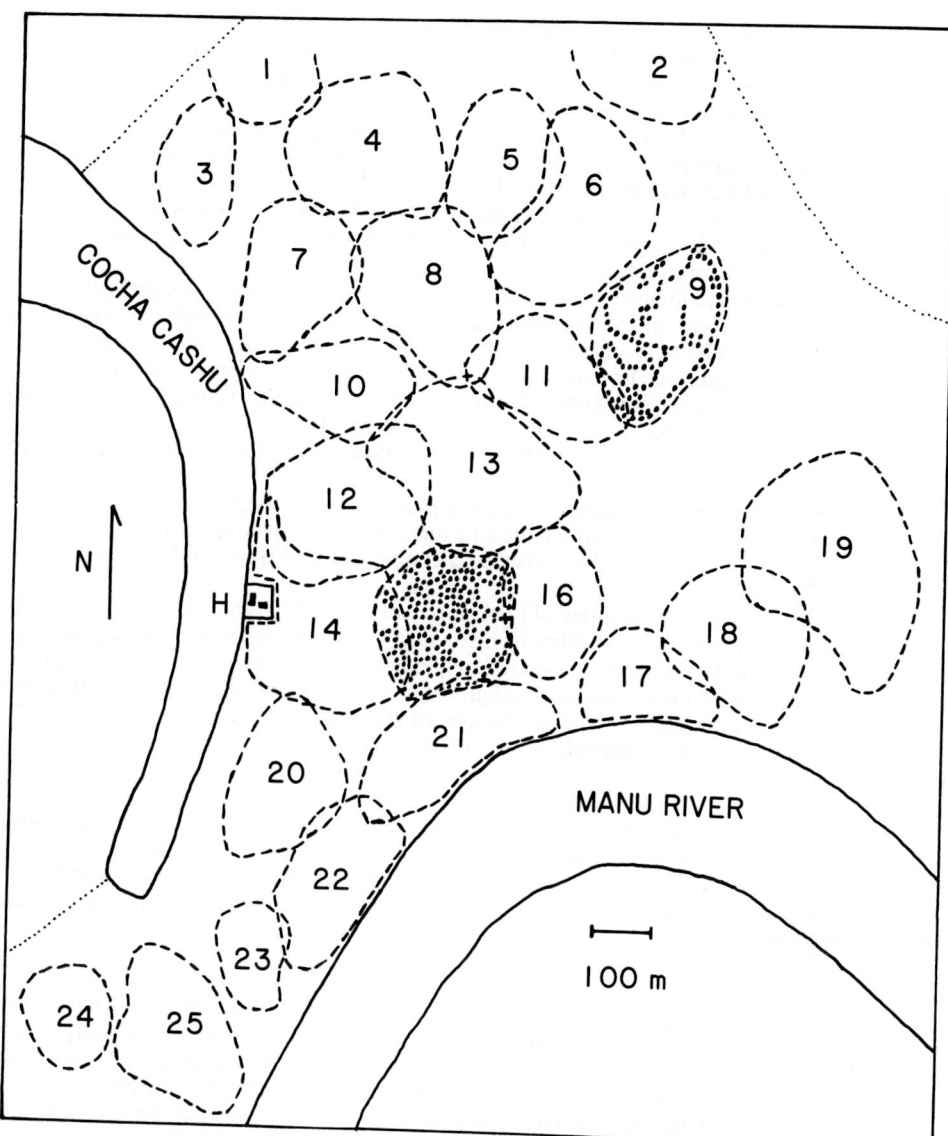

FIG. 1.   Territories of 25 different understory flocks in 1.8 km² of lowland forest at Cocha Cashu, Peru. "H" indicates the location of two houses in a clearing. The dots in territories 15 (unnumbered) and 9 represent locations where I sighted these flocks. The number of sightings of flock 15 is typical of the number for each near-house flock, while the number of sightings of flock 9 is typical of the number for each far-from-house flock. The gap north of flocks 16, 17, and 18 is a seasonally-inundated fig (*Ficus trigona*) swamp.

July–November 1982). Approximately two meters of rain fall annually at this site, mostly from December through May. Although my field seasons always spanned the driest months of July, August, and September, I observed no behavioral or structural changes in the flocks during the wetter months of October–January. Even in October 1982, during the heaviest rains and flooding on record for the site, the flocks functioned normally. Casual observations by my colleagues indicate that the flocks remain intact from February through May as well (J. Terborgh, C. Janson, pers. comm.). I studied understory flocks in all years, and canopy flocks only during 1981 and 1982. Understory birds were netted annually on long netlines along trails, while in 1982, canopy birds were captured using nets at up to 40 m above the ground. More than 950 individual birds of different flock-joining species were marked for the present study. The canopy nets, which will be described in a paper on techniques, captured 150 birds of 46 species in about 700 net-hours of use. Scientific and common names of all species considered are given in Appendix I.

I observed understory flocks with 7 × 42 binoculars and canopy flocks with tripod-mounted 15 × 60 binoculars, noting the presence of marked birds. Since 1976, I have followed understory and canopy flocks at the site for about 1500 h and 250 h, respectively. Birds in 1976–1981 were banded with numbered aluminum bands and celluloid color bands. In 1982 I used specially-prepared, loose-fitting celluloid color bands to avoid problems with fungal or virus infections on the legs of small birds (<17 g).

In addition, I was able to identify some individual birds by means of natural disfigurements of their facial areas and legs, or bizarre molting patterns. Finally, I also found it possible to identify individual birds in both understory and canopy flocks by virtue of consistent peculiarities in their vocalizations. For instance, each of the seven different male *Hylophilus hypoxanthus* in the seven different canopy flocks could be recognized. I tested this by listening to the greenlet in a canopy flock and predicting the canopy flock before actually looking at any marked birds. This same method worked with certain individual antshrikes, antwrens, and tanagers as well.

I used a sighting compass and counted paces to plot flock movements with reference to a network of marked trails. There is no location in the 1.8 km² study area that lies more than 150 m from a trail, and each time that I crossed a trail, I reoriented myself before setting off again into the forest.

To calculate the population densities of flock species for which all individuals in the study area were banded, I summed the number of birds that I knew to be in each flock (Table 1). For species whose populations were less completely banded, I estimated high and low values. For unbanded, or virtually unbanded species, I give crude high and low estimates accompanied by a question mark or a plus sign. For all of the species whose populations were not completely marked, I assumed that each unmarked bird used the same number of flocks as marked conspecifics (Table 1), and then I made low and high estimates based on the tendency of each species to travel with conspecifics in flocks.

In order to measure canopy height, I stopped every 25 m along 3.2 km of trails and chose the tallest tree within a circle of 10 m radius. I then measured the tree's height with a "Ranging 620" Rangefinder. If the top of the tree was obscured by vegetation, I estimated the height by measuring adjacent trees or by adding an estimated number of meters to the height of the highest visible branches.

RESULTS

After marking and initial observation of birds in canopy flocks, I noticed immediately that the canopy flocks shared the core structure of mated pairs that I had found previously in the understory flocks. Other sorts of similarities, however, only emerged after closer inspection of several marked canopy flocks. For instance, understory and canopy flocks each have their own set of species with population densities higher or lower than those of the core species. The lower density species switch back and forth between several different understory or canopy flocks, depending upon the species, whereas the higher density species join the appropriate type of flock for as long as a flock remains in their relatively small territory. Apparently, only the prior presence of a conspecific individual or pair prevents lower density species from entering a particular flock. Likewise, in both the understory and canopy flocks, a mated pair of higher density species apparently only drops out of the flock in response to aggression from conspecific neighbors that attempt to join the flock as it enters their territory.

CANOPY FLOCKS

Composition.—In 1981, daily study of two principal canopy flocks (7 and 4, Fig. 2) and intermittent study of flocks 5 and 6 confirmed that each flock contained only one mated pair of *Lanio versicolor*. One or both individuals in each of the four pairs was color-marked, and the daily presence of the same marked birds in flocks at particular locations suggested the possibility of permanent canopy flocks. In 1982, I banded 10 additional birds in canopy flocks 3, 4, and 7 (Fig. 2). Subsequent following in September through 14 November 1982 established that each of these three flocks always contained mated pairs of the following species: *Terenura humeralis, Tolmomyias assimilis, Lanio versicolor, Tachyphonus rufiventer, Tachyphonus luc-*

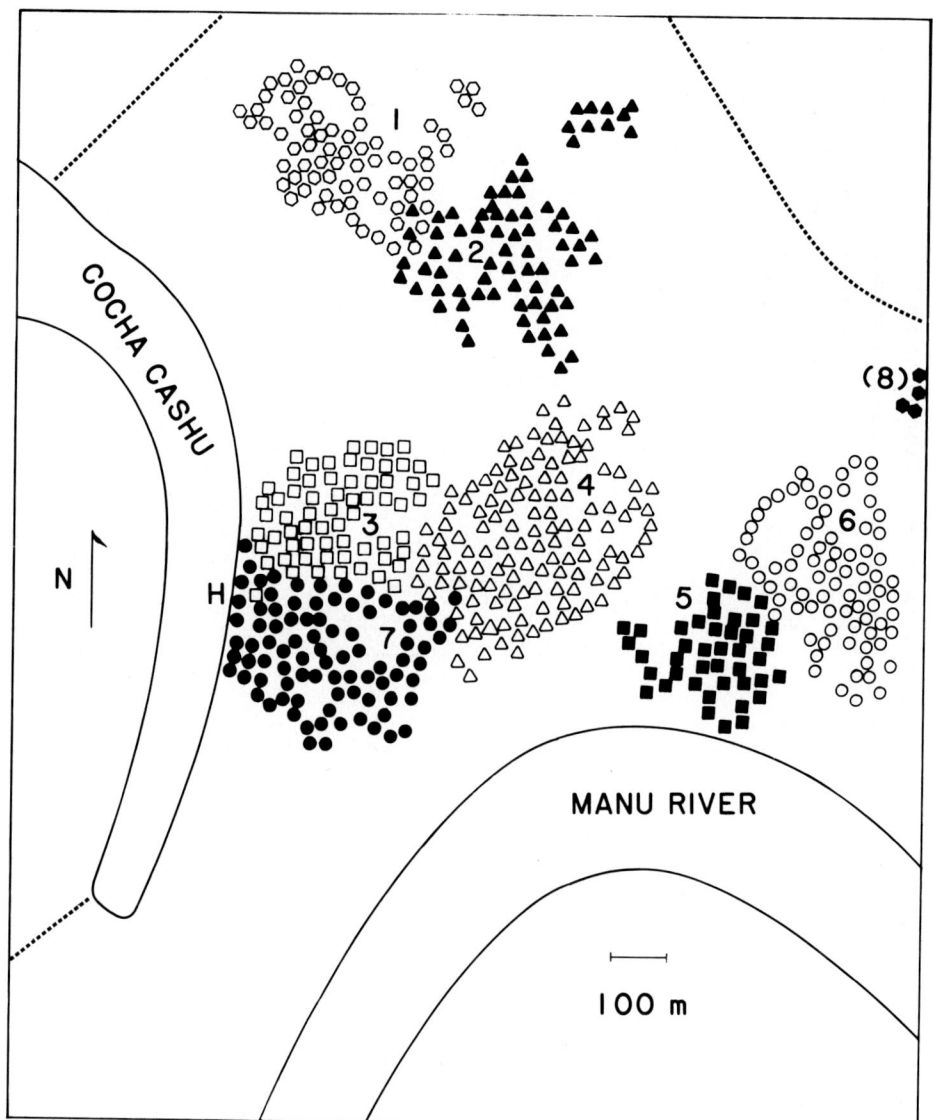

Fig. 2. Territories of seven canopy flocks in 1.8 km² of lowland forest at Cocha Cashu, Peru. "H" indicates the location of two houses. Symbols were placed wherever I followed individually-identifiable flocks during the 1981 and 1982 seasons.

tuosus, and *Hylophilus hypoxanthus.* The other four canopy flocks had similar cores of mated pairs of these six species. The same pairs associated with one another daily from soon after dawn until late afternoon, as did core pairs in the understory flocks. I never saw any of these individually-identifiable canopy birds in a flock other than their normal one. It seems probable that, like the marked *Lanio* pairs, the five other core species spend their entire lives in one canopy flock.

Individuals of 53 other species join these canopy flocks with varying frequency (Table 1). Some of these species have higher, and some lower population densities than the core species. Individuals of species with population densities higher than those of the core species tend to drop in and out of the local canopy flock as it moves through their territories, whereas individuals of species with lower population densities switch freely back and forth between several different flocks. Examples of the higher population density species include *Monasa*

TABLE 1

SPECIES COMPOSITION OF 32 FLOCKS AT COCHA CASHU AND POPULATION DENSITIES OF COMPONENT SPECIES[1]

Species	Understory flocks[2]																
	1	2	3	4	5	6	7	8	9	10	11	12	13	14	15	16	17
CUCULIDAE																	
Coccyzus erythrophthalmus																	
Coccyzus americanus												1					
Coccyzus melacoryphus												1					
Piaya cayana																	
GALBULIDAE																	
Galbula cyanescens														+	+	+	+
BUCCONIDAE																	
Monasa nigrifrons				+			+	+	+		+	+	+	+	+	+	
Monasa morphoeus		+				+	+										
CAPITONIDAE																	
Capito niger						+											
Eubucco richardsoni																	
PICIDAE																	
Picumnus rufiventris	0+1																0+1

TABLE 1
CONTINUED

Species	Understory Flocks[2]								Canopy flocks[2]							Trans.[3]	Total (density)[4]
	18	19	20	21	22	23	24	25	1	2	3	4	5	6	7		
CUCULIDAE																	
Coccyzus erythrophthalmus										1						+M	
Coccyzus americanus				1											1	+M	
Coccyzus melacoryphus													1			+	
Piaya cayana												1			3	+	
GALBULIDAE																	
Galbula cyanescens	+	+	+	2	+	+		+								+P	
BUCCONIDAE																	
Monasa nigrifrons	+	+		+	+			+	+	+	+	+	+	+	+	+	
Monasa morphoeus										2						+	
CAPITONIDAE																	
Capito niger				+					+	+		♂︎|♀︎	+	+	+	+	
Eubucco richardsoni										+	+	+			♂︎|♀︎|i·j	+	10–20? (6–11?)
PICIDAE																	
Picumnus rufiventris																+	

TABLE 1
CONTINUED

Species	Understory flocks[2]																
	1	2	3	4	5	6	7	8	9	10	11	12	13	14	15	16	17
Piculus chrysochloros																	
Veniliornis passerinus																	
Veniliornis affinis				♂ ♀					♂ ♀								
DENDROCOLAPTIDAE																	
Dendrocincla fuliginosa												1[a]/1	1[a]/1		1[a]/1		
Deconychura longicauda				1													
Sittasomus griseicapillus						1*/1	1	1*/1		1		1*/1	1*/1	1*/1	+	+	1
Glyphorynchus spirurus							1/1					1/1			1/1		
Dendrexetastes rufigula															1*	1*	
Xiphorhynchus obsoletus	1																
Xiphorhynchus ocellatus							1/1			1				1[a]/1 · 1j[b]/1 · 1[c]/1	1[a]/1 · 1j[b]/1 · 1[c]/1		
Xiphorhynchus spixii		2		1	2*	2*		1/1	2		1[a]/1 · 1[b]/1	1/1 · 1/1 · 1/1	1[a]/1 · 1[b]/1 · 1/1		1[d]/1 · 1[e]/1	1[d]/1 · 1[e]/1	1[e]/1

TABLE 1
CONTINUED

Species	Understory Flocks[2]								Canopy flocks[2]							Trans.[3]	Total (density)[4]
	18	19	20	21	22	23	24	25	1	2	3	4	5	6	7		
Piculus chrysochloros												♂				+	
Veniliornis passerinus								1				♀				+	
Veniliornis affinis				♀								♂ / ♀			♂ / ♀	+	
DENDROCOLAPTIDAE																	
Dendrocincla fuliginosa	+															+	
Deconychura longicauda				1												+	
Sittasomus griseicapillus[a]	1		1[a]	1	1[a]	1[a]			1	1*	1	1	1	1	1*	+	20–30?
[b] / †	1		1† / 1		1†	1†			1	1	1	1	1	1	1[b]		(11–17?)
Glyphorynchus spirurus								+								+	
Dendrexetastes rufigula																	
Xiphorhynchus obsoletus	1 / 1	1 / 1, / 1[d]	1[a] / 1j[b] / 1[c]	1[a]								1		1	1	+ / +P	1–4 (1–2)
Xiphorhynchus ocellatus				1[c]	1[c]												10–12 (6–7)
Xiphorhynchus spixii																	14–16 (8–9)

TABLE 1
CONTINUED

Species	Understory flocks[2]																
	1	2	3	4	5	6	7	8	9	10	11	12	13	14	15	16	17
Xiphorhynchus guttatus	1			1/1			1[a]/1	1[a]/1	1/1	1		1/1	1/1	1	1/1		
Lepidocolaptes albolineatus								1/1				1/1	1/1				
Campylorhamphus trochilirostris												1	1				
FURNARIIDAE																	
Cranioleuca gutturata																	
Hyloctistes subulatus				1	1	1		1[a]/1		+		1[b]/1	1[a]/1	1[b]		+	
Ancistrops strigilatus				+	2	1	1		1j	1[a]/1		1[a]/1	1[a]/1	1		+	
Philydor erythrocercus																	1*
Philydor rufus																	
Philydor erythropterus									1, 1j, 1			1[a]/1	1[a]/1	1[a]/1	+		1
Philydor ruficaudatus										1/1		1		1			
Automolus infuscatus	1										1[a]/1						
Automolus ochrolaemus			1/1			1/1						1/1	1/1				
Automolus rufipileatus																+	1[a]/1

TABLE 1
CONTINUED

Species	Understory Flocks[2]								Canopy flocks[2]							Trans.[3]	Total (density)[4]
	18	19	20	21	22	23	24	25	1	2	3	4	5	6	7		
Xiphorhynchus guttatus	1, / 1	1														+	30+ (17+)
Lepidocolaptes albolineatus				+*				+*	2		2	2		2	2	+	8–16 (4–9)
Campylorhamphus trochilirostris				1,[a] / 1													2 (1)
FURNARIIDAE																	
Cranioleuca guturata		1															1–2 (1)
Hyloctistes subulatus	1	1															8–12 (4–7)
Ancistrops strigilatus		1											1	1			2–4 (1–2)
Philydor erythrocercus	1*	1[b]	1										1*	1[b]			14–20 (8–11)
Philydor rufus				2*	2*	+*		1*									2 (1)
Philydor erythropterus									1	1	1	1	1	1	1		7–10 (4–6)
Philydor ruficaudatus			1[b]	1	1	1[b]		1[b] / 1	1	1	1[a] / 1	1[a] / 1	2	1	1		16–20 (9–11)
Automolus infuscatus													2[j]		1		8–12 (4–7)
Automolus ochrolaemus							+										6–8 (3–4)
Automolus rufipileatus			1[a] / 1j														2–3 (1–2)

TABLE 1
CONTINUED

Species	\multicolumn{17}{c}{Understory flocks[2]}																
	1	2	3	4	5	6	7	8	9	10	11	12	13	14	15	16	17
Xenops rutilans				1	1	1	1	1	1		1	1ᵃ	1	1ᵇ		+	+
Xenops minutus				1ᵃ 1		1	1ᵃ										
FORMICARIIDAE																	
Cymbilaimus lineatus	+			+			+	+	+		+	♂ ♀	♂ ♀	♂ ♀	♂ ♀ ♂	♂ ♀ ♂ ♀	♂ ♀ ♂ ♀
Thamnophilus schistaceus				+			♂ ♀	♂ ♀ ♂ ♂	+		♂ ♀	♂ ♀	♂ ♀ ?	♂ ?	♂ ♀ ♂	♂ ♀ ♂	♂ ♀ ♂
Pygiptila stellaris	+					+	+	♂ ♀ ♂	♂ ?	♂ ♀ ♂	+	♂ ♀		♂ ♀	+	♂ ♀	♂ ♀
Thamnomanes ardesiacus	+	♂ ♀	♂ ♀	♂ ♀	♂ ♀	♂ ♀	♂ ♀	♂ ♀ ♂	♂ ♀	♂	♂ ♀	♂ ♀ ♂ ♂	♂ ♀ ♂ ♂ ♂ ♀	♂ ♀ ♂ ♂	♂ ♀	♂ ♀	♂ ♀
Thamnomanes schistogynus	♂ ♀			♂ ♀ ♂		♂ ♀	♂ ♀	♂ ♀		♂ ♀ ♂	♂ ♀	♂ ♀ ♂ ♂	♂ ♀ ♂	♂	♂ ♀	♂ ♀ ♂	♂ ♀ ♂
Myrmotherula brachyura								+	+			+	+	+	+		+
Myrmotherula sclateri																	

TABLE 1
CONTINUED

Species	Understory Flocks[2]								Canopy flocks[2]							Trans.[3]	Total (density)[4]
	18	19	20	21	22	23	24	25	1	2	3	4	5	6	7		
Xenops rutilans	1					1*		1*	1	2		2	1	2	2		10–16 (6–9)
Xenops minutus			1ᵇ	1ᵇ	1												22–30 (12–17)
FORMICARIIDAE																	
Cymbilaimus lineatus		+			+						+	+	+	+	+	+P	16+ (9+)
Thamnophilus schistaceus	♂⌐, ♀⌐ ♀+⌐	?, ♀	♂, ♀	♂, ♀⌐ ♀	?, ♀											+P	40–80 (22–44?)
Pygiptila stellaris	♂⌐, ♀	♂		♂, ♀, ♂j	+			+	♂, ♀	♂, ♀, j	♂, ♀	♂⌐, ♀?, ♂, ♀	♂, ♀	♂, ♀, j	♂, ♀, j		50–90? (28–50?)
Thamnomanes ardesiacus	♂⌐, ♀+⌐, ♀j, ♀j	♂, ♀+⌐, ♂j, ♀j	♂⌐, ♀+⌐	♂ᵃ⌐													44 (24)
Thamnomanes schistogynus	♂⌐, ♀+⌐	♂⌐, ♀+⌐	♂, ♀+⌐	♂, ♀	♂⌐, ♀	♂⌐	♂, ♀	♂⌐, ♀									45 (25)
Myrmotherula brachyura				+	+			+	+	+	+	+	+	+	+	+P	40+ (22+)
Myrmotherula sclateri					+			+	♂, ♀	♂, ♀	♂, ♀	♂, ♀		♂	♂, ♀	+P	14–20 (8–11)

TABLE 1
CONTINUED

Species	Understory flocks[2]																
	1	2	3	4	5	6	7	8	9	10	11	12	13	14	15	16	17
Myrmotherula hauxwelli							+	♂♀|♀|					♂[a]♀|	♂♀|,♂[a]	♀|	♀+	♂♀| ♀
Myrmotherula leucophthalma					♂ ♀	♂ ♀		♂| ♀			♂ ♀	♂| ♀	♂| ♀|♂| ♀| ♀		♂ ♀|	♀	
Myrmotherula ornata																	
Myrmotherula axillaris	♂	♂ ♀	♂ ♂ ♀	♂ ♀ ♂| ♂ ♀	♂ ♀	♂| ♀|	♂| ♀||	♂| ♀||	♂| ♀	♂| ♀	♂| ♀	♂ ♀|	♂| ♀	♂| ♀	♂| ♀	♂| ♀	♂| ♀
Myrmotherula longipennis	♂|	♂ ♀ ♂			♂ ♂| ♀		♂| ♀||♂|	♂| ♀ ♂|		♂| ♀|	♂| ♀||♂| ♂| ♀|♂|	♂| ♀ ♂| ♀|	♂ ♀ ♂ ♀| ♀	♂| ♀|♀|		
Myrmotherula iheringi					·j							♂|* ♀|, ♀[c]|	♂|* ♀| ♂|[c]	♂|* ♀| ♂|[a] ♀||	♂|[a]♀| ♂||§	♂|[a]♀[b] ♂ ♀ ♀|		
Myrmotherula menetriesii	♂ ♀	♂ ♀	? ♀	♂ ?	♂ ♀	♂ ♀	♂ ♀	♂ ♀	♂ ♀	♂ ♀ ♂|	♂ ♀	♂ ♀ ♂|	♂ ♀ ♂|		♂ ♀|	♂ ♀ ♂|	♂| ♀
Microrhopias quixensis																	
Terenura humeralis																	
Hypocnemis cantator	+															+	♂| ♀|*

TABLE 1
CONTINUED

Species	Understory Flocks[2]								Canopy flocks[2]							Trans.[3]	Total (density)[4]
	18	19	20	21	22	23	24	25	1	2	3	4	5	6	7		
Myrmotherula hauxwelli	♂ ♀	♂ ♀														+	40+ (22+)
Myrmotherula leucophthalma																	18–20 (10–11)
Myrmotherula ornata																	0–1 (0–1)
Myrmotherula axillaris	♂│♀ ♂j	♂│♀ ♂j	♂ ♀	♂│♀	♂ ♀│	♂ ♀	♂ ♀	♂ ♀									58 (32)
Myrmotherula longipennis	♂ ♀│♀│ ♂j ♀j	♂│♀│ ♀j ♂j	♂│♀														46 (26)
Myrmotherula iheringi			♂ ♀ ♂j	♂[a] ♀[b] ♀j§													9 (5)
Myrmotherula menetriesii	♂ ♀	♂ ♀ ♂j	♂[b] ♀[d] ♂j[c] ♂ ♀	♂ ♀│	♂ ♀	♂ ♀	♂ ♀	♂ ♀	♂ ♀	♂ ♀	♂ ♀	♂ ♀ ♀	♂ ♀ ♀	♂[b] ♀[d] ♂j[c]	♂[a] ♀[c]│		56–66 (31–37)
Microrhopias quixensis			♂* ?		♂* ?												4 (2)
Terenura humeralis	♂ ♀								♂ ♀	♂ ♀	♂ ♀	♂ ♀│ ·j	♂ ♀	♂ ♀	♂ ♀│		16–17 (9)
Hypocnemis cantator				1*│												+	3–4 (2)

TABLE 1
CONTINUED

Species	Understory flocks[2]																
	1	2	3	4	5	6	7	8	9	10	11	12	13	14	15	16	17
TYRANNIDAE																	
Zimmerius gracilipes																	
Tyrannulus elatus							1										
Myiopagis gaimardii				+								+			+		
Mionectes oleaginea	+			+				+				1[a]	+	1[a]	1	1	
Leptopogon amaurocephalus							+										
Ornithion inerme																	
Tolmomyias assimilis																	
Onychorhynchus coronatus											+		+				
Attila spadiceus								1						1			
Sirystes sibilator																	
PIPRIDAE																	
Piprites chloris							+	+			+	+		+	1[a]		
Tyranneutes stolzmanni															1*		

TABLE 1
CONTINUED

Species	Understory Flocks[2]								Canopy flocks[2]							Trans.[3]	Total (density)[4]
	18	19	20	21	22	23	24	25	1	2	3	4	5	6	7		
TYRANNIDAE																	
Zimmerius gracilipes									1					1	1	+	8–12? (4–7?)
Tyrannulus elatus									+	+	+	+	+	+	+	+	
Myiopagis gaimardii					+			+	+	+	+	+	+	+	+	+	
Mionectes oleaginea	1															+	
Leptopogon amaurocephalus	1			1											+		6–8? (3–4?)
Ornithion inerme									+	+			+	+		+	8–10? (4–6)
Tolmomyias assimilis									+	+	+	+	+	+	+		14 (8)
Onychorhynchus coronatus										+	+	+				+	0–2 (0–1)
Attila spadiceus												+				+	
Sirystes sibilator															+	+	4–6 (2–3)
PIPRIDAE																	
Piprites chloris	+			1[a] 1*	+											+	14–20 (8–11)
Tyranneutes stolzmanni				+	+				+	+	1		+	+	1	+	8♂♂ (4♂♂)

TABLE 1
CONTINUED

Species	Understory flocks[2]																
	1	2	3	4	5	6	7	8	9	10	11	12	13	14	15	16	17
COTINGIDAE																	
Pachyramphus polychopterus																	
Pachyramphus marginatus														+			
Pachyramphus minor																	
TROGLODYTIDAE																	
Thryothorus genibarbis	+															1[a]/1[b]	
MUSCICAPIDAE																	
Ramphocaenus melanurus												1[a]					1*
EMBERIZIDAE																	
Euphonia xanthogaster																	
Euphonia chrysopasta																	
Euphonia minuta																	
Euphonia rufiventris																	
Tangara velia																	
Tangara callophrys																	

TABLE 1
CONTINUED

Species	Understory Flocks[2]								Canopy flocks[2]							Trans.[3]	Total (density)[4]
	18	19	20	21	22	23	24	25	1	2	3	4	5	6	7		
COTINGIDAE																	
Pachyramphus polychopterus					+						♂a/1	♂			♂a/1	+	2–4 (1–2)
Pachyramphus marginatus									♂ ♀	♂ ♀	♂/1 ♀	♂/1 ♀	♂ ♀	♂ ♀	♂ ♀	+	14 (8)
Pachyramphus minor					♂ ♀				♂ ♀		♂b/1 ♀/1		♂ ♀	♂ ♀	♂a/1 ♀/1	+	10–14 (6–8)
TROGLODYTIDAE																	
Thryothorus genibarbis				1a / 1b												+	4 (2)
MUSCICAPIDAE																	
Ramphocaenus melanurus				1*													1 (1)
EMBERIZIDAE																	
Euphonia xanthogaster									♂ ♀	♂ ♀		♂a/1 ♀			♂/1 ♀	+	4–8 (2–4)
Euphonia chrysopasta									♂ ♀	♂ ♀	♂ ♀	♂ ♀/1*	♂ ♀	♂ ♀	♂*	+	10–14 (6–8)
Euphonia minuta									♂ ♀	♂ ♀	♂ ♀*	♀+1*			♂*	+	4–8 (2–4)
Euphonia rufiventris									♂ ♀	♂ ♀	♂a/1 ♀	♂a/1 ♀	♂ ♀	♂	♂a/1	+	12–24? (7–13?)
Tangara velia									6*	6*	5*	4*	3*	4*	5*	+	10–20? (6–11?)
Tangara callophrys									5*	7*	4+	4+†	+†	+†	2+	+	10–20? (6–11?)

TABLE 1
Continued

Species	Understory flocks[2]																
	1	2	3	4	5	6	7	8	9	10	11	12	13	14	15	16	17
Tangara chilensis																	
Tangara schrankii				1			+	2				1[a]/1		1[a]/1	+		
Tangara nigrocincta																	
Tangara mexicana																	
Piranga rubra																	
Habia rubica		♂ ♀ j	♂ ♀		♂ ♂ ♀	♂│ ♀	♂│	♂ ♀	♂│			♂│ ♀│ ♂│	♂│ ♀│ ♂│	♂│ ♀│ ♂│	♂│ ♀		
Lanio versicolor								♂									
Tachyphonus rufiventer							♂ ♀										
Tachyphonus luctuosus							♂ ♀ ♂j			♂ ♀ ♂j				♂ ♂j ♀	♂ ♀		♂ ♀
Hemithraupis guira																	
Hemithraupis flavicollis																	
Lamprospiza melanoleuca																	

TABLE 1
CONTINUED

Species	Understory Flocks[2]								Canopy flocks[2]							Trans.[3]	Total (density)[4]
	18	19	20	21	22	23	24	25	1	2	3	4	5	6	7		
Tangara chilensis									+	+	+	+	+	+	+	+	20–40? (11–22?)
Tangara schrankii	1								2	2	1[b], 1, 1	1[b], 1[c], 1	1, 1[c]	1, 1	1, 1	+	30–60? (17–33?)
Tangara nigrocincta											1	1				+	0–1 (0–1)
Tangara mexicana									2	3	1[a], 1*	2	2	2	1[a], 1*	+	14–28? (8–16?)
Piranga rubra										2						M	1–3 (1–2)
Habia rubica	♂, ♀, ♀j	♂|, ♀|, ♀|	♂|														29 (16)
Lanio versicolor									♂, ♀	♂, ♀, ♂j	♂, ♀	♂|, ♀|·j	♂, ♀|	♂, ♀|, ♂j	♂|, ♀|		15–18 (8–10)
Tachyphonus rufiventer									♂, ♀	♂, ♀	♂, ♀|, ♀	♂, ♀|	♂, ♀	♂, ♀	♂, ♀, j		16 (9)
Tachyphonus luctuosus				♂, ♀, ♂j	♂, ♀, ♂j	♂, ♀, ♂j	+		♂, ♀	♂, ♀	♂, ♀	♂|, ♀	♂, ♀	♂, ♀, ♀	♂, ♀	+	25–39? (14–22?)
Hemithraupis guira								♂, ♀		♂[a], ♀*		♂[a], ♀*					4 (2)
Hemithraupis flavicollis										♂, ♀							2 (1)
Lamprospiza melanoleuca										5*	6*					+	6 (3)

TABLE 1
Continued

Species	Understory flocks[2]																
	1	2	3	4	5	6	7	8	9	10	11	12	13	14	15	16	17
Cyanerpes caeruleus																	
Dacnis cayana																	
Dacnis lineata																	
Dacnis flaviventer																	
Chlorophanes spiza																	
VIREONIDAE																	
Smaragdolanius leucotis							+										
Vireo olivaceus																	
Hylophilus hypoxanthus	+	2		2	+	+		1/1				1/1	1/1	1	1/1		
Hylophilus ochraceiceps								1/1				1/1	1j	1	1/1		
ICTERIDAE																	
Cacicus cela												+		+			
Icterus cayanensis																	

[1] Data from canopy flock 8 not included.

[2] Numbers indicate flocks in Figures 1 and 2. A "+" in a column under a flock number indicates that the species was present in the flock, but the number of individuals is not known. Numerals in flock columns indicate the number of individuals of that species known to be in that flock at one time. Underlined numerals or symbols indicate individuals recognizable by color-bands, voice, disfigurements, or molt patterns. A "?" indicates uncertainty about the presence of another individual or about the precise population density of a species; "j" indicates a juvenile. A comma after an entry indicates that birds below that entry were not present in the flock at the same time as those before the comma. Within a given species, entries marked with the same lettered superscript indicate the same individual present in different flocks; entries marked with the same symbol (i.e., *,

† indicate an unmarked bird suspected of switching flocks.

TABLE 1
CONTINUED

Species	Understory Flocks²								Canopy flocks²							Trans.³	Total (density)⁴
	18	19	20	21	22	23	24	25	1	2	3	4	5	6	7		
Cyanerpes caeruleus												1			1	+	0–2 (0–1)
Dacnis cayana ♂									♂	♂	♂ᵃ	♂	♂	♂	♂ᵃ	+	6–12 (3–7)
Dacnis cayana ♀									♀	♀	♀*	♀	♀	♀	♀*		
Dacnis lineata ♂									♂	♂	♂	♂	♂	♂	♂	+	6–16 (3–9)
Dacnis lineata ♀									♀	♀	♀	♀	♀	♀	♀		
Dacnis flaviventer															♂	+	0–1 (0–1)
Chlorophanes spiza ♂									♂	♂	♂ᵃ	♂	♂	♂	♂ᵃ	+	6–12 (3–7)
Chlorophanes spiza ♀									♀	♀	♀*	♀	♀	♀	♀*		
VIREONIDAE																	
Smaragdolanius leucotis			+		+				+	+	+	+	+	+	1 / 1 / 1j	+	18–30? (10–17?)
Vireo olivaceous					+				2	+	+	2	2	+	1 / 1	+	14–28 (8–16?)
Hylophilus hypoxanthus ♂				♂*		♂*											15–16 (8–9)
Hylophilus hypoxanthus ♀				♀†		♀†											
Hylophilus ochraceiceps ♂	1	1	1						♂1				♂1	♂1	♂1		25–27 (14–15)
Hylophilus ochraceiceps ♀	1	1	?						♀				♀	♀	♀		
		1j															
ICTERIDAE																	
Cacicus cela				+						+	+	+	+	+	+	+	
Icterus cayanensis ♂													♂	♂	♂1	+	4–6 (2–3)
Icterus cayanensis ♀													♀	♀	♀1		

³ Trans. = transient species. A "+" in this column indicates a species with a high population density whose individuals enter and leave flocks as the flocks pass through their territories, or a species that appears to have a lower population density than species in flocks and yet often forages out of flocks. M = a migrant species that appears at the study site during the northern hemisphere fall and winter. P = a species that occupies patchy habitats, such as thick vine tangles, and joins flocks as they pass through.

⁴ Density = number of individuals per km² (calculated from the total number of individuals per 1.8 km²).

nigifrons and *Pygiptila stellaris.* Since I captured and marked only eight *Monasa* during the study, and their upright posture and thick body feathers made bands hard to see, I used other methods to establish their high population density. First, they did not follow the canopy flocks in long distance movements, and yet *Monasa* appeared at many sites in each canopy flock territory. Second, while watching a canopy flock, I often heard two groups of *Monasa* singing simultaneously from sites 100 m or more apart, both within the territory of the flock. I observed *Pygiptila stellaris* individuals singing and chasing one another on three occasions in canopy flocks. On one of these occasions, the marked male from canopy flock 4 (Table 1) fought an unmarked male and subsequently flew from the flock with an unmarked female, presumably his mate. In their places were the unmarked male and an unmarked female. More marked birds are needed in order to understand territoriality in these high population density species.

Hemithraupis guira, Euphonia minuta, and *Philydor ruficaudatus* are lower population density species (Table 1). I saw one marked pair of *E. minuta* in two different canopy flocks (Table 1); the same was true for a marked pair of *H. guira* (Table 1). *Philydor ruficaudatus* appears to have a density similar to or greater than the core canopy species, but because this species joins understory as well as canopy flocks, and two marked individuals switched from one canopy flock to another and from canopy to understory flock, I classify them as low population density, switching species (Table 1). Moreover, I frequently saw only one *P. ruficaudatus* in a flock at one time (Table 1), which is also characteristic of low population density species.

Lead species.—Each of the seven canopy flocks was led by the resident pair of *Lanio versicolor* in long distance (>20 m) flock movements. On 20 occasions in 1982 I saw a pair of *L. versicolor* fly 20 to 50 m while giving chattering vocalizations apparently reserved for long distance moves, and the other flock species followed within seconds. Only twice did I observe other species flying first across a large clearing, and in one of those instances, the canopy flock contained no *L. versicolor.* I heard some chattering vocalizations from *Lanio* and saw birds flying on scores of other occasions, but vegetation generally obscured the flying order of the birds.

Habitat selection by canopy flocks.—The canopy is significantly lower in the gaps between territories of canopy flocks than it is within these territories ($\bar{X}$ = 27.4 m, s.d. = 9.43, N = 57 vs $\bar{X}$ = 31.7 m, s.d. = 6.84, N = 65; t = 2.89, P < 0.005, one-tailed test). In particular, the large gap between the territories of canopy flocks 1 and 2 and those of flocks 3 and 4 (Fig. 2) is littered with large, recently-fallen trees. Consequently, the area is now a mosaic of very dense, low vegetation, 10 to 20 m tall, and occasional remnant emergents. Vines and saplings dominate the patches of dense vegetation, which grow over the large fallen trees. Under the emergents, on the other hand, the understory is still very open and easily passable. While it is possible that these flocks prefer areas with a higher canopy, frequent observations of canopy flock 4 foraging in an area of pure fig trees (*Ficus trigona*) with a continuous canopy at only 20 to 25 m do not support this hypothesis. These observations suggest, rather, that the flocks may select areas of continuous canopy in preference to areas with broken canopy in order to avoid the potentially dangerous and energetically-expensive procedure of flying from isolated emergent to emergent or from emergent down to low vine tangles and back up to the next emergent. One measure of canopy continuity is the variance of canopy heights in the gap areas. Since the variance associated with the gap areas is significantly greater than that for the flock territory areas ($F_{.05}$ = 1.90; d.f. = 56,64; P < .05), it is possible that canopy flocks prefer areas with continuous canopy to areas of broken canopy, regardless of the canopy height.

Habitat requirements may not be the same among the core species in a canopy flock. For instance, *Terenura humeralis, Hylophilus hypoxanthus,* and *Tachyphonus luctuosus* lived in understory flocks 22 and 25 (Table 1, Fig. 1), perhaps because there was no *L. versicolor* to join in that area. *Hemithraupis guira* and *Xenops rutilans,* which are normally canopy flock species, also lived in these understory flocks. The habitat in that area was river edge second growth dominated by *Ficus insipida* and *Cecropia* sp., whereas complete canopy flocks mostly occupied areas of tall, mature primary forest. Whether the absence of *Lanio versicolor* and *Tachyphonus rufiventer* from this river edge second growth reflects habitat selection by these species or simply a temporary shortage of these two species is not certain. Nevertheless, the absence of these two species from the river edge from 1981 to 1983 suggests that habitat selection is the more likely explanation. On what basis these species select different habitats is not known.

Tanagers as members of canopy flocks.—Several species of largely frugivorous tanagers and honeycreepers sometimes formed temporary flocks that joined the canopy flocks. *Tangara chilensis* individuals seemed to lead these flocks, which contained any combination of the following species: *Tangara velia, T. callophrys, T. schrankii, T. nigrocincta, T. mexicana, Dacnis cayana, D. lineata, D. flaviventer,* and *Chlorophanes spiza.* These flocks often contained up to seven individuals of some of the *Tangara* spp., whereas the *Dacnis* spp. were represented by single mated pairs. While watching permanent canopy flocks, I regularly saw a temporary flock of tanagers and honeycreepers fly into an emergent tree above the canopy flock and forage for insects or preen for five to fifteen minutes before flying off in another direction. The extreme heights at which these birds travelled and their irregular vocalizations made them difficult to observe. Consequently, I noted their presence only when they were conspicuous. I could discern little pattern in their organization and schedule. Indeed, monospecific *Tangara* groups (*T. chilensis, T. velia, T. callophrys,* or *T. mexicana*) of two to eight individuals occasionally foraged in emergents more than 100 m from the nearest canopy flock, and I have even seen various of these flocks fly west across the lake (Cocha Cashu, Fig. 2) to forage in the forest on the other side. Single color-marked *Tangara mexicana, Chlorophanes spiza,* and *Dacnis cayana* frequently appeared with their unmarked mates on the edges of canopy flocks 3 and 7. Often, but not always, they appeared in conjunction with other species, such as *T. chilensis* and *T. schrankii,* typical of nonpermanent tanager flocks.

Combined canopy and understory flocks.—I discovered that certain canopy flocks spent many hours each day associated with certain understory flocks. In particular, canopy flock 3 (Fig. 2) spent most of each day with understory flock 12 (Fig. 1), while canopy flocks 1, 2, 4, 5, 6, and 7 each typically spent several hours daily with understory flocks 4, 6, 15, 18, 19, and 14, respectively. Periodically, nonpermanent tanager-honeycreeper flocks joined and left these combined canopy-understory flocks; when all three types of flocks were together, the combined flock consisted of 60 to 70 species of close to 100 individuals. These very diverse flocks regularly foraged for an hour or more in the crowns and understory beneath one or two large canopy trees before eventually moving off together.

Territory defense.—Several times during the 1981 and 1982 seasons, adjacent canopy flocks met at their common territorial boundary. On these occasions, conspecific pairs of *Hylophilus hypoxanthus, Lanio versicolor,* and *Terenura humeralis* in the two flocks restricted their territorial defense to countersinging. The core canopy species' apparent lack of physical aggression in territorial defense contrasts sharply with the displaying and chasing Munn and Terborgh (1979) described for understory species. I do not know why these canopy flock core species are apparently less aggressive than core species of understory flocks.

Population density estimates.—For 78 different flocking species, I estimated population densities in the 1.8 km² (180 ha) study tract. These are the first estimates of absolute population size for many of these species. The only comparable data for these flock species are from Karr's (1971) study in Panamanian lowland forest. He netted birds on a two hectare plot in the mainland forests of the Canal Zone and presented population density data for 56 resident species on the plot. Although Karr's Panamanian site and my Peruvian site share 10 species that appear in flocks at least occasionally, only one, *Myrmotherula axillaris,* for which we both have density data, is a core species. Karr reported a population density of 150 individuals per 100 ha, while I estimated only 32 individuals per 100 ha. Likewise, Karr obtained a density estimate of 150 individuals per 100 ha for *Myrmotherula fulviventris.* This species is the exact ecological replacement of the Peruvian antwren *Myrmotherula leucophthalma* (pers. observ.; Gradwohl and Greenberg 1982), for which I estimated 13 individuals per 100 ha. The densities in Karr's plot of two other flock species found both in Panama and Peru, *Automolus ochrolaemus* and *Xenops minutus,* were 60 and 40 individuals per 100 ha, respectively, which were much higher than my estimates of 3 to 4 and 12 to 17 individuals per 100 ha.

DISCUSSION

Canopy and understory flocks are similar in some respects and different in others. Both are permanent associations of mated pairs, each of different species. Furthermore, each is led by a noisy, sallying species, *Lanio versicolor* for the canopy flocks, and *Thamnomanes schistogynus* for the understory flocks (Munn and Terborgh 1979; Wiley 1980). Additionally, both types are joined by a variety of other species, some of which have higher population densities, and others that have lower population densities than those of the core species.

The differences between canopy and understory flocks include the following: (1) canopy flocks are much less common than understory flocks (in my 1.8 km² study area, 7 vs. 25, respectively); (2) canopy flock species tend to eat fruit as well as insects, whereas understory species are almost purely insectivorous; and (3) understory species seem to defend their territories more aggressively than canopy species.

Population densities.—In my study area of 1.8 km², each canopy flock core species was represented by about 14 individuals, while the most common understory flock core species had population sizes of 40 to 58. Since these core species are virtually always in these flocks, and since flocks are relatively easy to find compared with lone birds, the present method of color-marking, following, and direct counting seems simple and reliable. For the flock species whose populations were unmarked or incompletely marked, this method does not work as well. Nevertheless, it is possible to hear, for example, an unmarked pair of *Tolmomyias assimilis* in one canopy flock and then to walk several hundred meters and find a pair in another flock. I know the majority of the vocalizations of most flock species; they sing regularly every morning, so it was possible to census species such as *T. assimilis* by song as well as by sight.

The flock species found both in Panama and in Peru apparently were much more common in Panama (Karr 1971), but Karr's methods were different from mine and may have over-estimated the population densities. Specifically, his density estimates for moist forest are based on netting in a two-hectare tract, which may be too small to give an accurate result. Portions of the home ranges of many pairs of birds may have overlapped on the two-hectare plot, which would have biased density estimates upward. Furthermore, although Karr used color-marked birds to estimate population sizes, he does not say how he did this.

Although it is not clear how our different methods compare, the apparent large differences in the population densities of flock species in the two sites may be real. If so, then they may reflect the fact that Karr's forest contains many fewer resident species than does mine (56+ vs 228+). If both locales are approximately equally productive, and if diffuse interspecific competition for food is important in these bird communities, then populations of each species should be higher in the locale with fewer species, i.e., Panama ("density compensation," Crowell 1962; Brown 1975; Cody 1975). If the understory forest in Panama were more productive than that in southeastern Peru, then that would also contribute to higher densities of flock species in Panama. Data on the relative productivity of these two sites plus evidence of interspecific competition in these forests are necessary to test this hypothesis of density compensation.

Differences between canopy and understory flocks.—In my study area, canopy flocks are only one-third as common as understory flocks (Figs. 1, 2). I cannot account for this difference, but it may be related to the fact that the canopy species are more frugivorous than understory species. I have seen every flock-joining species of tanager (Emberizidae, Table 1) as well as *Eubucco richardsoni, Capito niger, Mionectes oleaginea, Tyranneutes stolzmanni, Pachyramphus minor, Vireo olivaceus, Cacicus cela,* and *Icterus cayanensis* regularly eat fruit in addition to insects. My observations of the rest of the flock species indicate that they are almost exclusively insectivorous. The majority of these partially frugivorous species join canopy and not understory flocks (Table 1). Possibly each canopy flock requires a minimum number of fruit trees of certain species in order to survive, but to test this hypothesis, data are required on the distribution and seasonality of insect and fruit resources in the canopy and on the pattern of use of these resources by canopy flock species. As far as I know, these data are unavailable.

Canopy flock species apparently do not chase and display towards conspecifics from adjacent flocks to the same degree as do understory flock species. This difference may be related to the dietary difference between the two kinds of flocks, since frugivores may not defend territories as aggressively as insectivores. Among the core species of the canopy flock, however, one might expect *Hylophilus hypoxanthus,* which is mostly or entirely insectivorous, to be more aggressive towards conspecific intruders than would be an apparently more frugivorous species such as *Tachyphonus luctuosus.* I do not know if this is true.

Topics for future research.—It is still not clear how certain high population density species are distributed in different flocks. For instance, *Pygiptila stellaris, Myrmotherula menetriesii,* and *Tachyphonus luctuosus* have population densities as high or higher than those of the core species of either the canopy or understory flocks, and all three species appear in both types of flocks (Table 1). Perhaps an individual pair joins the flock, regardless of type, that is closest to the center of its territory. On the other hand, it is also possible that particular pairs live permanently in either one or the other type of flock. The latter seems to be true for certain recognizable pairs of *T. luctuosus* that I found daily in particular understory or canopy flocks (Table 1). It is theoretically possible, therefore, that mated pairs of such species vertically overlap conspecific pairs, with some pairs permanently in canopy flocks and others permanently in understory flocks. Demonstration of such vertical stratification or territoriality within any of these three species will require additional observations of marked individuals.

I showed that gaps exist in the distribution of canopy flocks at my study site and that these areas tend to have a lower, less continuous canopy than the areas occupied by canopy flocks. This suggests that the number of canopy flocks is limited by the amount of suitable habitat. Consider, however, that between January and July 1982, a new canopy flock formed (canopy flock 3, Fig. 2) and occupied an area that had been devoid of canopy flocks in 1981. Since there was a vacancy, which was filled by this new flock, there may be other vacancies. Thus, it is possible that these canopy species are not yet at their carrying capacity. The gaps in their distribution in the area may, therefore, reflect not their habitat requirements so much as habitat preferences in a range of acceptable, but only partially-occupied habitats. Despite the difficulty of resolving these questions, at least the patterns of distribution shown by these species at Cocha Cashu are completely natural and not influenced by man. Research on spatial and temporal distribution of insect and fruit distributions as well as further observation of marked individuals of flock-joining species would contribute enormously to our knowledge of these complex mixed-species flocks and would allow us to compare these flocks with mixed-species flocks in other tropical and temperate locales.

ACKNOWLEDGMENTS

I thank my advisor, J. Terborgh, for suggesting mixed-species flocks as a potential subject for research. He introduced me to tropical forests in 1976 and has provided intellectual and logistical support, as well as friendship, ever since. M. Foster, J. Hoogland, H. Horn, E. Morton, G. Powell, S. Robinson, D. Rubenstein, and R. Zell provided helpful suggestions and criticism of this work, for which I am grateful. I thank the Peruvian Ministerio de Agricultura for permission to work in Manu National Park. The Frank M. Chapman Memorial Fund and the Princeton University Department of Biology generously supported my research. I thank my wife, Martha Brecht, for innumerable ingenious suggestions in the field, as well as for help with some of the more difficult aspects of the field work. Her sound scientific and personal judgement proved invaluable throughout.

LITERATURE CITED

BATES, H. W. 1863. The Naturalist on the River Amazons. Dent, London.

BROWN, J. H. 1975. Geographical ecology of desert rodents. Pp. 315–341, *In* M. L. Cody and J. M. Diamond (eds.), Ecology and Evolution of Communities. Harvard University Press, Cambridge, Massachusetts.

CODY, M. L. 1975. Towards a theory of continental species diversity. Pp. 214–257, *In* M. L. Cody and J. M. Diamond (eds.), Ecology and Evolution of Communities. Harvard University Press, Cambridge, Massachusetts.

CROWELL, K. 1962. Reduced interspecific competition among the birds of Bermuda. Ecology 43:75–88.

CROXALL, J. P. 1976. The composition and behaviour of some mixed-species bird flocks in Sarawak. Ibis 118:333–346.

FOGDEN, M. P. L. 1972. The seasonality and population dynamics of equatorial forest birds in Sarawak. Ibis 114:307–343.

GRADWOHL, J., AND R. GREENBERG. 1980. The formation of antwren flocks on Barro Colorado Island, Panama. Auk 97:385–395.

GRADWOHL, J., AND R. GREENBERG. 1982. The effect of a single species of avian predator on the arthropods of aerial leaf litter. Ecology 63:581–583.

JONES, S. E. 1978. Coexistence in mixed-species antwren flocks. Oikos 29:366–375.

KARR, J. R. 1971. Structure of avian communities in selected Panama and Illinois habitats. Ecol. Monogr. 41:207–233.

McCLURE, H. E. 1967. The composition of mixed species flocks in lowland and submontane forests of Malaya. Wilson Bull. 79:131–154.

MORSE, D. H. 1970. Ecological aspects of some mixed-species foraging flocks of birds. Ecol. Monogr. 40:119–168.

MORTON, E. S. 1980. Adaptations to seasonal changes by migrant land birds in the Panama Canal Zone. Pp. 437–453, *In* A. Keast and E. S. Morton (eds.), Migrant Birds in the Neotropics. Smithsonian Institution Press, Washington, D.C.

MOYNIHAN, M. 1962. The organization and probable evolution of some mixed species flocks of neo-tropical birds. Smithson. Misc. Collect. 143:1–140.

MUNN, C. A., AND J. W. TERBORGH. 1979. Multi-species territoriality in neotropical foraging flocks. Condor 81:338–347.

PARTRIDGE, L., AND R. ASHCROFT. 1976. Mixed-species flocks of birds in hill forest in Ceylon. Condor 78:449–453.

POWELL, G. V. N. 1979. Structure and dynamics of interspecific flocks in a neotropical mid-elevation forest. Auk 96:375–390.

RICHARDS, P. W. 1952. The Tropical Rain Forest. Cambridge University Press, Cambridge, England.

TERBORGH, J. 1983. Five New World Primates: A Study in Comparative Ecology. Princeton University Press, Princeton, New Jersey.

WILEY, R. H. 1980. Multispecies antbird societies in lowland forests of Surinam and Ecuador: stable membership and foraging differences. J. Zool. (Lond.) 191:127–145.

APPENDIX I
LIST OF SPECIES CONSIDERED IN THIS STUDY

Scientific name	Common name
Coccyzus erythrophthalmus	Black-billed Cuckoo
C. americanus	Yellow-billed Cuckoo
C. melacoryphus	Dark-billed Cuckoo
Piaya cayana	Squirrel Cuckoo
Galbula cyanescens	Bluish-fronted Jacamar
Monasa nigrifrons	Black-fronted Nunbird
M. morphoeus	White-fronted Nunbird
Capito niger	Black-spotted Barbet
Eubucco richardsoni	Lemon-throated Barbet
Picumnus rufiventris	Rufous-breasted Piculet
Piculus chrysochloros	Golden-green Woodpecker
Veniliornis passerinus	Little Woodpecker
V. affinis	Red-stained Woodpecker
Dendrocincla fuliginosa	Plain-brown Woodcreeper
Deconychura longicauda	Long-tailed Woodcreeper
Sittasomus griseicapillus	Olivaceous Woodcreeper
Glyphorynchus spirurus	Wedge-billed Woodcreeper
Dendrexetastes rufigula	Cinnamon-throated Woodcreeper
Xiphorhynchus obsoletus	Striped Woodcreeper
X. ocellatus	Ocellated Woodcreeper
X. spixii	Spix's Woodcreeper
X. guttatus	Buff-throated Woodcreeper
Lepidocolaptes albolineatus	Lineated Woodcreeper
Campylorhamphus trochilirostris	Red-billed Scythebill
Cranioleuca gutturata	Speckled Spinetail
Hyloctistes subulatus	Striped Woodhaunter
Ancistrops strigilatus	Chestnut-winged Hookbill
Philydor erythrocercus	Rufous-rumped Foliage-gleaner
P. rufus	Buff-fronted Foliage-gleaner
P. erythropterus	Chestnut-winged Foliage-gleaner
P. ruficaudatus	Rufous-tailed Foliage-gleaner
Automolus infuscatus	Olive-backed Foliage-gleaner
A. ochrolaemus	Buff-throated Foliage-gleaner
A. rufipileatus	Chestnut-crowned Foliage-gleaner
Xenops rutilans	Streaked Xenops
X. minutus	Plain Xenops
Cymbilaimus lineatus	Fasciated Antshrike
Thamnophilus schistaceus	Black-capped Antshrike
Pygiptila stellaris	Spot-winged Antshrike
Thamnomanes ardesiacus	Dusky-throated Antshrike
T. schistogynus	Bluish-slate Antshrike
Myrmotherula brachyura	Pygmy Antwren
M. sclateri	Sclater's Antwren
M. hauxwelli	Plain-throated Antwren
M. fulviventris	Checker-throated Antwren
M. leucophthalma	White-eyed Antwren
M. ornata	Ornate Antwren
M. axillaris	White-flanked Antwren
M. longipennis	Long-winged Antwren
M. iheringi	Ihering's Antwren
M. menetriesii	Gray Antwren
Microrhopias quixensis	Dot-winged Antwren
Terenura humeralis	Chestnut-shouldered Antwren
Hypocnemis cantator	Warbling Antbird
Zimmerius gracilipes	Slender-footed Tyrannulet
Tyrannulus elatus	Yellow-crowned Tyrannulet
Myiopagis gaimardii	Forest Elaenia
Mionectes oleaginea	Ochre-bellied Flycatcher
Leptopogon amaurocephalus	Sepia-capped Flycatcher
Ornithion inerme	White-lored Tyrannulet

APPENDIX I

CONTINUED

Scientific name	Common name
Tolmomyias assimilis	Yellow-margined Flycatcher
Onychorhynchus coronatus	Royal Flycatcher
Attila spadiceus	Bright-rumped Attila
Syristes sibilator	Syristes
Piprites chloris	Wing-barred Manakin
Tyranneutes stolzmanni	Dwarf Tyrant-manakin
Pachyramphus polychopterus	White-winged Becard
P. marginatus	Black-capped Becard
P. minor	Pink-throated Becard
Thryothorus genibarbis	Moustached Wren
Ramphocaenus melanurus	Long-billed Gnatwren
Euphonia xanthogaster	Orange-bellied Euphonia
E. chrysopasta	Golden-bellied Euphonia
E. minuta	White-vented Euphonia
E. rufiventris	Rufous-bellied Euphonia
Tangara velia	Opal-rumped Tanager
T. callophrys	Opal-crowned Tanager
T. chilensis	Paradise Tanager
T. schrankii	Green and Gold Tanager
T. nigrocincta	Masked Tanager
T. mexicana	Turquoise Tanager
Piranga rubra	Scarlet Tanager
Habia rubica	Red-crowned Ant-tanager
Lanio versicolor	White-winged Shrike-tanager
Tachyphonus rufiventer	Yellow-crested Tanager
T. luctuosus	White-shouldered Tanager
Hemithraupis guira	Guira Tanager
H. flavicollis	Yellow-backed Tanager
Lamprospiza melanoleuca	Red-billed Pied Tanager
Cyanerpes caeruleus	Purple Honeycreeper
Dacnis cayana	Blue Dacnis
D. lineata	Black-faced Dacnis
D. flaviventer	Yellow-bellied Dacnis
Chlorophanes spiza	Green Honeycreeper
Smaragdolanius leucotis	Slaty-capped Shrike-vireo
Vireo olivaceus	Red-eyed Vireo
Hylophilus hypoxanthus	Dusky-capped Greenlet
H. ochraceiceps	Tawny-crowned Greenlet
Cacicus cela	Yellow-rumped Cacique
Icterus cayanensis	Epaulet Oriole

SOCIOBIOLOGY AND ADAPTIVE SIGNIFICANCE OF INTERSPECIFIC FORAGING FLOCKS IN THE NEOTROPICS

GEORGE V. N. POWELL

National Audubon Society, Research Department, 115 Indian Mound Trail, Tavernier, Florida 33070 USA

ABSTRACT. The sociobiology of mixed species flocks is reviewed and used as a basis for evaluating the adaptive significance of flocking. Virtually all data currently available deal with mixed flocks of insectivores, so this review deals primarily with those flocks with only minimal reference to flocks of frugivores and granivores.

Most insectivorous species that participate in mixed flocks remain paired and defend exclusive feeding territories year round. Territoriality restricts intraspecific group sizes in flocks to single pairs or small family groups. Since territorial individuals are forced to leave a flock whenever it moves beyond their respective territory boundaries, there is a high turnover rate in flock composition. Some species reduce or eliminate turnover by aligning their respective territories so that they overlap completely. This alignment enables individuals of these species to remain permanently associated in the same mixed flock (except during some phases of the breeding cycle). Territory sizes of mixed flocking species are small (< 1–10 ha), and territory holders forage throughout their entire territories every few days. Foraging ranges of mixed flock participants remain relatively constant throughout the year with slight expansions due to increased trespassing during some periods of the year.

The foraging niches of mixed flocking insectivores include a wide array of behaviors. Analyses of foraging niches reveal little overlap among most species, although sympatric congeners of at least one genus of Formicariidae (*Myrmotherula*) appear to overlap more than expected.

Hypotheses concerning the adaptive significance of flocking are categorized as those predicting that flocking enhances foraging or reduces predation pressure. Analyses of spatial use and foraging niche characteristics of flocking species support some of both hypothesis categories. In general, however, the small intraspecific group sizes and home ranges resulting from permanent territoriality, and the high diversity of foraging niches tend to minimize the applicability of forage enhancement hypotheses. Limited cases of imitative foraging and foraging facilitation from the flushing of insects (beater effect) are reported. Predation reduction, through improvement of surveillance, appears to be an important function of mixed species flocking. Support for this conclusion is still largely indirect, but increasing field evidence is accumulating.

RESUMEN. Se revee la sociobiología de bandadas mixtas (compuestas por varias especies) y se usa como una base para evaluar el significado adaptativo de andar en bandadas. Casi toda la información disponible se refiere a bandadas mixtas de insectívoros, por lo que esta revisión se refiere en forma mínima a bandadas de frugívoros o granívoros.

La mayoría de las especies insectívoras que participan de bandadas mixtas se mantienen en pareja y defienden territorios de alimentación exclusivos durante todo el año. Esta territorialidad restringe el tamaño de los grupos intraespecíficos en una bandada a únicamente pares o pequeños grupos familiares. Debido a que individuos territorialistas están obligados a salir de la bandada cuando esta se translada fuera de los límites de sus territorios, hay un alto nivel de cambio en la composición de la bandada. Algunas especies disminuyen o eliminan el intercambio de individuos, alineando sus respectivos territorios, de manera que ellos se superpongan completamente. Esta alineación le permite a los individuos de esa especie quedar asociados permanentemente en la misma bandada mixta (excepto durante algunas fases del ciclo reproductivo). Las superficies de los territorios de varias especies participantes de bandadas mixtas son pequeñas (<1–10 ha) y los ciudadores de los territorios forrajean a lo largo de todos sus territorios cada pocos dias. Los rangos de forraje de miembros de bandadas mixtas son relativamente constantes a lo largo del año con pequeñas expansiones debido al aumento de las intrusiones durante ciertos períodos del año.

Los nichos de forrajeo de bandadas mixtas de insectívoros incluyen una amplia variedad de comportamientos. Análisis de nichos de forraje revelan poca superposición entre la mayoría de las especies, aunque congéneres simpátricos de al menos un género de Formicariidae (*Myrmotherula*) parecen superponerse más que los esperado.

Las hipótesis concernientes al significado adaptativo de andar en bandadas están clasificadas como aquellas que predicen que facilitan el forrajeo o que reducen las presiones de depredación. Análisis del uso espacial y de las características de los nichos de forraje de especies en bandadas, apoyan algo de ambas categorías hipotéticas. Sin embargo, en general, los grupos pequeños intraespecíficos y áreas de vida pequeñas debidas a territorios permanentes y la gran diversidad de nichos de forraje tienden a minimizar la aplicabilidad de la hipótesis de realce de forraje. Se informa sobre casos de imitación de forraje y facilitamiento de forraje debido a la desaparición de insectos ("beater effect"). El mejoramiento de la vigilancia parece ser una función importante de las bandadas compuestas por varias especies ya que disminuye la depredación. Los elementos que apoyan esta conclusión son muy indirectos, pero cada vez se están acumulando más evidencias desde el campo.

"One may pass several days without seeing many birds; but now and then the surrounding bushes and trees appear suddenly to swarm with them. There are scores, probably hundreds, of birds all moving about with the greatest activity.... The bustling crowd loses no time, and although moving in concert, each bird is occupied on its own account in searching bark, or leaf, or twig; .... In a few minutes, the host is gone, and the forest path remains deserted and silent as before."

Henry W. Bates 1863 (pp. 334–335)

Multispecific associations of birds occur throughout the neotropics in virtually every terrestrial habitat. These associations may be categorized as aggregations or flocks. Aggregations are groupings that incidentally form when individuals are drawn together by a factor of the environment. Generally, these factors are stationary concentrations of resources such as fruit (Diamond and Terborgh 1967; Leck 1971), but they may be mobile resource patches, such as army ant swarms (Willis 1967; Willis and Oniki 1978) or burning grasslands (Winterbottom 1949) that enhance the availability of arthropods. In contrast, flocks are groupings whose cohesion is dependent on members' responses to one another, i.e., the flock generates its own *raison d'etre*. Most recent studies of mixed species flocking have been directed toward identifying the function of flocking, which, presumably, is manifest through increased fitness. Hypothesized selective advantages of mixed flock formation are numerous. Flocking is thought to enhance foraging, decrease the likelihood of predation, or some combination of both. Flocking may enhance foraging by facilitating the location of food-rich areas (Moynihan 1962a) or areas with few competitors (Morse 1970). Flocks may make food more available by flushing prey (Belt 1874) or by enabling flock members to learn of new food sources through copying (Leck 1971). Flocking may permit more systematic resource harvesting (Cody 1971), reduce interspecific aggression (Austin and Smith 1972), or facilitate reduction of niche overlap (Morse 1967). Flocking may reduce susceptibility to predation by producing confusion, or by reducing a predator's capacity to surprise intended prey (Bates 1863). Birds in a flock also may physically threaten a predator (Olson 1964). Following flocks may allow birds to avoid areas that are likely to harbor predators (Moynihan 1962a).

To effectively evaluate these hypotheses, it is necessary to have a working knowledge of the structure, dynamics, and spatial organization of mixed species flocks. To this end, I summarize the information available on neotropical mixed species flocks, and then evaluate hypotheses about the adaptive significance of flocking. For the evaluation, I have incorporated relevant data from temperate zone mixed species flocks.

Mixed species flocks may be categorized on the basis of the foods selected by flock members. Flocks composed primarily of insectivores tend to be stable and to have small intraspecific group sizes and high species diversity (Table 1). In contrast, flocks composed primarily of frugivores tend to be unstable and to include larger intraspecific groups. In the field, it is frequently difficult to determine whether a group of fruit-eating birds is a flock that moves together or simply an aggregation of birds that have arrived at a tree independently. Typically, assemblages at fruiting trees consist of both components with birds tending to arrive and depart in single species flocks (Diamond and Terborgh 1967; Leck 1971). Flocks of granivorous

TABLE 1
GEOGRAPHIC VARIATION IN THE SIZE AND SPECIES RICHNESS OF MIXED FLOCKS OF INSECTIVOROUS BIRDS

	Nucleus[1] species	Average flock size[2]	Total no. of species	No. of regular species[3]	Source[4]
		Humid Tropical Forest—low elevation			
Amazonia	F	30–35 (1)*	48	16	a
	F	25–30 (3)*	35	14	b
Venezuela	V	ID	42	7	c
Panama	F	6 (44)	22	5	d
(Barro	F	ID	28	7	e
Colorado	F	8 (97)	40	7	f
Island)	F	7 (1)*	34	8	g
Mexico	F-V	ID	44	ID	h
Honduras	F-V	10–15 (166)	67	3	i
Southern Brazil	Th	ID	20	6	j
Costa Rica	F	ID	31	8	k
		Humid Tropical Forest—middle elevation			
Panama	Th	8–15 (3)*	21	8	l
Costa Rica	P	8 (1)*	43	5	m
Colombia	F	22 (16)	46	10	n
		Dry Tropical Forest			
Mexico	T	40 (5)	10	3	o
Brazil	P	ID	10	5	j

[1] F, Formicariidae; T, Troglodytidae; V, Vireonidae, P, Parulinae; Th, Thraupinae.

[2] Flock size was calculated from attendance frequency and intraspecific group size data. An * denotes values derived from repeated observations of color-marked flocks and number in parentheses gives number of independent flocks; in other cases, the number of independent flocks included in sample is unknown; ID = insufficient data.

[3] Because the Total Number of Species includes many species that rarely associate with mixed species flocks, the Number of Regular Species (species denoted as common, regular, or present >25% of observations except for Venezuela = >33%), is a better index of flock composition.

[4] Source a = Munn and Terborgh 1979; b = Powell, unpubl. data; c = Morton 1979; d = Wiley 1971; e = Willis 1972a; f = Jones 1977; g = Gradwohl and Greenberg 1980; h = S. Barrios M. and J. H. Rappole, unpubl. data; i = Willis 1960; j = Powell, pers. observ.; k = Slud 1960; l = Buskirk et al. 1972; m = Powell 1979; n = Willis 1966; o = Short 1961.

species are generally associated with grassland habitats; these flocks usually have few species with large intraspecific group sizes (Rubenstein et al. 1977).

With few exceptions, mixed flock studies in the neotropics have dealt with insectivorous flocks, and little information is available to characterize mixed flocks of frugivores and granivores. Therefore, the following review primarily analyzes insectivorous flocks and only briefly mentions other flock types. Mixed flocks of insectivores are typically largest and most species rich in the Amazon lowlands (Table 1). In middle elevation habitats, canopy and understory species tend to flock together, whereas in taller lowland forests, they usually form separate flocks. In drier tropical habitats, typically smaller mixed flocks of insectivores frequent gallery forests and to a lesser extent, xerophytic cerrado and thorn-scrub forests.

FLOCK CHARACTERISTICS

FLOCK COMPOSITION

Traditionally, species have been categorized according to their propensity to associate with mixed flocks and their impact on flock development. Nucleus species contribute significantly to the maintenance of flock cohesion (Winterbottom 1943) whereas circumference or attendant species (Rand 1954) do not. Moynihan (1962a) further defined nucleus species as either active, that is, tending to join or follow individuals of other species, or passive, tending to be joined or followed by other species. The nucleus-attendant classification is still used by behaviorists, whereas, the active-passive terminology has been largely dropped.

Species in mixed flocks seem to be designated as nucleus or attendant more by gestalt than by rigorous evaluation of their contribution to flock cohesion. Behavioral and physical characteristics of nucleus species have been identified, but even these require subjective evaluation. Nucleus species tend to be intraspecifically gregarious ($\bar{X}$ group size = 2.4, N = 31 species,

TABLE 2
AVERAGE INTRASPECIFIC GROUP SIZE OF TAXONOMIC FAMILIES THAT PARTICIPATE IN MIXED SPECIES FLOCKS

	Low elevation				Middle elevation		X̄
Nucleus							
Formicariidae	2.3 (6)[1]	2.5 (6)	2.3 (2)	1.5 (3)		1.8 (2)	2.1
Vireonidae	2.0 (1)	2.0 (1)	1.9 (1)	1.0 (1)	1.8 (1)	2.7 (1)	1.9
Parulidae					2.3 (2)	3.0 (1)	2.7
Thraupidae	2.0 (1)			3.1 (1)		4.2 (1)	3.1
Attendant							
Picidae		1.8 (1)		1.0 (1)	1.0 (1)	2.0 (3)	1.5
Dendrocolaptidae	1.0 (2)	1.5 (4)	1.1 (2)	1.4 (3)	1.2 (4)	1.3 (4)	1.3
Furnariidae	1.0 (3)	1.7 (3)	1.3 (1)	1.0 (1)	1.6 (5)	1.4 (5)	1.3
Tyrannidae		1.2 (4)	1.0 (1)	1.2 (4)	1.1 (4)	1.7 (3)	1.2
Troglodytidae	2.0 (1)	2.0 (1)	1.5 (2)		1.6 (2)	1.5 (2)	1.8
Parulinae							
Resident					1.4 (1)	1.7 (3)	1.6
Migrant			1.1 (1)	1.0 (8)	1.0 (5)	1.0 (4)	1.0
Thraupinae					2.0 (3)	2.6 (4)	2.3
Fringillidae		4.0 (2)		1.5 (1)	1.8 (1)	1.0 (3)	1.4
Source[2]	a	b	i	f	l	m	

[1] Numbers of species per taxonomic family that participated in mixed species flocks on each study site are given in parentheses.
[2] See Table 1.

Table 2), and/or have conspicuous color patterns, movements, or vocalizations (Moynihan 1962a). These characteristics are generally assumed to make the birds easy to follow. The ecological characteristics of a nucleus species have not been identified, but they probably depend on the pattern, rate, and height of foraging. These parameters presumably must be compatible with foraging behavior of potential attendant species. Whether species that function as nucleus species have evolved to enhance this role, or whether their attractiveness to other species is a by-product of selection to promote intraspecific gregariousness has not been determined. Based on Tinbergen's (1959) idea of adaptation for reducing individual distance, Moynihan (1960) proposed that as an adaptation for interspecific flock formation, nucleus species have evolved conspicuous patterns of dull colors (neutral coloration) to make them less threatening to other species while still remaining conspicuous enough to be followed. He further hypothesized that mixed species flock members, in general, have evolved similar colorations and vocalizations as a form of social mimicry (Moynihan 1960). The prediction of convergence in coloration and vocalizations among mixed species flock members has not been evaluated. Hamilton and Barth (1962) attempted to evaluate the neutral coloration hypothesis by correlating winter plumages of North American migrants and their propensity to join winter flocks. Their inability to eliminate other hypotheses that also predicted dull coloration, precluded resolution of the question.

Another proposed adaptation to enhance mixed species flocking is the evolution of specialized vocalizations to maintain flock cohesion. Antshrikes (*Thamnomanes* spp.) are believed to use special calls for assembling mixed flocks at the beginning of the day and when wind or rain interferes with maintenance of cohesion by intraspecific contact notes (Munn and Terborgh 1979; Wiley 1980). These vocalizations differ markedly from agonistic vocalizations, typical contact notes, and warning calls (Powell, unpubl. data).

Attendant species are those that do not contribute appreciably to the cohesion of mixed flocks. Generally, they are inconspicuous while foraging and rarely vocalize, though they may be present in pairs or family groups the same size as nucleus species. Mean intraspecific group sizes of attendant species were significantly smaller than mean group sizes of nucleus species (X̄ group size = 2.4, N = 31 spp. for nucleus spp.; X̄ group size = 1.4, N = 103 spp. for attendant spp.; *t*-test, $t = 7.0$, $P < 0.01$, N = 134; Table 2). In the larger, more complex flocks of the Amazon Basin, it is difficult to differentiate nucleus from attendant species because as many as six species contribute in differing degrees to flock cohesion.

FLOCK STRUCTURE

The structure and individual composition of mixed flocks are generally highly stable. Flocks are composed primarily of one pair or a pair with young of each species. Therefore, flock size is highly correlated with the number of species present. For example, in Costa Rican flocks the correlation coefficient for flock size versus number of species was 0.95 (Powell 1979). Essentially, flocks are enlarged by the addition of new species rather than more individuals of the same species. Mean intraspecific group size in six studies of insectivorous flocks averaged 1.7 birds per species (Table 2). Mean intraspecific group size may increase during the breeding season, reflecting the presence of young that remain with parents for up to a year. However, even when young are present, intraspecific group size increases less than expected, because adults with young may forage separately, each adult accompanied by a single young (Powell 1977).

In all studies involving color-marked birds, mixed flocks were composed almost exclusively of permanent residents. With most species, a resident pair defends its territory against trespass by all conspecifics, thereby restricting flock membership to a single pair or family group per species. Territories are defended throughout the year, generally with agonistic vocalizations and displays. Occasionally, agonistic displays are accompanied by chasing and supplanting, and rarely by actual physical fighting (Wiley 1971; Jones 1977; Powell 1977; Munn and Terborgh 1979; Gradwohl and Greenberg 1980).

The relative configurations of territories of flock participants have a major impact on flock membership and stability. As a flock moves through its range, species join and leave the group within the constraints of their own home ranges. Powell (1979) found that 65 percent of birds leaving flocks left at territory boundaries. The impact of territorial restriction on resident flocking species was inversely proportional to territory size (linear regression, $r^2 = 0.80$, $P < 0.01$, N = 10 spp.; Powell 1979). Some species have mitigated the impact of territorial limits by aligning their territories to conform with the home range of a flock. In middle elevation flocks in Costa Rica, Spotted Barbtails (*Premnoplex brunnescens*), had territories aligned with those of the nucleus species so that barbtails could associate with mixed flocks 75 percent of the time, almost twice as long as any other species (Powell 1977). Two antwren species (*Myrmotherula fulviventris* and *Microrhopias quixensis*), which are the nucleus species for insectivorous flocks in lowland Panama, shared completely overlapping territories and remained together most of the time (Gradwohl and Greenberg 1980). As many as eight species in interspecific flocks in the Amazon basin have matching territories (Munn and Terborgh 1979; Powell, unpubl. data). These include antshrikes (*Thamnomanes* spp.) and antwrens (*Myrmotherula* spp.), the nucleus component of the flocks, and woodcreepers (*Xiphorhynchus* sp.), foliage-gleaners (*Philydor* sp.), and greenlets (*Hylophilus* sp.), which are attendants. Pairs of these species have common territory boundaries and remain together as a permanent flock except while roosting and during some nesting activities. Although they share common boundaries, each species defends its territory independently. During confrontations between neighboring flocks, up to a dozen species may interact with agonistic vocalizations and displays, giving the appearance of a crowded aviary. Interflock interactions last up to 30 minutes before the flocks separate and move into their respective territories (Munn and Terborgh 1979; Gradwohl and Greenberg 1980). Flock members that are not confronted by trespassing conspecifics ignore the confrontation and continue foraging.

Flocking species with home ranges that differ in spatial configuration from that of the nucleus of the flock include those with larger home ranges (or lower population densities), those with smaller home ranges (higher densities), and transients. Birds with home ranges larger than those of a single flock nucleus associate with the two or more flocks that pass through their home ranges and frequently switch flocks during inter-flock encounters. They are rarely precluded from associating with flocks because one is usually present somewhere in their home range.

Species with ranges smaller than those typical of a flock nucleus frequently cannot associate with a flock because none is present in their home range. The likelihood that a species of this category will join a flock is most strongly correlated with the length of time the flock is in the individual's (or pair's) territory (linear regression, $r^2 = 0.8$, $P < 0.01$, N = 3 spp. and 644 flock joinings), and whether or not a conspecific is associated with the flock when it enters the bird's territory ($\chi^2 = 26.94$, $P < 0.01$, N = 898; Powell 1979). Birds join a flock sooner and closer to the edges of their territories when conspecifics are with the flock, presumably, in response to the intrusion by the conspecifics.

Transients or nonresident flock participants are species without restricted home ranges that can remain with flocks for extended periods. In middle elevation flocks, transients are typically elevational migrants that have wandered into the area (Powell 1979). Transients may associate with flocks throughout a day or several days in succession or for only a few minutes.

Flocks of frugivores tend to include many species and have large intraspecific group sizes. Membership appears to be highly flexible both in terms of species composition and numbers of individuals per species (Moynihan 1962a; Leck 1971). Flocks of granivores generally include few species and have large intraspecific group sizes and, like flocks of frugivores, highly variable membership (Moynihan 1962a; Rubenstein et al. 1977). Nothing is known of their spatial and structural organization.

MOVEMENT RATES AND PATTERNS

As is the case with other aspects of mixed flock sociobiology, virtually all data available on movement rates and patterns were collected from understory, insectivorous flocks. Therefore, what follows is by necessity a discussion of insectivorous flocks. Generally, they begin to assemble during the dawn chorus period and are fully formed by the time birds begin to forage about 0.5 h after daylight (Pearson 1977; Munn and Terborgh 1979; Powell, unpubl. data). During this period, the birds are active and vociferous, but individual activity is not integrated into flock activity. Neotropical flocks do not follow fixed courses as do flocks in Malaya (McClure 1967), but meander through the forest on courses that frequently cross, but rarely retrace previous paths (Willis 1972a; Munn and Terborgh 1979; Powell 1979; Gradwohl and Greenberg 1980). Flock movement is centered straight ahead; the frequency distribution of turn angles approximates a circular normal distribution (Batschelet 1965). At two sites where directional data were collected (Costa Rica and Amazonia), flock movement patterns approximated random walks (Powell 1977, unpubl. data). In these areas, flocks apparently do not avoid areas previously visited, nor do they exhibit movement patterns that emphasize resource-rich patches. Areas are not revisited more often than predicted for random walks of similar directional distributions, and directional probabilities do not depict doubling back to remain in food-rich patches (Powell 1977). In some cases, flocks show preferences for portions of their home ranges. Gradwohl and Greenberg (1980) attributed this to heterogeneity of the habitat and preference by the nucleus species for vine tangles, particularly those centrally located in the home range. Similarly, Powell (1977) found that a middle elevation mixed flock tended to favor an area of dense understory vegetation during one time of the year.

Flocks remain active throughout the daylight period, although foraging activity may be reduced during the middle of the day (Buskirk et al. 1972; Munn and Terborgh 1979; Powell 1979). Pearson (1977) did not observe a mid-day decrease in foraging activity at any of three locations in upper Amazonian forests, but he observed that birds tended to forage lower during mid-day. He concluded that antbird foraging activity declined slowly through the afternoon (13:00–16:00) although he presented no data to support that conclusion. At the onset of dusk, flock members may move to a water source for bathing (Munn and Terborgh 1979; Powell 1979, unpubl. data) before going to roost. At roost, flocks may either fragment (Munn and Terborgh 1979) or remain in close proximity (Gradwohl and Greenberg 1980). At least some flock participants roost solitarily in holes away from the rest of the flock (Skutch 1969, 1977).

The rate at which flocks move through the forest varies from 1 to 80 m/min, although flocks that are actively foraging consistently move about 5 m/min. Geographically, movement rate appears to be relatively consistent for understory flocks. In Central Amazonia, antshrike-antwren flocks moved an average (± s.d.) of 4.9 ± 1.3 m/min (Powell, unpubl. data), whereas middle elevation flocks in Costa Rica moved at a mean rate of 5.4 ± 2.4 m/min (Powell 1979). Understory flocks at the central Amazonian site moved significantly more slowly when associated with canopy flocks of insectivores ($\bar{X} = 3.8$ m/min) than when independent of them ($\bar{X} = 5.2$ m/min; t-test, $t = 5.5$, $P < 0.01$, N = 4 flocks followed during 21 days; Powell, unpubl. data).

FEEDING ECOLOGY

Much of the research on mixed flocking has focused on foraging behavior. Numerous attempts have been made to measure the level of interspecific overlap in food niche among flock members, and the impact of this overlap on foraging behavior. Again, most data were collected from insectivorous flocks.

Insectivores in mixed flocks capture insects with a wide array of methods from virtually all substrates and at all levels in the forest (for a thorough analysis of foraging methods see Buskirk 1972). MacArthur (1972) was among the first to quantify niche separation among mixed species flock participants. He concluded, based on data collected by J. W. Terborgh, that foraging height differences among four species of antwrens (*Myrmotherula*) were sufficient to preclude competition even when all four species were present in the same flock. However, these four species account for only part of the *Myrmotherula* complement, which may include as many as seven sympatric species. When foraging heights of the full complement of species were considered, differences were not sufficient to separate foraging niches (Pearson 1977; Wiley 1980). Munn and Terborgh (1979:341) analyzed foraging parameters of *Myrmotherula* more completely (including the same species considered by MacArthur) and concluded that while most of the sympatric species on their study area diverged in foraging height or substrate, some species appeared to "overlap broadly in foraging height, foraging technique and substrate." This apparent overlap in the *Myrmotherula* foraging niches, when considered in conjunction with their excessively large home ranges (Munn and Terborgh 1979; Powell, unpubl. data), may reflect a flock induced reduction of interspecific competition. Because the small antwrens (8–9 g) have home ranges that conform to those of larger antshrikes (18 g), the density of individual antwren species may be lower than the carrying capacity. The home range of *Myrmotherula axillaris* in Panama, where it flocks with the small antwrens, *Myrmotherula fulviventris* and *Microrhopias quixensis* (Gradwohl and Greenberg 1980), was less than half the size of the range it used in Amazonia. At the same time, woodcreepers (*Xiphorhynchus*) and foliage-gleaners (*Automolus, Xenops*) that join flocks in Amazonia have home ranges that are similar in size or smaller than those of congeners in Panamanian and Costa Rican flocks (Powell 1977; Gradwohl and Greenberg 1980).

Overlap in food niche may also result from specialization on arthropods that are flushed by the flock. Up to half of foraging moves by members of the genera *Thamnomanes* (antshrikes) and *Lanio* (shrike-tanagers) are in pursuit of prey that attempt to escape from other flock members (C. A. Munn, pers. comm.; Powell, unpubl. data). The escaping prey may be chased by an antshrike and/or a shrike-tanager, as well as the bird that flushed it. Both antshrikes and shrike-tanagers may enhance their success at obtaining flushed prey through the use of warning calls that distract the bird that flushed it.

Other species in Amazonian mixed flocks use different behaviors, substrates, and/or heights for foraging (Munn and Terborgh 1979; Powell, unpubl. data). All species in insectivorous mixed flocks outside the Amazon basin differed significantly in some foraging parameters, so their food niches are thought to overlap little if any (Buskirk 1972; Jones 1977).

Frugivorous mixed flock participants are generally considered to have largely overlapping food niches because they all feed on fruit at the same foraging sites. Different species, however, may harvest resources from the same species of fruiting tree in different ways, making the overlap less than expected (Willis 1966; Diamond and Terborgh 1967; Terborgh and Diamond 1970). Limited data available for granivorous flocks indicate a high degree of overlap in seed selection and, presumably, food niche, among flock participants (Moynihan 1962a; Rubenstein et al. 1977).

SEASONALITY

Mixed species flocks exist year-round in neotropical environments, although average flock size usually varies seasonally. At the only site where flocks have been monitored throughout the year, mean flock size varied seasonally by a factor of four (Powell 1979). Because mixed flocks are composed primarily of permanent residents, flocking cycles cannot be attributed to population changes. Rather, flocking cycles result from changes in flocking propensities of resident species. During the breeding season, birds generally continue to participate in flocks, but at a lower frequency. They may be precluded from participating when nest building, incubating, brooding, and sometimes while caring for nestlings. The impact of reproductive activities on flock participation is at least partially dependent on home range size. Birds with large home ranges are limited to foraging with flocks in the vicinity of the nest (Munn and Terborgh 1979). Nests of flocking species tend to be scattered through the flock range, probably in response to predation (Tinbergen 1952), so flocks are frequently beyond foraging distances of nesting birds. Species with small home ranges tend to work on the nest when the flock is in its vicinity and to make trips back and forth to the flock when feeding young (R. Greenberg, pers. comm.).

The decline in mixed flock size during the breeding season appears to be more pronounced for middle elevation mixed flocks than for lowland flocks. This is probably because middle elevation flocks have a single nucleus species that becomes ineffective as a cohesive force while breeding. In the absence of an alternative nucleus species, the flocks tend to disintegrate. In more complex flocks, the several nucleus species are typically slightly out of phase in breeding activity (R. O. Bierregaard, C. Strang, P. Polshek, pers. comm.; Powell, unpubl. data) so an active nucleus is always present. In Amazonian flocks, year-old offspring of one nucleus species, Cinereous Antshrike (*Thamnomanes caesias*) remain with the flock through the next breeding season filling the role of flock nucleus while their parents are breeding (Powell, unpubl. data).

ADAPTIVE SIGNIFICANCE OF MIXED SPECIES FLOCKS

Data that have been collected from mixed species foraging flocks of insectivores are sufficient to allow at least a qualitative evaluation of hypotheses that explain the adaptive significance of flocking. Data are insufficient to permit the same for other types of foraging flocks. Therefore, except for a few specific studies in which the relevance of hypotheses to frugivorous or granivorous flocks is evaluated, my discussion is restricted to insectivorous forms. Flocking may either enhance foraging efficiency or reduce predation. In most instances, the hypotheses are not mutually exclusive, and different species, even different individuals of the same species, may benefit from flocking in different ways.

FORAGING ENHANCEMENT HYPOTHESES

Disturbance created by flock members as a mechanism for flushing cryptic insects (Belt 1874).—The phenomenon in which disturbance causes cryptic invertebrates to move, thereby increasing their vulnerability to opportunistic predators, occurs regularly in special situations. Documented examples of the beater effect in terrestrial environments usually involve a source of disturbance extrinsic to the birds themselves. Known sources include fire (Winterbottom 1949), army ants (Willis 1967, 1972a), and large mammals (Siegfried 1971; Siegfried and Batt 1972). The significance of the beater effect appears to vary among mixed species flocks. In smaller flocks, the beater effect appears to be rare. Willis (1966) concluded that food was rarely flushed by one individual for another in ant-tanager (*Habia*) mixed flocks. During extensive observations of mixed flocks of insectivores on Barro Colorado Island, Panama, he observed only one incident of a bird frightening a prey that was captured by another (Willis 1972a). Buskirk (1972) reported one similar interaction in Costa Rican mixed flocks. Powell (1977) observed only one definite case of an individual functioning as a beater for another in 700 h of observation at the same site. Outside the neotropics, Greig-Smith (1978a) saw no "clear instances" of insect-flushing in African mixed flocks, whereas in Kashmir, MacDonald and Henderson (1977) found the beater effect an important flock function for some species, especially flycatchers. In larger mixed flocks, the beater effect appears to be more significant. Munn and Terborgh (1979) suggested that antshrikes (*Thamnomanes*) in Amazonian mixed flocks gained significantly from the beater effect, a prediction supported by data collected by Powell (unpubl. data). Almost 30 percent of 119 observed prey capture attempts by *Thamnomanes caesius* involved insects flushed by other flock members. Croxall (1976) concluded that the beater effect was significant in flocks in Sarawak, but did not present supporting data.

It is possible that the significance of the beater effect has been underestimated because flock disturbance causes arthropods to become more conspicuous to gleaning birds through minor shifts that are undetectable by human observers. However, mixed flock members generally forage several meters apart (Buskirk 1972; Wiley 1980) and rarely forage on substrates canvassed by other flock members, so beneficiaries of the beater effect must make flights of several meters to capture prey (C. A. Munn, pers. comm.; Powell, unpubl. data). To fully evaluate the significance of the beater effect to gleaning guild members of mixed flocks, it will be necessary to study the reactions of cryptic arthropods to bird presence.

Flocking as a means of transferring information about food resources (Ward 1965).—Transfer of information about food resources frequently occurs among birds in groups (Fisher and Hinde 1949; Klopfer 1957, 1959; Pinowski 1959; Turner 1965; Horn 1968; Sealy 1973). Such information transfer can enhance foraging success of birds in groups in experimental situations (Krebs et al. 1972; Sasvari 1979). Information could be transferred deliberately, i.e., active cooperation, or unintentionally, i.e., passive cooperation (Pinowski 1959). An active cooperater uses special signals to communicate the presence of resources (e.g., Frings et al. 1955); a passive cooperater draws attention to a resource by its efforts to harvest the resource. The distinction is relevant because active cooperation implies that information transfer is a primary function of flocking, whereas passive cooperation is likely to be a consequence of grouping [e.g., kleptoparasitism in seabirds (Fuchs 1977) and herons (Kushlan 1978)].

Information transfer in mixed flocks appears to be passive. Flocking birds seem to be forced to share food discoveries with other flock members (Emlen 1952; Hinde 1952; Lack 1954; Lockie 1956; Ward 1965; Crook and Goss-Custard 1972). Active communication of resource related information has not been observed in mixed species flocks. Moynihan (1962b) made detailed behavioral analyses of neotropical mixed flock participants but failed to recognize specialized food discovery vocalizations or behavioral cues. Despite thorough analyses of vocal repertoires of both tropical (Willis 1972a) and temperate (Odum 1942; Hinde 1952) mixed species flocks, special interspecific assembly calls to identify food concentrations have not been reported.

If information transfer were a primary selective advantage of flocking, the transfer should be reciprocal for all flock members. Otherwise, flock participants that are transmitters, but not recipients, of information should disassociate themselves from flocks (Trivers 1971; Gadil 1972; Baker 1978). Several characteristics of mixed flocks composed of territorial species are likely to limit reciprocity of foraging information transfer.

Flock participants probably vary in their abilities to locate food resources. Foraging capacities of neotropical mixed flock participants have not been measured, but some data are available for temperate zone species. Smith and Sweatman (1974) noted marked differences in performances of individual titmice (*Parus*) in experiments designed to test foraging behavior relative to varied food distribution. Sasvari (1979) noted interspecific differences in similar types of experiments with tits (*Parus* spp.). Partridge (1976) observed individual differences in foraging efficiencies of captive tits, and Vince (1956) found the same differences in wild ones. Foraging success has been shown to be age dependent in large species (Orians 1969; Recher and Recher 1969; Murton et al. 1971; Salt and Willard 1971; Buckley and Buckley 1974; Verbeek 1977). No one has attempted to evaluate the significance of age to foraging efficiency of birds in mixed flocking species, but it is reasonable to assume that, as in large species, experience influences foraging efficiency. If so, older or more experienced individuals would tend to be better at discovering food, and would provide foraging information more consistently.

Dominance hierarchies, which are typical of small monospecific flocks (Colquhoun 1942; Hamerstrom 1942; Brian 1949; Sabine 1949; Dixon 1965; Fretwell 1969; Murton et al. 1972; Feare et al. 1974) and probably exist in interspecific associations (see Morse 1974 for review), will inhibit reciprocation in transfer of foraging information (Lack 1954; Lockie 1956; Fretwell 1969; Woolfenden et al. 1976). Subordinates that discover food are likely to lose much of what they find (e.g., Murton et al. 1971; Krebs et al. 1972; Baker 1978), whereas dominants that locate a resource will invoke spatial dominance (e.g., Murton et al. 1971; Davies 1976).

Territorial behavior, which is characteristic of most mixed flock participants, severely limits the potential for the transfer of foraging information to conspecifics. Even young birds and floaters, classes with the greatest potential for gaining information by joining mixed flocks, are prevented from participating in flocks with conspecific residents. Where resident mixed species flocks of marked birds have been studied, migrants and young birds, other than offspring of the flock, have played minor roles (Munn and Terborgh 1979; Powell 1979; Gradwohl and Greenberg 1980). Instead, migrants tend to form flocks of their own in disturbed habitats where resident mixed flocks are relatively rare (Willis 1966; Chipley 1976).

Information regarding new types or concentrations of food could be transferred within mixed flocks; however, the restricted home ranges of and high food niche diversity among insectivorous mixed flock participants minimize the potential for such transfer. Continuous following of color-marked birds in flocks revealed that most flock attendants have small home ranges (0.7–10 ha) and visit all parts of their ranges frequently (Buskirk et al. 1972; Munn and Terborgh 1979; Powell 1979; Gradwohl and Greenberg 1980). Therefore, they can regularly monitor food availability and probably maintain a high degree of familiarity with food resource distributions within their respective home ranges. Because residents only follow flocks in their home ranges, flocking cannot provide information about food resources in unfamiliar areas. In addition, most studies of foraging niche diversity in mixed flocks have revealed very little food niche overlap (Wiley 1971; Buskirk 1972; Jones 1977; Munn and Terborgh 1979; Gradwohl and Greenberg 1980). Thus, information about new food resources for one species would be irrelevant to other participating species.

Despite these limitations on the potential for the transfer of food resource information among insectivorous flock members, information transfer, referred to as imitative foraging, has been observed in limited cases outside the neotropics (Greig-Smith 1978a; Morse 1978; Herrera 1979). These are all examples of one species altering its foraging niche presumably to copy feeding behavior of another flock species (Herrera had changes in niche breadth of several flocking species when data for all species were grouped, but only one species showed a significant change when species were analyzed individually). The only suspected case of imitative foraging in neotropical flocks involves a Paruline warbler (*Myioborus miniatus*) that switches its solitary behavior of hawking insects to foliage gleaning while with flocks (Buskirk 1972). This change in foraging behavior may also indicate that hawking for insects is not compatible with flock following because *Myioborus* can no longer remain in small understory openings as it does while foraging solitarily.

One variant of the information transfer hypothesis predicts that flocking enhances discovery of food rich patches (Ward 1965; Karr 1971; Zahavi 1971; Morton 1973). Several models have been generated to demonstrate the theoretical basis for this hypothesis (Charnov 1973; Thompson et al. 1974). These models are probably relevant to frugivorous and granivorous flocks, which concentrate on temporary food rich patches. The degree to which neotropical flocks of insectivores maximize the exploitation of food patches has not been measured. In a temperate zone forest, Nilsson (1979) identified resource rich patches and concluded that flocks did not concentrate in those areas. A flock's tendency to specialize on resource rich patches can be measured by analyzing the distribution of flock movement rates. Theoretically, flocks that specialize in locating and exploiting patches, will move rapidly between patches and slowly within patches (Charnov 1973). Individual birds experimentally presented with food distributed in patches followed this pattern (Croze 1970; Krebs 1973; O. Garton, pers. comm.). Frugivorous and granivorous flocks follow this pattern (Moynihan 1962a; Rubenstein et al. 1977), but flocks of insectivores do not. Movement rates obtained by following insectivorous flocks in middle elevation Costa Rica (Powell 1977) and in the Amazon basin (Powell, unpubl. data) were not serially correlated. Instead, the temporal distributions of flock movements indicate a fairly uniform use of space. This is consistent with the relatively uniform movement rates recorded for mixed flocks outside the neotropics (Odum 1942; Dixon 1965; McClure 1967; Nilsson 1979). The only flocks of insectivores known to move with bimodally distributed rates are flocks of Long-tailed Tits (*Aegithalos caudatus*, Gibb 1960; Gaston 1973).

The transfer of food resource information within frugivorous and granivorous mixed species flocks is potentially an important function of flocking. In these flocks, foraging niches appear to overlap considerably (Leck 1971; Howe 1977; Rubenstein et al. 1977), and food sources are concentrated in ephemeral patches that must constantly be discovered. Dominance hierarchies would be less likely to restrict food information transfer because food patches tend to be too large or ephemeral for defense by single individuals.

Mixed flocking lessens interspecific competition by reducing foraging niche overlap (Morse 1967, 1970) or spatial overlap (Miller 1922; Short 1961). —In fluctuating communities, such as those with large migratory components, flocking may allow members to monitor competition levels and to avoid areas with high competitor densities or to modify foraging behavior to reduce niche overlap. In contrast, in more stable communities with long-term residents, the need to constantly reassess competition levels and readjust foraging niche breadth is probably minimal. Most neotropical species that flock defend the same territory throughout the year, making relatively minor seasonal adjustments in their home ranges by means of increased trespassing (Willis 1972a; Munn and Terborgh 1979; Powell 1979). With this level of community stability, permanent residents with overlapping foraging niches could adjust niche breadth to reduce or eliminate overlap independent of mixed flocks (Austin and Smith 1972; Morse 1978; Alatalo 1981).

Most measures of the impact of flocking on niche overlap indicate that flocking results in a reduction in foraging niche diversity. Both Pearson (1971) and Buskirk (1972) found significant decreases in foraging height diversity when birds joined mixed species flocks. Buskirk (1972) found that one warbler (*Myioborus miniatus*) switched from its solitary foraging strategy of hawking insects to foliage gleaning, a foraging pattern used by other associates of the flocks. The only example of decreased foraging niche overlap in flocks involves the White-flanked Antwren (*Myrmotherula axillaris*), which forages higher when *M. longipennis* is present (Wiley 1980).

The second part of the overlap reduction hypothesis predicts that flocking reduces spatial overlap by allowing flock members to monitor locations used by flocks and subsequently to avoid those areas until food resources are restored or redistributed (e.g., Drent and Swierstra 1977). Interspecific flocks of territorial residents present a special case that restricts the applicability of this hypothesis. As already discussed, mixed flocking does not alter the level of intraspecific overlap since participating species forage in pairs, or pairs with young, in the same home ranges regardless of their flock affiliation. The potential for reducing spatial overlap among species is minimized by the constraints of permanent territoriality. Because most flock members are restricted to small home ranges and exhibit high return frequencies to all parts of their home ranges (Buskirk et al. 1972; Munn and Terborgh 1979; Powell 1979; Gradwohl and Greenberg 1980) the potential for allowing food resources to replenish between visits is eliminated.

PREDATION AVOIDANCE HYPOTHESES

The significance of predation to the evolution of social systems of small passerines is debated (Lack 1968; Cody 1971; Murton 1971; Greig-Smith 1978b). It has been suggested that predation on small birds may be too rare to exercise a pronounced impact on social behavior. Analyzing the importance of predation is complicated by the nature of predator-prey interactions themselves. In most studies, predation rates are negatively biased by the presence of an observer who generally frightens away unhabituated predators (Busse 1978). The impact of an observer on the behavior of a forest raptor was demonstrated by a study of a radio-tagged forest-falcon (*Micrastur gilvicollis*, L. Harper, unpubl. data). During a 22 day tracking period, the forest-falcon remained extremely wary and virtually impossible to observe. Efforts to habituate the bird failed, and attempts to approach the bird invariably caused it to flush. Given this level of response to a human presence, it seems unlikely that *Micrastur* spp. would attack a flock followed by an observer. Harper's radio-tagged *Micrastur* had a home range of less than 20 ha which was the same area used by several mixed flocks being followed concurrently by Powell (unpubl. data). Despite the overlap, Powell never observed the forest-falcon.

Morse (1978) suggested that historically, predation on mixed flocks may have been more frequent, but that recent declines in predator populations, particularly in the temperate zone where most studies of flocks have been conducted, may have reduced the rate of predation. Finally, even if predation frequency is presently low, this does not imply that predator-prey interactions have not had a major evolutionary impact on shaping behavior that now effectively minimizes predation (Macan 1965).

Evidence that predation promotes flocking is largely indirect. Willis (1972b, 1973) noted an absence of mixed flocks in forests of Hawaii and Puerto Rico where raptors that prey on birds are rare or absent. Buskirk (1976) evaluated the significance of predation with respect to flocking by intuitively analyzing the vulnerability of flocking and non-flocking species to predation. He concluded that morphological and behavioral characteristics of flocking species make them especially vulnerable to predation and the flocking is a mechanism to mitigate this vulnerability. Morton (1979) analyzed forest bird communities in Venezuela following Buskirk's format and also concluded that predation was the primary determinant of flocking propensity. Howe (1977) hypothesized that predation was a determinant of the patterns in which frugivores visited fruiting trees, but had no quantitative measure of predation pressure.

Although the significance to flock evolution of avian predator-prey interactions is debated, predation or attempted predation by raptors on birds has been observed hundreds of times. Many reports are incomplete or anecdotal, but virtually all describe the use of surprise by avian predators to capture bird prey (Tinbergen 1946; Rudebeck 1950, 1951; Pielowski 1959; Markgren 1960; Bengtson 1971; Morse 1973; Newton 1973; Schipper et al. 1975; Page and Whitacre 1975; MacDonald and Henderson 1977; Kenward 1978; Dekker 1980). This indicates that early detection of predators can reduce the likelihood of predator attack and success. The density of the vegetation in forests inhabited by species that associate in mixed flocks allows predators to unobtrusively stalk or "sit and wait" for small actively foraging birds to wander within the domain of danger (Hamilton 1971).

The wariness of birds participating in mixed flocks suggests that predation pressure is a selective force acting on behavior (Morse 1970; Balda et al. 1972; Willis 1972a). Flock participants frequently give warning calls in addition to exhibiting escape behavior in response to rapid descending flights by flock members and nonpredators in the vicinity of the flock (Willis 1972a; Powell 1977).

Observations of encounters between raptors and mixed species flocks in neotropical forests are relatively rare. Pearson (1977) observed three raptor attacks on flocks during 1500 h of observations; all were unsuccessful. During 1700 h of observations, Powell (1977) witnessed four unsuccessful attacks by raptors on mixed flocks in Costa Rica. Munn and Terborgh (1979) observed a single unsuccessful attack by *Micrastur* sp. on their flock in 440 h of observations in eastern Peru, and Powell (unpubl. data) observed one successful and two unsuccessful predatory attacks on mixed flocks during 380 h of observations in central Amazonia.

Increased effectiveness of surveillance against attacks from predators as a result of flocking (Bates 1863).—Surveillance can be improved either by pooling the independent efforts of flock members or by using sentinels. Sentinels, i.e., individuals that interrupt foraging for extended periods to watch for predators, have not been observed in mixed species flocks. However, flock participants vary in their capacities to maintain awareness while foraging. Therefore, it is possible that less efficient surveillants, such as woodcreepers and foliage-gleaners, may obtain a greater increase in surveillance by participating in mixed flocks than they would by forming intraspecific associations. At a Costa Rican study site, two species that initiated 75 percent of the mobbing episodes comprised only 15 percent of flocks (N = 16 mobbing episodes, Powell 1977). The two species, therefore, functioned as sentinels for other species in the flocks. In eastern Peru, *Thamnomanes* fills the role of the functional sentinel because its flycatching behavior allows it to be more vigilant than species that focus on proximal substrates (Munn and Terborgh 1979).

The hypothesis that groups of birds maintain greater predator awareness by pooling their independent efforts has been modeled by Koopman (1956) and Treisman (1975), and tested experimentally by Powell (1974) and Lazarus (1979a, b). Field observations of Starlings (Jennings and Evans 1980), Dark-eyed Juncos, *Junco hyemalis* (Goldman 1980), and Downy Woodpeckers, *Dendrocopos pubescens* (Sullivan 1984) have confirmed the laboratory experiments. Davis (1975) determined that awareness of danger is transferred among flock members by rapid, very subtle behavioral cues. There has been no experimentation with hawk models and neotropical flocking species. Indirect field evidence suggests that flocking is a mechanism to improve surveillance. Willis (1972a) observed that solitary birds were more wary of his presence than birds in flocks, and several cases of interspecific communication during attacks by raptors have been observed (Munn and Terborgh 1977; Powell 1977).

Flocking makes it difficult for raptors to attack potential prey (Miller 1922; Tinbergen 1946; Brock and Riffenburgh 1960).—Flocks may frustrate attacks by making it difficult for the predator to select a specific prey or by presenting the threat of physical damage from collision with other flock members. These hypotheses probably function primarily in large associations. Observations of successful or attempted predation on small flocks indicate that raptors select a specific prey prior to initiating attacks and before flock members respond to their presence (Rudebeck 1950, 1951; Pielowski 1959; Page and Whitacre 1975; MacDonald and Henderson 1977). Characteristics of mixed species flocks further reduce the potential for confusion. Intraspecific group sizes are small, and most species have distinct markings.

The risk of collision between a predator and fleeing flock members will be minimal during attacks on small mixed flocks. Whereas birds in large flocks frequently form compact aerial groupings to discourage predator attacks (Tinbergen 1951), birds of mixed flocks scatter to cover to escape (Willis 1972a; Morse 1973; Powell 1977; Munn and Terborgh 1979).

Flocking decreases the frequency of encounters between predators and prey (Koopman 1956; Olson 1964).—This hypothesis deals specifically with spatial aspects of social organization. It predicts that the frequency of encounters between predators and prey will decrease when birds form flocks because flocking increases predator search time. The impact of flocking on predator-prey encounter rate is dependent on the nature of the flock, the predator, and the habitat (Koopman 1956; Olson 1964). In the case of small resident mixed flocks, these parameters tend to minimize the benefits of grouping with respect to predator-prey encounters. The probability, P, that a predator with a detection range, R, moving at a velocity, V, for a time, T, will encounter a specific individual (or flock) with a home range of area, A, is given by the following equation (Olson 1964):

$$P = 1 - e^{2RVT/A}.$$

The probability of encountering an individual if it is in a flock differs in the substitution of R by R' where R' is the range from which the predator can detect a flock plus the radius of the flock.

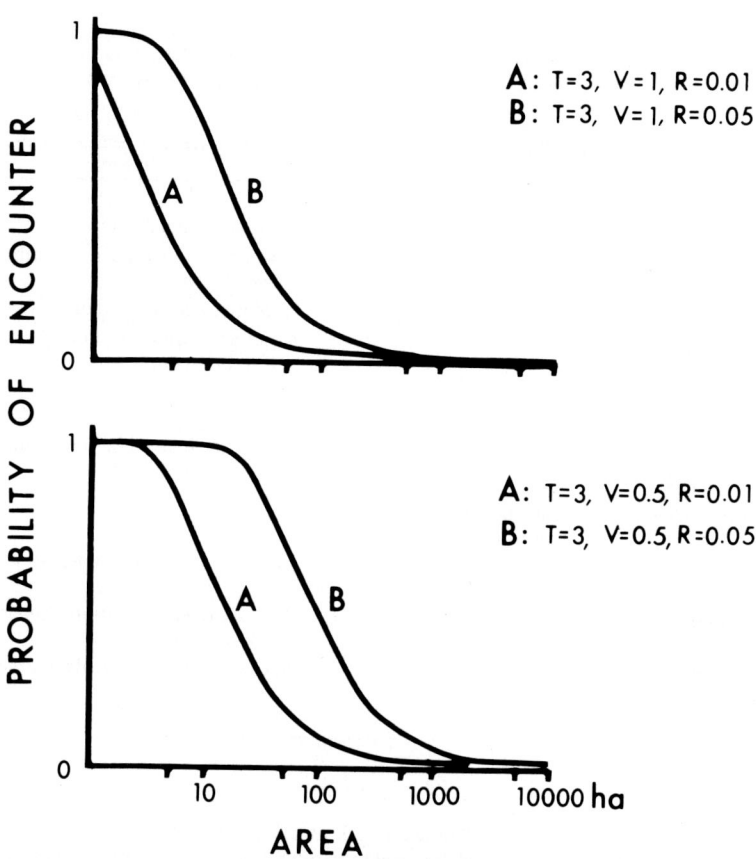

FIG. 1. Predicted frequency of predator-prey encounters between flocks and forest raptors based on Olson's (1964) model, $P = 1 - e^{2R'VT/A}$, where P = probability that a predator will encounter a flock, R' = the range of prey detection by the raptor, V = the raptor's velocity, T = hunting time per prey capture, and A = flock home range. The ranges of values used in the solutions were selected to be inclusive of realistic values estimated from field observations.

Actual values for the parameters in Olson's equation are largely unavailable, but reasonable estimates can be made based on characteristics of flocks and one of the principal forest understory bird predators, *Micrastur*. There are no quantitative measures of the relative detectability of flocks versus individuals or pairs, but most observers consider flocks to be more conspicuous. Therefore, R' (flock detection range) will be larger than R. Conservative values of 10 and 50 m were substituted for R and R' respectively. Ten meters is probably a reasonable minimum for detecting individuals, and 50 meters a maximum detection distance for foraging flocks. Greater detection distances, as during flock confrontations, would be offset by periods of relative quiet when flocks are inaudible at distances greater than 10 to 20 meters. The velocity at which *Micrastur* moves through the forest is not known. *Micrastur* attacks with short rapid flights, but L. Harper (pers. comm.) found that a radio-tagged *Micrastur gilvicollis* remained stationary most of the time. If the predator remains stationary, V would equal the flock's velocity or about 0.5 km/h. Allowing for occasional moves by the predator, V would be the sum of the two movement rates. For a conservative range of estimates, 0.5 and 1 km/h were substituted for V. A substitution value for foraging time, T, was based on data for similar sized raptors that consume the equivalent of one to two birds per day (Tin- bergen 1946). If *Micrastur* spent the entire daylight period foraging, it would have at least 6 h foraging time per capture. As a conservative estimate, allowing 50 percent of activity budget for non-hunting, the equation was solved with T = 3 h. Solving the equation with these values reveals that flock home range, A, must be very large to significantly decrease the frequency

TABLE 3

ATTRIBUTES OF MIXED FLOCKS OF INSECTIVOROUS SPECIES THAT ARE EITHER CONSISTENT OR INCONSISTENT WITH ASSUMPTIONS OF HYPOTHESES CONCERNING THE ADAPTIVE SIGNIFICANCE OF FLOCKING

	Consistent	Inconsistent
Foraging Enhancement		
Beater effect	mobile-cryptic prey	flock members dispersed
Information sharing		permanent territoriality eliminates intra-specific sharing
Food types		little niche overlap, food resources static
Food locations		small territories, food distribution appears homogeneous
Reduced competition		
Niche overlap		greater in flocks
Spatial overlap		greater in flocks, high revisitation rate
Predation Avoidance		
Improved surveillance	interspecific warning against predation	
Collision with clumped prey		flocks scatter when attacked
Decreased predator-prey encounters		prey spatial distribution unchanged, more conspicuous
Reduced probability of being selected as prey		flocks small, individuals distinctly marked

of predator-prey encounter (Fig. 1). Because resident mixed flocks typically have home ranges of less than 10 ha, the frequency of encounters between raptors and flocks will be high. Furthermore, most species that participate in flocks use the same home range irrespective of their affiliation with the flock, so A is constant when comparing predator-prey encounter rates for flocking and non-flocking individuals. Because the detection distance of a flock is greater than for individuals, flocking may actually increase the frequency of encounter between raptors and potential prey.

Flocking reduces the probability of a given individual being selected as prey during predator-prey encounters (Brock and Riffenburgh 1960; Hamilton 1971; Vine 1971).—Again, this hypothesis is most applicable to very large aggregations of organisms. Mixed species flocks are typically small, and most species remain in pairs regardless of their affiliation with these flocks. Assuming random selection by a predator, there is relatively little change in the probability of being selected in a predator-prey encounter. According to Olson's (1964) equation for predator-prey encounters, the probability of encounters between resident mixed flocks and forest raptors should be high so individuals affiliated with flocks will still be attacked frequently.

CONCLUSIONS

The determinants of mixed flock formation are a combination of spatial, temporal, climatic, and behavioral factors. For insectivorous flocks, at least, composition is controlled in large degree by intraspecific social organization. Permanent territoriality and pair bonding limit intraspecific group size to pairs or family groups, making interspecific diversity the principal determinant of flock size. Maximum potential flock size is, therefore, somewhat larger than twice the number of flocking species resident in the area. This maximum is rarely attained, however, and mean flock size is typically less than half the potential size. This discrepancy between predicted and observed values results, in part, from members constantly being forced to leave the flock when it crosses territorial boundaries. The likelihood that an owner of a newly entered territory will join a mixed flock is directly proportional to the time the flock is in its territory. The lag in joining appears to be an expression of the costs and benefits of joining the flock.

TABLE 4

ATTRIBUTES OF MIXED FLOCKS OF FRUGIVOROUS AND GRANIVOROUS SPECIES THAT ARE EITHER CONSISTENT OR INCONSISTENT WITH ASSUMPTIONS OF HYPOTHESES CONCERNING THE ADAPTIVE SIGNIFICANCE OF FLOCKING

	Consistent	Inconsistent
Foraging Enhancement		
Beater effect		immobile food
Information sharing		
Food types	ephemeral food high niche overlap	
Food locations	food patchy	
	(home range size unknown)	
Reduced competition		
Niche overlap	high niche overlap	
Spatial overlap	high renewal rates	
Predation Avoidance		
Improved surveillance	interspecific warning against predation	
Collision with clumped prey		flocks scatter when attacked
Decreased predator-prey encounters	(the impact of flocking on spatial distribution is unknown)	
Reduced probability of being selected as prey	flocks large	individuals distinctly marked

Assessing the benefits of flocking has proven elusive. In part, this is because early analyses tended to disregard the diversity of bird associations and attempted to assign a single adaptive significance to all flocks or flock members. It is probable that different types of flocks have evolved in response to different selective pressures and that mixed flocks of permanent residents that engage in year-round territoriality serve different functions from those of fluid flocks of transient birds. Furthermore, the adaptive significance of flocking probably differs for frugivorous or granivorous species and insectivorous species. Therefore, it is necessary to determine the relative importance of improved foraging efficiency or predator evasion for a specific flocking cohort. This problem is complicated by the fact that efficient foraging and predator evasion are probably interdependent. Increased protection against predation allows more efficient foraging and vice versa (Lazarus 1972). This relationship was demonstrated experimentally (Powell 1974) and in the field (Abramson 1979; Goldman 1980). Caraco (1980) further refined the foraging model by adding an energy parameter. He quantified the trade-offs between predation evasion and foraging efficiency by altering energy costs at food sites. Evaluating flocking hypotheses may be further complicated by the fact that advantages and disadvantages of flocking are probably balanced. This is evidenced by the low propensity of residents to join mixed flocks that are accessible. Typically, 50 percent or less of the individuals of flocking species participate in flocks at a given time. A partial explanation for this low attendance is the failure of individuals to join flocks that are within their home ranges, but not in their immediate vicinity (Powell 1979). Apparently, the interruption of foraging, the energetic costs of moving to the flock's location, and/or increased exposure to predation while moving to the flock's location offset the advantages accruing from flocking. This precision in the balance of parameters makes it exceedingly difficult to discriminate quantitatively among hypotheses.

On a qualitative basis, organizational and behavioral characteristics of resident insectivores, the major component of many neotropical mixed flocks, minimize the applicability of foraging efficiency hypotheses as explanations of the flock's adaptive significance. Intraspecific group size is unchanged by flocking so mixed flock formation cannot reduce intraspecific overlap.

The potential for reducing interspecific overlap is minimal since there appears to be little interspecific food niche overlap. Two benefits of flocking related to foraging that have been observed in limited cases are imitative foraging and the beater effect. Both benefits appear to be gained by a restricted subset of flocking species. Qualitative evaluation of predation hypotheses again indicates that spatial and group size restrictions imposed by permanent residency minimize the applicability of most predation reducing hypotheses (Table 3). Formation of these small flocks probably does not substantially reduce the frequency of predator-prey encounters. Intraspecific group sizes in these mixed flocks are small so potential prey are distinctly marked and predator confusion is unlikely. Because flock members are spread out while foraging and scatter when attacked, the probability of collisions between the predator and flock members is low. The predation hypothesis most relevant to these flocks is that flocking increases the quality of surveillance and decreases the likelihood of approaches by undetected predators.

As mentioned, an evaluation of the adaptive significance of flocking by frugivores and granivores is precluded by the absence of sufficient data. Speculatively, flocks of frugivores, and possibly granivores, have more attributes that fulfill the prerequisites of foraging hypotheses (Table 4). These flocks appear to have considerable overlap in foraging niche, so a potential for foraging enhancement through information sharing, imitative foraging and possibly manipulation of niche to spatial overlap exists. However, without measures of foraging niche parameters, renewal rates of food resources, and spatial organization, the respective foraging enhancement hypotheses cannot be tested. The same problem exists with respect to predation evasion hypotheses. Home ranges are not known, so the impact of flocking on the frequency of predator-prey encounters is unknown. The potential for improved surveillance and decreased likelihood of being selected by a predator would seem to be great because associations in fruiting trees tend to be large and individuals that are processing fruit can serve as sentinels for those searching the foraging substrate.

ACKNOWLEDGMENTS

Gene Eisenmann was responsible for arranging my first trip to the Neotropics for which I am forever grateful. My recent work in Amazonia was supported by Postdoctoral fellowship #BEGES 73698 from the Organization of American States and logistic support from the World Wildlife Fund US. I wish to thank H. Powell, D. Robertson, R. Greenberg, E. Morton, C. Munn, and M. Foster for their editorial reviews.

LITERATURE CITED

ABRAMSON, M. 1979. Vigilance as a factor influencing flock formation among curlews *Numenius arquata*. Ibis 121:213–216.

ALATALO, R. V. 1981. Interspecific competition in tits *Parus* spp. and the goldcrest *Regulus regulus*: foraging shifts in multispecific flocks. Oikos 37:335–344.

AUSTIN, G. T., AND E. L. SMITH. 1972. Winter foraging ecology of mixed insectivorous bird flocks in oak woodland in southern Arizona. Condor 74:17–24.

BAKER, M. C. 1978. Flocking and feeding in the great tit *Parus major*—an important consideration. Am. Nat. 115:779–781.

BALDA, R. P., G. C. BATEMAN, AND F. G. FOSTER. 1972. Flocking associates of the piñon jay. Wilson Bull. 84:60–76.

BATSCHELET, E. 1965. Statistical methods for the analysis of problems in animal orientation and certain biological rhythms. American Institute Biological Science, Washington, D.C.

BATES, H. W. 1863. The Naturalist on the River Amazons. Murray Press, London.

BELT, T. W. 1874. The Naturalist in Nicaragua. Murray Press, London.

BENGTSON, S. 1971. Hunting methods of Gyrfalcons *Falco rusticolus* at Myvatn in Northeast Iceland. Ibis 113:468–476.

BRIAN, A. D. 1949. Dominance in the great tit. Scott. Nat. 61:144–155.

BROCK, V. E., AND R. H. RIFFENBURGH. 1960. Fish schooling: a possible factor in reducing predation. J. Cons. Int. Explor. Mer 25:307–317.

BUCKLEY, F. G., AND P. A. BUCKLEY. 1974. Comparative feeding ecology of wintering adult and juvenile Royal Terns (Aves: Laridae, Sterninae). Ecology 55:1053–1063.

BUSKIRK, W. H. 1972. Ecology of bird flocks in a tropical forest. Unpubl. Ph.D. dissert., University of California, Davis.

BUSKIRK, W. H. 1976. Social systems in a tropical forest avifauna. Am. Nat. 110:293–310.

BUSKIRK, W. H., G. V. N. POWELL, J. R. WITTENBERGER, R. E. BUSKIRK, AND T. U. POWELL. 1972. Interspecific bird flocks in tropical highland Pamama. Auk 89:612–624.

BUSSE, C. D. 1978. Chimpanzee predation as a possible factor in the evolution of Red Colobus Monkey social organization. Evolution 31:907–911.

CARACO, T. 1980. Stochastic dynamics of avian foraging flocks. Am. Nat. 115:262–275.

CHARNOV, E. L. 1973. Optimal foraging: some theoretical explorations. Unpubl. Ph.D. dissert., University of Washington, Seattle.

CHIPLEY, R. M. 1976. Impact of wintering migrant wood warblers on resident insectivorous passerines in a subtropical Colombian oak woods. Living Bird 15:119–143.

CODY, M. L. 1971. Finch flocks in the Mojave Desert. Theor. Popul. Biol. 2:142–158.

COLQUHOUN, M. K. 1942. Notes on the social behavior of Blue Tits. Br. Birds 35:234–240.

CROOK, J. H., AND J. D. GOSS-CUSTARD. 1972. Social ethology. Ann. Rev. Psychol. 23:282–312.

CROXALL, J. P. 1976. The composition and behavior of some mixed-species bird flocks in Sarawak. Ibis 118:333–346.

CROZE, H. 1970. Searching image in Carrion Crows. Hunting strategy in a predator and some antipredator devices in camouflaged prey. Z. Tierpsychol. Beih. 5:1–86.

DAVIES, N. B. 1976. Food, flocking, and territorial behavior of the Pied Wagtail (Motacilla alba varrellii Gould) in winter. J. Anim. Ecol. 45:235–253.

DAVIS, J. M. 1975. Socially induced flight reactions in pigeon. Anim. Behav. 23:597–601.

DEKKER, D. 1980. Hunting success rates, foraging habits and prey selection of Peregrine Falcons migrating through Central Alberta. Can. Field-Nat. 94:371–382.

DIAMOND, J. M., AND J. W. TERBORGH. 1967. Observations on bird distribution and feeding assemblages along the Rio Callaria, Department of Loreto, Peru. Wilson Bull. 79:273–282.

DIXON, K. L. 1965. Dominance-subordination relationships in Mountain Chickadees. Condor 67:291–299.

DRENT, R., AND P. SWIERSTRA. 1977. Goose flocks and food finding: field experiments with Barnacle Geese in Winter. Wildfowl 28:15–20.

EMLEN, J. T. 1952. Flocking behavior in birds. Auk 69:160–170.

FEARE, C. J., G. M. DUNNET, AND I. J. PATTERSON. 1974. Ecological studies of the Rook (Corvus frugilegus L.) in north-east Scotland: food intake and feeding behavior. J. Appl. Ecol. 11:867–896.

FISHER, J., AND R. A. HINDE. 1949. The opening of milk bottles by birds. Br. Birds 42:347–358.

FRETWELL, S. D. 1969. Dominance behavior and winter habitat distribution in Juncos (Junco hyemalis). Bird-Banding 40:1–25.

FRINGS, H., M. FRINGS, B. COX, AND L. PEISSNER. 1955. Auditory and visual mechanisms in food-finding behavior of the Herring Gull. Wilson Bull. 67:155–170.

FUCHS, E. 1977. Kleptoparasitism of Sandwich Terns (Sterna sandvicensis) by Black-headed Gulls (Larus ridibundus). Ibis 119:183–190.

GADIL, M. 1972. The function of communal roosts: relevance of mixed roosts. Ibis 114:531–533.

GASTON, A. J. 1973. The ecology and behavior of the Long-tailed Tit. Ibis 115:330–351.

GIBB, J. 1960. Populations of tits and goldcrests and their food supply in pine plantations. Ibis:163–208.

GOLDMAN, P. 1980. Flocking as a possible predator defense in Dark-eyed Juncos. Wilson Bull. 92:88–95.

GRADWOHL, J., AND R. GREENBERG. 1980. The formation of antwren flocks on Barro Colorado Island, Panama. Auk 97:385–395.

GREIG-SMITH, P. W. 1978a. Imitative foraging in mixed species flocks of Seychelles birds. Ibis 120:233–235.

GREIG-SMITH, P. W. 1978b. The formation, structure and function of mixed-species insectivorous bird flocks in West African savanna woodland. Ibis 120:284–297.

HAMILTON, T. H., AND R. H. BARTH, JR. 1962. The biological significance of seasonal change in male plumage appearance in some new world migratory bird species. Am. Nat. 46:129–144.

HAMILTON, W. D. 1971. Geometry for the selfish herd. J. Theor. Biol. 31:295–311.

HAMERSTROM, F. N. 1942. Dominance in winter flocks of chickadees. Wilson Bull. 54:32–42.

HERRERA, C. M. 1979. Ecological aspects of heterospecific flock formation in a Mediterranean passerine bird community. Oikos 33:85–96.

HINDE, R. A. 1952. The behavior of the Great Tit (Parus major) and some other related species. Behav. Suppl. 2:1–201.

HORN, H. S. 1968. The adaptive significance of colonial nesting in the Brewer's Blackbird (Euphagus cyanocephalus). Ecology 49:682–690.

HOWE, H. F. 1977. Bird activity and seed dispersal of a tropical wet forest tree. Ecology 58:539–550.

JENNINGS, T., AND S. M. EVANS. 1980. Influence of position in the flock and flock size on vigilance in the starling, Sturnus vulgaris. Anim. Behav. 28:22–31.

JONES, S. E. 1977. Coexistence in mixed species antwren flocks. Oikos 29:366–375.

KARR, J. R. 1971. Structure of avian communities in selected Panama and Illinois habitats. Ecol. Monogr. 41:207–233.

KENWARD, R. E. 1978. Hawks and doves: factors affecting success and selection in Goshawk attacks on Wood Pigeons. J. Anim. Ecol. 47:449–460.

KLOPFER, P. H. 1957. Empathic learning in ducks. Am. Nat. 91:61–63.

KLOPFER, P. H. 1959. Social interactions in discrimination learning with special reference to feeding behavior in birds. Behaviour 14:282–299.

KOOPMAN, B. O. 1956. The theory of search, II. Target detection. Operations Res. 4:503–531.

KREBS, J. R. 1973. Social learning and the significance of mixed-species flocks of chickadees (Parus spp.). Can. J. Zool. 51:1275–1288.

KREBS, J. R., H. M. MACROBERTS, AND J. M. CULLEN. 1972. Flocking and feeding in the Great Tit Parus major—an experimental study. Ibis 114:507–530.

KUSHLAN, J. A. 1978. Feeding ecology of wading birds. Pp. 249–298, In A. Sprunt, IV, J. C. Ogden, and S. Winckler (eds.), Wading birds. National Audubon Society Res. Rep. No. 7.

LACK, D. 1954. The Natural Regulation of Animal Numbers. Oxford Press, London.

LACK, D. 1968. Ecological Adaptions for Breeding Birds. Methuen Press, London.

LAZARUS, J. 1972. Natural selection and the functions of flocking in birds: a reply to Murton. Ibis 114:556–558.

LAZARUS, J. 1979a. The early warning function of flocking in birds: an experimental study with captive Quelea. Anim. Behav. 27:855–865.

LAZARUS, J. 1979b. Flock size and behaviour in captive Red-billed Weaverbirds (Quelea quelea): Implications for social facilitation and the functions of flocking. Anim. Behav. 27:855–865.

LECK, C. F. 1971. Measurement of social attractions between tropical passerine birds. Wilson Bull. 83:278–283.

LOCKIE, J. D. 1956. Winter fighting in feeding flocks of Rooks, Jackdaws, and Carrion Crows. Bird Study 3:180–190.

MACAN, T. T. 1965. Predation as a factor in the ecology of water bugs. J. Anim. Ecol. 34:691–698.

MACARTHUR, R. 1972. Geographical Ecology: Patterns of Species Distribution. Harper Row, New York.

MACDONALD, D. W., AND D. G. HENDERSON. 1977. Aspects of the behavior and ecology of mixed-species bird flocks in Kashmir. Ibis 119:481–493.

MARKGREN, M. 1960. Fugitive reactions in avian behavior. Acta Vertebr. 2:1–160.

McCLURE, H. E. 1967. The composition of mixed species flocks in lowland and sub-montane forests of Malaya. Wilson Bull. 79:131–154.

MILLER, R. C. 1922. The significance of the gregarious habit. Ecology 3:122–126.

MORSE, D. H. 1967. Foraging relationships of Brown-headed Nuthatches and Pine Warblers. Ecology 48:94–103.

MORSE, D. H. 1970. Ecological aspects of some mixed-species foraging flocks of birds. Ecol. Monogr. 40:119–168.

MORSE, D. H. 1973. Interactions between tit flocks and Sparrowhawks (Accipiter nisus). Ibis 115:591–593.

MORSE, D. H. 1974. Niche breadth as a function of social dominance. Am. Nat. 108:818–830.

MORSE, D. H. 1978. Structure and foraging patterns of flocks of tits and associated species in an English woodland during the winter. Ibis 120:298–312.

MORTON, E. S. 1973. Evolutionary advantages and disadvantages of fruit eating in tropical birds. Am. Nat. 107:8–22.

MORTON, E. S. 1979. A comparative survey of avian social systems in northern Venezuelan habitats. Pp. 233–259, In J. F. Eisenberg (ed.), Ecology of the Northern Neotropics. Smithsonian Press, Washington, D.C.

MOYNIHAN, M. 1960. Some adaptations which help to promote gregariousness. Proc. 12th Int. Ornithol. Congr., pp. 523–541.

MOYNIHAN, M. 1962a. The organization and probable evolution of some mixed species flocks of neotropical birds. Smithson. Misc. Collect. 143:1–140.

MOYNIHAN, M. 1962b. Display patterns of tropical American "nine-primaried" songbirds I. Chlorospingus. Auk 79:310–344.

MUNN, C. A., AND J. W. TERBORGH. 1979. Multi-species territoriality in neotropical foraging flocks. Condor 81:338–347.

MURTON, R. K. 1971. Why do some bird species feed in flocks? Ibis 113:534–536.

MURTON, R. K., A. J. ISAACSON, AND N. J. WESTWOOD. 1971. The significance of gregarious feeding behavior and adrenal stress in a population of Wood-Pigeons (Columba palumbus). J. Zool. (Lond.) 165:53–84.

MURTON, R. K., C. F. B. COOMBS, AND R. J. P. THEARLE. 1972. Ecological studies of the feral pigeon Columba livia II. Flock behavior and social organization. J. Appl. Ecol. 9:875–889.

NEWTON, I. 1973. Studies of Sparrowhawks. Br. Birds 66:271–278.

NILSSON, S. G. 1979. Seed density, cover, predation and the distribution of birds in a beech wood in southern Sweden. Ibis 121:177–185.

ODUM, E. P. 1942. Annual cycle of the Black-capped Chickadee—3. Auk 59:499–531.

OLSON, F. C. W. 1964. The survival value of fish schooling. J. Cons. Int. Explor. Mer 29:115–116.

ORIANS, G. H. 1969. Age and hunting success in the Brown Pelican (Pelecanus occidentalis). Anim. Behav. 17:316–319.

PAGE, G., AND D. F. WHITACRE. 1975. Raptor predation on wintering shorebirds. Condor 77:73–83.

PARTRIDGE, L. 1976. Individual differences in feeding efficiencies and feeding preferences of captive Great Tits. Anim. Behav. 24:230–240.

PEARSON, D. L. 1971. Vertical stratification of birds in a tropical dry forest. Condor 73:46–55.

PEARSON, D. L. 1977. Ecological relationships of small antbirds in Amazonian bird communities. Auk 94:283–292.

PIELOWSKI, Z. 1959. Studies on the relationship: predator (Goshawk)-Prey (Pigeon). Pol. Akad. Nauk. Bull. Ser. Sci. Bio. 7:401–403.

PINOWSKI, J. 1959. Factors influencing the number of feeding Rooks (*Corvus frugilegus f.* L.) in various field environments. Ekol. Pol. Ser. A 7:434–482.

POWELL, G. V. N. 1974. Experimental analysis of the social value of flocking by Starlings (*Sturnus vulgaris*) in relation to predation and foraging. Anim. Behav. 22:501–505.

POWELL, G. V. N. 1977. Socioecology of mixed species flocks in a neotropical forest. Unpubl. Ph.D. dissert., University of California, Davis.

POWELL, G. V. N. 1979. Structure and dynamics of interspecific flocks in a mid-elevation neotropical forest. Auk 96:375–390.

RAND, A. L. 1954. Social feeding behavior of birds. Fieldiana Zool. 36:5–71.

RECHER, H. F., AND J. A. RECHER. 1969. Comparative foraging efficiency of adult and immature Little Blue Herons (*Florida caerulea*). Anim. Behav. 17:320–322.

RUBENSTEIN, D. I., R. J. BARNETT, R. S. RIDGELY, AND P. H. KLOPFER. 1977. Adaptive advantages of mixed-species feeding flocks among seed eating finches in Costa Rica. Ibis 119:10–21.

RUDEBECK, G. 1950. The choice of prey and modes of hunting of predatory birds with special reference to their selective effect 1. Oikos 3:65–88.

RUDEBECK, G. 1951. The choice of prey and modes of hunting of predatory birds with special reference to their selective effect 2. Oikos 4:200–231.

SABINE, W. S. 1949. Dominance in winter flocks of juncos and tree sparrows. Physiol. Zool. 22:64–85.

SALT, G. W., AND D. E. WILLARD. 1971. The hunting behavior and success of Forester's tern. Ecology 52:989–998.

SASVARI, L. 1979. Observational learning in Great, Blue, and Marsh Tits. Anim. Behav. 27:767–771.

SCHIPPER, W. J. A., L. S. BUURMA, AND P. H. BOSSENBROEK. 1975. Comparative study of hunting behavior in wintering Hen Harriers *Circus cyaneus* and Marsh Harriers *Circus aeruginosus*. Ardea 63:1–29.

SEALY, S. G. 1973. Interspecific feeding assemblages of marine birds off British Columbia. Auk 90:796–802.

SHORT, L. L., JR. 1961. Interspecies flocking of birds of montane forest in Oaxaca, Mexico. Wilson Bull. 73:341–347.

SIEGFRIED, W. R. 1971. Communal roosting of the cattle egret. Trans. R. Soc. S. Afr. 39:419–443.

SIEGFRIED, W. R., AND B. D. J. BATT. 1972. Wilson's Phalaropes forming feeding association with shovelers. Auk 89:667–668.

SKUTCH, A. F. 1969. Life Histories of Central American Birds III. Pac. Coast Avif. No. 35.

SKUTCH, A. F. 1977. A Birdwatcher in Tropical America. University Texas Press, Austin, Texas.

SLUD, P. 1960. The birds of finca "La Selva" Costa Rica: a tropical wet forest locality. Bull. Am. Mus. Nat. Hist. 121:53–148.

SMITH, J. N. M., AND H. P. A. SWEATMAN. 1974. Food searching behavior of titmice in patchy environments. Ecology 55:1216–1232.

SULLIVAN, K. A. 1984. Information exploitation in mixed species flocks. Anim. Behav. In press.

TERBORGH, J., AND J. N. DIAMOND. 1970. Niche overlap in feeding assemblages of New Guinea birds. Wilson Bull. 82:29–52.

THOMPSON, W. A., J. VERTINSKY, AND J. R. KREBS. 1974. The survival value of flocking in birds: a simulation model. J. Anim. Ecol. 43:785–820.

TINBERGEN, L. 1946. De sperwer als roofvijand van zangvogels. Ardea 34:1–213.

TINBERGEN, N. 1951. The Study of Instinct. Clarendon Press, Oxford, England.

TINBERGEN, N. 1952. "Derived" activities, their causation, biological significance, origin, and emancipation during evolution. Q. Rev. Biol. 27:1–32.

TINBERGEN, N. 1959. Comparative studies on the behavior of gulls (Laridae): a progress report. Behaviour 15:1–70.

TREISMAN, M. 1975. Predation and the evolution of gregariousness. I. Models for concealment and evasion. Anim. Behav. 23:779–800.

TRIVERS, R. L. 1971. The evolution of reciprocal altruism. Q. Rev. Biol. 46:35–57.

TURNER, E. R. A. 1965. Social feeding in birds. Behaviour 24:1–46.

VERBEEK, N. A. M. 1977. Comparative feeding of immature and adult herring gulls. Wilson Bull. 89:415–421.

VINCE, M. A. 1956. "String-pulling" in birds. I. Individual differences in wild adult Great Tits. Anim. Behav. 4:111–116.

VINE, I. 1971. Risk of visual detection and pursuit by a predator and the selective advantage of flocking behavior. J. Theor. Biol. 30:405–422.

WARD, P. 1965. Feeding ecology of the Black-faced Dioch (*Quelea quelea*) in Nigeria. Ibis 107:173–214.

WILEY, R. H. 1971. Cooperative roles in mixed flocks of antwrens (Formicariidae). Auk 88:881–892.

WILEY, R. H. 1980. Multispecies antbird societies in lowland forests of Surinam and Ecuador: stable membership and foraging differences. J. Zool. (Lond.) 191:127–145.

WILLIS, E. O. 1960. Red-crowned Ant-Tanagers, Tawny-crowned Greenlets, and forest flocks. Wilson Bull. 72:105–106.

WILLIS, E. O. 1966. Competitive exclusion and birds at fruiting trees in western Colombia. Auk 83:479–480.

WILLIS, E. O. 1967. The behavior of bicolored antbirds. Univ. Calif. Publ. Zool. 79:1–132.

WILLIS, E. O. 1972a. The behavior of spotted antbirds. Ornithol. Monogr. No. 10.

WILLIS, E. O. 1972b. Do birds flock in Hawaii, a land without predators? Calif. Birds 3:1–8.

WILLIS, E. O. 1973. Local distribution of mixed flocks in Puerto Rico. Wilson Bull. 85:75–77.

WILLIS, E. O., AND Y. ONIKI. 1978. Birds and army ants. Ann. Rev. Ecol. Syst. 9:243–263.

WINTERBOTTOM, J. M. 1943. On woodland bird parties in Northern Rhodesia. Ibis 85:437–442.

WINTERBOTTOM, J. M. 1949. Mixed bird parties in the tropics, with special reference to northern Rhodesia. Auk 66:258–263.

WOOLFENDEN, G. E., S. C. WHITE, R. L. MUMME, AND W. B. ROBERTSON, JR. 1976. Aggression among starving cattle egrets. Bird-Banding 47:48–53.

ZAHAVI, A. 1971. The function of pre-roost gatherings and communal roosts. Ibis 113:106–109.

COMMUNITY ORGANIZATION AND ECOLOGY OF BIRDS OF HIGH ELEVATION HUMID FOREST OF THE BOLIVIAN ANDES

J. V. Remsen, Jr.

Museum of Zoology, Louisiana State University, Baton Rouge, Louisiana 70893 USA

ABSTRACT. The avian community composition, and foraging behavior and social organization of birds of two humid forest sites near timberline in the Bolivian Andes were studied and compared to the same features of montane communities in other regions, particularly New Hampshire. Most of the difference in species richness between the tropical and temperate latitude sites can be explained by the addition of "tropical" resources not specially used by, or unavailable to, temperate birds, i.e., fruit, nectar, epiphytic vegetation, and bamboo thickets. When Andean species using these resources are eliminated, the overall species richness between the two regions is very similar, as are other community characteristics, such as use of foraging strata of different heights. The foraging maneuvers used by Andean birds are also discussed in detail. Almost 50 percent of the Andean species are regular participants in mixed-species flocks. In this respect, as well as with regard to the finer categories of social organization, the Andean community is very similar to one in montane Costa Rica but differs in important ways from others in Venezuela and New Hampshire.

RESUMEN. Se estudió la composición de la comunidad avícola y los comportamientos de forraje y organización social de aves en dos sitios de bósques húmedos cercanos al límite superior boscoso ("timberline") en los Andes bolivianos y se los comparó con los de otras comunidades montañosas similares en otras regiones, particularmente en Nueva Hampshire, E.E.U.U. La mayoría de las diferencias en la riqueza de especies entre las áreas de latitudes templadas y tropicales se pueden explicar por la presencia de recursos "tropicales" adicionales no especialmente usados o fuera del alcance de las aves de regiones templadas (por ej. frutas, néctar, vegetación epífita y cañaverales). Cuando se eliminan las especies andinas que usan esos recursos ls riqueza total de especies entre las dos regiones es muy similar, así como lo son otras características de la comunidad tales como el uso de estratos de forraje de deferentes alturas. También se discuten en detalle las maniobras de forraje usadas por las especies andinas. Casi el 50% de las especies andinas participan de las bandadas de varias especies de manera regular. En este aspecto, así como con respecto a las categorías más detalladas de organización social, la comunidad andina es muy similar a una en montañas costarricences, pero diferente en aspectos importantes de otras en Venezuela y Nueva Hampshire, en el noreste de E.E.U.U.

The problem of latitudinal differences in bird species diversity has often been investigated by comparing properties of land bird communities of tropical lowland regions to those of communities at temperate latitudes (Ricklefs 1966; Orians 1969; Howell 1971; Karr 1971; Faaborg 1980; Terborgh 1980; Greenberg 1981). Only comparisons by Stiles (1978) for alder habitats and by Fjeldså (1981) for wetlands have focused on communities in the Neotropics other than those of lowland areas. My purpose in this paper is to provide insight into latitudinal diversity gradients by comparing relatively low diversity bird communities in montane forests of tropical and temperate regions, using methods similar to those of Karr (1975) and Terborgh (1980). I also present information on the foraging ecology of birds of high elevation forest of the humid Andes and compare it with that available from similar habitats elsewhere.

STUDY SITES AND METHODS

Two localities in the upper "Temperate Zone" of the humid Andes of Dpto. La Paz, Bolivia were studied, 1 km south of Chuspipata, 3050 m (hereafter referred to as Chuspipata), and Cotapata, 4.5 km WNW Chuspipata, 3300 m (hereafter, Cotapata). Although these sites are within 5 km of one another, they differ in the following important ways, (1) air temperatures are substantially lower at the upper site (mean daily lows 2°C versus 8°C, mean daily highs 10°C versus 14°C); (2) average canopy height is lower at the upper site (5–12 m vs 14–17 m);

(3) tree species richness is much lower at the upper site, where only four species make up an estimated 90 percent of the trees more than 5 m tall.

Both sites are dominated by humid forest in which trees are heavily laden with moss and other epiphytes. The four common tree species at Cotapata are *Myrica pubescens* (Myricaceae), *Brunellia* sp. (Brunelliaceae), and two species of *Clusia* (Clusiaceae). Also present are two species of *Weinmannia* (Cunnoniaceae), *Freziera* sp. (Theaceae), several species of *Solanum* (Solanaceae), *Schefflera* sp. (Araliaceae), and *Gaiodendron* sp. (Loranthaceae). At Chuspipata, these latter species all become more common as *Myrica, Brunellia,* and *Clusia* become less dominant; *Tibouchina* and other Melastomaceae also become common. The dense undergrowth at both sites is dominated by bamboo (*Chusquea*), but the following shrubs and vines are also common, especially in edge situations: *Cavendishia* sp. (and many other Ericaceae), *Baccharis* sp. (Compositae), *Monnina* sp. (Polygalaceae), *Syphocampylus* sp. (Campanulaceae), *Gunnera* sp. (Haloragidaceae), and *Bomarea* sp. (Amaryllidaceae). At Chuspipata, *Munnozia* sp. and *Barnadesia* sp. (Compositae) are also common. The terrain at both sites was extremely steep. A dirt road bisected each site, creating a substantial edge effect. Between 1979 and 1982, both sites were altered considerably by human disturbance and deteriorated markedly in their suitability as study areas. Cotapata is no longer a tenable site, and Chuspipata will not last much longer.

In spite of the disturbance, the avifaunas at both sites seemed to be intact in 1980 and 1981. Birds sensitive to human disturbance, such as guans, raptors, and parrots, were present, and to our knowledge, no species that should have been absent if the area were disturbed, was missing. High elevation forest in the Andes is naturally disturbed by frequent landslides, and so even in its pristine state, the canopy is seldom continuous and the forest often patchy.

Typically, skies at Cotapata remained clear until 10:00 to 12:00 hours, after which clouds poured over the ridge from the east; low overcast or dense fog persisted the rest of the afternoon and into the evening. My visit, from 27 May to 25 June 1981, apparently coincided with the early part of the dry season; rain was recorded on only nine of 28 days, and on only one day did it rain hard. Nevertheless, dew soaked the vegetation each morning, and dense fog kept vegetation damp or wet during the afternoon. Skies were clear on only five days. Frost and ice covered the vegetation on many mornings.

My visit to Chuspipata, 22 July to 10 August 1982, coincided with the height of the dry season. Except for afternoon fog on the ridges several hundred meters above camp, skies were generally clear, although during the last six days a major southern storm brought light to moderate rain that lasted through most daylight hours.

During the months of my visits, most insectivorous and nectarivorous bird species were not breeding, but at least some fraction of the population of many frugivorous species were breeding (as indicated by gonad size data: Appendix I). Each site was sampled intensively by mist-netting; as many as 44 nets were operated simultaneously at each site, for a total of 11,221 daylight net hours at the upper site and 6366 at the lower site. Nets were not closed at night, and so an equivalent number of nighttime hours were accumulated. Species not prone to mist net capture were hunted selectively. Complete species lists for both localities are given in Appendix I; these are the first inventories available for high elevation Andean forests. Details on new distributional records for Bolivia and natural history observations on some poorly known species have been published elsewhere (Parker et al. 1980; Cardiff and Remsen 1981; Remsen 1981; Remsen et al. 1982; Schulenberg and Remsen 1982).

Foraging data and other natural history information were accumulated on a daily basis from dawn to dusk, giving approximately 450 man-hours observation time at the upper site and 625 at the lower site. Visual observations and mist net lanes encompassed elevations that extended approximately 50 m above and below the upper site and 100 m above and below the lower site. All data were gathered within a 0.5 km radius of the campsite. Each prey capture attempt by a bird was considered as a separate foraging observation. No more than three consecutive foraging observations for a given bird were analyzed. Sample sizes for foraging observations were low relative to most studies from temperate latitudes, partly because the precipitous terrain and impenetrable undergrowth precluded following birds away from trails and roadsides. Furthermore, the highly clumped spatial distribution (in flocks) of nearly half of the species resulted in "feast or famine" data collection; during the few minutes when a flock was visible, it was impossible to get data on all species or individuals, and intervening periods of an hour or more could be spent in search of another flock. Most species not in flocks were secretive or difficult to observe (e.g., antpittas, tapaculos, spinetails, cotingas,

TABLE 1
DIET CATEGORY MEMBERSHIP FOR BIRDS FROM STUDY SITES IN THE ANDES AND WHITE
MOUNTAINS

Diet	Number of species (% locality total)		
	Chuspipata	Cotapata	Mt. Moosilauke
Fruit	10 (13.5)	8 (16.7)	0
Fruit + insects	10 (13.5)	6 (12.5)	0
Fruit + insects + nectar	1 (1.4)	1 (2.1)	0
Nectar + insects	11 (14.9)	8 (14.6)	0
Insects	38 (51.4)	22 (16.7)	14 (82.3)
Insects + seeds	3 (4.1)	2 (4.2)	3 (17.7)
Seeds	1 (1.4)	1 (2.1)	0
Total species	74	48	17

and undergrowth tyrannids). Researchers accustomed to following passerines for extended periods in habitats at temperate latitudes will be discouraged by the low yield of data per hour in the Andes.

For general comparisons with a high elevation temperate community, I chose Sabo's (1980) data from the White Mountains of New Hampshire, primarily because of the volume of foraging information available for the insectivorous species. For specific comparisons, I used a hypothetical community constructed by S. R. Sabo (in litt.) for a similar census effort in an area of similar size at 1200 m (approximately the same number of meters below timberline as Cotapata in Bolivia). The 18 species at 1200 m were: Sharp-shinned Hawk (*Accipiter striatus*), Spruce Grouse (*Dendragapus canadensis*), Yellow-bellied Flycatcher (*Empidonax flaviventris*), Common Raven (*Corvus corax*), Boreal Chickadee (*Parus hudsonicus*), Red-breasted Nuthatch (*Sitta canadensis*), Winter Wren (*Troglodytes troglodytes*), Swainson's Thrush (*Catharus ustulatus*), Gray-cheeked Thrush (*C. minimus*), Golden-crowned Kinglet (*Regulus satrapa*), Ruby-crowned Kinglet (*R. calendula*), Nashville Warbler (*Vermivora ruficapilla*), Magnolia Warbler (*Dendroica magnolia*), Yellow-rumped Warbler (*D. coronata*), Blackpoll Warbler (*D. striata*), White-throated Sparrow (*Zonotrichia albicollis*), Dark-eyed Junco (*Junco hyemalis*), and Purple Finch (*Carpodacus purpureus*). The forest at 1200 m is dominated by two conifers (*Abies balsamea, Picea rubens*) and two broadleaf trees (*Betula papyrifera, Sorbus americana*); the canopy height is similar to that at the Bolivian sites (Sabo 1980).

COMMUNITY COMPOSITION, SPECIES DIVERSITY, AND FORAGING SITE SELECTION

Excluding raptors and nocturnal species, resident bird species at the Bolivian and New Hampshire localities were assigned to one of seven diet categories (Table 1; Appendices I and II for details). Four diet categories that involve heavy use of fruit and nectar are strongly represented in the tropical latitude communities but are absent from the temperate latitude community. These four categories encompass 43 percent (at 3050 m) and 48 percent (at 3300 m) of all resident species at the tropical latitude localities. Thus, nearly half of the species in the Andean communities are dependent on resources not used in and virtually unavailable to the White Mountains community (S. R. Sabo, pers. comm.). When species in these four diet categories are eliminated, the species richness of the temperate and tropical latitude communities is much more similar, although the temperate community is still less speciose.

A further breakdown of the insectivorous species into foraging substrate categories (Fig. 1) shows that the five substrate categories in New Hampshire either have the identical number of species in Bolivia or differ by only one species. The eight "extra" insectivores at Cotapata all use one of five substrate categories not used by or unavailable to birds at the Mt. Moosilauke (White Mountains) site. Two of these substrates ("air" and "leaves and air") are obviously available at the New Hampshire site but are not used. Aerial foraging swifts and swallows are only rare visitors at such elevations, and the Olive-sided Flycatcher (*Contopus borealis*) is a scarce and local resident.

The other three substrate categories ("bamboo," "moss," and "dead leaves, moss, and bark") are more "tropical" in character in being largely unavailable for specialized use by

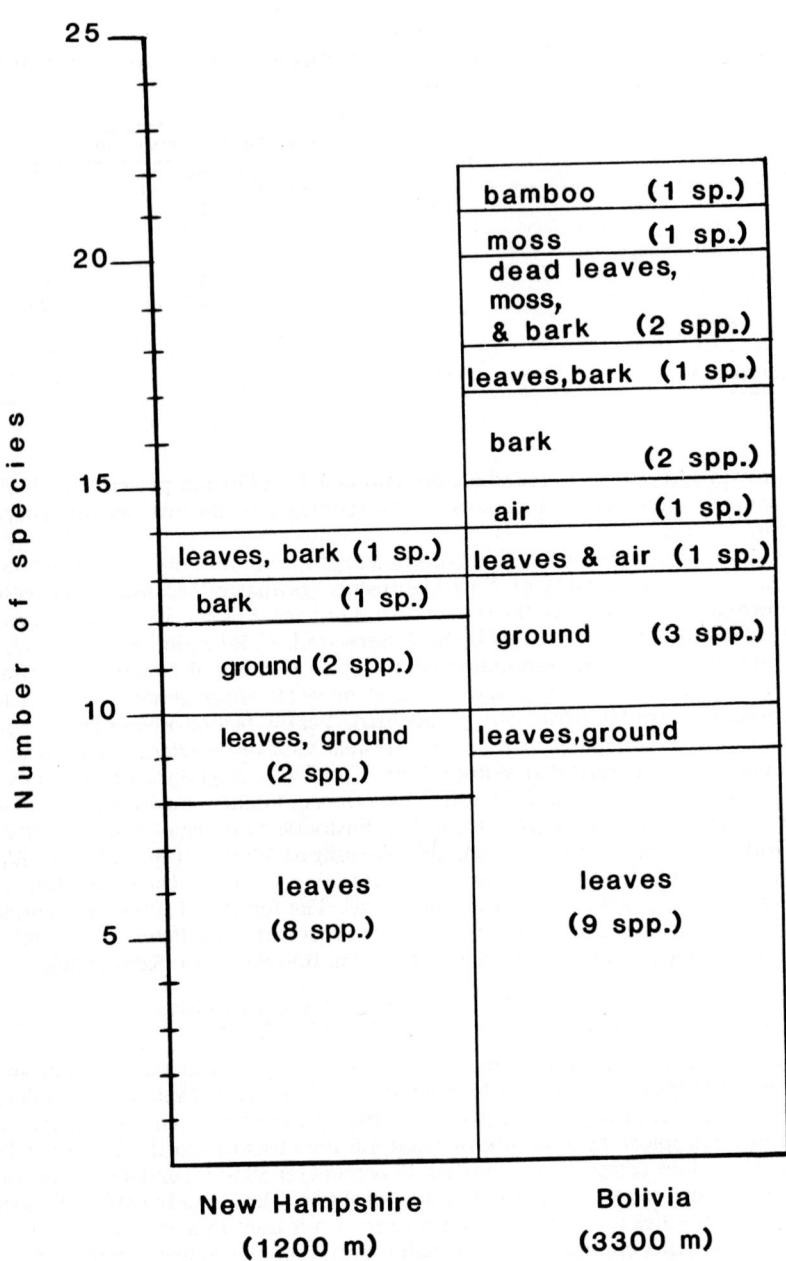

Substrate Use by Insectivores

FIG. 1.   Number of insectivorous species in various substrate use categories at near-timberline sites in the White Mountains and the Andes. Foraging substrates for each Andean species are those from Appendix II necessary to include at least 60% of a species' foraging observations; assignment to substrate category for White Mountains species was based on data from Sabo (1980). The "moss" category includes all epiphytic vegetation, but only one species, *Pseudocolaptes boissonneautii*, was recorded feeding in epiphytes other than mosses.

temperate latitude birds. Bamboo thickets throughout the Neotropics support a unique avifauna (Parker 1982; Parker and Remsen, unpubl. data), only one species of which, *Catamblyrhynchus diadema* (Hilty et al. 1979), occurs as high as 3300 m. *Hemispingus calophrys* also forages primarily in bamboo, as does *Cranioleuca albiceps* when gleaning foliage. The Andean subspecies of *Cacicus holosericeus* may also be a bamboo thicket specialist, but our observations are too few to make this certain. Although mosses and other epiphytes occur at most latitudes (lichen growth is heavy at Mt. Moosilauke; S. R. Sabo, pers. comm.), apparently only in the tropics do some birds specialize upon this substrate (Remsen and Parker, unpubl. data). At Cotapata, only one species, *Margarornis squamiger*, is a specialized moss-searcher, although *Delothraupis castaneoventris* forages almost exclusively on moss-covered branches when not feeding on fruit. At Chuspipata, three additional epiphyte-searching specialists are *Lepidocolaptes affinis*, *Pseudocolaptes boissonneautii*, and *Troglodytes solstitialis* (the latter two are rare visitors to Cotapata). *Chlorornis riefferii*, *Buthraupis montana*, *Tangara vassorii*, and *Cacicus chrysonotus* also search moss frequently at the two Bolivian sites, but none is exclusively insectivorous. Finally, two species, *Cranioleuca albiceps* and *Atlapetes rufinucha*, use a combination of substrates (dead leaves, moss, bark, and green leaves). The "tropical" nature of dead leaves suspended above ground is discussed elsewhere (Remsen and Parker 1984).

Once those species dependent on resources not available (at least on a year-round basis or to a degree that would allow specialization) at temperate latitudes are excluded, the timberline sites in Bolivia and New Hampshire are very similar in species richness. The New Hampshire site, however, contains both coniferous and broadleaf vegetation. These two types of vegetation are structurally very different, each with characteristic bird species, and so species richness would almost certainly be somewhat reduced at a site at 1200 m with only coniferous or only broadleaf vegetation. No such heterogeneity exists at the Bolivian sites, although, as S. R. Sabo (pers. comm.) has pointed out, frugivore diversity must be at least partly dependent on floristic diversity. Also, the analysis of the insectivorous species ignores those frugivorous and nectarivorous species that include insects in their diet. It it fair to exclude these species? For all but four of the 23 species involved at Cotapata, the answer is "probably." Except for *Cacicus chrysonotus*, the nectarivores are either hummingbirds or flowerpiercers, most of which take insects mainly from or near the flowers upon which they feed, or in the case of sallying and gleaning hummingbirds, capture insects that are presumably smaller than those used by other species at either the Andes or White Mountains sites. Among the six frugivores that also feed on insects, two species (*Buthraupis montanus* and *Delothraupis castaneoventris*) search for insects primarily in moss on tree branches, a substrate not used at the White Mountains site. This leaves only four frugivorous/insectivorous species (three flycatchers and a thrush) that search for insects in substrates (foliage, ground) used by White Mountains birds.

Why the pronounced difference in species richness between the 3300 and 3050 m sites in Bolivia (48 vs 74 resident species)? The species at the higher elevation site are mainly a subset of the species from the lower elevation site (with 16 of the residents at 3050 m also occurring as visitors to the 3300 site). Only six species (*Metallura aeneocauda*, *Myiophobus ochraceiventris*, *Turdus serranus*, *Hemispingus trifasciatus*, *Iridosornis jelskii*, and *Diglossa lafresnayii*) are residents at the 3300 m site but absent from or only visitors at 3050 m. *Turdus serranus* is common at elevations below 3050 m (LSUMZ specimens at 2575 m at Sacramento Alto, only a few km away), and so its rarity at 3050 m is puzzling. Of the five species truly restricted to the upper site, direct competition can possibly be invoked for restriction to upper elevations for only one, *Hemispingus trifasciatus*. At about 3100 m, this species is replaced by *H. superciliaris*, which is extremely similar in morphology and almost all aspects of foraging behavior, e.g., substrate selection (Appendix I), foraging maneuvers, and foraging position (see below). The only substantial difference between the two is in foraging height, and this is merely a consequence of differing canopy heights at the two elevations. Both species are seen in the same flocks at about 3100 m. For the remaining four species restricted to upper elevations, no congeneric ecological counterpart replaces them at the 3050 m site.

The proportional decrease in species in each of the major diet categories from 3050 to 3300 m is very similar to the overall proportional decrease, 34 percent, in species richness (Table 1). The decreases within diet categories range from 30 percent for "nectar and insects" to 42 percent for "insects." Thus, the depauperization of the avifauna is spread rather evenly among the diet categories, in contrast to the findings of Terborgh (1977) over a broader elevational gradient in the Peruvian Andes. A similar analysis for foraging substrate categories is made difficult by the fewer species per category. The decreases, however, in species in two categories,

TABLE 2

DEGREE OF SPECIALIZATION OF SOME COMMON INSECTIVORES ON UPPER OR LOWER LEAF
SURFACES AT ANDEAN STUDY SITES

	Percent of foraging on upper surface[1]
Undergrowth species	
Hemispingus calophrys	50 N.S.
Thlypopsis ruficeps	23***
Basileuterus luteoviridis	20***
Basileuterus signatus	26*
Canopy species	
Gleaners	
Hemispingus superciliaris	46 N.S.
Hemispingus trifasciatus	54 N.S.
Conirostrum sitticolor[2]	44 N.S.
Sallyers	
Mecocerculus leucophrys	24**
Mecocerculus stictopterus	21**
Myiophobus ochracieventris	47 N.S.
Myioborus melanocephalus	31**
Edge species	
Ochthoeca rufipectoralis	82**

[1] The percent of all foraging bouts observed directed at upper surface is followed by significance levels for a two-tailed Binomial Test: N.S. = $P > 0.05$; * = $P \le 0.05$; ** = $P \le 0.01$; *** = $P \le 0.001$.

[2] Moynihan (1963) considered this species to be an undergrowth species that frequently fed at flowers (in Ecuador).

"moss" and "air," do seem disproportionately high (75%). Not much significance can be attached to the difference in aerial foragers, because one is a nocturnal caprimulgid that could have been missed at the upper site, and two other species at the lower site were rare. The difference in moss-searchers, however, is more noteworthy, especially because as a substrate, moss is at least as extensive at the upper site as at the lower. Perhaps insects within moss become disproportionately depleted at the upper site.

Thus, except for moss-searchers, the decrease in species richness within various diet and foraging categories is more or less proportional to the overall decrease in species richness. This implies that the overall decline reflects general resource depletion (as a result of colder temperatures?) at upper elevations.

Greenberg and Gradwohl (1980) pointed out that many lowland tropical bird species distribute their search efforts unequally with respect to upper and lower surfaces of leaves, with a majority of species, particularly those of the undergrowth, specializing on lower surfaces. Their preliminary data from higher elevations showed a shift in insect abundance from lower to upper leaf surfaces, and their model consequently predicts a shift in searching effort by birds toward the upper surface as one moves up in elevation.

Comparison of upper surface versus lower surface searching by birds at the Bolivian sites (Appendix II, Table 2) with similar data for lowland species from Panama (Greenberg and Gradwohl 1980) shows that (1) undergrowth species at both sites have similar (low) percent upper surface values; the exception among the Andean species is Hemispingus calophrys, which forages primarily in bamboo, the leaves of which are very differently shaped and arranged from those of woody undergrowth plants; (2) sallying species at the Andean sites have higher percent upper surface values than Panamanian sallying species (= "hoverers" in terminology of Greenberg and Gradwohl 1980); (3) canopy foliage-gleaners at the Andean sites show no demonstrable upper or lower surface preference. Trends 2 and 3 are consistent with the hypothesis that insects themselves show less preference for the lower surface at high than at low elevations. Unfortunately, too few bird species are involved for statistical analysis.

FORAGING MANEUVERS

The maneuvers used by the birds in gathering food and the frequency with which they were used were recorded for each species (Table 3). Most of the discussion below pertains only to those 33 species for which more than 10 observations were obtained. Data from other species about which nothing has been published are also included in Table 3.

Six species (19%) used only one type of maneuver. Sixteen species (49%) used four or more maneuver types. Using Levins (1968) niche breadth formula ($NB = 1/\Sigma\ [p_{ij}]^2$), the three most generalized species in terms of foraging maneuvers were the flycatchers *Myiophobus ochraceiventris* ($NB = 2.5$), *Mecocerculus leucophrys* (2.6), and *M. stictopterus* (3.0). The latter was the only species to use more than 50 percent (6 of 10) of the maneuver types (but it also had one of the largest sample sizes of observations). That three flycatchers should rank the highest in foraging maneuver "niche breadth" will surprise many investigators accustomed to the stereotyped foraging behavior exhibited by most temperate latitude tyrannids. Unfortunately, the small sample sizes for observations make interpretations of the NB values tenuous.

The maneuvers were ranked on a subjective scale of increasing acrobatic complexity and accompanying requisite morphological specialization (Table 3, from left to right). At one end of the scale is "pick," the maneuver in which the bird picks its food from a substrate without extending its legs or wings. Seventeen species (51%) used this maneuver for 50 percent or more of the observations. Thus, most species use most frequently the maneuver that requires the lowest degree of agility and energy expenditure. Although 50 percent of the species used the "reach out/up" maneuver (fully extending body by extending legs and neck to reach prey), only one species, *Basileuterus signatus,* used this more than any other maneuver. Two other species of the dense undergrowth, *Basileuterus luteoviridis* and the warbler-like tanager *Thlypopsis ruficeps,* used this maneuver frequently.

Only the bamboo-stem-climbing *Catamblyrhynchus diadema* used the "pull/rip" maneuver frequently. This species feeds primarily by opening bamboo stem and leaf nodes (Hilty et al. 1979). The moss-and-epiphyte-searching furnariid, *Pseudocolaptes boissonneautii,* regularly used the "probe" maneuver (thrusting bill deep into substrate with extension of neck). Its allospecies *P. lawrenceii* exhibits similar behavior in Costa Rica (Skutch 1969). *Cacicus chrysonotus* may also use this maneuver regularly (but N = only 5 foraging records).

Only *Basileuterus luteoviridis* frequently used the "lunge" maneuver (rapid lateral movement that appears to combine rapid leg movements and some wing propulsion; differs from "reach out/up" in that the bird moves along a perch rather than remaining stationary). The "reach down" maneuver is one in which the bird leans down, stretching much of its body below the level of the perch to search a substrate. The feet still grasp the upperside of the perch. It is similar to the "reach out/up" maneuver but is directed downward. This maneuver is employed by more than half (52%) of the species examined and is the predominant maneuver used by the tanager *Buthraupis montana.* It is also heavily used by another mossy-branch-searching tanager, *Delothraupis castaneoventris,* and a hummingbird, *Metallura tyrianthina.* (*M. aeneocauda* and the several *Diglossa* species may also use this maneuver frequently, but sample sizes of observations were small.)

The "hang" maneuver differs from "reach down" in that most of the bird's body is suspended below the level of the feet, and the feet grasp the underside of the perch. Although this maneuver is used by more species (19) than "reach down," it is a prominent part of the foraging repertoire of only *Conirostrum sitticolor,* which uses this maneuver in the majority of its foraging maneuvers, primarily to reach the tips and curled portions of *Brunellia* leaves (c.f. Moynihan 1963), and *Margarornis squamiger,* which frequently hitches along the undersides of mossy branches.

The "sally" maneuver (flight from perch to pick object from substrate or out of the air) is used by fewer than half (46%) of the species, but it is the second most frequently used foraging maneuver after "pick" when mean percent use over all 33 species is calculated.

The "hover-glean" maneuver is the same as "sally" except that the bird hovers briefly in front of the target substrate before returning to a perch. Both in terms of number of species that use it and mean percent use over all 33 species, it is much less widely used than "sally," presumably because of the difficulty and high energy expenditure of hovering flight. Only the flycatchers *Mecocerculus leucophrys* and *Mionectes striaticollis* employ "hover-glean" frequently.

The "hawk" maneuver is the searching and pursuing of prey in continuous flight. Only a swift and two swallows, hawking specialists, use this maneuver (the only one that they use). Presumably, the extremely high wing length to body weight ratio required for hawking flight would make other foraging maneuvers very difficult.

TABLE 3
Foraging Maneuvers Used by Birds in the Humid Temperate Zone of the Bolivian Andes[1]

Species	Pick	Reach out/up	Pull/rip	Probe	Lunge	Reach down	Hang	Sally	Hover-glean	Mean sally or hover-glean distance (m)	Hawk	Hover	N
Cypseloides rutilus											1.00		19
Pterophanes cyanoptera												1.00	11
Coeligena violifer												1.00	37
Heliangelus amethysticollis						.12		.12				.76	17
Chalcostigma ruficeps						.20						.80	5
Metallura aeneocauda						.60						.40	5
M. tyrianthina	1.00												28
Lepidocolaptes affinis	.60	.10				.10	.20						11
Schizoeaca harterti	.92					.05		.03					20
Cranioleuca albiceps	.73						.27						36
Margarornis squamiger	.59			.38			.03						30
Pseudocolaptes boissonneautii	.44	.22					.22		.11				32
Phyllomyias uropygialis	.50						.08	.42		0.3			9
Elaenia albiceps	.13						.03	.51	.33	0.6			12
Mecocerculus leucophrys	.14	.06					.09	.51	.20	0.3			30
M. stictopterus	.56						.11		.33	0.6			35
Mionectes striaticollis	.12					.02	.08	.60	.16	0.6			18
Myiophobus ochraceiventris	.03	.02						.89	.08	1.0			51
Ochthoeca rufipectoralis						.20		.80		1.1			36
O. frontalis								1.00		2.0			5
Myiotheretes fuscorufus											1.00		3
Notiochelidon murina											1.00		128
N. flavipes	.69	.13				.19	.10						26
Troglodytes solstitialis	.61	.13				.10	.11	.06					16
Atlapetes rufinucha	.32		.57				.11						31
Catamblyrhynchus diadema	.64	.14				.07	.06	.04					19
Hemispingus calophrys	.73	.12				.09	.03						28
H. superciliaris	.87	.05				.03	.01	.02					34
H. trifasciatus	.56					.10	.06						58
Thlypopsis ruficeps	.22	.33				.54							40
Buthraupis montana	.73	.09				.06	.02						35
Anisognathus igniventris		.10	.09			.10	.08	.05					42
Delothraupis castaneoventris	.53					.39	.08						13

TABLE 3
CONTINUED

Species	Proportion total foraging maneuvers									Mean sally or hover-glean distance (m)	Hawk	Hover	N
	Pick	Reach out/up	Pull/rip	Probe	Lunge	Reach down	Hang	Sally	Hover-glean				
Tangara vassorii	.67					.33							6
Diglossa baritula	.20			.20		.40		.20					10
D. lafresnayii	.67	.17				.33							9
D. carbonaria						.33	.50						6
D. cyanea	.64					.18	.18						17
Conirostrum sitticolor	.24	.04				.11	.57	.04					54
Myioborus melanocephalus	.25	.05				.01		.61	.08				88
Basileuterus luteoviridis	.62	.23			.01			.10	.04	1.1			70
B. signatus	.26	.42						.32					19
Cacicus chrysonotus	.83			.17									5
Mean frequency for all species	.36	.05	.02	.02	<.01	.12	.07	.15	.03		.07	.11	
Mean frequency for species for which N > 10 observations	.40	.06	.02	.01	<.01	.08	.07	.13	.04		.09	.10	

[1] Data for 3050 m and 3300 m sites combined; see text for descriptions of maneuvers.

The "hover" maneuver is the searching for food while hovering continuously (in contrast to "hover-glean," in which actual hovering is only a small component of the total maneuver). Hummingbirds are the only species that use this maneuver.

Climbing and hitching up branches were not included among the maneuvers because they are used more for moving between foraging sites than for prey capture itself. *Lepidocolaptes affinis* and *Margarornis squamiger* are specialized bark-climbers, and *Pseudocolaptes boissonneautii* and *Cranioleuca albiceps* are almost exclusively "hitchers" that use a combination of climbing and hopping motions to ascend stems but not large vertical branches. *Catamblyrhynchus diadema* also uses the hitching technique frequently.

Sabo's (1980) foraging data were taken in a way differing sufficiently from that in the present study to preclude formal comparisons. Nevertheless, it seems clear that all maneuvers used by the tropical latitude birds are also used by those at the temperate latitude site, except "hawk" and "hover" (as defined above). These are absent from the Mt. Moosilauke birds' repertoires only because swifts, hummingbirds, and swallows do not occur at the White Mountains study site (Sabo, in litt.).

FORAGING POSITIONS

Foraging position here refers to the location of feeding sites with respect to the ground and the canopy (Table 4). Very few deep undergrowth or terrestrial foragers were considered because of difficulty in observing these species. The discussion that follows is restricted to species for which I had more than 10 observations unless stated otherwise.

Except for the swallows and swift, the three species with greatest mean foraging heights are *Hemispingus superciliaris, Delothraupis castaneoventris,* and *Troglodytes solstitialis.* The three species with the smallest mean distance-to-canopy measures are the bush-top flycatcher *Ochthoeca rufipectoralis* and two hummingbirds (*Pterophanes cyanopterus, Coeligena violifer*) that characteristically feed in flowers on the outer surface of shrubs, mostly at forest edges.

The relative position (RP) of most canopy species in the foliage column (height above ground/height above ground + distance to canopy) ranges from .75 to .90 (Table 4). The values for most undergrowth species are biased upward by the relative ease of detection of these species when in edge situations, where canopy height is reduced.

Species foraging in the densest foliage (Table 4) are *Elaenia albiceps,* which feeds in the densest sections of the canopy, and three small insectivores that maneuver through dense foliage, *Thlypopsis ruficeps, Hemispingus trifasciatus,* and *H. superciliaris.* Bias against observations in dense undergrowth is responsible for the unexpectedly low values (open areas) for some insectivores of dense foliage (e.g., *Schizoeaca harterti, Basileuterus luteoviridis, B. signatus, Hemispingus calophrys*). Other than species with a high aerial component to their foraging, the species with the lowest foliage density values are the bark-foraging *Lepidocolaptes affinis* and the bush-top hummingbirds *Pterophanes cyanopterus* and *Coeligena violifer.*

To determine if foraging sites of bird species in the tropical and temperate sites differed with respect to distance above ground, I assigned primarily insectivorous species to a category based on the heights most frequently used (Table 5). For the Andes, only data from the 3300 m site were used because canopy height at the 3050 m was substantially greater than that at the 3300 m site or at Mt. Moosilauke, preventing meaningful comparisons using absolute heights.

The only notable difference between the tropical and temperate sites (Table 5) was the number of species in the understory zone, five in the Andes versus none in the White Mountains (although 10 of 13 insectivores at the White Mountains site used the understory on occasion). Otherwise, the two sites are very similar, but sample sizes of species per category are too small for statistical comparison.

FORAGING SOCIAL SYSTEMS

Using Buskirk's (1976) categories, I characterized the foraging social system of each resident species at the two Andean sites (Appendix I). As expected, almost every species was seen solitarily on occasion, and many species characterized as solitary were occasionally seen to accompany mixed-species flocks. Thus, I considered only the most frequently observed system, although two or more designations are given for particularly flexible species.

At both sites, nearly half of the species participated regularly in mixed-species flocks (Table 6), and because the small passerines that predominate in such flocks are for the most part the most abundant birds at the sites, a large majority of the individual birds seen were found in

TABLE 4
Characteristics of Foraging Sites Used by Andean Forest Birds

	Height (m) above ground ± s.d.	Distance (m) to canopy ± s.d.	Relative position[1]	Foliage density[2]	N
Pterophanes cyanoptera	2.2 ± 2.0	0.3 ± 0.4	.88	2.0 ± 0.9	11
Coeligena violifer	2.2 ± 2.4	0.3 ± 0.6	.88	2.1 ± 1.0	37
Heliangelus amethysticollis	3.2 ± 3.0	0.6 ± 1.4	.84	1.8 ± 1.0	17
Chalcostigma ruficeps	3.7 ± 1.5	1.9 ± 3.3	.66	2.4 ± 0.6	5
Metallura aeneocauda	1.2 ± 0.3	1.5 ± 1.1	.44	2.4 ± 1.1	5
M. tyrianthina	1.7 ± 1.0	1.0 ± 1.6	.63	2.4 ± 0.6	29
Piculus rivolii	6.6 ± 1.5	0.8 ± 1.0	.89	2.2 ± 1.0	6
Lepidocolaptes affinis	8.2 ± 5.0	3.7 ± 3.4	.69	1.6 ± 0.5	12
Margarornis squamiger	4.9 ± 3.8	2.4 ± 1.4	.67	2.6 ± 1.1	31
Cranioleuca albiceps	2.5 ± 1.1	4.4 ± 2.0	.36	3.0 ± 0.9	37
Pseudocolaptes boissonneautii	8.5 ± 3.4	4.0 ± 2.3	.68	2.7 ± 0.7	34
Schizoeaca harterti	1.4 ± 1.0	1.9 ± 2.0	.42	2.9 ± 1.2	33
Ochthoeca rufipectoralis	2.6 ± 3.2	0.0 ± 0.5	1.00	1.3 ± 1.1	38
O. frontalis	1.6 ± 0.9	3.2 ± 1.2	.33	2.4 ± 0.5	7
Myiotheretes fuscorufus	12.2 ± 0.0	0.5 ± 0.9	1.04	0.7 ± 1.2	3
Myiophobus ochraceiventris	4.7 ± 1.8	0.8 ± 0.9	.85	2.2 ± 0.7	52
Phyllomyias uropygialis	6.2 ± 5.4	4.3 ± 2.9	.59	2.6 ± 0.7	9
Elaenia albiceps	6.3 ± 2.3	0.7 ± 0.3	.90	3.3 ± 0.7	12
Mecocerculus leucophrys	6.2 ± 2.3	1.5 ± 0.9	.81	2.4 ± 0.8	31
M. stictopterus	8.3 ± 2.9	2.4 ± 2.5	.78	2.9 ± 0.4	35
Mionectes striaticollis	5.3 ± 1.7	1.0 ± 1.0	.84	2.8 ± 0.6	19
Troglodytes solstitialis	9.6 ± 5.9	3.1 ± 2.5	.76	2.8 ± 0.7	17
Atlapetes rufinucha	4.9 ± 4.2	3.2 ± 2.6	.60	2.4 ± 1.0	31
Catamblyrhynchus diadema	2.0 ± 1.3	3.4 ± 1.7	.37	2.9 ± 0.8	19
Hemispingus calophrys	3.4 ± 1.3	3.0 ± 2.8	.53	2.9 ± 0.4	28
H. superciliaris	10.6 ± 4.7	2.0 ± 1.5	.84	3.1 ± 0.8	34
H. trifasciatus	6.0 ± 2.4	1.4 ± 1.3	.81	3.1 ± 0.7	58
Thlypopsis ruficeps	4.1 ± 2.8	1.9 ± 2.5	.68	3.2 ± 0.7	40
Buthraupis montana	6.6 ± 2.3	1.8 ± 1.2	.79	2.5 ± 1.0	35
Anisognathus igniventris	5.8 ± 2.9	1.5 ± 1.3	.80	3.0 ± 0.9	42
Delothraupis castaneoventris	10.1 ± 3.6	2.5 ± 1.4	.80	2.3 ± 1.0	14
Tangara vassorii	8.9 ± 3.5	1.6 ± 1.4	.85	2.8 ± 0.8	6
Diglossa baritula	6.9 ± 2.5	3.1 ± 3.3	.69	1.9 ± 1.5	10
D. lafresnayii	2.1 ± 1.2	0.8 ± 1.5	.75	2.6 ± 0.9	9
D. carbonaria	6.4 ± 1.1	0.5 ± 0.3	.93	3.2 ± 0.8	6
D. cyanea	3.9 ± 2.8	1.4 ± 1.7	.73	2.6 ± 0.9	18
Conirostrum sitticolor	5.7 ± 2.9	1.1 ± 1.6	.84	2.2 ± 0.6	54
Myioborus melanocephalus	7.0 ± 3.6	2.1 ± 3.9	.77	1.7 ± 1.3	88
Basileuterus luteoviridis	3.3 ± 2.5	3.0 ± 2.4	.52	2.6 ± 0.8	71
B. signatus	1.4 ± 2.0	1.8 ± 1.8	.44	2.8 ± 0.5	19
Cacicus chrysonotus	8.2 ± 5.5	3.9 ± 1.1	.68	2.0 ± 1.0	5
Notiochelidon murina	21.4 ± 14.2	+10.5 ± 7.2	1.49	0.0 ± 0.0	128
N. flavipes	18.3 ± 9.2	+3.3 ± 2.7	1.18	0.0 ± 0.0	26
Cypseloides rutilus	85.7 ± 34.5	+61.0 ± 29.3	1.71	0.0 ± 0.0	19

[1] Relative position in the foliage column is measured by dividing the height above ground at which a bird is recorded by the height of the canopy at that point; species with values close to 1.00 are high canopy foragers and those with values close to 0 are near-ground foragers.
[2] Foliage density was estimated for each observation on a subjective scale from 0 to 5. A value of 0 indicates no vegetation within a ½ m radius sphere around the foraging site and a value of 5 indicates extremely dense vegetation: no light passes through this sphere.

mixed-species flocks. One can often walk for an hour without seeing a bird until a mixed-species flock is encountered.

At the 3300 m site, four types of mixed-species flocks were noted, (A) small insectivores in the canopy, (B) small frugivores in the canopy, (C) small insectivores in the undergrowth, and (D) large omnivores in the canopy.

A typical Type A flock consisted of *Hemispingus trifasciatus* (4–8 individuals), *Myiophobus ochraceiventris* (4–5), *Myioborus melanocephalus* (2), *Conirostrum sitticolor* (2), *Margarornis squamiger* (2), and *Delothraupis castaneoventris* (1). A typical Type B flock included *Aniso-*

TABLE 5

FORAGING HEIGHT STRATA OF INSECTIVOROUS BIRDS AT STUDY SITES IN THE ANDES AND
WHITE MOUNTAINS

	No. of species (% locality total)	
Vertical zone[1]	Andes (3300 m)[2]	White Mtns.
Forest floor	3 (14)	3 (23)
Understory, 0.1–2 m	6 (29)	0
Midstory, 2–5 m	6 (29)	4 (31)
Midcanopy, 5–8 m	6 (29)	5 (38)
Upper canopy, 8 m	0 (0)	1 (8)
Total	21	13[3]

[1] See text for explanation of assignment of species to categories.
[2] "Visitors" (Appendix I) and aerial foragers were not included in the analysis, but several species not listed in Table 4 were included: *Grallaria squamigera, G. rufula,* and *Scytalopus magellanicus* in the "forest floor" category; *Anairetes parulus* and *Cacicus holosericeus* in the "understory" category; and *Veniliornis nigriceps* in the "midstory" category.
[3] *Corvus corax* was not included in the analysis because of lack of data.

gnathus igniventris (4–8), *Iridosornis jelskii* (2), *Diglossa cyanea* (1–4), *Mionectes striaticollis* (1), and often *Elaenia albiceps* (1–2). A typical Type C flock included *Basileuterus luteoviridis* (4), *Atlapetes rufinucha* (4), *Cranioleuca albiceps* (2), and *Catamblyrhynchus diadema* (2–4); these were often accompanied by *Ochthoeca rufipectoralis* (2) in edge situations, and *Hemispingus calophrys* (4–8). The boundaries between these first three flock types were not rigid, and two or more types often mingled temporarily, especially A and B in *Myrica* trees. Type D flocks, however, did not intermingle with the others and were much less frequently noted. A typical Type D flock included *Buthraupis montana* (6–12), *Cacicus chrysonotus* (5–10), *Piculus rivolii* (1), and often *Cyanolyca viridicyana* (2–4). All but *Piculus rivolii* were seen frequently in monospecific flocks as well.

Within A, C, and D flocks, ecological differences in feeding behavior between member species were obvious. For example, in Type A flocks, *Hemispingus trifasciatus* gleaned leaves in dense foliage, particularly *Myrica* trees; *Conirostrum sitticolor* hung acrobatically to reach leaves, particularly those of *Brunellia,* generally out of reach of a "normal" foliage-gleaner; *Myiophobus ochraceiventris* sallied to foliage, especially *Clusia* leaves, for insects and to dehiscent fruit clusters of *Brunellia; Margarornis squamiger* searched moss on branches of all sizes and inclinations; *Delothraupis castaneoventris* searched horizontal mossy branches; and *Myioborus melanocephalus* pursued flushed insects through the air, often for many meters. As expected, the relative abundance of each species within a flock seemed in proportion to the available volume of favored substrate. Thus, *Hemispingus trifasciatus,* the most common bird, foraged primarily in the most abundant tree, *Myrica,* and *Delothraupis castaneoventris,* which focused most of its effort upon a relatively scarce substrate, horizontal mossy branches, was the least common.

Ecological segregation within Type B flocks was less obvious, because all species except *Iridosornis jelskii* ate the small fruits of *Myrica.* The sallying tyrannids presumably were able to reach fruit that the perch-gleaning tanagers could not.

At the 3050 m site, the same four flock types were noted, with similar but richer species composition. The greatest differences in composition between the two sites were in Type A flocks, which at 3050 m included *Hemispingus superciliaris* (4–6), *H. xanthophthalmus* (2–4), *Myioborus melanocephalus* (2), *Mecocerculus leucophrys* (2), *M. stictopterus* (2), *Delothraupis castaneoventris* (1), *Troglodytes solstitialis* (1), and regularly, *Lepidocolaptes affinis* (2; below 3100 m only), *Margarornis squamiger* (2; above 3000 m only), *Pseudocolaptes boissonneautii* (1–2), *Chlorornis riefferii* (1), and *Veniliornis nigriceps* (2). A typical Type B flock was composed of *Anisognathus igniventris* (4–8), *Tangara vassorii* (2–4), *Diglossa cyanea* (1–4), and often *Thraupis cyanocephala* (4–6). A typical Type C flock was similar to one at 3300 m except that *Thlypopsis ruficeps* (1–2), and, often, in edge situations, *Myioborus melanocephalus* (2) were added. A typical Type D flock was the same as that at 3300 m, but usually more *Buthraupis montana* were present, sometimes as many as 25.

During the breeding season, when Sabo's (1980) Mt. Moosilauke study was conducted,

TABLE 6

SUMMARY OF AVIAN SOCIAL FORAGING SYSTEMS FOR STUDY SITES IN THE ANDES AND COASTAL
MOUNTAINS OF VENEZUELA

Interspecific[1]	Intraspecific[2]	Chuspipata	Cotapata	Costa Rica[3]	Guatopo[4]
N	N	18 (24.7)	14 (29.2)	24 (42.1)	64 (33.0)
N	N, P; P	7 (9.6)	4 (8.3)	2 (3.5)	32 (16.5)
N	N, S; P, S; N, P, S	7 (9.6)	5 (10.4)	0	0
N	S; L; S, L; P, L	7 (9.6)	2 (4.2)	4 (7.0)	35 (18.0)
F	N; P; N, P	16 (21.9)	9 (18.8)	18 (31.6)	38 (19.6)
F	N, S; P, S; P, L	3 (4.1)	4 (8.3)	2 (3.5)	1 (0.5)
F	S; L; S, L	8 (11.0)	4 (8.3)	7 (12.3)	16 (8.2)
N, F	N; N, P; P; N, S; P, S; N, P, S	4 (5.5)	4 (8.3)	0	6 (3.1)
N, F	S; L; S, L	3 (4.1)	2 (4.2)	0	2 (1.0)
	Total species	73	48	57	194

[1] Refers to interspecific sociality: N = does not form flocks with other species; F = flocks with other species.
[2] Refers to intraspecific sociality: N = not social, solitary; P = pairs; S = small flocks, 3–5 birds; L = large flocks, 6+ birds; see Appendix I.
[3] Data from Buskirk (1976).
[4] Data from Morton (1979).

virtually all species are paired and defending territories; thus, mixed-species flocks are nearly non-existent during the temperate summer, at least until the end of the breeding season.

Buskirk's (1976) analysis of social systems of the avifauna of a tropical montane locality in Costa Rica provides an excellent comparison for the Bolivian sites. The proportion of species that participate in mixed-species flocks at the Bolivian sites is virtually identical to that in Costa Rica. Eliminating rare transients, predators, and North American migrants from Buskirk's Monteverde list leaves 57 species, of which 27 (47%) are interspecifically gregarious; this compares with 34 (47%) and 23 (48%) at the 3050 m and 3300 m sites, respectively (Table 6). Differences between the sites in proportion of the avifauna within finer subdivisions of foraging system categories are probably due primarily to differences in scoring technique—I was more liberal in giving multiple categories for each species. Even so, the overall similarity between the Bolivian and Costa Rican sites is impressive, especially because the montane forest of the Costa Rican site differs from that in Bolivia in not being near timberline and because the sites share only two species and 18 of 53 genera.

In light of these ecological and taxonomic differences, and the tremendous geographic distances between the two areas, it is noteworthy that in 15 of 18 cases the shared resident congeners were independently characterized as having the same foraging social system (with allowance for my more liberal dispensing of multiple categories for a species). The three conflicting cases differed only in flock size category (*Columba, Amazona,* and *Chlorospingus*), and in two of these, the Bolivian species (*Amazona mercenaria* and *Chlorospingus ophthalmicus*) were only marginal community members at 3050 m and at lower elevations have larger flock sizes similar to their Costa Rican counterparts. Thus, the constancy of foraging social systems over geographic and taxonomic distance is impressive. This contrasts with the extensive intraspecific geographic variation perceived by Moynihan (1979) for Andean birds.

Morton (1979) presented similar data for foraging social systems of birds of the coastal mountains of northern Venezuela at Guatopo National Park, and I have summarized his findings with predators and North American migrants removed (Table 6). Of Morton's 194 species, 63 (33%) are regular participants in mixed-species flocks. This is significantly lower ($\chi^2 < 3.9$, $P < 0.05$) than the percentage of mixed-species flocking birds at the Costa Rican and Bolivian sites.

Inspection of the proportion of the avifauna in various foraging social system categories reveals that most of this difference is due to a higher proportion of the Guatopo avifauna in two categories: single-species flocks and non-interspecifically-gregarious species that forage in pairs (Table 6). The Guatopo avifauna has a higher proportion of flocking parrots, doves, and icterids than the other three sites; this is to be expected, because the habitat at Guatopo consists of much second-growth forest at low elevation, where such birds are more common. The ultimate reasons for the higher proportion of single-species flocking birds presumably lies with the greater patchiness of food distribution in more open, seasonal areas (Crook 1965), but pertinent data on dispersion patterns of food are not available. As for the higher proportion at Guatopo of non-interspecifically social species that forage in pairs, no plausible hypothesis is immediately evident.

Patterns of presumed food dispersion and vulnerability to predators with respect to foraging social system yield results very similar to and supportive of the analysis of Buskirk (1976), as would be expected from the great similarity in foraging social organization between the Costa Rican and Bolivian avifaunas (Table 6). Species that feed on clumped resources tend to be either in single-species flocks or sentinel foragers, species that feed on dispersed resources in the canopy tend to be in mixed-species flocks, and species with good anti-predator surveillance that feed on dispersed resources tend to be solitary.

As a consequence of the Costa Rican-Bolivian similarities, most differences between the foraging social organization of Costa Rican and Venezuelan avifaunas revealed by Morton's (1979) thorough analysis are mirrored in a comparison of the Bolivian and Venezuelan avifaunas, and will not be repeated here. Similar community-wide analyses in various habitats throughout the Neotropics are needed to test the generality of these interhabitat and interregional differences.

Because the categorization of foraging social systems presented here for several Andean species differs from that of Vuilleumier (1970, 1982) for a site in central Peru, I feel that it is necessary to address the discrepancies: (1) *Diglossa lafresnayii* is listed by Vuilleumier as a participant in mixed-species flocks; in Bolivia, this species is solitary, only occasionally and temporarily participating in mixed-species flocks; (2) *Buthraupis montana* and (3) *Thraupis cyanocephala* are listed by Vuilleumier as mixed-species flock members; in Bolivia both species are also commonly found in single-species flocks; (4) *Troglodytes solstitialis* is listed as a single-species flock member; in Bolivia this species is found primarily in pairs or solitarily in mixed-species flocks; (5) *Mionectes striaticollis* is listed among the solitary species; in Bolivia this species is a regular, frequent participant in mixed-species flocks; (7) *Mecocerculus stictopterus* is listed as a non-social species; in Bolivia this species is found primarily in mixed-species flocks, usually in pairs. Among species not found or not resident at the Bolivia sites, Vuilleumier lists *Entomodestes leucotis* as a mixed-species flock member; in Bolivia this species is solitary (Remsen, unpubl. data) as it is also in Peru (T. A. Parker, pers. comm.). *Saltator cinctus* (= "?*Arremon*" from the 1970 paper) is listed as a mixed-species flock member; because virtually nothing is known about this species in the field (O'Neill and Schulenberg 1979), this classification is premature at best. *Haplospiza rustica* is listed as a mixed-species flock member; in Peru, this species is found primarily as singles or in single-species flocks (T. A. Parker, pers. comm.).

At first one might explain these differences as simply due to geographic variation in foraging social systems (Moynihan 1979). However, I know of only two such examples; *Margarornis squamiger* and *Chlorornis riefferii* in central and northern Peru, Ecuador, and Colombia are frequently seen in small flocks within mixed-species flocks (T. A. Parker, R. S. Ridgely, T. S. Schulenberg, pers. comms.), but at my sites in Bolivia they occur primarily as pairs or singles. Vuilleumier's (1970) conclusions were based on three days of fieldwork and observation of eight flocks; at least three of these flocks had five or fewer birds and another included canopy species (*Tangara nigroviridis*) with undergrowth species (*Basileuterus luteoviridis*). Additionally, of the 24 species listed by Vuilleumier (1970), four were not positively identified and nine others were noted only once. Thus, Vuilleumier's categorizations and subsequent analyses must remain suspect.

GENERAL CONSIDERATIONS

When ornithologists discuss tropical species diversity, they almost always focus on tall, lowland tropical forest with its impressively large avifauna—as many as 550 species have been recorded in a two km^2 area after only a few years of fieldwork (Parker 1982). The bird diversity at tropical latitudes near timberline in the Andes in cold, fog-shrouded, stunted forest is almost as impressive. In the Bolivian Andes at 3300 m, an elevation at which only a handful of breeding birds are found at temperate latitudes, nearly 50 resident species are present in a 20 hectare area. This is at least 10 more species than coexist in some of the tallest, richest forests of eastern North America (Terborgh 1980). Even compared to tropical regions, the diversity of the Andean avifauna is impressive. For example, in forested montane New Guinea one must descend below 1800 m to reach an avifauna with 50 species (Diamond 1972). It also must be kept in mind that northern Bolivia is far south of the center of Andean bird diversity; an inventory in northern Peru similar to the one presented here would yield a community richer by another 10 species or more (LSUMZ collecting locality data).

As discussed in a previous section, most of this increased species richness of the Andes relative to temperate latitudes can be accounted for by the addition of "tropical" resources. Also, Stiles' (1978) data for alder forests in Washington and Costa Rica indicate that once nectarivores and epiphyte-searchers are removed from the species lists, species richness is very similar (17 vs 18) at the temperate and tropical sites. In contrast, Terborgh (1980) found that only 34 percent of the increase in species number in a lowland tropical forest in Peru could be accounted for in such a manner, with the remaining difference best explained by larger tropical niches (broader resource spectra within a guild) or increased species-packing within guilds. Karr (1975) found that approximately 70 percent of the "excess" species in a lowland forest in Panama could be accounted for by addition of "tropical" resources.

At least two hypotheses exist to explain the difference between the montane and lowland results. Terborgh (1980) is one of many (e.g., especially Haffer 1974; Karr 1976; Pearson 1977, 1982) who have invoked historical factors, primarily increased numbers of speciation events, to explain increased species-packing in the tropics. Because the Andes are geologically recent relative to the South American lowlands, one would predict that if historical factors are important, they would be more important in the lowlands than in the Andes. The results of the montane-lowland comparison are consistent with this hypothesis. Also predicted by this hypothesis, therefore, is a decrease in the importance of historical factors with increasing elevation along a lowland-to-highland gradient at a given latitude. This follows from Chapman's (1917, 1926) analysis of the origin of the forest avifauna of the Andes: avifaunas at higher elevations are more recent in origin, because the species at a given elevation are derived primarily from those of the next lowest elevational "zone." This hypothesis needs to be tested.

Although the influence of historical factors is undeniable, perhaps the importance of reduced extinction rates in the tropics, rather than increased speciation rates, should be emphasized. A speciation event can be considered analogous to reproduction by fission, and so with only two refugia, one parental species potentially can produce over 100 "off-spring" species in just seven "generations." Hypothetically, therefore, only a handful of allopatric speciation episodes are necessary to generate an avifauna far more diverse than currently exists in the forests of North America, and certainly, the number of geographic isolating events over the last 20 million years has been far, far greater than this minimum number. Proponents of the importance of speciation rates in determining species richness need to consider why current species numbers are so small relative to the hypothetical upper limit.

Obviously, time is needed between speciation events for the ecological and morphological changes required before syntopy of recently speciated forms can be achieved (and single-point diversity thereby increased). This period, however, may be much shorter than once thought. For example, the adaptive radiation of the Drepanidinae of Hawaii in just a few million years or so (Amadon 1950; Olson and James 1982; Sibley and Ahlquist 1982) attests to the speed with which secondary morphological changes can occur if ecologically stimulated by lack of competitors, i.e., if the resources are available.

The large number of speciation events that have apparently occurred in the Amazon Basin did nothing but fragment parental species into allopatric component species, i.e., produce superspecies, and this does not directly increase single-point diversity. Clearly, the proponents of the importance of historical factors do not take such a narrow view of the mechanism by which historical factors increase species diversity. Nevertheless, I think it is useful to take such a view briefly to emphasize that to increase single-point diversity, the resources that promote coexistence must be available. Thus, trying to separate historical from ecological factors may not be a realistic way of looking at the process of species enrichment.

If it is true that there has been time for far more speciating events and subsequent morphological change than needed to generate today's avifauna, then differential extinction rates become critical. Although no one any longer believes that the tropics are climatologically and ecologically stable, certainly their stability in these respects is greater than that of temperate regions, and this stability could reduce extinction rates, thereby "buying time" for ecological and morphological changes that promote survival of species and syntopy. On the other hand, as Darwin pointed out, species diversity in a stable equilibrium state may be reduced by competitive exclusion.

An alternative hypothesis to the importance of historical factors is that the degree of species-packing in the lowlands has been overestimated because our knowledge of the natural history of the birds is very incomplete. In other words, perhaps many of the species currently assigned to a guild also found at temperate latitudes are really specializing on different resources and should form their own separate guild. Could we be overlooking many examples of resource specialization? Lowland avifaunas are notoriously complex and inherently difficult to study. Many species exist in very low densities or are extremely difficult to observe. Our knowledge of diets through analysis of stomach contents is virtually nil. Certainly the conspicuous "tropical" resources, such as year-round fruit and nectar (Karr 1971; Morton 1973) and army ants (Willis and Oniki 1978), have already been identified and their contribution assessed. The contribution of other foraging substrates, habitats, and microhabitats, such as dead leaves suspended in the foliage (Terborgh 1980; Remsen and Parker, 1984), vine tangles (Terborgh 1980), tree falls (Schemske and Brokaw 1981), oxbow lakes and flooded forest (Remsen and Parker, 1983), bamboo thickets (Parker 1982; Parker and Remsen, unpubl. data), and epiphytic vegetation (Orians 1969; Remsen and Parker, unpubl. data), has only recently been identified or assessed. It thus seems likely that many more resources found in, or exploited only in, the tropics are yet to be discovered and that the degree to which tropical species diversity can be explained by additional or expanded resources is still to be resolved.

ACKNOWLEDGMENTS

Permission to do research in Bolivia was granted by the Dirección de Ciencia y Tecnologia through Gastón Bejarano, with the assistance of the Academia Nacional de Ciencias. I am grateful to the Groves Construction Company, T. and J. Heindel, A. Saavedra M., J. Solomon, and O. Suarez M. for their help in Bolivia. The fieldwork was generally supported by B. M. Odom, J. S. McIlhenny, H. I. Schweppe, and L. Schweppe. This study would not have been possible without the capable help of my assistants at Cotapata and Chuspipata: A. P. Capparella, C. S. Cardiff, S. W. Cardiff, L. S. Hale, S. M. Lanyon, M. Sánchez S., T. S. Schulenberg, and D. A. Wiedenfeld. M. S. Foster, T. A. Parker, T. S. Schulenberg, and an anonymous reviewer provided many useful comments on the manuscript. I am especially grateful to S. R. Sabo for sharing unpublished data and his careful review of the manuscript.

LITERATURE CITED

AMADON, D. 1950. The Hawaiian honeycreepers (Aves, Drepaniidae). Bull. Am. Mus. Nat. Hist. 95: 151–162.
BUSKIRK, W. H. 1976. Social systems in a tropical forest avifauna. Am. Nat. 110:293–310.
CARDIFF, S. W., AND J. V. REMSEN, JR. 1981. Three bird species new to Bolivia. Bull. Br. Ornithol. Club 101:304–305.
CHAPMAN, F. M. 1917. The distribution of bird-life in Colombia: a contribution to a biological survey of South America. Bull. Am. Mus. Nat. Hist. 36:1–729.
CHAPMAN, F. M. 1926. The distribution of bird-life in Ecuador. A contribution to a study of the origin of Andean bird-life. Bull. Am. Mus. Nat. Hist. 55:1–784.
CROOK, J. H. 1965. The adaptive significance of avian social organizations. Symp. Zool. Soc. Lond. 18: 237–258.
DIAMOND, J. M. 1972. Avifauna of the eastern highlands of New Guinea. Publ. Nuttall Ornithol. Club No. 12.
FAABORG, J. 1980. Patterns in the nonpasserine component of tropical avifaunas. Proc. 17th Int. Ornithol. Congr., pp. 979–983.
FJELDSÅ, J. 1981. A comparison of bird communities in temperate and subarctic wetlands in northern Europe and the Andes. Proc. 2nd Nord. Ornithol. Congr., pp. 101–108.
GREENBERG, R. 1981. Dissimilar bill shapes in New World tropical versus temperate forest foliage-gleaning birds. Oecologia (Berl.) 49:143–147.
GREENBERG, R., AND J. GRADWOHL. 1980. Leaf surface specializations of birds and arthropods in a Panamanian forest. Oecologia (Berl.) 46:115–124.
HAFFER, J. 1974. Avian speciation in tropical South America. Publ. Nuttall Ornithol. Club No. 14.
HILTY, S. L., T. A. PARKER, III, AND J. SILLIMAN. 1979. Observations on Plush-capped Finches in the Andes with a description of the juvenal and immature plumages. Wilson Bull. 91:145–148.
HOWELL, T. R. 1971. An ecological study of the birds of the lowland pine savanna and adjacent rain forest in northeastern Nicaragua. Living Bird 10:185–242.

KARR, J. R. 1971. The structure of avian communities in selected Panama and Illinois habitats. Ecol. Monogr. 41:207–233.

KARR, J. R. 1975. Production, energy pathways, and community diversity in forest birds. Pp. 161–176, In F. B. Golley and E. Medina (eds.), Tropical Ecological Systems. Springer-Verlag, New York.

KARR, J. 1976. Within- and between-habitat diversity in African and Neotropical lowland habitats. Ecol. Monogr. 46:457–481.

LEVINS, R. 1968. Evolution in changing environments. Monogr. Popul. Biol. No. 2.

MORTON, E. S. 1973. On the evolutionary advantages and disadvantages of fruit-eating in tropical birds. Am. Nat. 107:8–22.

MORTON, E. S. 1979. A comparative survey of avian social systems in northern Venezuelan habitats. Pp. 233–259, In J. Eisenberg (ed.), Vertebrate Ecology in the Northern Neotropics. Smithsonian Institution Press, Washington, D.C.

MOYNIHAN, M. 1963. Inter-specific relations between some Andean birds. Ibis 105:327–339.

MOYNIHAN, M. 1979. Geographic variation in social behavior and in adaptations to competition among Andean birds. Publ. Nuttall Ornithol. Club No. 18.

OLSON, S. L., AND H. F. JAMES. 1982. Prodromus of the fossil avifauna of the Hawaiian Islands. Smithson. Contrib. Zool. 365:1–59.

O'NEILL, J. P., AND T. S. SCHULENBERG. 1979. Notes on the Masked Saltator, *Saltator cinctus,* in Peru. Auk 96:610–613.

ORIANS, G. H. 1969. The number of bird species in some tropical forests. Ecology 50:783–801.

PARKER, T. A., III. 1982. Observations of some unusual rainforest and marsh birds in southeastern Peru. Wilson Bull. 94:477–493.

PARKER, T. A., III, J. V. REMSEN, JR., AND J. A. HEINDEL. 1980. Seven bird species new to Bolivia. Bull. Br. Ornithol. Club 100:160–162.

PEARSON, D. L. 1977. A pantropical comparison of bird community structure on six lowland forest sites. Condor 79:232–244.

PEARSON, D. L. 1982. Historical factors and bird species richness. Pp. 441–452, In G. T. Prance (ed.), Biological Diversification in the Tropics. Columbia University Press, New York.

REMSEN, J. V., JR. 1981. A new subspecies of *Schizoeaca harterti* with notes on taxonomy and natural history of *Schizoeaca* (Aves: Furnariidae). Proc. Biol. Soc. Wash. 94:1068–1075.

REMSEN, J. V., JR., AND T. A. PARKER, III. 1983. Contribution of river-created habitats to bird species richness in Amazonia. Biotropica 15:223–231.

REMSEN, J. V., JR., AND T. A. PARKER, III. 1984. Arboreal dead-leaf-searching birds of the neotropics. Condor 86:36–41.

REMSEN, J. V., JR., T. A. PARKER, III, AND R. S. RIDGELY. 1982. Natural history notes on some poorly known Bolivian birds. Gerfaut 72:77–87.

RICKLEFS, R. 1966. The temporal component of diversity among species of birds. Evolution 20:235–242.

SABO, S. R. 1980. Niche and habitat relations in subalpine bird communities of the White Mountains of New Hampshire. Ecol. Monogr. 50:241–259.

SCHEMSKE, D. W., AND N. BROKAW. 1981. Treefalls and the distribution of understory birds in a tropical forest. Ecology 62:938–945.

SCHULENBERG, T. S., AND J. V. REMSEN, JR. 1982. Eleven bird species new to Bolivia. Bull. Br. Ornithol. Club 102:52–57.

SIBLEY, C. G., AND J. E. AHLQUIST. 1982. The relationships of the Hawaiian honeycreepers (Drepaninini) as indicated by DNA-DNA hybridization. Auk 99:130–140.

SKUTCH, A. F. 1969. Life histories of Central American birds, III. Pac. Coast Avif. No. 35.

STILES, E. W. 1978. Avian communities in temperate and tropical alder forests. Condor 80:276–284.

TERBORGH, J. 1977. Bird species diversity on an Andean elevational gradient. Ecology 58:1007–1019.

TERBORGH, J. 1980. Causes of tropical species diversity. Proc. 17th Int. Ornithol. Congr., pp. 955–961.

VUILLEUMIER, F. 1970. L'organisation sociale des bandes vagabondes d'oiseaux dans les Andes du Pérou central. Rev. Suisse Zool. 77:209–235.

VUILLEUMIER, F. 1982. Ecological aspects of speciation in birds, with special reference to South American birds. Pp. 101–148, In O. A. Reig (ed.), Ecologia y Genetica de la Especiación Animal. Universidad Simon Bolivar, Caracas.

WILLIS, E. O., AND Y. ONIKI. 1978. Birds and army ants. Annu. Rev. Ecol. Syst. 9:243–263.

APPENDIX I

SPECIES FOUND AT EACH STUDY SITE WITH ECOLOGICAL AND MORPHOLOGICAL CHARACTERISTICS OF EACH SPECIES

	Relative[1] abundance		Foraging[2] system		Diet[3]	Breeding[4] condition	Mean body weight (g)[5]			
	3050 m	3300 m	Intersp.	Intrasp.			♂ (N)	♀ (N)	Juv. ♂ (N)	Juv. ♀ (N)
CATHARTIDAE										
Cathartes aura	(V)									
Vultur gryphus		(V)								
ACCIPITRIDAE										
Accipiter striatus		(V)								
Buteo leucorrhous		(V)								
?Buteo albigula	(V)									
Buteo polyosoma		(V)								740 (1)
FALCONIDAE										
Phalcoboenus megalopterus	(V)	V					105 (1)			
Falco sparverius	(V)	V								
CRACIDAE										
Chamaepetes goudotii	R	R	N	P, S	f	±, +, −	730 (3)	773 (3)		
PHASIANIDAE										
Odontophorus balliviani	(R)		N	?	(i, f)					
SCOLOPACIDAE										
Gallinago stricklandii		V					252 (1)			
COLUMBIDAE										
Columba fasciata	FC		N	L	f	+, −	265 (1)	280 (1)		
Geotrygon frenata	(V)									
PSITTACIDAE										
Hapalopsittaca melanotis	R		N	S, N	f	+, −	158 (2)			
Pionus tumultuosus	(U)		N	S, L	f					
Amazona mercenaria	(U?)		N	P, S	(f)					
STRIGIDAE										
?Otus albogularis	(V)									
?Ciccaba albitarsus	(V)		N	N	i (c)	−, +	63.7 (3)	73.5 (2)		
Glaudicium jardinii	FC									
CAPRIMULGIDAE										
?Uropsalis segmentata	(R)		N	?	(i)					

APPENDIX I
CONTINUED

	Relative[1] abundance		Foraging[2] system		Diet[3]	Breeding[4] condition	Mean body weight (g)[5]			
	3050 m	3300 m	Intersp.	Intrasp.			♂ (N)	♀ (N)	Juv. ♂ (N)	Juv. ♀ (N)
APODIDAE										
Cypseloides rutilus	(R)		N	L	i					
TROCHILIDAE										
Colibri corruscans	R	V	N	N	n, i	—	7.8 (1)	6.8 (2)		
C. thalassinus	U		N	N	n, i	—	5.0 (1)			
Pterophanes cyanopterus	R	U	N	N	n, (i)	—	10.2 (3)	8.4 (3)	10.1 (3)	
Ensifera ensifera	FC		N	N	n, i		10.2 (3)	8.4 (3)	10.1 (3)	
Coeligena torquata	C	V	N	N	n, i	—	7.3 (4)	6.4 (5)	6.6 (1)	
C. violifer	FC	FC	N	N	n, i		7.7 (44)	7.1 (23)	7.9 (1)	
Heliangelus amethysticollis	FC	FC	N	N	n, i	—, +	5.3 (15)	4.8 (11)		
Chalcostigma ruficeps	U	(V)	N	N	n, i	—, +	3.8 (1)	3.3 (2)		
Metallura aeneocauda	V	FC	N	N	n, i	—, +	5.3 (16)	5.0 (9)	5.3 (19)	
M. tyrianthina	FC	FC	N	N	n, i	+, —	3.5 (18)	3.4 (10)	3.5 (4)	
TROGONIDAE										
Trogon personatus	V						64.0 (1)			
RAMPHASTIDAE										
Aulacorhynchus coeruleicinctus	V	R	N	S, N	f	±, —	225 (1)			
Andigena cucullata	U	R					272 (4)			
PICIDAE										
Piculus rivolii	U	R	F	N	i, (f)	—		91.5 (2)		
Veniliornis nigriceps	U	R	F	P	i	—	46.0 (2)	42.8 (5)		
DENDROCOLAPTIDAE										
Lepidocolaptes affinis	(U)		F	P, N	i					
FURNARIIDAE										
Cinclodes fuscus	V	V	N	P	i		26.7 (4)	12.0 (3)		
Synallaxis azarae	FC		N	N, P	i	—	12.6 (5)	12.7 (21)		
Schizoeaca harterti	FC	C	F	P	i	—, ±	13.2 (11)	15.2 (8)		
Cranioleuca albiceps	FC	FC			i	—, +	15.0 (15)	16.0 (1)		
Asthenes urubambensis	V									
Margarornis squamiger	U	U	F	P	i	—	15.5 (8)	15.1 (6)		
Pseudocolaptes boissonneautii	U	V	F	N, P	i	—, +	45.5 (4)	42.7 (3)		13.0 (1)

APPENDIX I
CONTINUED

	Relative[1] abundance		Foraging[2] system		Diet[3]	Breeding[4] condition	Mean body weight (g)[5]			
	3050 m	3300 m	Intersp.	Intrasp.			δ (N)	♀ (N)	Juv. δ (N)	Juv. ♀ (N)
FORMICARIIDAE										
Chamaeza mollissima	U		N	N	g, i	—	71 (1)	65 (1)		
Grallaria squamigera	(R?)	R?	N	N	i	—		166 (1)		
G. erythrotis	R	R	N	N	i	—	53 (1)	61 (1)		
G. rufula	FC		N	N	i	—	35.3 (6)	34.7 (3)		
G. ferrugineipectus	R		N	N	i	—	17.0 (3)	16.0 (2)		
RHINOCRYPTIDAE										
Scytalopus magellanicus	FC	C	N	N	i	—	14.6 (7)	14.6 (4)		12.0 (1)
COTINGIDAE										
Ampelion rubrocristatus	(V)	R	N	N	f	—	69.9 (7)	65.6 (4)	60.0 (1)	61.0 (1)
Pipreola intermedia	R	V	N	N, P	f	—	52.8 (4)	57.0 (2)		
P. arcuata	U	U	N	N, P	f	+, —	117.3 (3)	111.3 (2)		
TYRANNIDAE										
Phyllomyias uropygialis	U	R	N	N	i	—	9.4 (11)	8.6 (8)		
Elaenia albiceps	U	U	N, F	N, S	f, i	—	16.2 (14)	16.3 (9)		
E. pallatangae	V	V				—	13.7 (1)			
Mecocerculus leucophrys	FC	R	F	P	i	±, —	12.2 (3)	10.0 (1)		
M. stictopterus	FC		F	P	i		8.8 (1)	8.7 (2)		
Anairetes parulus	R	R	N	P	i	—, ±	6.4 (5)	6.9 (1)		
Mionectes striaticollis	R	FC	F	N	f, (i)	—	15.2 (60)	12.6 (18)		
Pseudotriccus ruficeps	R		N	N	i	—	9.9 (2)	9.8 (4)		
Hemitriccus granadensis		V					7.7 (1)	6.6 (1)		
Myiophobus ochraceiventris	V	FC	F	S	i, f	—, ±	11.2 (11)	9.8 (3)	11.0 (1)	
Pyrrhomyias cinnamomea	(V)		N	N	i	—				
Ochthoeca frontalis	FC	FC	N	N	i		10.4 (34)	9.6 (36)		
O. pulchella	V					—		10.2 (1)		
O. rufipectoralis	FC	U	F, N	P	i		10.5 (4)	9.5 (10)		
O. fumicolor	V	V					13.4 (2)	13.5 (1)		
Myiotheretes fuscorufus	R	V	N	P	i	—	32.4 (2)	31.0 (3)		
Myiarchus (sp.)		(V)								

APPENDIX I
CONTINUED

	Relative[1] abundance		Foraging[2] system		Diet[3]	Breeding[4] condition	Mean body weight (g)[5]			
	3050 m	3300 m	Intersp.	Intrasp.			♂ (N)	♀ (N)	Juv. ♂ (N)	Juv. ♀ (N)
HIRUNDINIDAE										
Notiochelidon murina	R	C	N	L	i	—	11.8 (7)	11.6 (7)		10.4 (3)
N. cyanoleuca	(V)									
N. flavipes	(R)		N	L	i					
CORVIDAE										
Cyanolyca viridicyana	U	V	N	S	i	—, ±	93.8 (6)	89.0 (2)		
TROGLODYTIDAE										
Cinnycerthia peruana	R		N	S	i	—, +	18.0 (1)	15.1 (6)		14.0 (1)
Troglodytes solstitialis	FC	(V)	F	N, P	i	—	11.6 (4)			
TURDINAE										
Myadestes ralloides	(V)									
Entomodestes leucotis	V	V						69.0 (1)		
Catharus fuscater	V							28.0 (1)		
Turdus chiguanco							100 (1)			
T. fuscater	FC	FC	N	N, S	f, (i)	—, ±	126 (4)	139 (5)		118 (2)
T. serranus	V	R	N	N	f	—	92.3 (3)	90.5 (4)		76.0 (1)
EMBERIZINAE										
Zonotrichia capensis	C	FC	N	N, P, S	g	—	24.0 (3)	21.9 (2)	23.0 (2)	
Catamenia homochroa		V					12.1 (2)			
C. analis	V							12.7 (1)		
Atlapetes rufinucha	C	U	F	S, P	g, i, f	—	22.2 (9)	20.9 (16)		
A. torquatus	R		N	P, S	i, g	—	41.7 (3)			
CATAMBLYRYHNCHINAE										
Catamblyrhynchus diadema	FC	R	F	S	i, v	—	14.9 (16)	13.4 (8)	13.5 (2)	12.5 (1)
THRAUPINAE										
Chlorornis riefferii	R		F	N	f, i	—	55.0 (1)	54.0 (1)		
Chlorospingus ophthalmicus	R		F	S	i, f	—	18.0 (1)			
Hemispingus calophrys	FC	R	N, F	S	i	—, ±, +	17.0 (17)	16.4 (10)		15.0 (1)
H. superciliaris	FC	(V)	F	S	i	—	13.3 (2)	12.5 (1)		
H. xanthophthalmus	FC		F	S	i	—, ±, —	11.4 (3)	11.3 (3)	14.0 (5)	
H. trifasciatus	R	C	F	S	i	+, —	14.9 (10)	13.3 (11)		

APPENDIX I
CONTINUED

	Relative[1] abundance		Foraging[2] system		Diet[3]	Breeding[4] condition	Mean body weight (g)[5]			
	3050 m	3300 m	Intersp.	Intrasp.			δ (N)	♀ (N)	Juv. δ (N)	Juv. ♀ (N)
Thlypopsis ruficeps	FC	(V)	F	N	i	—	12.0 (1)	11.4 (5)		
Thraupis cyanocephala	FC	(V)	F	S	f, i	—, ±	40.0 (6)	37.0 (4)		
T. bonariensis	V								32.0 (1)	
Buthraupis montana	C	U	N, F	L	f, i	—, ±, +	79.3 (16)	77.5 (9)		
Anisognathus igniventris	C	U	F	S	f	—, ±, +	34.9 (37)	32.8 (17)	70.3 (3)	
Iridosornis jelskii		U	F	N, P	f	—, ±	19.3 (15)	19.2 (13)		
Delothraupis castaneoventris	U	U	F	N	i, f	—	26.4 (7)			16.0 (1)
Tangara vassorii	U	U	F	S	f, i	—	17.6 (3)	17.5 (4)	16.3 (1)	
Diglossa baritula	R	V	F	N	n, i	—	9.3 (1)	8.6 (1)		
D. carbonaria	FC	U	N	N	n, i		13.1 (12)	12.1 (9)	11.8 (16)	10.9 (16)
D. lafresnayii	V	FC	N	N	n, i	—, ±	16.5 (17)	15.0 (11)	14.9 (25)	14.2 (19)
D. cyanea	FC	FC	F	N, S	i, f, n	—, ±	17.6 (12)	16.6 (20)	16.5 (5)	15.5 (9)
CARDINALINAE										
Pheucticus aureoventris	U	V	N, F	N	f	—			47.0 (2)	
PARULIDAE										
Myioborus melanocephalus	FC	U	F	P	i	—	10.5 (9)	10.6 (3)		
Basileuterus luteoviridis	FC	FC	F	S, P	i	—	13.7 (18)	12.2 (11)		10.5 (1)
B. signatus	C	U	N, F	P, S	i	—	11.5 (21)	11.1 (16)		
Conirostrum sitticolor	R	U	F	P	i	—, ±	11.0 (2)	10.8 (3)		
ICTERIDAE										
Psarocolius atrovirens	(V)		N, F	S	i, n	—	99.0 (2)	63.3 (4)	80.5 (2)	
Cacicus chrysonotus	FC	U	N	N, P	i	—	58.5 (2)	49.5 (4)		
C. holosericeus	R	R								
FRINGILLIDAE										
Carduelis (sp.)		(V)								

[1] C = common (>10 noted/day), FC = fairly common (3–10/day), U = uncommon (1–2/day), R = rare (<1 detection/day), and V = visitor (only a few records; not part of the resident avifauna). When the relative abundance symbol is put in parentheses, no specimen was obtained from that site. For visitors, the only additional data presented are body weights, when available.

[2] Terminology for foraging social system follows Buskirk (1976): N = no sociality, F = flocks, P = pairs, S = small flocks (3–5 individuals), and L = large flock (>5 individuals).

[3] i = insectivore (includes all invertebrate prey), f = frugivore, n = nectarivore, v = unidentifiable vegetable matter, and g = granivore. Symbols in parentheses indicate that the diet category is inferred from morphology in the absence of feeding observations or stomach contents analysis. When more than one diet category is listed, the one observed most frequently is listed first.

[4] + = enlarged gonads (arbitrarily, testes >7 × 4 mm for birds under 50 g or >10 × 5 mm for birds over 50 g; largest ova >1 mm diameter for birds under 50 g or >2 mm for birds over 50 g); — = nonreproductive gonads (arbitrarily, testes <4 × 3 mm for birds under 50 g or <7 × 4 mm for birds over 50 g; largest ova ≤1 mm in diameter); ± = intermediate testis size. When more than one condition was noted, the most frequent is given first.

[5] Body weights taken only from specimens from the two study sites.

APPENDIX II

Foraging Substrate Use and Characteristics for Birds at Two Forested Sites in the Humid Temperate Zone of the Bolivian Andes[1]

Species	N	Frequency of substrate use								Characteristics of substrates					
		Green foliage	Dead foliage	Bark/stem	Moss[2]	Ground	Fruit	Flower	Air	% Upperside green leaves	x̄ leaf length (cm)	x̄ leaf width (cm)	x̄ diam. branch (cm)[3]	x̄ diam. fruit (cm)	x̄ corolla length (cm)
Cypseloides rutilus	19								1.00						
Pterophanes cyanopterus	11							1.00							5.0
Coeligena violifer	37							1.00							5.5
Heliangelus amethysticollis	17							.88	.12						2.3
Chalcostigma ruficeps	5							1.00							2.1
Metallura aeneocauda	5							1.00							2.7
M. tyrianthina	28							1.00							1.2
Piculus rivolii	6						.50								
Lepidocolaptes affinis	11			.33	.17					.90	8.5	2.1	10.4		
Schizoeaca harterti	20	.55		.18	.82					.25	8.6	2.9	3.2		
Cranioleuca albiceps	36	.33	.24	.40	.05								3.5		
Margarornis squamiger	30		.03	.27	.77										
Pseudocolaptes boissonneautii	32	.06	.06	.20	.82										
Phyllomyias uropygialis	9	.44		.22	.11			.22		.00					
Elaenia albiceps	12						1.00							0.3	
Mecocerculus leucophrys	30	.73		.10				.10	.07	.24	6.8	3.1			
M. stictopterus	35	.94		.03	.03					.21	7.8	4.8			
Mionectes striaticollis	18	.06		.02			.94							0.3	
Myiophobus ochraceiventris	51	.59					.35			.47	10.9	5.9		0.3	
Ochthoeca rufipectoralis	36	.47		.20		.36			.17	.82	8.0	3.4			
O. frontalis	5	.80								1.00					
Myiotheretes fuscorufus	3	.33							.67						
Notiochelidon murina	128								1.00						
N. flavipes	26								1.00						
Troglodytes solstitialis	16	.13		.19	.87					.63	9.7	3.7	11.0		
Allapetes rufinucha	31	.26	.03	.84	.26	.10	.03								
Catamblyrhynchus diadema	19	.16	.04	.11	.04			.10					0.6		
Hemispingus calophrys	28	.81								.50	11.8	2.6			
H. superciliaris	34	1.00								.46	5.6	2.7			
H. trifasciatus	58	.96	.03	.02				.02		.54	7.1	3.6			
Thlypopsis ruficeps	40	.86	.03	.03				.08		.23					
Buthraupis montana	35	.20			.40		.37						4.4	0.5	

APPENDIX II
CONTINUED

Species	N	Green foliage	Dead foliage	Bark/stem	Moss[2]	Ground	Fruit	Flower	Air	% Upper-side green leaves	x̄ leaf length (cm)	x̄ leaf width (cm)	x̄ diam. branch (cm)[3]	x̄ diam. fruit (cm)	x̄ corolla length (cm)
		Frequency of substrate use								Characteristics of substrates					
Anisognathus igniventris	42	.07					.91		.02					0.3	
Delothraupis castaneoventris	13	.15		.08	.77		.67						5.8	0.3	
Tangara vassorii	6				.33			.80							3.6
Diglossa baritula	10							.89	.20						0.8
D. lafresnayii	9						.11	1.00							3.9
D. carbonaria	6							.82							2.5
D. cyanea	17	.06			.06		.06	.02							
Conirostrum sitticolor	54	.96			.02			.05	.38	.44	13.0	5.3			
Myioborus melanocephalus	88	.49	.01	.02	.05				.01	.31	10.9	4.7			
Basileuterus luteoviridis	70	.98	.01							.20	10.4	4.9			
B. signatus	19	1.00								.26	14.4	3.6			
Cacicus chrysonotus	5			.20	.60			.20							

[1] Data for 3050 m and 3300 m sites combined.
[2] Includes other epiphytic vegetation.
[3] Branch diameters include observations from "moss" substrates, all of which were moss-covered branches, as well as "bark/stem" substrates (bare branches).

SEASONAL PATTERNS AND COEVOLUTION IN THE HUMMINGBIRD-FLOWER COMMUNITY OF A COSTA RICAN SUBTROPICAL FOREST

F. GARY STILES

Escuela de Biología, Universidad de Costa Rica, Ciudad Universitaria, Costa Rica

ABSTRACT. This two-year study seeks to quantify the relations of the breeding, molt, and population movements of hummingbirds, with the flowering seasons of their foodplants, at La Montura, a site in premontane rain forest at 1000 m on the Caribbean slope of Costa Rica. The community comprises 22 hummingbird species (9–10 residents, 5–6 regular seasonal visitors, the rest rare to accidental), that collectively visit at least 70 species of plants, ca. 50 of which they pollinate.

Patterns of flower visitation by the birds allow partitioning of the community into subcommunities, each with its own seasonal rhythm. These are: (a) Lancebill Subcommunity: the Green-fronted Lancebill (*Doryfera ludovicae*) and five species of large shrubby epiphytes, and probaby several bromeliads. Morphologically, the bird's long, slender, nearly straight bill and the flowers' correspondingly shaped corolla tubes characterize this subcommunity. The five principal foodplants bloomed in a nearly perfectly staggered sequence through the study period; breeding in *D. ludovicae* coincided with periods of greatest flower abundance, molt with intermediate flower availability. (b) Hermit Subcommunity: chiefly the Green Hermit (*Phaethornis guy*) with its long, decurved bill, and 19 understory and subcanopy flowers with correspondingly shaped corollas. Flower availability is high during breeding, peaks during the period of molt-breeding overlap, and declines during molt and quiescent periods; the species of *Heliconia* appear to supply critical nectar resources for breeding. *P. guy*'s foodplants show staggered blooming sequences when the bird's breeding and nonbreeding periods are analyzed separately. Also included here is the White-tipped Sicklebill (*Eutoxeres aquila*), whose visits to the study area are linked to the blooming of a single foodplant. (c) *Heliodoxa-Marcgravia* Subcommunity: the Green-fronted Brilliant (*Heliodoxa jacula*) and three species of sphingid- or bat-pollinated *Marcgravia*, about which the bird's breeding and molting cycles appear to be organized. The three *Marcgravia* species bloom in a staggered sequence that is exploited by *H. jacula*, but this cannot reflect a coevolutionary relationship as the hummingbird is a nectar thief. (d) "Generalized" Subcommunity: various small- to medium-sized, straight-billed hummingbirds, and flowers with straight, short- to medium-length corollas. Sufficient overlap in flower visitation exists within this assemblage to preclude further subdivision. The principal hummingbird species (*Lampornis hemileucus, Eupherusa nigriventris, Elvira cupreiceps*) breed in close association with the blooming of a series of canopy epiphytes, mostly in the family Ericaceae. Many individuals of the latter two species, plus *Colibri delphinae*, emigrate following breeding and at least the start of molt; their places are taken by several species of postbreeding seasonal visitors from other elevations, which mostly exploit the flowers of the abundant understory treelet *Cephaelis elata*.

Most La Montura hummingbirds breed from the mid- to late wet season, through the early dry season, roughly October through March, and molt from March through June, into the early rainy season. Males of many species molt one to two months ahead of females. The cycle of *H. jacula* is displaced one to two months earlier than those of most species, while that of *P. guy* is nearly the opposite: breeding April–September, molting July–November. Differences in seasonality in other hummingbird-flower communities reflect different climatic regimes and taxonomic affinities—e.g., in all areas Ericaceae bloom mainly in the wet season, *Heliconia* mostly June–August.

The various hummingbird-flower communities in the wet forests of Costa Rica show similar divisions into subcommunities, given different species richnesses and taxonomic affinities. In dry forest and second growth, most or all species pertain to the generalized subcommunity. The relation between plant-pollinator interactions and flowering seasons is reassessed in the light of changing nectar demands of the birds due to breeding and molt. It is concluded that competition for pollinators may be an important selective force in the spacing of flowering seasons, but that its intensity varies according to the annual cycles of the pollinators, and cannot be assessed properly without taking the pollinators' biology into account. Finally, based on ecological and biogeographical evidence, it is argued that many (but by no means all) of the bird-flower interactions in the La Montura community represent true coevolved mutualisms.

RESUMEN. Este estudio de dos años, pretende cuantificar la relación de la reproducción, muda y movimientos de poblaciones de colibríes con las épocas de floración de sus plantas alimenticias, en La Montura, un sitio de la selva lluviosa premontana a 1.000 m de altura en la ladera caribeña costarricense. La comunidad está compuesta por 22 especies de colibríes (9–10 residentes; 5–6 visitantes regulares estacionales y el resto raros o accidentales) que visitan en forma collectiva el menos 70 especies de plantas, de las cuales polinizan casi 50.

Los patrones de las aves para visitar las flores, permiten dividir la comunidad en subcomunidades, cada una con su propio ritmo estacional. Estas son: a—Subcomunidad Pico de Lanza: el pico de lanza mayor (*Doryfera ludovicae*) y cinco especies de arbustos epífitos grandes y probablemente varias bromelias. Esta subcomunidad está caracterizada morfológicamente por el pico largo fino y casi recto, que corresponde con la forma del tubo de las corolas de las flores. Las cinco plantas principales de alimentación florecen en secuencia casi perfecta a lo largo del período de estudio; la temporada de reproducción de *D. ludovicae* coincide con la gran abundancia de flores; la muda con disponibilidad intermedia de flores. b—Subcomunidad Ermitaño: Comandada por el ermitaño verde (*Phaethornis guy*) con su pico largo, curvado hacia abajo y 19 flores con sus corolas de forma similar del sotobosque ("understory") y del subdosel bajo las copas de los árboles ("subcanopy"). La disponibilidad de flores es alta durante la reproducción, teniendo su pico máximo durante el período en que la muda y reproducción se superponen, y declina durante la muda y los períodos de poca actividad ("quiescent"); las especies *Heliconia* parecen proveer recursos de néctar críticos para reproducir. Las plantas de las cuales se alimenta *P. guy* florecen en secuencia una tras otra cuando se analizan separadamente los períodos de reproducción y no-reproducción del ave. También se incluye acá al Pico de hoz coliolivia (*Eutoxeres aquila*) cuyas visitas al área de estudio entuvieron relacionadas al florecimineto de una sola planta de alimentación. c—Subcomunidad *Heliodoxa-Marcgravia*: el colibrí jacula (*Heliodoxa jacula*) y tres especies de *Marcgravia*, polinizadas por polillas de la familia Sphingidae o murciélagos, sobre las cuales parece estar basado el ciclo de reproducción y muda del ave. Las tres especies de *Marcgravia* florecen en una secuencia que es utilizada por *H. jacula*, pero esto no refleja una relación de coevolución, ya que el colibrí es un ladrón de néctar. d—Subcomunidad "Generalizada": varios picaflores de tamaño pequeño a mediano, de pico recto y flores de corola recta corta o mediana. En este grupo existe suficiente superposición en las visitas a las flores, lo cual imposibilita mayores subdivisiones. Las principales especies de picaflores (*Lampornis hemileucus, Eupherusa nigriventris, Elvira cupreiceps*) reproducen en asociación íntima con una serie de epífitas de las copas de los árboles mayormente de la familia Ericaceae. Muchos individuos de estas dos últimas especies, además de *Colibri delphinae* emigran luego de la reproducción o como máximo cuando comienza la muda; sus lugares son ocupados por varias especies de vistantes post-reproductivos de otras elevaciones, que mayormente utilizan las flores abundantes del arbolito *Cephaelis elata*, del nivel de vegetación inferior.

La mayoría de los picaflores de La Montura reproducen desde la mitad o final de la estación húmeda hasta el principio de la temporada seca, aproximadamente desde octubre hasta marzo y mudan desde marzo hasta junio, cuando comienza la estación de las lluvias. Machos de muchas especies mudan uno o dos meses antes que las hembras. El ciclo de *H. jacula* está desplazado de uno a dos meses unico con el de la mayoría de las especies, mientras que el de *P. guy* es casi lo opuesto: reproduce entre abril y septiembre y muda entre julio y noviembre. Las diferencias en las estaciones en otras comunidades de flores y picaflores refleja régimenes climáticos diferentes y afinidades taxonómicas por ejemplo en todas las áreas las Ericáceas florece mayormente en la estación húmeda y *Heliconia* de junio a agosto.

Las diversas comunidades de flores y picaflores en los bosques húmedos de Costa Rica muestran divisiones en subcomunidades similares, habiendo diferentes riquezas de especies y afinidades taxonómicas. En bosque seco y vegetación secundaria, la mayoría o todas las especies pertenecen a la subcomunidad generalizada. La relación entre la interacción planta-polinizador y temporada de floración está reevaluada a la luz de los cambios en las demandas de néctar de las aves debido a la reproducción y muda. Se concluye que la competencia por polinizadores puede ser una fuerza selectiva de importancia en la temporada de estaciones de floración, pero cuya intensidad varía de acuerdo con el ciclo anual de los polinizadores, y no puede ser propiamente determinado sin tomar en cuenta la biología de los polinizadores. Finalmente basándose en evidencias ecológicas y biogeográficas, se discute que muchas (pero por cierto no todas) las interacciones ave-flor en la comunidad de La Montura representan verdaderos mutualismos coevolucionados.

Communities of tropical birds have well-defined seasonal rhythms of reproduction, molt, and population movements just as do avifaunas of higher latitudes (Skutch 1950; Fogden 1972). In tropical communities, however, there is no overriding zeitgeber corresponding to the annual photoperiodic (and related temperature) cycle. Seasonal changes in the avifauna, as well as in the resources required by the various bird species, are thus related in large part to patterns of rainfall. Different resources (insects, flowers, fruit, etc.) often peak at different times, or show different patterns of availability in different years, reflecting variability in rainfall (Stiles 1978; Wolda 1978).

Interpretation of avian seasonal rhythms may be facilitated if the birds' critical resources can be specified and counted with relative precision. This is possible with hummingbirds, whose critical energetic resource, nectar, comes in conspicuous, stationary, countable packages with a predictable daily pattern of energy yield (cf. Wolf et al. 1972, 1975; Stiles 1975; Feinsinger 1976). Although small insects and spiders are nutritionally important to hummingbirds (e.g., Scheithauer 1966), there is no evidence that these resources are limiting for the performance of seasonal activities like breeding and molt, which have been shown to be strongly influenced by flowering patterns (Skutch 1950; Stiles 1978, 1980). Therefore, detailed data on flowering phenology and flower visitation by the birds are required to compare seasonal patterns in different hummingbird communities.

When such data exist for several hummingbird-flower communities in a single region, several interesting questions can be addressed. Are flowering patterns similar or offset in different communities? How does this affect breeding, molting, and seasonal movements of the hummingbirds? Does a given plant species or group respond similarly in different communities to a given set of environmental conditions (especially rainfall)? How much variation exists between communities in flower choice by hummingbirds or in plant-pollinator specificity? Are certain taxonomic or ecological groups of flowers of outstanding importance to the hummingbirds, and how do these vary with parameters such as elevation and moisture?

Previously I studied a tropical wet lowland hummingbird-flower community (Stiles 1978, 1980; Stiles and Wolf 1979), and participated in a study of a high-montane community (Wolf et al. 1976). Comparisons of the two communities were hindered, however, by the complete lack of taxonomic overlap between them. Accordingly, I initiated a study of hummingbird-flower communities at middle elevations. In this paper, I present data gathered at the first such site, La Montura (elev. = ca. 1000 m) on the Caribbean slope of the Cordillera Central of Costa Rica. I make preliminary comparisons between the seasonality of this hummingbird-flower community, and those of Finca La Selva (elev. = ca. 100 m, at the foot of this cordillera), and Cerro de la Muerte (elev. = ca. 3000 m at the crest of the Cordillera de Talamanca, the next mountain range to the south), as well as to the second-growth, middle-elevation community at Monteverde (elev. = ca. 1100 m in the Cordillera de Tilarán in northwestern Costa Rica), studied by Feinsinger (1976, 1977, 1978).

METHODS

STUDY AREA

The present study was carried out at La Montura (10°7'N, 83°58'W) in Parque Nacional Braulio Carrillo, a 32,000 ha reserve centered around Volcán Barba in the Cordillera Central of Costa Rica. The topography of the La Montura area is rugged, characterized by long steep-sided ridges that extend down from the continental divide. The tops of the ridges are relatively gently sloping, but their sides drop off precipitously into deep canyons containing narrow, rocky rivers that flow swiftly northward toward the Caribbean coastal plain (Fig. 1). The ridge of La Montura divides the watersheds of the Río Patria to the west and the Río La Hondura to the east.

Most of my observations were made along ca. 2.5 km of study trails, chiefly within a roughly rectangular area some 12 ha in extent, just east of the ridgetop. The study area has a mean elevation of ca. 1000 m, with extremes of 900 and 1070 m; its topography is varied, including

Fig. 1.   Above: General aspect of topography and vegetation at La Montura. View from upper part of the study area, looking east across the Río Patria drainage. Below: View of forest understory at La Montura. Note the dominance of dicot shrubs, large-leaved monocots (palms, *Heliconia,* etc.), and tree-ferns.

steep slopes and ravines, level areas, and both permanent and intermittent streams. Except for a small area along a now-abandoned road at one corner, the entire area is covered with primary forest. Perhaps the most striking feature of the La Montura forest is the abundance and diversity of epiphytes. Many trees support huge masses of moss, aroids, cyclanths, large

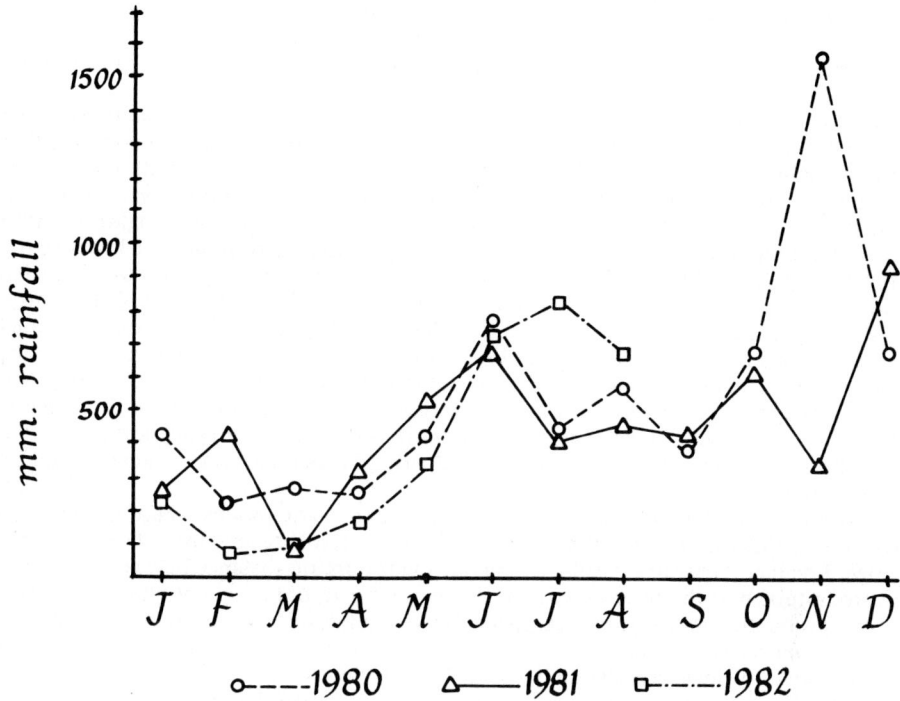

FIG. 2.   Rainfall recorded at Cariblanco, ca. 30 km E of La Montura, 1980–1982.

hemiepiphytes like *Clusia*, epiphytic shrubs (e.g., Ericaceae, Gesneriaceae), and vines (e.g., Marcgraviaceae). Bromeliads and orchids are also common but do not dominate the epiphyte flora to the extent that they do in some lowland forests (e.g., Gentry 1982). The forest canopy is quite irregular; large trees reach 35–40 m, but treefalls are frequent as a result of heavy rains followed by high winds. Light gaps of various sizes and ages are thus common in the understory, which is dominated by tall shrubs (notably Rubiaceae, Gesneriaceae), palms, treeferns, and large herbs such as *Heliconia* and *Calathea* (Fig. 1).

The area of La Montura is mapped as Premontane Rain Forest in the Holdridge Life Zone System (Tosi 1969), but unfortunately precise weather data for the area do not exist. The closest weather station is at Guápiles, ca. 20 km northeast but at a much lower elevation (300 m) at the edge of the coastal plain. The station with the most comparable weather is at Cariblanco, some 30 km to the northeast but similar in elevation and topographic situation. The Cariblanco rainfall data correspond well with my subjective impressions of La Montura rainfall, and I consider them representative of the latter for purposes of comparison. The driest months at Cariblanco are February and March, although only in 1982 was the dry season well marked. Over the rest of the year rainfall is heavy, especially from June to August and October to December. In both years of my study more than 6 m of rain probably fell at La Montura (Fig. 2). Average daytime temperatures reached ca. 24°C and nighttime temperatures averaged about 15°C at Cariblanco.

FIELD WORK

Between late June 1980 and late May 1982, I made 21 two- to seven-day visits to La Montura, at intervals of four (when possible) to six weeks. During each visit, I estimated the numbers of all species of flowers visited by hummingbirds within 10 m of all study trails (exact counts were impossible for many canopy species). I also noted whether flowering was increasing or decreasing, based upon the relative numbers of unopened buds, open flowers, and developing fruits. I had great difficulty in identifying and counting hummingbirds in the dense, epiphyte-laden canopy, especially under the foggy to drizzly conditions that often prevailed. I therefore concluded that an overall subjective estimation of hummingbird abundance, based upon mist-net data and all visual observations, would be little less accurate (and potentially far less misleading) than numbers derived from ground-based censuses. Abundance of each species was evaluated according to a semiquantitative scale running from 0 (= absent) and 1 (= rare) to 5 (= very abundant). I also recorded all hummingbird visits to flowers, where both bird and plant could be identified.

Mist-nets set 0 to 3 m above ground were used to sample hummingbird populations. I tried to accumulate ca. 300 net-hours per visit, but this was often impossible due to rainy weather. Each hummingbird was marked with a colored plastic tag (Stiles and Wolf 1973) or an aluminum leg band with photoengraved letters. Birds were weighed to the nearest 0.1 gm with a Pesola spring balance, and measured (exposed culmen, wing chord, tail length) to within 0.5 mm with a millimeter rule. Each individual was also checked for molt, which was scored according to the system of Stiles and Wolf (1974). Young birds were diagnosed by distinctive plumages (cf. Ridgway 1911) or by striations in the maxillary ramphotheca (Ortiz-Crespo 1972). Breeding females could be recognized as such by the presence of a brood patch, although this is less obvious in hummingbirds than in most birds, because few or no feathers are lost when it is formed. However, the skin does become noticeably thickened and crinkly, and more bluish or grayish in color. All available means were used to determine the breeding seasons of the La Montura hummingbirds: observations of females carrying nesting material or feeding fledglings, the nests themselves, the presence of brood patches on females, and bill striations on juveniles.

I attempted to collect vouchers for identification of all plant species whose flowers were visited by hummingbirds, but this was impossible for some epiphytes and trees. To characterize flower size, I defined "effective corolla length" as the minimum distance between the mouth of the corolla tube and the nectar chamber. I was unable to obtain quantitative samples of daily nectar production for most species, particularly epiphytes. However, by considering such parameters as flower size, dimensions of the nectar chamber, and data from related species, I usually was able to categorize the daily nectar flow as "low" (<20 μl/flower/day) or "high" (≥ 20 μl/flower/day). For evaluating blooming seasonality, I used either the actual flower counts or a simplified measure derived therefrom, the number of species in "good bloom," defined as those species attaining 50% or better of the maximum flower count in a given blooming season.

Vouchers of plants collected are deposited in the herbaria of the Universidad de Costa Rica and/or the Museo Nacional de Costa Rica. I also collected vouchers of most species and plumage types of hummingbirds, but I did not attempt to obtain series of specimens for gonad analysis. Bird specimens are deposited in the Museo de Zoología, Universidad de Costa Rica.

DATA ANALYSIS

To evaluate the relation between flowering parameters and hummingbird activities, I calculated non-parametric correlation coefficients (Spearman's r_s) between the number of flowers or the number of species in good bloom, and the total hummingbird abundance (according to the semi-quantitative system above) or the number of species breeding, for each census period. For single hummingbird species, I compared the numbers of flowers available on censuses during the bird's breeding, molting, and quiescent (neither breeding nor molting) periods with Kruskal-Wallis nonparametrical analysis of variance (Zar 1974).

I also tested statistically whether the periods of good bloom of the flowers used by particular hummingbird species were random, aggregated, or hyperdispersed. By interpolation from the flower counts (including my estimation of the direction of flowering—increasing or decreasing), I determined whether a given flower species was in good bloom during each 2-week interval of the study period. I then summed the number of intervals with 0, 1, . . . , N species in good bloom (N = the total number of species used by the hummingbird in question) and determined the mean ñ number of species in good bloom per interval. If blooming periods are randomly distributed, the number of intervals with 0, 1, . . . , ñ species in good bloom should approximate a Poisson distribution with mean ñ. Chi-square tests were used to test for goodness of fit between the expected (Poisson) and observed distributions. The direction of any deviation from randomness is given by the variance: mean ratio (Zar 1974).

While a detailed analysis of flower choice by the hummingbirds is beyond the scope of this paper, some means of comparing the visitation patterns of different species was required for the purpose of defining subcommunities. To this end I calculated overlap values using the formula of Feinsinger (1976) between each pair of resident hummingbird species over all flower species throughout the study period. If the mean overlap of any species or group of

TABLE 1

STATUS, ABUNDANCE, AND PHYSICAL CHARACTERISTICS OF LA MONTURA HUMMINGBIRDS[1]

	Status[2]	Max. abund.[3]	No. banded	$\bar{X}$ wt.[4]	$\bar{X}$ bill length[4]	Bill[5] curv.
Hermits (Phaethorninae)						
Eutoxeres aquila (Ea): White-tipped Sicklebill	R?	U	25	10.7	32.2	V
Phaethornis guy (Pg): Green Hermit	R	A	113	6.0	40.9	M
P. longuemareus (Pl): Little Hermit	V	R	3	2.6	21.2	S
Threnetes ruckeri (Tr): Band-tailed Barbthroat	V	X	1	5.6	30.1	S
Nonhermits (Trochilinae)						
Amazilia saucerrottei (As): Steely-vented Hummingbird	V	X	1	4.8	18.7	O
Campylopterus hemileucurus (Ch): Violet Sabrewing	N	U	5	10.7	31.4	S-M
Colibri delphinae (Cd): Brown Violet-ear	B	C	6	6.5	16.8	O
C. thalassinus (Ct): Green Violet-ear	N	U	8	5.4	21.3	O-S
Discosura conversii (Pc): Green Thorntail	R?	R?	2	3.0	12.0[6]	O
Doryfera ludovicae (Dl): Green-fronted Lancebill	R	U	17	5.7	34.4	S'
Elvira cupreiceps (Ec): Coppery-headed Emerald	R	C	52	3.3	14.1	S
Eupherusa nigriventris (En): Black-bellied Hummingbird	R	A	71	3.5	14.9	O
Florisuga mellivora (Fm): White-necked Jacobin	V	X	1	7.0	18.1	O
Heliodoxa jacula (Hj): Green-crowned Brilliant	R	A	63	8.6	21.9	O
Heliothryx barroti (Hb): Purple-crowned Fairy	R	U	6	5.6	17.2	O
Hylocharis eliciae (He): Blue-throated Goldentail	V	X	1	3.7	17.6	O
Klais guimeti (Kg): Violet-headed Hummingbird	V	R	2	2.8	13.2	O
Lampornis calolaema (Lc): Purple-throated Mountain-gem	N	U	6	5.4	19.4	O
L. hemileucus (Lh): White-bellied Mountain-gem	R	A	75	5.5	19.3	O
Lophornis helenae (Loh): Black-crested Coquette	V?	X	0	2.9	11.3[6]	O
Microchera albocoronata (Ma): Snowcap	V	R	2	2.5	11.9	O
Thalurania colombica (Tf): Crowned Wood-nymph	N	C	12	4.3	20.1	O

[1] Abbreviations for species (in parentheses) are used in figures.
[2] R = resident (most or all year); B = breeding resident; N = regular nonbreeding resident; V = vagrant or stray.
[3] A = abundant; C = common; U = uncommon; R = rare; X = fewer than three records.
[4] Mean of all birds banded; weights in grams; bill length in mm. Birds from other sites included if the sample size was very small.
[5] V = very strongly curved; M = moderately curved; S = slightly curved; O = straight; S' = slightly recurved.
[6] Data from Ridgway (1911).

species with all other species is less than 0.1, I consider that species (or group) to constitute a subcommunity. The value of 0.1, while arbitrary, is reasonably in accord for values between "coexisting specialists . . . whose specialties do not overlap extensively" (Feinsinger 1976: 282).

COMMUNITY COMPOSITION

HUMMINGBIRDS

To date, 22 species of hummingbirds have been recorded at La Montura (Table 1). Eight are rare to accidental visitors whose effects upon the dynamics of the community are negligible. Most of these are vagrants from lower elevations on the Caribbean slope, but one species, *Amazilia saucerrottei,* evidently strayed across the continental divide from the deforested Valle Central. Four species are regular visitors in good numbers; one of these (*Thalurania colombica*) moves uphill, while the remaining three move down from higher elevations. Taken as a whole, these visitors appear mainly in the early- to mid-rainy season, with peaking in June and July (Fig. 3). The presence of most of these species is associated with the flowering of *Cephaelis elata* (Rubiaceae), an abundant understory treelet. Individuals of these species netted in the study area were mostly immatures, with a few molting adults. This suggests that these species invade La Montura following their respective breeding seasons in other areas.

The remaining 10 species are present at La Montura for much or all of the year and are known or suspected to breed there (or nearby); they comprise the area's "nuclear" hummingbird community. Of these species, the status of *Discosura conversii* is least certain. This small species seems restricted to the upper canopy and may be more numerous than infrequent sightings suggest. It is very rarely captured in mist-nets, and I lack firm data on breeding and

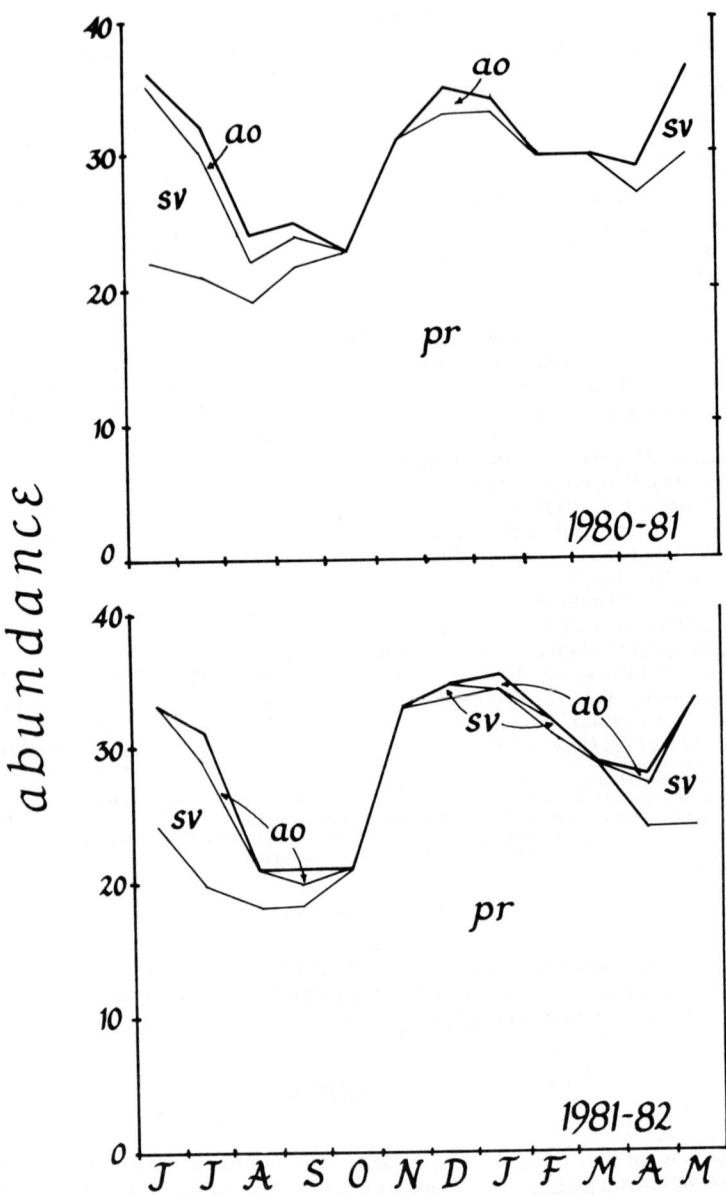

FIG. 3. Hummingbird abundance at La Montura during the two years of this study, according to the semiquantitative system presented in the text. pr = permanent residents and breeding residents; sv = nonbreeding seasonal visitors; ao = accidentals and occasional visitors.

molt. Nevertheless, it seems to be present in most months, and I have seen possible breeding displays by males in December and March. The other marginal species is *Eutoxeres aquila,* which reaches its upper elevational limit at La Montura. I found no evidence of its nesting on the study area, although a few individuals could be seen or netted at any time of year. Given the extremely powerful flight of this large species, it is possible that the birds of the study area, including females with brood patches, might be commuting from lower elevations on a daily basis.

The numerically dominant species of the community are *Phaethornis guy, Lampornis hemileucus,* and *Heliodoxa jacula* on a year-round basis, although each shows seasonal fluctuations in abundance. Species that are abundant at certain seasons but rare or absent at others include *Elvira cupreiceps* and *Eupherusa nigriventris*. Such a pattern is also shown by *Colibri delphinae,* which however, is never as numerous as the two preceding species. Present year-round in smaller numbers are *Heliothryx barroti* and *Doryfera ludovicae*. Morphologically, the nuclear community consists of two hermits, one very large and with a very strongly curved bill (*E. aquila*), and one medium-sized, with a less decurved but very long bill, and six nonhermits. One of the latter has a very long, slender, slightly upturned bill (*D. ludovicae*); four are of medium to rather large size with straight bills of rather intermediate lengths; three are small-sized and short-billed (Table 1). The implications of these patterns will be considered below.

Relatively few species are strictly confined to a particular forest stratum at La Montura, and most species occur in edge habitats (e.g., light gaps, streams). For example, *P. guy* ascends freely into the lower canopy at times to visit flowers, unlike its congener *P. superciliosus* at lower elevations (Stiles 1980). Individuals of species such as *E. nigriventris* move freely between canopy and understory, feeding largely in the former and singing or nesting in the latter. Often males tend to occur more in the canopy, and females in the understory, but both sexes (notably of *E. cupreiceps, L. hemileucus*) may mix at patches of flowers in either. For hummingbirds such as *H. jacula* and, to some extent, *E. cupreiceps* and *L. hemileucus,* the preferred foodplants themselves occur at a wide range of heights above ground (see below). These patterns doubtless reflect the broken, irregular canopy of the La Montura forest. Only *D. ludovicae, C. delphinae,* and *D. conversii* seem confined to the canopy in this forest; only *E. aquila* is restricted to the understory (Table 1).

FLOWERS

The flowers of well over 60 species of plants are more or less regularly visited by hummingbirds at La Montura (Appendix I). At least 50 species are pollinated primarily by hummingbirds. Fully half of these ornithophilous species are epiphytes. Shrubs and large terrestrial herbs are represented by 10 or more species each, while the remaining species include a few vines and (mostly small) trees. The list of non-ornithophilous flowers at which hummingbirds forage also includes a high proportion of epiphytes, and relatively more trees. This list is decidedly incomplete, due in part to the difficulty of obtaining specimens of many trees and epiphytes for identification. Many such species are also visited only infrequently or irregularly, depending upon what else is in bloom.

The great majority of the hummingbird-pollinated flowers of La Montura are concentrated in a few plant families: Musaceae (= Heliconiaceae of some authors), Bromeliaceae, Gesneriaceae, and Ericaceae (Appendix I). If one considers numbers of flowers, the Musaceae and Bromeliaceae decline in overall importance, while the Ericaceae is easily the most important family, followed by the Rubiaceae, Gesneriaceae, and Acanthaceae. Other parameters useful to consider are corolla length (and to a lesser extent curvature) and nectar production. Some families can be characterized by a particular combination of these characters (e.g., all Musaceae have relatively high nectar production and long corollas), whereas others (e.g., Gesneriaceae, Ericaceae) are much more heterogeneous. Taking the plant species visited (Appendix I) as a whole, there is a highly significant positive association between corolla length and nectar production, and a highly significant negative association between flower abundance and nectar production (Table 2). That is, the most abundant flowers tend to be those with short corollas and low nectar production; flowers with long corollas and high nectar production tend to be scarce overall, although they may occur in relatively large clumps (e.g., *Satyria warsewiczii*). Most species of hummingbird foodplants at La Montura have a single discrete flowering peak each year. Certain families tend to have more irregular patterns of flowering (especially the Bromeliaceae), but there is no obvious association between the type of blooming seasonality and either corolla length or nectar production (Table 2, Appendix I).

SEASONAL PATTERNS IN THE COMMUNITY

HUMMINGBIRD NUMBERS, BREEDING, AND MOLT

Overall hummingbird numbers were consistently high through most of the year, roughly from November through July, and rather sharply lower between about August and October (Fig. 3). However, patterns of abundance varied considerably among different members of the community. Some species were relatively common year-round, albeit less numerous in some months than others. For instance, *P. guy* was less numerous between about December and April; *L. hemileucus* declined somewhat between August and October. Seasonal fluctua-

TABLE 2

RELATION BETWEEN DAILY NECTAR SECRETION AND EFFECTIVE COROLLA LENGTH, PEAK
FLOWER ABUNDANCE AND BLOOMING SEASONALITY OF 40 SPECIES OF HUMMINGBIRD
FOODPLANTS AT LA MONTURA

Nectar secretion[1]	Flower characteristics				
	Effective corolla length (mm)[1]				
	<10	10–19	20–29	30–39	≥40
Low	4	14	2	1	0
High	0	0	4	11	4

χ^2 (combining species above and below 20 mm) = 26.78; $P < 0.0001$

	Maximum flower count in censuses[1]			
	<10	10–100	101–1000	>1000
Low	1	8	7	5
High	3	13	3	0

χ^2 (combining species above and below 100 flowers) = 5.63; $P < 0.05$

	Type of blooming seasonality[1]					
	A_1	A_2	B_1	B_2	C	D
Low	14	0	1	1	2	3
High	9	3	3	0	3	1

[1] As in Appendix I.

tions in the numbers of other species, notably *Elvira cupreiceps, Eupherusa nigriventris,* and
to some extent *Heliodoxa jacula,* were more marked. All of these species are consistently
abundant at certain times of year; the timing and extent of their periods of relative scarcity
seem more variable. For instance, *P. guy* decreased on the study area in late November in
1980–1981, but not until about mid-January in 1981–1982. Similarly, in 1981 the declines
in *E. nigriventris* and *E. cupreiceps* occurred about early- to mid-July; in 1982, they occurred
in April. *Eutoxeres aquila, D. ludovicae,* and *H. barroti,* were never very numerous, but were
evidently present year-round. The short-term visitors varied considerably in both numbers
and timing of their visits on a year-to-year basis. Both *T. colombica* and *C. hemileucurus*
were more numerous and stayed longer in the area in 1980 than in 1981; *L. calolaema* appeared
in April 1982, but not until June in 1981. Taken as a whole, the peak abundance of these
visitors was June–July. The overall abundance of hummingbirds in the community changed
rather little, however, as several resident species declined at this time. The August to October
low point of hummingbird abundance resulted from the gap between the departure of these
visitors, and the return to the area of large numbers of resident species such as *E. nigriventris*
and *E. cupreiceps* (Fig. 3).

The peak breeding season for most La Montura hummingbirds was about November to
March, with breeding commencing sometime between August and October; virtually all these
species ceased breeding by April (Fig. 4). The breeding season of *H. jacula* was slightly earlier,
roughly late July through December or January. *Phaethornis guy,* however, departs completely
from this pattern, with a breeding peak from about April through August. I lack adequate
information on breeding by *E. aquila, H. barroti* and *D. conversii.* At La Montura, a female
H. barroti with nesting material was seen in September 1980; female *Eutoxeres* with brood
patches were netted mostly from January through March (but also one in May and one in
July). At Finca La Selva in the lowlands, *E. aquila* definitely breeds between December and
February, and *H. barroti* between July and October (Stiles 1980, unpubl. data).

As is usual in hummingbirds, molt followed immediately after breeding in most or all
species at La Montura (Fig. 5). Most species began molt in February or March, *H. jacula* as
early as late November. Perhaps because males have lower energetic demands than females
during breeding, males of many species commenced molt one to two months ahead of females
(January vs March in *L. hemileucus,* February vs March in *E. nigriventris,* March vs April
in *E. cupreiceps,* November vs December or January in *H. jacula*). Only in *P. guy* was there
some temporal overlap between molt and breeding, with a few individuals commencing molt

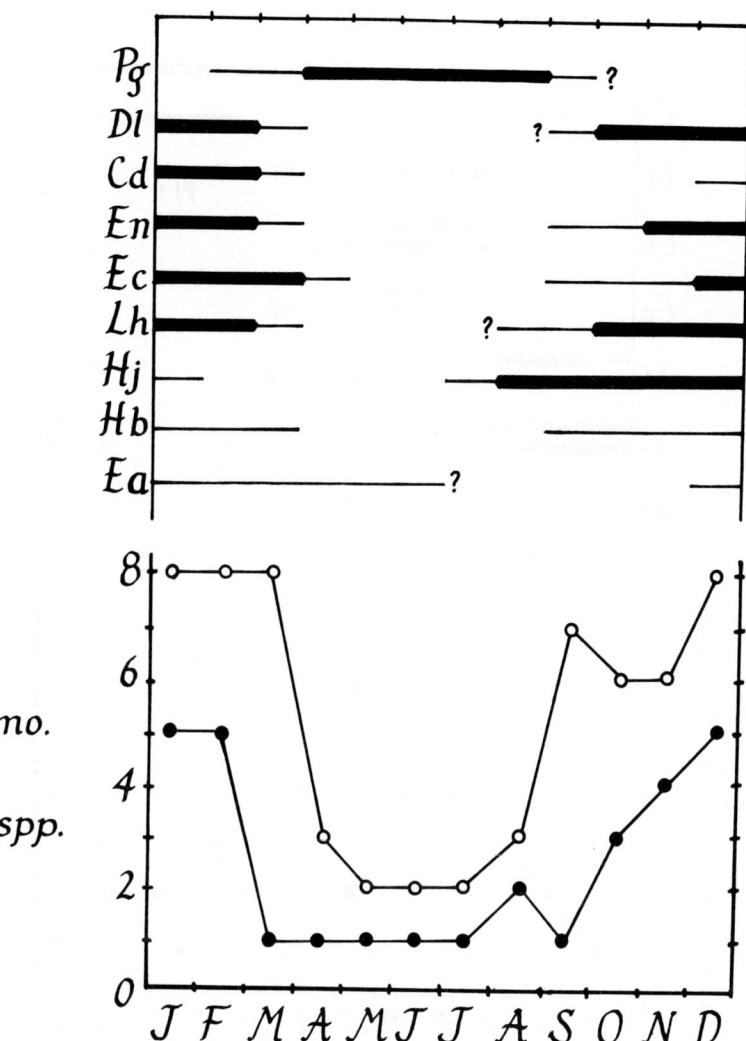

FIG. 4. Above: breeding seasons of La Montura hummingbirds; abbreviations for species as in Table 1. Heavy lines indicate peak of breeding season (insufficient data are available to determine breeding peaks in *E. aquila, H. barroti*). Below: number of species showing breeding activity (upper line) or at peak of breeding (lower line) in different months.

as early as late May or June. Overlap was at best slight for the bulk of the population, however, which showed a well-defined peak of molt from September to November–December.

I detected no significant between-year variation in the timing of molt for any species at La Montura. The timing of breeding did appear to vary somewhat from year to year, however. The breeding seasons of *L. hemileucus, E. cupreiceps, E. nigriventris,* and, perhaps, *H. jacula* all appeared to begin about a month earlier in 1980–1981 than in 1981–1982. In 1980, some *P. guy* continued breeding at least through October; in 1981, no breeding was detected after early- to mid-September. In general, more data are needed for a thorough evaluation of annual variation in breeding (or molting) seasonality in La Montura hummingbirds.

Population movements, particularly altitudinal migrations, seem to be characteristic of many hummingbirds at La Montura. The periods of lower abundance of many resident species result from partial or complete emigrations from the study area to lower elevations. For instance, *L. hemileucus* and *E. cupreiceps* regularly appear as low as 600 to 700 m during the

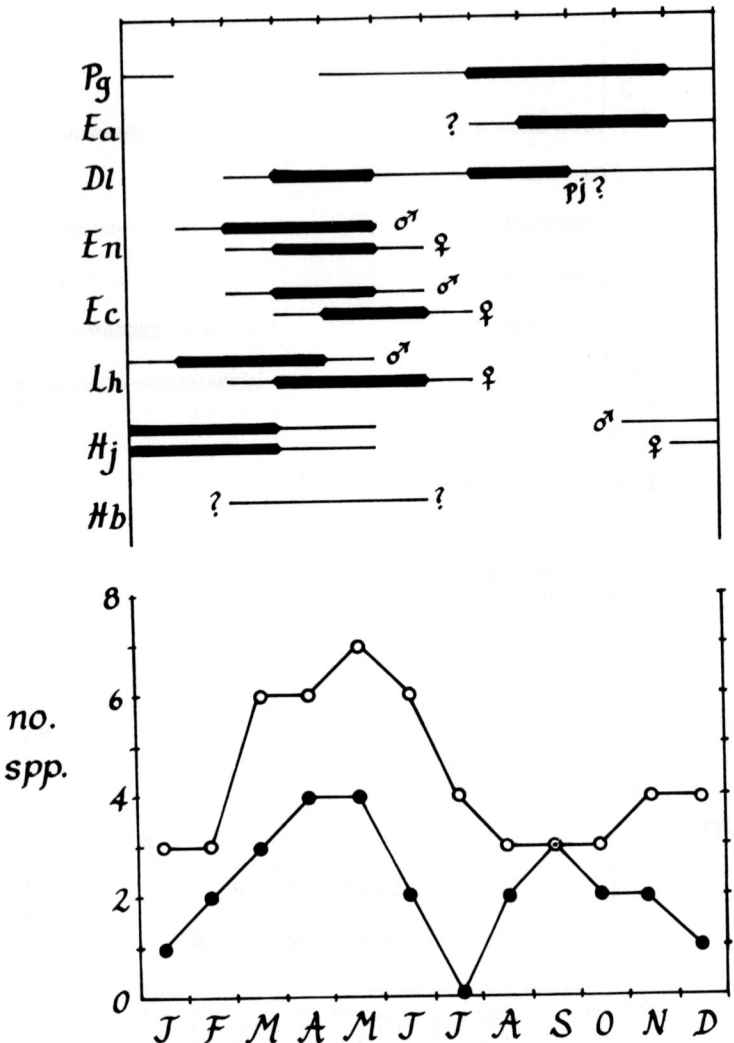

FIG. 5. Above: molting seasons of La Montura hummingbirds; abbreviations for species as in Table 1. Heavy line = peak of molting season (50% or more of all individuals mist-netted in molt); pj? = suspected postjuvenal molt. Lower: number of species showing some (upper line) or intense (lower line) molting activity in different months.

period when they are scarcest at La Montura. *Colibri delphinae* occurs at least occasionally as low as 75 m (Finca La Selva) in June and July, whereas *D. conversii* appears there in numbers from July through September of some years (Slud 1960, Stiles 1980). Small numbers of immature *P. guy* also appear at La Selva during these months; in view of the observed breeding season at La Montura, these birds probably represent post-fledging wanderers. Adult *P. guy* very rarely appear at such low elevations, and then only during the nonbreeding season (mostly December–February).

Influxes of several seasonal visitors definitely reflect altitudinal movements. The species that appear during the flowering of *Cephaelis elata* are drawn from several different elevational bands. *Thalurania colombica* breeds in the lowlands and foothills up to ca. 600 m, *C. thalassinus* in the mountains mostly above 2000 m (cf. Wolf et al. 1976), and *C. hemileucurus* at upper-middle elevations from ca. 1400 to 2300 m (pers. observ.). Elevational movements are strongly suspected, but not yet confirmed for some other species (e.g., *E. nigriventris*). Finally, movements of a few species seem to represent mainly nonaltitudinal shifts to other

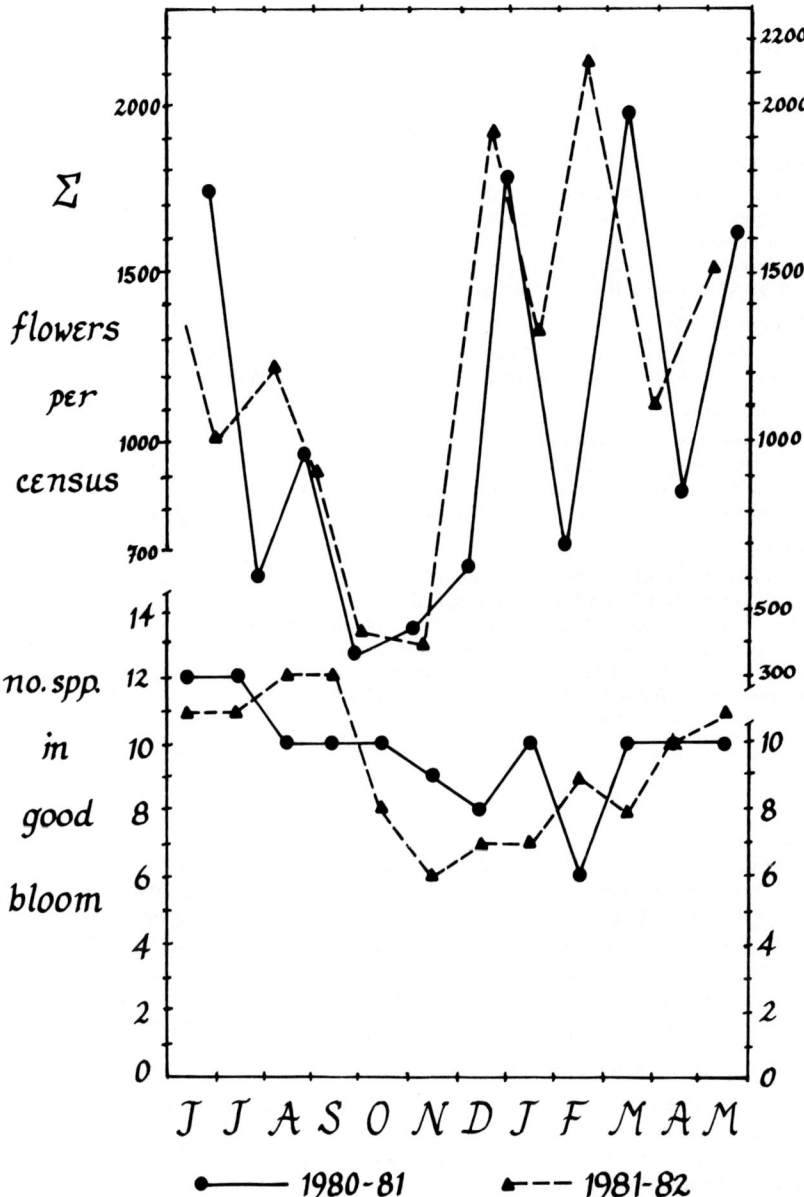

Fɪɢ. 6.   Numbers of hummingbird-pollinated flowers recorded on flowering censuses at La Montura.
Lower: number of ornithophilous plants in good bloom (see text) by month, June 1980–May 1982.

habitats where flowers not present on the study area occur. The scarcity of *H. jacula* in June
and July reflects its departure to riparian habitats where *Heliconia latispatha* is in good bloom;
it returns to the study area as blooming of this plant declines.

Bʟᴏᴏᴍɪɴɢ Sᴇᴀsᴏɴs ᴏғ Hᴜᴍᴍɪɴɢʙɪʀᴅ Fᴏᴏᴅᴘʟᴀɴᴛs

The numbers of ornithophilous plants in good bloom in the community were greatest in
May and September, and lowest sometime between November and February (Fig. 6). However,
the total number of hummingbird-pollinated flowers present at a given time showed a very
different pattern, being lowest around October–November, and relatively high, but widely
fluctuating, between about December and June (Fig. 6). The lack of correlation between these

TABLE 3

CORRELATIONS BETWEEN HUMMINGBIRD BREEDING AND ABUNDANCE, AND FLOWERING
PARAMETERS

	r_s^1
Entire Community	
Total no. of flowers vs total hummingbird abundance	.419*
No. spp. in good bloom vs total hummingbird abundance	−.250
Total no. of flowers vs no. breeding hummingbird spp.	.196
No. spp. in good bloom vs no. breeding hummingbird spp.	−.710**
No. spp. in good bloom vs total number of flowers	−.132
Lancebill Subcommunity[2]	
Total no. of flowers (including bromeliads) vs abundance of *D. ludovicae*	.366
No. flowers of 5 principal foodplants, only, vs abundance of *D. ludovicae*	.440*
Hermit Subcommunity[2]	
Total no. of flowers vs abundance of *P. guy*	.502*
No. of *Heliconia* flowers vs abundance of *P. guy*	.706**
No. flowers of *Centropogon* flowers vs abundance of *E. aquila*	.643**
No. *Centropogon* plus *Heliconia trichocarpa* flowers vs abundance of *E. aquila*	.652**
Generalized Subcommunity[2]	
Total no. of flowers vs total hummingbird abundance	.784**
Total no. of flowers vs no. breeding hummingbird spp.	.482**
No. flowers of epiphytes only vs abundance of breeding hummingbird spp.	.799**
No. flowers of epiphytes only vs no. breeding hummingbird spp.	.819**
No. flowers of *Cephaelis* vs abundance of seasonal visitors	.572**
No. spp. breeding vs abundance of breeding hummingbird spp.	.785**
Abundance of seasonal visitors vs abundance of breeding spp.	−.350

[1] r_s = Spearman nonparametric correlation coefficient. Total 21 censuses, thus 19 degrees of freedom for all tests. n.s. = $P > 0.05$; * = $P \leq 0.05$; ** = $P \leq 0.01$.
[2] Total for these tests refers to total numbers within the particular subcommunity.

parameters (Table 3) suggests their inadequacy as indicators of relative resource availability to hummingbirds. Moreover, only a weak correlation exists between total flower availability and hummingbird breeding activity, and a strong negative correlation between the latter and the number of species in good bloom (Table 3). This result, at first unexpected, indicates that the community as a whole might not be the appropriate unit for analysis, as different segments of it might have rather different seasonal rhythms. The situation is further complicated by the fact that overall flowering patterns differed between the two years of the study.

HUMMINGBIRD-FLOWER SUBCOMMUNITIES

Certain flowers seem to be characteristically associated with certain hummingbird species that are usually (but not invariably) their primary pollinators. Other flowers are visited and pollinated by a much wider variety of hummingbirds (Appendix I). Conversely, the different hummingbird species overlap to varying extents in their patterns of flower visitation. These overlap values permit subdivision of the community into sets of birds that exploit similar sets of flowers; seasonal patterns in the birds can then be related specifically to the set of flowers used. By these criteria, the nuclear hummingbird community of La Montura includes a group of five species among which overlaps are moderate to high, and four species that overlap relatively little with this group or with each other (Fig. 7).

The five broadly overlapping species (*L. hemileucus, E. nigriventris, E. cupreiceps, C. delphinae,* and *D. conversii*) are all small to medium-sized (3–6 g) with short to medium-length (11–20 mm), straight bills (Table 1). All visit a rather wide array of flowers that have short to moderate-length corolla tubes, many of which are entomophilous (Appendix I). I have recorded each species visiting 10 or more flower species (except for *D. conversii,* for which I have few data).

Closest to the preceding "generalized" group is *H. jacula,* which visits many of the same flowers during part of the year. Between about August and February or March, however, nearly all of my foraging records for this species are at three species of *Marcgravia* that are

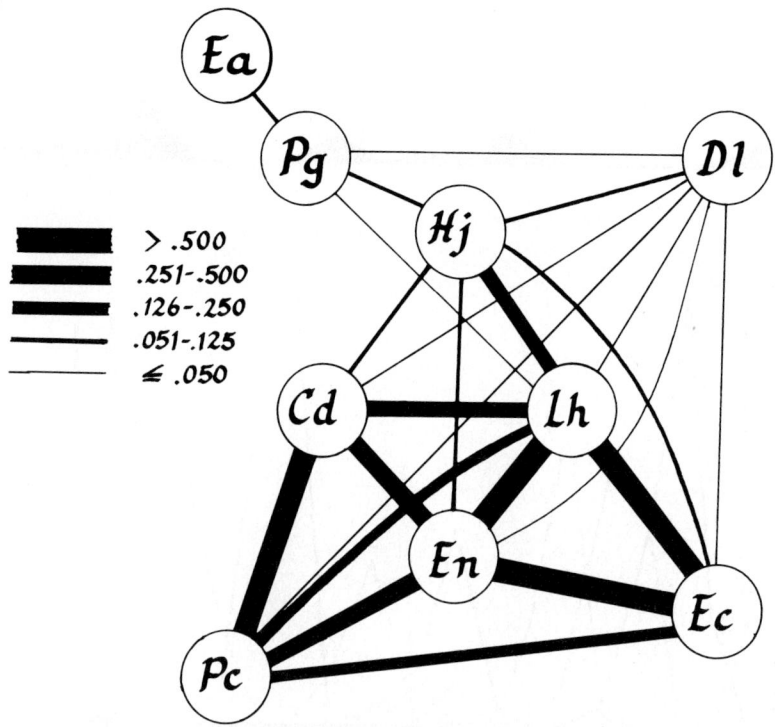

FIG. 7. Pairwise overlap values in flower visitation between all resident species of La Montura hummingbirds throughout the study period. For abbreviations of species see Table 1. Total number of flower visits for all species—786.

very rarely used by other hummingbirds. Thus, during these months one may define a *Heliodoxa-Marcgravia* subcommunity.

The two species of hermits clearly form a separate group. Overlap between them is low, because *E. aquila* visited only a very limited subset of the flowers visited by *P. guy.* However, since it visited no flowers not also visited by the latter, the two species are best included in the same subcommunity. *Doryfera ludovicae* is practically the sole visitor and pollinator of five species of canopy epiphytes with long corolla tubes. This hummingbird and its foodplants clearly constitute another recognizable subcommunity.

One species, the awl-billed *Heliothryx barroti,* was not included in the overlap analysis. This bird is a highly specialized, flower-piercing nectar thief. I know of no flower that it consistently pollinates. I have seen *H. barroti* rob flowers pollinated by hermits, lancebills, and members of the "generalized" group; thus, it could be considered to overlap all of these subcommunities. However, the subcommunities themselves presumably result from birds of similar morphologies (especially bill length and curvature) visiting flowers of corresponding corolla types (Snow and Snow 1972, 1980; Wolf et al. 1972). Since its feeding mode renders *H. barroti* essentially independent of floral morphology, I consider this species to exist outside of the system of subcommunities as such.

Lancebill subcommunity.—This clearly-defined grouping comprises the lancebill, *D. ludovicae,* and five species of epiphytic shrubs: the ericads *Satyria warsewiczii, Psammisia ramiflora, Cavendishia* sp., and *C. callista,* and the mistletoe *Psittacanthus nodosus.* Possible additional members of the subcommunity are five species of the bromeliad genus *Guzmania.* However, all of these latter are relatively scarce and bloom very irregularly, thus their importance to *D. ludovicae* is probably slight.

The most striking feature of the flowering of this subcommunity over the study period was its uniformity: major blooming peaks of the five principal foodplants were nearly perfectly staggered. In only five of the 48 sampling periods was no species in good bloom, and in only

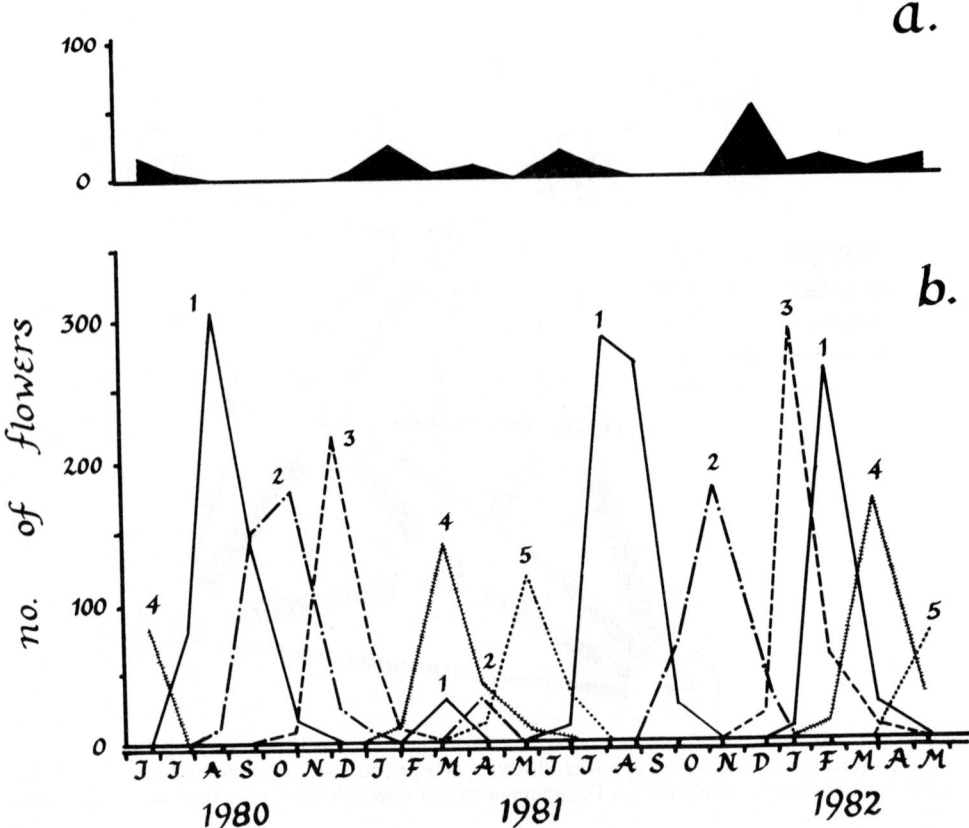

FIG. 8.   Numbers of flowers recorded on different lancebill (*D. ludovicae*) foodplants during the study period. a. total for five species of *Guzmania* bromeliads; b. five principal species:1 = *Satyria warsewiczii*; 2 = *Psittacanthus nodosus*; 3 = *Cavendishia* sp.; 4 = *Psammisia ramiflora*; 5 = *Cavendishia callista*.

seven were there as many as two (Fig. 8). This represents a highly significant deviation from a random distribution of flowering times, in the direction of overdispersion (Table 4). Because blooming peaks of different species involved different numbers of flowers, total flower availability to *D. ludovicae* fluctuated through the year. The breeding season generally coincided with periods during which flower availability was moderately (but not significantly) higher than it was during the nonbreeding season. A significant difference in flower availability did exist, however, between breeding, molting, and quiescent periods (Table 5). Flower availability was also significantly correlated with abundance of *D. ludovicae* on the study area (Table 3). Thus, it appears that the five epiphytic shrubs provide a dependable, staple year-round food supply to the lancebill, the major events of whose annual cycle are timed to fluctuations in this food supply.

Hermit subcommunity.—Of the two hummingbird species of this subcommunity, the Green Hermit is much more abundant, exploits many more species of flowers, and definitely breeds on the study area. Some 18–20 species of understory herbs and shrubs (including six species of *Heliconia*) are visited and pollinated principally or exclusively by *P. guy*. Although these foodplants collectively tend toward a uniform distribution of blooming seasons through the year, the difference from a random distribution is not significant (Table 4). Flower availability varies seasonally, with pronounced peaks in March to April and August to September (Fig. 9). These peaks are considerably less pronounced for nectar availability, however; the two species responsible for the high numbers of flowers, *Razisea spicata* and *Columnea macrophylla*, have less nectar per flower than other species of the subcommunity (Appendix I). Moreover, the flowers of *Razisea* in particular are often pierced and robbed by Bananaquits (*Coereba flaveola*) and other hummingbirds, notably *H. barroti*.

TABLE 4
MEAN AND VARIANCE OF NUMBERS OF FLOWER SPECIES PER TWO-WEEK SAMPLING INTERVAL IN GOOD OR PEAK BLOOM IN DIFFERENT SUBCOMMUNITIES

Subcommunity flowers and breeding period	No. sampling intervals	Measure of flowering	Mean no. spp. flowering	Variance : mean ratio	χ^2 [1]
Lancebill: 5 major flower spp., 1980–82	48	Peak bloom	0.83	0.27	31.90**
	48	Good bloom	1.04	0.26	30.81**
Heliodoxa-Marcgravia (3 spp.)					
1980–81	15	Good bloom	1.07	0.21	10.73**
1981–82	14	Good bloom	1.33	0.29	7.57*
Hermit: 18 spp. *P. guy* foodplants					
Entire 2 years	48	Peak bloom	2.31	0.81	2.37 n.s.
Entire 2 years	48	Good bloom	4.65	0.46	11.94 n.s.
Breeding-early molting periods only	35	Good bloom	5.14	0.28	20.25**
Late molt-nonbreeding periods only	13	Good bloom	3.00	0.18	9.19*
Generalized: 5 spp. canopy epiphytes during breeding season					
1980–81	13	Peak bloom	1.00	0.50	2.88 n.s.
	13	Good bloom	2.23	0.34	8.34*
1981–82	12	Peak bloom	1.42	0.83	0.71 n.s.
	12	Good bloom	2.42	0.34	4.97 n.s.

[1] χ^2 (Chi-square) test: Comparison of observed values with those expected according to Poisson distribution (see text). n.s. $= P > 0.05$; $* = P \le 0.05$; $** = P \le 0.01$.

Both abundance and breeding of *P. guy* on the study area are positively associated with overall flower abundance, but even more closely associated with the numbers of *Heliconia* flowers (Tables 3, 4). Total flower availability is high during breeding but maximal in the period of molt-breeding overlap, much lower when the birds are only molting or quiescent (Table 4). Breeding of *P. guy* is specifically associated with high numbers of *Heliconia* flowers, suggesting that this hummingbird times its breeding cycle to the blooming of these flowers. Molt is mostly finished before the period of flower scarcity that begins in November or December (Figs. 6, 10).

The Sicklebill, *E. aquila,* is never numerous at La Montura, where it visits only the flowers of *Centropogon granulosus* and *Heliconia trichocarpa.* The former is visited relatively infrequently by *P. guy,* which is the chief visitor and pollinator of the latter. At both species *E. aquila* feeds by perching, grasping the inflorescence with its powerful feet; *P. guy* always hovers to feed. In no other hermit foodplants at La Montura is it possible for *E. aquila* to perch and reach the nectar with its strongly curved bill. The abundance of *E. aquila* at La Montura is positively correlated with flowering by *C. granulosus*; this correlation is scarcely improved if the flowering of *H. trichocarpa* is also considered (Table 3). However, the flowering of these plants and the putative breeding season of *E. aquila* do not correspond (Table 5). This strongly suggests that *E. aquila* does not breed at La Montura, and that the critical floral resources for breeding are found elsewhere, most likely at elevations below ca. 800 m, where *Heliconia longa* and *H. pogonantha,* both important foodplants for *E. aquila,* grow (Stiles 1975, 1979a).

Heliodoxa-Marcgravia subcommunity. — During much of the year, most of the activity of *H. jacula* is centered around the inflorescences of three species of *Marcgravia.* Each candelabra-shaped inflorescence of long-stalked flowers is subtended by a tube- or funnel-shaped bract into which the nectar is secreted. *Heliodoxa,* which like *E. aquila* has very powerful feet, perches on the bract or the flower stalks to feed. It is highly unlikely that *H. jacula* pollinates these species: anthesis occurs at or shortly before dusk; and the anthers have mostly fallen by dawn. I have seen the flowers of *M. schippii* being visited, and probably pollinated, by sphinx moths (*Eumorpha* sp.) after nightfall. However, these moths evidently do not use up the nectar in the bracts, which is harvested by *H. jacula* from dawn through mid-morning, and again in late afternoon prior to anthesis of the next crop of flowers. The significantly staggered sequence of flowering peaks shown by the three species (Fig. 10, Table 5), collectively spanned the period July through January in 1980–1981, and August through March 1981–1982; these months comprise the breeding season and at least the start of the molting season of *H. jacula.*

TABLE 5

Comparisons of the Numbers of Flowers Available to Different Hummingbird Species Within Their Respective Subcommunities at Different Seasons[1]

	Breeding	Breeding-molt overlap	Molt	Quiescent	Statistic[2]
P. guy—all flowers	218.9 ± 94.2	380.8 ± 340.6	102.0 ± 51.6	115.6 ± 109.0	$H = 45.52$**
P. guy—*Heliconia* flowers only	51.1 ± 13.3	47.6 ± 19.8	8.4 ± 8.2	4.2 ± 2.0	$H = 78.71$**
D. ludovicae—5 principal spp.	190.0 ± 119.5	—	134.4 ± 78.1	36.7 ± 23.8	$H = 6.55$*

	Breeding	Nonbreeding	Statistic[2]
D. ludovicae—5 principal spp.	190.0 ± 119.5	115.6 ± 71.5	$U_s = 24$ n.s.
D. ludovicae—all spp.	190.0 ± 119.5	117.1 ± 86.9	$U_s = 28$ n.s.
P. guy—all flowers	365.5 ± 213.6	117.4 ± 80.4	$U_s = 16$*
P. guy—*Heliconia* flowers only	51.5 ± 15.3	8.8 ± 9.0	$U_s = 11$*
E. aquila—*Centropogon* flowers	37.8 ± 22.7	35.8 ± 19.5	$U_s = 42.5$ n.s.
E. aquila—*Centropogon* plus *H. trichocarpa* flowers	37.8 ± 22.7	45.4 ± 25.2	$U_s = 60$ n.s.

[1] Mean and standard deviation of numbers of flowers at different seasons.

[2] Statistics: H = Kruskal-Wallis nonparametric ANOVA; U_s = Mann-Whitney test. n.s. = $P > 0.05$; * = $P \leq 0.05$; ** = $P \leq 0.01$.

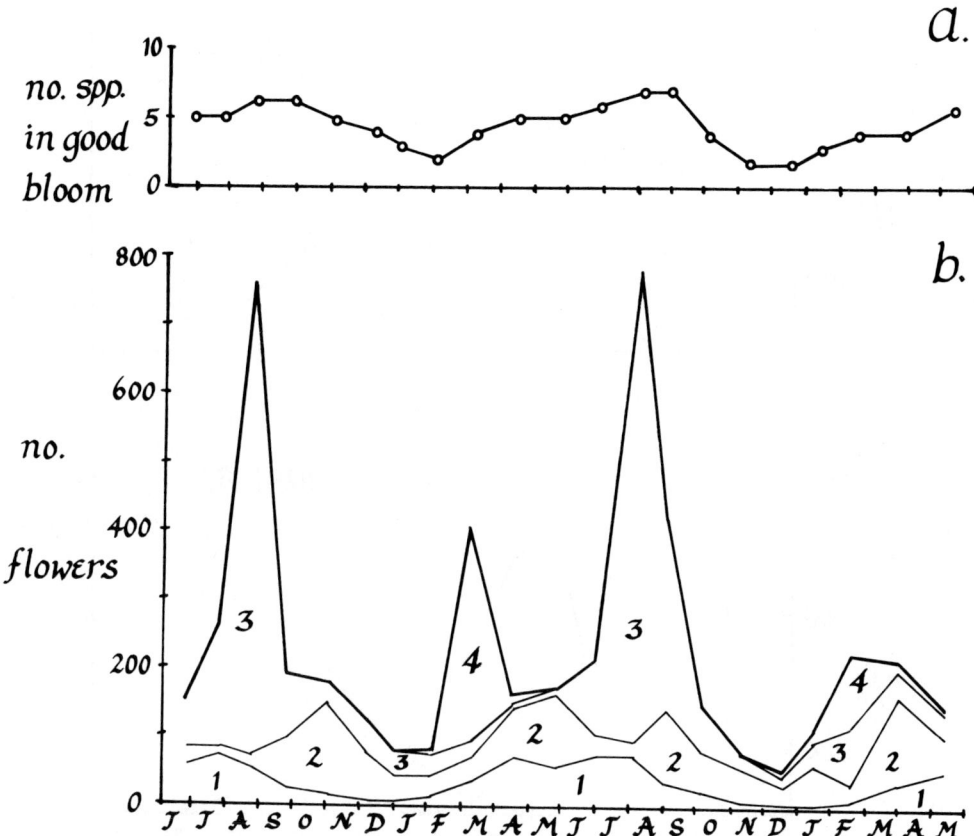

FIG. 9. Numbers of hermit (*Phaethornis guy*) foodplants in good bloom throughout the study period. 1 = *Heliconia* spp., *Costus* spp.; 2 = several dicot foodplants (Gesneriaceae, Labiatae, Malvaceae, Rubiaceae); 3 = *Razisea spicata*; 4 = *Columnea macrophylla*.

Were *H. jacula* the pollinator of *Marcgravia* spp., this would appear to be a fine example of bird-flower coevolution. In reality, the hummingbird is essentially a nectar-thief that seems to have organized its annual cycle at La Montura around such parasitism.

A variety of other flowers was visited, and in some cases pollinated, by *H. jacula* at La Montura. These were of secondary importance until *Marcgravia* flowers became scarce; then *H. jacula* often visited flowering trees (e.g., *Inga*, *Calliandra*), and sometimes defended territories at the abundant flowers of *Cephaelis elata* and (off the study area) *Heliconia latispatha*.

"*Generalized*" *subcommunity.*—As suggested above, this subcommunity represents something of a catch-all for small- to medium-sized hummingbirds, and flowers with short to medium-length corollas. No exclusive bird-flower associations are evident within this loose assemblage, in which flower choice by the birds more often reflects such factors as microhabitat preferences and aggressiveness. The number of hummingbird-pollinated flowers in good bloom in this subcommunity fluctuates irregularly during most of the year; few species were in good bloom in August–September 1980 or in October 1981. A somewhat different picture emerges if number of flowers is considered, since three species (the epiphytes *Cavendishia quereme* and *Thibaudia costaricensis,* and the treelet *Cephaelis elata*) produce many more flowers than the rest. The first two of these, along with *Columnea querceti* and two other species of *Cavendishia,* provided a rich supply of flowers in the forest canopy from about December through April (Fig. 11). This included the peak of the breeding season and at least the start of molt for *L. hemileucus, E. nigriventris, E. cupreiceps,* and *C. delphinae.* For these species, breeding was closely correlated with the abundance of flowers of these canopy epiphytes (Table 3). Blooming of these epiphytes tended to be uniformly distributed (but usually not significantly

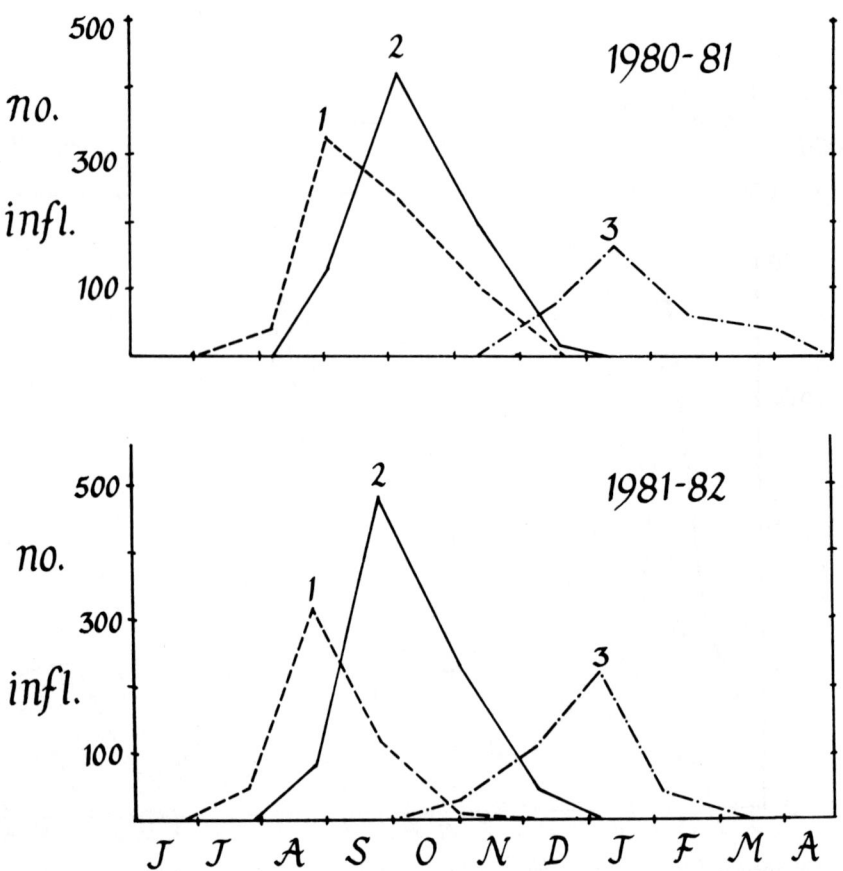

Fig. 10. Numbers of inflorescences recorded on three *Marcgravia* species used by *H. jacula*, 1980–1982. 1 = *M. schippii*; 2 = *M. affinis*; 3 = *M. pittierii*. Each inflorescence is subtended by one nectar-bearing bract.

so) during the birds' breeding seasons (Table 5). Annual variations in timing of breeding in these hummingbirds corresponded rather well with annual variations in timing of flowering, in particular the attainment of good bloom by *C. capitulata*, which occurred about a month earlier in 1980–1981. On the other hand, in both years *E. nigriventris*, *C. delphinae*, and to a lesser extent *E. cupreiceps* left the community during much of the flowering period of *Cephaelis*. This coincided approximately with an influx of *T. colombica*, *C. hemileucurus*, and *C. thalassinus*, and largely explains the negative correlation between the abundance of these two groups (Table 3).

A final group of flowers that is best discussed here includes those species that are visited, but not pollinated (at least not exclusively) by various species of hummingbirds. These species have mostly short-tubed or cup-shaped corollas or nectar-bracts; their often very abundant flowers generally produce little nectar. Included are several trees, especially of the Mimosaceae (*Inga* spp., *Pithecellobium arboreum*, and especially *Calliandra arborea*), two species of Marc-graviaceae (*Norantea sessilis*, *Souroubea* sp.), and a variety of other epiphytes (e.g., *Clusia* spp.), shrubs (e.g., *Palicourea* spp., *Witheringia warsewiczii*), and lianas (Bignoniaceae, Com-positae, Apocynaceae) (Appendix I). For most species of La Montura hummingbirds, these flowers are "supplementary," visited primarily when preferred, usually ornithophilous flowers are scarce or unavailable (due to defense by other, more aggressive hummingbirds). However, two hummingbird species seem to be more regularly associated with such plants (and are thus somewhat peripheral in terms of the community of hummingbird-pollinated plants *per se*): the fairly large *C. delphinae*, and the small *D. conversii*. The most important plant of this

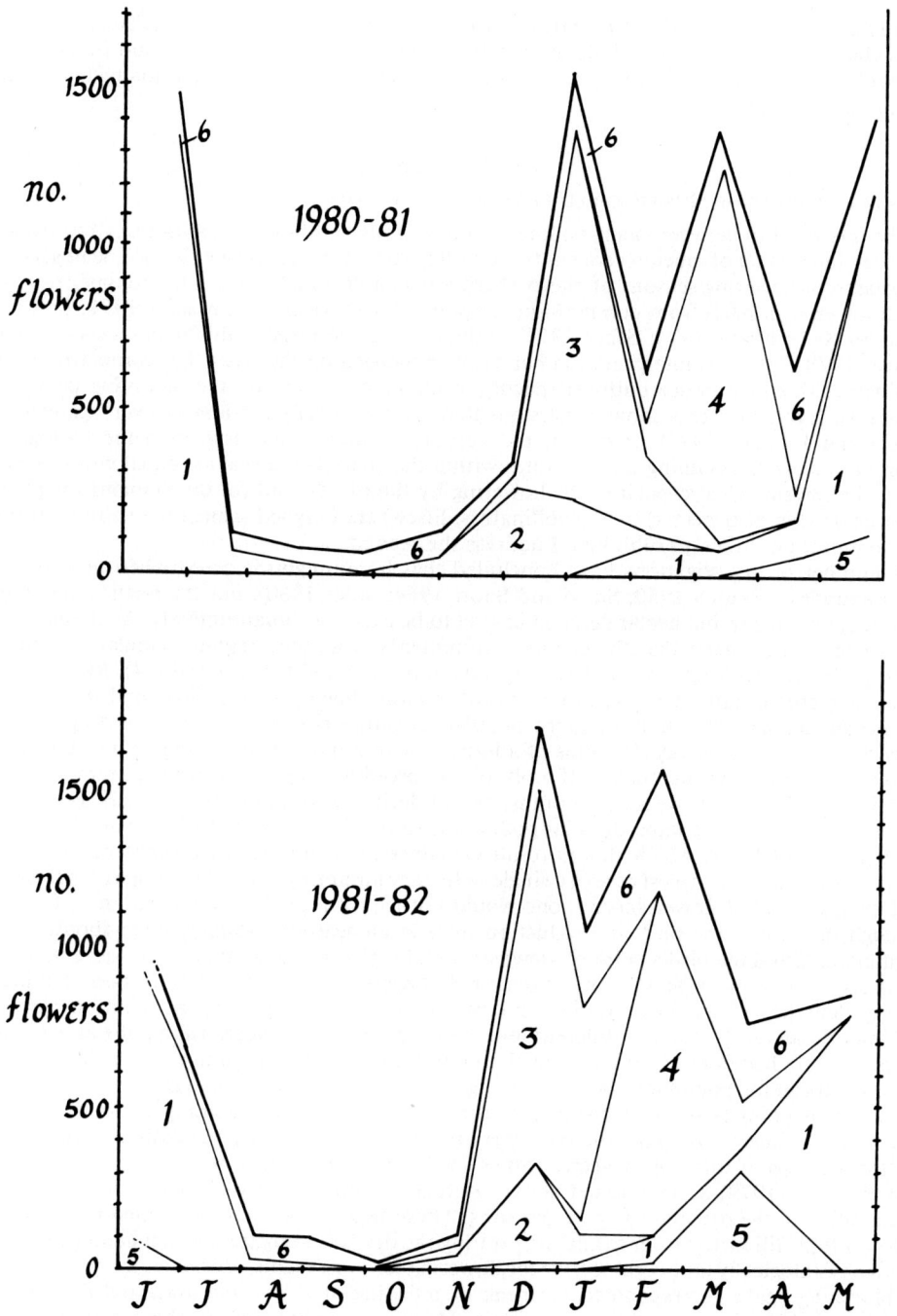

FIG. 11. Numbers of flowers recorded on foodplants used by hummingbirds of the generalized sub-community throughout the study period. 1 = *Cephaelis elata*; 2 = *Cavendishia capitulata*; 3 = *C. quereme*; 4 = *Thibaudia costaricensis*; 5 = *C. endresii*; 6 = other species (Rubiaceae, Cucurbitaceae, Gesneriaceae, etc.).

group for the hummingbird community as a whole is undoubtedly *Calliandra arborea* because of its abundance and year-round blooming (albeit with a pronounced peak around June–July). Probably pollinated mainly by sphingids or small bats, this tree is also a mainstay of the nectar-robbing Bananaquit.

DISCUSSION

SEASONAL ASPECTS OF HUMMINGBIRD-FLOWER INTERACTIONS

The relationship between plant-pollinator interactions and flowering seasons of the plants has been the subject of much recent controversy (see review in Feinsinger 1983). The observed dispersion of flowering seasons of plants sharing pollen vectors has often been considered to result (at least in part) from competition for pollinator services (e.g., Macior 1971; Frankie et al. 1974; Heithaus 1974; Stiles 1977), although specific tests of this hypothesis are few (Waser 1978a, b). This interpretation has been challenged on the basis that competition for pollinators should produce uniform spacing of blooming peaks, yet the blooming peaks of several such sets of plants show dispersions that are statistically random or even aggregated (Poole and Rathcke 1979). However, the validity of these tests rests upon the biological appropriateness of assuming that all times within the analysis interval are equal with respect to (a) the physiological capacities for flowering by the plants, and (b) the demand for plant rewards (in this case, nectar) by the pollinators. Elsewhere I argued against the validity of the first assumption (Stiles 1979b); here I address the second.

A number of investigations have concluded that hummingbirds breed when flowers are most abundant (Skutch 1950; Snow and Snow 1964; Stiles 1980), but the relation between flowering phenology and nectar demand has yet to be examined quantitatively. Available data and calculations suggest that the energy requirements of a hummingbird population will at least double in the course of the breeding season (Calder 1974; Ricklefs 1974). Molt might easily increase the daily energy requirements of an individual by up to 30% (King 1980). Thus, the nectar demand of a hummingbird population shoud rise through the breeding season (reflecting increased energy demands of adults, growth requirements of the young, and ultimately the population increment). If molt follows breeding directly, nectar requirements will remain high for several weeks or months, before declining to maintenance levels. From the flowers' point of view, competition for pollinators should be relaxed at this time due to higher pollinator availability, which should result in reduced selection for divergence in blooming seasons. Even in the simplest case of a single sedentary hummingbird interacting with a limited and exclusive set of flower species, one would not expect a uniform distribution of flowers through the year if the bird's reproduction were at all seasonal. Rather, there should be a predictable (from the bird's point of view) season of higher nectar availability—more flowers, more species of flowers, and/or more nectar-rich flowers—corresponding to the period of high energy demands of the bird. Within this period there might well be competition for the pollinator's services—but at a different level from the rest of the year. Taking the entire year as the unit of phenological analysis may therefore lead to misinterpretation of the significance of competition for pollinators in determining the spacing of blooming seasons.

A case in point is my conclusion that the dispersion of the blooming peaks of hermit (*Phaethornis superciliosus*)-pollinated flowers at Finca La Selva reflected possible competition for the services of this pollen vector (Stiles 1977). This conclusion was criticized by Poole and Rathcke (1979) as mentioned above. Actually, there were two broad peaks of flower availability, in the dry and early wet seasons. These two periods were separated by a period of lower (but still high) flower availability during the dry-wet transition, which I argued (1979) was a physiologically unfavorable time for flowering for most plants. However, the two clusters of blooming peaks correspond to the peak of breeding of *P. superciliosus,* and the end of breeding and the peak of molt, respectively. Moreover, within each of these periods taken separately, blooming peaks are overdispersed (Cole 1981). Thus, the observed pattern is precisely what would be expected to result from competition for pollinators, if the physiology of the plants and the nectar demand of the pollinators were taken into account.

Population movements also affect the energy demands of an area's hummingbird community, either accentuating or cushioning the amplitude of variations in the demands of sedentary individuals and species. A very constant nectar demand by one or several species over the year will most likely reflect emigration during the breeding season, immigration in the non-breeding season, or both. Conversely, such movements may be an integral part of a species' strategy for satisfying its energy requirements when the availability of rich nectar sources varies in space and time (Feinsinger 1980; Stiles 1980).

Many of these patterns are clearly reflected in the flowering of hummingbird foodplants at La Montura. The theoretical case of a single sedentary hummingbird with its limited, exclusive set of foodplants is rather closely approached by the Lancebill subcommunity. Superimposed upon a highly uniform distribution of flowering peaks is a period of higher flower abundance during which *D. ludovicae* breeds; intermediate flower availability prevails during molt. The lower flower abundance during molt, and the fivefold difference in this parameter between breeding and quiescent periods, evidently correspond to a partial postbreeding emigration.

The two members of the hermit subcommunity place different demands upon the study area's nectar resources. *E. aquila* varies in abundance with the flowering of *Centropogon,* but its energy demands for breeding and molt are apparently satisfied elsewhere. The phenologies of *P. guy* and its foodplants appear nearly as closely linked as in the Lancebill subcommunity. Blooming periods are not uniformly distributed over the entire year, but they tend toward uniform distributions within both breeding and quiescent periods considered separately. Peak flower availability corresponds with the period of molt-breeding overlap, and earlier in the breeding season flower availability averages twice that during late molt and quiescent periods. The specific association of breeding by *P. guy* and blooming of the high-nectar *Heliconia* flowers emphasizes the close ecological relationship that seems to exist between these two genera throughout their distributions (Stiles 1979a, 1981).

Such patterns are much less clearcut in the generalized subcommunity because of the number of species involved, the low degree of bird-plant specificity, and the manifest importance of population movements in the annual cycles of most of the hummingbird species. Certainly, periods of high flower availability on the one hand, and of high nectar demand (breeding and molt) by the pollinators on the other, generally correspond (Table 4). There are some suggestions of competitive interactions among certain foodplants, however. There was little or no overlap in blooming between the three really abundant plant species (*Cavendishia quereme, Thibaudia costaricensis, Cephaelis elata*). Blooming peaks of some rarer species may be "repulsed" by those of more abundant species, notably that of *Cavendishia endresii* by that of *Thibaudia*; as a result, the former may exhibit major changes in its blooming period from one year to the next. The almost complete separation of blooming periods of *Cavendishia capitulata* and *C. endresii* may reflect selection for genetic isolation, as these species are virtually identical in floral morphology. In general, however, specific interactions in this subcommunity are difficult to detect and more study is required, particularly with respect to the possible relation between the flowering of such canopy epiphytes, and hummingbird visitation of the various "supplementary" flowers of the area.

SEASONALITY IN OTHER HUMMINGBIRD-FLOWER COMMUNITIES

Seasonal activity patterns of birds and flowers have been studied in detail at two other Costa Rican sites: Finca La Selva, at 100 m in the Caribbean lowlands below La Montura (Stiles 1975, 1978, 1980; Stiles and Wolf 1979); and Cerro de la Muerte, at 3000 m on the crest of the Cordillera de Talamanca (Wolf 1969, 1976; Wolf and Stiles 1970; Colwell 1973; Wolf et al. 1976). In addition, Skutch (1950, 1966) discussed hummingbird breeding seasons in the El General Valley, at 650 m on the Pacific slope below Cerro de la Muerte, and fragmentary data are available for hummingbirds of the dry forests of Guanacaste, in northwestern Costa Rica (Stiles and Wolf 1970; Wolf 1970). Finally, Feinsinger (1976, 1977, 1978) carried out detailed studies of flowering and hummingbird foraging at Monteverde, at 1200 m on the Pacific face of the Cordillera de Tilarán in northern Costa Rica.

As at La Montura, primary habitats were emphasized in the La Selva study, but some secondary habitats were also included. The Cerro de la Muerte study was done in secondary habitats, but fairly extensive comparisons with primary forest were included. Both primary and secondary habitats were included in the Guanacaste and El General Valley studies, but the Monteverde work was done entirely in young second growth, in a transitional area between the major forest types of the region.

The hummingbird-flower community of La Selva is similar in size to that of La Montura (ca. 20 hummingbirds, 50 hummingbird-pollinated plants), but differs in composition: there are five resident hermit hummingbirds and eight to nine nonhermits, three regular seasonal visitors, and four rare to accidental species. Floristically, the hummingbird foodplants of the understory (many *Heliconia, Costus, Cephaelis elata, Besleria* spp.) are more similar to those of La Montura than are those of the canopy (bromeliads dominant, Ericaceae insignificant). Cerro de la Muerte has a smaller community (ca. five hummingbirds pollinating 20 flower species) with no hermits; all five hummingbird species breed there, but three emigrate for part of the year. Species of Ericaceae are important foodplants in the canopy as at La Montura, but the understory is distinct in terms of hummingbird foodplants. Practically all hummingbird species at La Selva breed between about January and June, and molt between June and September; all groups of hummingbird foodplants show flowering peaks in March–April and July–August, with a pronounced flower scarcity in November–December. The only hummingbird species that apparently departs from this pattern is the nectar-robbing, highly insectivorous *H. barroti,* which breeds from July or August to at least October. Thus, breeding and molting activity at La Selva is at its nadir when most species at La Montura are attaining their peak of breeding; the molting seasons of the La Montura species coincide with the main breeding season at La Selva. Only *P. guy* at La Montura overlaps the La Selva species in its breeding season, which begins 1 to 3 months later than that of its lowland congener, *P. superciliosus* (Stiles 1980). In the El General Valley the hermits also tend to show a breeding peak around July–August, and most nonhermits around January in the dry season (Skutch 1950). On Cerro de la Muerte, the main breeding season is roughly September through about January or February, but there is considerable variation among species and between years. The abundant *Panterpe insignis* may begin to breed as early as May and may finish in October; *Colibri thalassinus* may not commence breeding until December. Flowering patterns vary greatly from year to year, and breeding seasons may do likewise (Wolf et al. 1976). The hummingbirds of the Guatemalan mountains resemble those of the Cerro de la Muerte in breeding mostly at the coldest and wettest time of year, roughly October to January, since flowers are most abundant at this time (Skutch 1950). On the Cerro de la Muerte, the relevant flowers are epiphytic Ericaceae (as at La Montura), as well as species of *Centropogon, Fuchsia,* and *Salvia.*

Although detailed data are lacking, most species of Guanacaste hummingbirds seem to breed during the dry season and molt during the early wet season (at least *Amazilia* spp.; cf. Stiles and Wolf 1970). The seasonal pattern thus somewhat resembles that at La Selva, but the flowers most important for the breeding of most Guanacaste hummingbirds are pollinated by other agents, notably large bees. Wolf (1970) interpreted this to mean that the hummingbirds are relatively recent invaders of a basically insect (and bat)-pollinated plant community. Unfortunately, too little information is available on breeding and molt of Monteverde hummingbirds to relate these events to flowering patterns, and movements into and out of Feinsinger's study areas evidently reflected in large part flowering in adjacent forest habitats. Nevertheless, he found that flowers were available on these areas year-round, and that at some seasons hummingbird numbers tracked flower abundance fairly closely (Feinsinger 1977, 1978).

SUBCOMMUNITIES IN OTHER HUMMINGBIRD-FLOWER COMMUNITIES

My principal criterion for dividing the La Montura hummingbird-flower community into subcommunities was the degree of overlap in flower use by the birds (Fig. 7). The resulting division is correlated with bill length and curvature, since these largely determine a hummingbird's ability to extract nectar from a given flower. However, I neither require nor necessarily expect that other aspects of morphology and behavior will sort out neatly along subcommunity lines. A very different way of subdividing hummingbird guilds, based precisely upon foraging tactics and attempting to draw detailed morphological correlations, was advanced by Feinsinger and Colwell (1978). These authors classified hummingbirds according to their "ecological roles," which were defined not only by the flowers visited, but even more by how they are visited—i.e., whether they are defended or traplined, etc. A detailed critique of this approach will be presented elsewhere; here I explain my reasons for considering the Feinsinger-Colwell scheme inappropriate for the aims of this paper.

My major disagreement with Feinsinger and Colwell concerns the extent to which particular "ecological roles" characterize particular hummingbird species under all, or even most circumstances. Consider *D. ludovicae,* morphologically a classical "high-reward trapliner" in their scheme, because of its specialized bill and flower visitation. While many individuals do indeed trapline small clumps of epiphyte flowers, many males consistently defend rich clumps of *Satyria* and *Cavendishia* sp., while other males and females attempt to poach. A "classical" territorialist like male *L. hemileucus* may trapline flowers of low to high nectar content at any time of the year, or may mix such traplining with defense of rich clumps, even on a daily basis. Much the same could be said for most La Montura hummingbirds, especially in the "generalized" subcommunity. Even the traplining *P. guy* may show transient defense of large clumps of flowers against other hermits before moving on. This does not necessarily imply that "ecological roles" do not exist, but I think that they are applied more appropriately to sex-age groups or to individuals at particular times and at particular flowers, than to species in most cases—and to me, seasonal patterns are most meaningfully treated at the species level. Similarly, Feinsinger and Colwell considered large and small clumps of flowers within a given plant species as separate "resource states." This distinction is valid if foraging tactics are being considered, but it would only be useful for my purposes were blooming seasons to vary systematically with clump size. I have no evidence that this occurs in any plant species at La Montura. For all these reasons, I prefer to compare hummingbird-flower communities in terms of subcommunities rather than "ecological roles."

The division of the hummingbird-flower community into subcommunities is quite notable at La Selva. The hermit subcommunity is quite diverse and complex, with some members overlapping more with nonhermits in flower use than at La Montura. One member, *Threnetes ruckeri* pierces many flowers, and seems especially closely associated with those of the large-bee-pollinated genus *Calathea*. As at La Montura, *E. aquila* is rare and visits only one or two flower species; its center of distribution appears to be at ca. 300 to 600 m between the two sites. The hermit subcommunities of La Selva and La Montura resemble each other most closely in both taxonomic composition and seasonal behavior. The generalized subcommunity at La Selva resembles that of La Montura in consisting mainly of canopy species; the sharp difference in seasonality reflects the difference in blooming between dominant groups of canopy foodplants—the Bromeliaceae at La Selva, the Ericaceae at La Montura. This subcommunity at La Selva may be further subdivided into a group of small hummingbirds (2–3 g) and nectar-poor flowers, and a group of medium-sized (4–7 g) hummingbirds and flowers with longer corolla tubes and greater nectar flow. No real analogue of the *Heliodoxa-Marcgravia* association occurs at La Selva, and the only long-billed canopy specialist possibly analogous to *D. ludovicae,* is the rare and perhaps nonresident *Heliomaster longirostris.*

On Cerro de la Muerte, with only four to five species, the possibilities for defining subcommunities are decidedly limited. Nevertheless, the long-billed *Eugenes fulgens* and its long-tubed flowers do form a subunit apart. Females in particular visit scattered shrubs in the forest understory, thus becoming hermit analogues; the males do this and also visit certain long-tubed canopy flowers (*Columnea magnifica, Passiflora* sp.) in the manner of lancebills. The remaining species comprise the "generalized" subcommunity, within which a great deal of foraging overlap is noted. The small (2.8 g) *Selasphorus flammula* shows a limited divergence from the others in that it visits relatively more small, low-nectar, short-corolla flowers, but even here it overlaps widely with the larger generalist *Panterpe insignis* (5–6 g). Like *E. nigriventris* and *E. cupreiceps* at La Montura, it scarcely merits separate subcommunity status.

Detailed foraging data are lacking for Guanacaste dry-forest hummingbirds, but most species appear to constitute a large generalized grouping. Only the long-billed *Heliomaster constantii* may be separable on the basis of its visits to some long-tubed flowers, but even this requires confirmation. Much the same might be said of the second-growth community at Monteverde studied by Feinsinger (1976, 1977, 1978). This community is far simpler than that of La Montura at a fairly similar elevation, containing only two "principal" hummingbird species and ca. five common seasonal visitors, that together visit some 10 to 15 flower species. Most hummingbirds overlap sufficiently at most plant species to render clearcut division of the community on this basis difficult.

Three subcommunity types thus recur with some regularity in the hummingbird-flower communities of humid tropical forests: (1) long-billed hummingbirds and long-tubed flowers of the forest canopy; (2) long, usually curve-billed hummingbirds and understory flowers with corresponding corollas; and (3) a large, generalized subcommunity of small- to medium-sized, straight-billed hummingbirds and flowers with straight, short- to medium-length corollas. In dry forests and young second growth, this may be the only subcommunity present. This subcommunity may be further divisible on the basis of hummingbird size and bill length, and flower length and nectar flow, along the lines suggested by Feinsinger and Colwell (1978) if values of overlap in flower visitation between the putative subgroups are sufficiently low.

Finally, extrinsic to these systems may be one or more nectar-robbing species like *H. barroti* (and *Coereba*) at La Montura, as well as species that parasitize other plant-pollinator systems, such as *H. jacula* at *Marcgravia* spp.

Is The La Montura Hummingbird-flower Community Coevolved?

This study has demonstrated a close correspondence between the annual cycles of some La Montura hummingbirds and the flowering phenology of their foodplants. To what extent is this mutualism truly a result of plant-pollinator coevolution? Janzen (1980) presented a rigorous definition of coevolution, emphasizing the repeated, reciprocal evolutionary responses of each partner to selective pressures generated by the other. He also described cases that mimic coevolution, in which one participant enters the system and takes advantage of the scarcity or absence of (or displaces) one coevolved partner, then interacts with the other in a manner suggestive of a coevolved relationship. Another complication occurs when one or both coevolved partners actually consists of groups of species; coevolutionary responses are then much more diffuse and difficult to distinguish from fortuitous relationships.

Indirect evidence must normally be used to distinguish the results of bona-fide coevolution from fortuitous, recently-derived associations. Such evidence could include the complexity, degree of obligateness (in the case of mutualisms), and exclusiveness of the interaction itself, with the more complex, exclusive, and obligate associations being more likely to be truly coevolved. The degrees of precision and efficiency are also likely to be greater in a coevolved mutualism than in a fortuitous one. Historical and biogeographic considerations also apply: is there evidence that one or both partners are recent invaders or at the edge of their distributions, or that other possible participants are now extinct? Do present distributions and ecology suggest that both parties to the interaction share a long period of evolution and adaptation in the same localities and habitats?

These criteria are satisfied to different degrees by the different subcommunities at La Montura. The complex, precise, and relatively exclusive correspondences between bills and corollas in the hermit and lancebill subcommunities are certainly suggestive of bona-fide coevolved relationships (although the degree of dependence of *E. aquila* on *Centropogon* flowers for energy resources for breeding and molt is nil, at least on the study area). The generalized subcommunity is clearly an example of diffuse coevolution at best, and some of its members (e.g., *C. delphinae*, *D. conversii*) may in fact be fortuitous associates. Certainly the annual cycles of *L. hemileucus*, *E. nigriventris*, and *E. cupreiceps* seem much more closely correlated with blooming patterns of the principal ornithophilous plants than are those of *C. delphinae* or *D. conversii*.

The *Heliodoxa jacula-Marcgravia* association is clearly not a mutualism, and probably is not coevolved at all. This hummingbird has organized its annual cycle about its parasitism of *Marcgravia*, but because its cycle does not deviate markedly from those of other straight-billed species, the association may not be more than facultative; *H. jacula* is certainly quite capable of visiting many other flowers on the study area. On the other hand, the inflorescence structure of *Marcgravia* seems to deter hummingbirds that cannot cling strongly to the nectar bract: the strong feet of *H. jacula* probably represent a preadaptation (perhaps even an adaptation) for overcoming this defense. Selection on *Marcgravia* to reduce or counteract parasitism by *H. jacula* may be weak because much nectar is secreted at night, and is thus available to pollinators; also, the timing of anthesis precludes direct interference with pollen flow by hummingbirds.

Seasonal patterns provide another way of assessing whether many bird-flower associations at La Montura are fortuitous or coevolved: do the principal events of the annual cycles of birds and flowers occur at the most favorable times in terms of other parameters, such as physiology (cf. Herrera 1982)? It is significant that the main breeding season of the birds, and the greatest flowering of ornithophilous plants occur during the coolest and wettest time of year; both end in the early dry season. For most plant groups and in most areas, heavy rainfall is decidedly unfavorable for flowering (reviews in Frankie et al. 1974; Stiles 1978). Heavy rainfall may also restrict the time available to hummingbirds for foraging, and thus diminish energy reserves for nesting females (Calder 1974; Foster 1974). On physiological grounds, nesting by the birds and flowering by the plants are both occurring at a relatively improbable time of year. The reason for this is not well understood, but may reflect a lower abundance of nectar-robbing insects at this time. Direct data on insect abundance are lacking, but it is noteworthy that nearly all insectivorous birds of La Montura breed in the late dry and early wet seasons (Stiles, unpubl. data; cf. Skutch 1950). A further indication that this relationship is more than fortuitous is the fact that nearly all La Montura hummingbirds start to breed about a month *before* their respective flowers attain high levels of blooming; thus, the increase in flowering coincides closely with the period of rapid growth of nestlings in their first broods. This suggests a more intricate degree of coadaptation of timing mechanisms than simply having an abundance of flowers trigger breeding.

All of the major hummingbirds and flowers of the La Montura community seem to be characteristic components of primary subtropical forest ecosystems in Costa Rica (for plant distributions and ecology see Standley 1937–1938; Durkee 1978; Skog 1978; Wilbur and Luteyn 1978; Dwyer 1980; and Stiles 1982; for similar information on birds see Slud 1964; Ridgely 1975). Indeed, many of the birds and plants are endemic to the mountains of Costa Rica and Panama including the principal birds and flowers of the generalized subcommunity. The hummingbirds *E. aquila, D. ludovicae, D. conversii,* and *P. guy* reach their northern limits in Costa Rica, but well to the north of La Montura; all range south into South America, where their associations with flowers resemble those seen at La Montura (Snow and Snow 1972, 1980; B. A. Stein, pers. comm.). Thus, most species of birds and flowers of the La Montura community seem to share a long period of evolution and adaptation in the same region and habitats, thereby also satisfying this criterion for coevolved relationships.

Many bird-flower relationships of the primary forests of La Selva and Cerro de la Muerte probably also satisfy the requirements for being considered true coevolved mutualisms (cf. Wolf et al. 1976: Stiles 1978, 1980). The situation is far from clear in the second-growth communities studied by Feinsinger (1978). Although the second growth itself is obviously of recent anthropogenic origin, many of the plants and birds of such habitats are also characteristic of natural forest edge and light gaps, and may in fact have been associated for long periods (Stiles 1975). Determining the degree to which plant-pollinator patterns in secondary vegetation reflect fortuitous accidents or historical associations will certainly require extensive studies in the original primary habitats of the region (Wolf et al. 1976; Feinsinger 1983). This represents one more argument for the preservation of primary forest, in order to better understand the second-growth vegetation with which man is so rapidly replacing it.

ACKNOWLEDGMENTS

The Servicio de Parques Nacionales de Costa Rica, especially G. Flores, F. Cortés, and B. Madriz, provided permits and facilitated logistics for working in Parque Nacional Braulio Carrillo. I thank P. J. DeVries for much encouragement, and I. Chacón, C. Gómez, T. J. Lewis, and R. Campos for help in the field. The manuscript was greatly improved by the comments and criticisms of P. Buckley, F. Gill, Y. Linhart, and especially P. Feinsinger and R. Colwell. Hummingbird leg bands were obtained through W. A. Calder. J. Gómez Laurito supplied many plant identifications.

LITERATURE CITED

CALDER, W. A., III. 1974. Consequences of body size for avian energetics. Pp. 86–151, *In* R. A. Paynter (ed.), Avian Energetics. Publ. Nuttall Ornithol. Club No. 15

COLE, B. J. 1981. Overlap, regularity, and flowering phenologies. Am. Nat. 117:991–997.

COLWELL, R. K. 1973. Competition and coexistence in a simple tropical community. Am. Nat. 107: 737–760.

DURKEE, L. H. 1978. Flora of Panama, Pt. IX: Acanthaceae. Ann. Mo. Bot. Gard. 65 (1).

DWYER, J. D. 1980. Flora of Panama, Pt. IX: Rubiaceae. Ann. Mo. Bot. Gard. 67.

FEINSINGER, P. 1976. Organization of a tropical guild of nectivorous birds. Ecol. Monogr. 46:257–291.

FEINSINGER, P. 1977. Notes on the hummingbirds of Monteverde, Cordillera de Tilaran, Costa Rica. Wilson Bull. 89:159–164.

FEINSINGER, P. 1978. Ecological interactions between plants and hummingbirds in a successional tropical community. Ecol. Monogr. 48:269–287.

FEINSINGER, P. 1980. Asynchronous migration patterns and the coexistence of tropical hummingbirds. Pp. 411–419, *In* A. Keast and E. S. Morton (eds.), Migrant Birds in the Neotropics. Smithsonian Institution Press, Washington, D. C.

FEINSINGER, P. 1983. Coevolution and pollination. Pp. 282–310, *In* D. J. Futuyma and M. Slatkin (eds.), Coevolution. Sinauer, New York.

FEINSINGER, P., AND R. K. COLWELL. 1978. Community organization among neotropical nectar-feeding birds. Am. Zool. 18:779–795.

FOGDEN, M. P. L. 1972. The seasonality and population dynamics of equatorial forest birds in Sarawak. Ibis 114:307–343.

FOSTER, M. S. 1974. Rain, feeding behavior, and clutch size in tropical birds. Auk 91:722–726.

FRANKIE, G. W., H. G. BAKER, AND P. A. OPLER. 1974. Comparative phenological studies of trees in tropical wet and dry forests of Costa Rica. J. Ecol. 62:881–919.

GENTRY, A. H. 1982. Patterns of neotropical plant species diversity. Evol. Biol. 15:1–141.

HEITHAUS, R. E. 1974. The role of plant-pollinator interactions in determining community structure. Ann. Missouri Bot. Gard. 61:675–691.

HERRERA, C. M. 1982. Seasonal variation in quality of fruits and diffuse coevolution between plants and avian dispersers. Ecology 63:773–785.

JANZEN, D. H. 1980. When is it coevolution? Evolution 34:611–612.

KING, J. R. 1980. Energetics of avian molt. Acta XVII Int. Ornithol. Congr. I:312–320.

MACIOR, L. W. 1971. Co-evolution of plants and animals: systematic insight from plant-insect interactions. Taxon 20:17–28.

ORTIZ-CRESPO, F. I. 1972. A new method to separate immature and adult hummingbirds. Auk 89:851–857.

POOLE, R., AND B. J. RATHCKE. 1979. Regularity, randomness, and aggression in flowering phenologies. Science 203:469–470.

RICKLEFS, R. E. 1974. Energetics of reproduction in birds. Pp. 152–297, In R. E. Paynter (ed.), Avian energetics. Publ. Nuttall Ornithol. Club No. 15.

RIDGELY, R. S. 1975. A Guide to the Birds of Panama. Princeton University Press, Princeton, New Jersey.

RIDGWAY, R. 1911. Birds of North and Middle America, Pt. 5. Bull. U. S. Natl. Mus. No. 50.

SCHEITHAUER, W. 1966. Hummingbirds. Thomas Y. Crowell, New York.

SKOG, L. E. 1978. Flora of Panama, Pt. IX: Gesneriaceae. Ann. Mo. Bot. Gard. 65 (3).

SKUTCH, A. F. 1950. The nesting seasons of Central American birds in relation to climate and food supply. Ibis 92:182–222.

SKUTCH, A. F. 1966. A breeding bird census and nesting success in Central America. Ibis. 108:1–16.

SLUD, P. 1960. The birds of Finca "La Selva," a tropical wet forest locality. Bull. Am. Mus. Nat. Hist. 121:49–128.

SLUD, P. 1964. The birds of Costa Rica: distribution and ecology. Bull. Am. Mus. Nat. Hist. Vol. 128.

SNOW, B. K., AND D. W. SNOW. 1972. Feeding niches of hummingbirds in a Trinidad valley. J. Anim. Ecol. 41:471–485.

SNOW, D. W., AND B. K. SNOW. 1964. Breeding seasons and annual cycles of Trinidad land-birds. Zoologica 49:1–39.

SNOW, D. W., AND B. K. SNOW. 1980. Relationships between hummingbirds and flowers in the Andes of Colombia. Bull. Br. Mus. (Nat. Hist.) 38:105–139.

STANDLEY, P. C. 1937–1938. Flora of Costa Rica Pts. I–IV. Field Mus. Nat. Hist., Bot. Ser. 18 (1–4).

STILES, F. G. 1975. Ecology, flowering phenology, and hummingbird pollination of some Costa Rican Heliconia species. Ecology 56:285–301.

STILES, F. G. 1977. Coadapted competitors: the flowering seasons of hummingbird-pollinated plants in a tropical forest. Science 198:1177–1178.

STILES, F. G. 1978. Temporal organization of flowering among the hummingbird foodplants of a tropical forest. Biotropica 10:194–210.

STILES, F. G. 1979a. Notes on the natural history of Heliconia (Musaceae) in Costa Rica. Brenesia 15(suppl.):151–180.

STILES, F. G. 1979b. Regularity, randomness, and aggregation of flowering phenologies. Science 203: 470–471.

STILES, F. G. 1980. The annual cycle in a tropical wet forest hummingbird community. Ibis 122:322–343.

STILES, F. G. 1981. Geographical aspects of bird-flower coevolution, with special reference to Central America. Ann. Mo. Bot. Gard. 68:323–351.

STILES, F. G. 1982. Taxonomy and distribution of Costa Rican Heliconia (Musaceae), II: Parque Nacional Braulio Carrillo. Brenesia 19/20:221–230.

STILES, F. G., AND L. L. WOLF. 1970. Hummingbird territoriality at a tropical flowering tree. Auk 87: 465–491.

STILES, F. G., AND L. L. WOLF. 1973. Methods for color-marking hummingbirds. Condor 75:225–227.

STILES, F. G., AND L. L. WOLF. 1974. A possible circannual molt rhythm in a tropical hummingbird. Am. Nat. 108:341–354.

STILES, F. G., AND L. L. WOLF. 1979. Ecology and evolution of a lek mating system in the Long-tailed Hermit hummingbird. Ornithol. Monogr. No. 27.

TOSI, J. A., JR. 1969. Mapa Ecológico de Costa Rica. Tropical Science Center, San José, Costa Rica.

WASER, N. M. 1978a. Competition for hummingbird pollination and sequential flowering in two Colorado wildflowers. Ecology 59:943–944.

WASER, N. M. 1978b. Interspecific pollen transfer and competition between co-occurring plant species. Oecologia 36:223–236.

WILBUR, R. J., AND J. L. LUTEYN. 1978. Flora of Panama Pt. IX: Ericaceae. Ann. Mo. Bot. Gard. 65 (1).

WOLDA, H. 1978. Fluctuations in abundance of tropical insects. Am. Nat. 112:1017–1045.

WOLF, L. L. 1969. Female territoriality in a tropical hummingbird. Auk 86:490–504.

WOLF, L. L. 1970. Impact of seasonal flowering on the biology of some tropical hummingbirds. Condor 72:1–15.

WOLF, L. L. 1976. Birds of the Cerro de la Muerte region, Cordillera de Talamanca, Costa Rica. Am. Mus. Novit. No. 2606.

WOLF, L. L., F. R. HAINSWORTH, AND F. B. GILL. 1975. Foraging efficiencies and time budgets of nectar-feeding birds. Ecology 56:117–128.

WOLF, L. L., F. R. HAINSWORTH, AND F. G. STILES. 1972. Energetics of foraging: rate and efficiency of nectar extraction by hummingbirds. Science 186:1351–1352.

WOLF, L. L., AND F. G. STILES. 1970. Evolution of pair cooperation in a tropical hummingbird. Evolution 24:759–773.

WOLF, L. L., F. G. STILES, AND F. R. HAINSWORTH. 1976. The ecological organization of a highland tropical hummingbird community. J. Anim. Ecol. 32:349–379.

ZAR, J. H. 1974. Biostatistical Analysis. Prentice-Hall, Englewood Cliffs, New Jersey.

APPENDIX I

HUMMINGBIRD FOODPLANTS OF LA MONTURA

	Growth habit[1]	Bloom. seas.[2]	Corolla length[3]	Corolla curv.[4]	Nectar prod.[5]	Coloration[6]	Max. abund.[7]	Visitors[8]
A. Hummingbird-pollinated species								
Heliconia deflexa (Musaceae)	H	A1	37	M	H	yellow + green, red[1]	10	Pg
H. ignescens (Musaceae)	H	A1	35	M	H	yellow, red-orange[1]	32	Pg
H. mathiasae (Musaceae)	H	B1	34	S	H	yellow + red, red[1]	6	Pg
H. rodriguezii (Musaceae)	H	B1	35	M	H	yellow, red[1]	52	Pg
H. trichocarpa (Musaceae)	H	A1	33	M	H	yellow, red[1]	18	Pg, Ea
H. atropurpurea (Musaceae)	H	A1	32	S-M	H	white, red[1]	4	Pg
Costus pulverulentus (Zingiberaceae)	H	A1	42	S	H	red, red[1]	29	Pg
C. sp. (Zingiberaceae)	H	A1	30	S	H	yellow, red-orange[1]	10	Pg, Hj
Renealmia cernua (Zingiberaceae)	H	A1	14	O-S	L	orange, orange[1]	7	Lh, En, Ec, Tf
Alpinia sp. (Zingiberaceae)	H	A1	28	S	H	white, red[1]	27	Pg, Hj
Guzmania lingulata (Bromeliaceae)	E	D	40	O	–	yellow, red[1]	5	Dl, Lh
G. donnell-smithii (Bromeliaceae)	E	D	32	O	H	white, red[1]	7	Dl, Lh
G. plicatifolia (Bromeliaceae)	E	B1?	22	O	–	yellow, orange[1]	13	Lh, Hj, Dl
G. coryostachia (Bromeliaceae)	E	B2	16	O	–	white, green[1]	16	Lh, Hj, Dl
G. sp. (Bromeliaceae)	E	D	27	O	L?	yellow, red[1]	55	Dl, Lh, Hj
G. aff. scherziana (Bromeliaceae)	E	D	16	O	–	orange, green[1]	7	Lh, Dl, Hj
Pitcairnia valerii (Bromeliaceae)	E	A1	8	O	L	red, green[1]	8	Lh, Ec, En
Elleanthus aurantiacum (Orchidaceae)	E	A1	ca. 5	O	–	orange	13i	Lh, En, Ec
E. sp. (Orchidaceae)	E	A1	ca. 5	O	L	purple	15i	Lh, En, Ec
Spiranthes sp. (Orchidaceae)	H	A1	36	O-S	–	red	4	Pg
Psittacanthus nodosus (Loranthaceae)	E	B2	28	O	–	yellow + orange	180	Dl, Hj
Malvaviscus arboreus (Malvaceae)	S	C	38	O	H	red	26	Pg, others pierce
Cephaelis elata (Rubiaceae)	S-T	B1	11	S	L	white, red[1]	1350	Lh, En, Ec, Ch, Tf, Hj, Lc
Ravnia triflora (Rubiaceae)	E	A1	43	S	–	red	21	Pg
R. pittierii (Rubiaceae)	E	A1	41	S	–	red	5	Pg
Aphelandra tridentata (Acanthaceae)	S	B1	31	S	H?	red-orange	15	Pg
Razisea spicata (Acanthaceae)	S	C	40	S	H	red	685	Pg, others pierce
Scutellaria costaricensis (Labiatae)	S	C	40	S	H	orange + yellow	60	Pg, others pierce
Centropogon granulosus (Lobeliaceae)	S	A2	34	V	H	red + yellow	95	Ea, Pg
Alloplectus tetragonus (Gesneriaceae)	S	B2	40	S	L	red	51	Pg
Besleria solanoides (Gesneriaceae)	S	A1	15	O	L	orange	72	Ec, En, Lh
B. notabilis (Gesneriaceae)	S	B2	14	O	–	orange	90	Ec, En, Ct
B. laxiflora (Gesneriaceae)	S	A2	13	O	H	orange	34	Ec
Columnea purpurata (Gesneriaceae)	E	A2	25	O-S	L	yellow, orange[2]	15	Lh, Lc, En
C. macrophylla (Gesneriaceae)	E	A1	33	S	L	orange	310	Pg
C. querceti (Gesneriaceae)	E	A1	15	S	L	red + yellow	392	Lh, En, Cd

APPENDIX I
CONTINUED

	Growth habit[1]	Bloom. seas.[2]	Corolla length[3]	Corolla curv.[4]	Nectar prod.[5]	Coloration[6]	Max. abund.[7]	Visitors[8]
C. consanguinea (Gesneriaceae)	E	D	18	O	L	yellow	14	Lh, En
Drymonia conchocalyx (Gesneriaceae)	S-T	D?	28	O-S	–	purple	22	Hj
Cavendishia capitulata (Ericaceae)	E	A_1	12	O	L	purple + white, pink[1]	250	Lh, En, Ec
C. quereme (Ericaceae)	E	A_1	11	O	L	orange + white, pink[1]	1320	Lh, En, Ec
C. endresii (Ericaceae)	E	A_1	12	O	L	purple + white, pink[1]	361	Lh, En, Ec
C. calliste (Ericaceae)	E	A_1	32	O	–	white, pink[1]	92	Dl, Hj
C. sp. (Ericaceae)	E	A_1	30	O	–	pink + white, pink[1]	327	Dl
Satyria warsewiczii (Ericaceae)	E	A_2	31	O	H	red + white	295	Dl
Thibaudia costaricensis (Ericaceae)	E	A_1	10	O	L	white, pink[1]	1200	Lh, En, Ec, Cd
Psammisia ramiflora (Ericaceae)	E	A_1	27	O	H	red + white	145	Dl, Pg, Lh
Gurania costaricensis (Curcurbitaceae)	V	D	7	O	L	yellow, orange[1]	21	Ec, En, Lh, (Pg)
G. levyana (Curcurbitaceae)	V	D	9/17	O	–	yellow, orange[1]	10	Ec, En, Lh
G. sp. (Curcurbitaceae)	V	D	–	–	–	yellow, orange[1]	7	Ec, En, Lh
Symphonia globulifera (Guttiferae)	T	A_1	10±	O	–	red	ca. 1000	Cd, Lh, En
Erythrina lanceolata (Fabaceae)	T	A_1	28	O	H	red	26	Pg, (Dl?)
B. Species frequently visited by hummingbirds, but pollinated mostly by other agents								
Palicourea lasiorrachis (Rubiaceae)	S	A_1	10	O	L	yellow	150	Lh, En, Ec
P. sp. (Rubiaceae)	S	A_1	6	O	L	green	70	En, Ec
Coussarea sp. (Rubiaceae)	S-T	A_1	20+	O	L	white	580	En, Ec, Pl (pierce)
Marcgravia schippii (Marcgraviaceae)	E	A_1	ca. 20	O-S	–	brown, green[1]	335	Hj
M. pittierii (Marcgraviaceae)	E	A_1	ca. 15	O	–	brown, green[1]	215	Hj
M. affinis (Marcgraviaceae)	E	A_1	ca. 12	O	–	brown, green[1]	550	Hj, Lh
Norantea sessilis (Marcgraviaceae)	E	A_1	5	O	–	red, brown[1]	70i	Cd, En, Ec, Lh, Pc
Clusia spp. (Guttiferae)	E	(A_1)	3–6	O	–	white to cream	300+	Cd, En, Ec, Lh, Pc
Inga spp. (Mimosaceae)	T	(A_1)	8–14	O	L?	white	3000+	Cd, Lh, En
Calliandra arborea (Mimosaceae)	T	C	12	O	L	pink, white	5000+	Cd, Lh, En, Ec, Pc, Ct
Witheringia warsewiczii (Solanaceae)	S	C	5	O	L	yellowish	350	Ec, En, Lh
Hampea appendiculata (Tiliaceae)	T	A_1	8	O	L?	whitish-yellow	ca. 200	En, Loh, Ec

[1] Growth habit: H = terrestrial herb; S = shrub; T = tree; E = epiphyte or hemiepiphyte; V = vine.

[2] Blooming Seasonality: A_1 = one discrete peak of flowering per year; A_2 = 2 discrete peaks per year (may vary in intensity); B_1 = population blooms fairly continuously, 1 well-defined peak per year; B_2 = population blooms irregularly but 1 well-defined peak per year; C = population blooms fairly continuously, with irregular and variable peaks; D = irregular bursts of flowering (discrete), much variation between years.

[3] Corolla length: effective corolla length: corolla opening to nectar chamber.

[4] Corolla curvature: same symbols as for bill curvature, Table 1.

[5] Nectar production: H ≥ 20 μl/flower/day; L < 20 μl.

[6] Color: color given is for corolla, except 1 = bract or inflorescence, 2 = calyx.

[7] Max. abundance: high count in flower censuses (at peak bloom), i = number of inflorescences (many small flowers per inflorescence, not possible to count).

[8] Principal visitors: according to abbreviations, Table 1, approximately in order of importance; species italicized accounted for 80% or more of observed visits.

COMPARATIVE FEEDING ECOLOGY OF TWENTY-TWO TROPICAL PISCIVORES

DAVID E. WILLARD

Bird Division, Field Museum of Natural History, Chicago, Illinois 60605 USA

ABSTRACT. Twenty-two fish-eating predators on a Peruvian lake show differences in hunting technique, habitat use, prey sizes, and prey species. The swimming predators take mostly different prey species. Waders differ in their prey species and sizes, and in the habitats in which they hunt. Perchers and cruisers hunt from different heights and take prey of different sizes and species. Many of the same types of differences are evident between groups using different hunting techniques, although several species pairs have remarkably similar diets.

RESUMEN. Veintidos depredadores de peces en un lago de Perú muestran diferencias en las técnicas de pesca, uso del habitat, tamaño y especies de las presas. Los depredadores nadadores, atrapan mayormente diferentes especies de presas. Las aves zancudas difieren en las especies y tamaños de las presas así como en los habitats en los cuales pescan. Aves que se posan para pescar ("perchers") y otras que recorren mucha distancia en su búsqueda por presas ("cruisers") pescan desde diferentes alturas y atrapan presas de diferentes tamaños y especies. Son evidentes muchas diferencias del mismo tipo entre grupos que usan distintas técnicas de pesca, aunque varios pares de especies tienen dietas remarcablemente similares.

Most comparative studies of potentially competing species have considered only closely related species on the assumption that morphological differences between those more distant insure relatively great differences in resource use. Root (1967) suggested that functional rather than taxonomic relationships are more important and that all species using the same class of resources should be considered because very different species can use the same resources.

In lowland Peru, oxbow lakes attract many fish-eating species that use a variety of hunting techniques. Some of these piscivores wade, some swim and pursue prey under water, some dive from flight, others dive from perches, and one skims the surface. Little information has been published on many of these species, and even those that are well known in some parts of their ranges are poorly known in South America. In this paper, I compare the feeding habits of twenty bird, one mammal, and one reptile species that I saw capture fish, and I analyze behavioral and ecological relationships within this portion of the fish-eating community.

STUDY SITE AND METHODS

Observations were made at Cocha Cashu, an oxbow lake of the Río Manu, in Manu National Park, Department of Madre de Dios, Peru, elevation about 350 m. The lake is 2 km long and about 100 m wide. During the periods that I was present, mostly in the late dry season, the lake reached a maximum depth of about 3 m. The water is murky, and the bottom is soft and silty.

I made observations in July and August, 1975, and August through December, 1976. I watched the members of the fish-eating community that regularly bring their prey to the water's surface. For size reference, weights for birds other than kingfishers (from Haverschmidt 1968) and the otter (from Duplaix 1980) are included in the following list. My own measurements for kingfishers are in Table 3. The species for which I gathered information are: Olivaceous Cormorant (*Phalacrocorax olivaceus*; 1113–1400 g), Anhinga (*Anhinga anhinga*; 1115–1250 g), White-necked Heron (*Ardea cocoi*; 1465–1750 g), Great Egret (*Casmerodius albus*; 770–1022 g), Green-backed Heron (*Butorides striatus*; 142–214 g), Snowy Egret (*Egretta thula*; 277–335 g), Chestnut-bellied Heron (*Agamia agami*; 565 g), Rufescent Tiger-Heron (*Tigrisoma lineatum*; 630–980 g), Capped Heron (*Pilherodius pileatus*; 444–632 g), Boat-billed Heron (*Cochlearius cochlearius*; 577–642 g), Black-collared Hawk (*Busarellus nigricollis*; 710–829 g), Osprey (*Pandion haliaetus*; 1440–1600 g), Large-billed Tern (*Phaetusa simplex*; 219–275 g), Yellow-billed Tern (*Sterna superciliaris*; 176–226 g), Black Skimmer (*Rynchops niger*; 222–377 g), Ringed Kingfisher (*Ceryle torquata*), Amazon Kingfisher (*Chloroceryle*

TABLE 1
MAXIMUM DAILY COUNT OF FISH-EATERS ON COCHA CASHU[1]

Phalacrocorax	17	Pandion	2
Anhinga	4	Ceryle	6
Ardea	3	Chloroceryle amazona	4
Casmerodius	4	C. americana	5
Egretta	1	C. inda	13
Butorides	11	C. aenea	3
Agamia	10	Phaetusa	5
Pilherodius	1	Sterna	6
Tigrisoma	14	Rynchops	4
Cochlearius	3	Pteronura	5
Busarellus	2		

[1] July–December.

amazona), Green-and-rufous Kingfisher (*Chloroceryle inda*), Green Kingfisher (*Chloroceryle americana*), American Pygmy Kingfisher (*Chloroceryle aenea*), Giant Otter (*Pteronura brasiliensis*; 24–34 kg), and Black Caiman (*Melanosuchus niger*).

I watched these species through 9× binoculars from a kayak and a dugout canoe, and through a 20× telescope from shore. I generally watched species as I encountered them and terminated observations when feeding stopped or when the subject moved from sight. When there was a choice between species, I chose the one for which I had the least information. The lakeshore was divided into discrete areas of open forest, brushy shrubs, grassy marshes, and dense stands of *Heliconia*. I recorded the habitat of all hunting birds, and if a bird moved from one habitat to another, I recorded its occurrence in both. Observations on any given day either covered the whole lake or half of it, with halves countered by the other half on another day so that habitats were censused approximately in proportion to their occurrences. I estimated prey lengths by comparison to known bill lengths. Perch and cruising heights were estimated by comparison to carefully gauged, strategically located landmarks. I sampled fish qualitatively using minnow traps, seine, throw nets, and hook and line; this allowed field identification of some of the prey species as they were captured by the birds. I censused fish-eating species by mapping daily sightings of individuals, and by counting birds as they arrived at lakeshore roosts in the evening. Mist-netting on the lakeshore, in the forest, and at an inland marsh provided additional information on population sizes and habitat use. Bill lengths and weights were taken for all netted kingfishers.

RESULTS

ABUNDANCES

The most common species were *Phalacrocorax*, *Chloroceryle inda*, *Tigrisoma*, *Butorides*, and *Agamia* (Table 1). No accurate census was possible for caimans, but they were extremely numerous. Numbers of several species, particularly *Phalacrocorax*, *Butorides*, and *Agamia* were augmented by arrival of immatures. Anhingas were absent from mid-August to early November. *Sterna* left the area by late October. Most *Phaetusa* left by late November; their disappearance coincided with feeding independence of their young and with rising water levels as the rainy season approached. *Egretta* and *Pilherodius*, both common in the backwaters of the Río Manu, only appeared on the lake when the river level was high. Two families of otters occasionally used the lake. They generally remained for only a day or two, often with several weeks between visits.

HUNTING BEHAVIOR

Five general classes of hunting behavior were used by the piscivores in this study: (1) swimming and diving from the surface; (2) wading; (3) diving from a perch; (4) cruising and diving from flight; (5) skimming. Within these classes there were some differences between species in details of hunting style, in prey sizes and species captured (Fig. 1) and in habitat use (Fig. 2).

(1) Swimming.—Anhingas and caimans hunted solitarily. Cormorants hunted either alone

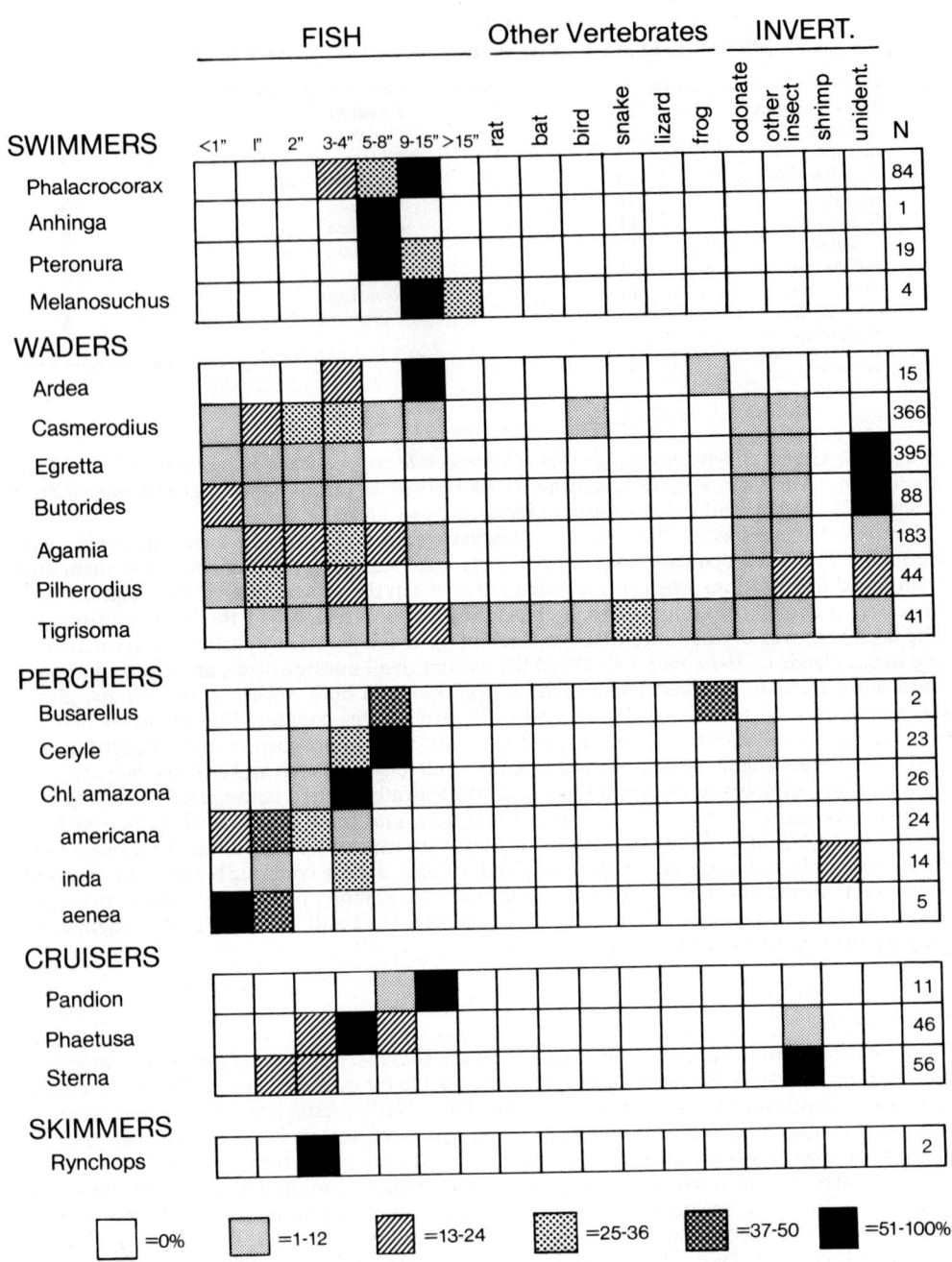

FIG. 1. Diets of Cocha Cashu fish-eaters.

or in small groups. Giant Otters were almost always in groups of 3 to 5. Anhingas, cormorants, and otters all pursued prey under water, whereas caimans waited quietly in shallow water, capturing fish by sudden ambush.

Cormorants took prey covering a broad range of sizes. Because I could not see prey that was swallowed under water or as a bird surfaced, my sample may be biased toward large and hard to handle fish. Sixty percent of the observed prey items were sedentary, shallow-water Loricariid catfish in the genera *Loricariicthys* and *Loricaria*. Most otter prey was large *Pla-*

HABITAT USE (%)

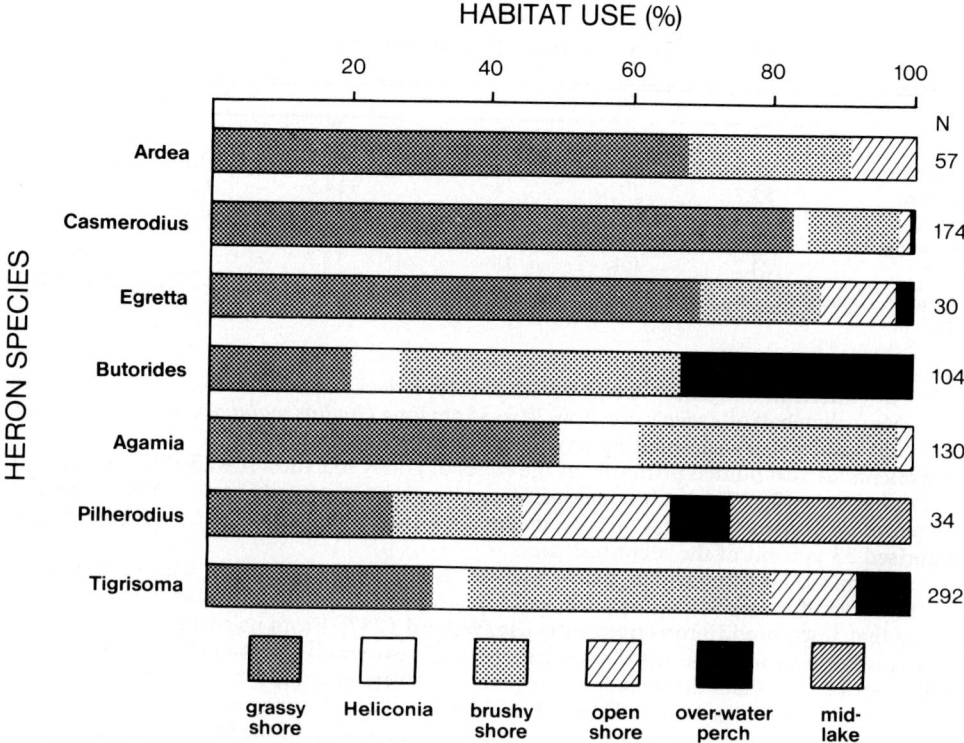

FIG. 2. Hunting habitats of Cocha Cashu herons.

gioscion squamosissimus (family Sciaenidae), a species I only captured in the deepest water of the lake. The four caiman prey recorded were Loricariid catfish in the genera *Hypostomus* and *Pterygoplicthys*. These were the largest fish prey observed in the study, and all were caught by caimans larger than 1 m. The total range of caiman prey must await observation of prey caught by the many small caimans also present.

(2) Wading. — *Tigrisoma* hunted by standing alert and motionless for long periods, moving only after a strike or a particularly long period of no activity. When prey was sighted, a bird pointed its bill toward it, and began slowly, almost imperceptibly bending toward it. Strikes were extremely powerful, often appearing to damage prey more severely than did strikes by other heron species. *Tigrisoma*'s diet was the most generalized of any species' in the study, with several snake species (individuals = 15–20 cm long) the most common prey. This heron was also one of the most generalized in habitat use.

Ardea largely restricted its hunting to the edges of grassy marshes. It hunted exclusively by slow wading mixed with long motionless periods, a hunting style almost identical to that of

TABLE 2

MEAN SEARCH TIME OF PERCHERS PER HUNTING PERCH[1]

	Mean	s.d.	N
Chloroceryle aenea	0.7	0.3	7
C. americana	1.5	1.9	48
C. inda	4.0	3.7	23
C. amazona	6.0	7.6	148
Ceryle torquata	12.0	12.9	101
Busarellus	59.0	15.0	4

[1] Time in minutes. All pairwise differences between means significant at $P < 0.025$ or less by Mann Whitney U-test.

TABLE 3
BILL LENGTH AND BODY WEIGHT OF KINGFISHERS[1]

	Bill			Weight		
	X̄	s.d.	N	X̄	s.d.	N
torquata	62.4	3.7	7	303.6	20.1	8
amazona	53.7	2.1	6	114.5	6.7	6
inda	41.5	1.7	24	53.6	4.1	33
americana	34.0	2.0	5	29.0	.8	7
aenea	24.2	1.4	14	13.3	.9	22

[1] Bill length = nostril to tip, in mm. Weight in grams. All pairwise differences between means significant at $P < 0.001$ by Mann Whitney U-test.

its North American congener, *Ardea herodias* (Meyerriecks 1960; Kushlan 1976a; Willard 1977). More than half of *Ardea*'s prey was 20 to 35 cm long *Hoplias malabaricus* (Erythrinidae), a fish that I did not see captured by any other predator in the study.

Casmerodius also hunted primarily at the edges of grassy marshes. It waded slowly, usually pausing only to strike. This technique was the same as that commonly used in North America (references as above). Prey was mostly in the 2 to 10 cm range. A cichlid (*Aequidens* sp.) comprised 33 percent of the identified prey.

Agamia mixed standing still and slow wading, usually in a deep crouch with belly feathers and the curve of the neck touching the water in a manner extremely similar to the crouched waiting that Tricolored Herons frequently use (Willard 1977). From its crouch, *Agamia* made strikes toward prey near the water's surface. *Agamia*'s extremely long bill (150 mm) appeared to allow strikes at greater distance from its body than were possible for similarly sized herons with shorter bills (as suggested by Hancock and Elliot 1978). This may be especially important for striking around, over, and under the many protruding twigs in the tangled, flooded brush where *Agamia* often hunted. *Agamia* took prey primarily of 2 to 10 cm, and 52 percent of the identified prey items were Characidae, particularly *Triportheus angulatus* and *Astyanax* sp., both commonly observed surface species.

Butorides, Egretta, Pilherodius, and probably *Cochlearius* were small prey specialists with a mixed diet of 1 to 10 cm fish and invertebrates, many of which were too small to identify before being swallowed. *Butorides* hunted either by wading in shallow water under dense brush or by perching on protruding branches over open water. It moved along perches, sometimes perching at water level and pecking for prey there, sometimes climbing short distances above the water and either hanging acrobatically from the perch to grab at prey or plunging into the water. Similar behavior for other races of the Green-backed Heron has been described by Meyerriecks (1960) and Snow (1974). The feeding style of *Egretta* was varied and generally active, with a mixture of slow walking, running, foot-stirring and wing-flicking (follows Kushlan 1976a). Its hunting was largely restricted to grassy shores. *Pilherodius*' primary hunting style on Cocha Cashu was haphazard rapid pecking with no apparent orientation toward individual prey items ("sandpiper-style pecking": Willard 1977). An extreme incident of this behavior involved 789 probing strikes in 56 min, with only seven strikes apparently successful. Remsen (in press) also observed this feeding style in Colombia. *Pilherodius* also hunted by slow walking and sometimes employed aerial hunting. Birds sallied out and hovered over the water, striking for fish and swallowing in flight. Occasionally birds landed on the water and struck while swimming. They also hunted from perches at shoreline, and I saw one hunt by bill-vibrating (Kushlan 1973). *Cochlearius* was a nocturnal hunter. Its activity coincided with a nightly emergence of insects that attracted myriads of small fish (*Moenkhausia*) to the surface. Neither the fish nor the insects were evident during the day when other herons were hunting. At Cocha Cashu, feeding of *Cochlearius* appeared to be tactile (Willard 1979).

(3) Perching.—Hunting behavior of the five kingfishers and *Busarellus* involved sitting on a perch and plunging into the water for prey. The kingfishers took prey with their bills and swallowed it whole, while the hawk captured prey with its feet. Its hooked bill, used for tearing, allowed it to take prey too large to swallow whole. The hunting styles of these six species differed primarily in activity level. The mean length of time per hunting perch increased with size of the birds (Tables 2, 3).

Mean prey sizes of the perchers increased from the tiny prey of *C. aenea* to the relatively

TABLE 4
PERCH AND CRUISING HEIGHT[1]

	Perch height				Cruising height		
	X̄	s.d.	N		X̄	s.d.	N
Ceryle torquata	10.0	4.7	203	*Sterna*	3.0	4.9	51
Chloroceryle amazona	6.9	3.8	207	*Phaetusa*	10.3	4.3	26
C. inda	4.9	4.2	111	*Pandion*	13.0	6.5	13
C. americana	2.9	4.7	110				
C. aenea	1.4	.9	24				

[1] Values given in meters. All pairwise differences between means within groups significant at $P < 0.01$ by Mann Whitney U-test.

large fish and frog recorded for *Busarellus*. *Ceryle* and *C. amazona* took the most similarly-sized prey of the perchers. Thirty-five percent of *Ceryle*'s prey was the cichlid *Aequidens* sp., which *C. amazona* rarely took, whereas 65 percent of *C. amazona*'s prey was Characidae, particularly *Triportheus angulatus* and *Astyanax* which were rare in *Ceryle*'s diet.

Ceryle, *Chloroceryle amazona*, and *C. americana* used open perches and dived into open water. *Chloroceryle inda* and *C. aenea* occasionally hunted in the open, but more generally perched in shoreline brush and dived for prey underneath. They also used swampy forest and heavily wooded streams where the other species rarely ventured. *Busarellus*' favored haunts were coves choked with water weeds. The only kingfisher regularly found in these spots was the tiny *C. aenea*.

The kingfishers show a regular size gradient (Table 3). Bill lengths increase approximately 10 mm from one species to the next, and weights approximately double until the somewhat larger jump between *C. amazona* and *Ceryle*. Although perch heights overlapped somewhat, mean perch height increased with increasing kingfisher size (Table 4). At over 700 g, *Busarellus*, which perched in the height range (1–3 m; X̄ = 1.3 m; N = 14) of the three smallest kingfishers, does not fit this sequence.

(4) Cruising.—The two terns and Osprey hunted by cruising over the lake, and hovering and plunging after spotting prey. The three species cruised at different heights above the water (Table 4). Like kingfisher perch height, cruising height increased with size of the bird.

Most Osprey prey items were fish larger than 20 cm, primarily *Prochilodus* sp. (Prochilodontidae). *Phaetusa* took 5 to 15 cm fishes, mostly *Triportheus angulatus* and *Astyanax*. David Duffy (pers. comm.) and I each once observed *Phaetusa* skimming the surface in a manner similar to Black Skimmer feeding. Remsen (in press) also observed this once in Colombia. *Phaetusa* occasionally hunts at night, and determining the extent to which it uses skimming will require extensive night observations. *Sterna* took unidentified fishes, mostly 5 cm or smaller. It also swooped to the surface and dipped its bill for surface prey (probably insects) and made sudden upward sallies for flying insects. Mixed bill-dipping and aerial hunting accounted for 59 percent of my feeding observations of *Sterna* (N = 131).

(5) Skimming.—Black Skimmers, like *Cochlearius*, hunted nocturnally when the small fish *Moenkhausia* was active at the surface. The only two prey items I could identify with certainty were both this fish. All observed hunting was by skimming the surface with bill, snapping it shut on contact with prey.

TABLE 5
PREY CAPTURE RATES[1]

Ardea	.01	(11/864)	*Ceryle*	.003 (4/1304)
Casmerodius	.16	(332/2067)	*Chloroceryle amazona*	.02 (15/965)
Egretta	1.17	(317/270)	*C. inda*	.02 (2/92)
Butorides	.15	(60/405)	*C. americana*	.09 (6/67)
Agamia	.08	(152/1875)	*C. aenea*	.7 (2/3)
Tigrisoma	.008	(15/1788)		
Pilherodius	.19	(18/96)		

[1] Prey items/minute hunting time.

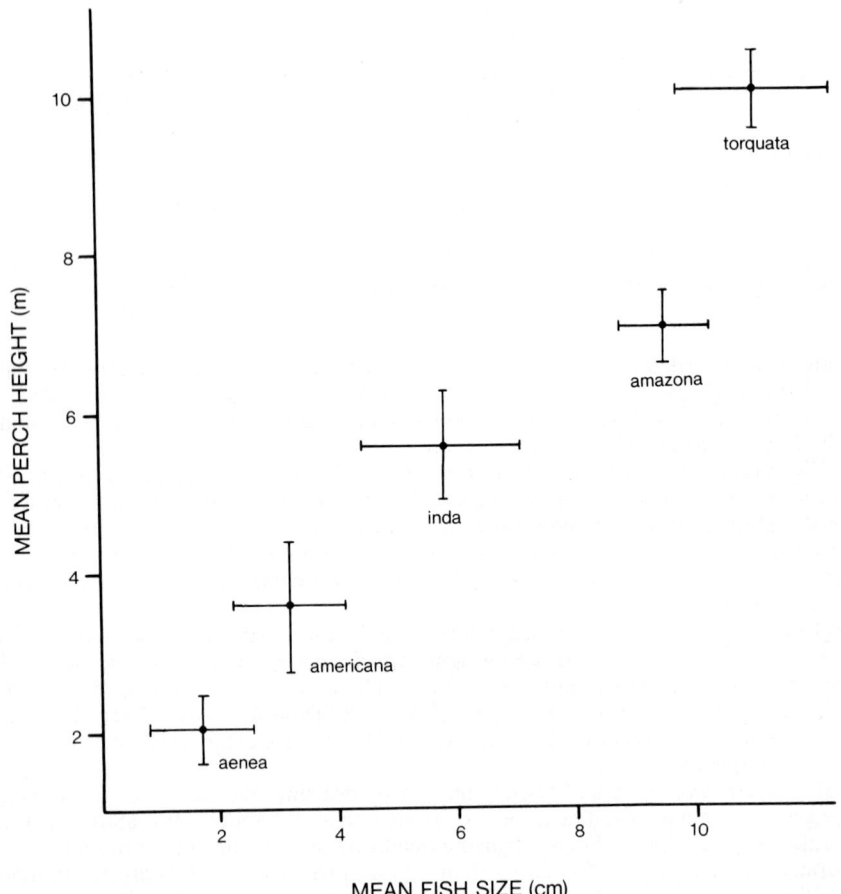

Fig. 3. Relationship between mean perch height and mean prey size of kingfishers. Vertical and horizontal lines indicate two standard errors on either side of the means.

Prey Capture Rate

Among herons and kingfishers, species that took larger prey had lower capture rates. Between species in these groups that took equivalently-sized prey, the heavier herons always had the greater capture rates (Table 5).

DISCUSSION

Perch and Cruising Height

MacArthur (1972) suggested that *Chloroceryle americana* perches near the water's surface in order to see the small fish that it is equipped to catch with its small bill. The larger *Ceryle*, because of its demands for greater absolute food intake, must perch higher to survey a greater area. On these high perches, it can no longer efficiently locate and capture small fish, but instead captures larger fish with its larger bill. MacArthur's argument is interesting in the context of the Cocha Cashu kingfisher community, where mean perch height and mean prey size are linearly related (Fig. 3).

The much larger *Busarellus*, while taking larger prey than any of the kingfishers, perched in the height range of the three smallest. It may be that the weedy coves where it habitually hunted cannot be effectively surveyed from higher perches. Although I saw only two prey items of *Busarellus*, I suspect that most of its prey is large because of the bird's large size and low capture rate (one prey item in 235 min of timed observation), and because it is not restricted to prey that it can swallow whole. This is supported by Remsen's (in press) observations of four captured fish ranging from 102 to 203 mm.

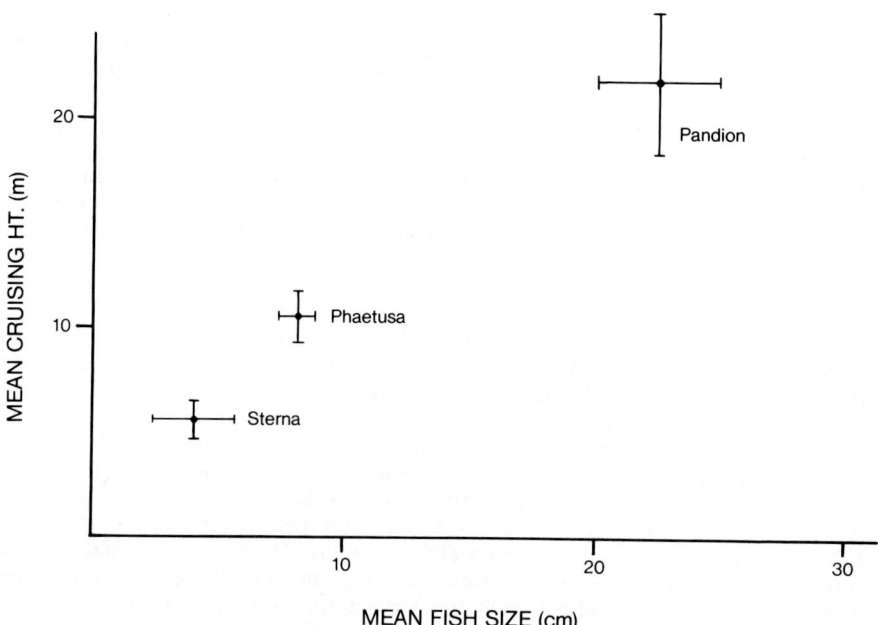

Fig. 4. Relationship between mean cruising height and mean prey size of terns and Osprey. Vertical and horizontal lines indicate two standard errors on either side of the means.

The relationship between cruising height and prey size in the two terns and Osprey (Fig. 4) shows a trend similar to that found for perch height and prey size in kingfishers.

SIMILARITY IN DIETS

Seven species (*Ardea, Tigrisoma, Pandion, Busarellus, Pteronura, Phalacrocorax*, and large *Melanosuchus*) took large prey, but showed virtually no overlap in their primary prey species. By contrast, the birds that took moderately-sized prey, had greater similarity of prey species. *Agamia, Chloroceryle amazona*, and *Phaetusa* all preyed heavily on the characins *Triportheus* and *Astyanax*; *Casmerodius* and *Ceryle* both took many of the cichlid *Aequidens*.

Populations of *Casmerodius* and *Agamia* may be near the carrying capacity of Cocha Cashu. Throughout this study, three *Casmerodius* hunted daily on the lake. I saw a fourth only once; each time it entered the water it was chased, even when the normal occupant was out roosting at the moment. *Agamia* populations also showed little fluctuation. The influx of young birds was short-lived, and, presumably, these were birds that had been fed by the adults normally present. If there were a superabundance of fish available for these herons, one would expect to see more fluctuation in their numbers.

Agamia captured prey four times as often as did *Chloroceryle amazona*, and *Casmerodius* captured prey 50 times as often as did *Ceryle*. If, in fact, Cocha Cashu heron populations are near carrying capacity, and if prey capture rates reflect food requirements of the birds, then even if diets, habitats, and hunting times were identical, the lake could support four *C. amazona* and 50 *Ceryle* with the food required by one additional *Agamia* or *Casmerodius*. The argument is obviously over-simplified, for there were some obvious dietary differences, particularly between *Casmerodius* and *Ceryle*. Nevertheless, the argument suggests the possibility of coexistence between two species with very similar diets, when one species consumes many fewer prey items, thus allowing several individuals of the less demanding species to survive where a single individual of the more demanding could not.

Phaetusa, while capturing many of the same prey species as *Agamia* and *Chloroceryle amazona*, often took them at mid-lake where they were not available to the other two. *Phaetusa*'s hunting style involved searching the whole lake for fish that happened to be at the surface at a given moment, whereas *Agamia* and *C. amazona* hunted smaller areas, waiting for prey to come within striking range.

The species that specialized on small prey had mixed diets of fish and invertebrates that I usually could not identify. *Sterna* hunted mostly at mid-lake where its prey was inaccessible to the others. The small kingfishers rarely hunted along the dense grassy shoreline that constituted *Egretta*'s main hunting area, probably because of lack of satisfactory perches there. *Pilherodius*' shoreline feeding resulted in smaller prey than much taken by *Chloroceryle inda* and *C. americana*; its mid-lake prey were not accessible to these low perching kingfishers. *Butorides,* which hunted in the open and amid dense brush, was similar to *C. inda* and *C. americana* both in its habitat and vertebrate prey sizes. It captured many more invertebrate prey than I saw for either kingfisher. *Chloroceryle aenea*'s small size allowed it to enter vegetation that was not open to any other Cocha Cashu fish-eater. Its total range of prey sizes was the smallest of any of the species.

OTHER FISH-EATING COMMUNITIES

Remsen (in press) concentrated primarily on kingfishers, but also observed other fish-eating birds. The regular fish-eaters in his Colombian and my Peruvian communities were the same, and, although our observations differed somewhat, particularly in kingfisher perch heights and prey sizes, our conclusions regarding resource use are very similar. We observed the same trends in prey size differences and many of the same habitat and behavioral differences.

Whitfield and Blaber (1978, 1979a, b) studied 10 members (three swimmers, four waders, and three cruisers) of a fish-eating community on a large East African lake. They did not mention what other fish-eaters were present, so a species by species comparison between their site and Cocha Cashu is not possible. Dietary differences based purely on prey sizes appeared to be more prevalent in the African community, although this might not be so if additional members were considered. Several common prey species appeared to be more generally used than was any single prey species at Cocha Cashu, where differences in prey types were more prevalent.

Whitfield and Blaber (1979a) suggested that tarsometatarsal length determined the fishing depths of the waders and, thus, indirectly, the sizes of fishes available to them. Other studies of wading bird communities (e.g., Kushlan 1976b; Willard 1977) have also suggested a fishing depth component of resource segregation. At Cocha Cashu, all herons restricted their hunting to very shallow water and had relatively minor hunting depth differences. In fact, the smallest heron, *Butorides,* hunted in the deepest water when it used protruding perches. One possible explanation for the differences between these studies is that the common presence of caimans at Cocha Cashu restricts herons to shallow water from which they can escape easily and in which large caimans cannot maneuver well. In deep water, with attention strongly focused on potential prey, even the largest herons may run the risk of becoming prey themselves.

ACKNOWLEDGMENTS

This work was supported by grants from the Frank M. Chapman Fund of the American Museum of Natural History for both years, and from Sigma Xi and Princeton University for the first year. J. Fitzpatrick, J. Terborgh, D. Duffy, D. Moskovits, R. Kiltie, G. Russell, C. Munn, C. Janson, H. Brokaw, and J. Weske all shared many observations with me and often led me to fish-eating occurrences and ideas I would not have found otherwise. J. Fitzpatrick did lakeshore mist-netting with me. He, H. Horn, J. V. Remsen, D. Stotz, D. Duffy, D. Mock, R. M. Erwin, M. Foster, and T. Custer made helpful comments on the manuscript. R. Horwitz identified the fish species. I wish to thank Ing. Carlos Ponce del Prado for permission to work at Manu National Park, and the park personnel for their assistance.

LITERATURE CITED

DUPLAIX, N. 1980. Observations on the ecology and behavior of the Giant River Otter *Pteronura brasiliensis* in Suriname. Rev. Ecol. (Terre Vie) 34:495–620.

HANCOCK, J., AND H. ELLIOTT. 1978. The Herons of the World. London Editions, London.

HAVERSCHMIDT, F. 1968. The Birds of Surinam. Oliver and Boyd, London.

KUSHLAN, J. A. 1973. Bill-vibrating. A prey attracting behavior of the Snowy Egret, *Leucophoyx thula*. Am. Midl. Nat. 89:509–512.

KUSHLAN, J. A. 1976a. Feeding behavior of North American herons. Auk 93:86–94.

KUSHLAN, J. A. 1976b. Wading bird predation in a seasonally fluctuating pond. Auk 93:464–476.

MACARTHUR, R. H. 1972. Geographical Ecology. Harper and Row, New York.

MEYERRIECKS, A. J. 1960. Comparative breeding behavior of four species of North American herons. Publ. Nuttall Ornithol. Club No. 2.

Remsen, J. V. In press. Geographical ecology of neotropical kingfishers. Univ. Calif. Publ. Zool.

Root, R. B. 1967. The niche exploitation pattern of the Blue-gray Gnatcatcher. Ecol. Monogr. 37:317–350.

Snow, B. K. 1974. The Plumbeous Heron of the Galapagos. Living Bird 13:51–72.

Whitfield, A. K., and S. J. M. Blaber. 1978. Feeding ecology of piscivorous birds at Lake St. Lucia, Part 1: Diving birds. Ostrich 49:185–198.

Whitfield, A. K., and S. J. M. Blaber. 1979a. Feeding ecology of piscivorous birds at Lake St. Lucia, Part 2: Wading birds. Ostrich 50:1–9.

Whitfield, A. K., and S. J. M. Blaber. 1979b. Feeding ecology of piscivorus birds at Lake St. Lucia, Part 3: Swimming birds. Ostrich 50:10–20.

Willard, D. E. 1977. The feeding ecology and behavior of five species of herons in southeastern New Jersey. Condor 79:462–470.

Willard, D. E. 1979. Comments on the feeding of the Boat-billed Heron (*Cochlearius cochlearius*). Biotropica 11:158.

BIRDS FORM TIGHTLY STRUCTURED COMMUNITIES IN THE PEARL ARCHIPELAGO, PANAMA

S. Joseph Wright,[1] John Faaborg,[2] and Claudia J. Campbell[1]

[1]Smithsonian Tropical Research Institute, APO, Miami, Florida, 34002 USA, and
Division of Biological Sciences, University of Missouri,
Columbia, Missouri, 65201 USA

ABSTRACT. We ask how disturbance affects the land birds of the Pearl Archipelago, Panama. We analyze 15 censuses conducted on 11 islands between 1971 and 1978. Avian abundances are highly predictable among islands and years. Two islands with unusual forests support unique bird communities. The nine remaining islands form two groups. The first includes large islands and their immediate satellites. These islands support several bird species that are not found on the second group of small, isolated islands. Community structure is similar among islands within each group and consistently different among islands drawn from different groups. These patterns are repeatable from year to year despite disturbances such as fire, windstorms, and drought, that occurred between 1971 and 1978. We conclude that disturbance has minimal impact on this community, and we suggest that disturbance will be more important to sessile organisms than to mobile organisms.

The number of Pearl Islands whose avifaunas are known is increased from four to 14. The species-area relation is exponential.

RESUMEN. Nos preguntamos en que manera las aves del Archipiélago de las Perlas son afectadas por perturbaciones. Analizamos 15 censos conducidos en 11 islas entre 1971 y 1978. Se puede predecir con relativa certeza la abundancia de aves a lo largo de las islas y los años. Dos islas que poseen bosques inusuales, soportan comunidades avícolas únicas. Las nueve islas restantes forman dos grupos. El primero incluye islas grandes y sus islas satélites que las circundan inmediatamente. Estas islas tienen varias especies de aves que no se encuentran en el segundo grupo de islas pequeñas y aisladas. La estructura de la comunidad es similar entre islas de un mismo grupo, y consistentemente diferente si se las compara con islas de grupos diferentes. Estos modelos se repiten año tras año, a pesar de perturbaciones tales como incendios, tormentas de viento y sequías, que han ocurrido entre 1971 y 1978. Nuestra conclusión es que las perturbaciones tienen un efecto mínimo en esta comunidad y sugerimos que serán mucho más importantes para organismos sésiles que para organismos móbiles.

El conocimiento de la avifauna de las islas del Archipiélago de las Perlas se incrementó de 4 a 14. La relación entre el número de especies y el área de la isla es exponencial.

Disturbances such as fire, severe storms, drought, and disease, affect the abundance of many species. If disturbances like these occur rarely, coexisting organisms may attain stable, equilibrial abundances determined by their interactions with one another and with the abiotic environment. On the other hand, if disturbances occur frequently, equilibria may not occur, and local patterns of abundance may depend upon the local history of disturbance. There is controversy among community ecologists over the impact of disturbance. Some minimize its role (e.g., Lack 1971; MacArthur 1972; Cody 1974, 1981; Schoener 1982, 1983) while others argue that disturbance and chance recolonization of disturbed sites are of paramount importance (e.g., Andrewartha and Birch 1954; Sale 1977; Wiens 1977; Connell 1978; Hubbell 1979; Huston 1979). We ask what impact disturbances have had on the terrestrial avifaunas of islands in the Pearl Archipelago, Panama. A single empirical study will not, of course, resolve this controversy. Empirical studies are needed, however, because resolution will require recognition of the fact that the role of disturbance varies from community to community. This is the first attempt to assess the effect of disturbance on a tropical bird community.

We examined censuses of terrestrial birds taken on islands in the Pearl Archipelago in 1971, 1976, and 1978. If disturbance has little impact on avian abundances, patterns of abundance should be stable among years and predictable among islands. On the other hand, if disturbance reigns, patterns of abundance should vary among years and islands. This is true for two reasons. First, significant disturbances occurred between census dates; these include fire, windstorms, and drought. Second, the study islands are separated by up to 40 km of open ocean and support independent bird communities. There is no reason to expect chance disturbances to generate patterns of abundance that are repeatable among independent communities.

We test the null hypothesis that the abundances of species vary independently from year to year and, after developing two *a priori* predictions, from island to island. Evidence that disturbance is relatively unimportant will be realized if the null hypotheses are rejected and if abundances are predictable among islands and years. We emphasize that if the null hypotheses are not rejected, we can draw no firm conclusions. We do not follow the dubious lead of Grossman et al. (1982) who performed a similar analysis, failed to reject the null hypothesis, and then asserted from acceptance of the null hypotheses that disturbance was important to a stream fish community.

The first *a priori* prediction results from an historical factor that affected current avian distributions. During the Wisconsin glaciation, sea levels in Panama Bay were 110 m below their present levels (Bartlett and Barghoorn 1973), and the Pearl Archipelago was connected to the mainland. While the land-bridge was intact, many bird species that do not cross open ocean had access to the archipelago-to-be. Some of these species persist in the archipelago today (MacArthur et al. 1972). We will show that these "land-bridge relicts" tend to be restricted to the three largest islands in the archipelago and to their immediate satellite islands, a possibility suggested by MacArthur et al. (1972). This provides two groups of islands with substantially different avifaunas. The larger islands and their satellites support "land-bridge relicts" while smaller, more isolated islands do not. We predict that the overall pattern of avian abundances should be similar on islands within each of these groups but dissimilar between groups. Differences are expected to result from substantially different interactions among species.

The second *a priori* prediction is related to habitat differences among islands. Forest structure and dominant plant species were similar on all study islands except San José and Pacheca (see below). Each bird species is likely to respond differently to the unique habitats of San José and Pacheca. Therefore, we predict that the pattern of avian abundances will also be unique on these islands.

The study islands grouped by *a priori* considerations are: (1) San José, (2) Pacheca, (3) isolated islands with "typical" forests (Saboga, Chapera, Bolaños, Platanales, and Bayoneta), and (4) large islands and their satellites with "typical" forests (Rey, Viveros, Cañas, Puercos). We predict that the pattern of avian abundances will be similar among islands drawn from each group and dissimilar among islands drawn from different groups.

STUDY AREA AND METHODS

THE PEARL ARCHIPELAGO

The Pearl Archipelago is located in Panama Bay (Fig. 1). The closest islands are 25 km from the mainland. Average annual rainfall is slightly more than 2.5 m (Portig 1976). Rainfall is highly seasonal, however, with less than 5 percent of the annual total falling between January 1 and April 30 (Fig. 2). In most years, little or no rain falls in February and March (Portig 1976; S. J. Wright, pers. observ.).

VEGETATION

We quantified forest structure on the 10 islands where we conducted avian censuses (listed below). On each island, we estimated canopy height at 20 points located at 12 m intervals along a line of mist nets (see below). At each point, we also estimated the horizontal distance at which half of a one meter square board would be obscured by vegetation at heights of 0.15, 0.6, 1.5, 3.1, and 6.1 m. We converted these distances into foliage densities and plotted foliage height profiles using methods developed by MacArthur and MacArthur (1961). The forest was taller than 6.1 m at each site, but our distance estimates became unreliable above this height.

We also noted dominant tree and shrub species (also see Erlanson 1947; Johnston 1949). On exposed, coastal areas and wherever the soil is thin, *Bombacopsis quinata, Pseudobombax septenatum, Bursera simaruba, Luehea seemannii, L. speciosa,* and *Ficus* sp. are common. Common species in more protected interior forests include a second *Ficus, Guarea* sp.,

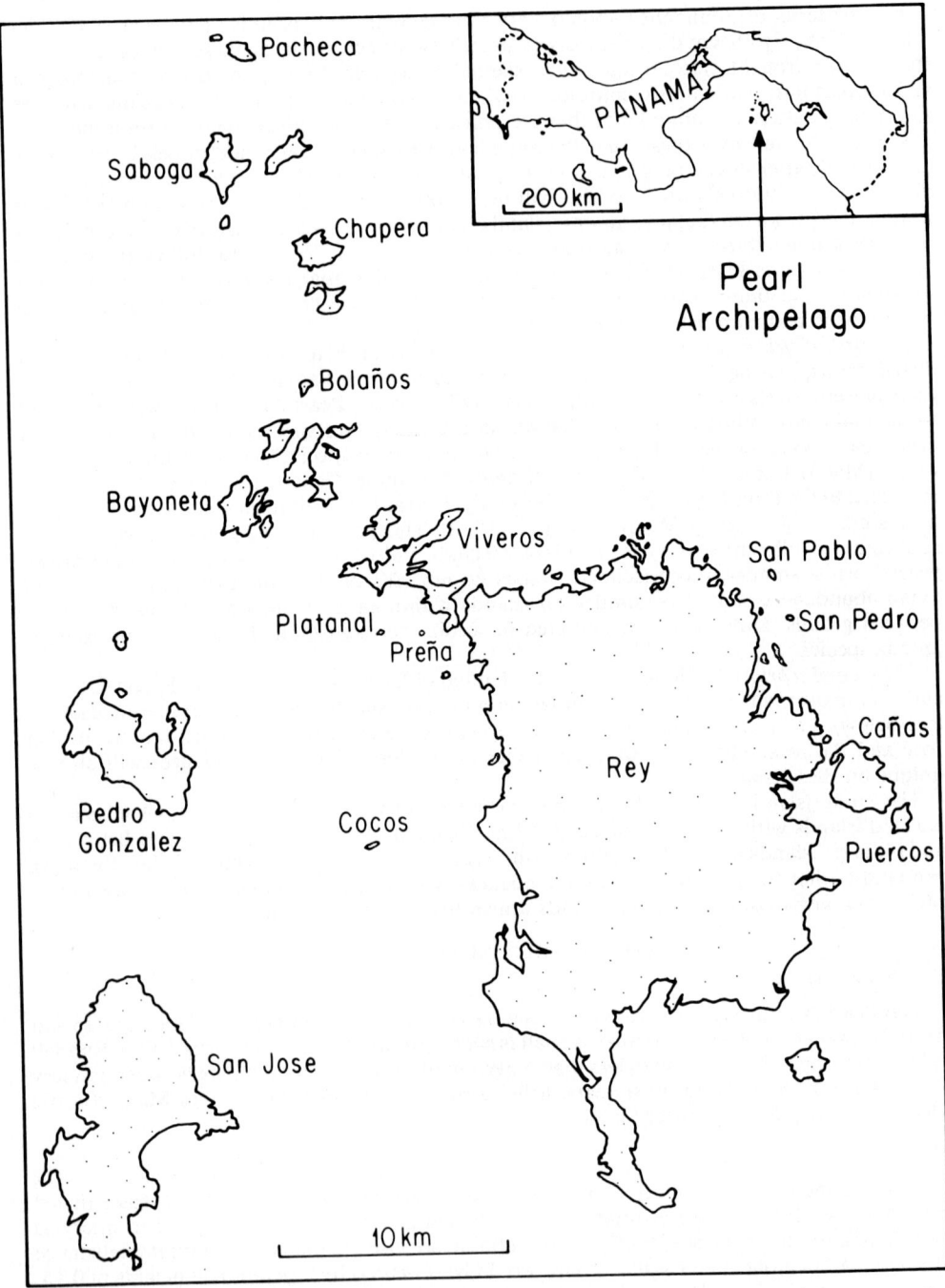

FIG. 1. Map of the Pearl Archipelago. Inset shows the position of the archipelago with respect to the Republic of Panama.

Schwartzia simplex, Pouteria campechiana, Cordia sp., *Manilkara chicle, Calycophyllum caudidissium, Nectandra gentlei,* and others. All study sites were located in the latter habitat.

BIRDS

We restrict our analysis to the resident, terrestrial avifauna of the Pearl Archipelago. Migrants that breed in North America, shorebirds, seabirds, and species that depend upon aquatic

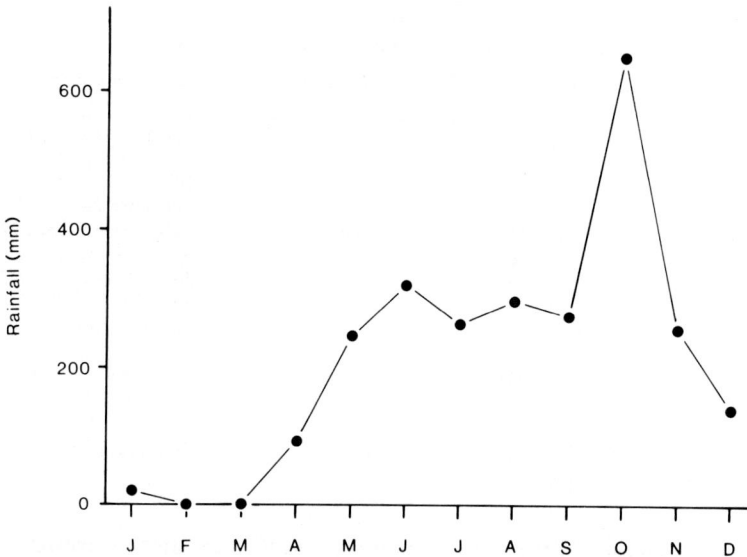

FIG. 2. Mean monthly rainfall on Rey. Data are from two stations and seven years (Portig 1976).

habitats and tidal mud flats are omitted. Fifty-five species remain. Nomenclature follows the 6th edition of the A.O.U. Check-list of North American Birds (American Ornithologists' Union 1983).

We studied this portion of the avifauna of 13 islands. During June, July and August 1976, S. J. Wright and J. Faaborg visited the following islands for from four to seven days each: Bayoneta, Bolaños, Chapera, Mina, Pacheca, Platanales, Saboga, and Viveros. During January, February, and March, 1978, S. J. Wright and C. J. Campbell visited the following islands for from three to 14 days each: Cañas, Chapera, Cocos, Pedro Gonzalez, Platanales, Rey, San José, and Viveros. We also include data collected by MacArthur et al. (1972, 1973) for Puercos (January 1971) and Cañas (December 1971, January 1972). Fragmentary distributional data, which we do not include, are available for several other islands [see Rendahl (1920) and MacArthur et al. (1972) for summaries].

We searched each island intensively to document avian distributions. Previous ornithologists had compiled faunal lists which proved to be complete or nearly complete for Rey (Bangs 1901; Thayer and Bangs 1905; Rendahl 1920), San José and Pedro Gonzalez (Wetmore 1947), and probably Puercos (MacArthur et al. 1972). We did not visit Puercos. We were either the first ornithologists to visit the remaining islands; or we doubled or tripled the number of known residents. Nevertheless, earlier investigators encountered several island populations that we did not. While it is possible that these populations have become extinct, the species involved are usually rare and/or secretive. To be conservative, we have assumed that these populations are still extant. They are noted in Table 2. For many of the more vagile species (raptors, parrots, swifts, swallows, and the pigeon, *Columba cayennensis*), presence on an island does not necessarily represent a breeding population.

We censused avian abundances with mist nets (Table 1). We placed nets end-to-end in a straight line. Each net measured 2.4 m × 12 m and had 36 mm mesh and four shelves. We opened nets at dawn and closed them at dusk for at least three consecutive days except at San José, where we opened nets for only two days. We estimated abundance separately for each species by dividing the number of captures by the number of net-days, e.g., 20 nets opened for three days equals 60 net-days. We only used captures from the first three days to estimate abundances (two days for San José). We banded each individual (clipped several rectrices from hummingbirds) so that recaptured birds could be identified and omitted from abundance estimates. We operated nets on Chapera, Platanales and Viveros in exactly the same locations in both 1976 and 1978 (Table 1). We will refer to these censuses by island and year, e.g., Viveros (1978).

MacArthur et al. (1972, 1973) also estimated avian abundances with mist nets. They used identical methods except adjacent nets were separated by an average gap of 19 m. As a result,

TABLE 1
ISLAND AREAS AND DETAILS OF CENSUSES

Island	Area (ha)[1]	No. of nets	Netting dates[2]
Rey	24,900	16	February 19–21, 1978
San José	4720	15	March 8–9, 1978
Viveros	649	20	July 14–16, 1976
		16	February 2–5, 1978
Cañas	522	19	February 11–13, 1978
Saboga	288	20	June 21–23, 1976
Bayoneta	226	20	June 29–July 1, 1976
Chapera	178	20	June 25–27, 1976
		15	January 23–25, 1978
Pacheca	62	18	June 16–18, 1976
Bolaños	21	20	July 19–21, 1976
Platanales	7	13	July 31–August 2, 1976
		13	January 27–29, 1978

[1] Taken from U.S. Topographic Sheets 4341-I, 4341-II, 4341-III, 4441-III, and 4340-I (scale 1:50,000) with a planimeter.
[2] Net days = no. days sampled × no. nets used.

the areas censused by each of their nets overlapped less than the areas censused by our nets, and we expect their abundance estimates to be systematically greater than ours. Therefore, we do not use their estimates in analyses of abundances of single species among islands. However, because the increase will be proportionately equal for all species, the correlation analyses to be described below are not affected, and the censuses of MacArthur et al. (1972, 1973) are included.

Not all birds are equally susceptible to capture in mist nets. The nets that we used are efficient for birds that weigh between 3 gm and 100 gm. The hummingbird, *Chlorostilbon canivetii,* and the dove, *Leptotila verreauxi,* push these limits and occasionally escaped after encountering a net. Also, the kingbird, *Tyrannus melancholicus,* and the honeycreeper, *Cyanerpes cyaneus,* tended to forage above mist nets. Low abundance estimates for these species could be an artifact of the census method. To be sure that this did not affect our conclusions, we repeated all analyses both with and without these four species.

Species that are much larger than 100 gm, species that do not enter forest, and crepuscular and nocturnal species were never captured and were omitted from all analyses of abundances. Species that are too large or that always forage above mist nets include a tinamou (*Cryturellus soui*), a vulture (*Coragyps atratus*), seven raptors, a chachalaca (*Ortalis cinereiceps*), a pigeon (*Columba cayennensis*), two parrots, a swift (*Chaetura vauxi*), a martin (*Progne chalybea*), and a gnatcatcher (*Polioptila plumbea*). Noctural and crepuscular species include two owls and a pauraque (*Nyctidromus albicollis*). Species that rarely or never entered forest include a ground-dove (*Columbina talpacoti*), an ani (*Crotophaga ani*), a grackle (*Quiscalus mexicanus*; captured twice), a grassquit (*Volatinia jacarina*), and a seedeater (*Sporophila nigricollis*). This left 28 species that we captured regularly in mist nets. We used the abundances of these 28 species to test hypotheses about community structure.

ANALYSES

To test the null hypothesis that abundances are independent among islands and years, we calculated Pearson product-moment correlation coefficients for all pairwise combinations of censuses. We calculated each correlation coefficient using the abundances of species known to be present on both islands. Species that had never been recorded on an island were omitted. Thus, sample size varied from correlation to correlation with minima and maxima of 10 and 24, respectively. The 15 censuses generated 105 pairwise correlation coefficients.

We performed a cluster analysis to determine whether similar patterns of abundance appeared as predicted in the introduction. The cluster analysis forms a dendrogram in which censuses with the most similar avian abundances are grouped together. We used the Pearson product-moment correlation coefficients described above to index similarity. The algorithm initially placed each of the 15 censuses in a separate group. The pair of censuses with the highest correlation coefficient was then linked. When groups included more than one census, the algorithm calculated the average correlation coefficient for all pairwise combinations of

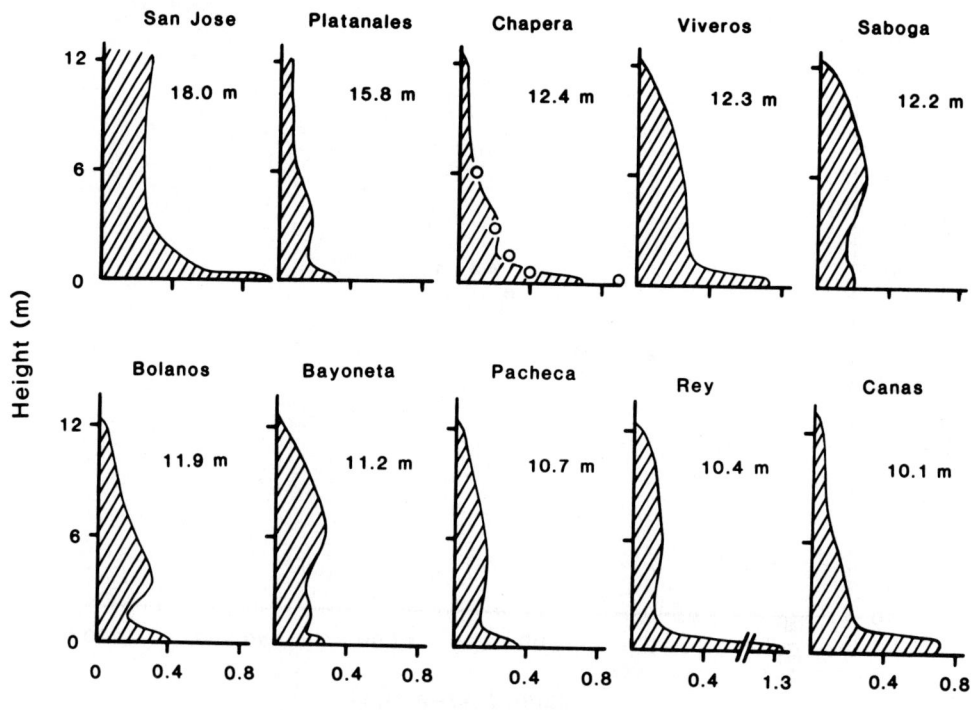

Foliage Density (m² of vegetation per m³ of space)

FIG. 3. Foliage-height profiles for islands where we operated mist nets. Foliage density is plotted on the abscissa; height above the ground on the ordinate. The name of the site and mean canopy height are specified above each graph. For Chapera, foliage densities in 1978 are represented by open circles, in 1976 by the hatched region. On this site, fire reduced mean canopy height from 12.4 m to 6.6 m sometime between 1976 and 1978.

censuses from different groups. The pair of groups with the highest average was then linked. The final dendrogram provides a visual representation of the pattern of similarities and dissimilarities among censuses. For further explication see Dixon (1981, BMDP1M, average similarity).

Finally, for widespread species, we tested the null hypothesis that abundances are equal on islands that lack and islands that support "land-bridge relicts." We used a nonparametric Kruskal-Wallis test for this analysis (Siegel 1956). We did not include the abundance estimates of MacArthur et al. (1972, 1973) for reasons given previously.

RESULTS

FOLIAGE HEIGHT PROFILES

Foliage height profiles were qualitatively similar at all sites except San José, where the canopy was higher and the vegetation denser (Fig. 3). The effect of forest structure on avian communities is well known (e.g., MacArthur and MacArthur 1961), and we therefore predicted that avian abundances on San José, would differ from those elsewhere. Forest species composition was also unique on San José; the dominant tree, *Tetragastris panamensis,* has not been found elsewhere in the archipelago (Erlanson 1947; Johnston 1949; S. J. Wright, pers. observ.). The only other noticeable differences among sites were the unusually dense herbaceous layers on Chapera, Viveros, Rey, and Cañas (Fig. 3). Because only one of the censused bird species, *Leptotila verreauxi,* is terrestrial this last difference is unlikely to have affected our analyses of avian abundances.

The suspected differences between the forests of Pacheca and the other islands cannot be detected from foliage height profiles (Fig. 3). Nevertheless, we believe that the difference is

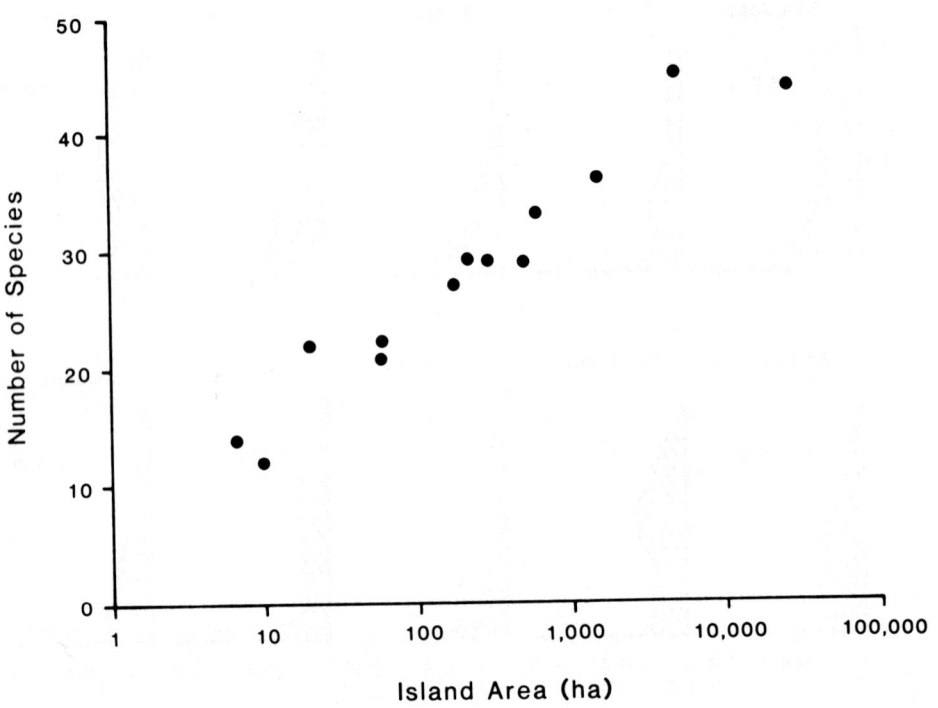

Island Area (ha)

Fig. 4. The exponential relation between species number (ordinate) and island area (abscissa) for the terrestrial avifauna of the Pearl Archipelago. Mina is excluded because its center is a brackish lagoon, and the forested area cannot be determined.

real. Pacheca has been home to tens of thousands of pelicans and cormorants and lesser numbers of frigate birds and boobies since at least 1882 (Bovallius 1886; Rendahl 1920). During the breeding season, each canopy tree supports tens to hundreds of their nests, and the ground and lower levels of the vegetation are spattered with guano, rotting fish, and bird carcasses. We believe that this input of nutrients must affect the flora and insects of Pacheca and, thereby, the terrestrial avifauna as well. We predicted, therefore, that avian community structure would also be unique on Pacheca.

Foliage height profiles were almost identical during both visits (1976 and 1978) to Viveros and Platanales. Few plants were deciduous in the interior forests that surrounded our study sites. On Chapera, our study site burned some time between 1976 and 1978. This reduced mean canopy height along our line of mist nets from 12.4 to 6.6 m (Fig. 3).

DISTRIBUTIONS WITHIN THE PEARL ARCHIPELAGO

Species number and island area are directly related, and an exponential function best describes the relationship (Fig. 4, $r = .97, P < 0.001$). Theoretically, the slope of this function is affected by the rates at which species immigrate and become extinct on islands of different area (MacArthur and Wilson 1967; Schoener 1976; Wright 1981). The slope observed for the Pearl Archipelago is unusually low for oceanic islands. This suggests that immigration rates are high and/or extinction rates are low. Direct observations will be needed to confirm this inference.

The slope of the species-area relationship would be even smaller were it not for the presence of relict species on the larger islands. MacArthur et al. (1972) tentatively concluded that 27 percent of the terrestrial bird species could not cross open water to reach the archipelago and that their presence must represent relict, land-bridge distributions. The species tentatively identified as relicts are noted in Table 2. Following MacArthur et al. (1972), we hypothesized that relict species would be restricted to the three largest islands in the Pearl Archipelago and to immediate satellites of these large islands. This is expected for two reasons: first, persistence times should be long on large islands because the probability of extinction is low (MacArthur and Wilson 1967), and second, immigration to satellite islands across water gaps of a few hundred meters should be possible for even the most sedentary birds. We have faunal lists for the three largest islands in the archipelago (Rey, San José, and Pedro Gonzalez), three satellite islands (Viveros, Cañas, and Puercos are less than 200 m from Rey or a steppingstone island) and seven smaller, more isolated islands. Sixteen species are restricted to the larger islands and their satellites while 37 species are more widespread (Table 2). Ten of the restricted species and four of the widespread species were tentatively identified as relicts by MacArthur et al. (1972). Relict species are overrepresented among species with restricted distributions ($\chi^2_c = 12.81$, $P < 0.001$). We conclude that within-archipelago distributions are consistent with the hypothesis that land-bridge relicts persist in the Pearl Archipelago.

These relict distributions are important for our community analyses because two distinct groups of islands emerge. Among islands and species for which we have censuses, Rey, Viveros, Puercos, and Cañas support a hummingbird, two or three antbirds, a wren, and possibly a gnatcatcher and flycatcher that are absent from the other sites (Table 2). We predict *a priori* that avian abundances will be similar within this group of islands and also within a second group of small, isolated islands that includes Bayoneta, Chapera, Saboga, Bolaños and Platanales. Again, we also predict that avian abundances will be unique on Pacheca and San José because they support unique habitats.

ABUNDANCES

The null hypothesis that avian abundances vary independently among sites is rejected. Correlations between the abundances of conspecific populations are significant at $P < 0.05$ for 36 percent and at $P < 0.01$ for 23 percent of the 105 pairwise comparisons of censuses (Table 3). All significant correlations are positive. Exclusion of the four species whose abundances might have been underestimated did not change the results. Thirty percent of the pairwise correlations are still significant ($P < 0.05$), and the patterns discussed below for the complete data set are robust.

The distribution of significant correlations is not random. The six censuses from Rey, Cañas (2 censuses), Puercos, and Viveros (2 censuses) generate 15 pairwise comparisons and 14 significant correlations (Table 3). The seven censuses from Bayoneta, Chapera (2 censuses), Bolaños, Saboga, and Platanales (2 censuses) generate 21 pairwise comparisons and 14 significant correlations (Table 3). Finally, there are 27 pairwise comparisons that involve Pacheca and/or San José, and none of these generates significant correlations (Table 3).

The dendrogram produced by the cluster analysis reflects this pattern (Fig. 5). The relative abundances of birds on Pacheca and San José are clearly different from one another and from those on all other islands. On Chapera, Bayoneta, Platanales, and Saboga, relative abundances are similar, and the same is true for Rey, Cañas, Puercos, Viveros, and Bolaños. The inclusion of Bolaños in the latter group provides the only deviation from the *a priori* predictions. The low average correlation between the subgroup that includes Puercos and Cañas (1971) and the subgroup that includes Rey, Cañas (1978), Viveros, and Bolaños (Fig. 5) reflects the very poor correlations between avian abundances on Bolaños and Puercos ($r = -.13$) and Bolaños and Cañas (1971) ($r = .04$).

For the three islands that we censused twice, the two replicate censuses are more similar to one another than to any other census (Fig. 5). Because the two censuses of a single site were separated by about 18 months (Table 1), many of the same birds were captured both times. On Viveros, Chapera, and Platanales, 15, 25 and 27 percent, respectively, of the birds captured in 1978 had been captured in 1976. Clearly, this would tend to make the replicate censuses similar to one another. Still, given that 73 to 85 percent of the individuals differed, the similarity between the wet season of 1976 and the dry season of 1978 is striking. The similarity between the two censuses on Chapera is especially striking because the study site burned some time between 1976 and 1978. The two censuses from Cañas are not as similar ($r = 0.63$, $P < 0.01$). However, these censuses were not conducted in exactly the same spot, and they were separated by six years rather than 18 months.

DISCUSSION

To facilitate discussion, we refer to the species of birds that we censused as a community, and we define community structure as the pattern of relative abundances of these species. The similarities in community structure among islands with similar forests and avifaunas and the

TABLE 2

Distributions of Terrestrial Birds and Numbers of Individuals Captured on the Pearl Islands, Panama[1]

Species	Rey	San José[2]	Viveros 1976/1978	Cañas	Saboga	Bayoneta	Chapera 1976/1978	Pacheca	Bolaños	Platanales[3] 1976/1978	Other islands[4]
Crypturellus soui[5]	+	+[6]	+								PG[6]
Coragyps atratus		+[6]									PG[6]
Chondrohierax uncinatus		+[6]			+[6]						PG
Ictinia plumbea	+[6]	+									PG, P
Buteo magnirostris	+	+		+			+				
Buteogallus anthracinus	+	+	+[6]	+		+					M, PG, P
Milvago chimachima	+		+		+[6]		+	+[6]		+/NP	M
Polyborus plancus			+				+	+[6]			PG
Ortalis cinereiceps[5]	+	+		+							M, PG, P
Columba cayennensis	+	+	+	+	+	+	+[6]	+	+	+	PG
Columbina talpacoti	+	+		+[6]	+[6]		+[6]	+	+		M, PG, P
Leptotila verreauxi	12	0	4/4	5	3	7	2/7	7	3	1/NP	M, PG, P
Amazona autumnalis	+		+			+	+		+		PG
A. ochrocephala	+	+									M, PG
Crotophaga ani	+	+	+		+	+	+		+		PG[6]
Otus choliba[5]	+	+	+								PG
Nyctidromus albicollis	+	+[6]									PG, P
Phaethornis anthophilus[5]	2	0	9/4	4	0	0		0	0		C, M, PG, P
Chlorostilbon canivetii	1	0	1/2	0	0	0	0/0			0/1	C, M, PG, P
Amazilia edward	4	1	4/2	1	+	3	7/5	1	4	1/0	M, PG
Chaetura vauxi		+	+		+	+	+				P
Melanerpes rubricapillus	3		4/3	2		4					P
Cercomacra nigricans[5]	4		7/2	9							PG, P
Thamnophilus doliatus[5,7]	12		8/9	15							PG
Formicivora grisea[5,8]	9	5	4/2								C, M, PG, P
Tyrannus melancholicus	0	0	0/0	0	3	0	3/1	2	3	3/0	C, M, PG, P
Myiarchus panamensis	6	8	2/0	11	17	8	10/8	19	8	22/8	C, M, PG, P
Myiodynastes maculatus[5]	0	1		0	1		1/0	0[6]			
Myiophobus fasciatus[5]	0	1		0[6]	0[6]		1/0				
Mionectes oleagineus[5]	1		0/0[6]								
Myiopagis viridicata	0	5	0/0	0	0	0	0/2	11	0	NP/2	
Elaenia flavogaster	0	9	3/0	1	1	2	5/1		4	1/0	C, PG
E. chiriquensis	1	17	1/0	1	1	2	3/0		2		M
Sublegatus modestus	3	0	3/3	3	5	2	7/7	2	6	10/4	C, M, PG, P
Camptostoma obsoletum	0	0	1/0	0	1	2	2/1	0[6]	1	NP/1	PG, P

TABLE 2
CONTINUED

Species	Rey	San José[2]	Viveros 1976/1978	Cañas	Saboga	Bayoneta	Chapera 1976/1978	Pacheca	Bolaños	Platanales[3] 1976/1978	Other islands[4]
Progne chalybea	+	+	+		+	+		+	+		M
Troglodytes aedon	0[6]	3									C, PG
Thryothorus leucotis[5]	5		16/10	13		2					P
Polioptila plumbea[5]	+[6]										
Vireo olivaceus											
Coereba flaveola	2	9	0/0	1	4	5	5/1	5	16		M, PG, P
Cyanerpes cyaneus	3	0	5/2	5	6	5	31/12	1	14	7/7	C, M, PG, P
Dendroica petechia	0	9		3	0					NP/0	PG, P
Quiscalus mexicanus	0	0	+	0	5	25	20/6	0	11	18/12	C, M, PG
Thraupis episcopus	+	+			+	+	+	+	+	+	C, M, PG
Ramphocelus dimidiatus[5]	0	0	0/0	0	2	10	7/7	48	4	1/1	C, M, PG, P
Saltator albicollis	12	2	10/5	11	2	15	14/10	7	28		PG
Volatinia jacarina	5	21	9/7	5	7	16	13/11	9	35	7/2	C, M, PG
Oryzoborus funereus[5]	+	+			0	+					PG
Sporophila nigricollis	0	0	4/1	1	+[6]	1	2/2	0	1	NP/1	

[1] Only individuals captured in the first three days of operation of mist nets (two days for San José) are included. + indicates presence of a species not susceptible to capture in mist nets.

[2] Seven species have been recorded in the Pearl Archipelago only from San José. They are *Falco rufigularis, Geotrygon montana, Pionus menstruus*, Tyto alba*, Nyctibius griseus, Myiozetetes similis,* and *Contopus cinereus* (Wetmore 1947; N. G. Smith, pers. comm.). Those denoted by asterisks are probably vagrants.

[3] Species present in 1976 or 1978, but not in both years, are denoted by NP. We are confident that we did not overlook populations on this 7 ha island.

[4] Reasonably complete faunal lists are available for four islands where we did not mist net. These are Pedro Gonzalez (PG, Wetmore 1947), Puercos (P. MacArthur et al. 1972), and Cocos (C) and Mina (M. S. J. Wright, pers. observ.). We omit fragmentary data for several other islands. We also omit *Trogon massena* (Rendahl 1920) from Pedro Gonzalez. Habitat destruction virtually precludes the presence of this species on Pedro Gonzalez today. Finally, we omit records for Chitre (MacArthur et al. 1972) because Chitre is broadly connected to Saboga during low tides.

[5] Tentatively identified as land-bridge relicts by MacArthur et al. (1972).

[6] We were not able to confirm the presence of these populations. The records were taken from Bangs (1901), Thayer and Bangs (1905), Rendahl (1920), Wetmore (1947), and Ridgely (1976).

[7] N. G. Smith (pers. comm.) now retracts records for Casaya, Chapera, and Bayoneta (Smith 1971).

[8] We omit a record attributed to Eisenmann for Saboga (Wetmore 1972:184) because it could not be confirmed in unpublished species accounts kept by Eisenmann. Other ornithologists that have visited Saboga and failed to record this species include: C. Bovallius in 1882, O. Bangs in 1904, A. Wetmore in 1960, R. S. Ridgely in 1970, SJW and JF in 1976, and SJW in 1981.

TABLE 3

SIGNIFICANT CORRELATIONS BETWEEN THE ABUNDANCES OF CONSPECIFIC POPULATIONS FOR ALL PAIRWISE COMBINATIONS OF CENSUSES[1]

	Cañas		Viveros		Puer-cos	Chapera		Bayon-eta	Platanales		Sa-boga	Bol-años	Pache-ca	San José
	1972	1978	1976	1978		1976	1978		1976	1978				
Rey	**	**	*	**	*	**								
Cañas 1972		**	*	**	**									
Cañas 1978			**	**	**	**					**	*		
Viveros 1976				**		*	**	**					**	
Viveros 1978					**	*							**	
Puercos														
Chapera 1976							**	*	**		*	*		
Chapera 1978								*	*			*	**	
Bayoneta									**				**	
Platanales 1976										**	**			
Platanales 1978												*		
Saboga														
Bolaños														
Pacheca														

[1] * = P ≤ 0.05; ** = P ≤ 0.01. The three boxed regions include all predictions. Patterns of abundance are predicted to be (1) similar on Rey, Cañas, Viveros, and Puercos; (2) similar on Chapera, Bayoneta, Platanales, Saboga, and Bolaños; and (3) unique on Pacheca and San José. See text for reasons.

predictable differences among islands with dissimilar forests and avifaunas suggest that avian abundances are determined primarily by predictable processes. Before considering the implications of this, we will consider three alternative hypotheses.

First, similar abundances could result from a common history of chance disturbances. The three primary sources of disturbance in the Pearl Archipelago are fire, wind, and fluctuations in rainfall. Fire and wind affect single islands. For example, much of the forest on Chapera burned between 1976 and 1978, and most of the large trees (dbh > 50 cm) on Cocos fell in a windstorm between 1978 and 1981. In both instances, nearby islands were unaffected and the abundances of resident birds on the disturbed island changed little if at all (Table 2 and Fig. 5 for Chapera; Wright, unpubl. data, for Cocos). The third source of disturbance, fluctuations in rainfall, would have archipelago-wide effects. On the adjacent mainland, the dry season of 1976–1977 was the most severe since 1920 (Panama Canal Company, Meteorological and Hydrographic Branch, unpubl. data). Two observations suggest that the avian community was little affected. First, patterns of abundance were similar in censuses taken before (1971 and 1976) and after (1978) the drought (Fig. 5). Second, predictable between-island differences remained despite the possible archipelago-wide effects of drought. We conclude that the two sources of local disturbance, fire and wind, have little effect on mobile birds, and large scale disturbances caused by fluctuations in rainfall did not occur despite a dry season more severe than any in the past 60 years.

A second alternative hypothesis is that similar abundances are an artifact of the census technique. This can be discounted for two reasons. First, the same census technique was used on each island, yet abundances differed systematically between islands with different forests and between islands where land-bridge relicts were present and islands from which they were absent (Table 3, Fig. 5). Second, the pattern of similarities and differences in community structure was not affected by omission of the species whose abundances were most likely to be misrepresented in the censuses.

The final alternative hypothesis that we consider is that movements of individuals between islands are so frequent that the entire Pearl Archipelago acts as a single avian community.

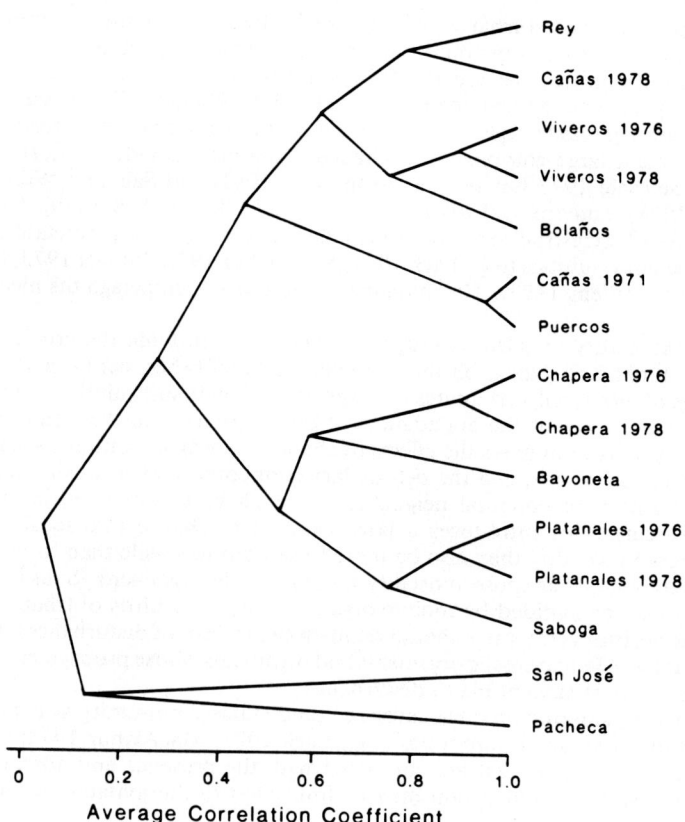

Rey

Cañas 1978

Viveros 1976

Viveros 1978

Bolaños

Cañas 1971

Puercos

Chapera 1976

Chapera 1978

Bayoneta

Platanales 1976

Platanales 1978

Saboga

San José

Pacheca

| 0 | 0.2 | 0.4 | 0.6 | 0.8 | 1.0 |

Average Correlation Coefficient

FIG. 5. Dendrogram of similarities among censuses. Pacheca and San José have unique forests and distinctive avian communities. The remaining censuses fall into two large groups. One includes small, isolated islands that lack land-bridge relicts: Chapera 1976 and 1978, Bayoneta, Platanales 1976 and 1978, and Saboga. The other group includes Bolaños and islands that support land-bridge relicts: Rey, Cañas 1971 and 1978, Viveros 1976 and 1978, and Puercos.

This hypothesis can be discounted for several reasons. First, most of the study islands are separated by 10 km or more of open ocean (Fig. 1). Second, differences in community structure among islands with dissimilar forests and avifaunas cannot be explained by this hypothesis. Third, the evidence indicates that inter-island movements are rare. The species censused tend to be sedentary, and, during eight months in the archipelago, we only recorded four individuals of these species in flight between islands. While these rare movements might have important consequences with regard to colonization and gene flow, their rarity indicates that local population dynamics are determined by local births and deaths and not by immigration and emigration. In sum, the alternative hypotheses cannot account for the predictable patterns in community structure.

We will now turn to the implications of predictable, highly structured communities. The local abundance of a species is affected by a myriad of predictable and unpredictable events. Today, there is a schism in ecological thought. One group of ecologists emphasizes the impact of unpredictable events [e.g., rolling boulders (Sousa 1979), treefall gaps (Connell 1978; Hubbell 1979), chance deaths (Sale 1977), and chance recruitment (Sale 1977; Chesson and Warner 1981)]. The other group emphasizes predictable effects, particularly competition for limiting resources (e.g., Lack 1971; MacArthur 1972; Cody 1974; Schoener 1982, 1983). Synthetic works acknowledge that both processes occur (e.g., Hutchinson 1959), but the schism is still real (e.g., Sale 1977 and Sale and Williams 1982 contra Anderson et al. 1981; Wiens 1977 contra Cody 1968, 1981). Our data indicate that the avifauna of the Pearl Archipelago is primarily affected by predictable events, but one example cannot resolve this controversy.

In fact, the importance of disturbance probably varies from community to community. We will now consider characteristics of communities that are affected primarily by deterministic interactions and communities that are affected primarily by chance disturbances. The latter are found in highly variable environments (Connell 1975; Wiens 1977 but see Cody 1981; Sousa 1979) and among sessile organisms whose abundances are limited by space and whose recruitment involves a large component of chance. Examples include territorial reef fishes with pelagic larvae (Sale 1977 but see Anderson et al. 1981, and Sale and Williams 1982), corals (Connell 1978), and tropical trees (e.g., Connell 1978; Hubbell 1979). On the other hand, communities of terrestrial vertebrates provide the paradigm for predictable, deterministically structured communities (e.g., Lack 1971; MacArthur 1972; Pianka 1973; Cody 1974; Brown 1975; but see Wiens 1977). The avifauna of the Pearl Archipelago fits nicely into this pattern.

Mobility and the ability to select microenvironment may provide the crucial differences between rodents, birds, and lizards on the one hand and reef fishes, corals, and trees on the other. Propagules of terrestrial vertebrates are cognizant of their surroundings and are able to incorporate information about the abundances of other species into their choice of habitat (Krebs et al. 1974). This minimizes the effects of chance. By way of contrast, seeds are at the mercy of their dispersal agents, and the pelagic larvae of corals and reef fishes do not assess the species composition of potential neighbors (although preferential settling on different substrates does occur). This introduces a large element of chance into local recruitment. Disturbance-induced mortality may also be more important to sessile than to mobile organisms. Local disturbances that cause mortality among sessile organisms (Sousa 1979; Paine and Levin 1981) may be avoided by mobile organisms (e.g., the birds of Chapera). In sum, the mobility of terrestrial vertebrates should dampen the impact of disturbances (Cody 1981; Karr and Freemark 1983), and sessile organisms and organisms whose propagules are passively dispersed should be most susceptible to disturbance.

We conclude by considering possible causes of predictable community structure. Competition influences many avian communities (e.g., Lack 1971; MacArthur 1972; Cody 1974), and differences among islands that are correlated with the presence and absence of "land-bridge relicts" suggest that competition may be important to the avifauna of the Pearl Archipelago.

Several of the land-bridge relicts are extremely successful. In particular, the three antbirds and the wren *Thryothorus leucotis* are consistently abundant wherever they are found (Table 2). Several widespread insectivores seem to respond to the presence of these four species. The mangrove warbler, *Dendroica petechia,* is common in forest on islands that lack the three antbirds and *T. leucotis.* Where these species are present, however, the mangrove warbler is restricted to mangroves. The wren, *Troglodytes aedon,* shows a similar pattern. It is absent from forest on islands that support *Thryothorus leucotis* and two or three antbirds, and moderately common in forest in their absence (Table 2; Thayer and Bangs 1905; Rendahl 1920; Wetmore 1947).

We compared the abundances of widespread species on islands where relict species are present and islands that lack relict species. In 23 percent of the comparisons, the widespread species was significantly ($P < 0.05$) less common in the presence of land-bridge relicts than in their absence. Again, this is consistent with the hypothesis that competition occurs. Nevertheless, while our data suggest that interspecific competition may be important, we cannot discount numerous alternative hypotheses. We are now carrying out introduction experiments to test the competition hypothesis. In addition, competition cannot account for the differences in community structure between Pacheca and San José and the other islands, for which we are seeking alternative explanations.

ACKNOWLEDGMENTS

A dissertation improvement grant from the National Science Foundation (DEB 77-07227) and travel funds from the University of California at Los Angeles supported SJW. A grant from the Research Council of the Graduate School of the University of Missouri at Columbia supported JF. Comments by M. S. Foster and an anonymous reviewer led to improvements in the manuscript.

LITERATURE CITED

AMERICAN ORNITHOLOGISTS' UNION. 1983. Check-list of North American birds. 6th ed. American Ornithologists' Union, Washington, D.C.

ANDERSON, G. R. V., A. H. EHRLICH, P. R. EHRLICH, J. D. ROUGHGARDEN, B. C. RUSSELL, AND F. H. TALBOT. 1981. The community structure of coral reef fishes. Am. Nat. 117:476–495.

ANDREWARTHA, H. G., AND L. C. BIRCH. 1954. The Distribution and Abundance of Animals. University Chicago Press, Chicago, Illinois.

BANGS, O. 1901. Birds of San Miguel Island, Panama. Auk 18:24–32.

BARTLETT, A. S., AND E. S. BARGHOORN. 1973. Phytogeographic history of the Isthmus of Panama during the past 12,000 years (A history of vegetation, climate, and sea-level change). Pp. 203–299, In A. Graham (ed.), Vegetation and Vegetational History of Northern Latin America. Elsevier Publishing Co., Amsterdam.

BOVALLIUS, C. 1886. En segling i Las Perlas-arkipelagen. Ymer 8:5–18.

BROWN, J. H. 1975. Geographical ecology of desert rodents. Pp. 315–341, In M. L. Cody and J. M. Diamond (eds.), Ecology and Evolution of Communities. Harvard University Press, Cambridge, Massachusetts.

CHESSON, P. L., AND R. R. WARNER. 1981. Environmental variability promotes coexistence in lottery competitive systems. Am. Nat. 117:923–943.

CODY, M. L. 1968. On the methods of resource division in grassland bird communities. Am. Nat. 102: 107–147.

CODY, M. L. 1974. Competition and the Structure of Bird Communities. Princeton University Press, Princeton, New Jersey.

CODY, M. L. 1981. Habitat selection in birds: the roles of vegetation structure, competitors, and productivity. Bioscience 31:107–113.

CONNELL, J. H. 1975. Some mechanisms producing structure in natural communities: a model and evidence from field experiments. Pp. 460–490, In M. L. Cody and J. M. Diamond (eds.), Ecology and Evolution of Communities. Harvard University Press, Cambridge, Massachusetts.

CONNELL, J. H. 1978. Diversity in tropical rain forests and coral reefs. Science 199:1302–1310.

DIXON, W. J. 1981. BMDP Statistical Software 1981. University of California Press, Berkeley, California.

ERLANSON, C. O. 1947. The vegetation of San José Island, Republic of Panama. Smithson. Misc. Collect. 106:1–12.

GROSSMAN, G. D., P. B. MOYLE, AND J. O. WHITAKER, JR. 1982. Stochasticity in structural and functional characteristics of an Indiana stream fish assemblage: a test of community theory. Am. Nat. 120: 423–454.

HUBBELL, S. P. 1979. Tree dispersion, abundance, and diversity in a tropical dry forest. Science 203: 1299–1309.

HUSTON, M. 1979. A general hypothesis of species diversity. Am. Nat. 113:81–101.

HUTCHINSON, G. E. 1959. Homage to Santa Rosalia, or why are there so many kinds of animals? Am. Nat. 93:145–159.

JOHNSTON, I. M. 1949. The botany of San José Island. Sargentia 8:1–298.

KARR, J. R., AND K. E. FREEMARK. 1983. Habitat selection and environmental gradients: dynamics in the 'stable' tropics. Ecology 64:1481–1494.

KREBS, J. R., J. RYAN, AND E. L. CHARNOV. 1974. Hunting by expectation or optimal foraging? A study of patch use by chickadees. Anim. Behav. 22:953–964.

LACK, D. 1971. Ecological Isolation in Birds. Harvard University Press, Cambridge, Massachusetts.

MACARTHUR, R. H. 1972. Geographical Ecology. Harper and Row, New York.

MACARTHUR, R. H., AND J. MACARTHUR. 1961. On bird species diversity. Ecology 42:594–598.

MACARTHUR, R. H., AND E. O. WILSON. 1967. The Theory of Island Biogeography. Princeton University Press, Princeton, New Jersey.

MACARTHUR, R. H., J. M. DIAMOND, AND J. R. KARR. 1972. Density compensation in island faunas. Ecology 53:330–342.

MACARTHUR, R. J. MACARTHUR, D. MACARTHUR, AND A. MACARTHUR. 1973. The effect of island area on population densities. Ecology 54:657–658.

PAINE, R. T., AND S. A. LEVIN. 1981. Intertidal landscapes: disturbances and the dynamics of pattern. Ecol. Monogr. 51:145–178.

PIANKA, E. R. 1973. The structure of lizard communities. Annu. Rev. Ecol. Syst. 4:53–74.

PORTIG, W. H. 1976. The climate of Central America. Pp. 405–478, In W. Schwerdtfeger (ed.), Climates of Central and South America. Elsevier Publishing Co., Amsterdam.

RENDAHL, H. 1920. A list of the birds of the Pearl Islands, Bay of Panama. Ark. Zool. 13:1–56.

RIDGELY, R. S. 1976. Birds of Panama. Princeton University Press, Princeton, New Jersey.

SALE, P. F. 1977. Maintenance of high diversity in coral reef fish communities. Am. Nat. 111:337–359.

SALE, P. F., AND D. McB. WILLIAMS. 1982. Community structure of coral reef fishes: Are the patterns more than those expected by chance? Am. Nat. 120:121–127.

SCHOENER, T. W. 1976. The species-area relation within archipelagos: models and evidence from island land birds. Proc. 16th Int. Ornithol. Congr., pp. 629–642.

SCHOENER, T. W. 1982. The controversy over interspecific competition. Am. Sci. 70:586–595.

SCHOENER, T. W. 1983. Field experiments on interspecific competition. Am. Nat. 122:240–285.

SIEGEL, S. 1956. Nonparametric Statistics. McGraw-Hill, New York.

SMITH, N. G. 1971. Biovarianism in an insular population of a neotropical bird. Am. Midl. Nat. 86: 238–241.

Sousa, W. P. 1979. Disturbance in marine intertidal boulder fields: the non-equilibrium maintenance of species diversity. Ecology 60:1225–1239.

Thayer, J. E., and O. Bangs. 1905. The mammals and birds of the Pearl Islands, Bay of Panama. Bull. Mus. Comp. Zool. 46:137–160.

Wetmore, A. 1947. The birds of San José and Pedro Gonzalez Islands, Republic of Panama. Smithson. Misc. Collect. 106:1–60.

Wetmore, A. 1972. The birds of the Republic of Panamá. Pt. 3. Smithson. Misc. Collect. 150(3).

Wiens, J. A. 1977. On competition and variable environments. Am. Sci. 65:590–597.

Wright, S. J. 1981. Intra-archipelago vertebrate distributions: the slope of the species-area relation. Am. Nat. 118:726–748.

EVOLUTIONARY AND BEHAVIORAL ECOLOGY

FOSTER, MERCEDES S. Pre-nesting Cooperation in Birds: Another Form of Helping Behavior 817

FRAGA, ROSENDO M. Host-Parasite Interactions Between Chalk-browed Mockingbirds and Shiny Cowbirds 829

GREENBERG, RUSSELL, AND JUDY GRADWOHL. A Comparative Study of the Social Organization of Antwrens on Barro Colorado Island, Panama 845

HOUSTON, DAVID C. Evolutionary Ecology of Afrotropical and Neotropical Vultures in Forests 856

MOERMOND, TIMOTHY C., AND JULIE SLOAN DENSLOW. Neotropical Avian Frugivores: Patterns of Behavior, Morphology, and Nutrition, with Consequences for Fruit Selection 865

ROBINSON, SCOTT K. The Yellow-rumped Cacique and its Associated Nest Pirates 898

SHERRY, THOMAS W. Adaptation to a Novel Environment: Food, Foraging, and Morphology of the Cocos Island Flycatcher 908

THOMAS, BETSY TRENT. Coexistence and Behavior Differences Among the Three Western Hemisphere Storks 921

PRE-NESTING COOPERATION IN BIRDS: ANOTHER FORM OF HELPING BEHAVIOR

MERCEDES S. FOSTER

Museum Section, U.S. Fish and Wildlife Service, National Museum of Natural History, Washington, D.C. 20560 USA

ABSTRACT. Reports of cooperative behavior among promiscuous species and cooperative performance of pre-nest reproductive activities such as mate attraction, courtship, and copulation are rare. Factors that influence the likelihood of helping behavior occurring in promiscuous species are considered. The occurrence of pre-nest helping behavior and factors that may have selected against its presence in most species and favored its evolution in the groups in which it is found are discussed. It is argued that help with pre-nest activities should be detrimental to both the helper and recipient in species with lasting pair bonds.

RESUMEN. Son raros los informes sobre comportamiento cooperativo entre especies promiscuas y sobre el desempeño cooperativo de actividades reproductivas pre-anidadoras, tales como atracción de pareja, cortejo y copulación. Se consideran los factores que influyen en la probabilidad de comportamiento de ayuda que se presenta en especies promiscuas. Se discute la presencia de comportamiento de ayuda pre-anidadora y los factores que pueden haber sido seleccionados en contra de su presencia en la mayoría de las especies y los que pueden haber favorecido su evolución en los grupos en que se presenta dicho comportamiento. Se discute que la ayuda en las actividades pre-anidadoras debe ser detrimente para ambos, el ayudante y el ayudado, en especies con vínculos de pareja perdurables.

The literature on cooperative breeding suggests that cooperation occurs much more frequently at some stages of the reproductive cycle than at others. In an overwhelming majority of species, the help provided involves feeding and care of the young, although in some, help with nest building, nest defense, nest sanitation, incubation, and territory defense is reported also. At the same time, the distribution of helping behavior among species with different types of mating systems is not uniform. So far, cooperative breeding has been reported primarily for monogamous species with Type A territories (i.e., those in which courtship, copulation, nesting, and feeding occur, Nice 1941), although helping in colonial nesters (e.g., Fry 1975; Dow 1977; Balda and Balda 1978; Emlen 1982) and polyandrous species (e.g., Maynard Smith and Ridpath 1972; Ridpath 1972; Mader 1979; Birkhead 1981) is also known. In all of these species males and females share duties at the nest.

It is not surprising that most help reported involves nest-related activities. When Skutch (1935) first reported the phenomenon, he dealt only with "helpers at the nest." In fact, there is no *a priori* reason why help could not be rendered with any reproductive process, including pre-nest activities, or in species with any type of mating system, yet, reports of cooperative behavior among promiscuous species, and cooperation with mate attraction, courtship, and copulation, are rare. Interestingly, help with these particular pre-nest activities has been reported only from promiscuous species, and promiscuous species (with one exception, Dow 1977) have been reported to exhibit only this type of help. In fact, the primary basis of this relationship may not be the type of mating system, *per se,* but rather, the associated characteristic of pair-bond length, or length of the male-female association.

In the present paper, I consider factors that influence the likelihood of various types of helping behavior occurring in promiscuous species as well as the likelihood of help with mate attraction, courtship, and copulation occurring in species with different types of mating systems. I argue that selection should favor help with pre-nest activities among forms with transient bonds (most promiscuous species), but that its occurrence should be detrimental to both the helper and recipient in species with lasting pair bonds (most bird species).

DEFINITIONS

I use Verner and Willson's (1966:143) definition of promiscuous species as "those in which a member of one sex copulates with more than one member of the other [sex] but no lasting

[pair] bond is formed." In birds with lasting pair bonds males and females are associated through at least one nesting cycle. In those with transient bonds the association persists for a few hours or a few days, the sexes being associated only during the period of copulation; one sex assumes all responsibility for incubation and rearing of the young.

I use cooperative breeding to refer to situations in which one individual assists another who is neither offspring nor mate with some activity directly related to reproduction, and in which the help provided is not simply a coincidental result of normal, non-cooperative behavior. Thus, assistance with the defense of a breeding site may be considered help; defense of a separate feeding site is not. Likewise lek males displaying simultaneously, but independently, on different courts are not considered to cooperate, since at any given time, each may display or not regardless of the activities of the other males. Any benefit accruing to a male as a result of activities in which a second male is engaged are an incidental consequence of the second male's activities. In contrast, males that engage in reproductive activities at the same court, particularly joint, simultaneous, coordinated displays, and in which the behavior of one male directly influences that of the other, are considered to cooperate.

THE POTENTIAL AMONG PROMISCUOUS SPECIES FOR COOPERATION WITH REPRODUCTIVE ACTIVITIES

Generally, females of promiscuous species are solely responsible for rearing the young, and male-female contact is limited to copulation. Characteristically, females breed their first year and so are unavailable to serve as helpers (e.g., Wiley 1974; Foster 1976). Even if environmental factors prevented females from breeding, however, it is unlikely that they would help with nest-related activities. If helpers could enhance nest success, then it is hard to image how selection would have favored male emancipation from nest-related duties in the first place. Likewise, it has been suggested for some species that the separation of the parents is necessary to minimize activity around the nest in order to decrease predation (D. W. Snow 1962; B. K. Snow 1970; Wrangham 1980). Helpers at the nest would negate this advantage.

Males of promiscuous species, in contrast, often delay breeding for one to several years and, thus, are potential helpers (e.g., Snow 1962; Wiley 1974; Lill 1976; Foster 1977). Selection would not be expected to favor males that help at the nest, for the same reasons given for females. Another factor with regard to male helpers is that they would be in a position to commit infanticide and thereby increase their probability of fathering the female's next brood. Such behavior would be more likely to occur in promiscuous species than in those with persistent pair bonds because helpers in the former species would be less likely to be closely related to the young in the nest. This would be extremely detrimental to the female who already has invested substantially in the nest, eggs, and incubation. Thus, selection should favor breeding females that do not tolerate helper males at the nest.

In addition, as adults, males will not contribute to the rearing of the young, so they are not apt to benefit from learning aspects of nestling or fledgling care. Nor is such behavior likely to improve their future abilities to obtain mating territories (which are separated from nesting areas), to obtain high positions in a dominance hierarchy, or to attract females for copulation. Virtually the only means by which benefit would accrue to helping males would be through kin selection, although potential benefits would be less than if the males, themselves, bred. If males of one brood were to stay with their mother and help to rear young of a subsequent brood, they would be assured only of helping either full or half sibs. Thus, their average relatedness, r, to the young helped would be $\frac{1}{2} \geq r \geq \frac{1}{4}$, less than the relatedness of helpers to their own offspring.

The only other way for non-breeding males to cooperate with reproduction would be to assist breeding males with pre-nest activities.

OCCURRENCE OF PRE-NESTING COOPERATION

Help can be provided with three types of pre-nesting behavior—mate attraction, courtship, and protection of the courting or copulating pair from disruption by rivals (Foster 1983). I have found five species, representing three families, for which such cooperative activities have been reported and four more in which it is likely to occur. Because the number of species is small and this type of help is poorly known, a brief summary of the activities of each species and the contexts in which they occur is in order.

MANAKINS

Three of the 51 species of Pipridae are known to exhibit such behavior, and it is suspected to occur in four others. As the behavior and mating systems of only a small number of manakin species are known well, it seems likely that such behavior will be found in other piprids, as well.

In the Long-tailed Manakin, *Chiroxiphia linearis,* pairs (or occasionally trios) of males establish bonds that persist throughout a reproductive season, and often from year to year (data for this species from Foster 1977, unpubl. data). Each pair occupies a court on an exploded lek (one on which males maintain auditory rather than visual contact; Gilliard 1963) where the males advertise for and court females. One male of a pair is dominant and performs all copulations. Within trios, the dominance hierarchy is linear. If the dominant male disappears, the beta male apparently assumes his place and acquires a new subordinate male partner. Females of this species and all other members of the genus are solely responsible for nest-building and rearing of the young (Foster 1976).

The males of a Long-tailed Manakin pair cooperate to attract females and advertise their readiness to display and copulate. They do this by means of a call that they give in synchrony continuously through the day and breeding season. It almost never is given by a single male. Because the calls are essentially synchronous, the area over which they are detectable, their active space (sensu Bradbury 1981), increases in direct proportion to the number of males calling (theoretically, doubles). The expansion of the active space around the two calling males, as opposed to one, increases the area from which they can draw females. If a female arrives, the males move with her to a display perch and perform a cooperative courtship display that cannot be performed by one male alone, but *requires* the participation of at least two individuals. When the display ends, the subordinate male moves to nearby vegetation while the dominant completes an additional precopulatory display for the female, and copulates. The cooperative display serves to excite the female and probably increases the likelihood that copulation will occur, but only by the dominant male who performs all copulations.

The calling and dance activities of the subordinate should benefit the dominant by increasing the number of females that he may court and enhancing his success at copulation. These activities incur some cost to the subordinate in terms of time and energy expended and, perhaps, in terms of increased vulnerability to predators. The subordinate has little or no opportunity to mate and accrues no immediate benefit.

Cooperative courtship also is found in the Swallow-tailed Manakin, *Chiroxiphia caudata* (data for this species from Foster 1981, unpubl. data), in which four to six males occupy a series of communal display courts. As in *C. linearis,* the male associations persist throughout a breeding season and between years. A linear dominance hierarchy exists among the males, each moving up a step with the elimination of a higher-ranking male. Only the dominant male advertises for females in this species. When one arrives at the court, however, the alpha and beta males perform a cooperative, precopulatory display like that of the Long-tailed Manakin. This requires participation by two individuals and, again, serves to excite the female, increasing the probability that copulation will occur. Should the alpha or beta male be absent when a female visits, the gamma male takes his place in the display, and so on down the length of the hierarchy. However, the alpha male was always observed to return before copulation occurred and was responsible for all copulations, which are preceded by a solo display as in the Long-tail. The subordinates, at some cost, benefit the dominant with their actions, but without the opportunity for any immediate benefit to themselves.

Although the mating systems of the other two *Chiroxiphia, C. pareola* and *C. lanceolata,* are not known, these species exhibit the same cooperative, coordinated displays described for *C. caudata* and *C. linearis* (Aldrich and Bole 1937; Friedmann and Smith 1955; Gilliard 1959; Snow 1963a). Thus, cooperative courtship display and cooperative mate attraction may exist in these species as well.

In the Band-tailed Manakin (*Pipra fasciicauda*) each territory on an arena is owned by a single male (data for this species from Robbins 1983, in press, pers. comm.). Closely associated with him is a subordinate or beta individual who spends less time on the territory and occasionally visits alpha males on adjacent territories. Below them in the dominance hierarchy are non-territorial males that make brief visits to many territories. The alpha and beta males, and sometimes the non-territorial birds, display on the courts, sometimes simultaneously, but usually independently. If a female arrives, the alpha male chases the beta male away from the main display perches and displays for the female alone. However, if a female has visited and displayed with the alpha male and then left without copulating, the male may perform a coordinated, joint display in an attempt to lure her back. If she does return, then the actions of the beta male have benefitted the alpha by providing another opportunity for him to copulate, but since again the alpha alone will display with her, the beta receives no immediate benefit from his presumably costly actions.

Also of benefit to the alpha male may be the presence and activity of the beta male at the court. Robbins (in press, pers. comm.) suggested that females may prefer territories with increased display activity as has been found for other lek species (e.g., Kruijt and Hogan 1967). The beta males seem analogous to the satellite males of the Ruff (*Philomachus pugnax*; discussed later), although whether or not the beta males have the opportunity to copulate has not been determined. Unlike the satellite males of the Ruff, beta male *P. fasciicauda* inherit dominance on a territory with the disappearance of the alpha male, and competition among non-territorial males for the alpha position is intense (Robbins, in press).

Males of both the Crimson-headed (*Pipra aureola*; Snow 1963b; Haverschmidt 1968) and Wire-tailed (*P. filicauda*; Schwartz and Snow 1979) Manakins, closely related members of a zoogeographical superspecies (Haffer 1970; Schwartz and Snow 1979) with *P. fasciicauda*, engage in coordinated, joint displays very similar to those of *P. fasciicauda*. The contexts in which the displays are given, their function, and the relationships of the males that perform them remain unknown, but also may turn out to be pre-nest stage helping.

TURKEY

In the Rio Grande subspecies of the Wild Turkey (*Meleagris gallopavo*, Phasianidae; data for this species from Watts 1968, Watts and Stokes 1971) males occur in groups consisting of individuals that were reared by the same hen, but are not necessarily kin related (Balph et al. 1980). Because some brood mixing occurs after hatching, and broods may combine to form large flocks, genetic relatedness of males in a group is not clear. In the late winter or early spring, both males and females visit the mating grounds where for about a month males court and copulate with females. The males continue to associate in their brood groups, a hierarchy existing both within groups and among them. Males of the most dominant group move about and display among the females at the arena, while those of subordinate groups display at the periphery. The males of each group display close together and in synchrony, which apparently provides for more rapid and intense stimulation of the female and increases the probability that copulation will occur. Usually, only the dominant male of the dominant group mates, disrupting attempts by subordinates of his own or other groups to copulate. Thus, synchronous display activities of subordinates contribute to the success of the dominant.

Copulation generally persists for four minutes or more, during which time members of the dominant's brood group fend off males from other groups, protecting him from disruption so that copulation proceeds undisturbed. Whether this last activity truly represents cooperation, however, has been questioned by Balph et al. (1980) who suggested that the "protection" provided by brood mates may be nothing more than a coincidental consequence of their defense of their own individual spaces.

During the latter months of the breeding season, male groups visit small groups of females occupying nesting sites. Because nesting sites are far more numerous than display grounds, some groups of females may be attended by only a single brood group. Thus, male groups that were subordinate on the display ground may be dominant at a nesting site, and the dominant males of many more groups may copulate. Again, subordinate members of the group display with the dominant and may protect him from disruption while he copulates.

Apparently, males never change brood groups, even when group size is reduced to one or two. With an average annual mortality of 40 percent among adult male turkeys, turnover in dominance position must be fairly rapid so that many males of a brood group may have the opportunity to copulate during their lives. However, among groups, dominance seems to be positively correlated with group size, so most males will have the opportunity to mate only on the nesting grounds and not at the display grounds. The dominance hierarchy that exists among the males of the brood group presumably determines the sequence of inheritance of the dominant position.

RUFFS

Males of the Ruff (Scolopacidae) occupy small territories grouped together on arenas where they court and copulate with females (data for this species from Hogan-Warburg 1966, van

Rhijn 1973). These males fall into three categories based on a variety of criteria, including differences in behavior. Resident males own and defend territories on the arena. Marginal males, who will not concern us, attempt to establish territories (usually at the margin of the arena) and sometimes succeed in becoming residents. Satellite males do not own territories but use those of resident males, often becoming associated with one (or a few) male(s) and persistently visiting his territory. Both males display on the territory, often simultaneously, but always independently.

The success of a resident male in achieving copulations is influenced significantly by the number of females that visit his residence in the absence of a satellite male. Females, however, preferentially visit territories where satellites are displaying, which means that a high incidence of satellite males on a territory results in a high number of female visits to that territory. Thus, it is advantageous to residents to tolerate, or even attract, satellites. Once females are present on a territory, however, the owner attacks the satellite(s) and attempts to drive him (them) out (satellites never respond aggressively toward residents). Females generally do not copulate when satellite males are present, so the departure of the satellite increases the probability that copulation will occur. In addition, the aggressive resident-satellite interaction increases the receptivity of females for copulation. With copulation completed and/or the females gone, the owner male again allows satellites onto his territory as a means of increasing his chances of attracting more females.

The activities of the satellites on the court of a resident contribute directly to the reproductive success of the resident by increasing the number of females that visit and by increasing the likelihood that copulation will occur once the females are there. If a resident male disappears, the satellite attaches himself to another territorial male. He does not become the owner of the territory and, so far as is known, remains a satellite throughout his life. Satellites benefit by visiting and, thus, intermittently sharing good territories on the lek, which sometimes provides them with an opportunity to display for females and copulate.

DISCUSSION

Costs, Benefits, and the Length of the Pair Bond

All of the cases reviewed involve species with transient pair bonds in which male helpers assist with pre-nest reproductive activities. Benefits accruing to recipients of this help seem fairly straight forward. Benefits to helpers are more difficult to ascertain, perhaps because they usually are delayed. Nevertheless, they are basically the same as those accruing to birds helping at any point in the reproductive cycle (i.e., increased survival, learning, acquisition of a display site or dominant position in a hierarchy, stolen copulations, increased inclusive fitness through kin selection). Why, then, does help at pre-nest stages appear to be confined to lek species, or, at least, to those with brief, transient, pair bonds? To answer this question, it is necessary to review the benefits and costs to both the donors and the recipients of the helping behavior and the ways in which these factors may be influenced by the length of the pair bond or male-female association.

Benefits to the recipient. — As a result of cooperative mate attraction, male advertising signals are broadcast over a greater area and, thus, should draw more females to the display ground. Females attracted to the arena and exposed to a cooperative courtship display are apparently more likely to mate. Likewise, if auxiliary males protect the mating pair from disruption by rival males, then copulation is more likely to succeed. All of these factors should directly increase the fitness of the breeding males. However, the magnitude of these benefits should decrease as the length of the pair bond increases (Fig. 1), for several reasons:

(1) When pair bonds are very short, they must be made rapidly and the female must be excited quickly for copulation. In conjunction with this, males of promiscuous species rank among those with the most elaborate and brightly colored plumages and the most elaborate displays (Sibley 1957). Addition of help by a subordinate can be considered an extension of this, analogous, perhaps, to the acquisition of an additional set of plumes or crests or to a functional increase in size of the displaying male. As the bonding/courtship period lengthens, the premium on speed of bonding decreases since the process may occur over a period of several days or weeks instead of just a few minutes. Thus, the addition of a second, helping male may not be favored by selection.

(2) As the length of the pair bond increases from mating systems of promiscuity to polygyny, and then to monogamy, the proportion of males who mate with only a single female will increase. Even in polygynous species, males generally have only a few mates. As females of

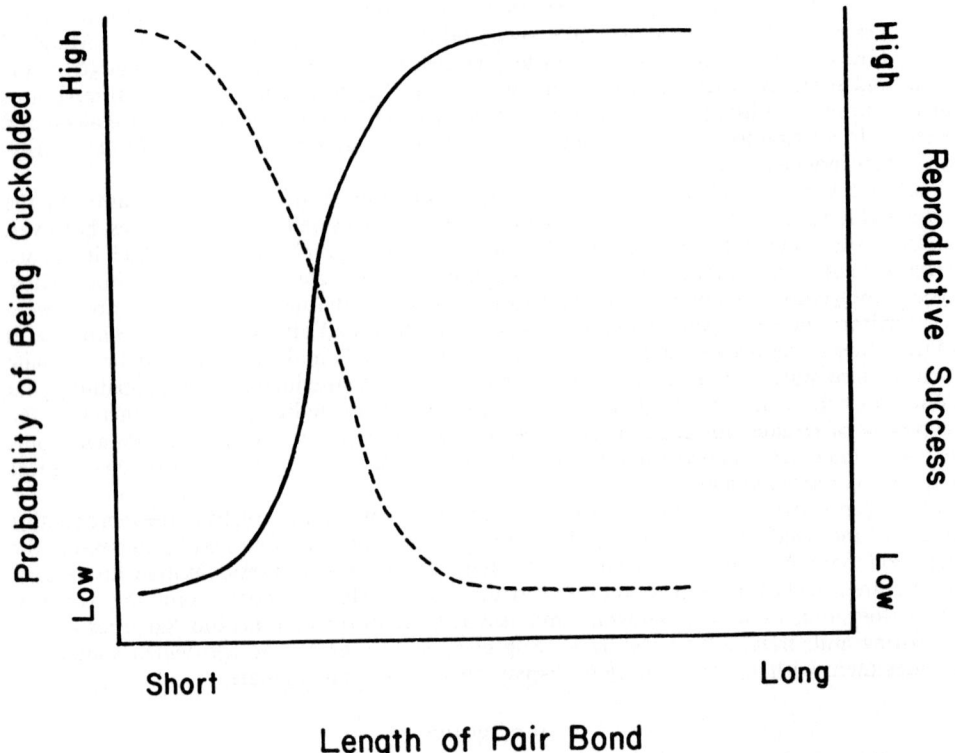

Fig. 1. The costs (probability of being cuckolded; solid line) and benefits (increased reproductive success through enhancement of mate attraction, more rapid mate excitation, and protection from disruption; dashed line) to male breeders of pre-nest cooperation with auxiliary males according to the length of the male–female bond.

promiscuous species are mated, the operational sex ratio (i.e., the ratio of fertilizable females to sexually active, potentially "eligible" males; Emlen and Oring 1977) will decrease. Thus, competition among males for females will increase continuously through the breeding season. In monogamous species, in contrast, the operational sex ratio should remain fixed as females become mated. In polygynous species, competition for females will increase, but at a slower rate than in promiscuous forms, while the operational sex ratio will show periodic, stepwise, decreases. In monogamous species with equal sex ratios, each male has a high probability of mating. In a promiscuous species with equal sex ratios, a few males will have a very high probability of mating, and most males will have a very low probability of mating. Thus, competition among males for mates will be intense, and anything that enhances a male's attractiveness to females, such as a helper, will be favored.

(3) Finally, the need for protection from disruption by rival males during copulation should be more common in birds with arena mating systems and associated short pair bonds because males of these species are usually closely grouped and in visual contact, and, thus, aware of visits by females to their rivals who are readily accessible for disruption. Birds with lasting pair bonds, on the other hand, tend to occupy large, Type A territories; neighbors are widely spaced and, perhaps, unaware when copulation occurs on an adjacent territory. Unmated males lacking territories should be more apt to disrupt, but the probability of their being present at the right moment for this should be very small.

Overall, a benefit curve should follow a sigmoid path (Fig. 1). Benefits should be very high during a brief courtship period and then drop rapidly to zero when the premium on rapid excitation of the female no longer exists and competition among males is less, as in mating systems with long pair bonds.

Costs to the recipient.—The costs to the recipient of helping behavior also will vary with the length of the pair bond, but now, directly as opposed to inversely (Fig. 1). Costs such as use of food or other resources that might be needed by the monogamous breeding pair or its offspring, or attraction of additional predators to the area, if anything, should increase because more birds will be occupying the territory, i.e., mate and young. In most promiscuous species, the birds do not feed on the mating territory, nor do females nest there, so attraction of predators to a nest is impossible. If helpers enhance activity at the mating site, however, they could attract more predators there.

The really significant cost to the recipient is the probability of being cuckolded. In species in which the male-female association is brief, a male can guard the female until she leaves the mating site. As the length of the courtship/mating period increases from a few minutes to a few days, however, the difficulty the dominant male will have guarding his mate will increase steeply as, therefore, will the probability of stolen copulation by a subordinate (Fig. 1). Although a male may stick with his mate closely throughout her receptive period, it is unlikely that he can be with her every instant. Important in conjunction with this is the proximity of a helping male to the mating site. A helper at courtship has access to that site; helper males at other stages do not.

If we consider these costs graphically, we see that the cost curve follows a sigmoid path (Fig. 1). Presumably, as long as courtship and copulation occur within a short period, the dominant male can be reasonably assured of protecting his mate from copulation with rivals. As the length of this period increases, the probability of his being cuckolded should rise steeply. At some point, however, his efficiency and the amount of time he spends with his mate per day should be constant, regardless of the length of the receptive period, and probability of being cuckolded should level.

A combination of high benefits and low costs for male recipients of pre-nest help would be expected only when the courtship/breeding period is very short, as in promiscuous (including arena-mating) species (to the left of the graph, Fig. 1). Because the slopes of both curves change steeply, the transition from net benefit to net loss should be relatively sharp. This means that males of non-promiscuous species should not tolerate the presence of helper males during the courtship/mating period and should actively drive them from the mating area or keep them away from the receptive female. This is what occurs in territorial males without help, and even in those species in which helping behavior occurs in post-copulatory stages (e.g., Andrews and Naik 1970; Woolfenden 1975; Zahavi 1976; Woolfenden and Fitzpatrick 1977).

Costs and benefits to the donors.—We also may consider the relative costs and benefits of pre-nesting help to the donor of the action, and the ways in which these factors may be influenced by the length of the pair bond. Benefits to helpers fall into five categories:

(1) Increased survival. Certain benefits associated with group living may enhance the survival of group members as opposed to solitary individuals (e.g., Stallcup and Woolfenden 1978). These should apply to birds with any social system. That they may be affected by stage in the life cycle or length of the breeding period is not evident.

(2) Increased experience or learning. In those species in which the pair bond is short, in which competition for mates is extreme, and in which speed and intensity of female excitation are at a premium (as in manakins), courtship behavior should be elaborate and intense. Thus, learning may be important, not necessarily with regard to performance of particular display elements, although this may occur, but with regard to tactics (i.e., use of particular display elements in response to particular female behaviors, as in Black Grouse [Kruijt and Hogan 1967]). If participation through helping is a form of learning, then one would expect to see this type of behavior. As the courtship period lengthens, and especially as courtship shifts from predominantly visual to predominantly auditory stimuli, the premium on learning should decrease. Thus, one would expect learning of courtship techniques to be most important in lek-displaying forms (Fig. 2).

(3) Increased probability of taking over a good display site or territory, or assuming the dominant position in a hierarchy, and (4) Increased probability of stealing copulations. Assumption of the dominant position in a hierarchy seems important in both manakins and turkeys. In those species for which data on marked individuals are available, dominance position on good sites is inherited in a linear sequence. Presumably, by helping, a male retains the opportunity to be present on the territory and to improve his status as males above him are eliminated. And, though position in a hierarchy and, thus, inheritance may be correlated with age, prior occupancy also may bestow an advantage in contests for dominance (e.g., Foster 1981). An auxiliary male also could enhance the quality of the territory in the eyes of

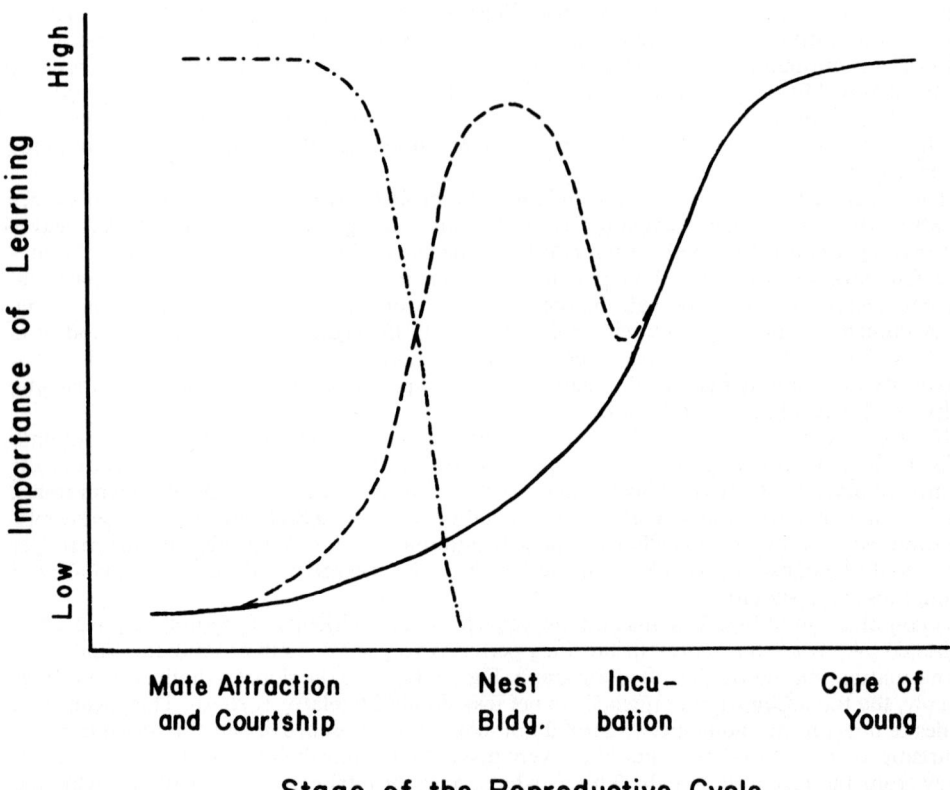

Fig. 2.   The importance of learning at different stages of the reproductive cycle for species with brief, transient pair bonds (dot-dash line), or extended, persistent pair bonds (solid line). For species with extended bonds, an additional period of learning (dashed line) may be important in those species with elaborate nests.

females who might return in subsequent years (when he is the dominant) to breed. Stolen copulations may occur among species with all types of mating systems (e.g., Bray et al. 1975; Fujioka and Yamagashi 1981), yet the closer and more frequently a male is at the site of couplation (e.g., Power and Doner 1980), the greater the probability that he will be able to take advantage of the opportunity should it arise. This surely must be a factor in both manakins and Ruffs and, perhaps, turkeys.

Increased opportunities for stolen copulation, and enhanced probability of taking over a good territory or assuming a dominant position in a hierarchy would seem to accrue to helpers regardless of the length of the pair bond. Thus, one would expect helpers to assist with courtship activities and, thereby, to increase their proximity to receptive females and the mating site, and, perhaps, to enhance their status, whenever the opportunity arose. That they do not in species with extended pair bonds probably results, at least in part, from the intolerance of the breeding males.

(5) Increased inclusive fitness through kin selection. Enhancement of inclusive fitness through kin selection requires first that the relatedness of helpers and recipients be greater than the relatedness of helpers and the general population. Second, it requires that the helpers be able in some way to insure that the probability that the individuals they help will be kin is greater than random. Very few data on the kin relationships of the species considered here are available. Various lines of evidence argue, however, that males of several of these species are not kin-related (Foster 1977; Balph et al. 1980) and that this factor is not important. Even so, from a theoretical point of view, especially since the sample of species so far reported to

TABLE 1

EXPECTED OCCURRENCE OF NET BENEFITS (+) OR COSTS (−) TO BREEDERS AND
HELPERS AS A RESULT OF COOPERATION WITH DIFFERENT REPRODUCTIVE ACTIVITIES, AS A
FUNCTION OF PAIR-BOND LENGTH

| | Pair-bond length | | | |
| | Long[1] | | Short[2] | |
Reproductive activity	Breeder	Helper	Breeder	Helper
Territory defense	+	+	+	+
Mate attraction	−	−	+	+
Courtship	−	−	+	+
Copulation	−	−	+	+
Nest-building	−/+	−/+	−	−
Incubation	−	−	−	−
Care of young[3]	+	+	−	−

[1] Monogamous and polygamous forms.
[2] Promiscuous forms.
[3] Includes nest defense, nest sanitation, brooding and feeding of young.

show pre-nest help is so small, it is useful to consider how the factor may be influenced by pair-bond length.

As indicated earlier, helpers at pre-nest stages may assist their fathers, to whom they are related by ½, brothers or half-brothers, to whom they are related on average by ½ or ¼, respectively, or various individuals of lesser kinship. Degree of relatedness should not be influenced by pair-bond length. The likelihood of association with related individuals, on the otherhand, definitely may be influenced by the persistence of the relationship between the helper's parents.

In monogamous species, the helpers most often are offspring of previous broods. From the time they hatch until they leave to breed on their own, they remain in the family territory or home range, associated with family members. Thus, the probability that the recipients of their help will be parents or sibs, as closely related to them as their own offspring, is high.

In promiscuous species, on the other hand, females alone rear young, usually in areas away from the display grounds where they copulated. Young males that delay breeding often undergo a period of pre-reproductive dispersal and may not become associated with a male as a helper until they are several years old (e.g., Foster 1977; Graves et al. 1983). These factors should greatly decrease the probability of a male becoming associated with close kin when he begins to help. The only other alternative would be for male sibs from the same brood to remain associated until they breed and then to assist one another. There is no evidence of this for any species so far described (Balph et al. 1980), and in some species, small clutch size, low reproductive success, and advanced age at which males begin helping (Foster 1976, 1977) argue against it.

Thus, one must conclude that kin selection is not likely to be a driving force in the evolution of pre-nest helping behavior in promiscuous species. In those species in which helpers and breeders share a close kin relationship, however, it may reinforce selective advantages that accure for other reasons.

ARENA VERSUS NON-ARENA FORMS

Males of species with promiscuous mating systems may be clumped on arenas or leks, loosely grouped on exploded leks, or uniformly dispersed throughout the habitat. Thus far, all reports of pre-nest helping behavior, as I have defined it, are for arena-occupying forms. This is not unexpected since the opportunity for the evolution of cooperative displays is greater in these species. Such displays often appear to have originated from male-male aggressive interactions related to disputes over display sites (Foster 1977, 1981) or territory ownership (satellite males). Such disputes would be expected to occur more frequently if males were closely grouped in a small area with a finite number of display perches than if they were widely dispersed through a habitat. The same argument applies to disruptions of copulations by rival males (Foster 1983).

OCCURRENCE OF HELP WITH OTHER REPRODUCTIVE ACTIVITIES

The arguments developed in an attempt to explain the rarity of pre-nest helping behavior and its apparent restriction to promiscuous species can be applied equally to help provided at other stages of reproduction. In Table 1, the major reproductive activities are listed along with the expected occurrence of net benefits or costs accruing as a result of cooperation with these activities among species having mating systems with long or short pair bonds. I will briefly consider the rationale behind these designations.

If acquisition of a territory or dominance in an area is one of the benefits accruing to helpers, then they should be expected to assist with territory defense. Territory owners, on the other hand, should welcome such help, even though it means that more individuals will occupy the territory, as long as the help ensures control of a larger and/or better quality area. These benefits should operate for both the donor and recipient of the help regardless of the length of the pair bond. Defense of territories by all members of a group has been reported for many monogamous species (e.g., Woolfenden and Fitzpatrick 1978; Grant and Grant 1979) and the promiscuous noisy miner (Dow 1979), although in the latter, unlike the situation in most promiscuous forms, males and females remain associated throughout the reproductive period.

For the reasons given above, promiscuous forms would be expected to cooperate with pre-nest stage but not nest-related activities (Table 1). In contrast, I have argued that in species with persistent pair bonds, individuals should not have pre-nest stage help (Table 1), but may be expected to cooperate with nest-related activities.

The major nest-oriented activities with which helpers of such species can cooperate are nest-building, incubation, and care of young (nest defense, nest sanitation, incubation, and feeding). As indicated earlier, helpers should maximize the likelihood of return on the help invested by helping at the latest reproductive stage possible and/or by helping at the most crucial stage. Nest-building does not fit either category except, perhaps, among species with complex nests, the building of which requires considerable effort or practice (e.g., Rowley 1978). In addition, male help with nest building generally should not be tolerated by breeding males if the helpers are sexually mature, since females usually are receptive during this period.

Cooperative incubation also should occur only rarely (Table 1) and involve species in which more than one female lays in a particular nest or in which more than one male mates with the laying female (Frith and Davies 1961; Ridpath 1972; Vehrencamp 1978; Koenig and Pitelka 1979; Mader 1979). All birds sharing in incubation would be parents (or potential parents). It is not likely that non-breeding helpers would practice infanticide if associated with a nest (at any stage) because of a high probability that the young would be full sibs. Breeders, however, might not welcome incubation by inexperienced helpers because exposure of eggs to temperatures outside the optimum range may be fatal or lead to developmental abnormalities (Drent 1975). It also may be that non-breeders are not physiologically equipped to incubate successfully (Vehrencamp 1982).

Finally, it would appear advantageous to both the donor and the recipient of help for birds to cooperate in feeding nestlings (Table 1). Inexperienced helpers may benefit from learning (Lawton and Guidon 1981), but, unlike the situation with incubation, an error (e.g., bringing too large a food item or eating it oneself) causes no direct damage to the nestling. Helpers also may benefit through kin selection because in most instances, their average relatedness to the nestlings they feed is ½. The breeders benefit from the help provided by producing more young (e.g., Rowley 1965; Woolfenden 1975), or by producing the average number of offspring, but at less cost to themselves (e.g., Krekorian 1978; Rowley 1978).

ACKNOWLEDGMENTS

The development of my ideas was enhanced by discussion with many colleagues, several of whom also read an early draft of the manuscript. For their helpful comments I thank B. Beehler, R. P. Balda, R. W. McDiarmid, E. S. Morton, M. B. Robbins, D. E. Wilson, and R. L. Zusi. I also thank G. Graves and M. Robbins, who kindly supplied copies of unpublished manuscripts for my use, and C. W. Angle, who prepared the figures.

LITERATURE CITED

ALDRICH, J. W., AND B. P. BOLE, JR. 1937. The birds and mammals of the western slope of the Azuero Peninsula [Republic of Panama]. Sci. Publ. Cleveland Mus. Nat. Hist. VII.

ANDREWS, M. I., AND R. M. NAIK. 1970. The biology of the Jungle Babbler. Pavo 8:1–34.

BALDA, R. P., AND J. H. BALDA. 1978. The care of young Piñon Jays (*Gymnorhinus cyanocephalus*) and their integration into the flock. J. Ornithol. 119:146–171.

BALPH, D. F., G. S. INNIS, AND M. H. BALPH. 1980. Kin selection in Rio Grande turkeys: a critical assessment. Auk 97:854–860.

BIRKHEAD, M. E. 1981. The social behaviour of the dunnock, *Prunella modularis*. Ibis 123:75–84.

BRADBURY, J. W. 1981. The evolution of leks. Pp. 138–169, *In* R. D. Alexander and D. W. Tinkle (eds.), Natural Selection and Social Behavior. Chiron Press, New York.

BRAY, O. E., J. J. KENNELLY, AND J. L. GUARINO. 1975. Fertility of eggs produced on territories of vasectomized Red-winged Blackbirds. Wilson Bull. 87:187–195.

DOW, D. D. 1977. Reproductive behavior of the Noisy Miner, a communally breeding honeyeater. Living Bird 16:163–185.

DOW, D. D. 1979. Agonistic and spacing behaviour of the Noisy Miner *Manorina melanocephala*, a communally breeding honeyeater. Ibis 121:423–436.

DRENT, R. 1975. Incubation. Pp. 333–420, *In* D. S. Farner and J. R. King (eds.), Avian Biology V. Academic Press, New York.

EMLEN, S. T. 1982. The evolution of helping. I. An ecological constraints model. Am. Nat. 119:29–39.

EMLEN, S. T., AND L. W. ORING. 1977. Ecology, sexual selection, and the evolution of mating systems. Science 197:215–223.

FOSTER, M. S. 1976. Nesting biology of the Long-tailed Manakin. Wilson Bull. 88:400–420.

FOSTER, M. S. 1977. Odd couples in manakins: a study of social organization and cooperative breeding in *Chiroxiphia linearis*. Am. Nat. 11:845–853.

FOSTER, M. S. 1981. Cooperative behavior and social organization of the Swallow-tailed Manakin (*Chiroxiphia caudata*). Behav. Ecol. Sociobiol. 9:167–177.

FOSTER, M. S. 1983. Disruption, dispersion, and dominance in lek-breeding birds. Am. Nat. 122:53–72.

FRIEDMANN, H., AND F. D. SMITH. 1955. A further contribution to the ornithology of northeastern Venezuela. Proc. U.S. Natl. Mus. 104:463–524.

FRITH, H. J., AND S. J. J. F. DAVIS. 1961. The ecology of the Magpie Goose. CSIRO Wildl. Res. 6:91–141.

FRY, C. H. 1975. Cooperative breeding in bee-eaters and longevity as an attribute of group-breeding birds. Emu 74 Suppl., pp. 308–309.

FUJIOKA, M., AND S. YAMAGISHI. 1981. Extramarital and pair copulations in the Cattle Egret. Auk 98:134–144.

GILLIARD, E. T. 1959. Notes on the courtship behavior of the Blue-backed Manakin (*Chiroxiphia pareola*). Am. Mus. Novit. No. 1942.

GILLIARD, E. T. 1963. The evolution of bowerbirds. Sci. Am. 209(2):38–46.

GRANT, P. R., AND N. GRANT. 1979. Breeding and feeding of Galapagos Mockingbirds, *Nesomimus parvulus*. Auk 96:723–736.

GRAVES, G. R., M. B. ROBBINS, AND J. V. REMSEN, JR. 1983. Age and sexual difference in spatial distribution and mobility in manakins (Pipridae): inferences from mist-netting. J. Field Ornithol. 54:407–412.

HAFFER, J. 1970. Art-Entstehung bei einiger Waldvogeln Amazoniens. J. Ornithol. 111:285–331.

HAVERSCHMIDT, F. 1968. Birds of Surinam. Oliver and Boyd, Edinburgh.

HOGAN-WARBURG, A. J. 1966. Social behavior of the Ruff, *Philomachus pugnax* (L.). Ardea 54:109–229.

KOENIG, W. D., AND F. A. PITELKA. 1979. Relatedness and inbreeding avoidance: counterploys in the communally nesting acorn woodpecker. Science 206:1103–1105.

KREKORIAN, C. O. 1978. Alloparental care in the Purple Gallinule. Condor 80:382–390.

KRUIJT, J. P., AND J. A. HOGAN. 1967. Social behavior on the lek in Black Grouse, *Lyrurus tetrix tetrix* (L.). Ardea 55:203–240.

LAWTON, M. F., AND C. F. GUINDON. 1981. Flock composition, breeding success, and learning in the Brown Jay. Condor 83:27–33.

LILL, A. 1976. Lek behavior in the Golden-headed Manakin, *Pipra erythrocephala* in Trinidad (West Indies). J. Ethol. Suppl. 18:1–84.

MADER, W. J. 1979. Breeding behavior of a polyandrous trio of Harris' Hawks in southern Arizona. Auk 96:776–788.

MAYNARD SMITH, J., AND M. G. RIDPATH. 1972. Wife sharing in the Tasmanian Native Hen, *Tribonyx mortierii*: a case of kin selection? Am. Nat. 106:447–452.

NICE, M. M. 1941. The role of territory in bird life. Am. Midl. Nat. 26:441–487.

POWER, H. W., AND C. G. P. DONER. 1980. Experiments on cuckoldry in the Mountain Bluebird. Am. Nat. 116:689–704.

RIDPATH, M. G. 1972. The Tasmanian Native Hen, *Tribonyx mortierii* II. The individual, the group, and the population. CSIRO Wildl. Res. 17:53–90.

ROBBINS, M. B. 1983. The display repertoire of the Band-tailed Manakin (*Pipra fasciicauda*). Wilson Bull. 95: 321–342.

ROBBINS, M. B. In press. Social organization of the Band-tailed Manakin (*Pipra fasciicauda*). Condor.

ROWLEY, I. 1965. The life history of the Superb Blue Wren *Malurus cyaneus*. Emu 64:251–297.

ROWLEY, I. 1978. Communal activities among White-winged Choughs *Corcorax melanorhamphus*. Ibis 12:178–197.

SCHWARTZ, P., AND D. W. SNOW. 1979. Display and related behavior of the Wire-tailed Manakin. Living Bird 17:51–78.

SIBLEY, C. G. 1957. The evolutionary and taxonomic significance of sexual dimorphism and hybridization in birds. Condor 59:166–191.

SKUTCH, A. F. 1935. Helpers at the nest. Auk 52:257–273.

SNOW, B. K. 1970. A field study of the Bearded Bellbird in Trinidad. Ibis 112:299–329.

SNOW, D. W. 1962. A field study of the Black and White Manakin, *Manacus manacus*, in Trinidad. Zoologica 47:65–104.

SNOW, D. W. 1963a. The display of the Blue-backed Manakin, *Chiroxiphia pareola*, in Tobago, WI. Zoologica 48:167–176.

SNOW, D. W. 1963b. The display of the Orange-headed Manakin. Condor 65:44–48.

STALLCUP, J. A., AND G. E. WOOLFENDEN. 1978. Family status and contributions to breeding by Florida Scrub Jays. Anim. Behav. 26:1144–1156.

VAN RHIJN, J. G. 1973. Behavioural dimorphism in male Ruffs, *Philomachus pugnax* (L.). Behav. 47: 153–229.

VEHRENCAMP, S. L. 1978. The adaptive significance of communal nesting in Groove-billed Anis (*Crotophaga sulcirostris*). Behav. Ecol. Sociobiol. 4:1–33.

VEHRENCAMP, S. L. 1982. Body temperature of incubating versus non-incubating roadrunners. Condor 84:203–207.

VERNER, J., AND M. F. WILLSON. 1966. The influence of habitats on mating systems in North American passerine birds. Ecology 47:143–147.

WATTS, C. R. 1968. Rio Grande turkeys in the mating season. Trans. North Am. Wildl. Nat. Res. Conf. 33:205–210.

WATTS, C. R., AND A. W. STOKES. 1971. The social order of turkeys. Sci. Am. 224(6):112–118.

WILEY, R. H. 1974. Evolution of social organization and life-history patterns among grouse. Q. Rev. Biol. 49:201–227.

WOOLFENDEN, G. E. 1975. Florida Scrub Jay helpers at the nest. Auk 92:1–15.

WOOLFENDEN, G. E., AND J. H. FITZPATRICK. 1977. Dominance in the Florida Scrub Jay. Condor 79: 1–12.

WOOLFENDEN, G. E., AND J. H. FITZPATRICK. 1978. The inheritance of territory in group-breeding birds. BioSci. 28:104–108.

WRANGHAM, R. W. 1980. Female choice of least costly males; a possible factor in the evolution of leks. Z. Tierpsychol. 54:357–367.

ZAHAVI, A. 1976. Co-operative nesting in Eurasian birds. Pp. 685–693, *In* H. J. Frith and J. H. Calaby (eds.), Proc. 16th Int. Ornithol. Congr. Vol. 2. Australian Academy Science, Canberra.

HOST-PARASITE INTERACTIONS BETWEEN CHALK-BROWED MOCKINGBIRDS AND SHINY COWBIRDS

Rosendo M. Fraga

Department of Biological Sciences, University of California, Santa Barbara,
Santa Barbara, California 93106 USA

ABSTRACT. The host-parasite interactions between a generalist brood parasite, the Shiny Cowbird (*Molothrus bonariensis*), and one of its hosts, the Chalk-browed Mockingbird (*Mimus saturninus*), were studied from 1972 to 1979 at a study site in Buenos Aires Province, Argentina. Particular attention was paid to the possible co-evolution of host defenses and parasite counterdefenses in the system. Field observations and experiments were used to estimate selective pressures and to detect correlative responses in both species.

The relationship between host and parasite is characterized by a high incidence of parasitism, as cowbird eggs were observed in 78.1 percent (50 of 65) of the mockingbird nests. Parasitism reduced the mean number of mockingbirds reared per nesting attempt by 69.9 percent (0.93 in non-parasitized nests, 0.28 in parasitized ones). Reproductive success of the parasite was low, as only six young cowbirds were reared from 102 eggs (5.9%); still, the lower success of the cowbird in nests of other local passerines makes mockingbirds a reasonably good host choice.

Possible mockingbird defenses against brood parasitism include early nesting (before the cowbird laying season) and nest guarding. Cowbird eggs in the study site are of two morphs, immaculate and spotted, occurring in roughly equal frequencies in nests of most host species. Mockingbirds ejected nearly all immaculate cowbird eggs, but few spotted ones.

Cowbird nestlings in mixed broods have a significantly lower chance of fledging than host nestlings: 37 percent starved in one to seven days. The smaller size and (presumably) the lower competitive ability of the parasitic nestling may explain this. This differential pattern of mortality resembles brood reduction; it may not have evolved as an antiparasitic defense of mockingbirds.

Shiny Cowbirds do not seem to have evolved any specific, successful adaptation to deal with this particular host, except perhaps gregariousness during nest searching; this may counteract nest guarding by mockingbirds. There was no evidence that cowbirds avoided laying immaculate eggs in nests of this host, although the success for this egg morph was almost zero. Female cowbirds that lay immaculate eggs have alternative, suitable hosts. No evidence exists of host fidelity among female Shiny Cowbirds, and probably most individuals that parasitize mockingbirds are reared by more abundant host species with fewer antiparasite defenses.

RESUMEN. Entre 1972 y 1979 en un sitio de estudio de la provincia de Buenos Aires, se estudió la interacción entre anfitrión y parásito en una especie de ave que parasita nidos de diferentes especies de aves, el *Molothrus bonariensis* y una de sus especies anfitrionas, *Mimus saturninus*. Se prestó particular atención a la posibilidad de coevolución de defensa del anfitrión y la contradefensa del parásito en el sistema. Se realizaron observaciones de campo y experimentos para determinar las presiones selectivas y detectar respuestas correlativas en ambas especies.

La relación entre la especie anfitriona y la parásita se caracteriza por una gran incidencia de parasitismo ya que los huevos de *Molothrus bonariensis* fueron observados en 78.1 por ciento (50 de 65) de los nidos de *Mimus saturninus*. La presencia del parásito redujo el promedio de polluelos de *saturninus* criados por nido en 69.9 por ciento (0.93 en nidos no parasitados y 0.28 en parasitados). El éxito reproductivo del parásito fue bajo, ya que sólo seis jovenes fueron criados de un total de 102 huevos (5.9%); pero de cualquier manera el poco éxito de los *bonariensis* en nidos de otros passeriformes, hace que sea rasonablemente buena la elección del *saturninus* como especie anfitriona. Es posible que las defensas del *saturninus* en contra del parasitismo de la nidada incluya anidación temprana (antes de que el *bonariensis* comience su época de desove) y cuidado del nido. Los huevos del *bonariensis*, en el área de estudio, fueron de dos diferentes formas, lisos o moteados, presentandose en frecuencia más o menos similar en los nidos de la especie anfitriona. Los *saturninus* echaron del nido casi todos los huevos lisos, pero sólo unos pocos de los moteados. Los polluelos de *bonariensis* en nidadas mixtas, tienen significativamente posibili-

829

dades más bajas de que sus crías sobrevivan que los de la especie anfitriona: 37 por
ciento murieron de hambre entre el primero y el séptimo día. El tamaño más pequeño
y (presumiblemente) la menor habilidad competitiva de los pichones parásitos pueden
explicar esto. Estos diferentes patrones de mortalidad se asemejan a la reducción de
la nidada; puede que no haya evolucionado como una defensa antiparásita.

Los *Molothrus bonariensis* no parecen haber desarrollado ninguna adaptación es-
pecífica y exitosa para sobreponerse a esta especie anfitriona "difícil," con excepción
quizás de su comportamiento gregario durante la búsqueda de nidos; esto puede ser
contrarrestado por el cuidado del nido del Mimidae. No hubo evidencias de que el
tordo evitará desovar huevos lisos en nidos de esta anfitriona, a pesar de que el éxito
de ese tipo de huevos fue casi cero. Hembras de los tordas que ponen huevos lisos
tienen alternativamente otras especies anfitrionas. No hay evidencias de que exista
una fidelidad hacia la especie anfitriona y es probable que la mayoría de los individuos
que parasitan sean criados por otras especies más abundantes con menor cantidad
de defensas antiparasitarias.

Interspecific brood parasitism is a breeding strategy found in about 90 species of birds (Lack
1968; Payne 1977), or in approximately 1 percent of all avian species. Parental investment
by brood parasites is usually reduced to the formation of eggs and sperm and the placement
of eggs in suitable host nests. All remaining aspects of avian parental care are provided by
the host. Most studies have shown that brood parasitism reduces the nesting success of the
host (data summarized in Payne 1977: table 1). If a negative impact on the breeding success
of the host exists, a host should evolve adaptations to avoid and/or counteract parasitism,
and the parasite should evolve counter-adaptations to overcome these host defenses.

In this paper I discuss the interaction between a generalist brood parasite, the Shiny Cowbird
(*Molothrus bonariensis*), and one of its hosts, the Chalk-browed Mockingbird (*Mimus saturninus*), at a study site in Buenos Aires, Argentina. I provide data showing the harmful effects
of cowbird parasitism on the breeding success of mockingbirds and later discuss potential
host and parasite adaptations related to parasitism.

The Shiny Cowbird is widespread through most of South America, with a major range gap
in the Amazonian rainforest; it has been spreading north and south during this century
(Johnson 1967; Post and Wiley 1977), being found today from the Antilles to 45°S latitude
in Argentina and Chile. The Chalk-browed Mockingbird is found only in eastern South
America, from scattered savannas north of the lower Amazon river in the Guianas and adjacent
states of Brazil south to 40°S latitude in Argentina. Its range seems to be completely included
within the range of the cowbirds, at least in Argentina (Olrog 1979). Both species occupy
similar nonforest habitats and, in the last century, have become adapted to living in modified
environments in Argentina.

Interactions between the Shiny Cowbird and the Chalk-browed Mockingbird have probably
existed for a considerable time. In the late eighteenth century Azara (1802: species numbers
61, 223) reported both species in Buenos Aires, Argentina, and Paraguay. The first recorded
cases of parasitism were observed by Gibson (*fide* Friedmann 1929) at Cape San Antonio,
Buenos Aires Province, probably around 1880. Barrows (1883) included this mockingbird
among the regular cowbird hosts in the province of Entre Rios. Parasitized nests of Chalk-
browed Mockingbirds have been found subsequently in the Argentinian provinces of Salta,
Tucumán, Santa Fé, and Córdoba as well as in Uruguay (Friedmann 1929; Friedmann et al.
1977; De la Peña 1983; Salvador 1983).

The Chalk-browed Mockingbird is not the only species of *Mimus* parasitized by the Shiny
Cowbird; the seven South American species in the genus have been recorded as hosts (Dinelli
1918; Friedmann 1963). At least in Argentina two other mockingbirds (*M. triurus* and *M.
patagonicus*) seem to be parasitized as frequently as *M. saturninus* (Pereyra 1937; pers. observ.)

Shiny Cowbird eggs are extremely variable. In eastern Argentina and neighboring parts of
South America, eggs are either spotted or immaculate (Hudson 1920; Friedmann 1929). Eggs
intermediate between the immaculate and spotted types are rare. The variation in egg morphology is, thus, essentially discontinuous and may constitute a true polymorphism (Ford
1964). Spotted eggs vary considerably and are hard to classify into discrete types, although
some authors have attempted to do so (Hoy and Ottow 1964). It has usually been assumed
(e.g., Sick 1958), although not demonstrated, that each female cowbird lays eggs of only one
morph.

TABLE 1
HOST AND PARASITE YOUNG IN BROODS CONTAINING SHINY COWBIRD NESTLINGS

No. young			Fate of cowbird nestlings		
Host	Parasite	No. nests	Starved	Preyed upon	Fledged
1	1	3	0	2	1
2	1	8	3	3	2
3	1	3	2	1	0
0	2	1	0	2	0
1	2	6	4	3	5
2	2	2	3	1	0
	Totals	23	12	12	8

STUDY AREA AND METHODS

Field data were gathered in a study site of about 60 ha (Fig. 1) centered in the main woodland of Estancia La Candelaria, Lobos, Buenos Aires Province, Argentina (35°15′S, 59°13′W). For descriptions of the study area see Fraga (1978, 1980, 1983). I resided in the study area from September to March (spring and summer) during 1972 to 1979. Periodic visits were made in the remaining months.

A total of 79 mockingbird nests were located and studied; nest contents were checked at intervals of one to two days. Eggs were marked with waterproof ink, measured to the nearest 0.1 mm with vernier calipers, and described. Nestlings were banded for individual recognition and weighed to the nearest 0.5 g with a spring scale. Of the 79 host nests, 69 were found before or during the egg stage, and 10 in the nestling stage (host and/or parasite). Complete records for the entire nesting cycle are available for 17 parasitized nests that were found during the nest-building period. Additionally, 53 adult cowbirds and five adult mockingbirds were mist-netted and color-banded for individual recognition.

My experiments with simulated parasitism followed the technique of Rothstein (1975a), with some modifications. I used only natural eggs. A total of 27 eggs (10 immaculate, 17 spotted; sizes in Appendix I) were introduced into 16 mockingbird nests. The sample of immaculate eggs consisted of eight immaculate Shiny Cowbird eggs plus two plain white eggs of Picui Doves (*Columbia picui*), which differed from the first in being less glossy and more elongate. The sample of spotted eggs included 12 spotted Shiny Cowbird eggs of variable aspect, plus five eggs of Screaming Cowbirds (*Molothrus rufoaxillaris*), four of which differed from most spotted Shiny Cowbird eggs in having a more reddish tone and more evenly distributed marks. One of the Screaming Cowbird eggs had a blue-green ground color, resembling a mockingbird egg. All experimental eggs were marked with waterproof ink.

Experiments were conducted between 1975 and 1979. Eggs were introduced between 07:00 and 09:30 during the laying period of the host or during the first three days following laying. Six experimental nests were continuously observed during the next 3 h from a car. In the other cases the nest contents were checked at intervals not exceeding 12 h. I did not remove eggs from the experimental nests.

Rothstein (1975a, b) regarded as "rejecter species" those that, within five days of the experimental introduction of cowbird eggs, removed, damaged or buried the eggs, or deserted the nests. Species not showing any of these behaviors were regarded as "accepters." The variability of Shiny Cowbird eggs in Argentina (Fraga 1978, 1983) requires a modification of these definitions, as some species may accept and/or reject only some types of eggs.

RESULTS

BEHAVIOR OF FEMALE SHINY COWBIRDS

Movements and spacing.—During the breeding season, I obtained three or more visual records each for five banded cowbird females; by plotting these points on a map I estimated minimal home range sizes of 21 to 48 ha. None of these females was observed for more than three consecutive weeks in the study area, although two returned the next fall.

Home ranges of breeding females overlap extensively. In a few cases I observed females threatening each other with "Bill Up" displays (Orians and Christman 1968); apparently the behavior is used for maintaining individual distances. Chasing or fighting was never observed among female Shiny Cowbirds, which are not territorial (Wiley and Wiley 1980).

Female Shiny Cowbirds show considerable gregariousness during the breeding season (Sick 1958). Groups of up to 16 were seen at feeding and roosting sites.

Nest searching.—Female Shiny Cowbirds seem to locate host nests chiefly by watching the behavior of host individuals (Sick 1958; Fraga 1978), but probably also by following other nest-searching females. In 17 percent (7 of 42) of the observed cowbird visits to mockingbird nests, two or more females were present simultaneously; twice I recorded visits by five females. Such high numbers were not seen around the nests of other hosts. Once, two females located a recently built mockingbird nest and perched in a nearby tree uttering series of vocalizations; within two minutes, three other females and three males were attracted to the site.

HOSTS OF THE SHINY COWBIRD

I found eggs of Shiny Cowbirds in the nests of 14 (40%) of the 35 species of passerines breeding in the study area and immediate surroundings (Fraga 1982).

In addition to studying parasitized nests, I censused cowbird fledglings in the study site between 1970 and 1979. A single host species, the Rufous-collared Sparrow, (*Zonotrichia capensis*) reared 46 percent of the young parasites. Similar results have been reported from other areas (Gibson 1880, 1918; Sick 1958; King 1973). Chalk-browed Mockingbirds reared 12 percent of the cowbird juveniles. As sparrows apparently do not rear more than one cowbird per nest (Fraga 1978, 1983), the 31 parasites reared by this host were probably fledged from 31 nests. The eight young cowbirds reared by mockingbirds were probably fledged from five nests.

The differences among host species in numbers of cowbirds reared is a function of several variables, including the relative number of nests of each species available to the parasite. I found one to four sparrow nests per hectare in the study site; thus, they were about 10 times more abundant than mockingbird nests. The estimated number of nests that produced young cowbirds was only six times higher in sparrows than in mockingbirds (31 vs 5); this suggests that mockingbirds may have been better hosts than sparrows.

BREEDING BIOLOGY OF CHALK-BROWED MOCKINGBIRDS

Although the Chalk-browed Mockingbird seems to be one of the most suitable cowbird hosts in terms of nestling diet and brood biomass, it was relatively scarce. Only 10 to 12 breeding pairs resided in the area during the years of my study (Fig. 1). Except one small plot of 2.35 ha where up to three pairs nested in relatively close proximity (Fraga 1979), most mockingbird nests were hundreds of meters apart. Mockingbirds remain in pairs throughout the year, forming family groups with the juveniles from their last broods during the non-breeding season. Adults are extremely sedentary, and banded individuals were invariably found in their territories during the winter, often roosting in the same shrubs they used for nesting.

The most favored nest sites are relatively isolated shrubs or small trees with dense foliage, a choice in part explained by the habits of the fledglings (see below). The nest is a large open cup of twigs lined with fibers.

Egg laying was recorded from 10 September to 9 January, and banded individuals attempted up to three successful broods per season. Clutch size ($\bar{X} \pm$ s.d.) was 3.57 ± 0.76 eggs in 14 non-parasitized nests (range 3–5 eggs). Eggs were usually laid at daily intervals. Mockingbird eggs were spotted and/or blotched in shades of brown over a blue-green ground color; the amount of variation in color and pattern was slight. Measurements ($\bar{X}$ length and width, $\pm$s.d.) of 62 eggs were $28.09 \pm 1.19 \times 20.72 \pm 0.52$ mm (ranges 25.7–30.3×19.6–21.7 mm). The mean weight of 16 fresh eggs was 6.7 g.

The incubation period was 13 to 15 days in three non-parasitized nests, with the first two or three eggs hatching on day 13. Hatching intervals of 36 to 72 hours were recorded at five nests with four or five eggs. The nestling period was usually 12 to 15 days (14 nestlings), but if disturbed, nestlings sometimes left the nest as early as day 9. Young leave the nest by climbing and hopping through nearby branches, and remain hidden and stationary in the nest shrub or in a nearby shrub for at least another week. During this period, fledglings do not fly even if closely approached. With their striped underparts and silent behavior, they are difficult to locate in shadowy foliage. After this cryptic period, juveniles start to fly, following the adults and frequently begging.

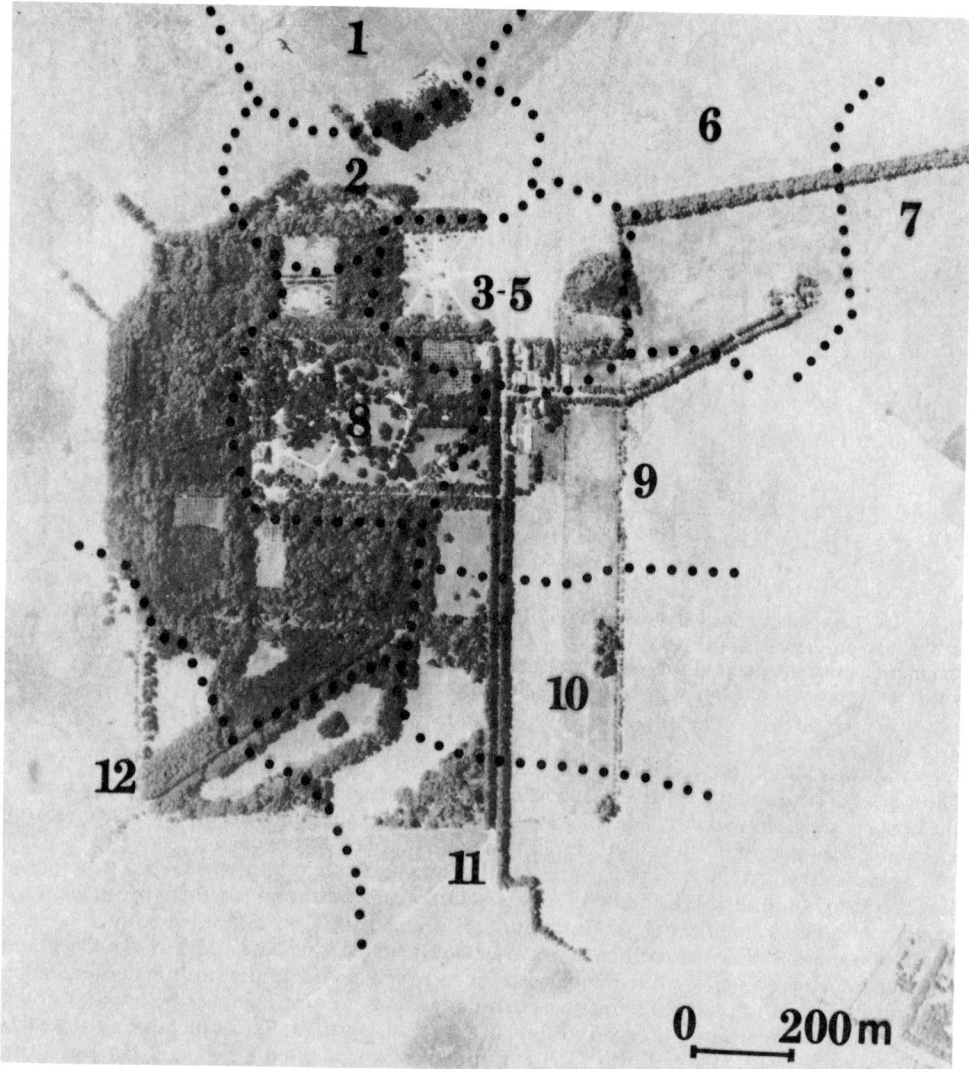

F<small>IG</small>. 1. Study area and approximate boundaries of Chalk-browed Mockingbird territories and nesting areas. In area 3–5, up to three pairs nested together in some breeding seasons. Pairs in areas 1, 11, and 12 sometimes nested outside the study site.

INCIDENCE OF PARASITISM

Cowbird eggs were observed in 50 of 69 (72.5%) mockingbird nests found before the nestling stage. As mockingbirds eject some parasitic eggs (see below), this figure may be a minimum estimate of the incidence of parasitism. In five of the 19 nests in which cowbird eggs were not detected, mockingbird eggs were punctured and/or removed, a typical result of cowbird parasitism (see below); fragments of cowbird eggshells were found on the ground 4 m from one of these nests. Probably these nests were parasitized, but through most of this paper they will be excluded from computations, giving an incidence of parasitism of 78.1 percent (50/64). Mason (1980) found a similar proportion (76.9%) of parasitized mockingbird nests in his study sites in Buenos Aires Province. Salvador (1983) found an even higher percent of parasitized nests (86.2%) in Córdoba. Very few studies of Shiny Cowbird hosts have reported higher incidences of parasitism (Post and Wiley 1977; Gochfeld 1979).

The egg-laying season of mockingbirds (10 September to 9 January, 121 days) was almost

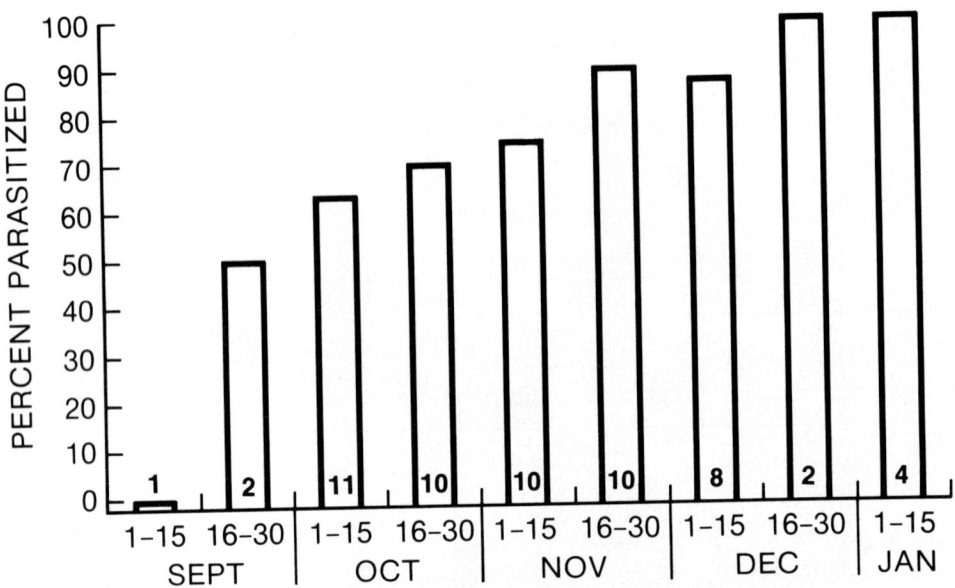

F<small>IG</small>. 2. Temporal incidence of parasitism. Thirty-six Chalk-browed Mockingbird nests are assigned to a period on the basis of the day egg-laying began (observed or estimated from hatching dates). The remaining nests (not found during egg-laying and not producing nestlings) are placed in the period in which they were found. Sample sizes shown inside histograms.

fully included within the cowbird laying season (26 September to 7 February, 134 days). The dates are the earliest and latest recorded during the nine years of study. Only 13.2 percent of the host laying season was outside the parasite breeding season. In terms of numbers of nesting attempts, only two of 64 host clutches (3.1%) were laid before the cowbird breeding season.

Early mockingbird nests were less frequently parasitized than later ones (Fig. 2). A phase-length test for time series (Kendall 1976) showed that the temporal trend in the incidence of parasitism departs significantly from random ($P < 0.001$, d.f. 7). In nests of Rufous-collared Sparrows (Fraga 1983), the incidence of parasitism reached a peak (87.5%) in late December to early January, showing coincidence with the mockingbird data. By nesting earlier, mockingbirds could avoid cowbird parasitism to some extent, a point I discuss below.

Nestling cowbirds were observed in four of 10 (40%) nests found in the nestling stage; the incidence of parasitism was lower in this sample, although the difference is not significant ($\chi^2 = 1.5435$, d.f. 1, $P > 0.05$).

C<small>ONTENTS OF</small> P<small>ARASITIZED</small> N<small>ESTS</small>

Host eggs.—Clutch size of mockingbirds averaged 3.57 eggs in 14 non-parasitized nests, and only 1.76 eggs (s.d. = 1.10) in 50 parasitized nests. Thus, mockingbird reproductive effort was reduced by 50.7 percent in parasitized sets; Mason (1980) found a reduction of 53 percent, and Salvador (1983, pers. comm.), of 51 percent. In my sample only 13 (26%) of the 50 parasitized nests contained "normal" mockingbird clutches of three or more eggs, and no host eggs remained in eight (16%) of these nests.

Observations show an association between egg puncturing and brood parasitism in Shiny Cowbirds. Male cowbirds were not recorded visiting nests. Reductions in the number of mockingbird eggs in the nest can be attributed to cowbird behavior (as suggested first by Hudson 1870) rather than to the laying of fewer eggs (as suggested by King 1973). In 17 parasitized mockingbird nests that I visited daily during the laying sequence of host and parasite, mockingbirds laid 58 eggs, giving a mean clutch size of 3.48 eggs; this does not differ significantly from mean clutch size of non-parasitized nests (3.57). During the process of parasitism, 14 eggs vanished, and 11 eggs were found punctured in the nests (43.1% of the total). One to three days later, four punctured host eggs vanished. The remaining damaged host eggs were left *in situ* in five nests that mockingbirds deserted. Yolk-stained eggs were often seen in nests from which mockingbird eggs vanished. Mockingbirds immediately eject shells from hatched eggs and probably some vanished host eggs were first punctured by the cowbirds and later removed by the mockingbirds.

Cowbird eggs.—I found spotted, immaculate, and intermediate egg morphs at the study site. Immaculate eggs are usually white, rarely pale bluish-white. Spotted eggs usually have a white, pale grey, pale blue or pale buff ground color and are marked with variable patterns of brown to reddish-brown blotches or spots. Often they have underlying marks in gray or pale lilac. Eggs intermediate between the immaculate and spotted types are rare, comprising one to two percent of the approximately 250 eggs I examined.

Cowbird eggs from mockingbird nests measured ($\bar{X}$ length and width, ±s.d.) 22.41 ± 1.20 × 18.06 ± 0.59 mm (N = 73 spotted) and 22.89 ± 0.97 × 18.16 ± 0.79 mm (N = 5, immaculate). Cowbird eggs from nests of Rufous-collared Sparrows (Fraga 1983) and mockingbirds do not differ significantly in size.

A total of 102 cowbird eggs was found in the 50 parasitized mockingbird nests, giving a mean (± s.d.) of 2.04 (±1.11) and a range of one to four cowbird eggs per nest. Multiple parasitism was more common than single parasitism. Twenty-eight nests were multiply parasitized (11 with two cowbird eggs, 10 with three, and seven with four), whereas only 22 contained a single egg. These figures are similar to those I obtained for 39 parasitized nests of Rufous-collared Sparrows in the study area ($\bar{X}$ = 1.95, range = 1–5 cowbird eggs).

Rufous-collared Sparrows have been shown experimentally to accept all types of cowbird eggs (Mason 1980; Fraga 1978, 1983). I found almost equal numbers of immaculate and spotted cowbird eggs in 39 parasitized sparrow nests observed between 1970 and 1979; 37 (46.7%) were immaculate, 38 (50%) were spotted, and one was intermediate. By contrast only five of 102 (4.9%) of the cowbird eggs in nests of mockingbirds were immaculate (χ^2 = 47.14, d.f. 1, P < 0.001). The different proportions of egg morphs were not due to spatial or temporal differences in nest distribution since the study area for mockingbirds included the areas where most of the other species also nested. The breeding season of mockingbirds was almost completely included within the breeding season of Rufous-collared Sparrows (Fraga 1983) and other species frequently parasitized with immaculate cowbird eggs. Cowbird egg morphs were not temporally segregated in the study area, and 30.8 percent of the parasitized sparrow nests contained immaculate and spotted eggs.

Immaculate and spotted cowbird eggs in nests of Rufous-collared Sparrows have similar chances of being punctured by cowbirds. In the three sparrow nests in which punctured cowbird eggs were found, two were immaculate and three were spotted; overall 5.4 percent (2 of 37) of the immaculate eggs and 7.9 percent (3 of 38) of the spotted eggs were damaged. As the sparrow accepts all cowbird eggs (Fraga 1978, 1983; Mason 1980), the puncturing was almost certainly done by cowbirds.

Spotted cowbird eggs in mockingbird nests are sometimes punctured. In 17 parasitized mockingbird nests with complete laying records, four of 32 (12.5%) spotted cowbird eggs were punctured, each in a different nest. Five immaculate cowbird eggs were observed in this sample of 17 nests; no eggs were punctured, but four disappeared.

Ejection of cowbird eggs by mockingbirds.—Cowbirds parasitized 11 nests into which I introduced additional cowbird eggs; the five remaining nests apparently were not parasitized.

The 10 immaculate eggs placed in the experimental nests (four of which were unaffected by natural parasitism) vanished in less than 48 h; in four cases, they vanished in less than 3 h. By contrast, 14 of 17 spotted eggs remained in the nest and were incubated for at least five days; five hatched. Nine of the spotted egg experiments were performed at nests where natural parasitism was not observed; one of these resulted in rejection. The results are significantly different in both the naturally parasitized and unparasitized subsamples of experimental nests (P < 0.05, Fisher Exact Test, Siegel 1956). Ejection occurred in one nest that was continuously observed from a car and that was only visited by mockingbirds.

No immaculate egg was found punctured, but two of the three spotted eggs regarded as rejected were punctured within 48 h; the third missing spotted egg simply vanished within four days. One of the punctured spotted eggs was found in a nest affected by natural parasitism.

Survival in parasitized and non-parasitized nests.—Eggs in parasitized nests had a decreased probability of reaching the nestling stage; some eggs (host and/or parasite) hatched in 10 of 14 non-parasitized nests, but only in 19 of 50 parasitized ones. The difference is significant (χ^2 = 4.93, d.f. 1, P < 0.05). This result probably explains the lower incidence of parasitism in the sample of nests found in the nestling stage. Desertion was not observed in non-parasitized nests, but it was frequent among parasitized ones: five of 17 parasitized nests with complete laying records were abandoned by the mockingbirds. All the deserted nests in this last sample contained conspicuously punctured mockingbird eggs not removed by the host or the parasite.

NESTLINGS AND FLEDGLINGS

Brood composition, nestling behavior, and nestling appearance.—Twenty-three cowbird eggs hatched in 16 naturally parasitized nests that reached the nestling stage. Four other nests found in the nestling stage contained six cowbird nestlings. During experiments with simulated parasitism, I left Shiny Cowbird nestlings in only three nests. I obtained data on 23 broods that included cowbird nestlings (Table 2). All but one of these also contained host nestlings.

Nestlings of mockingbirds and cowbirds differ in general aspect and behavior. Nestling mockingbirds have pink to flesh-colored skin, with dense tufts of blackish down; the oral flanges are pale yellow, and the mouth lining is bright yellow. Nestling cowbirds have a yellowish skin, with sparse tufts of pale gray down; the oral flanges range from white to yellow (Fraga 1978). Although the mouth lining is somewhat variable in color, it always has a reddish tinge.

Calls of nestlings and fledglings of each species are quite different (Fig. 3). Up to the age of four to five days, cowbirds utter short, loud "peeps," which in older nestlings and fledglings are transformed into longer, tremulous calls with a hissing quality. The frequency range of all these calls was 4 to 9 kHz. Calls of nestling and fledgling mockingbirds can be described as soft whistles, with an overall frequency range of 3 to 7 kHz.

Although the cowbird incubation period (usually 12 days, Fraga 1978) is at least 24 h shorter than the incubation period of mockingbirds, nestlings of both species hatch at a similar stage of development (closed eyes, no visible feather tracts). Cowbird nestlings usually exhibited a more vigorous begging behavior, but, mockingbirds had stronger tarsi and relatively longer necks and could stretch up closer to feeding adults. Feather growth started at a similar age (usually 7 days) for both species, and the nestling period was the same (14–15 days). When disturbed, mockingbirds were more likely to leave the nest prematurely.

Shiny Cowbirds did not have a stationary, cryptic post-fledging period, as mockingbirds do. Three successfully fledged young cowbirds from two broods remained with their mockingbird foster parents for at least 20 to 21 days after leaving the nest.

Nestling growth.—The weight of adult Shiny Cowbirds ($\bar{X} \pm$ s.d. = 49.4 $\pm$ 6.0 g, N = 27, both sexes combined) was 62.8 percent of the weight of adult mockingbirds (78.7 $\pm$ 4.7 g, N = 4, sexes combined). This difference is also found in egg weights; the average weight of cowbird eggs (4.0 $\pm$ 0.3 g, N = 19) was 59.7 percent of the mean weight of host eggs.

Cowbird nestlings apparently reached a maximum weight at day 8 (Fig. 4), with a slight decline in the late nestling period; mockingbird nestlings reached maximum weight at day 12 or later. The growth pattern of nestling Shiny Cowbirds in this study conforms to previous reports in the literature in showing little (if any) increase in weight after day 8 (King 1973: fig. 1, table 6; Fraga 1978: table 2, both with Rufous-collared Sparrows as hosts). The weight ratio of parasite nestlings to host nestlings followed an inverse U-shaped curve through time with maximum values of 88.5 to 87.0 percent at days 4 to 6, and a minimum value near 50.0 percent at the late nestling period, when cowbirds no longer gain weight.

Survival of host and parasite.—Cowbird nestlings in mixed broods had a significantly lower chance of fledging than mockingbird nestlings. Excluding broods destroyed by predators, the ratio of fledglings to nestlings was .40 for cowbirds versus .94 for mockingbirds (χ^2 = 12.48, d.f. 1, $P < 0.005$).

In the mixed broods, 12 cowbird nestlings that failed to fledge died or vanished one to seven days after hatching, with no sign of predation (i.e., other members of the brood survived). The mean age at death was 3.25 days. Four nestlings were found dead in the nest or died while I weighed them. Undernourishment and starvation probably caused death of the 12 nestlings, as all showed low weights (Fig. 4) and/or no increase (or even a decline) in weight during two to three consecutive days. By contrast, only one mockingbird nestling (reared with two cowbirds) did not increase in weight on two consecutive days and failed to fledge.

Post-fledging survival of parasites.—Eight cowbirds fledged from six parasitized nests. At three nests in which host nestlings were simultaneously and successfully fledged, no cowbird young fledged. In the other three nests the young mockingbirds vanished after leaving the nest, and four cowbird fledglings survived to independence.

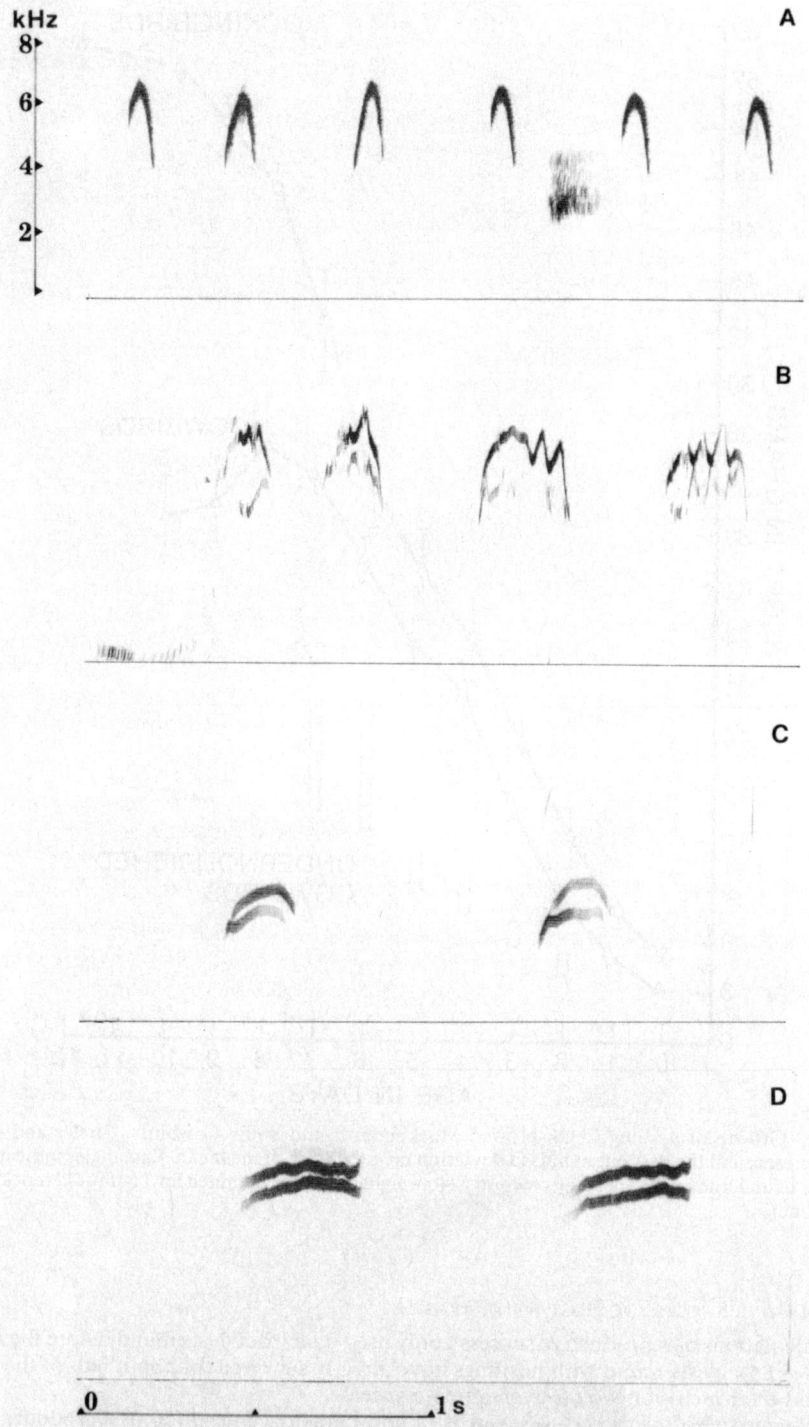

FIG. 3. Nestling vocalizations. A: Two-day old nestling Shiny Cowbird; B: Seven-day old nestling Shiny Cowbird; C: Two-day old nestling Chalk-browed Mockingbird; D: Seven-day old nestling Chalk-browed Mockingbird. Vertical axis: frequency in kHz; horizontal axis: time in seconds; scales applicable to all vocalizations.

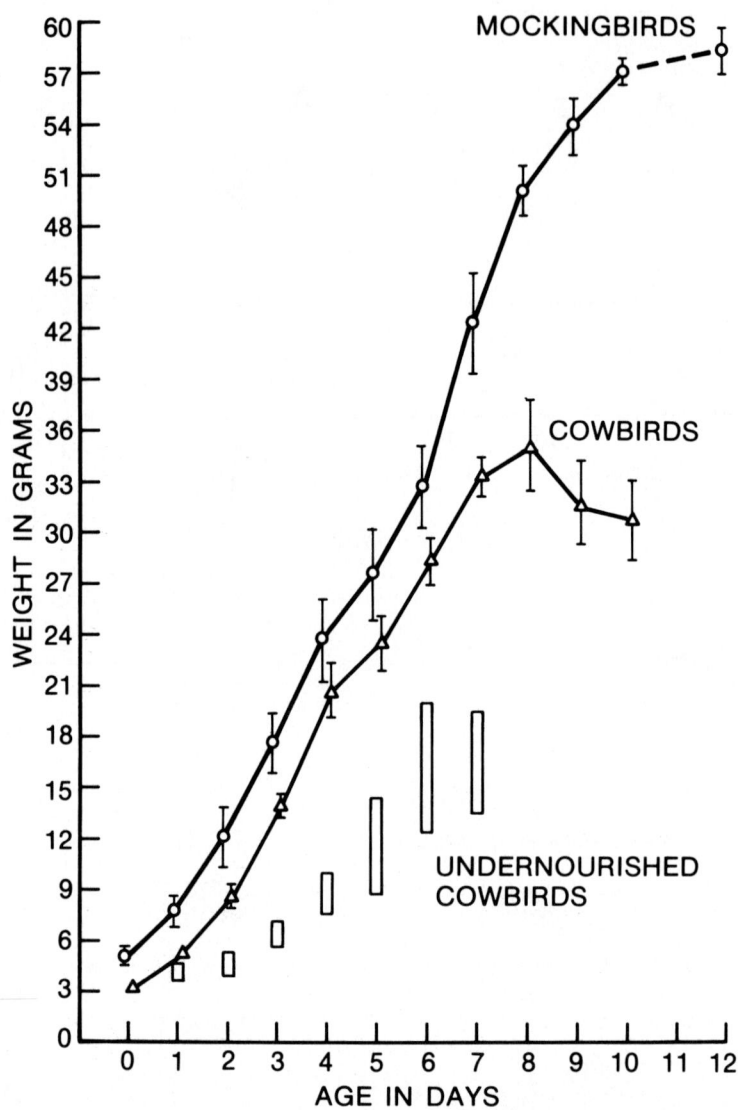

Fɪɢ. 4. Growth of nestling Chalk-browed Mockingbirds and Shiny Cowbirds. Circles and triangles indicate means, and the bars one standard deviation on either side of the mean. Rectangles indicate ranges of weights of undernourished nestling cowbirds. No weight data were obtained for 11-day old mockingbirds (dashed line).

REPRODUCTIVE SUCCESS OF HOST AND PARASITE

For calculations of reproductive success I only used data from nests found before the nestling stage (N = 64). Nests found with nestlings have already survived for about half of the nesting cycle, and their inclusion would overestimate success.

Mockingbird breeding success.—Seven of 14 non-parasitized nests (50.0%) produced at least one fledgling; the other nests were destroyed by predators. The ratio of nestlings hatched to eggs laid was 35 to 50 (70.0%), and the ratio of fledglings to nestlings was 17 to 35 (48.6%), giving a first estimate of total success of 34.0 percent. Only 13 of the 17 fledglings survived to independence; thus, the corrected estimate of success is 26.0 percent (13 of 50).

Only 10 of 50 parasitized nests (20%) produced at least one host fledgling. Breeding success can be calculated using the maximum number of host eggs observed (N = 127) or using the number of host eggs presumed to have been laid, assuming that mean host clutch size (3.57 eggs) was the same in parasitized and non-parasitized nests (N = 178). The first method underestimates the number of host eggs, because cowbirds destroy many of these before an observer can count them.

Using the first criterion, the ratio of nestlings to eggs is .236 (30 to 127) and the ratio of fledglings to nestlings .633 (19 to 30), giving a first estimate of 15.0 percent for total success. With 178 as total number of eggs, the ratio of nestlings to eggs is .168 (30 to 178), and the ratio of fledglings to nestlings remains unchanged, giving a total success of 10.6 percent. Only 14 fledglings survived to independence, giving a corrected estimate of success of either 11.0 percent or 7.9 percent.

Cowbird success.—Only natural cases of parasitism were used to evaluate success. Ninety-seven spotted eggs were found within active host nests. The ratio of nestlings to eggs was .237 (23 to 97), the ratio of fledglings to nestlings .261 (6 to 23). Total fledging success was 6.2 percent. As only three fledglings survived to independence, the corrected estimate of total success was 3.1 percent. The estimates show that in nests of mockingbirds, about 16 spotted cowbird eggs are needed to produce one fledgling, and from 32 to 35 eggs to produce one independent juvenile.

Only one of five immaculate eggs found within active nests was incubated by the mockingbirds, but it vanished in five to six days. Success for this egg morph was zero. Total fledging success for all cowbird eggs observed was 5.9 percent (6 of 102).

Breeding success of cowbirds with other hosts in the study area.—The success of the cowbird is highest when using the mockingbird as compared to four other species for which my samples consisted of at least 10 cowbird eggs (Fraga, unpubl. data). Success on the Rufous-collared Sparrow was similar to that on the mockingbird (5.6%, N = 76). Successful fledging was not observed on any of three other host species.

Rufous-collared Sparrows produced more cowbird fledglings than any other host. Estimates of cowbird success in nests of mockingbirds and sparrows did not differ significantly. There is no indication of a higher post-fledgling mortality of cowbirds when reared by mockingbirds.

DISCUSSION

The Chalk-browed Mockingbird is a host highly preferred by Shiny Cowbirds in many, if not all parts of its range. This host is even more heavily parasitized at sites other than the study area. Nests with nine to 14 cowbird eggs have been reported by several authors (Friedmann 1929:109–111; Hoy and Ottow 1964; S. Narosky *in* Friedmann et al. 1977:57, and pers. comm.; Mason 1980). These results could be explained by assuming a lower relative density of cowbirds in my study area, but reliable information is lacking. Alternatively, mockingbirds may have ejected a larger proportion of parasitic eggs in my study area. Mason (1980) and Salvador (1983) have both reported lower frequencies of the immaculate egg in nests of species that are known to accept all morphs of cowbird eggs.

EVOLUTION OF HOST DEFENSES AGAINST PARASITISM

Mockingbirds produced an average of 0.93 young per nesting attempt in the absence of cowbird parasitism but only 0.28 young when parasitized. Any trait that would reduce the incidence of parasitism will be favored by natural selection. Several defenses are possible.

Early nesting.—Parasitism of mockingbird nests increased as the breeding season advanced (Fig. 2). Thus, cowbird parasitism created a selective pressure favoring early breeding. Mockingbirds exhibited a tendency to nest earlier than other local hosts (pers. observ.). In the most extreme case (late winter of 1977) two pairs started laying at least one month in advance of the earliest cowbird laying date. However, some environmental parameters that affect nesting success may select against late winter breeding. For example, nest cover is probably an important factor; only pairs that had evergreen shrubs in their territories nested early. Mockingbirds continue to nest with an increased chance of parasitism in late spring and summer, but parasitized nests still produce some juveniles. The species is multiple brooded, and banded pairs or individuals attempted three successful broods per season.

Nest guarding.—Nest guarding and aggressive behavior have been regarded by some as defenses against brood parasites (e.g., Scott 1977). Mason (1980) tested the responses of Chalk-browed Mockingbirds to mounted specimens of Shiny Cowbirds and reported low levels of aggression and poor attentiveness at nests. My observations suggest otherwise.

In my study site mockingbirds spend considerable time guarding nests, usually from elevated perches; in territories 3, 4, and 5, extra individuals often joined the guarding pair. As a typical example, I visited one nest in the incubation stage in this area 15 times between 11 and 14 October 1977; invariably I observed one to three adults on elevated perches within 25 m of the site. I observed female cowbirds alighting in or near active nests 11 times; in eight cases they were immediately attacked and chased by the mockingbirds.

The high incidence of parasitism suggests that nest guarding was not totally successful, but without guarding, the number of cowbird eggs per nest might have been higher. The gregarious habits of female cowbirds while visiting nests may also weaken this defense.

Defenses against parasitic eggs and nestlings.—The main damage inflicted by cowbirds was the destruction of host eggs. The presence of parasitic eggs and nestlings could, however, reduce the chance of hatching and fledging of the remaining host eggs and nestlings. Several defenses against this are possible. One is the ejection of Shiny Cowbird eggs, but by ejecting only immaculate eggs, mockingbirds eliminate no more than half of the cowbird eggs. Nevertheless, my data indicate that selection favors ejection of the spotted cowbird eggs.

In the first place, the presence of parasitic eggs apparently reduced the chance of hatching of host eggs. Hatchability of mockingbird eggs was 75.0 percent (30 of 40) when cowbird eggs were present, versus 95.4 percent (21 of 22) in sets without cowbird eggs. Punctured host eggs are excluded from the first sample, but I may have included eggs damaged in a more subtle way by cowbirds. The significant difference ($\chi^2 = 4.069$, d.f. 1, $P < 0.05$) suggests, however, that cowbird eggs have a detrimental effect on hatchability; the fact that unhatched host eggs occurred chiefly in clutches containing six or more eggs of both species, or in nests in which cowbirds hatched earlier than mockingbirds, gives some plausibility to the idea. Friedmann (1963:21–22) discussed data that suggested that parasitized clutches may exceed the maximum volume of eggs that host species can successfully incubate. Post and Wiley (1977) reported an increased incidence of hatching failures in parasitized nests of Yellow-shouldered Blackbirds (*Agelaius xanthomus*) containing large numbers of eggs, even in areas where Shiny Cowbirds do not puncture eggs.

The chance of a mockingbird nestling fledging from a parasitized nest is 63.3 percent; the 20.4 percent decrease in hatchability thus reduced fledgling productivity of parasitized nests by 12.9 percent (= 0.204 × 0.633). This could be a first estimate of the selection coefficient for ejection of spotted cowbird eggs; additionally, as explained in the next section, cowbird nestlings have a slight negative effect on survival of mockingbird nestlings. The combination of both factors may raise the selection coefficient to 15 or 20 percent.

Mockingbirds may also discriminate against cowbird nestlings. This defense is far less efficient than ejection of the parasitic eggs and may be difficult to evolve since birds have been strongly selected for positive feeding responses toward nestlings (Hamilton and Orians 1965).

The differential mortality of cowbirds in mixed broods resembled brood reduction. This trait is found among mockingbirds. In two of 10 non-parasitized clutches that reached the nestling stage, the brood was reduced to three nestlings through starvation of the smaller and less competitive individuals.

The hypothesis that smaller relative size is the main factor causing cowbird mortality is supported by observations of mixed broods in Córdoba (S. Salvador, pers. comm.; pers. observ.): female cowbirds, cowbird eggs, and nestlings are significantly heavier there than in Buenos Aires, and preliminary observations indicate that cowbirds in mixed broods survive better.

EVOLUTION OF COWBIRD STRATEGIES

Although the antiparasitic defenses of mockingbirds seem limited, they still may be selecting for correlative counteradaptations in the cowbird. Gregariousness during nest searching may be advantageous to females that learn the location of nests from more experienced individuals. Alternatively, it may confer some immediate advantages to all the females involved. When dealing with potentially aggressive nest-guarding hosts, such as mockingbirds, individual females may have an improved chance of parasitizing a nest if they arrive in a group. In one case I observed three female cowbirds approaching a nest from different sides; although the mockingbirds supplanted them, one female reached the nest and punctured a host egg.

Groups of female cowbirds were observed more frequently and were larger around nests of mockingbirds than near nests of other hosts in my study sites; however, this could be an effect of the relative conspicuousness and location of the nests, rather than a trait favored by the nest guarding habits of mockingbirds.

Parasitism of a single mockingbird nest by several female cowbirds generates a complex situation as regards individual advantages and disadvantages. A cost of communal parasitism is the increased risk that cowbird eggs will be punctured by female cowbirds (above, and Fraga 1983).

Elimination of host eggs and nestlings.—Mockingbird nestlings are detrimental to the survival of cowbird nestlings. Cowbirds could double their reproductive output by replacing all mockingbird eggs with their own, or by eliminating the host nestlings. Cowbirds eliminated all host eggs in only eight (16%) of the parasitized nests. In only one of the eight cases, however, did mockingbirds accept and successfully incubate a clutch composed solely of cowbird eggs.

In several species of brood parasites in the families Cuculidae and Indicatoriidae, the parasitic nestlings eject or kill the host nestlings (reviews in Payne 1977; Morton and Farabaugh 1979; Salvador 1982), but these habits are not known to occur, at least regularly, in any species of cowbird. Only once did I record (in a nest of Fork-tailed Flycatchers *Tyrannus savana*) a possible case of ejection of a host nestling by a nestling Shiny Cowbird; this may indicate the presence of variation upon which selection may act.

Cowbird egg polymorphism.—I have assumed that female cowbirds lay eggs of one type, either immaculate or spotted, although no proof for this exists. Constancy in egg type has been shown to occur in some species of passerines with variable eggs (e.g., Victoria 1972). Within the study area, most of the unusual and distinctive variants of spotted eggs were found during restricted periods (seldom exceeding one month) and only during one particular breeding season.

Through most of their range, Shiny Cowbirds lay mainly spotted eggs (Friedmann 1929). Immaculate eggs may reach a frequency of five percent or more along the southwest coast of Ecuador (Marchant 1960) and probably the coast of Peru, and in parts of east Argentina, Uruguay, and southeast Brazil. In Argentina, immaculate eggs may reach a frequency near 50 percent in areas of Buenos Aires Province (Hudson 1920:81; Fraga 1978, 1983), but the incidence declines outside the pampean region in the provinces of Córdoba (Salvador 1983), Sante Fé (De la Peña 1983), and Chaco (Hartert and Venturi 1909).

Reaction of local host species to cowbird egg morphs.—The 14 species of passerines recorded as cowbird hosts at my study site can be divided into four categories according to their acceptance or rejection of both egg morphs. Most small passerines accept eggs of both morphs (Mason 1980; pers. observ.). Argentinian passerines that are known to eject both morphs of cowbird eggs (e.g., Tropical Kingbird, *Tyrannus melancholicus*, Fraga, unpubl. data; Rufous Hornero, *Furnarius rufus*, Fraga 1980, Mason 1980) are relatively larger.

Only one host species, the Yellow-winged Blackbird (*Agelaius thilius*), is thought to accept immaculate eggs and reject spotted ones (G. Orians *in* Friedmann et al. 1977; S. Narosky, pers. comm.). This bird is a regular breeder in marshes around the study area.

Species other than the mockingbird may eject immaculate eggs and accept spotted ones. The Brown-and-Yellow Marshbird (*Pseudoleistes virescens*) seems to behave in this way (Hudson 1920:124–125; G. Orians *in* Friedmann et al. 1977), and perhaps the Rufous-bellied Thrush (*Turdus rufiventris*) (Mason 1980). The marshbird does not breed in or near the study area, and the thrush is a rare and occasional breeder there. In the study area, Chalk-browed Mockingbirds are probably the only numerically important hosts in this category.

Cowbird egg morphs in mockingbird nests.—Probably the main selective force exerted by mockingbirds on cowbirds was related to polymorphism in egg color. The corrected estimate of success was 5.9 percent for spotted eggs, but zero for immaculate ones; thus, natural selection would operate against female cowbirds that lay immaculate eggs in nests of mockingbirds. The magnitude of the selection coefficients depends on the degree of host-specificity displayed by female cowbirds. In any case, the clear prediction is that laying of immaculate eggs in mockingbird nests should be rare if cowbirds behave adaptively (avoidance hypothesis).

The ejection behavior of mockingbirds makes it difficult to estimate the original proportion of eggs of each morph laid in nests of this host, but the available data give no support to the avoidance hypothesis. Immaculate eggs were found in 10 percent of the 50 parasitized mockingbird nests. The proportion of immaculate eggs was higher in the sample of 17 nests found before the incubation stage (5 of 37 or 13.5%) than in the remaining parasitized sets (0 of 65;

three immaculate eggs were found on the ground near nests in this last group). Without doubt, the incidence of immaculate eggs in the first sample was less influenced by mockingbird ejection behavior, and thus is closer to the true proportion laid by cowbirds.

No cowbird eggs were detected in five nests that were probably visited by cowbirds (host eggs punctured and/or removed). Fragments of immaculate cowbird eggs were found on the ground 4 m from one of the nests. In a sixth nest with a similar history, and deserted by the mockingbirds, one immaculate egg was subsequently laid. Probably all the five nests were parasitized with immaculate eggs, ejected almost immediately by the host.

From the data, I estimate that at least a third of the parasitized mockingbird nests received immaculate cowbird eggs. Even if the original proportion of eggs of this morph in nests of mockingbirds was significantly smaller than in nests of sparrows and other accepter species, a large number of immaculate eggs are lost every breeding season. This conclusion may also apply to other ejecters of this egg morph, such as Brown-and-Yellow Marshbirds (Hudson 1920:124–125). Brown-headed Cowbirds in North America also lose eggs by parasitizing ejecters or other unsuitable hosts (Rothstein 1976).

Ankney and Scott (1980) objected to Rothstein's (1976) idea that Brown-headed Cowbirds "waste" eggs by parasitizing ejecters and other unsuitable hosts; their arguments can also be applied to Shiny Cowbirds, as both parasites have similar breeding biologies and feeding habits. Female Brown-headed Cowbirds seem to produce large numbers of eggs each breeding season, but Ankney and Scott were unable to find any evidence of nutritional and energetic costs in laying females. When no suitable hosts are available, parasitism of ejecter species may not be maladaptive; the slight chance that an ejecter host will successfully rear a cowbird (the benefit) may still exceed the very low cost of producing an egg (for a contrary view on the cost of egg laying see Payne 1976).

These considerations do not seem to apply to the cowbird-mockingbird interaction. At no time were mockingbirds the only available host in my study area; besides, 60 percent of the immaculate eggs observed in or near mockingbird nests were laid in October–November, a period when all host species that accept this type of egg are breeding (e.g., for sparrows this is the peak of the breeding season, Fraga 1983). Scarcity of suitable hosts probably cannot account for the existence of such a high number of misplaced immaculate eggs.

Reproductive strategy of Shiny Cowbirds.—Shiny Cowbirds do not seem to have evolved any specific and successful adaptations, with the possible exception of gregariousness, to counteract mockingbird defenses. In the evolutionary and ecological literature, generalists, such as this parasite, are regarded as "jacks of all trades and masters of none" (e.g., MacArthur 1972:61). Generalist habits in other brood parasites, however, do not necessarily imply lack of adaptations for dealing with specific hosts. Several species of parasitic cuckoos seem to be subdivided into populations of host-specific females ("gentes"), which may display remarkable adaptations, especially egg mimicry, to a limited number of particular hosts (Southern 1954; Payne 1977). The Giant Cowbird (*Scaphidura oryzivora*) may have evolved a similar system (Haverschmidt 1966; Smith 1968).

The number of gentes of parasitic cuckoos that can coexist locally seems limited, up to four in the European Cuckoo (*Cuculus canorus*; Payne 1977), and three in the Didric Cuckoo (*Chrysococcyx caprius*) in South Africa (Jensen and Vernon 1970). Although a single gens in cuckoos may be adapted to parasitize two host species with similar eggs (Jensen and Vernon 1970), this system may be impossible for brood parasites with many local hosts, such as Shiny and Brown-headed Cowbirds (Rothstein 1976).

Probably the extreme variability found in the eggs of Shiny Cowbirds in the study area has been influenced by the behavior of its hosts. Immaculate and spotted morphs may occur as a result of selective pressures by ejecters. Further refinements are possible, for instance, one common type of spotted egg resembles in color and markings the eggs of three local hosts in the family Tyrannidae. At least one species in this group is an ejecter. However, nests of these species receive all kinds of cowbird eggs. The raw materials for evolution of mimicry are present, but there is no further adaptation in the way of host specialization.

No evidence for host fidelity among female Shiny Cowbirds exists. In other brood parasites host specificity may be maintained by imprinting (Payne 1977). Perhaps the low recruitment rate of cowbirds in nests of most local hosts (Table 1) precludes the continuous existence of gentes, which would have a high chance of random extinction. This is likely to be the case with mockingbirds; presumably most female cowbirds that parasitize this host in the study site are reared by sparrows. A hypothetical gens of Shiny Cowbirds specializing on mockingbirds and laying only spotted eggs would still have a low recruitment rate unless it evolved additional adaptations to increase the survival of parasitic nestlings, such as larger eggs and nestlings.

ACKNOWLEDGMENTS

This paper is based on a portion of an M.A. Thesis submitted to the University of California, Santa Barbara. I am particularly grateful to the members of my committee S. Rothstein, A. Kuris, S. Sweet, and R. Warner for their comments and advice. S. Rothstein provided encouragement through all my work with cowbirds. A. Garsd helped me on some statistical points. E. Stevens, D. Nakashima and D. Yokel answered some bothering questions during the preparation of the manuscript. S. Narosky and S. Salvador gave me access to unpublished data. P. Mason discussed several facets of cowbird parasitism with me, and also helped in the preparation of the manuscript.

The author was supported by the National Science Foundation grant DEB 8214999 to S. Rothstein during the writing of the final drafts of the manuscript.

LITERATURE CITED

ANKNEY, C. D., AND D. M. SCOTT. 1980. Changes in nutrient reserves and diet of breeding Brown-headed Cowbirds. Auk 97:684–696.

AZARA, F. DE. 1802. Apuntamientos para la historia natural de los páxaros del Paraguay y Río de la Plata. 3 vols. Vda. de Ibarra, Madrid.

BARROWS, W. B. 1883. Birds of the lower Uruguay. Bull. Nuttall Ornithol. Club 8:82–94.

BENT, A. C. 1958. Life histories of North American blackbirds, orioles, tanagers and allies. U.S. Natl. Mus. Bull. No. 211.

DE LA PEÑA, M. 1983. Hábitos parasitarios de algunas especies de aves. Hornero 12, No. Extraord.: 165–169.

DINELLI, L. 1918. Notas biológicas sobre aves del noroeste de la Rep. Argentina. Hornero 1:57–68.

FORD, E. B. 1964. Ecological Genetics. Methuen, London.

FRAGA, R. M. 1978. The Rufous-collared Sparrow as a host of the Shiny Cowbird. Wilson Bull. 90: 271–284.

FRAGA, R. M. 1979. Helpers at the nest in passerines from Buenos Aires Province, Argentina. Auk 96: 606–608.

FRAGA, R. M. 1980. The breeding of Rufous Horneros (Furnarius rufus). Condor 82:58–68.

FRAGA, R. M. 1982. Host-brood parasite interactions between Chalk-browed Mockingbirds and shiny cowbirds. Unpubl. M.A. thesis, University of California, Santa Barbara.

FRAGA, R. M. 1983. Parasitismo de cría del Renegrido Molothrus bonariensis sobre el Chingolo Zonotrichia capensis: nuevas observaciones y conclusiones. Hornero 12, No. Extraord.:245–255.

FRIEDMANN, H. 1929. The Cowbirds. C. C. Thomas, Springfield, Illinois.

FRIEDMANN, H. 1963. Host relations of the parasitic cowbirds. U.S. Natl. Mus. Bull. No. 223.

FRIEDMANN, H., L. F. KIFF, AND S. I. ROTHSTEIN. 1977. A further contribution to the knowledge of the host relations of the parasitic cowbirds. Smithson. Contrib. Zool. No. 235.

GIBSON, E. 1880. Ornithological notes from the neighbourhood of Cape San Antonio, Buenos Ayres, Part I. Ibis, Ser. 10, 6:363–415.

GIBSON, E. 1918. Further ornithological notes from the neighbourhood of Cape San Antonio, Province of Buenos Ayres. Ibis, Ser. 4, 4:1–38.

GOCHFELD, M. 1979. Brood parasite and host coevolution: interactions between Shiny Cowbirds and two species of meadowlarks. Am. Nat. 113:855–870.

HAMILTON, W. J., AND G. H. ORIANS. 1965. Evolution of brood parasitism in altricial birds. Condor 67:361–382.

HARTERT, E., AND S. VENTURI. 1909. Notes sur les oiseaux de la République Argentine. Nov. Zool. 16: 159–276.

HAVERSCHMIDT, F. 1966. The eggs of the Giant Cowbird. Bull. Br. Ornithol. Club 86:144–147.

HOY, G., AND H. OTTOW. 1964. Biological and oological studies of the molothrine cowbirds (Icteridae) of Argentina. Auk 81:186–203.

HUDSON, W. H. 1870. Letters on the Ornithology of Buenos Ayres. Reedited by D. Dewar, 1951. Cornell University Press, Ithaca, New York.

HUDSON, W. H. 1920. Birds of La Plata. 2 vols. J. M. Dent, London.

JENSEN, R. A. C., AND C. J. VERNON. 1970. On the biology of the Didric Cuckoo in Southern Africa. Ostrich 41:237–246.

JOHNSON, A. W. 1967. The Birds of Chile, Vol. 2. Platt, Buenos Aires.

KENDALL, M. G. 1976. Time Series. Griffin, London.

KING, J. R. 1973. Reproductive relationships of the Rufous-collared Sparrow and the Shiny Cowbird. Auk 90:19–34.

LACK, D. 1968. Ecological Adaptations for Breeding in Birds. Methuen, London.

MACARTHUR, R. H. 1972. Geographical Ecology. Harper and Row, New York.

MARCHANT, S. 1960. The breeding of some SW Ecuadorian birds. Ibis 104:584–599.

MASON, P. 1980. Ecological and evolutionary aspects of host selection in cowbirds. Unpubl. Ph.D. dissert., University of Texas, Austin.

MASON, P. 1985. The nesting biology of some passerines of Buenos Aires, Argentina. Pp. 954–972, *In* P. A. Buckley, M. S. Foster, E. S. Morton, R. S. Ridgely, and F. G. Buckley (eds.), Neotropical Ornithology. Ornithol. Monogr. No. 36.

MORTON, E. S., AND S. M. FARABAUGH. 1979. Infanticide and other adaptations of the nestling Striped Cuckoo, *Tapera naevia*. Ibis 121:212–213.

OLROG, C. C. 1979. Nueva Lista de la Avifauna Argentina. Inst. Lillo, Tucumán, Argentina.

ORIANS, G. H., AND G. M. CHRISTMAN. 1968. A comparative study of the behavior of Red-winged, Tricolored and Yellow-headed Blackbirds. Univ. Calif. Publ. Zool. 84:1–85.

PAYNE, R. B. 1976. The clutch size and number of eggs of Brown-headed Cowbirds: effects of latitude and breeding season. Condor 78:337–342.

PAYNE, R. B. 1977. The ecology of brood parasitism in birds. Ann. Rev. Ecol. Syst. 8:1–28.

PEREYRA, J. A. 1937. Contribución al estudio y observaciones ornithológicas de la zona norte de la Gobernación de La Pampa. Mem Jardín Zool. La Plata 7:198–326.

POST, W., AND J. W. WILEY. 1977. Reproductive interactions of the Shiny Cowbird and the Yellow-shouldered Blackbird. Condor 79:176–184.

ROTHSTEIN, S. I. 1975a. Evolutionary rates and host defenses against avian brood parasites. Am. Nat. 109:161–176.

ROTHSTEIN, S. I. 1975b. An experimental and teleonomic investigation of avian brood parasitism. Condor 77:250–271.

ROTHSTEIN, S. I. 1976. Cowbird parasitism of the Cedar Waxwing and its evolutionary implications. Auk 93:675–691.

SALVADOR, S. A. 1982. Estudio de parasitismo del Crespín *Tapera naevia chochi* (Vieillot) (Aves: Cuculidae). Hist. Nat. (Corrientes, Argentina) 2:65–70.

SALVADOR, S. A. 1983. Parasitismo de cría del Renegrido (*Molothrus bonariensis*) en Villa María, Córdoba, Argentina (Aves: Icteridae). Hist. Nat. (Corrientes, Argentina) 3:149–158.

SCOTT, D. M. 1977. Cowbird parasitism on the Gray Catbird at London, Ontario. Auk 94:18–27.

SICK, H. 1958. Notas biológicas sóbre o Gaudério *Molothrus bonariensis* (Gmelin) (Aves: Icteridae). Rev. Bras. Biol. 18:417–431.

SIEGEL, S. 1956. Nonparametric Statistics for the Behavioral Sciences. McGraw Hill, New York.

SMITH, N. G. 1968. The advantage of being parasitized. Nature 219:690–694.

SOUTHERN, H. N. 1954. Mimicry in cuckoo's eggs. Pp. 219–232, *In* J. Huxley, A. C. Hardy, and E. B. Ford (eds.), Evolution as a Process. Allen and Unwin, London.

VICTORIA, J. K. 1972. Clutch characteristics and egg discriminability of the African Village Weaverbird, *Ploceus cucullatus*. Ibis 114:367–376.

WILEY, R. H., AND M. S. WILEY. 1980. Spacing and timing in the nesting ecology of a tropical blackbird: comparison of populations in different environments. Ecol. Monogr. 50:153–178.

APPENDIX I

MEASUREMENTS OF EGGS USED IN EXPERIMENTS OF SIMULATED PARASITISM

Species	Egg type	$\bar{X} \pm$ s.d. length $\times$ width (mm) (range)	$\bar{X}$ weight (g)
Shiny Cowbird	7 immaculate[1]	22.84 ± 1.27 × 17.91 ± 0.56 (21.5–24.5 × 17.2–18.7)	4.0
	12 spotted	22.71 ± 1.29 × 18.01 ± 0.56 (20.5–24.9 × 17.2–18.9)	4.0
Picui Dove	2 immaculate	(23.2–23.4 × 17.9)	3.8
Screaming Cowbird	4 reddish	23.59 ± 0.64 × 17.64 ± 0.93 (22.9–24.2 × 16.3–18.3)	4.2
	1 blue-green	(24.2 × 18.3)	4.4

[1] One egg used in two experiments.

A COMPARATIVE STUDY OF THE SOCIAL ORGANIZATION OF ANTWRENS ON BARRO COLORADO ISLAND, PANAMA

Russell Greenberg and Judy Gradwohl

Department of Zoological Research, National Zoological Park, Washington, D.C. 20008 USA, and Smithsonian Tropical Research Institute, Apartado 2072, Balboa, Panama

ABSTRACT. For six years we studied the territorial systems and group composition of three species of antwren in the lowland tropical forest on Barro Colorado Island, Panama. Checker-throated Antwrens (*Myrmotherula fulviventris*) and Dot-winged Antwrens (*Microrhopias quixensis*) co-defended small (1.5 ha) territories that remained stable for the entire study. White-flanked Antwrens (*Myrmotherula axillaris*) defended larger (2–4 ha) more overlapping, and more fluctuating territories. We suggest that the territorial system of Dot-winged and Checker-throated antwrens is based on stable areas of dense vine-tangles, whereas White-flanked Antwrens are much less dependent on these areas.

White-flanked Antwrens and Dot-winged Antwrens occur in small family groups; young remain with parents for long periods, occasionally overlapping breeding efforts in the following rainy season. Checker-throated Antwren young disperse relatively quickly. No difference in demography or population dynamics accounts for the interspecific variation in group size. Foraging ecology is the best predictor of non-breeding season group size, not only in antwrens, but in understory insectivores in general. Foliage-gleaning species have the greatest tendency toward clan formation. We suggest that adults of foliage gleaning species are the least sensitive to disturbance from the addition of conspecifics (young) foraging in close proximity.

RESUMEN. Hemos estudiado los sistemas territoriales y la composición de grupos de tres especies de pájaros hormigueros en las tierras bajas del bosque tropical de la Isla de Barro Colorado, Panamá, durante seis años. El hormiguero de garganta ajedrezada (*Myrmotherula fulviventris*) y el matorralero (*Microrhopias quixensis*) co-defienden pequeños territorios (1,5 ha) que han permanecido estables a lo largo de todo el estudio. Los hormigueros de flancos blancos (*Myrmotherula axillaris*) defienden territorios mayores (2–4 ha) que se superponen más y son más fluctuantes. Sugerimos que los sistemas territoriales de las dos primeras especies estan basados en áreas de vida estables en vegetación enmarañada y densa, mientras que la última especie es mucho menos dependiente de esas áreas.

El matorralero y el hormiguero de flancos blancos, se encuentran en pequeños grupos familiares, los jóvenes permanecen junto a los padres por largos períodos de tiempo y ocasionalmente se superponen a los esfuerzos reproductivos de la siguiente temporada de lluvia. Los jóvenes del hormiguero de garganta ajedrezada se dispersan relativamente rápido. No se cuenta con diferencias en la demografía o en la dinámica de las poblaciones para explicar la variación interespecífica en el tamaño de los grupos. La ecología de forraje es el mejor elemento para predecir el tamaño de los grupos fuera de la temporada de reproducción no sólo en hormigueros, sinõ también de insectívoros en vegetación inferior en general. Las especies que picotean comida de las hojas "foliage-gleaning") tienen la mayor tendencia a la formación de clanes. Sugerimos que los adultos de las especies "foliage-gleaning" son menos sensibles a ser perturbados por la presencia adicional de con-específicos (los jóvenes) forrajeando muy cerca.

In this paper we undertake a comparative analysis of the social systems of three small arboreal antbirds, Checker-throated Antwren (*Myrmotherula fulviventris*), White-flanked Antwren (*M. axillaris*), and Dot-winged Antwren (*Microrhopias quixensis*), that are common in lowland tropical forests of Central America. We base our comparison on seven years of field work on Barro Colorado Island and other localities in Central Panama. From this comparison we suggest how features of social organization correlate with habitat use and foraging behavior in antwrens and other tropical forest birds.

Although antwrens, because of their sheer diversity, are an attractive group for such a comparative study, little has been published on their social behavior. Studies by Willis (1972), Wiley (1971, 1980), Jones (1977), Pearson (1977), Munn and Terborgh (1979), and Gradwohl and Greenberg (1980) concentrated on foraging ecology and interspecific gregariousness. Skutch (1967) provided a brief account of antwren social behavior and parental care. The data we present on group size, territoriality, and the demography of these Central American antwren species provide a hint of the patterns that may emerge when the myriad South American species are more closely examined.

Of the three species of antwrens we studied, the Checker-throated Antwren is the largest (11 g) and feeds by pulling arthropods from dead curled leaves that hang in the forest understory. White-flanked Antwren (8 g) is an active gleaner of live foliage of the low to mid-understory. The Dot-winged Antwren (8 g) is also an active foliage gleaning species, but it is more common in higher and denser foliage than the White-flanked Antwren (Wiley 1971; Greenberg 1984).

STUDY SITES

We studied antwrens for 23 months on Barro Colorado Island (BCI), Panama Canal Area from October 1977 to September 1983 (10/77–3/78, 10/78–9/79, 7/80–9/80, 7/81–9/81, 8/82–9/82, 8/83–9/83). BCI is a 15 km² island in Lake Gatun that is covered with both secondary and old growth tropical forest (Foster and Brokaw 1983). The annual rainfall averages about 260 cm and falls primarily between early May and mid-December. Most antwren breeding occurs during the rainy season (Willis and Eisenmann 1979; Gradwohl and Greenberg 1982a). We frequently visited other moist and wet forest sites in Central Panama (Rio Frijoles and Limbo Hunt Club) and Darien (Cerro Pirre) for comparative observations.

METHODS

Antwren group composition. — Each month we walked 15–25 km of the 40 km trail system on BCI; antwren groups encountered within 15 m of either side of the trail were followed for 10–30 min to ascertain the total composition. We counted groups along any particular trail only once each month to avoid duplication. This census was facilitated by the close association between young and parent antwrens as well as the species-specific begging notes of juveniles and young immatures. We classified young antwrens, on the basis of behavioral and morphological characteristics, into several crude age categories: young juveniles (<1 month old), old juveniles (1 ≤ 2 months old), and immatures (>2 months old). Young juveniles beg frequently, retain fluffy juvenile body feathers, and, in Checker-throated and Dot-winged antwrens have yellow mandible edges. In Dot-winged Antwrens young juveniles have short tails. Older juveniles continue to beg, retain some juvenile feathering and bill coloration, but they look essentially like adults. In Checker-throated and White-flanked antwrens males finish molting into their distinct adult plumage during this period. Immatures (except young male White-flanked Antwrens) are indistinguishable from adults. We estimated the natality of the three species based on the number of juveniles (0–2 months out of the nest) with antwren groups in late July through mid-September. All surviving young produced from nests initiated from May through early August should appear in this census. While this is not a complete census of the annual production of young antwrens, it probably includes the peak of the breeding effort (Willis and Eisenmann 1979; Gradwohl and Greenberg 1982a).

Focal flock following. — Employing methods similar to those described in Gradwohl and Greenberg (1980), we followed a focal Dot-winged Antwren–Checker-throated Antwren group for 60 hr in the dry season (February) and 60 hr in the rainy season (June) of 1980 to examine ranging patterns. Following was carried out in five 12-hr shifts, each shift (12:00–18:00, 06:00–12:00) on a gridded plot so that the location of the group could be recorded to the nearest 10 m unit each 15 minutes.

Intensive resightings. — We have maintained a marked population of antwrens, particularly White-flanked and Checker-throated, on a 6–12 ha study site on the BCI plateau since 1977; from 1977 to 1980 the study site was the 6 ha described in Gradwohl and Greenberg (1980); 6 ha were added between 1980 and 1981. We banded 77 Checker-throated, 37 White-flanked, and 6 Dot-winged Antwrens which includes individuals from eight Checker-throated and Dot-winged Antwren groups and two to six White-flanked Antwren groups. We frequently traversed the study area and recorded the locations of all marked antwrens. When a marked pair was located, we followed the group through the gridded study area for two hours. We used these "tracks," single point sightings, and net captures to generate territory maps. In addition, we were able to determine the number of marked territorial adults that disappeared each year. We searched the territories surrounding our plot for marked birds, so we are reasonably certain that the disappearance rate of adults is close to the true annual mortality rate.

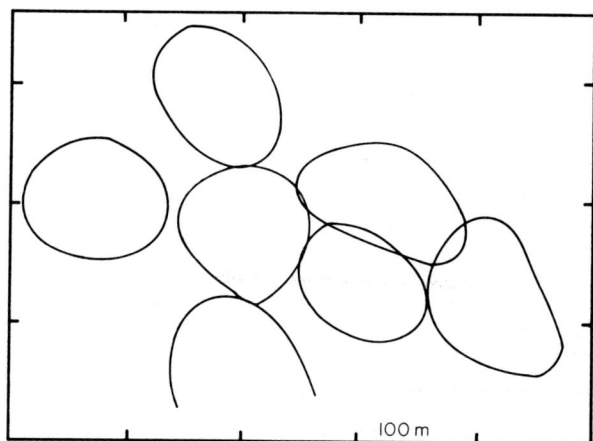

Fig. 1. Co-defended territories of Checker-throated and Dot-winged antwrens on the plateau study site. The plots are based on 532 sightings of marked Checker-throated Antwrens over the 6 years. The territories remained essentially the same from 1977 to 1983 and so are depicted only once.

Group size of understory insectivorous birds on BCI.—We recorded the group size of 26 other understory insectivorous birds on BCI for comparison with the data for antwrens. We censused group size during the late rainy and dry seasons of 1977–1978 and 1978–1979 (November through April) which is presumably the nadir of breeding activity for all of the species (Willis and Eisenmann 1979). The group size census was taken opportunistically while gathering similar data on antwrens (See Antwren Group Composition) and warblers (Greenberg 1984).

During the non-breeding seasons (November–April) of 1977–1978 and 1978–1979, we recorded the number of conspecifics within 5 m of a focal individual for 12 species of insectivorous birds (Greenberg 1984). Focal individuals were generally the first of that species located in a mixed species flock. The data were gathered over as many forest trails as possible in order to minimize the possibility of replicate observations on the same birds.

RESULTS

Territoriality

All three species of antwrens defend territories. Pairs or family groups forage over a home-range that is usually free of conspecifics, except for an occasional dispersing immature. Aggressive interactions between groups are lengthy, vociferous, and occur along consistent boundaries. These displays end with both groups retreating from apparent borders toward the centers of their territories.

Unlike territories of most small insectivorous birds, antwren territories are defended and advertised almost entirely through boundary interactions. With the possible exception of a brief period of dawn singing in White-flanked and Dot-winged antwrens, advertising songs are not used. Because of the mode of defense and the large size of the territories relative to the small size of the birds, interloping by neighboring groups and wandering through by dispersing immatures is not uncommon, particularly during the dry season. Trespassing groups will move a small distance into a neighboring territory before contact is made and an interaction centering on the border occurs.

Dot-winged Antwrens defend small territories that coincide with the territories of Checker-throated Antwren groups (Gradwohl and Greenberg 1980); the mean territory size is 1.5 ha (range 1.0–2.2 ha, N = 8; Fig. 1). Both species use dense vine-tangled portions of the forest as core areas of their territories. Home range size and use vary seasonally. In the dry season, Dot-winged and Checker-throated antwrens cover a larger total area and the birds move over their home range more quickly. The focal antwren group moved over a 3.2 ha area in the dry and 2.2 ha area in the rainy season. On a daily basis, the antwrens covered an average of 1.7 ha per 12 hr following (s.e. = 0.23) in the dry and 1.1 (s.e. = 0.11) in the rainy season ($t_{5,5}$ = 3.49, $P < 0.01$). Seasonal variation in home range size results from an increased rate of intrusion beyond the territory borders. From year to year the territories of these species are

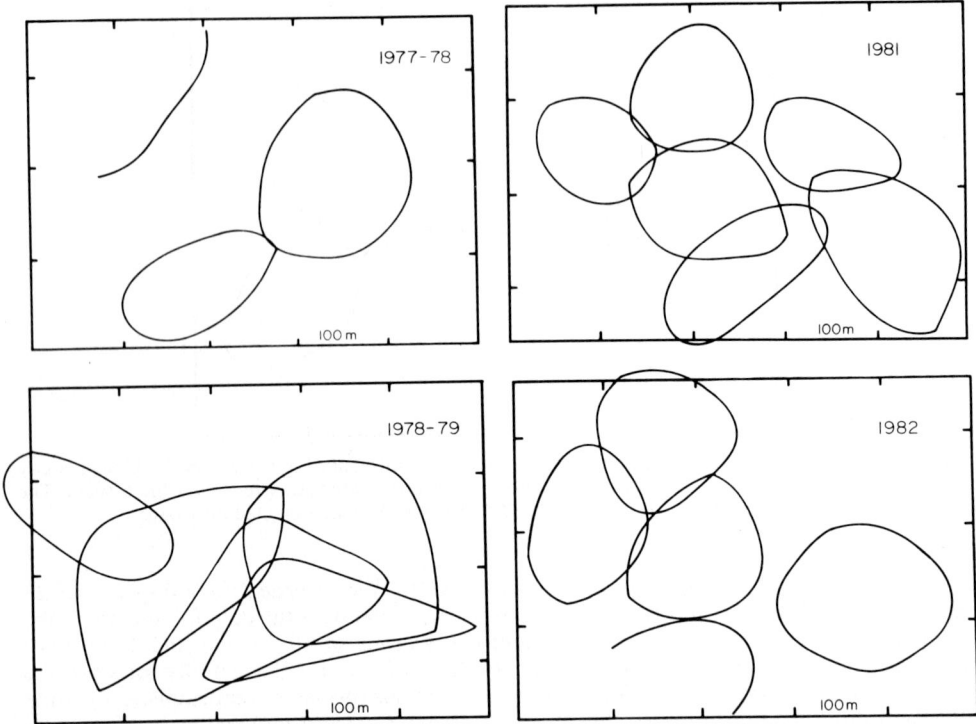

Fig. 2. Territories of White-flanked Antwren groups on the plateau study site: 1977–1978 (n = 62 sightings); 1978–1979 (267); 1981 (73); 1982 (48).

remarkably constant (Fig. 1). We have found territorial boundaries to remain essentially unchanged for seven rainy seasons.

Territorial boundary interactions are very common. We found that 21.5 percent of both Checker-throated and Dot-winged Antwren groups observed on trailside censuses were involved in such displays (N = 322, 334 group sightings, respectively). These figures probably overestimate the amount of time that antwren groups spend in such displays; censuses are conducted during periods of peak antwren activity, displaying groups are probably more conspicuous, and because up to 30 min are spent with a group, the censuses do not give a "spot" assessment of territorial participation. The focal Checker-throated Antwren group spent 6 percent of its waking time (N = 60 hr) in the dry season and 1 percent in the wet season (N = 60 hr) engaged in territorial interactions.

The displays are lengthy and energy demanding. Dot-winged Antwren displays lasted from a few minutes to three hours; the mean display period for the focal group was 22 min (N = 12, s.e. = 6.3). As groups approach the borders, males meet males and females meet females in posturing that usually results in chases, often 15 to 25 m long, in dense canopy foliage. Both sexes fluff contour feathers, exposing white back patches, fan out rectrices, droop wings, and hold their heads erect and bills up. In this position they chase, emitting a rasping buzz. Unlike the young of the other two species, juveniles and immatures may participate in displays and often sing from nearby trees.

Checker-throated Antwren displays are shorter and less vigorous than Dot-winged Antwren displays; the average border dispute involving the focal Checker-throated Antwren group lasted 10 min (N = 20, s.e. = 6.7). In contrast to male Dot-winged Antwrens, males display at border interactions more frequently than females (100% vs 63%, N = 30); females often continue foraging while males display. Checker-throated Antwrens approach the border exchanging series of chip notes antiphonally with the birds in the neighboring group. When the antwrens meet, both birds chip continuously at close range (5–25 cm), usually on the same or adjacent branches.

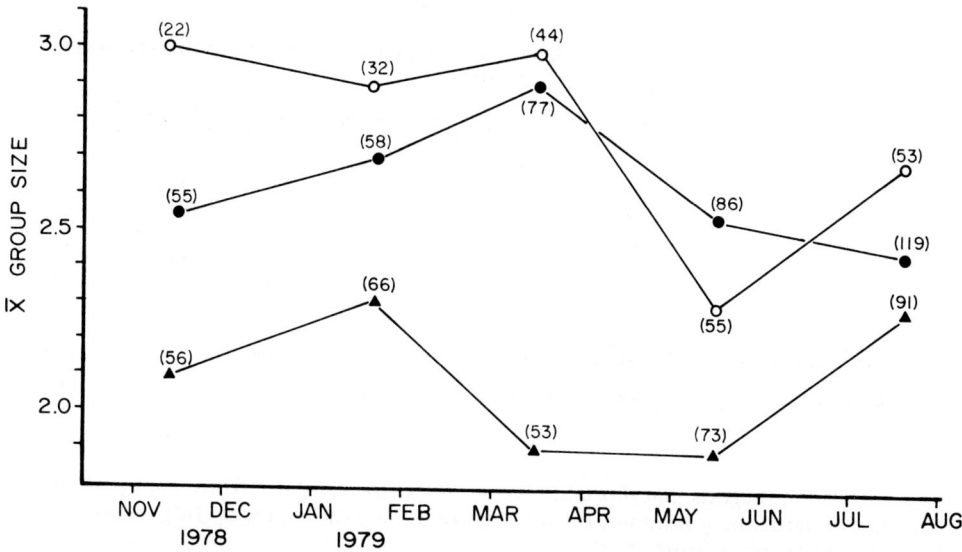

FIG. 3.   Mean group size of antwrens encountered on monthly censuses of Barro Colorado Island trails: White-flanked Antwren (clear circles), Dot-winged Antwren (dark circles), Checker-throated Antwren (triangles). The number of different groups censused each month is indicated in parentheses.

The territories of the White-flanked Antwren are the largest and most variable of the three species (Fig. 2). They range from 1.5–4.5 ha ($\bar{X}$ = 2.5 ha, N = 15); boundaries change considerably between years. One White-flanked Antwren group is commonly associated with one to three Dot-winged–Checker-throated antwren territories. Territorial defense differs markedly from that of other antwrens. Border interactions are uncommon; only 9 percent of the 208 groups encountered on the trailside censuses were involved in border disputes, which differs significantly from the 21.5 percent value for Dot-winged (χ^2 = 15.1, d.f. = 1, $P < 0.001$) and Checker-throated (χ^2 = 13.6, d.f. = 1, $P < 0.001$) Antwrens. The White-flanked Antwren displays usually consist of males exchanging calls from the relatively large distance of 10 to 20 m. Males give the "pew" alarm note, a low "whip-whip-whip" call, and a song consisting of a slow descending series of whistles. Only rarely do males engage in close interactions, and these result in rapid, long-distance chasing. Females and immatures rarely participate, and we have observed no equivalent of the extended physical displays of the other two species.

GROUP COMPOSITION

Antwren pairs generally forage close together throughout the year. Dot-winged Antwrens, in particular, maintain close contact, whereas Checker-throated and White-flanked antwren pair members can be 10 to 15 m apart. The close association of pair members throughout the year distinguishes them from many other monogamous bird species on BCI, including other antbirds, such as the Chestnut-backed Antbird (*Myrmeciza exsul*), and the Spotted Antbird (*Hylophylax naevioides*).

Young maintain a close association with parents, even when they are essentially independent. The length of the association varies among species (Fig. 3). Checker-throated Antwrens usually occur in pairs; young are present in groups, so that the average group size exceeds 2.0 from the mid-rainy to the early dry seasons. The breeding season for antwrens is primarily the mid-rainy season, so young are retained for a few months after breeding. This suggests that young disperse relatively quickly from natal groups, an inference confirmed by our observation of eight color-marked juveniles fledged on our study plot, that dispersed between one and three months after fledging.

In contrast, Dot-winged and White-flanked antwren groups ranged from two to six individuals for all months of the year (annual $\bar{X}$ = 2.7 for White-flanked, and 2.6 for Dot-winged Antwrens). The lack of any perceptible decline in group size during the dry season, when few Dot-winged Antwren and no White-flanked Antwren young are produced (Gradwohl and

TABLE 1

ANNUAL SURVIVORSHIP OF SETTLED ADULT CHECKER-THROATED AND WHITE-FLANKED
ANTWRENS[1]

Year	White-flanked Antwren	Checker-throated Antwren
1977–1978[2]	100 (3)	9 (11)
1978–1979	100 (3)	64 (11)
1979–1980	30 (10)	60 (12)
1980–1981	66 (3)	50 (14)
1981–1982	60 (10)	93 (15)
1982–1983	83 (6)	89 (19)
X̄ ± s.e.	74 ± 10	62 ± 12

[1] Based on color-marked birds on a 6–12 ha study plot on Barro Colorado Island. Values given are percent survivorship followed by the sample size in parentheses.
[2] Survivorship was measured from September–September, except in 1977–1978 when it was measured from November–November.

Greenberg 1982a), suggests that mortality is low, and dispersal slow. White-flanked Antwren groups show a particularly pronounced drop in size in May through early July corresponding to the peak of antwren nesting efforts.

Dot-winged Antwren groups showed a strong male bias, also noted by Skutch (1967) and Willis (1972). If we assume that each group has an adult male and female, 65 percent of the associated immatures were males. This proportion is significantly different from 50 percent ($\chi^2 = 16.6$, d.f. = 1, N = 149, $P < 0.001$; the analysis was done only on groups that we were certain were not censused twice and includes data from years other than 1978–1979 presented in Figure 3). Checker-throated Antwren groups also had more males than females, but the sample size of censused immatures is small (25) and the 60 percent value is not significantly different from 50:50 ($\chi^2 = 1.0$, d.f. = 1, n.s.). White-flanked Antwren groups had exactly an even sex ratio (N = 110). The most reasonable hypothesis to explain the skewed sex ratio in Dot-winged Antwren groups is that, like many other bird species (Emlen 1978), females disperse at an earlier age than males. Dot-winged Antwren juveniles (unlike the other two species) are dimorphic in color; we found that the sex ratio of nonindependent juveniles was slightly and insignificantly biased towards males (55%, $\chi^2 = 0.64$, d.f. = 1, N = 77, n.s.). On the other hand, the observed male to female ratio in juveniles is not significantly different from the ratio in immatures ($\chi^2 = 3.39$, d.f. = 1, N = 77,149, $P < 0.10$). A larger sample of censused immatures and juveniles will be required to satisfactorily reject the hypothesis that the sex ratio is biased towards males prior to dispersal.

We have observed immatures of all three species in groups where the adults are attempting or have completed a subsequent nesting effort. In Dot-winged Antwrens, however, most young apparently leave their natal group before eggs are laid. We found that the mean group size for Dot-winged Antwrens in the process of building a nest was 2.6 (N = 26, s.e. = 0.21), while it was 2.1 (N = 12, s.e. = 0.01) for groups that had eggs or young in a nest (Mann-Whitney $U_{26,12} = 67$, $P < 0.001$). We cannot perform a similar analysis for the other species of antwren because we found too few active nests. Of the 63 White-flanked Antwren groups we observed with juveniles, five (8%) had additional auxiliary immatures. In Checker-throated Antwrens only one of 181 groups (0.6%) with fledglings had additional older young as well (the difference between the species cannot be tested, as the expected values fall below 5 for 2 cells of the contingency table). Despite the presence of some auxiliary young during subsequent breeding efforts, we have observed no evidence that these immatures help the adults care for young.

ADULT MORTALITY

We had sufficient numbers of marked adult Checker-throated and White-flanked antwrens to estimate annual mortality (Table 1). The mean annual survival rate for Checker-throated Antwren was 62 percent (s.e. = 12) and for White-flanked Antwren, 74 percent (s.e. = 10). We tested the difference in survivorship with a Mann-Whitney U test based on the values obtained each year for each species and with a χ^2 test based on the number of total year to year transitions in which an adult antwren survived (this method is weighted more strongly toward years in which more birds were marked). We found no significant difference between the two species indicated using either method ($U_{5,5} = 12.5$, $P = 0.11$; $\chi^2 = 0$, N = 35, 74, d.f. = 1, n.s.). The overall survivorship based on the total year to year transitions was 74 percent for each species, which is within the range found by Willis (1974) for ant following birds on BCI.

NATALITY

While we do not have a complete census of the number of young produced each year, we can compare the number of juveniles produced during the peak of fledgling production. Although these numbers are incomplete, the interspecific comparison should be valid unless the species have different breeding seasons. Our census data from 1978–1979 indicate that the peak of breeding is similar for the three species, although the production of young is somewhat less synchronized in the Dot-winged Antwren. The mean number of young per group was 0.54 for White-flanked Antwren (N = 132), 0.38 for Checker-throated Antwren (N = 181) and 0.30 for Dot-winged Antwren (N = 279). Only the Dot-winged and White-flanked antwrens differ significantly (χ^2 = 20.9, d.f. = 1, $P < 0.001$).

GROUP SIZE IN UNDERSTORY INSECTIVOROUS BIRDS

Understory birds on BCI occur solitarily, in pairs, or small family groups (Willis 1972). Our data on the mean number of birds per group (Table 2) should distinguish which is the dominant social group for a given species. We assume that species that have an average group size of two are pair-forming (e.g., Checker-throated Antwren), and that those with a group size of substantially more than two are clan-forming. The censuses were performed during the non-breeding season for insectivorous birds, the period during which we found the maximum difference in group size between Checker-throated and the other two species of antwren. We have divided the 29 species into foraging guilds to correlate group size with species ecology. The major pattern that emerges is that the group size of foliage gleaning species (species mean = 2.96) is larger than for species in the other guilds (species mean = 1.39, Mann-Whitney $U_{6,23}$ = 0, $P < 0.001$). Willis (1972) suggested a similar correlation for BCI forest birds.

A consequence of the larger group size in foliage gleaning birds is that an individual is more likely to be foraging near a conspecific. The average number of conspecifics within 5 m of a foliage gleaning bird (the Grand Mean for 5 species) was 1.09, versus 0.38 for non-foliage gleaners ($U_{5,7}$ = 0, $P < 0.001$).

DISCUSSION

TERRITORIALITY

Dot-winged and Checker-throated antwrens share small, stable territories perhaps in response to a similar pattern of habitat use. Both species prefer to forage in vine-tangles (Gradwohl and Greenberg 1980) that tend to be concentrated in areas of disturbance in the forest (Foster and Brokaw 1982). Concentrations of dense vine-tangles are far from ephemeral in the BCI forest. Areas of dense vine-tangles that we located in 1977 (Gradwohl and Greenberg 1980) were still present in 1983 and have persisted as core areas of antwren territories. One large tree fall created a new area of viny vegetation; it occurred near an old core area and so had little effect on antwren activity. Other species that depend heavily upon areas of vine-tangles, such as the Slaty Antshrike, Long-billed Gnatwren, Southern Bentbill, and Chestnut-backed Antbird, appear to show rigid territorial behavior similar to Checker-throated and Dot-winged antwrens (pers. observ.). The stability of antwren territory size is reflected in the constancy of island-wide densities (Greenberg and Gradwohl, unpubl. data). White-flanked Antwren territories tend to be larger and more variable from year to year than those of the other species. White-flanked Antwren territories are not strongly associated with vine-tangled areas or any other identifiable feature of forest structure.

Habitat difference may account for the different modes of territorial defense observed; both Checker-throated and Dot-winged antwrens have an extended, ritualized, short-range display, whereas White-flanked Antwrens interact briefly and from a much greater distance. White-flanked Antwren groups encounter each other in low open understory where flashes of white flank patches and puffed white back patches are visible for greater distances than they would be in the dense vegetation in which the other species often display.

GROUP SIZE

The difference in group size between the pair-forming Checker-throated Antwren and the clan-forming Dot-winged and White-flanked antwrens may result from differences in demog-

TABLE 2
Average Group Size of Understory Insectivorous Birds[1]

Guild or species	N	Group size X̄ (s.e.)	Dominant group
Foliage Gleaners			
White-flanked Antwren (*Myrmotherula axillaris*)	98	2.93 (.08)	clan
Dot-winged Antwren (*Microrhopias quixensis*)	190	2.75 (.06)	clan
Lesser Greenlet (*Hylophilus decurtatus*)	93	2.83 (.09)	clan
Blue Dacnis (*Dacnis cayana*)	58	3.02 (.15)	clan
Red-throated Ant-Tanager (*Habia fuscicauda*)	7	3.63 (.53)	clan
White-shouldered Tanager (*Tachyphonus luctuosus*)	47	2.57 (.09)	clan
Dead-leaf Gleaner			
Checker-throated Antwren (*Myrmotherula fulviventris*)	174	2.05 (.06)	pair
Generalists (long-attack distance)			
Slaty Antshrike (*Thamnophilus punctatus*)	94	1.92 (.04)	pair
Bright-rumped Atilla (*Atilla spadiceus*)	10	1.00 (0)	solitary
Hover-gleaners			
Squirrel Cuckoo (*Piaya cayana*)	20	1.75 (.15)	pair
Black-throated Trogon (*Trogon rufus*)	22	1.48 (.16)	solitary-pair
White-whiskered Puffbird (*Malacoptila panamensis*)	13	1.32 (.18)	solitary
Spot-crowned Antvireo (*Dysithamnus puncticeps*)	30	1.90 (.07)	pair
Rufous Mourner (*Rhyipterna holerythra*)	7	1.00 (.00)	solitary
Ruddy-tailed Flycatcher (*Terenotriccus erythrurus*)	43	1.00 (.00)	solitary
Golden-crowned Spadebill (*Platyrinchus coronatus*)	10	1.9 (.23)	pair
Yellow-margined Flycatcher (*Tolmomyias assimilis*)	35	1.13 (.06)	solitary
Olivaceous Flatbill (*Rhynchocyclus olivaceus*)	15	1.21 (.11)	solitary
Southern Bentbill (*Oncostoma olivaceum*)	29	1.19 (.07)	solitary
Black-capped Pygmy-tyrant (*Myiornis atricapillus*)	4	1.60 (.30)	pair
Twig, Branch or Trunk Gleaners			
Tropical Gnatcatcher (*Polioptila plumbea*)	30	1.22 (.07)	solitary
Long-billed Gnatwren (*Ramphocaenus rufiventris*)	15	1.90 (.13)	pair
Wedge-billed Woodcreeper (*Glyphorynchus spirurus*)	35	1.34 (.09)	solitary
Buff-throated Woodcreeper (*Xiphorhynchus guttatus*)	20	1.00 (.00)	solitary
Black-striped Woodcreeper (*Xiphorhynchus lachrymosus*)	35	1.22 (.08)	solitary
Plain Xenops (*Xenops minutus*)	59	1.11 (.04)	solitary
Ground Gleaners			
Scaly-throated Leaftosser (*Sclerurus guatemalensis*)	8	1.00 (.00)	solitary
Spotted Antbird (*Hylophylax naevioides*)	24	1.50 (.10)	solitary-pair
Chestnut-backed Antbird (*Myrmeciza exsul*)	26	1.46 (.10)	solitary-pair

[1] November–April 1977–1979 on Barro Colorado Island.

raphy, differences in population stability, or the variation in costs and benefits of the young remaining on the natal territory that is associated with different foraging behaviors.

Demography.—It is possible that antwren group size reflects a simple equilibrium between the rate of formation of reproductive vacancies in the population versus the rate of production of young. Larger mean group size may result from increased natality or decreased adult mortality. We only have data on mortality in two species. These data suggest that the clan-forming White-flanked Antwren has a similar mortality rate to the pair-forming Checker-throated Antwren (26% for each species). The only significant difference we found in the number of juveniles produced in the early and mid-rainy season was between the two clan-forming antwrens.

Population stability.—The habitat saturation hypothesis of cooperative breeding in birds (see Emlen 1978) suggests that in populations where territories and densities are stable, and alternative suboptimal habitats are unavailable, young may be forced to remain in their natal group until a "reproductive opening" occurs. Therefore, species with similar demographies, but different population dynamics, may differ in group size. The BCI forest is a closed habitat for all three species; no alternative habitats are found on the island, and the antwrens rarely,

TABLE 3
NUMBERS OF CONSPECIFICS WITHIN 5 M OF FOCAL FORAGING INDIVIDUALS IN SPECIES OF UNDERSTORY INSECTIVOROUS BIRDS[1]

Species (guild[2])	N	X̄ (s.e.)
White-flanked Antwren (FG)	55	1.11 (0.16)
Dot-winged Antwren (FG)	65	0.91 (0.17)
Lesser Greenlet (FG)	65	1.65 (0.21)
Blue Dacnis (FG)	25	0.77 (0.20)
White-shouldered Tanager (FG)	65	0.82 (0.18)
Slaty Antshrike (Gen)	25	0.52 (0.14)
Spot-crowned Antvireo (HG)	25	0.50 (0.20)
Checker-throated Antwren (DLG)	25	0.74 (0.14)
Spot-crowned Antvireo (HG)	25	0.50 (0.20)
Ruddy-tailed Flycatcher (HG)	40	0.00 (0.00)
Southern Bentbill (HG)	30	0.40 (0.21)
Plain Xenops (BG)	30	0.04 (0.01)

[1] November–March 1977–78 and 1978–79 on Barro Colorado Island.
[2] Guilds are Foliage-gleaner (FG), Generalist (Gen), Hover-gleaner (HG), Dead Leaf Gleaner (DLG), Bark Gleaner (BG).

if ever, emigrate from the island. The degree to which young birds can establish themselves on new territories depends on the fluidity of the territorial system in the BCI forest. The two species that have the rigid territorial system are the clan-forming Dot-winged Antwren and the pair-forming Checker-throated Antwren. Population stability and habitat saturation alone do not seem to explain the variation in group size in antwren species.

Foraging ecology.—Differences in group size may result from variation in the dispersal strategy of young as it relates to foraging ecology. One line of evidence that supports this hypothesis is the correlation between group size and foraging behavior reported in insectivorous birds of neotropical forests (Table 2; see also Buskirk 1969 and Willis 1972). The major pattern that emerges in our analysis of understory insectivorous birds on BCI is that foliage gleaning species tend to occur in family groups in the non-breeding season, whereas other types of insectivores occur in pairs or solitarily. This analysis does not include the dominant ant followers (Willis 1974) that depend on a highly patchy and variable insect source. Buskirk (1969) presented a correlation between foraging microhabitat and sociality of a mid-elevation bird community (excluding ground foragers) in Costa Rica. He found that group size increased with decreasing perch size, and the small perched species were mainly foliage gleaners.

We suggest three possible reasons for the relationship between foraging ecology and group size in antwrens (and other neotropical understory insectivorous birds), (1) foraging competition, (2) anti-predator vigilance, and (3) learning.

(1) Competition. The possibility of competition is greater for foraging Checker-throated Antwrens because they forage from a scarce microhabitat (dead leaves) that is sensitive to disturbance. Because of this, Checker-throated Antwren parents may be at a greater disadvantage than the other antwren species if they keep their young on their territories (see also Buskirk 1969, Willis 1972). White-flanked and Dot-winged antwrens exploit an abundant microhabitat, live leaf surfaces, and the birds presumably forage by scanning large areas. Checker-throated Antwrens search dead leaves, which occur at densities of about one per 50 live leaves, and each leaf is searched individually. An adult Checker-throated Antwren would have a high probability of searching dead curled leaves that have been disturbed by the activities of their young. For this argument to apply it is not necessary that the antwrens actually deplete the numbers of dead leaf inhabiting arthropods; foraging visits must merely disturb the arthropods so that they are less likely to be captured by subsequently visiting antwrens. Checker-throated Antwrens do, in fact, severely depress the population of arthropods in dead leaves; an exclosure experiment showed that prey populations were lowered approximately 50 percent in a six week period (Gradwohl and Greenberg 1982b).

The competition argument is supported by the examination of the difference between foliage-gleaning birds and members of other guilds. A consequence of occurring in larger family groups is that foliage gleaners are more likely to have a conspecific foraging in close proximity (Table 3). As Buskirk (1969) suggested, species that forage on relatively rare substrates (trunks or large limbs) may be more sensitive to the disturbance caused by another bird foraging close by. Similarly, hover-gleaning species sit motionlessly by a large number of leaf surfaces which they can attack from a great distance. This makes them susceptible to interference from other hover-gleaners entering their large foraging spheres. Ground gleaners that do not depend upon arthropods flushed by antswarms should also be vulnerable to having their search paths crossed by attendant young. Song Wrens provide a notable exception to the lack of sociality in litter foraging birds (Willis 1972; Morton 1978). Morton suggested that Song Wrens are particularly adept at seizing flushed orthopterans and so may gain foraging advantages from the disturbance of conspecifics.

(2) Predation. Willis (1972) suggested that "intensive" foragers, those species that peer intently at a single foraging substrate, are less effective sentinels for predators. An adult Checker-throated Antwren would gain less by retaining its young than would a White-flanked Antwren, since the latter species peers extensively at foliage, whereas the former probes into dead curled leaves. This hypothesis, however, appears to contradict the observation of Willis (1972) and Wiley (1971) that Checker-throated Antwrens are the most reliable sentinels in antwren flocks. The predation hypothesis readily explains the small group sizes characteristic of bark gleaners which are also intensive foragers, but it is less useful in accounting for the general solitary existence of hover-gleaning species. Willis proposes that in these species, predator vigilance is so effective that adults gain little increased advantage by retaining their young. However, many hover-gleaners join mixed species flocks, presumably to take advantage of the anti-predatory vigilance of the other species (Willis 1972). It is difficult to invoke the efficacy of the predator detection as the sole factor selecting for increased group size in hover-gleaning birds without considering intraspecific competition as an additional factor (i.e., it is better to use individuals of other species that are less likely to disturb your foraging zone for a predator alarm system).

(3) Learning. While little is known about the learning process of juveniles of different species of insectivorous birds, it is tempting to speculate that gleaners of live foliage may face special problems in diet selection. This is at least defensible for the comparison of live foliage versus dead leaf foragers. Dead curled leaves contain a very depauperate arthropod fauna dominated by cryptically colored orthopterans and spiders; brightly colored arthropods that might be aposomatically colored comprise much less than 1 percent of the dead leaf fauna versus approximately 5 percent of the live leaf fauna (Greenberg and Gradwohl 1980; Gradwohl, unpubl. data). However, young Dot-winged and White-flanked antwrens often remain in natal groups far longer than is required for parental care. During the period after parental feeding ceases, it is unlikely that young can learn the quality of prey from adults. In addition, it is not clear that this line of reasoning would explain the longer retention of young foliage gleaners in comparison to certain other guilds, such as hover-gleaners.

There is a striking correlation between foraging ecology and the degree of sociality of insectivorous birds on Barro Colorado Island. At this point, the single strongest hypothesis explaining this relationship is that group size is smaller in species that are sensitive to foraging disturbance by conspecifics. Variation in predator-detection efficiency may also contribute to the net gain or loss in fitness experienced by adults that retain their young for longer periods. It may ultimately prove difficult to determine the relative importance of these two factors. It is currently difficult to see how learning or other forms of extended parental care could account for the variation in the stability of family units in insectivorous tropical birds.

ACKNOWLEDGMENTS

We thank M. S. Foster, C. Munn, and E. O. Willis for helpful comments on earlier drafts of this paper. E. Leigh, E. Morton, F. Pitelka, and N. Smith offered advice and encouragement for our research on BCI. Financial support was provided from the Smithsonian Tropical Research Institute, a doctoral dissertation improvement grant from the National Science Foundation, the Frank Chapman Fund, and the Environmental Sciences Program of the Smithsonian Institution. The manuscript was typed by T. Cummings.

LITERATURE CITED

BUSKIRK, W. H. 1969. Social systems in a tropical forest avifauna. Am. Nat. 110:293–310.
EMLEN, S. 1978. Cooperative breeding in birds. Pp. 245–282, In J. R. Krebs and N. B. Davies (eds.), Behavioural Ecology: an Evolutionary Approach. Blackwell, Oxford, England.
FOSTER, R., AND N. BROKAW. 1982. Structure and the history of vegetation on Barro Colorado Island, Panama. Pp. 67–83, In E. G. Leigh, A. S. Rand, and D. Windsor (eds.), The Ecology of a Tropical

Forest: Seasonal Rhythms and Long-term Changes. Smithsonian Institution Press, Washington, D.C.

GRADWOHL, J., AND R. GREENBERG. 1980. The formation of antwren flocks on Barro Colorado Island, Panama. Auk 97:385–395.

GRADWOHL, J., AND R. GREENBERG. 1982a. The breeding season of antwrens on Barro Colorado Island. Pp. 347–351, In E. G. Leigh, A. S. Rand, and D. Windsor (eds.), The Ecology of a Tropical Forest: Seasonal Rhythms and Long-term Changes. Smithsonian Institution Press, Washington, D.C.

GRADWOHL, J., AND R. GREENBERG. 1982b. The effect of a single species of avian predator on the arthropods of aerial leaf litter. Ecology 63:581–583.

GREENBERG, R. 1984. The winter exploitation systems of Bay-breasted and Chestnut-sided warblers in Panama. Univ. Calif. Publ. Zool. 116:1–107.

JONES, S. 1977. Coexistence in mixed species antwren flocks. Oikos 29:366–375.

MORTON, E. S. 1978. Reintroducing recently extirpated birds into a tropical forest reserve. Pp. 379–384, In S. A. Temple (ed.), Endangered Birds: Management Techniques for Preserving Threatened Species. University Wisconsin Press, Madison, Wisconsin.

MUNN, C., AND J. TERBORGH. 1979. Multi-species territoriality in neotropical foraging flocks. Condor 81:338–347.

PEARSON, D. L. 1977. Ecological relationships of small antbird communities. Auk 94:283–292.

SKUTCH, A. 1967. Life histories of Central American birds. Pac. Coast Avif. No. 35.

WILEY, R. H. 1971. Cooperative roles in mixed species flocks of antwrens (Formicariidae). Auk 88:881–892.

WILEY, R. H. 1980. Multispecies antbird societies in lowland forests of Surinam and Ecuador: stable membership and foraging differences. J. Zool. (Lond.) 191:127–145.

WILLIS, E. O. 1972. The behavior of Spotted Antbirds. Ornithol. Monogr. No. 10.

WILLIS, E. O. 1974. Populations and local extinctions of birds on Barro Colorado Island, Panama. Ecol. Monogr. 44:153–169.

WILLIS, E. O., AND E. EISENMANN. 1979. An annotated checklist of the birds of Barro Colorado Island. Smithson. Contrib. Zool. No. 292.

EVOLUTIONARY ECOLOGY OF AFROTROPICAL AND NEOTROPICAL VULTURES IN FORESTS

DAVID C. HOUSTON

Department of Zoology, University of Glasgow, G12 8QQ, Britain

ABSTRACT. Several species of vultures are commonly found in neotropical forests, but none has occupied the forested regions of the Old World. It is suggested that a greater abundance of food for large scavenging animals may exist in New World forests because of their higher mammalian biomass and the complexity of their invertebrate communities. Cathartid vultures are highly efficient at searching for food and may be the major scavengers of large mammal carcasses in the neotropical forest.

RESUMEN. Se encuentran varias especies de buítres en bosques neotropicales, pero ninguno ha ocupado regiones boscosas en el Viejo Mundo. Se sugiere que la mayor abundancia de comida para animales carroñeros, existente en los bosques del Nuevo Mundo se debería a su mayor biomasa de mamíferos y a la complejidad de su comunidad de invertebrados. Los buitres de la familia Cathartidae son muy eficientes buscando comida y pueden ser los mayores carroñeros de carcasas de mamíferos grandes en los bosques neotropicales.

Vultures differ from other meat eating birds in that they virtually never kill their prey, but obtain almost all food by scavenging from dead animals. This method of feeding has been exploited by two groups of birds, the New World vultures and the Old World vultures, which are quite unrelated. The 15 species of vultures that are found in Africa, Asia and southern Europe are classified in the family Accipitridae together with the eagles and hawks to which they are obviously closely related (Brown and Amadon 1968). The seven species of New World vultures, however, appear to show affinities to the storks, family Ciconiidae (Ligon 1967; Feduccia 1980); these Cathartid vultures are usually placed for convenience as a suborder within the Falconiformes (Brown and Amadon 1968), although it is widely recognized that they do not belong there (Voous 1973).

Despite their different ancestry, the two groups of modern vultures are extremely similar in superficial appearance. They are textbook examples of convergent evolution. Both groups have a similar diversity of species, and share many aspects of plumage pattern and a common dependence on soaring flight. In contrast, the habitats occupied by these two vulture groups are markedly different in one respect. The Old World vultures are confined to open habitats such as savannas, grasslands, and semideserts, and none of these species is found in any of the forested areas of Africa or Asia. Among the New World vultures, the two species of condors (*Gymnogyps californianus* and *Vultur gryphus*) live in mountainous areas. The five smaller Cathartid vultures, however, have centers of distribution in the neotropical forest region and several species live in dense forest conditions.

In this review I consider some possible reasons why neotropical forests are able to support several species of vultures, whereas the tropical forests of Africa and Asia support none; this implies that the scavenger food chains in the two regions of forest may differ significantly. I first briefly consider the evolutionary history of the two vulture groups, and then consider the potential size of the food supply for scavenging animals in forested areas, the potential competition among scavengers for this resource, and finally the comparative efficiency of vultures at locating food. Little information is available about the ecology of any of the Cathartid vultures in forested regions; thus many of these considerations are tentative, and much of what follows is speculation.

EVOLUTIONARY HISTORY OF VULTURES

The fossil record of both vulture groups is comparatively good (Brodkorb 1964; Rich 1983). Thus, we know that although today the New World vultures and the Old World vultures are clearly separated geographically, they did not develop in isolation; fossil New World species have been found in the Old World, and similarly fossil Old World species have been found in the New World. Ten Old World species from four genera have been discovered from the

Lower Miocene to the Upper Pleistocene of North America. Lower Miocene deposits in France also contained fossil species of Old World vultures, and so it is likely that from the Lower Miocene until the Late Pleistocene this group of scavenging birds was commonly found in both Old and New worlds. It became extinct in North America only relatively recently, toward the end of the Pleistocene.

The earliest of the New World vultures are recorded from Europe, a disputed fossil dating from Late Paleocene deposits in England, and several species are known from middle and late Eocene deposits in France and Germany. No fossil remains have been found in the Old World after the early Miocene, when the group may have become extinct in this region. New World vultures appear later in the fossil record in America, in the early Oligocene (earlier descriptions are probably incorrect, Feduccia 1980), and diverse genera flourished in the Pleistocene.

Both Old and New World species of vultures, therefore, coexisted in North America until the late Pleistocene and in Europe until the Miocene. Vultures have a long ancestry, and scavenging birds of this kind probably have occupied a similar role in wildlife communities since the earliest radiation of modern mammals after the late Paleocene. They do not appear to be groups that have undergone rapid evolutionary change; the modern Old World vulture genus, *Gyps,* has a fossil record back to the middle Pleistocene, and the modern New World genera *Vultur* and *Sarcoramphus* date from the middle Pliocene.

Both vulture groups play similar ecological roles and contain modern species that cover the same range in body size, from the Andean Condor (*Vultur gryphus*) and Himalayan Griffon Vulture (*Gyps himalayensis*), which are the heaviest species at about 12 kg to the species of *Cathartes, Coragyps, Necrosyrtes* and *Neophron,* which range from 1 to 2 kg (Brown and Amadon 1968).

There are, however, two obvious differences between the vulture groups. Firstly there is the distinction that is discussed in this paper, that only New World species live in forests. Secondly, New World vultures do not include any large species that are ecologically equivalent to the griffon vultures of the Old World, living in lowland grasslands or plains habitats. The Pleistocene fossil record, however, shows an abundance of large Cathartid vultures, and similar birds such as teratorns, which did live in such habitats feeding on the large mammal communities then present in the Americas (Feduccia 1980). These birds must have filled the same ecological role as the modern griffon vultures of Africa and Asia. The loss of these Cathartid species is almost certainly associated with the late Pleistocene extinctions of large mammals in America, on which these birds depended for food. Pleistocene mammal extinctions were far less extensive within Africa (Martin 1967).

FOOD SUPPLY FOR CARNIVOROUS ANIMALS IN FORESTED REGIONS

Savanna regions of Africa support a complex community of ungulates that live at high densities. It is clear that these mammals can provide a large food supply for scavenging animals (Houston 1979). The distributions and feeding ecologies of the various Old World vulture species differ considerably (Kruuk 1967; Pennycuick 1972; Houston 1976), but most of these birds depend on the large ungulate species for food. Contrary to popular belief, these vultures do not feed extensively from the remains of predator kills. Their food comes chiefly from ungulates that die from malnutrition, disease, accidents, or old age. In communities dominated by migratory ungulate herds, vultures probably consume more meat than any of the mammalian carnivores (Houston 1979). None of these Old World vulture species is found in tropical forest.

Among the American vultures, however, five species are found in regions of tropical rainforest: Greater Yellow-headed Vulture (*Cathartes melambrotus*), Lesser Yellow-headed Vulture (*Cathartes burrovianus*), Turkey Vulture (*Cathartes aura*), Black Vulture (*Coragyps atratus*), and King Vulture (*Sarcorhamphus papa*). Our knowledge of their habitat selection is poor, but my observations suggest that probably only one species, the Greater Yellow-headed, is largely confined to forest conditions. The Turkey Vulture and King Vulture occur extensively in forest, but their range also extends widely into savannas and open grasslands in South and Central America. The Lesser Yellow-headed Vulture is usually found in open savannas but will sometimes enter forest. The Black Vulture is the only species which is rarely found inside the forest, although it occurs along large waterways, clearings and other forest edges. The ranges of the Turkey and Black vultures are far more extensive than those of the other species and include much of North America where the birds are partial migrants. In many areas these

TABLE 1

ESTIMATES OF ANNUAL MORTALITY AMONG CLIMBING AND ARBOREAL MAMMAL SPECIES FROM
A NEOTROPICAL AND TWO AFRICAN FOREST SITES[1]

Species	No./km²	Weight (kg)	Estimated mortality rate	No. dying/ year/km²	Wt. dying/ year/km²
Barro Colorado Island, Panama					
Philander opossum (Four-eyed Opossum)	27	1.4	40	11	15
Didelphis marsupialis (Common Opossum)	70	1.0	40	28	28
Aotus trivirgatus (Douroucouli)	3	0.8	50	1.5	1
Alouatta palliata (Howler Monkey)	67	5.5	15	10	55
Cebus capucinus (Capuchin Monkey)	17	2.6	25	4	10
Ateles geoffroyi (Spider Monkey)	1	5.0	15	—	—
Saguinus geoffroyi (Geoffroy's Tamarin)	3	0.8	50	1.5	1
Bradypus infuscatus (Three Toed Sloth)	760	2.8	15	114	319
Choloepus hoffmanni (Two Toed Sloth)	187	3.5	15	28	98
Nasus nasua (Coati)	40	3.0	25	10	30
Total				208	557
Kibale Forest, Western Uganda					
Colobus badius (Red Colobus)	298	6.0	15	45	270
C. guereza (Abyssinian Colobus)	12	5.0	20	2.5	12
Cercopithecus ascanius (Red Tailed Guenon)	65	2.5	20	13	33
C. mitis (Diana Monkey)	42	3.0	20	8	25
C. lhoesti (L'hoest's Monkey)	5	2.5	20	1	2
Cercocebus albigena (Grey Cheeked Mangabey)	9	6.5	15	1	7
Pan troglodytes (Chimpanzee)	1	34.0	5	—	2
Total				71	351
Lombe Forest, Western Cameroon					
Colobus satanus (Black Colobus)	31	9.5	15	5	48
Cercopithecus erythrotis (Red Eared Guenon)	31	4.0	20	6	24
C. nictitans (White-nosed Monkey)	30	5.5	20	6	33
C. mona (Mona Monkey)	10	4.0	20	2	81
C. poganias (Crowned Guenon)	8	4.0	20	2	8
Cercocebus albigena (Cloaked Mangabey)	9	8.0	15	1	8
Total				22	129

[1] Estimated densities and body weight data for Panama from Glanz (1982) and Eisenberg and Thorington (1973), respectively, for Uganda, from Struhsaker (1975), and for Cameroon, from McKey and Clutton-Brock and Harvey (1977), respectively.

two species feed from road kills and rubbish dumps, and recent human activity has probably enabled them to extend their natural ranges. Nevertheless, all five species occur commonly in the neotropical forested region of Central and South America.

Most areas of rainforest appear superficially to have a low density of mammals. It is at first difficult to understand how vultures could obtain an adequate supply of carcasses, especially bearing in mind how extremely difficult it is to locate dead animals in dense forest. To consider why scavenging birds have come to occupy American tropical forests, but not those of the Old World, we should first consider the potential size of the food supply in these habitats.

Vultures feed chiefly on large herbivorous mammals. They will take almost any available carcasses, such as large birds, reptiles or predatory mammals, but such animals have low biomasses and so contribute a smaller food supply from the community than the herbivorous mammals. All these animals will, of course, eventually die. Some may be killed and eaten wholly or partly by predators. The remainder will die from other causes and will form the food supply for scavenging animals. It should be possible to estimate the number of carcasses that will become available as food for large meat eating animals if we know the population densities of forest mammal species and their annual mortality rates. Unfortunately, extremely little information has been obtained on mammal communities in tropical forests or on the population dynamics of individual species. Estimates of mammal density and biomass from a few sites in African and neotropical forest communities are, however, available (Table 1). I also estimated (Table 1) the annual mortality rate for each species, based on the assumption that a general relationship exists between body weight in mammals and their mortality rates, (Western 1979), bearing in mind that primate species have a considerably lower annual mortality rate than other mammals of comparable body size (Western 1979), a feature that I have assumed sloths to share because of their low metabolic rate (Goffart 1971). This enabled me to estimate the number and weight of animals dying per year per square kilometer in the various forests (Table 1).

In tropical forests almost all plant production occurs in the tree canopy (Fittkau and Klinge 1973), and herbivorous mammals are, therefore, concentrated there. Barro Colorado Island in Panama is the only area of neotropical forest for which detailed mammalian biomass data are available: Eisenberg and Thorington (1973) suggested a total mammalian biomass of 4400 kg/km^2, 70 percent of which is found in the canopy. I have only considered the arboreal part of this community because the surveys made in African forests are confined to this zone. It is reasonable to assume that African forests also have a low biomass of terrestrial mammals for the same reason; this is often suggested (Collins 1959; Bourliere 1972) although no information is available. The arboreal mammalian biomass on Barro Colorado Island appears to be about 3080 kg/km^2, compared to 2200 kg/km^2 for the Kibale Forest in Uganda, and 728 kg/km^2 for the Lombe Forest in Cameroon (Table 1). On Barro Colorado Island about 557 kg of canopy mammals are estimated to die each year per km^2. The mean body weight of these mammals is comparatively small and results in a food supply, on average, of one 2.3 kg dead animal every two days per km^2. Comparable calculations for the Kibale Forest in East Africa suggest that the food supply for carnivorous animals in this area is only about 63 percent that on Barro Colorado: about 350 kg of animals die each year per km^2, or, on average, about one 5 kg animal dies every 5.1 days per km^2. Lombe forests may have 129 kg of animals dying each year per km^2 (23 percent of that on Barro Colorado), which is equivalent to one animal weighing 6 kg dying every 17 days per km^2.

These calculations are, of course, based on many dubious assumptions. Also, no justification exists for assuming that biomass figures taken from one area will be representative of forests in a whole region: forests show considerable variation in soil type, elevation, and climate, all of which almost certainly influence mammalian density, although there are almost no data available on these factors. The figures presented (Table 1) are intended only to suggest the possible order of magnitude of the food supply. There are, however, major differences between the mammalian faunas of neotropical forests and the forests of the Old World which could cause the suggested differences in carcass availability. South American forests contain some mammalian groups that are absent from the Old World. Sloths live at high densities in neotropical forest and are far more abundant there than any of the primate species. This group of edentates has a specialized digestion and other adaptations that enable it to feed on the mature leaves of forest trees (Montgomery and Sunquist 1979), and no mammal fulfills the same ecological role in the forests of the Old World. In African forests, primates are the major mammalian herbivores, but they live at comparatively low densities, and most species cannot digest mature leaves. The digestive tract of colobus monkeys (*Colobus* spp.) does enable them to feed on leaves (Bauchop 1979). These monkeys are considerably larger than sloths, and only the Red Colobus (*C. badius*) seems to achieve high densities (Struthsaker 1975). The difference in biomass between Kibale and Lombe forests is largely due to the presence of this species at the former but not the latter site. Opossums are abundant omnivores in neotropical forests, but are another mammal group not found in the Old World forests. These differences in the mammalian community between New and Old World forests may result in neotropical forests sustaining a greater mammalian biomass. In addition, mean body size of herbivorous mammals is smaller in the neotropical than in the Old World region, probably resulting in a faster turnover of the population. The implications of these features for scavenging animals are that neotropical forests may provide a food supply larger than that in comparable areas of Old World forest, and also a more regular supply in the form of smaller carcasses.

These estimates suggest the general level of food available for all meat eating animals in the forest. No information on the proportions taken by predators and scavengers is available. Predation among arboreal herbivores, however, probably accounts for a far smaller proportion of deaths than in a terrestrial herbivore community, for the obvious reason that it is dangerous for predatory mammals to operate 30 m up in the forest canopy. No large mammalian carnivores live in the canopy. Large predators, such as Jaguar (*Panthera onca*), prey either on species on the forest floor or among low vegetation, and also live at very low densities (Schaller and Crawshaw 1980). A few birds of prey (e.g., Harpy Eagle, *Harpia harpyja*) will take monkeys and other arboreal herbivores, but they also live at extremely low densities and cannot be responsible for a significant proportion of herbivore mortality. Therefore, although there is no detailed information on mortality, we can conclude that very few animals are killed by predation, and most die from some other cause. Animals often are found dead on the forest floor. Mittermeier (1973), for example, while covering the 35 km of trails on Barro Colorado Island, once encountered 15 to 20 dead mammals on the forest floor during the late rainy season. Most arboreal herbivores that die are, therefore, not killed and eaten by predators, but fall to the ground and are then available to scavenging animals.

It is interesting to compare these figures with those obtained from large ungulate populations which form the typical food supply for many Old World vultures in the African savannas. Such areas support the highest mammalian biomass recorded from any natural community (Talbot and Talbot 1963). Figures for the Serengeti region of Tanzania (biomass about 9000 kg/km^2) suggest that about 1000 kg of animals die per km^2 every year. This supply of carcasses takes, on average, the form of one 75 kg mammal dying per km^2 every 26 days (Houston, in press). This is about double the overall food supply suggested from canopy mammals on Barro Colorado Island, but in the Serengeti about half of the potential food for meat eating animals is taken by the various mammalian carnivores, the remainder going largely to the vulture community (Houston 1979). Although neotropical forests, therefore, appear to be an unpromising habitat for vultures, they may, in fact, provide a food supply for scavenging animals that is as large as that of the Serengeti savannas.

COMPETITION AMONG SCAVENGERS

Many other scavengers compete with vultures for carcasses. Apart from other avian species and mammalian carnivores, many invertebrates feed on meat, and some Diptera and Coleoptera depend on carrion for their larval development. Microorganisms also rapidly develop in a carcass, and, although bacteria and fungi cannot move the food away, they can infuse it with toxins to prevent animals from feeding (Janzen 1977). The food supply available to vultures will depend partly on the speed and effectiveness of competition from these other scavenging animals. In this section I consider the invertebrate scavengers; I will comment later on mammalian scavengers.

Preliminary observations suggest that the activities of invertebrate carrion feeders in the New and Old worlds may differ in a major way, and that these could have important implications for the amount of food available for larger scavengers, such as vultures. Invertebrate scavengers are important competitors because of the speed with which they can consume carrion. *Lucilia illustris,* a European blowfly weighing 0.1 mg on hatching, completes larval growth in 50 hrs at 35°C to reach a final weight of 45 mg, with a production efficiency of 90 percent under optimum conditions (Hanski 1976a). These larvae feed by secreting digestive enzymes onto the meat, which is partially digested, producing fluids that are then taken into the body. When these larvae are present at a high density in the carcass of a small forest mammal the combined effects of their feeding cause the complete breakdown and liquifaction of the soft tissues within two days. Competition occurs within carcasses between different species of dipteran larvae, and selection for rapid hatching of eggs and short periods of larval development has been strong (Hanski 1976b).

In African forests insect larvae can consume small carcasses within a few days. I placed 10 meat baits (2–10 kg) in riverine forests at Kirawira in Tanzania, principally to determine if vultures could locate them. They did not, nor did any mammalian scavengers, but the meat was almost totally consumed by maggots within three days. This was not a true rainforest area. Hanski (pers. comm.), however, studied carrion insects in rainforest at Lombe, Liberia, and he also found that maggot communities developed rapidly and consumed all available food within a few days. Janzen and McKey (1975) simlarly found that in Cameroon rainforest trappers lost about half of their catch because carrion flies destroyed the carcasses before the trappers reached them.

Neotropical forests seem to have a different and more complex community of insects that exploit carrion than do the forests of the Old World. In a short study of the scavengers that were attracted to carrion at a forest site north of Manaus in Brazil (Houston, unpubl. data), I placed a series of 2 kg carcasses on the ground at least 1 km apart and visited each three times a day to record any animals that were present. The carcasses were observed for up to 17 days. The site was an area of disturbed forest, adjacent to a transamazon highway, where mammal density was artificially low (Smith 1976), and vultures were not commonly

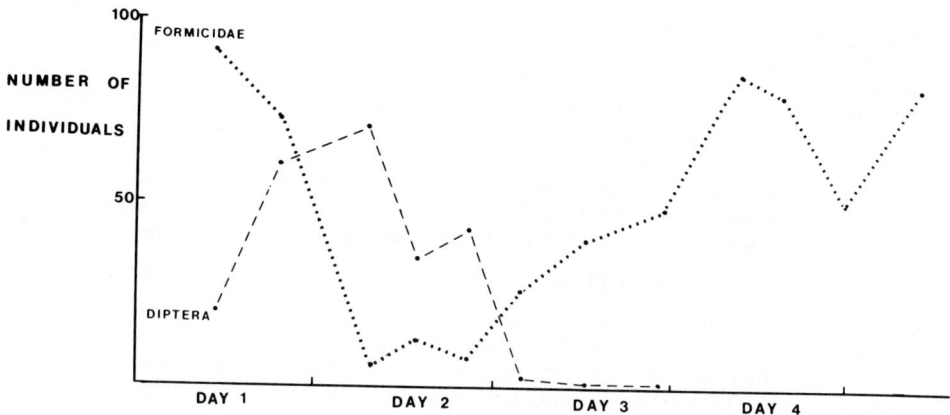

FIG. 1. The sequence of attendance of ants and adult carrion flies recorded at a carcass placed in a forest site 60 km north of Manaus, Brazil.

seen. A wide range of insect species, however, were recorded. Insects attending a carcass usually followed a typical progression (Fig. 1). Ants investigated a bait within a few minutes of its arrival, and most species fed off blood and soft pieces of tissue. Usually, only one species was present at each bait. Several species of diptera visited the bait in large numbers within the first two days to lay eggs on the meat. Some flies, such as *Lucilla exima,* laid 50 to 250 eggs. Up to 50 adults of this species were present at any time on a carcass during the first two days, and so tens of thousands of eggs could be laid on a single food source. These carrion flies were no longer attracted to bait, and no additional eggs were laid, after about day four. After a few days ant numbers increased. Several ant species were found to feed on the fly eggs and young maggots; ants of some of the larger species, such as *Camponotus* sp. and *Ectatomma* sp., carried off large maggots and caught and carried adult flies away. These species of ants used the carrion as a source of insect prey and removed large numbers of dipteran eggs and larvae. To consider the relative importance of ant predation on dipteran larvae, baits were placed in two situations along a small river. One group was placed within 20 m of the river bank, where access of ants to the food was unrestricted. The other group was placed on a small sand island in the river that ants could not reach, but which both diptera and hymenoptera could reach by flying only a few meters. The number of maggots present in carcasses to which ants had access was reduced by as much as 90 percent. This predation greatly extended the period during which the bait was available to other scavengers. Where development of dipteran larvae was unrestricted, it was usual for meat to be completely liquified and destroyed within 3 to 4 days, but where larvae were prevented from reaching a high density, the bait survived up to 17 days.

These results come from a short study in only one area. I have since carried out similar trials in different forest types in Venezuela and Barro Colorado Island, but did not find ant species behaving in this way. However, this effect of predation on carrion feeding insects has been recorded for other invertebrate scavenger communities. Hanski (1976b) reported from Finland that staphylinid beetles and their larvae through predation exert a major influence on the fly larvae community. Nuorteva (1970) found that histerid beetles act in a similar way in southern Finland.

It is known that the ant faunas of the neotropical and the Ethiopian and Oriental regions differ greatly (Brown 1972). The neotropics contain a diversity of old genera of ants which are often highly specialized in their feeding. For example, *Discothyrea* and *Proceratium* specialize at feeding on arthropod eggs (B. Bolton, pers. comm.). Africa, however, has a fauna that is dominated by more recent and less specialized genera such as *Crematogaster* and *Tetramorium* (B. Bolton, pers. comm.). Such a distinction between the New and Old World forest insect communities could further exaggerate the difference in the abundance of food for large scavenging animals in the two regions. For example, if the average number of animals dying per day per km^2 were two times greater in a neotropical region than that in a comparable African forest and in addition each carcass were available for about 10 days in a neotropical region, but only for two days in an African forest, then on any one day 10 times more carcasses would be available per km² for a large scavenger in the neotropics than in the Old World.

Food may, therefore, be available to large scavengers for longer periods in the neotropics, provided that vultures are willing to use it. Over a period of about a week, meat becomes somewhat dry and begins to smell bad, but it does not become unbearably foul and rancid. I took about 1 kg of meat which had remained for 10 days on the forest floor to a group of Black Vultures in a village: they ate some of it with obvious reluctance. I have also seen Turkey, Lesser Yellow-headed, and King Vultures feed readily on meat five days old. Presumably cathartid vultures have developed resistance to many bacterial toxins.

EFFICIENCY OF VULTURES AT LOCATING FOOD IN FORESTS

The food searching behavior of the Old World vultures is reasonably well known. All rely on vision to locate food, and none of the species uses a sense of smell. Flight behaviors differ between species (Pennycuick 1972), but the most abundant species, the griffon vultures, usually fly at altitudes from 100 to 1000 m, tending to concentrate at the lower altitudes where food is abundant (Houston 1974). They do not enter forested areas.

The Cathartid vultures do forage widely in forests and show a range of searching techniques. The Turkey Vulture has a well-developed sense of smell (Bang 1960, 1967; Stager 1964), and it is likely that the other *Cathartes* species also rely on olfaction to locate food. These birds fly comparatively low when searching for food, usually at the level of the tree canopy or just above it. They do not have a straight flight path, but usually bank and turn between emergent trees projecting from the canopy. They use rising air currents on the windward side of tall trees to gain lift, and rarely flap their wings. They are very agile and can maneuver skillfully between tall trees with a buoyant flight path. They only descend below the canopy after detecting food, when they move from branch to branch, or walk over the forest floor to locate the food. The Turkey Vulture seems to realy heavily on smell to locate food in the forest. In some recent field studies on Barro Colorado Island in Panama, I provided more than 70 carcasses in a wide range of situations. The amount of overhead cover made no difference to the time taken by vultures to locate carcasses and, rather surprisingly, carcasses that were invisible under a covering of leaf litter were located as efficiently as those in open positions.

The King Vulture may not be able to locate food by smell. Experiments with captive birds suggest that they cannot (Houston 1984), and their olfactory lobes are much smaller than those of the *Cathartes* species (Stager 1964). King Vultures do, however, feed in dense forest and descend to the forest floor to feed. This vulture flies at considerably higher altitudes than the *Cathartes* vultures when it is searching for food, usually above 300 m and far above the forest canopy. At such a height it seems unlikely that olfactory cues would be present. It is probable that this species locates most of its food by watching the activities of the *Cathartes* vultures, following them to carcasses and displacing them (Koster and Koster-Stoewesand 1978). I have seen King Vultures feeding on twelve occasions in forest conditions in Venezuela, and in every case the carcass had first been located by a *Cathartes* species.

Black Vultures are not usually found inside the forest unless they are in association with Turkey Vultures, who probably lead them to the food. Black Vultures are unable to locate food by smell (Stager 1964), have comparatively small olfactory lobes (Bang 1967), and are usually found near large villages or towns where they depend on garbage. During extensive travelling in Central America and northern South America, N. J. H. Smith (pers. comm.) observed that Black Vultures were confined in their distribution to areas near settlements of more than 3000 people. These birds were probably comparatively scarce before the European settlement of the neotropical forests and were then, perhaps, confined to feeding on stranded fish, oil palm seeds, and other items along large waterways and lakes shores.

In undisturbed forests the density of small cathartid vultures is comparatively high; *Cathartes* spp. are among the most abundant meat eating birds in neotropical forests. Chapman (1929, 1938), working on Barro Colorado Island in Panama, estimated that the island usually contained about 12 of these vultures, giving a density of about 1 bird per 1.25 km². My recent work on the island suggests that this is still a reasonable estimate. This is similar to the density of *Cathartes* spp. recorded along the banks of the Rio Negro in Brazil (Houston 1984). Birds are able to locate food with surprising speed. During some recent studies on Barro Colorado Island I found that more than 70 percent of 74 chicken carcasses that I provided were located and eaten by Turkey Vultures within 12 hrs.

Mammalian scavengers will also compete for carrion, but they do not seem to be nearly as efficient at locating food. In some field trials north of Manaus, Brazil no mammalian species located any of the carcasses I provided during 83 bait-days of observations. Mammalian density was, however, greatly reduced in this area. But on Barro Colorado Island, where scavenging mammals are abundant and opossums (*Didelphis marsupialis*) and Coatis (*Nasua nasua*) were seen feeding at night from carrion, of the 32 baits I provided only 8 percent were located by mammalian scavengers, and only small amounts of meat were removed by them (Houston, unpubl. data).

DISCUSSION

The aim of this short review has been to consider some of the possible reasons why forest scavenging birds should have developed in the New World, but not the Old. Some tentative estimates have suggested that neotropical forests may provide a greater food supply than equivalent areas of Old World forest because of higher mammalian biomass and smaller mean body size of herbivorous mammals. Carrion may also remain available for longer periods in some areas of neotropical forest because of the more complex ant communities in this region, which can prevent large maggot populations from developing in a carcass. Neotropical forests may, therefore, potentially have a good food supply for a large scavenging animal which is able to locate carcasses efficiently. Cathartid vultures, with their highly developed sense of smell, are astonishingly efficient at finding dead animals. There is no mammalian species that is an exclusive scavenger, but many omnivorous mammals scavenge meat whenever they have the opportunity. These species are far less efficient at locating carrion than the vultures, however, and take only a small proportion of the potential food. Presumably, this is because no mammal can search such a large area of forest each day as a vulture. The energy costs of widespread foraging for a terrestrial animal are probably so high as to make deliberate searching for carcasses an unprofitable activity, while the soaring flight of vultures enables them to forage at very low energy cost and thus to have become exclusive scavengers.

Forested regions in the Old World may not provide such a large food supply for scavenging birds, partly because mammal density may be lower. In addition, any carcasses that do become available must be located within two or three days, before they are destroyed by insect larvae. It seems possible that this food supply has been insufficient to enable any scavenging bird to specialize on it. It is also, of course, possible that Old World forests could supply adequate food to support scavenging birds, but that Old World vultures have never occupied these areas simply because these birds have a poorly developed sense of smell and cannot locate the carrion in forest. This seems unlikely for two reasons. Firstly, the fossil record shows that cathartid vultures were once widespread in the Old World and may have been expected to occupy such a role if the food resource were adequate. Secondly, we know that many unrelated groups of birds have developed good olfactory abilities (Wenzel 1973). I strongly suspect that Old World vultures would have developed their sense of smell if the selective advantage in doing so had been strong enough.

ACKNOWLEDGMENTS

I am grateful to the Royal Society for a Scientific Investigations Grant and exchange fellowship with CNPq Brazil, and to the Carnegie Trust for the Universities of Scotland for funding the fieldwork. Mr. B. Bolton of the British Museum (Natural History) kindly identified the insect specimens, and I am most grateful to Dr. N. G. Smith for his helpful comments on the text.

LITERATURE CITED

BANG, B. G. 1960. Anatomical evidence for olfactory function in some species of birds. Nature 188: 547–549.

BANG, B. G. 1967. The nasal organs of the Black and Turkey Vultures: a comparative study of the cathartid species. J. Morphol. 115:153–184.

BAUCHOP, T. 1979. Digestion of leaves in vertebrate arboreal folivores. Pp. 193–204, *In* G. G. Montgomery (ed.), The Ecology of Arboreal Folivores. Smithsonian Institution Press, Washington, D.C.

BRODKORB, P. 1964. Catalogue of fossil birds: Part 1. Bull. Fl. State Mus. Biol. Sci. 8:201–345.

BROWN, L. H., AND D. AMADON. 1968. Eagles, Hawks and Falcons of the World. County Life, London.

BROWN, W. L. 1972. A comparison of the Hylean and Congo-West African rain forest ant faunas. Pp. 161–177, *In* B. J. Meggers, E. Eyensu, and W. D. Duckworth (eds.), Tropical Forest Ecosystems in Africa and South America: a Comparative Review. Smithsonian Institution Press, Washington, D.C.

BOURLIERE, F. 1972. Comparative ecology of rainforest mammals. Pp. 279–292, *In* B. J. Meggers, E. Eyensu, and W. D. Duckworth (eds.), Tropical Forest Ecosystems in Africa and South America: a Comparative Review. Smithsonian Institution Press, Washington, D.C.

CHAPMAN, F. M. 1929. My Tropical Air Castle. Appleton-Century, New York.

CHAPMAN, F. M. 1938. Life in an Air Castle. Appleton-Century, New York.

COLLINS, W. B. 1959. The Perpetual Forest. Lippincott, Philadelphia, Pennsylvania.

CLUTTON-BROCK, T. H., AND P. H. HARVEY. 1977. Species differences in feeding and ranging behaviour in primates. Pp. 557–579, *In* T. H. Clutton-Brock (ed.), Primate Ecology. Academic Press, New York.

EISENBERG, J. F., AND R. W. THORINGTON. 1973. A preliminary analysis of a neotropical mammal fauna. Biotropica 5:150–161.

FEDUCCIA, A. 1980. The Age of Birds. Harvard University Press, Cambridge, Massachusetts.

FITTKAU, E. J., AND H. KLINGE. 1973. On biomass and trophic structure of the central Amazonian rain forest ecosystem. Biotropica 5:2–14.

GOFFART, M. 1971. Function and form in the sloth. Pergamon Press, New York.

GLANZ, W. E. 1982. The terrestrial mammal fauna of Barro Colorado Island: censuses and long-term changes. Pp. 455–468, *In* E. G. Leigh, A. S. Rand, and D. M. Windsor (eds.), The Ecology of a Tropical Forest: Seasonal Rhythms and Long-term Changes. Smithsonian Institution Press, Washington, D.C.

HANSKI, I. 1976a. Assimilation by *Lucilia illustris* (Diptera) larvae in constant and changing temperatures. Oikos 27:288–299.

HANSKI, I. 1976b. Breeding experiments with carrion flies (Diptera) in natural conditions. Ann. Entomol. Fenn. 43:108–115.

HOUSTON, D. C. 1974. Food searching in Griffon Vultures. East. Afr. Wildl. J. 12:63–77.

HOUSTON, D. C. 1976. Ecological isolation in African scavenging birds. Ardea 63:56–64.

HOUSTON, D. C. 1979. The adaptations of scavengers. Pp. 263–286, *In* A. R. E. Sinclair, and M. Norton Griffiths (eds.), Serengeti: Dynamics of an Ecosystem. University Chicago Press, Chicago, Illinois.

HOUSTON, D. C. 1984. Does the King Vulture *Sarcorhamphus papa* use a sense of smell to locate food? Ibis 126:67–69.

HOUSTON, D. C. In press. A comparison of the food supply of African and South American vultures. Proc. 4th Pan Afr. Cong.

JANZEN, D. H. 1977. Why fruits rot, seeds mold and meat spoils. Am. Nat. 111:691–713.

JANZEN, D. H., AND D. MCKEY. 1975. What the tropical trappers leave behind. Biotropica 7:7.

KOSTER, F., AND H. KOSTER-STOEWESAND. 1978. Konigsgeier Beobachtungen im Tayrona Nationalpark im Norden Kolumbiens, Sudamerika. Z. Koeln. Zoo 21:35–41.

KRUUK, H. 1967. Competition for food between vultures in East Africa. Ardea 55:171–193.

LIGON, J. D. 1967. Relationships of the cathartid vultures. Occas. Pap. Mus. Zool. Univ. Mich. No. 651.

MARTIN, P. S. 1967. Prehistoric overkill. Pp. 75–120, *In* P. S. Martin, and H. E. Wright (eds.), Pleistocene Extinctions, the Search for a Cause. Yale University Press, New Haven, Connecticut.

MCKEY, D. B. 1978. Plant chemical defences and the feeding and ranging behavior of Colobus monkeys in African rainforests. Unpubl. Ph.D. dissert. University of Michigan, Ann Arbor.

MITTERMEIER, R. A. 1973. Group activity and population dynamics of the howler monkey on Barro Colorado Island. Primates 14:1–19.

MONTGOMERY, G. G., AND M. E. SUNQUIST. 1979. Habitat selection and use by two toed and three toed sloths. Pp. 329–360, *In* G. G. Montgomery (ed.), Ecology of Arboreal Folivores. Smithsonian Institution Press, Washington, D.C.

NUORTEVA, P. 1970. Histerid beetles as predators of blowflies in Finland. Ann. Zool. Fenn. 7:195–198.

PENNYCUICK, C. J. 1972. Soaring behaviour and performance of some East African birds, observed from a motor-glider. Ibis 114:178–218.

RICH, P. V. 1983. The fossil history of vultures: a world perspective. Pp. 3–25, *In* S. R. Wilbur, and A. J. Jackson (eds.), Vulture Biology and Management. University of California Press. Berkeley.

SCHALLER, G. B., AND P. G. CRAWSHAW. 1980. Movement patterns of jaguar. Biotropica 12:161–168.

SMITH, N. J. H. 1976. Utilization of game along Brazil's transamazon highway. Acta Amazonica 6:455–466.

STAGER, K. E. 1964. The role of olfaction in food location by the Turkey Vulture. Los Ang. Cty. Mus. Contrib. Sci. No. 81.

STRUHSAKER, T. T. 1975. The Red Colobus Monkey. University Chicago Press, Chicago, Illinois.

TALBOT, L. M., AND M. H. TALBOT. 1963. The high biomass of wild ungulates in East Africa. Trans. 28th N. Am. Wildl. Nat. Res. Conf. pp. 465–476.

VOOUS, K. H. 1973. List of recent holarctic bird species. Ibis 115:612–638.

WENZEL, B. M. 1973. Chemoreception. Pp. 389–416, *In* D. S. Farner, and J. R. King (eds.), Avian Biology Vol. 3. Academic Press, New York.

WESTERN, D. 1979. Size, life history and ecology in mammals. Afr. J. Ecol. 17:185–204.

NEOTROPICAL AVIAN FRUGIVORES: PATTERNS OF BEHAVIOR, MORPHOLOGY, AND NUTRITION, WITH CONSEQUENCES FOR FRUIT SELECTION

Timothy C. Moermond[1] and Julie Sloan Denslow[2]

[1]Department of Zoology, and [2]Department of Botany, University of Wisconsin–Madison, Madison, Wisconsin 53706 USA

ABSTRACT. A large number of neotropical bird species in many families regularly eat fruit. We discuss the physiological, morphological, and behavioral adaptations associated with eating fruits. Fruits present a number of difficulties for the frugivores, such as the low protein to calorie ratio, the watery bulk, and the undigestible seed mass. Although we find no consistent morphological adaptations for digesting fruits, we show that important adaptations in digestive physiology are to be expected. Both large and small birds have similarly diverse diets, but small birds, such as tanagers and manakins, eat mainly small, carbohydrate-rich fruits, while large species, such as cotingas and toucans, eat a wider range of fruit sizes including both small, carbohydrate-rich fruits and large, lipid-rich fruits.

Most fruit-eaters have large gapes, and many birds swallow fruits whole, although a number of bird species, both small and large, may facultatively eat some large fruits piecemeal. Tanagers and some emberizid finches generally eat fruits by mashing the pulp, swallowing the pulp, juices and small seeds, but dropping medium and large seeds. Birds that take fruits on the wing usually have short, wide, flat bills, whereas those that reach fruits from perches often have long and narrow bills. Differences in bill structure affect the manner of fruit handling and may influence fruit selection.

Most species take fruits either primarily on the wing or primarily from perches. We divide these two general foraging methods into four foraging maneuvers on the wing (hovering, stalling, swooping, and snatching) and three foraging maneuvers from a perch (picking, reaching, and hanging). These foraging techniques differentially affect access to fruits. Differences in flight maneuvers are associated with differences in wing-loading, wing aspect ratio, and degree of slotting. Likewise differences in ability to reach fruit from a perch are correlated with differences in tarsus shape and leg musculature. Very few birds are able to reach fruits well and also pluck them expertly on the wing.

The dichotomy of frugivorous birds into specialists and generalists, with only a couple exceptions, is not supported by comparison of diet, behavior, morphology, or function as dispersers. The important differences among fruit-eating birds depend on the constraints of morphology on behavior. Differences in digestive systems, bill strength and shape, wing and leg morphology, and body weight differentially affect the accessibility of fruits, the ease with which they are handled, the rate at which they are eaten, the efficiency with which they are digested, and thus influence which fruits are selected. The presence of fruit-eating birds in so many families is not evidence for lack of requirements for eating fruits but rather represents adaptive specialization to exploit fruits displayed in different ways.

RESUMEN. En las zonas neotropicales se encuentran un gran numero de especies de aves pertenecientes a numerosas familias cuya dieta consiste de frutas. Presentamos una discusión sobre las adaptaciones fisiológicas, morfológicas y de comportamiento que están asociadas con la costumbre de comer frutas. El comer frutas le presenta una serie de problemas a aves frugívoras como por ejemplo: una relación baja entre proteínas y calorías, un gran volumen de materia acuosa y la presencia de semillas indigeribles. Aunque no encontramos adaptaciones digestivas morfológicas generales para comer frutas, demostramos que adaptaciones digestivas fisiológicas en aves frugívoras son de esperarse. Las aves frugívoras, tanto las grandes como las pequeñas, tienen dietas similares en cuanto a la diversidad de frutas. Sin embargo, las aves pequeñas, como las tangaras y los saltarines, se alimentan mayormente de frutas pequeñas que son ricas en carbohidratos; mientras que las aves de mayor tamaño, como las cotingas y los tucanes, se alimentan de frutas de diversos tamaños; tanto de frutas pequeñas ricas en carbohidratos, como de frutas grandes ricas en lípidos.

La mayoría de las especies frugívoras tienen el rictus ancho y muchas ingieren la fruta entera sin despedazarla. Sin embargo, algunas especies, tanto grandes como pequeñas, tienen la capacidad facultativa de comerze la fruta trozo a trozo. Las tangaras y algunos de los pinzones usualmente se comen la fruta después de aplastarla con el pico. Usando este método ellos ingieren la pulpa, el jugo y las semillas pequeñas, pero rechazan las semillas grandes y medianas. Las especies que obtienen la fruta volando, generalmente tienen picos cortos, anchos y de poca profundidad. Mientras que las que obtienen la fruta desde una percha comúnmente tienen picos largos y estrechos. Estas diferencias en la estructura del pico afectan la manera en que el ave manipula la fruta y pueden influenciar el proceso de selección entre diferentes frutas.

La mayoría de las especies usan uno de dos métodos principales para obtener las frutas, las toman volando o las toman desde una percha. Para las especies que obtienen las frutas volando, dividimos las maniobras de forrajeo en cuatro categorías: "hovering," "stalling," "snatching" y "swooping." Para las especies que obtienen la fruta desde una percha, dividimos las maniobras de forrajeo en tres categorias: "picking," "reaching" y "hanging." Todos estos métodos de forrajeo afectan de manera diferente el acceso que un ave pueda tener a una fruta. Las diferencias en maniobras de vuelo están asociadas con diferencias aerodinámicas como el "wing-loading," "wing aspect ratio" y el grado de "slotting." De igual manera, las diferencias entre las especies en la abilidad de alcanzar frutas desde una percha están asociadas con diferencias en la forma del tarso y la musculatura de las patas. Son muy pocas las especies que tienen la capacidad de obtener frutas desde una percha fácilmente y a la misma vez poder tomar frutas en vuelo con destreza.

La dicotomía existente entre aves frugívoras "especialistas" y aves frugívoras "generalistas" no está apoyada por nuestras comparasiones de las dietas, comportamientos, morfologías, o la función de las especies como dispersoras de semillas. Las diferencias mas importantes entre las aves frugívoras se basan en las restricciones impuestas por la morfología y el comportamiento del ave. Diferencias en el sistema digestivo, la fuerza y configuración del pico, la morfología de las patas y alas y el peso del ave afectan de la manera diferente la accesibilidad de las frutas, la facilidad con que se manipulan, la rapidez con que se ingieren y la eficiencia con que se digieren y por lo tanto, influyen de manera muy importante en el proceso de selección entre diferentes frutas. La presencia de tantas aves frugívoras pertenencientes a numerosas familias no representa una evidencia de que no se necesitan adaptaciones para comer frutas, si no demuestra numerosas especializaciones adaptivas para explotar frutas que están presentadas en maneras diferentes.

Fruit-eating among tropical birds is so wide-spread as to make definition of "frugivorous" difficult. Few species are so restricted to fruits that their nestlings are completely reared on fruits (Morton 1973). Oilbirds (*Steatornis caripensis*, Snow 1961, 1962c) and Bearded Bellbirds (*Procnias averano*, Snow 1970) are in this category. At the other extreme are the many primarily insectivorous birds that exploit the abundant fruits of some trees (e.g., Eisenmann 1961; Olson and Blum 1968; Morton 1971; Leck 1972a, 1973; Howe and DeSteven 1979) or take an occasional fruit while foraging for insects. Traylor and Fitzpatrick (1982) noted for flycatchers (Tyrannidae) that while fruit features heavily in the diets of only a few species, nearly all eat some fruit occasionally. The same is probably true of woodpeckers, and even rails have been observed taking fruits in Costa Rica (*Aramides cajanea* feeding on low growing Melastomataceae berries, D. J. Levey, pers. comm.; *Porphyrula* feeding on *Heliconia* fruits, Skutch 1933). Between these extremes are birds whose diets consist of some mixture of fruits and insects and occasionally vertebrates, snails, or leaves (Jenkins 1969; Morton 1978; Stiles 1983).

Here we use the word "frugivore" for birds whose diets include a substantial portion of fruit at least during some seasons, rather than confining it to birds that eat solely fruits. Nearly all of the so-called frugivorous birds eat other types of foods as well (usually insects or other invertebrates) and should be properly referred to as omnivores. However, the term "frugivore" has been commonly used for such frequent but not exclusive fruit-eaters as tanagers. Here we use the terms "frugivore" and "fruit-eater" interchangeably.

Adaptations of fruits that enhance seed dispersal by birds have been described in detail (Ridley 1930; van der Pijl 1969). The most obvious common characteristic is an energy or nutrient reward in the form of a soft, fleshy pulp surrounding the seed. In addition, bright colors and/or morphological modifications presumably advertise the presence of the reward (Morden-Moore and Willson 1982; Willson and Thompson 1982; Janson 1983). The seed itself is frequently provided with some form of protection that reduces the probability of destruction by the bird or discourages invertebrate seed-eaters, e.g., hard seed coat or toxins in the seed, seed coat, or fruit pulp (Herrera 1982). Given that fruits, especially small-seeded, sweet, berry-like fruits, are at times abundant, visible, and easily accessible, it is not surprising that many birds occasionally exploit this ready source of energy.

Morphological adaptations for fruit-eating in birds have not been so clearly identified. The most commonly cited examples concern a relatively large gape (Snow 1973) and lack of crop or muscular gizzard. Even these modifications of the gut may not be typical (see Morphology of the Digestive Tract below). Some species exhibit behavioral adaptations, such as regional or elevational migrations that track local fruit abundances (e.g., Fogden 1972; Crome 1975a, c; Morton 1977; Karr et al. 1982; Wheelwright 1983; F. G. Stiles, pers. comm.). Most fruit-eating birds do not specialize on one or a few species of fruits and dietary overlap among many fruit-eating birds is broad (e.g., Snow and Snow 1971; Leck 1971b; Crome 1975a), especially with respect to small-seeded fruits (Eisenmann 1961; Land 1963; Willis 1966; Olson and Blum 1968; Leck 1973; Howe 1981, 1982).

In contrast to this general picture is the persistent notion of frugivores arrayed on a continuum between specialists and generalists (McKey 1975; Howe and Estabrook 1977; Snow 1980). The idea is based on the assumption that reliable attraction of dispersers to large-seeded fruits entails provision of a high quality reward to the bird. Hence, large-seeded fruits are seen to be nutritious and closely linked evolutionarily with highly frugivorous birds that depend on fruits for most of their protein and energy needs (hence specialists) and whose gapes are large enough to handle large seeds (Snow 1971a; McKey 1975). These birds are also seen to provide high quality dispersal, i.e., they preferentially feed on these large fruits and disperse the seeds away from the parent plant. The seeds are not destroyed. Most small, fruit-eating birds are seen as opportunists (generalists) that take fruits nonselectively as they are encountered and are willing to settle for small, juicy, low-reward fruits (McKey 1975; Howe and Estabrook 1977).

However, there is no reason to expect that fruit specialists provide a higher quality dispersal than generalists (Wheelwright and Orians 1982), and we shall show that the diets of specialists and generalists often have much in common. Our own studies of Costa Rican fruit-eating birds have suggested that fruits displayed in different ways are differentially accessible to birds that forage in different ways (Moermond and Denslow 1983; Denslow and Moermond, in press). Here we compare frugivores with respect to fruit choices, fruit handling, and foraging behavior and suggest that these factors form a more functional basis on which to categorize this phylogenetically and morphologically diverse group of birds.

WHO EATS FRUITS

Neotropical fruit-eating birds are found among a broad variety of bird families. Very few eat only fruits as adults and nestlings. Most supplement their fruit diets with different, relatively protein-rich foods, and those that do feed almost exclusively on fruits as adults usually feed insects to their young. The nature of these alternative protein sources—seeds on one hand and insects or vertebrates on the other—provides a convenient basis for classifying frugivores.

Seed-eating fruvigores.—That relatively few species combine a diet of seeds and fruit pulp is not surprising in view of the different digestive apparati necessary for processing the two foods. Granivorous birds are characterized by large crops, muscular gizzards, long small intestines, and slow passage rates through the digestive tract in contrast to frugivores (Ziswiler and Farner 1972). Many fruit-eaters in this category are, in fact, feeding on both seed and pulp of fleshy fruits. They are often seed destroyers, rather than dispersers, and generally prefer green (unripe) or partially ripe fruits to fully ripe fruits in which the seed coats are likely to be fully hardened. Most also supplement their diets with young leaves, shoot tips, buds, and occasional insects.

Columba pigeons swallow seeds and berries whole; large seeds especially are undoubtedly destroyed in the large, muscular gizzard, although very small seeds, such as those of *Cecropia*, *Ficus*, or *Miconia*, may pass through intact. The bill is relatively weak and only very soft fruits can be eaten by biting or tearing out chunks (Goodwin 1970). New World parrots are also primarily seed-eaters that break up seeds with their bills before swallowing (Forshaw 1978; Howe 1981; Janzen 1981). The woody husks surrounding arillate seeds of many canopy tree fruits may offer some measure of protection from such seed predators.

The Cracidae also subsist largely on fruits and seeds. Curassows (*Crax, Nothocrax*) are primarily terrestrial foragers that feed on fallen or low growing, unripe fruits (Delacour and Amadon 1973). Crop and gizzard are well-developed, so many seeds are probably digested in addition to the fruit pulp. Chachalacas (*Ortalis*) and guans (*Penelope*) are primarily arboreal and forage on ripe fruits (Delacour and Amadon 1973). Long legs and neck facilitate reaching of fruits on small diameter, pendent twigs. The gizzard is thin-walled and the crop is poorly developed, so most seeds probably pass through the gut intact.

Finches in the subfamilies Emberizinae and Cardinalinae feed primarily on seeds, but some cardinalines, especially *Saltator* spp. and the grosbeaks (e.g., *Caryothraustes, Pytilus,* and *Passerina (Cyanocompsa) cyanoides*), eat a substantial amount of fruit as well. All include insects in their diets, and saltators are also known to feed on young leaves (Jenkins 1969; pers. observ.). The size and hardness of a seed probably influence how it is handled. Seeds of *Heliconia* spp. (Musaceae) are too hard for even the heavy-billed *Passerina cyanoides* to break (Stiles 1979, pers. observ.), but both *P. cyanoides* and *Caryothraustes poliogaster* will crack the smaller seeds of *Psychotria brachiata* (Rubiaceae). Saltators appear to digest very few seeds and supplement their diets primarily with insects (Jenkins 1969). The fruit-eaters in this group crush the fruit pulp before swallowing (Moermond 1983). Large seeds are frequently dropped in place if not eaten and, therefore, poorly dispersed, whereas small seeds may be swallowed intact (Moermond and Denslow, unpubl. data; D. J. Levey, pers. comm.).

Insect-eating frugivores.—Most fruit-eating birds eat some insects and provide them to young nestlings. Like frugivores, primarily insectivorous birds have thin-walled gizzards and small crops (Cvitanic 1970). Passage rates through the digestive tracts of the two groups are comparable (Ziswiler and Farner 1972; Herrera 1984). Incorporation of fruits into the diet of an insectivore is apparently done as readily as is the reverse. Frugivores frequently take advantage of locally superabundant insect populations (Eisenmann 1961), and some migrants, e.g., *Dendroica castanea, D. pensylvanica,* and *Vermivora peregrina* are insectivorous on their temperate breeding grounds but often highly frugivorous in the tropics (articles in Keast and Morton 1980). In fact north temperate migrants may be important dispersers of seeds of some neotropical trees (Leck 1972a; Howe and DeSteven 1979; Howe and Vande Kerckhove 1980; Greenberg 1981). Except during seasons of low fruit abundance when partially ripe fruits are taken (Foster 1977), insect-eating fruvigores prefer fully ripe fruits to partially ripe fruits (Moermond and Denslow 1983). Most take both watery, sweet fruits and dry or oily fruits. Fruit-eating birds in the following families supplement their fruit diets with insects and, occasionally, small vertebrates: Cotingidae, Pipridae, Tyrannidae, Muscicapidae (Turdinae), and Emberizidae (including Parulinae, Icterinae, Thraupinae, and Cardinalinae).

NUTRITION AND DIGESTION

FRUIT AS A FOOD RESOURCE

Pulp quality varies widely with respect to amount, caloric and nutrient contents, digestibility of the pulp (fiber, water content), seed size and total bulk of seeds, toughness of fruit coat, and presence of secondary compounds (e.g., phenols and alkaloids, Herrera 1982). Seeds ingested with the pulp may add considerably to the non-digestible bulk of the fruit (e.g., commonly more than 50 percent of the total fruit weight for few-seeded fruits but much less for fruits with many small seeds, White 1974; Snow 1981; Wheelwright et al. 1984). In contrast to both seeds and insects, fruits are generally low in protein and in lipid and may be high in carbohydrate (Table 1). A survey of families producing bird-dispersed seeds (Table 1) also reveals that in general fruits are either rich in lipid or in carbohydrate. Only in a few instances does a fruit contain more than 20 percent each of lipid and carbohydrate. The median protein content for 29 tropical families of bird-dispersed fruits is 8.4 percent of the dry weight of the fruit pulp (Table 1) compared to averages of approximately 12 percent for 12 species of cereal seeds (Jenkins 1969) and 66 percent for 19 species of insects (White 1974). Among these fruits, there is no correlation between lipid and protein content of fruit pulp (Spearman $R = 0.182$, n.s.).

The water content of fruits is generally very high, contributing to the impression that fruits are an extremely low quality food source. Experimental diets fed to granivores are low in moisture (about 11%) in comparison to fruits (50% to 94%, Table 1). On a dry weight basis, the protein contents of many fruits are only slightly less than those of seeds, and nutrients are easily assimilated from an aqueous solution. Fruit pulp is also generally low in fiber, requiring little mechanical reduction to render the cell contents accessible to digestive processes. Fruits thus are characterized by a bulky, dilute pulp coupled with a substantial portion of undigestible seed mass that must also be handled. It is apparent that efficient handling of a bulky food is one of the major constraints on a bird dependent on fruits for a large portion of its energetic and nutritional needs. One solution for the birds is a rapid gut passage time

TABLE 1
ENERGETIC AND NUTRIENT CONTENTS OF FRUITS IN TWENTY-NINE PLANT FAMILIES THAT USE BIRDS AS SEED-DISPERSERS[1]

	No. species sampled	Water %	Protein[2]	Lipid	Carbohydrate	kcal/g[3]	P/C[4]	Source[5]
Annonaceae	1	86	4.8	3.3	21.0	1.40	.034	8
Araliaceae	1	68	12.0	33.0	55.0	5.83	.020	7
Apocynaceae	2	73	11.0	17.3	4.5	2.35	.047	6
Boraginaceae	2	80	16.8	5.2	68.1	3.96	.042	3, 7
Burseraceae	4	85	11.0	3.9	65.0	3.53	.031	1, 7
Connaraceae	1	79	8.4	15.9	70.3	4.72	.018	7
Flacourtiaceae	1	90	5.7	.9	37.8	1.89	.030	8
Lauraceae	24	68	6.2	25.3	9.0	3.03	.020	1, 2, 8
Leguminosae	1	56	1.9	22.7	73.2	5.14	.004	7
Loranthaceae	1	55	5.8	52.7	38.1	6.73	.009	7
Malvaceae	1	56	3.1	0.3	35.2	1.60	.019	8
Melastomataceae	2	66	4.2	6.8	84.0	4.21	.010	7
Meliaceae	1	54	15.1	59.7	22.3	7.23	.020	9
Moraceae	4	78	6.4	4.3	79.0	3.89	.016	3, 7
Musaceae	1	74	10.6	39.2	40.4	5.81	.018	7
Myrsinaceae	2	87	6.0	2.7	53.8	2.71	.022	5, 8
Myristicaceae	1	55	2.5	63.0	9.0	6.35	.004	4
Myrtaceae	2	81	4.5	1.6	53.8	2.53	.018	7, 8
Ochnaceae	1	57	4.9	11.0	81.0	4.52	.011	7
Olacaceae	1	68	17.5	17.6				7
Palmae	4	85	9.5	20.8	66.6	4.88	.027	1, 7
Phytolaccaceae	1	80	1.9	2.4				7
Rosaceae	3	80	6.5	1.5	38.0	2.00	.032	3, 8
Rubiaceae	2	87	13.2	5.2	74.6	4.14	.032	7
Rutaceae	1	77	8.8	1.2				8
Sapindaceae	1	83	9.9	3.2	82.5	4.11	.024	7
Solanaceae	1	89	9.2	1.3	15.6	1.22	.075	8
Verbenaceae	3	80	2.2	2.2	38.4	1.86	.012	7, 8
Zingiberaceae	1	92	17.9	1.8	60.9	3.54	.051	7
Medians		79 (n = 29)	6.5 (n = 29)	5.2 (n = 29)	53.8 (n = 26)	3.92 (n = 26)	.020 (n = 26)	

[1] Median values are provided when data for more than one species per family are available. Except for water, all percents are based on dry weights.

[2] Protein was calculated as 4.4 × percent nitrogen (Milton and Dintzis 1981).

[3] The following conversion factors were used to estimate coloric content: protein 5.2 kcal/g; lipid 9.3 kcal/g; carbohydrate 4.0 kcal/g (Watt and Merrill 1963).

[4] Protein : calorie ratio is calculated as % protein/kcal·kg⁻¹ food.

[5] Sources: 1 = Snow 1962c; 2 = Snow 1981; 3 = Munzell et al. 1949; 4 = Howe and Vande Kerckhove 1980; 5 = M. Foster 1977; 6 = McDiarmid et al. 1977; 7 = White 1974; 8 = Wheelwright et al. 1984; 9 = Foster and McDiarmid 1983.

(Groebbels 1932; Walsberg 1975; Herrera 1984). Another solution is for the birds to select fruits with low proportions of undigestible matter. Studies by Howe and Vande Kerckhove (1980) and Herrera (1981) have shown that birds prefer fruits with a large pulp to seed ratio (at least among large-seeded fruits), and Sorensen (1984) correlated fruit preference in *Turdus merula* with fruits whose seeds are regurgitated rather than defecated. Regurgitated seeds have significantly less residence time in the gut.

Although yearly extremes of fruit abundance may commonly differ only two-fold (Fogden 1972) compared to many-fold in temperate habitats, seasons of fruit abundance may nevertheless be well-defined even in aseasonal wet tropical climates (Snow 1965; Janzen 1967; Smythe 1970; Frankie et al. 1974; Hilty 1980; Foster 1982a). Fruiting seasons may show marked year-to-year variation (Ward 1969a; Fogden 1972; Foster 1982a, b). Trees may fruit on a supraannual periodicity (Frankie et al. 1974), crops of some species may fail completely in some years (Crome 1975a; Foster 1982b; Faaborg and Terborgh 1980), and some species may not fruit consistently at the same time each year (Foster 1982a). Crop sizes, fruit characteristics, spatial distributions, and phenologies of fruit production differ between canopy and understory plants. Understory trees and shrubs are more likely to ripen few, small fruits over an extended season than are canopy trees (Frankie et al. 1974; Hilty 1980; Opler et al. 1980).

Thus, even though fruits appear to be common year round in rain forest, suitability of fruits for a particular frugivore may differ, and favored fruits may be patchily distributed in space (Snow 1962c; Fogden 1972) and poorly predictable in time. The tracking and exploitation of a locally variable but sometimes abundant food source (for example by flocking, Leck 1971a; Buskirk 1976) is then another important attribute of a frugivorous habit.

NUTRITIONAL CONSTRAINTS ON FRUGIVORES

Unfortunately, studies on the nutritional requirements of frugivorous birds are scarce. We are often limited to extrapolations from studies on granivores—chickens, domestic pigeons, and sparrows—the results of which may be of limited value in understanding digestive physiology of frugivores. Nevertheless, some general patterns are suggested.

Studies of temperate bird species suggest that ingestion is limited primarily by energetic requirements (Fisher 1972) and (to some extent) volumetric constraints (chickens, Hill and Dansky 1954; Tree Sparrows, *Spizella arborea,* Martin 1968). For primarily carnivorous, insectivorous, or granivorous birds, other nutritional requirements such as protein and minerals are likely to be largely satisfied at the same time that energetic requirements are met if there is variety in the diet.

Studies of adult birds suggest that a diet of 4 percent to 8 percent protein (fresh weight) is necessary for maintenance (chicken = 7%, Leveille and Fisher 1958; *Spizella* = 8%, Martin 1968; and papers reviewed by Berthold 1976a). Martin (1968) found that although a diet of 4 percent protein was minimal for survival of *Spizella arborea,* assimilation efficiencies of nitrogen and carbohydrate are significantly greater in diets containing at least 8 percent protein. Initial indications are, however, that protein requirements (or ability to extract sufficient protein from fruits) differ importantly among species. Berthold (1976a) showed clearly that several European passerines (*Sylvia atricapilla, S. borin, Turdus merula,* and *Erithacus rubecula*) were unable to maintain weight on a diet consisting solely of fruits. The birds consumed adequate amounts of calories but not of protein. In contrast, the waxwing (*Bombycilla garrulus*) maintained weight for 10 to 18 days on a diet consisting solely of berries (Berthold 1976b). In another study, the heavily frugivorous *Phainopepla nitens* maintained weight on a diet of mistletoe berries, but the house finch (*Carpodacus mexicanus*) did not (Walsberg 1975). These studies suggest that there are important constraints on the digestive physiology of birds dependent on fruits for the majority of their calorie and nutrient requirements. Although many species may exploit fruits on an occasional basis, dependence on fruits for longer periods of time may not be so easily accomplished.

For adults of many frugivores, protein and calorie levels of many fruits are adequate providing that sufficient quantity can be processed daily. We have kept tropical birds of families Emberizidae (tanagers and finches), Pipridae, Trogonidae, and Ramphastidae, and temperate birds in the Turdinae, Bombycillidae, and Mimidae on a synthetic diet for several months during which time their weights stayed close to or above capture weight. The content of this diet was richer than that of many tropical fruits: 86 percent water; 13 percent protein; 6 percent lipid; 78 percent carbohydrate by dry weight. Consistent with Walsberg's study, the three exceptions that required an additional protein supplement to maintain weight were all seed-eating finches—*Arremon aurantiirostris, Caryothraustes poliogaster,* and *Passerina* (*Cyanocompsa*) *cyanoides* (Moermond and Denslow, unpubl. data; D. J. Levey, pers. comm.). The other species did well on the diet and, although we do not know whether they could have reproduced, they were able to molt. Caged temperate frugivores maintained solely on our synthetic diet ingested between 1.8 and 2.2 g (fresh weight) of food per gram of body weight per day in comparison to 0.2 g food per gram of body weight per day in the case of the granivorous *Spizella* (Martin 1968). A 30 g frugivore must consume approximately 60 g of our artificial diet daily—the volume equivalent to 171 *Ardisia revoluta* fruits (6.0 mm diameter) or 20 *Cinnamomum laubautii* fruits (23.3 × 13.2 mm), including their seeds (data from Foster 1978). In caloric equivalents the numbers of fruits are 287 and 12.2 fruits respectively. A similar pattern has been found for a frugivorous neotropical bat (Morrison 1980).

It is apparent from these estimates that processing time (as opposed to searching or handling time) is likely to be an important component of the foraging equation for frugivorous birds. For birds handling large-seeded fruits with firmly attached pulp (e.g., Lauraceae and Palmae), separation of the pulp from the seed may be an important limitation to processing rate.

Resplendent Quetzals (*Pharomachrus mocinno*) may take an hour to regurgitate the seed from a single large *Beilschmiedia* (Lauraceae) fruit (N. T. Wheelwright, pers. comm.), and Collared Aracaris (*Pteroglossus torquatus*) may regurgitate a single *Ocotea* (Lauraceae) fruit and then reswallow it three times over a period of almost an hour before the pulp is removed from the seed (E. Santana C., pers. comm.). For birds feeding on small-seeded, sugary fruits, physiological specializations that permit rapid absorption of nutrients from dilute aqueous solution are likely to be more important. Studies on starlings (*Sturnus vulgaris*) suggest that unusually high rates of kidney function may be one component of the physiological adaptations of frugivores to their diets (B. Wentworth, pers. comm.).

Digestive adaptations for efficiently processing watery, sweet fruits may constrain the facility with which some frugivores are able to incorporate dry or oily fruits into their diets and at least partially explain the general preference of frugivores for ripe over partially ripe fruits (Foster 1977; Moermond and Denslow 1983). Preliminary data suggest that the watery texture of fully ripe fruits allows them to be more efficiently assimilated although unripe fruits may also contain higher levels of digestion inhibitors such as tannins. Caged Collared Araçaris thoroughly digested nearly all fully ripe *Hamelia patens* (Rubiaceae) fruits offered but digested a significantly lower proportion of partially ripe fruits (Santana C., Moermond, and Denslow, unpubl. data).

Differences in physiological treatments of proteins and carbohydrates impose further constraints on the nutrition of frugivores. Assimilated protein can be converted to fat, stored as protein in muscles (Fogden and Fogden 1979), or burned as an immediate source of energy. Nitrogen in excess of that required for body maintenance, feather production, or reproduction can be excreted. Assimilated lipid and carbohydrate may be burned for energy or stored as fat or glycogen but cannot be excreted. Unless frugivores are unusual in possessing some mechanism by which protein can be differentially assimilated relative to carbohydrate, protein and carbohydrate are probably assimilated with similar efficiencies (about 75%, White 1974).

For these reasons, the protein : calorie ratio (here reported as percent protein : Kcal/100 g food, White 1974) is an important indication of the quality of the diet. The few bird species for which dietary requirements have been determined appear to do best on dietary protein : calorie ratios of approximately .02 to .05 for adults (.039 for laying quail, Nestler et al. 1944; .038 to .052 for laying and breeding chickens, National Research Council 1960; .028 for *Spizella*, Martin 1968). Protein : calorie ratios have been calculated for several tropical fruits (Table 1). Some are considerably lower than the predicted dietary requirements, e.g., lipid-rich Myristicaceae .004; at best the protein : calorie requirements of adult birds can be met only minimally by a random selection of tropical fruits if frugivores have similar protein assimilation and utilization efficiencies as granivores.

The problem of lack of protein is compounded for nestlings and for reproductive and molting adults, which require protein for both maintenance and for new tissue production. During times in which protein demands are high, calories are also likely to be in excess of metabolic needs (Foster 1978). Thus, birds feeding predominantly on fruits must possess some mechanism to deal with excess calories in their diets. For example, young Oilbirds deposit large quantities of fat before leaving the nest even though dietary protein is so low as to prolong the fledgling period much beyond that usual for birds of that size (Snow 1961, 1962c; White 1974). Other neotropical frugivores (e.g., Cotingidae, Pipridae, Thraupinae) do not appear to store large quantities of oil and duration of the nestling period appears to be no longer (on average) than for non-frugivores (Ricklefs 1974).

Walsberg (1975) suggested that frugivores may be less efficient at assimilating calories than granivores, or, alternatively, that the daily energy budgets of birds feeding mainly on fruits may be higher than those for birds supplementing their diets with insects. This appears to be the pattern in bats (D. W. Thomas, pers. comm.), and White (1974) presented evidence that the efficiency of conversion of protein to tissue is about 63 percent in Oilbird nestlings in contrast to 6.8 percent for the insectivore nestlings he studied. Protein is used to meet only 14.2 percent of the metabolic needs of Oilbird nestlings in comparison to 75.5 percent of the needs of insectivorous nestlings. For most nestlings and many adult frugivorous birds, some protein is supplied from insects (Skutch 1954, 1960, 1969; Snow 1962a, b; Snow and Snow 1971; Morton 1980), and fruits are then used primarily as a source of calories. Thus, most birds that are highly or totally frugivorous as adults feed insects to very young nestlings; the proportion of fruits in the diet increases as the nestlings mature, and the ratio of protein to calories metabolized approaches that of adult birds (Morton 1973; White 1974; Skutch 1976;

Wheelwright 1983). White (1974) noted that the ratios of protein to calories metabolized by nestlings of five tropical species were all initially higher than that obtainable from fruits alone, but that this ratio was lower for birds feeding both fruits and insects to nestlings (*Elaenia flavogaster, Turdus grayi, Zarhynchus wagleri*) than for a total insectivore (*Iridoprocne albilinea*). Oilbirds only bring fruit to nestlings, but the nestlings receive food from the parents with a higher protein:calorie ratio than fruits provide. White (1974) suggested that adults provision the young nestlings with a high protein regurgitated liquid (Snow 1961) analagous to that produced by pigeons.

Nevertheless, for many species, fruits are, at times, an important energy source. In this sense even those species that feed primarily on insects may be obligate frugivores, because fruits provide critical calories, perhaps freeing insect proteins for egg or feather production. The demand for fruits is particularly noticeable during seasons in which few ripe fruits are available. During these times, birds may take green (unripe) fruits (Leck 1972b), greatly increasing the foraging time necessary to acquire sufficient food (Foster 1974, 1977). In a Costa Rican rain forest, we have recorded 15 species of birds eating green or partially ripe fruits during one season of particularly low fruit availability (Moermond and Denslow, unpubl. data).

Reproduction in an adult bird often depends on the level of protein in the diet (Ward 1969a; Fogden 1972; Crome 1975a, b; Fogden and Fogden 1979). Among tropical frugivores, breeding and molting seasons may be closely tied to the quality and abundance of the resource base both for protein and for energetic requirements. In Trinidad the breeding seasons of Oilbirds (Snow 1961) and Bearded Bellbirds (Snow 1970) coincide with the fruiting seasons of Lauraceae, Burseraceae, and Palmae (all of which produce particularly nutritious fruits). In Singapore and Sarawak, Yellow-vented Bulbuls (*Pycnonotus goiavier*) breed during seasons of fruit and insect abundance (Ward 1969b; Fogden 1972).

Fogden (1972) suggested that frugivores may not be sufficiently good at hunting insects to meet both daily metabolic protein requirements and to store sufficient protein for reproduction or feather production on a totally insect diet. Reproduction and molting are, therefore, closely dependent on energy obtained from fruits. Under such circumstances, all protein acquired from insects can then be used to meet protein rather than energetic demands. This thesis is supported in the breeding of *Phainopepla* in the deserts of the southwestern United States. Nesting coincides with abundance of both insects and mistletoe fruits. If either resource fails, nesting is abandoned or not initiated (Walsberg 1977). The breeding seasons of the Black-and-white and the Golden-headed Manakins (*M. manacus* and *Pipra erythrocephala*) in Trinidad coincide with fruiting peaks of Melastomataceae and Rubiaceae, neither of which have notable concentrations of either proteins or lipids (Snow 1962a, b, 1965).

MORPHOLOGY OF THE DIGESTIVE TRACT

We expect that the digestive tract of primarily frugivorous birds should reflect a diet of bulky, watery, non-fibrous foods that are generally low in protein, high in soluble carbohydrate, and variable in lipid. As noted above, it would be advantageous for frugivores to process fruit pulp quickly, absorbing only the easily available nutrients from large volumes of fruits. Thus, we expect to find little provision for food storage, short intestines (Pulliainen et al. 1981), non-muscular gizzards (because fruits require little mechanical reduction), and large livers relative to body size (detoxification of the secondary compounds frequently present in fruits (Herrera 1982) occurs primarily in the liver in vertebrates (Brattsten 1979)).

Comparative data on the morphology of the digestive apparatus of predominantly frugivorous birds are not extensive. Moreover, they are complicated by seasonal variation in diets (especially among temperate frugivores), differences in fruit and seed handling behavior, and the nature of alternative sources of protein (e.g., seeds, insects, vertebrates). As Herrera (1984) noted, functional and behavior modifications may be more important components of the evolution of frugivory than less-flexible morphological adaptations. Although some generalizations are apparent, there is considerable variation in the morphology of the digestive tract among fruit-eaters, and some of the generalizations suggested above are not supported. We must await considerably larger sample sizes that take into account diverse feeding ecologies of frugivores.

In most species there appears to be little provision for short-term storage of fruits (but see Ziswiler and Farner 1972). The crop (if present), esophagus, and/or proventriculus are small

(Wetmore 1914; Cadow 1933; Jenkins 1969 and papers described therein; Walsberg 1975). However, the proventriculus is enlarged in *Euphonia,* and the esophagus is extensible in swallow-tanagers (Schaefer 1953). Wheelwright (1983) also reported the presence of an expandable proventriculus in Resplendent Quetzals that presumably accommodates large lauraceous fruits. These birds have no crop, and the esophagus is not used to store food.

The soft, non-fibrous nature of fruit pulp is reflected by the non-muscular gizzard in many frugivores (Cadow 1933; Jenkins 1969 and papers described therein; Walsberg 1975). The quetzal has rings of muscles around the esophagus, presumably to facilitate regurgitation of large seeds, and also a muscular gizzard (Wheelwright 1983). Herrera (1984) found no difference in gizzard size among seed dispersers (all Muscicapidae) and non-frugivores in a Mediterranean woodland. In addition, Herrera (1984) reported no morphological differences between seed dispersers and non-frugivores with respect to liver size or intestine length in relation to body size.

Other studies also have failed to find a consistent pattern of intestine length in relation to body weight of frugivores. Although some authors have suggested that the intestine length in frugivores is short (e.g., Desselberger 1931; Docters van Leeuwen 1954; Ziswiler and Farner 1972; Walsberg 1975), waxwings apparently have long intestines (Cvitanic 1970; but see Walsberg 1977), and Jenkins (1969) reported no consistent pattern among species of Costa Rican frugivores he examined. Intestinal caecae are not common in frugivores, presumably reflecting the non-fibrous nature of their food; however, Wheelwright (1983) described paired caecal sacs packed with fruit skins in the Resplendent Quetzal.

It is apparent that some of the most important adaptations to a fruit-eating habit involve a digestive physiology that can accommodate the processing of a bulky, dilute food that is relatively high in calories in comparison to protein. To some degree these may be reflected in the morphology of the digestive tract, but this structure is also likely to retain elements that maintain the capability of processing other foods and that reflect the phylogenetic affinities of the frugivores, obscuring patterns associated primarily with frugivory. Moreover, fruits do not represent a homogeneous food resource. Important differences among them in source of calories (primarily lipid or carbohydrate), seed number and size, and the firmness with which the pulp is attached to the seed are likely to be reflected in the morphology of the digestive tract and the digestive physiology of birds feeding on fruits.

DIET BREADTH AND FRUIT SELECTION

With a few notable exceptions tropical frugivores exploit many species of fruits (Snow 1981; Stiles, in press; Wheelwright et al. 1984). Three species of Costa Rican saltators together included 189 fruit species in their diets (Jenkins 1969). Worthington (1982) recorded 38 species in the diet of *Manacus vitellinus* on Orchid Island, Panama, Snow (1962a, b) observed 66 and 43 species in the diets of two species of manakins in Trinidad, and N. T. Wheelwright (pers. comm.) recorded more than 96 species of fruits in the diets of Emerald Toucanets (*Aulacorhynchus prasinus*). Even Oilbirds, which take few if any insects, use fruits from 37 species of plants during their nesting season (Snow 1962c).

Such data, together with observations of many diverse bird species exploiting abundant crops of some fruiting trees, have been the basis for the hypothesis that some frugivores, especially the small birds that feed on small, berry-like fruits, are not highly selective but choose fruits opportunistically as they are encountered (Howe and Estabrook 1977; Fleming 1979). The hypothesis is supported by studies such as those of Sorensen (1981), who was unable to find a basis for selection of fruits by British thrushes and titmice, and Worthington (1982, pers. comm.) who concluded that manakins likely take fruits randomly as they are encountered.

The notion that some of these species are specialists on a few fruits while others are indiscriminate generalists on many fruits derives in part from the great abundance and diversity of small, sweet, watery, low quality fruits in comparison to the relatively fewer large, lipid-rich fruits. The apparent richness of the large fruits and their sheer size, which restricts the number of bird species that are able to eat them, offer a plausible basis for assuming that such fruits could provide a staple diet to the few bird species able to eat them and hence allow diet specialization (Snow 1971a; McKey 1975).

However, data on diets of a range of small and large frugivores reveal no particular evidence for marked differences in fruit specialization at the level of the plant family (Table 2). Fruits from only three or four families make up large percentages of the diets of all birds studied.

TABLE 2
THE THREE OR FOUR DOMINANT PLANT FAMILIES IN THE DIETS OF SOME NEOTROPICAL FRUIT-EATING BIRDS

	Body weight (g)	No. species in diet	% diet comprising 3–4 dominant families	Aquifoliaceae	Araceae	Araliaceae	Bromeliaceae	Burseraceae	Cactaceae	Caprifoliaceae	Chrysobalanaceae	Dilleniaceae	Euphorbiaceae	Flacourtiaceae	Guttiferae	Lauraceae	Loranthaceae	Malvaceae	Melastomataceae	Meliaceae	Moraceae	Palmae	Rubiaceae	Sapindaceae	Solanaceae	Theaceae	Ulmaceae	Sources[1]
Pipra erythrocephala	13.0	43	79			X							X						X								X	2
P. mentalis	14.4	29	65																X					X			X	4
Manacus manacus	17.0	66	61										X								X		X					3
M. vitellinus	17.6	38	59		X							X							X				X		X			5
Chiroxiphia linearis	21	37	51							X				X											X		X	1
Cyanerpes caeruleus	12.6	12	70										X		X												X	1
Dacnis cayana	14.1	26	89		X							X	X						X									1
Cyanerpes cyaneus	14.2	12	68				X																X					1
Euphonia violacea	14.7	19	62						X								X											1
Chlorophanes spiza	18.2	14	78										X						X								X	1
Tangara guttata	18.4	14	89										X						X								X	1
T. gyrola	20.7	32	76	X															X		X						X	1
T. mexicana	20.9	26	62				X												X		X						X	1
Ramphocelus carbo	28.5	39	79				X												X		X						X	1
Tachyphonus rufus	36.2	31	68		X														X		X						X	1
Thraupis episcopus	37.1	23	61			X													X		X							1
T. palmarum	38.6	24	68			X								X		X					X							5
Turdus plebejus	95	44	50[2]													X		X			X							5
Aulacorhynchus prasinus	162	96	70[2]			X					X					X				X					X			6
Querula purpurata	100	8[3]	90			X		X											X									7
Procnias averano	120	41[3]	62			X		X											X	X								8
Rupicola rupicola	140	26[3]	74											X		X										X		5
Procnias tricarunculata	210	29	82			X		X																				9
Perissocephalus tricolor	340	37[3]	82											X		X					X	X						5
Pharomachrus mocinno	206	43	99													X						X						10
Steatornis caripensis	415	37	99					X								X						X						

[1] Sources: 1 = Snow and Snow 1971; 2 = Snow 1962b; 3 = Snow 1962a; 4 = Worthington 1982; 5 = Wheelwright et al. 1984, Wheelwright, pers. comm.; 6 = Snow 1971b; 7 = Snow 1970; 8 = Snow 1971c; 9 = Snow 1972; 10 = Snow 1962c.

[2] Estimates only (N. T. Wheelwright, pers. comm.).

[3] Species recorded during breeding season only.

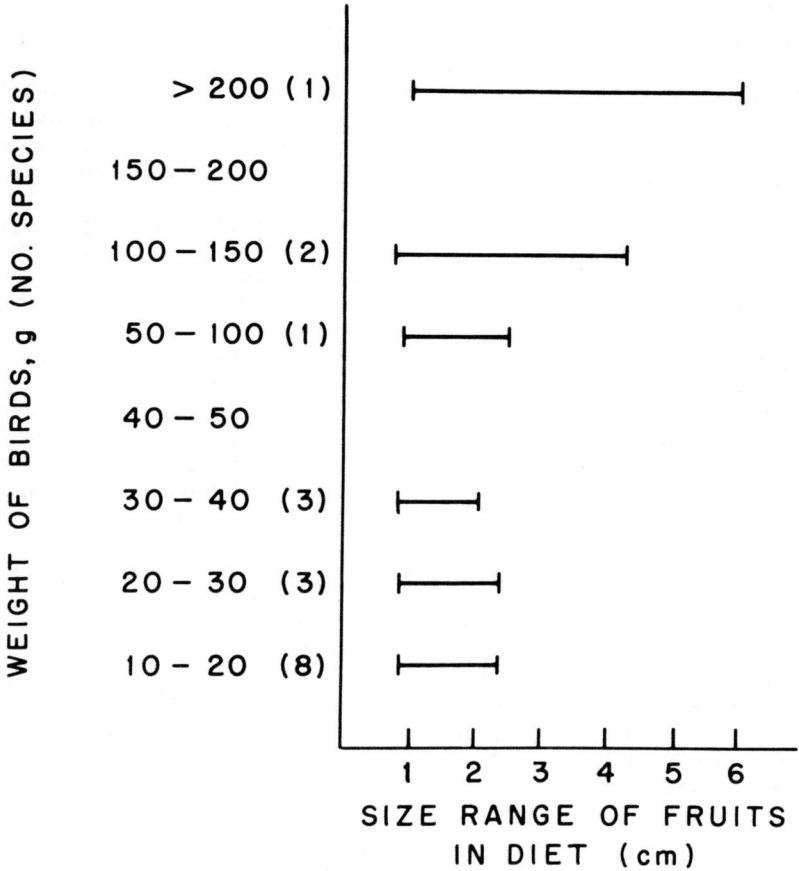

FIG. 1. Fruit sizes in the diets of neotropical frugivorous birds grouped according to body weight. Data summarized from sources listed in Table 2.

Although some large frugivores are closely dependent on a few species of fruiting trees, most take a diversity of fruits and concentrate on three or four plant families.

Even dependence at the family level is not rigid, as can be seen by comparing the diets of two congeneric pairs of manakins (considered specialists by Snow 1981). *Pipra mentalis* and *P. erythrocephala* are morphologically and ecologically very similar, and *Manacus manacus* and *M. vitellinus* are so similar that Snow (1979) described them as subspecies of *M. manacus*. On the large island of Trinidad, berries of the diverse family Melastomataceae figure prominantly in the diets of both species (*P. erythrocephala*—63 percent of fruits eaten; *M. manacus*—47 percent, Snow 1962a, b). On the very small Orchid Island, melastomes are much less diverse and apparently less common (Worthington 1982). In any case they are not particularly important to the two manakins there, both of which, however, include a high proportion of fruits of epiphytes in the family Araceae in their diets (*P. mentalis*—22 percent, *M. vitellinus*—18 percent).

Small birds are restricted in the diameters of fruits they are able to swallow by the size of their gapes. Data on the sizes of fruits in the diets of large birds shows that they take both large and small fruits, including the small fruits taken by small species (Fig. 1; Wheelwright, in press). Although a number of large species do feed extensively on the large fruits of lauraceous plants (Snow 1981), many of these take a variety of other fruits (Table 2; Wheelwright 1983; Wheelwright et al. 1984). Even when we consider cases where the degree of dependence between the bird and plant is assumed to be very strong, as with mistletoes and their dispersers (e.g., *Dicaeum* spp., Docters Van Leeuwen 1954; Salomonsen 1964; Euphonias, Snow 1981; *Phainopepla*, Walsberg 1975; *Zimmerius* (*Tyranniscus*), G. Stiles, pers. comm.; pers. observ.;

Ampelion (*Zaratornis*) *stresemanni,* Parker 1981), we find that the bird species span a wide range of size and families and that none of the birds feeds exclusively on the mistletoe berries.

With respect to the fruits for which data are available, both tanagers and manakins feed mainly on small, juicy fruits that are rich in carbohydrate but relatively low in lipid (Tables 1, 2). Larger birds (5 species of cotingas, a trogon, the Oilbird, and a toucanet) include both lipid rich and carbohydrate rich fruits in their diets (Tables 1, 2). But, with the exception of the Oilbird, median protein contents of fruits in families in the diets of frugivores were not different from the median for all fruits analyzed.

Also, no evidence suggests that highly frugivorous birds provide higher quality dispersal or that plants and their dispersers are tightly co-evolved (Wheelwright and Orians 1982). The same tree species is likely to be visited by different frugivores in different parts of its range (Howe and Primak 1975; Howe 1977), and frugivores may exploit different plant families in different parts of their ranges (Snow 1962a, b; Worthington 1982).

With regard to the selectivity of fruit-eating birds, experimental work with caged birds indicates strong feeding preferences among a wide variety of fruit-eating birds. Although all of the bases for these choices are not yet known, some patterns are becoming clear. Costa Rican tanagers, saltators, and manakins each show hierarchical preferences among fruits of different species, ripeness, size, and accessibility (Moermond and Denslow 1983). Preference for fruits with large pulp/seed ratios have been demonstrated for toucans (Howe and Vande Kerckhove 1980), *Sylvia atricapilla* (Herrera 1981), and araçaris and trogons (Santana C., Moermond, and Denslow, unpubl. data).

Responses of the birds to particular characteristics of the fruits themselves are not too surprising, but experimental aviary and field studies have also shown that the accessibility of fruits strongly influences fruit choice (Denslow and Moermond 1982, in press; Moermond and Denslow 1983). Numerous observations of responses of birds in the field to fruits differing in accessibility (e.g., D. W. Snow 1971b; Snow and Snow 1971; Howe 1977; B. K. Snow 1977; Simms 1978; Kantak 1979; Sorensen 1981; Denslow and Moermond 1982, in press) under-score the likely importance of this aspect of fruit selection. This may be particularly critical with regard to adaptations for frugivory, since it is the single characteristic of fruits most likely to influence the morphology of the bird.

BEHAVIOR AND MORPHOLOGY

METHODS OF HANDLING FRUIT

Fruits may be swallowed whole, eaten piecemeal, or mashed. Many birds actively handled fruits in their bills, rotating or moving the fruit with rapid movements of the mandibles. We refer to this action as mandibulation to distinguish this action on the item from the effect on the item (cf. Kear 1962; Snow and Snow 1971; Moermond 1983). Cases in which a fruit is mandibulated to such an extent that it is partially flattened or crushed we have termed "mashing" (Moermond 1983). When fruits are swallowed whole, the amount of mandibulation is usually minimal. Eating fruits in piecemeal fashion refers to cases in which birds bite or tear our chunks of pulp. These differences in fruit handling are potentially of great importance to the plants and are likely to have differing consequences to the birds' feeding and digestion rates.

The size of fruit swallowed by a bird is limited by the breadth of its gape, where fruit is defined as the unit of seed(s) and surrounding pulp that is ingested (Snow 1973; Wheelwright, in press). Small birds such as *Manacus* spp. can only handle fruits smaller than about 15 mm diameter; seeds of large fruits, such as those of many Lauraceae and Palmae, can be dispersed only by birds that have large gapes such as the larger cotingas and a number of the non-passeriform frugivores. Many fruits, however, consist of removable units containing one or a few seeds with the surrounding pulp within a woody capsule (e.g., *Guarea* spp., Meliaceae; *Glusia* spp., Guttiferae). As with berries and drupes, the frugivore is limited by the diameter of the unit to be swallowed. Seeds may be either defecated if small or regurgitated if large (Snow 1971a; Worthington 1982; Moermond and Denslow, unpubl. data). In both cases the seed is carried within the bird from several minutes to an hour or more before being dropped (pers. observ.; D. J. Levey, pers. comm.).

Mashing of fruits appears to be nearly completely restricted to the tanagers and emberizid finches (Moermond 1983). Other birds, such as manakins, certainly mandibulate their fruits to some extent, but such actions appear to serve primarily for positioning the fruit in preparation for swallowing and only on rare occasion is the fruit actually crushed (Moermond and Denslow, unpubl. data). All 20 species of tanagers we have observed crush the fruit pulp, often thoroughly, squeezing out and swallowing much of the fruit pulp and juice. In the case of very small fruits (<3 mm diameter), the mashing may be very rapid (<1 sec), and Snow and Snow (1971) reported that some species of *Tangara* swallow small fruits whole without noticeable mandibulation. Moermond (1983) has suggested that the "suction-drinking" capability of tanagers may be directly related to efficient swallowing of juice crushed out of watery fruits. In addition, large seeds, as well as the tough outer skin of some fruits, are frequently separated from the pulp and dropped during the mashing process. This reduces the bulk of indigestible material to be processed but also results in little dispersal of the seeds of fruits with even moderately large seeds (>3.5 mm diam., Moermond and Denslow, unpubl. data). Whether or not a seed is dropped depends not only on the size of the seed and the size of the bill but also on the bill shape and the handling technique of the bird. This was recently demonstrated in systematic experiments with captive birds of five tanager species and four other frugivorous species in Costa Rica (D. J. Levey, unpubl. data).

Small seeds, such as those found in *Phytolacca* (Phytolaccaceae), many *Psychotria*, and Melastomataceae, are usually swallowed whole by tanagers along with the crushed pulp and the juices. In some epiphytic plant species with small fruits, small seeds, and a tough fruit coat such as *Rhipsalis* (Cactaceae) and many species of *Anthurium* (Araceae), the fruit coat is nearly always dropped by tanagers, but the seeds are forced out of the fruit with the first drops of pulp and juice as the fruits are mashed and all are swallowed together before the seed coat is dropped (Moermond and Denslow, unpubl. data). This is likely an adaptation of the plant to insure ingestion of the seeds by fruit mashers such as *Euphonia* tanagers that are often the principal consumers of these fruits.

Eating fruits piecemeal is not the speciality of any particular species of frugivore but rather appears to be a facultative response to fruits that are large relative to the bird's gape. Birds that have strong bills such as toucans, barbets, tanagers, and finches are the most likely to eat piecemeal fruits too large to be easily swallowed whole (Snow and Snow 1971; Skutch 1981; pers. observ.); however, even relatively weak-billed birds such as the manakins occasionally eat pieces out of large soft fruits such as those of *Coussarea* (Rubiaceae; pers. observ.; D. J. Levey, pers. comm.) and *Henriettea* (Melastomataceae; Snow 1962b) or out of catkins or compound fruits (Snow 1962a). The ability of birds to eat fruits piecemeal increases the size range of exploitable fruits beyond those that are swallowed whole.

The fate of seeds from fruits eaten piecemeal depends on how and where they are eaten. Large cultivated fruits such as papaya (*Carica papaya*), banana (*Musa sapiens*), and the Pejiballe palm (*Bactris gasipaes*) are regularly eaten in piecemeal fashion. Pejiballe fruits attract many species of birds and mammals, but among the birds, only a few very large species, such as oropendolas (*Gymnostinops montezuma*), are actually able to carry the intact fruit from the tree (pers. observ.). Other fruits, such as *Cecropia* (Moraceae), *Clusia, Drymonia* (Gesneriaceae), *Piper* (Piperaceae), *Stemmadenia* (Apocynaceae), and *Xylopia* (Annonaceae), appear to be adapted to being eaten piecemeal in that the seeds are small and imbedded in the matrix of pulp which is eaten.

In other cases fruits eaten piecemeal are first carried away from the plant, and the seeds thus dispersed to some degree, e.g., *Tersina viridis* eating large fruits (Schaefer 1953), *Trogon massena* eating large lauraceous fruits in captivity (Santana C., Moermond, and Denslow, unpubl. data), *Ramphastos sulfuratus* and *Pteroglossus torquatus* eating *Swartzia cubensis* fruits, and *Euphonia gouldi* eating *Asterogyne martiana* fruits (pers. observ.). Whether or not the seeds of fruits eaten piecemeal are dispersed effectively bears on the coevolution between the plants and their dispersers and hence may influence the nature of food resources available to fruit-eating birds.

BILL MORPHOLOGY

Given the diverse set of species that regularly take fruits, the absence of common morphological adaptations for fruit-eating *per se* is not surprising (see Karr and James 1975; Ricklefs 1977). Even bills, the structures used directly for "capturing" and handling fruits, exhibit no common pattern, as can be seen in part by examining the diversity of bill shapes among tanagers (Storer 1969) and frugivorous tyrannids (Traylor and Fitzpatrick 1982). Nevertheless, if differences in the methods of taking and handling fruits are considered, a few general adaptive features of bill size and shape can be seen. A statistical comparison of a large set of temperate

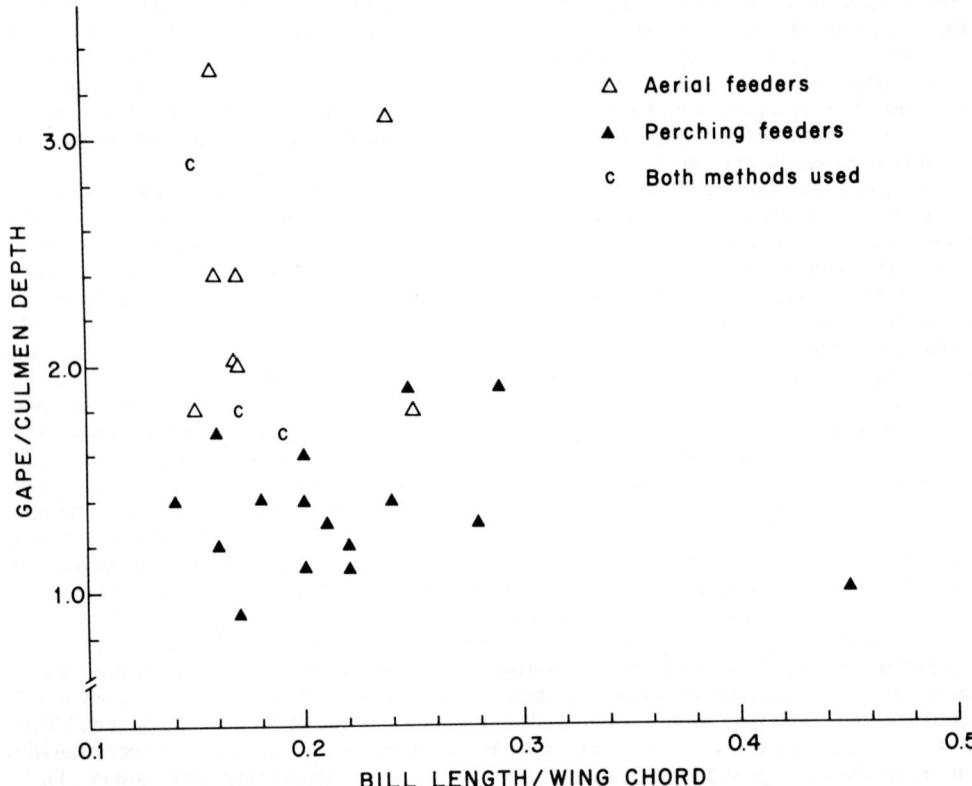

Fig. 2. Variation in bill size and shape associated with different methods of removing fruits. Gape/culmen depth is used to separate wide flat bills from narrow deep ones. Measurements are from skins in the University of Wisconsin Zoology Museum representing 28 species in 11 families or subfamilies of neotropical frugivorous birds. Culmen depth was taken at the base of the nares and bill length from the juncture with the skull.

species feeding on fruits in southern Spain demonstrated that those species (all muscicapids) that were actual seed dispersers (swallowing fruit whole) had flatter and wider bills than species that were seed and pulp predators (Herrera 1984). We compared bill shape to method of capturing fruits by plotting the ratio of gape width to culmen depth against relative bill size (bill length/wing chord) for a diverse set of species (Fig. 2). Species that predominantly use flight to take fruits have relatively short, wide, flat bills which separate them sharply from those picking or reaching fruit while perched. A similar pattern appears also among insect-eaters (Partridge 1976).

Frugivores from several families that have relatively wide, flat bills commonly take fruits on the wing (see next section). Even within families in which most species predominantly take fruits from perches, those species that take many fruits on the wing again have the widest, flattest bills, e.g., *Tersina viridis* (Thraupinae) and *Myadestes melanops* (Turdinae).

Although a wide bill is in part associated with aerial feeding *per se*, a wide bill and gape are also advantageous for handling and swallowing large fruits. Among African drongos (Dicruridae), a family of birds that take most of their food items on the wing, the species that are most heavily frugivorous have shorter, flatter bills (Karr and James 1975). Similarly, among the flycatchers, many of which take their food on the wing, some of the most heavily frugivorous species (e.g., *Myiozetetes* spp., *Megarhynchus pitangua*, *Myiodynastes* spp., *Legatus leucophaius*) have wider and flatter bills than average (Traylor and Fitzpatrick 1982). Other highly frugivorous flycatchers such as *Mionectes* spp., *Elaenia* spp., and *Zimmerius* (*Tyranniscus*) spp., that feed on relatively small fruits, have short, somewhat terete bills, but nevertheless still have relatively large gapes (F. G. Stiles, pers. comm.). A parallel pattern is found among cotingas, most of which are highly frugivorous and take most food on the wing. Those species such as *Procnias* and *Xipholena* that feed on relatively large fruits have very wide and short bills, whereas those species such as *Capornis* and *Pipreola* that eat relatively small fruits have short but narrower and less specialized bills (Snow 1982). The Sharpbill, *Oxyruncus cristatus* (recently placed in the Cotingidae, Sibley et al. 1984) which has a longer, narrower bill than most cotingas, takes fruits by reaching and hanging (F. G. Stiles, pers. comm.).

The advantage to a frugivore of a large gape is clear since gape size limits the size of fruit that can be swallowed (Snow 1973; Wheelwright, in press), but the bill length and strength may be associated with other aspects of fruit handling. Virtually all the aerial feeders we considered (Fig. 2) swallow fruits whole; however, the type of fruit handling employed by species with longer, deeper bills appears to depend on the presence of other structural features of the bills. For example, although wide bills need not also be long, a long bill may be advantageous for reaching fruits from perches. Woodpeckers, barbets, and toucans, which often swallow fruits whole, have long to extremely long, deep bills, which they use to reach fruits from perches. The toucans, in particular, reach much farther with their long bills than any other frugivore and are so dexterous that they are able to select and pick individual small fruits with the tip of the bill (Santana C., Moermond, and Denslow, unpubl. data).

Tanagers and finches, which also pick most of their fruits from perches, mash most of their fruits. These birds have short to relatively long beaks that are markedly narrower, deeper, and frequently much stronger than those of aerial fruit-eaters. Their bills have well-developed lateral ridges on the horny palate (Beecher 1951), which may facilitate ingestion of crushed pulp and juice as they mash fruits. Some species, such as the *Ramphocelus* tanagers, have expanded ramphothecae, which may act to prevent fruit juices from fouling the feathers (Storer 1969).

Within each of the groups discussed, modifications in bill shape and form may influence the method of fruit handling and the types of fruits taken. Although few appropriate data exist, two within-family comparisons will serve to illustrate consequences of relatively minor differences in bill shape. Within the Trogonidae, both *Trogon massena* and *Pharomachrus mocinno* are highly frugivorous, and both frequently eat large lauraceous fruits (Skutch 1972; Wheelwright 1983). Differences in bill size and strength, however, affect the ways in which the two species handle fruits. The bill of *T. massena* is narrower, deeper, relatively longer, and noticeably heavier than that of *Pharomachrus* (Fig. 3), even though *T. massena* is markedly smaller than the *P. mocinno* (141 g vs 206 g). In captivity, *T. massena* has been observed to eat large lauraceous fruits piecemeal (Santana C., Moermond, and Denslow, unpubl. data), whereas in a detailed field study Wheelwright never observed *P. mocinno* to eat fruits piecemeal (N. T. Wheelwright, pers. comm.). Similarly, among paleotropic fruit-eating pigeons, the strong-billed *Treron* spp. are able to bite chunks out of fruits whereas the weaker billed *Ducula* and *Ptilinopus* spp. swallow fruits whole (Goodwin 1970).

Such examples further reinforce the idea that bill shape and size are likely to influence which fruits can be taken most efficiently. We contend that bill features are more likely to be associated with particular fruit handling behavior than with the frugivorous habit in general.

METHODS OF TAKING FRUITS

Snow and Snow (1971) suggested that fruit can be plucked by birds in only a few different ways. We agree, but suggest that differences among those ways influence which fruits are most accessible and the benefit:cost ratio of taking those fruits. Although few quantitative data are available, we shall offer descriptions of techniques birds use to "capture" fruits and, with a few examples, propose how the capture techniques may influence fruit selection. Examination of both fruit capture methods and their morphological correlates (see next section) provides a basis for understanding adaptive divergence and specialization among frugivores.

Fruits may be taken on the wing or from a perched position (Herrera and Jordano 1981; Denslow and Moermond, in press). A comparison of the use frequencies of these two general techniques for 10 bird species feeding on the same tree (Fig. 4) illustrates a typical pattern: most species predominantly use only one of the two techniques, taking fruits either on the wing or from a perch. Only two of the species in this example, the *Tityra* and the *Catharus* thrush, commonly used both techniques. Despite the strong implication of stereotypy in fruit capture techniques among these species, nearly all of them occasionally used the other technique. If other fruit species were included in the comparison, the versatility in foraging

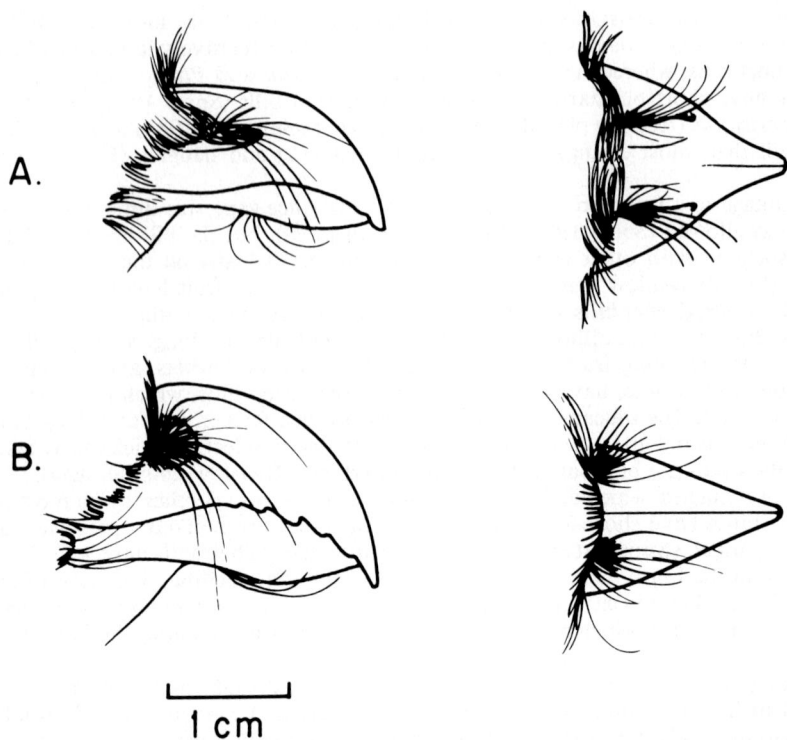

1 cm

Fig. 3.   Bill shapes (lateral [left] and dorsal [right] views) of male (A) *Pharomachrus mocinno* and (B) *Trogon massena.* The bills were drawn from specimens in the University of Wisconsin Zoology Museum.

techniques used by several bird species would be more obvious. For example, even though *Pteroglossus torquatus* rarely takes fruit on the wing, captive *Pteroglossus* were induced to take fruits on the wing repeatedly under exceptional circumstances in which highly desirable food items were inaccessible by any other capture method (Santana C., Moermond, and Denslow, unpubl. data). The *Querula* which clearly took most *Miconia multispicata* fruits on the wing, as is typical for this cotinga, was observed to regularly perch "to pluck the fruits of *Didymopanax morototoni* and *Guarea trichilioides,* both of which bear their fruits in large bunches on strong stalks that afford a foothold" (Snow 1971b:7). *Perissocephalus tricolor* is another larger cotinga that takes most fruits exclusively on the wing, except when feeding on *Didymopanax* fruits which it picks while perched (Snow 1972).

These two fruit capture categories can be further divided to reveal additional functional differences among frugivores. Unfortunately, few studies report fruit capture techniques, and those that do often use general categories that are inconsistent from study to study. The subtle functional differentiation of prey capture methods recently applied to some insect-eaters (Davies and Green 1976; Fitzpatrick 1980) has not yet been developed for fruit-eaters. The following descriptions are based on our own observations of over 80 species of neotropical birds in the field in numerous localities and on experiments with 23 species of six families in captivity.

We have observed four distinct flight maneuvers used to pluck fruit while on the wing: hovering, stalling, swooping, and snatching. A hovering bird pauses in front of the fruit while flapping its wings so as to maintain zero air speed. This method is used commonly by manakins, tyrannid flycatchers, and many small tanagers, especially those that weigh less than 20 g, such as *Tangara* and *Euphonia* spp. A stalling bird pauses briefly in front of the fruit by using a very steep wing attack angle allowing the bird to slow down and stall just in front of or below the fruit. Both hovering and stalling are functionally equivalent to "hover-gleaning" of Fitzpatrick (1980), but the flight motions employed and the morphological features associated

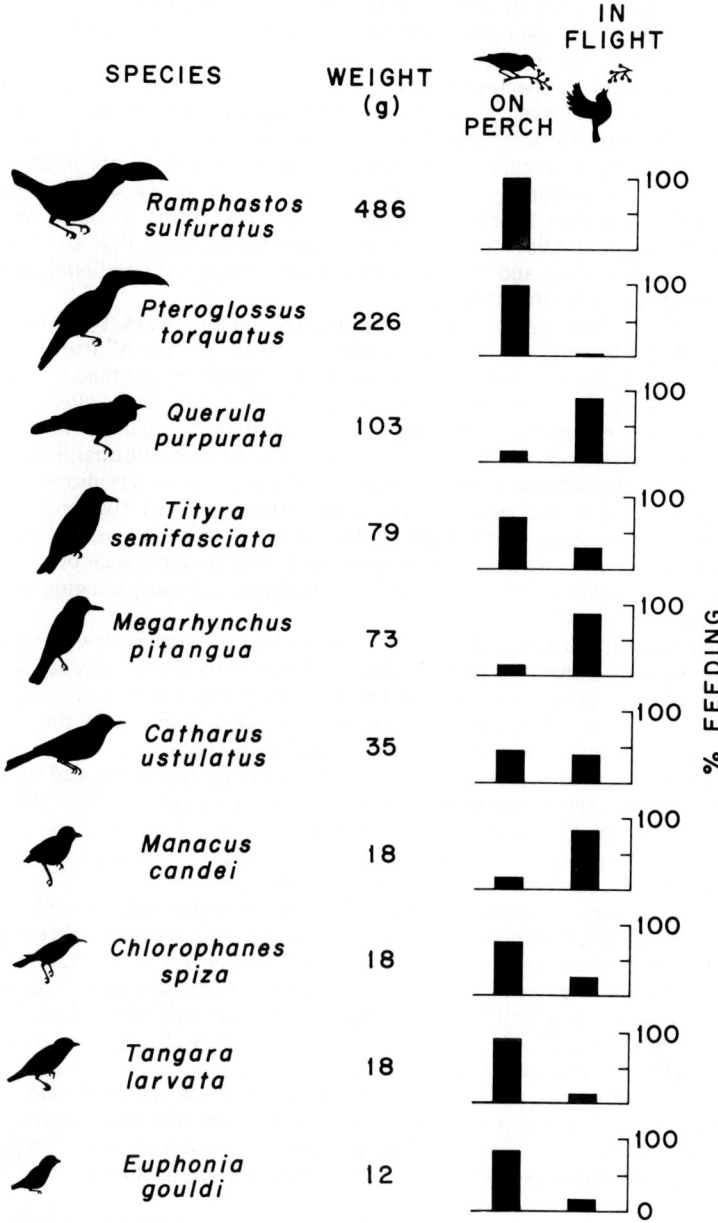

FIG. 4.   Feeding behavior of the most common visitors to *Miconia multispicata* (Melastomataceae). Histograms show percentages of total fruit removals (Denslow and Moermond, in press, with additions).

with each technique differ (see next section). *Trogon massena* was observed and photographed while stalling, and morphological examination of trogon species of three genera (*Trogon, Pharomachrus,* and *Harpactes*) suggest this technique may be generally employed by the Trogonidae and possibly *Phoenicircus* spp. (Cotingidae; Moermond and Santana C., unpubl. data). Swooping and snatching both involve continuous movement past the fruit as it is taken. In swooping, the wings are held out and the bird glides up to the fruit, whereas in snatching, the wings are flapped throughout. In both cases, some forward momentum is conserved in contrast to hovering or stalling in which the bird actually stops at the point of contact with the fruit. We have observed swooping used by several cotingids (*Querula purpurea, Carpodectes nitidus*) and by *Tityra* spp. (*Tityrinae*).

We suspect that nearly all species may at times snatch fruits, but the first three methods are used primarily by those species that use flight to capture most of their food items. The costs and skill of taking fruits by snatching and swooping are likely to differ from those associated with hovering and stalling. Hovering and possibly stalling are likely to be more costly than snatching or swooping (Pennycuick 1975; Rayner 1981), but the total distance flown when capturing a fruit is sometimes shorter for those using hovering vs snatching. Hovering and stalling may allow a more precise approach to fruits that are less accessible from a longer flight. Snatching and swooping may result in more "misses" and possibly more "collisions" with twigs and infructescences.

Birds that take fruits from perches may do so in three different ways, by picking, reaching, or hanging. While birds using all of these techniques "pick" or "reach" fruit in an ordinary sense, we have restricted use of these terms to describe the position assumed by the bird when taking the fruits. "Picking" refers to all cases where the birds take fruits close to their perch without extending their bodies or assuming special positions. "Reaching" refers to cases in which the bird extends its body out or down from the perch. Appropriate qualifiers or adjectives can be applied as necessary to distinguish the degree of reach (e.g., tarsus above or below level of perch, wings fluttered or not, reach horizontal or vertically down). Hanging refers to cases in which the bird's entire body and legs are under the perch with the ventral side up. Picking and reaching are often used by a wide variety of birds, but we have seen only woodpeckers (e.g., *Melanerpes pucherani* and *Campephilus guatemalensis*) regularly hanging below perches when eating fruits.

Although virtually all birds are able to pick fruits close to perches, few are able to reach very far below a perch. Some appear restricted to taking fruits within easy reach of a perch, while others can reach outward or downward close to the limit of their body length (Denslow and Moermond, in press). The ability to use reaching or hanging does not appear to be associated with the ability to take fruits on the wing. Some species that take fruits primarily on the wing, such as trogons, flycatchers, some manakins (e.g., *Manacus*), and some cotingids (e.g., *Querula, Cephalopterus*), are unable to reach well from a perch (Snow 1982; Denslow, Moermond, and Santana C., unpubl. data), but other primarily aerial feeders, such as *Tityra* spp., and several cotingids (e.g., *Procnias, Carpodectes*) regularly reach fruits from a perch (Snow 1977; Santana C. and Milligan 1984; pers. observ.).

Likewise, birds that take most of their fruits from perches also differ considerably in their ability to reach. Neotropical thrushes (Turdinae) and emberizine ground finches (e.g., *Arremon aurantiirostris* and *Arremonops conirostris*) are poor reachers. These species tend to take all their fruits by either picking or snatching (e.g., *Catharus ustulatus* in Fig. 4). Toucans, barbets, and most tanagers can reach well below perches. The tanagers are particularly interesting in this respect, showing considerable variation in reaching ability among similar-sized species. Some tanagers such as *Thraupis palmarum* (36 g) and *Euphonia gouldi* (12 g) are so adept at reaching that they can extend their entire bodies and legs below the perch and recover their original upright position without the aid of their wings (Denslow and Moermond, unpubl. data). Others such as *Habia fuscicauda* and *Ramphocelus passerinii* are relatively poor reachers as tested in aviary experiments (Denslow and Moermond, in press, unpubl. data; D. J. Levey, pers. comm.).

Differences in reaching ability observed in the field were described for tanagers and honeycreepers of Trinidad by Snow and Snow (1971). Their descriptions correspond to our observations on Costa Rican species and show that marked differences in such abilities may be found within the same genus. For example, *Thraupis palmarum* and *Tachyphonus delatrii* demonstrated greater reaching abilities than *Thraupis episcopus* and *Tachyphonus rufus*, respectively. Although such abilities have been little studied, they seem to be associated with some readily identifiable morphological traits (see next section) and, as with different flight techniques, they also likely influence the types of fruits that can be efficiently exploited.

We compared the limits of reach of two small tanagers to that of *Manacus candei* (Fig. 5). All three species were able to pick fruits above them at nearly full body lengths, but only the tanagers were able to reach full body lengths downward. Similar determinations of the distances fruits can be taken from a perch for *Pteroglossus torquatus* and *Trogon massena* demonstrated a similar pattern. *Pteroglossus* was able to reach a full body length below its perch, whereas

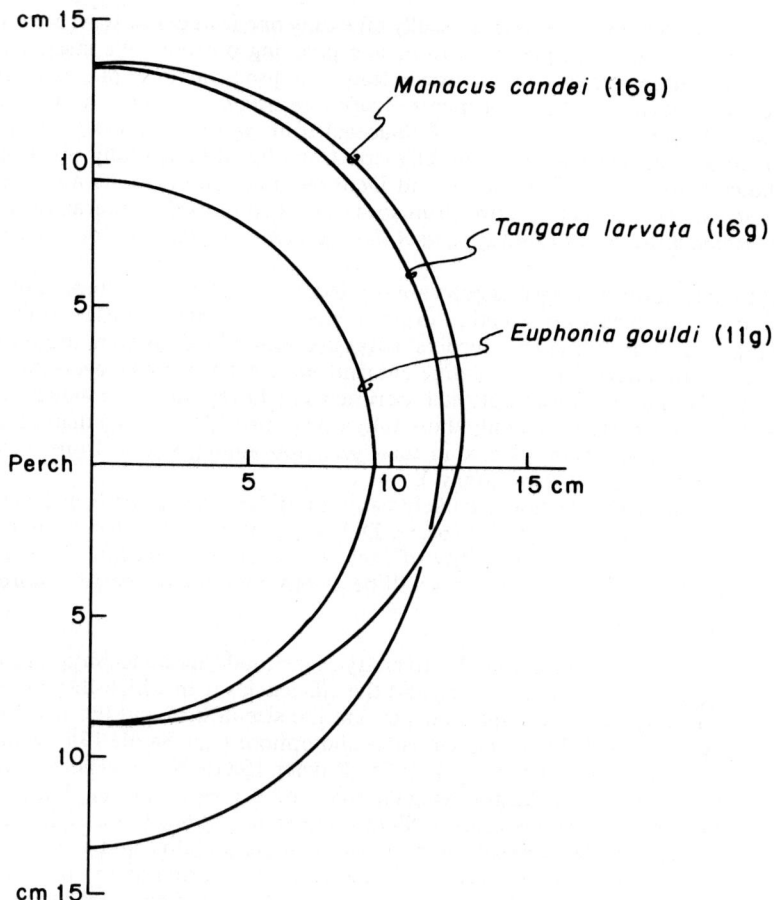

FIG. 5.   Distances of maximum reach in 3 directions from a sturdy perch for 3 small frugivores. *Manacus candei,* an aerial feeder, can reach only as far below the perch as the much smaller *Euphonia gouldi,* a perching feeder.

the trogon was able to pick fruits slightly below the perch only with difficulty and was unable to reach (Santana C., Moermond, and Denslow, unpubl. data). Clearly such differences are likely to restrict the number of fruits that can be taken by the manakin or trogon from a single perch relative to the number obtainable by the tanagers and toucanet. On the other hand, both the manakin and the trogon were adept at taking fruits on the wing and in aviary experiments were shown to select fruits hanging clear from perches (Denslow, Moermond, and Santana C., unpubl. data). In the field both species commonly take hanging fruits that other species take only rarely and with difficulty.

Species using different fruit capture techniques can exploit the same species of fruit (Fig. 4); however, detailed data are likely to show that species using different techniques take fruits from different portions of the tree as has been reported in other studies (Kantak 1979; Herrera and Jordano 1981; Santana C. and Milligan 1984). All fruits in a tree were not equivalent resources contrary to such a suggestion by Leck (1971b); fruits on an infructescence differ with respect to distances to a perch and thus to accessibility and cost of "capture" to different species of birds. Differences in feeding techniques and ability therefore may affect which fruit species are incorporated into a bird's diet.

A comparison of trogons, which rarely take fruits from a perch, to toucans which are rarely observed to take fruits on the wing (Wagner 1944; Skutch 1971, 1972; Bourne 1974; Wheelwright 1983; Santana C. and Milligan 1984), illustrates other possible consequences of differences in feeding technique. Trogons usually take only one fruit per sally whereas toucans frequently reach and take multiple fruits from one perching position. We suggest that the likely greater expenditure per fruit by the trogon leads it to use a narrower range of fruit types compared to toucans. In aviary experiments *Trogon massena* showed a nearly exclusive preference for fully ripe *Hamelia patens* (Rubiaceae) fruit; whereas *Pteroglossus torquatus* readily accepted several stages of less than fully ripe *Hamelia,* although fully ripe fruits were their first choice (Santana C., Moermond, and Denslow, unpubl. data). As a consequence, trogons are also expected to include fewer fruit species in their diets than toucans as has been shown for *Pharomachrus mocinno* (43 spp.) vs *Aulocorhynchus prasinus* (96 spp.; Wheelwright 1983).

When additional appropriate species pairs are studied in sufficient detail, these comparisons may be extended. Our own fruit selection experiments with manakins (aerial fruit feeders) and tanagers (primarily perching fruit feeders) suggested that manakins were more discriminating in choices between paired fruits of different fruit species than were tanagers (Moermond and Denslow 1983), and recently completed experiments by Levey show that manakins accept fewer stages of ripening berries (i.e., only those fully ripe or nearly fully ripe) than do tanagers and that the manakins are more precise in their selection among sets of fruits at different stages of ripeness (D. J. Levey, pers. comm.).

These differences in fruit selection are likely based on differences in the benefit : cost ratio associated with each technique of fruit capture. Differences in the costs depend not only on the technique used but also on the ability of each bird species to execute the maneuvers required to use the technique. Such abilities will be determined in large part by morphology.

LOCOMOTORY MORPHOLOGY

Morphology is assumed to determine the diversity of the food capture techniques employed by a given bird species, their relative costs, and the efficiencies with which they are used.

Wings.—Basic aerodynamic principles dictate that the size of bird and the size and shape of its wings strongly influence flight characteristics and options (e.g., Savile 1957; Pennycuick 1969, 1975; Kokshaysky 1973; Greenewalt 1975; Rayner 1981). A few general predictions can be made with regard to wing length, wing loading, and degree of slotting. For a bird of a given weight, long wings reduce the costs of flight (Pennycuick 1969, 1975; Hails 1979) and aid hovering (Norberg 1979), whereas short wings increase stability and maneuverability (Savile 1957; Kokshaysky 1973; Norberg 1981), increase acceleration due in part as a result of higher wing beat frequency (Kokshaysky 1973; Norberg 1981), and reduce the inertial forces on the wing skeleton (Norberg 1979, 1981). Low wing loading reduces the cost of flight (Pennycuick 1969, 1975; Greenewalt 1975) and, as a consequence, aids hovering and maneuverability (Norberg 1981). Increased slotting of the primaries allows a higher angle of attack of the wing which may increase lift, delay stalling, and increase acceleration (Savile 1957; Brown 1963; Alexander 1968; Kokshaysky 1973). The suggestion that slotting primarily serves to reduce the vortex at the wing tip (Savile 1957; Cone 1968) is not supported (Kokshaysky 1973). All of these predictions are constrained by body weight of the bird. Among geometrically similarly shaped birds, a decrease in size is associated with increases in the margin between power required for flight and the power available (Pennycuick 1969, 1975; Greenewalt 1975) and improvements in most aspects of aerial performance such as rate of acceleration, velocity range, and maneuverability (Norberg 1981; DeJong 1983).

Constraints on flight appear far less critical for birds below about 100 g body weight (Greenewalt 1975; DeJong 1983). Among frugivorous birds heavier than 100 g (Table 3), birds with wing loadings greater than 0.42 g/cm take fruits primarily by reaching, whereas those with lower wing loadings take fruits primarily on the wing. Such a correlation does not hold well for species weighing less than 100 g. Among these forms, some birds that reach well also appear able to hover easily (e.g., several *Tangara* spp. [Denslow and Moermond unpubl. data]), whereas *Manacus* spp., which take virtually all fruits on the wing, have unusually high wing loadings (Table 3, Fig. 6).

Fitzpatrick (1978) showed that large, aerial-feeding tyrannid flycatchers have long and often somewhat slotted wings, while small forest flycatchers have short, rounded wings presumably giving them greater maneuverability. A similar pattern holds for large sally-gleaning antshrikes versus perch-gleaning antshrikes (Schulenberg 1983). However, small frugivores, such as tanagers (*Euphonia* and *Tangara* spp.) and honeycreepers (*Cyanerpes, Dacnis,* and *Chlorophanes* spp.) generally have relatively longer and narrower wings ($\bar{X}$ aspect ratio of 15 species = 1.76)

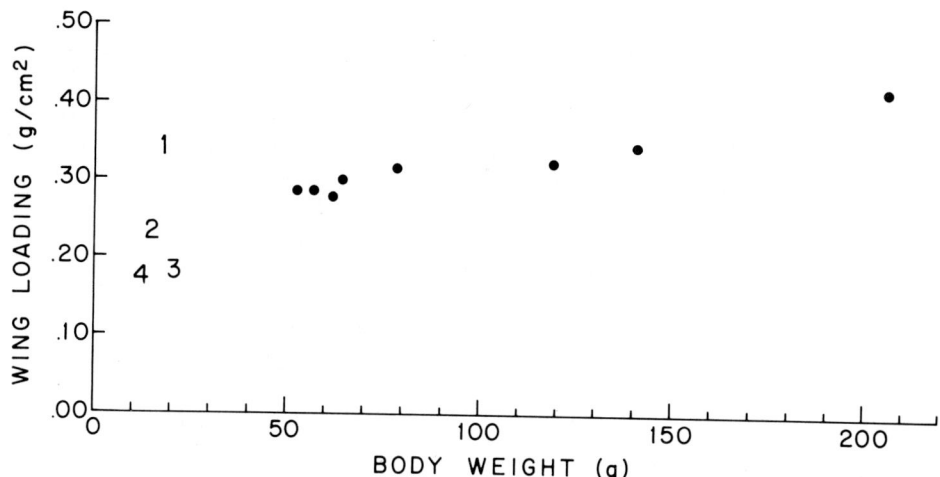

FIG. 6. Wing loading for manakins (numbers) and trogons (closed symbols) of different body weights (data from Hartman 1961); 1 = *Manacus vitellinus*, 2 = *Pipra mentalis*, 3 = *Chiroxiphia lanceolata*, 4 = *Corapipo leucorrhea*.

than small insectivorous tyrannids (19 species, $\bar{X}$ = 1.62) and formicariids (15 species, $\bar{X}$ = 1.46) (P < 0.01, median test; data from Hartman 1961). The short, rounded wings of the insectivores may be important in maneuverability and rapid acceleration in capturing insects. For the tanagers feeding on fruits, the longer wings may enhance their ability to hover for fruits and reduce the flight costs of commuting between fruiting plants. A consequence of longer, narrower wings may be the higher wing loadings of the tanagers compared to similarly sized tyrannids (.16 g/cm vs .22 g/cm; P < 0.01, median test; data from Hartman 1961).

This first order approach gives an unexpected result when applied to frugivorous manakins which take nearly all their food on the wing. The four manakin species we considered (Table 3) have wings with a mean aspect ratio (1.62) similar to that of the small flycatchers discussed above and distinctly lower than the small tanagers, but manakin wing loadings ($\bar{X}$ = .22 g/cm) are higher than that of the flycatchers but equivalent to that of the tanagers. The functional advantage of this type of wing for the manakins is not clear.

Assessments based on single factors such as wing loading or aspect ratio are unlikely to explain satisfactorily foraging differences among similar species (e.g., Partridge 1976, Levey et al. 1984). Wing loading is highly correlated with body weight for some homogeneous groups of birds such as trogons (8 species in 2 genera), but no clear relationship emerges from a similar plot for four species of manakins of four genera (Fig. 6). Similarly, wing length with respect to body weight varies considerably among tyrannids (Fitzpatrick 1978) and cotingids (Snow 1982), although, in general, larger birds have longer wings. A suite of wing characteristics must be considered to understand differences in foraging style among ecologically similar species (Norberg 1979).

A more detailed comparison of the four species of manakins (Fig. 6) reveals distinct differences in aspect ratio and degree of slotting. *Manacus vitellinus,* with the highest wing loading, has a relatively low aspect ratio and a high degree of slotting, which may compensate for the low wing area. *Pipra mentalis,* with intermediate wing loading, has the highest aspect ratio and virtually no slotting similar to the wings of some *Tangara* tanagers with similar wing loadings. *Chiroxiphia lanceolata* has an intermediate wing loading, relatively low aspect ratio, and an intermediate degree of slotting with fewer and shallower slots than *Manacus*. A similar pattern can be found within the Cotingidae. The Purple-throated Fruit-crow (*Querula purpurata*), (91–112 g) in the Guianas has much longer wings (165–195 mm chord) than the similar-sized Guianan Red Cotinga (*Phoenicircus carnifex,* 75–85 g, 90–110 mm wing chord; Snow 1982). Like *Manacus vitellinus, Phoenicircus* has highly slotted wings as compared to the little slotted wings of the fruitcrow (Fig. 7). Regardless of the effect of wing characteristics on courtship displays of these birds, wing morphology should strongly influence feeding behavior. Such parallel patterns in wing loading, slotting, and aspect ratios among divergent

TABLE 3

MORPHOLOGICAL[1] AND BEHAVIORAL CHARACTERISTICS OF SELECTED SPECIES OF NEOTROPICAL
FRUIT-EATING BIRDS

	Body weight (g)	Flight muscles (% body weight)	Leg muscles (% body weight)	Wing loading (g/cm²)	Aspect ratio (span/ chord)	FCM[3]	FHM[4]
Columbidae							
Columba speciosa	259	29.1	5.0	0.64	2.0	P	Pr
C. nigrirostris	144	28.7	7.0	0.46	1.8	P	Pr
Psittacidae							
Amazona autumnalis	416	18.7	8.0	0.60	2.0	P	Pr
Pionis senilis	213	20.8	7.2	0.46	2.2	P	Pr
Steatornithidae							
Steatornis caripensis[2]	415	15.6	—	0.29	3.2	A	S
Trogonidae							
Pharomachrus mocinno	206	22.2	3.4	0.42	1.9	A	S
Trogon massena	141	21.8	3.2	0.34	2.0	A	S
T. curucui	50	22.3	2.7	0.28	2.1	A	S
Momotidae							
Momotus momotus	133	20.5	6.2	0.34	1.7	A	S
Capitonidae							
Eubucco bourcierii	33	13.1	9.4	0.33	1.6	P	S
Ramphastidae							
Aulacorhynchus prasinus	155	144	11.0	0.53	1.7	P	S
Pteroglossus torquatus	226	13	9.9	0.52	1.7	P	S
Ramphastos swainsonii	480	14.2	13.0	0.56	1.6	P	S
Picidae							
Phloeoceastes guatemalensis	240	15.3	11.5	0.41	1.7	P	S
Melanerpes pucherani	54	17.9	7.1	0.28	2.0	P	S
Pipridae							
Pipra mentalis	15	19.9	5.4	0.23	1.7	A	S
Chiroxiphia lanceolata	20	18.4	6.5	0.18	1.6	A	S
Corapipo leucorrhoa	12	19.2	5.2	0.18	1.6	A	S
Manacus vitellinus	18	17.8	10.2	0.34	1.6	A	S
Cotingidae							
Cotinga ridgwayi	57	25.2	6.4	0.31	1.9	A/P	S
Lipaugus unirufus	80	21.1	—	0.29	1.9	A	S
Querula purpurata	103	17.1	5.0	0.15	1.7	A	S
Tyrannidae							
Tityra semifasciata	79	18.4	9.2	0.33	1.7	A/P	S
Tyrannus melancholichus	40	24.4	3.4	0.21	1.9	A	S
Legatus leucophaius	26	22.1	4.3	0.22	2.0	A	S
Myiodynastes maculatus	46	24.7	4.5	0.23	1.8	A	S
Megarhynchus pitangua	71	24.0	3.9	0.31	1.7	A	S
Myiozetetes granadensis	29	22.1	3.6	0.20	1.8	A	S
Elaenia flavogaster	24	19.2	6.0	0.21	1.6	A	S
E. frantzii	20	22.6	5.5	0.18	1.8	A	S
Tyranniscus vilissimus	9	18.9	4.3	0.16	1.7	A	S
Mionectes oleaginea	11	24.5	4.2	0.17	1.5	A	S
Mimidae							
Dumatella carolinensis	39	14.0	8.4	0.25	1.7	P	S
Mimus polyglottos	51	16.0	9.7	0.24	1.7	P	S

TABLE 3
Continued

	Body weight (g)	Flight muscles (% body weight)	Leg muscles (% body weight)	Wing loading (g/cm²)	Aspect ratio (span/chord)	FCM[3]	FHM[4]
Muscicapidae, Turdinae							
Turdus plebejus	87	22.0	7.6	0.30	1.9	P	S
Myadestes melanops	32	18.7	4.8	0.24	1.7	P	S
Catharus ustulatus	33	19.6	5.6	0.24	1.9	A/P	S
Bombycillidae							
Bombycilla cedrorum	33	22.6	4.0	0.29	2.1	P	S
Vireonidae							
Vireo olivaceus	17	18.6	5.9	0.18	1.9	P	S
Emberizidae, Parulinae							
Vermivora peregrina	9	17.1	6.6	0.15	1.9	P	S
Dendrioca pennsylvanica	10	15.2	7.0	0.14	1.9	P	S
Icterinae							
Zarhynchus wagleri	163	17.2	9.4	0.34	2.0	P	S
Cacicus uropygialis	59	15.6	9.9	0.26	1.7	P	S
Icterus galbula	34	18.0	8.1	0.24	1.8	P	S
Thraupinae							
Chlorophanes spiza	18	18.8	8.1	0.23	1.7	P	M
Cyanerpes lucidus	13	17.5	7.0	0.22	1.8	P	M
Dacnis venusta	16	22.1	6.3	0.21	1.8	P	M
Chlorophonia callophrys	25	18.3	7.2	0.30	1.8	P	M
Euphonia luteicapilla	12	18.2	5.2	0.25	1.8	P	M
E. imitans	13	17.6	6.1	0.20	1.7	P	M
Tangara larvata	19	21.5	6.1	0.21	1.7	P	M
T. guttata	20	17.8	9.3	0.26	1.7	P	M
Thraupis episcopus	32	20.9	6.9	0.25	1.8	P	M
T. palmarum	39	20.8	—	0.26	1.6	P	M
Piranga rubra	30	19.1	5.6	0.22	1.9	P	M
P. olivacea	31	20.1	5.7	0.19	2.0	P	M
Tachyphonus luctuosus	15	16.1	6.0	0.19	1.6	P	M
T. rufus	35	17.4	9.9	0.25	1.6	P	M
Ramphocelus dimidiatus	30	15.9	7.4	0.25	1.6	P	M
R. passerinii	32	18.3	8.4	0.30	1.6	P	M
Habia rubica	36	18.8	8.8	0.25	1.5	P	M
H. fuscicauda	39	16.4	8.7	0.22	1.6	P	M
Chlorospingus ophthalmicus	20	15.7	8.9	0.25	1.6	P	M
Cardinalinae							
Saltator maximus	48	18.4	8.2	0.26	1.5	P	M
Passerina (Cyanocompsa) cyanoides	32	19.0	6.8	0.31	1.5	P	M/Pr
Emberizinae							
Arremonops conirostris	40	14.5	13.0	0.34	1.5	P	M/Pr

[1] Data from Hartman 1961 except as noted. Values presented are averages for males and females where given separately by Hartman.
[2] Data for *Steatornis caripensis* from Snow 1961.
[3] FCM = fruit capture mode; A = in air; P = from perch.
[4] FHM = fruit handling mode; S = swallow whole; M = mash; Pr = seed predator.

groups of fruit-eating birds encourage the search for general functional relationships between wing morphology and feeding behavior.

If frugivores can be fit into a functional-morphological space (see James 1982), then we offer three examples from Costa Rica that represent three extremes of that space and that will serve to illustrate the morphological adaptations and behavioral consequences that characterize

Fig. 7.   Wing outlines of (A) *Querula purpurata,* (B) *Trogon massena,* and (C) *Pteroglossus torquatus* drawn to approximately same scale from photographs of birds in flight.

frugivores within that space. The Collared Aracari, the Slaty-tailed Trogon, *Trogon massena,* and the Purple-throated Fruit-crow represent three divergent types of wing morphology and feeding behavior of frugivores (Fig. 7). Because of the marked aerodynamic constraints imposed by increasing size (Pennycuick 1969; Kokshaysky 1973; Greenewalt 1975; Norberg 1981), these relatively large birds (>100 g) may better illustrate the limiting cases, at least as far as the aerial feeders are concerned.

Querula purpurata obtains most of its fruits by swooping, although snatching and picking are used (Snow 1971b; pers. observ.; Fig. 4). *Trogon massena* takes its fruits by stalling (we have never seen it hover); and *Pteroglossus torquatus* obtains nearly all its food by picking and reaching (rarely using snatching) (Santana C., Moermond, and Denslow, unpubl. data; Fig. 4). The morphological differences in the species correspond to differences in feeding behavior. *Querula* has the lowest wing loading of any bird of its size for which data are available (Table 3; Snow 1982: fig. 1). While all three birds have relatively long wings, *Querula* has unusually wide wings mainly due to the broad, long secondaries and shows a relatively small degree of slotting at the wing tips (Fig. 7). This very broad wing allows these birds to make long, relatively slow, swooping flights to pluck fruits, one at a time. By comparison, *Trogon massena* has a narrower wing (higher aspect ratio) with very prominent, deep square slots. This slotting pattern allows the trogon to fly very slowly before stalling. Its flight pattern consists of a rise, followed by an abrupt slowing until it stalls and drops. Fruits are plucked singly at the top of the rise at the point of stalling. This type of flight is likely facilitated by extreme slotting of the primaries as well as by well-developed flight muscles and heart (Table 3; Hartman 1961; Moermond and Santana C., unpubl. data). The unusually large, broad tail of the trogon (see Hartman 1961) is spread during stalling and may aid in braking or in controlling the precise moment of the stall.

In contrast to these two aerial feeders, *Pteroglossus torquatus* appears to have difficulty flying slowly. Its wing loading is high, and it has poorly developed V-shaped slots that may operate effectively only when the wing is under maximum stress, as in take-off (pers. observ.). These toucanets usually fly from tree to tree in a straight line with rapid wing beats and relatively high impact landings. In comparison to *Querula* or *Trogon,* the toucanet's velocity range is narrower and its horizontal cruising velocity faster, as expected for a bird with high wing loading (cf. Pennycuick 1969, 1975). Correlated with better reaching ability, leg musculature of *Pteroglossus* is considerably heavier than that of the trogon or fruit-crow (Table 3).

The morphological differences between the toucanet and the two aerial species are marked and correspond well with differences in fruit-taking behavior and fruit choice described in earlier sections; however, the strong morphological differences between the trogon and fruit-crow are more surprising given that both species take fruit primarily on the wing. Nonetheless, the differences in their feeding techniques likely have consequences for fruit selection. The trogon's flight style is characterized by expensive powered flapping; the fruit-crow's swooping glides may provide cheaper flight and lower costs of plucking its fruit. Trogons, however, appear better at negotiating small spaces in the understory and lower parts of tree crowns, while the fruit-crow forages more often (and presumably more efficiently) among the larger open spaces of the canopy and takes more fruits from the upper portions of trees.

Field observations of similar birds at Monte Verde, Costa Rica (Snow 1977; Wheelwright 1983; Santana C. and Milligan 1984) provide strong support for the patterns we observed in the lowland birds listed above. Observations of the Resplendent Quetzel (*Pharomachrus mocinno*), and the Three-wattled Bellbird (*Procnias tricarunculata*), a large, broad-winged cotinga, feeding in the same trees show that *Pharomachrus* take fruit primarily from the lower portions of the trees whereas the bellbirds take significantly more fruits from the upper portions of the trees (Santana C. and Milligan 1984). Compared to *Querula,* the bellbird does not appear to be as specialized for aerial flight, having a shorter wing, likely higher wing loading, and relatively longer (and stronger?) legs (Snow 1982). The bellbird often is able to pick and reach some fruit from a perch (Snow 1977; Santana C. and Milligan 1984) but frequently resorts to taking fruit on the wing if the fruits are not easily accessible or are difficult to remove (Snow 1977). The two toucans at Monte Verde, *Aulacorhynchus prasinus* and *Ramphastos sulfuratus,* pick or reach virtually all their food items. The *Ramphastos,* which are much heavier than the *Aulacorhynchus,* are "able to pluck every fruit they reach out for and grasp with their mandibles, so they hardly move at all while feeding" (Snow 1977:628).

The small wing area and deep slotting of both *Manacus* and *Phoenicircus* suggest wing movements and flight behavior similar to those described for *Trogon massena* above. The relatively longer wing of *Pipra mentalis* and other small manakins may improve their hovering abilities and lower other flight costs (cf. Norberg 1981). Other examples with relatively minor but, nevertheless, noticeable, differences in wing shape invite further analysis.

Legs. — Legs are the other major morphological feature influencing manner of capture and selection of fruits. Although muscle weights are seldom taken, Hartman's data (1961; Table 3) show considerable interspecific variation in the percent of body weight in leg muscles. The leg musculature of trogons accounts for only about 3 percent of their body weight (2.7–3.2 percent for 7 *Trogon* spp. and 3.4 percent for *Pharomachrus*), and their feet are so weak that the birds may be unable to turn around on a perch without using their wings (Santana C., Moermond, and Denslow, unpubl. data). Many other species, however, have two to four times as much of the body weight in leg muscles (e.g., Ramphastidae and tanagers, Table 3). In the absence of functional analyses, we assume that birds with a greater percentage of body weight in leg musculature have stronger legs.

Species with strong legs fall into two fairly distinct functional groups: ground feeding birds, such as most thrushes (e.g., *Catharus* spp. and *Turdus* spp.) and emberizines (e.g., *Arremon* and *Arremonops* spp.), and perching birds, such as many tanagers and honeycreepers, that reach food items from perches on narrow branches (see previous section).

Relative tarsal lengths differ in the two groups of strong-legged species. Those birds that often perch or feed on the ground have longer tarsi than similar species that feed primarily above the ground; this pattern has been observed within diverse groups of birds (Dilger 1956; Newton 1967; Fretwell 1969; Partridge 1976; Fitzpatrick 1978). The tarsi of birds that perch primarily on branches vary in length and diameter depending on their feeding behavior. Among tyrannid flycatchers, aerial feeders using perches for resting have short tarsi, while those that feed by gleaning from perches have long, narrow tarsi (Fitzpatrick 1978). The long legs of perch-gleaners are presumed to increase the visible scanning area and to facilitate clinging to angled perches (Fitzpatrick 1978). Few tyrannids, however, reach below their perch level.

Species that reach down from perches tend to have shorter and thicker tarsi (Grant 1965, 1966; Sturman 1968; Moermond, Howe, Bruskewitz, and Rusterholz, unpubl. data; see Partridge 1976 and references therein).

Among frugivores, species with the "ground" type legs are able to reach out or down from perches only with difficulty. Our aviary experiments with two such species (*Catharus minimus* and *Arremon aurantiirostris*) have shown that these birds would rather jump up to hanging fruits that reach down when fruits are further than a few centimeters below a perch (Denslow and Moermond, in press, unpubl. data). Species with strong legs of the "perching" type are usually agile on small diameter perches and able to reach below their perch to pluck fruits. Many of the smaller tanagers and honeycreepers appear able to reach well below a perch. As in hovering ability, small absolute size facilitates such behavior (Newton 1967). Nevertheless, not all small birds reach equally well. *Thraupis palmarum*, for example, can reach its full body length below a perch, while the somewhat smaller *Ramphocelus passerini* cannot do so without apparent difficulty (Denslow and Moermond, in press, unpubl. data). The *Thraupis*, a heavier bird, has a distinctly shorter but thicker tarsus than the *Ramphocelus*.

Additional morphological features may enhance the reaching abilities of "perchers" (Osterhaus 1962; Leisler and Thaler 1982). Snow and Snow (1971) noted aspects of the foot, such as grip strength and claw shape, that may enhance the reaching ability of *Thraupis palmarum*. They also noted that *T. palmarum* and *Tangara guttata*, which frequently foraged by clinging head-down, had longer tails than their sympatric congeners. We have also noted in our aviary experiments that the long tail of *Saltator maximus* appears to aid in balance, allowing this species to forage in hanging positions on thin stems or leaves.

Reaching ability may be constrained by perch diameter. We have observed that a small bird such as *Euphonia gouldi* will reach readily from a 3 mm diameter perch but reluctantly from a 12 mm diameter perch (Moermond and Denslow 1983). A study of the ability of two paruline warbler species to reach from a series of perches demonstrated a considerable difference in distance reached over a relatively small range of perch diameters (3 mm–24 mm) (Moermond, Howe, Bruskewitz, and Rusterholz, unpubl. data). Similar detailed studies of closely related species have shown that minor differences in leg dimensions are correlated with distinct differences in activity patterns and microhabitat (Dilger 1956 with thrushes; Pearson 1977 with antwrens; Leisler and Thaler 1982 with Kinglets). These studies suggest that the variations in tarsal length and leg muscle mass described for fruit-eating birds are associated with differences in ability to negotiate different substrates and to pluck fruits presented on infructescences of different structure.

Among medium to large tanagers (Table 3), species that forage in the canopy or outer edges of vegetation (*Piranga rubra*, *P. olivacea*, *Thraupis episcopus*, *T. palmarum*, *Tachyphonus luctuosus*) have lower wing loading, higher aspect ratios, a higher percent of body weight in flight muscles, and a lower percent of body weight in their legs muscles than most species that frequently forage on or near the ground and in low vegetation (*Tachyphonus rufus*, *Ramphocelus passerinii*, *R. icteronotus*, *Habia rubica*, *H. fuscicauda*). Among the canopy tanagers, the ones with the best reaching ability (*Thraupis* spp.) have the highest percent of body weight in their lower extremities, although the percentage is still below that shown by the "low vegetation and ground" species.

The suite of morphological features of a particular bird are expected to determine the relative accessibility of and the costs of taking available fruits and, thus, to influence strongly the bird's diet. While this is most clearly seen in such divergent birds as trogons and toucans, we believe it also contributes importantly to fruit choice among all fruit-eaters. Although species of diverse morphology often feed at the same fruiting tree (Eisenmann 1961; Land 1963; Willis 1966; Olson and Blum 1968; Leck 1969; Howe 1977; McDiarmid et al. 1977; Fig. 4), a discriminant function analysis of data from four of these studies was partially successful in separating groups of species on the basis of body, wing, and leg characteristics, although not of bill dimensions (Ricklefs 1977). Similarly, a multivariate analysis (Denslow and Moermond, in press) of 15 species of tanagers and honeycreepers studied by Snow and Snow (1971) demonstrated that these species could be separated by the species of fruits in their diets and that the differences among the bird groups appeared correlated with aspects of their morphology (bill size, percent body weight in leg muscles, wing loading). Likewise our aviary choice tests with 20 species of tanagers, manakins, thrushes, finches, and tyrannids have shown that all those tested exhibit clear, well-defined preferences for more accessible fruits with the differences in preferences associated with differences in morphology (Moermond and Denslow 1983; Denslow and Moermond, in press, unpubl. data; Levey, unpubl. data).

A phylogenetic survey of frugivores shows, not surprisingly, that the members of each family are usually similar in morphology and similar in the techniques they use to take fruits (Table 3). Where important variation in locomotory morphology exists in these families, it is often associated with different techniques of fruit capture. The adaptive values of these patterns are apparent if one assumes that fruits presented in different ways actually constitute different resources. If certain morphological characteristics allow a bird to take some of these fruits more cheaply than others, then selection on such morphological adaptations may be enhanced. The presence of frugivorous species in so many families is not evidence for lack of requirements for eating fruits but rather represents adaptive "specialization" to exploit fruits displayed in different ways.

SPECIALISTS, GENERALISTS, AND SEED DISPERSAL

Snow (1980) designated as fruit specialists those species that can survive entirely on fruit and rear their young mainly on fruit. Such species (Oilbirds, a few cotingids) represent a small subset of the total range of fruit-eating birds that still may be classed as "legitimate" frugivores, i.e., species that digest the fruit pulp or juices but not the seed (Snow 1980). A further division of this large group of "legitimate" frugivores into specialist and generalists would be difficult and beset with contradictions and inconsistencies. For example, although large, lipid-rich fruits are taken primarily by large frugivores, these same birds generally take many species of small fruits also. By virtue of their sizes, the large frugivores (including those usually considered specialists) are certainly capable of taking the largest size range of fruits and, therefore, the largest number of fruit species (Wheelwright in press; Table 2, Fig. 1). In addition, morphological specializations for fruit-eating have been noted for only a few species, but include both large and small birds (e.g., Oilbirds and euphonias). Specializations in bill, wing, and leg morphology that affect methods of removal of fruits can be identified in several species among both traditional specialists and generalists. Finally, the "high quality" seed dispersal said to be associated with specialists is not necessarily characteristic of all birds traditionally called specialists (Oilbirds leave many seeds on the floor of their roosting caves), and, in addition, many traditional generalists appear to be good seed dispersers (e.g., Leck 1972a; Howe and DeSteven 1979; Greenberg 1981). With the exception of a few species that appear to be total frugivores, there does not seem to be sufficient basis in behavior, morphology, or function on which to recognize generalists and specialists among frugivorous birds.

As Wheelwright and Orians (1982) pointed out, the quality of seed dispersal for a given plant is likely to depend on a variety of factors and is not easily predicted just from the species of birds feeding on the plant's fruits. Safe sites for seed germination are not likely recognizeable by the birds and in any case a "disperser" (in contrast to a pollinator) is "rewarded" for taking the seed but not for delivering it. Adaptive strategies for seed dispersal are more likely to involve simply moving as many seeds as possible from the immediate vicinity of the parent tree (Denslow 1980). Co-evolution among fruiting plants and their dispersers has very obviously occurred as evidenced in the characteristics of fruit displays (Van der Pijl 1969), but a better understanding of the interaction at the species level must take into account the factors influencing fruit selection by the dispersers.

The birds' choice of fruits is neither confined to specific fruits nor is it dictated only by chance of encounter. Experiments with a number of species of small and two species of large frugivores have shown that the birds make systematic decisions and that their choices depend on such factors as fruit size, pulp-to-seed ratio, ripeness, color, taste, and importantly, accessibility of the fruits from a perch and the distance between fruit clusters (Best 1981; Herrera 1981; Sorensen 1981; Moermond and Denslow 1983; Denslow and Moermond, in press; Levey et al. 1984; Santana C., Moermond, Denslow, unpubl. data).

Of most interest among the aviary experiments are choices in which a bird is induced to change an initial fruit preference by a change in one of the conditions under which the preferred fruit is offered (for example, a decrease in the accessibility of the preferred fruit or an increase in the flight distance to the preferred fruit). Such changes in preference have been observed in each of the frugivores we have studied so far, leading us to believe that fruit choice in the field is a dynamic process, i.e., the species and quantities of fruits taken are contingent upon the context within which each fruit is found (e.g., Best 1981; Levey et al. 1984).

Differences in fruits taken by different frugivores living in the same habitat may be attributed to differences in the birds' morphological abilities and physiological requirements that translate into differences in the benefit : cost balance associated with a given choice. The morphological and behavioral differences among frugivores are assumed to influence not only number and species of fruits eaten but also to influence the quality of dispersal of seeds. For example, very small seeds, such as those of Melastomataceae or *Ficus,* are easily passed through the digestive tract of large and small frugivores. Dispersal of larger seeds, however, is increasingly limited by their size. Not only may they be swallowed only by birds with sufficiently large gapes, but small and medium-sized birds (tanagers and finches) mash the fruits before swallowing them, often dropping the seeds at or near the parent plant. Thus, it may be advantageous for a tree producing medium to large seeds to display fruits in such a way as to lower the probability of removal by birds in this group. Because these birds are perching birds, display of fruits at the ends of long peduncles or at the tips of narrow flexible twigs, in effect, lowers their accessibility (Snow 1977; Denslow and Moermond, in press). Such fruit displays may also even enhance their probability of removal by aerial-feeding, fruit-swallowing birds (Denslow and Moermond 1982, in press; Santana C., Moermond, and Denslow, unpubl. data).

We suggest that the key to understanding such differences among fruit-eating birds lies in the constraints of morphology on behavior. Differences in digestive systems, bill strength and shape, wing and leg morphology, and body weight differentially affect the accessibility of fruits, the ease with which they are handled, the rates at which they are eaten, and the efficiency with which they are digested. Such constraints are expected to influence the birds' choices of fruits, the benefits and costs of taking those fruits, and ultimately, the fruiting patterns and display structures of plants competing for birds as dispersers.

ACKNOWLEDGMENTS

We thank D. J. Levey and E. Santana C. for assistance in the field, for unpublished observations and for valuable discussions. We thank J. G. Blake, E. DeVito, R. L. Knight, D. J. Levey, B. Loiselle, E. Santana C., D. W. Snow, F. G. Stiles, B. C. Wentworth, N. T. Wheelwright, and M. F. Willson for comments on portions of the manuscript. D. E. Stone, D. A. Clark, D. B. Clark of the Organization for Tropical Studies provided much needed logistic support at the La Selva Biological Station in Costa Rica. F. G. Stiles, J. R. Karr, and S. Flores P. made it possible to visit other areas of the neotropics. Researchers at La Selva as well as many others have added comments, advice, and observations for which we are grateful. The University of Wisconsin Zoology Museum provided permission to work with its collection. C. M. Hughes prepared the figures and graphs. The field work was supported in part by grants from the National Science Foundation, P. and M. Hess, the Nave Fund and the Graduate School of the University of Wisconsin. Assistance and support was also generously provided by the Departments of Zoology and Botany of the University of Wisconsin.

LITERATURE CITED

ALEXANDER, R. M. 1968. Animal Mechanics. University Washington Press, Seattle, Washington.

BEECHER, W. J. 1951. Convergence in the Coerebidae. Wilson Bull. 63:274–287.

BERTHOLD, P. 1976a. The control and significance of animal and vegetable nutrition in omnivorous songbirds. Ardea 64:140–154.

BERTHOLD, P. 1976b. Der Seidenschwanz *Bombycilla garrulus* als frugivorer Ernahrungsspezialist. Experientia 32:1445.

BEST, L. S. 1981. The effect of specific fruit and plant characteristics on seed dispersal. Unpubl. Ph.D. dissert., University of Washington, Seattle.

BOURNE, G. R. 1974. The Red-billed Toucan in Guyana. Living Bird 13:99–126.

BRATTSTEN, L. B. 1979. Biochemical defense mechanisms in herbivores against plant allelochemicals. Pp. 199–270, *In* G. A. Rosenthal and D. H. Janzen (eds.), Herbivores: Their Interaction with Secondary Plant Metabolites. Academic Press, New York.

BROWN, R. H. J. 1963. The flight of birds. Biol. Rev. 38:460–489.

BUSKIRK, W. H. 1976. Social systems in a tropical forest avifauna. Am. Nat. 110:293–310.

CADOW, G. 1933. Magen und Darm der Fruchttauben. J. Ornithol. 81:236–252.

CONE, C. D., JR. 1968. The aerodynamics of flapping flight. Virginia Inst. Marine Sci., Spec. Scient. Rep. No. 52.

CROME, F. H. J. 1975a. The ecology of fruit pigeons in tropical northern Queensland. Aust. Wildl. Res. 2:155–185.

CROME, F. H. J. 1975b. Notes on the breeding of the purple-crowned pigeon. Emu 75:172–174.

CROME, F. H. J. 1975c. Breeding, feeding, and status of the Torres Strait Pigeon on Low Isles, northeastern Queensland. Emu 75:189–198.

CVITANIC, A. 1970. The relationships between intestine and body length and the nutrition of some bird species. Larus 21–22:181–190.

DAVIES, N. B., AND R. E. GREEN. 1976. The development and ecological significance of feeding techniques in the reed warbler (Acrocephalus scirpaceus). Animal Behaviour 24:213–229.

DEJONG, M. J. 1983. Bounding flight in birds. Unpubl. Ph.D. dissert., University of Wisconsin–Madison, Madison.

DELACOUR, J., AND D. AMADON. 1973. Curassows and related birds. American Museum Natural History, New York.

DENSLOW, J. S. 1980. Gap partitioning among tropical rainforest trees. Biotropica 12 Suppl.:47–55.

DENSLOW, J. S., AND T. C. MOERMOND. 1982. The effect of accessibility on rates of fruit removal from neotropical shrubs: an experimental study. Oecologia 54:170–176.

DENSLOW, J. S., AND T. C. MOERMOND. in press. The interaction of fruit display and the foraging strategies of small frugivorous birds. In W. D'Arcy and M. D. Correa A. (eds.), The Botany and Natural History of Panama: la Botanica y Historia Natural de Panama. Missouri Botanical Garden, St. Louis, Missouri.

DESSELBERGER, H. 1931. Der Verdauungskanal der Dicaeiden nach Gestalt und Funktion. J. Ornithol. 79:353–370.

DILGER, W. C. 1956. Adaptive modifications and ecological isolating mechanisms in the thrush genera Catharus and Hylocichla. Wilson Bull. 68:171–199.

DOCTERS VAN LEEUWEN, W. M. 1954. On the biology of some Javanese Loranthaceae and the role birds play in their life-histories. Beaufortia 4:105–205.

EISENMANN, E. 1961. Favorite foods of neotropical birds: flying termites and Cecropia catkins. Auk 78: 636–637.

FAABORG, J. R., AND J. W. TERBORGH. 1980. Patterns of migration in the West Indies. Pp. 157–163, In A. Keast and E. S. Morton (eds.), Migrant Birds in the Neotropics. Smithsonian Institution Press, Washington, D.C.

FISHER, H. 1972. The nutrition of birds. Pp. 431–469, In D. S. Farner and J. R. King (eds.), Avian Biology Vol. II. Academic Press, New York.

FITZPATRICK, J. W. 1978. Foraging behavior and adaptive radiation in the avian family Tyrannidae. Unpubl. Ph.D. dissert., Princeton University, Princeton, New Jersey.

FITZPATRICK, J. W. 1980. Foraging behavior of neotropical tyrant flycatchers. Condor 82:43–57.

FLEMING, T. H. 1979. Do tropical frugivores compete for food? Am. Zool. 19:1157–1172.

FOGDEN, M. P. L. 1972. The seasonality and population dynamics of equatorial forest birds in Sarawak. Ibis 114:307–343.

FOGDEN, M. P. L., AND P. M. FOGDEN. 1979. The role of fat and protein reserves in the annual cycle of the Grey-backed Camaroptera in Uganda (Aves:Sylviidae). J. Zool. (Lond.) 189:233–258.

FORSHAW, J. M. 1978. Parrots of the World. David and Charles, London.

FOSTER, M. S. 1974. Rain, feeding behavior, and clutch size in tropical birds. Auk 91:722–726.

FOSTER, M. S. 1977. Ecological and nutritional effects of food scarcity on a tropical frugivorous bird and its fruit source. Ecology 58:73–85.

FOSTER, M. S. 1978. Total frugivory in tropical passerines: a reappraisal. Trop. Ecol. 19:131–154.

FOSTER, M. S., AND R. MCDIARMID. 1983. Nutritional value of the aril of Trichilia cuneata, a bird-dispersed fruit. Biotropica 15:26–31.

FOSTER, R. B. 1982a. The seasonal rhythm of fruit fall on Barro Colorado Island. Pp. 151–172, In E. G. Leigh, Jr., A. S. Rand, and D. Windsor (eds.), The Ecology of a Tropical Forest. Smithsonian Institution Press, Washington, D.C.

FOSTER, R. B. 1982b. Famine on Barro Colorado Island. Pp. 201–212, In E. G. Leigh, Jr., A. S. Rand, and D. Windsor (eds.), The Ecology of a Tropical Forest. Smithsonian Institution Press, Washington, D.C.

FRANKIE, G. W., H. G. BAKER, AND P. A. OPLER. 1974. Comparative phenological studies of trees in tropical wet and dry forests in the lowlands of Costa Rica. J. Ecol. 62:881–919.

FRETWELL, S. D. 1969. Ecotypic variation in the nonbreeding season in migratory populations: a study of tarsal length in some Fringillidae. Evolution 23:406–420.

GOODWIN, D. 1970. Pigeons and Doves of the World. British Museum (Natural History), London.

GRANT, P. R. 1965. The adaptive significance of some size trends in island birds. Evolution 19:355–367.

GRANT, P. R. 1966. Further information on the relative length of the tarsae in land birds. Postilla 98: 1–13.

GREENBERG, R. 1981. Frugivory in some migrant tropical forest wood warblers. Biotropica 13:215–223.

GREENEWALT, C. H. 1975. The flight of birds. Trans. Am. Phil. Soc., New Ser. 65:1–67.

GROEBBELS, F. 1932. Der Vogel. Bau, Funktion, Lebenserscheinung, Einpassung. Vol. 1. Borntraeger, Berlin.

HAILS, C. J. 1979. A comparison of flight energetics in hirundines and other birds. Comp. Biochem. Physiol. A 63:581–586.

HARTMAN, F. A. 1961. Locomotor mechanisms of birds. Smithson. Misc. Collect. 143:1–91.

HERRERA, C. M. 1981. Fruit variation and competition for dispersers in natural populations of Smilax aspera. Oikos 36:51–58.

HERRERA, C. M. 1982. Defense of ripe fruits from pests: its significance in relation to plant-disperser interactions. Am. Nat. 120:218–241.

HERRERA, C. M. 1984. Adaptation to frugivory of Mediterranean avian seed dispersers. Ecology 65: 609–617.

HERRERA, C. M., AND P. JORDANO. 1981. Prunus mahaleb and birds: the high-efficiency seed dispersal system of a temperate fruiting tree. Ecol. Monogr. 51:203–218.

HILL, F. W., AND L. M. DANSKY. 1954. Studies of the energy requirements of chickens I. The effect of dietary energy level on growth and feed consumption. Poult. Sci. 33:112–119.

HILTY, S. L. 1980. Flowering and fruiting periodicity in a premontane rain forest in Pacific Colombia. Biotropica 12:292–306.

HOWE, H. F. 1977. Bird activity and seed dispersal of a tropical wet forest tree. Ecology 58:539–550.

HOWE, H. F. 1981. Dispersal of a neotropical nutmeg (Virola sebifera) by birds. Auk 98:88–98.

HOWE, H. F. 1982. Fruit production and animal activity at two tropical trees. Pp. 189–199, In E. G. Leigh, Jr., A. S. Rand, and D. Windsor (eds.), The Ecology of a Tropical Forest. Smithsonian Institution Press, Washington, D.C.

HOWE, H. F., AND D. DESTEVEN. 1979. Fruit production, migrant bird visitation, and seed dispersal of Guarea glabra in Panama. Oecologia 39:185–196.

HOWE, H. F., AND G. F. ESTABROOK. 1977. On intraspecific competition for avian dispersers in tropical trees. Am. Nat. 111:817–832.

HOWE, H. F., AND R. B. PRIMAK. 1975. Differential seed dispersal by birds of the tree Casearia nitida (Flacourtiaceae). Biotropica 7:278–283.

HOWE, H. F., AND G. A. VANDE KERCKHOVE. 1980. Nutmeg dispersal by tropical birds. Science 210: 925–927.

JAMES, F. C. 1982. The ecological morphology of birds: a review. Ann. Zool. Fennici 19:265–275.

JANSON, C. 1983. Adaptation of fruit morphology to dispersal agents in a neotropical forest. Science 219:187–189.

JANZEN, D. H. 1967. Synchronization of sexual reproduction of trees within the dry season in Central America. Evolution 21:620–637.

JANZEN, D. H. 1981. Ficus ovalis seed predation by an orange-chinned parakeet (Brotogeris jugularis) in Costa Rica. Auk 98:841–844.

JENKINS, R. 1969. Ecology of three species of Saltators in Costa Rica with special reference to their frugivorous diet. Unpubl. Ph.D. dissert., Harvard University, Cambridge, Massachusetts.

KANTAK, G. E. 1979. Observations on some fruit-eating birds in Mexico. Auk 96:183–186.

KARR, J. R., AND F. C. JAMES. 1975. Eco-morphological configurations and convergent evolution in species and communities. Pp. 258–291. In M. L. Cody and J. M. Diamond (eds.), Ecology and Evolution of Communities. Harvard University Press, Cambridge, Massachusetts.

KARR, J. R., D. W. SCHEMSKE, AND N. BROKAW. 1982. Temporal variation in the undergrowth bird community of a tropical forest. Pp. 441–453. In E. G. Leigh, Jr., A. S. Rand, and D. Windsor (eds.), The Ecology of a Tropical Forest. Smithsonian Institution Press, Washington, D.C.

KEAR, J. 1962. Food selection in finches with special reference to interspecific differences. Proc. Zool. Soc. London 138:163–204.

KEAST, A., AND E. S. MORTON (EDS.). 1980. Migrant Birds in the Neotropics. Smithsonian Institution Press, Washington, D.C.

KOKSHAYSKY, N. V. 1973. Functional aspects of some details of bird wing configurations. Syst. Zool. 22:442–450.

LAND, H. C. 1963. A tropical feeding tree. Wilson Bull. 75:199–200.

LECK, C. F. 1969. Observations of birds exploiting a Central American fruit tree. Wilson Bull. 81:264–269.

LECK, C. F. 1971a. Measurement of social attractions between tropical passerine birds. Wilson Bull. 83:278–283.

LECK, C. F. 1971b. Overlap in the diet of some neotropical birds. Living Bird 10:89–106.

LECK, C. F. 1972a. The impact of some North American migrants at fruiting trees in Panama. Auk 89: 842–850.

LECK, C. F. 1972b. Seasonal changes in feeding pressures of fruit and nectar eating birds in the neotropics. Condor 74:54–60.

LECK, C. F. 1973. Observations of birds at Cecropia trees in Puerto Rico. Wilson Bull. 84:498–500.

LEISLER, B., AND E. THALER. 1982. Differences in morphology and foraging behaviour in the goldcrest Regulus regulus and the firecrest R. ignicapillus. Ann. Zool. Fennici 19:277–284.

LEVEY, D. J., T. C. MOERMOND, AND J. S. DENSLOW. 1984. Fruit choice in neotropical birds: the effect of distance between fruits on preference patterns. Ecology 65:844–850.

LEVEILLE, G. A., AND H. FISHER. 1958. The amino acid requirements for maintenance in the adult rooster. 1. Nitrogen and energy requirements in normal and protein-depleted animals receiving whole egg protein and amino acid diets. J. Nutr. 66:441–453.

MARTIN, E. W. 1968. The effects of dietary protein on the energy and nitrogen balance of the Tree Sparrow, (*Spizella arborea arborea*). Physiol. Zool. 41:313–331.

MCDIARMID, R. W., R. E. RICKLEFS, AND M. S. FOSTER. 1977. Dispersal of *Stemmadenia donnell-smithii* (Apocynaceae) by birds. Biotropica 9:9–25.

MCKEY, D. 1975. The ecology of coevolved seed dispersal systems. Pp. 159–191, *In* L. E. Gilbert and P. H. Raven (eds.), Coevolution of Animals and Plants. University Texas Press, Austin, Texas.

MILTON, K., AND F. R. DINTZIS. 1981. Nitrogen-to-protein conversion factors for tropical plant samples. Biotropica 13:177–181.

MOERMOND, T. C. 1983. Suction-drinking in tanagers and its relation to fruit handling. Ibis 125:545–549.

MOERMOND, T. C., AND J. S. DENSLOW. 1983. Fruit choice in neotropical birds: effects of fruit type and accessibility on selectivity. J. Anim. Ecol. 52:407–420.

MORDEN-MOORE, A. L., AND M. F. WILLSON. 1982. On the ecological significance of fruit color in *Prunus serotina* and *Rubus occidentalis:* field experiments. Can. J. Bot. 60:1554–1560.

MORRISON, D. W. 1980. Efficiency of food utilization by fruit bats. Oecologia 45:270–273.

MORTON, E. S. 1971. Food and migration habits of Eastern Kingbird in Panama. Auk 88:925–926.

MORTON, E. S. 1973. On the evolutionary advantages and disadvantages of fruit eating in tropical birds. Am. Nat. 107:8–22.

MORTON, E. S. 1977. Intratropical migration in the Yellow-green Vireo and Piratic Flycatcher. Auk 94:97–106.

MORTON, E. S. 1978. Avian arboreal folivores: Why not? Pp. 123–129, *In* G. G. Montgomery (ed.), The Ecology of Arboreal Folivores. Smithsonian Institution Press, Washington, D.C.

MORTON, E. S. 1980. Adaptations to seasonal changes by migrant land birds in the Panama Canal Zone. Pp. 437–453, *In* A. Keast and E. S. Morton (eds.), Migrant Birds in the Neotropics. Smithsonian Institution Press, Washington, D.C.

MUNZELL, H. E., L. D. WILLIAMS, L. P. GUILD, C. B. TROESCHER, G. NIGHTINGALE, AND R. S. HARRIS. 1949. Composition of food plants of Central America I. Honduras. Food Research 14:144–164.

NATIONAL RESEARCH COUNCIL. 1960. Nutrient requirements of domestic animals I. Nutrient requirements of poultry. Publ. 827.

NESTLER, R. B., W. W. BAILEY, L. M. LLEWELLYN, AND M. J. RENSBERGER. 1944. Winter protein requirements of bobwhite quail. J. Wildl. Manage. 8:218–222.

NEWTON, I. 1967. The adaptive radiation and feeding ecology of some British finches. Ibis 109:33–98.

NORBERG, U. M. 1979. Morphology of the wings, legs and tail of three coniferous forest tits, the goldcrest, and the treecreeper in relation to locomotor pattern and feeding station selection. Philos. Trans. R. Soc. Lond. B Biol. Sci. 287:131–165.

NORBERG, U. M. 1981. Flight, morphology and the ecological niche in some birds and bats. Symp. Zool. Soc. Lond. 48:173–197.

OLSON, S. L., AND K. E. BLUM. 1968. Avian dispersal of plants in Panama. Ecology 49:565–566.

OPLER, P. A., G. W. FRANKIE, AND H. G. BAKER. 1980. Comparative phenological studies of treelet and shrub species in tropical wet and dry forests in the lowlands of Costa Rica. J. Ecol. 68:167–188.

OSTERHAUS, M. B. 1962. Adaptive modifications in the leg structure of some North American warblers. Am. Midl. Nat. 68:474–486.

PARKER, T. A. 1981. Distribution and biology of the White-cheeked Cotinga *Zaratornis stresemanni*, a high Andean frugivore. Bull. Br. Ornithol. Club 101:256–265.

PARTRIDGE, L. 1976. Some aspects of the morphology of Blue Tits (*Parus caerulens*) and Coal Tits (*Parus ater*) in relation to their behaviour. J. Zool. (Lond.) 179:121–133.

PEARSON, D. L. 1977. Ecological relationships of small antbirds in Amazonian bird communities. Auk 94:283–292.

PENNYCUICK, C. J. 1969. The mechanics of bird migration. Ibis 111:525–556.

PENNYCUICK, C. J. 1975. Mechanics of flight. Pp. 1–75, *In* D. S. Farner and J. R. King (eds.), Avian Biology, Vol. 5. Academic Press, New York.

PULLAINEN, E., P. HELLE, AND P. TUNKKARI. 1981. Adaptive radiation of the digestive system, heart and wings of *Turdus pilaris, Bombycilla garrulus, Sturnus vulgaris, Pyrrhula pyrrhula, Pinicola enuclieator* and *Loxia pytyopsittacus*. Ornis Fenn. 58:21–28.

RAYNER, J. M. V. 1981. Flight adaptations of vertebrates. Symp. Zool. Soc. Lond. 48:137–172.

RICKLEFS, R. E. 1974. Energetics of reproduction in birds. Pp. 152–292, *In* R. A. Paynter (ed.), Avian Energetics. Publ. Nuttall Ornithol. Club No. 15.

RICKLEFS, R. E. 1977. A discriminant function analysis of assemblages of fruit-eating birds in Central America. Condor 79:228–231.

RIDLEY, H. N. 1930. The dispersal of plants throughout the world. L. Reeve and Co., Ashford, Kent, England.

SALOMONSEN, F. 1964. Flowerpeckers. Pp. 311–313, In A. Landsborough Thomson (ed.), A New Dictionary of Birds. Thomas Nelson, London.

SANTANA C., E., AND B. G. MILLIGAN. 1984. Behavior of toucanets, bellbirds, and quetzals feeding on Lauraceous fruits. Biotropica 16:152–154.

SAVILE, D. B. O. 1957. Adaptive evolution in the avian wing. Evolution 11:212–224.

SCHAEFER, E. 1953. Contribution to the life history of the swallow-tanager. Auk 70:403–460.

SCHULENBERG, T. S. 1983. Foraging behavior, eco-morphology, and systematics of some antshrikes (Formicariidae: *Thamnomanes*). Wilson Bull. 95:505–521.

SIBLEY, C. G., S. M. LANYON, AND J. E. ALQUIST. 1984. The relationships of the Sharpbill (*Oxyruncus cristatus*). Condor 86:48–52.

SIMMS, E. 1978. British Thrushes. Collins, London.

SKUTCH, A. F. 1933. The aquatic flowers of a terrestrial plant, *Heliconia bihai* L. Am. J. Bot. 20:535–544.

SKUTCH, A. F. 1954. Life histories of Central American Birds. Pacific Coast Avifauna No. 31.

SKUTCH, A. F. 1960. Life histories of Central American Birds Vol. II. Pacific Coast Avifauna No. 34.

SKUTCH, A. F. 1969. Life histories of Central American Birds Vol. III. Pacific Coast Avifauna No. 35.

SKUTCH, A. F. 1971. Life history of the Keel-billed Toucan. Auk 88:381–396.

SKUTCH, A. F. 1972. Studies of tropical American birds. Publ. Nuttall Ornithol. Club No. 10.

SKUTCH, A. F. 1976. Parent Birds and Their Young. University Texas Press, Austin, Texas.

SKUTCH, A. F. 1981. New studies of tropical American birds. Publ. Nuttall Ornithol. Club No. 19.

SMYTHE, N. 1970. Relations between fruiting seasons and seed dispersal methods in a neotropical forest. Am. Nat. 104:25–35.

SNOW, B. K. 1970. A field study of the Bearded Bellbird in Trinidad. Ibis 112:299–329.

SNOW, B. K. 1972. A field study of the Calfbird, *Perissocephalus tricolor*. Ibis 114:139–162.

SNOW, B. K. 1977. Territorial behavior and courtship of the Three-wattled Bellbird. Auk 94:623–645.

SNOW, B. K., AND D. W. SNOW. 1971. The feeding ecology of tanagers and honeycreepers in Trinidad. Auk 88:291–322.

SNOW, D. W. 1961. The natural history of the Oilbird, *Steatornis caripensis*, in Trinidad, W. I. Part 1. General behavior and breeding habits. Zoologica (N.Y.) 46:27–48.

SNOW, D. W. 1962a. A field study of the Black and White Manakin, *Manacus manacus*, in Trinidad. Zoologica (N.Y.) 47:65–104.

SNOW, D. W. 1962b. A field study of the Golden-headed Manakin, *Pipra erythrocephala*, in Trinidad. Zoologica (N.Y.) 47:183–198.

SNOW, D. W. 1962c. The natural history of the Oilbird, *Steatornis caripensis*, in Trinidad, W. I. Part 2. Population, breeding ecology, food. Zoologica (N.Y.) 47:199–221.

SNOW, D. W. 1965. A possible selective factor in the evolution of fruiting seasons in tropical forests. Oikos 15:274–281.

SNOW, D. W. 1971a. Evolutionary aspects of fruit-eating by birds. Ibis 113:194–202.

SNOW, D. W. 1971b. Observations on the Purple-throated Fruit-crow in Guyana. Living Bird 10:5–17.

SNOW, D. W. 1971c. Notes on the biology of the Cock-of-the-rock (*Rupicola rupicola*). J. Ornithol. 112:323–333.

SNOW, D. W. 1973. Distribution, ecology and evolution of the bellbirds (*Procnias*, Cotingidae). Br. Mus. (Nat. Hist.) Bull. Zool. 25:369–391.

SNOW, D. W. 1979. Family Pipridae. Pp. 245–280, In M. A. Traylor, Jr. (ed.), Check-list of Birds of the World, Vol. 3. Museum Comparative Zoology, Cambridge, Massachusetts.

SNOW, D. W. 1980. Regional differences between tropical floras and the evolution of frugivory. Acta XVII Congr. Int. Ornithol. 2:1192–1198.

SNOW, D. W. 1981. Tropical frugivorous birds and their food plants: a world survey. Biotropica 13:1–14.

SNOW, D. W. 1982. The Cotingas. Cornell University Press, Ithaca, New York.

SORENSEN, A. E. 1981. Interactions between birds and fruits in a temperate woodland. Oecologia 50:242–249.

SORENSEN, A. E. 1984. Nutrition, energy and passage time: experiments with fruit preference in European blackbirds (*Turdus merula*). J. Anim. Ecol. 53:545–557.

STILES, F. G. 1979. Notes on the natural history of *Heliconia* (Musaceae) in Costa Rica. Brenesia 15:151–180.

STILES, F. G. 1983. Birds. Pp. 502–530, In D. H. Janzen (ed.), Costa Rican Natural History. University Chicago Press, Chicago, Illinois.

STILES, F. G. In press. On the roles of birds in the dynamics of neotropical forests. In A. W. Diamond and T. Lovejoy (eds.), Conservation of Tropical Forest Birds. International Council for Bird Preservation, Cambridge, England.

STORER, R. W. 1969. What is a tanager? Living Bird 8:127–136.

STURMAN, W. A. 1968. The foraging ecology of *Parus atricapillus* and *P. rufescens* in the breeding season, with comparisons with other species of *Parus*. Condor 70:309–322.

TRAYLOR, M. A., JR., AND J. W. FITZPATRICK. 1982. A survey of the tyrant flycatchers. Living Bird 19: 7–50.

VAN DER PIJL, L. 1969. Principles of Dispersal in Higher Plants. Springer-Verlag, New York.

WAGNER, H. O. 1944. Notes on the life history of the emerald toucanet. Wilson Bull. 56:65–76.

WALSBERG, G. E. 1975. Digestive adaptations of *Phainopepla nitens* associated with the eating of mistletoe berries. Condor 77:169–174.

WALSBERG, G. E. 1977. Ecology and energetics of contrasting social systems in *Phainopepla nitens* (Aves: Ptilogonatidae). Univ. Calif. Publ. Zool. 108:1–63.

WARD, P. 1969a. Seasonal and diurnal changes in the fat content of an equatorial bird. Physiol. Zool. 42:85–95.

WARD, P. 1969b. The annual cycle of the Yellow-vented Bulbul *Pycnonotus goiavier* in a humid equatorial environment. J. Zool. (Lond.) 157:25–45.

WATT, B. K., AND A. L. MURIEL. 1963. Composition of foods. Agric. Handbk. No. 8. U.S. Dept. Agriculture, Washington, D.C.

WETMORE, A. 1914. The development of the stomach in the euphonias. Auk 31:458–461.

WHEELWRIGHT, N. T. 1983. Fruits and the ecology of Resplendent Quetzals. Auk 100:286–301.

WHEELWRIGHT, N. T. In press. Fruit size, gape width, and the diets of fruit-eating birds. Ecology.

WHEELWRIGHT, N. T., W. A. HABER, K. G. MURRAY, AND C. GUINDON. 1984. Tropical fruit-eating birds and their food plants: a survey of a Costa Rican lower montane forest. Biotropica 16:173–192.

WHEELWRIGHT, N. T., AND G. H. ORIANS. 1982. Seed dispersal by animals: contrasts with pollen dispersal, problems of terminology, and constraints on coevolution. Am. Nat. 119:402–413.

WHITE, S. C. 1974. Ecological aspects of growth and nutrition in tropical fruit-eating birds. Unpubl. Ph.D. dissert., University of Pennsylvania, Philadelphia.

WILLIS, E. O. 1966. Competitive exclusion and birds at fruiting trees in Western Columbia (sic.). Auk 83:479–480.

WILLSON, M. F., AND J. N. THOMPSON. 1982. Phenology and ecology of color in bird-dispersed fruits, or why some fruits are red when they are "green." Can. J. Bot. 60:701–713.

WORTHINGTON, A. 1982. Population sizes and breeding rhythms of two species of manakins in relation to food supply. Pp. 213–225, *In* E. G. Leigh, Jr., A. S. Rand, and D. M. Windsor (eds.), The Ecology of a Tropical Forest. Smithsonian Institution Press, Washington, D.C.

ZISWILER, V., AND D. W. FARNER. 1972. Digestion and the digestive system. Pp. 343–430, *In* D. S. Farner and J. R. King (eds.), Avian Biology, Vol. 2. Academic Press, New York.

THE YELLOW-RUMPED CACIQUE AND ITS ASSOCIATED NEST PIRATES

SCOTT K. ROBINSON

Department of Biology, Princeton University, Princeton, New Jersey 08544 USA;
present address: Illinois Natural History Survey, 607 East Peabody Drive,
Champaign, Illinois 61820 USA

ABSTRACT. I describe the interactions between Yellow-rumped Caciques (*Cacicus cela*) in southeastern Peru and three species that often pirate their nests or destroy their eggs and kill nestlings. Individuals of one of these species, the Piratic Flycatcher (*Legatus leucophaius*), harass caciques until they abandon their nests, which the flycatchers then use for their own eggs and young. The second species, the Russet-backed Oropendola (*Psarocolius angustifrons*), sometimes destroys the eggs or kills the young in cacique nests near its own in mixed colonies. The third species, the Troupial (*Icterus icterus*), exhibits both forms of behavior, it takes over a nest for its own use and then destroys the eggs and kills the young in adjacent cacique nests. The driving force behind nest destruction is unlikely to be interspecific competition for food because caciques, oropendolas, and Troupials forage in very different microhabitats. Instead, I hypothesize that the adaptive value of the destruction of neighboring nests is the creation of a maze of empty nests which makes it difficult for predators to find the occupied active nest. Direct observations of predator attacks on colonies indicate that such mazes are effective defenses against several frequent nest predators.

Because nest pirates are very rare compared with the cacique host, their impact on the distribution of host nests appears to be negligible. Rarity of the piratic species relative to the host may be a necessary condition for the evolution of nest piracy, as in mimicry.

RESUMEN. Describo la interacción entre la calandria (*Cacicus cela*) en el sureste de Perú y tres especies que comúnmente piratean sus nidos o destruyen sus huevos y matan los polluelos. Individuos de una de estas especie, el atrapa moscas pirata (*Legatus leucophaius*) hostigas a las calandrias hasta que ellas abandonan sus nidos, los cuales son usados por los atrapa moscas para sus propios huevos y polluelos. La segunda especie, el oropéndola de lomo rojizo (*Psarocolius angustifrons*), a veces destruye los huevos y mata a los polluelos en los nidos de calandrias cercanos a sus propios nidos en colonias mixtas. La tercera especie, el troupial (*Icterus icterus*), presenta ambas formas de comportamiento, se apodera del nido para su propio uso y luego destruye los huevos y polluelos de los nidos de calandrias circundantes. La fuerza que lleva a la destrucción de los nidos dificilmente sea competencia interespecifica, ya que las calandrias, los oropéndolas y los troupiales se alimentan en microhabitats muy diferentes. Mi hipótesis, en cambio, es que el valor adaptativo de la destrucción de los nidos vecinos es la creación de un laberinto de nidos vacios que dificulte a los depredadores, el encontrar el nido activo. Observaciones directas de ataques de depredadores a colonias, indican que semejantes laberintos son defensas efectivas contra ataques frecuentes de depredadores de nidos.

Debido a que los piratas de nidos son muy raros comparandolos con las calandrias anfitrionas, su impacto en la distribución de la especie anfitriona parece ser mínimo. La rareza en la abundancia de la especie pirata en relación con la especies pirateada, parece una condición necesaria para la evolución del comportamiento de pirata de nidos como en el caso de mímica.

A rare but extreme form of interspecific aggression is nest piracy in which one species usurps an active nest of another species and then uses it for its own eggs and young. Two well-documented examples of nest piracy exist for the neotropics. First, Piratic Flycatchers (*Legatus leucophaius*) take over the nests of many species that build enclosed nests, including the Yellow-rumped Cacique (*Cacicus cela*) (Skutch 1960; Morton 1977). Second, Troupials (*Icterus icterus*) pirate the domed nests of the Plain-fronted Thornbird (*Phacellodomus rufirons*) (Skutch 1969; Thomas 1983). The adaptive value of nest piracy is clear; the nest pirate is able to nest in relatively well-protected enclosed sites without having to build its own nest.

In effect, the nest pirate is parasitizing the specialized nest-building behaviors and morphologies of other species. Pearson (1974) hypothesized that the use of nests of other species may be an intermediate stage in the evolution of brood parasitism.

There are also species, however, that destroy the eggs and kill young in nests of other species, but do not then use the nests for their own eggs and young. This behavior is especially well-developed in wrens (see refs. in Picman 1977), which build their own nests, but also puncture the eggs of other species that nest within their territories. Two hypotheses have been advanced to explain the adaptive value of this behavior. First, it has been suggested that the destruction of eggs and young by the (Long-billed) Marsh Wren (*Cistothorus palustris*) reduces food competition in Marsh Wren territories (Verner 1965; Picman 1977). Marsh Wrens are especially aggressive toward Red-winged (*Agelaius phoeniceus*) (Picman 1980) and Yellow-headed Blackbirds (*Xanthocephalus xanthocephalus*) (Verner 1975), both of which search for insects in the vegetation of marshes and are, therefore, likely to be competitors. Second, for Sedge (Short-billed Marsh) Wrens (*C. platensis*), nest destruction may minimize interference near the nest (Picman and Picman 1980). Sedge and Marsh Wrens often attack nests of Least Bitterns (*Ixobrychus exilis*) (Picman 1977) and Cinnamon Teals (*Anas cyanoptera*) (Picman and Picman 1980), which are not likely to be food competitors, but which are large enough to damage Sedge Wren nests incidentally. Verner (1975) argued that the Yellow-headed Blackbird's tendency to destroy Marsh Wren nests near its own is a defense against Marsh Wren piracy. I suggest a third hypothesis, that destruction of eggs and young in nearby nests creates a maze of empty nests among which it may be difficult for predators to find active nests. Pearson (1974) used this argument to explain why Troupials often nest in abandoned colonies of the Yellow-rumped Cacique.

The Yellow-rumped Cacique and the species that nest with it are an ideal system for studying the adaptive value of nest piracy for three reasons. First, it is possible to study the adaptive value of both nest piracy and destruction of eggs and offspring in the same system and, even in the same species. As I describe in this paper, there is one species, the Piratic Flycatcher, which pirates single cacique nests whereas the Russet-backed Oropendola (*Psarocolius angustifrons*) sometimes destroys the eggs and kills the young in cacique nests surrounding its own. In addition, the Troupial pirates nests for its own eggs and then destroys the eggs and young in adjacent cacique nests. Second, because cacique colonies are regularly attacked by conspicuous, diurnal predators, it is possible to observe how each pirate species responds to predators and how predators search for nests. These data are essential for evaluating the importance of nest predators in selecting for the destruction of eggs and young in surrounding nests. And third, it is possible to observe how and where each pirate species searches for prey and what food items it brings its nestlings. Such data are essential for evaluating the role of food competition in selecting for interspecific nest destruction.

In this paper I describe the interactions between Yellow-rumped Caciques and their three nest pirates, evaluate the three hypothesized advantages of the destruction of the eggs and young of other species, compare the intensity of selection by nest predators and nest pirates, and discuss the conditions that may have led to the evolution of nest piracy.

STUDY AREA AND METHODS

This study was conducted in the Manu National Park in southeastern Peru at 71°19'W, 11°51'S in the Department of Madre de Dios. All observations were made within 4 km of the Cocha Cashu Biological Station, an area of undisturbed lowland floodplain forest of the Manu River (elev. = 400 m). This area includes two oxbow lakes, Cocha Cashu and Cocha Totora on which all observations were made. The biological station is 3 to 4 days by boat from the nearest road. For a detailed account of the study area, its climate and vegetation, see Terborgh (1983).

During a four year (1979–1982) study (Robinson, unpubl. data) of the mating system and foraging behavior of the Yellow-rumped Cacique, I also noted their interactions with other species that nest in the same sites, including the Troupial, Russet-backed Oropendola and Piratic Flycatcher. During all four years, I visited each colony on Cocha Cashu on a daily basis, and in 1982, I also visited each colony of Cocha Totora twice a week. I recorded the position of each cacique nest with respect to other nests, dates of nest initiation and completion, dates on which incubation and nestling-feeding began, and the dates on which young fledged or were preyed upon, or the nest was pirated or abandoned. I made all observations from a dugout canoe or a portable kayak moored at least 25 m from the colonies, which are usually located near water. In order to avoid disturbing the colonies, I did not look in nests. Whenever I observed nest takeover attempts, I recorded which nests were being attacked by which species and described the behaviors used by the attacker and by the defender. I also recorded which individual caciques defended the nests (715 caciques were color-banded).

In 1980, I gathered data on substrate use by foraging birds of both sexes of Russet-backed Oropendolas, Yellow-rumped Caciques, and Troupials. I recorded the substrate a foraging bird was searching when I first observed it (living leaves, dead leaves, palm fronds, or bark). For statistical analyses I used only one substrate from each independent foraging sequence.

Whenever I obtained a clear view of a nestling-feeding female returning to the colony, I recorded the kind of food she was carrying. Some large prey, such as katydids (Orthoptera: Tettigoniidae), 2 to 5 cm long caterpillars (presumably mostly Lepidopteran larvae), and spiders could be identified easily. However, birds of these species sometimes brought back several small prey items rather than single large ones. Small prey were usually crushed together and so were impossible to identify; they were classified as "unidentified small prey."

BREEDING BIOLOGY OF THE YELLOW-RUMPED CACIQUE

In southeastern Peru Yellow-rumped Caciques (adult males = 106.3 ± 7.5 (s.d.) g, N = 126; adult females = 68.9 ± 3.6 (s.d.) g, N = 225) breed from early July through late February. Cocha Cashu has a resident population of 50 to 75 females, 25 to 35 males, and a variable number of short-term residents. At any one time from August through January, usually 50 to 100 nests are active on Cocha Cashu. Cocha Totora has a resident population of about 15 to 20 females, 8 to 10 males, and an extremely variable number of short-term residents. At any one time usually from 15 to 50 nests are active on Cocha Totora. Caciques nest in sites, such as trees on islands and in marshes, that are isolated from the forest. They also nest around Polistine wasp nests, which provide protection against mammals (Robinson, unpubl. data) and botflies (Smith 1968). Colonies range in size from 2 to 240 nests of which as many as 100 may be active at any one time.

Cacique colonies are attacked by many predators of eggs and nestlings, including the Brown Capuchin (*Cebus apella*), White-fronted Capuchin (*C. albifrons*), Squirrel Monkey (*Saimiri sciureus*), Cuvier's Toucan (*Ramphastos cuvieri*), Black Caracara (*Daptrius ater*), Great Black-Hawk (*Buteogallus urubitinga*), and several species of snake. Clusters of nests on islands and around wasp nests suffer the least predation (Robinson, in press). Small colonies in bushes and trees in marshes and on branches of canopy trees that overhang the water are usually completely destroyed by predators.

Yellow-rumped Caciques have a polygynous mating system in which dominant males consort with females during the period immediately prior to the laying of the first of two eggs. The most dominant males consort with many females, whereas males in the bottom half of the dominance hierarchy occupy peripheral positions within a colony and seldom consort with females. Females build the deep enclosed pouch-like nests, incubate, and feed their nestlings with no help from the males. Males do mob predators, however.

INTERACTIONS BETWEEN CACIQUES AND NEST PIRATES

PIRATIC FLYCATCHER

The Piratic Flycatcher (family Tyrannidae) is a small (18–24 g, Haverschmidt 1968) intra-tropical migrant (Morton 1977) that is usually present in the Cocha Cashu area from July through November. In most years there are two pairs in the Cocha Cashu area and one pair in the Cocha Tortora area. Piratic Flycatchers usurp active cacique nests as well as those of oropendolas (Skutch 1960; N. G. Smith, pers. comm.), flycatchers of the genus *Myiozetetes* (Robinson, pers. observ.; N. G. Smith, pers. comm.) and the Scarlet-rumped Cacique (*C. uropygialis*) (N. G. Smith, pers. comm.). From 1979 to 1982 Piratic Flycatchers pirated only four of the 736 active cacique nests on Cocha Cashu. The rest of their nesting attempts were in abandoned cacique (three times) or Russet-backed Oropendola (one time) nests.

Piratic Flycatchers are monogamous; both sexes harass nest owners and young, and mob predators (Skutch 1960; Robinson, pers. observ.). They are extremely effective at mobbing and chasing away avian nest predators (N. G. Smith, pers. comm.). Adults feed almost exclusively on fruit and feed their nestlings both fruit and insects (Skutch 1960; Morton 1977).

Piratic Flycatchers take over active nests of other species by harassing the owners until they abandon them. The behaviors used have been described by Skutch (1960) and will be summarized only briefly here. Pairs of Piratic Flycatchers attack together. They select a nest and dive at the owners as they approach or leave the nest. When cacique females are inside the nest, the flycatchers hover in front of and sometimes perch on the nest entrance, a behavior that usually results in the cacique flying out and chasing the flycatcher. As soon as it leaves the nest, the other member of the flycatcher pair dives at and chases the cacique. By persistently harassing a female, the flycatchers sometimes cause her to abandon the nest. Skutch (1960) observed Piratic Flycatchers tossing out eggs of the original nest owner. After establishing a nest site, the flycatchers occasionally harass other caciques, but do not take over additional nests. In Panamá Piratic Flycatchers sometimes usurp two or three adjacent nests (N. G. Smith, pers. comm.).

Piratic Flycatchers only succeeded in taking over spatially isolated cacique nests, apparently because of the defensive behavior of caciques. Three of the nests they pirated were solitary nests. The fourth was isolated from the other four nests in a colony by a distance of about 1 m. The three times I observed Piratic Flycatchers attacking dense clusters of active nests, they were unsuccessful. Each time the Piratic Flycatcher hovered in front of a nest in a dense cluster, it was attacked by several female caciques including some from neighboring nests. Flycatchers do not single out particular nests in dense groups, but harass several caciques, thus reducing the effectiveness of the attack. On all three occasions, Piratic Flycatchers eventually used nests in these clusters, but only after the nests had been abandoned following predation. Thus, dense clumping of nests is an effective defense against Piratic Flycatchers. E. S. Morton (pers. comm.) and N. G. Smith (pers. comm.) have also observed that Piratic Flycatchers nest in peripheral positions in cacique colonies in Panamá.

RUSSET-BACKED OREPENDOLA

Russet-backed Oropendolas (males = 440 g, N = 1; females = 195 to 215 g, N = 6) nest at the same time and in the same kinds of sites as caciques, with which they sometimes form mixed colonies. Cocha Cashu has a resident population of 15 to 25 females and at least 7 to 8 males. No oropendolas nest at Cocha Totora. Even in mixed colonies, oropendolas usually, but not always, nest apart from caciques (Koepcke 1972). Often, oropendolas and caciques nest in adjacent, but separate trees in the Cocha Cashu lake bed. In one especially large nest tree that was occupied by both species, all of the oropendolas nested on a branch 10 to 15 m from the branch on which all of the caciques were nesting. Oropendolas are highly polygynous; males play no role in building nests or in feeding young (Smith 1968; Robinson, pers. observ.). Males do, however, attack predators and brood parasites such as the Giant Cowbird (*Scaphidura oryzivora*).

In 1979 and 1980, I observed no aggression between caciques and oropendolas. Indeed, in 1980, some caciques actually wove their nests into the sides of oropendola nests. However, in 1981 and 1982, at least one unmarked male oropendola persistently harassed and supplanted any cacique that got to within 2 m of any active oropendola nest. It also attacked five nests within this 2 m radius; all five had been built before the oropendolas switched to this colony. The male oropendola ripped a hole in the side of one nest, pulled out the nestling, and dropped it into the water. On two other occasions, it attacked very old nestlings as they sat in the nest entrance. One of these nestlings fell into the water and was eaten by a Black Caiman (*Melanosuchus niger*). The other flew weakly to a low branch where I caught it; it had a fresh 1.5 cm gash in its belly, presumably inflicted by the oropendola. The other two nests were abandoned before young could have fledged, possibly as a result of harassment by the oropendola. Each of the five nests was isolated from other active nests by at least 1 m.

Whenever male oropendolas attacked dense clusters of nests, they were very aggressively mobbed by male and female caciques. Mobbing was effective in preventing harm to the nestlings (although one nest fell from the tree during a fight), but only at a heavy cost to the cacique. On six occasions I saw a male oropendola catch and hold a cacique in its claws. The cacique was released, but only after other caciques intensified their mobbing. One female cacique lost her tail feathers during such a fight, and a male lost over 20 breast feathers, at least four secondaries and half its tail.

TROUPIALS

Troupials (males = 51–59 g, N = 3; females = 63 g, N = 1) are present in the Cocha Cashu area at least from July through December, and possibly for the rest of the year; they breed at the same time as caciques. Troupials are present at low population densities; I have never found more than one pair per lake in the Manu area. In 1979, 1980, and 1982 there was one pair on Cocha Cashu. In 1981, the female of the mated pair disappeared, and the male, who was color-marked, moved to Cocha Totora where it remained in 1982 with a new mate. In 1982, a new pair of Troupials moved into Cocha Cashu. Troupials were present on Cocha Totora all four years, but I was unable to observe their nesting attempts, except in 1982. T. A. Parker (pers. comm.) also found only one pair of Troupials per oxbow lake in the Rio Tambopata area of southeastern Peru, regardless of the size of the lake. The Troupial is also rare in eastern Ecuador where Pearson (1974) found only five pairs along 20 km of river.

In the Cocha Cashu area, three of the seven nesting attempts by Troupials that I observed were in abandoned cacique nests (see also Pearson 1974). The other four attempts involved the piracy of active cacique nests and the destruction of eggs and young in adjacent active nests.

Troupials attack caciques in pairs and, unlike Piratic Flycatchers, actually grapple with them. Sometimes while fighting, the Troupial and cacique fall into the water. During an aggressive encounter with a cacique, a Troupial crouches low, cocks its tail, waves its head from side to side and pumps its tail, sometimes while singing. When attacking, it lunges upward at the cacique with its sharp bill. Single Troupials are generally able to hold their own in fights against the larger caciques, but pairs of Troupials are always dominant to lone caciques. Thus, being monogamous gives Troupials a tremendous advantage over female caciques, which are seldom aided by males in nest defense against Troupials. In three of the four attacks by Troupials on cacique colonies, male caciques did not attack the troupials. In the fourth, one male grappled with the Troupials for about 10 minutes, but then left the colony.

Before choosing a particular group of nests, Troupials check many different colonies. Once they choose a group, they systematically fight each female and chase away any male that approaches. Troupials use a variety of different approaches to attack females in a nest. Sometimes they hang from the nest and peck incubating females through the nest wall. The pair of Troupials that nested on Cocha Totora in 1982 consecutively attacked seven females inside their nests. Each time, one member of the pair entered the nest while the other waited outside and chased away any other caciques that approached. Then, when the female cacique burst out of the nest, the Troupial that was outside either attacked and grappled with her or chased her from the colony tree. On two occasions when the cacique burst out of its nest, it left behind a cloud of feathers, all of which proved to be its own when I recovered them from the water. Presumably these feathers were pulled out by the Troupial, but it is also possible that some became tangled in the nest and were pulled out during the sudden exit. None of the feathers I recovered was from the Troupial. After evicting each female and tossing out fragments of her eggs, which they had presumably destroyed, the Troupials left. In two days this pair evicted the seven females nesting in the colony and had taken over a nest in the center of the group for its own use. Thomas (1983) has also observed Troupials attacking Plain-fronted Thornbirds inside their nest.

Another pair of Troupials that attacked a group of five nests used somewhat different tactics, attacking females as they flew in and out of their nests. Each time the Troupials chased a cacique from the tree, one entered its nest and 60 to 90 sec later threw out yolky shell fragments, which implies that the eggs had been destroyed. The caciques, including one male, attacked the Troupials as a group and kept them out for about 20 minutes. Eventually, the male left, and the Troupials were able to enter all five nests in the group. Again, each time a Troupial entered a nest, the other stood guard and grappled with any cacique that approached.

Later in the breeding season the same pair of Troupials attacked a group of five nests using the same tactics. One female resisted all efforts to evict her and, when I left, was still using her nest, the only one in the tree that had not been taken over by the Troupials. Each time the Troupials attacked, the female sat in the entrance of her nest and pecked them. The Troupials could neither enter the nest nor grapple with her outside the nest. Curiously, the Troupials did not go into her nest when she was off foraging.

Troupials kill nestlings as well as destroy eggs. On one occasion a pair of Troupials attacked a 16 to 17 day old nestling by tearing a hole in the side of a nest. For two days the Troupials pecked and pulled at the nestling through the hole. By the end of the second day, the nestling was clearly dead—it was hanging halfway out of the nest.

ADAPTIVE VALUE OF NEST PIRACY

The theft and use of the nest of another species would clearly benefit the pirate if it could usurp the nest with less effort than it takes to build a comparably well-protected site on its own. By taking over cacique nests, Piratic Flycatchers and Troupials are able to occupy newly-built, intricately woven enclosed nests constructed in sites that are well-protected from mammalian predators. In effect, nest pirates are parasitizing the specialized nest-building behaviors of the cacique. The energetic costs of piracy are difficult to gauge, however. Relative to the time it takes a cacique to build a nest, a Troupial needs very little time to take over a nest. One Troupial pair evicted 12 caciques from their nests in one afternoon. In contrast, caciques usually need eight to 20 days to build nests. Also, by waiting until the cacique nests are completed and, therefore, are too tightly woven to be robbed, Piratic Flycatchers and Troupials do not have to defend their nests against other caciques, which regularly steal each other's nest material. Piratic Flycatchers, however, are only capable of pirating isolated cacique nests; apparently the Piratic Flycatcher, which is much smaller than the cacique, is incapable of overcoming groups of caciques. Why the Troupial and Russet-backed Oropendola destroy eggs and kill young in cacique nests surrounding their own is less clear. Three explanations are possible:

REDUCED INTERSPECIFIC COMPETITION

By destroying eggs and young in cacique nests near their own, Troupials and oropendolas may be reducing the number of competitors for food in the area around the colony. Troupials, oropendolas, and caciques all eat essentially the same kinds of fruits (chiefly *Ficus* spp. and other Moraceous fruits such as *Coussapoa* spp.) and nectar (largely from *Combretum* spp. and *Quararibea* spp.). And while the nestling diets of the three species consist primarily of arthropods (Fig. 1), caciques feed their young more caterpillars and small unidentified arthropods relative to orthopterans than oropendolas (Fig. 1; $\chi^2 = 36.27$, d.f. = 4, $P \ll 0.001$). Troupials bring significantly more large orthopterans to their young than do caciques (Fig. 1; Fisher Exact Probability Test, $P = 0.0003$), but this is based on a small sample (N = 13 prey items for the Troupial). Thus, while the proportions of prey items differ, all three species bring in many large orthopterans.

However, these three species differ considerably in their insect foraging microhabitats. Troupials search significantly more often in dead leaves than caciques, which mostly search living leaves (Fig. 2; Fisher Exact Probability Test, $P \ll 0.001$). Female oropendolas, which differ significantly in their substrate use from male oropendolas (Fig. 2; $\chi^2 = 53.2$, d.f. = 3, $P \ll 0.001$), search significantly more in dead leaves and palm fronds than do caciques (Fig. 2; $\chi^2 = 122.3$, d.f. = 2, $P \ll 0.001$). Similarly, male oropendolas search bark on horizontal limbs of canopy trees significantly more often than do caciques, which search almost exclusively in outer foliage of trees and vines (Fig. 2; Fisher Exact Probability Test, $P \ll 0.001$). Troupials search almost exclusively in the dense vines along the edge of the lake, whereas caciques forage extensively inside the forest, away from the lake edge.

Because Troupials, Russet-backed Oropendolas, and Yellow-rumped Caciques search markedly different microhabitats and substrates for insects, it is unlikely that the destruction of cacique nests by Troupials and oropendolas would significantly reduce interspecific competition for arthropods. Likewise, the fruits and nectar eaten by these three species are abundant and are used also by other birds, primates, bats, and a variety of other mammals and insects (Janson et al. 1981; J. Terborgh, pers. comm.; Robinson, unpubl. data). The proportion of fruit and nectar saved by the destruction of cacique nests is therefore likely to be a very small proportion of that available. Also, most evicted females immediately renest in the same area and continue to use the same fruit and nectar sources.

Another indication that food competition is unlikely to be the driving force behind interspecific nest destruction is that both Troupials and oropendolas only attack neighboring nests, regardless of the number of active nests in that colony. In order to significantly reduce competition with caciques, Troupials would have to destroy more nests in larger colonies. Yet, Troupials on Cocha Cashu destroyed roughly the same number of nests in colonies that differed widely in size. For example in one colony of 60 active nests, a pair of Troupials only destroyed the eggs and killed the young in the five nests nearest the nest it had taken over for its own eggs and young. On two other occasions, a pair of Troupials pirated and destroyed eggs in all of the nests in a colony but made no attempt to evict caciques from adjacent colonies that were less than 20 m away.

REDUCED INTERSPECIFIC INTERFERENCE

The adaptive value of nest destruction may be that it reduces interference from caciques. Second-year male caciques chase and peck cacique females and fledglings, and at least occa-

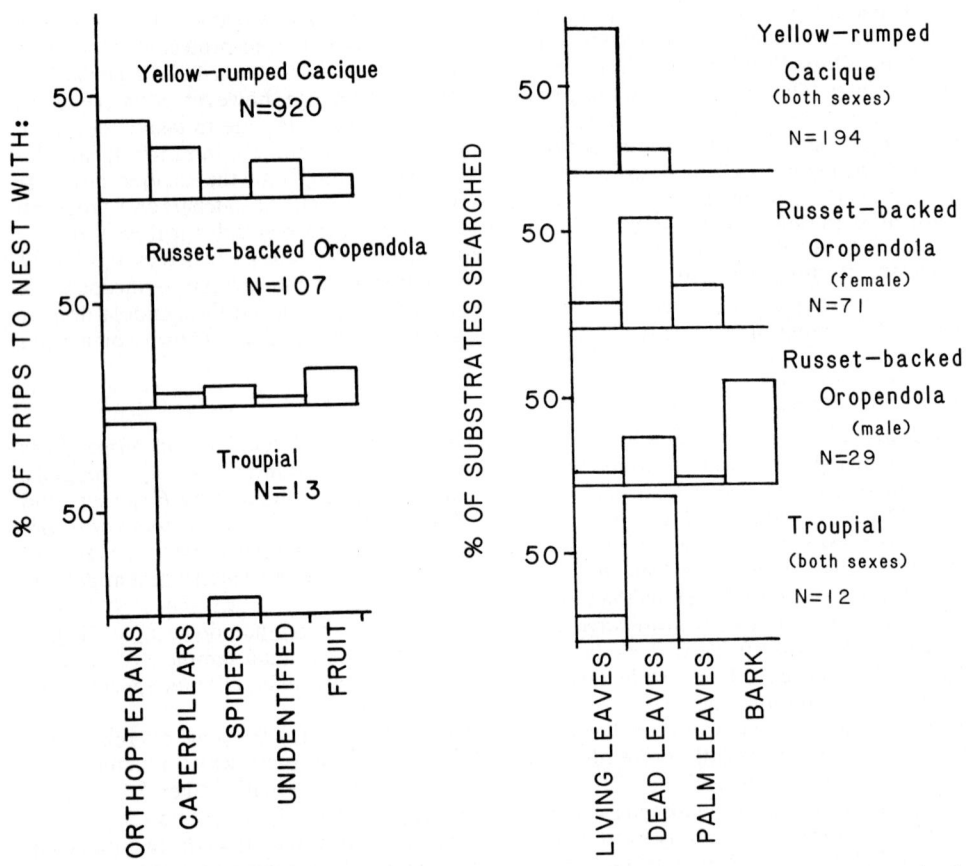

Fig. 1. Nestling diets of Yellow-rumped Caciques and two species, Troupial and Russet-backed Oropendola, that destroy eggs and kill young in cacique nests, August through November 1980. Data are from 48 female caciques, 10 female oropendolas and one pair of Troupials. Unidentified prey consists of several small caterpillars crushed together in most cases.

Fig. 2. Substrates searched by Yellow-rumped Caciques and two species, Troupial and Russet-backed Oropendola, that destroy eggs and kill young in cacique nests. N = number of independent sequences from which only the first substrate searched is used (see Methods).

sionally kill conspecific nestlings (Robinson, unpubl. data). However, I have never seen an adolescent male harass an oropendola or Troupial nestling or fledgling. Female caciques occasionally steal nest material from oropendolas, but most of the destruction of cacique nests by oropendolas occurs after the oropendolas have finished building their nests, and completed nests cannot be effectively robbed. Thus, there is no evidence that reduced interference is the primary adaptive value of nest destruction.

REDUCED CHANCES OF NEST PREDATION

I believe that the primary advantage of the destruction of eggs and young in cacique nests is that it creates a situation in which the active nest is hidden from a searching predator amid a maze of empty nests. Cacique females mob only when their own nests are being attacked by predators. Therefore, by destroying adjacent nests, the pirates do not necessarily decrease their protection from nest predators. Furthermore, caciques do not mob several of their most important predators, including primates, Great Black-Hawks and, presumably, snakes, which attack at night. The enclosed, pouch-like nests of the cacique are very difficult to search. Avian predators have to tear holes in the nests to see inside, whereas snakes and primates reach inside. Great Black-Hawks shake nests before opening them, a behavior that may reveal the presence of large nestlings inside. Great Black-Hawks generally give up after shaking or tearing open three or four empty nests. Cuvier's Toucans give up after tearing open one or two empty nests (Robinson, in press). I have even seen a Brown Capuchin Monkey give up after reaching in only one nest in a group of five, three of which had large nestlings. Thus, a maze of empty nests in a colony may cause Cuvier's Toucans, Great Black-Hawks, and, at least occasionally, primates to give up and try another colony or group within that colony. Because all of these predators return repeatedly to groups in which they were successful, it is particularly important to prevent them from finding young the first time they attack. Black Caracaras, unlike the above-mentioned predators, are able to distinguish between active and inactive nests, so they are less likely to be fooled by empty nests in a colony.

During one series of predator attacks on a colony, I actually saw a maze of nests provide protection for a Troupial nest. On four consecutive occasions, lone Cuvier's Toucans attacked a group of three nests in which Troupials were nesting. Two of these toucans opened two empty nests that had been destroyed by the Troupials (the Troupial's nest was the farthest out on the branch and, hence, the hardest to reach). The other two toucans left after only briefly searching the nests that had already been opened. The Troupial nest was the only nest in that colony, which included two other groups, to escape predation; every active cacique nest in that colony was found by toucans.

In this context, it is interesting that Troupials do not mob predators. When predators attack, Troupials skulk in dense cover near the colony. Caciques do not mob Great Black-Hawks or primates either, although they do mob Black Caracaras and Cuvier's Toucans. It is, therefore, possible that caciques also rely on the maze effect to deter some predators since most groups of cacique nests consist of a mixture of active and abandoned nests. Piratic Flycatchers, on the other hand, mob predators very effectively (N. G. Smith, pers. comm.) and only take over one nest. This suggests two strategies of nest defense. Troupials adopt a passive defense hiding their nests in a maze of empty nests which they have created. Piratic Flycatchers adopt an aggressive defense that requires no nest destruction. Finally, caciques adopt either strategy depending upon which predator is attacking. This may explain why groups of caciques often build their nests together in the middle of clusters of abandoned nests.

Why Russet-backed Oropendolas nest apart from caciques is less clear. Male oropendolas aggressively exclude all caciques from the area around their nests with behavior akin to interspecific territoriality. The only major vertebrate predators of the Russet-backed Oropendola are large snakes and Great Black-Hawks. Lone oropendolas can easily chase away Cuvier's Toucans and Black Caracaras, and oropendolas only nest in sites that cannot be reached by primates. Oropendolas have no defense against Great Black-Hawks, and most leave the colony when these hawks attack. However, I have never seen a Great Black-Hawk tear open an active oropendola nest. Snakes, however, are likely to be very important predators of oropendola nests. Most oropendola colonies that failed were abandoned over a period of several days during which the nests were apparently preyed upon at night, a pattern characteristic of snake attacks (Robinson, in press). On one occasion I found a large snake (species unknown) at the bottom of the tree. Spatial segregation between oropendolas and the far more numerous cacique nests may reduce the chances of a snake finding oropendola nests, especially if snakes preferentially attack dense clusters of nests. Indeed, both times I found a snake in a colony tree, cacique nests in the largest clusters were being abandoned at night. At present, however, I know too little about how snakes search for nests to evaluate their effect on nest distributions.

I believe that the destruction of surrounding nests by Russet-backed Oropendolas and Troupials acts as a predator deterrent by creating a maze around active nests. As Pearson (1974) proposed, Troupials could receive the same benefits from using nests in abandoned colonies. Abandoned nests, however, are usually old and may fall during strong winds and rain. The indiscriminate destruction of the nests of other species by *Cistothorus* Marsh Wrens (Picman 1977; Picman and Picman 1980) may serve the same function, deterring predators from searching an area that either has few active nests or that appears to have been attacked by a predator already. The numerous "dummy" nests built by Long-billed Marsh Wrens (Verner 1965) may also form a maze that deters predators. These nests are enclosed like cacique nests and, therefore, are difficult to search, which enhances their value as predator deterrents. Such predator defenses may be especially important in marshes, where nest predation is very high (Ricklefs 1969). The related behavior of the Yellow-backed Oriole (*Icterus chrysater*) in Panamá, which destroys eggs and kills young in groups of cacique nests but then builds its own nest (N.G. Smith, pers. comm.), may also serve the same function.

EFFECTS OF NEST PIRACY AND DESTRUCTION ON CACIQUE COLONY STRUCTURE

Oropendolas and Piratic Flycatchers destroy or pirate significantly more spatially isolated nests than would be expected if nests were selected at random. All eight nests pirated or destroyed by oropendolas and Piratic Flycatchers were at least 0.5 m from the nearest active cacique nest. Only 186 of the 845 cacique nests on Cocha Cashu and Cocha Totora were spatially isolated by at least 0.5 m (Fisher Exact Probability Test, $P \ll 0.001$). Neither pirate species, however, destroyed or stole a large enough proportion of the nests to have a significant impact on cacique nest distribution. In Panamá, where virtually every cacique colony is occupied by a pair of Piratic Flycatchers (E. S. Morton, pers. comm.; N. G. Smith, pers. comm.), the impact of the flycatchers on cacique nest distribution may be greater.

Troupials tend to take fewer isolated nests than expected if selection is random, but the difference is not significant ($\chi^2 = 3.30$, d.f. = 1, $P = 0.07$, N = 27 nests of which only two were more than 0.5 m from the nearest active cacique nest). However, compared with avian nest predators such as Cuvier's Toucans and the Black Caracara, which prey upon spatially isolated nests (Robinson, in press), the selection against clumped nests by Troupials may be negligible. Altogether, Troupials pirated and destroyed only 1.2 percent of the 736 nests on Cocha Cashu compared with the 5.3 percent preyed upon by Cuvier's Toucan and the 4.9 percent destroyed by the Black Caracara. On Cocha Totora where there are fewer cacique nests, the proportional impact of Troupials is considerably greater. In 1982, Troupials pirated or destroyed 15.2 percent of the 105 nests present.

THE EVOLUTION OF NEST PIRACY

Nest piracy and brood parasitism are similar to each other in that both involve parasitism of the reproductive efforts of another species. Hamilton and Orians (1965) developed a model that includes three conditions favoring the evolution of brood parasitism: (1) the potential parasite should be relatively rare compared with the host, (2) the host should be a colonial nester because these species show reduced territorial behavior and their nests are therefore easy to find, and (3) the potential parasite should be closely related to the host which should, in turn, increase the probability that the nestling diet will be appropriate. Pearson (1974) has indeed proposed that Troupials, which meet all three conditions and have a similar diet to that of the cacique, are an example of a species in the early stages of evolving brood parasitism. For the evolution of nest piracy, the third condition listed above is not relevant since the pirate feeds its own young. Instead, I would substitute the condition that the potential host species should nest in or near the preferred foraging areas of the potential nest pirate. A fourth condition for the evolution of nest piracy may be that the host builds well-protected nests in sites that are safe from predators.

Troupials and Piratic Flycatchers meet all of the above conditions for the evolution of nest piracy. First, Troupials and Piratic Flycatchers are very rare with respect to their host species and their impact on cacique social behavior is probably swamped by the effects of nest predators. If Troupials were any commoner, they might select for more effective defenses on the part of the caciques. Second, the host species, the Yellow-rumped Cacique, is colonial, their nests are extremely conspicuous, and most females only defend their own nests, thus making it easy for pairs of nest pirates to dominate them in aggressive encounters. The nests of the Plain-fronted Thornbird, which are pirated by several species (Thomas 1983) are also conspicuous and only weakly defended. Third, both Piratic Flycatchers and Troupials feed near the edge of lakes and in second growth, which is where most Yellow-rumped Cacique colonies are located. Fourth, caciques build large, enclosed nests that are difficult to search and are often situated around wasp nests and on islands, which are well protected against mammals. The rarity of nest piracy among birds may be due to the restrictive nature of most of these conditions.

ACKNOWLEDGMENTS

I would like to thank J. Terborgh, my thesis adviser, for encouragement and for the opportunity to work at Cocha Cashu. D. Rubenstein, H. Horn, and J. Hoogland provided advice at critical stages of my work. D. Rubenstein, P. Buckley, J. Terborgh, M. Foster, N. Smith, J. Picman, R. Ridgely, and H. Horn provided much constructive criticism of the manuscript. I thank the Ministry of Agriculture of the Peruvian Government for permission to work in the Manu National Park. I received financial support for this work from the Frank M. Chapman Memorial Fund of the American Museum of Natural History, Society of Sigma Xi, Department of Biology at Princeton University, and from the National Science Foundation (Dissertation Improvement Grant DEB-8025975).

LITERATURE CITED

HAMILTON, W. J., III, AND G. H. ORIANS. 1965. Evolution of brood parasitism in altricial birds. Condor 67:361–382.

HAVERSCHMIDT, F. 1968. Birds of Surinam. Oliver and Boyd, Edinburgh.

JANSON, C. H., J. TERBORGH, AND L. H. EMMONS. 1981. Non-flying mammals as pollinating agents in the Amazonian forest. Biotropica Suppl. 13:1–6.

KOEPCKE, M. 1972. Uber die Resistenzformen der Vogelnester in einem begrenzten Gebiet des tropischen Regenwaldes in Peru. J. Ornithol. 113:138–160.

MORTON, E. S. 1977. Intratropical migration in the Yellow-Green Vireo and Piratic Flycatcher. Auk 94:97–106.

PEARSON, D. L. 1974. Use of abandoned cacique nests by nesting Troupials (*Icterus icterus*): precursor to parasitism? Wilson Bull. 86:290–291.

PICMAN, J. 1977. Destruction of eggs by the Long-billed Marsh Wren (*Telmatodytes palustris palustris*). Can. J. Zool. 55:1914–1920.

PICMAN, J. 1980. Impact of marsh wrens on reproductive strategy of Red-winged Blackbirds. Can. J. Zool. 58:337–350.

PICMAN, J., AND A. K. PICMAN. 1980. Destruction of nests by the Short-billed Marsh Wren. Condor 82:176–179.

RICKLEFS, R. E. 1969. An analysis of nesting mortality in birds. Smithsonian Contrib. Zool. 9:1–48.

ROBINSON, S. K. In press. Coloniality as a defense against nest predators in the Yellow-rumped Cacique. Auk.

SKUTCH, A. F. 1960. Life histories of Central American birds, II. Pac. Coast Avif. 34:451–464.

SKUTCH, A. F. 1969. A study of the Rufous-fronted Thornbird and associated birds (Pts. 1, 2). Wilson Bull. 81:5–43, 123–139.

SMITH, N. G. 1968. The advantage of being parasitized. Nature 219:690–694.

TERBORGH, J. 1983. Five New World Primates: A Study in Comparative Ecology. Princeton University Press, Princeton, New Jersey.

THOMAS, B. T. 1983. The Plain-fronted Thornbird: nest construction, material choice and nest defense behavior. Wilson Bull. 95:106–117.

VERNER, J. 1965. Breeding history of the Long-billed Marsh Wren. Condor 67:6–30.

VERNER, J. 1975. Interspecific aggression between Yellow-headed Blackbirds and Long-billed Marsh Wrens. Condor 77:329–331.

ADAPTATION TO A NOVEL ENVIRONMENT: FOOD, FORAGING, AND MORPHOLOGY OF THE COCOS ISLAND FLYCATCHER

Thomas W. Sherry

Department of Biological Sciences, Dartmouth College, Hanover, New Hampshire 03755 USA

ABSTRACT. Cocos Island, Costa Rica, is a small and lushly forested island in the tropical eastern Pacific Ocean between Costa Rica and the Galapagos Archipelago. During two expeditions there, I quantified the stomach contents, available food (i.e., foliage-inhabiting arthropods sampled with sweep nets), foraging behavior, and morphology of the endemic Cocos Island Flycatcher (*Nesotriccus ridgwayi*, Tyrannidae). *Nesotriccus* individuals captured a diversity of arthropods in proportion to their availability (P $\gg$ 0.1), using diverse foraging tactics. Stages of the birds' annual cycles differed during the two expeditions, but diet and foraging behavior were remarkably consistent. Fulgoroid Homoptera dominated stomach contents (43–64% of prey individuals), and probably explain why *Nesotriccus* foraged regularly with acrobatic pursuits (11–12% of all feeding tactics) much like a mainland Homoptera specialist, *Terenotriccus erythrurus*. *Nesotriccus* is morphologically and behaviorally distinct from its primarily frugivorous mainland relatives, *Phaeomyias* and *Capsiempis*; its wings and tail are structurally convergent with those of *Terenotriccus*, but its bill is comparatively longer and probably evolved for the capture of non-homopteran insects.

Nesotriccus is a food specialist or generalist depending on one's frame of reference — available food, mainland insectivorous flycatchers, closest mainland relatives, or other resident land birds on the island. The diet and adaptations of *Nesotriccus*, in combination with other evidence, strongly support the hypothesis that insufficient abundance of many resource types precludes persistence in Cocos Island forests by virtually all but the endemic land birds. High endemism of the depauperate land bird fauna on the island appears to have resulted as much from this ecological impoverishment as from a lack of potential immigrants.

RESUMEN. La Isla del Coco es una isla costarricense pequeña y densamente arbolada en el océano Pacífico tropical, entre Costa Rica y el Archipiélago de las Galapagos. Durante dos expediciones a la isla cuantifiqué los contenidos estomacales, alimento disponible (es decir artrópodos habitantes de los follajes muestreados con redes a mano), comportamiento de forraje y morfología del atrapamoscas endémico (*Nesotriccus ridgwayi*, Tyrannidae) de la Isla del Coco. Individuos de *Nesotriccus* utilizan diversas tácticas de forraje para capturar una diversidad de artrópodos en proporción a disponibilidad (P $\gg$ 0.1). Las etapas del ciclo anual de las aves fueron diferentes durante las dos expediciones, pero la dieta y el comportamiento de forraje fueron consistentes de manera remarcable. Los contenidos estomacales estuvieron dominados por Homoptera (Fulgoroidea, 43–64% de los individuos presa) y probablemente eso explique porque *Nesotriccus* forrajea regularmente con cazas acrobáticas (11–12% de todas las tácticas de alimentación) de manera muy similar a la especialista en Homoptera del continente *Terenotriccus erythrurus*. *Nesotriccus* es morfológicamente y por su comportamiento, distinto de sus parientes del continente, primariamente frugívoros, *Phaeomyias* y *Capsiempis*; sus alas y cola son de estructura convergente con aquellas de *Terenotriccus* pero su pico es comparativamente más largo y probablemente ha evolucionado para capturar insectos no-homopteros.

Nesotriccus se alimenta como un especialista o no-especialista, dependiendo del marco de referencia en el que se lo ubique — comida disponible, los atrapamoscas insectívoros del continente, parientes más cercanos del continente u otras aves terrestres residentes en la isla. La dieta y adaptaciones de *Nesotriccus*, en combinación con otra evidencia, apoya fuertemente la hipótesis de que las aves, con excepción de las terrestres endémicas, al no disponer de manera abundante de muchos tipos de recursos se les hace imposible la permanencia en bosques de la Isla del Coco. El alto endemismo de la paupérrima fauna de aves terrestres de la isla, parece haber resultado tanto por su empobrecimiento ecológico como por la falta de inmigrantes potenciales.

Small, remote oceanic islands tend to have few resident species of land birds, and these are often endemic and broad-niched (MacArthur 1972; Lack 1976; Williamson 1981). Island biologists do not currently agree whether the most important cause of this is reduced dispersal or a poverty of ecological resources on such islands (Lack 1976; Abbott 1980; Williamson 1981). The difficulty in resolving this controversy arises from the lack of knowledge of available resources and how island birds use them (Abbott 1980). Many biologists have discussed the morphological characteristics of birds in island populations (e.g., Lack 1947; Van Valen 1965; Grant 1968, 1981; Keast 1968; Grant et al. 1976), but again, a lack of information about available resources and their use has limited interpretation of the data (Abbott 1980). Clearly, the need to study resource availability, resource use, and adaptations of island birds is compelling.

High endemism and low species turnover characterize the land birds of Cocos Island, Costa Rica, and challenge contemporary beliefs about island biogeography (Slud 1976; Abbott 1980). Ecological studies of the biota of this island are virtually nonexistent, probably due in part to its isolation, small size, and humid climate. In this paper I examine the diet, available resources, foraging behavior, and morphology of the Cocos Island Flycatcher (*Nesotriccus ridgwayi*, Tyrannidae), an endemic genus. Specifically, I address six questions: (1) How does *Nesotriccus* feed? (2) What is the nature of the food resource exploited relative to resources available? (3) How seasonal are the behavior and diet of *Nesotriccus*? (4) How has *Nesotriccus* become adapted morphologically to exploit available resources? (5) Is *Nesotriccus* a generalist or specialist? (6) Why is the land bird fauna of Cocos Island depauperate and largely endemic?

The Cocos Island Flycatcher is the second most abundant of four resident land bird species (Slud 1967; Sherry and T. K. Werner, unpubl. data). A study of *Nesotriccus* is timely because of new information concerning its phylogeny (Lanyon 1984; Sherry, in press), a conspicuous gap in recent discussions of tyrannid genera (Traylor 1977; Fitzpatrick 1980; Traylor and Fitzpatrick 1982), recent ecological studies of mainland tyrannids (Fitzpatrick 1980; Traylor and Fitzpatrick 1982; Sherry 1982, 1983, 1984), and recent studies of how other land birds on Cocos Island exploit available resources (Smith and Sweatman 1976; Werner and Sherry, unpubl. data).

STUDY SITE

Cocos Island, Costa Rica, lies in the tropical eastern Pacific Ocean (5°32'57"N, 86°59'17"W), approximately 630 km northeast of the Galapagos Archipelago and 500 km southwest of Costa Rica. The island is approximately 46.6 km² in area and rises from abrupt cliffs (with heights up to 180 m) around most of its perimeter to a maximum elevation of 573 m on Cerro Iglesias (Hertlein 1963; F. Cortés, pers. comm.). Geologically, Cocos Island, like the Galapagos Islands, arose volcanically during the Pliocene period and probably was never connected with the mainland by a land bridge (Dalrymple and Cox 1968). Typically, the eastward-flowing, equatorial counter-current includes the latitude of Cocos Island, and the warm waters of this current help create a humid climate (up to 8 m of rain annually, C. Hogue, pers. comm.), with the least rainfall between January and March. Air temperatures range between 20°C and 33°C (Hertlein 1963).

Cocos Island is densely forested throughout. It is floristically impoverished with only 155 vascular and 48 nonvascular plant species (Fournier 1966). Life forms are diverse and include epiphytes, palms, ferns, understory shrubs, lianas, vines, and trees. Most behavioral observations were made in three locations with differing vegetation: (1) *Hibiscus tiliaceus* and melastome thickets, ca. 8–10 m high, near the Bahia Wafer beach; (2) interior forest dominated by *Saccoglottis* trees, ca. 20–25 m tall, with an understory including melastomes and rubiaceous shrubs, and with generally little ground cover due to the activity of feral pigs; (3) Mirador cloud forest, ca. 460 m elevation, comprised of an epiphyte-laden canopy dominated by *Saccoglottis* (less than 20 m tall), an understory of tree ferns and occasional *Rooseveltia* palms, and dense ground cover of Piperaceae spp., with occasional ferns.

The island is also faunistically impoverished. At present, 363 arthropod species (Hogue and Miller 1981), two lizards, and 82 bird species (Slud 1967; T. W. Sherry, F. G. Stiles, and T. K. Werner, unpubl. data), but no snakes, lizards, amphibians, or native mammals are known to occur there. Deer, pigs, cats, and rats have established populations (Hogue and Miller 1981), and pigs occur virtually throughout the island (Sherry, unpubl. data). There are only four resident land bird species, of which three are endemic. Cocos Island has never had a permanent settlement and was declared a Costa Rican national park in 1978 (Hogue and Miller 1981).

METHODS

Most observations were made during two brief expeditions to Cocos Island (30 June–6 July, 1978, and 24 February–24 March, 1980); a few observations made during February to April 1984 are also reported. It rained every day during the 1978 expedition, as is typical for that season (Hertlein 1963), and the birds collected were all in a late stage of molt. It was drier during the 1980 expedition with rain about every third day; the birds were either in a late stage of the reproductive cycle, or beginning to molt (Sherry, in press).

Diet and Available Food

To facilitate dietary comparisons between *Nesotriccus* and mainland insectivorous flycatchers, the methods used to collect and analyze stomach samples were standardized (see Sherry 1984, for details). I collected actively foraging birds (five males, four females, and two of unknown sex in 1978; four males, 10 females, and one post-fledging male in 1980), mostly before noon, to ensure full stomachs. Birds were collected in *Hibiscus* thickets and interior forest (three widely separated locations) in 1978, and in all habitats but mostly interior forest and cloud forest in 1980. Stomachs were dissected from a bird immediately upon collection and stored in 90% ethanol. All recognizable food remains (arthropods and seeds) were later removed from the contents such that the minimum number of individuals or fruits present could be estimated. For arthropods I reconstructed individuals from body parts so as not to overcount individuals (Sherry 1984). Arthropods were identified to order or family, and species were often identifiable due to the low insect diversity on the island. A reference collection in the Natural History Museum of Los Angeles County (Hogue and Miller 1981) facilitated these identifications.

Because *Nesotriccus* fed extensively on foliage-inhabiting arthropods in 1978 (see below), available arthropods were sampled in 1980 by vigorously beating understory vegetation in 100-sweep units with a 38 cm-diameter beating net (Janzen 1973). Sweeping was conducted during the late morning to correspond as closely as possible with times that birds were collected. Arthropods were killed with cyanide, sorted from plant debris, then identified. Because *Nesotriccus* individuals rarely if ever eat arthropods less than 0.5 mm long (Sherry, unpubl. data), I made no effort to sort smaller arthropods from the debris (see Hespenheide 1979).

I used curves of dietary diversity plotted against sample size (number of stomachs) to determine adequacy of sample size and to estimate dietary diversity of the *Nesotriccus* population in a particular season (Hurtubia 1973; Pielou 1975; Sherry 1984). Although the Shannon-Wiener diversity index (H′) is appropriate and frequently used to estimate collection diversity from a sample, the Brillouin diversity measure,

$$H = (1/N)\log_e(N!/(n_{1!} \cdot n_{2!} \ldots n_{t!})),$$

where n_i = number of prey in taxon "i" and $N = \sum_i n_i$, is appropriate for the present application. Specifically, one uses the Brillouin index to estimate diversity from collections (stomachs, in this case) that are not necessarily random samples of the larger collection (Pielou 1975). The diversity of prey contents in stomach 1 is plotted, then the diversity in stomachs 1 and 2, then in stomachs 1 to 3, and so forth; the curve reaches a plateau if additional stomachs contribute no new information concerning the diversity of the pooled stomach contents.

To quantify the heterogeneity of prey taxa in different stomachs (= population dietary heterogeneity, PDH), I calculated $G/d.f.$, where G = the G-statistic for independence (Sokal and Rohlf 1969), and d.f. = degrees of freedom for the particular number of taxa and stomachs examined (see Sherry 1984). To avoid taking logarithms of zero cells I arbitrarily added 0.1 to each frequency value. To compare prey taxa eaten by different species or by one species in two seasons, I calculated Euclidian distances for all pairwise combinations of species or samples taken in different seasons. For this analysis I used $\log_e$ of (1 + % composition) of stomach contents, based on 15 prey taxa (defined by Sherry 1984), and I clustered species (or seasonal samples within species) with a complete-linkage algorithm (Johnson 1967; Holmes et al. 1979). To emphasize frequently eaten taxa I did not standardize prey taxa (to mean = 0, s.d. = 1) across bird species. Comparisons of prey taxa in stomachs with those available were made with the G-statistic (Sokal and Rohlf 1969), using original frequencies.

Original prey taxa eaten by mainland flycatchers clustered into eight statistically significant, biologically meaningful groups (Sherry 1984). *Nesotriccus* stomachs contained six of these: ants, Coleoptera plus Hemiptera, Homoptera plus Arachnida, Orthoptera plus Lepi-

TABLE 1

PERCENT COMPOSITION OF ARTHROPOD REMAINS, AND NUMBER OF FRUIT SEEDS IN STOMACHS OF *NESOTRICCUS RIDGWAYI*[1]

Arthropod taxon	1978	1980		
		Total	Mirador cloud forest	Forest interior
Orthoptera	12.4 (10)[2]	6.4 (10)	4.5 (2)	6.7 (8)
Heteroptera	–	2.0 (5)	6.7 (2)	1.2 (3)
Homoptera	43.5 (11)	62.3 (15)	51.1 (3)	64.3 (12)
Coleoptera	15.3 (9)	3.7 (5)	–	4.4 (5)
Lepidoptera				
Larvae	1.8 (2)	2.0 (6)	2.2 (1)	2.0 (5)
Adults	1.2 (1)	0.3 (1)	–	0.4 (1)
Diptera	4.1 (5)	5.1 (6)	13.3 (2)	3.6 (4)
Hymenoptera				
Formicidae	13.5 (7)	7.1 (10)	11.1 (3)	6.3 (7)
Other	2.9 (3)	5.7 (7)	2.2 (1)	6.3 (6)
Arachnida	5.3 (5)	5.4 (10)	8.9 (3)	4.8 (7)
Total no. fruit seeds	4 (1)[2]	150 (10)	10 (3)	140 (7)
Total no. arthropods	170	297	45	252
Total no. stomachs	11	15	3	12

[1] From birds collected in 1978 and 1980 on Cocos Island, Costa Rica.
[2] Number of stomachs in which taxon was found given in parentheses.

doptera larvae, parasitoid Hymenoptera plus Diptera, and adult Lepidoptera. I used these groups to calculate cumulative prey-taxon diversity and PDH, primarily for comparisons with mainland flycatchers. I further pooled adult Lepidoptera (the three such specimens were moths) with "parasitoid Hymenoptera plus Diptera" to avoid zero cell frequencies with *G*-tests.

FORAGING BEHAVIOR AND MORPHOLOGY

I observed *Nesotriccus* forage in *Hibiscus* thickets and interior forest in 1978 and in all three habitats in 1980. Most observations were made before noon both years, but some were made at other times through the day. I recognized six tactic types (Sherry 1982), hawk, snatch, hover, pounce, glean, and pursue. Cocos Island Flycatchers often pursued evasive prey acrobatically after initiating an attack with another tactic. I treated such cases as two behaviors (one of which was "pursue") when tallying tactic types used. I calculated the diversity of foraging tactics used with the Shannon-Wiener information-theory statistic ($H' = -\sum_{i=1}^{6}$ $P_i \log_e p_i$, where p_i = relative frequency of tactic type "i"), and the number of "equally common tactic types" as $e^{H'}$ (MacArthur 1972). To test whether tactics used were identical in 1978 and 1980, I used the *G*-statistic with original frequencies (Sokal and Rohlf 1969). In addition, I quantified attack distances (cm) and time intervals (sec) between successive attacks (Inter-attack-intervals, IAIs) to compare species with respect to food accessibility and feeding rate. Distributions of both attack distances (ADs) and IAIs were approximately normalized by a logarithmic transformation (Sherry 1982), and I used the geometric mean (approximately equal to the median) as the statistic to compare such distributions.

RESULTS

DIET

Homoptera dominated the arthropod remains in stomachs of *Nesotriccus ridgwayi*, comprising on average between 43 percent and 64 percent of prey individuals (Table 1). Homoptera was the one taxon found in every stomach examined. Fulgoroidea dominated the Homoptera (95%) and included Nogodinidae (at least two species), Cixiidae (at least one species), Flatidae (at least one species) and Tropiduchidae (one species). The remaining Homoptera were Cercopoidea (mostly one species of Cicadellidae). These same families and species of Homoptera were observed in both the 1978 and 1980 samples of stomachs. After Homoptera, Orthoptera (largely one species of Gryllidae) and Hymenoptera (Formicidae—at least several species)

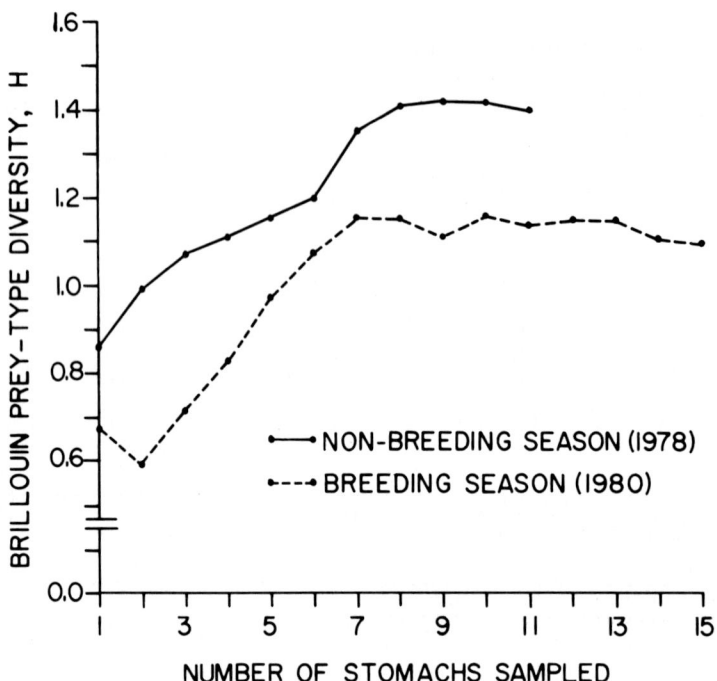

NON-BREEDING SEASON (1978)
BREEDING SEASON (1980)

FIG. 1. Curves of cumulative dietary diversity plotted against sample size (number of stomachs) for breeding-season (1980) and non-breeding-season (1978) samples of *Nesotriccus*.

were about equally important, taking into consideration percentages and frequency of occurrence; Arachnida (spiders) and Coleoptera (largely a zygopine curculionid and a species of Cerambycidae) were next in overall importance (Table 1; Sherry, unpubl. data). Fruit, represented by seeds, was moderately important in the 1980 sample of stomachs and present in only one stomach from the 1978 sample (Table 1). The large numbers of seeds in the 1980 forest-interior sample resulted from one stomach with 122 seeds, probably because far fewer many-seeded fruits were eaten. I never observed *Nesotriccus* eat fruit, but one individual netted near Bahia Wafer in February 1984 regurgitated one *Clusia rosea* seed. Fruit was a minor part of the contents of most, if not all, the stomachs examined.

Individual stomachs of *Nesotriccus* tended to contain just a sample of the arthropod types eaten by a particular population (Fig. 1), but a plateau in the curve of cumulative diet diversity against sample size was evident with both 1978 and 1980 data in samples of eight stomachs or more. The 1978 plateau indicates slightly greater diet breadth than the 1980 sample. Both plateau values of population diet diversity, 1.1 and 1.4 (Fig. 1), are well within the range (0 to >1.5) for mainland insectivorous flycatchers (Sherry 1984). Population dietary heterogeneity (PDH), which quantifies for a collection of stomachs how independent each stomach is from every other with respect to prey taxa, was slightly higher in the more diverse 1978 collection of stomachs (1.68) than the 1980 collection (1.56). Both values, however, fall within the range for mainland flycatchers and would place *Nesotriccus* in the second highest quartile of values observed in 16 mainland species (Sherry 1984). Canopy-inhabiting mainland tyrannids had PDHs most comparable to *Nesotriccus,* which is primarily a canopy forager (Sherry, unpubl. data). I conclude that *Nesotriccus* had a diet neither more nor less diverse (but see below) and neither more nor less heterogeneous than average mainland flycatchers that also feed primarily on arthropods.

I tested the null hypothesis that *Nesotriccus* ate the same arthropod taxa in the wet (1978) as in the dry (1980) season, as indicated by the stomach contents, and rejected this hypothesis (Table 1; $G = 28.0$, 5 d.f., $P < 0.001$). The 1980 collection of stomachs contained relatively more Homoptera, and fewer Orthoptera, Coleoptera, and Formicidae than did the 1978 collection (Table 1).

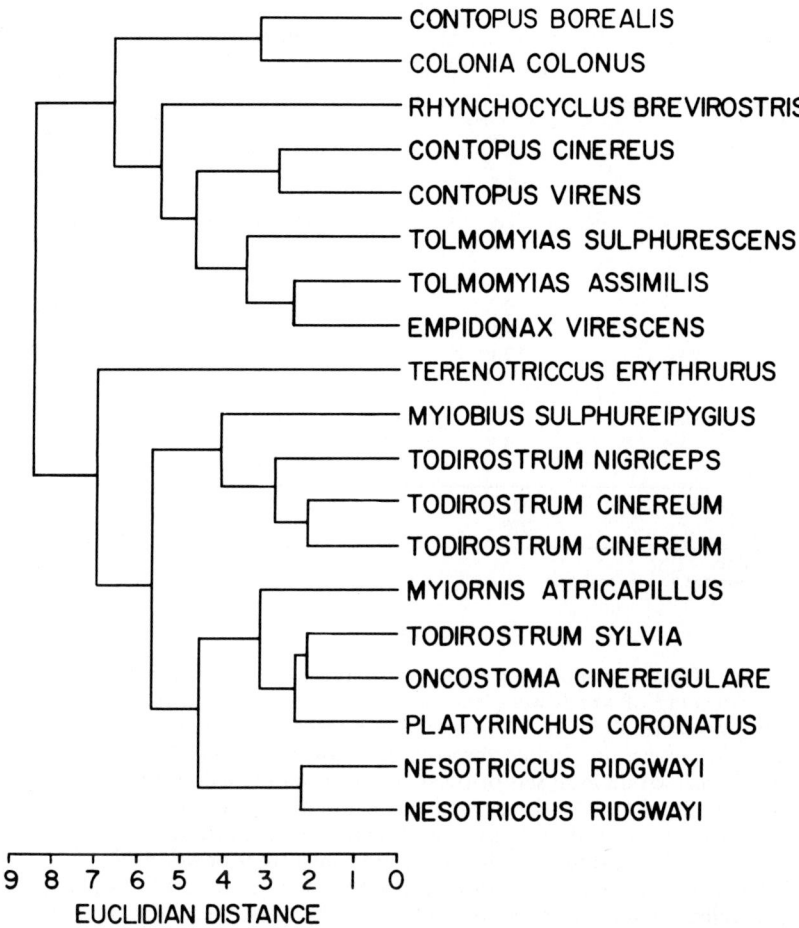

CONTOPUS BOREALIS
COLONIA COLONUS
RHYNCHOCYCLUS BREVIROSTRIS
CONTOPUS CINEREUS
CONTOPUS VIRENS
TOLMOMYIAS SULPHURESCENS
TOLMOMYIAS ASSIMILIS
EMPIDONAX VIRESCENS
TERENOTRICCUS ERYTHRURUS
MYIOBIUS SULPHUREIPYGIUS
TODIROSTRUM NIGRICEPS
TODIROSTRUM CINEREUM
TODIROSTRUM CINEREUM
MYIORNIS ATRICAPILLUS
TODIROSTRUM SYLVIA
ONCOSTOMA CINEREIGULARE
PLATYRINCHUS CORONATUS
NESOTRICCUS RIDGWAYI
NESOTRICCUS RIDGWAYI

9 8 7 6 5 4 3 2 1 0
EUCLIDIAN DISTANCE

FIG. 2. Clustering of tyrannid species based on the pooled arthropod taxa in stomachs. For *Nesotriccus ridgwayi* and *Todirostrum cinereum* both non-breeding and breeding-season samples of stomach contents were available; for all other species just non-breeding samples were available (Sherry 1984).

Dietarily, *Nesotriccus* clustered with a group of small-bodied "euscarthmine" flycatchers on the mainland (Fig. 2; Sherry 1982). Even though the 1978 and 1980 collections of *Nesotriccus* prey were significantly different, they clustered together when compared with mainland flycatchers (Fig. 2). This dendrogram (Fig. 2) contains one other species (*Todirostrum cinereum*) for which breeding and non-breeding season collections of stomachs were available, and again the taxa were not different enough to obscure a predator-specific identity with respect to arthropod taxa in stomachs. This result did not depend on the clustering algorithm used since primary clusters based on similarity matrices invariably combine samples that are each other's nearest neighbor, as were the *Todirostrum* and *Nesotriccus* collections from different seasons.

The final dietary character examined was the number of items per stomach. In 1978, the mean ($\pm$ 1 s.d.) was 15.45 ± 4.78 items per stomach (N = 11 stomachs), whereas in 1980 it was 19.8 ± 7.29 items (N = 15 stomachs). This difference was not statistically significant ($t = 1.72$, 24 d.f., $P > 0.05$).

PREY SELECTION

To understand prey selection, it is important to know what resources are available (e.g., Smith 1982). A total of 403 arthropods were collected with 200 sweeps of a beating net at

TABLE 2

PERCENT COMPOSITION BY TAXON OF UNDERSTORY, LEAF-INHABITING ARTHROPODS IN TWO
HABITATS ON COCOS ISLAND[1]

Arthropod taxon	Mirador cloud forest	Interior forest	Interior forest
Orthoptera	3.2	1.1	2.4
Heteroptera	1.2	0.9	2.0
Homoptera	52.6	6.4	14.1
Coleoptera	3.0	1.1	2.4
Lepidoptera			
Larvae	0.3	0.2	0.5
Adults	3.0	0.9	2.0
Diptera	17.9	18.5	40.9
Hymenoptera			
Formicidae	7.9	64.9	22.5[2]
Other	3.2	1.1	2.4
Arachnida	7.7	4.9	10.8
Total no. arthropods	403	1001	453
Total no. sweeps	200	400	(400)

[1] Data from March 1980.
[2] Ants <2 mm long excluded from the analysis.

the Mirador, and 1001 arthropods were collected with 400 sweeps in interior forest (Table 2). I excluded mites (all of which were less than 1 mm) from all analyses because they were never eaten by *Nesotriccus* (Table 1). Clearly, Homoptera dominated the Mirador sweeps, whereas ants and Diptera dominated forest-interior sweeps (Table 2). This difference was consistent, moreover, within local sets of 100-sweeps. Homoptera dominated both sets of 100 sweeps in Mirador cloud forest whereas ants and Diptera were the two most important taxa in all four 100-sweep samples of forest interior. I therefore examined the relationship between diets and available prey separately for Mirador and forest-interior samples.

Three *Nesotriccus* were collected in 1980 at the Mirador in the vicinity where arthropods were sweep-sampled. I could not reject the null hypothesis that these Mirador flycatchers ate arthropods (Table 1) in proportion to their availability as quantified by the two-hundred sweeps (Table 2; $G = 3.25$, 4 d.f., $P \gg 0.1$). I executed the test again with different groupings of prey, and in no case could I approach rejection of the null hypothesis.

Twelve flycatchers were collected in forest-interior habitat, four of these along Rio Genio in the vicinity of the forest-interior sweep-sample locations and eight in very similar forest in the island interior along ridges above Bahia Chatham. These flycatcher stomachs contained different arthropods (Table 1) from the sweep samples (Table 2; $G = 467.36$, 4 d.f., $P \ll 0.01$). A large number of small ants (<2 mm) in the forest-interior sweeps (Table 2), most of which were in one 100-sweep sample, provided the most obvious explanation for the discrepancy. I excluded ants less than 2 mm long from the forest-interior sweep sample and still observed a significant difference between arthropod taxa in stomachs (Table 1) and those "available" (Table 2, $G = 185.97$, 4 d.f., $P \ll 0.01$).

Two hypotheses may explain why prey taxa were selected in proportion to those available on the Mirador but not in the forest interior. First, flycatcher behavior (selectivity) may not be the same in the two habitats, or second, the sweeps of vegetation may not have sampled what was available to the birds in the forest-interior habitat. Two lines of evidence suggesting that behavior did not differ are (a) that the birds were at a comparable stage of breeding (pers. observ.), thus equalizing demand for food throughout the island, and more importantly, (b) that prey in Mirador and forest-interior stomachs were statistically indistinguishable in size (Sherry, unpubl. data) as well as proportions of taxa represented (Table 1; $G = 2.69$, 4 d.f., $P \gg 0.1$). Because *Nesotriccus* concentrates its foraging in the canopy, and only occasionally pursues an arthropod to the ground (pers. observ.), it forages farther from where I sampled available arthropods in the taller, forest-interior habitat, than in Mirador cloud forest habitat. I tentatively conclude that the Mirador sweeps were more representative of the arthropods

TABLE 3
Percent Composition of Foraging Tactics Used by *Nesotriccus*[1]

Year	Foraging tactic						(N)[2]	H'[3]
	Hawk	Snatch	Hover	Pounce	Glean	Pursue		
1978	15.5	43.1	6.5	8.1	15.5	11.3	123	1.57
1980	20.7	40.0	5.9	11.9	9.6	11.9	135	1.59

[1] Data taken in 1978 and 1980, on Cocos Island, Costa Rica.
[2] N = Sample size.
[3] H' = Shannon-Wiener diversity index (see text).

available to *Nesotriccus* within canopy throughout the island, and that *Nesotriccus* ate arthropods in proportions remarkably similar to those in which they were available.

FORAGING BEHAVIOR

I could not reject the null hypothesis that frequencies of foraging tactics were identical in 1978 and 1980 ($G = 9.04$, 5 d.f., $P > 0.1$). Cocos Island Flycatchers used more snatches to capture prey than any other foraging tactic (Table 3). Most of these were "upward-strikes" rather than "downward-strikes" (for definitions, see Fitzpatrick 1980), 43 versus 11. *Nesotriccus* also employed a variety of other foraging tactics frequently (Table 3) in contrast to many mainland genera of tyrannids that are stereotyped upward-strikers (Fitzpatrick 1980; Sherry 1982). The information-theory indices of diversity (H') were 1.57 and 1.59 for the 1978 and 1980 samples of foraging tactics, respectively (Table 3). These values are 88–89 percent of the maximum possible value (H'$_{max}$ = $\log_e 6 = 1.79$, for six foraging categories). Mainland insectivorous flycatchers, categorized for the same foraging tactics, gave an approximately normal distribution of H' values ($\bar{X} = 0.70$, $s^2 = 0.124$, n = 16 species; Fig. 3). The null hypothesis that *Nesotriccus* foraging tactics were not more diverse than those of ecologically similar mainland species was rejected ($t = 2.4$, 15 d.f., $P < 0.025$, one-tailed). This was a conservative result insofar as the mainland diversity values were inflated because I arbitrarily added 0.001 to each frequency value. The number of "equally-common foraging tactics" for *Nesotriccus* was $e^{1.57} = 4.81$ compared with $e^{0.7} = 2.01$ for an average mainland flycatcher. I conclude that *Nesotriccus* as a population used a more diverse array of foraging tactics (by a factor of greater than two) than mainland insectivorous flycatchers, and this resulted from both the equitability and total number of tactics used by *Nesotriccus*.

I compared inter-attack-intervals (in seconds) in the two seasons. Mean natural logarithm of IAIs ± 1 s.d. were 2.46 ± 0.915 (N = 56) and 2.37 ± 0.7798 (N = 73) in 1978 and 1980, respectively ($t = 0.59$, $P \gg 0.1$). I also compared attack distances (originally in cm). Mean natural logarithm of ADs ± 1 s.d. were 3.89 ± 0.789 (N = 83) and 3.69 ± 0.729 (N = 82), respectively ($t = 1.69$, $P > 0.05$). I conclude that foraging behavior of non-breeding *Nesotriccus* in 1978 was statistically indistinguishable from that of the late-breeding-stage population in 1980 in all respects measured.

DISCUSSION

DIET OF *NESOTRICCUS*

The most conspicuous characteristic of *Nesotriccus* stomach contents is the numerical dominance and consistent representation of Homoptera (Table 1). The high relative abundance of Homoptera where *Nesotriccus* feeds provides the simplest explanation. I cannot reject this hypothesis for the 1980 observations of stomach contents and available arthropods in Mirador cloud forest, and additional data suggest that this hypothesis may have applied in other habitats as well. A relative abundance of Homoptera throughout the island, if a general phenomenon, is consistent with the "disharmoniousness" of the Cocos Island entomofauna (Hogue and Miller 1981) and that of many islands (Williamson 1981). For example, grasshoppers, mantids, treehoppers, many beetle and fly families, most butterfly families, and bees have never been recorded on Cocos Island (Hogue and Miller 1981). A disproportionate representation of Homoptera was also observed on several Caribbean islands (Janzen 1973), although the explanation for this phenomenon, if a general one, remains obscure.

The breadth of prey types eaten is another characteristic of *Nesotriccus* stomach contents. Although the diversity of prey taxa (Fig. 1) was no more or less than that of mainland

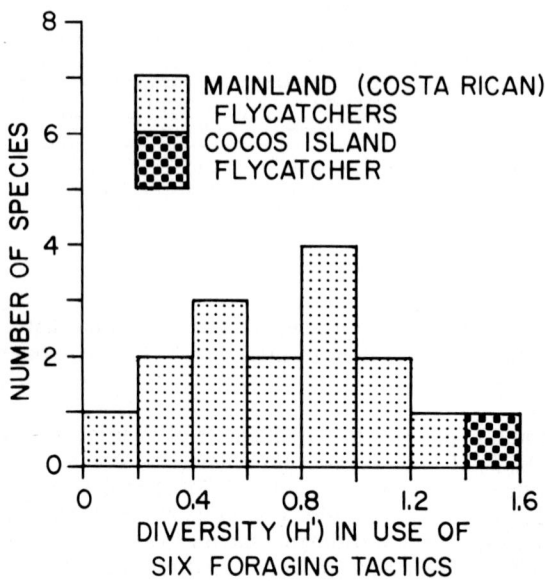

FIG. 3. Frequency distribution of diversity values (H') for the use 17 flycatcher species (*N. ridgewayi* and 16 species from mainland Costa Rica, Sherry 1982) made of six possible foraging tactics.

insectivorous flycatchers, the range of prey *types,* from the perspective of the skills required to detect or catch or handle prey, was greater for *Nesotriccus.* This assertion follows from two kinds of evidence. First, the prey taxa of *Nesotriccus* (Table 1) broadly spanned those of at least four mainland guilds (Sherry 1982): Homoptera plus Arachnida eaten by pursuing species of flycatcher, Orthoptera plus Lepidoptera larvae eaten by searching species, Coleoptera plus Hemiptera plus Formicidae eaten by several canopy or open-country species, and Diptera plus parasitoid Hymenoptera eaten by *Todirostrum* flycatchers. On the mainland, different flycatchers tended to eat one group of arthropod taxa to the exclusion of others depending on where and how the bird hunted, detected, or captured prey (Sherry 1984). Secondly, the diversity of foraging tactics *Nesotriccus* used was significantly greater than that of mainland species (Fig. 3). Furthermore, a factor-analysis of eight behavioral and nine morphological characters of insectivorous Neotropical flycatchers (including *Nesotriccus*) demonstrated that the latter broadly overlaps at least two mainland guilds insofar as it had close to the extreme scores on two multivariate axes ("pursuer" and "pouncer-gleaner," Sherry 1982).

SEASONAL VARIATION IN DIET AND BEHAVIOR

Although arthropod abundance was not quantified in 1978, I noted (field record, 1 July), "how *very few* insects I saw on the vegetation compared with what I would have seen in the lowland rainforest in Costa Rica proper." In 1980, by contrast, arthropods were abundant and easily captured especially in non-forested habitat. Considerations of foraging optimality (reviewed by Pyke et al. 1977) predict that *Nesotriccus* should have been relatively selective when food was abundant in 1980. The 1980 stomach samples indeed contained significantly different arthropod taxa from the 1978 samples and were more dominated by Homoptera. Correspondingly, the cumulative population diversity (Fig. 1) and PDH were lower in 1980 than 1978. Number of items per stomach was greater in 1980, and IAIs and ADs were both shorter (although none of these differences was significant), suggesting higher feeding rates on more accessible food. Availability of fruit (Table 1) may have contributed to overall food abundance. At present, however, I cannot exclude the alternative hypothesis that Homoptera were disproportionately abundant in 1980, compared with 1978, and eaten in proportion to their abundance.

Similarities of stomach contents and foraging behaviors in 1978 and 1980 were more striking than differences. This was so even though rainfall regimes on the island and stage of the *Nesotriccus* annual cycle differed considerably. The two sets of stomach contents were more similar to each other than either was to that of any other tropical flycatcher for which I had data (Fig. 2). The same observation applies to the stomachs of *Todirostrum cinereum*. Clearly, diets of these tropical flycatchers change seasonally, but not enough to obscure a species-specific dietary identity. Several ecologists have suggested that vertebrates are opportunistic, that available food fluctuates dramatically and unpredictably, and that species-specific segregation of diets with respect to prey types is minimal (Wiens 1977; Wiens and Rotenberry 1979; Bell 1980; Fenton and Thomas 1980; Rotenberry 1980; Fenton 1982; Rosenberg et al. 1982). This view does not apply to *Nesotriccus* and the insectivorous tropical flycatchers inhabiting the humid Caribbean lowlands of Costa Rica (Sherry 1984). These flycatchers illustrate considerable dietary specialization and/or homogeneity, and appropriate morphological adaptations (Sherry 1982), perhaps because resources in such environments do not fluctuate as dramatically as in seasonal tropical and higher-latitude environments. I infer that Homoptera and other diverse arthropod types (and correspondingly broad foraging tactics) are fundamental to *Nesotriccus* ecology. Subsequent discussion of morphological adaptation, specialization, and island colonization is based on this premise.

ADAPTATIONS OF *NESOTRICCUS*

The consistently broad range of prey types eaten by *Nesotriccus* prompts the question of the configuration of morphological characteristics "optimal" for a species performing in one environment the ecological role of many species, each adapted and specialized for a different role, in another environment. Specifically, if pursuing flycatchers have a different morphology from aerial-hawking species on the mainland (Sherry 1982), what can we predict about *Nesotriccus* which both pursues and aerially hawks? Fitzpatrick (1978) suggested that generalist flycatchers tend to have intermediate morphological characters.

Nesotriccus appears, instead, to have evolved in such a way that different morphological complexes (e.g., wings vs bill) are each appropriate to a different behavior or component of the diet and, thus, less well adapted to other components. Wing aspect-ratio (pointedness), and "effective" wing and tail area (i.e., effect of body mass removed statistically) of *Nesotriccus* are appropriate for acrobatic flight and are virtually indistinguishable from those of *Terenotriccus erythrurus*, a mainland Central and South American tyrannid (Sherry 1982, 1983). Both species pursue evasive prey acrobatically within vegetation, and *Terenotriccus* is a specialist on fulgoroid Homoptera (Sherry 1984). *Nesotriccus*, however, has a relatively long bill quite unlike that of *Terenotriccus*, and its "effective" bill size (length, width, and depth) is most similar to mainland euscarthmine flycatchers (Sherry 1982). The diet and foraging behavior of *Nesotriccus* broadly overlap those of euscarthmine species (Table 1; Sherry 1982, 1984), and I suggest that the bill of *Nesotriccus* has become adapted accordingly. Elongation of the bill of *Nesotriccus*, relative to its closest living relatives (see below), supports the above hypothesis, but wing and tail data are not presently available for similar comparisons.

Palate and syrinx morphology, nest and egg characteristics, song, and plumage strongly suggest that *Capsiempis flaveola* and especially *Phaeomyias murina* are the closest relatives of *Nesotriccus* (Lanyon 1984; Sherry, in press) and probably much like its ancestor. Both *Phaeomyias* and *Capsiempis* belong to a group of elaeniine Tyrannidae that are generalized frugivores/insectivores that mostly upward hover-glean (*Phaeomyias*) or perch-glean (*Capsiempis*, Fitzpatrick 1980). *Phaeomyias* relies heavily on mistletoe (Loranthaceae) fruits in canopy vegetation throughout much of the year (Fitzpatrick 1980, pers. comm.). If the *Nesotriccus* ancestor was at all like *Phaeomyias* or *Capsiempis*, then *Nesotriccus* has changed considerably since reaching the island. The ancestor would necessarily have switched from frugivory to insectivory and added several foraging tactics (aerial hawk, pounce, and acrobatic pursuit) to its repertoire. Morphologically, *Phaeomyias* and *Capsiempis* have short bills like other frugivorous Tyrannidae (Fitzpatrick 1978, pers. comm.; Traylor and Fitzpatrick 1982). The bill length of *Nesotriccus* is more than twice that of *Phaeomyias* (exposed culmen = 13.8 and 6.2 mm, respectively), and should be effective in catching active and agile arthropods such as Diptera (Sherry 1982) or Orthoptera (Greenberg 1981), but longer bills may also be effective for larger prey (Grant 1968; Hespenheide 1973, 1975). *Nesotriccus*, thus, has many morphological and behavioral characteristics expected for its diet, and this flycatcher has presumably changed considerably in these respects from its putative ancestor.

IS *NESOTRICCUS* A GENERALIST OR SPECIALIST?

One may make ecological inferences about *Nesotriccus* from its diet and feeding behavior, and evolutionary inferences from its ancestry and from heritable traits such as morphology. *Nesotriccus* is an ecological generalist compared with mainland insectivorous flycatchers in

TABLE 4

CLOSEST RELATIVE, HABITAT AND ABUNDANCE, AND DIET OF THE FOUR RESIDENT LAND BIRD
SPECIES OF COCOS ISLAND, COSTA RICA

Species	Closest relative	Habitat and abundance[1]	Diet[2]
Dendroica petechia aureola	D. petechia aureola[3]	Edge vegetation near beach, common	Small insects
Coccyzus ferrugineus	C. minor	Edge vegetation near beach, regular; rarer inland	Large arthropods (and small lizards?)
Nesotriccus ridgwayi	Phaeomyias (or Capsiempis)	All habitats, common	Diverse small Arthropods, some fruit
Pinaroloxias inornata	Geospizinae	All habitats, common–abundant	Arthropods (including crustaceans), fruit, nectar, seeds, small lizards (?)

[1] Based on Slud (1967) and Sherry and Werner (unpubl. data).
[2] Based on Slud (1967), Smith and Sweatman (1976), the present study, and Werner and Sherry (unpubl. data).
[3] Population inhabiting the Galapagos Archipelago.

that it catches diverse arthropod types with diverse feeding behaviors. The close match between diet and available foliage-inhabiting arthropods also classifies *Nesotriccus* as a generalist by the definition of some ecologists (e.g., Smith 1982). Evolutionarily, however, this flycatcher appears to have acquired specialized aerodynamic capabilities and a bill specialized to capture active and/or large-sized prey. Relative to *Phaeomyias*, *Nesotriccus* has ceased to rely on fruit and has become specialized largely on insects. It would be of considerable interest if *Nesotriccus* is more "specialized" than *Phaeomyias* since animals are expected to broaden their diets when they colonize competitor-poor environments. *Nesotriccus* is certainly both an evolutionary and ecological specialist in comparison with *Pinaroloxias inornata* (Geospizinae), the endemic genus of Darwin's Finch with which *Nesotriccus* shares all Cocos Island habitats (Table 4).

Nesotriccus seems to be both an ecological generalist and an evolutionary specialist. The foregoing discussion suggests moreover that generalization and specialization are meaningful concepts only in a comparative sense, and the conclusions depend on the reference used for comparison.

WHY IS THE RESIDENT LAND BIRD FAUNA OF COCOS ISLAND DEPAUPERATE AND LARGELY ENDEMIC?

Three lines of evidence suggest that Cocos Island is ecologically impoverished from the perspective of land birds (see Lack 1976; Abbott 1980), and that few immigrants are adapted to survive in the forest which covers much of the island. (1) Many arthropod taxa that birds encounter on the mainland are poorly represented or absent from the island, and few taxa, except perhaps Homoptera, compensate in population density for missing taxa. The ability of *Nesotriccus* to live and breed in Cocos Island forest (Slud 1967; Sherry and Werner, unpubl. data) probably reflects its adaptation for exploitation of particular arthropods for which it was not previously adapted. *Nesotriccus* also eats fruit, but the virtual absence of fruit in the 1978 stomach samples suggests that fruit was rare at that time. That fruit might be seasonally rare on a floristically impoverished island such as Cocos is not unexpected and further emphasizes the predicament of any frugivore that successfully disperses to the island (cf. Faaborg and Terborgh 1980).

(2) Evidence from the four resident land birds on Cocos Island suggests that evolutionary modification of the species was necessary before they could expand into forest habitat. Specifically, these species illustrate a trend in which greater endemism, from none to a level of subfamily, and greater morphological divergence from closest living relatives correspond with broader habitat use and diet, and with greater abundance in forest (Table 4). Scarcity of appropriate food is a more likely explanation than predation for the avoidance of forest by

certain Cocos species (Table 4) because at present the forest contains no snakes, raptors, or other obvious avian predators. Whether or not the residents depress resources enough to affect each other or other land birds is unknown.

(3) The distribution and ecology of migratory birds also suggest that interior forest provides poor foraging opportunities. Migratory land birds, many of which inhabit forested habitats elsewhere, occur overwhelmingly in edge habitats on Cocos Island, generally near shore (Slud 1967; F. G. Stiles, pers. comm.; Sherry, unpubl. data). The exception is the American Redstart (*Setophaga ruticilla*, Parulinae). One female that I collected on 29 February, 1980 in subcanopy on a forested ridge between Bahia Wafer and Bahia Chatham had been eating fulgoroid Homoptera (10 of 11 identifiable items in the stomach). A yearling male observed on 14 April, 1984 foraged much as *Nesotriccus* does in interior forest between Bahia Wafer and the Mirador. Redstarts eat Homoptera elsewhere and are convergent morphologically and behaviorally with tyrannid flycatchers that specialize on Homoptera (Robinson and Holmes 1982; Sherry 1982, 1983, unpubl. data).

Potential colonists have arrived on Cocos Island, both independently and with man's assistance (T. W. Sherry, T. K. Werner, and F. G. Stiles, unpubl. data), even though no new species have definitely established breeding populations since the avifauna was described late in the last century (Slud 1976). Thus, lack of immigrants cannot alone explain why Cocos Island has a depauperate and endemic avifauna. This conclusion does not mean that dispersal barriers have not also contributed to the small number of resident land birds on isolated oceanic islands (Williamson 1981), especially if birds tend to disperse better than the organisms upon which they rely for food.

ACKNOWLEDGMENTS

I thank F. G. Stiles for kindling my interest in the Cocos Island Flycatcher, C. L. Hogue for continued interest in Cocos Island, and both for loans of equipment. My wife, T. K. Werner, was an invaluable companion and colleague during the 1980 expedition. F. Torres and J. Sanchez helped with logistic arrangements. J. M. Rodriguez, G. Flores, M. Rojas, J. Barborak, and B. Carlson of Servicio de Parques Nacionales de Costa Rica and personnel of the Guardia Nacional de Costa Rica facilitated and permitted the observations reported here. I think P. A. Buckley, M. S. Foster, T. K. Werner, and C. Whelan for constructively criticizing the manuscript. For financial assistance I thank the National Science Foundation dissertation-improvement program, the Western Foundation of Vertebrate Zoology, and the Frank M. Chapman Memorial Fund of the American Museum of Natural History.

LITERATURE CITED

ABBOTT, I. 1980. Theories dealing with the ecology of landbirds on islands. Adv. Ecol. Res.11:329–371.

BELL, G. P. 1980. Habitat use and response to patches of prey by desert insectivorous bats. Can. J. Zool. 58:1876–1883.

DALRYMPLE, G. B., AND A. COX. 1968. Paleomagnetism, potassium-argon ages and petrology of some volcanic rocks. Nature 217:323–326.

FAABORG, J. R., AND J. W. TERBORGH. 1980. Patterns of migration in the West Indies. Pp. 157–163, *In* A. Keast and E. S. Morton (eds.), Migrant Birds in the Neotropics. Smithsonian Institution Press, Washington, D.C.

FENTON, M. B. 1982. Echolocation, insect hearing, and feeding ecology of insectivorous bats. Pp. 261–285, *In* T. H. Kunz (ed.), Ecology of Bats. Plenum Press, New York.

FENTON, M. B., AND D. W. THOMAS. 1980. Dry-season overlap in activity patterns, habitat use, and prey selection by sympatric African insectivorous bats. Biotropica 12:81–90.

FITZPATRICK, J. W. 1978. Foraging behavior and adaptive radiation in the avian family Tyrannidae. Unpubl. Ph.D. dissert., Princeton University, Princeton, New Jersey.

FITZPATRICK, J. W. 1980. Foraging behavior of Neotropical tyrant flycatchers. Condor 82:43–57.

FOURNIER, L. A. 1966. Botany of Cocos Island, Costa Rica. Pp. 183–186, *In* R. I. Bowman (ed.), The Galapagos. University California Press, Berkeley, California.

GRANT, P. R. 1968. Bill size, body size and the ecological adaptations of bird species to competitive situations on islands. Syst. Zool. 17:319–333.

GRANT, P. R. 1981. Speciation and the adaptive radiation of Darwin's finches. Am. Sci. 69:653–663.

GRANT, P. R., B. R. GRANT, J. N. M. SMITH, I. J. ABBOTT, AND L. K. ABBOTT. 1976. Darwin's finches: population variation and natural selection. Proc. Natl. Acad. Sci. U.S.A. 73:257–261.

GREENBERG, R. 1981. Dissimilar bill shapes in New World tropical versus temperate forest foliage-gleaning birds. Oecologia (Berl.) 49:143–147.

HERTLEIN, L. G. 1963. Contribution to the biogeography of Cocos Island, including a bibliography. Proc. Calif. Acad. Sci., Ser. 4, 32:219–289.

HESPENHEIDE, H. A. 1973. Ecological inferences from morphological data. Annu. Rev. Ecol. Syst. 4: 213–229.

HESPENHEIDE, H. A. 1975. Prey characteristics and predator niche width. Pp. 158–180, In M. L. Cody and J. M. Diamond (eds.), Ecology and Evolution of Communities. Belknap Press, Cambridge, Massachusetts.

HESPENHEIDE, H. A. 1979. Are there fewer parasitoids in the tropics? Am. Nat. 113:766–769.

HOGUE, C. L., AND S. E. MILLER. 1981. Entomofauna of Cocos Island, Costa Rica. Atoll Res. Bull. 250.

HOLMES, R. T., R. E. BONNEY JR., AND S. W. PACALA. 1979. Guild structure of the Hubbard Brook bird community: a multivariate approach. Ecology 60:512–520.

HURTUBIA, J. 1973. Trophic diversity measurement in sympatric predatory species. Ecology 54:885–890.

JANZEN, D. H. 1973. Sweep samples of tropical foliage insects: effects of seasons, vegetation types, elevation, time of day, and insularity. Ecology 54:687–708.

JOHNSON, S. C. 1967. Hierarchical clustering schemes. Psychometrika 32:241–254.

KEAST, A. 1968. Competitive interactions and the evolution of ecological niches as illustrated by the Australian honeyeater genus *Melitheptus* (Meliphagidae). Evolution 22:762–784.

LACK, D. 1947. Darwin's Finches. Cambridge University Press, Cambridge, England.

LACK, D. 1976. Island biology, Illustrated by the Birds of Jamaica. University California Press, Berkeley, California.

LANYON, W. E. 1984. The systematic position of the Cocos Flycatcher. Condor 86:42–47.

MACARTHUR, R. H. 1972. Geographical Ecology: Patterns in the Distribution of Species. Harper and Row, New York.

PIELOU, E. C. 1975. Ecological Diversity, John Wiley and Sons, New York.

PYKE, G. H., H. R. PULLIAM, AND E. L. CHARNOV. 1977. Optimal foraging: selective review of theory and tests. Q. Rev. Biol. 52:137–154.

ROBINSON, S. K., AND R. T. HOLMES. 1982. Foraging behavior of forest birds: the relationships among search tactics, diet, and habitat structure. Ecology 63:1918–1931.

ROSENBERG, K. V., R. D. OHMART, AND B. W. ANDERSON. 1982. Community organization of riparian breeding birds: response to an annual resource peak. Auk 99:260–274.

ROTENBERRY, J. T. 1980. Dietary relationships among shrubsteppe passerine birds: competition or opportunism in a variable environment? Ecol. Monogr. 50:93–110.

SHERRY, T. W. 1982. Ecological and evolutionary inferences from morphology, foraging behavior, and diet of sympatric insectivorous neotropical flycatchers (Tyrannidae). Unpubl. Ph.D. dissert., University California, Los Angeles.

SHERRY, T. W. 1983. *Terenotriccus erythrurus* (Mosquitero Colirrufo, Tontillo, Ruddy-tailed Flycatcher). Pp. 605–607, In D. H. Janzen (ed.), Costa Rican Natural History. University Chicago Press, Chicago, Illinois.

SHERRY, T. W. 1984. Comparative dietary ecology of sympatric insectivorous neotropical flycatchers (Tyrannidae). Ecol. Monogr. 54:313–338.

SHERRY, T. W. In press. Notes on nesting, seasonal cycle, and nematode parasites of the Cocos Island Flycatcher. Condor.

SLUD, P. 1967. The birds of Cocos Island (Costa Rica). Bull. Am. Mus. Nat. Hist. 134:263–295.

SLUD, P. 1976. Geographic and climatic relationships of avifaunas with special reference to comparative distributions in the Neotropics. Smithson. Contrib. Zool. 212:1–149.

SMITH, E. P. 1982. Niche breadth, resource availability, and inference. Ecology 63:1675–1681.

SMITH, J. N. M., AND H. P. A. SWEATMAN. 1976. Feeding habits and morphological variation in Cocos Finches. Condor 78:244–248.

SOKAL, R. R., AND F. J. ROHLF. 1969. Biometry. W. H. Freeman Co., San Francisco, California.

TRAYLOR, M. A. 1977. A classification of the tyrant flycatchers (Tyrannidae). Bull. Mus. Comp. Zool. 148:129–184.

TRAYLOR, M. A. JR., AND J. W. FITZPATRICK. 1982. A survey of the tyrant flycatchers. Living Bird 19: 7–50.

VAN VALEN, L. 1965. Morphological variation and width of ecological niche. Am. Nat. 99:377–390.

WIENS, J. A. 1977. On competition and variable environments. Am. Sci. 65:590–597.

WIENS, J. A., AND J. T. ROTENBERRY. 1979. Diet niche relationships among North American grassland and shrubsteppe birds. Oecologia (Berl.) 42:253–293.

WILLIAMSON, M. 1981. Island Populations. Oxford University Press, New York.

COEXISTENCE AND BEHAVIOR DIFFERENCES AMONG THE THREE WESTERN HEMISPHERE STORKS

BETSY TRENT THOMAS

Apartado 80844, Caracas 1080-A, Venezuela;
present address: 1 Wetsel Road, Troy, New York 12182 USA

ABSTRACT. The only storks in the New World, the Wood Stork (*Mycteria americana*), the Maguari Stork (*Ciconia maguari*), and the Jabiru (*Jabiru mycteria*), were studied in the llanos of Venezuela. Stork weights and linear measurements are given, and behaviors are described and compared. The storks use both different foraging-techniques and intraspecific foraging-associations; these may result in resource partitioning among the storks and with other waterbirds. Although their diets of mostly aquatic organisms overlap broadly, available data suggest that the percent of each food class taken, and prey-sizes, are not alike. The three storks also differ in morphology, resident status, pair bond length, breeding months, nest-site preference, and nesting dispersion. The timing and amount of annual rainfall influences the storks' breeding months through the availability and abundance of food.

RESUMEN. En los llanos de Venezuela fueron estudiadas las únicas especies de cigüeñas del Nuevo Mundo, la Cigüeña Gabán (*Mycteria americana*), la Cigüeña Americana (*Ciconia maguari*) y el Jabirú (*Jabiru mycteria*). Se brinda información sobre pesos y medidas lineales de las cigüeñas y se describen y comparan comportamientos. Las cigüeñas usan tanto técnicas de alimentación y diferentes como asociaciones de forraje interespecíficas; esas pueden resultar en la repartición de los recursos entre cigüeñas así como con otras aves acuáticas. Aunque sus dietas son mayormente de organismos acuáticos y se superponen ampliamente, la información disponible sugiere que el porcentaje de cada clase de comida consumida y el tamaño de las presas, no son similiares. Las tres cigüeñas también son diferentes en sus aspectos morfológicos, situación como residentes, duración de parejas, meses de reproducción, sitios preferidos para anidación y dispersión de anidación. La época y la cantidad de lluvias anuales, influyen en los meses de reproducción a través de la disponibilidad y abundancia de comida.

Only three of the world's 17 living species of storks are native to the Western Hemisphere (Kahl 1971a); they are the Wood Stork (*Mycteria americana*), the Maguari Stork (*Ciconia maguari = Euxenura galeata*), and the Jabiru (*Jabiru mycteria*). Probably because of their large size storks are the socially dominant species of the waterbird guild in the Venezuelan llanos. Current ecological theory holds that no two species occupy exactly the same niche (Moreau 1966); thus, it is of interest to examine some of the differences among these sympatric storks. In this paper I review some of the morphological and behavioral differences that may facilitate their coexistence.

When he revised the taxonomy of the world's storks, Kahl (1971a) maintained these three species in separate genera, although he changed the Maguari Stork to *Ciconia* as an indication of its close relationship to the White Stork (*Ciconia ciconia*). Even though the three species share many familial traits of morphology and behavior, their placement in separate genera suggests, correctly, that distinctions also exist.

STUDY AREA AND METHODS

Beginning in 1972, and continuing for 11 years, I studied storks in the llanos in Venezuela. My main study area was a cattle ranch, Fundo Pecuario Masaguaral, which lies roughly in the center of the llanos, in the state of Guárico. This area is classified, under the Holdridge system, as tropical dry forest (Ewel and Madriz 1968). The habitat consists of open grassland savannas with occasional clumps of trees and gallery forests. The area is generally below 100 m in elevation; thus, the climate is tropical with one wet season and one dry season each year. Vegetation of the study site is described in Troth (1979), and the species of birds found there are listed in Thomas (1979).

I also observed storks in other parts of the llanos, mainly in southern Guárico and throughout

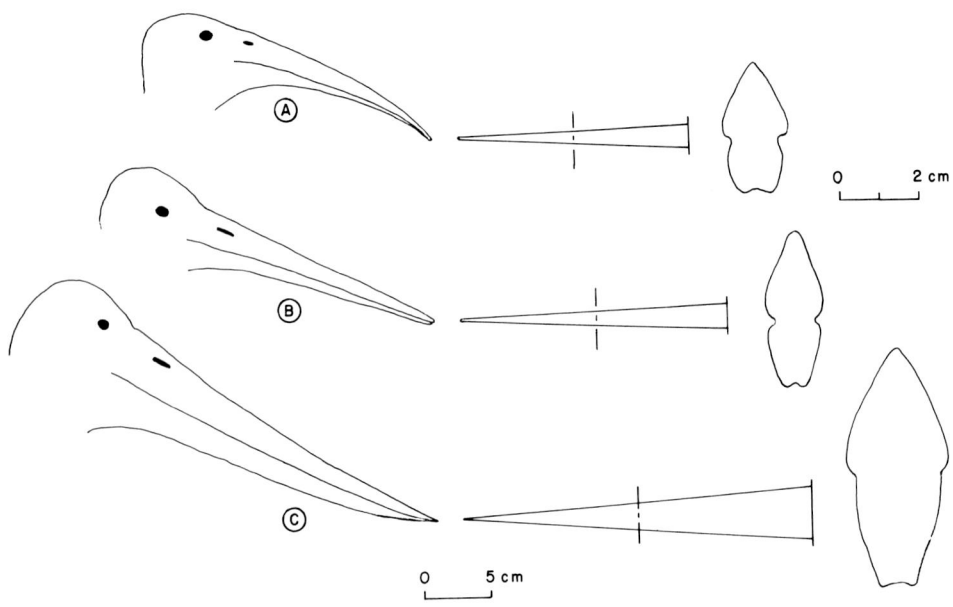

Fig. 1.   Comparison of bills of the (A) Wood Stork, (B) Maguari Stork, (C) Jabiru. Left: in profile, center: dorsal view. Right: cross sections at each bill's mid-point as indicated in the center figures. Note larger scale for the cross sections.

most of the state of Apure. The habitat of western Apure is described in Ojasti (1973). Census data there were taken along a 138 km highway, from El Saman de Apure to Bruzual, by driving slowly and stopping at every stork seen, recording the species, age-class (for Maguari Storks), usually the substrate on which the bird stood, stork behavior, intraspecific distance, and location. Later each stork was plotted on a census-route map. Two censuses were made 1 to 3 days apart near the middle of August, September, and November 1982. Six aerial surveys were made in January 1982 covering much of Apure and eastern Barinas where there are no roads. An additional aerial survey was made in October 1983, and observations of nesting Wood Storks were made at a remote llanos ranch in February 1984.

Mensural data of all birds are from museum specimens and the literature, from all localities, and from Venezuelan zoo birds. The zoo birds (N = 10) were weighed to the nearest 50 g and measured to the nearest 0.5 cm; they were controlled during handling by the use of special hoods (Thomas 1977). All other measurements were made with a vernier caliper graduated to the nearest 0.1 mm.

The Maguari Stork food-class percentages are based on 344 items brought to nestlings or eaten by adults; the fish sample was 41 measured prey. Jabiru foods are from 25 items either fed to nestlings or eaten by adults, of those 19 were fish. The length of each food item was either measured from collected samples or estimated to the nearest cm, and its weight calculated from weights of preserved samples of food species. The Jabiru nestling period was estimated from observations at a single nest (Thomas 1981). Sample sizes of visual versus tactile foraging were collected from 395 Wood Stork, 232 Maguari Stork, and 43 Jabiru observations in Venezuela, when the birds' behaviors could be determined.

Venezuelan rainfall data are from the Ministerio del Ambiente y de los Recoursos Naturales Renovables station 6 km south of Masaguaral, supplemented with daily records kept at Masaguaral. In addition I maintained a calibrated measuring stick at a blind in a marsh at the largest Maguari Stork nesting colony on the ranch, where I recorded the absolute marsh water depth from 1974 to 1982 during the stork breeding season.

MORPHOLOGY

It is axiomatic that similar birds sharing a habitat differ in some measure of size, often that of the bill. The three storks' bills generally differ in shape, length, width, curvature, and cross section (Fig. 1), although the bills of the Wood Stork and the Maguari Stork are similar in

TABLE 1

Mensural Data for the Three Species of Storks[1]

Species	Weight (g)	Culmen (cm)	Tarsus (cm)	Longest toe (cm)
Mycteria americana	♂ 2500 [2700–3090]	22.5 (20.5–23.0) [21.5–24.4]	22.5 (19.7–20.5) [19.4–21.3]	12.5 (12.0–13.2)
	♀ 2150 [2050–2180]	20.5 (20.0) [19.0–20.5]	18.5 (18.0) [17.5–19.2]	10.5 (11.0)
Ciconia maguari	♂ 4000	23.5 (22.5–25.5) [20.3–24.5]	26.0 (26.5–27.5) [26.5]	12.0 (10.0–12.5)
	♂ 4200	23.0	26.0	12.5
	♂ 4250	23.5	26.0	11.0
	♂ 4350	23.0	27.0	12.5
	♀ 3400	20.0 (20.5–22.7) [20.5–22.9]	24.0	11.0
	♀ 3500	20.5	23.5	10.0
Jabiru mycteria	♂ 6700	31.0 (33.0) [29.5–34.3]	31.0 (32.0) [28.5–33.0]	11.5 (12.5)
	♂ 7700	30.0	32.0	12.0
	♀ [?♀ 5470]	(30.5) [28.4–32.8]	(29.5)	(11.5)

[1] The first figures in each column are data from zoo birds. Data in parentheses are ranges from sexed museum skins: *Mycteria americana* N = 2 ♂♂, 1 ♀; *Ciconia maguari* N = 8 ♂♂, 6 ♀♀; *Jabiru mycteria* N = 1 ♂, 1 ♀. Data in brackets are ranges from the literature (Benedict and Fox 1927; Hartman 1961; Palmer 1962; Kahl 1964; Wetmore 1965; Haverschmidt 1968; Blake 1977).

exposed culmen length (Table 1). Fry (1980), after examining neotropical congeneric kingfishers, concluded that body weight is a better measure of size differences between sympatric species than linear measurements. Although they are not congeneric, body weights among the storks are more different than other morphological measurements. Weights of adult storks are rarely recorded. No museum has large numbers of specimens, of those only about half are sexed birds, and none that I examined had weight data. Capturing such large birds in the field is impractical; thus, I used weights of captive birds. The weights given in Table 1 were all of adult birds, three to nine years old, taken as nestlings in Venezuela and reared in captivity. While these weights may differ somewhat from those of wild birds, their relative weights more nearly express the size differences.

Tarsal measurements of the three species differ less. Their toes, as Chubb (1916) noted, are remarkably small for such large birds and, surprisingly, the lengths of the longest toe of each species are similar (Table 1). Unlike the other two species, the Wood Stork uses its hallux to grasp branches when standing and roosting in trees, a frequent behavior, whereas neither the Maguari Stork, nor the Jabiru grasp with the hallux. Both species roost on the ground, although both also nest in trees (Thomas 1981, 1984, unpubl. data).

In handling the living birds, I found the tactile dissimilarities of their head and neck surfaces very noticable. The Wood Stork's bare head and neck are rough and horny, being covered with irregular scaly plates; the Maguari Stork's head and neck are covered with semiplumes that function, when erected, in agonistic and sexual displays; in contrast, the Jabiru's head and neck are bare, except for a few coarse semiplumes on the back of the head, and its skin is smooth and exceedingly silky to touch.

In each of the three species males are larger than females (Table 1). Soft part colors of the storks differ and appear to function in intra-sexual recognition; however, sexual behavioral differences are less conspicuous than interspecific ones.

FORAGING AND DIET

More than 26 fish-eating species of birds, and two fish-eating reptiles, live in the llanos. They include large numbers of cormorants (1 species), anhingas (1), herons and egrets (10), spoonbills (1), ibises (2), raptors (2), larids (2), skimmers (1), and kingfishers (4), in addition to the three storks. At a western Apure study site Great Egrets (*Casmerodius albus*) took more than 50 percent of the fish eaten by birds (Pinowski et al. 1980). MacArthur (1971) pointed out that differences in size of sympatric birds may lead to foods of different size, and moreover, that sexual size dimorphism may also result in the use of different size prey. Sexual dimorphism

TABLE 2
Foods of the Three Storks[1]

Food	Mycteria americana	Ciconia maguari	Jabiru mycteria
Mammal	X	X	X
Bird	X	X	
Frog/tadpole	X	X	X
Reptile	X	X	X
Fish	X	X	X
Eel	X	X	X
Earthworm		X	
Mollusk	X		X
Crustacean	X	X	X
Insect	X	X	X
Plant/seed	X		

[1] Data for *Mycteria americana* taken from Palmer 1962, Kahl 1964, Ogden et al. 1976, and Thomas, this study; for *Ciconia maguari*, from Hudson 1920; Kahl 1971b; and Thomas 1984; and for *Jabiru mycteria*, from Bent 1926; Kahl 1971b; and Thomas, this study.

effects differential niche use (Selander 1966). Both of these ecological separating mechanisms may be operating among the storks and between the sexes.

The three storks eat a wide variety of mostly aquatic organisms (Table 2). However, most of the diet of Wood Storks in Florida is fish, where they selectively take larger fish, 1.5 to 13 cm long, than could be expected from those found in their feeding areas (Ogden et al. 1976). The Maguari Stork is more of a generalist. It ate or fed to nestlings frogs and tadpoles (32%), fish (28%), freshwater eels (16%), rats (12%), and crabs, insects, earthworms and a few other vertebrates (12%). These items (N = 344) are from five years, although percentages varied in different years probably with local abundance (Thomas 1984). Jabiru foods have been less studied. In the llanos of Venezuela I recorded fish (total length = 8–20 cm, N = 19), freshwater eels (total length = 30–80 cm, N = 4), and crabs (carapace width = 7 cm, N = 2), being eaten by adult Jabirus or fed to nestlings. Although I saw more than four times as many fish eaten as eels, the total estimated weight of the eels was 830 g compared with the total estimated 253 g weight of the fish. The length and curvature of the Jabiru bill suggests that it may specialize in digging large eels from deep mud (D. Mock, pers. comm.). Possibly the Jabiru and Wood Stork, because of bill differences, do not use the same size or species of fish in Venezuela. However, the Maguari Stork takes nine species of fish (total length = 3–18 cm, N = 85; Thomas 1984), overlapping in size the fish prey of the other two storks.

Once, a Jabiru ate pieces of cow dung, a behavior that I have observed also in Maguari Storks. Kahl (1964) saw three Wood Storks eating cow dung, and also he reported dung being eaten by the White Stork. The reason for this behavior remains obscure.

More apparent than interspecific differences in food classes and food sizes among the storks, are their foraging techniques. Wood Storks feed primarily by tactile location (Kahl 1964). In Venezuela, Wood Storks occasionally search visually (4% of 395 individuals observed), but more often they feed in large groups by walking, sometimes quickly, in water about 10 cm deep, holding their bills open and half submerged. At other times groups advance, on a wide front, swinging their open bills from side to side in the water. Wood Stork foraging has been described by Rechnitzer (1956), Kahl (1964), Ogden et al. (1976), and summarized by Kushlan (1978). Experimentally blind-folded Wood Storks located prey without using their eyes, and their bill-snap reflex is one of the fastest known among vertebrates (Kahl and Peacock 1963; Kahl 1964). Most of the foraging waters of Wood Storks in Venezuela are muddy, and the birds are attracted to drying ponds as they are in Florida.

The Maguari Stork is almost entirely a visual feeder (Kahl 1971b; pers. observ.). Rarely (3% of 232 individuals observed) did a bird grope with its bill in muddy water, and those that did were birds that had returned to their breeding area before sufficient rain had formed clear, freshwater marshes. The usual manner of Maguari hunting is to walk slowly in wet marshes having water 2 to 15 cm deep, with the bill lowered, but not in the water, and often open, poised to seize any prey encountered. Occasionally at the beginning or the end of the breeding season, members of migratory flocks search dry, short-grass fields for arthropods, examining bushes and lifting cow dung with their bills. Less commonly a Maguari Stork hunts visually in open water 10 to 30 cm deep.

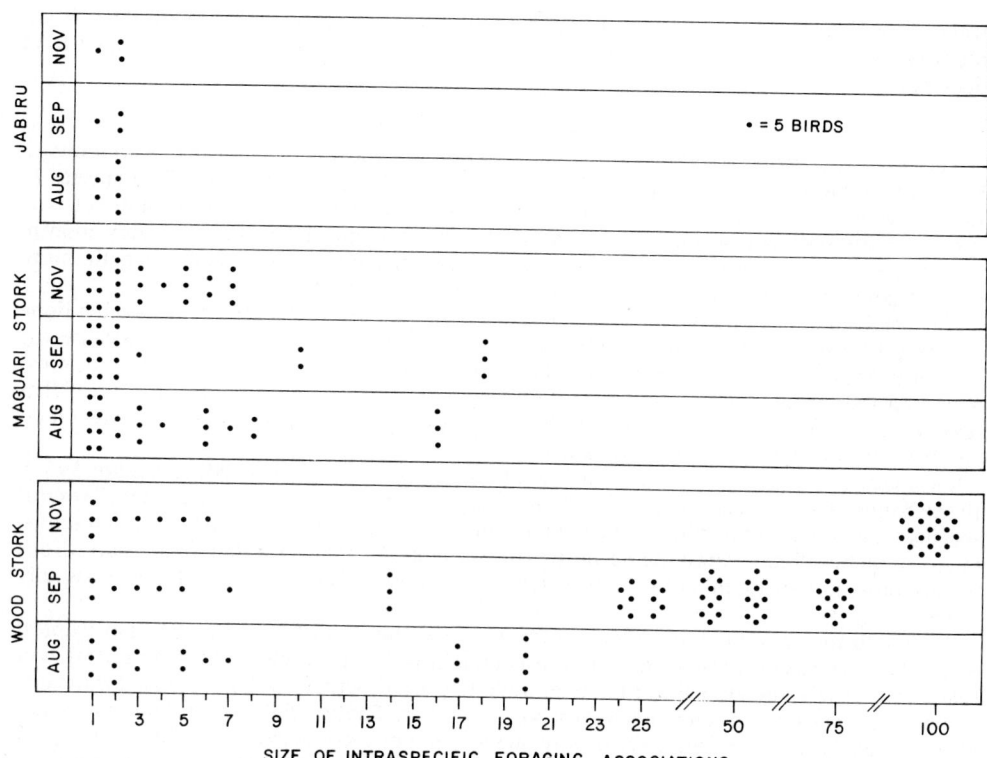

FIG. 2. Size and frequency of intraspecific foraging associations of storks. Each dot indicates five birds; rounded up or down to the nearest five. Abscissa shows the size of the foraging associations. For example, in August there were 12 solitary Jabirus and 20 Jabirus in pairs, in September 3 solitary Jabirus and 10 in pairs.

Jabirus forage by walking slowly in shallow muddy water, with the tips of their bills open and groping in the water. When they catch a fish they frequently carry it, crosswise in the bill, onto the shore where they pinch it 5 to 20 times before swallowing it whole. Jabirus hunt for eels buried deep in mud by plunging their bills into the mud. At other times they run four or five steps in shallow water and grab with their bills, or stalk slowly in pairs in water 20 to 40 cm deep 'biting' the water surface. They appear to find about half of their prey by tactile location (54% of 43 individuals observed), but sometimes they combine tactile and visual behaviors. For example, one pair of Jabirus stood beside a muddy pond watching the water surface, then hurried into the water, on some cue, to grab with their bills.

In the western part of the state of Apure the habitat is almost entirely grassy marsh interspersed with open water areas, during the wet season. Six censuses in this habitat showed that the storks there were sympatric, but their distribution reflected preferences for the same areas on all censuses. No storks were ever observed closer than 5 km to Mantecal, the only town on the census route. The size of intraspecific stork foraging associations are shown in Figure 2. Wood Storks were the most abundant. They foraged solitarily and in small groups, the group size increasing as the breeding season approached. The average number of Wood Storks observed per census was 89.83 (range = 32–180, s.d. = 55.83, N = 6). Maguari Storks, excluding a small breeding colony, averaged 51.66 birds per census (range = 37–62, s.d. = 10.15, N = 6). More Maguari Storks were solitary than the other two species, although there were some larger associations, but their mean inter-individual distance was 55 m (range = 20–300 m, N = 48). There were few Jabirus; the mean number of birds per census was 10.16 (range = 3–22, s.d. = 6.73, N = 6), and all birds were either solitary or in pairs. The visual impression offered by Figure 2 is similar to that given by the storks' distribution in the habitat.

In a study of the relationship of food distribution to social distribution, Morton (1979) found that the Maguari Stork was solitary or occurred in small groups of three to five birds, whereas the Jabiru occurred solitarily or in pairs. Both fed on dispersed food. The Wood Stork was not included in his analysis, but its gregariousness should reflect a clumped food distribution.

BREEDING MONTHS

The months of stork breeding can be understood only within the context of variations in both timing and quantity of rainfall. This is a particularly important point. Bird species in the tropics breed at discrete times of year, just as they do at higher latitudes; however, one species or another can be found breeding, at the same tropical location, in every month. Nevertheless, in the llanos, where the wet and dry seasons are roughly equal in length, more than 85 percent of all birds breed in the wet season (Thomas 1979).

It is confusing to call the wet season 'winter' and the dry season 'summer,' as is often done in tropical Latin America (fide E. Eisenmann). These terms should be avoided because the year in tropical latitudes is little differentiated by temperature variations. The time of most abundant food in the tropics, and most breeding, seems to be the rainy season. But if that season is called 'winter,' it conflicts with the meaning of 'winter' at higher latitudes when food and breeding are generally not abundant.

The onset of rain appears to stimulate low latitude, neotropical bird-breeding (Collins 1968). In the llanos of Venezuela the onset of the first rains varies from late March to June. Tropical species are adapted to this fluctuating regimen and can delay breeding for one or two months, in the equitable climate, until the optimum time occurs. The llanos rainfall, in all years, drops precipitously from September to December, and the months of December to March are usually rainless (Fig. 3).

Adding to the complication of unpredictably late or light rainfall, is the fact that tropical South America has exceedingly diverse rainfall patterns. For example, in Suriname, 1600 km east of Venezuela, the annual rainfall is bimodally peaked, with two dry seasons, a short one February to April and a long one August to November. The rainfall pattern 2000 km south of Venezuela is still different; there is only one wet and one dry season, but about 5° south of the equator the cycle is the reverse of that found in Venezuela. Therefore, the same species of birds often breed in entirely different months of the year, at different sites on the same continent. Snethlage (1928) examined this rainfall variation in relation to the breeding months of birds, giving examples from Guyana (6°N), to Para, Brazil (1°S), and Rio de Janeiro (22°S). He showed that breeding seasons change in the same progression as does the timing of rainfall, when the equator is crossed.

Because the onset of substantial Venezuelan rains, which form large freshwater, llanos marshes, is variable in both time and quantity (Fig. 3), I found it the most important variable controlling the beginning of the Maguari Stork breeding season. Later in the rainy season, the breeding of the other two storks is also related to the rainfall. I used egg-laying months as a measure of the breeding seasons of the three storks, since incubation for all three species is about one month. The nestling period for the Wood Stork is approximately eight weeks (Kahl 1962), about nine weeks for the Maguari Stork (Thomas 1984), and 10 to 11 weeks for the Jabiru (Thomas 1981).

The months of Maguari Stork egg-laying varied among years. One old, experienced pair laid a clutch in the last week of May 1981, following unusually heavy and early rains (Fig. 3). Conversely, in years of late rains younger breeders laid eggs as late as October (Thomas 1984; unpubl. data). Even so Maguari laying usually only spanned a period of about five to seven weeks in any one year, excluding replacement clutches. Furthermore, it generally followed that the years when Maguari Stork egg-laying was early, laying for the Jabiru and Wood Stork was also early. Thus the beginning of breeding for each of the storks appears to be rainfall dependent. The usual months of laying for each species in the llanos of Venezuela are shown in Figure 3.

While rainfall may be the proximate cue, the ultimate effect is the food that presumably becomes available for storks in freshwater marshes and ponds. The Maguari may breed first because during the early months of the rainy season the marsh water is clear for visual foraging; the Jabiru may breed next, because it uses both visual and tactile searching behavior that is appropriate to water that is changing from clear to muddy; finally, for the Wood Stork the opaque, muddy water of the drying habitat creates optimal feeding conditions.

Although no Wood Storks nested on Masaguaral during my study, they were present in all months, with numbers varying from one to 100 birds. Many of the Wood Storks were immature

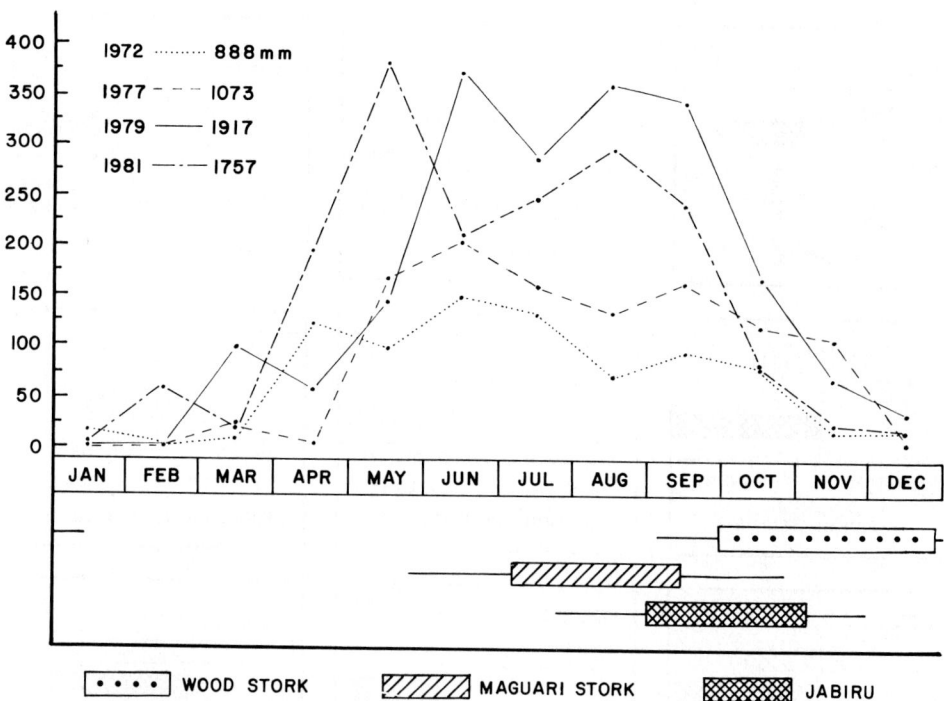

FIG. 3.   Rainfall during the two years of highest (1979, 1981) and lowest (1972, 1977) precipitation recorded during 1972–1983 on the principal study site (above). The main months of egg-laying are indicated (below) by wide bars, with extremes indicated by narrow lines.

individuals, and the largest groups occurred at the end of the rainy season and during the dry season. One marsh that was always dry from December to April, but reached a depth of 62 cm during some rainy seasons, was monitored. At the end of each wet season this marsh often dried as fast as 4 cm per day. Although this rapid drying might concentrate prey for the tactile-foraging Wood Storks, they rarely fed there, preferring the edges of permanent lagoons. Possibly, such rapid drying is too fast to sustain adequate local food for Wood Stork breeding, which may explain why non-breeding, highly mobile, immature birds used the area. In western Apure the land remains flooded longer, and suitable Wood Stork breeding conditions obviously occur as reflected by the presence of large breeding colonies of more than 1000 nests.

There is no breeding record for the Maguari Stork in Suriname, but Spaans (1975) found that the Jabiru and the Wood Stork breed there in the opposite calendar order from that of Venezuela. Wood Storks, that bred in only one of the three years of Spaans' study, initiated egg-laying in late September and continued through October, whereas Jabirus, that bred in two of the three years, laid eggs in late October through November. These dates show that the Wood Stork initiated nesting in the Suriname long dry season, just as it does in Venezuela and Florida (Kahl 1964). By beginning later in the year, the Jabiru fledged young in the short dry season; Jabiru young in Venezuela also fledge in the dry season. Nevertheless, egg-laying of the two storks in Suriname, appears to be temporally separate.

NEST DISPERSION AND PAIR BONDS

The intraspecific dispersions of nests of the three stork species differ. Wood Storks nest in large conspecific colonies; thousands of nests have been found together in Florida (Bent 1926; Palmer 1962). In Venezuela they nest in very large clumps of trees and dense high bushes (Ayarzaguena et al. 1981; Thomas, pers. observ.). The Maguaris nest in close small groups, in small clumps of bushes and in low, isolated trees (77% of nests are colonial, Thomas, unpub. data). Jabirus nest solitarily on the tops of palm trees and in open branches of large isolated trees (Kahl 1971b; Thomas 1981). Interestingly, these distinct social nesting patterns are reflected in the three species' intraspecific foraging associations (Fig. 2).

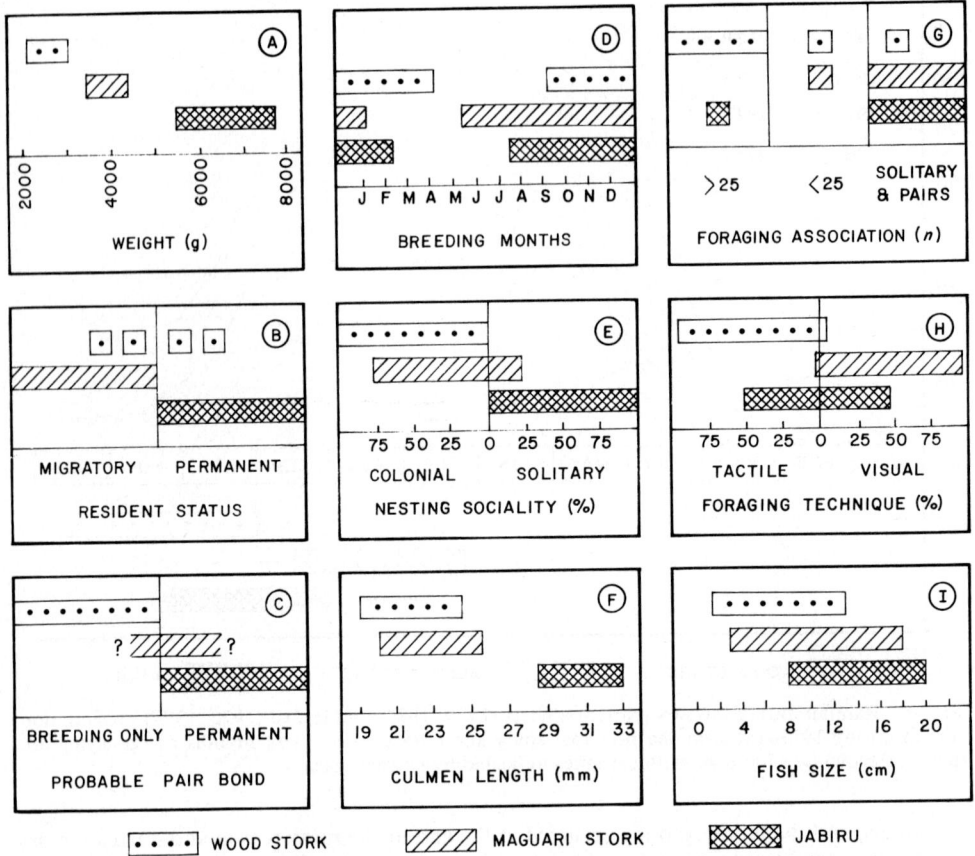

FIG. 4.   Differences among the three Western Hemisphere storks in Venezuela. Breeding months (D) are from 1972–1984 and cover the period from egg-laying through fledging. Fish prey size (I) data from Ogden et al. (1976) for 3000 prey of Wood Storks in Florida, and from Thomas (1984) for 41prey of Maguari Storks and 19 prey of Jabirus in Venezuela.

Wood Storks are highly mobile, often flying long distances to reach suitable, ephemeral, feeding sites (Browder 1976). They probably form a new pair bond at the beginning of each breeding season (Kahl 1964, 1972). The Maguari Stork is migratory, leaving its breeding area when marshes dry and its aquatic food disappears (Thomas 1984, unpubl. data). Some pairs bred together in succeeding years on the same nest-site or in the same area (Thomas, unpubl. data), but data from banded birds are too few to ascertain the extent of mate-fidelity in this species. Jabirus appear to be permanent residents, although no marked pairs have ever been studied. Kahl (1971b, 1973) concluded on the basis of the undemonstrative pre-breeding and nesting behavior of Jabirus that they probably form permanent pair bonds. Jabirus defend their general nest areas from conspecifics (Thomas 1981) and may also maintain exclusive territories at adjacent feeding sites.

INTERSPECIFIC BEHAVIOR

I have observed few incidences of overt interspecific interactions between the storks; in all cases a larger species attacked or supplanted a smaller one. Twice Jabirus flew at and chased Maguari fledglings that flew within 100 m of Jabiru nests or nest-sites. On another occasion, during a drought, I watched two aerial chases by a Jabiru directed at a Wood Stork that had just caught a 30 cm freshwater eel. Despite the considerable size difference of the birds, the Wood Stork landed, turned its back to the Jabiru, raised its wings defensively, and swallowed its prey. Once two fledgling Maguari Storks, together, chased and supplanted two or three Wood Storks at the edge of a lagoon where all were foraging. When mature pairs of Jabirus meet mature pairs of Maguari Storks, the birds all watch each other closely and walk past one another no closer than five to seven m.

Small groups of immature Wood Storks are often attracted to large trees overlooking breeding Maguari Storks, but they are always ignored; occasionally Wood Storks roosted in trees near active Maguari nests without incident. Generally, Wood Storks on my study site did not mingle with the other two species of storks, but spent their time standing in large trees or actively foraging in muddy ponds and lagoons.

CONCLUSIONS

Of the three species, the Wood Stork may be the most adaptable to varying environmental (pluvial) conditions. Kahl (1964) has shown that in Florida it deserts entire breeding colonies when sudden adverse climatic conditions occur and that it nests again only when favorable conditions are re-established. Wood Storks in Florida feed in both fresh and brackish water (Ogden et al. 1976), as they do in Venezuela. These more flexible behaviors may be the reason that the Wood Stork has the widest geographical range of the three storks (Blake 1977).

Resources, for which competition is most likely among sympatric species of birds, are nest-sites and food (Ashmole 1971). Many of the differences found among the New World storks suggest niche partitioning (Fig. 4). Stork size (Figs. 4A, F) probably influences both nest-site choice (Thomas 1984) and food selection. Wood Storks (Fig. 4B) are not clearly assignable to either resident or migratory categories because of their dispersal behavior and their inter-area movements in pursuit of concentrations of food. Yet they may not make long distance migrations like Maguari Storks, nor do they remain in one place like Jabirus.

Intraspecific foraging associations (Fig. 4G) are shown schematically, based not only on highway census data, but also on my occasional observations that Jabirus sometimes gather in large numbers (maximum 83) at temporarily rich food sources, along with Wood Storks and other wading birds. Data for prey species and prey sizes in Venezuela are generally lacking for Wood Storks and few for Jabirus. Differences in bill structure result in differences in pursuit and securing of prey, releasing sympatric fish-eating birds from direct competition (Owre 1967), and further, Orians (1971) suggested that different foraging behaviors among non-territorial and community feeding birds influence foraging success. The differences of the storks' bill structure (Figs. 1, 4F), and foraging associations and techniques (Fig. 4G, H) suggest that the storks in the Venezuelan llanos use different prey. Fish, however, is only one of the food classes taken by storks in the llanos and may be important only for Wood Storks, because the Maguari Stork is a food generalist, and possibly the Jabiru specializes on freshwater eels.

In summary, morphological and ecological differences among the Western Hemisphere storks, which may result in niche partitioning, appear to make their coexistence possible in the llanos of Venezuela.

ACKNOWLEDGMENTS

I am indebted to the following persons and institutions for generous assistance: T. Blohm for access to and frequent help at Masaguaral, J. B. Trent for financial support, C. C. Wicker and N. T. Wicker for the January aerial surveys, and Sociedad Conservacionista Audubon de Venezuela for the October aerial survey. E. O. Boede kindly gave me permission to handle the birds at Zoológico Las Delicias, Maracay, Venezuala, and along with C. Diaz, assisted me with the live birds. E. López de Caballos readily arranged for my visit to his ranch. Opportunity to examine specimens in their care was willingly given by W. H. Phelps, Jr., R. Avelado, and L. Pérez of the Coleccion Ornithológica Phelps, Caracas; H. Sick of the Museo Boa Vista, Rio de Janeiro, Brazil; G. Maussberger of the Humboldt Museum, Berlin, East Germany; W. E. Lanyon and J. Bull of the American Museum of Natural History, New York; G. Medina and D. Figueroa of the Rancho Grande Collection, Maracay, Venezuela. Others who helped me or kindly shared important data with me are E. Schüz, J. C. Ogden, E. Eisenmann, W. H. Mader, J. Pinowski, J. A. Browder, C. T. Collins, R. Savage, P. Trebbeau, M. Thiele, C. Ramo, A. Spaans, M. Howe, and J. Ingels. Important and useful comments on an early draft were contributed by D. Mock, J. Ogden, J. Ingels, C. Ramo and N. Wicker, as well as by reviewers M. Foster, P. Buckley, P. Kahl, and J. Kushlan.

LITERATURE CITED

ASHMOLE, N. P. 1971. Seabird ecology and the marine environment. Pp. 223–286, *In* D. S. Farner and J. R. King (eds.), Avian Biology, Vol. 1. Academic Press, New York.

AYARZAGUENA S., J., J. PEREZ T., AND C. RAMO H. 1981. Los Garceros del Llano. Cuadernos Lagoven, Caracas.

BENEDICT, F. G., AND E. L. FOX. 1927. The gaseous metabolism of large wild birds under aviary conditions. Proc. Am. Philos. Soc. 66:511–534.

BENT, A. C. 1926. Life histories of North American marsh birds. U.S. Natl. Mus. Bull. 135:1–392.

BLAKE, E. R. 1977. Manual of Neotropical Birds, Vol. 1. University Chicago Press, Chicago, Illinois.

BROWDER, J. A. 1976. Water, wetlands and Wood Storks in southwest Florida. Unpub. Ph.D. dissert. University Florida, Gainesville.

CHUBB, C. 1916. The Birds of British Guiana. Bernard Quaritch, London.

COLLINS, C. T. 1968. The comparative biology of two species of swifts in Trinidad, West Indies. Bull. Fl. State Mus. Biol. Sci. 11:257–320.

EWEL, J. J., AND A. MADRIZ. 1968. Zonas de Vida de Venezuela. Ministerio Agricultura y Cria, Caracas.

FRY, C. H. 1980. The evolutionary biology of kingfishers (Alcedinidae). Living Bird 18:113–160.

HARTMAN, F. A. 1961. Locomotor mechanisms of birds. Smithson. Misc. Collect. 143 No. 1.

HAVERSCHMIDT, F. 1968. Birds of Surinam. Oliver and Boyd, London.

HUDSON, W. H. 1920. Birds of La Plata. J. M. Dent, London.

KAHL, M. P. 1962. Bioenergetics of growth in nestling Wood Storks. Condor 64:169–183.

KAHL, M. P. 1964. Food ecology of the Wood Stork (*Mycteria americana*) in Florida. Ecol. Monogr. 34:97–117.

KAHL, M. P., 1971a. Social behavior and taxonomic relationships of the storks. Living Bird 10:151–170.

KAHL, M. P. 1971b. Observations on the Jabiru and Maguari Storks in Argentina, 1969. Condor 73:220–229.

KAHL, M. P. 1972. Comparative ethology of the Ciconiidae, the Wood-Storks (Genera *Mycteria* and *Ibis*). Ibis 114:15–29.

KAHL, M. P. 1973. Comparative ethology of the Ciconiidae. Part 6. The Blacknecked, Saddlebill, and Jabiru Storks (Genera *Xenorhynchus, Ephippiorhynchus,* and *Jabiru*). Condor 75:17–27.

KAHL, M. P., AND L. J. PEACOCK. 1963. The bill-snap reflex: a feeding mechanism in the American Wood Stork. Nature 199:505–506.

KUSHLAN, J. A. 1978. Feeding ecology of wading birds. Pp. 249–297, *In* A. Sprunt, J. C. Ogden, and S. Wickler (eds.), Wading Birds. National Audubon Society, New York.

MACARTHUR, R. 1971. Patterns of terrestrial bird communities. Pp. 189–221, *In* D. S. Farner and J. R. King (eds.), Avian Biology, Vol. 1. Academic Press, New York.

MOREAU, R. E. 1966. The Bird Faunas of Africa and its Islands. Academic Press, New York.

MORTON, E. S. 1979. A comparative survey of avian social systems in northern Venezuelan habitats. Pp. 233–259, *In* J. F. Eisenberg (ed.), Vertebrate Ecology of the Northern Neotropics. Smithsonian Institution Press, Washington, D.C.

OGDEN, J. C., J. A. KUSHLAN, AND J. T. TILMANT. 1976. Prey selection by the Wood Storks. Condor 78:324–330.

OJASTI, J. 1973. Estudio Biologico del Chigüire o Capibara. Fundo Nacional Investigaciones Agropecuarias, Caracas.

ORIANS, G. 1971. Ecological aspects of behavior. Pp. 513–546, *In* D. S. Farner and J. R. King (eds.), Avian Biology, Vol. 1. Academic Press, New York.

OWRE, O. T. 1967. Adaptions for locomotion and feeding in the Anhinga and the Double-crested Cormorant. Ornithol. Monogr. No. 6.

PALMER, R. S. 1962. Handbook of North American Birds, Vol. 1. Yale University Press, New Haven, Connecticut.

PINOWSKI, J., L. G. MORALES, J. PACHECO, K. A. DOBROWOLSKI, AND B. PINOWSKA. 1980. Estimation of the food consumption of fish-eating birds in the seasonally-flooded savannas (llanos) of Alto Apure, Venezuela. Bull. Acad. Pol. Sci. Ser. Sci. Biol. 28:163–170.

RECHNITZER, A. B. 1956. Foraging habits and local movements of the Wood Ibis in San Diego County, California. Condor 58:427–432.

SELANDER, R. K. 1966. Sexual dimorphism and differential niche utilization in birds. Condor 68:113–151.

SNETHLAGE, H. 1928. Meine Reise durch Nordostbrasilien. J. Ornithol. 76:503–581.

SPAANS, A. L. 1975. The status of the Wood Stork, Jabiru and Maguari Stork along the Surinam coast, South America. Ardea 63:116–130.

THOMAS, B. T. 1977. Hooding and other techniques for holding and handling nestling storks. North Am. Bird Bander 2:47–49.

THOMAS, B. T. 1979. The birds of a ranch in the Venezuelan llanos. Pp. 213–232, *In* J. F. Eisenberg (ed.), Vertebrate Ecology of the Northern Neotropics. Smithsonian Institution Press, Washington, D.C.

THOMAS, B. T. 1981. Jabiru nest, nest building and quintuplets. Condor 83:84–85.

THOMAS, B. T. 1984. Maguari Stork nesting: juvenile growth and behavior. Auk 101:812–823.

TROTH, R. G. 1979. Vegetational types of a ranch in the central llanos of Venezuela. Pp. 17–36, *In* J. F. Eisenberg (ed.), Vertebrate Ecology of the Northern Neotropics. Smithsonian Institution Press, Washington, D.C.

WETMORE, A. 1965. The Birds of the Republic of Panama. Pt. 1. Smithson. Misc. Collect. Vol. 150(1).

BREEDING BIOLOGY

ESCALANTE, RODOLFO. Taxonomy and Conservation of Austral-Breeding Royal Terns   935

HUMPHREY, P. S., AND B. C. LIVEZEY. Nest, Eggs, and Downy Young of the White-headed Flightless Steamer-duck   945

MASON, PAUL. The Nesting Biology of Some Passerines of Buenos Aires, Argentina   954

TAXONOMY AND CONSERVATION OF AUSTRAL-BREEDING ROYAL TERNS

RODOLFO ESCALANTE

Guayaquí 3425, Apartamiento 301, Montevideo, Uruguay

ABSTRACT. On the Atlantic coast of South America, populations of Royal Tern, *Sterna maxima*, show two different calendars of breeding activity. In French Guiana, boreal breeding occurs within the same period as in the northern hemisphere (April–August). In the southern Argentina population, austral reproduction occurs during the southern hemisphere spring and summer (October–March). This latter population migrates northward to winter in Uruguay and (principally) southern Brazil. No data are available on the status of this species between French Guiana (5°N) and Rio de Janeiro (23°S) on the northern and eastern coasts of Brazil. The plumage status of specimens from Brazil, Uruguay, and Argentina suggests tentatively that only a few puzzling specimens from Brazil are individuals from boreal breeding populations. On the other hand, no banded bird from North America has been recovered from Brazil, Uruguay, or Argentina. Nonetheless, these three populations are mensurally indistinguishable from each other and from those breeding in the southern United States and should, therefore, be treated as *S. m. maxima*.

RESUMEN. En la costa atlántica de Sudamérica las poblaciones de gaviotín real, *Sterna maxima*, muestran dos calendarios diferentes de actividad reproductora. En la Guayana Francesa, la reproducción boreal ocurre en el mismo período que en el hemisferio norte (abril–agosto). En la población del sur de Argentina, la reproducción austral ocurre durante la primavera y el verano del hemisferio sur (octubre–marzo). Esta última población migra hacia el norte para pasar el invierno en Uruguay y (principalmente) en el sur de Brasil. No se dispone de información sobre la situación de esta especie entre Guayana Francesa (5°N) y Rio de Janeiro (23°S) en las costas del norte y este de Brasil. Las condiciones del plumaje de los especímenes de Brasil, Uruguay y Argentina sugieren, tentativamente, que sólo unos pocos e intrigantes ejemplares de Brasil son individuos de poblaciones con reproducción boreal. Por otra parte ni en Brasil, Uruguay o Argentina se han recuperado gaviotines reales marcados en América del Norte. De cualquier manera, estas tres poblaciones no se pueden diferenciar entre si por sus medidas lineales, ni tampoco con la población que anida en el sur de los Estados Unidos de Norteamérica y por lo tanto deben ser tratadas como *S. m. maxima*.

The Royal Tern has traditionally been considered to comprise two races. *Sterna maxima maxima* breeds from May to July on the North American side of the Atlantic from Virginia to Mexico and on some Caribbean islands; it migrates in the boreal winter to the Caribbean, Central America, and northern South America, and to areas along the Pacific coast of South America as far south as Peru and along the Atlantic coast, to Brazil, Uruguay, and Argentina (Meyer de Schauensee 1966; American Ornithologists' Union 1983). The other population, *S. m. albidorsalis*, nests on the west coast of Africa from the Straits of Gibraltar to Angola (Blake 1977; Howard and Moore 1980).

Sixteen years ago evidence was presented that strongly suggested the existence of an additional breeding population along the coasts of Uruguay, probably Argentina, and perhaps Brazil (Escalante 1968, 1971); it was confirmed later, when nesting birds, eggs, chicks, and non-flying young were observed or collected in Argentina (Korschenewski 1969; Erize and Korschenewski *in* Boswall and Prytherch 1972; Daciuk 1972; Jehl et al. 1973; Devillers 1977; Olrog 1979; S. Narosky *in litt*).

Regarding the status of this bird on the Atlantic side of South America, Blake (1977:635–636) commented that "A breeding population may be resident on the coast of Uruguay and northern Argentina, and coastal Chubut," and that it ". . . winters . . . to Buenos Aires (occasionally Chubut) Argentina, in the east." Therefore, Blake (1977) considered that coastal Brazil is visited only by North American migrants, and he also asserted that these migrants winter in Argentina. Olrog (1979) expressed similar points of view. These authors, in general, seemed unaware that to date no evidence supports the idea that Royal Terns occurring on

TABLE 1
MEASUREMENTS OF *STERNA MAXIMA* FROM BRAZIL[1]

Specimen no.[2]	Date	Age[3]	Wing chord[4]	Tail length[4]	Culmen length	Gonys length	Tarsus length	Middle toe with claw
Males								
36663	4 Aug.	Ad. (N)	370	185	64.0	30.7	31.9	33.9
7923	28 Aug.	Ad. (N)	354	167	57.3	27.6	30.0	28.4
37152	31 Aug.	Ad. (?)	369	165	63.0	33.0	32.0	31.5
36664	4 Aug.	Sub. (?)	360+m	180+m	64.0	30.0	32.0	32.6
7925	28 Aug.	Sub. (?)	345	150	59.4	30.5	30.0	32.0
37151	31 Aug.	Sub. (?)	360	172	62.0	33.8	31.0	32.9
31476	16 Oct.	Sub. (?)	350m	135m	63.0	31.5	32.5	32.0
31477	18 Oct.	Sub. (?)	372m	120m	62.7	29.4	33.0	31.9
32972	9 Apr.	Juv.	357	125	54.7	25.3	34.0	33.5
37150	31 Aug.	(Juv.)	350++m	181++m	61.4	29.8	32.0	31.0
Females								
35391	8 Jul.	Ad. (N)	357+m	178	57.0	26.0	30.8	32.8
36666	4 Aug.	Ad. N	358	188	58.7	30.0	30.5	30.0
7924	28 Aug.	Ad. (N)	362	180	59.0	31.0	32.7	30.0
36665	4 Aug.	Sub. (?)	365	180	61.0	30.6	33.2	34.2

[1] Measurements in mm.
[2] Museu de Zoologia da Universidade de São Paulo. Localities of specimens: Casquerinho, Piassa Guerra (Santos, Estado de São Paulo), 7923–7925; Mun. Itanhaem (south of Santos), 31476–34177; Inst. Osvaldo Cruz (Rio de Janeiro, 32972; Praia Grande (Santos), 35391, 36663–36666, 37150–37152.
[3] Ad. = adult; N = full nuptial plumage; (N) = nuptial plumage with some white feathers on the anterior crown; Sub. = subadult; Juv. = juvenile in its first covering of true feathers; (Juv.) = juvenile in first postjuvenal molt; (?) = fully grown birds of unknown age.
[4] m = quills in postnuptial molt; + = worn feathers; ++ = very worn feathers.

the Atlantic coast south of Brazil are North American migrants, although the possibility that some North American birds may winter in the area cannot be dismissed. Furthermore, according to Ashmole and Tovar (1968) and especially Van Velzen (1968, 1971), recoveries of Royal Terns banded in the United States suggest a partial movement down the northern coast of South America (Venezuela, Trinidad), but principally across Central America to Colombia, Ecuador, and Peru.

In the present paper I attempt to clarify the status of *S. maxima* along the Atlantic coast of South America based, in part, on evidence hitherto unpublished.

MATERIALS AND METHODS

Plumages, molt, and measurements of Royal Tern specimens from Brazil (Museu de Zoologia da Universidade de São Paulo [MZUSP]; Museu Paraense "Emilio Goeldi," Para [MP]; British Museum [Natural History] [BMNH]); Uruguay (Museo Nacional de Historia Natural, Montevideo), and Argentina (Museo Argentino de Ciencias Naturales, Buenos Aires [MACN]; BMNH; U.S. National Museum of Natural History [USNM]) were compared with those of known breeders from North America and northern South America, and banding data supplied by the Bird Banding Laboratory (U.S. Fish and Wildlife Service, Laurel, Maryland, K. Klimkiewicz, *in litt.*) were analyzed.

RESULTS

Brazil.—Data were obtained from 14 specimens (all MZUSP) collected in the austral winter (April–October) along the coasts of São Paulo and Rio de Janeiro states between 23° and 25° (Table 1). Their stages of development are, April:1 fresh juvenile (32972); July:1 adult (35391) with worn plumage in postnuptial molt, forehead and crown black with some fresh white feathers, 4 innermost primaries (P-1 to P-4) renewed, P-5 lacking; August:1 adult (36666) in full and fresh nuptial plumage (complete black pileum); 3 adults (7923, 7924, 36663) in fresh plumage, black pileum with few white feathers on the anterior crown and forehead; 4 subadults (?) (7925, 36665, 37151, 37152) with new white feathers with black centers on the crown and fresh primaries; 1 subadult (36664) with white feathers with black centers on the crown, molting P-1 and P-2; 1 juvenile (37150) with dark cubital bands, white head with only limited black behind eyes and on nape, with worn outer primaries, P-1 through P-3 new, P-4 lacking

TABLE 2

COMPARATIVE MEASUREMENTS OF THE EXPOSED CULMEN IN GEOGRAPHICAL SAMPLES OF
AMERICAN ROYAL TERNS[1]

Locality	Sex	N	X̄	Range	s.d.	s.e.	C.V.	95% confid. interval
United States of North America,	♂♂	10	64.1	59.0–68.0	—	—	—	—
Mexico, Cuba	♀♀	18	62.7	57.5–67.0	—	—	—	—
Brazil (Rio de Janeiro, Sao	♂♂	8	61.9	57.3–64.0	2.37	0.83	3.82	±1.9
Paulo, 23°–25°S)	♀♀	4	58.9	57.0–61.0	1.63	0.81	2.76	±2.6
Uruguay, Argentina	♂♂	5	64.1	61.7–66.4	1.85	0.83	2.89	±2.3
(35°–42°30'S)	♀♀	10	62.7	58.7–69.0	3.30	1.04	5.26	±2.7

[1] Measurements in mm. Data for North American birds from Ridgway (1919), for Brazilian birds, from Table 1, and for Uruguayan and Argentinian birds, from Escalante (1968, 1970, 1971).

(postjuvenal molt); October:2 subadults (?) (31476, 31477) with worn plumage, molting P-3 and P-4. Bill measurements of these individuals are in Tables 1 and 2.

The author received from D. W. Snow and P. E. Colston (*in litt.*) a listing of Royal Terns in the British Museum (Natural History) that have been collected on the Atlantic coast of South America. Among nine Brazilian skins, five had sufficient data to be useful. Three specimens, collected on 18 July 1886, were from Rio de Janeiro, including 1 adult female that "is presumably a breeding bird which shows white feathers on forehead and crown," and 2 unsexed birds that are considered to be "a year plus, or in first summer plumage" ("portlandi-ca" plumage). One adult male "in breeding plumage, i.e., black pileum with many white feathers on forehead and crown" and fresh quills, was collected on 1 August in Santa Catharina (27°S, H. Saunders coll. No. 3614). The fifth specimen (30 April 1891, Brazil, BMNH No. 1891.5.16.1) is an adult male with a "black pileum with many white feathers on forehead and crown, indicative of a breeding bird."

There is one male skin from the state of Para (1°N–1°S) mentioned by Snethlage (1914). Fernando C. Novaes (*in litt.*) informed me that this specimen (MP 3992) was collected on 17 November 1905, but the label bears neither name of the collector nor the locality although the state has 600 km each of Atlantic and Amazon River shorelines (Escalante 1972). According to Novaes, the bird seems to be a juvenile in first winter plumage similar to that mentioned by Escalante (1968:245, fig. 1).

Uruguay.—Seventeen specimens from Uruguay in the Museo Nacional de Historia Natural have already been discussed by Escalante (1968). Exposed culmen measurements are shown in Table 2. The molt cycle of Uruguayan Royal Terns, indicated in this sample, has not been previously reported, and data on the renewal of the 10 primaries are given in Table 3. Feathers examined were from the right or left wing or both, depending on condition. In addition, I here report a photograph of part of a flock of 45 Royal Terns obtained on 1 August 1978 in the mouth of Arroyo Carrasco (Depto. Canelones, 15 km E of Montevideo), where most of the birds exhibited a solidly black pileum. This is the only photographic record to date of birds in nuptial plumage in southern South America at a probable prebreeding time.

Argentina.—Snow and Colston (*in litt.*) listed four skins in the British Museum, including three juveniles taken on 13 March; the fourth (male) was from Buenos Aires Province, collected in March 1910 (BMNH 1920.9.30.16), and was either one year old or a winter adult. A breeding plumage male in the U.S. National Museum of Natural History was collected near Cabo San Antonio (Buenos Aires Province) on 4 November 1920 (Escalante 1968), and Escalante (1971) recorded two more specimens in the Museo Argentino de Ciencias Naturales, one adult male with black pileum (MACN 8963) from Chubut Province, 14 November 1916, and one juvenile (MACN 4466) taken in Buenos Aires Province on 12 April 1937; their measurements are lumped in Table 2 with those of specimens from Uruguay and Argentina. There are no Argentinian winter specimens.

DISCUSSION

The status of Royal Tern for each country on the Atlantic coast of South America may now be summarized as follows:

Guyana: no specimens are known but there are undated sight records near Georgetown

TABLE 3

Molt Program of Primaries of Royal Terns from Uruguay

Month and specimen no.[1]	Sex and age[2]	Primary number[3]									
		1	2	3	4	5	6	7	8	9	10
January											
RE 574	♀ Ad.	N	G	—	O	O	O	O	O	O	O
February											
GB 487	♀ Ad.	N	N	—	—	O	O	O	O	O	O
GB 488	? Ad.	N	N	N	N	G	—	O	O	O	O
GB 100	♂ Juv.	first plumage of true feathers									
March											
RE 547	♀ Ad.	N	N	N	N	N	N	G	G	—	O
RE 613	? Ad.	N	N	N	N	N	N	G	B	O	O
May											
MHNM 1639	♂ Ad.	N	N	N	N	N	N	N	N	N	G
>MHNM 1641	♀ Ad.?	N	N	N	N	N	N	N	N	G	O
MHNM 1638	♂ Sub.	N	N	N	N	N	N	N	N	G	O
MHNM 1640	♀ Sub.	N	N	N	G	O	O	O	O	O	O
MHNM 1642	? Juv.	first winter plumage									
June											
MHNM 1643	♀ Ad.?	N	N	N	N	N	N	N	N	N	G
MHNM 1644	? Ad.	N	N	N	N	N	N	N	N	N	G
August											
R.E. 617	♀ Sub.	N	N	N	N	N	N	—	O	O	O
September											
ST 763	♂ Ad.	full and fresh nuptial plumage									
ST 458	♀ Sub.	N	N	N	N	N	N	N	—	O	O
ST 459	♀ Sub.	N	N	N	N	N	N	N	N	G	O

[1] Specimens from the Museo Nacional de Historia Natural (Montevideo). For complementary data see Escalante 1968 (table 1) and 1970 (table IX).

[2] Ad. = adult; Sub. = subadult or immature; Juv. = young or juvenile.

[3] N = new feather; G = growing or sheathed feather; O = old feather; B = brush feather; — = feather lacking.

(Snyder 1966). *Surinam:* banded specimens from the North American breeding population were recovered between 1972 and 1982 at approximately 6°N, 57°W (K. Klimkiewicz, *in litt.*). There are no other North American band recoveries east or south of Surinam. *French Guiana:* a population of unknown size breeds from May to September (5°N, 52°W); for example, from 1 to 7 August 1974, juveniles outnumbered eggs and chicks (Condamin 1978). The dispersal and winter range of these birds are unknown.

Brazil: virtually nothing is known about the status of Royal Tern along ±4000 km of the northern and eastern coasts of this country south from its boundary with French Guiana (5°N) to Rio de Janeiro (23°S). Nevertheless, it seems possible that May through August breeding birds may visit or nest on this coast near French Guiana. Analysis of British Museum and Museu de Sao Paulo specimens suggests, however, that the birds from Rio de Janeiro south belong to the southern or austral breeding population of *S. maxima*. They may be local breeders (Escalante 1980), migrants from Argentina (Escalante 1980; R. van Halewijn, *in litt.*), or a mixture of both. Although the birds of both samples were collected in the austral winter, Teixeira (*in litt.*) recently reported 16 birds in nonbreeding plumage in "Baia de Guanabara" (Rio de Janeiro) on 28 December 1982, so Royal Terns seem to occur the year round in that area. In Rio Grande do Sul (32°S), Vooren et al. (1982) considered *maxima* a winter visitor only (April–September).

Although Sick and Leao (1965) reported 30 Royal Terns on an island near Cabo Frio (Rio de Janeiro), among a big colony of *S. sandvicensis eurygnatha,* breeding remains unproven for Brazil. However, Junge and Voous (1955) believed that eggs (60 mm × 43 mm) from Brazil ascribed by Ihering (1900) to *S. s. eurygnatha* probably were those of *S. maxima;* this

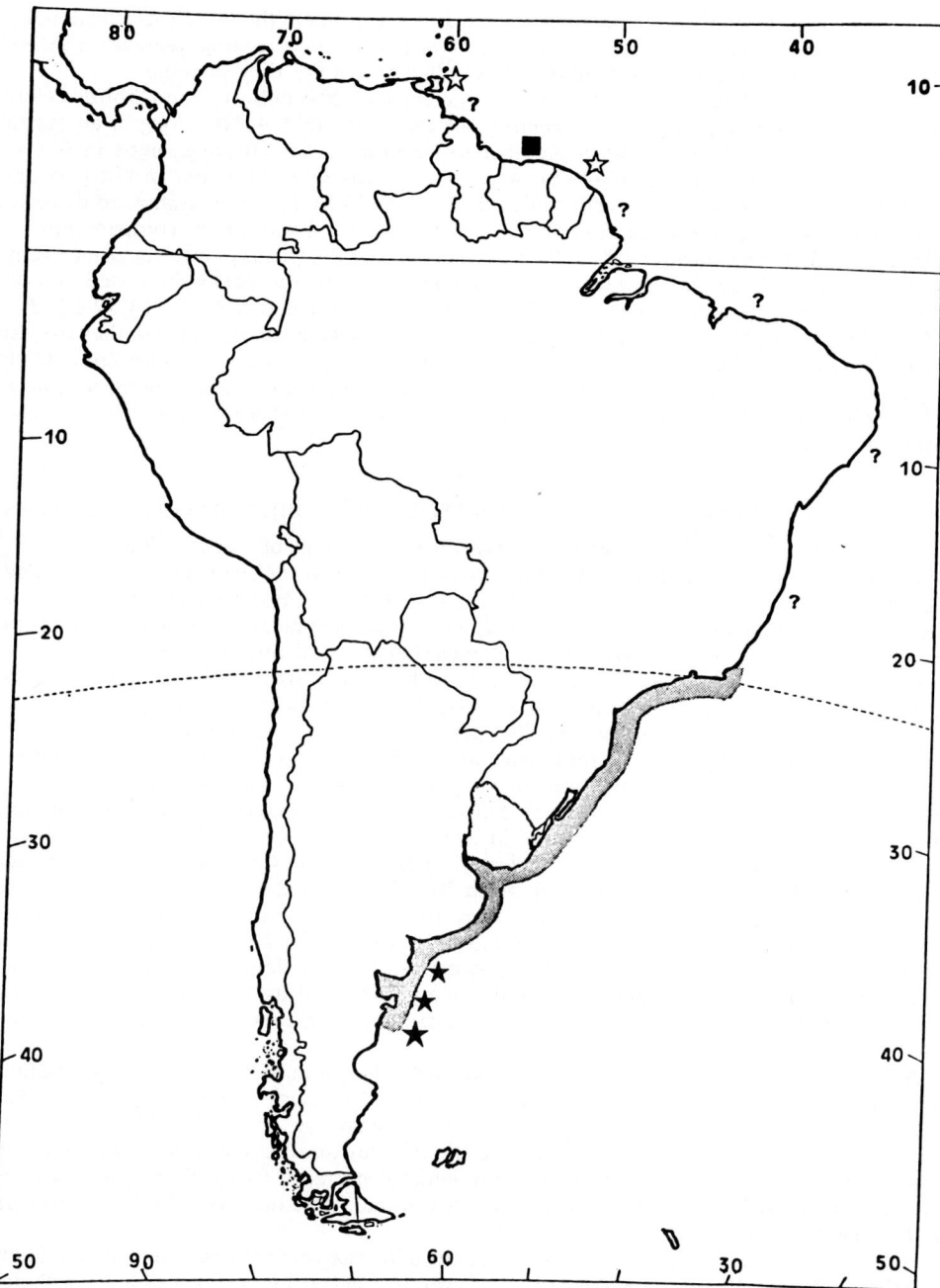

FIG. 1. Distribution of various Royal Tern populations on the east coast of South America. Stippled areas indicate the probable limits of the austral breeding population. Solid stars indicate known nesting places of the austral breeding population. Open stars indicate known nesting places of boreal breeding populations. The solid square marks the only locality at which banded specimens of the North American breeding population have been recovered. Question marks indicate areas for which the breeding status of the Royal Tern is unknown.

is surely the case. Complicating the picture are a July specimen in the middle of a postnuptial primary molt (MZUSP 35391) and an April specimen in breeding plumage (BMNH 1891.5.16.1). These may represent individuals with a boreal breeding calendar.

Uruguay: the plumages and molts of the 17 specimens (35°S) indicate that all birds belong to an austral breeding population. *Argentina:* all specimens (37°–44°S) belong to an austral population breeding from August–September to December–February (e.g., nests in Buenos Aires and Chubut Provinces, 40°–44°S). Argentinian birds seem to winter in Uruguay and southern Brazil, perhaps as far north as Rio de Janeiro (23°S), but their southward dispersal is unknown. Because of wear or molt of outermost primaries, I was able to compare only the culmens of adults and subadults from North America, Brazil, Uruguay, and Argentina (Table 2). The figures indicate that the birds of Rio de Janeiro-São Paulo are closely related to those of Uruguay and Argentina (males: $t = 1.754$, 11 d.f., $P > 0.1$; females: $t = 0.549$, 12 d.f., $P > 0.5$). Although only the means and ranges for North American birds are at hand, they overlap so much with these from South America that statistically significant differences are unlikely to be found. As plumage also seems not to differ among the various populations, all should tentatively be known as *Sterna maxima maxima,* regardless of location, or of austral versus boreal breeding seasons.

BIOLOGY AND CONSERVATION OF THE SOUTHERN BREEDING POPULATION

Nesting and ecology.—The Argentinian breeding population of *maxima* has so far only been recorded nesting clustered within large colonies of *Larus* (*belcheri*) *atlanticus, S. sandvicensis eurygnatha,* and *S. hirundinacea* (Korschenewski 1969; Boswall and Prytherch 1972; Daciuk 1972, Jehl et al. 1973; Devillers 1977), similar to the manner in which Atlantic coast *S. s. acuflavidus* nest only in Royal Tern colonies (Buckley and Buckley 1980).

Most of the birds observed in Uruguay seem to fish out of sight of the shoreline. Among those occasionally fishing inshore, none was seen skimming or catching crabs. The birds usually caught surface fishes by diving, principally "pejerreyes" (Atherinidae) and "anchoitas" (Engraulidae). Occasionally some bottom fishes were obtained on the beaches as fishing refuse; on 1 August 1978 I photographed birds picking up "roncaderas" and "burriquetas" (*Micropogon* sp. and *Ophioscion* sp.: Scianidae). The fish lengths were greater than 10 cm, and one tern I collected had a 17 cm "roncadera" in its gullet. On October 1964 I recorded a Royal Tern's behavior when it had caught a "pejerrey" perhaps 20 cm long. The bird first took the fish by the middle of the body, and then, when 20 m high, jerked its neck and tossed the fish in mid-air, recaptured it by the head, and then swallowed it on the wing; part of the fish's body and tail stuck out of the bill of the bird as it flew away. This almost exclusive preference for fish contrasts with descriptions of North American birds in Bent (1921) and Buckley and Buckley (1972), who mentioned birds taking many crabs and fishes only 5 to 10 cm long. Also, they did not record specimens from the Scianidae even though that family occurs on the U.S. Atlantic coast.

According to Buckley and Buckley (1972), Royal Terns appear to have few critical predators after the egg stage. However, Escalante (1969) recorded the killing of a Royal Tern by a Skua (*Catharacata* sp.) at Punta del Este (Maldonado, Uruguay) on 11 April 1968, so predation on adults is not unknown. More in line with the Buckley and Buckley (1972) observations, Daciuk (1972) believed that 40 to 50 percent, perhaps more, of Royal Tern eggs on Isla de los Pajaros, Golfo San Jose (42°25'S; Chubut, Argentina), were destroyed by *Larus scoresbii* and *L. dominicanus.*

Human disturbance.—Most of the breeding pairs of *S. maxima* have been recorded in Punta Tombo (44°S) and Peninsula Valdez (42°S), Argentina, according to Boswall and Prytherch (1972). Boswall (1973) pointed out that the uncontrolled wandering of tourists in Punta Tombo was a cause of great concern to local conservationists. In December, January, and February of 1971 and 1972 Daciuk (1972) reported that in Argentina, tourists on Isla de los Pajaros caused the death of chicks that fled into colonies of *L. dominicanus.* As pairs of austral breeding Royals nest adjacent to the gulls, their nests, eggs, and chicks are equally at risk. In Uruguay, the coasts of the Río de la Plata and the Atlantic Ocean are visited during the austral spring and summer by 200,000 to 400,000 tourists with boats and pets, and there are no protected places where seabirds can rest free of disturbance. The same problem seems to occur in Brazil. Olrog (1979) and Devillers (1977) reported, in addition, that eggs were frequently collected from colonies of *L.* (*b.*) *atlanticus* in Bahia San Blas and Bahia Anegada (40°S, Argentina).

Uncontrolled sewage and chemical residues, oil spillage from ships, tankers, and various devices for discharging fuels are increasing, with deleterious effects on seabirds in the Río de la Plata and neighboring oceanic waters.

Overfishing of inshore waters of the Río de la Plata along 100 km of coast occurred between 1973 and 1978 in Uruguay and probably had an adverse effect on Royal Terns and other seabirds. It presumably will continue to increase.

Conservation.—Few reports of austral Royal Tern breeding colonies, migrant flocks, or individuals seen or collected are available (Escalante 1968; Korschenewski 1969; Boswall and Prytherch 1972; Daciuk 1972; Devillers 1977). It is my impression that this probably small population has declined since 1970 and seems especially vulnerable to human disturbance. Quick and effective actions must be undertaken if we are to preserve this population (Escalante 1980, in press).

ACKNOWLEDGMENTS

The author expresses his gratitude to H. de A. Camargo, O. de O. Pinto, P. E. Vanzolini, M. A. Klappenbach, and to J. R. Navas who permitted him to study skins in the collections in their care. I am also indebted with the following persons for their generosity in supplying information or help in many other ways: J. Boswall, F. G. Buckley, P. Devillers, J. R. Jehl Jr., K. Klimkiewicz, K. Lambert, S. Narosky, F. C. Novaes, H. Sick, D. M. Teixeira, R. van Halewijn, and W. T. Van Velzen. Special mention goes to D. W. Snow and P. R. Colston for important information about skins in the British Museum, to E. Mayr for his systematics opinions, to F. G. Buckley and J. S. Weske for helpful criticism of earlier drafts of this paper, and to P. A. Buckley for aid in polishing my English. The visit of the author to the Brazilian museums was supported by a grant from the Frank M. Chapman Memorial Fund of the American Museum of Natural History.

LITERATURE CITED

AMERICAN ORNITHOLOGISTS' UNION. 1983. Check-list of North American Birds, 6th ed. American Ornithologists' Union, Washington, D.C.

ASHMOLE, N. P., AND H. TOVAR. 1968. Prolonged parental care in Royal Terns and other birds. Auk 85:90–100.

BENT, A. C. 1921. Life histories of North American gulls and terns. U.S. Natl. Mus. Bull. 113.

BLAKE, E. R. 1977. Manual of Neotropical Birds, Vol. 1. University Chicago Press, Chicago, Illinois.

BOSWALL, J. 1973. Supplementary notes on the birds of Point Tombo, Argentina. Bull. Br. Ornithol. Club. 93:33–36.

BOSWALL, J., AND R. J. PRYTHERCH. 1972. Some notes in the birds of Point Tombo, Argentina. Bull. Br. Ornithol. Club. 92:118–129.

BUCKLEY, F. G., AND P. A. BUCKLEY. 1972. The breeding ecology of Royal Terns Sterna (*Thalasseus*) *maxima maxima*. Ibis 114:344–359.

BUCKLEY, F. G., AND P. A. BUCKLEY. 1980. Habitat selection and marine birds. Pp. 69–112, *In* J. Burger, B. L. Olla, and H. E. Winn (eds.), Marine Birds, Vol. 4. Plenum Press, New York.

CONDAMIN, M. 1978. Nidification d'oiseaux de mer a Guyane. Oiseau Rev. Fr. Ornithol. 48:115–121.

DACIUK, J. 1972. Notas faunisticas y bioecologicas de Peninsula Valdes y Patagonia. XIV. Pequena colonia de nidificacion del Gaviotin Brasilero en "Isla de los Pajaros" (Golfo San Jose, Chubut, Rep. Argentina). Neotropica 18:103–106.

DEVILLERS, P. 1977. Observations at a breeding colony of Larus (*belcheri*) *atlanticus*. Gerfaut 67:22–43.

ESCALANTE, R. 1968. Notes on the Royal Tern in Uruguay. Condor 70:243–247.

ESCALANTE, R. 1969. Gaviotin Real apresado por una Gaviota Parda. Neotropica 15:64.

ESCALANTE, R. 1970. Aves marinas del Río de la Plata y aguas vecinas oceano Atlántico. Barreiro y Ramos, Montevideo, Uruguay.

ESCALANTE, R. 1971. El Gaviotin Real en la Argentina. Neotropica 17:101–104.

ESCALANTE, R. 1972. First Pomarine Jaeger specimen from Brazil. Auk 89:663–665.

ESCALANTE, R. 1980. Notas sobre algunas aves de la vertiente atlántica de Sud America (Rallidae, Laridae). Resumenes Letras Jornadas Cienc. Nat. Montevideo 1:33–34.

ESCALANTE, R. In press. Problemas en la conservacion de dos poblaciones de Laridos sobre la costa atlantica de Sud America (*Larus* (*b.*) *atlanticus* y *Sterna maxima*). Actas IIIa Reunion Ibero. Conserv. Zool. Vertebr. Buenos Aires, Argentina.

HOWARD, R., AND A. MOORE. 1980. A Complete Checklist of the Birds of the World. Oxford University Press, Oxford, England.

IHERING, H. VON. 1900. Catalogo critico-comparativo dos ninhos e ovos das aves do Brazil. Rev. Mus. Paulista 4:191–300.

JEHL, J. R., JR., M. A. E. RUMBOLL, AND J. P. WINTER. 1973. Winter bird populations of Golfo San Jose, Argentina. Bull. Br. Ornithol. Club 93:56–63.

JUNGE, G. C. A., AND K. H. VOOUS. 1955. The distribution and relationship of *Sterna eurygnatha* Saunders. Ardea 43:226–247.

KORSCHENEWSKI, P. 1969. Observations sobre aves del litoral patagonico. Hornero 11:48–52.

MEYER DE SCHAUENSEE, R. M. 1966. Birds of South America. Livingston Publ. Co., Narberth, Pennsylvania.

OLROG, C. C. 1979. Nueva lista de la avifauna argentina. Opera Lilloana 27:102.

RIDGWAY, R. 1919. The birds of North and Middle America. U.S. Natl. Mus. Bull. 50 (8).

SICK, H., AND A. P. A. LEAO. 1965. Breeding sites of *Sterna eurygnatha* and other seabirds off the Brazilian coast. Auk 82:507–508.

SNETHLAGE, E. 1914. Catalogo das aves amazonicas. Bol. Mus. Para. E. Goeldi Nov. ser. Hist. Nat. Etnol. 8:1–79.

SNYDER, D. E. 1966. The Birds of Guyana. Peabody Museum, Salem, Massachusetts.

VAN VELZEN, W. T. 1968. The status and dispersal of Virginia Royal Terns. Raven 39:55–60.

VAN VELZEN, W. T. 1971. Recoveries of Royal Terns banded in the Carolinas. Chat 35:64–66.

VOOREN, C. M., G. A. L. BRANDAO, A. FILIPPINI, W. S. FERREIRA, AND G. J. PEDRAS. 1982. Shore and seabirds of south Brazil. Int. Symp. Util. Coast. Ecosys. 22–27 November 1982, Rio Grande do Sul, Brazil.

PLATE VII.   Downy young of *Tachyeres*. From top to bottom: *T. leucocephalus* (KUMNH 79501), *T. brachypterus* (MCZ 70520), *T. patachonicus* (USNM 485600), *T. pteneres* (AMNH 443689). From a watercolor painting by Robert M. Mengel.

NEST, EGGS, AND DOWNY YOUNG OF THE WHITE-HEADED FLIGHTLESS STEAMER-DUCK

PHILIP S. HUMPHREY AND BRADLEY C. LIVEZEY

Museum of Natural History and Department of Systematics and Ecology, University of Kansas,
Lawrence, Kansas 66045 USA

ABSTRACT. Nest, eggs, and downy young of the recently described White-headed Flightless Steamer-Duck (*Tachyeres leucocephalus*) are described and compared with those of other species of steamer-duck. Nests of *T. leucocephalus,* like those of other steamer-ducks, are shallow bowls lined heavily with down, and typically located in dense woody shrubs on peninsulas and off-shore islands. Egg size of *T. leucocephalus* is most similar to that of *T. brachypterus,* but eggs of all species of *Tachyeres* overlap substantially in their dimensions. Downy young of *T. leucocephalus* differ from those of congeners in eyelid and back color, and facial pattern. These characters are used to construct a key to the downy young of the four species of *Tachyeres.*

RESUMEN. Los nidos, huevos, y pichones de *Tachyeres leucocephalus* estan descritos y comparados con los de las otras especies del género. Los nidos de *T. leucocephalus* semejan cuencos pocos profundos forrados con plumón y estan situados típicamente abajo de arbustos ubicados en islotes y peninsulas. Los huevos de *T. leucocephalus* tienen un tamaño mas semejante a los de *T. brachypterus* pero las dimensiones de los huevos de todas las especies de *Tachyeres* se superponen. Los pichones de *T. leucocephalus* se distinguen de otros del mismo género en el color de los párpados y la espalda, y el dibujo de la distribución de las manchas de la cabeza. Se presenta una clave para la identificación de los pichones de patos vapores.

The White-headed Flightless Steamer-Duck (*Tachyeres leucocephalus*) from the coast of southern Chubut, Argentina, is very similar in appearance to other species of steamer-duck and for this reason was not discovered until 1979 (Humphrey and Thompson 1981). During field work at Puerto Melo, Chubut, in December 1981, we found nests and downy young of this species. The downy young were of particular interest because of variation in the head patterns of downies of the other three species of steamer-duck and their possible utility as taxonomic characters (Delacour 1954). In this paper we describe the nest, eggs, and newly hatched young of the White-headed Flightless Steamer-Duck, and compare them to those of the other three known species of steamer-duck.

MATERIALS AND METHODS

NESTS, EGGS, AND BROODS

We found two active and three recently vacated nests of *T. leucocephalus* on Isolate Escobar at Puerto Melo, Chubut, Argentina, during 14 to 17 December 1981. For each nest, we collected the following data, if possible: date; nest attendance; deposition of down; diameter and depth of nest bowl; height and composition of cover; distances from nest to shore and to another nest; habitat type; clutch size; stage of incubation (Weller 1956); length and width of eggs. We also recorded the size and approximate ages of broods of *T. leucocephalus* encountered during field work. Age classes of broods and skin specimens were defined as follows: class I—downy young with no juvenal feathers visible; class II—young covered by a mixture of down and juvenal feathers; and class III—young birds completely covered with feathers, no down visible.

Our data for *T. leucocephalus* were supplemented by those collected by Boswall and Prytherch (1972), Boswall (1973), Boswall and MacIver (1979), and Daciuk (1976) at other localities in coastal Chubut. Eggs described in these papers were identified as *T. patachonicus* by their authors, but clearly pertain to *T. leucocephalus* based on a photograph of an incubating female in Daciuk (1976), distributional information, locomotor behavior, and, to a lesser extent, dimensions of the eggs. The data for *T. leucocephalus* were compared to published descriptions and our own information on nests and eggs of the Flying Steamer-Duck (*T. patachonicus*) and Magellanic Flightless Steamer-Duck (*T. pteneres*) and published descriptions of the Falkland Flightless Steamer-Duck (*T. brachypterus*).

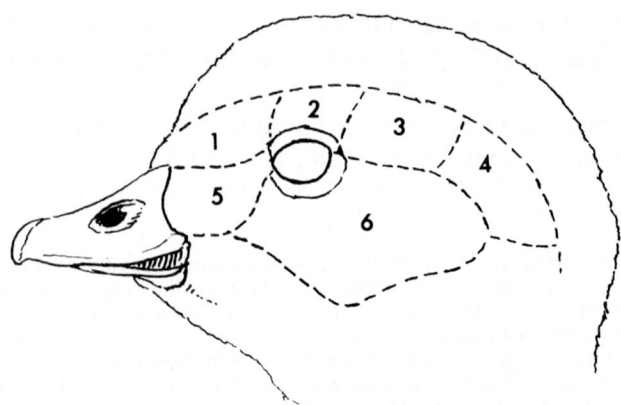

Fig. 1. Diagram of head of newly hatched downy steamer-duck showing: (1) supraloral patch; (2) supraocular patch; (3) anterior postocular streak; (4) posterior postocular streak; (5) lores; (6) cheek.

Downy Young

We collected six specimens of downy *T. leucocephalus* at Puerto Melo during December 1981, and compared these specimens with borrowed material representing the three other species of *Tachyeres*. These comparisons and the characters discussed by Lowe (1934) and Murphy (1936) formed the basis for a key to young downies of *Tachyeres*.

Nomenclature used for the regions of the head and neck is explained in Figure 1. We use Murphy's (1936) term, "yoke," to designate the upper back and scapular region. We follow Palmer (1962) for color nomenclature.

Several of the specimens examined are class-II downies with considerable body down and, for some, emerging tips of contour feathers. We based our descriptions only on downies which, because of their very small size, we determined to be roughly comparable in age to the specimens of *T. leucocephalus*.

Specimens examined (all class-I except as indicated) are as follows:

T. leucocephalus. 6 specimens from Puerto Melo, Chubut, Argentina, collected 12–14 December 1981: MACN 52698 (male, 97.5 g, no yolk), KUMNH 79624 (62.5 g), KUMNH 79625 (85.0 g), KUMNH 79501 (female, 83.0 g, yolk sac 20 mm. in diameter), KUMNH 79502 (male, 92.0 g, yolk sac 20 mm in diameter).

T. brachypterus. 5 specimens from the Falkland Islands, sex unknown: AMNH 419160 (7 December 1915), MCZ 70521 (15 December 1915), MCZ 70522 (figured *in* Murphy 1936), BM 1930. 12.18.1, BM (no number, 29 November 1936).

T. patachonicus. 3 specimens from Chile: USNM 485600 (female, 1 February 1964), FMNH 120521 (female, class-II, 28 January 1940), FMNH 120522 (class-II, 28 January 1940).

T. pteneres. 11 specimens from Chile: AMNH 443684 (male, 3 December 1914), AMNH 443685 (male, 4 December 1914), AMNH 443687 (female, 19 January 1915), AMNH 443689 (male, 30 November 1914), AMNH 443704, 443705, 443706, 443708 (3 males, 1 female, 11 January 1915), BM 80.11.18.561 (class-II), FMNH 62419 (male, class-II, 30 January 1923), FMNH H.B. Conover 2763 (male, class-II, 4 February 1923).

RESULTS AND DISCUSSION

Nests

Nesting season.—We found two active nests of *T. leucocephalus* at Puerto Melo on 14 December 1982; one contained downy young and the other contained five well incubated eggs. Other nest dates for *T. leucocephalus* are 22 October (Boswall 1973; Boswall and MacIver 1979), 16 November (Boswall and Prytherch 1972), and 29 January (Daciuk 1976). Daciuk (1976) estimated that nesting lasts from October to February; we saw class-I downies frequently in December and class-II broods occasionally during February.

These nest dates for *T. leucocephalus* conform to the nesting seasons of other continental species of *Tachyeres*. Nesting of *T. pteneres* (Germain 1860; Murphy 1936; Humphrey et al. 1970; Weller 1975) and continental populations of *T. patachonicus* (Housse 1945; Johnson

Fig. 2. A: nest cover on Islote Escobar, Puerto Melo, Chubut. B: nest and eggs of *T. leucocephalus* at same locality. Photographs by Livezey, 14 December 1981.

1965; Zapata 1967; Jehl and Rumboll 1976) occurs from late October through January. Steamer-ducks begin nesting earlier in the Falkland Islands, evidently because of the less severe maritime climate of the archipelago. There, *T. patachonicus* begins nesting in October or earlier (Vallentin 1924; Woods 1975), and *T. brachypterus* nests predominantly from mid-September through December, although nesting of *T. brachypterus* occurs infrequently throughout the rest of the year (Abbott 1861; Cobb 1910, 1933; Ramsay 1915; Brooks 1917; Vallentin 1924; Murphy 1936; Beck *in* Murphy 1936; Cawkell and Hamilton 1961; Pettingill 1965; Weller 1972; Woods 1975).

Locations of nests.—All six nest sites of *T. leucocephalus* for which we have detailed notes were located on a small peninsula of Islote Escobar, Puerto Melo (Fig. 2). The nests averaged 12.7 m (range 6–17) above the high-tide level on the island. All nests were situated within or beneath zampa brush (*Atriplex vulgatissima*) which averaged 77 cm (range 51–115) in height. One nest was located 1 m inside a large bush and could be reached only by a small tunnel through the base of the bush. Nest density on the small island was high; during two half-hour visits we found at least ten nests. Inter-nest distances for six nests averaged only 4.8 m (range 2–8.5), and in some cases sitting hens probably would be within sight of each other.

A single nest of *T. leucocephalus* found by Boswall and Prytherch (1972) was 10 m above the high-tide mark on a shingle beach at the base of a small bush. The nest described by Boswall (1973) and Boswall and MacIver (1979) at Punta Tombo was 2 m above high-tide level beside a dead branch. Daciuk (1976) described nest sites beneath jume brush (*Suaeda divaricata*), in dead brush 10 to 50 m from the shore, and a single nest placed in a pile of old sacks of guano.

Other species of *Tachyeres* typically nest very close to water, often only a few meters above high-tide level (Cunningham 1871; Vallentin 1904; Crawshay 1907; Cobb 1910, 1933; Beck 1918; Phillips 1925; Reynolds 1934; Murphy 1936; Beck *in* Murphy 1936; Pettingill 1965; Johnson 1965; Humphrey et al. 1970; Venegas C. and Jory H. 1979). Rarely, however, nests of *T. patachonicus* (Vallentin 1924), as well as of flightless *T. brachypterus* (Brooks 1917; Reynolds *in* Lowe 1934; Woods 1975; 1982) and *T. pteneres* (Johnson 1965), are found as far as 400 m from the sea. *T. patachonicus* also nests on inland freshwater lakes (Reynolds *in* Lowe 1934; Todd 1979; pers. observ.). Although Olrog (1948) stated that *T. pteneres* nests on Lago Fagnano (Tierra del Fuego), we know of no other evidence that flightless species nest there or on other inland freshwater lakes. Like *T. leucocephalus*, the other three species show a marked preference for nesting on small islands and peninsulas (Germain 1860; Oustalet 1891; Cobb 1933; Reynolds *in* Lowe 1934; Johnson 1965; Zapata 1967; Todd 1979). *T. patachonicus* and *T. pteneres* are known to nest syntopically on islands in the Beagle Channel (Humphrey et al. 1970).

TABLE 1

CLUTCH SIZES OF FOUR SPECIES OF STEAMER-DUCK

Species	Mean[1] (N)	Range	Sources[2]
T. leucocephalus	4.6 (5)	3–6	13, 15, 16, 17 present study
T. brachypterus	6.0 (2)	4–11	3, 4, 6, 7
T. patachonicus	6.2 (4)	5–9	2, 3, 5, 6, 10, 14
T. pteneres	6.6 (11)	4–8	1, 5, 6, 8, 9, 11, 12

[1] Means based only on actual "random" clutches, not on generalized modes or extremes.
[2] Sources: (1) Germain 1860; (2) Abbott 1861; (3) Vallentin 1924; (4) Cobb 1933; (5) Reynolds 1934; (6) Murphy 1936; (7) Beck *in* Murphy 1936; (8) Goodall et al. 1951; (9) Johnson 1965; (10) Zapata 1967; (11) Griswold 1968; (12) Humphrey et al. 1970; (13) Boswall and Prytherch 1972; (14) Boswall 1973; (15) Jehl and Rumboll 1976; (16) Daciuk 1976; (17) Boswall and MacIver 1979.

Selection of dense cover for nest sites also is typical of the entire genus, particularly use of dense tangles of woody brush (e.g., *Empetrum, Lomaria, Berberis*), tussock grasses, dead kelp (*Macrocystis pyrifera*), and other shoreline debris (Coppinger 1883; Brooks 1917; Vallentin 1924; Murphy 1936; Johnson 1965; Pettingill 1965; Woods 1975). Construction of nests with tunnel entrances in dense brush was noted as well for *T. brachypterus* (Cawkell and Hamilton 1961). Percy (*in* Phillips 1925) reported that steamer-ducks (probably *T. pteneres*) occasionally nested in inland caves on small islands off Chiloé, and that these caves may have been used for roosting by adults with broods. *T. brachypterus* regularly nests in abandoned burrows of Magellanic Penguins (*Spheniscus magellanicus*) (Cobb 1910, 1933; Murphy 1936; Woods 1975; Todd 1979), and a single nest was found inside an overturned boat at tideline (Cobb 1933). Not all nests of steamer-ducks are well concealed. Crawshay (1907) found a nest of either *T. pteneres* or *T. patachonicus* in kelp on the outskirts of a nesting colony of South American Terns (*Sterna hirundinacea*). Johnson (1965) described a nest of *T. pteneres* located in the open, a short distance from the landing stage for a steamboat.

Structure of nest bowls.—The two active and four old nests of *T. leucocephalus* found at Puerto Melo were simple depressions in the ground among branches of concealing shrubs. All were heavily lined with down mixed with little or no dry vegetation; one bowl contained an old plastic bag. Diameters of five nest bowls averaged 27.4 cm (range 25–31); mean depth of the bowls was 7.5 cm (range 6.4–8.9). Nests of *T. leucocephalus* described by Boswall and Prytherch (1972), Daciuk (1976), and Boswall and MacIver (1979) also contained much down and little or no vegetation. Boswall and MacIver (1979) reported a bowl diameter of 27 cm.

Nest bowls of other species of *Tachyeres* are similar to those of *T. leucocephalus*. The nests of the three other species are shallow depressions in the soil, lined thickly with down (Germain 1860; Crawshay 1907; Reynolds 1934; Murphy 1936; Korchenevsky 1969; Humphrey et al. 1970; Kear 1970) or a mixture of down and dry grasses or twigs (Cunningham 1871; Coppinger 1883; Vallentin 1904; Phillips 1925; Housse 1942, 1948; Woods 1975, 1982). Heavy deposition of down in nests of *T. brachypterus* prompted Bennett (1926) to suggest that the down, if cleaned, might be of commercial value. Four diameters of nest bowls for *T. pteneres* averaged 27.9 cm (Humphrey et al. 1970; pers. observ.). Reynolds (1934) found a nest of *T. patachonicus* constructed on top of an old nest from a previous year. This observation and the sturdy nests of *T. leucocephalus* that we found in large shrubs suggest that steamer-ducks may reuse favored nest sites in subsequent years.

EGGS

Clutch sizes.—Four clutches for *T. leucocephalus* averaged only 4.6 eggs (Table 1). This mean is somewhat smaller than those of other species of *Tachyeres,* but small sample sizes of actual clutches for most species prevent statistical comparisons. Extreme clutch sizes of 11 reported for *T. brachypterus* (Murphy 1936), 12–14 for *T. pteneres* (Des Murs 1847; Barros V. 1948), and 12–14 for *T. patachonicus* (Housse 1942, 1948) are uncommon and may pertain

TABLE 2
Measurements of Eggs of Four Species of Steamer-Duck

Species (N)	Egg length (mm)	Egg width (mm)	Sources[1]
T. leucocephalus (15)			9, 10, 11, present study
Mean	81.2	54.2	
Range	72.3–85.6	51.2–56.1	
T. brachypterus (11)			1, 4, 5
Mean	81.8	56.6	
Range	79–86	56–57	
T. patachonicus (40)			4, 5, 6, 7
Mean	77.1	52.2	
Range	73–84	51–55	
T. pteneres (32)			2, 3, 4, 5, 6, 8
Mean	82.7	56.5	
Range	78–88	52–61	

[1] Sources: (1) Gould 1859; (2) Coppinger 1883; (3) Schalow 1898; (4) Reynolds *in* Lowe 1934; (5) Murphy 1936; (6) Goodall et al. 1951; (7) Cawkell and Hamilton 1961; (8) Griswold 1968; (9) Boswall 1973; (10) Daciuk 1976; (11) Boswall and MacIver 1979.

to dump nests or multiple layings; indiscriminant laying of single eggs in exposed places has been observed in *T. patachonicus* (Reynolds *in* Lowe 1934).

Color of eggs.—Six eggs of *T. leucocephalus* from two clutches at Puerto Melo were "pale creamy tan" in color (field notes). Boswall (1973), Daciuk (1976) and Boswall and MacIver (1979) did not describe egg color for *T. leucocephalus.*

The few data on egg color for *T. leucocephalus* are difficult to compare with the varied descriptions of egg color for other species of *Tachyeres* because of differences in color terminology of the different authors, and in a few cases, uncertainty concerning species identifications. Eggs of steamer-ducks have been described variously as from white to buffy (Gould 1859; Cunningham 1871; Coppinger 1883; Oates 1902; Crawshay 1907; Vallentin 1904, 1924; Murphy 1936; Pereyra 1943; Housse 1942, 1945, 1948; Cawkell and Hamilton 1961; Johnson 1965; Korchenevsky 1969; Woods 1975, 1982). Murphy (1936) concluded that "The eggs of the three steamer ducks [*T. pteneres, T. brachypterus,* and *T. patachonicus*] look much alike." We feel that methodologically consistent comparison of fresh clutches of all four species may reveal interspecific differences in egg color.

Size of eggs.—Egg size of *T. leucocephalus* closely approximates that of *T. brachypterus,* and both are intermediate between the smaller eggs of *T. patachonicus* and the larger eggs of *T. pteneres* (Table 2). These interspecific differences are small, and ranges of egg sizes overlap broadly among all species. Based on these data, egg measurements are not sufficient for identification of species of *Tachyeres,* despite the belief of Yahgan Indians that ". . . the eggs of the *alacush* [*T. pteneres*] are double the size of the *tushca's* [*T. patachonicus*] . . ." (Bridges 1948). Shape of eggs may differ interspecifically, however; Wace (1921) concluded that the eggs of *T. patachonicus* were more pointed than those of *T. brachypterus.*

We lack data on thickness of egg shells for *T. leucocephalus.* Reynolds (*in* Lowe 1934), Murphy (1936), and Todd (1979) reported that egg shells of flightless steamer-ducks were considerably thicker than those of the flying species. Weights of six eggs of *T. leucocephalus* found by Boswall and MacIver (1979) averaged 136.7 g (range 130–145). Griswold (1968) reported that four fresh eggs of *T. pteneres* had a mean weight of 137.3 g (range 131–140).

Nesting Ecology and Broods

Placement of nests in dense shrubs and penguin burrows, covering of eggs with down by departing hens, and apparent preference for nesting on peninsulas and small islands indicate that nest predation is substantial for all species of steamer-duck. High nest densities on islands is especially suggestive in such a territorial and presumably sedentary species as *T. leucocephalus.* Uplands adjacent to highly tidal waterfronts are probably unattractive to flightless species for nesting because low tides would separate the nesting bird from the safety of water

TABLE 3

SIZES AND PARENTAL ATTENDANCE OF BROODS OF *T. LEUCOCEPHALUS*, BY AGE CLASS, AT
PUERTO MELO, 12 DECEMBER–19 FEBRUARY

		Brood size		No. (%) attended
Age class	N	X̄	s.d.	by both sexes[1]
I	10	2.80	1.93	9
II	7	3.00	1.41	6
III	16	3.38	1.50	14
All	33	3.12	1.60	29 (88)

[1] One class-I brood was accompanied only by an adult female, one class-II brood only by an adult male, and two class-III broods only by adult females.

by often extensive tidal flats. Weller (1972) observed that *T. brachypterus* pairs usually occupied territories with sloped gravel shores, rather than abrupt rocky shores, and suggested that such shorelines permitted the adults to walk from the water to their nests.

Incubation periods in *Tachyeres* exceed 30 days (Zapata 1967; Griswold 1968; Kear 1970), during which several probable avian (*Macronectes giganteus, Catharacta* spp., *Larus dominicanus*) and mammalian (*Dusicyon* spp., *Lyncodon* spp., *Conepatus* spp.) nest predators undoubtedly contribute to egg mortality. Beck (*in* Phillips 1925) saw a caracara (species not given) destroy three of a clutch of four eggs of a steamer-duck in the Magellan Straits. In addition, eggs of steamer-ducks are considered good eating, and egg collectors are responsible for much egg mortality of *T. leucocephalus* (F. Fauring, pers. comm.) and of *T. brachypterus*, or have been in the past (Abbott 1861; Cobb 1910; Phillips 1925).

Several avian species (*Macronectes giganteus, Catharacta* spp., *Larus dominicanus*) are known to prey at least occasionally on young steamer-ducks (Pettingill 1965; Woods 1975, 1982; Todd 1979). Weller (1972) estimated that hatchlings of *T. brachypterus* require 12 weeks to fledge. Griswold (1968) presented limited data on growth of young *T. pteneres*. During this comparatively long developmental period, we suspect that young steamer-ducks are especially vulnerable to predators.

Steamer-ducks are monogamous, and generally both adults attend the brood until fledging (Table 3; Beck *in* Phillips 1925; Murphy 1936; Pettingill 1965; Kear 1970; Weller 1972, 1976; Todd 1979). Data from Weller (1976), and our own, substantiate that biparental attendance of broods is typical in *T. pteneres* (79% of broods of all ages seen were accompanied by both adults, N = 19), *T. brachypterus* (91%, N = 72), and *T. patachonicus* (78%, N = 36). Attendance of steamer-duck broods by both adults may reflect, in part, the need for protection of ducklings from predators. Both sexes of *T. brachypterus* defend their ducklings from gulls (Pettingill 1965; Weller 1972). However, in *Tachyeres* only the female incubates or broods the young (Pettingill 1965; Kear 1970; Woods 1975; Todd 1979).

Our observations of broods of *T. leucocephalus* indicate a very slight increase in mean sizes of broods with age (Table 3). This trend, although not statistically significant, may indicate that sizes of clutches, and hence broods, decrease during the nesting season, or that fusion of older broods occurs occasionally in this species. Data from Weller (1972) for *T. brachypterus* also suggest a slight increase in brood size between age-classes I (X̄ = 4.36, N = 60) and II (X̄ = 4.73, N = 12). However, Cobb (1933), Cawkell and Hamilton (1961), and Pettingill (1965) concluded that brood size of *T. brachypterus* decreases with age of young, which Pettingill (1965) attributed to losses of ducklings to avian predators.

We suspect that the young of *T. leucocephalus* are driven from the parental territory by the adults shortly after fledging, and soon congregate with other immature birds in the large flocks of non-territorial steamer-ducks that we frequently observed at Puerto Melo. Late-season flocks of immature birds and parental eviction of fledglings have been observed in *T. brachypterus* (Vallentin 1904; Murphy 1936; Pettingill 1965; Weller 1972).

DESCRIPTION OF DOWNY YOUNG

Lowe (1934: plate IX) was the first to describe and figure downy young of *Tachyeres patachonicus* and *T. pteneres* (his "*brachypterus*"). Murphy (1936) described and figured the downy young of *T. brachypterus* and compared patterns of the downy young of the three species of steamer-duck recognized at that time. Later authors (Delacour 1954; Weller 1976) based their figures and descriptions of downies of the three species of *Tachyeres* on Murphy's excellent diagnoses.

The principal features used by Lowe (1934) and Murphy (1936) to characterize downy young of three species of steamer-duck were distribution and continuity of white in front of, above, and behind the eye, and color of the yoke or upper back. We have found additional diagnostic characters including coloration of eyelids and width of supraloral patch (Fig. 1). For this reason we redescribe the downies of *T. brachypterus*, *T. patachonicus*, and *T. pteneres* as a basis for comparing them with the downy young of *T. leucocephalus* (Plate VII) and preparing a key for the identification of very young downies of *Tachyeres*.

T. leucocephalus. Upper and lower eyelids whitish to pale pearl gray; cheeks from light to medium grayish-brown becoming paler ventrally and posteriorly in some individuals. A dark grayish-brown streak through the eye forms the dorsal margin of the cheeks and lores; lores light to medium grayish-brown; forehead, crown, and nape dark grayish-brown; supraloral patch broad, and continuous with the supraocular patch and postocular streak all of which are whitish or pale pearl gray; yoke, light grayish-brown; lower back and rump, dark grayish-fuscous.

T. brachypterus. Upper eyelid medium fuscous and lower eyelid whitish to pale fuscous; cheeks, lores, forehead, crown, and nape, medium fuscous; supraloral and supraocular patches very narrow and continuous with the broader postocular streak but narrower at divisions between the supraloral and supraocular patches and between the supraocular patch and post-ocular streak; anterior part of postocular streak constricted posteriorly as it continues into the posterior part; supraloral and supraocular patches and postocular streak very pale smoke gray, almost whitish; ventral margin of supraocular patch above upper eyelid, dark dusky brown; yoke, light fuscous; lower back and rump, medium to dark fuscous.

T. patachonicus. Lower eyelid whitish, and upper eyelid, cheeks, lores, forehead, crown, and nape, dark brownish olive; supraocular and supraloral patches very narrow, almost occluded anteriorly, and separated widely from the broad postocular streak; supraloral and supraocular patches and postocular streak, whitish; yoke light fuscous; lower back and rump dark brownish-olive.

T. pteneres. Lower eyelid whitish, and upper eyelid and most of the supraocular patch, blackish-brown; cheeks, lores, forehead, crown, and nape, dark brownish-olive becoming slightly lighter at the anterior part of forehead; supraloral patch very small or absent; supra-ocular region blackish-brown except for a whitish patch which may be absent, faintly indicated, or small and, when present, is always separate from the supraloral patch if present; postocular streak whitish and divided; upper parts of body, dark brownish-olive, with the yoke slightly lighter.

Small downy young of *T. leucocephalus* look very grayish compared to downies of the other three species all of which are brownish in overall coloration of skin specimens and living individuals (color slides). In addition, the whitish upper eyelid and broad, continuous whitish streak comprising the undivided supraloral and supraocular patches and postocular streak immediately distinguish small downy young of *T. leucocephalus* from small downies of the other three species.

KEY TO DOWNY YOUNG

Small downy young of the four described species of *Tachyeres* may be identified using a relatively small number of characters as set forth in the following key. We believe that most of these characters also are valid for class-II downies although we have not had an opportunity to test this for all species.

1a. Upper eyelid *whitish*; crown *darker* than cheeks; supraloral and supraocular patches wide and continuous with wide postocular streak .... *T. leucocephalus*
1b. Upper eyelid *dark*; crown *not darker* than cheeks 2
2a. Postocular streak *divided*; supraloral and supraocular patches very small or *absent*; *separate* when present .... *T. pteneres*
2b. Postocular streak *undivided*; supraloral and supraocular patches *present* 3
3a. Supraloral and supraocular patches *very narrow* (almost occluded anteriorly), con-*tinuous,* and *separated* from postocular streak .... *T. patachonicus*
3b. Supraloral and supraocular patches *narrow* and *continuous* with postocular streak
.... *T. brachypterus*

ACKNOWLEDGMENTS

Our studies of steamer-ducks would not have been possible without the generous interest, assistance, and hospitality of many people and organizations. We are grateful to the following: The authorities of the Field Museum of Natural History (FMNH), American Museum of Natural History (AMNH), National Museum of Natural History (USNM), Museum of Comparative Zoology (MCZ), Museo Argentino de Ciencias Naturales (MACN), and the British Museum (Natural History) (BM) for enabling us to examine specimens in their care; B. Mayer and F. Fauring for their boundless help, friendship, warm hospitality, and provision of field facilities at Puerto Melo, Chubut; and Mr. and Mrs. Ralph Gibson for being home and family in La Lucila; J. M. Gallardo, Director, MACN; J. Navas for space and facilities and generous assistance of many kinds; R. A. Bockel, Jefe de Servicios Publicos; and the administrative and library staff of the MACN. E. O. Gonzalez Ruiz, R. L. Cejas and their colleagues, Dirección Nacional de Fauna Sylvestre, and L. O. Saigg de Chialva, Directora de Protección Ambiental, Provincia de Chubut, for issuing permits and facilitating our field work; authorities of the Gendarmeria Nacional, province of Chubut, and particularly N. J. Black, Comandante Principal for making us welcome throughout our travels; B. de Ferradas, M. A. E. Rumboll, F. Erize, P. Canevari, R. Straneck, and T. Narosky for assistance of many kinds; M. E. Mulgura de Romero, Instituto de Botanica Darwinion, and P. Raven, R. Liesner, and N. R. Morin of the Missouri Botanical Garden for assistance in identifying a plant specimen; R. M. Mengel for preparing figure 1 and Plate VII; R. Patterson for helping us prepare for our field work; B. Padget and K. McManness for typing the manuscript, the authorities of the University of Kansas for awarding Humphrey research leave for two months, and M. L. Humphrey for her endless patience and enthusiastic support.

This study was supported by National Science Foundation Grant No. DEB-8012403 to Humphrey.

LITERATURE CITED

ABBOTT, C. C. 1861. Notes on the birds of the Falkland Islands. Ibis (Ser. 1) 3:149–167.

BARROS V., R. 1948. Anotaciones sobre las aves de Maullín. Rev. Univ. (Universidad Catolica, Santiago) 33:35–60.

BECK, R. M. 1918. Narrative of a bird quest in the vicinity of Cape Horn. Am. Mus. J. 18:1–16, 111–119.

BENNETT, A. G. 1926. A list of the birds of the Falkland Islands and dependencies. Ibis (Ser. 12) 2:306–333.

BOSWALL, J. Supplementary notes on the birds of Point Tombo, Argentina. Bull. Br. Ornithol. Club 93: 33–36.

BOSWALL, J., AND D. MACIVER. 1979. Nota sobre el pato vapor volador (Tachyeres patachonicus). Hornero 12:75–78.

BOSWALL, J. AND R. J. PRYTHERCH. 1972. Some notes on the birds of Point Tombo, Argentina. Bull. Br. Ornithol. Club 92:118–129.

BRIDGES, E. L. 1948. Uttermost Part of the Earth. E. P. Dutton and Co., New York.

BROOKS, W. S. 1917. Notes on some Falkland Island birds. Bull. Mus. Comp. Zool. 61:135–160.

CAWKELL, E. M., AND J. E. HAMILTON. 1961. The birds of the Falkland Islands. Ibis 103a:1–27.

COBB, A. F. 1910. Wild Life in the Falkland Islands. Gowans and Gray, Ltd., London.

COBB, A. F. 1933. Birds of the Falkland Islands. H. F. and G. Witherby, London.

COPPINGER, R. W. 1883. Cruise of the "Alert." Four years in Patagonian, Polynesian, and Mascarene waters. W. Swan Sonnenschein and Co., London.

CRAWSHAY, R. 1907. The Birds of Tierra del Fuego. Bernard Quaritch, London.

CUNNINGHAM, R. O. 1871. Notes on the Natural History of the Strait of Magellan and West Coast of Patagonia Made During the Voyage of H.M.S. "Nassau" in the Years 1866, 67, 68 and 69. Edmonston and Douglas, Edinburgh.

DACIUK, J. 1976. Notas faunísticas y bioecológicas de península Valdés y Patagonia. XVIII. Comportamiento del pato vapor volador observado durante el ciclo reproductivo en costas e islas de Chubut (Rep. Argentina)—(Anserif., Anatidae). Neotropica 22:27–29.

DELACOUR, J. 1954. The Waterfowl of the World, Vol. 1. Country Life, London.

DES MURS, SEÑOR. 1847. Aves. Pp. 183–494, In C. Gay, Historia Física y Política de Chile. Zoología, Vol. 1. Mus. Hist. Nat., Santiago.

GERMAIN, M. F. 1860. Notes upon the mode and place of nidification of some of the birds of Chili. Proc. Boston Soc. Nat. Hist. 7:308–316.

GOODALL, J. D., A. W. JOHNSON, AND R. A. PHILLIPI B. 1951. Las Aves de Chile, Vol. 2. Platt Establ. Graf., Buenos Aires.

GOULD, J. 1859. List of birds from the Falkland Islands, with descriptions of the eggs of some of the species, from specimens collected principally by Captain C. C. Abbott, of the Falkland Island Detachment. Proc. Zool. Soc. Lond. 27:93–98.

GRISWOLD, J. A. 1968. First breeding of the Magellanic Flightless Steamer Duck in captivity. Wildfowl 19:32.

HOUSSE, R. E. 1942. Les oiseaux des Andes. Ann. Sci. Nat. Zool. Biol. Anim. (Ser. 11) 4:138–238.

HOUSSE, R. E. 1945. Las Aves de Chile en su Clasificación Moderna. Ediciones Univ. Chile, Santiago.

HOUSSE, R. E. 1948. Les Oiseaux du Chili. Masson and Co., Paris.

HUMPHREY, P. S., AND M. C. THOMPSON. 1981. A new species of steamer-duck (*Tachyeres*) from Argentina. Occas. Pap. Mus. Nat. Hist. Univ. Kans. 95:1–12.

HUMPHREY, P. S., D. BRIDGE, P. W. REYNOLDS, AND R. T. PETERSON. 1970. Birds of Isla Grande (Tierra del Fuego). Smithsonian Institution Press, Washington, D.C.

JEHL, J. R., JR., AND M. A. E. RUMBOLL. 1976. Notes on the avifauna of Isla Grande and Patagonia, Argentina. Trans. San Diego Soc. Nat. Hist. 18:145–154.

JOHNSON, A. W. 1965. The birds of Chile and adjacent regions of Argentina, Bolivia and Peru, Vol. 1. Platt Establ. Graf., Buenos Aires.

KEAR, J. 1970. The adaptive radiation of parental care in waterfowl. Pp. 357–392, *In* J. H. Crook (ed.), Social Behaviour in Birds and Mammals. Academic Press, New York.

KORCHENEVSKY, P. 1969. Observaciones sobre aves del litoral patagónico. Hornero 11:48–52.

LOWE, P. R. 1934. On the evidence for the existence of two species of steamer duck (*Tachyeres*), and primary and secondary flightlessness in birds. Ibis (Ser. 13) 4:467–495.

MURPHY, R. C. 1936. Oceanic Birds of South America, Vol. 2. American Museum Natural History, New York.

OATES, E. W. 1902. Catalogue of Bird's Eggs in the British Museum, Vol. 2. Longman's, London.

OLROG, C. C. 1948. Observaciones sobre la avifauna de Tierra del Fuego y Chile. Acta Zool. Lilloana 5:437–531.

OUSTALET, E. 1891. Mission scientifique du Cape Horn, 1882–1983 (Oiseaux), Zoologie 6(B). Minis. L'instruct. Publ., Paris.

PALMER, R. S. (ed.). 1962. Handbook of North American Birds, Vol. 1. Yale University Press, New Haven, Connecticut.

PEREYRA, J. A. 1943. Nuestras aves. Minist. Obras Pub. Prov. Buenos Aires, La Plata, Argentina.

PETTINGILL, O. S., JR. 1965. Kelp Geese and flightless steamer ducks in the Falkland Islands. Living Bird 4:65–78.

PHILLIPS, J. C. 1925. A Natural History of the Ducks, Vol. 3. Houghton Mifflin, Boston, Massachusetts.

RAMSAY, L. N. G. 1915. Ornithology of the Scottish National Antarctic Expedition. Part XIV of Section III.—The Falkland Islands. Pp. 211–214, *In* W. S. Bruce (ed.), Report of the scientific results of the voyage of S.Y. "Scotia" during the years 1902, 1903, and 1904, Vol. IV.—Zoology. Scot. Oceanogr. Lab., Edinburgh.

REYNOLDS, P. W. 1934. Apuntes sobre aves de Tierra del Fuego. Hornero 5:339–353.

SCHALOW, H. 1898. Die Vögel der Sammlung Plate. Zool. Jahrb. Suppl. 4, Pt. 3, pp. 641–749.

TODD, F. S. 1979. Waterfowl: Ducks, Geese and Swans of the World. Harcourt-Brace Jovanovich, New York.

VALLENTIN, R. 1904. Notes on the Falkland Islands. Mem. Proc. Manchester Lit. Philos. Soc. 48: 23–45.

VALLENTIN, R. 1924. Zoology. Pp. 285–335, *In* V. F. Boyson, The Falkland Islands. Clarendon Press, Oxford, England.

VENEGAS C., C., AND J. JORY H. 1979. Guía de Campo para las Aves de Magallanes. Inst. Patagonia, Punta Arenas, Chile.

WACE, R. H. 1921. Lista de aves de las islas Falkland. Hornero 2:194–204.

WELLER, M. W. 1956. A simple field candler for waterfowl eggs. J. Wildl. Manage. 20:111–113.

WELLER, M. W. 1972. Ecological studies of Falkland Islands' waterfowl. Wildfowl 23:25–44.

WELLER, M. W. 1975. Habitat selection by waterfowl of Argentine Isla Grande. Wilson Bull. 87:83–90.

WELLER, M. W. 1976. Ecology and behaviour of steamer ducks. Wildfowl 27:45–53.

WOODS, R. W. 1975. The Birds of the Falkland Islands. Compton Press, Wiltshire, England.

WOODS, R. W. 1982. Falkland Island Birds. Anthony Nelson, Shropshire, England.

ZAPATA, A. R. P. 1967. Observaciones sobre aves de Puerto Deseado Provincia de Santa Cruz. Hornero 10:351–378.

THE NESTING BIOLOGY OF SOME PASSERINES OF BUENOS AIRES, ARGENTINA

PAUL MASON

Department of Zoology, University of Texas, Austin, Texas 78712 USA;
present address: Department of Biological Sciences,
University of California at Santa Barbara,
Santa Barbara, California 93106 USA

ABSTRACT. Nesting data were collected over two seasons at two study sites approximately 20 km apart in Buenos Aires Province, Argentina. Community composition differed between sites. The data include brief descriptions of nests, eggs, and nestlings. The duration of incubation and nestling periods are also presented. Maximum likelihood estimators of nest survivorship are calculated separately for nests with eggs and nests with nestlings. Nestlings of eight species were parasitized with botfly larvae (probably *Philornis* sp.).

Broadly speaking, species that nest in complex or closed nests have higher survivorship than species which build cup nests. However, a comparison of mortality of the Rufous Hornero and the Saffron Finch using nests of the former indicates that the behavior of the species contributes greatly to survivorship. Saffron Finches nesting in hornero nests show poorer survivorship than the horneros themselves. Nonetheless, Saffron Finches using these nest sites have superior survivorship to those which use other nests. The demographic data also permit the rejection of the hypothesis of prevention of total nest failure as the selective agent favoring asynchronous hatching in the Rufous Hornero.

RESUMEN. Se coleccionó información de anidación durante dos temporadas en dos sitios distantes 20 km entre sí, en la provincia de Buenos Aires, Argentina. La composición de la comunidad difiere entre ambos sitios. Los datos incluyen breves descripciones de nido, huevos y polluelos. También se presenta información respecto a la duración de la incubación y el período de anidación. Se calculan separadamente las estimaciones máximas de sobrevivencia, para nidos con huevos y nidos con polluelos. Polluelos de ocho especies estaban parasitados por larvas de la mosca (probablemente *Philornos* sp.).

Generalmente, las especies que anidan en nidos complejos o cerrados tienen mayor tasa de sobrevivencia que las especies cuyos nidos tienen forma de copa. Sin embargo una comparación de la mortalidad del hornero Rufous y el Pinzón Saffron usando nidos de hornero indican que la sobrevivencia está relacionada con el comportamiento de la especie. Pinzones Saffron que anidan en nidos de horneros, tienen menor sobrevivencia que los mismos horneros. A pesar de lo cual, los pinzones que usan nidos de hornero tienen mayor sobrevivencia que los pinzones que usan otros nidos. La información demográfica también permite rechazar la hipótesis de prevención de fracaso total de anidación como un agente selectivo favoreciendo la incubación asincrónica del hornero Rufous.

Studies of nesting biology contribute to the development of biological knowledge in a variety of ways. Descriptive natural history observations document the diversity of life, and the accumulation of such observations eventually allows the construction of general hypotheses. Some information can be interpreted immediately through its application to current theory. Many of the data collected at nests, such as egg size, clutch size, length of incubation and nestling periods, and nest survivorship are among the most fundamental of avian life-history attributes. The values of these parameters, which contribute to fitness and are subject to selection, are often unknown.

The purpose of this paper is twofold. First, it supplies a body of descriptive data on events that occur at nests of temperate, South American passerines. Second, it interprets appropriate portions of the demographic data. In particular, these analyses examine the relationship between nest survivorship and nest placement, both inter- and intraspecifically. Finally, a comparison of survival rates during the egg and nestling phases is used to test a hypothesis of asynchronous hatching in the single passerine species observed here to display this trait.

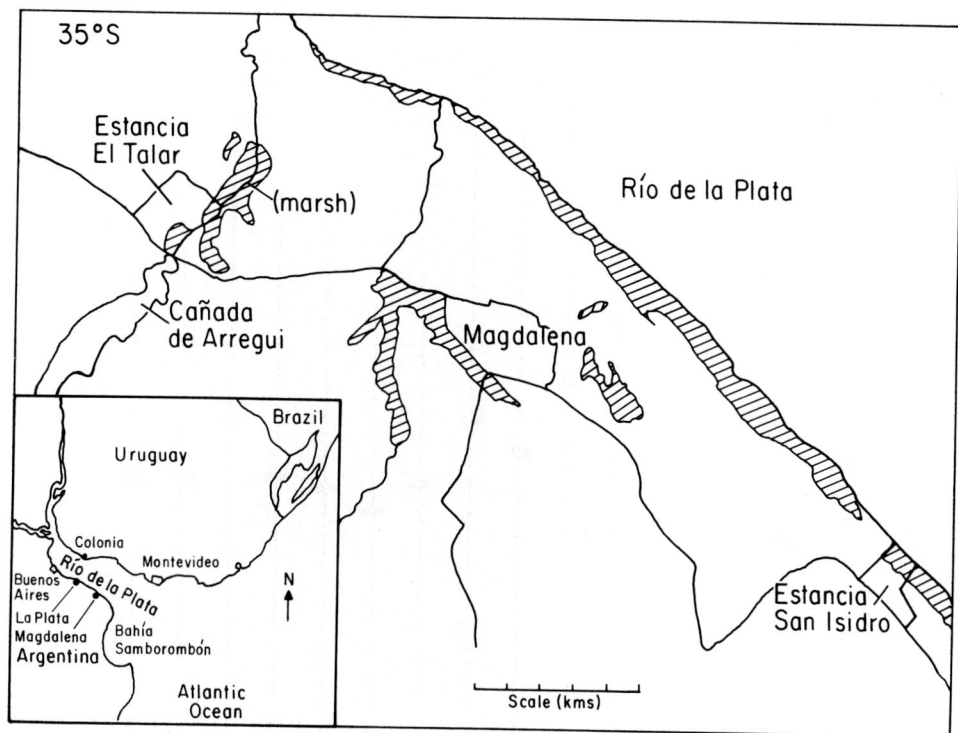

FIG. 1. Map of locations of the study sites.

The data presented here were obtained during two seasons of field work (1977–1978 and 1978–1979) at two sites approximately 20 km apart along the Río de La Plata, Buenos Aires Province, Argentina (approximately 35°S). Geological, climatological, and floral descriptions of the general area are available in Cabrera (1965) and Barbetti (1982). Surveys of the avifauna and natural histories of birds in the area have been presented by Hudson (1870, 1871, 1872a, b, c, 1873, 1874, 1876, 1920), Gibson (1880, 1918), Sclater and Hudson (1888, 1889), Grant (1911, 1912), Wetmore (1926), Pereyra (1938), Weller (1967), and Narosky (1978). Myers and Myers (1979) described the shorebird community on the Atlantic coast.

STUDY SITES

The two study sites were located close to the village of Magdalena, ca. 125 km ESE of the city of Buenos Aires. Fieldwork was performed at Estancia San Isidro during the 1977–1978 season, and at Estancia El Talar during the 1978–1979 season. The areas of the study sites were ca. 2 km^2 (Estancia San Isidro) and ca. 4 km^2 (Estancia El Talar). Both estancias are situated in a riparian zone, distinct from the pampas, which stretches approximately from the delta of the Río Paraná to the Bahía Samborombón. Estancia San Isidro is located on the shore of the Río de La Plata, whereas El Talar is some 4 km inland (Fig. 1).

The ground is nowhere more than a few meters above sea level. Drainage is poor and temporary pools may form after rain at any time of year. The habitat consists of marsh, pasture, and broken woodland. Rainfall records from Estancia San Isidro, maintained since 1939, indicate that there is no pronounced rainy season (Fig. 2). The climate is mild temperate and seasonal. Occasional frosts occur in winter. Summertime temperatures rarely exceed 30°C in the shade.

Six native tree species were abundant at San Isidro: *Erythrina crista-galli*, Leguminosae; *Phytolaca dioica*, Phytolacaceae; *Jodina rhombifolia*, Santalaceae; *Celtis spinosa*, Ulmaceae; *Scutia buxifolia*, Rhamnaceae; and *Schinus longifolius*, Anacardiaceae. *Erythrina crista-galli* grows in a narrow band just along the coast, whereas the others are found further inland. Only *Celtis spinosa* was abundant at El Talar, a site which was conspicuously more open and less

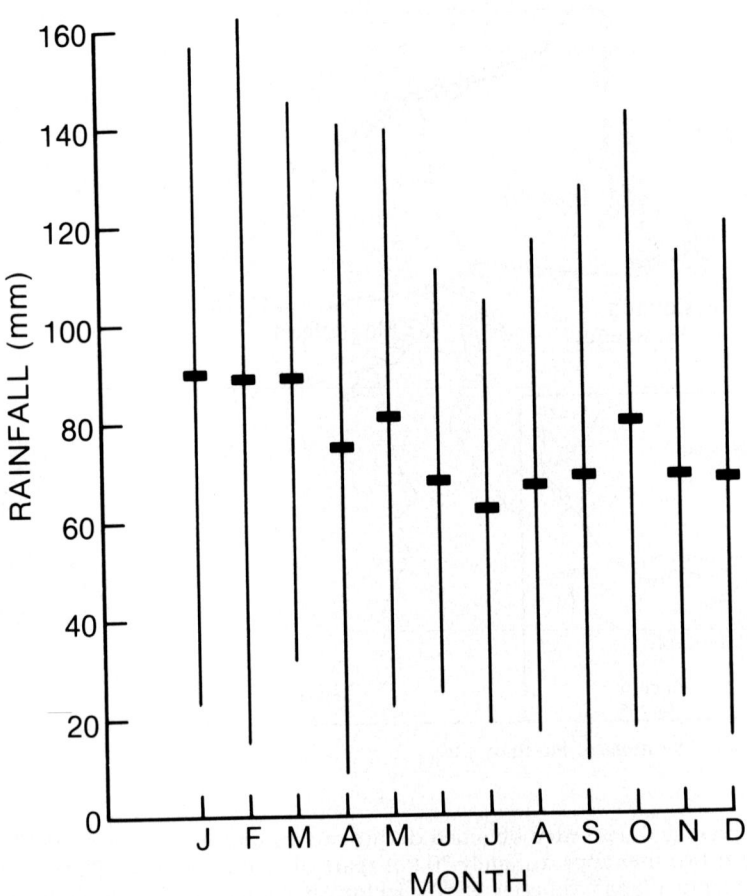

Fig. 2.   Rainfall at Estancia San Isidro, 1939–1977. Monthly means (solid block) and one standard deviation on either side of the mean (vertical lines) are indicated.

wooded than San Isidro. At both sites, ornamental and fruit trees have been planted around dwellings. Eucalyptus have been introduced as shade trees.

Marshes were present on both sites. Rushes (*Scirpus* sp.) and cattails (*Typha* sp.) predominated; stands of the deciduous *Solanum malacoxylon* were less common. Wet grasslands border the marshes. Plant distributions are given in Cabrera (1965) and Celulosa Argentina (1973).

Mammalian nest predators included the Argentine skunk (*Conepatus chinga*), Azara's fox (*Dusicyon gymnocercus*), and the white-eared oppossum (*Didelphis albiventris*). The last is the only important mammalian predator on tree nests. There were some domestic cats. Ground nests were subject to predation by a variety of snake species. These same nests may also be prey to a frog, the "escuerzo" (*Ceratophrys ornata*). Two avian predators on eggs and nestlings were abundant at both sites, the Chimango Caracara (*Milvago chimango*) and the Guira Cuckoo (*Guira guira*).

Observations were carried out on almost a daily basis. The period of observation extended from 20 October 1977 to 24 February 1978 at Estancia San Isidro, and from 21 September 1978 to 10 February 1979 at Estancia El Talar. Nomenclature (both scientific and common names) and arrangement follow Meyer de Schauensee (1970) unless otherwise noted.

Data were collected as part of a larger project examining the ecology of brood parasitism in two locally abundant brood parasites, the Shiny Cowbird (*Molothrus bonariensis*) and the Screaming Cowbird (*M. rufoaxillaris*). Most nests were artificially parasitized with either real or artificial cowbird eggs. I attempted to survey as many nests as possible, but even some abundant passerines yielded very little data because of factors such as nest placement or inconspicuousness.

METHODS

The composition of the avian community was determined by census. Once a week I assigned each species recorded into one of several broad categories of abundance typically used by bird watchers: abundant (>25 adults seen/day), common (5–25/day), uncommon (1–4/day), or rare (<1/day). I used these weekly determinations to estimate overall abundance for the breeding season.

I continually searched for nests, and attempted to visit those found, at least on alternate days. Visits sometimes occurred at longer intervals, because of weather or short absences from the study sites. Visits were often only a few seconds long, but longer when information other than the activity of the nest was recorded. I was rarely at a nest for more than 20 minutes. Eggs were numbered inconspicuously, measured to the nearest 0.1 mm with vernier calipers, and weighed to the nearest 0.1 g with a spring scale. Clutch sizes were recorded only from nests that were unparasitized by cowbirds. The fate of individual eggs and/or nestlings was recorded, and an attempt was made to identify important sources of mortality. Special techniques were required to observe the contents of complex furnariid nests. For domed mud nests of the Rufous Hornero (*Furnarius rufus*), I followed the procedure of Fraga (1980): an observation hole was cut into the wall of the nest chamber and later closed with a wooden plug (approximately five cm in diameter) and sealed with mud. Bulky stick nests with long, convoluted entrance tunnels (such as those of the Firewood-gatherer, *Anumbius annumbi*, and the thornbirds, *Phacellodomus* spp.) were opened in a similar manner, but the holes were plugged with oversized pieces of sponge. The tension exerted by the compressed sponge prevented deterioration of the structure by holding the sticks in place. Sponge plugs were also used to close observation holes in the tiny, delicate, woven nests of the Wren-like Rushbird (*Phleocryptes melanops*). At both sites, I hung several nest boxes, which measured ca. 30 cm on a side. One side was only half closed to serve as an entrance.

Nestlings were uniquely banded with pieces of plastic tape, weighed with spring scales (to the nearest 0.1 g if less than 10 g, otherwise to the nearest 0.5 g), and described. Samples of food brought to nestlings were collected on a few occasions by fitting nestlings with collars of short pieces of pipe cleaner. Data describing nestling weight increase over time were subjected to linear regression analysis to determine a mean growth rate for each species.

I separately calculated daily survivorship rates for nests with eggs and nests with nestlings using the technique of Hensler and Nichols (1981). Computationally, this is identical to the Mayfield (1975) method, but Hensler and Nichols demonstrate that the calculation is the maximum likelihood estimator of survivorship, and they also present equations describing the variance associated with this mean. I only considered sources of mortality that affected all nest occupants simultaneously and similarly, namely, losses to predation (or parasitism) when all young perished, and losses from severe weather.

The calculation of survivorship requires scoring each nest for the number of days of observation and its success or failure, when the lengths of the egg and nestling phases are known. Any nest whose outcome was unknown was scored as successful over the period of time it was observed to be active. This scoring procedure simply counts the number of "exposure" days (Mayfield 1975), and is legitimate because it gives information about the distribution of failures.

The production of young from nests is not presented for a number of reasons: the handling of eggs may have occasionally introduced additional mortality, eggs were occasionally collected, the presence of cowbird eggs (in almost all nests) and young affects production, rarely some young were strangled by collaring, and on one occasion an adult ejected nestlings with collars from the nest (*Zonotrichia capensis*).

To determine the average survivorship of the nest over each period (egg or nestling), daily mean survivorship must be raised to a power equal to the length of that period. The variance associated with mean survivorship over the interval can be described by the following equation (M. K. Uyenoyama, pers. comm.):

$$\mathrm{VAR}(\hat{P}^J) = \hat{P}^{2(J-2)}\hat{v}^2 J^2\left[\hat{P}^2 - \frac{(J-1)^2}{4}\hat{v}^2\right]$$

where $\hat{P}$ = daily survivorship, $\hat{v}^2$ = variance associated with $\hat{P}$, and J = the length of the interval (symbols are those of Hensler and Nichols 1981).

Although Hensler and Nichols (1981) used incubation period as the duration of the egg phase, this is not strictly true since the egg-laying period should also be included. The length of the egg phase was determined by subtracting 1 from the incubation period and adding the number of days required to lay an average-sized clutch. This assumes that incubation is initiated with the laying of the last egg and that hatching is synchronous. When hatching is asynchronous, the mean duration of asynchrony can be subtracted from the duration of the egg phase as calculated above to get the true duration of the egg phase. Nests with both eggs and nestlings were assumed to be in the nestling phase, in agreement with Clark and Wilson (1981). Only one species (*Furnarius rufus*) regularly showed asynchrony in hatching, although the spread of hatching was highly irregular.

The length of the intervals was estimated by assuming that hatching (or fledging) occurred at the midpoint between observation dates, unless other considerations indicated that some adjustment was necessary. These considerations are primarily judgements of hatching (or fledging) dates when observation dates were more than two days apart. These adjustments were made to be consistent with other nests of known age in the same cohort. Nests were scored as having successfully fledged young when the nest was empty on the next observation, and the young had reached a stage of development sufficient for them to leave the nest. Since I did not always actually see fledglings, I may have scored predations occurring very late in the nesting cycle as fledges. The results of this judgement would be an underestimate of the true interval length and an overestimate of survivorship. The method assumes no variation in the duration of either phase, a simplification whose effect on the reliability of the estimates is unknown.

Overall survivorship, the product of survivorship through the egg phase times survivorship through the nestling phase, yields the probability of a nest fledging at least one young. Subtracting the overall survivorship from unity gives a nest's probability of total failure.

I made two interspecific comparisions of survivorship rates to see what the effect of nest architecture might be on nesting success. First, I placed species into one of two categories: those nesting in complex, closed nests (including hole nesters) and those nesting in cup nests. The species were then ranked by survivorship and compared using the Mann-Whitney U-test (Siegel 1956). Second, I compared survivorship of nests of the Saffron Finch (*Sicalis flaveola*) built in nests of the Rufous Hornero to the survivorship of the hornero. This comparison allowed me to see if species-specific behavior affected survivorship independent of nest site selection.

Two intraspecific comparisons were performed. The first compared survivorship of Saffron Finches in hornero nests versus survivorship in other nests. This comparison gives a finer examination of the effect of nest site selection, by holding the species constant. The second comparison examined differences in the daily survivorship rates in the egg and nestling phases of the Rufous Hornero. This allowed me to test a proposed explanation for the evolution of asynchronous hatching (Clark and Wilson 1981). The technique of statistical comparison for daily survivorship rates is given by Hensler and Nichols (1981). Interval survivorship rates were compared using Student's t-test. Because the sample sizes and variances differed, correction was made for degrees of freedom (Hayes 1973).

RESULTS

Despite the close proximity of the study sites, their passerine communities differed (Table 1), presumably as a result of contrasting vegetational structures. San Isidro is more heavily and diversely wooded than El Talar. Agricultural practices at the two sites are probably responsible for the differences, although relative proximity to the Río de La Plata may play some role.

The data base consists of observations from 360 nests of 28 passerine species. Many of the descriptive data are self-explanatory and require discussion only when they are in conflict with or are supplemented by previous reports (See Species Accounts). Data on nesting chronology from both sites are pooled (Fig. 3). Table 2 describes the distribution of nests by site and gives nest heights. Table 3 is a summary of clutch sizes and egg measurements. Table 4 shows the duration of average nesting attempts for all species. Tables 5 and 6 present the nest survivorship data. Table 7 describes various aspects of nestling morphology and growth.

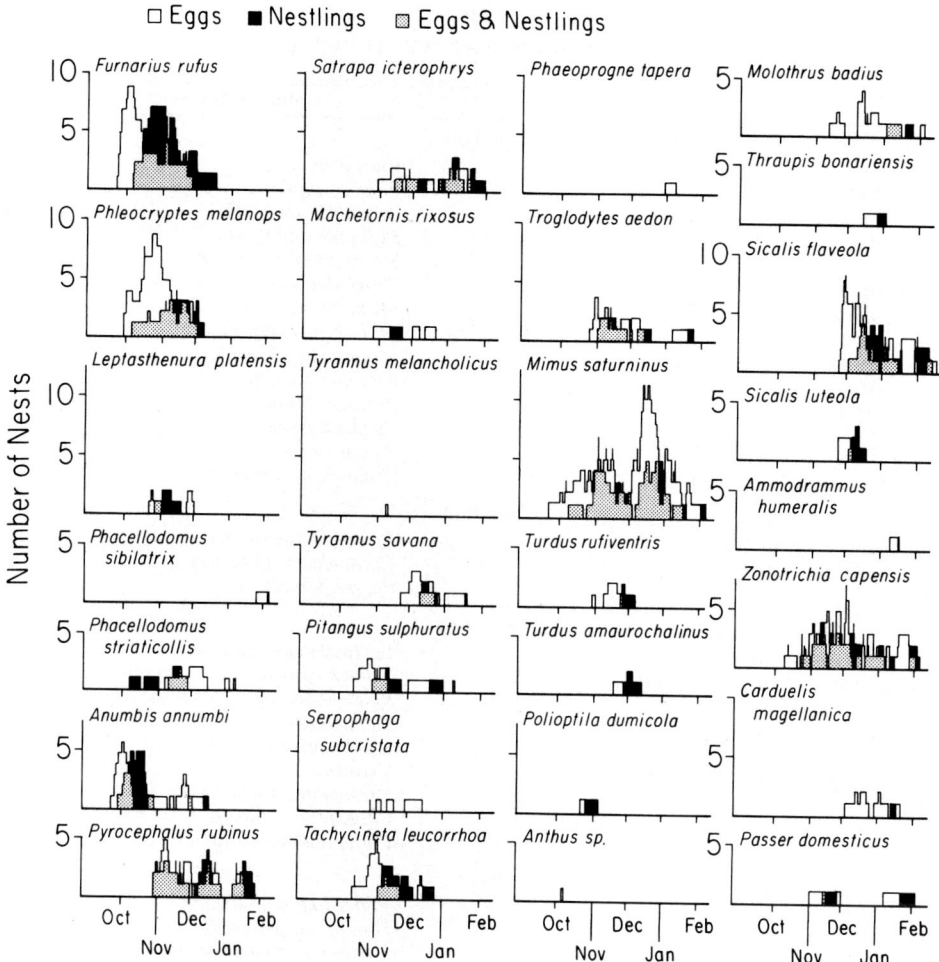

☐ Eggs   ■ Nestlings   ▨ Eggs & Nestlings

FIG. 3. Nesting chronology of all passerine nests found during the two field seasons. Open bars indicate nests with eggs, closed bars indicate nests with nestlings, shaded bars indicate both.

Curvilinear regressions of nestling weight gain over time (Ricklefs 1967) were not performed because of the heterogeneity of the sample and because the biological significance of the different growth equations is obscure. The original data from which the regressions were obtained can be found in Mason (1980).

Species using complex nests have higher survivorship than those using cup nests (Table 8), despite the diversity in the natural histories of species within each category. Different species nesting in hornero nests are characterized by different probabilities of survivorship (Table 9). Further statistical comparisons of nest survivorship were performed only for the Saffron Finch and the Rufous Hornero because of constraints of small sample sizes for other species. Horneros have superior survivorship to Saffron Finches nesting in hornero nests, but the latter have superior survivorship to finches nesting in other sites.

SPECIES ACCOUNTS

The following species accounts amplify previous natural history observations (cited appropriately in the text). Species for which I found no new information are not discussed.

Furnarius rufus. Rufous Hornero. My sample of nests examined was biased toward those at low heights (Table 2), since many were beyond my reach with the equipment available. Although nest-building can begin several months prior to egg-laying, the birds are capable of

TABLE 1
PASSERINE ABUNDANCE AT THE STUDY SITES[1]

Estancia San Isidro	Estancia El Talar
Abundant	
Furnarius rufus	*Furnarius rufus*
Phleocryptes melanops	*Phleocryptes melanops*
Pitangus sulphuratus	*Anumbius annumbi*
Tachycineta leucorrhoa	*Pitangus sulphuratus*
Troglodytes aedon	*Tachycineta leucorrhoa*
Mimus saturninus	*Troglodytes aedon*
Turdus rufiventris	*Mimus saturninus*
Molothrus bonariensis	*Molothrus bonariensis*
Molothrus rufoaxillaris	*Molothrus rufoaxillaris*
Molothrus badius	*Molothrus badius*
Agelaius thilius	*Agelaius thilius*
Pseudoleistes virescens	*Sicalis flaveola*
Sicalis flaveola	*Sicalis luteola*
Zonotrichia capensis	*Zonotrichia capensis*
Common	
Tyrannus savana	*Pyrocephalus rubinus*
Thraupis bonariensis	*Carduelis magellanica*
	Passer domesticus
Uncommon	
Leptasthenura platensis	*Leptasthenura platensis*
Phacellodomus striaticollis	*Schoeniophylax phryganophila*
Hymenops perspicillata	*Phacellodomus striaticollis*
Polioptila dumicola	*Satrapa icterophrys*
Sicalis luteola	*Machetornis rixosus*
Embernagra platensis	*Tyrannus savana*
	Serpophaga subcristata
	Phaeoprogne tapera
	Polioptila dumicola
Rare	
Lepidocolaptes angusitrostris	*Phacellodomus sibilatrix*
Limnornis curvirostris	*Hymenops perspicillata*
Schoeniophylax phryganophila	*Tyrannus melancholicus*
Phacellodomus sibilatrix	*Pseudocolopteryx flaviventris*
Thamnophilus ruficapillus	*Tachuris rubigastra*
Pyrocephalus rubinus	*Phytotoma rutila*
Satrapa icterophrys	*Turdus rufiventris*
Machetornis rixosus	*Anthus* sp.
Tyrannus melancholicus	*Icterus cayenensis*
Serpophaga subcristata	*Agelaius ruficapillus*
Elaenia parvirositris	*Pseudoleistes virescens*
Phytotoma rutila	*Leistes superciliaris*
Turdus amaurochalinus	*Paroaria coronata*
Cyclarhis gujanensis	*Ammodramus humeralis*
Icterus cayenensis	*Embernagra platensis*
Parula pitiayumi	
Stephanophorus diadematus	
Paroaria coronata	
Sporophila caerulescens	
Poospiza nigro-rufa	
Carduelis magellanica	
Passer domesticus	

[1] Species listed in taxonomic order within abundance categories.

TABLE 2
Nest Height and Distribution by Study Site[1]

Species	Number of nests			Nest height (m)		
	Site 1	Site 2	Total	X̄	s.d.	Range
Furnarius rufus[2]	1	16	17	2.6	1.2	1.2–4.9
Phleocryptes melanops	0	21	21	0.8	0.2	0.4–0.9
Leptasthenura platensis	1	3	4	2.3	0.4	1.8–2.6
Phacellodomus sibilatrix	1	0	1	2.0	—	—
Phacellodomus striaticollis	3	6	9	2.2	0.4	1.5–2.8
Anumbius annumbi[2]	0	13	13	2.7	0.8	2.1–4.6
Pyrocephalus rubinus	0	23	23	2.5	1.1	1.4–6.1
Satrapa icterophrys	3	8	11	2.6	1.1	1.1–5.0
Machetornis rixosus	1	2	3	3.0	1.1	1.8–3.9
Tyrannus melancholicus	1	0	1	3.0	—	—
Tyrannus savana	0	9	9	4.2	1.2	2.1–6.5
Pitangus sulphuratus[2]	0	7	7	3.4	0.6	2.3–4.2
Serpophaga subcristata	1	3	4	1.8	0.5	1.0–2.1
Tachycineta leucorrhoa	1	9	10	2.2	0.5	1.3–2.9
Phaeoprogne tapera	0	1	1	1.9	—	—
Troglodytes aedon[3]	3	13	16	1.3	0.5	0.4–2.0
Mimus saturninus	39	40	79	1.9	0.6	0.7–3.4
Turdus rufiventris	5	0	5	2.0	0.7	1.1–2.8
Turdus amaurochalinus	2	0	2	1.6	—	—
Polioptila dumicola	0	1	1	3.4	—	—
Molothrus badius	7	4	11	2.8	1.3	1.3–5.1
Thraupis bonariensis	1	0	1	2.1	—	—
Sicalis flaveola	14	20	36	2.4	1.0	1.0–4.2
Sicalis luteola	0	5	5	0	—	—
Ammodrammus humeralis	0	1	1	0	—	—
Zonotrichia capensis	40	19	59	0	—	—
Carduelis magellanica	0	7	7	2.9	0.7	2.2–3.8
Passer domesticus	0	3	3	3.4	0.8	2.5–4.0

[1] Site 1 is Estancia San Isidro; Site 2 is Estancia El Talar.
[2] Mean nest height underestimated because many nests were located beyond the reach of the investigator.
[3] Includes only heights of natural nests.

rapid construction. Two nests toppled by a severe thunderstorm were completely rebuilt and contained the first egg in 15 days. One nest had not yet received eggs, while the other had eggs which had been incubated for 13 days. White (1882) described a single instance of construction which required 10 days.

Eggs are pure white when laid, but accumulate mud smears as incubation progresses. Hudson (Sclater and Hudson 1888; Hudson 1920) listed clutch size as five, but this is outside the range I observed (Table 3) and has been recorded in only one clutch since that time (Narosky et al. 1983). Clutches of three or four eggs generally hatch asynchronously by about one day, but the hatching spread is variable. Fraga (1980) reported hatching spreads of 6 hrs to greater than 48 hrs. Food items collected from nestlings' mouths at one nest included: one nymphal mole cricket (Gryllotalpidae); two adult crickets of the same species (Gryllidae: Gryllinae); and one weevil (Curculionidae).

I was unable to weigh nestlings after about day 10, because they were very agile and could not be withdrawn easily through the observation hole. At this time, they were almost completely feathered and weighed 50 to 55 g, almost adult weight (Contreras 1979; Fraga 1980). Second nesting attempts may occur in the same nest (two observations). Of the 12 nests that successfully fledged young, I observed a single case of starvation. This occurred at a nest with four eggs, where hatching spread exceeded 48 hrs (greatest observed).

Phleocryptes melanops. Wren-like Rushbird. The closed, woven nest may be built in *Solanum malacoxylon* as well as in reeds. I never found a nest built in cattails. Nests are generally built 0.8 m above the surface of the water, but heavy rains can cause flooding. Differential growth of rushes supporting the nest can cause the inclination of the structure to change. In one case, this was sufficient to pour the eggs out of the nest. The diameter of the nest entrance

TABLE 3

CLUTCH SIZES AND EGG DIMENSIONS[1]

Species	Clutch size Mean, range (n)	Mean egg sizes		
		Length	Width	Weight
Furnarius rufus	3.38, 2–4 (16)	28.5	21.9 (29)	7.1 (29)
Phleocryptes melanops	2.71, 2–3 (21)	20.7	15.6 (13)	2.6 (10)
Leptasthenura platensis	3.33, 3–4 (3)	18.3	13.9 (9)	1.8 (9)
Phacellodomus sibilatrix	3.00, — (1)	20.1	16.3 (3)	2.9 (2)
Phacellodomus striaticollis	4.00, — (3)	22.4	16.5 (8)	3.2 (8)
Anumbius annumbi	3.91, 3–6 (11)	23.4	17.6 (32)	3.8 (32)
Pyrocephalus rubinus	2.89, 2–3 (18)	17.7	13.2 (45)	1.6 (22)
Satrapa icterophrys	3.00, 2–4 (10)	20.4	15.2 (19)	2.4 (19)
Machetornis rixosus	3.33, 3–4 (3)	23.8	17.5 (9)	3.9 (9)
Tyrannus savana	3.20, 3–4 (5)	21.6	16.4 (12)	3.0 (11)
Pitangus sulphuratus	4.67, 4–5 (3)	28.9	20.6 (19)	6.4 (19)
Serpophaga subcristata	2.67, 2–4 (3)	15.7	11.9 (3)	1.2 (5)
Tachycineta leucorrhoa	4.75, 4–6 (4)	20.0	14.0 (16)	2.0 (15)
Phaeoprogne tapera	5.00, — (1)	23.5	16.8 (3)	3.4 (3)
Troglodytes aedon	5.25, 4–6 (4)	17.5	13.2 (25)	1.7 (21)
Mimus saturninus	3.00, 2–4 (10)	28.4	20.4 (100)	6.0 (50)
Turdus rufiventris	3.50, 3–4 (2)	29.9	21.5 (4)	7.0 (4)
Turdus amaurochalinus	—	30.7	20.1 (1)	6.2 (1)
Polioptila dumicola	3.00, — (1)	15.8	11.8 (3)	1.0 (3)
Anthus sp.	3.00, — (1)	20.7	14.9 (3)	2.3 (3)
Molothrus bonariensis	—	23.6	18.6 (36)	4.5 (10)
Molothrus rufoaxillaris	—	23.2	17.6 (10)	4.2 (10)
Molothrus badius	3.44, 3–4 (9)	24.2	18.3 (12)	4.4 (12)
Thraupis bonariensis	2.00, — (1)	25.7	18.0 (2)	4.4 (2)
Sicalis flaveola	4.22, 3–6 (23)	19.5	14.2 (78)	2.1 (78)
Sicalis luteola	5.00, — (3)	18.1	13.7 (10)	1.8 (10)
Ammodramus humeralis	4.00, — (1)	19.8	15.7 (4)	2.6 (4)
Zonotrichia capensis	3.03, 2–4 (36)	20.1	15.3 (100)	2.4 (100)
Carduelis magellanica	3.67, 3–4 (3)	17.1	12.8 (10)	1.6 (10)
Passer domesticus	3.50, 3–4 (2)	22.7	15.5 (6)	2.9 (6)

[1] Egg dimensions in mm; egg weight in grams.

is about 25 to 30 mm. The birds require at least three days to build a nest and the delay between nest completion and laying is generally two to four days, although in one extreme case a completed nest remained empty for 17 days. No eggs are laid until the wet rushes of which the nest is built have dried. The first egg of the season (29 September) was perhaps laid by a female other than the builder of the nest. It was placed in a nest which was still moist and not yet enclosed. The next day, construction was complete, but the egg was absent.

Eggs are immaculate, and a deep greenish blue in color. Only animal food is brought to the nestlings. Samples from two nests included one naiad and one adult damselfly (Coenagrionidae), one adult male moth (Noctuidae: Amphipyrinae), and a set of balled up cocoons of parasitic Hymenoptera (probably Braconidae).

Leptasthenura platensis. Tufted Tit-spinetail. All nests examined had been built by filling old nests of the Rufous Hornero with grass. One hornero nest had been used by its builders earlier in the season. One spinetail nest was occupied and guarded for close to a month prior to the onset of egg-laying. Two nests were taken over and used as nests by mice, while the other two both successfully fledged young. Eggs are immaculate white and somewhat pointed.

Phacellodomus sibilatrix. Little Thornbird. The Little Thornbird, like the Rufous-fronted Thornbird (*P. rufifrons*, see Skutch 1969), begins its nest by piling sticks at the end of a slender horizontal branch. As the mass of interlocking twigs accumulates, the bough is bent downward until the nest is pendulous. I was only able to examine a single nest (at San Isidro) which hung down to a height of 2 m. Most others are substantially higher than this (but Barrows, *fide* Sclater and Hudson 1888, claimed that they often hang down to a height of only a meter). Barrows also noted that a single nest may be used in subsequent seasons, but that reconstruction on the same branch is more common. He listed the breeding season as lasting from 1 October to 1 January, but Pereyra (1938) found a nest with eggs in mid-September. Eggs are immaculate white. The young that hatched in the single nest I examined died from parasitism by botfly larvae, which were visible on the first day.

Phacellodomus striaticollis. Freckle-breasted Thornbird. The globular stick nest is built within the branches of a tree and is supported partially by several branches. The entrance faces outward and upward. The nest can be built in eight to 10 days. When heavy storms occur, the independent movement of the supporting twigs can damage the integrity of the structure. At least one and possibly two nests were lost to Bay-winged Cowbirds (*Molothrus badius*), the only nesting cowbird. Piracy of nests of other furnariids has been reported (Hudson 1920; Daguerre 1924; Friedmann 1929), but in neither of the cases I report here did Bay-winged Cowbirds subsequently occupy the nests. Eggs are immaculate white. Pereyra (1938) described the bird as double-brooded, with laying occurring from late September until December.

Anumbius annumbi. Firewood-gatherer. The bulky stick nests are generally built on one principal branch, but as the volume of the nest grows, many other smaller branches are typically incorporated into the nest structure. A long entrance tunnel extends from near the top of the structure to the nest chamber, located near the bottom. Average nest height recorded in this study (2.7 m) certainly underestimates the true mean, although low nest sites have been described by Wetmore (1926). Durnford (*fide* Sclater and Hudson 1888) reported nesting heights as great as 60 to 70 feet (18.5 to 21.5 m) in introduced poplars in Buenos Aires. Nest building seems to go on continuously throughout the breeding season. Second nesting attempts occur in the same nest (two observations), and the young of the first brood may accompany their parents during this time. I did not observe helping, but my observations are not sufficiently detailed to rule out this possibility. Hudson noted that young may stay with their parents for as long as three or four months, "going about and feeding in company, and roosting together in the same nest" (Sclater and Hudson 1888:189; Hudson 1920:169). In the same sources, Hudson reported clutch size as five, which probably is an overestimate (Table 3). Eggs are immaculate white. Young at four (of 13) nests were parasitized by botfly larvae, and infestation was sufficient to kill all the young at one nest.

Pyrocephalus rubinus. Vermillion Flycatcher. The natural history of this flycatcher in Argentina has been summarized by Fraga (1977). The nest, constructed of lichens and spider webs, is well hidden in the crotch of a small limb. The eggs are ovate with an off-white or creamish ground color. Spots and blotches of black and lilac are concentrated at the wide end. Hudson (Sclater and Hudson 1888; Hudson 1920) claimed a clutch size of four, which exceeds clutch sizes I observed for this species (Table 3).

Nestlings are melanic dorsally. They remain quiet and motionless, maintaining the nest's inconspicuousness. Food delivered to one nest included one mature female sawfly (Tenthredinidae), and one butterfly (Satyridae) with wings and legs removed. Young in one nest were eaten by ants. At another nest, one of two nestlings died at four to five days of age. Two days later, the remaining nestling was found dead but apparently uninjured in the nest. The adult male was found unable to fly moments before death, but without obvious injury. Fraga (1977) reported the nestling period as 16 to 17 days, but observations at five nests in this study indicate a mean duration of 12.8 days (Table 4). Perhaps my frequent visits to nests caused early departure of young. On the other hand, Marchant (1960) reported a nestling stage of 13 to 15 days in southwestern Ecuador.

Satrapa icterophrys. Yellow-browed Tyrant. This flycatcher builds a cup nest of a few fine twigs lined with grass and finally feathers. The structure is delicate and made with a minimal amount of material over a period of four to five days. Severe weather can damage the nest, and in one case the eggs began to sink through the floor of the nest. This nest was deserted. The eggs have a creamy ground color and the wide end is marked with large, chocolate-brown spots. Hudson (Sclater and Hudson 1888; Hudson 1920) reported clutch size as four, the largest and least frequent clutch size I observed (2 of 10). Young at one nest were parasitized by botflies. Food items brought to one nest included six spiders: one female wolf spider (Lycosidae), four Araneidae (three females, one male, in two species), and the abdomen of one additional, unidentified spider.

Tyrannus savana. Fork-tailed Flycatcher. Two of the nine nests of the Fork-tailed Flycatcher (formerly *Muscivora tyrannus,* Traylor 1979) were known to contain only eggs of the Shiny Cowbird. Both nests showed evidence of continual disturbance, and I was uncertain if incubation had been initiated at either. These two nests are not included in the data analysis. This

TABLE 4

LAYING RHYTHM AND DURATION OF NESTLING PHASES

Species	Laying rhythm[1]	Mean phase duration[2]		
		Incubation	Eggs[3]	Nestling[3]
Furnarius rufus	A	17.0	20.8[4]	24.4
Phleocryptes melanops	A	16.1	19.5	15.0
Leptasthenura platensis	C?	14.0	16.3	15.5
Phacellodomus striaticollis	A?	16.0	22.0	12.5
Anumbius annumbi	A	16.0	23.8	17.4
Pyrocephalus rubinus	C	15.0	16.9	12.8
Satrapa icterophrys	C	15.0	19.0	15.0
Machetornis rixosus	—	14.5?	—	—
Tyrannus savana	C	15.0	17.2	—
Pitangus sulphuratus	A	16.0	22.7	15.0?
Serpophaga subcristata	A	—	—	—
Tachycineta leucorrhoa	C	15.0	18.8	18.0
Phaeoprogne tapera	C	—	—	—
Troglodytes aedon	C	13.0	17.3	15.0?
Mimus saturninus	C	13.4	15.4	12.0
Turdus rufiventris	C	14.0	16.5	14.0
Turdus amaurochalinus	—	—	—	15.0
Molothrus badius	C	12.0	14.4	14.5
Sicalis flaveola	C	13.3	16.5	14.6
Sicalis luteola	C	11.0	15.0	—
Zonotrichia capensis	C	13.0	15.0	10.4
Carduelis magellanica	C	—	—	—
Passer domesticus	C	14.5	17.0	—

[1] Laying on consecutive or alternate days; ? = data equivocal.
[2] In days.
[3] Values used to calculate nest survivorship.
[4] Decreased by one day to account for asynchronous hatching. See methods for discussion of this procedure.

bird builds its grass cup nest high in trees. At least five days seem to be necessary for construction. Eggs have a white ground color and few large lilac and dark gray splotches at the wide end. Hudson (Sclater and Hudson 1888; Hudson 1920) presented the largest clutch size (four) as representative. I observed this clutch size once out of five observations. At one nest where only a single young hatched, the nestling showed numerous small lesions in the skin, particularly about the breast, by day 7. By day 9, it was dead and crawling with maggots. A similar pattern was repeated at a second nest; lesions were apparent in all three nestlings at day 7, but they looked more healthy by day 9. By day 11, however, the nestlings were gone.

Pitangus sulphuratus. Great Kiskadee. This aggressive and conspicuous tyrant weaves a retort-shaped nest of grass. In my experience, construction takes place over a period of about four weeks, although Hudson (1920) claimed that five or six weeks is more usual. Although this bird was abundant at both sites, I examined only a few nests because they are generally built high in the trees at the tips of branches. Observed mean nest height (3.4 m) certainly underestimates the true mean. Eggs are cream colored with large brown spots concentrated at the wide end. The young are provisioned with a variety of prey. Items collected at one nest included part of a glass lizard (Anguidae), one mature and one metamorphosing cricket frog (Hylidae), three leptodactylid frogs (Leptodactylidae, three species), a single mature butterfly (Satyridae), and a spider egg sac (probably Lycosidae).

Tachycineta leucorrhoa. White-rumped Swallow. Most nests were built inside old deserted nests of the Rufous Hornero, but others were placed in old woodpecker holes or other cavities in trees. The nest is made of coarse grass and lined with feathers. The eggs are elongated and immaculate white. At three nests, cracked eggs were not ejected although most passerines remove damaged eggs from the nest (Rothstein 1982; pers. observ.). Hudson (Sclater and Hudson 1888; Hudson 1920) gave the range in clutch size as five to seven, but four to six is the range that I observed (Table 3). Young may stay in the nest as long as 21 days before fledging. The birds left both study sites during the first week of January.

Troglodytes aedon. House Wren. Five of 16 nests were built in nest boxes (one at San Isidro,

TABLE 5
SURVIVORSHIP OVER THE EGG PHASE[1]

	Daily		Interval			
	Mean	s.d.	Mean	s.d.	No. nests	No. days
Furnarius rufus	.988	.007	.773	.114	15	244
Phleocryptes melanops	.967	.011	.518	.113	20	271
Leptasthenura platensis	.920	.054	.257	.220	4	25
Phacellodomus striaticollis	.917	.036	.147	.115	7	60
Anumbius annumbi	.975	.014	.547	.187	11	120
Pyrocephalus rubinus	.940	.019	.354	.120	22	151
Satrapa icterophrys	.979	.015	.667	.189	10	95
Machetornis rixosus	.938	.043	—	—	3	32
Tyrannus savana	.958	.024	.481	.199	7	72
Pitangus sulphuratus	.962	.019	.418	.178	7	106
Serpophaga subcristata	.852	.068	—	—	4	27
Tachycineta leucorrhoa	.966	.019	.525	.192	9	89
Troglodytes aedon, w/o boxes	.864	.052	.079	.071	9	44
Troglodytes aedon, w/ boxes	.923	.026	.250	.119	14	104
Mimus saturninus	.922	.012	.288	.058	59	477
Turdus rufiventris	.933	.046	.320	.238	3	30
Molothrus badius	.890	.035	.187	.101	10	82
Sicalis flaveola	.919	.019	.247	.082	29	209
Sicalis luteola	.960	.039	.542	.318	3	25
Zonotrichia capensis	.899	.020	.201	.066	45	227
Carduelis magellanica	.878	.047	—	—	7	49
Passer domesticus	.970	.030	.593	.300	3	33

[1] Dash = phase length unknown; survivorship cannot be calculated.

four at El Talar). The remainder were built in recesses such as knotholes, rotted fence posts, and stumps. Once, an old furnariid stick nest was used. Except for the Saffron Finch (one nest), wrens were the only species to use nest boxes. Nest boxes were bigger and more conspicuous than natural nests and had relatively larger entrances. Although the entrance holes of natural nests are often irregular, they are generally about 3 cm wide. Nests in boxes were subjected to extremely high levels of parasitism by Shiny Cowbirds, an effect which may be due to conspicuousness and entrance hole size. Survivorship estimates for the cohort nesting in boxes are presented separately from those nesting in natural cavities (Tables 5, 6). Nesting chronology is probably little affected by nest type, however, and all data were pooled (Fig. 3). In one case, nest construction (in a box) required 16 days. Eggs are densely, but evenly mottled with fine, brick-red spots. Hudson (Sclater and Hudson 1888; Hudson 1920) reported a clutch size of nine, far outside the range I observed (Table 3). Two of three broods at San Isidro were lost to botfly parasitism.

Mimus saturninus. Chalk-browed Mockingbird. Nests are built of small twigs, and then lined primarily with rootlets. Nest construction requires about five days (range: three to eight days). Nest sites are usually small, isolated trees surrounded by pasture. After nest failures, subsequent nests are generally built close to the original site, often in the same tree. Eggs have a blue background and are marked with large, reddish-brown spots. Entire broods at four nests (all at San Isidro) succumbed to botfly parasitism, and non-lethal parasitism was observed at one nest at El Talar. Young were eaten by ants at one nest. Food items collected from the mouths of nestlings at three nests include: three intact nymphal mole crickets (Gryllotalpidae); two intact adult crickets (Gryllidae); one scarab (Scarabeidae), legs on one side missing; one weevil (Curculionidae); and the anterior half of an earthworm (Lumbricidae). More on the natural history of this species may be found in Fraga (1985).

Turdus rufiventris. Rufous-bellied Thrush. The nest is a solid and deep cup of small twigs, vines, and/or coarse grass, plastered with mud, and then lined with fine grasses and rootlets. Nests are built in dense foliage at the tips of branches, sometimes near buildings. Eggs have a blue ground color, heavily mottled with reddish-brown. Excluded from analysis is one of five nests that contained two freshly pecked Shiny Cowbird eggs and not known to be active subsequently.

TABLE 6
SURVIVORSHIP OVER THE NESTLING PHASE[1]

	Daily		Interval			
	Mean	s.d.	Mean	s.d.	No. nests	No. days
Furnarius rufus	.996	.004	.918	.078	13	285
Phleocryptes melanops	.976	.014	.697	.144	11	126
Leptasthenura platensis	1.00	.180	1.00	*[2]	2	31
Phacellodomus striaticollis	.979	.021	.769	.200	4	48
Anumbius annumbi	.972	.016	.607	.173	8	106
Pyrocephalus rubinus	.937	.020	.435	.119	14	143
Satrapa icterophrys	.949	.025	.459	.175	8	79
Tyrannus savana	.857	.066	—	—	4	28
Pitangus sulphuratus	.949	.035	.454	.244	3	39
Tachycineta leucorrhoa	.986	.014	.772	.198	5	70
Troglodytes aedon, w/o boxes	1.00	.156	1.00	*	3	41
Troglodytes aedon, w/ boxes	.967	.023	.601	.213	6	60
Mimus saturninus	.953	.014	.563	.097	28	235
Turdus rufiventris	.938	.061	.405	.332	2	16
Turdus amaurochalinus	1.00	.224	1.00	*	2	20
Molothrus badius	1.00	.224	1.00	*	2	20
Sicalis flaveola	.978	.013	.720	.135	12	135
Sicalis luteola	.920	.054	—	—	3	25
Zonotrichia capensis	.901	.024	.337	.091	26	161
Passer domesticus	.963	.036	—	—	2	27

[1] Dash = phase length unknown; survivorship cannot be calculated.
[2] * = assumption of the model violated; see text p. 967 for discussion.

Turdus amaurochalinus. Creamy-bellied Thrush. The nests and eggs of this species resemble those of the previous species, but the Creamy-bellied Thrush apparently prefers to build its nest on branches several centimeters in diameter. The two nests found were built on the sloping trunks of trees, where the first major branches originated.

Polioptila dumicola. Masked Gnatcatcher. The single nest found was positioned within the crotch of a sapling, shaded by larger trees. The nest was covered with lichens, lined with fine grasses, thistle down, and hair. Eggs are light blue with evenly distributed reddish-brown spots. Nestlings are melanic dorsally. The gape and tongue are yellow, and the latter possesses two black spots at its base.

Molothrus badius. Bay-winged Cowbird. The natural history of the Bay-winged Cowbird in Argentina is complicated by its interaction with its specialized parasite, the Screaming Cowbird. All nests were parasitized, so mean clutch size (Table 3) is based on egg identification and may be biased. In addition, this is a social species with helpers at the nest (Fraga 1972; Orians 1980). Ten of eleven active nests were built either in old furnariid stick nests, or in the domed grass nests of Great Kiskadees. One nest was built beneath the loose bark of a eucalyptus tree. Nest piracy by Bay-winged Cowbirds is common (Sclater and Hudson 1888; Hudson 1920; Daguerre 1924). Eggs have a variable background color and are covered with brown spots. Eggs of the parasite can be very similar, and disginguishing between the eggs of the two congeners is probably impossible without a detailed knowledge of events occurring at the nest over a period of time (see Fraga 1983a for a more detailed description of the eggs of the two species). Identifications of eggs collected where the Bay-winged and Screaming Cowbirds are sympatric should be considered suspect. As a result of the misidentification of eggs, Hudson (Sclater and Hudson 1888; Hudson 1920) mistakenly claimed that several female Bay-winged Cowbirds laid in the same nest. He also reported clutch size as five, more than in my experience and Fraga's (1972). The nestlings of the two species are also quite similar, but do differ slightly at some periods of development (Fraga 1979). Food items collected at one nest include two mature female spiders (Araneidae), and one nymphal grasshopper (Acrididae).

Thraupis bonariensis. Blue-and-Yellow Tanager. The single nest found was a cup which contained only two eggs. The eggs have a blue background with black scrawls and spots on

TABLE 7

CHARACTERISTICS OF NESTLING

Species	Gape color	Rictal flange	Natal down	Hatching weight[1]	Fledging weight[1]	Growth rate[2]
Furnarius rufus	yellow	pale yellow	absent	6.0 (9)	>50.0 (13)	4.33 (34)
Phelocryptes melanops	yellow	deep yellow	dark gray, dense	2.2 (2)	17.3 (6)	1.48 (16)
Leptasthenura platensis	yellow	yellow	light gray, dense	1.5 (2)	18.0 (3)	1.47 (2)
Phacellodomus sibilatrix	yellow	white	gray, sparse	2.1	—	—
Phacellodomus striaticollis	yellow	white	gray, sparse	2.6 (1)	26.5 (6)	2.14 (9)
Anumbius annumbi	yellow	yellow	light gray, dense	3.5 (3)	39.4 (14)	2.42 (16)
Pyrocephalus rubinus	yellow	yellow	light gray, dense, short	1.3 (5)	17.8 (5)	1.32 (26)
Satrapa icterophrys	yellow	yellow	light gray, dense	2.1 (1)	20.5 (1)	1.64 (7)
Tryannus savana	yellow	yellow	light gray	2.8 (3)	—	2.42 (7)
Pitangus sulphuratus	yellow	yellow	gray, long	5.2 (4)	57.2 (3)	4.41 (12)
Tachycineta leucorrhoa	yellow	white	very light gray, sparse	1.6 (1)	—	2.28 (4)
Mimus saturninus	yellow	white	dark gray	4.3 (3)	56.0 (11)	5.19 (23)
Turdus rufiventris	yellow	pale yellow	yellow gray	—	64.0 (2)	—
Turdus amaurochalinus	yellow	pale	yellow gray	—	53.0 (1)	—
Polioptila dumicola	yellow	cream	—	—	7.9 (2)	0.85 (2)
Molothrus badius	red	white	gray	—	—	4.09 (2)
Sicalis flaveola	pink	white	light gray	1.6 (2)	16.3 (5)	1.46 (15)
Sicalis luteola	pink	white	very light gray	—	14.0 (2)	1.54 (6)
Zonotrichia capensis	pink	white	gray	1.9 (3)	19.0 (10)	2.03 (22)

[1] In grams; sample size in parentheses.
[2] In grams increase per day; sample size in parentheses.

the wide end. The incubating female was found mutilated in the nest after seven days of tending the nestling.

Sicalis flaveola. Saffron Finch. All 14 of the nests examined at San Isidro were built inside old hornero nests, but other furnariid stick nests were used as well. Nine of 22 nests found at El Talar were built in such sites, and a single nest was built in a nest box. Many nests had been used earlier in the season, both by the nests' builders (e.g., Rufous Horneros, Firewood-gatherers, etc.) or by other, subsequent occupants (e.g., Tufted Tit-spinetails, White-rumped Swallows, etc.). The nest is lined with fine grass. The eggs have a dark-gray background, overlayed with darker, dense maculations. The young are provisioned entirely with regurgitated seeds. Botfly larvae parasitized the nestlings at one nest.

Zonotrichia capensis. Rufous-collared Sparrow. This ground nesting sparrow is widely distributed in South and Central America, often abundant, and has been the subject of several studies. An adequate natural historical base is already available for this bird in this area (Fraga 1978, 1983b). Hudson (Sclater and Hudson 1888; Hudson 1920) reported clutch size as four to five, but in my experience four egg clutches are infrequent (8 of 36 clutches) and I did not observe a five-egg clutch. Botfly parasitism was observed at two nests (both at San Isidro), and killed all young in one case.

Carduelis magellanica. Hooded Siskin. The Hooded Siskin (formerly *Spinus magellanicus,* Howell et al. 1968) builds a small grass cup nest, whose construction is apparently finished two to three days before egg laying commences. Of the five nests found before laying began, one was apparently complete but without eggs for as long as 10 days. The eggs may be pure

TABLE 8

SPECIES RANKINGS BY OVERALL SURVIVORSHIP[1]

Rank	Species	Survivorship	Nest type
1.	*Furnarius rufus*	.710	complex
2.	*Tachycineta leucorrhoa*	.405	complex
3.	*Phleocryptes melanops*	.360	complex
4.	*Anumbius annumbi*	.332	complex
5.	*Satrapa icterophrys*	.305	cup
6.	*Leptasthenura platensis*	.257	complex
7.	*Pitangus sulphuratus*	.189	complex
8.	*Molothrus badius*	.187	complex
9.	*Sicalis flaveola*	.177	complex
10.	*Mimus saturninus*	.162	cup
11.	*Pyrocephalus rubinus*	.154	cup
12.	*Troglodytes aedon*, w/boxes	.150	complex
13.	*Turdus rufiventris*	.130	cup
14.	*Phacellodomus striaticollis*	.113	complex
15.	*Zonotrichia capensis*	.068	cup

[1] Overall survivorship = product of survivorships over the egg and nestling phases; groups defined by nest type differ significantly ($U = 11$, $P = 0.05$).

white (one clutch) as claimed by Hudson (Sclater and Hudson 1888; Hudson 1920), but typically the wide end is lightly marked with a few fine dark brown spots. Pereyra (1938) also described this variation. The first nest found held eggs that did not hatch for 21 days before predation occurred. One nest, found after incubation had been initiated, was observed for 13 days before hatching occurred. Seed loads were often visible in the esophagi of nestlings.

DISCUSSION

Species abundance.—The census method used to record differences in community composition is approximate and biased by conspicuousness. Nonetheless, real differences do exist between the sites (Table 1). The bias has its strongest effect on species described as uncommon or rare, but not all species in these categories are inconspicuous (three extreme examples: *Hymenops perspicillata*; *Paroaria coronata*; and *Passer domesticus*). Those species recorded at both sites are likely to be ranked realistically relative to each other. Species listed as abundant or common at one site could not be overlooked at the other. Two examples make this clear. The Rufous-bellied Thrush was seen every day at Estancia San Isidro, but heard singing and sighted on only a single occasion at Estancia El Talar. Conversely, only a single pair of Vermillion Flycatchers was recorded at Estancia San Isidro, but the species was seen daily at Estancia El Talar. Twelve nesting species were categorized as abundant at each site, 10 of which were common to both. The remaining two abundant species (at each site) were ranked in abundance at the other site as uncommon (one case), rare (two cases), or absent (one case).

The relationship between species abundance (Table 1) and the number of nests found at each site (Table 2) is unreliable. In many cases, this is due to small sample sizes. The techniques for opening furnariid nests were being developed during the first season (at Estancia San Isidro), so species using these nest sites are underrepresented. On the other hand, relative abundances are probably accurately described for species building easy-to-find cup nests or represented by larger cohorts (e.g., Vermillion Flycatcher, Yellow-browed Tyrant, Chalk-browed Mockingbird, Saffron Finch, and Rufous-collared Sparrow).

Clutch size.—The clutch size data presented here are often at odds with those presented by Hudson, who failed to provide data on the distribution of clutch sizes. His best known work, *Birds of La Plata* (1920) is an edited version of excerpts from *Argentine Ornithology* (Sclater and Hudson 1888, 1889). Much of this material in turn was taken from a series of letters and articles published by the Zoological Society of London (Hudson 1870, 1871, 1872a, b, c, 1873, 1874, 1876). I found few substantive changes between the sources listed above, but in the introduction to a recent Spanish language translation of the 1920 work (see Hudson 1920 citation), M. Victoria provides a letter by Hudson in which he admits to editing from memory (27 years after he left Argentina). In addition, at least one contemporary author (Gibson 1880)

TABLE 9
Survivorship of Species Using Hornero Nests[1]

	Daily		Interval			
	Mean	s.d.	Mean	s.d.	No. nests	No. days
Furnarius rufus						
Eggs	.988	.007	.773	.114	15	244
Nestlings	.996	.004	.918	.078	13	285
Leptasthenura platensis						
Eggs	.920	.054	.257	.220	4	25
Nestlings	1.00	.018	1.00	*[2]	2	31
Tachycineta leucorrhoa						
Eggs	.965	.024	.511	.236	6	57
Nestlings	.977	.022	.661	.268	3	44
Sicalis flaveola, in hornero nests						
Eggs	.956	.018	.475	.142	16	136
Nestlings	.982	.013	.767	.143	9	111
Sicalis flaveola, in other nesting sites						
Eggs	.842	.042	.059	.044	13	76
Nestlings	.958	.041	.537	.319	3	24
Passer domesticus						
Eggs	1.00	.186	1.00	*	2	29
Nestlings	.963	.036	—	—	2	27

[1] Dash = phase length unknown; survivorship cannot be calculated.
[2] * = assumption of model violated; see text p. 969 for discussion.

presented conflicting data on clutch size. Hudson may have presented the largest observed clutch size as if it were representative.

Parasitism by botfly larvae.—Parasitism of nestlings by botfly larvae occurred sporadically at the nests of seven species: Freckle-breasted Thornbird, Firewood-gatherer, Yellow-browed Tyrant, House Wren, Chalk-browed Mockingbird, Saffron Finch, and Rufous-collared Sparrow, and also was recorded for the brood-parasitic Shiny Cowbird. Such larvae have also been reported for nestlings of the Rufous-collared Sparrow and the Shiny Cowbird in Brazil (Sick 1958). Although adults were not reared for identification, the flies are probably members of the genus *Philornis*. Such flies are probably of common occurrence in nests of both passerines and non-passerines in Central and South America (Smith 1968; Hector 1982). Larvae attack nestlings before the feathers emerge. Nestlings may die when infected with six or more larvae, but the lethality of fly parasitism may depend on a variety of factors such as the age and size of the nestlings, the amount of food that adults can bring to the nest, and the thermal environment of the nest. I found no evidence that cowbird nestlings could preen larvae from other nestlings, as has been claimed for nestlings of the Giant Cowbird, *Scaphidura oryzivora* in Panama (Smith 1968). Continued research will doubtlessly expand the number of avian species used as hosts by these flies.

Survivorship.—The data presented here are primarily descriptive. Some species are represented by only a few nests, and most nests were not visited on a daily basis. Some care should be exercised, therefore, in accepting the validity of some of the estimates. Average durations of nesting attempts fall into this category, as do the survivorship estimates. An additional peculiarity emerged with the variance associated with mean survivorship at nests where no failures were observed. These variance estimates are typically much larger than those calculated for nests where failures did occur. In four cases for species represented by extremely small sample sizes (Tufted Tit-spinetail, House Wren, Creamy-bellied Thrush, and Bay-winged Cowbird), the equation describing the variance over the nestling interval resulted in negative values. These cases are marked with asterisks in Table 6. Evidently, this "impossible" result issues from a violation of the model for estimating nesting success that requires that failures be distributed normally over the interval in question. Despite these shortcomings, the use of

TABLE 10

COMPARISONS OF SURVIVORSHIP RATES[1]

I. Interval survivorship

Furnarius versus Sicalis in hornero nests

Eggs (.773 versus .475):	$t = 6.46$	30 d.f.	$P < 0.001$, 1-tailed
Nestlings (.918 versus .767):	$t = 2.88$	12 d.f.	$P < 0.01$, 1-tailed

Sicalis, in hornero nests versus other nest sites

Eggs (.475 versus .059):	$t = 10.9$	20 d.f.	$P < 0.001$, 1-tailed

II. Daily survivorship

Furnarius versus Sicalis, in hornero nests

Eggs (.988 versus .956):	$z = 1.71$	$F(z) = .956$	$P < 0.05$, 1-tailed
Nestlings (.996 versus .982):	$z = 1.03$	$F(z) = .848$	n.s., 1-tailed

Sicalis, in hornero nests versus other nest sites

Eggs (.956 versus .842):	$z = 2.49$	$F(z) = .994$	$P < 0.01$, 1-tailed

Furnarius, eggs versus nestlings

.988 versus .996	$z = 0.99$	$F(z) = .839$	n.s., 1-tailed

[1] Survivorship probabilities given in Table 9.

the maximum likelihood estimate of survivorship is probably the best estimate available, and the practice should become standard. Data in this form are of great value because they can be readily used to compare diverse species, and because they lend themselves well to use in analytic modelling within and between species (see below).

When species are assigned to one of two groups on the basis of nest architecture, species building complex nests, which are closed from above tend to have superior survivorship to those building simpler cup nests, which are open from above (Table 8). This trend has been reported for other bird species as well (Ricklefs 1969). The trend in survivorship reported here, significant at the 0.05 level for 15 species, disappears with removal of the only ground-nesting species in the table (Rufous-collared Sparrow).

Comparison of survivorship of several species using hornero nests indicates that the behavior of the birds contributes to survivorship independently of nest site selection (Tables 9, 10). Comparisons involving the Saffron Finch indicate that two effects can be seen in the same species. Although Saffron Finches nesting in hornero nests have superior survivorship to those nesting in other sites, the former still show lower survivorship than the hornero itself. Such comparisons of species that nest in identical sites are enlightening, but are not free from bias. The Rufous Hornero breeds very early in the season (Fig. 3), and season may affect survivorship.

Asynchronous hatching is sometimes better explained as an adaptation to prevent total nest failure than as a mechanism of brood reduction in variable environments (Clark and Wilson 1981). Many species achieve asynchronous hatching by initiating incubation on the penultimate egg. The Rufous Hornero conforms to this pattern and was the only species surveyed to do so. Clark and Wilson (1981, p. 258) predict this specific behavior when daily survivorship of the egg phase exceeds that of the nestling phase (but not extremely so). This hypothesis can be rejected for the hornero. Daily survivorship over the nestling phase exceeded that of the egg phase, but not significantly (Table 10).

ACKNOWLEDGMENTS

M. Christie helped me plan my initial trip to Argentina, introduced me to many members of the Argentine scientific community, and accompanied me in the field on several occasions. Dr. J. E. B. Ostrowsky of the Facultad de Veterinaria, Universidad de Buenos Aires, put me in touch with the Earnshaw family, owners of Estancia San Isidro. D. Earnshaw made the initial arrangements with A. Landa, owner of Estancia El Talar, for my second field season. I am grateful to these families for allowing me to use their property and for their hospitality. J. M. Gallardo and J. R. Navas, Museo Argentino de Ciencias Naturales, Buenos Aires, allowed me free access to the museum collections. M. Canevari was a source of intellectual stimulation and entertainment whenever I was in Buenos Aires. I am also indebted to M. K. Uyenoyama for deriving the equation to describe the variance associated with mean interval survivorship. J. Rawlins kindly identified the nestling food samples. R. M. Fraga shared with me his expertise, offered valuable suggestions, and reviewed an earlier copy of this manuscript. The manuscript was greatly improved by editorial comments by M. Foster, M. Nores, and L. Short.

Financial support was provided in part by the Frank M. Chapman Memorial Fund of the American Museum of Natural History. The author was supported by NSF grant BSR 8214999 to S. I. Rothstein during the preparation of the final draft of this paper.

LITERATURE CITED

BARBETTI, R. 1982. Algunas Plantas Autóctonas de Magdalena. Fundación Elsa Shaw de Pearson, Buenos Aires.

CABRERA, A. I. 1965. Flora de la Provincia de Buenos Aires, Vol. IV. Instituto Nacional Tecnología Agricultura, Buenos Aires.

CELULOSA ARGENTINA. 1973. Libro del Arbol: Essencias Forestales Indígenas de la Argentina de Aplicación Ornamental. Tomo 1. Ramos Mejía, Buenos Aires.

CLARK, A. B., AND D. S. WILSON. 1981. Avian breeding adaptation: hatching asynchrony, brood reduction, and nest failure. Q. Rev. Biol. 56:253–277.

CONTRERAS, J. R. 1979. Bird weights from northeastern Argentina. Bull. Br. Ornithol. Club 99:21–24.

DAGUERRE, J. B. 1924. Observaciones sobre la nidificación de los tordos, Molothrus brevirostris y M. badius. Hornero 3:285.

FRAGA, R. M. 1972. Cooperative breeding and a case of successive polyandry in the bay-winged cowbird. Auk 89:447–449.

FRAGA, R. M. 1977. Notas sobre la reprodución del churrinche, Pyrocephalus rubinus. Hornero 11: 380–383.

FRAGA, R. M. 1978. The rufous-collared sparrow as a host of the shiny cowbird. Wilson Bull. 90:271–284.

FRAGA, R. M. 1979. Differences between nestlings and fledglings of screaming and bay-winged cowbirds. Wilson Bull. 91:151–154.

FRAGA, R. M. 1980. The breeding of the rufous hornero (Furnarius rufus). Condor 82:58–68.

FRAGA, R. M. 1983a. The eggs of the parasitic screaming cowbird (Molothrus rufoaxillaris) and its host, the bay-winged cowbird (M. badius): is there evidence for mimicry? J. Ornithol. 124:187–194.

FRAGA, R. M. 1983b. Parasitismo de cría del Renegrido (Molothrus bonariensis) sobre el Chingolo (Zonotrichia capensis): nuevas observaciones y conclusiones. Hornero 12, No. Extraord.:245–255.

FRAGA, R. M. 1985. Host-parasite interactions between Chalk-browed Mockingbirds and Shiny Cowbirds. Pp. 829–844, In P. A. Buckley, M. S. Foster, E. S. Morton, R. S. Ridgely, and F. G. Buckley (eds.). Neotropical Ornithology. Ornithol. Monogr. No. 36.

FRIEDMANN, H. 1929. The Cowbirds: a Study in the Biology of Social Parasitism. Charles C Thomas, Baltimore, Maryland.

GIBSON, E. 1880. Ornithological notes from the neighborhood of Cape San Antonio, Buenos Aires. Ibis, 4th Ser., 4:15.

GIBSON, E. 1918. Further ornithological notes from the neighborhood of Cape San Antonio, Province of Buenos Aires. Part I, Passeres. Ibis, 10th Ser., 6:363–415.

GRANT, C. H. B. 1911. List of birds collected in Argentina, Paraguay, Bolivia, and southern Brazil, with field notes. Part I, Passeres. Ibis, 9th Ser., 5:104–131.

GRANT, C. H. B. 1912. Notes on some South American birds. Ibis, 10th Ser., 1:273–280.

HAYES, W. 1973. Statistics for the Social Sciences, 2nd ed. Holt, Rinehart and Winston, New York.

HECTOR, D. P. 1982. Botfly (Diptera:Muscidae) parasitism of nestling aplomado falcons. Condor 84: 443–444.

HENSLER, G. L., AND J. D. NICHOLS. 1981. The Mayfield method of estimating nesting success: a model, estimators, and simulation results. Wilson Bull. 93:42–53.

HOWELL, T. R., R. A. PAYNTER, JR., AND A. L. RAND. 1968. Subfamily Carduelinae, serins, goldfinches, linnets, rose finches, grosbeaks, and allies. Pp. 207–305, In R. A. Paynter, Jr. (ed.), Check-list of the birds of the world, Vol. XIV. Museum Comparative Zoology, Cambridge, Massachusetts.

HUDSON, W. H. 1870. [Untitled letters to the Secretary of the Zoological Society of London.] Proc. Zool. Soc. Lond.: Pt. I, 87–89; Pt. I, 112–114; Pt. I, 158–160; Pt. II, 332–334; Pt. II, 545–550; Pt. III, 671–673; Pt. III, 748–750; Pt. III, 798–802.

HUDSON, W. H. 1871. [Untitled letters to the Secretary of the Zoological Society of London.] Proc. Zool. Soc. Lond.: Pt. I, 4–7; Pt. II, 258–262; Pt. II, 326–329.

HUDSON, W. H. 1872a. Notes on the habits of the churrinche (Pyrocephalus rubinus). Proc. Zool. Soc. Lond., pp. 808–810.

HUDSON, W. H. 1872b. On the habits of the swallows of the genus Progne met with in the Argentine Republic. Proc. Zool. Soc. Lond., pp. 605–609.

HUDSON, W. H. 1872c. Further observations on the swallows of Buenos Ayres. Proc. Zool. Soc. Lond., pp. 844–846.

HUDSON, W. H. 1873. Notes on the habits of the pipit of the Argentine Republic. Proc. Zool. Soc. Lond., pp. 771–772.

HUDSON, W. H. 1874. Notes on the procreant instincts of the three species of *Molothrus* found in Buenos Ayres. Proc. Zool. Soc. Lond., pp. 153–174.

HUDSON, W. H. 1876. Notes on the rails of the Argentine Republic. Proc. Zool. Soc. Lond., pp. 102–109.

HUDSON, W. H. 1920. Birds of La Plata, Vol. I. E. P. Dutton, New York. [See also the Spanish language version, Aves de La Plata. 1974. H. C. Mangonnet de Gollan and J. S. Gollan, transl. Introduction by M. Victoria. Libros de Hispanoamerica, Buenos Aires.]

MARCHANT, S. 1960. The breeding of some S.W. Ecuadorian birds. Ibis 102:349–382.

MASON, P. 1980. Ecological and evolutionary aspects of host selection in cowbirds. Unpubl. Ph.D. dissert. University of Texas, Austin.

MAYFIELD, H. F. 1975. Suggestions for calculating nesting success. Wilson Bull. 78:456–466.

MEYER DE SCHAUENSEE, R. M. 1970. A Guide to the Birds of South America. Livingstone Publ. Co., Wynnewood, Pennsylvania.

MYERS, J. P., AND L. P. MYERS. 1979. Shorebirds of coastal Buenos Aires Province, Argentina. Ibis 121:186–200.

NAROSKY, S. 1978. Aves Argentinas: guía para el reconocimiento de la avifauna bonaerense. Asoc. Ornitol. Plata, Buenos Aires.

NAROSKY, S., R. M. FRAGA, AND M. DE LA PENA. 1983. Nidificacion de las Aves Argentinas (Dendrocolaptidae y Furnariidae). Asoc. Ornitol. Plata, Buenos Aires.

ORIANS, G. H. 1980. Some adaptations of marsh-nesting blackbirds. Monogr. Popul. Biol. No. 14.

PEREYRA, J. A. 1938. Aves de la zona riberena de la provincia de Buenos Aires. Mem. Jardin Zool. La Plata 9:1–304.

RICKLEFS, R. E. 1967. A graphical method of fitting equations to growth curves. Ecology 48:978–983.

RICKLEFS, R. E. 1969. An analysis of nesting mortality in birds. Smithson. Contrib. Zool. 9:1–48.

ROTHSTEIN, S. I. 1982. Successes and failures in avian and nestling recognition with comments on the utility of optimality reasoning. Am. Zool. 22:547–560.

SCLATER, P. L., AND W. H. HUDSON. 1888. Argentine Ornithology, Vol. 1: Passeres. R. H. Porter, London.

SCLATER, P. L., AND W. H. HUDSON. 1889. Argentine Ornithology, Vol. 2. R. H. Porter, London.

SICK, H. 1958. Notas biologicas sobre o gauderio, *Molothrus bonariensis* (Gmelin) (Icteridae, Aves). Rev. Bras. Biol. 18:417–431.

SIEGEL, S. 1956. Non-parametric Statistics for the Behavioral Sciences. McGraw-Hill, New York.

SKUTCH, A. F. 1969. A study of the rufous-fronted thornbird and associated birds. Wilson Bull. 81:5–43.

SMITH, N. G. 1968. The advantage of being parasitized. Nature 219:690–694.

TRAYLOR, M. A., JR. (ED.) 1979. Check-list of the Birds of the World, Vol. VIII. Museum Comparative Zoology, Cambridge, Massachusetts.

WELLER, M. W. 1967. Notes on some marshbirds of Cape San Antonio, Argentina. Ibis 109:391–411.

WETMORE, A. 1926. Observations on the birds of Argentina, Paraguay, Uruguay, and Chile. U.S. Natl. Mus. Bull. 133.

WHITE, E. W. 1882. Notes on birds collected in the Argentine Republic. Proc. Zool. Soc. Lond., pp. 591–629.

CONSERVATION

COATS, SADIE, AND WILLIAM H. PHELPS, JR. The Venezuelan Red Siskin: Case History
of an Endangered Species 977
FFRENCH, RICHARD. Changes in the Avifauna of Trinidad 986
GARRIDO, ORLANDO H. Cuban Endangered Birds 992
HILTY, STEVEN L. Distributional Changes in the Colombian Avifauna: A Preliminary
Blue List 1000
RAPPOLE, JOHN H., AND EUGENE S. MORTON. Effects of Habitat Alteration on a Tropical
Avian Forest Community 1013

THE VENEZUELAN RED SISKIN: CASE HISTORY OF
AN ENDANGERED SPECIES

SADIE COATS AND WILLIAM H. PHELPS, JR.

Sociedad Venezolana de Ciéncias Naturales, Caracas, Venezuela, and *Department of Ornithology,
American Museum of Natural History, Central Park West at 79th Street,
New York, New York, 10024 USA (*mailing address)

ABSTRACT. Excessive trapping for the cagebird trade during the early 20th century placed the Red Siskin, *Carduelis cucullata,* a Neotropical cardueline finch, in grave danger of extinction. 1981 estimates indicated only 600 to 800 individuals in scattered small populations in the western and central regions of the northern cordilleras of Venezuela, and trapping of these populations still continues. Capture, sale, and export of Red Siskins has been illegal in Venezuela since the 1940's, but protection has not succeeded because trapping occurs in remote regions difficult to police, and the birds are smuggled out of Venezuela to nearby Curaçao. Red Siskins are protected under CITES, but the Netherlands is not a signatory party, hence CITES regulations do not apply in Curaçao or Holland. Red Siskins are semi-nomadic, use many types of habitat including scrub woodland and evergreen forest, and feed on a variety of seeds and fruits. The main breeding season is May–July, with a secondary peak in November–December. Management recommendations include further field studies, establishment of national parks or reserves where Red Siskin populations are numerically strongest, and cooperation by the Netherlands and Curaçao governments in stopping the smuggling of endangered animals into Curaçao.

RESUMEN. La excesiva captura para el comercio de aves de jaula a comienzos del siglo XX puso al cardenalito, *Carduelis cucullata,* un pinzón neotropical, en grave peligro de extinción. Las estimaciones de 1981 indican que hay sólo entre 600 y 800 individuos dispersos en pequeñas poblaciones en las regiones del oeste y central de las cordilleras del norte de Venezuela y las capturas aún continuan en esas poblaciones. La captura, venta y exportación de *C. cucullata* en Venezuela es ilegal desde la década de 1940, pero la protección no ha tenido éxito debido a que las capturas continúan en regiones remotas difíciles de controlar para la policía y las aves son sacadas de contrabando de Venezuela hacia la cercana isla de Curaçao. *C. cucullata* está protegida por CITES (Convención para el Tráfico Internacional de Especies de Flora y Fauna en Peligro de Extinción) pero Holanda no es signataria de la convención, por lo que las regulaciones de CITES no se aplican a Curaçao u Holanda. *C. cucullata* es seminómade, usa muchos tipos de habitats incluyendo bosques arbustivos y perennes y se alimenta de una variedad de semillas y frutas. La principal temporada de reproducción es mayo–julio, con un pico secundario en noviembre–diciembre. Las recomendaciones de manejo incluyen más estudios de campo, establecimiento de parques nacionals y reservas donde las poblaciones de *C. cucullata* son numéricamente más grandes, y cooperación de los gobiernos de Curaçao y Holanda, para detener el contrabando hacia Curaçao, de esta ave en peligro de extinción.

Public awareness of the vulnerability of the natural environment has resulted from cases of extinction or near-extinction of wild species due to human activities. In presenting this case study of the Red Siskin (*Carduelis cucullata*), an endangered Neotropical cardueline finch, we hope to contribute to that awareness and to point out some persisting problems in the regulation of international trade in wildlife. We summarize what little is known about this species, briefly trace the history of its exploitation for the cagebird trade, describe the measures that have been taken to preserve it, and conclude with some recommendations for further action.

The Red Siskin's decline was caused by excessive trapping for the cagebird trade, which reflects the ability of this species to hybridize with the domestic canary (*Serinus canarius*) and produce fertile offspring of various reddish or coppery colors. As the supply of Red Siskins has diminished, international avicultural demand for the species has pushed the price to nearly one thousand dollars per individual in the world markets. Twenty-five years ago, William H. Phelps, Sr., noted (pers. comm.) that if a Red Siskin could be bought for one hundred dollars, and if there were one million canary fanciers in the world, then there was a pressure of one hundred million dollars on the bird. Today there are more canary breeders, the price has risen considerably, and the number of wild Red Siskins is very small. The pressure is now so great that as a wild species, this bird is probably doomed to extinction.

DISTRIBUTION

The limited geographical distribution of *C. cucullata* is an important factor in its vulnerability. Now essentially a Venezuelan endemic (Phelps and Phelps 1963), earlier it was also found on Monos Island (Chapman 1894), Gasparee Island, and Trinidad (Belcher and Smooker 1937; ffrench 1973), populations which apparently have been extirpated, and south of Cúcuta, Norte de Santander, Colombia (specimens in the Smithsonian Institution). It was introduced (presumably) into Cuba, where its present status is unknown, and into Puerto Rico, where a small, very local, breeding population persists (Raffaele 1983; Coats, pers. observ. 1982).

Within Venezuela, it is very patchily distributed in the piedmontane zone (ca. 280–1300 m), in three separate regions of the northern cordilleras (Fig. 1). The western region includes sites in the states of Lara, Falcón, Portuguesa, Yaracuy, Trujillo, and Mérida; the central region includes small areas in Guárico, Miranda, Anzoátegui, the Distrito Federal, and probably Aragua; the eastern region includes parts of Sucre and Monagas. Historically Red Siskins were much more continuously distributed and much more abundant in each of these regions. They are semi-nomadic, regularly occurring in different localities and habitats at different times of year.

CURRENT STATUS

This section summarizes a study made in Venezuela by the senior author in 1981–1982 (Coats 1982). Specific localities are omitted from published reports to try to protect the few surviving populations from increased exploitation.

In January–May 1981, in collaboration with Antonio Rivero M. of the Instituto Pedagógico, Barquisimeto, Eduardo Lara, Coats surveyed localities where Red Siskins were reported to have occurred. In all, over 70 localities in the western and central regions of Venezuela were visited; the species was found only at five of these localities during the survey period and at a sixth during August, all in the central region. These sightings probably represent six different local populations. Reports from birdcatchers indicate four additional local populations in this region. All are small; extrapolation from the population density in the area of the natural history study (see below) to the area available to each of these local populations gives an estimate of approximately 250 to 300 *C. cucullata* in the central region.

The species was not encountered (except in cages) in the western region during this survey, but Rivero later found it at three localities and received reports of its occurrence in other places. Its distribution in western Venezuela is still under study by Rivero, and it is likely that a few more local populations exist there than in the central region. Western Venezuela has a greater extent of suitable habitat than the central region, but it is also more heavily trapped. Our rough estimate for the western region in 1981 was 350 to 500 birds.

The eastern region has not been studied. Birdcatchers say that the species has been extirpated there, but the matter deserves investigation. Parts of Sucre and Monagas where Red Siskin habitat occurs remain relatively undeveloped at present.

NATURAL HISTORY

STUDY AREA

The natural history of *C. cucullata* was studied in the central region from June 1981 to March 1982. Observations were concentrated in a single area ca. 15 km by 10 km where the movements of the local population were mapped and its ecology studied. The center of the study area was about 65°55′W, 9°57′N, along the border of the states of Guárico and Miranda on the southern edge of the Serranía del Interior in the Cordillera de la Costa. This area encompassed two quite distinct habitat zones. Rising abruptly from the seasonally dry llanos (elev. = ca. 220 m), the southern escarpment of the Serranía is covered with dry deciduous woodland and shrubby grassland. Above approximately 650 m the wetter cordilleran climate dominates, resulting in mixed deciduous and evergreen forest, with a few cafetals, small gardens, and clearings for grazing stock. The narrow zone of transition between these two

FIG. 1. Distribution of *Carduelis cucullata* in South America. Stippled areas represent regions from which the species has been reported since 1940. Black square shows location of study area (Coats 1982).

climatically distinct zones has little of its original vegetation, being now a settlement with approximately 25 families spread along a road distance of 4 km.

HABITAT AND DIET

Carduelis cucullata uses a variety of habitats, including dry deciduous woodland, mixed deciduous forest, evergreen forest, and the savannah-forest ecotone. Breeding areas lie in the moist forests of the higher (750–1300 m) zone. During the post-breeding period, the birds travel many kilometers daily, often feeding in the lower zone but usually moving up the mountainsides (above 650 m) to communal roosts in the evenings. In this study in northern Guárico and southern Miranda, the local population, ca. 30 birds, used or traveled through some 5600 hectares of the 15,000 hectare study area.

Red Siskins were observed feeding on dry seeds and fleshy fruits of five species from three plant families: *Urera baccifera* (Urticaceae); *Cordia currasavica* (Boraginaceae); *Trixis divaricata, Eupatorium odoratum*, and *Wedelia caracasana* (Compositae). Twenty other species in fourteen families are said by local birdcatchers to be food plants (Coats 1982). Most of these fruits and seeds are available for only a limited period each year, and the birds feed on more than one plant species much of the time. When *Urera baccifera* is in fruit, it seems to be the preferred food.

REPRODUCTION AND MOLT

The main breeding period is from May through early July; a second period occurs in November and December, but many fewer juvenile birds are seen in January and February than in August and September. Nesting of wild birds was not observed, but data from captive breeding indicate that a single nesting takes about 45 days from the beginning of nest-building until the young birds begin to feed themselves. It is likely that only one brood is reared each breeding period. Birdcatchers say the nest is placed in clumps of *Tillandsia usneoides* hanging from tall trees (25 m or more). In captivity, the female constructs a cup-shaped nest in a partially enclosed nest chamber if that is available. The usual clutch is 3 to 5, incubation commencing with the laying of the last or next to last egg. The female incubates, and the male feeds her throughout the 11 to 13 day incubation period. The diet of wild-bred nestlings has not been observed, but many cardueline species rear nestlings primarily on vegetable foods (Newton 1973), and captive-bred Red Siskins develop normally when reared on a varied diet of seeds and fruit.

Fledging occurs 14 to 16 days post-hatching. Family groups appear to stay together to forage and roost for several weeks after the young have left the nest. During the first month after fledglings appeared in the population, the average number of immature birds per group for all apparent family groups (i.e., an adult pair or a single adult with one or more fledglings) was 1.4. This does not take into account those pairs that produced no young.

In the non-breeding season, *C. cucullata* is gregarious. The observed roost trees were in the lower part of the cordilleran zone, but other, undiscovered roosts were used on stormy evenings.

The annual pre-basic molt begins in late July and continues approximately 10 to 12 weeks. A pre-nuptual molt is unknown.

PREDATION

No predation on *C. cucullata* was observed, but 31 species of raptors (22 Falconiformes and 9 Strigiformes) were found on the study area. On the basis of known food habits, habitat preferences, and times of activity, 20 of these species are considered to be possible predators of Red Siskins, and five species very likely are: *Leptodon cayanensis, Buteo nitidus, Falco femoralis, Falco sparverius,* and *Glaucidium brasilianum.*

Sixteen species of snakes were found on the study area, of which seven, including *Boa constrictor, Epicrates cenchria,* and *Spilotes pullata* are known predators on birds and nests. Other snake species, many arboreal bird predators, also occur in the Red Siskin's breeding habitat, but were not observed during this study.

If the Siskins' nests are placed in clumps of *Tillandsia usneoides* hanging from high branches, as birdcatchers indicate, mammal predation is probably negligible.

COMPETITION

Potential competitors for food plants included 25 species of tanagers and fringillids, most of which were seen to feed on one or more plants also eaten by *C. cucullata.* The only overt agonistic interaction observed between Red Siskins and another species involved a male *Tachyphonus rufus* chasing a pair of adult Red Siskins from a stand of *Urera* and then taking their place feeding on the *Urera* fruits.

HISTORY OF EXPLOITATION

CAPTURE

Most information about birdcatching practices and about the distribution and abundance of *C. cucullata* in Venezuela during this century has come from birdcatchers, either directly to the authors or via notes made by W. H. Phelps, Sr. in the records of the Colección Ornitológica Phelps.

A standard technique is used to catch Red Siskins. A caged Red Siskin, the *pitador,* used to attract the wild birds with its calls, is put where wild Red Siskins are expected to come to forage or to drink, and sticks or stiff wires coated with sticky *Ficus* sap are stuck into the ground or vegetation nest to the cage. Small birds perch on these, become stuck, and are transferred to a cage.

Another, less favored, method is the use of a *trampajaula,* a trap-cage. A central compartment contains the *pitador,* and the spring-loaded tops of two side compartments are opened upward, propped by a delicately positioned piece of wood. When a bird perches on this piece of wood, the wood slips and the door springs shut, trapping the bird.

These methods have been quite successful. It is said that thousands of Red Siskins were trapped yearly during the first third of the twentieth century. It was necessary to catch many more than one planned to sell because of high mortality of captured birds. Many perished during the initial post-capture period, and many more during the long trip to foreign markets.

A veteran birdcatcher for more than 65 years (despite the fact that it has been illegal for many years), whose statements were echoed by other birdcatchers and by campesinos, described recent changes in the practice of hunting Red Siskins. During the early 20th century, Red Siskins were not usually hunted from the onset of the rainy season in March until after the main breeding season, because the areas used by the birds during this period were generally inaccessible. Most captures occurred from July to September when flocks of adults and fledglings fed in certain accessible areas at the base of the mountains. Young birds are preferred because they adapt to captivity more easily than adults, and because of their potentially longer life span. Juvenile Red Siskins that survive the first year after capture may live to 10 or 11 years in captivity. A bird that has been captured after it has acquired its full adult plumage cannot be aged, hence its remaining lifespan is unknown. Previously, most females captured were released. Males were preferred for their brilliant colors and more vigorous singing (although females are also excellent singers), and only males were hybridized with canaries.

In recent years new roads have made many breeding areas accessible, and birds are now taken at all times of year. Females are not released and are being hybridized with canaries, although not all Red Siskin hens will accept a canary male. Professional birdcatchers do not often hunt Red Siskins, because other birds can be caught in greater quantity to produce a steady income. Nevertheless, because of the siskins' high market value, it is worthwhile for bird dealers to buy them from those who trap them part time for extra income. For some of the latter, trapping Red Siskins is akin to sport, and the difficulty only heightens the challenge. With jeeps and pickup trucks, the trappers are mobile and visit many localities, following the predictable movements of the siskins. They also buy birds from campesinos to resell for high profits. Many of these part time birdcatchers are "Canarios," immigrants to Venezuela from the Canary Islands, where there is a long tradition of keeping Red Siskins to hybridize with canaries.

THE MARKET

We do not know if Amerindians in pre-colonial Venezuela kept this species, but it was a common cagebird there after the Spanish arrived, and almost certainly was first exported by ships traveling from Venezuela to Spain's other colonies, including the Canary Islands. Canaries were imported into Venezuela from early colonial times, and the initial discovery that these species would hybridize may have resulted from the Venezuelan practice of keeping them together in aviaries.

The Red Siskin was made known to science in 1820 when a color plate and description of a male adult *Carduelis cucullata* were published in *Zoological Illustrations, or Original Figures and Descriptions of New, Rare, or Interesting Animals* (Swainson 1820: pl. 7). *Cucullata* means hooded, hence, in England it was called the Hooded Siskin until recently. Swainson stated that he had seen only one example of the species, the possession of an Englishman who had received it along with other rarities from "the Spanish Main" (Swainson 1820: pl. 7).

The Red Siskin was generally unknown in the European cagebird trade during the 1800's. Although the British Museum (Natural History) has a specimen collected in 1867 from Carúpano, Venezuela, and one collected in 1857 labeled "Trinidad," there is little or no mention of this species in the English or American avicultural literature of the 19th century. A major encyclopedic work on aviculture, *The Illustrated Book of Canaries and Cage-birds, British and Foreign* (Blakston et al. 1877–1880), mentions neither the Red Siskin nor the "red canary" even though two chapters are devoted specifically to the topic of crossing canaries with other species, especially siskins, and there is an extensive discussion of color varieties. One of the authors later published a note (Wiener 1903) stating that he had acquired a live Red Siskin in 1877, but that he had not included it in his section on foreign cagebirds because it was so rare.

By the early 1800's, however, it was known in the Canary Islands where it was soon being crossed with domestic canary. In 1902, articles were published in which it was stated that on Tenerife this siskin was commonly crossed with the canary to produce a very attractive "mule" (Astley 1902a, b). Fertility of the hybrids was not mentioned. Small numbers of Red Siskins were soon being bred in captivity and crossed with canaries in many countries (Amsler 1912; Delacour 1917; Astley 1920; Hopkinson 1920); hence it was soon discovered that some of the F_1 males were fertile (Sich 1928).

Reproduction of cagebred birds was very low and did not meet the demand for breeding stock, which was provided by importation from Venezuela, often through the Canary Islands. Usually a siskin cock was crossed with a canary hen. Red Siskin hens were difficult to acquire; it was said that they were rarely exported from Venezuela. The price for a male Red Siskin in the Canary Islands in 1920 was £5 (Hopkinson 1920), but because this species was a poor traveler and difficult to keep alive through the long period of acclimatization to the northern regions, the price in London was undoubtedly much higher.

Among canary breeders, both commercial and amateur, there has long been great interest in producing new genetic varieties, especially "color canaries." Red canaries are highly prized, but so far as is known, they are produced only by the introduction of genetic red factors from the Red Siskin. During the 1920's, scientific experiments on the genetics of plumage color of the canary, using hybridization with species of *Carduelis,* including *cucullata,* showed that several new color varieties could be produced by using different color strains of canaries to make the crosses and back-crosses (Promatova 1930; Duncker 1930). The fertility of the hybrids was soon being discussed in avicultural journals (Lukes 1932; Amsler 1935; Bennett 1935; Ewins 1936). The hybrids tended to inherit the canary size and song, increasing their appeal to canary fanciers. One breeder, however, tried to use hybridization to produce a more cage-hardy Red Siskin rather than a red canary (Amsler 1935). It is a sad paradox that if the time comes when the Red Siskin exists only as a cage species, its natural genotypes may be lost because of such crosses, even though they spring from appreciation of the siskin.

In the early 20th century, interest in birdkeeping and commerce in cagebirds grew concomitantly. The popularity of the canary, with its many varieties, increased enormously when laws prohibiting the taking of native birds were passed in many European countries and in the United States. Thus, as Europe and the United States protected their native birds, an increased demand was created for the Red Siskin, which at that time had no such protection.

PROTECTION OF THE RED SISKIN

By the early 1940's the rarity of the Red Siskin and its persecution by birdcatchers had become a source of concern to Venezuelan ornithologists William Phelps, Sr. and Jr., who communicated their concern to colleagues in the Ornithology section of the Sociedad Venezolana de Ciéncias Naturales (SVCN). The matter was placed before the Ministerio de Agricultura y Cría (MAC), which at that time had governmental responsibility for wildlife; between 1944 and 1947, MAC issued resolutions prohibiting sale and export of the Red Siskin. These measures intended to protect the species had an unforeseen and unfortunate result. The world market was alerted to the precarious situation of the Red Siskin, and demand for them rose enormously, causing the price to increase manyfold, a stimulus to trappers and dealers. Because this was the first time a species had been protected under Venezuelan law, the skilled enforcement needed was lacking. Many ways were found to smuggle birds out of Venezuela, with the usual destination being the nearby island of Curaçao, a colony of the Netherlands. Because it was not native to any Netherlands territory, the Red Siskin was not regulated by law in Curaçao and could be openly shipped from there. It is difficult to verify the actual extent of smuggling, but birdcatchers in Venezuela claim hundreds, even thousands, of Red Siskins were sent to Curaçao annually in the 1940's and 1950's.

MAC continued its efforts to protect the Red Siskin in the late 1940's. Several communications were sent to the Ministry of Hacienda, urging vigilance in preventing illegal export of the birds, and to forestry officials of the state of Falcón, urging use of all legal means available to restrict the trapping of the Red Siskin. An Executive Resolution categorically forbidding the capture of birds that were endangered or threatened was prepared by MAC, but it was shelved for reconsideration because it would have been very unpopular (Muñoz-Tébar 1952). Venezuelans have a long tradition of keeping cagebirds, and it was certain that there would be a vigorous protest.

That resolution was of marginal importance anyway, because the biggest threat to the Red Siskin was not the domestic (Venezuelan) market, but the world market. It was clear that international protection was required. Therefore, the Species Survival Commission of the International Union for the Conservation of Nature and Natural Resources (IUCN) was invited to meet in Caracas to consider the problem. At this meeting in September 1952, William Phelps, Sr., addressed the General Assembly about the Red Siskin and its need for protection at an international level. The main markets for this siskin were assumed to be the United States, the Canary Islands, and Europe, where there were large numbers of canary breeders. International recognition of its endangered status, and enforcement of import prohibitions could greatly diminish the flow of Red Siskins into these countries, and as a result, the number being smuggled out of Venezuela might decrease. The IUCN agreed to include *Carduelis cucullata* on its list of endangered species, an important first step toward protection by international treaty.

A few months later, Ricardo Muñoz-Tébar, of SVCN and the Colección Ornitológica Phelps, published an article about the Red Siskin's situation and the efforts being made to protect it, in the magazine "El Farol," a Venezuelan publication of the Creole Petroleum Corporation (Muñoz-Tébar 1952). He stated that in past decades, the Red Siskin was common in bird shops in Venezuela and was often shipped to foreign markets in lots of 500, but that now it was exceedingly rare. This article was one of the first attempts to make an educated and influential sector of the public aware of a threat to part of Venezuela's avifauna.

Other efforts during the 1950's and 1960's to make the public aware of and sympathetic to the plight of the Red Siskin included the election of a National Bird, but the siskin lost to the Troupial (*Icterus icterus*).

The Venezuelan government underwent extensive reorganization during those two decades. The Ministerio del Ambiente y de los Recursos Naturales Renovables (MARNR) was created to manage the environment and was given responsibility for protection of the fauna and flora, regulation of hunting and commerce in wildlife, and maintenence and administration of National Parks. Preparation of new laws pertaining to wildlife was begun, culminating in the current Ley de la Proteccion de la Fauna Silvestre (Law for the Protection of Wildlife), issued in 1970. Under command of the Fuerzas Armadas de Cooperación, a new enforcement branch, the Guardería Ambiental y de los Recursos Naturales Renovables, was created to police National Parks and enforce laws pertaining to the environment.

Despite the new laws and new enforcement branch, Red Siskins continued to be smuggled out of Venezuela. The remote localities from which they were trapped were generally unknown except to birdcatchers and local campesinos, and laws regulating wildlife were virtually unenforceable (and often unknown) in these areas.

From 1950 to 1970 was also a time of great activity in the international conservation movement. A treaty regulating international commerce in endangered species was being prepared. This document, the Convention on International Trade in Endangered Species of Wild Fauna and Flora (CITES) was submitted to various countries for ratification in the early 1970's, and took effect for the signators in 1977. CITES had three categories for species covered by its regulations, listed in its Appendices 1–3. The most stringent controls were directed toward the species in danger of extinction, Appendix 1, to which *Carduelis cucullata* was added 1 July 1975.

CITES offers protection to endangered species only in the context of international transport and trade. Within the boundaries of a country, or between various territories of one country, CITES restrictions do not apply. Many countries, including the United States, have passed strict laws governing the internal transport and trade of endangered wildlife, native and nonnative. In June 1976, *Carduelis cucullata* was listed as endangered under the United States Endangered Species Act.

More than 70 countries have now signed the CITES treaty and have established regulatory procedures for import and export of species it covers. Unfortunately, the Netherlands is not among them. Neither in the Netherlands, nor in its overseas territories, is commerce in Red Siskins restricted. Birds smuggled out of Venezuela or Colombia can be taken openly into Curaçao and shipped from there to Holland where they are used to produce hybrid red canaries. The hybrids can be sent to other countries, subject only to regulations for cagebirds. It is said that large scale canary breeding operations in Holland can now make use of as many Red Siskins as can be smuggled out of Venezuela. Because of the proximity of Curaçao to the western region of Venezuela, from which most captive Red Siskins come, birds can be smuggled in ways that are very difficult to detect. They need only be hidden for the brief time that it takes to go through Venezuelan customs and for the airplane or ship to depart; then they can be removed from hiding places. They are also transported by small private boats that can cross from the Port of La Guaira to Curaçao in a few hours. So few of these birds are left in the wild that even this piecemeal form of smuggling constitutes a heavy burden for the species.

In addition to governmental action, efforts to help the Red Siskin have come from private organizations. International Council for Bird Preservation (ICPB) in conjunction with the Species Survival Commission of the IUCN publishes the Aves volume of the Red Data Book, in the second edition of which *Carduelis cucullata* was included (King 1979).

In 1978 FUDENA, the Venezuelan branch of the World Wildlife Fund, commissioned one of its investigators, Jose Laiz B., to gather data on captive breeding of Red Siskins in Venezuela. Although this was not a field study, Laiz asked birdbreeders in Venezuela about their sources of the Red Siskin, and his report (unpublished) was the first information on its present distribution in the wild.

The organization that has had the most significant role in the effort to preserve the Red

Siskin is SVCN, which, in conjunction with ornithologists working with Colección Ornitológica Phelps, brought the bird's peril to the attention of the Venezuelan government, the ICBP, and the IUCN. The President of SVCN for more than 25 years, Ramón Aveledo, has actively worked for greater public and governmental concern in Venezuela for endangered species, especially the Red Siskin. Despite concern about this species, it had never been studied and very little was known about it. Therefore, in 1980 SCVN, with a grant from MARNR, sponsored a 13 month field study (Coats 1982). Much remains to be learned about the Red Siskin in Venezuela, but it is hoped that the findings of that study will enhance protection efforts and will stimulate further investigations.

RECOMMENDATIONS FOR CONSERVATION

1. Creation of one or more national parks or protected natural reserves, of size and location such that the total area used by a population throughout the year would be protected, is crucial. Areas where the largest Red Siskin populations occur are rich in wildlife, hence other species, including jaguars and other spotted cats, caimans, parrots, and macaws also would benefit. Economic development and rapid increase in human population in northern Venezuela make it imperative to act quickly if these areas, with their rich fauna and flora, are to be preserved for posterity.

2. Further studies, especially radiotelemetric, are needed to provide more precise information on daily and seasonal movements of local populations, and to increase the autecological data base. These studies should be made in an area being considered for a reserve, in order to provide information necessary for setting adequate boundaries for the reserve.

3. Efforts in Venezuela to raise public concern and enlist support for protection of endangered native fauna should be continued and expanded.

4. The cooperation of the Netherlands government and its participation in CITES must be sought. As long as Curaçao can be used as a port of trade by unscrupulous animal dealers, it poses a clear threat to Red Siskins and other species endangered by the exotic pet trade, particularly because of its proximity to mainland sources of those animals.

ACKNOWLEDGMENTS

We wish to thank the following persons for their assistance in various ways during the research reported in this paper: R. Aveledo H., P. Fuster, R. Muñoz-Tébar, A. Rivero M., R. F. Smith, G. J. Nelson, R. S. Bottome, D. Conde, D. Bolívar, K. Phelps, M. L. Goodwin, B. Hicks, A. Arzola and family, F. Canache and family, L. Pérez, R. Pasquier, W. B. King. Financial support for the field study was provided through a grant to the Sociedad Venezolana de Ciéncias Naturales (SVCN) from the Venezuelan Ministerio del Ambiente y de los Recursos Naturales Renovables, and SVCN provided essential administrative support. The following institutions made available collections, libraries, and archives that were essential to this project: American Museum of Natural History, British Museum (Natural History), Coleccion Ornitologica Phelps, Smithsonian Institution, New York Botanical Gardens. M. Foster, R. Ohmart, and R. Pasquier provided very helpful criticism of an earlier draft of this paper.

LITERATURE CITED

AMSLER, M. 1912. Breeding of the Hooded Siskin, *Chrysomitris cucullata.* Avic. Mag. (Ser. 3), 4:51–54.
AMSLER, M. 1935. Fertility of the Hooded Siskin and canary hybrid. Avic. Mag. (Ser. 4), 13:229–232.
ASTLEY, H. D. 1902a. The Hooded Siskin and the wild canary. Avic. Mag. (Ser. 1), 8:123–124.
ASTLEY, H. D. 1902b. The Hooded Siskin. Avic. Mag. (Ser. 2), 1:47–51.
ASTLEY, H. D. 1920. Miscellanies. Avic. Mag. (Ser. 3), 11:116.
BELCHER, C., AND G. D. SMOOKER. 1937. Birds of the colony of Trinidad and Tobago, Part 6. Ibis (Ser. 14), 1:504–550.
BENNETT, C. B. 1935. Note on the red color of birds. Roller Canary Journal and Bird World, Feb. 1935, pp. 33 ff.
BLAKSTON, W. A., W. SWAYSLAND, AND A. F. WIENER. 1877–1880. The Illustrated Book of Canaries and Cage-Birds, British and Foreign. Cassell, Peter, Galpin, London.
CHAPMAN, F. 1894. On the birds of the island of Trinidad. Bull. Am. Mus. Nat. Hist. 6:1–86.
COATS, S. 1982. The distribution and natural history of the Cardenalito, *Carduelis cucullata,* in Venezuela. Unpubl. report submitted to Soc. Venez. Cienc. Nat., Caracas.
DELACOUR, J. 1917. Notes on my birds at Villers-Bretonneux in 1916. Avic. Mag. (Ser. 3), 8:69–75.
DUNCKER, H. 1930. Fettsfarbstoffvererbung bei Kanarienvoegeln im Lichte der Goldschmidtschen physiologischen Vererbungstheorie. Z. Indukt. Abstammungs. Vererbungsl. 54:267–271.

EWINS, J. 1936. Colored canaries. Aviculture (Ser. 3), 6:9–10.

FFRENCH, R. P. 1973. A Guide to the Birds of Trinidad and Tobago. Livingston Publ. Co., Wynnewood, Pennsylvania.

HOPKINSON, E. 1920. Hooded Siskin mules. Avic. Mag. (Ser. 3), 11:150.

KING, W. B. (comp.) 1979. Red Data Book, Vol. 2: Aves. Int. Union Conserv. Nat., Nat. Res., Morges, Switzerland.

LUKES, W. L. 1932. The thistle finches, pt. 2. Aviculture (Ser. 2), 4:249–254.

MUÑOZ-TÉBAR B., R. 1952. El Cardenalito: un ave en peligro. El Farol 43:20–22.

NEWTON, I. 1973. Finches. Taplinger Publ. Co., New York.

PHELPS, W. H., SR., AND W. H. PHELPS, JR. 1963. Lista de las aves de Venezuela y su distribucion, Vol. 1, Pt. 2, Passeriformes, 2nd ed. Bol. Soc. Venez. Cienc. Nat. 24:(104, 105).

PROMATOVA, A. N. 1930. Hybridization of fringillidae. Biol. Abstr. 4:640. (orig. published 1928 in Russian: J. Biol. Exp., Ser. A, 4:30–64.)

RAFFAELE, H. 1983. The raising of a ghost—*Spinus cucullatus* in Puerto Rico. Auk 100:737–739.

SICH, H. L. 1928. Fertile hybrids. Avic. Mag. (Ser. 4), 6:300–301.

SWAINSON, W. 1820. Zoological Illustrations, Vol. 1. Baldwin, Craddock, and Joy, London.

WIENER, A. F. 1903. The Hooded Siskin. Avic. Mag. (Ser. 2), 1:115–116.

CHANGES IN THE AVIFAUNA OF TRINIDAD

RICHARD FFRENCH

Texaco Trinidad Inc., Pointe-a-Pierre, Trinidad

ABSTRACT. During the 20th century, the economy of Trinidad has changed from an agricultural to an industrial one, with accompanying urbanization. Drainage of swampland, human encroachment into both marshland and foothills, the development of cattle ranches and the neglect of traditional cocoa and citrus estates have brought about significant ecological change. Human population increase has led to both pollution and some over-exploitation of wild areas through tourism.

In this century the Horned Screamer and the Blue-and-yellow Macaw have been extirpated from Trinidad, populations of six seed-eating species, mostly *Oryzoborus* and *Sporophila* spp., have been decimated by trappers, and the Scarlet Ibis, first encouraged to breed by the establishment of a sanctuary, has recently ceased to do so, possibly because of mismanagement of the sanctuary. Species showing range expansions include Cattle Egret, Southern Lapwing, Yellow-headed Caracara, White-tailed and Pearl Kites, Olivaceous Cormorant, Tropical Mockingbird, and Azure Gallinule. The wintering Dickcissel is present some years in enormous numbers, but entirely absent other years.

RESUMEN. Durante el siglo XX la economía de Trinidad ha cambiado de agrícola a industrial, acompañada por desarrollo urbano. El drenaje de tierras pantanosas, la invasión humana a las áreas pantanosas y a las colinas, el desarrollo de ranchos ganaderos y el descuído de las plantaciones de cacao y cítricos, ha producido cambios ecológicos significativos. El incremento de las poblaciones humanas ha producido cotaminación, así como sobreexplotación de áreas silvestres para el desarrollo turístico.

En este siglo el *Anhima cornuta* y el *Ara ararauna* han sido extirpados de Trinidad, poblaciones de seis comedores de semillas, mayormente *Oryzoborus* y *Sporophila* spp., han sido destruídas por los cazadores que usan trampas y en un principio se fomento la reproducción del Ibis (*Eudocimus ruber*) por medio del establecimiento de un sanctuario, pero recientemente ha dejado de reproducirse, probablemente debido al mal manejo del sanctuario. Las especies que muestran expansiones en sus rangos de distribución son la garcita bueyera (*Bulbulcus ibis*), *Vanellus chilensis*, *Milvago chimachima*, *Elanus leucurus*, *Gampsonyx swainsonii*, *Phalacrocorax olivaceus*, *Mimus gilbus*, y *Porphyrula flavirostris*. El invernante *Spiza americana* está presente durante algunos años en grandes cantidades, pero otros años está completamente ausente.

Considering its small size (4543 sq. km), the island of Trinidad in the West Indies has attracted abundant attention from ornithologists. The earliest contributors mainly provided lists of specimens collected (Taylor 1864; Leotaud 1866; Chapman 1894; Hellmayr 1906), but occasionally observations were added concerning the status of certain species.

Important contributions on other aspects of bird life histories came from Williams (1922), Belcher and Smooker (1934, 1937), and Roberts (1934). The first attempt at a synthesis of knowledge of Trinidad's birds came from Herklots (1961), whose work included descriptions of plumage and voice, and accounts of nesting, status, habitat, and range.

Since the late 1950's, a number of professional scientists (notably D. W. Snow, B. K. Snow, and C. T. Collins) have advanced our knowledge of the birds of the island considerably. In addition, the increasing influx of visitors interested in watching and studying birds has provided a wealth of observations which have assisted in determining the status and distribution of many species. My own experience with the local avifauna began in 1956, and the majority of bird records in Trinidad since that time have been either my own or have been sent to me by other observers for publication (ffrench and ffrench 1966; ffrench 1977, 1980, 1981, 1983; Ramcharan et al. 1982; ffrench and Manolis 1983).

Any account of individual species must acknowledge the possibility that impressions about abundance or scarcity may reflect the greater number of observers in recent, as opposed to earlier years, rather than a real change in species status. Nevertheless, several significant changes can be clearly discerned, and in some cases the causes of these changes are apparent.

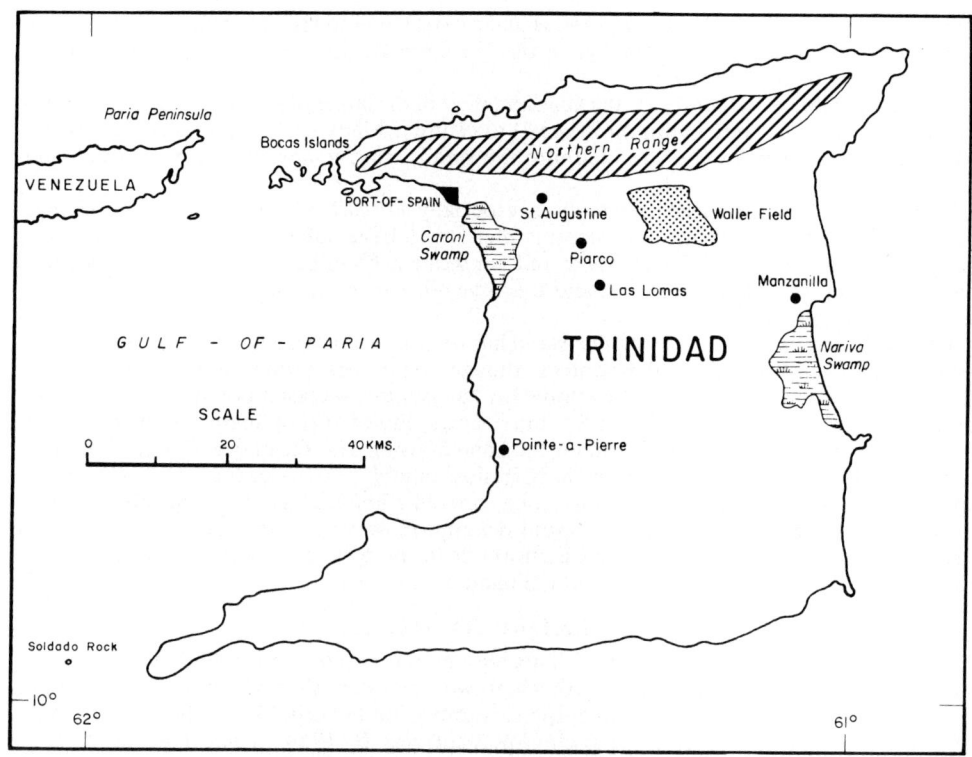

FIG. 1. The island of Trinidad, showing localities mentioned in the text.

ECOLOGICAL CHANGES IN TRINIDAD DURING THE 20TH CENTURY

The island has undergone very considerable changes during the past eighty years. The almost entirely agricultural economy has been replaced by an industrial one, accompanied by a vast population increase and consequent urbanization.

Attempts have been made, with varying success, to drain the two large swamps on the island (Fig. 1). The Caroni Swamp (present area 8120 hectares) extended in the 1920's well to the east of its present boundaries, and much of the adjoining marshland figured prominently in the area studied by Smooker as Nelson's Estate (Belcher and Smooker 1934). However, a reclamation scheme begun about 1921 and continuing fitfully into the 1940's resulted in the creation of a long dike separating the salt-water mangrove area from the fresh-water marsh to the east. This latter area was gradually drained, so that by the early 1960's it resembled a savannah. Interestingly, as a result of the inability of State authorities to maintain the dike and sluice-gates, the area has, during the past 12 years, again been subject to salt water infiltration, and the mangroves have recently extended their range considerably eastward.

While no similar drainage scheme has diminished the somewhat larger Nariva Swamp, it has nevertheless been affected by encroachment from local farmers with more modern agricultural techniques. Hunting and trapping of birds in the area has also been facilitated by improved transportation methods.

Another feature of development of the lowlands has been the extension of savannah and ranch land. Large areas of secondary forest and scrubland were cleared for airports at Waller Field and Piarco, and although the former was abandoned when the United States forces left soon after 1945, it was not long before most of the area was cleared again for cattle-raising. Other formerly cultivated estates at Las Lomas and Manzanilla have also been converted into pasture.

In the hilly districts forests have inevitably been opened by roads connecting villages formerly reached only by rural trails. Quarrying has taken a heavy toll in the foothills of the Northern Range, whereas lowland forests in south Trinidad have been much dissected by oilfield operations and the roads connecting them. Where roads go, shifting cultivators have followed, chipping gradually away at the island's extensive forests, especially in the foothills. Thus, the lower reaches of the valleys in the Northern Range are now largely occupied by small farms or secondary scrub.

Another notable development is the abandonment of the larger plantations planted in cocoa, coffee, or citrus, that were common in the 19th century. Many such estates have reverted to secondary forest or have been divided into the small, more temporary farms of local land-holders.

Urbanization has increased enormously, especially in lowland areas. Significant changes have occurred in waterways; large streams and rivers have suffered from increased silting because of the removal of forest cover and consequent flooding. Extensive new reservoirs have been created at three locations, and the large oil refinery at Pointe-a-Pierre is served by sizeable man-made lakes.

During the last twenty-five years, the island has been affected increasingly by the authorities' attempt to manage the natural resources, though sometimes serious mismanagement has resulted. For example, industrial development has greatly increased pollution of the waters draining into and through the Caroni Swamp near the city of Port of Spain. Conflicts between the needs of the increasing population for food and housing, and the desire of conservationists to protect the natural environment have been unavoidable. Tourism, focused largely on the spectacular Scarlet Ibis (*Eudocimus ruber*), has proved a financial asset that may have helped to preserve the ibis's habitat in the Caroni Swamp. However, mismanagement of tourists' visits to the swamp may have been a significant factor in driving the wary ibis away to breed in a less disturbed environment on the mainland.

EXTIRPATIONS

Anhima cornuta. Horned Screamer. This species was resident on Trinidad in the fresh-water Nariva Swamp at least until 1910, when two birds were shot by a hunter (Belcher and Smooker 1934). According to Smooker (pers. comm.), hunters still "knew the species" in the 1930's, one of them even describing to him a nest and eggs. By 1934 the screamer was "verging on extinction." I have no recent records of this species, except for an uncorroborated hunter's report from 1964. On the basis of extensive fieldwork in Nariva during 1981–1982, I conclude that the species is now extirpated from the island of Trinidad.

Ara ararauna. Blue-and-yellow Macaw. This species, confined to the vicinity of Nariva Swamp, has always been uncommon. However, I saw flocks of up to 15 birds in 1959, and pairs until about 1970. Reports during the 1970's have been uncorroborated, and it now seems certain that this spectacular bird has ceased to exist in the wild on Trinidad.

Throughout the last several decades, the pet trade in macaws has continued unabated with little more than token protest from the authorities. Many hundreds of these macaws have been imported illegally from Venezuela through Trinidad's southwestern peninsula, often to be re-exported abroad. Between October 1979 and June 1980, at least 125 Blue-and-yellow Macaws were exported to the United States (Roet et al. 1981). I strongly doubt if any of these were native to Trinidad, for the local population had been virtually eliminated by trappers some years earlier.

SPECIES ADVERSELY AFFECTED BY HUMAN ACTIVITY

Eudocimus ruber. Scarlet Ibis. Although the species seems to have been fairly numerous from early days, estimates of its numbers have varied from the euphoric, somewhat journalistic accounts of "many thousands darkening the sky" to much more conservative figures. For the first half of the century the ibis could be legally hunted in an open season, but, perhaps realizing the tourist potential, the government created a sanctuary in Caroni Swamp in 1953, following which the species bred successfully for 10 years. Careful counts of the birds (ffrench and Haverschmidt 1970) during 1963 to 1965 revealed an adult population of about 5000, and it is possible that at peak times since then, as many as 10,000 ibis have inhabited the Caroni Swamp.

However, breeding of the ibis in Trinidad became sporadic during the late 1960's and ceased altogether about 1972. At the same time the species changed its dispersal pattern. When the ibis bred on Trinidad, birds arrived on the island in February, nested some time between April and September, and dispersed to the continent thereafter. During the 1970's, the pattern was reversed. The ibis population clearly has been breeding on the mainland, probably in the Orinoco delta, during April through July. Thus, it now is scarce on Trinidad at that time.

After the breeding season, large numbers of immature birds begin to arrive, followed in due course by the adults. Thus, by December numbers have reached an annual peak; the average of nine "Christmas counts" during 1971 to 1980 is 5700, including immature birds.

The cause of the ibis's desertion of Trinidad as a breeding location is not certain. It may be significant that irregularity in breeding coincided with a pronounced increase in tourist traffic to Caroni Swamp beginning in the late 1960's. The Scarlet Ibis is very wary at the nest (ffrench and Haverschmidt 1970). Thus, large boat-loads of tourists, often noisy and knowing little of birds' habits, may well have prevented the ibis from breeding, even though the disturbance was not sufficient to frighten them from the nightly roost.

FINCHES AND SEEDEATERS

These form another group of birds that have suffered extensively from human persecution on Trinidad.

Carduelis cucullata. Red Siskin. This species has been practically unknown on the island since the 1920's. Never common, it was much prized by trappers and aviculturists and was rarely seen outside a cage. It has not been recorded at all during the last 20 years and is also exceedingly scarce on the mainland of South America (W. Phelps, Jr., pers. comm.).

Oryzoborus maximiliani. Great-billed Seed-Finch. This species and the Lesser Seed-Finch (*O. angolensis*) have also been much sought as cage-birds. The larger species was not common in Belcher and Smooker's day (1937), but nests were still found, and these authors termed *angolensis* a "widely distributed resident." At present, both are extremely rare, the former very possibly extirpated, in response to the high prices paid for caged specimens and to the gradual encroachment by humans into the remoter parts of their habitat. Both species are imported from the continent, a practice again ignored by the authorities. It cannot be said that *Oryzoborus* habitat on Trinidad has disappeared, so it seems likely that the scarcity of these two species is a direct result of uncontrolled trapping.

Sporophila spp. Five species of this genus are known from Trinidad; *S. schistacea* is a migrant from South America. *Sporophila intermedia, S. lineola,* and *S. nigricollis* were once fairly widespread (Belcher and Smooker 1937), but all are now quite rare. Numbers of the latter two may be increased by birds dispersing from the continent during the later months of the year, but no doubt exists that all these species have suffered from excessive persecution by bird-trappers. The smallest species, *S. minuta,* was quite common in 1957 and was almost never trapped, because it was "too small to bother with." Nowadays, trappers hunt for *minuta* on a regular basis, since they have few suitable alternatives. Attempts by conservation organizations within Trinidad to encourage action by the government to protect these finches and seedeaters have met with little success. Apparently cultural traditions in Trinidad favor the right of the rural people to exploit the animals in their environment. No heed is paid to the fact that the worst exploitation comes from the entrepreneur, whose sole motive is profit, or to the likelihood that without prompt action to control the situation several species will soon be extinct.

RANGE EXPANSIONS

Several species appear to have benefited from the opportunity to exploit new territory, resulting from human activity.

INCREASES ASSOCIATED WITH RANCHES OR DEFORESTATION

I will only allude in passing to the well-known population explosion of the immigrant Cattle Egret (*Bubulcus ibis*). This species seems to have arrived on Trinidad about 1951. It was common by the end of that decade, and is now the commonest and most conspicuous heron on the island. Undoubtedly it has benefited from the development of ranch-lands in the Waller Field area.

Another species that has exploited the ranch development is the Southern Lapwing (*Vanellus chilensis*). It was first recorded on Trinidad in 1961 (ffrench and ffrench 1966). More and more individuals were recorded, and finally breeding was confirmed at Waller Field in 1976. The species is now widespread in small numbers throughout Trinidad, usually associating with cattle on rough pasture land.

Similarly, there has been an increase in population of the Yellow-headed Caracara (*Milvago chimachima*). First recorded on Trinidad in 1942 (Herklots 1961), this caracara has been seen much more commonly since the early 1960's. Although it ranges more widely than the previous species, it has been found most regularly at the large cattle ranches in the eastern part of the island. Although breeding has not yet been confirmed, the presence of several immature birds indicates that birds probably do breed on the island.

Two other raptors seem to have extended their range somewhat to include Trinidad. The White-tailed Kite (*Elanus leucurus*) has apparently benefited from recent deforestation in Latin America (Eisenmann 1971). Although it cannot be said that deforestation in Trinidad has had a significant effect on the kite's available habitat here, it is a fact that the species, hitherto considered a rare visitor, has not only been more frequently recorded, especially during the last 10 years, but has also been found to breed on at least two occasions (Ramcharan et al. 1982).

The Pearl Kite (*Gampsonyx swainsonii*), another inhabitant of open woodlands and savannah, has also recently been found to breed on Trinidad on a regular basis (ffrench 1982). In this case the birds are exploiting an abundant food supply, consisting largely of lizards, that is available at an extensive suburban housing development at Pointe-a-Pierre, with well-kept gardens, a golf course, open grasslands, and many large ornamental trees. Before the 1960's, this species was considered an occasional visitor from the mainland.

INCREASES ASSOCIATED WITH URBANIZATION

One species to benefit from human settlement and development is the Olivaceous Cormorant (*Phalacrocorax olivaceus*). Whereas Belcher and Smooker (1934) found it an occasional visitor to the mangroves of Caroni Swamp, it was not recorded elsewhere on Trinidad. I have found it present in small numbers from December to July in the various mangrove swamps. At the Pointe-a-Pierre reservoirs, which cover some 85 hectares, this cormorant has increased its numbers enormously. During the late 1950's, when I made regular counts of roosting birds beside the reservoirs, the average total was 40. By 1972 average numbers were near 200; the present roosting population approximates 2000. By day the majority flies to the nearby coast to feed, but all return to roost in bamboo and trees bordering the reservoirs. Without doubt the oil company's policy of protecting wildlife has made the area more attractive for cormorants. During the breeding season, a few birds remain, whereas the majority migrate to the Orinoco delta region; it would not be surprising if cormorants were to breed at Pointe-a-Pierre in the near future.

Urbanization, at least where plenty of gardens adjoin the houses, also seems to suit the Tropical Mockingbird (*Mimus gilvus*). In the early years of this century it seems to have been unrecorded (Chapman 1894; Hellmayr 1906; Cherrie 1906, 1908; Williams 1922). Roberts (1934) found it restricted to St. Augustine, while Belcher and Smooker (1937) found it only in the populated strip between Port of Spain and St. Augustine. Twenty years later it had colonized residential areas in south Trinidad and had become common in urban Port of Spain and its environs. It is now found all over the island wherever sizeable human settlements occur, even several kilometers up the valleys of the Northern Range, along the east coast, and on the islands of the Bocas, off the northwestern peninsula.

OTHER INCREASES

Porphyrula flavirostris. Azure Gallinule. During the last five years, I have collected a considerable number of sight records (from responsible observers) of this species. All have come from two localities in the Nariva Swamp area. The species was not recorded on Trinidad before July 1978, although many ornithologists visited or worked in the Nariva Swamp during the previous twenty years. Blake (1977) mentioned that the species is spottily distributed, but locally abundant. It is known from Delta Amacuro, the nearest district of Venezuela to the south of Trinidad, about 80 km from the Nariva Swamp. There is no record of migration in this species, so this may represent a range expansion. However, its larger congener, *P. martinica*, is still the commonest gallinule in the Nariva Swamp.

Spiza americana. Dickcissel. This species is well-known for shifting its breeding-grounds from time to time (Gross 1956; A.O.U. Check-list 1957). This irregular behavior seems to be shown also in the species' wintering range (ffrench 1967). Although I found very large flocks on Trinidad between December and April from 1958 to 1966, earlier ornithologists such as Williams (1922), Belcher and Smooker (1937), and Junge and Mees (1958), did not encounter the species at all during the period 1920 to 1953. Moreover, Dickcissels began to diminish in numbers about 1972, and by 1975 seemed to be entirely absent through the winter. Since then, the only record of the species on Trinidad has been a flock of 50 birds present in February

1982. The occurrence of Dickcissels on Trinidad may depend on the extent of the dry season in Venezuela during the later months of the year, but this correlation has not been proven.

LITERATURE CITED

AMERICAN ORNITHOLOGISTS' UNION. 1957. Check-list of North American Birds, 5th ed. Port City Press, Baltimore, Maryland.

BELCHER, C., AND G. D. SMOOKER. 1934. Birds of the colony of Trinidad and Tobago. Pt. 1. Ibis (Ser. 13), 4:572–595.

BELCHER, C., AND G. D. SMOOKER. 1937. Birds of the colony of Trinidad and Tobago. Pt. 6. Ibis (Ser. 14), 1:504–550.

BLAKE, E. R. 1977. Manual of Neotropical Birds, Vol. 1. University Chicago Press, Chicago, Illinois.

CHAPMAN, F. M. 1894. On the birds of the island of Trinidad. Bull. Am. Mus. Nat. Hist. 6:1–86.

CHERRIE, G. K. 1906. A collection of birds from Aripo, Trinidad. Brooklyn Inst. Bull. 1:188–191.

CHERRIE, G. K. 1908. Further notes on a collection of birds from Trinidad. Brooklyn Inst. Bull. 1:353–370.

EISENMANN, E. 1971. Range expansion and population increase in North and Middle America of the White-tailed Kite. Am. Birds 25:529–536.

FFRENCH, R. P. 1967. The Dickcissel on its wintering grounds in Trinidad. Living Bird 6:123–140.

FFRENCH, R. P. 1977. Some interesting bird records from Trinidad and Tobago. J. Trinidad Field Nat. Club 1977:9–10.

FFRENCH, R. P. 1980. A Guide to the Birds of Trinidad and Tobago, 3rd ed. Harrowood Press, Newtown Square, Pennsylvania.

FFRENCH, R. P. 1981. Some recent additions to the avifauna of Trinidad and Tobago. J. Trinidad Field Nat. Club 1981:35–36.

FFRENCH, R. P. 1982. The breeding of the Pearl Kite in Trinidad. Living Bird 19:121–131.

FFRENCH, R. P. 1983. Further notes on the avifauna of Trinidad and Tobago. J. Trinidad Field Nat. Club 1983:32–34.

FFRENCH, R. P., AND M. FFRENCH. 1966. Recent records of birds in Trinidad and Tobago. Wilson Bull. 78:5–11.

FFRENCH, R. P., AND F. HAVERSCHMIDT. 1970. The Scarlet Ibis in Surinam and Trinidad. Living Bird 9:147–165.

FFRENCH, R. P., AND T. MANOLIS. 1983. Notes on some birds of Trinidad wetlands. J. Trinidad Field Nat. Club 1983:29–31.

GROSS, A. O. 1956. The recent reappearance of the Dickcissel (Spiza americana) in eastern North America. Auk 73:66–70.

HELLMAYR, C. E. 1906. On the birds of the island of Trinidad. Novit. Zool. 13:1–60.

HERKLOTS, G. A. C. 1961. The Birds of Trinidad and Tobago. Collins, London.

JUNGE, G. C. A., AND G. F. MEES. 1958. The avifauna of Trinidad and Tobago. Zool. Verh. (Leiden) No. 37.

LEOTAUD, A. 1866. Oiseaux de l'ile de la Trinidad. Chronicle Press, Port of Spain, Trinidad.

RAMCHARAN, E. K., G. DE SOUZA, AND R. P. FFRENCH. 1982. Inventory of the living resources of coastal wetlands in Trinidad. Inst. Marine Affairs, Chaguaramas, Trinidad.

ROBERTS, H. R. 1934. List of Trinidad birds with field notes. Trop. Agric. Trinidad 11:87–99.

ROET, E. C., D. S. MACK, AND N. DUPLAIX. 1981. Psittacines imported by the United States. (October 1979–June 1980). Pp. 21–45, In R. F. Pasquier (ed.), Conservation of New World Parrots. ICBP Tech. Publ. No. 1. Smithsonian Institution Press, Washington, D.C.

TAYLOR, E. C. 1864. Five months in the West Indies. Pt. 1. Ibis 6:73–97.

WILLIAMS, C. B. 1922. Notes on the food and habits of some Trinidad birds. Bull. Dep. Agric. Trinidad Tobago 20:123–185.

CUBAN ENDANGERED BIRDS

ORLANDO H. GARRIDO

Dirección Nacional de Flora y Fauna, Ministerio del Transporte, Habana, Cuba

ABSTRACT. Although Cuba has strict hunting regulations and protective laws, as well as an excellent system of preserves, some species and races of Cuban birds are endangered. Here, I discuss the current status of these and bring the Red Data Book up-to-date for Cuban birds. Most of the endangered races are those inhabiting keys off the coast of Cuba. Endemic Cuban species are, as a whole doing well and benefitting from newly established preserves. Two endemics, the Cuban Kite and the Zapata Wren are very rare and, perhaps, nearly extinct. The wren may recover now that its Zapata Swamp habitat is protected. Gundlach's Hawk is more common than reported by earlier workers. The Bachman's Warbler, although recently sighted, is essentially extinct as a non-breeding bird on Cuba.

RESUMEN. A pesar de que Cuba posee regulaciones estrictas para la caza y leyes de protección, así como un excelente sistema de reservas, algunas especies y razas de aves cubanas están en peligro de extinción. En el presente trabajo discuto la situación actual de estas aves y actualizo el Libro Rojo de Datos ("Red Data Book") para las aves cubanas. La mayoría de las razas que están en peligro de extinción son aquellas que habitan en los cayos fuera de la costa de Cuba. Las especies endémicas cubanas, en general, se encuentran bien, y aprovechando del nuevo sistema de reservas establecido. Dos especies endémicas, el Gavilán Caguarero y la Fermina son muy raros y posiblemente cercanos a su extinción. La población de la Fermina podría recuperarse ahora que la ciénaga de Zapata está protegida. El Gavilán Coliargo es más común ahora que lo que habían manifestado otros autores anteriormente. La Bijirita de Bachman, aunque ha sido vista recientemene, se le puede considerar virtualmente extinguida, aunque no cría en Cuba.

Most countries are proud of their natural heritage and are concerned about the conservation of their natural resources. Many countries have created programs to protect these natural resources, including the creation of sanctuaries, natural reservations, and national parks that provide refuges for plants and animals. Cuba, during the past two decades, has increased its efforts to protect its native flora and fauna. Several reservations have been created and many more are planned, supported by protective laws, forest guards, and staff that guarantee that these areas shall function effectively for conservation.

In Cuba, birds are the most diverse and important faunal element; the mammalian fauna is meager, and other groups are not as well known as are the birds. Garrido and García Montaña (1975) reported 380 avian forms for the Cuban Archipelago. Some small keys and islets harbor endemic races with limited distributions; many of these have been described only recently and do not appear in the Red Data Book for Birds (King 1981) even though some are more threatened than some mainland Cuban birds. Four compilations of endangered Cuban birds have been published in the last 21 years, mainly by Bond who has largely relied upon my information (Bond 1961, 1968, 1978; King 1981).

Here, I add to the list of known endangered birds, including those species or subspecies that are threatened with decline because they are very locally distributed or vulnerable through habitat loss. This information is derived from my own observations over the last 20 years. I follow the Catalogue of Cuban Birds (Garrido and García Montaña, 1975), updated to include an unpublished supplement covering the last seven years.

SPECIES ACCOUNTS

Pterodroma hasitata. Black-capped Petrel, Pajaro de las Brujas. A colony was discovered in December 1976 by Nicasio Vinas in the southern coastal slopes of Sierra Maestra, near a place called "Las Brujas." Some birds were collected as they approached the coast at dusk. A definite breeding record has not been obtained, but the noises made by the birds at night were probably from breeding individuals and nestlings. The status of the birds is unknown, but the population undoubtedly represents a new breeding colony for this rare petrel, which was also recently discovered breeding on Hispaniola (Bond 1956). The Cuban breeding site is almost unreachable by people, so the site is safe. This species may also breed on Trinidad, near Playa Yaguanabo, where George Reynard and I were told of nocturnal sounds emanating from a perfect hillside breeding habitat. In Cuba, the whole Sierra Maestra territory has been selected as a protected area.

Dendrocygna arborea. West Indian Tree Duck, Yaguasa. I include this species because it is listed in the Red Data Book (King 1981). In Cuba, this species has declined, but it now seems to be recovering in response to restrictions on hunting. About 15 years ago, tree ducks were shot in great numbers to protect rice plantations, but now *D. arborea* is completely protected and is found locally in swamps, along coasts, and even on keys, in good numbers.

Chondrohierax willsonii. Cuban Kite, Gavilan Caguarero. This species is the rarest of Cuban hawks and is declining. Formerly, it was more widely distributed; Gundlach (1876, 1893) reported it from central Cuba (Bahia de Cochinos), and the type came from western Cuba in Oriente Province. Now, it is only found between Baracoa and, perhaps, Sierra de Toa in eastern Cuba. Since its purported food, the beautiful snail *Polymita picta,* is restricted to eastern Cuba, the kite must have eaten other species of snail in central Cuba, from which *Polymita* is absent. I have only seen the Cuban Kite once, near Duaba Arriba (Baracoa), as it flew over a river, chased by Cuban Crows (*Corvus nasicus*). Since we know little about its habits, biology, or exact range, little can be done to protect it. Some consider the Cuban Kite conspecific with *Chondrohierax uncinatus.* I will consider *C. wilsonii* a full species until more is known about its biology.

Rostrhamus sociabilis plumbeus. Snail Kite, Gavilan Caracolero. The Snail Kite was much more common during the last century but gradually declined and disappeared from some areas due to unknown causes. It was always common around Treasure Lake (Zapata Swamp), where it is still common and surely breeds. I have seen this species in nearly every province: near Cortes in Pinar del Rio, El Laguito and Laguna de Ariguanabo in Havana, Treasure Lake in Matanzas, the suburbs of Finca El Dorado (Isabela de Sagua), near Sancti-Spiritus (El Jibaro), and on Cayo Santa Maria (Archipiélago de Sabana-Camaguey), Cayo Romano, and Cayo Coco. It is also found in several areas in the southeastern regions of Santiago de Cuba. The species is increasing in numbers now and is fully protected by law, as are all of other species of Falconiformes.

Accipiter striatus fringilloides. Sharp-shinned Hawk, Halconcito. This small hawk has never been considered endangered, but for the past twenty years at least, it has become increasingly rare. The Sharp-shinned inhabits forests at middle elevations; it only occasionally is seen in lowland forests. Most birds observed at sea level are probably of the North American race, *A. s. velox,* for which a few specimens exist. The race *A. s. fringilloides* lives in pairs. I have seen it at 600 m on Pico Turquino (Sierra Maestra) but not at higher elevations in these mountains. The eggs of this species have been found only once, near Monte Alto (Sagua La Grande). Jorge de la Cruz (pers. comm.) has seen it nesting in the mountains of Trinidad.

This species has suffered with the degradation of forests. Although it is reported from the Isle of Pines (Isla de la Juventud), I have not seen it there, but it may occur in the northern section on Sierra de Casas and Sierra de Caballos.

Accipiter gundlachi. Gundlach's Hawk, Gavilan Colilargo. Barbour (1943:32) considered this hawk one of the rarest in the world. I have not found this to be true and, in fact, believe that the species is not endangered. This is not a specialized hawk, for it frequents forest borders, swamps, wooded coasts, and mountains below 800 m elevation. Little is known about its life history. Villalba obtained the only egg known from the oviduct of a bird collected near Artemisa (Pinar del Rio; Bond 1936). Gundlach (1876) reported seeing nestlings in the Zapata Swamp, and James Clements (pers. comm.) found a nest high in a casuarina tree along the channel to Treasure Lake in the Zapata Swamp. Recently, Carlos Wotskow (pers. comm.) found a nest with two eggs in Sopillar, a wooded area near Playa Larga (Zapata Swamp). On a second visit the nest contained two nestlings. Rogelio Garcia Arencibia (pers. comm.), a resident of the area, told me he had seen the nest in the same place for three consecutive years. I have also seen and collected this hawk in Casilda (Trinidad), Gibara (west of Holguin), and Cupeyal (Sagua de Tanamo).

I was surprised to see Gundlach's Hawk on Cayo Cantiles, and recently, it was seen on Cayo Coco (Regalado 1981). James Bond (pers. comm.) believed that the Cantiles observation may have been of a migrant *Accipiter cooperii* rather than of Gundlach's Hawk. Gundlach's Hawk is closely related to the North American Cooper's Hawk (*A. cooperii*) but is larger with heavier feet and a different plumage pattern. I have examined two specimens that I consider to be migrant *A. cooperii,* one from Gibara collected by Joaquin de la Vara, and the other collected by Gundlach and in his collection. Gundlach's Hawk may be a subspecies of the northern Cooper's Hawk. Gundlach's Hawk flies low to the ground and often surprises an observer by gliding sideways away from him. It gives a repeated cackling near nest sites.

Cyanolimnas cerverai. Zapata Rail, Gallinuela de Santo Tomas. Although listed in the Red Data Book (King 1981), little has been learned about this elusive rail since its discovery by Fermin Cervera. Only James Bond succeeded in collecting a few specimens and in observing several others (pers. comm.). Bond (1974:70) described its voice, and George Reynard recorded the voice in the mid 1970's. Later, Eugene Morton obtained some additional calls. This is the only Cuban species that I have not observed in life, although I have heard it and followed tracks rustling in the sawgrass. Eugene Morton and I, in 1979, were watching a big tussock of grass where two birds were calling, and he was lucky enough to see one. A year later Hiram Gonzalez, our guide Antonio, and I cornered one bird in a large tussock of sawgrass, but the bird moved to another tussock. It passed right by Hiram's surprised eyes but, once again, my view was blocked by the grass!

The Zapata Rail was thought to be restricted to the portion of the Zapata Swamp located north of the village of Santo Tomás together with the swamp's other two famous endemics, the Zapata Wren (*Ferminia cerverai*) and the Zapata Sparrow (*Torreornis inexpectata*). When James Clements visited Treasure Lake, however, he found the rail nearby in one of the channels leading to the Guama Tourist Village. George Reynard and the guide, Rogelio García, also found the rail at Treasure Lake and photographed a young bird through binoculars. Thus, the Zapata Rail may be more widespread in the swamp than previously thought and, if no disastrous fires or droughts occur, the rail should not be in any danger. Formerly, the rail was more widely distributed than now, as is shown by the finding of fossils in Pinar del Rio (at Pica Pica) and in the northern Isle of Pines area (Olson 1974).

Grus canadensis nesiotes. Sandhill Crane, Grulla. The Sandhill Crane has always worried conservationists because of its conspicuousness, large size, and edibility (Gundlach 1893). Fortunately, Cuba has always protected the crane from hunting, and hunters seem to have abided by the law, for the crane is now more common than formerly. The Zapata Swamp harbors most cranes, with smaller flocks scattered in flat country in the northern section of the Isle of Pines (Walkinshaw and Baker 1946). I have never seen more than three birds together on that island. Small flocks have also been reported near Vinales, Playa de la Teja (Itabo), and scattered places in Camaguey (Cayo Romano, Berovides and Acosta, pers. comm.). It is rumored, but not confirmed, that the crane is found in northern Oriente Province. These scattered colonies suggest that the bird is spreading to new breeding grounds.

Ara tricolor. Cuban Macaw, Guacamayo. Bond (1956) suggested that this bird went extinct sometime between 1850 and 1925. Only Gundlach (1893) narrated some of his experiences with this macaw.

Melanerpes superciliaris. West Indian Woodpecker, Carpintero Jabado. *Melanerpes superciliaris* is polytypic, with the nominal form inhabiting mainland Cuba and Cayo Cantiles, *M. s. murceus* endemic to the Isle of Pines, *M. s. florentinoi* endemic to Cayo Largo del Sur, and *M. s. sanfelipensis* known only from Cayo Real. The first two races are common while the latter two are endangered, especially *M. s. sanfelipensis.*

Cayo Largo del Sur is a large key belonging to the populous Archipiélago de los Canarreos (Jardines y Jardinillos) located southeast of the Zapata Peninsula. The key has a diversity of habitats (Garrido 1966). Bond (1950) was unable to locate *M. s. florentinoi* there probably because he did not search in the eastern half of the key. Here, the Juraguano's Palm (*Cocothrinax* sp.) forms the dominant vegetation type, and this is preferred by this subspecies. Now that the key has become a tourist attraction, hotels and other accommodations have reduced the amount of habitat for the woodpecker. Another problem for the survival of *M. s. florentinoi* is the possible reinvasion of the key by the mainland or Cayo Cantiles populations, for this race is larger and less habitat specialized than individuals of *M. s. florentinoi.*

The subspecies *M. s. sanfelipensis* has been found only on Cayo Real, the largest and most wooded of San Felipe's keys, located southwest of La Coloma (Pinar del Rio Province). These keys are part of the western end of the Archipiélago de los Canarreos. Local residents of La Coloma have informed me that the woodpecker was more common previously, especially when coconuts were commonly grown on Cayo Real. When I last visited, few coconuts remained, although there were some *Cocothrinax.* This race, as with *M. s. florentinoi,* is smaller than the nominate race but has a color pattern different from those of *florentinoi* and *murceus.* Since this race does not occur on any adjacent keys, it is extremely vulnerable to extinction.

Colaptes fernandinae. Fernandina's Woodpecker, Carpintero Churrosa. The rarity of this species is not due to human disturbance but may be due to habitat specialization. Its populations seem stable in the limited areas where they occur. The species prefers forest edge and areas of palm mixed with other vegetation that supplies shade and shelter. It also dwells in middle elevation areas such as Sierra del Rosario (Nortey). It is more common in the center of Cuba than in the western and eastern sections, and has not been found in far western Pinar del Río nor in Zapata or the eastern end of Cuba near Baracoa. This is the only Cuban woodpecker that habitually walks on the ground as well as hops (Bond 1956). It is a terrestrial forager, searching in leaf litter among the scattered trees that offer safety. The nest hole is a meter or so above ground, usually in a dead palm. I have found these nests in Sierra del Rosario (Nortey) and near El Dorado (Isabel de Sagua). The voice is similar to that of the Cuban race of the Northern Flicker (*Colaptes auratus chrysocaulosus*) but can be differentiated.

Campephilus principalis bairdii. Ivory-billed Woodpecker, Carpintero Real. Seeing this bird is one of the great thrills still to be found on Cuba, but the species is on the verge of extinction. The Ivory-bill was more widely distributed on Cuba late in the last century, but was declining rapidly (Gundlach 1876, 1893). It was found in Pinar del Río (Cordillera de los Organos), Matanzas (Zapata Swamp), and was widespread in the eastern sections of the country. In the 1950's, it could still be found in some areas south of Moa, (e.g., La Melba), and during the 1940's, the entomologist Fernando de Zayas reported seeing an Ivory-bill nailed to a peasant's door in keeping with a local superstition. In 1978 an engineer reported (Garrido, unpubl. data) seeing an Ivory-bill near Nuevo Mundo, not far from La Melba in Oriente. I had visited there previously but had not found any sign of the bird. There have been recent rumors of the bird in Pinares de Mayari and in the Sierra Maestra, but no confirmations. Remarkably, in April 1982, a member of a group of birdwatchers led by John Rowlet and Theodore Parker, III (pers. comm.) was separated from the main group; he claimed to have seen and heard an Ivory-bill. This was near Soroa in Sierra del Rosario. The voice was recorded, and I heard it a day later after being notified of the find. I did not recognize the recording as that of any bird with which I am familiar. The Cuban guide said he had also heard this bird and even imitated its voice, succeeding in attracting the bird even closer to a companion about 200 meters away. I have explored this area before and never heard an Ivory-bill there, but the guide informed me that he had received recent news of this woodpecker from farmers long before the birdwatchers visited.

Thus far, the only official recent report of an Ivory-bill is mine. I saw a female in February 1968, a few kilometers northeast of Cupeyal near the common boundary of Sagua de Tanamo, Moa, and Guantanamo. I had climbed along a hillside path, and, when I entered an open area, I saw a bird on a dead tree ahead, pecking on the bark. I froze, and the bird stopped pecking and looked at me with its head turned sideways. It then gave a short call as it flew directly away to a wooded area on the hillside. There was little habitat for the bird in the immediate area so it must have been wandering. Not far away was a zone with pine trees (called Cayos de Pinos) that sometimes form closed canopy forests. I have also seen this habitat near Nibujon and Nuevo Mundo in the district of Moa where, according to the literature, this woodpecker occurs. Local residents told us that years ago, near Vega Grande (in the jurisdiction of Guantanamo), two birds were seen several times, then only one. There is an unexplored area in eastern Cuba northwest of Guantanamo, southwest of Moa, and west of Baracoa, where some birds may, hopefully, still exist.

Corvus palmarum. Palm Crow, Cao Pinalero. This bird is listed because it is very localized. Earlier, it was thought to exist (Bond 1956:119) near Mantua, Puerto Esperanza, and Guane (Pinar del Rio), and near Jimanayagua (Camaguey). I have not found Palm Crows in any of these areas nor in Mina Dora (Pinar del Río) where two crows were reported. The only place that I know the crow exists is around Sierra de Najasa near La Belen where I have seen four individuals. Local people there are familiar with the species, calling it "Cao Ronco" (raucous crow). Sierra de Najasa consists of low hills separated by flat land and valleys with palm groves. The Palm Crow lives in the flat lands; it is greatly outnumbered by the Cuban Crow (*Corvus nasicus*), but this is most common at higher elevations.

Ferminia cerverai. Zapata Wren, Ferminia. This is one of the three famous endemic birds of the Zapata Swamp near the town of Santo Tomás. It was discovered by Fermin Cervera, a Spanish naturalist resident in Cuba, who occassionally collected for Thomas Barbour. Barbour published on the new Zapata Swamp species collected by Cervera in 1926, but he never saw any of the Zapata endemics (*Ferminia, Torreornis inexpectata, Cyanolimnas cerverai*) in life.

After its discovery, the Zapata Wren was seldom observed or collected. Before 1960, only Stephan C. Bruner, Gaston S. Villalba, and James Bond had seen it (Bruner 1934; Bond 1936, 1956, pers. comm.). All felt the bird was common in its restricted habitat of sawgrass and woody hummocks in the swamp directly north of the small village of Santo Tomás. Barbour (1928) also commented that the bird was common there. The nest has never been found, Bond's (1936) description being based on that of a local resident or "cienaguero."

In 1962, with the late Florentino Garcia, I first visited the area, hoping to collect all three endemics. On the last day we finally collected one *Torreornis,* but on later expeditions we obtained several *Ferminia* and more *Torreornis. Ferminia* was then common and several birds could be heard singing from the channel, a narrow canal cut through the swamp from the Rio Hatiguanico to Santo Tomás. In 1974, I visited the swamp again to collect a specimen of *Ferminia.* I was surprised not to hear the bird singing as before. Only a long walk into the swamp yielded a single specimen. After a few more years, I returned to the swamp with George Reynard but could not locate a single bird. This was repeated the next year, as Reynard tried in vain to record the bird on tape. A few months later, in 1979, Eugene Morton and Storrs Olson, from the Smithsonian Institution, and I spent almost a week looking for *Ferminia.* We even ventured deeper into the swamp than any scientists had before, looking for the wren. When we again failed to find it, I told Morton that I thought the bird was extinct. In 1980, I returned to the swamp with Hiram Gonzalez to study *Torreornis* but still could not find a single *Ferminia;* I was now convinced the species was extinct. Then I was told the local people used to set fires in the swamp to collect turtles (*Chrysemys decussata*) for eating. This news provided a reason for the wren's disappearance. I also found evidence that the mongoose had recently invaded the region and saw three animals in the swamp. I wrote Bond that I was convinced the wren was gone, but he did not believe me. Fortunately, he was correct, for late in 1981, Hiram Gonzalez reported one bird, and he later tape-recorded one. In the spring of 1982, Carlos Wotzkow saw two birds during two separate trips.

Apparently, *Ferminia* withdrew because of the fires and is now slowly (hopefully) reoccupying its former haunts. It still consists of only two known birds, both probably males. We need much information on this wren before its relationships with other wren genera can be ascertained. I disagree with the Greenway's (1958) proposal that *Ferminia* is congeneric with *Thryomanes.*

Mimus g. gundlachii. Bahama Mockingbird, Sinsote Prieto. This species is only found on some keys, not on the mainland of Cuba. Gundlach (1893) reported it from Cayo Santa Maria, but I have not seen it there. I have found it on Cayo Lanzanillo and Cayo Tio Pepe, both north of Isabela de Sagua. It has also been found on Cayo Coco (Garrido 1976; Regalado 1981). Although this species is more common on the Bahama Islands and in Jamaica, on Cuba its small and scattered populations on keys makes it vulnerable, and it requires protection.

Myadestes elisabeth retrusus. Cuban Solitaire, Ruisenor de Isla de Pinos. The mainland race of this species is common in the mountains of both eastern and western Cuba, but the present race from the Isle of Pines is virtually extinct. Even when first described it was considered rare and restricted to the densest forests at Pasadita (Bangs and Zappey 1905). Todd (1916) believed that the bird might exist also at Hato in the southern part of the island, but I have not found it there. Bond (1956, pers. comm.) obtained his only specimen in 1934 from the edge of the Cienaga de Lanier. I searched this area and talked to local people but have not found it, although it was formerly there but never common. The only possible place for its existence is in southern highlands (Mogotes de Santa Isabel and San Juan), for the bird has not been found in any of its previously known haunts.

Vireo gundlachii sanfelipensis. Cuban Vireo, Juan Chivi. While the other three races of this vireo are common and stable, this form, restricted as it is to Cayo Real, is extremely vulnerable. *Vireo gundlachii sanfelipensis* lives only in the wooded center of the key and in coastal bushy zones. The population is small, and only three specimens were taken for the subspecific designation. It is particularly vulnerable to wood cutting and fires which could eliminate the form in a short time.

Vermivora bachmani. Bachman's Warbler, Bijirita de Pecho Negro. This species spends much more time in Cuba as a non-breeding resident than in its breeding ground in the United

States. Bachman's Warbler is considered the most endangered bird of North America but has not always been rare. Collectors could easily find Bachman's Warblers in the summer months of late July and August in the heart of Havana in a small forest area near the botanical garden during the 1930's. I did not see it there after 1960 despite many visits at all times of the year.

I have only seen this warbler three times, all before 1965. I saw a female with a flock of the Cuban endemic warbler *Teretistris fernandinae* near Santo Tomás at the Zapata Swamp in 1962. I saw another female near the mouth of Treasure Lake in the Zapata Swamp in February, 1964. My last sighting, in 1966, was in the hills of Soroa. I have not seen another, but S. Dillon Ripley (J. Bond, pers. comm.) reported one at the Zapata Swamp in May 1981.

Coereba flaveola bahamensis. Bahama Bananaquit, Reinita. This species, the Bahaman Mockingbird, and the Black-faced Grassquit (*Tiaris b. bicolor*) are Bahaman forms that have islolated populations on some of the northern Cuban keys. They have not differentiated from the races found on the Bahamas and, therefore, may only be vagrants and not a breeding population. Specimens have been collected from Cayo Tio Pepe an arid coastal area north of Isabela de Sagua and near Gibara. I suspect that some of the northern Cuban keys also may have small breeding populations.

Agelaius humeralis scopulus. Tawny-shouldered Blackbird, Mayito. While the nominate race is common on the Cuban mainland, this race exists in isolated small groups of no more than 10 individuals on Cayo Cantiles, a wooded key south of the Zapata Peninsula. Birds of this race are smaller and have longer and thinner bills than birds of the nominate form. The forests on this key are vulnerable to exploitation and should be preserved; two other endemic avian subspecies and several reptile races are found here (Garrido 1978a, b, unpubl. data).

Tiaris b. bicolor. Black-faced Grassquit, Tomeguin Prieto. This race is common on the Bahamas but in Cuba is found only on Cayo Tio Pepe. The population is small but it definitely breeds there. If discovered by local "pajareros," the populations will rapidly be eliminated by bird catchers for the cagebird trade.

Torreornis inexpectata. Zapata Sparrow, Cabrerito del la Cienaga. The Zapata Sparrow, known since 1926, has always been considered common but extremely local (Bond 1936, 1956; Ripley and Watson 1956; Garrido 1980). Nearly forty years later, a second population was discovered 750 km east of the swamp by the herpetologists Albert Swartz and Ronald Klinikowski who were collecting reptiles along the desert-like coast from Tortuguilla to Baitiquiri. Schwartz and Klinikowski (1963) collected more specimens on a return trip for a total of 10. I had difficulty finding the bird until my third trip to the area when I found it regularly.

The eastern population (*Torreronis i. sigmani*), being more difficult to find than the Zapata Swamp birds, was thought to be rarer and probably endangered (Bond 1956). However, the range of the eastern form is know known to extend farther east to Imias (La Chivera) and Cajobabo (Garrido and García Montaña, unpubl. data). This arid zone is not in danger of disturbance from farming, because it lacks aricultural potential. This dry habitat of *T. i. sigmani* is, however, in extreme danger from fire which destroys the brush which is then replaced by grasses uninhabitable by the bird (Morton and Gonzalez 1982).

A third race (*T. i. varonai*) is found on Cayo Coco (Regalado 1981). None has been found in other large keys such as Cayo Romano (Berovides and Acosta, pers. comm.). Thus, although this bird exists only in widely scattered populations, it is fairly common where found. S. L. Olson (pers. comm.) suggested that these are remnant populations and that the bird was of more widespread distribution in the pleistocene.

DISCUSSION

Most of the endangered forms of Cuban birds are those restricted to keys. A complete list is given by Silva Taboada (1974). Other rare forms are discussed in earlier papers (Garrido 1966, 1971a, b, 1973, 1978a, b; Garrido and Kreisel 1972). The keys also contain species of seabirds that were not discussed. Species nesting only on keys include *Larus atricilla, Sterna h. hirundo, S. d. dougalli, S. anaethetus recognita, S. f. fuscata, S. albifrons antillarum, S. m. maxima, S. sandwichensis acuflavidus,* and *Anous s. stolidus* (García Montaña and Garrido 1965). *Pelicanus occidentalis, Sula leucogaster, Catoptrophorus semipalmatus,* and *Coccyzus minor* also nest only on keys. The White-tailed Tropicbird (*Phaethon lepturus catesbyi*) nests on the mainland but in only two scattered colonies with few individuals each. One is in the cliffs of Cabo Cruz and the other in cliffs at Baitiquiri (Garrido and García Montaña, 1975).

The flamingo (*Phoenicopterus ruber*) has some large colonies on Cuba. I saw about 2000 birds in the Zapata Swamp at Salinas. Nesting grounds differ in location from feeding grounds and some nesting grounds were prone to human disturbance (near Nuevitas, Birama, Tunas de Zaza). The birds are now completely protected at these breeding sites.

Some species have benefitted by human intervention and have increased. These include *Bulbulcus ibis, Anas bahamensis, Dendrocygna bicolor, Colinus virginianus cubanensis, Rallus elegans, Zenaida macroura, Mimus polyglottus, Quiscalis niger, Dives atroviolaceus,* and *Agelaius h. humeralis.*

ACKNOWLEDGMENTS

First, I am very grateful to James Bond of the Academy of Natural Sciences of Philadelphia for all the advice, information, and recommendations that he has given me over the last twenty years. I also thank the editors for allowing me to contribute to Neotropical Ornithology. E. S. Morton and M. S. Foster rewrote the manuscript, but all statements are my own.

LITERATURE CITED

BANGS, O., AND W. R. ZAPPEY. 1905. Birds of the Isle of Pines. Am. Nat. 39:179–215.
BARBOUR, T. 1928. Notes on three Cuban birds. Auk 45:28–32.
BARBOUR, T. 1943. Cuban Ornithology. Publ. Nuttall Ornithol. Club No. 9.
BOND, J. 1936. Birds of the West Indies. Academy Natural Sciences, Philadelphia, Pennsylvania.
BOND, T. 1950. Results of the Catherwood-Chaplin West Indies Expedition, 1948. Part II. Birds of Cayo Largo (Cuba), San Andres and Providencia. Proc. Acad. Nat. Sci. Phila. 102:43–68.
BOND, T. 1956. Check-list of birds of the West Indies. 4th ed. Academy Natural Sciences, Philadelphia, Pennsylvania.
BOND, T. 1961. Extinct and near extinct birds of the West Indies. Pan Am. Int. Coun. Bird Preserv., Res. Rep. 4:1–6.
BOND, T. 1968. Thirteenth Supplement to the check-list of birds of the West Indies (1956). Academy Natural Sciences, Philadelphia, Pennsylvania.
BOND, T. 1974. Ninteenth Supplement to the check-list of birds of the West Indies (1956). Academy Natural Sciences, Philadelphia, Pennsylavania.
BOND, T. 1978. Twenty-second Supplement to the check-list of birds of the West Indies (1956). Academy of Natural Sciences, Philadelphia, Pennsylvania.
BRUNER, S. C. 1934. Observaciones sobre *Ferminia cerverai* (Aves: Troglodytidae). Mem. Soc. Cubana Hist. Nat. 8:97–102.
GARCÍA MONTAÑA, F., AND O. H. GARRIDO. 1965. Nuevos registros de nidificación de aves en Cuba. Poeyana, La Habana, Ser. A 9:1–3.
GARRIDO, O. H. 1966. Nueva subespecie del Carpintero Jabado, *Centurus superciliaris* (Aves:Picidae), para Cuba. Poeyana, La Habana, Ser. A 29:1–4.
GARRIDO, O. H. 1971a. Una nueva subespecie del *Vireo gundlachii* (Aves: Vireonidae) para Cuba. Poeyana, La Habana, Ser. A 81:1–8.
GARRIDO, O. H. 1971b. Variación del género monotípico *Xiphidiopicus* (Aves Picidae) en Cuba. Poeyana, La Habana, Ser. A 83:1–12.
GARRIDO, O. H. 1973. Anfibios, reptiles, y aves de Cayo Real (Cayos de San Felipe), Cuba. Poeyana, La Habana, Ser. A 119:1–50.
GARRIDO, O. H. 1976. Aves y reptiles de Cayo Coco, Cuba. Misc. Zool., Inst. Zool., La Habana 3: 1–4.
GARRIDO, O. H. 1978a. Nueva subespecie de Carpintero Verde (Aves: Picidae) para Cayo Coco, Cuba. Inf. Cient.-Tec., Inst. Zool., La Habana 67:1–6.
GARRIDO, O. H. 1978b. Nuevo Bobito Chico (Aves: Strigidae) para Cuba. Inf. Cient.-Tec., Inst. Zool., La Habana 68:3–6.
GARRIDO, O. H. 1980. Los vertebrados terrestres de la Peninsula de Zapata. Poeyana, La Habana 203: 1–49.
GARRIDO, O. H., AND F. GARCÍA MONTAÑA. 1975. Catálogo de las Aves de Cuba. Academia Ciencias Cuba, La Habana.
GARRIDO, O. H., AND H. KREISEL. 1972. 1st finding of a great northern diver *Gavia immer* on the Cuban coasts. Poeyana, La Habana, Ser. A 98:1–4.
GREENWAY, J. C. JR. 1958. Extinct and Vanishing Birds of the World. Am. Comm. Int. Wild Life Protec. Spec. Publ. No. 13.
GUNDLACH, J. 1876. Contribución a la ornitología Cubana. Imp. La Antilla, Habana.
GUNDLACH, J. 1893. Ornitología Cubana. Imp. La Habana, Habana.
KING, W. B. 1981. Endangered Birds of the World. The ICBP Red Data Book. Smithsonian Institution Press and International Council Bird Preservation, Washington, D.C.
MORTON, E. S., AND H. J. GONZALEZ ALONSO. 1982. The biology of *Torreornis inexpectata I.* A comparison of vocalizations in *T. i. inexpectata* and *T. i. sigmani.* Wilson Bull. 94:433–446.
OLSON, S. 1974. A new species of *Nesotrochis* from Hispaniola, with notes on other fossil rails from the West Indies (Aves: Rallidae). Proc. Biol. Soc. Wash. 87:439–450.

REGALADO, P. 1981. El género *Torreornis* (Aves: Fringillidae), descripción de una nueva subespecie en Cayo Coco, Cuba. Centro Agrícola 2:87–112.

RIPLEY, S. D., AND G. WATSON. 1956. Cuban bird notes. Postilla 26:1–6.

SCHWARTZ, A., AND R. KLINIKOWSKI. 1963. Observations on West Indian birds. Proc. Acad. Nat. Sci. Phila. 15:53–77.

SILVA TABOADA, G. 1974. Las especies amenazadas de vertebrados cubanos. Acad. Cienc. Cuba, Inst. Zool. 32 pp.

TODD, C. 1916. The birds of the Isle of Pines. Ann. Carnegie Mus. 10:146–296.

WALKINSHAW, L. H., AND B. W. BAKER. 1946. Notes of the birds of the Isle of Pines, Cuba. Wilson Bull. 58:133–142.

DISTRIBUTIONAL CHANGES IN THE COLOMBIAN AVIFAUNA: A PRELIMINARY BLUE LIST

STEVEN L. HILTY

5151 East Rosewood, Tucson, Arizona 85711 USA

ABSTRACT. Some important man-caused changes in the distributions of Colombian bird populations are discussed. Examination of the changes is based on a preliminary Blue List of 135 species with declining populations. Canopy frugivores dominate the list, followed by terrestrial and aquatic species, respectively. There are proportionately more highland species on the list than lowland species. Two regions, the central Colombian highlands, and the Cauca-Magdalena lowlands contain more declining species than other avifaunal regions. Almost two-thirds of Colombia's endemic species have been placed on the list. Some species have profited from deforestation or other human activities, and these are discussed.

RESUMEN. Se discuten algunos cambios importantes, causados por el hombre, en las distribuciones de poblaciones de aves colombianas. El reconocimiento de los cambios está basado en la Lista Azul preliminar de 135 especies con poblaciones en disminución. Los frugívoros del tope del bosque dominan la lista, seguidos respectivamente por especies terrestres y acuáticas. Proporcionalmente en la lista hay más especies de tierras altas que de tierras bajas. Dos regiones, las tierras altas del centro de Colombia y las tierras bajas de Cauca-Magdalena, presentan más especies en disminución que otras regiones de avifaunas. Casi dos tercios de las especies endémicas de Colombia están incluídas en la lista. Se discute también la ganancia que obtuvieron otras aves por la deforestación y otras actividades humanas.

In this paper I examine changes in the distributions of Colombian bird populations. Olivares (1970) and Lehmann (1970) discussed some effects of environmental change on birds in Colombia and mentioned examples of species that have profited or been harmed by these changes. However, neither author attempted to provide a comprehensive overview of the country's avifauna. Some of Colombia's endangered or threatened species are discussed in the International Council for Bird Preservation (ICBP) Bird Red Data Book (King 1981). Unfortunately, little other information has been compiled on distributional changes of Colombian birds. Furthermore, bird populations in Colombia (or most anywhere in developing regions of the world) are not systematically surveyed or evaluated like they are through the Blue List for North America (Arbib 1971), the Christmas Bird Counts, or more recently the natural heritage inventories of The Nature Conservancy. There is no systematic research on endangered Colombian species either by government or private agencies.

In order to examine distributional changes in Colombia's avifauna a preliminary Blue List was prepared. This list is similar in concept to that initiated in North America in 1971 by Robert Arbib (Arbib 1971). The latter list, updated annually, includes North American birds that are of special concern because they appear to be "... giving indications of non-cyclic population decline or range contractions, either locally or widespread" (Arbib 1971). Species on the list may be fairly common locally, or even widespread, but because of habitat reduction, chemical contamination, hunting pressure, or other causes, their populations appear to be in decline.

As noted by Arbib (1971, 1972) the Blue List is essentially an "early warning system" that focuses attention on problem species and may help alert the scientific community, government agencies, and the public to situations where action is needed. I include on the present list (Table 1) most species whose geographical ranges in Colombia are believed to have declined by at least 50 percent. Unlike the North American Blue List, I also include all endangered species listed by the ICBP Bird Red Data Book (King 1981) because information on these species is scarce. I exclude species known only from one or a few localities near the Colombian border, or species known only from "Bogotá" specimens without precise locality (Meyer de Schauensee 1948). I also exclude a considerable number of species whose ranges probably have contracted by 50 percent or more due to deforestation since pre-Colombian times but that still occur commonly wherever tracts of forestland remain. Examples of species in this

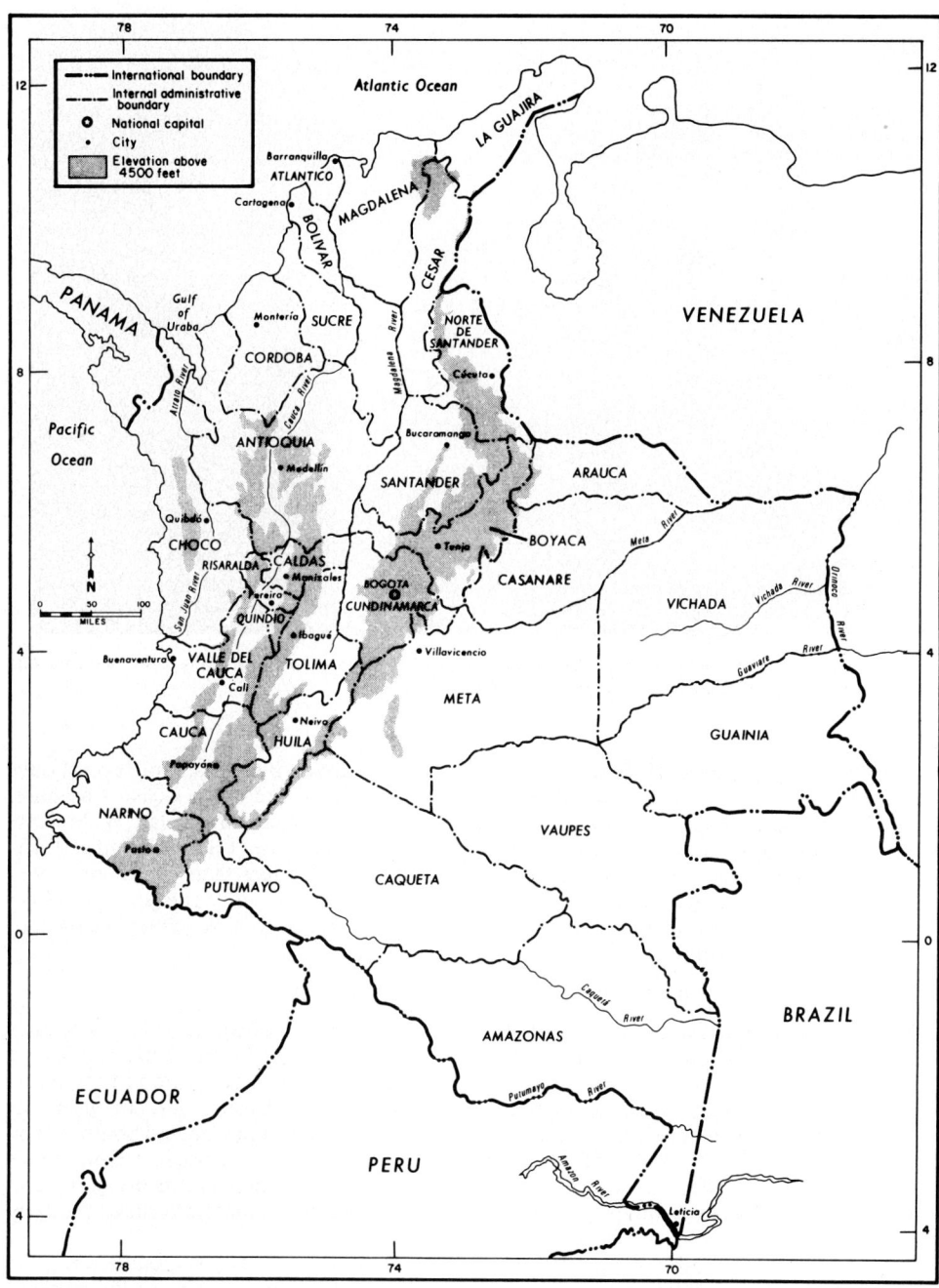

Fig. 1. Departmental map of Colombia (from American University 1977).

group are the Crimson-rumped Toucanet (*Aulacorhynchus haematopygus*), Golden-faced Tyrannulet (*Zimmerius viridiflavus*), Andean Solitaire (*Myadestes ralloides*), and Three-striped Warbler (*Basileuterus tristriatus*).

Data, photos, and locality lists used in the preparation of this paper have been compiled during the course of field work carried out in Colombia over 13 years. Initial work was undertaken from October 1971 to July 1973; subsequently I worked 1 to 5 months each year

FIG. 2.   Deforestation west of Cali, on the eastern slope of the Western Andes, 1100–1400 m elevation, Department of Valle, Colombia, 1973 (photo: S. Hilty).

from 1975 to 1984 (except 1983 when no field work was done). Field visits of several days to several weeks have been conducted in every Department in Colombia except Casanare, Risaralda, and Sucre (Fig. 1). Information from Casanare was provided by the late W. McKay. For logistical reasons more work has been conducted in Amazonas, Caquetá, Cauca, Cundinamarca, Huila, Magdalena, Nariño, Putumayo, Valle, and Vaupés than elsewhere.

Total numbers of lowland and montane species, and species in other categories used for comparison in this paper have been obtained by compiling totals from Meyer de Schauensee (1970) and Hilty and Brown (in press).

RECENT GEOGRAPHICAL HISTORY

Documentation of avian distributional changes in Colombia is of world-wide importance because Colombia has one of the largest bird faunas in the world (nearly 1700 species), it harbors more than 50 endemic species, and it is an important wintering ground for many north temperate and some south temperate zone breeding birds. Because of its geographical location, Colombia is an important contact zone between Middle American and South American avifaunas (Haffer 1974, 1975). Documentation is also important because Colombia's Andes and western valleys have been subject to greater human population pressures and habitat alteration than any other region in South America except southeastern Brazil (Ridgely 1982).

Topographically Colombia is complex. The western half of the country is mountainous, the eastern half flat. Most endemic birds, or those with restricted ranges, occur in the maze of mountains, valleys, and floodplains in the west. More than 85 percent of the total human population of nearly 30 million also resides in the mountainous west (American University 1977). In contrast, most birds east of the Andes are widely distributed, although not necessarily common, members of either the Amazonian, Guianian, or northern Llanos avifauna. Almost no birds east of the Andes in Colombia are confined to that country. Furthermore, because the number of people living east of the Andes is small, habitat alteration is minimal, although there are a few major exceptions including rapidly expanding human settlements in the Macarena-western Meta area and in western Caquetá and western Putumayo.

Human colonization of lowland tropical forest east of the Andes has been slow, but during

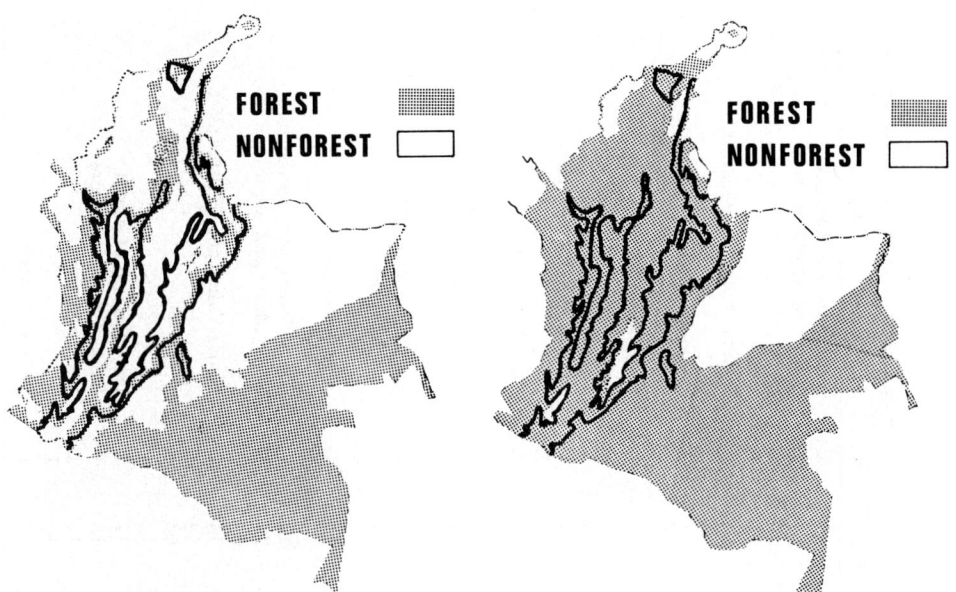

FIG. 3.   Approximate forest and nonforest zones in Colombia prior to Spanish settlement (right) and about 1980 (left). Heavy black lines are the 1000 m contour at the base of the Andes and Santa Marta Mountains.

the 1970's, the Colombian Institute for Agrarian Reform sponsored colonization projects in forested areas of Caquetá, Meta, Putumayo, and Vaupés (Shane 1980). Settlers now are occupying these regions in increasing numbers. All the lowland forest of western Caquetá from Meta to the Putumayo border has been replaced by pastureland and cattle. Deforestation in western Putumayo is less extensive so far, but the same processes are occurring. In extreme eastern Colombia deforestation is confined to limited areas near the few larger towns, e.g., Puerto Inírida, Mitú, San Jose del Guaviare, and Leticia, and also along rivers.

In the mountainous western half of Colombia the situation is very different and much more alarming. Environmental degradation, exacerbated by a high human population growth rate, has been severe for most of the twentieth century (Fig. 2). Photographs taken between 1911 and 1915 and comments in Chapman (1917) confirm that large areas of floodplain and slopes of the Cauca and Magdalena valleys, and the Eastern Andean highlands were already deforested by the beginning of the twentieth century. It is possible that significant changes may have been made by Amerindians even before the Spanish arrived in the 1500's (F. C. Lehmann, pers. comm.).

Nevertheless, areas of forest or nonforest cover have changed significantly between pre-Colombian times and the present (Fig. 3). LANDSAT photos were unsatisfactory for mapping and calculating forest and nonforest zones in Colombia because of heavy cloud cover (W. L. Brown, pers. comm.). Therefore, my estimates of forest cover (Fig. 3) are based on personal observation, reports in the literature, and Instituto Geografico Agustín Codazzi (Latorre 1967) maps from Bogotá.

COMMENTS ON DISTRIBUTIONAL CHANGE

My Blue List for Colombia (Table 1) includes 135 species of birds that have shown signs of serious population decline. Twenty-three of these are listed as endangered by the ICBP (King 1981). One of them, the endemic Cauca Guan (*Penelope perspicax*), is surely extinct. Another, the endemic Colombian Grebe (*Podiceps andinus*), may be near extinction; it has not been seen in several years. Several more are now very rare including the Southern Pochard (*Netta erythrophthalma*), Gorgeted Wood-Quail (*Odontophorus strophium*), and Yellow-eared Parrot (*Ognorhynchus icterotis*). Subspecies of several other species listed by ICBP, and many more subspecies not listed by them have certainly disappeared. Lehmann (1970) stated that

TABLE 1

PRELIMINARY BLUE LIST OF COLOMBIAN BIRDS

Taxon[1]	Distribution[2]	Elevation zone[3]	Foraging[4] guild	Body[5] size	Comments
Gray Tinamou, *Tinamus tao*	W	Tropical	6	L	very spotty distribution
Black Tinamou, *T. osgoodi*	Andes	Subtrop.	6	L	former range unknown
Highland Tinamou, *Nothocercus bonapartei*	Andes	Subtrop.	6	L	
Tawny-breasted Tinamou, *N. julius*	Andes	Subtrop.	6	L	*columbianus* and *saltuarius*
Red-legged Tinamou, *Crypturellus erythropus*[6]	Cauca-Magdal.	Tropical	6	S	
Choco Tinamou, *C. kerriae*	Chocó-Pacif.	Tropical	6	S	very rare or extinct(?)
COLOMBIAN GREBE, *Podiceps andinus*[6]	Eastern Andes	Temperate	1	L	may also occur in lowlands
Fasciated Tiger-Heron, *Tigrisoma fasciatum*	Mts.	Subtrop.	1	L	status east of Andes = ?
Jabiru, *Jabiru mycteria*	Cauca-Magdal.	Tropical	1	L	
Roseate Spoonbill, *Ajaia ajaja*	Cauca-Magdal.	Tropical	1	L	now mainly Guajira
Greater Flamingo, *Phoenicopterus ruber*	Cauca-Magdal.	Tropical	1	L	very local in Cauca Valley
Horned Screamer, *Anhima cornuta*	Cauca-Magdal.	Tropical	1	L	
Northern Screamer, *Chauna chavaria*	Cauca-Magdal.	Tropical	1	L	heavily hunted
Orinoco Goose, *Neochen jubata*	east of Andes	Tropical	1	L	
Torrent Duck, *Merganetta armata*	Andes	Subtrop./Temp.	1	L	
Southern Pochard, *Netta erythrophthalma*[6]	Andes	Temperate	1	L	probably extinct 1952
Cinnamon Teal, *Anas cyanoptera borreroi*	Andes	Temperate	1	L	also local east of Andes
and *A. c. tropicus*	(Cauca-Magdal.)	(Tropical)			
Yellow-billed Pintail, *A. georgica niceforoi*	Andes	Temperate	1	L	in serious decline
Comb Duck, *Sarkidiornis melanotos*	W	Tropical	1	L	
Muscovy Duck, *Cairina moschata*	Mts.	Temperate	14	L	
Andean Condor, *Vultur gryphus*	Mts.	Temperate	14	L	
Solitary Eagle, *Harpyhaliaetus solitarius*	Cauca-Magdal.	Subtrop.	14	L	Cauca Valley pop. extirpated(?)
Harris' Hawk, *Parabuteo unicinctus*	Mts.	Tropical	14	L	Santa Marta pop. secure
White-rumped Hawk, *Buteo leucorrhous*	Andes	Subtrop.	14	L	
White-throated Hawk, *B. albigula*	Cauca-Magdal.	Subtrop.	14	L	Cauca Valley pop. extirpated(?)
White-tailed Hawk, *B. albicaudatus*	W	Tropical	14	L	
Crested Eagle, *Morphnus guianensis*[6]	W	Tropical	14	L	
Harpy Eagle, *Harpia harpyja*[6]	Mts.	Subtrop.	14	L	
Black-and-chestnut Eagle, *Oroaetus isidori*	Andes	Subtrop.	12	L	common east of Andes
Variable Chachalaca, *Ortalis motmot columbiana*	Mts.	Subtrop.	12	L	Santa Marta pop. secure
Band-tailed Guan, *Penelope argyrotis*	Mts.	Subtrop.	12	L	
CAUCA GUAN, *P. perspicax*[6]	Andes	Subtrop.	12	L	perhaps extinct
Crested Guan, *Penelope purpurascens*	W	Tropical	12	L	
Andean Guan, *P. montagnii*	Andes	Temperate	12	L	common locally

TABLE 1
CONTINUED

Taxon[1]	Distribution[2]	Elevation zone[3]	Foraging[4] guild	Body[5] size	Comments
Wattled Guan, *Pipile aburri*	Andes	Subtrop.	12	L	quite local
Northern Helmeted Curassow, *Crax pauxi*	Catatumbo	Subtrop.	9	L	may not belong on list
Great Curassow, *C. rubra*	Chocó-Pacif.	Tropical	9	L	hunting/deforest. threaten
BLUE-BILLED CURASSOW, *C. alberti*[6]	Caribbean/Cauca-Mag.	Tropical	9	L	hunting/deforest. threaten
Yellow-knobbed Curassow, *C. daubentoni*	Catatumbo/E. Llanos	Tropical	9	L	Catatumbo heavily settled
CHESTNUT WOOD-QUAIL, *Odontophorus hyperythrus*	Andes	Subtrop.	9	S	
GORGETED WOOD-QUAIL, *O. strophium*[6]	Eastern Andes	Temperate	9	S	very rare and local
BOGOTA RAIL, *Rallus semiplumbeus*[6]	Eastern Andes	Temperate	1	S	very local
Large-billed Tern, *Phaetusa simplex*	eastern Colombia	Tropical	1	L	still common; breeding sites need protection
Yellow-billed Tern, *Sterna superciliaris*	eastern Colombia	Tropical	1	S	as previous species
Black Skimmer, *Rynchops nigra*	eastern Colombia	Tropical	1	L	as previous species
Maroon-chested Ground-Dove, *Claravis mondetoura*	Andes	Subtrop.	9	S	
TOLIMA DOVE, *Leptotila conoveri*[6]	Cauca-Magdal.	Subtrop.	9	S	little habitat remains
White-throated Quail-Dove, *Geotrygon frenata*	Andes	Subtrop.	9	S	
Blue-and-yellow Macaw, *Ara ararauna*	W	Tropical	12	L	problems esp. west of Andes
Great Green Macaw, *A. ambigua*	Chocó-Pacif.	Tropical	12	L	
Scarlet Macaw, *A. macao*	W	Tropical	12	L	problems esp. west of Andes
Red-and-green Macaw, *A. chloroptera*	W	Tropical	12	L	more numerous than *A. macao*
Chestnut-fronted Macaw, *A. severa*	W	Tropical	12	L	
Scarlet-fronted Parakeet, *Aratinga wagleri*	Andes	Subtrop.	12	L	formerly very common
Golden-plumed Parakeet, *Leptosittaca branickii*	Andes	Temperate	12	L	
Yellow-eared Parrot, *Ognorhynchus icterotis*[6]	Andes	Temperate	12	L	very rare and local
FLAME-WINGED PARAKEET, *Pyrrhura calliptera*	Eastern Andes	Temperate	12	L	still in numbers locally
Painted Parakeet, *P. picta subandina*	Cauca-Magdal.	Tropical	11	S	Perijá subsp. prob. secure
RUFOUS-FRONTED PARAKEET, *Bolborhynchus ferrugineifrons*[6]	Andes	Temperate	11	S	prob. never numerous
Spot-winged Parrotlet, *Touit stictoptera*	Andes	Subtrop.	11	S	deforestation threatens
Saffron-headed Parrot, *Pionopsitta pyrilia*	Chocó-Pac./Cauca-Mag.	Tropical	11	S	
Rusty-faced Parrot, *Hapalopsittaca amazonina*[6]	Andes	Temperate	11	S	both subspp. rare

TABLE 1
CONTINUED

Taxon[1]	Distribution[2]	Elevation zone[3]	Foraging[4] guild	Body[5] size	Comments
Bronze-winged Parrot, *Pionus chalcopterus*	Andes	Subtrop.	11	S	numerous locally
Speckle-faced Parrot, *P. seniloides*	Andes	Subtrop.	11	S	numerous locally
Red-lored Parrot, *Amazona autumnalis*	W	Tropical	12	L	deforestation threatens
Yellow-crowned Parrot, *A. ochrocephala*	W	Tropical	12	L	
Scaly-naped Parrot, *A. mercenaria*	Andes	Subtrop.	12	L	
Rufous-banded Owl, *Ciccaba albitarsus*	Andes	Subtrop.	14	S	
Buff-fronted Owl, *Aegolius harrisii*	Andes	Temperate	14	S	
Oilbird, *Steatornis caripensis*	Andes	Subtrop.	12	L	
Lyre-tailed Nightjar, *Uropsalis lyra*	Andes	Subtrop.	3	S	
WHITE-CHESTED SWIFT, *Cypseloides lemosi*	Cauca-Magdal.	Tropical	5	S	poorly known; may not belong on list
CHESTNUT-BELLIED HUMMINGBIRD, *Amazilia castaneiventris*	Cauca-Magdal.	Subtrop.	13	S	
BLOSSOMCROWN, *Anthocephala floriceps*	Santa Marta/Cauca-Mag.	Subtrop.	13	S	only Magdal. pop. threatened
Fawn-breasted Brilliant, *Heliodoxa rubinoides*	Andes	Subtrop.	13	S	
BLACK INCA, *Coeligena prunellei*[6]	Andes	Temperate	13	S	small pop. rediscovered in Boyaca, 1978
Golden-bellied Starfrontlet, *C. bonapartei*	Eastern Andes	Temperate	13	S	
Emerald-bellied Puffleg, *Eriocnemis alinae*	Andes	Subtrop.	13	S	
COLORFUL PUFFLEG, *E. mirabilis*	Western Andes	Subtrop.	13	S	known from one area
Wedge-billed Hummingbird, *Schistes geoffroyi*	Andes	Subtrop.	13	S	
Crested Quetzal, *Pharomachrus antisianus*	Andes	Subtrop.	12	L	
Golden-headed Quetzal, *P. auriceps*	Andes	Subtrop.	12	L	
WHITE-MANTLED BARBET, *Capito hypoleucus*	Cauca-Magdal.	Tropical	11	S	
Toucan Barbet, *Semnornis ramphastinus*[6]	Chocó-Pacif.	Subtrop.	11	S	much habitat remains but it is a popular cage bird
Gray-breasted Mountain-Toucan, *Andigena hypoglauca*	Central Andes	Temperate	12	L	
Black-billed Mountain-Toucan, *A. nigrirostris*	Andes	Subtrop.	12	L	
Citron-throated Toucan, *Ramphastos citreolaemus*	Cauca-Mag.	Tropical	12	L	
Powerful Woodpecker, *Campephilus pollens*	Andes	Subtrop.	15	L	
Tyrannine Woodcreeper, *Dendrocincla tyrannina*	Andes	Subtrop.	15	S	
Greater Scythebill, *Campylorhamphus pucheranii*	Andes	Temperate	15	S	

TABLE 1
CONTINUED

Taxon[1]	Distribution[2]	Elevation zone[3]	Foraging[4] guild	Body[5] size	Comments
Flammulated Treehunter, Thripadectes flammulatus	Mts.	Subtrop.	3	S	Santa Marta pop. secure
Recurve-billed Bushbird, Clytoctantes alixi	Cauca-Magdal.	Tropical	2	S	needs *dense* young regrowth
Western Antshrike, Thamnomanes occidentalis	Western Andes	Subtrop.	3	S	may not belong on list
Rufous-tailed Antthrush, Chamaeza ruficauda	Andes	Subtrop.	2	S	
Barred Antthrush, C. mollissima	Andes	Temperate	2	S	
Undulated Antpitta, Grallaria squamigera	Andes	Subtrop.	2	S	
Giant Antpitta, Grallaria gigantea	Central Andes	Temperate	2	S	no recent records
BICOLORED ANTPITTA, G. rufocinerea	Andes	Subtrop.	2	S	
MOUSTACHED ANTPITTA, G. alleni[6]	Central Andes	Subtrop.	2	S	
BROWN-BANDED ANTPITTA, G. milleri[6]	Central Andes	Subtrop.	2	S	
Rusty-breasted Antpitta, Grallaricula ferrugineipectus	Andes	Subtrop.	2	S	
Ocellated Tapaculo, Acropternis orthonyx	Andes	Subtrop.	2	S	
YELLOW-HEADED MANAKIN, Chloropipo flavicapilla	Andes	Subtrop.	10	S	
Andean Cock-of-the-Rock, Rupicola peruviana	Andes	Subtrop.	12	L	sought for cage trade
Chestnut-crested Cotinga, Ampelion rufaxilla	Andes	Subtrop.	11	S	
Red-ruffed Fruitcrow, Pyroderus scutatus	Andes	Subtrop.	12	L	
Long-wattled Umbrellabird, Cephalopterus penduliger[6]	Western Andes	Subtrop.	12	L	most habitat still remains; prob. never numerous
Bearded Tachuri, Polystictus pectoralis[6]	Andes	Temperate	16	S	common east of Andes
Yellow-throated Spadebill, Platyrinchus flavigularis	Andes	Subtrop.	3	S	Perijá pop. prob. secure
APOLINAR'S MARSH-WREN, Cistothorus apolinari[6]	Eastern Andes	Temperate	1	S	
Spot-breasted Wren, Thryothorus maculipectus	Andes	Subtrop.	3	S	
Niceforo's Wren, T. nicefori	Cauca-Magdal.	Tropical	3	S	
Crested Oropendola, Psarocolius decumanus melanterus	W	Tropical	12	L	not declining east of Andes
Black Oropendola, P. guatimozinus	Cauca-Magdal.	Tropical	12	L	
CHESTNUT-MANTLED OROPENDOLA, P. cassini	Chocó-Pacif.	Tropical	12	L	
RED-BELLIED GRACKLE, Hypopyrrhus pyrophpogaster	Cauca-Magdal.	Subtrop.	8	S	very local

TABLE 1
CONTINUED

Taxon[1]	Distribution[2]	Elevation zone[3]	Foraging[4] guild	Body[5] size	Comments
MOUNTAIN GRACKLE, *Macroagelaius subalaris*	Eastern Andes	Temperate	8	S	
Gray-throated Warbler, *Basileuterus cinereicollis*	East. Andes	Subtrop.	4	S	perhaps does not belong on list but
Black-throated Flower-Piercer, *Diglossa brunneiventris*	Andes	Temperate	13	S	habitat in range highly degraded
TURQUOISE DACNIS-TANAGER, *Pseudodacnis hartlaubi*	Andes	Subtrop.	8	S	
MULTICOLORED TANAGER, *Chlorochrysa nitidissima*	Cauca-Magdal.	Subtrop.	8	S	range highly fragmented
Masked Mountain-Tanager, *Buthraupis wetmorei*	Central Andes	Temperate	11	S	may not belong on list
BLACK-AND-GOLD TANAGER, *Bangsia melanochlamys*	Chocó-Pac./Cauca-Mag.	Subtrop.	11	S	Cauca-Mag. pop. perhaps extinct
GOLD-RINGED TANAGER, *B. aureocincta*	Chocó-Pacif.	Subtrop.	11	S	
Red-hooded Tanager, *Piranga rubriceps*	Andes	Subtrop.	8	S	
White-winged Tanager, *P. leucoptera*	Andes	Subtrop.	8	S	
Red-throated Ant-Tanager, *Habia fuscicauda*	Caribbean	Tropical	7	S	
SOOTY ANT-TANAGER, *H. gutturalis*	Cauca-Magdal.	Tropical	7	S	*erythrolaema* subsp.
CRESTED ANT-TANAGER, *H. cristata*	Western Andes	Subtrop.	7	S	
Rufous-crested Tanager, *Creurgops verticalis*	Andes	Subtrop.	8	S	
DUSKY-HEADED BRUSH-FINCH, *Atlapetes fuscoolivaceus*	Cauca-Magdal.	Subtrop.	7	S	does accept bushy young second growth
OLIVE-HEADED BRUSH-FINCH, *A. flaviceps*[6]	Cauca-Magdal.	Subtrop.	7	S	
TUMACO SEEDEATER, *Sporophila insulata*[6]	Chocó-Pac.	Tropical	16	S	sought for cage trade
Red Siskin, *Spinus cucullatus*[6]	Catatumbo	Tropical	16	S	local; Cauca Valley only
Grasshopper Sparrow, *Ammodramus savannarum caucae*	Cauca-Magdal.	Tropical	16	S	

[1] Subspecies listings denote the specific population that is declining. Names of endemic species capitalized.

[2] Andes = e slope E Andes to e slope W Andes above 1200 m elev.; Santa Marta = Santa Marta Mts. above 1200 m; Mts. = both Andes and Santa Marta Mts.; Cauca-Magdal. = Cauca, Nechí, and Magdalena river basins below 1200 m; Chocó-Pacif. = w slope W Andes and Pacific lowlands; Caribbean = dry coastal zone from Rio Sinú northeastward; Catatumbo = Catatumbo river basin, Norte de Santander; W = widespread range.

[3] Tropical = 0–1200 m elev.; Subtrop. = Subtropical, 1200–2400 m; Temperate = 2400–3600 m (to treeline).

[4] 1 = aquatic; 2 = terrestrial/semiterrestrial insectivore; 3 = lower story (below ca. 15 m) insectivore; 4 = upper story (above ca. 15 m) insectivore; 5 = aerial insectivore; 6 = terrestrial omnivore; 7 = lower story omnivore; 8 = upper story omnivore; 9 = terrestrial/semiterrestrial frugivore; 10 = lower story frugivore; 11 = small canopy frugivore; 12 = large canopy frugivore; 13 = nectivore; 14 = raptorial; 15 = bark surface gleaner (scansorial); 16 = unclassified (mostly open zone species).

[5] L = large, total length >300 mm; S = small, total length ≤300 mm (incl. *Uropsalis lyra* which has abnormally long tail).

[6] Listed in ICBP Bird Red Data Book (King 1981).

TABLE 2

LAND AREA AND NUMBER OF SPECIES AT RISK IN EACH ELEVATIONAL ZONE

Zone and elevation in Colombia	Land area (%)[1]	No. of species[2]
Tropical, 0–1200 m	83	43
Subtropical, 1200–2400 m	9	57
Temperate, 2400–3600 m	6	34
Paramo, snowfields, etc., 3600–5900 m	2	0

[1] Total land area of Colombia = approximately 1,138,555 km².
[2] Based on species in Table 1.

as many as 500 forms of Colombian birds may have been lost. This estimate seems very high, but it is clear (Fig. 3) that much of the subtropical forest between the western slope of the Eastern Andes and the eastern slope of the Western Andes has disappeared, and with it many forest birds of this zone. In fact, many of the famous "Bogotá" skins collected more than a century ago came from this region or the adjacent eastern slope of the Andes. A number of the birds in these collections have not been relocated in Colombia. Examples include Brown Tinamou (*Crypturellus obsoletus*), White-winged Potoo (*Nyctibius leucopterus*), and Turquoise-throated Puffleg (*Eriocnemis godini*). The Rufous-browed Tyrannulet (*Phylloscartes superciliaris*), once known only from "Bogotá," was located in northwestern Cundinamarca by M. A. Carriker Jr. in 1943 but has not been found since (Hilty and Brown 1983).

Human activities that have little to do with deforestation have also been disasterous to birds. Construction, in the late 1960's, of the isthmus highway connecting Barranquilla and Santa Marta impeded saline water flow and destroyed the mangrove and estuarine communities within Salamanca National Park. Increased human settlement as well as persecution probably drove the Greater Flamingo (*Phoenicopterus ruber*) and Jabiru (*Jabiru mycteria*) from the lower Magdalena region. Persistent egg robbing from large river island tern and skimmer colonies in Amazonia is almost certainly resulting in the decline of these species. I included a few species on the list because they have highly restricted ranges; the actual status of each is unknown. These include Northern Helmeted Curassow (*Crax pauxi*), Tolima Dove (*Leptotila conoveri*), White-chested Swift (*Cypseloides lemosi*), Western Antshrike (*Thamnomanes occidentalis*), Colorful Puffleg (*Eriocnemis mirabilis*), Moustached Antpitta (*Grallaria alleni*), Brown-banded Antpitta (*G. milleri*) and Tumaco Seedeater (*Sporophila insulata*).

The Blue List is dominated by aquatic species, terrestrial or semiterrestrial species (tinamous, curassows, wood-quail, antpittas), and canopy frugivores (parrots, toucans, cotingas, tanagers), represented by 19, 25, and 43 species, respectively (Table 1) out of about 102, 124, and 234 in each guild in Colombia. These Blue List totals account for about 20 percent of Colombia's terrestrial or semiterrestrial species, and more than 18 percent each of the aquatic residents and canopy frugivores. The small number of raptors (9 of 67 or about 13%) may reflect the fact that many are very widespread, or do well in open zones. Some have even benefited from human activities [e.g., some vultures and caracaras, Black-shouldered Kite (*Elanus caeruleus*), Pearl Kite (*Gampsonyx swainsonii*), Savanna Hawk (*Heterospizias meridionalis*) and American Kestrel (*Falco sparverius*)].

Large birds (more than 300 mm in total length) appear in the table at a rate of more than twice their overall representation in the avifauna. They account for about 47 percent of the species listed but less than 20 percent of the breeding avifauna. In the highlands, where deforestation problems have been particularly accute, large montane species are particularly vulnerable. They occur almost four times more frequently on the list (38% of all montane species listed, Table 1) than they do in the overall montane avifauna (only 9.8%). This disproportionate number of big birds is not surprising. It is well known that in tropical forests, among guilds of similarily foraging birds, large or specialized species with low densities are usually the first to disappear when habitat is reduced. Examples of the early loss of large forest species from Barro Colorado Island, Panamá, are documented by Willis (1974) and Willis and Eisenmann (1979); similar cases in Hawaii are cited by Wilson and Willis (1975).

The great number of montane species in all categories on the Blue List (68% are montane; Table 1) is distressing. For example, 72 percent of the terrestrial/semiterrestrial birds bluelisted are montane; 65 percent of the canopy frugivores and all of the nectivores are montane.

TABLE 3

Major Avifaunal Regions in Colombia and Number of Species at Risk[1]

Avifaunal region	No. of species
Caribbean[2]	1
Santa Marta	0
Chocó-Pacific	7
Cauca-Magdalena[3]	26
Central Colombian Mountains[4]	75
Catatumbo	3
Eastern Llanos	1
Amazonian	3

[1] Species with more than a 50% range loss, or those at risk because of human activities or unknown factors. Species identified with more than one avifaunal region are not included.
[2] Dry littoral from Rio Sinu northeastward.
[3] Drainage basins below 1200 m.
[4] Land from eastern slope of Eastern Andes to eastern slope of Western Andes above 1200 m elevation.

The list of 92 montane species represents almost one in five (an estimated 18.5%) of all montane species in Colombia. Furthermore, subtropical land area between 1200 and 2400 m elevation comprises only about 9 percent of the Colombian land surface (Table 2), whereas temperate zone land area, between about 2400 and 3600 m, accounts for only another 6 percent. Yet, 48 percent of the blue-listed birds occur primarily in the subtropics, and another 20 percent are principally temperate zone birds. These two montane zones make up only about 15 percent of Colombia's land surface but contain at least 68 percent of the species that are endangered or show evidence of population decline. These are also the regions most densely colonized by humans because of favorable soils and climate.

When the blue-listed species are grouped by avifaunal region (Table 3), the problem areas—the Central Andean highlands and the Cauca-Magdalena river basins—are confirmed. Deforestation of these two regions probably exceeds 60 percent; data are badly needed for the central mountain region from the western slope of the Eastern Andes to the eastern slope of the Western Andes, but there it may exceed 80 percent. I have placed 30 of Colombia's endemic species (about two-thirds of the total) on the Blue List. Almost all are confined to the Cauca-Magdalena river basin or the Andean slopes above these rivers. None of the 12 Santa Marta endemics is believed to face problems.

By far the most important generalization that can be made about changes in the distributions of Colombian birds is that at present forest species continue to decline while those favoring cleared land and disturbed areas are increasing, often rapidly. Newly deforested valleys on the Pacific slope and lowlands host many new invaders, such as Yellow-headed Caracara (*Milvago chimachima*), Spot-tailed Nightjar (*Caprimulgus maculicaudus*), Dwarf Cuckoo (*Coccyzus pumilus*), and Red-breasted Blackbird (*Leistes militaris*), to open areas. Increased second growth following lumbering operations in the Pacific lowlands of Valle and Chocó has benefited many species. Examples include Jet Antbird (*Cercomacra nigricans*), Fulvous-bellied Antpitta (*Hylopezus fulviventris*), Black-bellied Wren (*Thryothorus fasciatoventris*), Black-chested Jay (*Cyanocorax affinis*), and Crimson-backed Tanager (*Ramphocelus dimidiatus*), all of which appear to be increasing and several even expanding southward. Escaped cage birds are probably the source of range expansions of species such as Orange-chinned Parakeets (*Brotogeris jugularis*) in the lower Dagua Valley, and Saffron Finches (*Sicalis flaveola*) in Cali and Buenaventura. House Sparrows (*Passer domesticus*), first noted at Buenaventura in 1978, probably arrived as stowaways on a ship. In 1983 they were also found farther south on the Pacific coast at Guapí, Cauca (R. S. Ridgely, pers. comm.). The species that have benefited significantly from deforestation are listed in Appendix I.

Colombia's rich hummingbird fauna may have changed dramatically during the transformation of Andean forests to agriculture land over the past century. During the middle of the 19th century, "Bogotá" hummingbirds, collected without a definite Colombian locality, were shipped to Europe in numbers. These specimens were the original source of many new hummingbird species' descriptions (Meyer de Schauensee 1948). More than 20 of these forms, once thought to be species, are now believed to be hybrids, yet, curiously, hybrid hummingbirds from Colombia are lacking in large collections made during the last half century. "Bogotá" hybrids may reflect collecting done when rapid land clearing permitted contact between previously isolated siblings—a man-made evolutionary experiment accelerated by rapid deforestation. Or, these hybrids may reflect the more intensive collecting efforts made during that earlier period. Finally, some of these birds may represent species that have now disappeared, although recent field work (e.g., Snow and Snow 1980) and collecting have failed to "turn up" any of these old forms.

Present land use practices and high population growth rates in Colombia do not bode well for the future of Colombia's highland forest birds. Some larger species will continue to decline and disappear, and a small number of avian commensals of man will profit as the human struggle to domesticate the land continues. In the long run enlightened efforts will be needed to prevent major losses from this magnificent Andean ecosystem.

ACKNOWLEDGMENTS

The following persons have provided field notes and locality lists from which information has been drawn: S. Furniss (Meta), A. Gast (Cundinamarca; Vichada); S. Gniadek (Valle); P. Gertler (Huila); T. B. Johnson (Magdalena); T. Lemke and W. Lamar (Meta and Vichada); W. McKay (Meta; Casanare); J. V. Remsen (Amazonas); R. S. Ridgely (many areas in both eastern and western Colombia); J. R. Silliman (Cauca); and P. Silverstone (Chocó). Collecting journals of the late M. A. Carriker, Jr., which are housed at the U. S. National Museum of Natural History, Washington, D.C., gave information on many areas in western Colombia. S. Olson kindly allowed me to examine these journals. I have also profited from conversations with J. Hernández, V. Rodríquez, the late A. Olivares, and the late F. C. Lehmann. Initial work in Colombia was sponsored by a University of Arizona-U.S. Peace Corps Graduate Research Grant and the Corporación Autónoma del Valle del Cauca (CVC), in Cali, Colombia.

LITERATURE CITED

AMERICAN UNIVERSITY, FOREIGN AREA STUDIES. 1977. Area Handbook for Colombia. U.S. Government Printing Office, Washington, D.C.

ARBIB, R. 1971. Announcing—the blue list: an "early warning system" for birds. Am. Birds 25:948–949.

ARBIB, R. 1972. The blue list for 1972. Am. Birds 26:932–933.

CHAPMAN, F. M. 1917. The distribution of bird-life in Colombia: a contribution to a biological survey of South America. Bull. Am. Mus. Nat. Hist. 36:1–729.

HAFFER, J. 1974. Avian speciation in tropical South America. Publ. Nuttall Ornithol. Club No. 14.

HAFFER, J. 1975. Avifauna of northwestern Colombia, South America. Bonn. Zool. Monogr. No. 7.

HILTY, S. L., AND W. L. BROWN. 1983. Range extensions of Colombian birds as indicated by the M. A. Carriker, Jr. collection at the National Museum of Natural History, Smithsonian Institution. Bull. Br. Ornithol. Club 103:5–17.

HILTY, S. L., AND W. L. BROWN. In press. A Guide to the Birds of Colombia. Princeton University Press, Princeton, New Jersey.

KING, W. B. (ed.). 1981. Endangered Birds of the World. The ICBP Bird Red Data Book. Smithsonian Institution Press, Washington, D.C.

LATORRE, E. A. (ed.). 1967. Atlas de Colombia. Instituto Geografico Agustín Codazzi, Bogotá.

LEHMANN, F. C. 1970. Avifauna in Colombia. Pp. 88–92, In H. K. Buechner and J. H. Buechner (eds.), The Avifauna of Northern Latin America. Smithson. Contr. Zool. No. 26.

MEYER DE SCHAUENSEE, R. 1948. The birds of the Republic of Colombia. Part I. Caldasia 5:251–380.

MEYER DE SCHAUENSEE, R. 1970. A Guide to the Birds of South America. Livingston Publ. Co., Wynnewood, Pennsylvania.

OLIVARES, A. 1970. Effects of the environmental changes on the avifauna of the Republic of Colombia. Pp. 77–87, In H. K. Buechner and J. H. Buechner (eds.), The Avifauna of Northern Latin America. Smithson. Contr. Zool. No. 26.

RIDGELY, R. S. 1982. The distribution, status and conservation of neotropical mainland parrots. Vols. 1, 2. Unpubl. Ph.D. dissert., Yale University, New Haven, Connecticut.

SHANE, D. R. 1980. Hoofprints on the forest: an inquiry into the beef cattle industry in the tropical forest areas of Latin America. U.S. Dept. State, Office Environmental Affairs, Washington, D.C.

SNOW, D. W., AND B. K. SNOW. 1980. Relationships between hummingbirds and flowers in the Andes of Colombia. Bull. Br. Mus. Nat. Hist. 38:105–139.

WILLIS, E. O. 1974. Populations and local extinctions of birds on Barro Colorado Island, Panama. Ecol. Monogr. 44:153–169.

WILLIS, E. O., AND E. EISENMANN. 1979. A revised list of birds of Barro Colorado Island, Panama. Smithson. Contr. Zool. No. 291.

WILSON, E. O., AND E. O. WILLIS. 1975. Applied biogeography. Pp. 522–534, In M. L. Cody and J. M. Diamond (eds.), Ecology and Evolution of Communities. Belknap Press, Harvard University, Cambridge, Massachusetts.

APPENDIX I

Primary Avian Beneficiaries of Deforestation in Colombia[1]

Cattle Egret, *Bubulcus ibis*
Black Vulture, *Coragyps atratus*
White-tailed Kite, *Elanus caeruleus*
Pearl Kite, *Gampsonyx swainsonii*
Yellow-headed Caracara, *Milvago chimachima*
Crested Caracara, *Polyborus plancus*
American Kestrel, *Falco sparverius*
Crested Bobwhite, *Colinus cristatus*
White-throated Crake, *Laterallus albogularis*
Wattled Jacana, *Jacana jacana*
Southern Lapwing, *Vanellus chilensis*
Eared Dove, *Zenaida auriculata*
Ruddy Ground-Dove, *Columbina talpacoti*
White-tipped Dove, *Leptotila verreauxi*
Spectacled Parrotlet, *Forpus conspicillatus*
Smooth-billed Ani, *Crotophaga ani*
Striped Cuckoo, *Tapera naevia*
White-tailed Nightjar, *Caprimulgus cayennensis*
Sparkling Violet-ear, *Colibri coruscans*[2]
Blue-tailed Emerald, *Chlorostilbon mellisugus*
Rufous-tailed Hummingbird, *Amazilia tzacatl*
Azara's Spinetail, *Synallaxis azarae*[2]
Yellow-bellied Elaenia, *Elaenia flavogaster*
Common Tody-Flycatcher, *Todirostrum cinereum*
Vermilion Flycatcher, *Pyrocephalus rubinus*
Great Kiskadee, *Pitangus lictor*

Rusty-margined Flycatcher, *Myiozetetes cayanensis*
Tropical Kingbird, *Tyrannus melancholicus*
Blue-and-white Swallow, *Notiochelidon cyanoleuca*[2]
Southern Rough-winged Swallow, *Stelgidopteryx ruficollis*
House Wren, *Troglodytes aedon*
Tropical Mockingbird, *Mimus gilvus*
Black-billed Thrush, *Turdus ignobilis*
Shiny Cowbird, *Molothrus bonariensis*
Red-breasted Blackbird, *Leistes militaris*
Eastern Meadowlark, *Sturnella magna*
Bananaquit, *Coereba flaveola*
Scrub Tanager, *Tangara vitriolina*
Blue-gray Tanager, *Thraupis episcopus*
White-lined Tanager, *Tachyphonus rufus*
Streaked Saltator, *Saltator albicollis*
Yellow-faced Grassquit, *Tiaris olivacea*
Variable Seedeater, *Sporophila americana*
Yellow-bellied Seedeater, *S. nigricollis*
Ruddy-breasted Seedeater, *S. minuta*
Blue-black Grassquit, *Volatina jacarina*
Rufous-collared Sparrow, *Zonotrichia capensis*[2]
Lesser Goldfinch, *Spinus psaltria*[2]

---

[1] List limited to species now very widespread and common. Many others profit temporarily from partial deforestation and regrowth.
[2] Highland species found mainly above 1200 m elevation.

EFFECTS OF HABITAT ALTERATION ON A TROPICAL AVIAN FOREST COMMUNITY

JOHN H. RAPPOLE[1] AND EUGENE S. MORTON[2]

[1]Caesar Kleberg Wildlife Research Institute, Texas A&I University, Kingsville, Texas 78363 USA,
and [2]National Zoological Park, Smithsonian Institution, Washington, D.C. 20008 USA

ABSTRACT. Individuals of a tropical forest avian community at a site near Catemaco in southern Veracruz were captured, banded, and observed during field work in the winters of 1973–1974 and 1974–1975. The site was revisited and resampled in November–December 1980 and 1981. Vegetation on the site had been changed during the interim as a result of human activity, from mature forest to various forms of disturbed forest and second growth. Changes in avian populations associated with habitat alteration included drastic decreases in total numbers of forest-dwelling migrant and resident species and individuals on the site and changes in the behavior of individuals from territoriality to nomadism. We hypothesize that destruction of mature forest will cause declines in forest-dwelling tropical residents and wintering migrants that breed in temperate North America.

RESUMEN. Durante un estudio de campo llevado a cabo en los inviernos 1973–1974 y 1974–1975, se capturaron, anillaron y observaron individuos de una comunidad de aves de un bosque tropical en una localidad cerca de Catemaco en el sur de Veracruz. El sitio fue visitado nuevamente y se volvieron a juntar muestras en noviembre–diciembre de 1980 y de 1981. Como resultado de la actividad humana durante el interín, la vegetación del sitio ha cambiado de bosque maduro a varios tipos de bosque alterado y vegetación secundaria. Cambios en las poblaciones de aves, asociados con la alteración del habitat, incluyeron disminución drástica en el número total de especies que habitan en el bosque, tanto residentes como migratorias, ya sea para cantidad de especies o de individuos; así mismo hubo cambios en el comportamiento de individuos – anteriormente territoriales, ahora nómades. Nuestra hipótesis es que la destrucción del bosque maduro causará reducciones en las especies tropicales residentes que habitan en el bosque, así como en las aves migratorias que invernan en la región pero que reproducen en regiones templadas de América del Norte.

Habitats in many parts of the neotropics are being converted rapidly to agriculture (Myers 1980). The high conversion rates are cause for concern for the welfare of tropical resident avian species as well as migrants from Canada and the United States that winter in the Neotropics (Vogt 1970; Briggs and Criswell 1979; Terborgh 1980; Deis 1981). Here we report on the first study of changes in wintering bird populations at a tropical site converted from undisturbed mature forest to disturbed forest and second growth. We believe the results have important implications for our understanding of tropical avian ecology. Furthermore, the data may help resolve the question of relative suitability of mature versus disturbed forest habitats for forest-dwelling species.

The birds wintering at the site were studied originally during winters (November–March) over a two year period from 1973 to 1975 (Ramos and Warner 1980; Rappole and Warner 1980). Five years later we returned to renew our studies and found that the site had been changed from primary tropical forest to various forms of undisturbed, isolated forest, disturbed forest, pasture, second growth, and edge. This change allowed us to compare our netting and observational results from the different sampling periods with regard to changes in the species composition of the community and social behavior of selected species. We were also able to obtain a measure of relative habitat suitability for these species before and after human disturbance.

STUDY SITE AND METHODS

The study area is located in the Tuxtla Mountains of southern Veracruz, Mexico, roughly 30 km northeast of the town of Catemaco. The climax vegetation for the lowlands of the region is "selva alta perennifolia," or tall rainforest (Andrle 1964; Pennington and Sarukhan

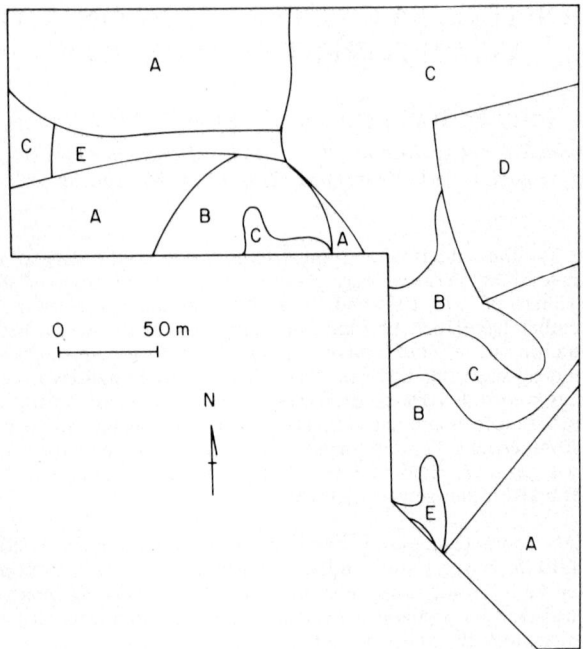

FIG. 1. Vegetation map of forest study site after modification. Letter designations correspond to those in Table 1.

1968; vegetation type C in Fig. 1 and Table 1). The study site covers 4.85 hectares. It was covered entirely with selva in 1975, but as of December 1980, 3.21 of the 4.85 hectares had been severely disturbed by cutting activities for pasture and agriculture. Only 1.64 hectares of undisturbed forest remained at that time. This section was essentially an island separated by 100 to 200 m from sections of untouched forest found on the steeper slopes. The main

TABLE 1

SELECTED VEGETATION PARAMETERS FOR HABITAT TYPES ON THE STUDY AREA AFTER HUMAN DISTURBANCE

	Habitats[1]				
Vegetation parameters estimated[2]	A (1.71 ha)[3]	B (0.70 ha)	C (1.64 ha)	D (0.57 ha)	E (0.23 ha)
Trees/ha, 8–15 cm DBH[4]	5	65	225	50 175	
Trees/ha, 15–23 cm DBH	5	30	85	30 100	
Trees/ha, 23–38 cm DBH	0	40	80	50	50
Trees/ha, 38–53 cm DBH	10	25	35	25	0
Trees/ha, 53–69 cm DBH	10	20	40	10	0
Trees/ha, 69–84 cm DBH	10	0	15	10	0
Trees/ha, 84–102 cm DBH	5	0	0	0	0
Shrub stems/ha	450	3500	19,500	750	7150
Percent ground cover	77	41	66	45	40
Percent canopy cover	42	66	85	16	85
Average canopy height (m)	20.4	21.0	18.4	11.4	16.0

[1] Letters used to designate habitats correspond to those used in Figure 1: A—open pasture, few large trees; B—forest from which most small to medium sized trees and undergrowth was cleared; C—selva; D—field of mixed cane, corn, bare dirt, scattered brush, large logs, a few large trees remaining; E—streamside vegetation, medium and small trees with dense undergrowth.
[2] Estimated using techniques described in James and Shugart 1970.
[3] Area of each vegetation type on the study site.
[4] DBH = Diameter at breast height.

FIG. 2. Photographs showing human-caused changes in vegetation at the Tuxtlas forest study site. The upper photograph was taken in 1974. The lower one was taken at the same location with the same aspect, in 1980.

structural characteristics of the different habitats in 1980 were assessed using the method of James and Shugart (1970; Table 1, Fig. 1).

We worked in the area from 9 November 1973–3 February 1974; 11 December 1974–13 March 1975; 15 November–9 December 1980; 19 November–6 December 1981. The site

TABLE 2

SAMPLING EFFORT AND BIRDS CAPTURED ON THE FOREST STUDY SITE, 1973–1975 VS. 1980–1981[1]

	Sampling periods					
	1973–1974	1974–1975	Total 1973–1975	1980	1981	Total 1980–1981
Net hours	3645	8755	12,400	3574	3139	6713
Total species	48	56	63	33	46	55
Total individuals	198	426	624[2]	138	119	257[2]
Total individuals/1000 net hrs	54.3	48.6	50.3	38.6	37.9	38.2
Forest-dwelling migrant species	12	16	16	10	13	13
Forest-dwelling migrant individuals	80	232	312[3]	63	50	113[3]
Forest-dwelling migrant individuals/1000 net hrs	21.9	26.5	25.2	17.6	15.9	16.8
Forest-dwelling resident species	28	26	32	11	15	19
Forest-dwelling resident individuals	106	157	263[4]	35	34	69[4]
Forest-dwelling resident individuals/1000 net hrs	29.1	17.9	21.2	9.8	10.8	10.2

[1] Based on data in Appendix I. Hummingbird data are omitted here for reasons mentioned in Appendix I.
[2] Significantly different from the expected values for this sample ($\chi^2 = 13.68$, 1 d.f., $P = 0.01$).
[3] $\chi^2 = 13.60$, 1 d.f., $P = 0.01$.
[4] $\chi^2 = 29.95$, 1 d.f., $P = 0.01$.

was mapped and divided into grids 25 m on a side in November 1973. Grid intersections were assigned a specific number-letter combination and marked with engineering tape. A mist net (12 m × 2.7 m; 30, 36, or 61 mm mesh) was placed in every other grid except where grids were located in open pasture.

We also worked in an overgrown pasture bordered by a hedgerow of trees from 10 to 30 m tall, and in streamside vegetation. The pasture was covered with *Solanum, Lantana,* and *Piper* (canopy height 1–2 m). These habitats were adjacent to our original study area. They, too were selva prior to clearing in 1968. We gridded these habitats in the same manner as described for the forest study area.

Each bird captured was given a special color band sequence for individual recognition in the field. Migrants were given Fish and Wildlife Service bands as well. Each capture was entered in a main catalogue and a species catalogue, and its capture point was plotted on a map of the study site kept for each species. Recaptures and resightings were also catalogued and mapped. The species composition of capture samples is given in Appendix I along with Latin names for all species mentioned in the text. The statistical significance of changes in the number of individuals within a given category (e.g., forest-dwelling migrants) was tested using a Chi-square, Goodness of Fit analysis (Sokal and Rohlf 1969:550–560) where the expected values for before and after cutting were calculated using the ratio of sample net hours to total net hours times the combined sample of individuals within a category (1973–1975 + 1980–1981).

Territory size was determined for the Hooded, Wilson's, and Magnolia Warblers through observation of marked birds and use of playback of call notes following procedures outlined in greater detail in Rappole and Warner (1980).

RESULTS

Structural analysis of the vegetation of the study area (summarized in Table 1), revealed a patchwork of vastly different habitat types where formerly there had been only selva and some ecotone. Figure 2 illustrates the changes caused by forest conversion at one specific site on the study area.

TABLE 3

NUMBER OF CAPTURES, TERRITORIES, AND RETURNS FOR SELECTED SPECIES FROM THE
PASTURE/HEDGEROW SITE

Species	Total captures (1980)	Estimated no. terr. (1980)	Returns[1]
Least Flycatcher	6	6	0
Gray Catbird	9	2	0
White-eyed Vireo	7	2	1
Worm-eating Warbler	1	0	0
Orange-crowned Warbler[2]	2	2	1
Yellow Warber	2	2	1
Magnolia Warbler	6	2	2
Kentucky Warbler	1	1	0
Common Yellowthroat	1	1	1
Gray-crowned Yellowthroat[2]	1	0	0
Yellow-breasted Chat	7	2	2
Hooded Warbler	1	1	0
Wilson's Warbler	13	2	2
Rufous-capped Warbler[3]	5	2	4
Indigo Bunting	2	0	0

[1] Birds originally captured in 1980, recaptured in 1981.
[2] Orange-crowned Warbler = *Vermivora celata*; Gray-crowned Yellowthroat = *Geothlypis poliocephala*.
[3] Probably remain in pairs or family groups for much of the year.

Despite the fact that one hectare of undisturbed forest remained on the site along with two hectares of various forms of disturbed forest, numbers of forest-dwelling species and individuals declined sharply (Table 2, Appendix I). This decrease was noted in numbers of total individuals, forest-dwelling migrants, and forest-dwelling residents at a high level of statistical significance (Table 2).

Changes were noted in territory size and social behavior for some forest-related species remaining on areas of disturbed forest, ecotone, or second growth. For example, 12 Hooded Warblers held territories on the study area in 1975, but only 5 held them in the same area in 1980 (Fig. 3). There were 5 territories on the site again in 1981 with 3 held by the same owners as in 1980.

Average territory size in the Hooded Warbler population on the site increased from 0.3 hectare in 1975 to 1.0 hectare in 1980. One individual (Red left) defended a territory more than 200 m long.

A similar situation was observed in Magnolia and Wilson's Warblers. In selva, members of these species defend territories of 0.2 to 0.4 hectare. However, a color banded individual of each of these species defended a territory more than 1.0 hectare in size and more than 500 m long in a hedgerow thicket bordering overgrown pasture.

The large size and small number of territories for these species was not due solely to a lack of birds. Nine Hooded Warblers were captured on the disturbed forest site in 1980, and 8 in 1981; and while many individuals of the Wilson's and Magnolia Warblers were captured or observed on the pasture/hedgerow site (Table 3), only one member of each species defended a territory. Both of these territories were in hedgerow trees, not in the short brush of the overgrown pasture. Non-territorial individuals generally were not resighted more than one or two days after capture.

DISCUSSION

A decline in the numbers and species of forest-associated birds is not an unexpected result of forest alteration; yet we did not foresee the degree of decline, particularly since much of the forest remains, although altered to varying degrees.

Two lines of evidence indicate that decreases in numbers and species of forest-dwelling birds were caused by the alteration of the forest habitat. First, overall populations of the migrant species studied as determined from breeding bird surveys from 1975–1979 have not declined (C. S. Robbins, pers. comm.). Second, Ramos and Warner (1980) showed that the wintering migrants in Veracruz are derived from a broad area of the breeding range. Thus,

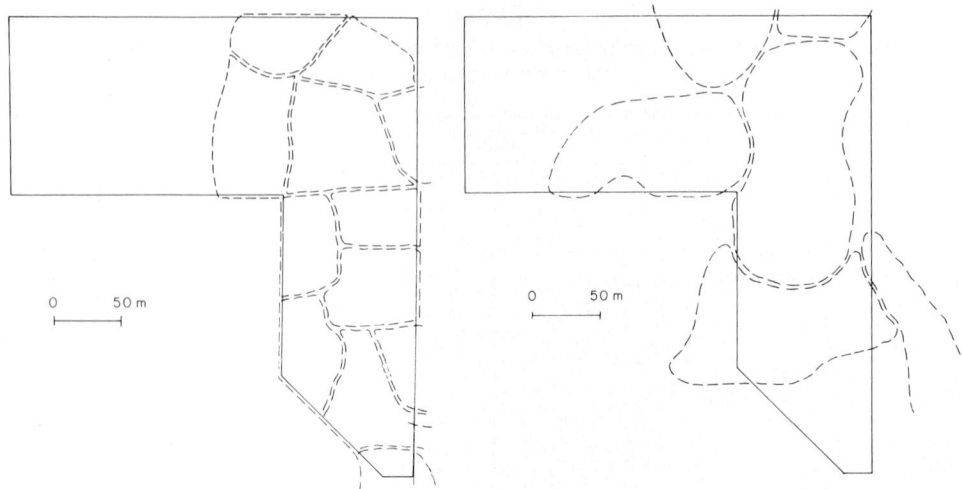

Fig. 3.   Maps of Hooded Warbler territories on the forest study site: upper—1975; lower—1980. Note that territories in the northeast section of the site were not determined in either year.

local changes in breeding habitat carrying capacity should not be reflected in reduced winter numbers in Veracruz.

We suggest that a major reason for declines is that the forest remnants at our site were isolated from larger tracts. This isolation and the small size of the remnants apparently made the forest patches unsuitable for use by multispecies foraging flocks. These flocks are important for many members of tropical forest avian communities (Willis 1966; Morton 1973; Buskirk 1976; Rappole et al. 1983). Several species occur only in association with these flocks while individuals of many others join a flock when it enters their home range or territory. Each flock has a foraging area of 2 to 3 hectares or more. This home range apparently does not overlap with that of other flocks (Buskirk et al. 1972).

Species that forage in family groups often form the nucleus of these flocks ("nuclear species," Moynihan 1962) while those that join these birds are termed "attendants." On our study site, Tawny-crowned Greenlets and Gray-headed Greenlets are important nuclear species. The Plain Xenops, Ivory-billed Woodcreeper, Chestnut-colored Woodpecker, Black-and-white Warbler, Sulphur-rumped Flycatcher, and Worm-eating Warbler are examples of attendants (usually one individual per flock). All of these species suffered declines or disappeared from the site (Appendix I).

Numbers and species of migratory birds as well as residents declined sharply, apparently as a result of forest removal. As mentioned above, species such as the Black-and-white and Worm-eating Warblers may have declined as a result of the absence of a multispecies flock on the site.

A second possible reason for declines in avian, forest-dwelling species is that a smaller amount of suitable habitat was available after cutting allowing fewer individuals to survive on the site. This explanation is often accepted for declines in resident species, but not for migrants. Yet migrants (e.g., the Wood Thrush, Hooded Warbler, Wilson's Warbler, and Magnolia Warbler) show declines as well. Nomads or "floaters" of these species create a false sense of the habitat's suitability for the bird. These individuals generally move through a specific degraded habitat site in a day or so. They do not use the areas on the same long-term basis as birds of the same species in selva.

Ecotones, second growth, and other forms of habitat perturbations can provide complex mixtures of vegetation types. These mixtures are often rich in avifauna derived from a variety of primary habitats. As a result, creation of edge sometimes is intimated as a means for increasing diversity or suitability of a habitat, especially for avian migrants in the Neotropics (Willis 1966:224; Morse 1971; Karr 1976). Our data, indicate however, that, unlike sedentary conspecifics in primary habitats, many individuals observed in edge and second growth habitats are nomads. Whether these birds suffer a lower survival rate than sedentary individuals is not known but seems likely given the additional problems, such as the lack of knowledge of stable food sources, local predator habits, best avenues of escape, and protected roost sites, with which a nomad must deal as compared with a sedentary conspecific.

Mature forests supported higher, more stable populations of forest-dwelling migrants and residents than disturbed forest, ecotones, and second growth on our study sites in Veracruz. Since tropical forest alteration is not a local phenomenon, our results provide some cause for alarm regarding the future of forest-dwelling species in the tropics.

ACKNOWLEDGMENTS

We thank the World Wildlife Fund for providing financial support for the 1980 and 1981 field work in Veracruz. In particular, T. Lovejoy III and N. Hammond were most helpful. We thank the Welder Wildlife Foundation for financial support of Rappole during the 1973–1975 field work and logistical support during the 1980 field work. A. Estrada, Director of the Estación de Biología "Los Tuxtlas," U.N.A.M., allowed us access to the University of Mexico's field station grounds and facilities, for which we are most grateful. S. Barrios, J. Vega, I. Valdovinos, M. Ramos, K. Ballard, L. Yoder, M. Yoder, M. Van der Vort, M. Wilson, A. Wilson, and E. Fisk all assisted with field work. J. Ballard assisted with preparation of the manuscript.

LITERATURE CITED

ANDRLE, R. F. 1964. A biogeographical investigation of the Sierra de Tuxtla in Veracruz, Mexico. Unpubl. Ph.D. dissert. Louisiana State University, Baton Rouge.

BRIGGS, S. A., AND J. H. CRISWELL. 1979. Gradual silencing of spring in Washington. Atl. Nat. 32:19–26.

BUSKIRK, W. H. 1976. Social systems in a tropical forest avifauna. Am. Nat. 110:293–310.

BUSKIRK, W. H., G. V. POWELL, J. F. WITTENBERGER, R. E. BUSKIRK, AND T. U. POWELL. 1972. Interspecific bird flocks in tropical highland Panama. Auk 89:612–624.

DEIS, R. 1981. Again silent spring. Defenders 56(2):6–10.

JAMES, F. C., AND H. H. SHUGART, JR. 1970. A quantitative method of habitat description. Am. Birds 24:727–736.

KARR, J. R. 1976. On the relative abundance of migrants from the northern temperate zone in tropical habitats. Wilson Bull. 88:433–458.

MORSE, D. H. 1971. The insectivorous bird as an adaptive strategy. Annu. Rev. Ecol. Syst. 2:177–200.

MORTON, E. S. 1973. On the evolutionary advantages and disadvantages of fruit eating in tropical birds. Am. Nat. 107:8–22.

MOYNIHAN, M. 1962. The organization and probable evolution of some mixed species flocks of neotropical birds. Smithson. Misc. Collect. 143:1–140.

MYERS, N. 1980. Conversion of Tropical Moist Forests. National Research Council, Committee on Research Priorities in Tropical Biology, National Academy of Sciences, Washington, D.C.

PENNINGTON, T. D., AND J. SARUKHAN. 1968. Arboles Tropicales de México. Instituto Nacional de Investigaciones Forestales, México, D.F.

RAMOS, M. A., AND D. W. WARNER. 1980. Analysis of North American subspecies of migrant birds wintering in Los Tuxtlas, southern Veracruz. Pp. 1973–180, In A. Keast and E. S. Morton (eds.), Migrant Birds in the Neotropics. Smithsonian Institution Press, Washington, D.C.

RAPPOLE, J. H., E. S. MORTON, T. J. LOVEJOY, III, AND J. L. RUOS. 1983. Nearctic Avian Migrants in the Neotropics. U.S. Fish Wildlife Service, Washington, D.C.

RAPPOLE, J. H., AND D. W. WARNER. 1980. Ecological apsects of migrant bird behavior in Veracruz, Mexico. Pp. 353–393, In A. Keast and E. S. Morton (eds.), Migrant Birds in the Neotropics. Smithsonian Institution Press, Washington, D.C.

SOKAL, R. R., AND F. J. ROHLF. 1969. Biometry. W. H. Freeman Co., San Francisco, California.

TERBOUGH, J. W. 1980. The conservation status of neotropical migrants: present and future. Pp. 21–30, In A. Keast and E. S. Morton (eds.), Migrant Birds in the Neotropics. Smithsonian Institution Press, Washington, D.C.

VOGT, W. 1970. The avifauna in a changing ecosystem. Pp. 8–16, In H. K. Buechner and J. H. Buechner (eds.), The Avifauna of Northern Latin America. Smithson. Inst. Contrib. Zool. Vol. 26.

WILLIS, E. O. 1966. The role of migrant birds at swarms of army ants. Living Bird 5:187–231.

APPENDIX I

SPECIES COMPOSITION OF CAPTURE SAMPLES FROM THE FOREST STUDY SITE IN VARIOUS YEARS[1]

Species	1973–1974	1974–1975	1980	1981
[2]*Accipiter striatus*, Sharp-shinned Hawk	0	0	0	1
Columbina inca, Inca Dove	0	0	0	1
C. talpacoti, Ruddy Ground-Dove	0	0	0	1
Claravis pretiosa, Blue Ground-Dove	0	0	0	1
[3]*Leptotila plumbeiceps*, Gray-headed Dove	1	2	0	4
[3]*Geotrygon montana*, Ruddy Quail-Dove	1	0	0	0
[3]*Piaya cayana*, Squirrel Cuckoo	1	0	0	0
Crotophaga sulcirostris, Groove-billed Ani	0	0	0	2
Glaucidium brasilianum, Ferruginous Pygmy Owl	1	0	1	0
[3,4]*Phaethornis superciliosus*, Long-tailed Hermit	9	15	—	13
[3,4]*P. longuemareus*, Little Hermit	2	3	—	0
[3,4]*Campylopterus curvipennis*, Wedge-tailed Sabrewing	3	5	—	3
[3,4]*C. hemileucurus*, Violet Sabrewing	1	7	—	0
[3,4]*Florisuga mellivora*, White-necked Jacobin	1	0	—	0
[3,4]*Amazilia candida*, White-bellied Emerald	2	23	—	3
[4]*A. tzacatl*, Rufous-tailed Hummingbird	1	7	—	3
[3]*Momotus momota*, Blue-crowned Motmot	0	2	1	0
[3]*Pteroglossus torquatus*, Collared Aracari	0	0	1	1
Melanerpes pucherani, Black-cheeked Woodpecker	0	0	0	1
M. aurifrons, Golden-fronted Woodpecker	0	0	2	1
[3]*Veniliornis fumigatus*, Smoky-brown Woodpecker	2	2	0	0
[3]*Celeus castaneus*, Chestnut-colored Woodpecker	0	1	0	0
[3]*Xenops minutus*, Plain Xenops	3	2	0	0
[3]*Dendrocincla anabatina*, Tawny-winged Woodcreeper	1	1	2	0
[3]*Glyphorhynchus spirurus*, Wedge-billed Woodcreeper	1	0	0	0
[3]*Xiphorhynchus flavigaster*, Ivory-billed Woodcreeper	1	4	1	0
[3]*Mionectes oleagineus*, Ochre-bellied Flycatcher	6	19	13	7
[3]*Rhynchocyclus brevirostris*, Eye-ringed Flatbill	1	4	2	1
[3]*Platyrinchus mystaceus*, White-throated Spadebill	4	9	2	0
[3]*Onychorhynchus mexicanus*, Royal Flycatcher	2	0	0	0
[3]*Myiobius sulphureipygius*, Sulphur-rumped Flycatcher	4	5	0	2
[2,3]*Empidonax flaviventris*, Yellow-bellied Flycatcher	3	13	5	9
[2]*E. minimus*, Least Flycatcher	3	2	2	3
[3]*Attila spadiceus*, Bright-rumped Attila	6	5	0	1
Megarhynchus pitangua, Boat-billed Flycatcher	1	1	0	0
Pachyramphus aglaiae, Rose-throated Becard	1	1	0	0
Tityra semifasciata, Masked Tityra	0	0	2	0
[3]*Pipra mentalis*, Red-capped Manakin	1	0	0	0
Campylorhynchus zonatus, Band-backed Wren	0	1	1	3
[3]*Thryothorus maculipectus*, Spot-breasted Wren	7	16	0	1
[3]*Henicorhina leucosticta*, White-breasted Wood-Wren	6	4	1	2
[3]*Ramphocaenus rufiventris*, Long-billed Gnatwren	1	3	0	0
[2,3]*Catharus ustulatus*, Swainson's Thrush	1	1	0	0
[2,3]*C. guttatus*, Hermit Thrush	0	1	0	0
[2,3]*Hylocichla mustelina*, Wood Thrush	16	74	14	6
[3]*Turdus assimilis*, White-throated Robin	0	4	0	0
[3]*T. grayi*, Clay-colored Robin	3	6	5	1
[2]*Dumetella carolinensis*, Gray Catbird	1	12	3	3
[2,3]*Vireo griseus*, White-eyed Vireo	4	12	6	1
[3]*Hylophilus ochraceiceps*, Tawny-crowned Greenlet	14	18	0	0
[3]*H. decurtatus*, Gray-headed Greenlet	1	4	0	0
[2]*Dendroica petechia*, Yellow Warbler	1	0	0	1
[2,3]*D. magnolia*, Magnolia Warbler	3	18	3	2
[2]*D. coronata*, Yellow-rumped Warbler	0	0	4	0
[2,3]*Mniotilta varia*, Black-and-white Warbler	6	14	0	0
[2,3]*Setophaga ruticilla*, American Redstart	0	1	4	1
[2,3]*Helmitheros vermivorus*, Worm-eating Warber	5	12	0	1
[2,3]*Seiurus aurocapillus*, Ovenbird	0	2	3	1
[2,3]*S. noveboracensis*, Northern Waterthrush	0	2	0	1

APPENDIX I

CONTINUED

Species	1973–1974	1974–1975	1980	1981
[2,3]*S. motacilla*, Louisiana Waterthrush	1	5	1	2
[2,3]*Oporornis formosus*, Kentucky Warbler	17	28	10	8
[2]*Geothlypis trichas*, Common Yellowthroat	0	1	2	0
[2,3]*Wilsonia citrina*, Hooded Warbler	15	30	9	8
[2,3]*W. pusilla*, Wilson's Warbler	8	16	8	9
[3]*Myioborus miniatus*, Slate-throated Redstart	1	0	0	0
[3]*Basileuterus culicivorus*, Golden-crowned Warber	4	3	0	0
B. rufifrons, Rufous-capped Warbler	0	2	2	4
[2]*Icteria virens*, Yellow-breasted Chat	0	1	0	1
Coereba flaveola, Bananaquit	0	2	0	1
[3]*Euphonia hirundinaceae*, Yellow-throated Euphonia	0	2	1	3
[3]*E. gouldi*, Olive-backed Euphonia	6	2	0	1
[3]*Eucometis penicillata*, Gray-headed Tanager	1	4	0	0
[3]*Habia rubica*, Red-crowned Ant-Tanager	12	19	0	2
[3]*H. gutturalis*, Red-throated Ant-Tanager	12	13	6	6
[2,3]*Piranga rubra*, Summer Tanager	1	3	0	1
[3]*Chlorospingus opthalmicus*, Common Bush Tanager	0	0	0	1
[3]*Cyanocompsa cyanoides*, Blue-black Grosbeak	3	3	0	1
C. parellina, Blue Bunting	3	2	0	4
[2]*Passerina cyanea*, Indigo Bunting	0	5	15	4
[2]*P. ciris*, Painted Bunting	0	1	5	1
Volatinia jacarina, Blue-black Grassquit	0	3	0	2
Sporophila torqueola, White-collared Seedeater	1	0	0	0
Tiaris olivacea, Yellow-faced Grassquit	0	3	0	0
Dives dives, Melodius Blackbird	0	0	1	0
Total birds	217	486	138	141

[1] Numbers in columns represent total individuals captured per sample period.

[2] Wintering migrants.

[3] Forest-dwelling birds (captured or observed regularly in undistrubed forest 50 m or more from another habitat type).

[4] Due to the difficulties involved in marking and subsequent separation of recaptures from new captures, numbers of hummingbird individuals shown must be considered as rough estimates. Numbers were not recorded during the 1980 field season.

OVERVIEW

PARKES, KENNETH C. Neotropical Ornithology—An Overview 1025

NEOTROPICAL ORNITHOLOGY—AN OVERVIEW

KENNETH C. PARKES

Carnegie Museum of Natural History, Pittsburgh, Pennsylvanis 15213 USA

ABSTRACT. The papers in this monograph are reviewed in the context of the progressive stages of ornithological knowledge and of current neotropical ornithological research in general. Active fields of study not represented in these pages are listed, to reflect more completely the breadth of research on birds of the neotropics in the 1980's.

RESUMEN. En esta monografía se repasan los artículos en los contextos de las etapas progresivas del conocimiento ornitológico y de las investigaciones de ornitología neotropical, en general. Se mencionan los campos de investigación activos que no están representados en estas páginas para reflejar de la manera más completa posible la amplitud de las investigaciones en aves neotropicales en la década de 1980.

The title of this final chapter could be construed as either an overview of the papers in this monograph or an overview of neotropical ornithology as a whole, as a scientific discipline. Obviously the latter interpretation, were it to be attempted in depth, would demand much more space than can be allotted here. Yet the question naturally arises: to what extent is the monograph itself, as a collection of papers by many of the active workers in the field, an "overview" of our present knowledge of neotropical birds? To answer that question I must attempt to synthesize, to some extent, the two approaches mentioned in my first sentence. Some readers may find this chapter a bit opinionated, but, as any good journalist will verify, strict objectivity is a myth.

Two recent publications are germane to this "overview." Haffer (1983)* has attempted, with notable success, a survey of "Results of Modern Ornithological Research in Tropical America" (title of English abstract). His stated purpose in publishing this article, in German, was to introduce the subject to Europeans unfamiliar with the predominantly American literature of neotropical ornithology. It will nevertheless prove to be a valuable summary for American readers fluent in German. Some of the most useful parts of the paper demand little linguistic facility, such as the lists of new species of birds described from South America between 1951 and 1981 (with a separate list for Peru!).

Somewhat similar to the present monograph is "Mammalian Biology in South America" (Mares and Genoways 1982; see also the comprehensive review by Patterson [1983]). Unlike its avian counterpart, the mammal volume consists of papers presented at a symposium, so that workers (especially those from South America) who could not attend the meeting are unrepresented. Furthermore, the focus of the symposium was continental South America (although some authors naturally mentioned other neotropical areas in passing), whereas one-fifth of the papers in this ornithological monograph concern themselves with birds of Middle America and the West Indies. Some readers may find it of interest to compare the kinds and levels of knowledge of neotropical birds and mammals; to mention only one obvious difference, karyotypes play a major role in the literature of mammalian systematics, but have been little used in ornithology (see Shields 1982).

There are three basic and successive stages of knowledge of birds that must precede all other aspects of the study of ornithology. These are (A) the *Inventory* stage: the kinds of birds that exist, (B) the *Classification* stage: how these birds are related to one another (and, thus, how we organize our information about them), and (C) the *Descriptive Zoogeography* stage: where these kinds of birds are found and where they are not found. Knowledge of these three factors, as complete as possible, must precede any analytical or interpretive studies. In these respects, our knowledge of neotropical birds lags far behind our knowledge of those of North American and western Europe. There, for example, our data base for category C is so thorough

---

* References with dates will be found in the Literature Cited at the end of this chapter; references without dates indicate papers in the present monograph.

that we can document in detail the dynamic range changes many species have undergone in the past century; in the neotropics we can do this for barely a handful of conspicuous species such as the Cattle Egret (*Bubulcus ibis*) or, in recent years, the Shiny Cowbird (*Molothrus bonariensis*; see Cruz et al.). All three of the initial stages of ornithological data-gathering are still very much under way in the neotropics, especially in large parts of South America. The Inventory stage approaches completion for many areas of the neotropics, but there are still relatively unexplored regions, notably in Peru, that continue to yield hitherto unknown species and even genera of birds (see Haffer 1983; Parker and O'Neill). Even from areas where the Inventory was thought to have been completed long ago, major surprises continue to appear, such as the recent discovery of the Elfin Woods Warbler (*Dendroica angelae*) on the well-studied Caribbean island of Puerto Rico (Kepler and Parkes 1972). Lowery and Tallman (1976) jolted our complacency about both the Inventory and Classification stages when they described the Pardusco (*Nephelornis oneilli*) but could not classify it with any precision beyond membership in the "New World nine-primaried oscine" complex. Certainly excellent studies can be done and are being done (as witness some of the papers in this volume) of single species or of ecological relationships in the neotropics. Any studies that attempt to use comparative numbers or distributions of avian taxa in the neotropics, however, run the risk of being based on relatively shaky data.

That subcategory of stages A and C that involves the study of geographic variation within species is still yielding much new information. That more such information is not being published is primarily a function of the small number of active workers in this field (Parkes 1982), as there are major collections of neotropical birds in such museums as the American Museum of Natural History, National Museum of Natural History, and Carnegie Museum of Natural History that contain much unstudied material. My own notes include partial revisions of dozens of neotropical species whose geographic variation has not been adequately described. Papers in this monograph include the designation as subspecies of newly discovered populations (Delgado, Stiles, Weske) as well as broader revisions of species (Dickerman, Storer and Getty, Traylor); although based chiefly on traditional museum studies, these use larger series of specimens and, often, more sophisticated statistical analyses, than hitherto available.

Field and museum studies are demonstrating that the number of avian species to be recognized cannot be considered settled, even leaving out of consideration those remaining to be discovered. Whether or not to "lump" allopatric forms has always been a matter of taxonomic outlook, but study of new or already existing specimens that demonstrate intergradation (and hence conspecificity) of populations has been going on for some time (Weske). More recently, the same fruitful combination of field and museum studies has led to proof that very similar forms long considered conspecific are actually good species (for example, *Turdus rufopalliatus* and *graysoni*, Phillips 1981; *Emberizoides herbicola, duidae*, and *ypirangus*, Eisenmann and Short 1982; *Cymbilaimus lineatus* and *sanctaemariae*, Pierpont and Fitzpatrick 1983). In such cases, additional collecting provides evidence of sympatry, while studies of vocal and other behavior reveal probable species-specific characters and isolating mechanisms. Braun and Parker go even farther in analyzing relationships among a group of synallaxine Furnariidae, using protein data to support the suggestion that *Schizoeaca helleri* and *S. harterti* are good species and should not be "lumped" with *S. fuliginosa*.

All over the world there are "problem" genera whose relationships have remained subject to debate for many years. I have the subjective and non-documented impression that there are more of these in the Indo-Australian region than elsewhere, but they surely are not lacking in the neotropics. Using traditional techniques of comparative anatomy that had not previously been applied to the taxa in question, Morony verifies the status of *Sericossypha* as a tanager, and Bock indicates that the flowerpiercer genus *Diglossa* is polyphyletic. Although Bock's conclusions are based on comparative anatomy, he brings to his analysis a more sophisticated functional/adaptive approach, based on modern evolutionary concepts, than has been used by earlier anatomists. Bock's warning about convergent evolution as a hazard in constructing classifications should be thoroughly absorbed by both cladists and pheneticists who tend largely to overlook the importance of functional/adaptive analyses.

The element of zoogeographic speculation in taxonomic papers has also advanced in recent years, as exemplified in this monograph by the papers of Traylor and Storer and Getty. These speculations, however, still rest on analyses of the distributions of traditional taxonomic characters of size and color.

At the level of genus and species, the Classification stage B is profiting by new kinds of data and new kinds of analyses. An especially good example of the application of these new techniques to a classification problem is the paper by Braun and Parker. They used a combination of traditional morphological characters [which had previously given results that were somewhat equivocal, although not as much so as would be implied by the treatment of Vaurie (1980)], more recent but by now virtually traditional evidence from comparisons of vocalizations, and new molecular data. Gratifyingly, all approaches gave the same answer, much more congruent than has been true of some other recent studies. Here again it is important to keep in mind my warning (Parkes 1982) that the protein molecules now accessible to genetical analyses are not known to be correlated in any way with the genes governing the morphological characters traditionally used in taxonomy, characters that appear to respond to selective influences, as beautifully illustrated by Grant's studies of Darwin's finches. Braun and Parker emphasize how few studies have been published that have combined traditional and molecular approaches; it will be a very long time before such combinations can be widely used, as the molecular techniques are time-consuming and expensive. Nevertheless, we may expect that these synthetic analyses will increase.

The lack of concordance between variation in those genetic characters susceptible to biochemical/genetical analysis and variation in traditional morphological characters is nowhere better illustrated than in the paper by Caparella and Lanyon. Two taxa of flycatchers so similar morphologically as to have been considered conspecific when first discriminated (but now well known to differ in vocalizations and other behavior, as is typical of many "sibling species") prove to be as well differentiated biochemically as are other congeneric species pairs that are also well differentiated morphologically. These findings support the suggested caution in deducing relationships among taxa that represent the converse situation, i.e., strong morphological and behavioral differentiation but weak differentiation among the few biochemical characters testable. The current fashionable term "genetic distance," it must always be remembered, is being used for only a tiny handful of loci whose relative importance to the overall adaptation of the organism is unknown.

Stage C, the plotting of the distribution of avian species, lags far behind (in the neotropics) the kind of knowledge exemplified by the many regional publications available for the nearctic, western palearctic, and even much of Africa. For many parts of South America in particular, the large series of study skins in museums are highly deceptive. All too often, they come from a few favorable, readily accessible localities (students of Argentine birds, for example, will recognize "km 10, Arroyo Uruguai-i, Misiones" as such a locality). Other "large" series represent material from a number of localities, but utterly inadequate when broken down by age, sex, and season (including molt stage). The heroic attempt of Haffer and Fitzpatrick to analyze geographic variation of Amazonian forest birds by new quantitative mapping techniques and analyses of regional patterns was confined to five species, and even of those, as the authors admit, the data base was "somewhat scanty," with some parts of their study area wholly unsampled.

The core of Cracraft's paper, although it contains a large element of historical speculation, is his lists of representative endemic taxa for each of 33 areas in South America. Although on a broad scale these lists are probably accurate and, indeed, form a potentially useful compilation, they, too, are subject to modification through additional information about ranges and changes in taxonomic status (for example, some subspecies used as examples of endemics by Cracraft, based on recognition in the standard literature, may prove upon further study with better material to be untenable). Such modifications will not greatly affect Cracraft's arguments.

Haffer's paper, although also dealing with endemism, goes into much more detail than was possible for Cracraft about taxonomic aspects of zoogeography. His approach to the study of endemism in the neotropics differs in several respects from that of Cracraft (as pointed out by the latter); Haffer (p. 133) quite rightly recommends "a pluralistic approach to zoogeographic analysis . . . in view of the diversity of phenomena causing the distribution patterns of any particular taxonomic group."

The papers by Fjeldså, Vuilleumier, and Snow also take an historical approach to zoogeographic analysis, but the limited geographical scope of their studies permits a somewhat more detailed presentation; the nature of the avifauna under discussion is such that Fjeldså's conclusions in particular are unlikely to be affected significantly by new taxonomic or distributional information.

With the deemphasis on museum collecting imposed by lack of interest and financial support from educational institutions and government agencies, as well as by bureaucratic red tape, it appears unlikely that the most important data base for zoogeographic (and, of course, many other) studies, i.e., specimen material, will improve markedly in the near future (and, as will be mentioned later, the very existence of the avifauna being analyzed is in jeopardy in many inadequately sampled or even ornithologically unknown regions). Nevertheless, a few papers still appear annually that provide additional distributional data for hitherto poorly known areas of the neotropics; papers in the present monograph that include such data and place these data in the context of previous information include those of Parker et al., Robbins et al., and Wiedenfeld et al. It is certainly not without significance that all of these contributions emanate from the same institution, Louisiana State University (LSU). Schulenberg's interesting paper on the taxonomic significance of an intergeneric conebill specimen (*Conirostrum × Oreomanes*) is a by-product of one of the LSU avifaunal collecting expeditions to Peru.

As exemplified by the papers in this collection, field research in the neotropics (as elsewhere) tends to fall into one of three basic categories, once the three initial stages have been essentially completed. These are: intensive studies of single species; studies of a small number of species, either comparative or with respect to interactions; and studies of ecological interrelationships among entire communities. All of these kinds of field studies provide material for interpretation, speculation, and theorizing—the testing of old theories or the formulation of new ones. In contrast to the literature of one or two decades ago, much of the interpretive and theoretical writing is now being done by the same individuals who conducted the field work, and who have a true "feel" for the environmental variables that must be taken into consideration before theoretical interpretations or predictions can be credible.

Also appearing from time to time in the current literature of neotropical ornithology are shorter papers, bringing up to date our knowledge of relatively well known avifaunas. These usually report both natural and human-caused changes. Examples in the present volume include the papers by Voous and ffrench; both, by coincidence, involve continental islands. Neither paper, however, concerns itself particularly with insularity *per se*. With the current intense interest in the evolution, distribution, and ecology of island birds and the consequent, much-needed influx of field data, it was to be expected that a symposium on neotropical ornithology would include several papers on these subjects. In the present volume, the contributions of Grant, Faaborg, Wright et al., and Sherry fall into this category.

Of the four papers, that of Faaborg is the most comprehensive. As a compilation of data based on personal field work plus a survey of available literature, it almost inevitably includes some outright errors (the weights of all West Indian *Dendroica* except *D. adelaidae* are given as 11 g, but the published weights of the four collected specimens of *D. angelae* ranged from 8.0 to 8.8 g) and some arguable generalizations (on Hispaniola, *Melanerpes striatus* is not merely a member of the bark gleaning guild, but is enough of a frugivore to be considered a serious pest by fruit-growers; on the same island, *Zonotrichia capensis* is not confined to pine forests and is, in fact, less common there than in other high elevation habitats). In connection with his classification of bird species into foraging guilds, I believe it is somewhat unorthodox to lump into "frugivores" those that are primarily graminivores (such as *Sporophila*, *Tiaris*, and *Columbina*); see Moermond and Denslow.

Nevertheless, Faaborg's data do show some remarkably consistent patterns of bird distribution in the West Indies, and he calls attention to those exceptions that merit further study. His is one of the first such analyses to acknowledge the significance of Pleistocene climatic shifts in explaining avian distribution in the West Indies, as demonstrated by Olson (1978) and Pregill and Olson (1981), but as yet the data from fossils, especially of small birds, are too sparse from most islands to influence conclusions drawn from present distributions. Such conclusions will inevitably be modified as fossil data accumulate, as has been the case with the Hawaiian avifauna (Olson and James 1982a, b).

Another conspicuous and deliberate omission from Faaborg's paper, acknowledged as such by the author, was any except brief passing reference to the structure of West Indian avian communities during the northern winter, when these are greatly augmented by migrant species. For a fuller picture, the reader must consult the admirable symposium volume *Migrant Birds in the Neotropics* (Keast and Morton 1980), which includes two papers on the West Indies coauthored by Faaborg. One paper in the present volume, that of Gochfeld, does address questions involving resident and migrant species in the West Indies. The principal conclusion of this paper, that he could find no evidence of negative or positive correlations (i.e., neither repulsion nor attraction) in his census data for these two groups, is perhaps less important than the caveats sprinkled through his pages. He reiterates the hazards of basing generalized conclusions on the results of a single field season (although he defends his publication of these single-season data by showing how closely they fit data reported by Lack from the same areas in Jamaica). He also cites as a personal communication the comments that I, too, received orally several times from Eugene Eisenmann, on the extreme skepticism Eisenmann exhibited about many studies of "species diversity and populations done in the neotropics" by hit-and-run observers from the north temperate zone (see Gochfeld's paper for details). It is undoubtedly symptomatic of a more recent appreciation of the difficulties of tropical field work that there are none of these formerly fashionable superficial papers, with more mathematics than data, in this Eisenmann memorial volume. I am sure, for example, that Eisenmann would have appreciated the depth of detail in Remsen's paper on high elevation forest bird populations of Bolivia, even though the two study sites were visited for only a month and 19 days, respectively (I once heard a paper given at a major ornithological meeting in which species diversity comparisons were made of two localities that had been mist-netted for a total of four days each!).

Two of the papers on island birds are of special interest in that, to some extent, they come to opposite conclusions. Wright et al., studying the avifauna of the Pearl Archipelago, Panama, argue that natural disturbances such as fire, windstorms, and drought, have minimal impact on bird communities of these islands (defined for the purposes of their paper as including only the 28 species the authors could mist-net). On the other hand, Grant, in this and previous papers cited in his bibliography, has clearly shown the sometimes drastic effects of short-term climatic fluctuations, especially drought (one of the "disturbances" listed by Wright et al.) on populations of Darwin's finches in the Galapagos archipelago. He extrapolates convincingly from these short-term fluctuations the influence of climate on the entire evolutionary history of these finches. The significance of this apparent disagreement is not clear, but the Pearl Island study obviously has not had the benefit of an extensive data base like that for the Galapagos available to Grant and his colleagues.

The fourth island paper, that by Sherry, is of quite a different genre. It concentrates on a single, little known species, the Cocos Island endemic flycatcher *Nesotriccus ridgwayi*, placing it in context with mainland flycatchers to which it is either taxonomically related or morphologically convergent. Although based on small sample sizes and limited observation periods, together with an awesome amount of mathematics and conjecture, Sherry's paper is, nevertheless, a useful addition to the literature of the Tyrannidae.

If any one taxonomic group of birds can be said to dominate the neotropical avifauna, it is the suboscine Passeriformes, which are represented elsewhere in the world by the limited number of species that have reached the nearctic region, and by a few relict families in the Old World. It is hardly surprising, therefore, that no fewer than nine papers in the present volume concern themselves wholly with some aspect of the suboscines. Several of these have already been mentioned.

Of the others, Lanyon's paper represents a natural expansion of his long-term studies of the flycatcher genus *Myiarchus*, examining the relationships among the tyrannine genera previously alleged to be or found by Lanyon to be related to *Myiarchus*. He uses a broad-spectrum approach in constructing his phylogeny, emphasizing especially the character states of two internal morphological complexes and of patterns of nesting behavior. The concordance of these characters is gratifying, and Lanyon's paper is one of several already published (see his bibliography) or forthcoming that also attempt a synthesis between morphological and behavioral characters in constructing phylogenies of the tyrant flycatchers.

Fitzpatrick's paper will undoubtedly take its place among the milestones in the development of our understanding of the Tyrannidae. Based on the author's detailed museum studies of form and his field studies of foraging behavior, this paper also invokes the "independently derived phylogenetic inferences made by Traylor (1977, 1979)" to create plausible evolutionary scenarios for various lineages of flycatchers.

Sibley and Ahlquist, using their now well known technique of DNA-DNA hybridization, present a phylogeny and resulting reclassification of all of the New World suboscines (mentioning the Old World groups in passing). The reclassification, as in other papers by these authors, is based on principles that are thoroughly explained. As would by now be expected, there are some major departures from traditional classifications of the suboscines. One surprise is the conclusion that a group of flycatcher genera is sufficiently distinct from the remaining

Tyrannoidea (including the rest of the flycatchers, cotingas, manakins, sharpbills, and plant-cutters combined into the single family Tyrannidae) to postulate its having differentiated prior to the origin of the above-listed taxa of Tyrannidae; this group is segregated as the family Mionectidae. The genus *Schiffornis,* generally placed with the manakins (Pipridae), but long recognized as aberrant, is tentatively placed in a monotypic tribe of the subfamily Tityrinae along with a second tribe for the tityras and becards.

It has been known for some time that the traditional antbird family Formicariidae contains two very distinct groups of genera (Heimerdinger and Ames 1967; Ames 1971). This division is supported by Sibley and Ahlquist; the chief surprise here is the postulated age of the separation of the "typical antbirds" (Thamnophilidae), which are placed as a separate "parvorder" Thamnophili, from the "parvorders" Tyranni and Furnarii, which between them include all of the rest of the New World suboscines.

Although Sibley and Ahlquist occasionally mention morphological or behavioral data, these references tend overwhelmingly to pertain to instances in which such data support the findings of DNA-DNA hybridization, and not when there is a conflict between the two sets of data. Examples of such unresolved conflicts include the status of *Conopophaga,* given its own family by Sibley and Ahlquist but found to be inseparable from the "ground antbird" group within the Formicariidae by Ames et al. (1968), and the question of the monophyly of the traditional family Tyrannidae as opposed to the regrouping resulting from the Sibley and Ahlquist separation of the "Mionectidae" (M. McKitrick, pers. comm.). Acceptance of the reclassification of the New World suboscines presented in this monograph will depend upon the continuing assessment of the validity of the methodology and the assumptions upon which all of the Sibley and Ahlquist *oeuvre* is based.

One more paper on suboscines *per se* remains to be mentioned. Greenberg and Gradwohl present the results of a comparative study of three sympatric antwrens on Barro Colorado island, Panama. Their paper, unlike most of the previous literature on this attractive group (their adjective), concentrates on social behavior, in the context of the better-known aspects of foraging ecology. Their findings have wider implications with respect to non-breeding group size, with foliage-gleaning species tending to remain in family groups, "whereas other types of insectivores occur in pairs or solitarily." It should be of much interest to avian ecologists to test this generalization in other areas with other taxa.

Foster's paper is, in fact, a generalized extension of her well known work on manakins, especially of the genus *Chiroxiphia.* After working out the social system in these birds, she is able to characterize a previously little-recognized behavior pattern; the attention paid by many workers to "helpers at the nest" now needs to be extended to search for additional examples of what Foster calls "pre-nesting cooperation."

Two papers concern brood parasitism by the Shiny Cowbird (*Molothrus bonariensis*), involving historical and recent victims. As shown by Fraga, this cowbird species has been parasitizing *Mimus saturninus* in Argentina for at least a century. The mockingbirds have evolved a defense (egg rejection) and also guard their nests against being parasitized in the first place, so they manage to fledge young in 20 percent of parasitized nests. As mentioned early in the present chapter, the Shiny Cowbird has been extending its range northward in historic times. Cruz et al. make important comparisons between the cowbird's impact on *Agelaius icterocephalus* on Trinidad, where the two species have long been sympatric, and on the closely related *A. xanthomus* on Puerto Rico, where the cowbird is a recent invader. That the Yellow-shouldered Blackbird of Puerto Rico is now considered endangered is attributed in large part to nest losses through cowbird parasitism. The Shiny Cowbird arrived in Hispaniola as recently as 1972, and was reported for the first time in Cuba in 1983 (Garrido 1984). The known impact of this species on vulnerable insular endemic songbirds is such that control measures similar to those practiced in Michigan against the Brown-headed Cowbird (*Molothrus ater*) where it seriously interfered with reproduction of Kirtland's Warbler (*Dendroica kirtlandii*) might well be considered in the Caribbean countries (Kelly and DeCapita 1982).

Another paper, that of Mason, had its genesis in a study of two brood parasites in Argentina, the Shiny Cowbird and the Screaming Cowbird (*Molothrus rufoaxillaris*), but the author also accumulated valuable data on the breeding biology of a number of other passerine (mostly host) species.

The monograph contains three single-species studies of neotropical passerines, each of which examines a relatively narrow aspect of its subject's biology. Two concern themselves with the neighbors of nesting birds. The long known association of bird nests with nests of colonial hymenopterans has generally been accepted both as advantageous to the birds and as occurring through deliberate nest-site selection, but there has been little quantification of the former and no test of the latter hypothesis. Wunderle and Pollock demonstrate conclusively that for Bananaquits (*Coereba flaveola*) on Grenada, reproductive success of pairs nesting near wasp nests is, indeed, significantly higher. On the other hand, they also demonstrate that the observed degree of Bananaquit-wasp nesting association *could* occur with nothing beyond random nest-site selection on the part of the female Bananaquit. The authors list twelve families (or subfamilies) of birds for which nesting near hymenopteran colonies has been reported, and it would appear that their study could be relatively easily replicated in other areas using different species. This would be highly desirable if for no other reason than the demonstrable tendency of authors of secondary literature to make broad generalized statements documented only by studies such as this (one species on one island).

Few papers in this volume treat the subject of predation on birds beyond brief mention. Robinson's contribution deals with a highly specialized kind of predation, piracy of the nests of the Yellow-rumped Cacique (*Cacicus cela*). With a color-banded sample of 715 (!) individuals of a highly conspicuous species nesting in highly conspicuous colonies, Robinson was in an ideal position to document nest predation. Piratic Flycatchers (*Legatus leucophaius*) harass caciques into abandoning their nests, which the flycatchers take over. Troupials (*Icterus icterus*) destroy eggs and young of caciques from several nests in addition to the one they usurp. The larger Russet-backed Oropendola (*Psarocolius angustifrons*), which often nests near or even in mixed colonies with the caciques, may destroy nearby cacique nests and their contents, but does not use the vacated nests. Robinson's field observations supply excellent data toward a plausible hypothesis of the evolution of nest piracy, and possibly, as suggested earlier by Pearson (1974), the evolution of brood parasitism.

The last of the single-species studies of a passerine is that of Coats and Phelps on the endangered Red Siskin (*Carduelis cucullata*) of Venezuela, appropriately placed in the "Conservation" section of this monograph. This attractive finch has long been in great demand among canary fanciers, as it produces fertile hybrids with canaries and introduces a desired red color factor into the genomes of the breeding stock. So little information is available about the Red Siskin as a wild bird that even after Coats's intensive nine-month field study, much of the knowledge of the species, especially its breeding biology, still rests on data obtained from captive birds and from the reports of bird catchers. It is to the latter that we owe the information that the Red Siskin nests in clumps of Spanish Moss hanging from tall trees, surely an unusual nest site for a cardueline finch! A gratifyingly international drive to preserve the Red Siskin has been under way for some years, led by the Sociedad Venezolana de Ciencias Naturales. Unfortunately, Curaçao is a major port of trade for unscrupulous animal dealers who export smuggled Red Siskins to Holland. The Netherlands government is conspicuous as a non-signatory to the Convention on International Trade in Endangered Species (CITES). As long as this situation persists, given the demand for Red Siskins from the Dutch canary breeders, the other conservation measures urged by Coats and Phelps will have minimal effect.

Two papers of the "comparative" genre in this volume were composed from the somewhat unusual viewpoint of workers personally familiar with both neotropical birds and their Old World equivalents. In the case of Houston's paper, the comparisons are between two convergent taxa, the New and Old World vultures. One of the most striking differences between the ecologies of the two groups is the absence of Old World vultures in tropical forests; five species of the New World Cathartidae are regularly found in forested areas (although the Black Vulture [*Coragyps atratus*] only marginally so). Houston assembles data to support some admittedly speculative but attractive explanations for this difference.

Lester Short, in contrast, discusses Old and New World radiations of two families found in the tropics of both hemispheres, the woodpeckers (Picidae) and barbets (Capitonidae); he also brings into the discussion two related families, the toucans (Ramphastidae) and honeyguides (Indicatoridae), each confined to the tropics of a single hemisphere. This massive paper is mostly descriptive, and the author himself admits that "we are left with the facts of the radiations as documented, and little in the way of explanations for them." Those familiar with Short's previous work, especially his monograph of the Picidae (Short 1982) may find the detailed accounts of the woodpecker radiations given in this paper to be somewhat redundant. He brings his own perspective to the Indicatoridae, some aspects of the biology of which have been extensively covered in various papers by Herbert Friedmann, and to the

Ramphastidae, acknowledging the probable validity of the zoogeographic scenarios of Haffer (1974). There has been no previous comparable survey of the Capitonidae, and this will undoubtedly be a seminal paper in the literature of that family. Incidentally, Short considers the toucans confamilial with the barbets, and even postulates that the toucans may be more closely related to the neotropical barbets than the latter are to Old World capitonids. This interesting suggestion deserves more than just a passing hint and should be tested with an appropriate cladistic study.

There is much current interest in ecological relationships within rather sharply circumscribed avifaunas or communities, some of which have already been mentioned in other contexts. Examples of such studies can be found in the papers of Stiles, Powell, and Munn for land birds (I will mention water birds later).

Stiles is well known for his long term studies of hummingbirds and their foodplants in Costa Rica. In his present paper he demonstrates the existence of rather well-defined hummingbird/ flower subcommunities, each with its own seasonal rhythm. His extensive data are placed in the context of other studies and especially of theories regarding "the relationship between plant-pollinator interactions and flowering seasons of the plants," and definitions of hummingbird/flower communities by other authors. Each of Stiles' subcommunities is carefully analyzed as to the likelihood of its representing a true coevolution. I predict that this will be a much-cited paper.

One of the most conspicuous communities of land birds in the neotropics is the interspecific foraging flock of insectivores in forests, vividly described by Bates as long ago as 1863 (and cited by both Powell and Munn). Powell gives a comprehensive survey of what is known and what is postulated about such flocks. However, he specifically mentions only briefly the existence of separate canopy flocks in tall forests; most, if not all of the field studies he cites and the generalizations he makes deal with understory or middle elevation flocks. This deficiency is admirably remedied in Munn's paper, which compares understory and canopy flocks in lowland forests of Amazonian Peru. He found that, as might have been expected, there is relatively little overlap of species in the two kinds of flocks, but the structures are virtually identical with respect to numbers of core and visiting species and the flocking behavior of these two groups. Munn's and Powell's papers should be read as a unit.

Although acknowledging that multispecies flocks of frugivores exist, Powell points out that it is uncertain whether these actually represent a community or merely a temporary association of individuals at a food source, and he consequently limits his discussion almost entirely to insectivores. Moermond and Denslow do not address the subject of flocking, but virtually every other aspect of the foraging of frugivores is covered in their paper. Although founded on their own work in Costa Rica, their paper appears to be a thorough survey of the literature of frugivory in neotropical birds, also citing studies from the Old World when appropriate; there appear to be few of these, and I do not know enough about the ecology of the Old World tropics to be able to guess whether this is due to a relative paucity of appropriate fruit trees, frugivorous birds, or interested ornithologists.

The two papers that show the most subject overlap in this volume are those of Skutch and Murray. Although Murray's title suggests a narrower field than that of Skutch, invoking only clutch size, he must, perforce, refer to nesting success and predation. Comparison of the two papers is quite instructive, as both cite many of the same previous studies but interpret them somewhat differently (for example, Snow and Snow's 1963 paper on three Trinidad *Turdus*). In the end, both reject the hypothesis, associated chiefly with the late David Lack, that clutch size is ultimately determined by the amount of food parents can bring to their young, with its concomitant explanations for the small clutch sizes of most tropical birds. Murray stresses the ability of tropical species to divide the minimum number of eggs necessary to assure parental replacement into several successive clutches during a longer breeding season than available to temperate birds. Skutch rightly points out that north temperate theoreticists like Lack have always tended to overestimate the uniformity of availability of food resources during the tropical year. Few quantitative data are actually available, but those that are suggest that the insects needed for feeding young (even in primarily frugivorous species) are *not* necessarily equally abundant the year around. Thus, the implication that any time during the tropical year is as good as any other for rearing young calls for renewed scrutiny. Skutch also calls attention to the possibility raised by Oniki (1979) that the long-accepted higher rate of nest predation in the tropics may in part be an artifact of man-caused alterations in the kind and proportion of predators in the compared areas.

In her own contribution to this volume, Oniki investigates other aspects of nesting, testing variations in egg color, nest construction, and nest placement against theories previously advanced for these variations. She found no evidence for any adaptive correlation other than anti-predator functions. Once more we are shown the value of field data in testing adaptive hypotheses.

To many temperate zone readers, mention of "neotropical birds" tends to conjure up an image of the multiplicity of frugivores and insectivores that have dominated the papers mentioned thus far. The water birds tend to be less associated in our minds with the neotropics, possibly because many of the species, although predominantly tropical in their total distribution, have ranges extending to our latitudes. The paper by Kushlan et al., indeed, compares and contrasts various parameters of ciconiiform faunas in the Florida Everglades and the Venezuelan Llanos. Of 26 species listed for the two areas combined, 11 (42%) occurred in both study areas, and much of the difference between the two lists is attributable to the presence of seven species of ibis in the Llanos, of which only two occur in Florida. Foraging differences between the two populations of conspecifics are minimal, arising principally from a larger spectrum of available prey items in the Llanos.

Willard's comparisons are much broader taxonomically. He studied 22 potentially competing piscivores at an oxbow lake in the Peruvian lowlands, including a mammal and a reptile. The avian species included members of five orders. Some of the descriptions of foraging methods, especially of relatively little-known tropical genera such as *Agamia* and *Pilherodius*, are contributions in their own right over and above their use as data for interspecies comparisons. I was especially interested in the observations of the Large-billed Tern (*Phaetusa simplex*) skimming like a *Rynchops,* as after my own 1961 observations of *Phaetusa* in Argentina (which did not include skimming) and examination of specimens, I formed a strong opinion that the ancestor of the Rynchopinae must have been very like a *Phaetusa.*

A major contribution to the literature of water birds is Thomas's paper comparing the only three storks in the New World, all of which are sympatric in her study area in Venezuela. Accusations of superficiality or incompleteness levied against some published papers on neotropical birds will never be applied to Thomas's work, which is based on 11 years of field work, plus studies of zoo birds and museum specimens.

Other than Schulenberg's paper on conebills, already mentioned, only one paper in this collection deals at length with a hybridization phenomenon. Jehl's study of the two overlapping oystercatcher taxa on the Pacific coast of Baja California is probably unique in ornithological literature in its historical perspective, demonstrating "the resumption of stability in the zone of hybridization after a period of dynamic change," the latter owing to the virtual extermination of some populations within the hybrid zone. Jehl presents field evidence, including of assortative mating, strongly supporting the treatment of the 6th edition of the American Ornithologists' Union Check-list of North American Birds (1983) of the oystercatchers as two species, *Haematopus palliatus* and *H. bachmani.*

Two of the water bird papers ask and answer questions with respect to migrants of northern shorebirds. Schneider found that invertebrate prey animals on the Bay of Panama were *not* depleted by visiting shorebirds, so that the movement of the birds out of the area must be associated with other factors. Myers and his team of Sanderling (*Calidris alba*) specialists examine wintering populations in California and in Peru and Chile. After a thorough analysis of their field-obtained data, they come to the reasonable conclusion that wintering resources are more favorable for Sanderlings at the far distant end of their wintering range, but point out carefully the additional studies that should be made. I enjoyed particularly the last paragraph of their Introduction, presenting an unassailable (to me) rationale for their long-term concentration on only one of a myriad of migratory shorebird species.

The "neotropics" as a concept is often construed loosely as everywhere south of the United States, or at least south of the Mexican Plateau. One paper in this volume, however, reminds us that there is also a non-tropical fauna at the other end—the south temperate Zone and the subantarctic. Humphrey and Livezey contribute another paper in their ongoing study of the unique steamer-ducks of southernmost South America, describing for the first time the nest, eggs, and downy young of a species discovered by Humphrey and his colleagues as recently as 1979.

I turn now to the last section of this monograph, that on conservation. As is repeatedly emphasized in a host of technical and popular publications, a substantial number of neotropical bird species or local populations are currently considered "endangered" through direct or indirect human influence. Direct influence in the past has been chiefly through hunting, but more recently (and for a quite different array of species) the cagebird trade has been responsible for much of the near-extermination of many local populations (see Coats and Phelps; Pasquier 1981 on Psittacidae). One important indirect influence on an avifauna is the introduction of predators, diseases, or competing species. The most pervasive indirect influence is, of course, habitat alteration, which affects not only birds but the entire biota.

Carnegie Museum of Natural History contains bird collections of major importance from several countries of northern South America (Colombia, Venezuela, French Guiana, Amazonian Brazil). With minor exceptions, none of these birds was collected after 1930. In studying geographic variation in the birds of northern South America, my chief source of comparative material has been the American Museum of Natural History; there, too, most of the collections from these countries were made prior to World War II. My point is that in outlining the ranges of subspecies, I and other workers have based our statements on specimen material most of which is at least fifty years old. These range statements are then duly incorporated in the secondary literature, such as the "Peters" check-lists. Most caveats about the use of old museum specimens concern postmortem color changes in plumage, but I do not think I have ever seen a warning that distributional statements based on such material may already be long obsolete when published today. I may say with some confidence that a subspecifically distinct population existed in a certain area *when those specimens were collected,* but without an intimate, up-to-date knowledge of that area I cannot say that the range as outlined is still valid.

Hilty's paper in this volume is the first that I have seen that attempts to analyze in detail the distributional changes effected by human activity in an area as large and diverse as Colombia. Local, detailed studies such as that by Rappole and Morton on a site in southern Veracruz, Mexico, are of great importance in analyzing manmade habitat changes, and constitute part of the data used in broader analyses. Hilty's paper should be emulated for as many neotropical countries as possible; made either by or with the cooperation of knowledgeable local ornithologists, such surveys should then be made available to individuals and government agencies in a position to slow or halt further environmental damage.

Garrido's report on the status of Cuban endangered birds is refreshingly optimistic about several species on mainland Cuba, but emphasizes once again the vulnerability of populations on small islands; these populations are often highly distinct and, in the case of the Cuban islands, only recently discriminated subspecies.

In thinking about bird conservation problems in tropical countries, it is all too easy for us, as concerned ornithologists, to be single-minded, and to overlook the complexities of human existence in these lands. It is obviously inappropriate, for example, to excoriate the rural farmer who clears a patch of forest for firewood or for cropland, even though by so doing he may be eliminating a local bird population. In very few instances is it as easy to designate a "villain" as it is in the case of the law-breaking smugglers of Red Siskins (Coats and Phelps). There are often many links between the proximate and ultimate responsibilities for actions detrimental to local avifaunas. Major programs of forest clearing are not uncommonly approved or even generated by official government agencies, which may be acting either in ignorance of or in defiance of soundly-based projections as to ecological consequences. This is not the place to insert a discussion of ecopolitical problems in Latin America, but we must never forget that they exist. Recommended background reading for this subject would include the papers presented at a Smithsonian Institution symposium in 1966 (Buechner and Buechner 1970); although some of the data in these papers are now obsolete, the reader will nevertheless develop an appreciation of conservation problems in Latin America.

It is now time to step back and look at the papers in this monograph in the total context of current research on neotropical birds. To what extent do these papers accurately reflect such research? It is especially unfortunate that, for various reasons, only a few Latin American ornithologists could be represented in this volume, which, therefore, does not suggest the true level of research activity in the Latin American countries. A quick check of the "Neotropical" portion of the AVES section of the Zoological Record for 1980 (the most recent volume available) reveals that approximately 50 workers resident in Argentina, Brazil, Chile, Colombia, Cuba, Mexico, Peru, Venezuela, and Uruguay, contributed approximately 120 titles to the ornithological literature that year.

Each of the categories into which the papers in this volume are grouped represents, obviously, an active field of research within neotropical ornithology in the 1980's. A quick survey of recent volumes of the four major North American ornithological journals suggests some other fields of current activity that would have to be included in a balanced presentation of neotropical ornithological research. In listing these below, I cite one or two recent papers to exemplify current research in each of the fields.

Although the papers in this volume cover a broad taxonomic spectrum of neotropical birds, two major groups are conspicuous by their absence. These are seabirds (Schreiber et al. 1981; Duffy 1983) and raptors (Mader 1982; Bierregaard 1984). General fields of ornithological research that might well have been included in this volume include (in no particular sequence) vocalizations (Morton and Gonzalez Alonso 1982; Stiles 1984), internal morphology (Bentz and Zusi 1982; Berman 1984), plumages and molts (Arnold et al. 1983), fossil birds (Olson 1982; Campbell and Tonni 1983), migration of neotropical birds (Lanyon 1982; McNeil 1982), eggs (Grant 1982), growth and development of young (Winterstein and Raitt 1983), diseases and parasites (Post 1981), and physiology (Bucher and Worthington 1982).

The papers that have found their way between the covers of this volume nevertheless constitute a most impressive contribution to our knowledge of the birds of the neotropics. Several of them will, I venture to say, be among the most frequently cited titles in their own specialties for years to come. Eugene Eisenmann, friend and colleague, could hardly have a more enduring and appropriate memorial.

LITERATURE CITED

AMERICAN ORNITHOLOGISTS' UNION. 1983. Check-list of North American birds, 6th ed. American Ornithologists' Union, Washington, D.C.

AMES, P. L. 1971. The morphology of the syrinx in passerine birds. Peabody Mus. Nat. Hist. Bull. 37.

AMES, P. L., M. A. HEIMERDINGER, AND S. L. WARTER. 1968. The anatomy and systematic position of the antpipits *Conopophaga* and *Corythopis*. Postilla 114.

ARNOLD, K. A., E. J. BOYD, AND C. T. COLLINS. 1983. Natal and juvenal plumages of the Blue-and-white Swallow (*Notiochelidon cyanoleuca*). Auk 100:203–205.

BATES, H. W. 1863. The naturalist on the River Amazons. Dent, London.

BENTZ, G. D., AND R. L. ZUSI. 1982. The humeroulnar pulley and its evolution in hummingbirds. Wilson Bull. 94:71–73.

BERMAN, S. L. 1984. The hindlimb musculature of the White-fronted Amazon (*Amazona albifrons*, Psittaciformes). Auk 101:74–92.

BIERREGAARD, R. O., JR. 1984. Observations of the nesting biology of the Guiana Crested Eagle (*Morphnus guianensis*). Wilson Bull. 96:1–5.

BUCHER, T. L., AND A. WORTHINGTON. 1982. Nocturnal hypothermia and oxygen consumption in manakins. Condor 84:327–331.

BUECHNER, H. K., AND J. H. BUECHNER (eds.). 1970. The avifauna of northern Latin America. Smithson. Contrib. Zool. 26.

CAMPBELL, K. E., JR., AND E. P. TONNI. 1983. Size and locomotion in teratorns (Aves: Teratornithidae). Auk 100:380–403.

DUFFY, D. C. 1983. The foraging ecology of Peruvian seabirds. Auk 100:800–810.

EISENMANN, E., AND L. S. SHORT. 1982. Systematics of the avian genus *Emberizoides* (Emberizidae). Am. Mus. Novit. No. 2740.

GARRIDO, O. H. 1984. *Molothrus bonariensis* (Aves: Icteridae), nuevo record para Cuba. Misc. Zool., Inst. Zool., La Habana 19:2–3.

GRANT, P. R. 1982. Variation in the size and shape of Darwin's finch eggs. Auk 99:15–23.

HAFFER, J. 1974. Avian speciation in tropical South America. Publ. Nuttall Ornith. Club No. 14.

HAFFER, J. 1983. Ergebnisse moderner ornithologischer Forschung im tropischen Amerika. Spixiana [München] Suppl. 9:117–166.

HEIMERDINGER, M. A., AND P. L. AMES. 1967. Variation in the sternal notches of suboscine passeriform birds. Postilla 105.

KEAST, A., AND E. S. MORTON (eds.). 1980. Migrant birds in the neotropics. Smithsonian Institution Press, Washington, D.C.

KELLY, S. T., AND M. E. DECAPITA. 1982. Cowbird control and its effect on Kirtland's Warbler reproductive success. Wilson Bull. 94:363–365.

KEPLER, C. B., AND K. C. PARKES. 1972. A new species of warbler (Parulidae) from Puerto Rico. Auk 89:1–18.

LANYON, W. E. 1982. Evidence for wintering and resident populations of Swainson's Flycatcher (*Myiarchus swainsoni*) in northern Suriname. Auk 99:581–582.

LOWERY, G. H., JR., AND D. A. TALLMAN. 1976. A new genus and species of nine-primaried oscine of uncertain affinities from Peru. Auk 93:415–428.

MADER, W. J. 1982. Ecology and breeding habits of the Savanna Hawk in the Llanos of Venezuela. Condor 84:261–271.

MARES, M. A., AND H. H. GENOWAYS (eds.). 1982. Mammalian biology in South America. Pymatuning Symposia Ecol. 6.

MCNEIL, R. 1982. Winter resident repeats and returns of austral and boreal migrant birds banded in Venezuela. J. Field Ornith. 53:125–132.

MORTON, E. S., AND H. J. GONZALEZ ALONSO. 1982. The biology of *Torreornis inexpectata* I. A comparison of vocalizations in *T. i. inexpectata* and *T. i. sigmani*. Wilson Bull. 94:433–446.

OLSON, S. L. 1978. A paleontological perspective of West Indian birds and mammals. Pp. 99–117, *In* F. B. Gill (ed.), Zoogeography in the Caribbean. Acad. Nat. Sci. Philadelphia Spec. Publ. 13.

OLSON, S. L. 1982. A new species of palm swift (*Tachornis*: Apodidae) from the Pleistocene of Puerto Rico. Auk 99:230–235.

OLSON, S. L., AND H. F. JAMES. 1982a. Fossil birds from the Hawaiian islands: evidence for wholesale extinction by man before western contact. Science 217:633–635.

OLSON, S. L., AND H. F. JAMES. 1982b. Prodromus of the fossil avifauna of the Hawaiian islands. Smithson. Contrib. Zool. 365.

ONIKI, Y. 1979. Is nesting success of birds low in the tropics? Biotropica 11:60–69.

PARKES, K. C. 1982. Subspecific taxonomy: unfashionable does not mean irrelevant. Auk 99:596–598.

PASQUIER, R. F. (ed.). 1981. Conservation of New World parrots. Int. Council Bird Preserv. Tech. Publ. 1.

PATTERSON, B. D. 1983. Review of Mares and Genoways 1982. Syst. Zool. 32:460–463.

PEARSON, D. L. 1974. Use of abandoned cacique nests by nesting Troupials (*Icterus icterus*): precursor to parasitism? Wilson Bull. 86:290–291.

PHILLIPS, A. R. 1981. Subspecies vs forgotten species: the case of Grayson's Robin (*Turdus graysoni*). Wilson Bull. 93:301–309.

PIERPONT, N., AND J. W. FITZPATRICK. 1983. Specific status and behavior of *Cymbilaimus sanctaemariae*, the Bamboo Antshrike, from southwestern Amazonia. Auk 100:607–620.

POST, W. 1981. The prevalence of some ectoparasites, diseases, and abnormalities in the Yellow-shouldered Blackbird. J. Field Ornithol. 52:16–22.

PREGILL, G. K., AND S. L. OLSON. 1981. Zoogeography of West Indian vertebrates in relation to Pleistocene climatic cycles. Annu. Rev. Ecol. Syst. 12:75–98.

SCHREIBER, R. W., D. W. BELITZKY, AND B. A. SORRIE. 1981. Notes on Brown Pelicans in Puerto Rico. Wilson Bull. 93:397–400.

SHIELDS, G. F. 1982. Comparative avian cytogenetics: a review. Condor 84:45–58.

SHORT, L. L. 1982. Woodpeckers of the world. Delaware Mus. Nat. Hist. Monogr. Ser. 4.

SNOW, D. W., AND B. K. SNOW. 1963. Breeding and the annual cycle in three Trinidad thrushes. Wilson Bull. 75:27–41.

STILES, F. G. 1984. The songs of *Microcerculus* wrens in Costa Rica. Wilson Bull. 96:99–103.

TRAYLOR, M. A., JR. 1977. A classification of the tyrant flycatchers. Bull. Mus. Comp. Zool. 148:129–184.

TRAYLOR, M. A., JR. 1979. Family Tyrannidae (part). Pp. 1–228, *In* M. A. Traylor, Jr. (ed.), Checklist of Birds of the World, Vol. 8. Mus. Comp. Zool., Cambridge, Massachusetts.

VAURIE, C. 1980. Taxonomy and geographical distribution of the Furnariidae (Aves, Passeriformes). Bull. Am. Mus. Nat. Hist. 166.

WINTERSTEIN, S. R., AND R. J. RAITT. 1983. Nestling growth and development and the breeding ecology of the Beechey Jay. Wilson Bull. 95:256–268.

INDEX

The "key words" listed below reflect the subject matter of the articles in this volume. These words and the names of contributors are indexed to the first page of each appropriate article; names of artists are indexed to pages on which their plates appear. New scientific names are set in boldface.

Adaptation, 908
Adaptive limitation of reproduction, 575
  radiation, 447, 559
Adjusted reproduction, 505
Afrotropical birds, 559
Agelaius icterocephalus, 607
  xanthomas, 607
Ahlquist, J. E., 396
Allen, S. E., 198
Amazonia, 147, 683
Amazonian birds, 536
Andean birds, 41, 85, 169, 733
  zoogeography, 305
Andean-Patagonian affinities, 233
Antwren, 845
Argentina, 829, 935, 954
Arid region avifauna, 169, 305
Aridity, 471
Aruba, 247
Atlantic coast, 935
Austral-breeding birds, 935
Avian abundance, 798
Avifaunal region, 1000
Baja California, 484
Bananaquit, 595
Barbet, 559
Behavioral ecology, 815
Benthic invertebrates, 546
Biogeography, 47, 49, 113, 147, 169, 238, 255,
  305, 621, 1000
Bird-wasp association, 595
Blue List, 1000
Bock, W. J., 319
Bolivia, 733
Bonaire, 247
Braun, M. J., 169, 333
Brazil, 233, 935
Breeding biology, 933
Brood parasitism, 607, 829
Cacicus cela, 898
Calidris alba, 520
Campbell, C. J., 798
Canopy flocks, 683
Capitonidae, 559
Capparella, A. P., 347
Census data, 654
Cerro Pirre, 198
Chalk-browed Mockingbird, 829
Chordeiles acutipennis inferior, 356
  acutipennis littoralis, 356

acutipennis micromeris, 356
acutipennis texensis, 356
Ciconia maguari, 921
Ciconiiformes, 663
CITES, 977
Classification, 396
Cloud forest birds, 41
Clutch size, 505, 575
Coats, S., 977
Cocos Island, 908
Coeligena torquata **eisenmanni**, 41
Coereba flaveola, 595
Coerebidae, 390
Coerebinae, 595
Coevolution, 757, 829
Collared Inca, 41
Colombia, 1000
Colonization, 607
Commensalism, 595
Community ecology, 605, 654
  structure, 621, 757, 798
Competition, 621, 654
Computer mapping, 147
Conebill, 390
Conirostrum, 390
Conservation, 975, 977, 992
Contour maps, 147
Convergence, 319
Cooperative breeding, 817
Corneous tongue, 319
Costa Rica, 23, 757, 908
Cotingidae, 383
Courtship, 817
Cracraft, J., 49
Cranial morphology, 383
  osteology, 319
Cranioleuca, 333
Cruz, A., 607
Cuba, 992
Curaçao, 247
Darién, 198
Darwin's Finches, 471
Dedication, vi
Delgado B., F. S., 17
Demography, 954
Dendroica magnolia, 1013
Density compensation, 683
Denslow, J. S., 865
Dickerman, R. W., 356
Diglossa, 319

Diglossopis, 319
Distribution, 47, 85, 169, 198, 621
DNA hybridization, 396
Dominican Republic, 607
Downy young, 945
Ecological change, 986
Egg color, 536, 829
  destruction, 898
  thermodynamics, 536
Eisenmann, Eugene, vii, 1
Endangered species, 977, 992
Endemism, 49, 113, 233, 255
Equatorial arid fauna, 305
Escalante, R., 935
Everglades, 663
Evolution, 113, 445, 471
Evolutionary ecology, 815
  genetics, 333
Extinct species, 986
Extinction, 471
Faaborg, J., 621, 978
Faunal history, 255
Feeding adaptations, 865
  behavior, 865
  ecology, 663, 788
ffrench, R., 986
Fiery-throated Hummingbird, 23
Fish-eating, 663, 788
Fitzpatrick, J. W., 147, 430, 447
Fjeldså, J., 85
Flock composition, 683, 713
Flocking behavior, 683, 713
  ecology, 683, 713
Florida, 663
Flowerpiercer, 319
Flycatcher, 347, 361, 431, 447, 908
Foraging behavior, 447, 908, 921
  ecology, 733
  enhancement hypothesis, 713
Formicariidae, 845
Foster, M. S., 817
Fraga, R. M., 829
Fringillidae, 977, 986
Frohring, P., 663
Frugivory, 865
Fruit selection, 865
Galápagos Islands, 471
Gardner, D., 22
Garrido, O. R., 992
Genetic divergence, 347
Geographic variation, 23, 31, 147
Getty, T., 31
Gochfeld, M., 654
Gradwohl, J., 845
Grant, P. R., 471
Graves, G. R., 169
Greater Antilles, 654
Grebe, 31
Greenberg, R., 845

Guild structure, 621
Habitat disturbance, 798, 1013
  fluctuation, 471, 798
Haematopodidae, 484
Haematopus bachmani, 484
  palliatus, 484
Haffer, J., 113, 147
Hatchability of eggs, 575
Hellmayrea, 333
Helping behavior, 817
Heron, 663, 788
Hilty, S. L., 1000
Hooded Warbler, 1013
Host-parasite coevolution, 829
Houston, D. C., 856
Howell, T. R., 1
Huancabamba region, 169
Human disturbance, 198
Hummingbird, 23, 41, 757
Humphrey, P. S., 945
Hybridization, 390, 484
Ibis, 663
Icterinae, 829
Icterus icterus, 898
Inca Wren, 9
Intergeneric hybrid, 390
Intertidal fauna, 546
Interspecific group, 713
Island biology, 908
Itatiaia, 233
Jabiru mycteria, 921
Jamaica, 654
Jaw musculature, 383
Jehl, J. R., Jr., 484
Kingfisher, 788
Kleinbaum, M., front cover, i
Kushlan, J., 663
Lack's hypothesis of clutch size, 505
Lanyon, S. M., 347
Lanyon, W. E., 361
Laridae, 935
Least Grebe, 31
Legatus leucophaius, 898
Lesser Nighthawk, 356
Livezey, B. C., 945
Llanos, 663, 921
Magnolia Warbler, 1013
Maguari Stork, 921
Manolis, T., 607
Maron, J. L., 520
Marshland, 663
Mason, P., 954
Mating system, 817
Mengel, R. M., 944
Mexico, 1013
Migration, 247, 520
Migratory birds, 247, 654, 1013
Mimidae, 829
Mimus saturninus, 829

Mionectes, 347
Mixed species flocks, 683, 713, 733
Moermond, T. C., 865
Molecular evolution, 333
Molothrus bonariensis, 607, 829
Montane region, 233
Morales, G., 663
Morony, J. J., Jr., 383
Morphology, 447
Morphometrics, 347
Morton, E. S., 1013
Munn, C. A., 683
Murray, B. G., Jr., 505
Mutualism, 757
Mycteria americana, 921
Myers, J. P., 520
Nasal capsule, 361
Natural history, 954
    selection, 471
Nectar-feeding, 319
Neotropical forest, 856
Nesotriccus, 908
Nest piracy, 898
    predation, 536, 575, 595, 898
    structure, 536
    success, 575
    thermodynamics, 536
Nesting behavior, 361
    biology, 945, 954
Nestling killing, 898
New species, 9
    subspecies, 9, 17, 23, 31, 41
    taxa, 7
Nine-primaried oscines, 383
Nomenclature, 356
Nothofagus forest, 255
Nutrition, 865
Ochthoeca, 431
O'Neill, J. P., iv, 9, 40
Oniki, Y., 536
Oreomanes, 390
Overview, 1025
Oystercatcher, 484
Painted Parakeet, 17
Pair bond, 817
Panama, 17, 198, 546, 798
*Panterpe insignis **eisenmanni**,* 23
    insignis insignis, 23
Parapatric model, 49
Parker, T. A., III, 9, 169, 198, 333
Parkes, K. C., 1025
Parulinae, 1013
Passerine, 954
Patagonia, 255
Pearl Islands, 798
Peterson, R. T., 382
Peru, 9, 41, 169, 305, 683, 788
Phelps, W. H., Jr., 977
Phenetic biogeography model, 49

Phylogeny, 361, 396
Picidae, 559
Piratic Flycatcher, 898
Pleistocene glaciation, 233
    refugia, 49
Pollock, K. H., 595
Polymorphism, 829
Population decline, 992
    density, 683
    ecology, 605
Powell, G. V. N., 713
Predation avoidance hypotheses, 713
Predator defense, 898
Promiscuity, 817
Protein electrophoresis, 347
Psarocolius angustifrons, 898
Psittacidae, 17
Puerto Rico, 607
*Pyrrhura picta **eisenmanni**,* 17
Range expansion, 607, 986
Rappole, J. H., 1013
Red Siskin, 977
Remsen, J. V., Jr., 733
Rhamphotheca, 319
Robbins, M. B., 198, 305
Robinson, S. K., 898
Royal Tern, 935
Russet-backed Oropendola, 898
Sallaberry, M., 520
Sanderling, 520
Savanna, 663
Scarlet Ibis, 986
Scavenger, 856
Schneider, D., 546
Schulenberg, T. S., 169, 305, 390
Scolopacidae, 520
Sericossypha, 383
Sherry, T. W., 908
Shiny Cowbird, 829
Shorebird, 520, 546
Short, L. L., 559
Sibley, C. G., 396
Sibling species, 347
Sick, H., 233
Singer, A. B., 16
Skutch, A. F., 575
Snow, D. W., 238
Social behavior, 683, 713, 817, 845
    organization, 683, 713
    systems, 733, 817
South Caribbean Islands, 247
Speciation, 85, 113, 255, 431
Species diversity, 654, 733
Spinetail, 333
Spinus cucullatus, 977
Steamer-duck, 945
Sterna maxima maxima, 935
Sterninae, 935
Stiles, F. G., 23, 757

Storer, R. W., 31
Stork, 663
Suboscine, 347, 361, 396, 431, 447, 908
Subspecies, 147
Superspecies, 431
Synallaxis, 333
Syrinx, 361
Systematics, 317, 361
Tachybaptus dominicus bangsi, 31
  dominicus brachyrhynchus, 31
  dominicus dominicus, 31
  dominicus eisenmanni, 31
  dominicus speciosa, 31
Tachyeres, 945
Taxonomy, 7, 9, 17, 23, 31, 41, 317, 319,
  333, 356, 361, 383, 390, 396, 431
Territoriality, 713, 845
Thomas, B. T., 921
Thraupinae, 319, 383
Thryothorus eisenmanni, 9
  euophrys atriceps, 9
  euophrys euophrys, 9
  euophrys longipes, 9
  euophrys schulenbergi, 9
Tobago, 238
Traylor, M. A., Jr., 431
Trinidad, 238, 607, 986
Trochilidae, 23, 41, 757
Troglodytidae, 9
Troupial, 898

Tumbes, 305
Tyrannidae, 347, 361, 431, 447, 908
Understory flocks, 683
Uruguay, 935
Venezuela, 663, 921
Vicariance, 49, 113
Voous, K. H., 247
Vuilleumier, F., 255
Vulture, 856
Wading birds, 663
Wasps, 595
Water bird, 85, 247, 788, 921
  nomadism, 247
Weske, J. S., 41
West Indies, 621
White-browed Spinetail, 333
Wiedenfeld, D. A., 305
Wiley, J. W., 607
Willard, D. E., 788
Williams, M. D., 360
Wilsonia citrina, 1013
  pusilla, 1013
Wilson's Warbler, 1013
Wood Stork, 921
Woodpeckers, 559
Wright, S. J., 798
Wunderle, J. M., Jr., 595
Yellow-rumped Cacique, 898
Zoogeography, 47, 49, 113, 147, 169, 238,
  255, 305, 621, 992

ORNITHOLOGICAL MONOGRAPHS

No. 1. **A Distributional Study of the Birds of British Honduras,** by Stephen M. Russell. 1964. $7.00 ($5.50 to AOU members).

No. 2. **A Comparative Study of Some Social Communication Patterns in the Pelecaniformes,** by Gerard Frederick van Tets. 1965. $3.50 ($2.50 to AOU members).

No. 3. **The Birds of Kentucky,** by Robert M. Mengel. 1965. $15.00 ($12.50 to AOU members).

No. 6. **Adaptations for Locomotion and Feeding in the Anhinga and the Double-crested Cormorant,** by Oscar T. Owre. 1967. $6.00 ($4.50 to AOU members).

No. 7. **A Distributional Survey of the Birds of Honduras,** by Burt L. Monroe, Jr. 1968. $14.00 ($11.00 to AOU members).

No. 9. **Mating Systems, Sexual Dimorphism, and the Role of Male North American Passerine Birds in the Nesting Cycle,** by Jared Verner and Mary F. Willson. 1969. $4.00 ($3.00 to AOU members).

No. 10. **The Behavior of Spotted Antbirds,** by Edwin O. Willis. 1972. $9.00 ($7.50 to AOU members).

No. 11. **Behavior, Mimetic Songs and Song Dialects, and Relationships of the Parasitic Indigobirds (*Vidua*) of Africa,** by Robert B. Payne. 1973. $12.50 ($10.00 to AOU members).

No. 12. **Intra-island Variation in the Mascarene White-eye *Zosterops borbonica*,** by Frank B. Gill. 1973. $3.50 ($2.50 to AOU members).

No. 13. **Evolutionary Trends in the Neotropical Ovenbirds and Woodhewers,** by Alan Feduccia. 1973. $3.50 ($2.50 to AOU members).

No. 14. **A Symposium on the House Sparrow (*Passer domesticus*) and European Tree Sparrow (*P. montanus*) in North America,** by S. Charles Kendeigh. 1973. $6.00 ($4.50 to AOU members).

No. 15. **Functional Anatomy and Adaptive Evolution of the Feeding Apparatus in the Hawaiian Honeycreeper Genus *Loxops* (Drepanididae),** by Lawrence P. Richards and Walter J. Bock. 1973. $9.00 ($7.50 to AOU members).

No. 16. **The Red-tailed Tropicbird on Kure Atoll,** by Robert R. Fleet. 1974. $5.50 ($4.50 to AOU members).

No. 17. **Comparative Behavior of the American Avocet and the Black-necked Stilt (Recurvirostridae),** by Robert Bruce Hamilton. 1975. $7.50 ($6.00 to AOU members).

No. 18. **Breeding Biology and Behavior of the Oldsquaw (*Clangula hyemalis* L.),** by Robert M. Alison. 1975. $3.50 ($2.50 to AOU members).

No. 19. **Bird Populations of Aspen Forests in Western North America,** by J. A. Douglas Flack. 1976. $7.50 ($6.00 to AOU members).

No. 20. **Sexual Size Dimorphism in Hawks and Owls of North America,** by Noel F. R. Snyder and James W. Wiley. 1976. $7.00 ($6.00 to AOU members).

No. 21. **Social Organization and Behavior of the Acorn Woodpecker in Central Coastal California,** by Michael H. MacRoberts and Barbara R. MacRoberts. 1976. $7.50 ($6.00 to AOU members).

No. 22. **Maintenance Behavior and Communication in the Brown Pelican,** by Ralph W. Schreiber. 1977. Price $6.50 ($5.00 to AOU members).

No. 23. **Species Relationships in the Avian Genus *Aimophila*,** by Larry L. Wolf. 1977. Price $12.00 ($10.50 to AOU members).

No. 24. **Land Bird Communities of Grand Bahama Island: The Structure and Dynamics of an Avifauna,** by John T. Emlen. 1977. Price $9.00 ($8.00 to AOU members).

No. 25. **Systematics of Smaller Asian Night Birds Based on Voice,** by Joe T. Marshall. 1978. Price $7.00 ($6.00 to AOU members).

No. 26. **Ecology and Behavior of the Prairie Warbler *Dendroica discolor*,** by Val Nolan, Jr. 1978. Price $29.50.

No. 27. **Ecology and Evolution of Lek Mating Behavior in the Long-tailed Hermit Hummingbird,** by F. Gary Stiles and Larry L. Wolf. 1979. Price $8.50 ($7.50 to AOU members).

No. 28. **The Foraging Behavior of Mountain Bluebirds with Emphasis on Sexual Foraging Differences,** by Harry W. Power. 1980. Price $8.50 ($7.50 to AOU members).

No. 29. **The Molt of Scrub Jays and Blue Jays in Florida,** by G. Thomas Bancroft and Glen E. Woolfenden. 1982. Price $8.00 ($6.50 to AOU members).

No. 30. **Avian Incubation: Egg Temperature, Nest Humidity, and Behavioral Thermoregulation in a Hot Environment,** by Gilbert S. Grant. 1982. Price $9.00 ($7.00 to AOU members).

No. 31. **The Native Forest Birds of Guam,** by J. Mark Jenkins. 1983. Price $9.00 ($7.00 to AOU members).

No. 32. **The Marine Ecology of Birds in the Ross Sea, Antarctica,** by David G. Ainley, Edmund F. O'Connor, and Robert J. Boekelheide. 1984. Price $9.00 ($8.00 to AOU members).

No. 33. **Sexual Selection, Lek and Arena Behavior, and Sexual Size Dimorphism in Birds,** by Robert B. Payne. 1984. Price $8.00 ($6.50 to AOU members).

No. 34. **Pattern, Mechanism, and Adaptive Significance of Territoriality in Herring Gulls (*Larus argentatus*),** by Joanna Burger. 1984. Price $9.00 ($7.00 to AOU members).

No. 35. **Ecogeographic Variation in Size and Proportions of Song Sparrows (*Melospiza melodia*),** by John W. Aldrich. 1984. Price $10.50 ($8.50 to AOU members).

No. 36. **Neotropical Ornithology,** P. A. Buckley, Mercedes S. Foster, Eugene S. Morton, Robert S. Ridgely, and Francine G. Buckley (eds.). 1985. Price. $70.00.

Like all other AOU publications, *Ornithological Monographs* are shipped prepaid. Make checks payable to "The American Ornithologists' Union." For the convenience of those who wish to maintain complete sets of *Ornithological Monographs* and to receive new numbers immediately upon issue, standing orders are encouraged.

Order from: **Frank R. Moore, Assistant to the Treasurer AOU, Department of Biology, University of Southern Mississippi, Southern Station Box 5018, Hattiesburg, Mississippi 39406.**